ALIGNMENT GUIDE FOR UPDATED

The Practice of Statistics for the AP® Exam, SIXTH EDITION

Find a complete correlation to the current AP® Statistics Curriculum Framework at
go.bfwpub.com/ap-course-updates

Prasit Chansareekorn/Moment/Getty Images

UPDATED

ANNOTATED TEACHER'S EDITION

Luke Wilcox
East Kentwood High School

for the AP® Exam

UPDATED The
Practice of
Statistics

SIXTH EDITION

Daren S. Starnes
The Lawrenceville School

Josh Tabor
Canyon del Oro High School

bedford, freeman & worth
high school publishers
New York

Updated Annotated Teacher's Edition
Updated *The Practice of Statistics* for the AP® Exam
Sixth Edition

Senior Vice President, Humanities & Social Sciences and High School: Charles Linsmeier
Senior Program Director, High School: Ann Heath
Development Editor: Donald Gecewicz
Editorial Assistant: Carla Duval
Marketing Manager: Thomas Menna
Marketing Coordinator: Kelly Noll
Senior Media Editor: Kim Morté
Director of Digital Production: Keri deManigold
Lead Media Project Manager: Jodi Isman
Director of Design, Content Management: Diana Blume
Cover and Interior Designer: Lumina Datamatics, Inc.
Director, Content Management Enhancement: Tracey Kuehn
Senior Managing Editor: Lisa Kinne
Senior Content Project Manager: Vivien Weiss
Senior Workflow Supervisor: Susan Wein
Illustrations: Lumina Datamatics, Inc.
Senior Photo Editor: Robin Fadool
Photo Researcher: Candice Cheesman
Art Manager: Matthew McAdams
Composition: Lumina Datamatics, Inc.
Printing and Binding: LSC Communications
Cover Photo: Prasit Chansareekorn/Moment/Getty Images

TI-84+™ screen shots are used with permission of the Publisher, © 1996, Texas Instruments Incorporated.

M&M'S is a registered trademark of Mars, Incorporated and its affiliates. This trademark is used with permission. Mars, Incorporated is not associated with Macmillan Learning. Images printed with permission of Mars, Incorporated.

Library of Congress Control Number: 2020930772
ISBN-13: 978-1-319-26932-6
ISBN-10: 1-319-26932-X
© 2020, 2018, 2015, 2012 by W. H. Freeman and Company

Printed in the United States of America
2   3   4   5   6   24   23   22

W. H. Freeman and Company
Bedford, Freeman & Worth
One New York Plaza
Suite 4600
New York, NY 10004-1562
highschool.bfwpub.com/catalog

What's Included in This Annotated Teacher's Edition

About the Author

Peter McDaniel

LUKE WILCOX is a National Board Certified Teacher who has spent his 19-year teaching career at East Kentwood High School, which is the most diverse public high school in the state of Michigan. His teaching has been recognized with the 2013 Presidential Award and the 2018 Michigan Teacher of the Year award, each giving him the opportunity to visit the White House to meet the president.

Luke is a veteran grader and table leader at the annual AP® Statistics Reading and spends his summers helping teachers at College Board AP® Summer Institutes around the country and speaking at local, national, and international conferences. To further help statistics teachers, Luke co-developed the website www.statsmedic.com, which is positively transforming statistics instruction in classrooms around the country. Over the past 14 years, he has guided more than 250 students to a 5 on the AP® Statistics exam, with a 96% overall pass rate. Luke leads his students with an "experience first, formalize later" approach to teaching statistics—in which students engage in the mathematical thinking and reasoning before being provided algorithmic formulas and definitions.

A Word from the Author

In 2013, I attended my very first National Council of Teachers of Mathematics Annual Conference in Denver. I was thrilled to find out that Daren Starnes, author of *The Practice of Statistics*, was presenting at a workshop. In my own classroom, we had just switched to this book, and it had transformed my teaching. After attending Daren's session, I approached him to ask some follow-up questions (and to get my AP® Statistics book signed!). I thanked him for all the work he put into *The Practice of Statistics* and divulged that my students' AP® scores had skyrocketed as a result of using it. Daren became interested in how a public school with such diversity could produce these scores, and several months later he was observing in my classroom. Now, seven years later, Daren is one of my greatest mentors, and I am truly honored to have worked with him (and Josh!) on this *Annotated Teacher's Edition*.

During the past 14 years, I have made it my mission to continuously improve the quality of instruction in my AP® Statistics classes. This quest has taken me to trainings, workshops, and conferences all over the country, affording me the opportunity to learn from legends. The community of AP® Statistics teachers is tight-knit, and the collaborative support of this group is unparalleled. To this day, I continue to find ways to make AP® Statistics more relevant, engaging, and understandable for my students. It is my goal with the *Annotated Teacher's Edition* to transform all these experiences into a set of real-life, usable teaching tips that can help new and experienced AP® Statistics teachers accelerate their improvement. I hope to give back to the community that has given so much to me.

In my first years teaching AP® Statistics, I didn't know what I didn't know. Now I realize there are infinite connections to be made in the curriculum as we lead students toward the eventual goal of AP® Statistics: inference. I now know there is a story to be told. And the story has a clear progression, with the details from the beginning of the story being essential to understanding the end of the story. The Teaching Tips labeled "Making Connections" and "Preparing for Inference" will help you build this story with your students.

Also, I wanted the *Annotated Teacher's Edition* (ATE) to be teacher friendly and effective. I have included only teaching tips that I have used myself and have seen work with my own students. Finally, I wanted the ATE to produce real results for student scores on the AP® Statistics exam. After all, one of the great benefits of the AP® program is that students will save money and can get ahead in achieving a college degree. Therefore, many of the teaching tips will allow you to help students maximize their AP® score at the end of the year.

I hope you find the ATE to be a helpful resource, whether you are a rookie or a veteran teacher of AP® Statistics. Creating this resource has been a rewarding experience for me, but I certainly didn't do it alone. Many thanks go to the following people:

- My beautiful wife Jamie for putting up with my 4:00 A.M. alarm clock in support of my ambitious efforts outside the classroom, all the while being an extraordinary mother and third-grade teacher.

- My children Reese and Trey for inspiring me to want to positively impact education on a scale greater than my own classroom.

- Daren Starnes, author, for being an iconic leader, but also so relatable. Your mentorship has challenged and improved my teaching practice and has given me the confidence and inspiration to help others improve theirs.

- Josh Tabor, author, for providing me with supportive guidance and feedback throughout the entire writing process. Your work ethic and productivity are rivaled only by Daren, and I strive to match this enthusiasm. Side note: I attended my first four-day workshop in 2007 with Josh as the instructor and have loved the dotplot ever since.

- Ann Heath, Publisher, for being a visionary leader and project manager. Your experience and wisdom in the publishing world have helped this project (and me!) reach the fullest potential.

- Don Gecewicz, Development Editor, for meticulous attention to details in the manuscript and valuable feedback that always improved the end product.

- Lindsey Gallas, colleague, for always challenging me to be more progressive in my instruction and for helping to create and grow Stats Medic.

- Al Reiff, for accuracy checking the Activity solutions and Alternate Examples.

- Kaitlyn Swygard and Carla Duval, Editorial Assistants, for all the behind-the-scenes work on manuscript preparation, Teacher's Resource Materials, and other tasks.

About the Program

Since the first edition of *The Practice of Statistics* (TPS) was published, teachers have lauded the ancillary materials that accompany the student textbook; our goal is to make them better with each revision. We aim to provide the best educational support through print and digital resources to help all teachers—new and experienced—teach the best possible AP® Statistics course.

For the updated sixth edition, we have consolidated all the best material into the *Updated Annotated Teacher's Edition* (ATE) and the Teacher's Resource Materials (TRM), which can be accessed in three ways: by clicking on the link in the Teacher's e-Book (TE-Book), opening the Teacher's Resource Flash Drive (TRFD), or accessing through the book's digital platform. Look for the Teacher's Resource Materials icon **TRM** throughout the ATE for reminders about the available resources.

The ATE includes:

- The student text
- The AP® Statistics Primer with general advice for teaching AP® Statistics
- "Blue pages" at the beginning of each chapter that precede the reduced student pages with wrap-around teacher's material. Blue pages provide an in-depth guide to the chapter content and support materials, including:

  - An overview of the chapter, featuring a discussion of the most important ideas and how the content relates to the AP® Statistics Course Framework
  - A summary of each section in the chapter, highlighting the main ideas
  - A list of resources, including a comprehensive list of AP® Free Response questions appropriate for that chapter
  - A pacing guide for the chapter, featuring Learning Targets and suggested homework assignments
  - An Alignment to the College Board's Fall 2019 AP® Statistics Course Framework that highlights the Course Skills, Learning Objectives, and Essential Knowledge Statements that are addressed in the chapter

- Answers (including many graphs) to all Exercises, Activities, Check Your Understanding problems, the FRAPPY! materials at the end of each chapter, and every chapter test and cumulative test. Care has been taken to make sure that the answer appears on the same page as the corresponding exercise. *Note:* In many cases, answers in the ATE are abbreviated to fit on the same page—or at least the same spread—as the exercise. Please click the link to Full Solutions in the Teacher's e-Book, or go to the TRFD or the digital platform for complete solutions to these exercises.

- Teaching Tips that are placed strategically throughout the chapter just when you need them. We have introduced three new types of teaching tips in the updated sixth edition to help you build the story of AP® Statistics:

  - Making Connections: An increasingly important goal of the course is to help students see and understand the big ideas in statistics and gain an appreciation for how they crosscut the content. These new tips will help teachers demonstrate connections to concepts from previous chapters, as well as how current concepts will connect to ideas in later chapters.

  - Preparing for Inference: One of the major content goals of AP® Statistics is to develop students' inferential thinking abilities. Inference is not an idea that should be saved for the end of the course. Rather, it is a big concept that must be built up slowly throughout the course. This feature will help teachers recognize and highlight the ideas in each chapter that will later lead to student success in inference.

  - ✚ Ask the StatsMedic: Blog posts have been created to help you become a more knowledgeable and effective AP® Statistics teacher. They can be found at www.statsmedic.com/askthestatsmedic and are linked from the *Annotated Teacher's Edition*. Topics include insights on understanding the content better, best practices for teaching for understanding, and activities that can be used in the classroom.

- Alternate Examples for each example in the student text coded to the appropriate AP® Skill from the new Course and Exam Description (CED). Almost all the Alternate Examples have been updated since the last edition to use contexts that are relevant to students. Students gain a much deeper understanding of statistical concepts when taught in an engaging and

memorable context. These examples are also available in Microsoft Word format for easy export into PowerPoint or class notes.

- AP® Exam Tips that discuss insights for maximizing student performance on the exam, including details from previous exams and rubrics.
- Common Student Errors highlight mistakes that students often make—and how your students can avoid them.
- Technology Tips to provide guidance when working with graphing calculators and other technology.
- Links to applets from the *Updated TPS6e* Student Site and many others.
- Links to valuable resources in the Teacher's Resource Materials, including video overviews of the chapter and sections, a quiz for each section, a test for every chapter, video introductions to all activities, printable Learning Targets grids, multiple worksheets, additional activities, FRAPPY! materials, and more.

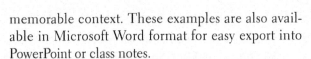

Resources Available with *Updated The Practice of Statistics*

Digital Resources

Sapling Plus/ACHIEVE with Online Homework

This comprehensive online platform combines the e-Book and its innovative high-quality teaching and learning features with our acclaimed online homework program. The platform gives students everything they need to prepare for class and exams, while giving teachers an extensive set of resources to set up and administer their courses efficiently and easily. Exercises from *Updated TPS* 6e as well as additional statistics problems may be assigned and graded using the online homework system, which includes expert guided feedback for students and useful analytics to help teachers guide individual students or the class as a whole. All student and teacher resources are available within the platform, and the integrated e-Book is mobile-ready for use on laptops, tablets, or smartphones. Learn more about our digital options at highschool.bfwpub.com/updatedtps6e.

Tutorial Videos

For Students

These videos are keyed to the student text with the play button icon ▶ and are accessible through a click in the student e-Book (and TE-Book) or by link from the book's digital platform. They feature the following:

- Worked examples: Experienced AP® teachers walk through key examples to give step-by-step instructions, just as teachers do in the classroom.
- Worked exercises: "For Practice, Try . . ." exercises are featured in step-by-step instructional videos to give students help when they need it.
- Chapter Review Exercises: New to the sixth edition, these videos offer additional insights and instruction as students assess their mastery of each chapter and prepare to take the Chapter Test.
- Technology Corners: Detailed, key-stroke-level instructional videos by Leigh Nataro show students how to use their TI-83/84 and TI-Nspire graphing calculators.

For Teachers

- PD videos: Brief chapter and section overviews by authors Daren Starnes and Josh Tabor offer advice on how to teach with the new Course Framework, what to emphasize in a given section, how to anticipate and overcome common pitfalls, as well as tips for helping students realize success in the course and on the AP® Statistics exam.
- Activity Overview videos: Beth Benzing or Doug Tyson gives an overview of each activity and guidance on how to use them most effectively in class.
- Mastering the FRQ with the FRAPPY video: Jason Molesky discusses the process of using the FRAPPY problem to help students prepare for Chapter Tests and the AP® Statistics exam.
- Getting the most from the ATE video: Luke Wilcox offers insights into how to teach a highly effective AP® Statistics course and how the ATE can help.
- Lesson plans: ATE author Luke Wilcox and his colleague, Lindsey Gallas, created daily lesson plans

to help you best teach using *The Updated Practice of Statistics*, Sixth Edition. The lessons and accompanying blog posts can be found at www.statsmedic.com under "150 Days of AP® Stats."

Updated Teacher's Resource Flash Drive (TRFD)
(ISBN: 1-319-26937-0)

The content referenced throughout the ATE by the **TRM** icon is available in digital form on the TRFD. The content may also be accessed through direct links embedded in the TE-Book and as part of the teacher's resources in the book's digital platform. The teacher's resources on the flash drive include:

- All videos included in the ATE that are identified by either a **PD** or play button ▶ icon.
- The TRM and other resources that work in tandem with the ATE, including additional problems, materials for activities, rubrics for projects, and review materials for the AP® Statistics exam.
- One prepared quiz for each section and one test per chapter in PDF format. You can also create your own quiz using the **ExamView® Assessment Suite** that is part of the *Updated TPS* 6e program. Questions are coded by Section, Learning Target, and level of difficulty to make it easy to build parallel quizzes using the Learning Target Keys.
- Word files for the Alternate Examples.
- **Supplemental chapters** in PDF format for use after the AP® Statistics exam:
  - Chapter 13: Analysis of Variance
  - Chapter 14: Multiple Linear Regression
  - Chapter 15: Logistic Regression
- Data files for the larger data sets.

Updated ExamView® Assessment Suite
(ISBN: 1-319-26935-4)

- The ExamView® Test Bank, written by experienced AP® Statistics Readers Erica Chauvet and James Bush (both of Waynesburg University), includes 1600 problems composed of more than 1100 multiple-choice questions and almost 500 free response questions. Each question has been tagged with the *Updated TPS* 6e section and Learning Target(s) to which it applies. The level

of difficulty and suggested points are also included. The test items have been checked for alignment with the text content and notation, overall usability, and accuracy. All questions can be found in the ExamView® Test Generator and may be printed, as desired.

- ExamView® Test Generator guides teachers through the process of creating online or paper tests and quizzes quickly and easily. Users may select from the bank of more than 1600 test questions or opt for the step-by-step tutorial to write their own questions. Questions may be scrambled to create different versions of tests. Tests may be printed in many different formats to provide maximum flexibility or may be administered online using the ExamView® Player. Student results flow to the ExamView® Test Manager to provide a comprehensive assessment management system for the teacher.

Updated Full Solutions

- Written by Erica Chauvet and accuracy-checked by Tonya Adkins, Kathleen Dickinsheets, Ann Cannon, and MathMadeVisible, complete worked solutions to each exercise in the student textbook are available as digital files in our digital platform, on the TRFD, or by clicking on the link in the Teacher's e-Book. Care has been taken to make sure that the language, methodology, and notation are consistent with that used by AP® Readers.

Updated Lecture Presentation Slides

- Created by Doug Tyson of Central York (PA) High School, the PowerPoint presentation slides are a great tool for both new and experienced teachers alike. Written by section, the slides highlight the important ideas in each section and include many examples from the book.

For Students

Updated Strive for a 5: Preparing for the AP® Statistics Examination
(ISBN: 1-319-20990-4)

- Written by Jason Molesky and Michael Legacy and updated by Erica Chauvet and Jeff Eicher, the *Strive for a 5 Guide* is designed for use with *The Practice of Statistics*,

Updated Sixth Edition. It is designed to help students evaluate their understanding of the material covered in the textbook, reinforce the key concepts, develop conceptual understanding and communication skills, and prepare students to succeed on the AP® Statistics exam. This book is divided into two sections: a study guide and a test preparation section. The study guide is written to be used throughout the course, while the prep section offers additional AP® test strategies and includes two full-length AP®-style practice exams—each with 40 multiple-choice questions, 5 free response questions, and an investigative task. Each exam has a full answer key with section references to relevant content in *Updated TPS* 6e and linked to the appropriate Learning Objective, Skill, and Unit from the College Board's Course Framework.

Student Site

- The textbook's Student Site, which may be accessed at highschool.bfwpub.com/updatedtps6e, includes the tutorial videos for the worked examples, "For Practice, Try. . ." exercises, chapter review exercises, and Technology Corners. The applets, data sets, and FRAPPY! Student Samples may also be accessed here.

About the Student Edition

Goals for the Updated Sixth Edition

After getting overwhelmingly positive feedback on the changes we made to the fifth edition of *The Practice of Statistics*, we set several new and continuing goals for the updated sixth edition.

- **Align with the AP® Statistics Course and Exam Description (CED).** This updated version of *TPS* 6e has been revised to perfectly align with the College Board's Fall 2019 Course and Exam Description. We reorganized the table of contents in *Updated TPS* 6e to match the 9 Unit sequence in the AP® Statistics Course Framework (CF). Every learning objective and essential knowledge statement in the CF is covered thoroughly in the text, including all new topics (e.g., mosaic plots, systematic random sampling). The book uses terms, notation, formulas,

and tables consistent with those found in the Course Framework and on the AP® Statistics exam formula sheet. Visit the book's website at highschool.bfwpub.com/updatedtps6e for a detailed alignment guide.

- **Maintain a clear focus on the AP® Statistics exam.** As the only textbook written exclusively for AP® Statistics, our primary goal is to help students achieve success on the AP® Statistics exam while gaining a solid foundation in statistics. Throughout the book, you will find recurring features that make exam preparation a clear priority: the Free Response AP® Problem, Yay! (FRAPPY!) at the end of each chapter, AP®-level multiple-choice questions in every set of section exercises, AP® Exam Tips throughout, model solutions in examples based on official scoring guidelines, an AP® Statistics Practice Test at the end of each chapter, and Cumulative AP® Practice Tests after Chapters 4, 7, 11, and 12. *Updated TPS* 6e also develops the Course Skills for AP® Statistics through frequent repetition. See the inside back cover of the student edition for more details.

- **Put all examples in problem/solution format, with model student solutions separated from teacher commentary.** All the integrated, narrative examples have been converted into problem/solution format, giving students many more examples of AP®-style model solutions. Furthermore, in each example, the voice of the teacher and the voice of the student have been separated, with model solutions in handwritten font and teacher talk in separate dialogue boxes next to the student solution.

- **Keep the book easy to read and easy to use.** We preserved the direct, student-friendly writing style of previous editions. In *Updated TPS* 6e, all key terms are called out in bold and carefully explained in a definition box. A summary of these key terms and definitions is provided in the Glossary/Glosario. "How To" boxes describe step-by-step procedures while Summary boxes recap important results developed in the narrative. Section summaries and the Chapter Reviews distill key vocabulary, concepts, and procedures. Cautions about common misconceptions appear in bold, red type and are labeled with a caution icon in the margin. Detailed answers are provided in a back-of-book appendix to all Check Your Understanding questions, odd-numbered section exercises, all Chapter

Review Exercises, and all Chapter and Cumulative AP® Statistics Practice Test questions.

- **Tighten the alignment between learning targets, examples, and exercises.** The learning targets at the beginning of each section link directly to the examples. Each "For Practice, Try . . ." odd-numbered exercise and its even-numbered counterpart in the Section Exercises mirror the associated example and include a page reference for that example. Tightening the alignment between these three elements encouraged us to carefully evaluate the purpose of each exercise, which resulted in a more robust set of exercises in each section. The tight alignment is reflected in the What Did You Learn? box at the end of each chapter, which maps the learning targets to the examples and to specific Chapter Review Exercises. This approach is extended to the ExamView® test bank, with each question being coded to the corresponding book section and Learning Target.

- **Emphasize that students learn statistics best by *doing* statistics.** As in previous editions, *Updated TPS* 6e includes hands-on activities in every chapter, as well as Check Your Understanding questions at strategic points in each section. Several new Chapter Projects give students an opportunity to practice multivariable thinking by analyzing larger data sets using statistical software.

- **Update examples and exercises with new contexts and current data.** More than 35% of the examples and exercises in *Updated TPS* 6e are new or revised.

- **Expand media resources.** With the growth of instructional videos on the Internet, we wanted to make sure that teachers and students using *Updated TPS* 6e would have a reliable source for video content that is aligned to the textbook and the AP® Statistics exam. We offer tutorial videos for every example, "For Practice, Try . . ." exercise, Technology Corner—in both TI-83/84 and Nspire format—and now for each Chapter Review exercise as well. All these videos were created by veteran AP® Statistics teachers and exam Readers to provide students with help when they need it most.

An enhanced e-Book version of the updated sixth edition is available for use on computers and mobile devices. A new online homework system is available for purchase by teachers who want to assign auto-graded exercises with guided hints and robust feedback provided to students.

- **Revamp external assessment resources.** We overhauled the tests, quizzes, and test bank to help address the challenge of teachers posting textbook assessments online. We developed model quizzes (one per section) and tests (one per chapter) that can be easily modified using items from the enhanced ExamView® test bank. All of the multiple choice and free response items in the test bank are coded to our Learning Targets so that teachers can easily swap out items and frustrate cheaters.

- **Preserve an appealing design.** Today's students are accustomed to operating in a visually stimulating environment. A textbook with an engaging look and feel is more likely to get used regularly. We have once again attempted to create visual appeal without making the pages too busy or distracting. And we kept the flamingos around for another edition!

Content and Structure of *Updated TPS* 6e

We made several important structural changes in the updated sixth edition to ensure tight alignment with the AP® Statistics Course and Exam Description.

- Transforming to achieve linearity moved from Section 12.2 in *TPS* 5e to Section 3.3 in *Updated TPS* 6e, so Chapter 3 now aligns with Unit 2: Exploring Two-Variable Data in the Course Framework.

- Sampling distribution of a difference in proportions moved from Section 10.1 in *TPS* 5e to Section 7.2 in *Updated TPS* 6e; sampling distribution of a difference in means moved from Section 10.2 in *TPS* 5e to Section 7.3 in *Updated TPS* 6e.

- Power moved from Section 9.3 in *TPS* 6e to Section 9.2 in *Updated TPS* 6e.

- Paired data moved from Section 9.3 in *TPS* 5e to Section 10.2 (confidence intervals) and Section 11.2 (significance tests) in *Updated TPS* 6e.

- Inference completely reorganized to match Course Framework units:

CF Unit	Title	*Updated TPS* 6e
6	Inference for Categorical Data: Proportions	Chapter 8 Estimating Proportions with Confidence Chapter 9 Testing Claims about Proportions

In the fifth edition, we introduced the 4-step State-Plan-Do-Conclude process in Chapter 1 as a framework for solving a statistical problem. Because this 4-step process is mainly intended to help students answer inference questions on the AP® Statistics exam, we have restricted its use to Chapters 8–12 in *Updated TPS* 6e. We also use a streamlined 2-step process for Normal distribution and binomial distribution calculations in the updated sixth edition that matches recent AP® exam rubrics.

Technology Corners in *Updated TPS* 6e feature the TI-83/84. TI-Nspire instructions are on the book's Student Site at highschool.bfwpub.com/updatedtps6e.

Following are specific content changes made in the updated sixth edition based on the AP® Statistics Course Framework.

Chapter 1

- Added mosaic plots.
- Added margin note discussing mean ±2 or 3 SD criterion for outliers.
- Emphasized terminology:
  - discrete versus continuous quantitative variables
  - descriptive statistics versus inferential statistics
  - statistic versus parameter
  - unimodal (single-peaked), bimodal (double-peaked), approximately uniform
  - variance versus standard deviation

Chapter 2

- Changed definition of percentile to *less than or equal to*, with equivalent changes to cumulative relative frequency graphs.
- Emphasized terminology:
  - empirical rule (68–95–99.7 rule)

Chapter 3

- Removed the term *outlier* in Section 3.1, as it has a more restrictive meaning in the Course Framework; replaced with the phrase *unusual value*.
- Changed the equation of the least-squares regression line to $\hat{y} = a + bx$.
- Introduced the term *high-leverage point* as one whose x value is far from $\bar{x}$. Also, restricted the use of *outlier* to points with big residuals. Both high-leverage points and outliers can be influential points.
- Emphasized terminology:
  - univariate versus bivariate data

Chapter 4

- Added systematic random sampling.
- Emphasized terminology:
  - homogeneous versus heterogeneous
  - prospective versus retrospective observational studies

Chapter 5

- Changed *chance process* to *random process*.
- Emphasized terminology:
  - trial
  - empirical versus theoretical probability
  - disjoint (mutually exclusive)
  - joint probability

Chapter 6

- Expanded discussion of 10% condition and independent observations.
- Increased coverage of geometric random variables to include shape, center (mean), and variability (SD) of the probability distribution.
- Emphasized terminology:
  - probability distribution versus population distribution; parameter
  - cumulative probability distribution
  - linear transformation versus linear combination

Chapter 7

- Increased emphasis on the 10% condition, including the idea that the traditional formulas for $\sigma_{\hat{p}}$ and $\sigma_{\bar{x}}$

overestimate the actual SD when sampling without replacement (but the difference is negligible when the 10% condition is met).

- Emphasized terminology:
  - point estimator
  - randomization distribution
  - independent observations versus independent samples

Chapters 8–11

- For each inference procedure, the conditions are introduced by stating that we need to check for independence in data collection methods (Random condition, 10% condition when sampling without replacement) and that the sampling distribution is approximately Normal (Large Counts, Normal/Large Sample).

Chapter 8

- Emphasized that a confidence interval is an interval estimate of plausible values for a parameter *based on sample data*.
- Emphasized that a confidence level represents the long-run capture rate *when conditions are met*.
- Modified generic formula for a confidence interval to use "standard error" rather than "standard deviation" to align with the new formula sheet.
- Emphasized that a confidence interval for a proportion can be used to create other intervals (e.g., for the total number of successes in a population).

Chapter 9

- Modified generic formula for a standardized test statistic to include standard deviation (error) of statistic in the denominator to align with the new formula sheet.
- Emphasized that the *P*-value is the probability of getting evidence for H_a as strong as or stronger than the observed evidence when H_0 is true, *assuming the probability model is valid (conditions are met)*.
- Adjusted the decision rule for rejecting H_0 to *P*-value $\leq \alpha$ instead of *P*-value $< \alpha$.
- For a two-sample z test of H_0: $p_1 - p_2 = 0$, the Large Counts condition now uses expected counts

calculated with the pooled (combined) proportion instead of observed counts.

Chapters 10 and 11

- Encouraged use of df from technology for two-sample t procedures.
- Placed paired data procedures immediately after two-sample procedures to help students differentiate these methods.

Chapter 12

- Emphasized that the chi-square statistic measures how far observed counts are from expected counts, *relative to expected counts*.
- Changed the equation of the population regression model to $\mu_Y = \alpha + \beta x$.
- Modified the Normal condition for inference about slope: At each x value, the distribution of y values is approximately Normal OR $n \geq 30$.
- Added mention of a one-parameter model (with y intercept = 0) that uses df = $n - 1$ (otherwise df = $n - 2$ when estimating both the slope and y intercept).

Why Did You Do That?

The organization of *Updated TPS* 6e now mirrors that of the College Board's Fall 2019 AP® Statistics Course Framework to support adopters' use of the College Board's Personal Progress Checks. There is no single best way to organize a first course in statistics. That said, our choices reflect thinking about both content and pedagogy, along with the unit structure of the AP® Statistics Course Framework. Here are comments on several "frequently asked questions" about the order and selection of material in our *Updated TPS* 6e program.

Why not begin with data production?

It is certainly reasonable to do so—the natural flow of a planned study is from study design to data analysis to inference. In fact, the book is structured so that you can start with Chapter 4 on designing studies if you wish. We put data analysis first because this placement matches Units 1 and 2 of the AP® Statistics Course Framework. Also, the graphical and

numerical techniques introduced there are useful even for data that are not produced according to a careful plan.

Why do Normal distributions appear so early in the book?

Density curves such as the Normal curves are just another tool to describe the distribution of a quantitative variable, along with dotplots, stemplots, histograms, and boxplots. We prefer not to suggest that Normal distributions are tied to probability, as the traditional order does. We also find it very helpful to break up the indigestible lump of probability that troubles students so much. Encountering Normal distributions early helps to do so and strengthens the "probability distributions are like data distributions" way of approaching probability. Furthermore, Normal distributions are included at the end of Unit 1 in the AP® Statistics Course Framework.

What happened to inference about a population mean μ when σ is known?

Early editions of the text introduced confidence intervals and significance tests about a population mean μ in an "idealized" setting where the population standard deviation σ is known. Doing so allows use of the standard Normal curve and z procedures. Unfortunately, this approach requires considerable time, is rarely used in practice, and sometimes confuses students later when we expect them to use z for proportions and t for means.

Why did you choose this inference sequence?

To match the AP® Statistics Course Framework! Students find the ideas and methods of estimating an unknown parameter much more straightforward than the logic of testing a claim about a parameter. That's a pedagogical argument for introducing confidence intervals first, and then discussing significance tests. We follow that approach in performing inference about one proportion and a difference in proportions, with confidence intervals in Chapter 8 and significance tests in Chapter 9. Then we use the same sequence in performing inference about one mean and a difference in means, with confidence intervals in Chapter 10 and significance tests in Chapter 11. Finally, we address inference about relationships between two categorical variables or two quantitative variables in Chapter 12.

Getting Started

Before You Begin Teaching AP® Statistics

Teaching AP® Statistics can be a tremendously rewarding experience, but it takes a lot of preparation. If you are like most teachers of AP® Statistics, you probably took few, if any, statistics courses in college. And even if you have some statistics courses in your background, they were probably very different from the AP® course.

Don't worry! The *Updated Annotated Teacher's Edition* includes an abundance of resources to help you teach AP® Statistics. There are also many other resources available from the College Board and other groups to make your life easier.

Here is a list of things you should do before you start teaching the course, if possible:

1. Attend a College Board–approved workshop. Ideally, you should attend a one-week summer institute. These institutes are held all over the country and are led by consultants endorsed by the College Board, including Daren, Josh, and Luke. There are also one- and two-day workshops offered by the College Board during the school year.

2. Download and read the AP® Statistics Course and Exam Description (CED), available at apcentral.collegeboard.org/pdf/ap-statistics -course-and-exam-description.pdf. This document contains the most important information you need to know, including an introduction to the course, an overview of course content, information about the AP® Statistics exam, and sample questions.

3. Participate in the AP® Teacher Community, an online forum for AP® Statistics teachers (apcommunity.collegeboard.org/web/apstatistics). Here, you can ask questions, get advice, give suggestions, and share resources. If you are like most other AP® Statistics teachers, you are the only one at your school who teaches statistics. The community is a great remedy for this problem. Plus, the authors are regular contributors.

4. Visit www.statsmedic.com regularly and sign up to be on its email list. This site offers regular blog posts about statistics content and pedagogy and includes resources specifically designed for *The Practice of Statistics*, Sixth Edition. Click on "150 Days of AP® Stats" to see daily lesson plans for *TPS* 6e—including an activity for every new lesson. Also consider joining the "Stats Medic Teacher Community" group on Facebook to collaborate with other AP® Statistics teachers who use the Experience First, Formalize Later teaching approach.

5. Complete the AP® Statistics Course Audit. See the AP® Coordinator at your school or go to cb.collegeboard.org/ap-course-audit/index .html for more information. A sample syllabus and pacing guide are included in the Teacher's Resource Materials.

Additional Internet Resources

In addition to the book's Student Site at highschool .bfwpub.com/updatedtps6e, AP® Central at apcentral .collegeboard.org/courses/ap-statistics, and the AP® Teacher Community at apcommunity.collegeboard.org/web /apstatistics, there are many other websites that AP® Statistics teachers have found useful. Here is a short list of some of the best:

1. Khan Academy has developed its own collection of student resources for AP® Statistics at khanacademy.org/math/ap-statistics. The site includes short videos for all the topics on the AP® Statistics exam, along with AP®-style practice questions.

2. On the last Thursday of the month, there is a Twitter chat using #statschat. Many of the leading statistics educators in the country participate.

3. The American Statistical Association (ASA) at amstat.org/ASA/Education/home.aspx has a number of great resources for teachers, including recordings of past education webinars and access to publications such as the *Journal of Statistics Education, Statistics Teacher, Significance Magazine,* and *Chance Magazine.* The ASA also sponsors a Poster and Project competition for K–12 students.

4. Census At School at amstat.org/censusatschool. Use their random selector to get data about a random sample of American or international students who have completed the Census At School survey. A great source for real data about students.

5. Gapminder at gapminder.org. Contains some amazing data and a collection of videos by the late Hans Rosling and others, including *The Joy of Stats.*

Note: Links to websites with applets are included in the front matter for each chapter and in margin notes throughout the ATE.

Sequencing and Pacing

With one possible exception (see below), we advise that you go through the chapters and sections in order. The decisions about the arrangement of topics in the updated sixth edition were made after careful thought and discussion, and they reflect the desire to keep broad topics together in chapter clusters (e.g., probability in Chapters 5–7) and the need to build new content on the foundation of earlier content.

Alternate Organization

Some experienced teachers choose to teach Chapter 4: Collecting Data at the start of the school year. Because data must be collected before they can be analyzed, it might make sense to learn how to properly collect data before learning about methods of analysis. This fits with the 4-step statistical process suggested by the *Guidelines for Assessment and Instruction in Statistics Education* (amstat.org/asa/education/Guidelines-for-Assessment -and-Instruction-in-Statistics-Education-Reports.aspx): Formulate questions, collect data, analyze data, interpret results.

Starting with Chapter 4 also sends a clear message to students that this isn't a typical math class—there is more reading, writing, and thinking than they might be accustomed to. On the other hand, some of the material in this chapter is challenging and many teachers choose to keep Chapter 4 in its place so students can be properly "warmed up" before tackling these difficult topics. Also,

if you typically have students adding courses during the first few weeks of school, starting with Chapter 1 might make more sense because this material should be easier to make up for students who join the course late. Finally, if you have students who aren't native speakers of English, Chapter 4 can be a challenge because of its more verbal nature. If you do choose to begin with Chapter 4, you will need to adjust assignment of the PPCs accordingly.

Pacing Guide

Pacing is often one of the biggest challenges for new (and experienced) teachers of AP® Statistics. Depending on your schedule and when the school year starts, you may not have much time to spare before the AP® Statistics exam in early May. The table below shows a suggested pacing guide based on a schedule with 110 class sessions before the AP® exam.

Chapter	Class sessions
1	9
2	7
3	10
4	11
5	8
6	9
7	8
8	7
9	9
10	6
11	7
12	9
Midterm	3
Review for AP® exam	7
Total	110

If you have more (or fewer) class sessions, you should adjust your schedule accordingly. Specific pacing guides with learning targets and suggested assignments are provided at the beginning of each chapter and in the Sample Syllabus and Pacing Guide found in the Teacher's Resource Materials. The Teacher's Resource Materials also include a pacing guide for a "semesterized" block schedule, courtesy of AP® Statistics teacher Doug Tyson.

Preparing Students for the AP® Statistics Exam

Unfortunately, there is no foolproof way to prepare students for success on the AP® Statistics exam. By choosing *The Updated Practice of Statistics*, however, you are off to a good start; it was written by high school teachers specifically for high school students in an AP® Statistics class. Here are some of the features of the program that are designed specifically to help students excel on the AP® exam:

- Every topic on the official AP® Statistics Course Framework is covered in detail and in the College Board's suggested order.

- For each chapter and section, teachers may access a video overview presented by the authors. Look for the PD icon throughout the *Annotated Teacher's Edition*.

- For every activity in the text, teachers may access a video created by an experienced AP® Statistics teacher and Reader offering insights and advice on how to use the activity effectively. Look for the gray "play" icon throughout the *Annotated Teacher's Edition*.

- Students can access short videos presented by veteran AP® Readers for all examples and corresponding exercises, plus all Technology Corners and Chapter Review exercises. Great care was taken to ensure that the solutions provided in the videos meet (and even exceed!) the standards on recent AP® exam rubrics. Look for the blue "play" icon throughout the student edition.

- Throughout the student text and *Annotated Teacher's Edition*, you will find AP® Exam Tips, including a cumulative set of tips in Appendix A.

- At the end of each chapter in the student textbook, there is an AP® Statistics Practice Test with multiple choice and free response questions like those found on the real AP® Statistics exam.

- At the end of each chapter in the student textbook, there is a FRAPPY! (Free Response AP® Problem, Yay!), which asks students to complete free response questions like those found on the AP® Statistics exam. After completing the questions, students can

look at sample responses on the book's Student Site and discuss the good and bad parts of each response. A model solution, scoring rubric, and commentary on the student samples are available for teachers in the Teacher's Resource Materials. For a more detailed explanation of the FRAPPY! process, watch the video from Jason Molesky.

- At the end of each set of exercises, we have provided "Recycle and Review" questions that integrate content from earlier chapters to reinforce what students learned previously and to help them make connections from chapter to chapter. In some cases, these exercises also foreshadow upcoming material.

- At the end of Chapters 4, 7, 11, and 12, there is a *Cumulative* AP® Statistics Practice Test with multiple choice and free response questions like those found on the real AP® exam.

- Additional helpful tools in the Teacher's Resource Materials include a worksheet that asks students to choose the correct inference procedure in a given setting and a set of 120 flash cards that review important topics from throughout the course. Make sure to use these!

Here are some additional strategies for preparing students for the AP® Statistics exam:

- Make reading and writing an integral part of your course. On the AP® exam, students are often surprised at how little they have to calculate. Most questions involve quite a bit of reading and writing. Make sure your students get plenty of practice throughout the year and show them previous AP® questions so they understand the high expectations of AP® Readers. Also, emphasize the correct use of statistical vocabulary—incorrect use of statistical terms can lower scores.

- Make sure your students understand the structure of the AP® Statistics exam.
  - Section I consists of 40 multiple choice questions to be answered in 90 minutes. This section accounts for 50% of the exam score. There is no penalty for guessing, so make sure your students answer every question.
  - Section II consists of six free response questions to be answered in 90 minutes. This section accounts for 50% of the exam score. The first five questions

should take about 13 minutes each and count for 75% of the Section II score. The last question, called the investigative task, should take about 25 minutes and is worth 25% of the Section II score. According to the AP® Statistics Course and Exam Description, "The investigative task assesses multiple skill categories and content areas, focusing on the application of skills and content in new contexts or in non-routine ways."

- Shortly before taking the AP® Statistics exam, have students come in after school or on the weekend to take a full-length AP® exam for practice. It is beneficial for students to experience a real exam, given under real conditions, so they can get a sense of how quickly they should work and what topics they should review. If possible, administer the practice exam in the same room in which students will take the actual exam. Currently, the complete 2002, 2007, and 2012 AP® Statistics exams are available from the College Board. Also, the international versions of the exam are available for recent years through your AP® audit account (look for the link to "Secure Documents" after logging in). These forms are secure, so students can't find them on AP® Central. Please help keep them secure by only using them in class and collecting them when you are done! The international version of the exam is different from the version administered to the majority of students in the United States and Canada. However, it is comparable both in quality and difficulty to other versions. There is also a "Practice Exam" available through your audit account, but this exam was never used with actual students and is no longer considered secure by most teachers. *Note:* It is best to decide at the beginning of the year which exams you will be using as your Practice Exam and Final Exam for your course, as you will want to avoid using any of these questions throughout the year. If you are using the College Board Question Bank to build assignments or assessments, be aware that all of these AP® questions are part of the pool of questions.

- Use as many actual AP® questions as possible. In addition to the correlation we provide that links AP® questions to chapters in the book, you can also search for AP® questions in a variety of ways using the Question

Bank in your AP® Classroom account. There are several ways to incorporate previous free response questions into your course. Some teachers include one or two AP® free response questions on each test. Other teachers assign them for homework or in cooperative groups during class. If you assign them for homework, be aware that students can access the solutions online. However, because there are more than 200 released free response questions, it's probably safe to include one or two on an exam. To help choose which questions to use, we have included a list of released AP® questions that are appropriate for each chapter in the *Annotated Teacher's Edition*. You can find the lists in the blue pages that precede each chapter. Also, the Teacher's Resource Materials contain two different sortable AP® Free Response Question Indices with all the questions listed in one place. This list will be updated each year on our digital platform and at statsmedic.com/free-response-questions. *Note:* Questions from the secure versions of the AP® Statistics exam are not included in these indices.

- Make sure students know how the AP® Statistics exam is graded. Show them the scoring guidelines (rubrics) and examples of student responses after you answer old AP® questions in class. The rubrics and examples of student responses (with commentary) are available at apcentral.collegeboard.org/courses/ap-statistics/exam.

- Early in the year, give students a copy of the formula sheets and tables that are provided on the AP® Statistics exam so they can familiarize themselves with them. You can find these materials in the AP® Statistics Course and Exam Description. Be sure that you are using the new formula sheet developed for the 2020 exam. Many teachers also copy these on colored card stock for students to use on chapter tests. It is better for students to spend their time learning which formula to use and how to use it, rather than spending time to memorize the formulas. We have also placed the formula sheets and tables in the back of the student text for easy reference.

- Give challenging assessments that mimic the AP® Statistics exam. Make sure to include multiple choice and free response questions and don't be afraid to ask difficult questions. Also, don't be afraid to be a strict grader—the rubrics on the real AP® Statistics

exam are often quite strict. To compensate for these practices, you will probably need to adjust the scores on your assessments. Experience will help you know how much to adjust them, but remember that getting a raw score of 70 out of 100 will almost certainly earn a student a 5 on the AP® Statistics exam.

- Give cumulative assessments. One of the hardest things for students to do on the AP® exam is to identify which method is appropriate for a specific problem. Giving students cumulative assessments will provide them with practice making these tough decisions and will keep the material fresh in their minds. Some teachers include cumulative questions on each exam, while others give cumulative exams every couple of chapters. The review questions at the end of each set of exercises and the cumulative AP® Practice Tests in the text should help as well.

- Have students use an AP® preparation book, such as *Strive for a 5 Guide*, which was written specifically for *The Practice of Statistics* by Jason Molesky and Michael Legacy and updated by Erica Chauvet and Jeff Eicher.

Assessments

Quizzes and Tests

Although we have provided a quiz for every section and a test for every chapter, you probably won't have time to use all of them. And while some teachers find time for quizzes, many teachers formally assess their students only at the end of each chapter. If you are new to teaching AP® Statistics, it is better to err on the side of fewer days of assessment so you have more time to cover the content. It will also save you time grading! Furthermore, don't forget our advice about being a strict grader and scaling the test scores like graders do on the real AP® Statistics exam.

The tests and quizzes in the Teacher's Resource Materials are designed to be ready to go and the *Updated TPS* 6e ExamView® Test Bank gives you the power to create many alternate assessments.

IMPORTANT: *Please do not post any of the questions or solutions online, because some students looking to cheat search online for questions and answers. We receive many emails and complaints from teachers who have discovered that their students*

found answers online. Once posted, they are readily available to all AP® Statistics students with access to the Internet.

Personal Progress Checks (PPC) and the Progress Dashboard

Teachers can assign the Personal Progress Checks at checkpoints throughout the year to assess student understanding. Students access these formative assessments through their College Board accounts. Teachers can then use the Progress Dashboard on AP® Classroom to access the results from these assessments, allowing teachers to pinpoint areas of strength and weakness in their students.

Question Bank

The College Board has created an extensive pool of questions, which includes released AP® Statistics exam multiple choice and free response questions, as well as many newly developed items. The questions are part of a searchable test bank, allowing teachers to create custom assignments and assessments to use with their students (online or printed). Before selecting items from the question bank, you should note where the item came from, reserving items from the most recent couple of secure exams so they can be used as a complete mock exam or a final exam in May.

Homework

The number of exercises available in the textbook should greatly exceed the number you would assign in any given school year. In fact, as teachers get more experience with AP® Statistics, they tend to assign fewer and fewer homework problems. It is better to have students spend time doing a few problems in depth, rather than giving shallow answers to a large set of problems. This philosophy is reflected in the suggested homework assignments provided at the beginning of each chapter.

The exercise sets are designed so that the problems generally come in pairs. For example, Problems 1 and 2 will cover the same concepts but in different contexts. This way, teachers have the flexibility to assign odd-numbered problems (with the solutions in the back of the student text) or to assign even-numbered problems. The suggested assignments generally include odd-numbered problems because we want students to get immediate feedback about their work and make corrections

before the next class. *Note:* In addition to answers to the odd-numbered section exercises, the back-of-book solutions include answers to *all* Check Your Understanding questions, Chapter Review exercises, Chapter Tests, and Cumulative AP® Practice Test exercises.

Finally, there is no consensus on how best to grade and count homework. Some teachers collect homework daily, while others collect it only occasionally and at random. Some teachers grade homework problems for correctness, while others grade it only for effort. Often, this decision is based on your own preferences and school culture. Using the new online homework platform that accompanies this text may solve some of your grading challenges.

Activities and Projects

One of the best parts about teaching AP® Statistics is that it is much easier to incorporate activities and projects than it is in other math classes. Activities and projects provide one of the best ways to help students learn, hands-on, the important concepts in AP® Statistics.

To emphasize the importance of activities, we often start chapters with a classroom-tested activity designed to introduce students to some of the most important ideas in that chapter. Included in the *Annotated Teacher's Edition* is an Activity Overview for each activity that provides teachers with tips for maximizing efficiency and learning. Although you may feel overwhelmed trying to get through the content, experienced teachers agree that getting students engaged in activities is time well spent.

Projects are a great way for students to put their knowledge to work. Suggestions for projects are included in the student edition at the end of Chapters 1, 3, 4, 11, and 12, with rubrics available in the Teacher's Resource Materials. Although it may not be possible to assign each of these projects, many teachers try to assign one project per semester.

Current Events

Another great part of teaching AP® Statistics is that there is no shortage of real-life applications. Almost every day, news articles describe current polls, studies, or experiments that you can discuss with your students. When students see that the ideas they are learning about get used in the real world, their motivation is sure to increase! A great resource for finding statistics related to current events is

the AP® Statistics Teacher Community (see "Before You Begin Teaching AP® Statistics" above). Several times a week, someone posts a link to an article that can be used in class. Alternatively, you can offer extra credit to students for finding articles related to statistics.

Technology

Technology is a vital part of AP® Statistics. Gone are the days when students in a statistics class spent most of their time calculating means, standard deviations, and correlations by hand. Instead, students in an AP® Statistics class are expected to use technology to do these calculations so they can focus on understanding and interpreting what they are calculating. There are a plethora of online applets that can perform all the calculations needed for this course.

Graphing Calculators

All students are expected to have a graphing calculator with the capability of making statistical graphs and performing calculations, such as finding the standard deviation of a set of data or the equation of a least-squares regression line. Students are allowed to use a graphing calculator on all parts of the AP® Statistics exam and should definitely be familiar with its uses by the time they take the exam.

Because the TI-83/84 is the most prominently used calculator in high school, *The Updated Practice of Statistics* includes Technology Corners throughout the student text, with step-by-step instructions for using this calculator. Instructions for the TI-Nspire can be found on the book's Student Site, highschool.bfwpub.com/updatedtps6e.

Although using technology is essential, teachers have different opinions about when to introduce students to that technology. For example, some teachers want students to have the experience of calculating things by hand before they show students how to calculate them with technology. In some cases, like calculating the standard deviation, having students do it by hand once or twice may be useful because the calculations reinforce what the standard deviation is all about (deviations from the mean).

The extent to which students should use the Normal table is another subject of debate regarding technology. Some teachers strongly prefer to have students calculate z-scores and use the Normal table, whereas other teachers quickly move to the normalcdf function on the calculator.

You can make this decision for yourself based on your own preferences. Make sure to pay attention to the AP® Exam Tips that often accompany the Technology Corners so that your students know how to properly show their work.

Computers

In many college statistics courses, students are asked to use statistical software, such as Minitab, rather than graphing calculators. Unlike high schools, colleges frequently have computer labs open to students at all times or have classes that meet in computer labs. Also unlike most high schools, they have the money to afford statistical software for their labs!

To help ensure that colleges continue to give credit to students who pass the AP® Statistics exam, questions on the AP® exam often include output from statistical software. In *Updated TPS* 6e, we have included examples of computer output throughout the book so students will be familiar with these types of questions.

Ideally, students in AP® Statistics will have the opportunity to use computer software on a regular basis. In fact, the projects following Chapters 1, 3, and 11 were designed to be completed using computer software. At a minimum, you should have a copy of at least one program that you can use for demonstrations in class. There are also some free statistical software packages and applets available on the Internet that you can use to create graphs and generate computer output.

After the AP® Statistics Exam

What to Teach

If you still have time with your students after the AP® Statistics exam, there are several options:

1. Assign students the Final Project introduced at the end of Chapter 12. This is an ideal time for students to bring together all the topics from the course. Ask them to design a study, conduct the study, analyze the data, and make the appropriate inferences. This project assignment and rubric are included in the Teacher's Resource Materials.

2. Learn some additional statistical topics. The Teacher's Resource Flash Drive and the textbook's website include three additional chapters:

- Chapter 13: Analysis of Variance (comparing more than two means)
- Chapter 14: Multiple Linear Regression (using more than one explanatory variable to predict a response variable)
- Chapter 15: Logistic Regression (predicting a categorical response variable)

3. Teach your students how to use computer software. Some statistical software can be used for free (some for a limited time) and providing students experience with computer software will be helpful to them in college. If you didn't already have students do the projects following Chapters 1, 3, and 11, this would be a great time to complete them.

Recruiting Students for Next Year

In many cases, students have no idea what a statistics class is all about. It is your job to promote the hands-on learning and real-world applications that make AP® Statistics a class worth taking! Make sure to check out the great resources at thisisstatistics.org for videos, fliers, and other information. To recruit students at your school, consider visiting upper-level math classes like Algebra II and above, and give the students a short presentation about taking statistics. When recruiting potential students, try to hit three main themes:

1. It is a fresh start: This isn't a typical math class. You don't have to remember how to factor a polynomial, graph a sine curve, or prove that triangles are congruent. However, this doesn't mean the class will be easy! You will get a chance to sharpen critical thinking and communication skills.

2. It is great preparation for college: Many majors require students to take a statistics course and be familiar with the methods of analysis in AP® Statistics. Also, AP® Statistics is taught over a full year (instead of a semester in college), so you have more time to learn the material. It helps if you can gather a list of majors at your local university that require a statistics course.

3. Students will *never* have to ask "When will we ever use this?" Problems are almost always based on real-world studies in a variety of fields. We will read current articles and learn how to think critically about claims made in the media.

I also try to give prospective students an example or two of the types of studies we design and analyze so they have a more concrete understanding of what actually happens in class. For example: Is there an association between sex and superpower preference? How does the wording of a survey question affect the responses? Is yawning contagious?

Many students will ask you how AP® Statistics compares to AP® Calculus and which class they should take. Here are some questions to consider discussing with these students:

1. *What credit or placement will you get for my AP® math class?* Each college/university has a policy about credit and/or placement awarded for specific AP® exams. At most schools that give credit, earning a "passing" score on the AP® Statistics or AP® Calculus exams results in one semester of college credit. The key might be which math course is required for your chosen major.

2. *What college major do you plan to pursue?* If you plan to pursue a degree in mathematics, physics, or chemistry, you will be expected to take calculus in college. If you select an English, history, fine arts, foreign language, life science, or social science major, you are more likely to need a statistics course. Business majors will probably have to take both calculus and statistics. You want to choose a course that sets you up well for your college math requirements.

3. *What is the difference between AP® Calculus and AP® Statistics?*

AP® Calculus	AP® Statistics
Graphical, numerical, and algebraic	Collecting and analyzing data
Builds on precalculus concepts	Computation de-emphasized
Computational proficiency helps	Focus on communication and interpretation
Emphasizes techniques, applications	Writing is critical

4. *Still can't decide?* Take both courses!

Student Edition Preface

UPDATED
to reflect the revised Course Framework

for the AP® Exam

The

Practice of
Statistics

SIXTH EDITION

Daren S. Starnes

The Lawrenceville School

Josh Tabor

Canyon del Oro High School

bedford, freeman & worth
high school publishers

Boston | New York

Contents

Additional Online Chapters

About the Authors

Ann Heath

DAREN S. STARNES is Mathematics Department Chair and holds the Robert S. and Christina Seix Dow Distinguished Master Teacher Chair in Mathematics at The Lawrenceville School near Princeton, New Jersey. He earned his MA in Mathematics from the University of Michigan and his BS in Mathematics from the University of North Carolina at Charlotte. Daren is also an alumnus of the North Carolina School of Science and Mathematics. Daren has led numerous one-day and weeklong AP® Statistics institutes for new and experienced teachers, and he has been a Reader, Table Leader, and Question Leader for the AP® Statistics exam since 1998. Daren is a frequent speaker at local, state, regional, national, and international conferences. He has written articles for *The Mathematics Teacher* and *CHANCE* magazine. From 2004 to 2009, Daren served on the ASA/NCTM Joint Committee on the Curriculum in Statistics and Probability (which he chaired in 2009). While on the committee, he edited the *Guidelines for Assessment and Instruction in Statistics Education* (GAISE) pre-K–12 report and coauthored (with Roxy Peck) *Making Sense of Statistical Studies*, a capstone module in statistical thinking for high school students. Daren is also coauthor of the popular on-level text *Statistics and Probability with Applications*.

Ann Heath

JOSH TABOR has enjoyed teaching on-level and AP® Statistics to high school students for more than 22 years, most recently at his alma mater, Canyon del Oro High School in Oro Valley, Arizona. He received a BS in Mathematics from Biola University, in La Mirada, California. In recognition of his outstanding work as an educator, Josh was named one of the five finalists for Arizona Teacher of the Year in 2011. He is a past member of the AP® Statistics Development Committee (2005–2009) and has been a Reader, Table Leader, and Question Leader at the AP® Statistics Reading since 1999. In 2013, Josh was named to the SAT® Mathematics Development Committee. Each year, Josh leads one-week AP® Summer Institutes and one-day College Board workshops around the country and frequently speaks at local, national, and international conferences. In addition to teaching and speaking, Josh has authored articles in *The American Statistician, The Mathematics Teacher, STATS Magazine,* and *The Journal of Statistics Education*. He is the author of the *Annotated Teacher's Edition* and *Teacher's Resource Materials* for *The Practice of Statistics,* Fourth Edition and Fifth Edition. Combining his love of statistics and love of sports, Josh teamed with Christine Franklin to write *Statistical Reasoning in Sports,* an innovative textbook for on-level statistics courses. Josh is also coauthor of the popular on-level text *Statistics and Probability with Applications*.

Content Advisory Board and Supplements Team

Ann Cannon, Cornell College, Mount Vernon, IA
Content Advisor, Accuracy Checker

Ann has served as Reader, Table Leader, Question Leader, and Assistant Chief Reader for the AP® Statistics exam for the past 17 years. She is also the 2017 recipient of the Mu Sigma Rho William D. Warde Statistics Education Award for a lifetime devotion to the teaching of statistics. Ann has taught introductory statistics at the college level for 25 years and is very active in the Statistics Education Section of the American Statistical Association, currently serving as secretary/treasurer. She is coauthor of *STAT2: Modeling with Regression and ANOVA* (W. H. Freeman and Company).

Luke Wilcox, East Kentwood High School, Kentwood, MI *Content Advisor, Teacher's Edition, Teacher's Resource Materials*

Luke has been a math teacher for 15 years and is currently teaching Intro Statistics and AP® Statistics. Luke recently received the Presidential Award for Excellence in Mathematics and Science Teaching and was named Michigan Teacher of the Year 2016–2017. He facilitates professional development for teachers in curriculum, instruction, assessment, and strategies for motivating students. Lindsey Gallas and Luke are the co-bloggers at The Stats Medic (www.statsmedic.com), a site dedicated to improving statistics education, which includes activities and lessons for this textbook.

Erica Chauvet, Waynesburg University, PA
Solutions, Tests and Quizzes, Test Bank, Online Homework, Strive for a 5 Guide

Erica has more than 15 years of experience in teaching high school and college statistics and has served as an AP® Statistics Reader for the past 10 years. Erica famously hosts the two most highly anticipated events at the Reading: the Fun Run and the Closing Ceremonies. She has also worked as a writer, consultant, and reviewer for statistics and calculus textbooks for the past 10 years.

Doug Tyson, Central York High School, York, PA
Content Advisor, Videos and Video Program Manager, Lecture Slide Presentations

Doug has taught mathematics and statistics to high school and undergraduate students for more than 25 years. He has taught AP® Statistics for 11 years and has been active as an AP® Reader and Table Leader for a decade. Doug is the coauthor of a curriculum module for the College Board and the *Teacher's Edition* for *Statistics and Probability with Applications*, Third Edition. He conducts student review sessions around the country and leads workshops on teaching statistics. Doug also serves on the NCTM/ASA Joint Committee on Curriculum in Statistics and Probability.

Beth Benzing, Strath Haven High School, Wallingford/Swarthmore School District, Wallingford, PA
Activity Videos

Beth has taught AP® Statistics since 2000 and has served as a Reader for the AP® Statistics exam for the past 7 years. She served as president, and is a current board member, of the regional affiliate for NCTM in the Philadelphia area and has been a moderator for an online course, Teaching Statistics with Fathom. Beth has an MA in Applied Statistics from George Mason University.

Paul Buckley, Gonzaga College High School, Washington, DC
Videos, Online Homework

Paul has taught high school math for 24 years and AP® Statistics for 16 years. He has been an AP® Statistics Reader for 10 years and a Table Leader for the past 4 years. Paul has presented at Conferences for AP®, NCTM, NCEA (National Catholic Education Association), and JSEA (Jesuit Secondary Education Association) and has served as a representative for the American Statistical Association at the American School Counselors Association annual conference.

James Bush, Waynesburg University, Waynesburg, PA
Test Bank, Videos

James has taught introductory and advanced courses in statistics for over 35 years. He is currently a Professor of Mathematics at Waynesburg University and is the recipient of the Lucas Hathaway Teaching Excellence Award. James serves as an AP® Table Leader, leads AP® Statistics preparation workshops through the National Math and Science Initiative, and has been a speaker at NCTM, USCOTS, and the Advance Kentucky Fall Forum.

Monica DeBold, Harrison High School, Harrison, NY
Videos

Monica has taught for 10 years at both the high school and college levels. She is experienced in probability and statistics, as well as AP® Statistics and International Baccalaureate math courses. Monica has served as a mentor teacher in her home district and, more recently, as an AP® Statistics Reader.

Lindsey Gallas, East Kentwood High School, Kentwood, MI
Videos

Lindsey has recently begun teaching AP® Statistics after spending many years teaching introductory statistics and algebra. Together with Luke Wilcox, Lindsey has created www.statsmedic.com, a site about how to teach high school statistics effectively—which includes daily lesson planning and activities for this textbook.

Vicki Greenberg, Atlanta Jewish Academy, Atlanta, GA
Videos

Vicki has taught mathematics and statistics to high school and undergraduate students for more than 18 years. She has taught AP® Statistics for 10 years and served as an AP® Reader for 7 years. She is the co-author of an AP® Statistics review book and conducts student review sessions and workshops for teachers. Her educational passion is making mathematics fun and relevant to enhance students' mathematical and statistical understanding.

DeAnna McDonald, University of Arizona, Tucson, AZ
Videos

DeAnna has taught introductory and AP® Statistics courses for 20 years. She currently teaches statistics as an adjunct instructor at the University of Arizona in the Mathematics Department and taught AP® Statistics at University High School in Tucson for many years. DeAnna has served as an AP® Statistics Reader for 12 years, including 4 years as a Table Leader.

Leigh Nataro, Moravian Academy, Bethlehem, PA
Technology Corner Videos, TI-Nspire Technology Corners

Leigh has taught AP® Statistics for 13 years and has served as an AP® Statistics Reader for the past 8 years. She enjoys

the challenge of writing multiple-choice questions for the College Board for use on the AP® Statistics exam. Leigh is a National Board Certified Teacher in Adolescence and Young Adulthood Mathematics and was previously named a finalist for the Presidential Award for Excellence in Mathematics and Science Teaching in New Jersey.

Jonathan Osters, The Blake School, Minneapolis, MN
Videos

Jonathan has taught high school mathematics for 12 years. He teaches AP® Statistics, Probability & Statistics, and Geometry and has been a reader for the AP® Statistics exam for 9 years. Jonathan writes a blog about teaching at experiencefirstmath.org and tweets at @callmejosters.

Tonya Adkins, Charlotte, NC
Accuracy Checker, Online Homework

Tonya has been teaching math and statistics courses for more than 20 years in high schools and colleges in Alabama and North Carolina. She taught AP® Statistics for 10 years and has served as an AP® Reader for the past four years. Tonya also works as a reviewer, consultant, and subject matter expert on mathematics and statistics projects for publishers.

Robert Lochel, Hatboro-Horsham High School, Horsham PA
Online Homework, Desmos Projects

Bob has served as a high school math teacher and curriculum coach in his district for 21 years, and has taught AP® Statistics for 13 years. He has been an AP® Statistics Reader for the past 6 years. Bob has a passion for developing lessons that leverage technology in math classrooms, and he has shared his ideas at national conferences for NCTM and ISTE. He has served as a section editor for NCTM Mathematics Teacher "Tech Tips" for the last 3 years.

Sandra Lowell, Brandeis High School, San Antonio, TX
Online Homework

Sandra was a software engineer for 8 years and has taught high school math for 24 years, serving as mathematics coordinator and lead AP® Statistics teacher. She has taught AP® Statistics for 19 years and has been an AP® Statistics Reader for 15 years, serving as a Table Leader for the last 3 years. Sandra is currently teaching at Brandeis High School and is an adjunct professor at Northwest Vista College.

Jason Molesky *Strive for a 5 Guide*

Jason served as an AP® Statistics Reader and Table Leader since 2006. After teaching AP® Statistics for 8 years and developing the FRAPPY system for AP® Statistics exam preparation, Jason served as the Director of Program Evaluation and Accountability for the Lakeville Area Public Schools. He has recently settled into his dream job as an educational consultant for Apple. Jason maintains the "Stats Monkey" website, a clearinghouse for AP® Statistics resources.

Michael Legacy, Greenhill School, Dallas, TX
Strive for a 5 Guide

Michael is a past member of the AP® Statistics Development Committee (2001–2005) and a former Table Leader at the Reading. He currently reads the Alternate Exam and is a presenter at many AP® Summer Institutes. Michael is the author of the 2007 College Board AP® Statistics Teacher's Guide and was named the Texas 2009–2010 AP® Math/Science Teacher of the Year by the Siemens Corporation.

Jeff Eicher, Jr., Classical Academy High School, Escondido, CA
Strive for a 5 Guide

Jeff has taught AP® Statistics for 9 years and has been an AP® Reader for 3 years. He has served on the AP® Statistics Instructional Design Team, creating resources for students and teachers. Jeff is also a NMSI consultant, mentoring teachers and hosting Saturday study sessions.

Dori Peterson, Northwest Vista College, San Antonio, TX
Online Homework

Dori taught high school math for 28 years and AP® Statistics for 12 years. She served as a mathematics coordinator and statistics lead instructor for 4 years. Dori is currently an adjunct professor of math and statistics at Northwest Vista College. She has been an AP® Statistics Reader for 9 years, serving as a Table Leader for the last 3 years. Dori is a member of the American Statistical Association and served as a project competition judge for 2 years.

Mary Simons, Midlothian, VA
Online Homework

Mary has taught high school math for 15 years and AP® Statistics for 9 years. She has been an AP® Statistics Reader for the past 5 years. Mary is a member of the American Statistical Association, serving as a project competition judge for the past 4 years. She has also worked as a member of the Delaware Department of Education Mathematics Assessment Committee, the Delaware Mathematics Coalition, and has served as a cooperating teacher for the University of Delaware.

Acknowledgments

First and foremost, we owe a tremendous debt of gratitude to David Moore and Dan Yates. Professor Moore reshaped the college introductory statistics course through publication of three pioneering texts: *Introduction to the Practice of Statistics (IPS)*, *The Basic Practice of Statistics (BPS)*, and *Statistics: Concepts and Controversies.* He was also one of the original architects of the AP® Statistics course. When the course first launched in the 1996–1997 school year, there were no textbooks written specifically for the high school student audience that were aligned to the AP® Statistics topic outline. Along came Dan Yates. His vision for such a text became reality with the publication of *The Practice of Statistics (TPS)* in 1998. Over a million students have used one of the first five editions of *TPS* for AP® Statistics! Dan also championed the importance of developing high-quality resources for AP® Statistics teachers, which were originally provided in a *Teachers' Resource Binder.* We stand on the shoulders of two giants in statistics education as we carry forward their visions in this and future editions.

The Practice of Statistics has continued to evolve, thanks largely to the support of our longtime editor and team captain, Ann Heath. Her keen eye for design is evident throughout the pages of the student and teacher's editions. More importantly, Ann's ability to oversee all of the complex pieces of this project while maintaining a good sense of humor is legendary. Ann has continually challenged everyone involved with *TPS* to innovate in ways that benefit AP® Statistics students and teachers. She is a good friend and an inspirational leader.

Teamwork is the secret sauce of *TPS*. We have been blessed to collaborate with many talented AP® Statistics teachers and introductory statistics college professors over the years we have been working on this project. We sincerely appreciate their willingness to give us candid feedback about early drafts of the student edition, and to assist with the development of an expanding cadre of resources for students and teachers.

On the sixth edition, we are especially grateful to the individuals who played lead roles in key components of the project. Ann Cannon did yeoman's work once again in reading, reviewing, and accuracy checking every line in the student edition. Her sage advice and willingness

to ask tough questions were much appreciated throughout the writing of *TPS 6e*. Luke Wilcox took on the herculean task of producing the *Teacher's Edition* (TE). We know teachers will appreciate his careful thinking about effective pedagogy and the importance of engaging students with relevant context throughout the TE chapters. Working with his colleague, Lindsey Gallas, Luke also oversaw creation of the fabulous "150 Days of AP® Statistics" resource for teachers at his StatsMedic site.

Erica Chauvet wrote all of the solutions for *TPS 6e* exercises. Her thorough attention to matching the details in worked examples was exceeded only by her remarkable speed in completing this burdensome task. Erica also agreed to manage a substantial revision of the test bank, including crafting prototype quizzes and tests, and has assisted with the online homework content.

Doug Tyson is overseeing production of the vast collection of *TPS 6e* tutorial videos for students and teachers. We are thankful for Doug's expertise in video creation and for his willingness to pitch in wherever we need him. Tonya Adkins kindly agreed to spearhead our new online homework system for this edition. Welcome to the team, Tonya!

Every member of the *TPS 6e* Content Advisory Board and Supplements Team is an experienced teacher with significant involvement in the AP® Statistics program. In addition to the individuals above, we offer our heartfelt thanks to the following list of superstars for their tireless work and commitment to excellence: Beth Benzing, Don Brechlin, Paul Buckley, James Bush, Monica Debold, Kathleen Dickensheets, Jeff Eicher, Lindsey Gallas, Vicki Greenberg, Michael Legacy, Bob Lochel, Sandra Lowell, DeAnna McDonald, Stephen Miller, Jason Molesky, Leigh Nataro, Jonathan Osters, Dori Peterson, Al Reiff, and Mary Simons.

Sincere gratitude also goes to everyone at Bedford, Freeman, and Worth (BFW) involved in *TPS 6e*. Don Gecewicz returned partway through the manuscript writing to offer helpful developmental edits. Vivien Weiss and Susan Wein oversaw the production process with their usual care and professionalism. Louise Ketz kept us clear and consistent with her thoughtful copyediting. Corrina Santos and Kaitlyn Swygard worked behind the scenes

to carefully prepare manuscript chapters for production. Diana Blume ensured that the design of the finished book exceeded our expectations. Special thanks go to all the dedicated people on the high school sales and marketing team at BFW who promote *TPS* 6e with enthusiasm. We also offer our thanks to Murugesh Rajkumar Namasivayam and the team at Lumina Datamatics for turning a complex manuscript into good-looking page proofs.

Thank you to all the reviewers who offered encouraging words and thoughtful suggestions for improvement in this and previous editions. And to the many outstanding statistics educators who have taken the time to share their questions and insights with us online, at conferences and workshops, at the AP® Reading, and in assorted other venues, we express our appreciation.

Daren Starnes and Josh Tabor

A final note from Daren: I feel extremely fortunate to have partnered with Josh Tabor in writing *TPS* 6e. He is a gifted teacher and talented author in his own right. Josh's willingness to take on half of the chapters in this edition pays tribute to his unwavering commitment to excellence. He enjoys exploring new possibilities, which ensures that *TPS* will keep evolving in future editions. Josh is a good friend and trusted colleague.

My biggest thank you goes to my wife, Judy. She has made incredible sacrifices throughout my years as a textbook author. For Judy's unconditional love and support, I would like to dedicate this edition to her. She is my inspiration.

A final note from Josh: I have greatly enjoyed working with Daren Starnes on this edition of *TPS*. No one I know works harder and holds himself to a higher standard than he does. His wealth of experience and vision for this edition made him an excellent writing partner. For your friendship, encouragement, and support—thanks!

I especially want to thank the two most important people in my life. To my wife, Anne, your patience while I spent countless hours working on this project is greatly appreciated. I couldn't have survived without your consistent support and encouragement. To my daughter, Jordan, I can't believe how quickly you are growing up. It won't be long until you are reading *TPS* as a student! I love you both very much.

Sixth Edition Survey Participants and Reviewers

Paul Bastedo, *Viewpoint School, Calabasas, CA*
Emily Beal, *Chagrin Falls High School, Chagrin Falls, OH*
Raquel Bocast, *Hamilton Union High School, Hamilton City, CA*
Lisa Bonar, *Fossil Ridge High School, Keller, TX*
Robert Boone, *First Coast High School, Jacksonville, FL*
John Bowman, *Hinsdale Central High School, Hinsdale, IL*
Brigette Brankin, *St. Viator High School, Arlington Heights, IL*
Chris Burke, *Hingham High School, Hingham, MA*
Kenny Contreras, *Wilcox High School, Santa Clara, CA*
Nancy Craft, *Chestnut Hill Academy, Philadelphia, PA*
Aimee Davenport, *Heritage High School, Frisco, TX*
Gabrielle Dedrick, *Ernest McBride High School, Long Beach, CA*
Michael Ditzel, *Hampden Academy, Hampden, ME*
Robin Dixon, *Panther Creek High School, Cary, NC*
Parisa Foroutan, *Renaissance School for the Arts, Long Beach, CA*
Rebecca Gaillot, *Metairie Park Country Day School, Metairie, LA*
Roger Gale, *Waukegan High School, Waukegan, IL*
Becky Gerek, *Abraham Lincoln High School, San Francisco, CA*
Lisa Haney, *Monticello High School, Charlottesville, VA*
Kellie Hodge, *Jordan High School, Long Beach, CA*
Susan Knott, *The Oakridge School, Arlington, TX*
Lauren Kriczky, *Clewiston High School, Clewiston, FL*
William Ladley, *Bell High School, Hurst, TX*
Cathy Lichodziejewski, *Fountain Valley High School, Fountain Valley, CA*
Veronica Lunde, *Apollo High School, Saint Cloud, MN*
Shannon McBriar, *Central High School, Macon, GA*
Maureen McMichael, *Seneca High School, Tabernacle, NJ*
Victor Mirrer, *Fairfield Ludlowe High School, Fairfield, CT*
Jose Molina, *St. Mary's Hall, San Antonio, TX*
Kevin Morgan, *Central Columbia High School, Bloomsburg, PA*
Karin Munro, *Marcus High School, Flower Mound, TX*
Leigh Nataro, *Moravian Academy, Bethlehem, PA*
Cindy Parliament, *Klein Oak High School, Spring, TX*
Juliet Pender, *New Egypt High School, New Egypt, NJ*
John Powers, *Cardinal Gibbons High School, Fort Lauderdale, FL*
Jessica Quinn, *Mayfield Senior School, Pasadena, CA*
Gary Remiker, *Cathedral Catholic High School, San Diego, CA*
Michael Rice, *Rainer Beach High School, Seattle, WA*
Laura Ringwood, *Westlake High School, Austin, TX*

Gina Ruth, *Woodrow Wilson High School, Beckley, WV*
Dan Schmidt, *Rift Valley Academy, Kijabe, Kenya*
Ned Smith, *Southwest High School, Fort Worth, TX*
Joseph Tanzosh, *Marian High School, Mishawaka, IN*
Rachael Thiele, *Polytechnic High School, Long Beach, CA*
Jenny Thom-Carroll, *West Essex Senior High School, North Caldwell, NJ*
Tara Truesdale, *Ben Lippen School, Columbia, SC*
Alethea Trundy, *Montachusett Regional Vocational Technical School, Fitchburg, MA*
Crystal Vesperman, *Prairie School, Racine, WI*
Kristine Witzel, *Duchesne High School, Saint Charles, MO*

Fifth Edition Survey Participants and Reviewers

Blake Abbott, *Bishop Kelley High School, Tulsa, OK*
Maureen Bailey, *Millcreek Township School District, Erie, PA*
Kevin Bandura, *Lincoln County High School, Stanford, KY*
Elissa Belli, *Highland High School, Highland, IN*
Jeffrey Betlan, *Yough School District, Herminie, PA*
Nancy Cantrell, *Macon County Schools, Franklin, NC*
Julie Coyne, *Center Grove High School, Greenwood, IN*
Mary Cuba, *Linden Hall, Lititz, PA*
Tina Fox, *Porter-Gaud School, Charleston, SC*
Ann Hankinson, *Pine View, Osprey, FL*
Bill Harrington, *State College Area School District, State College, PA*
Ronald Hinton, *Pendleton Heights High School, Pendleton, IN*
Kara Immonen, *Norton High School, Norton, MA*
Linda Jayne, *Kent Island High School, Stevensville, MD*
Earl Johnson, *Chicago Public Schools, Chicago, IL*
Christine Kashiwabara, *Mid-Pacific Institute, Honolulu, HI*
Melissa Kennedy, *Holy Names Academy, Seattle, WA*
Casey Koopmans, *Bridgman Public Schools, Bridgman, MI*
David Lee, *Sun Prairie High School, Sun Prairie, WI*
Carolyn Leggert, *Hanford High School, Richland, WA*
Jeri Madrigal, *Ontario High School, Ontario, CA*
Tom Marshall, *Kents Hill School, Kents Hill, ME*
Allen Martin, *Loyola High School, Los Angeles, CA*
Andre Mathurin, *Bellarmine College Preparatory, San José, CA*
Brett Mertens, *Crean Lutheran High School, Irvine, CA*
Sara Moneypenny, *East High School, Denver, CO*
Mary Mortlock, *The Harker School, San José, CA*
Mary Ann Moyer, *Hollidaysburg Area School District, Hollidaysburg, PA*
Howie Nelson, *Vista Murrieta High School, Murrieta, CA*
Shawnee Patry, *Goddard High School, Wichita, KS*
Sue Pedrick, *University High School, Hartford, CT*
Shannon Pridgeon, *The Overlake School, Redmond, WA*
Sean Rivera, *Folsom High, Folsom, CA*
Alyssa Rodriguez, *Munster High School, Munster, IN*

Sheryl Rodwin, *West Broward High School, Pembroke Pines, FL*
Sandra Rojas, *Americas High School, El Paso, TX*
Amanda Schneider, *Battle Creek Public Schools, Charlotte, MI*
Christine Schneider, *Columbia Independent School, Boonville, MO*
Steve Schramm, *West Linn High School, West Linn, OR*
Katie Sinnott, *Revere High School, Revere, MA*
Amanda Spina, *Valor Christian High School, Highlands Ranch, CO*
Julie Venne, *Pine Crest School, Fort Lauderdale, FL*
Dana Wells, *Sarasota High School, Sarasota, FL*
Luke Wilcox, *East Kentwood High School, Grand Rapids, MI*
Thomas Young, *Woodstock Academy, Putnam, CT*

Fourth Edition Focus Group Participants and Reviewers

Gloria Barrett, *Virginia Advanced Study Strategies, Richmond, VA*
David Bernklau, *Long Island University, Brookville, NY*
Patricia Busso, *Shrewsbury High School, Shrewsbury, MA*
Lynn Church, *Caldwell Academy, Greensboro, NC*
Steven Dafilou, *Springside High School, Philadelphia, PA*
Sandra Daire, *Felix Varela High School, Miami, FL*
Roger Day, *Pontiac High School, Pontiac, IL*
Jared Derksen, *Rancho Cucamonga High School, Rancho Cucamonga, CA*
Michael Drozin, *Munroe Falls High School, Stow, OH*
Therese Ferrell, *I. H. Kempner High School, Sugar Land, TX*
Sharon Friedman, *Newport High School, Bellevue, WA*
Jennifer Gregor, *Central High School, Omaha, NE*
Julia Guggenheimer, *Greenwich Academy, Greenwich, CT*
Dorinda Hewitt, *Diamond Bar High School, Diamond Bar, CA*
Dorothy Klausner, *Bronx High School of Science, Bronx, NY*
Robert Lochel, *Hatboro-Horsham High School, Horsham, PA*
Lynn Luton, *Duchesne Academy of the Sacred Heart, Houston, TX*
Jim Mariani, *Woodland Hills High School, Greensburgh, PA*
Stephen Miller, *Winchester Thurston High School, Pittsburgh, PA*
Jason Molesky, *Lakeville Area Public Schools, Lakeville, MN*
Mary Mortlock, *The Harker School, San José, CA*
Heather Nichols, *Oak Creek High School, Oak Creek, WI*
Jamis Perrett, *Texas A&M University, College Station, TX*
Heather Pessy, *Mount Lebanon High School, Pittsburgh, PA*
Kathleen Petko, *Palatine High School, Palatine, IL*
Todd Phillips, *Mills Godwin High School, Richmond, VA*
Paula Schute, *Mount Notre Dame High School, Cincinnati, OH*
Susan Stauffer, *Boise High School, Boise, ID*
Doug Tyson, *Central York High School, York, PA*
Bill Van Leer, *Flint High School, Oakton, VA*
Julie Verne, *Pine Crest High School, Fort Lauderdale, FL*
Steve Willot, *Francis Howell North High School, St. Charles, MO*
Jay C. Windley, *A. B. Miller High School, Fontana, CA*

To the Student

Statistical Thinking and You

The purpose of this book is to give you a working knowledge of the big ideas of statistics and of the methods used in solving statistical problems. Because data always come from a real-world context, doing statistics means more than just manipulating data. *The Practice of Statistics* (TPS), Sixth Edition, is full of data. Each set of data has some brief background to help you understand where the data come from. We deliberately chose contexts and data sets in the examples and exercises to pique your interest.

TPS 6e is designed to be easy to read and easy to use. This book is written by current high school AP® Statistics teachers, for high school students. We aimed for clear, concise explanations and a conversational approach that would encourage you to read the book. We also tried to enhance both the visual appeal and the book's clear organization in the layout of the pages.

Be sure to take advantage of all that *TPS* 6e has to offer. You can learn a lot by reading the text, but you will develop deeper understanding by doing the Activities and Projects and answering the Check Your Understanding questions along the way. The walkthrough guide on pages xvi–xxii gives you an inside look at the important features of the text.

You learn statistics best by doing statistical problems. This book offers many different types of problems for you to tackle.

- **Section Exercises** include paired odd- and even-numbered problems that test the same skill or concept from that section. There are also some multiple-choice questions to help prepare you for the AP® Statistics exam. Recycle and Review exercises at the end of each exercise set involve material you studied in preceding sections.

- **Chapter Review Exercises** consist of free-response questions aligned to specific learning targets from the chapter. Go through the list of learning targets summarized in the Chapter Review and be sure you can say of each item on the list, "I can do that." Then prove it by solving some problems.

- The **AP® Statistics Practice Test** at the end of each chapter will help you prepare for in-class exams. Each test has about 10 multiple-choice questions and 3 free-response problems, very much in the style of the AP® Statistics exam.

- Finally, the **Cumulative AP® Practice Tests** after Chapters 4, 7, 11, and 12 provide challenging, cumulative multiple-choice and free-response questions like those you might find on a midterm, final, or the AP® Statistics exam.

The main ideas of statistics, like the main ideas of any important subject, took a long time to discover and thus take some time to master. The basic principle of learning them is to be persistent. Once you put it all together, statistics will help you make informed decisions based on data in your daily life.

TPS and AP® Statistics

The Practice of Statistics (TPS) was the first book written specifically for the Advanced Placement (AP®) Statistics course. This updated version of *TPS* 6e is organized to closely follow the AP® Statistics Course Framework (CF). Every learning objective and essential knowledge statement in the CF is covered thoroughly in the text. Visit the book's website at highschool.bfwpub.com/updatedtps6e for a detailed alignment guide, including "The Nitty Gritty Alignment" guide to EKs and LOs in Updated *TPS* 6e. The few topics in the book that go beyond the AP® Statistics syllabus are marked with an asterisk (*).

Most importantly, *TPS* 6e is designed to prepare you for the AP® Statistics exam. The author team has been involved in the AP® Statistics program since its early days. We have more than 40 years' combined experience teaching AP® Statistics and grading the AP® exam! Both of us have served as Question Leaders for more than 10 years, helping to write scoring rubrics for free-response questions. Including our Content Advisory Board and Supplements Team (page vii), we have extensive knowledge of how the AP® Statistics exam is developed and scored.

TPS 6e will help you get ready for the AP® Statistics exam throughout the course by:

- **Using terms, notation, formulas, and tables consistent with those found in the Course Framework and on the AP® Statistics exam.** Key terms are shown in bold in the text, and they are defined in the Glossary. Key terms also are cross-referenced in the Index. See page F-1 to find "Formulas for the AP® Statistics Exam," as well as Tables A, B, and C in the back of the book for reference.

- **Following accepted conventions from AP® Statistics exam rubrics when presenting model solutions.** Over the years, the scoring guidelines for free-response questions have become fairly consistent. We kept these guidelines in mind when writing the solutions that appear throughout *TPS* 6e. For example, the four-step State–Plan–Do–Conclude process that we use to complete inference problems in Chapters 8–12 closely matches the four-point AP® scoring rubrics.

- **Including AP® Exam Tips in the margin where appropriate.** We place exam tips in the margins as "on-the-spot" reminders of common mistakes and how to avoid them. These tips are collected and summarized in the About the AP® Exam and AP® Exam Tips appendix.

- **Providing over 1600 AP®-style exercises throughout the book.** Each chapter contains a mix of free-response and multiple-choice questions that are similar to those found on the AP® Statistics exam. At the start of each Chapter Wrap-Up, you will find a FRAPPY (Free Response AP® Problem, Yay!). Each FRAPPY gives you the chance to solve an AP®-style free-response problem based on the material in the chapter. After you finish, you can view and critique

two example solutions from the book's Student Site (highschool.bfwpub.com/updatedtps6e). Then you can score your own response using a rubric provided by your teacher.

- **Developing the Course Skills for AP® Statistics** through frequent repetition. See the inside back cover for more details.

Turn the page for a tour of the text. See how to use the book to realize success in the course and on the AP® Statistics exam.

READ THE TEXT and use the book's features to help you grasp the big ideas.

Read the **LEARNING TARGETS** at the beginning of each section. Focus on mastering these skills and concepts as you work through the chapter.

Scan the margins for the green notes, which represent the "voice of the teacher" giving helpful hints for being successful in the course. Many of these notes include important reminders from the AP® Statistics Course Framework.

Read the **AP® EXAM TIPS.** They give advice on how to be successful on the AP® Statistics exam.

Watch for **CAUTION ICONS.** They alert you to common mistakes that students often make.

It's important to learn the language of statistics. Take note of the green **DEFINITION** boxes that explain important vocabulary. Flip back to them to review key terms and their definitions, or turn to the Glossary/Glosario at the back of the book.

Look for the boxes with the green bands. Some explain how to make graphs or set up calculations, while others recap important concepts.

SECTION 3.1 Scatterplots and Correlation

LEARNING TARGETS *By the end of the section, you should be able to:*

- Distinguish between explanatory and response variables for quantitative data.
- Make a scatterplot to display the relationship between two quantitative variables.
- Describe the direction, form, and strength of a relationship displayed in a scatterplot and identify unusual features.
- Interpret the correlation.
- Understand the basic properties of correlation, including how the correlation is influenced by unusual points.
- Distinguish correlation from causation.

A one-variable data set is sometimes called univariate data. *A data set that describes the relationship between two variables is sometimes called* bivariate data.

Most statistical studies examine data on more than one variable for a group of individuals. Fortunately, analysis of relationships between two variables builds on the same tools we used to analyze one variable. The principles that guide our work also remain the same:

- Plot the data, then look for overall patterns and departures from those patterns.
- Add numerical summaries.
- When there's a regular overall pattern, use a simplified model to describe it.

Explanatory and Response Variables

In the "Candy grab" activity, the number of candies is the **response variable.** Hand span is the **explanatory variable** because we anticipate that knowing a student's hand span will help us predict the number of candies that student can grab.

> **DEFINITION** Response variable, Explanatory variable
>
> A **response variable** measures an outcome of a study. An **explanatory variable** may help predict or explain changes in a response variable.

You will often see explanatory variables called independent variables *and response variables called* dependent variables. *Because the words* independent *and* dependent *have other meanings in statistics, we won't use them here.*

It is easiest to identify explanatory and response variables when we initially specify the values of one variable to see how it affects another variable. For instance, to study the effect of alcohol on body temperature, researchers gave several different amounts of alcohol to mice. Then they measured the change in each mouse's body temperature 15 minutes later. In this case, amount of alcohol is the explanatory variable, and change in body temperature is the response variable. When we don't specify the values of either variable before collecting the data, there may or may not be a clear explanatory variable.

AP® EXAM TIP

When you are asked to *describe* the association shown in a scatterplot, you are expected to discuss the direction, form, and strength of the association, along with any unusual features, *in the context of the problem.* This means that you need to use both variable names in your description.

HOW TO DESCRIBE A SCATTERPLOT

To describe a scatterplot, make sure to address the following four characteristics in the context of the data:

- **Direction:** A scatterplot can show a positive association, negative association, or no association.
- **Form:** A scatterplot can show a linear form or a nonlinear form. The form is linear if the overall pattern follows a straight line. Otherwise, the form is nonlinear.
- **Strength:** A scatterplot can show a weak, moderate, or strong association. An association is strong if the points don't deviate much from the form identified. An association is weak if the points deviate quite a bit from the form identified.

 Few relationships are linear for all values of the explanatory variable. Don't make predictions using values of *x* that are much larger or much smaller than those that actually appear in your data.

LEARN STATISTICS BY *DOING* STATISTICS

Every chapter begins with a hands-on **ACTIVITY** that introduces the content of the chapter. Many of these activities involve collecting data and drawing conclusions from the data. In other activities, you'll use dynamic applets to explore statistical concepts.

ACTIVITY   Candy grab

In this activity, you will investigate if students with a larger hand span can grab more candy than students with a smaller hand span.[1]

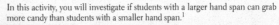

1. Measure the span of your dominant hand to the nearest half-centimeter (cm). Hand span is the distance from the tip of the thumb to the tip of the pinkie finger on your fully stretched-out hand.

2. One student at a time, go to the front of the class and use your dominant hand to grab as many candies as possible from the container. You must grab the candies with your fingers pointing down (no scooping!) and hold the candies for 2 seconds before counting them. After counting, put the candy back into the container.

3. On the board, record your hand span and number of candies in a table with the following headings:

Hand span (cm)	Number of candies

4. While other students record their values on the board, copy the table onto a piece of paper and make a graph. Begin by constructing a set of coordinate axes. Label the horizontal axis ⬚⬚⬚ the vertical axis "Number of candies." ⬚⬚⬚ scale for each axis and plot each point ⬚⬚⬚ as accurately as you can on the graph. ⬚⬚⬚ell you about the relationship between ⬚⬚ of candies? Summarize your observa-⬚⬚.

ACTIVITY   Investigating properties of the least-squares regression line

In this activity, you will use the *Correlation and Regression* applet to explore some properties of the least-squares regression line.

1. Launch the applet at highschool.bfwpub.com/updatedtps6e.

2. Click on the graphing area to add 10 points in the lower-left corner so that the correlation is about $r = 0.40$. Also, check the boxes to show the "Least-Squares Line" and the "Mean X & Y"

Chapters 1, 3, 4, 11, and 12 conclude with a **CHAPTER PROJECT**. Three of the projects (Chapters 1, 3, and 11) provide an opportunity to think like a statistician by analyzing larger data sets with multiple variables of interest. The other two (Chapters 4 and 12) are longer-term projects that require you to engage in the statistical problem-solving process: Ask Questions, Collect Data, Analyze Data, Interpret Results.

Chapter 3 Project   Investigating Relationships in Baseball

What is a better predictor of the number of wins for a baseball team, the number of runs scored by the team or the number of runs they allow the other team to score? What variables can we use to predict the number of runs a team scores? To predict the number of runs it allows the other team to score? In this project, you will use technology to help answer these questions by exploring a large set of data from Major League Baseball.

Part 1

1. Download the "MLB Team Data 2012–2016" Excel file from the book's website, along with the "Glossary for MLB Team Data file," which explains each of the variables included in the data set.[58] Import the data into th⬚ cal software package you prefer.

2. Create a scatterplot to investigate the relationship ⬚ runs scored per game (R/G) and wins (W). The ⬚ late the equation of the least-squares regression ⬚ standard deviation of the residuals, and r^2. *Note:* ⬚ the section for hitting statistics and W is in the se⬚ pitching statistics.

5. Because the number of wins a team has is dependent on both how many runs they score and how many runs they allow, we can use a combination of both variables to predict the number of wins. Add a column in your data table for a new variable, run differential. Fill in the values using the formula R/G – RA/G.

6. Create a scatterplot to investigate the relationship between run differential and wins. Then calculate the equation of the least-squares regression line, the standard deviation of the residuals, and r^2.

7. Is run differential a better predictor than the variable you chose in Question 4? Explain your reasoning.

CHECK YOUR UNDERSTANDING questions appear throughout the section. They help clarify definitions, concepts, and procedures. Be sure to check your answers in the back of the book.

CHECK YOUR UNDERSTANDING

In Exercises 3 and 7, we asked you to make and describe a scatterplot for the hiker data shown in the table.

Body weight (lb)	120	187	109	103	131	165	158	116
Backpack weight (lb)	26	30	26	24	29	35	31	28

1. Calculate the equation of the least-squares regression line.
2. Make a residual plot for the linear model in Question 1.
3. What does the residual plot indicate about the appropriateness of the linear model? Explain your answer.

EXAMPLES: Model statistical problems and how to solve them

Read through each **EXAMPLE**, and then try out the concept yourself by working the **FOR PRACTICE, TRY** exercise in the Section Exercises.

Need extra help? Examples and exercises marked with the **PLAY ICON** ▶ are supported by short video clips prepared by experienced AP® Statistics teachers. The video guides you through each step in the example and solution and provides additional explanation when you need it.

EXAMPLE

Old Faithful and fertility
Describing a scatterplot

PROBLEM: Describe the relationship in each of the following contexts.

(a) The scatterplot on the left shows the relationship between the duration (in minutes) of an eruption and the interval of time until the next eruption (in minutes) of Old Faithful during a particular month.

(b) The scatterplot on the right shows the relationship between the average income (gross domestic product per person, in dollars) and fertility rate (number of children per woman) in 187 countries.[4]

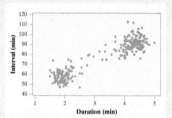
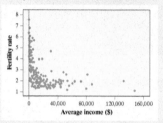

SOLUTION:

(a) There is a strong, positive linear relationship between the duration of an eruption and the interval of time until the next eruption. There are two main clusters of points: one cluster has durations around 2 minutes with in... has durations aro...

(b) There is a modera... between average... is a country outsi... $30,000 and a f...

Even with the clusters, the overall direction is still positive. In some cases, however, the points in a cluster...

Example: Old Faithful and fertility

Describe the relationship in each of the following contexts.

(a) The scatterplot on the left shows the relationship between the duration (in minutes) of an eruption and the interval of time until the next eruption (in minutes) of Old Faithful during a particular month.

Solution: There is a strong, positive linear relationship between the duration of an eruption and the interval of time until the next eruption. There are two main clusters of points: one cluster has durations around 2 minutes with intervals around 55 minutes, and the other cluster has durations around 4.5 minutes with intervals around 90 minutes.

EXAMPLE

Caffeine and pulse rates
How random assignment works

PROBLEM: A total of 20 students have agreed to participate in an experiment comparing the effects of caffeinated cola and caffeine-free cola on pulse rates. Describe how you would randomly assign 10 students to each of the two treatments:

(a) Using 20 identical slips of paper
(b) Using technology
(c) Using Table D

SOLUTION:

(a) On 10 slips of paper, write the letter "A"; on the remaining 10 slips, write the letter "B." Shuffle the slips of paper and hand out one slip of paper to each volunteer. Students who get an "A" slip receive the cola with caffeine and students who get a "B" slip receive the cola without caffeine.

When describing a method of random assignment, don't stop after creating the groups. Make sure to identify which group gets which treatment.

(b) Label each student with a different integer from 1 to 20. Then randomly generate 10 different integers from 1 to 20. The students with these labels receive the cola with caffeine. The remaining 10 students receive the cola without caffeine.

When using a random number generator or a table of random digits to assign treatments, make sure to account for the possibility of repeated numbers when describing your method.

(c) Label each student with a different integer from 01 to 20. Go to a line of Table D and read two-digit groups moving from left to right. The first 10 different labels between 01 and 20 identify the 10 students who receive cola with caffeine. The remaining 10 students receive the caffeine-free cola. Ignore groups of digits from 21 to 00.

FOR PRACTICE, TRY EXERCISE 63

The **SOLUTION** is presented in a special font and models the style, steps, and language that you should use to earn full credit on the AP® Statistics exam.

THE VOICE OF THE TEACHER. Study the worked examples and pay special attention to the carefully placed **"Teacher Talk" comment boxes** that guide you step by step through the solution. These comments offer lots of good advice—as if your teacher is working directly with you to solve a problem.

The blue page number icon next to an exercise points you back to the page on which the model example appears.

63. **Layoffs and "survivor guilt"** Workers who survive a layoff of other employees at their location may suffer from "survivor guilt." A study of survivor guilt and its effects used as subjects 120 students who were offered an opportunity to earn extra course credit by doing proofreading. Each subject worked in the same cubicle as another student, who was an accomplice of the experimenters. At a break midway through the work,

EXERCISES: Practice makes perfect!

Start by reading the **SECTION SUMMARY** to be sure that you understand the big ideas and key concepts.

Section 3.1 | Summary

- A **scatterplot** displays the relationship between two quantitative variables measured on the same individuals. Mark values of one variable on the horizontal axis (*x* axis) and values of the other variable on the vertical axis (*y* axis). Plot each individual's data as a point on the graph.
- If we think that a variable *x* may help predict, explain, or even cause changes in another variable *y*, we call *x* an **explanatory variable** and *y* a **response variable**. Always plot the explanatory variable on the *x* axis of a scatterplot. Plot the response variable on the *y* axis.
- When describing a scatterplot, look for an overall pattern (direction, form, strength) and departures from the pattern (unusual features) and always answer in cor...
  - Directio...
    variable t...

Practice! Work the **EXERCISES** assigned by your teacher. Compare your answers to those in the Solutions appendix at the back of the book. Short solutions to the exercises numbered in red are found in the appendix.

Section 3.1 | Exercises

1. **Coral reefs and cell phones** Identify the explanatory variable and the response variable for the following relationships, if possible. Explain your reasoning.
   pg 154
   (a) The weight gain of corals in aquariums where the water temperature is controlled at different levels
   (b) The number of text messages sent and the number of phone calls made in a sample of 100 students

2. **Teenagers and corn yield** Identify the explanatory variable and the response variable for the following relationships, if possible. Explain your reasoning.
   (a) The height and arm span of a sample of 50 teenagers
   (b) The yield of corn in bushels per acre and the amount of rain in the growing season

3. **Heavy backpacks** Ninth-grade students at the Webb Schools go on a backpacking trip each fall. Students are divided into hiking groups of size 8 by selecting names from a hat. Before leaving, students and their backpacks are weighed. The data here are from one hiking group. Make a scatterplot by hand that shows how backpack weight relates to body weight.
   pg 155

Body weight (lb)	120	187	109	103	131	165	158	116
Backpack weight (lb)	26	30	26	24	29	35	31	28

4. **Putting success** How well do professional golfers putt from various distances to the hole? The data show various distances to the hole (in feet) and the percent of putts mad...
   Make a scatt...
   of putts mad...

Distance (ft)		

Most of the **exercises are paired**, meaning that odd- and even-numbered exercises test the same skill or concept. If you answer an assigned exercise incorrectly, try to figure out your mistake. Then see if you can solve the paired exercise.

Look for **ICONS** that appear next to selected **EXERCISES**. They will guide you to
- the Example that models the exercise.
- videos that provide step-by-step instructions for solving the exercise.

Exercise: Heavy Backpacks

Make a scatterplot by hand that shows how backpack weight relates to body weight.

Track and Field team.[10] Describe the relationship between height and weight for these athletes.

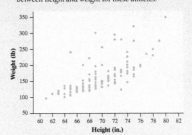

6. **Starbucks** The scatterplot shows the relationship between the amount of fat (in grams) and number of calories in products sold at Starbucks.[11] Describe the relationship between fat and calories for these products.

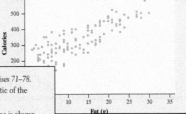

Multiple Choice: *Select the best answer for Exercises 71–78.*

71. Which of the following is *not* a characteristic of the least-squares regression line?

(a) The slope of the least-squares regression line is always between −1 and 1.

(b) The least-squares regression line always goes through the point $(\bar{x}, \bar{y})$.

(c) The least-squares regression line minimizes the sum of squared residuals.

(d) The slope of the least-squares regression line w... always have the same sign as the correlation.

(e) The least-squares regression line is not resistant... outliers.

Recycle and Review

79. **Fuel economy** (2.2) In its recent *Fuel Economy Guide*, the Environmental Protection Agency (EPA) gives data on 1152 vehicles. There are a number of outliers, mainly vehicles with very poor gas mileage or hybrids with very good gas mileage. If we ignore the outliers, however, the combined city and highway gas mileage of the other 1120 or so vehicles is approximately Normal with mean 18.7 miles per gallon (mpg) and standard deviation 4.3 mpg.

(a) The Chevrolet Malibu with a four-cylinder engine has a combined gas mileage of 25 mpg. What percent of the 1120 vehicles have worse gas mileage than the Malibu?

Various types of problems in the Section Exercises let you practice solving many different types of questions, including AP®-style **multiple-choice** and **free-response**. The **Recycle and Review** exercises refer back to concepts and skills learned in an earlier section, noted in purple after the problem title.

REVIEW and PRACTICE for quizzes and tests

Study the **CHAPTER REVIEW** to be sure that you understand the key concepts in each section.

Chapter 3 Wrap-Up

Chapter 3 Review

Section 3.1: Scatterplots and Correlation

In this section, you learned how to explore the relationship between two quantitative variables. As with distributions of a single variable, the first step is always to make a graph.

A scatterplot is the appropriate type of graph to investigate relationships between two quantitative variables. To describe a scatterplot, be sure to discuss four characteristics: direction, form, strength, and unusual features. The direction of a

Use the **WHAT DID YOU LEARN?** table that directs you to examples and exercises to verify your mastery of each **LEARNING TARGET**.

What Did You Learn?

Learning Target	Section	Related Example on Page(s)	Relevant Chapter Review Exercise(s)
Distinguish between explanatory and response variables for quantitative data.	3.1	154	R3.4
Make a scatterplot to display the relationship between two quantitative variables.	3.1	155	R3.4

SUMMARY TABLES in Chapters 8–12 review important details of each inference procedure, including conditions and formulas.

	Comparing confidence intervals for proportions	
	Confidence interval for p	**Confidence interval for $p_1 - p_2$**
Name (TI-83/84)	One-sample z interval for p (1-PropZInt)	Two-sample z interval for $p_1 - p_2$ (2-PropZInt)
Conditions	• **Random:** The data come from a random sample from the population of interest. ○ **10%:** When sampling without replacement, $n < 0.10N$. • **Large Counts:** Both $n\hat{p}$ and $n(1-\hat{p})$ are at least 10. That is, the number of successes and the number of failures in the sample are both at least 10.	• **Random:** The data come from two independent random samples or from two groups in a randomized experiment. ○ **10%:** When sampling without replacement, $n_1 < 0.10N_1$ and $n_2 < 0.10N_2$. • **Large Counts:** The counts of "successes" and "failures" in each sample or group— $n_1\hat{p}_1, n_1(1-\hat{p}_1), n_2\hat{p}_2, n_2(1-\hat{p}_2)$—are all at least 10.
Formula	$\hat{p} \pm z^* \sqrt{\dfrac{\hat{p}(1-\hat{p})}{n}}$	$(\hat{p}_1 - \hat{p}_2) \pm z^* \sqrt{\dfrac{\hat{p}_1(1-\hat{p}_1)}{n_1} + \dfrac{\hat{p}_2(1-\hat{p}_2)}{n_2}}$

Chapter 3 Review Exercises

These exercises are designed to help you review the important ideas and methods of the chapter.

R3.1 Born to be old? Is there a relationship between the gestational period (time from conception to birth) of an animal and its average life span? The figure shows a scatterplot of the gestational period and average life span for 43 species of animals.[5]

Tackle the **CHAPTER REVIEW EXERCISES** for practice in solving problems that test concepts from throughout the chapter. Need more help or just want additional insights before you take the practice test? Watch the **Chapter Review Exercise Videos**.

Review Exercise: Late bloomers?

(b) Use technology to calculate the correlation and the equation of the least-squares regression line. Interpret the slope and y intercept of the line in this setting.

The correlation is $r = -0.85$.

The equation of the LSRL is $\hat{y} = \underline{33.12} - \underline{4.69}x$, *intercept slope* where $\hat{y}$ represents the predicted number of days and x represents the average March temperature.

Chapter 3 AP® Statistics Practice Test

Section I: Multiple Choice *Select the best answer for each question.*

T3.1 A school guidance counselor examines how many extracurricular activities students participate in and their grade point average. The guidance counselor says, "The evidence indicates that the correlation between the number of extracurricular activities a student participates in and his or her grade point average is close to 0." Which of the following is the most appropriate conclusion?

(a) Students involved in many extracurricular activities tend to be students with poor grades.

(b) Students with good grades tend to be students who are not involved in many extracurricular activities.

(c) Students involved in many extracurricular activities are just as likely to get good grades as bad grades.

(d) Students with good grades tend to be students who are involved in many extracurricular activities.

(e) No conclusion should be made based on the correlation without looking at a scatterplot of the data.

Questions T3.3–T3.5 refer to the following setting. Scientists examined the activity level of 7 fish at different temperatures. Fish activity was rated on a scale of 0 (no activity) to 100 (maximal activity). The temperature was measured in degrees Celsius. A computer regression printout and a residual plot are provided. Notice that the horizontal axis on the residual plot is labeled "Fitted value," which means the same thing as "predicted value."

> Each chapter concludes with an **AP® STATISTICS PRACTICE TEST**. This test includes about 10 AP®-style multiple-choice questions and 3 free-response questions.

Cumulative AP® Practice Test 1

Section I: Multiple Choice *Choose the best answer for Questions AP1.1–AP1.14.*

AP1.1 You look at real estate ads for houses in Sarasota, Florida. Many houses have prices from $200,000 to $400,000. The few houses on the water, however, have prices up to $15 million. Which of the following statements best describes the distribution of home prices in Sarasota?

(a) The distribution is most likely skewed to the left, and the mean is greater than the median.

(b) The distribution is most likely skewed to the left, and the mean is less than the median.

(c) The distribution is roughly symmetric with a few high outliers, and the mean is approximately equal to the median.

(d) The distribution is most likely skewed to the right, and the mean is greater than the median.

(e) The distribution is most likely skewed to the right, and the mean is less than the median.

AP1.2 A child is 40 inches tall, which places her at the 90th percentile of all children of similar age. The heights for children of this age form an approximately Normal distribution with a mean of 38 inches. Based on this information, what is the standard deviation of the heights of all children of this age?

(a) 0.20 inch
(b) 0.31 inch
(c) 0.65 inch
(d) 1.21 inches
(e) 1.56 inches

> Four **CUMULATIVE AP® PRACTICE TESTS** simulate the real exam. They are placed after Chapters 4, 7, 11, and 12. The tests expand in length and content coverage as you work through the book. The last test models a full AP® Statistics exam.

FRAPPY! FREE RESPONSE AP® PROBLEM, YAY!

The following problem is modeled after actual AP® Statistics exam free response questions. Your task is to generate a complete, concise response in 15 minutes.

Directions: Show all your work. Indicate clearly the methods you use, because you will be scored on the correctness of your methods as well as on the accuracy and completeness of your results and explanations.

Two statistics students went to a flower shop and randomly selected 12 carnations. When they got home, the students prepared 12 identical vases with exactly the same amount of water in each vase. They put one tablespoon of sugar in 3 vases, two tablespoons of sugar in 3 vases, and three tablespoons of sugar in 3 vases. In the remaining 3 vases, they put no sugar. After the vases were prepared, the students randomly assigned 1 carnation to each vase and observed how many hours each flower continued to look fresh. A scatterplot of the data is shown below.

(a) Briefly describe the association shown in the scatterplot.

(b) The equation of the least-squares regression line for these data is $\hat{y} = 180.8 + 15.8x$. Interpret the slope of the line in the context of the study.

(c) Calculate and interpret the residual for the flower that had 2 tablespoons of sugar and looked fresh for 204 hours.

(d) Suppose that another group of students conducted a similar experiment using 12 flowers, but included different varieties in addition to carnations. Would you expect the value of r^2 for the second group's data to be greater than, less than, or about the same as the value of r^2 for the first group's data? Explain.

After you finish, you can view two example solutions on the book's website (highschool.bfwpub.com/updatedtps6e). Determine whether you think each solution is "complete," "substantial," "developing," or "minimal." If the solution is not complete, what improvements would you suggest to the student who wrote it? Finally, your teacher will provide you with a scoring rubric. Score your response and note what, if anything, you would do differently

> Learn how to answer free response questions successfully by working the **FRAPPY!**—the Free Response AP® Problem, Yay!—that begins the Chapter Wrap-Up in every chapter.

Use TECHNOLOGY to discover and analyze

3. Technology Corner — COMPUTING NUMERICAL SUMMARIES

TI-Nspire and other technology instructions are on the book's website at highschool.bfwpub.com/updatedtps6e.

Let's find numerical summaries for the boys' shoes data from the example on page 64. We'll start by showing you how to compute summary statistics on the TI-83/84 and then look at output from computer software.

I. **One-variable statistics on the TI-83/84**

 1. Enter the data in list L1.

 2. Find the summary statistics for the shoe data.
  - Press STAT (CALC); choose 1-VarStats.
   OS 2.55 or later: In the dialog box, press 2nd 1 (L1) and ENTER to specify L1 as the List. Leave FreqList blank. Arrow down to Calculate and press ENTER.
   Older OS: Press 2nd 1 (L1) and ENTER.
  - Press ▼ to see the rest of the one-variable statistics.

II. **Output from statistical software** We used Minitab statistical software to calculate descriptive statistics for the boys' shoes data. Minitab allows you to choose which numerical summaries are included in the output.

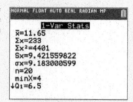

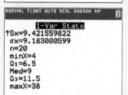

```
Descriptive Statistics: Shoes

Variable  N  Mean  StDev  Minimum  Q₁   Median  Q₃    Maximum
Shoes     20 11.65 9.42   4.00     6.25 9.00    11.75 38.00
```

Note: The TI-83/84 gives the first and third quartiles of the boys' shoes distribution as $Q_1 = 6.5$ and $Q_3 = 11.5$. Minitab reports that $Q_1 = 6.25$ and $Q_3 = 11.75$. What happened? Minitab and some other software use slightly different rules for locating quartiles. Results from the various rules are usually close to each other. Be aware of possible differences when calculating quartiles as they may affect more than just the IQR.

Use technology as a tool for discovery and analysis. **TECHNOLOGY CORNERS** give step-by-step instructions for using the TI-83/84 calculator. Instructions for the TI-Nspire and other calculators are on the book's Student Site (highschool.bfwpub.com/updatedtps6e) and in the e-Book platform.

Technology Corner videos are also available to walk you through the key strokes needed to perform each analysis.

Although the Technology Corners focus on the TI-83/84 graphing calculator, output from multiple programs—including Minitab and JMP—is used in the book's Examples and Exercises to help you become familiar with reading and interpreting many different kinds of statistical summaries.

3.2 Technology Corners

TI-Nspire and other technology instructions are on the book's website at highschool.bfwpub.com/updatedtps6e.

9. Calculating least-squares regression lines	Page 184
10. Making residual plots	Page 187

Find the Technology Corners easily by consulting the summary table at the end of each section or the complete table at the back of the book.

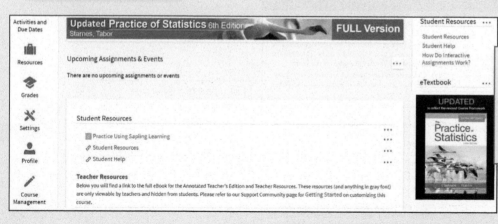

Read, practice, access the resources, and do homework assignments online with the new **Online Homework and e-Book Platform** that may be purchased to enhance your learning experience.

OVERVIEW:
What Is Statistics?

Does listening to music while studying help or hinder learning? If an athlete fails a drug test, how sure can we be that she took a banned substance? Does having a pet help people live longer? How well do SAT scores predict college success? Do most people recycle? Which of two diets will help obese children lose more weight and keep it off? Can a new drug help people quit smoking? How strong is the evidence for global warming?

These are just a few of the questions that statistics can help answer. But what is statistics? And why should you study it?

Statistics Is the Science of Learning from Data

istockphoto

Data are usually numbers, but they are not "just numbers." *Data are numbers with a context.* The number 10.5, for example, carries no information by itself. But if we hear that a family friend's new baby weighed 10.5 pounds at birth, we congratulate her on the healthy size of the child. The context engages our knowledge about the world and allows us to make judgments. We know that a baby weighing 10.5 pounds is quite large, and that a human baby is unlikely to weigh 10.5 ounces or 10.5 kilograms. The context makes the number meaningful.

In your lifetime, you will be bombarded with data and statistical information. Poll results, television ratings, music sales, gas prices, unemployment rates, medical study outcomes, and standardized test scores are discussed daily in the media. Using data effectively is a large and growing part of most professions. A solid understanding of statistics will enable you to make sound, data-based predictions, decisions, and conclusions in your career and everyday life.

Data Beat Personal Experiences

It is tempting to base conclusions on your own experiences or the experiences of those you know. But our experiences may not be typical. In fact, the incidents that stick in our memory are often the unusual ones.

Do Cell Phones Cause Brain Cancer?

Bloomberg via Getty Images

Italian businessman Innocente Marcolini developed a brain tumor at age 60. He also talked on a cellular phone up to 6 hours per day for 12 years as part of his job. Mr. Marcolini's physician suggested that the brain tumor may have been caused by cell-phone use. So Mr. Marcolini decided to file suit in the Italian court system. A court ruled in his favor in October 2012.

Several statistical studies have investigated the link between cell-phone use and brain cancer. One of the largest was conducted by the Danish Cancer Society.

Over 350,000 residents of Denmark were included in the study. Researchers compared the brain-cancer rate for the cell-phone users with the rate in the general population. The result: no statistical difference in brain-cancer rates.[1] In fact, most studies have produced similar conclusions. In spite of the evidence, many people (like Mr. Marcolini) are still convinced that cell phones can cause brain cancer.

In the public's mind, the compelling story wins every time. A statistically literate person knows better. *Data are more reliable than personal experiences because they systematically describe an overall picture, rather than focus on a few incidents.*

Where the Data Come from Matters

Are You Kidding Me?

The famous advice columnist Ann Landers once asked her readers, "If you had it to do over again, would you have children?" A few weeks later, her column was headlined "70% OF PARENTS SAY KIDS NOT WORTH IT." Indeed, 70% of the nearly 10,000 parents who wrote in said they would not have children if they could make the choice again. Do you believe that 70% of all parents regret having children?

You shouldn't. The people who took the trouble to write to Ann Landers are not representative of all parents. Their letters showed that many of them were angry with their children. All we know from these data is that there are some unhappy parents out there. A statistically designed poll, unlike Ann Landers's appeal, targets specific people chosen in a way that gives all parents the same chance to be asked. Such a poll showed that 91% of parents *would* have children again.

Where data come from matters a lot. If you are careless about how you get your data, you may announce 70% "No" when the truth is close to 90% "Yes."

Who Talks More—Women or Men?

According to Louann Brizendine, author of *The Female Brain*, women say nearly 3 times as many words per day as men. Skeptical researchers devised a study to test this claim. They used electronic devices to record the talking patterns of 396 university students from Texas, Arizona, and Mexico. The device was programmed to record 30 seconds of sound every 12.5 minutes without the carrier's knowledge. What were the results?

According to a published report of the study in *Scientific American*, "Men showed a slightly wider variability in words uttered. . . . But in the end, the sexes came out just about even in the daily averages: women at 16,215 words and men at 15,669."[2] When asked where she got her figures, Brizendine admitted that she used unreliable sources.[3]

The most important information about any statistical study is how the data were produced. Only carefully designed studies produce results that can be trusted.

Always Plot Your Data

Yogi Berra, a famous New York Yankees baseball player known for his unusual quotes, had this to say: "You can observe a lot just by watching." That's a motto for learning from data. *A carefully chosen graph helps us describe patterns in data and identify important departures from those patterns.*

Do People Live Longer in Wealthier Countries?

The Gapminder website, www.gapminder.org, provides loads of data on the health and well-being of the world's inhabitants. The graph below displays some data from Gapminder.[4] The individual points represent all the world's nations for which data are available. Each point shows the income per person and life expectancy for one country, along with the region (color of point) and population (size of point).

We expect people in richer countries to live longer. The overall pattern of the graph does show this, but the relationship has an interesting shape. Life expectancy rises very quickly as personal income increases and then levels off. People in very rich countries like the United States live no longer than people in poorer but not extremely poor nations. In some less wealthy countries, people live longer than in the United States. Several other nations stand out in the graph. What's special about each of these countries?

Graph of the life expectancy of people in many nations against each nation's income per person in 2015.

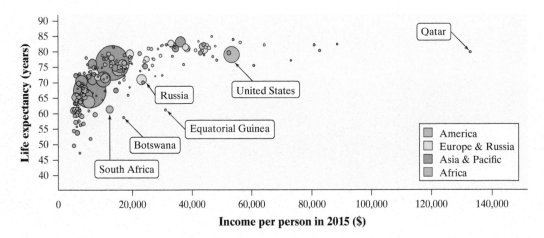

Variation Is Everywhere

Individuals vary. Repeated measurements on the same individual vary. Chance outcomes—like spins of a roulette wheel or tosses of a coin—vary. Almost everything varies over time. Statistics provides tools for understanding variation.

Have Most Students Cheated on a Test?

Researchers from the Josephson Institute were determined to find out. So they surveyed about 23,000 students from 100 randomly selected schools (both public and private) nationwide. The question was: "How many times have you cheated during a test at school in the past year?" Fifty-one percent said they had cheated at least once.[5]

If the researchers had asked the same question of *all* high school students, would exactly 51% have answered "Yes"? Probably not. If the Josephson Institute had selected a different sample of about 23,000 students to respond to the survey, they would probably have gotten a different estimate. *Variation is everywhere!*

Fortunately, statistics provides a description of how the sample results will vary in relation to the actual population percent. Based on the sampling method that this study used, we can say that the estimate of 51% is very likely to be within 1% of the true population value. That is, we can be quite confident that between 50% and 52% of *all* high school students would say that they have cheated on a test.

Because variation is everywhere, conclusions are uncertain. Statistics gives us the tools to quantify our uncertainty, allowing for valid, data-based predictions, decisions, and conclusions.

UPDATED
to reflect the revised Course Framework

for the AP® Exam

The
Practice of
Statistics

Chapter 1

Chapter 1

Data Analysis

Overview

If you haven't yet read the Introduction to the Annotated Teacher's Edition, please do this now. There are many great pieces of advice that apply to all chapters, including this one!

Chapters 1–3 present the principles and tools of basic data analysis. In Chapter 1, we introduce categorical and quantitative variables and the idea of a distribution. In Section 1.1, we show how to describe the distribution of a categorical variable and the relationship between two categorical variables. In Sections 1.2 and 1.3, we focus on graphing and summarizing distributions of a single quantitative variable.

Here are three principles for effectively teaching data analysis.

1. **Emphasize the Strategy, Not Just the Skills:** It is easy to treat data analysis as simply a collection of descriptive statistics such as the mean, median, and standard deviation. However, as part of a larger strategy for exploring data, we also introduce ways to visualize the data using dotplots, stemplots, histograms, and boxplots. We emphasize this strategy by following a consistent approach to data analysis:

   • Begin with a graph, move to numerical descriptions of specific aspects of the data, and (sometimes) to a compact mathematical model. Determining which graphs, numerical summaries, and models are appropriate depends on the context and type of data.

   • Look for an overall pattern and for striking departures from that pattern. Departures such as outliers may influence the choice of descriptive summaries, and the presence and clarity of the overall pattern suggest what models may be useful.

2. **Focus on Understanding, Not Just Mechanics:** Statistics exams would be much easier to grade if correctly graphing a histogram or calculating a standard deviation by hand showed mastery of data analysis. Unfortunately, students can possess these skills and still have no idea what data analysis is all about. Modern statistics courses use technology to quickly graph the data and calculate the associated numerical summaries, so the focus of the course is on what we can learn from these summaries. This includes knowing which summaries are appropriate in different situations and how to interpret the summaries. It also includes being able to communicate this knowledge. Students are often surprised with how much writing and how little "math" they have to do in AP® Statistics!

In Chapters 1–3, we present a variety of tools for data analysis. Typically, we start with "by hand" methods when introducing a new graph or numerical summary and then follow with a more technological approach. Some teachers move to the technology right away, while others wait until students feel comfortable with the "by hand" methods. In any case, it is much more important that students understand what information the summaries provide, rather than the mechanics of how they are calculated.

3. **Use Real Data:** Remember the mantra: Data are not just numbers; they are numbers with a context. The context enables students to communicate conclusions in words and to judge whether their conclusions are sensible. Data usually come with a little bit of background, though for an introductory course the background may not fully reflect the complexities of the real world. We're willing to simplify for the sake of clarity, but not to ask students to perform empty operations with mere numbers.

We have tried to provide small- and moderate-size data sets in adequate number for basic instruction, both in the student text and in the *Annotated Teacher's Edition*. All the medium to large data sets used in examples, exercises, and projects are available in a variety of formats in the book's platform, on the Teacher's Resource Flash Drive, and on the Student Site (highschool .bfwpub.com/updatedtps6e) for easy downloading. The Resources section in the Introduction to the *Annotated Teacher's Edition* also highlights additional places to look for good data sets. Finally, nothing is more interesting to students than data about themselves. The more data you can collect in class the better. Consider using Mr. Starnes's Infamous AP® Statistics Survey in the Teacher's Resource Materials to build up your collection of student data.

The Main Ideas

One of the challenges in teaching the AP® Statistics course is keeping students focused on the big picture and not just the details of each section. We outline the main ideas for the chapter here.

Chapter 1 Introduction, Statistics: The Science and Art of Data

In this brief section, we introduce several fundamental concepts that will be important throughout the course: the idea of a distribution and the distinction between quantitative and categorical variables. We also introduce a strategy for exploring data:

- Begin by examining each variable by itself. Then move on to study relationships between variables.
- Start with a graph or graphs. Then add numerical summaries.

Also included in this section is an excellent activity for the first day of class. We strongly encourage you to use it with your students. The activity introduces students to the major themes in AP® Statistics and gives them a taste of what the rest of the course will be like. In less than one class period, you can have your students identify the evidence for a claim, use simulation to create a sampling distribution, and analyze the distribution to make a decision using the same inferential reasoning that they will be learning later in the course.

Section 1.1 Analyzing Categorical Data

In Section 1.1, we use our data analysis strategy to explore categorical data. Starting with the distribution of a single categorical variable, we show students how to display the data with bar graphs, including segmented bar graphs and mosaic plots (and sometimes pie charts), and what to look for when describing these displays. Make sure students get in the habit of properly labeling their graphs. Failure to do this is a very common error on the AP® Statistics exam for which students can lose points. Students should also be able to recognize misleading graphs and be careful to avoid making misleading graphs themselves.

Next, we show students how to analyze the relationship between two categorical variables. After looking at the *marginal* relative frequencies of each variable separately, students should consider *conditional* relative frequencies of one variable for each category of the other variable. Graphing and comparing the distribution of these conditional relative frequencies will allow students to look for an association between the variables. If there is no association between the two variables, then graphs of the distribution of conditional relative frequencies will look the same. However, if there are differences in the distribution of conditional relative frequencies, an association exists between the variables. In the fourth edition of this book, we included optional material on Simpson's paradox in this section. This content has been moved to the Teacher's Resource Materials. There is one homework exercise that can be used to introduce Simpson's paradox to students (page 28).

Section 1.2 Displaying Quantitative Data with Graphs

In this section, we begin our study of quantitative variables by introducing three types of graphs: dotplots, stemplots, and histograms. Each of the graphs has its own distinct benefits, but all of them are good tools for examining the distribution of a quantitative variable. On the AP® Statistics exam, students will sometimes be expected to create these types of graphs, label them properly, and comment on their characteristics; but more often, they will be asked to do some analysis based on graphs provided.

Dotplots are typically the easiest graph to construct, and many students may already be familiar with this type of display. A dotplot is much less abstract than a histogram or boxplot, making it easier for students to focus on interpreting the plot rather than trying to understand what it represents. For this reason, we will continue to use dotplots throughout the book as one of our main ways to display distributions of quantitative data.

Stemplots are similar to dotplots in that they give a quick picture of the distribution and show each individual observation. They also have the added benefit of including the numerical values of the observations. However, this means that stemplots of large data sets can be cumbersome.

Histograms group numerical observations into intervals and then use bars to represent the number of observations in each interval. Histograms are the best choice when there is a large number of observations. Be sure to emphasize the connection between dotplots and histograms.

To examine the distribution of a quantitative variable, students should look at a graph for the overall pattern (shape, center, variability) and striking departures from that pattern (outliers). We introduce the acronym SOCV (shape, outliers, center, variability) to help students remember these four characteristics. Likewise, when comparing distributions, students should include explicit comparison words such as "the center of the distribution of male heights is *greater than* the center of the distribution of female heights." A very common error on the AP® Statistics exam is describing the characteristics of each distribution separately without making these explicit comparisons. When describing/comparing distribution(s), students should always include the context of the variable name.

In this section, we only briefly address center and variability and focus on shape. Because there will be a much more thorough discussion of center and variability in Section 1.3, we recommend that you spend most of your time in this section discussing the various shapes that distributions can take and save the details of center and variability for later.

Section 1.3 Describing Quantitative Data with Numbers

In this section, we introduce numerical summaries to describe the center and variability of a distribution of quantitative data and show how to identify outliers in a distribution. Although it can be helpful if students know how to calculate summary statistics by hand, more important is that students know when they are appropriate to use and how to interpret them. With this in mind, we also present instructions for using technology to compute these numerical summaries. By the end of this section, students should feel very comfortable comparing several distributions of quantitative data by explicitly discussing the similarities and differences in their shape, outliers, center, and variability.

To measure the center of a distribution of quantitative data, students should use the mean or median of the distribution. Students should understand that the median is a more resistant measure of center than the mean and should be able to choose which measure of center is more appropriate for various distribution shapes.

To measure the variability of a distribution of quantitative data, students can use the range, interquartile range, or standard deviation. The interquartile range (IQR) is a more resistant measure of variability than the range because it ignores the upper 25% and lower 25% of the distribution. The standard deviation is the most commonly used measure of variability and approximates the typical distance of the values in a distribution from the mean. The standard deviation is less resistant than the IQR because it uses every value in the distribution rather than just the middle 50%. Having students develop a solid understanding of what the standard deviation measures will definitely help them be successful in later chapters.

To identify outliers in a distribution of quantitative data, students are introduced to a rule using the quartiles and interquartile range of a distribution. We also introduce boxplots as a graphical way to summarize a distribution of quantitative data. Boxplots are great for comparing distributions because they identify the five-number summary, allowing students to compare center (median) and variability (range, IQR). However, boxplots do not reveal modes, clusters, or gaps in a distribution. As always, it is imperative that students use proper labels when creating a visual display of data!

A project (American Community Survey) has been added at the end of Chapter 1 that will give students a chance to work with a large data set. Students will need some sort of statistical software to properly analyze the data presented in this project. The Case Study featured in the previous two editions of the book has been moved to the Teacher's Resource Materials as an optional resource.

Chapter 1 Resources
Teacher's Resource Materials

The following resources, identified by the TRM in the annotated student pages, can be found by clicking on the link in the Teacher's e-Book (TE-Book), searching by category or chapter on the Teacher's Resource Flash Drive (TRFD), or logging into the book's digital platform and searching the Teacher's Resources menu (teacher log-in required).

- Alternate Examples: one file per section
- Lecture Presentation slides: one per section
- Additional First-Day Activities
  - Mr. Starnes's Infamous AP® Statistics Survey
  - Getting to Know Your Textbook
- Chapter 1 Learning Target Grid
- Smelling Parkinson's activity
- Simpson's paradox optional material
- "Motivating SOCV" activity
- "Do you know your geography?" activity
- "The memory game" activity
- FRAPPY! Materials
- Chapter 1 Project American Community Survey
  - Chapter 1 Project data
  - Chapter 1 Project code sheet
  - Chapter 1 Project solutions
- Chapter 1 Case Study
- Complete solutions for the Check Your Understanding problems, section exercises, review exercises, and practice test
- Quizzes: one per section
- Chapter 1 Test

Free Response Questions from Previous AP® Statistics Exams

Questions can be found on the AP® Central website: apcentral .collegeboard.org/courses/ap-statistics/exam

Students should be able to answer all the free response questions listed with material learned in this chapter. Questions that contain content from this chapter but also require content from later chapters are listed in the last chapter required to complete the entire question. This list will be updated after each AP® Statistics exam and will be posted to the Teacher's Resource section of the book's digital platform and to www.statsmedic.com/free-response-questions. Questions marked with an asterisk in later chapters are from exams with released multiple-choice questions.

Year	#	Content
2019	1	• Describing a bimodal histogram • Identifying outliers • Boxplots versus histograms
2018	5	• Medians from histograms • Mean from combined sample • Mean $+/-$ SD on histogram
2017	4	• Comparing boxplots • Using boxplots to classify
2016	1	• Describing a distribution • Effect of changing a value on mean and median
2015	1	• Comparing boxplots • Using boxplots to make decisions
2011B	1	• Estimating medians of histograms • Comparing histograms • Relationship between mean and median
2010B	1	• Comparing distributions (boxplots) • Stemplots • Comparing boxplots and stemplots
2007B	1	• Constructing a stemplot • Summarizing a distribution of univariate data (stemplot) • Bimodal distribution
2006	1	• Comparing distributions of univariate data (dotplots) • Comparing variability • Measuring center
2005B	1	• Describing shape of a stemplot • Mean vs. median • Midrange
2004	1	• Constructing parallel boxplots • Outliers • Properties of boxplots • Mean vs. median
2002B	5	• Constructing parallel boxplots • Comparing distributions of univariate data (boxplots)
2001	1	• Identifying outliers and unusual values
2000	3	• Graphing and comparing two frequency distributions

Applets

- The *Mean and Median* applet allows students to add, delete, and drag points to investigate the relationship between the mean and the median. Go to highschool .bfwpub.com/updatedtps6e, navigate to the Student Site, click on "Statistical Applets," and choose "Mean and Median."

- Other applets allow you to calculate summary statistics and make graphical displays of categorical and quantitative data, including dotplots, stemplots, histograms, and boxplots. They are available on the Student Site at highschool.bfwpub.com/updatedtps6e under the Extra Applets menu.

Chapter 1: Pacing Guide, Learning Targets, and Suggested Assignments

This pacing guide is based on a schedule with 110, 50-minute sessions before the AP® Statistics exam. If you have a different number of sessions before the AP® exam, you can modify the pacing guide to suit your needs. If you have additional time, consider incorporating quizzes, released AP® Statistics free response questions, additional activities, and projects. See the Resources section above for suggestions.

The suggested homework assignments list odd-numbered exercises, whenever possible, so students can check their answers against the back-of-book answers. If you would rather students not have access to the answers while doing homework, adding 1 to the exercise numbers usually will do the trick, because the homework exercises typically are paired. For example, Exercises 1 and 2 will generally cover the same topics, but in different contexts. You may also choose to include the Recycle and Review questions at the end of each section, which review topics from previous sections or chapters.

If your school is using the digital platform, Sapling Plus, that accompanies Updated TPS6, you will find these assignments prebuilt as online homework assignments for Chapter 1.

Day	Content	Learning Targets: Students will be able to . . .	Suggested Assignment (MC bold)
1	Chapter 1 Introduction	• Identify the individuals and variables in a set of data. • Classify variables as categorical or quantitative.	1, 3, 5, 7, **9, 10**
2	1.1 Bar Graphs and Pie Charts, Graphs: Good and Bad, Analyzing Data on Two Categorical Variables	• Make and interpret bar graphs for categorical data. • Identify what makes some graphs of categorical data misleading. • Calculate marginal and joint relative frequencies from a two-way table.	13, 15, 17, 19, 21, 23
3	1.1 Relationships Between Two Categorical Variables	• Calculate conditional relative frequencies from a two-way table. • Use bar graphs to compare distributions of categorical data. • Describe the nature of the association between two categorical variables.	27, 29, 33, 35, **40–43**
4	1.2 Dotplots, Stemplots, Histograms, Describing Shape	• Make and interpret dotplots, stemplots, and histograms of quantitative data. • Identify the shape of a distribution from a graph.	45, 49, 51, 59, 63
5	1.2 Describing Distributions, Comparing Distributions, Using Histograms Wisely	• Describe the overall pattern (shape, center, and variability) of a distribution and identify any major departures from the pattern (outliers). • Compare distributions of quantitative data using dotplots, stemplots, and histograms.	55, 65, 69, 77, **80–85**
6	1.3 Measuring Center: Mean and Median, Comparing the Mean and Median, Measuring Variability: Range, Standard Deviation and *IQR*	• Calculate measures of center (mean, median) for a distribution of quantitative data. • Calculate and interpret measures of variability (range, standard deviation, *IQR*) for a distribution of quantitative data. • Explain how outliers and skewness affect measures of center and variability.	87, 89, 91, 95, 97, 101, 103, 105, 121
7	1.3 Identifying Outliers, Making and Interpreting Boxplots, Comparing Distributions with Boxplots	• Identify outliers using the $1.5 \times IQR$ rule. • Make and interpret boxplots of quantitative data. • Use boxplots and numerical summaries to compare distributions of quantitative data.	109, 111, 113, 115, **123–126**
8	Chapter 1 Review/FRAPPY!		Chapter 1 Review Exercises
9	Chapter 1 Test		

Chapter 1 Alignment to the College Board's Fall 2019 AP® Statistics Course Framework*

Relationship to College Board Units

Chapter 1 in this book covers Topics 1.1–1.9 in Unit 1 of the College Board Course Framework. Students will be ready to take the Personal Progress Check for Unit 1 once they have completed Chapters 1 and 2.

Big Ideas and Enduring Understandings

Chapter 1 develops these Big Ideas and related Enduring Understandings outlined in the Course Framework:

- **Big Idea 1: Variation and Distribution (EU: VAR 1):** The distribution of measures for individuals within a sample or population describes variation. The value of a statistic varies from sample to sample. How can we determine whether differences between measures represent random variation or meaningful distinctions? Statistical methods based on probabilistic reasoning provide the basis for shared understandings about variation and about the likelihood that variation between and among measures, samples, and populations is random or meaningful.
- **Big Idea 2: Patterns and Uncertainty (EU: UNC 1):** Statistical tools allow us to represent and describe patterns in data and to classify departures from patterns. Simulation and probabilistic reasoning allow us to anticipate patterns in data and to determine the likelihood of errors in inference.

Course Skills

Chapter 1 helps students to develop the skills identified in the Course Framework.

- 1: **Selecting Statistical Methods** (1.A)
- 2: **Data Analysis** (2.A, 2.B, 2.C, 2.D)
- 4: **Statistical Argumentation** (4.B)

Learning Objectives and Essential Knowledge Statements

Section	Learning Objectives	Essential Knowledge Statements
Introduction	VAR-1.A, VAR-1.B, VAR-1.C, UNC-1.F	VAR-1.A.1, VAR-1.B.1, VAR-1.C.1, VAR-1.C.2, UNC-1.F.1, UNC-1.F.2
1.1	UNC-1.A, UNC-1.B, UNC-1.C, UNC-1.D, UNC-1.E, UNC-1.P, UNC-1.Q, UNC-1.R, VAR-1.D	UNC-1.A.1, UNC-1.B.1, UNC-1.B.2, UNC-1.C.1, UNC-1.C.2, UNC-1.C.3, UNC-1.D.1, UNC-1.E.1, UNC-1.P.1, UNC-1.P.2, UNC-1.P.3, UNC-1.P.4, UNC-1.Q.1, UNC-1.Q.2, UNC-1.R.1, VAR-1.D.1
1.2	UNC-1.G, UNC-1.H, UNC-1.N	UNC-1.G.1, UNC-1.G.2, UNC-1.G.3, UNC-1.G.5, UNC-1.H.1, UNC-1.H.2, UNC-1.H.3, UNC-1.H.4, UNC-1.H.5, UNC-1.H.6, UNC-1.H.7, UNC-1.N.1
1.3	UNC-1.I, UNC-1.J, UNC-1.K, UNC-1.L, UNC-1.M, UNC-1.O, VAR-2.A	UNC-1.I.1, UNC-1.I.2, UNC-1.I.3, UNC-1.I.4, UNC-1.J.1, UNC-1.J.2, UNC-1.J.3, UNC-1.K.1, UNC-1.K.2, UNC-1.L.1, UNC-1.L.2, UNC-1.M.1, UNC-1.M.2, UNC-1.O.1, VAR-2.A.1

A detailed alignment (The Nitty Gritty Guide) that can be sorted by Course Framework Unit, Topic, Learning Objective, Essential Knowledge Statement, or textbook section, is available on the TRFD and in the Teacher's Resources folder on Sapling Plus. **TRM**

*Should changes be made to the Course Framework in the future, an updated alignment will be placed on our AP® updates page at go.bfwpub.com/ap-course-updates.

UNIT 1
Exploring One-Variable Data

Chapter 1

Data Analysis

AP-Photo.com/Alamy Stock Photo

Preparing for Inference

One of the major content goals of AP® Statistics is developing students' inferential thinking. Inference is not an idea that should be saved for the end of the course. Rather, it is a big concept that must be built up slowly throughout the course. This feature will help teachers to recognize and highlight the ideas in each chapter that will lead to student success in formal inference.

Making Connections

The core skills and concepts in statistics build topic by topic to culminate in statistical inference. New ideas are always connected to those previously learned and set the stage for subsequent learning. This feature will help you to identify and demonstrate these connections to students so that you can build a cohesive story for the course.

Teaching Tip

Unit 1 in the College Board Course Framework aligns to Chapters 1 and 2 in this book. Students will be ready to take the Personal Progress Check #1 after completing Chapters 1 and 2.

PD **Chapter 1 Overview**

To watch the video overview of Chapter 1 (for teachers), click on the link in the TE-Book, look on the TRFD, or download from the Teacher's Resources on the book's digital platform.

TRM **Lecture Presentation Slides**

If you are new to teaching AP® Statistics or are short on time when preparing for class, you may find the Lecture Presentation Slides to be helpful. Experienced AP® Teacher Doug Tyson has created one slide presentation per section. You may use them as is, modify them to fit your needs, or share them with students who miss class. Find them on the TRFD and in the Teacher's Resources on the book's digital platform.

Teaching Tip

As you use this *Annotated Teacher's Edition*, be sure to pay close attention to three important features that appear in the margins. They are:

+ **Ask the StatsMedic**

Blog posts centered on high-quality statistics instruction are created at www.statsmedic.com and are linked to the Teacher's Edition. Topics include insights on understanding the content better, best practices for teaching for understanding, and using statistics in the real world. Many blog posts include links to other sites, pictures of student work, and ideas about motivating students for success.

INTRODUCTION Statistics: The Science and Art of Data

LEARNING TARGETS *By the end of the section, you should be able to:*

- Identify the individuals and variables in a set of data.
- Classify variables as categorical or quantitative.

We live in a world of *data*. Every day, the media report poll results, outcomes of medical studies, and analyses of data on everything from stock prices to standardized test scores to global warming. The data are trying to tell us a story. To understand what the data are saying, you need to learn more about **statistics.**

> **DEFINITION** **Statistics**
>
> **Statistics** is the science and art of collecting, analyzing, and drawing conclusions from data.

A solid understanding of statistics will help you make good decisions based on data in your daily life.

Organizing Data

Rudy Sulgan/Corbis Documentary/Getty Images

Every year, the U.S. Census Bureau collects data from over 3 million households as part of the American Community Survey (ACS). The table displays some data from the ACS in a recent year.

Household	Region	Number of people	Time in dwelling (years)	Response mode	Household income	Internet access?
425	Midwest	5	2–4	Internet	52,000	Yes
936459	West	4	2–4	Mail	40,500	Yes
50055	Northeast	2	10–19	Internet	481,000	Yes
592934	West	4	2–4	Phone	230,800	No
545854	South	9	2–4	Phone	33,800	Yes
809928	South	2	30+	Internet	59,500	Yes
110157	Midwest	1	5–9	Internet	80,000	Yes
999347	South	1	<1	Mail	8,400	No

Most data tables follow this format—each row describes an **individual** and each column holds the values of a **variable**.

Sometimes the individuals in a data set are called *cases* or *observational units*.

> **DEFINITION**   Individual, Variable
>
> An **individual** is an object described in a set of data. Individuals can be people, animals, or things.
>
> A **variable** is an attribute that can take different values for different individuals.

For the American Community Survey data set, the *individuals* are households. The *variables* recorded for each household are region, number of people, time in current dwelling, survey response mode, household income, and whether the dwelling has Internet access. Region, time in dwelling, response mode, and Internet access status are **categorical variables**. Number of people and household income are **quantitative variables**.

Note that household is *not* a variable. The numbers in the household column of the data table are just labels for the individuals in this data set. Be sure to look for a column of labels—names, numbers, or other identifiers—in any data table you encounter.

> **DEFINITION**   Categorical variable, Quantitative variable
>
> A **categorical variable** assigns labels that place each individual into a particular group, called a category.
>
> A **quantitative variable** takes number values that are quantities—counts or measurements.

 Not every variable that takes number values is quantitative. Zip code is one example. Although zip codes are numbers, they are neither counts of anything, nor measurements of anything. They are simply labels for a regional location, making zip code a categorical variable. Some variables—such as gender, race, and occupation—are categorical by nature. Time in dwelling from the ACS data set is also a categorical variable because the values are recorded as intervals of time, such as 2–4 years. If time in dwelling had been recorded to the nearest year for each household, this variable would be quantitative.

To make life simpler, we sometimes refer to *categorical data* or *quantitative data* instead of identifying the variable as categorical or quantitative.

EXAMPLE	**Census At School**
	Individuals and Variables

PROBLEM: Census At School is an international project that collects data about primary and secondary school students using surveys. Hundreds of thousands of students from Australia, Canada, Ireland, Japan, New Zealand, South Africa, South Korea, the United Kingdom, and the United States have taken part in the project. Data from the surveys are available online. We used the site's "Random Data Selector" to choose 10 Canadian students who completed the survey in a recent year. The table displays the data.

Garry Black/Alamy

Teaching Tip

Emphasize that "individuals" don't have to be people. For example, the individuals in a study could be the trees in an orchard or the M&M'S® chocolate candies in a bag.

Teaching Tip

Notice the categorical variable that was identified as "Internet access *status*," rather than simply "Internet access." Describing the variable as "Internet access" would be missing a reference to levels or status and is not sufficient.

Teaching Tip

A **statistical question** is one in which the answer is based on data that vary depending on the individual that is chosen.

Teaching Tip

Any quantitative variable can be made into a categorical variable by creating groups for certain intervals of values.

Teaching Tip

The icon at the top of this example (and many others) indicates that students can watch a video presentation of the example by clicking on the link in the e-Book or viewing the videos on the Student Site at highschool.bfwpub.com /updatedtps6e. If you have a projector and Internet access, we recommend that you show your students how to access the videos.

ALTERNATE EXAMPLE						Skill 2.A
Which car is best for Mr. Starnes? Individuals and variables						

Mr. Starnes is thinking about buying a new car. He has narrowed his choices down to 10 vehicles. Here is information about each vehicle:

Make	Model	Price ($)	Drivetrain	# doors	City MPG	Engine type
Ford	Focus	17,650	FWD	Four	26	Inline 4
Ford	Escape	27,745	4WD	Four	22	Inline 4
Ford	Fusion	23,485	FWD	Four	22	Inline 4
Honda	CR-V	26,045	4WD	Four	25	Inline 4
Honda	Civic	19,985	FWD	Two	28	Inline 4
Honda	Pilot	33,295	AWD	Four	18	V-6
Porsche	Cayman	53,650	RWD	Two	20	Flat-6
Toyota	Camry	23,935	FWD	Four	24	Inline 4
Toyota	Prius	25,550	FWD	Four	54	Electric/Inline 4
Toyota	RAV4	27,250	AWD	Four	22	Inline 4

(continues)

All data taken from www.caranddriver.com.

PROBLEM:
(a) Identify the individuals in this data set.
(b) What are the variables? Classify each as categorical or quantitative.

SOLUTION:
(a) The 10 vehicles that Mr. Starnes has identified for potential purchase.
(b) Categorical: Make, Model, Drivetrain, Number of doors, Engine type
Quantitative: Price ($), City MPG.

TRM Chapter 1 Introduction Alternate Examples

You can find the Alternate Examples for this section in Microsoft Word format by clicking on the link in the TE-Book, opening the TRFD, or downloading from the Teacher's Resources on the book's digital platform.

Making Connections

In Chapter 6, students will distinguish *random variables* as being **discrete** (with gaps between possible values) or **continuous** (no gaps between possible values).

Preparing for Inference

Categorical and quantitative variables each require different inference procedures. Identifying the type of variable will help students choose the correct inference procedure for a given set of data.

Province	Gender	Number of languages spoken	Handedness	Height (cm)	Wrist circumference (mm)	Preferred communication
Saskatchewan	Male	1	Right	175.0	180	In person
Ontario	Female	1	Right	162.5	160	In person
Alberta	Male	1	Right	178.0	174	Facebook
Ontario	Male	2	Right	169.0	160	Cell phone
Ontario	Female	2	Right	166.0	65	In person
Nunavut	Male	1	Right	168.5	160	Text messaging
Ontario	Female	1	Right	166.0	165	Cell phone
Ontario	Male	4	Left	157.5	147	Text messaging
Ontario	Female	2	Right	150.5	187	Text messaging
Ontario	Female	1	Right	171.0	180	Text messaging

(a) Identify the individuals in this data set.
(b) What are the variables? Classify each as categorical or quantitative.

SOLUTION:

(a) 10 randomly selected Canadian students who participated in the Census At School survey.

(b) Categorical: Province, gender, handedness, preferred communication method

Quantitative: Number of languages spoken, height (cm), wrist circumference (mm)

> We'll see in Chapter 4 why choosing at random, as we did in this example, is a good idea.

> There is at least one suspicious value in the data table. We doubt that the girl who is 166 cm tall really has a wrist circumference of 65 mm (about 2.6 inches). Always look to be sure the values make sense!

FOR PRACTICE, TRY EXERCISE 1

There are two types of quantitative variables: *discrete* and *continuous*. Most **discrete variables** result from counting something, like the number of languages spoken in the preceding example. **Continuous variables** typically result from measuring something, like height or wrist circumference. Be sure to report the units of measurement (like centimeters for height and millimeters for wrist circumference) for a continuous variable.

> **DEFINITION   Discrete variable, Continuous variable**
>
> A quantitative variable that takes a fixed set of possible values with gaps between them is a **discrete variable**.
>
> A quantitative variable that can take any value in an interval on the number line is a **continuous variable**.

The proper method of data analysis depends on whether a variable is categorical or quantitative. For that reason, it is important to distinguish these two types of variables. The type of data determines what kinds of graphs and which numerical summaries are appropriate.

ANALYZING DATA A variable generally takes values that vary (hence the name *variable*!). Categorical variables sometimes have similar counts in each category and sometimes don't. For instance, we might have expected similar numbers of

males and females in the Census At School data set. But we aren't surprised to see that most students are right-handed. Quantitative variables may take values that are very close together or values that are quite spread out. We call the pattern of variation of a variable its **distribution**.

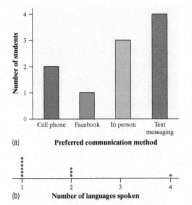

FIGURE 1.1 (a) Bar graph showing the distribution of preferred communication method for the sample of 10 Canadian students. (b) Dotplot showing the distribution of number of languages spoken by these students.

> **DEFINITION   Distribution**
>
> The **distribution** of a variable tells us what values the variable takes and how often it takes those values.

Let's return to the data for the sample of 10 Canadian students from the preceding example. Figure 1.1(a) shows the distribution of preferred communication method for these students in a *bar graph*. We can see how many students chose each method from the heights of the bars: cell phone (2), Facebook (1), in person (3), text messaging (4). Figure 1.1(b) shows the distribution of number of languages spoken in a *dotplot*. We can see that 6 students speak one language, 3 students speak two languages, and 1 student speaks four languages.

Section 1.1 begins by looking at how to describe the distribution of a single categorical variable and then examines relationships between categorical variables. Sections 1.2 and 1.3 and all of Chapter 2 focus on describing the distribution of a quantitative variable. Chapter 3 investigates relationships between two quantitative variables. In each case, we begin with graphical displays, then add numerical summaries for a more complete description.

HOW TO ANALYZE DATA

- Begin by examining each variable by itself. Then move on to study relationships among the variables.
- Start with a graph or graphs. Then add numerical summaries.

This process of exploratory data analysis is known as *descriptive statistics*.

CHECK YOUR UNDERSTANDING

Jake is a car buff who wants to find out more about the vehicles that his classmates drive. He gets permission to go to the student parking lot and record some data. Later, he does some Internet research on each model of car he found. Finally, Jake makes a spreadsheet that includes each car's license plate, model, number of cylinders, color, highway gas mileage, weight, and whether it has a navigation system.

1. Identify the individuals in Jake's study.
2. What are the variables? Classify each as categorical or quantitative.
3. Identify each quantitative variable as discrete or continuous.

An alternative activity that explores the same concepts is found in the "Smelling Parkinson's" handout created by Doug Tyson.

Preparing for Inference

This activity can be used on the first day of class. The context is accessible for all learners, and it gets students to think inferentially.

▶ ACTIVITY OVERVIEW

To help prepare for using this activity, watch the overview video by clicking on the link in the TE-Book, opening the TRFD, or downloading from the Teacher's Resources on the book's digital platform.

Time: 20 minutes

Materials: Bag with 25 beads (15 of one color and 10 of another) or 25 slips of paper (15 labeled "male" and 10 labeled "female") for each group of students

Teaching Advice: Please do this activity! It introduces students to the major themes of the course. Consider making the dotplot on chart paper or a poster and saving it to refer back to when you get to significance tests in Chapter 9. The theoretical probability of 5 or more females is about 0.128.

Extension: Would your advice change if the lottery had chosen 6 female (and 2 male) pilots? What about 7 female pilots? Explain.

Answers:

1–2. Answers will vary for each class.

3. Here is one possible dotplot:

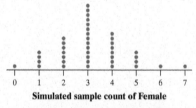

Simulated sample count of Female

4. Yes, it is believable that the lottery was fair because selecting 5 or more female pilots would not be that surprising (6/40 = 0.15). I would advise the male pilot not to file a grievance with the pilots' union.

From Data Analysis to Inference

Sometimes we're interested in drawing conclusions that go beyond the data at hand. That's the idea of *inferential statistics*. In the "Census At School" example, 9 of the 10 randomly selected Canadian students are right-handed. That's 90% of the *sample*. Can we conclude that exactly 90% of the *population* of Canadian students who participated in Census At School are right-handed? No.

If another random sample of 10 students were selected, the percent who are right-handed might not be exactly 90%. Can we at least say that the actual population value is "close" to 90%? That depends on what we mean by "close." The following activity gives you an idea of how statistical inference works.

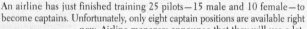

ACTIVITY Hiring discrimination—it just won't fly!

An airline has just finished training 25 pilots—15 male and 10 female—to become captains. Unfortunately, only eight captain positions are available right now. Airline managers announce that they will use a lottery to determine which pilots will fill the available positions. The names of all 25 pilots will be written on identical slips of paper. The slips will be placed in a hat, mixed thoroughly, and drawn out one at a time until all 8 captains have been identified.

A day later, managers announce the results of the lottery. Of the 8 captains chosen, 5 are female and 3 are male. Some of the male pilots who weren't selected suspect that the lottery was not carried out fairly. One of these pilots asks your statistics class for advice about whether to file a grievance with the pilots' union.

The key question in this possible discrimination case seems to be: *Is it plausible (believable) that these results happened just by chance?* To find out, you and your classmates will *simulate* the lottery process that airline managers said they used.

1. Your teacher will give you a bag with 25 beads (15 of one color and 10 of another) or 25 slips of paper (15 labeled "M" and 10 labeled "F") to represent the 25 pilots. Mix the beads/slips thoroughly. Without looking, remove 8 beads/slips from the bag. Count the number of female pilots selected. Then return the beads/slips to the bag.

2. Your teacher will draw and label a number line for a class *dotplot*. On the graph, plot the number of females you got in Step 1.

3. Repeat Steps 1 and 2 if needed to get a total of at least 40 simulated lottery results for your class.

4. Discuss the results with your classmates. Does it seem plausible that airline managers conducted a fair lottery? What advice would you give the male pilot who contacted you?

Our ability to do inference is determined by how the data are produced. Chapter 4 discusses the two main methods of data production—sampling and

Preparing for Inference

The hiring discrimination activity foreshadows several of the big ideas of formal inference:

Hypotheses: To start the activity, we assume that the lottery was carried out fairly. In Chapter 9, this claim will be called the **null hypothesis.**

P-value: Count the number of dots in the dotplot that represent 5 females or more. Convert this count to a percent by dividing by the total number of dots. In Chapter 9, this estimated probability will be called the **P-value.**

Conclusion: If the *P*-value is small, the observed outcome is unlikely to happen purely by chance if the null hypothesis is true, and we have convincing evidence that the lottery was unfair. What constitutes a *P*-value that is "small"? This depends on the context, but a generally accepted rule is that less than 5% is considered **statistically significant** (Chapter 4).

experiments—and the types of conclusions that can be drawn from each. As the activity illustrates, the logic of inference rests on asking, "What are the chances?" *Probability*, the study of chance behavior, is the topic of Chapters 5–7. We'll introduce the most common inference techniques in Chapters 8–12.

Introduction | Summary

- **Statistics** is the science and art of collecting, analyzing, and drawing conclusions from data.
- A data set contains information about a number of **individuals.** Individuals may be people, animals, or things. For each individual, the data give values for one or more **variables.** A variable describes some characteristic of an individual, such as a person's height, gender, or salary.
- A **categorical variable** assigns a label that places each individual in one of several groups, such as male or female. A **quantitative variable** has numerical values that count or measure some characteristic of each individual, such as number of siblings or height in meters.
- There are two types of quantitative variables: discrete and continuous. A **discrete variable** has a fixed set of possible numeric values with gaps between them. A **continuous variable** can take any value in an interval on the number line. Discrete variables usually result from counting something; continuous variables usually result from measuring something.
- The **distribution** of a variable describes what values the variable takes and how often it takes them.

Introduction | Exercises

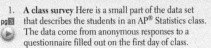

The solutions to all exercises numbered in red may be found in the Solutions Appendix, starting on page S-1.

1. **A class survey** Here is a small part of the data set that describes the students in an AP® Statistics class. The data come from anonymous responses to a questionnaire filled out on the first day of class.

Gender	Grade level	GPA	Children in family	Homework last night (min)	Android or iPhone?
F	9	2.3	3	0–14	iPhone
M	11	3.8	6	15–29	Android
M	10	3.1	2	15–29	Android
F	10	4.0	1	45–59	iPhone
F	10	3.4	4	0–14	iPhone
F	10	3.0	3	30–44	Android
M	9	3.9	2	15–29	iPhone
M	12	3.5	2	0–14	iPhone

(a) Identify the individuals in this data set.

(b) What are the variables? Classify each as categorical or quantitative.

2. **Coaster craze** Many people like to ride roller coasters. Amusement parks try to increase attendance by building exciting new coasters. The following table displays data on several roller coasters that were opened in a recent year.[1]

Roller coaster	Type	Height (ft)	Design	Speed (mph)	Duration (sec)
Wildfire	Wood	187.0	Sit down	70.2	120
Skyline	Steel	131.3	Inverted	50.0	90
Goliath	Wood	165.0	Sit down	72.0	105
Helix	Steel	134.5	Sit down	62.1	130
Banshee	Steel	167.0	Inverted	68.0	160
Black Hole	Steel	22.7	Sit down	25.5	75

(a) Identify the individuals in this data set.

(b) What are the variables? Classify each as categorical or quantitative.

3. **Hit movies** According to the Internet Movie Database, *Avatar* is tops based on box-office receipts worldwide as of January 2017. The following table displays data on several popular movies. Identify the individuals

1.3 The individuals are movies. The variables are year (quantitative), rating (categorical), time (min) (quantitative), genre (categorical), and box office ($) (quantitative). *Note:* Year might be considered categorical if we want to know how many of these movies were made each year, rather than the average year.

1.4 The individuals are the tallest buildings in the world as of February 2017. The variables are country (categorical), height (m) (quantitative), floors (quantitative), use (categorical), and year completed (quantitative).

1.5 The categorical variables are type of wood, type of water repellent, and paint color. The quantitative variables are paint thickness and weathering time.

1.6 The categorical variables are gender, race, and smoker status. The quantitative variables are age, systolic blood pressure, and level of calcium in the blood.

1.7 The discrete variables are number of siblings and how many books they have read in the past month. The continuous variables are the distance from their home to campus and how long it took them to complete an online survey.

1.8 The discrete variables are the number of times they visited the site and the number of likes received. The continuous variables are the time spent on the site and how long since they created a member profile.

1.9 b

1.10 c

and variables in this data set. Classify each variable as categorical or quantitative.

Movie	Year	Rating	Time (min)	Genre	Box office ($)
Avatar	2009	PG-13	162	Action	2,783,918,982
Titanic	1997	PG-13	194	Drama	2,207,615,668
Star Wars: The Force Awakens	2015	PG-13	136	Adventure	2,040,375,795
Jurassic World	2015	PG-13	124	Action	1,669,164,161
Marvel's The Avengers	2012	PG-13	142	Action	1,519,479,547
Furious 7	2015	PG-13	137	Action	1,516,246,709
The Avengers: Age of Ultron	2015	PG-13	141	Action	1,404,705,868
Harry Potter and the Deathly Hallows: Part 2	2011	PG-13	130	Fantasy	1,328,111,219
Frozen	2013	PG	108	Animation	1,254,512,386
Iron Man 3	2013	PG-13	129	Action	1,172,805,920

4. **Skyscrapers** Here is some information about the tallest buildings in the world as of February 2017. Identify the individuals and variables in this data set. Classify each variable as categorical or quantitative.

Building	Country	Height (m)	Floors	Use	Year completed
Burj Khalifa	United Arab Emirates	828.0	163	Mixed	2010
Shanghai Tower	China	632.0	121	Mixed	2014
Makkah Royal Clock Tower Hotel	Saudi Arabia	601.0	120	Hotel	2012
Ping An Finance Center	China	599.0	115	Mixed	2016
Lotte World Tower	South Korea	554.5	123	Mixed	2016
One World Trade Center	United States	541.0	104	Office	2013
Taipei 101	Taiwan	509.0	101	Office	2004
Shanghai World Financial Center	China	492.0	101	Mixed	2008
International Commerce Center	China	484.0	118	Mixed	2010
Petronas Tower 1	Malaysia	452.0	88	Office	1998

5. **Protecting wood** What measures can be taken, especially when restoring historic wooden buildings, to help wood surfaces resist weathering? In a study of this question, researchers prepared wooden panels and then exposed them to the weather. Some of the variables recorded were type of wood (yellow poplar, pine, cedar); type of water repellent (solvent-based, water-based); paint thickness (millimeters); paint color (white, gray, light blue); weathering time (months). Classify each variable as categorical or quantitative.

6. **Medical study variables** Data from a medical study contain values of many variables for each subject in the study. Some of the variables recorded were gender (female or male); age (years); race (Asian, Black, White, or other); smoker (yes or no); systolic blood pressure (millimeters of mercury); level of calcium in the blood (micrograms per milliliter). Classify each variable as categorical or quantitative.

7. **College life** A college admissions office collects data from each incoming freshman on several quantitative variables: distance from their home to campus, number of siblings, how many books they have read in the past month, and how long it took them to complete an online survey. Classify each variable as discrete or continuous.

8. **Social media** A social media company records data from each of its users on several quantitative variables: time spent on the site, how many times they visited the site, number of likes received, and how long since they created a member profile. Classify each variable as discrete or continuous.

Multiple Choice: *Select the best answer.*

Exercises 9 and 10 refer to the following setting. At the Census Bureau website www.census.gov, you can view detailed data collected by the American Community Survey. The following table includes data for 10 people chosen at random from the more than 1 million people in households contacted by the survey. "School" gives the highest level of education completed.

Weight (lb)	Age (years)	Travel to work (min)	School	Gender	Income last year ($)
187	66	0	Ninth grade	1	24,000
158	66	n/a	High school grad	2	0
176	54	10	Assoc. degree	2	11,900
339	37	10	Assoc. degree	1	6000
91	27	10	Some college	2	30,000
155	18	n/a	High school grad	2	0
213	38	15	Master's degree	2	125,000
194	40	0	High school grad	1	800
221	18	20	High school grad	1	2500
193	11	n/a	Fifth grade	1	0

9. The individuals in this data set are

(a) households.   (b) people.   (c) adults.

(d) 120 variables.   (e) columns.

10. This data set contains

(a) 7 variables, 2 of which are categorical.

(b) 7 variables, 1 of which is categorical.

(c) 6 variables, 2 of which are categorical.

(d) 6 variables, 1 of which is categorical.

(e) None of these.

SECTION 1.1 Analyzing Categorical Data

LEARNING TARGETS *By the end of the section, you should be able to:*

- Make and interpret bar graphs for categorical data.
- Identify what makes some graphs of categorical data misleading.
- Calculate marginal and joint relative frequencies from a two-way table.
- Calculate conditional relative frequencies from a two-way table.
- Use bar graphs to compare distributions of categorical data.
- Describe the nature of the association between two categorical variables.

Here are the data on preferred communication method for the 10 randomly selected Canadian students from the example on page 3:

In person	In person	Facebook	Cell phone	In person
Text messaging	Cell phone	Text messaging	Text messaging	Text messaging

We can summarize the distribution of this categorical variable with a **frequency table** or a **relative frequency table**.

Some people use the terms *frequency distribution* and *relative frequency distribution* instead.

DEFINITION Frequency table, Relative frequency table

A **frequency table** shows the number of individuals having each value.

A **relative frequency table** shows the proportion or percent of individuals having each value.

To make either kind of table, start by tallying the number of times that the variable takes each value. Note that the frequencies and relative frequencies listed in these tables are not data. The tables summarize the data by telling us how many (or what proportion or percent of) students in the sample said "Cell phone," "Facebook," "In person," and "Text messaging."

Preferred method	Tally
Cell phone	II
Facebook	I
In person	III
Text messaging	IIII

Frequency table

Preferred method	Frequency
Cell phone	2
Facebook	1
In person	3
Text messaging	4

Relative frequency table

Preferred method	Relative frequency
Cell phone	2/10 = 0.20 or 20%
Facebook	1/10 = 0.10 or 10%
In person	3/10 = 0.30 or 30%
Text messaging	4/10 = 0.40 or 40%

The same process can be used to summarize the distribution of a quantitative variable. Of course, it would be hard to make a frequency table or a relative frequency table for quantitative data that take many different values, like the ages of people attending a Major League Baseball game. We'll look at a better option for quantitative variables with many possible values in Section 1.2.

PD Section 1.1 Overview

The video overview of Section 1.1 (for teachers) can be found by clicking on the link in the TE-Book, opening the TRFD, or downloading from the Teacher's Resources on the book's digital platform.

TRM Section 1.1 Alternate Examples

You can find the Alternate Examples for this section in Microsoft Word format by clicking on the link in the TE-Book, opening the TRFD, or downloading from the Teacher's Resources on the book's digital platform.

Teaching Tip

Frequency tables show counts. Relative frequency tables show the counts *relative* to the total (proportion or percent).

Displaying Categorical Data: Bar Graphs and Pie Charts

A frequency table or relative frequency table summarizes a variable's distribution with numbers. To display the distribution more clearly, use a graph. You can make a **bar graph** or a **pie chart** for categorical data.

> Bar graphs are sometimes called *bar charts*. Pie charts are sometimes called *circle graphs*.

> **DEFINITION** Bar graph, Pie chart
>
> A **bar graph** shows each category as a bar. The heights of the bars show the category frequencies or relative frequencies.
>
> A **pie chart** shows each category as a slice of the "pie." The areas of the slices are proportional to the category frequencies or relative frequencies.

Figure 1.2 shows a bar graph and a pie chart of the data on preferred communication method for the random sample of Canadian students. Note that the percents for each category come from the relative frequency table.

Relative frequency table	
Preferred method	Relative frequency
Cell phone	2/10 = 0.20 or 20%
Facebook	1/10 = 0.10 or 10%
In person	3/10 = 0.30 or 30%
Text messaging	4/10 = 0.40 or 40%

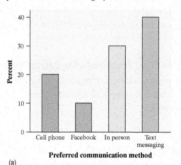
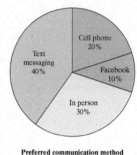

FIGURE 1.2 (a) Bar graph and (b) pie chart of the distribution of preferred communication method for a random sample of 10 Canadian students.

It is fairly easy to make a bar graph by hand. Here's how you do it.

> **HOW TO MAKE A BAR GRAPH**
>
> - **Draw and label the axes.** Put the name of the categorical variable under the horizontal axis. To the left of the vertical axis, indicate whether the graph shows the frequency (count) or relative frequency (percent or proportion) of individuals in each category.
> - **"Scale" the axes.** Write the names of the categories at equally spaced intervals under the horizontal axis. On the vertical axis, start at 0 and place tick marks at equal intervals until you exceed the largest frequency or relative frequency in any category.
> - **Draw bars above the category names.** Make the bars equal in width and leave gaps between them. Be sure that the height of each bar corresponds to the frequency or relative frequency of individuals in that category.

Making a graph is not an end in itself. The purpose of a graph is to help us understand the data. When looking at a graph, always ask, "What do I see?" We can see from both graphs in Figure 1.2 that the most preferred communication method for these students is text messaging.

EXAMPLE

What's on the radio?
Making and interpreting bar graphs

PROBLEM: Arbitron, the rating service for radio audiences, categorizes U.S. radio stations in terms of the kinds of programs they broadcast. The frequency table summarizes the distribution of station formats in a recent year.[2]

(a) Identify the individuals in this data set.

(b) Make a frequency bar graph of the data. Describe what you see.

Format	Number of stations	Format	Number of stations
Adult contemporary	2536	Religious	3884
All sports	1274	Rock	1636
Contemporary hits	1012	Spanish language	878
Country	2893	Variety	1579
News/Talk/Information	4077	Other formats	4852
Oldies	831	**Total**	**25,452**

SOLUTION:

(a) U.S. radio stations

(b)

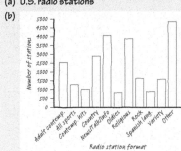

To make the bar graph:
- **Draw and label the axes.**
- **"Scale" the axes.** The largest frequency is 4852. So we choose a vertical scale from 0 to 5000, with tick marks 500 units apart.
- **Draw bars above the category names.**

On U.S. radio stations, the most frequent formats are Other (4852), News/talk/information (4077), and Religious (3884), while the least frequent are Oldies (831), Spanish language (878), and Contemporary hits (1012).

FOR PRACTICE, TRY EXERCISE 11

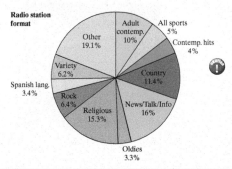

Here is a pie chart of the radio station format data from the preceding example. You can use a pie chart when you want to emphasize each category's relation to the whole. Pie charts are challenging to make by hand, but technology will do the job for you. Note that a pie chart must include all categories that make up a whole, which might mean adding an "other" category, as in the radio station example.

ALTERNATE EXAMPLE Skill 2.B

Who listens to the Beatles?
Making and interpreting bar graphs

PROBLEM:
Spotify tracks data about listeners. The frequency table summarizes the distribution of ages for a sample of 100 Beatles listeners.

Age (years)	Number of listeners
17 and under	5
18–24	25
25–34	33
35–44	16
45–54	10
55+	11

(a) Identify the individuals in this data set.
(b) Make a frequency bar graph of the data. Describe what you see.

SOLUTION:
(a) 100 Spotify users who listen to the Beatles.
(b)

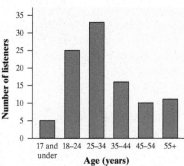

Of the sample of 100 Spotify users who listen to the Beatles, the most frequent ages are 25–34 years (33) and 18–24 years (25), while the least frequent are 17 years and under (5), and 45–54 years (10).

✓ Answers to CYU

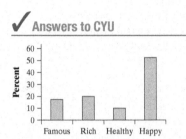

A slight majority (52.5%) of students in the sample said that they would prefer to be Happy. Rich and Famous were preferred about equally (20% and 17.5%). The least popular choice was Healthy (10%).

Teaching Tip

There are many good websites that illustrate deceptive graphs. For example, try www.datavis.ca/gallery/index.php or do an Internet search for "deceptive graphs." The classic book *How to Lie with Statistics* by Darrell Huff also includes some good examples.

▼ CHECK YOUR UNDERSTANDING

The American Statistical Association sponsors a web-based project that collects data about primary and secondary school students using surveys. We used the site's "Random Sampler" to choose 40 U.S. high school students who completed the survey in a recent year.[3] One of the questions asked:

Which would you prefer to be? Select one.

_____ Rich   _____ Happy   _____ Famous   _____ Healthy

Here are the responses from the 40 randomly selected students:

Famous	Healthy	Healthy	Famous	Happy	Famous	Happy	Happy	Famous
Rich	Happy	Happy	Rich	Happy	Happy	Happy	Rich	Happy
Famous	Healthy	Rich	Happy	Happy	Rich	Happy	Happy	Rich
Healthy	Happy	Happy	Rich	Happy	Happy	Rich	Happy	Famous
Famous	Happy	Happy	Happy					

Make a relative frequency bar graph of the data. Describe what you see.

Graphs: Good and Bad

Bar graphs are a bit dull to look at. It is tempting to replace the bars with pictures or to use special 3-D effects to make the graphs seem more interesting. Don't do it! Our eyes react to the area of the bars as well as to their height. When all bars have the same width, the area (width × height) varies in proportion to the height, and our eyes receive the right impression about the quantities being compared.

EXAMPLE	**Who buys iMacs?**
	Beware the pictograph!

PROBLEM: When Apple, Inc., introduced the iMac, the company wanted to know whether this new computer was expanding Apple's market share. Was the iMac mainly being bought by previous Macintosh owners, or was it being purchased by first-time computer buyers and by previous PC users who were switching over? To find out, Apple hired a firm to conduct a survey of 500

iMac customers. Each customer was categorized as a new computer purchaser, a previous PC owner, or a previous Macintosh owner. The table summarizes the survey results.[4]

Previous ownership	Count	Percent (%)
None	85	17.0
PC	60	12.0
Macintosh	355	71.0
Total	500	100.0

(a) To the right is a clever graph of the data that uses pictures instead of the more traditional bars. How is this pictograph misleading?

(b) Two possible bar graphs of the data are shown below. Which one could be considered deceptive? Why?

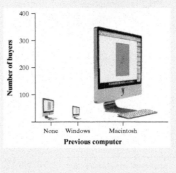

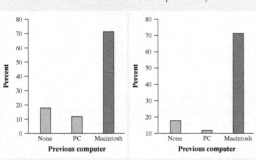

SOLUTION:

(a) The pictograph makes it look like the percentage of iMac buyers who are former Mac owners is at least 10 times larger than either of the other two categories, which isn't true.

(b) The bar graph on the right is misleading. By starting the vertical scale at 10 instead of 0, it looks like the percentage of iMac buyers who previously owned a PC is less than half the percentage who are first-time computer buyers, which isn't true.

> In part (a), the *heights* of the images are correct. But the *areas* of the images are misleading. The Macintosh image is about 6 times as tall as the PC image, but its area is about 36 times as large!

FOR PRACTICE, TRY EXERCISE 19

 There are two important lessons to be learned from this example: (1) beware the pictograph, and (2) watch those scales.

Analyzing Data on Two Categorical Variables

You have learned some techniques for analyzing the distribution of a single categorical variable. What should you do when a data set involves two categorical variables? For example, Yellowstone National Park staff surveyed a random sample of 1526 winter visitors to the park. They asked each person whether he or she belonged to an environmental club (like the Sierra Club). Respondents were also

---

Is Steph Curry the greatest of all time?
Beware the pictograph!

PROBLEM:
In the 2017–2018 regular season, basketball player Stephen Curry averaged 6.1 assists per game and 1.6 steals per game. Explain what is wrong with the following graph and how to make it better.

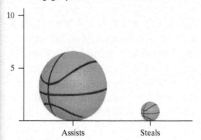

SOLUTION:
Although the heights of the basketballs are correct, our eyes respond to the *areas* of the basketballs. Curry averaged between 3 and 4 times as many assists as steals, but the area of the basketball representing assists is more than 10 times as big. It would be better to use equally wide bars for assists and steals so that only the height of the bars determines the area.

franz12/Shutterstock

asked whether they owned, rented, or had never used a snowmobile. The data set looks something like the following:

Respondent	Environmental club?	Snowmobile use
1	No	Own
2	No	Rent
3	Yes	Never
4	Yes	Rent
5	No	Never
⋮	⋮	⋮

The **two-way table** summarizes the survey responses.

		Environmental club member?	
		No	Yes
Snowmobile use	Never	445	212
	Rent	497	77
	Own	279	16

A two-way table is sometimes called a *contingency table*.

DEFINITION   Two-way table

A **two-way table** is a table of counts that summarizes data on the relationship between two categorical variables for some group of individuals.

It's easier to grasp the information in a two-way table if row and column totals are included, like the one shown here.

		Environmental club		
		No	Yes	Total
Snowmobile use	Never used	445	212	657
	Snowmobile renter	497	77	574
	Snowmobile owner	279	16	295
	Total	1221	305	1526

Now we can quickly answer questions like:

- What percent of people in the sample are environmental club members?

$$\frac{305}{1526} = 0.200 = 20.0\%$$

- What proportion of people in the sample never used a snowmobile?

$$\frac{657}{1526} = 0.431$$

These percents or proportions are known as **marginal relative frequencies** because they are calculated using values in the margins of the two-way table.

DEFINITION   Marginal relative frequency

A **marginal relative frequency** gives the percent or proportion of individuals that have a specific value for one categorical variable.

We could call this distribution the *marginal distribution* of environmental club membership.

We can compute marginal relative frequencies for the *column* totals to give the distribution of environmental club membership in the entire sample of 1526 park visitors:

$$\text{No: } \frac{1221}{1526} = 0.800 \text{ or } 80.0\% \qquad \text{Yes: } \frac{305}{1526} = 0.200 \text{ or } 20.0\%$$

We can compute marginal relative frequencies for the *row* totals to give the distribution of snowmobile use for all the individuals in the sample:

$$\text{Never: } \frac{657}{1526} = 0.431 \text{ or } 43.1\%$$

$$\text{Rent: } \frac{574}{1526} = 0.376 \text{ or } 37.6\%$$

$$\text{Own: } \frac{295}{1526} = 0.193 \text{ or } 19.3\%$$

We could call this distribution the *marginal distribution* of snowmobile use.

Note that we could use a bar graph or a pie chart to display either of these distributions.

A marginal relative frequency tells you about only *one* of the variables in a two-way table. It won't help you answer questions like these, which involve values of *both* variables:

- What percent of people in the sample are environmental club members and own snowmobiles?

$$\frac{16}{1526} = 0.010 = 1.0\%$$

- What proportion of people in the sample are not environmental club members and never use snowmobiles?

$$\frac{445}{1526} = 0.292$$

These percents or proportions are known as **joint relative frequencies**.

In the body, a joint is where two bones come together. In a two-way table, a *joint frequency* is shown where a row and column come together.

> **DEFINITION** **Joint relative frequency**
>
> A **joint relative frequency** gives the percent or proportion of individuals that have a specific value for one categorical variable and a specific value for another categorical variable.

Making Connections

In Chapter 5, we will use probability to describe these fractions. For a randomly selected member of the sample:

$\frac{16}{1526} = 0.010 = 1.0\%$ will be P(environmental club AND own snowmobile).

$\frac{445}{1526} = 0.292$ will be P(no environmental club AND never use snowmobile).

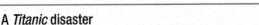

EXAMPLE

A *Titanic* disaster

Calculating marginal and joint relative frequencies

The New York Times.

TITANIC SINKS FOUR HOURS AFTER HITTING ICEBERG; 866 RESCUED BY CARPATHIA, PROBABLY 1250 PERISH; ISMAY SAFE, MRS. ASTOR MAYBE, NOTED NAMES MISSI

Blank Archives/Getty Images

PROBLEM: In 1912 the luxury liner *Titanic*, on its first voyage across the Atlantic, struck an iceberg and sank. Some passengers got off the ship in lifeboats, but many died. The two-way table gives information about adult passengers who survived and who died, by class of travel.

iPhone or Android?

Calculating marginal and joint relative frequencies

PROBLEM:

The Pew Research Center asked a random sample of 2024 adult cell-phone owners from the United States which type of cell phone they own: iPhone, Android, or other (including non-smartphones). Here are the results, broken down by age category:

		Age			
		18–34	35–54	55 +	Total
Type of phone	iPhone	169	171	127	467
	Android	214	189	100	503
	Other	134	277	643	1054
	Total	517	637	870	2024

(a) What proportion of the adult cell-phone owners have an iPhone?

(b) Find the distribution of ages for adult cell-phone owners using relative frequencies.

(c) What percent of adult cell-phone owners are 55 or older and own an iPhone?

SOLUTION:

(a) $\dfrac{467}{2024} = 0.231$

(b) 18–34: $\dfrac{517}{2024} = 0.255 = 25.5\%$

35–54: $\dfrac{637}{2024} = 0.315 = 31.5\%$

55 +: $\dfrac{870}{2024} = 0.430 = 43.0\%$

(c) $\dfrac{127}{2024} = 0.063 = 6.3\%$

(a) What proportion of adult passengers on the *Titanic* survived?

(b) Find the distribution of class of travel for adult passengers on the *Titanic* using relative frequencies.

(c) What percent of adult *Titanic* passengers traveled in third class and survived?

		Class of travel		
		First	Second	Third
Survival status	Survived	197	94	151
	Died	122	167	476

SOLUTION:

(a) $\dfrac{442}{1207} = 0.366$

(b) First: $\dfrac{319}{1207} = 0.264 = 26.4\%$

Second: $\dfrac{261}{1207} = 0.216 = 21.6\%$

Third: $\dfrac{627}{1207} = 0.519 = 51.9\%$

(c) $\dfrac{151}{1207} = 0.125 = 12.5\%$

Start by finding the marginal totals.

		Class of travel			
		First	Second	Third	Total
Survival status	Survived	197	94	151	442
	Died	122	167	476	765
	Total	319	261	627	1207

Remember that a distribution lists the possible values of a variable and how often those values occur.

Note that the three percentages for class of travel in part (b) do not add to exactly 100% due to roundoff error.

FOR PRACTICE, TRY EXERCISE 23

CHECK YOUR UNDERSTANDING

An article in the *Journal of the American Medical Association* reports the results of a study designed to see if the herb St. John's wort is effective in treating moderately severe cases of depression. The study involved 338 patients who were being treated for major depression. The subjects were randomly assigned to receive one of three treatments: St. John's wort, Zoloft (a prescription drug), or placebo (an inactive treatment) for an 8-week period. The two-way table summarizes the data from the experiment.[5]

		Treatment		
		St. John's wort	Zoloft	Placebo
Change in depression	Full response	27	27	37
	Partial response	16	26	13
	No response	70	56	66

1. What proportion of subjects in the study were randomly assigned to take St. John's wort? Explain why this value makes sense.

2. Find the distribution of change in depression for the subjects in this study using relative frequencies.

3. What percent of subjects took Zoloft and showed a full response?

✔ **Answers to CYU**

1. $113/338 = 0.334$; this value makes sense because there were three treatments, so we would expect about one-third of the subjects to be assigned to this treatment.

2. The distribution of change in depression is:
Full response: $91/338 = 0.269$
Partial response: $55/338 = 0.163$
No response: $192/338 = 0.568$

3. $27/338 = 0.08 = 8\%$

Relationships Between Two Categorical Variables

Let's return to the data from the Yellowstone National Park survey of 1526 randomly selected winter visitors. Earlier, we calculated marginal and joint relative frequencies from the two-way table. These values do not tell us much about the *relationship* between environmental club membership and snowmobile use for the people in the sample.

		Environmental club		
		No	Yes	Total
Snowmobile use	Never used	445	212	657
	Snowmobile renter	497	77	574
	Snowmobile owner	279	16	295
	Total	1221	305	1526

We can also use the two-way table to answer questions like:

- What percent of environmental club members in the sample are snowmobile owners?

$$\frac{16}{305} = 0.052 = 5.2\%$$

- What proportion of snowmobile renters in the sample are not environmental club members?

$$\frac{497}{574} = 0.866$$

These percents or proportions are known as **conditional relative frequencies**.

> **DEFINITION** **Conditional relative frequency**
>
> A **conditional relative frequency** gives the percent or proportion of individuals that have a specific value for one categorical variable among individuals who share the same value of another categorical variable (the condition).

EXAMPLE

A *Titanic* disaster
Conditional relative frequencies

PROBLEM: In 1912 the luxury liner *Titanic*, on its first voyage across the Atlantic, struck an iceberg and sank. Some passengers made it off the ship in lifeboats, but many died. The two-way table gives information about adult passengers who survived and who died, by class of travel.

		Class of travel			
		First	Second	Third	Total
Survival status	Survived	197	94	151	442
	Died	122	167	476	765
	Total	319	261	627	1207

Teaching Tip

If you are displaying the table with a projector or on the board, physically cover up the "No" column to answer the first question. In the second question, physically cover up the rows with "Never used" and "Snowmobile owner."

Preparing for Inference

The *P*-value of a significance test is a conditional probability. More specifically, it is the probability of getting evidence for the alternative hypothesis as strong as or stronger than the observed evidence, given that the null hypothesis is true.

Do older people own iPhones?
Conditional relative frequencies

PROBLEM:

The Pew Research Center asked a random sample of 2024 adult cell-phone owners from the United States which type of cell phone they own: iPhone, Android, or other (including non-smartphones). Here are the results, broken down by age category:

		Age			
		18–34	35–54	55 +	Total
Type of phone	iPhone	169	171	127	467
	Android	214	189	100	503
	Other	134	277	643	1054
	Total	517	637	870	2024

(a) What proportion of iPhone owners were 55 or older?

(b) What percent of 18- to 34-year-olds owned Android phones?

SOLUTION:

(a) $\dfrac{127}{467} = 0.272$

(b) $\dfrac{214}{517} = 0.414 = 41.4\%$

Making Connections

In Chapter 5, these proportions will be referred to as *conditional probabilities* and will have the following notation:

$P(\text{never} \mid \text{yes}) = 0.695$

$P(\text{rent} \mid \text{yes}) = 0.252$

$P(\text{own} \mid \text{yes}) = 0.052$

(a) What proportion of survivors were third-class passengers?

(b) What percent of first-class passengers survived?

SOLUTION:

(a) $\dfrac{151}{442} = 0.342$ (b) $\dfrac{197}{319} = 0.618 = 61.8\%$

> Note that a proportion is always a number between 0 and 1, whereas a percent is a number between 0 and 100. To get a percent, multiply the proportion by 100.

FOR PRACTICE, TRY EXERCISE 27

We can study the snowmobile use habits of environmental club members by looking only at the "Yes" column in the two-way table.

		Environmental club		
		No	Yes	Total
Snowmobile use	Never used	445	212	657
	Snowmobile renter	497	77	574
	Snowmobile owner	279	16	295
	Total	1221	305	1526

It is easy to calculate the proportions or percents of environmental club members who never use, rent, and own snowmobiles:

Never: $\dfrac{212}{305} = 0.695$ or 69.5%   Rent: $\dfrac{77}{305} = 0.252$ or 25.2%

Own: $\dfrac{16}{305} = 0.052$ or 5.2%

This is the distribution of snowmobile use among environmental club members.

> We could also refer to this distribution as the *conditional distribution* of snowmobile use among environmental club members.

We can find the distribution of snowmobile use among the survey respondents who are not environmental club members in a similar way. The table summarizes the conditional relative frequencies for both groups.

Snowmobile use	Not environmental club members	Environmental club members
Never	$\dfrac{445}{1221} = 0.364$ or 36.4%	$\dfrac{212}{305} = 0.695$ or 69.5%
Rent	$\dfrac{497}{1221} = 0.407$ or 40.7%	$\dfrac{77}{305} = 0.252$ or 25.2%
Own	$\dfrac{279}{1221} = 0.229$ or 22.9%	$\dfrac{16}{305} = 0.052$ or 5.2%

AP® EXAM TIP

When comparing groups of different sizes, be sure to use relative frequencies (percents or proportions) instead of frequencies (counts) when analyzing categorical data. Comparing only the frequencies can be misleading, as in this setting. There are many more people who never use snowmobiles among the non-environmental club members in the sample (445) than among the environmental club members (212). However, the *percentage* of environmental club members who never use snowmobiles is much higher (69.5% to 36.4%). Finally, make sure to avoid statements like "More club members never use snowmobiles" when you mean "A greater percentage of club members never use snowmobiles."

Figure 1.3 compares the distributions of snowmobile use for Yellowstone National Park visitors who are environmental club members and those who are not environmental club members with (a) a **side-by-side bar graph,** (b) a **segmented bar graph,** and (c) a **mosaic plot.** Notice that the segmented bar graph can be obtained by stacking the bars in the side-by-side bar graph for each of the two environmental club membership categories (no and yes). The bar widths in the mosaic plot are proportional to the number of survey respondents who are (305) and are not (1221) environmental club members.

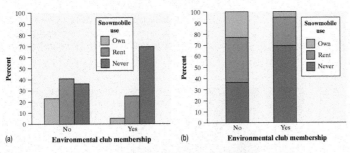

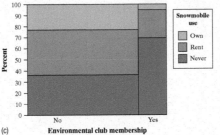

FIGURE 1.3 (a) Side-by-side bar graph, (b) segmented bar graph, and (c) mosaic plot displaying the distribution of snowmobile use among environmental club members and among non-environmental club members from the 1526 randomly selected winter visitors to Yellowstone National Park.

> **DEFINITION** Side-by side bar graph, Segmented bar graph, Mosaic plot
>
> A **side-by-side bar graph** displays the distribution of a categorical variable for each value of another categorical variable. The bars are grouped together based on the values of one of the categorical variables and placed side by side.
>
> A **segmented bar graph** displays the distribution of a categorical variable as segments of a rectangle, with the area of each segment proportional to the percent of individuals in the corresponding category.
>
> A **mosaic plot** is a modified segmented bar graph in which the width of each rectangle is proportional to the number of individuals in the corresponding category.

Teaching Tip

Segmented bar graphs and mosaic plots were not listed on the previous AP® Topic Outline, but they are now specifically mentioned in the new Course Framework.

Teaching Tip: Using Technology

Go to the Student Site at highschool .bfwpub.com/updatedtps6e and choose "Two Categorical Variables" from the Extra Applets. Type in the categories for each variable and the frequency of each. Click "Begin Analysis." Graph options here are the side-by-side bar graph and segmented bar graph.

Teaching Tip

Emphasize the concept of association between two variables. This will be a focus throughout the course. It is OK to use the word *relationship* as a synonym for *association,* but do not use the word *correlation.* It is reserved for a specific measure of the strength of a linear relationship between two quantitative variables.

Preparing for Inference

How do we calculate the percents for these hypothetical distributions? They are the marginal relative frequencies calculated earlier.

Never: $\frac{657}{1526} = 0.431$ or 43.1%

Rent: $\frac{574}{1526} = 0.376$ or 37.6%

Own: $\frac{295}{1526} = 0.193$ or 19.3%

These percents will be used in Chapter 12 to calculate the expected values for a chi-square test for independence. In this test, we will assume that there is no association between the two variables in the population.

Preparing for Inference

In Chapter 12, the chi-square test for independence will be used to test for an association between two categorical variables in a population. Two variables with no association are said to be independent, while two variables that have an association are not independent.

All three graphs in Figure 1.3 show a clear **association** between environmental club membership and snowmobile use in this random sample of 1526 winter visitors to Yellowstone National Park. The environmental club members were much less likely to rent (25.2% versus 40.7%) or own (5.2% versus 29.0%) snowmobiles than non-club-members and more likely to never use a snowmobile (69.5% versus 36.4%). Knowing whether or not a person in the sample is an environmental club member helps us predict that individual's snowmobile use.

> **DEFINITION   Association**
>
> There is an **association** between two variables if knowing the value of one variable helps us predict the value of the other. If knowing the value of one variable does not help us predict the value of the other, then there is no association between the variables.

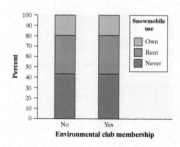

What would the graphs in Figure 1.3 look like if there was *no association* between environmental club membership and snowmobile use in the sample? The blue segments would be the same height for both the "Yes" and "No" groups. So would the green segments and the red segments, as shown in the graph at left. In that case, knowing whether a survey respondent is an environmental club member would *not* help us predict his or her snowmobile use.

Which distributions should we compare? Our goal all along has been to analyze the relationship between environmental club membership and snowmobile use for this random sample of 1526 Yellowstone National Park visitors. We decided to calculate conditional relative frequencies of snowmobile use among environmental club members and among non-club-members. Why? Because we wanted to see if environmental club membership helped us predict snowmobile use. What if we had wanted to determine whether snowmobile use helps us predict whether a person is an environmental club member? Then we would have calculated conditional relative frequencies of environmental club membership among snowmobile owners, renters, and non-users. *In general, you should calculate the distribution of the variable that you want to predict for each value of the other variable.*

Can we say that there is an association between environmental club membership and snowmobile use in the *population* of all winter visitors to Yellowstone National Park? Making this determination requires formal inference, which will have to wait until Chapter 12.

EXAMPLE

A *Titanic* disaster
Conditional relative frequencies and association

PROBLEM: In 1912 the luxury liner *Titanic,* on its first voyage across the Atlantic, struck an iceberg and sank. Some passengers made it off the ship in lifeboats, but many died. The two-way table gives information about adult passengers who survived and who died, by class of travel.

ALTERNATE EXAMPLE

Do older people own iPhones?   Conditional relative frequencies and association   Skills 2.C, 2.D

PROBLEM:
The Pew Research Center asked a random sample of 2024 adult cell-phone owners from the United States which type of cell phone they own: iPhone, Android, or other (including non-smartphones). Here are the results, broken down by age category:

| | | Age | | | |
		18–34	35–54	55 +	Total
Type of phone	iPhone	169	171	127	467
	Android	214	189	100	503
	Other	134	277	643	1054
	Total	517	637	870	2024

(continues)

(a) Find the distribution of survival status for each class of travel. Make a segmented bar graph to compare these distributions.

(b) Describe what the graph in part (a) reveals about the association between class of travel and survival status for adult passengers on the *Titanic*.

	Class of travel			
	First	Second	Third	Total
Survived	197	94	151	442
Died	122	167	476	765
Total	319	261	627	1207

Survival status (row label to the left of Survived/Died/Total)

SOLUTION:

(a) First class   Survived: $\frac{197}{319} = 0.618 = 61.8\%$ Died: $\frac{122}{319} = 0.382 = 38.2\%$

Second class   Survived: $\frac{94}{261} = 0.360 = 36.0\%$ Died: $\frac{167}{261} = 0.640 = 64.0\%$

Third class   Survived: $\frac{151}{627} = 0.241 = 24.1\%$ Died: $\frac{476}{627} = 0.759 = 75.9\%$

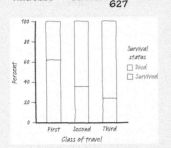

To make the segmented bar graph:
- **Draw and label the axes.** Put class of travel on the horizontal axis and percent on the vertical axis.
- **"Scale" the axes.** Use a vertical scale from 0 to 100%, with tick marks every 20%.
- **Draw bars.** Make each bar have a height of 100%. Be sure the bars are equal in width and leave spaces between them. Segment each bar based on the conditional relative frequencies you calculated. Use different colors or shading patterns to represent the two possible statuses—survived and died. Add a key to the graph that tells us which color (or shading) represents which status.

(b) Knowing a passenger's class of travel helps us predict his or her survival status. First class had the highest percentage of survivors (61.8%), followed by second class (36.0%), and then third class (24.1%).

FOR PRACTICE, TRY EXERCISE 29

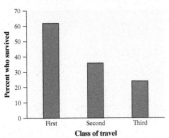

FIGURE 1.4 Bar graph comparing the percents of passengers who survived among each of the three classes of travel on the *Titanic*.

Because the variable "Survival status" has only two possible values, comparing the three distributions displayed in the segmented bar graph amounts to comparing the percent of passengers in each class of travel who survived. The bar graph in Figure 1.4 shows this comparison. Note that the bar heights do *not* add to 100%, because each bar represents a different group of passengers on the *Titanic*.

We offer a final caution about studying the relationship between two variables: association does not imply causation. It may be true that being in a higher class of travel on the *Titanic* increased a passenger's chance of survival. However, there isn't always a cause-and-effect relationship between two variables even if they are clearly associated. For example, a recent study proclaimed that people who are overweight are less likely to die within a few years than are people of normal

35–54:

iPhone: $\frac{171}{637} = 0.268 = 26.8\%$

Android: $\frac{189}{637} = 0.297 = 29.7\%$

Other: $\frac{277}{637} = 0.435 = 43.5\%$

55+:

iPhone: $\frac{127}{870} = 0.146 = 14.6\%$

Android: $\frac{100}{870} = 0.115 = 11.5\%$

Other: $\frac{643}{870} = 0.739 = 73.9\%$

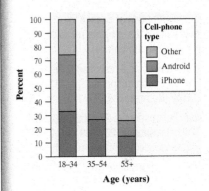

(b) Knowing an adult's age helps us predict his or her cell-phone type. A much higher percentage of people 18–34 years old have iPhones (32.7%) than those who are 35–54 years old (26.8%) or over 55 years old (14.6%). Also, a much higher percentage of people over 55 years old have "Other" phones (73.9%) than do people who are 35–54 years old (43.5%) or 18–34 years old (25.9%).

ALTERNATE EXAMPLE   (continued)

(a) Find the distribution of type of cell phone for each age group. Make a segmented bar graph to compare these distributions.

(b) Describe what the graph in (a) reveals about the association between age and type of cell phone for adult cell-phone owners.

SOLUTION:

(a) *18–34:*

iPhone: $\frac{169}{517} = 0.327 = 32.7\%$

Android: $\frac{214}{517} = 0.414 = 41.4\%$

Other: $\frac{134}{517} = 0.259 = 25.9\%$

Preparing for Inference

In Chapter 4, we will make the distinction between an observational study and an experiment (where a treatment is imposed). While an observational study can provide evidence of an association, a properly designed experiment is needed to show causation.

✓ Answers to CYU

1. $27/91 = 0.297$

2. $70/113 = 0.619 = 61.9\%$

3. The distribution of change in depression for the subjects receiving each of the three treatments is:

St. John's wort: Full response: $27/113 = 0.239$, Partial response: $16/113 = 0.142$, No response: $70/113 = 0.619$

Zoloft: Full response: $27/109 = 0.248$, Partial response: $26/109 = 0.239$, No response: $56/109 = 0.514$

Placebo: Full response: $37/116 = 0.319$, Partial response: $13/116 = 0.112$, No response: $66/116 = 0.569$

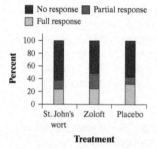

4. There does not appear to be a strong association between treatment and change in depression for these subjects because the distribution of response status is very similar for the three different treatments. Also, note that the treatment with the highest rate of "full response" was the placebo.

---

weight. Does this mean that gaining weight will cause you to live longer? Not at all. The study included smokers, who tend to be thinner and also much more likely to die in a given period than non-smokers. Smokers increased the death rate for the normal-weight category, making it appear as if being overweight is better.[6] The moral of the story: *beware other variables!*

CHECK YOUR UNDERSTANDING

An article in the *Journal of the American Medical Association* reports the results of a study designed to see if the herb St. John's wort is effective in treating moderately severe cases of depression. The study involved 338 subjects who were being treated for major depression. The subjects were randomly assigned to receive one of three treatments: St. John's wort, Zoloft (a prescription drug), or placebo (an inactive treatment) for an 8-week period. The two-way table summarizes the data from the experiment.

		Treatment		
		St. John's wort	Zoloft	Placebo
Change in depression	Full response	27	27	37
	Partial response	16	26	13
	No response	70	56	66

1. What proportion of subjects who showed a full response took St. John's wort?
2. What percent of subjects who took St. John's wort showed no response?
3. Find the distribution of change in depression for the subjects receiving each of the three treatments. Make a segmented bar graph to compare these distributions.
4. Describe what the graph in Question 3 reveals about the association between treatment and change in depression for these subjects.

---

1. Technology Corner   ANALYZING TWO-WAY TABLES

Statistical software will provide marginal relative frequencies, joint relative frequencies, and conditional relative frequencies for data summarized in a two-way table. Here is output from Minitab for the data on snowmobile use and environmental club membership. Use the information on cell contents at the bottom of the output to help you interpret what each value in the table represents.

Teaching Tip

Remind students that they can watch a video of each of the Technology Corners by clicking on the link in the e-Book or viewing the videos on the Student Site at highschool.bfwpub.com/updatedtps6e.

Teaching Tip: Using Technology

Make sure to point out the "key" at the bottom of the computer output. Almost all computer output showing two-way tables will have a key of some kind to explain what each value in the table represents.

Section 1.1 | Summary

- The distribution of a categorical variable lists the categories and gives the **frequency** (count) or **relative frequency** (percent or proportion) of individuals that fall in each category.
- You can use a **pie chart** or **bar graph** to display the distribution of a categorical variable. When examining any graph, ask yourself, "What do I see?"
- Beware of graphs that mislead the eye. Look at the scales to see if they have been distorted to create a particular impression. Avoid making graphs that replace the bars of a bar graph with pictures whose height and width both change.
- A **two-way table** of counts summarizes data on the relationship between two categorical variables for some group of individuals.
- You can use a two-way table to calculate three types of relative frequencies:
    - A **marginal relative frequency** gives the percent or proportion of individuals that have a specific value for one categorical variable. Use the appropriate row total or column total in a two-way table when calculating a marginal relative frequency.
    - A **joint relative frequency** gives the percent or proportion of individuals that have a specific value for one categorical variable and a specific value for another categorical variable. Use the value from the appropriate cell in the two-way table when calculating a joint relative frequency.
    - A **conditional relative frequency** gives the percent or proportion of individuals that have a specific value for one categorical variable among individuals who share the same value of another categorical variable (the condition). Use conditional relative frequencies to compare distributions of a categorical variable for two or more groups.
- Use a **side-by-side bar graph**, a **segmented bar graph**, or a **mosaic plot** to compare the distribution of a categorical variable for two or more groups.
- There is an **association** between two variables if knowing the value of one variable helps predict the value of the other. To see whether there is an association between two categorical variables, find the distribution of one variable for each value of the other variable by calculating an appropriate set of conditional relative frequencies.

1.1 Technology Corner

TI-Nspire and other technology instructions are on the book's website at highschool.bfwpub.com/updatedtps6e.

1. Analyzing two-way tables   Page 22

TRM Section 1.1 Quiz

There is one quiz available for this section in the Teacher's Resource Materials. Click on the link in the TE-Book, look on the TRFD, or download from the Teacher's Resources on the book's digital platform. You can also create your own quiz using the ExamView® Assessment Suite that is part of the TPS6 program. Questions are coded by Learning Target to make it easy to build parallel quizzes.

Answers to Section 1.1 Exercises

1.11 (a) The individuals are the babies. **(b)** It appears that births occurred with similar frequencies on weekdays (Monday through Friday), but with noticeably smaller frequencies on the weekend days (Saturday and Sunday).

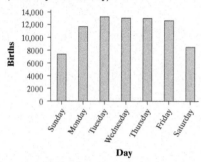

1.12 (a) The individuals are the elevators in New York City. **(b)** A large majority of the elevators in New York City are passenger elevators. All other types of elevators are much less common.

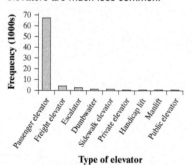

1.13 First, a relative frequency table must be constructed.

Camera brand	Relative frequency
Canon	23/45 = 51.1%
Sony	6/45 = 13.3%
Nikon	11/45 = 24.4%
Fujifilm	3/45 = 6.7%
Olympus	2/45 = 4.4%

The relative frequency bar graph is given below.

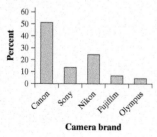

Section 1.1 | Exercises

11. Birth days The frequency table summarizes data on the numbers of babies born on each day of the week in the United States in a recent week.[7]

Day	Births
Sunday	7374
Monday	11,704
Tuesday	13,169
Wednesday	13,038
Thursday	13,013
Friday	12,664
Saturday	8459

(a) Identify the individuals in this data set.

(b) Make a frequency bar graph to display the data. Describe what you see.

12. Going up? As of 2015, there were over 75,000 elevators in New York City. The frequency table summarizes data on the number of elevators of each type.[8]

Type	Count
Passenger elevator	66,602
Freight elevator	4140
Escalator	2663
Dumbwaiter	1143
Sidewalk elevator	943
Private elevator	252
Handicap lift	227
Manlift	73
Public elevator	45

(a) Identify the individuals in this data set.

(b) Make a frequency bar graph to display the data. Describe what you see.

13. Buying cameras The brands of the last 45 digital single-lens reflex (SLR) cameras sold on a popular Internet auction site are listed here. Make a relative frequency bar graph for these data. Describe what you see.

Canon	Sony	Canon	Nikon	Fujifilm
Nikon	Canon	Sony	Canon	Canon
Nikon	Canon	Nikon	Canon	Canon
Canon	Nikon	Fujifilm	Canon	Nikon
Nikon	Canon	Canon	Canon	Canon
Olympus	Canon	Canon	Canon	Nikon
Olympus	Sony	Canon	Canon	Sony
Canon	Nikon	Sony	Canon	Fujifilm
Nikon	Canon	Nikon	Canon	Sony

14. Disc dogs Here is a list of the breeds of dogs that won the World Canine Disc Championships from 1975 through 2016. Make a relative frequency bar graph for these data. Describe what you see.

Whippet	Mixed breed	Australian shepherd
Whippet	Australian shepherd	Australian shepherd
Whippet	Border collie	Australian shepherd
Mixed breed	Australian shepherd	Border collie
Mixed breed	Mixed breed	Border collie
Other purebred	Mixed breed	Australian shepherd
Labrador retriever	Mixed breed	Border collie
Mixed breed	Border collie	Border collie
Mixed breed	Border collie	Other purebred
Border collie	Australian shepherd	Border collie
Mixed breed	Border collie	Border collie
Mixed breed	Australian shepherd	Border collie
Labrador retriever	Border collie	Mixed breed
Labrador retriever	Mixed breed	Australian shepherd

15. Cool car colors The most popular colors for cars and light trucks change over time. Silver advanced past green in 2000 to become the most popular color worldwide, then gave way to shades of white in 2007. Here is a relative frequency table that summarizes data on the colors of vehicles sold worldwide in a recent year.[9]

Color	Percent of vehicles	Color	Percent of vehicles
Black	19	Red	9
Blue	6	Silver	14
Brown/beige	5	White	29
Gray	12	Yellow/gold	3
Green	1	Other	??

(a) What percent of vehicles would fall in the "Other" category?

(b) Make a bar graph to display the data. Describe what you see.

(c) Would it be appropriate to make a pie chart of these data? Explain.

The most popular brand of camera among the 45 most recent purchases on the Internet auction site is Canon, followed by Nikon, Sony, Fujifilm, and Olympus. Canon is the overwhelming favorite, with over 50% of the customers purchasing this brand. Also noteworthy is that almost 25% of the customers purchased a Nikon camera.

1.14 First, a relative frequency table must be constructed.

Breed group	Relative frequency
Whippet	3/42 = 7.1%
Mixed breed	12/42 = 28.6%
Other purebred	2/42 = 4.8%
Labrador retriever	3/42 = 7.1%
Border collie	13/42 = 31.0%
Australian shepherd	9/42 = 21.4%

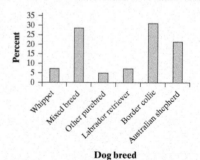

At the World Canine Disc Championships from 1975 to 2016, Border collie dogs won most often (31% of the time), followed by Mixed breed (28.6% of the time) and Australian shepherds (21.4% of the time). Whippets (7.1%), Labrador retrievers (7.1%), and other purebreds (4.8%) were the least frequent winners.

16. Spam Email spam is the curse of the Internet. Here is a relative frequency table that summarizes data on the most common types of spam:[10]

Type of spam	Percent
Adult	19
Financial	20
Health	7
Internet	7
Leisure	6
Products	25
Scams	9
Other	??

(a) What percent of spam would fall in the "Other" category?

(b) Make a bar graph to display the data. Describe what you see.

(c) Would it be appropriate to make a pie chart of these data? Explain.

17. Hispanic origins Here is a pie chart prepared by the Census Bureau to show the origin of the more than 50 million Hispanics in the United States in 2010.[11] About what percent of Hispanics are Mexican? Puerto Rican?

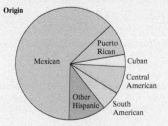

18. Which major? About 3 million first-year students enroll in colleges and universities each year. What do they plan to study? The pie chart displays data on the percent of first-year students who plan to major in several disciplines.[12] About what percent of first-year students plan to major in business? In social science?

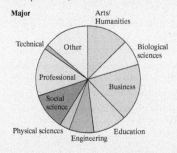

19. Going to school Students in a high school statistics class were given data about the main method of transportation to school for a group of 30 students. They produced the pictograph shown. Explain how this graph is misleading.

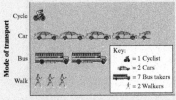

20. Social media The Pew Research Center surveyed a random sample of U.S. teens and adults about their use of social media. The following pictograph displays some results. Explain how this graph is misleading.

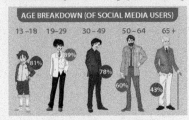

21. Binge-watching Do you "binge-watch" television series by viewing multiple episodes of a series at one sitting? A survey of 800 people who binge-watch were asked how many episodes is too many to watch in one viewing session. The results are displayed in the bar graph.[13] Explain how this graph is misleading.

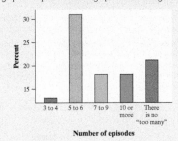

22. Support the court? A news network reported the results of a survey about a controversial court decision. The network initially posted on its website a bar graph of the data similar to the one that follows. Explain how this graph is misleading. (*Note*: When notified about

---

followed closely by Financial (20%), and Adult (19%). **(c)** It would be appropriate to make a pie chart of these data (including the other category) because the numbers in the table refer to parts of a single whole.

1.17 Estimates will vary, but should be close to 63% Mexican and 9% Puerto Rican.

1.18 Estimates will vary, but should be close to 17% Business and 10% Social science.

1.19 The areas of the pictures should be proportional to the numbers of students they represent. As drawn, it appears that most of the students arrived by car but, in reality, most came by bus (14 took the bus, 9 came in cars).

1.20 Based on the heights of the people, it appears that the peak of social media use is ages 50–64, but actually it is ages 19–29.

1.21 By starting the vertical scale at 12 instead of 0, it looks like the percent of binge-watchers who think that 5 to 6 episodes are too many to watch in one viewing session is almost 20 times higher than the percent of binge-watchers who think that 3 to 4 episodes are too many to watch in one viewing session. In truth, the percent of binge-watchers who think that 5 to 6 episodes are too many to watch in one viewing session (31%) is less than 3 times higher than the percent of binge-watchers who think that 3 to 4 episodes are too many to watch in one viewing session (13%). Similar arguments can be made for the relative sizes of the other categories represented in the bar graph.

Teaching Tip

Exercises 17 and 18 help students see that it is hard to determine relative frequencies from a pie chart. Bar graphs are much easier to use.

1.15 (a) 2%

(b)

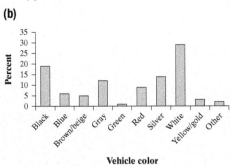

The most popular color of vehicles sold that year was white, followed by black, silver, and gray. It appears that a majority of car buyers that year preferred vehicles that were shades of black and white. **(c)** It would be appropriate to make a pie chart of these data (including the other category) because the numbers in the table refer to parts of a single whole.

1.16 (a) 7%

(b)

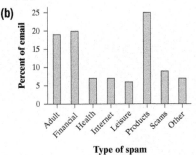

The bar graph shows that the most common type of spam is Products (25% of spam email),

1.22 By starting the vertical scale at 53 instead of 0, it looks as if the percent of Democrats who agree with the decision is almost 10 times higher than the percent of Republicans and Independents who agree. In truth, the percent of Democrats who agree with the decision (62%) is only slightly higher than the percent of Republicans (54%) and Independents (54%) who agree.

1.23 (a) 50/150 = 0.333 **(b)** 29/150 = 19.3% said they saw broken glass at the accident; 121/150 = 80.7% said they did not; 14% said they saw broken glass at the accident; 12% said they saw broken glass at the accident. **(c)** 10.67%

1.24 (a) 0.518 **(b)** Fly: 99/415 = 0.239 = 23.9%, Freeze time: 96/415 = 0.231 = 23.1%, Invisibility: 67/415 = 0.161 = 16.1%, Superstrength: 43/415 = 0.104 = 10.4%, and Telepathy: 110/415 = 0.265 = 26.5% **(c)** 10.6%

1.25 (a) 71.25% **(b)** 0.633 **(c)** 12.1%

1.26 (a) 55.6% **(b)** 0.690 **(c)** 28.9%

the misleading nature of its graph, the network posted a corrected version.)

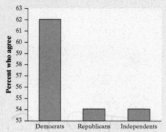

23. **A smash or a hit?** Researchers asked 150 subjects to pg15 recall the details of a car accident they watched on video. Fifty subjects were randomly assigned to be asked, "About how fast were the cars going when they smashed into each other?" For another 50 randomly assigned subjects, the words "smashed into" were replaced with "hit." The remaining 50 subjects—the control group—were not asked to estimate speed. A week later, all subjects were asked if they saw any broken glass at the accident (there wasn't any). The table shows each group's response to the broken glass question.[14]

		Treatment		
		"Smashed into"	"Hit"	Control
Response	Yes	16	7	6
	No	34	43	44

(a) What proportion of subjects were given the control treatment?

(b) Find the distribution of responses about whether there was broken glass at the accident for the subjects in this study using relative frequencies.

(c) What percent of the subjects were given the "smashed into" treatment and said they saw broken glass at the accident?

24. **Superpowers** A total of 415 children from the United Kingdom and the United States who completed a survey in a recent year were randomly selected. Each student's country of origin was recorded along with which superpower they would most like to have: the ability to fly, ability to freeze time, invisibility, superstrength, or telepathy (ability to read minds). The data are summarized in the following table.[15]

		Country	
		U.K.	U.S.
Superpower	Fly	54	45
	Freeze time	52	44
	Invisibility	30	37
	Superstrength	20	23
	Telepathy	44	66

(a) What proportion of students in the sample are from the United States?

(b) Find the distribution of superpower preference for the students in the sample using relative frequencies.

(c) What percent of students in the sample are from the United Kingdom and prefer telepathy as their superpower preference?

25. **Body image** A random sample of 1200 U.S. college students was asked, "What is your perception of your own body? Do you feel that you are overweight, underweight, or about right?" The two-way table summarizes the data on perceived body image by gender.[16]

		Gender		
		Female	Male	Total
Body image	About right	560	295	855
	Overweight	163	72	235
	Underweight	37	73	110
	Total	760	440	1200

(a) What percent of respondents feel that their body weight is about right?

(b) What proportion of the sample is female?

(c) What percent of respondents are males and feel that they are overweight or underweight?

26. **Python eggs** How is the hatching of water python eggs influenced by the temperature of the snake's nest? Researchers randomly assigned newly laid eggs to one of three water temperatures: hot, neutral, or cold. Hot duplicates the extra warmth provided by the mother python, and cold duplicates the absence of the mother. The two-way table summarizes the data on whether or not the eggs hatched.[17]

		Water temperature			
		Cold	Neutral	Hot	Total
Hatched?	Yes	16	38	75	129
	No	11	18	29	58
	Total	27	56	104	187

(a) What percent of eggs were randomly assigned to hot water?

(b) What proportion of eggs in the study hatched?

(c) What percent of eggs in the study were randomly assigned to cold or neutral water and hatched?

27. **A smash or a hit** Refer to Exercise 23.

pg17 (a) What proportion of subjects who said they saw broken glass at the accident received the "hit" treatment?

(b) What percent of subjects who received the "smashed into" treatment said they did not see broken glass at the accident?

28. **Superpower** Refer to Exercise 24.

(a) What proportion of students in the sample who prefer invisibility as their superpower are from the United States?

(b) What percent of students in the sample who are from the United Kingdom prefer superstrength as their superpower?

29. **A smash or a hit** Refer to Exercise 23.

pg 20

(a) Find the distribution of responses about whether there was broken glass at the accident for each of the three treatment groups. Make a segmented bar graph to compare these distributions.

(b) Describe what the graph in part (a) reveals about the association between response about broken glass at the accident and treatment received for the subjects in the study.

30. **Superpower** Refer to Exercise 24.

(a) Find the distribution of superpower preference for the students in the sample from each country (i.e., the United States and the United Kingdom). Make a segmented bar graph to compare these distributions.

(b) Describe what the graph in part (a) reveals about the association between country of origin and superpower preference for the students in the sample.

31. **Body image** Refer to Exercise 25.

(a) Of the respondents who felt that their body weight was about right, what proportion were female?

(b) Of the female respondents, what percent felt that their body weight was about right?

(c) The mosaic plot displays the distribution of perceived body image by gender. Describe what this graph reveals about the association between these two variables for the 1200 college students in the sample.

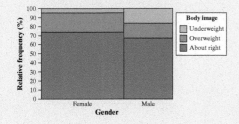

32. **Python eggs** Refer to Exercise 26.

(a) Of the eggs that hatched, what proportion were randomly assigned to hot water?

(b) Of the eggs that were randomly assigned to hot water, what percent hatched?

(c) The mosaic plot displays the distribution of hatching status by water temperature. Describe what this graph reveals about the association between these two variables for the python eggs in this experiment.

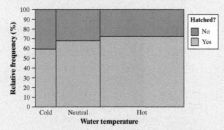

33. **Far from home** A survey asked first-year college students, "How many miles is this college from your permanent home?" Students had to choose from the following options: 5 or fewer, 6 to 10, 11 to 50, 51 to 100, 101 to 500, or more than 500. The side-by-side bar graph shows the percentage of students at public and private 4-year colleges who chose each option.[18] Write a few sentences comparing the distributions of distance from home for students from private and public 4-year colleges who completed the survey.

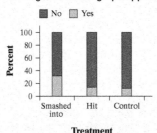

1.27 (a) 0.241 **(b)** 68%

1.28 (a) 0.552 **(b)** 10%

1.29 (a) The distributions of responses for the three treatment groups are:

"Smashed into"	"Hit"	Control
Yes: 16/50 = 32%	Yes: 7/50 = 14%	Yes: 6/50 = 12%
No: 34/50 = 68%	No: 43/50 = 86%	No: 44/50 = 88%

The segmented bar graph appears below.

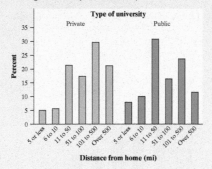

(b) The segmented bar graph reveals that there is an association between opinion about broken glass at the accident and treatment received for subjects in the study. Knowing which treatment a subject received helps us predict whether or not that person will respond that he or she saw broken glass at the accident.

1.30 (a) The distributions of superpower preference for the students in the sample from each country are:

United Kingdom	United States
Fly: 54/200 = 27%	Fly: 45/215 = 20.9%
Freeze time: 52/200 = 26%	Freeze time: 44/215 = 20.5%
Invisibility: 30/200 = 15%	Invisibility: 37/215 = 17.2%
Superstrength: 20/200 = 10%	Superstrength: 23/215 = 10.7%
Telepathy: 44/200 = 22%	Telepathy: 66/215 = 30.7%

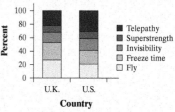

(b) There is an association between country of origin and superpower preference for the students in the sample. Students from the United States were more likely to choose the superpower of telepathy and less likely to choose the superpowers of flying or freezing time than were students from the United Kingdom.

1.31 (a) 0.655 **(b)** 73.7% **(c)** Based on the mosaic plot, there is an association between perceived body image and gender. Males are about 4 times as likely as females to perceive that they are underweight. Males are also less likely to perceive that they are overweight. However, the overwhelming majority of both genders perceives that their body image is about right.

1.32 (a) 0.581 **(b)** 72.1% **(c)** Based on the mosaic plot, there is an association between water temperature and whether the eggs hatched or not. The python eggs were least likely to hatch in cold water (59.3% hatched) and most likely to hatch in hot water (72.1% hatched), so the chance of hatching increased as the water temperature increased.

1.33 Answers may vary. Regardless of whether a student went to a private or public college, most students chose a school that was at least 11 miles from home. Those who went to a public university were most likely to choose a school that was 11 to 50 miles from home (about 30%), while those who went to a private university were most likely to choose a school that was 101 to 500 miles from home (about 29%).

1.34 Answers may vary. Asians overwhelmingly preferred white cars (about 41%). North Americans also preferred white cars, but not as strongly (about 26%). Also noteworthy is that Asians preferred gray and red cars about 7% and 6% of the time, respectively, while North Americans preferred those colors with approximately twice those frequencies.

1.35 (a) The graph reveals that as age increases, the percent that use smartphones for navigation decreases. **(b)** It would not be appropriate to make a pie chart of these data because the category percentages are not parts of the same whole.

1.36 (a) The graph reveals that as age increases, the percentage of people who attended a movie decreases. **(b)** It would not be appropriate to make a pie chart of these data because the category percentages are not parts of the same whole.

1.37 Answers will vary. Two possible tables are given below.

| 10 | 40 | | 30 | 20 |
| 50 | 0 | | 30 | 20 |

1.38 (a) The distributions of survival status for males and females within each class of travel are:

First class
Female survived: $140/144 = 97.2\%$
Female died: $4/144 = 2.8\%$
Male survived: $57/175 = 32.6\%$
Male died: $118/175 = 67.4\%$

Second class
Female survived: $80/93 = 86\%$
Female died: $13/93 = 14\%$
Male survived: $14/168 = 8.3\%$
Male died: $154/168 = 91.7\%$

Third class
Female survived: $76/165 = 46.1\%$
Female died: $89/165 = 53.9\%$
Male survived: $75/462 = 16.2\%$
Male died: $387/462 = 83.8\%$
Regardless of class of travel, women survived the disaster at higher rates than men. Of those who were traveling first class, women were about 3 times likely to survive than men. Of those who were traveling second class, women were about 10 times more likely to survive

34. **Popular car colors** Favorite car colors may differ among countries. The side-by-side bar graph displays data on the most popular car colors in a recent year for North America and Asia. Write a few sentences comparing the distributions.[19]

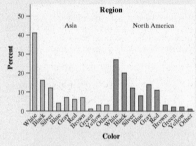

35. **Phone navigation** The bar graph displays data on the percent of smartphone owners in several age groups who say that they use their phone for turn-by-turn navigation.[20]

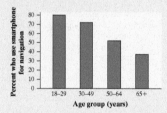

(a) Describe what the graph reveals about the relationship between age group and use of smartphones for navigation.

(b) Would it be appropriate to make a pie chart of the data? Explain.

36. **Who goes to movies?** The bar graph displays data on the percent of people in several age groups who attended a movie in the past 12 months.[21]

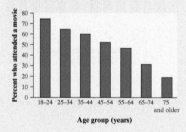

(a) Describe what the graph reveals about the relationship between age group and movie attendance.

(b) Would it be appropriate to make a pie chart of the data? Explain.

37. **Marginal totals aren't the whole story** Here are the row and column totals for a two-way table with two rows and two columns:

a	b	50
c	d	50
60	40	100

Find *two different* sets of counts *a*, *b*, *c*, and *d* for the body of the table that give these same totals. This shows that the relationship between two variables cannot be obtained from the two individual distributions of the variables.

38. **Women and children first?** Here's another table that summarizes data on survival status by gender and class of travel on the *Titanic*:

	Class of travel					
	First class		Second class		Third class	
Survival status	Female	Male	Female	Male	Female	Male
Survived	140	57	80	14	76	75
Died	4	118	13	154	89	387

(a) Find the distributions of survival status for males and for females within each class of travel. Did women survive the disaster at higher rates than men? Explain.

(b) In an earlier example, we noted that survival status is associated with class of travel. First-class passengers had the highest survival rate, while third-class passengers had the lowest survival rate. Does this same relationship hold for both males and females in all three classes of travel? Explain.

39. **Simpson's paradox** Accident victims are sometimes taken by helicopter from the accident scene to a hospital. Helicopters save time. Do they also save lives? The two-way table summarizes data from a sample of patients who were transported to the hospital by helicopter or by ambulance.[22]

		Method of transport		
		Helicopter	Ambulance	Total
Survival status	Died	64	260	324
	Survived	136	840	976
	Total	200	1100	1300

than men. Of those who were traveling third class, women were about 3 times more likely to survive than men. **(b)** For women, there is a clear association between class of travel and survival status. Women in first class were more likely to survive than women in second class, who were more likely to survive than women in third class. For men, the same relationship does not hold true. Although men traveling first class were more likely to survive than men traveling second class, the men traveling third class were more likely to survive than men traveling second class as well.

TRM **Simpson's Paradox Optional Materials**

Optional material, including exercises, on Simpson's paradox can be found in the TE-Book, on the TRFD, or in the Teacher's Resources on the book's digital platform. Although this is an interesting topic, it is not on the AP® Statistics Topic Outline. Exercise 39 can be used to introduce Simpson's paradox to students.

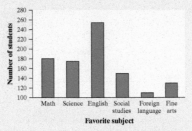

(a) What percent of patients died with each method of transport?

Here are the same data broken down by severity of accident:

Serious accidents

Survival status		Method of transport		
		Helicopter	Ambulance	Total
Survival status	Died	48	60	108
	Survived	52	40	92
	Total	100	100	200

Less serious accidents

Survival status		Method of transport		
		Helicopter	Ambulance	Total
Survival status	Died	16	200	216
	Survived	84	800	884
	Total	100	1000	1100

(b) Calculate the percent of patients who died with each method of transport for the serious accidents. Then calculate the percent of patients who died with each method of transport for the less serious accidents. What do you notice?

(c) See if you can explain how the result in part (a) is possible given the result in part (b).

Note: This is an example of *Simpson's paradox,* which states that an association between two variables that holds for each value of a third variable can be changed or even reversed when the data for all values of the third variable are combined.

Multiple Choice: *Select the best answer for Exercises 40–43.*

40. For which of the following would it be *inappropriate* to display the data with a single pie chart?

(a) The distribution of car colors for vehicles purchased in the last month

(b) The distribution of unemployment percentages for each of the 50 states

(c) The distribution of favorite sport for a sample of 30 middle school students

(d) The distribution of shoe type worn by shoppers at a local mall

(e) The distribution of presidential candidate preference for voters in a state

41. The following bar graph shows the distribution of favorite subject for a sample of 1000 students. What is the most serious problem with the graph?

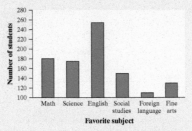

Favorite subject

(a) The subjects are not listed in the correct order.

(b) This distribution should be displayed with a pie chart.

(c) The vertical axis should show the percent of students.

(d) The vertical axis should start at 0 rather than 100.

(e) The foreign language bar should be broken up by language.

42. The Dallas Mavericks won the NBA championship in the 2010–2011 season. The two-way table displays the relationship between the outcome of each game in the regular season and whether the Mavericks scored at least 100 points.

		Points scored		
		100 or more	Fewer than 100	Total
Outcome of game	Win	43	14	57
	Loss	4	21	25
	Total	47	35	82

Which of the following is the best evidence that there is an association between the outcome of a game and whether or not the Mavericks scored at least 100 points?

(a) The Mavericks won 57 games and lost only 25 games.

(b) The Mavericks scored at least 100 points in 47 games and fewer than 100 points in only 35 games.

(c) The Mavericks won 43 games when scoring at least 100 points and only 14 games when scoring fewer than 100 points.

(d) The Mavericks won a higher proportion of games when scoring at least 100 points (43/47) than when they scored fewer than 100 points (14/35).

(e) The combination of scoring 100 or more points and winning the game occurred more often (43 times) than any other combination of outcomes.

1.39 (a) 64 of the 200 patients transported by helicopter, or 32% of patients died. 260 of the 1100 patients transported by ambulance, or 23.6% of patients died. **(b)** Of the patients who were in serious accidents, 48% who were transported by helicopter died and 60% who were transported by ambulance died. Of the patients who were in less serious accidents, 16% who were transported by helicopter died and 20% who were transported by ambulance died. Whether the patients were in a serious accident or less serious accident, the percentage who died was greater for those who were transported by ambulance. **(c)** Overall, a greater percentage of patients who were transported by helicopter died, but when broken down by seriousness of the accident, in both instances, a greater percentage of patients who were transported by ambulance died. This is because people in more serious accidents were also more likely to be transported by helicopter and were more likely to die. People who were involved in less serious accidents were less likely to be transported by helicopter and were less likely to die. Overall, this makes it appear that those who are transported by helicopters are more likely to die; but when the patients are broken out by seriousness of the accident, we can see that the driving factor for the overall death rates is the seriousness of the accident, not the method of transportation.

1.40 b

1.41 d

1.42 d

1.43 c

1.44 The individuals are the hotels. The variables in the data set are whether there is a pool (categorical), whether there is an exercise room (categorical), Internet cost ($/day) (quantitative), number of restaurants (quantitative), distance to site (mi) (quantitative), room service availability (categorical), and room rate ($/day) (quantitative).

43. The following partially completed two-way table shows the marginal distributions of gender and handedness for a sample of 100 high school students.

		Gender		
		Male	Female	Total
Dominant hand	Right	x		90
	Left			10
	Total	40	60	100

If there is no association between gender and handedness for the members of the sample, which of the following is the correct value of x?

(a) 20 (d) 45

(b) 30 (e) Impossible to determine without more information.

(c) 36

Recycle and Review

44. **Hotels** (Introduction) A high school lacrosse team is planning to go to Buffalo for a three-day tournament. The tournament's sponsor provides a list of available hotels, along with some information about each hotel. The following table displays data about hotel options. Identify the individuals and variables in this data set. Classify each variable as categorical or quantitative.

Hotel	Pool	Exercise room?	Internet ($/day)	Restaurants	Distance to site (mi)	Room service?	Room rate ($/day)
Comfort Inn	Out	Y	0.00	1	8.2	Y	149
Fairfield Inn & Suites	In	Y	0.00	1	8.3	N	119
Baymont Inn & Suites	Out	Y	0.00	1	3.7	Y	60
Chase Suite Hotel	Out	N	15.00	0	1.5	N	139
Courtyard	In	Y	0.00	1	0.2	Dinner	114
Hilton	In	Y	10.00	2	0.1	Y	156
Marriott	In	Y	9.95	2	0.0	Y	145

PD **Section 1.2 Overview**

To watch the video overview of Section 1.2, click on the link in the TE-Book, look on the TRFD, or download from the Teacher's Resources on the book's digital platform.

Teaching Tip

Emphasize the switch from categorical (Section 1.1) to quantitative variables (Section 1.2). It is important that students begin to use this distinction to classify the different methods of data analysis.

Displaying Quantitative Data with Graphs

LEARNING TARGETS *By the end of the section, you should be able to:*

- Make and interpret dotplots, stemplots, and histograms of quantitative data.
- Identify the shape of a distribution from a graph.
- Describe the overall pattern (shape, center, and variability) of a distribution and
- identify any major departures from the pattern (outliers).
- Compare distributions of quantitative data using dotplots, stemplots, and histograms.

To display the distribution of a categorical variable, use a bar graph or a pie chart. How can we picture the distribution of a quantitative variable? In this section, we present several types of graphs that can be used to display quantitative data.

Dotplots

One of the simplest graphs to construct and interpret is a **dotplot**.

DEFINITION **Dotplot**

A **dotplot** shows each data value as a dot above its location on a number line.

Preparing for Inference

Be aware of the power of the dotplot. Instructionally, the dotplot can be created very quickly to display class data. More importantly, we will use the dotplot later to display the distribution of a statistic calculated from many samples, which will lead us to an understanding of sampling distributions.

Note that goals scored is a discrete variable.

Here are data on the number of goals scored in 20 games played by the 2016 U.S. women's soccer team:

5 5 1 10 5 2 1 1 2 3 3 2 1 4 2 1 2 1 9 3

Figure 1.5 shows a dotplot of these data.

FIGURE 1.5 Dotplot of goals scored in 20 games by the 2016 U.S. women's soccer team.

It is fairly easy to make a dotplot by hand for small sets of quantitative data.

HOW TO MAKE A DOTPLOT

- **Draw and label the axis.** Draw a horizontal axis and put the name of the quantitative variable underneath. Be sure to include units of measurement for continuous variables.
- **Scale the axis.** Look at the smallest and largest values in the data set. Start the horizontal axis at a convenient number equal to or less than the smallest value and place tick marks at equal intervals until you equal or exceed the largest value.
- **Plot the values.** Mark a dot above the location on the horizontal axis corresponding to each data value. Try to make all the dots the same size and space them out equally as you stack them.

Remember what we said in Section 1.1: Making a graph is not an end in itself. When you look at a graph, always ask, "What do I see?" From Figure 1.5, we see that the 2016 U.S. women's soccer team scored 4 or more goals in $6/20 = 0.30$ or 30% of its games. That's quite an offense! Unfortunately, the team lost to Sweden on penalty kicks in the 2016 Summer Olympics.

EXAMPLE

Give it some gas!
Making and interpreting dotplots

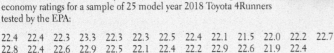

PROBLEM: The Environmental Protection Agency (EPA) is in charge of determining and reporting fuel economy ratings for cars. To estimate fuel economy, the EPA performs tests on several vehicles of the same make, model, and year. Here are data on the highway fuel economy ratings for a sample of 25 model year 2018 Toyota 4Runners tested by the EPA:

22.4 22.4 22.3 23.3 22.3 22.3 22.5 22.4 22.1 21.5 22.0 22.2 22.7
22.8 22.4 22.6 22.9 22.5 22.1 22.4 22.2 22.9 22.6 21.9 22.4

Teaching Tip: Using Technology

All the graphs in this section can be created with technology. Go to the Student Site and choose "One Quantitative Variable" from the Extra Applets. Type the variable name and then the values for the variable. Then click "Begin Analysis." Graph options include dotplot, stemplot, histogram, and boxplot.

Teaching Tip

It is more important that students can interpret graphs and use them to answer questions than it is for them to create graphs. Don't spend too much class time constructing graphs by hand. Technology can usually create the graphs much more quickly.

Teaching Tip: Using Technology

Go to the Student Site and choose "One Quantitative Variable" from the Extra Applets. Type in the variable name and the data values. Click "Begin Analysis." The default graph type is dotplot.

TRM Section 1.2 Alternate Examples

You can find the Alternate Examples for this section in Microsoft Word format by clicking on the link in the TE-Book, opening the TRFD, or downloading from the Teacher's Resources on the book's digital platform.

Fair soda cans?
Making and interpreting dotplots

PROBLEM:
Phoebe and Lisa bought a 12-pack of soda and measured the number of fluid ounces in each can. Here are the data:

12.01	12.05	11.98	12.12
12.08	12.05	12.01	12.09
12.11	12.05	12.09	11.97

(a) Make a dotplot of these data.
(b) The company claims that its cans contain 12 ounces of soda. Do these data give Phoebe and Lisa reason to doubt that claim?

SOLUTION:
(a)

Fluid ounces

(b) No; of the 12 cans tested, 10 had at least 12 ounces of soda.

Preparing for Inference

Inferential thinking requires students to use data to evaluate the evidence against a claim. At the end of this Example, ask students, "Is Toyota's claim plausible (believable) or do we have convincing evidence against the claim?"

Teaching Tip

The word *skew* has a looser meaning in the English language than in statistics class. In statistics, it should only be used to describe the shape of a distribution. Students should not say, for example, "The outlier *skews* the data" or, in later chapters, that "the sampling method *skews* the data."

(a) Make a dotplot of these data.
(b) Toyota reports the highway gas mileage of its 2018 model year 4Runners as 22 mpg. Do these data give the EPA sufficient reason to investigate that claim?

SOLUTION:
(a)

Highway fuel economy rating (mpg)

To make the dotplot:
• **Draw and label the axis.** Note the name and units for this continuous variable.
• **Scale the axis.** The smallest value is 21.5 and the largest value is 23.3. So we choose a scale from 21.5 to 23.5 with tick marks 0.1 units apart.
• **Plot the values.**

(b) No. 23 of the 25 cars tested had an estimated highway fuel economy of 22 mpg or greater.

FOR PRACTICE, TRY EXERCISE 45

Describing Shape

When you describe the shape of a dotplot or another graph of quantitative data, focus on the main features. Look for major *peaks*, not for minor ups and downs in the graph. Look for *clusters* of values and obvious *gaps*. Decide if the distribution is roughly **symmetric** or clearly **skewed**.

We could also describe a distribution with a long tail to the left as "skewed toward negative values" or "negatively skewed" and a distribution with a long right tail as "positively skewed."

DEFINITION   Symmetric and skewed distributions

A distribution is roughly **symmetric** if the right side of the graph (containing the half of observations with the largest values) is approximately a mirror image of the left side.

Roughly symmetric

A distribution is **skewed to the right** if the right side of the graph is much longer than the left side.

Skewed to the right

A distribution is **skewed to the left** if the left side of the graph is much longer than the right side.

Skewed to the left

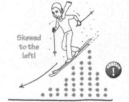

Skewed to the left!

For ease, we sometimes say "left-skewed" instead of "skewed to the left" and "right-skewed" instead of "skewed to the right." The direction of skewness is toward the long tail, not the direction where most observations are clustered. The drawing is a cute but corny way to help you keep this straight. To avoid danger, Mr. Starnes skis on the gentler slope—in the direction of the skewness.

Teaching Tip

Distributions seldom have perfect shapes. Tell students to use *-ly* words to modify their description of the shape. For example, say "roughly symmetric," or "slightly skewed to the right," or "strongly skewed to the left."

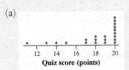

EXAMPLE

Quiz scores and die rolls
Describing shape

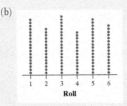

PROBLEM: The dotplots display two different sets of quantitative data. Graph (a) shows the scores of 21 statistics students on a 20-point quiz. Graph (b) shows the results of 100 rolls of a 6-sided die. Describe the shape of each distribution.

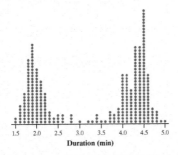

SOLUTION:

(a) The distribution of statistics quiz scores is skewed to the left, with a single peak at 20 (a perfect score). There are two small gaps at 12 and 16.

(b) The distribution of die rolls is roughly symmetric. It has no clear peak.

> We can describe the shape of the distribution in part (b) as *approximately uniform* because the frequencies are about the same for all possible rolls.

FOR PRACTICE, TRY EXERCISE 49

Some people refer to graphs with a single peak, like the dotplot of quiz scores in part (a) of the example, as *unimodal*. Figure 1.6 is a dotplot of the duration (in minutes) of 220 eruptions of the Old Faithful geyser. The graph has two distinct clusters and two peaks: one at about 2 minutes and one at about 4.5 minutes. We would describe this distribution's shape as roughly symmetric and *bimodal*. (Although we could continue the pattern with "trimodal" for three peaks and so on, it's more common to refer to distributions with more than two clear peaks as *multimodal*.) When you examine a graph of quantitative data, describe any pattern you see as clearly as you can.

FIGURE 1.6 Dotplot displaying duration (in minutes) of 220 Old Faithful eruptions. This graph has two distinct clusters and two clear peaks (bimodal).

ALTERNATE EXAMPLE   Skill 2.A

Pick a random number for pizza?
Describing shape

PROBLEM:
The dotplots below display two different sets of quantitative data. Graph (a) shows the results of 100 integers from a random digit generator. Graph (b) shows the number of grams of fiber in one serving of frozen cheese pizza for several different brands. Describe the shape of each distribution.

(a)

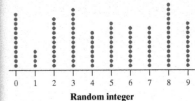

Random integer

(b)

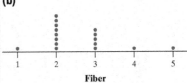

Fiber

SOLUTION:
(a) The distribution of random integers is roughly symmetric. It has no clear peak. We can describe the distribution as "approximately uniform."
(b) The distribution of the number of grams of fiber in one serving of frozen cheese pizza is skewed to the right with a peak at 2 grams of fiber.

AP® EXAM TIP

Student solutions on the free response section of the AP® Statistics exam should always include the context of the problem. Exam questions are unlikely to say "in context" because that is always the expectation. Establish early in the course the importance of including context in solutions. In the example on this page, a student solution for part (a) that says "skewed to the left" is insufficient; however, "the distribution of statistics quiz scores is skewed to the left" includes context.

AP® EXAM TIP

When describing shape, the first priority is to use the terms *symmetric*, *skewed to the right,* or *skewed to the left,* if appropriate. Next identify peaks and gaps, if appropriate.

Some quantitative variables have distributions with predictable shapes. Many biological measurements on individuals from the same species and gender—lengths of bird bills, heights of young women—have roughly symmetric distributions. Salaries and home prices, on the other hand, usually have right-skewed distributions. There are many moderately priced houses, for example, but the few very expensive mansions give the distribution of house prices a strong right skew.

CHECK YOUR UNDERSTANDING

Knoebels Amusement Park in Elysburg, Pennsylvania, has earned acclaim for being an affordable, family-friendly entertainment venue. Knoebels does not charge for general admission or parking, but it does charge customers for each ride they take. How much do the rides cost at Knoebels? The table shows the cost for each ride in a sample of 22 rides in a recent year.

Name	Cost	Name	Cost
Merry Mixer	$1.50	Looper	$1.75
Italian Trapeze	$1.50	Flying Turns	$3.00
Satellite	$1.50	Flyer	$1.50
Galleon	$1.50	The Haunted Mansion	$1.75
Whipper	$1.25	StratosFear	$2.00
Skooters	$1.75	Twister	$2.50
Ribbit	$1.25	Cosmotron	$1.75
Roundup	$1.50	Paratrooper	$1.50
Paradrop	$1.25	Downdraft	$1.50
The Phoenix	$2.50	Rockin' Tug	$1.25
Gasoline Alley	$1.75	Sklooosh!	$1.75

1. Make a dotplot of the data.
2. Describe the shape of the distribution.

Describing Distributions

Here is a general strategy for describing a distribution of quantitative data.

Variability is sometimes referred to as *spread*. We prefer variability because students sometimes think that spread refers only to the distance between the maximum and minimum value of a quantitative data set (the *range*). There are several ways to measure the variability (spread) of a distribution, including the range.

> ### HOW TO DESCRIBE THE DISTRIBUTION OF A QUANTITATIVE VARIABLE
>
> In any graph, look for the *overall pattern* and for clear *departures* from that pattern.
>
> - You can describe the overall pattern of a distribution by its **shape, center,** and **variability.**
> - An important kind of departure is an **outlier,** an observation that falls outside the overall pattern.

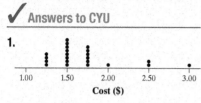

We will discuss more formal ways to measure center and variability and to identify outliers in Section 1.3. For now, just use the *median* (middle value in the ordered data set) when describing center and the *minimum* and *maximum* when describing variability.

Let's practice with the dotplot of goals scored in 20 games played by the 2016 U.S. women's soccer team.

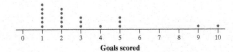

Goals scored

When describing a distribution of quantitative data, don't forget:
Statistical **O**pinions **C**an **V**ary
(**S**hape, **O**utliers, **C**enter, **V**ariability).

Shape: The distribution of goals scored is skewed to the right, with a single peak at 1 goal. There is a gap between 5 and 9 goals.

Outliers: The games when the team scored 9 and 10 goals appear to be outliers.

Center: The median is 2 goals scored.

Variability: The data vary from 1 to 10 goals scored.

➕ Ask the StatsMedic

Knix the Trix: Except for SOCV and DUFS

In this post, we discuss consequences for when students use acronyms to memorize procedures. DUFS will be introduced in Chapter 3.

EXAMPLE

Give it some gas!
Describing a distribution

PROBLEM: Here is a dotplot of the highway fuel economy ratings for a sample of 25 model year 2018 Toyota 4Runners tested by the EPA. Describe the distribution.

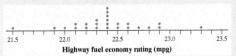

Highway fuel economy rating (mpg)

Daren Starnes

SOLUTION:

Shape: The distribution of highway fuel economy ratings is roughly symmetric, with a single peak at 22.4 mpg. There are two clear gaps: between 21.5 and 21.9 mpg and between 22.9 and 23.3 mpg.

Outliers: The cars with 21.5 mpg and 23.3 mpg ratings are possible outliers.

Center: The median rating is 22.4 mpg.

Variability: The ratings vary from 21.5 to 23.3 mpg.

> Be sure to include context by discussing the variable of interest, highway fuel economy ratings. And give the units of measurement: miles per gallon (mpg).

FOR PRACTICE, TRY EXERCISE 53

ALTERNATE EXAMPLE   Skill 2.A

How many calories in that slice of pizza?   Describing a distribution

PROBLEM:

Here are the calories per serving for 16 brands of frozen cheese pizza, along with a dotplot of the data. Describe the distribution.

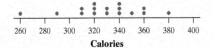

Calories

SOLUTION:

Shape: The distribution of number of calories per serving is roughly symmetric with peaks at 320 and 340 calories. There are three gaps: between 260 and 290 calories, between 290 and 310 calories, and between 360 and 380 calories.

Outliers: One pizza has an unusually small number of calories (260).

Center: The median number of calories is 330 calories.

Variability: The values vary from 260 to 380 calories.

Comparing Distributions

Some of the most interesting statistics questions involve comparing two or more groups. Which of two popular diets leads to greater long-term weight loss? Who texts more—males or females? As the following example suggests, you should always discuss shape, outliers, center, and variability whenever you compare distributions of a quantitative variable.

| **EXAMPLE** | Household size: U.K. versus South Africa |
| | Comparing distributions |

PROBLEM: How do the numbers of people living in households in the United Kingdom (U.K.) and South Africa compare? To help answer this question, we used Census At School's "Random Data Selector" to choose 50 students from each country. Here are dotplots of the household sizes reported by the survey respondents. Compare the distributions of household size for these two countries.

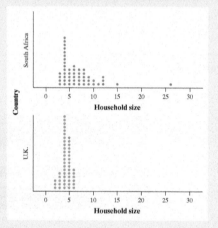

FrankvandenBergh/Getty Images

> **AP® EXAM TIP**
>
> When comparing distributions of quantitative data, it's not enough just to list values for the center and variability of each distribution. You have to explicitly *compare* these values, using words like "greater than," "less than," or "about the same as."

SOLUTION:

Shape: The distribution of household size for the U.K. sample is roughly symmetric, with a single peak at 4 people. The distribution of household size for the South Africa sample is skewed to the right, with a single peak at 4 people and a clear gap between 15 and 26.

Outliers: There don't appear to be any outliers in the U.K. distribution. The South African distribution seems to have two outliers: the households with 15 and 26 people.

Center: Household sizes for the South African students tend to be larger (median = 6 people) than for the U.K. students (median = 4 people).

Variability: The household sizes for the South African students vary more (from 3 to 26 people) than for the U.K. students (from 2 to 6 people).

> Don't forget to include context! It isn't enough to refer to the U.K. distribution or the South Africa distribution. You need to mention the variable of interest, household size.

FOR PRACTICE, TRY EXERCISE 55

How many shoes are too many shoes?
Comparing distributions

PROBLEM:
How many pairs of shoes does a typical teenager own? To find out, a group of statistics students surveyed separate random samples of 20 female students and 20 male students from their large high school. Here are dotplots of the number of pairs of shoes owned by the survey respondents. Compare the distributions of number of pairs of shoes for females and males.

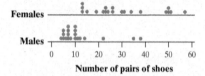

SOLUTION:

Shape: The distribution of numbers of pairs of shoes for the female students is slightly skewed right, with a single peak at 13 pairs of shoes. The distribution of numbers of pairs of shoes for the male students is also skewed to the right, with peaks at 7 and 10 pairs of shoes.

Outliers: There don't appear to be any outliers in the female student distribution. The male distribution seems to have three outliers: the male students with 22, 35, and 38 pairs of shoes.

Center: The numbers of pairs of shoes for female students tend to be greater (median = 26 pairs) than for the male students (median = 9 pairs).

Variability: The numbers of pairs of shoes for female students vary more (from 13 to 57 pairs) than for the males (from 4 to 38 pairs).

Teaching Tip

It is incorrect to say that the "center of South Africa" is higher than the "center of the United Kingdom." After all, the middle of South Africa is much farther south than the middle of the U.K.! It is better to say that the "center of *the distribution of household size* in South Africa is much greater than the center *of the distribution of household size* in the U.K."

Notice that in the preceding example, we discussed the distributions of household size only for the two *samples* of 50 students. We might be interested in whether the sample data give us convincing evidence of a difference in the *population* distributions of household size for South Africa and the United Kingdom. We'll have to wait a few chapters to decide whether we can reach such a conclusion, but our ability to make such an inference later will be helped by the fact that the students in our samples were chosen at random.

CHECK YOUR UNDERSTANDING

For a statistics class project, Jonathan and Crystal hosted an ice-cream-eating contest. Each student in the contest was given a small cup of ice cream and instructed to eat it as fast as possible. Jonathan and Crystal then recorded each contestant's gender and time (in seconds), as shown in the dotplots. Compare the distributions of eating times for males and females.

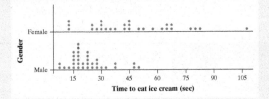

Stemplots

Another simple type of graph for displaying quantitative data is a **stemplot**.

A stemplot is also known as a *stem-and-leaf plot.*

> **DEFINITION Stemplot**
>
> A **stemplot** shows each data value separated into two parts: a *stem*, which consists of all but the final digit, and a *leaf*, the final digit. The stems are ordered from lowest to highest and arranged in a vertical column. The leaves are arranged in increasing order out from the appropriate stems.

Here are data on the resting pulse rates (beats per minute) of 19 middle school students:

71 104 76 88 78 71 68 86 70 90 74 76 69 68 88 96 68 82 120

Figure 1.7 shows a stemplot of these data.

```
 6 | 8889
 7 | 0114668      Key: 8|2 is a
 8 | 2688         student whose
 9 | 06           resting pulse
10 | 4            rate is 82 beats
11 |              per minute.
12 | 0
```

FIGURE 1.7 Stemplot of the resting pulse rates of 19 middle school students.

According to the American Heart Association, a resting pulse rate above 100 beats per minute is considered high for this age group. We can see that $2/19 = 0.105 = 10.5\%$

✓ **Answers to CYU**

Shape: The distribution of time to eat ice cream (in seconds) for the female sample is skewed to the right, with no clear peak. The distribution of time to eat ice cream (in seconds) for the male sample is skewed to the right, with a single peak at about 17.5 seconds.

Outliers: The female distribution appears to have one outlier: the female who took approximately 105 seconds to eat the ice cream. The male distribution does not appear to contain any outliers.

Center: The time it took female students to eat ice cream was generally longer (median $\approx$ 45 seconds) than for male students (median $\approx$ 20 seconds).

Variability: The ice-cream-eating times for female students vary more (from about 13 seconds to about 107 seconds) than for the male students (from about 5 seconds to about 50 seconds).

 COMMON STUDENT ERROR

The stem of 11 has no leaves. Students will often omit this stem. This is not appropriate, because doing so will hide the gap that exists in the distribution. Also, students will often forget the key or labels on a stemplot.

Teaching Tip: Using Technology

Go to the Student Site and choose "One Quantitative Variable" from the Extra Applets. Type in the variable name and the data values. Click "Begin Analysis." Change the graph option to stemplot.

of these students have high resting pulse rates by this standard. Also, the distribution of pulse rates for these 19 students is skewed to the right (toward the larger values).

Stemplots give us a quick picture of a distribution that includes the individual observations in the graph. It is fairly easy to make a stemplot by hand for small sets of quantitative data.

HOW TO MAKE A STEMPLOT

- **Make stems.** Separate each observation into a stem, consisting of all but the final digit, and a leaf, the final digit. Write the stems in a vertical column with the smallest at the top. Draw a vertical line at the right of this column. Do not skip any stems, even if there is no data value for a particular stem.
- **Add leaves.** Write each leaf in the row to the right of its stem.
- **Order leaves.** Arrange the leaves in increasing order out from the stem.
- **Add a key.** Provide a key that identifies the variable and explains what the stems and leaves represent.

ALTERNATE EXAMPLE   Skill 2.B

How many medals?

Making and interpreting stemplots

PROBLEM:
The table displays the total number of medals 28 countries won during the 2016 Summer Olympic Games in Rio de Janeiro.

Country	Medals
United States	121
China	70
Great Britain	67
Russia	56
Germany	42
France	42
Japan	41
Australia	29
Italy	28
Canada	22
South Korea	21
Netherlands	19
Brazil	19
Spain	18
New Zealand	18
Azerbaijan	18
Kazakhstan	17
Hungary	15
Denmark	15
Kenya	13
Uzbekistan	13
Jamaica	11
Cuba	11
Sweden	11
Ukraine	11
Poland	11
Croatia	10
South Africa	10

(a) Make a stemplot of these data.
(b) Describe the shape of the distribution. Are there any obvious outliers?

EXAMPLE   Wear your helmets!

Making and interpreting stemplots

Pete Saloutos/AGE Fotostock

PROBLEM: Many athletes (and their parents) worry about the risk of concussions when playing sports. A football coach plans to obtain specially made helmets for his players that are designed to reduce the chance of getting a concussion. Here are the measurements of head circumference (in inches) for the 30 players on the team:

23.0 22.2 21.7 22.0 22.3 22.6 22.7 21.5 22.7 25.6 20.8 23.0 24.2 23.5 20.8
24.0 22.7 22.6 23.9 22.5 23.1 21.9 21.0 22.4 23.5 22.5 23.9 23.4 21.6 23.3

(a) Make a stemplot of these data.
(b) Describe the shape of the distribution. Are there any obvious outliers?

SOLUTION:

(a)

```
20 | 88
21 | 05679
22 | 0234566777
23 | 001345599
24 | 02
25 | 6
```

Key: 23|5 is a player with a head circumference of 23.5 inches.

(b) The distribution of head circumferences for the 30 players on the team is roughly symmetric, with a single peak on the 22-inch stem. There are no obvious outliers.

To make the stemplot:
- **Make stems.** The smallest head circumference is 20.8 inches and the largest is 25.6 inches. We use the first two digits as the stem and the final digit as the leaf. So we need stems from 20 to 25.
- **Add leaves.**
- **Order leaves.**
- **Add a key.**

FOR PRACTICE, TRY EXERCISE 59

SOLUTION:

(a)
```
 1 | 0 0 1 1 1 1 1 3 3 5 5 7 8 8 8 9 9
 2 | 1 2 8 9
 3 |
 4 | 1 2 2
 5 | 6
 6 | 7
 7 | 0
 8 |
 9 |
10 |
11 |
12 | 1
```

Key: 4|1 is a country with 41 total medals

(b) The distribution of number of medals for the 28 countries is skewed to the right, with a single peak on the 10–19 medals stem. The United States, with 121 medals, appears to be an outlier.

We can get a better picture of the head circumference data by *splitting stems*. In Figure 1.8(a), leaf values from 0 to 9 are placed on the same stem. Figure 1.8(b) shows another stemplot of the same data. This time, values with leaves from 0 to 4 are placed on one stem, while those with leaves from 5 to 9 are placed on another stem. Now we can see the shape of the distribution even more clearly—including the possible outlier at 25.6 inches.

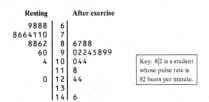

```
20 | 88
21 | 05679
22 | 02345566777          Key: 23|5 is a player with a head
23 | 001345599             circumference of 23.5 inches.
24 | 02
25 | 6
```

```
20 | 88
21 | 0
21 | 5679
22 | 0234
22 | 5566777
23 | 00134
23 | 5599
24 | 02
24 |
25 |
25 | 6
```

(a)   (b)

FIGURE 1.8 Two stemplots showing the head circumference data. The graph in (b) improves on the graph in (a) by splitting stems.

Here are a few tips to consider before making a stemplot:

- There is no magic number of stems to use. Too few or too many stems will make it difficult to see the distribution's shape. Five stems is a good minimum.
- If you split stems, be sure that each stem is assigned an equal number of possible leaf digits.
- When the data have too many digits, you can get more flexibility by rounding or truncating the data. See Exercises 61 and 62 for an illustration of rounding data before making a stemplot.

You can use a *back-to-back stemplot* with common stems to compare the distribution of a quantitative variable in two groups. The leaves are placed in order on each side of the common stem. For example, Figure 1.9 shows a back-to-back stemplot of the 19 middle school students' resting pulse rates and their pulse rates after 5 minutes of running.

```
  Resting        After exercise
     9888 |  6 |
  8664110 |  7 |
     8862 |  8 | 6788
       60 |  9 | 02245899       Key: 8|2 is a student
        4 | 10 | 044            whose pulse rate is
          | 11 | 8              82 beats per minute.
        0 | 12 | 44
          | 13 |
          | 14 | 6
```

FIGURE 1.9 Back-to-back stemplot of 19 middle school students' resting pulse rates and their pulse rates after 5 minutes of running.

COMMON STUDENT ERROR

When identifying the shape of each distribution using a back-to-back stemplot, students will often give the wrong direction of the skew for the distribution displayed on the left side of the back-to-back stemplot. For this example, the resting pulse rates are mostly in the 70s, with values extending farther to larger values than to smaller values (skewed to the right).

Answers to CYU

1. *Shape:* The distribution of resting pulse rates is skewed to the right, with a single peak on the 70s stem. The distribution of after-exercise pulse rates is also skewed to the right, with a single peak on the 90s stem. *Outliers:* The distribution of resting pulse rates appears to have one outlier: the student whose resting pulse rate was 120 bpm. The distribution of after-exercise pulse rates appears to have one outlier as well: the student whose after-exercise pulse rate was 146 bpm. *Center:* The students' resting pulse rates tended to be lower (median = 76 bpm) than their "after-exercise" pulse rates (median = 98 bpm). *Variability:* The "after-exercise" pulse rates vary more (from 86 bpm to 146 bpm) than the resting pulse rates (which vary from 68 bpm to 120 bpm).

2. b

3. e

4. c

CHECK YOUR UNDERSTANDING

1. Write a few sentences comparing the distributions of resting and after-exercise pulse rates in Figure 1.9.

Multiple Choice: *Select the best answer for Questions 2–4.*

Here is a stemplot of the percent of residents aged 65 and older in the 50 states and the District of Columbia:

```
  6 | 8
  7 |
  8 | 8
  9 | 79
 10 | 08
 11 | 15566
 12 | 012223444457888999
 13 | 01233333444899
 14 | 02666
 15 | 23
 16 | 8
```

Key: 8|8 represents a state in which 8.8% of residents are 65 and older.

2. The low outlier is Alaska. What percent of Alaska residents are 65 or older?
   (a) 0.68   (b) 6.8   (c) 8.8   (d) 16.8   (e) 68

3. Ignoring the outlier, the shape of the distribution is
   (a) skewed to the right.
   (b) skewed to the left.
   (c) skewed to the middle.
   (d) double-peaked (bimodal).
   (e) roughly symmetric.

4. The center of the distribution is close to
   (a) 11.6%.   (b) 12.0%.   (c) 12.8%.   (d) 13.3%.   (e) 6.8% to 16.8%.

Histograms

You can use a dotplot or stemplot to display quantitative data. Both graphs show every individual data value. For large data sets, this can make it difficult to see the overall pattern in the graph. We often get a clearer picture of the distribution by grouping together nearby values. Doing so allows us to make a new type of graph: a **histogram**.

> **DEFINITION** **Histogram**
>
> A **histogram** shows each interval of values as a bar. The heights of the bars show the frequencies or relative frequencies of values in each interval.

Figure 1.10 shows a dotplot and a histogram of the durations (in minutes) of 220 eruptions of the Old Faithful geyser. Notice how the histogram groups together nearby values.

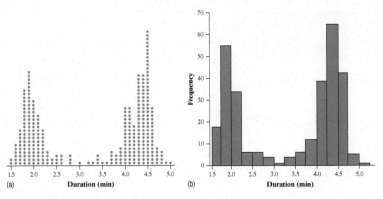

FIGURE 1.10 (a) Dotplot and (b) histogram of the duration (in minutes) of 220 eruptions of the Old Faithful geyser.

It is fairly easy to make a histogram by hand. Here's how you do it.

HOW TO MAKE A HISTOGRAM

- **Choose equal-width intervals** that span the data. Five intervals is a good minimum.
- **Make a table** that shows the frequency (count) or relative frequency (percent or proportion) of individuals in each interval. Put values that fall on an interval boundary in the interval containing larger values.
- **Draw and label the axes.** Draw horizontal and vertical axes. Put the name of the quantitative variable under the horizontal axis. To the left of the vertical axis, indicate whether the graph shows the frequency (count) or relative frequency (percent or proportion) of individuals in each interval.
- **Scale the axes.** Place equally spaced tick marks at the smallest value in each interval along the horizontal axis. On the vertical axis, start at 0 and place equally spaced tick marks until you exceed the largest frequency or relative frequency in any interval.
- **Draw bars** above the intervals. Make the bars equal in width and leave no gaps between them. Be sure that the height of each bar corresponds to the frequency or relative frequency of individuals in that interval. An interval with no data values will appear as a bar of height 0 on the graph.

It is possible to choose intervals of unequal widths when making a histogram. Such graphs are beyond the scope of this book.

How many medals?
Making and interpreting a histogram

PROBLEM:
The table displays the total number of medals 28 countries won during the 2016 Summer Olympic Games in Rio de Janeiro.

Country	Medals
United States	121
China	70
Great Britain	67
Russia	56
Germany	42
France	42
Japan	41
Australia	29
Italy	28
Canada	22
South Korea	21
Netherlands	19
Brazil	19
Spain	18
New Zealand	18
Azerbaijan	18
Kazakhstan	17
Hungary	15
Denmark	15
Kenya	13
Uzbekistan	13
Jamaica	11
Cuba	11
Sweden	11
Ukraine	11
Poland	11
Croatia	10
South Africa	10

(a) Make a frequency histogram to display the data.
(b) What percent of values in the graph are less than 30? Interpret this result in context.

SOLUTION:
(a) The data vary from 10 to 121 medals. I will choose intervals of width 10, starting at 10 medals.

Interval	Frequency
10 to <20	17
20 to <30	4
30 to <40	0
40 to <50	3
50 to <60	1
60 to <70	1
70 to <80	1
80 to <90	0
90 to <100	0
100 to <110	0
110 to <120	0
120 to <130	1

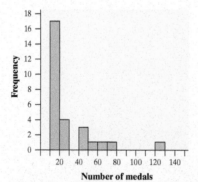

(b) 21/28 = 75%. 75% of the countries represented in the table won a total of fewer than 30 medals at the 2016 Summer Olympics.

EXAMPLE
How much tax?
Making and interpreting histograms

PROBLEM: Sales tax rates vary widely across the United States. Four states charge no state or local sales tax: Delaware, Montana, New Hampshire, and Oregon. The table shows data on the average total tax rate for each of the remaining 46 states and the District of Columbia.[23]

State	Tax rate (%)	State	Tax rate (%)	State	Tax rate (%)
Alabama	9.0	Louisiana	9.0	Oklahoma	8.8
Alaska	1.8	Maine	5.5	Pennsylvania	6.3
Arizona	8.3	Maryland	6.0	Rhode Island	7.0
Arkansas	9.3	Massachusetts	6.3	South Carolina	7.2
California	8.5	Michigan	6.0	South Dakota	5.8
Colorado	7.5	Minnesota	7.3	Tennessee	9.5
Connecticut	6.4	Mississippi	7.1	Texas	8.2
Florida	6.7	Missouri	7.9	Utah	6.7
Georgia	7.0	Nebraska	6.9	Vermont	6.2
Hawaii	4.4	Nevada	8.0	Virginia	5.6
Idaho	6.0	New Jersey	7.0	Washington	8.9
Illinois	8.6	New Mexico	7.5	West Virginia	6.2
Indiana	7.0	New York	8.5	Wisconsin	5.4
Iowa	6.8	North Carolina	6.9	Wyoming	5.4
Kansas	8.6	North Dakota	6.8	District of Columbia	5.8
Kentucky	6.0	Ohio	7.1		

(a) Make a frequency histogram to display the data.
(b) What percent of values in the distribution are less than 6.0? Interpret this result in context.

SOLUTION:
(a)

Interval	Frequency
1.0 to <2.0	1
2.0 to <3.0	0
3.0 to <4.0	0
4.0 to <5.0	1
5.0 to <6.0	6
6.0 to <7.0	15
7.0 to <8.0	11
8.0 to <9.0	9
9.0 to <10.0	4

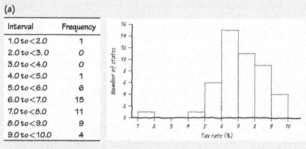

(b) 8/47 = 0.170 = 17.0%; 17% of the states (including the District of Columbia) have tax rates less than 6%.

To make the histogram:
• **Choose equal-width intervals** that span the data. The data vary from 1.8 percent to 9.5 percent. So we choose intervals of width 1.0, starting at 1.0%.
• **Make a table.** Record the number of states in each interval to make a frequency histogram.
• **Draw and label the axes.** Don't forget units (percent) for the variable (tax rate).
• **Scale the axes.**
• **Draw bars.**

Figure 1.11 shows two different histograms of the state sales tax data. Graph (a) uses the intervals of width 1% from the preceding example. The distribution has a single peak in the 6.0 to <7.0 interval. Graph (b) uses intervals half as wide: 1.0 to <1.5, 1.5 to <2.0, and so on. Now we see a distribution with more than one distinct peak. The choice of intervals in a histogram can affect the appearance of a distribution. Histograms with more intervals show more detail but may have a less clear overall pattern.

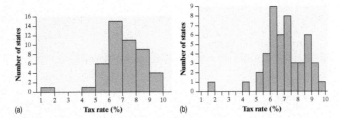

FIGURE 1.11 (a) Frequency histogram of the sales tax rate in the states that have local or state sales taxes and the District of Columbia with intervals of width 1.0%, from the preceding example. (b) Frequency histogram of the data with intervals of width 0.5%.

You can use a graphing calculator, statistical software, or an applet to make a histogram. The technology's default choice of intervals is a good starting point, but you should adjust the intervals to fit with common sense.

---

2. Technology Corner — MAKING HISTOGRAMS

TI-Nspire and other technology instructions are on the book's website at highschool.bfwpub.com/updatedtps6e.

1. Enter the data from the sales tax example in your Statistics/List Editor.
   - Press [STAT] and choose Edit...
   - Type the values into list L1.

2. Set up a histogram in the Statistics Plots menu.
   - Press [2nd] [Y=] (STAT PLOT).
   - Press [ENTER] or [1] to go into Plot1.
   - Adjust the settings as shown.

Teaching Tip: Using Technology

The *Histogram Bin-Width* applet allows students to explore how changing the classes in a histogram can greatly alter the apparent shape of a distribution. Go to www.rossmanchance.com/applets and choose "Histogram Bin-Width" and be sure to click on the JavaScript version.

Teaching Tip

Some people prefer "single-peaked" to "unimodal." In a histogram with one peak, we can't ascertain that there is one mode (because values are grouped together). "Unimodal" is too strong in many cases, but "single-peaked" is OK even if there is more than 1 mode (or no modes).

Teaching Tip: Using Technology

The screen shots we show for the TI-83/84 are from the TI-84 Plus CE. In many cases, the instructions are the same for older and newer operating systems.

Teaching Tip: Using Technology

Remind students that instructions for using the TI-Nspire are available on the Student Site.

Teaching Tip: Using Technology

Students may ask what the "Freq" setting is for. "Freq" stands for frequency, so when Freq = 1, each observation in L1 is counted once. The only time you might use something other than 1 is if you have data already in a frequency table. For example, suppose that the distribution of scores for your students on last year's AP® Statistics exam was

Score	Frequency
1	3
2	5
3	14
4	6
5	4

If you enter the values 1 to 5 in L1 and the frequencies in L2, you should use Freq = L2 when making a graph or calculating summary statistics.

3. Use ZoomStat to let the calculator choose intervals and make a histogram.
   - Press [ZOOM] and choose ZoomStat.
   - Press [TRACE] to examine the intervals.

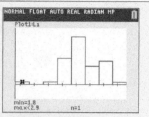

4. Adjust the intervals to match those in Figure 1.11(a), and then graph the histogram.
   - Press [WINDOW] and enter the values shown for Xmin, Xmax, Xscl, Ymin, Ymax, and Yscl.
   - Press [GRAPH].
   - Press [TRACE] to examine the intervals.

5. See if you can match the histogram in Figure 1.11(b).

> **AP® EXAM TIP**
>
> If you're asked to make a graph on a free-response question, be sure to label and scale your axes. Unless your calculator shows labels and scaling, don't just transfer a calculator screen shot to your paper.

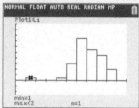

✔ **Answers to CYU**

1. One possible histogram is displayed below.

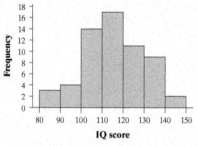

2. The distribution of IQ scores is roughly symmetric and bell-shaped with a single peak (unimodal). There are no obvious outliers. The typical IQ score appears to be between 110 and 120 (median = 114). The IQ scores vary from about 80 to 150.

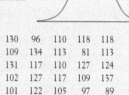

CHECK YOUR UNDERSTANDING

Many people believe that the distribution of IQ scores follows a "bell curve," like the one shown. But is this really how such scores are distributed? The IQ scores of 60 fifth-grade students chosen at random from one school are shown here.[24]

145	139	126	122	125	130	96	110	118	118
101	142	134	124	112	109	134	113	81	113
123	94	100	136	109	131	117	110	127	124
106	124	115	133	116	102	127	117	109	137
117	90	103	114	139	101	122	105	97	89
102	108	110	128	114	112	114	102	82	101

1. Construct a histogram that displays the distribution of IQ scores effectively.
2. Describe what you see. Is the distribution bell-shaped?

Using Histograms Wisely

We offer several cautions based on common mistakes students make when using histograms.

1. **Don't confuse histograms and bar graphs.** Although histograms resemble bar graphs, their details and uses are different. A histogram displays the distribution of a quantitative variable. Its horizontal axis identifies intervals of values that the variable takes. A bar graph displays the distribution of a categorical variable. Its horizontal axis identifies the categories. Be sure to draw bar graphs with blank space between the bars to separate the categories. Draw histograms with no space between bars for adjacent intervals. For comparison, here is one of each type of graph from earlier examples:

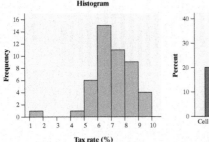

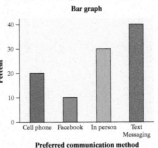

2. **Use percents or proportions instead of counts on the vertical axis when comparing distributions with different numbers of observations.** Mary was interested in comparing the reading levels of a biology journal and an airline magazine. She counted the number of letters in the first 400 words of an article in the journal and of the first 100 words of an article in the airline magazine. Mary then used statistical software to produce the histograms shown in Figure 1.12(a). This figure is misleading—it compares frequencies, but the two samples were of very different sizes (400 and 100). Using the same data, Mary's teacher produced the histograms in Figure 1.12(b). By using relative frequencies, this figure makes the comparison of word lengths in the two samples much easier.

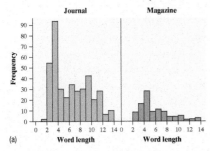

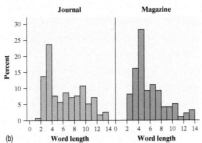

FIGURE 1.12 Two sets of histograms comparing word lengths in articles from a biology journal and from an airline magazine. In graph (a), the vertical scale uses frequencies. Graph (b) fixes the problem of different sample sizes by using percents (relative frequencies) on the vertical scale.

Teaching Tip: Using Technology

The bar graph on the left is not helpful because it doesn't allow us to identify the shape, center, or variability of the distribution of name length. To be a proper histogram, the variable should be on the horizontal axis and frequency (or relative frequency) should be on the vertical axis. Warn your students against using Excel for graphing quantitative variables unless they have a statistics add-on (which they probably don't). It is much better to use statistical software or applets designed for such a purpose.

✓ Answers to CYU

1. The distribution of word length for both the Journal and the Magazine have shapes that are skewed to the right and single-peaked. Neither distribution appears to have any outliers. The centers for both distributions are about the same at approximately 5–6 letters per word. Both distributions have similar variability, as the length of words in the Journal varies from 1 letter to 14 letters and the length of words in the Magazine varies from 2 letters to 14 letters.

2. This is a bar graph. It displays categorical data about first-year students' planned field of study.

3. No, because the variable is categorical and the categories could be listed in any order on the horizontal axis.

3. Just because a graph looks nice doesn't make it a meaningful display of data. The 15 students in a small statistics class recorded the number of letters in their first names. One student entered the data into an Excel spreadsheet and then used Excel's "chart maker" to produce the graph shown on the left. What kind of graph is this? It's a bar graph that compares the raw data values. But first-name length is a quantitative variable, so a bar graph is not an appropriate way to display its distribution. The histogram on the right is a much better choice because the graph makes it easier to identify the shape, center, and variability of the distribution of name length.

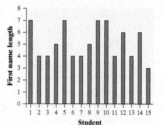

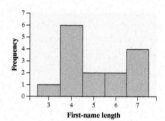

CHECK YOUR UNDERSTANDING

1. Write a few sentences comparing the distributions of word length shown in Figure 1.12(b).

Questions 2 and 3 refer to the following setting. About 3 million first-year students enroll in colleges and universities each year. What do they plan to study? The graph displays data on the percent of first-year students who plan to major in several disciplines.[25]

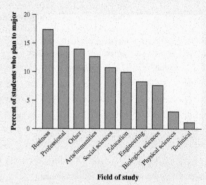

2. Is this a bar graph or a histogram? Explain.
3. Would it be correct to describe this distribution as right-skewed? Why or why not?

Section 1.2 | Summary

- You can use a **dotplot, stemplot,** or **histogram** to show the distribution of a quantitative variable. A dotplot displays individual values on a number line. Stemplots separate each observation into a stem and a one-digit leaf. Histograms plot the frequencies (counts) or relative frequencies (proportions or percents) of values in equal-width intervals.
- Some distributions have simple shapes, such as **symmetric, skewed to the left,** or **skewed to the right.** The number of peaks is another aspect of overall shape. So are distinct clusters and gaps.
- A single-peaked graph is sometimes called *unimodal,* and a double-peaked graph is sometimes called *bimodal.*
- When examining any graph of quantitative data, look for an *overall pattern* and for clear *departures* from that pattern. **Shape, center,** and **variability** describe the overall pattern of the distribution of a quantitative variable. **Outliers** are observations that lie outside the overall pattern of a distribution.
- When comparing distributions of quantitative data, be sure to compare shape, center, variability, and possible outliers.
- Remember: histograms are for quantitative data; bar graphs are for categorical data. Be sure to use relative frequencies when comparing data sets of different sizes.

1.2 Technology Corner

TI-Nspire and other technology instructions are on the book's website at highschool.bfwpub.com/updatedtps6e.

2. Making histograms   Page 43

Section 1.2 | Exercises

45. Feeling sleepy? Students in a high school statistics
pg 31 class responded to a survey designed by their teacher. One of the survey questions was "How much sleep did you get last night?" Here are the data (in hours):

9	6	8	7	8	8	6	6.5	7	7	9.0	4	3	4
5	6	11	6	3	7	6	10.0	7	8	4.5	9	7	7

(a) Make a dotplot to display the data.

(b) Experts recommend that high school students sleep at least 9 hours per night. What proportion of students in this class got the recommended amount of sleep?

46. Easy reading? Here are data on the lengths of the first 25 words on a randomly selected page from Toni Morrison's *Song of Solomon*:

2	3	4	10	2	11	2	8	4	3	7	2	7
5	3	6	4	4	2	5	8	2	3	4	4	

(a) Make a dotplot of these data.

(b) Long words can make a book hard to read. What percentage of words in the sample have 8 or more letters?

47. U.S. women's soccer—2016 Earlier, we examined data on the number of goals scored by the 2016 U.S. women's soccer team in 20 games played. The following dotplot displays the goal differential for those same games, computed as U.S. goals scored minus opponent goals scored.

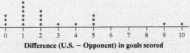

Difference (U.S. − Opponent) in goals scored

(a) Explain what the dot above 3 represents.

(b) What does the graph tell us about how well the team did in 2016? Be specific.

Answers to Section 1.2 Exercises

1.45 (a)

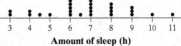

Amount of sleep (h)

(b) 0.179

1.46 (a)

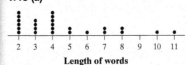

Length of words

(b) 16%

1.47 (a) The dot above 3 indicates that there was one game in which the U.S. women's soccer team scored 3 more goals than their opponent. **(b)** All 20 of the values are zero or more, which indicates that the U.S. women's soccer team had a very good season. They won $17/20 = 85\%$ of their games, tied with the other team in $3/20 = 15\%$ of their games, and never lost.

1.48 (a) The dot above −3 represents a car that got 3 mpg less on the highway than it did in city driving. **(b)** From the dotplot, we see that the EPA mpg rating is greater on the highway than in the city for all but one of the cars. These other 23 cars all averaged at least 7 mpg more on the highway than in the city.

1.49 The shape of the distribution is left-skewed with a peak between 90 and 100 years. There is a small gap around 70 years.

1.50 The shape of the distribution is roughly symmetric, with a single peak at 7.

1.51 The shape of the distribution is roughly symmetric, with a single peak at 7.

1.52 The shape of the distribution is skewed to the right, with peaks at 2 and 4 letters.

1.53 The distribution of difference (U.S. − Opponent) in goals scored is skewed to the right, with two potential outliers: when the U.S. team outscored their opponents by 9 and 10 goals. The median difference was 2 goals and the differences varied from 0 to 10 goals.

1.54 The distribution of difference (Highway − City) in EPA mileage rating is roughly symmetric, with a low outlier at −3. The median difference is 9 mpg and the differences vary from −3 to 12 miles per gallon.

1.55 The distribution of total family income for Indiana is roughly symmetric, while the distribution of total family income for New Jersey is slightly skewed right. The value of $125,000 may be an outlier in the Indiana distribution. There are no obvious outliers in the New Jersey distribution. The median for both distributions is about the same, approximately $49,000. The distribution of total family income in Indiana is less variable than the New Jersey distribution. The incomes in Indiana vary from $0 to about $125,000. The incomes in New Jersey vary from $0 to about $170,000.

1.56 The distributions of nitrate concentration for Stony Brook and for Mill Brook are roughly symmetric. The values 18 mg/L and 20 mg/L in the Mill Brook distribution may be outliers. There do not appear to be any outliers in the Stony Brook distribution. The typical nitrate concentration in Stony Brook (median ≈ 5 mg/L) is less

48. **Fuel efficiency** The dotplot shows the difference (Highway − City) in EPA mileage ratings, in miles per gallon (mpg) for each of 24 model year 2018 cars.

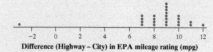

Difference (Highway − City) in EPA mileage rating (mpg)

(a) Explain what the dot above −3 represents.

(b) What does the graph tell us about fuel economy in the city versus on the highway for these car models? Be specific.

49. **Getting older** How old is the oldest person you know? pg 33 Prudential Insurance Company asked 400 people to place a blue sticker on a huge wall next to the age of the oldest person they have ever known. An image of the graph is shown here. Describe the shape of the distribution.

50. **Pair-o-dice** The dotplot shows the results of rolling a pair of fair, six-sided dice and finding the sum of the up-faces 100 times. Describe the shape of the distribution.

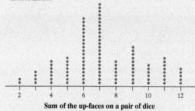

Sum of the up-faces on a pair of dice

51. **Feeling sleepy?** Refer to Exercise 45. Describe the shape of the distribution.

52. **Easy reading?** Refer to Exercise 46. Describe the shape of the distribution.

53. **U.S. women's soccer—2016** Refer to Exercise 47. pg 35 Describe the distribution.

54. **Fuel efficiency** Refer to Exercise 48. Describe the distribution.

55. **Making money** The parallel dotplots show the pg 36 total family income of randomly chosen individuals from Indiana (38 individuals) and New Jersey

(44 individuals). Compare the distributions of total family incomes in these two samples.

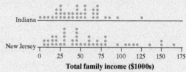

Total family income ($1000s)

56. **Healthy streams** Nitrates are organic compounds that are a main ingredient in fertilizers. When those fertilizers run off into streams, the nitrates can have a toxic effect on fish. An ecologist studying nitrate pollution in two streams measures nitrate concentrations at 42 places on Stony Brook and 42 places on Mill Brook. The parallel dotplots display the data. Compare the distributions of nitrate concentration in these two streams.

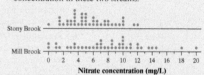

Nitrate concentration (mg/L)

57. **Enhancing creativity** Do external rewards—things like money, praise, fame, and grades—promote creativity? Researcher Teresa Amabile recruited 47 experienced creative writers who were college students and divided them at random into two groups. The students in one group were given a list of statements about external reasons (E) for writing, such as public recognition, making money, or pleasing their parents. Students in the other group were given a list of statements about internal reasons (I) for writing, such as expressing yourself and enjoying wordplay. Both groups were then instructed to write a poem about laughter. Each student's poem was rated separately by 12 different poets using a creativity scale.[26] These ratings were averaged to obtain an overall creativity score for each poem. Parallel dotplots of the two groups' creativity scores are shown here.

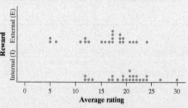

Average rating

(a) Is the variability in creativity scores similar or different for the two groups? Justify your answer.

(b) Do the data suggest that external rewards promote creativity? Justify your answer.

than the typical nitrate concentration in Mill Brook (median ≈ 8 mg/L). The distribution of nitrate concentration for Stony Brook is less variable than the Mill Brook distribution. The nitrate concentrations in the Stony Brook distribution vary from 0 to 12 mg/L. The nitrate concentrations in the Mill Brook distribution vary from 0 to 20 mg/L.

1.57 (a) Both distributions have about the same amount of variability. The "external reward" distribution varies from 5 to about 24. And the "internal reward" distribution varies from about 12 to 30. **(b)** The center of the internal distribution is greater than the center of the external distribution, indicating that external rewards do not promote creativity.

58. **Healthy cereal?** Researchers collected data on 76 brands of cereal at a local supermarket.[27] For each brand, the sugar content (grams per serving) and the shelf in the store on which the cereal was located (1 = bottom, 2 = middle, 3 = top) were recorded. A dotplot of the data is shown here.

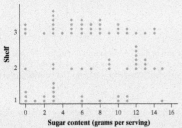

(a) Is the variability in sugar content of the cereals on the three shelves similar or different? Justify your answer.

(b) Critics claim that supermarkets tend to put sugary cereals where kids can see them. Do the data from this study support this claim? Justify your answer. (Note that Shelf 2 is at about eye level for kids in most supermarkets.)

59. **Snickers® are fun!** Here are the weights (in grams) of 17 Snickers Fun Size bars from a single bag:

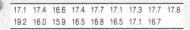

| 17.1 | 17.4 | 16.6 | 17.4 | 17.7 | 17.1 | 17.3 | 17.7 | 17.8 |
| 19.2 | 16.0 | 15.9 | 16.5 | 16.8 | 16.5 | 17.1 | 16.7 |

(a) Make a stemplot of these data.

(b) What interesting feature does the graph reveal?

(c) The advertised weight of a Snickers Fun Size bar is 17 grams. What proportion of candy bars in this sample weigh less than advertised?

60. **Eat your beans!** Beans and other legumes are a great source of protein. The following data give the protein content of 30 different varieties of beans, in grams per 100 grams of cooked beans.[28]

7.5	8.2	8.9	9.3	7.1	8.3	8.7	9.5	8.2	9.1
9.0	9.0	9.7	9.2	8.9	8.1	9.0	7.8	8.0	7.8
7.0	7.5	13.5	8.3	6.8	10.6	8.3	7.6	7.7	8.1

(a) Make a stemplot of these data.

(b) What interesting feature does the graph reveal?

(c) What proportion of these bean varieties contain more than 9 grams of protein per 100 grams of cooked beans?

61. **South Carolina counties** Here is a stemplot of the areas of the 46 counties in South Carolina. Note that the data have been rounded to the nearest 10 square miles (mi²).

```
 3 | 9999
 4 | 0116689
 5 | 01115566778
 6 | 47899
 7 | 01245579
 8 | 0011
 9 | 13
10 | 8
11 | 233
12 | 2
```

Key: 6|4 represents a county with an area of 635 to 644.99 square miles.

(a) What is the area of the largest South Carolina county?

(b) Describe the distribution of area for the 46 South Carolina counties.

62. **Shopping spree** The stemplot displays data on the amount spent by 50 shoppers at a grocery store. Note that the values have been rounded to the nearest dollar.

```
0 | 399
1 | 1345677889
2 | 000123455668888
3 | 25699
4 | 1345579
5 | 0359
6 | 1
7 | 0
8 | 366
9 | 3
```

Key: 9|3 = $92.50 to $93.49 spent

(a) What was the smallest amount spent by any of the shoppers?

(b) Describe the distribution of amount spent by these 50 shoppers.

63. **Where do the young live?** Here is a stemplot of the percent of residents aged 25 to 34 in each of the 50 states:

```
11 | 44
11 | 66778
12 | 0134
12 | 666778888
13 | 0000001111444
13 | 7788999
14 | 0044
14 | 567
15 | 11
15 |
16 | 0
```

(a) Why did we split stems?

(b) Give an appropriate key for this stemplot.

(c) Describe the shape of the distribution. Are there any outliers?

others (13.5 grams per 100 grams of cooked beans). **(c)** 0.233

1.61 (a) The area of the largest South Carolina county is 1220 square miles (rounded to the nearest 10 mi²). **(b)** The distribution of the area for the 46 South Carolina counties is right-skewed, with distinct peaks on the 500 mi² and 700 mi² stems. There are no clear outliers. A typical South Carolina county has an area of about 655 square miles. The area of the counties varies from about 390 square miles to 1220 square miles.

1.62 (a) The smallest amount spent by any of the shoppers is between $2.50 and $3.49. **(b)** The distribution of the amount of money spent by shoppers at this supermarket is skewed to the right, with distinct peaks on the $20 and $40 stems. There are some potential outliers on the high end ($83, $86, and $93). A typical shopper spent about $28. The amount spent by the 50 shoppers varies from about $3 to about $93.

1.63 (a) If we had not split the stems, most of the data would appear on just a few stems, making it hard to identify the shape of the distribution. **(b)** Key: 16 | 0 means that 16.0% of that state's residents are aged 25 to 34. **(c)** The distribution of percent of residents aged 25–34 is roughly symmetric, with a possible outlier at 16.0%.

1.58 (a) All three distributions of sugar content have about the same amount of variability, as they all vary from about 0 grams per serving to about 15 grams per serving. **(b)** Yes! The distribution of sugar content for shelf 2, which is right at eye level for most kids, has the largest center indicating that supermarkets tend to put sugary kids' cereals where the kids are most likely to see them.

1.59 (a)
```
15 | 9
16 | 0 5 5 6 7 8
17 | 1 1 1 3 4 4 7 7 8
18 |
19 | 2
```
Key: 15 | 9 = 15.9 grams

(b) The graph reveals that there was one Fun Size Snickers® bar that is "gigantic"! It weighs 19.2 grams. **(c)** 0.412

1.60 (a)
```
 6 | 8
 7 | 0 1 5 5 6 7 8 8
 8 | 0 1 1 2 2 3 3 3 7 9 9
 9 | 0 0 0 1 2 3 5 7
10 | 6
11 |
12 |
13 | 5
```
Key 6 | 8 = 6.8 grams per 100 grams of cooked beans

(b) The graph reveals that there was one variety of bean that has much more protein than the

1.64 (a) If we had not split the stems, most of the data would appear on just a few stems, making it hard to identify the shape of the distribution. **(b)** Key: 1 | 5 = a soft drink for which an 8-oz serving has 15 mg of caffeine. **(c)** This distribution is slightly skewed to the right, with two distinct peaks, one on the 25–29 stem and the other on the 35–39 stem. There do not appear to be any outliers.

1.65 The distribution of acorn volume for the Atlantic coast is skewed to the right. The distribution of acorn volume for California is roughly symmetric, with one high outlier of 17.1 cubic centimeters. The distribution of volume of acorn for the Atlantic coast has 3 potential outliers: 8.1, 9.1, and 10.5 cubic centimeters. The typical acorn volume for Atlantic coast oak tree species (median = 1.7 cubic centimeters) is less than the typical acorn volume for California oak tree species (median = 4.1 cubic centimeters). The Atlantic coast distribution (with acorn volumes from 0.3 to 10.5 cubic centimeters) varies less than the California distribution (with acorn volumes from 0.4 to 17.1 cubic centimeters).

1.66 The distributions of study time for men and for women are both roughly symmetric. Both distributions have one upper outlier: one woman who studied 360 minutes and one man who studied 300 minutes. The women (median = 175 minutes) tended to study more than the men (median = 120 minutes). The variability of the two study time distributions is about the same: the women's study times varied from 60 to 360 minutes and the men's study times varied from 0 to 300 minutes.

1.67 (a)

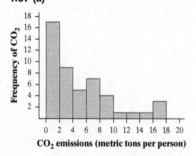

(b) The distribution of amount of carbon dioxide emissions per person in these 48 countries is right-skewed. Visually, none of the countries appear to be outliers.

64. **Watch that caffeine!** The U.S. Food and Drug Administration (USFDA) limits the amount of caffeine in a 12-ounce can of carbonated beverage to 72 milligrams. That translates to a maximum of 48 milligrams of caffeine per 8-ounce serving. Data on the caffeine content of popular soft drinks (in milligrams per 8-ounce serving) are displayed in the stemplot.

```
1 | 556
2 | 033344
2 | 55667778888899
3 | 113
3 | 55567778
4 | 33
4 | 77
```

(a) Why did we split stems?

(b) Give an appropriate key for this graph.

(c) Describe the shape of the distribution. Are there any outliers?

65. **Acorns and oak trees** Of the many species of oak trees in the United States, 28 grow on the Atlantic Coast and 11 grow in California. The back-to-back stemplot displays data on the average volume of acorns (in cubic centimeters) for these 39 oak species.[29] Write a few sentences comparing the distributions of acorn size for the oak trees in these two regions.

```
Atlantic Coast |    | California
       998643  | 0  | 4
  88864211111  | 1  | 06
           50  | 2  | 06
         6640  | 3  |
            8  | 4  | 1
               | 5  | 59
            8  | 6  | 0
               | 7  | 1
            1  | 8  |
            1  | 9  |
            5  | 10 |
               | 11 |
               | 12 |
               | 13 |
               | 14 |
               | 15 |
               | 16 |
               | 17 | 1
```

Key: 2|6 = An oak species whose acorn volume is 2.6 cm³.

66. **Who studies more?** Researchers asked the students in a large first-year college class how many minutes they studied on a typical weeknight. The back-to-back stemplot displays the responses from random samples of 30 women and 30 men from the class, rounded to the nearest 10 minutes. Write a few sentences comparing the male and female distributions of study time.

Women		Men
	0	03333
96	0	56668999
22222222	1	02222222
888888888875555	1	558
4440	2	00344
	2	
	3	0
6	3	

Key: 2|3 = 225–234 minutes

67. **Carbon dioxide emissions** Burning fuels in power plants and motor vehicles emits carbon dioxide (CO_2), which contributes to global warming. The table displays CO_2 emissions per person from countries with populations of at least 20 million.[30]

pg 42

(a) Make a histogram of the data using intervals of width 2, starting at 0.

(b) Describe the shape of the distribution. Which countries appear to be outliers?

Country	CO₂	Country	CO₂
Algeria	3.3	Mexico	3.8
Argentina	4.5	Morocco	1.6
Australia	16.9	Myanmar	0.2
Bangladesh	0.4	Nepal	0.1
Brazil	2.2	Nigeria	0.5
Canada	14.7	Pakistan	0.9
China	6.2	Peru	2.0
Colombia	1.6	Philippines	0.9
Congo	0.5	Poland	8.3
Egypt	2.6	Romania	3.9
Ethiopla	0.1	Russia	12.2
France	5.6	Saudi Arabia	17.0
Germany	9.1	South Africa	9.0
Ghana	0.4	Spain	5.8
India	1.7	Sudan	0.3
Indonesia	1.8	Tanzania	0.2
Iran	7.7	Thailand	4.4
Iraq	3.7	Turkey	4.1
Italy	6.7	Ukraine	6.6
Japan	9.2	United Kingdom	7.9
Kenya	0.3	United States	17.6
Korea, North	11.5	Uzbekistan	3.7
Korea, South	2.9	Venezuela	6.9
Malaysia	7.7	Vietnam	1.7

68. **Traveling to work** How long do people travel each day to get to work? The following table gives the average travel times to work (in minutes) for workers in each state and the District of Columbia who are at least 16 years old and don't work at home.[31]

AL	23.6	LA	25.1	OH	22.1
AK	17.7	ME	22.3	OK	20.0
AZ	25.0	MD	30.6	OR	21.8
AR	20.7	MA	26.6	PA	25.0
CA	26.8	MI	23.4	RI	22.3
CO	23.9	MN	22.0	SC	22.9
CT	24.1	MS	24.0	SD	15.9
DE	23.6	MO	22.9	TN	23.5
FL	25.9	MT	17.6	TX	24.6
GA	27.3	NE	17.7	UT	20.8
HI	25.5	NV	24.2	VT	21.2
ID	20.1	NH	24.6	VA	26.9
IL	27.9	NJ	29.1	WA	25.2
IN	22.3	NM	20.9	WV	25.6
IA	18.2	NY	30.9	WI	20.8
KS	18.5	NC	23.4	WY	17.9
KY	22.4	ND	15.5	DC	29.2

(a) Make a histogram to display the travel time data using intervals of width 2 minutes, starting at 14 minutes.

(b) Describe the shape of the distribution. What is the most common interval of travel times?

69. **DRP test scores** There are many ways to measure the reading ability of children. One frequently used test is the Degree of Reading Power (DRP). In a research study on third-grade students, the DRP was administered to 44 students.[32] Their scores were as follows.

40	26	39	14	42	18	25	43	46	27	19
47	19	26	35	34	15	44	40	38	31	46
52	25	35	35	33	29	34	41	49	28	52
47	35	48	22	33	41	51	27	14	54	45

Make a histogram to display the data. Write a few sentences describing the distribution of DRP scores.

70. **Country music** The lengths, in minutes, of the 50 most popular mp3 downloads of songs by country artist Dierks Bentley are given here.

4.2	4.0	3.9	3.8	3.7	4.7
3.4	4.0	4.4	5.0	4.6	3.7
4.6	4.4	4.1	3.0	3.2	4.7
3.5	3.7	4.3	3.7	4.8	4.4
4.2	4.7	6.2	4.0	7.0	3.9
3.4	3.4	2.9	3.3	4.0	4.2
3.2	3.4	3.7	3.5	3.4	3.7
3.9	3.7	3.8	3.1	3.7	3.6
4.5	3.7				

Make a histogram to display the data. Write a few sentences describing the distribution of song lengths.

71. **Returns on common stocks** The return on a stock is the change in its market price plus any dividend payments made. Return is usually expressed as a percent of the beginning price. The figure shows a histogram of the distribution of monthly returns for the U.S. stock market over a 273-month period.[33]

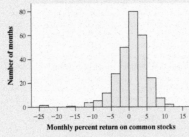

Monthly percent return on common stocks

(a) Describe the overall shape of the distribution of monthly returns.

(b) What is the approximate center of this distribution?

(c) Explain why you cannot find the exact value for the minimum return. Between what two values does it lie?

(d) A return less than 0 means that stocks lost value in that month. About what percent of all months had returns less than 0?

72. **Healthy cereal?** Researchers collected data on calories per serving for 77 brands of breakfast cereal. The histogram displays the data.[34]

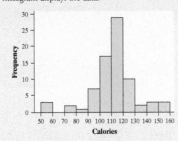

Calories

(a) Describe the overall shape of the distribution of calories.

(b) What is the approximate center of this distribution?

(c) Explain why you cannot find the exact value for the maximum number of calories per serving for

1.70 The data vary from 2.9 to 7.0 minutes. We chose intervals of width 0.5, beginning at 2.5.

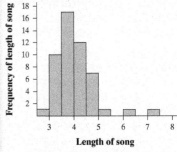

Length of song

(b) The distribution of song lengths of the most popular mp3 downloads from country artist Dierks Bentley is skewed to the right, with a single peak in the 3.5 to < 4.0 minute interval. There are gaps in the 5.5 to < 6.0 minute interval and the 6.5 to < 7.0 interval. The songs with lengths of 6.2 and 7 minutes are possible outliers. Song lengths vary from 2.9 to 7 minutes.

1.71 (a) The shape of the distribution is slightly skewed to the left, with a single peak. **(b)** The center is between 0% and 2.5% return on common stocks. **(c)** The exact value for the minimum return cannot be identified because we have a histogram of only the returns, not the actual data. The lowest return was in the interval −22.5% to −25%. **(d)** About 37% of these months (102 out of 273) had negative returns.

1.72 (a) The shape of the distribution is roughly symmetric, with a single peak. **(b)** The center of the distribution is between 110 and 120 calories. **(c)** The exact value for the maximum number of calories per serving cannot be identified because we have a histogram of only the calories, not the actual data. The maximum number of calories per serving is in the interval 150 to 160 calories.

1.68 (a)

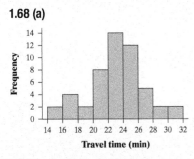

Travel time (min)

(b) The distribution of travel times is roughly symmetric. The most common interval of travel times is 22 to 24 minutes.

1.69 The data vary from 14 to 54. We chose intervals of width 6, beginning at 12.

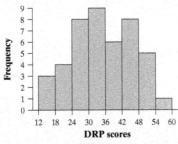

DRP scores

The distribution of DRP scores is roughly symmetric. There do not appear to be any outliers. The center of the DRP score distribution is between 30 and 36 (with median = 35). The DRP scores vary from 14 to 54.

1.73 It is difficult to effectively compare the salaries of the two teams with these two histograms because the scale on the horizontal axis is very different from one graph to the other. It also does not help that the scales on the *y* axis differ as well.

1.74 The distribution of salary for both teams is skewed to the right. It appears that the Yankees' distribution has an outlier somewhere between $32 and $36 million, while the Phillies' distribution does not appear to have any outliers. The salaries for the Yankees have a higher center of about $16 million than the Phillies' center of about $4 million. The distribution of salaries for the Yankees also has more variability as their salaries vary from $0 to $36 million and the Phillies' salaries vary from $0 to $16 million.

1.75 (a) No, it would not be appropriate to use frequency histograms instead of relative frequency histograms in this setting because there were many more graduates surveyed (314) than non-graduates (57). **(b)** The distribution of total personal income for each group is skewed to the right and single-peaked. There are some possible high outliers in the graduate distribution. There do not appear to be any outliers in the non-graduate distribution. The center of the personal income distribution is larger for graduates than non-graduates, indicating that graduates typically have higher incomes than non-graduates in this sample. The incomes for graduates vary a lot more (from $0 to $150,000) than non-graduates (from $0 to $60,000).

1.76 (a) Yes, it would be all right to use frequency histograms instead of relative frequency histograms in this setting because the sample sizes were similar. The study was based on 30 Bounty paper towels and 30 generic paper towels. **(b)** The distributions of the number of quarters the Bounty and generic paper towels could hold until breaking are roughly symmetric. Neither distribution appears to contain any outliers. The Bounty paper towels tended to hold more quarters before breaking, as their center was around 115 quarters, while the generic brand tended to only hold about 90 quarters before breaking. The variability of both brands was about the same as the number of quarters held by the Bounty paper towels varied from approximately 100 to 130 quarters and the number of quarters held by the generic paper towels varied from about 75 to 105.

these cereal brands. Between what two values does it lie?

(d) About what percent of the cereal brands have 130 or more calories per serving?

73. **Paying for championships** Does paying high salaries lead to more victories in professional sports? The New York Yankees have long been known for having Major League Baseball's highest team payroll. And over the years, the team has won many championships. This strategy didn't pay off in 2008, when the Philadelphia Phillies won the World Series. Maybe the Yankees didn't spend enough money that year. The figure shows histograms of the salary distributions for the two teams during the 2008 season. Why can't you use these graphs to effectively compare the team payrolls?

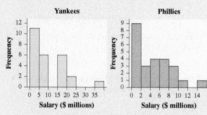

74. **Paying for championships** Refer to Exercise 73. Here is a better graph of the 2008 salary distributions for the Yankees and the Phillies. Write a few sentences comparing these two distributions.

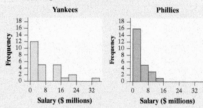

75. **Value of a diploma** Do students who graduate from high school earn more money than students who do not? To find out, we took a random sample of 371 U.S. residents aged 18 and older. The educational level and total personal income of each person were recorded. The data for the 57 non-graduates (No) and the 314 graduates (Yes) are displayed in the relative frequency histograms.

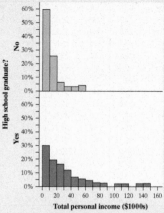

(a) Would it be appropriate to use frequency histograms instead of relative frequency histograms in this setting? Explain why or why not.

(b) Compare the distributions of total personal income for the two groups.

76. **Strong paper towels** In commercials for Bounty paper towels, the manufacturer claims that they are the "quicker picker-upper," but are they also the stronger picker-upper? Two of Mr. Tabor's statistics students, Wesley and Maverick, decided to find out. They selected a random sample of 30 Bounty paper towels and a random sample of 30 generic paper towels and measured their strength when wet. To do this, they uniformly soaked each paper towel with 4 ounces of water, held two opposite edges of the paper towel, and counted how many quarters each paper towel could hold until ripping, alternating brands. The data are displayed in the relative frequency histograms. Compare the distributions.

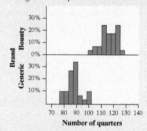

(a) Would it be appropriate to use frequency histograms instead of relative frequency histograms in this setting? Explain why or why not.

(b) Compare the distributions of number of quarters until breaking for the two paper towel brands.

77. **Birth months** Imagine asking a random sample of 60 students from your school about their birth months. Draw a plausible (believable) graph of the distribution of birth months. Should you use a bar graph or a histogram to display the data?

78. **Die rolls** Imagine rolling a fair, six-sided die 60 times. Draw a plausible graph of the distribution of die rolls. Should you use a bar graph or a histogram to display the data?

79. **AP® exam scores** The table gives the distribution of grades earned by students taking the AP® Calculus AB and AP® Statistics exams in 2016.[35]

			Grade			
	5	**4**	**3**	**2**	**1**	**Total**
Calculus AB	76,486	53,467	53,533	30,017	94,712	308,215
Statistics	29,627	44,884	51,367	32,120	48,565	206,563

(a) Make an appropriate graphical display to compare the grade distributions for AP® Calculus AB and AP® Statistics.

(b) Write a few sentences comparing the two distributions of exam grades.

Multiple Choice: *Select the best answer for Exercises 80–85.*

80. Here are the amounts of money (cents) in coins carried by 10 students in a statistics class: 50, 35, 0, 46, 86, 0, 5, 47, 23, 65. To make a stemplot of these data, you would use stems

(a) 0, 2, 3, 4, 6, 8.

(b) 0, 1, 2, 3, 4, 5, 6, 7, 8.

(c) 0, 3, 5, 6, 7.

(d) 00, 10, 20, 30, 40, 50, 60, 70, 80, 90.

(e) None of these.

81. The histogram shows the heights of 300 randomly selected high school students. Which of the following is the best description of the shape of the distribution of heights?

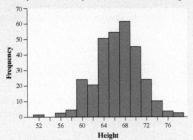

(a) Roughly symmetric and single-peaked (unimodal)

(b) Roughly symmetric and double-peaked (bimodal)

(c) Roughly symmetric and multi-peaked (multimodal)

(d) Skewed to the left

(e) Skewed to the right

82. You look at real estate ads for houses in Naples, Florida. There are many houses ranging from $200,000 to $500,000 in price. The few houses on the water, however, are priced up to $15 million. The distribution of house prices will be

(a) skewed to the left.

(b) roughly symmetric.

(c) skewed to the right.

(d) single-peaked.

(e) approximately uniform.

83. The histogram shows the distribution of the percents of women aged 15 and over who have never married in each of the 50 states and the District of Columbia. Which of the following statements about the histogram is correct?

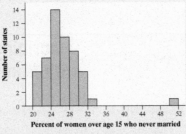

Percent of women over age 15 who never married

(a) The center (median) of the distribution is about 36%.

(b) There are more states with percentages above 32 than there are states with percentages less than 24.

(c) It would be better if the values from 34 to 50 were deleted on the horizontal axis so there wouldn't be a large gap.

(d) There was one state with a value of exactly 33%.

(e) About half of the states had percentages between 24% and 28%.

84. When comparing two distributions, it would be best to use relative frequency histograms rather than frequency histograms when

(a) the distributions have different shapes.

(b) the distributions have different amounts of variability.

(c) the distributions have different centers.

(d) the distributions have different numbers of observations.

(e) at least one of the distributions has outliers.

85. Which of the following is the best reason for choosing a stemplot rather than a histogram to display the distribution of a quantitative variable?

(a) Stemplots allow you to split stems; histograms don't.

(b) Stemplots allow you to see the values of individual observations.

(c) Stemplots are better for displaying very large sets of data.

(d) Stemplots never require rounding of values.

(e) Stemplots make it easier to determine the shape of a distribution.

1.79 A score earned on the AP® Statistics exam is quantitative, so histograms are shown below using relative frequencies since there are many more students who take the AP® Calculus AB exam.

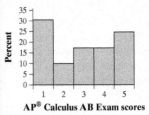

AP® Calculus AB Exam scores

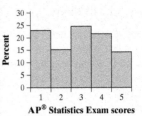

AP® Statistics Exam scores

The shapes of the two distributions are very different. The distribution of scores on the AP® Calculus AB exam has a peak at 1 and another slightly lower peak at 5. The distribution of scores on the AP® Statistics exam, however, is more uniform, with scores of 1, 3, and 4 being the most frequent and scores of 5 being the least frequent. Neither distribution has any outliers. The center of both distributions is 3. Although scores on both exams vary from 1 to 5, there are more scores close to the center on the AP® Statistics exam and more scores at the extremes on the AP® Calculus exam.

1.80 b	**1.82** c	**1.84** d
1.81 a	**1.83** e	**1.85** b

1.77 A bar graph should be used because birth month is a categorical variable. A possible bar graph is given below.

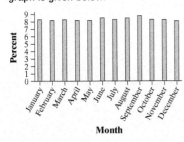

Month

1.78 A histogram should be used because die roll is a quantitative variable. A possible histogram is given below.

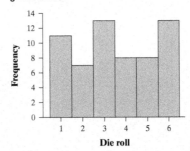

Die roll

1.86 (a) 8.3% were elite soccer players; 5% developed arthritis of the hip or knee. **(b)** 14.1% developed arthritis; 23.3% were elite soccer players. **(c)** Of the elite soccer players, 14.1% developed arthritis later in life. Of the non-elite soccer players, 4.2% developed arthritis later in life. Of those that did not play, 4.2% developed arthritis later in life. The data do appear to confirm the suspicion that more serious soccer players were more likely to develop arthritis later in life.

PD   Section 1.3 Overview

The video overview can be found by clicking on the link in the TE-Book, opening the TRFD, or downloading from the Teacher's Resources on the book's digital platform.

TRM   Section 1.3 Alternate Examples

You can find the Alternate Examples for this section in Microsoft Word format by clicking on the link in the TE-Book, opening the TRFD, or downloading from the Teacher's Resources on the book's digital platform.

Making Connections

In earlier grades, most students learn mean, median, and mode together and assume that all of them are measures of center. The mode may not be a measure of center, as there is no guarantee that the mode is near the center of a distribution. The mode can be the smallest value in the data set.

Recycle and Review

86. **Risks of playing soccer** (1.1) A study in Sweden looked at former elite soccer players, people who had played soccer but not at the elite level, and people of the same age who did not play soccer. Here is a two-way table that classifies these individuals by whether or not they had arthritis of the hip or knee by their mid-fifties:[36]

		Soccer level		
		Elite	Non-elite	Did not play
Whether person developed arthritis	Yes	10	9	24
	No	61	206	548

(a) What percent of the people in this study were elite soccer players? What percent of the people in this study developed arthritis?

(b) What percent of the elite soccer players developed arthritis? What percent of those who got arthritis were elite soccer players?

(c) Researchers suspected that the more serious soccer players were more likely to develop arthritis later in life. Do the data confirm this suspicion? Calculate appropriate percentages to support your answer.

SECTION 1.3   Describing Quantitative Data with Numbers

LEARNING TARGETS   *By the end of the section, you should be able to:*

- Calculate measures of center (mean, median) for a distribution of quantitative data.
- Calculate and interpret measures of variability (range, standard deviation, *IQR*) for a distribution of quantitative data.
- Explain how outliers and skewness affect measures of center and variability.
- Identify outliers using the $1.5 \times IQR$ rule.
- Make and interpret boxplots of quantitative data.
- Use boxplots and numerical summaries to compare distributions of quantitative data.

How much offense did the 2016 U.S. women's soccer team generate? The dotplot (reproduced from Section 1.2) shows the number of goals the team scored in 20 games played.

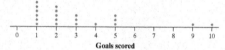

Goals scored

The distribution is right-skewed and single-peaked (unimodal). The games in which the team scored 9 and 10 goals appear to be outliers. How can we describe the center and variability of this distribution?

The *mode* of a data set is the most frequently occurring value. For the distribution of goals scored by the 2016 U.S. women's soccer team, the mode is 1. As you can see, the mode is rarely a good way to describe the center of the distribution.

Measuring Center: The Mean

The most common measure of center is the **mean**.

DEFINITION   The mean

The **mean** of a distribution of quantitative data is the average of all the individual data values. To find the mean, add all the values and divide by the total number of data values.

If the n observations are $x_1, x_2, \ldots, x_n$, the sample mean $\bar{x}$ (pronounced "x-bar") is given by the formula

$$\bar{x} = \frac{\text{sum of data values}}{\text{number of data values}} = \frac{x_1 + x_2 + \cdots + x_n}{n} = \frac{\sum x_i}{n}$$

The $\sum$ (capital Greek letter sigma) in the formula is short for "add them all up." The subscripts on the observations x_i are just a way of keeping the n data values distinct. They do not necessarily indicate order or any other special facts about the data.

AP® EXAM TIP

The formula for the sample mean $\bar{x}$ is included on the AP® Statistics formula sheet that can be used on both parts of the AP® Statistics exam. We recommend that you point out the Formula sheet in the appendix (page F-1) and give each student a copy right away so he or she can get accustomed to referring to it. Consider making laminated copies for students to use during tests and quizzes.

EXAMPLE

How many goals?
Calculating the mean

PROBLEM: Here are the data on the number of goals scored in 20 games played by the 2016 U.S. women's soccer team:

5 5 1 10 5 2 1 1 2 3 3 2 1 4 2 1 2 1 9 3

(a) Calculate the mean number of goals scored per game by the team. Show your work.
(b) The earlier description of these data (page 35) suggests that the games in which the team scored 9 and 10 goals are possible outliers. Calculate the mean number of goals scored per game by the team in the other 18 games that season. What do you notice?

SOLUTION:

(a) $\bar{x} = \dfrac{5 + 5 + 1 + 10 + 5 + 2 + 1 + 1 + 2 + 3 + 3 + 2 + 1 + 4 + 2 + 1 + 2 + 1 + 9 + 3}{20}$

$= \dfrac{63}{20} = 3.15 \text{ goals}$

$\boxed{\bar{x} = \dfrac{\sum x_i}{n}}$

(b) The mean for the other 18 games is

$\bar{x} = \dfrac{5 + 5 + 1 + 5 + 2 + 1 + 1 + 2 + 3 + 3 + 2 + 1 + 4 + 2 + 1 + 2 + 1 + 3}{18}$

$= \dfrac{44}{18} = 2.44 \text{ goals}$

These two games increased the team's mean number of goals scored per game by 0.71 goals.

FOR PRACTICE, TRY EXERCISE 87

The notation $\bar{x}$ refers to the mean of a *sample*. Most of the time, the data we encounter can be thought of as a sample from some larger population. When we need to refer to a *population mean*, we'll use the symbol μ (Greek letter mu, pronounced "mew"). If you have the entire population of data available, then you calculate μ in just the way you'd expect: add the values of all the observations, and divide by the number of observations.

We call $\bar{x}$ a **statistic** and μ a **parameter**. Remember **s** and **p**: **s**tatistics come from **s**amples and **p**arameters come from **p**opulations.

Preparing for Inference

In Chapter 4, we will reinforce that a value calculated from a sample (a statistic) is used to estimate a value for a population (a parameter). We use different notation for statistics than we do for parameters so readers will know if the reported value is an estimate or the true value.

ALTERNATE EXAMPLE   Skill 2.C

How many likes on Instagram for ASA?
Calculating the mean

PROBLEM:
The American Statistical Association (www.amstat.org) has an Instagram account (@amstatnews) to post updates on new statistical publications and adorable normal distribution plushies. Here are the number of Instagram likes for 10 posts selected at random:

16	4	8	7	8
6	15	2	9	5

(a) Calculate the mean number of Instagram likes for these 10 posts. Show your work.
(b) The posts with 15 and 16 likes are possible outliers. Calculate the mean number of Instagram likes in the other 8 posts. What do you notice?

SOLUTION:

(a) $\bar{x} = \dfrac{\sum x_i}{n} = \dfrac{16 + 4 + 8 + 7 + 8 + 6 + 15 + 2 + 9 + 5}{10} = \dfrac{80}{10} = 8 \text{ likes}$

(b) The mean for the other 8 posts is

$\bar{x} = \dfrac{\sum x_i}{n} = \dfrac{4 + 8 + 7 + 8 + 6 + 2 + 9 + 5}{8} = \dfrac{49}{8} = 6.1 \text{ likes}$

These two posts increased the mean number of Instagram likes per post by 1.9 likes.

> **DEFINITION** **Statistic, Parameter**
>
> A **statistic** is a number that describes some characteristic of a sample.
> A **parameter** is a number that describes some characteristic of a population.

 The preceding example illustrates an important weakness of the mean as a measure of center: the mean is sensitive to extreme values in a distribution. These may be outliers, but a skewed distribution that has no outliers will also pull the mean toward its long tail. We say that the mean is not a **resistant** measure of center.

> **DEFINITION** **Resistant**
>
> A statistical measure is **resistant** if it isn't sensitive to extreme values.

The mean of a distribution also has a physical interpretation, as the following activity shows.

ACTIVITY   Mean as a "balance point"

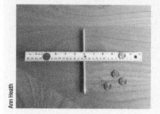

In this activity, you'll investigate an important property of the mean.

1. Stack 5 pennies on top of the 6-inch mark on a 12-inch ruler. Place a pencil under the ruler to make a "seesaw" on a desk or table. Move the pencil until the ruler balances. What is the relationship between the location of the pencil and the mean of the five data values 6, 6, 6, 6, and 6?

2. Move one penny off the stack to the 8-inch mark on your ruler. Now move one other penny so that the ruler balances again without moving the pencil. Where did you put the other penny? What is the mean of the five data values represented by the pennies now?

3. Move one more penny off the stack to the 2-inch mark on your ruler. Now move both remaining pennies from the 6-inch mark so that the ruler still balances with the pencil in the same location. Is the mean of the data values still 6?

4. Discuss with your classmates: Why is the mean called the "balance point" of a distribution?

The activity gives a physical interpretation of the mean as the balance point of a distribution. For the data on goals scored in each of 20 games played by the 2016 U.S. women's soccer team, the dotplot balances at $\bar{x} = 3.15$ goals.

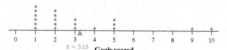

$\bar{x} = 3.15$
Goals scored

Teaching Tip

Resistant = doesn't want to change

 ACTIVITY OVERVIEW

To prepare for using this activity, watch the overview video by clicking on the link in the TE-Book, opening the TRFD, or downloading from the Teacher's Resources on the book's digital platform.

Time: 10 minutes

Materials: A foot-long ruler, pencil, and 5 pennies per student or group of students

Teaching Advice: Students can work in pairs or small groups if it is difficult to balance the pennies and keep the pencil from moving. This activity helps students recognize that the mean is the balancing point/center of mass/centroid of the distribution. Using pencils with hexagonal (not round) shafts will make this activity easier to complete

Answers:

1. The pencil is at the mean value, 6.

2. The other penny should balance the ruler when it is at 4. The mean is still 6.

3. The remaining two pennies should be at 8, for a total of three pennies. The mean is still 6.

4. Answers will vary. One possible answer is that the sum of the distances to the right of 6 will "balance out" the sum of the distances to the left of 6.

Teaching Tip

If you don't have the time to do the ruler activity, you can illustrate the basic concepts quickly on the board using dotplots. Instead of placing 5 pennies on the 6-inch mark, place a stack of 5 dots above 6 on a dotplot running from 0 to 12. Then "move" one of the dots from 6 to 8 and ask students what needs to happen for the mean to remain equal to 6. When they suggest "moving" a dot from 6 to 4, point out that the deviations from 6 (the mean) are in balance. Now "move" another dot from 6 to 2 and have students discuss how to keep the mean balanced at 6.

Measuring Center: The Median

We could also report the value in the "middle" of a distribution as its center. That's the idea of the **median**.

> **DEFINITION Median**
>
> The **median** is the midpoint of a distribution, the number such that about half the observations are smaller and about half are larger.
>
> To find the median, arrange the data values from smallest to largest.
>
> - If the number n of data values is odd, the median is the middle value in the ordered list.
> - If the number n of data values is even, use the average of the two middle values in the ordered list as the median.

The median is easy to find by hand for small sets of data. For instance, here are the data from Section 1.2 on the highway fuel economy ratings for a sample of 25 model year 2018 Toyota 4Runners tested by the EPA:

22.4 22.4 22.3 23.3 22.3 22.3 22.5 22.4 22.1 21.5 22.0 22.2 22.7
22.8 22.4 22.6 22.9 22.5 22.1 22.4 22.2 22.9 22.6 21.9 22.4

Start by sorting the data values from smallest to largest:

21.5 21.9 22.0 22.1 22.1 22.2 22.2 22.3 22.3 22.3 22.4 22.4 **22.4**
22.4 22.4 22.4 22.5 22.5 22.6 22.6 22.7 22.8 22.9 22.9 23.3

There are $n = 25$ data values (an odd number), so the median is the middle (13th) value in the ordered list, the bold **22.4**.

EXAMPLE

How many goals?
Finding the median

Icon Sports Wire/Getty Images

PROBLEM: Here are the data on the number of goals scored in 20 games played by the 2016 U.S. women's soccer team:

5 5 1 10 5 2 1 1 2 3 3 2 1 4 2 1 2 1 9 3

Find the median.

SOLUTION:

1 1 1 1 1 1 2 2 2 ② ② 3 3 3 4 5 5 5 9 10

The median is $\dfrac{2+2}{2} = 2$.

> To find the median, sort the data values from smallest to largest. Because there are $n = 20$ data values (an even number), the median is the average of the middle two values in the ordered list.

FOR PRACTICE, TRY EXERCISE 89

Teaching Tip

In *TPS* 6e, we do not use a symbol to represent the median the way we do for the mean ($\bar{x}$ or μ). Although some people use the symbol M (e.g., $M = 20$), we have found it clearer for students to simply write "median = 20."

Making Connections

Many students have learned the "crossing out method" for finding a median, where they successively cross out values from the ends of the distribution, until they get to the center.

ALTERNATE EXAMPLE   Skill 2.C

How many likes on Instagram for ASA?
Finding the median

PROBLEM:
The American Statistical Association (www.amstat.org) has an Instagram account (@amstatnews). Here are the number of Instagram likes for 10 posts selected at random:

16	4	8	7	8
6	15	2	9	5

Find the median.

SOLUTION:

2 4 5 6 (7 | 8) 8 9 15 16

The median is $\dfrac{7+8}{2} = 7.5$ likes.

Comparing the Mean and the Median

Which measure—the mean or the median—should we report as the center of a distribution? That depends on both the shape of the distribution and whether there are any outliers.

- **Shape:** Figure 1.13 shows the mean and median for dotplots with three different shapes. Notice how these two measures of center compare in each case. The mean is pulled in the direction of the long tail in a skewed distribution.

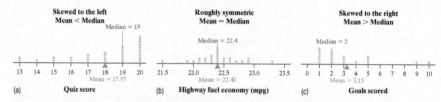

FIGURE 1.13 Dotplots that show the relationship between the mean and median in distributions with different shapes: (a) Scores of 30 statistics students on a 20-point quiz, (b) highway fuel economy ratings for a sample of 25 model year 2018 Toyota 4Runners, and (c) number of goals scored in 20 games played by the 2016 U.S. women's soccer team.

You can compare how the mean and median behave by using the *Mean and Median* applet at the book's website, highschool.bfwpub.com/updatedtps6e.

- **Outliers:** We noted earlier that the mean is sensitive to extreme values. If we remove the two possible outliers (9 and 10) in Figure 1.13(c), the mean number of goals scored per game decreases from 3.15 to 2.44. The median number of goals scored is 2 whether we include these two games or not. The median is a resistant measure of center, but the mean is not.

EFFECT OF SKEWNESS AND OUTLIERS ON MEASURES OF CENTER

- If a distribution of quantitative data is roughly symmetric and has no outliers, the mean and median will be similar.
- If the distribution is strongly skewed, the mean will be pulled in the direction of the skewness but the median won't. For a right-skewed distribution, we expect the mean to be greater than the median. For a left-skewed distribution, we expect the mean to be less than the median.
- The median is resistant to outliers but the mean isn't.

The mean and median measure center in different ways, and both are useful. In Major League Baseball (MLB), the distribution of player salaries is strongly skewed to the right. Most players earn close to the minimum salary (which was $507,500 in 2016), while a few earn more than $20 million. The median salary for MLB players in 2016 was about $1.5 million—but the mean salary was about $4.4 million. Clayton Kershaw, Miguel Cabrera, John Lester, and several other highly paid superstars pulled the mean up but that did not affect the median.

The median gives us a good idea of what a "typical" MLB salary is. If we want to know the total salary paid to MLB players in 2016, however, we would multiply the mean salary by the total number of players: ($4.4 million)(862) ≈ $3.8 billion!

CHECK YOUR UNDERSTANDING

Some students purchased pumpkins for a carving contest. Before the contest began, they weighed the pumpkins. The weights in pounds are shown here, along with a histogram of the data.

3.6   4.0   9.6   14.0   11.0   12.4   13.0   2.0   6.0   6.6   15.0   3.4

12.7   6.0   2.8   9.6   4.0   6.1   5.4   11.9   5.4   31.0   33.0

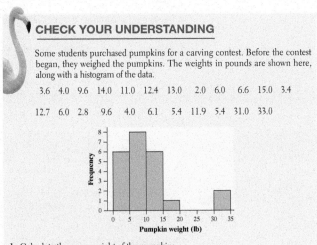

1. Calculate the mean weight of the pumpkins.
2. Find the median weight of the pumpkins.
3. Would you use the mean or the median to summarize the typical weight of a pumpkin in this contest? Explain.

Measuring Variability: The Range

Being able to describe the shape and center of a distribution is a great start. However, two distributions can have the same shape and center, but still look quite different.

Figure 1.14 shows comparative dotplots of the length (in millimeters) of separate random samples of PVC pipe from two suppliers, A and B.[37] Both distributions are roughly symmetric and single-peaked (unimodal), with centers at about 600 mm, but the variability of these two distributions is quite different. The sample of pipes from Supplier A has much more consistent lengths (less variability) than the sample from Supplier B.

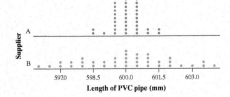

FIGURE 1.14 Comparative dotplots of the length of PVC pipes in separate random samples from Supplier A and Supplier B.

Teaching Tip

To introduce measures of variability, draw two dotplots with the same center and the same shape but very different amounts of variability. Hopefully, this will convince students that simply describing the shape and center does not give a complete picture of a distribution.

 **Answers to CYU**

1. The mean weight of the pumpkins is $\bar{x} = \dfrac{3.6 + 4.0 + 9.6 + \ldots + 5.4 + 31 + 33}{23} = 9.935$ pounds.

2. First we must put the weights in order: 2.0, 2.8, 3.4, 3.6, 4.0, 4.0, 5.4, 5.4, 6.0, 6.0, 6.1, **6.6**, 9.6, 9.6, 11.0, 11.9, 12.4, 12.7, 13.0, 14.0, 15.0, 31.0, and 33.0. Because there are 23 weights, the median is the 12th weight in this ordered list. The median weight of the pumpkins is 6.6 pounds.

3. I would use the median to summarize the typical weight of a pumpkin in this contest because the distribution of pumpkin weights is skewed to the right with two possible upper outliers: 31.0 and 33.0 pounds.

There are several ways to measure the variability of a distribution. The simplest is the **range**.

Here are the data on the number of goals scored in 20 games played by the 2016 U.S. women's soccer team, along with a dotplot:

5   5   1   10   5   2   1   1   2   3   3   2   1   4   2   1   2   1   9   3

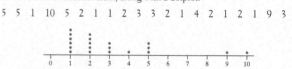

 The range of this distribution is $10 - 1 = 9$ goals. Note that the range of a data set is a single number. In everyday language, people sometimes say things like, "The data values range from 1 to 10." A correct statement is "The number of goals scored in 20 games played by the 2016 U.S. women's soccer team varies from 1 to 10, a range of 9 goals."

The range is *not* a resistant measure of variability. It depends on only the maximum and minimum values, which may be outliers. Look again at the data on goals scored by the 2016 U.S. women's soccer team. Without the possible outliers at 9 and 10 goals, the range of the distribution would decrease to $5 - 1 = 4$ goals.

The following graph illustrates another problem with the range as a measure of variability. The parallel dotplots show the lengths (in millimeters) of a sample of 11 nails produced by each of two machines.[38] Both distributions are centered at 70 mm and have a range of $72 - 68 = 4$ mm. But the lengths of the nails made by Machine B clearly vary more from the center of 70 mm than the nails made by Machine A.

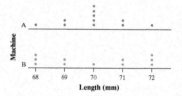

Measuring Variability: The Standard Deviation

If we summarize the center of a distribution with the mean, then we should use the **standard deviation** to describe the variation of data values around the mean. To obtain the standard deviation, we start by calculating the **variance**.

Teaching Tip

Insist that students provide a single number when discussing the range of a distribution. The range is the distance between the minimum and maximum value in the distribution.

Teaching Tip

It is not always possible to calculate the standard deviation if we do not have all the individual values in the data set (e.g., if we are given a histogram). In this case, the estimated range may be the best measure of variability.

Teaching Tip

The average of the absolute values of the deviations from the mean is called the mean absolute deviation, or MAD. This is not on the AP® Statistics exam, but you can tell students that it is a legitimate way to measure the variability of a distribution. However, it is not nearly as useful as the standard deviation for many reasons. For example, the Normal distributions are determined by their mean and standard deviation, not the mean absolute deviation.

DEFINITION Standard deviation, Variance

The **standard deviation** measures the typical distance of the values in a distribution from the mean. It is calculated by finding an average of the squared deviations and then taking the square root. This average squared deviation is called the **variance**. If the values in a data set are $x_1, x_2, \ldots, x_n$, the sample variance s_x^2 is given by the formula

$$s_x^2 = \frac{(x_1 - \overline{x})^2 + (x_2 - \overline{x})^2 + \cdots + (x_n - \overline{x})^2}{n-1} = \frac{\sum(x_i - \overline{x})^2}{n-1}$$

The sample standard deviation s_x is the square root of the variance:

$$s_x = \sqrt{\frac{\sum(x_i - \overline{x})^2}{n-1}}$$

AP® EXAM TIP

The formula sheet provided with the AP® Statistics exam gives the sample standard deviation in the equivalent form

$$s_x = \sqrt{\frac{1}{n-1}\sum(x_i - \overline{x})^2}$$

How do we calculate the standard deviation s_x of a quantitative data set with n values? Here are the steps.

HOW TO CALCULATE THE SAMPLE STANDARD DEVIATION s_x

- Find the mean of the distribution.
- Calculate the *deviation* of each value from the mean:
  deviation = value – mean
- Square each of the deviations.
- Add all the squared deviations, then divide by $n-1$ to get the sample variance.
- Take the square root to return to the original units.

The population standard deviation σ is calculated by dividing the sum of squared deviations by the population size N (not $N-1$) before taking the square root.

The notation s_x refers to the standard deviation of a *sample*. When we need to refer to the standard deviation of a population, we'll use the symbol σ (Greek lowercase sigma). We often use the sample statistic s_x to estimate the population parameter σ.

EXAMPLE

How many friends?
Calculating and interpreting standard deviation

PROBLEM: Eleven high school students were asked how many "close" friends they have. Here are their responses, along with a dotplot:

1 2 2 2 3 3 3 3 4 4 6

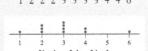

Number of close friends

LaraBelova/Getty Images

Calculate the sample standard deviation. Interpret this value.

Teaching Tip

Why square each deviation? To make all values positive. *Why square root?* Because we squared each deviation in Step 3. We are undoing that process so we can return to the original unit of measurement. *Why divide by $n-1$?* The standard deviation we are calculating is just an estimate of the true standard deviation—the "average" distance all the values in the population are from the mean of the population. Because we don't know the mean of the population (μ), we can't calculate the deviations from μ. So we use the mean of the sample ($\overline{x}$) instead. However, because $\overline{x}$ is the exact balancing point of the data, the data will almost always be closer to $\overline{x}$, on average, than they will be to μ. Thus, the sum of the squared deviations from $\overline{x}$ will underestimate the sum of the squared deviations from μ. To correct for this, we divide by a slightly smaller number when finding the average ($n-1$ instead of n). We will demonstrate this with a simulation in Chapter 7.

Preparing for Inference

In Chapter 4, we will reinforce that a value calculated from a sample (a statistic) is used to estimate a value for a population (a parameter). We use different notation for statistics than we do for parameters, so readers will know if the reported value is an estimate or the true value.

AP® EXAM TIP

The equation for standard deviation is on the formula sheet provided on both parts of the AP® Statistics exam, but it is unlikely that students will be asked to perform this calculation. It is more important that they can interpret the standard deviation as the typical distance the values in a distribution are from the mean.

Teaching Tip: AP® Connections

AP® Psychology students are taught to divide by n for the standard deviation calculation. This is not necessarily wrong, but it presupposes that the data given are the population data.

Teaching Tip

Going through a single example of calculating standard deviation by hand is worthwhile in helping students to understand the correct interpretation of standard deviation as the "typical distance from the mean" as well as to understand some of the properties of standard deviation.

SOLUTION:

$$\bar{x} = \frac{1+2+2+2+3+3+3+3+4+4+6}{11} = 3$$

x_i	$x_i - \bar{x}$	$(x_i - \bar{x})^2$
1	$1-3=-2$	$(-2)^2 = 4$
2	$2-3=-1$	$(-1)^2 = 1$
2	$2-3=-1$	$(-1)^2 = 1$
2	$2-3=-1$	$(-1)^2 = 1$
3	$3-3=0$	$0^2 = 0$
3	$3-3=0$	$0^2 = 0$
3	$3-3=0$	$0^2 = 0$
3	$3-3=0$	$0^2 = 0$
4	$4-3=1$	$1^2 = 1$
4	$4-3=1$	$1^2 = 1$
6	$6-3=3$	$3^2 = 9$
		Sum $= 18$

To calculate the sample standard deviation:
- Find the mean of the distribution.
- Calculate the *deviation* of each value from the mean:
  deviation = value − mean
- Square each of the deviations.
- Add all the squared deviations, then divide by $n-1$ to get the sample variance.
- Take the square root to return to the original units.

$$s_x^2 = \frac{18}{11-1} = 1.80$$

$$s_x = \sqrt{1.80} = 1.34 \text{ close friends}$$

$$s_x = \sqrt{\frac{\sum (x_i - \bar{x})^2}{n-1}}$$

Interpretation: The number of close friends these students have typically varies by about 1.34 close friends from the mean of 3 close friends.

FOR PRACTICE, TRY EXERCISE 99

In the preceding example, the sample variance is $s_x^2 = 1.80$. Unfortunately, the units are "squared close friends." Because variance is measured in squared units, it is not a very helpful way to describe the variability of a distribution.

Think About It

WHY IS THE STANDARD DEVIATION CALCULATED IN SUCH A COMPLEX WAY? Add the deviations from the mean in the preceding example. You should get a sum of 0. Why? Because the mean is the balance point of the distribution. We square the deviations to avoid the positive and negative deviations balancing each other out and adding to 0. It might seem strange to "average" the squared deviations by dividing by $n-1$. We'll explain the reason for doing this in Chapter 7. It's easier to understand why we take the square root: to return to the original units (close friends).

ALTERNATE EXAMPLE   Skill 2.C

How many likes on Instagram for ASA?
Calculating and interpreting standard deviation

PROBLEM:

The American Statistical Association (www.amstat.org) has an Instagram account (@amstatnews). Here are the number of Instagram likes for 10 posts selected at random:

2	4	5	6	7
8	8	9	15	16

Calculate the sample standard deviation. Interpret this value in context.

SOLUTION:

$$\bar{x} = \frac{\sum x_i}{n} = \frac{2+4+5+6+7+8+8+9+15+16}{10} = \frac{80}{10} = 8 \text{ likes}$$

x_i	$x_i - \bar{x}$	$(x_i - \bar{x})^2$
2	$2-8=-6$	$(-6)^2 = 36$
4	$4-8=-4$	$(-4)^2 = 16$
5	$5-8=-3$	$(-3)^2 = 9$
6	$6-8=-2$	$(-2)^2 = 4$
7	$7-8=-1$	$(-1)^2 = 1$
8	$8-8=0$	$0^2 = 0$
8	$8-8=0$	$0^2 = 0$
9	$9-8=1$	$1^2 = 1$
15	$15-8=7$	$7^2 = 49$
16	$16-8=8$	$8^2 = 64$

Sum $= 180$

$$s_x = \sqrt{\frac{180}{10-1}} = 4.47 \text{ likes}$$

Interpretation: The number of Instagram likes for each ASA post typically varies by about 4.47 likes from the mean of 8 likes.

Making Connections

Why do we bother with variance? You will find out in Chapter 6 when we combine random variables and the *variances* will add up (not the standard deviations!).

More important than the details of calculating s_x are the properties of the standard deviation as a measure of variability:

- s_x **is always greater than or equal to 0.** $s_x = 0$ only when there is no variability, that is, when all values in a distribution are the same.
- **Larger values of s_x indicate greater variation** from the mean of a distribution. The comparative dotplot shows the lengths of PVC pipe in random samples from two different suppliers. Supplier A's pipe lengths have a standard deviation of 0.681 mm, while Supplier B's pipe lengths have a standard deviation of 2.02 mm. The lengths of pipes from Supplier B are typically farther from the mean than the lengths of pipes from Supplier A.

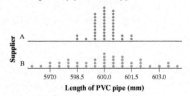

- s_x **is not a resistant measure of variability.** The use of squared deviations makes s_x even more sensitive than $\bar{x}$ to extreme values in a distribution. For example, the standard deviation of the number of goals scored in 20 games played by the 2016 U.S women's soccer team is 2.58 goals. If we omit the possible outliers of 9 and 10 goals, the standard deviation drops to 1.46 goals.

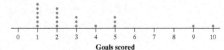

- s_x **measures variation about the mean.** It should be used only when the mean is chosen as the measure of center.

In the close friends example, 11 high school students had an average of $\bar{x} = 3$ close friends with a standard deviation of $s_x = 1.34$. What if a 12th high school student was added to the sample who had 3 close friends? The mean number of close friends in the sample would still be $\bar{x} = 3$. How would s_x be affected? Because the standard deviation measures the typical distance of the values in a distribution from the mean, s_x would *decrease* because this 12th value is at a distance of 0 from the mean. In fact, the new standard deviation would be

$$s_x = \sqrt{\frac{\sum(x_i - \bar{x})^2}{n-1}} = \sqrt{\frac{18}{12-1}} = 1.28 \text{ close friends}$$

Measuring Variability: The Interquartile Range (*IQR*)

We can avoid the impact of extreme values on our measure of variability by focusing on the middle of the distribution. Start by ordering the data values from smallest to largest. Then find the **quartiles**, the values that divide the distribution into four

Teaching Tip

Use the formula for standard deviation

$$s_x = \sqrt{\frac{1}{n-1}\sum(x_i - \bar{x})^2}$$ to help explain

each of these properties.

Teaching Tip

Point out that the definition for quartiles in this book implies that the median is never included when computing quartiles.

Teaching Tip: Using Technology

Statistical software packages may differ in how they calculate quartiles. However, the differences are usually small, especially when there is a large number of individuals in the data set.

Teaching Tip

Just as with range, be careful to identify *IQR* as a single value and not an interval of values. For example, the "*IQR* is 5 pairs of shoes" is correct, while the "*IQR* is from 6.5 to 11.5 shoes" is not correct.

groups of roughly equal size. The **first quartile** Q_1 lies one-quarter of the way up the list. The second quartile is the median, which is halfway up the list. The **third quartile** Q_3 lies three-quarters of the way up the list. The first and third quartiles mark out the middle half of the distribution.

For example, here are the amounts collected each hour by a charity at a local store: $19, $26, $25, $37, $31, $28, $22, $22, $29, $34, $39, and $31. The dotplot displays the data. Because there are 12 data values, the quartiles divide the distribution into 4 groups of 3 values.

First quartile Q_1 Second quartile (median) Third quartile Q_3

Amount collected ($)

DEFINITION   Quartiles, First quartile Q_1, Third quartile Q_3

The **quartiles** of a distribution divide the ordered data set into four groups having roughly the same number of values. To find the quartiles, arrange the data values from smallest to largest and find the median.

The **first quartile** Q_1 is the median of the data values that are to the left of the median in the ordered list.

The **third quartile** Q_3 is the median of the data values that are to the right of the median in the ordered list.

The **interquartile range** (*IQR*) measures the variability in the middle half of the distribution.

DEFINITION   Interquartile range (*IQR*)

The **interquartile range** *(IQR)* is the distance between the first and third quartiles of a distribution. In symbols:

$$IQR = Q_3 - Q_1$$

Notice that the *IQR* is simply the range of the "middle half" of the distribution.

EXAMPLE

Boys and their shoes?
Finding the *IQR*

PROBLEM: How many pairs of shoes does a typical teenage boy own? To find out, two AP® Statistics students surveyed a random sample of 20 male students from their large high school and recorded the number of pairs of shoes that each boy owned. Here are the data, along with a dotplot:

14 7 6 5 12 38 8 7 10 10 10 11 4 5 22 7 5 10 35 7

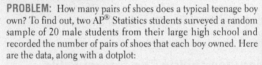

Number of pairs of shoes

Skill 2.C

ALTERNATE EXAMPLE

How many Instagram likes for ASA?   Finding the *IQR*

PROBLEM:
The American Statistical Association (www.amstat.org) has an Instagram account (@amstatnews). Here are the number of Instagram likes for 10 of their posts selected at random, along with a dotplot:

2	4	5	6	7
8	8	9	15	16

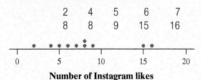

Number of Instagram likes

Find the interquartile range.

SOLUTION:

2   4   5   6   (7 | 8)   8   9   15   16
Median = 7.5

[2   4   (5)   6   7] [8   8   (9)   15   16]

$Q_1 = 5$ $Q_3 = 9$
IQR = 9 − 5 = 4 likes

Find the interquartile range.

SOLUTION:

4 5 5 5 6 7 7 7 7 ⑧ ⑩ 10 10 10 11 12 14 22 35 38

Median = 9

> Sort the data values from smallest to largest and find the median.

4 5 5 5 ⑥ ⑦ 7 7 7 8 ¦ 10 10 10 10 ⑪ ⑫ 14 22 35 38

$Q_1 = 6.5$ Median $Q_3 = 11.5$

$IQR = 11.5 - 6.5 = 5$ pairs of shoes

> Find the first quartile Q_1 and the third quartile Q_3.

> $IQR = Q_3 - Q_1$

FOR PRACTICE, TRY EXERCISE 105

The quartiles and the interquartile range are *resistant* because they are not affected by a few extreme values. For the shoe data, Q_3 would still be 11.5 and the *IQR* would still be 5 if the maximum were 58 rather than 38.

Be sure to leave out the median when you locate the quartiles. In the preceding example, the median was not one of the data values. For the earlier close friends data set, we ignore the circled median of 3 when finding Q_1 and Q_3.

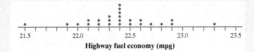

1   2   ②   2   3 ¦ ③ ¦ 3   3   ④   4   6

Q_1 Median Q_3

CHECK YOUR UNDERSTANDING

Here are data on the highway fuel economy ratings for a sample of 25 model year 2018 Toyota 4Runners tested by the EPA, along with a dotplot:

22.4   22.4   22.3   23.3   22.3   22.3   22.5   22.4   22.1   21.5   22.0   22.2   22.7
22.8   22.4   22.6   22.9   22.5   22.1   22.4   22.2   22.9   22.6   21.9   22.4

Highway fuel economy (mpg)

1. Find the range of the distribution.
2. The mean and standard deviation of the distribution are 22.404 mpg and 0.363 mpg, respectively. Interpret the standard deviation.
3. Find the interquartile range of the distribution.
4. Which measure of variability would you choose to describe the distribution? Explain.

Numerical Summaries with Technology

Graphing calculators and computer software will calculate numerical summaries for you. Using technology to perform calculations will allow you to focus on choosing the right methods and interpreting your results.

Teaching Tip: Using Technology

We recommend that you start allowing students to calculate numerical summaries with technology at this point. Students will be expected to use their calculators on the AP® Statistics exam, which allows the writers of the AP® exam to ask more conceptual questions about the application and interpretation of the numerical summaries. Make sure your students get plenty of practice with these types of questions.

Making Connections

There is no frequency list because we want each entry in L1 to be used once. In Chapter 6, students will encounter distributions where each value is weighted according to a probability. In this scenario, we will use another list in the calculator to show the probabilities (or relative frequencies).

Teaching Tip

"Ruler" as in a measuring tool. Not a king.

---

3. Technology Corner COMPUTING NUMERICAL SUMMARIES

TI-Nspire and other technology instructions are on the book's website at highschool.bfwpub.com/updatedtps6e.

Let's find numerical summaries for the boys' shoes data from the example on page 64. We'll start by showing you how to compute summary statistics on the TI-83/84 and then look at output from computer software.

I. **One-variable statistics on the TI-83/84**

1. Enter the data in list L1.

2. Find the summary statistics for the shoe data.

- Press STAT (CALC); choose 1-Var Stats.
  OS 2.55 or later: In the dialog box, press 2nd 1 (L1) and ENTER to specify L1 as the List. Leave FreqList blank. Arrow down to Calculate and press ENTER.
  Older OS: Press 2nd 1 (L1) and ENTER.

- Press ▼ to see the rest of the one-variable statistics.

II. **Output from statistical software** We used Minitab statistical software to calculate descriptive statistics for the boys' shoes data. Minitab allows you to choose which numerical summaries are included in the output.

```
Descriptive Statistics: Shoes

Variable   N   Mean  StDev  Minimum   Q₁  Median   Q₃  Maximum
Shoes     20  11.65   9.42     4.00  6.25    9.00  11.75   38.00
```

Note: The TI-83/84 gives the first and third quartiles of the boys' shoes distribution as $Q_1 = 6.5$ and $Q_3 = 11.5$. Minitab reports that $Q_1 = 6.25$ and $Q_3 = 11.75$. What happened? Minitab and some other software use slightly different rules for locating quartiles. Results from the various rules are usually close to each other. Be aware of possible differences when calculating quartiles as they may affect more than just the *IQR*.

Identifying Outliers

Besides serving as a measure of variability, the interquartile range (*IQR*) is used as a "ruler" for identifying outliers.

There are other rules for determining outliers, such as "any value that is more than 2 (or 3) standard deviations from the mean." We always use the 1.5 × *IQR* rule in this book because it is based on statistics that are resistant to outliers, unlike the mean and standard deviation.

HOW TO IDENTIFY OUTLIERS: THE 1.5 × *IQR* RULE

Call an observation an outlier if it falls more than $1.5 \times IQR$ above the third quartile or below the first quartile. That is,

$$\text{Low outliers} < Q_1 - 1.5 \times IQR \qquad \text{High outliers} > Q_3 + 1.5 \times IQR$$

Here are sorted data on the highway fuel economy ratings for a sample of 25 model year 2018 Toyota 4Runners tested by the EPA, along with a dotplot:

21.5 21.9 22.0 22.1 22.1 22.2 22.2 22.3 22.3 22.3 22.4 22.4 22.4
22.4 22.4 22.4 22.5 22.5 22.6 22.6 22.7 22.8 22.9 22.9 23.3

Highway fuel economy (mpg)

Does the $1.5 \times IQR$ rule identify any outliers in this distribution? If you did the preceding Check Your Understanding, you should have found that $Q_1 = 22.2$ mpg, $Q_3 = 22.6$ mpg, and $IQR = 0.4$ mpg. For these data,

$$\text{High outliers} > Q_3 + 1.5 \times IQR = 22.6 + 1.5 \times 0.4 = 23.2$$

and

$$\text{Low outliers} < Q_1 - 1.5 \times IQR = 22.2 - 1.5 \times 0.4 = 21.6$$

The cars with estimated highway fuel economy ratings of 21.5 and 23.3 are identified as outliers.

> **AP® EXAM TIP**
>
> You may be asked to determine whether a quantitative data set has any outliers. Be prepared to state and use the rule for identifying outliers.

EXAMPLE

How many goals?
Identifying outliers

PROBLEM: Here are sorted data on the number of goals scored in 20 games played by the 2016 U.S women's soccer team, along with a dotplot:

1 1 1 1 1 1 2 2 2 2 2 3 3 3 4 5 5 5 9 10

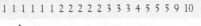

Goals scored

Identify any outliers in the distribution. Show your work.

SOLUTION:

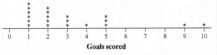

$IQR = Q_3 - Q_1 = 4.5 - 1 = 3.5$

Low outliers $< 1 - 1.5 \times 3.5 = -4.25$

High outliers $> 4.5 + 1.5 \times 3.5 = 9.75$

There are no data values less than -4.25, but the game in which the team scored 10 goals is an outlier.

> Low outliers $< Q_1 - 1.5 \times IQR$
> High outliers $> Q_3 + 1.5 \times IQR$

> The game in which the team scored 9 goals is not identified as an outlier by the $1.5 \times IQR$ rule.

FOR PRACTICE, TRY EXERCISE 107

ALTERNATE EXAMPLE   Skill 4.B

How many likes on Instagram for ASA?
Identifying outliers

PROBLEM:

The American Statistical Association (www.amstat.org) has an Instagram account (@amstatnews). Here are the number of Instagram likes for 10 of their posts selected at random, along with a dotplot:

2	4	5	6	7
8	8	9	15	16

Number of Instagram likes

Identify any outliers in the distribution. Show your work.

SOLUTION:

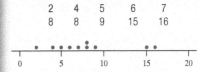

$Q_1 = 5$ Median $= 7.5$ $Q_3 = 9$

$IQR = 9 - 5 = 4$

Low outliers $< Q_1 - 1.5 \times IQR$
$= 5 - 1.5 \times 4 = -1$

High outliers $> Q_3 + 1.5 \times IQR$
$= 9 + 1.5 \times 4 = 15$

There are no data values less than -1, but the post that has 16 likes is an outlier ($16 > 15$). Note that 15 is *not* an outlier because $15 = Q_3 + 1.5 \times IQR$, and a high outlier must be *greater than* $Q_3 + 1.5 \times IQR$.

It is important to identify outliers in a distribution for several reasons:

1. **They might be inaccurate data values.** Maybe someone recorded a value as 10.1 instead of 101. Perhaps a measuring device broke down. Or maybe someone gave a silly response, like the student in a class survey who claimed to study 30,000 minutes per night! Try to correct errors like these if possible. If you can't, give summary statistics with and without the outlier.
2. **They can indicate a remarkable occurrence.** For example, in a graph of net worth, Bill Gates is likely to be an outlier.
3. **They can heavily influence the values of some summary statistics,** like the mean, range, and standard deviation.

Making and Interpreting Boxplots

You can use a dotplot, stemplot, or histogram to display the distribution of a quantitative variable. Another graphical option for quantitative data is a **boxplot**. A boxplot summarizes a distribution by displaying the location of 5 important values within the distribution, known as its **five-number summary**.

> A boxplot is sometimes called a *box-and-whisker* plot.

> **DEFINITION** Five-number summary, Boxplot
>
> The **five-number summary** of a distribution of quantitative data consists of the minimum, the first quartile Q_1, the median, the third quartile Q_3, and the maximum.
>
> A **boxplot** is a visual representation of the five-number summary.

Figure 1.15 illustrates the process of making a boxplot. The dotplot in Figure 1.15(a) shows the data on EPA estimated highway fuel economy ratings for a sample of 25 model year 2018 Toyota 4Runners. We have marked the first quartile, the median, and the third quartile with vertical green lines. The process of testing for outliers with the $1.5 \times IQR$ rule is shown in red. Because the values of 21.5 mpg and 23.3 mpg are outliers, we mark these separately. To get the finished boxplot in Figure 1.15(b), we make a box spanning from Q_1 to Q_3 and then draw "whiskers" to the smallest and largest data values that are not outliers

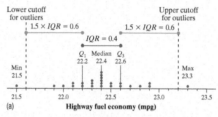
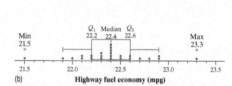

FIGURE 1.15 A visual illustration of how to make a boxplot for the Toyota 4Runner highway gas mileage data. (a) Dotplot of the data with the five-number summary and $1.5 \times IQR$ marked. (b) Boxplot of the data with outliers identified (*).

As you can see, it is fairly easy to make a boxplot by hand for small sets of data. Here's a summary of the steps.

 COMMON STUDENT ERROR

Many students refer to the box in a boxplot as the *IQR*. For example, they will say "the median is in the middle of the *IQR*." This is incorrect because the *IQR* is a number (the length of the box), not the box itself.

 COMMON STUDENT ERROR

When describing the shape of a boxplot, many students will describe a symmetric boxplot as "approximately Normal." This is incorrect because a boxplot does not reveal the location of the modes or any gaps for a distribution. It is sometimes difficult to assess the shape of a distribution using only a boxplot, because boxplots don't show peaks, gaps, or clusters. See 2019 free response #1 for a good example of this idea.

HOW TO MAKE A BOXPLOT

- **Find the five-number summary** for the distribution.
- **Identify outliers** using the $1.5 \times IQR$ rule.
- **Draw and label the axis.** Draw a horizontal axis and put the name of the quantitative variable underneath, including units if applicable.
- **Scale the axis.** Look at the minimum and maximum values in the data set. Start the horizontal axis at a convenient number equal to or below the minimum and place tick marks at equal intervals until you equal or exceed the maximum.
- **Draw a box** that spans from the first quartile (Q_1) to the third quartile (Q_3).
- **Mark the median** with a vertical line segment that's the same height as the box.
- **Draw whiskers**—lines that extend from the ends of the box to the smallest and largest data values that are *not* outliers. Mark any outliers with a special symbol such as an asterisk (*).

We see from the boxplot in Figure 1.15 that the distribution of highway gas mileage ratings for this sample of model year 2018 Toyota 4Runners is roughly symmetric with one high outlier and one low outlier.

AP® EXAM TIP

Some textbooks and the TI graphing calculators make the distinction between a boxplot that doesn't show outliers and a boxplot that shows outliers with a special symbol, such as an asterisk. Students can assume that any boxplot they encounter on the AP® Statistics exam displays outliers (if there are outliers).

EXAMPLE

Picking pumpkins
Making and interpreting boxplots

PROBLEM: Some students purchased pumpkins for a carving contest. Before the contest began, they weighed the pumpkins. The weights in pounds are shown here.

| 3.6 | 4.0 | 9.6 | 14.0 | 11.0 | 12.4 | 13.0 | 2.0 | 6.0 | 6.6 | 15.0 | 3.4 |
| 12.7 | 6.0 | 2.8 | 9.6 | 4.0 | 6.1 | 5.4 | 11.9 | 5.4 | 31.0 | 33.0 | |

(a) Make a boxplot of the data.
(b) Explain why the median and IQR would be a better choice for summarizing the center and variability of the distribution of pumpkin weights than the mean and standard deviation.

SOLUTION:

(a)

Min .... Q_1 .... Median

2.0 2.8 3.4 3.6 4.0 (4.0) 5.4 5.4 6.0 6.0 6.1 (6.6)
9.6 9.6 11.0 11.9 12.4 (12.7) 13.0 14.0 15.0 31.0 33.0
.... Q_3 .... Max

$IQR = Q_3 - Q_1 = 12.7 - 4.0 = 8.7$

Low outliers $< Q_1 - 1.5 \times IQR = 4.0 - 1.5 \times 8.7 = -9.05$

High outliers $> Q_3 + 1.5 \times IQR = 12.7 + 1.5 \times 8.7 = 25.75$

The pumpkins that weighed 31.0 and 33.0 pounds are outliers.

> To make the boxplot:
> - Find the five-number summary.
> - Identify outliers.
> - Draw and label the axis.
> - Scale the axis.
> - Draw a box.
> - Mark the median.
> - Draw whiskers to the smallest and largest data values that are *not* outliers. Mark outliers with an asterisk.

ALTERNATE EXAMPLE

Skill 2.B

How many likes on Instagram for BFW?   Making and interpreting boxplots

PROBLEM:
Bedford, Freeman & Worth (highschool.bfwpub.com/catalog), the publisher of this textbook, has an Instagram account (@bfwhighschool). Here are the number of Instagram likes for 10 posts selected at random:

| 19 | 21 | 18 | 14 | 15 |
| 18 | 8 | 16 | 11 | 38 |

(a) Make a boxplot of the data.
(b) Explain why the median and IQR would be a better choice for summarizing the center and variability of the distribution of number of Instagram likes than the mean and standard deviation.

SOLUTION:

8 11 (14) 15 16 | 18 18 (19) 21 38

$IQR = Q_3 - Q_1 = 19 - 14 = 5$
Outliers $< Q_1 - 1.5 \times IQR$
$= 14 - 1.5 \times 5 = 6.5$
Outliers $> Q_3 + 1.5 \times IQR$
$= 19 + 1.5 \times 5 = 26.5$
The post that has 38 likes is an outlier $(38 > 26.5)$.

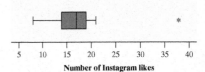

Number of Instagram likes

(b) The distribution of number of likes on Instagram has one high outlier. Because the mean and standard deviation are sensitive to outliers, it would be better to use the resistant median and IQR.

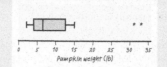

(b) The distribution of pumpkin weights is skewed to the right with two high outliers. Because the mean and standard deviation are sensitive to outliers, it would be better to use the median and IQR, which are resistant.

We know the distribution is skewed to the right because the left half of the distribution varies from 2.0 to 6.6 pounds, while the right half of the distribution (excluding outliers) varies from 6.6 to 15.0 pounds.

FOR PRACTICE, TRY EXERCISE 111

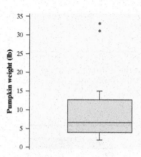

Boxplots provide a quick summary of the center and variability of a distribution. The median is displayed as a line in the central box, the interquartile range is the length of the box, and the range is the length of the entire plot, including outliers. Note that some statistical software orients boxplots vertically. At left is a vertical boxplot of the pumpkin weight data from the preceding example. You can see that the graph is skewed toward the larger values.

Boxplots do not display each individual value in a distribution. And boxplots don't show gaps, clusters, or peaks. For instance, the dotplot below left displays the duration, in minutes, of 220 eruptions of the Old Faithful geyser. The distribution of eruption durations is clearly double-peaked (*bimodal*). But a boxplot of the data hides this important information about the shape of the distribution.

CHECK YOUR UNDERSTANDING

Ryan and Brent were curious about the amount of french fries they would get in a large order from their favorite fast-food restaurant, Burger King. They went to several different Burger King locations over a series of days and ordered a total of 14 large fries. The weight of each order (in grams) is as follows:

165 163 160 159 166 152 166 168 173 171 168 167 170 170

1. Make a boxplot to display the data.
2. According to a nutrition website, Burger King's large fries weigh 160 grams, on average. Ryan and Brent suspect that their local Burger King restaurants may be skimping on fries. Does the boxplot in Question 1 support their suspicion? Explain why or why not.

Comparing Distributions with Boxplots

Boxplots are especially effective for comparing the distribution of a quantitative variable in two or more groups, as seen in the following example.

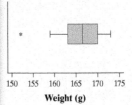

EXAMPLE

Which company makes better tablets?
Comparing distributions with boxplots

PROBLEM: In a recent year, *Consumer Reports* rated many tablet computers for performance and quality. Based on several variables, the magazine gave each tablet an overall rating, where higher scores indicate better ratings. The overall ratings of the tablets produced by Apple and Samsung are given here, along with parallel boxplots and numerical summaries of the data.[39]

| Apple | 87 | 87 | 87 | 87 | 86 | 86 | 86 | 86 | 84 | 84 |
| | 84 | 84 | 83 | 83 | 83 | 83 | 81 | 79 | 76 | 73 |

| Samsung | 88 | 87 | 87 | 86 | 86 | 86 | 86 | 86 | 84 | 84 | 83 | 83 |
| | 77 | 76 | 76 | 75 | 75 | 75 | 75 | 75 | 74 | 71 | 62 | |

	$\bar{x}$	s_x	Min	Q_1	Median	Q_3	Max	IQR
Apple	83.45	3.762	73	83	84	86	87	3
Samsung	79.87	6.74	62	75	83	86	88	11

Compare the distributions of overall rating for Apple and Samsung.

Answers to CYU

1. $M = \dfrac{166 + 167}{2} = 166.5$ grams;

$Q_1 = 163$ grams; $Q_3 = 170$ grams.
The $IQR = 170 - 163 = 7$ grams.
An outlier is any value below
$Q_1 - 1.5 \times IQR = 163 - 1.5 \times 7 = 152.5$ grams or above
$Q_3 + 1.5 \times IQR = 170 + 1.5 \times 7 = 180.5$ grams. This means that the value 152 grams is an outlier. The boxplot is displayed below.

2. No, the graph does not support their suspicion. The first quartile of the distribution of weight is 163 grams, which means that at least 75% of the large fries that they purchased weighed more than the advertised weight of 160 grams.

Skill 2.D

ALTERNATE EXAMPLE

Who has more likes on Instagram?   Comparing distributions with boxplots

PROBLEM:
The American Statistical Association (ASA) has an Instagram account (@amstatnews). Bedford, Freeman & Worth (BFW), the publisher of this textbook, also has an Instagram account (@bfwhighschool). Here are the number of Instagram likes for 10 randomly selected posts from each account, along with parallel boxplots and numerical summaries of the data:

| ASA | 16 | 4 | 8 | 7 | 8 | 6 | 15 | 2 | 9 | 5 |
| BFW | 19 | 21 | 18 | 14 | 15 | 18 | 8 | 16 | 11 | 38 |

Group name	n	mean	SD	min	Q_1	med	Q_3	max
1: ASA	10	8	4.472	2	5	7.5	9	16
2: BFW	10	17.8	8.08	8	14	17	19	38

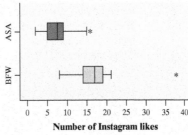

Compare the distributions of number of Instagram likes for ASA and BFW.

(continues)

ALTERNATE EXAMPLE   (continued)

SOLUTION:
Shape: Both distributions of number of Instagram likes are slightly skewed to the right.
Outliers: There is one high outlier in the ASA distribution: the post with 16 likes. There is also one high outlier in the BFW distribution: the post with 38 likes.
Center: The BFW Instagram account had a higher median number of likes (17) than the ASA Instagram account (7.5). More importantly, 100% of the BFW posts had a number of likes greater than the median for the ASA account.
Variability: There is more variation in number of likes among the BFW posts than the ASA posts. The *IQR* for the BFW posts (5) is larger than the *IQR* for the ASA posts (4).

SOLUTION:
Shape: Both distributions of overall ratings are skewed to the left.
Outliers: There are two low outliers in the Apple tablet distribution: overall ratings of 73 and 76. The Samsung tablet distribution has no outliers.
Center: The Apple tablets had a slightly higher median overall rating (84) than the Samsung tablets (83). More importantly, about 75% of the Apple tablets had overall ratings that were greater than or equal to the median for the Samsung tablets.
Variability: There is much more variation in overall rating among the Samsung tablets than the Apple tablets. The *IQR* for Samsung tablets (11) is almost four times larger than the *IQR* for Apple tablets (3).

> Remember to compare shape, outliers, center, and variability!

> Because of the strong skewness and outliers, use the median and *IQR* instead of the mean and standard deviation when comparing center and variability.

FOR PRACTICE, TRY EXERCISE 115

AP® EXAM TIP

Use statistical terms carefully and correctly on the AP® Statistics exam. Don't say "mean" if you really mean "median." Range is a single number; so are Q_1, Q_3, and *IQR*. Avoid poor use of language, like "the outlier *skews* the mean" or "the median is in the middle of the *IQR*." Skewed is a shape and the *IQR* is a single number, not a region. If you misuse a term, expect to lose some credit.

Here's an activity that gives you a chance to put into practice what you have learned in this section.

TRM Do You Know Your Geography?

Consider doing the activity, "Do you know your geography?," found in the Teacher's Resource Materials. The activity has each student answer a couple of questions about geography. However, there are two versions of the "geography quiz." Comparing the results with parallel boxplots demonstrates how the results of a survey can be manipulated by changing the way the questions are asked. Thanks to Sanderson Smith for sharing this great activity.

TRM The Memory Game

Consider doing "The memory game" activity found in the Teacher's Resource Materials. The activity has each student listen to some statements and then take a quiz about how much he or she remembers. However, one version of the quiz provides additional instructions. Comparing the results with parallel boxplots demonstrates how the results of a study can differ when given additional directions. Thanks to Adam Shrager for sharing this great activity.

ACTIVITY   Team challenge: Did Mr. Starnes stack his class?

In this activity, you will work in a team of three or four students to resolve a dispute.

Mr. Starnes teaches AP® Statistics, but he also does the class scheduling for the high school. There are two AP® Statistics classes—one taught by Mr. Starnes and one taught by Ms. McGrail. The two teachers give the same first test to their classes and grade the test together. Mr. Starnes's students earned an average score that was 8 points higher than the average for Ms. McGrail's class. Ms. McGrail wonders whether Mr. Starnes might have "adjusted" the class rosters from the computer scheduling program. In other words, she thinks he might have "stacked" his class. He denies this, of course.

To help resolve the dispute, the teachers collect data on the cumulative grade point averages and SAT Math scores of their students. Mr. Starnes provides the GPA data from his computer. The students report their SAT Math scores. The following table shows the data for each student in the two classes.

Did Mr. Starnes stack his class? Give appropriate graphical and numerical evidence to support your conclusion. Be prepared to defend your answer.

▶ ACTIVITY OVERVIEW

To prepare for using this activity, watch the overview video by clicking on the link in the TE-Book, opening the TRFD, or downloading from the Teacher's Resources on the book's digital platform.

Time: 30 minutes

Materials: A calculator or computer for each group of students

Teaching Advice: Students should use calculators, computer software, or an Extra Applet (highschool.bfwpub.com/updatedtps6e)

to analyze the data. At the end of the activity, pose the following question: If students were randomly assigned to the two teachers, how likely is it that there would be a difference in average test scores of 8 points? If the difference is so large that it is unlikely to happen purely by chance, we have convincing evidence that Mr. Starnes stacked the classes. In Chapter 11, this probability will be the *P*-value for a two-sample *t* test for the difference of means.

Student	Teacher	GPA	SAT-M
1	Starnes	2.900	670
2	Starnes	2.860	520
3	Starnes	2.600	570
4	Starnes	3.600	710
5	Starnes	3.200	600
6	Starnes	2.700	590
7	Starnes	3.100	640
8	Starnes	3.085	570
9	Starnes	3.750	710
10	Starnes	3.400	630
11	Starnes	3.338	630
12	Starnes	3.560	670
13	Starnes	3.800	650
14	Starnes	3.200	660
15	Starnes	3.100	510

Student	Teacher	GPA	SAT-M
16	McGrail	2.900	620
17	McGrail	3.300	590
18	McGrail	3.980	650
19	McGrail	2.900	600
20	McGrail	3.200	620
21	McGrail	3.500	680
22	McGrail	2.800	500
23	McGrail	2.900	502.5
24	McGrail	3.950	640
25	McGrail	3.100	630
26	McGrail	2.850	580
27	McGrail	2.900	590
28	McGrail	3.245	600
29	McGrail	3.000	600
30	McGrail	3.000	620
31	McGrail	2.800	580
32	McGrail	2.900	600
33	McGrail	3.200	600

You can use technology to make boxplots, as the following Technology Corner illustrates.

4. Technology Corner   MAKING BOXPLOTS

TI-Nspire and other technology instructions are on the book's website at highschool.bfwpub.com/updatedtps6e.

The TI-83/84 can plot up to three boxplots in the same viewing window. Let's use the calculator to make parallel boxplots of the overall rating data for Apple and Samsung tablets.

1. Enter the ratings for Apple tablets in list L1 and for Samsung in list L2.

2. Set up two statistics plots: Plot1 to show a boxplot of the Apple data in list L1 and Plot2 to show a boxplot of the Samsung data in list L2. The setup for Plot1 is shown. When you define Plot2, be sure to change L1 to L2.

Note: The calculator offers two types of boxplots: one that shows outliers and one that doesn't. We'll always use the type that identifies outliers.

3. Use the calculator's Zoom feature to display the parallel boxplots. Then Trace to view the five-number summary.

- Press ZOOM and select ZoomStat.
- Press TRACE .

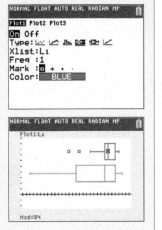

OVERVIEW (continued)

Answers:

I. *The GPA Data*

Here are boxplots and summary statistics comparing the GPA data for McGrail and Starnes:

Variable	N	Mean	Median	StDev
Starnes GPA	15	3.2129	3.2000	0.3642
McGrail GPA	18	3.1347	3.0000	0.3580

Variable	Min	Max	Q_1	Q_3
Starnes GPA	2.6000	3.8000	2.9000	3.5600
McGrail GPA	2.8000	3.9800	2.9000	3.2450

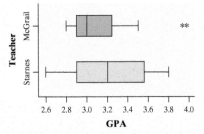

GPA

The GPA distribution in Mr. Starnes's class is roughly symmetric, whereas the distribution for Ms. McGrail's class is right-skewed with two high outliers. The median GPA is higher in Mr. Starnes's class. Even though the range of GPAs is about the same for students in both

classes, the *IQR* is much larger for students in Mr. Starnes's class, mostly because the value of Q_3 is much higher for Mr. Starnes's class. Because the median and third quartile of the distribution of GPAs in Mr. Starnes's class are much higher than in Ms. McGrail's class, there is evidence that he stacked his class.

II. *The SAT Math Data*

Here are boxplots and summary statistics comparing the SAT Math data for Starnes and McGrail:

Variable	N	Mean	Median	StDev
Starnes SAT-M	15	622.0	630.0	61.3
McGrail SAT-M	18	600.1	600.0	44.1

Variable	Min	Max	Q_1	Q_3
Starnes SAT-M	510.0	710.0	570.0	670.0
McGrail SAT-M	500.0	680.0	590.0	620.0

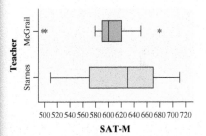

SAT-M

McGrail's distribution is somewhat symmetric, though there are two low outliers and one high outlier. Starnes's distribution of SAT Math scores is slightly skewed to the left. The median SAT Math score is higher for Starnes's class. The SAT Math scores vary more in Starnes's class, mostly due to the larger values for Q_3 and the maximum in Starnes's distribution. Again, because the median, Q_3, and maximum are all higher for students in Mr. Starnes's class, it appears that Mr. Starnes stacked his class.

Teaching Tip: Using Technology

When making calculator boxplots, always have your students use the boxplot option that marks the outliers. The TI-Nspire automatically graphs boxplots this way.

Section 1.3 | Summary

- A numerical summary of a distribution should include measures of **center** and **variability.**
- The **mean** and the **median** describe the center of a distribution in different ways. The mean is the average of the observations. In symbols, the sample mean $\bar{x} = \frac{\sum x_i}{n}$. The median is the midpoint of the distribution, the number such that about half the observations are smaller and half are larger.
- A **statistic** is a number that describes a sample. A **parameter** is a number that describes a population. We often use statistics (like the sample mean $\bar{x}$) to estimate parameters (like the population mean μ).
- The simplest measure of variability for a distribution of quantitative data is the **range,** which is the distance from the maximum value to the minimum value.
- When you use the mean to describe the center of a distribution, use the **standard deviation** to describe the distribution's variability. The standard deviation gives the typical distance of the values in a distribution from the mean. In symbols, the sample standard deviation $s_x = \sqrt{\frac{\sum(x_i - \bar{x})^2}{n-1}}$. The standard deviation s_x is 0 when there is no variability and gets larger as variability from the mean increases.
- The **variance** is the average of the squared deviations from the mean. The sample variance s_x^2 is the square of the sample standard deviation.
- When you use the median to describe the center of a distribution, use the **interquartile range** (*IQR*) to describe the distribution's variability. The **first quartile** Q_1 has about one-fourth of the observations below it, and the **third quartile** Q_3 has about three-fourths of the observations below it. The interquartile range measures variability in the middle half of the distribution and is found using $IQR = Q_3 - Q_1$.
- The median is a **resistant** measure of center because it is relatively unaffected by extreme observations. The mean is not resistant. Among measures of variability, the *IQR* is resistant, but the standard deviation and range are not.
- According to the **$1.5 \times IQR$ rule,** an observation is an outlier if it is less than $Q_1 - 1.5 \times IQR$ or greater than $Q_3 + 1.5 \times IQR$.
- The **five-number summary** of a distribution consists of the minimum, Q_1, the median, Q_3, and the maximum. A **boxplot** displays the five-number summary, marking outliers with a special symbol. The box shows the variability in the middle half of the distribution. Whiskers extend from the box to the smallest and the largest observations that are not outliers. Boxplots are especially useful for comparing distributions.

1.3 Technology Corners

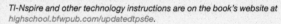

TI-Nspire and other technology instructions are on the book's website at highschool.bfwpub.com/updatedtps6e.

3. Computing numerical summaries	Page 66
4. Making boxplots	Page 73

Section 1.3 | Exercises

87. Quiz grades Joey's first 14 quiz grades in a marking period were as follows:

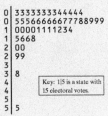

| 86 | 84 | 91 | 75 | 78 | 80 | 74 |
| 87 | 76 | 96 | 82 | 90 | 98 | 93 |

(a) Calculate the mean. Show your work.

(b) Suppose Joey has an unexcused absence for the 15th quiz, and he receives a score of 0. Recalculate the mean. What property of the mean does this illustrate?

88. Pulse rates Here are data on the resting pulse rates (in beats per minute) of 19 middle school students:

| 71 | 104 | 76 | 88 | 78 | 71 | 68 | 86 | 70 | 90 |
| 74 | 76 | 69 | 68 | 88 | 96 | 68 | 82 | 120 | |

(a) Calculate the mean. Show your work.

(b) The student with a 120 pulse rate has a medical issue. Find the mean pulse rate for the other 18 students. What property of the mean does this illustrate?

89. Quiz grades Refer to Exercise 87.

(a) Find the median of Joey's first 14 quiz grades.

(b) Find the median of Joey's quiz grades after his unexcused absence. Explain why the 0 quiz grade does not have much effect on the median.

90. Pulse rates Refer to Exercise 88.

(a) Find the median pulse rate for all 19 students.

(b) Find the median pulse rate excluding the student with the medical issue. Explain why this student's 120 pulse rate does not have much effect on the median.

91. Electing the president To become president of the United States, a candidate does not have to receive a majority of the popular vote. The candidate does have to win a majority of the 538 Electoral College votes. Here is a stemplot of the number of electoral votes in 2016 for each of the 50 states and the District of Columbia:

```
0 | 3333333344444
0 | 555566666677788999
1 | 00001111234
1 | 5668
2 | 00
2 | 99
3 |
3 | 8
4 |               Key: 1|5 is a state with
4 |               15 electoral votes.
5 |
5 | 5
```

(a) Find the median.

(b) Without doing any calculations, explain how the mean and median compare.

(c) Is the value you found in part (a) a statistic or a parameter? Justify your answer.

92. Birthrates in Africa One of the important factors in determining population growth rates is the birthrate per 1000 individuals in a population. The dotplot shows the birthrates per 1000 individuals (rounded to the nearest whole number) for all 54 African nations.

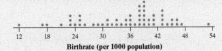

Birthrate (per 1000 population)

(a) Find the median.

(b) Without doing any calculations, explain how the mean and median compare.

(c) Is the value you found in part (a) a statistic or a parameter? Justify your answer.

93. House prices The mean and median selling prices of existing single-family homes sold in September 2016 were $276,200 and $234,200.[40] Which of these numbers is the mean and which is the median? Explain your reasoning.

94. Mean salary? Last year a small accounting firm paid each of its five clerks $32,000, two junior accountants $60,000 each, and the firm's owner $280,000.

(a) What is the mean salary paid at this firm? How many of the employees earn less than the mean? What is the median salary?

(b) Write a sentence to describe how an unethical recruiter could use statistics to mislead prospective employees.

95. Do adolescent girls eat fruit? We all know that fruit is good for us. Here is a histogram of the number of servings of fruit per day claimed by 74 seventeen-year-old girls in a study in Pennsylvania:[41]

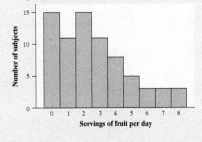

Servings of fruit per day

Answers to Section 1.3 Exercises

1.87 (a) The mean of Joey's first 14 quiz scores is 85. **(b)** Including a 15th quiz score of 0, Joey's mean would be 79.3. This illustrates the property of nonresistance. The mean is not resistant. It is sensitive to extreme values.

1.88 (a) The mean of the resting pulse rates for the 19 middle school students is 81.21 beats per minute. **(b)** Deleting the student from the list with a resting pulse rate of 120 beats per minute, the new mean is 79.06 beats per minute. This illustrates the property of nonresistance. The mean is not resistant. It is sensitive to extreme values.

1.89 (a) 85 **(b)** Since there are now 15 quiz scores, the median is 84. Notice that the median did not change much. This shows that the median is resistant to outliers.

1.90 (a) The median pulse rate for all 19 students is 76 beats per minute. **(b)** 76 beats per minute; notice that the median did not change. This shows that the median is resistant to outliers.

1.91 (a) 8 electoral votes **(b)** Since the distribution of number of electoral votes is skewed to the right, the mean of this distribution is greater than its median. **(c)** The median, calculated in part (a), is a parameter because it is a number that describes the population of all 50 states and the District of Columbia.

1.92 (a) The median is 37.5 births per 1000 individuals in the population. **(b)** Since the distribution of number of births per 1000 individuals in the population is slightly skewed to the left, the mean of this distribution is less than its median. **(c)** The median, calculated in part (a), is a parameter because it is a number that describes the population of all 54 African nations.

1.93 The mean house price is $276,200 and the median is $234,200. The distribution of house prices is likely to be quite skewed to the right because of a few very expensive homes, some of which may be outliers. When a distribution is skewed to the right, the mean is bigger than the median.

1.94 (a) The mean salary is $70,000. Seven of the eight employees (everyone but the owner) earned less than the mean. The median salary is $32,000. **(b)** An unethical recruiter would report the mean salary as the "typical" or "average" salary. However, the median is a more accurate depiction of a "typical" employee's earnings, because it is not influenced by the outlier ($280,000).

1.95 (a) The median is 2 servings of fruit per day. **(b)** $\bar{x} = 194/74 = 2.62$ servings of fruit per day

1.96 (a) The 50th percentile corresponds to a word length of 4. **(b)** The mean is $413/100 = 4.13$. Alternatively, we can estimate the mean by finding the balance point of the distribution, which appears to be slightly larger than 4.

1.97 (a) Range = Max − Min = 98 − 74 = 24; the range of Joey's quiz grades after his unexcused absence is 98 − 0 = 98. **(b)** The range may not be the best way to describe variability for a distribution of quantitative data because the range can be heavily affected by outliers.

1.98 (a) Range = 52 beats per minute; the range of the resting pulse rates excluding the student with the medical issue is 104 − 68 = 36 beats per minute. **(b)** The range may not be the best way to describe variability for a distribution of quantitative data because the range can be heavily affected by outliers.

1.99 The standard deviation is 2.52 cm. The foot lengths of these 14-year-olds from the U.K. typically vary by about 2.52 cm from the mean of 24 cm.

1.100 The standard deviation is 1.673 hours. The amount of sleep that these 6 students got last night typically varies by about 1.673 hours from the mean of 8 hours.

1.101 (a) The size of these 18 files typically varies by about 1.9 megabytes from the mean of 3.2 megabytes. **(b)** If the music file that takes up 7.5 megabytes of storage space is replaced with another version of the file that takes up only 4 megabytes, the mean would decrease slightly. The standard deviation would decrease as well because a file of size 4 megabytes will be closer to the new mean than the file of 7.5 megabytes was to the former mean.

1.102 (a) The fat content of these 12 hamburgers typically varies by about 9.06 grams from the mean of 22.83 grams. **(b)** If the restaurant replaces the burger that has 22 grams of fat with a burger that has 35 grams of fat, the mean would increase slightly. The standard deviation would increase as well because a hamburger with 35 grams of fat will be farther from the new mean than the hamburger of 22 grams was from the former mean.

1.103 Variable B has a smaller standard deviation because more of the observations have values closer to the mean than in Variable A's distribution.

(a) Find the median number of servings of fruit per day from the histogram. Explain your method clearly.

(b) Calculate the mean of the distribution. Show your work.

96. **Shakespeare** The histogram shows the distribution of lengths of words used in Shakespeare's plays.[42]

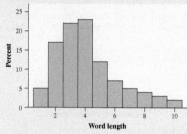

(a) Find the median word length in Shakespeare's plays from the histogram. Explain your method clearly.

(b) Calculate the mean of the distribution. Show your work.

97. **Quiz grades** Refer to Exercise 87.

(a) Find the range of Joey's first 14 quiz grades and the range of Joey's quiz grades after his unexcused absence.

(b) Explain what part (a) suggests about using the range as a measure of variability for a distribution of quantitative data.

98. **Pulse rates** Refer to Exercise 88.

(a) Find the range of the pulse rates for all 19 students and the range of the pulse rates excluding the student with the medical issue.

(b) Explain what part (a) suggests about using the range as a measure of variability for a distribution of quantitative data.

99. **Foot lengths** Here are the foot lengths (in centimeters)
pg 61 for a random sample of seven 14-year-olds from the United Kingdom:

| 25 | 22 | 20 | 25 | 24 | 24 | 28 |

Calculate the sample standard deviation. Interpret this value.

100. **Well rested?** A random sample of 6 students in a first-period statistics class was asked how much sleep (to the nearest hour) they got last night. Their responses were 6, 7, 7, 8, 10, and 10. Calculate the sample standard deviation. Interpret this value.

101. **File sizes** How much storage space does your music use? Here is a dotplot of the file sizes (to the nearest tenth of a megabyte) for 18 randomly selected files on Nathaniel's mp3 player:

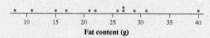

(a) The distribution of file size has a mean of $\bar{x} = 3.2$ megabytes and a standard deviation of $s_x = 1.9$ megabytes. Interpret the standard deviation.

(b) Suppose the music file that takes up 7.5 megabytes of storage space is replaced with another version of the file that only takes up 4 megabytes. How would this affect the mean and the standard deviation? Justify your answer.

102. **Healthy fast food?** Here is a dotplot of the amount of fat (to the nearest gram) in 12 different hamburgers served at a fast-food restaurant:

(a) The distribution of fat content has a mean of $\bar{x} = 22.83$ grams and a standard deviation of $s_x = 9.06$ grams. Interpret the standard deviation.

(b) Suppose the restaurant replaces the burger that has 22 grams of fat with a new burger that has 35 grams of fat. How would this affect the mean and the standard deviation? Justify your answer.

103. **Comparing SD** Which of the following distributions has a smaller standard deviation? Justify your answer.

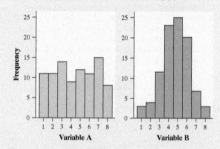

That is, the typical distance from the mean is smaller for Variable B than for Variable A.

1.104 Machine B has a larger standard deviation because more of the observations have values farther from the mean than in Machine A's distribution.

1.105 $Q_1 = 1.9$ megabytes; $Q_3 = 4.7$; $IQR = Q_3 - Q_1 = 4.7 - 1.9 = 2.8$ megabytes

1.106 $Q_1 = 16$ grams; $Q_3 = 28$ grams; $IQR = Q_3 - Q_1 = 28 - 16 = 12$ grams

1.107 An outlier is any value below $Q_1 - 1.5(IQR) = 1.9 - 1.5(2.8) = -2.3$ megabytes or above $Q_3 + 1.5(IQR) = 4.7 + 1.5(2.8) = 8.9$ megabytes. No files in the data set have less than −2.3 megabytes or more than 8.9 megabytes; there are no outliers.

1.108 An outlier is any value below $Q_1 - 1.5(IQR) = 16 - 1.5(12) = -2$ grams of fat or above $Q_3 + 1.5(IQR) = 28 + 1.5(12) = 46$ grams of fat. No hamburgers in the data set have fewer than −2 grams of fat or more than 46 grams of fat, so there are no outliers in the distribution.

1.109 (a) The distribution is skewed to the right because the mean is much larger than the median. Also, Q_3 is much farther from the median than Q_1. **(b)** The amount of money spent typically varies by about $21.70 from the mean of $34.70. **(c)** $Q_1 = 19.27$; $Q_3 = 45.40$; so the IQR is 45.40 − 19.27 = 26.13. Any points below $19.27 - 1.5(26.13) = -19.925$ or above $45.40 + 1.5(26.13) = 84.595$ are outliers. Because the maximum of 93.34 is greater than 84.595, there is at least one outlier.

104. Comparing SD The parallel dotplots show the lengths (in millimeters) of a sample of 11 nails produced by each of two machines. Which distribution has the larger standard deviation? Justify your answer.

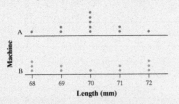

105. File sizes Refer to Exercise 101. Find the interquartile pg 64 range of the file size distribution shown in the dotplot.

106. Healthy fast food? Refer to Exercise 102. Find the interquartile range of the fat content distribution shown in the dotplot.

107. File sizes Refer to Exercises 101 and 105. Identify any pg 67 outliers in the distribution. Show your work.

108. Healthy fast food? Refer to Exercises 102 and 106. Identify any outliers in the distribution. Show your work.

109. Shopping spree The figure displays computer output for data on the amount spent by 50 grocery shoppers.

	$\bar{x}$	s_x	Min	Q_1	Med	Q_3	Max
Amount spent	34.70	21.70	3.11	19.27	27.86	45.40	93.34

(a) What would you guess is the shape of the distribution based only on the computer output? Explain.

(b) Interpret the value of the standard deviation.

(c) Are there any outliers? Justify your answer.

110. C-sections A study in Switzerland examined the number of cesarean sections (surgical deliveries of babies) performed in a year by samples of male and female doctors. Here are summary statistics for the two distributions:

	$\bar{x}$	s_x	Min	Q_1	Med	Q_3	Max
Male doctors	41.333	20.607	20	27	34	50	86
Female doctors	19.1	10.126	5	10	18.5	29	33

(a) Based on the computer output, which distribution would you guess has a more symmetrical shape? Explain your answer.

(b) Explain how the *IQR*s of these two distributions can be so similar even though the standard deviations are quite different.

(c) Does either distribution have any outliers? Justify your answer.

111. Don't call me According to a study by Nielsen pg 69 Mobile, "Teenagers ages 13 to 17 are by far the most prolific texters, sending 1742 messages a month." Mr. Williams, a high school statistics teacher, was skeptical about the claims in the article. So he collected data from his first-period statistics class on the number of text messages they had sent in the past 24 hours. Here are the data:

0	7	1	29	25	8	5	1	25	98	9	0	26
8	118	72	0	92	52	14	3	3	44	5	42	

(a) Make a boxplot of these data.

(b) Use the boxplot you created in part (a) to explain how these data seem to contradict the claim in the article.

112. Acing the first test Here are the scores of Mrs. Liao's students on their first statistics test:

93	93	87.5	91	94.5	72	96		95	93.5	93.5	73
82	45	88	80	86	85.5	87.5	81	78	86	89	
92	91	98	85	82.5	88	94.5	43				

(a) Make a boxplot of these data.

(b) Use the boxplot you created in part (a) to describe how the students did on Mrs. Liao's first test.

113. Electing the president Refer to Exercise 91. Here are a boxplot and some numerical summaries of the electoral vote data:

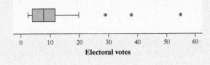

Electoral votes

Variable	N	Mean	SD	Min	Q_1	Median	Q_3	Max
Electoral votes	51	10.55	9.69	3	4	8	12	55

(a) Explain why the median and *IQR* would be a better choice for summarizing the center and variability of the distribution of electoral votes than the mean and standard deviation.

(b) Identify an aspect of the distribution that the stemplot in Exercise 91 reveals that the boxplot does not.

(b) The article claims that teens send 1742 texts a month, or about 58 texts a day (assuming a 30-day month). Nearly all (21 of 25) sent fewer than 58 texts per day, which seems to contradict the claim in the article.

1.112 (a) The median is 87.75, $Q_1 = 82$; $Q_3 = 93$. The *IQR* is $93 - 82 = 11$. Any observation above $Q_3 + 1.5IQR = 93 + 1.5(11) = 109.5$ or below $Q_1 - 1.5IQR = 82 - 1.5(11) = 65.5$ is considered an outlier. Thus, the scores 43 and 45 are outliers. The boxplot is shown below.

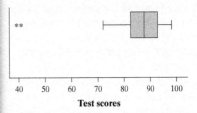

Test scores

(b) Most students did quite well. In fact, more than 75% of the class scored higher than 80. Only two of the students did very poorly.

1.113 (a) The median and *IQR* would be a better choice for summarizing the center and variability of the distribution of electoral votes than the mean and standard deviation because the boxplot reveals that there are three outliers in the data set. The mean and standard deviation are not resistant measures of center and variability, so their values are sensitive to these extreme values. **(b)** The stemplot reveals that the distribution has a single peak, which cannot be discerned from the boxplot. Also, the stemplot reveals that there are actually *four* upper outliers rather than three. The value of 29, which is an outlier, gives the number of electoral votes for *two* states. In the boxplot, this appears as one asterisk. However, there are two states that have that many electoral votes, not one.

1.110 (a) The distribution for the female doctors is more likely to be symmetric since the mean and median are relatively close together (19.1 and 18.5, respectively). The mean and median for the male doctors are quite far apart (41.333 and 34). **(b)** The *IQR* measures the range of the middle 50% of the data. The standard deviation, uses every observation and is not resistant to outliers. So, while the middle 50% of the data sets may have similar spreads, if the lower and upper 25% of one distribution are more variable than the other distribution, the distribution will have a larger standard deviation. **(c)** *Male distribution:* Any points below $27 - 1.5(23) = -7.5$ or above $50 + 1.5(23) = 84.5$ are outliers. Because 86 is greater than 84.5, there is at least one upper outlier. There are no lower

outliers. *Female distribution:* Any points below $10 - 1.5(19) = -18.5$ or above $29 + 1.5(19) = 57.5$ are outliers. This distribution does not have any outliers.

1.111 (a) The median is 9, the first quartile is 3, and the third quartile is 43. The *IQR* is $43 - 3 = 40$. An outlier would be any value below $3 - 1.5(40) = -57$ or above $43 + 1.5(40) = 103$. This means that the value of 118 is an outlier. The boxplot is shown below.

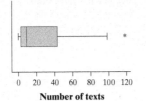

Number of texts

1.114 (a) The median and *IQR* would be a better choice for summarizing the center and variability of the distribution of birthrates in African countries because the boxplot is left-skewed. We can confirm the left skew of this distribution by noticing that the distance from median to minimum is 23.5, while the distance from median to maximum is 15.5. Also, the mean is less than the median. **(b)** The dotplot reveals that there is a single peak at 38 births per 1000 individuals in the population. This cannot be discerned from the boxplot.

1.115 *Shape:* The distribution of energy cost (in dollars) for top freezers looks roughly symmetric. The distribution of energy cost (in dollars) for side freezers looks roughly symmetric. The distribution of energy cost (in dollars) for bottom freezers looks skewed to the right. *Outliers:* There are no outliers for the top or side freezers. There are at least two bottom freezers with unusually high energy costs (over $140 per year). *Center:* The typical energy cost for the side freezers (median ≈ $75) is greater than the typical cost for the bottom freezers (median ≈ $69), which is greater than the typical cost for the top freezers (median ≈ $56). *Variability:* There is much more variability in the energy costs for bottom freezers (*IQR* ≈ $20), than for side freezers (*IQR* ≈ $12), than for top freezers (*IQR* ≈ $8).

1.116 *Shape:* The distributions of annual household income for Maine and Massachusetts are both skewed to the right, while the distribution for Connecticut is fairly symmetric. *Outliers:* There is one outlier in the distribution for Maine at ≈ $160,000, and one outlier in the distribution for Massachusetts at ≈ $430,000. There are no outliers in the Connecticut distribution. *Center:* The center of the distribution is lowest for Maine (median ≈ $43,000), followed by Connecticut (median ≈ $60,000), and then Massachusetts (median ≈ $95,000). *Variability:* The variability of the household income distribution is similar for Maine (*IQR* ≈ $47,000) and Connecticut (*IQR* ≈ $50,000), but much more variable for Massachusetts (*IQR* ≈ $130,000).

1.117 (a)

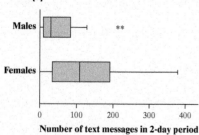

Number of text messages in 2-day period

---

114. **Birthrates in Africa** Refer to Exercise 92. Here are a boxplot and some numerical summaries of the birth-rate data:

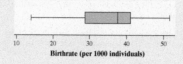

Birthrate (per 1000 individuals)

Variable	N	Mean	SD	Min	Q_1	Median	Q_3	Max
Birthrate	54	34.91	8.57	14.00	29.00	37.50	41.00	53.00

(a) Explain why the median and *IQR* would be a better choice for summarizing the center and variability of the distribution of birthrates in African countries than the mean and standard deviation.

(b) Identify an aspect of the distribution that the dotplot in Exercise 92 reveals that the boxplot does not.

115. **Energetic refrigerators** *Consumer Reports* magazine **pg 71** rated different types of refrigerators, including those with bottom freezers, those with top freezers, and those with side freezers. One of the variables they measured was annual energy cost (in dollars). The following boxplots show the energy cost distributions for each of these types. Compare the energy cost distributions for the three types of refrigerators.

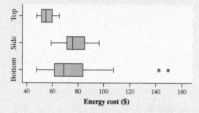

Energy cost ($)

116. **Income in New England** The following boxplots show the total income of 40 randomly chosen households each from Connecticut, Maine, and Massachusetts, based on U.S. Census data from the American Community Survey. Compare the distributions of annual incomes in the three states.

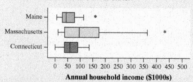

Annual household income ($1000s)

117. **Who texts more?** For their final project, a group of AP® Statistics students wanted to compare the texting habits of males and females. They asked a random

---

sample of students from their school to record the number of text messages sent and received over a two-day period. Here are their data:

Males	127	44	28	83	0	6	78	6
	5	213	73	20	214	28	11	
Females	112	203	102	54	379	305	179	24
	127	65	41	27	298	6	130	0

(a) Make parallel boxplots of the data.

(b) Use your calculator to compute separate numerical summaries for the males and for the females. Are these values statistics or parameters? Explain your answer.

(c) Do these data suggest that males and females at the school differ in their texting habits? Use the results from parts (a) and (b) to support your answer.

118. **SSHA scores** Here are the scores on the Survey of Study Habits and Attitudes (SSHA) for a random sample of 18 first-year college women:

154	109	137	115	152	140	154	178	101
103	126	126	137	165	165	129	200	148

Here are the SSHA scores for a random sample of 20 first-year college men:

108	140	114	91	180	115	126
92	169	146	109	132	75	88
113	151	70	115	187	104	

Note that high scores indicate good study habits and attitudes toward learning.

(a) Make parallel boxplots of the data.

(b) Use your calculator to compute separate numerical summaries for the women and for the men. Are these values statistics or parameters? Explain your answer.

(c) Do these data support the belief that men and women differ in their study habits and attitudes toward learning? Use your results from parts (a) and (b) to support your answer.

119. **Income and education level** Each March, the Bureau of Labor Statistics compiles an Annual Demographic Supplement to its monthly Current Population Survey.[43] Data on about 71,067 individuals between the ages of 25 and 64 who were employed full-time were collected in one of these surveys. The parallel boxplots compare the distributions of income for people with five levels of education. This figure is a variation of the boxplot idea: because large data sets often contain very extreme observations, we omitted the individuals in each category with the top 5% and bottom 5% of incomes. Also, the whiskers are drawn all the way to the maximum and minimum values of the remaining data for each distribution.

(b) Numerical summaries are given below.

	n	$\bar{X}$	s_X	Min	Q_1	Median	Q_3	Max
Male	15	62.4	71.4	0	6	28	83	214
Female	16	128.3	116.0	0	34	107	191	379

These values are statistics because they are numbers that describe the sample of students. **(c)** The data give very strong evidence that male and female texting habits differ considerably at the school. A typical female sends and receives about 79 more text messages in 2 days than a typical male. The males as a group are also much more consistent in their texting frequency than the females.

1.118 (a)

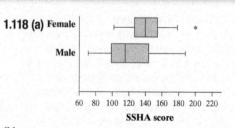

SSHA score

(b)

Variable	n	$\bar{X}$	s_X	Min	Q_1	Median	Q_3	Max
Women	18	141.06	26.44	101.00	126.00	138.50	154.00	200.0
Men	20	121.25	32.85	70.00	98.00	114.50	143.00	187.0

(c) The data give very strong evidence that first-year male and female college students' study habits and attitudes differ considerably. A typical first-year female earns, on average, about 20 more points on the SSHA than a typical male first-year student.

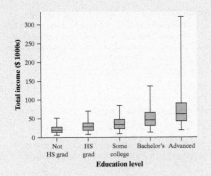

Use the graph to help answer the following questions.

(a)   What shape do the distributions of income have?

(b)   Explain how you know that there are outliers in the group that earned an advanced degree.

(c)   How does the typical income change as the highest education level reached increases? Why does this make sense?

(d)   Describe how the variability in income changes as the highest education level reached increases.

120. **Sleepless nights** Researchers recorded data on the amount of sleep reported each night during a week by a random sample of 20 high school students. Here are parallel boxplots comparing the distribution of time slept on all 7 nights of the study:[44]

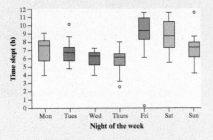

Use the graph to help answer the following questions.

(a)   Which distributions have a clear left-skewed shape?

(b)   Which outlier stands out the most, and why?

(c)   How does the typical amount of sleep that the students got compare on these seven nights?

(d)   On which night was there the most variation in how long the students slept? Justify your answer.

121. **SD contest** This is a standard deviation contest. You must choose four numbers from the whole numbers 0 to 10, with repeats allowed.

(a)   Choose four numbers that have the smallest possible standard deviation.

(b)   Choose four numbers that have the largest possible standard deviation.

(c)   Is more than one choice possible in either part (a) or (b)? Explain.

122. **What do they measure?** For each of the following summary statistics, decide (i) whether it could be used to measure center or variability and (ii) whether it is resistant.

(a) $\dfrac{Q_1 + Q_3}{2}$ (b) $\dfrac{\text{Max} - \text{Min}}{2}$

Multiple Choice: *Select the best answer for Exercises 123–126.*

123.   If a distribution is skewed to the right with no outliers, which expression is correct?

(a)   mean < median   (d)   mean > median

(b)   mean ≈ median   (e)   We can't tell without examining the data.

(c)   mean = median

124.   The scores on a statistics test had a mean of 81 and a standard deviation of 9. One student was absent on the test day, and his score wasn't included in the calculation. If his score of 84 was added to the distribution of scores, what would happen to the mean and standard deviation?

(a)   Mean will increase, and standard deviation will increase.

(b)   Mean will increase, and standard deviation will decrease.

(c)   Mean will increase, and standard deviation will stay the same.

(d)   Mean will decrease, and standard deviation will increase.

(e)   Mean will decrease, and standard deviation will decrease.

1.119 (a) All five income distributions are skewed to the right. This tells me that within each level of education, there is much more variability in incomes for the upper 50% of the distribution than for the lower 50% of the distribution. **(b)** Even though the boxplots are not modified (i.e., do not show outliers), it is clear that the upper whisker for the advanced degree boxplot is much longer than 1.5 times its *IQR,* indicating that there is at least one upper outlier in the group that earned an advanced degree. **(c)** As education level rises, the median, quartiles, and extremes rise as well—that is, every value in the 5-number summary gets larger. This makes sense because one would expect that individuals with more education tend to attain jobs that yield more income. **(d)** The variability in income increases as the education level increases. Both the width of the box (the *IQR*) and the distance from one extreme to the other increase as education levels increase.

1.120 (a) Monday, Wednesday, and Thursday nights' distributions have a clear left-skewed shape. This shape tells us that, on those nights, there is much more variability in the lower 50% of sleep amounts than the upper 50% of sleep amounts. **(b)** The outlier of 0 hours on sleep on Friday night stands out the most. It is very unusual for a student to get 0 hours of sleep! **(c)** Students tended to get more sleep on Friday and Saturday nights than the other five nights of the week. **(d)** The most variation seems to exist in the number of hours that the students slept on Friday night. That evening, the number of hours the student slept varied from 0 to 12 hours, which is greater than every other night.

1.121 (a) One possible answer is 1, 1, 1, and 1. **(b)** 0, 0, 10, 10 **(c)** For part (a), any set of four identical numbers will have $s_x = 0$. For part (b), however, there is only 1 possible answer. We want the values to be as far from the mean as possible, so the squared deviations from the mean can be as big as possible. Our best choice is two values at each extreme, which makes all four squared deviations equal to 5^2.

1.122 (a) This could be used to measure the center since we are averaging the 25th and 75th percentiles, effectively finding a middle point between these positions. It would be resistant to outliers, because any outliers would occur below Q_1 or above Q_3. **(b)** This could be used as a measure of variability since it finds the distance between the smallest and largest values and then divides by 2, giving half of the range. This measure, however, would not be resistant to outliers. If outliers exist, they would, by definition, include the max, the min, or both.

1.123 d

1.124 b

1.125 e

1.126 a

1.127 A histogram is given below.

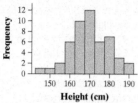

Height (cm)

This distribution is roughly symmetric, with a single peak at 170 cm. There do not appear to be any outliers. The center of the distribution of heights can be described by the mean of 169.88 cm or the median of 169.5 cm. The heights vary from 145.5 to 191 cm, so the range is 45.5 cm. The standard deviation of heights is 9.687 cm and the *IQR* is $177 - 163 = 14$ cm.

1.128 Women appear to be more likely to engage in behaviors that are indicative of good "habits of mind." They are especially more likely to revise papers to improve their writing (about 55% of females report this as opposed to about 37% of males). The difference is a little less for seeking feedback on their work. In that case, about 49% of the females did this as opposed to about 38% of males.

125. The stemplot shows the number of home runs hit by each of the 30 Major League Baseball teams in a single season. Home run totals above what value should be considered outliers?

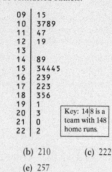

```
09 | 15
10 | 3789
11 | 47
12 | 19
13 |
14 | 89
15 | 34445
16 | 239
17 | 223
18 | 356
19 | 1
20 | 3
21 | 0
22 | 2
```

Key: 14|8 is a team with 148 home runs.

(a) 173   (b) 210   (c) 222

(d) 229   (e) 257

126. Which of the following boxplots best matches the distribution shown in the histogram?

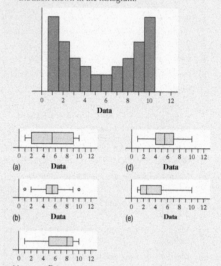

Data

Recycle and Review

127. **How tall are you?** (1.2) We used Census At School's "Random Data Selector" to choose a sample of 50 Canadian students who completed a survey in a recent year. Here are the students' heights (in centimeters):

166.5	170.0	178.0	163.0	150.5	169.0	173.0	169.0	171.0	166.0
190.0	183.0	178.0	161.0	171.0	170.0	191.0	168.5	178.5	173.0
175.0	160.5	166.0	164.0	163.0	174.0	160.0	174.0	182.0	167.0
166.0	170.0	170.0	181.0	171.5	160.0	178.0	157.0	165.0	187.0
168.0	157.5	145.5	156.0	182.0	168.5	177.0	162.5	160.5	185.5

Make an appropriate graph to display these data. Describe the shape, center, and variability of the distribution. Are there any outliers?

128. **Success in college** (1.1) The Freshman Survey asked first-year college students about their "habits of mind"—specific behaviors that college faculty have identified as being important for student success. One question asked students, "How often in the past year did you revise your papers to improve your writing?" Another asked, "How often in the past year did you seek feedback on your academic work?" The figure is a bar graph comparing the percent of males and females who answered "frequently" to these two questions.[45]

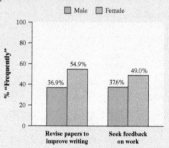

What does the graph reveal about the habits of mind of male and female college freshmen?

Chapter 1 Wrap-Up

FRAPPY! FREE RESPONSE AP® PROBLEM, YAY!

The following problem is modeled after actual AP® Statistics exam free response questions. Your task is to generate a complete, concise response in 15 minutes.

Directions: Show all your work. Indicate clearly the methods you use, because you will be scored on the correctness of your methods as well as on the accuracy and completeness of your results and explanations.

Using data from the 2010 census, a random sample of 348 U.S. residents aged 18 and older was selected. Among the variables recorded were gender (male or female), housing status (rent or own), and marital status (married or not married).

The two-way table below summarizes the relationship between gender and housing status.

	Male	Female	Total
Own	132	122	254
Rent	50	44	94
Total	182	166	348

(a) What percent of males in the sample own their home?

(b) Make a graph to compare the distribution of housing status for males and females.

(c) Using your graph from part (b), describe the relationship between gender and housing status.

(d) The two-way table below summarizes the relationship between marital status and housing status.

	Married	Not Married	Total
Own	172	82	254
Rent	40	54	94
Total	212	136	348

For the members of the sample, is the relationship between marital status and housing status stronger or weaker than the relationship between gender and housing status that you described in part (c)? Justify your choice using the data provided in the two-way tables.

After you finish, you can view two example solutions on the book's website (highschool.bfwpub.com/updatedtps6e). Determine whether you think each solution is "complete," "substantial," "developing," or "minimal." If the solution is not complete, what improvements would you suggest to the student who wrote it? Finally, your teacher will provide a scoring rubric. Score your response and note what, if anything, you would do differently to improve your own score.

Chapter 1 Review

Introduction: Data Analysis: Making Sense of Data

In this brief section, you learned several fundamental concepts that will be important throughout the course: the idea of a distribution and the distinction between quantitative and categorical variables. You also learned a strategy for exploring data:

- Begin by examining each variable by itself. Then move on to study relationships between variables.
- Start with a graph or graphs. Then add numerical summaries.

Section 1.1: Analyzing Categorical Data

In this section, you learned how to display the distribution of a single categorical variable with bar graphs and pie charts and what to look for when describing these displays. Remember to properly label your graphs! Poor labeling is an easy way to lose points on the AP® Statistics exam. You should also be able to recognize misleading graphs and be careful to avoid making misleading graphs yourself.

81

➕ **Ask the StatsMedic**
How to Write a Great Test for AP® Statistics

This post details the proper format of a great AP® Statistics test and also provides resources for finding questions.

➕ **Ask the StatsMedic**
How to Grade Your AP® Statistics Test

This post gives suggestions on the best way to give students feedback on their tests.

TRM Chapter 1 FRAPPY! Materials

Please consult the Teacher's Resource Materials for sample student responses, a scoring rubric, and a printable version of the original question with space for students to write their responses. We present a model solution here.

Answers:

(a) $132/182 = 0.725 = 72.5\%$

(b) A segmented bar graph is shown below.

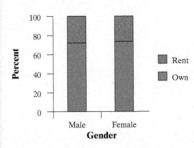

(c) There doesn't seem to be much of an association between gender and housing status. The percent of each gender that owns their home is roughly the same, although it is slightly higher for females (73.5% vs. 72.5%).

(d) A graph of housing status for married and not married respondents is shown below. Because the percent of married respondents who own their home (81.1%) is quite a bit larger than the percent of not married respondents (60.3%), there seems to be an association between marital status and housing status. Also, this association is stronger than the association between gender and housing status because the difference in percents who own a home is greater for marital status $(81.1 - 60.3 = 20.8\%)$ than for gender $(73.5 - 72.5 = 1\%)$.

Next, you learned how to investigate the relationship between two categorical variables. Using a two-way table, you learned how to calculate and display marginal and joint relative frequencies. Conditional relative frequencies and side-by-side bar graphs, segmented bar graphs, or mosaic plots allow you to look for an association between the variables. If there is no association between the two variables, comparative bar graphs of the distribution of one variable for each value of the other variable will be identical. If differences in the corresponding conditional relative frequencies exist, there is an association between the variables. That is, knowing the value of one variable helps you predict the value of the other variable.

Section 1.2: Displaying Quantitative Data with Graphs

In this section, you learned how to create three different types of graphs for a quantitative variable: dotplots, stemplots, and histograms. Each of the graphs has distinct benefits, but all of them are good tools for examining the distribution of a quantitative variable. Dotplots and stemplots are handy for small sets of data. Histograms are the best choice when there are a large number of observations. On the AP® exam, you will be expected to create each of these types of graphs, label them properly, and comment on their characteristics.

When you are describing the distribution of a quantitative variable, you should look at its graph for the overall pattern (shape, center, variability) and striking departures from that pattern (outliers). Use the acronym SOCV (shape, outliers, center, variability) to help remember these four characteristics. When comparing distributions, you should include explicit comparison words for center and variability such as "is greater than" or "is approximately the same as." When asked to compare distributions, a very common mistake on the AP® exam is describing the characteristics of each distribution separately without making these explicit comparisons.

Section 1.3: Describing Quantitative Data with Numbers

To measure the center of a distribution of quantitative data, you learned how to calculate the mean and the median of a distribution. You also learned that the median is a resistant measure of center, but the mean isn't resistant because it can be greatly affected by skewness or outliers.

To measure the variability of a distribution of quantitative data, you learned how to calculate the range, standard deviation, and interquartile range. The standard deviation is the most commonly used measure of variability and approximates the typical distance of a value in the data set from the mean. The standard deviation is not resistant—it is heavily affected by extreme values. The interquartile range (IQR) is a resistant measure of variability because it ignores the upper 25% and lower 25% of the distribution, but the range isn't resistant because it uses only the minimum and maximum value.

To identify outliers in a distribution of quantitative data, you learned the $1.5 \times IQR$ rule. You also learned that boxplots are a great way to visually summarize a distribution of quantitative data. Boxplots are helpful for comparing the center (median) and variability (range, IQR) for multiple distributions. Boxplots aren't as useful for identifying the shape of a distribution because they do not display peaks, clusters, gaps, and other interesting features.

What Did You Learn?

Learning Target	Section	Related Example on Page(s)	Relevant Chapter Review Exercise(s)
Identify the individuals and variables in a set of data.	Intro	3	R1.1
Classify variables as categorical or quantitative.	Intro	3	R1.1
Make and interpret bar graphs for categorical data.	1.1	11	R1.2
Identify what makes some graphs of categorical data misleading.	1.1	12	R1.3
Calculate marginal and joint relative frequencies from a two-way table.	1.1	15	R1.4
Calculate conditional relative frequencies from a two-way table.	1.1	17	R1.4, R1.5
Use bar graphs to compare distributions of categorical data.	1.1	20	R1.5
Describe the nature of the association between two categorical variables.	1.1	20	R1.5
Make and interpret dotplots, stemplots, and histograms of quantitative data.	1.2	Dotplots: 31 Stemplots: 38 Histograms: 42	R1.6, R1.7

Teaching Tip

Be sure to discuss the "What Did You Learn?" table with your students. Ask them to read each learning target and self-assess whether or not they can do each one. For each target, the grid shows the section in which it was covered, page references for examples that illustrate the target, and relevant chapter review exercises that students can use to assess their understanding of that target. If you didn't distribute the Learning Target grid at the beginning of the chapter, you may do so now. Encourage your students to use this grid as part of their preparations for the chapter test.

Learning Target	Section	Related Example on Page(s)	Relevant Chapter Review Exercise(s)
Identify the shape of a distribution from a graph.	1.2	33	R1.6
Describe the overall pattern (shape, center, and variability) of a distribution and identify any major departures from the pattern (outliers).	1.2	35	R1.6
Compare distributions of quantitative data using dotplots, stemplots, and histograms.	1.2	36	R1.8
Calculate measures of center (mean, median) for a distribution of quantitative data.	1.3	Mean: 55 Median: 57	R1.6
Calculate and interpret measures of variability (range, standard deviation, IQR) for a distribution of quantitative data.	1.3	SD: 61 IQR: 64	R1.9
Explain how outliers and skewness affect measures of center and variability.	1.3	69	R1.9
Identify outliers using the 1.5 × IQR rule.	1.3	67	R1.7, R1.9
Make and interpret boxplots of quantitative data.	1.3	69	R1.7
Use boxplots and numerical summaries to compare distributions of quantitative data.	1.3	71	R1.10

Chapter 1 Review Exercises

These exercises are designed to help you review the important ideas and methods of the chapter.

R1.1 Who buys cars? A car dealer keeps records on car buyers for future marketing purposes. The table gives information on the last 4 buyers.

Buyer's name	Zip code	Gender	Buyer's distance from dealer (mi)	Car model	Model year	Price
P. Smith	27514	M	13	Fiesta	2018	$26,375
K. Ewing	27510	M	10	Mustang	2015	$39,500
L. Shipman	27516	F	2	Fusion	2016	$38,400
S. Reice	27243	F	4	F-150	2016	$56,000

(a) Identify the individuals in this data set.

(b) What variables were measured? Classify each as categorical or quantitative.

R1.2 I want candy! Mr. Starnes bought some candy for his AP® Statistics class to eat on Halloween. He offered the students an assortment of Snickers®, Milky Way®, Butterfinger®, Twix®, and 3 Musketeers® candies. Each student was allowed to choose one option. Here are the data on the type of candy selected. Make a relative frequency bar graph to display the data. Describe what you see.

Twix	Snickers	Butterfinger
Butterfinger	Snickers	Snickers
3 Musketeers	Snickers	Snickers
Butterfinger	Twix	Twix
Twix	Twix	Twix
Snickers	Snickers	Twix
Snickers	Milky Way	Twix
Twix	Twix	Butterfinger
Milky Way	Butterfinger	3 Musketeers
Milky Way	Butterfinger	Butterfinger

TRM **Full Solutions to Chapter 1 Review Exercises**

The full Solutions can be found by clicking on the link in the TE-Book, opening the TRFD, or downloading from the Teacher's Resources on the book's digital platform.

Teaching Tip

Recommend that your students watch the Chapter Review Exercise videos. They feature an experienced AP® Statistics teacher walking step-by-step through the solution and offering valuable tips and advice for the AP® Statistics exam. The videos are available by clicking on the link in the e-Book or viewing them on the Student Site at highschool.bfwpub.com/updatedtps6e.

Answers to Chapter 1 Review Exercises

R1.1 (a) The individuals are buyers.
(b) The variables are zip code (categorical), gender (categorical), buyer's distance from the dealer (in miles) (quantitative), car model (categorical), model year (categorical), and price (quantitative).

R1.2 First, a relative frequency table must be constructed.

Candy selected	Relative frequency
Snickers®	8/30 = 26.7%
Milky Way	3/30 = 10%
Butterfinger	7/30 = 23.3%
Twix	10/30 = 33.3%
3 Musketeers	2/30 = 6.7%

The relative frequency bar graph is given below.

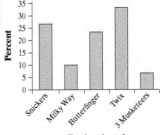

Candy selected

Students preferred Twix the most (one-third of the students chose this candy), followed by Snickers (27% relative frequency), then Butterfinger (23% relative frequency), then Milky Way (10% relative frequency), and lastly 3 Musketeers (7% relative frequency).

R1.3 (a) The graph is misleading because the "bars" are different widths. For example, the bar for "send receive text messages" should be roughly twice the size of the bar for "camera" when it is actually much more than twice as large in area. **(b)** It would not be appropriate to make a pie chart for these data because they do not describe parts of the same whole. Students were free to answer in more than one category.

R1.4 (a) $148/219 = 0.676 = 67.6\%$ were Facebook users.
(b) $67/219 = 0.306 = 30.6\%$ were aged 28 and over.
(c) $21/219 = 0.096 = 9.6\%$ were older Facebook users.
(d) $78/148 = 0.527 = 52.7\%$ of the Facebook users were younger students.

R1.5 (a)

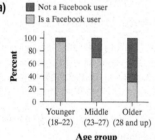

(b) From both the table and the graph, we can see that there is an association between age and Facebook usage. As age increases, the percent of Facebook users decreases. For younger students, about 95% use Facebook. That drops to 70% for 23- to 27-year-old students and drops even further to 31.3% for older students.

R1.6 (a) A stemplot is shown below.

```
48 | 8
49 |
50 | 7
51 | 0
52 | 6799
53 | 04469
54 | 2467
55 | 03578
56 | 12358
57 | 59
58 | 5
```

Key: 48 | 8 = 4.88

(b) The distribution is roughly symmetric, with one possible outlier at 4.88. The center of the distribution is between 5.4 and 5.5; the densities vary from 4.88 to 5.85. **(c)** The mean of the distribution of Cavendish's 29 measurements is 5.45. So these estimates suggest that the earth's density is about 5.45 times the density

R1.3 I'd die without my phone! In a survey of over 2000 U.S. teenagers by Harris Interactive, 47% said that "their social life would end or be worsened without their cell phone."[46] One survey question asked the teens how important it is for their phone to have certain features. The following figure displays data on the percent who indicated that a particular feature is vital.

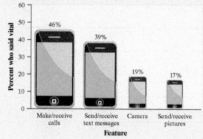

(a) Explain how the graph gives a misleading impression.

(b) Would it be appropriate to make a pie chart to display these data? Why or why not?

R1.4 Facebook and age Is there a relationship between Facebook use and age among college students? The following two-way table displays data for the 219 students who responded to the survey.[47]

		Age		
		Younger (18–22)	Middle (23–27)	Older (28 and up)
Facebook user?	Yes	78	49	21
	No	4	21	46

(a) What percent of the students who responded were Facebook users?

(b) What percent of the students in the sample were aged 28 or older?

(c) What percent of the students who responded were older Facebook users?

(d) What percent of the Facebook users in the sample were younger students?

R1.5 Facebook and age Refer to the preceding exercise.
(a) Find the distribution of Facebook use for each of the three age groups. Make a segmented bar graph to compare these distributions.

(b) Describe what the graph in part (a) reveals about the association between age and Facebook use.

R1.6 Density of the earth In 1798, the English scientist Henry Cavendish measured the density of the earth several times by careful work with a torsion balance. The variable recorded was the density of the earth as a multiple of the density of water. Here are Cavendish's 29 measurements:[48]

5.50	5.61	4.88	5.07	5.26	5.55	5.36	5.29	5.58	5.65
5.57	5.53	5.62	5.29	5.44	5.34	5.79	5.10	5.27	5.39
5.42	5.47	5.63	5.34	5.46	5.30	5.75	5.68	5.85	

(a) Make a stemplot of the data.

(b) Describe the distribution of density measurements.

(c) The currently accepted value for the density of earth is 5.51 times the density of water. How does this value compare to the mean of the distribution of density measurements?

R1.7 Guinea pig survival times Here are the survival times (in days) of 72 guinea pigs after they were injected with infectious bacteria in a medical experiment.[49] Survival times, whether of machines under stress or cancer patients after treatment, usually have distributions that are skewed to the right.

43	45	53	56	56	57	58	66	67	73	74	79
80	80	81	81	81	82	83	83	84	88	89	91
91	92	92	97	99	99	100	100	101	102	102	102
103	104	107	108	109	113	114	118	121	123	126	128
137	138	139	144	145	147	156	162	174	178	179	184
191	198	211	214	243	249	329	380	403	511	522	598

(a) Make a histogram of the data. Does it show the expected right skew?

(b) Now make a boxplot of the data.

(c) Compare the histogram from part (a) with the boxplot from part (b). Identify an aspect of the distribution that one graph reveals but the other does not.

R1.8 Household incomes Rich and poor households differ in ways that go beyond income. Here are histograms that compare the distributions of household size (number of people) for low-income and high-income households.[50] Low-income households had annual incomes less than $15,000, and high-income households had annual incomes of at least $100,000.

of water. The currently accepted value for the density of the earth (5.51 times the density of water) is slightly larger than the mean of the distribution of density measurements.

R1.7 (a)

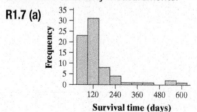

The survival times are right-skewed, as expected.

(b)

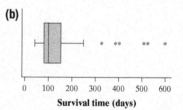

(c) While both graphs clearly show that the distribution is strongly skewed to the right, the boxplot shows that there are several high outliers.

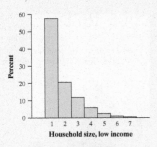

Household size, low income

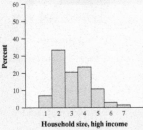

Household size, high income

(a) About what percent of each group of households consisted of four or more people?

(b) Describe the similarities and differences in these two distributions of household size.

Exercises R1.9 and R1.10 refer to the following setting. Do you like to eat tuna? Many people do. Unfortunately, some of the tuna that people eat may contain high levels of mercury. Exposure to mercury can be especially hazardous for pregnant women and small children. How much mercury is safe to consume? The Food and Drug Administration will take action (like removing the product from store shelves) if the mercury concentration in a 6-ounce can of tuna is 1.00 ppm (parts per million) or higher.

What is the typical mercury concentration in cans of tuna sold in stores? A study conducted by Defenders of Wildlife set out to answer this question. Defenders collected a sample of 164 cans of tuna from stores across the United States. They sent the selected cans to a laboratory that is often used by the Environmental Protection Agency for mercury testing.[51]

R1.9 Mercury in tuna Here are a dotplot and numerical summaries of the data on mercury concentration in the sampled cans (in parts per million, ppm):

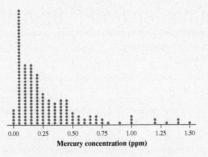

Mercury concentration (ppm)

Variable	N	Mean	SD	Min
Mercury	164	0.285	0.300	0.012
Variable	Q_1	Med	Q_3	Max
Mercury	0.071	0.180	0.380	1.500

(a) Interpret the standard deviation.

(b) Determine whether there are any outliers.

(c) Explain why the mean is so much larger than the median of the distribution.

R1.10 Mercury in tuna Is there a difference in the mercury concentration of light tuna and albacore tuna? Use the parallel boxplots and the computer output to write a few sentences comparing the two distributions.

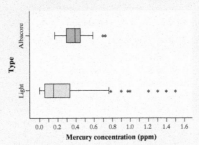

Mercury concentration (ppm)

Type	N	Mean	SD	Min
Albacore	20	0.401	0.152	0.170
Light	144	0.269	0.312	0.012
Type	Q_1	Med	Q_3	Max
Albacore	0.293	0.400	0.460	0.730
Light	0.059	0.160	0.347	1.500

R1.8 (a) About 11% of low-income and 40% of high-income households consisted of four or more people. **(b)** The shapes of both distributions are skewed to the right. However, the skewness is much stronger in the distribution for low-income households. On average, household size is larger for high-income households. In fact, the majority of low-income households consist of only one person. Only about 7% of high-income households consist of one person. One-person households might have less income because they would include many young single people who have no job or retired single people with a fixed income.

R1.9 (a) The amount of mercury per can of tuna will typically vary by about 0.3 ppm from the mean of 0.285 ppm. **(b)** The *IQR* = 0.380 − 0.071 = 0.309, so any point below 0.071 − 1.5(0.309) = −0.393 or above 0.38 + 1.5(0.309) = 0.8435 would be considered an outlier. Because the smallest value is 0.012, there are no low outliers. According to the histogram, some values fall above 0.8435, so there are several high outliers. **(c)** The mean is much larger than the median of the distribution because the distribution is strongly skewed to the right and there are several high outliers.

R1.10 The distribution for light tuna is skewed to the right with several high outliers. The distribution for albacore tuna is more symmetric with just a couple of high outliers. The albacore tuna generally has more mercury. Its minimum, first quartile, median, and third quartile are all greater than the respective values for light tuna. But that doesn't mean light tuna is always better. It has much larger variation in mercury concentration, with some cans having as much as twice the amount of mercury as the largest amount in the albacore tuna.

TRM Chapter 1 Test

There is one Chapter Test available in the Teacher's Resource Materials. Click on the link in the TE-Book, open in the TRFD, or download from the Teacher's Resources on the book's digital platform. You can also create your own test using the ExamView® Assessment Suite that is part of the TPS6 program. Questions are coded by Learning Target to make it easy to build parallel tests.

Answers to Chapter 1 AP® Statistics Practice Test

T1.1 c

T1.2 d

T1.3 b

T1.4 b

T1.5 c

T1.6 e

T1.7 c

Chapter 1 AP® Statistics Practice Test

Section I: Multiple Choice *Select the best answer for each question.*

T1.1 An airline records data on several variables for each of its flights: model of plane, amount of fuel used, time in flight, number of passengers, and whether the flight arrived on time. The number and type of variables recorded are

(a) 1 categorical, 4 quantitative (1 discrete, 3 continuous)

(b) 1 categorical, 4 quantitative (2 discrete, 2 continuous)

(c) 2 categorical, 3 quantitative (1 discrete, 2 continuous)

(d) 2 categorical, 3 quantitative (2 discrete, 1 continuous)

(e) 3 categorical, 2 quantitative (1 discrete, 1 continuous)

T1.2 The students in Mr. Tyson's high school statistics class were recently asked if they would prefer a pasta party, a pizza party, or a donut party. The following bar graph displays the data.

This graph is misleading because

(a) it should be a histogram, not a bar graph.

(b) there should not be gaps between the bars.

(c) the bars should be arranged in decreasing order by height.

(d) the vertical axis scale should start at 0.

(e) preferred party should be on the vertical axis and number of students should be on the horizontal axis.

T1.3 Forty students took a statistics test worth 50 points. The dotplot displays the data. The third quartile is

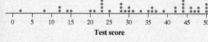

(a) 45.

(b) 44.

(c) 43.

(d) 32.

(e) 23.

Questions T1.4–T1.6 refer to the following setting. Realtors collect data in order to serve their clients more effectively.

In a recent week, data on the age of all homes sold in a particular area were collected and displayed in this histogram.

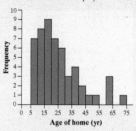

T1.4 Which of the following could be the median age?

(a) 19 years   (b) 24 years   (c) 29 years

(d) 34 years   (e) 39 years

T1.5 Which of the following is most likely true?

(a) mean > median, range < IQR

(b) mean < median, range < IQR

(c) mean > median, range > IQR

(d) mean < median, range > IQR

(e) mean = median, range > IQR

T1.6 The standard deviation of the distribution of house age is about 16 years. Interpret this value.

(a) The age of all houses in the sample is within 16 years of the mean.

(b) The gap between the youngest and oldest house is 16 years.

(c) The age of all the houses in the sample is 16 years from the mean.

(d) The gap between the first quartile and the third quartile is 16 years.

(e) The age of the houses in the sample typically varies by about 16 years from the mean age.

T1.7 The mean salary of all female workers at a company is $35,000. The mean salary of all male workers at the company is $41,000. What must be true about the mean salary of all workers at the company?

(a) It must be $38,000.

(b) It must be larger than the median salary.

(c) It could be any number between $35,000 and $41,000.

(d) It must be larger than $38,000.

(e) It cannot be larger than $40,000.

Questions T1.8 and T1.9 refer to the following setting. A survey was designed to study how business operations vary by size. Companies were classified as small, medium, or large. Questionnaires were sent to 200 randomly selected businesses of each size. Because not all questionnaires are returned, researchers decided to investigate the relationship between the response rate and the size of the business. The data are given in the following two-way table.

		Business size		
		Small	Medium	Large
Response?	Yes	125	81	40
	No	75	119	160

T1.8 What percent of all small companies receiving questionnaires responded?

(a) 12.5%   (b) 20.8%   (c) 33.3%
(d) 50.8%   (e) 62.5%

T1.9 Which of the following conclusions seems to be supported by the data?

(a) There are more small companies than large companies in the survey.

(b) Small companies appear to have a higher response rate than medium or big companies.

(c) Exactly the same number of companies responded as didn't respond.

(d) Overall, more than half of companies responded to the survey.

(e) If we combined the medium and large companies, then their response rate would be equal to that of the small companies.

T1.10 An experiment was conducted to investigate the effect of a new weed killer to prevent weed growth in onion crops. Two chemicals were used: the standard weed killer (S) and the new chemical (N). Both chemicals were tested at high and low concentrations on 50 test plots. The percent of weeds that grew in each plot was recorded. Here are some boxplots of the results.

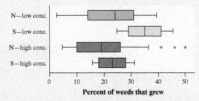

Which of the following is *not* a correct statement about the results of this experiment?

(a) At both high and low concentrations, the new chemical results in better weed control than the standard weed killer.

(b) For both chemicals, a smaller percentage of weeds typically grew at higher concentrations than at lower concentrations.

(c) The results for the standard weed killer are less variable than those for the new chemical.

(d) High and low concentrations of either chemical have approximately the same effects on weed growth.

(e) Some of the results for the low concentration of weed killer show a smaller percentage of weeds growing than some of the results for the high concentration.

Section II: Free Response *Show all your work. Indicate clearly the methods you use, because you will be graded on the correctness of your methods as well as on the accuracy and completeness of your results and explanations.*

T1.11 You are interested in how many contacts older adults have in their smartphones. Here are data on the number of contacts for a random sample of 30 elderly adults with smartphones in a large city:

7	20	24	25	25	28	28	30	32	35
42	43	44	45	46	47	48	48	50	51
72	75	77	78	79	83	87	88	135	151

(a) Construct a histogram of these data.

(b) Are there any outliers? Justify your answer.

(c) Would it be better to use the mean and standard deviation or the median and *IQR* to describe the center and variability of this distribution? Why?

T1.12 A study among the Pima Indians of Arizona investigated the relationship between a mother's diabetic status and the number of birth defects in her children. The results appear in the two-way table.

		Diabetic status		
		Nondiabetic	Prediabetic	Diabetic
Number of birth defects	None	754	362	38
	One or more	31	13	9

(a) What proportion of the women in this study had a child with one or more birth defects?

(b) What percent of the women in this study were diabetic or prediabetic, and had a child with one or more birth defects?

T1.12 (a) $53/1207 = 0.044$
(b) $22/1207 = 1.8\%$ **(c)** The conditional probabilities are shown in the table below.

Number of birth defects	Diabetic status		
	Nondiabetic	Prediabetic	Diabetic
None	96.1%	96.5%	80.9%
One or more	3.9%	3.5%	19.1%

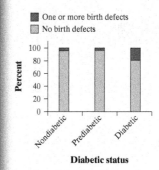

(d) There is an association between diabetic status and number of birth defects for the women in this study. Nondiabetics and prediabetics appear to have babies with birth defects at about the same rate. However, those with diabetes have a much higher rate of babies with birth defects.

T1.8 e

T1.9 b

T1.10 d

T1.11 (a)

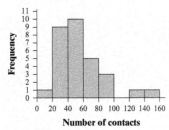

(b) The first quartile is 30 contacts; the third quartile is 77. *IQR* = 77 − 30 = 47; any value below 30 − 1.5(47) = −40.5 or above 77 + 1.5(47) = 147.5 is an outlier. So the observation of 151 contacts is an outlier. **(c)** It would be better to use the median and *IQR* to describe the center and variability because the distribution of number of contacts is skewed to the right and has a high outlier.

T1.13 (a) The longest that any battery lasted was between 550 and 559 hours. **(b)** Someone might prefer to use Brand X because it has a higher minimum lifetime or because its lifetimes are more consistent (less variable). **(c)** Someone might prefer Brand Y because it has a higher median lifetime.

T1.14 (a) These numerical summaries are statistics because they are numbers that describe the samples. **(b)** The distribution of reaction time for the "Athlete" group is slightly skewed to the right. The distribution of reaction time for the "Other" group is roughly symmetric, with two high outliers. It appears that the "Athlete" distribution has one high outlier, while the "Other" distribution has two high outliers. The reaction times for the students who have not been varsity athletes tended to be slower (median = 292.0 milliseconds) than for the athletes (median = 261.0 milliseconds). The distribution of reaction time for the "Other" group also has more variability as their reaction times have an *IQR* of 70 milliseconds and the athletes' reaction times have an *IQR* of 64 milliseconds.

Teaching Tip

If you are going to use the Chapter Project, plan ahead because students will need several days (or weeks) to complete it. Consider assigning the project at the beginning of the chapter and giving students a due date around the time of the chapter's completion. This project can be completed using many different software packages, but the most popular are Minitab, SPSS, SAS, JMP, R, and Fathom. The analysis can also be done using the Extra Applets on the Student Site at highschool.bfwpub.com/updatedtps6e.

(c) Make a segmented bar graph to display the distribution of number of birth defects for the women with each of the three diabetic statuses.

(d) Describe the nature of the association between mother's diabetic status and number of birth defects for the women in this study.

T1.13 The back-to-back stemplot shows the lifetimes of several Brand X and Brand Y batteries.

Brand X		Brand Y
	1	
	1	7
	2	2
	2	6
2110	3	
99775	3	
3221	4	223334
	4	56889
4	5	0
5	5	

Key: 4|2 represents 420–429 hours.

(a) What is the longest that any battery lasted?

(b) Give a reason someone might prefer a Brand X battery.

(c) Give a reason someone might prefer a Brand Y battery.

T1.14 Catherine and Ana suspect that athletes (i.e., students who have been on at least one varsity team) typically have a faster reaction time than other students. To test this theory, they gave an online reflex test to separate random samples of 33 varsity athletes and 29 other students at their school. Here are parallel boxplots and numerical summaries of the data on reaction times (in milliseconds) for the two groups of students.

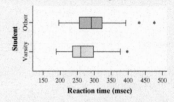

Student	n	Mean	StDev	Min	Q_1	Med	Q_3	Max
Other	29	297.3	65.9	197.0	255.0	292.0	325.0	478.0
Athlete	33	270.1	57.7	189.6	236.0	261.0	300.0	398.0

(a) Are these numerical summaries statistics or parameters? Explain your answer.

(b) Write a few sentences comparing the distribution of reaction time for the two types of students.

Chapter 1 Project American Community Survey

Each month, the U.S. Census Bureau selects a random sample of about 300,000 U.S. households to participate in the American Community Survey (ACS). The chosen households are notified by mail and invited to complete the survey online. The Census Bureau follows up on any uncompleted surveys by phone or in person. Data from the ACS are used to determine how the federal government allocates over $400 billion in funding for local communities.

The file **acs survey ch1 project.xls** can be accessed from the book's website at highschool.bfwpub.com/updatedtps6e. It contains data for 3000 randomly selected households in one month's ACS survey. Download the file to a computer for further analysis using the application specified by your teacher.

Each row in the spreadsheet describes a household. A serial number that identifies the household is in the first column. The other columns contain values of several variables. See the code sheet on the book's website for details on how each variable is recorded. Note that all the categorical variables have been coded to have numerical values in the spreadsheet.

Use the files provided to answer the following questions.

1. How many variables are recorded? Classify each one as categorical or quantitative.

2. Examine the distribution of location (division or region) for the households in the sample. Make a bar graph to display the data. Then calculate numerical summaries (counts, percents, or proportions). Describe what you see.

3. Explore the relationship between two categorical variables of interest to you. Summarize the data in a two-way table. Then calculate appropriate conditional relative frequencies and make a side-by-side or segmented bar graph. Write a few sentences comparing the distributions.

4. Analyze the distribution of household income (HINCP) using appropriate graphs and numerical summaries.

5. Compare the distribution of a quantitative variable that interests you in two or more groups. For instance, you might compare the distribution of number of people in a family (NPF) by region. Make appropriate graphs and calculate numerical summaries. Then write a few sentences comparing the distributions.

TRM **Chapter 1 Project American Community Survey Data**

The data file and supporting resources for this project can be found by clicking on the link in the TE-Book, opening the TRFD, or downloading from the Teacher's Resources on the book's digital platform. The data files are also available on the Student Site.

TRM **Chapter 1 Case Study**

The Case Study feature that was found in the previous two editions of TPS has been moved to the Teacher's Resource Materials. Click on the link in the TE-Book, open in the TRFD, or download from the Teacher's Resources on the book's digital platform.

Chapter 2

Chapter 2

Modeling Distributions of Quantitative Data

PD Overview

In Chapter 2, we continue exploring distributions of quantitative variables. Consider each of the following situations:

- A student gets a test back with a score of 78 marked clearly at the top.
- A middle-aged man goes to his doctor to have his cholesterol checked. His total cholesterol reading is 210 mg/dl.
- An employee in a large company earns an annual salary of $42,000.
- An 11th-grader scores 650 on the SAT Math test.

Isolated numbers don't always provide enough information. It is human nature to ask, "Where do I stand?" when presented with numbers like these. This chapter focuses on ways to describe an individual's position within a distribution.

We begin by learning two different ways to measure an individual's location in a distribution: percentiles and standardized scores (z-scores). This leads to a discussion of how various transformations affect the characteristics of a distribution.

Then we extend our strategy for data analysis by suggesting that when a distribution has a clear overall pattern, we can model that pattern with a smooth curve, called a *density curve*. The most commonly used density curve is the Normal curve, a symmetric, bell-shaped curve that is a good model for the distribution of many different quantitative variables. We will learn how to use Normal curves to identify percentiles in a symmetric, bell-shaped distribution and how to identify the value of a variable at a given percentile.

The Main Ideas

One of the challenges in teaching the AP® Statistics course is keeping students focused on the big picture, not just the details of each section. We outline the main ideas for the chapter here.

Chapter 2 Introduction

In Chapter 1, we learned to describe characteristics of distributions such as shape, center, variability, and striking departures such as outliers. In this chapter, we focus on identifying the location of individuals within a distribution. The activity on page 90 is a great way to get students thinking about where they stand in a distribution.

Section 2.1 Describing Location in a Distribution

In this section, we focus on describing the location of individuals in a distribution. Percentiles are perhaps the most common way to identify where a value is located in a distribution. Students are probably already familiar with this concept, because almost all standardized tests report a student's performance with a percentile. For example, if Anne scored at the 97th percentile, she scored better than or equal to 97% of people who took the test.

A cumulative relative frequency graph is handy for identifying percentiles in a distribution. You can use it to estimate the percentile for a particular value of a variable or the value of a variable at a particular percentile.

Another very common way to measure the positions of individuals in a distribution is with a standardized score, also called a z-score. A standardized score measures how many standard deviations an observation is above or below the mean of the distribution. Using z-scores is also a great way to compare observations from different distributions because they are on a standardized scale.

To find the standardized score for a particular observation, we transform the value by subtracting the mean and dividing the difference by the standard deviation. This is just one example in which we transform data from one scale to another. Another common example is converting measurements

from one unit to another, such as inches to centimeters. Adding a positive constant to (or subtracting it from) each value in a data set changes the measures of location, but not the shape or variability of the distribution. Multiplying each value in a data set by a positive constant changes the measures of location and measures of variability, but not the shape of the distribution.

Section 2.2 Density Curves and Normal Distributions

We begin this section by introducing density curves as a way to model distributions of data with a smooth curve. In many cases, using a density curve is more convenient than working with the actual data. The most common type of density curve is the Normal curve, which will be the focus of most of this section.

Some people may find it a bit unusual to present density curves and Normal distributions as part of data analysis. However, Normal curves have long been used as approximations of distributions of data that are not necessarily connected with probability theory. And giving students a brief taste of "probability" here will make it easier for them when we formally introduce random variables in Chapter 6. We think of data analysis as progressing from graphs to numerical summaries of specific aspects of the data to (when appropriate) a mathematical model for the overall pattern. That's the role Normal distributions play in data analysis for a single quantitative variable.

After introducing the basic properties of the Normal distributions and the reasons we should study them in a statistics course, we present the empirical rule, sometimes called the 68–95–99.7 *rule*. This handy rule allows students to tie together the ideas of standardized scores and Normal distributions by providing approximations for what proportion of the observations should fall within 1, 2, and 3 standard deviations of the mean.

Unfortunately, the empirical rule doesn't tell us much about what proportion of the observations fall within 1.5 standard deviations of the mean or what proportion of the observations

are more than 2.73 standard deviations above the mean. The standard Normal distribution and Table A come to the rescue, providing the proportion of observations below each standardized score from -3.49 to 3.49.

Students then use the standard Normal distribution and Table A to answer many questions that involve Normal distributions, such as the proportion of golf balls going a certain distance after being "crushed" by Jordan Spieth. We also learn how to "go backward" and find the distance that corresponds to a specific percentile of the distribution.

In this section, we also present methods for performing Normal calculations on the TI-83/84. Instructions for the TI-Nspire are available on the book's website. Whether your students use Table A or their calculators, make sure they follow the two-step process that we detail on page 122. It is very important that students always identify the distribution (including the parameters) and the values of interest, show work, and answer the question. If students choose to use technology, make sure they pay close attention to the AP® Exam Tips that provide guidelines for receiving full credit.

Finally, we spend time showing students how to investigate whether a distribution of data is approximately Normal. Knowing how to decide if a sample could have come from a Normal population will be a very important skill when doing inference in Chapters 10–12. We introduce it here because it nicely ties together our data analysis strategy:

- Graph the data.
- Look for an overall pattern and departures from this pattern.
- Calculate numerical summaries.
- If the data follow a regular overall pattern, describe it with a smooth curve.

We also introduce a special type of graph designed specifically for assessing Normality: the Normal probability plot. Although it is not on the AP® Statistics Course Framework, it is a useful tool and easy to create with technology. The more linear the Normal probability plot, the more Normal the distribution of data.

Chapter 2 Resources

Teacher's Resource Materials

The following resources, identified by the **TRM** in the annotated student pages, can be found by clicking on the link in the Teacher's e-Book (TE-Book), searching by category or chapter on the Teacher's Resource Flash Drive (TRFD), or logging into the book's digital platform and searching the Teacher's Resources menu (teacher log-in required).

- Alternate Examples: one file per section
- Lecture Presentation slides: one per section
- Chapter 2 Learning Targets Grid
- Data Exploration: The speed of light
- "Investigating Normal distributions" activity
- Extra practice with the Normal distribution
- FRAPPY! Materials
- Chapter 2 Case Study
- Complete solutions for the Check Your Understanding problems, section exercises, review exercises, and practice test.
- Quizzes: one per section
- Chapter 2 Test

Free Response Questions from Previous AP® Statistics Exams

Questions can be found on the AP® Central website: apcentral .collegeboard.org/courses/ap-statistics/exam.

Students should be able to answer all the free response questions listed below with material learned in this chapter. Questions that contain content from this chapter but also require content from later chapters are listed in the last chapter required to complete the entire question. This list will be updated after each AP® Statistics exam and will be posted to the Teacher's Resource section of the book's digital platform

and to www.statsmedic.com/free-response-questions. Questions marked with an asterisk are from exams with released multiple-choice questions. You may want to save these questions until the end of the year so you can give your students a complete released exam for practice.

Year	#	Content
2011	1	• Assessing Normality from summary statistics • Calculating and interpreting a *z*-score • Using *z*-scores to make a comparison
2009B	1	• Estimating median and *IQR* from a boxplot • Linear transformations of data
2008	1	• Comparing distributions with boxplots • Linear transformations of data • Effect of shape on the relationship between mean and median
2006B	1	• Interpreting cumulative relative frequency graphs
1997	1	• Interpreting cumulative relative frequency graphs • Finding the median and *IQR* from a cumulative relative frequency graph • Comparing center and variability

Applets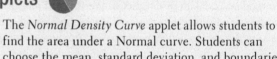

- The *Normal Density Curve* applet allows students to find the area under a Normal curve. Students can choose the mean, standard deviation, and boundaries. Go to the Student Site at highschool.bfwpub.com /updatedtps6e (click on Statistical Applets and select "Normal Density Curve").

- The *Normal Distribution* applet allows students to find areas under a Normal curve and to work backward to find boundary values. Students can enter the mean and standard deviation, choose which direction to shade, and specify the boundary value or area. Go to the Extra Applets on the Student Site at highschool.bfwpub.com /updatedtps6e (click on the *Probability* applet and select "Normal Distribution").

Chapter 2: Pacing Guide, Learning Targets, and Suggested Assignments

This pacing guide is based on a schedule with 110, 50-minute sessions before the AP® Statistics exam. If you have a different number of sessions before the AP® exam, you can modify the pacing guide to suit your needs. If you have additional time, consider incorporating quizzes, released AP® Statistics free response questions, or additional activities. See the Resources section above for suggestions.

The suggested homework assignments list odd-numbered exercises, whenever possible, so students can check their answers against the back-of-book answers. If you would rather students not have access to the answers while doing homework, adding 1 to the exercise numbers usually will do the trick, because the homework exercises typically are paired. For example, Exercises 1 and 2 will generally cover the same topics, but in different contexts. You may also choose to include the Recycle and Review questions at the end of each section, which review topics from previous sections or chapters. If your school is using the digital platform that accompanies this program, you will find these assignments pre-built as online homework assignments for Chapter 2.

Day	Content	Learning Targets: Students will be able to . . .	Suggested Assignment (MC bold)
1	2.1 Measuring Location: Percentiles, Cumulative Relative Frequency Graphs, Measuring Location: Standardized Scores	• Find and interpret the percentile of an individual value within a distribution of data. • Estimate percentiles and individual values using a cumulative relative frequency graph. • Find and interpret the standardized score (z-score) of an individual value within a distribution of data.	1, 3, 7, 9, 11, 13, 15, 19
2	2.1 Transforming Data	• Describe the effect of adding, subtracting, multiplying by, or dividing by a constant on the shape, center, and variability of a distribution of data.	21, 25, 29, 31, **33–38**
3	2.2 Density Curves, Describing Density Curves, Normal Distributions, The Empirical Rule	• Use a density curve to model distributions of quantitative data. • Identify the relative locations of the mean and median of a distribution from a density curve. • Use the empirical rule to estimate (i) the proportion of values in a specified interval, or (ii) the value that corresponds to a given percentile in a Normal distribution.	41, 45, 47, 49, 51
4	2.2 Finding Areas in a Normal Distribution, Working Backward: Finding Values from Areas	• Find the proportion of values in a specified interval in a Normal distribution using Table A or technology. • Find the value that corresponds to a given percentile in a Normal distribution using Table A or technology.	53, 55, 57, 59, 61, 63
5	2.2 Assessing Normality	• Determine whether a distribution of data is approximately Normal from graphical and numerical evidence.	73, 75, 77, 79, 81, **85–90**
6	Chapter 2 Review/FRAPPY!		Chapter 2 Review Exercises
7	Chapter 2 Test		

Chapter 2 Alignment to the College Board's Fall 2019 AP® Statistics Course Framework*

Relationship to College Board Units

Chapter 2 in this book covers Topics 1.5–1.10 in Unit 1 of the College Board Course Framework. Students will be ready to take the Personal Progress Check for Unit 1 once they have completed Chapters 1 and 2.

Big Ideas and Enduring Understandings

Chapter 2 develops these Big Ideas and related Enduring Understandings outlined in the Course Framework:

- **Big Idea 1: Variation and Distribution** (EU: VAR 2,6): The distribution of measures for individuals within a sample or population describes variation. The value of a statistic varies from sample to sample. How can we determine whether differences between measures represent random variation or meaningful distinctions? Statistical methods based on probabilistic reasoning provide the basis for shared understandings about variation and about the likelihood that variation between and among measures, samples, and populations is random or meaningful.
- **Big Idea 2: Patterns and Uncertainty** (EU: UNC 1): Statistical tools allow us to represent and describe patterns in data and to classify departures from patterns. Simulation and probabilistic reasoning allow us to anticipate patterns in data and to determine the likelihood of errors in inference.

Course Skills

Chapter 2 helps students develop the skills identified in the Course Framework.

- **2: Data Analysis** (2.B, 2.C, 2.D)
- **3: Using Probability and Simulation** (3.A, 3.C)

Learning Objectives and Essential Knowledge Statements

Section	Learning Objectives	Essential Knowledge Statements
2.1	VAR-2.B, VAR-2.C, UNC-1.G, UNC-1.I, UNC-1.J	VAR-2.B.1, VAR-2.B.2, VAR-2.C.1, UNC-1.G.4, UNC-1.I.5, UNC-1.J.4
2.2	VAR-2.A, VAR-2.B, VAR-6.B, VAR-6.C	VAR-2.A.2, VAR-2.A.3, VAR-2.A.4, VAR-2.B.3, VAR-2.B.4, VAR-6.B.1, VAR-6.B.2, VAR-6.C.1

A detailed alignment (The Nitty Gritty Guide) that can be sorted by Course Framework Unit, Topic, Learning Objective, Essential Knowledge Statement, or textbook section is available on the TRFD and in the Teacher's Resources folder on Sapling Plus. **TRM**

*Should changes be made to the Course Framework in the future, an updated alignment will be placed on our AP® updates page at go.bfwpub.com/ap-course-updates.

Chapter 2

Modeling Distributions of Quantitative Data

AtriPics.com/Alamy

Teaching Tip

Unit 1 in the College Board Course Framework aligns to Chapters 1 and 2 in this book. Students will be ready to take the Personal Progress Check for Unit 1 once they have completed Chapters 1 and 2.

PD Chapter 2 Overview

To watch the video overview of Chapter 2 (for teachers), click on the link in the TE-Book, look on the TRFD, or download from the Teacher's Resources on the book's digital platform.

TRM Lecture Presentation Slides

If you are new to teaching AP® Statistics or are short on time when preparing for class, you may find the Lecture Presentation Slides to be helpful. Experienced AP® Statistics Teacher Doug Tyson has created one slide presentation per section. You may use them as is, modify them to fit your needs, or share them with students who miss class. Find them on the TRFD and in the Teacher's Resources on the book's digital platform.

Preparing for Inference

Students will use Normal distributions in Chapters 8 and 9 to construct confidence intervals and perform significance tests. Building a good basic understanding of Normal distribution calculations in this chapter will set them up for later success.

 ACTIVITY OVERVIEW

To prepare for using this activity, watch the overview video by clicking on the link in the TE-Book, opening the TRFD, or downloading from the Teacher's Resources on the book's digital platform.

Time: 15 minutes

Materials: Masking tape (to make a number line on the floor)

Teaching Advice: Use this activity as a way to review big ideas from Chapter 1 (describing a distribution using SOCV plus context) and also to preview some big ideas in Chapter 2 (percentile, z-score, transforming data). Use the calculator or one of the Extra Applets at highschool .bfwpub.com/updatedtps6e to find the mean and standard deviation of the heights.

Answers:

1–6. Answers will vary for each student.

7. The shape of the distribution of the class's heights would stay the same. The center, variability, and measures of location would all be multiplied by 2.54.

INTRODUCTION

Suppose Emily earns 43 out of 50 points on a statistics test. Should she be satisfied or disappointed with her performance? That depends on how her score compares with the scores of the other students who took the test. If 43 is the highest score, Emily might be very pleased. Maybe her teacher will "scale" the grades so that Emily's 43 becomes an "A." But if Emily's 43 falls below the class average, she may not be so happy.

Section 2.1 focuses on describing the location of an individual within a distribution of quantitative data. We begin by discussing a familiar measure of location: *percentiles*. Next, we introduce a new type of graph that is useful for displaying percentiles. Then we consider another way to describe an individual's location that is based on the mean and standard deviation. In the process, we examine the effects of transforming data on the shape, center, and variability of a distribution.

Sometimes it is helpful to use graphical models called *density curves* to estimate an individual's location in a distribution, rather than relying on actual data values. Such models are especially helpful when data fall in a bell-shaped pattern called a *Normal distribution*. Section 2.2 examines the properties of Normal distributions and shows you how to perform useful calculations with them.

ACTIVITY **Where do I stand?**

In this activity, you and your classmates will explore ways to describe where you stand (literally!) within a distribution.

1. Your teacher will mark out a number line on the floor with a scale running from about 58 to 78 inches.
2. Make a human dotplot. Each member of the class should stand at the appropriate location along the number line scale based on height (to the nearest inch).
3. Your teacher will make a copy of the dotplot on the board for your reference. Describe the class's distribution of heights.
4. What percent of the students in the class have heights less than or equal to yours? This *percentile* is one way to measure your location in the distribution of heights.
5. Calculate the mean and standard deviation of the class's distribution of height from the dotplot. Confirm these values with your classmates.
6. Does your height fall above or below the mean? How far above or below the mean is it? How many standard deviations above or below the mean is it? This *standardized score* (also called a *z-score*) is another way to measure your location in the class's distribution of heights.
7. *Class discussion:* What would happen to the class's distribution of height if you converted each data value from inches to centimeters? (There are 2.54 centimeters in 1 inch.) How would this change of units affect the shape, center, variability, and the measures of location (percentile and *z*-score) that you calculated?

Want to know more about where you stand—in terms of height, weight, or even body mass index? Do a web search for "Clinical Growth Charts" at the National Center for Health Statistics site, www.cdc.gov/nchs.

SECTION 2.1 Describing Location in a Distribution

LEARNING TARGETS *By the end of the section, you should be able to:*

- Find and interpret the percentile of an individual value in a distribution of data.
- Estimate percentiles and individual values using a cumulative relative frequency graph.
- Find and interpret the standardized score (*z*-score) of an individual value in a distribution of data.
- Describe the effect of adding, subtracting, multiplying by, or dividing by a constant on the shape, center, and variability of a distribution of data.

There are 25 students in Mr. Tabor's statistics class. He gives them a first test worth 50 points. Here are the students' scores:

35 18 37 38 42 41 25 37 36 32 12 43 31
29 32 48 44 45 38 40 45 38 38 40 22

The score marked in red is Emily's 43. How did she perform on this test relative to her classmates?

Figure 2.1 displays a dotplot of the class's test scores, with Emily's score marked in red. The distribution is skewed to the left with some possible low outliers. From the dotplot, we can see that Emily's score is above the mean (balance point) of the distribution. We can also see that Emily did better on the test than most other students in the class.

FIGURE 2.1 Dotplot of scores (out of 50 points) on Mr. Tabor's first statistics test. Emily's score of 43 is marked in red.

Measuring Location: Percentiles

One way to describe Emily's location in the distribution of test scores is to calculate her **percentile**.

> Some people define the *p*th percentile of a distribution as the value with *p*% of observations *less than it*. This distinction matters for discrete variables (like score on Mr. Tabor's first test), but not for continuous variables.

> **DEFINITION Percentile**
>
> The *p*th **percentile** of a distribution is the value with *p*% of observations less than or equal to it.

 Using the dotplot, we see that four students in the class earned test scores greater than Emily's 43. Because 21 of the 25 observations (84%) are less than or equal to her score, Emily is at the 84th percentile in the class's test score distribution.

Be careful with your language when describing percentiles. Percentiles are specific locations in a distribution, so an observation isn't "in" the 84th percentile. Rather, it is "at" the 84th percentile.

PD Section 2.1 Overview

To watch the video overview of Section 2.1, click on the link in the TE-Book, look on the TRFD, or download from the Teacher's Resources on the book's digital platform.

TRM Learning Targets Grid

At the beginning of each section, we present the relevant learning targets. Point these out to your students and refer back to the targets when you cover them in class. There is a PDF version of the grid with an additional column that students can use to keep track of their progress. Find it in the Teacher's Resource Materials located in the TE-Book, on the TRFD, or in the Teacher's view in the book's digital platform.

AP® EXAM TIP

Depending on the reference source, percentile can be defined using "less than" or "less than or equal to." The new College Board Course Framework has chosen "less than or equal to" for AP® Statistics.

Teaching Tip

Remind students that the "percent correct" on a test does not measure the same thing as a "percentile."

TRM Section 2.1 Alternate Examples

You can find the Alternate Examples for this section in Microsoft Word format by clicking on the link in the TE-Book, opening the TRFD, or downloading from the Teacher's Resources on the book's digital platform.

Teaching Tip

The icon at the top of this example (and many others) indicates that students can watch a video presentation of the example by clicking on the link in the e-Book or viewing the videos at the Student Site highschool.bfwpub.com/updatedtps6e.

ALTERNATE EXAMPLE   Skill 2.C

Wins in Major League Baseball
Finding and interpreting percentiles

PROBLEM:
The stemplot below shows the number of wins for each of the 30 Major League Baseball teams in 2016.

```
 5 | 9
 6 |
 6 | 8  8  8  8  9  9
 7 | 1  3  4
 7 | 5  8  8  9
 8 | 1  4  4
 8 | 6  6  6  7  7  9  9
 9 | 1  3  4
 9 | 5  5
10 | 3
```
Key: 9|1 = 91 wins

(a) Find the percentile for the Detroit Tigers, who had 86 wins.
(b) The number of wins for the Cleveland Indians is at the 90th percentile of the distribution. Interpret this value in context. How many wins did the Cleveland Indians have in 2016?

SOLUTION:
(a) $20/30 = 0.67$, so the Detroit Tigers were at the 67th percentile for this season. *Note:* There are three teams with 86 wins, and all of them are at the 67th percentile.
(b) 90% of the teams have a number of wins less than or equal to the Cleveland Indians. Because $(0.90)(30) = 27$, the number of wins for the Cleveland Indians was greater than or equal to that for 27 of the 30 teams. The Cleveland Indians had 94 wins in the 2016 season.

Teaching Tip

Other examples where a high percentile is not desired include: time to run a race, golf score, and student debt.

EXAMPLE

Mr. Tabor's first test
Finding and interpreting percentiles

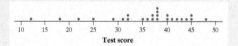

PROBLEM: Refer to the dotplot of 25 scores on Mr. Tabor's first statistics test.

Test score

(a) Find the percentile for Jacob, who scored 18 on the test.
(b) Maria's test score is at the 48th percentile of the distribution. Interpret this value in context. What score did Maria earn?

SOLUTION:

(a) $2/25 = 0.08$, so Jacob scored at the 8th percentile on this test.

> Only 2 of the 25 scores in the class are less than or equal to Jacob's 18.

(b) 48% of students in the class earned a test score less than or equal to Maria's. Because $(0.48)(25) = 12$, Maria scored greater than or equal to 12 of the 25 students. Maria earned a 37 on the test.

> Two students in the class scored a 37 on the test. Both these students' scores are at the 48th percentile because 12 of the 25 students in the class earned equal or lower scores.

FOR PRACTICE, TRY EXERCISE 1

> The three quartiles Q_1, Q_2 (median), and Q_3 divide a distribution into four groups of roughly equal size. Similarly, the 99 percentiles should divide a distribution into 100 groups of roughly equal size. This concept of percentile makes sense for only very large sets of quantitative data.

Using the alternate definition of percentile (see margin note on page 91), it is possible for an individual to fall at the 0th percentile. If we used this definition, Jacob's score of 18 would fall at the 4th percentile (1 of 25 scores were less than 18). Calculating percentiles is not an exact science, especially with small data sets!

The median of a distribution is roughly the 50th percentile because about half the observations are less than or equal to the median. The first quartile Q_1 is roughly the 25th percentile of a distribution because it separates the lowest 25% of values from the upper 75%. Likewise, the third quartile Q_3 is roughly the 75th percentile.

A high percentile is not always a good thing. For example, a man whose cholesterol level is at the 90th percentile for his age group may need treatment for his high cholesterol!

Cumulative Relative Frequency Graphs

There are some interesting graphs that can be made with percentiles. One of the most common graphs starts with a frequency table for a quantitative variable. For instance, the frequency table on the next page summarizes the ages of the first 45 U.S. presidents when they took office.

Teaching Tip

If two observations have the same value, they will be at the same percentile. To find the percentile, calculate the percent of the values in the distribution that are less than or equal to the given value (including both instances of the given value).

AP® EXAM TIP

On the AP® Statistics exam, students may use any reasonable definition of "percentile" on the free response section. On the multiple-choice section, the answer choices will be designed so that any reasonable method of finding percentiles will yield the correct answer. When working with large data sets, the different definitions will give very similar values for the percentiles of a distribution.

Age	Frequency
40–45	2
45–50	7
50–55	13
55–60	12
60–65	7
65–70	3
70–75	1

Note that age at inauguration is a *continuous* variable. For instance, Donald Trump was 70 years, 7 months, 7 days (about 70.603 years) old when he took office. The intervals in the frequency table are given as 40–45, 45–50, and so on because each president's inauguration age falls in exactly one of these intervals on the number line.

Let's expand this table to include columns for relative frequency, *cumulative frequency*, and *cumulative relative frequency*.

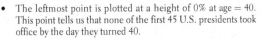

Age	Frequency	Relative frequency	Cumulative frequency	Cumulative relative frequency (percentile)
40–45	2	$2/45 = 0.044 = 4.4\%$	2	$2/45 = 0.044 = 4.4\%$
45–50	7	$7/45 = 0.156 = 15.6\%$	9	$9/45 = 0.200 = 20.0\%$
50–55	13	$13/45 = 0.289 = 28.9\%$	22	$22/45 = 0.489 = 48.9\%$
55–60	12	$12/45 = 0.267 = 26.7\%$	34	$34/45 = 0.756 = 75.6\%$
60–65	7	$7/45 = 0.156 = 15.6\%$	41	$41/45 = 0.911 = 91.1\%$
65–70	3	$3/45 = 0.067 = 6.7\%$	44	$44/45 = 0.978 = 97.8\%$
70–75	1	$1/45 = 0.022 = 2.2\%$	45	$45/45 = 1.000 = 100\%$

The table reveals that 20.0% of U.S. presidents took office by the time they turned 50. In other words, the 20th percentile of the distribution of inauguration age is 50.000 years. We can display the percentiles from the table in a **cumulative relative frequency graph.**

Some people refer to cumulative relative frequency graphs as *ogives* (pronounced "o-jives") or as *percentile plots.*

> **DEFINITION   Cumulative relative frequency graph**
>
> A **cumulative relative frequency graph** plots a point corresponding to the percentile of a given value in a distribution of quantitative data. Consecutive points are then connected with a line segment to form the graph.

Figure 2.2 shows the completed cumulative relative frequency graph for the presidential age at inauguration data. Notice the following details:

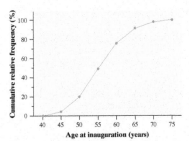

FIGURE 2.2 Cumulative relative frequency graph of the ages of U.S. presidents when they took office.

- The leftmost point is plotted at a height of 0% at age = 40. This point tells us that none of the first 45 U.S. presidents took office by the day they turned 40.
- The next point to the right is plotted at a height of 4.4% at age = 45. This point tells us that 4.4% of these presidents were inaugurated by their 45th birthday.
- The graph grows very gradually at first because few presidents were inaugurated when they were in their 40s. Then the graph gets very steep beginning at age 50 because most U.S. presidents were in their 50s when they took office. The rapid growth in the graph slows at age 60.
- The rightmost point on the graph is plotted above age 75 and has cumulative relative frequency 100%. That's because 100% of these U.S. presidents took office by age 75.

A cumulative relative frequency graph can be used to describe the position of an individual value in a distribution or to locate a specified percentile of the distribution.

State median household incomes
Interpreting a cumulative relative frequency graph

PROBLEM:
The cumulative relative frequency graph shows the distribution of median household incomes for the 50 states and the District of Columbia in a recent year.

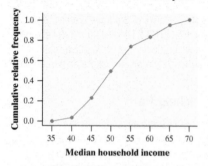

Use the cumulative relative frequency graph for the state income data to answer each question.
(a) At what percentile is California, with a median household income of $57,445?
(b) Estimate and interpret the first quartile of this distribution.

SOLUTION:
(a) California is at about the 78th percentile for household income.

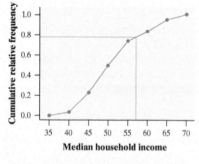

(b) The first quartile of this distribution is the 25th percentile. About 25% of states have median incomes less than or equal to $45,000.

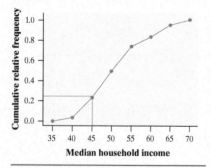

EXAMPLE

Ages of U.S. presidents
Interpreting a cumulative relative frequency graph

PROBLEM: Use the graph in Figure 2.2 to help you answer each question.
(a) Was Barack Obama, who was first inaugurated at age 47 years, 169 days (about 47.463 years), unusually young?
(b) Estimate and interpret the 65th percentile of the distribution.

SOLUTION:
(a)

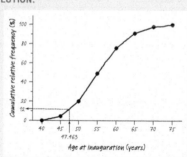

To find President Obama's location in the distribution, draw a vertical line up from his age (47.463) on the horizontal axis until it meets the graph. Then draw a horizontal line from this point to the vertical axis.

Barack Obama's inauguration age places him at about the 12th percentile. About 12% of the first 45 U.S. presidents took office by the time they were 47.463 years old. So Obama was fairly young, but not unusually young, when he took office.

(b)

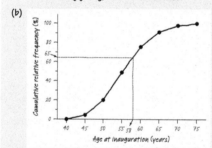

The 65th percentile of the distribution is the age with cumulative relative frequency (percentile) 65%. To find this value, draw a horizontal line across from the vertical axis at a height of 65% until it meets the graph. Then draw a vertical line from this point down to the horizontal axis.

The 65th percentile is about 58 years old. About 65% of the first 45 U.S. presidents took office by the time they turned 58.

FOR PRACTICE, TRY EXERCISE 9

Teaching Tip

To check for understanding, ask your students to make a boxplot of the data using the example, "Ages of U.S. presidents." They should be able to estimate the first quartile, median, and third quartile by finding the values of the 25th, 50th, and 75th percentiles. The min and max values cannot be determined exactly from the cumulative frequency plot, but the minimum can't be less than 40 and the maximum can't be more than 75.

CHECK YOUR UNDERSTANDING

1. *Multiple choice: Select the best answer.* Mark receives a score report detailing his performance on a statewide test. On the math section, Mark earned a raw score of 39, which placed him at the 68th percentile. This means that
   (a) Mark did the same as or better than about 39% of those who took the test.
   (b) Mark did worse than about 39% of those who took the test.
   (c) Mark did the same as or better than about 68% of those who took the test.
   (d) Mark did worse than about 68% of those who took the test.
   (e) Mark got more than half of the questions correct on this test.

2. Mrs. Munson is concerned about how her daughter's height and weight compare with those of other girls of the same age. She uses an online calculator to determine that her daughter is at the 87th percentile for weight and the 67th percentile for height. Explain to Mrs. Munson what these values mean.

Questions 3 and 4 relate to the following setting. The graph displays the cumulative relative frequency of the lengths of phone calls made from the mathematics department office at Gabalot High last month.

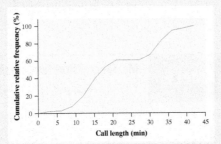

3. About what percent of calls lasted less than or equal to 30 minutes? More than 30 minutes?

4. Estimate Q_1, Q_3, and the IQR of the distribution of phone call length.

Measuring Location: Standardized Scores

A percentile is one way to describe the location of an individual in a distribution of quantitative data. Another way is to give the **standardized score (z-score)** for the observed value.

> **DEFINITION Standardized score (z-score)**
>
> The **standardized score (z-score)** for an individual value in a distribution tells us how many standard deviations from the mean the value falls, and in what direction. To find the standardized score (z-score), compute
>
> $$z = \frac{\text{value} - \text{mean}}{\text{standard deviation}}$$

Preparing for Inference

When we get to significance tests in Chapter 9, the z-score will become the standardized test statistic. We will use the z-score to calculate a P-value and make a decision about the null hypothesis. It is important for students to start thinking about whether a value is far away from what is expected (mean).

Teaching Tip

Emphasize that a z-score is directional. If the z-score is positive, it means the observation is above the mean. If the z-score is negative, it means the observation is below the mean.

✓ **Answers to CYU**

1. c

2. For girls of the same age, 87% weigh ≤ Mrs. Munson's daughter and 67% of girls are ≤ her daughter's height.

3. About 65% of calls lasted less than or equal to 30 minutes. This means that about 35% of calls lasted more than 30 minutes.

4. $Q_1 = 13$ minutes; $Q_3 = 32$ minutes; $IQR = 32 - 13 = 19$ minutes

Making Connections

In Section 2.2, students will use a z-score and Table A to find a percentile in a Normal distribution.

COMMON STUDENT ERROR

Many students will incorrectly assume that converting a value into a z-score means the data follow a Normal distribution. This is not the case, as shown with Mr. Tabor's test scores. Any value from any distribution can be converted into a z-score. Only when the value comes from a Normal distribution can we then use that z-score with Table A to find an area.

Teaching Tip

Make sure to discuss why knowing that Emily is 7.56 points above average doesn't tell you much about her location in the distribution. Depending on the variability of the distribution, Emily might be just barely above average or really far above average. This is why we have to incorporate a measure of variability (standard deviation) to have a good understanding of how far she is above the mean.

Teaching Tip

Point out that a z-score is not measured in the same units as the variable. It would be incorrect to say Emily was 0.86 points above average. Instead, we say she is 0.86 standard deviations above average.

➕ Ask the StatsMedic

Interpret the z-Score (Like It's Your Job)

In this post, we discuss the importance of students being able to interpret the z-score and how you can assign one student in your class to be responsible for this task.

Values larger than the mean have positive z-scores. Values smaller than the mean have negative z-scores.

Let's return to the data from Mr. Tabor's first statistics test. Figure 2.3 shows a dotplot of the data, along with numerical summaries.

> The relationship between the mean and the median is about what you'd expect in this left-skewed distribution.

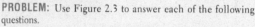

n	$\bar{x}$	s_x	Min	Q_1	Med	Q_3	Max
25	35.44	8.77	12	31.5	38	41.5	48

FIGURE 2.3 Dotplot and summary statistics of the scores on Mr. Tabor's first statistics test. Emily's score of 43 is marked in red on the dotplot.

Where does Emily's 43 (marked in red on the dotplot) fall in the distribution? Her standardized score (z-score) is

$$z = \frac{\text{value} - \text{mean}}{\text{standard deviation}} = \frac{43 - 35.44}{8.77} = 0.86$$

That is, Emily's test score is 0.86 standard deviations above the mean score of the class.

EXAMPLE

Mr. Tabor's first test, again
Finding and interpreting z-scores

PROBLEM: Use Figure 2.3 to answer each of the following questions.
(a) Find and interpret the z-score for Jacob, who earned an 18 on the test.
(b) Tamika had a standardized score of 0.292. Find Tamika's test score.

SOLUTION:

(a) $z = \dfrac{18 - 35.44}{8.77} = -1.99$

Jacob's test score is 1.99 standard deviations below the class mean of 35.44.

(b) $0.292 = \dfrac{\text{value} - 35.44}{8.77}$

$0.292(8.77) + 35.44 = \text{value}$

$38 = \text{value}$

Tamika's test score was 38.

> Be sure to interpret the magnitude (number of standard deviations) and direction (greater than or less than the mean) of a z-score in context.

> Be sure to show your work when finding the value that corresponds to a given z-score.

FOR PRACTICE, TRY EXERCISE 13

ALTERNATE EXAMPLE   Skill 3.A

Who rules the Windy City?   Finding and interpreting z-scores

PROBLEM:

In 2016, the mean number of wins for teams in Major League Baseball was 81 wins with a standard deviation of 10.7 wins.
(a) The Chicago Cubs broke the Curse of the Billy Goat by winning the World Series in 2016. Find and interpret the z-score for the Chicago Cubs, who had 103 wins in 2016.
(b) The Chicago White Sox had a z-score of −0.28. Find the number of wins for the Chicago White Sox for 2016.

SOLUTION:

(a) $z = \dfrac{103 - 81}{10.7} = 2.06$. The number of wins by the Chicago Cubs in 2016 is 2.06 standard deviations above the Major League Baseball mean of 81 wins.

(b) $-0.28 = \dfrac{\text{value} - 81}{10.7}$

$-0.28(10.7) + 81 = \text{value}$
$78 = \text{value}$
The Chicago White Sox had 78 wins in the 2016 season.

We often standardize observed values to express them on a common scale. For example, we might compare the heights of two children of different ages or genders by calculating their *z*-scores.

- At age 9, Jordan is 55 inches tall. Her height puts her at a *z*-score of 1. That is, Jordan is 1 standard deviation above the mean height of 9-year-old girls.
- Zayne's height at age 11 is 58 inches. His corresponding *z*-score is 0.5. In other words, Zayne is 1/2 standard deviation above the mean height of 11-year-old boys.

Even though Zayne's height is larger, Jordan is taller relative to girls her age than Zayne is relative to boys his age. The standardized heights tell us where each child stands (pun intended!) in the distribution for his or her age group.

CHECK YOUR UNDERSTANDING

1. Mrs. Navard's statistics class has just completed the first three steps of the "Where do I stand?" activity (page 90). The figure shows a dotplot of the distribution of height for the class, along with summary statistics from computer output.

Height (in.)

Variable	n	$\bar{x}$	s_x	Min	Q_1	Med	Q_3	Max
Height	25	67	4.29	60	63	66	70	75

   Lynette, a student in the class, is 62 inches tall. Find and interpret her *z*-score.

2. Brent is a member of the school's basketball team and is 74 inches tall. The mean height of the players on the team is 76 inches. Brent's height translates to a standardized score of −0.85 in the team's distribution of height. What is the standard deviation of the team members' heights?

Transforming Data

To find the standardized score (*z*-score) for an individual observation, we transform this data value by subtracting the mean and dividing the difference by the standard deviation. Transforming converts the observation from the original units of measurement (e.g., inches) to a standardized scale.

There are other reasons to transform data. We may want to change the units of measurement for a data set from kilograms to pounds (1 kg ≈ 2.2 lb), or from Fahrenheit to Celsius $\left[°C = \dfrac{5}{9}(°F - 32)\right]$. Or perhaps a measuring device is calibrated wrong, so we have to add a constant to each data value to get accurate measurements. What effect do these kinds of transformations—adding

Teaching Tip

The example of heights of Jordan and Zayne could be used as a context to motivate the reason why we might want to standardize values using *z*-scores.

✓ Answers to CYU

1. $z = \dfrac{62 - 67}{4.29} = -1.166$. *Interpretation:* Lynette's height is 1.166 standard deviations below the mean height of the class.

2. Because Brent's *z*-score is −0.85, we know that $-0.85 = \dfrac{74 - 76}{\sigma}$. Solving for σ, we find that $\sigma = 2.35$ inches.

or subtracting; multiplying or dividing—have on the shape, center, and variability of a distribution?

EFFECT OF ADDING OR SUBTRACTING A CONSTANT Recall that Mr. Tabor gave his class of 25 statistics students a first test worth 50 points. Here is a dotplot of the students' scores along with some numerical summaries.

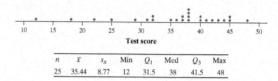

n	$\bar{x}$	s_x	Min	Q_1	Med	Q_3	Max
25	35.44	8.77	12	31.5	38	41.5	48

Suppose Mr. Tabor was nice and added 5 points to each student's test score. How would this affect the distribution of scores? Figure 2.4 shows graphs and numerical summaries for the original test scores and adjusted scores.

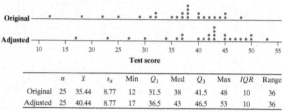

FIGURE 2.4 Dotplots and summary statistics for the original scores and adjusted scores (with 5 points added) on Mr. Tabor's statistics test.

	n	$\bar{x}$	s_x	Min	Q_1	Med	Q_3	Max	IQR	Range
Original	25	35.44	8.77	12	31.5	38	41.5	48	10	36
Adjusted	25	40.44	8.77	17	36.5	43	46.5	53	10	36

From both the graph and summary statistics, we can see that the measures of center (mean and median) and location (min, Q_1, Q_3, and max) increased by 5 points. The shape of the distribution did not change. Neither did the variability of the distribution—the range, the standard deviation, and the interquartile range (IQR) all stayed the same.

As this example shows, adding the same positive number to each value in a data set shifts the distribution to the right by that number. Subtracting a positive constant from each data value would shift the distribution to the left by that constant.

THE EFFECT OF ADDING OR SUBTRACTING A CONSTANT

Adding the same positive number a to (subtracting a from) each observation:

- Adds a to (subtracts a from) measures of center and location (mean, five-number summary, percentiles)
- Does not change measures of variability (range, IQR, standard deviation)
- Does not change the shape of the distribution

EXAMPLE

How wide is this room?
Effect of adding/subtracting a constant

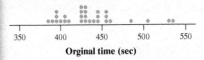

PROBLEM: Soon after the metric system was introduced in Australia, a group of students was asked to guess the width of their classroom to the nearest meter. A dotplot of the data and some numerical summaries are shown.[1]

The actual width of the room was 13 meters. We can examine the distribution of students' errors by defining a new variable as follows: error = guess − 13. Note that a negative value for error indicates that a student's guess for the width of the room was too small.

(a) What shape does the distribution of error have?

(b) Find the mean and the median of the distribution of error.

(c) Find the standard deviation and interquartile range (IQR) of the distribution of error.

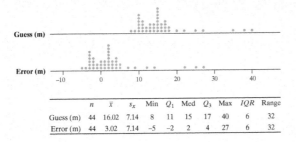

Guess (m)

	n	$\bar{x}$	s_x	Min	Q_1	Med	Q_3	Max	IQR	Range
Guess (m)	44	16.02	7.14	8	11	15	17	40	6	32

SOLUTION:

(a) The same shape as the original distribution of guesses: skewed to the right with two distinct peaks.

(b) Mean: $16.02 - 13 = 3.02$ meters; Median: $15 - 13 = 2$ meters

(c) The same as for the original distribution of guesses:
Standard deviation = 7.14 meters, IQR = 6 meters

> It is not a surprise that the mean is greater than the median in this right-skewed distribution.

FOR PRACTICE, TRY EXERCISE 21

Figure 2.5 confirms the results of the example.

Guess (m)

Error (m)

FIGURE 2.5 Dotplots and summary statistics for the Australian students' guesses of classroom width and the errors in their guesses, in meters.

	n	$\bar{x}$	s_x	Min	Q_1	Med	Q_3	Max	IQR	Range
Guess (m)	44	16.02	7.14	8	11	15	17	40	6	32
Error (m)	44	3.02	7.14	−5	−2	2	4	27	6	32

What about outliers? You can check that the four highest guesses—which are 27, 35, 38, 40 meters—are outliers by the $1.5 \times IQR$ rule. The same individuals will still be outliers in the distribution of error, but each of their values will be decreased by 13 meters: 14, 22, 25, and 27 meters.

ALTERNATE EXAMPLE   Skill 2.C

How fast is the track team?
Effect of adding/subtracting a constant

PROBLEM:

All 30 girls on a high school track team ran a 1-mile race, and their times were recorded. A dotplot and numerical summaries are shown.

Orginal time (sec)

n	mean	SD	min	Q_1	med	Q_3	max
30	439.197	37.876	387.8	409.9	432.85	455.8	535.7

It was later discovered that the person in charge of the stopwatch did not start the time until 10 seconds into the race. This means that the times must be corrected by taking the original time and adding 10 seconds to each one. Put another way:

corrected time = original time + 10

(a) What shape would the distribution of corrected time have?

(b) Find the mean and median of the distribution of corrected time.

(c) Find the standard deviation and interquartile range of the distribution of corrected time.

SOLUTION:

(a) The same shape as the distribution of original time: skewed to the right with one peak.

(b) Mean: $439.197 + 10 = 449.197$ seconds; median: $432.85 + 10 = 442.85$ seconds

(c) The same as for the distribution of original time: Standard deviation = 37.876 seconds, IQR = $455.8 - 409.9 = 45.9$ seconds.

EFFECT OF MULTIPLYING OR DIVIDING BY A CONSTANT Suppose that Mr. Tabor wants to convert his students' adjusted test scores to percents. Because the test was out of 50 points, he should multiply each score by 2 to be counted out of 100 points instead. Here are graphs and numerical summaries for the adjusted scores and the doubled scores:

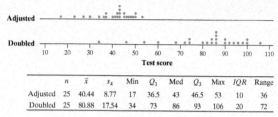

	n	$\bar{x}$	s_x	Min	Q_1	Med	Q_3	Max	IQR	Range
Adjusted	25	40.44	8.77	17	36.5	43	46.5	53	10	36
Doubled	25	80.88	17.54	34	73	86	93	106	20	72

From the graphs and summary statistics, we can see that the measures of center, location, and variability have all doubled, just like the individual observations. But the shape of the two distributions is the same.

> It is not common to multiply (or divide) each observation in a data set by a *negative* number b. Doing so would multiply (or divide) the measures of variability by the *absolute value* of b. We can't have a negative amount of variability! Multiplying or dividing by a negative number would also affect the shape of the distribution, as all values would be reflected over the y axis.

EFFECT OF MULTIPLYING OR DIVIDING BY A CONSTANT

Multiplying (or dividing) each observation by the same positive number b:

- Multiplies (divides) measures of center and location (mean, five-number summary, percentiles) by b
- Multiplies (divides) measures of variability (range, IQR, standard deviation) by b
- Does not change the shape of the distribution

EXAMPLE

How far off were our guesses?

Effect of multiplying/dividing by a constant

PROBLEM: Refer to the preceding example. The graph and numerical summaries describe the distribution of the Australian students' guessing errors (in meters) when they tried to estimate the width of their classroom.

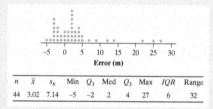

n	$\bar{x}$	s_x	Min	Q_1	Med	Q_3	Max	IQR	Range
44	3.02	7.14	−5	−2	2	4	27	6	32

Because the students are having some difficulty with the metric system, it may not be helpful to tell them that their guesses tended to be about 2 meters too high. Let's convert the error data to feet before we report back to them.

Teaching Tip

What happens to the variance? Because the variance is the square of the standard deviation, the variance is multiplied by b^2.

Teaching Tip

You might want to show your students this short "proof" for how the IQR changes when multiplying each observation by the same number b:

Original $IQR = Q_3 - Q_1$
New $IQR = Q_3 \cdot b - Q_1 \cdot b$
$= b(Q_3 - Q_1)$
$= b(\text{Original } IQR)$

ALTERNATE EXAMPLE   Skill 2.C

Now how fast did the track team run?

Effect of multiplying/dividing by a constant

PROBLEM:

All 30 girls on a high school track team ran a 1-mile race, and their times were recorded. After a slight timing error was fixed, we have the corrected times. A dotplot and numerical summaries are shown.

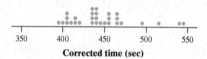

Corrected time (sec)

n	mean	SD	min	Q_1	med	Q_3	max
30	449.197	37.876	397.8	419.9	442.85	465.8	545.7

The athletes wondered what these times would look like if they were given in minutes instead of seconds. They took each one of the times (in seconds) and divided by 60 to get a new distribution of times (in minutes).
(a) What shape would the resulting distribution have?
(b) Find the median of the new distribution in minutes.
(c) Find the interquartile range of the new distribution in minutes.

SOLUTION:
(a) The same shape as the original distribution of corrected time: skewed to the right and unimodal.
(b) Median $= 442.85/60 = 7.38$ minutes
(c) Original distribution:
$IQR = 465.8 - 419.9 = 45.9$ seconds; new distribution: $IQR = 45.9/60 = 0.765$ minutes

There are roughly 3.28 feet in a meter. For the student whose error was −5 meters, that translates to

$$-5 \text{ meters} \times \frac{3.28 \text{ feet}}{1 \text{ meter}} = -16.4 \text{ feet}$$

To change the units of measurement from meters to feet, we multiply each of the error values by 3.28.

(a) What shape does the resulting distribution of error have?

(b) Find the median of the distribution of error in feet.

(c) Find the interquartile range (IQR) of the distribution of error in feet.

SOLUTION:

(a) The same shape as the original distribution of guesses: skewed to the right with two distinct peaks.

(b) Median = 2 × 3.28 = 6.56 feet

(c) IQR = 6 × 3.28 = 19.68 feet

FOR PRACTICE, TRY EXERCISE 25

Figure 2.6 confirms the results of the example.

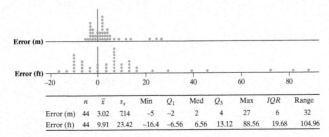

FIGURE 2.6 Dotplots and summary statistics for the errors in Australian students' guesses of classroom width, in meters and in feet.

	n	$\bar{x}$	s_x	Min	Q_1	Med	Q_3	Max	IQR	Range
Error (m)	44	3.02	7.14	−5	−2	2	4	27	6	32
Error (ft)	44	9.91	23.42	−16.4	−6.56	6.56	13.12	88.56	19.68	104.96

PUTTING IT ALL TOGETHER: ADDING/SUBTRACTING AND MULTIPLYING/DIVIDING What happens if we transform a data set by both adding or subtracting a constant and multiplying or dividing by a constant? We just use the facts about transforming data that we've already established and the order of operations.

EXAMPLE

Too cool at the cabin?
Analyzing the effects of transformations

PROBLEM: During the winter months, the temperatures at the Starnes's Colorado cabin can stay well below freezing (32°F, or 0°C) for weeks at a time. To prevent the pipes from freezing, Mrs. Starnes sets the thermostat at 50°F. She also buys a digital thermometer that records the indoor temperature each night at midnight. Unfortunately, the thermometer is programmed to measure the temperature in degrees Celsius. Following are a dotplot and numerical summaries of the midnight temperature readings for a 30-day period.

PROBLEM:
Taking an Uber ride in New York City has an initial fee of $2.55 with an additional charge of $1.75 per mile (we will ignore the small per minute waiting fee). In equation form, cost = 2.55 + 1.75(miles). A local New York City resident records the number of miles for his first 25 rides with Uber. The mean distance of his rides is 5.6 miles with a standard deviation of 1.2 miles.
(a) Find the mean cost of the 25 trips.
(b) Calculate the standard deviation of the cost of the 25 trips. Interpret this value in context.

SOLUTION:
(a) Mean = (1.75)(5.6) + 2.55 = $12.35
(b) SD = (1.75)(1.2) = $2.10; the costs of an Uber ride for this resident typically vary from the mean by about $2.10.

Making Connections

Not all transformations maintain the shape of a distribution of data. In Chapter 3, we will use logarithms and other mathematical operations to transform data, which will also change the shape of a distribution. And in Chapter 10, we will find that multiplying (or dividing) the values in a distribution by a *variable* changes the shape (*t* distributions).

Making Connections

In Section 2.2, standardizing a Normal distribution will allow us to use Table A to find areas for any Normal distribution.

Use the fact that $°F = (9/5)°C + 32$ to help you answer the following questions.
(a) Find the mean temperature in degrees Fahrenheit. Does the thermostat setting seem accurate?
(b) Calculate the standard deviation of the temperature readings in degrees Fahrenheit. Interpret this value in context.

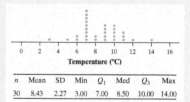

n	Mean	SD	Min	Q_1	Med	Q_3	Max
30	8.43	2.27	3.00	7.00	8.50	10.00	14.00

SOLUTION:
(a) Mean = (9/5)(8.43) + 32 = 47.17°F. The thermostat doesn't seem to be very accurate. It is set at 50°F, but the mean temperature over the 30-day period is about 47°F.
(b) SD = (9/5)(2.27) = 4.09°F. The temperature readings typically vary from the mean by about 4°F. That's a lot of variation!

> Multiplying each observation by 9/5 multiplies the standard deviation by 9/5. However, adding 32 to each observation doesn't affect the variability.

FOR PRACTICE, TRY EXERCISE 29

Many other types of transformations can be very useful in analyzing data. We have only studied what happens when you transform data by adding, subtracting, multiplying, or dividing by a constant.

CONNECTING TRANSFORMATIONS AND *z*-SCORES What happens if we standardize *all* the values in a distribution of quantitative data? Here is a dotplot of the original test scores for the 25 students in Mr. Tabor's statistics class, along with some numerical summaries:

n	$\bar{x}$	s_x
25	35.44	8.77

We calculate the standardized score for each student using

$$z = \frac{\text{score} - 35.44}{8.77}$$

In other words, we subtract 35.44 from each student's test score and then divide by 8.77. What effect do these transformations have on the shape, center, and variability of the distribution?

Here is a dotplot of the class's *z*-scores. Let's describe the distribution.

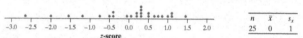

n	$\bar{x}$	s_x
25	0	1

- **Shape:** *The shape of the distribution of z-scores is the same as the shape of the original distribution of test scores*—skewed to the left. Neither subtracting a constant nor dividing by a constant changes the shape of the graph.
- **Center:** *The mean of the distribution of z-scores is 0.* Subtracting 35.44 from each test score would reduce the mean from 35.44 to 0. Dividing each of

these new data values by 8.77 would divide the new mean of 0 by 8.77, which still yields a mean of 0.

- **Variability:** *The standard deviation of the distribution of z-scores is 1.* Subtracting 35.44 from each test score does not affect the standard deviation. However, dividing all of the resulting values by 8.77 would divide the original standard deviation of 8.77 by 8.77, yielding 1.

CHECK YOUR UNDERSTANDING

Knoebels Amusement Park in Elysburg, Pennsylvania, has earned acclaim for being an affordable, family-friendly entertainment venue. Knoebels does not charge for general admission or parking, but it does charge customers for each ride they take. How much do the rides cost at Knoebels? The figure shows a dot-plot of the cost for each of 22 rides in a recent year, along with summary statistics.

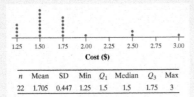

n	Mean	SD	Min	Q_1	Median	Q_3	Max
22	1.705	0.447	1.25	1.5	1.5	1.75	3

1. Suppose you convert the cost of the rides from dollars to cents ($1 = 100 cents). Describe the shape, mean, and standard deviation of the distribution of ride cost in cents.
2. Knoebels' managers decide to increase the cost of each ride by 25 cents. How would the shape, center, and variability of this distribution compare with the distribution of cost in Question 1?
3. Now suppose you convert the increased costs from Question 2 to z-scores. What would be the shape, mean, and standard deviation of this distribution? Explain your answers.

Section 2.1 | Summary

- Two ways of describing an individual value's location in a distribution are **percentiles** and **standardized scores (z-scores).** The *p*th percentile of a distribution is the value with *p* percent of the observations less than or equal to it.
- To standardize any data value, subtract the mean of the distribution and then divide the difference by the standard deviation. The resulting z-score

$$z = \frac{\text{value} - \text{mean}}{\text{standard deviation}}$$

measures how many standard deviations the data value lies above or below the mean of the distribution. We can also use percentiles and z-scores to compare the relative location of individuals in different distributions.
- A **cumulative relative frequency graph** allows us to examine location in a distribution. The completed graph allows you to estimate the percentile for an individual value, and vice versa.

Making Connections

Highlight the fact that standardizing a distribution will give a mean of 0 and standard deviation of 1 for any distribution. In Section 2.2, we will standardize Normal distributions, giving us a *standard Normal distribution* that has a mean of 0 and standard deviation of 1.

 **Answers to CYU**

1. Converting the cost of the rides from dollars to cents will not change the shape. However, it will multiply the mean and the standard deviation by 100.

2. Adding 25 cents to the cost of each ride will not change the shape of the distribution, nor will it change the variability. It will, however, add 25 cents to the measures of center (mean, median).

3. Converting the costs to z-scores will not change the shape of the distribution. It will change the mean to 0 and the standard deviation to 1.

TRM Data Exploration: The Speed of Light

The Data Exploration feature that was found in the previous two editions of *The Practice of Statistics* student edition has been moved to the Teacher's Resource Materials. Click on the link in the TE-Book, open in the TRFD, or download from the Teacher's Resources on the book's digital platform.

TRM Section 2.1 Quiz

There is one quiz available for this section in the Teacher's Resource Materials. Click on the link in the TE-Book, look on the TRFD, or download from the Teacher's Resources on the book's digital platform. You can also create your own quiz using the ExamView® Assessment Suite that is part of the TPS6 program. Questions are coded by Learning Target to make it easy to build parallel quizzes.

Teaching Tip

Make sure to point out the two icons next to Exercise 1. The "pg 92" icon reminds students that the example on page 92 is very similar to this exercise. The "play" icon reminds students that there is a video solution available in the student e-Book or on the Student Site at highschool.bfwpub .com/updatedtps6e.

TRM Full Solutions to Section 2.1 Exercises

Click on the link in the TE-Book, open the TRFD, or download from the Teacher's Resources on the book's digital platform.

Answers to Section 2.1 Exercises

2.1 (a) Because 18 of the 20 students (90%) own ≤ the number of pairs of shoes that Jackson owns (22 pairs of shoes), Jackson is at the 90th percentile in the number of pairs of shoes distribution. **(b)** 45% of the boys own ≤ the number of pairs of shoes that Raul owns. Raul is at the 45th percentile, meaning that 45% of the 20 boys, or 9 boys, have the same number or fewer pairs of shoes. Therefore, Raul's response is the 9th value in the ordered list. He owns 7 pairs of shoes.

2.2 (a) Because 4 of the 50 states (10%) have smaller values for the percent of residents aged 65 and older that are ≤ that of Colorado, Colorado is at the 10th percentile in the distribution. 10% of states have a percent of residents aged 65 and older that is ≤ that of Colorado. **(b)** 80% of states have a percent of residents aged 65 and older that is ≤ that of Rhode Island. Rhode Island is at the 80th percentile, meaning that 80% of the 50 states, or 40 states, have a percent of residents aged 65 and older that is ≤ that of Rhode Island. Rhode Island is the 40th value in the ordered list. 13.8% of Rhode Island's residents are aged 65 and older.

2.3 (a) Because 11 of the 30 observations (36.7%) are ≤ Antawn's head circumference (22.4 inches), Antawn is at the 36.7th percentile in the head circumference distribution. **(b)** 90% of the 30 players (27) will have a head circumference that is ≤ this player's. The player at the 90th percentile will have a head circumference that is the

- It is necessary to **transform data** when changing units of measurement.
  - When you add a positive constant a to (subtract a from) all the values in a data set, measures of center and location—mean, five-number summary, percentiles—increase (decrease) by a. Measures of variability—range, IQR, SD—do not change.
  - When you multiply (divide) all the values in a data set by a positive constant b, measures of center, location, and variability are multiplied (divided) by b.
  - Neither of these transformations changes the shape of the distribution.

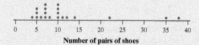

Section 2.1   Exercises

1. **Shoes** How many pairs of shoes does a typical teenage boy own? To find out, two AP® Statistics students surveyed a random sample of 20 male students from their large high school. Then they recorded the number of pairs of shoes that each boy owned. Here is a dotplot of the data:

   Number of pairs of shoes

   (a) Find the percentile for Jackson, who reported owning 22 pairs of shoes.

   (b) Raul's reported number of pairs of shoes is at the 45th percentile of the distribution. Interpret this value. How many pairs of shoes does Raul own?

2. **Old folks** Here is a stemplot of the percents of residents aged 65 and older in the 50 states:

   ```
    7 | 0
    8 | 8
    9 | 8
   10 | 019
   11 | 16777
   12 | 01122456778999
   13 | 0001223344455689
   14 | 023568
   15 | 24
   16 | 9
   ```

   Key: 15|2 means 15.2% of this state's residents are 65 or older.

   (a) Find the percentile for Colorado, where 10.1% of the residents are aged 65 and older.

   (b) Rhode Island is at the 80th percentile of the distribution. Interpret this value. What percent of Rhode Island's residents are aged 65 and older?

3. **Wear your helmet!** Many athletes (and their parents) worry about the risk of concussions when playing sports. A football coach plans to obtain specially made helmets for his players that are designed to reduce the chance of getting a concussion. Here are the

measurements of head circumference (in inches) for the players on the team:

23.0	22.2	21.7	22.0	22.3	22.6	22.7	21.5
22.7	25.6	20.8	23.0	24.2	23.5	20.8	24.0
22.7	22.6	23.9	22.5	23.1	21.9	21.0	22.4
23.5	22.5	23.9	23.4	21.6	23.3		

(a) Antawn, the team's starting quarterback, has a head circumference of 22.4 inches. What is Antawn's percentile?

(b) Find the head circumference of the player at the 90th percentile of the distribution.

4. **Don't call me** According to a study by Nielsen Mobile, "Teenagers ages 13 to 17 are by far the most prolific texters, sending 1742 messages a month." Mr. Williams, a high school statistics teacher, was skeptical about the claims in the article. So he collected data from his first-period statistics class on the number of text messages they had sent over the past 24 hours. Here are the data:

0	7	1	29	25	8	5	1	25	98	9	0	26
8	118	72	0	92	52	14	3	3	44	5	42	

(a) Sunny was the student who sent 42 text messages. What is Sunny's percentile?

(b) Find the number of text messages sent by Joelle, who is at the 20th percentile of the distribution.

5. **Setting speed limits** According to the *Los Angeles Times*, speed limits on California highways are set at the 85th percentile of vehicle speeds on those stretches of road. Explain to someone who knows little statistics what that means.

6. **Blood pressure** Larry came home very excited after a visit to his doctor. He announced proudly to his wife, "My doctor says my blood pressure is at the 90th percentile among men like me. That means I'm better off than about 90% of similar men."

27th value in the ordered list. The player with a head circumference of 23.9 inches is at the 90th percentile of the distribution.

2.4 (a) Because 19 of the 25 students (76%) sent less than or equal to the number of text messages sent by Sunny (who sent 42 texts), Sunny is at the 76th percentile in the distribution of number of text messages sent in the past 24 hours. **(b)** Joelle is at the 20th percentile. This means that 20% of the 25 students in the class, or 5 students, sent ≤ the number of text messages than she sent. Therefore, the number

of texts Joelle sent will be the 5th value in the ordered list. Joelle sent 1 text message in the past 24 hours.

2.5 This means that the speed limit is set at such a speed that 85% of the vehicle speeds are ≤ the posted speed.

2.6 Larry's wife should tell him that being at the 90th percentile for blood pressures is *not* a good thing. It means that 90% of men like him have a blood pressure that is ≤ his! When it comes to blood pressure, a high number is not desirable. Larry may need treatment for his high blood pressure.

How should his wife, who is a statistician, respond to Larry's statement?

7. **Growth charts** We used an online growth chart to find percentiles for the height and weight of a 16-year-old girl who is 66 inches tall and weighs 118 pounds. According to the chart, this girl is at the 48th percentile for weight and the 78th percentile for height. Explain what these values mean in plain English.

8. **Track star** Peter is a star runner on the track team. In the league championship meet, Peter records a time that would fall at the 80th percentile of all his race times that season. But his performance places him at the 50th percentile in the league championship meet. Explain how this is possible. (Remember that shorter times are better in this case!)

9. **Run fast!** As part of a student project, high school pg 84 students were asked to sprint 50 yards, and their times (in seconds) were recorded. A cumulative relative frequency graph of the sprint times is shown here.

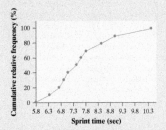

(a) One student ran the 50 yards in 8.05 seconds. Is a sprint time of 8.05 seconds unusually slow?

(b) Estimate and interpret the 20th percentile of the distribution.

10. **Household incomes** The cumulative relative frequency graph describes the distribution of median household incomes in the 50 states in a recent year.[2]

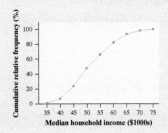

(a) The median household income in North Dakota that year was $55,766. Is North Dakota an unusually wealthy state?

(b) Estimate and interpret the 90th percentile of the distribution.

11. **Foreign-born residents** The cumulative relative frequency graph shows the distribution of the percent of foreign-born residents in the 50 states.

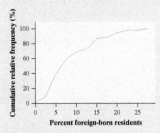

(a) Estimate the interquartile range (IQR) of this distribution. Show your method.

(b) Arizona had 15.1% foreign-born residents that year. Estimate its percentile.

(c) Explain why the graph is fairly flat between 20% and 27.5%.

12. **Shopping spree** The figure is a cumulative relative frequency graph of the amount spent by 50 consecutive grocery shoppers at a store.

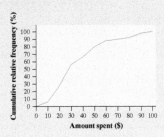

(a) Estimate the interquartile range (IQR) of this distribution. Show your method.

(b) One shopper spent $19.50. Estimate this person's percentile.

(c) Explain why the graph is steepest between $10 and $30.

2.11 (a) The first quartile is the 25th percentile. Find 25 on the y axis, read over to the line and then down to the x axis to get about $Q_1 = 4\%$. The 3rd quartile is the 75th percentile. Find 75 on the y axis and then read down to the x axis to get about $Q_3 = 14\%$. The IQR is approximately $14 - 4 = 10\%$. **(b)** Arizona, which had 15.1% foreign-born residents that year, is approximately at the 85th percentile. **(c)** The graph is fairly flat between 20% and 27.5% foreign-born residents because there were very few states that had 20% to 27.5% foreign-born residents that year.

(d)

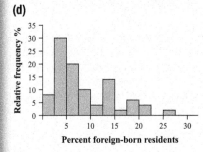

2.12 (a) The first quartile is the 25th percentile. Find 25 on the y axis and then read down to the x axis to get about $Q_1 = \$19$. The 3rd quartile is the 75th percentile. Find 75 on the y axis and then read down to the x axis to get about $Q_3 = \$46$. The IQR is approximately $\$46 - \$19 = \$27$. **(b)** The person who spent $19.50 is just above the value we determine for the 25th percentile. It appears that $19.50 is at about the 26th percentile. **(c)** The graph is steepest between $10 and $30 because more shoppers spent amounts in this interval.

(d)

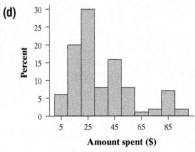

2.7 48% of girls her age weigh $\le$ her weight and 78% of girls her age are $\le$ her height. Because she is taller than 78% of girls, but only weighs $\ge$ 48% of girls, she is probably fairly thin.

2.8 Peter's time was $\le$ 80% of his previous race times that season, but it was $\le$ only 50% of the racers at the meet. Because this time was relatively slow for Peter but at the median for the runners in the league championship, Peter must be a good runner.

2.9 (a) No, a sprint time of 8.05 seconds is not unusually slow. A student with an 8-second sprint is at the 75th percentile, so 25% of the students took that long or longer.

(b) The 20th percentile of the distribution is approximately 6.7 seconds. 20% of the students completed the 50-yard sprint in 6.7 seconds or less.

2.10 (a) No; North Dakota, with a median household income of $55,766, is not an unusually wealthy state. North Dakota is at the 65th percentile of median household income. That means 35% of states have a median household income that is larger than that of North Dakota. **(b)** The 90th percentile of the distribution of median household income is approximately $65,000. 90% of the states have a median household income of $65,000 or less.

2.13 (a) The z-score for Montana is
$$z = \frac{1.9 - 8.73}{6.12} = -1.12.$$ Montana's
percent of foreign-born residents is 1.12
standard deviations below the mean
percent of foreign-born residents for all
states. **(b)** If we let x denote the percent
of foreign-born residents in New York
at that time, then we can solve for x in
the equation $2.10 = \dfrac{x - 8.73}{6.12}$. Thus,
x = 21.582% foreign-born residents.

2.14 (a) The z-score for North Carolina
is $z = \dfrac{41,553 - 51,742.44}{8,210.64} = -1.24.$
North Carolina's median household
income is 1.24 standard deviations below
the average median household income
for all states. **(b)** If we let x denote the
median household income for New
Jersey, then we can solve for x in the
equation $1.82 = \dfrac{x - 51,742.44}{8,210.64}$. Thus,
x = $66,685.80 is the median household
income for New Jersey.

2.15 (a) The number of pairs of shoes
owned by Jackson is 1.10 standard
deviations above the average number of
pairs of shoes owned by the students
in the sample. **(b)** If we let $\bar{x}$ denote the
mean number of pairs of shoes owned by
students in the sample, then we can solve
for $\bar{x}$ in the equation $1.10 = \dfrac{22 - \bar{x}}{9.42}$.
Thus, $\bar{x}$ = 11.64 is the mean number of
pairs of shoes owned by the students in
the sample.

2.16 (a) The number of texts sent by
Alejandro is 1.89 standard deviations
above the average number of texts sent
by the students in the sample. **(b)** If we
let $\bar{x}$ denote the mean number of texts
sent in the sample, then we can solve for
$\bar{x}$ in the equation $1.89 = \dfrac{92 - \bar{x}}{34.15}$. Thus,
$\bar{x}$ = 27.46 is the mean number of texts
sent by the students in the sample.

2.17 (a) The fact that your standardized
score is negative indicates that your bone
density is below the average for your
peer group. In fact, your bone density
is about 1.5 standard deviations below
average among 25-year-old women.
(b) If we let σ denote the standard
deviation of the bone density in
Judy's reference population, then
we can solve for σ in the equation
$-1.45 = \dfrac{948 - 956}{\sigma}$. Thus,
σ = 5.52 g/cm².

13. **Foreign-born residents** Refer to Exercise 11. Here
    are summary statistics for the percent of foreign-born
    residents in the 50 states:

n	Mean	SD	Min	Q_1	Med	Q_3	Max
50	8.73	6.12	1.3	4.1	6.2	13.4	27.1

(a) Find and interpret the z-score for Montana, which had
1.9% foreign-born residents.

(b) New York had a standardized score of 2.10. Find the
percent of foreign-born residents in New York at that time.

14. **Household incomes** Refer to Exercise 10. Here are
summary statistics for the state median household
incomes:

n	Mean	SD	Min	Q_1	Med	Q_3	Max
50	51,742.44	8210.64	36,641	46,071	50,009	57,020	71,836

(a) Find and interpret the z-score for North Carolina, with
a median household income of $41,553.

(b) New Jersey had a standardized score of 1.82. Find New
Jersey's median household income for that year.

15. **Shoes** Refer to Exercise 1. Jackson, who reported owning
22 pairs of shoes, has a standardized score of $z = 1.10$.

(a) Interpret this z-score.

(b) The standard deviation of the distribution of number of
pairs of shoes owned in this sample of 20 boys is 9.42.
Use this information along with Jackson's standardized
score to find the mean of the distribution.

16. **Don't call me** Refer to Exercise 4. Alejandro, who sent
92 texts, has a standardized score of $z = 1.89$.

(a) Interpret this z-score.

(b) The standard deviation of the distribution of number
of text messages sent over the past 24 hours by the
students in Mr. Williams's class is 34.15. Use this
information along with Alejandro's standardized score
to find the mean of the distribution.

17. **Measuring bone density** Individuals with low bone
density (osteoporosis) have a high risk of broken bones
(fractures). Physicians who are concerned about low
bone density in patients can refer them for specialized
testing. Currently, the most common method for
testing bone density is dual-energy X-ray absorptiometry
(DEXA). The bone density results for a patient who
undergoes a DEXA test usually are reported in grams
per square centimeter (g/cm²) and in standardized units.
Judy, who is 25 years old, has her bone density
measured using DEXA. Her results indicate bone
density in the hip of 948 g/cm² and a standardized
score of $z = -1.45$. The mean bone density in the hip
is 956 g/cm² in the reference population of 25-year-old
women like Judy.[3]

(a) Judy has not taken a statistics class in a few years.
Explain to her in simple language what the
standardized score reveals about her bone density.

(b) Use the information provided to calculate the standard
deviation of bone density in the reference population.

18. **Comparing bone density** Refer to Exercise 17. Judy's
friend Mary also had the bone density in her hip
measured using DEXA. Mary is 35 years old. Her
bone density is also reported as 948 g/cm², but her
standardized score is $z = 0.50$. The mean bone density
in the hip for the reference population of 35-year-old
women is 944 grams/cm².

(a) Whose bones are healthier for her age: Judy's or
Mary's? Justify your answer.

(b) Calculate the standard deviation of the bone density in
Mary's reference population. How does this compare
with your answer to Exercise 17(b)? Are you surprised?

19. **SAT versus ACT** Eleanor scores 680 on the SAT
Mathematics test. The distribution of SAT Math scores
is symmetric and single-peaked with mean 500 and
standard deviation 100. Gerald takes the American
College Testing (ACT) Mathematics test and scores
29. ACT scores also follow a symmetric, single-peaked
distribution—but with mean 21 and standard deviation
5. Find the standardized scores for both students.
Assuming that both tests measure the same kind of
ability, who has the higher score?

20. **Comparing batting averages** Three landmarks of
baseball achievement are Ty Cobb's batting average
of 0.420 in 1911, Ted Williams's 0.406 in 1941, and
George Brett's 0.390 in 1980. These batting averages
cannot be compared directly because the distribution
of major league batting averages has changed over the
years. The distributions are quite symmetric, except for
outliers such as Cobb, Williams, and Brett. While the
mean batting average has been held roughly constant
by rule changes and the balance between hitting and
pitching, the standard deviation has dropped over time.
Here are the facts:

Decade	Mean	Standard deviation
1910s	0.266	0.0371
1940s	0.267	0.0326
1970s	0.261	0.0317

Find the standardized scores for Cobb, Williams, and
Brett. Who had the best performance for the decade he
played?[4]

21. **Long jump** A member of a track team was practicing
the long jump and recorded the distances (in
centimeters) shown in the dotplot. Some numerical
summaries of the data are also provided.

2.18 (a) Mary's z-score (0.5) indicates that her
bone density score is about half a standard
deviation above the average score for all women
her age. Because her z-score is higher than
Judy's ($z = -1.45$), Mary's bones are healthier
in comparison to other women in their age
groups. **(b)** If we let σ denote the standard
deviation of the bone density in Mary's reference
population, then we can solve for σ in the
equation $0.5 = \dfrac{948 - 944}{\sigma}$. Thus, $\sigma = 8$ g/cm².
There is more variability in the bone densities
for older women. This isn't surprising because,
as women get older, there is more time for their
good or bad health habits to have an effect,
creating a wider range of bone densities.

2.19 Eleanor's standardized score,
$$z = \frac{680 - 500}{100} = 1.8$$ is higher than Gerald's
standardized score, $z = \dfrac{29 - 21}{5} = 1.6$.

2.20 The standardized batting averages
(z-scores) for these three outstanding hitters
are Cobb $= z = \dfrac{0.420 - 0.266}{0.0371} = 4.15$;
Williams $= z = \dfrac{0.406 - 0.267}{0.0326} = 4.26$;
Brett $= z = \dfrac{0.390 - 0.261}{0.0317} = 4.07$. All three
hitters were at least 4 standard deviations above
their peers, but Williams's z-score is the highest.

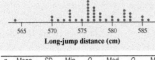

Long-jump distance (cm)

n	Mean	SD	Min	Q_1	Med	Q_3	Max
40	577.3	4.713	564	574.5	577	581.5	586

After chatting with a teammate, the jumper realized that he measured his jumps from the back of the board instead of the front. Thus, he had to subtract 20 centimeters from each of his jumps to get the correct measurement for each jump.

(a) What shape would the distribution of corrected long-jump distance have?

(b) Find the mean and median of the distribution of corrected long-jump distance.

(c) Find the standard deviation and interquartile range (*IQR*) of the distribution of corrected long-jump distance.

22. **Step right up!** A dotplot of the distribution of height for Mrs. Navard's class is shown, along with some numerical summaries of the data.

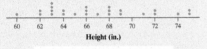

Height (in.)

n	$\bar{x}$	s_x	Min	Q_1	Med	Q_3	Max
25	67	4.29	60	63	66	70	75

Suppose that Mrs. Navard has the entire class stand on a 6-inch-high platform and then asks the students to measure the distance from the top of their heads to the ground.

(a) What shape would this distribution of distance have?

(b) Find the mean and median of the distribution of distance.

(c) Find the standard deviation and interquartile range (*IQR*) of the distribution of distance.

23. **Teacher raises** A school system employs teachers at salaries between $38,000 and $70,000. The teachers' union and school board are negotiating the form of next year's increase in the salary schedule. Suppose that every teacher is given a $1000 raise. What effect will this raise have on each of the following characteristics of the resulting distribution of salary?

(a) Shape

(b) Mean and median

(c) Standard deviation and interquartile range (*IQR*)

24. **Used cars, cheap!** A used-car salesman has 28 cars in his inventory, with prices ranging from $11,500 to $25,000. For a Labor Day sale, he reduces the price of each car by $500. What effect will this reduction have on each of the following characteristics of the resulting distribution of price?

(a) Shape

(b) Mean and median

(c) Standard deviation and interquartile range (*IQR*)

25. **Long jump** Refer to Exercise 21. Suppose that the corrected long-jump distances are converted from centimeters to meters (note that 100 cm = 1 m).

(a) What shape would the resulting distribution have? Explain your answer.

(b) Find the mean of the distribution of corrected long-jump distance in meters.

(c) Find the standard deviation of the distribution of corrected long-jump distance in meters.

26. **Step right up!** Refer to Exercise 22. Suppose that the distances from the tops of the students' heads to the ground are converted from inches to feet (note that 12 in. = 1 ft).

(a) What shape would the resulting distribution have? Explain your answer.

(b) Find the mean of the distribution of distance in feet.

(c) Find the standard deviation of the distribution of distance in feet.

27. **Teacher raises** Refer to Exercise 23. Suppose each teacher receives a 5% raise instead of a $1000 raise. What effect will this raise have on each of the following characteristics of the resulting salary distribution?

(a) Shape

(b) Median

(c) Interquartile range (*IQR*)

28. **Used cars, cheap!** Refer to Exercise 24. Suppose each car's price is reduced by 10% instead of $500. What effect will this discount have on each of the following characteristics of the resulting price distribution?

(a) Shape

(b) Median

(c) Interquartile range (*IQR*)

29. **Cool pool?** Coach Ferguson uses a thermometer to measure the temperature (in degrees Fahrenheit) at 20 different locations in the school swimming pool. An analysis of the data yields a mean of 77°F and a standard deviation of 3°F. (Recall that $°C = \frac{5}{9}°F - \frac{160}{9}$.)

2.21 (a) The shape of the distribution of corrected long-jump distance will be the same as the original distribution of long-jump distance: roughly symmetric with a single peak. **(b)** The mean is $577.3 - 20 = 557.3$ centimeters. The median is $577 - 20 = 557$ centimeters. **(c)** The standard deviation is the same at 4.713 centimeters. The *IQR* is the same as the *IQR* of the distribution of long-jump distance, 7 centimeters.

2.22 (a) The shape of the distribution of distance will be the same as the original distribution of heights: slightly skewed right with several peaks. **(b)** The mean is $67 + 6 = 73$ inches. The median is $66 + 6 = 72$ inches. **(c)** The standard deviation of the distribution of distance is the same as the SD of the distribution of heights, 4.29 inches. The *IQR* is the same as the *IQR* of the distribution of heights, 7 inches.

2.23 (a) The shape of the new salary distribution will be the same as the shape of the original salary distribution. **(b)** The mean and median salaries will each increase by $1000. **(c)** The standard deviation and *IQR* of the new salary distribution will each be the same as they were for the original salary distribution.

2.24 (a) The shape of the new and original price distribution will be the same. **(b)** The mean and median of the new price distribution will each be $500 lower than for the original price distribution. **(c)** The standard deviation and *IQR* of the new and original price distribution will be the same.

2.25 (a) The shape of the distribution of corrected long-jump distance will be the same in meters and in centimeters: roughly symmetric with a single peak. **(b)** The mean of the distribution of corrected long-jump distance, in meters, is $577.3 - 20 = 557.3$ cm $\div$ 100 cm/m $= 5.573$ meters. **(c)** The standard deviation of the distribution of corrected long-jump distance, in meters, is 4.713 centimeters $\div$ 100 $= 0.04713$ meters.

2.26 (a) The shape of the distribution of distance, in feet, will be the same as the original distribution of heights: slightly skewed right with several peaks. **(b)** The mean of the distribution of distance, in feet, is $67 + 6 = 73$ inches $\div$ 12 in./ft $= 6.083$ feet. **(c)** The standard deviation of the distribution of distance, in feet, is 4.29 inches $\div$ 12 $= 0.358$ feet.

2.27 (a) The shape of the resulting salary distribution will be the same as the original distribution of salaries. **(b)** The median will increase by 5% because each value in the distribution is multiplied by 1.05. **(c)** The *IQR* will increase by 5% because each value in the distribution is multiplied by 1.05.

2.28 (a) The shape of the resulting price distribution will be the same as the shape of the original price distribution. **(b)** The median of the new price distribution will be 0.90 times the median for the original price distribution because each value in the distribution is multiplied by 0.90. **(c)** The *IQR* of the new price distribution will be 0.90 times the *IQR* for the original price distribution because each value in the distribution is multiplied by 0.90.

2.29 (a) The mean temperature reading in degrees Celsius $= \frac{5}{9}(77) - \frac{160}{9} = 25$ degrees Celsius. **(b)** The standard deviation of the temperature reading in degrees Celsius $= \frac{5}{9}(3) = 1.667$ degrees Celsius.

2.30 (a) To find the correct measurement in centimeters, we subtract the 0.2 inch that Clarence mistakenly added to each value. Then to transform that measurement to centimeters, we multiply by 2.54. So the mean of the corrected measurements, in centimeters, is $(3.2 - 0.2)(2.54) = 7.62$ cm. **(b)** To calculate the standard deviation of the corrected measurements in centimeters, we just multiply the old standard deviation by 2.54 because subtracting a constant from each value in a distribution does not affect the standard deviation. The standard deviation of the corrected measurements is $0.1(2.54) = 0.254$ cm.

2.31 To determine the mean of the lengths of his cab rides in miles, we substitute \$15.45 for the mean fare and solve for the mean number of miles.

$15.45 = 2.85 + 2.7$ Mean(miles)
$\rightarrow$ Mean(miles) = 4.667 miles

To determine the standard deviation of the lengths of his cab rides in miles, we use the equation:

SD(Fare) = 2.7 SD(miles) $\rightarrow$ 10.20 =
2.7 SD(miles) $\rightarrow$ SD(miles) = 3.778 miles

The mean and standard deviation of the lengths of his cab rides, in miles, are 4.667 miles and 3.778 miles, respectively.

2.32 To make the standard deviation increase from 3 to 12, multiply each score by 4. This will make the standard deviation equal to $4(3) = 12$ and the mean equal to $4(12) = 48$. To make the mean increase from 48 to 75, add 27 to each adjusted score. Adding 27 does not change the variability, so the mean will be $48 + 27 = 75$ and the standard deviation will still be 12.

2.33 c

2.34 a

2.35 d

2.36 b

2.37 c

2.38 d

(a) Find the mean temperature reading in degrees Celsius.

(b) Calculate the standard deviation of the temperature readings in degrees Celsius.

30. **Measure up** Clarence measures the diameter of each tennis ball in a bag with a standard ruler. Unfortunately, he uses the ruler incorrectly so that each of his measurements is 0.2 inch too large. Clarence's data had a mean of 3.2 inches and a standard deviation of 0.1 inch. (Recall that 1 in. = 2.54 cm.)

(a) Find the mean of the corrected measurements in centimeters.

(b) Calculate the standard deviation of the corrected measurements in centimeters.

31. **Taxi!** In 2016, taxicabs in Los Angeles charged an initial fee of \$2.85 plus \$2.70 per mile. In equation form, Fare = $2.85 + 2.7$(miles). At the end of a month, a businessman collects all his taxicab receipts and calculates some numerical summaries. The mean fare he paid was \$15.45 with a standard deviation of \$10.20. What are the mean and standard deviation of the lengths of his cab rides in miles?

32. **Quiz scores** The scores on Ms. Martin's statistics quiz had a mean of 12 and a standard deviation of 3. Ms. Martin wants to transform the scores to have a mean of 75 and a standard deviation of 12. What transformations should she apply to each test score? Explain your answer.

Multiple Choice: *Select the best answer for Exercises 33–38.*

33. Jorge's score on Exam 1 in his statistics class was at the 64th percentile of the scores for all students. His score falls

(a) between the minimum and the first quartile.

(b) between the first quartile and the median.

(c) between the median and the third quartile.

(d) between the third quartile and the maximum.

(e) at the mean score for all students.

34. When Sam goes to a restaurant, he always tips the server \$2 plus 10% of the cost of the meal. If Sam's distribution of meal costs has a mean of \$9 and a standard deviation of \$3, what are the mean and standard deviation of his tip distribution?

(a) \$2.90, \$0.30   (d) \$11.00, \$2.00

(b) \$2.90, \$2.30   (e) \$2.00, \$0.90

(c) \$9.00, \$3.00

35. Scores on the ACT college entrance exam follow a bell-shaped distribution with mean 21 and standard deviation 5. Wayne's standardized score on the ACT was −0.6. What was Wayne's actual ACT score?

(a) 3   (b) 13   (c) 16

(d) 18   (e) 24

36. George's average bowling score is 180; he bowls in a league where the average for all bowlers is 150 and the standard deviation is 20. Bill's average bowling score is 190; he bowls in a league where the average is 160 and the standard deviation is 15. Who ranks higher in his own league, George or Bill?

(a) Bill, because his 190 is higher than George's 180.

(b) Bill, because his standardized score is higher than George's.

(c) Bill and George have the same rank in their leagues, because both are 30 pins above the mean.

(d) George, because his standardized score is higher than Bill's.

(e) George, because the standard deviation of bowling scores is higher in his league.

Exercises 37 and 38 refer to the following setting. The number of absences during the fall semester was recorded for each student in a large elementary school. The distribution of absences is displayed in the following cumulative relative frequency graph.

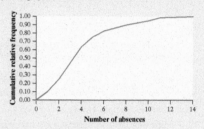

37. What is the interquartile range (IQR) for the distribution of absences?

(a) 1   (b) 2   (c) 3

(d) 5   (e) 14

38. If the distribution of absences was displayed in a histogram, what would be the best description of the histogram's shape?

(a) Symmetric   (d) Skewed right

(b) Uniform   (e) Cannot be determined

(c) Skewed left

Recycle and Review *Exercises 39 and 40 refer to the following setting.* We used Census At School's Random Data Selector to choose a sample of 50 Canadian students who completed a survey in a recent year.

39. **Travel time** (1.2) The dotplot displays data on students' responses to the question "How long does it usually take you to travel to school?" Describe the distribution.

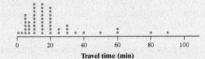

40. **Lefties** (1.1) Students were asked, "Are you right-handed, left-handed, or ambidextrous?" The responses (R = right-handed, L = left-handed, A = ambidextrous) are shown here.

R	R	R	R	R	R	R	R	R	R	R	L	R	R
R	R	R	R	R	R	R	R	R	R	R	R	R	A
R	R	R	R	A	R	R	L	R	R	R	R	L	A
R	R	R	R	R	R	R	R						

(a) Make an appropriate graph to display these data.

(b) Over 10,000 Canadian high school students took the Census At School survey that year. What percent of this population would you estimate is left-handed? Justify your answer.

2.39 The distribution is skewed to the right. The two largest values appear to be outliers. The data are centered roughly around a median of 15 minutes and the *IQR* is approximately 10 minutes.

2.40 (a)

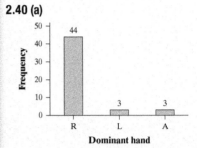

(b) Because $3/50 = 6\%$ of the sample was left-handed, our best estimate of the percentage of the population that is left-handed is 6%.

PD Section 2.2 Overview

To watch the video overview of Section 2.2, click on the link in the TE-Book, look on the TRFD, or download from the Teacher's Resources on the book's digital platform.

SECTION 2.2 # Density Curves and Normal Distributions

LEARNING TARGETS *By the end of the section, you should be able to:*

- Use a density curve to model a distribution of quantitative data.
- Identify the relative locations of the mean and median of a distribution from a density curve.
- Use the empirical rule to estimate (i) the proportion of values in a specified interval, or (ii) the value that corresponds to a given percentile in a Normal distribution.

- Find the proportion of values in a specified interval in a Normal distribution using Table A or technology.
- Find the value that corresponds to a given percentile in a Normal distribution using Table A or technology.
- Determine whether a distribution of data is approximately Normal from graphical and numerical evidence.

In Chapter 1, we developed graphical and numerical tools for describing distributions of quantitative data. Our work gave us a clear strategy for exploring data on a single quantitative variable.

- Always plot your data: make a graph—usually a dotplot, stemplot, or histogram.
- Look for the overall pattern (shape, center, variability) and for striking departures such as outliers.
- Calculate numerical summaries to describe center and variability.

In this section, we add one more step to this strategy.

- When there's a regular overall pattern, use a simplified model called a *density curve* to describe it.

Making Connections

In Chapter 3, we will continue using this strategy when analyzing relationships between quantitative variables. Instead of describing a distribution with a smooth curve, we will describe a linear relationship between two variables with a line.

Density Curves

Selena works at a bookstore in the Denver International Airport. She takes the airport train from the main terminal to get to work each day. The airport just opened a new walkway that would allow Selena to get from the main terminal to the bookstore in 4 minutes. She wonders if it will be faster to walk or take the train to work.

Figure 2.7(a) shows a dotplot of the amount of time it has taken Selena to get to the bookstore by train each day for the last 1000 days she worked. To estimate the percent of days on which it would be quicker for her to take the train, we could find the percent of dots (marked in red) that represent journey times of less than 4 minutes. Surely, there's a simpler way than counting all those dots!

Figure 2.7(b) shows the dotplot modeled with a **density curve.** You might wonder why the density curve is drawn at a height of 1/3. That's so the area under the density curve between 2 minutes and 5 minutes is equal to

$$3 \times 1/3 = 1.00 = 100\%$$

representing 100% of the observations in the distribution shown in Figure 2.7(a).

DEFINITION   Density curve

A **density curve** models the distribution of a quantitative variable with a curve that

- Is always on or above the horizontal axis
- Has area exactly 1 underneath it

The area under the curve and above any interval of values on the horizontal axis estimates the proportion of all observations that fall in that interval.

The red shaded area under the density curve in Figure 2.7(b) provides a good approximation for the proportion or percent of red dots. Because the shaded region is rectangular,

$$\text{area} = \text{base} \times \text{height} = 2 \times 1/3 = 2/3 = 0.667 = 66.7\%$$

So we estimate that it would be quicker for Selena to take the train to work on about 66.7% of days. In fact, on 669 of the 1000 days, Selena's journey from the terminal to the bookstore took less than 4 minutes. That's $669/1000 = 0.669 = 66.9\%$ — very close to the estimate we got using the density curve.

> Recall from Chapter 1 that we can describe the distribution of journey times in Figure 2.7(a) as *approximately uniform.* The density curve in Figure 2.7(b) is called a *uniform density curve* because it has constant height.

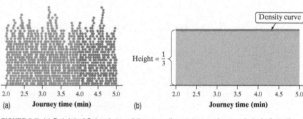

FIGURE 2.7 (a) Dotplot of Selena's travel time over the past several years via train from the Denver airport main terminal to the bookstore where she works. The red dots indicate times when it took her less than 4 minutes to get to work. (b) Density curve modeling the dotplot in part (a). The red shaded area estimates the proportion of times that it took Selena less than 4 minutes to get to work.

Preparing for Inference

The most commonly used density curve is the Normal curve, which we will learn about in a few pages. We will use other density curves, such as t curves and χ^2 curves, to model the distributions of the t and χ^2 test statistics in later chapters.

Teaching Tip

The discussion about the *uniform* density curve is setting the stage for students to understand the Normal distribution as a density curve, where area under the curve represents the proportion of values within a certain interval. In the end, the Normal distribution will be much more valuable for students than the uniform distribution.

Teaching Tip

Whenever your students look at a density curve, tell them to imagine that it is tracing the distribution of dots on a dotplot. Dotplots are much less abstract than density curves. This should help them understand what the density curve represents and how to approximate the median and mean.

No set of quantitative data is exactly described by a density curve. The curve is an approximation that is easy to use and accurate enough in most cases. The density curve simply smooths out the irregularities in the distribution.

EXAMPLE

That's so random!
Density curves

PROBLEM: Suppose you use a calculator or computer random number generator to produce a number between 0 and 2 (like 0.84522 or 1.1111119). The random number generator will spread its output uniformly across the entire interval from 0 to 2 as we allow it to generate a long sequence of random numbers.

(a) Draw a density curve to model this distribution of random numbers. Be sure to include scales on both axes.

(b) About what percent of the randomly generated numbers will fall between 0.87 and 1.55?

(c) Estimate the 65th percentile of this distribution of random numbers.

SOLUTION:

(a)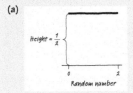

(b)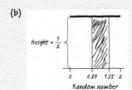

(c)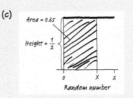

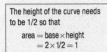
The height of the curve needs to be 1/2 so that

area = base × height
= 2 × 1/2 = 1

Area = (1.55 − 0.87) × 1/2 = 0.34 = 34%

0.65 = (x − 0) × 1/2
0.65 = (1/2)x
1.30 = x

FOR PRACTICE, TRY EXERCISE 41

Describing Density Curves

Density curves come in many shapes. As with the distribution of a quantitative variable, we start by looking for rough symmetry or clear skewness. Then we identify any clear peaks. Figure 2.8 shows three density curves with distinct shapes.

Skewed to the left, unimodal

Roughly symmetric, bimodal

Skewed to the right, unimodal

FIGURE 2.8 Density curves with different shapes.

Teaching Tip

Notice from the above example that a random number generator produces a density curve that is uniform. This means there is not some interval of values that is more likely to occur than another equally sized interval of values. This will not be true for a Normal distribution.

ALTERNATE EXAMPLE   Skill 3.A

I am your density
Density curves

PROBLEM:
A certain density curve is defined by the line segment that connects the points (0, 0) and (4, 0.5).

(a) Draw a picture of the density curve.

(b) What percent of values will fall between 0 and 2?

(c) Estimate the median of the distribution.

SOLUTION:

(a)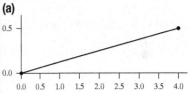

(b)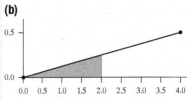

$$\text{Area} = \frac{1}{2} \cdot 2 \cdot 0.25 = 0.25 = 25\%$$

(c) We will find the x value such that the area of the triangle equals 0.5. The height of the triangle is the y value of the line given by the equation $y = 1/8\, x$.

$$\text{Area} = \frac{1}{2} \cdot \text{base} \cdot \text{height} = 0.5$$

$$= \frac{1}{2} \cdot x \cdot \left(\frac{1}{8}x\right) = 0.5$$

so $x = \sqrt{8} \approx 2.83$. The median of the distribution is about 2.83.

Teaching Tip

Tell students to imagine that the density curve is made out of wood. Ask them where they would put a fulcrum so that the piece of wood will balance. It also helps to remind them of the balancing-point activity with the ruler and pennies from Chapter 1.

Our measures of center and variability apply to density curves as well as to distributions of quantitative data. Recall that the mean is the balance point of a distribution. Figure 2.9 illustrates this idea for the **mean of a density curve**.

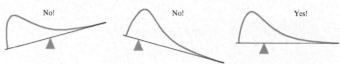

FIGURE 2.9 The mean of a density curve is its balance point.

The median of a distribution of quantitative data is the point with half the observations on either side. Similarly, the **median of a density curve** is the point with half of the area on each side.

> **DEFINITION**   **Mean of a density curve, Median of a density curve**
>
> The **mean of a density curve** is the point at which the curve would balance if made of solid material.
>
> The **median of a density curve** is the equal-areas point, the point that divides the area under the curve in half.

A symmetric density curve balances at its midpoint because the two sides are identical. So the mean and median of a symmetric density curve are equal, as in Figure 2.10(a). It isn't so easy to spot the equal-areas point on a skewed density curve. We used technology to locate the median in Figure 2.10(b). The mean is greater than the median because the balance point of the distribution is pulled toward the long right tail.

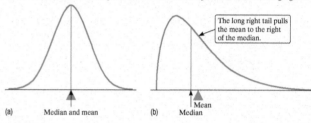

FIGURE 2.10 (a) Both the median and mean of a symmetric density curve lie at the point of symmetry. (b) In a right-skewed density curve, the mean is pulled away from the median toward the long tail.

The long right tail pulls the mean to the right of the median.

(a) Median and mean   (b) Mean Median

Making Connections

Recall from Chapter 1 that the mean is sensitive to outliers and skewness, while the median is resistant.

EXAMPLE   **What does the left skew do?**
Mean versus median

PROBLEM: The density curve that models a distribution of quantitative data is shown here. Identify the location of the mean and median by letter. Justify your answers.

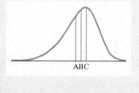

ABC

SOLUTION:

Median = B, Mean = A. B is the equal-areas point of the distribution. The mean will be less than the median due to the left-skewed shape.

> Even though C is directly under the peak of the curve, more than half of the area is to its left, so C cannot be the median.

FOR PRACTICE, TRY EXERCISE 45

ALTERNATE EXAMPLE   Skill 2.A

What does the right skew do?   Mean versus median

PROBLEM:

The density curve that models a distribution of quantitative data is shown. Identify the location of the mean and median by letter. Justify your answers.

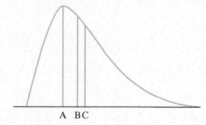

A BC

SOLUTION:

Median = B, mean = C; B is the equal-areas point of the distribution. The mean will be greater than the median due to the right-skewed shape. Even though A is directly under the peak of the curve, more than half of the area is to its right, so it cannot be the median.

A density curve is an idealized model for the distribution of a quantitative variable. As a result, we label the mean of a density curve as μ and the standard deviation of a density curve as σ. This is the same notation we used for the population mean and standard deviation in Chapter 1. In both cases, we refer to μ and σ as *parameters.*

We can roughly locate the mean μ of any density curve by eye, as the balance point. No easy way exists to estimate the standard deviation for density curves in general. But there is one family of density curves for which we can estimate the standard deviation by eye.

Preparing for Inference

Correct notation is very important in this course. We often will use a sample to estimate a population value. Remind students that values calculated from a sample are called statistics (e.g., $\overline{X}$, s_x), while values for the population are called parameters (e.g., μ, σ).

Normal Distributions

When we examine a distribution of quantitative data, how does it compare with an idealized density curve? Figure 2.11(a) shows a histogram of the scores of all seventh-grade students in Gary, Indiana, on the vocabulary part of the Iowa Test of Basic Skills (ITBS).[5] The scores are grade-level equivalents, so a score of 6.3 indicates that the student's performance is typical for a student in the third month of grade 6. The histogram is roughly symmetric, and both tails fall off smoothly from a single center peak. There are no large gaps or obvious outliers.

The density curve drawn through the tops of the histogram bars in Figure 2.11(b) is a good description of the overall pattern of the ITBS score distribution. We call it a **Normal curve.** The distributions described by Normal curves are called **Normal distributions.** In this case, the ITBS vocabulary scores of Gary, Indiana, seventh-graders are approximately Normally distributed.

Teaching Tip

Get accustomed to using the word *approximately* when identifying a distribution as Normal, because a real-life distribution would very seldom be *exactly* Normal.

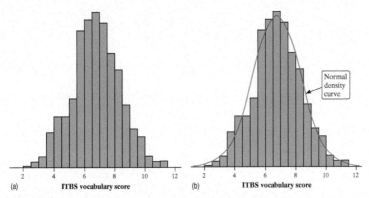

FIGURE 2.11 (a) Histogram of the Iowa Test of Basic Skills (ITBS) vocabulary scores of all seventh-grade students in Gary, Indiana. (b) The Normal density curve shows the overall shape of the distribution.

Normal distributions play a large role in statistics, but they are rather special and not at all "normal" in the sense of being usual or typical. We capitalize Normal to remind you that these density curves are special.

Look at the two Normal distributions in Figure 2.12. They illustrate several important facts:

- **Shape:** All Normal distributions have the same overall shape: symmetric, single-peaked (unimodal), and bell-shaped.
- **Center:** The mean μ is located at the midpoint of the symmetric density curve and is the same as the median.
- **Variability:** The standard deviation σ measures the variability (width) of a Normal distribution.

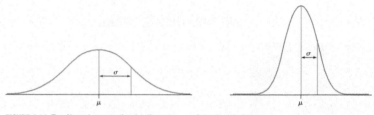

FIGURE 2.12 Two Normal curves, showing the mean μ and standard deviation σ.

You can estimate σ by eye on a Normal curve. Here's how. Imagine that you are skiing down a mountain that has the shape of a Normal distribution. At first, you descend at an increasingly steep angle as you go out from the peak:

Fortunately, before you find yourself going straight down, the slope begins to get flatter rather than steeper as you go out and down:

The points at which this change of curvature takes place are located at a distance σ on either side of the mean μ. (Advanced math students know these as "inflection points.") You can feel the change as you run a pencil along a Normal curve, which will allow you to estimate the standard deviation.

DEFINITION   Normal distribution, Normal curve

A **Normal distribution** is described by a symmetric, single-peaked, bell-shaped density curve called a **Normal curve**. Any Normal distribution is completely specified by two parameters: its mean μ and standard deviation σ.

The distribution of ITBS vocabulary scores for seventh-grade students in Gary, Indiana, is modeled well by a Normal distribution with mean $\mu = 6.84$ and standard deviation $\sigma = 1.55$. The figure shows this distribution with

Teaching Tip

When the standard deviation σ increases, the Normal curve is more spread out. This also means that the Normal curve will be shorter, because the area under the curve must remain equal to 1.

Teaching Tip: AP® Connections

AP® Calculus students learn that the inflection points of a function are the points at which concavity changes. These points can be found algebraically (if the function is defined) by setting the second derivative equal to zero and solving the equation.

Teaching Tip

Some teachers refer to the concave-up portions of the Normal curve as "happy" faces and the concave-down portion as a "sad" face. This may help the students who haven't taken calculus and have never heard of inflection points.

the points 1, 2, and 3 standard deviations from the mean labeled on the horizontal axis.

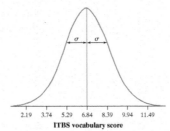

ITBS vocabulary score

You will be asked to make reasonably accurate sketches of Normal distributions to model quantitative data sets like the ITBS vocabulary scores. The best way to learn is to practice.

EXAMPLE

Stop the car!

Sketching a Normal distribution

PROBLEM: Many studies on automobile safety suggest that when automobile drivers make emergency stops, the stopping distances follow an approximately Normal distribution. Suppose that for one model of car traveling at 62 mph under typical conditions on dry pavement, the mean stopping distance is $\mu = 155$ ft with a standard deviation of $\sigma = 3$ ft. Sketch the Normal curve that approximates the distribution of stopping distance. Label the mean and the points that are 1, 2, and 3 standard deviations from the mean.

SOLUTION:

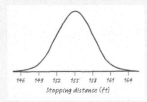

Stopping distance (ft)

The mean (155) is at the midpoint of the bell-shaped density curve. The standard deviation (3) is the distance from the center to the change-of-curvature points on either side. Label the mean and the points that are 1, 2, and 3 SDs from the mean:

1 SD: $155 - 1(3) = 152$ and $155 + 1(3) = 158$
2 SD: $155 - 2(3) = 149$ and $155 + 2(3) = 161$
3 SD: $155 - 3(3) = 146$ and $155 + 3(3) = 164$

FOR PRACTICE, TRY EXERCISE 47

 Remember that μ and σ alone do not specify the appearance of most distributions. The shape of density curves, in general, does not reveal σ. These are special properties of Normal distributions.

ALTERNATE EXAMPLE   Skill 2.D

Chapter 1 Test scores

Sketching a Normal distribution

PROBLEM:

Chapter 1 Test scores from Mrs. Gallas's first-hour class follow an approximately Normal distribution with a mean of 81 and standard deviation of 6. Sketch the Normal curve that approximates the distribution of Chapter 1 test scores. Label the mean and the points that are 1, 2, and 3 standard deviations from the mean.

SOLUTION:

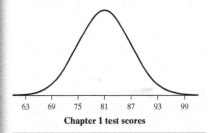

Chapter 1 test scores

Why are Normal distributions important in statistics? Here are three reasons.

1. Normal distributions are good descriptions for some distributions of real data. Distributions that are often close to Normal include:
   - Scores on tests taken by many people (such as SAT exams and IQ tests)
   - Repeated careful measurements of the same quantity (like the diameter of a tennis ball)
   - Characteristics of biological populations (such as lengths of crickets and yields of corn)

2. Normal distributions are good approximations to the results of many kinds of chance outcomes, like the proportion of heads in many tosses of a fair coin.

3. Many of the inference methods in Chapters 8–12 are based on Normal distributions.

Normal curves were first applied to data by the great mathematician Carl Friedrich Gauss (1777–1855). He used them to describe the small errors made by astronomers and surveyors in repeated careful measurements of the same quantity. You will sometimes see Normal distributions labeled "Gaussian" in honor of Gauss. His image was even featured on a previous German DM 10 bill, along with a sketch of the Normal distribution.

The Empirical Rule

Earlier, we saw that the distribution of Iowa Test of Basic Skills (ITBS) vocabulary scores for seventh-grade students in Gary, Indiana, is approximately Normal with mean $\mu = 6.84$ and standard deviation $\sigma = 1.55$. How unusual is it for a Gary seventh-grader to get an ITBS score less than 3.74? The figure shows the Normal density curve for this distribution with the area of interest shaded. Note that the boundary value, 3.74, is exactly 2 standard deviations below the mean.

Calculating the shaded area isn't as easy as multiplying base × height, but it's not as hard as you might think. The following activity shows you how to do it.

2.19   3.74   5.29   6.84   8.39   9.94   11.49
ITBS vocabulary score

ACTIVITY **What's so special about Normal distributions?**

In this activity, you will use the *Normal Density Curve* applet at *highschool .bfwpub.com/updatedtps6e* to explore an interesting property of Normal distributions.

Change the mean to 6.84 and the standard deviation to 1.55, and click on "UPDATE." (These are the values for the distribution of ITBS vocabulary scores of seventh-graders in Gary, Indiana.) A figure like the one that follows should appear:

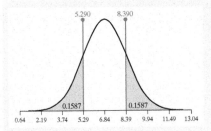

Use the applet to help you answer the following questions.

1. (a) What percent of the area under the Normal curve lies within 1 standard deviation of the mean? That is, about what percent of Gary, Indiana, seventh-graders have ITBS vocabulary scores between 5.29 and 8.39?

(b) What percent of the area under the Normal curve lies within 2 standard deviations of the mean? Interpret this result in context.

(c) What percent of the area under the Normal curve lies within 3 standard deviations of the mean? Interpret this result in context.

2. The distribution of IQ scores in the adult population is approximately Normal with mean $\mu = 100$ and standard deviation $\sigma = 15$. Adjust the applet to display this distribution. About what percent of adults have IQ scores within 1, 2, and 3 standard deviations of the mean?

> When you hear the phrase "standard Normal distribution," think standardized scores (z-scores), which have a mean of 0 and a standard deviation of 1.

3. Adjust the applet to have a mean of 0 and a standard deviation of 1. Then click "UPDATE." The resulting density curve describes the *standard Normal distribution*. What percent of the area under this Normal density curve lies within 1, 2, and 3 standard deviations of the mean?

4. Summarize by completing this sentence: "For any Normal distribution, the area under the Normal curve within 1, 2, and 3 standard deviations of the mean is about ___%, ___%, and ___%."

Although there are many Normal distributions, they all have properties in common. In particular, all Normal distributions obey the **empirical rule**.

> *Empirical* means "learned from experience or by observation."

DEFINITION   The empirical rule

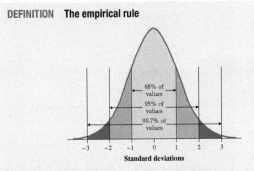

In a Normal distribution with mean μ and standard deviation σ:

- Approximately **68%** of the observations fall within σ of the mean μ.
- Approximately **95%** of the observations fall within 2σ of the mean μ.
- Approximately **99.7%** of the observations fall within 3σ of the mean μ.

This result is known as the **empirical rule**.

Preparing for Inference

When we get to confidence intervals for proportions (Chapter 8), you will have to reveal the truth: 95% of the values in a Normal distribution fall within 1.96 σ of the mean (not 2!). This will be the critical value (z^*) for a 95% confidence interval.

Teaching Tip

You may notice that we ask questions about the "proportion" or "percent" of observations that are between two boundaries, rather than the "probability" that a randomly selected observation is between two boundaries. While we will spend plenty of time using Normal distributions to calculate probabilities in Chapter 6 and beyond, we prefer for now to use the Normal distribution as a model for data.

Teaching Tip

Some students benefit from physically seeing this reasoning. Hold your hands up with your palms touching at the center of an imaginary Normal curve. Then move each hand 1 and then 2 standard deviations away from the mean. Ask students: "What percent of the data is between my hands?" Then "What percent of the data is outside my hands?" Then "So what percent is on each end?"

ALTERNATE EXAMPLE   Skill 2.D

Scoring the Chapter 1 test scores
Using the empirical rule

PROBLEM:
Chapter 1 test scores from Mrs. Gallas's first-hour class follow an approximately Normal distribution with a mean of 81 and standard deviation of 6.
(a) About what percent of students scored greater than 69 on the Chapter 1 test? Show your method clearly.
(b) A student who scored a 69 would be at about what percentile of the distribution? Justify your answer.

SOLUTION:
(a)

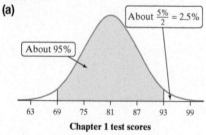

About 95 + 2.5 = 97.5% of students scored greater than 69.
(b) About the 2.5th percentile because about 100% − 97.5% = 2.5% of students scored less than or equal to 69.

Recall from Chapter 1 that there are other rules for determining outliers, such as "any value that is more than 2 (or 3) standard deviations from the mean." This rule makes better sense when we are discussing a Normal distribution!

Some people refer to the empirical rule as the *68–95–99.7 rule*. By remembering these three numbers, you can quickly estimate proportions or percents of observations (areas) using Normal distributions and recognize when an observation is unusual.

Earlier, we asked how unusual it would be for a Gary seventh-grader to get an ITBS score less than 3.74. Figure 2.13 gives the answer in graphical form. By the empirical rule, about 95% of these students have ITBS vocabulary scores between 3.74 and 9.94, which means that about 5% of the students have scores less than 3.74 or greater than 9.94. Due to the symmetry of the Normal distribution, about 5% / 2 = 2.5% of students have scores less than 3.74. So it is quite unusual for a Gary, Indiana, seventh-grader to get an ITBS vocabulary score below 3.74.

FIGURE 2.13 Using the empirical rule to estimate the percent of Gary, Indiana, seventh-graders with ITBS vocabulary scores less than 3.74.

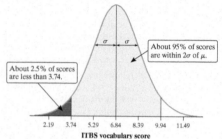
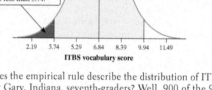
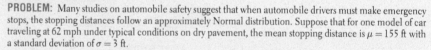

How well does the empirical rule describe the distribution of ITBS vocabulary scores for Gary, Indiana, seventh-graders? Well, 900 of the 947 scores are between 3.74 and 9.94. That's 95.04%, which is very accurate indeed. Of the remaining 47 scores, 20 are below 3.74 and 27 are above 9.94. The number of values in each tail is not quite equal, as it would be in an exactly Normal distribution. Normal distributions often describe real data better in the center of the distribution than in the extreme high and low tails. As famous statistician George Box once noted, "All models are wrong, but some are useful!"

EXAMPLE   **Stop the car!**
Using the empirical rule

PROBLEM: Many studies on automobile safety suggest that when automobile drivers must make emergency stops, the stopping distances follow an approximately Normal distribution. Suppose that for one model of car traveling at 62 mph under typical conditions on dry pavement, the mean stopping distance is $\mu = 155$ ft with a standard deviation of $\sigma = 3$ ft.
(a) About what percent of cars of this model would take more than 158 feet to make an emergency stop? Show your method clearly.
(b) A car of this model that takes 158 feet to make an emergency stop would be at about what percentile of the distribution? Justify your answer.

SOLUTION:

(a)

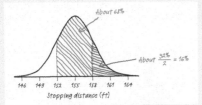

About 68%

About $\frac{32\%}{2}$ = 16%

146 149 152 155 158 161 164
Stopping distance (ft)

> Start by sketching a Normal curve and labeling the values 1, 2, and 3 standard deviations from the mean. Then shade the area of interest.

> Use the empirical rule and the symmetry of the Normal distribution to find the desired area.

About 16% of cars of this model would take more than 158 feet to make an emergency stop.

(b) About the 84th percentile because about 100% − 16% = 84% of cars of this model would stop in less than or equal to 158 feet.

FOR PRACTICE, TRY EXERCISE 51

 Note that the empirical rule applies *only* to Normal distributions. Is there a similar rule that would apply to *any* distribution? Sort of. A result known as *Chebyshev's inequality* says that in any distribution, the proportion of observations falling within k standard deviations of the mean is *at least* $1 - \frac{1}{k^2}$. If $k = 2$, for example, Chebyshev's inequality tells us that at least $1 - \frac{1}{2^2} = 0.75$ of the observations in *any* distribution are within 2 standard deviations of the mean. For Normal distributions, we know that this proportion is much higher than 0.75. In fact, it's approximately 0.95.

> Chebyshev's inequality is an interesting result, but it is not required for the AP® Statistics exam.

CHECK YOUR UNDERSTANDING

 The distribution of heights of young women aged 18 to 24 is approximately Normal with mean $\mu = 64.5$ inches and standard deviation $\sigma = 2.5$ inches.

1. Sketch the Normal curve that approximates the distribution of young women's height. Label the mean and the points that are 1, 2, and 3 standard deviations from the mean.
2. About what percent of young women have heights less than 69.5 inches? Show your work.
3. Is a young woman with a height of 62 inches unusually short? Justify your answer.

Finding Areas in a Normal Distribution

Let's return to the distribution of ITBS vocabulary scores among all Gary, Indiana, seventh-graders. Recall that this distribution is approximately Normal with mean $\mu = 6.84$ and standard deviation $\sigma = 1.55$. What proportion of

---

> **Preparing for Inference**
>
> Finding area in a Normal distribution will be used to calculate a *P*-value for significance tests for proportions in Chapter 9.

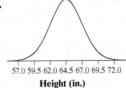

COMMON STUDENT ERROR

Students sometimes try to apply the empirical rule to non-Normal distributions. Make sure to emphasize that this rule works reasonably well only for distributions that are approximately Normal.

Teaching Tip

Chebyshev's inequality is not a topic on the AP® Statistics exam. We mention it here as a contrast to the empirical rule, which applies only to approximately Normal distributions.

✓ **Answers to CYU**

1.

57.0 59.5 62.0 64.5 67.0 69.5 72.0
Height (in.)

2. Because 69.5 inches is 2 standard deviations above the mean, approximately $\frac{100\% - 95\%}{2} = 2.5\%$ of young women have heights greater than 69.5 inches. Therefore, 97.5% of young women have heights less than 69.5 inches.

3. Because 62 is 1 standard deviation below the mean, approximately $\frac{100\% - 68\%}{2} = 16\%$ of young women have heights below 62 inches. This is not unusually short.

these seventh-graders have vocabulary scores that are below sixth-grade level? Figure 2.14 shows the Normal curve with the area of interest shaded. We can't use the empirical rule to find this area because the boundary value of 6 is not exactly 1, 2, or 3 standard deviations from the mean.

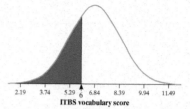

FIGURE 2.14 Normal curve we would use to estimate the proportion of Gary, Indiana, seventh-graders with ITBS vocabulary scores that are less than 6—that is, below sixth-grade level.

As the empirical rule suggests, all Normal distributions are the same if we measure in units of size σ from the mean μ. Changing to these units requires us to standardize, just as we did in Section 2.1:

$$z = \frac{\text{value} - \text{mean}}{\text{standard deviation}} = \frac{x - \mu}{\sigma}$$

Recall that subtracting a constant and dividing by a constant don't change the shape of a distribution. If the quantitative variable we standardize has an approximately Normal distribution, then so does the new variable z. This new distribution of standardized values can be modeled with a Normal curve having mean $\mu = 0$ and standard deviation $\sigma = 1$. It is called the **standard Normal distribution**.

> **DEFINITION** **Standard Normal distribution**
>
> The **standard Normal distribution** is the Normal distribution with mean 0 and standard deviation 1.

Because all Normal distributions are the same when we standardize, we can find areas under any Normal curve using the standard Normal distribution. Table A in the back of the book gives areas under the standard Normal curve. The table entry for each z-score is the area under the curve *to the left* of z.

For the ITBS test score data, we want to find the area to the left of 6 under the Normal distribution with mean 6.84 and standard deviation 1.55. See Figure 2.15(a). We start by standardizing the boundary value $x = 6$:

$$z = \frac{\text{value} - \text{mean}}{\text{standard deviation}} = \frac{6 - 6.84}{1.55} = -0.54$$

Figure 2.15(b) shows the standard Normal distribution with the area to the left of $z = -0.54$ shaded. Notice that the shaded areas in the two graphs are the same.

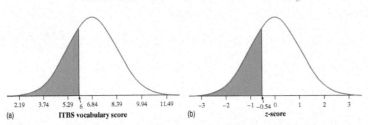

FIGURE 2.15 (a) Normal distribution estimating the proportion of Gary, Indiana, seventh-graders who earn ITBS vocabulary scores less than sixth-grade level. (b) The corresponding area in the standard Normal distribution.

z	.03	.04	.05
−0.6	.2643	.2611	.2578
−0.5	.2981	.2946	.2912
−0.4	.3336	.3300	.3264

To find the area to the left of $z = -0.54$ using Table A, locate -0.5 in the left-hand column, then locate the remaining digit 4 as .04 in the top row. The entry to the right of -0.5 and under .04 is .2946. This is the area we seek. We estimate that about 29.46% of Gary, Indiana, seventh-grader scores fall below the sixth-grade level on the ITBS vocabulary test. *Note that we have made a connection between z-scores and percentiles when the shape of a distribution is approximately Normal.*

It is also possible to find areas under a Normal curve using technology.

---

5. Technology Corner   FINDING AREAS FROM VALUES IN A NORMAL DISTRIBUTION

TI-Nspire and other technology instructions are on the book's website at highschool.bfwpub.com/updatedtps6e.

The normalcdf command on the TI-83/84 can be used to find areas under a Normal curve. The syntax is normalcdf(lower bound, upper bound, mean, standard deviation). Let's use this command to calculate the proportion of ITBS vocabulary scores in Gary, Indiana, that are less than 6. Note that we can do the area calculation using the standard Normal distribution or the Normal distribution with mean 6.84 and standard deviation 1.55.

(i) *Using the standard Normal distribution:* What proportion of observations in a standard Normal distribution are less than $z = -0.54$? Recall that the standard Normal distribution has mean $\mu = 0$ and standard deviation $\sigma = 1$.

- Press 2nd VARS (Distr) and choose normalcdf(.
  OS 2.55 or later: In the dialog box, enter these values: lower: -1000, upper: -0.54, μ:0, σ:1, choose Paste, and then press ENTER.
  Older OS: Complete the command normalcdf$(-1000,-0.54,0,1)$ and press ENTER.

  Note: We chose -1000 as the lower bound because it's many, many standard deviations less than the mean.

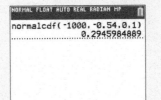

```
NORMAL FLOAT AUTO REAL RADIAN MP
normalcdf(-1000,-0.54,0,1)
                   0.2945984889
```

Teaching Tip

Consider using some notation to help keep track of all the details. Label the first graph as N(6.84, 1.55) and the second graph as N(0, 1). **Caution:** There is some disagreement in the statistics field as to whether it is best to include standard deviation or variance as the second number using this notation.

Teaching Tip: Using Technology

Remind students that they can watch a video of this Technology Corner on the e-Book or at the Student Site at highschool.bfwpub.com/updatedtps6e.

Teaching Tip: Using Technology

If your students are using a TI-84 with OS 2.55 or later, make sure they turn on the Stat Wizards by pressing Mode and scrolling down to the second page. If students are using a TI-83 Plus or TI-84 with an older operating system, they can run an app called *Catalog Help* (ctlghelp) to remind them what values to enter for certain commands. Press the APPS button to see if it is already loaded. If it is, press "Enter" and read the directions. If not, download the app at education.ti.com or copy it from another calculator.

Teaching Tip: Using Technology

If the last two arguments (mean and standard deviation) are omitted, the calculator will automatically assume a standard Normal distribution (with a mean of 0 and standard deviation of 1). This means that the calculation shown here could have been entered as normalcdf $(-1000, -0.54)$.

Teaching Tip: Using Technology

We use -1000 as a lower bound when finding areas to the left in a standard Normal distribution because it is virtually impossible to be 1000 standard deviations below the mean in a Normal distribution. In theory, the lower bound is negative infinity, but many calculators do not have an infinity button. When doing Normal calculations for Normal distributions other than the standard Normal distribution, you should pick a different lower (or upper) boundary, depending on the scale of the numbers involved.

Teaching Tip: Using Technology

In addition to the normalcdf command, the calculator also lists a normalpdf command. It is not important that your students know how to use this command, but some students may ask what normalpdf does. Tell them that it gives the height of a Normal curve for a particular value of x, mean μ, and standard deviation σ. For example, normalpdf (x value: 0, μ: 0, σ: 1) = 0.3989 gives the height of the standard Normal curve at its peak ($x = 0$).

Teaching Tip: Using Technology

Go to the Student Site at highschool .bfwpub.com/updatedtps6e and choose the *Probability* applet from the Extra Applets to calculate area under a Normal distribution using an applet.

Ask the StatsMedic

Table A or normalcdf?

Should students use *z*-scores and Table A or should you recommend normalcdf? There is no correct answer, but this post has some good discussion about the advantages and disadvantages of each.

AP® EXAM TIP

Have students draw a picture for *every* Normal distribution calculation. They can then use the picture to confirm that their answer makes sense. The picture often saves students from a careless mistake (like forgetting to subtract the area from 1 when finding area to the *right* of a value when using Table A).

(ii) *Using the unstandardized Normal distribution*: What proportion of observations in a Normal distribution with mean $\mu = 6.84$ and standard deviation $\sigma = 1.55$ are less than $x = 6$?

- Press [2nd] [VARS] (Distr) and choose normalcdf(.
  OS 2.55 or later: In the dialog box, enter these values: lower: −1000, upper:6, μ:6.84, σ:1.55, choose Paste, and then press [ENTER].
  Older OS: Complete the command normalcdf(−1000,6,6.84,1.55) and press [ENTER].

```
NORMAL FLOAT AUTO REAL RADIAN MP
normalcdf(-1000,6,6.84,1.▶
                 0.2939314473
```

This answer differs slightly from the one we got using the standard Normal distribution because we rounded the standardized score to two decimal places: $z = -0.54$.

The following box summarizes the process of finding areas in a Normal distribution. In Step 2, each method of performing calculations has some advantages, so check with your teacher to see which option will be used in your class.

HOW TO FIND AREAS IN ANY NORMAL DISTRIBUTION

Step 1: Draw a Normal distribution with the horizontal axis labeled and scaled using the mean and standard deviation, the boundary value(s) clearly identified, and the area of interest shaded.

Step 2: Perform calculations—show your work! Do one of the following:

(i) Standardize each boundary value and use Table A or technology to find the desired area under the standard Normal curve; or

(ii) Use technology to find the desired area without standardizing.

Be sure to answer the question that was asked.

AP® EXAM TIP

Students often do not get full credit on the AP® Statistics exam because they use option (ii) with "calculator-speak" to show their work on Normal calculation questions—for example, normalcdf(−1000,6,6.84,1.55). This is *not* considered clear communication. To get full credit, follow the two-step process above, making sure to carefully label each of the inputs in the calculator command if you use technology in Step 2: normalcdf(lower: −1000, upper: 6, mean: 6.84, SD:1.55).

EXAMPLE

Stop the car!

Finding area to the left

PROBLEM: As noted in the preceding example, studies on automobile safety suggest that stopping distances follow an approximately Normal distribution. For one model of car traveling at 62 mph, the mean stopping distance is $\mu = 155$ ft with a standard deviation of $\sigma = 3$ ft. Danielle is driving one of these cars at 62 mph when she spots a wreck 160 feet in front of her and needs to make an emergency stop. About what percent of cars of this model when going 62 mph would be able to make an emergency stop in less than 160 feet? Is Danielle likely to stop safely?

SOLUTION:

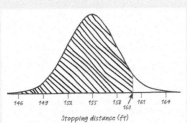

Stopping distance (ft)

> 1. **Draw a normal distribution.** Be sure to:
>   - Scale the horizontal axis.
>   - Label the horizontal axis with the variable name, including units of measurement.
>   - Clearly identify the boundary value(s).
>   - Shade the area of interest.

(i) $z = \dfrac{160 - 155}{3} = 1.67$

Using Table A: Area for $z < 1.67$ is 0.9525.
Using technology: normalcdf(lower: −1000, upper: 1.67, mean: 0, SD: 1) = 0.9525

(ii) normalcdf(lower: −1000, upper: 160, mean: 155, SD: 3) = 0.9522

> 2. **Perform calculations—show your work!**
>   (i) Standardize the boundary value and use Table A or technology to find the desired probability; or
>   (ii) Use technology to find the desired area without standardizing.

z	.06	.07	.08
1.5	.9406	.9418	.9429
1.6	.9515	.9525	.9535
1.7	.9608	.9616	.9625

About 95% of cars of this model would be able to make an emergency stop within 160 feet. So Danielle is likely to be able to stop safely.

> Be sure to answer the question that was asked.

FOR PRACTICE, TRY EXERCISE 53

What percent of cars of this model would be able to make an emergency stop in less than 140 feet? The standardized score for $x = 140$ is

$$z = \frac{140 - 155}{3} = -5.00$$

Table A does not go beyond $z = -3.50$ and $z = 3.50$ because it is highly unusual for a value to be more than 3.5 standard deviations from the mean in a Normal distribution. For practical purposes, we can act as if there is approximately zero probability outside the range of Table A. So there is almost no chance that a car of this model going 62 mph would be able to make an emergency stop within 140 feet.

FINDING AREAS TO THE RIGHT IN A NORMAL DISTRIBUTION What proportion of Gary, Indiana, seventh-grade scores on the ITBS vocabulary test are *at least* 9? Start with a picture. Figure 2.16(a) on the next page shows the Normal distribution with mean $\mu = 6.84$ and standard deviation $\sigma = 1.55$ with the area of interest shaded. Next, standardize the boundary value:

$$z = \frac{9 - 6.84}{1.55} = 1.39$$

Figure 2.16(b) shows the standard Normal distribution with the area to the right of $z = 1.39$ shaded. Again, notice that the shaded areas in the two graphs are the same.

Teaching Tip

How far does a value have to be away from the mean to be considered *unusual*? Most would say that any value that is more than 2 standard deviations away from the mean is considered *unusual*.

ALTERNATE EXAMPLE   Skill 3.A

What is so Normal about SAT math scores?

Finding area to the left

PROBLEM:

In the class of 2016, more than 1.6 million students took the SAT. The distribution of scores on the math section (out of 800) follows an approximately Normal distribution with a mean of 508 and standard deviation of 110. About what percent of students who took the SAT scored less than 350 on the math section?

SOLUTION:

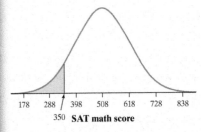

350   **SAT math score**

(i) $z = \dfrac{350 - 508}{110} = -1.44$

Table A: Area for $z < -1.44$ is 0.0749.
Tech: normalcdf(lower: −1000, upper: −1.44, mean: 0, SD: 1) = 0.0749

(ii) normalcdf(lower: −1000, upper: 350, mean: 508, SD: 110) = 0.0754

About 7.5% of students from the class of 2016 who took the SAT scored less than 350 on the math section.

Teaching Tip

In this alternate example, consider requiring that students draw the unstandardized distribution (as shown), as well as the standardized distribution (not shown). This will help students see the rationale for using Table A. You can relax this requirement for Normal distribution calculations in later chapters. An alternative would be to put a second scale on the original curve.

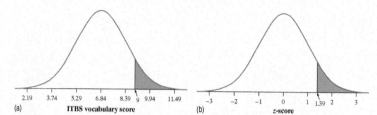

FIGURE 2.16 (a) Normal distribution estimating the proportion of Gary, Indiana, seventh-graders who earn ITBS vocabulary scores at the ninth-grade level or higher. (b) The corresponding area in the standard Normal distribution.

z	.07	.08	.09
1.2	.8980	.8997	.9015
1.3	.9147	.9162	.9177
1.4	.9292	.9306	.9319

To find the area to the right of $z = 1.39$, locate 1.3 in the left-hand column of Table A, then locate the remaining digit 9 as .09 in the top row. The entry to the right of 1.3 and under .09 is .9177. However, this is the area *to the left* of $z = 1.39$. We can use the fact that the total area in the standard Normal distribution is 1 to find that the area *to the right* of $z = 1.39$ is $1 - 0.9177 = 0.0823$. We estimate that about 8.23% of Gary, Indiana, seventh-graders earn scores at the ninth-grade level or above on the ITBS vocabulary test.

A common student mistake is to look up a z-score in Table A and report the entry corresponding to that z-score, regardless of whether the problem asks for the area to the left or to the right of that z-score. This mistake can usually be prevented by drawing a Normal distribution and shading the area of interest. Look to see if the area should be closer to 0 or closer to 1. In the preceding example, for instance, it should be obvious that 0.9177 is *not* the correct area.

Are you worthy to be a Wolverine?
Finding area to the right

PROBLEM:
The distribution of scores on the math section of the SAT (out of 800) follows an approximately Normal distribution with a mean of 508 and standard deviation of 110. The University of Michigan has a recommended math SAT score of at least 730. What percent of students who took the math SAT meet this requirement?

SOLUTION:

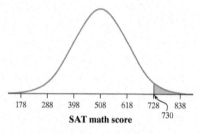

(i) $z = \dfrac{730 - 508}{110} = 2.02$

Table A: Area for $z < 2.02$ is 0.9783; area for $z \geq 2.02$ is $1 - 0.9783 = 0.0217$.
Tech: normalcdf(lower: 2.02, upper: 1000, mean: 0, SD: 1) = 0.0217
(ii) normalcdf(lower: 730, upper: 10,000, mean: 508, SD: 110) = 0.0218
Note: The reason these answers differ (by 0.0001) is that the z-score of 2.02 was rounded in order to use Table A. However, each answer would receive full credit on the AP® Statistics exam.

About 2.2% of students who took the SAT scored at least 730 on the math section. It appears that the University of Michigan is a tough school to get into.

EXAMPLE

Can Spieth clear the trees?
Finding area to the right

PROBLEM: When professional golfer Jordan Spieth hits his driver, the distance the ball travels can be modeled by a Normal distribution with mean 304 yards and standard deviation 8 yards. On a specific hole, Jordan would need to hit the ball at least 290 yards to have a clear second shot that avoids a large group of trees. What percent of Spieth's drives travel at least 290 yards? Is he likely to have a clear second shot?

SOLUTION:

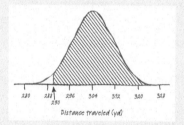

1. Draw a Normal distribution.

Teaching Tip

To make Normal curves less abstract, have students mentally "fill" each Normal curve with dots. Explain that the Normal curve is tracing what a dotplot would look like after many observations. In this example, you can even suggest that students fill the Normal curve with little golf balls, each one representing the distance that Jordan Spieth drives a ball.

(i) $z = \dfrac{290 - 304}{8} = -1.75$

Using Table A: Area for $z < -1.75$ is 0.0401. Area for $z \geq -1.75$ is $1 - 0.0401 = 0.9599$.

Using technology: normalcdf(lower: −1.75, upper: 1000, mean: 0, SD: 1) = 0.9599

(ii) normalcdf(lower: 290, upper: 1000, mean: 304, SD: 8) = 0.9599

About 96% of Jordan Spieth's drives travel at least 290 yards. So he is likely to have a clear second shot.

> **2. Perform calculations—show your work!**
> (i) Standardize and use Table A or technology; or
> (ii) Use technology without standardizing.

> Be sure to answer the question that was asked.

FOR PRACTICE, TRY EXERCISE 55

Think About It

WHAT PROPORTION OF JORDAN SPIETH'S DRIVES GO EXACTLY 290 YARDS? There is no area under the Normal density curve in the preceding example directly above the point 290.000000000. . . . So the answer to our question based on the Normal distribution is 0. One more thing: the areas under the curve with $x \geq 290$ and $x > 290$ are the same. According to the Normal model, the proportion of Spieth's drives that travel at least 290 yards is the same as the proportion that travel more than 290 yards.

FINDING AREAS BETWEEN TWO VALUES IN A NORMAL DISTRIBUTION How do you find the area in a Normal distribution that is between two values? For instance, suppose we want to estimate the proportion of Gary, Indiana, seventh-graders with ITBS vocabulary scores between 6 and 9. Figure 2.17(a) shows the desired area under the Normal curve with mean $\mu = 6.84$ and standard deviation $\sigma = 1.55$. We can use Table A or technology to find the desired area.

Option (i): If we standardize each boundary value, we get

$$z = \frac{6 - 6.84}{1.55} = -0.54, \qquad z = \frac{9 - 6.84}{1.55} = 1.39$$

Figure 2.17(b) shows the corresponding area of interest in the standard Normal distribution.

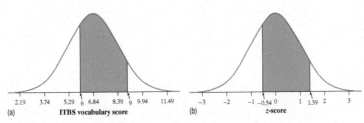

(a)   2.19   3.74   5.29   6   6.84   8.39   9   9.94   11.49   **ITBS vocabulary score**

(b)   −3   −2   −1   −0.54   0   1   1.39   2   3   **z-score**

FIGURE 2.17 (a) Normal distribution approximating the proportion of seventh-graders in Gary, Indiana, with ITBS vocabulary scores between 6 and 9. (b) The corresponding area in the standard Normal distribution.

Making Connections

The fact that the proportion of observations with $x \geq 290$ is the same as the proportion of observations with $x > 290$ is true when using any density curve, not just a Normal curve. Students will have to recall this fact later in the course when we discuss the differences between discrete random variables and continuous random variables in Chapter 6.

Teaching Tip: AP® Connections

AP® Calculus students will understand that the area under a continuous function for one specific value of x is 0.

Teaching Tip

There are three types of questions for finding area under a Normal curve: (1) area to the left, (2) area to the right, and (3) area between two values. Make sure your students are proficient in all three types.

Using Table A: The table makes this process a bit trickier because it only shows areas to the left of a given z-score. The visual shows one way to think about the calculation.

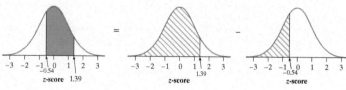

Area between $z = -0.54$ and $z = 1.39$
= Area to the left of $z = 1.39$ − Area to the left of $z = -0.54$
= $0.9177 - 0.2946$
= 0.6231

Using technology: normalcdf(lower: −0.54, upper: 1.39, mean: 0, SD: 1) = 0.6231

Option (ii): normalcdf(lower: 6, upper: 9, mean: 6.84, SD: 1.55) = 0.6243.

About 62% of Gary, Indiana, seventh-graders earned grade-equivalent scores between 6 and 9.

ALTERNATE EXAMPLE   Skill 3.A

SAT math scores

Finding areas between two values

PROBLEM:
The distribution of scores on the math section of the SAT (out of 800) follows an approximately Normal distribution with a mean of 508 and standard deviation of 110. What percent of students score in the 500s?

SOLUTION:

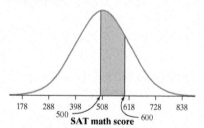

(i) $z = \dfrac{600 - 508}{110} = 0.84$

$z = \dfrac{500 - 508}{110} = -0.07$

Table A: $0.7995 - 0.4721 = 0.3274$

Tech: normalcdf(lower: −0.07, upper: 0.84, mean: 0, SD: 1) = 0.3274

(ii) normalcdf(lower: 500, upper: 600, mean: 508, SD: 110) = 0.3275

About 33% of students who took the SAT scored in the 500s on the math section.

EXAMPLE

Can Spieth reach the green?
Finding areas between two values

PROBLEM: When professional golfer Jordan Spieth hits his driver, the distance the ball travels can be modeled by a Normal distribution with mean 304 yards and standard deviation 8 yards. On another golf hole, Spieth has the opportunity to drive the ball onto the green if he hits the ball between 305 and 325 yards. What percent of Spieth's drives travel a distance that falls in the interval? Is he likely to get the ball on the green with his drive?

SOLUTION:

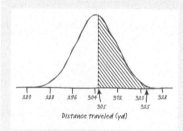

1. **Draw a Normal distribution.**

(i) $z = \dfrac{305 - 304}{8} = 0.13$ $z = \dfrac{325 - 304}{8} = 2.63$

2. **Perform calculations—show your work!**
   (i) Standardize and use Table A or technology; or
   (ii) Use technology without standardizing.

Using Table A: $0.9957 - 0.5517 = 0.4440$
Using technology: normalcdf(lower:0.13, upper:2.63, mean:0, SD:1) = 0.4440

(ii) normalcdf(lower:305, upper:325, mean:304, SD:8) = 0.4459

About 45% of Spieth's drives travel between 305 and 325 yards. He has a fairly good chance of getting the ball on the green—assuming he hits the shot straight.

> Be sure to answer the question that was asked.

FOR PRACTICE, TRY EXERCISE 57

CHECK YOUR UNDERSTANDING

High levels of cholesterol in the blood increase the risk of heart disease. For 14-year-old boys, the distribution of blood cholesterol is approximately Normal with mean $\mu = 170$ milligrams of cholesterol per deciliter of blood (mg/dl) and standard deviation $\sigma = 30$ mg/dl.[6]

1. Cholesterol levels higher than 240 mg/dl may require medical attention. What percent of 14-year-old boys have more than 240 mg/dl of cholesterol?

2. People with cholesterol levels between 200 and 240 mg/dl are at considerable risk for heart disease. What proportion of 14-year-old boys have blood cholesterol between 200 and 240 mg/dl?

Working Backward: Finding Values from Areas

So far, we have focused on finding areas in Normal distributions that correspond to specific values. What if we want to find the value that corresponds to a particular area? For instance, suppose we want to estimate the 90th percentile of the distribution of ITBS vocabulary scores for Gary, Indiana, seventh-graders. Figure 2.18(a) shows the Normal curve with mean $\mu = 6.84$ and standard deviation $\sigma = 1.55$ that models this distribution. We're looking for the ITBS score x that has 90% of the area under the curve less than or equal to it. Figure 2.18(b) shows the standard Normal distribution with the corresponding area shaded.

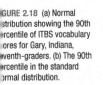

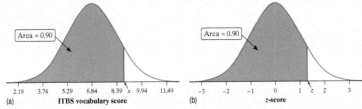

FIGURE 2.18 (a) Normal distribution showing the 90th percentile of ITBS vocabulary scores for Gary, Indiana, seventh-graders. (b) The 90th percentile in the standard Normal distribution.

We can use Table A or technology to find the z-score with an area of 0.90 to its left. Because Table A gives the area to the left of a specified z-score, all we have to do is find the value closest to 0.90 in the middle of the table. From the

1. (i) $z = \dfrac{240 - 170}{30} = 2.33$

Table A: The proportion of z-scores above 2.33 is $1 - 0.9901 = 0.0099$.
Tech: normalcdf(lower: 2.33, upper: 1000, mean: 0, SD: 1) = 0.0099.

(ii) *Tech:* normalcdf(lower: 240, upper: 1000, mean: 170, SD: 30) = 0.0098. About 1% of 14-year-old boys have cholesterol above 240 mg/dl.

2. (i) $z = \dfrac{200 - 170}{30} = 1$ and

$z = \dfrac{240 - 170}{30} = 2.33$

Table A: The proportion of z-scores below $z = 1.00$ is 0.8413 and the proportion of z-scores below 2.33 is 0.9901. Thus, the proportion of z-scores between 1 and 2.33 is $0.9901 - 0.8413 = 0.1488$.
Tech: normalcdf(lower: 1, upper: 2.33, mean: 0, SD: 1) = 0.1488.

(ii) *Tech:* normalcdf(lower: 200, upper: 240, mean: 170, SD: 30) = 0.1488. About 15% of 14-year-old boys have cholesterol between 200 and 240 mg/dl.

Teaching Tip

When starting a Normal distribution problem with students, start by asking, "Are you given a value and asked to find an area OR given an area and asked to find a value?"

Teaching Tip

There is a way to do this problem without algebra by thinking about the interpretation of a *z*-score. The *z*-score of 1.28 means that the value *x* is 1.28 standard deviations above the mean. All that is needed is to add 1.28 standard deviations to the mean:

Mean + 1.28 standard deviations = 6.84 + 1.28(1.55) = 8.824

Notice that this solution is perfectly parallel with the algebraic solution.

Teaching Tip: Using Technology

Go to highschool.bfwpub.com/updatedtps6e and choose the *Probability* applet from the Extra Applets to find a value for a given area under a Normal distribution.

Making Connections

In earlier problems, we were given a value and asked to find an area. In these problems, we are given an area and asked to find a value. This sounds like an *inverse* function from algebra class.

z	.07	.08	.09
1.1	.8790	.8810	.8830
1.2	.8980	.8997	.9015
1.3	.9147	.9162	.9177

reproduced portion of Table A, you can see that the desired value is $z = 1.28$. Then we "unstandardize" to get the corresponding ITBS vocabulary score *x*.

$$z = \frac{x - \mu}{\sigma}$$

$$1.28 = \frac{x - 6.84}{1.55}$$

$$1.28(1.55) + 6.84 = x$$

$$8.824 = x$$

So the 90th percentile of the distribution of ITBS vocabulary scores for Gary, Indiana, seventh-graders is 8.824.

It is also possible to find the 90th percentile of either distribution in Figure 2.18 using technology.

---

6. Technology Corner   FINDING VALUES FROM AREAS IN A NORMAL DISTRIBUTION

TI-Nspire and other technology instructions are on the book's website at highschool.bfwpub.com/updatedtps6e.

The TI-83/84 invNorm command calculates the value corresponding to a given percentile in a Normal distribution. The syntax is invNorm(area to the left, mean, standard deviation). Let's use this command to confirm the 90th percentile for the ITBS vocabulary scores in Gary, Indiana. Note that we can do the calculation using the standard Normal distribution or the Normal distribution with mean 6.84 and standard deviation 1.55.

(i) *Using the standard Normal distribution:* What is the 90th percentile of the standard Normal distribution?

- Press [2nd] [VARS] (Distr) and choose invNorm(.
  OS 2.55 or later: In the dialog box, enter these values: area:0.90, μ:0, σ:1, choose Paste, and then press [ENTER].
  Older OS: Complete the command invNorm(0.90,0,1) and press [ENTER].

Note: The most recent TI-84 Plus CE OS has added an option for specifying area in the LEFT, CENTER, or RIGHT of the distribution. Choose LEFT in this case.

This result matches what we got using Table A. Now "unstandardize" as shown preceding the Technology Corner to get $x = 8.824$.

(ii) *Using the unstandardized Normal distribution:* What is the 90th percentile of a Normal distribution with mean $\mu = 6.84$ and standard deviation $\sigma = 1.55$?

- Press [2nd] [VARS] (Distr) and choose invNorm(.
  OS 2.55 or later: In the dialog box, enter these values: area:0.90, μ:6.84, σ:1.55, choose Paste, and then press [ENTER].
  Older OS: Complete the command invNorm(0.90,6.84,1.55) and press [ENTER].

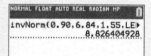

Note: The most recent TI-84 Plus CE OS has added an option for specifying area in the LEFT, CENTER, or RIGHT of the distribution. Choose LEFT in this case.

The following box summarizes the process of finding a value corresponding to a given area in a Normal distribution. In Step 2, each method of performing calculations has some advantages, so check with your teacher to see which option will be used in your class.

HOW TO FIND VALUES FROM AREAS IN ANY NORMAL DISTRIBUTION

Step 1: Draw a Normal distribution with the horizontal axis labeled and scaled using the mean and standard deviation, the area of interest shaded and labeled, and unknown boundary value clearly marked.

Step 2: Perform calculations—show your work! Do one of the following:

(i) Use Table A or technology to find the value of z with the indicated area under the standard Normal curve, then "unstandardize" to transform back to the original distribution; or

(ii) Use technology to find the desired value without standardizing.

Be sure to answer the question that was asked.

AP® EXAM TIP

As noted previously, to make sure that you get full credit on the AP® Statistics exam, do not use "calculator-speak" alone—for example, invNorm(0.90,6.84,1.55). This is *not* considered clear communication. To get full credit, follow the two-step process above, making sure to carefully label each of the inputs in the calculator command if you use technology in Step 2: invNorm(area: 0.90, mean: 6.84, SD:1.55).

EXAMPLE

How tall are 3-year-old girls?
Finding a value from an area

PROBLEM: According to www.cdc.gov/growthcharts/, the heights of 3-year-old females are approximately Normally distributed with a mean of 94.5 centimeters and a standard deviation of 4 centimeters. Seventy-five percent of 3-year-old girls are taller than what height?

SOLUTION:

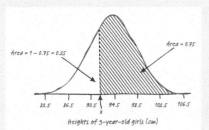

Heights of 3-year-old girls (cm)

If 75% of 3-year-old girls are taller than a certain height, then 25% of 3-year-old girls are shorter than that height. So we just need to find the 25th percentile of this distribution of height.

1. Draw a Normal distribution.
From the empirical rule, we know that about 16% of the observations in a Normal distribution will fall more than 1 standard deviation less than the mean. So the 25th percentile will be located slightly to the right of 90.5, as shown.

Great Scott!
Finding a value from an area

PROBLEM:
After accelerating for 20 seconds, a DeLorean sports car has a wide range of speeds that it can achieve, depending on traction. The distribution of speed follows an approximately Normal distribution with a mean of 80 mph and standard deviation of 7.7 mph. Marty wants the next acceleration run to be in the fastest 15% of all possible speeds. How fast will the car have to go?

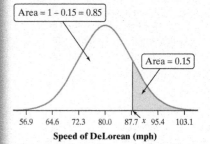

Area = 1 − 0.15 = 0.85

Area = 0.15

| 56.9 | 64.6 | 72.3 | 80.0 | 87.7 | 95.4 | 103.1 |

x

Speed of DeLorean (mph)

SOLUTION:
(i) *Table A:* 0.85 area to the left $\rightarrow z = 1.04$
Tech: invNorm(area: 0.85, mean: 0, SD: 1) = 1.04

$$1.04 = \frac{x - 80}{7.7}$$

$$1.04(7.7) + 80 = x$$

$$88.0 = x$$

(ii) invNorm(area: 0.85, mean: 80, SD: 7.7) = 88.0

Marty will have to achieve a speed in the DeLorean of at least 88 mph.

TRM Extra Practice with the Normal Distribution

Find additional worksheets for students to practice Normal distribution calculations. Click on the link in the TE-Book, open in the TRFD, or download from the Teacher's Resources on the book's digital platform.

(i) *Using Table A:* 0.25 area to the left → z = −0.67

Using technology: invNorm(area: 0.25, mean: 0, SD: 1) = −0.67

$$-0.67 = \frac{x - 94.5}{4}$$

$$-0.67(4) + 94.5 = x$$

$$91.82 = x$$

(ii) invNorm (area: 0.25, mean: 94.5, SD: 4) = 91.80

About 75% of 3-year-old girls are taller than 91.80 centimeters.

> **2. Perform calculations—show your work!**
> (i) Use Table A or technology to find the value of z with the indicated area under the standard Normal curve, then "unstandardize" to transform back to the original distribution; or
> (ii) Use technology to find the desired value without standardizing.

> Be sure to answer the question that was asked.

FOR PRACTICE, TRY EXERCISE 63

Here's an activity that gives you a chance to apply what you have learned so far in this section.

ACTIVITY   Team challenge: The vending machine problem

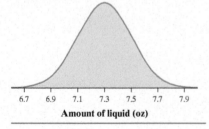
Ed Honowitz/The Image Bank/Getty Images

In this activity, you will work in a team of three or four students to resolve a real-world problem.

Have you ever purchased a hot drink from a vending machine? The intended sequence of events goes something like this: You insert your money into the machine and select your preferred beverage. A cup falls out of the machine, landing upright. Liquid pours out until the cup is nearly full. You reach in, grab the piping-hot cup, and drink happily.

Except sometimes, things go wrong. The machine might swipe your money. Or the cup might fall over. More frequently, everything goes smoothly until the liquid begins to flow. It might stop flowing when the cup is only half full. Or the liquid might keep coming until your cup overflows. Neither of these results leaves you satisfied.

The vending machine company wants to keep its customers happy. So it has decided to hire you as a statistical consultant. The company provides you with the following summary of important facts about the vending machine:

- Cups will hold 8 ounces.
- The amount of liquid dispensed varies according to a Normal distribution centered at the mean μ that is set in the machine.
- $\sigma = 0.2$ ounces.

If a cup contains too much liquid, a customer may get burned from a spill. This could result in an expensive lawsuit for the company. On the other hand, customers may be irritated if they get a cup with too little liquid from the machine.

Given these issues, what mean setting for the machine would you recommend? Provide appropriate graphical and numerical evidence to support your conclusion. Be prepared to defend your answer.

▶ ACTIVITY OVERVIEW

To prepare for using this activity, watch the overview video by clicking on the link in the TE-Book, opening the TRFD, or downloading from the Teacher's Resources on the book's digital platform.

Time: 20 minutes

Materials: None

Teaching Advice: This activity has a very open-ended question. There is no single desired correct answer. It is much more important that students can clearly communicate their reasoning. Clear communication should include pictures, calculations, and written descriptions.

Answers:

Answers will vary. Here is one possible answer: We recommend a mean setting for the machine of 7.3 ounces. With this setting, 99.7% of the cups will get fluid ounce values that are within 3 standard deviations of the mean. Since $7.3 - 3(0.2) = 6.7$ and $7.3 + 3(0.2) = 7.9$, we know that 99.7% of the cups will receive between 6.7 and 7.9 ounces. At 6.7 ounces or more, customers should feel they are getting enough liquid. Also, we don't want a setting of more than 7.9 ounces, because customers may spill hot liquid on themselves.

```
6.7  6.9  7.1  7.3  7.5  7.7  7.9
          Amount of liquid (oz)
```

> ### CHECK YOUR UNDERSTANDING
>
> High levels of cholesterol in the blood increase the risk of heart disease. For 14-year-old boys, the distribution of blood cholesterol is approximately Normal with mean $\mu = 170$ milligrams of cholesterol per deciliter of blood (mg/dl) and standard deviation $\sigma = 30$ mg/dl. What cholesterol level would place a 14-year-old boy at the 10th percentile of the distribution?

Assessing Normality

Normal distributions provide good models for some distributions of quantitative data. Examples include SAT and IQ test scores, the highway gas mileage of 2018 Corvette convertibles, weights of 9-ounce bags of potato chips, and heights of 3-year-old girls (see Figure 2.19).

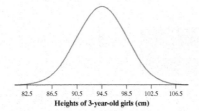

Heights of 3-year-old girls (cm)

FIGURE 2.19 The heights of 3-year-old girls are approximately Normally distributed with a mean of 94.5 centimeters and standard deviation of 4 centimeters.

The distributions of other quantitative variables are skewed and therefore distinctly non-Normal. Examples include single-family home prices in a certain city, survival times of cancer patients after treatment, and number of siblings for students in a statistics class.

While experience can suggest whether or not a Normal distribution is a reasonable model in a particular case, it is risky to assume that a distribution is approximately Normal without first analyzing the data. As in Chapter 1, we start with a graph and then add numerical summaries to assess the Normality of a distribution of quantitative data.

If a graph of the data is clearly skewed, has multiple peaks, or isn't bell-shaped, that's evidence the distribution is not Normal. Here is a dotplot of the number of siblings reported by each student in a statistics class. This distribution is skewed to the right and therefore not approximately Normal.

Number of siblings

✔ **Answers to CYU**

(i) *Table A:* Look in the body of Table A for the value closest to 0.10. A *z*-score of -1.28 gives the closest value (0.1003). *Tech:* invNorm(area: 0.10, mean: 0, SD: 1) $= -1.28$

$$-1.28 = \frac{x - 170}{30}$$
$$-38.4 = x - 170$$
$$x = 131.6$$

(ii) *Tech:* invNorm(area: 0.10, mean: 170, SD: 30) $= 131.6$. About 10% of 14-year-old boys have cholesterol levels that are less than or equal to 131.6, so a 14-year-old boy who has a cholesterol level of 131.6 would be at the 10th percentile of the distribution.

Teaching Tip

As a warm-up activity for this lesson, give students all the scores (without names) from your Chapter 1 test. Ask them the open-ended question "Are these data approximately Normal?" Then let them try their own approaches. Most will start with a graph, many will add numerical summaries, and some will even utilize the empirical rule.

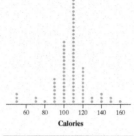

n   Mean   SD   Min Q_1 Med Q_3 Max

77 106.883 19.484   50   100   110   110   160

FIGURE 2.20 Dotplot and summary statistics for data on calories per serving in 77 different brands of breakfast cereal.

Even if a graph of the data looks roughly symmetric and bell-shaped, we shouldn't assume that the distribution is approximately Normal. The empirical rule can give additional evidence in favor of or against Normality.

Figure 2.20 shows a dotplot and numerical summaries for data on calories per serving in 77 brands of breakfast cereal.[7] The graph is roughly symmetric, single-peaked (unimodal), and somewhat bell-shaped. Let's count the number of data values within 1, 2, and 3 standard deviations of the mean:

Mean $\pm$ 1 SD: 106.883 $\pm$ 1(19.484)   87.399 to 126.367   63 out of 77 = 81.8%

Mean $\pm$ 2 SD: 106.883 $\pm$ 2(19.484)   67.915 to 145.851   71 out of 77 = 92.2%

Mean $\pm$ 3 SD: 106.883 $\pm$ 3(19.484)   48.431 to 165.335   77 out of 77 = 100.0%

In a Normal distribution, about 68% of the values fall within 1 standard deviation of the mean. For the cereal data, almost 82% of the brands had between 87.399 and 126.367 calories per serving. These two percentages are far apart. So this distribution of calories in breakfast cereals is not approximately Normal.

PROBLEM:
Here are the Chapter 1 test scores for the 32 students in Mrs. Gallas's first-hour AP® Statistics class:

68   70   73   74   74   75   75   77

77   77   78   78   78   79   80   80

81   82   83   83   83   83   84   85

85   86   88   88   90   91   92   96

A histogram and summary statistics for the data are shown. Is this distribution of Chapter 1 test scores approximately Normal? Justify your answer based on the graph and the empirical rule.

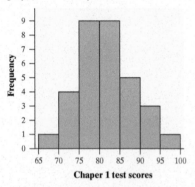

n	mean	SD	min	Q_1	med	Q_3	max
32	81.031	6.508	68	77	80.5	85	96

EXAMPLE

Are IQ scores Normally distributed?
Assessing Normality

PROBLEM: Many people believe that the distribution of IQ scores follows a Normal distribution. Is that really the case? To find out, researchers obtained the IQ scores of 60 randomly selected fifth-grade students from one school. Here are the data:[8]

81	82	89	90	94	96	97	100	101	101	101
102	102	102	103	105	106	108	109	109	109	110
110	110	112	112	113	113	114	114	114	115	116
117	117	117	118	118	122	122	123	124	124	124
125	126	127	127	128	130	131	133	134	134	136
137	139	139	142	145						

A histogram and summary statistics for the data are shown. Is this distribution of IQ scores of fifth-graders at this school approximately Normal? Justify your answer based on the graph and the empirical rule.

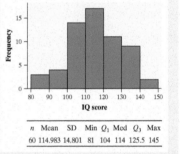

n   Mean   SD   Min Q_1 Med Q_3 Max

60 114.983 14.801   81   104   114   125.5   145

SOLUTION:
The histogram looks roughly symmetric, single-peaked, and somewhat bell-shaped. The percents of values within 1, 2, and 3 standard deviations of the mean are

Mean $\pm$ 1 SD: 81.031 $\pm$ 6.508   74.523 to 87.539   21 out of 32 = 65.6%

Mean $\pm$ 2 SD: 81.031 $\pm$ 2(6.508)   68.015 to 94.047   30 out of 32 = 93.8%

Mean $\pm$ 3 SD: 81.031 $\pm$ 3(6.508)   61.507 to 100.555   32 out of 32 = 100.0%

These percents are very close to the 68%, 95%, and 99.7% targets for a Normal distribution. The graphical and numerical evidence suggests that this distribution of Chapter 1 test scores is approximately Normal.

SOLUTION:

The histogram looks roughly symmetric, single-peaked, and somewhat bell-shaped. The percents of values within 1, 2, and 3 standard deviations of the mean are

Mean ± 1 SD: 114.983 ± 1(14.801)   100.182 to 129.784   41 out of 60 = 68.3%
Mean ± 2 SD: 114.983 ± 2(14.801)   85.381 to 144.585   57 out of 60 = 95.0%
Mean ± 3 SD: 114.983 ± 3(14.801)   70.580 to 159.386   60 out of 60 = 100.0%

These percents are very close to the 68%, 95%, and 99.7% targets for a Normal distribution. The graphical and numerical evidence suggests that this distribution of IQ scores is approximately Normal.

FOR PRACTICE, TRY EXERCISE 75

> **AP® EXAM TIP**
>
> Never say that a distribution of quantitative data *is* Normal. Real-world data always show at least slight departures from a Normal distribution. The most you can say is that the distribution is "approximately Normal."

Because the IQ data come from a random sample, we can use the sample mean IQ score to make an inference about the population mean IQ score of all fifth-graders at the school. As you will see in later chapters, the methods for inference about a population mean work best when the population distribution is Normal. Because the distribution of IQ scores in the sample is approximately Normal, it is reasonable to believe that the population distribution is approximately Normal.

NORMAL PROBABILITY PLOTS A graph called a **Normal probability plot** (or a *Normal quantile plot*) provides a good assessment of whether or not a distribution of quantitative data is approximately Normal.

> **DEFINITION Normal probability plot**
>
> A **Normal probability plot** is a scatterplot of the ordered pair (data value, expected *z*-score) for each of the individuals in a quantitative data set. That is, the *x*-coordinate of each point is the actual data value and the *y*-coordinate is the expected *z*-score corresponding to the percentile of that data value in a standard Normal distribution.

> Some software plots the data values on the horizontal axis and the expected *z*-scores on the vertical axis, while other software does just the reverse. The TI-83/84 gives you both options. We prefer the data values on the horizontal axis, which is consistent with other types of graphs we have made.

> **AP® EXAM TIP**
>
> Normal probability plots are not included in the AP® Statistics Course Framework. However, these graphs are very useful for assessing Normality. You may use them on the AP® Statistics exam if you wish—just be sure that you know what you're looking for.

Technology Corner 7 at the end of this subsection shows you how to make a Normal probability plot. For now, let's focus on how to interpret Normal probability plots.

> **Preparing for Inference**
>
> In Chapters 10–11, students will have to check the Normal/Large Sample condition for performing inference. If the shape of the population distribution is unknown and the sample size is small, students will have to graph the sample data and use the graph to assess the potential Normality of the population distribution. This skill will also be used in Chapter 12 when checking the condition that the distribution of residuals is approximately Normal. In these cases, the fastest way to assess Normality is a dotplot or histogram of the sample data or a Normal probability plot.

Figure 2.21 shows dotplots and Normal probability plots for each of the data sets in this subsection.

- Panel (a): We confirmed earlier that the distribution of IQ scores is approximately Normal. Its Normal probability plot has a *linear* form.
- Panel (b): The distribution of number of siblings is clearly right-skewed. Its Normal probability plot has a curved form.
- Panel (c): We determined earlier that the distribution of calories in breakfast cereals is *not* approximately Normal, even though the graph looks roughly symmetric and is somewhat bell-shaped. Its Normal probability plot has a different kind of nonlinear form.

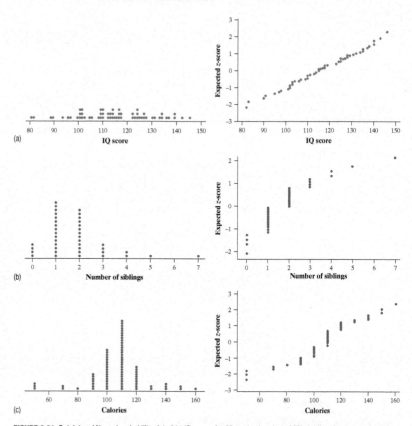

FIGURE 2.21 Dotplot and Normal probability plot of (a) IQ scores for 60 randomly selected fifth-graders from one school, (b) Number of siblings for each student in a college statistics class, and (c) Calories per serving in 77 brands of breakfast cereal. The distribution of IQ scores in (a) is approximately Normal because the Normal probability plot has a linear form. The nonlinear Normal probability plots in (b) and (c) confirm that neither of these distributions is approximately Normal.

Teaching Tip

Students will often say, "It is Normal, because it is linear." Remind them that "it" is inappropriate language to use in statistics class and will earn them a detention after school. A better interpretation is "The distribution of IQ scores is approximately Normal because the Normal probability plot is fairly linear."

✚ Ask the StatsMedic

Ban "IT" from Your Statistics Classroom?

In this post, we discuss the value of using precise statistical vocabulary to clearly communicate understanding of statistical concepts.

Teaching Tip

We prefer putting the *x*-values on the horizontal axis so that when there are values to the right of what we expect, we know the distribution is skewed to the right. You can also point out that if all the points on a Normal probability plot suddenly fell down to the horizontal axis, the result would be a dotplot of the data.

HOW TO ASSESS NORMALITY WITH A NORMAL PROBABILITY PLOT

If the points on a Normal probability plot lie close to a straight line, the data are approximately Normally distributed. A nonlinear form in a Normal probability plot indicates a non-Normal distribution.

When examining a Normal probability plot, look for shapes that show clear departures from Normality. Don't overreact to minor wiggles in the plot. We used a TI-84 to generate three different random samples of size 20 from a Normal distribution. The screen shots show Normal probability plots for each of the samples. Although none of the plots is perfectly linear, it is reasonable to believe that each sample came from a Normal population.

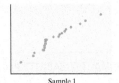

Sample 1

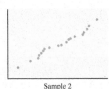

Sample 2

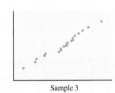

Sample 3

EXAMPLE

How Normal are survival times?

Interpreting Normal probability plots

PROBLEM: Researchers recorded the survival times in days of 72 guinea pigs after they were injected with infectious bacteria in a medical experiment.[9] A Normal probability plot of the data is shown. Use the graph to determine if the distribution of survival times is approximately Normal.

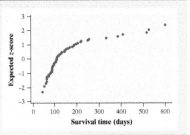

SOLUTION:

The Normal probability plot is clearly curved, indicating that the distribution of survival time for the 72 guinea pigs is not approximately Normal.

FOR PRACTICE, TRY EXERCISE 79

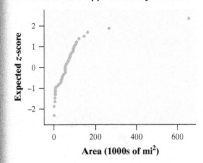

Think About It

HOW CAN WE DETERMINE SHAPE FROM A NORMAL PROBABILITY PLOT? Look at the Normal probability plot of the guinea pig survival data in the example. Imagine all the points falling down onto the horizontal axis. The resulting dotplot would have many values stacked up between 50 and 150 days, and fewer values that are further spread apart from 150 to about 600 days. The distribution would be skewed to the right due to the greater variability in the upper half of the data set. The dotplot of the data confirms our answer.

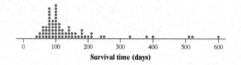

Survival time (days)

7. Technology Corner   MAKING NORMAL PROBABILITY PLOTS

TI-Nspire and other technology instructions are on the book's website at highschool.bfwpub.com/updatedtps6e.

Let's use the TI-83/84 to make a Normal probability plot for the IQ score data (page 132).

1. Enter the data values in list L1.

  • Press STAT and choose Edit....

  • Type the values into list L_1.

2. Set up a Normal probability plot in the statistics plots menu.

  • Press 2nd Y= (STAT PLOT).

  • Press ENTER or 1 to go into Plot1.

  • Adjust the settings as shown.

3. Use ZoomStat to see the finished graph.

Interpretation: The Normal probability plot is quite linear, which confirms our earlier belief that the distribution of IQ scores is approximately Normal.

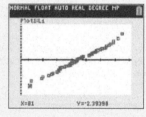

Section 2.2 | Summary

- We can describe the overall pattern of the distribution of a quantitative variable with a **density curve.** A density curve always remains on or above the horizontal axis and has total area 1 underneath it. An area under a density curve estimates the proportion of observations that fall in an interval of values.
- A density curve is an idealized description of a distribution that smooths out any irregularities. We write the **mean of a density curve** as μ and the **standard deviation of a density curve** as σ. The values of μ and σ are the parameters of the density curve.
- The mean and the median of a density curve can be located by eye. The mean μ is the balance point of the curve. The median divides the area under the curve in half. The standard deviation σ cannot be located by eye on most density curves.
- The mean and median are equal for symmetric density curves. The mean of a skewed density curve is located farther toward the long tail than is the median.
- **Normal distributions** are described by a special family of bell-shaped, symmetric density curves, called **Normal curves.** The mean μ and standard deviation σ completely specify a Normal distribution. The mean is the center of the curve, and σ is the distance from μ to the change-of-curvature points on either side.
- The **empirical rule** describes what percent of observations in any Normal distribution fall within 1, 2, and 3 standard deviations of the mean: about 68%, 95%, and 99%, respectively.
- All Normal distributions are the same when observations are standardized. If x follows a Normal distribution with mean μ and standard deviation σ, we can standardize using

$$z = \frac{x - \mu}{\sigma}$$

Then the variable z has the **standard Normal distribution** with mean 0 and standard deviation 1.
- **Table A** in the back of the book gives percentiles for the standard Normal distribution. You can use Table A or technology to determine area for given values of the variable or the value that corresponds to a given percentile in any Normal distribution.
- To find the area in a Normal distribution corresponding to given values:

    Step 1: Draw a Normal distribution with the horizontal axis labeled and scaled using the mean and standard deviation, the boundary value(s) clearly identified, and the area of interest shaded.

    Step 2: Perform calculations—show your work! Do one of the following:

    (i) Standardize each boundary value and use Table A or technology to find the desired area under the standard Normal curve; or

    (ii) Use technology to find the desired area without standardizing.

    Be sure to answer the question that was asked.

TRM Section 2.2 Quiz

There is one quiz available for this section in the Teacher's Resource Materials. Click on the link in the TE-Book, look on the TRFD, or download from the Teacher's Resources on the book's digital platform. You can also create your own quiz using the ExamView® Assessment Suite that is part of the TPS6 program. Questions are coded by Learning Target to make it easy to build parallel quizzes.

Teaching Tip

Make sure to point out the two icons next to Exercise 41. The "pg 111" icon reminds students that the example on page 111 is much like this exercise. The "play" icon reminds students that a video solution is available in the student e-Book or at the Student Site at highschool.bfwpub.com/updatedtps6e.

TRM **Full Solutions to Section 2.2 Exercises**

Click on the link in the TE-Book, open the TRFD, or download from the Teacher's Resources on the book's digital platform.

Answers to Section 2.2 Exercises

2.41 (a)

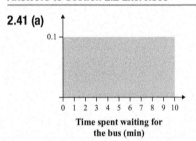

Time spent waiting for the bus (min)

(b) Area = (base)(height) = $(5.3 - 2.5)(0.1) = 0.28 = 28\%$. On about 28% of days, Sally waits between 2.5 and 5.3 minutes for the bus.
(c) The 70th percentile will have 70% of the wait times to the left of it, so the 70th percentile of Sally's wait times is 7 minutes.

2.42 (a)

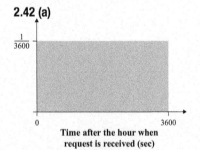

Time after the hour when request is received (sec)

(b) Area = (base)(height) =

$$(300 - 0)\left(\frac{1}{3600}\right) = 0.083$$

The proportion of requests that are received within the first 5 minutes after the hour is 300 out of 3600, or 0.083.

(c) The third quartile is the 75th percentile: Area = (base)(height); $0.75 = (Q_3 - 0)(1/3600)$; $Q_3 = 2700$ seconds. The first quartile is the 25th percentile: Area = (base)(height); $0.25 = (Q_1 - 0)(1/3600)$; $Q_1 = 900$ seconds; $IQR = 2700 - 900 = 1800$ seconds.

- To find the value in a Normal distribution corresponding to a given area (percentile):

Step 1: Draw a Normal distribution with the horizontal axis labeled and scaled using the mean and standard deviation, the area of interest shaded and labeled, and unknown boundary value clearly marked.

Step 2: Perform calculations—show your work! Do one of the following:

(i) Use Table A or technology to find the value of z with the indicated area under the standard Normal curve, then "unstandardize" to transform back to the original distribution; or

(ii) Use technology to find the desired area without standardizing.

Be sure to answer the question that was asked.

- To assess Normality for a given set of quantitative data, we first observe the shape of a dotplot, stemplot, or histogram. Then we can check how well the data fit the empirical rule for Normal distributions. Another good method for assessing Normality is to construct a **Normal probability plot.** If the Normal probability plot has a linear form, then we can say that the distribution is approximately Normal.

2.2 Technology Corners

TI-Nspire and other technology instructions are on the book's website at highschool.bfwpub.com/updatedtps6e.

5. Finding areas from values in a Normal distribution	Page 121
6. Finding values from areas in a Normal distribution	Page 128
7. Making Normal probability plots	Page 136

Section 2.2 | Exercises

41. Where's the bus? Sally takes the same bus to work every morning. The amount of time (in minutes) that she has to wait for the bus can be modeled by a uniform distribution on the interval from 0 minutes to 10 minutes.

(a) Draw a density curve to model the amount of time that Sally has to wait for the bus. Be sure to include scales on both axes.

(b) On about what percent of days does Sally wait between 2.5 and 5.3 minutes for the bus?

(c) Find the 70th percentile of Sally's wait times.

42. Still waiting for the server? How does your web browser get a file from the Internet? Your computer sends a request for the file to a web server, and the web server sends back a response. For one particular web server, the time (in seconds) after the start of an hour at which a request is received can be modeled by a uniform distribution on the interval from 0 to 3600 seconds.

(a) Draw a density curve to model the amount of time after an hour at which a request is received by the web server. Be sure to include scales on both axes.

(b) About what proportion of requests are received within the first 5 minutes (300 seconds) after the hour?

(c) Find the interquartile range of this distribution.

43. Quick, click! An Internet reaction time test asks subjects to click their mouse button as soon as a light flashes on the screen. The light is

---

2.43 (a) The density curve must have the height 0.25 because the area must equal 1: (base)(height) = $(4)(0.25) = 1$.
(b) Area = (base)(height) = $(5 - 3.75)(0.25) = 0.3125 = 31.25\%$. About 31.25% of the time, the light will flash more than 3.75 seconds after the subject clicks "Start." **(c)** The 38th percentile of this distribution is the time for which 38% of the observations are below it: Area = (base)(height); $0.38 = (x - 1)(0.25)$; $1.52 = x - 1$; $x = 2.52$. Therefore, the 38th percentile of this distribution is 2.52 seconds; 38 percent of the time, the light will flash in 2.52 seconds or less after the subject clicks "Start."

programmed to go on at a randomly selected time after the subject clicks "Start." The density curve models the amount of time the subject has to wait for the light to flash.

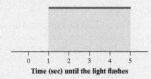

Time (sec) until the light flashes

(a) What height must the density curve have? Justify your answer.

(b) About what percent of the time will the light flash more than 3.75 seconds after the subject clicks "Start"?

(c) Calculate and interpret the 38th percentile of this distribution.

44. **Class is over!** Mr. Shrager does not always let his statistics class out on time. In fact, he seems to end class according to his own "internal clock." The density curve models the distribution of the amount of time after class ends (in minutes) when Mr. Shrager dismisses the class. (A negative value indicates he ended class early.)

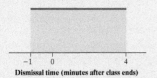

Dismissal time (minutes after class ends)

(a) What height must the density curve have? Justify your answer.

(b) About what proportion of the time does Mr. Shrager dismiss class within 1 minute of its scheduled end time?

(c) Calculate and interpret the 20th percentile of the distribution.

45. **Mean and median** The figure displays two density curves that model different distributions of quantitative data. Identify the location of the mean and median by letter for each graph. Justify your answers.

(a) A B C   (b) A   B   C

46. **Mean and median** The figure displays two density curves that model different distributions of quantitative data. Identify the location of the mean and median by letter for each graph. Justify your answers.

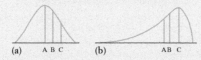

(a) A B C   (b) AB C

47. **Potato chips** The weights of 9-ounce bags of a particular brand of potato chips can be modeled by a Normal distribution with mean $\mu = 9.12$ ounces and standard deviation $\sigma = 0.05$ ounce. Sketch the Normal density curve. Label the mean and the points that are 1, 2, and 3 standard deviations from the mean.

48. **Batter up!** In baseball, a player's batting average is the proportion of times the player gets a hit out of his total number of times at bat. The distribution of batting averages in a recent season for Major League Baseball players with at least 100 plate appearances can be modeled by a Normal distribution with mean $\mu = 0.261$ and standard deviation $\sigma = 0.034$. Sketch the Normal density curve. Label the mean and the points that are 1, 2, and 3 standard deviations from the mean.

49. **Normal curve** Estimate the mean and standard deviation of the Normal density curve below.

50. **Normal curve** Estimate the mean and standard deviation of the Normal density curve below.

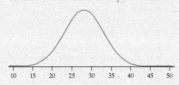

51. **Potato chips** Refer to Exercise 47. Use the empirical rule to answer the following questions.

(a) About what percent of bags weigh less than 9.02 ounces? Show your method clearly.

(b) A bag that weighs 9.07 ounces is at about what percentile in this distribution? Justify your answer.

2.48 Normal density curve with mean 0.261 and standard deviation 0.034

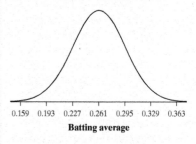

Batting average

2.49 The curve approaches the horizontal axis around 3 standard deviations from the mean. On the positive side of the curve, the curve shown seems to approach the horizontal axis around the value 16, so 16 should be approximately 3 standard deviations above the mean. Because $16 - 10 = 6$ units, we estimate 3 standard deviations to be 6 units and therefore 1 standard deviation would be about 2 units.

2.50 We know that the curve approaches the horizontal axis around 3 standard deviations from the mean. On the positive side of the curve, the curve shown seems to approach the horizontal axis around the value 43, so 43 should be approximately 3 standard deviations above the mean. Because $43 - 28 = 15$ units, we estimate 3 standard deviations to be 15 units and therefore 1 standard deviation would be about 5 units.

2.51 (a) The value 9.02 is 2 standard deviations below the mean, so

approximately $\dfrac{100\% - 95\%}{2} = 2.5\%$

of bags weigh less than 9.02 ounces.
(b) The value 9.07 is 1 standard deviation below the mean. About

$\dfrac{100\% - 68\%}{2} = 16\%$ of the bags

weigh less than 9.07 ounces. So 9.07 is approximately the 16th percentile of the weights of these potato chip bags.

2.44 (a) The density curve must have the height 0.2 because the area must equal 1: (base)(height) = (5)(0.2) = 1.
(b) Area = (base)(height) = $(1 - (-1))(0.2) = 0.4 = 40\%$. About 40% of the time, Mr. Shrager dismisses class within 1 minute of the scheduled end time.
(c) The 20th percentile of this distribution is the time for which 20% of the observations are below it: Area = (base)(height); $0.2 = (x - (-1))(0.2)$; $1 = x + 1$; $x = 0$. Therefore, the 20th percentile of this distribution is right at dismissal (time 0); 20 percent of the time, Mr. Shrager will dismiss class at or before the time the class ends.

2.45 (a) Mean is C, median is B (the right skew pulls the mean to the right of the median).
(b) Mean is B, median is B (this distribution is symmetric, so mean = median).

2.46 (a) Mean is A, median is A (the distribution is symmetric, so mean = median). **(b)** Mean is A, median is B (the left skew pulls the mean to the left of the median).

2.47 The Normal density curve with mean 9.12 and standard deviation 0.05

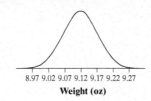

8.97 9.02 9.07 9.12 9.17 9.22 9.27
Weight (oz)

2.52 (a) 0.363 is 3 SDs above the mean, so approximately $(100\% - 99.7\%)/2 = 0.15\%$ of Major League Baseball players with 100 plate appearances had a batting average of 0.363 or higher. **(b)** The value 0.227 is 1 SD below the mean. About $(100\% - 68\%)/2 = 16\%$ of Major League Baseball players with 100 plate appearances had a batting average that is less than 0.227. So 0.227 is approximately the 16th percentile of the batting averages.

2.53 $z = (9 - 9.12)/0.05 = -2.40$; *Table A:* 0.0082; *Tech:* normalcdf(lower: −1000, upper: 9, mean: 9.12, SD: 0.05) = 0.0082. This is not likely to pose a problem because the percentage of bags that weigh less than the advertised amount is very small.

2.54 $z = (0.200 - 0.261)/0.034 = -1.79$; *Table A:* 0.0367; *Tech:* normalcdf(lower: −1000, upper: 0.2, mean: 0.261, SD: 0.034) = 0.0364. About 3.6% of players are at risk of sitting on the bench during important games.

2.55 $z = (2400 - 2000)/500 = 0.80$; *Table A:* 0.2119; *Tech:* normalcdf(lower: 2400, upper: 1000000, mean: 2000, SD: 500) = 0.2119. About 21.19%.

2.56 $z = (100 - 70)/20 = 1.50$; *Table A:* 0.0668; *Tech:* normalcdf(lower: 100, upper: 1000, mean: 70, SD: 20) = 0.0668. About 0.0668 of adults.

2.57 $z = (1200 - 2000)/500 = -1.60$ and $z = (1800 - 2000)/500 = -0.40$; *Table A:* $0.3446 - 0.0548 = 0.2898$; *Tech:* normalcdf(lower: 1200, upper: 1800, mean: 2000, SD: 500) = 0.2898. About 28.98%.

2.58 $z = (80 - 70)/20 = 0.50$ and $z = (90 - 70)/20 = 1$; *Table A:* $0.8413 - 0.6915 = 0.1498$; *Tech:* normalcdf(lower: 80, upper: 90, mean: 70, SD: 20) = 0.1499. About 14.99%.

2.59 (a) *Table A:* $1 - 0.0485 = 0.9515$; *Tech:* normalcdf(lower: −1.66, upper: 1000, mean: 0, SD: 1) = 0.9515.
(b) *Table A:* $0.9978 - 0.0485 = 0.9493$; *Tech:* normalcdf(lower: −1.66, upper: 2.85, mean: 0, SD: 1) = 0.9494.

2.60 (a) *Table A:* 0.0069; *Tech:* normalcdf(lower: −1000, upper: −2.46, mean: 0, SD: 1) = 0.0069.
(b) *Table A:* $0.9931 - 0.8133 = 0.1798$; *Tech:* normalcdf(lower: 0.89, upper: 2.46, mean: 0, SD: 1) = 0.1798.

2.61 (a) $z = (3 - 5.3)/0.9 = -2.56$; *Table A:* 0.0052; *Tech:* normalcdf (lower: −1000, upper: 3, mean: 5.3, SD: 0.9) = 0.0053.

52. **Batter up!** Refer to Exercise 48. Use the empirical rule to answer the following questions.
   (a) About what percent of Major League Baseball players with 100 plate appearances had batting averages of 0.363 or higher? Show your method clearly.
   (b) A player with a batting average of 0.227 is at about what percentile in this distribution? Justify your answer.

53. **Potato chips** Refer to Exercise 47. About what pg 122 percent of 9-ounce bags of this brand of potato chips weigh less than the advertised 9 ounces? Is this likely to pose a problem for the company that produces these chips?

54. **Batter up!** Refer to Exercise 48. A player with a batting average below 0.200 is at risk of sitting on the bench during important games. About what percent of players are at risk?

55. **Watch the salt!** A study investigated about pg 124 3000 meals ordered from Chipotle restaurants using the online site Grubhub. Researchers calculated the sodium content (in milligrams) for each order based on Chipotle's published nutrition information. The distribution of sodium content is approximately Normal with mean 2000 mg and standard deviation 500 mg.[10] About what percent of the meals ordered exceeded the recommended daily allowance of 2400 mg of sodium?

56. **Blood pressure** According to a health information website, the distribution of adults' diastolic blood pressure (in millimeters of mercury) can be modeled by a Normal distribution with mean 70 and standard deviation 20. A diastolic pressure above 100 for an adult is classified as very high blood pressure. About what proportion of adults have very high blood pressure according to this criterion?

57. **Watch the salt!** Refer to Exercise 55. About what pg 126 percent of the meals ordered contained between 1200 mg and 1800 mg of sodium?

58. **Blood pressure** Refer to Exercise 56. According to the same health information website, a diastolic blood pressure between 80 and 90 indicates borderline high blood pressure. About what percent of adults have borderline high blood pressure?

59. **Standard Normal areas** Find the proportion of observations in a standard Normal distribution that satisfies each of the following statements.
   (a) $z > -1.66$
   (b) $-1.66 < z < 2.85$

60. **Standard Normal areas** Find the proportion of observations in a standard Normal distribution that satisfies each of the following statements.
   (a) $z < -2.46$
   (b) $0.89 < z < 2.46$

61. **Sudoku** Mrs. Starnes enjoys doing Sudoku puzzles. The time she takes to complete an easy puzzle can be modeled by a Normal distribution with mean 5.3 minutes and standard deviation 0.9 minute.
   (a) What proportion of the time does Mrs. Starnes finish an easy Sudoku puzzle in less than 3 minutes?
   (b) How often does it take Mrs. Starnes more than 6 minutes to complete an easy puzzle?
   (c) What percent of easy Sudoku puzzles take Mrs. Starnes between 6 and 8 minutes to complete?

62. **Hit an ace!** Professional tennis player Novak Djokovic hits the ball extremely hard. His first-serve speeds can be modeled by a Normal distribution with mean 112 miles per hour (mph) and standard deviation 5 mph.
   (a) How often does Djokovic hit his first serve faster than 120 mph?
   (b) What percent of Djokovic's first serves are slower than 100 mph?
   (c) What proportion of Djokovic's first serves have speeds between 100 and 110 mph?

63. **Sudoku** Refer to Exercise 61. Find the 20th percentile pg 129 of Mrs. Starnes's Sudoku times for easy puzzles.

64. **Hit an ace!** Refer to Exercise 62. Find the 85th percentile of Djokovic's first-serve speeds.

65. **Deciles** The deciles of any distribution are the values at the 10th, 20th, ..., 90th percentiles. The first and last deciles are the 10th and the 90th percentiles, respectively. What are the first and last deciles of the standard Normal distribution?

66. **Outliers** The percent of the observations that are classified as outliers by the $1.5 \times IQR$ rule is the same in any Normal distribution. What is this percent? Show your method clearly.

67. **IQ test scores** Scores on the Wechsler Adult Intelligence Scale (an IQ test) for the 20- to 34-year-old age group are approximately Normally distributed with $\mu = 110$ and $\sigma = 25$.
   (a) What percent of people aged 20 to 34 have IQs between 125 and 150?

(b) $z = (6 - 5.3)/0.9 = 0.78$; *Table A:* $1 - 0.7823 = 0.2177$; *Tech:* normalcdf(lower: 6, upper: 1000, mean: 5.3, SD: 0.9) = 0.2183.
(c) $z = (6 - 5.3)/0.9 = 0.78$ and $z = (8 - 5.3)/0.9 = 3$; *Table A:* $0.9987 - 0.7823 = 0.2164$; *Tech:* normalcdf(lower: 6, upper: 8, mean: 5.3, SD: 0.9) = 0.2170.

2.62 (a) $z = (120 - 112)/5 = 1.60$; *Table A:* $1 - 0.9452 = 0.0548$; *Tech:* normalcdf(lower: 120, upper: 1000, mean: 112, SD: 5) = 0.0548.
(b) $z = (100 - 112)/5 = -2.40$; *Table A:* 0.0082; *Tech:* normalcdf(lower: −1000, upper: 100, mean: 112, SD: 5) = 0.0082.
(c) $z = (100 - 112)/5 = -2.40$ and $z = (110 - 112)/5 = -0.40$; *Table A:* $0.3446 - 0.0082 = 0.3364$; *Tech:* normalcdf(lower: 100, upper: 110, mean: 112, SD: 5) = 0.3364.

2.63 *Table A:* Solving $-0.84 = (x - 5.3)/0.9$ gives $x = 4.544$ minutes; *Tech:* invNorm(area: 0.2, mean: 5.3, SD: 0.9) = 4.543 minutes.

2.64 *Table A:* Solving $1.04 = (x - 112)/5$ gives $x = 117.2$ mph; *Tech:* invNorm(area: 0.85, mean: 112, SD: 5) = 117.18 mph.

2.65 *Table A:* The value $z = -1.28$ has an area of 0.1003 to the left and the value $z = 1.28$ has an area of 0.8997 to the left. *Tech:* invNorm(area: 0.10, mean: 0, SD: 1) = −1.282 and invNorm(area: 0.90, mean: 0, SD: 1) = 1.282.

2.66 The value $z = -0.67$ has an area of 0.2514 to the left and the value $z = 0.67$ has an area of 0.7486 to the left. Thus, $Q_1 = -0.67$, $Q_3 = 0.67$, and $IQR = 0.67 - (-0.67) = 1.34$. Outliers are values less than $-0.67 - 1.5(1.34) = -2.68$

(b) MENSA is an elite organization that admits as members people who score in the top 2% on IQ tests. What score on the Wechsler Adult Intelligence Scale would an individual aged 20 to 34 have to earn to qualify for MENSA membership?

68. **Post office** A local post office weighs outgoing mail and finds that the weights of first-class letters are approximately Normally distributed with a mean of 0.69 ounce and a standard deviation of 0.16 ounce.

(a) Estimate the 60th percentile of first-class letter weights.

(b) First-class letters weighing more than 1 ounce require extra postage. What proportion of first-class letters at this post office require extra postage?

Exercises 69 and 70 refer to the following setting. At some fast-food restaurants, customers who want a lid for their drinks get them from a large stack near the straws, napkins, and condiments. The lids are made with a small amount of flexibility so they can be stretched across the mouth of the cup and then snugly secured. When lids are too small or too large, customers can get very frustrated, especially if they end up spilling their drinks. At one particular restaurant, large drink cups require lids with a "diameter" of between 3.95 and 4.05 inches. The restaurant's lid supplier claims that the diameter of its large lids follows a Normal distribution with mean 3.98 inches and standard deviation 0.02 inch. Assume that the supplier's claim is true.

69. **Put a lid on it!**

(a) What percent of large lids are too small to fit?

(b) What percent of large lids are too big to fit?

(c) Compare your answers to parts (a) and (b). Does it make sense for the lid manufacturer to try to make one of these values larger than the other? Why or why not?

70. **Put a lid on it!** The supplier is considering two changes to reduce to 1% the percentage of its large-cup lids that are too small. One strategy is to adjust the mean diameter of its lids. Another option is to alter the production process, thereby decreasing the standard deviation of the lid diameters.

(a) If the standard deviation remains at $\sigma = 0.02$ inch, at what value should the supplier set the mean diameter of its large-cup lids so that only 1% are too small to fit?

(b) If the mean diameter stays at $\mu = 3.98$ inches, what value of the standard deviation will result in only 1% of lids that are too small to fit?

(c) Which of the two options in parts (a) and (b) do you think is preferable? Justify your answer. (Be sure to consider the effect of these changes on the percent of lids that are too large to fit.)

71. **Flight times** An airline flies the same route at the same time each day. The flight time varies according to a Normal distribution with unknown mean and standard deviation. On 15% of days, the flight takes more than an hour. On 3% of days, the flight lasts 75 minutes or more. Use this information to determine the mean and standard deviation of the flight time distribution.

72. **Brush your teeth** The amount of time Ricardo spends brushing his teeth follows a Normal distribution with unknown mean and standard deviation. Ricardo spends less than 1 minute brushing his teeth about 40% of the time. He spends more than 2 minutes brushing his teeth 2% of the time. Use this information to determine the mean and standard deviation of this distribution.

73. **Normal highway driving?** The dotplot shows the EPA highway gas mileage estimates in miles per gallon (mpg) for a random sample of 21 model year 2018 midsize cars.[11] Explain why this distribution of highway gas mileage is not approximately Normal.

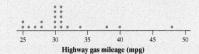

Highway gas mileage (mpg)

74. **Normal to be foreign born?** The histogram displays the percent of foreign-born residents in each of the 50 states.[12] Explain why this distribution of the percent of foreign-born residents in the states is not approximately Normal.

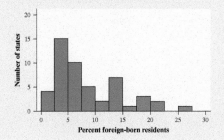

Percent foreign-born residents

proportion of lids that are too small, rather than too big. If lids are too small, customers will just try another lid. But if lids are too large, the customer may not notice and then spill the drink.

2.70 (a) *Table A:* Solving $-2.33 = (3.95 - \mu)/0.02$ gives $\mu = 4.00$ inches. *Tech:* invNorm(area: 0.01, mean: 0, SD: 1) gives $z = -2.33$. Solving $-2.33 = (3.95 - \mu)/0.02$ gives $\mu = 4.00$ inches. **(b)** *Table A:* Solving $-2.33 = (3.95 - 3.98)/\sigma$ gives $\sigma = 0.013$ inch. *Tech:* invNorm(area: 0.01, mean: 0, SD: 1) gives $z = -2.33$. Solving $-2.33 = (3.95 - 3.98)/\sigma$ gives $\sigma = 0.013$ inch. **(c)** We prefer reducing the SD, as in part (b). This will reduce the number of lids that are too small and the number of lids that are too big. If we make the mean a little larger, as in part (a), we will reduce the number of lids that are too small, but we will increase the number of lids that are too big.

2.71 $z = 1.04$ has an area of 0.15 to the right and $z = 1.88$ has an area of 0.03 to the right. Solving the system of equations $1.04 = \dfrac{60 - \mu}{\sigma}$ and $1.88 = \dfrac{75 - \mu}{\sigma}$ gives $\sigma = 17.86$ minutes and $\mu = 41.43$ minutes.

2.72 $z = -0.25$ has an area of 0.40 to the left and $z = 2.05$ has an area of 0.02 to the right. Solving the system of equations $-0.25 = \dfrac{1 - \mu}{\sigma}$ and $2.05 = \dfrac{2 - \mu}{\sigma}$ gives $\sigma = 0.4348$ minutes and $\mu = 1.1087$ minutes.

2.73 Because the distribution is skewed to the right.

2.74 Because the distribution is skewed to the right.

and greater than $0.67 + 1.5(1.34) = 2.68$. Finally, the area to the left of -2.68 is 0.0037 and to the right of 2.68 is $1 - 0.9963 = 0.0037$. Thus, $0.0037 + 0.0037 = 0.74\%$ are outliers. *Tech:* invNorm(area: 0.25, mean: 0, SD: 1) gives $Q_1 = -0.674$ and invNorm(area: 0.75, mean: 0, SD: 1) gives $Q_3 = 0.674$. Thus, $IQR = 0.674 - (-0.674) = 1.348$. Outliers are values less than $-0.674 - 1.5(1.348) = -2.696$ and greater than $0.674 + 1.5(1.348) = 2.696$. Finally, 1-normalcdf(lower: -2.696, upper: 2.696, mean: 0, SD: 1) $= 1 - 0.9930 = 0.0070$. Thus, 0.70% are outliers.

2.67 (a) $z = (125 - 110)/25 = 0.6$ and $z = (150 - 110)/25 = 1.6$; *Table A:* $0.9452 - 0.7257 = 0.2195$; *Tech:* normalcdf(lower: 125, upper: 150,

mean: 110, SD: 25) $= 0.2195$. **(b)** *Table A:* Solving $2.05 = (z - 110)/25$ gives $x = 161.25$; *Tech:* invNorm(area: 0.98, mean: 110, SD: 25) $= 161.34$.

2.68 (a) *Table A:* Solving $0.25 = (x - 0.69)/0.16$ gives $x = 0.73$; *Tech:* invNorm(area: 0.6, mean: 0.69, SD: 0.16) $= 0.73$. **(b)** $z = (1 - 0.69)/0.16 = 1.94$; *Table A:* $1 - 0.9738 = 0.0262$; *Tech:* normalcdf(lower: 1, upper: 1000, mean: 0.69, SD: 0.16) $= 0.0263$.

2.69 (a) $z = (3.95 - 3.98)/0.02 = -1.5$; *Table A:* 0.0668; *Tech:* normalcdf(lower: -1000, upper: 3.95, mean: 3.98, SD: 0.02) $= 0.0668$. **(b)** $z = (4.05 - 3.98)/0.02 = 3.5$; *Table A:* approximately 0; *Tech:* normalcdf(lower: 4.05, upper: 1000, mean: 3.98, SD: 0.02) $= 0.0002$. **(c)** It makes more sense to have a larger

2.75 The histogram of these data is roughly symmetric and bell-shaped. The mean and standard deviation of these data are $\bar{x} = 15.825$ cubic feet and $s_x = 1.217$ cubic feet.

- $\bar{x} \pm 1s_x = (14.608, 17.042)$; 24 of 36 observations, or 66.7% of the observations, are within 1 standard deviation of the mean.
- $\bar{x} \pm 2s_x = (13.391, 18.259)$; 34 of 36 observations, or 94.4% of the observations, are within 2 standard deviations of the mean.
- $\bar{x} \pm 3s_x = (12.174, 19.476)$; 36 of 36 observations, or 100% of the observations, are within 3 standard deviations of the mean.

These percentages are quite close to what we would expect based on the empirical rule. Combined with the graph, this gives good evidence that the distribution is approximately Normal.

2.76 The distribution of shark lengths is roughly symmetric and somewhat bell-shaped. The mean and standard deviation of these data are $\bar{x} = 15.586$ feet and $s_x = 2.55$ feet.

- $\bar{x} \pm 1s_x = (13.036, 18.136)$; 30 of 44 observations, or 68.2% of the observations, are within 1 standard deviation of the mean.
- $\bar{x} \pm 2s_x = (10.486, 20.686)$; 42 of 44 observations, or 95.5% of the observations, are within 2 standard deviations of the mean.
- $\bar{x} \pm 3s_x = (7.936, 23.236)$; 44 of 44 observations, or 100% of the observations, are within 3 standard deviations of the mean.

These percentages are quite close to what we would expect based on the empirical rule. Combined with the graph, this gives good evidence that the distribution is approximately Normal.

2.77 The distribution of tuitions in Michigan is not approximately Normal. If it was Normal, then the minimum value should be around 3 standard deviations below the mean. However, the actual minimum has a z-score of just

$$z = \frac{1873 - 10{,}614}{8049} = -1.09.$$ Also, if the distribution was Normal, the minimum and maximum should be about the same distance from the mean. However, the maximum is much farther from the mean $(30{,}823 - 10{,}614 = 20{,}209)$ than the minimum $(10{,}614 - 1873 = 8741)$.

75. Refrigerators *Consumer Reports* magazine pg 132 collected data on the usable capacity (in cubic feet) of a sample of 36 side-by-side refrigerators. Here are the data:

12.9	13.7	14.1	14.2	14.5	14.5	14.6	14.7	15.1	15.2	15.3	15.3
15.3	15.3	15.5	15.6	15.6	15.8	16.0	16.0	16.2	16.2	16.3	16.4
16.5	16.6	16.6	16.6	16.8	17.0	17.0	17.2	17.4	17.4	17.9	18.4

A histogram of the data and summary statistics are shown here. Is this distribution of refrigerator capacities approximately Normal? Justify your answer based on the graph and the empirical rule.

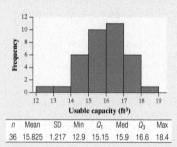

Usable capacity (ft³)

n	Mean	SD	Min	Q_1	Med	Q_3	Max
36	15.825	1.217	12.9	15.15	15.9	16.6	18.4

76. Big sharks Here are the lengths (in feet) of 44 great white sharks:[13]

18.7	12.3	18.6	16.4	15.7	18.3	14.6	15.8	14.9	17.6	12.1
16.4	16.7	17.8	16.2	12.6	17.8	13.8	12.2	15.2	14.7	12.4
13.2	15.8	14.3	16.6	9.4	18.2	13.2	13.6	15.3	16.1	13.5
19.1	16.2	22.8	16.8	13.6	13.2	15.7	19.7	18.7	13.2	16.8

A dotplot of the data and summary statistics are shown below. Is this distribution of shark length approximately Normal? Justify your answer based on the graph and the empirical rule.

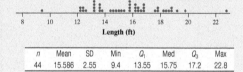

Length (ft)

n	Mean	SD	Min	Q_1	Med	Q_3	Max
44	15.586	2.55	9.4	13.55	15.75	17.2	22.8

77. Is Michigan Normal? We collected data on the tuition charged by colleges and universities in Michigan. Here are some numerical summaries for the data:

Mean	SD	Min	Max
10,614	8049	1873	30,823

Based on the relationship between the mean, standard deviation, minimum, and maximum, is it reasonable to believe that the distribution of Michigan tuitions is approximately Normal? Explain your answer.

78. Are body weights Normal? The heights of people of the same gender and similar ages follow Normal distributions reasonably closely. How about body weights? The weights of women aged 20 to 29 have mean 141.7 pounds and median 133.2 pounds. The first and third quartiles are 118.3 pounds and 157.3 pounds. Is it reasonable to believe that the distribution of body weights for women aged 20 to 29 is approximately Normal? Explain your answer.

79. Runners' heart rates The following figure is a Normal pg 133 probability plot of the heart rates of 200 male runners after 6 minutes of exercise on a treadmill.[14] Use the graph to determine if this distribution of heart rates is approximately Normal.

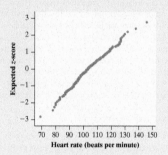

Heart rate (beats per minute)

80. Carbon dioxide emissions The following figure is a Normal probability plot of the emissions of carbon dioxide (CO_2) per person in 48 countries.[15] Use the graph to determine if this distribution of CO_2 emissions is approximately Normal.

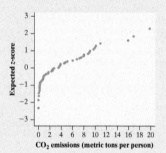

CO_2 emissions (metric tons per person)

2.78 The distribution of weights of women aged 20–29 is not approximately Normal. In a Normal distribution, Q_1 and Q_3 should be about the same distance from the median. However, the distance from Q_1 to the median $(133.2 - 118.3 = 14.9)$ is smaller than the distance from the median to Q_3 $(157.3 - 133.2 = 24.1)$.

2.79 The distribution is approximately Normal because the Normal probability plot is nearly linear.

2.80 The sharp curve in the Normal probability plot suggests that the distribution is right-skewed. This can be seen in the steep, nearly vertical section in the lower left. These numbers were much closer to the mean than would be expected in a Normal distribution, meaning the values that would be in the left tail are piled up close to the center of the distribution.

81. **Normal states?** The Normal probability plot displays data on the areas (in thousands of square miles) of each of the 50 states. Use the graph to determine if the distribution of land area is approximately Normal.

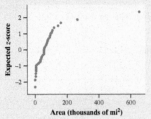

Area (thousands of mi²)

82. **Density of the earth** In 1798, the English scientist Henry Cavendish measured the density of the earth several times by careful work with a torsion balance. The variable recorded was the density of the earth as a multiple of the density of water. A Normal probability plot of the data is shown.[16] Use the graph to determine if this distribution of density measurement is approximately Normal.

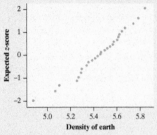

Density of earth

83. **Refrigerators** Refer to Exercise 75.

(a) Use your calculator to make a Normal probability plot of the data. Sketch this graph on your paper.

(b) What does the graph in part (a) imply about whether the distribution of refrigerator capacity is approximately Normal? Explain.

84. **Big sharks** Refer to Exercise 76.

(a) Use your calculator to make a Normal probability plot of the data. Sketch this graph on your paper.

(b) What does the graph in part (a) imply about whether the distribution of shark length is approximately Normal? Explain.

Multiple Choice: *Select the best answer for Exercises 85–90.*

85. Two measures of center are marked on the density curve shown. Which of the following is correct?

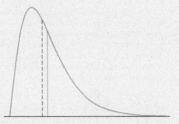

(a) The median is at the yellow line and the mean is at the red line.

(b) The median is at the red line and the mean is at the yellow line.

(c) The mode is at the red line and the median is at the yellow line.

(d) The mode is at the yellow line and the median is at the red line.

(e) The mode is at the red line and the mean is at the yellow line.

Exercises 86–88 refer to the following setting. The weights of laboratory cockroaches can be modeled with a Normal distribution having mean 80 grams and standard deviation 2 grams. The following figure is the Normal curve for this distribution of weights.

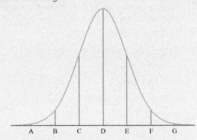

86. Point C on this Normal curve corresponds to

(a) 84 grams. (b) 82 grams. (c) 78 grams.

(d) 76 grams. (e) 74 grams.

87. About what percent of the cockroaches have weights between 76 and 84 grams?

(a) 99.7% (b) 95% (c) 68%

(d) 47.5% (e) 34%

88. About what proportion of the cockroaches will have weights greater than 83 grams?

(a) 0.0228 (b) 0.0668 (c) 0.1587

(d) 0.9332 (e) 0.0772

2.84 (a) A Normal probability plot is given below.

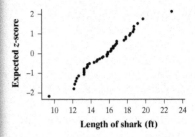

Length of shark (ft)

(b) Except for one small shark and one large shark, the plot is fairly linear, indicating that the distribution of shark lengths is approximately Normal.

2.85 b

2.86 c

2.87 b

2.88 b

2.81 The sharp curve in the Normal probability plot suggests that the distribution is right-skewed. This can be seen in the steep, nearly vertical section in the lower left. These numbers were much closer to the mean than would be expected in a Normal distribution, meaning the values that would be in the left tail are piled up close to the center of the distribution.

2.82 The distribution is approximately Normal because the Normal probability plot is nearly linear.

2.83 (a) A Normal probability plot is given below.

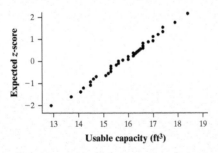

Usable capacity (ft³)

(b) The plot is fairly linear, indicating that the distribution of usable capacity is approximately Normal.

2.89 a

2.90 d

2.91 37 of the 38 family incomes in Indiana are at or below $95,000. Because $37/38 = 0.974$, or 97.4%, this individual's income is at the 97.4th percentile. 36 of the 44 family incomes in New Jersey are at or below $95,000. Because $36/44 = 0.818$, or 81.8%, this individual's income is at about the 82nd percentile.

The individual from Indiana has standardized income of 1.62, while the individual from New Jersey has standardized income of 0.88:

$$z = \frac{95,000 - 47,400}{29,400} = 1.62$$

$$z = \frac{95,000 - 58,100}{41,900} = 0.88$$

The individual from Indiana has an income that is 1.62 standard deviations above the mean of $47,400. The individual from New Jersey has an income that is 0.88 standard deviation above the mean of $58,100.

The individual from Indiana has a higher income, relative to others in his or her state because he or she had a higher percentile (97.4th versus 81.8th) and had a higher z-score (1.62 vs. 0.88) than the individual from New Jersey.

2.92 (a) The range of the distribution of total family incomes for individuals from Indiana is approximately range = max − min = $160,000 − $0 = $160,000. The range of the distribution of total family incomes for individuals from New Jersey is approximately range = max − min = $165,000 − $5,000 = $160,000. The two distributions have the same range. **(b)** The standard deviation of the total family incomes in the New Jersey sample is so much larger than for the Indiana sample because there are many large observations in the New Jersey sample that are far above the mean of the total family income for New Jersey, whereas in the Indiana sample, with the exception of one high outlier, the observations are quite close to the mean total family income for Indiana.

89. A different species of cockroach has weights that are approximately Normally distributed with a mean of 50 grams. After measuring the weights of many of these cockroaches, a lab assistant reports that 14% of the cockroaches weigh more than 55 grams. Based on this report, what is the approximate standard deviation of weight for this species of cockroach?

(a) 4.6
(b) 5.0
(c) 6.2
(d) 14.0
(e) Cannot determine without more information.

90. The following Normal probability plot shows the distribution of points scored for the 551 players in a single NBA season.

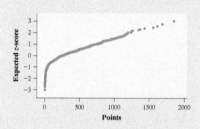

If the distribution of points was displayed in a histogram, what would be the best description of the histogram's shape?

(a) Approximately Normal

(b) Symmetric but not approximately Normal

(c) Skewed left

(d) Skewed right

(e) Cannot be determined

Recycle and Review

91. **Making money** (2.1) The parallel dotplots show the total family income of randomly chosen individuals from Indiana (38 individuals) and New Jersey (44 individuals). Means and standard deviations are given below the dotplots.

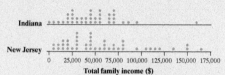

	Mean	Standard deviation
Indiana	$47,400	$29,400
New Jersey	$58,100	$41,900

Consider individuals in each state with total family incomes of $95,000. Which individual has a higher income, relative to others in his or her state? Use percentiles and z-scores to support your answer.

92. **More money** (1.3) Refer to Exercise 91.

(a) How do the ranges of the two distributions compare? Justify your answer.

(b) Explain why the standard deviation of the total family incomes in the New Jersey sample is so much larger than for the Indiana sample.

Chapter 2 Wrap-Up

FRAPPY! FREE RESPONSE AP® PROBLEM, YAY!

The following problem is modeled after actual AP® Statistics exam free response questions. Your task is to generate a complete, concise response in 15 minutes.

Directions: Show all your work. Indicate clearly the methods you use, because you will be scored on the correctness of your methods as well as on the accuracy and completeness of your results and explanations.

The distribution of scores on a recent test closely followed a Normal distribution with a mean of 22 points and a standard deviation of 4 points.

  (a) What proportion of the students scored at least 25 points on this test?

  (b) What is the 31st percentile of the distribution of test scores?

  (c) The teacher wants to transform the test scores so that they have an approximately Normal distribution with a mean of 80 points and a standard deviation of 10 points. To do this, she will use a formula in the form

$$new\ score = a + b(old\ score)$$

    Find the values of a and b that the teacher should use to transform the distribution of test scores.

  (d) Before the test, the teacher gave a review assignment for homework. The maximum score on the assignment was 10 points. The distribution of scores on this assignment had a mean of 9.2 points and a standard deviation of 2.1 points. Would it be appropriate to use a Normal distribution to calculate the proportion of students who scored below 7 points on this assignment? Explain your answer.

After you finish, you can view two example solutions on the book's website (highschool.bfwpub.com/updatedtps6e). Determine whether you think each solution is "complete," "substantial," "developing," or "minimal." If the solution is not complete, what improvements would you suggest to the student who wrote it? Finally, your teacher will provide a scoring rubric. Score your response and note what, if anything, you would do differently to improve your own score.

Chapter 2 Review

Section 2.1: Describing Location in a Distribution

In this section, you learned two different ways to describe the location of individuals in a distribution: percentiles and standardized scores (z-scores). Percentiles describe the location of an individual value in a distribution by measuring what percent of the observations are less than or equal to that value. A cumulative relative frequency graph is a handy tool for identifying percentiles in a distribution. You can use it to estimate the percentile for a particular value of a variable or to estimate the value of the variable at a particular percentile.

Standardized scores (z-scores) describe the location of an individual in a distribution by measuring how many standard deviations the individual is above or below the mean. To find the standardized score for a particular observation, transform the value by subtracting the mean and then dividing the difference by the standard deviation. Besides describing the location of an individual in a distribution, you can also use z-scores to compare observations from different distributions—standardizing the values puts them on a standard scale.

(a)

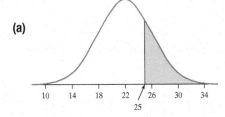

$$z = \frac{25 - 22}{4} = 0.75$$

(i) *Table A:* $1 - 0.7734 = 0.2266$
(ii) *Tech:* normalcdf(lower:25,upper:1000, μ:22,σ:4) = 0.2266. The proportion of students who scored at least 25 points is about 0.23.

(b)

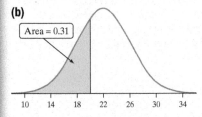

(i) *Table A:* 0.31 area to the left
$\rightarrow z = -0.50$. *Tech:* invNorm(area:0.31, mean:0, SD:1) $= -0.50$. Solving

$$-0.50 = \frac{x - 22}{4} \text{ gives } x = 20.$$

(ii) invNorm(area:0.31, mean:22, SD:4) $= 20.02$. The 31st percentile is about 20 points.

(c) Because adding a constant does not affect measures of variability,

$$new\ SD = b\,(old\ SD)$$
$$10 = b\,(4)$$
$$so\ b = 2.5.$$

Because adding a constant and multiplying by a constant affect measures of center,

$$new\ mean = a + b\,(old\ mean)$$
$$80 = a + 2.5\,(22)$$
$$so\ a = 25$$

The linear transformation should be old score = 25 + 2.5 (old score).

(d) If the distribution was Normal, scores should go at least 2 standard deviations above the mean. However, in this distribution, the highest possible score is only $z = (10 - 9.2)/2.1 = 0.38$ standard deviation above the mean. Therefore, it would be inappropriate to use a Normal distribution to do calculations.

There is one additional Chapter Test available for this section in the Teacher's Resource Materials. Click on the link in the TE-Book, open in the TRFD, or download from the Teacher's Resources on the book's digital platform. You can also create your own test using the TPS quiz and test builder (ExamView).

Ask the StatsMedic

How to Write a Great Test for AP® Statistics

This post details the proper format of a great AP® Statistics test and also provides resources for finding questions.

Ask the StatsMedic

How to Grade Your AP® Statistics Test

This post gives suggestions on the best way to give students feedback on their tests.

Teaching Tip

Be sure to discuss the "What Did You Learn?" table with your students. Ask them to read each learning target and self-assess whether or not they can do each one. For each target, the grid shows the section in which it was covered, page references for examples that illustrate the target, and relevant chapter review exercises that students can use to assess their understanding of that target. Encourage your students to use this grid as part of their preparations for the chapter test.

You also learned to describe the effects on the shape, center, and variability of a distribution when transforming data from one scale to another. Adding a positive constant to (or subtracting it from) each value in a data set changes the measures of center and location, but not the shape or variability of the distribution. Multiplying or dividing each value in a data set by a positive constant changes the measures of center and location and measures of variability, but not the shape of the distribution.

Section 2.2: Density Curves and Normal Distributions

In this section, you learned how density curves are used to model distributions of quantitative data. An area under a density curve estimates the proportion of observations that fall in a specified interval of values. The total area under a density curve is 1, or 100%.

The most commonly used density curve is called a Normal curve. The Normal curve is symmetric, single-peaked, and bell-shaped with mean μ and standard deviation σ. For any distribution of data that is approximately Normal in shape, about 68% of the observations will be within 1 standard deviation of the mean, about 95% of the observations will be within 2 standard deviations of the mean, and about 99.7% of the observations will be within 3 standard deviations of the mean. This handy result is known as the empirical rule.

When observations do not fall exactly 1, 2, or 3 standard deviations from the mean, you learned how to use Table A or technology to identify the proportion of values in any specified interval under a Normal curve. You also learned how to use Table A or technology to determine the value of an individual that falls at a specified percentile in a Normal distribution. On the AP® Statistics exam, it is extremely important that you clearly communicate your methods when answering questions that involve a Normal distribution. Shading a Normal curve with the mean, standard deviation, and boundaries clearly identified is a great start. If you use technology to perform calculations, be sure to label the inputs of your calculator commands.

Finally, you learned how to determine if a distribution of data is approximately Normal using graphs (dotplots, stemplots, histograms) and the empirical rule. You also learned that a Normal probability plot is a great way to determine whether the shape of a distribution is approximately Normal. The more linear the Normal probability plot, the more Normal the distribution of the data.

What Did You Learn?

Learning Target	Section	Related Example on Page(s)	Relevant Chapter Review Exercise(s)
Find and interpret the percentile of an individual value in a distribution of data.	2.1	92	R2.1, R2.3(c)
Estimate percentiles and individual values using a cumulative relative frequency graph.	2.1	94	R2.2
Find and interpret the standardized score (z-score) of an individual value in a distribution of data.	2.1	96	R2.1
Describe the effect of adding, subtracting, multiplying by, or dividing by a constant on the shape, center, and variability of a distribution of data.	2.1	99, 100, 101	R2.3
Use a density curve to model a distribution of quantitative data.	2.2	111	R2.4
Identify the relative locations of the mean and median of a distribution from a density curve.	2.2	112	R2.4
Use the empirical rule to estimate (i) the proportion of values in a specified interval, or (ii) the value that corresponds to a given percentile in a Normal distribution.	2.2	118	R2.5
Find the proportion of values in a specified interval in a Normal distribution using Table A or technology.	2.2	122, 124, 126	R2.5, R2.6, R2.7
Find the value that corresponds to a given percentile in a Normal distribution using Table A or technology.	2.2	129	R2.5, R2.6, R2.7
Determine whether a distribution of data is approximately Normal from graphical and numerical evidence.	2.2	132, 135	R2.8, R2.9

Teaching Tip

Now that you have completed Chapters 1 and 2 in this book, students can login to their College Board accounts and take the **Personal Progress Check for Unit 1**.

Chapter 2 Review Exercises

These exercises are designed to help you review the important ideas and methods of the chapter.

R2.1 Is Paul tall? According to the National Center for Health Statistics, the distribution of heights for 15-year-old males has a mean of 170 centimeters (cm) and a standard deviation of 7.5 cm. Paul is 15 years old and 179 cm tall.

(a) Find the *z*-score corresponding to Paul's height. Explain what this value means.

(b) Paul's height puts him at the 85th percentile among 15-year-old males. Explain what this means to someone who knows no statistics.

R2.2 Computer use Mrs. Causey asked her students how much time they had spent watching television during the previous week. The figure shows a cumulative relative frequency graph of her students' responses.

(a) At what percentile is a student who watched TV for 7 hours last week?

(b) Estimate from the graph the interquartile range (*IQR*) for time spent watching TV.

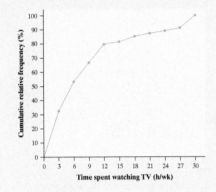

R2.3 Aussie, Aussie, Aussie A group of Australian students were asked to estimate the width of their classroom in feet. Use the dotplot and summary statistics to answer the following questions.

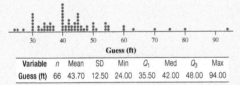

Variable	n	Mean	SD	Min	Q_1	Med	Q_3	Max
Guess (ft)	66	43.70	12.50	24.00	35.50	42.00	48.00	94.00

(a) Suppose we converted each student's guess from feet to meters (3.28 ft = 1 m). How would the shape of the distribution be affected? Find the mean, median, standard deviation, and *IQR* for the transformed data.

(b) The actual width of the room was 42.6 feet. Suppose we calculated the error in each student's guess as follows: guess − 42.6. Find the mean and standard deviation of the errors in feet.

(c) Find the percentile for the student who estimated the classroom width as 63 feet.

R2.4 Density curves The following figure is a density curve that models a distribution of quantitative data. Trace the curve onto your paper.

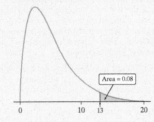

(a) What percent of observations have values less than 13? Justify your answer.

(b) Mark the approximate location of the median. Explain your choice of location.

(c) Mark the approximate location of the mean. Explain your choice of location.

R2.5 Low-birth-weight babies Researchers in Norway analyzed data on the birth weights of 400,000 newborns over a 6-year period. The distribution of birth weights is approximately Normal with a mean of 3668 grams and a standard deviation of 511 grams.[17] Babies that weigh less than 2500 grams at birth are classified as "low birth weight."

(a) Fill in the blanks: About 99.7% of the babies had birth weights between ____ and ____ grams.

(b) What percent of babies will be identified as low birth weight?

(c) Find the quartiles of the birth weight distribution.

---

63 feet or less, so the student who estimated the width as 63 feet is at the $62/66 = 0.939 = 93.94$th percentile.

R2.4 (a) The percent of observations that have values less than 13 is $1 - 0.08 = 0.92 = 92\%$. **(b)** Answers will vary, but the line indicating the median (line A in the graph below) should be slightly to the right of the main peak with half of the area to the left and half to the right. **(c)** Answers will vary, but the line indicating the mean (line B in the graph below) should be slightly to the right of the line for the median at the balance point.

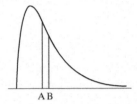

R2.5 (a) About 99.7% of the observations will fall within 3 standard deviations of the mean. About 97.7% of the babies had birth weights between <u>2135</u> and <u>5201</u> grams.

(b) (i) $z = \dfrac{2500 - 3668}{511} = -2.29$

The proportion of *z*-scores below -2.29 is 0.0110. (ii): normalcdf(lower: -1000, upper: 2500, mean: 3668, SD: 511) $= 0.0111$. About 1% of babies will be identified as low birth weight.
(c) (i) The first quartile is the boundary value with 25% of the area to its left $\rightarrow z = -0.67$. The third quartile is the boundary value with 75% of the area to its left $\rightarrow z = 0.67$. Solving

$$-0.67 = \frac{x - 3668}{511}$$ gives $Q_1 = 3325.63$.

Solving $0.67 = \dfrac{x - 3668}{511}$ gives

$Q_3 = 4010.37$. (ii) invNorm(area: 0.25, mean: 3668, SD: 511) gives $Q_1 = 3323.34$; invNorm(area: 0.75, mean: 3668, SD: 511) gives $Q_3 = 4012.66$. The quartiles are $Q_1 = 3323.34$ grams and $Q_3 = 4012.66$ grams.

TRM Full Solutions to Chapter 2 Review Exercises

The full solutions can be found by clicking on the link in the TE-Book, opening the TRFD, or downloading from the Teacher's Resources on the book's digital platform.

Answers to Chapter 2 Review Exercises

R2.1 (a) $z = \dfrac{179 - 170}{7.5} = 1.20$

Paul's height is 1.20 standard deviations above the average male height for his age. **(b)** 85% of boys Paul's age are the same height as or shorter than Paul.

R2.2 (a) 7 hours corresponds to about the 58th percentile. **(b)** Q_1 is approximately

2.5 hours. Q_3 is approximately 11 hours. Thus, $IQR = 11 - 2.5 = 8.5$ hours per week.

R2.3 (a) Converting from feet to meters would not change the shape of the distribution. The

new mean would be $\dfrac{43.7}{3.28} = 13.32$ meters,

median $= \dfrac{42}{3.28} = 12.80$ meters, standard

deviation $= \dfrac{12.5}{3.28} = 3.81$ meters, and the

$IQR = \dfrac{12.5}{3.28} = 3.81$ meters.

(b) The mean error would be $43.7 - 42.6 = 1.1$ feet. The standard deviation of the errors would be the same as the standard deviation of the guesses, 12.5 feet. **(c)** 62 of the 66 students estimated the width of their classroom to be

R2.6 (a) (i) $z = \dfrac{500 - 694}{112} = -1.73$,

$z = \dfrac{900 - 694}{112} = 1.84$; the proportion of z-scores below -1.73 is 0.0418. The proportion of z-scores above 1.84 is $1 - 0.9671 = 0.0329$. (ii) normalcdf(lower: -1000, upper: 500, mean: 694, SD: 112) = 0.0418 and normalcdf(lower: 900, upper: 100000, mean: 694, SD: 112) = 0.0329. The percent of test takers who earn a score less than 500 or greater than 900 on the GRE Chemistry test is $4.18\% + 3.29\% = 7.47\%$. **(b)** (i) 99% area to the left of $z \to z = 2.33$. Solving

$2.33 = \dfrac{x - 694}{112}$ gives $x = 954.96$.

(ii) invNorm(area: 0.99, mean: 694, SD: 112) = 954.55; the 99th percentile score on the GRE Chemistry test is 954.55.

R2.7 (a) (i) $z = \dfrac{1.2 - 1.05}{0.08} = 1.88$,

$z = \dfrac{1 - 1.05}{0.08} = -0.63$; the proportion of z-scores between -0.63 and 1.88 is $0.9699 - 0.2643 = 0.7056$. (ii) normalcdf(lower: 1, upper: 1.2, mean: 1.05, SD: 0.08) = 0.7036; about 70% of the time, the dispenser will put between 1 and 1.2 ounces of ketchup on a burger. **(b)** Because the mean of 1.1 is in the middle of the interval from 1 to 1.2, we are looking for the middle 99% of the distribution. This leaves 0.5% in each tail. (i) 0.005 area to the left of $z \to z - 2.58$.

Solving $-2.58 = \dfrac{1 - 1.1}{\sigma}$ gives

$\sigma = 0.039$; a standard deviation of at most 0.039 ounce will result in at least 99% of burgers getting between 1 and 1.2 ounces of ketchup.

R2.8 The distribution of percent of residents aged 65 and older in the 50 states and the District of Columbia is roughly symmetric and somewhat bell-shaped. The mean and standard deviation of these data are $\bar{x} = 13.255\%$ and $s_x = 1.668\%$.

- $\bar{x} \pm 1s_x = (11.587, 14.923)$; 40 of 51 observations, or 78.4% of the observations, are within 1 standard deviation of the mean.
- $\bar{x} \pm 2s_x = (9.919, 16.591)$; 48 of 51 observations, or 94.1% of the observations, are within 2 standard deviations of the mean.

R2.6 Acing the GRE The Graduate Record Examinations (GREs) are widely used to help predict the performance of applicants to graduate schools. The scores on the GRE Chemistry test are approximately Normal with mean = 694 and standard deviation = 112.

(a) Approximately what percent of test takers earn a score less than 500 or greater than 900 on the GRE Chemistry test?

(b) Estimate the 99th percentile score on the GRE Chemistry test.

R2.7 Ketchup A fast-food restaurant has just installed a new automatic ketchup dispenser for use in preparing its burgers. The amount of ketchup dispensed by the machine can be modeled by a Normal distribution with mean 1.05 ounces and standard deviation 0.08 ounce.

(a) If the restaurant's goal is to put between 1 and 1.2 ounces of ketchup on each burger, about what percent of the time will this happen?

(b) Suppose that the manager adjusts the machine's settings so that the mean amount of ketchup dispensed is 1.1 ounces. How much does the machine's standard deviation have to be reduced to ensure that at least 99% of the restaurant's burgers have between 1 and 1.2 ounces of ketchup on them?

R2.8 Where the old folks live Here are a stemplot and numerical summaries of the percents of residents aged 65 and older in the 50 states and the District of Columbia. Is this distribution of the percent of state residents who are age 65 and older approximately Normal? Justify your answer based on the graph and the empirical rule.

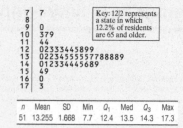

n	Mean	SD	Min	Q_1	Med	Q_3	Max
51	13.255	1.668	7.7	12.4	13.5	14.3	17.3

R2.9 Assessing Normality Catherine and Ana gave an online reflex test to 33 varsity athletes at their school. The following Normal probability plot displays the data on reaction times (in milliseconds) for these students. Is the distribution of reaction times for these athletes approximately Normal? Why or why not?

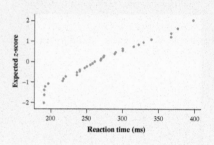

Chapter 2 AP® Statistics Practice Test

Section I: Multiple Choice *Select the best answer for each question.*

T2.1 Many professional schools require applicants to take a standardized test. Suppose that 1000 students take such a test. Several weeks after the test, Pete receives his score report: he got a 63, which placed him at the 73rd percentile. This means that

(a) Pete's score was below the median.

(b) Pete did worse than about 63% of test takers.

(c) Pete did worse than about 73% of test takers.

(d) Pete did the same as or better than about 63% of test takers.

(e) Pete did the same as or better than about 73% of test takers.

T2.2 For the Normal distribution shown, the standard deviation is closest to

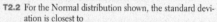

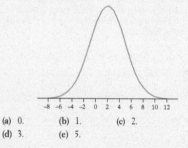

(a) 0.   (b) 1.   (c) 2.
(d) 3.   (e) 5.

- $\bar{x} \pm 3s_x = (8.251, 18.259)$; 50 of 51 observations, or 98.0% of the observations, are within 3 standard deviations of the mean.

These percentages are close to what we would expect based on the empirical rule. Combined with the graph, this gives good evidence that the distribution is approximately Normal.

R2.9 The curve in the Normal probability plot suggests that the data are slightly right-skewed. This can be seen in the steep, nearly vertical section in the lower left. These numbers were much closer to the mean than would be expected in a Normal distribution, meaning the values that would be in the left tail are piled up close to the center of the distribution.

TRM Full Solutions to Chapter 2 AP® Statistics Practice Test

The full solutions can be found by clicking on the link in the TE-Book, opening the TRFD, or downloading from the Teacher's Resources on the book's digital platform.

Answers to Chapter 2 AP® Statistics Practice Test

T2.1 e

T2.2 d

T2.3 Rainwater was collected in water containers at 30 different sites near an industrial complex, and the amount of acidity (pH level) was measured. The mean and standard deviation of the values are 4.60 and 1.10, respectively. When the pH meter was recalibrated back at the laboratory, it was found to be in error. The error can be corrected by adding 0.1 pH unit to all of the values and then multiplying the result by 1.2. What are the mean and standard deviation of the corrected pH measurements?

(a) 5.64, 1.44   (d) 5.40, 1.32

(b) 5.64, 1.32   (e) 5.64, 1.20

(c) 5.40, 1.44

T2.4 The figure shows a cumulative relative frequency graph of the number of ounces of alcohol consumed per week in a sample of 150 adults who report drinking alcohol occasionally. About what percent of these adults consume between 4 and 8 ounces per week?

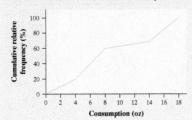

(a) 20%   (b) 40%   (c) 50%

(d) 60%   (e) 80%

T2.5 The average yearly snowfall in Chillyville is approximately Normally distributed with a mean of 55 inches. If the snowfall in Chillyville exceeds 60 inches in 15% of the years, what is the standard deviation?

(a) 4.83 inches   (d) 8.93 inches

(b) 5.18 inches   (e) The standard deviation cannot

(c) 6.04 inches   be computed from the given information.

T2.6 The figure shown is the density curve of a distribution. Seven values are marked on the density curve. Which of the following statements is true?

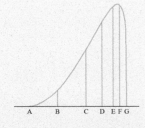

(a) The mean of the distribution is E.

(b) The area between B and F is 0.50.

(c) The median of the distribution is C.

(d) The 3rd quartile of the distribution is D.

(e) The area under the curve between A and G is 1.

T2.7 If the heights of a population of men are approximately Normally distributed, and the middle 99.7% have heights between 5′0″ and 7′0″, what is the standard deviation of the heights in this population?

(a) 1″   (b) 3″   (c) 4″

(d) 6″   (e) 12″

T2.8 The distribution of the time it takes for different people to solve a certain crossword puzzle is strongly skewed to the right with a mean of 30 minutes and a standard deviation of 15 minutes. The distribution of z-scores for those times is

(a) Normally distributed with mean 30 and SD 15.

(b) skewed to the right with mean 30 and SD 15.

(c) Normally distributed with mean 0 and SD 1.

(d) skewed to the right with mean 0 and SD 1.

(e) skewed to the right, but the mean and standard deviation cannot be determined without more information.

T2.9 The Environmental Protection Agency (EPA) requires that the exhaust of each model of motor vehicle be tested for the level of several pollutants. The level of oxides of nitrogen (NOX) in the exhaust of one light truck model was found to vary among individual trucks according to an approximately Normal distribution with mean $\mu = 1.45$ grams per mile driven and standard deviation $\sigma = 0.40$ gram per mile. Which of the following best estimates the proportion of light trucks of this model with NOX levels greater than 2 grams per mile?

(a) 0.0228   (b) 0.0846   (c) 0.4256

(d) 0.9154   (e) 0.9772

T2.10 Until the scale was changed in 1995, SAT scores were based on a scale set many years ago. For Math scores, the mean under the old scale in the early 1990s was 470 and the standard deviation was 110. In 2016, the mean was 510 and the standard deviation was 103. Gina took the SAT in 1994 and scored 500. Her cousin Colleen took the SAT in 2016 and scored 530. Who did better on the exam, and how can you tell?

(a) Colleen—she scored 30 points higher than Gina.

(b) Colleen—her standardized score is higher than Gina's.

(c) Gina—her standardized score is higher than Colleen's.

(d) Gina—the standard deviation was larger in 2016.

(e) The two cousins did equally well—their z-scores are the same.

T2.3 b

T2.4 b

T2.5 a

T2.6 e

T2.7 c

T2.8 d

T2.9 b

T2.10 c

T2.11 (a) 27 of the 40 sale prices were less than or equal to the house indicated in red on the dotplot so that home is at the $27/40 = 0.675 = 67.5$th percentile.

(b) *Interpretation:* The sale price for this home is 0.35 standard deviation above the average sale price for the homes in the sample.

T2.12 (a) (i) $z = \dfrac{6 - 7.11}{0.74} = -1.5$;

The proportion of z-scores below -1.5 is 0.0668; (ii) normalcdf(lower: -1000, upper: 6, mean: 7.11, SD: 0.74) = 0.0668. About 6.68% of the students ran the mile in less than 6 minutes, which means that $(0.0668)(12000) = 801.6$, or about 802, students ran the mile in less than 6 minutes. **(b)** (i): 0.90 area to the left of

$z \to z = 1.28$. Solving $1.28 = \dfrac{x - 7.11}{0.74}$

gives $x = 8.06$ minutes.

(ii) invNorm(area: 0.90, mean: 7.11, SD: 0.74) = 8.06 minutes. It took about 8.06 minutes for the slowest 10% of students to run the mile. **(c)** If the mile run times were converted from minutes to seconds, the mean would be $(7.11)(60) = 426.6$ seconds and the standard deviation would be $(0.74)(60) = 44.4$ seconds.

(i) $z = \dfrac{400 - 426.6}{44.4} = -0.60$,

$z = \dfrac{500 - 426.6}{44.4} = 1.65$; the difference

of z-scores between -0.60 and 1.65 is $0.9505 - 0.2743 = 0.6762$. (ii) normalcdf(lower: 400, upper: 500, mean: 426.6, SD: 44.4) = 0.6763. About 67.6% of students who ran the mile had times between 400 and 500 seconds.

T2.13 No, these data do not seem to follow a Normal distribution. First, there is a large difference between the mean and the median. In a Normal distribution, the mean and median are the same, but in this distribution, the mean is 48.25 and the median is 37.80. Second, the distance between the minimum and the median is $37.80 - 2 = 35.80$, but the distance between the median and the maximum is $204.90 - 37.80 = 167.10$. In a Normal distribution, these distances should be approximately the same. Because the mean is larger than the median and the distance from the median to the maximum is larger than the distance from the minimum to the median, these data appear to be skewed to the right.

Section II: Free Response *Show all your work. Indicate clearly the methods you use, because you will be graded on the correctness of your methods as well as on the accuracy and completeness of your results and explanations.*

T2.11 The dotplot gives the sale prices for 40 houses in Ames, Iowa, sold during a recent month. The mean sale price was $203,388 with a standard deviation of $87,609.

Price ($1000s)

(a) Find the percentile of the house indicated in red on the dotplot.

(b) Calculate and interpret the standardized score (z-score) for the house indicated by the red dot, which sold for $234,000.

T2.12 A study of 12,000 able-bodied male students at the University of Illinois found that their times for the mile run were approximately Normally distributed with mean 7.11 minutes and standard deviation 0.74 minute.[18]

(a) About how many students ran the mile in less than 6 minutes?

(b) Approximately how long did it take the slowest 10% of students to run the mile?

(c) Suppose that these mile run times are converted from minutes to seconds. Estimate the percent of students who ran the mile in between 400 and 500 seconds.

T2.13 A study recorded the amount of oil recovered from the 64 wells in an oil field, in thousands of barrels. Here are descriptive statistics for that set of data from statistical software.

Descriptive Statistics: Oilprod

Variable	N	Mean	Median	StDev	Min	Max	Q_1	Q_3
Oilprod	64	48.25	37.80	40.24	2.00	204.90	21.40	60.75

Based on the summary statistics, is the distribution of amount of oil recovered from the wells in this field approximately Normal? Justify your answer.

TRM Chapter 2 Case Study

The Case Study feature that was found in the previous two editions of TPS has been moved to the Teacher's Resource Materials. Click on the link in the TE-Book, open in the TRFD, or download from the Teacher's Resources on the book's digital platform.

Chapter 3

Chapter 3

Exploring Two-Variable Quantitative Data

PD Overview

Investigating the relationships between variables is central to what we do in statistics. In Section 1.1, we explored relationships between two categorical variables, such as environmental club membership and snowmobile use for visitors to Yellowstone National Park. We used two-way tables to summarize the data and bar graphs to visually display the relationship between these variables. We also introduced the idea of association: two variables have an association if knowing the value of one variable makes it easier to predict the value of the other variable. For visitors to Yellowstone National Park, we learned that environmental club members were less likely to rent or own a snowmobile.

In Sections 1.2 and 1.3, we explored the relationship between quantitative and categorical variables, such as the overall ratings of tablets and brand (Apple or Samsung). Although we didn't explicitly use the terms *relationship* or *association* in these sections, we used summary statistics (like the mean and median) and graphs (like dotplots and boxplots) to compare the two brands. We learned that Apple tablets typically had higher overall ratings with much less variability than the overall ratings for the Samsung tablets.

In Chapter 2, we learned that when data follow a regular overall pattern, we can use a simplified model to describe the pattern we see. For distributions of a single quantitative variable, we can use a Normal distribution to model the data if the distribution is symmetric and bell-shaped.

In this chapter, we consider relationships between two quantitative variables, such as the hand span of a person and the number of Starburst^{TM} candies he or she can grab. We will use a scatterplot to display the relationship between two variables, correlation to measure the strength and direction of a linear association, and a least-squares regression line to model a linear relationship. For a scatterplot that shows a curved relationship, we will transform one or both of the variables using logarithms or powers with the hope that we can straighten out the form of the relationship and fit a least-squares regression line.

The Main Ideas

One of the challenges in teaching the AP® Statistics course is keeping students focused on the big picture, not just the details of each section. We outline the main ideas for the chapter here.

Chapter 3 Introduction

Students should understand that knowing the distribution of a single variable often is not particularly useful. As the Old Faithful example illustrates, knowing the distribution of interval times isn't very helpful if you are trying to predict when the next eruption will occur. Instead, we look for relationships among variables to help explain patterns—and to make predictions.

The activity in this section asks students to investigate the relationship between hand span of a person and the number of Starburst^{TM} candies he or she can grab. This context is engaging for students (as long as they get some candy at the end of the activity) and will be used in several examples throughout the rest of Chapter 3.

Section 3.1 Scatterplots and Correlation

In this section, we revisit the principles of data analysis introduced in Chapter 1:

- Plot the data, then add numerical summaries.
- Look for overall patterns and departures from those patterns.
- When there is a regular overall pattern, use a simplified model to describe it.

Students will often lose sight of these principles as they get more involved with the specific content of the chapter. It is a good idea to remind students of the big picture by occasionally showing them a scatterplot and asking, "What do you see?"

In Chapter 1, we learned four important characteristics to address when describing the distribution of a single quantitative variable: shape, outliers, center, and variability, (SOCV). When describing the association between two variables in a scatterplot, students also should address four important characteristics:

- Direction of the association
- Unusual features
- Form of the association
- Strength of the association

You might help students remember these by using an acronym such as DUFS. Be sure that students include the context of the two variables involved in the association.

Students often find that describing the strength of an association in a scatterplot is challenging. The distinction between a moderately strong and a moderately weak association is much harder to recognize than the distinction between a positive and a negative association, for example. Fortunately, correlation is a numerical summary that can provide a more specific measure of the strength of a linear association. Instead of going straight to the formula, which can be difficult for some students to understand, we have chosen to focus on the properties of correlation. Once your students understand what correlation is all about, consider looking more carefully at the formula. Knowing how correlation is calculated can give insight into its properties. However, understanding the formula and calculating correlation by hand are not required for success on the AP® Statistics exam and will take valuable time away from other important topics.

Just as outliers can greatly influence the mean and standard deviation of a distribution, unusual points in a scatterplot can influence the correlation. Doing the activity on page 163 using the *Correlation and Regression* applet nicely illustrates the effects of unusual points on the correlation.

Section 3.2 Least-Squares Regression

Once we determine that the data fall into a regular, overall pattern, we seek a simplified model to describe the pattern. In this section, we use least-squares regression lines as models for relationships between variables that have a linear association. We will learn how to create models for nonlinear relationships in Section 3.3

As was the case with correlation, we de-emphasize formulas and hand calculations and focus instead on understanding and interpreting numerical summaries such as the slope, standard deviation of the residuals, and r^2. When data are provided, students should use technology to compute the least-squares

regression line and summary statistics. Hand computation is time-consuming, tedious, and not required on the AP® Statistics exam.

As you teach this section, emphasize the difference between the actual data and the model used to describe the data. When we substitute a value for x in the equation of a least-squares line, the result is a *predicted* value of y (denoted $\hat{y}$), not an actual value of y. Likewise, if the slope of a least-squares regression line is 3, we should say, "For each 1 unit increase in x, the linear model *predicts* that the value of y will increase by 3." Neglecting to include the word *predicts* suggests a deterministic relationship between x and y that lets us predict y exactly for a given value of x. To emphasize that the model provides only predicted values, least-squares regression lines are always expressed in terms of $\hat{y}$ instead of y.

The difference between the observed value of y and the predicted value of y is called a *residual* (residual = $y - \hat{y}$). Residuals are the key to understanding most of the material in this section. To find the equation of the least-squares regression line, we find the line that minimizes the sum of the squared residuals. To see if a linear model is appropriate, we make a residual plot. To assess how well our line models the data, we calculate the standard deviation of the residuals s to estimate the size of a typical prediction error. We also calculate r^2, which measures the percent reduction in the sum of squared residuals when using the least-squares regression line to make predictions rather than the mean value of y. In many cases, closely analyzing the residuals can give us insights about the data, such as why individuals with certain characteristics regularly exceed our predictions.

Like correlation in Section 3.1, students should understand the influence that outliers and high-leverage points have on the equation of the least-squares regression line, the standard deviation of the residuals s, and the value of r^2.

Students are also expected to use computer output to obtain the equation of the least-squares regression line and other numerical summaries. Ideally, students will have the opportunity to use a statistical software program to generate their own computer output. However, if you do not have the time or resources to have students use computer software for themselves, we have included examples of computer output in the narrative, examples, and exercises. There are many other examples in the Teacher's Resource Materials and released AP® Statistics exam questions.

Emphasize that we cannot conclude that a cause-and-effect relationship exists between two variables just because there is an association between them. Although changes in one variable may accompany changes in the other, we cannot be certain of the cause unless we account for other variables that may have an effect on the association.

Users of the first three editions of *TPS* should note that we have omitted the distinction between *confounding* and *common response* in this section. We found that focusing on this distinction caused students to miss the big picture. We will address the issue of confounding more thoroughly in Chapter 4, when we learn how to establish cause-and-effect relationships using experiments. We have also eliminated use of the term *lurking variable* in this chapter and throughout the book, as the term doesn't have a commonly accepted meaning in the statistics world.

Section 3.3 Transforming to Achieve Linearity

In the first two sections of Chapter 3, we spent most of our time analyzing linear relationships between two quantitative variables. However, many pairs of quantitative variables have nonlinear relationships, such as the income per person and life expectancy in countries around the world.

There are two basic strategies for modeling associations that are nonlinear:

1. Find a nonlinear function to model the association.
2. Transform the data so a linear model is appropriate.

Both strategies are reasonable, yet we will spend our time learning the second strategy. Not only does the AP® Statistics Course Framework specifically include transformations to achieve linearity, but understanding how to transform data is also a useful skill in other contexts. We have already had experience performing linear transformations of data in Chapter 2.

We begin by using powers and roots to transform data that follow a power model in the form $y = ax^p$. The goal is to make the form of the scatterplot as linear as possible. We know we have found the right transformation when the residual plot from a linear regression analysis has no leftover patterns.

If transformations with powers and roots don't accomplish the goal of straightening out the data, we can also transform one or both variables using logarithms, such as a base-10 logarithm (log) or the natural logarithm (ln). If a graph of log y versus x has a linear form, the original data follow an exponential model in the form $y = ab^x$. If a graph of log y versus log x has a linear form, the original data follow a power model in the form $y = ax^p$.

It's easy for us, as math teachers, to get carried away with all the interesting mathematical and statistical connections that can be made in this section. Focus on logarithmic transformations and don't worry about re-expressing the linear model in $\hat{y} = ab^x$ or $\hat{y} = ax^p$ form. Students will be expected only to predict values

of y for specific values of x using the least-squares regression line and transforming the predicted value back into the original units.

Chapter 3 Resources
Teacher's Resource Materials

The following resources, identified by the **TRM** in the annotated student pages, can be found by clicking on the link in the Teacher's e-Book (TE-Book), searching by category or chapter on the Teacher's Resource Flash Drive (TRFD), or logging into the book's digital platform (highschool.bfwpub.com/updatedtps6e) and searching the Teacher's Resources menu (teacher log-in required).

- Alternate Examples: one file per section
- Lecture Presentation Slides: one per section
- CSI Stats
- Chapter 3 Learning Targets Grid
- Additional Content: Time plots
- Data Exploration: The SAT essay: Is longer better?
- Examples of Computer Output
- Regression to the mean in sports
- Additional Transformation Problems
- FRAPPY! Materials
- Chapter 3 Project: Investigating Relationships in Baseball
- Chapter 3 Case Study
- Complete solutions for the Check Your Understanding problems, section exercises, review exercises, and practice test.
- Quizzes: one per section
- Chapter 3 Test

Free Response Questions from Previous AP® Statistics Exams

Questions can be found on the AP® Central website: apcentral.collegeboard.org/courses/ap-statistics/exam

Students should be able to answer all the free response questions listed below with material learned in this chapter. Questions that contain content from this chapter but also require content from later chapters are listed in the last chapter

required to complete the entire question. This list will be updated after each AP® Statistics exam and will be posted to the Teacher's Resource section of the book's digital platform and to www.statsmedic.com/free-response-questions. You may want to save these questions until the end of the year so you can give your students a complete released exam for practice. Notice how consistently the AP® exam has tested content from Chapter 3 in recent years.

Year	#	Content
2018	1	• Interpret intercept and r^2 • Identify outlier in scatterplot
2017	1	• Explaining "positive, linear, strong" • Interpreting slope • Calculating actual value from residual
2016	6	• Describing scatterplots • Interpreting slope from computer output • Estimating medians from a scatterplot • Accounting for a third variable
2015	5	• Describing a scatterplot • Classifying observations • Making a prediction
2014	6	• Calculate, interpret, and identify residuals • Comparing associations • Multiple regression and variable selection
2013	6	• Comparing distributions • Describing trends in a scatterplot (time plot) • Moving averages
2012	1	• Describing a scatterplot with nonlinear association • Influential points • Determining which points meet a consumer's criterion
2007B	4	• Graphing a least-squares regression line • Calculating a residual • Influential points
2005	3	• Assessing linearity with residual plots • Understanding and interpreting slope • Interpreting r^2 • Extrapolation
2004B	1	• Describing a scatterplot • Interpreting r^2 • Interpreting a residual plot for a least-squares regression line using transformed data
2003B	1	• Influential points
2002B	1	• Making a scatterplot • Interpreting correlation • Assessing linearity • Interpreting r^2

Year	#	Content
2002	4	• Using regression output to state the equation of a least-squares regression line • Finding and interpreting the correlation from computer output • Clusters and influential points
2000	1	• Describing scatterplots
1999	1	• Using a residual plot to assess linearity • Identifying slope and y intercept from computer output • Interpreting slope and y intercept • Making a prediction using a least-squares regression line • Using residuals to estimate actual values
1998	2	• Making a histogram of one variable from a scatterplot • Describing a histogram • Describing a scatterplot
1998	4	• Using regression output to state the equation of a least-squares regression line • Analyzing patterns in a residual plot (*Note:* The residual plot uses predicted values on the horizontal axis instead of the values of the explanatory variable.)
1997	6	• Making predictions using least-squares regression lines, including transformed data • Determining if models are appropriate • Creating a better model

Applets

- The *Correlation and Regression* applet allows students to add, delete, and drag points to investigate how outliers can influence the correlation and least-squares regression line. It also allows students to draw their own line and compare it with the actual least-squares regression line. Go to highschool.bfwpub.com/updatedtps6e (click on Statistical Applets and select "Correlation and Regression").
- The *Guess the Correlation* applet asks students to guess the value of the correlation and keeps track of their results. Go to www.rossmanchance.com/applets (select "Guess the Correlation").
- The *Least-Squares Regression* applet lets students fit a line to data and compare it with the least-squares regression line. They can also show the residuals and squared residuals for either line. Students can also add, delete, and drag points around. Go to www.rossmanchance.com/applets (select "Least-Squares Regression").
- The *Two-Quantitative Variables* applet in the Extra Applets on the Student Site at highschool.bfwpub.com/updatedtps6e allows you to enter your own two-variable data, make a scatterplot, find the correlation and the least-squares regression line, calculate summary statistics, and make residual plots.

Chapter 3: Pacing Guide, Learning Targets, and Suggested Assignments

This pacing guide is based on a schedule with 110, 50-minute sessions before the AP® Statistics exam. If you have a different number of sessions before the AP® exam, you can modify the pacing guide to suit your needs. If you have additional time, consider incorporating quizzes, released AP® free response questions, or additional activities. See the Resources section above for suggestions.

The suggested homework assignments list odd-numbered exercises, whenever possible, so students can check their answers against the back-of-book answers. If you would rather students not have access to the answers while doing homework, adding 1 to the exercise numbers usually will do the trick, because the homework exercises typically are paired. For example, Exercises 1 and 2 will generally cover the same topics, but in different contexts. You may also choose to include the Recycle and Review questions at the end of each section, which review topics from previous sections or chapters. If your school is using the digital platform that accompanies TPS6, you will find these assignments pre-built as online homework assignments for Chapter 3.

Day	Content	Learning Targets: Students will be able to . . .	Suggested Assignment (MC bold)
1	3.1 Explanatory and Response Variables, Displaying Relationships: Scatterplots, Describing a Scatterplot	• Distinguish between explanatory and response variables for quantitative data. • Make a scatterplot to display the relationship between two quantitative variables. • Describe the direction, form, and strength of a relationship displayed in a scatterplot and identify unusual features.	1, 3, 5, 9, 11
2	3.1 Measuring Linear Association: Correlation, Cautions about Correlation, Calculating Correlation, Additional Facts about Correlation	• Interpret the correlation. • Understand the basic properties of correlation, including how the correlation is influenced by unusual points. • Distinguish correlation from causation.	13, 15, 17, 19, 23, **29–34**
3	3.2 Prediction, Residuals, Interpreting a Regression Line	• Make predictions using regression lines, keeping in mind the dangers of extrapolation. • Calculate and interpret a residual. • Interpret the slope and y intercept of a least-squares regression line.	37, 39, 41, 43, 45
4	3.2 The Least-Squares Regression Line, Determining if a Linear Model Is Appropriate: Residual Plots	• Determine the equation of a least-squares regression line using technology or computer output. • Construct and interpret residual plots to assess whether a regression model is appropriate.	47, 49, 51, 53
5	3.2 How Well the Line Fits the Data: The Role of s and r^2 in Regression, Interpreting Computer Regression Output	• Interpret the standard deviation of the residuals and r^2 and use these values to assess how well the least-squares regression line models the relationship between two variables. • Describe how the slope, y intercept, standard deviation of the residuals, and r^2 are influenced by unusual points.	55, 57, 59, 67
6	3.2 Regression to the Mean, Correlation and Regression Wisdom	• Find the slope and y intercept of the least-squares regression line from the means and standard deviations of x and y and their correlation.	63, 65, **71–78**
7	3.3 Transforming with Powers and Roots; Transforming with Logarithms: Power Models	• Use transformations involving powers, roots, or logarithms to create a linear model that describes the relationship between two quantitative variables, and use the model to make predictions.	81, 83, 85, 87
8	3.3 Transforming with Logarithms: Exponential Models; Putting It All Together: Which Transformation Should We Choose?	• Determine which of several models does a better job of describing the relationship between two quantitative variables.	89, 91, 93, **95–96**
9	Chapter 3 Review/FRAPPY!		Chapter 3 Review Exercises
10	Chapter 3 Test		

Chapter 3 Alignment to the College Board's Fall 2019 AP® Statistics Course Framework*

Relationship to College Board Units

Chapter 3 in this book covers the topics addressed in Unit 2 of the College Board Course Framework, except for Topics 2.2 and 2.3, which were previously covered in Section 1.1. Students will be ready to take the Personal Progress Check for Unit 2 once they have completed Chapter 3.

Big Ideas and Enduring Understandings

Chapter 3 develops these Big Ideas and related Enduring Understandings outlined in the Course Framework:

- **Big Idea 2: Patterns and Uncertainty (EU: UNC 1):** Statistical tools allow us to represent and describe patterns in data and to classify departures from patterns. Simulation and probabilistic reasoning allow us to anticipate patterns in data and to determine the likelihood of errors in inference.
- **Big Idea 3: Data-Based Predictions, Decisions, and Conclusions (EU: DAT 1):** Data-based regression models describe relationships between variables and are a tool for making predictions for values of a response variable. Collecting data using random sampling or randomized experimental design means that findings may be generalized to the part of the population from which the selection was made. Statistical inference allows us to make data-based decisions.

Course Skills

Chapter 3 helps students to develop the skills identified in the Course Framework.

- **2: Data Analysis** (2.A, 2.B, 2.C)
- **4: Statistical Argumentation** (4.B)

Learning Objectives and Essential Knowledge Statements

Section	Learning Objectives	Essential Knowledge Statements
3.1	UNC-1.S, DAT-1.A, DAT-1.B, DAT-1.C	UNC-1.S.1, UNC-1.S.2, UNC-1.S.3, DAT-1.A.1, DAT-1.A.2, DAT-1.A.3, DAT-1.A.4, DAT-1.A.5, DAT-1.A.6, DAT-1.B.1, DAT-1.B.2, DAT-1.B.3, DAT-1.C.1, DAT-1.C.2
3.2	DAT-1.D, DAT-1.E, DAT-1.F, DAT-1.G, DAT-1.H, DAT-1.I	DAT-1.D.1, DAT-1.D.2, DAT-1.D.3, DAT-1.E.1, DAT-1.E.2, DAT-1.F.1, DAT-1.F.2, DAT-1.G.1, DAT-1.G.2, DAT-1.G.3, DAT-1.G.4, DAT-1.H.1, DAT-1.H.2, DAT-1.H.3, DAT-1.I.1, DAT-1.I.2, DAT-1.I.3
3.3	DAT-1.J	DAT-1.J.1, DAT-1.J.2

A detailed alignment that can be sorted by Course Framework Unit, Topic, Learning Objective, Essential Knowledge Statement, or textbook section, is available on the TRFD and in the Teacher's Resources folder on Sapling Plus. **TRM**

*Should changes be made to the Course Framework in the future, an updated alignment will be placed on our AP® updates page at go.bfwpub.com/ap-course-updates.

UNIT 2
Exploring Two-Variable Data

Chapter 3

Exploring Two-Variable Quantitative Data

AtriPics.com/Alamy Stock Photo

Teaching Tip

Unit 2 in the College Board Course Framework aligns with Chapter 3 in this book (except for Topics 2.2 and 2.3, which were covered in Section 1.1). Students will be ready to take the Personal Progress Check for Unit 2 once they have completed Chapter 3.

PD Chapter 3 Overview

To watch the video overview of Chapter 3 (for teachers), click on the link in the TE-Book, look on the TRFD, or download from the Teacher's Resources on the book's digital platform.

Teaching Tip

Some teachers cover Chapter 3 at the end of the course to make a more seamless connection between Chapter 3 and Section 12.3. This approach saves time by not requiring review of the content of Chapter 3 before starting Section 12.3.

TRM Lecture Presentation Slides

If you are new to teaching AP® Statistics or are short on time when preparing for class, you may find the Lecture Presentation Slides to be helpful. Experienced AP® Statistics Teacher Doug Tyson has created one slide presentation per section. You may use them as is, modify them to fit your needs, or share them with students who miss class. Find them on the TRFD and in the Teacher's Resources on the book's digital platform.

Making Connections

Relationships between two *categorical* variables were discussed in Section 1.1. Relationships between two *quantitative* variables are the focus of Chapter 3. In Section 12.3, we will extend the content from this chapter to doing inference for the relationship between two quantitative variables.

✚ Ask the StatsMedic

Barbie Bungee

This post outlines an alternative activity that you can use to collect two-variable data. The Barbie Bungee context will be referenced in several of the alternate examples later in this chapter.

▶ ACTIVITY OVERVIEW

To prepare for using this activity, watch the overview video by clicking on the link in the TE-Book, opening the TRFD, or downloading from the Teacher's Resources on the book's digital platform.

Time: 15 minutes

Materials: Large container, candy (Tootsie Rolls, Starbursts™, Dum Dum suckers, or any candy that is individually wrapped)

Teaching Advice: Be sure that you select a candy that each student can pick up at least 10 of at once. Students can make a scatterplot by hand or can use the *Two-Quantitative Variables* applet in the Extra Applets section of the Student Site at highschool.bfwpub.com/updatedtps6e.

Answers:

1–4. Answers will vary for each student.

5. It is likely that there will be a strong positive, linear relationship between hand span and number of candies.

INTRODUCTION

Investigating relationships between variables is central to what we do in statistics. When we understand the relationship between two variables, we can use the value of one variable to help us make predictions about the other variable. In Section 1.1, we explored relationships between *categorical* variables, such as membership in an environmental club and snowmobile use for visitors to Yellowstone National Park. The association between these two variables suggests that members of environmental clubs are less likely to own or rent snowmobiles than nonmembers.

In this chapter, we investigate relationships between two *quantitative* variables. What can we learn about the price of a used car from the number of miles it has been driven? What does the length of a fish tell us about its weight? Can students with larger hands grab more candy? The following activity will help you explore the last question.

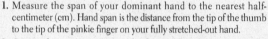

ACTIVITY   Candy grab

In this activity, you will investigate if students with a larger hand span can grab more candy than students with a smaller hand span.[1]

1. Measure the span of your dominant hand to the nearest half-centimeter (cm). Hand span is the distance from the tip of the thumb to the tip of the pinkie finger on your fully stretched-out hand.

2. One student at a time, go to the front of the class and use your dominant hand to grab as many candies as possible from the container. You must grab the candies with your fingers pointing down (no scooping!) and hold the candies for 2 seconds before counting them. After counting, put the candy back into the container.

3. On the board, record your hand span and number of candies in a table with the following headings:

Hand span (cm)	Number of candies

4. While other students record their values on the board, copy the table onto a piece of paper and make a graph. Begin by constructing a set of coordinate axes. Label the horizontal axis "Hand span (cm)" and the vertical axis "Number of candies." Choose an appropriate scale for each axis and plot each point from your class data table as accurately as you can on the graph.

5. What does the graph tell you about the relationship between hand span and number of candies? Summarize your observations in a sentence or two.

TRM CSI Stats

In this activity, a valuable item has been stolen from the classroom, but the thief has left a handprint. Students work to predict the height of the thief from the size of the handprint so that they can identify the culprit. Click on the link in the TE-Book, look on the TRFD, or download from the Teacher's Resources on the book's digital platform.

SECTION 3.1 Scatterplots and Correlation

LEARNING TARGETS *By the end of the section, you should be able to:*

- Distinguish between explanatory and response variables for quantitative data.
- Make a scatterplot to display the relationship between two quantitative variables.
- Describe the direction, form, and strength of a relationship displayed in a scatterplot and identify unusual features.

- Interpret the correlation.
- Understand the basic properties of correlation, including how the correlation is influenced by unusual points.
- Distinguish correlation from causation.

> A one-variable data set is sometimes called *univariate data*. A data set that describes the relationship between two variables is sometimes called *bivariate data*.

Most statistical studies examine data on more than one variable for a group of individuals. Fortunately, analysis of relationships between two variables builds on the same tools we used to analyze one variable. The principles that guide our work also remain the same:

- Plot the data, then look for overall patterns and departures from those patterns.
- Add numerical summaries.
- When there's a regular overall pattern, use a simplified model to describe it.

Explanatory and Response Variables

In the "Candy grab" activity, the number of candies is the **response variable**. Hand span is the **explanatory variable** because we anticipate that knowing a student's hand span will help us predict the number of candies that student can grab.

> **DEFINITION** Response variable, Explanatory variable
>
> A **response variable** measures an outcome of a study. An **explanatory variable** may help predict or explain changes in a response variable.

> You will often see explanatory variables called *independent variables* and response variables called *dependent variables*. Because the words *independent* and *dependent* have other meanings in statistics, we won't use them here.

It is easiest to identify explanatory and response variables when we initially specify the values of one variable to see how it affects another variable. For instance, to study the effect of alcohol on body temperature, researchers gave several different amounts of alcohol to mice. Then they measured the change in each mouse's body temperature 15 minutes later. In this case, amount of alcohol is the explanatory variable, and change in body temperature is the response variable. When we don't specify the values of either variable before collecting the data, there may or may not be a clear explanatory variable.

> **AP® EXAM TIP**
>
> Students might be asked to identify the explanatory variable in a context like this one. The response of "alcohol" is incomplete. The best answer is "amount of alcohol." A variable should have values that are capable of varying. "Alcohol" doesn't vary, but the "amount of alcohol" does have variability. Ask students to think about exactly what they are measuring and recording.

PD **Section 3.1 Overview**

To watch the video overview of Section 3.1 (for teachers), click on the link in the TE-Book, look on the TRFD, or download from the Teacher's Resources on the book's digital platform.

TRM **Learning Targets Grid**

At the beginning of each section, we present the relevant learning targets. Point these out to your students and refer back to the targets when you cover them in class. There is a PDF version of the grid with an additional column that students can use to keep track of their progress. Find it in the Teacher's Resource Materials located in the TE-Book, on the TRFD, or in the Teacher's view in the book's digital platform.

TRM **Section 3.1 Alternate Examples**

You can find the Alternate Examples for this section in Microsoft Word format by clicking on the link in the TE-Book, opening the TRFD, or downloading from the Teacher's Resources on the book's digital platform.

Making Connections

Remind students how we used these principles in previous chapters.
- Make a graph such as a dotplot or histogram.
- Add numerical summaries such as the mean and standard deviation or the median and interquartile range (*IQR*).
- Look for outliers and explore their effects on the numerical summaries.
- Check for Normality—if the shape of the data is approximately Normal, use a Normal distribution as a model.

Teaching Tip

You can also say that the explanatory variable "accounts for" changes in the response variable. This wording will come back later in our interpretation of r^2.

Teaching Tip

The "play" icon at the top of this example (and many others) indicates that students can watch a video presentation of the example by clicking on the link in the e-Book or viewing the videos at the Student Site highschool.bfwpub.com/updatedtps6e.

ALTERNATE EXAMPLE   Skill 2.B

Studying and final exam scores
Explanatory or response?

PROBLEM:

Identify the explanatory variable and response variable for the following relationships, if possible. Explain your reasoning.
(a) The score on a statistics final exam and the number of hours studied for a sample of students
(b) The final exam score for statistics and the final exam score for biology for a sample of students taking both courses

SOLUTION:

(a) Explanatory: number of hours studied. Response: score on the statistics final exam. The number of hours studied helps explain the score on the statistics final exam.
(b) Either variable can be the explanatory variable because each one can be used to predict or explain the other.

Teaching Tip

Point out that scatterplots are the only choice for displaying the relationship between two quantitative variables. For a single quantitative variable, there are many choices for displaying its distribution, including dotplots, histograms, boxplots, and stemplots.

Teaching Tip

If two individuals have identical values for the explanatory and response variables, another option is to jitter the two values so they don't overlap. This means, however, that neither dot will be in exactly the correct location.

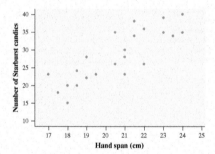

EXAMPLE   **Diamonds and the SAT**
Explanatory or response?

PROBLEM: Identify the explanatory variable and response variable for the following relationships, if possible. Explain your reasoning.

(a) The weight (in carats) and the price (in dollars) for a sample of diamonds.
(b) The SAT math score and the SAT evidence-based reading and writing score for a sample of students.

SOLUTION:

(a) Explanatory: weight; Response: price. The weight of a diamond helps explain how expensive it is.
(b) Either variable could be the explanatory variable because each one could be used to predict or explain the other.

FOR PRACTICE, TRY EXERCISE 1

In many studies, the goal is to show that changes in one or more explanatory variables actually *cause* changes in a response variable. However, other explanatory–response relationships don't involve direct causation. In the alcohol and mice study, alcohol actually *causes* a change in body temperature. But there is no cause-and-effect relationship between SAT math and evidence-based reading and writing scores.

Displaying Relationships: Scatterplots

Although there are many ways to display the distribution of a single quantitative variable, a **scatterplot** is the best way to display the relationship between two quantitative variables.

DEFINITION   Scatterplot

A **scatterplot** shows the relationship between two quantitative variables measured on the same individuals. The values of one variable appear on the horizontal axis, and the values of the other variable appear on the vertical axis. Each individual in the data set appears as a point in the graph.

Figure 3.1 shows a scatterplot that displays the relationship between hand span (cm) and number of Starburst™ candies for the 24 students in Mr. Tyson's class

FIGURE 3.1 Scatterplot of hand span (cm) and number of Starburst candies grabbed by 24 students. Only 23 points appear because two students had hand spans of 19 cm and grabbed 28 Starburst candies.

Teaching Tip

Notice that the bottom left value of the scatterplot (Figure 3.1) is not the origin (0, 0). This could be made clearer by providing a break or squiggly at the beginning of the x and y axis. The AP® Statistics exam does not typically include the axis break.

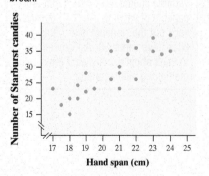

Making Connections

We will investigate correlation and causation in more detail at the end of this Section 3.1.

---

who did the "Candy grab" activity. As you can see, students with larger hand spans were typically able to grab more candies.

After collecting bivariate quantitative data, it is easy to make a scatterplot.

HOW TO MAKE A SCATTERPLOT

- **Label the axes.** Put the name of the explanatory variable under the horizontal axis and the name of the response variable to the left of the vertical axis. If there is no explanatory variable, either variable can go on the horizontal axis.
- **Scale the axes.** Place equally spaced tick marks along each axis, beginning at a convenient number just below the smallest value of the variable and continuing until you exceed the largest value.
- **Plot individual data values.** For each individual, plot a point directly above that individual's value for the variable on the horizontal axis and directly to the right of that individual's value for the variable on the vertical axis.

The following example illustrates the process of constructing a scatterplot.

EXAMPLE

Buying wins
Making a scatterplot

PROBLEM: Do baseball teams that spend more money on players also win more games? The table shows the payroll (in millions of dollars) and number of wins for each of the 30 Major League Baseball teams during the 2016 regular season.[2] Make a scatterplot to show the relationship between payroll and wins.

Team	Payroll	Wins	Team	Payroll	Wins
Arizona Diamondbacks	103	69	Milwaukee Brewers	75	73
Atlanta Braves	122	68	Minnesota Twins	112	59
Baltimore Orioles	157	89	New York Mets	150	87
Boston Red Sox	215	93	New York Yankees	227	84
Chicago Cubs	182	103	Oakland Athletics	98	69
Chicago White Sox	141	78	Philadelphia Phillies	117	71
Cincinnati Reds	114	68	Pittsburgh Pirates	106	78
Cleveland Indians	114	94	San Diego Padres	127	68
Colorado Rockies	121	75	San Francisco Giants	181	87
Detroit Tigers	206	86	Seattle Mariners	155	86
Houston Astros	118	84	St. Louis Cardinals	167	86
Kansas City Royals	145	81	Tampa Bay Rays	71	68
Los Angeles Angels	181	74	Texas Rangers	169	95
Los Angeles Dodgers	274	91	Toronto Blue Jays	159	89
Miami Marlins	81	79	Washington Nationals	163	95

Teaching Tip

Students may be asked to produce a scatterplot on the AP® Statistics exam. More often, they are asked to interpret a given scatterplot.

Teaching Tip: Using Technology

Use the *Two Quantitative Variables* applet at in the Extra Applets on the Student Site at highschool.bfwpub.com/updatedtps6e to make a scatterplot.

ALTERNATE EXAMPLE   Skill 2.B

Track and field day!
Making a scatterplot

Each member of a small statistics class ran a 40-yard sprint and then did a long jump (with a running start). The table below shows the sprint time (in seconds) and the long-jump distance (in inches):

Sprint time (sec)	Long-jump distance (in.)
5.41	171
5.05	184
7.01	90
7.17	65
6.73	78
5.68	130
5.78	173
6.31	143
6.44	92
6.50	139
6.80	120
7.25	110

PROBLEM:
Make a scatterplot of the relationship between sprint time and long-jump distance.

SOLUTION:

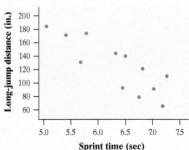

SOLUTION:

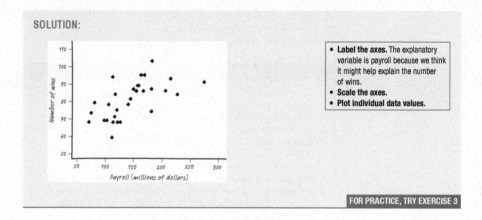

- **Label the axes.** The explanatory variable is payroll because we think it might help explain the number of wins.
- **Scale the axes.**
- **Plot individual data values.**

FOR PRACTICE, TRY EXERCISE 3

Describing a Scatterplot

To describe a scatterplot, follow the basic strategy of data analysis from Chapter 1: look for patterns and important departures from those patterns.

The scatterplot in Figure 3.2(a) shows a **positive association** between wins and payroll for MLB teams in 2016. That is, teams that spent more money typically won more games. Other scatterplots, such as the one in Figure 3.2(b), show a **negative association**. Teams that allow their opponents to score more runs typically win *fewer* games.

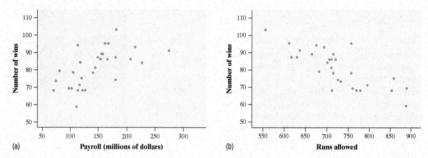

FIGURE 3.2 Scatterplots using data from the 30 Major League Baseball teams in 2016. (a) There is a positive association between payroll (in millions of dollars) and number of wins. (b) There is a negative association between runs allowed and number of wins.

In some cases, there is **no association** between two variables. For example, the following scatterplot shows the relationship between height (in centimeters) and the typical amount of sleep on a non-school night (in hours) for a sample

of students.[3] Knowing the height of a student doesn't help predict how much he or she likes to sleep on Saturday night!

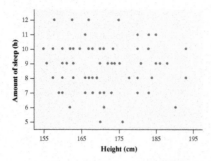

Recall from Section 1.1 that two variables have an *association* if knowing the value of one variable helps us predict the value of the other variable.

> **DEFINITION**   **Positive association, Negative association, No association**
>
> Two variables have a **positive association** when values of one variable tend to increase as the values of the other variable increase.
>
> Two variables have a **negative association** when values of one variable tend to decrease as the values of the other variable increase.
>
> There is **no association** between two variables if knowing the value of one variable does not help us predict the value of the other variable.

Making Connections

An alternative definition for a positive association states that *above-average* values of the explanatory variable (positive z_x-scores) tend to be paired with *above-average* values of the response variable (positive z_y-scores).

Identifying the direction of an association in a scatterplot is a good start, but there are several other characteristics that we need to address when describing a scatterplot.

AP® EXAM TIP

When you are asked to *describe* the association shown in a scatterplot, you are expected to discuss the direction, form, and strength of the association, along with any unusual features, *in the context of the problem*. This means that you need to use both variable names in your description.

> **HOW TO DESCRIBE A SCATTERPLOT**
>
> To describe a scatterplot, make sure to address the following four characteristics in the context of the data:
>
> - **Direction:** A scatterplot can show a positive association, negative association, or no association.
> - **Form:** A scatterplot can show a linear form or a nonlinear form. The form is linear if the overall pattern follows a straight line. Otherwise, the form is nonlinear.
> - **Strength:** A scatterplot can show a weak, moderate, or strong association. An association is strong if the points don't deviate much from the form identified. An association is weak if the points deviate quite a bit from the form identified.
> - **Unusual features:** Look for individual points that fall outside the overall pattern and distinct clusters of points.

Teaching Tip

It is possible to have a strong nonlinear relationship. Imagine a curved scatterplot where all the points lie very close to the identified curved pattern.

Teaching Tip

Unusual features might be points that are far from the other points in the *x* direction, far from the other points in the *y* direction, or both. They can also be any point that doesn't fit in with the pattern of the rest of the points.

Making Connections

Recall the acronym SOCV + context from Chapter 1 (**s**hape, **o**utliers, **c**enter, and **v**ariability). Use SOCV to describe the distribution of a single quantitative variable. Use the acronym DUFS + context to remember how to describe the relationship between two quantitative variables (**d**irection, **u**nusual features, **f**orm, **s**trength).

 **Ask the StatsMedic**

Nix the Trix: Except for SOCV and DUFS

In this post, we discuss consequences for when students use acronyms to memorize procedures.

Making Connections

In this chapter, we have intentionally avoided the term "outlier" until students have been introduced to the concept of a residual in Section 3.2. Then students learn that an outlier is a point that does not follow the pattern of the data and *has a large residual*.

ALTERNATE EXAMPLE   Skill 2.A

World records for sprints and marathons
Describing a scatterplot

PROBLEM:
Describe the relationship in each of the following contexts.

(a) The scatterplot shows the relationship between the years since 1900 and the 100-meter sprint record time (in seconds) for the years 1983 to 2010.

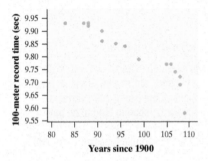

(b) The scatterplot shows the relationship between the year and the world record time for the marathon (hours).

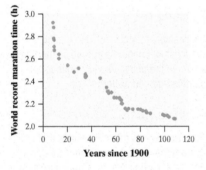

SOLUTION:
(a) There is a fairly strong, negative linear relationship between the years since 1900 and the 100-meter sprint record time. Usain Bolt's 9.58-second performance in 2009 is a clear unusual point.
(b) There is a strong, negative nonlinear relationship between years since 1900 and world record marathon time. There are no obvious unusual points.

Even though they have opposite directions, both associations in Figure 3.2 on page 156 have a linear form. However, the association between runs allowed and wins is stronger than the relationship between payroll and wins because the points in Figure 3.2(b) deviate less from the linear pattern. Each scatterplot has one unusual point: In Figure 3.2(a), the Los Angeles Dodgers spent $274 million and had "only" 91 wins. In Figure 3.2(b), the Texas Rangers gave up 757 runs but had 95 wins.

Even when there is a clear relationship between two variables in a scatterplot, the direction of the association describes only the overall trend—not an absolute relationship. For example, even though teams that spend more generally have more wins, there are plenty of exceptions. The Minnesota Twins spent more money than six other teams, but had fewer wins than any team in the league.

EXAMPLE   **Old Faithful and fertility**
Describing a scatterplot

PROBLEM: Describe the relationship in each of the following contexts.

(a) The scatterplot on the left shows the relationship between the duration (in minutes) of an eruption and the interval of time until the next eruption (in minutes) of Old Faithful during a particular month.

(b) The scatterplot on the right shows the relationship between the average income (gross domestic product per person, in dollars) and fertility rate (number of children per woman) in 187 countries.[4]

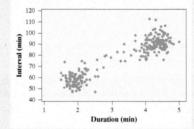
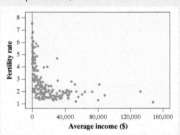

SOLUTION:

(a) There is a strong, positive linear relationship between the duration of an eruption and the interval of time until the next eruption. There are two main clusters of points: one cluster has durations around 2 minutes with intervals around 55 minutes, and the other cluster has durations around 4.5 minutes with intervals around 90 minutes.

> Even with the clusters, the overall direction is still positive. In some cases, however, the points in a cluster go in the opposite direction of the overall association.

(b) There is a moderately strong, negative nonlinear relationship between average income and fertility rate in these countries. There is a country outside this pattern with an average income around $30,000 and a fertility rate around 4.7.

> The association is called "nonlinear" because the pattern of points is clearly curved.

FOR PRACTICE, TRY EXERCISE 5

CHECK YOUR UNDERSTANDING

Is there a relationship between the amount of sugar (in grams) and the number of calories in movie-theater candy? Here are the data from a sample of 12 types of candy.[5]

Name	Sugar (g)	Calories	Name	Sugar (g)	Calories
Butterfinger Minis	45	450	Reese's Pieces	61	580
Junior Mints	107	570	Skittles	87	450
M&M'S®	62	480	Sour Patch Kids	92	490
Milk Duds	44	370	SweeTarts	136	680
Peanut M&M'S®	79	790	Twizzlers	59	460
Raisinets	60	420	Whoppers	48	350

1. Identify the explanatory and response variables. Explain your reasoning.
2. Make a scatterplot to display the relationship between amount of sugar and the number of calories in movie-theater candy.
3. Describe the relationship shown in the scatterplot.

8. Technology Corner   MAKING SCATTERPLOTS

TI-Nspire and other technology instructions are on the book's website at highschool.bfwpub.com/updatedtps6e.

Making scatterplots with technology is much easier than constructing them by hand. We'll use the MLB data from page 155 to show how to construct a scatterplot on a TI-83/84.

1. Enter the payroll values in L1 and the number of wins in L2.

   • Press STAT and choose Edit....

   • Type the values into L1 and L2.

2. Set up a scatterplot in the statistics plots menu.

   • Press 2nd Y= (STAT PLOT).

   • Press ENTER or 1 to go into Plot1.

   • Adjust the settings as shown.

Answers to CYU

1. The explanatory variable is the amount of sugar (in grams). The response variable is the number of calories. The amount of sugar helps to explain, or predict, the number of calories in movie-theater candy.

2. A scatterplot is shown below.

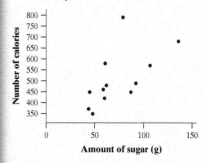

3. There is a moderately strong, positive linear relationship between the amount of sugar contained in movie-theater candy and the number of calories in the candy. The point for peanut M&M'S® is an unusual point with 79 grams of sugar and 790 calories.

Teaching Tip: Using Technology

Remind students that they can watch a video of this technology corner on the e-Book or at the Student Site.

Teaching Tip: Using Technology

Students can also use the *Two Quantitative Variables* applet in the Extra Applets on the Student Site at highschool .bfwpub.com/updatedtps6e to make a scatterplot.

TRM Additional Content: Time Plots

Time plots are scatterplots, but always with time on the *x* axis. Often the dots in these scatterplots are connected to show trends over time. You can find this extra content by clicking on the link in the TE-Book, opening the TRFD, or downloading from the Teacher's Resources on the book's digital platform.

✚ Ask the StatsMedic

Graphing Calculators Are the New Slide Rule

Fancy calculators allow our statistics students to do a whole lot of statistics that would be very tedious to do by hand. Unfortunately, the user interface with calculators has changed very little over the last 20 years and is now far behind other technology.

3. Use ZoomStat to let the calculator choose an appropriate window.

  • Press ZOOM and choose ZoomStat.

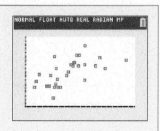

AP® EXAM TIP

If you are asked to make a scatterplot, be sure to label and scale both axes. *Don't* just copy an unlabeled calculator graph directly onto your paper.

Measuring Linear Association: Correlation

A scatterplot displays the direction, form, and strength of a relationship between two quantitative variables. Linear relationships are particularly important because a straight line is a simple pattern that is quite common. A linear relationship is considered strong if the points lie close to a straight line and is considered weak if the points are widely scattered about the line. Unfortunately, our eyes are not the most reliable tools when it comes to judging the strength of a linear relationship. When the association between two quantitative variables is linear, we can use the **correlation** *r* to help describe the strength and direction of the association.

Some people refer to *r* as the "correlation coefficient."

DEFINITION Correlation *r*

For a linear association between two quantitative variables, the **correlation** *r* measures the direction and strength of the association.

Here are some important properties of the correlation *r*:

• The correlation *r* is always a number between −1 and 1 ($-1 \leq r \leq 1$).
• The correlation *r* indicates the direction of a linear relationship by its sign: $r > 0$ for a positive association and $r < 0$ for a negative association.
• The extreme values $r = -1$ and $r = 1$ occur *only* in the case of a perfect linear relationship, when the points lie exactly along a straight line.
• If the linear relationship is strong, the correlation *r* will be close to 1 or −1. If the linear relationship is weak, the correlation *r* will be close to 0.

 It is only appropriate to use the correlation to describe strength and direction for a linear relationship. This is why the word *linear* kept appearing in the list above!

Figure 3.3 shows six scatterplots that correspond to various values of *r*. To make the meaning of *r* clearer, the standard deviations of both variables in these plots are equal, and the horizontal and vertical scales are the same. The correlation *r* describes the direction and strength of the linear relationship in each scatterplot.

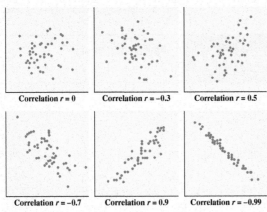

Correlation *r* = 0   Correlation *r* = −0.3   Correlation *r* = 0.5

Correlation *r* = −0.7   Correlation *r* = 0.9   Correlation *r* = −0.99

FIGURE 3.3 How correlation measures the strength and direction of a linear relationship. When the dots are tightly packed around a line, the correlation will be close to 1 or −1.

ACTIVITY   Guess the correlation

In this activity, we will have a class competition to see who can best guess the correlation.

1. Load the *Guess the Correlation* applet at www.rossmanchance.com/applets.

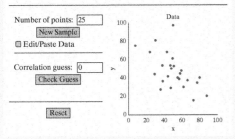

Correlation Guessing Game

Number of points: 25
New Sample
☐ Edit/Paste Data

Correlation guess: 0
Check Guess

Reset

2. The teacher will press the "New Sample" button to see a "random" scatterplot. As a class, try to guess the correlation. Type the guess in the "Correlation guess" box and press "Check Guess" to see how the class did. Repeat several times to see more examples. For the competition, there will be two rounds.

3. Starting on one side of the classroom and moving in order to the other side, the teacher will give each student *one* new sample and have him or her guess the correlation. The teacher will then record how far off the guess was from the true correlation.

4. Once every student has made an attempt, the teacher will give each student a second sample. This time, the students will go in reverse order so that the student who went first in Round 1 will go last in Round 2. The student who has the closest guess in either round wins a prize!

The following example illustrates how to interpret the correlation.

ALTERNATE EXAMPLE   Skill 4.B

Back to the track!
Interpreting a correlation

PROBLEM:
Here is a scatterplot that shows the relationship between the years since 1900 and the 100-meter sprint record time (in seconds) for the years 1983 to 2010. For these data, $r = -0.927$. Interpret the value of r.

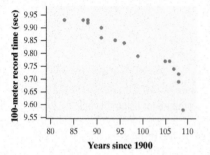

SOLUTION:
The correlation of $r = -0.927$ confirms that the linear association between years since 1900 and 100-meter record time is strong and negative.

✔ **Answers to CYU**

The correlation of $r = -0.838$ confirms that the linear association between dash time and long-jump distance is strong and negative.

EXAMPLE   Payroll and wins
Interpreting a correlation

PROBLEM: Here is the scatterplot showing the relationship between payroll (in millions of dollars) and wins for MLB teams in 2016. For these data, $r = 0.613$. Interpret the value of r.

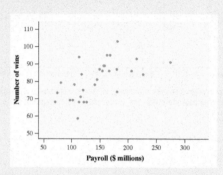

SOLUTION:

The correlation of $r = 0.613$ confirms that the linear association between payroll and number of wins is moderately strong and positive.

> Always include context by using the variable names in your answer.

FOR PRACTICE, TRY EXERCISE 15

CHECK YOUR UNDERSTANDING

The scatterplot shows the 40-yard-dash times (in seconds) and long-jump distances (in inches) for a small class of 12 students. The correlation for these data is $r = -0.838$. Interpret this value.

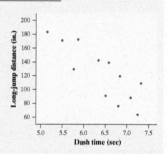

Cautions about Correlation

While the correlation is a good way to measure the strength and direction of a linear relationship, it has several limitations.

Correlation doesn't imply causation. In many cases, two variables might have a strong correlation, but changes in one variable are very unlikely to cause changes in the other variable. Consider the following scatterplot showing total revenue generated by skiing facilities in the United States and the number of people who died by becoming tangled in their bedsheets in 10 recent years.[6] The correlation for these data is $r = 0.97$. Does an increase in skiing revenue *cause* more people to die by becoming tangled in their bedsheets? We doubt it!

Even though we shouldn't automatically conclude that there is a cause-and-effect relationship between two variables when they have an association, in some cases there might actually be a cause-and-effect relationship. You will learn how to distinguish these cases in Chapter 4.

The following activity helps you explore some additional limitations of the correlation.

ACTIVITY   Correlation and Regression applet

In this activity, you will use an applet to investigate some important properties of the correlation. Go to the book's website at highschool.bfwpub.com/updatedtps6e and launch the *Correlation and Regression* applet.

1. You are going to use the *Correlation and Regression* applet to make several scatterplots that have correlation close to 0.7.

   (a) Start by putting two points on the graph. What's the value of the correlation? Why does this make sense?

   (b) Make a lower-left to upper-right pattern of 10 points with correlation about $r = 0.7$. You can drag points up or down to adjust r after you have 10 points.

   (c) Make a new scatterplot, this time with 9 points in a vertical stack at the left of the plot. Add 1 point far to the right and move it until the correlation is close to 0.7.

   (d) Make a third scatterplot, this time with 10 points in a curved pattern that starts at the lower left and rises to the right. Adjust the points up or down until you have a smooth curve with correlation close to 0.7.

Summarize: If you know only that the correlation between two variables is $r = 0.7$, what can you say about the form of the relationship?

2. Click on the scatterplot to create a group of 7 points in a U shape so that there is a strong nonlinear association. What is the correlation?

Summarize: If you know only that the correlation between two variables is $r = 0$, what can you say about the strength of the relationship?

Teaching Tip

Another way to demonstrate the influence of a single point on the correlation is to use the "oval method" described in an earlier Teaching Tip. If including the point makes the oval longer and skinnier, then that point makes the correlation get closer to ±1. If including the point makes the oval become more circular, that point makes the correlation get closer to 0.

✚ Ask the StatsMedic

Beware *r* by Itself!

This post outlines a quick demonstration that you can use with students to reinforce the idea that an *r*-value close to 1 does not guarantee a strong linear relationship.

Teaching Tip

Remember that an association is strong if the points don't deviate much from the form identified. In this scatterplot, the form is a nonlinear curve, and the points almost perfectly fit the curve. Thus, the relationship is a strong one.

3. Click on the scatterplot to create a group of 10 points in the lower-left corner of the scatterplot with a strong linear pattern (correlation about 0.9).

(a) Add 1 point at the upper right that is in line with the first 10. How does the correlation change?

(b) Drag this last point straight down. How small can you make the correlation? Can you make the correlation negative?

Summarize: What did you learn from Step 3 about the effect of an unusual point on the correlation?

 The activity highlighted some important cautions about correlation. Correlation does not measure form. Here is a scatterplot showing the speed (in miles per hour) and the distance (in feet) needed to come to a complete stop when a motorcycle's brake was applied.[7] The association is clearly curved, but the correlation is quite large: $r = 0.98$. In fact, the correlation for this *nonlinear* association is much greater than the correlation of $r = 0.613$ for the MLB payroll data, which had a clear linear association.

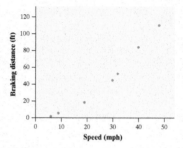

 Correlation should only be used to describe linear relationships. The association displayed in the following scatterplot is extremely strong, but the correlation is $r = 0$. This isn't a contradiction because correlation doesn't measure the strength of nonlinear relationships.

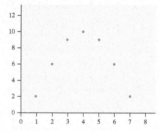

Making Connections

While we can only use correlation to describe the strength and direction for *linear* relationships, there is a way to assess the strength and direction for a nonlinear relationship (exponential, power, logarithmic). To do so, the original nonlinear relationship must be transformed into a linear relationship, and the correlation *r* is calculated for the transformed data. A strong linear relationship for the transformed data informs us about the strength and direction of the nonlinear model for the original data. We will discuss nonlinear relationships in Section 3.3.

The correlation is not a resistant measure of strength. In the following scatterplot, the correlation is $r = -0.13$. But when the unusual point in the lower right corner is excluded, the correlation becomes $r = 0.72$.

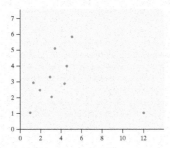

Like the mean and the standard deviation, the correlation can be greatly influenced by unusual points.

EXAMPLE

Nobel chocolate
Cautions about correlation

PROBLEM: Most people love chocolate for its great taste. But does it also make you smarter? A scatterplot like this one recently appeared in the *New England Journal of Medicine*.[8] The explanatory variable is the chocolate consumption per person for a sample of countries. The response variable is the number of Nobel Prizes per 10 million residents of that country.

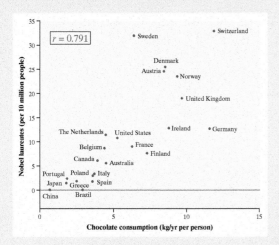

ALTERNATE EXAMPLE   Skill 4.B

Does playing video games help you get a degree?
Cautions about correlation

PROBLEM:
Computer science majors are often gamers, frequenting the local arcades. Are the arcade games helping them toward earning a doctorate in computer science? Below is a scatterplot created using U.S. data for the years 2000 to 2009. The explanatory variable is the arcade revenue (in millions of dollars) and the response variable is the number of doctoral degrees awarded in computer science.

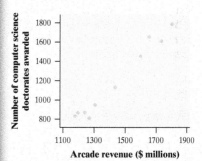

(a) Will playing lots of video games at the arcade make you more likely to get a computer science doctorate? Explain your answer.

(b) What effect does the year 2008 ($1803 million in arcade revenue and 1787 computer science doctorates awarded) have on the correlation? Explain your answer.

SOLUTION:
(a) No; even though there is a strong correlation between arcade revenue and the number of computer science doctorates awarded, causation should not be inferred. It may be that both these variables are changing due to another variable, such as overall progress of technology.

(b) Because the point for 2008 is in the positive linear pattern of the rest of the points, it makes the correlation closer to 1.

(a) If people in the United States started eating more chocolate, could we expect more Nobel Prizes to be awarded to residents of the United States? Explain.

(b) What effect does Switzerland have on the correlation? Explain.

SOLUTION

(a) No; even though there is a strong correlation between chocolate consumption and the number of Nobel laureates in a country, causation should not be inferred. It is possible that both of these variables are changing due to another variable, such as per capita income.

> Not all questions about cause and effect include the word *cause*. Make sure to read questions—and reports in the media—very carefully.

(b) When Switzerland is included with the rest of the points, it makes the association stronger because it doesn't vary much from the linear pattern. This makes the correlation closer to 1.

FOR PRACTICE, TRY EXERCISES 17 AND 19

Calculating Correlation

Now that you understand the meaning and limitations of the correlation, let's look at how it's calculated.

HOW TO CALCULATE THE CORRELATION r

Suppose that we have data on variables x and y for n individuals. The values for the first individual are x_1 and y_1, the values for the second individual are x_2 and y_2, and so on. The means and standard deviations of the two variables are $\bar{x}$ and s_x for the x-values, and $\bar{y}$ and s_y for the y-values. The correlation r between x and y is

$$r = \frac{1}{n-1}\left[\left(\frac{x_1 - \bar{x}}{s_x}\right)\left(\frac{y_1 - \bar{y}}{s_y}\right) + \left(\frac{x_2 - \bar{x}}{s_x}\right)\left(\frac{y_2 - \bar{y}}{s_y}\right) + \cdots + \left(\frac{x_n - \bar{x}}{s_x}\right)\left(\frac{y_n - \bar{y}}{s_y}\right)\right]$$

or, more compactly,

$$r = \frac{1}{n-1}\sum\left(\frac{x_i - \bar{x}}{s_x}\right)\left(\frac{y_i - \bar{y}}{s_y}\right)$$

The formula for the correlation r is a bit complex. It helps us understand some properties of correlation, but in practice you should use your calculator or software to find r. Exercises 21 and 22 ask you to calculate a correlation step by step from the definition to solidify its meaning.

Figure 3.4 shows the relationship between the payroll (in millions of dollars) and the number of wins for the 30 MLB teams in 2016. The red dot on the right represents the Los Angeles Dodgers, whose payroll was $274 million and who won 91 games.

FIGURE 3.4 Scatterplot showing the relationship between payroll (in millions of dollars) and number of wins for 30 Major League Baseball teams in 2016. The point representing the Los Angeles Dodgers is highlighted in red.

The Los Angeles Dodgers had a payroll of $274 million and won 91 games.

The formula for r begins by standardizing the observations. The value

$$\frac{x_i - \bar{x}}{s_x}$$

in the correlation formula is the standardized payroll (z-score) of the ith team. In 2016, the mean payroll was $\bar{x} = \$145.033$ million with a standard deviation of $s_x = \$46.879$ million. For the Los Angeles Dodgers, the corresponding z-score is

$$z_x = \frac{274 - 145.033}{46.879} = 2.75$$

The Dodgers' payroll is 2.75 standard deviations above the mean.
Likewise, the value

$$\frac{y_i - \bar{y}}{s_y}$$

in the correlation formula is the standardized number of wins for the ith team. In 2016, the mean number of wins was $\bar{y} = 80.9$ with a standard deviation of $s_y = 10.669$. For the Los Angeles Dodgers, the corresponding z-score is

$$z_y = \frac{91 - 80.9}{10.669} = 0.95$$

The Dodgers' number of wins is 0.95 standard deviation above the mean.

Multiplying the Dodgers' two z-scores, we get a product of $(2.75)(0.95) = 2.6125$. The correlation r is an "average" of the products of the standardized scores for all the teams. Just as in the case of the standard deviation s_x, we divide by 1 fewer than the number of individuals to find the average. Finishing the calculation reveals that $r = 0.613$ for the 30 MLB teams.

To understand what correlation measures, consider the graphs in Figure 3.5 on the next page. At the left is a scatterplot of the MLB data with two lines added—a vertical line at the mean payroll and a horizontal line at the mean number of wins. Most of the points fall in the upper-right or lower-left "quadrants" of the graph. Teams with above-average payrolls tend to have above-average numbers of wins, like the Dodgers. Teams with below-average payrolls tend to have numbers of wins that are below average. This confirms the positive association between the variables.

Some people like to write the correlation formula as

$$r = \frac{1}{n-1}\sum z_x z_y$$

to emphasize the product of standardized scores in the calculation.

Teaching Tip

If you are going to work through a calculation of the correlation with students, consider making a table to organize the work. Here are the headers for each of the 5 columns: x, y, z_x, z_y, $z_x z_y$.

Teaching Tip

Note that the Los Angeles Dodgers have an above-average payroll paired with an above-average number of wins. This corresponds to the alternative definition for a positive association presented in an earlier teaching tip.

Below on the right is a scatterplot of the standardized scores. To get this graph, we transformed both the *x*- and the *y*-values by subtracting their mean and dividing by their standard deviation. As we saw in Chapter 2, standardizing a data set converts the mean to 0 and the standard deviation to 1. That's why the vertical and horizontal lines in the right-hand graph are both at 0.

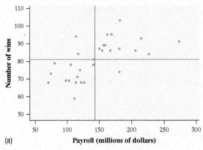 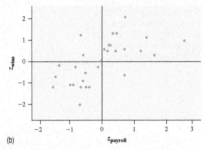

FIGURE 3.5 (a) Scatterplot showing the relationship between payroll (in millions of dollars) and number of wins for 30 Major League Baseball teams in 2016, with lines showing the mean of each variable. (b) Scatterplot showing the relationship between the standardized values of payroll and the standardized values of number of wins for the same 30 teams.

For the points in the upper-right quadrant and the lower-left quadrant, the products of the standardized values will be positive. Because most of the points are in these two quadrants, the sum of the *z*-score products will also be positive, resulting in a positive correlation *r*.

What if there was a negative association between two variables? Most of the points would be in the upper-left and lower-right quadrants and their *z*-score products would be negative, resulting in a negative correlation.

Additional Facts about Correlation

Now that you have seen how the correlation is calculated, here are some additional facts about correlation.

1. *Correlation requires that both variables be quantitative,* so that it makes sense to do the arithmetic indicated by the formula for *r*. We cannot calculate a correlation between the incomes of a group of people and what city they live in because city is a categorical variable. When one or both of the variables are categorical, use the term *association* rather than *correlation*.

2. *Correlation makes no distinction between explanatory and response variables.* When calculating the correlation, it makes no difference which variable you call *x* and which you call *y*. Can you see why from the formula?

$$r = \frac{1}{n-1}\sum \left(\frac{x_i - \bar{x}}{s_x}\right)\left(\frac{y_i - \bar{y}}{s_y}\right)$$

Teaching Tip

Be careful with the word *correlation*. Correlation is a numerical measure of the direction and strength of the relationship between two quantitative variables. When describing the relationship between two categorical variables (as in Section 1.1) or the relationship between a categorical and quantitative variable, use the word *association* instead.

Teaching Tip

Use the formula for correlation to help explain these additional facts:

$$r = \frac{1}{n-1}\sum \left(\frac{x_i - \bar{x}}{s_x}\right)\left(\frac{y_i - \bar{y}}{s_y}\right)$$

$$= \frac{1}{n-1}\sum z_x z_y$$

1. We must have numbers to plug into the formula, so our variables must be quantitative.

2. If we decided to switch our explanatory and response variables, the z_x and z_y would swap places in the formula. Because multiplication is commutative, we will get the same correlation value ($z_x z_y = z_y z_x$). You might also mention that if there is a strong correlation between height and weight, a strong correlation will also exist between weight and height.

3. Because r uses the standardized values of the observations, r *does not change when we change the units of measurement of x, y, or both.* The correlation between height and weight won't change if we measure height in centimeters rather than inches and measure weight in kilograms rather than pounds.

4. *The correlation r has no unit of measurement.* It is just a number.

EXAMPLE

Long strides
More about correlation

PROBLEM: The following scatterplot shows the height (in inches) and number of steps needed for a random sample of 36 students to walk the length of a school hallway. The correlation is $r = -0.632$.

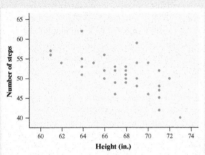

(a) Explain why it isn't correct to say that the correlation is -0.632 steps per inch.
(b) What would happen to the correlation if number of steps was used as the explanatory variable and height was used as the response variable?
(c) What would happen to the correlation if height was measured in centimeters instead of inches? Explain.

SOLUTION:

(a) *Because correlation is calculated using standardized values, it doesn't have units.*

(b) *The correlation would be the same because correlation doesn't make a distinction between explanatory and response variables.*

(c) *The correlation would be the same. Because r is calculated using standardized values, changes of units don't affect correlation.*

> Although it is unlikely that you will need to calculate the correlation by hand, understanding how the formula works makes it easier to answer questions like these.

> Changing from inches to centimeters won't change the locations of the points, only the numbers on the horizontal scale.

FOR PRACTICE, TRY EXERCISE 23

Teaching Tip

Use the formula for correlation to help explain these additional facts:

$$r = \frac{1}{n-1} \sum \left(\frac{x_i - \bar{x}}{s_x} \right) \left(\frac{y_i - \bar{y}}{s_y} \right)$$

$$= \frac{1}{n-1} \sum z_x z_y$$

3. z-scores tell us the number of standard deviations above or below the mean. This value does not depend on the units. If a value measured in inches is 1.78 standard deviations above the mean, then the same value measured in centimeters will still be 1.78 standard deviations above the mean. Also, remind students that changing the units doesn't change the way the scatterplot looks; it only changes the numbers on the scales.

4. Suppose the explanatory variable is measured in inches. Then $(x - \bar{x})$ would be measured in inches, and when divided by the standard deviation (which is also measured in inches), the units would cancel out. Thus, r has no units.

**TRM Data Exploration:
The SAT Essay: Is Longer Better?**

The Data Exploration feature that was found in the previous two editions of *The Practice of Statistics* student edition has been moved to the Teacher's Resource Materials. Click on the link in the TE-Book, open in the TRFD, or download from the Teacher's Resources on the book's digital platform.

ALTERNATE EXAMPLE

Skill 4.B

Arm span versus height   Additional facts about correlation

PROBLEM:
The following scatterplot shows the arm span (in centimeters) and height (in centimeters) for a random sample of 19 twelfth graders. The correlation is $r = 0.902$.

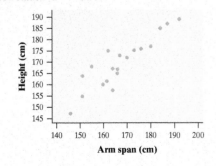

(a) Explain why it is incorrect to say that the correlation is 0.902 centimeter.
(b) What would happen to the correlation if height was measured in inches instead of centimeters? Explain your answer.
(c) What would happen to the correlation if height was used as the explanatory variable and arm span was used as the response variable?

SOLUTION:
(a) Because correlation is calculated using standardized values, it doesn't have units.
(b) The correlation would be the same. Because it is calculated using standardized values, changes of units don't affect correlation.
(c) The correlation would be the same because correlation doesn't make a distinction between explanatory and response variables.

Section 3.1 | Summary

- A **scatterplot** displays the relationship between two quantitative variables measured on the same individuals. Mark values of one variable on the horizontal axis (x axis) and values of the other variable on the vertical axis (y axis). Plot each individual's data as a point on the graph.

- If we think that a variable x may help predict, explain, or even cause changes in another variable y, we call x an **explanatory variable** and y a **response variable.** Always plot the explanatory variable on the x axis of a scatterplot. Plot the response variable on the y axis.

- When describing a scatterplot, look for an overall pattern (direction, form, strength) and departures from the pattern (unusual features) and always answer in context.
  - **Direction:** A relationship has a **positive association** when values of one variable tend to increase as the values of the other variable increase, a **negative association** when values of one variable tend to decrease as the values of the other variable increase, or **no association** when knowing the value of one variable doesn't help predict the value of the other variable.
  - **Form:** The form of a relationship can be linear or nonlinear (curved).
  - **Strength:** The strength of a relationship is determined by how close the points in the scatterplot lie to a simple form such as a line.
  - **Unusual features:** Look for individual points that fall outside the pattern and distinct clusters of points.

- For linear relationships, the **correlation r** measures the strength and direction of the association between two quantitative variables x and y.

- Correlation indicates the direction of a linear relationship by its sign: $r > 0$ for a positive association and $r < 0$ for a negative association. Correlation always satisfies $-1 \le r \le 1$ with stronger linear associations having values of r closer to 1 and -1. Correlations of $r = 1$ and $r = -1$ occur only when the points on a scatterplot lie exactly on a straight line.

- Remember these limitations of r: Correlation does not imply causation. The correlation is not resistant, so unusual points can greatly change the value of r. The correlation should only be used to describe linear relationships.

- Correlation ignores the distinction between explanatory and response variables. The value of r does not have units and is not affected by changes in the unit of measurement of either variable.

3.1 Technology Corner

TI-Nspire and other technology instructions are on the book's website at highschool.bfwpub.com/updatedtps6e.

8. Making scatterplots Page 159

Section 3.1 | Exercises

1. **Coral reefs and cell phones** Identify the explanatory
pg 154 variable and the response variable for the following
relationships, if possible. Explain your reasoning.

(a) The weight gain of corals in aquariums where the
water temperature is controlled at different levels

(b) The number of text messages sent and the number of
phone calls made in a sample of 100 students

2. **Teenagers and corn yield** Identify the explanatory
variable and the response variable for the following
relationships, if possible. Explain your reasoning.

(a) The height and arm span of a sample of 50 teenagers

(b) The yield of corn in bushels per acre and the amount
of rain in the growing season

3. **Heavy backpacks** Ninth-grade students at the Webb
pg 156 Schools go on a backpacking trip each fall. Students
are divided into hiking groups of size 8 by selecting
names from a hat. Before leaving, students and their
backpacks are weighed. The data here are from one
hiking group. Make a scatterplot by hand that shows
how backpack weight relates to body weight.

Body weight (lb)	120	187	109	103	131	165	158	116
Backpack weight (lb)	26	30	26	24	29	35	31	28

4. **Putting success** How well do professional golfers putt
from various distances to the hole? The data show
various distances to the hole (in feet) and the percent
of putts made at each distance for a sample of golfers.[9]
Make a scatterplot by hand that shows how the percent
of putts made relates to the distance of the putt.

Distance (ft)	Percent made	Distance (ft)	Percent made
2	93.3	12	25.7
3	83.1	13	24.0
4	74.1	14	31.0
5	58.9	15	16.8
6	54.8	16	13.4
7	53.1	17	15.9
8	46.3	18	17.3
9	31.8	19	13.6
10	33.5	20	15.8
11	31.6		

5. **Olympic athletes** The scatterplot shows the relation-
pg 158 ship between height (in inches) and weight (in pounds)
for the members of the U.S. 2016 Olympic Track and

Field team.[10] Describe the relationship between height
and weight for these athletes.

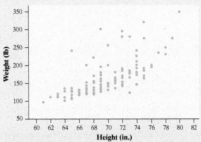

6. **Starbucks** The scatterplot shows the relationship
between the amount of fat (in grams) and number
of calories in products sold at Starbucks.[11] Describe
the relationship between fat and calories for these
products.

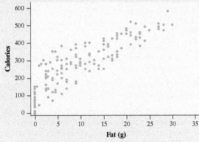

7. **More heavy backpacks** Refer to your graph from
Exercise 3. Describe the relationship between body
weight and backpack weight for this group of hikers.

8. **More putting success** Refer to your graph from
Exercise 4. Describe the relationship between distance
from hole and percent of putts made for the sample of
professional golfers.

9. **Does fast driving waste fuel?** How does the fuel
consumption of a car change as its speed increases?
Here are data for a British Ford Escort. Speed is measured in kilometers per hour, and fuel consumption is
measured in liters of gasoline used per 100 kilometers
traveled.[12]

Teaching Tip

Make sure to point out the two icons next to
Exercise 1. The "pg 154" icon reminds students
that the example on page 154 is very similar to
this exercise. The "play" icon reminds students
that there is a video solution available in the
student e-Book or at the Student Site.

TRM **Full Solutions to Section 3.1
Exercises**

Click on the link in the TE-Book, open the TRFD,
or download from the Teacher's Resources on the
book's digital platform.

Answers to Section 3.1 Exercises

3.1 (a) Water temperature is the
explanatory variable and weight gain is
the response variable. Water temperature
may help predict or explain changes in the
response variable, weight gain. Weight
gain measures the outcome of the study.
(b) Either variable could be the explanatory
variable because each one could be used
to predict or explain the other.

3.2 (a) Either variable, height or arm
span, could be the explanatory variable,
because each could be used to predict
or explain the other. **(b)** Amount of rain is
the explanatory variable and yield of corn
is the response variable. Amount of rain
may help predict or explain changes in the
response variable, yield of corn. Yield of
corn measures the outcome of the study.

3.3

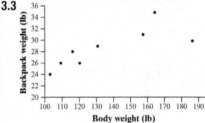

3.4

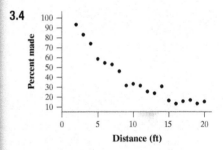

3.5 There is a moderately strong positive,
linear relationship between height and
weight for these athletes. A few athletes
weigh much more than other athletes of
the same height.

3.6 There is a moderately strong positive,
linear relationship between amount of fat and
number of calories in Starbucks products.

3.7 There is a moderately strong positive,
linear association between backpack
weight and body weight for these
students. There is an unusual point in
the graph—the hiker with body weight
187 pounds and pack weight 30 pounds.
This hiker makes the form appear to be
nonlinear for weights above 140 pounds.

3.8 There is a strong negative, nonlinear
relationship between distance and percent
of putts made for this sample of golfers.
There is an unusual point with a distance
of 14 feet and 31% of putts made.

3.9 (a) A scatterplot with speed as the explanatory variable is shown below.

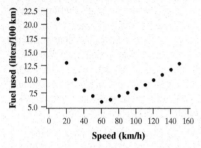

(b) There is a strong nonlinear relationship between speed and amount of fuel used. The relationship is negative for speeds up to 60 km/h and positive for speeds beyond 60 km/h. There are no unusual points.

3.10 (a) A scatterplot with mass as the explanatory variable is shown below.

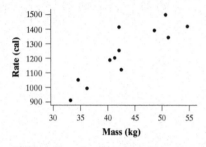

(b) There is a strong positive, linear association between lean body mass and metabolic rate. There are no unusual points.

3.11 For both groups of athletes, there is a moderately strong positive, linear association between height and weight; however, athletes who participate in the shot put, discus throw, and hammer throw tend to weigh more than other track and field athletes of the same height.

3.12 For both food products and beverage products, there is a moderately strong positive, linear association between amount of fat and number of calories; however, the majority of high-fat items are food products and the majority of low-fat items are drink products.

3.13 The relationship is positive, so $r > 0$. Also, r is closer to 1 than 0 because the relationship is strong.

3.14 The relationship appears to be slightly negative, so $r < 0$. Also, r is much closer to 0 than to -1 because the relationship is very weak.

Speed (km/h)	Fuel used (L/100 km)	Speed (km/h)	Fuel used (L/100 km)
10	21.00	90	7.57
20	13.00	100	8.27
30	10.00	110	9.03
40	8.00	120	9.87
50	7.00	130	10.79
60	5.90	140	11.77
70	6.30	150	12.83
80	6.95		

(a) Make a scatterplot to display the relationship between speed and fuel consumption.

(b) Describe the relationship between speed and fuel consumption.

10. **Do muscles burn energy?** Metabolic rate, the rate at which the body consumes energy, is important in studies of weight gain, dieting, and exercise. We have data on the lean body mass and resting metabolic rate for 12 women who are subjects in a study of dieting. Lean body mass, given in kilograms, is a person's weight leaving out all fat. Metabolic rate is measured in calories burned per 24 hours. The researchers believe that lean body mass is an important influence on metabolic rate.

Mass	36.1	54.6	48.5	42.0	50.6	42.0	40.3	33.1	42.4	34.5	51.1	41.2
Rate	995	1425	1396	1418	1502	1256	1189	913	1124	1052	1347	1204

(a) Make a scatterplot to display the relationship between lean body mass and metabolic rate.

(b) Describe the relationship between lean body mass and metabolic rate.

11. **More Olympics** Athletes who participate in the shot put, discus throw, and hammer throw tend to have different physical characteristics than other track and field athletes. The scatterplot shown here enhances the scatterplot from Exercise 5 by plotting these athletes with blue squares. How are the relationships between height and weight the same for the two groups of athletes? How are the relationships different?

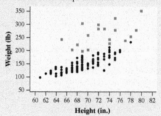

12. **More Starbucks** How do the nutritional characteristics of food products differ from drink products at

Starbucks? The scatterplot shown here enhances the scatterplot from Exercise 6 by plotting the food products with blue squares. How are the relationships between fat and calories the same for the two types of products? How are the relationships different?

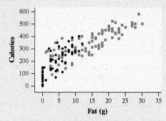

13. **Manatees** Manatees are large, gentle, slow-moving sea creatures found along the coast of Florida. Many manatees are injured or killed by boats. Here is a scatterplot showing the relationship between the number of boats registered in Florida (in thousands) and the number of manatees killed by boats for the years 1977 to 2015.[13] Is $r > 0$ or $r < 0$? Closer to $r = 0$ or $r = \pm 1$? Explain your reasoning.

14. **Windy city** Is it possible to use temperature to predict wind speed? Here is a scatterplot showing the average temperature (in degrees Fahrenheit) and average wind speed (in miles per hour) for 365 consecutive days at O'Hare International Airport in Chicago.[14] Is $r > 0$ or $r < 0$? Closer to $r = 0$ or $r = \pm 1$? Explain your reasoning.

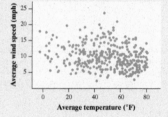

15. Points and turnovers Here is a scatterplot showing the
pg 162 relationship between the number of turnovers and the
number of points scored for players in a recent NBA
season.[15] The correlation for these data is $r = 0.92$.
Interpret the correlation.

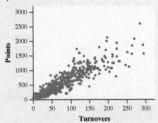

16. Oh, that smarts! Infants who cry easily may be more
easily stimulated than others. This may be a sign of
higher IQ. Child development researchers explored
the relationship between the crying of infants 4 to
10 days old and their IQ test scores at age 3 years. A
snap of a rubber band on the sole of the foot caused the
infants to cry. The researchers recorded the crying and
measured its intensity by the number of peaks in the
most active 20 seconds. The correlation for these data
is $r = 0.45$.[16] Interpret the correlation.

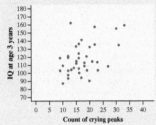

17. More turnovers? Refer to Exercise 15. Does the fact
pg 168 that $r = 0.92$ suggest that an increase in turnovers will
cause NBA players to score more points? Explain your
reasoning.

18. More crying? Refer to Exercise 16. Does the fact that
$r = 0.45$ suggest that making an infant cry will increase
his or her IQ later in life? Explain your reasoning.

19. Hot dogs Are hot dogs that are high in calories also
pg 169 high in salt? The following scatterplot shows the
calories and salt content (measured in milligrams of
sodium) in 17 brands of meat hot dogs.[17]

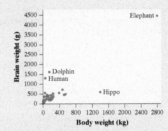

(a) The correlation for these data is $r = 0.87$. Interpret this
value.

(b) What effect does the hot dog brand with the smallest
calorie content have on the correlation? Justify your answer.

20. All brawn? The following scatterplot plots the average
brain weight (in grams) versus average body weight
(in kilograms) for 96 species of mammals.[18] There
are many small mammals whose points overlap at the
lower left.

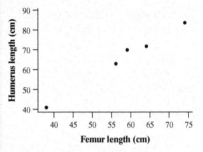

(a) The correlation between body weight and brain weight
is $r = 0.86$. Interpret this value.

(b) What effect does the human have on the correlation?
Justify your answer.

21. Dem bones Archaeopteryx is an extinct beast that
had feathers like a bird but teeth and a long bony tail
like a reptile. Only six fossil specimens are known to
exist today. Because these specimens differ greatly in
size, some scientists think they are different species
rather than individuals from the same species. If the
specimens belong to the same species and differ in size
because some are younger than others, there should
be a positive linear relationship between the lengths
of a pair of bones from all individuals. A point outside
the pattern would suggest a different species. Here are
data on the lengths (in centimeters) of the femur (a leg

3.20 (a) The correlation of 0.86 indicates
that the linear relationship between body
weight and brain weight of mammals
is strong and positive. **(b)** The human
decreases the correlation. The human
has an exceptionally high brain weight
compared to his body weight relative to
the other species of mammals.

3.21 (a) There is a strong positive, linear
relationship between the femur lengths and
humerus lengths. It appears that all five
specimens come from the same species.

(b) The femur has mean of 58.2 cm
and standard deviation of 13.2 cm.
The humerus has mean of 66 cm and
standard deviation of 15.89 cm. The table
shows the standardized measurements
(labeled z_{femur} and $z_{humerus}$) and the product
($z_{femur} \times z_{humerus}$) of the standardized
measurements.

Femur	Humerus	z_{femur}	$z_{humerus}$	Product
38	41	−1.53030	−1.57332	2.40765
56	63	−1.16667	−0.18880	0.03147
59	70	0.06061	0.25173	0.01526
64	72	0.43939	0.37760	0.16591
74	84	1.19697	1.13279	1.35591

The sum of the products is 3.97620,
so the correlation coefficient is

$$r = \frac{1}{4}(3.97620) = 0.9941.$$ The very

high value of the correlation confirms
the strong positive, linear association
between femur length and humerus
length in the scatterplot from part (a).

3.15 The correlation of 0.92 indicates that the
linear relationship between number of turnovers
and number of points scored for players in the
2017 NBA season is strong and positive.

3.16 The correlation of 0.45 indicates that the
linear relationship between the count of crying
peaks and IQ at age 3 years old is somewhat
weak and positive.

3.17 Probably not; although there is a strong positive
association, an increase in turnovers is not likely
to cause an increase in points for NBA players. It is
likely that both these variables are changing due to
several other variables, such as time played.

3.18 Probably not; although there is a somewhat
weak positive association, an increase in the
number of times an infant cries is not likely to
cause higher IQ scores later in life. It is likely that
both these variables are changing due to several
other variables, such as the baby's alertness or
irritability.

3.19 (a) The correlation of 0.87 indicates that the
linear relationship between amount of sodium
and number of calories is strong and positive.
(b) The hot dog with the lowest calorie content
increases the correlation. It falls in the linear
pattern of the rest of the data.

3.22 (a) A moderate positive, linear association, so r should be positive, but not close to 1.

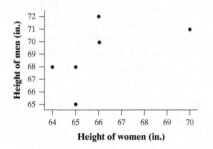

(b) For women, the mean is 66 inches and standard deviation is 2.098 inches. For the men, the mean is 69 inches and standard deviation is 2.53 inches. The table shows the standardized measurements (labeled z_{women} and z_{men}) and the product ($z_{women} \times z_{men}$) of the standardized measurements.

Women	Men	z_{women}	z_{men}	Product
66	72	0	1.18577	0
64	68	−0.95329	−0.39526	0.37679
66	70	0	0.39526	0
65	68	−0.47664	−0.39526	0.18840
70	71	1.90658	0.79051	1.50718
65	65	−0.47664	−1.58103	0.75359

The sum of the products is 2.82596, so the correlation coefficient is

$r = \dfrac{1}{5}(2.82596) = 0.5652$. The positive

correlation provides evidence that taller women tend to date taller men (and shorter women date shorter men). Because the correlation is not close to 1, the association is not strong.

3.23 (a) Correlation is unitless.
(b) The correlation would stay the same. Correlation makes no distinction between explanatory and response variables.
(c) If sodium was measured in grams instead of milligrams, the correlation would still be 0.87. Because r uses the standardized values of the observations, r does not change when we change the units of measurement of x, or y, or both.

3.24 (a) Correlation is unitless. **(b)** The correlation would stay the same. Correlation makes no distinction between explanatory and response variables.
(c) The correlation of brain weight would be 0.86 whether measured in kilograms or grams. Because r uses the standardized values, r does not change when we change the units of measurement of x, or y, or both.

bone) and the humerus (a bone in the upper arm) for the five specimens that preserve both bones:[19]

Femur (x)	38	56	59	64	74
Humerus (y)	41	63	70	72	84

(a) Make a scatterplot. Do you think that all five specimens come from the same species? Explain.

(b) Find the correlation r step by step, using the formula on page 166. Explain how your value for r matches your graph in part (a).

22. **Data on dating** A student wonders if tall women tend to date taller men than do short women. She measures herself, her dormitory roommate, and the women in the adjoining dorm rooms. Then she measures the next man each woman dates. Here are the data (heights in inches):

Women (x)	66	64	66	65	70	65
Men (y)	72	68	70	68	71	65

(a) Make a scatterplot of these data. Describe what you see.

(b) Find the correlation r step by step, using the formula on page 166. Explain how your value for r matches your description in part (a).

23. **More hot dogs** Refer to Exercise 19.

 pg 169

(a) Explain why it isn't correct to say that the correlation is 0.87 mg/cal.

(b) What would happen to the correlation if the variables were reversed on the scatterplot? Explain your reasoning.

(c) What would happen to the correlation if sodium was measured in grams instead of milligrams? Explain your reasoning.

24. **More brains** Refer to Exercise 20.

(a) Explain why it isn't correct to say that the correlation is 0.86 g/kg.

(b) What would happen to the correlation if the variables were reversed on the scatterplot? Explain your reasoning.

(c) What would happen to the correlation if brain weight was measured in kilograms instead of grams? Explain your reasoning.

25. **Rank the correlations** Consider each of the following relationships: the heights of fathers and the heights of their adult sons, the heights of husbands and the heights of their wives, and the heights of women at age 4 and their heights at age 18. Rank the correlations between these pairs of variables from largest to smallest. Explain your reasoning.

26. **Teaching and research** A college newspaper interviews a psychologist about student ratings of the teaching of faculty members. The psychologist says, "The evidence indicates that the correlation between the research productivity and teaching rating of faculty members is close to zero." The paper reports this as "Professor McDaniel said that good researchers tend to be poor teachers, and vice versa." Explain why the paper's report is wrong. Write a statement in plain language (don't use the word *correlation*) to explain the psychologist's meaning.

27. **Correlation isn't everything** Marc and Rob are both high school English teachers. Students think that Rob is a harder grader, so Rob and Marc decide to grade the same 10 essays and see how their scores compare. The correlation is $r = 0.98$, but Rob's scores are always lower than Marc's. Draw a possible scatterplot that illustrates this situation.

28. **Limitations of correlation** A carpenter sells handmade wooden benches at a craft fair every week. Over the past year, the carpenter has varied the price of the benches from $80 to $120 and recorded the average weekly profit he made at each selling price. The prices of the bench and the corresponding average profits are shown in the table.

Price	$80	$90	$100	$110	$120
Average profit	$2400	$2800	$3000	$2800	$2400

(a) Make a scatterplot to show the relationship between price and profit.

(b) The correlation for these data is $r = 0$. Explain how this can be true even though there is a strong relationship between price and average profit.

Multiple Choice: *Select the best answer for Exercises 29–34.*

29. You have data for many years on the average price of a barrel of oil and the average retail price of a gallon of unleaded regular gasoline. If you want to see how well the price of oil predicts the price of gas, then you should make a scatterplot with _____ as the explanatory variable.

(a) the price of oil

(b) the price of gas

(c) the year

(d) either oil price or gas price

(e) time

3.25 We would expect the height of girls at age 4 and their height at age 18 to be the highest correlation—it is reasonable to expect taller children to become taller adults. The next highest would be the correlation between the heights of male parents and their adult children because they share genes. The lowest correlation would be between husbands and their wives. Some tall men may prefer to marry tall women, but this isn't always the case.

3.26 The correlation of $r \approx 0$ means that there is no linear association between research productivity and teaching rating. Knowledge of a professor's research productivity will not help predict her teaching rating.

3.27 Answers will vary. Here is one possibility:

Marc's score	65	70	75	80	85	90	92	93	94	95
Rob's score	55	58	65	68	77	79	85	87	87	83

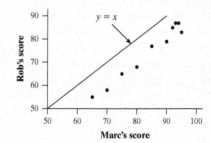

30. In a scatterplot of the average price of a barrel of oil and the average retail price of a gallon of gas, you expect to see

(a) very little association.

(b) a weak negative association.

(c) a strong negative association.

(d) a weak positive association.

(e) a strong positive association.

31. The following graph plots the gas mileage (in miles per gallon) of various cars from the same model year versus the weight of these cars (in thousands of pounds). The points marked with red dots correspond to cars made in Japan. From this plot, we may conclude that

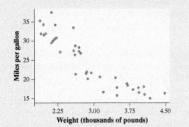

(a) there is a positive association between weight and gas mileage for Japanese cars.

(b) the correlation between weight and gas mileage for all the cars is close to 1.

(c) there is little difference between Japanese cars and cars made in other countries.

(d) Japanese cars tend to be lighter in weight than other cars.

(e) Japanese cars tend to get worse gas mileage than other cars.

32. If women always married men who were 2 years older than themselves, what would be the correlation between the ages of husband and wife?

(a) 2   (b) 1   (c) 0.5

(d) 0   (e) Can't tell without seeing the data

33. The scatterplot shows reading test scores against IQ test scores for 14 fifth-grade children. What effect does the point at IQ = 124 and reading score = 10 have on the correlation?

(a) It makes the correlation closer to 1.

(b) It makes the correlation closer to 0 but still positive.

(c) It makes the correlation equal to 0.

(d) It makes the correlation negative.

(e) It has no effect on the correlation.

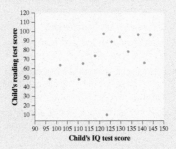

34. If we leave out this point, the correlation for the remaining 13 points in the preceding figure is closest to

(a) −0.95.   (b) −0.65.   (c) 0.

(d) 0.65.   (e) 0.95.

Recycle and Review

35. Big diamonds (1.2) Here are the weights (in milligrams) of 58 diamonds from a nodule carried up to the earth's surface in surrounding rock. These data represent a population of diamonds formed in a single event deep in the earth.[20]

13.8	3.7	33.8	11.8	27.0	18.9	19.3	20.8	25.4	23.1	7.8
10.9	9.0	9.0	14.4	6.5	7.3	5.6	18.5	1.1	11.2	7.0
7.6	9.0	9.5	7.7	7.6	3.2	6.5	5.4	7.2	7.8	3.5
5.4	5.1	5.3	3.8	2.1	2.1	4.7	3.7	3.8	4.9	2.4
1.4	0.1	4.7	1.5	2.0	0.1	0.1	1.6	3.5	3.7	2.6
4.0	2.3	4.5								

Make a histogram to display the distribution of weight. Describe the distribution.

36. Fruit fly thorax lengths (2.2) Fruit flies are used frequently in genetic research because of their quick reproductive cycle. The length of the thorax (in millimeters) for male fruit flies is approximately Normally distributed with a mean of 0.80 mm and a standard deviation of 0.08 mm.[21]

(a) What proportion of male fruit flies have a thorax length greater than 1 mm?

(b) What is the 30th percentile for male fruit fly thorax lengths?

3.35 One possible histogram is shown below.

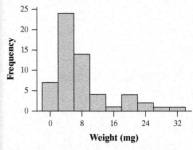

The distribution of weight is skewed to the right, with several possible high outliers. The median weight is 5.4 mg and the *IQR* is 5.5 mg.

3.36 (a) $z = \dfrac{1 - 0.8}{0.08} = 2.50$. *Table A:*

The proportion of *z*-scores above 2.50 is 0.0062. *Tech:* normalcdf(lower: 1, upper: 1000, mean: 0.8, SD: 0.08) = 0.0062. About 6 out of 1000 (or a proportion of 0.0062) male fruit flies have a thorax length greater than 1 mm. **(b)** *Table A:* Look in the body of Table A for the value closest to 0.30. A *z*-score of −0.52 gives the closest value (0.3015). Solving

$-0.52 = \dfrac{x - 0.8}{0.08}$ gives $x = 0.76$ mm.

Tech: invNorm(area: 0.3, mean: 0.8, SD: 0.08) = 0.76 mm. The 30th percentile of male fruit fly thorax lengths is about 0.76 mm.

3.28 (a)

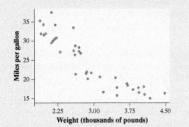

(b) It is possible that the correlation is $r = 0$, even though there is a strong relationship between price and average profit, because correlation measures the strength of a *linear* association between two quantitative variables. This plot shows a nonlinear relationship between price and average profit.

3.29 a

3.30 e

3.31 d

3.32 b

3.33 b

3.34 d

Teaching Tip

Students are most familiar with slope intercept form for equations of lines ($y = mx + b$). So why might statisticians prefer equations in the form $y = a + bx$ or $y = b_0 + b_1x$? We discuss this briefly on page 181, but the answer is that most often in the real world, there is more than one explanatory variable that can help predict the response variable. Statisticians might create a model with three explanatory variables (x_1, x_2, x_3) that looks like this: $\hat{y} = b_0 + b_1x_1 + b_2x_2 + b_3x_3$. The y intercept is a starting point for making a prediction for the response variable; each time we add one more explanatory variable, we are refining that prediction. This process is called *multiple regression* (not part of the AP® Statistics course). Also, this format resembles the commonly used form of the equation for exponential growth/decay, $y = ab^x$, where the starting value is first.

Teaching Tip

The original 6th edition of this book used the notation $\hat{y} = b_0 + b_1x$, which was consistent with the old formula sheet for the AP® Statistics exam. In this updated edition, we have changed to the notation $\hat{y} = a + bx$ to be consistent with the new formula sheet.

Teaching Tip

Get students in the habit of including the "hat" (caret) on the response variable to clearly indicate we are predicting that variable. This will help later when we are calculating residuals and interpreting slope and y intercept.

SECTION 3.2 **Least-Squares Regression**

LEARNING TARGETS *By the end of the section, you should be able to:*

- Make predictions using regression lines, keeping in mind the dangers of extrapolation.
- Calculate and interpret a residual.
- Interpret the slope and y intercept of a regression line.
- Determine the equation of a least-squares regression line using technology or computer output.
- Construct and interpret residual plots to assess whether a regression model is appropriate.
- Interpret the standard deviation of the residuals and r^2 and use these values to assess how well a least-squares regression line models the relationship between two variables.
- Describe how the least-squares regression line, standard deviation of the residuals, and r^2 are influenced by unusual points.
- Find the slope and y intercept of the least-squares regression line from the means and standard deviations of x and y and their correlation.

Linear (straight-line) relationships between two quantitative variables are fairly common. In the preceding section, we found linear relationships in settings as varied as Major League Baseball, geysers, and Nobel prizes. Correlation measures the strength and direction of these relationships. When a scatterplot shows a linear relationship, we can summarize the overall pattern by drawing a line on the scatterplot. A **regression line** models the relationship between two variables, but only in a specific setting: when one variable helps explain the other. Regression, unlike correlation, requires that we have an explanatory variable and a response variable.

> Sometimes regression lines are referred to as *simple linear regression models*. They are called "simple" because they involve only one explanatory variable.

DEFINITION **Regression line**

A **regression line** is a line that models how a response variable y changes as an explanatory variable x changes. Regression lines are expressed in the form $\hat{y} = a + bx$ where $\hat{y}$ (pronounced "y-hat") is the predicted value of y for a given value of x.

It is common knowledge that cars and trucks lose value the more they are driven. Can we predict the price of a used Ford F-150 SuperCrew 4 × 4 truck if we know how many miles it has on the odometer? A random sample of 16 used Ford F-150 SuperCrew 4 × 4s was selected from among those listed for sale at autotrader.com. The number of miles driven and price (in dollars) were recorded for each of the trucks.[22] Here are the data:

Miles driven	70,583	129,484	29,932	29,953	24,495	75,678	8359	4447
Price ($)	21,994	9500	29,875	41,995	41,995	28,986	31,891	37,991

Miles driven	34,077	58,023	44,447	68,474	144,162	140,776	29,397	131,385
Price ($)	34,995	29,988	22,896	33,961	16,883	20,897	27,495	13,997

Figure 3.6 is a scatterplot of these data. The plot shows a moderately strong, negative linear association between miles driven and price. There are two distinct clusters of trucks: a group of 12 trucks between 0 and 80,000 miles driven and a group of 4 trucks between 120,000 and 160,000 miles driven. The correlation is $r = -0.815$. The line on the plot is a regression line for predicting price from miles driven.

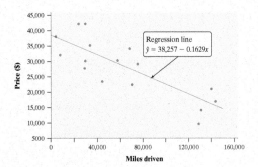

FIGURE 3.6 Scatterplot showing the price and miles driven of used Ford F-150s, along with a regression line.

Prediction

We can use a regression line to predict the value of the response variable for a specific value of the explanatory variable. For the Ford F-150 data, the equation of the regression line is

$$\widehat{\text{price}} = 38{,}257 - 0.1629 \,(\text{miles driven})$$

When we want to refer to the predicted value of a variable, we add a hat on top. Here, price refers to the predicted price of a used Ford F-150.

If a used Ford F-150 has 100,000 miles driven, substitute $x = 100{,}000$ in the equation. The predicted price is

$$\widehat{\text{price}} = 38{,}257 - 0.1629(100{,}000) = \$21{,}967$$

This prediction is illustrated in Figure 3.7.

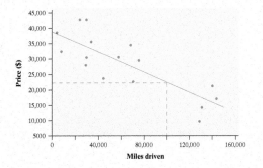

FIGURE 3.7 Using the regression line to predict price for a Ford F-150 with 100,000 miles driven.

Teaching Tip

In the real world, most prediction models include many explanatory variables (multiple regression). For this course, students need only understand simple linear regression. To learn more about multiple regression, check out Chapter 14 in the additional online chapters.

Teaching Tip

Always have students use context rather than x and y when writing out a regression equation. Also, make sure they don't forget the "hat" on the response variable, or the word *predicted* in front: *predicted price* = 38,257 − 0.1629 (*miles driven*).

Teaching Tip

Be sure to connect the algebraic calculation of a prediction to the representation on the graph (see dotted lines). This will help prepare students for understanding a residual on the graph, as well as residual plots.

Even though the value $\hat{y} = \$21,967$ is unlikely to be the actual price of a truck that has been driven 100,000 miles, it's our best guess based on the linear model using $x =$ miles driven. We can also think of $\hat{y} = \$21,967$ as the average price for a sample of trucks that have each been driven 100,000 miles.

Can we predict the price of a Ford F-150 with 300,000 miles driven? We can certainly substitute 300,000 into the equation of the line. The prediction is

$$\widehat{\text{price}} = 38,257 - 0.1629(300,000) = -\$10,613$$

The model predicts that we would need to *pay* $10,613 just to have someone take the truck off our hands!

A negative price doesn't make much sense in this context. Look again at Figure 3.7. A truck with 300,000 miles driven is far outside the set of *x* values for our data. We can't say whether the relationship between miles driven and price remains linear at such extreme values. Predicting the price for a truck with 300,000 miles driven is an **extrapolation** of the relationship beyond what the data show.

> **DEFINITION   Extrapolation**
>
> **Extrapolation** is the use of a regression line for prediction outside the interval of *x* values used to obtain the line. The further we extrapolate, the less reliable the predictions.

 Few relationships are linear for all values of the explanatory variable. Don't make predictions using values of *x* that are much larger or much smaller than those that actually appear in your data.

EXAMPLE

How much candy can you grab?
Prediction

PROBLEM: The scatterplot below shows the hand span (in cm) and number of Starburst™ candies grabbed by each student when Mr. Tyson's class did the "Candy grab" activity. The regression line $\hat{y} = -29.8 + 2.83x$ has been added to the scatterplot.

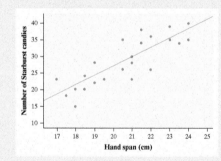

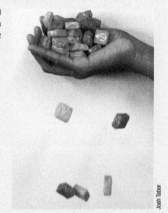

Josh Tabor

(a) Andres has a hand span of 22 cm. Predict the number of Starburst™ candies he can grab.

(b) Mr. Tyson's young daughter McKayla has a hand span of 12 cm. Predict the number of Starburst candies she can grab.

(c) How confident are you in each of these predictions? Explain.

SOLUTION:

(a) $\hat{y} = -29.8 + 2.83(22)$
$\hat{y} = 32.46$ Starburst candies

> Don't worry that the predicted number of Starburst candies isn't an integer. Think of 32.46 as the average number of Starburst candies that a group of students, each with a hand span of 22 cm, could grab.

(b) $\hat{y} = -29.8 + 2.83(12)$
$\hat{y} = 4.16$ Starburst candies

(c) The prediction for Andres is believable because $x = 22$ is within the interval of x-values used to create the model. However, the prediction for McKayla is not trustworthy because $x = 12$ is far outside the x-values used to create the regression line. The linear form may not extend to hand spans this small.

FOR PRACTICE, TRY EXERCISE 37

Residuals

In most cases, no line will pass exactly through all the points in a scatterplot. Because we use the line to predict y from x, the prediction errors we make are errors in y, the vertical direction in the scatterplot.

Figure 3.8 shows a scatterplot of the Ford F-150 data with a regression line added. The prediction errors are marked as bold vertical segments in the graph. These vertical deviations represent "leftover" variation in the response variable after fitting the regression line. For that reason, they are called **residuals**.

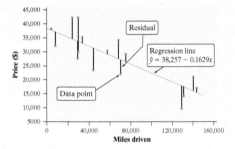

FIGURE 3.8 Scatterplot of the Ford F-150 data with a regression line added. A good regression line should make the residuals (shown as bold vertical segments) as small as possible.

DEFINITION   Residual

A **residual** is the difference between the actual value of y and the value of y predicted by the regression line. That is,

$$\text{residual} = \text{actual } y - \text{predicted } y$$
$$= y - \hat{y}$$

Making Connections

Emphasize the connection between residuals and deviations from the mean. When calculating the standard deviation in Chapter 1, we used the quantity $x_i - \bar{x}$, which was the difference between the actual value of x and the mean (expected) value of x. Likewise, in Chapter 2, when we calculated standardized scores, we found the difference between the actual value and the mean (expected) value. Looking ahead, we will see this same formulation again when we calculate standardized test statistics in Chapters 9, 11, and 12.

AP® EXAM TIP

Students will often mix up the order in a calculation of a residual by taking (predicted y − actual y). An easy way to remember the correct order of subtraction is to think **AP** = **A**ctual − **P**redicted. They should be able to remember this one.

PROBLEM:
Mrs. Gallas's class performed the "Barbie Bungee" activity. In this activity, students made a chain of rubber bands, connecting them one at a time to Barbie's feet and then measuring the distance that Barbie travels on her bungee jump. The distance is measured from the edge of the jumping platform to the lowest point that Barbie's head reaches. Here is the scatterplot of data from one of the groups with the regression line $\hat{y} = 27.42 + 7.21x$ added to the scatterplot:

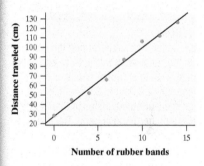

(a) Students did not collect data for 7 rubber bands. Predict the distance that Barbie will travel for 7 rubber bands.
(b) A small package of rubber bands comes with 30 rubber bands. Predict the distance that Barbie would travel if you used all these rubber bands.
(c) How confident are you in each of these predictions? Explain your reasoning.

SOLUTION:
(a) $\hat{y} = 27.42 + 7.21(7)$
$\hat{y} = 77.89$ cm
(b) $\hat{y} = 27.42 + 7.21(30)$
$\hat{y} = 243.72$ cm

(c) The prediction for 7 rubber bands is believable because $x = 7$ is within the interval of x-values used to create the model. However, the prediction for 30 rubber bands is not trustworthy because $x = 30$ is far outside the x-values used to create the regression line. The linear form may not extend to bungee jumps that go this far. Also, we don't want Barbie crashing to the ground.

In Figure 3.8 above, the highlighted data point represents a Ford F-150 that had 70,583 miles driven and a price of $21,994. The regression line predicts a price of

$$\widehat{price} = 38{,}257 - 0.1629(70{,}583) = \$26{,}759$$

for this truck, but its actual price was $21,994. This truck's residual is

$$
\begin{aligned}
\text{residual} &= \text{actual } y - \text{predicted } y \\
&= y - \hat{y} \\
&= 21{,}994 - 26{,}759 = -\$4765
\end{aligned}
$$

The actual price of this truck is $4765 *less* than the cost predicted by the regression line with $x =$ miles driven. Why is the actual price less than predicted? There are many possible reasons. Perhaps the truck needs body work, has mechanical issues, or has been in an accident.

EXAMPLE

Can you grab more than expected?
Calculating and interpreting a residual

PROBLEM: Here again is the scatterplot showing the hand span (in cm) and number of Starburst™ candies grabbed by each student in Mr. Tyson's class. The regression line is $\hat{y} = -29.8 + 2.83x$.

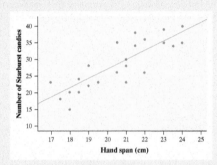

Josh Tabor

Find and interpret the residual for Andres, who has a hand span of 22 cm and grabbed 36 Starburst candies.

SOLUTION:
$\hat{y} = -29.8 + 2.83(22) = 32.46$ Starburst candies
Residual $= 36 - 32.46 = 3.54$ Starburst candies

Andres grabbed 3.54 more Starburst candies than the number predicted by the regression line with $x =$ hand span.

> Residual = actual y − predicted y
> $= y - \hat{y}$

FOR PRACTICE, TRY EXERCISE 39

ALTERNATE EXAMPLE | Skill 2.B

Barbie dares to bungee jump again
Calculating and interpreting a residual

PROBLEM:
Mrs. Gallas's class performed the "Barbie Bungee" activity. In this activity, students made a chain of rubber bands, connecting them one at a time to Barbie's feet and then measuring the distance that Barbie travels on her bungee jump. The distance is measured from the edge of the jumping platform to the lowest point that Barbie's head reaches. Here is the scatterplot of data from one of the groups with the regression line $\hat{y} = 27.42 + 7.21x$ added to the scatterplot:

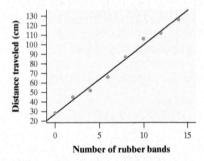

Find and interpret the residual for the trial with 4 rubber bands, where students measured that Barbie traveled 52 centimeters.

SOLUTION:
$\hat{y} = 27.42 + 7.21(4) = 56.26$ cm
Residual $= 52 - 56.26 = -4.26$ cm

The students' measurement for 4 rubber bands was 4.26 cm less than the length predicted by the regression line with $x =$ number of rubber bands.

CHECK YOUR UNDERSTANDING

Some data were collected on the weight of a male white laboratory rat for the first 25 weeks after its birth. A scatterplot of y = weight (in grams) and x = time since birth (in weeks) shows a fairly strong, positive linear relationship. The regression equation $\hat{y} = 100 + 40x$ models the data fairly well.

1. Predict the rat's weight at 16 weeks old.
2. Calculate and interpret the residual if the rat weighed 700 grams at 16 weeks old.
3. Should you use this line to predict the rat's weight at 2 years old? Use the equation to make the prediction and discuss your confidence in the result. (There are 454 grams in a pound.)

Interpreting a Regression Line

A regression line is a *model* for the data, much like the density curves of Chapter 2. The y **intercept** and **slope** of the regression line describe what this model tells us about the relationship between the response variable y and the explanatory variable x.

<aside>
The data used to calculate a regression line typically come from a sample. The statistics a and b in the sample regression model estimate the y intercept and slope parameters of the population regression model. You'll learn more about how this works in Chapter 12.
</aside>

> **DEFINITION**   **y intercept, Slope**
>
> In the regression equation $\hat{y} = a + bx$:
>
> - a is the **y intercept,** the predicted value of y when $x = 0$
> - b is the **slope,** the amount by which the predicted value of y changes when x increases by 1 unit

You are probably accustomed to the form $y = mx + b$ for the equation of a line from algebra. Statisticians have adopted a different form for the equation of a regression line. Some use $\hat{y} = b_0 + b_1x$. We prefer the form $\hat{y} = a + bx$ for three reasons: (1) it's simpler, (2) your calculator uses this form, and (3) the formula sheet provided on the AP® exam uses this form. Just remember that the slope is the coefficient of x, no matter what form is used.

Let's return to the Ford F-150 data. The equation of the regression line for these data is $\hat{y} = 38{,}257 - 0.1629x$, where x = miles driven and y = price. The slope $b = -0.1629$ tells us that the *predicted* price of a used Ford F-150 goes down by \$0.1629 (16.29 cents) for each additional mile that the truck has been driven. The y intercept $a = 38{,}257$ is the *predicted* price (in dollars) of a used Ford F-150 that has been driven 0 miles.

The slope of a regression line is an important numerical description of the relationship between the two variables. Although we need the value of the y intercept to draw the line, it is statistically meaningful only when the explanatory variable can actually take values close to 0, as in the Ford F-150 data. In other cases, using the regression line to make a prediction for $x = 0$ is an extrapolation.

<aside>
AP® EXAM TIP

When asked to interpret the slope or y intercept, it is very important to include the word *predicted* (or equivalent) in your response. Otherwise, it might appear that you believe the regression equation provides actual values of y.
</aside>

Making Connections

In previous math classes, students learned various definitions and formulas for slope. Try to connect their prior knowledge of slope.

$$\text{slope} = \frac{\text{rise}}{\text{run}} = \frac{\text{change in } y}{\text{change in } x} = \frac{\Delta y}{\Delta x}$$

➕ Ask the StatsMedic

How to Interpret the Slope of the Line of Best Fit

In this post, we discuss some "tricks" for helping students write precise interpretations for the slope of a line of best fit.

Preparing for Inference

If the data for a scatterplot come from a sample, the y intercept of the sample regression line (a) is a statistic that is being used to estimate the parameter α = true y intercept of the population regression model. The slope of the sample regression line (b) is a statistic that is being used to estimate the parameter β = true slope of the population regression model. Students will learn this formally in Section 12.3.

Teaching Tip

For the F-150 data, start by writing the slope as

$$\frac{-0.1629}{1}.$$

Then connect to students' prior knowledge about slope.

$$\text{slope} = \frac{\text{rise}}{\text{run}} = \frac{\Delta y}{\Delta x} = \frac{\text{change in predicted price}}{\text{change in miles}}$$
$$= \frac{-0.1629}{1}$$

The interpretation flows nicely from this description of slope.

Preparing for Inference

If we were to take a different sample of 16 F-150s and make a line of best fit, would we get exactly the same slope? Not likely, so instead of giving a single value as our estimate of the slope of the population regression line, it might be better to give an interval of plausible values for the slope. We will do this in Section 12.3 when we calculate a *t interval for slope.*

ALTERNATE EXAMPLE   Skill 4.B

Barbie dares to jump one more time
Interpreting the slope and *y* intercept

PROBLEM:
Mrs. Gallas's class performed the "Barbie Bungee" activity. In this activity, students made a chain of rubber bands, connecting them one at a time to Barbie's feet and then measuring the distance that Barbie travels on her bungee jump. The distance is measured from the edge of the jumping platform to the lowest point that Barbie's head reaches. Here is the scatterplot of data from one of the groups with the regression line $\hat{y} = 27.42 + 7.21x$ added to the scatterplot:

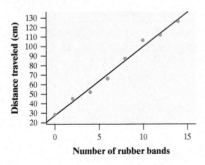

(a) Interpret the slope of the regression line.
(b) Does the value of the *y* intercept have meaning in this context? If so, interpret the *y* intercept. If not, explain why.

SOLUTION:
(a) The predicted distance Barbie's head travels goes up by 7.21 centimeters for each additional rubber band that is added.
(b) The *y* intercept does have meaning in this case. It is the predicted distance that Barbie's head travels when there are 0 rubber bands. This is Barbie's predicted height!

Teaching Tip

In the "Grabbing more candy" example on this page, be careful when trying to identify the *y* intercept using the graph. In this case, the vertical axis of the graph is not the *y* axis because the *x* axis does not start at 0. You can't always see the *y* intercept on the graph, especially if it is an extrapolation.

EXAMPLE

Grabbing more candy
Interpreting the slope and *y* intercept

PROBLEM: The scatterplot shows the hand span (in cm) and number of Starburst™ candies grabbed by each student in Mr. Tyson's class, along with the regression line $\hat{y} = -29.8 + 2.83x$.

(a) Interpret the slope of the regression line.
(b) Does the value of the *y* intercept have meaning in this context? If so, interpret the *y* intercept. If not, explain why.

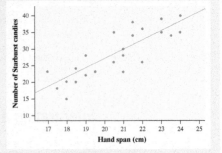

SOLUTION:

(a) The predicted number of Starburst candies grabbed goes up by 2.83 for each increase of 1 cm in hand span.

(b) The *y* intercept does not have meaning in this case, as it is impossible to have a hand span of 0 cm.

> Remember that the slope describes how the *predicted* value of *y* changes, not the actual value of *y*.

> Predicting the number of Starburst candies when *x* = 0 is an extrapolation—and results in an unrealistic prediction of −29.8.

FOR PRACTICE, TRY EXERCISE 41

For the Ford F-150 data, the slope $b = -0.1629$ is very close to 0. This does *not* mean that change in miles driven has little effect on price. The size of the slope depends on the units in which we measure the two variables. In this setting, the slope is the predicted change in price (in dollars) when the distance driven increases by 1 mile. There are 100 cents in a dollar. If we measured price in cents instead of dollars, the slope would be 100 times steeper, $b = -16.29$. *You can't say how strong a relationship is by looking at the slope of the regression line.*

CHECK YOUR UNDERSTANDING

Some data were collected on the weight of a male white laboratory rat for the first 25 weeks after its birth. A scatterplot of *y* = weight (in grams) and *x* = time since birth (in weeks) shows a fairly strong, positive linear relationship. The regression equation $\hat{y} = 100 + 40x$ models the data fairly well.

1. Interpret the slope of the regression line.
2. Does the value of the *y* intercept have meaning in this context? If so, interpret the *y* intercept. If not, explain why.

✓ **Answers to CYU**

1. The slope is 40. *Interpretation:* The predicted weight goes up by 40 grams for each increase of 1 week in the rat's age.

2. Yes, the *y* intercept has meaning in this context. The *y* intercept, 100, is predicted weight (in grams) for a rat at birth (*x* = 0 weeks).

Preparing for Inference

Perhaps there is not really a negative linear relationship between miles driven and predicted price for all F-150s and we got a negative slope value purely by chance because our random sample had certain values in it. We will investigate claims like this one in Chapter 12 when we do a *t test for slope*.

The Least-Squares Regression Line

There are many different lines we could use to model the association in a particular scatterplot. A *good* regression line makes the residuals as small as possible.

In the F-150 example, the regression line we used is $\hat{y} = 38,257 - 0.1629x$. How does this line make the residuals "as small as possible"? Maybe this line minimizes the *sum* of the residuals. If we add the prediction errors for all 16 trucks, the positive and negative residuals cancel out, as shown in Figure 3.9(a). That's the same issue we faced when we tried to measure deviation around the mean in Chapter 1. We'll solve the current problem in much the same way—by squaring the residuals.

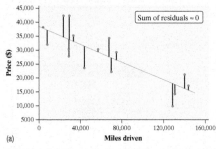

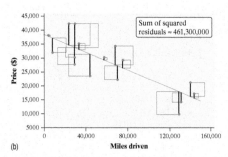

FIGURE 3.9 Scatterplots of the Ford F-150 data with the regression line added. (a) The residuals will add to approximately 0 when using a good regression line. (b) A good regression line should make the sum of squared residuals as small as possible.

A good regression line will have a sum of residuals near 0. But the regression line we prefer is the one that minimizes the sum of the squared residuals. That's what the line shown in Figure 3.9(b) does for the Ford F-150 data, which is why we call it the **least-squares regression line**. No other regression line would give a smaller sum of squared residuals.

> In addition to minimizing the sum of squared residuals, the least-squares regression line always goes through the point $(\bar{x}, \bar{y})$.

DEFINITION Least-squares regression line

The **least-squares regression line** is the line that makes the sum of the squared residuals as small as possible.

Your calculator or statistical software will give the equation of the least-squares line from data that you enter. Then you can concentrate on understanding and using the regression line.

AP® EXAM TIP

When displaying the equation of a least-squares regression line, the calculator will report the slope and intercept with much more precision than we need. There is no firm rule for how many decimal places to show for answers on the AP® Statistics exam. Our advice: decide how much to round based on the context of the problem you are working on.

Teaching Tip: Using Technology

On older operating systems for the TI-83/84, the default lists for LinReg(a+bx) are L1 and L2, so if you do not enter any lists, the calculator will automatically use L1 and L2. Discourage your students from using this shortcut, because there will be times when they may not use L1 and L2 and forget to enter the correct lists.

Teaching Tip: Using Technology

In the CALC menu, there are two choices for a line: LinReg(ax + b) and LinReg(a + bx). We choose the latter to be consistent with the general preference of statisticians.

9. Technology Corner   CALCULATING LEAST-SQUARES REGRESSION LINES

TI-Nspire and other technology instructions are on the book's website at highschool.bfwpub.com/updatedtps6e.

Let's use the Ford F-150 data to show how to find the equation of the least-squares regression line on the TI-83/84. Here are the data again:

Miles driven	70,583	129,484	29,932	29,953	24,495	75,678	8359	4447
Price ($)	21,994	9500	29,875	41,995	41,995	28,986	31,891	37,991
Miles driven	34,077	58,023	44,447	68,474	144,162	140,776	29,397	131,385
Price ($)	34,995	29,988	22,896	33,961	16,883	20,897	27,495	13,997

1. Enter the miles driven data into L1 and the price data into L2.

2. To determine the least-squares regression line, press [STAT]; choose CALC and then LinReg(a+bx).

- **OS 2.55 or later:** In the dialog box, enter the following: Xlist:L1, Ylist:L2, FreqList (leave blank), Store RegEQ (leave blank), and choose Calculate.
- **Older OS:** Finish the command to read LinReg(a+bx) L1,L2 and press [ENTER].

Note: If r^2 and r do not appear on the TI-83/84 screen, do this one-time series of keystrokes:

- **OS 2.55 or later:** Press [MODE] and set STAT DIAGNOSTICS to ON. Then redo Step 2 to calculate the least-squares line. The r^2 and r values should now appear.
- **Older OS:** Press [2nd] [0] (CATALOG), scroll down to DiagnosticOn, and press [ENTER]. Press [ENTER] again to execute the command. The screen should say "Done." Then redo Step 2 to calculate the least-squares line. The r^2 and r values should now appear.

To graph the least-squares regression line on the scatterplot:

1. Set up a scatterplot (see Technology Corner 8 on page 159).

2. Press [Y =] and enter the equation of the least-squares regression line in Y1.

3. Press [ZOOM] and choose ZoomStat to see the scatterplot with the least-squares regression line.

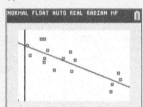

Note: When you calculate the equation of the least-squares regression line, you can have the calculator store the equation to Y1. When setting up the calculation, enter Y1 for the StoreRegEq prompt blank (OS 2.55 or later) or use the following command (older OS): LinReg(a+bx) L1,L2,Y1. Y1 is found by pressing [VARS] and selecting Y-VARS, then Function, then Y1.

Determining if a Linear Model Is Appropriate: Residual Plots

One of the first principles of data analysis is to look for an overall pattern and for striking departures from the pattern. A regression line describes the overall pattern of a linear relationship between an explanatory variable and a response variable. We see departures from this pattern by looking at a **residual plot**.

> *Some software packages prefer to plot the residuals against the predicted values $\hat{y}$ instead of against the values of the explanatory variable. The basic shape of the two plots is the same because $\hat{y}$ is linearly related to x.*

> **DEFINITION   Residual plot**
>
> A **residual plot** is a scatterplot that displays the residuals on the vertical axis and the explanatory variable on the horizontal axis.

Residual plots help us assess whether or not a linear model is appropriate. In Figure 3.10(a), the scatterplot shows the relationship between the average income (gross domestic product per person, in dollars) and fertility rate (number of children per woman) in 187 countries, along with the least-squares regression line. The residual plot in Figure 3.10(b) shows the average income for each country and the corresponding residual.

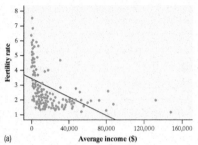
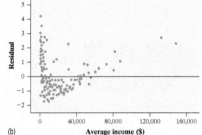

(a)   (b)

FIGURE 3.10 The (a) scatterplot and (b) residual plot for the linear model relating fertility rate to average income for a sample of countries.

The least-squares regression line clearly doesn't fit this association very well! For most countries with average incomes under $5000, the actual fertility rates are greater than predicted, resulting in positive residuals. For countries with average incomes between $5000 and $60,000, the actual fertility rates tend to be smaller than predicted, resulting in negative residuals. Countries with average incomes above $60,000 all have fertility rates greater than predicted, again resulting in positive residuals. This U-shaped pattern in the residual plot indicates that the linear form of our model doesn't match the form of the association. A curved model might be better in this case.

In Figure 3.11(a), the scatterplot shows the Ford F-150 data, along with the least-squares regression line. The corresponding residual plot is shown in Figure 3.11(b).

Looking at the scatterplot, the line seems to be a good fit for this relationship. You can "see" that the line is appropriate by the lack of a leftover curved pattern in the residual plot. In fact, the residuals look randomly scattered around the residual = 0 line.

Teaching Tip

Make sure to point out the title of this subsection. The purpose of a residual plot is to assess if the form of the model we are using matches the form of the association. If a residual plot indicates that a linear model is appropriate, we can ask the follow-up question found in the next subsection: How well does the line fit the data?

Teaching Tip

For residual plots, software packages put the predicted values (sometimes called *fitted values*) on the horizontal axis because they are built to do multiple regression. In multiple regression, many different explanatory variables are combined into one model to predict a response variable. Instead of picking one of the explanatory variables to put on the horizontal axis, software selects the predicted values, which are a function of all the explanatory variables.

Teaching Tip

Often students will suggest that we can just take the scatterplot and tilt it so the least-squares regression line is horizontal and we have created a residual plot. While this is close to correct, this method incorrectly shows the residual as the *perpendicular* distance from the point to the line. The true residual plot shows the *vertical* distance from each point to the line.

Making Connections

In Section 3.3, we will investigate nonlinear models, such as exponential, power, and logarithmic.

■ AP® EXAM TIP

When asked if a linear model is appropriate, students will sometimes incorrectly use the correlation to justify linearity. However, a strong correlation doesn't mean an association is linear. An association can be clearly nonlinear and still have a correlation close to ± 1. Only a residual plot can adequately address whether a line is an appropriate model for the data by showing the pattern of deviations from the line. For example, graphing the function $y = x^2$ for the integers 1 to 10 yields a correlation of $r = 0.97$, but the residual plot shows an obvious pattern.

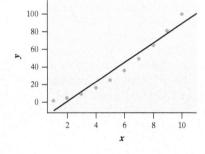
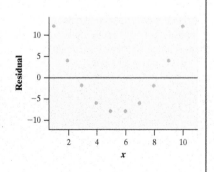

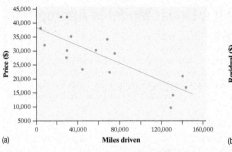
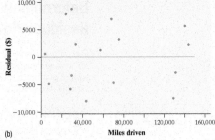

(a)   (b)

FIGURE 3.11 The (a) scatterplot and (b) residual plot for the linear model relating price to miles driven for Ford F-150s.

HOW TO INTERPRET A RESIDUAL PLOT

To determine whether the regression model is appropriate, look at the residual plot.
- If there is no leftover curved pattern in the residual plot, the regression model is appropriate.
- If there is a leftover curved pattern in the residual plot, consider using a regression model with a different form.

EXAMPLE

Pricing diamonds
Interpreting a residual plot

PROBLEM: Is a linear model appropriate to describe the relationship between the weight (in carats) and price (in dollars) of round, clear, internally flawless diamonds with excellent cuts? We calculated a least-squares regression line using x = weight and y = price and made the corresponding residual plot shown.[23] Use the residual plot to determine if the linear model is appropriate.

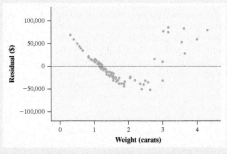

SOLUTION:

The linear model relating price to carat weight is not appropriate because there is a U-shaped pattern left over in the residual plot.

FOR PRACTICE, TRY EXERCISE 47

ALTERNATE EXAMPLE   Skill 2.A

Calories worth the cost?
Interpreting a residual plot

PROBLEM:

Is a linear model appropriate to describe the relationship between the number of calories and cost for sandwiches at Wendy's? Using a random sample of 11 sandwiches at Wendy's, we calculated a least-squares regression line using x = number of calories and y = price ($) and made the residual plot shown. Use the residual plot to determine if the linear model is appropriate.

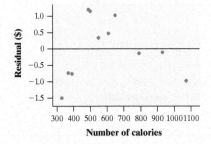

SOLUTION:
The linear model relating number of calories to price is not appropriate because there is an upside-down U-shaped pattern left over in the residual plot.

Teaching Tip

In the above example, require that students use precise language. A student may write, "It is not linear because there is a pattern." We don't know what "it" refers to. Stated more precisely, this becomes "The best model for the original carat weight and price data is not linear because there is a pattern in the residual plot."

Think About It

WHY DO WE LOOK FOR PATTERNS IN RESIDUAL PLOTS? The word *residual* comes from the Latin word *residuum*, meaning "left over." When we calculate a residual, we are calculating what is left over after subtracting the predicted value from the actual value:

$$\text{residual} = \text{actual } y - \text{predicted } y$$

Likewise, when we look at the form of a residual plot, we are looking at the form that is left over after subtracting the form of the model from the form of the association:

$$\text{form of residual plot} = \text{form of association} - \text{form of model}$$

When there is a leftover form in the residual plot, the form of the association and form of the model are not the same. However, if the form of the association and form of the model are the *same*, the residual plot should have no form, other than random scatter.

10. Technology Corner   MAKING RESIDUAL PLOTS

TI-Nspire and other technology instructions are on the book's website at highschool.bfwpub.com/updatedtps6e.

Let's continue the analysis of the Ford F-150 miles driven and price data from Technology Corner 9 (page 184). You should have already made a scatterplot, calculated the equation of the least-squares regression line, and graphed the line on the scatterplot. Now, we want to calculate residuals and make a residual plot. Fortunately, your calculator has already done most of the work. Each time the calculator computes a regression line, it computes the residuals and stores them in a list named RESID.

1. Set up a scatterplot in the statistics plots menu.

   • Press [2nd] [Y=] (STAT PLOT).

   • Press [ENTER] or [1] to go into Plot1.

   • Adjust the settings as shown. The RESID list is found in the List menu by pressing [2nd] [STAT]. *Note:* You have to calculate the equation of the least-squares regression line using the calculator *before* making a residual plot. Otherwise, the RESID list will include the residuals from a different least-squares regression line.

2. Use ZoomStat to let the calculator choose an appropriate window.

   • Press [ZOOM] and choose 9: ZoomStat.

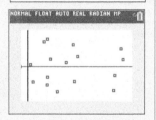

Note: If you want to see the values of the residuals, you can have the calculator put them in L3 (or any list). In the list editor, highlight the heading of L3, choose the RESID list from the LIST menu, and press [ENTER].

Preparing for Inference

Preparing for Inference

If the residual plot shows increasing spread moving from left to right, but otherwise looks randomly scattered, a line is the right model to use. When residual plots have this pattern, the precision of predictions will vary depending on the value of *x*. Also, the presence of increasing spread will cause problems in Section 12.3 when we perform significance tests and construct confidence intervals for the slope of a least-squares regression line.

Teaching Tip: Using Technology

Use the *Two Quantitative Variables* applet in the Extra Applets section of the Student Site at highschool.bfwpub.com /updatedtps6e to create a residual plot.

Teaching Tip: Using Technology

Emphasize the note on this page— students must calculate the equation of the least-squares regression line before using the RESID list. Otherwise, the residuals in the RESID list will be from whatever line they calculated most recently.

✓ Answers to CYU

1. The equation of the least-squares regression line is $\hat{y} = 16.2649 + 0.0908x$, where x = body weight and $\hat{y}$ = the predicted backpack weight.

2. A residual plot is given below.

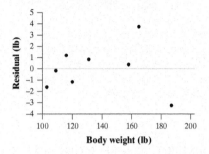

Body weight (lb)	Backpack weight (lb)	Prediction	Residual
120	26	27.1609	−1.1609
187	30	33.2444	−3.2444
109	26	26.1621	−0.1621
103	24	25.6173	−1.6173
131	29	28.1597	0.8403
165	35	31.2468	3.7532
158	31	30.6112	0.3888
116	28	26.7977	1.2023

3. Because there appears to be a negative-positive-negative pattern in the residual plot, a linear model is not appropriate for these data.

Teaching Tip

Again, make sure to point out the title of this subsection. The standard deviation of the residuals and r^2 do *not* reveal if a linear model is appropriate. Rather, once a residual plot reveals that a linear model is appropriate, the values of s and r^2 measure how good our predictions will be using the linear model.

Making Connections

When introducing the idea of the standard deviation of the residuals, remind students of the standard deviation we learned about earlier. In Chapter 1, the standard deviation was the typical distance of the actual values from the mean. In this chapter, the standard deviation is the typical distance of the actual values from the least-squares regression line. In general, a standard deviation is the typical distance that the actual values are from their expected values.

CHECK YOUR UNDERSTANDING

In Exercises 3 and 7, we asked you to make and describe a scatterplot for the hiker data shown in the table.

Body weight (lb)	120	187	109	103	131	165	158	116
Backpack weight (lb)	26	30	26	24	29	35	31	28

1. Calculate the equation of the least-squares regression line.
2. Make a residual plot for the linear model in Question 1.
3. What does the residual plot indicate about the appropriateness of the linear model? Explain your answer.

How Well the Line Fits the Data: The Role of *s* and *r²* in Regression

We use a residual plot to determine if a least-squares regression line is an appropriate model for the relationship between two variables. Once we determine that a least-squares regression line is appropriate, it makes sense to ask a follow-up question: How well does the line work? That is, if we use the least-squares regression line to make predictions, how good will these predictions be?

THE STANDARD DEVIATION OF THE RESIDUALS We already know that a residual measures how far an actual y value is from its corresponding predicted value $\hat{y}$. Earlier in this section, we calculated the residual for the Ford F-150 with 70,583 miles driven and price $21,994. As shown in Figure 3.12, the residual was −$4765, meaning that the actual price was $4765 *less* than we predicted.

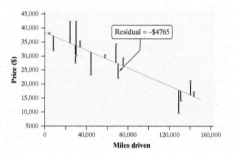

FIGURE 3.12 Scatterplot of the Ford F-150 data with a regression line added. Residuals for each truck are shown with vertical line segments.

To assess how well the line fits *all* the data, we need to consider the residuals for each of the trucks, not just one. Here are the residuals for all 16 trucks:

−4765	−7664	−3506	8617	7728	3057	−5004	458
2289	1183	−8121	6858	2110	5572	−5973	−2857

Teaching Tip

The standard deviation of the residuals is listed in Topic 9.2 of the Course Framework. We introduce it here because it is a great way to assess how well a regression line fits the data.

Using these residuals, we can estimate the "typical" prediction error when using the least-squares regression line. To do this, we calculate the **standard deviation of the residuals** *s*.

> ### DEFINITION Standard deviation of residuals *s*
> The **standard deviation of the residuals** *s* measures the size of a typical residual. That is, *s* measures the typical distance between the actual *y* values and the predicted *y* values.

To calculate *s*, use the following formula:

$$s = \sqrt{\frac{\text{sum of squared residuals}}{n-2}} = \sqrt{\frac{\sum(y_i - \hat{y}_i)^2}{n-2}}$$

For the Ford F-150 data, the standard deviation of the residuals is

$$s = \sqrt{\frac{(-4765)^2 + (-7664)^2 + \cdots + (-2857)^2}{16-2}} = \sqrt{\frac{461,264,136}{14}} = \$5740$$

Interpretation: The actual price of a Ford F-150 is typically about $5740 away from the price predicted by the least-squares regression line with *x* = miles driven. If we look at the residual plot in Figure 3.11, this seems like a reasonable value. Although some of the residuals are close to 0, others are close to $10,000 or −$10,000.

> ### Think About It
>
> **DOES THE FORMULA FOR *s* LOOK SLIGHTLY FAMILIAR?** It should. In Chapter 1, we defined the standard deviation of a set of quantitative data as
>
> $$s_x = \sqrt{\frac{\sum(x_i - \bar{x})^2}{n-1}}$$
>
> We interpreted the resulting value as the "typical" distance of the data points from the mean. In the case of two-variable data, we're interested in the typical (vertical) distance of the data points from the regression line. We find this value in much the same way: first add up the squared deviations, then average them (again, in a funny way), and take the square root to get back to the original units of measurement.

THE COEFFICIENT OF DETERMINATION r^2 There is another numerical quantity that tells us how well the least-squares line predicts values of the response variable *y*. It is r^2, the **coefficient of determination**. Some computer packages call it "R-sq." You may have noticed this value in some of the output that we showed earlier. Although it's true that r^2 is equal to the square of the correlation *r*, there is much more to this story.

Teaching Tip

Students often struggle with the interpretation of r^2. It might help to start by saying something like, "The sum of squared residuals has been reduced by 66.4% when using the linear model relating asking price to number of miles." This interpretation follows nicely from the explanation on this page. Then move to the more traditional interpretation: "_____% of the variation in (*y* variable) is accounted for by the linear model relating (*y* variable) to (*x* variable)." In this interpretation, the "variation in *y*" is measured by the sum of squared residuals.

Teaching Tip

If you have data that fall on a perfectly horizontal line, your calculator will say that r and r^2 are undefined. The formula for r uses the standard deviation of *y* in the denominator (0 in the case of a horizontal line). In practice, however, it would be very unusual to have a response variable with no variability.

Teaching Tip

Some books give the following formula:
$$r^2 = \frac{SSM - SSE}{SSM},$$
where *SSM* is the sum of the squares about the mean and *SSE* is the sum of squares for error.

Some people interpret r^2 as the proportion of variation in the response variable that is explained by the explanatory variable in the model.

> **DEFINITION** The coefficient of determination r^2
>
> The **coefficient of determination** r^2 measures the percent reduction in the sum of squared residuals when using the least-squares regression line to make predictions, rather than the mean value of *y*. In other words, r^2 measures the percent of the variability in the response variable that is accounted for by the least-squares regression line.

Suppose we wanted to predict the price of a particular used Ford F-150, but we didn't know how many miles it had been driven. Our best guess would be the average cost of a used Ford F-150, $\bar{y} = \$27,834$. Of course, this prediction is unlikely to be very good, as the prices vary quite a bit from the mean ($s_y = \$9570$). If we knew how many miles the truck had been driven, we could use the least-squares regression line to make a better prediction. How much better are predictions that use the least-squares regression line with $x =$ miles driven, rather than predictions that use only the average price? The answer is r^2.

The scatterplot in Figure 3.13(a) shows the squared residuals along with the sum of squared residuals (approximately 1,374,000,000) when using the average price as the predicted value. The scatterplot in Figure 3.13(b) shows the squared residuals along with the sum of squared residuals (approximately 461,300,000) when using the least-squares regression line with $x =$ miles driven to predict the price. Notice that the squares in part (b) are quite a bit smaller.

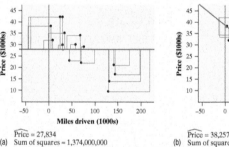
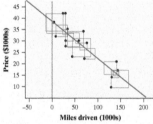

(a) $\widehat{\text{Price}} = 27,834$ Sum of squares = 1,374,000,000

(b) $\widehat{\text{Price}} = 38,257 - 0.1629$ Miles driven; $r^2 = 0.66$ Sum of squares $\approx 461,300,000$

FIGURE 3.13 (a) The sum of squared residuals is about 1,374,000,000 if we use the mean price as our prediction for all 16 trucks. (b) The sum of squared residuals from the least-squares regression line is about 461,300,000.

To find r^2, calculate the percent reduction in the sum of squared residuals:

$$r^2 = \frac{1,374,000,000 - 461,300,000}{1,374,000,000} = \frac{912,700,000}{1,374,000,000} = 0.66$$

The sum of squared residuals has been reduced by 66%.

Teaching Tip

Remind students that if they are given the value of r^2 and are asked to find the correlation, they have to consider the direction of the association. That is, $r = \pm\sqrt{r^2}$, where the $+$ or $-$ is determined by the direction of the association.

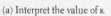

Interpretation: About 66% of the variability in the price of a Ford F-150 is accounted for by the least-squares regression line with $x =$ miles driven. The remaining 34% is due to other factors, including age, color, and condition.

If all the points fall directly on the least-squares line, the sum of squared residuals is 0 and $r^2 = 1$. Then all the variation in y is accounted for by the linear relationship with x. In the worst-case scenario, the least-squares line does no better at predicting y than $y = \bar{y}$ does. Then the two sums of squared residuals are the same and $r^2 = 0$.

It's fairly remarkable that the coefficient of determination r^2 is actually the square of the correlation. This fact provides an important connection between correlation and regression. When you see a linear association, square the correlation to get a better feel for how well the least-squares line fits the data.

Think About It

WHAT'S THE RELATIONSHIP BETWEEN s AND r^2? Both s and r^2 are calculated from the sum of squared residuals. They also both measure how well the line fits the data. The standard deviation of the residuals reports the size of a typical prediction error, in the same units as the response variable. In the truck example, $s = \$5740$. The value of r^2, however, does not have units and is usually expressed as a percentage between 0% and 100%, such as $r^2 = 66\%$. Because these values assess how well the line fits the data in different ways, we recommend you follow the example of most statistical software and report both.

Knowing how to interpret s and r^2 is much more important than knowing how to calculate them. Consequently, we typically let technology do the calculations.

EXAMPLE

Grabbing candy, again
Interpreting s and r^2

PROBLEM: The scatterplot shows the hand span (in centimeters) and number of Starburst™ candies grabbed by each student in Mr. Tyson's class, along with the regression line $\hat{y} = -29.8 + 2.83x$. For this model, technology gives $s = 4.03$ and $r^2 = 0.697$.

(a) Interpret the value of s.
(b) Interpret the value of r^2.

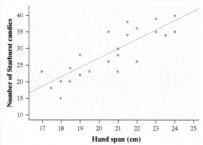

Teaching Tip

In the interpretation of r^2, "is accounted for" can be replaced with "is explained by," although we want to be careful not to imply causation here.

➕ Ask the StatsMedic

An Activity to Help Students Interpret r^2

Unfortunately for students, interpreting r^2 often becomes a memorized fill-in-the-blank exercise: "___% of the variability in (response variable) can be accounted for by the least-squares regression lines with $x =$ (explanatory variable)." This post shows how you can use the "Can you guess my IQ?" activity to help motivate an understanding of the interpretation.

ALTERNATE EXAMPLE Skill 2.C

The last Barbie bungee jump
Interpreting s and r^2

PROBLEM:
Mrs. Gallas's class performed the "Barbie Bungee" activity. They connected rubber bands one at a time in a chain to Barbie's feet and then measured the distance that Barbie travels on her (last) bungee jump. The distance is measured from the edge of the jumping platform to the lowest point that Barbie's head reaches. Here is the scatterplot of data from one of the groups with the regression line $\hat{y} = 27.42 + 7.21x$. For this model, technology gives $s = 4.11$ and $r^2 = 0.989$.

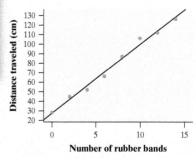

(a) Interpret the value of s.
(b) Interpret the value of r^2.

SOLUTION:
(a) The distance traveled by Barbie is typically about 4.11 cm away from the distance predicted by the least-squares regression line with $x =$ number of rubber bands.

(b) About 98.9% of the variability in distance traveled by Barbie is accounted for by the least-squares regression line with $x =$ number of rubber bands.

AP® EXAM TIP

Students will not be required to know how to use Minitab or JMP software, but they will be required to interpret output that comes from such software.

Teaching Tip

For our purposes, we will always use R-sq (RSquare), not R-sq(adj) (RSquare Adj). The adjusted r^2 value is only for situations in which there is more than one explanatory variable (multiple regression).

Preparing for Inference

In Section 12.3, we will learn that in the row containing the slope, the value of "T (t Ratio)" will be the standardized test statistic for a t test for slope and that "P (Prob > |t|)" will be the P-value for the t test for slope (two-sided).

SOLUTION:

(a) The actual number of Starburst™ candies grabbed is typically about 4.03 away from the number predicted by the least-squares regression line with x = hand span.

(b) About 69.7% of the variability in number of Starburst candies grabbed is accounted for by the least-squares regression line with x = hand span.

FOR PRACTICE, TRY EXERCISE 55

Interpreting Computer Regression Output

Figure 3.14 displays the basic regression output for the Ford F-150 data from two statistical software packages: Minitab and JMP. Other software produces very similar output. Each output records the slope and y intercept of the least-squares line. The software also provides information that we don't yet need, although we will use much of it later. Be sure that you can locate the slope, the y intercept, and the values of s (called *root mean square error* in JMP) and r^2 on both computer outputs. *Once you understand the statistical ideas, you can read and work with almost any software output.*

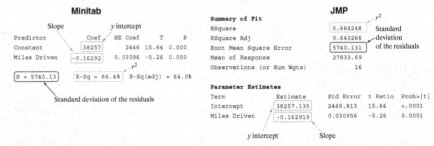

FIGURE 3.14 Least-squares regression results for the Ford F-150 data from Minitab and JMP statistical software. Other software produces similar output.

EXAMPLE   **Using feet to predict height**
Interpreting regression output

PROBLEM: A random sample of 15 high school students was selected from the U.S. Census At School database. The foot length (in centimeters) and height (in centimeters) of each student in the sample were recorded. Here are a scatterplot with the least-squares regression line added, a residual plot, and some computer output:

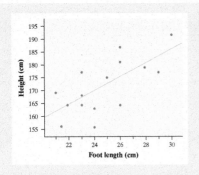

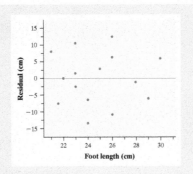

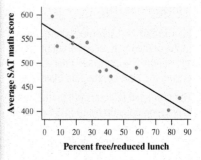

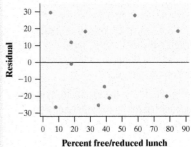

Predictor	Coef	SE Coef	T	P
Constant	103.4100	19.5000	5.30	0.000
Foot length	2.7469	0.7833	3.51	0.004

S = 7.95126 R-Sq = 48.6% R-Sq(adj) = 44.7%

(a) Is a line an appropriate model to use for these data? Explain how you know.

(b) Find the correlation.

(c) What is the equation of the least-squares regression line that models the relationship between foot length and height? Define any variables that you use.

(d) By about how much do the actual heights typically vary from the values predicted by the least-squares regression line with $x =$ foot length?

SOLUTION:

(a) *Because the scatterplot shows a linear association and the residual plot has no obvious leftover curved patterns, a line is an appropriate model to use for these data.*

(b) $r = \sqrt{0.486} = 0.697$

> The correlation r is the square root of r^2, where r^2 is a value between 0 and 1. Because the square root function on your calculator will always give a positive result, make sure to consider whether the correlation is positive or negative. If the slope is negative, so is the correlation.

(c) $\widehat{height} = 103.41 + 2.7469 \, (foot\,length)$

(d) *$s = 7.95$, so the actual heights typically vary by about 7.95 cm from the values predicted by the regression line with $x = $ foot length.*

> We could also write the equation as $\hat{y} = 103.41 + 2.7469x$, where $\hat{y} = $ predicted height (cm) and $x = $ foot length (cm).

FOR PRACTICE, TRY EXERCISE 59

Predictor	Coef	SE Coef	T	P
Constant	577.9	12.5	46.16	0.000
Foot length	−1.993	0.276	−7.22	0.000

S = 23.3168 R-Sq = 85.29% R-Sq(adj) = 83.66%

(a) Is a line an appropriate model to use for these data? Explain how you know the answer.

(b) Find the correlation.

(c) What is the equation of the least-squares regression line that describes the relationship between percent free/reduced lunch and average SAT math score? Define any variables that you use.

(d) By about how much do the actual average SAT math scores typically vary from the values predicted by the least-squares regression line with $x = $ percent free/reduced lunch?

SOLUTION:

(a) Because the scatterplot shows a linear association and the residual plot has no obvious leftover curved patterns, a line is an appropriate model to use for these data.

(b) $r = \pm\sqrt{0.8529} = \pm0.924$, and because the relationship is negative, we know $r = -0.924$.

(c) The equation is $\hat{y} = 577.9 - 1.993x$, where $\hat{y} = $ predicted average SAT math score and x is percent free/reduced lunch.

(d) $s = 23.3168$, so the actual average SAT math scores typically vary by about 23.3168 from the values predicted by the regression line using $x = $ percent free/reduced lunch.

ALTERNATE EXAMPLE

Skills 2.A, 2.C

Can we predict a school's average SAT math score? Interpreting regression output

PROBLEM:

A random sample of 11 high schools was selected from all the high schools in Michigan. The percent of students who are eligible for free/reduced lunch and the average SAT math score of each high school in the sample were recorded. Students with household income below a certain threshold are eligible for free/reduced lunch. Here are a scatterplot with the least-squares regression line added, a residual plot, and some computer output:

✓ Answers to CYU

1. $\hat{y} = 33.347 + 13.2854x$, where $x =$ duration of the most recent eruption (in minutes) and $\hat{y} =$ predicted interval of time until the next eruption (in minutes).

2. The predicted interval of time until the next eruption goes up by 13.2854 minutes for each increase of 1 minute in the duration of the most recent eruption.

3. The actual interval of time until the next eruption (in minutes) is typically about 6.49 minutes away from the time predicted by the least-squares regression line with $x =$ duration of the most recent eruption (in minutes).

4. $r^2 = 85.4\%$ of the variability in interval is accounted for by the least-squares regression line with $x =$ duration.

AP® EXAM TIP

Remind students that these two formulas are on the formula sheet provided on both parts of the AP® Statistics exam. However, the formula for the y intercept is solved for $\bar{y}$, not a:

$$\bar{y} = a + b\bar{x}$$

Teaching Tip

Remind students that because of rounding errors, the slope and y intercept may be a little different from what technology provides.

CHECK YOUR UNDERSTANDING

In Section 3.1, you read about the Old Faithful geyser in Yellowstone National Park. The computer output shows the results of a regression of $y =$ interval of time until the next eruption (in minutes) and $x =$ duration of the most recent eruption (in minutes) for each eruption of Old Faithful in a particular month.

```
Summary of Fit
RSquare                          0.853725
RSquare Adj                      0.853165
Root Mean Square Error           6.493357
Mean of Response                77.543730
Observations (or Sum Wgts)    263.000000

Parameter Estimates
Term        Estimate    Std Error   t Ratio   Prob>|t|
Intercept   33.347442   1.201081    27.76     <.0001*
Duration    13.285406   0.340393    39.03     <.0001*
```

1. What is the equation of the least-squares regression line that models the relationship between interval and duration? Define any variables that you use.
2. Interpret the slope of the least-squares regression line.
3. Identify and interpret the standard deviation of the residuals.
4. What percent of the variability in interval is accounted for by the least-squares regression line with $x =$ duration?

Regression to the Mean

Using technology is often the most convenient way to find the equation of a least-squares regression line. It is also possible to calculate the equation of the least-squares regression line using only the means and standard deviations of the two variables and their correlation. Exploring this method will highlight an important relationship between the correlation and the slope of a least-squares regression line—and reveal why we include the word *regression* in the expression *least-squares regression line*.

HOW TO CALCULATE THE LEAST-SQUARES REGRESSION LINE USING SUMMARY STATISTICS

We have data on an explanatory variable x and a response variable y for n individuals and want to calculate the least-squares regression line $\hat{y} = a + bx$. From the data, calculate the means $\bar{x}$ and $\bar{y}$ and the standard deviations s_x and s_y of the two variables and their correlation r. The **slope** is:

$$b = r\frac{s_y}{s_x}$$

Because the least-squares regression line passes through the point $(\bar{x}, \bar{y})$, the **y intercept** is:

$$a = \bar{y} - b\bar{x}$$

The formula for the slope reminds us that the distinction between explanatory and response variables is important in regression. Least-squares regression makes the distances of the data points from the line small only in the y direction. If we reverse the roles of the two variables, the values of s_x and s_y will reverse in the slope formula, resulting in a different least-squares regression line. This is *not* true for correlation: switching x and y does *not* affect the value of r.

The formula for the y intercept comes from the fact that the least-squares regression line always passes through the point $(\overline{x}, \overline{y})$. Once we know the slope (b) and that the line goes through the point $(\overline{x}, \overline{y})$, we can use algebra to solve for the y intercept. Substituting $(\overline{x}, \overline{y})$ into the equation $\hat{y} = a + bx$ produces the equation $\overline{y} = a + b\overline{x}$. Solving this equation for a gives the equation shown in the definition box, $a = \overline{y} - b\overline{x}$. To see how these formulas work in practice, let's look at an example.

EXAMPLE

More about feet and height
Calculating the least-squares regression line

PROBLEM: In the preceding example, we used data from a random sample of 15 high school students to investigate the relationship between foot length (in centimeters) and height (in centimeters). The mean and standard deviation of the foot lengths are $\overline{x} = 24.76$ and $s_x = 2.71$. The mean and standard deviation of the heights are $\overline{y} = 171.43$ and $s_y = 10.69$. The correlation between foot length and height is $r = 0.697$. Find the equation of the least-squares regression line for predicting height from foot length.

panpote/Getty Images

SOLUTION:

$$b = 0.697 \frac{10.69}{2.71} = 2.75$$

$\begin{aligned} b &= r\dfrac{s_y}{s_x} \\ a &= \overline{y} - b\overline{x} \end{aligned}$

$$a = 171.43 - 2.75(24.76) = 103.34$$

The equation of the least-squares regression line is $\hat{y} = 103.34 + 2.75x$.

FOR PRACTICE, TRY EXERCISE 63

There is a close connection between the correlation and the slope of the least-squares regression line. The slope is

$$b = r\frac{s_y}{s_x} = \frac{r \cdot s_y}{s_x}$$

This equation says that along the regression line, a change of 1 standard deviation in x corresponds to a change of r standard deviations in y. When the variables are perfectly correlated ($r = 1$ or $r = -1$), the change in the predicted response $\hat{y}$ is the same (in standard deviation units) as the change in x. For example, if $r = 1$ and x is 2 standard deviations above $\overline{x}$, then the corresponding value of $\hat{y}$ will be 2 standard deviations above $\overline{y}$.

Teaching Tip

If students are provided an explanatory variable (x), a response variable (y), and a least-squares regression line and then are asked to predict an x-value from a y-value, they might simply plug the y-value into the regression line and work backward to solve for x. This is not correct. A new least-squares regression line must be calculated using y as the explanatory variable and x as the response variable: $\hat{x} = c + dy$. Then simply plug the given y-value into this new equation. The first equation ($\hat{y} = a + bx$) is created by minimizing the sum of the squares of the *vertical* distances of the points to the lines. The second equation $\hat{x} = c + dy$ is created by minimizing the sum of the squares of the *horizontal* distances of the points to the lines.

ALTERNATE EXAMPLE   Skill 2.C

SAT math scores again
Calculating the least-squares regression line

PROBLEM:
In the preceding alternate example, we used data from a random sample of 11 high schools in Michigan to investigate the relationship between the percent of students who are eligible for free/reduced lunch and the average SAT math score. The mean and standard deviation of the percent of students on free/reduced lunch are $\overline{x} = 37.55$ and $s_x = 26.37$. The mean and standard deviation of the average SAT math scores are $\overline{y} = 503.04$ and $s_y = 57.68$. The correlation between percent free/reduced lunch and average SAT math score is $r = -0.9236$. Find the equation of the least-squares regression line for predicting average SAT math scores from percent free/reduced lunch. Show your work.

SOLUTION:

$$b = -0.9236\frac{57.68}{26.73} = -1.993$$

$a = 503.04 - (-1.993)(37.55) = 577.9$
The equation of the least-squares regression line is $\hat{y} = 577.9 - 1.993x$.

Teaching Tip

There are several other ways to illustrate regression to the mean.

- Read the *New York Times* article "In Climbing Income Ladder, Location Matters" at http://tinyurl.com/climbingincomelocationmatters and use its interactive feature to compare the income percentiles for parents and their children. For example, a child in Tucson, Arizona, whose parents' income is at the 10th percentile is predicted to have an income in the 36th percentile (still below the mean, but closer to the mean than the parents). Likewise, a child whose parents' income is at the 90th percentile is predicted to have an income in the 58th percentile (still above the mean, but closer to the mean than the parents).

- Make a scatterplot of your students' grades on the first two tests and include the line $y = x$. Typically, students who do well on the first exam don't do quite as well on the second. Likewise, students who do very poorly on the first exam tend to do better the second time around. Make sure to try this before you do it in front of your students, just in case.

- Assign and discuss Exercise 65 or 66. In these exercises, above-average values of x lead to not-quite-as-far-above-average predicted values for y. Likewise, below-average values of x lead to not-quite-as-far-below-average predicted values for y.

TRM **Regression to the Mean in Sports**

This alternate activity, found in the Teacher's Resource Materials, is another nice way to illustrate regression to the mean. Find it by clicking on the link in the TE-Book, opening the TRFD, or downloading from the Teacher's Resources on the book's digital platform.

However, if the variables are not perfectly correlated ($-1 < r < 1$), the change in $\hat{y}$ is *less than* the change in x, when measured in standard deviation units. To illustrate this property, let's return to the foot length and height data from the preceding example.

Figure 3.15 shows the scatterplot of height versus foot length and the regression line $\hat{y} = 103.34 + 2.75x$. We have added four more lines to the graph: a vertical line at the mean foot length $\bar{x}$, a vertical line at $\bar{x} + s_x$ (1 standard deviation above the mean foot length), a horizontal line at the mean height $\bar{y}$, and a horizontal line at $\bar{y} + s_y$ (1 standard deviation above the mean height).

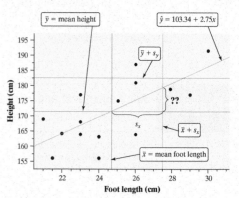

FIGURE 3.15 Scatterplot showing the relationship between foot length and height for a sample of students, along with lines showing the means of x and y and the values 1 standard deviation above each mean.

When a student's foot length is 1 standard deviation above the mean foot length $\bar{x}$, the predicted height $\hat{y}$ is above the mean height $\bar{y}$—but not an entire standard deviation above the mean. How far above the mean is the value of $\hat{y}$?

From the graph, we can see that

$$b = \text{slope} = \frac{\text{change in } y}{\text{change in } x} = \frac{??}{s_x}$$

From earlier, we know that

$$b = \frac{r \cdot s_y}{s_x}$$

Setting these two equations equal to each other, we have

$$\frac{??}{s_x} = \frac{r \cdot s_y}{s_x}$$

Thus, $\hat{y}$ must be $r \cdot s_y$ above the mean $\bar{y}$.

In other words, for an increase of 1 standard deviation in the value of the explanatory variable x, the least-squares regression line predicts an increase of *only r* standard deviations in the response variable y. When the correlation isn't $r = 1$ or $r = -1$, the predicted value of y is closer to its mean $\bar{y}$ than the value of x is to its mean $\bar{x}$. *This is called regression to the mean, because the values of y "regress" to their mean.*

Sir Francis Galton (1822–1911) is often credited with discovering the idea of regression to the mean. He looked at data on the heights of children versus the heights of their parents. He found that taller-than-average parents tended to have

children who were taller than average but not quite as tall as their parents. Likewise, shorter-than-average parents tended to have children who were shorter than average but not quite as short as their parents. Galton used the symbol r for the correlation because of its important relationship to regression.

Correlation and Regression Wisdom

Correlation and regression are powerful tools for describing the relationship between two variables. When you use these tools, you should be aware of their limitations.

CORRELATION AND REGRESSION LINES DESCRIBE ONLY LINEAR RELATIONSHIPS You can calculate the correlation and the least-squares line for any relationship between two quantitative variables, but the results are useful only if the scatterplot shows a linear pattern. *Always plot your data first!*

The following four scatterplots show very different relationships. Which one do you think shows the greatest correlation?

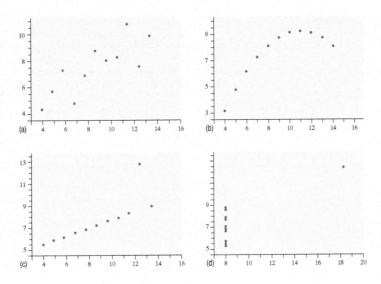

Answer: All four have the same correlation, $r = 0.816$. Furthermore, the least-squares regression line for each relationship is exactly the same, $\hat{y} = 3 + 0.5x$. These four data sets, developed by statistician Frank Anscombe, illustrate the importance of graphing data before doing calculations.[24]

CORRELATION AND LEAST-SQUARES REGRESSION LINES ARE NOT RESISTANT You already know that the correlation r is not resistant. One unusual point in a scatterplot can greatly change the value of r. Is the least-squares line resistant? The following activity will help you answer this question.

Making Connections

When starting with a nonlinear relationship between two variables, we can often transform the data into linear form by using logarithms. Once the data are transformed, we can use all the regression analysis from this section. We will investigate nonlinear relationships in Section 3.3.

Teaching Tip

Remind students that correlation measures direction and strength, but not form. Knowing that the correlation is $r = 0.816$ doesn't tell us anything about the form of the association, as illustrated in the scatterplots on this page.

Making Connections

Recall from Chapter 1 that mean and standard deviation are not resistant, because the introduction of an outlier can have a large effect on these values.

ACTIVITY OVERVIEW

To prepare for using this activity, watch the overview video by clicking on the link in the TE-Book, opening the TRFD, or downloading from the Teacher's Resources on the book's digital platform.

Time: 15 minutes

Materials: Computers, laptops, or devices with Internet access

Teaching Advice: Students could work individually or in pairs for this activity. Encourage students to play around with dragging points on the applet. The biggest takeaway from this activity is for students to understand that unusual points can have a large effect on the least-squares regression line.

Answers:

1–2. Answers will vary.

3. The least-squares regression line did not change. The value of r^2 increases (gets closer to 1).

4. The least-squares regression line is dragged toward the point. When the rightmost point is at the mean of y, the value of r^2 is close to zero. The value of r^2 increases as the point moves away from the mean of y (either up or down).

5. The slope of the least-squares regression line does not change, but the y intercept moves up and down with the point. When the point is at the mean of y, the value of r^2 is the greatest. The value of r^2 decreases as the point moves away from the mean of y (either up or down).

6. Unusual points can greatly affect the least-squares regression line, with unusual points in the horizontal direction having more influence than unusual points in the vertical direction.

ACTIVITY   Investigating properties of the least-squares regression line

In this activity, you will use the *Correlation and Regression* applet to explore some properties of the least-squares regression line.

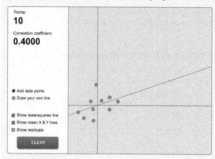

1. Launch the applet at highschool.bfwpub.com/updatedtps6e.

2. Click on the graphing area to add 10 points in the lower-left corner so that the correlation is about $r = 0.40$. Also, check the boxes to show the "Least-Squares Line" and the "Mean X & Y" lines as in the screen shot. Notice that the least-squares regression line goes though the point $(\bar{x}, \bar{y})$.

3. If you were to add a point on the least-squares regression line at the right edge of the graphing area, what do you think would happen to the least-squares regression line? To the value of r^2? (Remember that r^2 is the square of the correlation coefficient, which is provided by the applet.) Add the point to see if you were correct.

4. Click on the point you just added, and drag it up and down along the right edge of the graphing area. What happens to the least-squares regression line? To the value of r^2?

5. Now, move this point so that it is on the vertical $\bar{x}$ line. Drag the point up and down on the $\bar{x}$ line. What happens to the least-squares regression line? To the value of r^2?

6. Briefly summarize how unusual points influence the least-squares regression line.

As you learned in the activity, unusual points may or may not have an influence on the least-squares regression line and the coefficient of determination r^2. The same is true for the correlation r and the standard deviation of the residuals s. Here are four scatterplots that summarize the possibilities. In all four scatterplots, the 8 points in the lower left are the same.

Case 1: No unusual points

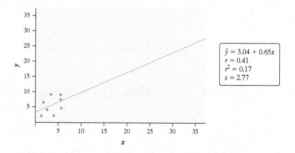

$\hat{y} = 3.04 + 0.65x$
$r = 0.41$
$r^2 = 0.17$
$s = 2.77$

Case 2: A point that is far from the other points in the *x* direction, but in the same pattern.

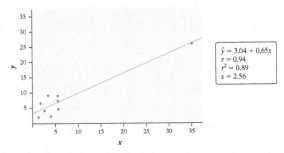

$\hat{y} = 3.04 + 0.65x$
$r = 0.94$
$r^2 = 0.89$
$s = 2.56$

Compared to Case 1, the equation of the least-squares regression line remained the same, but the values of *r* and r^2 greatly increased. The standard deviation of the residuals got a bit smaller because the additional point has a very small residual.

Case 3: A point that is far from the other points in the *x* direction, and not in the same pattern.

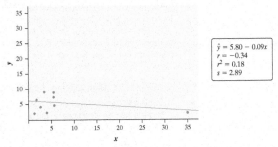

$\hat{y} = 5.80 - 0.09x$
$r = -0.34$
$r^2 = 0.18$
$s = 2.89$

Compared to Case 1, the equation of the least-squares regression line is much different, with the slope going from positive to negative and the *y* intercept increasing. The value of *r* is now negative while the value of r^2 stayed almost the same. Even though the new point has a relatively small residual, the standard deviation of the residuals got a bit larger because the line doesn't fit the remaining points nearly as well.

Teaching Tip

To help students understand how an unusual point influences the equation of the least-squares regression line, imagine that each point is connected to the line with a vertical spring. When the unusual point in Case 2 moved down in Case 3, its spring pulled the "right side" of the line down.

Case 4: A point that is far from the other points in the *y* direction, and not in the same pattern.

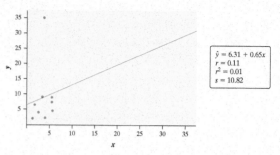

$\hat{y} = 6.31 + 0.65x$
$r = 0.11$
$r^2 = 0.01$
$s = 10.82$

Compared to Case 1, the slope of the least-squares regression line is the same, but the *y* intercept is a little larger as the line appears to have shifted up slightly. The values of *r* and r^2 are much smaller than before. Because the new point has such a large residual, the standard deviation of the residuals is much larger.

In Cases 2 and 3, the unusual point had a much bigger *x* value than the other points. Points whose *x* values are much smaller or much larger than the other points in a scatterplot have **high leverage**. In Case 4, the unusual point had a very large residual. Points with large residuals are called **outliers**. All three of these unusual points are considered **influential points** because adding them to the scatterplot substantially changed either the equation of the least-squares regression line or one or more of the other summary statistics (r, r^2, s).

DEFINITION   High leverage, Outlier, Influential point

Points with **high leverage** in regression have much larger or much smaller *x* values than the other points in the data set.

An **outlier** in regression is a point that does not follow the pattern of the data and has a large residual.

An **influential point** in regression is any point that, if removed, substantially changes the slope, *y* intercept, correlation, coefficient of determination, or standard deviation of the residuals.

 Outliers and high-leverage points are often influential in regression calculations! The best way to investigate the influence of such points is to do regression calculations with and without them to see how much the results differ. Here is an example that shows what we mean.

Does the age at which a child begins to talk predict a later score on a test of mental ability? A study of the development of young children recorded the age in months at which each of 21 children spoke their first word and their Gesell Adaptive Score, the result of an aptitude test taken much later.[25] A scatterplot of the data appears in Figure 3.16, along with a residual plot, and computer output. Two points, child 18 and child 19, are labeled on each plot.

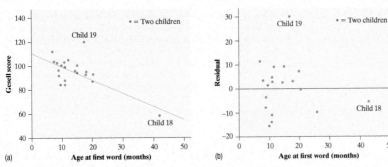

FIGURE 3.16 (a) Scatterplot of Gesell Adaptive Scores versus the age at first word for 21 children, along with the least-squares regression line. (b) Residual plot for the linear model. The point for Child 18 has high leverage and the point for Child 19 is an outlier. Each purple point in the graphs stands for two individuals.

The point for Child 18 has high leverage because its x value is much larger than the x values of other points. The point for Child 19 is an outlier because it falls outside the pattern of the other points and has a very large residual. How do these two points affect the regression? Figure 3.17 shows the results of removing each of these points on the equation of the least-squares regression line, the standard deviation of the residuals, and r^2.

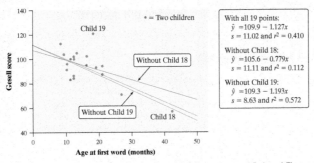

FIGURE 3.17 Three least-squares regression lines of Gesell score on age at first word. The green line is calculated from all the data. The dark blue line is calculated leaving out only Child 18. The red line is calculated leaving out only Child 19.

You can see that removing the point for Child 18 moves the line quite a bit. Because of Child 18's extreme position on the age (x) scale, removing this high-leverage point makes the slope closer to 0 and the y intercept smaller. Removing Child 18 also increases the standard deviation of the residuals because its small residual was making the typical distance from the regression line smaller. Finally, removing Child 18 also decreases r^2 (and makes the correlation closer to 0) because the linear association is weaker without this point.

Teaching Tip

Remind students that we are deleting individuals only to investigate their influence on the least-squares regression line and other summary measures. In practice, it would be wrong to delete inconvenient observations without a good justification.

Child 19's Gesell score was far above the least-squares regression line, but this child's age (17 months) is very close to $\bar{x} = 14.4$ months, making this point an outlier with low leverage. Thus, removing Child 19 has very little effect on the least-squares regression line. The line shifts down slightly from the original regression line, but not by much. Child 19 has a bigger influence on the standard deviation of the residuals: without Child 19's big residual, the size of the typical residual goes from $s = 11.02$ to $s = 8.63$. Likewise, without Child 19, the strength of the linear association increases and r^2 goes from 0.410 to 0.572.

Think About It

WHAT SHOULD WE DO WITH UNUSUAL POINTS? The strong influence of Child 18 makes the original regression of Gesell score on age at first word misleading. The original data have $r^2 = 0.41$. That is, the least-squares line with $x =$ age at which a child begins to talk accounts for 41% of the variability in Gesell score. This relationship is strong enough to be interesting to parents. If we leave out Child 18, r^2 drops to only 11%. The apparent strength of the association was largely due to a single influential observation.

What should the child development researcher do? She must decide whether Child 18 is so slow to speak that this individual should not be allowed to influence the analysis. If she excludes Child 18, much of the evidence for a connection between the age at which a child begins to talk and later ability score vanishes. If she keeps Child 18, she needs data on other children who were also slow to begin talking, so the analysis no longer depends as heavily on just one child.

Teaching Tip

Remind students that there is no rule for determining outliers or influential points in scatterplots. Students should be able to explain what effect any point has on the correlation, slope, *y* intercept, standard deviation of the residuals, and r^2.

ALTERNATE EXAMPLE   Skill 2.A

SAT math scores one more time
Outliers and high-leverage points

PROBLEM:
The scatterplot below shows the percent of students who are eligible for free/reduced lunch and the average SAT math score for 11 randomly selected high schools in Michigan in 2016, along with the least-squares regression line. The points highlighted in red are Northville High School (upper left) and East Kentwood High School (right middle).

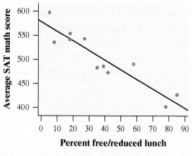

(a) Describe the influence the point representing Northville High School has on the equation of the least-squares regression line. Explain your reasoning.
(b) Describe the influence the point representing East Kentwood High School has on the standard deviation of the residuals and r^2. Explain your reasoning.

EXAMPLE

Dodging the pattern?
Outliers and high-leverage points

PROBLEM: The scatterplot shows the payroll (in millions of dollars) and number of wins for Major League Baseball teams in 2016, along with the least-squares regression line. The points highlighted in red represent the Los Angeles Dodgers (far right) and the Cleveland Indians (upper left).

(a) Describe what influence the point representing the Los Angeles Dodgers has on the equation of the least-squares regression line. Explain your reasoning.
(b) Describe what influence the point representing the Cleveland Indians has on the standard deviation of the residuals and r^2. Explain your reasoning.

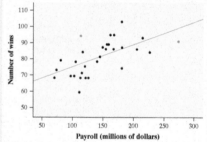

SOLUTION:
(a) Because the point for Northville High School is on the far left and above the least-squares regression line, it is making the slope of the line steeper (farther from 0) and the *y* intercept greater. If the point for Northville High School was removed, the line would be less steep.
(b) Because the point for East Kentwood High School has a large residual, it is making the standard deviation of the residuals greater and the value of r^2 smaller.

SOLUTION:

(a) Because the point for the Los Angeles Dodgers is on the right and below the least-squares regression line, it is making the slope of the line closer to 0 and the y intercept greater. If the Dodgers' point was removed, the line would be steeper.

> The point for the Dodgers has high leverage because its x value is much larger than the others.

(b) Because the point for the Cleveland Indians has a large residual, it is making the standard deviation of the residuals greater and the value of r^2 smaller.

> The point for the Indians is an outlier because it has a large residual.

FOR PRACTICE, TRY EXERCISE 67

ASSOCIATION DOES NOT IMPLY CAUSATION When we study the relationship between two variables, we often hope to show that changes in the explanatory variable *cause* changes in the response variable. A strong association between two variables is not enough to draw conclusions about cause and effect. Sometimes an observed association really does reflect cause and effect. A household that heats with natural gas uses more gas in colder months because cold weather requires burning more gas to stay warm. In other cases, an association is explained by other variables, and the conclusion that x causes y is not valid.

A study once found that people with two cars live longer than people who own only one car.[26] Owning three cars is even better, and so on. There is a substantial positive association between number of cars x and length of life y. Can we lengthen our lives by buying more cars? No. The study used number of cars as a quick indicator of wealth. Well-off people tend to have more cars. They also tend to live longer, probably because they are better educated, take better care of themselves, and get better medical care. The cars have nothing to do with it. There is no cause-and-effect link between number of cars and length of life.

> Remember: It only makes sense to talk about the *correlation* between two *quantitative* variables. If one or both variables are categorical, you should refer to the *association* between the two variables. To be safe, use the more general term *association* when describing the relationship between any two variables.

Section 3.2	Summary

- A **regression line** models how a response variable y changes as an explanatory variable x changes. You can use a regression line to **predict** the value of y for any value of x by substituting this x value into the equation of the line.

- The **slope** b of a regression line $\hat{y} = a + bx$ describes how the predicted value of y changes for each increase of 1 unit in x.

- The **y intercept** a of a regression line $\hat{y} = a + bx$ is the predicted value of y when the explanatory variable x equals 0. This prediction does not have a logical interpretation unless x can actually take values near 0.

- Avoid **extrapolation**, using a regression line to make predictions using values of the explanatory variable outside the values of the data from which the line was calculated.

- The most common method of fitting a line to a scatterplot is least squares. The **least-squares regression line** is the line that minimizes the sum of the squares of the vertical distances of the observed points from the line.

+ Ask the StatsMedic

Correlation Does Not Mean Causation

In this post, several examples are presented that provide good contexts for discussion with students.

Making Connections

In Chapter 4, we will see that a properly designed experiment (with two or more treatments and random assignment) is required to establish cause and effect. An observational study is only strong enough to identify an association between two variables.

TRM Section 3.2 Quiz

There is one quiz available for this section in the Teacher's Resource Materials. Click on the link in the TE-Book, look on the TRFD, or download from the Teacher's Resources on the book's digital platform. You can also create your own quiz using the ExamView® Assessment Suite that is part of the TPS6 program. Questions are coded by Learning Target to make it easy to build parallel quizzes.

- You can examine the fit of a regression line by studying the **residuals,** which are the differences between the actual values of y and predicted values of y: Residual $= y - \hat{y}$. Be on the lookout for curved patterns in the **residual plot,** which indicate that a linear model may not be appropriate.

- The **standard deviation of the residuals** s measures the typical size of a residual when using the regression line.

- The **coefficient of determination** r^2 is the percent of the variation in the response variable that is accounted for by the least-squares regression line using a particular explanatory variable.

- The least-squares regression line of y on x is the line with slope $b = r\frac{s_y}{s_x}$ and intercept $a = \bar{y} - b\bar{x}$. This line always passes through the point $(\bar{x}, \bar{y})$.

- **Influential points** can greatly affect correlation and regression calculations. Points with x values far from $\bar{x}$ have **high leverage** and can be very influential. Points with large residuals are called **outliers** and can also affect correlation and regression calculations.

- Most of all, be careful not to conclude that there is a cause-and-effect relationship between two variables just because they are strongly associated.

3.2 Technology Corners

TI-Nspire and other technology instructions are on the book's website at highschool.bfwpub.com/updatedtps6e.

9. Calculating least-squares regression lines   Page 184
10. Making residual plots   Page 187

Section 3.2   Exercises

37. **Predicting wins** Earlier we investigated the relationship between x = payroll (in millions of dollars) and y = number of wins for Major League Baseball teams in 2016. Here is a scatterplot of the data, along with the regression line $\hat{y} = 60.7 + 0.139x$:

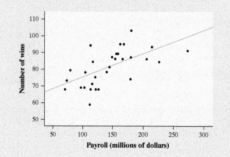

(a) Predict the number of wins for a team that spends $200 million on payroll.

(b) Predict the number of wins for a team that spends $400 million on payroll.

(c) How confident are you in each of these predictions? Explain your reasoning.

38. **How much gas?** Joan is concerned about the amount of energy she uses to heat her home. The scatterplot (on page 205) shows the relationship between x = mean temperature in a particular month and y = mean amount of natural gas used per day (in cubic feet) in that month, along with the regression line $\hat{y} = 1425 - 19.87x$.

(a) Predict the mean amount of natural gas Joan will use per day in a month with a mean temperature of 30°F.

Answers to Section 3.2 Exercises

3.37 (a) $\hat{y} = 60.7 + 0.139(200) = 88.5$ **(b)** $\hat{y} = 60.7 + 0.139(400) = 116.3$ **(c)** I am more confident in the prediction in part (a) than the prediction in part (b). The payroll for the teams varied from about $75 million to about $275 million. Although $200 million is in this interval of payrolls, $400 million is not, which makes the prediction in part (b) an extrapolation.

3.38 (a) $\hat{y} = 1425 - 19.87(30) = 828.9$ cubic feet **(b)** $\hat{y} = 1425 - 19.87(65) = 133.45$ cubic feet **(c)** I am more confident in the prediction in part (a) than the prediction in part (b). The average temperatures varied from about 27°F to 57°F. Although 30°F is in this interval of temperatures, 65°F is not, which makes the prediction in part (b) an extrapolation.

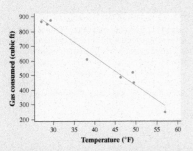

(b) Predict the mean amount of natural gas Joan will use per day in a month with a mean temperature of 65°F.

(c) How confident are you in each of these predictions? Explain your reasoning.

39. Residual wins Refer to Exercise 37. The Chicago
pg180 Cubs won the World Series in 2016. They had 103 wins and spent $182 million on payroll. Calculate and interpret the residual for the Cubs.

40. Residual gas Refer to Exercise 38. During March, the average temperature was 46.4°F and Joan used an average of 490 cubic feet of gas per day. Calculate and interpret the residual for this month.

41. More wins? Refer to Exercise 37.
pg182
(a) Interpret the slope of the regression line.

(b) Does the value of the y intercept have meaning in this context? If so, interpret the y intercept. If not, explain why.

42. Less gas? Refer to Exercise 38.

(a) Interpret the slope of the regression line.

(b) Does the value of the y intercept have meaning in this context? If so, interpret the y intercept. If not, explain why.

43. Long strides The scatterplot shows the relationship between x = height of a student (in inches) and y = number of steps required to walk the length of a school hallway, along with the regression line $\hat{y} = 113.6 - 0.921x$.

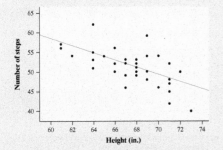

(a) Calculate and interpret the residual for Kiana, who is 67 inches tall and took 49 steps to walk the hallway.

(b) Matthew is 10 inches taller than Samantha. About how many fewer steps do you expect Matthew to take compared to Samantha?

44. Crickets chirping The scatterplot shows the relationship between x = temperature in degrees Fahrenheit and y = chirps per minute for the striped ground cricket, along with the regression line $\hat{y} = -0.31 + 0.212x$.[27]

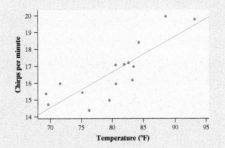

(a) Calculate and interpret the residual for the cricket who chirped 20 times per minute when the temperature was 88.6°F.

(b) About how many additional chirps per minute do you expect a cricket to make if the temperature increases by 10°F?

45. More Olympic athletes In Exercises 5 and 11, you described the relationship between height (in inches) and weight (in pounds) for Olympic track and field athletes. The scatterplot shows this relationship, along with two regression lines. The regression line for the shotput, hammer throw, and discus throw athletes (blue squares) is $\hat{y} = -115 + 5.13x$. The regression line for the remaining athletes (black dots) is $\hat{y} = -297 + 6.41x$.

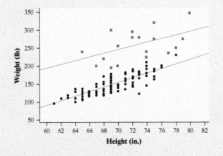

3.43 (a) The predicted number of steps for Kiana, who is 67 inches tall, is $\hat{y} = 113.6 - 0.921(67) = 51.893$ steps. The residual = actual y − predicted y = 49 − 51.893 = −2.893. *Interpretation:* Kiana took about 2.893 fewer steps than the number of steps predicted by the regression line with x = 67 inches. **(b)** Since Matthew is 10 inches taller than Samantha, I expect Matthew to take (10)(−0.921) = 9.21 fewer steps than Samantha.

3.44 (a) The predicted number of chirps per minute when the outside temperature is 88.6°F is $\hat{y} = -0.31 + 0.212(88.6) = 18.473$ chirps per minute. The residual = actual y − predicted y = 20 − 18.473 = 1.527. *Interpretation:* This cricket chirped about 1.527 more times per minute than the number of chirps predicted by the regression line with x = 88.6°F. **(b)** If the temperature increases by 10°F, I expect a cricket to make an additional (10)(0.212) = 2.12 chirps per minute.

3.39 The predicted number of wins for the Chicago Cubs, who spent $182 million on payroll, is $\hat{y} = 60.7 + 0.139(182) = 85.998$ wins. The residual = actual y − predicted y = 103 − 85.998 = 17.002. *Interpretation:* The Chicago Cubs won 17.002 more games than the number of games predicted by the regression line with x = $182 million.

3.40 The predicted amount of gas usage when the average temperature was 46.4°F is $\hat{y} = 1425 - 19.87(46.4) = 503.032$ cubic feet. The residual = actual y − predicted y = 490 − 503.032 = −13.032. *Interpretation:* The actual mean amount of gas consumed in the month of March was 13.032 cubic feet less than predicted by the regression line with x = 46.4°F.

3.41 (a) The slope is 0.139. *Interpretation:* The predicted number of wins goes up by 0.139 for each increase of $1 million in payroll. **(b)** The y intercept does not have meaning in this context. It is not reasonable for a team to have a payroll of $0.

3.42 (a) The slope is −19.87. *Interpretation:* The predicted mean amount of gas consumed in Joan's home decreases by 19.87 cubic feet for each additional 1 degree increase in the mean monthly temperature. **(b)** The y intercept is 1425 cubic feet. When the mean monthly temperature is 0°F, the predicted mean gas consumption for Joan's home is 1425 cubic feet. This prediction is an extrapolation because the data only included months with an average temperature of more than 20°F.

3.45 (a) The regression lines are nearly parallel, but the *y* intercept is much greater for the throwers. **(b)** Discus thrower: $\hat{y} = -115 + 5.13(72) = 254.36$ pounds. 72-inch sprinter: $\hat{y} = -297 + 6.41(72) = 164.52$ pounds. Based on the least-squares regression lines computed from the data, we expect a 72-inch discus thrower to weigh about 89.84 pounds more than a 72-inch sprinter.

3.46 (a) The least-squares regression line based on the drink products has a smaller *y* intercept and steeper slope than the least-squares regression line based on the food products. **(b)** Food item: $\hat{y} = 170 + 11.8(5) = 229$ calories. Drink item: $\hat{y} = 88 + 24.5(5) = 210.5$ calories. Based on the least-squares regression lines computed from the data, we expect a food item with 5 grams of fat to contain about 18.5 more calories than a drink item with 5 grams of fat.

3.47 No; there is an obvious negative-positive-negative pattern in the residual plot, so a linear model is not appropriate for these data. A curved model would be better.

3.48 No; there is an obvious positive-negative-positive pattern in the residual plot, so a linear model is not appropriate for these data. A curved model would be better.

3.49 The predicted mean weight of infants in Nahya who are 1 month old is $\hat{y} = 4.88 + 0.267(1) = 5.147$ kg. From the residual plot, the mean weight of 1-month-old infants is about 0.85 kg less than predicted. So the actual mean weight of the infants when they were 1 month old is about $5.147 - 0.85 = 4.297$ kg.

3.50 The predicted fuel consumption of the car driving 20 kilometers per hour is $\hat{y} = 11.058 - 0.01466(20) = 10.765$ liters. From the residual plot, the actual fuel consumption is about 2.4 liters more than predicted. So the actual fuel consumption of the car driving 20 kilometers per hour is about $10.765 + 2.4 = 13.165$ liters.

(a) How do the regression lines compare?

(b) How much more do you expect a 72-inch discus thrower to weigh than a 72-inch sprinter?

46. **More Starbucks** In Exercises 6 and 12, you described the relationship between fat (in grams) and the number of calories in products sold at Starbucks. The scatterplot shows this relationship, along with two regression lines. The regression line for the food products (blue squares) is $\hat{y} = 170 + 11.8x$. The regression line for the drink products (black dots) is $\hat{y} = 88 + 24.5x$.

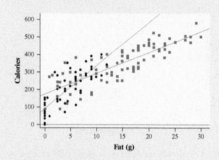

(a) How do the regression lines compare?

(b) How many more calories do you expect to find in a food item with 5 grams of fat compared to a drink item with 5 grams of fat?

47. **Infant weights in Nahya** A study of nutrition in developing countries collected data from the Egyptian village of Nahya. Researchers recorded the mean weight (in kilograms) for 170 infants in Nahya each month during their first year of life. A hasty user of statistics enters the data into software and computes the least-squares line without looking at the scatterplot first. The result is weight = $4.88 + 0.267$ (age). Use the residual plot to determine if this linear model is appropriate.

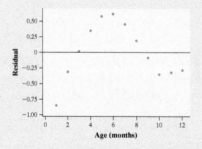

48. **Driving speed and fuel consumption** Exercise 9 (page 171) gives data on the fuel consumption *y* of a car at various speeds *x*. Fuel consumption is measured in liters of gasoline per 100 kilometers driven, and speed is measured in kilometers per hour. A statistical software package gives the least-squares regression line $\hat{y} = 11.058 - 0.01466x$. Use the residual plot to determine if this linear model is appropriate.

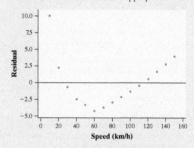

49. **Actual weight** Refer to Exercise 47. Use the equation of the least-squares regression line and the residual plot to estimate the *actual* mean weight of the infants when they were 1 month old.

50. **Actual consumption** Refer to Exercise 48. Use the equation of the least-squares regression line and the residual plot to estimate the *actual* fuel consumption of the car when driving 20 kilometers per hour.

51. **Movie candy** Is there a relationship between the amount of sugar (in grams) and the number of calories in movie-theater candy? Here are the data from a sample of 12 types of candy:

Name	Sugar (g)	Calories	Name	Sugar (g)	Calories
Butterfinger Minis	45	450	Reese's Pieces	61	580
Junior Mints	107	570	Skittles	87	450
M&M'S®	62	480	Sour Patch Kids	92	490
Milk Duds	44	370	SweeTarts	136	680
Peanut M&M'S®	79	790	Twizzlers	59	460
Raisinets	60	420	Whoppers	48	350

(a) Sketch a scatterplot of the data using sugar as the explanatory variable.

(b) Use technology to calculate the equation of the least-squares regression line for predicting the number of calories based on the amount of sugar. Add the line to the scatterplot from part (a).

(c) Explain why the line calculated in part (b) is called the "least-squares" regression line.

3.51 (a) See part (b). **(b)** $\hat{y} = 300.04 + 2.829x$, where $\hat{y} = $ the predicted number of calories and $x = $ the amount of sugar (grams).

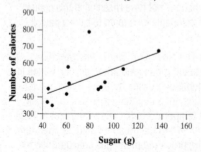

(c) The line calculated in part (b) is called the *least-squares* regression line because this line makes the sum of the squares of the residuals as small as possible.

52. Long jumps Here are the 40-yard-dash times (in seconds) and long-jump distances (in inches) for a small class of 12 students:

Dash time (sec)	5.41	5.05	7.01	7.17	6.73	5.68
Long-jump distance (in.)	171	184	90	65	78	130
Dash time (sec)	5.78	6.31	6.44	6.50	6.80	7.25
Long-jump distance (in.)	173	143	92	139	120	110

(a) Sketch a scatterplot of the data using dash time as the explanatory variable.

(b) Use technology to calculate the equation of the least-squares regression line for predicting the long-jump distance based on the dash time. Add the line to the scatterplot from part (a).

(c) Explain why the line calculated in part (b) is called the "least-squares" regression line.

53. More candy Refer to Exercise 51. Use technology to create a residual plot. Sketch the residual plot and explain what information it provides.

54. More long jumps Refer to Exercise 52. Use technology to create a residual plot. Sketch the residual plot and explain what information it provides.

55. Longer strides In Exercise 43, we modeled the
pg 191 relationship between x = height of a student (in inches) and y = number of steps required to walk the length of a school hallway, with the regression line $\hat{y} = 113.6 - 0.921x$. For this model, technology gives $s = 3.50$ and $r^2 = 0.399$.

(a) Interpret the value of s.

(b) Interpret the value of r^2.

56. Crickets keep chirping In Exercise 44, we modeled the relationship between x = temperature in degrees Fahrenheit and y = chirps per minute for the striped ground cricket, with the regression line $\hat{y} = -0.31 + 0.212x$. For this model, technology gives $s = 0.97$ and $r^2 = 0.697$.

(a) Interpret the value of s.

(b) Interpret the value of r^2.

57. Olympic figure skating For many people, the women's figure skating competition is the highlight of the Olympic Winter Games. Scores in the short program x and scores in the free skate y were recorded for each of the 24 skaters who competed in both rounds during the 2010 Winter Olympics in Vancouver, Canada.[28] Here is a scatterplot with least-squares regression line $\hat{y} = -16.2 + 2.07x$. For this model, $s = 10.2$ and $r^2 = 0.736$.

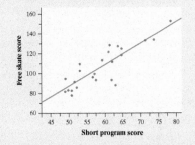

(a) Calculate and interpret the residual for the 2010 gold medal winner Yu-Na Kim, who scored 78.50 in the short program and 150.06 in the free skate.

(b) Interpret the slope of the least-squares regression line.

(c) Interpret the standard deviation of the residuals.

(d) Interpret the coefficient of determination.

58. Age and height A random sample of 195 students was selected from the United Kingdom using the Census At School data selector. The age x (in years) and height y (in centimeters) were recorded for each student. Here is a scatterplot with the least-squares regression line $\hat{y} = 106.1 + 4.21x$. For this model, $s = 8.61$ and $r^2 = 0.274$.

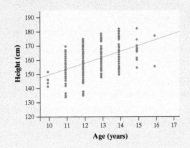

(a) Calculate and interpret the residual for the student who was 141 cm tall at age 10.

(b) Interpret the slope of the least-squares regression line.

(c) Interpret the standard deviation of the residuals.

(d) Interpret the coefficient of determination.

3.52 (a) See part (b). **(b)** $\hat{y} = 414.79 - 45.743x$, where $\hat{y}$ = the predicted long-jump distance (inches) and x = dash time (seconds).

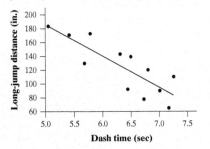

(c) The line calculated in part (b) is called the *least-squares* regression line because this line makes the sum of the squares of the residuals as small as possible.

3.53

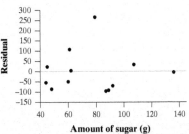

The linear model relating the amount of sugar to the number of calories is appropriate because there is no leftover pattern in the residual plot. The residuals look randomly scattered around the residual = 0 line.

3.54

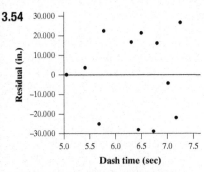

The linear model relating the dash time to the long-jump distance is appropriate because there is no leftover pattern in the residual plot. The residuals look randomly scattered around the residual = 0 line.

3.55 (a) The actual number of strides required to walk the length of a school hallway is typically about 3.50 away from the number predicted by the least-squares regression line with x = height of a student (in inches). **(b)** About 39.9% of the variability in number of steps required to walk the length of a school hallway is accounted for by the least-squares regression line with x = height of a student (in inches).

3.56 (a) The actual number of chirps per minute is typically about 0.97 away from the number predicted by the least-squares regression line with x = temperature in degrees Fahrenheit. **(b)** About 69.7% of the variability in number of chirps is accounted for by the least-squares regression line with x = temperature in degrees Fahrenheit.

3.57 (a) The predicted free skate score is $\hat{y} = -16.2 + 2.07(78.5) = 146.295$. The residual is $y - \hat{y} = 150.06 - 146.295 = 3.765$. *Interpretation:* Yu-Na Kim's free skate score was 3.765 points higher than predicted based on her short program score. **(b)** The slope is 2.07. *Interpretation:* The predicted free skate score increases by 2.07 points for each additional 1-point increase in the short program score. **(c)** The actual free skate score is typically about 10.2 points away from the score predicted by the least-squares regression line with x = short program score. **(d)** About 73.6% of the variability in the free skate score is accounted for by the least-squares regression line with x = short program score.

3.58 (a) The predicted height is $\hat{y} = 106.1 + 4.21(10) = 148.2$. The residual is $y - \hat{y} = 141 - 148.2 = -7.2$. This student's height was 7.2 cm less than predicted based on the student's age. **(b)** The slope is 4.21. *Interpretation:* The predicted height increases by 4.21 centimeters for each additional 1-year increase in the student's age. **(c)** The actual height is typically about 8.61 centimeters away from the height predicted by the least-squares regression line with x = age (years). **(d)** About 27.4% of the variability in height is accounted for by the least-squares regression line with x = age (years).

3.59 (a) Yes; there is no leftover pattern in the residual plot, so a linear model is appropriate for these data. **(b)** $r^2 =$ 60.21% and the slope is positive, so the correlation is $r = \sqrt{0.6021} = 0.776$. **(c)** $\hat{y} = 1.0021 + 0.0708x$, where x is the number of Mentos and $\hat{y}$ is the predicted amount expelled. **(d)** The value of $s = 0.067$ ml. *Interpretation:* The actual amount expelled is typically about 0.067 ml away from the amount predicted by the least-squares regression line with $x =$ number of Mentos. The value of r^2 is 60.21%. *Interpretation:* About 60.21% of the variability in the amount expelled is accounted for by the least-squares regression line with $x =$ number of Mentos.

3.60 (a) Yes; there is no leftover pattern in the residual plot, so a linear model is appropriate for these data. **(b)** $r^2 = 85.38\%$ and the slope is negative, so the correlation is $r = -\sqrt{0.8538} = -0.924$. **(c)** $\hat{y} = 106.36 - 2.635x$, where x is the tapping time (in seconds) and $\hat{y}$ is the predicted amount expelled. **(d)** The value of $s = 5.0$ ml. *Interpretation:* The actual amount expelled is typically about 5 ml away from the amount predicted by the least-squares regression line with $x =$ tapping time (in seconds). The value of r^2 is 85.38%. *Interpretation:* About 85.38% of the variability in the amount expelled is accounted for by the least-squares regression line with $x =$ tapping time (in seconds).

59. **More mess?** When Mentos are dropped into a newly opened bottle of Diet Coke, carbon dioxide is released from the Diet Coke very rapidly, causing the Diet Coke to be expelled from the bottle. To see if using more Mentos causes more Diet Coke to be expelled, Brittany and Allie used twenty-four 2-cup bottles of Diet Coke and randomly assigned each bottle to receive either 2, 3, 4, or 5 Mentos. After waiting for the fizzing to stop, they measured the amount expelled (in cups) by subtracting the amount remaining from the original amount in the bottle.[29] Here is computer output from a linear regression of $y =$ amount expelled on $x =$ number of Mentos:

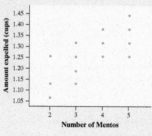

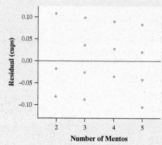

Term	Coef	SE Coef	T-Value	P-Value
Constant	1.0021	0.0451	22.21	0.000
Mentos	0.0708	0.0123	5.77	0.000

S = 0.06724   R-Sq = 60.21% R-Sq(adj) = 58.40%

(a) Is a line an appropriate model to use for these data? Explain how you know.

(b) Find the correlation.

(c) What is the equation of the least-squares regression line? Define any variables that you use.

(d) Interpret the values of s and r^2.

60. **Less mess?** Kerry and Danielle wanted to investigate whether tapping on a can of soda would reduce the amount of soda expelled after the can has been shaken. For their experiment, they vigorously shook 40 cans of soda and randomly assigned each can to be tapped for 0 seconds, 4 seconds, 8 seconds, or 12 seconds. After waiting for the fizzing to stop, they measured the amount expelled (in milliliters) by subtracting the amount remaining from the original amount in the can.[30] Here is computer output from a linear regression of $y =$ amount expelled on $x =$ tapping time:

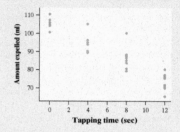

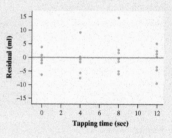

Term	Coef	SE Coef	T-Value	P-Value
Constant	106.360	1.320	80.34	0.000
Tapping_time	-2.635	0.177	-14.90	0.000

S = 5.00347   R-Sq = 85.38% R-Sq(adj) = 84.99%

(a) Is a line an appropriate model to use for these data? Explain how you know.

(b) Find the correlation.

(c) What is the equation of the least-squares regression line? Define any variables that you use.

(d) Interpret the values of s and r^2.

61. **Temperature and wind** The average temperature (in degrees Fahrenheit) and average wind speed (in miles per hour) were recorded for 365 consecutive days at Chicago's O'Hare International Airport. Here is computer output for a regression of y = average wind speed on x = average temperature:

```
Summary of Fit
RSquare                        0.047874
RSquare Adj                    0.045251
Root Mean Square Error         3.655950
Mean of Response               9.826027
Observations (or Sum Wgts)     365

Parameter Estimates
Term        Estimate  Std Error  t Ratio  Prob>|t|
Intercept   11.897762 0.521320   22.82    <.0001*
Avg temp    -0.041077 0.009615   -4.27    <.0001*
```

(a) Calculate and interpret the residual for the day where the average temperature was 42°F and the average wind speed was 2.2 mph.

(b) Interpret the slope.

(c) By about how much do the actual average wind speeds typically vary from the values predicted by the least-squares regression line with x = average temperature?

(d) What percent of the variability in average wind speed is accounted for by the least-squares regression line with x = average temperature?

62. **Beetles and beavers** Do beavers benefit beetles? Researchers laid out 23 circular plots, each 4 meters in diameter, in an area where beavers were cutting down cottonwood trees. In each plot, they counted the number of stumps from trees cut by beavers and the number of clusters of beetle larvae. Ecologists believe that the new sprouts from stumps are more tender than other cottonwood growth, so beetles prefer them. If so, more stumps should produce more beetle larvae.[31] Here is computer output for a regression of y = number of beetle larvae on x = number of stumps:

```
Summary of Fit
RSquare                        0.839144
RSquare Adj                    0.831484
Root Mean Square Error         6.419386
Mean of Response               25.086960
Observations (or Sum Wgts)     23

Parameter Estimates
Term       Estimate   Std Error  t Ratio  Prob>|t|
Intercept  -1.286104  2.853182   -0.45    0.6568
Number of
stumps     11.893733  1.136343   10.47    <.0001*
```

(a) Calculate and interpret the residual for the plot that had 2 stumps and 30 beetle larvae.

(b) Interpret the slope.

(c) By about how much do the actual number of larvae typically vary from the values predicted by the least-squares regression line with x = number of stumps?

(d) What percent of the variability in number of larvae is accounted for by the least-squares regression line with x = number of stumps?

63. **Husbands and wives** The mean height of married American women in their early 20s is 64.5 inches and the standard deviation is 2.5 inches. The mean height of married men the same age is 68.5 inches with standard deviation 2.7 inches. The correlation between the heights of husbands and wives is about $r = 0.5$.

(a) Find the equation of the least-squares regression line for predicting a husband's height from his wife's height for married couples in their early 20s.

(b) Suppose that the height of a randomly selected wife was 1 standard deviation below average. Predict the height of her husband *without* using the least-squares line.

64. **The stock market** Some people think that the behavior of the stock market in January predicts its behavior for the rest of the year. Take the explanatory variable x to be the percent change in a stock market index in January and the response variable y to be the change in the index for the entire year. We expect a positive correlation between x and y because the change during January contributes to the full year's change. Calculation from data for an 18-year period gives

$$\bar{x} = 1.75\% \quad s_x = 5.36\% \quad \bar{y} = 9.07\%$$
$$s_y = 15.35\% \quad r = 0.596$$

(a) Find the equation of the least-squares line for predicting full-year change from January change.

(b) Suppose that the percent change in a particular January was 2 standard deviations above average. Predict the percent change for the entire year *without* using the least-squares line.

65. **Will I bomb the final?** We expect that students who do well on the midterm exam in a course will usually also do well on the final exam. Gary Smith of Pomona College looked at the exam scores of all 346 students who took his statistics class over a 10-year period.[32] Assume that both the midterm and final exam were scored out of 100 points.

(a) State the equation of the least-squares regression line if each student scored the same on the midterm and the final.

3.61 (a) $\hat{y} = 11.898 - 0.041(42) = 10.176$ mph. The residual = actual y − predicted $y = 2.2 - 10.176 = -7.976$ mph. *Interpretation:* The actual average wind speed was 7.976 mph less than the average wind speed predicted by the regression line with $x = 42°F$. **(b)** The slope is -0.041. *Interpretation:* The predicted average wind speed decreases by 0.041 mph for each additional 1-degree increase in average temperature (in degrees Fahrenheit). **(c)** The actual average wind speeds typically vary by $s = 3.66$ mph from the values predicted by the least-squares regression line using x = average temperature. **(d)** $r^2 = 4.8\%$ of the variability in average wind speed is accounted for by the least-squares regression line using x = average temperature.

3.62 (a) $\hat{y} = -1.286 + 11.894(2) = 22.502$ clusters of beetle larvae. The residual = actual y − predicted $y = 30 - 22.502 = 7.498$ clusters of beetle larvae. *Interpretation:* The actual number of clusters of beetle larvae was 7.498 more than the number of clusters of beetle larvae predicted by the regression line with x = 2 stumps. **(b)** The slope is 11.894. *Interpretation:* The predicted number of clusters of beetle larvae increases by 11.894 for each additional 1 stump on the plot of land. **(c)** The actual number of clusters of beetle larvae typically vary by $s = 6.419$ clusters from the values predicted by the least-squares regression line using x = number of stumps. **(d)** $r^2 = 83.9\%$ of the variability in number of clusters of beetle larvae is accounted for by the least-squares regression line using x = number of stumps.

3.63 (a) The slope is
$$b_1 = 0.5\left(\frac{2.7}{2.5}\right) = 0.54.$$
The y intercept is $b_0 = 68.5 - 0.54(64.5) = 33.67$. So the equation for predicting y = husband's height from x = wife's height is $\hat{y} = 33.67 + 0.54x$. **(b)** If the value of x is 1 standard deviation below $\bar{x}$, the predicted value of y will be r standard deviations of y below $\bar{y}$. So the predicted value for the husband is $68.5 - 0.5(2.7) = 67.15$ inches.

3.64 (a) The slope is
$$b_1 = 0.596\left(\frac{15.35}{5.36}\right) = 1.707.$$
The y intercept is $b_0 = 9.07 - 1.707(1.75) = 6.083$. So the equation for predicting y = percent change in the index for the entire year from x = percent change in the index in January is $\hat{y} = 6.083 + 1.707x$. **(b)** If the value of x is 2 standard deviations above $\bar{x}$, the predicted value of y will be $2r$ standard deviations of y above $\bar{y}$. So the predicted value for the percent change for the entire year is $9.07 + 2(0.596)(15.35) = 27.4\%$.

3.65 (a) $\hat{y} = x$, where $\hat{y}$ = predicted grade on final and x = grade on midterm. **(b)** A student with a score of 50 on the midterm is predicted to score $\hat{y} = 46.6 + 0.41(50) = 67.1$ on the final. A student with a score of 100 on the midterm is predicted to score $\hat{y} = 46.6 + 0.41(100) = 87.6$ on the final. **(c)** These predictions illustrate regression to the mean because the student who did poorly on the midterm (50) is predicted to do better on the final (closer to the mean), whereas the student who did very well on the midterm (100) is predicted to do worse on the final (closer to the mean).

3.66 (a) $\hat{y} = x$, where $\hat{y}$ = predicted rest-of-season batting average and x = first-month batting average. **(b)** A player with a first-month batting average of 0.200 is predicted to have a rest-of-season batting average of $\hat{y} = 0.245 + 0.109(0.200) = 0.267$. A player with a first-month batting average of 0.400 is predicted to have a rest-of-season batting average of $\hat{y} = 0.245 + 0.109(0.400) = 0.289$. **(c)** These predictions illustrate regression to the mean because the player who hit poorly in the first month (0.200) is predicted to hit better the rest of the season (closer to the mean), whereas the player who hit very well in the first month (0.400) is predicted to do worse (closer to the mean).

3.67 (a) Because Jacob has an above-average height and an above-average vertical jump, his point increases the positive slope of the least-squares regression line and decreases the y intercept. **(b)** Jacob's vertical jump is farther from the least-squares regression line than the other students' vertical jumps. Because Jacob's point has such a large residual, it increases the standard deviation of the residuals. Also, because Jacob's vertical jump is farther from the least-squares regression line than the other students' vertical jumps, the value of r^2 decreases. The linear association is weaker because of the presence of this point.

3.68 (a) Because the mixer from Walmart was lighter and cheaper than the other mixers, this point may slightly increase the positive slope of the least-squares regression line and slightly decrease the y intercept. However, it probably doesn't have much effect on the line. **(b)** Because the mixer from Walmart is a little closer to the least-squares regression line than the rest of the points in the data set, this mixer will slightly decrease the standard deviation of the residuals. The value of r^2 may increase

(b) The actual least-squares line for predicting final-exam score y from midterm-exam score x was $\hat{y} = 46.6 + 0.41x$. Predict the score of a student who scored 50 on the midterm and a student who scored 100 on the midterm.

(c) Explain how your answers to part (b) illustrate regression to the mean.

66. It's still early We expect that a baseball player who has a high batting average in the first month of the season will also have a high batting average the rest of the season. Using 66 Major League Baseball players from a recent season,[33] a least-squares regression line was calculated to predict rest-of-season batting average y from first-month batting average x. Note: A player's batting average is the proportion of times at bat that he gets a hit. A batting average over 0.300 is considered very good in Major League Baseball.

(a) State the equation of the least-squares regression line if each player had the same batting average the rest of the season as he did in the first month of the season.

(b) The actual equation of the least-squares regression line is $\hat{y} = 0.245 + 0.109x$. Predict the rest-of-season batting average for a player who had a 0.200 batting average the first month of the season and for a player who had a 0.400 batting average the first month of the season.

(c) Explain how your answers to part (b) illustrate regression to the mean.

67. Who's got hops? Haley, Jeff, and Nathan measured the height (in inches) and vertical jump (in inches) of 74 students at their school.[34] Here is a scatterplot of the data, along with the least-squares regression line. Jacob (highlighted in red) had a vertical jump of nearly 3 feet!

pg 202

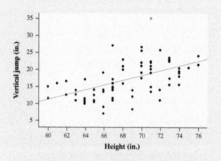

(a) Describe the influence that Jacob's point has on the equation of the least-squares regression line.

(b) Describe the influence that Jacob's point has on the standard deviation of the residuals and r^2.

68. Stand mixers The scatterplot shows the weight (in pounds) and cost (in dollars) of 11 stand mixers.[35] The mixer from Walmart (highlighted in red) was much lighter—and cheaper—than the other mixers.

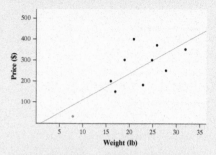

(a) Describe what influence the highlighted point has on the equation of the least-squares regression line.

(b) Describe what influence the highlighted point has on the standard deviation of the residuals and r^2.

69. Managing diabetes People with diabetes measure their fasting plasma glucose (FPG, measured in milligrams per milliliter) after fasting for at least 8 hours. Another measurement, made at regular medical checkups, is called HbA. This is roughly the percent of red blood cells that have a glucose molecule attached. It measures average exposure to glucose over a period of several months. The table gives data on both HbA and FPG for 18 diabetics five months after they had completed a diabetes education class.[36]

Subject	HbA (%)	FPG (mg/ml)	Subject	HbA (%)	FPG (mg/ml)
1	6.1	141	10	8.7	172
2	6.3	158	11	9.4	200
3	6.4	112	12	10.4	271
4	6.8	153	13	10.6	103
5	7.0	134	14	10.7	172
6	7.1	95	15	10.7	359
7	7.5	96	16	11.2	145
8	7.7	78	17	13.7	147
9	7.9	148	18	19.3	255

(a) Make a scatterplot with HbA as the explanatory variable. Describe what you see.

(b) Subject 18 has an unusually large x value. What effect do you think this subject has on the correlation? What effect do you think this subject has on the equation of the least-squares regression line? Calculate the correlation and equation of the least-squares regression line with and without this subject to confirm your answer.

slightly because the linear association is stronger due to the presence of this point.

3.69 (a)

400	
350	•
300	
FBG mg/ml 250	•
200	•
150	• • • •
100	• • • •
50	
	5.0 7.5 10.0 12.5 15.0 17.5 20.0
	HbA %

There is a moderate positive, linear association between HbA and FBG. Subject 15 (near the top) is a possible outlier; Subject 18 (far right) is a high-leverage point. **(b)** Subject 18 is a high-leverage point. Because the point for subject 18 is in the positive, linear pattern, it will make the correlation

closer to 1; the point is likely to be below the least-squares regression line and will "pull down" the line on the right side, making the slope closer to 0. Without this point, the correlation decreases from $r = 0.4819$ to $r = 0.3837$ as expected. Likewise, without this point, the equation of the line changes from $\hat{y} = 66.4 + 10.4x$ to $\hat{y} = 52.3 + 12.1x$. **(c)** Subject 15 is an outlier because it has a large residual. The point for subject 15 makes the correlation closer to 0 because it decreases the strength of what would otherwise be a moderately strong positive association. Because this point's x coordinate is very close to $\bar{x}$, it won't influence the slope very much. However, it will make the y intercept increase because its y coordinate is so large compared to the rest of the values. Without the outlier, the correlation increases from $r = 0.4819$ to $r = 0.5684$, the slope changes from 10.4 to 8.92, and the y intercept increases from 66.4 to 69.5.

(c) Subject 15 has an unusually large y value. What effect do you think this subject has on the correlation? What effect do you think this subject has on the equation of the least-squares regression line? Calculate the correlation and equation of the least-squares regression line with and without this subject to confirm your answer.

70. **Rushing for points** What is the relationship between rushing yards and points scored in the National Football League? The table gives the number of rushing yards and the number of points scored for each of the 16 games played by the Jacksonville Jaguars in a recent season.[37]

Game	Rushing yards	Points scored	Game	Rushing yards	Points scored
1	163	16	9	141	17
2	112	3	10	108	10
3	128	10	11	105	13
4	104	10	12	129	14
5	96	20	13	116	41
6	133	13	14	116	14
7	132	12	15	113	17
8	84	14	16	190	19

(a) Make a scatterplot with rushing yards as the explanatory variable. Describe what you see.

(b) Game 16 has an unusually large x value. What effect do you think this game has on the correlation? On the equation of the least-squares regression line? Calculate the correlation and equation of the least-squares regression line with and without this game to confirm your answers.

(c) Game 13 has an unusually large y value. What effect do you think this game has on the correlation? On the equation of the least-squares regression line? Calculate the correlation and equation of the least-squares regression line with and without this game to confirm your answers.

Multiple Choice: *Select the best answer for Exercises 71–78.*

71. Which of the following is *not* a characteristic of the least-squares regression line?

(a) The slope of the least-squares regression line is always between −1 and 1.

(b) The least-squares regression line always goes through the point $(\bar{x}, \bar{y})$.

(c) The least-squares regression line minimizes the sum of squared residuals.

(d) The slope of the least-squares regression line will always have the same sign as the correlation.

(e) The least-squares regression line is not resistant to outliers.

72. Each year, students in an elementary school take a standardized math test at the end of the school year. For a class of fourth-graders, the average score was 55.1 with a standard deviation of 12.3. In the third grade, these same students had an average score of 61.7 with a standard deviation of 14.0. The correlation between the two sets of scores is $r = 0.95$. Calculate the equation of the least-squares regression line for predicting a fourth-grade score from a third-grade score.

(a) $\hat{y} = 3.58 + 0.835x$

(b) $\hat{y} = 15.69 + 0.835x$

(c) $\hat{y} = 2.19 + 1.08x$

(d) $\hat{y} = -11.54 + 1.08x$

(e) Cannot be calculated without the data.

73. Using data from the LPGA tour, a regression analysis was performed using x = average driving distance and y = scoring average. Using the output from the regression analysis shown below, determine the equation of the least-squares regression line.

```
Predictor           Coef      SE Coef      T        P
Constant         87.974000   2.391000   36.78    0.000
Driving Distance -0.060934   0.009536   -6.39    0.000

S = 1.01216    R-Sq = 22.1%    R-Sq(adj) = 21.6%
```

(a) $\hat{y} = 87.974 + 2.391x$

(b) $\hat{y} = 87.974 + 1.01216x$

(c) $\hat{y} = 87.974 - 0.060934x$

(d) $\hat{y} = -0.060934 + 1.01216x$

(e) $\hat{y} = -0.060934 + 87.947x$

Exercises 74 to 78 refer to the following setting. Measurements on young children in Mumbai, India, found this least-squares line for predicting y = height (in cm) from x = arm span (in cm):[38]

$$\hat{y} = 6.4 + 0.93x$$

74. By looking at the equation of the least-squares regression line, you can see that the correlation between height and arm span is

(a) greater than zero.

(b) less than zero.

(c) 0.93.

(d) 6.4.

(e) Can't tell without seeing the data.

to 0 because it decreases the strength of what would otherwise be a moderately strong positive association. Because this point's x coordinate is very close to $\bar{x}$, it won't influence the slope very much. However, it will make the y intercept increase because its y coordinate is so large compared to the rest of the values. Without the outlier, the correlation increases from $r = 0.10$ to $r = 0.32$, the slope changes from 0.031 to 0.050, and the y intercept drops from 11.4 to 7.2.

3.71 a

3.72 a

3.73 c

3.74 a

3.70 (a) There appears to be a very weak positive association between points scored and rushing yards. With the exception of the outlier at 116 yards and 41 points, the association looks fairly linear.

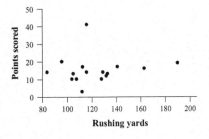

(b) Game 16 is a high-leverage point. Because this point is in the positive, linear pattern formed by most of the data values, it will make the correlation closer to 1. Also, because the point is likely to be above the least-squares regression line, it will "pull up" the line on the right side, making the slope a little steeper. The correlation with the point is $r = 0.10$; without this point, the correlation drops to $r = 0.02$. Likewise, with the point, the equation of the least-squares regression line is $\hat{y} = 11.4 + 0.031x$. Without the point, the equation changes to $\hat{y} = 14 + 0.008x$. **(c)** Game 13 is an outlier because it has a large residual. This point brings the correlation closer

3.75 d

3.76 a

3.77 b

3.78 e

3.79 (a) $z = \dfrac{25 - 18.7}{4.3} = 1.47$. *Table A:*
The proportion of z-scores below 1.47 is 0.9292. *Tech:* normalcdf(lower: -1000, upper: 25, mean: 18.7, SD: 4.3) = 0.9286. About 93% of vehicles get worse combined mileage than the Chevrolet Malibu. **(b)** *Table A:* Look in the body of Table A for the value closest to 0.90. A z-score of 1.28 gives the closest value (0.8997). Solving $1.28 = \dfrac{x - 18.7}{4.3}$ gives $x = 24.2$. *Tech:* invNorm(area: 0.9, mean: 18.7, SD: 4.3) = 24.2. The top 10% of all vehicles get at least 24.2 mpg.

3.80 (a) There is evidence of an association between accident rate and marijuana use. Those people who use marijuana more often are more likely to have caused accidents.

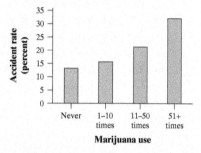

(b) Even if there is a strong association between two variables, we should not conclude that changes in one variable necessarily cause changes in the other variable. It could be that drivers who use marijuana more often are more willing to take risks than other drivers and that the willingness to take risks is what is causing the higher accident rate.

---

75. In addition to the regression line, the report on the Mumbai measurements says that $r^2 = 0.95$. This suggests that

(a) although arm span and height are correlated, arm span does not predict height very accurately.

(b) height increases by $\sqrt{0.95} = 0.97$ cm for each additional centimeter of arm span.

(c) 95% of the relationship between height and arm span is accounted for by the regression line.

(d) 95% of the variation in height is accounted for by the regression line with $x =$ arm span.

(e) 95% of the height measurements are accounted for by the regression line with $x =$ arm span.

76. One child in the Mumbai study had height 59 cm and arm span 60 cm. This child's residual is

(a) -3.2 cm.   (b) -2.2 cm.   (c) -1.3 cm.

(d) 3.2 cm.   (e) 62.2 cm.

77. Suppose that a tall child with arm span 120 cm and height 118 cm was added to the sample used in this study. What effect will this addition have on the correlation and the slope of the least-squares regression line?

(a) Correlation will increase, slope will increase.

(b) Correlation will increase, slope will stay the same.

(c) Correlation will increase, slope will decrease.

(d) Correlation will stay the same, slope will stay the same.

(e) Correlation will stay the same, slope will increase.

78. Suppose that the measurements of arm span and height were converted from centimeters to meters by dividing each measurement by 100. How will this conversion affect the values of r^2 and s?

(a) r^2 will increase, s will increase.

(b) r^2 will increase, s will stay the same.

(c) r^2 will increase, s will decrease.

(d) r^2 will stay the same, s will stay the same.

(e) r^2 will stay the same, s will decrease.

Recycle and Review

79. **Fuel economy** (2.2) In its recent *Fuel Economy Guide*, the Environmental Protection Agency (EPA) gives data on 1152 vehicles. There are a number of outliers, mainly vehicles with very poor gas mileage or hybrids with very good gas mileage. If we ignore the outliers, however, the combined city and highway gas mileage of the other 1120 or so vehicles is approximately Normal with mean 18.7 miles per gallon (mpg) and standard deviation 4.3 mpg.

(a) The Chevrolet Malibu with a four-cylinder engine has a combined gas mileage of 25 mpg. What percent of the 1120 vehicles have worse gas mileage than the Malibu?

(b) How high must a vehicle's gas mileage be in order to fall in the top 10% of the 1120 vehicles?

80. **Marijuana and traffic accidents** (1.1) Researchers in New Zealand interviewed 907 drivers at age 21. They had data on traffic accidents and they asked the drivers about marijuana use. Here are data on the numbers of accidents caused by these drivers at age 19, broken down by marijuana use at the same age:[39]

	Marijuana use per year			
	Never	1–10 times	11–50 times	51+ times
Number of drivers	452	229	70	156
Accidents caused	59	36	15	50

(a) Make a graph that displays the accident rate for each category of marijuana use. Is there evidence of an association between marijuana use and traffic accidents? Justify your answer.

(b) Explain why we can't conclude that marijuana use *causes* accidents based on this study.

SECTION 3.3 Transforming to Achieve Linearity

LEARNING TARGETS *By the end of the section, you should be able to:*

- Use transformations involving powers, roots, or logarithms to create a linear model that describes the relationship between two quantitative variables, and use the model to make predictions.

- Determine which of several models does a better job of describing the relationship between two quantitative variables.

In Section 3.2, we learned how to analyze relationships between two quantitative variables that showed a linear pattern. When bivariate data show a curved relationship, we must develop new techniques for finding an appropriate model. This section describes several simple *transformations* of data that can straighten a nonlinear pattern. Once the data have been transformed to achieve linearity, we can use least-squares regression to generate a useful model for making predictions.

The Gapminder website (www.gapminder.org) provides loads of data on the health and well-being of the world's inhabitants. Figure 3.18 shows a scatterplot of data from Gapminder.[40] The individuals are all the world's nations for which data were available in 2015. The explanatory variable, income per person, is a measure of how rich a country is. The response variable is life expectancy at birth.

We expect people in richer countries to live longer because they have better access to medical care and typically lead healthier lives. The overall pattern of the scatterplot does show this, but the relationship is not linear. Life expectancy rises very quickly as income per person increases and then levels off. People in

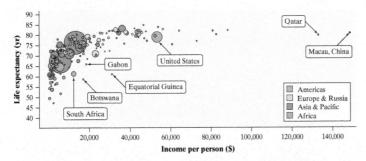

FIGURE 3.18 Scatterplot of the life expectancy of people in many nations against each nation's income per person. The color of each circle indicates the geographic region in which that country is located. The size of each circle is based on the population of the country—bigger circles indicate larger populations.

very rich countries such as the United States live no longer than people in poorer but not extremely poor nations. In some less wealthy countries, people live longer than in the United States.

Four African nations are outliers. Their life expectancies are much smaller than would be expected based on their income per person. Gabon and Equatorial Guinea produce oil, and South Africa and Botswana produce diamonds. It may be that income from mineral exports goes mainly to a few people and so pulls up income per person without much effect on either the income or the life expectancy of ordinary citizens. That is, income per person is a mean, and we know that mean income can be much higher than median income.

The scatterplot in Figure 3.18 shows a curved pattern. We can straighten things out using logarithms. Figure 3.19 plots the logarithm of income per person against life expectancy for these same countries. The effect is remarkable. This graph has a clear, linear form.

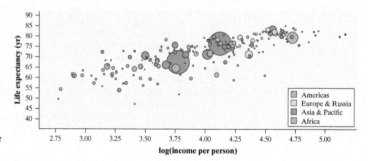

FIGURE 3.19 Scatterplot of life expectancy against log(income per person) for many nations.

Applying a function such as the logarithm or square root to a quantitative variable is called *transforming the data*. Transforming data amounts to changing the scale of measurement that was used when the data were collected. In Chapter 2, we discussed *linear transformations*, such as converting temperature in degrees Fahrenheit to degrees Celsius or converting distance in miles to kilometers. However, linear transformations cannot straighten a curved relationship between two variables. To do that, we resort to functions that are not linear. The logarithm function, applied in the income and life expectancy example, is a nonlinear function. We'll return to transformations involving logarithms later.

Transforming with Powers and Roots

When you visit a pizza parlor, you order a pizza by its diameter—say, 10 inches, 12 inches, or 14 inches. But the amount you get to eat depends on the area of the pizza. The area of a circle is π times the square of its radius r. So the area of a round pizza with diameter x is

$$\text{area} = \pi r^2 = \pi \left(\frac{x}{2}\right)^2 = \pi \left(\frac{x^2}{4}\right) = \frac{\pi}{4}x^2$$

This is a *power model* of the form $y = ax^p$ with $a = \pi/4$ and $p = 2$.

Making Connections

Now that the original data have been transformed into a relationship that is fairly linear, we can use the strategies from Sections 3.1 and 3.2. More specifically, we can find a least-squares regression line for the transformed data and then use it to make predictions for life expectancy. It is important to note that the equation for the least-squares regression line will have the form $\hat{y} = a + b \log x$ because $\log x$ is on the x axis.

Teaching Tip

Notice that the power model we present here has only one term. There are other multi-term models that can be used for nonlinear associations, such as quadratic models in the form $y = ax^2 + bx + c$. These more complicated models are not included on the AP® Statistics Topic Outline, however.

When we are dealing with things of the same general form, whether circles or fish or people, we expect area to go up with the square of a dimension such as diameter or height. Volume should go up with the cube of a linear dimension. That is, geometry tells us to expect power models in some settings. There are other physical relationships between two variables that are described by power models. Here are some examples from science.

- The distance that an object dropped from a given height falls is related to time since release by the model

$$distance = a(time)^2$$

- The time it takes a pendulum to complete one back-and-forth swing (its period) is related to its length by the model

$$period = a\sqrt{length} = a(length)^{1/2}$$

- The intensity of a light bulb is related to distance from the bulb by the model

$$intensity = \frac{a}{distance^2} = a(distance)^{-2}$$

Although a power model of the form $y = ax^p$ describes the nonlinear relationship between x and y in each of these settings, there is a *linear* relationship between x^p and y. If we transform the values of the explanatory variable x by raising them to the p power, and graph the points (x^p, y), the scatterplot should have a linear form. The following example shows what we mean.

Teaching Tip

Having trouble thinking about how the graph of distance = a(time)^2 can be linear? Think of distance as the y variable and (time)^2 as the x variable, so now $y = ax$. This is a simple line with slope a and y intercept of 0. This means we need to graph the values for (time)^2 on the x axis and the values for distance on the y axis. Thus, the transformation needed to make the relationship linear is to square all the values for time.

EXAMPLE

Go fish!
Transforming with powers

PROBLEM: Imagine that you have been put in charge of organizing a fishing tournament in which prizes will be given for the heaviest Atlantic Ocean rockfish caught. You know that many of the fish caught during the tournament will be measured and released. You are also aware that using delicate scales to try to weigh a fish that is flopping around in a moving boat will probably not yield very accurate results. It would be much easier to measure the length of the fish while on the boat. What you need is a way to convert the length of the fish to its weight.

You contact the nearby marine research laboratory, and it provides reference data on the length (in centimeters) and weight (in grams) for Atlantic Ocean rockfish of several sizes.[41] Here is a scatterplot of the data. Note the clear curved form.

Length	5.2	8.5	11.5	14.3	16.8	19.2	21.3	23.3	25.0	26.7
Weight	2	8	21	38	69	117	148	190	264	293
Length	28.2	29.6	30.8	32.0	33.0	34.0	34.9	36.4	37.1	37.7
Weight	318	371	455	504	518	537	651	719	726	810

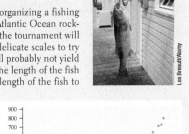

A scatterplot of the data follows. Note the clear curved shape.

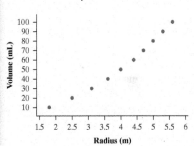

Because radius is one-dimensional and volume is three-dimensional, a power model of the form volume = a(radius)^3 should describe the relationship. But the height of the oil spill is constant for various volumes, so the relationship will actually be quadratic. A power model of the form volume = a(radius)^2 should describe the relationship. Here is a scatterplot of volume versus radius^2:

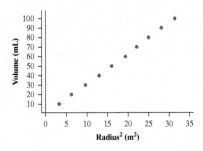

ALTERNATE EXAMPLE

Containing oil spills   Transforming with powers

Skill 2.C

PROBLEM:
Oil spills often occur in lakes and oceans, creating a thin layer of oil on the surface of the water in the shape of a circle (ignoring wind, current, and tides). Aerial photography makes it easy to determine the radius of the circle created by the oil spill. An environmental protection company wants to be able to estimate the volume of an oil spill from the known radius of the spill. To create a prediction model, the company conducts a study in which various amounts of oil are spilled into a controlled testing pool of water and the radius of the spill and the volume for each amount are recorded. Here are the results:

Radius (m)	1.8	2.5	3.1	3.6	4	4.4	4.7	5	5.3	5.6
Volume (mL)	10	20	30	40	50	60	70	80	90	100

(continues)

ALTERNATE EXAMPLE **(continued)**

Because the transformation made the association roughly linear, we used computer software to perform a linear regression analysis of $y = $ volume versus $x = $ radius^2.

Regression Analysis: Volume versus Radius^2				
Predictor	Coef	SE Coef	T	P
Constant	−0.852	0.500	−1.70	0.127
Radius^2	3.2106	0.0256	125.53	0.000
S = 0.72338	R-Sq = 99.95%	R-Sq(adj) = 99.94%		

(a) Give the equation of the least-squares regression line. Define any variables you use.

(b) Suppose an oil spill created a circle with a radius of 2.9 meters. Use the model from part (a) to predict the volume of the oil.

SOLUTION:

(a) $\widehat{\text{volume}} = -0.852 + 3.2106 \,(\text{radius})^2$

(b) $\widehat{\text{volume}} = -0.852 + 3.2106 \,(2.9)^2 = 26.15$ mL

Teaching Tip

In part (b) of the student example on this page, students will often forget to cube the length when making the prediction. Direct their attention to the scatterplot above, which graphs weight versus length^3. Therefore, in the equation $\hat{y} = a + bx$, we must replace x with length^3 and y with weight. This is another reason we prefer that students use contextual variables rather than x and y.

Teaching Tip

In the alternate example on this page, we transformed the data by squaring the radius of the oil spill. Another way to transform the data to achieve linearity is to take the square root of the volume and graph $\sqrt{\text{volume}}$ versus the radius. See the next alternate example.

Because length is one-dimensional and weight (like volume) is three-dimensional, a power model of the form weight $= a(\text{length})^3$ should describe the relationship. Here is a scatterplot of weight versus length^3:

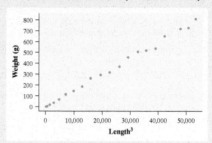

Because the transformation made the association roughly linear, we used computer software to perform a linear regression analysis of $y = $ weight versus $x = $ length^3.

Regression Analysis: Weight versus Length^3				
Predictor	Coef	SE Coef	T	P
Constant	4.066	6.902	0.59	0.563
Length^3	0.0146774	0.0002404	61.07	0.000
S = 18.8412	R-Sq = 99.5%	R-Sq(adj) = 99.5%		

(a) Give the equation of the least-squares regression line. Define any variables you use.

(b) Suppose a contestant in the fishing tournament catches an Atlantic Ocean rockfish that's 36 centimeters long. Use the model from part (a) to predict the fish's weight.

SOLUTION:

(a) $\widehat{\text{weight}} = 4.066 + 0.0146774 \,(\text{length})^3$

(b) $\widehat{\text{weight}} = 4.066 + 0.0146774 \,(36)^3$

$\widehat{\text{weight}} = 688.9 \, \text{grams}$

> If you write the equation as $\hat{y} = 4.066 + 0.0146774x^3$ make sure to define $y = $ weight and $x = $ length.

FOR PRACTICE, TRY EXERCISE 81

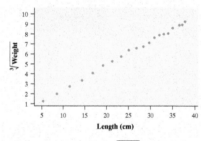

FIGURE 3.20 The scatterplot of $\sqrt[3]{\text{weight}}$ versus length is linear.

There's another way to transform the data in the "Go fish!" example to achieve linearity. We can take the cube root of the weight values and graph $\sqrt[3]{\text{weight}}$ versus length. Figure 3.20 shows that the resulting scatterplot has a linear form.

Why does this transformation work? Start with weight $= a(\text{length})^3$ and take the cube root of both sides of the equation:

$$\sqrt[3]{\text{weight}} = \sqrt[3]{a(\text{length})^3}$$

$$\sqrt[3]{\text{weight}} = \sqrt[3]{a}(\text{length})$$

That is, there is a linear relationship between length and $\sqrt[3]{\text{weight}}$.

Teaching Tip

You may need to help your students see how $\sqrt[3]{\text{weight}} = \sqrt[3]{a}\,(\text{length})$ is in a linear form. Remind them that an equation is linear if it is in the form

variable = constant + constant(variable)

In this case, $\sqrt[3]{\text{weight}}$ is the variable on the left side, 0 is the first constant, $\sqrt[3]{a}$ is the second constant, and length is the second variable.

EXAMPLE

Go fish!
Transforming with roots

Doug Wilson/Alamy

PROBLEM: Figure 3.20 shows that the relationship between length and $\sqrt[3]{\text{weight}}$ is roughly linear for Atlantic Ocean rockfish. Here is computer output from a linear regression analysis of $y = \sqrt[3]{\text{weight}}$ versus $x = $ length:

```
Regression Analysis: ∛Weight versus Length
Predictor    Coef     SE Coef      T       P
Constant   −0.02204   0.07762    −0.28   0.780
Length      0.246616  0.002868    86.00   0.000
S = 0.124161    R-Sq = 99.8%    R-Sq(adj) = 99.7%
```

(a) Give the equation of the least-squares regression line. Define any variables you use.

(b) Suppose a contestant in the fishing tournament catches an Atlantic Ocean rockfish that's 36 centimeters long. Use the model from part (a) to predict the fish's weight.

SOLUTION:

(a) $\widehat{\sqrt[3]{\text{weight}}} = -0.02204 + 0.246616\,(\text{length})$

(b) $\widehat{\sqrt[3]{\text{weight}}} = -0.02204 + 0.246616\,(36) = 8.856$

$\widehat{\text{weight}} = 8.856^3 = 694.6\,\text{grams}$

> If you write the equation as $\widehat{\sqrt[3]{y}} = -0.02204 + 0.246616x$, make sure to define $y = $ weight and $x = $ length.

> The least-squares regression line gives the predicted value of the *cube root* of weight. To get the predicted weight, reverse the cube root by raising the result to the third power.

FOR PRACTICE, TRY EXERCISE 83

When experience or theory suggests that a bivariate relationship is described by a power model of the form $y = ax^p$ where p is known, there are two methods for transforming the data to achieve linearity.

1. Raise the values of the explanatory variable x to the p power and plot the points (x^p, y).
2. Take the pth root of the values of the response variable y and plot the points $(x, \sqrt[p]{y})$.

What if you have no idea what value of p to use? You could guess and test until you find a transformation that works. Some technology comes with built-in sliders that allow you to dynamically adjust the power and watch the scatterplot change shape as you do.

It turns out that there is a much more efficient method for linearizing a curved pattern in a scatterplot. Instead of transforming with powers and roots, we use logarithms. This more general method works when the data follow an unknown power model or any of several other common mathematical models.

Containing more oil spills
Transforming with roots

PROBLEM:
In the preceding alternate example, an environmental protection company discovered a quadratic relationship between the radius of an oil spill and the volume of oil in the spill. Another way to transform the original data into a roughly linear relationship is to take the square root of the volume. Here is the scatterplot:

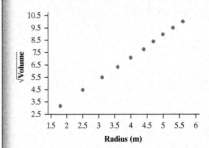

Here is computer output from a linear regression analysis of $y = \sqrt{\text{volume}}$ versus $x = $ radius:

```
Regression Analysis: √Volume versus
                     Radius
Predictor    Coef    SE Coef     T       P
Constant   −0.0891   0.0550    −1.62   0.144
Radius      1.7985   0.0132   136.43   0.000
S = 0.0492537  R-Sq = 99.96%  R-Sq(adj) = 99.95%
```

(a) Give the equation of the least-squares regression line. Define any variables you use.

(b) Suppose an oil spill created a circle with a radius of 2.9 meters. Use the model from part (a) to predict the volume of the oil.

SOLUTION:

(a) $\widehat{\sqrt{\text{volume}}} = -0.0891 + 1.7985\,(\text{radius})$

(b) $\widehat{\sqrt{\text{volume}}} = -0.0891 + 1.7985\,(2.9)$

$\widehat{\text{volume}} = 5.12655^2 = 26.28\,\text{mL}$

Teaching Tip

Students will likely groan at the mention of a logarithm. Assure them that we will not spend much time developing the meaning and properties of logarithms. Rather, we will use logarithms simply as a tool for transforming data.

Teaching Tip

In the linear graph of log y = log a + p log x the variable on the x axis is log x, the variable on the y axis is log y, the y intercept is log a and the slope is p.

Teaching Tip

Make sure that students are comfortable using both the natural log and the base-10 log when transforming data. Both transformations could appear on the AP® Statistics exam and we use both throughout this section.

ALTERNATE EXAMPLE   Skill 2.C

Containing oil spills

Transforming with logarithms: Power models

PROBLEM:
In the previous examples, an environmental protection company used powers and roots to find a model for predicting the volume of an oil spill from the radius of the circular spill created on the surface of the water. We still expect a power model of the form volume = a(radius)^2 based on geometry. Here once again is a scatterplot of the data from the study:

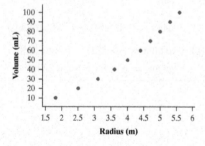

We took the natural logarithm (base e) of the values for both variables. Some computer output from a linear regression analysis on the transformed data is shown.

Transforming with Logarithms: Power Models

To achieve linearity from a power model, we apply the logarithm transformation to *both* variables. Here are the details:

1. A power model has the form $y = ax^p$, where a and p are constants.
2. Take the logarithm of both sides of this equation. Using properties of logarithms, we get

$$\log y = \log(ax^p) = \log a + \log(x^p) = \log a + p \log x$$

The equation

$$\log y = \log a + p \log x$$

shows that taking the logarithm of both variables results in a linear relationship between log x and log y. *Note:* You can use base-10 logarithms or natural (base-e) logarithms to straighten the association.

3. Look carefully: the *power p* in the power model becomes the *slope* of the straight line that links log y to log x.

If a power model describes the relationship between two variables, a scatterplot of the logarithms of both variables should produce a linear pattern. Then we can fit a least-squares regression line to the transformed data and use the linear model to make predictions. Here's an example.

EXAMPLE

Go fish!
Transforming with logarithms:
Power models

PROBLEM: In the preceding examples, we used powers and roots to find a model for predicting the weight of an Atlantic Ocean rockfish from its length. We still expect a power model of the form weight = a(length)^3 based on geometry. Here once again is a scatterplot of the data from the local marine research lab:

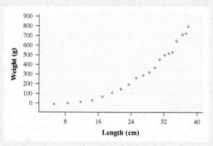

We took the logarithm (base 10) of the values for both variables. Here is some computer output from a linear regression analysis of the transformed data.

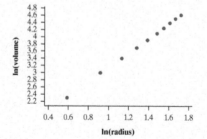

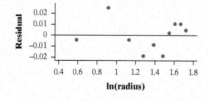

```
       Regression Analysis: ln(Volume)
              versus ln(Radius)

Predictor    Coef     SE Coef     T        P

Constant     1.1188   0.0184      60.87    0.000
ln(Radius)   2.0209   0.0133      151.38   0.000

S = 0.0145241  R-Sq = 99.97%  R-Sq(adj) = 99.96%
```

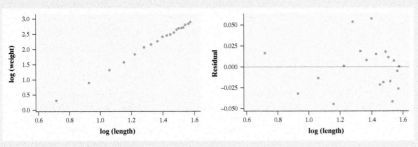

```
Regression Analysis: log(Weight) versus log(Length)
Predictor       Coef     SE Coef        T         P
Constant     -1.89940    0.03799    -49.99     0.000
log(Length)   3.04942    0.02764    110.31     0.000
S = 0.0281823      R-Sq = 99.9%      R-Sq(adj) = 99.8%
```

(a) Based on the output, explain why it would be reasonable to use a power model to describe the relationship between weight and length for Atlantic Ocean rockfish.

(b) Give the equation of the least-squares regression line. Be sure to define any variables you use.

(c) Suppose a contestant in the fishing tournament catches an Atlantic Ocean rockfish that's 36 centimeters long. Use the model from part (b) to predict the fish's weight.

SOLUTION:

(a) *The scatterplot of log(weight) versus log(length) has a linear form, and the residual plot shows a fairly random scatter of points about the residual = 0 line. So a power model seems reasonable here.*

> If a power model describes the relationship between two variables x and y, then a *linear* model should describe the relationship between log x and log y.

(b) $\widehat{\log(\text{weight})} = -1.89940 + 3.04942 \log(\text{length})$

> If you write the equation as $\widehat{\log(y)} = -1.89940 + 3.04942 \log(x)$, make sure to define $y =$ weight and $x =$ length.

(c) $\widehat{\log(\text{weight})} = -1.89940 + 3.04942 \log(36)$
$\widehat{\log(\text{weight})} = 2.8464$
$\widehat{\text{weight}} = 10^{2.8464} \approx 702.1 \text{ grams}$

> The least-squares regression line gives the predicted value of the base-10 *logarithm* of weight. To get the predicted weight, undo the logarithm by raising 10 to the 2.8464 power.

FOR PRACTICE, TRY EXERCISE 85

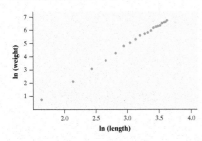

On the TI-83/84, you can "undo" the logarithm using the [2nd] function keys. To solve log(weight) = 2.8464, press [2nd] [LOG] 2.8464 [ENTER].

In addition to base-10 logarithms, you can also use natural (base-*e*) logarithms to transform the variables. Using the same Atlantic Ocean rockfish data, here is a scatterplot of ln(weight) versus ln(length).

The least-squares regression line for these data is

$$\widehat{\ln(\text{weight})} = -4.3735 + 3.04942 \ln(\text{length})$$

ALTERNATE EXAMPLE (continued)

(a) Based on the output, explain why it would be reasonable to use a power model to describe the relationship between volume and radius for oil spills.

(b) Give the equation of the least-squares regression line. Be sure to define any variables you use.

(c) Suppose an oil spill created a circle with a radius of 2.9 meters. Use the model from part (b) to predict the volume of the oil.

SOLUTION:

(a) The scatterplot of ln(volume) versus ln(radius) has a linear form, and the residual plot shows a fairly random scatter of points about the residual = 0 line. So a power model seems reasonable here.

(b) $\widehat{\ln(\text{volume})} = 1.1188 + 2.0209 \ln(\text{radius})$

(c) $\widehat{\ln(\text{volume})} = 1.1188 + 2.0209 \ln(2.9)$
$\widehat{\ln(\text{volume})} = 3.2705$
$\widehat{\text{volume}} = e^{3.2705} \approx 26.32 \text{ mL}$

Teaching Tip

Students must know that the inverse operation for a standard logarithm is "10 raised to the power of" and that the inverse operation for a natural logarithm is "e raised to the power of." On the TI-83/84 calculator, this can be done with 2nd LOG or 2nd LN.

Transformation (log or ln)

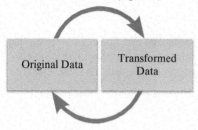

Inverse Transformation (10^ or e^)

Teaching Tip

When using least-squares regression lines in the form $\log y = a + b \log x$, the slope b corresponds to the exponent in a power model. In this example, the slope of transformed data (3.04942) becomes the exponent for the power model.

To predict the weight of an Atlantic Ocean rockfish that is 36 centimeters, we start by substituting 36 for length.

$$\widehat{\ln(\text{weight})} = -4.3735 + 3.04942 \ \ln(36) = 6.55415$$

To get the predicted weight, we then undo the natural logarithm by raising e to the 6.55415 power.

$$\widehat{\text{weight}} = e^{6.55415} = 702.2 \text{ grams}$$

On the TI-83/84, you can "undo" the natural logarithm using the 2nd function keys. To solve ln(weight) = 6.55415, press 2nd LN 6.55415 ENTER.

Your calculator and most statistical software will calculate the logarithms of all the values of a variable with a single command. The important thing to remember is that if a bivariate relationship is described by a power model, then we can linearize the relationship by taking the logarithm of *both* the explanatory and response variables.

Think About It

HOW DO WE FIND THE POWER MODEL FOR PREDICTING *Y* FROM *X*? The least-squares line for the transformed rockfish data is

$$\widehat{\log(\text{weight})} = -1.89940 + 3.04942 \ \log(\text{length})$$

If we use the definition of the logarithm as an exponent, we can rewrite this equation as

$$\widehat{\text{weight}} = 10^{-1.89940 + 3.04942 \ \log(\text{length})}$$

Using properties of exponents, we can simplify this as follows:

$$\widehat{\text{weight}} = 10^{-1.89940} \cdot 10^{3.04942 \log(\text{length})} \qquad \text{using the fact that } b^m b^n = b^{m+n}$$
$$\widehat{\text{weight}} = 10^{-1.89940} \cdot 10^{\log(\text{length})^{3.04942}} \qquad \text{using the fact that } p \log x = \log x^p$$
$$\widehat{\text{weight}} = 0.0126(\text{length})^{3.04942} \qquad \text{using the fact that } 10^{\log x} = x$$

This equation is now in the familiar form of a power model $y = ax^p$ with $a = 0.0126$ and $p = 3.04942$. Notice how close the power is to 3, as expected from geometry.

We could use the power model to predict the weight of a 36-centimeter-long Atlantic Ocean rockfish:

$$\widehat{\text{weight}} = 0.0126(36)^{3.04942} \approx 701.76 \text{ grams}$$

This is roughly the same prediction we got earlier. Here is the scatterplot of the original rockfish data with the power model added. Note how well this model fits the association!

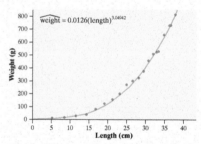

Teaching Tip: Using Technology

Using the PwrReg option on the TI-83/84 calculator will yield the same power model produced here: $\widehat{\text{weight}} = 0.0126(\text{length})^{3.04942}$.

The calculator will also report the correlation value r for the model. This is a bit confusing, as r only measures the strength of the linear relationship between two variables. The r value that the calculator reports is the correlation for the *transformed* data.

Transforming with Logarithms: Exponential Models

A linear model has the form $y = a + bx$. The value of y increases (or decreases) at a constant rate as x increases. The slope b describes the constant rate of change of a linear model. That is, for each 1-unit increase in x, the model predicts an increase of b units in y. You can think of a linear model as describing the repeated addition of a constant amount. Sometimes the relationship between y and x is based on repeated *multiplication* by a constant factor. That is, each time x increases by 1 unit, the value of y is multiplied by b. An *exponential model* of the form $y = ab^x$ describes such growth by multiplication.

Populations of living things tend to grow exponentially if not restrained by outside limits such as lack of food or space. More pleasantly (unless we're talking about credit card debt!), money also displays exponential growth when interest is compounded each time period. Compounding means that the last period's income earns income in the next period. Figure 3.21 shows the balance of a savings account where $100 is invested at 6% interest, compounded annually (assuming no additional deposits or withdrawals). After x years, the account balance y is given by the exponential model $y = 100(1.06)^x$.

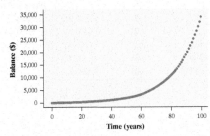

FIGURE 3.21 Scatterplot of the exponential growth of a $100 investment in a savings account paying 6% interest, compounded annually.

An exponential model of the form $y = ab^x$ describes the relationship between x and y, where a and b are constants. We can use logarithms to produce a linear relationship. Start by taking the logarithm of each side (we'll use base 10, but the natural logarithm ln using base e would work just as well). Then use algebraic properties of logarithms to simplify the resulting expressions. Here are the details:

$\log y = \log(ab^x)$	taking the logarithm of both sides
$\log y = \log a + \log(b^x)$	using the property $\log(mn) = \log m + \log n$
$\log y = \log a + x \log b$	using the property $\log m^p = p \log m$

We can then rearrange the final equation as

$$\log y = \log a + (\log b)x$$

Notice that $\log a$ and $\log b$ are constants because a and b are constants. So the equation gives a linear model relating the explanatory variable x to the transformed variable $\log y$. Thus, if the relationship between two variables follows an exponential model, a scatterplot of the logarithm of y against x should show a roughly linear association.

Teaching Tip

You can also use models in the form $y = ab^x$ to model exponential decay, as in the alternate example on the next page. When the value of y decreases as the value of x increases, the base b will be between 0 and 1.

Teaching Tip

In the linear graph of $\log y = \log a + (\log b)x$, the variable on the x axis is x, the variable on the y axis is $\log y$, the y intercept is $\log a$ and the slope is $\log b$.

ALTERNATE EXAMPLE   Skill 2.C

Back to the future
Transforming with logarithms:
Exponential models

PROBLEM:

A mad scientist buys 1020 milligrams of Plutonium-239 for a science experiment. Plutonium-239 is a special isotope of plutonium that can be used as fuel for nuclear reactions (needed for time travel). Plutonium is subject to natural radioactive decay, so the amount of plutonium decreases over time. The mad scientist expects that it will take him 12 years to build the science experiment, and he is concerned that he won't have the 500 milligrams needed for the experiment upon completion of the project. He collects the following data over the next 10 years.

Time (yr)	Amount (mg)
0	1020
1	965
2	930
3	887
4	839
5	805
6	759
7	722
8	695
9	665
10	628

Here are a scatterplot and residual plot for the data. An exponential decay model should describe the relationship between the variables.

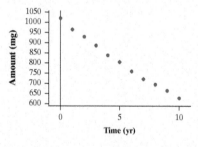

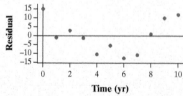

(a) A scatterplot of the logarithm (base 10) of the amount of plutonium versus time is shown. Based on this graph, explain why it would be reasonable to use an exponential model to describe the relationship between amount of plutonium and time since purchase.

EXAMPLE

Moore's law and computer chips

Transforming with logarithms:
Exponential models

PROBLEM: Gordon Moore, one of the founders of Intel Corporation, predicted in 1965 that the number of transistors on an integrated circuit chip would double every 18 months. This is Moore's law, one way to measure the revolution in computing. Here are data on the dates and number of transistors for Intel microprocessors:[42]

Processor	Year	Transistors	Processor	Year	Transistors
Intel 4004	1971	2,300	Pentium III Tualatin	2001	45,000,000
Intel 8008	1972	3,500	Itanium 2 McKinley	2002	220,000,000
Intel 8080	1974	4,500	Itanium 2 Madison 6M	2003	410,000,000
Intel 8086	1978	29,000	Itanium 2 with 9 MB cache	2004	592,000,000
Intel 80286	1982	134,000	Dual-core Itanium 2	2006	1,700,000,000
Intel 80386	1985	275,000	Six-core Xeon 7400	2008	1,900,000,000
Intel 80486	1989	1,180,235	8-core Xeon Nehalem-EX	2010	2,300,000,000
Pentium	1993	3,100,000	10-core Xeon Westmere-EX	2011	2,600,000,000
Pentium Pro	1995	5,500,000	61-core Xeon Phi	2012	5,000,000,000
Pentium II Klamath	1997	7,500,000	18-core Xeon Haswell-E5	2014	5,560,000,000
Pentium III Katmai	1999	9,500,000	22-core Xeon Broadwell-E5	2016	7,200,000,000
Pentium 4 Willamette	2000	42,000,000			

Here is a scatterplot that shows the growth in the number of transistors on a computer chip from 1971 to 2016. Notice that we used "years since 1970" as the explanatory variable. We'll explain this on page 224. If Moore's law is correct, then an exponential model should describe the relationship between the variables.

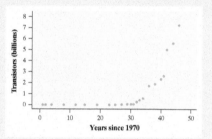

(a) Here is a scatterplot of the natural (base-e) logarithm of the number of transistors on a computer chip versus years since 1970. Based on this graph, explain why it would be reasonable to use an exponential model to describe the relationship between number of transistors and years since 1970.

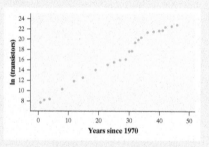

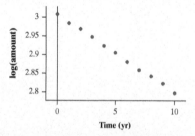

(b) Here is some computer output from a linear regression analysis on the transformed data. Give the equation of the least-squares regression line. Be sure to define any variables you use.

```
Regression Analysis: log(Amount) versus Time

Predictor     Coef      SE Coef      T        P
Constant    3.00832   0.00134   2243.65   0.000
Time       -0.020926  0.000227   -92.330   0.000
S = 0.0023770   R-Sq = 99.89%   R-Sq(adj) = 99.88%
```

(c) Use your model from part (b) to predict the amount of plutonium 12 years after purchase.

SOLUTION:
(a) The scatterplot of log(amount) versus time since purchase has a fairly linear pattern. So an exponential model seems reasonable here.

(b) $\widehat{\log(\text{amount})} = 3.00832 - 0.020926(\text{time})$

(c) $\widehat{\log(\text{amount})} = 3.00832 - 0.020926(12)$
$= 2.757$

$\widehat{\text{amount}} = 10^{2.757} = 571.5$ milligrams

This model predicts that the amount of Plutonium-239 after 12 years will be about 571.5 milligrams. The mad scientist should have enough Plutonium-239 for his science experiment.

(b) Here is some computer output from a linear regression analysis of the transformed data. Give the equation of the least-squares regression line. Be sure to define any variables you use.

Predictor	Coef	SE Coef	T	P
Constant	7.2272	0.3058	23.64	0.000
Years since 1970	0.3542	0.0102	34.59	0.000

S = 0.6653   R-Sq = 98.2%   R-Sq(adj) = 98.2%

(c) Use your model from part (b) to predict the number of transistors on an Intel computer chip in 2020.

SOLUTION:

(a) The scatterplot of ln(transistors) versus years since 1970 has a fairly linear pattern. So an exponential model seems reasonable here.

> If an exponential model describes the relationship between two variables x and y, we expect a scatterplot of (x, ln y) to be roughly linear.

(b) $\overline{\ln(\text{transistors})} = 7.2272 + 0.3542(\text{years since 1970})$

> If you write the equation as $\widehat{\ln(y)} = 7.2272 + 0.3542x$, make sure to define y = number of transistors and x = years since 1970.

(c) $\overline{\ln(\text{transistors})} = 7.2272 + 0.3542(50) = 24.9372$
$\overline{\text{transistors}} = e^{24.9372} = 67,622,053,360$

> 2020 is 50 years since 1970.

This model predicts that an Intel chip made in 2020 will have about 68 billion transistors.

> The least-squares regression line gives the predicted value of ln(transistors). To get the predicted number of transistors, undo the logarithm by raising e to the 24.9372 power.

FOR PRACTICE, TRY EXERCISE 89

Here is a residual plot for the linear regression in part (b) of the example:

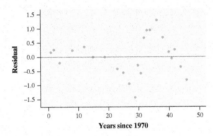

The residual plot shows a leftover pattern, with the residuals going from positive to negative to positive to negative as we move from left to right. However, the residuals are small in size relative to the transformed y-values, and the scatterplot of the transformed data is much more linear than the original scatterplot. We feel reasonably comfortable using this model to make predictions about the number of transistors on a computer chip.

Let's recap this big idea: When an association follows an exponential model, the transformation to achieve linearity is carried out by taking the logarithm of the response variable. The crucial property of the logarithm for our purposes is that *if a variable grows exponentially, its logarithm grows linearly*.

Teaching Tip

In part (c) of this example, students often forget to undo the natural logarithm in the final step. Direct their attention to the scatterplot on the bottom of the previous page, which graphs ln(transistors) versus years since 1970. Therefore, in the equation $\hat{y} = a + bx$, we must replace x with years since 1970 and y with ln(transistors). This is another reason we prefer that students use contextual variables rather than x and y.

Teaching Tip

This residual plot seems to show that the model is near perfect at predicting the number of transistors, as the residual values are all very small. But remember that these residuals do not indicate how far the predicted number of transistors is from the actual number of transistors. They indicate how far the predicted ln(transistors) is from the actual ln(transistors).

Teaching Tip

Even though the residual plot has a leftover pattern, the model still has value for making predictions. Remind students of the quote from George Box: "All models are wrong, but some are useful."

Teaching Tip: Using Technology

Using the ExpReg option on the TI-83/84 calculator will yield the same exponential model produced here:

$\overline{\text{transistors}} = 1376.2(1.4250)^{(\text{years since } 1970)}$.

The calculator will also report the correlation value *r* for the model. This is a bit confusing, because *r* only measures the strength of the linear relationship between two variables. The *r* value that the calculator reports is the correlation for the *transformed* data.

Think About It

HOW DO WE FIND THE EXPONENTIAL MODEL FOR PREDICTING Y FROM X? The least-squares line for the transformed data in the computer chip example is

$$\overline{\ln(\text{transistors})} = 7.2272 + 0.3542(\text{years since } 1970)$$

If we use the definition of the logarithm as an exponent, we can rewrite this equation as

$$\overline{\text{transistors}} = e^{7.2272 + 0.3542(\text{years since } 1970)}$$

Using properties of exponents, we can simplify this as follows:

$\overline{\text{transistors}} = e^{7.2272} \cdot e^{0.3542(\text{years since } 1970)}$ using the fact that $b^m b^n = b^{m+n}$

$\overline{\text{transistors}} = e^{7.2272} \cdot (e^{0.3542})^{(\text{years since } 1970)}$ using the fact that $(b^m)^n = b^{mn}$

$\overline{\text{transistors}} = 1376.4 \cdot (1.4250)^{(\text{years since } 1970)}$ simplifying

This equation is now in the familiar form of an exponential model $y = ab^x$ with $a = 1376.4$ and $b = 1.4250$. Here is the scatterplot of the original transistor data with the exponential model added:

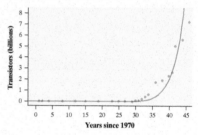

We could use the exponential model to predict the number of transistors on an Intel chip in 2020: $\overline{\text{transistors}} = 1376.4(1.4250)^{50} \approx 6.7529 \cdot 10^{10}$. This is roughly the same prediction we obtained earlier.

The calculation at the end of the Think About It feature might give you some idea of why we used years since 1970 as the explanatory variable in the example. To make a prediction, we substituted the value $x = 50$ into the equation for the exponential model. This value is the exponent in our calculation. If we had used year as the explanatory variable, our exponent would have been 2020. Such a large exponent can lead to overflow errors on a calculator.

Putting It All Together: Which Transformation Should We Choose?

Suppose that a scatterplot shows a curved relationship between two quantitative variables x and y. How can we decide whether a power model or an exponential model better describes the relationship?

Teaching Tip

Here is a quick summary:
Power model: Plot log *y* versus log *x* to transform the association to linear.
Exponential model: Plot log *y* versus *x* to transform the association to linear.

Teaching Tip

Students might ask what model is represented when graphing *y* versus log *x* shows a linear relationship. This is a logarithmic model:
$\hat{y} = a + b \log x$

HOW TO CHOOSE A MODEL

When choosing between different models to describe a relationship between two quantitative variables:

- Choose the model whose residual plot has the most random scatter.
- If there is more than one model with a randomly scattered residual plot, choose the model with the largest coefficient of determination, r^2.

It is not advisable to use the standard deviation of the residuals s to help choose a model, as the y values for the different models might be on different scales.

The following example illustrates the process of choosing the most appropriate model for a curved relationship.

EXAMPLE

Stop that car!
Choosing a model

PROBLEM: How is the braking distance for a car related to the amount of tread left on the tires? Researchers collected data on the braking distance (measured in car lengths) for a car making a panic stop in standing water, along with the tread depth of the tires (in 1/32 inch).[43]

Here is linear regression output for three different models, along with a residual plot for each model. Model 1 is based on the original data, while Models 2 and 3 involve transformations of the original data.

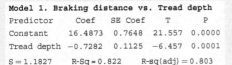

Model 1. Braking distance vs. Tread depth				
Predictor	Coef	SE Coef	T	P
Constant	16.4873	0.7648	21.557	0.0000
Tread depth	−0.7282	0.1125	−6.457	0.0001
S = 1.1827	R-Sq = 0.822	R-sq(adj) = 0.803		

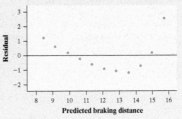

Model 2. ln(braking distance) vs. ln(tread depth)				
Predictor	Coef	SE Coef	T	P
Constant	2.9034	0.0051	566.34	0.0000
ln(tread depth)	−0.2690	0.0029	−91.449	0.0000
S = 0.007	R-sq = 0.999	R-sq(adj) = 0.999		

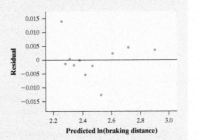

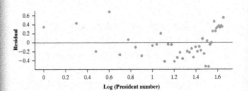

Model 2. Log(percent America) vs. log(President number)				
Predictor	Coef	SE Coef	T	P
Constant	−1.8979	0.1618	−11.54	0.0000
log(President number)	0.9372	0.1236	7.584	0.0000
S = 0.3118	R-Sq = 0.584	R-Sq(adj) = 0.574		

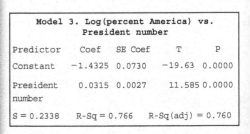

Model 3. Log(percent America) vs. President number				
Predictor	Coef	SE Coef	T	P
Constant	−1.4325	0.0730	−19.63	0.0000
President number	0.0315	0.0027	11.585	0.0000
S = 0.2338	R-Sq = 0.766	R-Sq(adj) = 0.760		

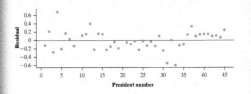

Skill 2.C

ALTERNATE EXAMPLE

Make "America" great Choosing a model

PROBLEM:

As part of the American Presidency Project, the annual State of the Union address has been analyzed going back to the first U.S. president. It turns out that the president's use of the word *America* has increased in relative frequency over the years. Here is linear regression output for three different models, along with a residual plot for each model. Model 1 is based on the original data, while Models 2 and 3 involve transformations of the original data.

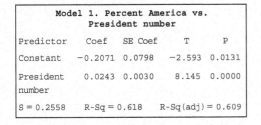

Model 1. Percent America vs. President number				
Predictor	Coef	SE Coef	T	P
Constant	−0.2071	0.0798	−2.593	0.0131
President number	0.0243	0.0030	8.145	0.0000
S = 0.2558	R-Sq = 0.618	R-Sq(adj) = 0.609		

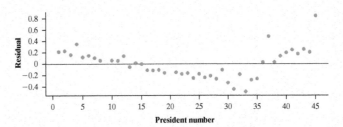

(continues)

(a) Which model does the best job of summarizing the relationship between president number and percent *America*? Explain your reasoning.

(b) Use the model chosen in part (a) to calculate and interpret the residual for the 45th president, Donald Trump, whose State of the Union address contained 1.73 percent *America*.

SOLUTION:

(a) Because the model that uses $x =$ president number and $y = \log$(percent America) produces the most randomly scattered residual plot with no leftover curved pattern, it is the most appropriate model.

(b) $\widehat{\log(\text{percent America})} = -1.4325 + 0.0315 \text{ (president number)}$

$\widehat{\log(\text{percent America})} = -1.4325 + 0.0315(45) = -0.015$

$\widehat{(\text{percent America})} = 10^{-0.015} = 0.97\%$
Residual $= 1.73 - 0.97 = 0.76$
Donald Trump's (the 45th president) use of the word *America* in the State of the Union address was 0.76% higher than the percent predicted by the model using $x =$ president number and $y = \log$(percent America).

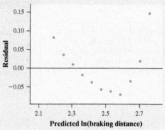

Model 3. ln(braking distance) vs. Tread depth				
Predictor	Coef	SE Coef	T	P
Constant	2.8169	0.0461	61.077	0.0000
Tread depth	−0.0569	0.0068	−8.372	0.0000
S = 0.071	R-sq = 0.886		R-sq(adj) = 0.874	

(a) Which model does the best job of summarizing the relationship between tread depth and braking distance? Explain your reasoning.

(b) Use the model chosen in part (a) to calculate and interpret the residual for the trial when the tread depth was 3/32 inch and the stopping distance was 13.6 car lengths.

SOLUTION:

(a) Because the model that uses $x = \ln$(tread depth) and $y = \ln$(braking distance) produced the most randomly scattered residual plot with no leftover curved pattern, it is the most appropriate model.

> Note that the value of r^2 is also closest to 1 for the model that uses $x = \ln$(tread depth) and $y = \ln$(braking distance).

(b) $\widehat{\ln(\text{braking distance})} = 2.9034 - 0.2690\ln(3) = 2.608$

$\widehat{\text{braking distance}} = e^{2.608} = 13.57 \text{ car lengths}$

Residual $= 13.6 - 13.57 = 0.03$

> The residual calculated here is on the original scale (car lengths), while the residuals shown in the residual plot for this model are on a logarithmic scale.

When the tread depth was 3/32 inch, the car traveled 0.03 car length farther than the distance predicted by the model using $x = \ln$(tread depth) and $y = \ln$(braking distance).

FOR PRACTICE, TRY EXERCISE 91

In the preceding example, the residual plots used the predicted values on the horizontal axis rather than the values of the explanatory variable. Plotting the residuals against the predicted values is common in statistical software. Because software allows for multiple explanatory variables in a single model, it makes sense to use a combination of the explanatory variables (the predicted values) on the horizontal axis rather than using only one of the explanatory variables. In the case of simple linear regression (one explanatory variable), we interpret a residual plot in the same way, whether the explanatory variable or the predicted values are used on the horizontal axis: the more randomly scattered, the more appropriate the model.

We have used statistical software to do all the transformations and linear regression analysis in this section so far. Now let's look at how the process works on a graphing calculator.

AP® EXAM TIP

Students should be comfortable interpreting residual plots with the explanatory variable on the *x*-axis as well as the predicted (fitted) values on the *x*-axis, as both versions have shown up on the AP® Statistics exam in the past.

11. Technology Corner | TRANSFORMING TO ACHIEVE LINEARITY

TI-Nspire and other technology instructions are on the book's website at highschool.bfwpub.com/updatedtps6e.

We'll use the Atlantic Ocean rockfish data to illustrate a general strategy for performing transformations with logarithms on the TI-83/84. A similar approach could be used for transforming data with powers and roots.

- Enter the values of the explanatory variable in L1 and the values of the response variable in L2. Make a scatterplot of *y* versus *x* and confirm that there is a curved pattern.

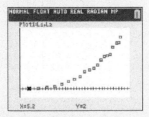

- Define L3 to be the natural logarithm (ln) of L1 and L4 to be the natural logarithm of L2. To see whether a power model fits the original data, make a plot of ln *y* (L4) versus ln *x* (L3) and look for linearity. To see whether an exponential model fits the original data, make a plot of ln *y* (L4) versus *x* (L1) and look for linearity.

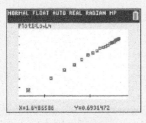 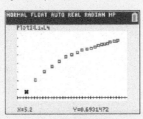

- If a linear pattern is present, calculate the equation of the least-squares regression line. For the Atlantic Ocean rockfish data, we executed the command LinReg(a + bx)L3, L4.

- Construct a residual plot to look for any left-over curved patterns. For Xlist, enter the list you used as the explanatory variable in the linear regression calculation. For Ylist, use the RESID list stored in the calculator. For the Atlantic Ocean rockfish data, we used L3 as the Xlist.

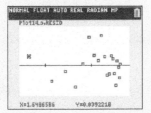

Teaching Tip: Using Technology

We strongly recommend that students use the RESID list on their calculators during this section. Although understanding how to calculate residuals is a valuable skill, there are so many other details in this section that we believe it worthwhile to let the calculator handle this step.

Teaching Tip: Using Technology

Notice that the residual plot graphs L3 on the *x* axis (ln length) rather than L1 (length). This is because the least-squares regression line was calculated with L3 as the explanatory variable.

CHECK YOUR UNDERSTANDING

One sad fact about life is that we'll all die someday. Many adults plan ahead for their eventual passing by purchasing life insurance. Many different types of life insurance policies are available. Some provide coverage throughout an individual's life (whole life), while others last only for a specified number of years (term life). The policyholder makes regular payments (premiums) to the insurance company in return for the coverage. When the insured person dies, a payment is made to designated family members or other beneficiaries.

How do insurance companies decide how much to charge for life insurance? They rely on a staff of highly trained actuaries—people with expertise in probability, statistics, and advanced mathematics—to establish premiums. For an individual who wants to buy life insurance, the premium will depend on the type and amount of the policy as well as personal characteristics like age, sex, and health status.

The table shows monthly premiums for a 10-year term-life insurance policy worth $1,000,000.[44]

Age (years)	Monthly premium
40	$29
45	$46
50	$68
55	$106
60	$157
65	$257

The output shows three possible models for predicting monthly premium from age. Option 1 is based on the original data, while Options 2 and 3 involve transformations of the original data. Each set of output includes a scatterplot with a least-squares regression line added and a residual plot.

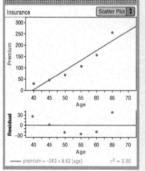 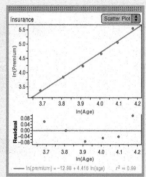 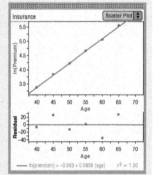

1. Use each model to predict how much a 58-year-old would pay for such a policy.
2. Which model does the best job summarizing the relationship between age and monthly premium? Explain your answer.

✓ Answers to CYU

1. *Option 1:* $\widehat{\text{premium}} = -343 + 8.63(58) = \$157.54$

   Option 2: $\widehat{\ln(\text{premium})} = -12.98 + 4.416(\ln 58) = 4.9509 \rightarrow \hat{y} = e^{4.9509} = \$141.30$

   Option 3: $\widehat{\ln(\text{premium})} = -0.063 + 0.0859(58) = 4.9192 \rightarrow \hat{y} = e^{4.9192} = \$136.89$

2. The exponential model (Option 3) best describes the relationship because this model produced the most randomly scattered residual plot with no leftover curved pattern. This model also has the greatest r^2 value of the three options.

Section 3.3 | Summary

- Curved relationships between two quantitative variables can sometimes be changed into linear relationships by **transforming** one or both of the variables. Once we transform the data to achieve linearity, we can fit a least-squares regression line to the transformed data and use this linear model to make predictions.
- When theory or experience suggests that the relationship between two variables follows a **power model** of the form $y = ax^p$, transformations involving powers and roots can linearize a curved pattern in a scatterplot.
  - **Option 1**: Raise the values of the explanatory variable x to the power p, then look at a graph of (x^p, y).
  - **Option 2**: Take the pth root of the values of the response variable y, then look at a graph of $(x, \sqrt[p]{y})$.
- Another useful strategy for straightening a curved pattern in a scatterplot is to take the **logarithm** of one or both variables. When a power model describes the relationship between two variables, a plot of log y versus log x (or ln y versus ln x) should be linear.
- For an **exponential model** of the form $y = ab^x$, the predicted values of the response variable are multiplied by a factor of b for each increase of 1 unit in the explanatory variable. When an exponential model describes the relationship between two variables, a plot of log y versus x (or ln y versus x) should be linear.
- To decide between competing models, choose the model with the most randomly scattered residual plot. If it is difficult to determine which residual plot is the most randomly scattered, choose the model with the largest value of r^2.

3.3 Technology Corner

TI-Nspire and other technology instructions are on the book's website at highschool.bfwpub.com/updatedtps6e.

11. Transforming to achieve linearity   Page 227

Section 3.3 | Exercises

81. **The swinging pendulum** Mrs. Hanrahan's precalculus
pg 215 class collected data on the length (in centimeters) of a pendulum and the time (in seconds) the pendulum took to complete one back-and-forth swing (called its period). The theoretical relationship between a pendulum's length and its period is

$$\text{period} = \frac{2\pi}{\sqrt{g}} \sqrt{\text{length}}$$

where g is a constant representing the acceleration due to gravity (in this case, $g = 980 \text{ cm/s}^2$). Here is a graph of period versus $\sqrt{\text{length}}$, along with output from a linear regression analysis using these variables.

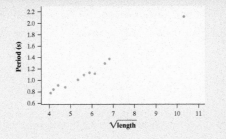

3.81 (a) $\hat{y} = -0.08594 + 0.21\sqrt{x}$, where y is the period and x is the length.
(b) $\hat{y} = -0.08594 + 0.21\sqrt{80} = 1.792$ seconds

3.82 (a) $\hat{y} = 0.3677 + 15.8994\left(\dfrac{1}{x}\right)$, where y is the pressure and x is the volume.

(b) $\hat{y} = 0.3677 + 15.8994\left(\dfrac{1}{17}\right) = 1.303$ atmospheres.

3.83 (a) $\widehat{y^2} = -0.15465 + 0.0428x$, where y is the period and x is the length.
(b) $\widehat{y^2} = -0.15465 + 0.0428(80) = 3.269$, so $\hat{y} = \sqrt{3.269} = 1.808$ seconds.

3.84 (a) $\widehat{\dfrac{1}{y}} = 0.1002 + 0.0398x$, where y is the pressure and x is the volume.

(b) $\widehat{\dfrac{1}{y}} = 0.1002 + 0.0398(17) = 0.7768$, so the predicted pressure is
$\hat{y} = \dfrac{1}{0.7768} = 1.287$ atmospheres.

```
Regression Analysis: (√length, period)
Predictor      Coef     SE Coef     T      P
Constant    -0.08594   0.05046   -1.70  0.123
sqrt(length) 0.209999 0.008322   25.23  0.000
S = 0.0464223   R-Sq = 98.6%   R-Sq(adj) = 98.5%
```

(a) Give the equation of the least-squares regression line. Define any variables you use.

(b) Use the model from part (a) to predict the period of a pendulum with length 80 cm.

82. Boyle's law If you have taken a chemistry or physics class, then you are probably familiar with Boyle's law: for gas in a confined space kept at a constant temperature, pressure times volume is a constant (in symbols, $PV = k$). Students in a chemistry class collected data on pressure and volume using a syringe and a pressure probe. If the true relationship between the pressure and volume of the gas is $PV = k$, then

$$P = k\frac{1}{V}$$

Here is a graph of pressure versus $\dfrac{1}{\text{volume}}$, along with output from a linear regression analysis using these variables:

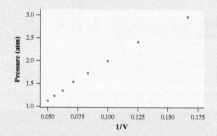

```
Regression Analysis: (  1   ,pressure)
                       volume
Predictor   Coef    SE Coef    T      P
Constant  0.36774  0.04055    9.07  0.000
1/V      15.8994   0.4190    37.95  0.000
S = 0.044205   R-Sq = 99.6%   R-Sq(adj) = 99.5%
```

(a) Give the equation of the least-squares regression line. Define any variables you use.

(b) Use the model from part (a) to predict the pressure in the syringe when the volume is 17 cubic centimeters.

83. The swinging pendulum Refer to Exercise 81. pg 217 Here is a graph of period² versus length, along with output from a linear regression analysis using these variables.

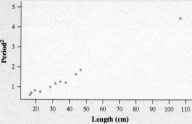

```
Regression Analysis: (length, period²)
Predictor    Coef      SE Coef     T      P
Constant   -0.15465   0.05802   -2.67  0.026
Length      0.042836  0.001320   32.46  0.000
S = 0.105469   R-Sq = 99.2%   R-Sq(adj) = 99.1%
```

(a) Give the equation of the least-squares regression line. Define any variables you use.

(b) Use the model from part (a) to predict the period of a pendulum with length 80 centimeters.

84. Boyle's law Refer to Exercise 82. Here is a graph of $\dfrac{1}{\text{pressure}}$ versus volume, along with output from a linear regression analysis using these variables:

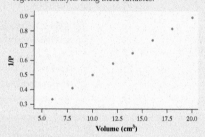

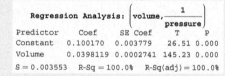

```
Regression Analysis: (volume,   1    )
                              pressure
Predictor   Coef      SE Coef    T      P
Constant  0.100170  0.003779   26.51  0.000
Volume    0.0398119 0.0002741 145.23  0.000
S = 0.003553   R-Sq = 100.0%   R-Sq(adj) = 100.0%
```

(a) Give the equation of the least-squares regression line. Define any variables you use.

(b) Use the model from part (a) to predict the pressure in the syringe when the volume is 17 cubic centimeters.

85. The swinging pendulum Refer to Exercise 81. We pg 218 took the logarithm (base 10) of the values for both length and period. Here is some computer output from a linear regression analysis of the transformed data.

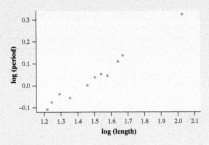

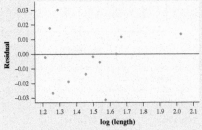

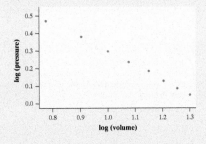

**Regression Analysis: log(Period)
versus log(Length)**

Predictor	Coef	SE Coef	T	P
Constant	−0.73675	0.03808	−19.35	0.000
log(Length)	0.51701	0.02511	20.59	0.000

S = 0.0185568 R-Sq = 97.9% R-Sq(adj) = 97.7%

(a) Based on the output, explain why it would be reasonable to use a power model to describe the relationship between the length and period of a pendulum.

(b) Give the equation of the least-squares regression line. Be sure to define any variables you use.

(c) Use the model from part (b) to predict the period of a pendulum with length 80 cm.

86. **Boyle's law** Refer to Exercise 82. We took the logarithm (base 10) of the values for both volume and pressure. Here is some computer output from a linear regression analysis of the transformed data.

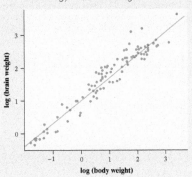

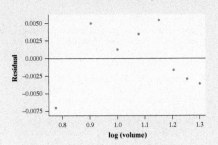

**Regression Analysis: log(Pressure)
versus log(Volume)**

Predictor	Coef	SE Coef	T	P
Constant	1.11116	0.01118	99.39	0.000
log(Volume)	−0.81344	0.01020	−79.78	0.000

S = 0.00486926 R-Sq = 99.9% R-Sq(adj) = 99.9%

(a) Based on the output, explain why it could be reasonable to use a power model to describe the relationship between pressure and volume.

(b) Give the equation of the least-squares regression line. Be sure to define any variables you use.

(c) Use the model from part (b) to predict the pressure in the syringe when the volume is 17 cubic centimeters.

87. **Brawn versus brain** How is the weight of an animal's brain related to the weight of its body? Researchers collected data on the brain weight (in grams) and body weight (in kilograms) for 96 species of mammals.[45] The following figure is a scatterplot of the logarithm of brain weight against the logarithm of body weight for all 96 species. The least-squares regression line for the transformed data is

$$\widehat{\log y} = 1.01 + 0.72 \log x$$

3.85 (a) The scatterplot of log(period) versus log(length) is roughly linear. Also, the residual plot shows no obvious leftover curved patterns. **(b)** $\widehat{\log y} = -0.73675 + 0.51701 \log(x)$, where y is the period and x is the length. **(c)** $\widehat{\log y} = -0.73675 + 0.51701 \log(80) = 0.24717$, so $\hat{y} = 10^{0.24717} = 1.77$ seconds.

3.86 (a) Even though the residual plot shows a leftover curved pattern, it is reasonable to use a power model here because the scatterplot of log(pressure) versus log(volume) is roughly linear. **(b)** $\widehat{\log y} = 1.11116 - 0.81344 \log(x)$, where y is the pressure in the syringe and x is volume. **(c)** $\widehat{\log y} = 1.11116 - 0.81344 \log(17) = 0.110264$. Thus, the predicted pressure is $\hat{y} = 10^{0.110264} = 1.28903$ atmospheres.

3.87 $\widehat{\log y} = 1.01 + 0.72 \log(127) = 2.525$, so $\hat{y} = 10^{2.525} = 334.97$ grams is the predicted brain weight of Bigfoot.

3.88 $\widehat{\ln y} = -2.00 + 2.42 \ln x$, where x is the diameter at breast height in cm and y is the aboveground biomass in kg. If a tree is $x = 30$ cm in diameter, then $\widehat{\ln y} = -2.00 + 2.42 \ln(30) = 6.231$. This means that $\hat{y} = e^{6.231} = 508.263$ kg is the predicted total aboveground biomass of the tree.

3.89 (a) Because the scatterplot of ln(count) versus time is fairly linear, an exponential model would be reasonable. **(b)** $\widehat{\ln y} = 5.97316 - 0.218425x$, where y is the count of surviving bacteria (in hundreds) and x is time in minutes. **(c)** $\widehat{\ln y} = 5.97316 - 0.218425(17) = 2.26$, so $\hat{y} = e^{2.26} = 9.58$ or 958 bacteria.

Based on footprints and some other sketchy evidence, some people believe that a large ape-like animal, called Sasquatch or Bigfoot, lives in the Pacific Northwest. Bigfoot's weight is estimated to be about 127 kilograms (kg). How big do you expect Bigfoot's brain to be?

88. **Determining tree biomass** It is easy to measure the diameter at breast height (in centimeters) of a tree. It's hard to measure the total aboveground biomass (in kilograms) of a tree, because to do this, you must cut and weigh the tree. The biomass is important for studies of ecology, so ecologists commonly estimate it using a power model. The following figure is a scatterplot of the natural logarithm of biomass against the natural logarithm of diameter at breast height (DBH) for 378 trees in tropical rain forests.[46] The least-squares regression line for the transformed data is

$$\widehat{\ln y} = -2.00 + 2.42 \ln x$$

Use this model to estimate the biomass of a tropical tree 30 cm in diameter.

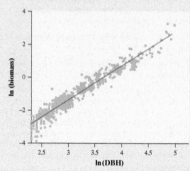

89. **Killing bacteria** Expose marine bacteria to X-rays for time periods from 1 to 15 minutes. Here is a scatterplot showing the number of surviving bacteria (in hundreds) on a culture plate after each exposure time:[47]

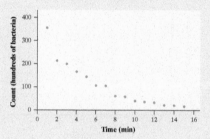

(a) Below is a scatterplot of the natural logarithm of the number of surviving bacteria versus time. Based on this graph, explain why it would be reasonable to use an exponential model to describe the relationship between count of bacteria and time.

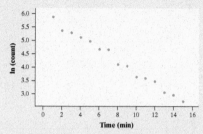

(b) Here is output from a linear regression analysis of the transformed data. Give the equation of the least-squares regression line. Be sure to define any variables you use.

Predictor	Coef	SE Coef	T	P
Constant	5.97316	0.05978	99.92	0.000
Time	−0.218425	0.006575	−33.22	0.000

S = 0.110016 R-Sq = 98.8% R-Sq(adj) = 98.7%

(c) Use your model to predict the number of surviving bacteria after 17 minutes.

90. **Light through water** Some college students collected data on the intensity of light at various depths in a lake. Here is a scatterplot of their data:

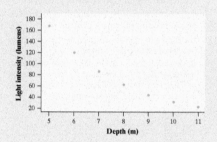

(a) At top right is a scatterplot of the natural logarithm of light intensity versus depth. Based on this graph, explain why it would be reasonable to use an exponential model to describe the relationship between light intensity and depth.

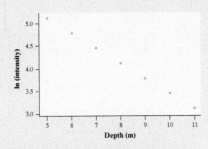

(b) Here is computer output from a linear regression analysis of the transformed data. Give the equation of the least-squares regression line. Be sure to define any variables you use.

Predictor	Coef	SE Coef	T	P
Constant	6.78910	0.00009	78575.46	0.000
Depth (m)	-0.333021	0.000010	-31783.44	0.000

S = 0.000055   R-Sq = 100.0%   R-Sq(adj) = 100.0%

(c) Use your model to predict the light intensity at a depth of 12 meters.

91. **Putting success** How well do professional golfers putt from different distances? Researchers collected data on the percent of putts made for various distances to the hole (in feet).[48]

pg 225

Here is linear regression output for three different models, along with a residual plot for each model. Model 1 is based on the original data, while Models 2 and 3 involve transformations of the original data.

Model 1. Percent made vs. Distance

Predictor	Coef	SE Coef	T	P
Constant	83.6081	4.7206	17.711	0.0000
Distance	-4.0888	0.3842	-10.64	0.0000

S = 9.17   R-sq = 0.870   R-Sq(adj) = 0.862

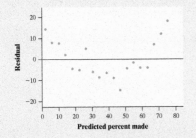

Model 2. ln(percent made) vs. ln(distance)

Predictor	Coef	SE Coef	T	P
Constant	5.5047	0.1628	33.821	0.0000
ln(distance)	-0.9154	0.0702	-13.04	0.0000

S = 0.196   R-sq = 0.909   R-sq(adj) = 0.904

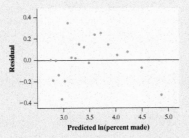

Model 3. ln(percent made) vs. Distance

Predictor	Coef	SE Coef	T	P
Constant	4.6649	0.0825	56.511	0.0000
Distance	-0.1091	0.0067	-16.24	0.0000

S = 0.160   R-sq = 0.939   R-sq(adj) = 0.936

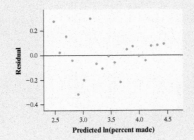

(a) Which model does the best job of summarizing the relationship between distance and percent made? Explain your reasoning.

(b) Using the model chosen in part (a), calculate and interpret the residual for the point where the golfers made 31% of putts from 14 feet away.

92. **Counting carnivores** Ecologists look at data to learn about nature's patterns. One pattern they have identified relates the size of a carnivore (body mass in kilograms) to how many of those carnivores exist in an area. A good measure of "how many" (abundance) is to count carnivores per 10,000 kg of their prey in the area. Researchers collected data on the abundance and body mass for 25 carnivore species.[49]

3.90 (a) Because the scatterplot of ln(intensity) versus depth is fairly linear, an exponential model would be reasonable. **(b)** $\widehat{\ln y} = 6.789 - 0.333x$, where y is the light intensity (lumens) and x is the depth (meters). **(c)** $\widehat{\ln y} = 6.789 - 0.333(12) = 2.793$, so the predicted light intensity at a depth of 12 meters is $\hat{y} = e^{2.793} = 16.33$ lumens.

3.91 (a) Model 3, which uses $x =$ distance and $y = $ ln(percent made), because this model produced the most randomly scattered residual plot with no leftover curved pattern. Also, Model 3 has the largest value of r^2.

(b) $\widehat{\ln(\text{percent made})} = 4.6649 - 0.1091(14) = 3.1375$

$\widehat{\text{percent made}} = e^{3.1375} = 23.05$ percent of putts made from 14 feet away.

Residual $= 31\% - 23.05\% = 7.95\%$

When the putting distance was 14 feet, the golfers' percent made is 7.95 greater than the percentage predicted by the model using $x =$ distance and $y = $ ln(percent made).

3.92 (a) The most appropriate model is Model 2, which uses $x = \ln(\text{body mass})$ and $y = \ln(\text{abundance})$, because this model produced the most randomly scattered residual plot with no leftover curved pattern. Also, Model 2 has the largest value of r^2.

(b)

$\widehat{\ln(\text{abundance})} = 4.4907 - 1.0481 \ln(13)$

$\ln(\text{abundance}) = 1.802$

Therefore, $\widehat{\text{abundance}} = e^{1.802} = 6.06$ carnivores per 10,000 kg of prey in the area

Residual $= 11.65 - 6.06 = 5.59$

The coyote's abundance is 5.59 carnivores per 10,000 kg of prey in the area greater than that predicted by the model using $x = \ln(\text{body mass})$ and $y = \ln(\text{abundance})$.

3.93 (a)

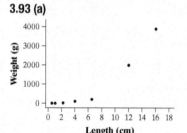

There is a strong, positive curved relationship between heart weight and length of left ventricle for mammals.

(b)

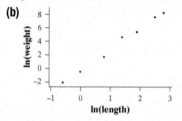

Because the relationship between ln(weight) and ln(length) is roughly linear, heart weight and length seem to follow a power model. An exponential model would not be appropriate because the relationship between ln(weight) and length is clearly curved.

(c) $\widehat{\ln y} = -0.314 + 3.1387 \ln x$, where y is the weight of the heart and x is the length of the cavity of the left ventricle. **(d)** $\widehat{\ln y} = -0.314 + 3.1387 \ln(6.8) = 5.703$, so $\hat{y} = e^{5.703} = 299.77$ grams.

Here is linear regression output for three different models, along with a residual plot for each model. Model 1 is based on the original data, while Models 2 and 3 involve transformations of the original data.

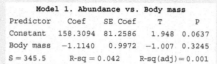

```
Model 1. Abundance vs. Body mass
Predictor    Coef    SE Coef      T      P
Constant   158.3094  81.2586   1.948  0.0637
Body mass   -1.1140   0.9972  -1.007  0.3245
S = 345.5     R-sq = 0.042    R-sq(adj)=0.001
```

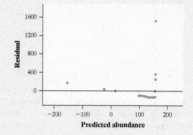

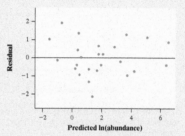

```
Model 2. ln(abundance) vs. ln(body mass)
Predictor          Coef   SE Coef      T       P
Constant         4.4907   0.3091  14.531  0.0000
ln(body mass)   -1.0481   0.0980 -10.693  0.0000
S = 0.975     R-sq = 0.833    R-sq(adj) = 0.825
```

```
Model 3. ln(abundance) vs. Body mass
Predictor    Coef    SE Coef      T      P
Constant   2.6375   0.4843   5.447  0.0000
Body mass  -0.0166  0.0059  -2.791  0.0104
S = 2.059     R-sq = 0.253    R-sq(adj) = 0.220
```

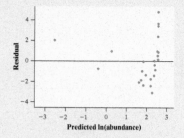

(a) Which model does the best job of summarizing the relationship between body mass and abundance? Explain your reasoning.

(b) Using the model chosen in part (a), calculate and interpret the residual for the coyote, which has a body mass of 13.0 kg and an abundance of 11.65.

93. Heart weights of mammals Here are some data on the hearts of various mammals:[50]

Mammal	Length of cavity of left ventricle (cm)	Heart weight (g)
Mouse	0.55	0.13
Rat	1.00	0.64
Rabbit	2.20	5.80
Dog	4.00	102.00
Sheep	6.50	210.00
Ox	12.00	2030.00
Horse	16.00	3900.00

(a) Make an appropriate scatterplot for predicting heart weight from length. Describe what you see.

(b) Use transformations to linearize the relationship. Does the relationship between heart weight and length seem to follow an exponential model or a power model? Justify your answer.

(c) Perform least-squares regression on the transformed data. Give the equation of your regression line. Define any variables you use.

(d) Use your model from part (c) to predict the heart weight of a human who has a left ventricle 6.8 cm long.

3.94 (a)

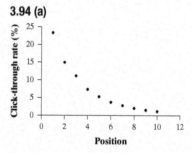

There is a strong, negative curved relationship between position and click-through rate (%) for these websites.

(continues)

94. Click-through rates Companies work hard to have their website listed at the top of an Internet search. Is there a relationship between a website's position in the results of an Internet search (1 = top position, 2 = 2nd position, etc.) and the percentage of people who click on the link for the website? Here are click-through rates for the top 10 positions in searches on a mobile device:[51]

Position	Click-through rate (%)
1	23.53
2	14.94
3	11.19
4	7.47
5	5.29
6	3.80
7	2.79
8	2.11
9	1.57
10	1.18

(a) Make an appropriate scatterplot for predicting click-through rate from position. Describe what you see.

(b) Use transformations to linearize the relationship. Does the relationship between click-through rate and position seem to follow an exponential model or a power model? Justify your answer.

(c) Perform least-squares regression on the transformed data. Give the equation of your regression line. Define any variables you use.

(d) Use your model from part (c) to predict the click-through rate for a website in the 11th position.

Multiple Choice: Select the best answer for Exercises 95 and 96.

95. Students in Mr. Handford's class dropped a kickball beneath a motion detector. The detector recorded the height of the ball (in feet) as it bounced up and down several times. Here is computer output from a linear regression analysis of the transformed data of log(height) versus bounce number. Predict the highest point the ball reaches on its seventh bounce.

```
Predictor    Coef      SE Coef     T       P
Constant     0.45374   0.01385    32.76   0.000
Bounce      -0.117160  0.004176  -28.06   0.000

S = 0.0132043   R-Sq = 99.6%   R-Sq(adj) = 99.5%
```

(a) 0.35 feet   (b) 0.37 feet   (c) 0.43 feet
(d) 2.26 feet   (e) 2.32 feet

96. A scatterplot of y versus x shows a positive, nonlinear association. Two different transformations are attempted to try to linearize the association: using the logarithm of the y-values and using the square root of the y-values. Two least-squares regression lines are calculated, one that uses x to predict log(y) and the other that uses x to predict $\sqrt{y}$. Which of the following would be the best reason to prefer the least-squares regression line that uses x to predict log(y)?

(a) The value of r^2 is smaller.

(b) The standard deviation of the residuals is smaller.

(c) The slope is greater.

(d) The residual plot has more random scatter.

(e) The distribution of residuals is more Normal.

Recycle and Review

97. **Shower time** (1.3, 2.2) Marcella takes a shower every morning when she gets up. Her time in the shower varies according to a Normal distribution with mean 4.5 minutes and standard deviation 0.9 minute.

(a) Find the probability that Marcella's shower lasts between 3 and 6 minutes on a randomly selected day.

(b) If Marcella took a 7-minute shower, would it be classified as an outlier by the 1.5IQR rule? Justify your answer.

98. **NFL weights** (1.2, 1.3) Players in the National Football League (NFL) are bigger and stronger than ever before. And they are heavier, too.[52]

(a) Here is a boxplot showing the distribution of weight for NFL players in a recent season. Describe the distribution.

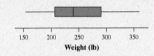
Weight (lb)

(b) Now, here is a dotplot of the same distribution. What feature of the distribution does the dotplot reveal that wasn't revealed by the boxplot?

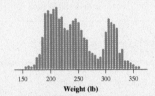
Weight (lb)

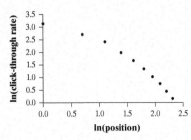

Answers:

(a) There is a fairly strong, positive linear relationship between amount of sugar and freshness.

(b) For each increase of 1 tablespoon of sugar, the predicted freshness increases by 15.8 hours.

(c) With 2 tablespoons of sugar, the predicted freshness is $\hat{y} = 180.8 + 15.8(2) = 212.4$ hours. Thus, the residual $= y - \hat{y} = 204 - 212.4 = -8.4$ hours. This carnation stayed fresh for 8.4 *fewer* hours than predicted based on the amount of sugar it received.

(d) The fact that other flowers will probably stay fresh for a longer or shorter period than carnations is an additional source of unaccounted-for variability in freshness. Because there will be less variation in freshness that is accounted for by the amount of sugar, the value of r^2 will decrease.

Chapter 3 Wrap-Up

FRAPPY! FREE RESPONSE AP® PROBLEM, YAY!

The following problem is modeled after actual AP® Statistics exam free response questions. Your task is to generate a complete, concise response in 15 minutes.

Directions: Show all your work. Indicate clearly the methods you use, because you will be scored on the correctness of your methods as well as on the accuracy and completeness of your results and explanations.

Two statistics students went to a flower shop and randomly selected 12 carnations. When they got home, the students prepared 12 identical vases with exactly the same amount of water in each vase. They put one tablespoon of sugar in 3 vases, two tablespoons of sugar in 3 vases, and three tablespoons of sugar in 3 vases. In the remaining 3 vases, they put no sugar. After the vases were prepared, the students randomly assigned 1 carnation to each vase and observed how many hours each flower continued to look fresh. A scatterplot of the data is shown below.

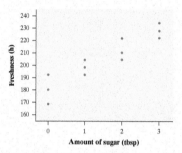

(a) Briefly describe the association shown in the scatterplot.

(b) The equation of the least-squares regression line for these data is $\hat{y} = 180.8 + 15.8x$. Interpret the slope of the line in the context of the study.

(c) Calculate and interpret the residual for the flower that had 2 tablespoons of sugar and looked fresh for 204 hours.

(d) Suppose that another group of students conducted a similar experiment using 12 flowers, but included different varieties in addition to carnations. Would you expect the value of r^2 for the second group's data to be greater than, less than, or about the same as the value of r^2 for the first group's data? Explain.

After you finish, you can view two example solutions on the book's website (highschool.bfwpub.com/updatedtps6e). Determine whether you think each solution is "complete," "substantial," "developing," or "minimal." If the solution is not complete, what improvements would you suggest to the student who wrote it? Finally, your teacher will provide you with a scoring rubric. Score your response and note what, if anything, you would do differently to improve your own score.

Chapter 3 Review

Section 3.1: Scatterplots and Correlation

In this section, you learned how to explore the relationship between two quantitative variables. As with distributions of a single variable, the first step is always to make a graph.

236

A scatterplot is the appropriate type of graph to investigate relationships between two quantitative variables. To describe a scatterplot, be sure to discuss four characteristics: direction, form, strength, and unusual features. The direction of a

relationship might be positive, negative, or neither. The form of a relationship can be linear or nonlinear. A relationship is strong if it closely follows a specific form. Finally, unusual features include points that clearly fall outside the pattern of the rest of the data and distinct clusters of points.

The correlation r is a numerical summary for linear relationships that describes the direction and strength of the association. When $r > 0$, the association is positive, and when $r < 0$, the association is negative. The correlation will always take values between -1 and 1, with $r = -1$ and $r = 1$ indicating a perfectly linear relationship. Strong linear relationships have correlations that are near 1 or -1, while weak linear relationships have correlations near 0. It isn't possible to determine the form of a relationship from only the correlation. Strong nonlinear relationships can have a correlation close to 1 or a correlation close to 0. You also learned that unusual points can greatly affect the value of the correlation and that correlation does not imply causation. That is, we can't assume that changes in one variable cause changes in the other variable, just because they have a correlation close to 1 or -1.

Section 3.2: Least-Squares Regression

In this section, you learned how to use least-squares regression lines as models for relationships between two quantitative variables that have a linear association. It is important to understand the difference between the actual data and the model used to describe the data. To emphasize that the model only provides predicted values, least-squares regression lines are always expressed in terms of $\hat{y}$ instead of y. Likewise, when you are interpreting the slope of a least-squares regression line, describe the change in the *predicted* value of y.

The difference between the actual value of y and the predicted value of y is called a residual. Residuals are the key to understanding almost everything in this section. To find the equation of the least-squares regression line, find the line that minimizes the sum of the squared residuals. To see if a linear model is appropriate, make a residual plot. If there is no leftover curved pattern in the residual plot, you know the model is appropriate. To assess how well a line fits the data, calculate the standard deviation of the residuals s to estimate the size of a typical prediction error. You can also calculate r^2, which measures the percent of the variation

in the y variable that is accounted for by the least-squares regression line.

You also learned how to obtain the equation of a least-squares regression line from computer output and from summary statistics (the means and standard deviations of two variables and their correlation). As with the correlation, the equation of the least-squares regression line and the values of s and r^2 can be greatly affected by influential points, such as outliers and points with high leverage. Make sure to plot the data and note any unusual points before making any calculations.

Section 3.3: Transforming to Achieve Linearity

When the association between two variables is nonlinear, transforming one or both of the variables can result in a linear association.

If the association between two variables follows a power model in the form $y = ax^p$, there are several transformations that will result in a linear association.

- Raise the values of x to the power of p and plot y versus x^p.
- Calculate the pth root of the y-values and plot $\sqrt[p]{y}$ versus x.
- Calculate the logarithms of the x-values and the y-values, and plot $\log(y)$ versus $\log(x)$ or $\ln(y)$ versus $\ln(x)$.

If the association between two variables follows an exponential model in the form $y = ab^x$, transform the data by computing the logarithms of the y-values and plot $\log(y)$ versus x or $\ln(y)$ versus x.

Once you have achieved linearity, calculate the equation of the least-squares regression line using the transformed data. Remember to include the transformed variables when you are writing the equation of the line. Likewise, when using the line to make predictions, make sure that the prediction is in the original units of y. If you transformed the y variable, you will need to undo the transformation after using the least-squares regression line.

To decide which of two or more models is most appropriate, choose the one that produces the most linear association and whose residual plot has the most random scatter. If more than one residual plot is randomly scattered, choose the model with the value of r^2 closest to 1.

What Did You Learn?

Learning Target	Section	Related Example on Page(s)	Relevant Chapter Review Exercise(s)
Distinguish between explanatory and response variables for quantitative data.	3.1	154	R3.4
Make a scatterplot to display the relationship between two quantitative variables.	3.1	155	R3.4

TRM Full Solutions to Chapter 3 Review Exercises

The full solutions can be found by clicking on the link in the TE-Book, opening the TRFD, or downloading from the Teacher's Resources on the book's digital platform.

Answers to Chapter 3 Review Exercises

R3.1 (a) There is a moderate positive, linear association between gestation and life span. Without the unusual points at the top and in the upper right, the association appears moderately strong positive, curved. **(b)** The hippopotamus makes the correlation closer to 0 because it decreases the strength of what would otherwise be a moderately strong positive association. Because this point's x coordinate is very close to $\overline{x}$, it won't influence the slope very much. However, it makes the y intercept higher because its y coordinate is so large compared to the values below it. Because it has such a large residual, it increases the standard deviation of the residuals. **(c)** Because the Asian elephant is in the positive, linear pattern formed by most of the data values, it will make the correlation closer to 1. Also, because the point is likely to be above the least-squares regression line, it will "pull up" the line on the right side, making the slope larger and the y intercept smaller. Because this point is likely to have a small residual, it decreases the standard deviation of the residuals.

Learning Target	Section	Related Example on Page(s)	Relevant Chapter Review Exercise(s)
Describe the direction, form, and strength of a relationship displayed in a scatterplot and identify unusual features.	3.1	158	R3.1, R3.2
Interpret the correlation.	3.1	162	R3.3
Understand the basic properties of correlation, including how the correlation is influenced by unusual points.	3.1	165, 169	R3.1, R3.2
Distinguish correlation from causation.	3.1	165	R3.6
Make predictions using regression lines, keeping in mind the dangers of extrapolation.	3.2	178	R3.4, R3.5
Calculate and interpret a residual.	3.2	180	R3.3, R3.4
Interpret the slope and y intercept of a regression line.	3.2	182	R3.4
Determine the equation of a least-squares regression line using technology or computer output.	3.2	192	R3.3, R3.4
Construct and interpret residual plots to assess whether a regression model is appropriate.	3.2	186	R3.3, R3.4
Interpret the standard deviation of the residuals and r^2 and use these values to assess how well a least-squares regression line models the relationship between two variables.	3.2	191	R3.3, R3.5
Describe how the least-squares regression line, standard deviation of the residuals, and r^2 are influenced by unusual points.	3.2	202	R3.1
Find the slope and y intercept of the least-squares regression line from the means and standard deviations of x and y and their correlation.	3.2	195	R3.5
Use transformations involving powers, roots, or logarithms to create a linear model that describes the relationship between two quantitative variables, and use the model to make predictions.	3.3	215, 217, 218, 222	R3.7
Determine which of several models does a better job of describing the relationship between two quantitative variables.	3.3	225	R3.7

Chapter 3 Review Exercises

These exercises are designed to help you review the important ideas and methods of the chapter.

R3.1 Born to be old? Is there a relationship between the gestational period (time from conception to birth) of an animal and its average life span? The figure shows a scatterplot of the gestational period and average life span for 43 species of animals.[53]

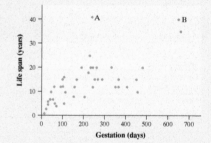

(a) Describe the relationship shown in the scatterplot.

(b) Point A is the hippopotamus. What effect does this point have on the correlation, the equation of the least-squares regression line, and the standard deviation of the residuals?

(c) Point B is the Asian elephant. What effect does this point have on the correlation, the equation of the least-squares regression line, and the standard deviation of the residuals?

R3.2 Penguins diving A study of king penguins looked for a relationship between how deep the penguins dive to seek food and how long they stay under water.[54] For all but the shallowest dives, there is an association between x = depth (in meters) and y = dive duration (in minutes) that is different for each penguin. The study gives a scatterplot for one penguin titled "The Relation of Dive Duration (y) to Depth (x)." The scatterplot shows an association that is positive, linear, and strong.

(a) Explain the meaning of the term *positive association* in this context.

(b) Explain the meaning of the term *linear association* in this context.

(c) Explain the meaning of the term *strong association* in this context.

(d) Suppose the researchers reversed the variables, using x = dive duration and y = depth. Would this change the correlation? The equation of the least-squares regression line?

R3.3 Stats teachers' cars A random sample of AP® Statistics teachers was asked to report the age (in years) and mileage of their primary vehicles. Here are a scatterplot, a residual plot, and other computer output:

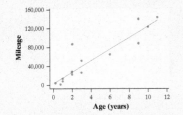

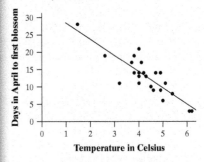

Predictor	Coef	SE Coef	T	P
Constant	3704	8268	0.45	0.662
Age	12188	1492	8.17	0.000

S = 20870.5   R-Sq = 83.7%   R-Sq(adj) = 82.4%

(a) Is a linear model appropriate for these data? Explain how you know this.

(b) What's the correlation between car age and mileage? Interpret this value in context.

(c) Give the equation of the least-squares regression line for these data. Identify any variables you use.

(d) One teacher reported that her 6-year-old car had 65,000 miles on it. Find and interpret its residual.

(e) Interpret the values of s and r^2.

R3.4 Late bloomers? Japanese cherry trees tend to blossom early when spring weather is warm and later when spring weather is cool. Here are some data on the average March temperature (in degrees Celsius) and the day in April when the first cherry blossom appeared over a 24-year period:[55]

Temperature (°C)	4.0	5.4	3.2	2.6	4.2	4.7	4.9	4.0	4.9	3.8	4.0	5.1
Days in April to first blossom	14	8	11	19	14	14	14	21	9	14	13	11

Temperature (°C)	4.3	1.5	3.7	3.8	4.5	4.1	6.1	6.2	5.1	5.0	4.6	4.0
Days in April to first blossom	13	28	17	19	10	17	3	3	11	6	9	11

(a) Make a well-labeled scatterplot that's suitable for predicting when the cherry trees will blossom from the temperature. Which variable did you choose as the explanatory variable? Explain your reasoning.

(b) Use technology to calculate the correlation and the equation of the least-squares regression line. Interpret the slope and y intercept of the line in this setting.

(c) Suppose that the average March temperature this year was 8.2°C. Would you be willing to use the equation in part (b) to predict the date of first blossom? Explain your reasoning.

R3.4 (a) Average March temperature is the explanatory variable because changes in March temperature probably have an effect on the date of first blossom. Also, we are predicting the date of first blossom from temperature.

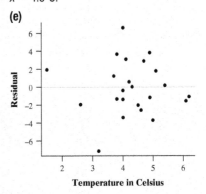

(b) The correlation is $r = -0.85$; $\hat{y} = 33.12 - 4.69x$, where x represents the average March temperature and $\hat{y}$ represents the predicted number of days. The slope is -4.69. *Interpretation:* The predicted number of days in April to the first blossom decreases by 4.69 days for each additional 1-degree increase in average March temperature (in degrees Celsius). The y intercept tells us that if the average March temperature was 0°C, the predicted number of days in April to first blossom is 33.12 (May 3). However, $x = 0$ is outside of the range of data, so this prediction is an extrapolation and may not be trustworthy. **(c)** No, $x = 8.2$ is well beyond the values of x in the data set (1.5 to 6.2). This prediction would be an extrapolation. **(d)** The predicted number of days until first blossom when the average March temperature was 4.5°C is $\hat{y} = 33.12 - 4.69(4.5) = 12.015$. The residual is $y - \hat{y} = 10 - 12.015 = -2.015$. *Interpretation:* The actual number of days until first blossom was 2.015 days less than the number of days predicted by the regression line with $x = 4.5$°C.

(e)

There is no leftover pattern in the residuals, indicating that a linear model is appropriate.

R3.2 (a) A positive relationship means that dives with larger values of depth also tend to have larger values of duration. **(b)** A linear relationship means that when depth increases by 1 meter, dive duration tends to change by a constant amount, on average. **(c)** A strong relationship means that the (dive depth, dive duration) data points fall close to a line. **(d)** If the variables are reversed, the correlation will remain the same. However, the slope and y intercept will be different.

R3.3 (a) A linear model is appropriate for these data because there is no leftover pattern in the residual plot. **(b)** Because $r^2 = 0.837$ and the slope is positive, the correlation is $r = +\sqrt{0.837} = 0.915$. *Interpretation:* The correlation of $r = 0.915$ confirms that the linear association between the age of cars and their mileage is strong and positive.

(c) $\hat{y} = 3704 + 12{,}188x$, where x represents the age and $\hat{y}$ represents the predicted mileage of the cars. **(d)** For a 6-year-old car, the predicted mileage is $\hat{y} = 3704 + 12{,}188(6) = 76{,}832$; the residual is $y - \hat{y} = 65{,}000 - 76{,}832 = -11{,}832$. *Interpretation:* The actual number of miles driven was 11,832 less than the number of miles predicted by the regression line with $x = 6$ years. **(e)** The value of $s = 20{,}870.5$ miles. *Interpretation:* The actual number of miles is about 20,870.5 miles away from the number of miles predicted by the least-squares regression line with x = age (in years). The value of $r^2 = 83.7\%$. *Interpretation:* About 83.7% of the variability in mileage is accounted for by the least-squares regression line with x = age (in years).

R3.5 (a) $b_1 = 0.6\left(\dfrac{8}{30}\right) = 0.16$; $b_0 =$
$75 - 0.16(280) = 30.2$;
$\hat{y} = 30.2 + 0.16x$, where $\hat{y} =$ the predicted final exam score and $x =$ total score before the final examination.
(b) $\hat{y} = 30.2 + 0.16(300) = 78.2$
(c) The least-squares regression line is the line that minimizes the sum of squared distances between the actual exam scores and predicted exam scores. **(d)** Because $r^2 = 0.36$, only 36% of the variability in the final exam scores is accounted for by the linear model relating final exam scores to total score before the final exam. More than half (64%) of the variation in final exam scores is *not* accounted for by the least-squares regression line, so Julie has reason to think this is not a good estimate.

R3.6 Even though there is a high correlation between number of calculators and math achievement, we shouldn't conclude that increasing the number of calculators will *cause* an increase in math achievement. It is possible that students who are more serious about school have better math achievement and also have more calculators.

R3.7 (a) The predictions of the price of a diamond of this type that weighs 2 carats are:

Model 1: $\widehat{\text{price}} = -98666 + 105932(2)$
$= \$113{,}198$.

Model 2:
$\widehat{\ln(\text{price})} = 9.7062 + 2.2913\ln(2)$
$\widehat{\ln(\text{price})} = 11.2944$
Therefore, $\widehat{\text{price}} = e^{11.2944} \approx \$80{,}370$.

Model 3: $\widehat{\ln(\text{price})} = 8.2709 + 1.3791(2)$
$= 11.0291$, therefore
$\widehat{\text{price}} = e^{11.0291} \approx \$61{,}642$.

(b) The model that does the best job of summarizing the relationship between weight and price is Model 2, which uses $x = \ln(\text{price})$ and $y = \ln(\text{weight})$, because this model produced the most randomly scattered residual plot with no leftover curved pattern. Also, Model 2 has the largest value of r^2.

(d) Calculate and interpret the residual for the year when the average March temperature was 4.5°C.

(e) Use technology to help construct a residual plot. Describe what you see.

R3.5 What's my grade? In Professor Friedman's economics course, the correlation between the students' total scores prior to the final examination and their final exam scores is $r = 0.6$. The pre-exam totals for all students in the course have mean 280 and standard deviation 30. The final exam scores have mean 75 and standard deviation 8. Professor Friedman has lost Julie's final exam but knows that her total before the exam was 300. He decides to predict her final exam score from her pre-exam total.

(a) Find the equation for the least-squares regression line Professor Friedman should use to make this prediction.

(b) Use the least-squares regression line to predict Julie's final exam score.

(c) Explain the meaning of the phrase "least squares" in the context of this question.

(d) Julie doesn't think this method accurately predicts how well she did on the final exam. Determine r^2. Use this result to argue that her actual score could have been much higher (or much lower) than the predicted value.

R3.6 Calculating achievement The principal of a high school read a study that reported a high correlation between the number of calculators owned by high school students and their math achievement. Based on this study, he decides to buy each student at his school two calculators, hoping to improve their math achievement. Explain the flaw in the principal's reasoning.

R3.7 Diamonds! Diamonds are expensive, especially big ones. To create a model to predict price from size, the weight (in carats) and price (in dollars) was recorded for each of 94 round, clear, internally flawless diamonds with excellent cuts.[56]

Here is linear regression output for three different models, along with a residual plot for each model. Model 1 is based on the original data, while Models 2 and 3 involve transformations of the original data.

(a) Use each of the three models to predict the price of a diamond of this type that weighs 2 carats.

(b) Which model does the best job of summarizing the relationship between weight and price? Explain your reasoning.

Model 1. Price vs. Weight

Predictor	Coef	SE Coef	T	P
Constant	−98666	7594.1	−12.992	0.0000
Weight	105932	4219.5	25.105	0.0000

S = 34073 R-sq = 0.873 R-sq(adj) = 0.871

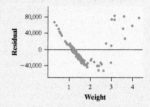

Model 2. ln(price) vs. ln(weight)

Predictor	Coef	SE Coef	T	P
Constant	9.7062	0.0209	465.102	0.0000
ln(weight)	2.2913	0.0332	68.915	0.0000

S = 0.171 R-sq = 0.981 R-sq(adj) = 0.981

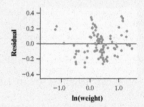

Model 3. ln(price) vs. Weight

Predictor	Coef	SE Coef	T	P
Constant	8.2709	0.0988	83.716	0.0000
Weight	1.3791	0.0549	25.123	0.0000

S = 0.443 R-sq = 0.873 R-sq(adj) = 0.871

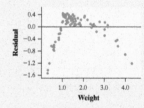

Chapter 3   AP® Statistics Practice Test

TRM Full Solutions to Chapter 3
AP® Statistics Practice Test

The full solutions can be found by clicking on the link in the TE-Book, opening the TRFD, or downloading from the Teacher's Resources on the book's digital platform.

**Answers to Chapter 3
AP® Statistics Practice Test**

T3.1 e

T3.2 d

T3.3 e

T3.4 a

T3.5 c

Section I: Multiple Choice *Select the best answer for each question.*

T3.1 A school guidance counselor examines how many extracurricular activities students participate in and their grade point average. The guidance counselor says, "The evidence indicates that the correlation between the number of extracurricular activities a student participates in and his or her grade point average is close to 0." Which of the following is the most appropriate conclusion?

(a) Students involved in many extracurricular activities tend to be students with poor grades.

(b) Students with good grades tend to be students who are not involved in many extracurricular activities.

(c) Students involved in many extracurricular activities are just as likely to get good grades as bad grades.

(d) Students with good grades tend to be students who are involved in many extracurricular activities.

(e) No conclusion should be made based on the correlation without looking at a scatterplot of the data.

T3.2 An AP® Statistics student designs an experiment to see whether today's high school students are becoming too calculator-dependent. She prepares two quizzes, both of which contain 40 questions that are best done using paper-and-pencil methods. A random sample of 30 students participates in the experiment. Each student takes both quizzes—one with a calculator and one without—in a random order. To analyze the data, the student constructs a scatterplot that displays a linear association between the number of correct answers with and without a calculator for the 30 students. A least-squares regression yields the equation

$$\widehat{\text{Calculator}} = -1.2 + 0.865\,(\text{Pencil}) \qquad r = 0.79$$

Which of the following statements is/are true?

I. If the student had used Calculator as the explanatory variable, the correlation would remain the same.

II. If the student had used Calculator as the explanatory variable, the slope of the least-squares line would remain the same.

III. The standard deviation of the number of correct answers on the paper-and-pencil quizzes was smaller than the standard deviation on the calculator quizzes.

(a) I only

(b) II only

(c) III only

(d) I and III only

(e) I, II, and III

Questions T3.3–T3.5 refer to the following setting.
Scientists examined the activity level of 7 fish at different temperatures. Fish activity was rated on a scale of 0 (no activity) to 100 (maximal activity). The temperature was measured in degrees Celsius. A computer regression printout and a residual plot are provided. Notice that the horizontal axis on the residual plot is labeled "Fitted value," which means the same thing as "predicted value."

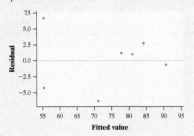

Predictor	Coef	SE Coef	T	P
Constant	148.62	10.71	13.88	0.000
Temperature	−3.2167	0.4533	−7.10	0.001

S = 4.78505   R-Sq = 91.0%   R-Sq(adj) = 89.2%

T3.3 What is the correlation between temperature and fish activity?

(a) 0.95 (b) 0.91 (c) 0.45

(d) −0.91 (e) −0.95

T3.4 What was the actual activity level rating for the fish at a temperature of 20°C?

(a) 87 (b) 84 (c) 81

(d) 66 (e) 3

T3.5 Which of the following gives a correct interpretation of *s* in this setting?

(a) For every 1°C increase in temperature, fish activity is predicted to increase by 4.785 units.

(b) The typical distance of the temperature readings from their mean is about 4.785°C.

(c) The typical distance of the activity level ratings from the least-squares line is about 4.785 units.

(d) The typical distance of the activity level readings from their mean is about 4.785 units.

(e) At a temperature of 0°C, this model predicts an activity level of 4.785 units.

T3.6 b

T3.7 e

T3.8 b

T3.9 e

T3.10 d

T3.6 Which of the following statements is *not* true of the correlation r between the lengths (in inches) and weights (in pounds) of a sample of brook trout?

(a) r must take a value between -1 and 1.

(b) r is measured in inches.

(c) If longer trout tend to also be heavier, then $r > 0$.

(d) r would not change if we measured the lengths of the trout in centimeters instead of inches.

(e) r would not change if we measured the weights of the trout in kilograms instead of pounds.

T3.7 When we standardize the values of a variable, the distribution of standardized values has mean 0 and standard deviation 1. Suppose we measure two variables X and Y on each of several subjects. We standardize both variables and then compute the least-squares regression line. Suppose the slope of the least-squares regression line is -0.44. We may conclude that

(a) the intercept will also be -0.44.

(b) the intercept will be 1.0.

(c) the correlation will be $1/-0.44$.

(d) the correlation will be 1.0.

(e) the correlation will also be -0.44.

T3.8 There is a linear relationship between the number of chirps made by the striped ground cricket and the air temperature. A least-squares fit of some data collected by a biologist gives the model $\hat{y} = 25.2 + 3.3x$, where x is the number of chirps per minute and $\hat{y}$ is the estimated temperature in degrees Fahrenheit. What is the predicted increase in temperature for an increase of 5 chirps per minute?

(a) $3.3°F$ (b) $16.5°F$ (c) $25.2°F$

(d) $28.5°F$ (e) $41.7°F$

T3.9 The scatterplot shows the relationship between the number of people per television set and the number of people per physician for 40 countries, along with the least-squares regression line. In Ethiopia, there were 503 people per TV and 36,660 people per doctor. Which of the following is correct?

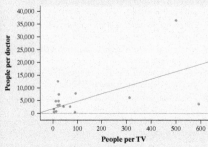

(a) Increasing the number of TVs in a country will attract more doctors.

(b) The slope of the least-squares regression line is less than 1.

(c) The correlation is greater than 1.

(d) The point for Ethiopia is decreasing the slope of the least-squares regression line.

(e) Ethiopia has more people per doctor than expected, based on how many people it has per TV.

T3.10 The scatterplot shows the lean body mass and metabolic rate for a sample of 5 adults. For each person, the lean body mass is the subject's total weight in kilograms less any weight due to fat. The metabolic rate is the number of calories burned in a 24-hour period.

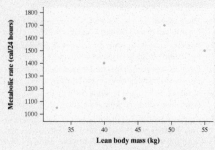

Because a person with no lean body mass should burn no calories, it makes sense to model the relationship with a direct variation function in the form $y = kx$. Models were tried using different values of k ($k = 25$, $k = 26$, etc.) and the sum of squared residuals (SSR) was calculated for each value of k. Here is a scatterplot showing the relationship between SSR and k:

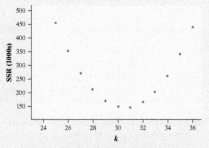

According to the scatterplot, what is the ideal value of k to use for predicting metabolic rate?

(a) 24   (b) 25   (c) 29

(d) 31   (e) 36

T3.11 We record data on the population of a particular country from 1960 to 2010. A scatterplot reveals a clear curved relationship between population and year. However, a different scatterplot reveals a strong linear relationship between the logarithm (base 10) of the population and the year. The least-squares regression line for the transformed data is

$$\overline{\log(\text{population})} = -13.5 + 0.01(\text{year})$$

Based on this equation, which of the following is the best estimate for the population of the country in the year 2020?

(a) 6.7

(b) 812

(c) 5,000,000

(d) 6,700,000

(e) 8,120,000

Section II: Free Response *Show all your work. Indicate clearly the methods you use, because you will be graded on the correctness of your methods as well as on the accuracy and completeness of your results and explanations.*

T3.12 Sarah's parents are concerned that she seems short for her age. Their doctor has kept the following record of Sarah's height:

Age (months)	36	48	51	54	57	60
Height (cm)	86	90	91	93	94	95

(a) Make a scatterplot of these data using age as the explanatory variable. Describe what you see.

(b) Using your calculator, find the equation of the least-squares regression line.

(c) Calculate and interpret the residual for the point when Sarah was 48 months old.

(d) Would you be confident using the equation from part (b) to predict Sarah's height when she is 40 years old? Explain.

T3.13 Drilling down beneath a lake in Alaska yields chemical evidence of past changes in climate. Biological silicon, left by the skeletons of single-celled creatures called diatoms, is a measure of the abundance of life in the lake. A rather complex variable based on the ratio of certain isotopes relative to ocean water gives an indirect measure of moisture, mostly from snow. As we drill down, we look further into the past. Here is a scatterplot of data from 2300 to 12,000 years ago:

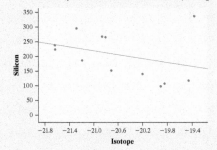

(a) Identify the unusual point in the scatterplot and estimate its *x* and *y* coordinates.

(b) Describe the effect this point has on

i.   the correlation.

ii. the slope and *y* intercept of the least-squares line.

iii. the standard deviation of the residuals.

T3.14 Long-term records from the Serengeti National Park in Tanzania show interesting ecological relationships. When wildebeest are more abundant, they graze the grass more heavily, so there are fewer fires and more trees grow. Lions feed more successfully when there are more trees, so the lion population increases. Researchers collected data on one part of this cycle, wildebeest abundance (in thousands of animals), and the percent of the grass area burned in the same year. The results of a least-squares regression on the data are shown here.[57]

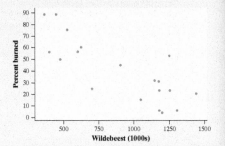

T3.11 c

T3.12 (a)

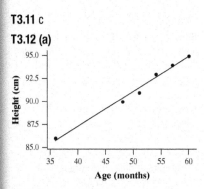

(b) The regression line for predicting *y* = height from *x* = age is $\hat{y} = 71.95 + 0.3833x$. **(c)** At age 48 months, we predict Sarah's height to be $\hat{y} = 71.95 + 0.3833(48) = 90.348$ cm. The residual for Sarah is $y - \hat{y} = 90 - 90.348 = -0.348$. *Interpretation:* Sarah's actual height was 0.348 cm less than the height predicted by the regression line with *x* = 48 months. **(d)** No; obviously, the linear trend will not continue until she is 40 years old. Our data were based only on the first 5 years of life and predictions should only be made for ages 0–5.

T3.13 (a) The unusual point is the one in the upper-right-hand corner with an isotope value about −19.3 and silicon value about 345. This point is unusual in that it has such a high silicon value for the given isotope value. **(b)** (i) If the point were removed, the correlation would get closer to −1 because it does not follow the linear pattern of the other points. (ii) Because this point is "pulling up" the line on the right side of the plot, removing it will make the slope steeper (more negative) and the *y* intercept smaller. Note that the *y* axis is to the *right* of the points in the scatterplot. (iii) Because this point has a large residual, removing it will make the size of the typical residual (*s*) a little smaller.

T3.14 (a) Yes; because there is no obvious leftover pattern in the residual plot, a linear model is appropriate for describing the relationship between wildebeest abundance and percent of grass area burned. **(b)** $\hat{y} = 92.29 - 0.05762x$, where x represents the number of wildebeest and $\hat{y}$ represents the predicted percent of the grass burned. **(c)** The slope $= -0.05762$. *Interpretation:* The predicted percent of grassy area burned decreases by about 0.058% for each additional 1000 wildebeest. The y intercept does not have meaning in this context, as making a prediction for 0 wildebeest is a big extrapolation. It is impossible to know what would happen going from some wildebeest to no wildebeest. **(d)** The value of $s = 15.988\%$. *Interpretation:* The actual percentage of burned area is typically about 15.988% away from the percent predicted by the least-squares regression line with $x =$ number of wildebeest (1000s). The value of $r^2 = 64.6\%$. *Interpretation:* About 64.6% of the variability in percentage of burned area is accounted for by the least-squares regression line with $x =$ number of wildebeest (1000s).

T3.15 (a) There is clear curvature evident in both the scatterplot and the residual plot.

(b) Option 1: $\hat{y} = 2.078 + 0.0042597(30)^3$ $= 117.09$ board feet

Option 2: $\widehat{\ln y} = 1.2319 + 0.113417(30)$ $= 4.63441$ and $\hat{y} = e^{4.63441} = 102.967$ board feet

(c) The residual plot for Option 1 is much more scattered, while the residual plot for Option 2 is curved, meaning that the model relating the amount of usable lumber to cube of the diameter is more appropriate. Thus, the prediction of 117.09 board feet seems more reliable.

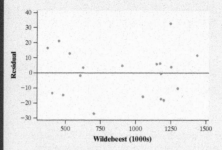

Wildebeest (1000s)

Predictor	Coef	SE Coef	T	P
Constant	92.29	10.06	9.17	0.000
Wildebeest (1000s)	-0.05762	0.01035	-5.56	0.001

S = 15.9880   R-Sq = 64.6%   R-Sq(adj) = 62.5%

(a) Is a linear model appropriate for describing the relationship between wildebeest abundance and percent of grass area burned? Explain.

(b) Give the equation of the least-squares regression line. Be sure to define any variables you use.

(c) Interpret the slope. Does the value of the y intercept have meaning in this context? If so, interpret the y intercept. If not, explain why.

(d) Interpret the standard deviation of the residuals and r^2.

T3.15 Foresters are interested in predicting the amount of usable lumber they can harvest from various tree species. They collect data on the diameter at breast height (DBH) in inches and the yield in board feet of a random sample of 20 Ponderosa pine trees that have been harvested. (Note that a board foot is defined as a piece of lumber 12 inches by 12 inches by 1 inch.) Here is a scatterplot of the data.

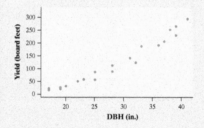

DBH (in.)

(a) Here is some computer output and a residual plot from a least-squares regression on these data. Explain why a linear model may not be appropriate in this case.

Predictor	Coef	SE Coef	T	P
Constant	-191.12	16.98	-11.25	0.000
DBH (inches)	11.0413	0.5752	19.19	0.000

S = 20.3290   R-Sq = 95.3%   R-Sq(adj) = 95.1%

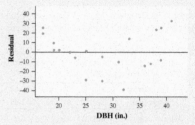

DBH (in.)

The foresters are considering two possible transformations of the original data: (1) cubing the diameter values or (2) taking the natural logarithm of the yield measurements. After transforming the data, a least-squares regression analysis is performed. Here is some computer output and a residual plot for each of the two possible regression models:

Option 1: Cubing the diameter values

Predictor	Coef	SE Coef	T	P
Constant	2.078	5.444	0.38	0.707
DBH^3	0.0042597	0.0001549	27.50	0.000

S = 14.3601   R-Sq = 97.7%   R-Sq(adj) = 97.5%

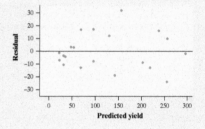

Predicted yield

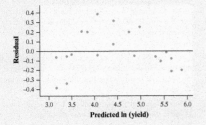

```
       Option 2: Taking natural logarithm
                of yield measurements
Predictor     Coef    SE Coef      T       P
Constant    1.2319     0.1795    6.86   0.000
DBH        0.113417  0.006081   18.65   0.000
S = 0.214894  R-Sq = 95.1%  R-Sq(adj) = 94.8%
```

(b) Use both models to predict the amount of usable lumber from a Ponderosa pine with diameter 30 inches.

(c) Which of the predictions in part (b) seems more reliable? Give appropriate evidence to support your choice.

TRM Chapter 3 Project:
Investigating Relationships
in Baseball

The data file and supporting resources for this project can be found by clicking on the link in the TE-Book, opening the TRFD, or downloading from the Teacher's Resources on the book's digital platform. The data files are also available on the Student Site.

Consider modifying the topic if you can obtain a reasonably large data set. Or, to make it more open-ended, have students pick a topic and find their own data. Possible topics include: Which is a better predictor of battery life in tablets: weight or cost? Which is a better predictor of the cost of a house: square footage, number of bedrooms, or age? Which is a better predictor of winning percentage in football: points scored or points allowed?

Chapter 3 Project Investigating Relationships in Baseball

What is a better predictor of the number of wins for a baseball team, the number of runs scored by the team or the number of runs they allow the other team to score? What variables can we use to predict the number of runs a team scores? To predict the number of runs it allows the other team to score? In this project, you will use technology to help answer these questions by exploring a large set of data from Major League Baseball.

Part 1

1. Download the "MLB Team Data 2012–2016" Excel file from the book's website, along with the "Glossary for MLB Team Data file," which explains each of the variables included in the data set.[58] Import the data into the statistical software package you prefer.

2. Create a scatterplot to investigate the relationship between runs scored per game (R/G) and wins (W). Then calculate the equation of the least-squares regression line, the standard deviation of the residuals, and r^2. *Note:* R/G is in the section for hitting statistics and W is in the section for pitching statistics.

3. Create a scatterplot to investigate the relationship between runs *allowed* per game (RA/G) and wins (W). Then calculate the equation of the least-squares regression line, the standard deviation of the residuals, and r^2. *Note:* Both of these variables may be found in the section for pitching statistics.

4. Compare the two associations. Is runs scored or runs allowed a better predictor of wins? Explain your reasoning.

5. Because the number of wins a team has is dependent on both how many runs they score and how many runs they allow, we can use a combination of both variables to predict the number of wins. Add a column in your data table for a new variable, run differential. Fill in the values using the formula R/G – RA/G.

6. Create a scatterplot to investigate the relationship between run differential and wins. Then calculate the equation of the least-squares regression line, the standard deviation of the residuals, and r^2.

7. Is run differential a better predictor than the variable you chose in Question 4? Explain your reasoning.

Part 2

It is fairly clear that the number of games a team wins is dependent on both runs scored and runs allowed. But what variables help predict runs scored? Runs allowed?

1. Choose either runs scored (R) or runs allowed (RA) as the response variable you will try to model.

2. Choose at least three different explanatory variables (or combinations of explanatory variables) that might help predict the response variable you chose in Question 1. Create a scatterplot using each explanatory variable. Then calculate the equation of the least-squares regression line, the standard deviation of the residuals, and r^2 for each relationship.

3. Which explanatory variable from Question 2 is the best? Explain your reasoning.

TRM **Chapter 3 Case Study**

The Case Study feature that was found in the previous two editions of TPS has been moved to the Teacher's Resource Materials. Click on the link in the TE-Book, open in the TRFD, or download from the Teacher's Resources on the book's digital platform.

Chapter 4

Chapter 4

Collecting Data

PD Overview

In this chapter, we cover every topic in Unit 3 of the Course Framework, Collecting Data. The topics in this section are often new to students and include a lot of vocabulary and difficult concepts. Between 12 and 15% of the questions on the AP® Statistics exam come from this unit, so make sure to spend adequate time here. As you can see in the pacing guide, we suggest spending 11 of the 110 class sessions covering this material, 2 more sessions than any other individual chapter.

Many teachers, including two of the authors, choose to teach Chapter 4 at the beginning of the school year. Because studies have to be designed to produce data to analyze, teaching this chapter first fits the order described in the Guidelines for Assessment and Instruction in Statistics Education (www.amstat.org/education/gaise/; see both the PreK–12 Report, especially Level C, and the College Report), endorsed by the American Statistical Association. This document suggests that "statistical problem solving is an investigative process that involves four components: Formulate Questions, Collect Data, Analyze Data, and Interpret Results."

A second benefit of this approach is that students understand immediately that AP® Statistics is very different from any other "math" class they may have encountered in the past. Right away, students are confronted with real-world applications that show the relevance of what they are going to learn in the course. They will also quickly realize that there is much more reading and writing than in other math courses.

It is important to move beyond written examples and exercises and have students perform some surveys and experiments. We have included several suggested activities in the student text as well as here, in the Annotated Teacher's Edition, to get you started. Our students find it much easier to remember complicated concepts like stratified random sampling and blocking if they have taken part in studies that use these strategies. Taking time to do these activities is another reason we suggest spending more time on this chapter than others.

Two final suggestions: First, read the module called "Planning and Conducting a Study," available at AP® Central (under Classroom Resources: Sampling and Experimentation). This is an outstanding resource written by first-rate statisticians. It goes into more depth than students need, but it helps build a strong foundation for teachers. Second, have your students do the "Response bias"

activity on page 320. It is a great way for students to implement the principles of a good study while learning an important lesson about how easy it is to manipulate the results of a survey. There is a sample rubric for this activity in the Teacher's Resource Materials.

The Main Ideas

One of the challenges of teaching the AP® Statistics course is keeping students focused on the big picture, not just the details of each section. We outline the main ideas for the chapter here.

Chapter 4 Introduction

The introduction gives several brief examples of the types of studies we will learn about in this chapter, including sample surveys, observational studies, and experiments.

Section 4.1 Sampling and Surveys

In this section, we emphasize the difference between a population and a sample. This is an extremely important distinction that will come up over and over again, especially in Chapters 7–12. In almost all cases, we take a sample from a population rather than conduct a census when we want to know something about a population. Unfortunately, there are many things that can go wrong along the way. Selecting people out of convenience or letting people choose whether or not to be included in the sample are definite problems, because these groups are not likely to be representative of any larger population. These methods of obtaining a sample are called *biased* because they will provide estimates that are consistently too big or consistently too small, depending on the context.

To avoid these biased methods of obtaining a sample, the members of the sample should be chosen at random. One way to do this is with a simple random sample, which is equivalent to putting the name of each member of the population on a slip of paper, mixing all the slips in a hat, and choosing the sample by selecting names from the hat one at a time. The Federalist Papers activity on page 250 is a great way to illustrate the benefits of using a simple random sample.

Next, we introduce three other sampling techniques: stratified random sampling, cluster sampling, and systematic random

sampling. Stratified random samples are a little more complicated to obtain and analyze than simple random samples but can produce estimates that have greater precision. Cluster samples can be easier to obtain than simple random samples, but if the clusters are too homogeneous, there is a chance that certain groups in the population might be over- or underrepresented just by chance. Students often get these first two sampling techniques confused, because both of them start by dividing the population into smaller subgroups. However, the strategy for forming the groups and selecting the sample is quite different. Finally, systematic random sampling is a practical method for choosing a sample from an ordered arrangement of the population.

We conclude this section with a discussion of other things that can go wrong in sample surveys—and how to avoid these problems. Undercoverage results when the sampling method systematically over- or underrepresents one part of the population. Nonresponse occurs when answers cannot be obtained from individuals chosen to be in the sample. Undercoverage and nonresponse can both lead to bias— consistently underestimating or overestimating the value we want to know. Even when the sample is chosen well and everyone in the sample responds, the members of the sample don't always respond honestly. Bias can be introduced by the wording of the question, characteristics of the interviewer, or other factors.

Section 4.2 Experiments

In this section, we begin by contrasting observational studies and experiments. Observational studies look at relationships between two variables but cannot show cause and effect because of potential confounding. Two variables are confounded when it is impossible to determine which of them is causing a change in the response variable. Experiments, on the other hand, deliberately impose treatments (conditions) to see whether there is a cause-and-effect relationship between two variables.

Next, we introduce the basic vocabulary of experiments and give an example of a poorly designed experiment about caffeine and pulse rates. To fix the problems in this experiment, we address four principles of experimental design: comparison, random assignment, control, and replication.

Most well-designed experiments include at least two treatments so that different values of the explanatory variable can be compared. To create groups of experimental units that are roughly equivalent at the beginning of the study, we randomly assign the units to the treatments. This random assignment should balance out the effects of other variables so that the only systematic difference between the groups is the treatment imposed. Once the treatments are assigned to experimental units, it is important that groups be treated in exactly the same manner, except for the treatments assigned to them. This is the principle of control: making sure all other variables stay the same for all units during the experiment. Without control, we can introduce confounding and additional variability, making it much harder to draw a conclusion about cause and effect. Finally, the principle of replication says that we should use enough experimental units in each treatment group so that a difference in the response can be attributed to the treatments, not the chance variation in the random assignment.

To avoid a poorly designed experiment, the placebo effect should be taken into account. Whenever possible, both the subjects and the person measuring the response should not know who is receiving which treatment. The Physicians' Health Study on page 282 is a perfect example of a randomized, placebo-controlled, double-blind experiment.

Finally, we introduce an additional principle that can enhance the design of some experiments: blocking. Blocking in experiments is similar to stratifying in sampling. To form blocks, group the experimental units that are similar with respect to variables that are most strongly associated with the response. Then randomly assign the units within each block to each treatment. When done correctly, an experiment with blocking will produce a more precise estimate than a completely randomized design would. Blocks that have only two experimental units are called *matched pairs*.

Section 4.3 Using Studies Wisely

We briefly introduce inference for sampling—using information from a sample to draw conclusions about a population. At this point, we want students to understand the overall idea of inference and begin developing the concept of sampling

variability. That is, we want students to realize that different random samples from the same population will produce different estimates of a population parameter, just by chance. We also show students how increasing sample size decreases the variability of the estimates. We will expand on these ideas in much more depth beginning in Chapter 7.

Next, we introduce inference for experiments. The results of an experiment are considered statistically significant if the difference in the response is too large to be accounted for by the random assignment of experimental units to treatments. The "Analyzing the caffeine experiment" activity on page 301 illustrates this reasoning nicely.

We emphasize that the types of conclusions we can draw depend on how the data are produced. When samples are selected at random, we can make inferences about the population from which the sample was drawn. When treatments are applied to groups formed by random assignment, we can conclude cause and effect.

Drawing a conclusion about cause and effect is often difficult because it is impossible or unethical to impose certain treatments, such as smoking cigarettes. We briefly discuss how to justify a conclusion about cause and effect in these circumstances and discuss some ethical issues related to collecting data. In a nutshell, studies should be approved by an institutional review board, subjects should give informed consent, and individual data must be kept confidential.

Chapter 4 Resources

Teacher's Resource Materials

The following resources, identified by the **TRM** in the annotated student pages, can be found by clicking on the link in the Teacher's e-Book (TE-Book), searching by category or chapter on the Teacher's Resource Flash Drive (TRFD), or logging into the book's digital platform and searching the Teacher's Resources menu (teacher log-in required).

- Alternate Examples: one file per section
- Lecture Presentation slides: one per section
- "See No Evil, Hear No Evil" activity
- "Do You Know Your Geography?" activity
- "The Memory Game" activity
- Chapter 4 Learning Targets Grid
- "The Federalist Papers" activity
- "Sampling Sunflowers" activity
- Template for Experiments
- FRAPPY! Materials
- Chapter 4 Project Response Bias
- Chapter 4 Case Study
- Complete solutions for the Check Your Understanding problems, section exercises, review exercises, practice test, and the Cumulative AP® Practice Test 1.
- Quizzes: one per section
- Chapter 4 Test

Free Response Questions from Previous AP® Statistics Exams

Questions can be found on the AP® Central website: apcentral .collegeboard.org/courses/ap-statistics/exam.

Students should be able to answer all the free response questions listed below with material learned in this chapter. Questions that contain content from this chapter but also require content from later chapters are listed in the last chapter required to complete the entire question. This list will be updated after each AP® Statistics exam and will be posted to the Teacher's Resource section of the book's digital platform and to www .statsmedic.com/free-response-questions.

Year	#	Content
2019	2	• Components of an experiment • Control group • Random assignment
2016	3	• Explanatory/response variables • Experiment vs. observational study • Confounding
2014	4	• Mean vs. median • Sampling methods and bias
2013	2	• Convenience sampling and bias • Selecting an SRS with a random number generator • Stratified sampling
2011B	2	• Observational study vs. experiment • Scope of inference • Purpose of random assignment
2011	3	• Cluster sampling • Stratified sampling
2010	1	• Treatments • Experimental units • Response variable • Scatterplots and linearity
2010B	2	• Simple random sampling • Stratified random sampling
2009	3	• Random assignment • Non-random assignment
2008	2	• Nonresponse bias
2007	2	• Control groups • Random assignment • Blocking
2007B	3	• Blocking • Randomization
2006	5	• Treatments • Randomization • Sources of variability • Generalizability
2006B	5	• Response variable • Treatments • Experimental units • Randomization • Replication • Confounding

Year	#	Content
2005	1	• Comparing distributions (stemplots) • Generalizability of results • Sampling variability
2004	2	• Blocking • Random assignment within blocks
2004B	2	• Selection and response bias
2003	4	• Random assignment • Control groups • Generalizability
2002	2	• Matched pairs experiment • Double-blind
2002B	3	• Designing experiment • Blocking
2001	4	• Blocking • Purpose of randomization
2000	5	• Designing experiment • Blocking • Double-blind
1999	3	• Experiment vs. observational study • Confounding • Cause and effect
1997	2	• Designing experiment • Blocking

Applets

- The *Simple Random Sample* applet allows students to randomly select a simple random sample from a population with labels from 1 to N. Go to highschool.bfwpub.com/updatedtps6e (click on Statistical Applets and select "Simple Random Sample").

- Use the *One Quantitative Variable* applet at highschool.bfwpub.com/updatedtps6e to perform the simulation in the "Caffeine and pulse rate" activity in Section 4.3. Use the *One Categorical Variable* applet for the Distracted Driver example.

Other Resources

- *Making Sense of Statistical Studies* by Roxy Peck and Daren Starnes, published by the American Statistical Association. This module consists of 15 hands-on investigations that provide students with valuable experience in designing and analyzing statistical studies.
- The "Planning and Conducting a Study Module" at AP® Central is an excellent resource that introduces teachers to the important concepts of data collection. It also goes just beyond the AP® curriculum and illustrates how to analyze the different types of designs encountered in AP® Statistics: https://apcentral.collegeboard.org/pdf /ap-statistics-module-planning-and-conducting-study.pdf.
- Newspapers: The health sections of major newspapers often include stories reporting the results of different studies: https://www.nytimes.com/section/health, https:// www.latimes.com/news/health/, and https://www.usatoday .com/news/health.

Chapter 4: Pacing Guide, Learning Targets, and Suggested Assignments

This pacing guide is based on a schedule with 110, 50-minute sessions before the AP® Statistics exam. If you have a different number of sessions before the AP® exam, you can modify the pacing guide to suit your needs. If you have additional time, consider incorporating quizzes, released AP® Statistics free response questions, or additional activities. See the Resources section above for suggestions.

The suggested homework assignments list odd-numbered exercises, whenever possible, so students can check their answers against the back-of-book answers. If you would rather students not have access to the answers while doing homework, adding 1 to the exercise numbers usually will do the trick, because the homework exercises typically are paired. For example, Exercises 1 and 2 will generally cover the same topics, but in different contexts. You may also choose to include the Recycle and Review questions at the end of each section, which review topics from previous sections or chapters.

If your school is using the digital platform that accompanies TPS6, you will find these assignments prebuilt as online homework assignments for Chapter 4.

Day	Content	Learning Targets Students will be able to . . .	Suggested Assignment (MC bold)
1	4.1 The Idea of a Sample Survey, How to Sample Badly, How to Sample Well: Random Sampling	• Identify the population and sample in a statistical study. • Identify voluntary response sampling and convenience sampling and explain how these sampling methods can lead to bias. • Describe how to select a simple random sample using slips of paper, technology, or a table of random digits.	1, 3, 5, 7, 11, 13
2	4.1 Other Random Sampling Methods	• Describe how to select a sample using stratified random sampling, cluster sampling, and systematic random sampling, and explain whether a particular sampling method is appropriate in a given situation.	15, 17, 19, 21, 22, 23
3	4.1 Sample Surveys: What Else Can Go Wrong?	• Explain how undercoverage, nonresponse, wording of questions, and other aspects of a sample survey can lead to bias.	25, 27, 29, 31, 33, **35–40**
4	4.2 Observational Studies Versus Experiments, The Language of Experiments	• Explain the concept of confounding and how it limits the ability to make cause-and-effect conclusions. • Distinguish between an observational study and an experiment, and identify the explanatory and response variables in each type of study. • Identify the experimental units and treatments in an experiment.	43, 45, 47, 49, 51, 53
5	4.2 Designing Experiments: Blinding and the Placebo Effect, Designing Experiments: Random Assignment	• Describe the placebo effect and the purpose of blinding in an experiment. • Describe how to randomly assign treatments in an experiment using slips of paper, technology, or a table of random digits.	57, 59, 61, 63
6	4.2 Designing Experiments: Comparison, Control, Replication, and Putting It All Together; Completely Randomized Designs	• Explain the purpose of comparison, random assignment, control, and replication in an experiment. • Describe a completely randomized design for an experiment.	55, 65, 67, 69
7	4.2 Randomized Block Designs	• Describe a randomized block design and a matched pairs design for an experiment and explain the purpose of blocking in an experiment.	71, 75, 77, 79, **83–90**
8	4.3 Inference for Sampling, Inference for Experiments	• Explain the concept of sampling variability when making an inference about a population and how sample size affects sampling variability. • Explain the meaning of "statistically significant" in the context of an experiment and use simulation to determine if the results of an experiment are statistically significant.	93, 95, 97, 99
9	4.3 The Scope of Inference: Putting It All Together, The Challenges of Establishing Causation, Data Ethics (optional)	• Identify when it is appropriate to make an inference about a population and when it is appropriate to make an inference about cause and effect. • Evaluate if a statistical study has been carried out in an ethical manner.	103, 105, 107, **117–118** (109, 111, 113, 115 optional)
10	Chapter 4 Review/FRAPPY!		Chapter 4 Review Exercises
11	Chapter 4 Test		Cumulative AP® Practice Test 1

Chapter 4 Alignment to the College Board's Fall 2019 AP® Statistics Course Framework*

Relationship to College Board Units

Chapter 4 in this book covers the topics addressed in Unit 3 of the College Board Course Framework. Students will be ready to take the Personal Progress Check for Unit 3 once they have completed Chapter 4.

Big Ideas and Enduring Understandings

Chapter 4 develops these Big Ideas and related Enduring Understandings outlined in the Course Framework:

- **Big Idea 1: Variation and Distribution (EU: VAR 1,3):** The distribution of measures for individuals within a sample or population describes variation. The value of a statistic varies from sample to sample. How can we determine whether differences between measures represent random variation or meaningful distinctions? Statistical methods based on probabilistic reasoning provide the basis for shared understandings about variation and about the likelihood that variation between and among measures, samples, and populations is random or meaningful.
- **Big Idea 3: Data-Based Predictions, Decisions, and Conclusions (EU: DAT 2):** Data-based regression models describe relationships between variables and are a tool for making predictions for values of a response variable. Collecting data using random sampling or randomized experimental design means that findings may be generalized to the part of the population from which the selection was made. Statistical inference allows us to make data-based decisions.

Course Skills

Chapter 4 helps students to develop the skills identified in the Course Framework.

- **1: Selecting Statistical Methods** (1.A, 1.B, 1.C)
- **4: Statistical Argumentation** (4.A, 4.B)

Learning Objectives and Essential Knowledge Statements

Section	Learning Objectives	Essential Knowledge Statements
4.1	DAT-2.A, DAT-2.C, DAT-2.D, DAT-2.E, VAR-1.E	DAT-2.A.1, DAT-2.A.2, DAT-2.C.1, DAT-2.C.2, DAT-2.C.3, DAT-2.C.4, DAT-2.C.5, DAT-2.C.6, DAT-2.D.1, DAT-2.E.1, DAT-2.E.2, DAT-2.E.3, DAT-2.E.4, DAT-2.E.5, DAT-2.E.6, VAR-1.E.1
4.2	DAT-2.A, VAR-3.A, VAR-3.B, VAR-3.C, VAR-3.D	DAT-2.A.3, DAT-2.A.4, VAR-3.A.1, VAR-3.A.2, VAR-3.A.3, VAR-3.A.4, VAR-3.B.1, VAR-3.C.1, VAR-3.C.2, VAR-3.C.3, VAR-3.C.4, VAR-3.C.5, VAR-3.C.6, VAR-3.C.7, VAR-3.C.8, VAR-3.C.9, VAR-3.D.1
4.3	DAT-2.B, VAR-3.E	DAT-2.B.1, DAT-2.B.2, DAT-2.B.3, VAR-3.E.1, VAR-3.E.2, VAR-3.E.3, VAR-3.E.4

A detailed alignment that can be sorted by Course Framework Unit, Topic, Learning Objective, Essential Knowledge Statement, or textbook section, is available on the TRFD and in the Teacher's Resources folder on Sapling Plus. **TRM**

*Should changes be made to the Course Framework in the future, an updated alignment will be placed on our AP® updates page at go.bfwpub.com/ap-course-updates.

UNIT 3
Collecting Data

Chapter 4

Collecting Data

AfriPics.com/Alamy Stock Photo

Teaching Tip

Unit 3 in the College Board Course Framework aligns to Chapter 4 in this book. Students will be ready to take the Personal Progress Check for Unit 3 once they have completed Chapter 4.

PD Chapter 4 Overview

To watch the video overview of Chapter 4 (for teachers), click on the link in the TE-Book, look on the TRFD, or download from the Teacher's Resources on the book's digital platform.

Teaching Tip

Many teachers start their course with Chapter 4. If you are doing so, consider starting with the "Hiring discrimination" activity from Chapter 1, because this activity highlights many of the key ideas of the course (see Chapter 1 Introduction). Then proceed with covering content from Chapter 4.

TRM Lecture Presentation Slides

If you are new to teaching AP® Statistics or are short on time when preparing for class, you may find the Lecture Presentation Slides to be helpful. Experienced AP® Statistics Teacher Doug Tyson has created one slide presentation per section. You may use them as is, modify them to fit your needs, or share them with students who miss class. Find them on the TRFD and in the Teacher's Resources on the book's digital platform.

INTRODUCTION

You can hardly go a day without hearing the results of a statistical study. Here are some examples:

- The National Highway Traffic Safety Administration (NHTSA) reports that seat belt use in passenger vehicles increased from 88.5% in 2015 to 90.1% in 2016.[1]
- According to a survey, U.S. teens aged 13 to 18 use entertainment media (television, Internet, social media, listening to music, etc.) nearly 9 hours a day, on average.[2]
- A study suggests that lack of sleep increases the risk of catching a cold.[3]
- For their final project, two AP® Statistics students showed that listening to music while studying decreased subjects' performance on a memory task.[4]

Can we trust these results? As you'll learn in this chapter, the answer depends on how the data were produced. Let's take a closer look at where the data came from in each of these studies.

Each year, the NHTSA conducts an *observational study* of seat belt use in vehicles. The NHTSA sends trained observers to record the behavior of people in vehicles at randomly selected locations across the country. The idea of an observational study is simple: you can learn a lot just by watching or by asking a few questions, as in the survey of teens' media habits. Common Sense Media conducted this survey using a random sample of 1399 U.S. 13- to 18-year-olds. Both of these studies use information from a *sample* to draw conclusions about some larger *population*. Section 4.1 examines the issues involved in sampling and surveys.

In the sleep and catching a cold study, 153 volunteers answered questions about their sleep habits over a two-week period. Then researchers gave them a virus and waited to see who developed a cold. This was a complicated observational study. Compare this with the *experiment* performed by the AP® Statistics students. They recruited 30 students and divided them into two groups of 15 by drawing names from a hat. Students in one group tried to memorize a list of words while listening to music. Students in the other group tried to memorize the same list of words while sitting in silence. Section 4.2 focuses on designing experiments.

In Section 4.3, we revisit two key ideas from Sections 4.1 and 4.2: drawing conclusions about a population based on a random sample and drawing conclusions about cause and effect based on a randomized experiment. In both cases, we will focus on the role of randomization in our analysis.

SECTION 4.1 Sampling and Surveys

LEARNING TARGETS *By the end of the section, you should be able to:*

- Identify the population and sample in a statistical study.
- Identify voluntary response sampling and convenience sampling and explain how these sampling methods can lead to bias.
- Describe how to select a simple random sample using slips of paper, technology, or a table of random digits.
- Describe how to select a sample using stratified random sampling, cluster sampling, and systematic random sampling, and explain whether a particular sampling method is appropriate in a given situation.
- Explain how undercoverage, nonresponse, question wording, and other aspects of a sample survey can lead to bias.

Suppose we want to find out what percent of young drivers in the United States text while driving. To answer the question, we will survey 16- to 20-year-olds who live in the United States and drive. Ideally, we would ask them all by conducting a **census**. But contacting every driver in this age group wouldn't be practical: it would take too much time and cost too much money. Instead, we pose the question to a **sample** chosen to represent the entire **population** of young drivers.

DEFINITION Population, Census, Sample

The **population** in a statistical study is the entire group of individuals we want information about. A **census** collects data from every individual in the population.

A **sample** is a subset of individuals in the population from which we collect data.

The distinction between population and sample is basic to statistics. To make sense of any sample result, you must know what population the sample represents.

EXAMPLE

Sampling monitors and voters
Populations and samples

PROBLEM: Identify the population and the sample in each of the following settings.

(a) The quality control manager at a factory that produces computer monitors selects 10 monitors from the production line each hour. The manager inspects each monitor for defects in construction and performance.

(b) Prior to an election, a news organization surveys 1000 registered voters to predict which candidate will be elected as president.

Cancan Chu/Getty Images

Teaching Tip

Use the following visual representation for ideas in Section 4.1. Build it with the class and refer back to it throughout the chapter. Recall the terms *parameter* and *statistic* from Chapter 1.

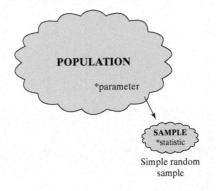

Simple random sample

ALTERNATE EXAMPLE   Skill 1.C

Cars and Twitter
Populations and samples

PROBLEM:

Identify the population and the sample in each of the following settings.

(a) An assembly line at a factory produces about 500 cars a day. Each day, quality control managers inspect 25 cars at the factory and perform an in-depth review of each car.

(b) A politician uses a Twitter poll to find out whether his followers agree with a recent bill that was passed and 432 people respond to the poll.

SOLUTION:

(a) The population is all the cars produced in this factory on a given day. The sample is the 25 cars selected from the assembly line.

(b) The population is all of the politician's Twitter followers. The sample is the 432 people who responded.

There are printable versions of this activity in the Teacher's Resource Materials located in the TE-Book, on the TRFD, or in the Teacher's view on the book's digital platform.

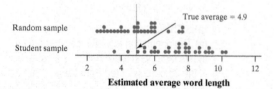

▶ ACTIVITY OVERVIEW

To prepare for using this activity, watch the overview video by clicking on the link in the TE-Book, opening the TRFD, or downloading from the Teacher's Resources on the book's digital platform.

Time: 15 minutes

Materials: Poster board or chart paper, sticker dots

Teaching Advice: Ask students to put sticker dots on a poster board or chart paper. Post the completed poster board in the classroom so that you can refer to this activity throughout the rest of the course. In the first part of the activity, rush students into choosing their 5 words. This will make them more likely to choose the larger words and to overestimate the true mean. The true mean word length of this paragraph is 4.9. A possible extension for this activity is to have students take a random sample of 10 words and find the sample mean to put on a third dotplot. This demonstrates a big idea that will become important later in the course: increasing the sample size decreases the variability of the sampling distribution.

Answers:

1–4. Answers will vary. Here is an example of what the dotplots might look like:

[Random sample / Student sample dotplot: True average = 4.9, x-axis "Estimated average word length" from 2 to 12]

5. Both dotplots of average word length will be fairly symmetric, with similar variability. The dotplot using the student sample will have a center that is greater than the center when using a random number generator. Students' eyes are naturally drawn to the larger words when they are choosing a sample.

SOLUTION:

(a) The population is all the monitors produced in this factory that hour. The sample is the 10 monitors selected from the production line.

(b) The population is all registered voters. The sample is the 1000 registered voters surveyed.

> Because the sample came from 1 hour's production at this factory, the population is the monitors produced that hour in this factory—not all monitors produced in the world or even all monitors produced by this factory.

FOR PRACTICE, TRY EXERCISE 1

The Idea of a Sample Survey

We often draw conclusions about a population based on a sample. Have you ever tasted a sample of ice cream and ordered a cone because the sample tastes good? Because ice cream is fairly uniform, the single taste represents the whole. Choosing a representative sample from a large and varied population (like all young U.S. drivers) is not so easy. The first step in planning a **sample survey** is to decide *what population* we want to describe. The second step is to decide *what we want to measure*.

> **DEFINITION** **Sample survey**
>
> A **sample survey** is a study that collects data from a sample to learn about the population from which the sample was selected.

By our definition, the individuals in a sample survey can be people, animals, or things. Some people use the terms *survey* or *sample survey* to refer only to studies in which people are asked questions, like the news organization survey in the preceding example. We'll avoid this more restrictive terminology.

The final step in planning a sample survey is to decide how to choose a sample from the population. Here is an activity that illustrates the process of conducting a sample survey.

ACTIVITY   Who wrote the Federalist Papers?

The Federalist Papers are a series of 85 essays supporting the ratification of the U.S. Constitution. At the time they were published, the identity of the authors was a secret known to only a few people. Over time, however, the authors were identified as Alexander Hamilton, James Madison, and John Jay. The authorship of 73 of the essays is fairly certain, leaving 12 in dispute. However, thanks in some part to statistical analysis, most scholars now believe that the 12 disputed essays were written by Madison alone or in collaboration with Hamilton.[5]

There are several ways to use statistics to help determine the authorship of a disputed text. One method is to estimate the average word length in a disputed text and compare it to the average word lengths of works where the authorship is not in dispute.

✚ Ask the StatsMedic

The Gettysburg Address

Does Beyoncé Write Her Own Lyrics?

These are two alternate activities to the "Who wrote the Federalist Papers?" activity. They are parallel activities with the same learning outcomes, so you need only choose one to use with students.

The following passage is the opening paragraph of Federalist Paper No. 51,[6] one of the disputed essays. The theme of this essay is the separation of powers between the three branches of government.

> To what expedient, then, shall we finally resort, for maintaining in practice the necessary partition of power among the several departments, as laid down in the Constitution? The only answer that can be given is, that as all these exterior provisions are found to be inadequate, the defect must be supplied, by so contriving the interior structure of the government as that its several constituent parts may, by their mutual relations, be the means of keeping each other in their proper places. Without presuming to undertake a full development of this important idea, I will hazard a few general observations, which may perhaps place it in a clearer light, and enable us to form a more correct judgment of the principles and structure of the government planned by the convention.

1. Choose 5 words from this passage. Count the number of letters in each of the words you selected, and find the average word length.
2. Your teacher will draw and label a horizontal axis for a class dotplot. Plot the average word length you obtained in Step 1 on the graph.
3. Your teacher will show you how to use a random number generator to select a random sample of 5 words from the 130 words in the opening passage. Count the number of letters in each of the selected words, and find the average word length.
4. Your teacher will draw and label another horizontal axis with the same scale for a comparative dotplot. Plot the average word length you obtained in Step 3 on the graph.
5. How do the dotplots compare? Can you think of any reasons why they might be different? Discuss with your classmates.

How to Sample Badly

Suppose we want to know how long students at a large high school spent doing homework last week. We might go to the school library and ask the first 30 students we see about the amount of time they spend on their homework. This method is known as a **convenience sampling**.

> **DEFINITION** **Convenience sampling**
>
> **Convenience sampling** selects individuals from the population who are easy to reach.

 Convenience sampling often produces unrepresentative data. Consider our sample of 30 students from the school library. It's unlikely that this convenience sample accurately represents the homework habits of all students at the high

school. In fact, if we were to repeat this sampling process day after day, we would almost always overestimate the average homework time in the population. Why? Because students who hang out in the library tend to be more studious. This is **bias**: using a method that systematically favors certain outcomes over others.

DEFINITION Bias

The design of a statistical study shows **bias** if it is very likely to underestimate or very likely to overestimate the value you want to know.

AP® EXAM TIP

If you're asked to describe how the design of a sample survey leads to bias, you're expected to do two things: (1) describe how the members of the sample might respond differently from the rest of the population, and (2) explain how this difference would lead to an underestimate or overestimate. Suppose you were asked to explain how using your statistics class as a sample to estimate the proportion of all high school students who own a graphing calculator could result in bias. You might respond, "This is a convenience sample. It would probably include a much higher proportion of students with a graphing calculator than in the population at large because a graphing calculator is required for the statistics class. So this method would probably lead to an overestimate of the actual population proportion."

 Bias is not just bad luck in one sample. It's the result of a bad study design that will consistently miss the truth about the population in the same way. Convenience sampling will almost always result in bias. So will **voluntary response sampling**.

Because voluntary response sampling typically gives a sample that is unrepresentative of the population we want to know about, the bias that results is called *voluntary response bias.*

DEFINITION Voluntary response sampling

Voluntary response sampling allows people to choose to be in the sample by responding to a general invitation.

Most Internet polls, along with call-in, text-in, and write-in polls, rely on voluntary response sampling. *People who self-select to participate in such surveys are usually not representative of the population of interest.* Voluntary response sampling attracts people who feel strongly about an issue, and who often share the same opinion. That leads to bias.

EXAMPLE **Boaty McBoatface**
Biased sampling methods

PROBLEM: In 2016, Britain's Natural Environment Research Council invited the public to name its new $300 million polar research ship. To vote on the name, people simply needed to visit a website and record their choice. Ignoring names suggested by the council, over 124,000 people voted for "Boaty McBoatface," which ended up having more than 3 times as many votes as the second-place finisher.[7]

ALTERNATE EXAMPLE   Skill 1.C

What is the average GPA?
Biased sampling methods

An AP® Statistics teacher was curious about the average grade point average (GPA) of students at his school. He used the 32 students in his second-period AP® Statistics class as a sample and concluded that the average GPA of students at his school is about 3.85.

PROBLEM:
What type of sampling did the teacher use? Explain how bias in this sampling method could have affected the results.

SOLUTION:
The teacher used convenience sampling: the second-period AP® Statistics class was an easy way to collect the data. Because the students in second-period AP® Statistics were all taking an Advanced Placement class, they are probably more dedicated to their school work and to learning than the population of all students at the school—and thus are more likely to have a higher GPA. The average GPA from the sample is likely to be greater than the average GPA of all students at the school.

What type of sampling did the council use in their poll? Explain how bias in this sampling method could have affected the poll results.

SOLUTION:

The council used voluntary response sampling: people chose to go online and respond. The people who chose to be in the sample were probably less serious about science than the British population as a whole—and more likely to prefer a funny name. The proportion of people in the sample who prefer the name Boaty McBoatface is likely to be greater than the proportion of all British residents who would choose this name.

> Remember to describe how the responses from the members of the sample might differ from the responses from the rest of the population *and* how this difference will affect the estimate.

FOR PRACTICE, TRY EXERCISE 5

 CHECK YOUR UNDERSTANDING

For each of the following situations, identify the sampling method used. Then explain how bias in the sampling method could affect the results.

1. A farmer brings a juice company several crates of oranges each week. A company inspector looks at 10 oranges from the top of each crate before deciding whether to buy all the oranges.

2. The ABC program *Nightline* once asked if the United Nations should continue to have its headquarters in the United States. Viewers were invited to call one telephone number to respond "Yes" and another to respond "No." There was a charge for calling either number. More than 186,000 callers responded, and 67% said "No."

How to Sample Well: Random Sampling

In convenience sampling, the researcher chooses easy-to-reach members of the population. In voluntary response sampling, people decide whether to join the sample. Both sampling methods suffer from bias due to personal choice. As you discovered in The Federalist Papers activity, a good way to avoid bias is to let chance choose the sample. That's the idea of **random sampling**.

> In everyday life, some people use the word *random* to mean "haphazard," as in "That's so random." In statistics, random means "using chance." Don't say that a sample was chosen at random if a chance process wasn't used to select the individuals.

DEFINITION Random sampling

Random sampling involves using a chance process to determine which members of a population are included in the sample.

For example, to choose a random sample of 6 students from a class of 30, start by writing each of the 30 names on a separate slip of paper, making sure the slips are all the same size. Then put the slips in a hat, mix them well, and pull out slips one at a time until you have identified 6 different students. An alternative approach would be to give each member of the population a distinct number and to use the "hat method" with these numbers instead of people's names. Note that this version would work just as well if the population consisted of animals or things. The resulting sample is called a **simple random sample**, or **SRS** for short.

Answers to CYU

1. Convenience sample; this could lead him to overestimate the quality if the farmer puts the best oranges on top or if the oranges at the bottom of the crate are damaged from the weight on top of them.

2. Voluntary response sample; those who are happy that the United Nations has its headquarters in the U.S. already have what they want and so are less likely to worry about responding to the question. This means that the proportion who answered "No" in the sample is likely to be higher than the true proportion in the U.S. who would answer "No."

Teaching Tip

Add to the visual that was started at the beginning of the chapter. This time, you can explain the three different sampling methods. A simple random sample (mini-cloud) is most likely to "look like" or be representative of the population (big cloud). The other sampling methods will not likely "look like" or be representative of the population.

Preparing for Inference

The fact that a simple random sample tends to be representative of the population means that we can take our conclusions from the sample results and generalize them to the whole population. In Chapters 8–12, we will check for a random sample in the Random condition so that we can generalize our results to the whole population.

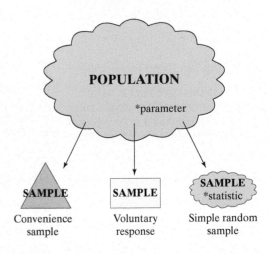

POPULATION

*parameter

SAMPLE
Convenience sample

SAMPLE
Voluntary response

SAMPLE
*statistic
Simple random sample

Teaching Tip

To help illustrate the definition of an SRS, describe how to select a sample of size 4 from your class. To do this, write each student's name on a slip of paper, mix the slips in a hat, and select 4 of them. Try this a few times and comment on the number of girls and boys in each sample. Then suggest that we may want to force the sample to have 2 girls and 2 boys. To do this, separate the names of the boys and girls into two hats and select 2 names from each hat (later, we will call this a stratified random sample). This is not an SRS because samples of 4 that do not contain 2 girls and 2 boys have no chance of being selected. In an SRS, each possible *sample of 4* has the same chance of occurring.

Teaching Tip: Using Technology

On an iPhone or Apple watch, you can request random numbers from Siri: "Give me a random number between 1 and ____ ." Siri will select a random number for you. On an Android phone, say, "Okay, Google, give me a random number between 1 and ____ ." Random.org also has an app that will generate random integers within a certain range.

Making Connections

In Chapter 6, one of the requirements for a variable to be considered a binomial random variable is that each trial is independent: knowing the outcome of one trial does not affect the probability of the next. When sampling without replacement from a finite population, the probabilities for each trial do depend on the outcome of the previous trial. But when the sample size is less than 10% of the population, the probability changes so little that we can view the trials as independent. This will later be called the 10% condition.

Teaching Tip: Using Technology

Remind students that they can watch a video presentation of Technology Corner 12 by clicking on the link in the e-Book or viewing the videos at the Student Site.

DEFINITION   Simple random sample (SRS)

A **simple random sample (SRS)** of size *n* is chosen in such a way that every group of *n* individuals in the population has an equal chance to be selected as the sample.

An SRS gives every possible sample of the desired size an equal chance to be chosen. Picture drawing 20 slips of paper (the sample) from a hat containing 200 identical slips (the population). Any set of 20 slips has the same chance as any other set of 20 to be chosen. This also means that each individual has the same chance to be chosen in an SRS. However, giving each individual the same chance to be selected is not enough to guarantee that a sample is an SRS. Some other random sampling methods give each member of the population an equal chance to be selected, but not each possible sample. We'll look at some of these methods later.

HOW TO CHOOSE A SIMPLE RANDOM SAMPLE The hat method won't work well if the population is large. Imagine trying to take a simple random sample of 1000 registered voters in the United States using a hat! In practice, most people use random numbers generated by technology to choose samples.

HOW TO CHOOSE AN SRS WITH TECHNOLOGY

- **Label.** Give each individual in the population a distinct numerical label from 1 to N, where N is the number of individuals in the population.
- **Randomize.** Use a random number generator to obtain *n different* integers from 1 to N, where *n* is the sample size.
- **Select.** Choose the individuals that correspond to the randomly selected integers.

When choosing an SRS, we **sample without replacement**. That is, once an individual is selected for a sample, that individual cannot be selected again. Many random number generators **sample with replacement**, so it is important to explain that repeated numbers should be ignored when using technology to select an SRS.

DEFINITION   Sampling without replacement, Sampling with replacement

When **sampling without replacement**, an individual from a population can be selected only once.

When **sampling with replacement**, an individual from a population can be selected more than once.

The following Technology Corner shows you how to select an SRS using a graphing calculator.

12. Technology Corner   CHOOSING AN SRS

TI-Nspire and other technology instructions are on the book's website at highschool.bfwpub.com/updatedtps6e.

Let's use a graphing calculator to select an SRS of 10 students from a population of students numbered 1 to 1750.

1. Check that your calculator's random number generator is working properly.

- Press MATH, then select PROB (PRB) and choose randInt(.
  Newer OS: In the dialogue box, enter these values: lower: 1, upper: 1750, n: 1, choose Paste, and press ENTER.

Teaching Tip: Using Technology

When brand new, all TI-83/84/89/Nspire calculators will generate the same random integers if given the same prompts. To ensure that all your students get different random integers, you must have them "seed" their calculators. To seed the random number generator on the TI-83/84, students must enter a unique number on their home screen, such as their phone number or student ID number. Then ask them to press the STO → button (just above the ON button). Finally, have them press the MATH button, scroll to the PROB menu, choose the first option, rand, and press ENTER. To seed the random number generator on the TI-89, from the home screen, press CATALOG and scroll down to RandSeed. Press ENTER to select the command. Then enter a unique number and press ENTER.

For the Nspire, press the MENU button, choose #5 Probability, then #4 Random, then #6 Seed to pull up the RandSeed function. Follow RandSeed with a unique number and press ENTER.

Older OS: Complete the command randInt(1,1750) and press ENTER.

NORMAL FLOAT AUTO REAL RADIAN CL

```
randInt(1,1750)
                                139.
randInt(1,1750)
                               1126.
randInt(1,1750)
                                920.
randInt(1,1750)
                               1089.
```

- Compare your results with those of your classmates. If several students got the same number, you'll need to seed your calculator's random integer generator with different numbers before you proceed. Directions for doing this are given in the *Teacher's Edition*.

2. Randomly generate 10 distinct numbers from 1 to 1750 by pressing ENTER until you have chosen 10 different labels.

Note: If you have a TI-84 Plus CE, use the command RandIntNoRep(1,1750,10) to get 10 distinct integers from 1 to 1750. If you have a TI-84 with OS 2.55 or later, use the command RandIntNoRep(1,1750) to sort the numbers from 1 to 1750 in random order. The first 10 numbers listed give the labels of the chosen students.

There are many random number generators available on the Internet, including those at www.random.org. You can also use the random number generator on your calculator.

If you don't have technology handy, you can use a table of random digits to choose an SRS. We have provided a table of random digits at the back of the book (Table D). Here is an excerpt:

Table D Random digits

LINE							
101	19223	95034	05756	28713	96409	12531	42544	82853
102	73676	47150	99400	01927	27754	42648	82425	36290
103	45467	71709	77558	00095	32863	29485	82226	90056

You can think of this table as the result of someone putting the digits 0 to 9 in a hat, mixing, drawing one, replacing it, mixing again, drawing another, and so on. The digits have been arranged in groups of five within numbered rows to make the table easier to read. The groups and rows have no special meaning—Table D is just a long list of randomly chosen digits. As with technology, there are three steps in using Table D to choose a random sample.

HOW TO CHOOSE AN SRS USING TABLE D

- **Label.** Give each member of the population a distinct numerical label with the same number of digits. Use as few digits as possible.
- **Randomize.** Read consecutive groups of digits of the appropriate length from left to right across a line in Table D. Ignore any group of digits that wasn't used as a label or that duplicates a label already in the sample. Stop when you have chosen *n* different labels.
- **Select.** Choose the individuals that correspond to the randomly selected labels.

Always use the shortest labels that will cover your population. For instance, you can label up to 100 individuals with two digits: 01, 02, . . . , 99, 00. As standard practice, we recommend that you begin with label 1 (or 01 or 001 or 0001, as needed). Reading groups of digits from the table gives all individuals the same chance to be chosen because all labels of the same length have the same chance to be found in the table. For example, any pair of digits in the table is equally likely to be any of the 100 possible labels 01, 02, . . . , 99, 00. Here's an example that shows how this process works.

Teaching Tip: Using Technology

It is also possible to generate sets of random numbers at one time. For example, the command randInt(1,28,4) will generate four random integers from 1 to 28. However, because each number is independently generated, you may not get four *unique* numbers with this command.

AP® EXAM TIP

Occasionally, students are asked to use random digits on the AP® Statistics exam, so make sure they use a table of random digits from time to time.

AP® EXAM TIP

Table D is not a universal designation. On the AP® Statistics exam, students should refer to a "Table of random digits" rather than to "Table D."

AP® EXAM TIP

When working with a table of random digits, it is very important that each label have the same number of digits. For example, if you need 50 labels, use 01–50 rather than 1–50. The single digit "1" has a 10% chance of occurring, but the two-digit combination "01" has the proper 2% chance of occurring. Also, students often lose credit for failing to address what they will do with a repeated label.

Teaching Tip

Teachers often have to select a sample of students to offer answers, to write on the board, to share good news, and so on. Consider the following methods for selecting a sample: On "voluntary response Mondays," students get to self-select; on "SRS Friday," students are randomly selected. Each Friday, use a different method for selecting a random sample (names in a hat, Table D, RandInt, random.org, Siri).

Good news
Choosing an SRS with Table D

PROBLEM:
To promote positive classroom culture, Mr. Wilcox often asks his students to share "Good News." Because he doesn't have time to let every student share each day, he takes a sample of students who will share.

(a) Describe how to use a random number generator to select an SRS of 5 students from the following list of 29 students.

Allison	Amari	Benjamin
Danijal	Kevin D.	Kevin H.
Damario	Emiley	Kayla
Tessa	Geneva	Micaela
Gabe L.	Anh	Sean
Kirah	Thai	Harrison
Turner	Bernard	Daejynae
Brandon	Jarrod	Kim
Emily	Jenny	Jackelyn
Gabe Y.	Luz	

(b) The random number generator at www.random.org was used to get the following random integers between 1 and 29. Use these integers to choose the sample.

4　14　21　19　14　12　25　2

SOLUTION:
(a) Label the students from 1 to 29 in the order that they are written (across rows). Use a random number generator to obtain 5 different integers from 1 to 29 (ignore repeats). Choose the students who correspond to the randomly selected integers to share "Good News."

(b) 4-select, 14-select, 21-select, 19-select, 14-skip (repeat), 12-select
The five students are 4: Danijal, 14: Anh, 21: Daejynae, 19: Turner, and 12: Micaela.

AP® EXAM TIP

When describing how to select a sample using a random number generator or Table D, many students forget to address what to do with repeated numbers. Students must explicitly state that repeated numbers should be ignored or say that they will generate random numbers until they get n different numbers in the specified range.

EXAMPLE

Attendance audit
Choosing an SRS with Table D

Holly Albrecht

PROBLEM: Each year, the state Department of Education randomly selects three schools from each district and conducts a detailed audit of their attendance records.

(a) Describe how to use a table of random digits to select an SRS of three schools from this list of 19 schools.

Amphitheater High School	Keeling Elementary School
Amphitheater Middle School	La Cima Middle School
Canyon del Oro High School	Mesa Verde Elementary School
Copper Creek Elementary School	Nash Elementary School
Coronado K-8 School	Painted Sky Elementary School
Cross Middle School	Prince Elementary School
Donaldson Elementary School	Rio Vista Elementary School
Harelson Elementary School	Walker Elementary School
Holaway Elementary School	Wilson K-8 School
Ironwood Ridge High School	

(b) Use the random digits here to choose the sample.

62081　64816　87374　09517　84534　06489　87201　97245

SOLUTION:

(a) Label the schools from 01 to 19 in alphabetical order. Move along a line of random digits from left to right, reading two-digit numbers, until three different numbers from 01 to 19 have been selected (ignoring repeated numbers and the numbers 20–99, 00). Audit the three schools that correspond with the numbers selected.

> Remember to include all three steps:
> • Label
> • Randomize
> • Select

(b) 62-skip, 08-select, 16-select, 48-skip, 16-repeat, 87-skip, 37-skip, 40-skip, 95-skip, 17-select.

The three schools are 08: Harelson Elementary School, 16: Prince Elementary School, and 17: Rio Vista Elementary School.

FOR PRACTICE, TRY EXERCISE 11

CHECK YOUR UNDERSTANDING

A furniture maker buys hardwood in batches that each contain 1000 pieces. The supplier is supposed to dry the wood before shipping (wood that isn't dry won't hold its size and shape). The furniture maker chooses five pieces of wood from each batch and tests their moisture content. If any piece exceeds 12% moisture content, the entire batch is sent back. Describe how to select a simple random sample of 5 pieces using each of the following:

1. A random number generator
2. A table of random digits

✓ Answers to CYU

1. Number the pieces of wood from 1 to 1000. Use the command randInt(1,1000) to select 5 different integers from 1 to 1000. Inspect the 5 pieces of wood that correspond with the numbers selected.

2. Number the pieces of wood from 000 to 999. Move along a line of random digits from left to right, reading three-digit numbers, until 5 different numbers between 000 and 999 have been selected. Inspect the corresponding 5 pieces of wood.

➕ Ask the StatsMedic
How Much Do Fans Love Justin Timberlake?

This blog post provides an alternative activity to the "Sampling sunflowers" activity found on the next page. In the Justin Timberlake activity, students take samples to determine which variable works best for setting up a stratified random sample.

Other Random Sampling Methods

One of the most common alternatives to simple random sampling is called **stratified random sampling**. This method involves dividing the population into non-overlapping groups (**strata**) of individuals who are expected to have similar responses, sampling from each of these groups, and combining these "subsamples" to form the overall sample.

> The singular form of *strata* is *stratum*.

DEFINITION   Strata, Stratified random sampling

Strata are groups of individuals in a population who share characteristics thought to be associated with the variables being measured in a study. **Stratified random sampling** selects a sample by choosing an SRS from each stratum and combining the SRSs into one overall sample.

Stratified random sampling works best when the individuals within each stratum are similar (homogeneous) with respect to what is being measured and when there are large differences between strata. For example, in a study of sleep habits on school nights, the population of students in a large high school might be divided into freshman, sophomore, junior, and senior strata. After all, it is reasonable to think that freshmen have different sleep habits than seniors. The following activity illustrates the benefit of choosing appropriate strata.

ACTIVITY   Sampling sunflowers

Li Ding/Alamy

A British farmer grows sunflowers for making sunflower oil. Her field is arranged in a grid pattern, with 10 rows and 10 columns as shown in the figure. Irrigation ditches run along the top and bottom of the field. The farmer would like to estimate the number of healthy plants in the field so she can project how much money she'll make from selling them. It would take too much time to count the plants in all 100 squares, so she'll accept an estimate based on a sample of 10 squares.

1. Use Table D or technology to take a simple random sample of 10 grid squares. Record the location (e.g., B6) of each square you select.

2. This time, select a stratified random sample using the *rows* as strata. Use Table D or technology to randomly select one square from each horizontal row. Record the location of each square—for example, Row 1: G, Row 2: B, and so on.

3. Now, take a stratified random sample using the *columns* as strata. Use Table D or technology to randomly select one square from each vertical column. Record the location of each square—for example, Column A: 4, Column B: 1, and so on.

4. Your teacher will provide the actual number of healthy sunflowers in each grid square. Use that information to calculate your estimate of the mean number of healthy sunflowers per square in the entire field for each of your samples in Steps 1, 2, and 3.

5. Make comparative dotplots showing the mean number of healthy sunflowers obtained using the three different sampling methods for all members of the class. Describe any similarities and differences you see.

Teaching Tip

Strata is the plural form of *stratum,* just as *data* is the plural form of *datum*.

Teaching Tip

Imagine the population (big cloud) split into several strata. Take an SRS within each stratum. Use the following visual representation.

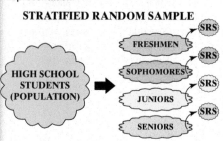

STRATIFIED RANDOM SAMPLE

Teaching Tip

How many individuals should be selected from each stratum? It depends on several factors, including the relative sizes of the strata and the variability within each stratum. One simple method is to choose samples of the same size from each stratum. Another method is to keep the sample size within each stratum roughly proportional to the size of the strata in the population. For example, if 30% of the students in a school are freshmen and we want a random sample of 50 students, we would select $(0.30)(50) = 15$ students from the freshmen stratum. Either of these is appropriate for AP® Statistics. However, there are other methods beyond the scope of AP® Statistics that can be used to increase the precision of the estimate.

ACTIVITY OVERVIEW

To prepare for this activity, watch the overview video by clicking on the link in the TE-Book, opening the TRFD, or downloading from the Teacher's Resources on the book's digital platform. Also provided is a document with the actual number of healthy sunflowers per square in the field.

Time: 15 minutes

Materials: Poster board or chart paper, sticker dots

Teaching Advice: After completing Step 3, ask students which method they believe will give the best estimate. Make sure they give a justification! In short, students should recognize that the irrigation ditches will be a source of variability in the number of sunflowers. Stratifying by row

will give the best estimates because it will take a sample from each different distance from the ditches. In the other two methods, it is possible that the selected squares will all be close to the ditches or far from the ditches. After calculating means for their samples, students should put sticker dots on a poster board or chart paper. Post the completed poster board in the classroom so that you can refer to this activity throughout the rest of the course. One possible extension would be to ask students how to take a cluster sample. For this sampling method, each column would be considered a cluster, and we would randomly select one column. Every plot of land in that column would be sampled. In this context, a cluster sample works as well as stratifying by row, but this is not usually the case. Most often, a cluster sample will not work as well as stratifying properly.

Answers:

1–4. Answers will vary. See dotplots on page 258 for examples.

5. All three dotplots of the mean number of plants are fairly symmetric with similar centers. The dotplot with rows for strata has less variability.

Teaching Tip

Make sure that students can explain how to choose a variable for stratification (the one that best predicts the response) *and* why stratified random samples are sometimes better than simple random samples (less variability of the estimates).

Teaching Tip

Imagine taking the population (big cloud) and dividing it into many clusters. Then take a random sample of a few of the clusters, and sample every individual in the selected clusters. Ideally, each of these clusters would be a mini-cloud that is representative of the population. This is not often the case, because clusters are based on proximity and tend to be homogeneous rather than heterogeneous.

The following dotplots show the mean number of healthy plants in 100 samples using each of the three sampling methods in the activity: simple random sampling, stratified random sampling with rows of the field as strata, and stratified random sampling with columns of the field as strata. Notice that all three distributions are centered at about 102.5, the true mean number of healthy plants in all squares of the field. That makes sense because random sampling tends to yield accurate estimates of unknown population means.

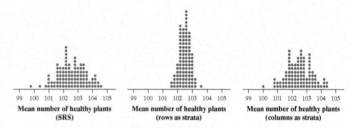

One other detail stands out in the graphs: there is much less variability in the estimates when we use the rows as strata. The table provided by your teacher shows the actual number of healthy sunflowers in each grid square. Notice that the squares within each row are relatively homogeneous because they contain a similar number of healthy plants, but that there are big differences between rows. *When we can choose strata that have similar responses (e.g., number of healthy plants) within strata but different responses between strata, stratified random samples give more precise estimates than simple random samples of the same size.*

Why didn't using the columns as strata reduce the variability of the estimates in a similar way? Because the numbers of healthy plants vary a lot within each column and aren't very different from other columns.

Both simple random sampling and stratified random sampling are hard to use when populations are large and spread out over a wide area. In that situation, we might prefer to use **cluster sampling**. This method involves dividing the population into non-overlapping groups (**clusters**) of individuals that are "near" one another, then randomly selecting whole clusters to form the overall sample.

> **DEFINITION** Clusters, Cluster sampling
>
> A **cluster** is a group of individuals in the population that are located near each other. **Cluster sampling** selects a sample by randomly choosing clusters and including each member of the selected clusters in the sample.

Cluster sampling is often used for practical reasons, like saving time and money. Ideally, the individuals within each cluster are heterogeneous (mirroring the population) and clusters are similar to each other in their composition. Imagine a large high school that assigns students to homerooms alphabetically by last name, in groups of 25. Administrators want to survey 200 randomly selected students about a proposed schedule change. It would be difficult to track down an SRS of 200 students, so the administration opts for a cluster sample of homerooms. The principal (who knows some statistics) selects an SRS of 8 homerooms and gives the survey to all 25 students in each of the selected homerooms.

CLUSTER SAMPLE

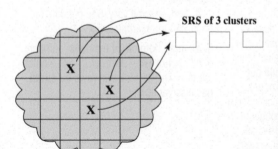

Teaching Tip

Emphasize that efficiency is the main benefit of cluster sampling. However, try to avoid the word *convenient* when describing this aspect of cluster sampling, as students may associate it (incorrectly) with a convenience sample. Cluster sampling is valid if the clusters are selected at random.

Be sure you understand the difference between strata and clusters. We want each stratum to contain similar individuals and for large differences to exist between strata. For a cluster sample, we'd *like* each cluster to look just like the population, but on a smaller scale. Unfortunately, cluster samples don't offer the statistical advantage of better information about the population that stratified random samples do. Here's an example that compares stratified random sampling and cluster sampling.

EXAMPLE

Sampling at a school assembly
Other sampling methods

PROBLEM: The student council wants to conduct a survey about use of the school library during the first five minutes of an all-school assembly in the auditorium. There are 800 students present at the assembly. Here is a map of the auditorium. Note that students are seated by grade level and that the seats are numbered from 1 to 800.

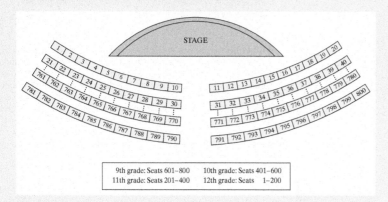

9th grade: Seats 601–800	10th grade: Seats 401–600
11th grade: Seats 201–400	12th grade: Seats 1–200

(a) Describe how to obtain a sample of 80 students using stratified random sampling. Explain your choice of strata and why this method might be preferred to simple random sampling.

(b) Describe how to obtain a sample of 80 students using cluster sampling. Explain your choice of clusters and why this method might be preferred to simple random sampling.

SOLUTION:

(a) *Because students' library use might be similar within grade levels but different across grade levels, use the grade-level seating areas as strata. For the 9th grade, generate 20 different random integers from 601 to 800 and give the survey to the students in those seats. Do the same for sophomores, juniors, and seniors using their corresponding seat numbers. Stratification by grade level should result in more precise estimates of student opinion than a simple random sample of the same size.*

Teaching Tip

In both stratified random sampling and cluster sampling, we take the population and split it into non-overlapping groups (strata or clusters). The big difference is that the individuals within a single stratum should be homogeneous (similar) with respect to the variable of interest, and individuals within a single cluster are ideally heterogeneous (different) with respect to the variable of interest.

✚ **Ask the StatsMedic**
Stratified Random Sampling Versus Cluster Sampling

In this post, we discuss the similarities and differences between these two sampling methods.

Teaching Tip

Here is another way to distinguish stratified sampling from cluster sampling: in a stratified sample, we divide the population into strata and take "some from all"; whereas in a cluster sample, we divide the population into clusters and take "all from some."

ALTERNATE EXAMPLE   Skill 1.C

A good read
Other sampling methods

PROBLEM:
A school librarian wants to know the average number of pages in all the books in the library. The library has 20,000 books, arranged by type (fiction, biography, history, etc.) in shelves that hold about 50 books each. You want to select a random sample of 500 books.
(a) Explain how to select a stratified random sample of 500 books. Justify your choice of strata. Why might the librarian want to choose a stratified random sample?
(b) Explain how to select a cluster sample of 500 books. Justify your choice of clusters. Why might the librarian want to choose a cluster sample?

SOLUTION:
(a) Stratify by type because different types of books may be longer (or shorter) than other types. This will provide a more precise estimate of the average page length than a simple random sample would. To select the sample, take an appropriately sized SRS of each type of book and combine the books selected from each type to form the sample. For example, if there are 1000 biographies (5% of 20,000), select an SRS of 25 biographies (5% of 500) using the method described in part (a).
(b) Clusters are formed by grouping books that are located near each other, making it easier for the librarian to select the random sample. We can use each shelf of 50 books as a cluster and randomly select 10 shelves to obtain the 500 books for our sample. Number the shelves from 1 to 400 and choose an SRS of 10 shelves. Then use all the books on the 10 selected shelves as the sample.

(b) Each column of seats from the stage to the back of the auditorium could be used as a cluster because it would be relatively easy to hand out the surveys to an entire column. Label the columns from 1 to 20 starting at the left side of the stage, generate 2 different integers from 1 to 20, and give the survey to the 80 students sitting in these two columns. Cluster sampling is much more efficient than finding 80 seats scattered about the auditorium, as required by simple random sampling.

> Note that each cluster contains students from all four grade levels, so each should represent the population fairly well. Randomly selecting 4 rows as clusters would also be easy, but this may over- or under-represent one grade level.

FOR PRACTICE, TRY EXERCISE 21

Another way to select a random sample of 80 students from the 800 students at the assembly is with **systematic random sampling**.

DEFINITION Systematic random sampling

Systematic random sampling selects a sample from an ordered arrangement of the population by randomly selecting one of the first k individuals and choosing every kth individual thereafter.

Because there are 800 students at the assembly and our desired sample size is 80, we want a systematic random sample that selects every $800/80 = 10$th student. For simplicity, we could survey students as they come through the door into the assembly. To choose a starting point, we randomly select a number from 1 to 10. Suppose we get the number 6. We would then survey the 6th student to enter, the 16th student to enter, the 26th student to enter, and so on until the 796th student enters the assembly.

 We could also use a systematic random sample with every 10th seat inside the auditorium. That is, we could survey the student sitting in seat 6, the student sitting in seat 16, and so on. There is a drawback to this method, however. It is possible that our starting number could have been 1 or 10, which would result in selecting only students who are sitting in aisle seats. If these seats are filled primarily by tardy students who are less responsible, the results of the survey might not provide an accurate estimate of library use for all students. If there are patterns in the way the population is ordered that coincide with the pattern in a systematic sample, the sample may not be representative of the population.

Systematic random sampling is particularly useful in certain contexts, such as exit polling at a polling place on Election Day. Because an unknown number of voters will come to the polling place that day, it isn't practical to select a simple random sample. However, it would be quite easy to generate a random number from 1 to 20 and to survey the chosen voter and every 20th voter thereafter.

Most large-scale sample surveys use *multistage sampling*, which combines two or more sampling methods. For example, the U.S. Census Bureau carries out a monthly Current Population Survey (CPS) of about 60,000 households. Researchers start by choosing a stratified random sample of neighborhoods in 756 of the 2007 geographical areas in the United States. Then they divide each neighborhood into clusters of four nearby households and select a cluster sample to interview.

Analyzing data from sampling methods other than simple random sampling takes us beyond basic statistics. But the SRS is the building block of more elaborate methods, and the principles of analysis remain much the same for these other methods.

CHECK YOUR UNDERSTANDING

A factory runs 24 hours a day, producing 15,000 wood pencils per day over three 8-hour shifts—day, evening, and overnight. In the last stage of manufacturing, the pencils are packaged in boxes of 10 pencils each. Each day a sample of 300 pencils is selected and inspected for quality.

1. Describe how to select a stratified random sample of 300 pencils. Explain your choice of strata.
2. Describe how to select a cluster sample of 300 pencils. Explain your choice of clusters.
3. Describe how to select a systematic random sample of 300 pencils.
4. Explain a benefit of each of these three methods in this context.

Sample Surveys: What Else Can Go Wrong?

As we have learned, the use of bad sampling methods (convenience or voluntary response) often leads to bias. Researchers can avoid these methods by using random sampling to choose their samples. Other problems in conducting sample surveys are more difficult to avoid.

Sampling is sometimes done using a list of individuals in the population, called a *sampling frame*. Such lists are seldom accurate or complete. The result is **undercoverage**.

> Because undercoverage typically results in a sample that is unrepresentative of the population we want to know about, the bias that occurs is sometimes called *undercoverage bias*.

DEFINITION Undercoverage

Undercoverage occurs when some members of the population are less likely to be chosen or cannot be chosen in a sample.

Most samples suffer from some degree of undercoverage. A sample survey of households, for example, will miss not only homeless people but also prison inmates and students in dormitories. An opinion poll conducted by calling landline telephone numbers will miss households that have only cell phones as well as households without a phone. The results of sample surveys may not be accurate if the people who are undercovered differ from the rest of the population in ways that affect their responses.

Even if every member of the population is equally likely to be selected for a sample, not all members of the population are equally likely to provide a response. Some people are never at home and cannot be reached by pollsters on the phone or in person. Other people see an unfamiliar phone number on their caller ID and never pick up the phone or quickly hang up when they don't recognize the voice of the caller. These are examples of **nonresponse**, another major source of bias in surveys.

> Because nonresponse typically results in a sample that is unrepresentative of the population we want to know about, the bias that occurs is sometimes called *nonresponse bias*.

DEFINITION Nonresponse

Nonresponse occurs when an individual chosen for the sample can't be contacted or refuses to participate.

Nonresponse leads to bias when the individuals who can't be contacted or refuse to participate would respond differently from those who do participate. Consider a landline telephone survey that asks people how many hours of television they watch per day. People who are selected but are out of the house won't

---

1. Because the quality of the pencils might be the same within each of the shifts, but differ across the shifts, use the shifts as strata. At the end of each 8-hour shift, label all the pencils produced during that shift 1 to 5000, where *N* is the total number of pencils produced on that shift. Generate 100 different random integers from 1 to 5000 and select those pencils for inspection.

2. The boxes of 10 pencils could be used as a cluster sample because it would be relatively easy to select boxes of 10 pencils. At the end of the day, label all the boxes of pencils 1 to 1500. Generate 30 different random integers from 1 to 1500 and inspect all the pencils in the selected boxes.

3. Select every 15,000/300 = 50th pencil that comes off the Start by randomly selecting a number from 1 to 50. Select that pencil and every 50th pencil thereafter until 300 pencils have been selected.

4. *Stratified:* We are guaranteed to inspect 100 pencils from each of the three shifts. This will lead to a more precise estimate of overall quality if quality is consistent within each shift but differs between the three shifts. *Cluster:* Simplifies the sampling process. Rather than having to label every pencil produced, only the packages of 10 pencils would need to be labeled. *Systematic:* It is easier to find the selected pencils by selecting them as they come off the production line, which will guarantee that samples of pencils that were manufactured at regular intervals over the course of the day are selected. This method will also allow us to address any manufacturing issues before too many faulty pencils are manufactured.

Teaching Tip

When discussing any type of bias, ask students to speculate on the direction of the bias. For example, if a survey about unemployment was conducted over the phone, nonresponse will lead to bias. It is likely that the estimated proportion of the unemployed will be too high because people who are unemployed are more likely to be at home and available for the survey.

Teaching Tip

Use the following visual representation for undercoverage. Notice that the sample doesn't "look like" the population.

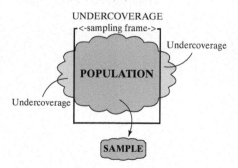

Teaching Tip

Nate Silver is a statistician who is famous for correctly predicting the outcome of the 2012 presidential election in all 50 states. He is author of the book *The Signal and the Noise* and maintains the website www.fivethirtyeight.com. The website is a blog that focuses on poll analysis, elections, and sports blogging. Statistics teachers should regularly visit the site for great articles about using data to make intelligent conclusions.

AP® EXAM TIP

Many students lose credit when describing what can go wrong in sample surveys because they use incorrect terminology. While it is important for students to understand and use the vocabulary of statistics correctly, they are rarely required to use specific vocabulary in their responses on the AP® Statistics exam. To be safe, tell your students not to worry about naming a specific problem or source of bias. Instead, have them clearly describe the problem and its consequences in the context of the question.

AP® EXAM TIP

If you're asked to describe how issues with the collection of survey data lead to bias, you're expected to address two ideas: (1) describe how the members of the sample might respond differently from the rest of the population, and (2) explain how this difference would lead to an underestimate or overestimate.

Teaching Tip

You may want to assign students the Chapter 4 Project on Response Bias while they are being introduced to the concept. To complete this project, students will need a good understanding of experimental design, which is covered in Section 4.2.

be able to respond. Because these people probably watch less television than the people who are at home when the phone call is made, the mean number of hours obtained in the sample is likely to be greater than the mean number of hours of TV watched in the population.

How bad is nonresponse? According to polling guru Nate Silver, "Response rates to political polls are dismal. Even polls that make every effort to contact a representative sample of voters now get no more than 10 percent to complete their surveys—down from about 35 percent in the 1990s."[8] In contrast, the Census Bureau's American Community Survey (ACS) has the lowest nonresponse rate of any poll we know: only about 1% of the households in the sample refuse to respond. The overall nonresponse rate, including "never at home" and other causes, is just 2.5%.[9]

Some students misuse the term *voluntary response* to explain why certain individuals don't respond in a sample survey. Their belief is that participation in the survey is optional (voluntary), so anyone can refuse to take part. What the students are describing is *nonresponse*. Think about it this way: nonresponse can occur only after a sample has been selected. In a voluntary response sample, every individual has opted to take part, so there won't be any nonresponse.

The wording of questions has an important influence on the answers given to a sample survey. Confusing or leading questions can introduce *question wording bias*. Even a single word can make a difference. In a recent Quinnipiac University poll, half of the respondents were asked if they support "stronger gun laws" and the other half were asked if they support "stronger gun control laws." In the first group, 52% of respondents supported stronger laws, but when the word *control* was added to the question, only 46% of respondents supported stronger laws.[10]

The gender, age, ethnicity, or behavior of the interviewer can also affect people's responses. People may lie about their age, income, or drug use. They may misremember how many hours they spent on the Internet last week. Or they might make up an answer to a question that they don't understand. All these issues can lead to **response bias**.

> **DEFINITION** Response bias
>
> **Response bias** occurs when there is a systematic pattern of inaccurate answers to a survey question.

EXAMPLE

Wash your hands!
Response bias

PROBLEM: What percent of Americans wash their hands after using the bathroom? It depends on how you collect the data. In a telephone survey of 1006 U.S. adults, 96% said they always wash their hands after using a public restroom. An observational study of 6028 adults in public restrooms told a different story: only 85% of those observed washed their hands after using the restroom. Explain why the results of the two studies are so different.[11]

SOLUTION:

When asked in person, many people may lie about always washing their hands because they want to appear to have good hygiene. When people are only observed and not asked directly, the percent who wash their hands will be smaller—and much closer to the truth.

FOR PRACTICE, TRY EXERCISE 29

ALTERNATE EXAMPLE   Skill 1.C

Trumping the polls   Response bias

PROBLEM:
Sometimes bias occurs when people respond to interview questions dishonestly because of the pressure of giving socially acceptable answers. In the 2016 presidential election, most polls predicted Hillary Clinton as the eventual winner. Donald Trump surprised many people when he was elected as the 45th president of the United States. Explain how this idea might have affected the poll's predictions.

SOLUTION:
It may be that the people interviewed in the polls responded that they intended to vote for Hillary Clinton because they thought that this choice was socially acceptable. When it was time to vote, however, the actual percentage who voted and cast their ballots for Hillary Clinton was less than what the polls predicted.

Even the order in which questions are asked is important. For example, ask a sample of college students these two questions:

- "How happy are you with your life in general?" (Answer on a scale of 1 to 5.)
- "How many dates did you have last month?"

There is almost no association between responses to the two questions when asked in this order. It appears that dating has little to do with happiness. Reverse the order of the questions, however, and a much stronger association appears: college students who say they had more dates tend to give higher ratings of happiness about life. The lesson is clear: the order in which questions are asked can influence the results.

CHECK YOUR UNDERSTANDING

1. Each of the following is a possible source of bias in a sample survey. Name the type of bias that could result.
   (a) The sample is chosen at random from a telephone directory.
   (b) Some people cannot be contacted in five calls.
   (c) Interviewers choose people walking by on the sidewalk to interview.
2. A survey paid for by makers of disposable diapers found that 84% of the sample opposed banning disposable diapers. Here is the actual question:

   It is estimated that disposable diapers account for less than 2% of the trash in today's landfills. In contrast, beverage containers, third-class mail, and yard wastes are estimated to account for about 21% of the trash in landfills. Given this, in your opinion, would it be fair to ban disposable diapers?[12]

   Do you think the estimate of 84% is less than, greater than, or about equal to the percent of all people in the population who would oppose banning disposable diapers? Explain your reasoning.

Section 4.1 | Summary

- A **census** collects data from every individual in the **population.**
- A **sample survey** selects a **sample** from the population of all individuals about which we desire information. The goal of a sample survey is to draw conclusions about the population based on data from the sample.
- **Convenience sampling** chooses individuals who are easiest to reach. In **voluntary response sampling,** individuals choose to join the sample in response to an open invitation. Both these sampling methods usually lead to **bias:** they will be very likely to underestimate or very likely to overestimate the value you want to know.
- **Random sampling** uses a chance process to select a sample.
- A **simple random sample (SRS)** gives every possible sample of a given size the same chance to be chosen. Choose an SRS by labeling the members of the population and using slips of paper, a table of random digits, or technology to select the sample. Make sure to use **sampling without replacement** when selecting an SRS.
- To use **stratified random sampling,** divide the population into non-overlapping groups of individuals (**strata**) that are similar in some way that might affect their responses. Then choose a separate SRS from each stratum and

➕ Ask the StatsMedic

What Could Students Do for a Response Bias Project?

In this post, we show several examples of student projects for collecting data that show a response bias.

✓ Answers to CYU

1. (a) Undercoverage

(b) Nonresponse

(c) Convenience

2. The estimate of 84% is greater than the percent of all people in the population who would oppose banning disposable diapers. By making it sound as if diapers are not a problem in the landfill, this question will result in fewer people suggesting that we should ban disposable diapers.

TRM Section 4.1 Quiz

There is one quiz available for this section in the Teacher's Resource Materials. Click on the link in the TE-Book, look on the TRFD, or download from the Teacher's Resources on the book's digital platform. You can also create your own quiz using the ExamView® Assessment Suite that is part of the TPS6 program. Questions are coded by Learning Target to make it easy to build parallel quizzes.

Teaching Tip

Make sure to point out the two icons next to Exercise 1. The "pg 249" icon reminds students that the example on page 249 is much like this exercise. The "play" icon reminds students that there is a video solution available in the student e-Book or at the Student Site.

combine these SRSs to form the sample. When strata are "similar within (homogeneous) but different between," stratified random samples tend to give more precise estimates of unknown population values than do simple random samples.

- To use **cluster sampling,** divide the population into non-overlapping groups of individuals that are located near each other, called **clusters.** Randomly select some of these clusters. All the individuals in the chosen clusters are included in the sample. Ideally, clusters are "different within (heterogeneous) but similar between." Cluster sampling saves time and money by collecting data from entire groups of individuals that are close together.

- To use **systematic random sampling,** select a value of k based on the population size and desired sample size, randomly select a value from 1 to k to identify the first individual in the sample, and choose every kth individual thereafter. If there are patterns in the way the population is ordererd that coincide with the value of k, the sample may not be represenative of the population. Otherwise, systematic random sampling can be easier to conduct than other sampling methods.

- Random sampling helps avoid bias in choosing a sample. Bias can still occur in the sampling process due to **undercoverage,** which happens when some members of the population are less likely to be chosen or cannot be chosen for the sample.

- Other serious problems in sample surveys can occur after the sample is chosen. The single biggest problem is **nonresponse:** when people can't be contacted or refuse to answer. Untruthful answers by respondents, poorly worded questions, and other problems can lead to **response bias.**

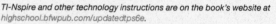

4.1 Technology Corner

TI-Nspire and other technology instructions are on the book's website at highschool.bfwpub.com/updatedtps6e.

12. Choosing an SRS   Page 254

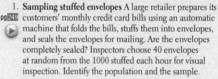

Section 4.1 | Exercises

Answers to Section 4.1 Exercises

4.1 Population: The 1000 envelopes stuffed during a given hour. Sample: The 40 randomly selected envelopes.

4.2 Population: All the artifacts discovered at the dig. Sample: Those artifacts (2% of the population) that are randomly chosen for inspection.

4.3 Population: All local businesses. Sample: The 73 businesses that return the questionnaire.

4.4 The population is all 45,000 people who made credit card purchases. The sample is the 137 people who returned the survey form.

1. **Sampling stuffed envelopes** A large retailer prepares its customers' monthly credit card bills using an automatic machine that folds the bills, stuffs them into envelopes, and seals the envelopes for mailing. Are the envelopes completely sealed? Inspectors choose 40 envelopes at random from the 1000 stuffed each hour for visual inspection. Identify the population and the sample.

2. **Student archaeologists** An archaeological dig turns up large numbers of pottery shards, broken stone tools, and other artifacts. Students working on the project classify each artifact and assign a number to it. The counts in different categories are important for understanding the site, so the project director chooses 2% of the artifacts at random and checks the students' work. Identify the population and the sample.

3. **Students as customers** A high school's student newspaper plans to survey local businesses about the importance of students as customers. From an alphabetical list of all local businesses, the newspaper staff chooses 150 businesses at random. Of these, 73 return the questionnaire mailed by the staff. Identify the population and the sample.

4. **Customer satisfaction** A department store mails a customer satisfaction survey to people who make credit card purchases at the store. This month, 45,000 people made

credit card purchases. Surveys are mailed to 1000 of these people, chosen at random, and 137 people return the survey form. Identify the population and the sample.

5. **Sleepless nights** How much sleep do high school students get on a typical school night? A counselor designed a survey to find out. To make data collection easier, the counselor surveyed the first 100 students to arrive at school on a particular morning. These students reported an average of 7.2 hours of sleep on the previous night.

(a) What type of sample did the counselor obtain?

(b) Explain why this sampling method is biased. Is 7.2 hours probably greater than or less than the true average amount of sleep last night for all students at the school? Why?

6. **Online polls** *Parade* magazine posed the following question: "Should drivers be banned from using all cell phones?" Readers were encouraged to vote online at parade.com. The subsequent issue of *Parade* reported the results: 2407 (85%) said "Yes" and 410 (15%) said "No."

(a) What type of sample did the *Parade* survey obtain?

(b) Explain why this sampling method is biased. Is 85% probably greater than or less than the true percent of all adults who believe that all cell phone use while driving should be banned? Why?

7. **Online reviews** Many websites include customer reviews of products, restaurants, hotels, and so on. The manager of a hotel was upset to see that 26% of reviewers on a travel website gave the hotel "1 star"—the lowest possible rating. Explain how bias in the sampling method could affect the estimate.

8. **Funding for fine arts** The band director at a high school wants to estimate the percentage of parents who support a decrease in the budget for fine arts. Because many parents attend the school's annual musical, the director surveys the first 30 parents who arrive at the show. Explain how bias in the sampling method could affect the estimate.

9. **Explain it to the congresswoman** You are on the staff of a member of Congress who is considering a bill that would provide government-sponsored insurance for nursing-home care. You report that 1128 letters have been received on the issue, of which 871 oppose the legislation. "I'm surprised that most of my constituents oppose the bill. I thought it would be quite popular," says the congresswoman. Are you convinced that a majority of the voters oppose the bill? How would you explain the statistical issue to the congresswoman?

10. **Sampling mall shoppers** You may have seen the mall interviewer, clipboard in hand, approaching people passing by. Explain why even a large sample of mall shoppers would not provide a trustworthy estimate of the current unemployment rate in the city where the mall is located.

11. **Do you trust the Internet?** You want to ask a sample of high school students the question "How much do you trust information about health that you find on the Internet—a great deal, somewhat, not much, or not at all?" You try out this and other questions on a pilot group of 5 students chosen from your class.

(a) Explain how you would use a line of Table D to choose an SRS of 5 students from the following list.

(b) Use line 107 to select the sample. Show how you use each of the digits.

Anderson	Drasin	Kim	Rider
Arroyo	Eckstein	Molina	Rodriguez
Batista	Fernandez	Morgan	Samuels
Bell	Fullmer	Murphy	Shen
Burke	Gandhi	Nguyen	Tse
Cabrera	Garcia	Palmiero	Velasco
Calloway	Glaus	Percival	Wallace
Delluci	Helling	Prince	Washburn
Deng	Husain	Puri	Zabidi
De Ramos	Johnson	Richards	Zhao

12. **Apartment living** You are planning a report on apartment living in a college town. You decide to select three apartment complexes at random for in-depth interviews with residents.

(a) Explain how you would use a line of Table D to choose an SRS of 3 complexes from the following list.

(b) Use line 117 to select the sample. Show how you use each of the digits.

Ashley Oaks	Country View	Mayfair Village
Bay Pointe	Country Villa	Nobb Hill
Beau Jardin	Crestview	Pemberly Courts
Bluffs	Del-Lynn	Peppermill
Brandon Place	Fairington	Pheasant Run
Briarwood	Fairway Knolls	Richfield
Brownstone	Fowler	Sagamore Ridge
Burberry	Franklin Park	Salem Courthouse
Cambridge	Georgetown	Village Manor
Chauncey Village	Greenacres	Waterford Court
Country Squire	Lahr House	Williamsburg

13. **Sampling the forest** To gather data on a 1200-acre pine forest in Louisiana, the U.S. Forest Service laid a grid of 1410 equally spaced circular plots over a map of the forest. A ground survey visited a sample of 10% of the plots.[13]

(a) Explain how you would use a random number generator to choose an SRS of 141 plots. Your description should be clear enough for a classmate to carry out your plan.

4.11 (a) Number the 40 students from 01 to 40 alphabetically. Moving left to right along a line from the random digit table, record two-digit numbers, skipping any numbers that are not between 01 and 40 and any repeated numbers, until you have 5 different numbers between 01 and 40. Select the corresponding 5 students. **(b)** Johnson (20), Drasin (11), Washburn (38), Rider (31), and Calloway (07).

4.12 (a) Number the 33 complexes from 01 to 33 alphabetically. Go to the random number table and pick a starting point. Moving left to right, record two-digit numbers, skipping any that are not between 01 and 33 and any repeated numbers, until you have 3 different numbers between 01 and 33. Select the 3 complexes corresponding to these numbers. **(b)** Fairington (16), Waterford Court (32), and Fowler (18)

4.13 (a) Number the plots from 1 to 1410. Use the command randInt(1,1410) to select 141 different integers from 1 to 1410. Select the corresponding 141 plots.

4.5 (a) A convenience sample **(b)** The estimate of 7.2 hours is probably less than the true average because students who arrive first to school had to wake up earlier and may have gotten less sleep than those students who are able to sleep in.

4.6 (a) A voluntary response sample **(b)** It is biased toward readers who feel most strongly about the issue. The reported value of 85% is probably greater than the true percent. Readers who have been involved in an accident caused by cell-phone use are more likely to respond to the poll and say "Yes."

4.7 This is a voluntary response sample. It is likely that those customers who volunteered to leave reviews feel strongly about the hotel, often due to a negative experience. As a result, the 26% from the sample is likely greater than the true percentage of all the hotel's customers who would give the hotel 1 star.

4.8 This is a convenience sample. It is likely that the first 30 parents to arrive at the show are very strong supporters of the fine arts. As a result, the proportion of parents in the sample who support the decrease in the budget will be less than the true proportion of all parents in the school who support the decrease.

4.9 Voluntary response sample. It is likely that the true proportion of constituents who oppose the bill is less than 871/1128.

4.10 This is a convenience sample. The sample is likely to overestimate the true unemployment rate because people without jobs have more time to be at the mall than those who are employed. It is also possible that the sample will underestimate the true unemployment rate if people who are unemployed do not have the money to shop at the mall.

4.13 (b) Answers will vary.

4.14 (a) Number the gravestones from 1 to 55,914. Use the command randInt(1,55914) to select 395 different integers from 1 to 55,914. Select the corresponding 395 plots. **(b)** Answers will vary.

4.15 Stratified random sampling may be preferable because the opinion of the employees might be the same within type of employee (servers, kitchen staff) but differ across different employee types. Using type of employee as strata may provide a more precise estimate of the overall proportion who approve of the policy. Select an SRS of 15 servers and 15 kitchen workers from the employees to form the sample.

4.16 Stratified random sampling may be preferable because the opinion of the students might be the same for those who live on campus but differ for those who commute. Using housing status (on or off campus) as strata may provide a more precise estimate of the overall proportion of undergraduate students who regularly park on campus. Select an SRS of 50 students who live on campus and 50 who do not live on campus. Combine these two SRSs into one sample.

4.17 No; in an SRS, each possible sample of 250 engineers is equally likely to be selected. The method described restricts the sample to exactly 200 males and 50 females.

4.18 No; in an SRS, each possible sample of 5 students is equally likely to be selected. The method described restricts the sample to exactly 3 students over 21 and 2 students under 21.

4.19 (a) This is cluster sampling. **(b)** The cable company could have chosen this method to save time and money. In an SRS, the company would have to visit individual homes all over the rural subdivision. With the cluster sampling method, the company only has to visit 5 locations.

4.20 (a) This is cluster sampling. **(b)** The lumber company could have chosen this method to save time and money. In an SRS, the company would have to inspect trees all over the entire forest. With the cluster sampling method, the company only has to visit 20 locations.

4.21 (a) Satisfaction likely varies depending on location of room, so we should stratify by floor and view. From

(b) Use your method from part (a) to choose the first 3 plots.

14. **Sampling gravestones** The local genealogical society in Coles County, Illinois, has compiled records on all 55,914 gravestones in cemeteries in the county for the years 1825 to 1985. Historians plan to use these records to learn about African Americans in Coles County's history. They first choose an SRS of 395 records to check their accuracy by visiting the actual gravestones.[14]

(a) Explain how you would use a random number generator to choose the SRS. Your description should be clear enough for a classmate to carry out your plan.

(b) Use your method from part (a) to choose the first 3 gravestones.

15. **No tipping** The owner of a large restaurant is considering a new "no tipping" policy and wants to survey a sample of employees. The policy would add 20% to the cost of food and beverages and the additional revenue would be distributed equally among servers and kitchen staff. Describe how to select a stratified random sample of approximately 30 employees. Explain your choice of strata and why stratified random sampling might be preferred in this context.

16. **Parking on campus** The director of student life at a university wants to estimate the proportion of undergraduate students who regularly park a car on campus. Describe how to select a stratified random sample of approximately 100 students. Explain your choice of strata and why stratified random sampling might be preferred in this context.

17. **SRS of engineers?** A corporation employs 2000 male and 500 female engineers. A stratified random sample of 200 male and 50 female engineers gives every individual in the population the same chance to be chosen for the sample. Is it an SRS? Explain your answer.

18. **SRS of students?** At a party, there are 30 students over age 21 and 20 students under age 21. You choose at random 3 of those over 21 and separately choose at random 2 of those under 21 to interview about their attitudes toward alcohol. You have given every student at the party the same chance to be interviewed. Is your sample an SRS? Explain your answer.

19. **High-speed Internet** Laying fiber-optic cable is expensive. Cable companies want to make sure that if they extend their lines to less dense suburban or rural areas, there will be sufficient demand so the work will be cost-effective. They decide to conduct a survey to determine the proportion of households in a rural subdivision that would buy the service. They select a simple random sample of 5 blocks in the subdivision and survey each family that lives on the selected blocks.

(a) What is the name for this kind of sampling method?

(b) Give a possible reason why the cable company chose this method.

20. **Timber!** A lumber company wants to estimate the proportion of trees in a large forest that are ready to be cut down. They use an aerial map to divide the forest into 200 equal-sized rectangles. Then they choose a random sample of 20 rectangles and examine every tree in the selected rectangles.

(a) What is the name for this kind of sampling method?

(b) Give a possible reason why the lumber company chose this method.

21. **How is your room?** A hotel has 30 floors with 40 rooms per floor. The rooms on one side of the hotel face the water, while rooms on the other side face a golf course. There is an extra charge for the rooms with a water view. The hotel manager wants to select 120 rooms and survey the registered guest in each of the selected rooms about his or her overall satisfaction with the property.
pg 259

(a) Describe how to obtain a sample of 120 rooms using stratified random sampling. Explain your choice of strata and why this method might be preferred to simple random sampling.

(b) Describe how to obtain a sample of 120 rooms using cluster sampling. Explain your choice of clusters and why this method might be preferred to simple random sampling.

22. **Go Blue!** Michigan Stadium, also known as "The Big House," seats over 100,000 fans for a football game. The University of Michigan Athletic Department wants to survey fans about concessions that are sold during games. Tickets are most expensive for seats on the sidelines. The cheapest seats are in the end zones (where one of the authors sat as a student). A map of the stadium is shown.

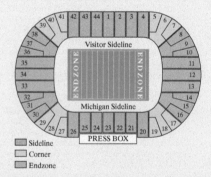

each floor, randomly select 2 rooms with each view. Using a stratified random sample would assure the manager of opinions from each type of room, thereby providing a more precise estimate of customer satisfaction. **(b)** Using floors as clusters, survey the registered guest in every room on each of 3 randomly selected floors. This would be simpler because the manager would survey guests on only three floors instead of throughout the hotel.

4.22 (a) Use the three types of seats (sideline, corner, and end zone) as the three strata because ticket prices are alike within each stratum but differ between the three. People who can afford more expensive tickets probably have different

opinions about concessions than people who can only afford the cheaper tickets. Select an SRS of fans from each of the three. Combine these SRSs into one sample. Stratification by seat type should result in more precise estimates of student opinion than a simple random sample of the same size would. **(b)** Each numbered section could be used as a cluster because it is easier to hand out surveys to everyone in a numbered section. Randomly select several numbered sections and survey all the fans in those sections. Cluster sampling is more efficient than finding an SRS of fans scattered about the stadium.

(a) Describe how to obtain a sample using stratified random sampling. Explain your choice of strata and why this method might be preferred to simple random sampling.

(b) Describe how to obtain a sample using cluster sampling. Explain your choice of clusters and why this method might be preferred to simple random sampling.

23. **Dead trees** In Rocky Mountain National Park, many mature pine trees along Highway 34 are dying due to infestation by pine beetles. Scientists would like to use a sample of size 200 to estimate the proportion of the approximately 5000 pine trees along the highway that have been infested.

(a) Explain why it wouldn't be practical for scientists to obtain an SRS in this setting.

(b) A possible alternative would be to use the first 200 pine trees along the highway as you enter the park. Why isn't this a good idea?

(c) Describe how to select a systematic random sample of 200 pine trees along Highway 34.

24. **iPhones** Suppose 1000 iPhones are produced at a factory today. Management would like to ensure that the phones' display screens meet the company's quality standards before shipping them to retail stores. Because it takes about 10 minutes to inspect an individual phone's display screen, managers decide to inspect a sample of 20 phones from the day's production.

(a) Explain why it would be difficult for managers to inspect an SRS of 20 iPhones that are produced today.

(b) An eager employee suggests that it would be easy to inspect the last 20 iPhones that were produced today. Why isn't this a good idea?

(c) Describe how to select a systematic random sample of 20 phones from the day's production.

25. **Eating on campus** The director of student life at a small college wants to know what percent of students eat regularly in the cafeteria. To find out, the director selects an SRS of 300 students who live in the dorms. Describe how undercoverage might lead to bias in this study. Explain the likely direction of the bias.

26. **Immigration reform** A news organization wants to know what percent of U.S. residents support a "pathway to citizenship" for people who live in the United States illegally. The news organization randomly selects registered voters for the survey. Describe how undercoverage might lead to bias in this study. Explain the likely direction of the bias.

27. **Reporting weight loss** A total of 300 people participated in a free 12-week weight-loss course at a community health clinic. After one year, administrators emailed each of the 300 participants to see how much weight they had lost since the end of the course. Only 56 participants responded to the survey. The mean

weight loss for this sample was 13.6 pounds. Describe how nonresponse might lead to bias in this study. Explain the likely direction of the bias.

28. **Nonresponse** A survey of drivers began by randomly sampling from all listed residential telephone numbers in the United States. Of 45,956 calls to these numbers, 5029 were completed. The goal of the survey was to estimate how far people drive, on average, per day. Describe how nonresponse might lead to bias in this study. Explain the likely direction of the bias.

29. **Running red lights** An SRS of 880 drivers was asked: "Recalling the last ten traffic lights you drove through, how many of them were red when you entered the intersections?" Of the 880 respondents, 171 admitted that at least one light had been red.[15] A practical problem with this survey is that people may not give truthful answers. Explain the likely direction of the bias.
pg 262

30. **Seat belt use** A study in El Paso, Texas, looked at seat belt use by drivers. Drivers were observed at randomly chosen convenience stores. After they left their cars, they were invited to answer questions that included questions about seat belt use. In all, 75% said they always used seat belts, yet only 61.5% were wearing seat belts when they pulled into the store parking lots.[16] Explain why the two percentages are so different.

31. **Boys don't cry?** Two female statistics students asked a random sample of 60 high school boys if they have ever cried during a movie. Thirty of the boys were asked directly and the other 30 were asked anonymously by means of a "secret ballot." When the responses were anonymous, 63% of the boys said "Yes," whereas only 23% of the other group said "Yes." Explain why the two percentages are so different.

32. **Weight? Wait what?** Marcos asked a random sample of 50 mall shoppers for their weight. Twenty-five of the shoppers were asked directly and the other 25 were asked anonymously by means of a "secret ballot." The mean reported weight was 13 pounds heavier for the anonymous group. Explain why the two means are so different.[17]

33. **Wording bias** Comment on each of the following as a potential sample survey question. Is the question clear? Is it slanted toward a desired response?

(a) "Some cell phone users have developed brain cancer. Should all cell phones come with a warning label explaining the danger of using cell phones?"

(b) "Do you agree that a national system of health insurance should be favored because it would provide health insurance for everyone and would reduce administrative costs?"

(c) "In view of escalating environmental degradation and incipient resource depletion, would you favor economic incentives for recycling of resource–intensive consumer goods?"

4.23 (a) Cluster sampling **(b)** Select every 5000/200 = 25th pine tree. Select every 25th pine tree after selecting a randomly selected starting point between 1 and 25.

4.24 (a) Difficult to locate the 20 phones from the 1000 produced that day and, assuming none of the phones can be shipped until after being inspected, choosing a random sample of 20 phones could delay the shipping process. **(b)** Not wise to inspect the last 20 iPhones as they may not be representative. **(c)** We would like to select every 1000/20 = 50th iPhone. Choose starting point between 1 and 50, then select every 50th iPhone until 20 are selected.

4.25 Students who do not live in the dorms cannot be part of the sample. Some of these students would live off campus, and are less likely to eat on campus than students who live in

the dorm, meaning the director's estimate for the percent of students who eat regularly on campus will likely be too high.

4.26 Undercoverage is a problem because the U.S. residents who are not registered voters cannot be part of the sample. Residents who are not registered voters are more likely to support a "pathway to citizenship" because this group includes some people who came to the U.S. illegally. So the news organization's estimate for the percent of U.S. residents who support a "pathway to citizenship" will likely be too low.

4.27 People who did not lose much weight (or who gained weight) after participating in the course may be less likely to respond. This will likely produce an estimated weight loss that is too large, because those who responded probably lost more weight than those who did not respond.

4.28 This survey will yield a biased result because people with long commutes are less likely to be at home and be in the sample. The estimate is likely to be too small, as people who are at home to answer the survey probably have shorter commutes (or none at all).

4.29 We would not expect many people to claim they have run red lights when they have not, but some people will deny running red lights when they have. So the proportion of drivers obtained in the sample who admitted to running a red light is likely to be less than the proportion of drivers who have run a red light.

4.30 When asked in person, people may lie about always wearing seat belts, or are embarrassed or ashamed to say they do not always wear them. When people are observed and not asked directly, the percent who wear a seat belt will be smaller and much closer to the actual percentage.

4.31 When asked in person, boys may claim they have never cried (when in reality, they have) because they are embarrassed or ashamed to admit it. Boys given an anonymous survey are more likely to be honest about their experiences.

4.32 When asked in person, the mall shoppers might be embarrassed or ashamed and lie about their true weight or refuse to answer. Mall shoppers given an anonymous survey are more likely to be honest about their weight.

4.33 (a) The wording is clear, but the question is slanted in favor of warning labels because of the first sentence about cell-phone users developing brain cancer. **(b)** The question is clear, but it is slanted in favor of national health insurance by asserting it would reduce administrative costs and not providing any counterarguments. The phrase "do you agree" also pushes respondents toward the desired response. **(c)** Not clear due to non-layman's terminology; for those who do understand the question, it is slanted because it suggests reasons why one should support recycling. It could be rewritten as "Do you support economic incentives to promote recycling?"

4.34 (a) The question is clear, but the two options presented are too extreme; no middle position on gun control is allowed. Also, the wording pushes respondents to choose option 2. The language used in option 2 is from the Constitution and people might avoid option 1 because they do not like the idea of government confiscating personal property. **(b)** The question is so complicated that it is not clear. The phrasing of this question will tend to make people respond in favor of a nuclear freeze because only one side of the issue is presented.

4.35 c

4.36 c

4.37 d

4.38 e

4.39 d

34. **Checking for bias** Comment on each of the following as a potential sample survey question. Is the question clear? Is it slanted toward a desired response?

(a) Which of the following best represents your opinion on gun control?

   i. The government should confiscate our guns.

   ii. We have the right to keep and bear arms.

(b) A freeze in nuclear weapons should be favored because it would begin a much-needed process to stop everyone in the world from building nuclear weapons now and reduce the possibility of nuclear war in the future. Do you agree or disagree?

Multiple Choice *Select the best answer for Exercises 35–40.*

35. A popular website places opinion poll questions next to many of its news stories. Simply click your response to join the sample. One of the questions was "Do you plan to diet this year?" More than 30,000 people responded, with 68% saying "Yes." Which of the following is true?

(a) About 68% of Americans planned to diet.

(b) The poll used a convenience sample, so the results tell us little about the population of all adults.

(c) The poll uses voluntary response, so the results tell us little about the population of all adults.

(d) The sample is too small to draw any conclusion.

(e) None of these.

36. To gather information about the validity of a new standardized test for 10th-grade students in a particular state, a random sample of 15 high schools was selected from the state. The new test was administered to every 10th-grade student in the selected high schools. What kind of sample is this?

(a) A simple random sample

(b) A stratified random sample

(c) A cluster sample

(d) A systematic random sample

(e) A voluntary response sample

37. Your statistics class has 30 students. You want to ask an SRS of 5 students from your class whether they use a mobile device for the online quizzes. You label the students 01, 02, . . . , 30. You enter the table of random digits at this line:

14459 26056 31424 80371 65103 62253 22490 61181

Your SRS contains the students labeled

(a) 14, 45, 92, 60, 56. (d) 14, 03, 10, 22, 06.

(b) 14, 31, 03, 10, 22. (e) 14, 03, 10, 22, 11.

(c) 14, 03, 10, 22, 22.

38. Suppose that 35% of the voters in a state are registered as Republicans, 40% as Democrats, and 25% as Independents. A newspaper wants to select a sample of 1000 registered voters to predict the outcome of the next election. If it randomly selects 350 Republicans, randomly selects 400 Democrats, and randomly selects 250 Independents, did this sampling procedure result in a simple random sample of registered voters from this state?

(a) Yes, because each registered voter had the same chance of being chosen.

(b) Yes, because random chance was involved.

(c) No, because not all registered voters had the same chance of being chosen.

(d) No, because a different number of registered voters was selected from each party.

(e) No, because not all possible groups of 1000 registered voters had the same chance of being chosen.

39. A local news agency conducted a survey about unemployment by randomly dialing phone numbers during the work day until it gathered responses from 1000 adults in its state. In the survey, 19% of those who responded said they were not currently employed. In reality, only 6% of the adults in the state were not currently employed at the time of the survey. Which of the following best explains the difference in the two percentages?

(a) The difference is due to sampling variability. We shouldn't expect the results of a random sample to match the truth about the population every time.

(b) The difference is due to response bias. Adults who are employed are likely to lie and say that they are unemployed.

(c) The difference is due to undercoverage bias. The survey included only adults and did not include teenagers who are eligible to work.

(d) The difference is due to nonresponse bias. Adults who are employed are less likely to be available for the sample than adults who are unemployed.

(e) The difference is due to voluntary response. Adults are able to volunteer as a member of the sample.

40. A simple random sample of 1200 adult Americans is selected, and each person is asked the following question: "In light of the huge national deficit, should the government at this time spend additional money to send humans to Mars?" Only 39% of those responding answered "Yes." This survey

(a) is reasonably accurate because it used a large simple random sample.

(b) needs to be larger because only about 24 people were drawn from each state.

(c) probably understates the percent of people who favor sending humans to Mars.

(d) is very inaccurate but neither understates nor overstates the percent of people who favor sending humans to Mars. Because simple random sampling was used, it is unbiased.

(e) probably overstates the percent of people who favor sending humans to Mars.

Recycle and Review

41. **Don't turn it over** (3.2) How many points do turnovers cost teams in the NFL? The scatterplot shows the relationship between $x =$ number of turnovers and $y =$ number of points scored by teams in the NFL during 2015, along with the least-squares regression line $\hat{y} = 460.2 - 4.084x$.

Number of turnovers

(a) Interpret the slope of the regression line in context.

(b) For this regression line, $s = 57.3$. Interpret this value.

(c) Calculate and interpret the residual for the San Francisco 49ers, who had 17 turnovers and scored 238 points.

(d) How does the point for the 49ers affect the least-squares regression line and standard deviation of the residuals? Explain your answer.

42. **Internet charges** (2.1) Some Internet service providers (ISPs) charge companies based on how much bandwidth they use in a month. One method that ISPs use to calculate bandwidth is to find the 95th percentile of a company's usage based on samples of hundreds of 5-minute intervals during a month.

(a) Explain what "95th percentile" means in this setting.

(b) Is it possible to determine the z-score for a usage total that is at the 95th percentile? If so, find the z-score. If not, explain why not.

SECTION 4.2 Experiments

LEARNING TARGETS *By the end of the section, you should be able to:*

- Explain the concept of confounding and how it limits the ability to make cause-and-effect conclusions.

- Distinguish between an observational study and an experiment, and identify the explanatory and response variables in each type of study.

- Identify the experimental units and treatments in an experiment.

- Describe the placebo effect and the purpose of blinding in an experiment.

- Describe how to randomly assign treatments in an experiment using slips of paper, technology, or a table of random digits.

- Explain the purpose of comparison, random assignment, control, and replication in an experiment.

- Describe a completely randomized design for an experiment.

- Describe a randomized block design and a matched pairs design for an experiment and explain the purpose of blocking in an experiment.

4.40 c

4.41 (a) The predicted number of points scored decreases by 4.084 points for each additional turnover. **(b)** $s = 57.3$; the actual number of points scored is typically about 57.3 points away from the number of points predicted by the least-squares regression line with $x =$ number of turnovers. **(c)** $\hat{y} = 460.2 - 4.084(17) = 390.772$, so the residual is $y - \hat{y} = 238 - 390.772 = -152.772$. The number of points scored by the San Francisco 49ers was 152.772 points less than predicted based on their number of turnovers. **(d)** Because the 49ers' point falls below the line and is to the left of the mean number of turnovers, their point decreases the y intercept and increases the slope, making the slope closer to 0 (less negative). Because the 49ers' point is farther from the line than the rest of the points in the data set, this point increases the standard deviation of the residuals.

4.42 (a) If a bandwidth measurement is at the 95th percentile, then 95% of the other bandwidth measurements will be less than this amount. **(b)** It is not possible to determine the z-score for a usage total that is at the 95th percentile because it is unknown if bandwidth measurements follow a Normal distribution, a uniform distribution, or some other distribution.

PD **Section 4.2 Overview**

To watch the video overview of Section 4.2 (for teachers), click on the link in the TE-Book, look on the TRFD, or download from the Teacher's Resources on the book's digital platform.

TRM **Section 4.2 Alternate Examples**

You can find the Alternate Examples for this section in Microsoft Word format by clicking on the link in the TE-Book, opening the TRFD, or downloading from the Teacher's Resources on the book's digital platform.

TRM **Template for Experiments**

There is a handy template for experiments available by clicking on the link in the TE-Book, opening the TRFD, or downloading from the Teacher's Resources on the book's digital platform.

A sample survey aims to gather information about a population without disturbing the population in the process. Sample surveys are one kind of **observational study**. Other observational studies record the behavior of animals in the wild or track the medical history of volunteers to look for associations between variables such as type of diet, amount of exercise, and blood pressure. Observational studies that examine existing data for a sample of individuals are called *retrospective*. Observational studies that track individuals into the future are called *prospective*.

> **DEFINITION Observational study**
>
> An **observational study** observes individuals and measures variables of interest but does not attempt to influence the responses.

Section 4.2 is about statistical designs for experiments, a very different way to produce data.

Observational Studies Versus Experiments

Is taking a vitamin D supplement good for you? Hundreds of observational studies have looked at the relationship between vitamin D concentration in a person's blood and various health outcomes.[18] In one prospective observational study, researchers found that teenage girls with higher vitamin D intakes were less likely to suffer broken bones.[19] Other observational studies have shown that people with higher vitamin D concentration have less cardiovascular disease, better cognitive function, and less risk of diabetes than people with lower concentrations of vitamin D.

In the observational studies involving vitamin D and diabetes, the **explanatory variable** is vitamin D concentration in the blood and the **response variable** is diabetes status—whether or not the person developed diabetes.

> **DEFINITION Response variable, Explanatory variable**
>
> A **response variable** measures an outcome of a study. An **explanatory variable** may help explain or predict changes in a response variable.

Unfortunately, it is very difficult to show that taking vitamin D *causes* a lower risk of diabetes using an observational study. As shown in the table, there are many possible differences between the group of people with high vitamin D concentration and the group of people with low vitamin D concentration. Any of these differences could be causing the difference in diabetes risk between the two groups of people.

Variable	Group 1	Group 2
Vitamin D concentration (explanatory)	**High vitamin D concentration**	**Low vitamin D concentration**
Quality of diet	Better diet	Worse diet
Amount of exercise	More exercise	Less exercise
⋮	⋮	⋮
Amount of vitamin supplementation	More likely to take other vitamins	Less likely to take other vitamins
Diabetes status (response)	**Less likely to have diabetes**	**More likely to have diabetes**

For example, it is possible that people who have healthier diets eat lots of foods that are high in vitamin D. Likewise, it is possible that people with healthier diets are less likely to develop diabetes. Vitamin D concentration may not have anything to do with diabetes status, even though there is an association between the two variables. In this case, we say there is **confounding** between vitamin D concentration and diet because we cannot tell which variable is causing the change in diabetes status.

> Some people call a variable that results in confounding, like diet in this case, a *confounding variable*.

DEFINITION Confounding

Confounding occurs when two variables are associated in such a way that their effects on a response variable cannot be distinguished from each other.

> **AP® EXAM TIP**
> If you are asked to identify a possible confounding variable in a given setting, you are expected to explain how the variable you choose (1) is associated with the explanatory variable and (2) is associated with the response variable.

Likewise, because sun exposure increases vitamin D concentration, it is possible that people who exercise a lot outside have higher concentrations of vitamin D. If people who exercise a lot are also less likely to get diabetes, then amount of exercise and vitamin D concentration are confounded—we can't say which variable is the cause of the smaller diabetes risk. In other words, exercise is a confounding variable because it is related to both vitamin D concentration and diabetes status.

Here is an example that describes a retrospective observational study.

EXAMPLE

Smoking and ADHD
Confounding

PROBLEM: In a study of more than 4700 children, researchers from Cincinnati Children's Hospital Medical Center found that those children whose mothers smoked during pregnancy were more than twice as likely to develop ADHD as children whose mothers had not smoked.[20] Explain how confounding makes it unreasonable to conclude that a mother's smoking during pregnancy causes an increase in the risk of ADHD in her children based on this study.

SOLUTION:

It is possible that the mothers who smoked during pregnancy were also more likely to have unhealthy diets. If people with unhealthy diets are also more likely to have children with ADHD, then it could be that unhealthy diets caused the increase in ADHD risk, not smoking.

> Notice that the solution describes how diet might be associated with the explanatory variable (smoking status) and with the response variable (ADHD status).

FOR PRACTICE, TRY EXERCISE 43

Observational studies of the effect of an explanatory variable on a response variable often fail because of confounding between the explanatory variable and one or more other variables. In contrast to observational studies, **experiments** don't just observe individuals or ask them questions. They actively impose some *treatment* to measure the response. Experiments can answer questions like "Does aspirin reduce the chance of a heart attack?" and "Can yoga help dogs live longer?"

DEFINITION Experiment

An **experiment** deliberately imposes treatments (conditions) on individuals to measure their responses.

Teaching Tip

In many other textbooks, including previous editions of *The Practice of Statistics*, the term *lurking variable* is introduced in the discussion about confounding. As we have taught the course, however, we've discovered that the term doesn't help student understanding. In fact, having an additional label for certain variables tended to make students more confused. We still think it is vital that students consider other variables that may affect the response variable, however.

| ALTERNATE EXAMPLE | Skill 1.C |

Alcohol and GPA
Confounding

PROBLEM:
In a recent study of about 13,900 college freshman, a researcher found that the more time students spent drinking alcohol, the lower their grade point averages (GPA). Explain how confounding makes it unreasonable to conclude that spending more time drinking causes a decrease in GPA for college freshman.

SOLUTION:
It is possible that the students who spend more time drinking are also less motivated toward academic success. If students with lower motivation have lower GPAs, it could be that the students' motivation level caused the GPAs to go down, not the time spent drinking alcohol.

Teaching Tip

To help students with the example on this page, use the following table:

Variable	Group 1	Group 2
Smoking status (explanatory)	Did not smoke during pregnancy	Smoked during pregnancy
Quality of diet	Healthy diet	Unhealthy diet
ADHD status (response)	Less likely to have ADHD	More likely to have ADHD

To determine if taking vitamin D actually causes a reduction in diabetes risk, researchers in Norway performed an experiment. The researchers randomly assigned 500 people with pre-diabetes to either take a high dose of vitamin D or to take a **placebo**—a pill that looked exactly like the vitamin D supplement but contained no active ingredient. After 5 years, about 40% of the people in each group were diagnosed with diabetes.[21] In other words, the association between vitamin D concentration and diabetes status disappeared when comparing two groups that were roughly the same to begin with.

> **DEFINITION   Placebo**
>
> A **placebo** is a treatment that has no active ingredient, but is otherwise like other treatments.

The experiment in Norway avoided confounding by letting chance decide who took vitamin D and who didn't. That way, people with healthier diets were split about evenly between the two groups. So were people who exercise a lot and people who take other vitamins. *When our goal is to understand cause and effect, experiments are the only source of fully convincing data.* For this reason, the distinction between observational study and experiment is one of the most important in statistics.

EXAMPLE
Facebook and financial incentives
Observational studies and experiments

PROBLEM: In each of the following settings, identify the explanatory and response variables. Then determine if each is an experiment or an observational study. Explain your reasoning.

(a) In a study conducted by researchers at the University of Texas, people were asked about their social media use and satisfaction with their marriage. Of the heavy social media users, 32% had thought seriously about leaving their spouse. Only 16% of non–social media users had thought seriously about leaving their spouse.[22]

(b) In a diet study using 100 overweight volunteers, 50 volunteers were randomly assigned to receive weight-loss counseling, monthly weigh-ins, and a three-month gym pass. The other 50 volunteers were given financial incentives (earning $20 for losing 4 pounds in a month or paying $20 otherwise) along with the counseling, weigh-ins, and gym pass. The group with the financial incentives lost 6.7 more pounds, on average.[23]

SOLUTION:

(a) Explanatory variable: Frequency of social media use. Response variable: Marital satisfaction. This is an observational study because people weren't assigned to use social media or not.

(b) Explanatory variable: Whether or not financial incentives were given. Response variable: Amount of weight lost. This is an experiment because researchers gave some volunteers financial incentives and did not give financial incentives to the other volunteers.

> In part (a), the response variable is not the *percent* who thought about leaving their spouse. This percent is a summary of all the responses. Likewise, in part (b), the response variable is not the *average* weight loss. This average is a summary of all the responses.

FOR PRACTICE, TRY EXERCISE 45

In part (a) of the example, it would be incorrect to conclude that using social media causes marital dissatisfaction. It could be that other variables are confounded with social media use—or even that marital dissatisfaction is causing increased social media use. In part (b), it is reasonable to conclude that the financial incentives caused the increase in weight loss because this was a well-designed experiment, and the difference in average weight loss between the two groups was too large to be explained by chance alone.

CHECK YOUR UNDERSTANDING

1. Does reducing screen brightness increase battery life in laptop computers? To find out, researchers obtained 30 new laptops of the same brand. They chose 15 of the computers at random and adjusted their screens to the brightest setting. The other 15 laptop screens were left at the default setting—moderate brightness. Researchers then measured how long each machine's battery lasted. Was this an observational study or an experiment? Justify your answer.

Questions 2–4 refer to the following setting. Does eating dinner with their families improve students' academic performance? According to an ABC News article, "Teenagers who eat with their families at least five times a week are more likely to get better grades in school."[24] This finding was based on a sample survey conducted by researchers at Columbia University.

2. Was this an observational study or an experiment? Justify your answer.

3. What are the explanatory and response variables?

4. Explain clearly why such a study cannot establish a cause-and-effect relationship. Suggest a variable that may be confounded with whether families eat dinner together.

The Language of Experiments

An experiment is a statistical study in which we actually do something (a **treatment**) to people, animals, or objects (the **experimental units** or **subjects**) to observe the response.

DEFINITION   Treatment, Experimental unit, Subjects

A specific condition applied to the individuals in an experiment is called a **treatment**. If an experiment has several explanatory variables, a treatment is a combination of specific values of these variables. An **experimental unit** is the object to which a treatment is randomly assigned. When the experimental units are human beings, they are often called **subjects**.

The best way to learn the language of experiments is to practice using it.

AP® EXAM TIP

Students often have trouble identifying the experimental units in an experiment. In the example on page 274, the experimental units are the schools, not the students. The decision about which treatment to apply was made for each school, not for each individual student. On the 2019 AP® Statistics exam question #2, the experimental units are the containers of insects, not the insects. The decision about which treatment to apply was made for each container, not for each individual insect.

ALTERNATE EXAMPLE — Skill 1.C

The best test scores
Vocabulary of experiments

PROBLEM:
Several AP® Statistics students wondered whether caffeine could improve test scores. They randomly assigned 30 student volunteers to either drink regular coffee or decaffeinated coffee the morning of the students' next test. At the end of the experiment, they recorded test scores for each student volunteer. Identify the treatments and the experimental units in this experiment.

SOLUTION:
This experiment compares two treatments: (1) regular coffee and (2) decaffeinated coffee. The experimental units are the 30 student volunteers.

ALTERNATE EXAMPLE — Skill 1.C

Growing the best tomatoes
Experiments with multiple explanatory variables

Does adding fertilizer affect the productivity of tomato plants? How about the amount of water given to the plants? To answer these questions, a gardener plants 24 similar tomato plants in identical pots in his greenhouse. He will add fertilizer to the soil in half the pots. Also, he will water 8 of the plants with 0.5 gallon of water per day, 8 of the plants with 1 gallon of water per day, and the remaining 8 plants with 1.5 gallons of water per day. At the end of 3 months, he will record the total weight of tomatoes produced by each plant.

PROBLEM:
(a) List the factors in this experiment and the number of levels for each factor.
(b) If the researchers used every possible combination of levels to form the treatments, how many treatments were included in the experiment?
(c) List all the treatments.

SOLUTION:
(a) Whether fertilizer is applied (2 levels) and the amount of water (3 levels).
(b) $2 \times 3 = 6$ different treatments
(c) (1) fertilizer, 0.5 gallon; (2) fertilizer, 1 gallon; (3) fertilizer, 1.5 gallons; (4) no fertilizer, 0.5 gallon; (5) no fertilizer, 1 gallon; (6) no fertilizer, 1.5 gallons

---

EXAMPLE

How can we prevent malaria?
Vocabulary of experiments

PROBLEM: Malaria causes hundreds of thousands of deaths each year, with many of the victims being children. Will regularly screening children for the malaria parasite and treating those who test positive reduce the proportion of children who develop the disease? Researchers worked with children in 101 schools in Kenya, randomly assigning half of the schools to receive regular screenings and follow-up treatments and the remaining schools to receive no regular screening. Children at all 101 schools were tested for malaria at the end of the study.[25] Identify the treatments and the experimental units in this experiment.

SOLUTION:
This experiment compares two treatments: (1) regular screenings and follow-up treatments and (2) no regular screening. The experimental units are 101 schools in Kenya.

> Note that the experimental units are the schools, not the students. The decision about who to screen was made school by school, not student by student. All students at the same school received the same treatment.

FOR PRACTICE, TRY EXERCISE 49

In the malaria experiment, there was one explanatory variable: screening status. In other experiments, there are multiple explanatory variables. Sometimes, these explanatory variables are called **factors**. In an experiment with multiple factors, the treatments are formed using the various **levels** of each of the factors. When there is only one factor, the levels are equivalent to the treatments.

> **DEFINITION** Factor, Levels
> In an experiment, a **factor** is an explanatory variable that is manipulated and may cause a change in the response variable. The different values of a factor are called **levels**.

Here's an example of a multifactor experiment.

EXAMPLE

The five-second rule
Experiments with multiple explanatory variables

PROBLEM: Have you ever dropped a tasty piece of food on the ground, then quickly picked it up and eaten it? If so, you probably thought about the "five-second rule," which states that a piece of food is safe to eat if it has been on the floor less than 5 seconds. The rule is based on the belief that bacteria need time to transfer from the floor to the food. But does it work?

Researchers from Rutgers University put the five-second rule to the test. They used four different types of food: watermelon, bread, bread with butter, and gummy candy. They dropped the food onto four different surfaces: stainless steel, ceramic tile, wood, and carpet. And they waited for four different lengths of time: less than 1 second, 5 seconds, 30 seconds, and 300 seconds. Finally, they used bacteria prepared two different ways: in a tryptic soy broth or peptone buffer. Once the bacteria were ready, the researchers spread them out on the different surfaces and started dropping food.[26]

(a) List the factors in this experiment and the number of levels for each factor.

(b) If the researchers used every possible combination of levels to form the treatments, how many treatments were included in the experiment?

(c) List two of the treatments.

SOLUTION:

(a) *Type of food (4 levels), type of surface (4 levels), amount of time (4 levels), method of bacterial preparation (2 levels)*

(b) $4 \times 4 \times 4 \times 2 = 128$ *different treatments*

(c) *Watermelon/stainless steel/less than 1 second/tryptic soy broth; gummy candy/wood/ 300 seconds/peptone buffer*

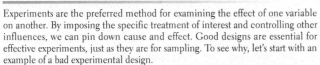

FOR PRACTICE, TRY EXERCISE 51

Teaching Tip

To determine the number of treatments, multiply the number of levels for each factor in the experiment. In the five-second rule example, the factor type of food had 4 levels, the factor type of surface had 4 levels, the factor amount of time had 4 levels and the factor method of bacterial preparation had 2 levels. Thus, there are $4 \times 4 \times 4 \times 2 = 128$ treatments.

Teaching Tip: AP® Connections

In AP® Biology, two treatments that work together to enhance their effect on a response variable are said to have synergy.

What did the researchers discover? The wetter foods had greater bacterial transfer and food dropped on carpet had the least bacterial transfer. There was greater bacterial transfer the longer the food was on the surface, although there was some transfer that happened almost instantaneously. Overall, the researchers concluded that the type of food and type of surface were at least as important as the amount of time the food remained on the surface.

This example shows how experiments allow us to study the combined effect of several factors. The interaction of several factors can produce effects that could not be predicted from looking at the effect of each factor alone. For example, although longer time was associated with more bacterial transfer in general, this relationship might not be true for very moist food.

Designing Experiments: Comparison

Experiments are the preferred method for examining the effect of one variable on another. By imposing the specific treatment of interest and controlling other influences, we can pin down cause and effect. Good designs are essential for effective experiments, just as they are for sampling. To see why, let's start with an example of a bad experimental design.

Does caffeine affect pulse rate? Many students regularly consume caffeine to help them stay alert. So it seems plausible that taking caffeine might increase an individual's pulse rate. Is this true? One way to investigate this claim is to ask volunteers to measure their pulse rates, drink some cola with caffeine, measure their pulse rates again after 10 minutes, and calculate the increase in pulse rate.

This experiment has a very simple design. A group of subjects (the students) were exposed to a treatment (the cola with caffeine), and the outcome (change in pulse rate) was observed. Here is the design:

Students → Cola with caffeine → Change in pulse rate

Unfortunately, even if the pulse rate of every student went up, we couldn't attribute the increase to caffeine. Perhaps the excitement of being in an experiment made their pulse rates increase. Maybe it was the sugar in the cola and not the caffeine. Perhaps their teacher told them a funny joke during the 10-minute

waiting period and made everyone laugh. In other words, there are many other variables that are potentially confounded with taking caffeine.

Many laboratory experiments use a design like the one in the caffeine example:

$$\text{Experimental units} \rightarrow \text{Treatment} \rightarrow \text{Measure response}$$

In the lab environment, simple designs often work well. Field experiments and experiments with animals or people deal with more varied conditions. *Outside the lab, badly designed experiments often yield worthless results because of confounding.*

The remedy for the confounding in the caffeine example is to do a comparative experiment with two groups: one group that receives caffeine and a **control group** that does not receive caffeine.

> **DEFINITION   Control group**
>
> In an experiment, a **control group** is used to provide a baseline for comparing the effects of other treatments. Depending on the purpose of the experiment, a control group may be given an inactive treatment (placebo), an active treatment, or no treatment at all.

In all other aspects, these groups should be treated exactly the same so that the only difference is the caffeine. That way, if there is convincing evidence of a difference in the average increase in pulse rates, we can safely conclude it was *caused* by the caffeine. This means that one group could get regular cola with caffeine, while the control group gets caffeine-free cola. Both groups would get the same amount of sugar, so sugar consumption would no longer be confounded with caffeine intake. Likewise, both groups would experience the same events during the experiment, so what happens during the experiment won't be confounded with caffeine intake either.

EXAMPLE

Preventing malaria
Control groups

Alexander Joe/Getty Images

PROBLEM: In an earlier example, we described an experiment in which researchers randomly assigned 101 schools in Kenya to either receive regular malaria screenings and follow-up treatments or to receive no regular screenings. Explain why it was necessary to include a control group of schools that didn't receive regular screenings.

SOLUTION:

The purpose of the control group is to provide a baseline for comparing the effect of the regular screenings and follow-up treatments. Otherwise, researchers wouldn't be able to determine if a decrease in malaria rates was due to the treatment or some other change that occurred during the experiment (like a drought that killed off mosquitos, slowing the spread of malaria).

FOR PRACTICE, TRY EXERCISE 55

Teaching Tip

If an experiment has more than one treatment and the researchers only want to know which treatment is best, a control group with no treatment (or with a placebo) might not be necessary. The important idea is that we have multiple groups to be able to *compare*.

Teaching Tip

For many years, coffee had a bad reputation—until scientists realized that smoking cigarettes was a confounding variable. People who drink coffee also are more likely to smoke. This is similar to the caffeine and sugar consumption scenario.

ALTERNATE EXAMPLE   Skill 1.B

Growing the best corn
Control groups

PROBLEM:
A group of AP® Biology students randomly assigned 100 corn seeds to be planted either at a depth of 1.5 inches or a depth of 3 inches. Explain why it was not necessary to include a control group of seeds that were not planted in soil.

SOLUTION:
The purpose of a control group is to provide a baseline so that the results of a treatment can be compared. In this experiment, there is no need for a control group because the two treatments can be compared. The AP® Biology students are interested in knowing which treatment is more effective when compared to the other treatment.

A control group was essential in the malaria experiment to determine if screening was effective. However, *not all experiments include a control group*—as long as comparison takes place. In the experiment about the five-second rule, there were 128 different treatments being compared and no control group. A control group wasn't essential in this experiment because researchers were interested in comparing different amounts of time on the floor, different types of food, and different types of surfaces.

Designing Experiments: Blinding and the Placebo Effect

In the caffeine experiment, we used comparison to help prevent confounding. But even when there is comparison, confounding is still possible. If the subjects in the experiment know what type of soda they are receiving, the expectations of the two groups will be different. The knowledge that a subject is receiving caffeine may increase his or her pulse rate, apart from the caffeine itself. This is an example of the **placebo effect**.

> **DEFINITION Placebo effect**
>
> The **placebo effect** describes the fact that some subjects in an experiment will respond favorably to any treatment, even an inactive treatment.

In one study, researchers zapped the wrists of 24 test subjects with a painful jolt of electricity. Then they rubbed a cream with no active medicine on subjects' wrists and told them the cream should help soothe the pain. When researchers shocked them again, 8 subjects said they experienced significantly less pain.[27] When the ailment is psychological, like depression, some experts think that the placebo effect accounts for about three-quarters of the effect of the most widely used drugs.[28]

Because of the placebo effect, it is important that subjects don't know what treatment they are receiving. It is also better if the people interacting with the subjects and measuring the response variable don't know which subjects are receiving which treatment. When neither group knows who is receiving which treatment, the experiment is **double-blind**. Other experiments are **single-blind**.

> **DEFINITION Double-blind, Single-blind**
>
> In a **double-blind** experiment, neither the subjects nor those who interact with them and measure the response variable know which treatment a subject is receiving.
>
> In a **single-blind** experiment, either the subjects or the people who interact with them and measure the response variable don't know which treatment a subject is receiving.

The idea of a double-blind design is simple. Until the experiment ends and the results are in, only the study's statistician knows for sure which treatment a subject is receiving. However, some experiments cannot be carried out in a

Teaching Tip

Here is a great TEDEd video with Emma Bryce that introduces the idea of the placebo effect:
https://ed.ted.com/lessons/the-power-of-the-placebo-effect-emma-bryce.

Teaching Tip

Placebos work even if subjects know that they are fake:
http://www.smithsonianmag.com/science-nature/why-i-take-fake-pills-180962765/.
On test day, offer students a "smart pill" before the test (just a Tic-Tac). You should see scores go up.

Teaching Tip

In the electricity experiment, it is not just that the subjects reported less pain—brain activity was lower in the pain center of the brain. See http://www.sciencechannel.com/video-topics/strange-creatures/weird-connections-power-of-the-placebo/.

Teaching Tip

Another form of control involves "blinding" the subjects, as well as those who interact with them, and then measuring the results. In this case, we are keeping the expectations the same for all treatment groups.

Teaching Tip

Suppose an experiment is comparing two treatments, one that uses a pill and one that uses an injection. At first glance, it seems that this experiment can't be double-blind because the subjects and the experimenters would certainly know which treatment was being given. The solution is for one experimental group to get a real pill and a fake injection (placebo), and for the other group to get a fake pill (placebo) and a real injection.

> **AP® EXAM TIP**
>
> Many students believe that an experiment cannot be double-blind because "someone has to know who gets which treatment." Although there is always someone who knows which subjects are receiving which treatment, as long as this person doesn't interact with the subjects or measure the response, the experiment can still be considered double-blind.

double-blind manner. For example, if researchers are comparing the effects of exercise and dieting on weight loss, then subjects will know which treatment they are receiving. Such an experiment can still be single-blind if the individuals who are interacting with the subjects and measuring the response variable don't know who is dieting and who is exercising. In other single-blind experiments, the subjects are unaware of which treatment they are receiving, but the people interacting with them and measuring the response variable do know.

ALTERNATE EXAMPLE   Skill 1.C

The first double-blind experiment
Blinding and the placebo effect

PROBLEM:

W. H. R. Rivers is credited with creating the first double-blind experiment in 1907. Rivers was interested in finding out if caffeine has an effect on work output. He quickly discovered the psychological effect on subjects who knew they were receiving a treatment. This led him to create a *placebo* for each experiment, which looked and tasted just like the caffeine treatment. Rivers would have a friend prepare the caffeine drinks and the placebo drinks and would then give the drinks to the subjects.

(a) Explain what it means for this experiment to be double-blind.

(b) Why was it important for this experiment to be double-blind?

SOLUTION:

(a) Neither the subjects nor the researcher (Rivers) knew which subjects were given the caffeine drink and which subjects were given the placebo drink.

(b) If subjects knew they were drinking caffeine, Rivers wouldn't be able to determine whether any improvement in work output was due to the caffeine or to the subjects' expectation of working harder (the placebo effect). If Rivers knew which subjects received which treatments, he might have treated one group of subjects differently from the other group. This would make it difficult to know if the caffeine was the cause of any improvement in work output.

EXAMPLE

Do magnets repel pain?
Blinding and the placebo effect

PROBLEM: Early research showed that magnetic fields affected living tissue in humans. Some doctors have begun to use magnets to treat patients with chronic pain. Scientists wondered if this type of therapy really worked. They designed a double-blind experiment to find out. A total of 50 patients with chronic pain were recruited for the study. A doctor identified a painful site on each patient and asked him or her to rate the pain on a scale from 0 (mild pain) to 10 (severe pain). Then the doctor selected a sealed envelope containing a magnet at random from a box with a mixture of active and inactive magnets. The chosen magnet was applied to the site of the pain for 45 minutes. After being treated, each patient was again asked to rate the level of pain from 0 to 10.[29]

(a) Explain what it means for this experiment to be double-blind.

(b) Why was it important for this experiment to be double-blind?

SOLUTION:

(a) Neither the subjects nor the doctors applying the magnets and recording the pain ratings knew which subjects had the active magnets and which had the inactive magnets.

(b) If subjects knew they were receiving an active treatment, researchers wouldn't know if any improvement was due to the magnets or to the expectation of getting better (the placebo effect). If the doctors knew which subjects received which treatments, they might treat one group of subjects differently from the other group. This would make it difficult to know if the magnets were the cause of any improvement.

FOR PRACTICE, TRY EXERCISE 59

CHECK YOUR UNDERSTANDING

A new analysis is casting doubt on a claimed benefit of omega-3 fish oil. For years, doctors have been recommending eating fish and taking fish oil supplements to prevent heart disease. But the new analysis reviewed 20 previous studies and showed that the effects of omega-3 aren't as great as once suspected. One reason is that an early trial of omega-3 supplements was conducted as an open-label study.[30] In this type of study, both patients and researchers know who is receiving which treatment.

1. Describe a potential problem with an open-label study in this context.

2. Describe how you can fix the problem identified in Question 1.

✓ Answers to CYU

1. The patients and the researchers know who is receiving which treatment. For some people, this knowledge could motivate them to take other measures (e.g., to exercise more or to eat better in general) that would also influence their heart health.

2. Use a double-blind experiment. The patients would not know which treatment they received, nor would the researchers know what treatment each patient received.

Designing Experiments: Random Assignment

Comparison alone isn't enough to produce results we can trust. If the treatments are given to groups that differ greatly when the experiment begins, confounding will result. If we allow students to choose what type of cola they will drink in the caffeine experiment, students who consume caffeine on a regular basis might be more likely to choose the regular cola. Due to their caffeine tolerance, these students' pulse rates might not increase as much as other students' pulse rates. In this case, caffeine tolerance would be confounded with the amount of caffeine consumed, making it impossible to conclude cause and effect.

To create roughly equivalent groups at the beginning of an experiment, we use **random assignment** to determine which experimental units get which treatment.

> **DEFINITION   Random assignment**
>
> In an experiment, **random assignment** means that experimental units are assigned to treatments using a chance process.

Let's look at how random assignment can be used to improve the design of the caffeine experiment.

EXAMPLE

Caffeine and pulse rates
How random assignment works

PROBLEM: A total of 20 students have agreed to participate in an experiment comparing the effects of caffeinated cola and caffeine-free cola on pulse rates. Describe how you would randomly assign 10 students to each of the two treatments:

(a) Using 20 identical slips of paper
(b) Using technology
(c) Using Table D

Zbia Snively

SOLUTION:

(a) On 10 slips of paper, write the letter "A"; on the remaining 10 slips, write the letter "B." Shuffle the slips of paper and hand out one slip of paper to each volunteer. Students who get an "A" slip receive the cola with caffeine and students who get a "B" slip receive the cola without caffeine.

> When describing a method of random assignment, don't stop after creating the groups. Make sure to identify which group gets which treatment.

(b) Label each student with a different integer from 1 to 20. Then randomly generate 10 different integers from 1 to 20. The students with these labels receive the cola with caffeine. The remaining 10 students receive the cola without caffeine.

> When using a random number generator or a table of random digits to assign treatments, make sure to account for the possibility of repeated numbers when describing your method.

(c) Label each student with a different integer from 01 to 20. Go to a line of Table D and read two-digit groups moving from left to right. The first 10 different labels between 01 and 20 identify the 10 students who receive cola with caffeine. The remaining 10 students receive the caffeine-free cola. Ignore groups of digits from 21 to 00.

FOR PRACTICE, TRY EXERCISE 63

AP® EXAM TIP

Many students lose credit on the AP® Statistics exam for failing to adequately describe how they assign the treatments to experimental units in an experiment. Most important, the method the students use must be *random*. Also, the method must be described in *sufficient detail* so that two knowledgeable users of statistics can follow the student's description and carry out the method in exactly the same way. For example, saying, "Assign students to the two groups using random digits" isn't sufficiently detailed because there are many ways to use random digits. If a student chooses to use random digits, he or she must use labels of the same length (e.g., 01–30, not 1–30). Students must also address how they would deal with repeated numbers that come up when using a random digit table or random number generator. For example, they can say "ignoring repeats" or state that they will generate 25 "different" numbers from 1 to 50.

Preparing for Inference

The fact that random assignment creates two roughly equivalent groups means that if we notice a significant difference between the two groups at the end of the experiment, it is likely due to the treatment. In other words, the treatment *caused* the difference in the results. In Chapters 8–12, we will check random assignment for the Random condition so that we can make conclusions about causation.

ALTERNATE EXAMPLE   Skill 1.C

Is it better to learn geometry online or in a class?
How random assignment works

PROBLEM:
Do students learn geometry better from an online course or in class with a teacher? To find out, a large high school set up an experiment with 500 student volunteers. The school randomly assigned half the students to take the geometry course online, watching videos to inform their learning. The other half took a more traditional course with lectures by a teacher. Describe how you would randomly assign 500 students to each of the two treatments:
(a) Using 500 identical slips of paper
(b) Using technology
(c) Using Table D

SOLUTION:
(a) On 250 slips of paper, write "Online"; on the remaining 250 slips, write "Teacher." Shuffle the slips of paper and hand out one slip of paper to each student. Students who get an "Online" slip will take the online geometry course and students who get a "Teacher" slip will take the geometry course taught by a classroom teacher.
(b) Label each student with a different integer from 1 to 500. Then randomly generate 250 different integers from 1 to 500. The students with these labels will take the online geometry course. The remaining 250 students will take the geometry course taught by a classroom teacher.
(c) Label each student with a different integer from 001 to 500. Go to a line of Table D and read three-digit groups moving from left to right. The first 250 different labels between 001 and 500 identify the 250 students who will take the online geometry course. The remaining 250 students will take the geometry course taught by a classroom teacher. Ignore groups of digits from 501 to 999 and also 000.

Teaching Tip

Don't say (or let your students say) that random assignment *eliminates* the effects of other variables. It doesn't. It simply balances their effects among the treatment groups.

Teaching Tip

For the variables that we do have control over (amount of sugar, experiences of subject during the experiment), we keep these variables constant for all experimental units to control their effects. Doing so reduces the variability in the response variable, making it easier to identify when a significant difference is present. For the variables that we do *not* have control over (caffeine tolerance, metabolism, and body size), we use random assignment to balance the effects. Later, we will use blocking to account for the variability in the response variable, making it easier to identify when a significant difference is present.

Preparing for Inference

In Chapter 9, we will see that the power of a significance test is the ability of a test to correctly find convincing evidence that one treatment is more effective, when there really is a difference. Controlling other variables increases the power of a significance test by reducing the variability in the response variable. To see how this works, consider the equation of the two-sample *t* statistic.

$$t = \frac{(\bar{x}_1 - \bar{x}_2) - (\mu_1 - \mu_2)}{\sqrt{\dfrac{s_1^2}{n_1} + \dfrac{s_2^2}{n_2}}}$$

Controlling other variables makes the values of s_1 and s_2 smaller, resulting in a larger *t* statistic and smaller *P*-value. In other words, more power!

Preparing for Inference

Is it plausible that the observed difference in average pulse rates between the two groups happened purely by chance because of the way that the 20 students were randomly assigned to the two groups? In Section 4.3, we will answer this question in the "Analyzing the caffeine experiment" activity. Later in Chapter 11, we will use a *two-sample t test for* $\mu_1 - \mu_2$ to help us make a conclusion for a scenario like this one.

Random assignment should distribute the students who regularly consume caffeine in roughly equal numbers to each group. It should also balance out the students with high metabolism and those with larger body sizes in the caffeine and caffeine-free groups. Random assignment helps ensure that the effects of other variables (e.g., caffeine tolerance, metabolism, or body size) are spread evenly among the two groups, making it easier to attribute a difference in mean pulse rate change to the caffeine.

Designing Experiments: Control

Although random assignment should create two groups of students that are roughly equivalent to begin with, we still have to ensure that the only consistent difference between the groups during the experiment is the type of cola they receive. We can **control** the effects of some variables by keeping them the same for both groups. For example, we should make both treatments contain the same amount of sugar. If one group got regular cola and the other group got caffeine-free *diet* cola, then the amount of sugar would be confounded with the amount of caffeine—we wouldn't know if it was the sugar or the caffeine that was causing a change in pulse rates.

> **DEFINITION** **Control**
>
> In an experiment, **control** means keeping other variables constant for all experimental units.

We also want to control other variables to reduce the variability in the response variable. Suppose we let volunteers in both groups choose how much cola they want to drink. In that case, the changes in pulse rate would be more variable than if we made sure each subject drank the same amount of soda. Letting the amount of cola vary will make it harder to determine if caffeine is really having an effect on pulse rates.

The dotplots on the left show the results of an experiment in which the amount of cola consumed was the same for all participating students. Because there is so little overlap in these graphs, it seems clear that caffeine increases pulse rates. The dotplots on the right show the results of an experiment in which the students were permitted to choose how much or how little cola they consumed. Notice that the centers of the distributions haven't changed, but the distributions are much more variable. The increased overlap in the graphs makes the evidence supporting the effect of caffeine less convincing.

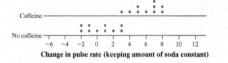

Change in pulse rate (keeping amount of soda constant)

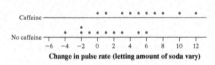

Change in pulse rate (letting amount of soda vary)

After randomly assigning treatments and controlling other variables, the two groups should be about the same, except for the treatments. Then a difference in the average change in pulse rate must be due either to the treatments themselves—or to the random assignment. We can't say that *any* difference

between the average pulse rate changes for students in the two groups must be caused by the difference in caffeine. There would be *some* difference, even if both groups received the same type of cola, because the random assignment is unlikely to produce two groups that are exactly equivalent with respect to every variable that might affect pulse rate.

Designing Experiments: Replication

Would you trust an experiment with just one student in each group? No, because the results would depend too much on which student was assigned to the caffeinated cola. However, if we randomly assign many subjects to each group, the effects of chance will balance out, and there will be little difference in the average responses in the two groups—unless the treatments themselves cause a difference. This is the idea of **replication**.

> There must be at least 2 experimental units receiving each treatment to achieve replication.

> **DEFINITION   Replication**
>
> In an experiment, **replication** means giving each treatment to enough experimental units so that a difference in the effects of the treatments can be distinguished from chance variation due to the random assignment.

 In statistics, replication means "use enough subjects." In other fields, the term *replication* has a different meaning. In these fields, replication means conducting an experiment in one setting and then having other investigators conduct a similar experiment in a different setting. That is, replication means repeatability.

Experiments: Putting It All Together

The following box summarizes the four key principles of experimental design: comparison, random assignment, control, and replication.

> **PRINCIPLES OF EXPERIMENTAL DESIGN**
>
> The basic principles for designing experiments are as follows:
>
> 1. **Comparison.** Use a design that compares two or more treatments.
> 2. **Random assignment.** Use chance to assign experimental units to treatments (or treatments to experimental units). Doing so helps create roughly equivalent groups of experimental units by balancing the effects of other variables among the treatment groups.
> 3. **Control.** Keep other variables the same for all groups, especially variables that are likely to affect the response variable. Control helps avoid confounding and reduces variability in the response variable.
> 4. **Replication.** Giving each treatment to enough experimental units so that any differences in the effects of the treatments can be distinguished from chance differences between the groups.

Teaching Tip

To illustrate the benefits of replication, ask your students to imagine using 6 subjects in the caffeine experiment, 2 of whom are regular coffee drinkers. There is a 50% chance that these 2 subjects will end up in the same treatment group. However, if we had 60 students in the experiment and 20 of them were regular coffee drinkers, there is practically no chance that they would all end up in the same treatment group. The more replication, the more balanced the treatment groups will be after the random assignment.

Teaching Tip: AP® Connections

AP® Chemistry and AP® Biology students will be familiar with the alternative meaning of replication: when an experiment is independently conducted in a different location by different investigators.

Let's see how these principles were used in designing a famous medical experiment.

Online learning versus teacher taught
Principles of experimental design

PROBLEM:
In a previous example, a school set up an experiment in which 500 geometry students were randomly assigned to take the course online or with lectures by a teacher. The same teacher was used for both courses, and both met for 58 minutes per day. At the end of the year, the success of each program was measured by students' scores on the geometry final exam, which was the same for both courses.

(a) Explain how this experiment used comparison.
(b) Explain the purpose of randomly assigning the students to the two treatments.
(c) Name two variables that were controlled in this experiment and why it was beneficial to control these variables.
(d) Explain how this experiment used replication. What is the purpose of replication in this context?

SOLUTION:
(a) Researchers used a design that compared two treatments: an online course and a teacher-taught course.
(b) Random assignment helped ensure that the two groups of students were roughly equivalent at the beginning of the experiment.
(c) The experiment used the same teacher for both courses, each course met for the same amount of time, and both courses used the same final exam. Keeping these variables constant helps reduce the variability in the response variable.
(d) There were 250 students in each treatment group. This helps ensure that the difference in scores on the geometry final exam is due to the type of course, not to chance variation in the random assignment.

EXAMPLE   **The Physicians' Health Study**
Principles of experimental design

PROBLEM: Does regularly taking aspirin help protect people against heart attacks? The Physicians' Health Study was a medical experiment that helped answer this question. In fact, the Physicians' Health Study looked at the effects of two drugs: aspirin and beta-carotene. Researchers wondered if beta-carotene would help prevent some forms of cancer. The subjects in this experiment were 21,996 male physicians. There were two explanatory variables (factors), each having two levels: aspirin (yes or no) and beta-carotene (yes or no). Combinations of the levels of these factors form the four treatments shown in the diagram. One-fourth of the subjects were assigned at random to each of these treatments.

On odd-numbered days, the subjects took either a tablet that contained aspirin or a placebo that looked and tasted like the aspirin but had no active ingredient. On even-numbered days, they took either a capsule containing beta-carotene or a placebo. There were several response variables—the study looked for heart attacks, several kinds of cancer, and other medical outcomes. After several years, 239 of the placebo group but only 139 of the aspirin group had suffered heart attacks. This difference is large enough to give good evidence that taking aspirin does reduce heart attacks.[31] It did not appear, however, that beta-carotene had any effect on preventing cancer.

(a) Explain how this experiment used comparison.
(b) Explain the purpose of randomly assigning the physicians to the four treatments.
(c) Name two variables that were controlled in this experiment and why it was beneficial to control these variables.
(d) Explain how this experiment used replication. What is the purpose of replication in this context?

SOLUTION:
(a) Researchers used a design that compared each of the active treatments to a placebo.
(b) Random assignment helped ensure that the four groups of physicians were roughly equivalent at the beginning of the experiment.
(c) The experiment used subjects of the same gender and same occupation. Using only male physicians helps to reduce the variability in the response variables.
(d) There were over 5000 subjects in each treatment group. This helped ensure that the difference in heart attacks was due to the aspirin and not to chance variation in the random assignment.

If women and people with other occupations were included, the results might be more variable, making it harder to determine the effects of aspirin and beta-carotene. However, using only male physicians means we don't know how females or other males would respond to these treatments.

FOR PRACTICE, TRY EXERCISE 67

Preparing for Inference

The difference in number of heart attacks between the placebo group and aspirin group was so large that it is not likely to have happened purely by chance due to random assignment. In Section 4.3, we will learn that the results are *statistically significant*.

The difference in number of heart attacks between the placebo group and beta-carotene group was small enough that it is likely to have happened purely by chance due to random assignment. In Section 4.3, we will say that the difference in results is *not statistically significant*.

The Physicians' Health Study shows how well-designed experiments can yield good evidence that differences in the treatments cause the differences we observe in the response.

CHECK YOUR UNDERSTANDING

Many utility companies have introduced programs to encourage energy conservation among their customers. An electric company considers placing small digital displays in households to show current electricity use and what the cost would be if this use continued for a month. Will the displays reduce electricity use? One cheaper approach is to give customers a chart and information about monitoring their electricity use from their outside meter. Would this method work almost as well? The company decides to conduct an experiment using 60 households to compare these two approaches (display, chart) with a group of customers who receive information about energy consumption but no help in monitoring electricity use.

1. Explain why it was important to have a control group that didn't get the display or the chart.
2. Describe how to randomly assign the treatments to the 60 households.
3. What is the purpose of randomly assigning treatments in this context?

Completely Randomized Designs

The diagram in Figure 4.1 presents the details of the caffeine experiment: random assignment, the sizes of the groups and which treatment they receive, and the response variable. This type of design is called a **completely randomized design**.

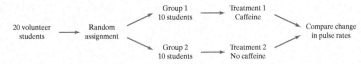

FIGURE 4.1 Outline of a completely randomized design to compare caffeine and no caffeine.

> **DEFINITION   Completely randomized design**
>
> In a **completely randomized design**, the experimental units are assigned to the treatments completely at random.

Although there are good statistical reasons for using treatment groups that are about equal in size, the definition of a completely randomized design does not require that each treatment be assigned to an equal number of experimental units. It does specify that the assignment of treatments must occur completely at random.

Making Connections

A completely randomized design in an experiment is somewhat analogous to an SRS when selecting a sample. In an SRS, there are no restrictions on who can be included in the sample, as is the case with a stratified random sample, a cluster sample, or a systematic random sample. Likewise, in a completely randomized design, there are no restrictions on who can be assigned to each treatment. There are other designs (to be discussed shortly) that do place restrictions on who can be assigned to each group.

Answers to CYU

1. A control group would show how much electricity customers tend to use naturally. This would serve as a baseline to determine how much less electricity is used in each of the treatment groups.

2. Number the houses from 1 to 60. Write the numbers 1 to 60 on the slips of paper. Shuffle well. Draw out 20 slips of paper (without replacement); those households will receive a display. Draw out another 20 slips of paper (without replacement); those households will receive a chart. The remaining 20 households will receive only information about energy consumption.

3. To create groups of households that are roughly equivalent at the beginning of the experiment. This will ensure that the effects of other variables (e.g., the thrifty inclination of some households) are spread evenly among the three groups.

Teaching Tip

When drawing these experiment outlines, students will be tempted to skip the step that says "Group 1" and "Group 2" and jump right to the treatments. There is an important reason for this intermediate step. At "Group 1" and "Group 2," the groups are roughly equivalent (because of random assignment). Consider having students add the ≈ symbol between Group 1 and Group 2. Only when we get to "Treatment 1" and "Treatment 2" do the groups become different, and the only consistent difference between the groups is the treatments. If there is a significant difference in pulse rates at the end of the experiment, we have strong evidence that the difference was *caused* by the treatments. Also, be sure that in the last step students clearly identify the response variable being measured and compared, rather than simply stating "Compare results."

Preparing for Inference

Having equal group sizes results in slightly more power when we perform significance tests. That is, if one treatment is really more effective than the other, we will be more likely to find convincing evidence of this when the group sizes are the same.

ALTERNATE EXAMPLE Skill 1.C

ALTERNATE EXAMPLE Skill 1.C
The Hawthorne effect
Completely randomized design

PROBLEM:

A Harvard researcher was once conducting some experiments at Western Electric's Hawthorne Works to see if certain changes in conditions would improve worker productivity. In one part of the study, a group of workers was provided additional lighting and was compared to a group with no additional lighting. The group with additional lighting showed significant improvements in worker productivity.

(a) Explain why it isn't reasonable to conclude that the additional lighting is effective for increasing worker productivity based on this study.

(b) To test the effectiveness of the additional lighting, you recruit 20 similar companies that have agreed to have employees participate in an experiment. Write a few sentences describing a completely randomized design for this experiment.

SOLUTION:

(a) It is possible that the group who received the additional lighting improved because they knew they were being measured, not because of the additional lighting. *Note:* This phenomenon is now commonly known as the Hawthorne effect, which describes how worker productivity can increase simply because workers know they are being measured.

(b) Number the companies from 1 to 20. Use a random number generator to produce 10 different random integers from 1 to 20 and increase the lighting at the facilities of the companies with these numbers. Leave lighting as is for the remaining 10 companies. Compare the increase in worker productivity between the two groups.

AP® EXAM TIP

The idea discussed in the AP® Exam Tip on this page was the focus of the Investigative Task question #6 on the 2017 AP® Statistics exam.

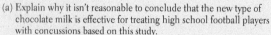
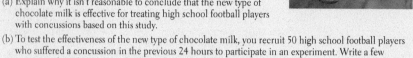

EXAMPLE
Chocolate milk and concussions
Completely randomized design

PROBLEM: "Concussion-Related Measures Improved in High School Football Players Who Drank New Chocolate Milk" announced a recent headline.[33] In the study, researchers compared a group of concussed football players given a new type of chocolate milk with a group of concussed football players who received no treatment.

(a) Explain why it isn't reasonable to conclude that the new type of chocolate milk is effective for treating high school football players with concussions based on this study.

(b) To test the effectiveness of the new type of chocolate milk, you recruit 50 high school football players who suffered a concussion in the previous 24 hours to participate in an experiment. Write a few sentences describing a completely randomized design for this experiment.

SOLUTION:

(a) It is possible that the group who received the new type of chocolate milk improved because they knew they were being treated and expected to get better, not because of the new chocolate milk.

(b) Number the players from 1 to 50. Use a random number generator to produce 25 different integers from 1 to 50 and give the new type of chocolate milk to the players with these numbers. Give regular chocolate milk to the remaining 25 players. Compare the concussion-related measures for the two groups.

FOR PRACTICE, TRY EXERCISE 69

AP® EXAM TIP

If you are asked to describe a completely randomized design, stay away from flipping coins. For example, suppose we ask each student in the caffeine experiment to toss a coin. If it's heads, then the student will drink the cola with caffeine. If it's tails, then the student will drink the caffeine-free cola. As long as all 20 students toss a coin, this is still a completely randomized design. Of course, the two groups are unlikely to contain exactly 10 students because it is unlikely that 20 coin tosses will result in a perfect 50-50 split between heads and tails.

The problem arises if we try to force the two groups to have equal sizes. Suppose we continue to have students toss coins until one of the groups has 10 students and then place the remaining students in the other group. In this case, the last two students in line are very likely to end up in the same group. However, in a completely randomized design, the last two subjects should only have a 50% chance of ending up in the same group.

Randomized Block Designs

Completely randomized designs are the simplest statistical designs for experiments. They illustrate clearly the principles of comparison, random assignment, control, and replication. But just as with sampling, there are times when the simplest method doesn't yield the most precise results. When a population consists of groups of individuals that are "similar within but different between," a stratified random sample gives a better estimate than a simple random sample. This same logic applies in experiments.

Suppose that a mobile phone company is considering two different keyboard designs (A and B) for its new smartphone. The company decides to perform an experiment to compare the two keyboards using a group of 10 volunteers. The response variable is typing speed, measured in words per minute.

How should the company address the fact that four of the volunteers already use a smartphone, whereas the remaining six volunteers do not? They could use a completely randomized design and hope that the random assignment distributes the smartphone users and non-smartphone users about evenly between the group using keyboard A and the group using keyboard B. Even so, there might be a lot of variability in typing speed within both treatment groups because some members of each treatment group are more familiar with smartphones than the others. This additional variability might make it difficult to detect a difference in the effectiveness of the two keyboards. What should the researchers do?

Because the company knows that experience with smartphones will affect typing speed, they could start by separating the volunteers into two groups—one with experienced smartphone users and one with inexperienced smartphone users. Each of these groups of similar subjects is known as a **block**. Within each block, the company could then randomly assign half of the subjects to use keyboard A and the other half to use keyboard B. To control other variables, each subject should be given the same passage to type while in a quiet room with no distractions. This **randomized block design** helps account for the variation in typing speed that is due to experience with smartphones.

> ### DEFINITION Block, Randomized block design
>
> A **block** is a group of experimental units that are known before the experiment to be similar in some way that is expected to affect the response to the treatments.
>
> In a **randomized block design**, the random assignment of experimental units to treatments is carried out separately within each block.

Figure 4.2 outlines the randomized block design for the smartphone experiment. The subjects are first separated into blocks based on their experience with smartphones. Then the two treatments are randomly assigned within each block.

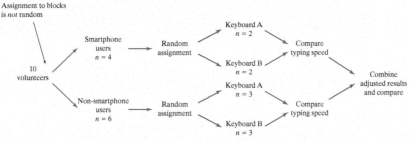

FIGURE 4.2 Outline of a randomized block design for the smartphone experiment. The blocks consist of volunteers who have used smartphones and volunteers who have not used smartphones. The treatments are keyboard A and keyboard B.

Using a randomized block design allows us to account for the variation in the response that is due to the blocking variable of smartphone experience. This makes it easier to determine if one treatment is really more effective than the other.

To see how blocking helps, let's look at the results of the smartphone experiment. In the block of 4 smartphone users, 2 were randomly assigned to use keyboard A and the other 2 were assigned to use keyboard B. Likewise, in the block of 6 non-smartphone users, 3 were randomly assigned to use keyboard A and the

other 3 were assigned to use keyboard B. Each of the 10 volunteers typed the same passage and the typing speed was recorded. Here are the results:

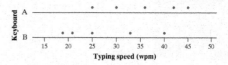

There is some evidence that keyboard A results in higher typing speeds, but the evidence isn't that convincing. Enough overlap occurs in the two distributions that the differences might simply be due to the chance variation in the random assignment.

If we compare the results for the two keyboards within each block, however, a different story emerges. Among the 4 smartphone users (indicated by the red squares), keyboard A was the clear winner. Likewise, among the 6 non-smartphone users (indicated by the black dots), keyboard A was also the clear winner.

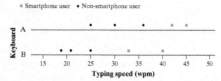

The overlap in the first set of dotplots was due almost entirely to the variation in smartphone experience—smartphone users were generally faster than non-smartphone users, regardless of which keyboard they used. In fact, the average typing speed for the smartphone users was 40, while the average typing speed for non-smartphone users was only 26, a difference of 14 words per minute. To account for the variation created by the difference in smartphone experience, let's subtract 14 from each of the typing speeds in the block of smartphone users to "even the playing field." Here are the results:

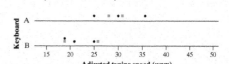

Because we accounted for the variation due to the difference in smartphone experience, the variation in each of the distributions has been reduced. There is now much less overlap between the two distributions, meaning that the evidence in favor of keyboard A is much more convincing. *When blocks are formed wisely, it is easier to find convincing evidence that one treatment is more effective than another.*

> **AP® EXAM TIP**
>
> Don't mix the language of experiments and the language of sample surveys or other observational studies. You will lose credit for saying things like "use a randomized block design to select the sample for this survey" or "this experiment suffers from nonresponse because some subjects dropped out during the study."

Preparing for Inference

In Chapter 9, we will see that the power of a significance test is the ability of a test to correctly find convincing evidence that one treatment is more effective, when there really is a difference. Blocking increases the power of a significance test by accounting for the variability in the response variable.

AP® EXAM TIP

Confusing stratified random sampling with a randomized block design is another way that students mix up the language of sample surveys and experiments. These two ideas are alike in some ways, because they both involve forming groups of similar subjects before random selection or random assignment. Both methods also help to account for the variability created by other variables. However, stratified random sampling is done only when taking a sample from a population. Likewise, blocking happens only when assigning units to treatments in an experiment.

The idea of blocking is an important additional principle of experimental design. A wise experimenter will form blocks based on the most important unavoidable sources of variation among the experimental units. In other words, the experimenter will form blocks using the variables that are the best predictors of the response variable. Random assignment will then average out the effects of the remaining other variables and allow a fair comparison of the treatments. The moral of the story is: *control what you can, block on what you can't control, and randomize to create comparable groups.*

EXAMPLE

Should I use the popcorn button?
Blocking in an experiment

PROBLEM: A popcorn lover wants to determine if it is better to use the "popcorn button" on her microwave oven or use the amount of time recommended on the bag of popcorn. To measure how well each method works, she will count the number of unpopped kernels remaining after popping. To obtain the experimental units, she goes to the store and buys 10 bags each of 4 different varieties of microwave popcorn (butter, cheese, natural, and kettle corn), for a total of 40 bags.

(a) Describe a randomized block design for this experiment. Justify your choice of blocks.

(b) Explain why a randomized block design might be preferable to a completely randomized design for this experiment.

SOLUTION:

(a) Form blocks based on variety, because the number of unpopped kernels is likely to differ by variety. Randomly assign 5 bags of each variety to the popcorn button treatment and 5 to the timed treatment by placing all 10 bags of a particular variety in a large box. Shake the box, pick 5 bags without looking, and assign them to be popped using the popcorn button. The remaining 5 bags will be popped using the instructions on the bags. Repeat this process for the remaining 3 varieties. After popping each of the 40 bags in random order, count the number of unpopped kernels in each bag and compare the results within each variety. Then combine the results from the 4 varieties after accounting for the difference in average response for each variety.

> It is important to pop the bags in random order so that changes over time (e.g., temperature, humidity) aren't confounded with the explanatory variable. For example, if the 20 "popcorn button" bags are popped last when the room temperature is greater, we wouldn't know if using the popcorn button or the warmer temperature was the cause of a difference in the number of unpopped kernels.

(b) A randomized block design accounts for the variability in the number of unpopped kernels created by the different varieties of popcorn (butter, cheese, natural, kettle). This makes it easier to determine if using the microwave button is more effective for reducing the number of unpopped kernels.

FOR PRACTICE, TRY EXERCISE 71

Another way to address the variability in unpopped kernels created by the different varieties is to use only one variety of popcorn in the experiment. Because variety of popcorn is no longer a variable, it will not be a source of variability. Of course, this means that the results of the experiment only apply to that one variety of popcorn—not ideal for the popcorn lover in the example!

MATCHED PAIRS DESIGN A common type of randomized block design for comparing two treatments is a **matched pairs design**. The idea is to create blocks by matching pairs of similar experimental units. The random assignment of subjects to treatments is done within each matched pair. Just as with other forms of blocking, matching helps account for the variation due to the variable(s) used to form the pairs.

ALTERNATE EXAMPLE | Skill 1.C

Geometry class on blocks
Blocking in an experiment

PROBLEM:
Do students learn geometry better from an online course or in class with a teacher? To find out, a large high school is setting up an experiment with the 500 students (100 freshmen, 400 sophomores) who signed up for geometry next year. The school's guidance counselors randomly assign half the students to take the geometry course online, watching videos. The other half of the students will take a geometry course based on teacher lectures. At the end of the year, the success of each program will be measured by students' scores on the geometry final exam, which will be the same for both courses.
The guidance counselors are convinced that these two learning formats will show different results for freshman and sophomore students.

(a) Describe a randomized block design for this experiment. Justify your choice of blocks.

(b) Explain why a randomized block design may be preferable to a completely randomized design for this experiment.

SOLUTION:

(a) Form blocks based on grade level, because the scores on the geometry final exam are likely to vary by grade level. Freshmen taking geometry tend to be more advanced in their math coursework and would likely have higher final exam scores. Randomly assign 50 of the freshmen to take the online geometry course and 50 of the freshmen to take the geometry course taught by a teacher. Randomly assign 200 of the sophomores to take the online geometry course and 200 of the sophomores to take the geometry course taught by a teacher.

(b) A randomized block design accounts for the variability in the scores on the geometry final exam created by the different grade levels. This makes it easier to determine which learning format is more effective for increasing scores on the geometry final exam.

Preparing for Inference

In Chapters 10 and 11, students will analyze data from a matched pairs design by finding the mean of the differences between the pairs. They will use the mean difference to construct a *paired t interval for a mean difference* and perform a *paired t test for a mean difference*.

Teaching Tip

In the first design of the classical music experiment, each student is "matched" with someone of very similar mathematical ability.

Teaching Tip

In the second design of the classical music experiment, each of the students is perfectly "matched" to themselves. Not only are they perfectly matched for the variable of mathematical ability, but they are also matched on all other variables (e.g., work ethic and willingness to study).

ALTERNATE EXAMPLE	Skill 1.C

Clockwise or counterclockwise?
Matched pairs design

PROBLEM:
A track coach wants to know whether his long-distance runners are faster running the track clockwise or counterclockwise. Design an experiment that uses a matched pairs design to investigate this question. Explain your method of pairing.

SOLUTION:
Have each long-distance runner race 1 mile in each direction. Some runners are faster than others, so using each runner as his or her own "pair" accounts for variation in 1-mile race times among the runners. For each runner, randomly assign the order in which the treatments (clockwise and counterclockwise) are assigned—by flipping a coin. Heads indicates the runner will race clockwise first and counterclockwise second; tails indicates the runner will race counterclockwise first and clockwise second. Allow adequate recovery time between the races. For each runner, record the 1-mile race times for each direction.

DEFINITION   **Matched pairs design**

A **matched pairs design** is a common experimental design for comparing two treatments that uses blocks of size 2. In some matched pairs designs, two very similar experimental units are paired and the two treatments are randomly assigned within each pair. In others, each experimental unit receives both treatments in a random order.

Suppose we want to investigate if listening to classical music while taking a math test affects performance. A total of 30 students in a math class volunteer to take part in the experiment. The difference in mathematical ability among the volunteers is likely to create additional variation in the test scores, making it harder to see the effect of classical music. To account for this variation, we could pair the students by their grade in the class—the two students with the highest grades are paired together, the two students with the next highest grades are paired together, and so on. Within each pair, one student is randomly assigned to take a math test while listening to classical music and the other member of the pair is assigned to take the math test in silence.

Sometimes, each "pair" in a matched pairs design consists of just one experimental unit that gets both treatments in random order. In the experiment about the effect of listening to classical music, we could have each student take a math test in both conditions. To decide the order, we might flip a coin for each student. If the coin lands on heads, the student takes a math test with classical music playing today and a similar math test without music playing tomorrow. If it lands on tails, the student does the opposite—no music today and classical music tomorrow.

Randomizing the order of treatments is important to avoid confounding. Suppose everyone did the classical music treatment on the first day and the no-music treatment on the second day, but the air conditioner wasn't working on the second day. We wouldn't know if any difference in mean test score was due to the difference in treatment or the difference in room temperature.

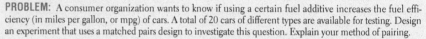

EXAMPLE	**Will an additive improve my mileage?** Matched pairs design

PROBLEM: A consumer organization wants to know if using a certain fuel additive increases the fuel efficiency (in miles per gallon, or mpg) of cars. A total of 20 cars of different types are available for testing. Design an experiment that uses a matched pairs design to investigate this question. Explain your method of pairing.

SOLUTION:
Give each car both treatments. It is reasonable to think that some cars are more fuel efficient than others, so using each car as its own "pair" accounts for the variation in fuel efficiency in the experimental units. For each car, randomly assign the order in which the treatments are assigned by flipping a coin. Heads indicates using the additive first and no additive second. Tails indicates using no additive first and then the additive second. For each car, record the fuel efficiency (mpg) after using each treatment.

FOR PRACTICE, TRY EXERCISE 77

In the preceding example, it is also possible to form pairs of two similar cars. For instance, we could pair together the two most fuel-efficient cars, the next two most fuel-efficient cars, and so on. This is less ideal, however, because there will still be some

differences between the members of each pair that may cause additional variation in the results. Using the same car twice creates perfectly matched "pairs," and it also doubles the number of pairs used in the experiment. Both these features make it easier to find convincing evidence that the gas additive is effective, if it really is effective.

CHECK YOUR UNDERSTANDING

Researchers would like to design an experiment to compare the effectiveness of three different advertisements for a new television series featuring the works of Jane Austen. There are 300 volunteers available for the experiment.

1. Describe a completely randomized design to compare the effectiveness of the three advertisements.
2. Describe a randomized block design for this experiment. Justify your choice of blocks.
3. Why might a randomized block design be preferable in this context?

Section 4.2 | Summary

- Statistical studies often try to show that changing one variable (the **explanatory variable**) causes changes in another variable (the **response variable**). Variables are **confounded** when their effects on a response variable can't be distinguished from each other.

- We can produce data to answer specific questions using **observational studies** or **experiments**. Observational studies that examine existing data for a sample of individuals are called *retrospective*. Observational studies that track individuals into the future are called *prospective*. Experiments actively do something to people, animals, or objects in order to measure their response. Experiments are the best way to show cause and effect.

- In an experiment, we impose one or more **treatments** on a group of **experimental units** (sometimes called **subjects** if they are human). Each treatment is a combination of the **levels** of the explanatory variables (also called **factors**).

- Some experiments give a **placebo** (fake treatment) to a **control group**. That helps prevent confounding due to the **placebo effect**, whereby some patients get better because they expect the treatment to work.

- Many behavioral and medical experiments are **double-blind**. That is, neither the subjects nor those interacting with them and measuring their responses know who is receiving which treatment. If one group knows and the other doesn't, then the experiment is **single-blind.**

- The basic principles of experimental design are:
  - **Comparison:** Use a design that compares two or more treatments.
  - **Random assignment:** Use chance (slips of paper, a random number generator, a table of random digits) to assign experimental units to treatments. This helps create roughly equivalent groups before treatments are imposed.
  - **Control:** Keep other variables the same for all groups. Control helps avoid confounding and reduces the variation in responses, making it easier to decide if a treatment is effective.

TRM Section 4.2 Quiz

There is one quiz available for this section in the Teacher's Resource Materials. Click on the link in the TE-Book, look on the TRFD, or download from the Teacher's Resources on the book's digital platform. You can also create your own quiz using the ExamView® Assessment Suite that is part of the TPS6 program. Questions are coded by Learning Target to make it easy to build parallel quizzes.

Preparing for Inference

In Chapter 9, we will see that the power of a significance test is the ability of a test to correctly find convincing evidence that one treatment is more effective, when there really is a difference. A matched pairs design increases the power of a significance test by accounting for the variability in the response variable.

✓ Answers to CYU

1. Number the volunteers 1 to 300. Use a random number generator to produce 100 different random integers from 1 to 300, and show the first advertisement to the volunteers with those numbers. Generate 100 additional random integers from 1 to 300, and show the second advertisement to the volunteers with those numbers. The remaining 100 volunteers will view the third advertisement. Compare the effectiveness of the advertisements for the three groups.

2. Because the effectiveness of the ads may depend on a volunteer's familiarity with Jane Austen, block by whether or not the individuals are familiar with the works of Jane Austen. The volunteers in each block would be numbered and randomly assigned to one of three treatment groups: One-third of the volunteers in each block would view the first advertisement, one-third the second advertisement, and one-third the third advertisement. After viewing the advertisements, the researchers would gauge the effectiveness of the advertisements.

3. A randomized block design accounts for the variability in effectiveness that is due to subjects' familiarity with the works of Jane Austen. This makes it easier to determine the effectiveness of the three different advertisements.

Answers to Section 4.2 Exercises

4.43 Although eating seafood may decrease the risk of colon cancer, it is possible that the male physicians who ate seafood were also more likely to exercise. Because exercise might decrease colon cancer risk, perhaps it was the exercise that caused the decrease in colon cancer risk, not eating seafood.

4.44 Although getting good grades may improve one's health, it is possible that the high school students who got good grades were also more informed about healthy life choices. Because being informed about healthy choices early in life is likely to lead to good health in later life, perhaps it was the knowledge about healthy life choices that caused the improved health, not getting good grades.

4.45 Explanatory: Type of program the people watched. Response: Number of calories consumed. This was an experiment because the treatments (20 minutes of *The Island*; 20 minutes of *The Island*, but without sound; and 20 minutes of *Charlie Rose*) were deliberately imposed on the students.

4.46 Explanatory: The instruction method. Response: The increase in biology test scores. This was an experiment because the treatments (computer software or textbook presentation) were deliberately imposed on the students to measure the response.

4.47 (a) Explanatory: Amount of time in child care from birth to age 4½. Response: Ratings of their behavior. **(b)** A prospective observational study. No treatments were assigned and the researchers followed the children through their 6th year in school, asking adults to rate their behavior several times along the way. **(c)** No, this study is an observational study, so we cannot make a cause-and-effect conclusion. For example, children who spend more time in child care probably have less time with their parents and get less instruction about proper behavior.

- **Replication:** Impose each treatment on enough experimental units so that the effects of the treatments can be distinguished from chance differences between the groups.

- In a **completely randomized design,** the experimental units are assigned to the treatments completely at random.

- A **randomized block design** forms groups (**blocks**) of experimental units that are similar with respect to a variable that is expected to affect the response. Treatments are assigned at random within each block. Responses are then compared within each block and combined with the responses of other blocks after accounting for the differences between the blocks. When blocks are chosen wisely, it is easier to determine if one treatment is more effective than another.

- A **matched pairs design** is a common form of randomized block design for comparing two treatments. In some matched pairs designs, each subject receives both treatments in a random order. In others, two very similar subjects are paired, and the two treatments are randomly assigned within each pair.

Section 4.2   Exercises

43. **Good for the gut?** Is fish good for the gut? Researchers tracked 22,000 male physicians for 22 years. Those who reported eating seafood of any kind at least 5 times per week had a 40% lower risk of colon cancer than those who said they ate seafood less than once a week. Explain how confounding makes it unreasonable to conclude that eating seafood causes a reduction in the risk of colon cancer, based on this study.[34]

44. **Straight A's now, healthy later** A study by Pamela Herd of the University of Wisconsin–Madison found a link between high school grades and health. Analyzing data from the Wisconsin Longitudinal Study, which has tracked the lives of thousands of Wisconsin high school graduates from the class of 1957, Herd found that students with higher grade-point averages were more likely to say they were in excellent or very good health in their early 60s. Explain how confounding makes it unreasonable to conclude that people will live healthier lives if they increase their GPA, based on this study.[35]

45. **Snacking and TV** Does the type of program people watch influence how much they eat? A total of 94 college students were randomly assigned to one of three treatments: watching 20 minutes of a Hollywood action movie (*The Island*), watching the same 20-minute excerpt of the movie with no sound, and watching 20 minutes of an interview program. While watching, participants were given snacks (M&M'S®, cookies, carrots, and grapes) and allowed to eat as much as they wanted. Subjects who watched the highly stimulating excerpt from *The Island* ate 65% more calories than subjects who watched the interview show. Participants who watched the silent version of *The Island* ate 46%

more calories than those who watched the interview show.[36] Identify the explanatory and response variables in this study. Then determine if it is an experiment or an observational study. Explain your reasoning.

46. **Learning biology with computers** An educator wants to compare the effectiveness of computer software for teaching biology with that of a textbook presentation. She gives a biology pretest to each student in a group of high school juniors, then randomly divides them into two groups. One group uses the computer, and the other studies the text. At the end of the year, she tests all the students again and compares the increase in biology test scores in the two groups. Identify the explanatory and response variables in this study. Then determine if it is an experiment or an observational study. Explain your reasoning.

47. **Child care and aggression** A study of child care enrolled 1364 infants and followed them through their sixth year in school. Later, the researchers published an article in which they stated that "the more time children spent in child care from birth to age 4½, the more adults tended to rate them, both at age 4½ and at kindergarten, as less likely to get along with others, as more assertive, as disobedient, and as aggressive."[37]

(a) What are the explanatory and response variables?

(b) Is this a prospective observational study, a retrospective observational study, or an experiment? Justify your answer.

(c) Does this study show that child care makes children more aggressive? Explain your reasoning.

48. **Chocolate and happy babies** A University of Helsinki (Finland) study wanted to determine if chocolate

consumption during pregnancy had an effect on infant temperament at age 6 months. Researchers began by asking 305 healthy pregnant women to report their chocolate consumption. Six months after birth, the researchers asked mothers to rate their infants' temperament using the traits of smiling, laughter, and fear. The babies born to women who had been eating chocolate daily during pregnancy were found to be more active and "positively reactive"—a measure that the investigators said encompasses traits like smiling and laughter.[38]

(a) What are the explanatory and response variables?

(b) Is this a prospective observational study, a retrospective observational study, or an experiment? Justify your answer.

(c) Does this study show that eating chocolate regularly during pregnancy helps produce infants with good temperament? Explain your reasoning.

49. **Growing in the shade** The ability to grow in shade may help pine trees found in the dry forests of Arizona to resist drought. How well do these pines grow in shade? Investigators planted pine seedlings in a greenhouse in either full light, light reduced to 25% of normal by shade cloth, or light reduced to 5% of normal. At the end of the study, they dried the young trees and weighed them. Identify the experimental units and the treatments.

50. **Sealing your teeth** Many children have their molars sealed to help prevent cavities. In an experiment, 120 children aged 6–8 were randomly assigned to a control group, a group in which sealant was applied and reapplied periodically for 36 months, or a group in which fluoride varnish was applied and reapplied periodically for 42 months. After 9 years, the percent of initially healthy molars with cavities was calculated for each group.[39] Identify the experimental units and the treatments.

51. **Improving response rate** How can we reduce the rate of refusals in telephone surveys? Most people who answer at all listen to the interviewer's introductory remarks and then decide whether to continue. One study made telephone calls to randomly selected households to ask opinions about the next election. In some calls, the interviewer gave her name; in others, she identified the university she was representing; and in still others, she identified both herself and the university. For each type of call, the interviewer either did or did not offer to send a copy of the final survey results to the person interviewed.

(a) List the factors in this experiment and state how many levels each factor has.

(b) If the researchers used every possible combination of levels to form the treatments, how many treatments were included in the experiment?

(c) List two of the treatments.

52. **Fabric science** A maker of fabric for clothing is setting up a new line to "finish" the raw fabric. The line will use either metal rollers or natural-bristle rollers to raise the surface of the fabric; a dyeing-cycle time of either 30 or 40 minutes; and a temperature of either 150°C or 175°C. Three specimens of fabric will be subjected to each treatment and scored for quality.

(a) List the factors in this experiment and state how many levels each factor has.

(b) If the researchers used every possible combination of levels to form the treatments, how many treatments were included in the experiment?

(c) List two of the treatments.

53. **Want a snack?** Can snacking on fruit rather than candy reduce later food consumption? Researchers randomly assigned 12 women to eat either 65 calories of berries or 65 calories of candy. Two hours later, all 12 women were given an unlimited amount of pasta to eat. The researchers recorded the amount of pasta consumed by each subject. The women who ate the berries consumed 133 fewer calories, on average. Identify the explanatory and response variables, the experimental units, and the treatments.

54. **Pricey pizza?** The cost of a meal might affect how customers evaluate and appreciate food. To investigate, researchers worked with an Italian all-you-can-eat buffet to perform an experiment. A total of 139 subjects were randomly assigned to pay either $4 or $8 for the buffet and then asked to rate the quality of the pizza on a 9-point scale. Subjects who paid $8 rated the pizza 11% higher than those who paid only $4.[40] Identify the explanatory and response variables, the experimental units, and the treatments.

55. **Oils and inflammation** The extracts of avocado and soybean oils have been shown to slow cell inflammation in test tubes. Will taking avocado and soybean unsaponifiables (called ASU) help relieve pain for subjects with joint stiffness due to arthritis? In an experiment, 345 men and women were randomly assigned to receive either 300 milligrams of ASU daily for three years or a placebo daily for three years.[41] Explain why it was necessary to include a control group in this experiment.

56. **Supplements for testosterone** As men age, their testosterone levels gradually decrease. This may cause a reduction in energy, an increase in fat, and other undesirable changes. Do testosterone supplements reverse some of these effects? A study in the Netherlands assigned 237 men aged 60 to 80 with low or low-normal testosterone levels to either a testosterone

---

4.52 (a) The factors are (1) roller type, which had 2 levels; (2) dyeing cycle time, which had 2 levels; and (3) temperature, which had 2 levels. **(b)** If the researchers used every possible combination to form the treatments, there would be $2 \times 2 \times 2 = 8$ treatments included in the experiment. **(c)** Answers may vary. The 8 treatments are:

(1) metal, 30 minutes, 150 degrees
(2) natural, 30 minutes, 150 degrees
(3) metal, 40 minutes, 150 degrees
(4) natural, 40 minutes, 150 degrees
(5) metal, 30 minutes, 175 degrees
(6) natural, 30 minutes, 175 degrees
(7) metal, 40 minutes, 175 degrees
(8) natural, 40 minutes, 175 degrees

4.53 Explanatory: Type of snack eaten (berries or candy). Response: Amount of pasta consumed (measured by the number of calories consumed). Experimental units: Women. Treatments: 65 calories of berries or 65 calories of candy.

4.54 Explanatory: The cost of the meal. Response: The rating of the quality of the pizza. Experimental units: The subjects (customers). Treatments: Paying either $4 or $8 for the buffet.

4.55 The control group can be used to show how changes in pain and joint stiffness progress over the 3 years naturally. If a control group was not used, the treatment could be deemed ineffective at improving the symptoms, when in reality it may be effective at preventing the symptoms from worsening.

4.56 A control group is used to provide a baseline for comparing the effects of other treatments. The control group allows the researchers to measure the effect of testosterone on energy level, fat content, and other changes, compared to no treatment at all. If a control group was not used, the treatment could be deemed effective for increasing energy, reducing fat, and other changes, when in reality the men believed these positive changes occurred simply because they were taking something that was *supposed* to have this effect.

4.48 (a) Explanatory: The mother's chocolate consumption. Response: The baby's temperament. **(b)** This is a prospective observational study. No treatments were assigned and the researchers asked the mothers to report their chocolate consumption while pregnant and later asked them about their babies' temperament at age 6 months. **(c)** No, this is an observational study, so we cannot make a cause-and-effect conclusion. It is possible that other variables are influencing the response. For example, women who eat chocolate daily may have less stressful lives, and that lack of stress might help their babies have better temperaments.

4.49 Experimental units: Pine seedlings. Treatments: Full light, 25% light, and 5% light.

4.50 Experimental units: 120 children. Treatments: No treatment, sealant, and fluoride varnish.

4.51 (a) The factors are (1) information provided by interviewer, which had 3 levels, and (2) whether caller offered survey results, which had 2 levels. **(b)** 6 treatments **(c)** Answers may vary. Here are two treatments: (1) giving name/no survey results; (2) identifying university/no survey results.

4.57 There was no control group. We do not know if this was a placebo effect or if the flavonols affected the blood flow. To make a cause-and-effect conclusion possible, we must randomly assign some subjects to get flavonols and others to get a placebo.

4.58 There was no control group. Over a year, many things can change. To make a cause-and-effect conclusion possible, we must randomly assign some people to receive the $500 bonus offer and other people to receive nothing.

4.59 Yes, if the treatment (ASU or placebo) assigned to a subject was unknown to both the subject and those monitoring the effectiveness of that treatment. If subjects knew they were receiving the placebo, their expectations would differ from those who received the ASU. Then it would be impossible to know if an improvement in pain was due to the difference in expectations or to the ASU. It is important for the experimenters to be "blind" so that they will be unbiased in their interactions with subjects.

4.60 Yes; this experiment could be double-blind if the treatment (testosterone or placebo) assigned to a subject was unknown to both the subject and those monitoring the effectiveness of that treatment. It is important for the experimenters to be "blind" so that they will be unbiased in their interactions with subjects.

4.61 Because the experimenter knew which subjects had learned the meditation techniques, he or she is not blind. If the experimenter believed that meditation was beneficial, he or she may subconsciously rate subjects in the meditation group as being less anxious.

4.62 Because the researchers know that a side effect of the new treatment is red skin, they may be able to identify which subjects are getting the old versus new treatment and thus are no longer blind to the treatment. If the researchers have high hopes for the new treatment, this could influence the researcher's judgment of the amount of acne present after the treatment.

4.63 (a) Write each name on a slip of paper, put the slips in a container, and mix thoroughly. Draw 40 slips of paper and assign these subjects to Treatment 1. Then draw 40 more slips and assign these subject to Treatment 2. Assign the remaining 40 subjects to Treatment 3. **(b)** Assign the students numbers from 1 to 120. Using RandInt(1,120), generate

supplement or a placebo.[42] Explain why it was necessary to include a control group in this experiment.

57. **Cocoa and blood flow** A study of blood flow involved 27 healthy people aged 18 to 72. Each subject consumed a cocoa beverage containing 900 milligrams of flavonols daily for 5 days. Using a finger cuff, blood flow was measured on the first and fifth days of the study. After 5 days, researchers measured what they called "significant improvement" in blood flow and the function of the cells that line the blood vessels.[43] What flaw in the design of this experiment makes it impossible to say if the cocoa really caused the improved blood flow? Explain your answer.

58. **Reducing unemployment** Will cash bonuses speed the return to work of unemployed people? A state department of labor notes that last year 68% of people who filed claims for unemployment insurance found a new job within 15 weeks. As an experiment, this year the state offers $500 to people filing unemployment claims if they find a job within 15 weeks. The percent who do so increases to 77%. What flaw in the design of this experiment makes it impossible to say if the bonus really caused the increase? Explain your answer.

59. **More oil and inflammation** Refer to Exercise 55. pg 278 Could blinding be used in this experiment? Explain your reasoning. Why is blinding an important consideration in this context?

60. **More testosterone** Refer to Exercise 56. Could blinding be used in this experiment? Explain your reasoning. Why is blinding an important consideration in this context?

61. **Meditation for anxiety** An experiment that claimed to show that meditation lowers anxiety proceeded as follows. The experimenter interviewed the subjects and rated their level of anxiety. Then the subjects were randomly assigned to two groups. The experimenter taught one group how to meditate and they meditated daily for a month. The other group was simply told to relax more. At the end of the month, the experimenter interviewed all the subjects again and rated their anxiety level. The meditation group now had less anxiety. Psychologists said that the results were suspect because the ratings were not blind. Explain what this means and how lack of blindness could affect the reported results.

62. **Side effects** Even if an experiment is double-blind, the blinding might be compromised if side effects of the treatments differ. For example, suppose researchers at a skin-care company are comparing their new acne treatment against that of the leading competitor. Fifty subjects are assigned at random to each treatment, and the company's researchers will rate the improvement for each of the 100 subjects. The researchers aren't

told which subjects received which treatments, but they know that their new acne treatment causes a slight reddening of the skin. How might this knowledge compromise the blinding? Explain why this is an important consideration in the experiment.

63. **Layoffs and "survivor guilt"** Workers who survive a pg 279 layoff of other employees at their location may suffer from "survivor guilt." A study of survivor guilt and its effects used as subjects 120 students who were offered an opportunity to earn extra course credit by doing proofreading. Each subject worked in the same cubicle as another student, who was an accomplice of the experimenters. At a break midway through the work, one of three things happened:

Treatment 1: The accomplice was told to leave; it was explained that this was because she performed poorly.

Treatment 2: It was explained that unforeseen circumstances meant there was only enough work for one person. By "chance," the accomplice was chosen to be laid off.

Treatment 3: Both students continued to work after the break.

The subjects' work performance after the break was compared with their performance before the break. Overall, subjects worked harder when told the other student's dismissal was random.[44] Describe how you would randomly assign the subjects to the treatments

(a) using slips of paper.

(b) using technology.

(c) using Table D.

64. **Precise offers** People often use round prices as first offers in a negotiation. But would a more precise number suggest that the offer was more reasoned and informed? In an experiment, 238 adults played the role of a person selling a used car. Each adult received one of three initial offers: $2000, $1865 (a precise under-offer), and $2135 (a precise over-offer). After hearing the initial offer, each subject made a counter-offer. The difference in the initial offer and counter-offer was the largest in the group that received the $2000 offer.[45] Describe how the researchers could have randomly assigned the subjects to the treatments

(a) using slips of paper.

(b) using technology.

(c) using Table D.

65. **Stronger players** A football coach hears that a new exercise program will increase upper-body strength better than lifting weights. He is eager to test this new program in the off-season with the players on his high

40 unique integers from 1 to 120 and assign the corresponding students to Treatment 1. Then generate 40 more unique integers from 1 to 120 and assign the corresponding students to Treatment 2. The remaining 40 students are assigned to Treatment 3. **(c)** Assign the students numbers from 001 to 120. Pick a spot on Table D and read off the first 40 unique numbers between 001 and 120 and assign those students to Treatment 1. The students corresponding to the next 40 unique numbers are assigned to Treatment 2. The remaining 40 students get Treatment 3.

4.64 (a) Write each name on a slip of paper, put the slips in a container, and mix thoroughly. Draw 80 slips of paper and assign these subjects to

Treatment 1. Draw 79 more slips of paper and assign those subjects to Treatment 2. Assign the remaining 79 subjects to Treatment 3. **(b)** Assign the adults numbers from 1 to 238. Using RandInt(1, 238) on the calculator, generate 80 unique numbers from 1 to 238 and assign the corresponding adults to Treatment 1 ($2000). Generate 79 more unique integers and assign the corresponding adults to Treatment 2 ($1865). Assign the remaining 40 adults to Treatment 3 ($2135). **(c)** Assign numbers from 001 to 238. Pick a spot on Table D, read off the first 80 unique numbers between 001 and 238, and assign them to Treatment 1. The adults with the next 79 unique numbers are assigned to Treatment 2. Assign the remaining 79 adults to Treatment 3.

school team. The coach decides to let his players choose which of the two treatments they will undergo for 3 weeks—exercise or weight lifting. He will use the number of push-ups a player can do at the end of the experiment as the response variable. Which principle of experimental design does the coach's plan violate? Explain how this violation could lead to confounding.

66. **Killing weeds** A biologist would like to determine which of two brands of weed killer, X or Y, is less likely to harm the plants in a garden at the university. Before spraying near the plants, the biologist decides to conduct an experiment using 24 individual plants. Which of the following two plans for randomly assigning the treatments should the biologist use? Why?

Plan A: Choose the 12 healthiest-looking plants. Then flip a coin. If it lands heads, apply Brand X weed killer to these plants and Brand Y weed killer to the remaining 12 plants. If it lands tails, do the opposite.

Plan B: Choose 12 of the 24 plants at random. Apply Brand X weed killer to those 12 plants and Brand Y weed killer to the remaining 12 plants.

67. **Boosting preemies** Do blood-building drugs help brain development in babies born prematurely? Researchers randomly assigned 53 babies, born more than a month premature and weighing less than 3 pounds, to one of three groups. Babies either received injections of erythropoietin (EPO) three times a week, darbepoetin once a week for several weeks, or no treatment. Results? Babies who got the medicines scored much better by age 4 on measures of intelligence, language, and memory than the babies who received no treatment.[46]

(a) Explain how this experiment used comparison.

(b) Explain the purpose of randomly assigning the babies to the three treatments.

(c) Name two variables that were controlled in this experiment and why it was beneficial to control these variables.

(d) Explain how this experiment used replication. What is the purpose of replication in this context?

68. **The effects of day care** Does preschool help low-income children stay in school and hold good jobs later in life? The Carolina Abecedarian Project (the name suggests the ABCs) has followed a group of 111 children for over 40 years. Back then, these individuals were all healthy but low-income black infants in Chapel Hill, North Carolina. All the infants received nutritional supplements and help from social workers. Half were also assigned at random to an intensive preschool program. Results? Children who were assigned to the preschool program had higher IQ's, higher standardized test scores, and were less likely to repeat a grade in school.[47]

(a) Explain how this experiment used comparison.

(b) Explain the purpose of randomly assigning the infants to the two treatments.

(c) Name two variables that were controlled in this experiment and why it was beneficial to control these variables.

(d) Explain how this experiment used replication. What is the purpose of replication in this context?

69. **Treating prostate disease** A large study used records from Canada's national health care system to compare the effectiveness of two ways to treat prostate disease. The two treatments are traditional surgery and a new method that does not require surgery. The records described many patients whose doctors had chosen what method to use. The study found that patients treated by the new method were more likely to die within 8 years.[48]

(a) Further study of the data showed that this conclusion was wrong. The extra deaths among patients who were treated with the new method could be explained by other variables. What other variables might be confounded with a doctor's choice of surgical or nonsurgical treatment?

(b) You have 300 prostate patients who are willing to serve as subjects in an experiment to compare the two methods. Write a few sentences describing a completely randomized design for this experiment.

70. **Diet soda and pregnancy** A large study of 3000 Canadian children and their mothers found that the children of mothers who drank diet soda daily during pregnancy were twice as likely to be overweight at age 1 than children of mothers who avoided diet soda during pregnancy.[49]

(a) A newspaper article about this study had the headline "Diet soda, pregnancy: Mix may fuel childhood obesity." This headline suggests that there is a cause-and-effect relationship between diet soda consumption during pregnancy and the weight of the children 1 year after birth. However, this relationship could be explained by other variables. What other variables might be confounded with a mother's consumption of diet soda during pregnancy?

(b) You have 300 pregnant mothers who are willing to serve as subjects in an experiment that compares three treatments during pregnancy: no diet soda, one diet soda per day, and two diet sodas per day. Write a few sentences describing a completely randomized design for this experiment.

71. **A fruitful experiment** A citrus farmer wants to know which of three fertilizers (A, B, and C) is most effective for increasing the number of oranges on his trees. He

each group makes it easier to rule out the chance variation in random assignment as a possible explanation for the differences observed.

4.68 (a) Researchers used a design that compared children in an intensive preschool program to children who were not in one. **(b)** Random assignment helps create two groups of children who are roughly equivalent at the start of the study. This ensures that the effects of other variables are spread evenly among the two groups of children. **(c)** Answers may vary. **(d)** There were more than 50 subjects in each treatment group. Having a large number of subjects in each group makes it easier to rule out chance variation as a possible explanation for the difference in outcomes.

4.69 (a) Expense and condition of the patient. If a patient is in very poor health, a doctor might choose not to recommend surgery. Then, if the non-surgery treatment has a higher death rate, we will not know if it is because of the treatment or because the initial health of the subjects was worse. **(b)** Write the names of all 300 patients on identical slips of paper, put the slips in a hat, and mix well. Draw out 150 slips and assign the corresponding subjects to receive surgery. The remaining 150 subjects receive the new method. At the end of the study, count how many patients survived in each group.

4.70 (a) Other variables that might be confounded may include the mother's weight, whether mothers who are overweight are more inclined to drink diet soda, and whether she exercises or not. Therefore, we cannot know if the diet soda, the mother's weight, or the lack of exercise was the cause of the higher rate of overweight children. **(b)** Write the names of all 300 mothers on identical slips of paper, put the slips in a hat, and mix well. Draw 100 slips and assign to no diet soda. Draw 100 more slips and assign to one diet soda per day. Assign the remaining 100 subjects to two diet sodas per day. At the end of the study, weigh the children and compare the proportion of overweight children in each group.

4.65 More motivated players may choose the new method. If they improve more, the coach cannot be sure whether the exercise program or player motivation caused the improvement.

4.66 The biologist should use Plan B. If the biologist uses Plan A, the brand assigned to the healthier plants might appear safer. We will not know if this is due to the brand or to the plants being healthier to begin with. With Plan B, the two groups of plants should be roughly equivalent at the start, so that a difference in response can be attributed to the type of weed killer.

4.67 (a) Researchers used a design that compared infants assigned to one of three treatments. **(b)** Random assignment helps create three groups of infants who are roughly equivalent with respect to the required conditions at the start of the study so that effects of other variables are spread evenly among the three groups. **(c)** Birthweight and whether or not the baby was born prematurely. It is beneficial to control these variables because otherwise they may become sources of variability, making it harder to determine the treatments' effectiveness. **(d)** Having about 17 infants in

4.71 (a) Because smaller trees are likely to have fewer oranges than medium or large trees, form blocks based on the size of the trees (small, medium, large). Then randomly assign the trees in each block to the three fertilizers. In the end, measure each tree's orange production. **(b)** Blocking helps account for the variability in number of oranges that is due to the differences in tree size, making it easier to determine if one fertilizer is better.

4.72 (a) Form blocks based on the rows because fertility differs between rows. Then randomly assign one plot in each row to each of the 5 varieties of corn. In the end, measure the yield of all 25 plots. **(b)** Blocking helps account for the variability in yield that is due to the differences in fertility in the field, making it easier to determine if one variety is better.

4.73 (a) The blocks are the different diagnoses because the type of care was assigned to patients within each diagnosis. **(b)** Blocking allows us to account for the variability in the response variables due to differences in diagnosis, making it easier to determine which type of care is more effective. **(c)** Advantage: there would be no variability in the response variables introduced by differences in diagnosis. However, we would be able to make conclusions about health and satisfaction with their medical care only for diabetic patients like the ones in the experiment.

4.74 (a) The blocks are the sexes because researchers will assign all three therapies to patients of each sex. **(b)** Blocking allows us to account for the variability in the effectiveness of the therapies that is due to differences in sex, making it easier to determine which therapy is more effective. **(c)** Advantage: there would be no variability in the effectiveness of the therapies introduced by differences in sex. However, we would be able to make conclusions about how the treatments work only for males like the ones in the experiment.

4.75 (a) We would not know if the weight gain was because of the diet or because of genetics and initial health. Initial health is a confounding variable because it is related to both diet and weight gain. **(b)** Number the rats in the first litter from 1 to 10. Use the command randInt(1,10) to select 5 different integers from 1 to 10. Select the corresponding 5 rats and assign them to Diet A. The other 5 rats in this litter will be assigned to Diet B. Repeat the same process for the second litter. **(c)** Answers will vary.

is willing to use 30 mature trees of various sizes from his orchard in an experiment with a randomized block design.

(a) Describe a randomized block design for this experiment. Justify your choice of blocks.

(b) Explain why a randomized block design might be preferable to a completely randomized design for this experiment.

72. **In the cornfield** An agriculture researcher wants to compare the yield of 5 corn varieties: A, B, C, D, and E. The field in which the experiment will be carried out increases in fertility from north to south. The researcher therefore divides the field into 25 plots of equal size, arranged in 5 east–west rows of 5 plots each, as shown in the diagram.

(a) Describe a randomized block design for this experiment. Justify your choice of blocks.

(b) Explain why a randomized block design might be preferable to a completely randomized design for this experiment.

73. **Doctors and nurses** Nurse-practitioners are nurses with advanced qualifications who often act much like primary-care physicians. Are they as effective as doctors at treating patients with chronic conditions? An experiment was conducted with 1316 patients who had been diagnosed with asthma, diabetes, or high blood pressure. Within each condition, patients were randomly assigned to either a doctor or a nurse-practitioner. The response variables included measures of the patients' health and of their satisfaction with their medical care after 6 months.[50]

(a) Which are the blocks in this experiment: the different diagnoses (asthma, diabetes, or high blood pressure) or the type of care (nurse or doctor)? Why?

(b) Explain why a randomized block design is preferable to a completely randomized design in this context.

(c) Suppose the experiment used only diabetes patients, but there were still 1316 subjects willing to participate. What advantage would this offer? What disadvantage?

74. **Comparing cancer treatments** The progress of a type of cancer differs in women and men. Researchers want to design an experiment to compare three therapies for this cancer. They recruit 500 male and 300 female patients who are willing to serve as subjects.

(a) Which are the blocks in this experiment: the three cancer therapies or the two sexes? Why?

(b) What are the advantages of a randomized block design over a completely randomized design using these 800 subjects?

(c) Suppose the researchers had 800 male and no female subjects available for the study. What advantage would this offer? What disadvantage?

75. **Aw, rats!** A nutrition experimenter intends to compare the weight gain of newly weaned male rats fed Diet A with that of rats fed Diet B. To do this, she will feed each diet to 10 rats. She has available 10 rats from one litter and 10 rats from a second litter. Rats in the first litter appear to be slightly healthier.

(a) If the 10 rats from Litter 1 were fed Diet A, then initial health would be a confounding variable. Explain this statement carefully.

(b) Describe how to randomly assign the rats to treatments using a randomized block design with litters as blocks.

(c) Use technology or Table D to carry out the random assignment.

76. **Comparing weight-loss treatments** A total of 20 overweight females have agreed to participate in a study of the effectiveness of four weight-loss treatments: A, B, C, and D. The researcher first calculates how overweight each subject is by comparing the subject's actual weight with her "ideal" weight. The subjects and their excess weights in pounds are as follows:

Birnbaum	35	Hernandez	25	Moses	25	Smith	29
Brown	34	Jackson	33	Nevesky	39	Stall	33
Brunk	30	Kendall	28	Obrach	30	Tran	35
Cruz	34	Loren	32	Rodriguez	30	Wilansky	42
Deng	24	Mann	28	Santiago	27	Williams	22

The response variable is the weight lost after 8 weeks of treatment.

(a) If the 5 most overweight women were assigned Treatment A, the next 5 most overweight women were assigned Treatment B, and so on, then the amount overweight would be a confounding variable. Explain this statement carefully.

(b) Describe how to randomly assign the women to treatments using a randomized block design. Use blocks of size 4 formed by the amount overweight.

(c) Use technology or Table D to carry out the random assignment.

4.76 (a) We would not know if the amount of weight lost is due to the effects of the weight-loss treatment or to the fact that the women had more (or less) to lose. **(b)** Arrange the women in order based on the number of excess pounds. The four women with the most excess weight form one block, the four women with the next most excess weight form the next block, and so on. For each block, number the subjects from 1 to 4. For each block, generate a random number using randInt(1, 4). Assign the corresponding person to Treatment A. Generate a different random number from 1 to 4 and assign this person to Treatment B, and so on. **(c)** Answers will vary.

77. **SAT preparation** A school counselor wants to compare the effectiveness of an online SAT preparation program with an in-person SAT preparation class. For an experiment, the counselor recruits 30 students who have already taken the SAT once. The response variable will be the improvement in SAT score.

(a) Design an experiment that uses a completely randomized design to investigate this question.

(b) Design an experiment that uses a matched pairs design to investigate this question. Explain your method of pairing.

(c) Which design do you prefer? Explain your answer.

78. **Valve surgery** Medical researchers want to compare the success rate of a new non-invasive method for replacing heart valves using a cardiac catheter with traditional open-heart surgery. They have 40 male patients, ranging in age from 55 to 75, who need valve replacement. One of several response variables will be the percentage of blood that flows backward—in the wrong direction—through the valve on each heartbeat.

(a) Design an experiment that uses a completely randomized design to investigate this question.

(b) Design an experiment that uses a matched pairs design to investigate this question. Explain your method of pairing.

(c) Which design do you prefer? Explain your answer.

79. **Look, Ma, no hands!** Does talking on a hands-free cell phone distract drivers? Researchers recruit 40 student subjects for an experiment to investigate this question. They have a driving simulator equipped with a hands-free phone for use in the study. Each subject will complete two sessions in the simulator: one while talking on the hands-free phone and the other while just driving. The order of the two sessions for each subject will be determined at random. The route, driving conditions, and traffic flow will be the same in both sessions.

(a) What type of design did the researchers use in their study?

(b) Explain why the researchers chose this design instead of a completely randomized design.

(c) Why is it important to randomly assign the order of the treatments?

(d) Explain how and why researchers controlled for other variables in this experiment.

80. **Chocolate gets my heart pumping** Cardiologists at Athens Medical School in Greece wanted to test if chocolate affects blood vessel function. The researchers recruited 17 healthy young volunteers, who were each given a 3.5-ounce bar of dark chocolate, either bittersweet or fake chocolate. On another day, the volunteers received the other treatment. The order in which subjects received the bittersweet and fake chocolate was determined at random. The subjects had no chocolate outside the study, and investigators didn't know if a subject had eaten the real or the fake chocolate. An ultrasound was taken of each volunteer's upper arm to observe the functioning of the cells in the walls of the main artery. The researchers found that blood vessel function was improved when the subjects ate bittersweet chocolate, and that there were no such changes when they ate the placebo (fake chocolate).[51]

(a) What type of design did the researchers use in their study?

(b) Explain why the researchers chose this design instead of a completely randomized design.

(c) Why is it important to randomly assign the order of the treatments for the subjects?

(d) Explain how and why researchers controlled for other variables in this experiment.

81. **Got deodorant?** A group of students wants to perform an experiment to determine whether Brand A or Brand B deodorant lasts longer. One group member suggests the following design: Recruit 40 student volunteers—20 male and 20 female. Separate by gender, because male and female bodies might respond differently to deodorant. Give all the males Brand A deodorant and all the females Brand B. Have the principal judge how well the deodorant is still working at the end of the school day on a 0 to 10 scale. Then compare ratings for the two treatments.

(a) Identify any flaws you see in the proposed design for this experiment.

(b) Describe how you would design the experiment. Explain how your design addresses each of the problems you identified in part (a).

82. **Close shave** Which of two brands (X or Y) of electric razor shaves closer? Researchers want to design and carry out an experiment to answer this question using 50 adult male volunteers. Here's one idea: Have all 50 subjects shave the left sides of their faces with the Brand X razor and shave the right sides of their faces with the Brand Y razor. Then have each man decide which razor gave the closer shave and compile the results.

(a) Identify any flaws you see in the proposed design for this experiment.

(b) Describe how you would design the experiment. Explain how your design addresses each of the problems you identified in part (a).

4.77 (a) Write the names of all 30 students on identical slips of paper and mix well in a hat. Draw 15 slips and assign the corresponding students to take the online program. The remaining 15 students will take the in-person class. Record the SAT improvement. **(b)** Pair the two students with the highest initial SAT score together, pair the two students with the next highest initial SAT scores, and so on. Within each pair, one student is randomly assigned to the online class and the other is assigned to the in-person class. Record the SAT improvement. **(c)** Matched pairs because we can account for the variability in improvement that is due to variability in student ability, making it easier to determine which program is more effective.

4.78 (a) Write the names of all 40 patients on identical slips of paper and mix well in a hat. Draw 20 slips and assign the corresponding patients to receive the new method. The remaining 20 patients will receive the traditional surgery. Record percentage of blood that flows backward. **(b)** Pair the two patients with the best overall heart health, pair the two patients with the next best overall heart health, and so on. Within each pair, one patient is randomly assigned to the new method and the other is assigned to surgery. Record percentage of blood that flows backward. **(c)** Matched pairs because we can account for the variability in the percentage of blood that flows backward that is due to overall heart health of the patients, making it easier to determine which treatment is more effective.

4.79 (a) Matched pairs because each subject was assigned both treatments. **(b)** In a matched pairs design, each student is compared with himself (or herself), so the variability in response due to differences between students (some students are more distractible) is accounted for, making it easier to see if one treatment is more effective. **(c)** If all the students used the hands-free phone during the first session and performed worse, we would not know if the better performance during the second session is due to the lack of phone or to learning from their mistakes. **(d)** The simulator, route, driving conditions, and traffic flow were all kept the same for both sessions. Researchers are preventing these variables from adding variability to the response.

4.80 (a) Matched pairs because each volunteer was assigned both treatments. **(b)** In a matched pairs design, each volunteer is compared with himself (or herself), so the differences between volunteers (some have better blood vessel function than others) are accounted for, making it easier to see if one treatment is more effective. **(c)** If everyone has the fake chocolate first and the bittersweet chocolate second, the explanatory variable will be confounded with other variables that vary between the two days. **(d)** The method of measurement and the amount of chocolate were kept the same for all subjects. Researchers are preventing these factors from adding variability to the response.

4.81 (a) They will not know if any difference in smell is due to gender or to the deodorant. **(b)** Each student should have one armpit randomly assigned to receive A and the other to receive B. This prevents the problem in part (a) and also accounts for the variability in smell between individuals, making it easier to see a difference in the effectiveness of the two deodorants.

4.82 (a) If Y appears better, we will not know if the difference is due to the brand of razor or because most men can shave better on the right side (because most are right-handed). **(b)** Have each man shave half his face with X and the other half with Y, randomly assigning which half receives which brand of razor. This prevents the problem in part (a) and also accounts for the variability between individuals, making it easier to see a difference in the effectiveness of the two razors.

4.83 c

4.84 b

4.85 b

4.86 d

4.87 c

4.88 d

4.89 b

Multiple Choice *Select the best answer for Exercises 83–90.*

83. Can a vegetarian or low-salt diet reduce blood pressure? Men with high blood pressure are assigned at random to one of four diets: (1) normal diet with unrestricted salt; (2) vegetarian with unrestricted salt; (3) normal with restricted salt; and (4) vegetarian with restricted salt. This experiment has

(a) one factor, the type of diet.

(b) two factors, high blood pressure and type of diet.

(c) two factors, normal/vegetarian diet and unrestricted/restricted salt.

(d) three factors, men, high blood pressure, and type of diet.

(e) four factors, the four diets being compared.

84. In the experiment of the preceding exercise, the subjects were randomly assigned to the different treatments. What is the most important reason for this random assignment?

(a) Random assignment eliminates the effects of other variables such as stress and body weight.

(b) Random assignment balances the effects of other variables such as stress and body weight among the four treatment groups.

(c) Random assignment makes it possible to make a conclusion about all men.

(d) Random assignment reduces the amount of variation in blood pressure.

(e) Random assignment prevents the placebo effect from ruining the results of the study.

85. To investigate if standing up while studying affects performance in an algebra class, a teacher assigns half of the 30 students in his class to stand up while studying and assigns the other half to not stand up while studying. To determine who receives which treatment, the teacher identifies the two students who did best on the last exam and randomly assigns one to stand and one to not stand. The teacher does the same for the next two highest-scoring students and continues in this manner until each student is assigned a treatment. Which of the following best describes this plan?

(a) This is an observational study.

(b) This is an experiment with blocking.

(c) This is a completely randomized experiment.

(d) This is a stratified random sample.

(e) This is a systematic random sample.

86. A gardener wants to try different combinations of fertilizer (none, 1 cup, 2 cups) and mulch (none, wood chips, pine needles, plastic) to determine which combination produces the highest yield for a variety of green beans. He has 60 green-bean plants to use in the experiment. If he wants an equal number of plants to be assigned to each treatment, how many plants will be assigned to each treatment?

(a) 1   (b) 3   (c) 4

(d) 5   (e) 12

87. Corn variety 1 yielded 140 bushels per acre last year at a research farm. This year, corn variety 2, planted in the same location, yielded only 110 bushels per acre. Based on these results, is it reasonable to conclude that corn variety 1 is more productive than corn variety 2?

(a) Yes, because 140 bushels per acre is greater than 110 bushels per acre.

(b) Yes, because the study was done at a research farm.

(c) No, because there may be other differences between the two years besides the corn variety.

(d) No, because there was no use of a placebo in the experiment.

(e) No, because the experiment wasn't double-blind.

88. A report in a medical journal notes that the risk of developing Alzheimer's disease among subjects who regularly opted to take the drug ibuprofen was about half the risk of those who did not. Is this good evidence that ibuprofen is effective in preventing Alzheimer's disease?

(a) Yes, because the study was a randomized, comparative experiment.

(b) No, because the effect of ibuprofen is confounded with the placebo effect.

(c) Yes, because the results were published in a reputable professional journal.

(d) No, because this is an observational study. An experiment would be needed to confirm (or not confirm) the observed effect.

(e) Yes, because a 50% reduction can't happen just by chance.

89. A farmer is conducting an experiment to determine which variety of apple tree, Fuji or Gala, will produce more fruit in his orchard. The orchard is divided into 20 equally sized square plots. He has 10 trees of each variety and randomly assigns each tree to a separate plot in the orchard. What are the experimental unit(s) in this study?

(a) The trees   (b) The plots   (c) The apples

(d) The farmer   (e) The orchard

90. Two essential features of all statistically designed experiments are

(a) comparing two or more treatments; using the double-blind method.

(b) comparing two or more treatments; using chance to assign subjects to treatments.

(c) always having a placebo group; using the double-blind method.

(d) using a block design; using chance to assign subjects to treatments.

(e) using enough subjects; always having a control group.

Recycle and Review

91. Seed weights (2.2) Biological measurements on the same species often follow a Normal distribution quite closely. The weights of seeds of a variety of winged bean are approximately Normal with mean 525 milligrams (mg) and standard deviation 110 mg.

(a) What percent of seeds weigh more than 500 mg?

(b) If we discard the lightest 10% of these seeds, what is the smallest weight among the remaining seeds?

92. Comparing rainfall (1.3) The boxplots summarize the distributions of average monthly rainfall (in inches) for Tucson, Arizona, and Princeton, New Jersey.[52] Compare these distributions.

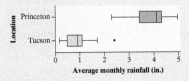

SECTION 4.3   Using Studies Wisely

LEARNING TARGETS *By the end of the section, you should be able to:*

- Explain the concept of sampling variability when making an inference about a population and how sample size affects sampling variability.

- Explain the meaning of statistically significant in the context of an experiment and use simulation to determine if the results of an experiment are statistically significant.

- Identify when it is appropriate to make an inference about a population and when it is appropriate to make an inference about cause and effect.

- Evaluate if a statistical study has been carried out in an ethical manner.*

Researchers who conduct statistical studies often want to draw conclusions that go beyond the data they produce. Here are two examples.

- The U.S. Census Bureau carries out a monthly Current Population Survey of about 60,000 households. Their goal is to use data from these randomly selected households to estimate the percent of unemployed individuals in the population.

- Scientists performed an experiment that randomly assigned 21 volunteer subjects to one of two treatments: sleep deprivation for one night or unrestricted sleep. The scientists hoped to show that sleep deprivation causes a decrease in performance two days later.[53]

What conclusions can be drawn from a particular study? The answer depends on how the data were collected.

---

*This is an important topic, but it is not required for the AP® Statistics exam.

4.90 b

4.91 (a) $z = \dfrac{500 - 525}{110} = -0.23$

Table A: $1 - 0.4090 = 0.5910$
Tech: normalcdf(lower: 500, upper: 10000, mean: 525, SD: 110) = 0.5899

(b) *Table A:* Solving $-1.28 = \dfrac{x - 525}{110}$

gives $x = 384.2$.
Tech: invNorm(area: 0.10, mean: 525, SD: 110) = 384.0

4.92 The distribution of average monthly rainfall for Tucson, Arizona, is slightly skewed to the right with one high outlier. The distribution of average monthly rainfall for Princeton, New Jersey, is skewed left with no outliers. The center of the distribution of average monthly rainfall for Tucson, Arizona, was considerably less (median ≈ 0.95 inch) than for Princeton, New Jersey (median ≈ 4.15 inches). The distribution of average monthly rainfall for Tucson, Arizona, has less variability (*IQR* ≈ 0.6 inch) than the average monthly rainfall for Princeton, New Jersey (*IQR* ≈ 0.9 inch).

PD **Section 4.3 Overview**

To watch the video overview of Section 4.3 (for teachers), click on the link in the TE-Book, look on the TRFD, or download from the Teacher's Resources on the book's digital platform.

TRM **Section 4.3 Alternate Examples**

You can find the Alternate Examples for this section in Microsoft Word format by clicking on the link in the TE-Book, opening the TRFD, or downloading from the Teacher's Resources on the book's digital platform.

Teaching Tip

In previous editions of this text, Section 4.3 covered scope of inference and data ethics. In this edition, we have added the discussion about sampling variability and statistical significance (the first two learning targets).

Inference for Sampling

When the members of a sample are selected at random from a population, we can use the sample results to *infer* things about the population. That is, we can make *inferences* about the population from which the sample was randomly selected. Inference from convenience samples or voluntary response samples would be misleading because these methods of choosing a sample are biased. In these cases, we are almost certain that the sample does *not* fairly represent the population.

Even when making an inference from a random sample, it would be surprising if the estimate from the sample was exactly equal to the truth about the population. For example, in a random sample of 1399 U.S. teens aged 13–18, 26% reported more than 8 hours of entertainment media use per day. Because of **sampling variability**, it would be surprising if exactly 26% of *all* U.S. teens aged 13–18 reported more than 8 hours of entertainment media use per day. Why? Because different samples of 1399 U.S. teens aged 13–18 will include different sets of people and produce different estimates.

> **DEFINITION** Sampling variability
>
> **Sampling variability** refers to the fact that different random samples of the same size from the same population produce different estimates.

The following activity explores the idea of sampling variability.

ACTIVITY Exploring sampling variability

When making an inference about a population from a random sample, we shouldn't expect the estimate to be exactly correct. But how much do sample results vary? Your teacher has prepared a large population of beads, where 30% have a certain color (e.g., red) so you can explore this question.

G. Curt Fiedler/Getty Images

1. In a moment, you will select a random sample of 20 beads. Do you expect that your sample will contain exactly 30% red beads? Explain your reasoning.

2. Mix the beads thoroughly, select a random sample of 20 beads from the population, calculate the percent of red beads in the sample, and replace the beads in the population.

3. After all students have selected a sample, make a class dotplot showing each student's estimate for the percent of red beads. Where is the graph centered? How much does the percent of red beads vary?

4. Imagine that you repeated Steps 2 and 3 with random samples of size 100. How do you expect the dotplots would compare? *If there's time, select random samples of size 100 to confirm your answer.*

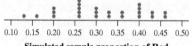

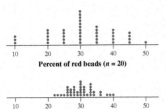

Percent of red beads (n = 20)

Percent of red beads (n = 100)

When Mrs. Storrs's class of 40 students did the red bead activity, they produced the dotplot shown at left (top) for samples of size 20. The dotplot is centered around 30%, the true percent of red beads. This shouldn't be surprising because random sampling helps avoid bias. Notice also that the estimates varied from 10% to 50% and that only 11 of the 40 estimates were equal to exactly 30%.

To see the effect of increasing the sample size, we simulated 40 random samples of size 100 from the same population and recorded the percent of red beads in each sample. Notice that the graph is still centered at 30%, but there is much less variability.

SAMPLING VARIABILITY AND SAMPLE SIZE

Larger random samples tend to produce estimates that are closer to the true population value than smaller random samples. In other words, estimates from larger samples are more precise.

EXAMPLE

Weighing football players

Inference for sampling

PROBLEM: How much do National Football League (NFL) players weigh, on average? In a random sample of 50 NFL players, the average weight is 244.4 pounds.

(a) Do you think that 244.4 pounds is the true average weight of all NFL players? Explain your answer.

(b) Which would be more likely to give an estimate close to the true average weight of all NFL players: a random sample of 50 players or a random sample of 100 players? Explain your answer.

SOLUTION:

(a) No. Different samples of size 50 would produce different average weights. So it would be surprising if this estimate is equal to the true average weight of all NFL players.

(b) A random sample of 100 players, because estimates tend to be closer to the truth when the sample size is larger.

FOR PRACTICE, TRY EXERCISE 93

Estimates from random samples often come with a *margin of error* that allows us to create an interval of plausible values for the true population value. In the preceding example about NFL players, the margin of error for the estimate of 244.4 pounds is 14.2 pounds. Based on this margin of error, it wouldn't be surprising if the true average weight for all NFL players was as small as 244.4 − 14.2 = 230.2 pounds or as large as 244.4 + 14.2 = 258.6 pounds.

You will learn how to calculate the margin of error for a population mean in Chapter 10. For now, make sure to remember the effect of sampling variability when using data from a random sample to make an inference about a population.

ALTERNATE EXAMPLE

Skill 4.B

How many likes for Selena Gomez? Inference for sampling

PROBLEM:
Selena Gomez was the most followed celebrity on Instagram in 2016 with 103 million followers. How many likes did she get for each Instagram post in 2016, on average? In a random sample of 30 posts, the average number of likes was 3.1 million.

(a) Do you think that 3.1 million is the true average number of likes for all Instagram posts made by Selena Gomez in 2016? Explain your reasoning.

(b) Which would be more likely to give an estimate closer to the true average number of likes for all Instagram posts made by Selena Gomez in 2016, a random sample of 30 posts or a random sample of 100 posts? Explain your reasoning.

SOLUTION:

(a) No; different samples of size 30 would produce different average numbers of likes. So it would be surprising if this estimate is equal to the true average number of likes for all Instagram posts made by Selena Gomez in 2016.

(b) A random sample of 100 posts, because estimates tend to be closer to the truth when the sample size is larger.

Preparing for Inference

From our 40 estimates, we can see that most of the samples will give an estimate that is within 15% of the truth. That 15% is known as a *margin of error* that describes how far we expect estimates to vary from the truth, at most.

Teaching Tip

When increasing sample size, it is as if all the dots in the dotplot are being squeezed toward the center (like taking a mound of sand and making it taller and skinnier). Physically demonstrate this for students by opening a gap between your hands and then closing that gap (like pushing the sand inward).

Preparing for Inference

In Chapter 7, we will discover that the standard deviation of the sampling distribution (for a sample proportion and a sample mean) has an inverse relationship with the square root of the sample size. It is easy to see in these formulas that as the sample size increases, the variability will decrease:

$$\sigma_{\hat{p}} = \sqrt{\frac{p(1-p)}{n}} \text{ and } \sigma_{\bar{x}} = \frac{\sigma}{\sqrt{n}}$$

Teaching Tip

Mention to students that the word *error* in the phrase *margin of error* does not mean that a mistake has been made. The margin of error compensates for the variability that results from taking a random sample from a population. It does not account for a mistake made during data collection.

Preparing for Inference

Margin of error is based on a confidence level. In this example, we have used a 95% confidence level to create a 95% confidence interval of (230.2, 258.6) pounds. In Chapter 10, the margin of error formula for estimating a mean is given by $ME = t^* \frac{s_x}{\sqrt{n}}$, where t^* depends on the confidence level, s is the standard deviation of the sample, and n is the sample size.

Preparing for Inference

We will learn much more about statistical significance in later chapters. For now, emphasize the connection between statistical significance and the variability that arises from random sampling and random assignment. Because no random sample will perfectly represent the population, and no random assignment will create perfectly equivalent treatment groups at the beginning of an experiment, there will likely be differences in the response variable just by chance. Results are deemed statistically significant only if the difference in the response is bigger than what would be expected due to chance.

Teaching Tip

We can discuss statistical significance for both observational studies and experiments. For observational studies, observed results that are too unusual to be explained by chance variation in the *random sampling* are considered statistically significant. For experiments, observed results that are too unusual to be explained by chance variation in the *random assignment* are considered statistically significant.

✚ Ask the StatsMedic

Tell the Whole Story: Evidence for *H*ₐ

In this post, guest blogger Josh Tabor (author!) offers a strategy for helping students to understand the purpose of a significance test.

Inference for Experiments

Well-designed experiments allow for inferences about cause and effect. But we should only conclude that changes in the explanatory variable cause changes in the response variable if the results of an experiment are **statistically significant**.

> **DEFINITION   Statistically significant**
>
> When the observed results of a study are too unusual to be explained by chance alone, the results are called **statistically significant**.

Mr. Wilcox and his students decided to perform the caffeine experiment from the preceding section. In their experiment, 10 student volunteers were randomly assigned to drink cola with caffeine and the remaining 10 students were assigned to drink caffeine-free cola. The table and graph show the change in pulse rate for each student (Final pulse rate – Initial pulse rate), along with the mean change for each group.

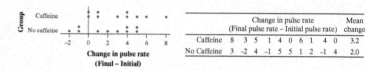

	Change in pulse rate (Final pulse rate – Initial pulse rate)										Mean change
Caffeine	8	3	5	1	4	0	6	1	4	0	3.2
No Caffeine	3	–2	4	–1	5	5	1	2	–1	4	2.0

The dotplots provide some evidence that caffeine has an effect on pulse rates. The mean change for the 10 students who drank cola with caffeine was 3.2, which is 1.2 greater than for the group who drank caffeine-free cola. But are the results statistically significant?

Recall that the purpose of random assignment in this experiment was to create two groups that were roughly equivalent at the beginning of the experiment. Subjects with high caffeine tolerance should be split up in about equal numbers, subjects with high metabolism should be split up in about equal numbers, and so on.

Of course, the random assignment is unlikely to produce groups that are exactly equivalent. One group might get more "favorable" subjects just by chance. That is, the caffeine group might end up with a few extra subjects who were likely to have a pulse rate increase, just due to chance variation in the random assignment.

There are two ways to explain why the mean change in pulse rate was 1.2 greater for the caffeine group:

1. Caffeine does *not* have an effect on pulse rates, and the difference of 1.2 happened because of chance variation in the random assignment.

2. Caffeine increases pulse rates.

If it is plausible to get a difference of 1.2 or more simply due to the chance variation in random assignment, the results of the experiment are not statistically

Preparing for Inference

In later chapters, when we perform significance tests, these two possible explanations will be the hypotheses:

1. Null hypothesis
2. Alternative hypothesis

significant. But if it is very unlikely to get a difference of 1.2 or more by chance alone, we rule out Explanation 1 and say the results are statistically significant—and that caffeine increases pulse rates.

How can we determine if a difference of 1.2 is statistically significant? You'll find out in the following activity.

ACTIVITY Analyzing the caffeine experiment

In the experiment performed by Mr. Wilcox's class, the mean change in pulse rate for the caffeine group was 1.2 greater than the mean change for the no-caffeine group. This provides some evidence that caffeine increases pulse rates. Is this evidence convincing? Or is it plausible that a difference of 1.2 would arise just due to chance variation in the random assignment? In this activity, we'll investigate by seeing what differences typically occur just by chance, assuming caffeine doesn't affect pulse rates. That is, we'll assume that the change in pulse rate for a particular student would be the same regardless of what treatment he or she was assigned.

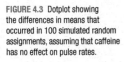

	Change in pulse rate (Final pulse rate – Initial pulse rate)										Mean change
Caffeine	8	3	5	1	4	0	6	1	4	0	3.2
No Caffeine	3	–2	4	–1	5	5	1	2	–1	4	2.0

1. Gather 20 index cards to represent the 20 students in this experiment. On each card, write one of the 20 outcomes listed in the table. For example, write "8" on the first card, "3" on the second card, and so on.

2. Shuffle the cards and deal two piles of 10 cards each. This represents randomly assigning the 20 students to the two treatments, *assuming that the treatment received doesn't affect the change in pulse rate.* The first pile of 10 cards represents the caffeine group, and the second pile of 10 cards represents the no-caffeine group.

3. Find the mean change for each group and subtract the means (Caffeine – No caffeine). *Note:* It is possible to get a negative difference.

4. Your teacher will draw and label an axis for a class dotplot. Plot the difference you got in Step 3 on the graph.

5. In Mr. Wilcox's class, the observed difference in means was 1.2. Is a difference of 1.2 statistically significant? Discuss with your classmates.

We used technology to perform 100 trials of the simulation described in the activity. The dotplot in Figure 4.3 shows that getting a difference of 1.2 isn't that unusual. In 19 of the 100 trials, we obtained a difference of 1.2 or more simply due to chance variation in the random assignment.

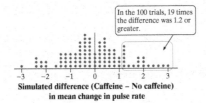

In the 100 trials, 19 times the difference was 1.2 or greater.

FIGURE 4.3 Dotplot showing the differences in means that occurred in 100 simulated random assignments, assuming that caffeine has no effect on pulse rates.

Simulated difference (Caffeine – No caffeine) in mean change in pulse rate

Preparing for Inference

Is the difference of 1.2 too unusual to be explained by the chance variation in the random assignment? In Chapter 11, we will make this decision based on the *P*-value, which is the probability of getting a difference of 1.2 or larger purely by chance (assuming the two treatments cause the same change in pulse rates). For this activity, we estimated the *P*-value by counting the number of dots at 1.2 or greater and dividing by the total number of dots. If it is small (less than 5%), the observed difference of 1.2 is statistically significant.

▶ ACTIVITY OVERVIEW

To prepare for using this activity, watch the overview video by clicking on the link in the TE-Book, opening the TRFD, or downloading from the Teacher's Resources on the book's digital platform.

Time: 15 minutes

Materials: Poster board or chart paper, sticker dots, index cards

Teaching Advice: To save time, prepare the index cards before class. This activity develops the thinking and reasoning students will need later to perform a significance test. Start class by drawing the tree diagram of the experiment on the board so students can see how this simulation matches the experiment. After students have completed the random assignment with index cards, use technology to extend the simulation. Go to the Extra Applets on the Student Site at highschool.bfwpub.com/updatedtps6e and choose "One Quantitative Variable," call the variable "Change in pulse rate," make 2 groups (caffeine and no caffeine), and input the observed values from the table. Click "Begin analysis," then under Perform Inference choose "Simulate difference in two means." The applet will also calculate the number/percent of dots that are 1.2 or greater.

Answers:

1–4. Answers will vary. Below is a possible dotplot.

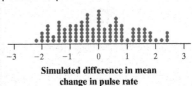

Simulated difference in mean change in pulse rate

5. No; in the above dotplot, there were 20 simulated differences (out of 100) of 1.2 or above. This difference could have occurred purely by chance due to the random assignment.

Making Connections

This dotplot is the result of many possible random assignments to two groups, rather than many possible random samples. Therefore, it is a representation of a *randomization distribution,* rather than a *sampling distribution.* This distinction is not important to point out to students, but will come up again in a later chapter.

Preparing for Inference

Does this mean that caffeine doesn't increase pulse rates? No, we just don't have convincing evidence that caffeine does increase pulse rates. This is the difference between "failing to reject H_0" and "accepting H_0."

ALTERNATE EXAMPLE Skill 4.B

Is yawning contagious?
Inference for experiments

PROBLEM:

According to the popular TV show *Mythbusters,* the answer is "Yes." The *Mythbusters* team conducted an experiment involving 50 subjects. Each subject was placed in a booth for an extended period of time and monitored by hidden camera. Thirty-four subjects were given a "yawn seed" by one of the experimenters; that is, the experimenter yawned in the subject's presence before leaving the room. The remaining 16 subjects were given no yawn seed. What happened in the experiment? The table below shows the results:

		Yawn seed?		
		Yes	No	Total
Subject yawned?	Yes	10	4	14
	No	24	12	36
	Total	34	16	50

(a) Calculate the difference (Yawn seed – No yawn seed) in the proportion of subjects who yawned in the two groups. A total of 100 trials of a simulation were performed to see what differences in proportions would occur due only to chance variation in the random assignment, assuming that the yawn seed doesn't affect whether a subject yawns. That is, 14 "yawners" and 36 "non-yawners" were randomly assigned to groups of 34 and 16.

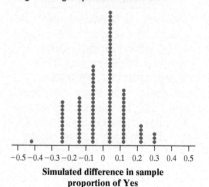

Simulated difference in sample proportion of Yes

Because a difference of 1.2 or greater is somewhat likely to occur by chance alone, the results of Mr. Wilcox's class experiment aren't statistically significant. Based on this experiment, there isn't convincing evidence that caffeine increases pulse rates.

EXAMPLE

Distracted driving
Inference for experiments

PROBLEM: Is talking on a cell phone while driving more distracting than talking to a passenger? David Strayer and his colleagues at the University of Utah designed an experiment to help answer this question. They used 48 undergraduate students as subjects. The researchers randomly assigned half of the subjects to drive in a simulator while talking on a cell phone, and the other half to drive in the simulator while talking to a passenger. One response variable was whether or not the driver stopped at a rest area that was specified by researchers before the simulation started. The table shows the results.[54]

(a) Calculate the difference (Passenger – Cell phone) in the proportion of students who stopped at the rest area in the two groups.

One hundred trials of a simulation were performed to see what differences in proportions would occur due only to chance variation in the random assignment, assuming that the type of distraction did not affect whether a subject stopped at the rest area. That is, 33 "stoppers" and 15 "non-stoppers" were randomly assigned to two groups of 24.

(b) There are three dots at 0.29. Explain what these dots mean in this context.

(c) Use the results of the simulation to determine if the difference in proportions from part (a) is statistically significant. Explain your reasoning.

		Treatment		
		Cell phone	Passenger	Total
Response	Stopped at rest area	12	21	33
	Didn't stop	12	3	15
	Total	24	24	48

Simulated difference (Passenger – Cell phone) in proportion of students who stopped

SOLUTION:

(a) Difference in proportions = 21/24 − 12/24 = 0.875 − 0.500 = 0.375

(b) When we assumed that the type of distraction doesn't matter, there were three simulated random assignments where the difference in the proportion of students who stopped at the rest area was 0.29.

(c) Because a difference of 0.375 or greater never occurred in the simulation, the difference is statistically significant. It is extremely unlikely to get a difference this big simply due to chance variation in the random assignment.

> Because the difference is statistically significant, we can make a cause-and-effect conclusion: talking on a cell phone is more distracting than talking with a passenger—at least for subjects like those in the experiment.

FOR PRACTICE, TRY EXERCISE 99

(b) There are five dots at 0.23. Explain what these dots mean in this context.

(c) Use the results of the simulation to determine if the difference in proportions from part (a) is statistically significant. Explain your reasoning.

SOLUTION:
(a) Difference in proportions = 10/34 − 4/16 = 0.29 − 0.25 = 0.04
(b) When we assumed that yawning is not contagious, there were 5 simulated random assignments where the difference in the proportion of people who yawned was 0.23.
(c) Because a difference of 0.04 or greater occurred 56/100 times in the simulation, the difference is not statistically significant. It is quite plausible to get a difference this big simply due to chance variation in the random assignment.

In the caffeine example, we said that a difference in means of 1.2 was not unusual because a difference that big or bigger occurred 19% of the time by chance alone. In the distracted drivers example, we said that a difference in proportions of 0.375 was unusual because a difference this big or bigger occurred 0% of the time by chance alone. So the boundary between "not unusual" and "unusual" must be somewhere between 0% and 19%. For now, we recommend using a boundary of 5% so that differences that would occur less than 5% of the time by chance alone are considered statistically significant.

The Scope of Inference: Putting it All Together

The type of conclusion (inference) that can be drawn from a study depends on how the data in the study were collected.

In the example about average weight in the NFL, the players were *randomly selected* from all NFL players. As you learned in Section 4.1, random sampling helps to avoid bias and produces reliable estimates of the truth about the population. Because the mean weight in the sample of players was 244.4 pounds, our best guess for the mean weight in the population of all NFL players is 244.4 pounds. Even though our estimates are rarely exactly correct, when samples are selected at random, we can make an *inference about the population*.

In the distracted driver experiment, subjects were *randomly assigned* to talk on a cell phone or talk to a passenger. As you learned in Section 4.2, random assignment helps ensure that the two groups of subjects are as alike as possible before the treatments are imposed. If the group assigned to talk with a passenger remembers to stop at the rest area more often than the group assigned to talk on a cell phone, and the difference is too large to be explained by chance variation in the random assignment, it must be due to the treatments. In that case, the researchers could safely conclude that talking on a cell phone is more distracting than talking to a passenger. That is, they can make an *inference about cause and effect*. However, because the experiment used volunteer subjects, the scientists can only apply this conclusion to subjects like the ones in their experiment.

Let's recap what we've learned about the scope of inference in a statistical study.

THE SCOPE OF INFERENCE
• Random selection of individuals allows inference about the population from which the individuals were chosen.
• Random assignment of individuals to groups allows inference about cause and effect.

Both random sampling and random assignment introduce chance variation into a statistical study. When performing inference, statisticians use the laws of probability to describe this chance variation. You'll learn how this works later in the book.

Preparing for Inference

In Chapter 9, this 5% boundary value will be our significance level α. While 5% is the most generally accepted significance level, there are circumstances where different significance values are appropriate.

AP® EXAM TIP

Students often confuse the two types of inferences we can make: inferences about a population and inferences about cause and effect. Over the years, there have been many questions on the AP® Statistics exam that ask students to discuss what types of inferences (conclusions) are appropriate based on the design of the study. Make sure students understand how the role of randomization in a study helps them decide which type of inference, if any, is appropriate.

Teaching Tip

Refer back to the visual from the beginning of the chapter. A simple random sample (mini-cloud) should be representative of the population (big cloud) so conclusions made from the sample can be generalized to the population.

Preparing for Inference

In Chapters 8–12, we will construct confidence intervals and perform significance tests, and students will be asked to check the Random condition. They will have to consider both random sampling and random assignment to decide if this condition is met.

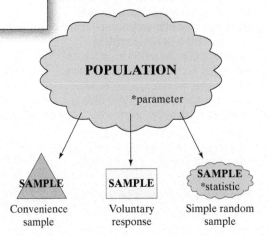

The following chart summarizes the possibilities.[55]

| | | Were individuals randomly assigned to groups? | |
		Yes	No
Were individuals randomly selected?	Yes	Inference about the population: YES Inference about cause and effect: YES	Inference about the population: YES Inference about cause and effect: NO
	No	Inference about the population: NO Inference about cause and effect: YES	Inference about the population: NO Inference about cause and effect: NO

Well-designed experiments randomly assign individuals to treatment groups. However, most experiments don't select experimental units at random from the larger population. That limits such experiments to inference about cause and effect for individuals like those who received the treatments. Observational studies don't randomly assign individuals to groups, which makes it challenging to make an inference about cause and effect. But an observational study that uses random sampling can make an inference about the population.

EXAMPLE

When will I ever use this stuff?
The scope of inference

PROBLEM: Researchers at the University of North Carolina were concerned about the increasing dropout rate in the state's high schools, especially for low-income students. Surveys of recent dropouts revealed that many of these students had started to lose interest during middle school. They said they saw little connection between what they were studying in school and their future plans. To change this perception, researchers developed a program called CareerStart. The central idea of the program is that teachers show students how the topics they're learning about can be applied to specific careers.

To test the effectiveness of CareerStart, the researchers recruited 14 middle schools in Forsyth County to participate in an experiment. Seven of the schools, determined at random, used CareerStart along with the district's standard curriculum. The other 7 schools just followed the standard curriculum. Researchers followed both groups of students for several years, collecting data on students' attendance, behavior, standardized test scores, level of engagement in school, and whether or not the students graduated from high school.

Results: Students at schools that used CareerStart generally had significantly better attendance and fewer discipline problems, earned higher test scores, reported greater engagement in their classes, and were more likely to graduate.[56]

What conclusion can we draw from this study? Explain your reasoning.

SOLUTION:
Because treatments were randomly assigned and the results were significant, we can conclude that using the CareerStart curriculum caused better attendance, fewer discipline problems, higher test scores, greater engagement, and increased graduation rates. However, these results only apply to schools like those in the study because the schools were not randomly selected from any population.

> With no random selection, the results of the study should be applied only to schools like those in the study. With random assignment, it is possible to make an inference about cause and effect.

FOR PRACTICE, TRY EXERCISE 103

SOLUTION:
With random selection, the results of the study can be applied to the entire population—in this case, all the students at this school. With no random assignment, however, we should not conclude anything about cause and effect. All we can conclude is that students at this school who listen to music while studying have lower GPAs than those who do not listen to music while studying. We don't know why their GPAs are lower, however.

CHECK YOUR UNDERSTANDING

When an athlete suffers a sports-related concussion, does it help to remove the athlete from play immediately? Researchers recruited 95 athletes seeking care for a sports-related concussion at a medical clinic and followed their progress during recovery. Researchers found statistically significant evidence that athletes who were removed from play immediately recovered more quickly, on average, than athletes who continued to play.[57] What conclusion can we draw from this study? Explain your answer.

The Challenges of Establishing Causation

A well-designed experiment can tell us that changes in the explanatory variable cause changes in the response variable. More precisely, it tells us that this happened for specific individuals in the specific environment of this specific experiment. In some cases, it isn't practical or even ethical to do an experiment. Consider these important questions:

- Does going to church regularly help people live longer?
- Does smoking cause lung cancer?

To answer these cause-and-effect questions, we just need to perform a randomized comparative experiment. Unfortunately, we can't randomly assign people to attend church or to smoke cigarettes. The best data we have about these and many other cause-and-effect questions come from observational studies.

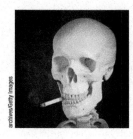

Doctors had long observed that most lung cancer patients were smokers. Comparison of smokers and similar nonsmokers showed a very strong association between smoking and death from lung cancer. Could the association be due to some other variable? Is there some genetic factor that makes people both more likely to become addicted to nicotine and to develop lung cancer? If so, then smoking and lung cancer would be strongly associated even if smoking had no direct effect on the lungs. Or maybe confounding is to blame. It might be that smokers live unhealthy lives in other ways (diet, alcohol, lack of exercise) and that some other habit confounded with smoking is a cause of lung cancer. Still, it is sometimes possible to build a strong case for causation in the absence of experiments. The evidence that smoking causes lung cancer is about as strong as nonexperimental evidence can be.

There are several criteria for establishing causation when we can't do an experiment:

- *The association is strong.* The association between smoking and lung cancer is very strong.
- *The association is consistent.* Many studies of different kinds of people in many countries link smoking to lung cancer. That reduces the chance that some other variable specific to one group or one study explains the association.
- *Larger values of the explanatory variable are associated with stronger responses.* People who smoke more cigarettes per day or who smoke over a longer period get lung cancer more often. People who stop smoking reduce their risk.

archives/Getty Images

- *The alleged cause precedes the effect in time.* Lung cancer develops after years of smoking. The number of men dying of lung cancer rose as smoking became more common, with a lag of about 30 years. Lung cancer kills more men than any other form of cancer. Lung cancer was rare among women until women began to smoke. Lung cancer in women rose along with smoking, again with a lag of about 30 years, and has passed breast cancer as the leading cause of cancer death among women.
- *The alleged cause is plausible.* Experiments with animals show that tars from cigarette smoke do cause cancer.

Medical authorities do not hesitate to say that smoking causes lung cancer. The U.S. Surgeon General states that cigarette smoking is "the largest avoidable cause of death and disability in the United States."[58] The evidence for causation is overwhelming—but it is not as strong as the evidence provided by well-designed experiments. Conducting an experiment in which some subjects were forced to smoke and others were not allowed to would be unethical. In cases like this, observational studies are our best source of reliable information.

Data Ethics*

There are some potential experiments that are clearly unethical. In other cases, the boundary between "ethical" and "unethical" isn't as clear. Decide if you think each of the following studies is ethical or unethical:

- A promising new drug has been developed for treating cancer in humans. Before giving the drug to human subjects, researchers want to administer the drug to animals to see if there are any potentially serious side effects.
- Are companies discriminating against some individuals in the hiring process? To find out, researchers prepare several equivalent résumés for fictitious job applicants, with the only difference being the gender of the applicant. They send the fake résumés to companies advertising positions and keep track of the number of males and females who are contacted for interviews.
- Will people try to stop someone from driving drunk? A television news program hires an actor to play a drunk driver and uses a hidden camera to record the behavior of individuals who encounter the driver.

The most complex issues of data ethics arise when we collect data from people. The ethical difficulties are more severe for experiments that impose some treatment on people than for sample surveys that simply gather information. Trials of new medical treatments, for example, can do harm as well as good to their subjects.

Here are some basic standards of data ethics that must be obeyed by all studies that gather data from human subjects, both observational studies and experiments. The law requires that studies carried out or funded by the federal government obey these principles.[59] But neither the law nor the consensus of experts is completely clear about the details of their application.

---

*This is an important topic, but it is not required for the AP® Statistics exam.

Teaching Tip

The section on data ethics is optional, because it is not covered on the AP® Statistics exam. However, it is a very important part of statistics. We recommend that you spend some time discussing the topics in this section.

Teaching Tip: AP® Connections

The AP® Psychology course outline has two sections related to data ethics:
1. Ethics in Research
2. Ethics and Standards in Testing

BASIC DATA ETHICS

- All planned studies must be reviewed in advance by an *institutional review board* charged with protecting the safety and well-being of the subjects.
- All individuals who are subjects in a study must give their *informed consent* before data are collected.
- All individual data must be kept *confidential*. Only statistical summaries for groups of subjects may be made public.

Institutional Review Boards The purpose of an *institutional review board* is not to decide whether a proposed study will produce valuable information or if it is statistically sound. The board's purpose is, in the words of one university's board, "to protect the rights and welfare of human subjects (including patients) recruited to participate in research activities." The board reviews the plan of the study and can require changes. It reviews the consent form to be sure that subjects are informed about the nature of the study and about any potential risks. Once research begins, the board monitors its progress at least once a year.

Informed Consent Both words in the phrase *informed consent* are important, and both can be controversial. Subjects must be informed in advance about the nature of a study and any risk of harm it may bring. In the case of a questionnaire, physical harm is not possible. But a survey on sensitive issues could result in emotional harm. The participants should be told what kinds of questions the survey will ask and roughly how much of their time it will take. Experimenters must tell subjects the nature and purpose of the study and outline possible risks. Subjects must then consent in writing.

Confidentiality It is important to protect individuals' privacy by keeping all data about them *confidential*. The report of an opinion poll may say what percent of the 1200 respondents believed that legal immigration should be reduced. It may not report what *you* said about this or any other issue. Confidentiality is not the same as *anonymity*. Anonymity means that individuals are anonymous—their names are not known even to the director of the study. Anonymity is rare in statistical studies. Even where anonymity is possible (mainly in surveys conducted by mail), it prevents any follow-up to improve nonresponse or inform individuals of results.

Section 4.3 | Summary

- **Sampling variability** refers to the idea that different random samples of the same size from the same population produce different estimates. Reduce sampling variability by increasing the sample size.
- When the observed results of a study are too unusual to be explained by chance alone, we say that the results are **statistically significant.**
- Most studies aim to make inferences that go beyond the data produced.

TRM Section 4.3 Quiz

There is one quiz available for this section in the Teacher's Resource Materials. Click on the link in the TE-Book, look on the TRFD, or download from the Teacher's Resources on the book's digital platform. You can also create your own quiz using the ExamView® Assessment Suite that is part of the TPS6 program. Questions are coded by Learning Target to make it easy to build parallel quizzes.

Answers to Section 4.3 Exercises

4.93 (a) Because different random samples will include different students and produce different estimates, it is unlikely that the sample result will be the same as the proportion of all students at the school who use Twitter. **(b)** An SRS of 100; estimates tend to be closer to the truth when the sample size is larger.

4.94 (a) Because different random samples will include different students and produce different estimates, it is unlikely that the sample result will be the same as the mean distance that all students live from campus. **(b)** An SRS of 100 students is more likely to get a sample result close to the true population value. Estimates tend to be closer to the truth when the sample size is larger.

4.95 (a) Yes, the true proportion could be as small as $0.37 - 0.031 = 0.339$ or as large as $0.37 + 0.031 = 0.401$, and 0.50 is not in this interval. **(b)** Increase the number of adults in the sample.

4.96 (a) No, this would not be surprising. According to the sample, the true proportion could be as small as $0.24 - 0.03 = 0.21$ or as large as $0.24 + 0.03 = 0.27$. Therefore, any proportion between 0.21 and 0.27 is plausible (which includes $p = 0.26$). **(b)** The traffic analyst could decrease the margin of error by increasing the number of cars in the sample.

4.97 (a) If we repeatedly take random samples of size 124 from a population of couples that have no preference for which way they kiss, the number of couples who kiss the "right way" in a sample varies from about 45 to 76. **(b)** Yes; in the study, 83 couples kissed the "right way." This is much higher than what we would expect to happen by chance alone. In the simulations, the largest number of couples kissing the "right way" was 76.

Section 4.3   Exercises

- **Inference about a population** requires that the individuals taking part in a study be randomly selected from the population.
- A well-designed experiment that randomly assigns experimental units to treatments allows **inference about cause and effect**.
- In the absence of an experiment, good evidence of causation requires a strong association that appears consistently in many studies, a clear explanation for the alleged causal link, and careful examination of other variables.
- Studies involving humans must be screened in advance by an **institutional review board**. All participants must give their **informed consent** before taking part. Any information about the individuals in the study must be kept **confidential.**

93. **Tweet, tweet!** What proportion of students at your school use Twitter? To find out, you survey a simple random sample of students from the school roster.

pg 299

(a) Will your sample result be exactly the same as the true population proportion? Explain your answer.

(b) Which would be more likely to produce a sample result closer to the true population value: an SRS of 50 students or an SRS of 100 students? Explain your answer.

94. **Far from home?** A researcher wants to estimate the average distance that students at a large community college live from campus. To find out, she surveys a simple random sample of students from the registrar's database.

(a) Will the researcher's sample result be exactly the same as the true population mean? Explain your answer.

(b) Which would be more likely to produce a sample result closer to the true population value: an SRS of 100 students or an SRS of 50 students? Explain your answer.

95. **Football on TV** A Gallup poll conducted telephone interviews with a random sample of 1000 adults aged 18 and older. Of these, 37% said that football is their favorite sport to watch on television. The margin of error for this estimate is 3.1 percentage points.

(a) Would you be surprised if a census revealed that 50% of adults in the population would say their favorite sport to watch on TV is football? Explain your answer.

(b) Explain how Gallup could decrease the margin of error.

96. **Car colors in Miami** Using a webcam, a traffic analyst selected a random sample of 800 cars traveling on I-195 in Miami on a weekday morning. Among the 800 cars in the sample, 24% were white. The margin of error for this estimate is 3.0 percentage points.

(a) Would you be surprised if a census revealed that 26% of cars on I-195 in Miami on a weekday morning were white? Explain your answer.

(b) Explain how the traffic analyst could decrease the margin of error.

97. **Kissing the right way** According to a newspaper article, "Most people are kissing the 'right way.'" That is, according to a study, the majority of couples prefer to tilt their heads to the right when kissing. In the study, a researcher observed a random sample of 124 kissing couples and found that 83/124 (66.9%) of the couples tilted to the right.[60] To determine if these data provide convincing evidence that couples are more likely to tilt their heads to the right, 100 simulated SRSs were selected.

Each dot in the graph shows the number of couples that tilt to the right in a simulated SRS of 124 couples, assuming that each couple has a 50% chance of tilting to the right.

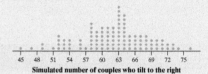

Simulated number of couples who tilt to the right

(a) Explain how the graph illustrates the concept of sampling variability.

(b) Based on the data from the study and the results of the simulation, is there convincing evidence that couples prefer to kiss the "right way"? Explain your answer.

98. **Weekend birthdays** Over the years, the percentage of births that are planned caesarean sections has been rising. Because doctors can schedule these deliveries, there might be more children born during the week and fewer born on the weekend than if births were uniformly distributed throughout the week. To investigate, Mrs. McDonald and her class selected an SRS of 73 people born since 1993. Of these people, 24 were born on Friday, Saturday, or Sunday.

To determine if these data provide convincing evidence that fewer than 43% (3/7) of people born since 1993 were born on Friday, Saturday, or Sunday, 100 simulated SRSs were selected. Each dot in the graph shows the number of people that were born on Friday, Saturday, or Sunday in a simulated SRS of 73 people, assuming that each person had a 43% chance of being born on one of these three days.

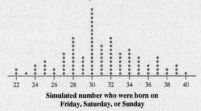

Simulated number who were born on
Friday, Saturday, or Sunday

(a) Explain how the dotplot illustrates the concept of sampling variability.

(b) Based on the data from the study and the results of the simulation, is there convincing evidence that fewer than 43% of people born since 1993 were born on Friday, Saturday, or Sunday? Explain your answer.

99. **I work out a lot** Are people influenced by what others say? Michael conducted an experiment in front of a popular gym. As people entered, he asked them how many days they typically work out per week. As he asked the question, he showed the subjects one of two clipboards, determined at random. Clipboard A had the question and many responses written down, where the majority of responses were 6 or 7 days per week. Clipboard B was the same, except most of the responses were 1 or 2 days per week. The mean response for the Clipboard A group was 4.68 and the mean response for the Clipboard B group was 4.21.[61]

(a) Calculate the difference (Clipboard A – Clipboard B) in the mean number of days for the two groups.

One hundred trials of a simulation were performed to see what differences in means would occur due only to chance variation in the random assignment, assuming that the responses on the clipboard don't matter. The results are shown in the dotplot.

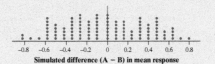

Simulated difference (A – B) in mean response

(b) There is one dot at 0.72. Explain what this dot means in this context.

(c) Use the results of the simulation to determine if the difference in means from part (a) is statistically significant. Explain your reasoning.

100. **A louse-y situation** A study published in the *New England Journal of Medicine* compared two medicines to treat head lice: an oral medication called ivermectin and a topical lotion containing malathion. Researchers studied 812 people in 376 households in seven areas around the world. Of the 185 households randomly assigned to ivermectin, 171 were free from head lice after 2 weeks, compared with only 151 of the 191 households randomly assigned to malathion.[62]

(a) Calculate the difference (Ivermectin – Malathion) in the proportion of households that were free from head lice in the two groups.

One hundred trials of a simulation were performed to see what differences in proportions would occur due only to chance variation in the random assignment, assuming that the type of medication doesn't matter. The results are shown in the dotplot.

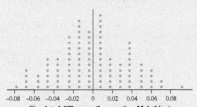

Simulated difference (Ivermectin – Malathion)
in proportion who were free from head lice

(b) There is one dot at 0.09. Explain what this dot means in this context.

(c) Use the results of the simulation to determine if the difference in proportions from part (a) is statistically significant. Explain your reasoning.

101. **Acupuncture and pregnancy** A study sought to determine if the ancient Chinese art of acupuncture could help infertile women become pregnant.[63] A total of 160 healthy women undergoing assisted reproductive therapy were recruited for the study. Half of the subjects were randomly assigned to receive acupuncture treatment 25 minutes before embryo transfer and

4.98 (a) If we repeatedly take random samples of size 73 from a population of people who are equally likely to be born on any day of the week, the number of people born on a weekend in a sample varies from about 22 to 40. **(b)** No, there is not convincing evidence that less than 43% of people born since 1993 were born on Friday, Saturday, or Sunday. In Mrs. McDonald's sample, 24 people were born on the weekend out of the sample of 73. One sample answer is: Of the 100 trials of a simulation, 24 or fewer people born on the weekend happened 6 times. This suggests that these results may have happened purely by chance.

4.99 (a) 4.68 − 4.21 = 0.47 days **(b)** When we assumed that the type of clipboard does not matter, there was one simulated random assignment where the difference (A − B) in the mean number of days for the two groups was 0.72. **(c)** Because a difference of means of 0.47 or higher occurred 16 out of 100 times in the simulation, the difference is not statistically significant. It is quite plausible to get a difference this big simply due to chance variation in the random assignment.

4.100 (a) The difference (Ivermectin − Malathion) in the proportion of households free from head lice in the two groups is $\frac{171}{185} - \frac{151}{191} = 0.92 - 0.79 = 0.13$. **(b)** When we assumed that the type of medicine does not matter, there was one simulated random assignment where the difference (Ivermectin − Malathion) in the proportion of households free from head lice in the two groups was 0.09. **(c)** Because a difference of proportions of 0.13 or higher never occurred in the simulation of 100 trials, the difference is statistically significant. It is extremely unlikely to get a difference this big simply due to chance variation in the random assignment.

4.101 (a) To make sure that the two groups were as similar as possible before the treatments were administered. **(b)** The difference in the percent of women who received acupuncture and became pregnant and those who lay still and became pregnant was large enough to conclude that the difference was not likely due to the chance variation created by the random assignment to treatments. **(c)** Because the women were aware of which treatment they received, we do not know if their expectations or the treatment was the cause of the increase in pregnancy rates.

4.102 (a) Researchers randomly assigned participants to diets to make sure that the two groups were as similar as possible before the treatments were administered. **(b)** The difference in weight loss observed was large enough to conclude that the difference was not likely due to the chance variation created by the random assignment to treatments. **(c)** Even though the low-carb dieters lost an average of 2 kg more than the low-fat group over the year, this difference was small enough that it could be due just to chance variation in the random assignment. In other words, it is plausible that more disciplined people were assigned to the low-carb group just by chance, and the weight-loss difference of 2 kg is due to the imbalance created by the random assignment.

4.103 Because this study involved random assignment to the treatments (foster or institutional), we can infer that the difference between foster care and institutional care caused the difference in response. However, the results should be applied only to children like the ones in the study, because these 136 children were not randomly selected from a larger population.

4.104 Because this study involved random assignment to the treatments (freezer or room temperature), we can infer that being stored in the freezer caused the increase in average charge. Also, because the batteries were randomly chosen from the warehouse, we can generalize this result to the entire population of batteries in the warehouse.

4.105 Because the subjects were not randomly assigned to attend religious services (or not), we cannot infer cause and effect. However, this study involved a random sample of adults, so we *can* make an inference about the population of adults. It appears that adults who attend religious services regularly have a lower risk of dying, but we do not know that attending religious services is the cause of the lower risk.

again 25 minutes after the transfer. The remaining 80 subjects were instructed to lie still for 25 minutes after the embryo transfer. *Results:* In the acupuncture group, 34 women became pregnant. In the control group, 21 women became pregnant.

(a) Why did researchers randomly assign the subjects to the two treatments?

(b) The difference in the percent of women who became pregnant in the two groups is statistically significant. Explain what this means to someone who knows little statistics.

(c) Explain why the design of the study prevents us from concluding that acupuncture caused the difference in pregnancy rates.

102. **Do diets work?** Dr. Linda Stern and her colleagues recruited 132 obese adults at the Philadelphia Veterans Affairs Medical Center in Pennsylvania. Half the participants were randomly assigned to a low-carbohydrate diet and the other half to a low-fat diet. Researchers measured each participant's change in weight and cholesterol level after six months and again after one year. Subjects in the low-carb diet group lost significantly more weight than subjects in the low-fat diet group during the first six months. At the end of a year, however, the average weight loss for subjects in the two groups was not significantly different.[64]

(a) Why did researchers randomly assign the subjects to the diet treatments?

(b) Explain to someone who knows little statistics what "lost significantly more weight" means.

(c) The subjects in the low-carb diet group lost an average of 5.1 kg in a year. The subjects in the low-fat diet group lost an average of 3.1 kg in a year. Explain how this information could be consistent with the fact that weight loss in the two groups was not significantly different.

103. **Foster care versus orphanages** Do abandoned children placed in foster homes do better than similar children placed in an institution? The Bucharest Early Intervention Project found statistically significant evidence that they do. The subjects were 136 young children abandoned at birth and living in orphanages in Bucharest, Romania. Half of the children, chosen at random, were placed in foster homes. The other half remained in the orphanages.[65] (Foster care was not easily available in Romania at the time and so was paid for by the study.) What conclusion can we draw from this study? Explain your reasoning.

104. **Frozen batteries** Will storing batteries in a freezer make them last longer? To find out, a company that produces batteries takes a random sample of 100 AA batteries from its warehouse. The company statistician randomly assigns 50 batteries to be stored in the freezer and the other 50 to be stored at room temperature for 3 years. At the end of that time period, each battery's charge is tested. *Result:* Batteries stored in the freezer had a significantly higher average charge. What conclusion can we draw from this study? Explain your reasoning.

105. **Attend church, live longer?** One of the better studies of the effect of regular attendance at religious services gathered data from a random sample of 3617 adults. The researchers then measured lots of variables, not just the explanatory variable (religious activities) and the response variable (length of life). A news article said: "Churchgoers were more likely to be nonsmokers, physically active, and at their right weight. But even after health behaviors were taken into account, those not attending religious services regularly still were significantly more likely to have died."[66] What conclusion can we draw from this study? Explain your reasoning.

106. **Rude surgeons** Is a friendly surgeon a better surgeon? In a study of more than 32,000 surgical patients from 7 different medical centers, researchers classified surgeons by the number of unsolicited complaints that had been recorded about their behavior. The researchers found that surgical complications were significantly more common in patients whose surgeons had received the most complaints about their behavior, compared with patients whose surgeons had received the fewest complaints.[67] What conclusion can we draw from this study? Explain your reasoning.

107. **Berry good** Eating blueberries and strawberries might improve heart health, according to a long-term study of 93,600 women who volunteered to take part. These berries are high in anthocyanins due to their pigment. Women who reported consuming the most anthocyanins had a significantly smaller risk of heart attack compared to the women who reported consuming the least. What conclusion can we draw from this study? Explain your reasoning.[68]

108. **Exercise and memory** A study of strength training and memory randomly assigned 46 young adults to two groups. After both groups were shown 90 pictures, one group had to bend and extend one leg against heavy resistance 60 times. The other group stayed relaxed, while the researchers used the same exercise machine to bend and extend their legs with no resistance. Two days later, each subject was shown 180 pictures—the original 90 pictures plus 90 new pictures and asked to identify which pictures were shown two days earlier. The resistance group was significantly more successful in identifying these pictures than was the relax group. What conclusions can we draw from this study? Explain your reasoning.[69]

4.106 Because surgeons were not randomly assigned to be friendly or unfriendly, we cannot infer that surgeon friendliness caused a decrease in complications. Also, this study did not involve a random sample of patients, so we should apply the results only to patients like the ones in this study.

4.107 Because this study does not involve random assignment to the treatments (amount of anthocyanins consumed), we cannot infer that the difference in blueberry and strawberry intake caused the difference in heart attack risk. Also, we should apply the results only to women like the ones in this study because these 93,600 women were not randomly selected from a larger population.

4.108 Because this study involved random assignment to the treatments (resistance or no resistance), we can infer that the difference between resistance and no resistance caused the difference in success rate in identifying pictures. However, we should apply these results only to people like the ones in the study because these 46 young adults were not randomly selected from a larger population.

109.* **Minimal risk?** You have been invited to serve on a college's institutional review board. You must decide whether several research proposals qualify for lighter review because they involve only minimal risk to subjects. Federal regulations say that "minimal risk" means the risks are no greater than "those ordinarily encountered in daily life or during the performance of routine physical or psychological examinations or tests." That's vague. Which of these do you think qualifies as "minimal risk"?

(a) Draw a drop of blood by pricking a finger to measure blood sugar.

(b) Draw blood from the arm for a full set of blood tests.

(c) Insert a tube that remains in the arm so that blood can be drawn regularly.

110.* **Who reviews?** Government regulations require that institutional review boards consist of at least five people, including at least one scientist, one nonscientist, and one person from outside the institution. Most boards are larger, but many contain just one outsider.

(a) Why should review boards contain people who are not scientists?

(b) Do you think that one outside member is enough? How would you choose that member? (For example, would you prefer a medical doctor? A religious leader? An activist for patients' rights?)

111.* **Facebook emotions** In cooperation with researchers from Cornell University, Facebook randomly selected almost 700,000 users for an experiment in "emotional contagion." Users' news feeds were manipulated (without their knowledge) to selectively show postings from their friends that were either more positive or more negative in tone, and the emotional tone of their own subsequent postings was measured. The researchers found evidence that people who read emotionally negative postings were more likely to post messages with a negative tone, whereas those who read positive messages were more likely to post messages with a positive tone. The research was widely criticized for being unethical. Explain why.[70]

112.* **No consent needed?** In which of the circumstances listed here would you allow collecting personal information without the subjects' consent?

(a) A government agency takes a random sample of income tax returns to obtain information on the average income of people in different occupations. Only the incomes and occupations are recorded from the returns, not the names.

(b) A social psychologist attends public meetings of a religious group to study the behavior patterns of its members.

(c) A social psychologist pretends to be converted to membership in a religious group and attends private meetings to study the behavior patterns of its members.

113.* **Anonymous? Confidential?** One of the most important nongovernment surveys in the United States is the National Opinion Research Center's General Social Survey (GSS). The GSS regularly monitors public opinion on a wide variety of political and social issues. Interviews are conducted in person in the subject's home. Are a subject's responses to GSS questions anonymous, confidential, or both? Explain your answer.

114.* **Anonymous? Confidential?** Texas A&M, like many universities, offers screening for HIV, the virus that causes AIDS. Students may choose either anonymous or confidential screening. An announcement says, "Persons who sign up for screening will be assigned a number so that they do not have to give their name." They can learn the results of the test by telephone, still without giving their name. Does this describe anonymous or confidential screening? Why?

115.* **The Willowbrook hepatitis studies** In the 1960s, children entering the Willowbrook State School, an institution for the intellectually disabled on Staten Island in New York, were deliberately infected with hepatitis. The researchers argued that almost all children in the institution quickly became infected anyway. The studies showed for the first time that two strains of hepatitis existed. This finding contributed to the development of effective vaccines. Despite these valuable results, the Willowbrook studies are now considered an example of unethical research. Explain why, according to current ethical standards, useful results are not enough to allow a study.

116.* **Unequal benefits** Researchers on aging proposed to investigate the effect of supplemental health services on the quality of life of older people. Eligible patients on the rolls of a large medical clinic were to be randomly assigned to treatment and control groups. The treatment group would be offered hearing aids, dentures, transportation, and other services not available without charge to the control group. The review board believed that providing these services to some but not other persons in the same institution raised ethical questions. Do you agree?

*Exercises 109–116: This is an important topic, but it is not required for the AP® Statistics exam.

4.109 Answers will vary.

4.110 Answers will vary. Possible answers include: (a) A non-scientist will be more likely to consider the subjects as people and not be influenced by the scientific results that might be discovered. (b) You might consider at least two outside members. A religious leader might be chosen because we would expect him or her to help lead the committee in ethical and moral discussions. You might also choose a patient advocate to speak for the subjects involved.

4.111 Those Facebook users involved in the study did not know that they were going to be subjected to treatments, and they did not provide informed consent before the study was conducted.

4.112 Answers will vary. Possible answers include: (a) Many would consider this to be an appropriate use of collecting data without participants' knowledge because the data are, in effect, anonymous. (b) Many would consider this to be appropriate because the meetings are public and the psychologist is not misleading the participants. (c) Most would consider this to be inappropriate because the psychologist is misleading the other participants and attending private meetings.

4.113 The responses to the GSS are confidential. The people giving the surveys know who is answering the questions because they are at that person's home, but they will not share the individual results with anyone else.

4.114 This describes the anonymous screening. The patient never gives a name, but rather is just assigned a number. No one at the clinic can put the results together with a name because the name was never given.

4.115 In this case, the subjects were not able to give informed consent. They did not know what was happening to them and they were not old enough to understand the ramifications in any event.

4.116 Answers will vary. One possible answer is: Yes, providing these potentially life-changing services to some but not all seniors in the study is unethical. We cannot withhold important services from some people.

4.117 d

4.118 b

4.119 (a)

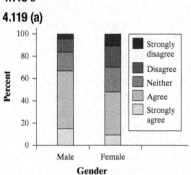

(b) Yes, there is an association between gender and opinion. Men are more likely to view animal testing as justified if it might save human lives: over two-thirds of men agree or strongly agree with this statement, compared to slightly less than half of women. The percentages who disagree or strongly disagree tell a similar story: 16% of men versus 30% of women.

4.120 This could happen because the median is resistant to outliers, whereas the mean is not. Also, the smallest change in stock price is −100%, whereas there is no upper limit for an increase in stock price. It is likely that there were several companies that were very high outliers, which had a big effect on the mean.

Multiple Choice *Select the best answer for Exercises 117 and 118.*

117. Do product labels influence customer perceptions? To find out, researchers recruited more than 500 adults and asked them to estimate the number of calories, amount of added sugar, and amount of fat in a variety of food products. Half of the subjects were randomly assigned to evaluate products with the word "Natural" on the label, while the other half were assigned to evaluate the same products without the "Natural" label. On average, the products with the "Natural" label were judged to have significantly fewer calories. Based on this study, is it reasonable to conclude that including the word "Natural" on the label causes a reduction in estimated calories?

(a) No, because the adults weren't randomly selected from the population of all adults.

(b) No, because there wasn't a control group for comparison.

(c) No, because association doesn't imply causation.

(d) Yes, because the adults were randomly assigned to the treatments.

(e) Yes, because there were a large number of adults involved in the study.

118. Some news organizations maintain a database of customers who have volunteered to share their opinions on a variety of issues. Suppose that one of these databases includes 9000 registered voters in California. To measure the amount of support for a controversial ballot issue, 1000 registered voters in California are randomly selected from the database and asked their opinion. Which of the following is the largest population to which the results of this survey should be generalized?

(a) The 1000 people in the sample

(b) The 9000 registered voters from California in the database

(c) All registered voters in California

(d) All California residents

(e) All registered voters in the United States

Review and Recycle

119. **Animal testing** (1.1) "It is right to use animals for medical testing if it might save human lives." The General Social Survey asked 1152 adults to react to this statement. Here is the two-way table of their responses:

		Gender		
		Male	Female	Total
Opinion about using animals for medical testing	Strongly agree	76	59	135
	Agree	270	247	517
	Neither agree nor disagree	87	139	226
	Disagree	61	123	184
	Strongly disagree	22	68	90
	Total	516	636	1152

(a) Construct segmented bar graphs to display the distribution of opinion for males and for females.

(b) Is there an association between gender and opinion for the members of this sample? Explain your answer.

120. **Initial public offerings** (1.3) The business magazine *Forbes* reports that 4567 companies sold their first stock to the public between 1990 and 2000. The *mean* change in the stock price of these companies since the first stock was issued was +111%. The *median* change was −31%.[71] Explain how this difference could happen.

Chapter 4 Wrap-Up

FRAPPY! FREE RESPONSE AP® PROBLEM, YAY!

The following problem is modeled after actual AP® Statistics exam free-response questions. Your task is to generate a complete, concise response in 15 minutes.

Directions: Show all your work. Indicate clearly the methods you use, because you will be scored on the correctness of your methods as well as on the accuracy and completeness of your results and explanations.

In a recent study, 166 adults from the St. Louis area were recruited and randomly assigned to receive one of two treatments for a sinus infection. Half of the subjects received an antibiotic (amoxicillin) and the other half received a placebo.[72]

(a) Describe how the researchers could have assigned treatments to subjects if they wanted to use a completely randomized design.

(b) All the subjects in the experiment had moderate, severe, or very severe symptoms at the beginning of the study. Describe one statistical benefit and one statistical drawback for using subjects with

moderate, severe, or very severe symptoms instead of just using subjects with very severe symptoms.

(c) At different stages during the next month, all subjects took the sino-nasal outcome test. After 10 days, the difference in average test scores was *not* statistically significant. In this context, explain what it means for the difference to be not statistically significant.

(d) One possible way that researchers could have improved the study is to use a randomized block design. Explain how the researchers could have incorporated blocking in their design.

After you finish, you can view two example solutions on the book's website (highschool.bfwpub.com/updatedtps6e). Determine whether you think each solution is "complete," "substantial," "developing," or "minimal." If the solution is not complete, what improvements would you suggest to the student who wrote it? Finally, your teacher will provide you with a scoring rubric. Score your response and note what, if anything, you would do differently to improve your own score

Chapter 4 Review

Section 4.1: Sampling and Surveys

In this section, you learned that a population is the group of all individuals that we want information about. A sample is the subset of the population that we use to gather this information. The goal of most sample surveys is to use information from the sample to draw conclusions about the population. Choosing people for a sample because they are located nearby or letting people choose whether or not to be in the sample are poor ways to select a sample. Because convenience sampling and voluntary response sampling will produce estimates that are likely to underestimate or likely to overestimate the value you want to know, these methods of choosing a sample are biased.

To avoid bias in the way the sample is formed, the members of the sample should be chosen at random. One way to do this is with a simple random sample (SRS), which is equivalent to selecting well-mixed slips of paper from a hat without replacement. It is often more convenient to select an SRS using technology or a table of random digits.

Three other random sampling methods are stratified sampling, cluster sampling, and systematic sampling. To obtain a stratified random sample, divide the population into non-overlapping groups (strata) of individuals that are likely to have similar responses, take an SRS from each stratum, and combine the chosen individuals to form the sample. Stratified random samples can produce estimates with much

313

TRM FRAPPY! Materials

Please consult the Teacher's Resource Materials for sample student responses, a scoring rubric, and a printable version of the original question with space for students to write their responses. We present a model solution here.

Answers:

(a) Label 83 note cards with "A" and 83 note cards with "B." Shuffle the cards well and hand one card to each subject at random. Subjects with "A" cards will receive the antibiotic and subjects with "B" cards will receive the placebo. **(b)** *Benefit:* We can make conclusions about a wider variety of subjects. If we used only subjects with very severe symptoms, we couldn't make conclusions about subjects with less severe symptoms. *Drawback:* Because these subjects varied in their initial conditions, the responses to the treatments will vary more than if all the subjects had the same initial condition. This additional variability will make it harder to determine if the antibiotic is really better. **(c)** If the difference is not statistically significant, the difference wasn't large enough to rule out random chance as a plausible explanation. That is, the difference could be due to the random assignment and not to the effects of the treatments. **(d)** To incorporate blocking, form blocks based on the initial conditions of the patients. That is, put all the patients with very severe symptoms into one block, and so on. Then within each block, randomly assign the subjects to treatments as in part **(a)**. Blocking by initial severity will help us account for the additional variability in test scores caused by the differences in severity.

TRM Chapter 4 Test

There is one Chapter Test available for this section in the Teacher's Resource Materials. Click on the link in the TE-Book, open in the TRFD, or download from the Teacher's Resources on the book's digital platform. You can also create your own Test using the TPS quiz and test builder (ExamView).

greater precision than simple random samples. To obtain a cluster sample, divide the population into non-overlapping groups (clusters) of individuals that are in similar locations, randomly select clusters, and use every individual in the chosen clusters. Cluster samples are easier to obtain than simple random samples or stratified random samples, but they may not produce very precise estimates. To obtain a systematic random sample, choose a value of k based on the population size and desired sample size, randomly select a number from 1 to k to determine which member of the population to survey first, and then survey every kth member thereafter.

Finally, you learned about other issues in sample surveys that can lead to bias: undercoverage occurs when the sampling method systematically underrepresents one part of the population. Nonresponse describes when answers cannot be obtained from some people that were chosen to be in the sample. Bias can also result when some people in the sample don't give accurate responses due to question wording, interviewer characteristics, or other factors.

Section 4.2: Experiments

In this section, you learned about the difference between observational studies and experiments. Experiments deliberately impose treatments to see if there is a cause-and-effect relationship between two variables. Observational studies look at relationships between two variables, but make it difficult to show cause and effect because other variables may be confounded with the explanatory variable. Variables are confounded when it is impossible to determine which of the variables is causing a change in the response variable.

A common type of comparative experiment uses a completely randomized design. In this type of design, the experimental units are assigned to the treatments completely at random. With random assignment, the treatment groups should be roughly equivalent at the beginning of the experiment. Replication means giving each treatment to as many experimental units as possible. This makes it easier to see the effects of the treatments because the effects of other variables are more likely to be balanced among the treatment groups.

During an experiment, it is important that other variables be controlled (kept the same) for each experimental unit. Doing so helps avoid confounding and removes a possible source of variation in the response variable. Also, beware of the placebo effect—the tendency for people to improve because they expect to, not because of the treatment they are receiving. One way to make sure that all experimental units have the same expectations is to make them blind—unaware of which treatment they are receiving. When the people interacting with the subjects and measuring the response variable are also blind, the experiment is called double-blind.

Blocking in experiments is similar to stratifying in sampling. To form blocks, group together experimental units that are similar with respect to a variable that is associated with the response. Then randomly assign the treatments within each block. A randomized block design that uses blocks with two experimental units is called a matched pairs design. Blocking helps us estimate the effects of the treatments more precisely because we can account for the variability introduced by the variables used to form the blocks.

Section 4.3: Using Studies Wisely

In this section, you learned that the types of conclusions we can draw depend on how the data are produced. When samples are selected at random, we can make inferences about the population from which the sample was drawn. However, the estimates we calculate from sample data rarely equal the true population value because of sampling variability. We can reduce sampling variability by increasing the sample size.

When treatments are applied to groups formed at random in an experiment, we can make an inference about cause and effect. Making a cause-and-effect conclusion is often difficult because it is impossible or unethical to perform certain types of experiments. Good data ethics requires that studies should be approved by an institutional review board, subjects should give informed consent, and individual data must be kept confidential.

Finally, the results of a study are statistically significant if they are too unusual to occur by chance alone.

What Did You Learn?

Learning Target	Section	Related Example on Page(s)	Relevant Chapter Review Exercise(s)
Identify the population and sample in a statistical study.	4.1	249	R4.1
Identify voluntary response sampling and convenience sampling and explain how these sampling methods can lead to bias.	4.1	252	R4.2
Describe how to select a simple random sample using slips of paper, technology, or a table of random digits.	4.1	256	R4.2
Describe how to select a sample using stratified random sampling, cluster sampling, and systematic random sampling, and explain whether a particular sampling method is appropriate in a given situation.	4.1	259	R4.2, R4.3
Explain how undercoverage, nonresponse, question wording, and other aspects of a sample survey can lead to bias.	4.1	262	R4.4

Teaching Tip

At the end of each chapter, a "What Did You Learn?" grid lists all the targets for the chapter. Make sure to discuss this grid with your students. Ask them to read each learning target and self-assess whether or not they can do each one. For each target, the grid shows the section in which it was covered, page references for examples that illustrate the target, and relevant chapter review exercises that students can use to assess their understanding of that target. Encourage your students to use this grid as part of their preparations for the chapter test.

Teaching Tip

Now that you have completed Chapter 4 in this book, students can login to their College Board accounts and take the **Personal Progress Check for Unit 3.**

Learning Target	Section	Related Example on Page(s)	Relevant Chapter Review Exercise(s)
Explain the concept of confounding and how it limits the ability to make cause-and-effect conclusions.	4.2	271	R4.5
Distinguish between an observational study and an experiment, and identify the explanatory and response variables in each type of study.	4.2	272	R4.5
Identify the experimental units and treatments in an experiment.	4.2	274	R4.6
Describe the placebo effect and the purpose of blinding in an experiment.	4.2	278	R4.8
Describe how to randomly assign treatments in an experiment using slips of paper, technology, or a table of random digits.	4.2	279	R4.9
Explain the purpose of comparison, random assignment, control, and replication in an experiment.	4.2	276, 282	R4.6, R4.8
Describe a completely randomized design for an experiment.	4.2	284	R4.6, R4.9
Describe a randomized block design and a matched pairs design for an experiment and explain the purpose of blocking in an experiment.	4.2	287, 288	R4.6, R4.9
Explain the concept of sampling variability when making an inference about a population and how sample size affects sampling variability.	4.3	299	R4.1
Explain the meaning of statistically significant in the context of an experiment and use simulation to determine if the results of an experiment are statistically significant.	4.3	302	R4.8
Identify when it is appropriate to make an inference about a population and when it is appropriate to make an inference about cause and effect.	4.3	304	R4.7
Evaluate if a statistical study has been carried out in an ethical manner.*	4.3		R4.10

*This is an important topic, but it is not required for the AP® Statistics exam.

Chapter 4 Review Exercises

R4.1 Nurses are the best A recent random sample of $n = 805$ adult U.S. residents found that the proportion who rated the honesty and ethical standards of nurses as high or very high is 0.85. This is 0.15 higher than the proportion recorded for doctors, the next highest-ranked profession.[73]

(a) Identify the sample and the population in this setting.

(b) Do you think that the proportion of all U.S. residents who would rate the honesty and ethical standards of nurses as high or very high is exactly 0.85? Explain your answer.

(c) What is the benefit of increasing the sample size in this context?

R4.2 Parking problems The administration at a high school with 1800 students wants to gather student opinion about parking for students on campus. It isn't practical to contact all students.

(a) Give an example of a way to obtain a voluntary response sample of students. Explain how this method could lead to bias.

(b) Give an example of a way to obtain a convenience sample of students. Explain how this method could lead to bias.

(c) Describe how to select an SRS of 50 students from the school. Explain how using an SRS helps avoid the biases you described in parts (a) and (b).

(d) Describe how to select a systematic random sample of 50 students from the school. What advantage does this method have over an SRS?

R4.3 Surveying NBA fans The manager of a sports arena wants to learn more about the financial status of the people who are attending an NBA basketball game. He would like to give a survey to a representative sample of about 10% of the fans in attendance. Ticket prices for the game vary a

---

Answers to Chapter 4 Review Exercises

R4.1 (a) Population: All adult U.S. residents. Sample: The 805 adult U.S. residents interviewed. **(b)** Even though the sample size is very large, it is unlikely that the percentage in the entire population would be exactly the same as the percentage in the sample because of sampling variability. **(c)** A larger random sample is more likely to get a sample result close to the true population value.

R4.2 (a) One possible answer: Announce in the daily bulletin that, for students who want to respond, there is a survey concerning student parking available in the main office. Because voluntary response surveys are generally responded to by only those who feel strongly about the issue, the opinions of respondents may differ from the population as a whole, resulting in an inaccurate estimate. **(b)** One possible answer: Personally interview a group of students as they come in from the parking lot. Because these students use the parking lot, their opinions may differ from the population as a whole, resulting in an inaccurate estimate. **(c)** Write the names of all 1800 students on identical pieces of paper. Place the slips of paper into a hat. Shuffle well. Draw out 50 names. Those students will form the SRS of 50 students from the school. **(d)** There are 1800 students and our desired sample size is 50, so select every 1800/50 = 36th student. If students enter the building at one location, select every 36th student who entered by randomly selecting a random number between 1 and 36. Then select every 36th student thereafter. The advantage of a systematic random sample over an SRS is that a systematic random sample selects the students on the spot as they enter the building, which is more efficient than tracking down students all over the school, as in an SRS.

R4.3 (a) You would have to identify 10% of the seats, go to those seats, and find the people sitting there. **(b)** The lettered rows are better as the strata because each lettered row is the same distance from the court and would contain only seats with the same (or nearly the same) ticket price. Within sections, ticket prices vary more. **(c)** Survey all fans in several randomly selected sections (clusters). People in a particular numbered section are in roughly the same location, making it easy to administer the survey. Furthermore, the people in each cluster reflect the variability found in the population, which is ideal.

R4.4 (a) The extra information, "Box-office revenues are at an all-time high," may lead listeners to believe they contributed to this fact and be more likely to overestimate the number of movies they have seen in the past 12 months. Eliminate this sentence. **(b)** A sample that only uses residential phone numbers is likely to underrepresent younger adults who use only cell phones. If younger adults go to movies more often than older adults, the estimated mean will be too small. **(c)** People who do not go to the movies often might be more likely to respond to the poll because they are at home. Because the frequent moviegoers will not be at home to respond, the estimated mean will be too small.

R4.5 (a) The data were collected after the anesthesia was administered. Hospital records were used to "observe" the death rates, rather than imposing different anesthetics on the subjects. **(b)** Explanatory: The type of anesthetic. Response: Whether or not a patient died. **(c)** One variable that might be confounded with choice of anesthetic is type of surgery. If anesthetic C is used more often with a type of surgery that has a higher death rate, we would not know if the death rate was higher due to the anesthesia or the type of surgery.

R4.6 (a) Experimental units: Potatoes. The factors are the storage method (3 levels) and time from slicing until cooking (2 levels). There are six treatments: (1) fresh picked and cooked immediately, (2) fresh picked and cooked after an hour, (3) stored at room temperature and cooked immediately, (4) stored at room temperature and cooked after an hour, (5) stored in refrigerator and cooked immediately, (6) stored in refrigerator and cooked after an hour. **(b)** Using 300 identical slips of paper, write "1" on 50 of them, write "2" on

great deal: seats near the court cost over $200 each, while seats in the top rows of the arena cost $50 each. The arena is divided into 50 numbered sections, from 101 to 150. Each section has rows of seats labeled with letters from A (nearest the court) to ZZ (top row of the arena).

(a) Explain why it might be difficult to give the survey to an SRS of fans.

(b) Explain why it would be better to select a stratified random sample using the lettered rows rather than the numbered sections as strata. What is the benefit of using a stratified sample in this context?

(c) Explain how to select a cluster sample of fans. What is the benefit of using a cluster sample in this context?

R4.4 Been to the movies? An opinion poll calls 2000 randomly chosen residential telephone numbers, then asks to speak with an adult member of the household. The interviewer asks, "Box office revenues are at an all-time high. How many movies have you watched in a movie theater in the past 12 months?" In all, 1131 people responded. The researchers used the responses to estimate the mean number of movies adults had watched in a movie theater over the past 12 months.

(a) Describe a potential source of bias related to the wording of the question. Suggest a change that would help fix this problem.

(b) Describe how using only residential phone numbers might lead to bias and how this will affect the estimate.

(c) Describe how nonresponse might lead to bias and how this will affect the estimate.

R4.5 Are anesthetics safe? The National Halothane Study was a major investigation of the safety of anesthetics used in surgery. Records of over 850,000 operations performed in 34 major hospitals showed the following death rates for four common anesthetics:[74]

Anesthetic	A	B	C	D
Death rate	1.7%	1.7%	3.4%	1.9%

There seems to be a clear association between the anesthetic used and the death rate of patients. Anesthetic C appears to be more dangerous.

(a) Explain why we call the National Halothane Study an observational study rather than an experiment, even though it compared the results of using different anesthetics in actual surgery.

(b) Identify the explanatory and response variables in this study.

(c) When the study looked at other variables that are related to a doctor's choice of anesthetic, it found that Anesthetic C was not causing extra deaths. Explain the concept of confounding in this context and identify a variable that might be confounded with the doctor's choice of anesthetic.

R4.6 Ugly fries Few people want to eat discolored french fries. To prevent spoiling and to preserve flavor, potatoes are kept refrigerated before being cut for french fries. But immediate processing of cold potatoes causes discoloring due to complex chemical reactions. The potatoes must therefore be brought to room temperature before processing. Researchers want to design an experiment in which tasters will rate the color and flavor of french fries prepared from several groups of potatoes. The potatoes will be freshly picked or stored for a month at room temperature or stored for a month refrigerated. They will then be sliced and cooked either immediately or after an hour at room temperature.

(a) Identify the experimental units, the factors, the number of levels for each factor, and the treatments.

(b) Describe a completely randomized design for this experiment using 300 potatoes.

(c) A single supplier has made 300 potatoes available to the researchers. Describe a statistical benefit and a statistical drawback of using potatoes from only one supplier.

(d) The researchers decided to do a follow-up experiment using potatoes from several different suppliers. Describe how they should change the design of the experiment to account for the addition of other suppliers.

R4.7 Don't catch a cold! A recent study of 1000 students at the University of Michigan investigated how to prevent catching the common cold. The students were randomly assigned to three different cold prevention methods for 6 weeks. Some wore masks, some wore masks and used hand sanitizer, and others took no precautions. The two groups who used masks reported 10–50% fewer cold symptoms than those who did not wear a mask.[75]

(a) Does this study allow for inference about a population? Explain your answer.

(b) Does this study allow for inference about cause and effect? Explain your answer.

R4.8 An herb for depression? Does the herb St. John's wort relieve major depression? Here is an excerpt from the report of one study of this issue: "Design: Randomized, Double-Blind, Placebo-Controlled Clinical Trial."[76] The study concluded that the difference in effectiveness of St. John's wort and a placebo was not statistically significant.

(a) Describe the placebo effect in this context. How did the design of this experiment account for the placebo effect?

(b) Explain the purpose of the random assignment.

(c) Why is a double-blind design a good idea in this setting?

(d) Explain what "not statistically significant" means in this context.

50 of them, and so on. Put the papers in a hat and mix well. Then select a potato and randomly select a slip from the hat to determine which treatment that potato will receive. Repeat this process for the remaining 299 potatoes, making sure not to replace the slips of paper into the hat. **(c)** Benefit: The quality of the potatoes should be fairly consistent, reducing a source of variability. Drawback: The results of the experiment could then be applied only to potatoes that come from that one supplier, rather than to potatoes in general. **(d)** Use a randomized block design with the suppliers as the blocks. For each supplier, randomly assign potatoes to the 6 treatments. Doing so would allow the researchers to account for the variability in color and flavor due to differences in the initial quality of the potatoes from different suppliers, making it easier to

estimate how the treatments affect color and flavor of the French fries.

R4.7 (a) No; the 1000 students were not randomly selected from any larger population, so we should apply the results only to students like those in the study. **(b)** Yes; the students were randomly assigned to the three treatments, so we can conclude that the reduction in cold symptoms was caused by the masks.

R4.8 (a) If all the patients received the St. John's wort, the researchers would not know if any improvement was due to the St. John's wort or to the expectations of the subjects (the placebo effect). By giving some patients a treatment that should have no effect at all—but that looks, tastes, and feels like the St. John's wort—the researchers can account

R4.9 How long did I work? A psychologist wants to know if the difficulty of a task influences our estimate of how long we spend working at it. She designs two sets of mazes that subjects can work through on a computer. One set has easy mazes and the other has difficult mazes. Subjects work until told to stop (after 6 minutes, but subjects do not know this). They are then asked to estimate how long they worked. The psychologist has 30 students available to serve as subjects.

(a) Describe an experiment using a completely randomized design to learn the effect of difficulty on estimated time. Make sure to carefully explain your method of assigning treatments.

(b) Describe a matched pairs experimental design using the same 30 subjects.

(c) Which design would be more likely to detect a difference in the effects of the treatments? Explain your answer.

R4.10* Deceiving subjects Students sign up to be subjects in a psychology experiment. When they arrive, they are told that interviews are running late and are taken to a waiting room. The experimenters then stage the theft of a valuable object left in the waiting room. Some subjects are alone with the thief, and others are present in pairs—these are the treatments being compared. Will the subject report the theft?

(a) The students had agreed to take part in an unspecified study, and the true nature of the experiment is explained to them afterward. Does this meet the requirement of informed consent? Explain your answer.

(b) What two other ethical principles should be followed in this study?

---

*This is an important topic, but it is not required for the AP® Statistics exam.

Chapter 4 AP® Statistics Practice Test

Section I: Multiple Choice *Select the best answer for each question.*

T4.1 When we take a census, we attempt to collect data from
(a) a stratified random sample.
(b) every individual chosen in a simple random sample.
(c) every individual in the population.
(d) a voluntary response sample.
(e) a convenience sample.

T4.2 You want to take a simple random sample (SRS) of 50 of the 816 students who live in a dormitory on campus. You label the students 001 to 816 in alphabetical order. In the table of random digits, you read the entries

95592 94007 69769 33547 72450 16632 81194 14873

The first three students in your sample have labels
(a) 955, 929, 400.   (d) 929, 400, 769.
(b) 400, 769, 769.   (e) 400, 769, 335.
(c) 559, 294, 007.

T4.3 A study of treatments for angina (pain due to low blood supply to the heart) compared bypass surgery, angioplasty, and use of drugs. The study looked at the medical records of thousands of angina patients whose doctors had chosen one of these treatments. It found that the average survival time of patients given drugs was the highest. What do you conclude?

(a) This study proves that drugs prolong life and should be the treatment of choice.
(b) We can conclude that drugs prolong life because the study was a comparative experiment.
(c) We can't conclude that drugs prolong life because the patients were volunteers.
(d) We can't conclude that drugs prolong life because the groups might differ in ways besides the treatment.
(e) We can't conclude that drugs prolong life because no placebo was used.

T4.4 Tonya wanted to estimate the average amount of time that students at her school spend on Facebook each day. She gets an alphabetical roster of students in the school from the registrar's office and numbers the students from 1 to 1137. Then Tonya uses a random number generator to pick 30 distinct labels from 1 to 1137. She surveys those 30 students about their Facebook use. Tonya's sample is a simple random sample because

(a) it was selected using a chance process.
(b) it gave every individual the same chance to be selected.
(c) it gave every possible sample of size 30 an equal chance to be selected.
(d) it doesn't involve strata or clusters.
(e) it is guaranteed to be representative of the population.

---

for the placebo effect by comparing the results for the two groups. **(b)** To create two groups of subjects that are roughly equivalent at the beginning of the experiment. **(c)** So that researchers can account for the placebo effect, subjects should not know which treatment they are getting. Also, the researchers should

be unaware of which subjects received which treatment, so they cannot influence the results. **(d)** Here, "not statistically significant" means that the difference in improvement between the St. John's wort and placebo groups was not large enough to rule out the variability caused by the random assignment as the explanation.

R4.9 (a) Use 30 identical slips of paper and write the name of each subject on a slip. Mix the slips in a hat and select 15 of them at random. These subjects will be assigned to Group 1. The remaining 15 will be assigned to Group 2. After the experiment, compare the time estimates of Group 1 with those of Group 2. **(b)** Each student does the activity twice, once with the easy mazes, and once with the hard mazes. For each student, randomly determine which set of mazes is used first. To do this, flip a coin for each subject. If the coin lands on heads, the subject will do the easy mazes followed by the hard mazes. If the coin lands on tails, the subject will do the hard mazes followed by the easy mazes. After the experiment, compare each student's "easy" and "hard" time estimate. **(c)** The matched pairs design would be more likely to detect a difference because it accounts for the variability between subjects.

R4.10 (a) No, because the subjects did not know the nature of the experiment before they agreed to participate. **(b)** All individual data should be kept confidential and the experiment should go before an institutional review board before being implemented.

TRM Full Solutions to Chapter 4 AP® Statistics Practice Test

The full solutions can be found by clicking on the link in the TE-Book, opening the TRFD, or downloading from the Teacher's Resources on the book's digital platform.

Answers to Chapter 4 AP® Statistics Practice Test

T4.1 c

T4.2 e

T4.3 d

T4.4 c

T4.5 d

T4.6 d

T4.7 a

T4.8 d

T4.9 d

T4.10 e

T4.5 Consider an experiment to investigate the effectiveness of different insecticides in controlling pests and their impact on the productivity of tomato plants. What is the best reason for randomly assigning treatment levels (spraying or not spraying) to the experimental units (farms)?

(a) Random assignment eliminates the effects of other variables, like soil fertility.

(b) Random assignment eliminates chance variation in the responses.

(c) Random assignment allows researchers to generalize conclusions about the effectiveness of the insecticides to all farms.

(d) Random assignment will tend to average out all other uncontrolled factors such as soil fertility so that they are not confounded with the treatment effects.

(e) Random assignment helps avoid bias due to the placebo effect.

T4.6 Researchers randomly selected 1700 people from Canada and rated the happiness of each person. Ten years later, the researchers followed up with each person and found that people who were initially rated as happy were less likely to have a heart problem.[77] Which of the following is the most appropriate conclusion based on this study?

(a) Happiness causes better heart health for all people.

(b) Happiness causes better heart health for Canadians.

(c) Happiness causes better heart health for the 1700 people in the study.

(d) Happier people in Canada are less likely to have heart problems.

(e) Happier people in the study were less likely to have heart problems.

T4.7 The sales force for a publishing company is constantly on the road trying to sell books. As a result, each salesperson accumulates many travel-related expenses that he or she charges to a company-issued credit card. To prevent fraud, management hires an outside company to audit a sample of these expenses. For each salesperson, the auditor prints out the credit card statements for the entire year, randomly chooses one of the first 20 expenses to examine, and then examines every 20th expense from that point on. Which type of sampling method is the auditor using for each salesperson?

(a) Convenience sampling

(b) Simple random sampling

(c) Stratified random sampling

(d) Cluster sampling

(e) Systematic random sampling

T4.8 *Bias* in a sampling method is

(a) any difference between the sample result and the truth about the population.

(b) the difference between the sample result and the truth about the population due to using chance to select a sample.

(c) any difference between the sample result and the truth about the population due to practical difficulties such as contacting the subjects selected.

(d) any difference between the sample result and the truth about the population that tends to occur in the same direction whenever you use this sampling method.

(e) racism or sexism on the part of those who take the sample.

T4.9 You wonder if TV ads are more effective when they are longer or repeated more often or both. So you design an experiment. You prepare 30-second and 60-second ads for a camera. Your subjects all watch the same TV program, but you assign them at random to four groups. One group sees the 30-second ad once during the program; another sees it three times; the third group sees the 60-second ad once; and the last group sees the 60-second ad three times. You ask all subjects how likely they are to buy the camera. Which of the following best describes the design of this experiment?

(a) This is a randomized block design, but not a matched pairs design.

(b) This is a matched pairs design.

(c) This is a completely randomized design with one explanatory variable (factor).

(d) This is a completely randomized design with two explanatory variables (factors).

(e) This is a completely randomized design with four explanatory variables (factors).

T4.10 Can texting make you healthier? Researchers randomly assigned 700 Australian adults to either receive usual health care or usual heath care plus automated text messages with positive messages, such as "Walking is cheap. It can be done almost anywhere. All you need is comfortable shoes and clothing." The group that received the text messages showed a statistically significant increase in physical activity.[78] What is the meaning of "statistically significant" in this context?

(a) The results of this study are very important.

(b) The results of this study should be generalized to all people.

(c) The difference in physical activity for the two groups is greater than 0.

(d) The difference in physical activity for the two groups is very large.

(e) The difference in physical activity for the two groups is larger than the difference that could be expected to happen by chance alone.

T4.11 You want to know the opinions of American high school teachers on the issue of establishing a national proficiency test as a prerequisite for graduation from high school. You obtain a list of all high school teachers belonging to the National Education Association (the country's largest teachers' union) and mail a survey to a random sample of 2500 teachers. In all, 1347 of the teachers return the survey. Of those who responded, 32% say that they favor some kind of national proficiency test. Which of the following statements about this situation is true?

(a) Because random sampling was used, we can feel confident that the percent of all American high school teachers who would say they favor a national proficiency test is close to 32%.

(b) We cannot trust these results, because the survey was mailed. Only survey results from face-to-face interviews are considered valid.

(c) Because over half of those who were mailed the survey actually responded, we can feel fairly confident that the actual percent of all American high school teachers who would say they favor a national proficiency test is close to 32%.

(d) The results of this survey may be affected by undercoverage and nonresponse.

(e) The results of this survey cannot be trusted due to voluntary response bias.

Section II: Free Response *Show all your work. Indicate clearly the methods you use, because you will be graded on the correctness of your methods as well as on the accuracy and completeness of your results and explanations.*

T4.12 Elephants sometimes damage trees in Africa. It turns out that elephants dislike bees. They recognize beehives in areas where they are common and avoid them. Can this information be used to keep elephants away from trees? Researchers want to design an experiment to answer these questions using 72 acacia trees and three treatments: active hives, empty hives, and no hives.[79]

(a) Identify the experimental units in this experiment.

(b) Explain why it is beneficial to include some trees that have no hives.

(c) Describe how the researchers could carry out a completely randomized design for this experiment. Include a description of how the treatments should be assigned.

T4.13 Google and Gallup teamed up to survey a random sample of 1673 U.S. students in grades 7–12. One of the questions was "How confident are you that you could learn computer science if you wanted to?" Overall, 54% of students said they were very confident.[80]

(a) Identify the population and the sample.

(b) Explain why it was better to randomly select the students rather than putting the survey question on a website and inviting students to answer the question.

(c) Do you expect that the percent of all U.S. students in grades 7–12 who would say "very confident" is exactly 54%? Explain your answer.

(d) The report also broke the results down by gender. For this question, 62% of males and 48% of females said they were very confident. Which of the three estimates (54%, 62%, 48%) do you expect is closest to the value it is trying to estimate? Explain your answer.

T4.14 Many people start their day with a jolt of caffeine from coffee or a soft drink. Most experts agree that people who consume large amounts of caffeine each day may suffer from physical withdrawal symptoms if they stop ingesting their usual amounts of caffeine. Researchers recruited 11 volunteers who were caffeine dependent and who were willing to take part in a caffeine withdrawal experiment. The experiment was conducted on two 2-day periods that occurred one week apart. During one of the 2-day periods, each subject was given a capsule containing the amount of caffeine normally ingested by that subject in one day. During the other study period, the subjects were given placebos. The order in which each subject received the two types of capsules was randomized. The subjects' diets were restricted during each of the study periods. At the end of each 2-day study period, subjects were evaluated using a tapping task in which they were instructed to press a button 200 times as fast as they could.[81]

(a) Identify the explanatory and response variables in this experiment.

(b) How was blocking used in the design of this experiment? What is the benefit of blocking in this context?

(c) Researchers randomized the order of the treatments to avoid confounding. Explain how confounding might occur if the researchers gave all subjects the placebo first and the caffeine second. In this context, what problem does confounding cause?

(d) Could this experiment have been carried out in a double-blind manner? Explain your answer.

T4.11 d

T4.12 (a) Experimental units: acacia trees. **(b)** This allows the researchers to measure the effect of active hives and empty hives on tree damage compared to no hives at all. **(c)** Assign the trees numbers from 01 to 72 and use a random number table to pick 24 different 2-digit numbers between 01 and 72. Those trees will get the active beehives. The trees associated with the next 24 different 2-digit numbers will get the empty beehives and the remaining 24 trees will remain empty. Compare the damage caused by elephants to the trees with active beehives, those with empty beehives, and those with no beehives.

T4.13 (a) Population: All U.S. students in grades 7–12. Sample: The 1673 U.S. students in grades 7–12 surveyed. **(b)** Random selection reduces the effects of bias due to self-selection and also allows the results to be inferred to a larger population. In this case, students who respond to an online survey might be more interested in computer science, leading to an estimate that is too high. **(c)** No, because of sampling variability. **(d)** The estimate of 54%. This estimate was based on a sample size of 1673, while the other two results were based on smaller sample sizes. Larger sample sizes should yield results that are closer to the truth about the population.

T4.14 (a) Explanatory: The treatment given (caffeine capsule or placebo). Response: The time it takes to complete the tapping task. **(b)** Each of the 11 individuals will be a block in a matched pairs design. Each participant will take the caffeine tablets on one of the two-day sessions and the placebo on the other. The blocking was done to account for individual differences in dexterity. **(c)** After the first trial, subjects might practice the tapping task and do better the second time. If all the subjects got caffeine the second time, the researchers would not know if the increase was due to the practice or the caffeine. **(d)** Yes, if neither the subjects nor the people who come in contact with them during the experiment (including those who record the number of taps) have knowledge of the order in which the caffeine or placebo was administered.

Chapter 4 Project Response Bias

TRM Chapter 4 Project Response Bias

You can find a description of this project as well as rubrics to grade this project by clicking on the link in the TE-Book, opening the TRFD, or downloading from the Teacher's Resources on the book's digital platform.

✚ Ask the StatsMedic

What Could Students Do for a Response Bias Project?

In this post, we show several examples of student projects for collecting data that show a response bias.

TRM Chapter 4 Case Study

The Case Study feature that was found in the previous two editions of TPS has been moved to the Teacher's Resource Materials. Click on the link in the TE-Book, open in the TRFD, or download from the Teacher's Resources on the book's digital platform.

TRM Full Solutions to Cumulative AP® Practice Test 1

The full solutions can be found by clicking on the link in the TE-Book, opening the TRFD, or downloading from the Teacher's Resources on the book's digital platform.

Answers to Cumulative AP® Practice Test 1

AP1.1 d

AP1.2 e

In this project, your team will design and conduct an experiment to investigate the effects of response bias in surveys.[82] You may choose the topic for your surveys, but you must design your experiment so that it can answer at least one of the following questions.

- Does the wording of a question affect the response?
- Do the characteristics of the interviewer affect the response?
- Does anonymity change the responses to sensitive questions?
- Does manipulating the answer choices affect the response?
- Can revealing other peoples' answers to a question change the response?

1. Write a proposal describing the design of your experiment. Be sure to include the following items:

   (a) Your chosen topic and which of the above questions you'll try to answer.

   (b) A detailed description of how you will obtain your subjects (minimum of 50). Your plan must be practical!

   (c) An explanation of the treatments in your experiment and how you will determine which subjects get which treatment.

   (d) A clear explanation of how you will implement your design.

   (e) Precautions you will take to collect data ethically.

   Here are two examples of successful student experiments.

   "Make-Up," by Caryn S. and Trisha T. (all questions asked to males):

   i. "Do you find females who wear makeup attractive?" (Questioner wearing makeup: 75% answered "Yes.")

   ii. "Do you find females who wear makeup attractive?" (Questioner not wearing makeup: 30% answered "Yes.")

   "Cartoons" by Sean W. and Brian H.:

   i. "Do you watch cartoons?" (90% answered "Yes.")

   ii. "Do you *still* watch cartoons?" (60% answered "Yes.")

2. Once your teacher has approved your design, carry out the experiment. Record your data in a table.

3. Analyze your data. What conclusion do you draw? Provide appropriate graphical and numerical evidence to support your answer.

4. Prepare a report that includes the data you collected, your analysis from Step 3, and a discussion of any problems you encountered and how you dealt with them.

Cumulative AP® Practice Test 1

Section I: Multiple Choice *Choose the best answer for Questions AP1.1–AP1.14.*

AP1.1 You look at real estate ads for houses in Sarasota, Florida. Many houses have prices from $200,000 to $400,000. The few houses on the water, however, have prices up to $15 million. Which of the following statements best describes the distribution of home prices in Sarasota?

(a) The distribution is most likely skewed to the left, and the mean is greater than the median.

(b) The distribution is most likely skewed to the left, and the mean is less than the median.

(c) The distribution is roughly symmetric with a few high outliers, and the mean is approximately equal to the median.

(d) The distribution is most likely skewed to the right, and the mean is greater than the median.

(e) The distribution is most likely skewed to the right, and the mean is less than the median.

AP1.2 A child is 40 inches tall, which places her at the 90th percentile of all children of similar age. The heights for children of this age form an approximately Normal distribution with a mean of 38 inches. Based on this information, what is the standard deviation of the heights of all children of this age?

(a) 0.20 inch

(b) 0.31 inch

(c) 0.65 inch

(d) 1.21 inches

(e) 1.56 inches

AP1.3 A large set of test scores has mean 60 and standard deviation 18. If each score is doubled, and then 5 is subtracted from the result, the mean and standard deviation of the new scores are

(a) mean 115 and standard deviation 31.

(b) mean 115 and standard deviation 36.

(c) mean 120 and standard deviation 6.

(d) mean 120 and standard deviation 31.

(e) mean 120 and standard deviation 36.

AP1.4 For a certain experiment, the available experimental units are eight rats, of which four are female (F1, F2, F3, F4) and four are male (M1, M2, M3, M4). There are to be four treatment groups, A, B, C, and D. If a randomized block design is used, with the experimental units blocked by gender, which of the following assignments of treatments is impossible?

(a) A → (F1, M1), B → (F2, M2), C → (F3, M3), D → (F4, M4)

(b) A → (F1, M2), B → (F2, M3), C → (F3, M4), D → (F4, M1)

(c) A → (F1, M2), B → (F3, F2), C → (F4, M1), D → (M3, M4)

(d) A → (F4, M1), B → (F2, M3), C → (F3, M2), D → (F1, M4)

(e) A → (F4, M1), B → (F1, M4), C → (F3, M2), D → (F2, M3)

AP1.5 For a biology project, you measure the weight in grams (g) and the tail length in millimeters (mm) of a group of mice. The equation of the least-squares line for predicting tail length from weight is

$$\text{predicted tail length} = 20 + 3 \times \text{weight}$$

Which of the following is *not* correct?

(a) The slope is 3, which indicates that a mouse's predicted tail length should increase by about 3 mm for each additional gram of weight.

(b) The predicted tail length of a mouse that weighs 38 grams is 134 millimeters.

(c) By looking at the equation of the least-squares line, you can see that the correlation between weight and tail length is positive.

(d) If you had measured the tail length in centimeters instead of millimeters, the slope of the regression line would have been 3/10 = 0.3.

(e) Mice that have a weight of 0 grams will have a tail of length 20 mm.

AP1.6 The figure shows a Normal density curve. Which of the following gives the best estimates for the mean and standard deviation of this Normal distribution?

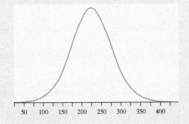

(a) $\mu = 200, \sigma = 50$

(b) $\mu = 200, \sigma = 25$

(c) $\mu = 225, \sigma = 50$

(d) $\mu = 225, \sigma = 25$

(e) $\mu = 225, \sigma = 275$

AP1.7 The owner of a chain of supermarkets notices that there is a positive correlation between the sales of beer and the sales of ice cream over the course of the previous year. During seasons when sales of beer were above average, sales of ice cream also tended to be above average. Likewise, during seasons when sales of beer were below average, sales of ice cream also tended to be below average. Which of the following would be a valid conclusion from these facts?

(a) Sales records must be in error. There should be no association between beer and ice cream sales.

(b) Evidently, for a significant proportion of customers of these supermarkets, drinking beer causes a desire for ice cream or eating ice cream causes a thirst for beer.

(c) A scatterplot of monthly ice cream sales versus monthly beer sales would show that a straight line describes the pattern in the plot, but it would have to be a horizontal line.

(d) There is a clear negative association between beer sales and ice cream sales.

(e) The positive correlation is most likely a result of the variable temperature; that is, as temperatures increase, so do both beer sales and ice cream sales.

AP1.3 b

AP1.4 c

AP1.5 e

AP1.6 c

AP1.7 e

AP1.8 e

AP1.9 d

AP1.10 d

AP1.11 d

AP1.12 b

AP1.8 Here are the IQ scores of 10 randomly chosen fifth-grade students:

| 145 | 139 | 126 | 122 | 125 | 130 | 96 | 110 | 118 | 118 |

Which of the following statements about this data set is *not* true?

(a) The student with an IQ of 96 is considered an outlier by the $1.5 \times IQR$ rule.

(b) The five-number summary of the 10 IQ scores is 96, 118, 123.5, 130, 145.

(c) If the value 96 were removed from the data set, the mean of the remaining 9 IQ scores would be greater than the mean of all 10 IQ scores.

(d) If the value 96 were removed from the data set, the standard deviation of the remaining 9 IQ scores would be less than the standard deviation of all 10 IQ scores.

(e) If the value 96 were removed from the data set, the IQR of the remaining 9 IQ scores would be less than the IQR of all 10 IQ scores.

AP1.9 Before he goes to bed each night, Mr. Kleen pours dishwasher powder into his dishwasher and turns it on. Each morning, Mrs. Kleen weighs the box of dishwasher powder. From an examination of the data, she concludes that Mr. Kleen dispenses a rather consistent amount of powder each night. Which of the following statements is true?

I. There is a high positive correlation between the number of days that have passed since the box of dishwasher powder was opened and the amount of powder left in the box.

II. A scatterplot with days since purchase as the explanatory variable and total amount of dishwasher powder used as the response variable would display a strong positive association.

III. The correlation between the amount of powder left in the box and the amount of powder used should be very close to -1.

(a) I only

(b) II only

(c) III only

(d) II and III only

(e) I, II, and III

AP1.10 The General Social Survey (GSS), conducted by the National Opinion Research Center at the University of Chicago, is a major source of data on social attitudes in the United States. Once each year, 1500 adults are interviewed in their homes all across the country. The subjects are asked their opinions about sex and marriage; attitudes toward women, welfare, foreign policy; and many other issues. The GSS begins by selecting a sample of counties from the 3000 counties in the country. The counties are divided into urban, rural, and suburban; a separate sample of counties is chosen at random from each group. This is a

(a) simple random sample.

(b) systematic random sample.

(c) cluster sample.

(d) stratified random sample.

(e) voluntary response sample.

AP1.11 You are planning an experiment to determine the effect of the brand of gasoline and the weight of a car on gas mileage measured in miles per gallon. You will use a single test car, adding weights so that its total weight is 3000, 3500, or 4000 pounds. The car will drive on a test track at each weight using each of Amoco, Marathon, and Speedway gasoline. Which is the best way to organize the study?

(a) Start with 3000 pounds and Amoco and run the car on the test track. Then do 3500 and 4000 pounds. Change to Marathon and go through the three weights in order. Then change to Speedway and do the three weights in order once more.

(b) Start with 3000 pounds and Amoco and run the car on the test track. Then change to Marathon and then to Speedway without changing the weight. Then add weights to get 3500 pounds and go through the three gasolines in the same order. Then change to 4000 pounds and do the three gasolines in order again.

(c) Choose a gasoline at random, and run the car with this gasoline at 3000, 3500, and 4000 pounds in order. Choose one of the two remaining gasolines at random and again run the car at 3000, then 3500, then 4000 pounds. Do the same with the last gasoline.

(d) There are nine combinations of weight and gasoline. Run the car several times using each of these combinations. Make all these runs in random order.

(e) Randomly select an amount of weight and a brand of gasoline, and run the car on the test track. Repeat this process a total of 30 times.

AP1.12 A linear regression was performed using the following five data points: A(2, 22), B(10, 4), C(6, 14), D(14, 2), E(18, −4). The residual for which of the five points has the largest absolute value?

(a) A

(b) B

(c) C

(d) D

(e) E

AP1.13 The frequency table summarizes the distribution of time that 140 patients at the emergency room of a small-city hospital waited to receive medical attention during the last month.

Waiting time	Frequency
Less than 10 minutes	5
At least 10 but less than 20 minutes	24
At least 20 but less than 30 minutes	45
At least 30 but less than 40 minutes	38
At least 40 but less than 50 minutes	19
At least 50 but less than 60 minutes	7
At least 60 but less than 70 minutes	2

Which of the following represents possible values for the median and *IQR* of waiting times for the emergency room last month?

(a) median = 27 minutes and *IQR* = 15 minutes
(b) median = 28 minutes and *IQR* = 25 minutes
(c) median = 31 minutes and *IQR* = 35 minutes
(d) median = 35 minutes and *IQR* = 45 minutes
(e) median = 45 minutes and *IQR* = 55 minutes

AP1.14 Boxplots of two data sets are shown.

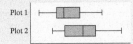

Based on the boxplots, which of the following is true?
(a) The range of both plots is about the same.
(b) The means of both plots are approximately equal.
(c) Plot 2 contains more data points than Plot 1.
(d) The medians are approximately equal.
(e) Plot 1 is more symmetric than Plot 2.

Section II: Free Response *Show all your work. Indicate clearly the methods you use, because you will be graded on the correctness of your methods as well as on the accuracy and completeness of your results and explanations.*

AP1.15 The manufacturer of exercise machines for fitness centers has designed two new elliptical machines that are meant to increase cardiovascular fitness. The two machines are being tested on 30 volunteers at a fitness center near the company's headquarters. The volunteers are randomly assigned to one of the machines and use it daily for two months. A measure of cardiovascular fitness is administered at the start of the experiment and again at the end. The following stemplot contains the differences (After − Before) in the two scores for the two machines. Note that greater differences indicate larger gains in fitness.

Machine A		Machine B
	0	2
54	1	0
876320	2	159
97411	3	2489
61	4	257
	5	359

Key: 2 | 1 represents a difference (After − Before) of 21 in fitness scores.

(a) Write a few sentences comparing the distributions of cardiovascular fitness gains from the two elliptical machines.

(b) Which machine should be chosen if the company wants to advertise it as achieving the highest overall gain in cardiovascular fitness? Explain your reasoning.

(c) Which machine should be chosen if the company wants to advertise it as achieving the most consistent gain in cardiovascular fitness? Explain your reasoning.

(d) Give one reason why the advertising claims of the company (the scope of inference) for this experiment would be limited. Explain how the company could broaden that scope of inference.

AP1.16 Those who advocate for monetary incentives in a work environment claim that this type of incentive has the greatest appeal because it allows the winners to do what they want with their winnings. Those in favor of tangible incentives argue that money lacks the emotional appeal of, say, a weekend for two at a romantic country inn or elegant hotel, or a weeklong trip to Europe.

A few years ago a national tire company, in an effort to improve sales of a new line of tires, decided to test which method—offering cash incentives or offering non-cash prizes such as vacations—was more successful in increasing sales. The company had 60 retail sales districts of various sizes across the country and data on the previous sales volume for each district.

(a) Describe a completely randomized design using the 60 retail sales districts that would help answer this question.

record the change in sales for each district and compare the mean change for each of the treatment groups.
(b) Number the 60 retail sales districts with a two-digit number from 01 to 60. Using a table of random digits, read two-digit numbers until 30 unique numbers from 01 to 60 have been selected. These 30 districts are assigned to the monetary incentives group and the remaining 30 to the tangibles incentives group. Using the digits provided, the districts labeled 07, 51, and 18 are the first three to be assigned to the monetary incentives group. **(c)** Matching the districts based on their size accounts for the variation among the experimental units due to their size on the response variable—sales volume. Pair the two largest districts in size, the next two largest, down to the two smallest districts. For each pair, pick one of the districts and flip a coin. If the flip is "heads," this district is assigned to the monetary incentives group. If it is "tails," this district is assigned to the tangible incentives group. The other district in the pair is assigned to the other group. After a specified period of time, record the change in sales for each district and compare within each pair.

AP1.13 a

AP1.14 a

AP1.15 (a) The distribution of gains for subjects using Machine A is roughly symmetric, while the distribution of gains for subjects using Machine B is skewed to the left (toward the smaller values). Neither distribution appears to contain any outliers. The center of the distribution of gains for subjects using Machine B (median = 38) is greater than the center of the distribution of gains for subjects using Machine A (median = 28). The distribution of gains for subjects using Machine B (range = 57, *IQR* = 22) is more variable than the distribution of gains for subjects using Machine A (range = 32, *IQR* = 15). **(b)** B; the median gain for Machine B (38) is greater than it is for Machine A (28), as is the mean

($\bar{x}_B$ = 35.4 versus $\bar{x}_A$ = 28.9). **(c)** A; Machine A exhibits less variation in gains than does machine B. The *IQR* for Machine A (15) is less than the *IQR* for Machine B (22). Additionally, the standard deviation for Machine A (9.38) is less than the standard deviation for Machine B (16.19). **(d)** The experiment was conducted at only one fitness center. Results may vary at other fitness centers in this city and in other cities. If the company wants to broaden its scope of inference, it should randomly select people from the population about whom it would like to draw an inference.

AP1.16 (a) Randomly assign 30 retail sales districts to the monetary incentives treatment and the remaining 30 retail sales districts to the tangible incentives treatment [see part (b) for method]. After a specified period of time,

AP1.17 (a) There is a very strong, positive, linear association between sales and shelf length. **(b)** $\hat{y} = 317.94 + 152.68x$, where $\hat{y}$ = predicted weekly sales (in dollars) and x = shelf length (in feet). **(c)** $\hat{y} = 317.94 + 152.68(5) = 1081.34$ **(d)** The actual weekly sales (in dollars) is typically about $22.9212 away from the weekly sales predicted by the least-squares regression line with x = shelf length (in feet). **(e)** About 98.2% of the variation in weekly sales revenue can be accounted for by the least-squares regression line with x = shelf length (in feet).

AP1.18 (a) $z = \dfrac{-5 - 0}{22.92} = -0.22$ and
$z = \dfrac{5 - 0}{22.92} = 0.22$

Table A: $0.5871 - 0.4219 = 0.1652$
Tech: normalcdf(lower: −5, upper: 5, mean: 0, SD: 22.92) = 0.1727
(b) *Table A:*

Solving $-1.96 = \dfrac{x - 0}{22.92}$ gives

$x = -\$44.92$.

Solving $1.96 = \dfrac{x - 0}{22.92}$ gives

$x = \$44.92$.

Tech: invNorm(area: 0.025, mean: 0, SD: 22.92) = −$44.92
and invNorm(area: 0.975, mean: 0, SD: 22.92) = $44.92

The middle 95% of residuals should be between −$44.92 and $44.92.

If 5 linear feet are allocated to the store's brand of men's grooming products, the weekly sales revenue can be expected to be between $1036.08 and $1125.92 (1081 +/−44.92).

(b) Explain how you would use the following excerpt from the table of random digits to do the random assignment that your design requires. Then use your method to make the first three assignments.

| 07511 | 88915 | 41267 | 16853 | 84569 | 79367 | 32337 | 03316 |
| 81486 | 69487 | 60513 | 09297 | 00412 | 71238 | 27649 | 39950 |

(c) One of the company's officers suggested that it would be better to use a matched pairs design instead of a completely randomized design. Explain how you would change your design to accomplish this.

AP1.17 In retail stores, there is a lot of competition for shelf space. There are national brands for most products, and many stores carry their own line of in-house brands, too. Because shelf space is not infinite, the question is how many linear feet to allocate to each product and which shelf (top, bottom, or somewhere in the middle) to put it on. The middle shelf is the most popular and lucrative, because many shoppers, if undecided, will simply pick the product that is at eye level.

A local store that sells many upscale goods is trying to determine how much shelf space to allocate to its own brand of men's personal-grooming products. The middle shelf space is randomly varied between 3 and 6 linear feet over the next 12 weeks, and weekly sales revenue (in dollars) from the store's brand of personal-grooming products for men is recorded. Here is some computer output from the study, along with a scatterplot:

Predictor	Coef	SE Coef	T	P
Constant	317.940	31.32	10.15	0.000
Shelf length	152.680	6.445	23.69	0.000
S = 22.9212	R-Sq = 98.2%	R-Sq(adj) = 98.1%		

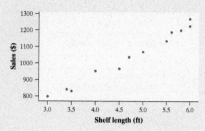

(a) Describe the relationship between shelf length and sales.

(b) Write the equation of the least-squares regression line. Be sure to define any variables you use.

(c) If the store manager were to decide to allocate 5 linear feet of shelf space to the store's brand of men's grooming products, what is the best estimate of the weekly sales revenue?

(d) Interpret the value of s.

(e) Identify and interpret the coefficient of determination.

AP1.18 The manager of the store in the preceding exercise calculated the residual for each point in the scatterplot and made a dotplot of the residuals. The distribution of residuals is roughly Normal with a mean of $0 and standard deviation of $22.92.

(a) What percent of the actual sales amounts do you expect to be within $5 of their expected sales amount?

(b) The middle 95% of residuals should be between which two values? Use this information to give an interval of plausible values for the weekly sales revenue if 5 linear feet are allocated to the store's brand of men's grooming products.

Chapter 5

Chapter 5

Probability

PD Overview

This chapter begins a three-chapter section about probability. It also begins the transition from the descriptive methods we learned about in Chapters 1–3 to the inferential methods we will learn about in Chapters 8–12.

Some of the concepts in this chapter will be familiar to students, but there will be very important new ideas as well. In Section 5.1, we introduce simulation as a useful way to estimate probabilities. In Section 5.2, we introduce the idea of a probability model and some basic probability rules. Also, we revisit two-way tables as a good way to investigate the relationship between two events. In Section 5.3, we introduce two concepts that we will use throughout the rest of the book: conditional probability and independence.

You may notice that many traditional probability topics such as counting rules, permutations, and combinations are not included in this chapter (or any other chapter). The reasons are twofold: (1) these topics are not on the AP® Statistics Course Framework, and (2) this is a course about statistics, so we spend time addressing only the probability topics that will help us do statistics.

When students read probability questions, they often try to identify the correct formula to use, so they can plug in numbers and get the right answer without thinking about the question being asked. Such an approach is dangerous, because this rush to an "answer" often causes them to sidestep the critical thinking and reasoning they will need to employ on the AP® Statistics exam. Challenge students to use other strategies to help them think about probability, such as simulation, sample space, two-way tables, Venn diagrams, and tree diagrams.

The Main Ideas

One of the challenges in teaching the AP® Statistics course is keeping students focused on the big picture, not just the details of each section. We outline the main ideas for the chapter here.

Chapter 5 Introduction

Chance is all around us, whether we realize it or not. We introduce this chapter with several quick examples of how chance shows up in our daily lives. The mathematics of chance is called probability, and probability is the basis for the confidence intervals and significance tests we will learn about in Chapters 8–12. This makes probability a vital bridge between designing studies and drawing conclusions from them.

The activity in this section highlights several important ideas, including simulation as a powerful tool to estimate probabilities. It also continues to foreshadow the logic of inference, which we will encounter in later chapters.

Section 5.1 Randomness, Probability, and Simulation

In this section, we introduce the idea of probability and show how simulations can be used to estimate probabilities and make decisions.

The probability of an outcome in a random process describes the proportion of times the outcome would occur in a very long series of repetitions. For example, when we say that the probability of getting heads when flipping a coin is 0.5, we do not mean that exactly 1 out of every 2 flips will result in heads. Instead, a probability of 0.5 means that if we were to flip the coin millions of times, the proportion of heads should be very close to 0.5. The fact that the proportion of heads will get very close to the true probability of heads after millions of flips is called the law of large numbers.

Many students (and adults!) have misconceptions about randomness, including a belief in short-term regularity. That is, streaks of the same outcome are sometimes viewed with suspicion even though they occur quite often just by chance. Also, many people mistakenly believe in the "law of averages," which suggests that outcomes will "even out" in a short number of trials. For example, upon seeing three heads in a row, many people think the next flip is more likely to be tails because there should be an equal number of heads and tails.

Simulations are powerful tools we can use to imitate random processes and estimate probabilities. Simulation is especially useful when it is difficult to calculate a probability theoretically. It is also a versatile tool that we will continue to use throughout the book to illustrate many different statistical concepts.

In previous editions of this book, students used the four-step process (State, Plan, Do, Conclude) to describe simulations. We have removed that requirement in this edition and will use the four-step process only for formal inference procedures (Chapters 8–12) where it is the most powerful.

Section 5.2 Probability Rules

In this section, we introduce probability models. Just as some distributions of data can be described by a Normal model and some relationships between quantitative variables can be described by a linear model, chance behavior can be described by a probability model. Probability models have two parts: a list of possible outcomes and the probability of each outcome.

We also introduce some basic rules of probability, including that probabilities must be between 0 and 1 and that the probabilities of all the outcomes in the sample space must add to 1. Other basic rules include the complement rule and the addition rule for mutually exclusive events.

A helpful way to organize information involving two events is a two-way table. We first saw two-way tables in Section 1.1, when we learned about relationships between categorical variables. Working with two-way tables leads to the general addition rule for two events, which works whether or not the events are mutually exclusive. We also present another useful way to organize information involving two events: a Venn diagram.

Section 5.3 Conditional Probability and Independence

In this section, we introduce two very important ideas: conditional probability and independence. These ideas will appear in every remaining chapter of the book, so make sure your students develop a solid understanding of them. A conditional probability describes the probability that an event occurs given that another event has already occurred. If knowing that event A occurs doesn't change the probability that event B occurs, events A and B are independent.

There is a connection between conditional probability and the conditional relative frequencies we learned about in Chapter 1. In Section 1.1, we looked at the distribution of one categorical variable for a specific value of another categorical variable. In this chapter, we look at the probability of a specific event occurring given that another event has already occurred.

Two-way tables and tree diagrams are two very useful ways to organize the information provided in a conditional probability problem. Two-way tables are best when the problem describes the number or proportion of cases with certain characteristics. Tree diagrams are best when the problem provides conditional probabilities of different events or describes events that are sequential.

In this section, we also introduce three formulas: the general multiplication rule, the multiplication rule for independent events, and the conditional probability formula. As always, it is important for students to recognize which formula to use in a particular situation and to be able to justify their choice.

Finally, an important type of conditional probability question involves going "backward" in a tree diagram. For example, we may know the probability that a person who has a disease should receive a positive result when given a test for the disease, but what is the probability that a person who receives a positive result actually has the disease? These two probabilities are not the same.

Chapter 5: Resources

Teacher's Resource Materials

The following resources, identified by the **TRM** in the annotated student pages, can be found by clicking on the link in the Teacher's e-Book (TE-Book), searching by category or chapter on the Teacher's Resource Flash Drive (TRFD), or logging into the book's digital platform and searching the Teacher's Resources menu (teacher log-in required).

- Alternate Examples: one file per section
- Lecture Presentation Slides: one per section
- Chapter 5 Learning Targets Grid
- Additional Simulation Problems
  - Feel the Power
  - The Duck Hunters
  - Airline Overbooking
- FRAPPY! Materials
- Chapter 5 Case Study
- Complete solutions for the Check Your Understanding problems, section exercises, review exercises, and practice test.
- Quizzes: one per section
- Chapter 5 Test

Free Response Questions from Previous AP® Statistics Exams

Questions can be found on the AP® Central website: apcentral .collegeboard.org/courses/ap-statistics/exam.

Students should be able to answer all the free response questions listed below with material learned in this chapter. Questions that contain content from this chapter but also require content from later chapters are listed in the last chapter required to complete the entire question. This list will be updated after each AP® Statistics exam and will be posted to the Teacher's Resource section of the book's digital platform and to www.statsmedic.com/free-response-questions.

Year	#	Content
2017	3	• Normal probability • General multiplication rule (tree diagram) • Conditional probability
2017	6	• Probabilities of different random assignments (coin vs. chip) • Which method of random assignment is best
2014	2	• General multiplication rule • Informal inference • Simulation design
2011	2	• Conditional probability from a two-way table • Independence of two events • Segmented bar charts and independence
2009B	2	• Conditional probability • Multiplication rule
2003B	2	• Two-way tables • Conditional probability • Independence
2001	3	• Simulation
1997	3	• Conditional probability

Applets

- The *Probability* applet simulates the flip of a coin and keeps track of the proportion of heads. It also allows you to change the probability of heads. Go to the Student Site at highschool.bfwpub.com/updatedtps6e click on Statistical Applets and select "Probability").

- The *Random Babies* applet simulates delivering four babies at random to four houses and recording the number of correct deliveries. The applet displays a histogram of the results and shows how the probability of each outcome changes as more trials are run. Go to www.rossmanchance.com/applets.

Chapter 5: Pacing Guide, Learning Targets, and Suggested Assignments

This pacing guide is based on a schedule with 110, 50-minute sessions before the AP® Statistics exam. If you have a different number of sessions before the AP® exam, you can modify the pacing guide to suit your needs. If time permits, consider incorporating quizzes, released AP® free-response questions, or additional activities. See the Resources section above for suggestions.

The suggested homework assignments list odd-numbered exercises, whenever possible, so students can check their answers against the back-of-book answers. If you would rather students not have access to the answers while doing homework, adding 1 to the exercise numbers usually will do the trick, because the homework exercises typically are paired. For example, Exercises 1 and 2 will generally cover the same topics, but in different contexts. You may also choose to include the Recycle and Review questions at the end of each section, which review topics from previous sections or chapters. If your school is using the digital platform that accompanies TPS6, you will find these assignments pre-built as online homework assignments for Chapter 5.

Day	Content	Learning Targets: Students will be able to . . .	Suggested Assignment (MC bold)
1	5.1 The Idea of Probability	• Interpret probability as a long-run relative frequency.	1, 3, 5, 7
2	5.1 Simulation	• Use simulation to model a random process.	9, 11, 15, 21, **23–28**
3	5.2 Probability Models, Basic Probability Rules	• Give a probability model for a random process with equally likely outcomes and use it to find the probability of an event. • Use basic probability rules, including the complement rule and the addition rule for mutually exclusive events.	31, 33, 35, 37, 39
4	5.2 Two-Way Tables, Probability, and the General Addition Rule; Venn Diagrams and Probability	• Use a two-way table or Venn diagram to model a random process and calculate probabilities involving two events. • Apply the general addition rule to calculate probabilities.	41, 47, 49, 51, 53, **55–58**
5	5.3 What Is Conditional Probability?, Conditional Probability and Independence, The General Multiplication Rule	• Calculate and interpret conditional probabilities. • Determine whether two events are independent. • Use the general multiplication rule to calculate probabilities.	61, 63, 65, 67, 69, 71, 77, 79
6	5.3 Tree Diagrams and Conditional Probability, The Multiplication Rule for Independent Events	• Use a tree diagram to model a random process involving a sequence of outcomes and to calculate probabilities. • When appropriate, use the multiplication rule for independent events to calculate probabilities.	81, 83, 87, 89, 91, 93, 99, **103–106**
7	Chapter 5 Review/FRAPPY!		Chapter 5 Review Exercises
8	Chapter 5 Test		

Chapter 5 Alignment to College Board's Fall 2019 AP® Statistics Course Framework*

Relationship to College Board Units

Chapter 5 in this book covers Topics 4.1–4.6 in Unit 4 of the College Board Course Framework. Students will be ready to take the Personal Progress Check for Unit 4 once they have completed Chapters 5 and 6.

Big Ideas and Enduring Understandings

Chapter 5 develops these Big Ideas and related Enduring Understandings outlined in the Course Framework:

- **Big Idea 1: Variation and Distribution (EU: VAR 1,4):** The distribution of measures for individuals within a sample or population describes variation. The value of a statistic varies from sample to sample. How can we determine whether differences between measures represent random variation or meaningful distinctions? Statistical methods based on probabilistic reasoning provide the basis for shared understandings about variation and about the likelihood that variation between and among measures, samples, and populations is random or meaningful.
- **Big Idea 2: Patterns and Uncertainty (EU: UNC 2):** Statistical tools allow us to represent and describe patterns in data and to classify departures from patterns. Simulation and probabilistic reasoning allow us to anticipate patterns in data and to determine the likelihood of errors in inference.

Course Skills

Chapter 5 helps students to develop the skills identified in the Course Framework.

- **1: Selecting Statistical Methods** (1.A)
- **3: Using Probability and Simulation** (3.A)
- **4: Statistical Argumentation** (4.B)

Learning Objectives and Essential Knowledge Statements

Section	Learning Objectives	Essential Knowledge Statements
5.1	VAR-1.F, VAR-4.A, VAR-4.B, UNC-2.A	VAR-1.F.1, VAR-4.A.3, VAR-4.B.1, UNC-2.A.1, UNC-2.A.4, UNC-2.A.5, UNC-2.A.6
5.2	VAR-4.A, VAR-4.C, VAR-4.E, UNC-2.A	VAR-4.A.1, VAR-4.A.2, VAR-4.A.4, VAR-4.C.1, VAR-4.C.2, VAR-4.E.3, VAR-4.E.4, UNC-2.A.2, UNC-2.A.3
5.3	VAR-4.D, VAR-4.E	VAR-4.D.1, VAR-4.D.2, VAR-4.E.1, VAR-4.E.2

A detailed alignment that can be sorted by Course Framework Unit, Topic, Learning Objective, Essential Knowledge Statement, or textbook section, is available on the TRFD and in the Teacher's Resources folder on Sapling Plus. **TRM**

*Should changes be made to the Course Framework in the future, an updated alignment will be placed on our AP® updates page at go.bfwpub.com/ap-course-updates.

UNIT 4

Probability, Random Variables, and Probability Distributions

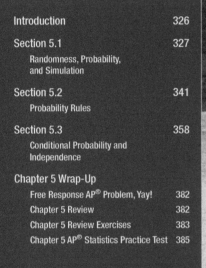

Chapter 5

Probability

Stefano Paterna-Alamy

Teaching Tip

Unit 4 in the College Board Course Framework aligns to Chapters 5 and 6 in this book. Students will be ready to take the Personal Progress Check for Unit 4 once they have completed Chapters 5 and 6.

PD Chapter 5 Overview

To watch the video overview of Chapter 5 (for teachers), click on the link in the TE-Book, look on the TRFD, or download from the Teacher's Resources on the book's digital platform.

Preparing for Inference

There will be several opportunities in this chapter to enhance students' thinking about significance tests. The "1 in 6 wins" activity, the NASCAR example, and the golden ticket example each provide a context for previewing hypotheses, evidence against a claim, and the *P*-value.

TRM Lecture Presentation Slides

If you are new to teaching AP® Statistics or are short on time when preparing for class, you may find the Lecture Presentation Slides to be helpful. Experienced AP® Teacher Doug Tyson has created one slide presentation per section. You may use them as is, modify them to fit your needs, or share them with students who miss class. Find them on the TRFD and in the Teacher's Resources on the book's digital platform.

ACTIVITY OVERVIEW

To prepare for using this activity, watch the overview video by clicking on the link in the TE-Book, opening the TRFD, or downloading from the Teacher's Resources on the book's digital platform.

Time: 15 minutes

Materials: Poster board or chart paper, sticker dots, dice

Teaching Advice: Ask students to put sticker dots on a poster board or chart paper. Post the completed poster board in the classroom so that you can refer to this activity throughout the rest of the course. This activity will lead students to the understanding that probability is a long-run relative frequency. The long run would be buying many, many sets of 30 sodas. This activity can be simulated using the *One Categorical Variable* applet in the Extra Applets on the Student Site at www.highschool.com/updatedtps6e. Put in the class data as 2 "Winner" and 28 "Nonwinner." At the bottom of the applet, you can "Simulate sample count." Use 0.167 as the hypothesized proportion. We suggest having students physically perform this simulation several times with a die before jumping to technology.

Answers:

1–3. Here is one possible answer.

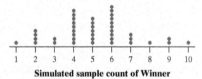

Simulated sample count of Winner

4. Answers will vary. Assuming the company is telling the truth, there is a 5/40 = 12.5% chance of getting two or fewer prizes out of 30 bottles of soda. Since this percentage is not small, it seems plausible that the company is telling the truth and the class was unlucky.

Making Connections

In Chapter 6, we will define the conditions necessary for a binomial distribution. In this activity, the number of winners is a binomial random variable with $n = 30$ and $p = 1/6$.

INTRODUCTION

Chance is all around us. You and your friend play rock-paper-scissors to determine who gets the last slice of pizza. A coin toss decides which team gets to receive the ball first in a football game. Many adults regularly play the lottery, hoping to win a big jackpot with a few lucky numbers. Others head to casinos or racetracks, hoping that some combination of luck and skill will pay off. People young and old play games of chance involving cards or dice or spinners. The traits that children inherit—hair and eye color, blood type, handedness, dimples, whether they can roll their tongues—are determined by the chance involved in which genes their parents pass along.

The mathematics of chance behavior is called *probability*. Probability is the topic of this chapter. Here is an activity that gives you some idea of what lies ahead.

ACTIVITY   The "1 in 6 wins" game

In this activity, you and your classmates will use simulation to test whether a company's claim is believable.

As a special promotion for its 20-ounce bottles of soda, a soft drink company printed a message on the inside of each bottle cap. Some of the caps said, "Please try again!" while others said, "You're a winner!" The company advertised the promotion with the slogan "1 in 6 wins a prize." The prize is a free 20-ounce bottle of soda.

Grayson's statistics class wonders if the company's claim holds true for the bottles at a nearby convenience store. To find out, all 30 students in the class go to the store and each student buys one 20-ounce bottle of the soda. Two of them get caps that say, "You're a winner!" Does this result give convincing evidence that the company's 1-in-6 claim is inaccurate?

For now, let's assume that the company is telling the truth, and that every 20-ounce bottle of soda it fills has a 1-in-6 chance of getting a cap that says, "You're a winner!" We can model the status of an individual bottle with a six-sided die: let 1 through 5 represent "Please try again!" and 6 represent "You're a winner!"

1. Roll your die 30 times to imitate the process of the students in Grayson's statistics class buying their sodas. How many of them won a prize?

2. Your teacher will draw and label axes for a class dotplot. Plot on the graph the number of prize winners you got in Step 1.

3. Repeat Steps 1 and 2, if needed, to get a total of at least 40 trials of the simulation for your class.

4. Discuss the results with your classmates. What percent of the time did the simulation yield two or fewer prize winners in a class of 30 students just by chance? Does it seem plausible (believable) that the company is telling the truth, but that the class just got unlucky? Or is there convincing evidence that the 1-in-6 claim is wrong? Explain your reasoning.

Preparing for Inference

This activity gets students thinking about the strength of evidence against a claim. In a significance test, the claim that the company is telling the truth is the *null hypothesis*. We will assume this claim is true, then find the probability of getting our result (2 out of 30 winners) or less purely by chance (*P-value*). If the probability is less than 5% (α, *alpha*), we have convincing evidence that the soda company is not telling the truth. If the probability is greater than 5%, it is plausible that the company is telling the truth and the class results happened purely by chance.

Teaching Tip: Using Technology

From the Extra Applets section on the Student Site at highschool.bfwpub.com/updatedtps6e, choose the *Probability* applet. Select "Binomial distribution" and set $n = 30$ and $p = 1/6$. You can then use the applet to find the probability of getting at most 2 successes (0.1028).

As the activity shows, *simulation* is a powerful method for modeling random behavior. Section 5.1 begins by examining the idea of probability and then illustrates how simulation can be used to estimate probabilities. In Sections 5.2 and 5.3, we develop the basic rules and techniques of probability.

Probability calculations are the basis for inference. When we produce data by random sampling or randomized comparative experiments, the laws of probability answer the question "What would happen if we repeated the random sampling or random assignment process many times?" Many of the examples, exercises, and activities in this chapter focus on the connection between probability and inference.

SECTION 5.1 Randomness, Probability, and Simulation

LEARNING TARGETS *By the end of the section, you should be able to:*

- Interpret probability as a long-run relative frequency.
- Use simulation to model a random process.

Imagine tossing a coin 10 times. How likely are you to get a run of 3 or more consecutive heads? An airline knows that a certain percent of customers who purchase tickets will not show up for a flight. If the airline overbooks a particular flight, what are the chances that they'll have enough seats for the passengers who show up? A couple plans to have children until they have at least one boy and one girl. How many children should they expect to have? To answer these questions, you need a better understanding of how random behavior operates.

The Idea of Probability

In football, a coin toss helps determine which team gets the ball first. Why do the rules of football require a coin toss? Because tossing a coin seems a "fair" way to decide. What does that mean exactly? The following activity should help shed some light on this question.

ACTIVITY What is probability?

If you toss a fair coin, what's the probability that it shows heads? It's 1/2, or 0.5, right? But what does probability 1/2 really mean? In this activity, you will investigate by flipping a coin several times.

1. Toss your coin once. Record whether you get heads or tails.
2. Toss your coin a second time. Record whether you get heads or tails. What proportion of your first two tosses is heads?

To prepare for using this activity, watch the overview video by clicking on the link in the TE-Book, opening the TRFD, or downloading from the Teacher's Resources on the book's digital platform.

Time: 10 minutes

Materials: Coins, devices with Internet access

Teaching Advice: It is important that students physically toss coins and calculate a few proportions by hand before jumping to technology to simulate. Without the coin tosses, students can lose sight of what is being measured (proportion of coin tosses that are heads) when they move to the technology.

Answers:

1–4. Answers will vary.

5. The probability seems to be approaching 0.5.

6. Answers will vary but most likely are close to 0.5.

7. The proportion of heads gets closer and closer to 0.5.

8. If you take a very large random sample of coin tosses, about 50% of them will be heads.

9. No, just because there are two outcomes does not mean there is a 50% probability of each. To find out the true probability, toss a thumbtack many, many times and record the proportion of times it lands point up.

---

3. Toss your coin 8 more times so that you have 10 tosses in all. Record whether you get heads or tails on each toss in a table like the one that follows.

4. Calculate the overall proportion of heads after each toss and record these values in the bottom row of the table. For instance, suppose you got tails on the first toss and heads on the second toss. Then your overall proportion of heads would be $0/1 = 0.00$ after the first toss and $1/2 = 0.50$ after the second toss.

Toss	1	2	3	4	5	6	7	8	9	10
Result (H or T)										
Proportion of heads										

5. Let's use technology to speed things up. Launch the *Idea of Probability* applet at highschool.bfwpub.com/updatedtps6e. Set the number of tosses to 10 and click toss. What proportion of the tosses were heads? Click "Reset" and toss the coin 10 more times. What proportion of heads did you get this time? Repeat this process several more times. What do you notice?

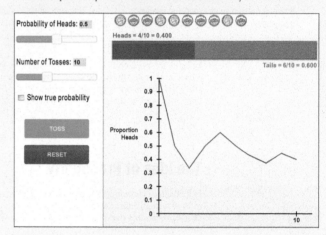

6. What if you toss the coin 100 times? Reset the applet and have it do 100 tosses. Is the proportion of heads exactly equal to 0.5? Close to 0.5?

7. Keep on tossing without hitting "Reset." What happens to the proportion of heads?

8. As a class, discuss what the following statement means: "If you toss a fair coin, the probability of heads is 0.5."

9. If you toss a coin, it can land heads or tails. If you "toss" a thumbtack, it can land with the point sticking up or with the point down. Does that mean the probability of a tossed thumbtack landing point up is 0.5? How can you find out? Discuss with your classmates.

Teaching Tip

To help students think about Question 9, bring up the context of the lottery. Someone who plays the lottery will either win money or lose money. Does this mean the player has a 50% chance of winning?

Figure 5.1 shows some results from the preceding activity. The proportion of tosses that land heads varies from 0.30 to 1.00 in the first 10 tosses. As we make more and more tosses, however, the proportion of heads gets closer to 0.5 and stays there.

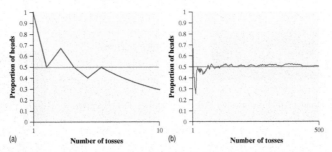

FIGURE 5.1 (a) The proportion of heads in the first 10 tosses of a coin. (b) The proportion of heads in the first 500 tosses of a coin.

When we watch coin tosses or the results of random sampling and random assignment closely, a remarkable fact emerges: A *random process is unpredictable in the short run but has a regular and predictable pattern in the long run.* This is the basis for the idea of **probability**.

> **DEFINITION   Random process, Probability**
>
> A **random process** generates outcomes that are determined purely by chance.
>
> The **probability** of any outcome of a random process is a number between 0 and 1 that describes the proportion of times the outcome would occur in a very long series of trials.

A trial is one repetition of a random process.

Outcomes that never occur have probability 0. An outcome that always occurs has probability 1. An outcome that happens half the time in a very long series of trials has probability 0.5.

The fact that the proportion of heads in many tosses eventually closes in on 0.5 is guaranteed by the **law of large numbers**. You can see this in Figure 5.1(b). The horizontal line represents the probability, and the proportion of heads in the simulation approaches this value as the number of trials becomes large.

Some people distinguish between tossing a coin several times to estimate the probability of a head (*empirical probability*) and using a computer applet to simulate this random process.

> **DEFINITION   Law of large numbers**
>
> The **law of large numbers** says that if we observe more and more trials of any random process, the proportion of times that a specific outcome occurs approaches its probability.

Probability gives us a language to describe the long-term regularity of a random process that is guaranteed by the law of large numbers. The outcome of a coin toss and the sex of the next baby born in a local hospital are both random. So is the result of a random sample or a random assignment. Even life insurance is

Preparing for Inference

Students should have a solid understanding of probability (and conditional probability) to be able to comprehend the *P*-value in Chapter 9. The *P*-value for a significance test is the probability of getting an observed result or more extreme, given that some claim is true.

Teaching Tip: AP® Connections

AP® Calculus students will recognize this idea as a limit. As the number of trials approaches infinity, the proportion of successes approaches a single value.

Teaching Tip

The "play" icon at the top of this example (and many others) indicates that students can watch a video presentation of the example by clicking on the link in the e-Book or viewing the videos on the Student Site at highschool.bfwpub.com/updatedtps6e.

PROBLEM:
The Chicago Cubs play their home games at Wrigley Field, located in the Lakeview neighborhood of Chicago. A recent *New York Times* study concluded that the probability that a randomly selected Lakeview resident is a Cubs fan is 0.44.
(a) Interpret this probability as a long-run relative frequency.
(b) If a researcher randomly selects 100 Lakeview residents, will exactly 44 of them be Cubs fans? Explain your answer.

SOLUTION:
(a) If you take a very large random sample of Lakeview residents, about 44% of them will be Cubs fans.
(b) Probably not; with only 100 randomly selected Lakeview residents, the number who are Cubs fans may not be very close to 44.

based on the fact that deaths occur at random among many individuals. Because men are more likely to die at a younger age than women, insurance companies sometimes charge a man up to 3 times more for a life insurance policy than a woman of the same age.

EXAMPLE   **Who drinks coffee?**
Interpreting probability

PROBLEM: According to *The Book of Odds*, the probability that a randomly selected U.S. adult drinks coffee on a given day is 0.56.
(a) Interpret this probability as a long-run relative frequency.
(b) If a researcher randomly selects 100 U.S. adults, will exactly 56 of them drink coffee that day? Explain your answer.

SOLUTION:
(a) If you take a very large random sample of U.S. adults, about 56% of them will drink coffee that day.
(b) Probably not. With only 100 randomly selected adults, the number who drink coffee that day may not be very close to 56.

> The random process consists of randomly selecting a U.S. adult and recording whether or not the person drinks coffee that day.

> Probability describes what happens in many, many repetitions (way more than 100) of a random process.

FOR PRACTICE, TRY EXERCISE 1

Life insurance companies, casinos, and others who make important decisions based on probability rely on the long-run predictability of random behavior.

UNDERSTANDING RANDOMNESS The idea of probability seems straightforward. It answers the question "What would happen if we did this many times?" Understanding random behavior is important for making decisions, especially when our data collection process includes random sampling or random assignment. But understanding randomness isn't always easy, as the following activity illustrates.

ACTIVITY   **Investigating randomness**

In this activity, you and your classmates will test your ability to imitate a random process.

1. Pretend that you are flipping a fair coin. Without actually flipping a coin, *imagine* the first toss. Write down the result you see in your mind, heads (H) or tails (T).
2. Imagine a second coin flip. Write down the result.

ACTIVITY OVERVIEW

To prepare for using this activity, watch the overview video by clicking on the link in the TE-Book, opening the TRFD, or downloading from the Teacher's Resources on the book's digital platform.

Time: 10 minutes

Materials: Imaginary coins, real coins

Teaching Advice: In general, students will not allow long runs in their imagined coin tosses because they think that such runs would not happen by chance. The random tosses will reveal that long runs happen quite often.

There are many ways to generate a list of 50 fair coin flips:

1. Toss a coin 50 times.

2. Using Table D, let even numbers represent heads and odd numbers represent tails. Ask each student to start at a different line.

3. On a graphing calculator, use randInt(1,2), where 1 is heads and 2 is tails.

4. Use https://www.random.org/coins/.

Be sure that your two dotplots line up vertically and have the same scale.

Answers:
1–7. Answers will vary.

8. The dotplot of longest run from the imagined tosses has a center less than the dotplot of longest run from the random tosses.

3. Keep doing this until you have recorded the results of 50 imaginary flips. Write your results in groups of 5 to make them easier to read, like this: HTHTH TTHHT, and so on.

4. A *run* is when the same result occurs in consecutive outcomes. In the example in Step 3, there is a run of two tails followed by a run of two heads in the first 10 coin flips. Read through your 50 imagined coin flips and find the longest run.

5. Your teacher will draw and label a number line for a class dotplot. Plot on the graph the length of the longest run you got in Step 4.

6. Use an actual coin, Table D, or technology to generate a similar list of 50 coin flips. Find the longest run that you have.

7. Your teacher will draw and label a number line with the same scale immediately above or below the one in Step 5. Plot on the new dotplot the length of the longest run you got in Step 6.

8. Compare the distributions of longest run from imagined tosses and random tosses. What do you notice?

The idea of probability is that randomness is predictable in the long run. Unfortunately, our intuition about random behavior tries to tell us that randomness should also be predictable in the short run. When it isn't, we look for some explanation other than chance variation.

Suppose you toss a coin 6 times and get TTTTTT. Some people think that the next toss must be more likely to give a head. It's true that in the long run, heads will appear half the time. What is a myth is that future outcomes must make up for an imbalance like six straight tails.

Coins and dice have no memories. A coin doesn't know that the first 6 outcomes were tails, and it can't try to get a head on the next toss to even things out. Of course, things do even out in the long run. That's the law of large numbers in action. After 10,000 tosses, the results of the first six tosses don't matter. They are overwhelmed by the results of the next 9994 tosses.

When asked to predict the sex—boy (B) or girl (G)—of the next seven babies born in a local hospital, most people will guess something like B-G-B-G-B-G-G. Few people would say G-G-G-B-B-B-G because this sequence of outcomes doesn't "look random." In fact, these two sequences of births are equally likely. "Runs" consisting of several of the same outcome in a row are surprisingly common in a random process. Many students are not aware of this fact when they imagine a sequence of 50 coin tosses in the "Investigating randomness" activity!

Is there such a thing as a "hot hand" in basketball? Belief that runs must result from something other than "just chance" influences behavior. If a basketball player makes several consecutive shots, both the fans and her teammates believe that she has a "hot hand" and is more likely to make the next shot. Several early studies of the hot hand theory showed that runs of baskets made or missed are no more frequent in basketball than would be expected if the result of each shot is unrelated to the outcomes of the player's previous shots. If a player makes half her shots in the long run, her made shots and misses behave just like tosses of a coin—which means that runs of makes and misses are more common than our intuition expects.[1]

Some people use the phrase *law of averages* to refer to the misguided belief that the results of a random process have to even out in the *short run*.

✔ Answers to CYU

1. *Interpretation:* If you take a very large random sample of Pedro's commutes, about 55% of the time the light will be red when Pedro reaches the light.

2. (a) This probability is 0. If an outcome can never occur, it will occur in 0% of the trials.
(b) This probability is 1. If an outcome will occur on every trial, it will occur in 100% of the trials.
(c) This probability is 0.001. An outcome that occurs in 0.1% of the trials is very unlikely, but will occur every once in a while.
(d) This probability is 0.6. An outcome that occurs in 60% of the trials will happen more than half of the time. Also, 0.6 is a better choice than 0.99 because the wording suggests that the event occurs often but not nearly every time.

3. The doctor is wrong because the sex of the next baby born is a random phenomenon that is unpredictable in the short run, even though it has a regular predictable pattern in the long run. So while approximately 50% of all babies born will be male, even after a couple has seven girls in a row, the probability of the eighth child being a girl is still 50%.

Teaching Tip

In this chapter, students will learn several strategies for approaching probability questions:

1. **Simulation**
2. Sample space
3. Two-way tables
4. Venn diagrams
5. Tree diagrams
6. Formulas

Teaching Tip

In previous editions of this book, students used the four-step process (State, Plan, Do, Conclude) to describe simulations. We have removed that requirement in this edition and will use the four-step process only for formal inference procedures (Chapters 8–12), where it is the most powerful.

Two more recent studies provide some evidence that there is a small hot hand effect for basketball players. These studies also suggest that "hot" shooters take riskier shots, which then masks the hot hand effect.[2]

CHECK YOUR UNDERSTANDING

1. Pedro drives the same route to work on Monday through Friday. His route includes one traffic light. According to the local traffic department, there is a 55% probability that the light will be red when Pedro reaches the light. Interpret the probability.

2. Probability is a measure of how likely an outcome is to occur. Match one of the probabilities that follow with each statement. Be prepared to defend your answer.

$$0 \quad 0.001 \quad 0.3 \quad 0.6 \quad 0.99 \quad 1$$

   (a) This outcome is impossible. It can never occur.
   (b) This outcome is certain. It will occur on every trial.
   (c) This outcome is very unlikely, but it will occur once in a while in a long sequence of trials.
   (d) This outcome will occur more often than not.

3. A husband and wife decide to have children until they have at least one child of each sex. The couple has had seven girls in a row. Their doctor assures them that they are much more likely to have a boy next. Explain why the doctor is wrong.

Simulation

We can model random behavior and estimate probabilities with a **simulation**.

DEFINITION   Simulation

A **simulation** imitates a random process in such a way that simulated outcomes are consistent with real-world outcomes.

You already have some experience with simulations. In the "Hiring discrimination—it just won't fly!" activity in Chapter 1 (page 6), you drew beads or slips of paper to imitate a random lottery to choose which pilots would become captains. The "Analyzing the caffeine experiment" activity in Chapter 4 (page 301) asked you to shuffle and deal piles of cards to mimic the random assignment of subjects to treatments. The "1 in 6 wins" game that opened this chapter had you roll a die several times to simulate buying 20-ounce sodas and looking under the cap.

The goal in each of these cases was to use simulation to answer a question of interest about some random process. Different random "devices" were used to perform the simulations—beads, slips of paper, cards, or dice. But the same basic strategy was followed each time.

THE SIMULATION PROCESS

- Describe how to use a random process to perform one trial of the simulation. Tell what you will record at the end of each trial.
- Perform many trials of the simulation.
- Use the results of your simulation to answer the question of interest.

For the 1-in-6 wins game, we wanted to estimate the probability of getting two or fewer prize winners in a class of 30 students if the company's 1-in-6 wins claim is true.

- We rolled a six-sided die 30 times to determine the outcome for each person's bottle of soda: 6 = wins a prize, 1 to 5 = no prize, and recorded the number of winners. The dotplot shows the number of winners in 40 trials of this simulation.

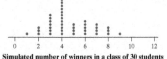

Simulated number of winners in a class of 30 students

- In 4 of the 40 trials, two or fewer of the students won a prize. So our estimate of the probability is 4/40 = 0.10 = 10%. According to these results, getting 2 winners isn't very likely, but it isn't unusual enough to conclude that the company is lying.

EXAMPLE

NASCAR cards and cereal boxes
Performing simulations

PROBLEM: In an attempt to increase sales, a breakfast cereal company decides to offer a NASCAR promotion. Each box of cereal will contain a collectible card featuring one of the following NASCAR drivers: Joey Lagano, Kevin Harvick, Chase Elliott, Danica Patrick, or Jimmie Johnson. The company claims that each of the 5 cards is equally likely to appear in any box of cereal. A NASCAR fan decides to keep buying boxes of the cereal until she has all 5 drivers' cards. She is surprised when it takes her 23 boxes to get the full set of cards. Does this outcome provide convincing evidence that the 5 cards are not equally likely? To help answer this question, we want to perform a simulation to estimate the probability that it will take 23 or more boxes to get a full set of 5 NASCAR collectible cards.

(a) Describe how to use a random number generator to perform one trial of the simulation.

The dotplot shows the number of cereal boxes it took to get all 5 drivers' cards in 50 trials.

(b) Explain what the dot at 20 represents.

Simulated number of boxes required

(c) Use the results of the simulation to estimate the probability that it will take 23 or more boxes to get a full set of cards. Does this outcome provide convincing evidence that the 5 cards are not equally likely?

The birthday problem Performing simulations

Skill 3.A

PROBLEM:
In a certain AP® Statistics class of 24 students, two of the students discovered they share the same birthday. Surprised by these results, the students decide to perform a simulation to estimate the probability that a class of 24 students has at least two students with the same birthday.
(a) Assume that birthdays are randomly distributed throughout the year (and ignore leap years). Describe how you would use a random number generator to carry out this simulation.

The dotplot shows the number of students who share a birthday with another student in a class of 24 students in 50 trials. There may be multiple sets of matching birthdays in each simulated class.

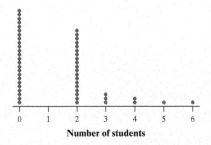

Number of students

(b) Explain what the dot at 5 represents.
(c) Use the results of the simulation to estimate the probability that a class of 24 students has at least two students with the same birthday. Were the results from this class surprising or unusual? Explain your answer.

SOLUTION:
(a) Label each day of the year with a number from 1 to 365. Use the random number generator to create a list of 24 numbers between 1 and 365, which will represent the birthdays of the 24 students. Record the number of students in the class who share a birthday with a fellow student.
(b) The dot at 5 represents one trial in which 5 students in the class shared a birthday with a fellow student.

(c) Probability $\approx \dfrac{26}{50} = 0.52$, so there's

about a 52% chance that a class of 24 students will have at least two students with the same birthday, assuming birthdays are randomly distributed. Because this probability isn't small, we should not be surprised by this class's results.

Preparing for Inference

The NASCAR example gets students thinking about the strength of evidence against a claim. In a significance test, the claim that the company is telling the truth is the *null hypothesis*. We will assume this claim is true, then find the probability of getting our result (23 cereal boxes) or more purely by chance (*P-value*). If the probability is less than 5% (α, *alpha*), we have convincing evidence that the company is not telling the truth. If the probability is greater than 5%, the NASCAR fan's results could have happened purely by chance (even if the company was being truthful). The decision that the company is not being truthful could be incorrect. In Chapter 9, we will see that this is a *Type I Error* (the null hypothesis is true and we incorrectly reject it).

ALTERNATE EXAMPLE   Skill 3.A

Names in a hat   Performing simulations

PROBLEM:
Suppose I want to choose a simple random sample of size 6 from a group of 60 seniors and 30 juniors. I want to be sure that my sample has 4 seniors and 2 juniors. To do this, I write each person's name on a slip of equal-size paper and mix the slips in a large hat. I will select names one at a time from the bag until I get 4 seniors and 2 juniors (i.e., after 4 seniors are selected, the remaining seniors will be discarded, and after 2 juniors are selected, the remaining juniors will be discarded). It takes 14 name selections to satisfy the requirement of getting 4 seniors and 2 juniors. Does this outcome provide convincing evidence that the names were not properly shuffled? To help answer this question, we want to perform a simulation to estimate the probability that it will take 14 or more selections to get 4 seniors and 2 juniors.
(a) Describe how to use Table D to perform one trial of the simulation.
(b) Perform 2 trials of the simulation using the random digits below. Make your procedure clear so that someone can follow what you did.

15185   39519   12834   15114
51978   76110   90435   66442

(c) In 50 trials of the simulation, there was 1 instance for which the number of name selections was 14 or greater. Do these results give convincing evidence that the names were not properly shuffled?

SOLUTION:

(a) Let 1 = Joey Lagano, 2 = Kevin Harvick, 3 = Chase Elliott, 4 = Danica Patrick, and 5 = Jimmie Johnson. Generate a random integer from 1 to 5 to simulate buying one box of cereal and looking at which card is inside. Keep generating random integers until all five labels from 1 to 5 appear. Record the number of boxes it takes to get all 5 cards.

> Describe how to use a random process to perform one trial of a simulation. Tell what you will record at the end of each trial.

(b) One trial where it took 20 boxes to get all 5 drivers' cards.

(c) Probability $\approx 0/50 = 0$, so there's about a 0% chance it would take 23 or more boxes to get a full set. Because it is so unlikely that it would take 23 or more boxes to get a full set when the drivers' cards are equally likely, this result provides convincing evidence that the 5 NASCAR drivers' cards are not equally likely to appear in each box of cereal.

FOR PRACTICE, TRY EXERCISE 9

It took our NASCAR fan 23 boxes to complete the set of 5 cards. Does that mean the company didn't tell the truth about how the cards were distributed? Not necessarily. Our simulation says that it's very unlikely for someone to have to buy 23 boxes to get a full set *if* each card is equally likely to appear in a box of cereal. The evidence suggests that the company's statement is incorrect. It's still possible, however, that the NASCAR fan was just very unlucky.

Here's one more example that shows the simulation process in action.

EXAMPLE

Golden ticket parking lottery
Performing simulations

PROBLEM: At a local high school, 95 students have permission to park on campus. Each month, the student council holds a "golden ticket parking lottery" at a school assembly. The two lucky winners are given reserved parking spots next to the school's main entrance. Last month, the winning tickets were drawn by a student council member from the AP® Statistics class. When both golden tickets went to members of that same class, some people thought the lottery had been rigged. There are 28 students in the AP® Statistics class, all of whom are eligible to park on campus. We want to perform a simulation to estimate the probability that a fair lottery would result in two winners from the AP® Statistics class.

(a) Describe how you would use a table of random digits to carry out this simulation.

(b) Perform 3 trials of the simulation using the random digits given. Make your procedure clear so that someone can follow what you did.

70708   41098   55181   94904   43563   56934   48394   51719

(c) In 9 of the 100 trials of the simulation, both golden tickets were won by members of the AP® Statistics class. Do these results give convincing evidence that the lottery was not carried out fairly? Explain your reasoning.

SOLUTION:

(a) Label the 60 seniors 01–60 and the 30 juniors 61–90. Use pairs of digits from Table D where repeats, 00, and 91–99 are skipped. Moving left to right across a row, we'll look at pairs of digits until we have 4 *different* labels from 01–60 and 2 *different* labels from 61–90. Then we will count how many *different* labels from 01–90 we looked at.

(b)

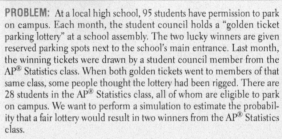

15	18	53	95	19	12	83	41	51	14	51	97	87	61	10	90	43	56	64	42
Sr	Sr	Sr	skip	Sr	Sr	Jr	Sr	Sr	Sr	skip	skip	Jr	Jr	Sr	Jr	Sr	Sr	Jr	Sr

10 selections | 7 selections

There were 0 trials of the first 2 in which the number of selections was 14 or greater.

(c) Yes. There's about a $\frac{1}{50} = 2\%$ chance that it will take 14 or more selections if the names are properly shuffled. Because this probability is small, we have convincing evidence that the names were not properly shuffled.

SOLUTION:

(a) Label the students in the AP® Statistics class from 01 to 28, and label the remaining students from 29 to 95. Numbers from 96 to 00 will be skipped. Moving left to right across a row, look at pairs of digits until we come across two *different* labels from 01 to 95. The two students with these labels win the prime parking spaces. Record whether or not both winners come from the AP® Statistics class. Perform many simulated lotteries. See what percent of the time both winners come from this statistics class.

> Describe how to use a random process to perform one trial of the simulation. Tell what you will record at the end of each trial.

(b)

> Perform many trials.

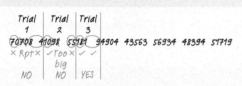

Trial 1	Trial 2	Trial 3					
70708	41098	55181	94904	43563	56934	48394	51719
× Rpt ×	✓Too× big	× ✓ ✓					
NO	NO	YES					

There was one trial out of the first 3 in which both golden parking tickets went to members of the AP® Statistics class.

> Use the results of your simulation to answer the question of interest.

(c) No; there's about a 9% chance of getting both winners from the AP® Statistics class in a fair lottery. Because this probability isn't very small, we don't have convincing evidence that the lottery was unfair. Outcomes like this could occur by chance alone in a fair lottery.

> Does that mean the lottery *was* conducted fairly? Not necessarily.

FOR PRACTICE, TRY EXERCISE 11

In the golden ticket lottery example, we ignored repeated numbers from 01 to 95 within a given trial. That's because the random process involved sampling students *without* replacement. In the NASCAR example, we allowed repeated numbers from 1 to 5 in a given trial. That's because we were selecting a small number of cards from a very large population of cards in thousands of cereal boxes. So the probability of getting, say, a Danica Patrick card in the next box of cereal was still very close to 1/5 even if we had already selected a Danica Patrick card.

CHECK YOUR UNDERSTANDING

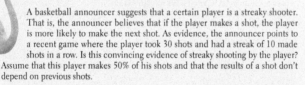

A basketball announcer suggests that a certain player is a streaky shooter. That is, the announcer believes that if the player makes a shot, the player is more likely to make the next shot. As evidence, the announcer points to a recent game where the player took 30 shots and had a streak of 10 made shots in a row. Is this convincing evidence of streaky shooting by the player? Assume that this player makes 50% of his shots and that the results of a shot don't depend on previous shots.

1. Describe how you would carry out a simulation to estimate the probability that a 50% shooter who takes 30 shots in a game would have a streak of 10 or more made shots.

Preparing for Inference

The golden ticket example gets students thinking about the strength of evidence against a claim. In a significance test, the claim that the lottery was conducted fairly is the *null hypothesis*. We will assume the claim is true, then find the probability of getting our result (2 AP® Statistics students chosen) purely by chance (*P-value*). If the probability is less than 5% (α, *alpha*), we have convincing evidence that the lottery was not fair. If the probability is greater than 5%, the results could have happened purely by chance (even if the lottery was fair). In this example, we did not have convincing evidence that the lottery was unfair. It could be the case that the lottery was unfair and we just made an incorrect decision based on our results. In Chapter 9, we will see that this constitutes a *Type II Error* (the alternative hypothesis is true and we fail to reject the null hypothesis).

Teaching Tip

Notice in this example that we state the probability is *about* 9% rather than saying the probability *is* 9%. When performing a simulation, emphasize that the probability is always an estimation.

✓ Answers to CYU

1. To carry out a simulation estimating the probability that a 50% shooter who takes 30 shots in a game would have a streak of 10 or more shots, begin by assigning digits to the outcomes. Let 1 = make a shot and let 2 = miss a shot. Generate 30 random integers from 1 to 2 to simulate taking 30 shots. Record whether or not there are a series of at least 10 "makes" in a row among the 30 shots. Perform many trials of this simulation. See what percent of the time there is a streak of at least 10 "makes" among the 30 shots.

Answers to CYU (continued)

2. Two trials resulted in 9 consecutive "makes" among the 30 shots.

3. Based on the dotplot, there was only one trial out of 50 simulated games in which the player had at least 10 "makes" in a row among the 30 shots. If the player's shot percentage truly is 50% and if the result of a shot truly does not depend on previous shots, then in 50 trials we would expect this player to have a streak of 10 or more makes only about 2% of the time (1 out of 50). Because this phenomenon is unlikely to happen strictly due to chance alone, and the announcer observed the player making 10 shots in a row in a recent game, the announcer is justified in claiming that the player is streaky.

TRM Additional Simulation Problems

See the Teacher's Resource Materials for some additional simulation problems you can use with your class. Find them in the Teacher's Resource Materials located in the TE-Book, on the TRFD, or in the Teacher's view in the book's digital platform.

TRM Section 5.1 Quiz

There is one quiz available for this section in the Teacher's Resource Materials. Click on the link in the TE-Book, look on the TRFD, or download from the Teacher's Resources on the book's digital platform. You can also create your own quiz using the ExamView® Assessment Suite that is part of the TPS6 program. Questions are coded by Learning Target to make it easy to build parallel quizzes.

Teaching Tip

Make sure to point out the two icons next to Exercise 1. The blue rectangle reminds students that the example on page 330 is very similar to this exercise. The "play" icon reminds students that there is a video solution available in the student e-Book or at the Student Site.

TRM Full Solutions to Section 5.1 Exercises

Click on the link in the TE-Book, open the TRFD, or download from the Teacher's Resources on the book's digital platform.

The dotplot displays the results of 50 simulated games in which this player took 30 shots.

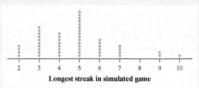

Longest streak in simulated game

2. Explain what the two dots above 9 indicate.

3. What conclusion would you draw about whether this player was streaky? Explain your answer.

Section 5.1 | Summary

- A **random process** generates outcomes that are determined purely by chance. Random behavior is unpredictable in the short run but has a regular and predictable pattern in the long run.
- The long-run relative frequency of an outcome after many trials of a random process is its **probability.** A probability is a number between 0 (never occurs) and 1 (always occurs).
- The **law of large numbers** says that in many trials of the same random process, the proportion of times that a particular outcome occurs will approach its probability.
- **Simulation** can be used to imitate a random process and to estimate probabilities. To perform a simulation:
  - Describe how to use a random process to perform one trial of the simulation. Tell what you will record at the end of each trial.
  - Perform many trials of the simulation.
  - Use the results of your simulation to answer the question of interest.

Section 5.1 | Exercises

1. **Another commercial** If Aaron tunes into his favorite radio station at a randomly selected time, there is a 0.20 probability that a commercial will be playing.

   (a) Interpret this probability as a long-run relative frequency.

   (b) If Aaron tunes into this station at 5 randomly selected times, will there be exactly 1 time when a commercial is playing? Explain your answer.

2. **Genetics** There are many married couples in which the husband and wife both carry a gene for cystic fibrosis but don't have the disease themselves. Suppose we select one of these couples at random. According to the laws of genetics, the probability that their first child will develop cystic fibrosis is 0.25.

   (a) Interpret this probability as a long-run relative frequency.

   (b) If researchers randomly select 4 such couples, is one of these couples guaranteed to have a first child who develops cystic fibrosis? Explain your answer.

3. **Mammograms** Many women choose to have annual mammograms to screen for breast cancer after age 40. A mammogram isn't foolproof. Sometimes the test suggests that a woman has breast cancer when she really doesn't (a "false positive"). Other times the test says that

---

Answers to Section 5.1 Exercises

5.1 (a) If you take a very large random sample of times that Aaron tunes into his favorite radio station, about 20% of the time a commercial will be playing. **(b)** No, chance behavior is unpredictable in the short run, but has a regular and predictable pattern in the long run. The value 0.20 describes the proportion of times that a commercial will be playing when Aaron tunes in to the station in a very long series of trials.

5.2 (a) If you take a very large random sample of couples in which the husband and wife both carry a gene for cystic fibrosis but do not have the disease themselves, about 25% of such couples will find that their first child develops cystic fibrosis. **(b)** No; chance behavior is unpredictable in the short run, but has a regular and predictable pattern in the long run. The value 0.25 describes the proportion of times that such couples will have a first child who develops cystic fibrosis in a very long series of trials.

a woman doesn't have breast cancer when she actually does (a "false negative"). Suppose the false negative rate for a mammogram is 0.10.

(a) Explain what this probability means.

(b) Which is a more serious error in this case: a false positive or a false negative? Justify your answer.

4. **Liar, liar!** Sometimes police use a lie detector test to help determine whether a suspect is telling the truth. A lie detector test isn't foolproof—sometimes it suggests that a person is lying when he or she is actually telling the truth (a "false positive"). Other times, the test says that the suspect is being truthful when he or she is actually lying (a "false negative"). For one brand of lie detector, the probability of a false positive is 0.08.

(a) Explain what this probability means.

(b) Which is a more serious error in this case: a false positive or a false negative? Justify your answer.

5. **Three pointers** The figure shows the results of a basketball player attempting many 3-point shots. Explain what this graph tells you about random behavior in the short run and long run.

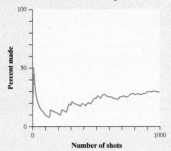

Number of shots

6. **Keep on tossing** The figure shows the results of two different sets of 5000 coin tosses. Explain what this graph tells you about random behavior in the short run and the long run.

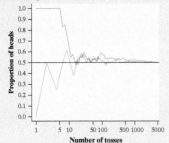

Number of tosses

7. **An unenlightened gambler**

(a) A gambler knows that red and black are equally likely to occur on each spin of a roulette wheel. He observes that 5 consecutive reds have occurred and bets heavily on black at the next spin. Asked why, he explains that "black is due." Explain to the gambler what is wrong with this reasoning.

(b) After hearing you explain why red and black are still equally likely after 5 reds on the roulette wheel, the gambler moves to a card game. He is dealt 5 straight red cards from a standard deck with 26 red cards and 26 black cards. He remembers what you said and assumes that the next card dealt in the same hand is equally likely to be red or black. Explain to the gambler what is wrong with this reasoning.

8. **Due for a hit** A very good professional baseball player gets a hit about 35% of the time over an entire season. After the player failed to hit safely in six straight at-bats, a TV commentator said, "He is due for a hit." Explain why the commentator is wrong.

9. **Will Luke pass the quiz?** Luke's teacher has assigned each student in his class an online quiz, which is made up of 10 multiple-choice questions with 4 options each. Luke hasn't been paying attention in class and has to guess on each question. However, his teacher allows each student to take the quiz three times and will record the highest of the three scores. A passing score is 6 or more correct out of 10. We want to perform a simulation to estimate the score that Luke will earn on the quiz if he guesses at random on all the questions.

pg 333

(a) Describe how to use a random number generator to perform one trial of the simulation.

The dotplot shows Luke's simulated quiz score in 50 trials of the simulation.

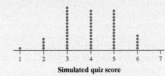

Simulated quiz score

(b) Explain what the dot at 1 represents.

(c) Use the results of the simulation to estimate the probability that Luke passes the quiz.

(d) Doug is in the same class and claims to understand some of the material. If he scored 8 points on the quiz, is there convincing evidence that he understands some of the material? Explain your answer.

10. **Double fault!** A professional tennis player claims to get 90% of her second serves in. In a recent match, the

---

5.7 (a) The wheel is not affected by its past outcomes—it has no memory. So on any one spin, black and red remain equally likely. **(b)** Removing a card changes the composition of the remaining deck. If you hold 5 red cards, the deck now contains 5 fewer red cards, so the probability of being dealt another red card decreases.

5.8 No, the TV commentator is incorrectly applying the law of large numbers to a small number of at-bats for the player. Chance behavior is unpredictable in the short run, but it has a regular and predictable pattern in the long run. The value 35% describes the percent of the time the player gets a hit in a very long series of trials.

5.9 (a) To carry out a simulation estimating the score that Luke will earn on the quiz, begin by assigning digits to the outcomes. For each question, let 1 = guessed correctly and let 2, 3, and 4 = guessed incorrectly. Generate 10 random integers from 1 to 4 to simulate the result of Luke's guess for each of the 10 questions. Record the number of correct guesses (1s). Repeat this 3 times and record the highest of the three scores. **(b)** One trial resulted in a highest quiz score of 1 correct response out of 10. **(c)** Based on the simulation, the probability that Luke passes the quiz (i.e., scores at least 6 out of 10) is $5/50 = 10\%$. Five of the 50 trials show a score of at least 6. **(d)** Yes; it is unlikely that Doug would score higher than 6 out of 10 based on randomly guessing the answers. Since Doug scored 8 points out of 10, there is reason to believe that Doug does understand some of the material.

5.3 (a) If you take a very large random sample of women with breast cancer, about 10% of the time the mammogram will indicate that the woman does not have breast cancer, when, in fact, she does have breast cancer. **(b)** A false negative is a more serious error because a woman who doesn't know she has breast cancer will not get the potentially life-saving treatment she needs. A false positive would result in temporary stress until a more thorough examination is performed.

5.4 (a) If you take a very large random sample of truthful people and give them a lie detector test, about 8% of the time the test will indicate that the truthful person is lying. **(b)** Answers will vary. A false positive would mean that a person telling the truth would be found to be lying. A false negative would mean that a person lying would be found to be telling the truth. An assumption of innocence until proven guilty beyond a reasonable doubt implies that a false positive would be worse—that is, saying someone is guilty (lying) when he is innocent is worse than finding someone to be truthful when he is guilty (lying).

5.5 In the short run, the percentage of made 3-point shots varied quite a bit. However, the percentage of made 3-point shots became less variable and approached about 0.30 as the number of shots increased.

5.6 In the short run, the proportion of heads varied a lot. In the long run, this proportion became less variable and settled down around 0.50 for both sets of 5000 tosses.

5.10 (a) To estimate the probability that the tennis player would miss > 5 of her first 20 second serves, assign digits to the outcomes. Let 0 = miss the second serve and let 1–9 = make the second serve. Generate 20 random integers from 0 to 9 to simulate making or missing each of her first 20 second serves. Record the second serve misses (the 0s). **(b)** One trial = 6/20 missed second serves. **(c)** A 10% chance of missing each second serve = missing ≥ 5 of her first 20 second serves, so 7/100 = 7%. Seven of the 100 trials show at least 5 missed second serves. **(d)** There is not convincing evidence at the 5% significance level that the player's claim is false.

5.11 (a) To use a table of random digits, number the first-class passengers 01–12 and the other passengers 13–76. Moving left to right across a row, look at pairs of digits until you have 10 unique numbers (no repetitions). Count the number of two-digit numbers between 01 and 12. Perform many trials of this simulation. Determine what percent of the trials showed no first-class passengers selected for screening. **(b)** The numbers, read in pairs, are **71 48 70** 99 84 **29 07** 71 48 **63 61 68 34** 70 **52**. The bold numbers indicate people who have been selected. The other numbers are either too large (over 76) or have already been selected. There is one person among the 10 selected who is in first class in this sample (underlined). **(c)** There is not convincing evidence that the TSA officers did not carry out a truly random selection. If the selection is truly random, there is about a 15% chance that no one in first class will be selected. So it is not surprising that a single random selection would contain no first-class passengers.

5.12 (a) Let the numbers 01–42 represent the vowels, 43–98 represent the consonants, and 99 and 00 represent the blank tiles. Moving left to right across a row, identify 7 unique numbers. Record whether all 7 numbers are between 01 and 42. Perform many trials of this simulation. Determine what percent of the trials had all 7 numbers between 01 and 42. **(b)** The numbers, read in pairs, are 00 69 **40** 59 77 **19** 66. The two numbers in bold are vowels, so the sample is not all vowels. **(c)** There is convincing evidence that the bag of tiles was not well mixed. If the bag was mixed well, there is about a 0.2% chance of getting 7 tiles that are all

---

player missed 5 of her 20 second serves. Is this a surprising result if the player's claim is true? Assume that the player has a 0.10 probability of missing each second serve. We want to carry out a simulation to estimate the probability that she would miss 5 or more of her 20 second serves.

(a) Describe how to use a random number generator to perform one trial of the simulation.

The dotplot displays the number of second serves missed by the player out of 20 second serves in 100 simulated matches.

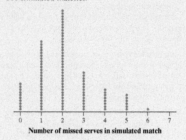

Number of missed serves in simulated match

(b) Explain what the dot at 6 represents.

(c) Use the results of the simulation to estimate the probability that the player would miss 5 or more of her 20 second serves in a match.

(d) Is there convincing evidence that the player misses more than 10% of her second serves? Explain your answer.

11. **Airport security** The Transportation Security Administration (TSA) is responsible for airport safety. On some flights, TSA officers randomly select passengers for an extra security check prior to boarding. One such flight had 76 passengers—12 in first class and 64 in coach class. Some passengers were surprised when none of the 10 passengers chosen for screening were seated in first class. We want to perform a simulation to estimate the probability that no first-class passengers would be chosen in a truly random selection.

(a) Describe how you would use a table of random digits to carry out this simulation.

(b) Perform one trial of the simulation using the random digits that follow. Copy the digits onto your paper and mark directly on or above them so that someone can follow what you did.

71487 09984 29077 14863 61683 47052 62224 51025

(c) In 15 of the 100 trials of the simulation, none of the 10 passengers chosen was seated in first class. Does this result provide convincing evidence that the TSA officers did not carry out a truly random selection? Explain your answer.

12. **Scrabble** In the game of Scrabble, each player begins by randomly selecting 7 tiles from a bag containing 100 tiles. There are 42 vowels, 56 consonants, and 2 blank tiles in the bag. Cait chooses her 7 tiles and is surprised to discover that all of them are vowels. We want to perform a simulation to determine the probability that a player will randomly select 7 vowels.

(a) Describe how you would use a table of random digits to carry out this simulation.

(b) Perform one trial of the simulation using the random digits given. Copy the digits onto your paper and mark directly on or above them so that someone can follow what you did.

00694 05977 19664 65441 20903 62371 22725 53340

(c) In 2 of the 1000 trials of the simulation, all 7 tiles were vowels. Does this result give convincing evidence that the bag of tiles was not well mixed?

13. **Bull's-eye!** In a certain archery competition, each player continues to shoot until he or she misses the center of the target twice. Quinn is one of the archers in this competition. Based on past experience, she has a 0.60 probability of hitting the center of the target on each shot. We want to design a simulation to estimate the probability that Quinn stays in the competition for at least 10 shots. Describe how you would use each of the following chance devices to perform one trial of the simulation.

(a) Slips of paper

(b) Random digits table

(c) Random number generator

14. **Free-throw practice** At the end of basketball practice, each player on the team must shoot free throws until he makes 10 of them. Dwayne is a 70% free-throw shooter. That is, his probability of making any free throw is 0.70. We want to design a simulation to estimate the probability that Dwayne makes 10 free throws in at most 12 shots. Describe how you would use each of the following chance devices to perform one trial of the simulation.

(a) Slips of paper

(b) Random digits table

(c) Random number generator

---

vowels. Because it is very unlikely to get a sample like this by chance, we have reason to believe the bag of tiles was not well mixed.

5.13 (a) Label 10 identically sized pieces of paper 0–9. Let 0–5 represent hitting the center of the target and let 6–9 represent not hitting the center. Place the slips of paper into a hat and mix well. Draw out one slip of paper at a time and record the digit. Replace the paper and mix well. Repeat this process until Quinn misses the center of the target twice. Determine whether she remained in the competition for at least 10 shots. **(b)** Using a random digits table, let 0–5 represent hitting the center of the target and let 6–9 represent

not hitting the center of the target. Moving left to right across a row, read single digits and record whether she hit the center of the target or not. Repeat this process until Quinn misses the center of the target twice. Determine whether she remained in the competition for at least 10 shots. **(c)** Using a random number generator, let 0–5 represent hitting the center of the target and let 6–9 represent not hitting the center of the target. Generate a random integer from 0–9 and record whether she hit the center of the target or not. Repeat this process until Quinn misses the center of the target twice. Determine whether she remained in the competition for at least 10 shots.

In Exercises 15–18, determine whether the simulation design is valid. Justify your answer.

15. **Smartphone addiction?** A media report claims that 50% of U.S. teens with smartphones feel addicted to their devices.[3] A skeptical researcher believes that this figure is too high. She decides to test the claim by taking a random sample of 100 U.S. teens who have smartphones. Only 40 of the teens in the sample feel addicted to their devices. Does this result give convincing evidence that the media report's 50% claim is too high? To find out, we want to perform a simulation to estimate the probability of getting 40 or fewer teens who feel addicted to their devices in a random sample of size 100 from a very large population of teens with smartphones in which 50% feel addicted to their devices.

    Let 1 = feels addicted and 2 = doesn't feel addicted. Use a random number generator to produce 100 random integers from 1 to 2. Record the number of 1's in the simulated random sample. Repeat this process many, many times. Find the percent of trials on which the number of 1's was 40 or less.

16. **Lefties** A website claims that 10% of U.S. adults are left-handed. A researcher believes that this figure is too low. She decides to test this claim by taking a random sample of 20 U.S. adults and recording how many are left-handed. Four of the adults in the sample are left-handed. Does this result give convincing evidence that the website's 10% claim is too low? To find out, we want to perform a simulation to estimate the probability of getting 4 or more left-handed people in a random sample of size 20 from a very large population in which 10% of the people are left-handed.

    Let 00 to 09 indicate left-handed and 10 to 99 represent right-handed. Move left to right across a row in Table D. Each pair of digits represents one person. Keep going until you get 20 different pairs of digits. Record how many people in the simulated sample are left-handed. Repeat this process many, many times. Find the proportion of trials in which 4 or more people in the simulated sample were left-handed.

17. **Notebook check** Every 9 weeks, Mr. Millar collects students' notebooks and checks their homework. He randomly selects 4 different assignments to inspect for all of the students. Marino is one of the students in Mr. Millar's class. Marino completed 20 homework assignments and did not complete 10 assignments. He is surprised when Mr. Millar only selects 1 assignment that he completed. Should he be surprised? To find out, we want to design a simulation to estimate the probability that Mr. Millar will randomly select 1 or fewer of the homework assignments that Marino completed.

    Get 30 identical slips of paper. Write "N" on 10 of the slips and "C" on the remaining 20 slips. Put the slips into a hat and mix well. Draw 1 slip without looking to represent the first randomly selected homework assignment, and record whether Marino completed it. Put the slip back into the hat, mix again, and draw another slip representing the second randomly selected assignment. Record whether Marino completed this assignment. Repeat this process two more times for the third and fourth randomly selected homework assignments. Record the number out of the 4 randomly selected homework assignments that Marino completed in this trial of the simulation. Perform many trials. Find the proportion of trials in which Mr. Millar randomly selects 1 or fewer of the homework assignments that Marino completed.

18. **Random assignment** Researchers recruited 20 volunteers—8 men and 12 women—to take part in an experiment. They randomly assigned the subjects into two groups of 10 people each. To their surprise, 6 of the 8 men were randomly assigned to the same treatment. Should they be surprised? We want to design a simulation to estimate the probability that a proper random assignment would result in 6 or more of the 8 men ending up in the same group.

    Get 20 identical slips of paper. Write "M" on 8 of the slips and "W" on the remaining 12 slips. Put the slips into a hat and mix well. Draw 10 of the slips without looking and place into one pile representing Group 1. Place the other 10 slips in a pile representing Group 2. Record the largest number of men in either of the two groups from this simulated random assignment. Repeat this process many, many times. Find the percent of trials in which 6 or more men ended up in the same group.

19. **Color-blind men** About 7% of men in the United States have some form of red–green color blindness. Suppose we randomly select one U.S. adult male at a time until we find one who is red–green color-blind. Should we be surprised if it takes us 20 or more men? Describe how you would carry out a simulation to estimate the probability that we would have to randomly select 20 or more U.S. adult males to find one who is red–green color-blind. *Do not perform the simulation.*

20. **Taking the train** According to New Jersey Transit, the 8:00 A.M. weekday train from Princeton to New York City has a 90% chance of arriving on time. To test this claim, an auditor chooses 6 weekdays at random during a month to ride this train. The train arrives late on 2 of those days. Does the auditor have convincing evidence that the company's claim is false? Describe how you would carry out a simulation to estimate the probability that a train with a 90% chance of arriving on time each day would be late on 2 or more of 6 days. *Do not perform the simulation.*

5.18 Legitimate simulation. By using 8 slips of paper labeled "M" and 12 slips of paper labeled "W," the simulation would model the fact that there were 8 men and 12 women who are being randomly assigned to two groups of 10.

5.19 To conduct a simulation estimating the probability that we would have to randomly select 20 or more U.S. adult males to find one who is red–green color-blind, let 0–6 = color-blind and let 7–99 = not color-blind. Use technology to pick an integer from 0 to 99 until we get a number between 0 and 6. Count how many numbers are in the sample. Repeat many times. Determine the percent of the time we would have to randomly select 20 or more U.S. adult males to find one who is red–green color-blind.

5.20 To conduct a simulation estimating the probability that a train with a 90% chance of arriving on time each day would be late on 2 or more of 6 days, let the digits 1–9 represent days that the train arrives on time and let the digit 0 represent days when the train does not arrive on time. Use technology to pick an integer from 0–9. Repeat this 6 times. Record whether there were at least two 0s among the 6 digits. Repeat this process many times. Determine the probability that a train with a 90% chance of arriving on time each day would be late on 2 or more of 6 days.

5.14 (a) Label 10 identically sized pieces of paper 0–9. Let 0–6 represent Dwayne making the shot and let 7–9 represent missing the shot. Place the slips of paper into a hat, mix well, draw one slip of paper at a time, and record whether he made or missed the shot. Replace the paper and mix well. Repeat 12 times. Determine whether Dwayne made 10 free throws within 12 shots. **(b)** Using a random digits table, let 0–6 represent Dwayne making the shot and let 7–9 represent Dwayne missing the shot. Moving left to right across a row, read single digits and record whether he made or missed the shot. Repeat this process 12 times. Determine whether Dwayne made 10 free throws within 12 shots. **(c)** Using a random number generator, let 0–6 represent Dwayne making the shot and let 7–9 represent Dwayne missing the shot. Generate a random integer from 0–9 and record whether he made or missed the shot. Repeat this process 12 times. Determine whether Dwayne made 10 free throws within 12 shots.

5.15 Legitimate simulation. By letting 1 = feel addicted and 2 = doesn't feel addicted, the simulation would model the claim that 50% of U.S. teens with smartphones feel addicted to their devices.

5.16 Not a legitimate simulation. As written, the simulation calls for "20 different pairs of digits" to be selected from Table D. Because the digits in the simulation represent left-handedness or right-handedness, the digits should be selected with replacement .

5.17 Not a legitimate simulation. When simulating the selection of homework assignments, the selection process would occur without replacement.

5.21 (a) In the simulation, 43 of the 200 samples yielded a sample proportion of at least 0.55. Obtaining a sample proportion of 0.55 or higher is not particularly unusual when 50% of all students recycle. **(b)** Only 1 of the 200 samples yielded a sample proportion of at least 0.63. This means that if 50% of all students recycle, we would see a sample proportion of at least 0.63 in only about 0.5% of samples. Because getting a sample proportion of at least 0.63 is very unlikely, we have convincing evidence that the percentage of all students who recycle is larger than 50%.

5.22 (a) If 27 out of 60 say they leave the water running, this means the sample proportion of students who leave the water running is 0.45. In the simulation, 44 of the 200 resulted in a proportion of 0.45 or less that leave the water running. Obtaining a sample proportion of 0.45 or lower is not particularly unusual when 50% of all people brush with the water off. **(b)** However, if 18 out of 60 say they leave the water running, this means the sample proportion of students who leave the water running is 0.30. In the simulation, no samples had a proportion this low. Because obtaining a sample proportion of at most 0.30 is very unlikely, we have strong evidence that fewer than 50% of the school's students brush their teeth with the water off.

5.23 c

5.24 a

5.25 b

---

21. **Recycling** Do most teens recycle? To find out, an AP® Statistics class asked an SRS of 100 students at their school whether they regularly recycle. In the sample, 55 students said that they recycle. Is this convincing evidence that more than half of the students at the school would say they regularly recycle? The dotplot shows the results of taking 200 SRSs of 100 students from a population in which the true proportion who recycle is 0.50.

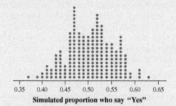

Simulated proportion who say "Yes"

(a) Explain why the sample result (55 out of 100 said "Yes") does not give convincing evidence that more than half of the school's students recycle.

(b) Suppose instead that 63 students in the class's sample had said "Yes." Explain why this result would give convincing evidence that a majority of the school's students recycle.

22. **Brushing teeth, wasting water?** A recent study reported that fewer than half of young adults turn off the water while brushing their teeth. Is the same true for teenagers? To find out, a group of statistics students asked an SRS of 60 students at their school if they usually brush with the water off. In the sample, 27 students said "Yes." The dotplot shows the results of taking 200 SRSs of 60 students from a population in which the true proportion who brush with the water off is 0.50.

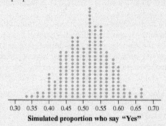

Simulated proportion who say "Yes"

(a) Explain why the sample result (27 of the 60 students said "Yes") does not give convincing evidence that fewer than half of the school's students brush their teeth with the water off.

(b) Suppose instead that 18 of the 60 students in the class's sample had said "Yes." Explain why this result would give convincing evidence that fewer than 50% of the school's students brush their teeth with the water off.

Multiple Choice: *Select the best answer for Exercises 23–28.*

23. You read in a book about bridge that the probability that each of the four players is dealt exactly 1 ace is approximately 0.11. This means that

(a) in every 100 bridge deals, each player has 1 ace exactly 11 times.

(b) in 1 million bridge deals, the number of deals on which each player has 1 ace will be exactly 110,000.

(c) in a very large number of bridge deals, the percent of deals on which each player has 1 ace will be very close to 11%.

(d) in a very large number of bridge deals, the average number of aces in a hand will be very close to 0.11.

(e) If each player gets an ace in only 2 of the first 50 deals, then each player should get an ace in more than 11% of the next 50 deals.

24. If I toss a fair coin 5 times and the outcomes are TTTTT, then the probability that tails appears on the next toss is

(a) 0.5.   (d) 0.

(b) less than 0.5.   (e) 1.

(c) greater than 0.5.

Exercises 25 to 27 refer to the following setting. A basketball player claims to make 47% of her shots from the field. We want to simulate the player taking sets of 10 shots, assuming that her claim is true.

25. To simulate the number of makes in 10 shot attempts, you would perform the simulation as follows:

(a) Use 10 random one-digit numbers, where 0–4 are a make and 5–9 are a miss.

(b) Use 10 random two-digit numbers, where 00–46 are a make and 47–99 are a miss.

(c) Use 10 random two-digit numbers, where 00–47 are a make and 48–99 are a miss.

(d) Use 47 random one-digit numbers, where 0 is a make and 1–9 are a miss.

(e) Use 47 random two-digit numbers, where 00–46 are a make and 47–99 are a miss.

26. A total of 25 trials of the simulation were performed. The number of makes in each set of 10 simulated shots was recorded on the dotplot. What is the approximate probability that a 47% shooter makes 5 or more shots in 10 attempts?

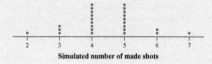

Simulated number of made shots

(a) 5/10   (b) 3/10   (c) 12/25

(d) 3/25   (e) 47/100

27. Suppose this player attempts 10 shots in a game and makes only 3 of them. Does this provide convincing evidence that she is less than a 47% shooter?

(a) Yes, because 3/10 (30%) is less than 47%.

(b) Yes, because she never made 47% of her shots in the simulation.

(c) No, because it is plausible (believable) that she would make 3 or fewer shots by chance alone.

(d) No, because the simulation was only repeated 25 times.

(e) No, because more than half of the simulated results were less than 47%.

28. Ten percent of U.S. households contain 5 or more people. You want to simulate choosing a household at random and recording "Yes" if it contains 5 or more people. Which of these is a correct assignment of digits for this simulation?

(a) Odd = Yes; Even = No

(b) 0 = Yes; 1–9 = No

(c) 0–5 = Yes; 6–9 = No

(d) 0–4 = Yes; 5–9 = No

(e) None of these

Recycle and Review

29. **AARP and Medicare** (4.1) To find out what proportion of Americans support proposed Medicare legislation to help pay medical costs, the AARP conducted a survey of their members (people over age 50 who pay membership dues). One of the questions was: "Even if this plan won't affect you personally either way,

do you think it should be passed so that people with low incomes or people with high drug costs can be helped?" Of the respondents, 75% answered "Yes."[4]

(a) Describe how undercoverage might lead to bias in this study. Explain the likely direction of the bias.

(b) Describe how the wording of the question might lead to bias in this study. Explain the likely direction of the bias.

30. **Waiting to park** (1.3, 4.2) Do drivers take longer to leave their parking spaces when someone is waiting? Researchers hung out in a parking lot and collected some data. The graphs and numerical summaries display information about how long it took drivers to exit their spaces.

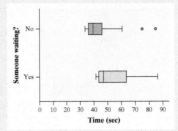

Descriptive Statistics: Time								
Waiting	n	Mean	StDev	Min	Q_1	Median	Q_3	Max
No	20	44.42	14.10	33.76	35.61	39.56	48.48	84.92
Yes	20	54.11	14.39	41.61	43.41	47.14	66.44	85.97

(a) Write a few sentences comparing these distributions.

(b) Can we conclude that having someone waiting causes drivers to leave their spaces more slowly? Why or why not?

SECTION 5.2   Probability Rules

LEARNING TARGETS   *By the end of the section, you should be able to:*

- Give a probability model for a random process with equally likely outcomes and use it to find the probability of an event.

- Use basic probability rules, including the complement rule and the addition rule for mutually exclusive events.

- Use a two-way table or Venn diagram to model a random process and calculate probabilities involving two events.

- Apply the general addition rule to calculate probabilities.

5.26 c

5.27 c

5.28 b

5.29 (a) This survey illustrates undercoverage because people who are less than 50 years old did not have a chance to participate in the survey. Younger people generally don't take as many prescription drugs and might be less willing to support a program like this than older people. The 75% who answered yes is likely an overestimate of the true proportion of Americans who support the program. **(b)** Including the additional information of "can be helped" might have encouraged respondents to say "Yes" to the program because they liked the idea of helping people. This would cause the 75% estimate to be higher than if the question were worded more neutrally.

5.30 (a) Both distributions are skewed to the right. There were no outliers for those with someone waiting, but there were two high outliers for those with no one waiting. The drivers generally took longer to leave when someone was waiting for the space (median = 47 seconds) than when no one was waiting (median = 40 seconds). There is more variability for the drivers with someone waiting as well—they had an *IQR* of 23.03 seconds as opposed to an *IQR* of 12.87 seconds for those with no one waiting. **(b)** No; the researchers merely observed what was happening. They did not randomly assign the treatments of either having a person waiting or not to the drivers of the cars leaving the lot, so we cannot conclude that having someone waiting causes drivers to leave their spaces more slowly.

PD   **Section 5.2 Overview**

To watch the video overview of Section 5.2 (for teachers), click on the link in the TE-Book, look on the TRFD, or download from the Teacher's Resources on the book's digital platform.

TRM   **Section 5.2 Alternate Examples**

You can find the Alternate Examples for this section in Microsoft Word format by clicking on the link in the TE-Book, opening the TRFD, or downloading from the Teacher's Resources on the book's digital platform.

The idea of probability rests on the fact that random behavior is predictable in the long run. In Section 5.1, we used simulation to imitate a random process. Do we always need to repeat a random process—tossing coins, rolling dice, drawing slips from a hat—many times to determine the probability of a particular outcome? Fortunately, the answer is no.

Probability Models

In Chapter 2, we saw that a Normal density curve could be used to model some distributions of quantitative data. In Chapter 3, we modeled linear relationships between two quantitative variables with a least-squares regression line. Now we're ready to develop a model for random behavior.

Many board games involve rolling dice. Imagine rolling two fair, six-sided dice—one that's red and one that's blue. How do we develop a **probability model** for this random process? Figure 5.2 displays the **sample space** of 36 possible outcomes. Because the dice are fair, each of these outcomes will be equally likely and have probability 1/36.

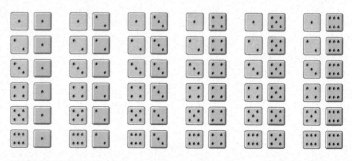

FIGURE 5.2 The 36 possible outcomes from rolling two six-sided dice, one red and one blue. Each of these equally likely outcomes has probability 1/36.

> **DEFINITION   Probability model, Sample space**
>
> A **probability model** is a description of some random process that consists of two parts: a list of all possible outcomes and the probability for each outcome.
>
> The list of all possible outcomes is called the **sample space**.

A sample space can be very simple or very complex. If we toss a coin once, there are only two possible outcomes in the sample space, heads and tails. When Gallup takes a random sample of 1523 U.S. adults and asks a survey question, the sample space consists of all possible sets of responses from 1523 of the over 240 million adults in the country.

A probability model does more than just assign a probability to each outcome. It allows us to find the probability of an **event**.

Teaching Tip

Some teachers prefer to write out sample spaces using set notation. For example, the sample space for rolling two fair, six-sided dice would be

S = {(1, 1), (1, 2), (1, 3), (1, 4), (1, 5), (1, 6),
(2, 1), (2, 2), (2, 3), (2, 4), (2, 5), (2, 6),
(3, 1), (3, 2), (3, 3), (3, 4), (3, 5), (3, 6),
(4, 1), (4, 2), (4, 3), (4, 4), (4, 5), (4, 6),
(5, 1), (5, 2), (5, 3), (5, 4), (5, 5), (5, 6),
(6, 1), (6, 2), (6, 3), (6, 4), (6, 5), (6, 6)}

Teaching Tip

In this chapter, students will learn several strategies for approaching probability questions:

1. Simulation
2. **Sample space**
3. Two-way tables
4. Venn diagrams
5. Tree diagrams
6. Formulas

Teaching Tip

There are 1.8×10^{8575} possible samples of size 1523 when choosing from 240 million adults, according to www.wolframalpha.com.

Making Connections

In Chapter 6, we will use a table to describe a probability model of all possible values for a random variable X and the probability for each value. We will call it the probability distribution of X.

X					
Probability					

DEFINITION Event

An **event** is any collection of outcomes from some random process.

Events are usually designated by capital letters, like A, B, C, and so on. For rolling two six-sided dice, we can define event A as getting a sum of 5. We write the probability of event A as P(A) or P(sum is 5).

It is fairly easy to find the probability of an event in the case of equally likely outcomes. There are 4 outcomes in event A:

The probability that event A occurs is therefore

$$P(A) = \frac{\text{number of outcomes with sum of 5}}{\text{total number of outcomes when rolling two dice}} = \frac{4}{36} = 0.111$$

FINDING PROBABILITIES: EQUALLY LIKELY OUTCOMES

If all outcomes in the sample space are equally likely, the probability that event A occurs can be found using the formula

$$P(A) = \frac{\text{number of outcomes in event A}}{\text{total number of outcomes in sample space}}$$

EXAMPLE

Spin the spinner
Probability models: Equally likely outcomes

PROBLEM: A spinner has three equal sections: red, blue, and yellow. Suppose you spin the spinner two times.
(a) Give a probability model for this random process.
(b) Define event A as spinning blue at least once. Find P(A).

SOLUTION:

(a) *Sample space: RR RB RY BR BB BY YR YB YY.*
Because the spinner has equal sections, each of these outcomes will be equally likely and have probability 1/9.

Remember: A probability model consists of a list of all possible outcomes and the probability of each outcome.

(b) *There are 5 outcomes with at least one blue:*
RB BR BB BY YB. So $P(A) = \frac{5}{9} = 0.556$.

If all outcomes in the sample space are equally likely,
$$P(A) = \frac{\text{number of outcomes in event A}}{\text{total number of outcomes in sample space}}$$

FOR PRACTICE, TRY EXERCISE 31

ALTERNATE EXAMPLE   Skill 3.A

The Perfect Outfit Probability models: Equally likely outcomes

PROBLEM:
The principal at a high school has three shirts to choose from (white, blue, and black) and four ties to choose from (red, white, green, and blue). One morning after a storm, there is no electricity and he chooses his outfit from the closet in the dark. In other words, he is randomly picking a shirt and tie.
(a) Give a probability model for the principal's random process.
(b) Define event A as the outfit has a matching-color shirt and tie. Find P(A).

SOLUTION:
(a) Sample space: white/red, white/white, white/green, white/blue, blue/red, blue/white, blue/green, blue/blue, black/red, black/white, black/green, black/blue. Because the principal is randomly choosing the shirt and tie, each of these outcomes will be equally likely and have probability 1/12.
(b) There are 2 outcomes with a matching-color shirt and tie: white/white, blue/blue.

So $P(A) = \dfrac{2}{12} = 0.17$.

Basic Probability Rules

Our work so far suggests three commonsense rules that a valid probability model must obey:

1. **If all outcomes in the sample space are equally likely, the probability that event A occurs is**

$$P(A) = \frac{\text{number of outcomes in event A}}{\text{total number of outcomes in sample space}}$$

2. **The probability of any event is a number between 0 and 1.** This rule follows from the definition of probability in Section 5.1: the proportion of times the event would occur in many trials of the random process. A proportion is a number between 0 and 1, so any probability is also a number between 0 and 1.

3. **All possible outcomes together must have probabilities that add up to 1.** Because some outcome must occur on every trial of a random process, the sum of the probabilities for all possible outcomes must be exactly 1.

Here's another rule that follows from the previous two:

4. **The probability that an event does not occur is 1 minus the probability that the event does occur.** If an event occurs in (say) 70% of all trials, it fails to occur in the other 30%. The probability that an event occurs and the probability that it does not occur always add to 100%, or 1. Earlier, we found that the probability of getting a sum of 5 when rolling two fair, six-sided dice is 4/36. What's the probability that the sum is *not* 5?

$$P(\text{sum is not 5}) = 1 - P(\text{sum is 5}) = 1 - \frac{4}{36} = \frac{32}{36} = 0.889$$

We refer to the event "not A" as the **complement** of A and denote it by A^C. For that reason, this handy result is known as the **complement rule**. Using the complement rule in this setting is much easier than counting all 32 possible ways to get a sum that isn't 5.

> Another common notation for the complement of event A is A′.

DEFINITION   **Complement rule, Complement**

The **complement rule** says that $P(A^C) = 1 - P(A)$, where A^C is the **complement** of event A; that is, the event that A does not occur.

Let's consider one more event involving the random process of rolling two fair, six-sided dice: getting a sum of 6. The outcomes in this event are

So $P(\text{sum is 6}) = \frac{5}{36}$. What's the probability that we get a sum of 5 *or* a sum of 6?

$$P(\text{sum is 5 or sum is 6}) = P(\text{sum is 5}) + P(\text{sum is 6}) = \frac{4}{36} + \frac{5}{36} = \frac{9}{36} = 0.25$$

Why does this formula work? Because the events "getting a sum of 5" and "getting a sum of 6" have no outcomes in common—that is, they can't both happen

at the same time. We say that these two events are **mutually exclusive (disjoint)**. As a result, this intuitive formula is known as the **addition rule for mutually exclusive events**.

> **DEFINITION**   **Mutually exclusive (disjoint), Addition rule for mutually exclusive events**
>
> Two events A and B are **mutually exclusive (disjoint)** if they have no outcomes in common and so can never occur together—that is, if $P(A$ and $B) = 0$.
>
> The **addition rule for mutually exclusive events** A and B says that
>
> $$P(A \text{ or } B) = P(A) + P(B)$$

Note that this rule works only for mutually exclusive events. We will soon develop a more general rule for finding $P(A$ or $B)$ that works for any two events.

We can summarize the basic probability rules more concisely in symbolic form.

> ### BASIC PROBABILITY RULES
>
> - For any event A, $0 \leq P(A) \leq 1$.
> - If S is the sample space in a probability model, $P(S) = 1$.
> - In the case of *equally likely* outcomes,
>
> $$P(A) = \frac{\text{number of outcomes in event A}}{\text{total number of outcomes in sample space}}$$
>
> - **Complement rule:** $P(A^C) = 1 - P(A)$.
> - **Addition rule for mutually exclusive events:** If A and B are mutually exclusive, $P(A$ or $B) = P(A) + P(B)$.

The earlier dice-rolling and spin the spinner settings involved equally likely outcomes. Here's an example that illustrates use of the basic probability rules when the outcomes of a random process are not equally likely.

We will see later in the chapter that this formula is really just a special case of the General Addition Formula: $P(A$ or $B) = P(A) + P(B) - P(A$ and $B)$, where $P(A$ and $B) = 0$ because events A and B are mutually exclusive.

Making Connections

EXAMPLE

Avoiding blue M&M'S®
Basic probability rules

PROBLEM: Suppose you tear open the corner of a bag of M&M'S® Milk Chocolate Candies, pour one candy into your hand, and observe the color. According to Mars, Inc., the maker of M&M'S, the probability model for a bag from its Cleveland factory is:

Color	Blue	Orange	Green	Yellow	Red	Brown
Probability	0.207	0.205	0.198	0.135	0.131	0.124

(a) Explain why this is a valid probability model.

(b) Find the probability that you don't get a blue M&M.

(c) What's the probability that you get an orange or a brown M&M?

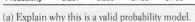

ALTERNATE EXAMPLE

Skill 3.A

Wing night   Basic probability rules

PROBLEM:

Buffalo Wild Wings ran a promotion called the Blazin' Bonus, in which every $25 gift card purchased also received a "Bonus" gift card for $5, $15, $25, or $100. According to the company, here are the probabilities for each Bonus gift card:

Blazin' Bonus	$5	$15	$25	$100
Probability	0.890	0.098	0.010	0.002

(a) Explain why this is a valid probability model.
(b) Find the probability that you don't get a $5 Bonus card.
(c) What's the probability that you get a $25 or $100 Bonus card?

SOLUTION:

(a) The probability of each outcome is a number between 0 and 1. The sum of the probabilities is $0.890 + 0.098 + 0.010 + 0.002 = 1$.

(b) $P(\text{not } \$5) = 1 - P(\$5) = 1 - 0.890 = 0.110$

(c) $P(\$25 \text{ or } \$100) = P(\$25) + P(\$100) = 0.010 + 0.002 = 0.012$

SOLUTION:

(a) The probability of each outcome is a number between 0 and 1. The sum of the probabilities is
$$0.207 + 0.205 + 0.198 + 0.135 + 0.131 + 0.124 = 1.$$

(b) $P(\text{not blue}) = 1 - P(\text{blue}) = 1 - 0.207 = 0.793$

> Using the complement rule: $P(A^c) = 1 - P(A)$.

(c) $P(\text{orange or brown}) = P(\text{orange}) + P(\text{brown})$
$$= 0.205 + 0.124 = 0.329$$

> Using the addition rule for mutually exclusive events because an M&M® can't be both orange and brown.

FOR PRACTICE, TRY EXERCISE 35

For part (b) of the example, we could also use an expanded version of the addition rule for mutually exclusive events:

$$P(\text{not blue}) = P(\text{orange or green or yellow or red or brown})$$
$$= P(\text{orange}) + P(\text{green}) + P(\text{yellow}) + P(\text{red}) + P(\text{brown})$$
$$= 0.205 + 0.198 + 0.135 + 0.131 + 0.124$$
$$= 0.793$$

Using the complement rule is much simpler than adding 5 probabilities together!

CHECK YOUR UNDERSTANDING

Suppose we choose an American adult at random. Define two events:

A = the person has a cholesterol level of 240 milligrams per deciliter of blood (mg/dl) or above (high cholesterol)

B = the person has a cholesterol level of 200 to < 240 mg/dl (borderline high cholesterol)

According to the American Heart Association, $P(A) = 0.16$ and $P(B) = 0.29$.

1. Explain why events A and B are mutually exclusive.
2. Say in plain language what the event "A or B" is. Then find $P(A \text{ or } B)$.
3. Let C be the event that the person chosen has a cholesterol level below 200 mg/dl (normal cholesterol). Find $P(C)$.

Two-Way Tables, Probability, and the General Addition Rule

So far, you have learned how to model random behavior and some basic rules for finding the probability of an event. What if you're interested in finding probabilities involving two events that are not mutually exclusive? For instance, a survey of all residents in a large apartment complex reveals that 68% use Facebook, 28% use Instagram, and 25% do both.[5] Suppose we select a resident at random. What's the probability that the person uses Facebook *or* uses Instagram?

There are two different uses of the word *or* in everyday life. In a restaurant, when you are asked if you want "soup or salad," the waiter wants you to choose

---

Answers to CYU

1. Events A and B are mutually exclusive because a person cannot have a cholesterol level of both 240 or above and between 200 and less than 240 at the same time.
2. The event "A or B" is a person that has either a cholesterol level of 240 or above or a cholesterol level between 200 and less than 240. $P(A \text{ or } B) = P(A) + P(B) = 0.16 + 0.29 = 0.45$
3. $P(C) = 1 - P(A \text{ or } B) = 1 - 0.45 = 0.55$

Teaching Tip

In this chapter, students will learn several strategies for approaching probability questions:

1. Simulation
2. Sample space
3. **Two-way tables**
4. Venn diagrams
5. Tree diagrams
6. Formulas

➕ Ask the StatsMedic

Applebee's vs. Starbucks "OR"

This post provides a story that can be used to demonstrate the different uses of the word *or* in everyday life.

one or the other, but not both. However, when you order coffee and are asked if you want "cream or sugar," it's OK to ask for one or the other or both. The same issue arises in statistics.

Mutually exclusive events A and B cannot both happen at the same time. For such events, "A or B" means that *only* event A happens or *only* event B happens. You can find $P(A \text{ or } B)$ with the addition rule for mutually exclusive events:

$$P(A \text{ or } B) = P(A) + P(B)$$

How can we find $P(A \text{ or } B)$ when the two events are *not* mutually exclusive? Now we have to deal with the fact that "A or B" means one or the other *or both*. For instance, "uses Facebook or uses Instagram" in the scenario just described includes U.S. adults who do both.

When you're trying to find probabilities involving two events, like $P(A \text{ or } B)$, a two-way table can display the sample space in a way that makes probability calculations easier.

Making Connections

As in Chapter 1, when there is an explanatory variable, we put the values of this variable across the top of a two-way table and the values of the response variable down the left side. This matches how we display the relationship between two quantitative variables on a scatterplot: explanatory variable on the horizontal axis and response variable on the vertical axis.

EXAMPLE

Who has pierced ears?
Two-way tables and probability

PROBLEM: Students in a college statistics class wanted to find out how common it is for young adults to have their ears pierced. They recorded data on two variables—gender and whether or not the student had a pierced ear—for all 178 people in the class. The two-way table summarizes the data.

		Gender		
		Male	Female	Total
Pierced ear	Yes	19	84	103
	No	71	4	75
	Total	90	88	178

Suppose we choose a student from the class at random. Define event A as getting a male student and event B as getting a student with a pierced ear.
(a) Find $P(B)$.
(b) Find $P(A \text{ and } B)$. Interpret this value in context.
(c) Find $P(A \text{ or } B)$.

SOLUTION:

(a) $P(B) = P(\text{pierced ear}) = \dfrac{103}{178} = 0.579$

(b) $P(A \text{ and } B) = P(\text{male and pierced ear}) = \dfrac{19}{178} = 0.107$. There's about an 11% chance that a randomly selected student from this class is male and has a pierced ear.

(c) $P(A \text{ or } B) = P(\text{male or pierced ear})$

$$= \frac{71 + 19 + 84}{178} = \frac{174}{178} = 0.978$$

> In statistics, *or* means "one the other or both." So "male or pierced ear" includes (i) male but no pierced ear; (ii) pierced ear but not male; and (iii) male and pierced ear.

FOR PRACTICE, TRY EXERCISE 41

Teaching Tip

When using the "Subject preference and gender" alternate example, consider using 25 index cards to represent the students in the class. For example, on 8 cards write "Male" on one side and "Math" on the other side. Then, on 12 cards write "Female" on one side and "Math" on the other side, and so on. Now you can physically show students the cards for each of the probabilities calculated in the example.

ALTERNATE EXAMPLE   Skills 3.A, 4.B

Subject preference and gender
Two-way tables and probability

PROBLEM:
Do males and females have a different preference for math or English classes? The two-way table summarizes data about gender and subject preference for a class of 25 AP® Statistics students.

		Gender		
		Male	Female	Total
Preferred subject	Math	8	12	20
	English	2	3	5
	Total	10	15	25

Suppose we choose a student from the class at random. Define event A as getting a male student and event B as getting a student who prefers math classes.
(a) Find $P(A)$.
(b) Find $P(A \text{ and } B)$. Interpret this value in context.
(c) Find $P(A \text{ or } B)$.

SOLUTION:

(a) $P(A) = P(\text{male}) = \dfrac{10}{25} = 0.40$

(b) $P(A \text{ and } B) = P(\text{male and math}) = \dfrac{8}{25}$
$= 0.32$

There's about a 32% chance that a randomly selected student from this class is male and prefers math.

(c) $P(A \text{ or } B) = P(\text{male or math})$

$$= \frac{2 + 8 + 12}{25} = \frac{22}{25} = 0.88$$

Making the index cards for the "Subject preference and gender" alternate example on page 347 can physically show how the addition rule for mutually exclusive events does *not* work in this case. The 8 males who preferred math are double-counted by attempting to separate the cards into a pile for "Male" and a pile for "Math." The 8 cards for the males who prefer math can't be put in both piles, but they *are* included in the totals on the table. This will help lead students to the general addition rule.

✚ Ask the StatsMedic
Mutually Exclusive Events? Ask Turner

This post demonstrates an activity whereby students can physically see how certain outcomes are double-counted when finding the *P*(A or B) for events that are not mutually exclusive. It is important that students experience the general addition rule in a context first before revealing the formula.

▌ AP® EXAM TIP

The general addition rule is included on the formula sheet that the students are provided on both sections of the AP® Statistics exam. However, the formula on the AP® exam uses the union (∪) and intersection (∩) symbols instead of the words *or* and *and*:

$$P(A \cup B) = P(A) + P(B) - P(A \cap B)$$

We will introduce these symbols shortly, in our discussion of Venn diagrams.

When we found *P*(male and pierced ear) in part (b) of the example, we could have described this as either *P*(A and B) or *P*(B and A). Why? Because "male and pierced ear" describes the same event as "pierced ear and male." Likewise, *P*(A or B) is the same as *P*(B or A). Don't get so caught up in the notation that you lose sight of what's really happening!

Part (c) of the example reveals an important fact about finding the probability *P*(A or B): we can't use the addition rule for mutually exclusive events unless events A and B have no outcomes in common. In this case, there are 19 outcomes that are shared by events A and B—the students who are male and have a pierced ear. If we did add the probabilities of A and B, we'd get 90/178 + 103/178 = 193/178. This is clearly wrong because the probability is bigger than 1! As Figure 5.3 illustrates, outcomes common to both events are counted twice when we add the probabilities of these two events.

The probability *P*(A and B) that events A and B both occur is called a *joint probability*. That's consistent with our description of 19/178 as a *joint relative frequency* in Chapter 1. If the joint probability is 0, the events are disjoint.

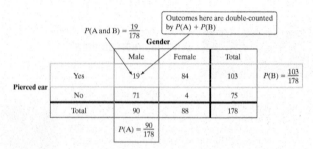

FIGURE 5.3 Two-way table showing events A and B from the pierced-ear example. These events are *not* mutually exclusive, so we can't find *P*(A or B) by just adding the probabilities of the two events.

We can fix the double-counting problem illustrated in the two-way table by subtracting the probability *P*(male and pierced ear) from the sum. That is,

P(male or pierced ear) = *P*(male) + *P*(pierced ear) − *P*(male and pierced ear)

$$= 90/178 + 103/178 - 19/178$$

$$= 174/178$$

This result is known as the **general addition rule**.

DEFINITION General addition rule

If A and B are any two events resulting from some random process, the **general addition rule** says that

$$P(A \text{ or } B) = P(A) + P(B) - P(A \text{ and } B)$$

Sometimes it's easier to label events with letters that relate to the context, as the following example shows.

✚ Ask the StatsMedic
Probability: 5 Strategies to Try Before Using a Formula

Students often lose their understanding of probability concepts when they try to immediately jump to using a formula. This post discusses 5 strategies that will promote student understanding better than formulas.

EXAMPLE

Facebook versus Instagram
General addition rule

PROBLEM: A survey of all residents in a large apartment complex reveals that 68% use Facebook, 28% use Instagram, and 25% do both. Suppose we select a resident at random. What's the probability that the person uses Facebook *or* uses Instagram?

SOLUTION:
Let event F = uses Facebook and I = uses Instagram.

$$P(F \text{ or } I) = P(F) + P(I) - P(F \text{ and } I)$$
$$= 0.68 + 0.28 - 0.25$$
$$= 0.71$$

FOR PRACTICE, TRY EXERCISE 47

What happens if we use the general addition rule for two mutually exclusive events A and B? In that case, $P(A \text{ and } B) = 0$, and the formula reduces to

$$P(A \text{ or } B) = P(A) + P(B) - P(A \text{ and } B) = P(A) + P(B) - 0 = P(A) + P(B)$$

In other words, the addition rule for mutually exclusive events is just a special case of the general addition rule.

You might be wondering if there is also a rule for finding $P(A \text{ and } B)$. There is, but it's not quite as intuitive. Stay tuned for that later.

CHECK YOUR UNDERSTANDING

Yellowstone National Park staff surveyed a random sample of 1526 winter visitors to the park. They asked each person whether they belonged to an environmental club (like the Sierra Club). Respondents were also asked whether they owned, rented, or had never used a snowmobile. The two-way table summarizes the survey responses.

		Environmental club	
		No	Yes
Snowmobile experience	Never used	445	212
	Renter	497	77
	Owner	279	16

Suppose we choose one of the survey respondents at random.

1. What's the probability that the person is an environmental club member?
2. Find *P*(not a snowmobile renter).
3. What's *P*(environmental club member and not a snowmobile renter)?
4. Find the probability that the person is not an environmental club member or is a snowmobile renter.

Teaching Tip

In this chapter, students will learn several strategies for approaching probability questions:

1. Simulation
2. Sample space
3. Two-way tables
4. **Venn diagrams**
5. Tree diagrams
6. Formulas

✚ Ask the StatsMedic

Best Friends Forever: Two-Way Tables and Venn Diagrams

This post identifies the close relationship between two-way tables and Venn diagrams.

Teaching Tip

If you made the index cards for the "Subject preference and gender" alternate example on page 347, separate them into 4 groups (Male/Math, Male/English, Female/Math, Female/English) and physically put them in the appropriate places on a Venn diagram.

Teaching Tip

Consider referring to these four regions as (1) the football, (2) the left Pac-Man, (3) the right Pac-Man, and (4) the outside.

Venn Diagrams and Probability

We have seen that two-way tables can be used to illustrate the sample space of a random process involving two events. So can **Venn diagrams**, like the one shown in Figure 5.4.

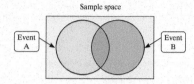

FIGURE 5.4 A typical Venn diagram that shows the sample space and the relationship between two events A and B.

> **DEFINITION   Venn diagram**
>
> A **Venn diagram** consists of one or more circles surrounded by a rectangle. Each circle represents an event. The region inside the rectangle represents the sample space of the random process.

In an earlier example, we looked at data from a survey on gender and ear piercings for a large group of college students. The random process was selecting a student in the class at random. Our events of interest were A: is male and B: has a pierced ear. Here is the two-way table that summarizes the data:

		Gender		
		Male	Female	Total
Pierced ear	Yes	19	84	103
	No	71	4	75
	Total	90	88	178

The Venn diagram in Figure 5.5 displays the sample space in a slightly different way. There are four distinct regions in the Venn diagram. These regions correspond to the four (non-total) cells in the two-way table as follows.

Region in Venn diagram	In words	Count
In the intersection of two circles	Male and pierced ear	19
Inside circle A, outside circle B	Male and no pierced ear	71
Inside circle B, outside circle A	Female and pierced ear	84
Outside both circles	Female and no pierced ear	4

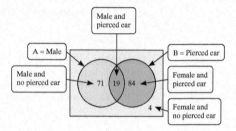

FIGURE 5.5 The completed Venn diagram for the large group of college students. The circles represent the two events A = male and B = has a pierced ear.

Because Venn diagrams have uses in other branches of mathematics, some standard vocabulary and notation have been developed that will make our work with Venn diagrams a bit easier.

- We introduced the *complement* of an event earlier. In Figure 5.6(a), the complement A^C contains the outcomes that are not in A.
- Figure 5.6(b) shows the event "A and B." You can see why this event is also called the **intersection** of A and B. The corresponding notation is $A \cap B$.
- The event "A or B" is shown in Figure 5.6(c). This event is also known as the **union** of A and B. The corresponding notation is $A \cup B$.

FIGURE 5.6 The green shaded region in each Venn diagram shows: (a) the *complement* A^C of event A, (b) the *intersection* of events A and B, and (c) the *union* of events A and B.

 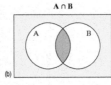 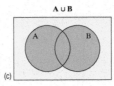

Here's a way to keep the symbols straight: $\cup$ for union; $\cap$ for intersection.

DEFINITION **Intersection, Union**

The event "A and B" is called the **intersection** of events A and B. It consists of all outcomes that are common to both events, and is denoted $A \cap B$.

The event "A or B" is called the **union** of events A and B. It consists of all outcomes that are in event A or event B, or both, and is denoted $A \cup B$.

With this new notation, we can rewrite the general addition rule in symbols as

$$P(A \cup B) = P(A) + P(B) - P(A \cap B)$$

This Venn diagram shows why the formula works in the pierced-ear example.

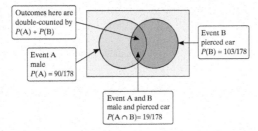

Outcomes here are double-counted by $P(A) + P(B)$

Event A male $P(A) = 90/178$

Event B pierced ear $P(B) = 103/178$

Event A and B male and pierced ear $P(A \cap B) = 19/178$

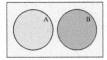

For mutually exclusive (disjoint) events A and B, the *joint probability* $P(A \cap B) = 0$ because the two events have no outcomes in common. So the corresponding Venn diagram consists of two non-overlapping circles. You can see from the figure at left why, in this special case, the general addition rule reduces to

$$P(A \cup B) = P(A) + P(B)$$

AP® EXAM TIP

The formula sheet provided on the AP® Statistics exam includes this formula (with this notation):

$$P(A \cup B) = P(A) + P(B) - P(A \cap B)$$

Teaching Tip

It's easy to illustrate the concept of mutual exclusivity by using a Venn diagram. But what does it "look like" on a two-way table? Here is an example of a survey of 30 high school students. The events "is a freshman" and "is taking AP® Statistics" are mutually exclusive because there are 0 students who are in both events.

		High school class		
		Fresh-man	Not a freshman	Total
In AP® Statistics?	Yes	0	3	3
	No	8	19	27
	Total	8	22	30

ALTERNATE EXAMPLE   Skill 3.A

Pandora or Spotify?
Venn diagrams and probability

PROBLEM:
According to a recent report, Pandora and Spotify are the most used music-streaming apps. A group of AP® Statistics students surveyed all the seniors in their school and found that 68% use Pandora, 38% use Spotify, and 24% use both. Suppose we select a senior at random.
(a) Make a Venn diagram to display the sample space of this random process using the events P: uses Pandora and S: uses Spotify.
(b) Find the probability that the person uses neither Pandora nor Spotify.

SOLUTION:
(a)

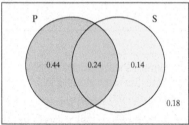

(b) P(no Pandora and no Spotify) =
1 − (0.44 + 0.24 + 0.14) = 0.18

Teaching Tip

In the "Facebook versus Instagram" example, some students will incorrectly put 0.68 for Facebook only and 0.28 for Instagram only. This is clearly incorrect, as the total of all probabilities would be well above 1. Give these students a 4-hour detention.

EXAMPLE   Facebook versus Instagram
Venn diagrams and probability

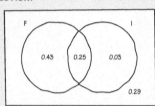

PROBLEM: A survey of all residents in a large apartment complex reveals that 68% use Facebook, 28% use Instagram, and 25% do both. Suppose we select a resident at random.
(a) Make a Venn diagram to display the sample space of this random process using the events F: uses Facebook and I: uses Instagram.
(b) Find the probability that the person uses neither Facebook nor Instagram.

SOLUTION:
(a)

F   0.43   0.25   0.03   I

0.29

- Start with the intersection: $P(F \cap I) = 0.25$.
- We know that 68% of residents use Facebook. That figure includes the 25% who also use Instagram. So 68% − 25% = 43% = 0.43 only use Facebook.
- We know that 28% of residents use Instagram. That figure includes the 25% who also use Facebook. So 28% − 25% = 3% = 0.03 only use Instagram.
- A total of 0.43 + 0.25 + 0.03 = 0.71 = 71% of residents use at least one of Facebook or Instagram. By the complement rule, 1 − 0.71 = 0.29 = 29% use neither Facebook nor Instagram.

(b) P(no Facebook and no Instagram) = 0.29

FOR PRACTICE, TRY EXERCISE 51

In the preceding example, the event "uses neither Facebook nor Instagram" is the complement of the event "uses at least one of Facebook or Instagram." To solve part (b) of the problem, we could have used our answer from the example on page 349 and the complement rule:

P(neither Facebook nor Instagram)
= 1 − P(at least one of Facebook or Instagram)
= 1 − 0.71 = 0.29

As you'll see in Section 5.3, the fact that "none" is the opposite of "at least 1" comes in handy for a variety of probability questions.

An alternate solution to the example uses a two-way table. Here is a partially completed table with the information given in the problem statement. Do you see how we can fill in the missing entries?

		Facebook use		
		Yes	No	Total
Instagram use	Yes	25%		28%
	No			
	Total	68%		100%

- $100\% - 68\% = 32\%$ of residents do not use Facebook.
- $100\% - 28\% = 72\%$ of residents do not use Instagram.
- $68\% - 25\% = 43\%$ of residents use Facebook but do not use Instagram.
- $28\% - 25\% = 3\%$ of residents use Instagram but do not use Facebook.
- $32\% - 3\% = 72\% - 43\% = 29\%$ of residents do not use Facebook and do not use Instagram.

The completed table is shown here. We can see the desired probability marked in bold in the table: P(neither Facebook nor Instagram) $= 0.29$.

		Facebook use		
		Yes	No	Total
Instagram use	Yes	25%	3%	28%
	No	43%	**29%**	72%
	Total	68%	32%	100%

Section 5.2 | Summary

- A **probability model** describes a random process by listing all possible outcomes in the **sample space** and giving the probability of each outcome. A valid probability model requires that all possible outcomes have probabilities that add up to 1.
- An **event** is a collection of possible outcomes from the sample space. The probability of any event is a number between 0 and 1.
- The event "A or B" is known as the **union** of A and B, denoted $A \cup B$. It consists of all outcomes in event A, event B, or both.
- The event "A and B" is known as the **intersection** of A and B, denoted $A \cap B$. It consists of all outcomes that are common to both events.
- To find the probability that an event occurs, we use some basic rules:
  - If all outcomes in the sample space are equally likely,

  $$P(A) = \frac{\text{number of outcomes in event A}}{\text{total number of outcomes in sample space}}$$

  - **Complement rule:** $P(A^C) = 1 - P(A)$, where A^C is the **complement** of event A; that is, the event that A does not happen.
  - **General addition rule:** For any two events A and B,

  $$P(A \cup B) = P(A) + P(B) - P(A \cap B)$$

  - **Addition rule for mutually exclusive events:** Events A and B are **mutually exclusive (disjoint)** if they have no outcomes in common. If A and B are mutually exclusive, $P(A \text{ or } B) = P(A) + P(B)$.
- A **two-way table** or **Venn diagram** can be used to display the sample space and to help find probabilities for a random process involving two events.

Answers to Section 5.2 Exercises

5.31 (a) The possible outcomes in the sample space will be equally likely and have probability $\frac{1}{16}$.

| | | First roll | | |
		1	2	3	4
Second roll	1	(1, 1)	(2, 1)	(3, 1)	(4, 1)
	2	(1, 2)	(2, 2)	(3, 2)	(4, 2)
	3	(1, 3)	(2, 3)	(3, 3)	(4, 3)
	4	(1, 4)	(2, 4)	(3, 4)	(4, 4)

(b) $P(A) = 0.25$; there are 4 ways to get a sum of 5 from these two dice: (1, 4), (2, 3), (3, 2), (4, 1) with a probability of getting a 5 = $P(A) = \frac{4}{16} = 0.25$.

5.32 (a) The sample space is {HHH, HHT, HTH, THH, HTT, THT, TTH, TTT}. Each of the 8 outcomes will be equally likely and have probability $\frac{1}{8}$. **(b)** $P(B) = 0.50$; there are 4 ways to get more heads than tails: HHH, HHT, HTH, THH with a probability of $P(B) = \frac{4}{8} = 0.50$.

5.33 (a) The sample space is: Connor/Declan, Connor/Lucas, Connor/Piper, Connor/Sedona, Connor/Zayne, Declan/Lucas, Declan/Piper, Declan/Sedona, Declan/Zayne, Lucas/Piper, Lucas/Sedona, Lucas/Zayne, Piper/Sedona, Piper/Zayne, and Sedona/Zayne. Each of these 15 outcomes will be equally likely and have probability 1/15. **(b)** There are 9 outcomes in which Piper or Sedona (or both) get to go to the show: Connor/Piper, Connor/Sedona, Declan/Piper, Declan/Sedona, Lucas/Piper, Lucas/Sedona, Piper/Sedona, Piper/Zayne, Sedona/Zayne. Define event A as Piper or Sedona (or both) get to go to the show. Then $P(A) = 9/15 = 0.60$.

5.34 (a) The sample space is: Abigail/Bobby, Abigail/Carlos, Abigail/DeAnna, Abigail/Emily, Bobby/Carlos, Bobby/DeAnna, Bobby/Emily, Carlos/DeAnna, Carlos/Emily, DeAnna/Emily. Each will be equally likely and have probability 1/10. **(b)** There are 7 outcomes in which Carlos or DeAnna (or both) end up paying for lunch: Abigail/Carlos, Abigail/DeAnna, Bobby/Carlos, Bobby/DeAnna,

---

Section 5.2 | Exercises

31. Four-sided dice A four-sided die is a pyramid whose four faces are labeled with the numbers 1, 2, 3, and 4 (see image). Imagine rolling two fair, four-sided dice and recording the number that is showing at the base of each pyramid. For instance, you would record a 4 if the die landed as shown in the figure.

(a) Give a probability model for this random process.

(b) Define event A as getting a sum of 5. Find $P(A)$.

32. Tossing coins Imagine tossing a fair coin 3 times.

(a) Give a probability model for this random process.

(b) Define event B as getting more heads than tails. Find $P(B)$.

33. Grandkids Mr. Starnes and his wife have 6 grandchildren: Connor, Declan, Lucas, Piper, Sedona, and Zayne. They have 2 extra tickets to a holiday show, and will randomly select which 2 grandkids get to see the show with them.

(a) Give a probability model for this random process.

(b) Find the probability that at least one of the two girls (Piper and Sedona) get to go to the show.

34. Who's paying? Abigail, Bobby, Carlos, DeAnna, and Emily go to the bagel shop for lunch every Thursday. Each time, they randomly pick 2 of the group to pay for lunch by drawing names from a hat.

(a) Give a probability model for this random process.

(b) Find the probability that Carlos or DeAnna (or both) ends up paying for lunch.

35. Mystery box Ms. Tyson keeps a Mystery Box in her classroom. If a student meets expectations for behavior, she or he is allowed to draw a slip of paper without looking. The slips are all of equal size, are well mixed, and have the name of a prize written on them. One of the "prizes"—extra homework—isn't very desirable! Here is the probability model for the prizes a student can win:

Prize	Pencil	Candy	Stickers	Homework pass	Extra homework
Probability	0.40	0.25	0.15	0.15	0.05

(a) Explain why this is a valid probability model.

(b) Find the probability that a student does not win extra homework.

(c) What's the probability that a student wins candy or a homework pass?

36. Languages in Canada Canada has two official languages, English and French. Choose a Canadian at random and ask, "What is your mother tongue?" Here is the distribution of responses, combining many separate languages from the broad Asia/Pacific region:[6]

Language	English	French	Asian/Pacific	Other
Probability	0.63	0.22	0.06	0.09

(a) Explain why this is a valid probability model.

(b) What is the probability that the chosen person's mother tongue is not English?

(c) What is the probability that the chosen person's mother tongue is one of Canada's official languages?

37. Household size In government data, a household consists of all occupants of a dwelling unit. Choose an American household at random and count the number of people it contains. Here is the assignment of probabilities for the outcome. The probability of finding 3 people in a household is the same as the probability of finding 4 people.

Number of people	1	2	3	4	5	6	7+
Probability	0.25	0.32	?	?	0.07	0.03	0.01

(a) What probability should replace "?" in the table? Why?

(b) Find the probability that the chosen household contains more than 2 people.

38. When did you leave? The National Household Travel Survey gathers data on the time of day when people begin a trip in their car or other vehicle. Choose a trip at random and record the time at which the trip started.[7] Here is an assignment of probabilities for the outcomes:

Time of day	10 P.M.–12:59 A.M.	1 A.M.–5:59 A.M.	6 A.M.–8:59 A.M.
Probability	0.040	0.033	0.144
Time of day	9 A.M.–12:59 P.M.	1 P.M.–3:59 P.M.	4 P.M.–6:59 P.M.
Probability	0.234	0.208	?
Time of day	7 P.M.–9:59 P.M.		
Probability	0.123		

Carlos/DeAnna, Carlos/Emily, DeAnna/Emily. Define event A as Carlos or DeAnna (or both) pay for lunch. Then $P(A) = 7/10 = 0.7$.

5.35 (a) This is a valid probability model because each probability is between 0 and 1 and the probabilities sum to 1. **(b)** P(student won't win extra homework) $= 1 - 0.05 = 0.95$; there is a 0.95 probability that a student will not win extra homework. **(c)** P(candy or homework pass) $= 0.25 + 0.15 = 0.40$; there is a 0.40 probability that a student wins candy or a homework pass.

5.36 (a) Valid because each probability is between 0 and 1 and the probabilities sum to 1. **(b)** P(not English) $= 1 - 0.63 = 0.37$; there is a 0.37 probability that a Canadian's mother tongue is not English. **(c)** P(English or French) $= 0.63 + 0.22 = 0.85$; there is a 0.85

probability that a Canadian's mother tongue is either English or French.

5.37 (a) The probabilities of all the possible outcomes must add to 1. The given probabilities add up to $0.25 + 0.32 + 0.07 + 0.03 + 0.01 = 0.68$. This leaves a probability of $1 - 0.68 = 0.32$ for P(3 or 4). Because the probability of finding 3 people in a household is the same as the probability of finding 4 people (and they are mutually exclusive), each probability must be $0.32/2 = 0.16$. **(b)** P(more than 2 people) $= 1 - P$(1 or 2 people) $= 1 - (0.25 + 0.32) = 0.43$; this could also be found using the addition rule for mutually exclusive events. P(more than 2 people) $= P$(3 or 4 or 5 or 6 or 7 +) $= 0.16 + 0.16 + 0.07 + 0.03 + 0.01 = 0.43$

(a) What probability should replace "?" in the table? Why?

(b) Find the probability that the chosen trip did not begin between 9 A.M. and 12:59 P.M.

39. **Education among young adults** Choose a young adult (aged 25 to 29) at random. The probability is 0.13 that the person chosen did not complete high school, 0.29 that the person has a high school diploma but no further education, and 0.30 that the person has at least a bachelor's degree.

(a) What must be the probability that a randomly chosen young adult has some education beyond high school but does not have a bachelor's degree? Why?

(b) Find the probability that the young adult completed high school. Which probability rule did you use to find the answer?

(c) Find the probability that the young adult has further education beyond high school. Which probability rule did you use to find the answer?

40. **Preparing for the GMAT** A company that offers courses to prepare students for the Graduate Management Admission Test (GMAT) has collected the following information about its customers: 20% are undergraduate students in business, 15% are undergraduate students in other fields of study, and 60% are college graduates who are currently employed. Choose a customer at random.

(a) What must be the probability that the customer is a college graduate who is not currently employed? Why?

(b) Find the probability that the customer is currently an undergraduate. Which probability rule did you use to find the answer?

(c) Find the probability that the customer is not an undergraduate business student. Which probability rule did you use to find the answer?

41. **Who eats breakfast?** Students in an urban school were curious about how many children regularly eat breakfast. They conducted a survey, asking, "Do you eat breakfast on a regular basis?" All 595 students in the school responded to the survey. The resulting data are shown in the two-way table.[8]

		Gender		
		Male	Female	Total
Eats breakfast regularly	Yes	190	110	300
	No	130	165	295
	Total	320	275	595

Suppose we select a student from the school at random. Define event F as getting a female student and event B as getting a student who eats breakfast regularly.

(a) Find $P(B^C)$.

(b) Find $P(F \text{ and } B^C)$. Interpret this value in context.

(c) Find $P(F \text{ or } B^C)$.

42. **Is this your card?** A standard deck of playing cards (with jokers removed) consists of 52 cards in four suits—clubs, diamonds, hearts, and spades. Each suit has 13 cards, with denominations ace, 2, 3, 4, 5, 6, 7, 8, 9, 10, jack, queen, and king. The jacks, queens, and kings are referred to as "face cards." Imagine that we shuffle the deck thoroughly and deal one card. Define events F: getting a face card, and H: getting a heart. The two-way table summarizes the sample space for this random process.

		Card		
		Face card	Nonface card	Total
Suit	Heart	3	10	13
	Nonheart	9	30	39
	Total	12	40	52

(a) Find $P(H^C)$.

(b) Find $P(H^C \text{ and } F)$. Interpret this value in context.

(c) Find $P(H^C \text{ or } F)$.

43. **Cell phones** The Pew Research Center asked a random sample of 2024 adult cell-phone owners from the United States their age and which type of cell phone they own: iPhone, Android, or other (including non-smartphones). The two-way table summarizes the data.

		Age			
		18–34	35–54	55+	Total
Type of cell phone	iPhone	169	171	127	467
	Android	214	189	100	503
	Other	134	277	643	1054
	Total	517	637	870	2024

Suppose we select one of the survey respondents at random. What's the probability that:

(a) The person is not age 18 to 34 and does not own an iPhone?

(b) The person is age 18 to 34 or owns an iPhone?

44. **Middle school values** Researchers carried out a survey of fourth-, fifth-, and sixth-grade students in Michigan. Students were asked whether good grades, athletic ability, or being popular was most important to them. The two-way table summarizes the survey data.[9]

		Grade			
		4th grade	5th grade	6th grade	Total
Most important	Grades	49	50	69	168
	Athletic	24	36	38	98
	Popular	19	22	28	69
	Total	92	108	135	335

5.38 (a) The probabilities of all the possible outcomes must add to 1, so $P(4 \text{ P.M.} - 6:59 \text{ P.M.}) = 1 - (0.040 + 0.033 + 0.144 + 0.234 + 0.208 + 0.123) = 1 - 0.782 = 0.218$.
(b) $P(\text{not between } 9 \text{ A.M.} - 12:59 \text{ P.M.}) = 1 - P(\text{between } 9 \text{ A.M.} - 12:59 \text{ P.M.}) = 1 - 0.234 = 0.766$

5.39 (a) The given probabilities have a sum of 0.72 and the sum of all probabilities should be 1. Thus, the probability that a randomly chosen young adult has some education beyond high school but does not have a bachelor's degree is $1 - 0.72 = 0.28$. There is a 0.28 probability that a young adult has some education beyond high school but does not have a bachelor's degree. **(b)** Using the complement rule, $P(\text{at least a high school education}) = 1 - P(\text{has}$

not finished high school$) = 1 - 0.13 = 0.87$. There is a 0.87 probability that a young adult has completed high school. **(c)** $P(\text{young adult has further education beyond high school}) = P(\text{young adult has some education beyond high school but does not have a bachelor's degree}) + P(\text{young adult has at least a bachelor's degree}) = 0.28 + 0.30 = 0.58$; the probability rule used is the addition rule for mutually exclusive events.

5.40 (a) The given probabilities have a sum of 0.95 and the sum of all probabilities should be 1. The probability that the customer is an unemployed college graduate is $1 - 0.95 = 0.05$. **(b)** $P(\text{undergraduate}) = P(\text{undergraduate student in business}) + P(\text{undergraduate student in other fields}) = 0.20 + 0.15 = 0.35$; there is a 0.35 probability

that the customer is currently an undergraduate. The probability rule used is the addition rule for mutually exclusive events. **(c)** $P(\text{not an undergraduate business student}) = 1 - P(\text{undergraduate business student}) = 1 - 0.20 = 0.80$; there is a 0.80 probability that the customer is not an undergraduate business student. The probability rule used is the complement rule.

5.41 (a) $P(B^C) = P(\text{student does not eat breakfast regularly}) = \dfrac{295}{595} = 0.496$

(b) $P(F \text{ and } B^C) = P(\text{student is a female and does not eat breakfast regularly}) = \dfrac{165}{595} = 0.277$ *Interpretation:* If we select a student from the school at random, the probability that the student will be female and does not eat breakfast regularly is 0.277.

(c) $P(F \text{ or } B^C) = P(\text{student is a female or does not eat breakfast regularly}) = \dfrac{110 + 165 + 130}{595} = 0.681$

5.42 (a) $P(H^C) = P(\text{do not get a heart}) = \dfrac{39}{52} = 0.75$

(b) $P(H^C \text{ and } F) = P(\text{do not get a heart and get a face card}) = \dfrac{9}{52} = 0.173$
Interpretation: If the deck is shuffled thoroughly and one card is dealt, the probability that the dealt card is not a heart and is a face card is 0.173.

(c) $P(H^C \text{ or } F) = P(\text{do not get a heart or get a face card}) = \dfrac{3 + 9 + 30}{52} = 0.808$

5.43 (a) $P(\text{not age 18 to 34 and not own an iPhone}) = \dfrac{189 + 100 + 277 + 643}{2024} = \dfrac{1209}{2024} = 0.597$

(b) $P(\text{age 18 to 34 or owns iPhone}) = \dfrac{169 + 214 + 134 + 171 + 127}{2024} = \dfrac{815}{2024} = 0.403$

5.44 (a) P(sixth grader or good grades) =
$$\frac{69 + 38 + 28 + 49 + 50}{335} = \frac{234}{335} = 0.699$$
(b) P(not sixth grader and did not rate good grades important) =
$$\frac{24 + 36 + 19 + 22}{335} = \frac{101}{335} = 0.301$$

5.45 (a)

Color

		Black	Not black	Total
	Yes	10	10	20
Even?	No	8	10	18
	Total	18	20	38

(b) $P(B) = \dfrac{18}{38} = 0.474$; $P(E) = \dfrac{20}{38} = 0.526$

(c) The event "B and E" would be that the ball lands in a spot that is black and even:
$$P(B \text{ and } E) = \frac{10}{38} = 0.263.$$

(d) The probability of the event "B or E" means the probability of landing in a spot that is either black, even, or both. If we add the probabilities of landing in a black spot and landing in an even spot, the spots that are black and even will be double-counted because events B and E are not mutually exclusive:
$$P(B \text{ or } E) = \frac{18}{38} + \frac{20}{38} - \frac{10}{38} = \frac{28}{38} = 0.737$$

5.46 (a)

Disk number

		Number 9	Not 9	Total
	Yes	1	8	9
Red?	No	3	24	27
	Total	4	32	36

(b) $P(R) = P(\text{red}) = 9/36 = 0.25$; $P(N) = P(\text{number } 9) = 4/36 = 0.111$

(c) The event "R and N" would be drawing one disk that is red and has the number 9 on it: $P(R \text{ and } N) = 1/36 = 0.028.$
(d) The events "disk is red" and "disk is the number 9" are not mutually exclusive (i.e., these two events can occur at the same time). There is 1 outcome that is shared by events R and N. If we add the probabilities of R and N, we would be double-counting this outcome. Using the general addition rule, $P(R \text{ or } N) = P(R) + P(N) - P(R \text{ and } N) = 9/36 + 4/36 - 1/36 = 12/36 = 0.333.$

Suppose we select one of these students at random. What's the probability of each of the following?

(a) The student is a sixth-grader or rated good grades as important.

(b) The student is not a sixth-grader and did not rate good grades as important.

45. **Roulette** An American roulette wheel has 38 slots with numbers 1 through 36, 0, and 00, as shown in the figure. Of the numbered slots, 18 are red, 18 are black, and 2—the 0 and 00—are green. When the wheel is spun, a metal ball is dropped onto the middle of the wheel. If the wheel is balanced, the ball is equally likely to settle in any of the numbered slots. Imagine spinning a fair wheel once. Define events B: ball lands in a black slot, and E: ball lands in an even-numbered slot. (Treat 0 and 00 as even numbers.)

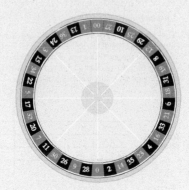

(a) Make a two-way table that displays the sample space in terms of events B and E.

(b) Find $P(B)$ and $P(E)$.

(c) Describe the event "B and E" in words. Then find the probability of this event.

(d) Explain why $P(B \text{ or } E) \neq P(B) + P(E)$. Then use the general addition rule to compute $P(B \text{ or } E)$.

46. **Colorful disks** A jar contains 36 disks: 9 each of four colors—red, green, blue, and yellow. Each set of disks of the same color is numbered from 1 to 9. Suppose you draw one disk at random from the jar. Define events R: get a red disk, and N: get a disk with the number 9.

(a) Make a two-way table that describes the sample space in terms of events R and N.

(b) Find $P(R)$ and $P(N)$.

(c) Describe the event "R and N" in words. Then find the probability of this event.

(d) Explain why $P(R \text{ or } N) \neq P(R) + P(N)$. Then use the general addition rule to compute $P(R \text{ or } N)$.

47. **Dogs and cats** In one large city, 40% of all households own a dog, 32% own a cat, and 18% own both. Suppose we randomly select a household. What's the probability that the household owns a dog or a cat?

pg 349

48. **Reading the paper** In a large business hotel, 40% of guests read the *Los Angeles Times*. Only 25% read the *Wall Street Journal*. Five percent of guests read both papers. Suppose we select a hotel guest at random and record which of the two papers the person reads, if either. What's the probability that the person reads the *Los Angeles Times* or the *Wall Street Journal*?

49. **Mac or PC?** A recent census at a major university revealed that 60% of its students mainly used Macs. The rest mainly used PCs. At the time of the census, 67% of the school's students were undergraduates. The rest were graduate students. In the census, 23% of respondents were graduate students and used a Mac as their main computer. Suppose we select a student at random from among those who were part of the census. Define events G: is a graduate student, and M: primarily uses a Mac.

(a) Find $P(G \cup M)$. Interpret this value in context.

(b) Consider the event that the randomly selected student is an undergraduate student and primarily uses a PC. Write this event in symbolic form and find its probability.

50. **Gender and political party** In January 2017, 52% of U.S. senators were Republicans and the rest were Democrats or Independents. Twenty-one percent of the senators were females, and 47% of the senators were male Republicans. Suppose we select one of these senators at random. Define events R: is a Republican, and M: is male.

(a) Find $P(R \cup M)$. Interpret this value in context.

(b) Consider the event that the randomly selected senator is a female Democrat or Independent. Write this event in symbolic form and find its probability.

51. **Dogs and cats** Refer to Exercise 47.

(a) Make a Venn diagram to display the outcomes of this random process using events D: owns a dog and C: owns a cat.

pg 352

(b) Find $P(D \cap C^C)$.

5.47 P(own a dog or a cat) = P(own a dog) + P(own a cat) − P(own a dog and a cat) = $0.40 + 0.32 - 0.18 = 0.54$

5.48 P(read *Los Angeles Times* or read *Wall Street Journal*) = P(read *Los Angeles Times*) + P(read *Wall Street Journal*) − P(read *Los Angeles Times* and read *Wall Street Journal*) = $0.40 + 0.25 - 0.05 = 0.60$

5.49 (a) We are given the following information: $P(M) = 0.6$, $P(G^C) = 0.67$, $P(G \text{ and } M) = 0.23$. Therefore, $P(G) = 1 - 0.67 = 0.33$; $P(G \cup M) = P(G) + P(M) - P(G \cap M) = 0.33 + 0.6 - 0.23 = 0.70$. *Interpretation:* The probability of randomly selecting a student from among those who were part of the census who is a graduate student or uses a Mac is 0.70.

(b) $P(G^C \cap M^C) = 1 - P(G \cup M) = 1 - 0.7 = 0.3$

5.50 (a) We are given the following information: $P(R) = 0.52$, $P(\text{female}) = 0.21$, $P(R \cap M) = 0.47$. Therefore, $P(M) = 1 - 0.21 = 0.79$; $P(R \cup M) = P(R) + P(M) - P(R \cap M) = 0.52 + 0.79 - 0.47 = 0.84$. *Interpretation:* The probability of randomly selecting a senator who is a Republican or male is 0.84.

(b) $P(R^C \cap M^C) = 1 - P(R \cup M) = 1 - 0.84 = 0.16$

5.51 (a)

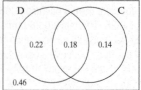

(b) $P(D \cap C^C) = 0.22$

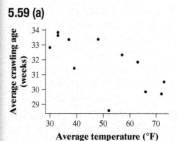

52. Reading the paper Refer to Exercise 48.

(a) Make a Venn diagram to display the outcomes of this random process using events L: reads the *Los Angeles Times* and W: reads the *Wall Street Journal*.

(b) Find $P(L^C \cap W)$.

53. Union and intersection Suppose A and B are two events such that $P(A) = 0.3$, $P(B) = 0.4$, and $P(A \cup B) = 0.58$. Find $P(A \cap B)$.

54. Union and intersection Suppose C and D are two events such that $P(C) = 0.6$, $P(D) = 0.45$, and $P(C \cup D) = 0.75$. Find $P(C \cap D)$.

Multiple Choice: *Select the best answer for Exercises 55–58.*

55. The partially completed table that follows shows the distribution of scores on the 2016 AP® Statistics exam.

Score	1	2	3	4	5
Probability	0.235	0.155	0.249	0.217	?

Suppose we randomly select a student who took this exam. What's the probability that he or she earned a score of at least 3?

(a) 0.249   (b) 0.361   (c) 0.390
(d) 0.466   (e) 0.610

56. In a sample of 275 students, 20 say they are vegetarians. Of the vegetarians, 9 eat both fish and eggs, 3 eat eggs but not fish, 1 eats fish but not eggs, and 7 eat neither. Choose one of the vegetarians at random. What is the probability that the chosen student eats fish or eggs?

(a) 9/20   (b) 13/20   (c) 22/20
(d) 9/275   (e) 22/275

Exercises 57 and 58 refer to the following setting. The casino game craps is based on rolling two dice. Here is the assignment of probabilities to the sum of the numbers on the up-faces when two dice are rolled:

Outcome	2	3	4	5	6	7	8	9	10	11	12
Probability	1/36	2/36	3/36	4/36	5/36	6/36	5/36	4/36	3/36	2/36	1/36

57. The most common bet in craps is the "pass line." A pass line bettor wins immediately if either a 7 or an 11 comes up on the first roll. This is called a *natural*. What is the probability that a natural does *not* occur?

(a) 2/36   (b) 6/36   (c) 8/36
(d) 16/36   (e) 28/36

58. If a player rolls a 2, 3, or 12, it is called *craps*. What is the probability of getting craps or an even sum on one roll of the dice?

(a) 4/36   (b) 18/36   (c) 20/36
(d) 22/36   (e) 32/36

Recycle and Review

59. Crawl before you walk (3.1, 3.2, 4.3) At what age do babies learn to crawl? Does it take longer to learn in the winter, when babies are often bundled in clothes that restrict their movement? Perhaps there might even be an association between babies' crawling age and the average temperature during the month they first try to crawl (around 6 months after birth). Data were collected from parents who brought their babies to the University of Denver Infant Study Center to participate in one of a number of studies. Parents reported the birth month and the age at which their child was first able to creep or crawl a distance of 4 feet within one minute. Information was obtained on 414 infants (208 boys and 206 girls). Crawling age is given in weeks, and average temperature (in degrees Fahrenheit) is given for the month that is 6 months after the birth month.[10]

Birth month	Average crawling age	Average temperature
January	29.84	66
February	30.52	73
March	29.70	72
April	31.84	63
May	28.58	52
June	31.44	39
July	33.64	33
August	32.82	30
September	33.83	33
October	33.35	37
November	33.38	48
December	32.32	57

(a) Make an appropriate graph to display the relationship between average temperature and average crawling age. Describe what you see.

Some computer output from a linear regression analysis of the data is shown.

Term	Coef	SE Coef	T-Value	P-Value
Constant	35.68	1.32	27.08	0.000
Average temperature	−0.0777	0.0251	−3.10	0.011

S = 1.31920   R-Sq = 48.96%   R-Sq(adj) = 43.86%

(b) What is the equation of the least-squares regression line that describes the relationship between average temperature and average crawling age? Define any variables that you use.

(c) Interpret the slope of the regression line.

(d) Can we conclude that warmer temperatures 6 months after babies are born causes them to crawl sooner? Justify your answer.

5.59 (a)

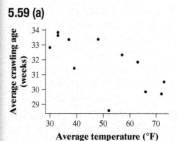

In this scatterplot, there appears to be a moderately strong, negative linear relationship between average temperature and average crawling age. **(b)** The equation for the least-squares regression line is $\hat{y} = 35.68 - 0.0777x$, where $\hat{y}$ is the predicted average crawling age and x is the average temperature in degrees Fahrenheit. **(c)** The slope $= -0.0777$. *Interpretation:* The predicted average crawling age decreases by 0.0777 week for each additional 1-degree increase in the average temperature in degrees Fahrenheit. **(d)** We cannot conclude that warmer temperatures 6 months after babies are born causes them to crawl sooner because this was an observational study and not an experiment. We cannot draw conclusions of cause and effect from observational studies.

5.52 (a)

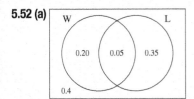

(b) $P(L^C \cap W) = 0.20$

5.53 The general addition rule states that $P(A \cup B) = P(A) + P(B) - P(A \cap B)$. We know that $P(A) = 0.3$, $P(B) = 0.4$, and $P(A \cup B) = 0.58$. Therefore, $0.58 = 0.3 + 0.4 - P(A \cap B)$. Solving for $P(A \cap B)$ gives $P(A \cap B) = 0.12$.

5.54 The general addition rule states that $P(C \cup D) = P(C) + P(D) - P(C \cap D)$. We know that $P(C) = 0.6$, $P(D) = 0.45$, and $P(C \cup D) = 0.75$. Therefore, $0.75 = 0.6 + 0.45 - P(C \cap D)$. Solving for $P(C \cap D)$ gives $P(C \cap D) = 0.30$.

5.55 e

5.56 b

5.57 e

5.58 c

5.60 (a) Label the women from 1 to 1649. Using a random number generator, output 825 different random integers from 1 to 1649. Assign the women with those numbers to take strontium ranelate. The other 824 women will take a placebo. At the end of 3 years, compare the number of new fractures for each group. **(b)** It is important to keep the calcium supplements and medical care the same for all the women in the experiment, to minimize that additional source of variation. In order to determine if strontium ranelate is effective in reducing the number of new fractures, it is important to control other factors (e.g., whether or not the women took calcium supplements and received adequate medical care) that might affect the response variable. **(c)** The fact that the women who took strontium ranelate had statistically significantly fewer new fractures, on average, than the women who took a placebo over a 3-year period means that the number of fractures observed in the strontium ranelate group were substantially fewer than what we would expect to happen by chance alone if the treatment were not effective.

PD Section 5.3 Overview

To watch the video overview of Section 5.3 (for teachers), click on the link in the TE-Book, look on the TRFD, or download from the Teacher's Resources on the book's digital platform.

TRM Section 5.3 Alternate Examples

You can find the Alternate Examples for this section in Microsoft Word format by clicking on the link in the TE-Book, opening the TRFD, or downloading from the Teacher's Resources on the book's digital platform.

60. **Treating low bone density** (4.2, 4.3) Fractures of the spine are common and serious among women with advanced osteoporosis (low mineral density in the bones). Can taking strontium ranelate help? A large medical trial was conducted to investigate this question. Researchers recruited 1649 women with osteoporosis who had previously had at least one fracture for an experiment. The women were assigned to take either strontium ranelate or a placebo each day. All the women were taking calcium supplements and receiving standard medical care. One response variable was the number of new fractures over 3 years.[11]

(a) Describe a completely randomized design for this experiment.

(b) Explain why it is important to keep the calcium supplements and medical care the same for all the women in the experiment.

(c) The women who took strontium ranelate had statistically significantly fewer new fractures, on average, than the women who took a placebo over a 3-year period. Explain what this means to someone who knows little statistics.

SECTION 5.3 # Conditional Probability and Independence

LEARNING TARGETS *By the end of the section, you should be able to:*

- Calculate and interpret conditional probabilities.
- Determine if two events are independent.
- Use the general multiplication rule to calculate probabilities.
- Use a tree diagram to model a random process involving a sequence of outcomes and to calculate probabilities.
- When appropriate, use the multiplication rule for independent events to calculate probabilities.

The probability of an event can change if we know that some other event has occurred. For instance, suppose you toss a fair coin twice. The probability of getting two heads is 1/4 because the sample space consists of the 4 equally likely outcomes

HH   HT   TH   TT

Suppose that the first toss lands tails. Now what's the probability of getting two heads? It's 0. Knowing that the first toss is a tail changes the probability that you get two heads.

This idea is the key to many applications of probability.

What Is Conditional Probability?

Let's return to the college statistics class from Section 5.2. Earlier, we used the two-way table shown on the next page to find probabilities involving events A: is male and B: has a pierced ear for a randomly selected student.

	Gender		
	Male	Female	Total
Pierced ear Yes	19	84	103
No	71	4	75
Total	90	88	178

Here is a summary of our previous results:

$P(A) = P(\text{male}) = 90/178$ $P(A \cap B) = P(\text{male and pierced ear}) = 19/178$

$P(B) = P(\text{pierced ear}) = 103/178$ $P(A \cup B) = P(\text{male or pierced ear}) = 174/178$

Now let's turn our attention to some other interesting probability questions.

	Gender		
	Male	Female	Total
Pierced ear Yes	19	84	103
No	71	4	75
Total	90	88	178

1. **If we know that a randomly selected student has a pierced ear, what is the probability that the student is male?** There are 103 students in the class with a pierced ear. We can restrict our attention to this group, since we are told that the chosen student has a pierced ear. Because there are 19 males among the 103 students with a pierced ear, the desired probability is

$$P(\text{male } given \text{ pierced ear}) = 19/103 = 0.184$$

	Gender		
	Male	Female	Total
Pierced ear Yes	19	84	103
No	71	4	75
Total	90	88	178

2. **If we know that a randomly selected student is male, what's the probability that the student has a pierced ear?** This time, our attention is focused on the males in the class. Because 19 of the 90 males in the class have a pierced ear,

$$P(\text{pierced ear } given \text{ male}) = 19/90 = 0.211$$

These two questions sound alike, but they actually ask two very different things. Each of these probabilities is an example of a **conditional probability**. The name comes from the fact that we are trying to find the probability that one event will happen under the *condition* that some other event is already known to have occurred. We often use the phrase "given that" to signal the condition.

> **DEFINITION   Conditional probability**
>
> The probability that one event happens given that another event is known to have happened is called a **conditional probability**. The conditional probability that event A happens given that event B has happened is denoted by $P(A|B)$.

With this new notation available, we can restate the answers to the two questions just posed as

$$P(\text{male}|\text{pierced ear}) = P(A|B) = 19/103$$
$$\text{and}$$
$$P(\text{pierced ear}|\text{male}) = P(B|A) = 19/90$$

Here's an example that illustrates how conditional probability works in a familiar setting.

Teaching Tip

When introducing the idea of conditional probability, display the entire two-way table and then physically cover up the rows or columns that are not part of the "given." Students can do this on paper problems with two index cards. Once they identify the given condition (which will often be one row or one column in the two-way table), ask them to cover up the rest of the two-way table with the index cards. They must then calculate the probability with only the numbers they can see.

Preparing for Inference

In Chapter 9 and beyond, we will do significance tests to evaluate the strength of evidence against a claim. To decide if we have convincing evidence, we must calculate a *P*-value, which is the probability of getting our observed results or more extreme purely by chance, *assuming a null hypothesis is true*. The *P*-value is a conditional probability: *P*-value = *P*(evidence as strong or stronger | H_0 is true).

Teaching Tip

Emphasize that order matters when asking conditional probability questions, just as it does with other mathematical operations such as subtraction or division.

ALTERNATE EXAMPLE   Skill 3.A

Olympic medals

Two-way tables and conditional probabilities

PROBLEM:

In the 2016 Summer Olympics, the United States and China won the most medals. Suppose we randomly select a medal from the 191 that are represented in the two-way table. Define events G: gold medal, U: United States, and B: bronze medal.

Medals

Country		Gold	Silver	Bronze	Total
	United States	46	37	38	121
	China	26	18	26	70
	Total	72	55	64	191

(a) Find $P(B \mid U)$. Interpret this value in context.

(b) Given that the chosen medal is not a gold medal, what's the probability that it came from the United States? Write your answer as a probability statement using correct symbols for the events.

SOLUTION:

(a) $P(B \mid U) =$ $P(\text{bronze medal} \mid \text{United States}) =$ $38/121 = 0.314$; given that the randomly chosen medal is from the United States, there is about a 31.4% chance that it is a bronze medal.

(b) $P(\text{United States} \mid \text{not gold}) =$

$$P(U \mid G^C) = \frac{37 + 38}{55 + 64} = \frac{75}{119} = 0.630$$

EXAMPLE

A *Titanic* disaster
Two-way tables and conditional probabilities

PROBLEM: In 1912, the luxury liner *Titanic*, on its first voyage across the Atlantic, struck an iceberg and sank. Some passengers got off the ship in lifeboats, but many died. The two-way table gives information about adult passengers who survived and who died, by class of travel.

Images Group/REX/Shutterstock

		Class of travel			
		First	Second	Third	Total
Survival status	Survived	197	94	151	442
	Died	122	167	476	765
	Total	319	261	627	1207

Suppose we randomly select one of the adult passengers from the *Titanic*. Define events F: first-class passenger, S: survived, and T: third-class passenger.

(a) Find $P(T \mid S)$. Interpret this value in context.

(b) Given that the chosen person is not a first-class passenger, what's the probability that she or he survived? Write your answer as a probability statement using correct symbols for the events.

SOLUTION:

(a) $P(T \mid S) = P(\text{third-class passenger} \mid \text{survived}) = 151/442 = 0.342$. Given that the randomly chosen person survived, there is about a 34.2% chance that she or he was a third-class passenger.

> To answer part (a), only consider values in the "Survived" row.

(b) $P(\text{survived} \mid \text{not first-class passenger})$

$$= P(S \mid F^C) = \frac{94 + 151}{261 + 627} = \frac{245}{888} = 0.276$$

> To answer part (b), only consider values in the "Second class" and "Third class" columns.

FOR PRACTICE, TRY EXERCISE 61

Is there is a connection between conditional probability and conditional relative frequency from Chapter 1? Yes! In part (a) of the example, we found the conditional probability $P(\text{third-class passenger} \mid \text{survived}) = 151/442 = 0.342$. In Chapter 1, we asked, "What proportion of survivors were third-class passengers?" Our answer was also $151/442 = 0.342$. This is a conditional relative frequency because we are finding the percent or proportion of third-class passengers among those who survived. So a conditional probability is just a conditional relative frequency that comes from a random process—in this case, randomly selecting an adult passenger.

Let's look more closely at how conditional probabilities are calculated using the data from the college statistics class. From the two-way table that follows, we see that

$$P(\text{male} \mid \text{pierced ear}) = \frac{19}{103} = \frac{\text{number of students who are male and have a pierced ear}}{\text{number of students with a pierced ear}}$$

		Gender		
		Male	Female	Total
Pierced ear	Yes	19	84	103
	No	71	4	75
	Total	90	88	178

What if we focus on probabilities instead of numbers of students? Notice that

$$\frac{P(\text{male and pierced ear})}{P(\text{pierced ear})} = \frac{\frac{19}{178}}{\frac{103}{178}} = \frac{19}{103} = P(\text{male} | \text{pierced ear})$$

This observation leads to a general formula for computing a conditional probability.

CALCULATING CONDITIONAL PROBABILITIES

To find the conditional probability $P(A | B)$, use the formula

$$P(A|B) = \frac{P(A \text{ and } B)}{P(B)} = \frac{P(A \cap B)}{P(B)} = \frac{P(\text{both events occur})}{P(\text{given event occurs})}$$

By the same reasoning,

$$P(B|A) = \frac{P(B \text{ and } A)}{P(A)} = \frac{P(B \cap A)}{P(A)}$$

EXAMPLE

Facebook or Instagram?
Calculating conditional probability

PROBLEM: A survey of all residents in a large apartment complex reveals that 68% use Facebook, 28% use Instagram, and 25% do both. Suppose we select a resident at random. Given that the person uses Facebook, what's the probability that she or he uses Instagram?

SOLUTION:

$$P(\text{Instagram} | \text{Facebook}) = P(I | F) = \frac{P(I \cap F)}{P(F)} = \frac{0.25}{0.68} = 0.368$$

FOR PRACTICE, TRY EXERCISE 69

AP® EXAM TIP

You can write statements like $P(A | B)$ if events A and B are clearly defined in a problem. Otherwise, it's probably easier to use contextual labels, like $P(I | F)$ in the preceding example. Or you can just use words: $P(\text{Instagram} | \text{Facebook})$.

Refer back to the example. If the person chosen is an Instagram user, what is the probability that he or she uses Facebook? By the conditional probability formula, it's

$$P(\text{Facebook} | \text{Instagram}) = P(F | I) = \frac{P(F \cap I)}{P(I)} = \frac{0.25}{0.28} = 0.893$$

If the chosen resident uses Instagram, it is extremely likely that he or she uses Facebook. However, if the chosen resident uses Facebook, he or she is not nearly so likely to use Instagram.

AP® EXAM TIP

The formula sheet provided on the AP® Statistics exam includes this formula: $P(A | B) = \dfrac{P(A \cap B)}{P(B)}$

Teaching Tip

In the formula for calculating a conditional probability, the numerator is the probability of "both" and the denominator is the probability of the "given."

ALTERNATE EXAMPLE   Skill 3.A

Pandora or Spotify? Replay.
Calculating conditional probability

PROBLEM:
According to a recent report, Pandora and Spotify are the most used music-streaming apps. A group of AP® Statistics students surveyed all the seniors in the school and found that 68% use Pandora, 38% use Spotify, and 24% use both. Suppose we select a senior at random. Given that the person uses Pandora, what's the probability that she or he uses Spotify?

SOLUTION:

$P(\text{Spotify} | \text{Pandora}) =$

$$\frac{P(\text{Spotify} \cap \text{Pandora})}{P(\text{Pandora})} = \frac{0.24}{0.68} = 0.353$$

Teaching Tip

Writing probability in terms of "A" and "B" pushes students toward using memorized formulas. Writing probability in terms of the context, like "Facebook" and "Instagram," keeps students thinking about the meaning of probability.

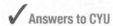

Answers to CYU

1. $P(N \mid E) = \dfrac{212}{305} = 0.695$; given that a survey respondent belongs to an environmental organization, there is about a 69.5% chance that he or she never used a snowmobile.

2. $P(E \mid S^C) = \dfrac{212 + 77}{657 + 574} = \dfrac{289}{1231} = 0.235$

3. $P(E^C \mid N) = \dfrac{445}{657} = 0.677$;

$P(E^C \mid \text{renter}) = \dfrac{497}{574} = 0.866$;

$P(E^C \mid \text{owner}) = \dfrac{279}{295} = 0.946$; the chosen person is more likely to not be an environmental club member if he or she owns a snowmobile.

Ask the StatsMedic

Using Two Aces to Introduce Conditional Probability

This post describes a game that will help students realize why conditional probabilities are needed for dependent events.

You could also use a two-way table to help you find these conditional probabilities. Here's the table that we made for this setting in Section 5.2 (page 353). It is easy to see that

$$P(\text{Instagram} \mid \text{Facebook}) = \frac{0.25}{0.68} = 0.368 \quad \text{and} \quad P(\text{Facebook} \mid \text{Instagram}) = \frac{0.25}{0.28} = 0.893$$

		Facebook use		
		Yes	No	Total
Instagram use	Yes	25%	3%	28%
	No	43%	29%	72%
	Total	68%	32%	100%

CHECK YOUR UNDERSTANDING

Yellowstone National Park surveyed a random sample of 1526 winter visitors to the park. They asked each person whether he or she owned, rented, or had never used a snowmobile. Respondents were also asked whether they belonged to an environmental organization (like the Sierra Club). The two-way table summarizes the survey responses.

		Environmental club		
		No	Yes	Total
Snowmobile experience	Never used	445	212	657
	Renter	497	77	574
	Owner	279	16	295
	Total	1221	305	1526

Suppose we randomly select one of the survey respondents. Define events E: environmental club member, S: snowmobile owner, and N: never used.

1. Find $P(N \mid E)$. Interpret this value in context.
2. Given that the chosen person is not a snowmobile owner, what's the probability that she or he is an environmental club member? Write your answer as a probability statement using correct symbols for the events.
3. Is the chosen person more likely to not be an environmental club member if he or she has never used a snowmobile, is a snowmobile owner, or a snowmobile renter? Justify your answer.

Conditional Probability and Independence

Suppose you toss a fair coin twice. Define events A: first toss is a head, and B: second toss is a head. We know that $P(A) = 1/2$ and $P(B) = 1/2$.

- What's $P(B \mid A)$? It's the conditional probability that the second toss is a head given that the first toss was a head. The coin has no memory, so $P(B \mid A) = 1/2$.
- What's $P(B \mid A^C)$? It's the conditional probability that the second toss is a head given that the first toss was not a head. Getting a tail on the first toss

does not change the probability of getting a head on the second toss, so $P(B \mid A^C) = 1/2$.

In this case, $P(B \mid A) = P(B \mid A^C) = P(B)$. Knowing the outcome of the first toss does not change the probability that the second toss is a head. We say that A and B are **independent events**.

DEFINITION **Independent events**

A and B are **independent events** if knowing whether or not one event has occurred does not change the probability that the other event will happen. In other words, events A and B are independent if

$$P(A \mid B) = P(A \mid B^C) = P(A)$$

Alternatively, events A and B are independent if

$$P(B \mid A) = P(B \mid A^C) = P(B)$$

Let's contrast the coin-toss scenario with our earlier pierced-ear example. In that case, the random process involved randomly selecting a student from a college statistics class. The events of interest were A: is male, and B: has a pierced ear. Are these two events independent?

| | **Gender** | | |
	Male	Female	Total
Pierced ear Yes	19	84	103
No	71	4	75
Total	90	88	178

- Suppose that the chosen student is male. We can see from the two-way table that $P(\text{pierced ear} \mid \text{male}) = P(B \mid A) = 19/90 = 0.211$.
- Suppose that the chosen student is female. From the two-way table, we see that $P(\text{pierced ear} \mid \text{female}) = P(B \mid A^C) = 84/88 = 0.955$.

Knowing that the chosen student is a male changes (greatly reduces) the probability that the student has a pierced ear. So these two events are not independent.

Another way to determine if two events A and B are independent is to compare $P(A \mid B)$ to $P(A)$ or $P(B \mid A)$ to $P(B)$. For the pierced-ear setting,

$$P(\text{pierced ear} \mid \text{male}) = P(B \mid A) = 19/90 = 0.211$$

The unconditional probability that the chosen student has a pierced ear is

$$P(\text{pierced ear}) = P(B) = 103/178 = 0.579$$

Again, knowing that the chosen student is male changes (reduces) the probability that the individual has a pierced ear. So these two events are not independent.

Making Connections

To help students understand the meaning of independence, sketch a scatterplot that displays no association. When there is no association between the variables, knowing the value of the x variable won't help you predict the value of the y variable. When two events are independent, knowing that one event occurred won't help you predict if the other event will occur. We will be talking about independence in every remaining chapter of the book, so emphasize what it means. The sooner students are comfortable with the idea of independence, the better.

Preparing for Inference

In this example, we showed that "male" and "pierced ear" are not independent for students in this statistics class. We made this conclusion because we had data for the whole population (this statistics class). In Chapter 12, we will use data from a sample to make an inference about whether two variables are independent for a whole population (chi-square test of independence).

Subject preference and gender, again
Checking for independence

PROBLEM:

Is there a relationship between gender and subject preference (math or English)? The two-way table summarizes the relationship between gender and subject preference for a class of 25 AP® Statistics students.

		Gender		
		Male	Female	Total
Preferred subject	Math	8	12	20
	English	2	3	5
	Total	10	15	25

Suppose we choose one of the students in the sample at random. Are the events "male" and "math" independent? Justify your answer.

SOLUTION:

$P(\text{math}|\text{male}) = 8/10 = 0.80$

$P(\text{math}|\text{female}) = 12/15 = 0.80$

Because these probabilities are equal, the events "male" and "math" are independent. Knowing that the student is male doesn't change the probability that the student prefers math.

Teaching Tip

If you made the index cards for the data in the first "Subject preference and gender" alternate example, divide the cards into two stacks: Male and Female. Hold up each pile and ask whether the probability of selecting a "Math" card is the same for both stacks.

Teaching Tip

Notice the last sentence in the example and alternate example. Although it isn't essential to include this contextual conclusion, we believe it helps students understand the two probabilities they are comparing.

EXAMPLE

Gender and handedness
Checking for independence

PROBLEM: Is there a relationship between gender and handedness? To find out, we used Census At School's Random Data Selector to choose an SRS of 100 Australian high school students who completed a survey. The two-way table summarizes the relationship between gender and dominant hand for these students.

		Gender		
		Male	Female	Total
Dominant hand	Right	39	51	90
	Left	7	3	10
	Total	46	54	100

Suppose we choose one of the students in the sample at random. Are the events "male" and "left-handed" independent? Justify your answer.

SOLUTION:

$P(\text{left-handed}|\text{male}) = 7/46 = 0.152$

$P(\text{left-handed}|\text{female}) = 3/54 = 0.056$

Because these probabilities are not equal, the events "male" and "left-handed" are not independent. Knowing that the student is male increases the probability that the student is left-handed.

FOR PRACTICE, TRY EXERCISE 71

In the example, we could have also determined that the two events are not independent by showing that

$$P(\text{left-handed}|\text{male}) = 7/46 = 0.152 \neq P(\text{left-handed}) = 10/100 = 0.100$$

Or we could have focused on whether knowing that the chosen student is left-handed changes the probability that the person is male. Because

$$P(\text{male}|\text{left-handed}) = 7/10 = 0.70 \neq P(\text{male}|\text{right-handed}) = 39/90 = 0.433$$

the events "male" and "left-handed" are not independent.

You might have thought, "Surely there's no connection between gender and handedness. The events 'male' and 'left-handed' are bound to be independent." As the example shows, you can't use your intuition to check whether events are independent. To be sure, you have to calculate some probabilities.

Is there a connection between independence of events and association between two variables? Yes! In the preceding example, we found that the events "male" and "left-handed" were not independent for the sample of 100 Australian high school students. Knowing a student's gender helped us predict his or her dominant hand. By what you learned in Chapter 1, there is an association between gender and handedness *for the students in the sample*. The segmented bar graph shows the association in picture form.

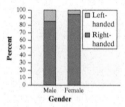

Does that mean an association exists between gender and handedness in the larger population? Maybe. If there is no association between the variables in the population, it would be surprising to choose a random sample of 100 students for which P(left-handed | male), P(left-handed | female), and P(left-handed) were *exactly* equal. But these probabilities should be close to equal if there's no association between the variables in the population. How close is close? We'll discuss this issue further in Chapter 12.

CHECK YOUR UNDERSTANDING

For each random process given, determine whether the events are independent. Justify your answer.

1. Shuffle a standard deck of cards, and turn over the top card. Put it back in the deck, shuffle again, and turn over the top card. Define events A: first card is a heart, and B: second card is a heart.
2. Shuffle a standard deck of cards, and turn over the top two cards, one at a time. Define events A: first card is a heart, and B: second card is a heart.
3. The 28 students in Mr. Tabor's AP® Statistics class took a quiz on conditional probability and independence. The two-way table summarizes the class's quiz results based on gender and whether the student got an A. Choose a student from the class at random. The events of interest are "female" and "got an A."

		Gender	
		Female	Male
Got an A	Yes	3	1
	No	18	6

The General Multiplication Rule

Suppose that A and B are two events resulting from the same random process. We can find the probability P(A or B) with the general addition rule:

$$P(A \text{ or } B) = P(A) + P(B) - P(A \text{ and } B)$$

How do we find the probability that both events happen, P(A and B)?

Consider this situation: about 55% of high school students participate in a school athletic team at some level. Roughly 6% of these athletes go on to play on a college team in the NCAA.[12] What percent of high school students play a sport in high school *and* go on to play on an NCAA team? About 6% of 55%, or roughly 3.3%.

Let's restate the situation in probability language. Suppose we select a high school student at random. What's the probability that the student plays a sport in high school and goes on to play on an NCAA team? The given information suggests that

$$P(\text{high school sport}) = 0.55 \text{ and } P(\text{NCAA team} | \text{high school sport}) = 0.06$$

By the logic just stated,

$$P(\text{high school sport and NCAA team})$$
$$= P(\text{high school sport}) \cdot P(\text{NCAA team} | \text{high school sport})$$
$$= (0.55)(0.06) = 0.033$$

✔ **Answers to CYU**

1. Events A and B are independent. Because we are putting the first card back and shuffling the cards before drawing the second card, knowing what the first card was will not help us predict what the second card will be.

2. Events A and B are not independent. Once we know the suit of the first card, the probability of getting a heart on the second card will change depending on what the first card was.

3. The two events "female" and "got an A" are independent. Knowing that the chosen person is female does not help us predict if she got an A or not. Overall, 4/28 or 1/7 of the students got an A on the quiz. And, among the females, 3/21 or 1/7 got an A. So P(got an A) $= P$(got an A | female).

Teaching Tip

Start with a simple context like this one where students can reason their way to an answer (6% of 55% is about 3.3%). Then have them formalize the process by identifying precisely what 55% and 6% represent in terms of probability. From this example, students can uncover the general multiplication rule.

Teaching Tip

Unlike the addition rule for mutually exclusive events and the general addition rule, we introduce the general multiplication rule before the special multiplication rule for independent events. We do this because the general multiplication rule flows very nicely from the conditional probability formula.

AP® EXAM TIP

The general multiplication rule is *not* on the formula sheets provided on the AP® Statistics exam. However, the conditional probability formula we introduced earlier *is* on the formula sheet. Fortunately, the formula for the general multiplication rule is actually the same as the conditional probability formula, just rearranged.

Teaching Tip

This formula can be extended for three or more events:

P(A and B and C)
$\;\;= P(A \cap B \cap C)$
$\;\;= P(A) \cdot P(B \mid A) \cdot P(C \mid A \cap B)$

ALTERNATE EXAMPLE   Skill 3.A

Hot coffee
The general multiplication rule

PROBLEM:
Students who work at a local coffee shop recorded the drink orders of all the customers on a Saturday. They found that 64% of customers ordered a hot drink, and 80% of these customers added cream to their drink. Find the probability that a randomly selected Saturday customer orders a hot drink and adds cream to the drink.

SOLUTION:

P(hot drink and adds cream)
$\;\;= P$(hot drink) $\cdot P$(adds cream | hot drink)
$\;\;= (0.64)(0.80)$
$\;\;= 0.512$

This is an example of the **general multiplication rule**.

> **DEFINITION   General multiplication rule**
> For any random process, the probability that events A and B both occur can be found using the **general multiplication rule**:
> $$P(\text{A and B}) = P(A \cap B) = P(A) \cdot P(B \mid A)$$

The general multiplication rule says that for both of two events to occur, first one must occur. Then, given that the first event has occurred, the second must occur. To confirm that this result is correct, start with the conditional probability formula

$$P(B \mid A) = \frac{P(B \cap A)}{P(A)}$$

The numerator gives the probability we want because $P(B \cap A)$ is the same as $P(A \cap B)$. Multiply both sides of the previous equation by $P(A)$ to get

$$P(A) \cdot P(B \mid A) = P(A \cap B)$$

EXAMPLE

Teens and social media
The general multiplication rule

PROBLEM: The Pew Internet and American Life Project reported that 79% of teenagers (ages 13 to 17) use social media, and that 39% of teens who use social media feel pressure to post content that will be popular and get lots of comments or likes.[13] Find the probability that a randomly selected teen uses social media and feels pressure to post content that will be popular and get lots of comments or likes.

SOLUTION:

P(use social media and feel pressure) $= P$(use social media) $\cdot P$(feel pressure | use social media)
$\;\;= (0.79)(0.39) = 0.308$

FOR PRACTICE, TRY EXERCISE 77

Tree Diagrams and Conditional Probability

Shannon hits the snooze button on her alarm on 60% of school days. If she hits snooze, there is a 0.70 probability that she makes it to her first class on time. If she doesn't hit snooze and gets up right away, there is a 0.90 probability that she makes it to class on time. Suppose we select a school day at random and record whether Shannon hits the snooze button and whether she arrives in class on time. Figure 5.7 shows a **tree diagram** for this random process.

Teaching Tip

In this chapter, students will learn several strategies for approaching probability questions:

1. Simulation
2. Sample space
3. Two-way tables
4. Venn diagrams
5. **Tree diagrams**
6. Formulas

Section 5.3 Conditional Probability and Independence 367

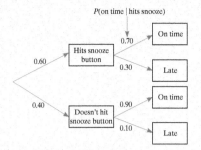

$P(\text{on time} \mid \text{hits snooze})$

FIGURE 5.7 A tree diagram displaying the sample space of randomly choosing a school day and noting if Shannon hits the snooze button or not and whether she gets to her first class on time.

Teaching Tip

When students are constructing a tree diagram, consider asking them to calculate the probability for each of the possible outcomes. In this case, there are four possible outcomes. Then ask students to check that the sum of the probabilities for all the outcomes is 1. Calculating all these values should help them answer any question that could be asked.

$(0.60)(0.70) = 0.42$

$(0.60)(0.30) = 0.18$

$(0.40)(0.90) = 0.36$

$(0.40)(0.10) = 0.04$

$0.42 + 0.18 + 0.36 + 0.04 = 1$

There are only two possible outcomes at the first "stage" of this random process: Shannon hits the snooze button or she doesn't. The first set of branches in the tree diagram displays these outcomes with their probabilities. The second set of branches shows the two possible results at the next "stage" of the process—Shannon gets to her first class on time or arrives late—and the probability of each result based on whether or not she hit the snooze button. Note that the probabilities on the second set of branches are *conditional* probabilities, like $P(\text{on time} \mid \text{hits snooze}) = 0.70$.

Teaching Tip

Students will often misinterpret the 0.70 as the $P(\text{on time and hits snooze})$. This is incorrect because 0.70 is the conditional probability $P(\text{on time} \mid \text{hits snooze})$. The $P(\text{on time and hits snooze})$ is given by $(0.60)(0.70) = 0.42$.

> **DEFINITION** **Tree diagram**
>
> A **tree diagram** shows the sample space of a random process involving multiple stages. The probability of each outcome is shown on the corresponding branch of the tree. All probabilities after the first stage are conditional probabilities.

We can ask some interesting questions related to the tree diagram:

- **What is the probability that Shannon hits the snooze button and is late for class on a randomly selected school day?** The general multiplication rule provides the answer:

$$P(\text{hits snooze and late}) = P(\text{hits snooze}) \cdot P(\text{late} \mid \text{hits snooze})$$
$$= (0.60)(0.30)$$
$$= 0.18$$

There is an 18% chance that Shannon hits the snooze button and is late for class. Note that the previous calculation amounts to multiplying probabilities along the branches of the tree diagram.

- **What's the probability that Shannon is late to class on a randomly selected school day?** Figure 5.8 on the next page illustrates two ways this can happen: Shannon hits the snooze button and is late or she doesn't hit snooze and is late. Because these outcomes are mutually exclusive,

$$P(\text{late}) = P(\text{hits snooze and late}) + P(\text{doesn't hits snooze and late})$$

Teaching Tip

Students will sometimes struggle with choosing the right strategy for solving a probability question. Here are some clues. Most conditional probability questions can be solved using a tree diagram or two-way table. However, if the problem provides conditional probabilities or describes a sequence of events, it is often easier to use a tree diagram. If the problem provides counts or proportions of people in different categories, a two-way table (or a Venn diagram) is often easier to use.

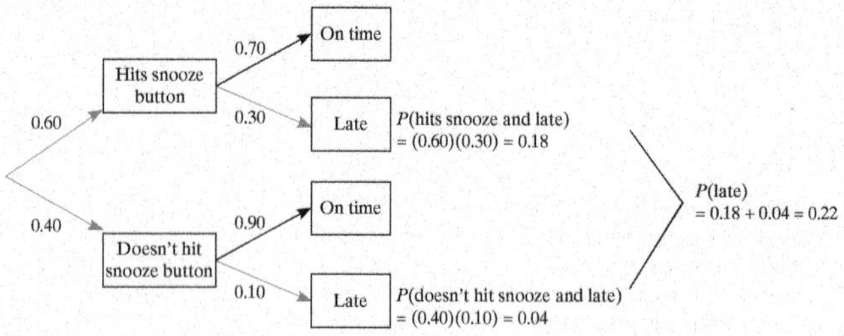

FIGURE 5.8 Tree diagram showing the two possible ways that Shannon can be late to class on a randomly selected day.

The general multiplication rule tells us that

$$P(\text{doesn't hit snooze and late}) = P(\text{doesn't hit snooze}) \cdot P(\text{late} \mid \text{doesn't hit snooze})$$
$$= (0.40)(0.10)$$
$$= 0.04$$

So $P(\text{late}) = 0.18 + 0.04 = 0.22$. There is a 22% chance that Shannon will be late to class.

- **Suppose that Shannon is late for class on a randomly chosen school day. What is the probability that she hit the snooze button that morning?** To find this probability, we start with the given information that Shannon is late, which is displayed on the second set of branches in the tree diagram, and ask whether she hit the snooze button, which is shown on the first set of branches. We can use the information from the tree diagram and the conditional probability formula to do the required calculation:

$$P(\text{hit snooze button} \mid \text{late}) = \frac{P(\text{hit snooze button and late})}{P(\text{late})}$$
$$= \frac{0.18}{0.22}$$
$$= 0.818$$

Given that Shannon is late for school on a randomly selected day, there is a 0.818 probability that she hit the snooze button.

Some interesting conditional probability questions—like this one about P(hit snooze button | late)—involve "going in reverse" on a tree diagram. Note that we just use the conditional probability formula and plug in the appropriate values to answer such questions.

This method for solving conditional probability problems that involve "going backward" in a tree diagram is sometimes referred to as *Bayes's theorem*. It was developed by the Reverend Thomas Bayes in the 1700s.

Teaching Tip

Students don't really need the formula to answer this question. Ask them to calculate the probability of each of the 4 possible outcomes. Next, they should circle the two events that represent the given condition of "late." Then ask students to put a star by the outcome we are looking for here (hits snooze button and late). The conditional probability is given by the starred value divided by the sum of the circled values.

$(0.60)(0.70) = 0.42$

$(0.60)(0.30) = 0.18*$

$(0.40)(0.90) = 0.36$

$(0.40)(0.10) = 0.04$

EXAMPLE

Do people read more ebooks or print books?
Tree diagrams and probability

PROBLEM: Recently, Harris Interactive reported that 20% of millennials, 25% of Gen Xers, 21% of baby boomers, and 17% of matures (age 68 and older) read more ebooks than print books. According to the U.S. Census Bureau, 34% of those 18 and over are millennials, 22% are Gen Xers, 30% are baby boomers, and 14% are matures. Suppose we select one U.S. adult at random and record which generation the person is from and whether she or he reads more ebooks or print books.

(a) Draw a tree diagram to model this random process.

(b) Find the probability that the person reads more ebooks than print books.

(c) Suppose the chosen person reads more ebooks than print books. What's the probability that she or he is a millennial?

SOLUTION:

(a)

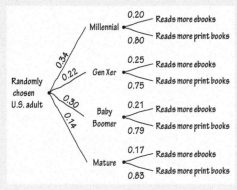

(b) $P(\text{reads more ebooks}) = (0.34)(0.20) + (0.22)(0.25) + (0.30)(0.21) + (0.14)(0.17)$
$$= 0.0680 + 0.0550 + 0.0630 + 0.0238$$
$$= 0.2098$$

(c) $P(\text{millennial} \mid \text{reads more ebooks}) = \dfrac{P(\text{millennial and reads more ebooks})}{P(\text{reads more ebooks})}$

$$\boxed{P(A \mid B) = \frac{P(A \cap B)}{P(B)}}$$

$$= \frac{0.068}{0.2098}$$
$$= 0.3241$$

FOR PRACTICE, TRY EXERCISE 81

Who prefers pop music?
Tree diagrams and conditional probability

PROBLEM:
In 2015, Spotify revealed that about 47% of its users are aged 13–24, 25% are aged 25–34, and 28% are 35 or older. Suppose that for the 13- to 24-year-olds, 85% identified pop as their favorite music genre, 59% of the 25- to 34-year-olds identified pop as their favorite music genre, and 23% of the users 35 or older identified pop as their favorite music genre. Suppose we select one 2015 Spotify user at random and record his or her age and whether his or her favorite music genre is pop.

(a) Draw a tree diagram to model this random process.

(b) Find the probability that the person identifies his or her favorite music genre as pop.

(c) Suppose the chosen person identifies his or her favorite music genre as pop. What's the probability that he or she is aged 13–24?

SOLUTION:

(a)

```
                        0.85  → Pop
                13-24 <
         0.47  /        0.15  → Not pop
              /
Spotify  0.25         0.59  → Pop
user  <------ 25-34 <
              \        0.41  → Not pop
         0.28  \
                35+  <  0.23  → Pop
                        0.77  → Not pop
```

(b) $P(\text{Pop}) = (0.47)(0.85) + (0.25)(0.59)$
$$+ (0.28)(0.23)$$
$$= 0.3995 + 0.1475 + 0.0644$$
$$= 0.6114$$

(c) $P(13 - 24 \mid \text{Pop})$

$$= \frac{P(13 - 24 \text{ and Pop})}{P(\text{Pop})}$$

$$= \frac{0.3995}{0.6114}$$

$$= 0.6534$$

PROBLEM:

In a 2012 interview with CBS News, Tom Hanks said that "80% of the population are really great, caring people who will help you and tell the truth. And I think 20% of the population are crooks and liars." A new lie detector suit has been tested and was shown to correctly identify truthful people 88.9% of the time and correctly identify liars 75.6% of the time. A positive test is one in which the person is identified as a liar and a negative test is one in which the person is identified as truthful. Sometimes, the test suggests that a truthful person is a liar (a "false positive"); other times, the test indicates that a liar is being truthful (a "false negative").

Assume that Tom Hanks and the company that makes the lie detector suit are *telling the truth.* A randomly selected person from the population tests positive for being a liar. Find the probability that this person is a liar.

```
               0.111    Positive test
        Truthful
   0.80           0.889   Negative test
Person
   0.20           0.756    Positive test
        Liar
               0.244    Negative test
```

$$P(\text{liar} \mid \text{positive test}) = \frac{P(\text{liar and positive test})}{P(\text{positive test})}$$

$$= \frac{(0.20)(0.756)}{(0.80)(0.111) + (0.20)(0.756)}$$

$$= \frac{0.1512}{0.2400} = 0.63$$

One of the most important applications of conditional probability is in the area of drug and disease testing.

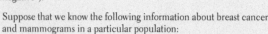

EXAMPLE

Mammograms
Tree diagrams and conditional probability

PROBLEM: Many women choose to have annual mammograms to screen for breast cancer after age 40. A mammogram isn't foolproof. Sometimes the test suggests that a woman has breast cancer when she really doesn't (a "false positive"). Other times, the test says that a woman doesn't have breast cancer when she actually does (a "false negative").

Suppose that we know the following information about breast cancer and mammograms in a particular population:

- One percent of the women aged 40 or over in this population have breast cancer.
- For women who have breast cancer, the probability of a negative mammogram is 0.03.
- For women who don't have breast cancer, the probability of a positive mammogram is 0.06.

A randomly selected woman aged 40 or over from this population tests positive for breast cancer in a mammogram. Find the probability that she actually has breast cancer.

SOLUTION:

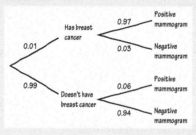

Start by making a tree diagram to summarize the possible outcomes.

- Because 1% of women in this population have breast cancer, 99% don't have breast cancer.
- Of those women who do have breast cancer, 3% would test negative on a mammogram. The remaining 97% would (correctly) test positive.
- Among the women who don't have breast cancer, 6% would test positive on a mammogram. The remaining 94% would (correctly) test negative.

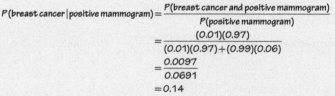

$$P(\text{breast cancer} \mid \text{positive mammogram}) = \frac{P(\text{breast cancer and positive mammogram})}{P(\text{positive mammogram})}$$

$$= \frac{(0.01)(0.97)}{(0.01)(0.97) + (0.99)(0.06)}$$

$$= \frac{0.0097}{0.0691}$$

$$= 0.14$$

FOR PRACTICE, TRY EXERCISE 83

Are you surprised by the final result of the example—given that a randomly selected woman from the population in question has a positive mammogram, there is only about a 14% chance that she has breast cancer? Most people are. Sometimes a two-way table that includes counts is more convincing.

Preparing for Inference

In Chapter 9, we will discuss Type I and Type II errors. In the example on this page, the null hypothesis is "The woman does not have breast cancer," so a Type I error would be a "false positive" (concluding she has cancer when she does not) and a Type II error would be a "false negative" (concluding she does not have cancer when she does). Clearly, a false negative has more serious consequences than a false positive. In Chapter 9, we will find that increasing the alpha level for a significance test will increase the probability of a Type I error and decrease the probability of a Type II error, which would be desired in this situation.

To make calculations simple, we'll suppose that there are exactly 10,000 women aged 40 or over in this population, and that exactly 100 have breast cancer (that's 1% of the women).

- How many of those 100 would have a positive mammogram? It would be 97% of 100, or 97 of them. That leaves 3 who would test negative.
- How many of the 9900 women who don't have breast cancer would get a positive mammogram? Six percent of them, or $(9900)(0.06) = 594$ women. The remaining $9900 - 594 = 9306$ would test negative.
- In total, $97 + 594 = 691$ women would have positive mammograms and $3 + 9306 = 9309$ women would have negative mammograms.

This information is summarized in the two-way table.

		Has breast cancer?		
		Yes	No	Total
Mammogram result	Positive	97	594	691
	Negative	3	9306	9309
	Total	100	9900	10,000

Given that a randomly selected woman has a positive mammogram, the two-way table shows that the conditional probability is

$$P(\text{breast cancer} \mid \text{positive mammogram}) = 97/691 = 0.14$$

This example illustrates an important fact when considering proposals for widespread testing for serious diseases or illegal drug use: if the condition being tested is uncommon in the population, many positives will be false positives. The best remedy is to retest any individual who tests positive.

CHECK YOUR UNDERSTANDING

A computer company makes desktop, laptop, and tablet computers at factories in two states: California and Texas. The California factory produces 40% of the company's computers and the Texas factory makes the rest. Of the computers made in California, 25% are desktops, 30% are laptops, and the rest are tablets. Of those made in Texas, 10% are desktops, 20% are laptops, and the rest are tablets. All computers are first shipped to a distribution center in Missouri before being sent out to stores. Suppose we select a computer at random from the distribution center and observe where it was made and whether it is a desktop, laptop, or tablet.[14]

1. Construct a tree diagram to model this random process.
2. Find the probability that the computer is a tablet.
3. Given that a tablet computer is selected, what is the probability that it was made in California?

✔ **Answers to CYU**

1.

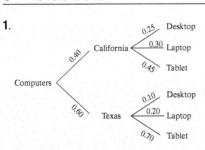

2. $P(\text{tablet}) = P(\text{California} \cap \text{tablet}) + P(\text{Texas} \cap \text{tablet}) = (0.40)(0.45) + (0.60)(0.70) = 0.60$; the probability of selecting a tablet is 0.60.

3. $P(\text{California} \mid \text{tablet}) = \dfrac{P(\text{California} \cap \text{tablet})}{P(\text{tablet})} = \dfrac{(0.4)(0.45)}{0.60} = \dfrac{0.18}{0.60} = 0.30$; given that a tablet was selected, there is a 0.30 probability that it was made in California.

The Multiplication Rule for Independent Events

What happens to the general multiplication rule in the special case when events A and B are independent? In that case, $P(B \mid A) = P(B)$. We can simplify the general multiplication rule as follows:

$$P(A \text{ and } B) = P(A \cap B) = P(A) \cdot P(B \mid A)$$
$$= P(A) \cdot P(B)$$

This result is known as the **multiplication rule for independent events**.

> This rule gives us another way to determine whether two events are independent. If $P(A \cap B) = P(A) \cdot P(B)$, then A and B are independent events.

> **DEFINITION   Multiplication rule for independent events**
>
> If A and B are independent events, the probability that A and B both occur is
>
> $$P(A \text{ and } B) = P(A \cap B) = P(A) \cdot P(B)$$

 Note that this rule applies only to independent events.

Suppose that Pedro drives the same route to work on Monday through Friday. His route includes one traffic light. The probability that the light will be green when Pedro arrives is 0.42, yellow is 0.03, and red is 0.55.

1. **What's the probability that the light is green on Monday and red on Tuesday?** Let event A be green light on Monday and event B be red light on Tuesday. These two events are independent because knowing that the light was green on Monday doesn't help us predict the color of the light on Tuesday. By the multiplication rule for independent events,

$$P(\text{green on Monday and red on Tuesday}) = P(A \text{ and } B)$$
$$= P(A) \cdot P(B)$$
$$= (0.42)(0.55)$$
$$= 0.231$$

There's about a 23% chance that the light will be green on Monday and red on Tuesday.

2. **What's the probability that Pedro finds the light red on Monday through Friday?** We can extend the multiplication rule for independent events to more than two events:

$P(\text{red Monday } and \text{ red Tuesday } and \text{ red Wednesday } and \text{ red Thursday } and \text{ red Friday})$
$= P(\text{red Monday}) \cdot P(\text{red Tuesday}) \cdot P(\text{red Wednesday}) \cdot P(\text{red Thursday}) \cdot P(\text{red Friday})$
$= (0.55)(0.55)(0.55)(0.55)(0.55)$
$= (0.55)^5$
$= 0.0503$

There is about a 5% chance that Pedro will encounter a red light on all five days in a work week.

EXAMPLE

The *Challenger* disaster
Multiplication rule for independent events

PROBLEM: On January 28, 1986, the space shuttle *Challenger* exploded on takeoff. All seven crew members were killed. Following the disaster, scientists and statisticians helped analyze what went wrong. They determined that the failure of O-ring joints in the shuttle's booster rockets was to blame. Under the cold conditions that day, experts estimated that the probability that an individual O-ring joint would function properly was 0.977. But there were six of these O-ring joints, and all six had to function properly for the shuttle to launch safely. Assuming that O-ring joints succeed or fail independently, find the probability that the shuttle would launch safely under similar conditions.

SOLUTION:

P(O-ring 1 OK and O-ring 2 OK and O-ring 3 OK and O-ring 4 OK and O-ring 5 OK and O-ring 6 OK)

$= P$(O-ring 1 OK) · P(O-ring 2 OK) · P(O-ring 3 OK) · P(O-ring 4 OK) · P(O-ring 5 OK) · P(O-ring 6 OK)

$= (0.977)(0.977)(0.977)(0.977)(0.977)(0.977)$

$= (0.977)^6$

$= 0.870$

FOR PRACTICE, TRY EXERCISE 89

The multiplication rule for independent events can also be used to help find P(at least one). In the preceding example, the shuttle would *not* launch safely under similar conditions if 1 or 2 or 3 or 4 or 5 or all 6 O-ring joints fail—that is, if *at least one* O-ring fails. The only possible number of O-ring failures excluded is 0. So the events "at least one O-ring joint fails" and "no O-ring joints fail" are complementary events. By the complement rule,

$$P(\text{at least one O-ring fails}) = 1 - P(\text{no O-ring fails})$$
$$= 1 - 0.87$$
$$= 0.13$$

That's a very high chance of failure! As a result of this analysis following the *Challenger* disaster, NASA made important safety changes to the design of the shuttle's booster rockets.

EXAMPLE

Rapid HIV testing
Finding the probability of "at least one"

PROBLEM: Many people who visit clinics to be tested for HIV, the virus that causes AIDS, don't come back to learn their test results. Clinics now use "rapid HIV tests" that give a result while the client waits. In a clinic in Malawi, for example, use of rapid tests increased the percentage of clients who learned their test results from 69% to over 99%.

Guess on every question
Multiplication rule for independent events

PROBLEM:
Mrs. Johnson decides to give her AP® Statistics class a pop quiz with 4 multiple-choice questions. Each multiple-choice question has 5 possible answer choices. John did not do the homework or do any studying, so he used the RandInt function on his calculator to guess on all 4 questions. What is the probability that John guesses correctly on all 4 questions?

SOLUTION:

P(#1 correct and #2 correct and #3 correct and #4 correct)

$= P$(#1 correct) · P(#2 correct) · P(#3 correct) · P(#4 correct)

$= (0.20)(0.20)(0.20)(0.20)$

$= (0.20)^4$

$= 0.0016$

Making Connections

Without the complement rule here, we would have to calculate the probability of exactly 1 failure, exactly 2 failures, exactly 3 failures, and so on. This would require significantly more work (but we could do it with binomial random variables in Chapter 6!).

First Trimester Screen Finding the probability of "at least one"

PROBLEM:
The First Trimester Screen is a noninvasive test given during the first trimester of pregnancy to determine if there are specific chromosomal abnormalities in the fetus. According to a study published in the *New England Journal of Medicine*, approximately 5% of normal pregnancies will receive a positive result.

If 100 women with normal pregnancies are tested with the First Trimester Screen, what is the probability that at least 1 woman will receive a positive result? Assume that test results for different individuals are independent.

SOLUTION:

P(no false positives) $= P$(all 100 tests negative)

$= (0.95)(0.95) \ldots (0.95)$

$= 0.95^{100}$

$= 0.0059$

P(at least one false positive) $= 1 - 0.0059 = 0.9941$

The trade-off for fast results is that rapid tests are less accurate than slower laboratory tests. Applied to people who have no HIV antibodies, one rapid test has a probability of about 0.004 of producing a false positive (i.e., of falsely indicating that antibodies are present).[15]

If a clinic tests 200 randomly selected people who are free of HIV antibodies, what is the probability that at least one false positive will occur? Assume that test results for different individuals are independent.

SOLUTION:

$$P(\text{no false positives}) = P(\text{all 200 tests negative})$$
$$= (0.996)(0.996) \cdots (0.996)$$
$$= 0.996^{200}$$
$$= 0.4486$$
$$P(\text{at least one false positive}) = 1 - 0.4486 = 0.5514$$

> Start by finding P(no false positives).

> The probability that any individual test result is negative is $1 - 0.004 = 0.996$.

FOR PRACTICE, TRY EXERCISE 91

Teaching Tip

There are only two "big idea" formulas in this chapter, and a special case for each one:

(1) General addition rule:
$P(A \text{ or } B) = P(A) + P(B) - P(A \text{ and } B)$
If the two events are *mutually exclusive,* then $P(A \text{ and } B) = 0$ and the formula becomes $P(A \text{ or } B) = P(A) + P(B)$.

(2) General multiplication rule:
$P(A \text{ and } B) = P(A) \cdot P(B \mid A)$
If the two events are *independent,* then $P(B \mid A) = P(B)$ and the formula becomes $P(A \text{ and } B) = P(A) \cdot P(B)$.
Remember that formulas can be almost completely avoided in this chapter if students are fluent with the other probability strategies.

Teaching Tip

Students will often confuse "mutually exclusive" and "independent" events, or even think that they are interchangeable. For each probability question, they must carefully consider the definition of each of these terms. Exercise 97 on page 380 offers a great context for this discussion.

USING THE MULTIPLICATION RULE FOR INDEPENDENT EVENTS WISELY
The multiplication rule $P(A \text{ and } B) = P(A) \cdot P(B)$ holds if A and B are *independent* but not otherwise. The addition rule $P(A \text{ or } B) = P(A) + P(B)$ holds if A and B are *mutually exclusive* but not otherwise. Resist the temptation to use these simple rules when the conditions that justify them are not met.

Hagar the Horrible

EXAMPLE

Watch the weather!
Beware lack of independence!

PROBLEM: Hacienda Heights and La Puente are two neighboring suburbs in the Los Angeles area. According to the local newspaper, there is a 50% chance of rain tomorrow in Hacienda Heights and a 50% chance of rain in La Puente. Does this mean that there is a $(0.5)(0.5) = 0.25$ probability that it will rain in both cities tomorrow?

ALTERNATE EXAMPLE   Skill 3.A

All blondes   Beware lack of independence!

PROBLEM:
In the United States, 28% of the population has hair that is blonde. A couple in the United States is expecting quadruplets. Does this mean there is a $(0.28)(0.28)(0.28)(0.28) = 0.0006$ probability that the quadruplets will all have hair that is blonde?

SOLUTION:
No; it is not appropriate to multiply the four probabilities, because each child's hair color is not independent of the hair colors of the other children. Knowing the hair color of one child will help predict the color of the other children because hair color is a trait inherited from parents' genes.

SOLUTION: No; it is not appropriate to multiply the two probabilities, because "raining tomorrow in Hacienda Heights" and "raining tomorrow in La Puente" are not independent events. If it is raining in one of these locations, there is a high probability that it is raining in the other location because they are geographically close to each other.

FOR PRACTICE, TRY EXERCISE 93

Is there a connection between mutually exclusive and independent? Let's start with a new random process. Choose a U.S. adult at random. Define event A: the person is male, and event B: the person is pregnant. It's fairly clear that these two events are mutually exclusive (can't happen together)! Are they also independent?

If you know that event A has occurred, does this change the probability that event B happens? Of course! If we know the person is male, then the chance that the person is pregnant is 0. But the probability of selecting *someone* who is pregnant is greater than 0. Because $P(B \mid A) \neq P(B)$, the two events are not independent. Two mutually exclusive events (with nonzero probabilities) can *never* be independent, because if one event happens, the other event is guaranteed not to happen.

CHECK YOUR UNDERSTANDING

Questions 1 and 2 refer to the following setting. New Jersey Transit claims that its 8:00 A.M. train from Princeton to New York has probability 0.9 of arriving on time. Assume that this claim is true.

1. Find the probability that the train arrives late on Monday but on time on Tuesday.
2. What's the probability that the train arrives late at least once in a 5-day week?
3. Government data show that 8% of adults are full-time college students and that 15% of adults are age 65 or older. If we randomly select an adult, is *P*(full-time college student and age 65 or older) = (0.08)(0.15)? Why or why not?

Section 5.3 | Summary

- A **conditional probability** describes the probability that one event happens given that another event is already known to have happened.
- One way to calculate a conditional probability is to use the formula

$$P(A \mid B) = \frac{P(A \text{ and } B)}{P(B)} = \frac{P(A \cap B)}{P(B)} = \frac{P(\text{both events occur})}{P(\text{given event occurs})}$$

- When knowing whether or not one event has occurred does not change the probability that another event happens, we say that the two events are independent. Events A and B are independent if

$$P(A \mid B) = P(A \mid B^C) = P(A)$$

or, alternatively, if

$$P(B \mid A) = P(B \mid A^C) = P(B)$$

- Use the **general multiplication rule** to calculate the probability that events A and B both occur:

$$P(A \text{ and } B) = P(A \cap B) = P(A) \cdot P(B \mid A)$$

TRM **Section 5.3 Quiz**

There is one quiz available for this section in the Teacher's Resource Materials. Click on the link in the TE-Book, look on the TRFD, or download from the Teacher's Resources on the book's digital platform. You can also create your own quiz using the ExamView® Assessment Suite that is part of the TPS6 program. Questions are coded by Learning Target to make it easy to build parallel quizzes.

Teaching Tip

Many students confuse when two events are independent and when two events are mutually exclusive. Here are some special cases to illustrate the difference.

- Select one card from a standard deck and define the events A: the card is red and B: the card is a club. Because there are no red clubs, events A and B are mutually exclusive. However, because $P(A) = 0.5$ and $P(A \mid B) = 0$, these events are not independent.

- Select one card from a standard deck and define the events A: the card is red and B: the card is a 7. Because there are two red 7s, events A and B are not mutually exclusive. However, because $P(A) = 0.5$ and $P(A \mid B) = 0.5$, these events are independent.

- Select one card from a standard deck and define the events A: the card is red and B: the card is a heart. Because all the hearts are red, events A and B are not mutually exclusive. Also, because $P(A) = 0.5$ and $P(A \mid B) = 1$, these events are not independent.

Overall, because of its prominence in later chapters, it is much more important that students understand the idea of independence than the idea of mutual exclusivity.

✓ **Answers to CYU**

1. P(train arrives late on Monday and on time on Tuesday) = P (Arrives late on Monday) · P(Arrives on time on Tuesday) = $(0.10)(0.90) = 0.09$

2. P(train arrives late at least once in a 5-day week) = $1 - P$ (train never arrives late in a 5-day week) = $1 - (0.90)^5 = 0.41$

3. P(full-time college student and age 65 or older) $\neq (0.08)(0.15)$ because "full time college student" and "adults age 65 or older" are not independent events. Knowing that a person is a full-time college student decreases the likelihood that he or she is an adult age 65 or older.

Answers to Section 5.3 Exercises

5.61 (a) $P(T|E) = \dfrac{44}{200} = 0.22$; given that the child is from England, there is a 0.22 probability that the student selected the superpower of telepathy.

(b) $P(E|S^C) = \dfrac{54 + 52 + 30 + 44}{99 + 96 + 67 + 110} = \dfrac{180}{372} = 0.484$; given that the child did not choose superstrength, there is a 0.484 probability that the child is from England.

5.62 (a) $P(G|M) = \dfrac{758}{2459} = 0.308$; given that the survey respondent is a male, there is a 0.308 probability that the person responded "a good chance."

(b) $P(M^C|N^C) = $
$\dfrac{426 + 696 + 663 + 486}{712 + 1416 + 1421 + 1083} = \dfrac{2271}{4632} = 0.490$; given that the respondent did not choose "almost no chance," there is a 0.490 probability that the respondent is a female.

5.63 (a) $P(\text{female}|\text{about right}) = \dfrac{560}{560 + 295} = 0.655$; given that the person perceived his or her body image as about right, there is a 0.655 probability that the person is a female.

(b) $P(\text{not overweight}|F) = $
$\dfrac{560 + 37}{560 + 163 + 37} = \dfrac{597}{760} = 0.786$; given that the person selected is female, there is a 0.786 probability that she did not perceive her body image as overweight.

5.64 (a) $P(\text{hatched}|\text{assigned to hot water}) = \dfrac{75}{75 + 29} = \dfrac{75}{104} = 0.721$; given that the chosen egg was assigned to hot water, there is a 0.721 probability that it hatched.

(b) $P(\text{not assigned to hot water}|\text{hatched}) = \dfrac{16 + 38}{16 + 38 + 75} = \dfrac{54}{129} = 0.419$; given that the chosen egg hatched, there is a 0.419 probability that it was not assigned to hot water.

- When a random process involves multiple stages, a **tree diagram** can be used to display the sample space and to help answer questions involving conditional probability.
- In the special case of independent events, the multiplication rule becomes
$$P(A \text{ and } B) = P(A \cap B) = P(A) \cdot P(B)$$

Section 5.3 | Exercises

61. Superpowers A random sample of 415 children from England and the United States who completed a survey in a recent year was selected. Each student's country of origin was recorded along with which superpower they would most like to have: the ability to fly, ability to freeze time, invisibility, superstrength, or telepathy (ability to read minds). The data are summarized in the two-way table.

		Country		
		England	U.S.	Total
Superpower	Fly	54	45	99
	Freeze time	52	44	96
	Invisibility	30	37	67
	Superstrength	20	23	43
	Telepathy	44	66	110
	Total	200	215	415

Suppose we randomly select one of these students. Define events E: England, T: telepathy, and S: superstrength.

(a) Find $P(T|E)$. Interpret this value in context.

(b) Given that the student did not choose superstrength, what's the probability that this child is from England? Write your answer as a probability statement using correct symbols for the events.

62. Get rich A survey of 4826 randomly selected young adults (aged 19 to 25) asked, "What do you think are the chances you will have much more than a middle-class income at age 30?" The two-way table summarizes the responses.[16]

		Gender		
		Female	Male	Total
Opinion	Almost no chance	96	98	194
	Some chance but probably not	426	286	712
	A 50-50 chance	696	720	1416
	A good chance	663	758	1421
	Almost certain	486	597	1083
	Total	2367	2459	4826

Choose a survey respondent at random. Define events G: a good chance, M: male, and N: almost no chance.

(a) Find $P(G|M)$. Interpret this value in context.

(b) Given that the chosen survey respondent didn't say "almost no chance," what's the probability that this person is female? Write your answer as a probability statement using correct symbols for the events.

63. Body image A random sample of 1200 U.S. college students was asked, "What is your perception of your own body? Do you feel that you are overweight, underweight, or about right?" The two-way table below summarizes the data on perceived body image by gender.[17]

		Gender	
		Female	Male
Body image	About right	560	295
	Overweight	163	72
	Underweight	37	73

Suppose we randomly select one of the survey respondents.

(a) Given that the person perceived his or her body image as about right, what's the probability that the person is female?

(b) If the person selected is female, what's the probability that she did not perceive her body image as overweight?

64. Temperature and hatching How is the hatching of water python eggs influenced by the temperature of a snake's nest? Researchers randomly assigned newly laid eggs to one of three water temperatures: cold, neutral, or hot. Hot duplicates the extra warmth provided by the mother python, and cold duplicates the absence of the mother.

		Nest temperature		
		Cold	Neutral	Hot
Hatching status	Hatched	16	38	75
	Didn't hatch	11	18	29

Suppose we select one of the eggs at random.

(a) Given that the chosen egg was assigned to hot water, what is the probability that it hatched?

(b) If the chosen egg hatched, what is the probability that it was not assigned to hot water?

65. Foreign-language study Choose a student in grades 9 to 12 at random and ask if he or she is studying a language other than English. Here is the distribution of results:

Language	Spanish	French	German	All others	None
Probability	0.26	0.09	0.03	0.03	0.59

(a) What's the probability that the student is studying a language other than English?

(b) What is the probability that a student is studying Spanish given that he or she is studying some language other than English?

66. Income tax returns Here is the distribution of the adjusted gross income (in thousands of dollars) reported on individual federal income tax returns in a recent year:

Income	< 25	25–49	50–99	100–499	≥ 500
Probability	0.431	0.248	0.215	0.100	0.006

(a) What is the probability that a randomly chosen return shows an adjusted gross income of $50,000 or more?

(b) Given that a randomly chosen return shows an income of at least $50,000, what is the conditional probability that the income is at least $100,000?

67. Tall people and basketball players Select an adult at random. Define events T: person is over 6 feet tall, and B: person is a professional basketball player. Rank the following probabilities from smallest to largest. Justify your answer.

$P(T)$ $P(B)$ $P(T|B)$ $P(B|T)$

68. Teachers and college degrees Select an adult at random. Define events D: person has earned a college degree, and T: person's career is teaching. Rank the following probabilities from smallest to largest. Justify your answer.

$P(D)$ $P(T)$ $P(D|T)$ $P(T|D)$

69. Dogs and cats In one large city, 40% of all households own a dog, 32% own a cat, and 18% own both. Suppose we randomly select a household and learn that the household owns a cat. Find the probability that the household owns a dog.
pg 361

70. Mac or PC? A recent census at a major university revealed that 60% of its students mainly used Macs. The rest mainly used PCs. At the time of the census, 67% of the school's students were undergraduates. The rest were graduate students. In the census, 23% of respondents were graduate students and used a Mac as their main computer. Suppose we select a student at random from among those who were part of the census and learn that the person mainly uses a Mac. Find the probability that the person is a graduate student.

71. Who owns a home? What is the relationship between educational achievement and home ownership? A random sample of 500 U.S. adults was selected. Each member of the sample was identified as a high school graduate (or not) and as a homeowner (or not). The two-way table summarizes the data.
pg 364

		High school graduate		
		Yes	No	Total
Homeowner	Yes	221	119	340
	No	89	71	160
	Total	310	190	500

Suppose we choose 1 member of the sample at random. Are the events "homeowner" and "high school graduate" independent? Justify your answer.

72. Is this your card? A standard deck of playing cards (with jokers removed) consists of 52 cards in four suits—clubs, diamonds, hearts, and spades. Each suit has 13 cards, with denominations ace, 2, 3, 4, 5, 6, 7, 8, 9, 10, jack, queen, and king. The jacks, queens, and kings are referred to as "face cards." Imagine that we shuffle the deck thoroughly and deal one card. The two-way table summarizes the sample space for this random process based on whether or not the card is a face card and whether or not the card is a heart.

		Type of card		
		Face card	Nonface card	Total
Suit	Heart	3	10	13
	Nonheart	9	30	39
	Total	12	40	52

Are the events "heart" and "face card" independent? Justify your answer.

73. Cell phones The Pew Research Center asked a random sample of 2024 adult cell-phone owners from the United States their age and which type of cell phone they own: iPhone, Android, or other (including non-smartphones). The two-way table summarizes the data.

		Age			
		18–34	35–54	55+	Total
Type of cell phone	iPhone	169	171	127	467
	Android	214	189	100	503
	Other	134	277	643	1054
	Total	517	637	870	2024

Suppose we select one of the survey respondents at random.

(a) Find $P(\text{iPhone} | 18\text{–}34)$.

(b) Use your answer from part (a) to help determine if the events "iPhone" and "18–34" are independent.

of college-educated among all adults, so $P(T|D) < P(D)$.

5.69 *Method 1:*
$$P(D|C) = \frac{P(D \cap C)}{P(C)} = \frac{0.18}{0.32} = 0.563;$$
given that a household owns a cat, there is a 0.563 probability that the household owns a dog.

Method 2: Use a Venn diagram;
$$P(D|C) = \frac{0.18}{0.18 + 0.14} = \frac{0.18}{0.32} = 0.563.$$

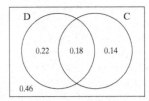

5.70 *Method 1:* $P(G|M) = \dfrac{P(G \cap M)}{P(M)} = \dfrac{0.23}{0.60} = 0.383$; given that a student mainly uses a Mac, there is a 0.383 probability that the person is a graduate student.

Method 2: Use a Venn diagram;
$$P(G|M) = \frac{0.23}{0.37 + 0.23} = \frac{0.23}{0.60} = 0.383.$$

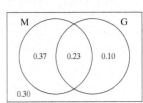

5.71 $P(\text{homeowner}) = 340/500 = 0.68$; $P(\text{homeowner} | \text{high school graduate}) = 221/310 = 0.713$. Because these probabilities are not equal, the events "homeowner" and "high school graduate" are not independent. Knowing that the person is a high school graduate increases the probability that the person is a homeowner.

5.72 $P(\text{heart}) = 13/52 = 0.25$; $P(\text{heart} | \text{face card}) = 3/12 = 0.25$. Because these probabilities are equal, the events "heart" and "face card" are independent. Knowing that a face card is dealt does not change the probability that the card is a heart.

5.73 (a) $P(\text{iPhone} | 18\text{–}34) = 169/517 = 0.327$; given that the adult is aged 18–34, there is a 0.327 probability that the person owns an iPhone. **(b)** First, we determine $P(\text{iPhone}) = 467/2024 = 0.231$. Because these probabilities are not equal, the events "own an iPhone" and "aged 18–34" are not independent. Knowing that the person is aged 18–34 increases the probability that the person owns an iPhone.

5.65 (a) $P(\text{is studying other than English}) = 1 - P(\text{none}) = 1 - 0.59 = 0.41$; there is a 0.41 probability that the student is studying a language other than English.

(b) $P(\text{Spanish} | \text{other than English}) = \dfrac{0.26}{0.41} = 0.6341$; given that the student is studying a language other than English, there is a 0.6341 probability that he or she is studying Spanish.

5.66 (a) $P(\$50,000 \text{ or more}) = 0.215 + 0.100 + 0.006 = 0.321$; there is a 0.321 probability that a randomly chosen return shows an adjusted gross income of $50,000 or more.

(b) $P(\text{at least } 100,000 | \text{at least } 50,000) = \dfrac{0.106}{0.321} = 0.3302$; given that the return shows an income of at least $50,000, there is a 0.3302 probability that the income is at least $100,000.

5.67 $P(B) < P(B|T) < P(T) < P(T|B)$; there are very few professional basketball players, so $P(B)$ should be the smallest probability. If you are a professional basketball player, it is quite likely that you are tall, so $P(T|B)$ should be the largest probability. Finally, it's much more likely to be over 6 feet tall (B) than it is to be a professional basketball player if you're over 6 feet tall (B|T).

5.68 $P(T) < P(T|D) < P(D) < P(D|T)$; nearly everyone whose career is teaching has a college degree, so $P(D|T)$ will have the largest probability (close to 1). A higher percentage of the general adult population will have college degrees than will be teachers, so $P(T) < P(D)$. Finally, the proportion of teachers among the college-educated is smaller than the proportion

5.74 (a) $P(\text{athletic} \mid \text{5th grade}) = 36/108 = 0.333$; given that the student is in 5th grade, there is a 0.333 probability that being athletic is most important to the student. **(b)** First, we determine $P(\text{athletic}) = 98/335 = 0.293$. Because these probabilities are not equal, the events "5th grade" and "athletic" are not independent. Knowing that the student is in 5th grade increases the probability that being athletic is most important to the student.

5.75 There are 36 different possible outcomes: (1, 1), (1, 2), . . . , (6, 6). If the second die is the green die, there are 6 ways for it to show a 4: (1, 4), (2, 4), (3, 4), (4, 4), (5, 4), (6, 4). Of those, only one way sums to 7, so $P(\text{sum of 7} \mid \text{green is 4}) = 1/6 = 0.1667$. Overall, there are 6 ways to get a 7: (1, 6), (2, 5), (3, 4), (4, 3), (5, 2), (6, 1). So $P(\text{sum of 7}) = 6/36 = 0.1667$. Because these two probabilities are the same, the events "sum of 7" and "green die shows a 4" are independent. Knowing that the green die shows a 4 does not change the probability that the sum is 7.

5.76 There are 36 different possible outcomes: (1, 1), (1, 2), . . . , (6, 6). If the second die is the green die, there are 6 ways for it to show a 4: (1, 4), (2, 4), (3, 4), (4, 4), (5, 4), (6, 4). Of those, only one way sums to 8, so $P(\text{sum of 8} \mid \text{green is 4}) = 1/6 = 0.1667$. Overall, there are 5 ways to get an 8: (2, 6), (3, 5), (4, 4), (5, 3), (6, 2). So $P(\text{sum of 8}) = 5/36 = 0.1389$. Because these two probabilities are not the same, the events "sum of 8" and "green die shows a 4" are not independent. Knowing that the green die shows a 4 changes the probability that the sum is 8.

5.77 $P(\text{download music} \cap \text{don't care}) = P(\text{download music}) \cdot P(\text{don't care} \mid \text{download music}) = (0.29)(0.67) = 0.1943 = 19.43\%$

5.78 $P(\text{belong to health club} \cap \text{go twice a week}) = P(\text{belong}) \cdot P(\text{go twice/week} \mid \text{belong}) = (0.40)(0.10) = 0.04 = 4\%$. About 4% of adults belong to a health club and go at least twice a week.

5.79 $P(\text{all three candies have soft centers}) = \frac{14}{20} \cdot \frac{13}{19} \cdot \frac{12}{18} = \frac{2184}{6840} = 0.319$

5.80 $P(\text{all three students are female}) = \frac{20}{30} \cdot \frac{19}{29} \cdot \frac{18}{28} = \frac{6840}{24360} = 0.281$

---

74. Middle school values Researchers carried out a survey of fourth-, fifth-, and sixth-grade students in Michigan. Students were asked whether good grades, athletic ability, or being popular was most important to them. The two-way table summarizes the survey data.[18]

		Grade			
		4th	5th	6th	Total
Most Important	Grades	49	50	69	168
	Athletic	24	36	38	98
	Popular	19	22	28	69
	Total	92	108	135	335

Suppose we select one of these students at random.

(a) Find $P(\text{athletic} \mid \text{5th grade})$.

(b) Use your answer from part (a) to help determine if the events "5th grade" and "athletic" are independent.

75. **Rolling dice** Suppose you roll two fair, six-sided dice—one red and one green. Are the events "sum is 7" and "green die shows a 4" independent? Justify your answer. (See Figure 5.2 on page 342 for the sample space of this random process.)

76. **Rolling dice** Suppose you roll two fair, six-sided dice—one red and one green. Are the events "sum is 8" and "green die shows a 4" independent? Justify your answer. (See Figure 5.2 on page 342 for the sample space of this random process.)

77. **Free downloads?** Illegal music downloading is a big problem: 29% of Internet users download music files, and 67% of downloaders say they don't care if the music is copyrighted.[19] Find the probability that a randomly selected Internet user downloads music and doesn't care if it's copyrighted.

78. **At the gym** Suppose that 10% of adults belong to health clubs, and 40% of these health club members go to the club at least twice a week. Find the probability that a randomly selected adult belongs to a health club and goes there at least twice a week.

79. **Box of chocolates** According to Forrest Gump, "Life is like a box of chocolates. You never know what you're gonna get." Suppose a candymaker offers a special "Gump box" with 20 chocolate candies that look alike. In fact, 14 of the candies have soft centers and 6 have hard centers. Suppose you choose 3 of the candies from a Gump box at random. Find the probability that all three candies have soft centers.

80. **Sampling students** A statistics class with 30 students has 10 males and 20 females. Suppose you choose 3 of the students in the class at random. Find the probability that all three are female.

---

81. **Fill 'er up!** In a certain month, 88% of automobile drivers filled their vehicles with regular gasoline, 2% purchased midgrade gas, and 10% bought premium gas.[20] Of those who bought regular gas, 28% paid with a credit card; of customers who bought midgrade and premium gas, 34% and 42%, respectively, paid with a credit card. Suppose we select a customer at random.

(a) Draw a tree diagram to model this random process.

(b) Find the probability that the customer paid with a credit card.

(c) Suppose the chosen customer paid with a credit card. What's the probability that the customer bought premium gas?

82. **Media usage and good grades** The Kaiser Family Foundation released a study about the influence of media in the lives of young people aged 8–18.[21] In the study, 17% of the youth were classified as light media users, 62% were classified as moderate media users, and 21% were classified as heavy media users. Of the light users who responded, 74% described their grades as good (A's and B's), while only 68% of the moderate users and 52% of the heavy users described their grades as good. Suppose that we select one young person from the study at random.

(a) Draw a tree diagram to model this random process.

(b) Find the probability that this person describes his or her grades as good.

(c) Suppose the chosen person describes his or her grades as good. What's the probability that he or she is a heavy user of media?

83. **First serve** Tennis great Andy Murray made 60% of his first serves in a recent season. When Murray made his first serve, he won 76% of the points. When Murray missed his first serve and had to serve again, he won only 54% of the points.[22] Suppose you randomly choose a point on which Murray served. You get distracted before seeing his first serve but look up in time to see Murray win the point. What's the probability that he missed his first serve?

84. **Lactose intolerance** Lactose intolerance causes difficulty in digesting dairy products that contain lactose (milk sugar). It is particularly common among people of African and Asian ancestry. In the United States (not including other groups and people who consider themselves to belong to more than one race), 82% of the population is White, 14% is Black, and 4% is Asian. Moreover, 15% of Whites, 70% of Blacks, and 90% of Asians are lactose intolerant.[23] Suppose we select a U.S. person at random and find that the person is lactose intolerant. What's the probability that she or he is Asian?

85. **HIV testing** Enzyme immunoassay (EIA) tests are used to screen blood specimens for the presence of

---

5.81 (a)

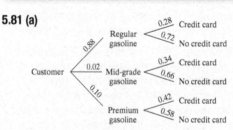

(b) $P(\text{credit card}) = (0.88)(0.28) + (0.02)(0.34) + (0.10)(0.42) = 0.295$

(c) $P(\text{premium gasoline} \mid \text{credit card}) = \frac{(0.10)(0.42)}{0.295} = \frac{0.042}{0.295} = 0.142$

antibodies to HIV, the virus that causes AIDS. Antibodies indicate the presence of the virus. The test is quite accurate but is not always correct. A false positive occurs when the test gives a positive result but no HIV antibodies are actually present in the blood. A false negative occurs when the test gives a negative result but HIV antibodies are present in the blood. Here are approximate probabilities of positive and negative EIA outcomes when the blood tested does and does not actually contain antibodies to HIV: [24]

		Test result	
		+	−
Truth	Antibodies present	0.9985	0.0015
	Antibodies absent	0.0060	0.9940

Suppose that 1% of a large population carries antibodies to HIV in their blood. Imagine choosing a person from this population at random. If the person's EIA test is positive, what's the probability that the person has the HIV antibody?

86. **Metal detector** A boy uses a homemade metal detector to look for valuable metal objects on a beach. The machine isn't perfect—it beeps for only 98% of the metal objects over which it passes, and it beeps for 4% of the nonmetallic objects over which it passes. Suppose that 25% of the objects that the machine passes over are metal. Choose an object from this beach at random. If the machine beeps when it passes over this object, find the probability that the boy has found a metal object.

87. **Fundraising by telephone** Tree diagrams can organize problems having more than two stages. The figure shows probabilities for a charity calling potential donors by telephone.[25] Each person called is either a recent donor, a past donor, or a new prospect. At the next stage, the person called either does or does not pledge to contribute, with conditional probabilities that depend on the donor class to which the person belongs. Finally, those who make a pledge either do or don't actually make a contribution. Suppose we randomly select a person who is called by the charity.

(a) What is the probability that the person contributed to the charity?

(b) Given that the person contributed, find the probability that he or she is a recent donor.

88. **HIV and confirmation testing** Refer to Exercise 85. Many of the positive results from EIA tests are false positives. It is therefore common practice to perform a second EIA test on another blood sample from a person whose initial specimen tests positive. Assume that the false positive and false negative rates remain the same for a person's second test. Find the probability that a person who gets a positive result on both EIA tests has HIV antibodies.

89. **Merry and bright?** A string of Christmas lights contains 20 lights. The lights are wired in series so that if any light fails, the whole string will go dark. Each light has probability 0.98 of working for a 3-year period. The lights fail independently of each other. Find the probability that the string of lights will remain bright for 3 years.

90. **Get rid of the penny** Harris Interactive reported that 29% of all U.S. adults favor abolishing the penny. Assuming that responses from different individuals are independent, what is the probability of randomly selecting 3 U.S. adults who all say that they favor abolishing the penny?

91. **Is the package late?** A shipping company claims that 90% of its shipments arrive on time. Suppose this claim is true. If we take a random sample of 20 shipments made by the company, what's the probability that at least 1 of them arrives late?

92. **On a roll** Suppose that you roll a fair, six-sided die 10 times. What's the probability that you get at least one 6?

93. **Who's pregnant?** According to the Current Population Survey (CPS), 27% of U.S. females are older than 55. The Centers for Disease Control and Prevention (CDC) report that 6% of all U.S. females are pregnant. Suppose that these results are accurate. If we randomly select a U.S. female, is $P(\text{pregnant and over 55}) = (0.06)(0.27) = 0.0162$? Why or why not?

94. **Late flights** An airline reports that 85% of its flights arrive on time. To find the probability that a random sample of 4 of this airline's flights into LaGuardia Airport in New York City on the same night all arrive on time, can we multiply $(0.85)(0.85)(0.85)(0.85)$? Why or why not?

95. **Fire or medical?** Many fire stations handle more emergency calls for medical help than for fires. At one fire station, 81% of incoming calls are for medical help. Suppose we choose 4 incoming calls to the station at random.

5.82 (a)

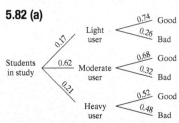

(b) $P(\text{good grades}) = (0.17)(0.74) + (0.62)(0.68) + (0.21)(0.52) = 0.657$

(c) $P(\text{heavy user} \mid \text{good grades}) =$
$$\frac{(0.21)(0.52)}{0.657} = \frac{0.1092}{0.657} = 0.166$$

5.83

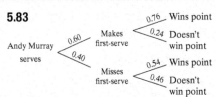

$P(\text{missed his first serve} \mid \text{won the point}) =$
$$\frac{(0.40)(0.54)}{(0.60)(0.76) + (0.40)(0.54)} = \frac{0.216}{0.672} = 0.32$$

5.84

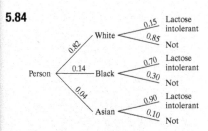

$P(\text{Asian} \mid \text{lactose intolerant})$

$$= \frac{(0.04)(0.90)}{(0.82)(0.15) + (0.14)(0.70) + (0.04)(0.90)}$$

$$= \frac{0.036}{0.257} = 0.14$$

5.85

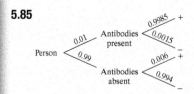

$P(\text{antibody} \mid \text{positive}) =$

$$\frac{(0.01)(0.9985)}{(0.01)(0.9985) + (0.99)(0.006)} = 0.6270$$

5.86

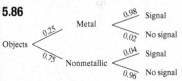

$P(\text{metal} \mid \text{signal}) =$

$$\frac{(0.25)(0.98)}{(0.25)(0.98) + (0.75)(0.04)} = 0.891$$

5.87 (a) $P(\text{contributed}) = (0.5)(0.4)(0.8) + (0.3)(0.3)(0.6) + (0.2)(0.1)(0.5) = 0.224$

(b) $P(\text{recent donor} \mid \text{contribute}) =$
$$\frac{0.16}{0.224} = 0.7143$$

5.88 $P(\text{has HIV antibody} \mid 2 \text{ positive tests}) =$

$$\frac{(0.01)(0.9985)(0.9985)}{(0.01)(0.9985)(0.9985) + (0.99)(0.006)(0.006)} = 0.9964$$

5.89 $(0.98)^{20} = 0.6676$

5.90 $(0.29)^3 = 0.024$

5.91 $P(\text{none are late}) = (0.90)^{20} = 0.1216$
$P(\text{at least 1 late}) = 1 - 0.1216 = 0.8784$

5.92 $P(\text{none are 6}) = (5/6)^{10} = 0.1615$
$P(\text{at least one 6}) = 1 - 0.1615 = 0.8385$

5.93 No; if a woman is over 55 years old, it is unlikely that she is pregnant. These events are not independent.

5.94 No; it is likely that if one flight is late, whatever is causing it to be late will also be affecting the other three flights. These four events are not independent.

5.95 (a) P(all 4 calls are for medical help) = P(1st is medical and 2nd is medical and 3rd is medical and 4th is medical) = P(1st is medical) $\cdot$ P(2nd is medical) $\cdot$ P(3rd is medical) $\cdot$ P(4th is medical) = $(0.81)(0.81)(0.81)(0.81) = (0.81)^4 = 0.430$. There is a 0.430 probability that all 4 calls are for medical help.

(b) P(at least 1 not for medical help) = $1 - P$(all 4 calls are for medical help) = $1 - 0.430 = 0.570$ probability that at least 1 of the 4 calls is not for medical help. **(c)** The calculation in part (a) might not be valid because the 4 consecutive calls being medical are not independent events. Knowing that the first call is medical might make it more likely that the next call is medical.

5.96 (a) P(all 7 references still work two years later)
= P(1st still works and 2nd still works . . . and 7th still works)
= P(1st still works) $\cdot$ P(2nd still works) $\cdot \cdot \cdot \cdot \cdot$ P(7th still works)
= $(0.87)(0.87)(0.87)(0.87)(0.87)(0.87)(0.87)$
= $(0.87)^7 = 0.377$
(b) P(at least 1 reference does not work two years later) = $1 - P$(all 7 references still work two years later) = $1 - 0.377 = 0.623$
(c) If 7 Internet references are chosen from one issue of the same journal, the calculation in part (a) might not be valid because the references working or not are not independent events. It is likely that the references from one issue of the same journal will continue working (or not) together.

5.97 (a)

Gender

		Male	Female	Total
Eye color	Blue	0	10	10
	Brown	20	20	40
	Total	20	30	50

The events "student is male" and "student has blue eyes" are mutually exclusive because they don't occur at the same time.

(b)

Gender

		Male	Female	Total
Eye color	Blue	4	6	10
	Brown	16	24	40
	Total	20	30	50

(a) Find the probability that all 4 calls are for medical help.

(b) What's the probability that at least 1 of the calls is not for medical help?

(c) Explain why the calculation in part (a) may not be valid if we choose 4 consecutive calls to the station.

96. **Broken links** Internet sites often vanish or move so that references to them can't be followed. In fact, 87% of Internet sites referred to in major scientific journals still work within two years of publication.[26] Suppose we randomly select 7 Internet references from scientific journals.

(a) Find the probability that all 7 references still work two years later.

(b) What's the probability that at least 1 of them doesn't work two years later?

(c) Explain why the calculation in part (a) may not be valid if we choose 7 Internet references from one issue of the same journal.

97. **Mutually exclusive versus independent** The two-way table summarizes data on the gender and eye color of students in a college statistics class. Imagine choosing a student from the class at random. Define event A: student is male, and event B: student has blue eyes.[27]

Gender

		Male	Female	Total
Eye color	Blue			10
	Brown			40
	Total	20	30	50

(a) Copy and complete the two-way table so that events A and B are mutually exclusive.

(b) Copy and complete the two-way table so that events A and B are independent.

(c) Copy and complete the two-way table so that events A and B are not mutually exclusive and not independent.

98. **Independence and association** The two-way table summarizes data from an experiment comparing the effectiveness of three different diets (A, B, and C) on weight loss. Researchers randomly assigned 300 volunteer subjects to the three diets. The response variable was whether each subject lost weight over a 1-year period.

Diet

		A	B	C	Total
Lost weight?	Yes		60		180
	No		40		120
	Total	90	100	110	300

(a) Suppose we randomly select one of the subjects from the experiment. Show that the events "Diet B" and "Lost weight" are independent.

(b) Copy and complete the table so that there is no association between type of diet and whether a subject lost weight.

(c) Copy and complete the table so that there is an association between type of diet and whether a subject lost weight.

99. **Checking independence** Suppose A and B are two events such that $P(A) = 0.3$, $P(B) = 0.4$, and $P(A \cap B) = 0.12$. Are events A and B independent? Justify your answer.

100. **Checking independence** Suppose C and D are two events such that $P(C) = 0.6$, $P(D) = 0.45$, and $P(C \cap D) = 0.3$. Are events C and D independent? Justify your answer.

101. **The geometric distributions** You are tossing a pair of fair, six-sided dice in a board game. Tosses are independent. You land in a danger zone that requires you to roll doubles (both faces showing the same number of spots) before you are allowed to play again.

(a) What is the probability of rolling doubles on a single toss of the dice?

(b) What is the probability that you do not roll doubles on the first toss, but you do on the second toss?

(c) What is the probability that the first two tosses are not doubles and the third toss is doubles? This is the probability that the first doubles occurs on the third toss.

(d) Do you see the pattern? What is the probability that the first doubles occurs on the kth toss?

102. **Matching suits** A standard deck of playing cards consists of 52 cards with 13 cards in each of four suits: spades, diamonds, clubs, and hearts. Suppose you shuffle the deck thoroughly and deal 5 cards face-up onto a table.

(a) What is the probability of dealing five spades in a row?

(b) Find the probability that all 5 cards on the table have the same suit.

Multiple Choice: *Select the best answer for Exercises 103–106.*

103. An athlete suspected of using steroids is given two tests that operate independently of each other. Test A has probability 0.9 of being positive if steroids have been used. Test B has probability 0.8 of being positive if steroids have been used. What is the probability that neither test is positive if the athlete has used steroids?

(a) 0.08   (b) 0.28   (c) 0.02

(d) 0.38   (e) 0.72

If the event "student is male" and the event "student has blue eyes" are independent, then

$P(\text{male}) = P(\text{male} \,|\, \text{blue}) \rightarrow \dfrac{20}{50} = \dfrac{x}{10} \rightarrow x = 4$.

(c) Answers may vary.

5.98 (a) The events "Diet B" and "lost weight" are independent because $P(\text{Diet B}) = P(\text{Diet B} \,|\, \text{lost weight})$.
$P(\text{Diet B}) = 100/300 = 0.333$.
$P(\text{Diet B} \,|\, \text{lost weight}) = 60/180 = 0.333$

(b)

Diet

		A	B	C	Total
Lost weight?	Yes	54	60	66	180
	No	36	40	44	120
	Total	90	100	110	300

(c) Answers may vary.

104. In an effort to find the source of an outbreak of food poisoning at a conference, a team of medical detectives carried out a study. They examined all 50 people who had food poisoning and a random sample of 200 people attending the conference who didn't get food poisoning. The detectives found that 40% of the people with food poisoning went to a cocktail party on the second night of the conference, while only 10% of the people in the random sample attended the same party. Which of the following statements is appropriate for describing the 40% of people who went to the party? (Let F = got food poisoning and A = attended party.)

(a) $P(F \mid A) = 0.40$
(b) $P(A \mid F^C) = 0.40$
(c) $P(F \mid A^C) = 0.40$
(d) $P(A^C \mid F) = 0.40$
(e) $P(A \mid F) = 0.40$

105. Suppose a loaded die has the following probability model:

Outcome	1	2	3	4	5	6
Probability	0.3	0.1	0.1	0.1	0.1	0.3

If this die is thrown and the top face shows an odd number, what is the probability that the die shows a 1?

(a) 0.10
(b) 0.17
(c) 0.30
(d) 0.50
(e) 0.60

106. If $P(A) = 0.24$, $P(B) = 0.52$, and A and B are independent events, what is P(A or B)?

(a) 0.1248
(b) 0.28
(c) 0.6352
(d) 0.76
(e) The answer cannot be determined from the information given.

Recycle and Review

107. **BMI** (2.2, 5.2, 5.3) Your body mass index (BMI) is your weight in kilograms divided by the square of your height in meters. Online BMI calculators allow you to enter weight in pounds and height in inches. High BMI is a common but controversial indicator of being overweight or obese. A study by the National Center for Health Statistics found that the BMI of American young women (ages 20 to 29) is approximately Normally distributed with mean 26.8 and standard deviation 7.4.[28]

(a) People with BMI less than 18.5 are often classed as "underweight." What percent of young women are underweight by this criterion?

(b) Suppose we select two American young women in this age group at random. Find the probability that at least one of them is classified as underweight.

108. **Snappy dressers** (4.2, 4.3) Matt and Diego suspect that people are more likely to agree to participate in a survey if the interviewers are dressed up. To test this idea, they went to the local grocery store to survey customers on two consecutive Saturday mornings at 10 A.M. On the first Saturday, they wore casual clothing (tank tops and jeans). On the second Saturday, they dressed in button-down shirts and nicer slacks. Each day, they asked every fifth person who walked into the store to participate in a survey. Their response variable was whether or not the person agreed to participate. Here are their results:

		Clothing	
		Casual	Nice
Participation	Agreed	14	27
	Declined	36	23

(a) Calculate the difference (Casual – Nice) in the proportion of subjects that agreed to participate in the survey in the two groups.

(b) Assume the study design is equivalent to randomly assigning shoppers to the "casual" or "nice" groups. A total of 100 trials of a simulation were performed to see what differences in proportions would occur due only to chance variation in this random assignment. Use the results of the simulation in the following dotplot to determine if the difference in proportions from part (a) is statistically significant. Explain your reasoning.

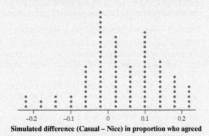

Simulated difference (Casual – Nice) in proportion who agreed

(c) What flaw in the design of this experiment would prevent Matt and Diego from drawing a cause-and-effect conclusion about the impact of an interviewer's attire on nonresponse in a survey?

5.99 Two events are independent if $P(A \cap B) = P(A) \cdot P(B)$. It is given that $P(A) = 0.3$, $P(B) = 0.4$, and $P(A \cap B) = 0.12$. Because $0.12 = (0.3)(0.4)$, events A and B are independent.

5.100 Two events are independent if $P(C \cap D) = P(C) \cdot P(D)$. It is given that $P(C) = 0.6$, $P(D) = 0.45$, and $P(C \cap D) = 0.3$. Because $0.3 \neq (0.6)(0.45) = 0.27$, events C and D are not independent.

5.101 This exercise previews the geometric distribution, to be covered in Section 6.3.
(a) There are 6 ways to get doubles out of 36 possibilities, so P(doubles) = 6/36 = 0.167. There is a 0.167 probability of getting doubles on a single toss of the dice. (b) Because the rolls are independent, we can use the multiplication rule for independent events:

P(no doubles first ∩ doubles second) = P(no doubles first) · P(doubles second) = (30/36)(6/36) = 0.139

There is a 0.139 probability of not getting doubles on the first toss and getting doubles on the second toss.

(c) P(first doubles on third roll) = P(no doubles) · P(no doubles) · P(doubles) =
$$\frac{5}{6}\left(\frac{5}{6}\right)\left(\frac{1}{6}\right) = \frac{25}{216} = 0.116$$

There is a 0.116 probability that the first doubles occurs on the third roll.
(d) For the first doubles on the fourth roll, the probability is $\left(\frac{5}{6}\right)^3\left(\frac{1}{6}\right)$. For the first doubles on the fifth roll, the probability is $\left(\frac{5}{6}\right)^4\left(\frac{1}{6}\right)$.

The probability that the first doubles are rolled on the kth roll is $\left(\frac{5}{6}\right)^{k-1}\left(\frac{1}{6}\right)$.

5.102 (a) Because the probability of dealing a spade changes each time a card is dealt, use the general multiplication rule. P(5 ♠ in a row) = $\frac{13}{52} \cdot \frac{12}{51} \cdot \frac{11}{50} \cdot \frac{10}{49} \cdot \frac{9}{48} = 0.000495$
(b) Because there are 4 suits, the probability of dealing 5 cards with the same suit is 4(0.000495) = 0.00198. There is a 0.00198 probability of dealing 5 cards with the same suit.

5.103 c

5.104 e

5.105 e

5.106 c

5.107 (a) $z = \frac{18.5 - 26.8}{7.4} = -1.12$
(i) The proportion of z-scores below −1.12 is 0.1314.
(ii) normalcdf(lower: −1000, upper: 18.5, mean: 26.8, SD: 7.4) = 0.1314
About 13.14% of young women are underweight by this criterion.
(b) Note that P(not underweight) = 1 − P(underweight) = 1 − 0.131 = 0.869.
P(at least one is underweight) = 1 − P(none are underweight) = 1 − 0.869² = 0.2448; there is a 0.2448 probability that at least one of the two women will be classified as underweight.

5.108 (a) The difference (Casual − Nice) in the proportions of subjects who agreed to participate in the survey in the two groups = 14/50 − 27/50 = 0.28 − 0.54 = −0.26. **(b)** Because a difference of proportions of −0.26 or lower never occurred in the 100 trials of the simulation, the difference is statistically significant. It is extremely unlikely to get a difference this big simply due to chance variation in the random assignment. **(c)** An experiment using random assignment is called for to determine cause-and-effect relationship. The treatments (casual clothes and nice clothes) were not randomly assigned to each subject because all the subjects on the first Saturday received casual clothes and all the subjects on the second Saturday received nice clothes.

Answers

(a) Here is a two-way table that organizes the given information:

		Gender		
		Male	Female	Total
Completed assignment?	Yes	11	19	30
	No	6	4	10
	Total	17	23	40

$P(\text{complete}) = 30/40 = 0.75$; there is a 0.75 probability that the student completed the assignment.

(b) No; $P(\text{complete}\,|\,\text{female}) = 19/23 = 0.826$ is not equal to $P(\text{complete}\,|\,\text{male}) = 11/17 = 0.647$. Knowing that the student is a female increases the probability that the assignment was completed.

(c) Let $01-30 =$ completed assignment and $31-40 =$ didn't complete assignment. Skip 00, $41-99$. Going left to right on the random digit table, choose 4 *different* two-digit numbers from $01-40$. Count the number of students among the four chosen who completed the assignment (i.e., count how many of the numbers are $01-30$). Repeat this process many times and divide the number of trials where 2 or fewer completed the assignment by the total number of trials.

(d)

Trial 1: 12 9̶7̶ 5̶1̶ 32 5̶8̶ 13 04 (3 of 4 completed)

Trial 2: 8̶4̶ 5̶1̶ 44 7̶2̶ 32 18 19 40 (2 of 4 completed)

Trial 3: 0̶0̶ 36 0̶0̶ 24 28 9̶6̶ 7̶6̶ 7̶3̶ 5̶9̶ 6̶4̶ 23 (3 of 4 completed)

$P(\text{2 or fewer completed assignment}) \approx 1/3$

Chapter 5 Wrap-Up

FRAPPY! FREE RESPONSE AP® PROBLEM, YAY!

The following problem is modeled after actual AP® Statistics exam free response questions. Your task is to generate a complete, concise response in 15 minutes.

Directions: Show all your work. Indicate clearly the methods you use, because you will be scored on the correctness of your methods as well as on the accuracy and completeness of your results and explanations.

A statistics teacher has 40 students in his class, 23 females and 17 males. At the beginning of class on a Monday, the teacher planned to spend time reviewing an assignment due that day. Unknown to the teacher, only 19 of the females and 11 of the males had completed the assignment. The teacher plans to randomly select students to do problems from the assignment on the whiteboard.

(a) What is the probability that a randomly selected student has completed the assignment?

(b) Are the events "selecting a female" and "selecting a student who completed the assignment" independent? Justify your answer.

Suppose that the teacher randomly selects 4 students to do a problem on the whiteboard and only 2 of the students had completed the assignment.

(c) Describe how to use a table of random digits to estimate the probability that 2 or fewer of the 4 randomly selected students completed the assignment.

(d) Complete three trials of your simulation using the random digits below and use the results to estimate the probability described in part (c).

12975	13258	13048	45144	72321	81940	00360	02428
96767	35964	23822	96012	94951	65194	50842	55372
37609	59057	66967	83401	60705	02384	90597	93600

After you finish, you can view two example solutions on the book's website (highschool.bfwpub.com/updatedtps6e). Determine whether you think each solution is "complete," "substantial," "developing," or "minimal." If the solution is not complete, what improvements would you suggest to the student who wrote it? Finally, your teacher will provide you with a scoring rubric. Score your response and note what, if anything, you would do differently to improve your own score.

Chapter 5 Review

Section 5.1: Randomness, Probability, and Simulation

In this section, you learned about the idea of probability. The law of large numbers says that when you repeat a random process many, many times, the relative frequency of an outcome will approach a single number. This single number is called the probability of the outcome—how often we expect the outcome to occur in a very large number of trials of the random process. Be sure to remember the "large" part of the law of large numbers. Although clear patterns emerge in a large number of trials, we shouldn't expect such regularity in a small number of trials.

Simulation is a powerful tool that we can use to imitate a random process and estimate a probability. To conduct a simulation, describe how to use a random process to perform one trial of the simulation. Tell what you will record at the end of each trial. Then perform many trials, and use the results of your simulation to answer the question of interest. If you are using random digits to perform your simulation, be sure to consider whether digits can be repeated within each trial.

382

Section 5.2: Probability Rules

In this section, you learned that random behavior can be described by a probability model. Probability models have two parts, a list of possible outcomes (the sample space) and a probability for each outcome. The probability of each outcome in a probability model must be between 0 and 1, and the probabilities of all the outcomes in the sample space must add to 1.

An event is a collection of possible outcomes from the sample space. The complement rule says the probability that an event occurs is 1 minus the probability that the event doesn't occur. In symbols, the complement rule says that $P(A) = 1 - P(A^C)$. Given two events A and B from some random process, use the general addition rule to find the probability that event A or event B occurs:

$$P(A \text{ or } B) = P(A \cup B) = P(A) + P(B) - P(A \cap B)$$

If the events A and B have no outcomes in common, use the addition rule for mutually exclusive events: $P(A \cup B) = P(A) + P(B)$.

Finally, you learned how to use two-way tables and Venn diagrams to display the sample space for a random process involving two events. Using a two-way table or a Venn diagram is a helpful way to organize information and calculate probabilities involving the union $(A \cup B)$ and the intersection $(A \cap B)$ of two events.

Section 5.3: Conditional Probability and Independence

In this section, you learned that a conditional probability describes the probability of an event occurring given that another event is known to have already occurred. To calculate the probability that event A occurs given that event B has occurred, use the formula

$$P(A \mid B) = \frac{P(A \cap B)}{P(B)} = \frac{P(A \text{ and } B)}{P(B)}$$

Two-way tables and tree diagrams are useful ways to organize the information provided in a conditional probability problem. Two-way tables are best when the problem describes the number or proportion of cases with certain characteristics. Tree diagrams are best when the problem provides the conditional probabilities of different events or describes a sequence of events.

Use the general multiplication rule for calculating the probability that event A and event B both occur:

$$P(A \text{ and } B) = P(A \cap B) = P(A) \cdot P(B \mid A)$$

If knowing whether or not event B occurs doesn't change the probability that event A occurs, then events A and B are independent. That is, events A and B are independent if $P(A \mid B) = P(A \mid B^C) = P(A)$. If events A and B are independent, use the multiplication rule for independent events to find the probability that events A and B both occur: $P(A \cap B) = P(A) \cdot P(B)$.

What Did You Learn?

Learning Target	Section	Related Example on Page(s)	Relevant Chapter Review Exercise(s)
Interpret probability as a long-run relative frequency.	5.1	330	R5.1
Use simulation to model a random process.	5.1	333, 334	R5.2
Give a probability model for a random process with equally likely outcomes and use it to find the probability of an event.	5.2	343	R5.3
Use basic probability rules, including the complement rule and the addition rule for mutually exclusive events.	5.2	345	R5.4
Use a two-way table or Venn diagram to model a random process and calculate probabilities involving two events.	5.2	347, 352	R5.5
Apply the general addition rule to calculate probabilities.	5.2	349	R5.5
Calculate and interpret conditional probabilities.	5.3	360, 361	R5.4, R5.5, R5.7
Determine if two events are independent.	5.3	364	R5.6
Use the general multiplication rule to calculate probabilities.	5.3	366	R5.6, R5.7
Use a tree diagram to model a random process involving a sequence of outcomes and to calculate probabilities.	5.3	369, 370	R5.7
When appropriate, use the multiplication rule for independent events to calculate probabilities.	5.3	373, 374, 375	R5.8

Chapter 5 Review Exercises

These exercises are designed to help you review the important ideas and methods of the chapter.

R5.1 Butter side down Researchers at Manchester Metropolitan University in England determined that if a piece of toast is dropped from a 2.5-foot-high table, the probability that it lands butter side down is 0.81.

(a) Explain what this probability means.

(b) Suppose that the researchers dropped 4 pieces of toast, and all of them landed butter side down. Does that make it more likely that the next piece of toast will land with the butter side up? Explain your answer.

R5.2 Butter side down Refer to the preceding exercise. Maria decides to test this probability and drops 10 pieces of toast from a 2.5-foot table. Only 4 of them

Answers to Chapter 5 Review Exercises

R5.1 (a) If you take a very large random sample of pieces of buttered toast and dropped them from a 2.5-foot-high table, about 81% of them will land butter side down. **(b)** No; if 4 dropped pieces of toast all landed butter side down, it does not make it more likely that the next piece of toast will land with the butter side up. Chance behavior is unpredictable in the short run, but it has a regular and predictable pattern in the long run. The value 0.81 describes the proportion of times that toast will land butter side down in a very long series of trials.

R5.2 (a) Using a table of random digits, let 00–80 = butter side down and 81–99 = butter side up. Moving left to right across a row, look at pairs of digits to simulate dropping one piece of toast. Read 10 such pairs of two-digit numbers to

simulate dropping 10 pieces. Record the number, out of 10 that land butter side down. Determine if 4 or fewer pieces of toast out of 10 landed butter side down. Repeat this process many times. **(b)** Bold numbers = butter side down *Trial 1:* **29 07 71 48 63 61 68 34 70 52**. Did 4 or fewer pieces of toast land butter side down? **No.** *Trial 2:* **62 22 45 10 25** 95 **05 29 09 08**. Did 4 or fewer pieces of toast land butter side down? **No.** *Trial 3:* **73 59 27 51** 86 87 **13 69 57 61**. Did 4 or fewer pieces of toast land butter side down? **No. (c)** Assuming the probability that each piece of toast lands butter side down is 0.81, the estimate of the probability that dropping 10 pieces of toast yields 4 or less butter side down is $1/50 = 0.02$. There is convincing evidence that the 0.81 claim is false. These results are unlikely to have occurred purely by chance.

R5.3 (a) The sample space is: rock/rock, rock/paper, rock/scissors, paper/rock, paper/paper, paper/scissors, scissors/rock, scissors/paper, and scissors/scissors. Because each player is equally likely to choose any of the three, each of these 9 outcomes will be equally likely and have probability 1/9. **(b)** There are 3 outcomes—rock/scissors, paper/rock, and scissors/paper—where Player 1 wins on the first play. P(Player 1 wins) = $3/9 = 0.33$.

R5.4 (a) The probability that the vehicle is a crossover is $1 - 0.46 - 0.15 - 0.10 - 0.05 = 0.24$. The sum of the probabilities must add to 1. **(b)** P(vehicle is not an SUV or a minivan) = $0.46 + 0.15 + 0.24 = 0.85$ **(c)** P(pickup truck | not a passenger car) =

$$\frac{0.15}{0.15 + 0.10 + 0.24 + 0.05} = \frac{0.15}{0.54} = 0.278$$

R5.5 (a) P(drives an SUV) = $39/120 = 0.325$ **(b)** P(drives a sedan or exercises) =

$$\frac{25 + 20 + 15 + 12}{120} = 0.60$$

(c) P(does not drive a truck | exercises) =

$$\frac{25 + 15}{25 + 15 + 12} = \frac{40}{52} = 0.769$$

R5.6 (a) Events "thick-crust pizza" and "pizza with mushrooms" are not mutually exclusive. A pizza can be thick-crust and have mushrooms.

(b)

		Type of crust	
		Thick	Thin	Total
Mushrooms?	Yes	2	4	6
	No	1	2	3
	Total	3	6	9

P(mushrooms) = $6/9 = 0.667$ and P(mushrooms | thick crust) = $2/3 = 0.667$; because P(mushrooms) = P(mushrooms | thick crust), the events "mushrooms" and "thick crust" are independent. **(c)** Use the general multiplication rule: P(both randomly selected pizzas have mushrooms) = P(first has mushrooms) · P(second has mushrooms | first has mushrooms) = $(6/9)(5/8) = 0.417$.

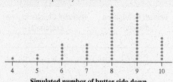

land butter side down. Maria wants to perform a simulation to estimate the probability that 4 or fewer pieces of toast out of 10 would land butter side down if the researchers' 0.81 probability value is correct.

(a) Describe how you would use a table of random digits to perform the simulation.

(b) Perform 3 trials of the simulation using the random digits given. Copy the digits onto your paper and mark directly on or above them so that someone can follow what you did.

29077	14863	61683	47052	62224	51025
95052	90908	73592	75186	87136	95761
27102	56027	55892	33063	41842	81868

(c) The dotplot displays the results of 50 simulated trials of dropping 10 pieces of toast. Is there convincing evidence that the researchers' 0.81 probability value is incorrect? Explain your answer.

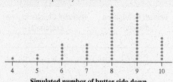

Simulated number of butter side down

R5.3 **Rock smashes scissors** Almost everyone has played the game rock-paper-scissors at some point. Two players face each other and, at the count of 3, make a fist (rock), an extended hand, palm side down (paper), or a "V" with the index and middle fingers (scissors). The winner is determined by these rules: rock smashes scissors; paper covers rock; and scissors cut paper. If both players choose the same object, then the game is a tie. Suppose that Player 1 and Player 2 are both equally likely to choose rock, paper, or scissors.

(a) Give a probability model for this random process.

(b) Find the probability that Player 1 wins the game on the first throw.

R5.4 **What kind of vehicle?** Randomly select a new vehicle sold in the United States in a certain month.[29] The probability model for the type of vehicle chosen is given here.

Vehicle type	Passenger car	Pickup truck	SUV	Crossover	Minivan
Probability	0.46	0.15	0.10	?	0.05

(a) What is the probability that the vehicle is a crossover? How do you know?

(b) Find the probability that the vehicle is not an SUV or a minivan.

(c) Given that the vehicle is not a passenger car, what is the probability that it is a pickup truck?

R5.5 **Drive to exercise?** The two-way table summarizes the responses of 120 people to a survey in which they were asked, "Do you exercise for at least 30 minutes 4 or more times per week?" and "What kind of vehicle do you drive?"

			Car type	
		Sedan	SUV	Truck
Exercise?	Yes	25	15	12
	No	20	24	24

Suppose one person from this sample is randomly selected.

(a) Find the probability that the person drives an SUV.

(b) Find the probability that the person drives a sedan or exercises for at least 30 minutes 4 or more times per week.

(c) Find the probability that the person does not drive a truck, given that she or he exercises for at least 30 minutes 4 or more times per week.

R5.6 **Mike's pizza** You work at Mike's pizza shop. You have the following information about the 9 pizzas in the oven: 3 of the 9 have thick crust and 2 of the 3 thick-crust pizzas have mushrooms. Of the remaining 6 pizzas, 4 have mushrooms. Suppose you randomly select one of the pizzas in the oven.

(a) Are the events "thick-crust pizza" and "pizza with mushrooms" mutually exclusive? Justify your answer.

(b) Are the events "thick-crust pizza" and "pizza with mushrooms" independent? Justify your answer.

Now suppose you randomly select 2 of the pizzas in the oven.

(c) Find the probability that both have mushrooms.

R5.7 **Does the new hire use drugs?** Many employers require prospective employees to take a drug test. A positive result on this test suggests that the prospective employee uses illegal drugs. However, not all people who test positive use illegal drugs. The test result could be a false positive. A negative test result could be a false negative if the person really does use illegal drugs. Suppose that 4% of prospective employees use drugs, and that the drug test has a false positive rate of 5%, and a false negative rate of 10%.[30] Imagine choosing a prospective employee at random.

(a) Draw a tree diagram to model this random process.

(b) Find the probability that the drug test result is positive.

(c) If the prospective employee's drug test result is positive, find the probability that she or he uses illegal drugs.

R5.8 **Lucky penny?** Harris Interactive reported that 33% of U.S. adults believe that finding and picking up a penny is good luck. Assuming that responses from different individuals are independent, what is the probability of randomly selecting 10 U.S. adults and finding at least 1 person who believes that finding and picking up a penny is good luck?

R5.7 (a)

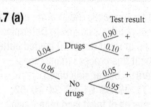

(b) P(drug test result is positive) =

$(0.04)(0.90) + (0.96)(0.05) = 0.084$

(c) P(uses illegal drugs | positive result) =

$$\frac{(0.04)(0.90)}{0.084} = 0.429$$

R5.8 P(at least 1 believes that finding and picking up a penny is good luck) = $1 - P$(none of the 10 people believe this) = $1 - (0.67)^{10} = 0.98177$

Chapter 5 AP® Statistics Practice Test

Section I: Multiple Choice *Select the best answer for each question.*

Questions T5.1 to T5.3 refer to the following setting. A group of 125 truck owners were asked what brand of truck they owned and whether or not the truck has four-wheel drive. The results are summarized in the two-way table below. Suppose we randomly select one of these truck owners.

		Four-wheel drive?	
		Yes	No
	Ford	28	17
Brand of truck	Chevy	32	18
	Dodge	20	10

T5.1 What is the probability that the person owns a Dodge or has four-wheel drive?

(a) 20/80   (b) 20/125   (c) 80/125
(d) 90/125   (e) 110/125

T5.2 What is the probability that the person owns a Chevy, given that the truck has four-wheel drive?

(a) 32/50   (b) 32/80   (c) 32/125
(d) 50/125   (e) 80/125

T5.3 Which one of the following is true about the events "Owner's truck is a Chevy" and "Owner's truck has four-wheel drive"?

(a) These two events are mutually exclusive and independent.

(b) These two events are mutually exclusive, but not independent.

(c) These two events are not mutually exclusive, but they are independent.

(d) These two events are neither mutually exclusive nor independent.

(e) These two events are mutually exclusive, but we do not have enough information to determine if they are independent.

T5.4 A spinner has three equally sized regions: blue, red, and green. Jonny spins the spinner 3 times and gets 3 blues in a row. If he spins the spinner 297 more times, how many more blues is he most likely to get?

(a) 97   (b) 99   (c) 100   (d) 101   (e) 103

Questions T5.5 and T5.6 refer to the following setting. Wilt is a fine basketball player, but his free-throw shooting could use some work. For the past three seasons, he has made only 56% of his free throws. His coach sends him to a summer clinic to work on his shot, and when he returns, his coach has him step to the free-throw line and take 50 shots. He makes 34 shots. Is this result convincing evidence that Wilt's free-throw shooting has improved? We want to perform a simulation to

estimate the probability that a 56% free-throw shooter would make 34 or more in a sample of 50 shots.

T5.5 Which of the following is a correct way to perform the simulation?

(a) Let integers from 1 to 34 represent making a free throw and 35 to 50 represent missing a free throw. Generate 50 random integers from 1 to 50. Count the number of made free throws. Repeat this process many times.

(b) Let integers from 1 to 34 represent making a free throw and 35 to 50 represent missing a free throw. Generate 50 random integers from 1 to 50 with no repeats allowed. Count the number of made free throws. Repeat this process many times.

(c) Let integers from 1 to 56 represent making a free throw and 57 to 100 represent missing a free throw. Generate 50 random integers from 1 to 100. Count the number of made free throws. Repeat this process many times.

(d) Let integers from 1 to 56 represent making a free throw and 57 to 100 represent missing a free throw. Generate 50 random integers from 1 to 100 with no repeats allowed. Count the number of made free throws. Repeat this process many times.

(e) None of the above is correct.

T5.6 The dotplot displays the number of made shots in 100 simulated sets of 50 free throws by someone with probability 0.56 of making a free throw.

Simulated number of made shots

Which of the following is an appropriate statement about Wilt's free-throw shooting based on this dotplot?

(a) If Wilt were still only a 56% shooter, the probability that he would make at least 34 of his shots is about 0.03.

(b) If Wilt were still only a 56% shooter, the probability that he would make at least 34 of his shots is about 0.97.

(c) If Wilt is now shooting better than 56%, the probability that he would make at least 34 of his shots is about 0.03.

(d) If Wilt is now shooting better than 56%, the probability that he would make at least 34 of his shots is about 0.97.

(e) If Wilt were still only a 56% shooter, the probability that he would make at least 34 of his shots is about 0.01.

Answers to Chapter 5 AP® Statistics Practice Test

T5.1 d

T5.2 b

T5.3 c

T5.4 b

T5.5 c

T5.6 a

T5.7 b

T5.8 a

T5.9 e

T5.10 e

T5.11 (a) *P*(student eats regularly in the cafeteria and is not a 10th grader) =
$$\frac{130 + 122 + 68}{805} = \frac{320}{805} = 0.398$$
There is a probability of 0.398 that a randomly selected student eats regularly in the cafeteria and is not a 10th grader.
(b) *P*(10th grader | eats regularly in the cafeteria) = 175/495 = 0.354; given that the student eats regularly in the cafeteria, there is a 0.354 probability that he or she is a 10th grader.
(c) *P*(10th grader) = 209/805 = 0.260; the events "10th grader" and "eats regularly in the cafeteria" are not independent because *P*(10th grader | eats regularly in the cafeteria) ≠ *P*(10th grader).

T5.12

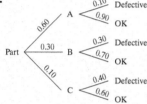

(a) To get the probability that a part randomly chosen from all parts produced in this factory is defective, add the probabilities from all branches in the tree that end in a defective part: *P*(defective) = (0.60)(0.10) + (0.30)(0.30) + (0.10)(0.40) = 0.06 + 0.09 + 0.04 = 0.19
(b) *P*(Machine B | defective) =
$$\frac{(0.30)(0.30)}{0.19} = \frac{0.09}{0.19} = 0.474;$$ given
that a part is inspected and found to be defective, there is a 0.474 probability that it was produced by Machine B.

T5.13 (a) *P*(customer will pay less than the usual cost of the buffet) = 23/36 = 0.639; there is a 0.639 probability that the customer will pay less than the usual cost of the buffet. **(b)** *P*(all 4 friends end up paying less than the usual cost of the buffet) = (23/36)^4 = 0.167; there is a 0.167 probability that all 4 of these friends end up paying less than the usual cost of the buffet. **(c)** *P*(at least 1 of the 4 friends ends up paying more than the usual cost of the buffet) = 1 − 0.167 = 0.833; there is a 0.833 probability that at least 1 of the 4 friends ends up paying more than the usual cost of the buffet.

T5.7 The partially complete table that follows shows the distribution of scores on the AP® Statistics exam for a class of students.

Score	1	2	3	4	5
Probability	0.10	0.20	???	0.25	0.15

Select a student from this class at random. If the student earned a score of 3 or higher on the AP® Statistics exam, what is the probability that the student scored a 5?

(a) 0.150   (b) 0.214   (c) 0.300   (d) 0.428   (e) 0.700

T5.8 In a class, there are 18 girls and 14 boys. If the teacher selects two students at random to attend a party with the principal, what is the probability that the two students are the same sex?

(a) 0.49   (b) 0.50   (c) 0.51   (d) 0.52   (e) 0.53

Section II: Free Response *Show all your work. Indicate clearly the methods you use, because you will be graded on the correctness of your methods as well as on the accuracy of your results and explanations.*

T5.11 The two-way table summarizes data on whether students at a certain high school eat regularly in the school cafeteria by grade level.

		Grade				
		9th	10th	11th	12th	Total
Eat in	Yes	130	175	122	68	495
cafeteria?	No	18	34	88	170	310
	Total	148	209	210	238	805

(a) If you choose a student at random, what is the probability that the student eats regularly in the cafeteria and is not a 10th-grader?

(b) If you choose a student at random who eats regularly in the cafeteria, what is the probability that the student is a 10th-grader?

(c) Are the events "10th-grader" and "eats regularly in the cafeteria" independent? Justify your answer.

T5.12 Three machines—A, B, and C—are used to produce a large quantity of identical parts at a factory. Machine A produces 60% of the parts, while Machines B and C produce 30% and 10% of the parts, respectively. Historical records indicate that 10% of the parts produced by Machine A are defective, compared with 30% for Machine B and 40% for Machine C. Suppose we randomly select a part produced at the factory.

(a) Find the probability that the part is defective.

(b) If the part is inspected and found to be defective, what's the probability that it was produced by Machine B?

T5.13 At Dicey Dave's Diner, the dinner buffet usually costs $12.99. Once a month, Dave sponsors "lucky buffet" night. On that night, each patron can either pay the usual price or roll two fair, six-sided dice and pay a number of dollars equal to the product of the numbers showing on the two faces. The table shows the sample space of this random process.

T5.9 Suppose that a student is randomly selected from a large high school. The probability that the student is a senior is 0.22. The probability that the student has a driver's license is 0.30. If the probability that the student is a senior or has a driver's license is 0.36, what is the probability that the student is a senior and has a driver's license?

(a) 0.060   (b) 0.066   (c) 0.080   (d) 0.140   (e) 0.160

T5.10 The security system in a house has two units that set off an alarm when motion is detected. Neither one is entirely reliable, but one or both *always* go off when there is motion anywhere in the house. Suppose that for motion in a certain location, the probability that detector A goes off and detector B does not go off is 0.25, and the probability that detector A does not go off is 0.35. What is the probability that detector B goes off?

(a) 0.1   (b) 0.35   (c) 0.4   (d) 0.65   (e) 0.75

		First die					
		1	2	3	4	5	6
	1	1	2	3	4	5	6
	2	2	4	6	8	10	12
Second	3	3	6	9	12	15	18
die	4	4	8	12	16	20	24
	5	5	10	15	20	25	30
	6	6	12	18	24	30	36

(a) A customer decides to play Dave's "lucky buffet" game. Find the probability that the customer will pay less than the usual cost of the buffet.

(b) A group of 4 friends comes to Dicey Dave's Diner to play the "lucky buffet" game. Find the probability that all 4 of these friends end up paying less than the usual cost of the buffet.

(c) Find the probability that at least 1 of the 4 friends ends up paying *more* than the usual cost of the buffet.

T5.14 Based on previous records, 17% of the vehicles passing through a tollbooth have out-of-state plates. A bored tollbooth worker decides to pass the time by counting how many vehicles pass through until he sees two with out-of-state plates. We would like to perform a simulation to estimate the average number of vehicles it takes to find two with out-of-state plates.[31]

(a) Describe how you would use a table of random digits to perform the simulation.

(b) Perform 3 trials of the simulation using the random digits given here. Copy the digits onto your paper and mark directly on or above them so that someone can follow what you did.

41050	92031	06449	05059	59884	31880
53115	84469	94868	57967	05811	84514
84177	06757	17613	15582	51506	81435

T5.14 (a) To carry out the simulation using a table of random digits, let 00–16 = cars with out-of-state plates and let 17–99 = other cars. Moving left to right across a row, look at pairs of digits from a random number table until you have found two numbers between 00 and 16 (repeats are allowed). Record how many two-digit numbers you had to read in order to get two numbers between 00 and 16. Repeat this process many times. **(b)** The first sample is 41 **05 09**. The bold numbers represent cars with out-of-state plates. In this sample, it took 3 cars to find 2 with out-of-state plates. The second sample is 20 31 **06** 44 90 50 59 59 88 43 18 80 53 **11**. In this sample, it took 14 cars to find 2 with out-of-state plates. The third sample is 58 44 69 94 86 85 79 67 **05** 81 18 45 **14**. In this sample, it took 13 cars to find 2 with out-of-state plates.

TRM **Chapter 5 Case Study**

The Case Study feature that was found in the previous two editions of *The Practice of Statistics* student edition has been moved to the Teacher's Resource Materials. Click on the link in the TE-Book, open in the TRFD, or download from the Teacher's Resources on the book's digital platform.

Chapter 6

Chapter 6

Random Variables and Probability Distributions

PD Overview

In this chapter, we continue our study of probability with a focus on random variables. A random variable is a numerical outcome of a random process, such as the height of a randomly selected student, the number of heads in three flips of a coin, or the difference between the diameter of a randomly chosen cup and the diameter of a randomly chosen lid.

Many students have a tough time with this chapter, possibly because of the large number of rules and formulas that are introduced. We suggest that you allow your students to use the formula sheet provided on the AP® Statistics exam during your classroom quizzes and tests. First, it allows students to spend less time memorizing and more time learning how and when to use the formulas. Second, because students will be allowed to use the formula sheets on the real AP® exam, it helps students to practice using them.

Besides learning how to calculate probabilities of events involving random variables, we will also focus on identifying the shape, center, and variability of the probability distribution of a random variable. Being able to anticipate the shape, center, and variability of a distribution of a random variable or combination of random variables will be extremely important as we proceed into Chapter 7 and beyond. To estimate a population proportion with 95% confidence, we need to know how the random variable (sample proportion) will behave in repeated sampling. Likewise, to investigate whether a difference in sample means is statistically significant, we need to know how the difference in these random variables (sample means) will behave when the true means are really the same.

Note: Consider starting the "Penny for your thoughts" activity from Chapter 7 during this chapter. Make sure to read the Teaching Tips that go along with the activity on page 468 for suggestions.

The Main Ideas

One of the challenges in teaching the AP® Statistics course is keeping students focused on the big picture, not just the details of each section. We outline the main ideas for the chapter here.

Chapter 6 Introduction

This chapter begins by introducing the idea of a random variable, a numerical outcome of a random process. In the "Bottled water versus tap water" activity, students identify which of three cups contains bottled water. This activity is a nice way to introduce random variables and a great way to review how to conduct a simulation. It also foreshadows one of the basic ideas of statistical inference—comparing the results obtained in a study with what might happen by chance alone.

Section 6.1 Discrete and Continuous Random Variables

In this section, we introduce probability distributions for random variables. These distributions give the possible values of a random variable and their probabilities. This is essentially the same definition for *distribution* that we used in Chapter 1, where we said that a distribution tells us what values a variable takes and how often it takes these values.

Random variables are divided into two categories, discrete and continuous. Discrete random variables take on a fixed set of values with gaps between them, whereas continuous random variables take on all values in an interval of numbers. Most discrete random variables result from counting something, such as the number of pencils that students have in their backpacks.

Continuous random variables typically result from measuring something, such as the lengths of pencils or weights of backpacks.

As in Chapter 1, we are often interested in the shape, center, and variability of a probability distribution. The shape can be identified by graphing a histogram of the probability distribution, with the height of each bar representing the probability of that outcome. The center of the distribution is usually identified by the mean (expected value) of the random variable. The expected value is the average value of the random variable if the random process could be repeated many, many times. The variability of a probability distribution is usually identified by the standard deviation of the random variable, which describes how much the values of the variable typically differ from the mean value.

Continuous probability distributions, such as uniform or Normal distributions, describe the distribution of continuous random variables. A density curve is used to graph a continuous probability distribution rather than a histogram, and probabilities are determined by finding the area under the density curve within the values of interest.

Section 6.2 Transforming and Combining Random Variables

In this section, we introduce rules for transforming and combining random variables. To do this, we include several examples from a fun context (Pete's Jeep Tours) to discover how certain transformations affect the shape, center, and variability of the distribution of a random variable. We then continue with the same context to discover how we can predict the shape, center, and variability of a combination of two or more random variables. We encourage you to allow students to discover the rules within a context, rather than starting with the rules and then moving to the context.

Linear transformations have the same effect on the probability distribution of a random variable as they have on a distribution of data. In Chapter 2, we learned that linear transformations do not change the shape, as long as we don't multiply by a negative number. However, linear transformations can change the center, variability, or both depending on the type of transformation. Multiplying each value by a constant b multiplies measures of center by b and most measures of variability by $|b|$. Adding a constant a to each value adds a to measures of center but doesn't change measures of variability.

Combining two or more random variables is a new and very important topic. Many of the methods we use later in this book involve differences in random variables, such as the difference in two means ("Is there a difference in the mean salaries for men and women in a certain industry?") or a difference in two proportions ("Does a new drug have a higher survival rate than the currently used drug?"). To analyze questions like these, students need to be able to anticipate what the distribution of a combination of random variables will look like.

We conclude this section by looking at several examples in which Normal random variables are combined and the resulting distribution is used to calculate a probability.

Because students can feel overwhelmed by the quantity of rules and formulas presented in this section, it is important that you help distinguish linear transformations (taking the values of *one* random variable and multiplying them by a constant and/or adding a constant) and combinations of random variables (adding or subtracting *two or more* random variables).

Section 6.3 Binomial and Geometric Random Variables

In this section, we introduce two common types of discrete random variables: binomial random variables and geometric random variables. Pay special attention to binomial random variables, because questions about binomial random variables often show up on the AP® Statistics exam. Also, the binomial distribution is closely related to the sampling distribution of a sample proportion, which we will learn more about in Chapter 7.

Binomial random variables count the number of successes in a fixed number of trials, whereas geometric random variables count the number of trials it takes to get a success. Otherwise, the binomial and geometric settings have the same conditions: there must be two possible outcomes for each trial (success or failure), the trials must be independent, and the probability of success must stay the same throughout all trials.

A very common application of the binomial distribution in statistics is counting the number of times a particular outcome occurs in a random sample from some population (e.g., the number of defective flash drives in a sample of size 10 from a shipment of 10,000). In cases like this, the sampling is almost always done without replacement. This means that the trials are no longer independent. However, if the sample is a small fraction of the population, the lack of independence has a very small effect on the probabilities we calculate.

Also included in this section is a discussion about using a Normal distribution to approximate a binomial distribution. This topic lays the foundation for using a Normal distribution to model the sampling distribution of a sample proportion in Chapter 7.

Chapter 6 Resources
Teacher's Resource Materials

The following resources, identified by the **TRM** in the annotated student pages, can be found by clicking on the link in the Teacher's e-Book (TE-Book), searching by category or chapter on the Teacher's Resource Flash Drive (TRFD), or logging into the book's digital platform and searching the Teacher's Resources menu (teacher log-in required).

- Alternate Examples: one file per section
- Lecture Presentation Slides: one per section
- Chapter 6 Learning Targets Grid
- The Casino Lab
- Airline Overbooking, Part II
- A Fair Coin?
- FRAPPY! Materials
- Chapter 6 Case Study
- Complete solutions for the Check Your Understanding problems, section exercises, review exercises, and practice test
- Quizzes: one per section
- Chapter 6 Test

Free Response Questions from Previous AP® Statistics Exams

Questions can be found on the AP® Central website: apcentral.collegeboard.org/courses/ap-statistics/exam.

Students should be able to answer all the free response questions listed below with material learned in this chapter. Questions that contain content from this chapter but also require content from later chapters are listed in the last chapter required to complete the entire question. This list will be updated after each AP® Statistics exam and will be posted to the Teacher's Resource section of the book's digital platform and to www.statsmedic.com/free-response-questions.

Year	#	Content
2019	3	• Two-way table probabilities • Justifying independence • Binomial probability
2019	5	• Percentile from Normal distribution • Probability from Normal distribution • Expected value calculation
2018	3	• Multiplication rule • Conditional probability • Binomial probability
2016	4	• Multiplication rule • Geometric probability calculation • Informal P-value and conclusion
2015	3	• Discrete probability distributions • Expected value • Conditional probability • Conditional expected value
2013	3	• Normal probability calculation • Mean and standard deviation of a sum of random variables
2012	2	• Discrete probability distributions • Expected value of a discrete random variable • Application of expected value • Normal probability calculation
2011B	3	• Geometric probability • Binomial probability • Cumulative binomial probability
2010B	3	• Binomial distribution • Expected value • Binomial calculations
2010	4	• Mean and standard deviation of a binomial distribution • Binomial calculations • Stratified sampling
2008	3	• Expected value • Basic probability rules
2008B	5	• Combining normal random variables • Normal calculations
2006B	3	• Normal calculations • Binomial calculations • Inverse normal calculations
2005	2	• Expected value • Median of a discrete random variable • Relationship of mean and median

Year	#	Content
2005B	2	• Mean and standard deviation of a discrete random variable • Combining independent random variables • Linear transformations of a random variable
2004	3	• Binomial conditions • Multiplication rule • Interpreting probability • Generalizability
2004	4	• Conditional probability • Expected value
2003	3	• Normal calculations • Binomial calculations
2002B	2	• Addition rule • Expected value • Conditional probability
2002	3	• Normal calculations • Combining independent random variables
2001	2	• Expected value
1999	4	• Normal calculations • Binomial calculations • Outlier rules
1999	5	• Sample space • Expected value
1998	6	• Normal calculations • Simulation • Expected value

Applets

- The *Law of Large Numbers* applet allows students to roll dice repeatedly and keep track of the mean of their rolls. After many repetitions, the mean of their rolls should approach the true mean, or expected value. Go to the Student Site at highschool.bfwpub.com/updatedtps6e, click on Statistical Applets, and select "Law of Large Numbers."

- The *Normal Density Curve* applet allows students to find the area under a Normal curve. Students can choose the mean, standard deviation, and boundaries. From the Statistical Applets selection on the Student Site at highschool.bfwpub.com/updatedtps6e, choose "Normal Density Curve."

- The *Normal Approximation to Binomial* applet allows students to try different combinations of n and p to see when it would be reasonable to use a Normal curve to model a binomial probability distribution. From the Statistical Applets selection on the Student Site at highschool.bfwpub.com/updatedtps6e, choose "Normal Approximation to Binomial."

- The *Probability* applet allows students to graph discrete probability distributions and find summary statistics, do Normal distribution calculations, and graph and analyze binomial distributions. Go to the Extra Applets on the Student Site and select "Probability."

Chapter 6: Pacing Guide, Learning Targets, and Suggested Assignments

This pacing guide is based on a schedule with 110, 50-minute sessions before the AP® Statistics exam. If you have a different number of sessions before the AP® exam, you can modify the pacing guide to suit your needs. If you have additional time, consider incorporating quizzes, released AP® free response questions, or additional activities. See the Resources section above for suggestions.

The suggested homework assignments list odd-numbered exercises, whenever possible, so students can check their answers against the back-of-book answers. If you would rather students not have access to the answers while doing homework, adding 1 to the exercise numbers usually will do the trick, because the homework exercises typically are paired. For example, Exercises 1 and 2 will generally cover the same topics, but in different contexts. You may also choose to include the Recycle and Review questions at the end of each section, which review topics from previous sections or chapters. If your school is using the digital platform that accompanies TPS6, you will find these assignments pre-built as online homework assignments for Chapter 6.

Day	Content	Learning Targets: Students will be able to . . .	Suggested Assignment (MC bold)
1	6.1 Discrete Random Variables, Analyzing Discrete Random Variables: Describing Shape, Measuring Center: The Mean (Expected Value) of a Discrete Random Variable	• Use the probability distribution of a discrete random variable to calculate the probability of an event. • Make a histogram to display the probability distribution of a discrete random variable and describe its shape. • Calculate and interpret the mean (expected value) of a discrete random variable.	1, 3, 5, 7, 9, 11
2	6.1 Measuring Variability: The Standard Deviation (and Variance) of a Discrete Random Variable, Continuous Random Variables	• Calculate and interpret the standard deviation of a discrete random variable. • Use the probability distribution of a continuous random variable (uniform or Normal) to calculate the probability of an event.	13, 19, 21, 23, 27, 29, **31–34**
3	6.2 Transforming a Random Variable	• Describe the effect of adding or subtracting a constant or multiplying or dividing by a constant on the probability distribution of a random variable.	37, 39, 41, 43, 47
4	6.2 Combining Random Variables, Standard Deviation of the Sum or Difference of Two Random Variables, Combining Normal Random Variables	• Calculate the mean and standard deviation of the sum or difference of random variables. • Find probabilities involving the sum or difference of independent Normal random variables.	49, 51, 55, 57, 59, 65, 67, **73–74**
5	6.3 Binomial Settings and Binomial Random Variables, Calculating Binomial Probabilities	• Determine whether the conditions for a binomial setting are met. • Calculate and interpret probabilities involving binomial distributions.	77, 79, 81, 83, 85, 89
6	6.3 Describing a Binomial Distribution: Shape, Center, and Variability; Binomial Distributions in Statistical Sampling, The Normal Approximation to Binomial Distributions	• Calculate the mean and standard deviation of a binomial random variable. Interpret these values. • When appropriate, use the Normal approximation to the binomial distribution to calculate probabilities.	91, 93, 95, 99, 101, 103, 105, 106, 117
7	6.3 Geometric Random Variables	• Calculate and interpret probabilities involving geometric random variables. • Calculate the mean and standard deviation of a geometric distribution. Interpret these values.	107, 109, 111, 113, **115–119**
8	Chapter 6 Review/FRAPPY!		Chapter 6 Review Exercises
9	Chapter 6 Test		

Chapter 6 Alignment to the College Board's Fall 2019 AP® Statistics Course Framework*

Relationship to College Board Units

Chapter 6 in this book covers Topics 4.7–4.12 in Unit 4 of the College Board Course Framework. Students will be ready to take the Personal Progress Check for Unit 4 once they have completed Chapters 5 and 6.

Big Ideas and Enduring Understandings

Chapter 6 develops these Big Ideas and related Enduring Understandings outlined in the Course Framework:

- **Big Idea 1: Variation and Distribution (EU: VAR 5,6):** The distribution of measures for individuals within a sample or population describes variation. The value of a statistic varies from sample to sample. How can we determine whether differences between measures represent random variation or meaningful distinctions? Statistical methods based on probabilistic reasoning provide the basis for shared understandings about variation and about the likelihood that variation between and among measures, samples, and populations is random or meaningful.
- **Big Idea 2: Patterns and Uncertainty (EU: UNC 3):** Statistical tools allow us to represent and describe patterns in data and to classify departures from patterns. Simulation and probabilistic reasoning allow us to anticipate patterns in data and to determine the likelihood of errors in inference.

Course Skills

Chapter 6 helps students to develop the skills identified in the Course Framework.

- **2: Data Analysis** (2.B)
- **3: Using Probability and Simulation** (3.A, 3.B, 3.C)
- **4: Statistical Argumentation** (4.B)

Learning Objectives and Essential Knowledge Statements

Section	Learning Objectives	Essential Knowledge Statements
6.1	VAR-5.A, VAR-5.B, VAR-5.C, VAR-5.D, VAR-6.A, VAR-6.B, VAR-6.C, UNC-3.A	VAR-5.A.1, VAR-5.A.2, VAR-5.A.3, VAR-5.A.4, VAR-5.B.1, VAR-5.C.1, VAR-5.C.2, VAR-5.C.3, VAR-5.D.1, VAR-6.A.1, VAR-6.A.2, VAR-6.A.3, VAR-6.B.1, VAR-6.B.2, VAR-6.C.1, UNC-3.A.1
6.2	VAR-5.E, VAR-5.F	VAR-5.E.1, VAR-5.E.2, VAR-5.E.3, VAR-5.F.1
6.3	UNC-3.A, UNC-3.B, UNC-3.C, UNC-3.D, UNC-3.E, UNC-3.F, UNC-3.G	UNC-3.A.2, UNC-3.B.1, UNC-3.C.1, UNC-3.D.1, UNC-3.E.1, UNC-3.E.2, UNC-3.F.1, UNC-3.G.1

A detailed alignment that can be sorted by Course Framework Unit, Topic, Learning Objective, Essential Knowledge Statement, or textbook section, is available on the TRFD and in the Teacher's Resources folder on Sapling Plus. **TRM**

*Should changes be made to the Course Framework in the future, an updated alignment will be placed on our AP® updates page at go.bfwpub.com/ap-course-updates.

Notes

Chapter 6

Random Variables and Probability Distributions

Stefano Paterna/Alamy

ACTIVITY OVERVIEW

To prepare for using this activity, watch the overview video by clicking on the link in the TE-Book, opening the TRFD, or downloading from the Teacher's Resources on the book's digital platform.

Time: 15 minutes

Materials: Cups, tap water, bottled water, index cards, sticker dots, poster or chart paper

Teaching Advice: Label three cups with A, B, and C for each student in the class and set up three water stations. At Station 1, fill the cups labeled B with bottled water and the others with tap water. At Station 2, fill the cups labeled A with bottled water and the others with tap water. At Station 3, fill the cups labeled C with bottled water and the others with tap water. Prepare an index card for each student, with roughly equal numbers labeled "Station 1," "Station 2," and "Station 3." Shuffle the cards and hand them out at random when students arrive to class.

For the simulation in Step 8, use the randInt feature on your calculator to generate *n* values (*n* = number of students who participated) from 1 to 3, letting 1 represent a correct guess and 2 and 3 represent an incorrect guess. Count the number of 1s and record this value. Ask students to put sticker dots on a poster board or chart paper. Post the completed poster board in the classroom so that you can refer to this activity throughout the rest of the course.

For example, in Mr. Hogarth's class (see the paragraph following the activity), we would randomly generate 21 integers from 1 to 3 and count the number of 1s. Here is a dotplot showing 100 trials of the simulation for his class. Not once in 100 trials were there 13 or more correct guesses. This gives convincing evidence that the students in Mr. Hogarth's class did better than random guessing.

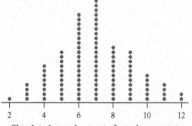

Simulated sample count of number correct

Answers:

1–6. Answers will vary.

7. 1/3

8. Answers will vary.

INTRODUCTION

Do you drink bottled water or tap water? According to an online news report, about 75% of people drink bottled water regularly. Some people do so because they believe bottled water is safer than tap water. (There's little evidence to support this belief.) Others say they prefer the taste of bottled water. Can people really tell the difference?

ACTIVITY   Bottled water versus tap water

This activity will give you and your classmates a chance to discover whether or not you can taste the difference between bottled water and tap water.

1. Before class begins, your teacher will prepare numbered stations with cups of water. You will be given an index card with a station number on it.

2. Go to the corresponding station. Pick up three cups (labeled A, B, and C) and take them back to your seat.

3. Your task is to determine which one of the three cups contains the bottled water. Drink all the water in Cup A first, then the water in Cup B, and finally the water in Cup C. Write down the letter of the cup that you think held the bottled water. Do not discuss your results with any of your classmates yet!

4. While you're tasting, your teacher will make a chart on the board like this one:

Station number	Bottled water cup?	Truth

5. When you are told to do so, go to the board and record your station number and the letter of the cup you identified as containing bottled water.

6. Your teacher will now reveal the truth about the cups of drinking water. How many students in the class identified the bottled water correctly? What percent of the class is this?

7. Let's assume that no one in your class can distinguish tap water from bottled water. In that case, students would just be guessing which cup of water tastes different. If so, what's the probability that an individual student would guess correctly?

8. How many correct identifications would you need to provide convincing evidence that the students in your class aren't just guessing? With your classmates, design and carry out a simulation to answer this question. What do you conclude about your class's ability to distinguish tap water from bottled water?

When Mr. Hogarth's class did the preceding activity, 13 out of 21 students made correct identifications. If we assume that the students in his class can't tell tap water from bottled water, then each one is basically guessing, with a 1/3 chance of being correct. So we'd expect about one-third of Mr. Hogarth's 21

Teaching Tip: Using Technology

This simulation in the activity can be done using the *One Categorical Variable* applet found in the Extra Applets section on the Student Site at highschool.bfwpub.com/updatedtps6e. Put in the class data for "number correct" and "number incorrect." At the bottom of the applet, you can "Simulate sample count." Use 0.333 as the hypothesized proportion.

Teaching Tip

There are many ways to modify the "Bottled water versus tap water" activity to suit the interests of your students. For example, using an identical setup, you can investigate whether students can correctly tell the difference between name-brand potato chips and store-brand potato chips or the difference between regular cola and diet cola.

students (i.e., about 7 students) to guess correctly. How likely is it that 13 or more of the 21 students would guess correctly? To answer this question without a simulation, we need a different kind of probability model from the ones we saw in Chapter 5.

Section 6.1 introduces the concept of a *random variable*, a numerical outcome of some random process (like the number of students who correctly guess the type of water). Each random variable has a *probability distribution* that gives us information about the likelihood that a specific event happens (like 13 or more correct guesses out of 21) and about what's expected to happen if the random process is repeated many times. Section 6.2 examines the effect of transforming and combining random variables on the shape, center, and variability of their probability distributions. In Section 6.3, we'll look at two random variables with probability distributions that are used enough to have their own names—*binomial* and *geometric*.

SECTION 6.1 Discrete and Continuous Random Variables

LEARNING TARGETS *By the end of the section, you should be able to:*

- Use the probability distribution of a discrete random variable to calculate the probability of an event.
- Make a histogram to display the probability distribution of a discrete random variable and describe its shape.
- Calculate and interpret the mean (expected value) of a discrete random variable.
- Calculate and interpret the standard deviation of a discrete random variable.
- Use the probability distribution of a continuous random variable (uniform or Normal) to calculate the probability of an event.

A probability model describes the possible outcomes of a random process and the likelihood that those outcomes will occur. For example, suppose you toss a fair coin 3 times. The sample space for this random process is

$$\text{HHH} \quad \text{HHT} \quad \text{HTH} \quad \text{THH} \quad \text{HTT} \quad \text{THT} \quad \text{TTH} \quad \text{TTT}$$

Because there are 8 equally likely outcomes, the probability is 1/8 for each possible outcome.

Define the **random variable** $X =$ the number of heads obtained in 3 tosses. The value of X will vary from one set of tosses to another, but it will always be one of the numbers 0, 1, 2, or 3. How likely is X to take each of those values? It will be easier to answer this question if we group the possible outcomes by the number of heads obtained:

$$X = 0: \text{TTT} \quad X = 1: \text{HTT THT TTH} \quad X = 2: \text{HHT HTH THH} \quad X = 3: \text{HHH}$$

Preparing for Inference

How likely is it that 13 or more of the 21 students would guess correctly? Using a binomial distribution, the answer is 0.007. This is the *P*-value for a significance test. Since this probability is small, we have convincing evidence that students were not simply guessing.

PD Section 6.1 Overview

To watch the video overview of Section 6.1 (for teachers), click on the link in the TE-Book, look on the TRFD, or download from the Teacher's Resources on the book's digital platform.

TRM Learning Targets Grid

At the beginning of each section, we present the relevant learning targets. Point these out to your students and refer back to the targets when you cover them in class. There is a PDF version of the grid with an additional column that students can use to keep track of their progress. Find it in the Teacher's Resource Materials located in the TE-Book, on the TRFD, or in the Teacher's view in the book's digital platform.

TRM Section 6.1 Alternate Examples

You can find the Alternate Examples for this section in Microsoft Word format by clicking on the link in the TE-Book, opening the TRFD, or downloading from the Teacher's Resources on the book's digital platform.

Teaching Tip

An alternate example could investigate the possible outcomes for a family that plans on having three children: BBB BBG BGB GBB BGG GBG GGB GGG.

Making Connections

Remind students that we could estimate this probability distribution using the simulation methods from Section 5.1. Looking ahead, we will also learn how to create this probability distribution using the binomial distributions in Section 6.3.

We can summarize the **probability distribution** of X in a table:

Value	0	1	2	3
Probability	1/8	3/8	3/8	1/8

DEFINITION Random variable, Probability distribution

A **random variable** takes numerical values that describe the outcomes of a random process.

The **probability distribution** of a random variable gives its possible values and their probabilities.

Making Connections

Emphasize that this use of the word *distribution* is the same as it has been since Chapter 1. A distribution describes the possible values a variable can take and how often it takes those values.

Recall the two types of quantitative variables from Chapter 1. *Discrete variables* typically result from counting something, while *continuous variables* typically result from measuring something.

We use capital, italic letters (like X or Y) to designate random variables and lowercase, italic letters (like x or y) to designate specific values of those variables. There are two main types of probability distributions, corresponding to two types of random variables: *discrete* and *continuous*.

Discrete Random Variables

The random variable X in the coin-tossing setting is a **discrete random variable**.

DEFINITION Discrete random variable

A **discrete random variable** X takes a fixed set of possible values with gaps between them.

Teaching Tip

Some students will assume that a discrete random variable must only take integer values. This is not the case. Shoe size can be 6, 7, or 8 but also 6.5, 7.5, and so on. Shoe size is a discrete random variable because there is a fixed set of possible values with gaps between them.

To illustrate the idea of "gaps between" the possible values, draw two number lines: one representing possible shoe sizes and one representing possible foot lengths. On the shoe size number line, draw solid dots at each half size (e.g., 6, 6.5, 7, 7.5, etc.) to represent the possible values. On the foot length number line, shade an entire interval (e.g., 6 to 12 inches) to represent the possible values. Shoe size is discrete because there are gaps between shoe sizes—there are no sizes between 6 and 6.5, for example. Foot length is continuous because there are no gaps between values—there are many possible foot lengths between 6 and 6.5 inches.

We can list the possible values of X = the number of heads in 3 tosses of a coin as 0, 1, 2, 3. Note that there are gaps between these values on a number line. For instance, a gap exists between $X = 1$ and $X = 2$ because X cannot take values such as 1.2 or 1.84.

The probability distribution of X is

Value	0	1	2	3
Probability	1/8	3/8	3/8	1/8

This probability distribution is valid because all the probabilities are between 0 and 1, and their sum is

$$1/8 + 3/8 + 3/8 + 1/8 = 1$$

PROBABILITY DISTRIBUTION FOR A DISCRETE RANDOM VARIABLE

The probability distribution of a discrete random variable X lists the values x_i and their probabilities p_i:

Value	x_1	x_2	x_3	...
Probability	p_1	p_2	p_3	...

For the probability distribution to be valid, the probabilities p_i must satisfy two requirements:

1. Every probability p_i is a number between 0 and 1, inclusive.
2. The sum of the probabilities is 1: $p_1 + p_2 + p_3 + \cdots = 1$.

The AP® Statistics exam formula sheet uses the notation $P(x_i)$ for the probability that $X = x_i$. We prefer the simpler notation p_i to represent $P(X = x_i)$.

Making Connections

In Chapter 5, our basic probability rules stated:

- The probability of any event is a number between 0 and 1.
- All possible outcomes together must have probabilities that add up to 1.

These now become our criteria for defining a valid probability distribution.

Teaching Tip

Even though a discrete random variable takes a fixed set of possible values, the set of possible values can be infinite. For example, if X = number of rolls of a fair die needed to get a 6, there is no upper limit for the values of X.

We can use the probability distribution of a discrete random variable to find the probability of an event. For instance, what's the probability that we get at least one head in three tosses of the coin? In symbols, we want to find $P(X \geq 1)$. We know that

$$P(X \geq 1) = P(X = 1 \text{ or } X = 2 \text{ or } X = 3)$$

Because the events $X = 1$, $X = 2$, and $X = 3$ are mutually exclusive, we can add their probabilities to get the answer:

$$P(X \geq 1) = P(X = 1) + P(X = 2) + P(X = 3)$$
$$= 3/8 + 3/8 + 1/8 = 7/8$$

Or we could use the complement rule from Chapter 5:

$$P(X \geq 1) = 1 - P(X < 1) = 1 - P(X = 0)$$
$$= 1 - 1/8 = 7/8$$

EXAMPLE

Apgar scores: Babies' health at birth
Discrete random variables

PROBLEM: In 1952, Dr. Virginia Apgar suggested five criteria for measuring a baby's health at birth: skin color, heart rate, muscle tone, breathing, and response when stimulated. She developed a 0-1-2 scale to rate a newborn on each of the five criteria. A baby's Apgar score is the sum of the ratings on each of the five scales, which gives a whole-number value from 0 to 10. Apgar scores are still used today to evaluate the health of newborns. Although this procedure was later named for Dr. Apgar, the acronym APGAR also represents the five scales: Appearance, Pulse, Grimace, Activity, and Respiration.

What Apgar scores are typical? To find out, researchers recorded the Apgar scores of over 2 million newborn babies in a single year.[1] Imagine selecting a newborn baby at random. (That's our random process.) Define the random variable X = Apgar score of a randomly selected newborn baby. The table gives the probability distribution of X.

Value x_i	0	1	2	3	4	5	6	7	8	9	10
Probability p_i	???	0.006	0.007	0.008	0.012	0.020	0.038	0.099	0.319	0.437	0.053

(a) Write the event "the baby has an Apgar score of 0" in terms of X. Then find its probability.

(b) Doctors decided that Apgar scores of 7 or higher indicate a healthy newborn baby. What's the probability that a randomly selected newborn baby is healthy?

SOLUTION:

(a) $P(X = 0) = 1 - (0.006 + 0.007 + \cdots + 0.053)$
$= 1 - 0.999$
$= 0.001$

> Use the complement rule:
> $P(X = 0) = 1 - P(X \neq 0)$

(b) $P(X \geq 7) = 0.099 + 0.319 + 0.437 + 0.053$
$= 0.908$

> The probability of choosing a healthy baby is
> $P(X \geq 7) = P(X = 7) + P(X = 8) + $
> $P(X = 9) + P(X = 10)$

FOR PRACTICE, TRY EXERCISE 1

ALTERNATE EXAMPLE   Skill 3.A

Will we get a snow day?   Discrete random variables

PROBLEM:
High school students in Michigan sometimes get "snow days" in the winter when the roads are so bad that school is canceled for the day. Define the random variable X = the number of snow days at a certain high school in Michigan for a randomly selected school year. Suppose the table below gives the probability distribution of X.

Value x_i	0	1	2	3	4	5	6	7	8	9	10
Probability p_i	???	0.19	0.14	0.10	0.07	0.05	0.04	0.04	0.02	0.01	0.01

(a) Write the event "the school year has 0 snow days" in terms of X. Then find its probability.
(b) At this high school, if more than 5 snow days are used over the course of the year, students are required to make up the time at the end of the school year. What's the probability that a randomly selected school year will require make-up days?

SOLUTION:
(a) $P(X = 0) = 1 - (0.19 + 0.14 + \cdots + 0.01) = 1 - 0.67 = 0.33$
(b) $P(X > 5) = 0.04 + 0.04 + 0.02 + 0.01 + 0.01 = 0.12$

Note that the probability of randomly selecting a newborn whose Apgar score is *at least* 7 is not the same as the probability that the baby's Apgar score is *greater than* 7. The latter probability is

$$P(X > 7) = P(X = 8) + P(X = 9) + P(X = 10)$$
$$= 0.319 + 0.437 + 0.053$$
$$= 0.809$$

 The outcome $X = 7$ is included in "at least 7" but is not included in "greater than 7." Be sure to consider whether to include the boundary value in your calculations when dealing with discrete random variables.

Analyzing Discrete Random Variables: Describing Shape

When we analyzed distributions of quantitative data in Chapter 1, we made it a point to discuss their shape, center, and variability. We'll do the same with probability distributions of random variables.

For the discrete random variable X = Apgar score of a randomly selected newborn baby, the probability distribution is

Value x_i	0	1	2	3	4	5	6	7	8	9	10
Probability p_i	0.001	0.006	0.007	0.008	0.012	0.020	0.038	0.099	0.319	0.437	0.053

We can display the probability distribution with a histogram. Values of the variable go on the horizontal axis and probabilities go on the vertical axis. There is one bar in the histogram for each value of X. The height of each bar gives the probability for the corresponding value of the variable.

Figure 6.1 shows a histogram of the probability distribution of X. This distribution is skewed to the left and unimodal, with a single peak at an Apgar score of 9.

We also discussed outliers when describing distributions of quantitative data. Outliers are generally of less interest for random variables because their probability distributions specify which values are likely and which are unlikely.

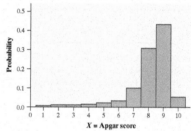

FIGURE 6.1 Histogram of the probability distribution for the random variable X = Apgar score of a randomly selected newborn baby.

There's another way to think about the graph displayed in Figure 6.1. The probability distribution of the random variable X models the *population distribution* of the quantitative variable "Apgar score of a newborn baby." So we can interpret our earlier result, $P(X > 7) = 0.809$, as saying that about 81% of all newborn babies have Apgar scores greater than 7. We also know that the shape of the population distribution is left-skewed with a single peak at 9.

EXAMPLE

Pete's Jeep Tours
Displaying a probability distribution

PROBLEM: Pete's Jeep Tours offers a popular day trip in a tourist area. There must be at least 2 passengers for the trip to run, and the vehicle will hold up to 6 passengers. Pete charges $150 per passenger. Let C = the total amount of money that Pete collects on a randomly selected trip. The probability distribution of C is given in the table.

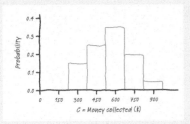

Total collected c_i	300	450	600	750	900
Probability p_i	0.15	0.25	0.35	0.20	0.05

Make a histogram of the probability distribution. Describe its shape.

SOLUTION:

> *Remember:* Values of the variable go on the horizontal axis and probabilities go on the vertical axis. Don't forget to properly label and scale each axis!

The graph is roughly symmetric and has a single peak at $600.

FOR PRACTICE, TRY EXERCISE 5

Notice the use of the label C (collects) for the random variable in the example. Sometimes we prefer contextual labels like this to the more generic X and Y.

CHECK YOUR UNDERSTANDING

Indiana University Bloomington posts the grade distributions for its courses online.[2] Suppose we choose a student at random from a recent semester of this university's Business Statistics course. The student's grade on a 4-point scale (with A = 4) is a random variable X with this probability distribution:

Value	0	1	2	3	4
Probability	0.011	0.032	???	0.362	0.457

1. Write the event "the student got a C" using probability notation. Then find this probability.
2. Explain in words what $P(X \geq 3)$ means. What is this probability?
3. Make a histogram of the probability distribution. Describe its shape.

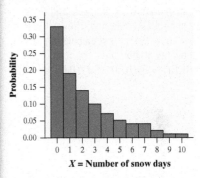
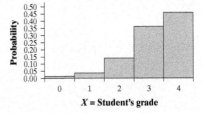

Teaching Tip

If your grading structure uses weighted grades, consider the following example: In a certain AP® Statistics class, homework is worth 10%, quizzes 30%, and tests 60%. Bernard has scores of 85, 78, 71, 76, and 75 on the first five tests. What is his test average?

$$\frac{85 + 78 + 71 + 76 + 75}{5} = 77$$

Bernard has a 98 average on homework, an 86 average on quizzes, and a 77 average on tests. Does this mean that Bernard's overall grade is $\frac{98 + 86 + 77}{3} = 87$?

Students will know that this is incorrect because homework, quizzes, and tests are not equally weighted. They will also likely know the correct calculation to find the overall grade:
$98(0.10) + 86(0.30) + 77(0.60) = 81.8$

Making Connections

The following example can help students see that the formula for calculating the mean in Chapter 1 is a special case of the formula on this page.

Bernard has scores of 85, 78, 71, 76, and 75 on the first five tests. What is his test average?

$$\frac{85 + 78 + 71 + 76 + 75}{5} = 77$$

which can be written equivalently as

$$85\left(\frac{1}{5}\right) + 78\left(\frac{1}{5}\right) + 71\left(\frac{1}{5}\right) + 76\left(\frac{1}{5}\right) + 75\left(\frac{1}{5}\right) = 77$$

Because each of the 5 tests is equally weighted, the probability is 1/5 for each, and now the calculation looks like the formula on this page.

Measuring Center: The Mean (Expected Value) of a Discrete Random Variable

In Chapter 1, you learned how to summarize the center of a distribution of quantitative data with either the median or the mean. For random variables, the mean is typically used to summarize the center of a probability distribution. Because a probability distribution can model the population distribution of a quantitative variable, we label the mean of a random variable X as μ_X. Like any parameter, μ_X is a single, fixed value.

To find the mean of a quantitative data set, we compute the sum of the individual observations and divide by the total number of data values. How do we find the *mean of a discrete random variable?*

Consider the random variable C = the total amount of money that Pete collects on a randomly selected jeep tour from the previous example. The probability distribution of C is given in the table.

Total collected c_i	300	450	600	750	900
Probability p_i	0.15	0.25	0.35	0.20	0.05

What's the average amount of money that Pete collects on his jeep tours?

Imagine a hypothetical 100 trips. Pete will collect \$300 on 15 of these trips, \$450 on 25 trips, \$600 on 35 trips, \$750 on 20 trips, and \$900 on 5 trips. Pete's average amount collected for these trips is

$$\frac{300 \cdot 15 + 450 \cdot 25 + 600 \cdot 35 + 750 \cdot 20 + 900 \cdot 5}{100}$$
$$= \frac{300 \cdot 15}{100} + \frac{450 \cdot 25}{100} + \frac{600 \cdot 35}{100} + \frac{750 \cdot 20}{100} + \frac{900 \cdot 5}{100}$$
$$= 300(0.15) + 450(0.25) + 600(0.35) + 750(0.20) + 900(0.05)$$
$$= 562.50$$

The third line of the calculation is just the values of the random variable C times their corresponding probabilities.

That is, the **mean of the discrete random variable** C is $\mu_C = \$562.50$. This is also known as the **expected value** of C, denoted by $E(C)$.

The mean (expected value) of any discrete random variable is found in a similar way. It is an average of the possible outcomes, but a weighted average in which each outcome is weighted by its probability.

DEFINITION Mean (expected value) of a discrete random variable

The **mean (expected value) of a discrete random variable** is its average value over many, many trials of the same random process.

Suppose that X is a discrete random variable with probability distribution

Value	x_1	x_2	x_3	...
Probability	p_1	p_2	p_3	...

To find the mean (expected value) of X, multiply each possible value of X by its probability, then add all the products:

$$\mu_X = E(X) = x_1p_1 + x_2p_2 + x_3p_3 + \cdots = \sum x_ip_i$$

The AP® Statistics exam formula sheet gives the formula for the mean (expected value) of a discrete random variable as $\sum x_i \cdot P(x_i)$.

AP® EXAM TIP

Let your students know that the formula for the mean (expected value) of a discrete random variable is included on the formula sheet provided to students on both sections of the AP® Statistics exam. *Note:* The revised formula sheet uses $P(x_i)$ rather than p_i to designate the probability of the value x_i. We use the p_i notation found in the previous formula sheet.

An alternate interpretation of $\mu_C = \$562.50$ comes from thinking of the probability distribution of C as a model for the *population distribution* of money collected: For all of his jeep tours, the average amount that Pete collects on a trip is about $562.50.

Recall that the mean is the balance point of a distribution. For Pete's distribution of money collected on a randomly selected jeep tour, the histogram balances at $\mu_C = 562.50$. How do we interpret this parameter? If we randomly select many, many jeep tours, Pete will make an average of about $562.50 per trip.

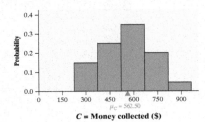

C = Money collected ($)

EXAMPLE

Apgar scores: What's typical?
Finding and interpreting the mean

PROBLEM: Earlier, we defined the random variable X to be the Apgar score of a randomly selected newborn baby. The table gives the probability distribution of X once again.

Value x_i	0	1	2	3	4	5	6	7	8	9	10
Probability p_i	0.001	0.006	0.007	0.008	0.012	0.020	0.038	0.099	0.319	0.437	0.053

Calculate and interpret the expected value of X.

SOLUTION:

$E(X) = \mu_X = (0)(0.001) + (1)(0.006) + \cdots + (10)(0.053) = 8.128$

$\boxed{E(X) = \mu_X = \sum x_i p_i}$

If many, many newborns are randomly selected, their average Apgar score will be about 8.128.

FOR PRACTICE, TRY EXERCISE 7

AP® EXAM TIP

If the mean of a random variable has a non-integer value but you report it as an integer, your answer will not get full credit.

Notice that the mean Apgar score, 8.128, is not a possible value of the random variable X because it is not a whole number between 0 and 10. The non-integer value of the mean shouldn't bother you if you think of the mean (expected value) as a long-run average over many trials of the random process.

How can we find the *median* of a discrete random variable? In Chapter 1, we defined the median as "the midpoint of a distribution, the number such that about half the observations are smaller and about half are larger." The median of a discrete random variable is the 50th percentile of its probability distribution. We can find the median from a *cumulative probability distribution*, like the one shown here for the random variable C = total amount of money Pete collects on a randomly selected jeep tour. We see from the table that $P(C \leq 300) = 0.15$ and that $P(C \leq 450) = 0.40$.

So $300 is the 15th percentile and $450 is the 40th percentile of the probability distribution of C. The median of a discrete random variable is the smallest value for which the cumulative probability equals or exceeds 0.5. So the median amount of money Pete collects on a randomly selected jeep tour is $600.

Total collected c_i	300	450	600	750	900
Probability p_i	0.15	0.25	0.35	0.20	0.05
Cumulative probability	0.15	0.40	0.75	0.95	1.00

AP® EXAM TIP

Emphasize the AP® Exam Tip on this page. Many students incorrectly believe that the *expected* value of a random variable must be equal to one of the *possible* values of the variable. This is not the case. In the Apgar score example, the expected value was 8.128, even though this was not a possible value for the score.

Teaching Tip

A cumulative probability distribution can be represented as a table or function. Each value in the table here gives the probability of being less than or equal to each value of the random variable (percentile).

COMMON STUDENT ERROR

In the example on this page, students may take this final answer of 8.128 and want to divide by 10 (as we would do in a mean calculation for a list of numbers). Remind them that there is no need to divide by 10 because they have already "accounted for this idea" when they multiplied each X value by its probability.

ALTERNATE EXAMPLE   Skills 3.B, 4.B

Blazin' Bonus
Finding and interpreting the mean

PROBLEM:
Buffalo Wild Wings ran a promotion called the Blazin' Bonus, in which every $25 gift card purchased also received a "Bonus" gift card for $5, $15, $25, or $100. According to the company, here are the probabilities for each Bonus gift card. Let X be the amount of money won on the Bonus gift card.

Value x_i	$5	$15	$25	$100
Probability p_i	0.890	0.098	0.010	0.002

Calculate and interpret the expected value of X.

SOLUTION:

$\mu_X = E(X) = (5)(0.890) + (15)(0.098) + (25)(0.010) + (100)(0.002) = \$6.37$

If many, many Bonus cards are randomly selected, the average amount of money that is won per card will be about $6.37.

Teaching Tip

In the histogram of the probability distribution, the median occurs where the sum of the areas of the rectangles equals or exceeds 0.5 as you move from left to right.

Measuring Variability: The Standard Deviation (and Variance) of a Discrete Random Variable

With the mean as our measure of center for a discrete random variable, it shouldn't surprise you that we'll use the standard deviation as our measure of variability. In Chapter 1, we defined the standard deviation of a distribution of quantitative data as the typical distance of the values in the data set from the mean. To get the standard deviation, we started by "averaging" the squared deviations from the mean to get the variance and then took the square root.

We can modify this approach to calculate the **standard deviation of a discrete random variable** X. Start by finding a weighted average of the squared deviations $(x_i - \mu_X)^2$ of the values of the variable X from its mean μ_X. The probability distribution gives the appropriate weight for each squared deviation. We call this weighted average of squared deviations the **variance** of X. Then take the square root to get the standard deviation. Because a probability distribution can model the population distribution of a quantitative variable, we label the variance of a random variable X as σ_X^2 and the standard deviation as σ_X. Like any parameter, σ_X is a single, fixed value.

> The AP® Statistics exam formula sheet gives the formula for the standard deviation of a discrete random variable as $\sqrt{\sum(x_i - \mu_X)^2 \cdot P(x_i)}$.

DEFINITION **Standard deviation of a discrete random variable, Variance**

The **standard deviation of a discrete random variable** measures how much the values of the variable typically vary from the mean in many, many trials of the random process.

Suppose that X is a discrete random variable with probability distribution

Value	x_1	x_2	x_3	$\cdots$
Probability	p_1	p_2	p_3	$\cdots$

and that μ_x is the mean of X. The **variance** of X is

$$\sigma_X^2 = (x_1 - \mu_X)^2 p_1 + (x_2 - \mu_X)^2 p_2 + (x_3 - \mu_X)^2 p_3 + \cdots$$
$$= \sum (x_i - \mu_X)^2 p_i$$

The standard deviation of X is the square root of the variance:

$$\sigma_X = \sqrt{\sigma_X^2} = \sqrt{(x_1 - \mu_X)^2 p_1 + (x_2 - \mu_X)^2 p_2 + (x_3 - \mu_X)^2 p_3 + \cdots}$$
$$= \sqrt{\sum (x_i - \mu_X)^2 p_i}$$

Let's return to the random variable C = the total amount of money that Pete collects on a randomly selected jeep tour. The left two columns of the following table give the probability distribution. Recall that the mean of C is $\mu_C = 562.50$. The third column of the table shows the squared deviation of each value from the mean. The fourth column gives the weighted squared deviations.

Making Connections

Show students the close relationship between the formula for standard deviation of a sample from Chapter 1 and the formula for standard deviation of a discrete random variable in Chapter 6.

$$s_x = \sqrt{\frac{(x_1 - \bar{x})^2 + (x_2 - \bar{x})^2 + (x_3 - \bar{x})^2 + \cdots}{n - 1}}$$

$$= \sqrt{\frac{\sum (x_i - \bar{x})^2}{n - 1}}$$

$$\sigma_X = \sqrt{(x_1 - \mu_X)^2 p_1 + (x_2 - \mu_X)^2 p_2 + (x_3 - \mu_X)^2 p_3 + \cdots}$$

$$= \sqrt{\sum (x_i - \mu_X)^2 p_i}$$

The biggest difference in the formulas is that instead of dividing the sum of the squared deviations by $n - 1$, we are now weighting each squared deviation individually.

Total collected c_i	Probability p_i	Squared deviation from the mean $(c_i - \mu_C)^2$	Weighted squared deviation $(c_i - \mu_C)^2 p_i$
300	0.15	$(300 - 562.50)^2$	$(300 - 562.50)^2 (0.15) =$ 10335.94
450	0.25	$(450 - 562.50)^2$	$(450 - 562.50)^2 (0.25) =$ 3164.06
600	0.35	$(600 - 562.50)^2$	$(600 - 562.50)^2 (0.35) =$ 492.19
750	0.20	$(750 - 562.50)^2$	$(750 - 562.50)^2 (0.20) =$ 7031.25
900	0.05	$(900 - 562.50)^2$	$(900 - 562.50)^2 (0.05) =$ 5695.31
			Sum = 26,718.75

Adding the weighted average of the squared deviations in the fourth column gives the variance of C:

$$\sigma_C^2 = \sum (c_i - \mu_C)^2 p_i$$
$$= (300 - 562.50)^2(0.15) + (450 - 562.50)^2(0.25) + \cdots + (900 - 562.50)^2(0.05)$$
$$= 10335.94 + 3164.06 + \cdots + 5695.31$$
$$= 26{,}718.75 \text{ (squared dollars)}$$

Because the probability distribution of C is a model for the *population distribution* of money collected, we can also interpret the parameter σ_C in this way: "For all of his jeep tours, the amount of money that Pete collects on a trip typically varies by about $163.46 from the mean of $562.50."

The standard deviation of C is the square root of the variance:

$$\sigma_C = \sqrt{26{,}718.75} = \$163.46$$

How do we interpret this parameter? If many, many jeep tours are randomly selected, the amount of money that Pete collects typically varies by about $163.46 from the mean of $562.50.

EXAMPLE

How much do Apgar scores vary?
Finding and interpreting the standard deviation

PROBLEM: Earlier, we defined the random variable X to be the Apgar score of a randomly selected newborn baby. The table gives the probability distribution of X once again. In the last example, we calculated the mean Apgar score of a randomly chosen newborn to be $\mu_x = 8.128$.

Value x_i	0	1	2	3	4	5	6	7	8	9	10
Probability p_i	0.001	0.006	0.007	0.008	0.012	0.020	0.038	0.099	0.319	0.437	0.053

Calculate and interpret the standard deviation of X.

SOLUTION:

$$\sigma_X^2 = (0 - 8.128)^2(0.001) + (1 - 8.128)^2(0.006) + \cdots + (10 - 8.128)^2(0.053) = 2.066$$
$$\sigma_X = \sqrt{2.066} = 1.437$$

If many, many newborns are randomly selected, the babies' Apgar scores will typically vary by about 1.437 units from the mean of 8.128.

> Start by calculating the variance
> $$\sigma_X^2 = \sum (x_i - \mu_X)^2 p_i.$$

FOR PRACTICE, TRY EXERCISE 13

ALTERNATE EXAMPLE

Skills 3.B, 4.B

Blazin' Bonus, again   Finding and interpreting the standard deviation

PROBLEM:
Buffalo Wild Wings ran a promotion called the Blazin' Bonus, in which every $25 gift card purchased also received a "Bonus" gift card for $5, $15, $25, or $100. According to the company, here are the probabilities for each Bonus gift card. Let X be the amount of money that is won on the Bonus gift card. Recall from the previous example that $\mu_x = \$6.37$.

Value x_i	$5	$15	$25	$100
Probability p_i	0.890	0.098	0.010	0.002

Calculate and interpret the standard deviation of X.

SOLUTION:

$$\sigma_X^2 = (5 - 6.37)^2(0.890) +$$
$$(15 - 6.37)^2(0.098) + (25 - 6.37)^2(0.010) +$$
$$(100 - 6.37)^2(0.002) = 29.97$$
$$\sigma_X = \sqrt{29.97} = 5.47$$

If many, many Blazin' Bonus cards are randomly selected, the amount of money that is won will typically vary by about $5.47 from the mean of $6.37.

You can use your calculator to graph the probability distribution of a discrete random variable and to calculate measures of center and variability, as the following Technology Corner illustrates.

13. Technology Corner   ANALYZING DISCRETE RANDOM VARIABLES

TI-Nspire and other technology instructions are on the book's website at highschool.bfwpub.com/updatedtps6e.

Let's explore what the TI-83/84 can do using the random variable X = Apgar score of a randomly selected newborn.

1. Enter the values of the random variable in list L1 and the corresponding probabilities in list L2.

2. To graph a histogram of the probability distribution:

   • In the statistics plot menu, define Plot 1 to be a histogram with Xlist: L1 and Freq: L2.

   • Adjust your window settings as follows: Xmin = −1, Xmax = 11, Xscl = 1, Ymin = −0.1, Ymax = 0.5, Yscl = 0.1.

   • Press GRAPH.

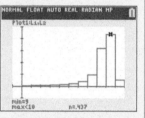

3. To calculate the mean and standard deviation of the random variable, use one-variable statistics with the values in L1 and the probabilities (relative frequencies) in L2. Press STAT, arrow to the CALC menu, and choose 1-Var Stats.

   OS 2.55 or later: In the dialog box, specify List: L1 and FreqList: L2. Then choose Calculate.

   Older OS: Execute the command 1-Var Stats L1,L2.

Note: If you leave Freq: L2 and try to calculate summary statistics for a quantitative data set that does not include frequencies, you will likely get an error message. Be sure to clear Freq when you are done with calculations to avoid this issue.

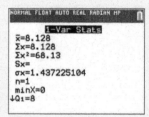

 The calculator's notation for the mean of the random variable X is incorrect. We should write $\mu_X = 8.128$. Fortunately, the notation for the standard deviation is correct: $\sigma_X = 1.437$.

AP® EXAM TIP

If you are asked to calculate the mean or standard deviation of a discrete random variable on a free response question, you must show numerical values substituted into the appropriate formula, as in the previous two examples. Feel free to use ellipses (…) if there are many terms in the summation, as we did. You may then use the method described in Technology Corner 13 to perform the calculation with 1-Var Stats. Writing only 1-Var Stats L1, L2 and then giving the correct values of the mean and standard deviation will *not* earn credit for showing work. Also, be sure to avoid incorrect notation when labeling these parameters.

CHECK YOUR UNDERSTANDING

Indiana University Bloomington posts the grade distributions for its courses online.[2] Suppose we choose a student at random from a recent semester of this university's Business Statistics course. The student's grade on a 4-point scale (with A = 4) is a random variable X with this probability distribution:

Value	0	1	2	3	4
Probability	0.011	0.032	0.138	0.362	0.457

1. Find the mean of X. Interpret this parameter.
2. Find the standard deviation of X. Interpret this parameter.

Continuous Random Variables

When we use Table D of random digits to select an integer from 0 to 9, the result is a discrete random variable (call it X). The probability distribution assigns probability 1/10 to each of the 10 equally likely values of X.

Suppose we want to choose a number at random between 0 and 9, allowing *any* number between 0 and 9 as the outcome (like 0.84522 or 7.1111119). Calculator and computer random number generators will do this. The sample space of this random process is the entire interval of values between 0 and 9 on the number line. If we define Y = randomly generated number between 0 and 9, then Y is a **continuous random variable**.

> **DEFINITION   Continuous random variable**
>
> A **continuous random variable** can take any value in an interval on the number line.

Most discrete random variables result from counting something, like the number of siblings that a randomly selected student has. Continuous random variables typically result from measuring something, like the height of a randomly selected student or the time it takes that student to run a mile.

How can we find the probability $P(3 \leq Y \leq 7)$ that the random number generator produces a value between 3 and 7? As in the case of selecting a random digit, we would like all possible outcomes to be equally likely. But we cannot assign probabilities to each individual value of Y and then add them, because there are infinitely many possible values.

The probability distribution of a continuous random variable is described by a density curve. Recall from Chapter 2 that any density curve has area exactly 1 underneath it, corresponding to a total probability of 1. We use areas under the density curve to assign probabilities to events.

For the continuous random variable Y = randomly generated number between 0 and 9, its probability distribution is a uniform density curve with constant

Making Connections

Just as the total area of the rectangles in a histogram for a discrete random variable is 1, the total area under a density curve for a continuous random variable is 1.

Teaching Tip: AP® Connections

AP® Calculus students will recognize that calculating the probability of an event for a continuous random variable can be done by taking a definite integral of the probability density function.

Teaching Tip

Another way to illustrate why continuous probability models assign probability 0 to every individual outcome is to remember that each outcome is just one of an infinite number of possible outcomes, so the probability of any particular outcome is $\frac{1}{\infty}$.

height 1/9 on the interval from 0 to 9. Note that this probability distribution is valid because the total area under the density curve is

$$Area = base \times height = 9 \times 1/9 = 1$$

Figure 6.2 shows the probability distribution of Y with the area of interest shaded. The area under the density curve between 3 and 7 is

$$Area = base \times height = 4 \times 1/9 = 4/9$$

So $P(3 \le Y \le 7) = 4/9 = 0.444$.

FIGURE 6.2 The probability distribution of the continuous random variable $Y =$ randomly generated number between 0 and 9. The shaded area represents $P(3 \le Y \le 7)$.

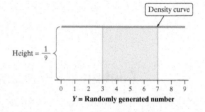

$Y =$ **Randomly generated number**

HOW TO FIND PROBABILITIES FOR A CONTINUOUS RANDOM VARIABLE

The probability of any event involving a continuous random variable is the area under the density curve and directly above the values on the horizontal axis that make up the event.

Consider a specific outcome from the random number generator setting, such as $P(Y = 7)$. The probability of this event is the area under the density curve that's above the point 7.0000 . . . on the horizontal axis. But this vertical line segment has no width, so the area is 0. In fact, all continuous probability distributions assign probability 0 to every individual outcome. For that reason,

$$P(3 \le Y \le 7) = P(3 \le Y < 7) = P(3 < Y \le 7) = P(3 < Y < 7) = 0.444$$

Remember: The probability distribution for a continuous random variable assigns probabilities to *intervals* of outcomes rather than to individual outcomes.

EXAMPLE	Will it be quicker to walk to work?
	Continuous random variables

PROBLEM: Selena works at a bookstore in the Denver International Airport. She takes the airport train from the main terminal to get to work each day. The airport just opened a new walkway that would allow Selena to get from the main terminal to the bookstore in 4 minutes. She wonders if it will be faster to walk or take the train to work. Let $Y =$ Selena's journey time to work (in minutes) by train on a randomly selected day. The probability distribution of Y can be modeled by a uniform density curve on the interval from 2 to 5 minutes. Find the probability that it will be quicker for Selena to take the train than to walk that day.

ALTERNATE EXAMPLE   Skill 3.A

Extra credit   Continuous random variables

PROBLEM:

A certain AP® Statistics teacher is feeling generous one day and decides that each student deserves some extra credit. The teacher assigns each student a random extra credit value between 0 and 5 (decimals included) by using 5*rand on the calculator.

Let $X =$ amount of extra credit for a randomly selected student. The probability distribution of X can be modeled by a uniform density curve on the interval from 0 to 5. Find the probability that a randomly selected student will get more than 3 points of extra credit.

SOLUTION:

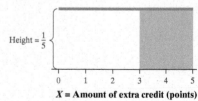

$X =$ **Amount of extra credit (points)**

$$Area = base \times height = 2 \times 1/5 = 2/5$$

$$P(X > 3) = 2/5 = 0.40$$

SOLUTION:

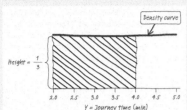

Start by drawing the density curve with the area of interest shaded. The height of the curve needs to be 1/3 so that
Area = base × height = 3 × 1/3 = 1

$$Shaded\ area = base \times height = 2 \times 1/3 = 2/3$$

$$P(Y < 4) = 2/3 = 0.667$$

There is a 66.7% chance that it will be quicker for Selena to take the train to work on a randomly selected day.

FOR PRACTICE, TRY EXERCISE 23

Density curves are probability distributions. They can also model populations of quantitative data, like Selena's train journey times in the preceding example.

EXAMPLE

Young women's heights
Normal probability distributions

PROBLEM: The heights of young women can be modeled by a Normal distribution with mean $\mu = 64$ inches and standard deviation $\sigma = 2.7$ inches. Suppose we choose a young woman at random and let $Y =$ her height (in inches). Find $P(68 \le Y \le 70)$. Interpret this value.

SOLUTION:

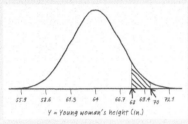

1. Draw a Normal distribution.
2. Perform calculations—show your work!
(i) Standardize and use Table A or technology; or
(ii) Use technology without standardizing.
Be sure to answer the question that was asked.

(i) $z = \dfrac{68-64}{2.7} = 1.48$

$z = \dfrac{70-64}{2.7} = 2.22$

Using Table A: $0.9868 - 0.9306 = 0.0562$

$P(68 \le Y \le 70) = P(1.48 \le Z \le 2.22)$

Using technology: normalcdf(lower:1.48, upper:2.22, mean:0, SD:1) = 0.0562

(ii) normalcdf(lower:68, upper:70, mean:64, SD:2.7) = 0.0561

The probability that a randomly selected young woman has a height between 68 and 70 inches is about 0.06.

FOR PRACTICE, TRY EXERCISE 27

Preparing for Inference

Students first performed Normal distribution calculations in Chapter 2. We revisit those calculations here in Chapter 6 in the context of continuous random variables. In Chapter 7, we will use Normal distributions to do probability calculations for sampling distributions. In Chapter 8, we use them to get z^* values for confidence intervals and again in Chapter 9 to calculate P-values. It is important to student success that they become fluent in Normal distribution calculations.

ALTERNATE EXAMPLE Skills 3.A, 4.B

Weights of 3-year-old females
Normal probability distributions

PROBLEM:
The weights of 3-year-old females closely follow a Normal distribution with a mean of $\mu = 30.7$ pounds and a standard deviation of 3.6 pounds. Suppose we randomly choose a 3-year-old female and call her weight X. What is the probability that she weighs at least 30 pounds?

SOLUTION:

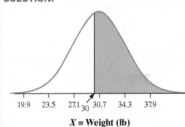

$X =$ **Weight (lb)**

(i) $z = \dfrac{30 - 30.7}{3.6} = -0.19$

Table A: $1 - 0.4247 = 0.5753$

Tech: normalcdf(lower: -0.19, upper: 1000, mean: 0, SD: 1) = 0.5753

(ii) normalcdf(lower: 30, upper: 1000, mean: 30.7, SD: 3.6) = 0.5771

There is about a 58% chance that the randomly selected 3-year-old female will weigh at least 30 pounds.

Teaching Tip

When reading this example with your students, mention that $P(68 \le X \le 70)$ is really the same as $P(68 < X < 70)$, because there is no area directly above $x = 68$ or $x = 70$.

Teaching Tip

Ask students to draw a picture for *every* Normal distribution calculation so that they can confirm their answer makes sense. The picture will often save students from making a careless mistake. Also, remind students that a z-score indicates the number of standard deviations above or below the mean that a value falls in a distribution. Finally, encourage students to label the axis with the random variable (e.g., $Y =$ young woman's height).

➕ Ask the StatsMedic
Table A vs. normalcdf

Should students use *z*-scores and Table A or should they use normalcdf? There is no correct answer here, but this post contains some good discussion of the pros and cons for each approach.

Teaching Tip: AP® Connections

AP® Calculus students may be interested in the formulas for the mean and variance of a continuous random variable. If $f(x)$ is the probability distribution function:

$$\mu_x = \int_{-\infty}^{\infty} x \cdot f(x)dx$$

$$\sigma_x^2 = \int_{-\infty}^{\infty} (x - \mu_x)^2 \cdot f(x)dx$$

TRM Section 6.1 Quiz

There is one quiz available for this section in the Teacher's Resource Materials. Click on the link in the TE-Book, look on the TRFD, or download from the Teacher's Resources on the book's digital platform. You can also create your own quiz using the ExamView® Assessment Suite that is part of the TPS6 program. Questions are coded by Learning Target to make it easy to build parallel quizzes.

> **■ AP® EXAM TIP**
>
> Students often do not get full credit on the AP® Statistics exam because they only use option (ii) with "calculator-speak" to show their work on Normal calculation questions—for example, normalcdf(68,70,64,2.7). This is not considered clear communication. To get full credit, follow the two-step process above, making sure to carefully label each of the inputs in the calculator command if you use technology in Step 2: normalcdf(lower:68, upper:70, mean:64, SD: 2.7).

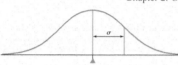

The calculation in the preceding example is the same as those we did in Chapter 2. Only the language of probability is new. By thinking of the density curve as a model for the population distribution of a quantitative variable, we can interpret the result as: "About 6% of all young women have heights between 68 and 70 inches."

What about the mean μ and standard deviation σ for continuous random variables? We interpret these parameters in the same way as we did for discrete random variables. Chapter 2 showed us how to find the mean of the distribution: it is the point at which the area under the density curve would balance if it were made out of solid material. The mean lies at the center of symmetric density curves such as Normal curves. We can locate the standard deviation of a Normal distribution from its inflection points. Exact calculation of the mean and standard deviation for most continuous random variables requires advanced mathematics.[3]

Section 6.1 | Summary

- A **random variable** takes numerical values determined by the outcome of a random process. The **probability distribution** of a random variable gives its possible values and their probabilities. There are two types of random variables: *discrete* and *continuous*.
- A **discrete random variable** has a fixed set of possible values with gaps between them.
  - A valid probability distribution assigns each of these values a probability between 0 and 1 such that the sum of all the probabilities is exactly 1.
  - The probability of any event is the sum of the probabilities of all the values that make up the event.
  - We can display the probability distribution as a histogram, with the values of the variable on the horizontal axis and the probabilities on the vertical axis.
- A **continuous random variable** can take any value in an interval on the number line.
  - A valid probability distribution for a continuous random variable is described by a density curve with area 1 underneath.
  - The probability of any event is the area under the density curve directly above the values on the horizontal axis that make up the event.
- A probability distribution can also be used to model the population distribution of a quantitative variable.
- We can describe the *shape* of a probability distribution histogram or density curve in the same way as we did a distribution of quantitative data—by identifying symmetry or skewness and any major peaks.

- Use the mean to summarize the *center* of a probability distribution. The **mean of a random variable** μ_X is the balance point of the probability distribution histogram or density curve.
  - The mean is the long-run average value of the variable after many, many trials of the random process. It is also known as the **expected value** of the random variable, $E(X)$.
  - If X is a discrete random variable, the mean is the average of the values of X, each weighted by its probability:

$$\mu_X = E(X) = \sum x_i p_i = x_1 p_1 + x_2 p_2 + x_3 p_3 + \cdots$$

- Use the standard deviation to summarize the variability of a probability distribution. The **standard deviation of a random variable** σ_X measures how much the values of the variable typically vary from the mean in many, many trials of the random process.
  - If X is a discrete random variable, the **variance** of X is the "average" squared deviation of the values of the variable from their mean:

$$\sigma_X^2 = \sum (x_i - \mu_X)^2 p_i = (x_1 - \mu_X)^2 p_1 + (x_2 - \mu_X)^2 p_2 + (x_3 - \mu_X)^2 p_3 + \cdots$$

The standard deviation σ_X is the square root of the variance.

6.1 Technology Corner

TI-Nspire and other technology instructions are on the book's website at highschool.bfwpub.com/updatedtps6e.

13. Analyzing discrete random variables   Page 398

Section 6.1 | Exercises

1. **Kids and toys** In an experiment on the behavior of young children, each subject is placed in an area with five toys. Past experiments have shown that the probability distribution of the number X of toys played with by a randomly selected subject is as follows:

Number of toys x_i	0	1	2	3	4	5
Probability p_i	0.03	0.16	0.30	0.23	0.17	???

(a) Write the event "child plays with 5 toys" in terms of X. Then find its probability.

(b) What's the probability that a randomly selected subject plays with at most 3 toys?

2. **Spell-checking** Spell-checking software catches "nonword errors," which result in a string of letters that is not a word, as when "the" is typed as "teh." When undergraduates are asked to write a 250-word essay (without spell-checking), the number Y of nonword errors in a randomly selected essay has the following probability distribution.

Value y_i	0	1	2	3	4
Probability p_i	0.1	???	0.3	0.3	0.1

(a) Write the event "one nonword error" in terms of Y. Then find its probability.

(b) What's the probability that a randomly selected essay has at least two nonword errors?

3. **Get on the boat!** A small ferry runs every half hour from one side of a large river to the other. The probability distribution for the random variable Y = money collected (in dollars) on a randomly selected ferry trip is shown here.

Money collected	0	5	10	15	20	25
Probability	0.02	0.05	0.08	0.16	0.27	0.42

(a) Find $P(Y < 20)$. Interpret this result.

(b) Express the event "at least $20 is collected" in terms of Y. What is the probability of this event?

Teaching Tip

Make sure to point out the two icons next to Exercise 1. The "pg 391" icon reminds students that the example on page 391 is very similar to this exercise. The "play" icon reminds students that there is a video solution available in the student e-Book or at the Student Site.

Answers to Section 6.1 Exercises

6.1 (a) $X = 5$; $P(X = 5) = 0.11$
(b) $P(X \leq 3) = 0.72$

6.2 (a) $Y = 1$; $P(Y = 1) = 0.2$
(b) $P(Y \geq 2) = 0.7$

6.3 (a) $P(Y < 20) = 0.31$; there is a 0.31 probability that the amount of money collected on a randomly selected ferry trip is less than $20.
(b) $Y \geq 20$; $P(Y \geq 20) = 0.69$

6.4 (a) $P(X > 20) = 0.41$; there is a 0.41 probability that Ana scores more than 20 points on a randomly selected roll of the ball. **(b)** $X \le 20$; $P(X \le 20) = 0.59$

6.5

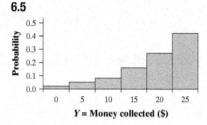

Y = Money collected ($)

Skewed to the left with a single peak at $25 collected.

6.6

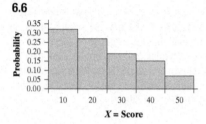

X = Score

Skewed to the right with a single peak at a score of 10 points.

6.7 $\mu_Y = 19.35$; if many, many ferry trips are randomly selected, the average amount of money collected would be about $19.35.

6.8 $\mu_Y = 23.8$; if Ana rolls many, many Skee Balls, the average number of points scored would be about 23.8 points.

6.9 (a) Right-skewed distribution; the most likely first digit is 1, and each subsequent digit is less likely than the previous digit. **(b)** $\mu_X = 3.441$; if many, many legitimate records are randomly selected, the average of the first digit would be about 3.441.

6.10 (a) Right-skewed distribution with a peak at 0 days. **(b)** $\mu_Y = 1.03$; if many people aged 19–25 are randomly selected, the average number of days they went to an exercise or fitness center or worked out in the past week would be about 1.03 days.

4. Skee Ball Ana is a dedicated Skee Ball player (see photo) who always rolls for the 50-point slot. The probability distribution of Ana's score X on a randomly selected roll of the ball is shown here.

Score	10	20	30	40	50
Probability	0.32	0.27	0.19	0.15	0.07

Stan Rohrer/Getty Images

(a) Find $P(X > 20)$. Interpret this result.

(b) Express the event "Ana scores at most 20" in terms of X. What is the probability of this event?

pg 393 **5. Get on the boat!** Refer to Exercise 3. Make a histogram of the probability distribution. Describe its shape.

6. Skee Ball Refer to Exercise 4. Make a histogram of the probability distribution. Describe its shape.

7. Get on the boat! Refer to Exercise 3. Find the mean of Y.
pg 395 Interpret this parameter.

Money collected	0	5	10	15	20	25
Probability	0.02	0.05	0.08	0.16	0.27	0.42

8. Skee Ball Refer to Exercise 4. Find the mean of X. Interpret this parameter.

Score	10	20	30	40	50
Probability	0.32	0.27	0.19	0.15	0.07

9. Benford's law Faked numbers in tax returns, invoices, or expense account claims often display patterns that aren't present in legitimate records. Some patterns, like too many round numbers, are obvious and easily avoided by a clever crook. Others are more subtle. It is a striking fact that the first digits of numbers in legitimate records often follow a model known as Benford's law.[4] Call the first digit of a randomly chosen legitimate record X for short. The probability distribution for X is shown here (note that a first digit cannot be 0).

First digit x_i	1	2	3	4	5	6	7	8	9
Probability p_i	0.301	0.176	0.125	0.097	0.079	0.067	0.058	0.051	0.046

(a) A histogram of the probability distribution is shown. Describe its shape.

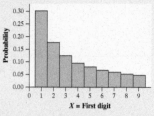

X = First digit

(b) Calculate and interpret the expected value of X.

10. Working out Choose a person aged 19 to 25 years at random and ask, "In the past seven days, how many times did you go to an exercise or fitness center or work out?" Call the response Y for short. Based on a large sample survey, here is the probability distribution of Y.[5]

Days y_i	0	1	2	3	4	5	6	7
Probability p_i	0.68	0.05	0.07	0.08	0.05	0.04	0.01	0.02

(a) A histogram of the probability distribution is shown. Describe its shape.

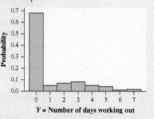

Y = Number of days working out

(b) Calculate and interpret the expected value of Y.

11. Get on the boat! A small ferry runs every half hour from one side of a large river to the other. The probability distribution for the random variable Y = money collected on a randomly selected ferry trip is shown here. From Exercise 7, $\mu_Y = \$19.35$.

Money collected	0	5	10	15	20	25
Probability	0.02	0.05	0.08	0.16	0.27	0.42

(a) Construct the cumulative probability distribution for Y.

(b) Use the cumulative probability distribution to find the median of Y.

(c) Compare the mean and median. Explain why this relationship makes sense based on the probability distribution.

6.11 (a)

Money collected	0	5	10	15	20	25
Probability	0.02	0.05	0.08	0.16	0.27	0.42
Cumulative probability	0.02	0.07	0.15	0.31	0.58	1

The median of Y is $20. **(b)** The mean of Y is less than the median of Y because the probability distribution is skewed to the left.

12. Skee Ball Ana is a dedicated Skee Ball player (see photo in Exercise 4) who always rolls for the 50-point slot. The probability distribution of Ana's score X on a randomly selected roll of the ball is shown here. From Exercise 8, $\mu_X = 23.8$.

Score	10	20	30	40	50
Probability	0.32	0.27	0.19	0.15	0.07

(a) Construct the cumulative probability distribution for X.

(b) Use the cumulative probability distribution to find the median of X.

(c) Compare the mean and median. Explain why this relationship makes sense based on the probability distribution.

13. Get on the boat! A small ferry runs every half hour from one side of a large river to the other. The probability distribution for the random variable Y = money collected on a randomly selected ferry trip is shown here. From Exercise 7, $\mu_Y = \$19.35$. Calculate and interpret the standard deviation of Y.

Money collected	0	5	10	15	20	25
Probability	0.02	0.05	0.08	0.16	0.27	0.42

14. Skee Ball Ana is a dedicated Skee Ball player (see photo in Exercise 4) who always rolls for the 50-point slot. The probability distribution of Ana's score X on a randomly selected roll of the ball is shown here. From Exercise 8, $\mu_X = 23.8$. Calculate and interpret the standard deviation of X.

Score	10	20	30	40	50
Probability	0.32	0.27	0.19	0.15	0.07

15. Benford's law Exercise 9 described how the first digits of numbers in legitimate records often follow a model known as Benford's law. Call the first digit of a randomly chosen legitimate record X for short. The probability distribution for X is shown here (note that a first digit can't be 0). From Exercise 9, $E(X) = 3.441$. Find the standard deviation of X. Interpret this parameter.

First digit x_i	1	2	3	4	5	6	7	8	9
Probability p_i	0.301	0.176	0.125	0.097	0.079	0.067	0.058	0.051	0.046

16. Working out Exercise 10 described a large sample survey that asked a sample of people aged 19 to 25 years, "In the past seven days, how many times did you go to an exercise or fitness center or work out?" The response Y for a randomly selected survey respondent has the probability distribution shown here. From Exercise 10, $E(Y) = 1.03$. Find the standard deviation of Y. Interpret this parameter.

Days y_i	0	1	2	3	4	5	6	7
Probability p_i	0.68	0.05	0.07	0.08	0.05	0.04	0.01	0.02

17. Life insurance A life insurance company sells a term insurance policy to 21-year-old males that pays $100,000 if the insured dies within the next 5 years. The probability that a randomly chosen male will die each year can be found in mortality tables. The company collects a premium of $250 each year as payment for the insurance. The amount Y that the company profits on a randomly selected policy of this type is $250 per year, less the $100,000 that it must pay if the insured dies. Here is the probability distribution of Y:

Age at death	21	22	23
Profit y_i	−$99,750	−$99,500	−$99,250
Probability p_i	0.00183	0.00186	0.00189

Age at death	24	25	26 or more
Profit y_i	−$99,000	−$98,750	$1250
Probability p_i	0.00191	0.00193	0.99058

(a) Explain why the company suffers a loss of $98,750 on such a policy if a client dies at age 25.

(b) Calculate the expected value of Y. Explain what this result means for the insurance company.

(c) Calculate the standard deviation of Y. Explain what this result means for the insurance company.

18. Fire insurance Suppose a homeowner spends $300 for a home insurance policy that will pay out $200,000 if the home is destroyed by fire in a given year. Let P = the profit made by the company on a single policy. From previous data, the probability that a home in this area will be destroyed by fire is 0.0002.

(a) Make a table that shows the probability distribution of P.

(b) Calculate the expected value of P. Explain what this result means for the insurance company.

(c) Calculate the standard deviation of P. Explain what this result means for the insurance company.

19. Size of American households In government data, a household consists of all occupants of a dwelling unit, while a family consists of two or more persons who live together and are related by blood or marriage. So all families form households, but some households are not families. Here are the distributions of household size and family size in the United States:

	Number of persons						
	1	2	3	4	5	6	7
Household probability	0.25	0.32	0.17	0.15	0.07	0.03	0.01
Family probability	0	0.42	0.23	0.21	0.09	0.03	0.02

Let H = the number of people in a randomly selected U.S. household and F = the number of people in a randomly selected U.S. family.

6.17 (a) The company has collected $1250 and has to pay out $100,000. The company earns $1250 − $100,000 = −$98,750.
(b) $\mu_Y = \$303.35$; if many 21-year-old males are insured by this company, the average amount the company would make, per person, would be about $303.35.
(c) $\sigma_Y = \$9707.57$; if many, many policies are randomly selected, the amount that the company earns will typically vary by about $9,707.57 from the mean of $303.35.

6.18 (a)

Profit	$300	−$199,700
Probability	0.9998	0.0002

(b) $\mu_P = \$260$; if many, many homes are insured by this company, the average amount the company would make, per home, would be about $260.
(c) $\sigma_P = \$2828.14$; if many, many policies are randomly selected, the amount that the company earns will typically vary by about $2,828.14 from the mean of $260.

6.12 (a)

Score	10	20	30	40	50
Probability	0.32	0.27	0.19	0.15	0.07
Cumulative probability	0.32	0.59	0.78	0.93	1

The median of X is 20 points. **(b)** The mean of X is greater than the median of X because the probability distribution is skewed to the right.

6.13 $\sigma_Y = 6.429$; if many, many ferry trips are randomly selected, the cost will typically vary by about $6.43 from the mean of $19.35.

6.14 $\sigma_X = 12.632$; if many, many Skee Ball rolls are randomly selected, the point value will typically vary by about 12.632 from the mean of 23.8.

6.15 $\sigma_X = 2.462$; if many, many records are randomly selected, the value of the first digit of the record will typically vary by about 2.462 from the mean of 3.441.

6.16 $\sigma_Y = 1.769$; if many, many respondents are randomly selected, the responses will typically vary by about 1.769 from the mean of 1.03.

6.19 (a) Both distributions are skewed to the right. The center for the "household" distribution is less than the center for the "family" distribution, but the variability of the "household" distribution is greater than the variability of the "family" distribution. Also, the event $X = 1$ has a much higher probability in the "household" distribution. **(b)** $\mu_H = 2.6$ and $\mu_F = 3.14$; the household distribution has a slightly smaller mean than the family distribution. **(c)** The standard deviations are: $\sigma_H = 1.421$ and $\sigma_F = 1.249$; the standard deviation for the household distribution is slightly larger than that for the family distribution.

6.20 (a) The distribution of the number of rooms is roughly symmetric for owners and skewed to the right for renters. Renter-occupied units tend to have fewer rooms (center about 4) than owner-occupied units (center about 6). There is more variability in the number of rooms for owner-occupied units. **(b)** $\mu_X = 6.284$; $\mu_Y = 4.187$. The renter-occupied distribution has a smaller mean than the owner-occupied distribution. **(c)** $\sigma_X = 1.6399$; $\sigma_Y = 1.3077$. The renter-occupied distribution has a smaller standard deviation than the owner-occupied distribution.

6.21 (a) $P(Y > 6) = 0.333$; $P(X > 6) = 0.155$. If an expense report contains more than a proportion of 0.155 of first digits that are greater than 6, it may be a fake expense report. **(b)** The mean is 5 because this distribution is symmetric. **(c)** To detect a fake expense report, compute the sample mean of the first digits and see if it is closer to 5 (suggesting a fake report) or near 3.441 (consistent with a truthful report).

6.22 (a) $\sigma_Y = 2.58$ **(b)** $\sigma_X^2 = 6.0605$, so $\sigma_X = 2.4618$. Comparing standard deviations would not be the best way to tell the difference between a fake and a real expense report because the standard deviations are similar.

(a) Here are histograms comparing the probability distributions of H and F. Describe any differences that you observe.

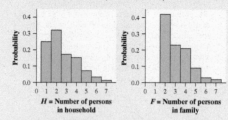

H = Number of persons in household   F = Number of persons in family

(b) Find the expected value of each random variable. Explain why this difference makes sense.

(c) The standard deviations of the two random variables are $\sigma_H = 1.421$ and $\sigma_F = 1.249$. Explain why this difference makes sense.

20. **Housing in San José** How do rented housing units differ from units occupied by their owners? Here are the distributions of the number of rooms for owner-occupied units and renter-occupied units in San José, California:[6]

	Number of rooms									
	1	2	3	4	5	6	7	8	9	10
Owned	0.003	0.002	0.023	0.104	0.210	0.224	0.197	0.149	0.053	0.035
Rented	0.008	0.027	0.287	0.363	0.164	0.093	0.039	0.013	0.003	0.003

Let X = the number of rooms in a randomly selected owner-occupied unit and Y = the number of rooms in a randomly chosen renter-occupied unit.

(a) Here are histograms comparing the probability distributions of X and Y. Describe any differences you observe.

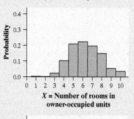

X = Number of rooms in owner-occupied units

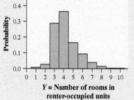

Y = Number of rooms in renter-occupied units

(b) Find the expected number of rooms for both types of housing unit. Explain why this difference makes sense.

(c) The standard deviations of the two random variables are $\sigma_X = 1.640$ and $\sigma_Y = 1.308$. Explain why this difference makes sense.

Exercises 21 and 22 examine how Benford's law (Exercise 9) can be used to detect fraud.

21. **Benford's law and fraud** A not-so-clever employee decided to fake his monthly expense report. He believed that the first digits of his expense amounts should be equally likely to be any of the numbers from 1 to 9. In that case, the first digit Y of a randomly selected expense amount would have the probability distribution shown in the histogram.

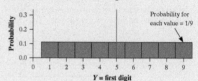

Probability for each value = 1/9

Y = first digit

(a) What's $P(Y > 6)$? According to Benford's law (see Exercise 9), what proportion of first digits in the employee's expense amounts should be greater than 6? How could this information be used to detect a fake expense report?

(b) Explain why the mean of the random variable Y is located at the solid red line in the figure.

(c) According to Benford's law, the expected value of the first digit is $\mu_X = 3.441$. Explain how this information could be used to detect a fake expense report.

22. **Benford's law and fraud**

(a) Using the graph from Exercise 21, calculate the standard deviation σ_Y. This gives us an idea of how much variation we'd expect in the employee's expense records if he assumed that first digits from 1 to 9 were equally likely.

(b) The standard deviation of the first digits of randomly selected expense amounts that follow Benford's law is $\sigma_X = 2.46$. Would using standard deviations be a good way to detect fraud? Explain your answer.

23. **Still waiting for the server?** How does your web pg 400 browser get a file from the Internet? Your computer sends a request for the file to a web server, and the web server sends back a response. Let Y = the amount of time (in seconds) after the start of an hour at which a randomly selected request is received by a particular web server. The probability distribution of Y can be

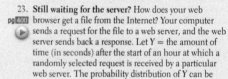

modeled by a uniform density curve on the interval from 0 to 3600 seconds. Find the probability that the request is received by this server within the first 5 minutes (300 seconds) after the hour.

24. **Where's the bus?** Sally takes the same bus to work every morning. Let $X =$ the amount of time (in minutes) that she has to wait for the bus on a randomly selected day. The probability distribution of X can be modeled by a uniform density curve on the interval from 0 minutes to 8 minutes. Find the probability that Sally has to wait between 2 and 5 minutes for the bus.

25. **Class is over!** Mr. Shrager does not always let his statistics class out on time. In fact, he seems to end class according to his own "internal clock." The density curve here models the distribution of Y, the amount of time after class ends (in minutes) when Mr. Shrager dismisses the class on a randomly selected day. (A negative value indicates he ended class early.)

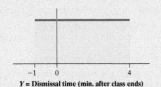

$Y =$ Dismissal time (min. after class ends)

(a) Find and interpret $P(-1 \le Y \le 1)$.

(b) What is μ_Y? Explain your answer.

(c) Find the value of x that makes this statement true: $P(Y \ge x) = 0.25$.

26. **Quick, click!** An Internet reaction time test asks subjects to click their mouse button as soon as a light flashes on the screen. The light is programmed to go on at a randomly selected time after the subject clicks "Start." The density curve models the amount of time Y (in seconds) that the subject has to wait for the light to flash.

$Y =$ Time until light flashes (sec)

(a) Find and interpret $P(Y > 3.75)$.

(b) What is μ_Y? Explain your answer.

(c) Find the value of x that makes this statement true: $P(Y \ge x) = 0.62$.

27. **Running a mile** A study of 12,000 able-bodied male students at the University of Illinois found that their times for the mile run were approximately Normal with mean 7.11 minutes and standard deviation 0.74 minute.[7] Choose a student at random from this group and call his time for the mile Y. Find $P(Y < 6)$. Interpret this value.

28. **Give me some sugar!** Machines that fill bags with powdered sugar are supposed to dispense 32 ounces of powdered sugar into each bag. Let $X =$ the weight (in ounces) of the powdered sugar dispensed into a randomly selected bag. Suppose that X can be modeled by a Normal distribution with mean 32 ounces and standard deviation 0.6 ounce. Find $P(X \le 31)$. Interpret this value.

29. **Horse pregnancies** Bigger animals tend to carry their young longer before birth. The length of horse pregnancies from conception to birth varies according to a roughly Normal distribution with mean 336 days and standard deviation 6 days. Let $X =$ the length of a randomly selected horse pregnancy.

(a) Write the event "pregnancy lasts between 325 and 345 days" in terms of X. Then find its probability.

(b) Find the 80th percentile of the distribution.

30. **Ace!** Professional tennis player Novak Djokovic hits the ball extremely hard. His first-serve speeds follow an approximately Normal distribution with mean 115 miles per hour (mph) and standard deviation 6 mph. Choose one of Djokovic's first serves at random. Let $Y =$ its speed, measured in miles per hour.

(a) Write the event "speed is between 100 and 120 miles per hour" in terms of Y. Then find its probability.

(b) Find the 15th percentile of the distribution.

Multiple Choice *Select the best answer for Exercises 31–34.*

Exercises 31–33 refer to the following setting. Choose an American household at random and let the random variable X be the number of cars (including SUVs and light trucks) they own. Here is the probability distribution if we ignore the few households that own more than 5 cars:

Number of cars	0	1	2	3	4	5
Probability	0.09	0.36	0.35	0.13	0.05	0.02

31. What's the expected number of cars in a randomly selected American household?

(a) 1.00

(b) 1.75

(c) 1.84

(d) 2.00

(e) 2.50

6.23 Length $= 3600$; height $= 1/3600$; probability $= 300(1/3600) = 0.083$. The probability that the request is received by this server within the first 5 minutes (300 seconds) after the hour $= 0.083$.

6.24 Length $= 8$; height $= 1/8$; probability $= (5 - 2)(1/8) = 0.375$. There is about a 37.5% chance that Sally has to wait between 2 and 5 minutes for the bus.

6.25 (a) $P(-1 \le Y \le 1) = 0.4$; the probability that Mr. Shrager dismisses class within a minute of the end of class is 0.4. **(b)** $\mu_Y = 1.5$; the mean is 1.5 because this distribution is symmetric. **(c)** $k = 2.75$; to find this value, we solve the equation $(4 - k)(1/5) = 0.25$ for k.

6.26 (a) $P(Y > 3.75) = 0.3125$; the probability that the subject has to wait more than 3.75 seconds for the light to flash is 0.3125. **(b)** $\mu_Y = 3$; the mean is 3 because this distribution is symmetric. **(c)** $k = 2.52$; to find this value, we solve the equation $(k - 1)(1/4) = 0.38$ for k.

6.27 (i) $z = -1.50$

Table A: $P(Z < -1.50) = 0.0668$

(ii) *Tech:* $P(Y < 6) = $ normalcdf(lower: -1000, upper: 6, mean: 7.11, SD: 0.74) $= 0.0668$. There is about a 6.68% chance that this student will run the mile in under 6 minutes.

6.28 (i) $z = -1.67$

Table A: $P(Z < -1.67) = 0.0475$

(ii) *Tech:* $P(X \le 31) = $ normalcdf(lower: -1000, upper: 31, mean: 32, SD: 0.6) $= 0.0478$. There is about a 4.78% chance that a randomly selected bag of powdered sugar will weigh at most 31 ounces.

6.29 (a) $P(325 < X < 345)$

(i) $z = -1.83$ and $z = 1.5$

Table A: $P(-1.83 < Z < 1.5) = 0.9332 - 0.0336 = 0.8996$

(ii) *Tech:* $P(325 < X < 345) = $ normalcdf(lower: 325, upper: 345, mean: 336, SD: 6) $= 0.8998$. There is about an 89.98% chance that a randomly selected horse pregnancy will last between 325 and 345 days.

(b) (i) 0.20 area to the right of $c \rightarrow z = 0.84$

Table A:

$$0.84 = \frac{c - 336}{6}$$
$$5.04 = c - 336$$
$$c = 341.04$$

(ii) *Tech:* invNorm(area: 0.80, mean: 336, SD: 6) $= 341.05$. About 20% of horses will have pregnancies that last more than 341.05 days.

6.30 (a) $P(100 < Y < 120)$

(i) $z = -2.5$ and $z = 0.83$

Table A: $P(-2.5 < Z < 0.83) = 0.7967 - 0.0062 = 0.7905$

(ii) *Tech:* $P(100 < Y < 120) = $ normalcdf(lower: 100, upper: 120, mean: 115, SD: 6) $= 0.7915$. There is a 0.7915 probability that a randomly selected first serve by Djokovic will be between 100 and 120 miles per hour.

(b) (i) *Table A:* Look in the body of Table A for the value closest to 0.15, $z = -1.04$.

$$-1.04 = \frac{c - 115}{6}$$
$$-6.24 = c - 115$$
$$c = 108.76$$

(ii) *Tech:* invNorm(area: 0.15, mean: 115, SD: 6) $= 108.76$. About 15% of Djokovic's first serves will be at most 108.76 miles per hour.

6.31 b

6.32 b

6.33 c

6.34 c

6.35 (a) Yes; if we look at the differences (Post − Pre) in the scores, the mean difference was 5.38 and the median difference was 3. This means that at least half of the students (though less than three-quarters because Q_1 was negative) improved their reading scores. **(b)** No; we do not have a control group that did not participate in the chess program for comparison. It may be that children of this age improve their reading scores for other reasons (e.g., regular school) and that the chess program had nothing to do with their improvement.

32. The standard deviation of X is $\sigma_X = 1.08$. If many households were selected at random, which of the following would be the best interpretation of the value 1.08?

(a) The mean number of cars would be about 1.08.

(b) The number of cars would typically be about 1.08 from the mean.

(c) The number of cars would be at most 1.08 from the mean.

(d) The number of cars would be within 1.08 from the mean about 68% of the time.

(e) The mean number of cars would be about 1.08 from the expected value.

33. About what percentage of households have a number of cars within 2 standard deviations of the mean?

(a) 68%   (d) 95%

(b) 71%   (e) 98%

(c) 93%

34. A deck of cards contains 52 cards, of which 4 are aces. You are offered the following wager: Draw one card at random from the deck. You win $10 if the card drawn is an ace. Otherwise, you lose $1. If you make this wager very many times, what will be the mean amount you win?

(a) About −$1, because you will lose most of the time.

(b) About $9, because you win $10 but lose only $1.

(c) About −$0.15; that is, on average, you lose about 15 cents.

(d) About $0.77; that is, on average, you win about 77 cents.

(e) About $0, because the random draw gives you a fair bet.

Recycle and Review

Exercises 35 and 36 refer to the following setting. Many chess masters and chess advocates believe that chess play develops general intelligence, analytical skill, and the ability to concentrate. According to such beliefs, improved reading skills should result from study to improve chess-playing skills. To investigate this belief, researchers conducted a study. All the subjects in the study participated in a comprehensive chess program, and their reading performances were measured before and after the program. The graphs and numerical summaries that follow provide information on the subjects' pretest scores, posttest scores, and the difference (Post − Pre) between these two scores.

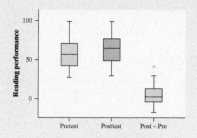

Descriptive Statistics: Pretest, Posttest, Post − Pre

Variable	N	Mean	Median	StDev	Min	Max	Q_1	Q_3
Pretest	53	57.70	58.00	17.84	23.00	99.00	44.50	70.50
Posttest	53	63.08	64.00	18.70	28.00	99.00	48.00	76.00
Post − Pre	53	5.38	3.00	13.02	−19.00	42.00	−3.50	14.00

35. Better readers? (1.3, 4.3)

(a) Did students tend to have higher reading scores after participating in the chess program? Justify your answer.

(b) If the study found a statistically significant improvement in the average reading score, could you conclude that playing chess causes an increase in reading skills? Justify your answer.

Some graphical and numerical information about the relationship between pretest and posttest scores is provided here.

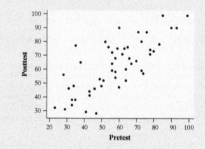

Regression Analysis: Posttest Versus Pretest				
Predictor	Coef	SE Coef	T	P
Constant	17.897	5.889	3.04	0.004
Pretest	0.78301	0.09758	8.02	0.000
S = 12.55	R-Sq = 55.8%		R-Sq(adj) = 54.9%	

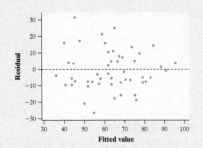

36. Predicting posttest scores (3.2)

(a) What is the equation of the least-squares regression line relating posttest and pretest scores? Define any variables used.

(b) Is a linear model appropriate for describing this relationship? Justify your answer.

(c) If we use the least-squares regression line to predict students' posttest scores from their pretest scores, how far off will our predictions typically be?

6.36 (a) $\hat{y} = 17.897 + 0.78301x$, where $\hat{y}$ is the predicted posttest score and x is the pretest score. **(b)** This linear model is appropriate because the residual plot does not show any leftover curved patterns. **(c)** $s = 12.55$; the actual posttest score is typically about 12.55 points away from the posttest score predicted by the least-squares regression line with $x =$ pretest score.

SECTION 6.2 Transforming and Combining Random Variables

LEARNING TARGETS *By the end of the section, you should be able to:*

- Describe the effect of adding or subtracting a constant or multiplying or dividing by a constant on the probability distribution of a random variable.

- Calculate the mean and standard deviation of the sum or difference of random variables.

- Find probabilities involving the sum or difference of independent Normal random variables.

In Section 6.1, we looked at several examples of random variables and their probability distributions. We also saw that the parameters μ_X and σ_X give us important information about a random variable.

Consider this new setting. An American roulette wheel has 38 slots numbered 1 through 36, plus 0 and 00. Half of the slots from 1 to 36 are red; the other half are black. Both the 0 and 00 slots are green. Suppose that a player places a $1 bet on red. If the ball lands in a red slot, the player gets the original dollar back, plus an extra dollar for winning the bet. If the ball lands in a different-colored slot, the player loses the $1 bet. Let $X =$ the net gain on a single $1 bet on red. Because there is an 18/38 chance that the ball lands in a red slot, the probability distribution of X is as shown in the table.

Value x_i	−$1	$1
Probability p_i	20/38	18/38

The mean of X is

$$\mu_X = (-1)\left(\frac{20}{38}\right) + (1)\left(\frac{18}{38}\right) = -\$0.05$$

That is, a player can expect to lose an average of 5 cents per $1 bet if he plays many, many games. You can verify that the standard deviation is $\sigma_X = \$1.00$. If

---

PD Section 6.2 Overview

To watch the video overview of Section 6.2 (for teachers), click on the link in the TE-Book, look on the TRFD, or download from the Teacher's Resources on the book's digital platform.

TRM Section 6.2 Alternate Examples

You can find the Alternate Examples for this section in Microsoft Word format by clicking on the link in the TE-Book, opening the TRFD, or downloading from the Teacher's Resources on the book's digital platform.

Teaching Tip

Be careful to distinguish between the distribution of $X =$ net gain (given in this example) and another random variable $Y =$ amount of winnings. The possible values for Y are $0 and $2. The difference between these two random variables is whether or not we are including the $1 cost to play the game.

Teaching Tip

Adding (or subtracting) a constant: Ask students to think about a histogram for some random variable. If we added 12 to each value, this would simply slide the entire histogram 12 units to the right, increasing the measures of center and location by 12—but it would not change the variability or shape.

Teaching Tip

Multiplying or dividing by a constant: Ask students to think about a histogram for some random variable that takes values between 0 and 8. If we multiplied each value by 10, the new histogram would go from 0 to 80. This would multiply the measures of center, location, and the variability by 10, but it would not change the shape.

Teaching Tip

For transforming random variables, consider having students work through a contextual example and then generalize to the rule or formula. Students can make predictions about the rules and then check their predictions empirically. Here is what it would look like using Pete's Jeep Tours as a context.

1. Give students the probability distributions for C = total money collected. Ask them to find the mean, standard deviation, and variance of C (calculate by hand).
2. Define a new variable for profit $V = C - 100$. Ask students to predict the mean, standard deviation, and variance for V.
3. Ask students to create the probability distribution for V. Use calculators or technology to find the mean, standard deviation, and variance for V. Verify or refute student predictions.
4. Define a new variable for profit (in pesos) $P = 20V$. Ask students to predict the mean, standard deviation, and variance for P.
5. Ask students to create the probability distribution for P. Use calculators or technology to find the mean, standard deviation, and variance for P. Verify or refute student predictions.
6. Generalize the results from this context to the rules and formulas.

the player only plays a few games, his actual net gain could be much better or worse than this expected value.

Would the player be better off playing one game of roulette with a $2 bet on red or playing two games and betting $1 on red each time? To find out, we need to compare the probability distributions of the random variables Y = gain from a $2 bet and T = total gain from two $1 bets. Which random variable (if either) has the higher expected gain in the long run? Which has the larger variability? By the end of this section, you'll be able to answer questions like these.

Transforming a Random Variable

In Chapter 2, we studied the effects of transformations on the shape, center, and variability of a distribution of quantitative data. Here's what we discovered:

1. *Adding (or subtracting) a constant:* Adding the same positive number a to (subtracting a from) each observation:
   - Adds a to (subtracts a from) measures of center and location (mean, median, quartiles, percentiles).
   - Does not change measures of variability (range, *IQR*, standard deviation).
   - Does not change the shape of the distribution.

2. *Multiplying or dividing by a constant:* Multiplying (or dividing) each observation by the same positive number b:
   - Multiplies (divides) measures of center and location (mean, median, quartiles, percentiles) by b.
   - Multiplies (divides) measures of variability (range, *IQR*, standard deviation) by b.
   - Does not change the shape of the distribution.

   How are the probability distributions of random variables affected by similar transformations?

EFFECT OF ADDING OR SUBTRACTING A CONSTANT Let's return to a familiar setting from Section 6.1. Pete's Jeep Tours offers a popular day trip in a tourist area. There must be at least 2 passengers for the trip to run, and the vehicle will hold up to 6 passengers. Pete charges $150 per passenger. Let C = the total amount of money that Pete collects on a randomly selected trip. The probability distribution of C is shown in the table and the histogram.

Total collected c_i	300	450	600	750	900
Probability p_i	0.15	0.25	0.35	0.20	0.05

Earlier, we calculated the mean of C as $\mu_C = \$562.50$ and the standard deviation of C as $\sigma_C = \$163.46$. We can describe the probability distribution of C as follows:

Shape: Roughly symmetric with a single peak

Center: $\mu_C = \$562.50$

Variability: $\sigma_C = \$163.46$

It costs Pete $100 to buy permits, gas, and a ferry pass for each day trip. The amount of profit V that Pete makes on a randomly selected trip is the total amount of money C that he collects from passengers minus $100. That is, $V = C - 100$. The probability distribution of V is

Profit v_i	200	350	500	650	800
Probability p_i	0.15	0.25	0.35	0.20	0.05

A histogram of this probability distribution is shown here.

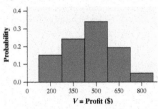

V = Profit ($)

We can see that the probability distribution of V has the same shape as the probability distribution of C. The mean of V is

$$\mu_V = (200)(0.15) + (350)(0.25) + (500)(0.35) + (650)(0.20) + (800)(0.05)$$
$$= \$462.50$$

On average, Pete will make a profit of $462.50 from the trip. That's $100 less than μ_C, his mean amount of money collected per trip. The standard deviation of V is

$$\sigma_V = \sqrt{(200 - 462.50)^2(0.15) + (350 - 462.50)^2(0.25) + \cdots + (800 - 462.50)^2(0.05)}$$
$$= \$163.46$$

That's the same as the standard deviation of C.

It's fairly clear that subtracting 100 from the values of the random variable C just shifts the probability distribution to the left by 100. This transformation decreases the mean by 100 (from $562.50 to $462.50), but it doesn't change the standard deviation ($163.46) or the shape. These results can be generalized for any random variable.

THE EFFECT OF ADDING OR SUBTRACTING A CONSTANT ON A PROBABILITY DISTRIBUTION

Adding the same positive number a to (subtracting a from) each value of a random variable:

- Adds a to (subtracts a from) measures of center and location (mean, median, quartiles, percentiles).
- Does not change measures of variability (range, IQR, standard deviation).
- Does not change the shape of the probability distribution.

Teaching Tip

All the probabilities for V are the same as the corresponding probabilities for C. To understand this, consider $V = 200$. To get a profit of $200, the total amount of money collected has to be $300. Since the probability of collecting $300 is 0.15, so is the probability of getting a profit of $200.

Making Connections

Emphasize that these results are exactly the same as we discovered in Chapter 2.

 COMMON STUDENT ERROR

When asked about the effect of a transformation on summary statistics, some students forget that adding (subtracting) a constant to (from) every value in the distribution has *no effect* on measures of variability, including the standard deviation, variance, range, and *IQR*.

Everyone gets a bonus
Effect of adding/subtracting a constant

PROBLEM:
A large corporation has thousands of employees. The distribution of annual earnings for the employees is skewed to the right, with a mean of $68,000 and a standard deviation of $18,000. Because business has been good this year, the CEO of the company decides that every employee will receive a $5000 bonus. Let X be the current annual earnings of a randomly selected employee before the bonus and Y be the employee's earnings after the bonus. Describe the shape, center, and variability of the probability distribution of Y.

SOLUTION:
Shape: Skewed right
Center: $\mu_Y = \mu_X + 5000 = 68,000 + 5000 = \$73,000$
Variability: $\sigma_Y = \sigma_X = \$18,000$

Teaching Tip

Students might argue that the shape of these two distributions is not the same (the second distribution is shorter and wider!). Have them consider changing the scale on the *x*- and *y*-axes. By squeezing the scale on the *x*-axis and stretching the scale on the *y*-axis, we could make the shape of the second distribution look identical to the first. Also note that "wider" should be used to describe variability, not shape.

Note that adding or subtracting a constant affects the distribution of a quantitative variable and the probability distribution of a random variable in exactly the same way.

EXAMPLE

Scaling test scores
Effect of adding/subtracting a constant

PROBLEM: In a large introductory statistics class, the score X of a randomly selected student on a test worth 50 points can be modeled by a Normal distribution with mean 35 and standard deviation 5. Due to a difficult question on the test, the professor decides to add 5 points to each student's score. Let Y be the scaled test score of the randomly selected student. Describe the shape, center, and variability of the probability distribution of Y.

SOLUTION:
Shape: Approximately Normal
Center: $\mu_Y = \mu_X + 5 = 35 + 5 = 40$
Variability: $\sigma_Y = \sigma_X = 5$

> Notice that $Y = X + 5$. Adding a constant doesn't affect the shape or the standard deviation of the probability distribution.

FOR PRACTICE, TRY EXERCISE 37

EFFECT OF MULTIPLYING OR DIVIDING BY A CONSTANT The professor in the preceding example decides to convert his students' scaled test scores Y to percentages. Because the test was scored out of 50 points, the professor multiplies each student's scaled score by 2 to convert to a percent score W. That is, $W = 2Y$. Figure 6.3 displays the probability distributions of the random variables Y and W. From the graphs, we can see that the measures of center, location, and variability have all doubled—just like the individual students' scores. But the shape of the two distributions is the same.

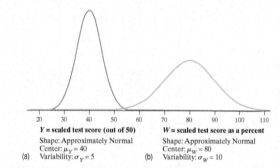

(a) Y = scaled test score (out of 50)
Shape: Approximately Normal
Center: $\mu_Y = 40$
Variability: $\sigma_Y = 5$

(b) W = scaled test score as a percent
Shape: Approximately Normal
Center: $\mu_W = 80$
Variability: $\sigma_W = 10$

FIGURE 6.3 Probability distribution of (a) Y = a randomly selected statistics student's scaled test score out of 50 and (b) W = the student's scaled test score as a percent.

It is not common to multiply (or divide) a random variable by a negative number *b*. Doing so would multiply (or divide) the measures of variability by |b|. Multiplying or dividing by a negative number would also affect the shape of the probability distribution, as all values would be reflected over the *y* axis.

THE EFFECT OF MULTIPLYING OR DIVIDING BY A CONSTANT ON A PROBABILITY DISTRIBUTION

Multiplying (or dividing) each value of a random variable by the same positive number *b*:

- Multiplies (divides) measures of center and location (mean, median, quartiles, percentiles) by *b*.
- Multiplies (divides) measures of variability (range, *IQR*, standard deviation) by *b*.
- Does not change the shape of the distribution.

Teaching Tip

Emphasize that these results are exactly the same as we discovered in Chapter 2.

Teaching Tip

While the standard deviation is multiplied by *b*, the variance is multiplied by b^2. See the Think About It on page 414.

Once again, multiplying or dividing by a constant has the same effect on the probability distribution of a random variable as it does on a distribution of quantitative data.

EXAMPLE

How much does college cost?

Effect of multiplying/dividing by a constant

PROBLEM: El Dorado Community College considers a student to be full-time if he or she is taking between 12 and 18 units. The number of units X that a randomly selected El Dorado Community College full-time student is taking in the fall semester has the following distribution.

Number of units	12	13	14	15	16	17	18
Probability	0.25	0.10	0.05	0.30	0.10	0.05	0.15

At right is a histogram of the probability distribution. The mean is $\mu_X = 14.65$ and the standard deviation is $\sigma_X = 2.056$.

At El Dorado Community College, the tuition for full-time students is \$50 per unit. That is, if T = tuition charge for a randomly selected full-time student, $T = 50X$.

(a) What shape does the probability distribution of T have?

(b) Find the mean of T.

(c) Calculate the standard deviation of T.

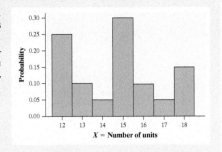
X = Number of units

SOLUTION:

(a) The same shape as the probability distribution of X: roughly symmetric with three peaks.

Multiplying by a constant doesn't change the shape.

(b) $\mu_T = 50\mu_X = 50(14.65) = \$732.50$

(c) $\sigma_T = 50\sigma_X = 50(2.056) = \$102.80$

FOR PRACTICE, TRY EXERCISE 41

ALTERNATE EXAMPLE

Selling cars   Effect of multiplying/dividing by a constant

Skill 3.C

PROBLEM:
Employees selling refrigerators at an appliance store make money on commission based on how many refrigerators they sell. The number of refrigerators R sold in a randomly selected hour has the following probability distribution:

Number of refrigerators	0	1	2	3	4	5
Probability	0.22	0.31	0.12	0.25	0.08	0.02

Here is a histogram of the probability distribution along with the mean and standard deviation.

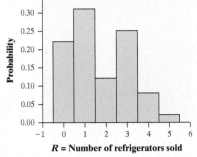
R = Number of refrigerators sold

$\mu_R = 1.72$

$\sigma_R = 1.36$

At this appliance store, the commission earned is \$30 for each refrigerator sold. That is, if C = total commission earned for a randomly selected hour, $C = 30R$.
(a) What shape does the probability distribution of C have?
(b) Find the mean of C.
(c) Calculate the standard deviation of C.

SOLUTION:
(a) The same shape as the probability distribution of R: slightly skewed right with two peaks.
(b) $\mu_C = 30\mu_R = 30(1.72) = \$51.60$
(c) $\sigma_C = 30\sigma_R = 30(1.36) = \$40.80$

✓ **Answers to CYU**

1. The probability distribution of X and the probability distribution of Y are shown below.

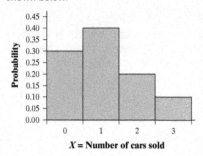

X = Number of cars sold

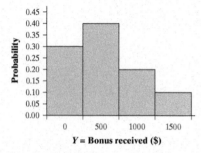

Y = Bonus received ($)

Both distributions are slightly skewed to the right with a single peak. Their shapes are identical.

2. $\mu_Y = 500(\mu_X) = 500(1.1) = \$550$

3. $\sigma_Y = 500(0.943) = \$471.50$; if many, many Fridays are randomly selected, the bonus earned in the first hour of business will typically vary by about $\$471.50$ from the mean of $\$550$.

4. Note that $T = Y - 75$.

Shape: The shape of the probability distribution of T will be the same as the shape of the probability distribution of Y: skewed right with a single peak.

Center:
$\mu_T = \mu_Y - 75 = 550 - 75 = \$475$

Variability: $\sigma_T = \sigma_Y = \$471.50$

Teaching Tip

Ask questions about other summary statistics such as the median and *IQR* to emphasize that these rules don't just apply to means and standard deviations.

Think About It

HOW DOES MULTIPLYING BY A CONSTANT AFFECT THE VARIANCE? For El Dorado Community College, the variance of the number of units that a randomly selected full-time student takes is $\sigma_X^2 = 4.2275$. The variance of the tuition charge for such a student is $\sigma_T^2 = 10{,}568.75$. That's $(2500)(4.2275)$. So $\sigma_T^2 = 2500\sigma_X^2$. Where did 2500 come from? It's just $(50)^2$. In other words, $\sigma_T^2 = (50)^2 \sigma_X^2$. Multiplying a random variable by a constant b multiplies the variance by b^2.

♦ **CHECK YOUR UNDERSTANDING**

A large auto dealership keeps track of sales made during each hour of the day. Let $X =$ the number of cars sold during the first hour of business on a randomly selected Friday. Based on previous records, the probability distribution of X is as follows:

Cars sold	0	1	2	3
Probability	0.3	0.4	0.2	0.1

The random variable X has mean $\mu_X = 1.1$ and standard deviation $\sigma_X = 0.943$. Suppose the dealership's manager receives a $500 bonus from the company for each car sold. Let $Y =$ the bonus received from car sales during the first hour on a randomly selected Friday.

1. Sketch a graph of the probability distribution of X and a separate graph of the probability distribution of Y. How do their shapes compare?
2. Find the mean of Y.
3. Calculate and interpret the standard deviation of Y.

The manager spends $75 to provide coffee and doughnuts to prospective customers each morning. So the manager's net profit T during the first hour on a randomly selected Friday is $75 less than the bonus earned.

4. Describe the shape, center, and variability of the probability distribution of T.

PUTTING IT ALL TOGETHER: ADDING/SUBTRACTING AND MULTIPLYING/ DIVIDING What happens if we transform a random variable by both adding or subtracting a constant and multiplying or dividing by a constant? Let's return to the preceding example.

El Dorado Community College charges each student a $100 fee per semester in addition to tuition charges. We can calculate a randomly selected full-time student's total charges Y for the fall semester directly from the number of units X the student is taking, using the equation $Y = 50X + 100$ or, equivalently, $Y = 100 + 50X$. This *linear transformation* of the random variable X includes two different transformations: (1) multiplying by 50 and (2) adding 100. Because neither of these transformations affects shape, the probability distribution of Y will have the same shape as the probability distribution of X. To get the mean of Y, we multiply the mean of X by 50, then add 100:

$$\mu_Y = 100 + 50\mu_X = 100 + 50(14.65) = \$832.50$$

> Can you see why this is called a "linear" transformation? The equation describing the sequence of transformations has the form $Y = a + bX$, which you should recognize as a linear equation.

To get the standard deviation of Y, we multiply the standard deviation of X by 50 (adding 100 doesn't affect SD):

$$\sigma_Y = 50\sigma_X = 50(2.056) = \$102.80$$

This logic generalizes to any linear transformation.

THE EFFECT OF A LINEAR TRANSFORMATION ON A RANDOM VARIABLE

If $Y = a + bX$ is a linear transformation of the random variable X,

- The probability distribution of Y has the same shape as the probability distribution of X if $b > 0$.
- $\mu_Y = a + b\mu_X$.
- $\sigma_Y = |b|\sigma_X$ (because b could be a negative number).

Note that these results apply to both discrete and continuous random variables.

EXAMPLE

The baby and the bathwater
Analyzing the effect of transformations

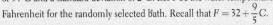

PROBLEM: One brand of baby bathtub comes with a dial to set the water temperature. When the "babysafe" setting is selected and the tub is filled, the temperature X of the water in a randomly selected bath follows a Normal distribution with a mean of 34°C and a standard deviation of 2°C. Let Y be the water temperature in degrees Fahrenheit for the randomly selected bath. Recall that $F = 32 + \frac{9}{5}C$.

(a) Find the mean of Y.
(b) Calculate and interpret the standard deviation of Y.

SOLUTION:

Michael DeLeon/Getty Images

Note that $Y = 32 + \frac{9}{5}X$.

(a) $\mu_Y = 32 + \frac{9}{5}(34) = 93.2°F$

$\mu_Y = a + b\mu_X$

(b) $\sigma_Y = \frac{9}{5}(2) = 3.6°F$

$\sigma_Y = |b|\sigma_X$

 If we randomly select many days when the dial is set on "babysafe," the temperature of the bath typically varies about 3.6°F from the mean of 93.2°F.

FOR PRACTICE, TRY EXERCISE 47

The probability distribution of Y = water temperature on a randomly selected day when the "babysafe" setting is used is Normal because the original distribution is Normal, and adding a constant and multiplying by a constant don't affect shape. We can use this Normal distribution to find probabilities as we did in Section 6.1. For instance, according to one source, the temperature of a baby's bathwater should be between 90°F and 100°F. What's $P(90 \leq Y \leq 100)$? Figure 6.4 shows the desired probability as an area under a Normal curve.

Making Connections

In the alternate example, the z-scores have a mean of 0 and a standard deviation of 1. In Chapter 2, we discovered that standardizing *any* Normal distribution will lead to a standard Normal distribution with a mean of 0 and a standard deviation of 1.

ALTERNATE EXAMPLE   Skill 3.C

What z-score did I get?
Analyzing the effect of transformations

PROBLEM:
In a large introductory statistics class, the distribution of scores on a recent test follows an approximately Normal distribution with a mean of 82.4 and a standard deviation of 5.1. Let X be the test score for a randomly selected student. To test the students' grasp of the material, the teacher cleverly decides to convert all the scores to z-scores and then report these values to students. If Z is the z-score for the randomly selected student, then

$$Z = \frac{X - 82.4}{5.1}$$

(a) Find the mean of Z.
(b) Calculate and interpret the standard deviation of Z.

SOLUTION:

(a) $\mu_Z = \dfrac{\mu_X - 82.4}{5.1} = \dfrac{82.4 - 82.4}{5.1} = 0$

(b) $\sigma_Z = \dfrac{\sigma_X}{5.1} = \dfrac{5.1}{5.1} = 1$

If many, many students are randomly selected, the z-score will typically vary by about 1 from the mean of 0.

Teaching Tip

Remind students that a *z*-score tells us the number of standard deviations above or below the mean that a value falls in a distribution.

Teaching Tip

For combining random variables, consider having students work through a contextual example and then generalize to the rule or formula. Students can make predictions about the rules, and then they can check their predictions empirically. Here is what it would look like using Pete's and Erin's Jeep Tours as a context.

1. Give students the probability distributions for *X* = # passengers for Pete and *Y* = # passengers for Erin. Give them the mean, standard deviation, and variance of *X* and *Y*. Inform students that *X* and *Y* are independent.

2. Define a new variable for the sum $S = X + Y$. Ask students to predict the mean, standard deviation, and variance for *S*.

3. Provide or ask students to create the probability distribution for *S*, which is given on page 419. Use calculators or technology to find the mean, standard deviation, and variance for *S*. Verify or refute student predictions.

4. Define a new variable for difference $D = X - Y$. Ask students to predict the mean, standard deviation, and variance for *D*.

5. Ask students to create the probability distribution for *D*, which is provided on page 420. Use calculators or technology to find the mean, standard deviation, and variance for *D*. Verify or refute student predictions.

6. Generalize the results from this context to formulas.

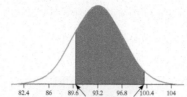

FIGURE 6.4 The Normal probability distribution of the random variable Y = the temperature (in degrees Fahrenheit) of the bathwater when the dial is set on "babysafe." The shaded area is the probability that the water temperature is between 90°F and 100°F.

To find the probability, we can either (i) standardize the boundary values and use Table A or technology; or (ii) use technology without standardizing.

(i) $z = \dfrac{90 - 93.2}{3.6} = -0.89$ $z = \dfrac{100 - 93.2}{3.6} = 1.89$

Using Table A: $0.9706 - 0.1867 = 0.7839$

Using technology: normalcdf(lower: −0.89, upper:1.89, mean:0, SD:1) = 0.7839

(ii) normalcdf(lower:90, upper:100, mean:93.2, SD:3.6) = 0.7835

When set on "babysafe" mode, there's about a 78% probability that the water temperature meets the recommendation for a randomly selected bath.

Combining Random Variables

So far, we have looked at settings that involved a single random variable. Many interesting statistics problems require us to combine two or more random variables.

Let's return to the familiar setting of Pete's Jeep Tours. Earlier, we focused on the amount of money *C* that Pete collects on a randomly selected day trip. This time we'll consider a different but related random variable: *X* = the number of passengers on a randomly selected trip. Here is its probability distribution:

Number of passengers x_i	2	3	4	5	6
Probability p_i	0.15	0.25	0.35	0.20	0.05

You can use what you learned earlier to confirm that $\mu_X = 3.75$ passengers and $\sigma_X = 1.0897$ passengers.

Pete's sister Erin runs jeep tours in another part of the country on the same days as Pete in her slightly smaller vehicle, under the name Erin's Adventures. The number of passengers *Y* on a randomly selected trip has the following probability distribution. You can confirm that $\mu_Y = 3.10$ passengers and $\sigma_Y = 0.943$ passengers.

Number of passengers y_i	2	3	4	5
Probability p_i	0.3	0.4	0.2	0.1

Here are two questions that we would like to answer based on this scenario:

- What is the distribution of the sum $S = X + Y$ of the number of passengers Pete and Erin will have on their tours on a randomly selected day?
- What is the distribution of the difference $D = X - Y$ in the number of passengers Pete and Erin will have on their tours on a randomly selected day?

As this setting suggests, we want to investigate what happens when we add or subtract random variables.

MEAN (EXPECTED VALUE) OF THE SUM OR DIFFERENCE OF TWO RANDOM VARIABLES How many total passengers *S* can Pete and Erin expect to have on their tours on a randomly selected day? Because Pete averages $\mu_X = 3.75$ passengers per trip and Erin averages $\mu_Y = 3.10$ passengers per trip, they will average a total of $\mu_S = 3.75 + 3.10 = 6.85$ passengers per day. We can generalize this result for any two random variables.

MEAN (EXPECTED VALUE) OF A SUM OF RANDOM VARIABLES

For any two random variables X and Y, if $S = X + Y$, the mean (expected value) of S is

$$\mu_S = \mu_{X+Y} = \mu_X + \mu_Y$$

In other words, the mean of the sum of two random variables is equal to the sum of their means.

What's the mean of the difference $D = X - Y$ in the number of passengers that Pete and Erin have on their tours on a randomly selected day? Because Pete averages $\mu_X = 3.75$ passengers per trip and Erin averages $\mu_Y = 3.10$ passengers per trip, the mean difference is $\mu_D = 3.75 - 3.10 = 0.65$ passengers. That is, Pete averages 0.65 more passengers per day than Erin does. Once again, we can generalize this result for any two random variables.

MEAN (EXPECTED VALUE) OF A DIFFERENCE OF RANDOM VARIABLES

For any two random variables X and Y, if $D = X - Y$, the mean (expected value) of D is

$$\mu_D = \mu_{X-Y} = \mu_X - \mu_Y$$

In other words, the mean of the difference of two random variables is equal to the difference of their means.

 The order of subtraction is important. If we had defined $D = Y - X$, then $\mu_D = \mu_Y - \mu_X = 3.10 - 3.75 = -0.65$. In other words, Erin averages 0.65 fewer passengers than Pete does on a randomly chosen day.

EXAMPLE

How much do Pete and Erin make?

Mean of a sum or difference of random variables

PROBLEM: Pete charges $150 per passenger and Erin charges $175 per passenger for a jeep tour. Let $C =$ the amount of money that Pete collects and $E =$ the amount of money that Erin collects on a randomly selected day. From our earlier work, we know that $\mu_C = 562.50$ and it is easy to show that $\mu_E = 542.50$. Define $S = C + E$. Calculate and interpret the mean of S.

SOLUTION:

$\mu_S = \mu_C + \mu_E = 562.50 + 542.50 = \$1105.00$

Pete and Erin expect to collect a total of $1105 per day, on average, over many randomly selected days.

FOR PRACTICE, TRY EXERCISE 49

How did we calculate μ_C and μ_E in the example? Earlier, we defined $X =$ the number of passengers that Pete has and $Y =$ the number of passengers that Erin has on a randomly selected day trip. Recall that $\mu_X = 3.75$ and $\mu_Y = 3.10$. Because Pete charges $150 per passenger, the amount of money that he collects

---

ALTERNATE EXAMPLE Skills 3.B, 4.B

Hoop Fever

Mean of a sum or difference of random variables

PROBLEM:
Hoop Fever is an arcade basketball game in which a player has 60 seconds to make as many baskets as possible. Morgan and Tim play head-to-head every Tuesday. Let $M =$ the number of baskets made by Morgan and $T =$ the number of baskets made by Tim in a randomly selected match. Based on previous matches, we know that $\mu_M = 39.8$ and $\mu_T = 31.2$. Let $D = M - T$. Calculate and interpret mean of D.

SOLUTION:
$\mu_D = \mu_M - \mu_T = 39.8 - 31.2 = 8.6$ baskets

The difference (Morgan − Tim) in the number of baskets made is 8.6 baskets, on average, over many randomly selected matches.

on a randomly selected day is $C = 150X$. Multiplying a random variable by a constant multiplies the value of the mean by the same constant:

$$\mu_C = 150\mu_X = 150(3.75) = \$562.50$$

Because Erin charges \$175 per passenger, $E = 175Y$ and

$$\mu_E = 175\mu_Y = 175(3.10) = \$542.50$$

The expression $aX + bY$ is called a *linear combination* of the random variables X and Y.

We can also write the total amount of money collected as $S = 150X + 175Y$. The discussion here shows that $\mu_S = 150\mu_X + 175\mu_Y$. More generally, if $S = aX + bY$, then $\mu_S = a\mu_X + b\mu_Y$.

What's the mean of the difference $D = C - E$ in the amounts that Pete and Erin collect on a randomly chosen day? It's

$$\mu_D = \mu_C - \mu_E = 562.50 - 542.50 = \$20.00$$

On average, Pete collects \$20 more per day than Erin does.

Standard Deviation of the Sum or Difference of Two Random Variables

How much variation is there in the total number of passengers $S = X + Y$ who go on Pete's and Erin's tours on a randomly chosen day? Here are the probability distributions of X and Y once again. Let's think about the possible values of S. The number of passengers X on Pete's tour is between 2 and 6, and the number of passengers Y on Erin's tour is between 2 and 5. So the total number of passengers S is between 4 and 11. That is, there's more variability in the values of S than in the values of X or Y alone. This makes sense, because the variation in X and the variation in Y both contribute to the variation in S.

Pete's Jeeps						Erin's Adventures				
Number of passengers x_i	2	3	4	5	6	Number of passengers y_i	2	3	4	5
Probability p_i	0.15	0.25	0.35	0.20	0.05	Probability p_i	0.3	0.4	0.2	0.1

$$\mu_X = 3.75 \qquad \sigma_X = 1.0897 \qquad\qquad \mu_Y = 3.10 \qquad \sigma_Y = 0.943$$

What's the standard deviation of $S = X + Y$? If we had the probability distribution of S, then we could calculate σ_S. Let's try to construct this probability distribution starting with the smallest possible value, $S = 4$. The only way to get a total of 4 passengers is if Pete has $X = 2$ passengers and Erin has $Y = 2$ passengers. We know that $P(X = 2) = 0.15$ and that $P(Y = 2) = 0.3$. If the events $X = 2$ and $Y = 2$ are *independent*, we can use the multiplication rule for independent events to find $P(X = 2 \text{ and } Y = 2)$. Otherwise, we're stuck. In fact, we can't calculate the probability for any value of S unless X and Y are **independent random variables**.

> **DEFINITION   Independent random variables**
>
> If knowing the value of X does not help us predict the value of Y, then X and Y are **independent random variables**. In other words, two random variables are independent if knowing the value of one variable does not change the probability distribution of the other variable.

It's reasonable to treat the random variables X = number of passengers on Pete's trip and Y = number of passengers on Erin's trip on a randomly chosen day

Making Connections

Remind students what we learned about two independent events in Chapter 5—knowing that one event occurred doesn't change the probability that the other event occurs. You can also remind students how we described "no association" in a scatterplot—knowing the value of one variable doesn't help us predict the value of the other variable. Many of the formulas and methods we learn assume that the random variables involved are independent, so make sure students understand the concept of independence.

as independent, because the siblings operate their trips in different parts of the country. Because X and Y are independent,

$$P(S = 4) = P(X = 2 \text{ and } Y = 2)$$
$$= (0.15)(0.3) = 0.045$$

There are two ways to get a total of $S = 5$ passengers on a randomly selected day: $X = 3$, $Y = 2$ or $X = 2$, $Y = 3$. So

$$P(S = 5) = P(X = 2 \text{ and } Y = 3) + P(X = 3 \text{ and } Y = 2)$$
$$= (0.15)(0.4) + (0.25)(0.3)$$
$$= 0.06 + 0.075 = 0.135$$

We can construct the probability distribution by listing all combinations of X and Y that yield each possible value of S and adding the corresponding probabilities. Here is the result:

Sum s_i	4	5	6	7	8	9	10	11
Probability p_i	0.045	0.135	0.235	0.265	0.190	0.095	0.030	0.005

The mean of S is

$$\mu_S = \sum s_i p_i = (4)(0.045) + (5)(0.135) + \cdots + (11)(0.005) = 6.85$$

Our calculation confirms that

$$\mu_S = \mu_X + \mu_Y = 3.75 + 3.10 = 6.85$$

The variance of S is

$$\sigma_S^2 = \sum (s_i - \mu_S)^2 p_i$$
$$= (4 - 6.85)^2 (0.045) + (5 - 6.85)^2 (0.135) + \cdots + (11 - 6.85)^2 (0.005)$$
$$= 2.0775$$

The variances of X and Y are $\sigma_X^2 = (1.0897)^2 = 1.1875$ and $\sigma_Y^2 = (0.943)^2 = 0.89$. Notice that

$$\sigma_X^2 + \sigma_Y^2 = 1.1875 + 0.89 = 2.0775 = \sigma_S^2$$

In other words, the variance of a sum of two independent random variables is the sum of their variances. To find the standard deviation of S, take the square root of the variance:

$$\sigma_S = \sqrt{2.0775} = 1.441$$

Over many randomly selected days, the total number of passengers on Pete's and Erin's trips typically varies by about 1.441 passengers from the mean of 6.85 passengers.

STANDARD DEVIATION OF THE SUM OF TWO INDEPENDENT RANDOM VARIABLES

The formula $\sigma_S^2 = \sigma_X^2 + \sigma_Y^2$ is sometimes referred to as the "Pythagorean theorem of statistics." It certainly looks similar to $c^2 = a^2 + b^2$! Just as the real Pythagorean theorem only applies to right triangles, the formula $\sigma_S^2 = \sigma_X^2 + \sigma_Y^2$ only applies if X and Y are independent random variables.

For any two *independent* random variables X and Y, if $S = X + Y$, the variance of S is

$$\sigma_S^2 = \sigma_{X+Y}^2 = \sigma_X^2 + \sigma_Y^2$$

To get the standard deviation of S, take the square root of the variance:

$$\sigma_S = \sigma_{X+Y} = \sqrt{\sigma_X^2 + \sigma_Y^2}$$

➕ Ask the StatsMedic

Using Memes to Teach Statistics

This post gives some examples of Internet memes that can help students remember important statistical concepts, including "One does not simply add standard deviations," which is a common student error.

Making Connections

We use both the general addition rule and the general multiplication rule from Chapter 5 here. We multiply (0.15)(0.4) because $X = 2$ and $Y = 3$ are independent, and we add 0.06 + 0.075 because ($X = 2$ and $Y = 3$) and ($X = 3$ and $Y = 2$) are mutually exclusive.

Teaching Tip

If students want to see all possible combinations of X and Y for Pete's Jeep Tours and Erin's Adventures, use the table below. It lists each combination, along with its sum and associated probability.

x_i	p_i	y_i	p_i	$s_i = x_i + y_i$	p_i
2	0.15	2	0.3	4	0.045
2	0.15	3	0.4	5	0.060
2	0.15	4	0.2	6	0.030
2	0.15	5	0.1	7	0.015
3	0.25	2	0.3	5	0.075
3	0.25	3	0.4	6	0.100
3	0.25	4	0.2	7	0.050
3	0.25	5	0.1	8	0.025
4	0.35	2	0.3	6	0.105
4	0.35	3	0.4	7	0.140
4	0.35	4	0.2	8	0.070
4	0.35	5	0.1	9	0.035
5	0.20	2	0.3	7	0.060
5	0.20	3	0.4	8	0.080
5	0.20	4	0.2	9	0.040
5	0.20	5	0.1	10	0.020
6	0.05	2	0.3	8	0.015
6	0.05	3	0.4	9	0.020
6	0.05	4	0.2	10	0.010
6	0.05	5	0.1	11	0.005

Teaching Tip

When we constructed the probability distribution for S, we had to assume that X and Y were independent so that we could multiply probabilities—which means that this formula for the variance of the sum of random variables works only if the random variables are independent.

You might be wondering whether there's a formula for computing the variance or standard deviation of the sum of two random variables that are *not* independent. There is, but it's beyond the scope of this course.

 When we add two independent random variables, their variances add. Standard deviations do not add. For Pete's and Erin's passenger totals,

$$\sigma_X + \sigma_Y = 1.0897 + 0.943 = 2.0327$$

This is very different from $\sigma_S = 1.441$.

Can you guess what the variance of the *difference* of two independent random variables will be? If you were thinking something like "the difference of their variances," think again! Here are the probability distributions of X and Y from the jeep tours scenario once again:

Pete's Jeeps					
Number of passengers x_i	2	3	4	5	6
Probability p_i	0.15	0.25	0.35	0.20	0.05

$\mu_X = 3.75$ $\sigma_X = 1.0897$

Erin's Adventures				
Number of passengers y_i	2	3	4	5
Probability p_i	0.3	0.4	0.2	0.1

$\mu_Y = 3.10$ $\sigma_Y = 0.943$

By following the process we used earlier with the random variable $S = X + Y$, you can build the probability distribution of $D = X - Y$.

Value d_i	−3	−2	−1	0	1	2	3	4
Probability p_i	0.015	0.055	0.145	0.235	0.260	0.195	0.080	0.015

You can use the probability distribution to confirm that:

1. $\mu_D = 0.65 = 3.75 - 3.10 = \mu_X - \mu_Y$
2. $\sigma_D^2 = 2.0775 = 1.1875 + 0.89 = \sigma_X^2 + \sigma_Y^2$
3. $\sigma_D = \sqrt{2.0775} = 1.441$

Result 2 shows that, just as with addition, when we subtract two independent random variables, variances add. There's more variability in the values of the difference D than in the values of X or Y alone. This should make sense, because the variation in X and the variation in Y both contribute to the variation in D.

STANDARD DEVIATION OF THE DIFFERENCE OF TWO INDEPENDENT RANDOM VARIABLES

For any two *independent* random variables X and Y, if $D = X - Y$, the variance of D is

$$\sigma_D^2 = \sigma_{X-Y}^2 = \sigma_X^2 + \sigma_Y^2$$

To get the standard deviation of D, take the square root of the variance:

$$\sigma_D = \sigma_{X-Y} = \sqrt{\sigma_X^2 + \sigma_Y^2}$$

Let's put this new rule to use in a familiar setting.

EXAMPLE

How much do Pete's and Erin's earnings vary?
SD of a sum or difference of random variables

PROBLEM: Pete charges $150 per passenger and Erin charges $175 per passenger for a jeep tour. Let C = the amount of money that Pete collects and E = the amount of money that Erin collects on a randomly selected day. From our earlier work, it is easy to show that $\sigma_C = \$163.46$ and $\sigma_E = \$165.03$. You may assume that these two random variables are independent. Define $D = C - E$. Earlier, we found that $\mu_D = \$20$. Calculate and interpret the standard deviation of D.

SOLUTION:

$D = C - E$. Because C and E are independent random variables,

$\sigma_D^2 = \sigma_C^2 + \sigma_E^2 = (163.46)^2 + (165.03)^2 = 53{,}954.07$

$\sigma_D = \sqrt{53{,}954.07} = \$232.28$

> Note that variances add when you are dealing with the sum *or* difference of independent random variables.

Over many randomly selected days, the difference (Pete − Erin) in the amount collected on their jeep tours typically varies by about $232.28 from the mean difference of $20.

FOR PRACTICE, TRY EXERCISE 57

How did we calculate σ_C and σ_E in the example? Earlier, we defined X = the number of passengers that Pete has and Y = the number of passengers that Erin has on a randomly selected day trip. Recall that $\sigma_X = 1.0897$ and $\sigma_Y = 0.943$. Because Pete charges $150 per passenger, the amount of money that he collects on a randomly selected day is $C = 150X$. Multiplying a random variable by a constant multiplies the value of the standard deviation by the same constant:

$$\sigma_C = 150\sigma_X = 150(1.0897) = \$163.46$$

Because Erin charges $175 per passenger, $E = 175Y$ and

$$\sigma_E = 175\sigma_Y = 175(0.943) = \$165.03$$

We can write the difference in the amount of money collected as $D = 150X - 175Y$. The discussion here shows that

$$\sigma_D^2 = (150\sigma_X)^2 + (175\sigma_Y)^2 = 150^2\sigma_X^2 + 175^2\sigma_Y^2$$

Recall that $aX + bY$ is a *linear combination* of the random variables X and Y.

More generally, if $S = aX + bY$, then $\sigma_S^2 = a^2\sigma_X^2 + b^2\sigma_Y^2$ for independent random variables X and Y.

MEAN AND STANDARD DEVIATION OF A LINEAR COMBINATION OF RANDOM VARIABLES

If $aX + bY$ is a linear combination of the random variables X and Y,

- Its mean is $a\mu_X + b\mu_Y$.
- Its standard deviation is $\sqrt{a^2\sigma_X^2 + b^2\sigma_Y^2}$ if X and Y are independent.

Note that these results apply to both discrete and continuous random variables.

ALTERNATE EXAMPLE Skills 3.B, 4.B

More Hoop Fever SD of a sum or difference of random variables

PROBLEM:
In Hoop Fever, a player has 60 seconds to make as many baskets as possible. Morgan and Tim play head-to-head every Tuesday. Let M = the number of baskets made by Morgan and T = the number of baskets made by Tim in a randomly selected match. Based on previous matches, we know that $\sigma_M = 5.7$ and $\sigma_T = 10.3$. Assume that these two random variables are independent. Define $D = M - T$. Earlier, we found that $\mu_D = 8.6$. Calculate and interpret the standard deviation of D.

SOLUTION:
Because M and T are independent random variables,

$\sigma_D^2 = \sigma_M^2 + \sigma_T^2 = 5.7^2 + 10.3^2 = 138.58$

$\sigma_D = \sqrt{138.58} = 11.77$ baskets

If many, many games are randomly selected, the difference (Morgan − Tim) in the number of baskets made will typically vary by about 11.77 baskets from the mean difference of 8.6 baskets.

Teaching Tip

We use subscripts on the X's because the value of X isn't always the same from game to game. In other words, X is a variable! This is different from how we use "variables" in algebra. In the algebraic equation $2x + x = 9$, x is a "variable," but there is only one value of x that makes the equation true. For more on this topic, do an online search for the article "Random Variables vs. Algebraic Variables," which can be found on the AP® Central website.

Teaching Tip

As an alternate example, ask students to consider the results of rolling a single die twice ($X_1 + X_2$), or rolling the die once and doubling the value ($2X$). It seems reasonable here that both distributions would have the same mean of 7. But when rolling a die twice, it is much less likely to get values at the extremes (2 and 12) than when doubling the value of a single die roll. Therefore, the variability of $X_1 + X_2$ will be less than for $2X$.

✓ **Answers to CYU**

1. $\mu_T = \mu_X + \mu_Y = 1.1 + 0.7 = 1.8$; if many, many Fridays are randomly selected, this dealership expects to sell or lease about 1.8 cars in the first hour of business, on average, over many randomly selected Fridays.

2. Because X and Y are independent, $\sigma_T^2 = \sigma_X^2 + \sigma_Y^2 = (0.943)^2 + (0.64)^2 = 1.2988$, so $\sigma_T = \sqrt{1.2988} = 1.14$. If many, many Fridays are randomly selected, the total number of cars sold or leased in the first hour will typically vary by about 1.14 cars from the mean of 1.8 cars.

3. The total bonus is $B = 500X + 300Y$, which means that $\mu_B = 500\mu_X + 300\mu_Y = 500(1.1) + 300(0.7) = \$760$. Because X and Y are independent, $\sigma_B^2 = (500\sigma_X)^2 + (300\sigma_Y)^2 = (500)^2(0.943)^2 + (300)^2(0.64)^2 = 259{,}176.25$. Therefore, $\sigma_B = \sqrt{259{,}176.25} = \$509.09$.

COMBINING VERSUS TRANSFORMING RANDOM VARIABLES We can extend our rules for combining random variables to situations involving repeated observations of the same random process. Let's return to the gambler we met at the beginning of this section. Suppose he plays two games of roulette, each time placing a $1 bet on red. What can we say about his total gain (or loss) from playing two games? Earlier, we showed that if X = the amount gained on a single $1 bet on red, then $\mu_X = -\$0.05$ and $\sigma_X = \$1.00$. Because we're interested in the player's total gain over two games, we'll define X_1 as the amount he gains from the first game and X_2 as the amount he gains from the second game. Then his total gain $T = X_1 + X_2$. Both X_1 and X_2 have the same probability distribution as X and, therefore, the same mean ($-\$0.05$) and standard deviation ($\$1.00$). The player's expected gain in two games is

$$\mu_T = \mu_{X_1} + \mu_{X_2} = (-\$0.05) + (-\$0.05) = -\$0.10$$

Because knowing the result of one game tells the player nothing about the result of the other game, X_1 and X_2 are independent random variables. As a result,

$$\sigma_T^2 = \sigma_{X_1}^2 + \sigma_{X_2}^2 = (1.00)^2 + (1.00)^2 = 2.00$$

and the standard deviation of the player's total gain is

$$\sigma_T = \sqrt{2.00} = \$1.41$$

At the beginning of the section, we asked whether a roulette player would be better off placing two separate $1 bets on red or a single $2 bet on red. We just showed that the expected total gain from two $1 bets is $\mu_T = -\$0.10$ with a standard deviation of $\sigma_T = \$1.41$. Now think about what happens if the gambler places a $2 bet on red in a single game of roulette. Because the random variable X represents a player's gain from a $1 bet, the random variable $Y = 2X$ represents his gain from a $2 bet.

What's the player's expected gain from a single $2 bet on red? It's

$$\mu_Y = 2\mu_X = 2(-\$0.05) = -\$0.10$$

That's the same as his expected gain from playing two games of roulette with a $1 bet each time. But the standard deviation of the player's gain from a single $2 bet is

$$\sigma_Y = 2\sigma_X = 2(\$1.00) = \$2.00$$

 Compare this result to $\sigma_T = \$1.41$. There's more variability in the gain from a single $2 bet than in the total gain from two $1 bets. Bottom line: $X_1 + X_2$ is not the same as 2X.

CHECK YOUR UNDERSTANDING

A large auto dealership keeps track of sales and lease agreements made during each hour of the day. Let X = the number of cars sold and Y = the number of cars leased during the first hour of business on a randomly selected Friday. Based on previous records, the probability distributions of X and Y are as follows:

Cars sold x_i	0	1	2	3
Probability p_i	0.3	0.4	0.2	0.1

Cars leased y_i	0	1	2
Probability p_i	0.4	0.5	0.1

$$\mu_X = 1.1 \quad \sigma_X = 0.943 \qquad \mu_Y = 0.7 \quad \sigma_Y = 0.64$$

Define $T = X + Y$. Assume that X and Y are independent.

1. Find and interpret μ_T.
2. Calculate and interpret σ_T.
3. The dealership's manager receives a $500 bonus for each car sold and a $300 bonus for each car leased. Find the mean and standard deviation of the manager's total bonus B.

TRM **Casino Lab**

At this point in the chapter, you might consider doing the Casino Lab, which includes activities exploring craps, roulette, blackjack, the Monty Hall problem, and how to make your own casino game. Find it in the Teacher's Resource Materials located in the TE-Book, on the TRFD, or in the Teacher's view in the book's digital platform.

Combining Normal Random Variables

So far, we have concentrated on developing rules for means and variances of random variables. If a random variable is Normally distributed, we can use its mean and standard deviation to compute probabilities. What happens if we combine two *independent* Normal random variables?

We used software to simulate separate random samples of size 1000 for each of two independent, Normally distributed random variables, X and Y. Their means and standard deviations are as follows:

$$\mu_X = 3, \sigma_X = 0.9 \qquad \mu_Y = 1, \sigma_Y = 1.2$$

Figure 6.5(a) shows the results. What do we know about the sum and difference of these two random variables? The histograms in Figure 6.5(b) came from adding and subtracting the corresponding values of X and Y for the 1000 randomly generated observations from each probability distribution.

FIGURE 6.5 (a) Histograms showing the results of randomly selecting 1000 values of two independent, Normal random variables X and Y. (b) Histograms of the sum and difference of the 1000 randomly selected values of X and Y.

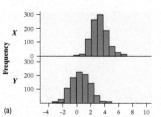
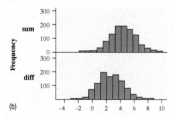

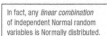 In fact, any *linear combination* of independent Normal random variables is Normally distributed.

As the simulation illustrates, *any sum or difference of independent Normal random variables is also Normally distributed*. The mean and standard deviation of the resulting Normal distribution can be found using the appropriate rules for means and standard deviations:

	Sum $X + Y$	Difference $X - Y$
Mean	$\mu_{X+Y} = \mu_X + \mu_Y = 3 + 1 = 4$	$\mu_{X-Y} = \mu_X - \mu_Y = 3 - 1 = 2$
SD	$\sigma^2_{X+Y} = \sigma^2_X + \sigma^2_Y = 0.9^2 + 1.2^2 = 2.25$ $\sigma_{X+Y} = \sqrt{2.25} = 1.5$	$\sigma^2_{X-Y} = \sigma^2_X + \sigma^2_Y = 0.9^2 + 1.2^2 = 2.25$ $\sigma_{X-Y} = \sqrt{2.25} = 1.5$

EXAMPLE

Will the lid fit?
Combining Normal random variables

PROBLEM: The diameter C of the top of a randomly selected large drink cup at a fast-food restaurant follows a Normal distribution with a mean of 3.96 inches and a standard deviation of 0.01 inch. The diameter L of a randomly selected large lid at this restaurant follows a Normal distribution with mean 3.98 inches and standard deviation 0.02 inch. Assume that L and C are independent random variables. Let the random variable $D = L - C$ be the difference between the lid's diameter and the cup's diameter.

(a) Describe the distribution of D.

(b) For a lid to fit on a cup, the value of L has to be bigger than the value of C, but not by more than 0.06 inch. Find the probability that a randomly selected lid will fit on a randomly selected cup. Interpret this value.

karandaev/Getty Images

Who will win?   Combining Normal random variables

PROBLEM:
A match of Hoop Fever gives each player 60 seconds to make as many baskets as possible. Morgan and Tim play head-to-head every Tuesday. Suppose that M = the number of baskets made by Morgan in a randomly selected match follows an approximately Normal distribution with $\mu_M = 39.8$ and $\mu_M = 5.7$. Suppose that T = the number of baskets made by Tim in a randomly selected match follows an approximately Normal distribution with $\mu_M = 31.2$. and $\sigma_T = 10.3$. Assume that these two random variables are independent and define $D = M - T$.
(a) Describe the distribution of D.
(b) What is the probability that Morgan will make more baskets than Tim in a randomly selected match?

SOLUTION:
(a) *Shape:* Approximately Normal
Center: $\mu_D = 39.8 - 31.2 = 8.6$ baskets
Variability: $\sigma_D = \sqrt{5.7^2 + 10.3^2} = 11.77$ baskets
(b) Morgan will win the match if $D = M - T > 0$.

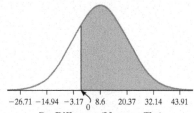

$-26.71 \quad -14.94 \quad -3.17 \quad 8.6 \quad 20.37 \quad 32.14 \quad 43.91$
0
D = Difference (Morgan – Tim)
in number of baskets made

(i) $z = \dfrac{0 - 8.6}{11.77} = -0.73$

Table A: $1 - 0.2327 = 0.7673$
Tech: normalcdf(lower: −0.73, upper: 10,000, mean: 0, SD: 1) = 0.7673
(ii) normalcdf(lower: 0, upper: 1000, mean: 8.6, SD: 11.77) = 0.7675. There's about a 77% chance that Morgan will make more baskets than Tim in a randomly selected match.

Preparing for Inference

In Chapter 7, we will look at the difference between two sample proportions and the difference between two sample means. Quite often, the sampling distribution of each proportion or each mean will be approximately Normal. We will want to know how to combine these Normal random variables to describe the shape, center, and variability for the sampling distributions of the differences in proportions or differences in means.

Preparing for Inference

If $\hat{p}_1$ and $\hat{p}_2$ are independent and the sampling distribution of each is approximately Normal, the sampling distribution of $\hat{p}_1 - \hat{p}_2$ is approximately Normal. If $\bar{x}_1$ and $\bar{x}_2$ are independent and the sampling distribution of each is approximately Normal, the sampling distribution of $\bar{x}_1 - \bar{x}_2$ is approximately Normal.

SOLUTION:

(a) **Shape:** Normal

Center: $\mu_D = 3.98 - 3.96 = 0.02$ inch

Variability: $\sigma_D = \sqrt{(0.02)^2 + (0.01)^2} = 0.0224$ inch

(b) The lid will fit if $0 < L - C \leq 0.06$, that is, if $0 < D \leq 0.06$.

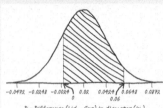

D = Difference (Lid – Cup) in diameter (in.)

D is the difference of two independent Normal random variables.

$$\mu_D = \mu_L - \mu_C$$

$$\sigma_D = \sqrt{\sigma_L^2 + \sigma_C^2}$$

1. Draw a Normal distribution.
2. Perform calculations—show your work!
(i) Standardize and use Table A or technology; or
(ii) Use technology without standardizing.

Be sure to answer the question that was asked.

(i) $z = \dfrac{0 - 0.02}{0.0224} = -0.89 \quad z = \dfrac{0.06 - 0.02}{0.0224} = 1.79$

Using Table A: $0.9633 - 0.1867 = 0.7766$

$\boxed{P(0 < D \leq 0.06) = P(-0.89 < Z \leq 1.79)}$

Using technology: normalcdf(lower: −0.89, upper: 1.79, mean: 0, SD: 1) = 0.7765

(ii) normalcdf(lower:0, upper:0.06, mean:0.02, SD:0.0224) = 0.7770

There's about a 78% chance that a randomly selected lid will fit on a randomly selected cup.

FOR PRACTICE, TRY EXERCISE 65

Preparing for Inference

In Chapter 7, we will find that the standard deviation of the sampling distribution of $\bar{x}$ is given by $\sigma_{\bar{x}} = \dfrac{\sigma_x}{\sqrt{n}}$.

This formula can be derived using ideas on this page. Think about the mean as the average of the results from n independent observations of the same random variable X:

$$\bar{x} = \frac{\sum X_i}{n} = \frac{1}{n}(X_1 + X_2 + \cdots + X_n)$$

The standard deviation of $\bar{x}$ is given by

$$\sigma_{\bar{x}} = \frac{1}{n}\sqrt{\sigma_{X_1}^2 + \sigma_{X_2}^2 + \cdots + \sigma_{X_n}^2}$$

$$= \frac{1}{n}\sqrt{n\sigma_X^2} = \frac{\sigma_X}{\sqrt{n}}$$

domin_domin/Getty Images

We can extend what we have learned about combining independent Normal random variables to settings that involve repeated observations from the same probability distribution. Consider this scenario. Mr. Starnes likes sugar in his hot tea. From experience, he needs between 8.5 and 9 grams of sugar in a cup of tea for the drink to taste right. While making his tea one morning, Mr. Starnes adds four randomly selected packets of sugar. Suppose the amount of sugar in these packets follows a Normal distribution with mean 2.17 grams and standard deviation 0.08 gram. What's the probability that Mr. Starnes's tea tastes right?

Let X = the amount of sugar in a randomly selected packet. Then X_1 = amount of sugar in Packet 1, X_2 = amount of sugar in Packet 2, X_3 = amount of sugar in Packet 3, and X_4 = amount of sugar in Packet 4. Each of these random variables has a Normal distribution with mean 2.17 grams and standard deviation 0.08 grams. We're interested in the total amount of sugar that Mr. Starnes puts in his tea: $T = X_1 + X_2 + X_3 + X_4$.

The random variable T is a sum of four independent Normal random variables. So T follows a Normal distribution with mean

$$\mu_T = \mu_{X_1} + \mu_{X_2} + \mu_{X_3} + \mu_{X_4} = 2.17 + 2.17 + 2.17 + 2.17 = 8.68 \text{ grams}$$

and variance

$$\sigma_T^2 = \sigma_{X_1}^2 + \sigma_{X_2}^2 + \sigma_{X_3}^2 + \sigma_{X_4}^2 = (0.08)^2 + (0.08)^2 + (0.08)^2 + (0.08)^2 = 0.0256$$

The standard deviation of T is

$$\sigma_T = \sqrt{0.0256} = 0.16 \text{ gram}$$

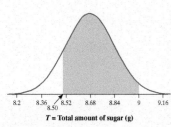

FIGURE 6.6 Normal distribution of the total amount of sugar in Mr. Starnes's tea.

We want to find the probability that the total amount of sugar in Mr. Starnes's tea is between 8.5 and 9 grams. Figure 6.6 shows this probability as the area under a Normal curve.

To find this area, we can use either of our two familiar methods:

(i) Standardize the boundary values and use Table A or technology:

$$z = \frac{8.5 - 8.68}{0.16} = -1.13 \quad \text{and} \quad z = \frac{9 - 8.68}{0.16} = 2.00$$

Using Table A: $P(-1.13 \le Z \le 2.00) = 0.9772 - 0.1292 = 0.8480$

Using technology: normalcdf(lower: -1.13, upper:2.00, mean:0, SD:1) $= 0.8480$

(ii) Use technology to find the desired area without standardizing.

normalcdf(lower:8.5, upper:9, mean:8.68, SD:0.16) $= 0.8470$

There's about an 85% probability that Mr. Starnes's tea will taste right.

Making Connections

In Chapter 7, we will use the sampling distribution of the sample mean to solve problems like this one. For this problem, we would want to know the probability that the *average* amount of sugar in each packet is between $8.5/4 = 2.125$ and $9/4 = 2.25$ grams.

Section 6.2 | Summary

- Adding a positive constant a to (subtracting a from) a random variable increases (decreases) measures of center and location by a, but does not affect measures of variability (range, *IQR*, standard deviation) or the shape of its probability distribution.
- Multiplying (dividing) a random variable by a positive constant b multiplies (divides) measures of center and location by b and multiplies (divides) measures of variability (range, *IQR*, standard deviation) by b, but does not change the shape of its probability distribution.
- If $Y = a + bX$ is a linear transformation of the random variable X with $b > 0$,
  - The probability distribution of Y has the same shape as the probability distribution of X.
  - $\mu_Y = a + b\mu_X$
  - $\sigma_Y = b\sigma_X$
- If X and Y are *any* two random variables,

  $\mu_{X+Y} = \mu_X + \mu_Y$: The mean of the sum of two random variables is the sum of their means.

  $\mu_{X-Y} = \mu_X - \mu_Y$: The mean of the difference of two random variables is the difference of their means.
- If X and Y are **independent random variables**, then knowing the value of one variable does not change the probability distribution of the other variable. In that case, variances add:

  $\sigma^2_{X+Y} = \sigma^2_X + \sigma^2_Y$: The variance of the sum of two independent random variables is the sum of their variances.

  $\sigma^2_{X-Y} = \sigma^2_X + \sigma^2_Y$: The variance of the difference of two independent random variables is the sum of their variances.
- To get the standard deviation of the sum or difference of two independent random variables, calculate the variance and then take the square root:

$$\sigma_{X+Y} = \sigma_{X-Y} = \sqrt{\sigma^2_X + \sigma^2_Y}$$

TRM **Section 6.2 Quiz**

There is one quiz available for this section in the Teacher's Resource Materials. Click on the link in the TE-Book, look on the TRFD, or download from the Teacher's Resources on the book's digital platform. You can also create your own quiz using the ExamView® Assessment Suite that is part of the TPS6 program. Questions are coded by Learning Target to make it easy to build parallel quizzes.

Answers to Section 6.2 Exercises

6.37 Approximately Normal

$\mu_T = 20$ minutes

$\sigma_T = 6.5$ minutes

6.38 Approximately Normal

$\mu_P = \$495$

$\sigma_P = \$185$

6.39 (a)

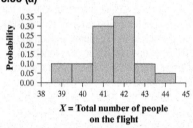

$X =$ **Total number of people on the flight**

The distribution of total number of people on the flight is roughly symmetric, with a single peak at 42 people. **(b)** $\mu_X = 41.4$; if many, many flights are randomly selected, the average of the total number of people on the flight will be about 41.4 people. **(c)** $\sigma_X = 1.24$; if many, many flights are randomly selected, the total number of people on the flight will typically vary by about 1.24 people from the mean of 41.4.

6.40 (a)

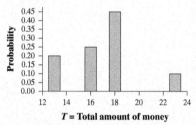

$T =$ **Total amount of money**

The distribution is roughly symmetric, with a single peak at \$18. **(b)** $\mu_T = \$17$; if many, many days are randomly selected, the average amount of money that Victoria pays for parking will be about \$17. **(c)** $\sigma_T = \$2.74$; if many, many days are randomly selected, the total amount of money Victoria pays for parking will typically vary by about \$2.74 from the mean of \$17.

- If $aX + bY$ is a linear combination of the random variables X and Y,
  - Its mean is $a\mu_X + b\mu_Y$.
  - Its standard deviation is $\sqrt{a^2\sigma_X^2 + b^2\sigma_Y^2}$ if X and Y are independent.
- A linear combination of independent Normal random variables is a Normal random variable.

Section 6.2 | Exercises

37. Driving to work The time X it takes Hattan to drive to pg 412 work on a randomly selected day follows a distribution that is approximately Normal with mean 15 minutes and standard deviation 6.5 minutes. Once he parks his car in his reserved space, it takes 5 more minutes for him to walk to his office. Let $T =$ the total time it takes Hattan to reach his office on a randomly selected day, so $T = X + 5$. Describe the shape, center, and variability of the probability distribution of T.

38. Toy shop sales Total gross profits G on a randomly selected day at Tim's Toys follow a distribution that is approximately Normal with mean \$560 and standard deviation \$185. The cost of renting and maintaining the shop is \$65 per day. Let $P =$ profit on a randomly selected day, so $P = G - 65$. Describe the shape, center, and variability of the probability distribution of P.

39. Airline overbooking Airlines typically accept more reservations for a flight than the number of seats on the plane. Suppose that for a certain route, an airline accepts 40 reservations on a plane that carries 38 passengers. Based on experience, the probability distribution of $Y =$ the number of passengers who actually show up for a randomly selected flight is given in the following table. You can check that $\mu_Y = 37.4$ and $\sigma_Y = 1.24$.

Number of passengers y_i	35	36	37	38	39	40
Probability p_i	0.10	0.10	0.30	0.35	0.10	0.05

There is also a crew of two flight attendants and two pilots on each flight. Let $X =$ the total number of people (passengers plus crew) on a randomly selected flight.

(a) Make a graph of the probability distribution of X. Describe its shape.

(b) Find and interpret μ_X.

(c) Calculate and interpret σ_X.

40. City parking Victoria parks her car at the same garage every time she goes to work. Because she stays at work for different lengths of time each day, the fee the parking garage charges on a randomly selected day is a random variable, G. The table gives the probability distribution of G. You can check that $\mu_G = \$14$ and $\sigma_G = \$2.74$.

Garage fee g_i	\$10	\$13	\$15	\$20
Probability p_i	0.20	0.25	0.45	0.10

In addition to the garage's fee, the city charges a \$3 use tax each time Victoria parks her car. Let $T =$ the total amount of money she pays on a randomly selected day.

(a) Make a graph of the probability distribution of T. Describe its shape.

(b) Find and interpret μ_T.

(c) Calculate and interpret σ_T.

41. Get on the boat! A small ferry runs every half hour pg 413 from one side of a large river to the other. The number of cars X on a randomly chosen ferry trip has the probability distribution shown here with mean $\mu_X = 3.87$ and standard deviation $\sigma_X = 1.29$.

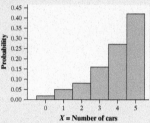

$X =$ **Number of cars**

The cost for the ferry trip is \$5. Define $M =$ money collected on a randomly selected ferry trip.

(a) What shape does the probability distribution of M have?

(b) Find the mean of M.

(c) Calculate the standard deviation of M.

42. Skee Ball Ana is a dedicated Skee Ball player who always rolls for the 50-point slot. Ana's score X on a randomly selected roll of the ball has the probability distribution shown here with mean $\mu_X = 23.8$ and standard deviation $\sigma_X = 12.63$.

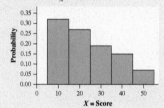

$X =$ **Score**

6.41 (a) The distribution of M has the same shape as the distribution of X: skewed to the left. **(b)** $\mu_M = \$19.35$ **(c)** $\sigma_M = \$6.45$

A player receives one ticket from the game for every 10 points scored. Define T = number of tickets Ana gets on a randomly selected roll.

(a) What shape does the probability distribution of T have?

(b) Find the mean of T.

(c) Calculate the standard deviation of T.

43. **Still waiting for the server?** How does your web browser get a file from the Internet? Your computer sends a request for the file to a web server, and the web server sends back a response. Let Y = the amount of time (in seconds) after the start of an hour at which a randomly selected request is received by a particular web server. The probability distribution of Y can be modeled by a uniform density curve on the interval from 0 to 3600 seconds. Define the random variable $W = Y/60$.

(a) Explain what W represents.

(b) What probability distribution does W have?

44. **Where's the bus?** Sally takes the same bus to work every morning. Let X = the amount of time (in minutes) that she has to wait for the bus on a randomly selected day. The probability distribution of X can be modeled by a uniform density curve on the interval from 0 minutes to 8 minutes. Define the random variable $V = 60X$.

(a) Explain what V represents.

(b) What probability distribution does V have?

Exercises 45 and 46 refer to the following setting. Ms. Hall gave her class a 10-question multiple-choice quiz. Let X = the number of questions that a randomly selected student in the class answered correctly. The computer output gives information about the probability distribution of X. To determine each student's grade on the quiz (out of 100), Ms. Hall will multiply his or her number of correct answers by 5 and then add 50. Let G = the grade of a randomly chosen student in the class.

Mean	Median	StDev	Min	Max	Q_1	Q_3
7.6	8.5	1.32	4	10	8	9

45. **Easy quiz**

(a) Find the median of G.

(b) Find the interquartile range (IQR) of G.

46. **More easy quiz**

(a) Find the mean of G.

(b) Find the range of G.

47. **Too cool at the cabin?** During the winter months, the pg416 temperatures at the Starneses' Colorado cabin can stay well below freezing (32°F or 0°C) for weeks at a time. To prevent the pipes from freezing, Mrs. Starnes sets the thermostat at 50°F. She also buys a digital thermometer that records the indoor temperature each night at midnight. Unfortunately, the thermometer is programmed to measure the temperature in degrees Celsius. Based on several years' worth of data, the temperature T in the cabin at midnight on a randomly selected night can be modeled by a Normal distribution with mean 8.5°C and standard deviation 2.25°C. Let Y = the temperature in the cabin at midnight on a randomly selected night in degrees Fahrenheit (recall that $F = (9/5)C + 32$).

(a) Find the mean of Y.

(b) Calculate and interpret the standard deviation of Y.

(c) Find the probability that the midnight temperature in the cabin is less than 40°F.

48. **How much cereal?** A company's single-serving cereal boxes advertise 1.63 ounces of cereal. In fact, the amount of cereal X in a randomly selected box can be modeled by a Normal distribution with a mean of 1.70 ounces and a standard deviation of 0.03 ounce. Let Y = the *excess* amount of cereal beyond what's advertised in a randomly selected box, measured in grams (1 ounce = 28.35 grams).

(a) Find the mean of Y.

(b) Calculate and interpret the standard deviation of Y.

(c) Find the probability of getting at least 1 gram more cereal than advertised.

49. **Community college costs** El Dorado Community pg417 College has a main campus in the suburbs and a downtown campus. The amount X spent on tuition by a randomly selected student at the main campus has mean $732.50 and standard deviation $103. The amount Y spent on tuition by a randomly selected student at the downtown campus has mean $825 and standard deviation $126.50. Suppose we randomly select one full-time student from each of the two campuses. Calculate and interpret the mean of the sum $S = X + Y$.

50. **Essay errors** Typographical and spelling errors can be either "nonword errors" or "word errors." A nonword error is not a real word, as when "the" is typed as "teh." A word error is a real word, but not the right word, as when "lose" is typed as "loose." When students are asked to write a 250-word essay (without spell-checking), the number of nonword errors X in a randomly selected essay has mean 2.1 and standard deviation 1.136. The number of word errors Y in the essay has mean 1.0 and standard deviation 1.0. Calculate and interpret the mean of the sum $S = X + Y$.

51. **Study habits** The Survey of Study Habits and Attitudes (SSHA) is a psychological test that measures academic motivation and study habits. The SSHA score F of a randomly selected female student at a large university has mean 120 and standard deviation 28, and the SSHA score M of a randomly selected male student at the university has mean 105 and standard deviation 35. Suppose we select one male student and one female student at random from this university and give them the SSHA test. Calculate and interpret the mean of the difference $D = F - M$ in their scores.

6.48 (a) $\mu_Y = 28.35(1.70) - 28.35(1.63) = 1.985$ **(b)** $\sigma_Y = 28.35(0.03) = 0.8505$; if many, many boxes are randomly selected, the excess amount of cereal beyond what is advertised will typically vary by about 0.8505 gram from the mean of 1.985 grams. **(c) (i)** $z = -1.16$; $P(Z \geq -1.16) = 0.8770$. **(ii)** $P(Y \geq 1) =$ normalcdf(lower: 1, upper: 1000, mean: 1.985, SD: 0.8505) $= 0.8766$. There is a 0.8766 probability of getting at least 1 gram more cereal than advertised.

6.49 $\mu_S = 732.50 + 825 = \$1,557.50$; the average of the sum of the tuitions would be about $1557.50 for many, many randomly selected pairs of students from each of the campuses.

6.50 $\mu_S = 2.1 + 1.0 = 3.1$; the average of the sum of the word errors and non-word errors would be about 3.1 errors for many, many randomly selected 250-word essays.

6.51 $\mu_{F-M} = 120 - 105 = 15$; the average of the difference would be about 15 points, if you were to repeat the process of selecting a single male student, selecting a single female student, and finding the difference (Female − Male) in their scores many times.

6.42 (a) The distribution of T has the same shape as the distribution of X: skewed to the right. **(b)** $\mu_T = 2.38$ **(c)** $\sigma_T = 1.263$

6.43 (a) W represents the amount of time (in minutes) after the start of an hour at which a randomly selected request is received by a particular Web server. **(b)** W has a uniform distribution on the interval from 0 to 60 minutes.

6.44 (a) V represents the amount of time (in seconds) that Sally has to wait for the bus on a randomly selected day. **(b)** The random variable V has a probability distribution that can be modeled by a uniform density curve on the interval from 0 to 480 seconds.

6.45 (a) $G = 5X + 50$; median$_G = 5(8.5) + 50 = 92.5$ **(b)** $IQR_G = 5(9 - 8) = 5$

6.46 (a) $G = 5X + 50$; $\mu_G = 5(7.6) + 50 = 88$ **(b)** $Range_G = 5(10 - 4) = 30$

6.47 (a) $\mu_Y = \dfrac{9}{5}(8.5) + 32 = 47.3$

(b) $\sigma_Y = \dfrac{9}{5}(2.25) = 4.05$; if many, many nights are randomly selected, the temperature in the cabin at midnight will typically vary by about 4.05°F from the mean of 47.3°F. **(c) (i)** $z = -1.80$; $P(Z < -1.80) = 0.0359$. **(ii)** $P(Y < 40) =$ normalcdf(lower: −1000, upper: 40, mean: 47.3, SD: 4.05) $= 0.0357$. There is a 0.0357 probability that the midnight temperature in the cabin is below 40°F.

6.52 $\mu_D = 12 - 16 = -4$; the average of the difference (Bus − Walk) in commute times would be about −4 minutes for many, many randomly selected days.

6.53 $\mu_{X/50 - Y/55} = \mu_{X/50} - \mu_{Y/55} = 14.65 - 15 = -0.35$

6.54 $\mu_{3X + 2Y} = \mu_{3X} + \mu_{2Y} = 6.3 + 2.0 = 8.3$ points

6.55 (a) Yes; the mean of a sum is always equal to the sum of the means. **(b)** No; the variance of the sum is not equal to the sum of the variances, because it is not reasonable to assume that X and Y are independent.

6.56 (a) Yes; the mean of a sum is always equal to the sum of the means. **(b)** No; the variance of the sum is not equal to the sum of the variances, because it is not reasonable to assume that X and Y are independent.

6.57 Because X and Y are independent, $\sigma_S = \sqrt{103^2 + 126.50^2} = \$163.13$. If many, many pairs of students are randomly and independently selected from each of the campuses, the sum of the tuitions will typically vary by about $\$163.13$ from the mean of $\$1557.50$.

6.58 Because X and Y are independent, $\sigma_S = \sqrt{1.136^2 + 1.0^2} = 1.513$. If many, many essays are randomly and independently selected, the sum of the number of word and non-word errors will typically vary by about 1.513 errors from the mean of 3.1 errors.

6.59 (a) Independence of F and M means that knowing the value of one student's score does not help us predict the value of the other student's score.
(b) $\sigma_{F-M} = \sqrt{28^2 + 35^2} = 44.822$; if many, many pairs of male and female college students are randomly and independently selected, the difference (Female − Male) of their SSHA scores will typically vary by about 44.822 points from the mean of 15 points. **(c)** No, we do not know the shapes of the distributions. We cannot assume that the distributions are Normal without additional information.

52. **Commuting to work** Sulé's job is just a few bus stops away from his house. While it can be faster to take the bus to work than to walk, the travel time is more variable due to traffic. The commute time B if Sulé takes the bus to work on a randomly selected day has mean 12 minutes and standard deviation 4 minutes. The commute time W if Sulé walks to work on a randomly selected day has mean 16 minutes and standard deviation 1 minute. Calculate and interpret the mean of the difference $D = B - W$ in the time it would take Sulé to get to work on a randomly selected day.

53. **Community college costs** Refer to Exercise 49. At the main campus, full-time students pay $50 per unit. At the downtown campus, full-time students pay $55 per unit. Find the mean of the difference D (Main − Downtown) in the number of units that the two randomly selected students take.

54. **Essay scores** Refer to Exercise 50. An English professor deducts 3 points from a student's essay score for each nonword error and 2 points for each word error. Find the mean of the total score deductions T for a randomly selected essay.

55. **Rainy days** Imagine that we randomly select a day from the past 10 years. Let X be the recorded rainfall on this date at the airport in Orlando, Florida, and Y be the recorded rainfall on this date at Disney World just outside Orlando. Suppose that you know the means μ_X and μ_Y and the variances σ_X^2 and σ_Y^2 of both variables.
    (a) Can we calculate the mean of the total rainfall $X + Y$ to be $\mu_X + \mu_Y$? Explain your answer.
    (b) Can we calculate the variance of the total rainfall to be $\sigma_X^2 + \sigma_Y^2$? Explain your answer.

56. **His and her earnings** Researchers randomly select a married couple in which both spouses are employed. Let X be the income of the husband and Y be the income of the wife. Suppose that you know the means μ_X and μ_Y and the variances σ_X^2 and σ_Y^2 of both variables.
    (a) Can we calculate the mean of the total income $X + Y$ to be $\mu_X + \mu_Y$? Explain your answer.
    (b) Can we calculate the variance of the total income to be $\sigma_X^2 + \sigma_Y^2$? Explain your answer.

57. **Community college costs** Refer to Exercise 49. pg 421 Note that X and Y are independent random variables because the two students are randomly selected from each of the campuses. Calculate and interpret the standard deviation of the sum $S = X + Y$.

58. **Essay errors** Refer to Exercise 50. Assume that the number of non-word errors X and word errors Y in a randomly selected essay are independent random variables. Calculate and interpret the standard deviation of the sum $S = X + Y$.

59. **Study habits** Refer to Exercise 51.
    (a) Assume that F and M are independent random variables. Explain what this means in context.
    (b) Calculate and interpret the standard deviation of the difference $D = F - M$ in their scores.
    (c) From the information given, can you find the probability that the randomly selected female student has a higher SSHA score than the randomly selected male student? Explain why or why not.

60. **Commuting to work** Refer to Exercise 52.
    (a) Assume that B and W are independent random variables. Explain what this means in context.
    (b) Calculate and interpret the standard deviation of the difference D (Bus − Walk) in the time it would take Sulé to get to work on a randomly selected day.
    (c) From the information given, can you find the probability that it will take Sulé longer to get to work on the bus than if he walks on a randomly selected day? Explain why or why not.

61. **Community college costs** Refer to Exercise 49. Note that X and Y are independent random variables because the two students are randomly selected from each of the campuses. At the main campus, full-time students pay $50 per unit. At the downtown campus, full-time students pay $55 per unit. Suppose we randomly select one full-time student from each of the two campuses. Find the standard deviation of the difference D (Main − Downtown) in the number of units that the two randomly selected students take.

62. **Essay scores** Refer to Exercise 50. Assume that the number of nonword errors X and word errors Y in a randomly selected essay are independent random variables. An English professor deducts 3 points from a student's essay score for each nonword error and 2 points for each word error. Find the standard deviation of the total score deductions T for a randomly selected essay.

Exercises 63 and 64 refer to the following setting. In Exercise 17 of Section 6.1, we examined the probability distribution of the random variable $X =$ the amount a life insurance company earns on a randomly chosen 5-year term life policy. Calculations reveal that $\mu_X = \$303.35$ and $\sigma_X = \$9707.57$.

63. **Life insurance** The risk of insuring one person's life is reduced if we insure many people. Suppose that we randomly select two insured 21-year-old males, and that their ages at death are independent. If X_1 and X_2 are the insurer's income from the two insurance policies, the insurer's average income W on the two policies is

$$W = \frac{X_1 + X_2}{2}$$

Find the mean and standard deviation of W. (You see that the mean income is the same as for a single policy, but the standard deviation is less.)

6.60 (a) Independence of B and W means that knowing the walking commute time does not help us predict the bus commute time.
(b) $\sigma_D = \sqrt{4^2 + 1^2} = 4.12$; if many, many days are randomly selected, the difference (Bus − Walk) in commute times will typically vary by about 4.12 minutes from the mean of −4 minutes. **(c)** No, we do not know the shapes of the distributions. We cannot assume that the distributions are Normal without additional information.

6.61 $\sigma_D = \sigma_{X/50 - Y/55} =$

$$\sqrt{\left(\frac{103}{50}\right)^2 + \left(\frac{126.50}{55}\right)^2} = 3.088;$$

the standard deviation of the difference D (Main − Downtown) in the number of units that two randomly selected students take is 3.088.

6.62 $\sigma_T = \sigma_{2X + 3Y} = \sqrt{(3 \cdot 1.136)^2 + (2 \cdot 1.0)^2} = 3.952$; the standard deviation of the total deductions T for a randomly selected essay is 3.952.

6.63 $\mu_{X_1 + X_2} = \$606.70$, and because the variables are independent, $\sigma_{X_1 + X_2} = \$13,728.58$. $W = \frac{1}{2}(X_1 + X_2)$, so $\mu_W = \$303.35$ and $\sigma_W = \$6864.29$.

64. **Life insurance** If we randomly select four insured 21-year-old men, the insurer's average income is

$$V = \frac{X_1 + X_2 + X_3 + X_4}{4}$$

where X_i is the income from insuring one man. Assuming that the amount of income earned on individual policies is independent, find the mean and standard deviation of V. (If you compare with the results of Exercise 63, you should see that averaging over more insured individuals reduces risk.)

65. **Time and motion** A time-and-motion study measures the time required for an assembly-line worker to perform a repetitive task. The data show that the time X required to bring a part from a bin to its position on an automobile chassis follows a Normal distribution with mean 11 seconds and standard deviation 2 seconds. The time Y required to attach the part to the chassis follows a Normal distribution with mean 20 seconds and standard deviation 4 seconds. The study finds that the times required for the two steps are independent.

(a) Describe the distribution of the total time required for the entire operation of positioning and attaching a randomly selected part.

(b) Management's goal is for the entire process to take less than 30 seconds. Find the probability that this goal will be met for a randomly selected part.

66. **Ohm-my!** The design of an electronic circuit for a toaster calls for a 100-ohm resistor and a 250-ohm resistor connected in series so that their resistances add. The resistance X of a 100-ohm resistor in a randomly selected toaster follows a Normal distribution with mean 100 ohms and standard deviation 2.5 ohms. The resistance Y of a 250-ohm resistor in a randomly selected toaster follows a Normal distribution with mean 250 ohms and standard deviation 2.8 ohms. The resistances X and Y are independent.

(a) Describe the distribution of the total resistance of the two components in series for a randomly selected toaster.

(b) Find the probability that the total resistance for a randomly selected toaster lies between 345 and 355 ohms.

67. **Yard work** Lamar and Hareesh run a two-person lawn-care service. They have been caring for Mr. Johnson's very large lawn for several years, and they have found that the time L it takes Lamar to mow the lawn on a randomly selected day is approximately Normally distributed with a mean of 105 minutes and a standard deviation of 10 minutes. The time H it takes Hareesh to use the edger and string trimmer on a randomly selected day is approximately Normally distributed

with a mean of 98 minutes and a standard deviation of 15 minutes. Assume that L and H are independent random variables. Find the probability that Lamar and Hareesh finish their jobs within 5 minutes of each other on a randomly selected day.

68. **Hit the track** Andrea and Barsha are middle-distance runners for their school's track team. Andrea's time A in the 400-meter race on a randomly selected day is approximately Normally distributed with a mean of 62 seconds and a standard deviation of 0.8 second. Barsha's time B in the 400-meter race on a randomly selected day is approximately Normally distributed with a mean of 62.8 seconds and a standard deviation of 1 second. Assume that A and B are independent random variables. Find the probability that Barsha beats Andrea in the 400-meter race on a randomly selected day.

69. **Swim team** Hanover High School has the best women's swimming team in the region. The 400-meter freestyle relay team is undefeated this year. In the 400-meter freestyle relay, each swimmer swims 100 meters. The times, in seconds, for the four swimmers this season are approximately Normally distributed with means and standard deviations as shown. Assume that the swimmer's individual times are independent. Find the probability that the total team time in the 400-meter freestyle relay for a randomly selected race is less than 220 seconds.

Swimmer	Mean	StDev
Wendy	55.2	2.8
Jill	58.0	3.0
Carmen	56.3	2.6
Latrice	54.7	2.7

70. **Toothpaste** Ken is traveling for his business. He has a new 0.85-ounce tube of toothpaste that's supposed to last him the whole trip. The amount of toothpaste Ken squeezes out of the tube each time he brushes can be modeled by a Normal distribution with mean 0.13 ounce and standard deviation 0.02 ounce. If Ken brushes his teeth six times on a randomly selected trip and the amounts are independent, what's the probability that he'll use all the toothpaste in the tube?

71. **Auto emissions** The amount of nitrogen oxides (NOX) present in the exhaust of a particular model of old car varies from car to car according to a Normal distribution with mean 1.4 grams per mile (g/mi) and standard deviation 0.3 g/mi. Two randomly selected cars of this model are tested. One has 1.1 g/mi of NOX; the other has 1.9 g/mi. The test station attendant finds this difference in emissions between two similar cars surprising. If the NOX levels for two randomly chosen cars of this type are independent, find the probability that the difference is greater than 0.8 or less than −0.8.

6.64 $\mu_{X_1+X_2+X_3+X_4} = \$1213.40$, and because the variables are independent, $\sigma_{X_1+X_2+X_3+X_4} = \$19,415.14$.

$V = \frac{1}{4}(X_1 + X_2 + X_3 + X_4)$, so $\mu_V = \$303.35$ and $\sigma_V = \$4853.785$. The standard deviation of V is less than the standard deviation of W, and exactly half the standard deviation of X.

6.65 (a) Normal distribution with mean $= 11 + 20 = 31$, and because variables are independent, standard deviation $= \sqrt{2^2 + 4^2} = 4.4721$. **(b) (i)** $z = -0.22$; $P(Z < -0.22) = 0.4129$ **(ii)** $P(X + Y < 30) =$ normalcdf(lower: −1000, upper: 30, mean: 31, SD: 4.4721) = 0.4115. There is a 0.4115 probability of completing the process in less than 30 seconds for a randomly selected part.

6.66 (a) Normal distribution with mean $100 + 250 = 350$, and because the

variables are independent, standard deviation $= \sqrt{2.5^2 + 2.8^2} = 3.7537$. **(b) (i)** $z = -1.33$ and $z = 1.33$; $P(-1.33 < Z < 1.33) = 0.9082 - 0.0918 = 0.8164$. **(ii)** $P(345 < X + Y < 355) =$ normalcdf(lower: 345, upper: 355, mean: 350, SD: 3.7537) = 0.8171. There is a 0.8171 probability that the total resistance is between 345 and 355 for a randomly selected toaster.

6.67 Let $D = L - H$. $\mu_D = 105 - 98 = 7$; because the random variables are independent, $\sigma_D = \sqrt{(10)^2 + (15)^2} = 18.03$. D is approximately Normally distributed with a mean of 7 and a standard deviation of 18.03; $P(-5 < L - H < 5)$ or $P(-5 < D < 5)$.

(i) $z = -0.11$ and $z = -0.67$; $P(-0.67 < Z < -0.11) = 0.4562 - 0.2514 = 0.2048$ **(ii)** $P(-5 < D < 5) =$ normalcdf(lower: −5, upper: 5, mean: 7,

SD: 18.03) = 0.2030. There is a 0.203 probability that Lamar and Hareesh finish their jobs within 5 minutes of each other on a randomly selected day.

6.68 Let $D = A - B$. $\mu_D = \mu_A - \mu_B = 62 - 62.8 = -0.8$ second; because the random variables are independent, $\sigma_D = \sqrt{\sigma_A^2 + \sigma_B^2} = \sqrt{(0.8)^2 + (1)^2} = 1.28$ seconds. D is approximately Normally distributed with a mean of −0.8 and a standard deviation of 1.28; $P(D > 0)$.

(i) $z = 0.63$; $P(Z > 0.63) = 0.2643$ **(ii)** $P(D > 0) =$ normalcdf(lower: 0, upper: 1000, mean: −0.8, SD: 1.28) = 0.2660. There is a 0.2660 probability that Barsha beats Ashley in the 400-meter race on a randomly selected day.

6.69 Let $T = X_1 + X_2 + X_3 + X_4$. $\mu_T = 55.2 + 58.0 + 56.3 + 54.7 = 224.2$; because the random variables are independent, $\sigma_T = \sqrt{(2.8)^2 + (3.0)^2 + (2.6)^2 + (2.7)^2} = 5.56$. T is Normally distributed with a mean of 224.2 and a standard deviation of 5.56; $P(T < 220)$.

(i) $z = -0.76$; $P(Z < -0.76) = 0.2236$ **(ii)** $P(T < 220) =$ normalcdf(lower: −1000, upper: 220, mean: 224.2, SD: 5.56) = 0.2250. There is a 0.2250 probability that the total team time is less than 220 seconds in a randomly selected race.

6.70 Let $T = X_1 + X_2 + X_3 + X_4 + X_5 + X_6$. $\mu_T = 0.78$ and $\sigma_T = \sqrt{0.02^2 + 0.02^2 + 0.02^2 + 0.02^2 + 0.02^2 + 0.02^2} = 0.049$ ounces. T has the Normal distribution with a mean of 0.78 and a standard deviation of 0.049; $P(T \geq 0.85)$.

(i) $z = 1.43$; $P(Z > 1.43) = 0.0764$ **(ii)** $P(T > 0.85) =$ normalcdf(lower: 0.85, upper: 1000, mean: 0.78, SD: 0.049) = 0.0766. There is a 0.0766 probability that Ken will use the entire tube of toothpaste on a randomly selected trip.

6.71 Let $D = X_1 - X_2$. $\mu_D = 1.4 - 1.4 = 0$; because the random variables are independent, $\sigma_{X_1 - X_2} = \sqrt{0.3^2 + 0.3^2} = 0.4243$. D has the Normal distribution with a mean of 0 and a standard deviation of 0.4243; $P(D < -0.8$ or $D > 0.8)$.

(i) $z = 1.89$ and $z = -1.89$; $P(Z < -1.89$ or $Z > 1.89) = 0.0588$

(ii) $P(D < -0.8$ or $D > 0.8) = 1 - $ normalcdf(lower: −0.8, upper: 0.8, mean: 0, SD: 0.4243) = $1 - 0.9406 = 0.0594$. There is a 0.0594 probability that difference is at least as large as the attendant observed.

6.72 Let $D = X_1 - X_2$. The mean is $\mu_D = 24 - 24 = 0$; because the random variables are independent, $\sigma_{X_1 - X_2} = \sqrt{2^2 + 2^2} = 2.83$. D is Normally distributed with a mean of 0 and a standard deviation of 2.83; $P(D \leq -5 \text{ or } D \geq 5)$.

(i) $z = 1.77$ and $z = -1.77$; $P(Z \leq -1.77 \text{ or } Z \geq 1.77) = 0.0768$

(ii) $P(D \leq -5 \text{ or } D \geq 5) = 1 - \text{normalcdf}$ (lower: -5, upper: 5, mean: 0, SD: 2.83) $= 0.0773$. There is a 0.0773 probability that the scores differ by 5 or more points in either direction.

6.73 c

6.74 d

6.75 (a) Yes; the girls who were assigned to receive the fluoride varnish may expect to get fewer cavities and, as a result, might take better care of their teeth. **(b)** This experiment could be double-blind if the girls who are not selected to receive the fluoride varnish received a placebo varnish that would not help or hurt their teeth. Also, the dental hygienist who checks the girls' teeth for cavities at the end of the 4 years should not know which girls received which treatment. **(c)** The random assignment should help to make the two groups as similar as possible before treatments are administered.

6.76 (a)

Amount	$562.50	$975	$1690
Probability	0.25	0.50	0.25

The probability that the stock is worth more than the $1000 paid for it, after two days, is 0.25 (the probability of it being worth $1690). **(b)** $\mu = 562.50(0.25) + 975(0.5) + 1690(0.25) = \$1,050.63$

PD **Section 6.3 Overview**

To watch the video overview of Section 6.3 (for teachers), click on the link in the TE-Book, look on the TRFD, or download from the Teacher's Resources on the book's digital platform.

TRM **Section 6.3 Alternate Examples**

You can find the Alternate Examples for this section in Microsoft Word format by clicking on the link in the TE-Book, opening the TRFD, or downloading from the Teacher's Resources on the book's digital platform.

72. **Loser buys the pizza** Leona and Fred are friendly competitors in high school. Both are about to take the ACT college entrance examination. They agree that if one of them scores 5 or more points better than the other, the loser will buy the winner a pizza. Suppose that, in fact, Fred and Leona have equal ability so that each score on a randomly selected test varies Normally with mean 24 and standard deviation 2. (The variation is due to luck in guessing and the accident of the specific questions being familiar to the student.) The two scores are independent. What is the probability that the scores differ by 5 or more points in either direction?

Multiple Choice *Select the best answer for Exercises 73 and 74, which refer to the following setting.*

The number of calories in a 1-ounce serving of a certain breakfast cereal is a random variable with mean 110 and standard deviation 10. The number of calories in a cup of whole milk is a random variable with mean 140 and standard deviation 12. For breakfast, you eat 1 ounce of the cereal with 1/2 cup of whole milk. Let T be the random variable that represents the total number of calories in this breakfast.

73. The mean of T is

(a) 110.   (b) 140.   (c) 180.

(d) 195.   (e) 250.

74. The standard deviation of T is

(a) 22.   (b) 16.   (c) 15.62.

(d) 11.66.   (e) 4.

Recycle and Review

75. **Fluoride varnish** (4.2) In an experiment to measure the effect of fluoride "varnish" on the incidence of tooth cavities, thirty-four 10-year-old girls whose parents volunteered them for the study were randomly assigned to two groups. One group was given fluoride varnish annually for 4 years, along with standard dental hygiene; the other group followed only the standard dental hygiene regimen. The mean number of cavities in the two groups was compared at the end of the 4 years.

(a) Are the participants in this experiment subject to the placebo effect? Explain.

(b) Describe how you could alter this experiment to make it double-blind.

(c) Explain the purpose of the random assignment in this experiment.

76. **Buying stock** (5.3, 6.1) You purchase a hot stock for $1000. The stock either gains 30% or loses 25% each day, each with probability 0.5. Its returns on consecutive days are independent of each other. You plan to sell the stock after two days.

(a) What are the possible values of the stock after two days, and what is the probability for each value? What is the probability that the stock is worth more after two days than the $1000 you paid for it?

(b) What is the mean value of the stock after two days?

Comment: You see that these two criteria give different answers to the question "Should I invest?"

SECTION 6.3 **Binomial and Geometric Random Variables**

LEARNING TARGETS *By the end of the section, you should be able to:*

- Determine whether the conditions for a binomial setting are met.
- Calculate and interpret probabilities involving binomial random variables.
- Calculate the mean and standard deviation of a binomial distribution. Interpret these values.
- When appropriate, use the Normal approximation to the binomial distribution to calculate probabilities.
- Calculate and interpret probabilities involving geometric random variables.
- Calculate the mean and standard deviation of a geometric distribution. Interpret these values.

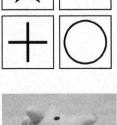

Alex Clark/Alamy

When the same random process is repeated several times, we are often interested in whether a particular outcome does or doesn't happen on each trial. Here are some examples:

- To test whether someone has extrasensory perception (ESP), choose one of four cards at random—a star, wave, cross, or circle. Ask the person to identify the card without seeing it. Do this a total of 50 times and see how many cards the person identifies correctly. *Random process*: choose a card at random. *Outcome of interest*: person identifies card correctly. *Random variable*: X = number of correct identifications.

- A shipping company claims that 90% of its shipments arrive on time. To test this claim, take a random sample of 100 shipments made by the company last month and see how many arrived on time. *Random process*: randomly select a shipment and check when it arrived. *Outcome of interest*: arrived on time. *Random variable*: Y = number of on-time shipments.

- In the game of Pass the Pigs, a player rolls a pair of pig-shaped dice. On each roll, the player earns points according to how the pigs land. If the player gets a "pig out," in which the two pigs land on opposite sides, she loses all points earned in that round and must pass the pigs to the next player. A player can choose to stop rolling at any point during her turn and to keep the points that she has earned before passing the pigs. *Random process*: roll the pig dice. *Outcome of interest*: pig out. *Random variable*: T = number of rolls it takes the player to pig out.

Some random variables, like X and Y in the first two bullets, count the number of times the outcome of interest occurs in a fixed number of trials. They are called *binomial random variables*. Other random variables, like T in the Pass the Pigs setting, count the number of trials of the random process it takes for the outcome of interest to occur. They are known as *geometric random variables*. These two special types of discrete random variables are the focus of this section.

Binomial Settings and Binomial Random Variables

Let's start with an activity that involves repeating a random process several times.

ACTIVITY Pop quiz!

It's time for a pop quiz! We hope you are ready. The quiz consists of 10 multiple-choice questions. Each question has five answer choices, labeled A through E. Now for the bad news: you will not get to see the questions. You just have to guess the answer for each one!

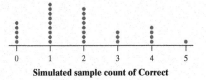

Preparing for Inference

For the activity, here are the probabilities of getting a certain number of correct answers, assuming the person is guessing:

P(4 or more correct) = 0.1209

P(5 or more correct) = 0.0328

P(6 or more correct) = 0.0064

P(7 or more correct) = 0.0009

P(8 or more correct) = 0.0001

P(9 or more correct) $\approx$ 0

P(10 or more correct) $\approx$ 0

These probabilities are P-values for a significance test. If the P-value is less than 5%, we have convincing evidence that the person has special powers. Of course, in a large enough class, someone is bound to get 5 or more correct by chance alone.

Teaching Tip

The term *trials* can be used interchangeably with the term *observations*.

Teaching Tip

Tell students that a "success" does not always mean something awesome happened. For example, a "success" might be defined as a faulty part or a person being diabetic.

Teaching Tip

When describing the independent condition, try to avoid using statements like "one trial doesn't affect another." When students hear this, they may think that the independence condition is violated only when there is a cause-and-effect relationship between the trials. Instead, use the wording that we give here and elsewhere: knowing the outcome of one trial tells us nothing about the outcomes of other trials.

1. Get out a blank sheet of paper. Write your name at the top. Number your paper from 1 to 10. Then guess the answer to each question: A, B, C, D, or E. Do not look at anyone else's paper! You have 2 minutes.

2. Now it's time to grade the quizzes. Exchange papers with a classmate. Your teacher will display the answer key. The correct answer for each of the 10 questions was determined randomly so that A, B, C, D, or E was equally likely to be chosen.

3. How did you do on your quiz? Make a class dotplot that shows the number of correct answers for each student in your class. As a class, describe what you see.

In the "Pop quiz" activity, each student is performing repeated *trials* of the same random process: guessing the answer to a multiple-choice question. We're interested in the number of times that a specific event occurs: getting a correct answer (which we'll call a "success"). Knowing the outcome of one question (right or wrong guess) tells us nothing about the outcome of any other question. That is, the trials are independent. The number of trials is fixed in advance: $n = 10$. And a student's probability of getting a "success" is the same on each trial: $p = 1/5 = 0.2$. When these conditions are met, we have a **binomial setting**.

DEFINITION Binomial setting

A **binomial setting** arises when we perform n independent trials of the same random process and count the number of times that a particular outcome (called a "success") occurs.

The four conditions for a binomial setting are:

- **B**inary? The possible outcomes of each trial can be classified as "success" or "failure."
- **I**ndependent? Trials must be independent. That is, knowing the outcome of one trial must not tell us anything about the outcome of any other trial.
- **N**umber? The number of trials n of the random process must be fixed in advance.
- **S**ame probability? There is the same probability of success p on each trial.

The boldface letters in the definition box give you a helpful way to remember the conditions for a binomial setting: just check the BINS!

When checking the binary condition, note that there can be more than two possible outcomes per trial—in the "Pop Quiz" Activity, each question (trial) had five possible answer choices: A, B, C, D, or E. If we define "success" as guessing the correct answer to a question, then "failure" occurs when the student guesses any of the four incorrect answer choices.

■ AP® EXAM TIP

On many questions involving binomial settings, students do not recognize that using a binomial distribution is appropriate. In fact, free response questions about the binomial distribution are often among the lowest-scoring questions on the exam. Make sure to spend plenty of time on how to identify a binomial distribution. You can also suggest that students check to see if it is a binomial setting whenever they aren't sure how to answer a probability question.

EXAMPLE

From blood types to aces
Identifying binomial settings

PROBLEM: Determine whether the given scenario describes a binomial setting. Justify your answer.

(a) Genetics says that the genes children receive from their parents are independent from one child to another. Each child of a particular set of parents has probability 0.25 of having type O blood. Suppose these parents have 5 children. Count the number of children with type O blood.

(b) Shuffle a standard deck of 52 playing cards. Turn over the first 10 cards, one at a time. Record the number of aces you observe.

(c) Shuffle a deck of cards. Turn over the top card. Put the card back in the deck, and shuffle again. Repeat this process until you get an ace. Count the number of cards you had to turn over.

SOLUTION:

(a) • Binary? "Success" = has type O blood. "Failure" = doesn't have type O blood.

> Check the BINS! A trial consists of observing the blood type for one of these parents' children.

• Independent? Knowing one child's blood type tells you nothing about another child's because they inherit genes independently from their parents.

• Number? $n = 5$

• Same probability? $p = 0.25$

This is a binomial setting.

> All the conditions are met and we are counting the number of successes (children with type O blood).

(b) • Binary? "Success" = get an ace. "Failure" = don't get an ace.

> Check the BINS! A trial consists of turning over a card from the deck and observing what's on the card.

• Independent? No. If the first card you turn over is an ace, then the next card is less likely to be an ace because you're not replacing the top card in the deck. If the first card isn't an ace, the second card is more likely to be an ace.

> To check for independence, you could also write
> P(2nd card ace | 1st card ace) = 3/51
> P(2nd card ace | 1st card not ace) = 4/51
> Because the two probabilities are not equal, the trials are not independent.

This is not a binomial setting because the independent condition is not met.

(c) • Binary? "Success" = get an ace. "Failure" = don't get an ace.

> Check the BINS! A trial consists of turning over a card from the shuffled deck of 52 cards and observing what's on the card.

• Independent? Yes. Because you are replacing the card in the deck and shuffling each time, the result of one trial doesn't tell you anything about the outcome of any other trial.

• Number? No. The number of trials is not fixed in advance.

> There's another clue that this is not a binomial setting: you're counting the number of trials to get a success and not the number of successes in a fixed number of trials.

Because there is no fixed number of trials, this is not a binomial setting.

FOR PRACTICE, TRY EXERCISE 77

ALTERNATE EXAMPLE **Skill 3.A**

Dice, cars, and hoops
Identifying binomial settings

PROBLEM:
Determine whether the random variables below have a binomial distribution. Justify your answer.
(a) Roll a fair die 10 times and let X = the number of 6s.
(b) Shoot a basketball 20 times from various distances on the court. Let Y = number of shots made.
(c) Observe the next 100 cars that go by and let C = color of each car.

SOLUTION:
(a) *Binary?* Yes; success = 6, failure = not a 6. *Independent?* Yes; knowing the outcomes of past rolls tells you nothing about the outcomes of future rolls. *Number?* Yes; there are $n = 10$ trials. *Same probability?* Yes; the probability of success is always $p = 1/6$. This is a binomial setting. The number of 6s, X, is a binomial random variable with $n = 10$ and $p = 1/6$.
(b) *Binary?* Yes; success = make the shot, failure = miss the shot. *Independent?* Yes; evidence suggests that it is reasonable to assume that knowing the outcome of one shot does not change the probability of making the next shot. *Number?* Yes; there are $n = 20$ trials. *Same probability?* No; the probability of success changes because the shots are taken from various distances. Because the probability of success is not constant, Y is not a binomial random variable.
(c) *Binary?* No; there are more than two possible colors. *Independent?* Yes; knowing the color of one car tells you nothing about the color of other cars. *Number?* Yes; there are $n = 100$ trials. *Same probability?* A success hasn't been defined, so we cannot determine if the probability of success is always the same. Because there are more than two possible outcomes, C is not a binomial random variable. Additionally, C is not even a random variable because it does not take numerical values.

Teaching Tip

In part (b) of the above example, the independent condition was violated because of sampling without replacement. Any time there is sampling without replacement, the trials will not be independent. But consider a large shipment of 10,000 M&M'S®, of which 2000 are yellow. Randomly choose 5 M&M'S and record whether or not they are yellow.

P(2nd is yellow | 1st is yellow) = 1999/9999 = 0.19992

P(2nd is yellow | 1st is not yellow) = 2000/9999 = 0.20002

While these two probabilities are not equal (and the independent condition is not met), they are close enough that we could still use a binomial distribution to closely approximate probabilities for this context. See the 10% condition later in this chapter.

Making Connections

In part (c) of the above example, we decided against the binomial setting. We will find out later in the chapter that this is a geometric setting because we are counting the number of trials until a success occurs.

Teaching Tip

Emphasize that the binomial random variable is the number (count) of successes in a fixed number of trials by having them define the variable each time they do a binomial calculation. This will help students recognize when a binomial distribution is appropriate.

Teaching Tip

Some teachers use $B(n, p)$ as the notation to identify a binomial distribution and its two parameters. This is similar to the $N(\mu, \sigma)$ notation for Normal distributions.

✓ **Answers to CYU**

1. *Binary?* "Success" = get an ace; "Failure" = don't get an ace. *Independent?* Because you are replacing the card in the deck and shuffling each time, the result of one trial does not tell you anything about the outcome of any other trial. *Number?* $n = 10$. *Same probability?* $p = 4/52$; this is a binomial setting, and X has a binomial distribution with $n = 10$ and $p = 4/52$.

2. *Binary?* "Success" = over 6 feet; "Failure" = not over 6 feet. *Independent?* Because we are selecting without replacement from a small number of students, the observations are not independent. *Number?* $n = 5$. *Same probability?* The (unconditional) probability of success will not change from trial to trial. Because the trials are not independent, this is not a binomial setting.

3. *Binary?* "Success" = roll a 5; "Failure" = don't roll a 5. *Independent?* Because you are rolling a die, the outcome of any one trial does not tell you anything about the outcome of any other trial. *Number?* $n = 100$. *Same probability?* No; the probability of success changes when the corner of the die is chipped off. Because the (unconditional) probability of success changes from trial to trial, this is not a binomial setting.

The Independent condition involves *conditional* probabilities. In part (b) of the example,

$$P(\text{2nd card ace} \mid \text{1st card ace}) = 3/51 \neq P(\text{2nd card ace} \mid \text{1st card not ace}) = 4/51$$

so the trials are not independent. The Same probability of success condition is about *unconditional* probabilities. Because

$$P(x\text{th card in a shuffled deck is an ace}) = 4/52$$

this condition is met in part (b) of the example. Be sure you understand the difference between these two conditions. When sampling is done without replacement, the Independent condition is violated.

The blood type scenario in part (a) of the example is a binomial setting. If we let $X =$ the number of children with type O blood, then X is a **binomial random variable**. The probability distribution of X is called a **binomial distribution**.

DEFINITION   **Binomial random variable, Binomial distribution**

The count of successes X in a binomial setting is a **binomial random variable**. The possible values of X are 0, 1, 2, ..., n.

The probability distribution of X is a **binomial distribution**. Any binomial distribution is completely specified by two numbers: the number of trials n of the random process and the probability p of success on each trial.

In the "Pop quiz" activity at the beginning of the lesson, $X =$ the number of correct answers is a binomial random variable with $n = 10$ and $p = 0.2$.

CHECK YOUR UNDERSTANDING

For each of the following situations, determine whether or not the given random variable has a binomial distribution. Justify your answer.

1. Shuffle a deck of cards. Turn over the top card. Put the card back in the deck, and shuffle again. Repeat this process 10 times. Let $X =$ the number of aces you observe.
2. Choose 5 students at random from your class. Let $Y =$ the number who are over 6 feet tall.
3. Roll a fair die 100 times. Sometime during the 100 rolls, one corner of the die chips off. Let $W =$ the number of 5s you roll.

Calculating Binomial Probabilities

How can we calculate probabilities involving binomial random variables? Let's return to the scenario from part (a) of the preceding example:

> Genetics says that the genes children receive from their parents are independent from one child to another. Each child of a particular set of parents has probability 0.25 of having type O blood. Suppose these parents have 5 children. Count the number of children with type O blood.

In this binomial setting, a child with type O blood is a "success" (S) and a child with another blood type is a "failure" (F). The count X of children with type O blood is a binomial random variable with $n = 5$ trials and probability $p = 0.25$ of success on each trial.

- What's $P(X = 0)$? That is, what's the probability that *none* of the 5 children has type O blood? The probability that any one of this couple's children doesn't have type O blood is $1 - 0.25 = 0.75$ (complement rule). By the multiplication rule for independent events (Section 5.3),

$$P(X=0) = P(FFFFF) = (0.75)(0.75)(0.75)(0.75)(0.75) = (0.75)^5 = 0.2373$$

- How about $P(X = 1)$? There are several different ways in which exactly 1 of the 5 children could have type O blood. For instance, the first child born might have type O blood, while the remaining 4 children don't have type O blood. The probability that this happens is

$$P(SFFFF) = (0.25)(0.75)(0.75)(0.75)(0.75) = (0.25)^1(0.75)^4$$

Alternatively, Child 2 could be the one that has type O blood. The corresponding probability is

$$P(FSFFF) = (0.75)(0.25)(0.75)(0.75)(0.75) = (0.25)^1(0.75)^4$$

There are three more possibilities to consider—those in which Child 3, Child 4, and Child 5 are the only ones to inherit type O blood. Of course, the probability will be the same for each of those cases. In all, there are five different ways in which exactly 1 child would have type O blood, each with the same probability of occurring. As a result,

$$P(X = 1) = P(\text{exactly 1 child with type O blood})$$

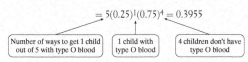

$$= 5(0.25)^1(0.75)^4 = 0.3955$$

| Number of ways to get 1 child out of 5 with type O blood | 1 child with type O blood | 4 children don't have type O blood |

The pattern of this calculation works for any binomial probability:

$$P(X = x) = (\text{\# of ways to get } x \text{ successes in } n \text{ trials})(\text{success probability})^x(\text{failure probability})^{n-x}$$

To use this formula, we must count the number of arrangements of x successes in n trials. This number is called the **binomial coefficient**. We use the following fact to do the counting without actually listing all the arrangements.

DEFINITION   Binomial coefficient

The number of ways to arrange x successes among n trials is given by the **binomial coefficient**

$$\binom{n}{x} = \frac{n!}{x!(n-x)!}$$

for $x = 0, 1, 2, \ldots, n$ where $n!$ (read as "n factorial") is given by

$$n! = n(n-1)(n-2) \cdot \cdots \cdot (3)(2)(1)$$

and $0! = 1$.

Teaching Tip

How many different ways are there of getting 5 failures? Only 1: FFFFF.

Making Connections

Students should be familiar with the binomial coefficient (or combinations) from their algebra classes. They likely used it to find the coefficients for expanding binomials. Students may be familiar with a slightly different wording of the definition: the number of ways of choosing x items among n items (where order doesn't matter).

Teaching Tip

When reading $n!$ to the class, be sure to speak in a louder voice to clearly express the exclamation point.

Teaching Tip

Don't spend too much time developing the formula for the binomial coefficient. Think of it as a tool that students will use to help them calculate binomial probabilities.

AP® EXAM TIP

The formula sheet for the AP® Statistics exam uses the notation $\binom{n}{x}$ rather than $_nC_x$.

The larger of the two factorials in the denominator of a binomial coefficient will cancel much of the $n!$ in the numerator. For example, the binomial coefficient we need to find the probability that exactly 2 of the couple's 5 children inherit type O blood is

$$\binom{5}{2} = \frac{5!}{2!3!} = \frac{(5)(4)(\cancel{3})(\cancel{2})(\cancel{1})}{(2)(1)(\cancel{3})(\cancel{2})(\cancel{1})} = \frac{(5)(4)}{(2)(1)} = 10$$

The binomial coefficient $\binom{5}{2}$ is not related to the fraction $\frac{5}{2}$. A helpful way to remember its meaning is to read it as "5 choose 2"—as in, how many ways are there to choose which 2 children have type O blood in a family with 5 children? Binomial coefficients have many uses, but we are interested in them only as an aid to finding binomial probabilities. If you need to compute a binomial coefficient, use your calculator.

> Some people prefer the notation $_5C_2$ instead of $\binom{5}{2}$ for the binomial coefficient.

14. Technology Corner   CALCULATING BINOMIAL COEFFICIENTS

TI-Nspire and other technology instructions are on the book's website at highschool.bfwpub.com/updatedtps6e.

To calculate a binomial coefficient like $\binom{5}{2}$ on the TI-83/84, proceed as follows:

- Type 5, press MATH, arrow over to PROB, choose nCr, and press ENTER. Then type 2 and press ENTER again to execute the command 5 nCr 2 (which displays as $_5C_2$ on devices with pretty print).

```
NORMAL FLOAT AUTO REAL RADIAN MP
5C2
                                    10
```

The binomial coefficient $\binom{n}{x}$ counts the number of different ways in which x successes can be arranged among n trials. The binomial probability $P(X = x)$ is this count multiplied by the probability of any one specific arrangement of the x successes.

> A function (like the binomial probability formula) can be used to specify the probability distribution of a random variable, in addition to a table or a graph.

BINOMIAL PROBABILITY FORMULA

Suppose that X is a binomial random variable with n trials and probability p of success on each trial. The probability of getting exactly x successes in n trials $(x = 0, 1, 2, \ldots, n)$ is

$$P(X = x) = \binom{n}{x} p^x (1-p)^{n-x}$$

where

$$\binom{n}{x} = \frac{n!}{x!(n-x)!}$$

With our formula in hand, we can now calculate any binomial probability.

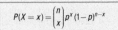

EXAMPLE

Inheriting blood type
Calculating a binomial probability

PROBLEM: Genetics says that the genes children receive from their parents are independent from one child to another. Each child of a particular set of parents has probability 0.25 of having type O blood. Suppose these parents have 5 children. Let $X =$ the number of children with type O blood. Find $P(X = 3)$. Interpret this value.

SOLUTION:

X is a binomial random variable with $n = 5$ and $p = 0.25$.

$$P(X = x) = \binom{n}{x} p^x (1-p)^{n-x}$$

$$P(X = 3) = \binom{5}{3}(0.25)^3 (0.75)^2$$

$$= 10(0.25)^3 (0.75)^2$$

$$= 0.08789$$

```
NORMAL FLOAT AUTO REAL RADIAN MP
₅C₃
                                    10
```

There is about a 9% probability that exactly 3 of the 5 children have type O blood.

FOR PRACTICE, TRY EXERCISE 81

There are times when we want to calculate a probability involving more than one value of a binomial random variable. As the following example illustrates, we can just use the binomial probability formula for each value of interest.

EXAMPLE

Inheriting blood type
Calculating binomial probabilities

PROBLEM: The preceding example tells us that each child of a particular set of parents has probability 0.25 of having type O blood. Suppose these parents have 5 children. Should the parents be surprised if more than 3 of their children have type O blood? Calculate an appropriate probability to support your answer.

SOLUTION:

Let $X =$ the number of children with type O blood. X has a binomial distribution with $n = 5$ and $p = 0.25$.

$$P(X > 3) = P(X = 4) + P(X = 5)$$

$$= \binom{5}{4}(0.25)^4 (0.75)^1 + \binom{5}{5}(0.25)^5 (0.75)^0$$

$$= 5(0.25)^4 (0.75)^1 + 1(0.25)^5 (0.75)^0$$

$$= 0.01465 + 0.00098$$

$$= 0.01563$$

Because there's only about a 1.5% probability of having more than 3 children with type O blood, the parents should definitely be surprised if this happens.

FOR PRACTICE, TRY EXERCISE 85

Teaching Tip

In both Examples on this page, the student solution starts out by defining the random variable and parameters for the distribution of that random variable. This serves two purposes. First, it helps students to identify the type of random variable that is needed to answer the question. Second, it might help students' scores on the AP® Statistics exam, as defining the type of distribution and its parameters has been part of rubrics in the past.

ALTERNATE EXAMPLE Skills 3.A, 4.B

Snake eyes
Calculating a binomial probability

PROBLEM:
When rolling two fair six-sided dice, getting a pair of 1s is called "snake eyes." The probability of getting "snake eyes" on any roll is 1/36. Suppose that a game player rolls the two dice 80 times. What is the probability that the player gets exactly 2 snake eyes? Interpret this value in context.

SOLUTION:
Let $X =$ the number of rolls that result in "snake eyes." X is a binomial random variable with $n = 80$ and $p = 1/36$.

$$P(X = 2) = \binom{80}{2}\left(\frac{1}{36}\right)^2\left(\frac{35}{36}\right)^{78}$$

$$= 3160\left(\frac{1}{36}\right)^2\left(\frac{35}{36}\right)^{78}$$

$$= 0.27089$$

There is about a 27% probability that exactly 2 of the 80 rolls of the two dice will result in "snake eyes."

ALTERNATE EXAMPLE Skills 3.A, 4.B

Not enough snake eyes
Calculating a binomial probability

PROBLEM:
See the preceding alternate example. Should the player be surprised if fewer than 2 of the rolls result in snake eyes? Calculate an appropriate probability to support your answer.

SOLUTION:
Let $X =$ the number of rolls that result in "snake eyes." X has a binomial distribution with $n = 80$ and $p = 1/36$.

$$P(X < 2) = P(X = 0) + P(X = 1)$$

$$= \binom{80}{0}\left(\frac{1}{36}\right)^0\left(\frac{35}{36}\right)^{80}$$

$$+ \binom{80}{1}\left(\frac{1}{36}\right)^1\left(\frac{35}{36}\right)^{79}$$

$$= 0.2400 + 0.1050 = 0.345$$

Because there's about a 34.5% probability of getting fewer than 2 "snake eyes" in 80 rolls, the player should not be surprised if this happens.

We can also use the calculator's binompdf and binomcdf commands to perform the calculations in the previous two examples. The following Technology Corner shows how to do it.

15. Technology Corner   CALCULATING BINOMIAL PROBABILITIES

TI-Nspire and other technology instructions are on the book's website at highschool.bfwpub.com/updatedtps6e.

There are two handy commands on the TI-83/84 for finding binomial probabilities: binompdf and binomcdf. The inputs for both commands are the number of trials n, the success probability p, and the values of interest for the binomial random variable X.

$$\text{binompdf}(n,p,x) \text{ computes } P(X = x)$$
$$\text{binomcdf}(n,p,x) \text{ computes } P(X \leq x)$$

Let's use these commands to confirm our answers in the previous two examples.

1. Find $P(X = 3)$.

- Press [2nd] [VARS] (DISTR) and choose binompdf(.

   OS 2.55 or later: In the dialog box, enter these values: trials:5, p:0.25, x value:3, choose Paste, and then press [ENTER].

   Older OS: Complete the command binompdf (5,0.25,3) and press [ENTER].

```
NORMAL FLOAT AUTO REAL RADIAN MP
binompdf(5,.25,3)
                      0.087890625
```

These results agree with our previous answer using the binomial probability formula: 0.08789.

2. Should the parents be surprised if more than 3 of their children have type O blood? To find $P(X > 3)$, use the complement rule:

$$P(X > 3) = 1 - P(X \leq 3) = 1 - \text{binomcdf}(5, 0.25, 3)$$

- Press [2nd] [VARS] (DISTR) and choose binomcdf(.

   OS 2.55 or later: In the dialog box, enter these values: trials:5, p:0.25, x value:3, choose Paste, and then press [ENTER]. Subtract this result from 1 to get the answer.

   Older OS: Complete the command binomcdf(5,0.25,3) and press [ENTER]. Subtract this result from 1 to get the answer.

```
NORMAL FLOAT AUTO REAL RADIAN MP
binomcdf(5,.25,3)
                          0.984375
1-Ans
                          0.015625
```

This result agrees with our previous answer using the binomial probability formula: 0.01563.

We could also have done the calculation for part (b) as $P(X > 3) = P(X = 4) + P(X = 5)$
$= \text{binompdf}(5, 0.25, 4) + \text{binompdf}(5, 0.25, 5) = 0.01465 + 0.00098 = 0.01563.$

Note the use of the complement rule to find $P(X > 3)$ in the Technology Corner: $P(X > 3) = 1 - P(X \leq 3)$. This is necessary because the calculator's

binomcdf(n,p,x) command computes the probability of getting *x or fewer* successes in *n* trials. Remember that a *cumulative probability distribution* always gives $P(X \leq x)$.

Students often have trouble identifying the correct third input for the binomcdf command when a question asks them to find the probability of getting less than, more than, or at least so many successes. Here's a helpful tip to avoid making such a mistake: write out the possible values of the variable, circle the ones you want to find the probability of, and cross out the rest. In the preceding example, *X* can take values from 0 to 5 and we want to find $P(X > 3)$:

$$\cancel{0} \quad \cancel{1} \quad \cancel{2} \quad \cancel{3} \quad \boxed{4 \quad 5}$$

Crossing out the values from 0 to 3 shows why the correct calculation is $1 - P(X \leq 3)$.

Take another look at the solutions in the two blood-type examples. The structure is much like the one we used when doing Normal calculations. Here is a summary box that describes the process.

HOW TO FIND BINOMIAL PROBABILITIES

Step 1: State the distribution and the values of interest. Specify a binomial distribution with the number of trials *n*, success probability *p*, and the values of the variable clearly identified.

Step 2: Perform calculations—show your work! Do one of the following:

(i) Use the binomial probability formula to find the desired probability; or

(ii) Use the binompdf or binomcdf command and label each of the inputs.

Be sure to answer the question that was asked.

Here's an example that shows the method at work.

HOW TO FIND AREAS IN ANY NORMAL DISTRIBUTION

Step 1: Draw a Normal distribution with the horizontal axis labeled and scaled using the mean and standard deviation, the boundary value(s) clearly identified, and the area of interest shaded.

Step 2: Perform calculations—show your work! Do one of the following:

(i) Standardize each boundary value and use Table A or technology to find the desired area under the standard Normal curve; or

(ii) Use technology to find the desired area without standardizing. Label the inputs of your calculator command.

Be sure to answer the question that was asked.

EXAMPLE

Free lunch?
Calculating a cumulative binomial probability

PROBLEM: A local fast-food restaurant is running a "Draw a three, get it free" lunch promotion. After each customer orders, a touchscreen display shows the message "Press here to win a free lunch." A computer program then simulates one card being drawn from a standard deck. If the chosen card is a 3, the customer's order is free. Otherwise, the customer must pay the bill.

(a) On the first day of the promotion, 250 customers place lunch orders. Find the probability that fewer than 10 of them win a free lunch.

(b) In fact, only 9 customers won a free lunch. Does this result give convincing evidence that the computer program is flawed?

eldatcariv/Getty Images

Skills 3.A, 4.B

Aussie instant lottery   Calculating binomial probabilities

PROBLEM:
The Australian Official Lottery has scratch-off instant lottery tickets that can be purchased for $1. The probability of winning a prize is 1 in 4.
(a) Mr. Urban is feeling lucky one day and decides to purchase 100 of the scratch-off instant lottery tickets. Find the probability that fewer than 20 tickets are winners.
(b) In fact, Mr. Urban won a prize for only 19 of the tickets. Does this result give convincing evidence that the probability of winning is less than 1 in 4?

SOLUTION:
(a) Let *Y* = the number of tickets that win a prize. *Y* has a binomial distribution with $n = 100$ and $p = 1/4$.

$P(Y < 20) = P(Y \leq 19) =$ binomcdf(trials: 100, p: 1/4, x value: 19) = 0.09953

(b) If the probability of winning a prize is 1 in 4, there is a 0.09953 probability that fewer than 20 of his 100 tickets win a prize. Because it is plausible that Mr. Urban would win on only 19 tickets purely by chance, we do not have convincing evidence that the probability of winning is less than 1 in 4.

Preparing for Inference

This example demonstrates the reasoning of a significance test from Chapter 9. The null hypothesis claim is that the computer program is running properly. Assuming the null hypothesis is true, the probability of getting fewer than 10 successes in 250 trials (0.00613) is the *P*-value. Because the *P*-value is small, we have convincing evidence that the computer program is flawed.

✓ **Answers to CYU**

1. *Binary?* "Success" = question answered correctly; "Failure" = question not answered correctly. *Independent?* Mr. Miller randomly determined correct answers to the questions, so knowing the result of one trial (question) should not tell you anything about the result on any other trial. *Number?* $n = 10$. *Same probability?* $p = 0.20$; this is a binomial setting and X has a binomial distribution with $n = 10$ and $p = 0.20$.

2. $P(X = 3) = \binom{10}{3}(0.2)^3(0.8)^7 = 0.2013$; there is about a 20% chance that Hannah will answer exactly 3 questions correctly.

3. $P(X \geq 6) = 1 - P(X < 6) =$ $1 - P(X \leq 5) = 1 -$ binomcdf(trials:10, p: 0.2, x value:5) $= 1 - 0.9936 = 0.0064$; there is only a 0.0064 probability that a student would get 6 or more correct answers, so we would be quite surprised if Hannah was able to pass.

SOLUTION:

(a) Let Y = the number of customers who win a free lunch. Y has a binomial distribution with $n = 250$ and $p = 4/52$.

$$P(Y < 10) = P(Y \leq 9)$$
$$= \text{binomcdf(trials: 250, }p\text{: 4/52, }x\text{ value: 9)}$$
$$= 0.00613$$

(b) There is only a 0.006 probability that fewer than 10 customers would win a free lunch if the computer program is working properly. Because only 9 customers won a free lunch on this day, we have convincing evidence that the computer program is flawed.

Step 1: State the distribution and the values of interest.

The values of Y that interest us are

⓪ 1 2 3 4 5 6 7 8 ⑨ 10 11 12 ... 250

Step 2: Perform calculations—show your work!

(i) Use the binomial probability formula to find the desired probability; or

(ii) use the binompdf or binomcdf command and label each of the inputs.

To use the binomial formula, you would have to add the probabilities for $Y = 0, 1, \ldots, 9$. That's too much work!

FOR PRACTICE, TRY EXERCISE 89

CHECK YOUR UNDERSTANDING

To introduce his class to binomial distributions, Mr. Miller does the "Pop quiz" activity at the beginning of this section (page 431). Each student in the class guesses an answer from A through E on each of the 10 multiple-choice questions. Mr. Miller determines the "correct" answer for each of the 10 questions randomly so that A, B, C, D, or E was equally likely to be chosen. Hannah is one of the students in this class. Let X = the number of questions that Hannah answers correctly.

1. What probability distribution does X have? Justify your answer.
2. Use the binomial probability formula to find $P(X = 3)$. Interpret this result.
3. To get a passing score on the quiz, a student must answer at least 6 questions correctly. Would you be surprised if Hannah earned a passing score? Calculate an appropriate probability to support your answer.

Describing a Binomial Distribution: Shape, Center, and Variability

What does the probability distribution of a binomial random variable look like? The table shows the possible values and corresponding probabilities for X = the number of children with type O blood from two previous examples. This is a binomial random variable with $n = 5$ and $p = 0.25$.

Value x_i	0	1	2	3	4	5
Probability p_i	0.23730	0.39551	0.26367	0.08789	0.01465	0.00098

Figure 6.7 shows a histogram of the probability distribution. This binomial distribution with $n = 5$ and $p = 0.25$ has a clear right-skewed shape. Why? Because

the probability that any one of the couple's children inherits type O blood is 0.25, it's quite likely that 0, 1, or 2 of the children will have type O blood. Larger values of X are much less likely.

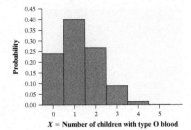

FIGURE 6.7 Histogram showing the probability distribution of the binomial random variable X = number of children with type O blood in a family with 5 children.

You can use technology to graph a binomial probability distribution like the one shown in Figure 6.7.

---

16. Technology Corner   GRAPHING BINOMIAL PROBABILITY DISTRIBUTIONS

TI-Nspire and other technology instructions are on the book's website at highschool.bfwpub.com/updatedtps6e.

To graph the binomial probability distribution for $n = 5$ and $p = 0.25$:

- Type the possible values of the random variable X into list L_1: 0, 1, 2, 3, 4, and 5.
- Highlight L_2 with your cursor. Enter the command binompdf(5,0.25) and press ENTER.
- Make a histogram of the probability distribution using the method shown in Technology Corner 13 (page 398).

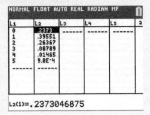

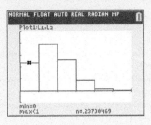

The binomial distribution with $n = 5$ and $p = 0.25$ is skewed to the right. Figure 6.8 on the next page shows two more binomial distributions with different shapes. The binomial distribution with $n = 5$ and $p = 0.51$ in Figure 6.8(a) is roughly symmetric. The binomial distribution with $n = 5$ and $p = 0.8$ in Figure 6.8(b) is skewed to the left. In general, when n is small, the probability distribution of a binomial random variable will be roughly symmetric if p is close to 0.5, right-skewed if p is much less than 0.5, and left-skewed if p is much greater than 0.5.

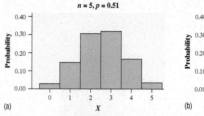

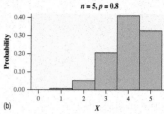

FIGURE 6.8 (a) Probability histogram for the binomial random variable X with $n = 5$ and $p = 0.51$. This binomial distribution is roughly symmetric. (b) Probability histogram for the binomial random variable X with $n = 5$ and $p = 0.8$. This binomial distribution has a left-skewed shape.

Coin tosses
Describing a binomial distribution

PROBLEM:
Mrs. Patenaude asked each student in her AP® Statistics class to toss a coin and record the result. There were 28 students in the class. Assume that all the coins are fair. Let X = the number of coins that landed heads. Here is a histogram of the probability distribution of X:

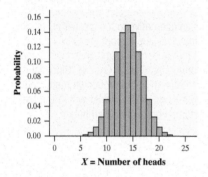

(a) What probability distribution does X have? Justify your answer.
(b) Describe the shape of the probability distribution.

SOLUTION:
(a) A trial consists of a student in the class recording the result of a coin toss.

- *Binary?* Success = heads; failure = tails
- *Independent?* Knowing the result of one student's coin toss does not help us predict the result of another student's coin toss.
- *Number?* $n = 28$
- *Same probability?* $p = 1/2$

X has a binomial distribution with $n = 28$ and $p = 1/2$.
(b) The probability distribution of X is symmetric, with a single peak at $X = 14$.

EXAMPLE

Bottled water versus tap water
Describing a binomial distribution

PROBLEM: Mr. Hogarth's AP® Statistics class did the activity on page 388. There were 21 students in the class. If we assume that the students in his class could *not* tell tap water from bottled water, then each one is guessing, with a 1/3 probability of being correct. Let X = the number of students who correctly identify the cup containing bottled water. Here is a histogram of the probability distribution of X:

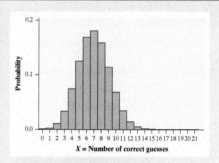

(a) What probability distribution does X have? Justify your answer.
(b) Describe the shape of the probability distribution.

SOLUTION:

(a) A trial consists of a student in the class trying to guess which of three cups contained bottled water.

- Binary? Success = correct guess; failure = incorrect guess
- Independent? Knowing whether one student guessed correctly does not help us predict whether another student guessed correctly.
- Number? $n = 21$
- Same probability? $p = 1/3$

X has a binomial distribution with $n = 21$ and $p = 1/3$.

(b) The probability distribution of X looks roughly symmetric with a single peak at $X = 7$.

> Check the BINS!

> You could also say that the graph is slightly right-skewed due to the long tail that extends out to $X = 21$.

FOR PRACTICE, TRY EXERCISE 91

MEAN AND STANDARD DEVIATION OF A BINOMIAL RANDOM VARIABLE

The random variable X = the number of children with type O blood from the previous two examples has a binomial distribution with $n = 5$ and $p = 0.25$. Its probability distribution is shown in the table.

Value x_i	0	1	2	3	4	5
Probability p_i	0.23730	0.39551	0.26367	0.08789	0.01465	0.00098

Because X is a discrete random variable, we can calculate its mean using the formula

$$\mu_X = E(X) = \sum x_i p_i = x_1 p_1 + x_2 p_2 + x_3 p_3 + \cdots$$

from Section 6.1. We get

$$\mu_X = (0)(0.23730) + (1)(0.39551) + \cdots + (5)(0.00098) = 1.25$$

So the expected number of children with type O blood in families like this one with 5 children is 1.25.

Did you think about why the mean is $\mu_X = 1.25$? Because each child has a 0.25 chance of inheriting type O blood, we'd expect one-fourth of the 5 children to have this blood type. In other words,

$$\mu_X = 5(0.25) = 1.25$$

This method can be used to find the mean of any binomial random variable.

MEAN (EXPECTED VALUE) OF A BINOMIAL RANDOM VARIABLE

If a count X of successes has a binomial distribution with number of trials n and probability of success p, the mean (expected value) of X is

$$\mu_X = E(X) = np$$

To calculate the standard deviation of X, we start by finding the variance.

$$\sigma_X^2 = \sum (x_i - \mu_X)^2 p_i$$
$$= (0 - 1.25)^2 (0.23730) + (1 - 1.25)^2 (0.39551) + \cdots + (5 - 1.25)^2 (0.00098)$$
$$= 0.9375$$

So the standard deviation of X is

$$\sigma_X = \sqrt{0.9375} = 0.968$$

The number of children with type O blood will typically vary by about 0.968 from the mean of 1.25 in families like this one with 5 children.

There is a simple formula for the standard deviation of a binomial random variable, but it isn't easy to explain (see the Think About It on page 445). For our family with $n = 5$ children and $p = 0.25$ of type O blood, the *variance* of X is

$$5(0.25)(0.75) = 0.9375$$

To get the standard deviation, we just take the square root:

$$\sigma_X = \sqrt{5(0.25)(0.75)} = \sqrt{0.9375} = 0.968$$

Teaching Tip: Using Technology

To find the mean and standard deviation of a binomial distribution, go to the Extra Applets on the Student Site at highschool .bfwpub.com/updatedtps6e and click on the *Probability* applet. In the dropdown menu, choose "Binomial distribution" and enter values for *n* and *p*.

AP® EXAM TIP

Let your students know that the formulas for the mean and standard deviation of a binomial distribution are included on the formula sheet provided to students on both sections of the AP® Statistics exam.

This method works for any binomial random variable.

STANDARD DEVIATION OF A BINOMIAL RANDOM VARIABLE

If a count X of successes has a binomial distribution with number of trials n and probability of success p, the standard deviation of X is

$$\sigma_X = \sqrt{np(1-p)}$$

 Remember that these formulas for the mean and standard deviation work only for binomial distributions. The interpretation of the parameters μ and σ is the same as for any discrete random variable.

EXAMPLE

Bottled water versus tap water
Describing a binomial distribution

PROBLEM: Assume that each of the 21 students in Mr. Hogarth's AP® Statistics class who did the bottled water versus tap water activity was just guessing, so there was a 1/3 chance of each student identifying the cup containing bottled water correctly. Let $X =$ the number of students who make a correct identification. At right is a histogram of the probability distribution of X.

(a) Calculate and interpret the mean of X.

(b) Calculate and interpret the standard deviation of X.

SOLUTION:

(a) $\mu_X = np = 21(1/3) = 7$

If all the students in Mr. Hogarth's class were just guessing and repeated the activity many times, the average number of students who guess correctly would be about 7.

(b) $\sigma_X = \sqrt{np(1-p)} = \sqrt{21(1/3)(2/3)} = 2.16$

If all the students in Mr. Hogarth's class were just guessing and repeated the activity many times, the number of students who guess correctly would typically vary by about 2.16 from the mean of 7.

FOR PRACTICE, TRY EXERCISE 95

Of the 21 students in Mr. Hogarth's class, 13 made correct identifications. Are you convinced that some of Mr. Hogarth's students could tell bottled water from tap water? The class's result corresponds to $X = 13$, a value that's nearly 3 standard

---

PROBLEM:

Mrs. Patenaude asked each of the 28 students in her AP® Statistics class to toss a coin and record the result. Assume that all the coins are fair. Let $X =$ the number of coins that landed heads. Here is a histogram of the probability distribution of X:

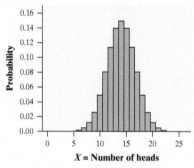

(a) Calculate and interpret the mean of X.
(b) Calculate and interpret the standard deviation of X.

SOLUTION:

(a) $\mu_X = np = 28(1/2) = 14$

If the coin tosses were repeated many times with groups of 28 students, the average number of students in the class who get heads would be about 14.

(b) $\sigma_X = \sqrt{np(1-p)} = \sqrt{28(1/2)(1/2)} = 2.65$

If the coin tosses were repeated many times with groups of 28 students, the number of students in the class who get heads would typically vary by about 2.65 from the mean of 14.

deviations above the mean. How likely is it that 13 or more of Mr. Hogarth's students would guess correctly? It's

$$P(X \geq 13) = 1 - P(X \leq 12)$$
$$= 1 - \text{binomcdf(trials:21, p:1/3, xvalue:12)}$$
$$= 1 - 0.9932$$
$$= 0.0068$$

The students had less than a 1% chance of getting so many correct identifications if they were all just guessing. This result gives convincing evidence that some of the students in the class could tell bottled water from tap water.

Think About It

WHERE DO THE BINOMIAL MEAN AND VARIANCE FORMULAS COME FROM? We can derive the formulas for the mean and variance of a binomial distribution using what we learned about combining random variables in Section 6.2. Let's start with the random variable B that's described by the following probability distribution.

Value b_i	0	1
Probability p_i	$1-p$	p

You can think of B as representing the result of a single trial of some random process. If a success occurs (probability p), then $B = 1$. If a failure occurs, then $B = 0$. Notice that the mean of B is

$$\mu_B = \sum b_i p_i = (0)(1-p) + (1)(p) = p$$

and that the variance of B is

$$\sigma_B^2 = \sum(b_i - \mu_B)^2 p_i = (0-p)^2(1-p) + (1-p)^2 p$$
$$= p^2(1-p) + (1-p)^2 p$$
$$= p(1-p)[p + (1-p)]$$
$$= p(1-p)$$

Now consider the random variable $X = B_1 + B_2 + \cdots + B_n$. We can think of X as counting the number of successes in n independent trials of this random process, with each trial having success probability p. In other words, X is a binomial random variable. By the rules from Section 6.2, the mean of X is

$$\mu_X = \mu_{B_1} + \mu_{B_2} + \cdots + \mu_{B_n} = p + p + \cdots + p = np$$

and the variance of X is

$$\sigma_X^2 = \sigma_{B_1}^2 + \sigma_{B_2}^2 + \cdots + \sigma_{B_n}^2$$
$$= p(1-p) + p(1-p) + \cdots + p(1-p)$$
$$= np(1-p)$$

The standard deviation of X is therefore

$$\sigma_X = \sqrt{np(1-p)}$$

Preparing for Inference

This example demonstrates the reasoning of a significance test from Chapter 9. The null hypothesis claim is that students were simply guessing to identify bottled water. Assuming the null hypothesis is true, the probability of getting 13 or more successes in 21 trials (0.0068) is the *P*-value. Because the *P*-value is small, we have convincing evidence that some of the students in the class could tell bottled water from tap water.

AP® EXAM TIP

The derivations of these two formulas are interesting, but they are not required for students to understand for the AP® Statistics exam.

Making Connections

We just derived the formulas for the mean and standard deviation for the distribution of the *count* of successes in n trials. In Chapter 7, we will use this formula to derive the mean and standard deviation for the distribution of the *proportion* of successes in n trials (by dividing the number of successes by n). This will give us the mean and standard deviation of the sampling distribution of a sample proportion:

$$\mu_{\hat{p}} = p \text{ and } \sigma_{\hat{p}} = \sqrt{\frac{p(1-p)}{n}}$$

✔ Answers to CYU

1.

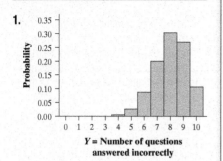

Y = Number of questions answered incorrectly

The probability distribution is skewed to the left with a single peak at $Y = 8$.

2. $\mu_Y = np = 10(0.80) = 8$; if many students took the quiz, the average number of questions students would answer incorrectly is about 8 questions.

3. $\sigma_Y = \sqrt{np(1-p)} =$ $\sqrt{10(0.80)(0.20)} = 1.265$; if many students took the quiz, we would expect individual students' scores to typically vary by about 1.265 incorrect answers from the mean of 8 incorrect answers.

4. The shape of X is skewed right and the shape of Y is skewed left, both with a single peak. The center (mean) of the distribution of X is 2, which is less than the center (mean) of the distribution of Y, which is 8. (Note that the sum of the means is 10 because $X + Y = 10$.) The variability of both probability distributions is the same (standard deviation = 1.265).

TRM Airline Overbooking, Part II

TRM A Fair Coin?

There are two additional resources you can use when teaching the binomial distribution. The "Airline Overbooking, Part II" problem is a continuation of the Airline Overbooking problem from Chapter 5, but you can use this one even if you didn't do the problem in the previous chapter. The "A Fair Coin?" problem reviews binomial probabilities and foreshadows the ideas of Type I and Type II errors. Find these in the Teacher's Resource Materials located in the TE-Book, on the TRFD, or in the Teacher's view on the book's digital platform.

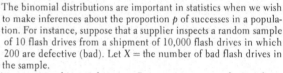

CHECK YOUR UNDERSTANDING

To introduce his class to binomial distributions, Mr. Miller does the "Pop quiz" activity at the beginning of this section (page 431). Each student in the class guesses an answer from A through E on each of the 10 multiple-choice questions. Mr. Miller determines the "correct" answer for each of the 10 questions randomly so that A, B, C, D, or E was equally likely to be chosen. Hannah is one of the students in this class. Let $Y =$ the number of questions that Hannah answers *incorrectly*.

1. Use technology to make a histogram of the probability distribution of Y. Describe its shape.
2. Calculate and interpret the mean of Y.
3. Calculate and interpret the standard deviation of Y.
4. On page 440, we defined $X =$ the number of *correct* answers that Hannah got on the quiz. How do the shape, center, and variability of the probability distribution of X compare to your answers for Questions 1 to 3?

Binomial Distributions in Statistical Sampling

The binomial distributions are important in statistics when we wish to make inferences about the proportion p of successes in a population. For instance, suppose that a supplier inspects a random sample of 10 flash drives from a shipment of 10,000 flash drives in which 200 are defective (bad). Let $X =$ the number of bad flash drives in the sample.

This is not quite a binomial setting. Because we are sampling without replacement, the independence condition is violated. The conditional probability that the second flash drive chosen is bad changes when we know whether the first is good or bad: $P(\text{second is bad} \mid \text{first is good}) = 200/9999 = 0.0200$ but $P(\text{second is bad} \mid \text{first is bad}) = 199/9999 = 0.0199$. These probabilities are very close because removing 1 flash drive from a shipment of 10,000 changes the makeup of the remaining 9999 flash drives very little. The distribution of X is very close to the binomial distribution with $n = 10$ and $p = 0.02$.

To illustrate this, let's compute the probability that none of the 10 flash drives is defective. Using the binomial distribution, it's

$$P(X = 0) = \binom{10}{0}(0.02)^0(0.98)^{10} = 0.8171$$

The actual probability of getting no defective flash drives is

$$P(\text{no defectives}) = \frac{9800}{10,000} \times \frac{9799}{9999} \times \frac{9798}{9798} \times \cdots \times \frac{9791}{9991} = 0.8170$$

Those two probabilities are quite close!

Almost all real-world sampling, such as taking an SRS from a population of interest, is done without replacement. As the preceding example illustrates, sampling without replacement leads to a violation of the Independent condition.

However, the flash drives context shows how we can use binomial distributions in the statistical setting of selecting a random sample. When the population is much larger than the sample, a count of successes in an SRS of size n has approximately the binomial distribution with n equal to the sample size and p equal to the proportion of successes in the population. What counts as "much larger"? In practice, the binomial distribution gives a good approximation as long as we sample less than 10% of the population. We refer to this as the **10% condition**.

> **DEFINITION   10% condition**
>
> When taking a random sample of size n from a population of size N, we can treat individual observations as independent when performing calculations as long as $n < 0.10N$.

Here's a scenario that shows why it's important to check the 10% condition before calculating a binomial probability. You might recognize the setting from the first activity in the book (page 6).

An airline has just finished training 25 pilots— 15 male and 10 female—to become captains. Unfortunately, only 8 captain positions are available right now. Airline managers announce that they will use a lottery to determine which pilots will fill the available positions. One day later, managers reveal the results of the lottery: Of the 8 captains chosen, 5 are female and 3 are male. Some of the male pilots who weren't selected suspect that the lottery was not carried out fairly.

What's the probability of choosing 5 female pilots in a fair lottery? Let $X =$ the number of female pilots selected in a random sample of size $n = 8$ from the population of $N = 25$ pilots. Notice that the sample size is almost 1/3 of the population size. If we ignore this fact and use a binomial probability calculation, we get

$$P(X = 5) = \binom{8}{5}(0.40)^5(0.60)^3 = 0.124$$

The correct probability, however, is 0.106. You can see that the binomial probability is off by about 17% (0.018/0.106) from the correct answer.

EXAMPLE

Teens and debit cards
Binomial distributions and sampling

PROBLEM: In a survey of 500 U.S. teenagers aged 14 to 18, subjects were asked a variety of questions about personal finance.[8] One question asked whether teens had a debit card. Suppose that exactly 12% of teens aged 14 to 18 have debit cards. Let $X =$ the number of teens in a random sample of size 500 who have a debit card.

Making Connections

Because all real-world sampling is done without replacement, we will always have to check the 10% condition when sampling without replacement in Chapters 6–12.

Teaching Tip

Some students mistakenly believe that the 10% condition implies that we want *small* samples. This is almost never the case. If we have a sample that is larger than 10% of the population, it means just that we shouldn't use the binomial distribution. We should use the hypergeometric distribution instead (which isn't covered on the AP® Statistics exam).

Teaching Tip

The correct probability of choosing 5 females and 3 males is given by

$$\frac{_{10}C_5 \cdot {}_{15}C_3}{_{25}C_8} = 0.106$$

ALTERNATE EXAMPLE   Skill 3.A

Teens and Facebook   Binomial distributions and sampling

PROBLEM:
A recent report from the Pew Research Center estimates that 71% of teenagers aged 13–17 use Facebook. Assume this claim is true. Suppose that some researchers are going to contact a random sample of 300 teenagers to find out if they use Facebook. Let $X =$ the number of teens in a random sample of size 300 who use Facebook.
(a) Explain why X can be modeled by a binomial distribution even though the sample was selected without replacement.
(b) Use a binomial distribution to estimate the probability that 200 or more teens in the sample use Facebook.

SOLUTION:
(a) 300 is less than 10% of all U.S. teenagers aged 13–17.
(b) X is approximately binomial with $n = 300$ and $p = 0.71$.
$P(X \geq 200) = 1 - P(X \leq 199) = 1 -$ binomcdf(trials: 300, p: 0.71, x value: 199)
$\qquad = 1 - 0.044 = 0.956$

(a) Explain why X can be modeled by a binomial distribution even though the sample was selected without replacement.

(b) Use a binomial distribution to estimate the probability that 50 or fewer teens in the sample have debit cards.

SOLUTION:

(a) 500 is less than 10% of all U.S. teenagers aged 14 to 18.

(b) X is approximately binomial with n = 500 and p = 0.12.

$$P(X \leq 50) = \text{binomcdf}(\text{trials: } 500, p: 0.12, x \text{ value: } 50)$$
$$= 0.0932$$

> Check the 10% condition:
> $n < 0.10N$

FOR PRACTICE, TRY EXERCISE 99

The Normal Approximation to Binomial Distributions

As you saw earlier, the shape of a binomial distribution can be skewed to the right, skewed to the left, or roughly symmetric. Something interesting happens to the shape as the number of trials *n* increases. You can investigate the relationship between *n* and *p* yourself using the *Normal Approximation to Binomial Distributions* applet at the book's website, highschool.bfwpub.com/updatedtps6e.

Figure 6.9 shows histograms of binomial distributions for different values of *n* and *p*. As the number of observations *n* becomes larger, the binomial distribution gets close to a Normal distribution.

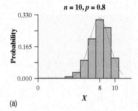
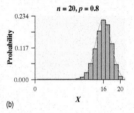
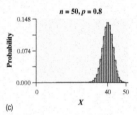

FIGURE 6.9 Histograms of binomial distributions with (a) $n = 10$ and $p = 0.8$, (b) $n = 20$ and $p = 0.8$, and (c) $n = 50$ and $p = 0.8$. As *n* increases, the shape of the probability distribution gets closer and closer to Normal.

When *n* is large, we can use Normal probability calculations to approximate binomial probabilities. To see if *n* is large enough, check the **Large Counts condition**.

> **DEFINITION   Large Counts condition**
>
> Suppose that a count *X* of successes has the binomial distribution with *n* trials and success probability *p*. The **Large Counts condition** says that the probability distribution of *X* is approximately Normal if
>
> $$np \geq 10 \quad \text{and} \quad n(1-p) \geq 10$$
>
> That is, the expected numbers (counts) of successes and failures are both at least 10.

This condition is called "large counts" because np is the expected (mean) count of successes and $n(1-p)$ is the expected (mean) count of failures in a binomial setting. Why do we require that both these values be at least 10? Look back at Figure 6.9. It is clear that a Normal distribution does not approximate the probability distributions in parts (a) or (b) very well. This isn't surprising because the Large Counts condition is not met in either case. For graph (a), both $np = 10(0.8) = 8$ and $n(1-p) = 10(0.2) = 2$ are less than 10. For graph (b), $np = 20(0.8) = 16 \geq 10$, but $n(1-p) = 20(0.2) = 4$ is less than 10. The Normal curve in graph (c) appears to model the binomial probability distribution well. This time, the Large Counts condition is met: $np = 50(0.8) = 40 \geq 10$ and $n(1-p) = 50(0.2) = 10 \geq 10$.

The accuracy of the Normal approximation improves as the sample size n increases. It is most accurate for any fixed n when p is close to 1/2 and least accurate when p is near 0 or 1. This is why the Large Counts condition depends on p as well as n.

EXAMPLE

Teens and debit cards
Normal approximation to a binomial distribution

PROBLEM: In a survey of 500 U.S. teenagers aged 14 to 18, subjects were asked a variety of questions about personal finance. One question asked whether teens had a debit card. Suppose that exactly 12% of teens aged 14 to 18 have debit cards. Let $X =$ the number of teens in a random sample of size 500 who have a debit card.

(a) Justify why X can be approximated by a Normal distribution.

(b) Use a Normal distribution to estimate the probability that 50 or fewer teens in the sample have debit cards.

SOLUTION:

(a) X is approximately binomial with $n = 500$ and $p = 0.12$. Because $np = 500(0.12) = 60 \geq 10$ and $n(1-p) = 500(0.88) = 440 \geq 10$, we can approximate X with a Normal distribution.

(b) $\mu_X = np = 500(0.12) = 60$ and
$\sigma_X = \sqrt{np(1-p)} = \sqrt{500(0.12)(0.88)} = 7.266$

> Start by calculating the mean and standard deviation of the binomial random variable X.

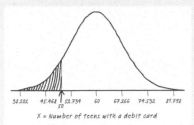
X = Number of teens with a debit card

> 1. Draw a Normal distribution.
> 2. Perform calculations—show your work!
>   (i) Standardize and use Table A or technology; or
>   (ii) Use technology without standardizing.
> Be sure to answer the question that was asked.

(i) $z = \dfrac{50-60}{7.266} = -1.38$

Using Table A: 0.0838

Using technology: normalcdf(lower: -1000, upper: -1.38, mean: 0, SD: 1) $= 0.0838$

> $P(X \leq 50) = P(Z \leq -1.38)$

(ii) normalcdf(lower: 0, upper: 50, mean: 60, SD: 7.266) $= 0.0844$

FOR PRACTICE, TRY EXERCISE 103

ALTERNATE EXAMPLE　　　Skill 3.A

Teens still on Facebook
Normal approximation to a binomial distribution

PROBLEM:
Recently, the Pew Research Center estimated that 71% of teenagers aged 13–17 use Facebook. Assume this claim is true. Suppose that some researchers are going to contact a random sample of 300 teenagers to find out if they use Facebook. Let $X =$ the number of teens in a random sample of size 300 who use Facebook.
(a) Justify why X can be approximated by a Normal distribution.
(b) Use a Normal distribution to estimate the probability that 200 or more teens in the sample use Facebook.

SOLUTION:
(a) $np = 300(0.71) = 213 \geq 10$ and $n(1-p) = 300(0.29) = 87 \geq 10$

(b) $\mu_X = np = 300(0.71) = 213$ and $\sigma_X = \sqrt{np(1-p)} = \sqrt{300(0.71)(0.29)} = 7.859$

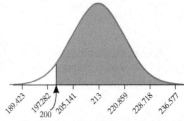
X = **Number of teens who use Facebook**

(i) $z = \dfrac{200-213}{7.859} = -1.65$

Table A: $1 - 0.0495 = 0.9505$
Tech: normalcdf(lower: -1.65, upper: 1000, mean: 0, SD: 1) $= 0.9505$
(ii) normalcdf(lower: 200, upper: 1000, mean: 213, SD: 7.859) $= 0.9510$

Notice the answer here (0.9510) is very close to the answer we got using the binomial distribution (0.956) in the preceding Alternate Example.

Teaching Tip

The geometric setting is similar to the binomial setting in that outcomes are binary, trials are independent, and the probability of success is the same for each trial. The important distinction is that while in a binomial setting the number of trials is fixed, a geometric setting does not have a fixed number of trials. A geometric random variable counts the number of trials it takes to get one success. This is why a geometric distribution is often referred to as a "wait time" distribution, because we are waiting for a success to occur.

 ACTIVITY OVERVIEW

To prepare for using this activity, watch the overview video by clicking on the link in the TE-Book, opening the TRFD, or downloading from the Teacher's Resources on the book's digital platform.

Time: 5 minutes

Materials: None

Teaching Advice: This is a good activity for introducing the idea of a geometric random variable, which measures the number of trials it takes to get a success (guessing the Lucky Day). In this setting, unlike the binomial setting, there is not a fixed number of trials, because we don't know how many trials it will take to find the first success.

Answers:

1. The class should choose to "gamble" because the expected number of homework problems with this strategy is less than 10.

2. Answers will vary.

The probability from the Normal approximation in this example, 0.0844, misses the exact binomial probability of 0.0932 from the preceding example by about 0.0088.

Geometric Random Variables

In a binomial setting, the number of trials n is fixed in advance, and the binomial random variable X counts the number of successes. The possible values of X are 0, 1, 2, . . . , n. In other situations, the goal is to repeat a random process *until a success occurs*:

- Roll a pair of dice until you get doubles.
- In basketball, attempt a 3-point shot until you make one.
- Keep placing a \$1 bet on the number 15 in roulette until you win.

These are all examples of a **geometric setting**.

> **DEFINITION** **Geometric setting**
>
> A **geometric setting** arises when we perform independent trials of the same random process and record the number of trials it takes to get one success. On each trial, the probability p of success must be the same.

Here's an activity your class can try that involves a geometric setting.

ACTIVITY Is this your lucky day?

Your teacher is planning to give you 10 problems for homework. As an alternative, you can agree to play the Lucky Day game. Here's how it works. A student will be selected at random from your class and asked to pick a day of the week (e.g., Thursday). Then your teacher will use technology to randomly choose a day of the week as the "lucky day." If the student picks the correct day, the class will have only one homework problem. If the student picks the wrong day, your teacher will select another student from the class at random. The chosen student will pick a day of the week and your teacher will use technology to choose a "lucky day." If this student gets it right, the class will have two homework problems. The game continues until a student correctly picks the lucky day. Your teacher will assign a number of homework problems that is equal to the total number of picks made by members of your class. Are you ready to play the Lucky Day game?

1. Decide as a class whether to "gamble" on the number of homework problems you will receive. You have 30 seconds.

2. Play the Lucky Day game and see what happens!

In a geometric setting, if we define the random variable X to be the number of trials needed to get the first success, then X is called a **geometric random variable**. The probability distribution of X is a **geometric distribution**.

DEFINITION   Geometric random variable, Geometric distribution

The number of trials X that it takes to get a success in a geometric setting is a **geometric random variable**. The probability distribution of X is a **geometric distribution** with probability p of success on any trial. The possible values of X are 1, 2, 3, ...

As with binomial random variables, it's important to be able to distinguish situations in which a geometric distribution does and doesn't apply. Let's consider the Lucky Day game. The random variable of interest in this game is $X =$ the number of picks it takes to correctly match the lucky day. Each pick is one trial of the random process. Knowing the result of one student's pick tells us nothing about the result of any other pick. On each trial, the probability of a correct pick is $1/7$. This is a geometric setting. Because X counts the number of trials to get the first success, it is a geometric random variable with $p = 1/7$.

What is the probability that the first student picks correctly and wins the Lucky Day game? It's $P(X = 1) = 1/7$. That's also the class's chance of having only one homework problem assigned. For the class to have two homework problems assigned, the first student selected must pick an incorrect day of the week and the second student must pick the lucky day correctly. The probability that this happens is

$$P(X = 2) = (6/7)(1/7) = 0.1224$$

Likewise,

$$P(X = 3) = (6/7)(6/7)(1/7) = 0.1050$$

In general, the probability that the first correct pick comes on the xth trial is

$$P(X = x) = (6/7)^{x-1}(1/7)$$

Let's summarize what we've learned about calculating a geometric probability.

GEOMETRIC PROBABILITY FORMULA

If X has the geometric distribution with probability p of success on each trial, the possible values of X are 1, 2, 3, ... If x is any one of these values,

$$P(X = x) = (1 - p)^{x-1} p$$

With the geometric probability formula in hand, we can now compute any geometric probability.

EXAMPLE

The Lucky Day game
Calculating geometric probabilities

PROBLEM: Mr. Lochel's class decides to play the Lucky Day game. Let $X =$ the number of homework problems that the class receives.
(a) Find the probability that the class receives exactly 10 homework problems as a result of playing the Lucky Day game.
(b) Find $P(X < 10)$ and interpret this value.

Skills 3.A, 4.B

Play until you win   Calculating geometric probabilities

PROBLEM:
The Australian Official Lottery has scratch-off instant lottery tickets that can be purchased for \$1. The probability of winning a prize is 1 in 4. Mr. Urban decides that he will buy lottery tickets until he finds a winner, and then he will quit.
(a) Find the probability Mr. Urban buys exactly 8 lottery tickets in order to get the first winner.
(b) Find the probability that Mr. Urban will buy at least 5 lottery tickets to get a winner.

SOLUTION:
Let $X =$ the number of lottery tickets needed to get the first winner. X has a geometric distribution with $p = 1/4$.
(a) $P(X = 8) = (3/4)^{7}(1/4) = 0.0334$

(b) $P(X \le 4) = P(X = 1) + P(X = 2) + P(X = 3) + P(X = 4)$

$= \left(\dfrac{1}{4}\right) + \left(\dfrac{3}{4}\right)^{1}\left(\dfrac{1}{4}\right) + \left(\dfrac{3}{4}\right)^{2}\left(\dfrac{1}{4}\right) +$

$\left(\dfrac{3}{4}\right)^{3}\left(\dfrac{1}{4}\right) = 0.6836$

$P(X \ge 5) = 1 - P(X \le 4) = 1 - 0.6836 = 0.3164$

There's about a 32% probability that Mr. Urban will buy at least 5 lottery tickets to get a winner. *Note:* This calculation can also be done by noticing that, for $X \ge 5$, the first four trials must be failures: $\left(\dfrac{3}{4}\right)^{4} = 0.3164$.

Students will often incorrectly introduce the binomial coefficient:

$$P(Y = 10) = \binom{10}{1}(6/7)^9(1/7) = 0.0357$$

The binomial coefficient is not required here because there is only one way to have 9 failures followed by 1 success (FFFFFFFFFS).

Teaching Tip

If students are feeling overwhelmed by calculator commands, you can skip over the geometpdf and geometcdf commands and tell them to use just the complement rule and multiplication rule for independent events. If you do teach these commands, remind students of the difference between pdf and cdf (the *p* stands for "probability" and the *c* stands for "cumulative"). Also, remind students that they must label the inputs of their calculator command if they want to avoid losing points.

SOLUTION:

X has a geometric distribution with $p = 1/7$.

(a) $P(X = 10) = (6/7)^9(1/7) = 0.0357$

(b) $P(X < 10) = P(X = 1) + P(X = 2) + P(X = 3) + \cdots + P(X = 9)$
$= 1/7 + (6/7)(1/7) + (6/7)^2(1/7) + \cdots + (6/7)^8(1/7)$
$= 0.7503$

There's about a 75% probability that the class will get fewer than 10 homework problems by playing the Lucky Day game.

FOR PRACTICE, TRY EXERCISE 107

There's a clever alternative approach to finding the probability in part (b) of the example. By the complement rule, $P(X < 10) = 1 - P(X \geq 10)$. What's the probability that it will take at least 10 picks for Mr. Lochel's class to win the Lucky Day game? It's the chance that the first 9 picks are all incorrect: $\left(\frac{6}{7}\right)^9 = 0.250$. So the probability that the class will win the Lucky Day game in fewer than 10 picks (and therefore have fewer than 10 homework problems assigned) is

$$P(X < 10) = 1 - P(X \geq 10) = 1 - 0.250 = 0.750$$

As you probably guessed, we can use technology to calculate geometric probabilities. The following Technology Corner shows how to do it.

17. Technology Corner CALCULATING GEOMETRIC PROBABILITIES

TI-Nspire and other technology instructions are on the book's website at highschool.bfwpub.com/updatedtps6e.

There are two handy commands on the TI-83/84 for finding geometric probabilities: geometpdf and geometcdf. The inputs for both commands are the success probability *p* and the value(s) of interest for the geometric random variable *X*.

$$\text{geometpdf }(p,x) \text{ computes } P(X = x)$$
$$\text{geometcdf }(p,x) \text{ computes } P(X \leq x)$$

Let's use these commands to confirm our answers in the previous example.

(a) Find the probability that the class receives exactly 10 homework problems as a result of playing the Lucky Day game.

- Press [2nd] [VARS] (DISTR) and choose geometpdf(.

  OS 2.55 or later: In the dialog box, enter these values: p:1/7, x value:10, choose Paste, and then press [ENTER].

  Older OS: Complete the command geometpdf(1/7,10) and press [ENTER].

```
NORMAL FLOAT AUTO REAL RADIAN MP
geometpdf(1/7,10)
            0.0356763859
```

These results agree with our previous answer using the geometric probability formula: 0.0357.

(b) Find $P(X < 10)$ and interpret this value. To find $P(X < 10)$, use the geometcdf command:

$$P(X < 10) = P(X \le 9) = \text{geometcdf}\,(1/7,9)$$

- Press [2nd] [VARS] (DISTR) and choose geometcdf.

  OS 2.55 or later: In the dialog box, enter these values: p:1/7, x value:9, choose Paste, and then press [ENTER].

  Older OS: Complete the command geometcdf(1/7,9) and press [ENTER].

These results agree with our previous answer using the geometric probability formula: 0.7503.

DESCRIBING A GEOMETRIC DISTRIBUTION: SHAPE, CENTER, AND VARIABILITY The table shows part of the probability distribution of $X =$ the number of picks it takes to match the lucky day. We can't show the entire distribution because the number of trials it takes to get the first success could be a very large number.

Value x_i	1	2	3	4	5	6	7	8	9	...
Probability p_i	0.143	0.122	0.105	0.090	0.077	0.066	0.057	0.049	0.042	

Figure 6.10 is a histogram of the probability distribution for values of X from 1 to 26. Let's describe what we see.

Shape: Skewed to the right. Every geometric distribution has this shape. That's because the most likely value of a geometric random variable is 1. The probability of each successive value decreases by a factor of $(1 - p)$.

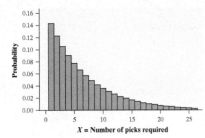

FIGURE 6.10 Histogram showing the probability distribution of the geometric random variable $X =$ number of trials needed for students to pick correctly in the Lucky Day game.

Center: The mean (expected value) of X is $\mu_X = 7$. If the class played the Lucky Day game many times, they would receive an average of 7 homework problems. It's no coincidence that $p = 1/7$ and $\mu_X = 7$. With probability of success 1/7 on each trial, we'd expect it to take an average of 7 trials to get the first success. That is, $\mu_X = 1/(1/7) = 7$.

Variability: The standard deviation of X is $\sigma_X = 6.48$. If the class played the Lucky Day game many times, the number of homework problems they receive would typically vary by about 6.5 problems from the mean of 7. That could mean a lot of homework! There is a simple formula for the standard deviation of a geometric random variable, but it isn't easy to explain. For the Lucky Day game,

$$\sigma_X = \frac{\sqrt{1 - 1/7}}{1/7} = 6.48$$

We can generalize these results for the mean and standard deviation of a geometric random variable.

Making Connections

Students have likely studied arithmetic and geometric sequences in a prior math class. A geometric sequence is defined as having a common ratio. The common ratio here is $(1 - p)$. Another interesting connection is thinking about the sum of all the probabilities in a geometric distribution. Students might know the sum of an infinite geometric series is $\dfrac{a_1}{1 - r}$, which in this context is

$$\frac{p}{1 - (1 - p)} = \frac{p}{p} = 1.$$

Teaching Tip

The factor of $(1 - p)$ here will always be a value less than 1, meaning that each successive bar in the histogram will be shorter than the previous one. Thus, the graph of a probability distribution of a geometric random variable will always be unimodal and skewed to the right, with the peak at $X = 1$. Binomial distributions, on the other hand, can be skewed to the right, skewed to the left, or roughly symmetric.

ALTERNATE EXAMPLE Skills 3.B, 4.B

Play some more

Mean and SD of a geometric distribution

PROBLEM:

At the beginning of class, Mrs. Gallas plays a simple game in which students earn extra credit. Each round, she rolls a single die. If the die shows a 1–5, the game continues to the next round. If the die shows a 6, the game is over, and the class earns as many extra credit points as the number of rounds played.

(a) Calculate the mean number of extra credit points a class earns on a randomly selected day. Interpret this value.

(b) Calculate the standard deviation of the number of extra credit points a class earns on a randomly selected day. Interpret this value.

SOLUTION:

Let X = the number of extra credit points a class earns on a randomly selected day. X has a geometric distribution with $p = 1/6$.

(a) $\mu_X = \dfrac{1}{(1/6)} = 6$

If Mrs. Gallas plays this game on many days, the average number of points a class earns would be about 6.

(b) $\sigma_X = \dfrac{\sqrt{1-1/6}}{1/6} = 5.48$

If Mrs. Gallas plays this game on many days, the number of points a class earns would typically vary by about 5.48 from the mean of 6.

MEAN (EXPECTED VALUE) AND STANDARD DEVIATION OF A GEOMETRIC RANDOM VARIABLE

If X is a geometric random variable with probability of success p on each trial, then its mean (expected value) is $\mu_X = E(X) = \dfrac{1}{p}$ and its standard deviation is

$$\sigma_X = \frac{\sqrt{1-p}}{p}.$$

We interpret the parameters μ and σ in the same way as for any discrete random variable.

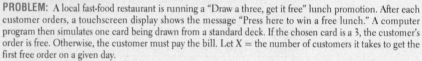

EXAMPLE

Waiting for a free lunch

Mean and SD of a geometric distribution

PROBLEM: A local fast-food restaurant is running a "Draw a three, get it free" lunch promotion. After each customer orders, a touchscreen display shows the message "Press here to win a free lunch." A computer program then simulates one card being drawn from a standard deck. If the chosen card is a 3, the customer's order is free. Otherwise, the customer must pay the bill. Let X = the number of customers it takes to get the first free order on a given day.

(a) Calculate and interpret the mean of X.

(b) Calculate and interpret the standard deviation of X.

SOLUTION:

X has a geometric distribution with $p = 4/52$.

(a) $\mu_X = \dfrac{1}{(4/52)} = 13$

If the restaurant runs this lunch promotion on many days, the average number of customers it takes to get the first free order would be about 13.

(b) $\sigma_X = \dfrac{\sqrt{1-4/52}}{4/52} = 12.49$

If the restaurant runs this lunch promotion on many days, the number of customers it takes to get the first free order would typically vary by about 12.5 from the mean of 13.

FOR PRACTICE, TRY EXERCISE 111

CHECK YOUR UNDERSTANDING

Suppose you roll a pair of fair, six-sided dice until you get doubles. Let T = the number of rolls it takes. Note that the probability of getting doubles on any roll is $6/36 = 1/6$.

1. Show that T is a geometric random variable.
2. Find $P(T = 3)$. Interpret this result.
3. In the game of Monopoly, a player can get out of jail free by rolling doubles within 3 turns. Find the probability that this happens.
4. Calculate the mean and standard deviation of T. Interpret these parameters.

Answers to CYU

1. Die rolls are independent; the probability of getting doubles is the same on each roll (1/6), and we are repeating the random process until we get a success (doubles). This is a geometric setting and T is a geometric random variable with $p = \dfrac{1}{6}$.

2. $P(T = 3) = \left(\dfrac{5}{6}\right)^2\left(\dfrac{1}{6}\right) = 0.1157$; the probability is 11.57% that you will get the first set of doubles on the third roll of the dice.

3. $P(T \leq 3) = \dfrac{1}{6} + \left(\dfrac{5}{6}\right)\left(\dfrac{1}{6}\right) + \left(\dfrac{5}{6}\right)^2\left(\dfrac{1}{6}\right) = 0.4213$; the probability is 42.13% of getting doubles in three or fewer rolls.

Section 6.3 | Summary

- A **binomial setting** arises when we perform n independent trials of the same random process and count the number of times that a particular outcome (a "success") occurs. The conditions for a binomial setting are:
  - **B**inary? The possible outcomes of each trial can be classified as "success" or "failure."
  - **I**ndependent? Trials must be independent. That is, knowing the result of one trial must not tell us anything about the result of any other trial.
  - **N**umber? The number of trials n of the random process must be fixed in advance.
  - **S**ame probability? There is the same probability of success p on each trial.

  Remember to check the BINS!
- The count of successes X in a binomial setting is a special type of discrete random variable known as a **binomial random variable.** Its probability distribution is a **binomial distribution.** Any binomial distribution is completely specified by two numbers: the number of trials n of the random process and the probability of success p on any trial. The possible values of X are the whole numbers $0, 1, 2, \ldots, n$.
- Use the binomial probability formula to calculate the probability of getting exactly x successes in n trials:

$$P(X = x) = \binom{n}{x} p^x (1-p)^{n-x}$$

  - The **binomial coefficient**

$$\binom{n}{x} = \frac{n!}{x!(n-x)!}$$

    counts the number of ways x successes can be arranged among n trials.
  - The factorial of n is
$$n! = n(n-1)(n-2) \cdots (3)(2)(1)$$
    for positive whole numbers n, and $0! = 1$.
- You can also use technology to calculate binomial probabilities. The TI-83/84 command binompdf(n,p,x) computes $P(X = x)$. The TI-83/84 command binomcdf(n,p,x) computes the cumulative probability $P(X \le x)$.
- A binomial distribution can have a shape that is roughly symmetric, skewed to the right, or skewed to the left.
- The mean and standard deviation of a binomial random variable X are

$$\mu_X = np \text{ and } \sigma_X = \sqrt{np(1-p)}$$

- The binomial distribution with n trials and probability p of success gives a good approximation to the count of successes in a random sample of size n from a large population containing proportion p of successes. This is true as long as the sample size n is less than 10% of the population size N (the **10% condition**). When the 10% condition is met, we can view individual observations as independent.

Answers to Section 6.3 Exercises

6.77 *Binary?* "Success" = survive and "Failure" = does not survive. *Independent?* Yes, knowing the outcome of one elk shouldn't tell us anything about the outcomes of other elk. *Number?* $n = 7$. *Same probability?* $p = 0.44$; X has a binomial distribution with $n = 7$ and $p = 0.44$.

6.78 *Binary?* "Success" = name has more than 6 letters and "Failure" = name has 6 letters or fewer. *Independent?* No, we are selecting without replacement from a small number of students. *Number?* $n = 4$. *Same probability?* The (unconditional) probability that a randomly chosen student's name has more than 6 letters is constant. X does not have a binomial distribution.

6.79 *Binary?* "Success" = hits and "Failure" = does not hit. *Independent?* Yes, the outcome of one shot does not tell us anything about the outcome of other shots. *Number?* No, there is not a fixed number of trials. Y does not have a binomial distribution.

- The Normal approximation to the binomial distribution says that if X is a count of successes having the binomial distribution with n trials and success probability p, then when n is large, X is approximately Normally distributed. You can use this approximation when $np \geq 10$ and $n(1-p) \geq 10$ (the **Large Counts condition**).

- A **geometric setting** consists of repeated trials of the same random process in which the probability p of success is the same on each trial, and the goal is to count the number of trials it takes to get one success. If $X =$ the number of trials required to obtain the first success, then X is a **geometric random variable**. Its probability distribution is called a **geometric distribution.**

- If X has the geometric distribution with probability of success p, the possible values of X are the positive integers 1, 2, 3, ... The probability that it takes exactly x trials to get the first success is given by

$$P(X = x) = (1-p)^{x-1}p$$

- You can also use technology to calculate geometric probabilities. The TI-83/84 command geometpdf(p,x) computes $P(X = x)$. The TI-83/84 command geometcdf(p,x) computes the cumulative probability $P(X \leq x)$.

- The mean and standard deviation of a geometric random variable X are

$$\mu_X = \frac{1}{p} \text{ and } \sigma_X = \frac{\sqrt{1-p}}{p}$$

6.3 Technology Corners

TI-Nspire and other technology instructions are on the book's website at highschool.bfwpub.com/updatedtps6e.

Section 6.3 | Exercises

In Exercises 77–80, determine whether the given scenario describes a binomial setting. Justify your answer.

77. **Baby elk** Biologists estimate that a randomly selected baby elk has a 44% chance of surviving to adulthood. Assume this estimate is correct. Suppose researchers choose 7 baby elk at random to monitor. Let $X =$ the number that survive to adulthood.

78. **Long or short?** Put the names of all the students in your statistics class in a hat. Mix up the names, and draw 4 without looking. Let $X =$ the number whose last names have more than six letters.

79. **Bull's-eye!** Lawrence likes to shoot a bow and arrow in his free time. On any shot, he has about a 10% chance of hitting the bull's-eye. As a challenge one day, Lawrence decides to keep shooting until he gets a bull's-eye. Let $Y =$ the number of shots he takes.

80. **Taking the train** According to New Jersey Transit, the 8:00 A.M. weekday train from Princeton to New York City has a 90% chance of arriving on time on a randomly selected day. Suppose this claim is true. Choose 6 days at random. Let $Y =$ the number of days on which the train arrives on time.

6.80 *Binary?* "Success" = on time and "Failure" = late. *Independent?* Because the days were randomly selected, the arrival times are independent. *Number?* $n = 6$. *Same probability?* $p = 0.90$; Y has a binomial distribution with $n = 6$ and $p = 0.90$.

6.81 X has a binomial distribution with $n = 7$ and $p = 0.44$;

$$P(X = 4) = \binom{7}{4}(0.44)^4(0.56)^3 = 0.2304.$$

Tech: $P(Y = 4) =$ binompdf(trials: 7, p: 0.44, x value: 4) $= 0.2304$. There is a 23.04% probability that exactly 4 of the 7 elk survive to adulthood.

81. Baby elk Refer to Exercise 77. Use the binomial
pg 437 probability formula to find $P(X = 4)$. Interpret this
value.

82. Taking the train Refer to Exercise 80. Use the
binomial probability formula to find $P(Y = 4)$.
Interpret this value.

83. Take a spin An online spinner has two colored
regions—blue and yellow. According to the website,
the probability that the spinner lands in the blue
region on any spin is 0.80. Assume for now that this
claim is correct. Suppose we
spin the spinner 12 times and
let X = the number of times it
lands in the blue region.

(a) Explain why X is a binomial
random variable.

(b) Find the probability that exactly
8 spins land in the blue region.

84. Red light! Pedro drives the
same route to work on Monday through Friday. His
route includes one traffic light. According to the local
traffic department, there is a 55% chance that the light
will be red on a randomly selected work day. Suppose
we choose 10 of Pedro's work days at random and let
Y = the number of times that the light is red.

(a) Explain why Y is a binomial random variable.

(b) Find the probability that the light is red on exactly 7 days.

85. Baby elk Refer to Exercise 77. How surprising would
pg 437 it be for more than 4 elk in the sample to survive to
adulthood? Calculate an appropriate probability to
support your answer.

86. Taking the train Refer to Exercise 80. Would you be
surprised if the train arrived on time on fewer than
4 days? Calculate an appropriate probability to support
your answer.

87. Take a spin Refer to Exercise 83. Calculate and
interpret $P(X \le 7)$.

88. Red light! Refer to Exercise 84. Calculate and
interpret $P(Y \ge 7)$.

89. The last kiss Do people have a preference for the
pg 439 last thing they taste? Researchers at the University of
Michigan designed a study to find out. The researchers
gave 22 students five different Hershey's Kisses (milk
chocolate, dark chocolate, crème, caramel, and
almond) in random order and asked the student to rate
each one. Participants were not told how many Kisses
they would be tasting. However, when the 5th and final
Kiss was presented, participants were told that it would
be their last one.[9] Assume that the participants in the

study don't have a special preference for the last thing
they taste. That is, assume that the probability a person
would prefer the last Kiss tasted is $p = 0.20$.

(a) Find the probability that 14 or more students would
prefer the last Kiss tasted.

(b) Of the 22 students, 14 gave the final Kiss the highest rating.
Does this give convincing evidence that the participants
have a preference for the last thing they taste?

90. 1 in 6 wins As a special promotion for its 20-ounce
bottles of soda, a soft drink company printed a message
on the inside of each bottle cap. Some of the caps said,
"Please try again!" while others said, "You're a winner!"
The company advertised the promotion with the
slogan "1 in 6 wins a prize." Grayson's statistics class
wonders if the company's claim holds true at a nearby
convenience store. To find out, all 30 students in the
class go to the store and each buys one 20-ounce bottle
of the soda.

(a) Find the probability that two or fewer students would
win a prize if the company's claim is true.

(b) Two of the students in Grayson's class got caps that say,
"You're a winner!" Does this result give convincing
evidence that the company's 1-in-6 claim is false?

91. Bag check Thousands of travelers pass through the
pg 442 airport in Guadalajara, Mexico, each day. Before leaving
the airport, each passenger must go through the customs
inspection area. Customs agents want to be sure that
passengers do not bring illegal items into the country.
But they do not have time to search every traveler's
luggage. Instead, they require each person to press a
button. Either a red or a green bulb lights up. If the red
light flashes, the passenger will be searched by customs
agents. A green light means "go ahead." Customs agents
claim that the light has probability 0.30 of showing red
on any push of the button. Assume for now that this
claim is true. Suppose we watch 20 passengers press the
button. Let R = the number who get a red light. Here is
a histogram of the probability distribution of R:

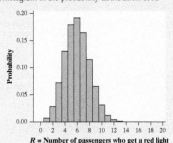

R = **Number of passengers who get a red light**

6.82 Y has a binomial distribution with $n = 6$
and $p = 0.90$;

$$P(Y = 4) = \binom{6}{4}(0.90)^4(0.10)^2 = 0.0984.$$

Tech: $P(Y = 4)$ = binompdf(trials: 6, p: 0.90,
x value: 4) = 0.0984. There is about a 9.84%
probability that exactly 4 of the 6 trains arrive
on time.

6.83 (a) *Binary?* "Success" = spinner lands in
the blue region; "Failure" = spinner does not
land in the blue region. *Independent?* Knowing if
one spin lands in the blue region tells you nothing
about whether or not another spin will. *Number?*
$n = 12$. *Same probability?* $p = 0.80$; X has a
binomial distribution with $n = 12$ and $p = 0.80$.

(b) $P(X = 8) = \binom{12}{8}(0.80)^8(0.20)^4 = 0.1329$;

Tech: $P(Y = 8)$ = binompdf(trials: 12, p: 0.80,
x value: 8) = 0.1329. There is about a 13.29%
probability that the spinner will land in the blue
region exactly 8 of the 12 times.

6.84 (a) *Binary?* "Success" = red; "Failure" =
not red. *Independent?* Knowing if the light is
red on one randomly selected workday tells
you nothing about whether it is red on another
randomly selected workday. *Number?* $n = 10$.
Same probability? $p = 0.55$; Y has a binomial
distribution with $n = 10$ and $p = 0.55$.
(b) $P(Y = 7) = 0.1665$; *Tech:* $P(Y = 7)$ =
binompdf(trials: 10, p: 0.55, x value: 7) =
0.1665. There is a probability of about 16.65%
that the light is red on exactly 7 of the 10 days.

6.85 X has a binomial distribution with
$n = 7$ and $p = 0.44$; $P(X > 4) = 0.1402$.

Tech: $P(X > 4) = 1 -$ binomcdf(trials: 7,
p: 0.44, x value: 4) = 0.1402. This
probability isn't very small. It is not surprising
for more than 4 elk to survive to adulthood.

6.86 Y has a binomial distribution
with $n = 6$ and $p = 0.90$;
$P(Y < 4) = P(Y = 0) + P(Y = 1) +$
$P(Y = 2) + P(Y = 3) = 0.0159.$ *Tech:*
$P(Y < 4)$ = binomcdf(trials: 6, p: 0.90, x
value: 3) = 0.0159. This probability is very
small. It would be surprising if the train
arrived on time fewer than 4 days.

6.87 X has a binomial distribution with
$n = 12$ and $p = 0.80$. *Tech:* $P(X \le 7)$ =
binomcdf(trials: 12, p: 0.80, x value: 7) =
0.0726. Assuming the website's claim is
true, there is a probability of about 7.26%
that there would be 7 or fewer spins
landing in the blue region.

6.88 X has a binomial distribution with
$n = 10$ and $p = 0.55$. *Tech:* $P(X \ge 7)$ =
$1 -$ binomcdf(trials: 10, p: 0.55, x value: 6) =
0.2660. Assuming the department's claim
is true, there is a probability of about
26.6% that there would be 7 or more red
lights out of 10 days.

6.89 (a) X has a binomial distribution with
$n = 22$ and $p = 0.20$. *Tech:* $P(X \ge 14)$ =
$1 -$ binomcdf(trials: 22, p: 0.20, x value:
13) = 0.00001. Assuming participants don't
have a special preference for the last thing
they taste, there is an almost 0% probability
that there would be at least 14 people
who choose the last Kiss. **(b)** Because this
outcome is very unlikely, we have convincing
evidence that participants have a preference
for the last thing they taste.

6.90 (a) X has a binomial distribution
with $n = 30$ and $p = 1/6$. *Tech:*
$P(X \le 2)$ = binomcdf(trials: 30, p: 1/6,
x value: 2) = 0.1028. Assuming the "1 in
6 wins" claim is true, there is a probability
of about 10.28% that 2 or fewer students
would win a prize. **(b)** No; if the "1 in 6
wins" claim is true, there is a 10.28%
probability that 2 or fewer students would
win a prize. Because this outcome is
not unlikely, we do not have convincing
evidence that the company's claim
is false.

6.91 (a) *Binary?* "Success" = red;
"Failure" = not red. *Independent?*
Knowing whether or not the light is red
for one randomly selected passenger
tells you nothing about whether or not the
light is red for another randomly selected
passenger. *Number?* $n = 20$. *Same
probability?* $p = 0.30$; R has a binomial
distribution with $n = 20$ and $p = 0.30$.
(b) The shape is fairly symmetric with a
single peak at $R = 6$ passengers.

6.92 (a) *Binary?* "Success" = starts on the first push of the button; "Failure" = does not start on the first push. *Independent?* Knowing if the mower starts on the first push one day tells nothing about how the mower starts on another day. *Number?* $n = 30$. *Same probability?* $p = 0.90$; T has a binomial distribution with $n = 30$ and $p = 0.90$. **(b)** The shape is skewed left, with a single peak at $T = 27$ times.

6.93

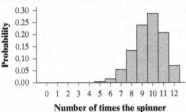

Number of times the spinner lands in the blue region

The shape is skewed left, with a single peak at $X = 10$.

6.94

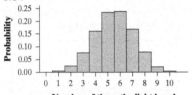

Number of times the light is red

The graph is roughly symmetric, with a single peak at $Y = 6$.

6.95 (a) $\mu_R = 20(0.3) = 6$; if many groups of 20 passengers were selected, the average number of passengers who would get a red light would be about 6. **(b)** $\sigma_R = \sqrt{20(0.3)(0.7)} = 2.049$; if many groups of 20 were selected, the number who would get a red light would typically vary from the mean (6) by about 2.049.

6.96 (a) $\mu_T = 30(0.9) = 27$; if many rounds of 30 attempts were completed, the average number of times the mower would start on the first button push is about 27 times. **(b)** $\sigma_T = \sqrt{30(0.9)(0.1)} = 1.643$; if many rounds of 30 attempts were completed, the number of times the mower would start would typically vary from the mean (27) by about 1.643.

6.97 (a) Let Y = number of calls *not* completed. Y has a binomial distribution with $n = 15$ and $p = 0.91$. *Tech:* $P(Y > 12) = 1 - \text{binomcdf}(15, 0.91, 12) = 0.8531$. There is an 85.31% probability that more than 12 calls are not completed. **(b)** $\mu_X = 15(0.09) = 1.35$; if the machine made many sets of 15 calls, the average number completed would be about 1.35. **(c)** $\sigma_X = \sqrt{15(0.09)(0.91)} = 1.11$; if the machine made many sets of 15 calls, the expected number of completed calls would typically vary from the mean (1.35) by about 1.11 calls.

(a) What probability distribution does R have? Justify your answer.

(b) Describe the shape of the probability distribution.

92. Easy-start mower? A company has developed an "easy-start" mower that cranks the engine with the push of a button. The company claims that the probability the mower will start on any push of the button is 0.9. Assume for now that this claim is true. On the next 30 uses of the mower, let T = the number of times it starts on the first push of the button. Here is a histogram of the probability distribution of T:

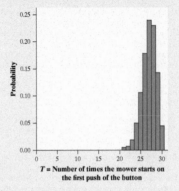

T = Number of times the mower starts on the first push of the button

(a) What probability distribution does T have? Justify your answer.

(b) Describe the shape of the probability distribution.

93. Take a spin An online spinner has two colored regions—blue and yellow. According to the website, the probability that the spinner lands in the blue region on any spin is 0.80. Assume for now that this claim is correct. Suppose we spin the spinner 12 times and let X = the number of times it lands in the blue region. Make a graph of the probability distribution of X. Describe its shape.

94. Red light! Pedro drives the same route to work on Monday through Friday. His route includes one traffic light. According to the local traffic department, there is a 55% chance that the light will be red on a randomly selected work day. Suppose we choose 10 of Pedro's work days at random and let Y = the number of times that the light is red. Make a graph of the probability distribution of Y. Describe its shape.

95. Bag check Refer to Exercise 91.

(a) Calculate and interpret the mean of R.

(b) Calculate and interpret the standard deviation of R.

96. Easy-start mower Refer to Exercise 92.

(a) Calculate and interpret the mean of T.

(b) Calculate and interpret the standard deviation of T.

97. Random digit dialing When a polling company calls a telephone number at random, there is only a 9% chance that the call reaches a live person and the survey is successfully completed.[10] Suppose the random digit dialing machine makes 15 calls. Let X = the number of calls that result in a completed survey.

(a) Find the probability that more than 12 calls are *not* completed.

(b) Calculate and interpret μ_X.

(c) Calculate and interpret σ_X.

98. Lie detectors A federal report finds that lie detector tests given to truthful persons have probability 0.2 of suggesting that the person is deceptive.[11] A company asks 12 job applicants about thefts from previous employers, using a lie detector to assess their truthfulness. Suppose that all 12 answer truthfully. Let Y = the number of people whom the lie detector indicates are being deceptive.

(a) Find the probability that the lie detector indicates that at least 10 of the people are being *honest*.

(b) Calculate and interpret μ_Y.

(c) Calculate and interpret σ_Y.

99. Lefties A total of 11% of students at a large high
pg 447   school are left-handed. A statistics teacher selects a random sample of 100 students and records L = the number of left-handed students in the sample.

(a) Explain why L can be modeled by a binomial distribution even though the sample was selected without replacement.

(b) Use a binomial distribution to estimate the probability that 15 or more students in the sample are left-handed.

100. In debt? According to financial records, 24% of U.S. adults have more debt on their credit cards than they have money in their savings accounts. Suppose that we take a random sample of 100 U.S. adults. Let D = the number of adults in the sample with more debt than savings.

(a) Explain why D can be modeled by a binomial distribution even though the sample was selected without replacement.

(b) Use a binomial distribution to estimate the probability that 30 or more adults in the sample have more debt than savings.

6.98 (a) Let X = number of people the lie detector test indicates as being honest. X has a binomial distribution with $n = 12$ and $p = 0.80$. *Tech:* $P(X \geq 10) = 1 - \text{binomcdf}(12, 0.80, 9) = 0.5583$. There is a 55.83% probability that at least 10 of the people are found to be honest. **(b)** $\mu_X = 12(0.20) = 2.4$; we would expect about 2.4 people to be declared deceptive. **(c)** $\sigma_X = \sqrt{12(0.20)(0.80)} = 1.39$; we would expect the number of people declared deceptive to typically vary by about 1.39 from the mean (2.4).

6.99 (a) As long as $n = 100$ is less than 10% of the size of the population, L can be modeled by a binomial distribution, even though the sample was selected without replacement. **(b)** *Tech:* $P(L \geq 15) \approx 1 - \text{binomcdf}(100, 0.11, 14) = 0.1330$. There is a 13.3% probability that ≥ 15 sampled students are left-handed.

6.100 (a) Because $n = 100$ is less than 10% of the size of the population (all U.S. adults), D can be modeled by a binomial distribution, even though the sample was selected without replacement. **(b)** *Tech:* $P(D \geq 30) \approx 1 - \text{binomcdf}(100, 0.24, 29) = 0.1009$. There is a 10.09% probability that 30 or more adults in the sample have more debt than savings.

6.101 No; sampling without replacement; sample size (10) > 10% of the population size (76). Not independent.

6.102 Yes; use the binomial distribution because the sample size (7) < 10% of the population size (100).

6.103 (a) $n = 100$ and $p = 0.11$, so $np = 11 \geq 10$ and $n(1 - p) = 89 \geq 10$. Since the expected number of successes and failures are both 10 or more, L can be approximated by a Normal distribution.

101. **Airport security** The Transportation Security Administration (TSA) is responsible for airport safety. On some flights, TSA officers randomly select passengers for an extra security check before boarding. One such flight had 76 passengers—12 in first class and 64 in coach class. Some passengers were surprised when none of the 10 passengers chosen for screening were seated in first class. Should we use a binomial distribution to approximate this probability? Justify your answer.

102. **Scrabble** In the game of Scrabble, each player begins by drawing 7 tiles from a bag containing 100 tiles. There are 42 vowels, 56 consonants, and 2 blank tiles in the bag. Cait chooses her 7 tiles and is surprised to discover that all of them are vowels. Should we use a binomial distribution to approximate this probability? Justify your answer.

103. **Lefties** Refer to Exercise 99.
   pg 449
   (a) Justify why L can be approximated by a Normal distribution.
   (b) Use a Normal distribution to estimate the probability that 15 or more students in the sample are left-handed.

104. **In debt?** Refer to Exercise 100.
   (a) Justify why D can be approximated by a Normal distribution.
   (b) Use a Normal distribution to estimate the probability that 30 or more adults in the sample have more debt than savings.

105. **10% condition** To use a binomial distribution to approximate the count of successes in an SRS, why do we require that the sample size n be less than 10% of the population size N?

106. **Large Counts condition** To use a Normal distribution to approximate binomial probabilities, why do we require that both np and $n(1-p)$ be at least 10?

107. **Cranky mower** To start her old lawn mower, Rita has
   pg 451
   to pull a cord and hope for some luck. On any particular pull, the mower has a 20% chance of starting.
   (a) Find the probability that it takes her exactly 3 pulls to start the mower.
   (b) Find the probability that it takes her more than 6 pulls to start the mower.

108. **1-in-6 wins** Alan decides to use a different strategy for the 1-in-6 wins game of Exercise 90. He keeps buying one 20-ounce bottle of the soda at a time until he gets a winner.
   (a) Find the probability that he buys exactly 5 bottles.
   (b) Find the probability that he buys at most 6 bottles. Show your work.

109. **Geometric or not?** Determine whether each of the following scenarios describes a geometric setting. If so, define an appropriate geometric random variable.
   (a) A popular brand of cereal puts a card bearing the image of 1 of 5 famous NASCAR drivers in each box. There is a 1/5 chance that any particular driver's card ends up in any box of cereal. Buy boxes of the cereal until you have all 5 drivers' cards.
   (b) In a game of 4-Spot Keno, Lola picks 4 numbers from 1 to 80. The casino randomly selects 20 winning numbers from 1 to 80. Lola wins money if she picks 2 or more of the winning numbers. The probability that this happens is 0.259. Lola decides to keep playing games of 4-Spot Keno until she wins some money.

110. **Geometric or not?** Determine whether each of the following scenarios describes a geometric setting. If so, define an appropriate geometric random variable.
   (a) Shuffle a standard deck of playing cards well. Then turn over one card at a time from the top of the deck until you get an ace.
   (b) Billy likes to play cornhole in his free time. On any toss, he has about a 20% chance of getting a bag into the hole. As a challenge one day, Billy decides to keep tossing bags until he gets one in the hole.

111. **Using Benford's law** According to Benford's law
   pg 454
   (Exercise 9, page 404), the probability that the first digit of the amount of a randomly chosen invoice is an 8 or a 9 is 0.097. Suppose you examine randomly selected invoices from a vendor until you find one whose amount begins with an 8 or a 9. Let X = the number of invoices examined.
   (a) Calculate and interpret the mean of X.
   (b) Calculate and interpret the standard deviation of X.

112. **Roulette** Marti decides to keep placing a \$1 bet on number 15 in consecutive spins of a roulette wheel until she wins. On any spin, there's a 1-in-38 chance that the ball will land in the 15 slot. Let Y = the number of spins it takes for Marti to win.
   (a) Calculate and interpret the mean of Y.
   (b) Calculate and interpret the standard deviation of Y.

113. **Using Benford's Law** Refer to Exercise 111. Would you be surprised if it took 40 or more invoices to find the first one with an amount that starts with an 8 or 9? Calculate an appropriate probability to support your answer.

114. **Roulette** Refer to Exercise 112. Would you be surprised if Marti won in 3 or fewer spins? Compute an appropriate probability to support your answer.

(b) $P(L \geq 15)$; $\mu_L = (100)(0.11) = 11$ and $\sigma_L = \sqrt{(100)(0.11)(0.89)} = 3.129$; **(i)** $z = 1.28$; $P(Z \geq 1.28) = 1 - 0.8997 = 0.1003$ **(ii)** $P(L \geq 15)$ = normalcdf(lower: 15, upper: 1000, mean: 11, SD: 3.129) = 0.1006 probability that 15 or more students sampled are left-handed.

6.104 (a) $np = (100)(0.24) = 24 \geq 10$ and $n(1-p) = 100(1-0.24) = 76 \geq 10$. D can be approximated by a Normal distribution. **(b)** $P(D \geq 30)$; $\mu_D = (100)(0.24) = 24$ and $\sigma_D = \sqrt{(100)(0.24)(0.76)} = 4.271$. **(i)** $z = 1.41$; *Table A:* $P(Z \geq 1.41) = 0.0793$. **(ii)** *Tech:* $P(D \geq 30)$ = normalcdf(lower: 30, upper: 1000, mean: 24, SD: 4.271) = 0.0800. There is a 0.08 probability that 30 or more adults in the sample have more debt than savings.

6.105 When sampling without replacement, the trials are not independent because knowing the outcomes of previous trials makes it easier to predict what will happen in future trials. However, if the sample is less than 10% of the population, the lack of independence isn't an issue.

6.106 The histogram for a binomial distribution is symmetric when $p = 0.5$. As p moves farther from 0.5, the probability distribution for a fixed n becomes more skewed. However, if we fix p (not 0.5) and increase n, the distribution becomes less skewed. When $np \geq 10$ and $n(1-p) \geq 10$, we know that the combination of n and p is such that the binomial distribution is close enough to Normal to use the Normal approximation.

6.107 (a) X is a geometric random variable with $p = 0.20$; $P(X = 3) = (0.8)^2(0.2) = 0.128$. *Tech:* geometpdf($p$: 0.20, x value: 3) = 0.128. There is a 0.128 probability that it takes Rita exactly 3 pulls to start the mower. **(b)** $P(X > 6) = 1 - [0.20 + (0.8)(0.2) + (0.8)^2(0.2) + \cdots + (0.8)^5(0.2)] = 0.2621$

Tech: $1 - $ geometcdf(p: 0.20, x value: 6) $= 0.2621$. There is a 0.2621 probability that it takes > 6 pulls to start the mower.

6.108 (a) X is a geometric random variable with $p = 1/6$; $P(X = 5) = 0.0804$. *Tech:* geometpdf(p: 1/6, x value: 5) $= 0.0804$. There is a 0.0804 probability that he buys exactly 5 bottles. **(b)** $P(X \leq 6) = 0.6651$. *Tech:* geometcdf(p: 1/6, x value: 6) $= 0.6651$. There is a 0.6651 probability that he buys at most 6 bottles.

6.109 (a) Not a geometric setting; we can't classify the possible outcomes on each trial (card) as a "success" or "failure" and no cards are selected until we get 1 success. **(b)** Games of 4-Spot Keno are independent—that is, the probability of winning is the same in each game ($p = 0.259$)—and Lola is repeating a random process until she gets a success. X is a geometric random variable with $p = 0.259$.

6.110 (a) Not a geometric setting; trials are not independent. **(b)** Assuming his shots are independent, this is a geometric setting because the probability of success is the same for each toss ($p = 0.20$) and he continues to toss until he gets a bag in the hole. X is a geometric random variable with $p = 0.20$.

6.111 (a) X is a geometric random variable with $p = 0.097$; $\mu_X = 10.31$. We would expect to examine about 10.31 invoices to find the first 8 or 9. **(b)** *Tech:* $P(X \geq 40) = 1 - $ geometcdf (p: 0.097, x value: 39) $= 0.0187$. Because the probability of not getting an 8 or 9 before the 40th invoice is small, we worry that the invoice amounts are fraudulent.

6.112 (a) Y is a geometric random variable with $p = 1/38$; $\mu_Y = \dfrac{1}{p} = \dfrac{1}{\frac{1}{38}} = 38$. We would expect it to take about 38 games for the ball to land in the 15 slot. If many, many games are played, the average number of games we expect to play for the ball to land in the 15 slot is about 38.

(b) $\sigma_Y = \dfrac{\sqrt{1 - (1/38)}}{1/38} = 37.497$

If many games are played, the number it will take for the ball to land in the 15 slot will typically vary by about 37.497 games from the mean of 38 games.

6.113 *Tech:* $P(X \geq 40) = 1 - P(X \leq 39) = 1 - $ geometcdf(p: 0.097, x value: 39) $= 1 - 0.9813 = 0.0187$. Because the probability of not getting an 8 or 9 before the 40th invoice is small, we may worry that the invoices are fraudulent.

6.114 *Tech:* geometcdf(p: 1/38, x value: 3) $= 0.0769$. Because the probability is not that small, I wouldn't be surprised if Marti won in 3 or fewer spins.

6.115 b

6.116 c

6.117 d

6.118 d

6.119 c

6.120 (a) This is an experiment because a treatment was deliberately imposed on the students. **(b)** Explanatory variable: The type of login box (genuine or not), which is categorical. Response variable: The student's action (logging in or not), which is also categorical.

6.121 The standard deviations, from smallest to largest, are B, C, A.

Multiple Choice: *Select the best answer for Exercises 115–119.*

115. Joe reads that 1 out of 4 eggs contains salmonella bacteria. So he never uses more than 3 eggs in cooking. If eggs do or don't contain salmonella independently of each other, the number of contaminated eggs when Joe uses 3 eggs chosen at random has the following distribution:

(a) binomial; $n = 4$ and $p = 1/4$

(b) binomial; $n = 3$ and $p = 1/4$

(c) binomial; $n = 3$ and $p = 1/3$

(d) geometric; $p = 1/4$

(e) geometric; $p = 1/3$

Exercises 116 and 117 refer to the following setting. A fast-food restaurant runs a promotion in which certain food items come with game pieces. According to the restaurant, 1 in 4 game pieces is a winner.

116. If Jeff gets 4 game pieces, what is the probability that he wins exactly 1 prize?

(a) 0.25

(b) 1.00

(c) $\binom{4}{1}(0.25)^1(0.75)^3$

(d) $\binom{4}{1}(0.25)^3(0.75)^1$

(e) $(0.75)^3(0.25)^1$

117. If Jeff keeps playing until he wins a prize, what is the probability that he has to play the game exactly 5 times?

(a) $(0.25)^5$

(b) $(0.75)^4$

(c) $(0.75)^5$

(d) $(0.75)^4(0.25)$

(e) $\binom{5}{1}(0.75)^4(0.25)$

118. Each entry in a table of random digits like Table D has probability 0.1 of being a 0, and the digits are independent of one another. Each line of Table D contains 40 random digits. The mean and standard deviation of the number of 0s in a randomly selected line will be approximately

(a) mean = 0.1, standard deviation = 0.05.

(b) mean = 0.1, standard deviation = 0.1.

(c) mean = 4, standard deviation = 0.05.

(d) mean = 4, standard deviation = 1.90.

(e) mean = 4, standard deviation = 3.60.

119. In which of the following situations would it be appropriate to use a Normal distribution to approximate probabilities for a binomial distribution with the given values of n and p?

(a) $n = 10, p = 0.5$

(b) $n = 40, p = 0.88$

(c) $n = 100, p = 0.2$

(d) $n = 100, p = 0.99$

(e) $n = 1000, p = 0.003$

Recycle and Review

120. **Spoofing** (4.2) To collect information such as passwords, online criminals use "spoofing" to direct Internet users to fraudulent websites. In one study of Internet fraud, students were warned about spoofing and then asked to log into their university account starting from the university's home page. In some cases, the log-in link led to the genuine dialog box. In others, the box looked genuine but, in fact, was linked to a different site that recorded the ID and password the student entered. The box that appeared for each student was determined at random. An alert student could detect the fraud by looking at the true Internet address displayed in the browser status bar, but most just entered their ID and password.

(a) Is this an observational study or an experiment? Justify your answer.

(b) What are the explanatory and response variables? Identify each variable as categorical or quantitative.

121. **Standard deviations** (6.1) Continuous random variables A, B, and C all take values between 0 and 10. Their density curves, drawn on the same horizontal scales, are shown here. Rank the standard deviations of the three random variables from smallest to largest. Justify your answer.

FRAPPY! FREE RESPONSE AP® PROBLEM, YAY!

The following problem is modeled after actual AP® Statistics exam free response questions. Your task is to generate a complete, concise response in 15 minutes.

Directions: Show all your work. Indicate clearly the methods you use, because you will be scored on the correctness of your methods as well as on the accuracy and completeness of your results and explanations.

Buckley Farms produces homemade potato chips that it sells in bags labeled "16 ounces." The total weight of each bag follows an approximately Normal distribution with a mean of 16.15 ounces and a standard deviation of 0.12 ounce.
(a) If you randomly selected 1 bag of these chips, what is the probability that the total weight is less than 16 ounces?
(b) If you randomly selected 10 bags of these chips, what is the probability that exactly 2 of the bags will have a total weight less than 16 ounces?

(c) Buckley Farms ships its chips in boxes that contain 6 bags. The empty boxes have a mean weight of 10 ounces and a standard deviation of 0.05 ounce. Calculate the mean and standard deviation of the total weight of a box containing 6 bags of chips.
(d) Buckley Farms decides to increase the mean weight of each bag of chips so that only 5% of the bags have weights that are less than 16 ounces. Assuming that the standard deviation remains 0.12 ounce, what mean weight should Buckley Farms use?

After you finish, you can view two example solutions on the book's website (highschool.bfwpub.com/updatedtps6e). Determine whether you think each solution is "complete," "substantial," "developing," or "minimal." If the solution is not complete, what improvements would you suggest to the student who wrote it? Finally, your teacher will provide you with a scoring rubric. Score your response and note what, if anything, you would do differently to improve your own score.

Chapter 6 Review

Section 6.1: Discrete and Continuous Random Variables

A random variable assigns numerical values to the outcomes of a random process. The probability distribution of a random variable describes its possible values and their probabilities. There are two types of random variables: discrete and continuous. Discrete random variables take on a fixed set of values with gaps in between. Continuous random variables can take on any value in an interval of numbers.

As in Chapter 1, we are often interested in the shape, center, and variability of a probability distribution. The shape of a discrete probability distribution can be identified by graphing a probability histogram, with the height of each bar representing the probability of a single value. The center is usually identified by the mean (expected value) of the random variable. The mean (expected value) is the average value of the random variable if the random process is repeated many times. The variability of a probability distribution is usually identified by the standard deviation, which describes how much the values of a random variable typically vary from the mean value, in many trials of the random process.

The probability distribution of a continuous random variable is described by a density curve. Probabilities for continuous random variables are determined by finding the area under the density curve and above the values of interest.

Section 6.2: Transforming and Combining Random Variables

In this section, you learned how linear transformations of a random variable affect the shape, center, and variability

(c) Let B = weight of box and C = combined weight of box and 6 bags of chips. Then $C = B + X_1 + X_2 + X_3 + X_4 + X_5 + X_6$. The mean is $\mu_C = 10 + 16.15 + 16.15 + 16.15 + 16.15 + 16.15 + 16.15 = 106.9$ ounces. $\sigma_C^2 = 0.05^2 + 0.12^2 + 0.12^2 + 0.12^2 + 0.12^2 + 0.12^2 + 0.12^2 = 0.1364$, so the standard deviation is $\sigma_C = \sqrt{0.1364} = 0.369$ ounce.

(d) Let X = the total weight of a bag of chips. X follows a Normal distribution with mean M and standard deviation 0.12 ounce. We want to find the value of M such that $P(X < 16) = 0.05$.

Look in the body of Table A for a value closest to 0.05. A z-score of -1.64 gives the closest value (0.0505).

Solving $-1.64 = \dfrac{16 - M}{0.12}$ gives

M = 16.1968 ounces. *Tech:* The command invNorm(area: 0.05, μ: 0, σ: 1) gives a value of -1.645.

Solving $-1.645 = \dfrac{16 - M}{0.12}$ gives

M = 16.1974 ounces. Buckley Farms should use a mean weight of about 16.2 ounces to make sure less than 5% of bags are underweight.

461

Answers

(a) Let X = the total weight of a bag of chips. X follows a Normal distribution with mean 16.15 ounces and standard deviation 0.12 ounce. We want to find $P(X < 16)$.

The standardized score for the boundary value is $z = \dfrac{16 - 16.15}{0.12} = -1.25$.

Table A: $P(Z < -1.25) = 0.1057$. *Tech:* The command normalcdf(lower: -1000, upper: 16, μ: 16.15, σ: 0.12) gives an area of 0.1057. There is about an 11% probability of randomly selecting a bag that weighs less than 16 ounces.

(b) Let Y = the number of bags that weigh less than 16 ounces. Y is binomial with $n = 10$ and $p = 0.1057$. We want to find $P(Y = 2)$.

$$P(Y = 2) = \binom{10}{2}(0.1057)^2(0.8943)^8 = 0.2057.$$

Tech: The command binompdf(trials: 10, p: 0.1057, x value: 2) gives 0.2057. There is about a 21% probability that exactly 2 out of 10 randomly selected bags weigh less than 16 ounces.

of its probability distribution. Similar to what you learned in Chapter 2, adding a positive constant to (or subtracting it from) each value of a random variable changes the measures of center and location, but not the shape or variability of the probability distribution. Multiplying or dividing each value of a random variable by a positive constant changes the measures of center and location and measures of variability, but not the shape of the probability distribution. A linear transformation $Y = a + bX$ (with $b > 0$) does not change the shape of the probability distribution. The mean and standard deviation of Y are:

$$\mu_Y = a + b\mu_X \quad \text{and} \quad \sigma_Y = b\sigma_X$$

You also learned how to calculate the mean and standard deviation for a combination of two or more random variables. The mean of a sum or difference of any two random variables X and Y is given by

$$\mu_{X+Y} = \mu_X + \mu_Y \quad \text{and} \quad \mu_{X-Y} = \mu_X - \mu_Y$$

If X and Y are any two *independent* random variables, variances add:

$$\sigma_{X+Y}^2 = \sigma_X^2 + \sigma_Y^2 \quad \text{and} \quad \sigma_{X-Y}^2 = \sigma_X^2 + \sigma_Y^2$$

To find the standard deviation, just take the square root of the variance. Recall that X and Y are independent if knowing the value of one variable does not change the probability distribution of the other variable. Also, if independent random variables X and Y are both Normally distributed, then their sum $X + Y$ and difference $X - Y$ are both Normally distributed as well.

The linear combination $aX + bY$ has mean $a\mu_X + b\mu_Y$. Its standard deviation is $\sqrt{a^2\sigma_X^2 + b^2\sigma_Y^2}$ if X and Y are independent.

Section 6.3: Binomial and Geometric Random Variables

In this section, you learned about two common types of discrete random variables, binomial random variables and geometric random variables. Binomial random variables count the number of successes in a fixed number of trials (n) of the same random process, whereas geometric random variables count the number of trials needed to get one success. Otherwise, the binomial and geometric settings have the same conditions:

there must be two possible outcomes for each trial (success or failure), the trials must be independent, and the probability of success p must stay the same throughout all trials.

To calculate probabilities for a binomial distribution with n trials and probability of success p on each trial, use technology or the binomial probability formula

$$P(X = x) = \binom{n}{x}p^x(1-p)^{n-x}$$

The mean and standard deviation of a binomial random variable X are

$$\mu_X = np \quad \text{and} \quad \sigma_X = \sqrt{np(1-p)}$$

The shape of a binomial distribution depends on both the number of trials n and the probability of success p. When the number of trials is large enough that both np and $n(1-p)$ are at least 10, the probability distribution of the binomial random variable X can be modeled with a Normal density curve. Be sure to check the Large Counts condition before using a Normal approximation to a binomial distribution.

A common application of the binomial distribution is when we count the number of times a particular outcome occurs in a random sample from some population. Because sampling is almost always done without replacement, the Independent condition is violated. However, if the sample size is a small fraction of the population size (less than 10%), we can view individual observations as independent. Be sure to check the 10% condition when sampling is done without replacement before using a binomial distribution.

Finally, to calculate probabilities for a geometric distribution with probability of success p on each trial, use technology or the geometric probability formula

$$P(X = x) = (1-p)^{x-1}p$$

A geometric distribution is always skewed to the right and unimodal, with a single peak at 1. The mean and standard deviation of a geometric random variable X are

$$\mu_X = 1/p \quad \text{and} \quad \sigma_X = \frac{\sqrt{1-p}}{p}$$

What Did You Learn?

Learning Target	Section	Related Example on Page(s)	Relevant Chapter Review Exercise(s)
Use the probability distribution of a discrete random variable to calculate the probability of an event.	6.1	391	R6.1
Make a histogram to display the probability distribution of a discrete random variable and describe its shape.	6.1	393	R6.3
Calculate and interpret the mean (expected value) of a discrete random variable.	6.1	395	R6.1, R6.3
Calculate and interpret the standard deviation of a discrete random variable.	6.1	397	R6.1, R6.3

Learning Target	Section	Related Example on Page(s)	Relevant Chapter Review Exercise(s)
Use the probability distribution of a continuous random variable (uniform or Normal) to calculate the probability of an event.	6.1	400, 401	R6.4
Describe the effect of adding or subtracting a constant or multiplying or dividing by a constant on the probability distribution of a random variable.	6.2	412, 413, 415	R6.2, R6.3
Calculate the mean and standard deviation of the sum or difference of random variables.	6.2	417, 421	R6.3, R6.4
Find probabilities involving the sum or difference of independent Normal random variables.	6.2	423	R6.4
Determine whether the conditions for a binomial setting are met.	6.3	433, 442	R6.5
Calculate and interpret probabilities involving binomial random variables.	6.3	437, 439, 447	R6.5
Calculate the mean and standard deviation of a binomial distribution. Interpret these values.	6.3	444	R6.6
When appropriate, use the Normal approximation to the binomial distribution to calculate probabilities.	6.3	449	R6.8
Calculate and interpret probabilities involving geometric random variables.	6.3	451	R6.7
Calculate the mean and standard deviation of a geometric random variable. Interpret these values.	6.3	454	R6.7

Chapter 6 Review Exercises

These exercises are designed to help you review the important ideas and methods of the chapter.

R6.1 Knees Patients receiving artificial knees often experience pain after surgery. The pain is measured on a subjective scale with possible values of 1 (low) to 5 (high). Let Y be the pain score for a randomly selected patient. The following table gives the probability distribution for Y.

Value	1	2	3	4	5
Probability	0.1	0.2	0.3	0.3	??

(a) Find $P(Y = 5)$. Interpret this value.

(b) Find the probability that a randomly selected patient has a pain score of at most 2.

(c) Calculate the expected pain score and the standard deviation of the pain score.

R6.2 A glass act In a process for manufacturing glassware, glass stems are sealed by heating them in a flame. Let X be the temperature (in degrees Celsius) for a randomly chosen glass. The mean and standard deviation of X are $\mu_X = 550°C$ and $\sigma_X = 5.7°C$.

(a) Is temperature a discrete or continuous random variable? Explain your answer.

(b) The target temperature is 550°C. What are the mean and standard deviation of the number of degrees off target, $D = X - 550$?

(c) A manager asks for results in degrees Fahrenheit. The conversion of X into degrees Fahrenheit is given by $Y = \frac{9}{5}X + 32$. What are the mean μ_Y and the standard deviation σ_Y of the temperature of the flame in the Fahrenheit scale?

R6.3 Keno In a game of 4-Spot Keno, the player picks 4 numbers from 1 to 80. The casino randomly selects 20 winning numbers from 1 to 80. The table shows the possible outcomes of the game and their probabilities, along with the amount of money (Payout) that the player wins for a $1 bet. If X = the payout for a single $1 bet, you can check that $\mu_X = \$0.70$ and $\sigma_X = \$6.58$.

Matches	0	1	2	3	4
Payout x_i	$0	$0	$1	$3	$120
Probability p_i	0.308	0.433	0.213	0.043	0.003

(a) Make a graph of the probability distribution. Describe what you see.

(b) Interpret the values of μ_X and σ_X.

TRM Full Solutions to Chapter 6 Review Exercises

The full solutions can be found by clicking on the link in the TE-Book, opening the TRFD, or downloading from the Teacher's Resources on the book's digital platform.

Answers to Chapter 6 Review Exercises

R6.1 (a) $P(Y = 5) = 0.1$; there is a 0.1 probability that a randomly selected patient would rate their pain as a 5 on a scale of 1 to 5. **(b)** $P(Y \le 2) = 0.3$ **(c)** $\mu_X = 3.1$; $\sigma_X = 1.136$

R6.2 (a) Temperature is a continuous random variable because it takes all values in an interval of numbers. **(b)** $\mu_D = 550 - 550 = 0°C$, and the standard deviation stays the same, $\sigma_D = 5.7°C$, because subtracting a constant does not change the variability. **(c)** $\mu_Y = \frac{9}{5}(550) + 32 = 1022°F$ and

$$\sigma_Y = \left(\frac{9}{5}\right)(5.7) = 10.26°F.$$

R6.3 (a)

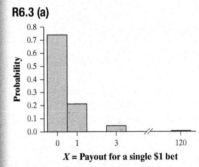

X = Payout for a single $1 bet

The distribution is skewed to the right, with a single peak at the $0 payout. The mean of the distribution is $0.70 and the standard deviation is $6.58. **(b)** Payout averages $0.70 per game; if you were to play many games of 4-Spot Keno, the payout amounts would vary, on average, by about $6.58 from the mean ($0.70). **(c)** $Y = 5X$; $\mu_Y = 5(0.70) = \$3.50$ and $\sigma_Y = 5(6.58) = \$32.90$. **(d)** Let W be the amount of Marla's payout. $W = X_1 + X_2 + X_3 + X_4 + X_5$; $\mu_W = 5(0.70) = \$3.50$ and $\sigma_W = \$14.71$.

R6.4 (a) (i) $z = 0.83$; $P(Z > 0.83) =$ $1 - P(Z \leq 0.83) = 0.2023$
(ii) $P(C > 11) =$ normalcdf(lower: 11, upper: 1000, mean: 10, SD: 1.2) $=$ 0.2023. There is a 0.2023 probability that a randomly selected cap has strength greater than 11 inch-pounds. **(b)** It is reasonable to assume the cap strength and torque are independent because the machine that makes the caps and the machine that applies the torque are not the same. **(c)** $C - T$ is Normal with mean $10 - 7 = 3$ inch-pounds and standard deviation $\sqrt{0.9^2 + 1.2^2} = 1.5$ inch-pounds. **(d)** (i) $z = -2$; $P(Z < -2) = 0.0228$. (ii) $P(C - T < 0) =$ normalcdf(lower: -1000, upper: 0, mean: 3, SD: 1.5) $=$ 0.0228. There is a 0.0228 probability that a randomly selected cap will break when being fastened by the machine.

R6.5 (a) *Binary?* "Success" = orange and "Failure" = not orange. *Independent?* The sample of size $n = 8$ is less than 10% of the large bag, so we can assume the outcomes of trials are independent. *Number?* We are choosing a fixed sample of $n = 8$ candies. *Success?* The probability of success remains constant at $p = 0.205$. This is a binomial setting, so X has a binomial distribution with $n = 8$ and $p = 0.205$. **(b)** $P(X = 3) = 0.1532$; *Tech:* $P(X = 3) =$ binompdf(trials: 8, p: 0.205, x value: 3) $= 0.1532$. There is a 15.32% probability that exactly 3 of the 8 M&M'S® will be orange. **(c)** *Tech:* $P(X \geq 4) = 1 -$ binomcdf(trials: 8, p: 0.205, x value: 3) $= 0.0610$. There is a 6.1% probability that at least 4 of the 8 M&M'S will be orange. **(d)** No; it is not very unusual to receive 4 or more orange M&M'S in a sample of 8 M&M'S when 20.5% of M&M'S produced are orange. This happens 6.1% of the time by chance alone, so there is not convincing evidence that the claim is false.

R6.6 (a) $\mu_X = 8(0.205) = 1.64$; if we selected many random samples of size 8, we would expect about 1.64 orange M&M'S. **(b)** $\sigma_X = \sqrt{8(0.205)(0.795)} =$ 1.14; if we were to select many random samples of size 8, the number of orange M&M'S would typically vary by about 1.14 from the mean of 164.

R6.7 (a) Y is a geometric random variable with $p = \dfrac{3}{12} = 0.25$; $P(Y \leq 3) = 0.5781$. *Tech:* geometcdf(p: 0.25, x value: 3) $=$ 0.5781.

(b) $\mu_Y = \dfrac{1}{0.25} = 4$. If this random process is repeated many times, the average number of spins it would take to get the first wasabi bomb is 4 spins.

(c) $\sigma_Y = \dfrac{\sqrt{1 - 0.25}}{0.25} = 3.464$
If repeated many times, the average number of spins needed would typically vary by about 3.464 spins from the mean of 4 spins.

R6.8 (a) Although the trials are binary ("success" = use and "failure" = does not use) and there is a fixed number of trials ($n = 200$), the residents were selected without replacement, violating the condition of independence. However, because the sample size ($n = 200$) is less than 10% of the population size (all residents in a large city), T has approximately a binomial distribution with $n = 200$ and $p = 0.34$.

(b) $n = 200$ and $p = 0.34$, so $np = 68 \geq 10$ and $n(1 - p) = 132 \geq 10$. Because the expected number of successes and failures are both 10 or more, T can be approximated by a Normal distribution.
(c) $P(T \leq 60)$; $\mu_T = (200)(0.34) = 68$ and $\sigma_T = \sqrt{(200)(0.34)(0.66)} = 6.699$ (i) $z = -1.19$; *Table A:* $P(Z \leq -1.19) = 0.1170$ (ii) *Tech:* $P(T \leq 60) =$ normalcdf(lower: -1000, upper: 60, mean: 68, SD: 6.699) $= 0.1162$. There is a 0.1162 probability that at most 60 residents in the sample use public transportation at least once per week.

Answers to Chapter 6
AP® Statistics Practice Test

T6.1 b

(c) Jerry places a single $5 bet on 4-Spot Keno. Find the expected value and the standard deviation of his winnings.

(d) Marla plays five games of 4-Spot Keno, betting $1 each time. Find the expected value and the standard deviation of her total winnings.

R6.4 Applying torque A machine fastens plastic screw-on caps onto containers of motor oil. If the machine applies more torque than the cap can withstand, the cap will break. Both the torque applied and the strength of the caps vary. The capping-machine torque T follows a Normal distribution with mean 7 inch-pounds and standard deviation 0.9 inch-pound. The cap strength C (the torque that would break the cap) follows a Normal distribution with mean 10 inch-pounds and standard deviation 1.2 inch-pounds.

(a) Find the probability that a randomly selected cap has a strength greater than 11 inch-pounds.

(b) Explain why it is reasonable to assume that the cap strength and the torque applied by the machine are independent.

(c) Let the random variable $D = C - T$. Find its mean and standard deviation.

(d) What is the probability that a randomly selected cap will break while being fastened by the machine?

Exercises R6.5 and R6.6 refer to the following setting. According to Mars, Incorporated, 20.5% of the M&M'S® Milk Chocolate Candies made at its Cleveland factory are orange. Assume that the company's claim is true. Suppose you take a random sample of 8 candies from a large bag of M&M'S. Let $X =$ the number of orange candies you get.

R6.5 Orange M&M'S

(a) Explain why it is reasonable to use the binomial distribution for probability calculations involving X.

(b) What's the probability that you get 3 orange M&M'S?

(c) Calculate $P(X \geq 4)$. Interpret this result.

(d) Suppose that you get 4 orange M&M'S in your sample. Does this result provide convincing evidence that Mars's claim about its M&M'S is false? Justify your answer.

R6.6 Orange M&M'S

(a) Find and interpret the expected value of X.

(b) Find and interpret the standard deviation of X.

R6.7 Sushi Roulette In the Japanese game show *Sushi Roulette*, the contestant spins a large wheel that's divided into 12 equal sections. Nine of the sections show a sushi roll, and three have a "wasabi bomb." When the wheel stops, the contestant must eat whatever food applies to that section. Then the game show host replaces the item of food on the wheel. To win the game, the contestant must eat one wasabi bomb. Let $Y =$ the number of spins required to get a wasabi bomb.

(a) Find the probability that it takes 3 or fewer spins for the contestant to get a wasabi bomb.

(b) Find the mean of Y. Interpret this parameter.

(c) Show that $\sigma_Y = 3.464$. Interpret this parameter.

R6.8 Public transportation In a large city, 34% of residents use public transportation at least once per week. Suppose the city's mayor selects a random sample of 200 residents. Let $T =$ the number in the sample who use public transportation at least once per week.

(a) What type of probability distribution does T have? Justify your answer.

(b) Explain why T can be approximated by a Normal distribution.

(c) Calculate the probability that at most 60 residents in the sample use public transportation at least once per week.

Chapter 6 AP® Statistics Practice Test

Section I: Multiple Choice *Select the best answer for each question.*

Questions T6.1–T6.3 refer to the following setting. A psychologist studied the number of puzzles that subjects were able to solve in a 5-minute period while listening to soothing music. Let X be the number of puzzles completed successfully by a randomly chosen subject. The psychologist found that X had the following probability distribution.

Value	1	2	3	4
Probability	0.2	0.4	0.3	0.1

T6.1 What is the probability that a randomly chosen subject completes more than the expected number of puzzles in the 5-minute period while listening to soothing music?

(a) 0.1   (b) 0.4   (c) 0.8

(d) 1   (e) Cannot be determined

T6.2 The standard deviation of X is 0.9. Which of the following is the best interpretation of this value?

(a) About 90% of subjects solved 3 or fewer puzzles.

(b) About 68% of subjects solved between 0.9 puzzles less and 0.9 puzzles more than the mean.

(c) The typical subject solved an average of 0.9 puzzles.

(d) The number of puzzles solved by subjects typically differed from the mean by about 0.9 puzzles.

(e) The number of puzzles solved by subjects typically differed from one another by about 0.9 puzzles.

T6.3 Let D be the difference in the number of puzzles solved by two randomly selected subjects in a 5-minute period. What is the standard deviation of D?

(a) 0

(b) 0.81

(c) 0.9

(d) 1.27

(e) 1.8

T6.4 Suppose a student is randomly selected from your school. Which of the following pairs of random variables are most likely independent?

(a) X = student's height; Y = student's weight

(b) X = student's IQ; Y = student's GPA

(c) X = student's PSAT Math score; Y = student's PSAT Verbal score

(d) X = average amount of homework the student does per night; Y = student's GPA

(e) X = average amount of homework the student does per night; Y = student's height

T6.5 A certain vending machine offers 20-ounce bottles of soda for $1.50. The number of bottles X bought from the machine on any day is a random variable with mean 50 and standard deviation 15. Let the random variable Y equal the total revenue from this machine on a randomly selected day. Assume that the machine works properly and that no sodas are stolen from the machine. What are the mean and standard deviation of Y?

(a) $\mu_Y = \$1.50, \sigma_Y = \$22.50$

(b) $\mu_Y = \$1.50, \sigma_Y = \$33.75$

(c) $\mu_Y = \$75, \sigma_Y = \$18.37$

(d) $\mu_Y = \$75, \sigma_Y = \$22.50$

(e) $\mu_Y = \$75, \sigma_Y = \$33.75$

T6.6 The weight of tomatoes chosen at random from a bin at the farmer's market follows a Normal distribution with mean $\mu = 10$ ounces and standard deviation $\sigma = 1$ ounce. Suppose we pick four tomatoes at random from the bin and find their total weight T. The random variable T is

(a) Normal, with mean 10 ounces and standard deviation 1 ounce.

(b) Normal, with mean 40 ounces and standard deviation 2 ounces.

(c) Normal, with mean 40 ounces and standard deviation 4 ounces.

(d) binomial, with mean 40 ounces and standard deviation 2 ounces.

(e) binomial, with mean 40 ounces and standard deviation 4 ounces.

T6.7 Which of the following random variables is geometric?

(a) The number of times I have to roll a single die to get two 6s

(b) The number of cards I deal from a well-shuffled deck of 52 cards to get a heart

(c) The number of digits I read in a randomly selected row of the random digits table to get a 7

(d) The number of 7s in a row of 40 random digits

(e) The number of 6s I get if I roll a die 10 times

T6.8 Seventeen people have been exposed to a particular disease. Each one independently has a 40% chance of contracting the disease. A hospital has the capacity to handle 10 cases of the disease. What is the probability that the hospital's capacity will be exceeded?

(a) 0.011

(b) 0.035

(c) 0.092

(d) 0.965

(e) 0.989

T6.9 The figure shows the probability distribution of a discrete random variable X. Which of the following best describes this random variable?

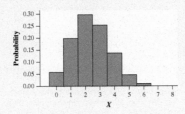

(a) Binomial with $n = 8$, $p = 0.1$

(b) Binomial with $n = 8$, $p = 0.3$

(c) Binomial with $n = 8$, $p = 0.8$

(d) Geometric with $p = 0.1$

(e) Geometric with $p = 0.2$

T6.2 d

T6.3 d

T6.4 e

T6.5 d

T6.6 b

T6.7 c

T6.8 b

T6.9 b

TRM Full Solutions to Chapter 6 AP® Statistics Practice Test

The full solutions can be found by clicking on the link in the TE-Book, opening the TRFD, or downloading from the Teacher's Resources on the book's digital platform.

T6.10 c

T6.11 (a) $P(Y \le 2) = 0.96$; there is a 96% chance that at least 10 eggs are unbroken in a randomly selected carton of "store-brand" eggs. **(b)** $\mu_Y = 0.38$; if we were to randomly select many cartons of eggs, we would expect an average of about 0.38 eggs to be broken. **(c)** $\sigma_Y = 0.8219$; if we were to randomly select many cartons of eggs, the number of broken eggs would typically vary by about 0.8219 egg from the mean of 0.38. **(d)** X is a geometric random variable with $p = 0.11$. We are looking for $P(X \le 3)$. *Tech:* geometcdf(p: 0.11, x value: 3) $= 0.2950$.

T6.12 (a) *Binary?* "Success" = greets dog first and "Failure" = does not greet dog first. *Independent?* We are sampling without replacement, but 12 is less than 10% of all dog owners. *Number?* $n = 12$. *Same probability?* The probability of success is constant for all trials ($p = 0.66$). X is a binomial random variable with $n = 12$ and $p = 0.66$. **(b)** $P(X = 6) = 0.1180$ **(c)** $P(X \le 4) = 0.0213$; *Tech:* binomcdf(trials: 12, p: 0.66, x value: 4) = 0.0213. Because this probability is very small, it is unlikely to have only 4 or fewer owners greet their dogs first by chance alone. This gives convincing evidence that the claim by *Ladies' Home Journal* is incorrect.

T6.13 (a) Let $D = A - E$. $\mu_D = 50 - 25 = 25$ minutes; because the amount of time they spend on homework is independent of each other, $\sigma_D = 11.18$ minutes. **(b)** (i) $z = -2.24$; *Table A:* $P(Z < -2.24) = 0.0125$. *Tech:* $P(D < 0) = $ normalcdf(lower: -1000, upper: 0, mean: 25, SD: 11.18) $= 0.0127$. There is a 0.0127 probability that Ed spent longer on his assignment than Adelaide did on hers.

T6.14 (a) X has a binomial distribution with $n = 1200$ and $p = 0.13$; $\mu_X = 1200(0.13) = 156$ and $\sigma_X = \sqrt{1200(0.13)(0.87)} = 11.6499$. **(b)** If the sample contains 10% Hispanics, there were $1200(0.10) = 120$ Hispanics in the sample; $P(X \le 120) = 0.00083$. *Tech:* binomcdf(trials: 1200, $p = 0.13$, x value: 120) = 0.00083. If we use the Normal approximation to the binomial distribution, $z = -3.09$ and $P(Z \le -3.09) = 0.001$. Because this probability is small, it is unlikely to select 120 or fewer Hispanics in the sample just by chance. This gives us reason to be suspicious of the sampling process.

T6.10 A test for extrasensory perception (ESP) involves asking a person to tell which of 5 shapes—a circle, star, triangle, diamond, or heart—appears on a hidden computer screen. On each trial, the computer is equally likely to select any of the 5 shapes. Suppose researchers are testing a person who does not have ESP and so is just guessing on each trial. What is the probability that the person guesses the first 4 shapes incorrectly but gets the fifth one correct?

(a) $1/5$

(b) $\left(\frac{4}{5}\right)^4$

(c) $\left(\frac{4}{5}\right)^4 \cdot \left(\frac{1}{5}\right)$

(d) $\binom{5}{1} \cdot \left(\frac{4}{5}\right)^4 \cdot \left(\frac{1}{5}\right)$

(e) $4/5$

Section II: Free Response *Show all your work. Indicate clearly the methods you use, because you will be graded on the correctness of your methods as well as on the accuracy and completeness of your results and explanations.*

T6.11 Let Y denote the number of broken eggs in a randomly selected carton of one dozen "store brand" eggs at a local supermarket. Suppose that the probability distribution of Y is as follows.

Value y_i	0	1	2	3	4
Probability p_i	0.78	0.11	0.07	0.03	0.01

(a) What is the probability that at least 10 eggs in a randomly selected carton are *unbroken*?

(b) Calculate and interpret μ_Y.

(c) Calculate and interpret σ_Y.

(d) A quality control inspector at the store keeps looking at randomly selected cartons of eggs until he finds one with at least 2 broken eggs. Find the probability that this happens in one of the first three cartons he inspects.

T6.12 *Ladies Home Journal* magazine reported that 66% of all dog owners greet their dog before greeting their spouse or children when they return home at the end of the workday. Assume that this claim is true. Suppose 12 dog owners are selected at random. Let $X = $ the number of owners who greet their dogs first.

(a) Explain why it is reasonable to use the binomial distribution for probability calculations involving X.

(b) Find the probability that exactly 6 owners in the sample greet their dogs first when returning home from work.

(c) In fact, only 4 of the owners in the sample greeted their dogs first. Does this give convincing evidence against the *Ladies Home Journal* claim? Calculate $P(X \le 4)$ and use the result to support your answer.

T6.13 Ed and Adelaide attend the same high school but are in different math classes. The time E that it takes Ed to do his math homework follows a Normal distribution with mean 25 minutes and standard deviation 5 minutes. Adelaide's math homework time A follows a Normal distribution with mean 50 minutes and standard deviation 10 minutes. Assume that E and A are independent random variables.

(a) Randomly select one math assignment of Ed's and one math assignment of Adelaide's. Let the random variable D be the difference in the amount of time each student spent on their assignments: $D = A - E$. Find the mean and the standard deviation of D.

(b) Find the probability that Ed spent longer on his assignment than Adelaide did on hers.

T6.14 According to the Census Bureau, 13% of American adults (aged 18 and over) are Hispanic. An opinion poll plans to contact an SRS of 1200 adults.

(a) What is the mean number of Hispanics in such samples? What is the standard deviation?

(b) Should we be suspicious if the sample selected for the opinion poll contains 10% or less Hispanic people? Calculate an appropriate probability to support your answer. **466**

TRM Chapter 6 Case Study

The Case Study feature that was found in the previous two editions of TPS has been moved to the Teacher's Resource Materials. Click on the link in the TE-Book, open in the TRFD, or download from the Teacher's Resources on the book's digital platform.

Chapter 7

Sampling Distributions

PD Overview

Sampling distributions are the culmination of the first three major content areas of the course—exploring data, collecting data, and probability—and the final piece needed for the formal study of statistical inference. Typically, sampling distributions are also very difficult for students to understand. To help them through this difficult topic, we have included an extra share of activities and applets that shed light on the important concepts. Make sure to do as many of these as you can.

We spend quite a bit of time in the Introduction and in Section 7.1 developing the idea of a sampling distribution. In short, a sampling distribution describes the possible values of a statistic and how often it takes those values. A statistic is any quantity that can be calculated from a sample, such as a sample mean $\bar{x}$ or a sample proportion $\hat{p}$.

Finally, we highly recommend that you download and read the Special Focus materials on Sampling Distributions available on the AP® Central website: https://apcentral.collegeboard.org/pdf/statistics-sampling-distributions-sf.pdf. These materials include an overview of sampling distributions, along with some great activities.

The Main Ideas

One of the challenges in teaching the AP® Statistics course is keeping students focused on the big picture, not just the details of each section. We outline the main ideas for the chapter here.

Chapter 7 Introduction

Two of the best activities of the year appear in this chapter, so make sure to take the time to do them with your students. The "A penny for your thoughts?" activity is an excellent way for students to discover properties of the sampling distribution of a sample proportion and of the sampling distribution of a sample mean. "The craft stick problem" activity is a wonderful way to show how statisticians can use creativity—and sampling distributions—to make good estimates.

Section 7.1 What Is a Sampling Distribution?

In this section, we introduce the "big ideas" of sampling distributions. The first big idea is the difference between a statistic and a parameter. A parameter is a number that describes some characteristic of a population. A statistic estimates the value of a parameter using a sample from the population. Although it may seem simple to students when introduced, the concept is a crucial distinction that students often fail to make when they perform statistical inference.

The second big idea is that statistics vary. For example, the mean height in a sample of high school students is a variable that will change from sample to sample. Parameters, while usually unknown, are constants. This is why statistics have distributions, whereas parameters do not.

The third big idea is the difference between the population distribution, the distribution of the sample, and the sampling distribution of a sample statistic. "The candy machine" activity and Central Limit Theorem applet activity illustrate this distinction well.

The fourth big idea is how to describe a sampling distribution. To describe a sampling distribution adequately, we have to address its shape, center, and variability. If the mean of the sampling distribution is the same as the value of the parameter being estimated, the statistic is called an unbiased estimator. Ideally, the variability of a sampling distribution will be very small, meaning that the statistic is very precise. Larger sample sizes result in sampling distributions with less variability.

Section 7.2 Sample Proportions

In this section, we focus on the sampling distribution of a sample proportion and the sampling distribution of a difference of sample proportions. "The candy machine" activity is a great way to start this section—and well worth the time it takes to do in class.

A sample proportion is simply the number of successes in the sample divided by the sample size. And because the number of successes

in a sample often follows a binomial distribution, the sampling distribution of a sample proportion should already be somewhat familiar to students. In fact, the sampling distribution of a proportion is just a rescaled (by a factor of $1/n$) version of a binomial distribution.

This makes it easy to calculate the mean and standard deviation of the sampling distribution of a sample proportion if we know the sample size n and the proportion of successes in the population p. Also, if the sample size is large enough and the value of p isn't too close to 0 or 1, the shape of the sampling distribution of $\hat{p}$ is approximately Normal, allowing us to do probability calculations.

We also explore the sampling distribution of a difference between two proportions. Understanding how the difference between two sample proportions varies due to the chance involved in random selection (or random assignment) is essential for drawing conclusions based on a difference in proportions.

Section 7.3 Sample Means

In this section, we focus on the sampling distribution of a sample mean and the sampling distribution of a difference of sample means. To introduce and reinforce the important lessons of this section, we highly recommend the activities on pages 505 and 509.

The sampling distribution of a sample mean behaves in much the same way as the sampling distribution of a sample proportion. Both the sample mean and sample proportion are unbiased estimators, so their sampling distributions are always centered at the value of the population parameter they are trying to estimate. Also, as the sample size increases, both sampling distributions become less variable and more Normal. The very useful fact that the distribution of $\bar{x}$ becomes more Normal as the sample size increases is called the *central limit theorem*. However, in the special case of a Normally distributed population, the sampling distribution of a sample mean will always be Normal for any sample size.

We also explore the sampling distribution of a difference between two means. Understanding how the difference between two sample means varies due to the chance involved in random selection (or random assignment) is essential for drawing conclusions based on a difference in means observed in a study.

Chapter 7: Resources

Teacher's Resource Materials

The following resources, identified by the **TRM** in the annotated student pages, can be found by clicking on the link in the Teacher's e-Book (TE-Book), searching by category or chapter on the Teacher's Resource Flash Drive (TRFD), or logging into the book's digital platform and searching the Teacher's Resources menu (teacher log-in required).

- Alternate Examples: one file per section
- Lecture Presentation slides: one per section
- Chapter 7 Learning Targets Grid
- German Tank Problem
- Sampling Distribution of a Sample Proportion
- Sampling Distribution of a Sample Mean
- FRAPPY! Materials
- Chapter 7 Case Study
- Complete solutions for the Check Your Understanding problems, section exercises, review exercises, practice test, and cumulative practice test.
- Quizzes: one per section
- Chapter 7 Test

Free Response Questions from Previous AP® Statistics Exams

Questions can be found on the AP® Central website: apcentral.collegeboard.org/courses/ap-statistics/exam.

Students should be able to answer all the free response questions listed below with material learned in this chapter. Questions that contain content from this chapter but also require content from later chapters are listed in the last chapter required to complete the entire question. This list will be updated after each AP® Statistics exam and will be posted to the Teacher's Resource section of the book's digital platform and to www.statsmedic.com/free-response-questions.

Year	#	Content
2015	6	• Choosing a sampling method • Describing distribution of a sample • Describing distribution of a sample mean • Comparing variability of sampling distributions
2014	3	• Normal probability calculation • Sampling distribution of $\bar{x}$ • Probability rules
2012	6	• Selecting an SRS • Standard error of the mean for a simple random sample • Standard error of the mean for a stratified random sample • How stratified random sampling reduces variability
2010	2	• Sampling distribution of the sample mean • Probability calculation for a total
2009	2	• Inverse Normal calculation • Binomial probability calculation • Probability calculation for the sample mean
2008B	2	• Properties of estimators: bias and variability
2007B	2	• Addition rule • Binomial probability calculation • Sampling distribution of the sample mean
2007	3	• Sampling distribution of the sample mean • Probability calculation for the sample mean • Central limit theorem
2006	3	• Normal probability calculation • Binomial probability calculation • Probability calculation for a sample mean
2004B	3	• Normal probability calculation and interpretation • Probability calculation and interpretation for a sample mean
1998	1	• Sampling distribution of the sample mean • Effect of sample size on shape of sampling distribution

Applets

- The *Reese's Pieces* applet at www.rossmanchance.com/applets allows students to investigate the sampling distribution of a sample proportion of orange candies. Students can change the sample size and proportion of successes to see how these factors affect the sampling distribution of the sample proportion.

- The *Normal Approximation to Binomial Distributions* applet at highschool.bfwpub.com/updatedtps6e allows students to use sliders to discover that the Normal approximation gets better as the sample size increases and the proportion of successes gets closer to 0.5 and farther from 0 and 1.

- The *Sampling Distribution* applet at http://onlinestatbook.com/stat_sim/sampling_dist/index.html allows students to take samples of different sizes from populations of different shapes to explore how these factors affect the sampling distribution of the sample mean.

Chapter 7: Pacing Guide, Learning Targets, and Suggested Assignments

This pacing guide is based on a schedule with 110, 50-minute sessions before the AP® Statistics exam. If you have a different number of sessions before the AP® exam, you can modify the pacing guide to suit your needs. If you have additional time, consider incorporating quizzes, released AP® free response questions, or additional activities. See the Resources section above for suggestions.

The suggested homework assignments list odd-numbered exercises, whenever possible, so students can check their answers against the back-of-book answers. If you would rather students not have access to the answers while doing homework, adding 1 to the exercise numbers usually will do the trick, because the homework exercises typically are paired. For example, Exercises 1 and 2 will generally cover the same topics, but in different contexts. You may also choose to include the Recycle and Review questions at the end of each section, which review topics from previous sections or chapters. If your school is using the digital platform that accompanies TPS6, you will find these assignments pre-built as online homework assignments for Chapter 7.

Day	Content	Learning Targets: Students will be able to...	Suggested Assignment (MC bold)
1	Chapter 7 Introduction, 7.1 Parameters and Statistics, The Idea of a Sampling Distribution	• Distinguish between a parameter and a statistic. • Create a sampling distribution using all possible samples from a small population.	1, 3, 5, 7, 9
2	7.1 The Idea of a Sampling Distribution, Describing Sampling Distributions	• Use the sampling distribution of a statistic to evaluate a claim about a parameter. • Distinguish among the distribution of a population, the distribution of a sample, and the sampling distribution of a statistic. • Determine if a statistic is an unbiased estimator of a population parameter. • Describe the relationship between sample size and the variability of a statistic.	11, 13, 15, 19, 21, 25, **26–30**
3	7.2 The Sampling Distribution of $\hat{p}$, Using the Normal Approximation for $\hat{p}$	• Calculate the mean and standard deviation of the sampling distribution of a sample proportion $\hat{p}$ and interpret the standard deviation. • Determine if the sampling distribution of $\hat{p}$ is approximately Normal. • If appropriate, use a Normal distribution to calculate probabilities involving $\hat{p}$ or $\hat{p}_1 - \hat{p}_2$.	35, 37, 41, 43
4	7.2 The Sampling Distribution of a Difference Between Two Proportions	• Calculate the mean and the standard deviation of the sampling distribution of a difference in sample proportions $\hat{p}_1 - \hat{p}_2$ and interpret the standard deviation. • Determine if the sampling distribution of $\hat{p}_1 - \hat{p}_2$ is approximately Normal. • If appropriate, use a Normal distribution to calculate probabilities involving $\hat{p}$ or $\hat{p}_1 - \hat{p}_2$.	49, 51, **53–56**
5	7.3 The Sampling Distribution of $\bar{x}$, Sampling from a Normal Population, The Central Limit Theorem	• Calculate the mean and standard deviation of the sampling distribution of a sample mean $\bar{x}$ and interpret the standard deviation. • Explain how the shape of the sampling distribution of $\bar{x}$ is affected by the shape of the population distribution and the sample size. • If appropriate, use a Normal distribution to calculate probabilities involving $\bar{x}$ or $\bar{x}_1 - \bar{x}_2$.	59, 61, 63, 65, 69, 71, 73
6	7.3 The Sampling Distribution of a Difference Between Two Means	• Calculate the mean and the standard deviation of the sampling distribution of a difference in sample means $\bar{x}_1 - \bar{x}_2$ and interpret the standard deviation. • Determine if the sampling distribution of $\bar{x}_1 - \bar{x}_2$ is approximately Normal. • If appropriate, use a Normal distribution to calculate probabilities involving $\bar{x}$ or $\bar{x}_1 - \bar{x}_2$.	75, 77, 83, 85, **87–90**
7	Chapter 7 Review/FRAPPY!		Chapter 7 Review Exercises
8	Chapter 7 Test		Cumulative AP® Practice Test 2

Chapter 7 Alignment to the College Board's Fall 2019 AP® Statistics Course Framework*

Relationship to College Board Units

Chapter 7 in this book covers the topics addressed in Unit 5 of the College Board Course Framework. Students will be ready to take the Personal Progress Check for Unit 5 once they have completed Chapter 7.

Big Ideas and Enduring Understandings

Chapter 7 develops these Big Ideas and related Enduring Understandings outlined in the Course Framework:

- **Big Idea 1: Variation and Distribution (EU: VAR 1):** The distribution of measures for individuals within a sample or population describes variation. The value of a statistic varies from sample to sample. How can we determine whether differences between measures represent random variation or meaningful distinctions? Statistical methods based on probabilistic reasoning provide the basis for shared understandings about variation and about the likelihood that variation between and among measures, samples, and populations is random or meaningful.
- **Big Idea 2: Patterns and Uncertainty (EU: UNC 3):** Statistical tools allow us to represent and describe patterns in data and to classify departures from patterns. Simulation and probabilistic reasoning allow us to anticipate patterns in data and to determine the likelihood of errors in inference.

Course Skills

Chapter 7 helps students to develop the skills identified in the Course Framework.

- **1: Selecting Statistical Methods** (1.A)
- **3: Using Probability and Simulation** (3.B, 3.C)
- **4: Statistical Argumentation** (4.B)

Learning Objectives and Essential Knowledge Statements

Section	Learning Objectives	Essential Knowledge Statements
7.1	VAR-1.G, UNC-3.H, UNC-3.I, UNC-3.J	VAR-1.G.1, UNC-3.H.1, UNC-3.H.4, UNC-3.H.5, UNC-3.I.1, UNC-3.J.1, UNC-3.J.2
7.2	UNC-3.K, UNC-3.L, UNC-3.M, UNC-3.N, UNC-3.O, UNC-3.P	UNC-3.K.1, UNC-3.K.2, UNC-3.L.1, UNC-3.M.1, UNC-3.N.1, UNC-3.N.2, UNC-3.O.1, UNC-3.P.1
7.3	UNC-3.H, UNC-3.Q, UNC-3.R, UNC-3.S, UNC-3.T, UNC-3.U, UNC-3.V	UNC-3.H.2, UNC-3.H.3, UNC-3.Q.1, UNC-3.Q.2, UNC-3.R.1, UNC-3.R.2, UNC-3.S.1, UNC-3.T.1, UNC-3.T.2, UNC-3.U.1, UNC-3.U.2, UNC-3.V.1

A detailed alignment (The Nitty Gritty Guide) that can be sorted by Course Framework Unit, Topic, Learning Objective, Essential Knowledge Statement, or textbook section, is available on the TRFD and in the Teacher's Resources folder on Sapling Plus. **TRM**

*Should changes be made to the Course Framework in the future, an updated alignment will be placed on our AP® updates page at go.bfwpub.com/ap-course-updates.

UNIT 5
Sampling Distributions

Chapter 7

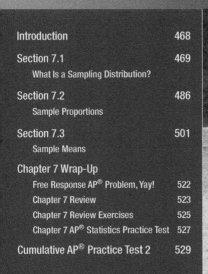

Sampling Distributions

Stefano Paterna/Alamy

Teaching Tip

Unit 5 in the College Board Course Framework aligns to Chapter 7 in this book. Students will be ready to take the Personal Progress Check for Unit 5 once they have completed Chapter 7.

PD Chapter 7 Overview

To watch the video overview of Chapter 7 (for teachers), click on the link in the TE-Book, look on the TRFD, or download from the Teacher's Resources on the book's digital platform.

Teaching Tip

We highly recommend that you download and read the Special Focus materials on Sampling Distributions available on AP® Central at https://apcentral.collegeboard .org/pdf/statistics-sampling-distributions-sf .pdf. These materials include an overview of sampling distributions, along with some great activities.

TRM Lecture Presentation Slides

If you are new to teaching AP® Statistics or are short on time when preparing for class, you may find the Lecture Presentation Slides to be helpful. Experienced AP® Statistics Teacher Doug Tyson has created one slide presentation per section. You may use them as is, modify them to fit your needs, or share them with students who miss class. Find them on the TRFD and in the Teacher's Resources on the book's digital platform.

Making Connections

Refer to this illustration used in Chapter 4. Remind students that it is rare to have data for a whole population. Most often, we have data from a sample.

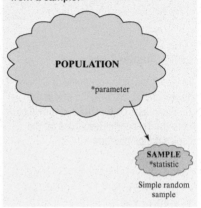

INTRODUCTION

In this chapter, we will return to a key idea about statistical inference from Chapter 4—making conclusions about a population based on data from a sample. Here are a few examples of statistical inference in practice:

- Each month, the Current Population Survey (CPS) interviews a random sample of individuals in about 60,000 U.S. households. The CPS uses the proportion of unemployed people in the sample $\hat{p}$ to estimate the national unemployment rate p.

- To estimate how much gasoline prices vary in a large city, a reporter records the price per gallon of regular unleaded gasoline at a random sample of 10 gas stations in the city. The range (Maximum − Minimum) of the prices in the sample is 25 cents. What can the reporter say about the range of gas prices at all the city's stations?

- A battery manufacturer wants to make sure that the AA batteries it produces each hour meet certain standards. Quality control inspectors collect data from a random sample of 50 AA batteries produced during one hour and use the sample mean lifetime $\bar{x}$ to estimate the unknown population mean lifetime μ for all batteries produced that hour.

Let's look at the battery example a little more closely. To make an inference about the batteries produced in the given hour, we need to know how close the sample mean $\bar{x}$ is likely to be to the population mean μ. After all, different random samples of 50 batteries from the same hour of production would yield different values of $\bar{x}$. How can we describe this *sampling distribution* of possible $\bar{x}$ values? We can think of $\bar{x}$ as a random variable because it takes numerical values that describe the outcomes of the random sampling process. As a result, we can examine its probability distribution using what we learned in Chapter 6.

The following activity will help you get a feel for the distribution of two very common statistics, the sample mean $\bar{x}$ and the sample proportion $\hat{p}$.

▶ ACTIVITY OVERVIEW

To prepare for using this activity, watch the overview video by clicking on the link in the TE-Book, opening the TRFD, or downloading from the Teacher's Resources on the book's digital platform.

Time: 60 minutes

Materials: Pennies (variety of old and new), poster board or chart paper, sticker dots

Teaching Advice: Some teachers spread out this activity over the month preceding Chapter 7. Each student does one sample per day as he or she arrives, increasing the sample size each week.

You will need a "population" of pennies for this activity. Consider having each student bring in 10 pennies for the activity. Ask them to put sticker dots on a poster board (they can write X or $\bar{x}$ or $\hat{p}$ on the sticker). Post the completed poster board in the classroom so that you can refer to this activity throughout the rest of the course.

Give each dotplot a title. They should be "Population distribution of year," "Simulated sampling distribution of $\bar{x}$," and "Simulated sampling distribution of $\hat{p}$." Point at a specific "dot" in each dotplot and ask students what it represents. For the distribution of X, they should reply, "One individual from the population." For the distribution of $\bar{x}$ and $\hat{p}$, they should reply, "One sample of 5 or 20 pennies

ACTIVITY A penny for your thoughts?

In this activity, your class will investigate how the mean year $\bar{x}$ and the proportion of pennies from the 2000s $\hat{p}$ vary from sample to sample, using a large population of pennies of various ages.[1]

1. Each member of the class should randomly select 1 penny from the population and record the year of the penny with an "X" on the dotplot provided by your teacher. Return the penny to the population. Repeat this process until at least 100 pennies have been selected and recorded. This graph gives you an idea of what the population distribution of penny years looks like.

2. Each member of the class should then select an SRS of 5 pennies from the population and note the year on each penny.
   - Record the average year of these 5 pennies (rounded to the nearest year) with an "$\bar{x}$" on a new class dotplot. Make sure this dotplot is on the same scale as the dotplot in Step 1.
   - Record the proportion of pennies from the 2000s with a "$\hat{p}$" on a different dotplot provided by your teacher.

and a mean (or proportion) calculated from that sample." This thinking will help students develop the concept of a population distribution and a sampling distribution.

Answers:

1. It is likely that this distribution is skewed left.

2–3. Answers will vary.

4. All three distributions have a similar center. The shape of the distribution of X is skewed left, while the shape of the distribution of $\bar{x}$ becomes closer to Normal when moving from sample size 5 to sample size 20. Compared to the variability of X, the distribution of $\bar{x}$ has less variability, with lower variability for samples of size 20 than size 5. Increasing sample size

makes the shape of the distribution of $\bar{x}$ more Normal and less variable, with the same center.

5. Both distributions of $\hat{p}$ have a similar center, but the shape of the distribution for samples of size 20 is more Normal and less variable than the distribution for samples of size 5. Increasing sample size makes the shape of the distribution of $\hat{p}$ more Normal and less variable, with the same center.

Return the pennies to the population. Repeat this process until there are at least 100 $\bar{x}$'s and 100 $\hat{p}$'s.

3. Repeat Step 2 with SRSs of size $n = 20$. Make sure these dotplots are on the same scale as the corresponding dotplots from Step 2.

4. Compare the distribution of X (year of penny) with the two distributions of $\bar{x}$ (mean year). How are the distributions similar? How are they different? What effect does sample size seem to have on the shape, center, and variability of the distribution of $\bar{x}$?

5. Compare the two distributions of $\hat{p}$. How are the distributions similar? How are they different? What effect does sample size seem to have on the shape, center, and variability of the distribution of $\hat{p}$?

Sampling distributions are the foundation of inference when data are produced by random sampling. Because the results of random samples include an element of chance, we can't guarantee that our inferences are correct. What we can guarantee is that our methods usually give correct answers. The reasoning of statistical inference rests on asking, "How often would this method give a correct answer if I used it many times?" If our data come from random sampling, the laws of probability help us answer this question. These laws also allow us to determine how far our estimates typically vary from the truth and what values of a statistic should be considered unusual.

Section 7.1 presents the basic ideas of sampling distributions. The most common applications of statistical inference involve proportions and means. Section 7.2 focuses on sampling distributions involving proportions. Section 7.3 investigates sampling distributions involving means.

<table>
<tr><td>SECTION 7.1</td><td></td></tr>
</table>

SECTION 7.1   What Is a Sampling Distribution?

LEARNING TARGETS   *By the end of the section, you should be able to:*

- Distinguish between a parameter and a statistic.
- Create a sampling distribution using all possible samples from a small population.
- Use the sampling distribution of a statistic to evaluate a claim about a parameter.

- Distinguish among the distribution of a population, the distribution of a sample, and the sampling distribution of a statistic.
- Determine if a statistic is an unbiased estimator of a population parameter.
- Describe the relationship between sample size and the variability of a statistic.

Because of some very large incomes, the mean total income ($73,750) was much larger than the median total income ($55,071).

What is the average income of U.S. residents with a college degree? Each March, the government's Current Population Survey (CPS) asks detailed questions about income. The random sample of about 70,000 U.S. college grads contacted in March 2016 had a mean "total money income" of $73,750 in 2015.[2] That $73,750 describes the sample, but we use it to estimate the mean income of all college grads in the United States.

Making Connections

Remind students of this image from Chapter 4. In this chapter, we will focus on two types of statistics: means and proportions.

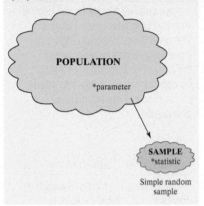

POPULATION

*parameter

SAMPLE
*statistic

Simple random sample

Teaching Tip

Suggest that students identify the sample and the statistic first. These are usually easier to identify and help clarify the population and the parameter. The population is the collection of individuals that could have been selected for the sample.

Preparing for Inference

At this point in the course, be tenacious about using correct notation. Don't let students get away with using μ when it really should be $\bar{x}$ or p when it should be $\hat{p}$. Being picky about notation now will help students be successful when we get to the formulas for confidence intervals and significance tests in Chapters 8–12.

Parameters and Statistics

As we begin to use sample data to draw conclusions about a larger population, we must be clear about whether a number describes a sample or a population. For the sample of college graduates contacted by the CPS, the mean income was $\bar{x} = \$73,750$. The number $73,750 is a **statistic** because it describes this one CPS sample. The population that the poll wants to draw conclusions about is the nearly 100 million U.S. residents with a college degree. In this case, the **parameter** of interest is the mean income μ of all these college graduates. We don't know the value of this parameter, but we can estimate it using data from the sample.

A sample statistic is sometimes called a *point estimator* of the corresponding population parameter because the estimate—$73,750 in this case—is a single point on the number line.

> **DEFINITION**　**Statistic, Parameter**
>
> A **statistic** is a number that describes some characteristic of a sample.
>
> A **parameter** is a number that describes some characteristic of a population.

Recall our hint from Chapter 1 about **s** and **p**: statistics come from **s**amples, and **p**arameters come from **p**opulations. As long as we were doing data analysis, the distinction between population and sample rarely came up. Now that we are focusing on statistical inference, however, it is essential. The notation we use should reflect this distinction. The table shows three commonly used statistics and their corresponding parameters.

It is common practice to use Greek letters for parameters and Roman letters for statistics. In that case, the population proportion would be π (pi, the Greek letter for "p") and the sample proportion would be p. We'll stick with the notation that's used on the AP® exam, however.

Sample statistic		Population parameter
$\bar{x}$ (the sample mean)	estimates	μ (the population mean)
$\hat{p}$ (the sample proportion)	estimates	p (the population proportion)
s_x (the sample SD)	estimates	σ (the population SD)

EXAMPLE

From ghosts to cold cabins
Parameters and statistics

PROBLEM: Identify the population, the parameter, the sample, and the statistic in each of the following settings.

(a) The Gallup Poll asked 515 randomly selected U.S. adults if they believe in ghosts. Of the respondents, 160 said "Yes."[3]

(b) During the winter months, the temperatures outside the Starneses' cabin in Colorado can stay well below freezing for weeks at a time. To prevent the pipes from freezing, Mrs. Starnes sets the thermostat at 50°F. She wants to know how low the temperature actually gets in the cabin. A digital thermometer records the indoor temperature at 20 randomly chosen times during a given day. The minimum reading is 38°F.

RyersonClark/Getty Images

ALTERNATE EXAMPLE　　　　　　　　　　　　　　　　　　Skill 1.C

From college tuition to German tanks　Parameters and statistics

PROBLEM:

Identify the population, parameter, sample, and statistic in each of the following settings.

(a) A high school student was interested in finding the mean annual tuition at a 4-year U.S. college. The student randomly selected 23 U.S. colleges and found a mean annual tuition of $19,800.

(b) During World War II, the United States captured several tanks from the German army. Based on the serial numbers on the tanks, statisticians estimated that the German army produced 7168 tanks during the war.

SOLUTION:

(a) *Population:* All 4-year U.S. colleges. *Parameter:* μ = the mean annual tuition. *Sample:* The 23 selected colleges. *Statistic:* $\bar{x}$ = the mean annual tuition in the sample = $19,800.

(b) *Population:* All German tanks produced during World War II. *Parameter:* The true total number of German tanks. *Sample:* The several tanks that were captured. *Statistic:* The estimated total number of German tanks based on the sample = 7168.

SOLUTION:

(a) *Population: all U.S. adults. Parameter: $p =$ the proportion of all U.S. adults who believe in ghosts. Sample: the 515 people who were interviewed in this Gallup Poll. Statistic: $\hat{p} =$ the proportion in the sample who say they believe in ghosts $= 160/515 = 0.31$.*

(b) *Population: all times during the day in question. Parameter: the true minimum temperature in the cabin at all times that day. Sample: the 20 randomly selected times. Statistic: the sample minimum temperature, 38°F.*

> Not all parameters and statistics have their own symbols. To distinguish parameters and statistics in these cases, use descriptors like "true" and "sample."

FOR PRACTICE, TRY EXERCISE 1

> **AP® EXAM TIP**
>
> Many students lose credit on the AP® Statistics exam when defining parameters because their description refers to the sample instead of the population or because the description isn't clear about which group of individuals the parameter is describing. When defining a parameter, we suggest including the word *all* or the word *true* in your description to make it clear that you aren't referring to a sample statistic.

The Idea of a Sampling Distribution

The students in Mrs. Gallas's class did the "Penny for your thoughts" activity at the beginning of the chapter. Figure 7.1 shows their "dotplot" of the sample mean year for 50 samples of size $n = 5$.

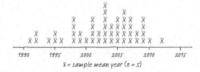

FIGURE 7.1 Distribution of the sample mean year of penny for 50 samples of size $n = 5$ from Mrs. Gallas's population of pennies.

It shouldn't be surprising that the statistic $\bar{x}$ is a random variable. After all, different samples of $n = 5$ pennies will produce different means. As you learned in Section 4.3, this basic fact is called **sampling variability**.

> **DEFINITION   Sampling variability**
>
> **Sampling variability** refers to the fact that different random samples of the same size from the same population produce different values for a statistic.

Knowing how statistics vary from sample to sample is essential when making an inference about a population. Understanding sampling variability reminds us that the value of a statistic is unlikely to be exactly equal to the value of the parameter it is trying to estimate. It also lets us say how much we expect an estimate to vary from its corresponding parameter.

Mrs. Gallas's class took only 50 random samples of 5 pennies. However, there are many, many possible random samples of size 5 from Mrs. Gallas's large population of pennies. If the students took every one of those possible samples, calculated the value of $\bar{x}$ for each, and graphed all those $\bar{x}$ values, then we'd have a **sampling distribution**.

Teaching Tip

When introducing the idea of a sampling distribution, emphasize that every statistic has a sampling distribution. That is, anything that can be calculated from a sample has its own distribution that can be approximated with simulation.

Preparing for Inference

Sampling variability is the reason why statistics are often reported with a *margin of error*. Because the value calculated from our sample is unlikely to be exactly equal to the value of the parameter, we give a *margin of error* to describe how far off our estimate could be from the true value, at most. Students will formally encounter *margin of error* in Chapter 8 when we introduce confidence intervals.

Making Connections

In Section 4.3, students learned about sampling variability in the "Red bead" activity on page 298. In this activity, students took several samples of size 20 and recorded the proportion of red beads for each sample. The proportion varied from sample to sample because of sampling variability.

AP® EXAM TIP

On the 2019 AP® Statistics exam, question #6 part (c) asked students to describe a sampling distribution. For full credit on the rubric, students needed to state that "all possible samples" must be taken to arrive at the theoretical sampling distribution; responding with merely "100 samples," "many samples," and "many, many samples" was not enough for full credit.

Teaching Tip

How many samples of size *n* are possible from a population of size *N*? We could use combinations to find out:

$$_NC_n = \binom{N}{n} = \frac{N!}{n!(N-n)!}$$

Teaching Tip

These activities and examples, in which we create a sampling distribution by taking many, many samples, are simply a thinking exercise for developing the concept of a sampling distribution. In reality, we will only take one sample from the population and then use an estimate from that one sample to make inferences about the population. To know how much variability an estimate could have, we must understand sampling distributions.

ALTERNATE EXAMPLE   Skill 3.C

Pick three
Creating a sampling distribution

PROBLEM:
Mrs. Chauvet has an interesting approach to assigning grades in her statistics class. Of the 5 tests students take throughout the semester, Mrs. Chauvet selects a random sample of 3, finds the average score of these tests, and records this average as the student's final grade. Joe's test scores are as follows: 93, 87, 96, 78, 90.
(a) List all 10 possible samples of size 3.
(b) Calculate the mean of each sample and display the sampling distribution of the sample mean using a dotplot.
(c) Calculate the range of each sample and display the sampling distribution of the sample range using a dotplot.

SOLUTION:
(a) Sample

93, 87, 96
93, 87, 78
93, 87, 90
93, 96, 78
93, 96, 90
93, 78, 90
87, 96, 78
87, 96, 90
87, 78, 90
96, 78, 90

Remember that a distribution describes the possible values of a variable and how often these values occur. Thus, a sampling distribution shows the possible values of a *statistic* and how often these values occur.

DEFINITION   Sampling distribution
The **sampling distribution** of a statistic is the distribution of values taken by the statistic in all possible samples of the same size from the same population.

For large populations, it is too difficult to take all possible samples of size *n* to obtain the exact sampling distribution of a statistic. Instead, we can approximate a sampling distribution by taking many samples, calculating the value of the statistic for each of these samples, and graphing the results. Because the students in Mrs. Gallas's class didn't take all possible samples of 5 pennies, their dotplot of $\bar{x}$'s in Figure 7.1 is called an *approximate sampling distribution*.

The following example demonstrates how to construct a complete sampling distribution using a small population.

EXAMPLE

Sampling heights
Creating a sampling distribution

PROBLEM: John and Carol have four grown sons. Their heights (in inches) are 71, 75, 72, and 68.
(a) List all 6 possible samples of size 2.
(b) Calculate the mean of each sample and display the sampling distribution of the sample mean using a dotplot.
(c) Calculate the range of each sample and display the sampling distribution of the sample range using a dotplot.

SOLUTION:

(a)
Sample
71, 75
71, 72
71, 68
75, 72
75, 68
72, 68

(b)
Sample	Sample mean
71, 75	73.0
71, 72	71.5
71, 68	69.5
75, 72	73.5
75, 68	71.5
72, 68	70.0

(dotplot) $\bar{x}$ = sample mean height (in.)

(c)
Sample	Sample range
71, 75	4
71, 72	1
71, 68	3
75, 72	3
75, 68	7
72, 68	4

(dotplot) Sample range of height (in.)

FOR PRACTICE, TRY EXERCISE 7

Being able to construct (or approximate) the sampling distribution of a statistic allows us to determine the values of the statistic that are likely to occur by chance alone—and the values that should be considered unusual. The following example shows how we can use a sampling distribution to evaluate a claim.

(b, c)

Sample	Sample mean	Sample range
93, 87, 96	92	9
93, 87, 78	86	15
93, 87, 90	90	6
93, 96, 78	89	18
93, 96, 90	93	6
93, 78, 90	87	15
87, 96, 78	87	18
87, 96, 90	91	9
87, 78, 90	85	12
96, 78, 90	88	18

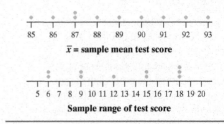

$\bar{x}$ = sample mean test score

Sample range of test score

EXAMPLE	Reaching for chips
	Using a sampling distribution to evaluate a claim

PROBLEM: To determine how much homework time students will get in class, Mrs. Lin has a student select an SRS of 20 chips from a large bag. The number of red chips in the SRS determines the number of minutes in class students get to work on homework. Mrs. Lin claims that there are 200 chips in the bag and that 100 of them are red. When Jenna selected a random sample of 20 chips from the bag (without looking), she got 7 red chips. Does this provide convincing evidence that less than half of the chips in the bag are red?

(a) What is the evidence that less than half of the chips in the bag are red?

(b) Provide two explanations for the evidence described in part (a).

We used technology to simulate choosing 500 SRSs of size $n = 20$ from a population of 200 chips, 100 red and 100 blue. The dotplot shows $\hat{p}$ = the sample proportion of red chips for each of the 500 samples.

(c) There is one dot on the graph at 0.80. Explain what this value represents.

(d) Would it be surprising to get a sample proportion of $\hat{p} = 7/20 = 0.35$ or smaller in an SRS of size 20 when $p = 0.5$? Justify your answer.

(e) Based on your previous answers, is there convincing evidence that less than half of the chips in the large bag are red? Explain your reasoning.

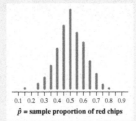

$\hat{p}$ = sample proportion of red chips

SOLUTION:

(a) Jenna's sample proportion was $\hat{p} = 7/20 = 0.35$, which is less than 0.50.

(b) It is possible that Mrs. Lin is telling the truth and Jenna got a $\hat{p}$ less than 0.50 because of sampling variability. It is also possible that Mrs. Lin is lying and less than half of the chips in the bag are red.

(c) In one simulated SRS of 20 chips, there were 16 red chips. So $\hat{p} = 16/20 = 0.80$ for this sample.

(d) No; there were many simulated samples that had $\hat{p}$ values less than or equal to 0.35.

(e) Because it isn't surprising to get a $\hat{p}$ less than or equal to 0.35 by chance alone when $p = 0.50$, there isn't convincing evidence that less than half of the chips in the bag are red.

FOR PRACTICE, TRY EXERCISE 13

When we simulate a sampling distribution using assumed values for the parameters, like in the chips example, the resulting distribution is sometimes called a *randomization distribution*.

Suppose that Jenna's sample included only 3 red chips, giving $\hat{p} = 3/20 = 0.15$. Would this provide convincing evidence that less than half of the chips in the bag are red? Yes. According to the simulated sampling distribution in the example, it would be very unusual to get a $\hat{p}$ value this small when $p = 0.50$. Therefore, sampling variability would not be a plausible explanation for the outcome of Jenna's sample. The only plausible explanation for a $\hat{p}$ value of 0.15 is that less than half of the chips in the bag are red.

Figure 7.2 (on the next page) illustrates the process of choosing many random samples of 20 chips from a population of 100 red chips and 100 blue chips and finding the sample proportion of red chips $\hat{p}$ for each sample. Follow the flow of the figure from the population distribution on the left, to choosing an SRS, graphing the distribution of sample data, and finding the $\hat{p}$ for that particular sample, to collecting together the $\hat{p}$'s from many samples. The first sample has $\hat{p} = 0.40$. The second sample is a different group of chips, with $\hat{p} = 0.55$, and so on.

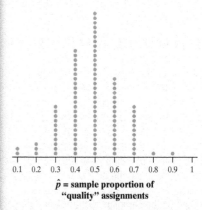

$\hat{p}$ = sample proportion of "quality" assignments

(c) There is one dot on the graph at 0.90. Explain what this value represents.

(d) Would it be surprising to get a sample proportion of $\hat{p} = 9/10 = 0.90$ or greater in an SRS of size 10 when $p = 0.5$? Justify your answer.

(e) Based on your previous answers, is there convincing evidence that more than half of Mr. Osters's students are turning in "quality" homework?

SOLUTION:

(a) Mr. Osters's sample proportion was $\hat{p} = 9/10 = 0.90$, which is more than 0.50.

(b) It is possible that only half of students produce "quality" assignments and Mr. Osters got a $\hat{p}$ greater than 0.50 because of sampling variability. It is also possible that Mr. Osters's claim is correct and more than half of his students are producing "quality" assignments.

(c) In one simulated SRS of 10 students, 9 turned in "quality" assignments. So $\hat{p} = 9/10 = 0.90$ for this simulated sample.

(d) Yes; only 1 simulated sample of 100 had a $\hat{p}$ of a value greater than or equal to 0.90.

(e) Because it is surprising to get a $\hat{p}$ greater than or equal to 0.90 by chance alone when $p = 0.50$, there is convincing evidence that more than half of Mr. Osters's students are producing "quality" homework.

ALTERNATE EXAMPLE

Do your homework Using a sampling distribution to evaluate a claim

Skill 4.B

PROBLEM:
A principal complained that only half of all students turn in "quality" homework. However, Mr. Osters believes that more than half of his 120 AP® Statistics students produce "quality" homework. To test his claim, he randomly selected 10 students to hand in their assignments one day. He determined that 9 of the 10 assignments could be considered "quality." Does this provide convincing evidence that more than half of Mr. Osters's students are producing "quality" homework?

(a) What is the evidence that more than half of Mr. Osters's students produced "quality" homework?

(b) Provide two explanations for the evidence described in part (a).

We used technology to simulate choosing 100 SRSs of size $n = 10$ from a population of 120 students, where half produce "quality" assignments and half do not. The dotplot shows $\hat{p}$ = the sample proportion of "quality" assignments for each of the 100 samples.

Preparing for Inference

The example on this page gets students thinking about the strength of evidence against a claim. In a significance test, the claim that half the chips in the bag are red is the *null hypothesis*. We will assume this claim is true, then find the probability of getting our result (7 out of 20 red) or less purely by chance (*P-value*). If the probability is less than 5% (α), we have convincing evidence that less than half the chips in the bag are red. If the probability is greater than 5%, it is plausible that half the chips in the bag are red and Jenna's results happened purely by chance.

Teaching Tip

Make sure your students completely understand the figure on this page. Students often have trouble making the distinction between the population distribution, the distribution of sample data, and the distribution of the sample statistic. One way to help make this distinction is to ask who the individuals are in each distribution. In the population distribution and the distribution of sample data, the individuals are chips (red or blue). In the sampling distribution of the sample proportion, the individuals are $\hat{p}$'s. That is, each dot in the dotplot represents the $\hat{p}$ obtained from a sample.

Preparing for Inference

When checking conditions for inference procedures involving sample proportions, many students incorrectly state that the population or the sample follows a Normal distribution. When this happens, revisit the figure on this page. When the data are categorical, neither the population nor the sample can be approximately Normal—there is just a bar for the successes and a bar for the failures.

AP® EXAM TIP

Students will often refer to the "sample" distribution when they mean "sampling" distribution. Remind them that a "sample" distribution represents one sample, whereas a "sampling" distribution represents many, many samples. In the text, we avoid the term *sample distribution*, opting instead for *distribution of sample data*.

The dotplot at the right of the figure shows the distribution of the values of $\hat{p}$ from 500 separate SRSs of size 20. This is the *approximate sampling distribution* of the statistic $\hat{p}$.

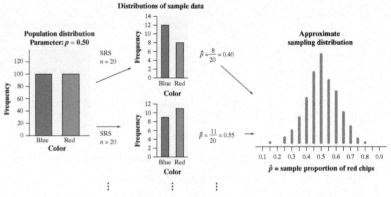

FIGURE 7.2 The idea of a sampling distribution is to take many samples from the same population, collect the value of the statistic from all the samples, and display the distribution of the statistic. The dotplot shows the approximate sampling distribution of $\hat{p}$ = the sample proportion of red chips.

AP® EXAM TIP

Terminology matters. Never just say "the distribution." Always say "the distribution of [blank]," being careful to distinguish the distribution of the population, the distribution of sample data, and the sampling distribution of a statistic. Likewise, don't use ambiguous terms like "sample distribution," which could refer to the distribution of sample data or to the sampling distribution of a statistic. You will lose credit on free response questions for misusing statistical terms.

As Figure 7.2 shows, there are three distinct distributions involved when we sample repeatedly and calculate the value of a statistic.

- The *population distribution* gives the values of the variable for all individuals in the population. In this case, the individuals are the 200 chips and the variable we're recording is color. Our parameter of interest is the proportion of red chips in the population, $p = 0.50$.
- The *distribution of sample data* shows the values of the variable for the individuals in a sample. In this case, the distribution of sample data shows the values of the variable color for the 20 chips in the sample. For each sample, we record a value for the statistic $\hat{p}$, the sample proportion of red chips.
- The *sampling distribution of the sample proportion* displays the values of $\hat{p}$ from all possible samples of the same size.

Remember that a sampling distribution describes how a *statistic* (e.g., $\hat{p}$) varies in many samples from the population. However, the population distribution and the distribution of sample data describe how *individuals* (e.g., chips) vary.

 CHECK YOUR UNDERSTANDING

Mars,® Inc. says that the mix of colors in its M&M'S® Milk Chocolate Candies from its Hackettstown, NJ, factory is 25% blue, 25% orange, 12.5% green, 12.5% yellow, 12.5% red, and 12.5% brown. Assume that the company's claim is true and that you will randomly select 50 candies to estimate the proportion that are orange.

AP® EXAM TIP

On the AP® Statistics exam, a common error is to write an ambiguous statement such as "the variability decreases when the sample size increases." This statement may be true for the sampling distribution, but it almost certainly is *not* true for the distribution of the sample. The correct statement here would be "the variability *of the sampling distribution* decreases when the sample size increases."

1. Identify the population, the parameter, the sample, and the statistic in this setting.
2. Graph the population distribution.
3. Imagine taking a random sample of 50 M&M'S® Milk Chocolate Candies. Make a graph showing a possible distribution of the sample data. Give the value of the statistic for this sample.
4. Which of these three graphs could be the approximate sampling distribution of the statistic? Explain your choice.

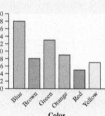
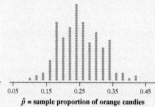
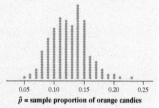

Describing Sampling Distributions

The fact that statistics from random samples have definite sampling distributions allows us to answer the question "How trustworthy is a statistic as an estimate of a parameter?" To get a complete answer, we will consider the shape, center, and variability of the sampling distribution. For reasons that will be clear later, we'll save shape for Sections 7.2 and 7.3.

Here is an activity that gets you thinking about the center and variability of a sampling distribution.

ACTIVITY   The craft stick problem

In this activity, you will create a statistic for estimating the total number of craft sticks in a bag (N). The sticks are numbered 1, 2, 3, ..., N. Near the end of the activity, your teacher will select a random sample of $n = 7$ sticks and read the number on each stick to the class. The team that has the best estimate for the total number of sticks will win a prize.

1. Form teams of three or four students. Each team will be given a statistic to begin their investigation.

2. For now, assume that there are $N = 100$ sticks in the bag and that you will be selecting a sample of size $n = 7$. To investigate the quality of the statistic you were given, generate a simulated sampling distribution:

(a) Using your TI-84 calculator, select an SRS of size 7 using the command RandIntNoRep(lower:1, upper:100, n:7). Calculate and record the value of your statistic for this simulated sample. [With a TI-83 or older TI-84 OS, use the command RandInt(lower:1, upper:100, n:7) and verify that there are no repeated numbers.]

1. The population is all M&M'S® Milk Chocolate Candies produced by the factory in Hackettstown, NJ. The parameter is $p =$ the proportion of all M&M'S Milk Chocolate Candies produced by the factory in Hackettstown, NJ, that are orange. The parameter is claimed to be $p = 0.25$. The sample is the 50 M&M'S Milk Chocolate Candies selected. The statistic is the proportion of the sample of 50 M&M'S that are orange, $\hat{p}$.

2.

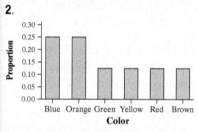

3. The graph below shows a possible distribution of sample data. For this sample, there are 11 orange M&M'S, so $\hat{p} = \dfrac{11}{50} = 0.22$.

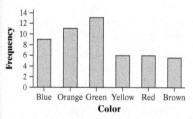

4. The middle graph is the approximate sampling distribution of $\hat{p}$. The statistic measures the proportion of orange candies in samples of 50 M&M'S. Assuming the company is correct, 25% of the M&M'S are orange, so the center of the distribution of $\hat{p}$ should be at approximately 0.25. The first graph shows the distribution of the colors for one sample, rather than the distribution of $\hat{p}$ from many samples, and the third graph is centered at 0.125, rather than 0.25.

Teaching Tip

If you are using the craft stick problem, be sure to read the Activity Overview on the next page.

TRM German Tank Problem

An alternate activity that can be used here is the German tank problem. In this scenario, U.S. intelligence is trying to estimate the total number of German tanks. The tanks that were captured are considered the sample and the serial numbers on the tanks reveal that they are essentially numbered 1, 2, 3, ..., N. We deliberately created a new context for this example so that students can't Google "German Tank Problem" to come up with a solution.

ACTIVITY OVERVIEW

To prepare for using this activity, watch the overview video by clicking on the link in the TE-Book, opening the TRFD, or downloading from the Teacher's Resources on the book's digital platform.

Time: 35 minutes

Materials: Craft sticks

Teaching Advice: Ask students to identify the parameter and statistic in this context; the parameter is the true total number of sticks in the bag (*N*), and the statistic is the estimate they come up with based on their sample data (they can call it $\hat{N}$). Then follow these steps for the questions in the activity.

1. The teacher provides the following 3 statistics: max, mean + 3SD, 2 × median. Multiple groups will likely have the same statistic. Don't give out any additional statistics; that way, students have a chance to create new statistics later in the activity.

2. For students to know which of the three statistics provides the best estimate, they must test them with a population in which they know the true value of the number of sticks (100). If you have the time (or a student assistant), prepare several bags of sticks numbered 1 to 100 so that students can take samples of sticks rather than using calculators to simulate this scenario. At the front of the room, prepare three parallel axes (0 to 200) for dotplots—one for each statistic.

3. Use the dotplots to guide the discussion.

4. Make students aware that many values can be calculated from the sample (mean, median, range, *IQR*, standard deviation) that may help in estimating the total.

5. Prepare a bag with craft sticks that are labeled 1, 2, 3, . . . , *N*. You can decide what value to use for *N*, but make it more than 50 and don't make *N* = 100, like students used in question #2. Remember, the winning team may have the best estimate, but not the best estimator.

Answers:

2. Here is an example of what the sampling distributions might look like for each statistic:

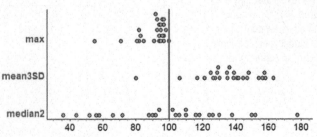

(b) Repeat the previous step at least 9 more times, recording the value of your statistic each time.

(c) Display the simulated sampling distribution of your statistic on the dotplot provided by your teacher.

3. As a class, discuss the quality of each of these statistics. Do any of them consistently overestimate or consistently underestimate the truth? Are some of the statistics more variable than others?

4. In your group, spend about 10–15 minutes creating a few additional statistics that could be used to estimate the total number of sticks. Create a simulated sampling distribution for each one (as in Step 2) to determine which statistic you will use for the competition.

5. It is time for the final competition. Your teacher will select a random sample of 7 sticks from the bag and read out the stick numbers. On a sheet of paper, write the names of your group, the statistic you think is best (a formula), and the value of the statistic calculated from the sample provided by the teacher. The closest estimate wins!

CENTER: BIASED AND UNBIASED ESTIMATORS Figure 7.3 shows the simulated sampling distribution of $\hat{p}$ = proportion of red chips when selecting samples of size *n* = 20 from a population where *p* = 0.5.

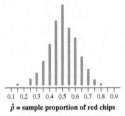

$\hat{p}$ = **sample proportion of red chips**

FIGURE 7.3 The distribution of $\hat{p}$ = the sample proportion of red chips in 500 SRSs of size *n* = 20 from a population where *p* = 0.5.

Notice that the center of this distribution is very close to 0.5, the parameter value. In fact, if we took all possible samples of 20 chips from the population, calculated $\hat{p}$ for each sample, and then found the mean of all those $\hat{p}$-values, we'd get *exactly* 0.5. For this reason, we say that $\hat{p}$ is an **unbiased estimator** of *p*.

> **DEFINITION   Unbiased estimator**
>
> A statistic used to estimate a parameter is an **unbiased estimator** if the mean of its sampling distribution is equal to the value of the parameter being estimated.

In a particular sample, the value of an unbiased estimator might be greater than the value of the parameter or it might be less than the value of the parameter. However, because the sampling distribution of the statistic is centered at the true value, the statistic will not consistently overestimate or consistently underestimate the parameter. This fits with our definition of bias from Chapter 4. The design of a statistical study shows bias if it is very likely to underestimate or very likely to overestimate the value we want to know.

---

3. Using the maximum value consistently underestimates the truth, while mean + 3SD consistently overestimates the truth. 2 × median is much more variable than the other two statistics.

4. Reasonable statistics include: maximum + minimum, $\bar{x} + 2s_x$, $2\bar{x}$, 2 × median, $\left(\dfrac{n+1}{n}\right)$ × maximum, $Q_1 + Q_3$, 2*IQR*, $\bar{x}$ + *median*, and *IQR* + *median*.

5. The best statistic is $\hat{N} = m + \dfrac{m}{n} - 1$, where *m* is the sample maximum and *n* is the sample size.

Making Connections

Remind students of "The Federalist Papers" activity from Section 4.1 and show them the dotplot posters that were produced in the activity. In this activity, we tried to estimate the average word length for some text. When students quickly circled a sample of words, they were likely to overestimate the true mean word length. Because of the sampling method, the sample mean was a *biased estimator* of the population mean.

We will confirm mathematically in Section 7.2 that the sample proportion $\hat{p}$ is an unbiased estimator of the population proportion p. This is a very helpful result if we're dealing with a categorical variable (like color). With quantitative variables, we might be interested in estimating the population mean, median, minimum, maximum, Q_1, Q_3, variance, standard deviation, IQR, or range. Which (if any) of these are unbiased estimators?

Let's revisit the "Sampling heights" example with John and Carol's four sons to investigate one of these statistics. Recall that the heights of the four sons are 71, 75, 72, and 68 inches. Here again is the sampling distribution of the sample mean $\bar{x}$ for samples of size 2:

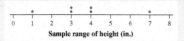

$\bar{x}$ = sample mean height (in.)

To determine if the sample mean is an unbiased estimator of the population mean, we need to compare the mean of the sampling distribution to the value we are trying to estimate—the mean of the population μ.

The mean of the sampling distribution of $\bar{x}$ is

$$\mu_{\bar{x}} = \frac{69.5 + 70 + 71.5 + 71.5 + 73 + 73.5}{6} = 71.5$$

The mean of the population is

$$\mu = \frac{71 + 75 + 72 + 68}{4} = 71.5$$

Because these values are equal, this example suggests that the sample mean $\bar{x}$ is an unbiased estimator of the population mean μ. We will confirm this fact in Section 7.3.

EXAMPLE

Estimating the range
Biased and unbiased estimators

PROBLEM: In the "Sampling heights" example, we created the sampling distribution of the sample range for samples of size $n = 2$ from the population of John and Carol's four sons with heights of 71, 75, 72, and 68 inches tall. Is the sample range an unbiased estimator of the population range? Explain your answer.

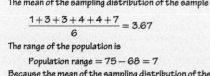

Sample range of height (in.)

SOLUTION:
The mean of the sampling distribution of the sample range is

$$\frac{1 + 3 + 3 + 4 + 4 + 7}{6} = 3.67$$

The range of the population is

Population range = $75 - 68 = 7$

Because the mean of the sampling distribution of the sample range (3.67) is not equal to the value it is trying to estimate (7), the sample range is not an unbiased estimator of the population range.

Brian Miller

FOR PRACTICE, TRY EXERCISE 19

ALTERNATE EXAMPLE

Pick three of Joe's scores   Biased and unbiased estimators

PROBLEM:
In the "Pick three" Alternate Example, we created the sampling distribution of the sample mean for samples of size $n = 3$ from the population of Joe's test scores of 93, 87, 96, 78, and 90. Is the sample mean an unbiased estimator of the population mean? Explain your answer.

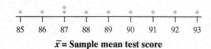

$\bar{x}$ = Sample mean test score

SOLUTION:
The mean of the sampling distribution is

$$\frac{92 + 86 + 90 + 89 + 93 + 87 + 87 + 91 + 85 + 88}{10} = 88.8$$

The mean of the population is

$$\text{Population mean} = \frac{93 + 87 + 96 + 78 + 90}{5} = 88.8$$

Because the mean of the sampling distribution of the sample mean (88.8) is equal to the value it is trying to estimate (88.8), the sample mean is an unbiased estimator of the population mean.

Because the sample range is consistently smaller than the population range, the sample range is a *biased estimator* of the population range.

Think About It

WHY DO WE DIVIDE BY $n - 1$ WHEN CALCULATING THE SAMPLE STANDARD DEVIATION? Now that we know about sampling distributions and unbiased estimators, we can finally answer this question. In Chapter 1, you learned that the formula for the sample standard deviation is $s_x = \sqrt{\dfrac{\sum(x_i - \bar{x})^2}{n-1}}$. You also learned that the value obtained before taking the square root in the standard deviation calculation is known as the variance. That is, the sample variance is $s_x^2 = \dfrac{\sum(x_i - \bar{x})^2}{n-1}$.

In an inference setting involving a quantitative variable, we might be interested in estimating the variance σ^2 of the population distribution. The most logical choice for our estimator is the sample variance s_x^2. We used technology to select 500 SRSs of size $n = 4$ from a population where the population variance is $\sigma^2 = 9$. For each sample, we recorded the value of two statistics:

$$\text{var}(n-1) = \frac{\sum(x_i - \bar{x})^2}{n-1} = s_x^2 \qquad \text{var}(n) = \frac{\sum(x_i - \bar{x})^2}{n}$$

Figure 7.4 shows the approximate sampling distributions of these two statistics. The blue vertical lines mark the means of these two distributions.

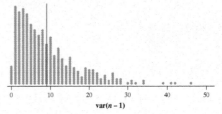

var(n − 1)

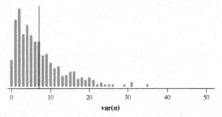

var(n)

FIGURE 7.4 Results from a simulation of 500 SRSs of size $n = 4$ from a population with variance $\sigma^2 = 9$. The sample variance s_x^2 (labeled "var($n-1$)" in the figure) is an unbiased estimator of the population variance. The "var(n)" statistic is a biased estimator of the population variance.

We can see that "var(n)" is a *biased* estimator of the population variance. The mean of its approximate sampling distribution (marked with a blue line segment) is clearly less than the value of the population parameter, $\sigma^2 = 9$. However, the statistic "var($n-1$)" (otherwise known as the sample variance s_x^2) is an unbiased estimator. Its values are centered at $\sigma^2 = 9$. That's why we divide by $n - 1$ when calculating the sample variance—and when calculating the sample standard deviation.

VARIABILITY: SMALLER IS BETTER!   To get a trustworthy estimate of an unknown population parameter, start by using a statistic that's an unbiased estimator. This ensures that you won't consistently overestimate or consistently underestimate the parameter. Unfortunately, using an unbiased estimator doesn't guarantee that the value of your statistic will be close to the actual parameter value.

To investigate the variability of a statistic, let's consider the proportion of people in a random sample who have ever watched the show *Survivor*. According to Nielsen ratings, *Survivor* was one of the most-watched television shows in the United States every week that it aired. Suppose that the true proportion of U.S. adults who have ever watched *Survivor* is $p = 0.37$.

$\hat{p}$ = sample proportion who have watched *Survivor* (*n* = 100)

The top dotplot in Figure 7.5 shows the results of drawing 400 SRSs of size $n = 100$ from a large population with $p = 0.37$ and recording the value of $\hat{p}$ = the sample proportion who have ever watched *Survivor*. We see that a sample of 100 people often gave a $\hat{p}$ quite far from the population parameter, $p = 0.37$.

Let's repeat our simulation, this time taking 400 SRSs of size $n = 1000$ from a large population with proportion $p = 0.37$ who have watched *Survivor*. The bottom dotplot in Figure 7.5 displays the distribution of the 400 values of $\hat{p}$ from these larger samples. Both graphs are drawn on the same horizontal scale to make comparison easy.

We can see that the variability shown in the top dotplot in Figure 7.5 is much greater than the variability shown in the bottom dotplot. With samples of size 100, the standard deviation of these $\hat{p}$ values is about 0.047. Using SRSs of size 1000, the standard deviation of these $\hat{p}$ values is about 0.016. This confirms what we learned in Section 4.3: larger random samples tend to produce estimates that are closer to the true population value.

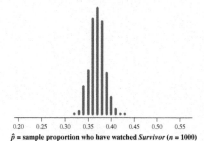

$\hat{p}$ = sample proportion who have watched *Survivor* (*n* = 1000)

FIGURE 7.5 The approximate sampling distribution of the sample proportion $\hat{p}$ from SRSs of size $n = 100$ and $n = 1000$ chosen from a large population with proportion $p = 0.37$. Both dotplots show the results of 400 SRSs.

One important and surprising fact is that the variability of a statistic does *not* depend very much on the size of the population, as long as the sample size is less than 10% of the population size. Suppose that in a small town of 25,000 people, 37% of the population have watched *Survivor*. That is, $p = 0.37$. Let's simulate taking 400 SRSs of size 1000 from this small town and compute $\hat{p}$, the sample proportion who have watched *Survivor*. The results are shown in Figure 7.6.

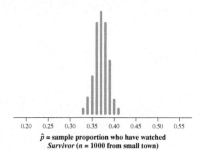

$\hat{p}$ = sample proportion who have watched
Survivor (*n* = 1000 from small town)

FIGURE 7.6 The approximate sampling distribution of the sample proportion $\hat{p}$ from 400 SRSs of size $n = 1000$ chosen from a population of 25,000 individuals with proportion $p = 0.37$.

Teaching Tip

Many students think that larger populations require larger samples to get good estimates. They think that to get a good estimate, a certain percentage of the population must be sampled. This is not the case. The precision of the estimate is based mostly on the number of individuals in the sample (not the percent of the population).

 Answers to CYU

1. The simulation does provide evidence that the sample median is a biased estimator of the population median because the mean of the sampling distribution of the sample median (73.5) is not equal to the value it is trying to estimate, the population median (75).

2. Increasing the sample size from 10 to 20 will decrease the variability of the sampling distribution. Larger samples provide more precise estimates, because larger samples include more information about the population distribution.

3. The sampling distribution of the sample median is skewed to the left and single-peaked.

Sergey Skleznev/Shutterstock

Notice that the distribution of $\hat{p}$ looks nearly the same when sampling from a small town of 25,000 residents as when sampling from the entire United States. In fact, the standard deviation for each sampling distribution is approximately 0.016.

Why does the size of the population have little influence on the behavior of statistics from random samples? Imagine sampling harvested corn by thrusting a scoop into a large sack of corn kernels. The scoop doesn't know if it is surrounded by a bag of corn or by an entire truckload. As long as the corn is well mixed (so that the scoop selects a random sample), the variability of the result depends mostly on the size of the scoop.

CHECK YOUR UNDERSTANDING

The histogram on the left shows the interval (in minutes) between eruptions of the Old Faithful geyser for all 222 recorded eruptions during a particular month. For this population, the median is 75 minutes. We used technology to take 500 SRSs of size 10 from the population. The 500 values of the sample median are displayed in the histogram on the right. The mean of these 500 values is 73.5.

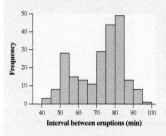

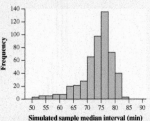

1. Does the simulation provide evidence that the sample median is a biased estimator of the population median? Justify your answer.
2. Suppose we had taken samples of size 20 instead of size 10. Would the variability of the sampling distribution of the sample median be larger, smaller, or about the same? Justify your answer.
3. Describe the shape of the sampling distribution of the sample median.

CHOOSING AN ESTIMATOR In many cases, it is obvious which statistic should be used as an estimator of a population parameter. If we want to estimate the population mean μ, use the sample mean $\bar{x}$. If we want to estimate a population proportion p, use the sample proportion $\hat{p}$. However, in other cases, there isn't an obvious best choice. When trying to estimate the population maximum in the "Craft stick" activity, there were many different estimators that could be used.

To decide which estimator to use when there are several choices, consider both bias and variability. Imagine the true value of the population parameter as the

bull's-eye on a target and the sample statistic as an arrow fired at the target. Both bias and variability describe what happens when we take many shots at the target.

Bias means that our aim is off and we consistently miss the bull's-eye in the same direction. Our sample statistics do not center on the population parameter. In other words, our estimates are not *accurate*. High variability means that repeated shots are widely scattered on the target. Repeated samples do not give very similar results. In other words, our estimates are not very *precise*. Figure 7.7 shows this target illustration.

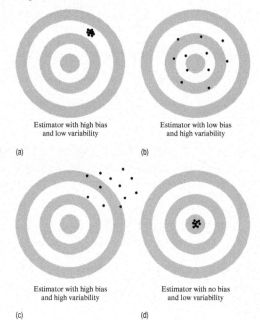

(a) Estimator with high bias and low variability

(b) Estimator with low bias and high variability

(c) Estimator with high bias and high variability

(d) Estimator with no bias and low variability

FIGURE 7.7 Bias and variability of four different estimators. (a) High bias, low variability. (b) Low bias, high variability. (c) High bias, high variability. (d) The ideal estimator: no bias, low variability.

Notice that an estimator with low variability can also have high bias, as in Figure 7.7(a). And an estimator with low or no bias can be quite variable, as in Figure 7.7(b). Ideally, we'd like to use an estimator that is unbiased with low variability, as in Figure 7.7(d).

AP® EXAM TIP

Make sure to understand the difference between accuracy and precision when writing responses on the AP® Statistics Exam. Many students use "accurate" when they really mean "precise." For example, a response that says "increasing the sample size will make an estimate more accurate" is incorrect. It should say that increasing the sample size will make an estimate more precise. If you can't remember which term to use, don't use either of them. Instead, explain what you mean without using statistical vocabulary.

Teaching Tip: AP® Connections

Students from AP® Chemistry and AP® Physics have likely encountered the discussion about accuracy versus precision.

Teaching Tip

Accurate means that the statistic has low or no bias. Bias is determined by the sampling method (see the Federalist Papers activity in Section 4.1) or the formula used to calculate the statistic (see the craft stick activity from earlier in this section).

Teaching Tip

Precise means that the statistic has low variability. Variability is most strongly influenced by the sample size (see the "A penny for your thoughts?" activity from earlier in this section) but can also be affected by sampling design. Stratified random samples and blocking in experiments often reduce variability.

TRM **Section 7.1 Quiz**

There is one quiz available for this section in the Teacher's Resource Materials. Click on the link in the TE-Book, look on the TRFD, or download from the Teacher's Resources on the book's digital platform. You can also create your own quiz using the ExamView® Assessment Suite that is part of the TPS6 program. Questions are coded by Learning Target to make it easy to build parallel quizzes.

TRM **Full Solutions to Section 7.1 Exercises**

Click on the link in the TE-Book, open the TRFD, or download from the Teacher's Resources on the book's digital platform.

Answers to Section 7.1 Exercises

7.1 *Population:* All people who signed a card saying they intend to quit smoking. *Parameter:* $p =$ the proportion of the population who had not smoked over the past 6 months. *Sample:* The 1000 people selected at random. *Statistic:* $\hat{p} =$ the proportion in the sample who had not smoked over the past 6 months $= 0.21$.

7.2 *Population:* All U.S. adults. *Parameter:* $p =$ the proportion of all U.S. adults who were unemployed in October 2016. *Sample:* The 60,000 randomly selected U.S. adults. *Statistic:* $\hat{p} =$ the proportion in the sample who were unemployed in October 2016 $= 0.049$.

7.3 *Population:* All dental practices in California. *Parameter:* The *IQR* of the price to fill a cavity for all dental practices in California. *Sample:* The 10 randomly selected dental practices that provided the price they charge to fill a cavity. *Statistic:* *IQR* $=$ the interquartile range of the price to fill a cavity for the 10 selected dental practices $= \$74$.

7.4 *Population:* All points in the turkey. *Parameter:* Minimum temperature in all points of the turkey. *Sample:* Four randomly chosen locations in the turkey. *Statistic:* Minimum temperature in the sample of four locations $= 170°F$.

7.5 *Population:* All bottles of Arizona Iced Tea produced that day. *Parameter:* $\mu =$ the average number of ounces per bottle in all bottles of Arizona Iced Tea produced that day. *Sample:* The 50 bottles of tea selected at random from the day's production. *Statistic:* $\bar{x} =$ the average number of ounces of tea contained in the 50 bottles $= 19.6$ ounces.

7.6 *Population:* All ball bearings in the production run. *Parameter:* $\mu =$ the average diameter of all ball bearings in the production run. *Sample:* The 100 ball bearings selected at random from the run. *Statistic:* $\bar{x} =$ the average diameter of the 100 ball bearings selected $= 2.5009$ cm.

7.7

1: Abigail(10), Bobby(5)	$\bar{x} = 7.5$
2: Abigail(10), Carlos(10)	$\bar{x} = 10$
3: Abigail(10), DeAnna(7)	$\bar{x} = 8.5$
4: Abigail(10), Emily(9)	$\bar{x} = 9.5$
5: Bobby(5), Carlos(10)	$\bar{x} = 7.5$
6: Bobby(5), DeAnna(7)	$\bar{x} = 6$
7: Bobby(5), Emily(9)	$\bar{x} = 7$
8: Carlos(10), DeAnna(7)	$\bar{x} = 8.5$
9: Carlos(10), Emily(9)	$\bar{x} = 9.5$
10: DeAnna(7), Emily(9)	$\bar{x} = 8$

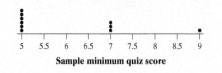

$\bar{x} =$ Sample mean quiz score

Section 7.1 Summary

- A **parameter** is a number that describes a population. To estimate an unknown parameter, use a **statistic** calculated from a sample.
- The **population distribution** of a variable describes the values of the variable for all individuals in a population. The **sampling distribution** of a statistic describes the values of the statistic in all possible samples of the same size from the same population. Don't confuse the sampling distribution with a **distribution of sample data,** which gives the values of the variable for all individuals in a particular sample.
- A statistic can be an **unbiased estimator** or a **biased estimator** of a parameter. A statistic is an unbiased estimator if the center (mean) of its sampling distribution is equal to the true value of the parameter.
- The **variability** of a statistic is described by the spread of its sampling distribution. Larger samples result in sampling distributions with less variability.
- When trying to estimate a parameter, choose a statistic with low or no bias and minimum variability.

Section 7.1 Exercises

For Exercises 1–6, identify the population, the parameter, the sample, and the statistic in each setting.

1. **Healthy living** From a large group of people who signed a card saying they intended to quit smoking, 1000 people were selected at random. It turned out that 210 (21%) of these individuals had not smoked over the past 6 months. *[pg 470]*

2. **Unemployment** Each month, the Current Population Survey interviews about 60,000 randomly selected U.S. adults. One of their goals is to estimate the national unemployment rate. In October 2016, 4.9% of those interviewed were unemployed.

3. **Fillings** How much do prices vary for filling a cavity? To find out, an insurance company randomly selects 10 dental practices in California and asks for the cash (non-insurance) price for this procedure at each practice. The interquartile range is $74.

4. **Warm turkey** Tom is cooking a large turkey breast for a holiday meal. He wants to be sure that the turkey is safe to eat, which requires a minimum internal temperature of 165°F. Tom uses a thermometer to measure the temperature of the turkey meat at four randomly chosen points. The minimum reading is 170°F.

5. **Iced tea** On Tuesday, the bottles of Arizona Iced Tea filled in a plant were supposed to contain an average of 20 ounces of iced tea. Quality control inspectors selected 50 bottles at random from the day's production. These bottles contained an average of 19.6 ounces of iced tea.

6. **Bearings** A production run of ball bearings is supposed to have a mean diameter of 2.5000 centimeters (cm). An inspector chooses 100 bearings at random from the run. These bearings have mean diameter 2.5009 cm.

Exercises 7–10 refer to the small population of 5 students in the table.

Name	Gender	Quiz score
Abigail	Female	10
Bobby	Male	5
Carlos	Male	10
DeAnna	Female	7
Emily	Female	9

7. **Sample means** List all 10 possible SRSs of size $n = 2$, calculate the mean quiz score for each sample, and display the sampling distribution of the sample mean on a dotplot. *[pg 472]*

8. **Sample minimums** List all 10 possible SRSs of size $n = 3$, calculate the minimum quiz score for each sample, and display the sampling distribution of the sample minimum on a dotplot.

9. **Sample proportions** List all 10 possible SRSs of size $n = 2$, calculate the proportion of females for each sample, and display the sampling distribution of the sample proportion on a dotplot.

10. **Sample medians** List all 10 possible SRSs of size $n = 3$, calculate the median quiz score for each sample, and display the sampling distribution of the sample median on a dotplot.

7.8

1: Abigail(10), Bobby(5), Carlos(10)	Min = 5
2: Abigail(10), Bobby(5), DeAnna(7)	Min = 5
3: Abigail(10), Bobby(5), Emily(9)	Min = 5
4: Abigail(10), Carlos(10), DeAnna(7)	Min = 7
5: Abigail(10), Carlos(10), Emily(9)	Min = 9
6: Abigail(10), DeAnna(7), Emily(9)	Min = 7
7: Bobby(5), Carlos(10), DeAnna(7)	Min = 5
8: Bobby(5), Carlos(10), Emily(9)	Min = 5
9: Bobby(5), DeAnna(7), Emily(9)	Min = 5
10: Carlos(10), DeAnna(7), Emily(9)	Min = 7

Sample minimum quiz score

7.9

1: Abigail(F), Bobby(M)	$\hat{p} = 0.50$
2: Abigail(F), Carlos(M)	$\hat{p} = 0.50$
3: Abigail(F), DeAnna(F)	$\hat{p} = 1$
4: Abigail(F), Emily(F)	$\hat{p} = 1$
5: Bobby(M), Carlos(M)	$\hat{p} = 0$
6: Bobby(M), DeAnna(F)	$\hat{p} = 0.50$
7: Bobby(M), Emily(F)	$\hat{p} = 0.50$
8: Carlos(M), DeAnna(F)	$\hat{p} = 0.50$
9: Carlos(M), Emily(F)	$\hat{p} = 0.50$
10: DeAnna(F), Emily(F)	$\hat{p} = 1$

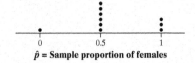

$\hat{p} =$ Sample proportion of females

11. **Doing homework** A school newspaper article claims that 60% of the students at a large high school completed their assigned homework last week. Assume that this claim is true for the 2000 students at the school.

(a) Make a bar graph of the population distribution.

(b) Imagine one possible SRS of size 100 from this population. Sketch a bar graph of the distribution of sample data.

12. **Tall girls** According to the National Center for Health Statistics, the distribution of height for 16-year-old females is modeled well by a Normal density curve with mean $\mu = 64$ inches and standard deviation $\sigma = 2.5$ inches. Assume this claim is true for the three hundred 16-year-old females at a large high school.

(a) Make a graph of the population distribution.

(b) Imagine one possible SRS of size 20 from this population. Sketch a dotplot of the distribution of sample data.

13. **More homework** Some skeptical AP® Statistics students want to investigate the newspaper's claim in Exercise 11, so they choose an SRS of 100 students from the school to interview. In their sample, 45 students completed their homework last week. Does this provide convincing evidence that less than 60% of all students at the school completed their assigned homework last week?

(a) What is the evidence that less than 60% of all students completed their assigned homework last week?

(b) Provide two explanations for the evidence described in part (a).

We used technology to simulate choosing 250 SRSs of size $n = 100$ from a population of 2000 students where 60% completed their assigned homework last week. The dotplot shows $\hat{p}$ = the sample proportion of students who completed their assigned homework last week for each of the 250 simulated samples.

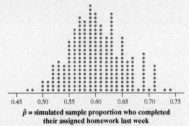

$\hat{p}$ = simulated sample proportion who completed their assigned homework last week

(c) There is one dot on the graph at 0.73. Explain what this value represents.

(d) Would it be surprising to get a sample proportion of $\hat{p} = 0.45$ or smaller in an SRS of size 100 when $p = 0.60$? Justify your answer.

(e) Based on your previous answers, is there convincing evidence that less than 60% of all students at the school completed their assigned homework last week? Explain your reasoning.

14. **Tall girls?** To see if the claim made in Exercise 12 is true at their high school, an AP® Statistics class chooses an SRS of twenty 16-year-old females at the school and measures their heights. In their sample, the mean height is 64.7 inches. Does this provide convincing evidence that 16-year-old females at this school are taller than 64 inches, on average?

(a) What is the evidence that the average height of all 16-year-old females at this school is greater than 64 inches, on average?

(b) Provide two explanations for the evidence described in part (a).

We used technology to simulate choosing 250 SRSs of size $n = 20$ from a population of three hundred 16-year-old females whose heights follow a Normal distribution with mean $\mu = 64$ inches and standard deviation $\sigma = 2.5$ inches. The dotplot shows $\bar{x}$ = the sample mean height for each of the 250 simulated samples.

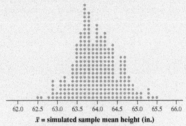

$\bar{x}$ = simulated sample mean height (in.)

(c) There is one dot on the graph at 62.5. Explain what this value represents.

(d) Would it be surprising to get a sample mean of $\bar{x} = 64.7$ or larger in an SRS of size 20 when $\mu = 64$ inches and $\sigma = 2.5$ inches? Justify your answer.

(e) Based on your previous answers, is there convincing evidence that the average height of all 16-year-old females at this school is greater than 64 inches? Explain your reasoning.

15. **Even more homework** Refer to Exercises 11 and 13. Suppose that the sample proportion of students who did all their assigned homework last week is $\hat{p} = 57/100 = 0.57$. Would this sample proportion provide convincing evidence that less than 60% of all

7.12 (a)

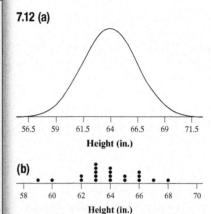

Height (in.)

(b)

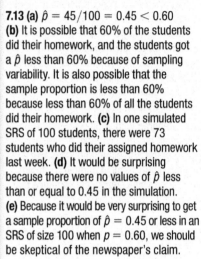

Height (in.)

7.13 (a) $\hat{p} = 45/100 = 0.45 < 0.60$ **(b)** It is possible that 60% of the students did their homework, and the students got a $\hat{p}$ less than 60% because of sampling variability. It is also possible that the sample proportion is less than 60% because less than 60% of all the students did their homework. **(c)** In one simulated SRS of 100 students, there were 73 students who did their assigned homework last week. **(d)** It would be surprising because there were no values of $\hat{p}$ less than or equal to 0.45 in the simulation. **(e)** Because it would be very surprising to get a sample proportion of $\hat{p} = 0.45$ or less in an SRS of size 100 when $p = 0.60$, we should be skeptical of the newspaper's claim.

7.14 (a) $\bar{x} = 64.7 > 64$ **(b)** It is possible that the average height of all 16-year-old females at this school is 64 inches, but the students got a sample mean $\bar{x}$ greater than 64 inches because of sampling variability. It is also possible that the sample mean is greater than 64 inches because the average height of all 16-year-old females at this school is greater than 64 inches. **(c)** In one simulated SRS of 20 students, the average height was $\bar{x} = 62.5$ inches. **(d)** It would not be surprising because about 12% of values of $\bar{x}$ were 64.7 or greater in the simulation. **(e)** Because it isn't surprising to get a sample mean of $\bar{x} = 64.7$ or greater in an SRS of size 20 from a Normal population with $\mu = 64$ and $\sigma = 2.5$, we do not have convincing evidence that the population mean height at the school is different from $\mu = 64$.

7.10

1: Abigail(10), Bobby(5), Carlos(10)	median = 10
2: Abigail(10), Bobby(5), DeAnna(7)	median = 7
3: Abigail(10), Bobby(5), Emily(9)	median = 9
4: Abigail(10), Carlos(10), DeAnna(7)	median = 10
5: Abigail(10), Carlos(10), Emily(9)	median = 10
6: Abigail(10), DeAnna(7), Emily(9)	median = 9
7: Bobby(5), Carlos(10), DeAnna(7)	median = 7
8: Bobby(5), Carlos(10), Emily(9)	median = 9
9: Bobby(5), DeAnna(7), Emily(9)	median = 7
10: Carlos(10), DeAnna(7), Emily(9)	median = 9

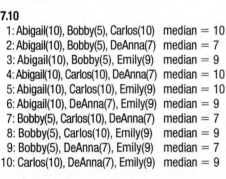

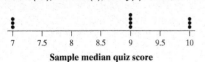

Sample median quiz score

7.11 (a)

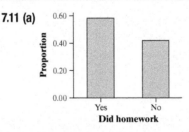

Did homework

(b)

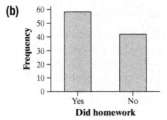

Did homework

7.15 In the simulation, $\hat{p} = 0.57$ is not an unusual value. Therefore, a sample proportion of $\hat{p} = 0.57$ does not provide convincing evidence that the population proportion of students who did all their assigned homework is less than $\hat{p} = 60\%$.

7.16 In the simulation, an $\overline{x} = 65.8$ would be unusually large. Therefore, a sample mean of $\overline{x} = 65.8$ does provide convincing evidence that the population mean height at the school is greater than $\mu = 64$.

7.17 A sample standard deviation of 5°F is quite large compared with what we would expect by chance alone, because none of the 300 simulated SRSs had a standard deviation that large. A sample standard deviation of 5°F provides convincing evidence that the manufacturer's claim is false and that the thermostat actually has more variability than claimed.

7.18 A sample minimum of 40°F is quite low compared with what we would expect to happen by chance alone, because only 2 of the 300 simulated SRSs had a minimum that small. A sample minimum of 40°F provides convincing evidence that the manufacturer's claim is false.

7.19 The 6 possible SRSs of size $n = 2$ and the sample proportion for each sample are:

1: Red, White	$\hat{p} = 0.5$
2: Red, Silver	$\hat{p} = 0.5$
3: Red, Red	$\hat{p} = 1$
4: White, Silver	$\hat{p} = 0$
5: White, Red	$\hat{p} = 0.5$
6: Silver, Red	$\hat{p} = 0.5$

$\hat{p}$ = **Sample proportion of red cars**

The population proportion of red cars is $p = 2/4 = 0.5$. Because the mean of the sampling distribution is 0.5, the sample proportion is an unbiased estimator of the population proportion.

7.20 The 6 possible SRSs of size $n = 2$ and the sample minimum for each sample are:

1: Red(1), White(5)	Min = 1
2: Red(1), Silver(8)	Min = 1
3: Red(1), Red(20)	Min = 1
4: White(5), Silver(8)	Min = 5
5: White(5), Red(20)	Min = 5
6: Silver(8), Red(20)	Min = 8

Sample minimum car age

The minimum age of the population of 4 cars is 1 year. The sample minimum is not an unbiased estimator of the population minimum. The mean of the sample minimums is 3.5.

---

students at the school completed all their assigned homework last week? Explain your reasoning.

16. **Even more tall girls** Refer to Exercises 12 and 14. Suppose that the sample mean height of the twenty 16-year-old females is $\overline{x} = 65.8$ inches. Would this sample mean provide convincing evidence that the average height of all 16-year-old females at this school is greater than 64 inches? Explain your reasoning.

Exercises 17 and 18 refer to the following setting. During the winter months, outside temperatures at the Starneses' cabin in Colorado can stay well below freezing (32°F, or 0°C) for weeks at a time. To prevent the pipes from freezing, Mrs. Starnes sets the thermostat at 50°F. The manufacturer claims that the thermostat allows variation in home temperature that follows a Normal distribution with $\sigma = 3$°F. To test this claim, Mrs. Starnes programs her digital thermometer to take an SRS of $n = 10$ readings during a 24-hour period. Suppose the thermostat is working properly and that the actual temperatures in the cabin vary according to a Normal distribution with mean $\mu = 50$°F and standard deviation $\sigma = 3$°F.

17. **Cold cabin?** The dotplot shows the results of taking 300 SRSs of 10 temperature readings from a Normal population with $\mu = 50$ and $\sigma = 3$ and recording the sample standard deviation s_x each time. Suppose that the standard deviation from an actual sample is $s_x = 5$°F. What would you conclude about the thermostat manufacturer's claim? Explain your reasoning.

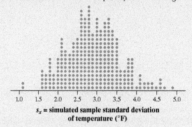

s_x = **simulated sample standard deviation of temperature (°F)**

18. **Really cold cabin** The dotplot shows the results of taking 300 SRSs of 10 temperature readings from a Normal population with $\mu = 50$ and $\sigma = 3$ and recording the sample minimum each time. Suppose that the minimum of an actual sample is 40°F. What would you conclude about the thermostat manufacturer's claim? Explain your reasoning.

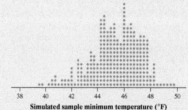

Simulated sample minimum temperature (°F)

Exercises 19–22 refer to the small population of 4 cars listed in the table.

Color	Age (years)
Red	1
White	5
Silver	8
Red	20

19. **Sample proportions** List all 6 possible SRSs of size $n = 2$, calculate the proportion of red cars in each sample, and display the sampling distribution of the sample proportion on a dotplot. Is the sample proportion an unbiased estimator of the population proportion? Explain your answer.

20. **Sample minimums** List all 6 possible SRSs of size $n = 2$, calculate the minimum age for each sample, and display the sampling distribution of the sample minimum on a dotplot. Is the sample minimum an unbiased estimator of the population minimum? Explain your answer.

21. **More sample proportions** List all 4 possible SRSs of size $n = 3$, calculate the proportion of red cars in the sample, and display the sampling distribution of the sample proportion on a dotplot with the same scale as the dotplot in Exercise 19. How does the variability of this sampling distribution compare with the variability of the sampling distribution from Exercise 19? What does this indicate about increasing the sample size?

22. **More sample minimums** List all 4 possible SRSs of size $n = 3$, calculate the minimum age for each sample, and display the sampling distribution of the sample minimum on a dotplot with the same scale as the dotplot in Exercise 20. How does the variability of this sampling distribution compare with the variability of the sampling distribution from Exercise 20? What does this indicate about increasing the sample size?

23. **A sample of teens** A study of the health of teenagers plans to measure the blood cholesterol levels of an SRS of 13- to 16-year-olds. The researchers will report the mean $\overline{x}$ from their sample as an estimate of the mean cholesterol level μ in this population. Explain to someone who knows little about statistics what it means to say that $\overline{x}$ is an unbiased estimator of μ.

24. **Predict the election** A polling organization plans to ask a random sample of likely voters who they plan to vote for in an upcoming election. The researchers will report the sample proportion $\hat{p}$ that favors the incumbent as an estimate of the population proportion p that favors the incumbent. Explain to someone who knows little about statistics what it means to say that $\hat{p}$ is an unbiased estimator of p.

7.21 The 4 possible SRSs of size $n = 3$ and the sample proportion for each sample are:

1: Red, White, Silver	$\hat{p} = 1/3$
2: Red, White, Red	$\hat{p} = 2/3$
3: Red, Silver, Red	$\hat{p} = 2/3$
4: White, Silver, Red	$\hat{p} = 1/3$

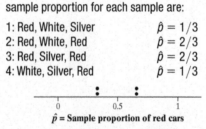

$\hat{p}$ = **Sample proportion of red cars**

The variability in this sampling distribution is less than the variability in the sampling distribution from Exercise 19. Increasing the sample size decreases the variability in the sampling distribution.

7.22 The 4 possible SRSs of size $n = 3$ and the sample minimum for each sample are:

1: Red(1), White(5), Silver(8)	Min = 1
2: Red(1), White(5), Red(20)	Min = 1

3: Red(1), Silver(8), Red(20)	Min = 1
4: White(5), Silver(8), Red(20)	Min = 5

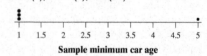

Sample minimum car age

The variability in this sampling distribution, as measured by the standard deviation, is less than the variability in the sampling distribution from Exercise 20.

7.23 If we chose many SRSs and calculated the sample mean $\overline{x}$ for each sample, the distribution of $\overline{x}$ would be centered at the value of μ.

7.24 If we chose many random samples and calculated the sample proportion $\hat{p}$ for each sample, the distribution of $\hat{p}$ would be centered at the value of p.

25. **Bias and variability** The figure shows approximate sampling distributions of 4 different statistics intended to estimate the same parameter.

(i)
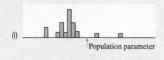
Population parameter

(ii)
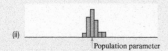
Population parameter

(iii)

Population parameter

(iv)
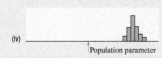
Population parameter

(a) Which statistics are unbiased estimators? Justify your answer.

(b) Which statistic does the best job of estimating the parameter? Explain your answer.

Multiple Choice: *Select the best answer for Exercises 26–30.*

26. At a particular college, 78% of all students are receiving some kind of financial aid. The school newspaper selects a random sample of 100 students and 72% of the respondents say they are receiving some sort of financial aid. Which of the following is true?

(a) 78% is a population and 72% is a sample.

(b) 72% is a population and 78% is a sample.

(c) 78% is a parameter and 72% is a statistic.

(d) 72% is a parameter and 78% is a statistic.

(e) 72% is a parameter and 100 is a statistic.

27. A statistic is an unbiased estimator of a parameter when

(a) the statistic is calculated from a random sample.

(b) in a single sample, the value of the statistic is equal to the value of the parameter.

(c) in many samples, the values of the statistic are very close to the value of the parameter.

(d) in many samples, the values of the statistic are centered at the value of the parameter.

(e) in many samples, the distribution of the statistic has a shape that is approximately Normal.

28. In a residential neighborhood, the distribution of house values is unimodal and skewed to the right, with a median of $200,000 and an *IQR* of $100,000. For which of the following sample sizes is the sample median most likely to be above $250,000?

(a) $n = 10$

(b) $n = 50$

(c) $n = 100$

(d) $n = 1000$

(e) Impossible to determine without more information.

29. Increasing the sample size of an opinion poll will reduce the

(a) bias of the estimates made from the data collected in the poll.

(b) variability of the estimates made from the data collected in the poll.

(c) effect of nonresponse on the poll.

(d) variability of opinions in the sample.

(e) variability of opinions in the population.

30. The math department at a small school has 5 teachers. The ages of these teachers are 23, 34, 37, 42, and 58. Suppose you select a random sample of 4 teachers and calculate the sample minimum age. Which of the following shows the sampling distribution of the sample minimum age?

(a)
```
    ·
 |--+----+----+----+----+--
   20   30   40   50   60
```

(b)
```
    ·
    ·
    ·
    ·
 |--+----+----+----+----+--
   20   30   40   50   60
```

(c)
```
    ·
    ·
    ·        ·
 |--+----+----+----+----+--
   20   30   40   50   60
```

(d)
```
    ·    · ·  ·       ·
 |--+----+----+----+----+--
   20   30   40   50   60
```

(e) None of these.

Recycle and Review

31. **Dem bones** (2.2) Osteoporosis is a condition in which the bones become brittle due to loss of minerals. To diagnose osteoporosis, an elaborate apparatus measures bone mineral density (BMD). BMD is usually reported in standardized form. The standardization is based on a population of healthy young adults. The World Health Organization (WHO) criterion

7.31 (a) We are looking for $P(z \leq -2.5)$.
(i) $P(z \leq -2.5) = 0.0062$
(ii) $P(z \leq -2.5) = $ normalcdf(lower: -1000, upper: -2.50, mean: 0, SD: 1) $= 0.0062$
Less than 1% of healthy young adults have osteoporosis.

(b) X follows a Normal distribution with a mean of -2 and a standard deviation of 1; $P(X \leq -2.5)$.
(i) $z = -0.5$; $P(z \leq -0.5) = 0.3085$
(ii) *Tech*: $P(X \leq -2.5) = $ normalcdf (lower: -1000, upper: -2.5, mean: -2, SD: 1) $= 0.3085$
About 30.85% of women aged 70–79 have osteoporosis.

7.32 (a) The linear model is appropriate for these data because there is no leftover pattern in the residual plot. **(b)** The equation for the least-squares regression line is $\hat{y} = 1.4146 + 0.4399x$, where $\hat{y}$ is the predicted average number of offspring per female and x is the index of the abundance of pinecones. **(c)** $r^2 = 57.2\%$ of the variability in the average number of offspring is accounted for by the least-squares regression line with $x = $ cone index; $s = 0.600309$. The actual number of offspring per female is typically about 0.600309 away from the amount predicted by the least-squares regression line with $x = $ cone index.

PD **Section 7.2 Overview**

To watch the video overview of Section 7.2 (for teachers), click on the link in the TE-Book, look on the TRFD, or download from the Teacher's Resources on the book's digital platform.

TRM **Section 7.2 Alternate Examples**

You can find the Alternate Examples for this section in Microsoft Word format by clicking on the link in the TE-Book, opening the TRFD, or downloading from the Teacher's Resources on the book's digital platform

for osteoporosis is a BMD score that is 2.5 standard deviations below the mean for young adults. BMD measurements in a population of people similar in age and gender roughly follow a Normal distribution.

(a) What percent of healthy young adults have osteoporosis by the WHO criterion?

(b) Women aged 70 to 79 are, of course, not young adults. The mean BMD in this age group is about -2 on the standard scale for young adults. Suppose that the standard deviation is the same as for young adults. What percent of this older population has osteoporosis?

32. **Squirrels and their food supply** (3.2) Animal species produce more offspring when their supply of food goes up. Some animals appear able to anticipate unusual food abundance. Red squirrels eat seeds from pinecones, a food source that sometimes has very large crops. Researchers collected data on an index of the abundance of pinecones and the average number of offspring per female over 16 years.[4] Computer output from a least-squares regression on these data and a residual plot are shown here.

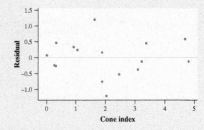

```
Predictor    Coef    SE Coef    T       P
Constant     1.4146  0.2517     5.62    0.000
Cone index   0.4399  0.1016     4.33    0.001
S = 0.600309    R-Sq = 57.2%    R-Sq(adj) = 54.2%
```

(a) Is a linear model appropriate for these data? Explain.

(b) Give the equation for the least-squares regression line. Define any variables you use.

(c) Interpret the values of r^2 and s in context.

SECTION 7.2 **Sample Proportions**

LEARNING TARGETS *By the end of the section, you should be able to:*

- Calculate the mean and standard deviation of the sampling distribution of a sample proportion $\hat{p}$ and interpret the standard deviation.

- Determine if the sampling distribution of $\hat{p}$ is approximately Normal.

- Calculate the mean and the standard deviation of the sampling distribution of a difference in

sample proportions $\hat{p}_1 - \hat{p}_2$ and interpret the standard deviation.

- Determine if the sampling distribution of $\hat{p}_1 - \hat{p}_2$ is approximately Normal.

- If appropriate, use a Normal distribution to calculate probabilities involving $\hat{p}$ or $\hat{p}_1 - \hat{p}_2$.

What proportion of U.S. teens know that 1492 was the year in which Columbus "discovered" America? A Gallup Poll found that 210 out of a random sample of 501 American teens aged 13 to 17 knew this historically important date.[5] The sample proportion

$$\hat{p} = \frac{210}{501} = 0.42$$

is the statistic that we use to gain information about the unknown population proportion p. Because another random sample of 501 teens would likely result in a different estimate, we can only say that "about" 42% of all U.S. teenagers know that Columbus discovered America in 1492. In this section, we'll use sampling distributions to clarify what "about" means.

FOXTROT©2005 Bill Amend. Andrews McMeel Syndicate.

Teaching Tip

Some books use π to represent the population proportion instead of p. This is consistent with the use of Greek letters for population parameters (μ for mean, σ for standard deviation, etc.). However, we use p for the population proportion to match the notation used on the formula sheet provided on the AP® Statistics exam.

The Sampling Distribution of $\hat{p}$

When Mrs. Gallas's class did the "Penny for your thoughts" activity at the beginning of the chapter, her students produced the "dotplot" in Figure 7.8. This graph approximates the **sampling distribution of the sample proportion** of pennies from the 2000s for samples of size $n = 20$ from Mrs. Gallas's population of pennies.

> **DEFINITION   Sampling distribution of the sample proportion**
>
> The **sampling distribution of the sample proportion** $\hat{p}$ describes the distribution of values taken by the sample proportion $\hat{p}$ in all possible samples of the same size from the same population.

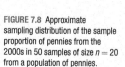

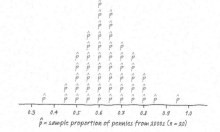

FIGURE 7.8 Approximate sampling distribution of the sample proportion of pennies from the 2000s in 50 samples of size $n = 20$ from a population of pennies.

This distribution is roughly symmetric with a mean of about 0.65 and a standard deviation of about 0.10. By the end of this section, you should be able to anticipate the shape, center, and variability of distributions like this one without getting your hands dirty in a jar of pennies.

ACTIVITY   The candy machine

Imagine a very large candy machine filled with orange, brown, and yellow candies. When you insert money, the machine dispenses a sample of candies. In this activity, you will use an applet to investigate the shape, center, and variability of the sampling distribution of the sample proportion $\hat{p}$.

TRM **Sampling Distribution of a Sample Proportion**

This assignment is built around the *Reese's Pieces* applet found at www.rossmanchance.com/applets and previews many of the examples in the section. It helps students explore the shape, center, and variability of the sampling distribution of $\hat{p}$ for different sample sizes and population proportions. You can assign it to students at the beginning of the section to preview the examples or assign it after discussing the examples as a way to reinforce what was covered in class. Find it in the Teacher's Resource Materials located in the TE-Book, on the TRFD, or in the Teacher's view on the book's digital platform.

ACTIVITY OVERVIEW

To prepare for using this activity, watch the overview video by clicking on the link in the TE-Book, opening the TRFD, or downloading from the Teacher's Resources on the book's digital platform.

Time: 30 minutes

Materials: Computers with Internet access

Teaching Advice: Use this activity to remind students of the difference between a population distribution, the distribution of sample data, and a sampling distribution. The population distribution is the candy machine at the top, the distribution of sample data is the one sample of 25 candies, and the (approximate) sampling distribution is the dotplot, which represents the sample proportion from many, many samples. Notice that the population distribution and the distribution of sample data are both made of candies, but the sampling distribution is made of $\hat{p}$'s. Show several individual samples with animation before jumping to the sampling distribution with 1000 samples. Students sometimes fail to grasp a concept if they have to move too quickly into using technology.

Answers:

1–2. Answers will vary.

3. The mean is close to the true proportion of 0.5. The standard deviation is around 0.10.

4. The shape is approximately Normal. The center is close to the true proportion of 0.5. The standard deviation is around 0.10.

5. The center of the sampling distribution of $\hat{p}$ is always the true proportion of orange candies. The shape of the sampling distribution of $\hat{p}$ is symmetric when the true proportion of orange candies is 0.5 and becomes more skewed and slightly less variable as the true proportion moves away from 0.5.

6. As the number of orange candies selected increases, the shape of the sampling distribution of $\hat{p}$ becomes more Normal and less variable, while the center remains unaffected.

7. The sampling distribution is more Normal when n is large and when p is closer to 0.5. If $np \geq 10$ and $n(1 - p) \geq 10$, then the sampling distribution appears approximately Normal.

1. Launch the *Reese's Pieces*® applet at www.rossmanchance.com/applets. Make sure the Probability of orange = 0.5, Number of candies (sample size) = 25, and Number of samples = 1. Choose "Proportion of orange" as the statistic to be calculated and check the box for Summary stats to be calculated, as shown in the screen shot for Step 2.

2. Click on the "Draw Samples" button. An animated simple random sample of $n = 25$ candies should be dispensed. The following screen shot shows the results of one such sample where the sample proportion of orange candies was $\hat{p} = 0.360$. How far was your sample proportion of orange candies from the actual population proportion, $p = 0.50$?

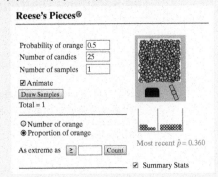

3. Click "Draw Samples" 9 more times so that you have a total of 10 sample proportions. Look at the dotplot of your $\hat{p}$ values. What is the mean of your 10 sample proportions? What is their standard deviation?

4. To take many more samples quickly, enter 990 in the "Number of samples" box. Click on the "Animate" box to turn the animation off. Then click "Draw Samples." You have now taken a total of 1000 samples of 25 candies from the machine. Describe the shape, center, and variability of the approximate sampling distribution of $\hat{p}$ shown in the dotplot.

5. How does the sampling distribution of $\hat{p}$ change if the proportion of orange candies in the machine is different from $p = 0.5$? Use the applet to investigate this question. Then write a brief summary of what you learned.

6. How does the sampling distribution of $\hat{p}$ change if the machine dispenses a different number of candies (n)? Use the applet to investigate this question. Then write a brief summary of what you learned.

7. For what combinations of n and p is the sampling distribution of $\hat{p}$ approximately Normal? Use the applet to investigate this question. Then write a brief summary of what you learned.

The graphs in Figure 7.9 show approximate sampling distributions of $\hat{p}$ for different combinations of p (population proportion) and n (sample size).

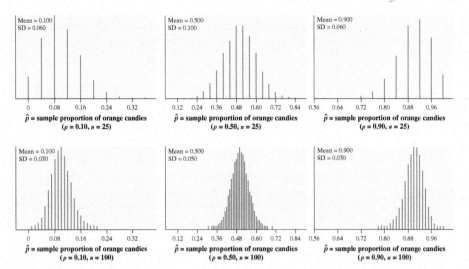

FIGURE 7.9 Approximate sampling distributions of $\hat{p}$ = sample proportion of orange candies for different combinations of p (population proportion) and n (sample size).

What do these graphs teach us about the sampling distribution of $\hat{p}$?

Shape: When n is small and p is close to 0, the sampling distribution of $\hat{p}$ is skewed to the right. When n is small and p is close to 1, the sampling distribution of $\hat{p}$ is skewed to the left. Finally, the sampling distribution of $\hat{p}$ becomes more Normal when p is closer to 0.5 or n is larger (or both).

Center: The mean of the sampling distribution of $\hat{p}$ is equal to the population proportion p. This makes sense because the sample proportion $\hat{p}$ is an *unbiased estimator* of p.

Variability: The value of $\sigma_{\hat{p}}$ depends on both n and p. For a specific sample size, the standard deviation $\sigma_{\hat{p}}$ is larger for values of p close to 0.5 and smaller for values of p close to 0 or 1. For a specific value of p, the standard deviation $\sigma_{\hat{p}}$ gets smaller as n gets larger. *Specifically, multiplying the sample size by 4 cuts the standard deviation in half.*

Here's a summary of the important facts about the sampling distribution of $\hat{p}$.

SAMPLING DISTRIBUTION OF A SAMPLE PROPORTION $\hat{p}$

Choose an SRS of size n from a population of size N with proportion p of successes. Let $\hat{p}$ be the sample proportion of successes. Then:

- The **mean** of the sampling distribution of $\hat{p}$ is $\mu_{\hat{p}} = p$.

AP® EXAM TIP

Let your students know that the formulas for the mean and standard deviation of the sampling distribution of a sample proportion are included on the formula sheet provided to students on both sections of the AP® Statistics exam.

Making Connections

In Chapter 6, we discovered that the shape of a binomial distribution was more Normal when the true proportion p was closer to 0.5 and with larger sample sizes. These ideas also hold true for the sampling distribution of the sample proportion $\hat{p}$. This makes sense because the sampling distribution of $\hat{p}$ is a scaled version of the binomial distribution $\hat{p} = (\text{binomial } X)/n$.

Teaching Tip: Using Technology

Consider having students explore the *Normal Approximation to Binomial Distributions* applet at highschool.bfwpub.com/updatedtps6e. Using the sliders helps students to see that the Normal approximation gets better as the sample size increases and as the proportion of successes gets closer to 0.5 and farther from 0 and 1.

Teaching Tip

Students are sometimes confused by the symbols $\mu_{\hat{p}}$ and $\sigma_{\hat{p}}$ These are simply the mean and standard deviation of the sampling distribution of the sample proportion. To illustrate, show your students the dotplot of sample proportions you made in the "A penny for your thoughts?" activity from the beginning of the chapter, and remind them that each of the values plotted on the graph is a $\hat{p}$. Then tell them that the way we denote the mean of the $\hat{p}$ distribution is $\mu_{\hat{p}}$ and the way we denote the standard deviation of the $\hat{p}$ distribution is $\sigma_{\hat{p}}$.

- The **standard deviation** of the sampling distribution of $\hat{p}$ is approximately

$$\sigma_{\hat{p}} = \sqrt{\frac{p(1-p)}{n}}$$

  as long as the *10% condition* is satisfied: $n < 0.10N$. The value $\sigma_{\hat{p}}$ measures the typical distance between a sample proportion $\hat{p}$ and the population proportion p.

- The sampling distribution of $\hat{p}$ is **approximately Normal** as long as the *Large Counts condition* is satisfied: $np \geq 10$ and $n(1-p) \geq 10$.

The two conditions mentioned in the preceding box are very important.

> We call it the "Large Counts" condition because *np* is the expected *count* of successes in the sample and $n(1-p)$ is the expected *count* of failures in the sample.

- *Large Counts condition:* If we assume that the sampling distribution of $\hat{p}$ is approximately Normal when it isn't, any calculations we make using a Normal distribution will be flawed.

- *10% condition:* When we sample *with* replacement, the standard deviation of the sampling distribution of $\hat{p}$ is exactly $\sigma_{\hat{p}} = \sqrt{\dfrac{p(1-p)}{n}}$. When we sample *without* replacement, the observations are not independent, and the actual standard deviation of the sampling distribution of $\hat{p}$ is smaller than the value given by the formula. If the sample size is less than 10% of the population size, however, the value given by the formula is nearly correct.

Because larger random samples give better information, it sometimes makes sense to sample more than 10% of a population. In such a case, there is an adjustment we can make to the formula for $\sigma_{\hat{p}}$ that correctly reduces the standard deviation. The adjustment is called a *finite population correction (FPC)*. We'll avoid situations that require the FPC in this text.

EXAMPLE

Backing the pack
The sampling distribution of $\hat{p}$

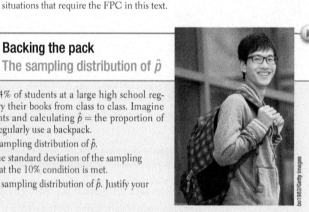

PROBLEM: Suppose that 84% of students at a large high school regularly use a backpack to carry their books from class to class. Imagine taking an SRS of 100 students and calculating $\hat{p}$ = the proportion of students in the sample who regularly use a backpack.

(a) Identify the mean of the sampling distribution of $\hat{p}$.
(b) Calculate and interpret the standard deviation of the sampling distribution of $\hat{p}$. Verify that the 10% condition is met.
(c) Describe the shape of the sampling distribution of $\hat{p}$. Justify your answer.

ALTERNATE EXAMPLE   Skills 3.B, 3.C

Business is major   The sampling distribution of $\hat{p}$

PROBLEM:
According to the National Center for Education Statistics, 20% of students who recently earned bachelor's degrees were business majors. Imagine taking an SRS of 300 students who recently earned bachelor's degrees and calculating $\hat{p}$ = the proportion of students in the sample who majored in business.
(a) Identify the mean of the sampling distribution of $\hat{p}$.
(b) Calculate and interpret the standard deviation of the sampling distribution of $\hat{p}$. Verify that the 10% condition is met.
(c) Describe the shape of the sampling distribution of $\hat{p}$. Justify your answer.

SOLUTION:
(a) $\mu_{\hat{p}} = 0.20$
(b) Assuming that $n = 300$ students is less than 10% of all students who recently earned a bachelor's degree,

$$\sigma_{\hat{p}} = \sqrt{\frac{0.20(1-0.20)}{300}} = 0.0231.$$ In SRSs of

size 300, the sample proportion of students who recently earned bachelor's degrees and majored in business will typically vary by about 0.0231 from the true proportion of $p = 0.20$.
(c) Because $300(0.20) = 60 \geq 10$ and $300(1-0.20) = 240 \geq 10$, the sampling distribution of $\hat{p}$ is approximately Normal.

SOLUTION:

(a) $\mu_{\hat{p}} = 0.84$

(b) Assuming that $n = 100$ students is less than 10% of students in a large high school, the standard deviation is approximately

$$\sigma_{\hat{p}} = \sqrt{\frac{0.84(1-0.84)}{100}} = 0.0367$$

In SRSs of size 100, the sample proportion of students who regularly use a backpack will typically vary by about 0.0367 from the true proportion of $p = 0.84$.

(c) Because $100(0.84) = 84 \geq 10$ and $100(1-0.84) = 16 \geq 10$, the sampling distribution of $\hat{p}$ is approximately Normal.

| $\mu_{\hat{p}} = p$ |

| When $n < 0.10N$, $\sigma_{\hat{p}} = \sqrt{\dfrac{p(1-p)}{n}}$ |

| When $np \geq 10$ and $n(1-p) \geq 10$, the sampling distribution of $\hat{p}$ is approximately Normal. |

FOR PRACTICE, TRY EXERCISE 33

Think About It

HOW IS THE SAMPLING DISTRIBUTION OF $\hat{p}$ RELATED TO THE BINOMIAL COUNT X? From Chapter 6, we know that the mean and standard deviation of a binomial random variable X are

$$\mu_X = np \quad \text{and} \quad \sigma_X = \sqrt{np(1-p)}$$

The sample proportion of successes is closely related to X:

$$\hat{p} = \frac{\text{count of successes in sample}}{\text{sample size}} = \frac{X}{n}$$

Because $\hat{p} = X/n = (1/n)X$, we're just multiplying the random variable X by a constant $(1/n)$ to get the random variable $\hat{p}$. Recall from Chapter 6 that multiplying a random variable by a constant multiplies both the mean and the standard deviation by that constant. We have

$$\mu_{\hat{p}} = \frac{1}{n}\mu_X = \frac{1}{n}(np) = p$$

$$\sigma_{\hat{p}} = \frac{1}{n}\sigma_X = \frac{1}{n}\sqrt{np(1-p)} = \sqrt{\frac{np(1-p)}{n^2}} = \sqrt{\frac{p(1-p)}{n}}$$

What about shape? Multiplying a random variable by a positive constant doesn't change the shape of the probability distribution. So the sampling distribution of $\hat{p}$ will have the same shape as the distribution of the binomial random variable X. If you remember the subsection in Chapter 6 about the Normal approximation to a binomial distribution, you already know that a Normal distribution can be used to approximate the sampling distribution of $\hat{p}$ whenever both np and $n(1-p)$ are at least 10.

CHECK YOUR UNDERSTANDING

Suppose that 75% of young adult Internet users (ages 18 to 29) watch online videos. A polling organization contacts an SRS of 1000 young adult Internet users and calculates the proportion $\hat{p}$ in this sample who watch online videos.

1. Identify the mean of the sampling distribution of $\hat{p}$.
2. Calculate and interpret the standard deviation of the sampling distribution of $\hat{p}$. Check that the 10% condition is met.

continues on next page

Making Connections

In the "A penny for your thoughts?" activity at the beginning of the chapter, consider asking students to make dotplots for both $X =$ number of pennies from the 2000s as well as $\hat{p} =$ the proportion of pennies from the 2000s. Save the dotplots (or take pictures) and use them here to make the connections between the binomial distribution and the sampling distribution of $\hat{p}$.

✔ Answers to CYU

1. $\mu_{\hat{p}} = p = 0.75$

2. Because $1000 < 10\%$ of all young adult Internet users, the 10% condition has been met. The standard deviation of the sampling distribution of $\hat{p}$ is approximately

$$\sigma_{\hat{p}} = \sqrt{\frac{p(1-p)}{n}} = \sqrt{\frac{0.75(0.25)}{1000}} =$$

0.0137. In SRSs of size 1000, the sample proportion of young adult Internet users who watch online videos will typically vary by about 0.0137 from the true proportion of $p = 0.75$.

3. Because $np = 1000(0.75) = 750 \geq 10$ and $n(1-p) = 1000(0.25) = 250 \geq 10$, the sampling distribution of $\hat{p}$ is approximately Normal.

4. If the sample size were 9000 instead of 1000, the sampling distribution would still be approximately Normal with mean 0.75. However, the standard deviation of the sampling distribution would be smaller by a factor of 3. In this case, the standard deviation is approximately

$$\sigma_{\hat{p}} = \sqrt{\frac{0.75(0.25)}{9000}} = 0.0046.$$

3. Is the sampling distribution of $\hat{p}$ approximately Normal? Check that the Large Counts condition is met.

4. If the sample size were 9000 rather than 1000, how would this change the sampling distribution of $\hat{p}$?

Using the Normal Approximation for $\hat{p}$

Inference about a population proportion p is based on the sampling distribution of $\hat{p}$. When the sample size is large enough for np and $n(1-p)$ to both be at least 10 (the *Large Counts condition*), the sampling distribution of $\hat{p}$ is approximately Normal. In that case, we can use a Normal distribution to estimate the probability of obtaining an SRS in which $\hat{p}$ lies in a specified interval of values. Here is an example.

EXAMPLE	Going to college
	Normal calculations involving $\hat{p}$

PROBLEM: A polling organization asks an SRS of 1500 first-year college students how far away their home is. Suppose that 35% of all first-year students attend college within 50 miles of home. Find the probability that the random sample of 1500 students will give a result within 2 percentage points of the true value.

SOLUTION:

Let $\hat{p}$ = sample proportion of all first-year college students who attend college within 50 miles of home.

$\mu_{\hat{p}} = 0.35$

Assuming that $1500 < 10\%$ of all first-year college students,

$\sigma_{\hat{p}} = \sqrt{\dfrac{(0.35)(0.65)}{1500}} = 0.0123$

Because $np = 1500(0.35) = 525 \geq 10$ and $n(1-p) = 1500(0.65) = 975 \geq 10$, the distribution of $\hat{p}$ is approximately Normal.

> Calculate the mean and standard deviation of the sampling distribution of $\hat{p}$.
>
> $\mu_{\hat{p}} = p \qquad \sigma_{\hat{p}} = \sqrt{\dfrac{p(1-p)}{n}}$

> Justify that the distribution of $\hat{p}$ is approximately Normal using the Large Counts condition.

> **1. Draw a Normal distribution.**

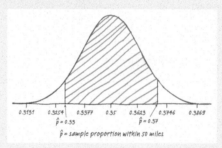

0.3131   0.3254   0.3377   0.35   0.3623   0.3746   0.3869
$\hat{p} = 0.33$ $\hat{p} = 0.37$

$\hat{p}$ = sample proportion within 50 miles

ALTERNATE EXAMPLE	Skill 4.B

Strive for 5
Normal calculations involving $\hat{p}$

PROBLEM:

In 2016, more than 200,000 students took the AP® Statistics exam. Of these students, 14% earned the highest score of 5. Find the probability that a random sample of 300 students who took the 2016 AP® Statistics exam would indicate that 15% or higher scored a 5.

SOLUTION:

$\mu_{\hat{p}} = 0.14$

We know that $300 < 10\%$ of all AP® Statistics exam takers (200,000+), so

$\sigma_{\hat{p}} = \sqrt{\dfrac{(0.14)(0.86)}{300}} = 0.02$. Because

$np = 300(0.14) = 42 \geq 10$ and $n(1-p) = 300(0.86) = 258 \geq 10$, the distribution of $\hat{p}$ is approximately Normal.

(i) $z = \dfrac{0.15 - 0.14}{0.02} = 0.50$

Table A: $P(\hat{p} \geq 0.15) = P(z \geq 0.50) = 1 - 0.6915 = 0.3085$

Tech: normalcdf (lower: 0.50, upper: 1000, mean: 0, SD: 1) = 0.3085

(ii) normalcdf (lower: 0.15, upper: 1, mean: 0.14, SD: 0.02) = 0.3085

(i) $z = \dfrac{0.33 - 0.35}{0.0123} = -1.63$ and $z = \dfrac{0.37 - 0.35}{0.0123} = 1.63$

Using Table A: $P(0.33 \le \hat{p} \le 0.37) = P(-1.63 \le z \le 1.63)$
$= 0.9484 - 0.0516 = 0.8968$

Using technology: normalcdf (lower: −1.63, upper: 1.63, mean: 0, SD: 1)
$= 0.8969$

(ii) normalcdf (lower: 0.33, upper: 0.37, mean: 0.35, SD: 0.0123) $= 0.8961$

> **2. Perform calculations.**
> (i) Standardize and use Table A or technology; or
> (ii) Use technology without standardizing.
> Be sure to answer the question that was asked.

FOR PRACTICE, TRY EXERCISE 43

Preparing for Inference

Have students consider the formula that was used in this example to calculate the z-scores:

$$z = \frac{\hat{p} - p}{\sqrt{\dfrac{p(1 - p)}{n}}}.$$ We will use this

formula to calculate the standardized test statistic for a *one-sample z-test for a proportion* in Chapter 9.

In the preceding example, about 90% of all SRSs of size 1500 from this population will give a result within 2 percentage points of the truth about the population. This result also suggests that in about 90% of all SRSs of size 1500 from this population, the true proportion will be within 2 percentage points of the sample proportion. This fact will become very important in Chapter 8 when we use sample data to create an interval of plausible values for a population parameter.

The Sampling Distribution of a Difference Between Two Proportions

Are males or females more likely to use Twitter? Many statistical questions involve comparing the proportion of individuals with a certain characteristic in two populations. Let's call these parameters of interest p_1 and p_2. The preferred strategy is to take a separate random sample from each population and to compare the sample proportions $\hat{p}_1$ and $\hat{p}_2$ with that characteristic.

Which of two treatments is more successful for helping people quit smoking? A randomized experiment can be used to answer this question. This time, the parameters p_1 and p_2 that we want to compare are the true proportions of successful outcomes for each treatment. We use the proportions of successes in the two treatment groups, $\hat{p}_1$ and $\hat{p}_2$, to make the comparison. Here's a table that summarizes these two situations:

Population or treatment	Parameter	Statistic	Sample size
1	p_1	$\hat{p}_1$	n_1
2	p_2	$\hat{p}_2$	n_2

Preparing for Inference

In Chapter 8, we will identify 2% as the *margin of error* for a 90% confidence interval. More specifically, we are 90% confident that the true proportion for the population will be within 2 percentage points of the sample proportion from an SRS of size 1500.

Making Connections

Notation matters! A statistic is used to estimate a parameter.

- $\hat{p}$ from the sample is used to estimate the population proportion p.
- $\hat{p}_1 - \hat{p}_2$ from two samples is used to estimate the difference of population proportions $p_1 - p_2$.

MachineHeadz/Getty Images

We compare the populations or treatments by doing inference about the difference $p_1 - p_2$ between the parameters. The statistic that estimates this difference is the difference between the two sample proportions, $\hat{p}_1 - \hat{p}_2$. To explore the sampling distribution of $\hat{p}_1 - \hat{p}_2$, let's start with two populations having a known proportion of successes.

Suppose there are two large high schools—each with more than 2000 students—in a certain town. At School 1, 70% of students did their homework last night ($p_1 = 0.70$). Only 50% of the students at School 2 did their homework last night ($p_2 = 0.50$). The counselor at School 1 selects an SRS of 100 students and records the proportion $\hat{p}_1$ that did the homework. School 2's counselor selects an SRS of 200 students and records the proportion $\hat{p}_2$ that did the homework. What can we say about the difference $\hat{p}_1 - \hat{p}_2$ in the sample proportions?

Earlier in this section, we saw that the sampling distribution of a sample proportion $\hat{p}$ has the following properties:

Shape: Approximately Normal if $np \geq 10$ and $n(1-p) \geq 10$

Center: $\mu_{\hat{p}} = p$

Variability: $\sigma_{\hat{p}} = \sqrt{\dfrac{p(1-p)}{n}}$ if $n < 0.10N$

For the sampling distributions of $\hat{p}_1$ and $\hat{p}_2$ in this case:

	Sampling distribution of $\hat{p}_1$	Sampling distribution of $\hat{p}_2$
Shape	Approximately Normal; $n_1 p_1 = 100(0.70) = 70 \geq 10$ and $n_1(1 - p_1) = 100(0.30) = 30 \geq 10$	Approximately Normal; $n_2 p_2 = 200(0.50) = 100 \geq 10$ and $n_2(1 - p_2) = 200(0.50) = 100 \geq 10$
Center	$\mu_{\hat{p}_1} = p_1 = 0.70$	$\mu_{\hat{p}_2} = p_2 = 0.50$
Variability	$\sigma_{\hat{p}_1} = \sqrt{\dfrac{p_1(1 - p_1)}{n_1}} = \sqrt{\dfrac{0.7(0.3)}{100}}$ $= 0.0458$ because $100 < 10\%$ of all students at School 1.	$\sigma_{\hat{p}_2} = \sqrt{\dfrac{p_2(1 - p_2)}{n_2}} = \sqrt{\dfrac{0.5(0.5)}{200}}$ $= 0.0354$ because $200 < 10\%$ of all students at School 2.

What about the sampling distribution of $\hat{p}_1 - \hat{p}_2$? We used software to randomly select 100 students from School 1 and 200 students from School 2. Our first set of samples gave $\hat{p}_1 = 0.72$ and $\hat{p}_2 = 0.47$, resulting in a difference of $\hat{p}_1 - \hat{p}_2 = 0.72 - 0.47 = 0.25$. A red dot for this value appears in Figure 7.10. The dotplot shows the results of repeating this process 500 times.

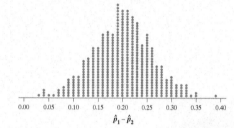

FIGURE 7.10 Simulated sampling distribution of the difference in sample proportions $\hat{p}_1 - \hat{p}_2$ in 500 SRSs of size $n_1 = 100$ from a population with $p_1 = 0.70$ and 500 SRSs of size $n_2 = 200$ from a population with $p_2 = 0.50$.

The figure suggests that the sampling distribution of $\hat{p}_1 - \hat{p}_2$ has an approximately Normal shape. This makes sense from what you learned in Section 6.2 because we are subtracting two independent random variables, $\hat{p}_1$ and $\hat{p}_2$, that have approximately Normal distributions.

The mean of the sampling distribution is 0.20. The true proportion of students who did last night's homework at School 1 is $p_1 = 0.70$ and at School 2 is $p_2 = 0.50$. We expect the difference $\hat{p}_1 - \hat{p}_2$ to center on the actual difference in the population proportions, $p_1 - p_2 = 0.70 - 0.50 = 0.20$. The standard deviation of the sampling distribution is 0.058. It can be found using the formula

$$\sqrt{\dfrac{p_1(1 - p_1)}{n_1} + \dfrac{p_2(1 - p_2)}{n_2}} = \sqrt{\dfrac{0.7(0.3)}{100} + \dfrac{0.5(0.5)}{200}} = 0.058$$

That is, the difference (School 1 – School 2) in the sample proportions of students at the two schools who did their homework last night typically varies by about 0.058 from the true difference in proportions of 0.20.

THE SAMPLING DISTRIBUTION OF $\hat{p}_1 - \hat{p}_2$

Choose an SRS of size n_1 from Population 1 with proportion of successes p_1 and an independent SRS of size n_2 from Population 2 with proportion of successes p_2. Then:

- The sampling distribution of $\hat{p}_1 - \hat{p}_2$ is **approximately Normal** if the *Large Counts condition* is met for both samples: $n_1 p_1 \geq 10$, $n_1(1 - p_1) \geq 10$, $n_2 p_2 \geq 10$, and $n_2(1 - p_2) \geq 10$.
- The **mean** of the sampling distribution of $\hat{p}_1 - \hat{p}_2$ is $\mu_{\hat{p}_1 - \hat{p}_2} = p_1 - p_2$.
- The **standard deviation** of the sampling distribution of $\hat{p}_1 - \hat{p}_2$ is approximately

$$\sigma_{\hat{p}_1 - \hat{p}_2} = \sqrt{\frac{p_1(1 - p_1)}{n_1} + \frac{p_2(1 - p_2)}{n_2}}$$

as long as the *10% condition* is met for both samples: $n_1 < 0.10 N_1$ and $n_2 < 0.10 N_2$.

Note that the formula for the standard deviation is exactly correct only when we have two types of independence:

- Independent samples, so that we can add the variances of $\hat{p}_1$ and $\hat{p}_2$.
- Independent observations within each sample. When sampling without replacement, the actual value of the standard deviation is smaller than the formula suggests. However, if the 10% condition is met for both samples, the difference is negligible.

The standard deviation of the sampling distribution tells us how much the difference in sample proportions will typically vary from the difference in the population proportions if we repeat the random sampling process many times.

Think About It

WHERE DO THE FORMULAS FOR THE MEAN AND STANDARD DEVIATION OF THE SAMPLING DISTRIBUTION OF $\hat{p}_1 - \hat{p}_2$ COME FROM? Both $\hat{p}_1$ and $\hat{p}_2$ are random variables. That is, their values would vary in repeated independent SRSs of size n_1 and n_2. Independent random samples yield independent random variables $\hat{p}_1$ and $\hat{p}_2$. The statistic $\hat{p}_1 - \hat{p}_2$ is the difference of these two independent random variables.

In Chapter 6, we learned that for any two random variables X and Y,

$$\mu_{X-Y} = \mu_X - \mu_Y$$

For the random variables $\hat{p}_1$ and $\hat{p}_2$, we have

$$\mu_{\hat{p}_1 - \hat{p}_2} = \mu_{\hat{p}_1} - \mu_{\hat{p}_2} = p_1 - p_2$$

We also learned in Chapter 6 that for *independent* random variables X and Y,

$$\sigma^2_{X-Y} = \sigma^2_X + \sigma^2_Y$$

AP® EXAM TIP

The formula for the standard deviation of the sampling distribution of $\hat{p}_1 - \hat{p}_2$ is provided on the formula sheet for the AP® Statistics exam.

Making Connections

In Chapter 6, we learned that we should add variances when combining random variables, but only if the two random variables are independent. To use the formula for standard deviation on this page, we must be convinced that the two sample proportions are independent of each other. We also need observations to be independent within each sample, which is satisfied if we are sampling with replacement. If we are sampling without replacement, we must check the 10% condition to be comfortable viewing observations within each sample as independent.

Making Connections

Students shouldn't be expected to derive the formulas for the mean and standard deviation of a difference in sample proportions, as we do here. However, reading through this Think About It is a great way to review the rules for combining random variables that we learned in Chapter 6 and the properties of the sampling distribution of a sample proportion that we learned earlier in this section.

For the random variables $\hat{p}_1$ and $\hat{p}_2$, we have

$$\sigma^2_{\hat{p}_1-\hat{p}_2} = \sigma^2_{\hat{p}_1} + \sigma^2_{\hat{p}_2} = \left(\sqrt{\frac{p_1(1-p_1)}{n_1}}\right)^2 + \left(\sqrt{\frac{p_2(1-p_2)}{n_2}}\right)^2$$

$$= \frac{p_1(1-p_1)}{n_1} + \frac{p_2(1-p_2)}{n_2}$$

So $\sigma_{\hat{p}_1-\hat{p}_2} = \sqrt{\dfrac{p_1(1-p_1)}{n_1} + \dfrac{p_2(1-p_2)}{n_2}}$.

When the conditions are met, we can use the Normal density curve shown in Figure 7.11 to model the sampling distribution of $\hat{p}_1 - \hat{p}_2$. Note that this would allow us to calculate probabilities involving $\hat{p}_1 - \hat{p}_2$ with a Normal distribution.

> When we analyzed the results of randomized experiments in Section 4.3, we used simulation to create a *randomization distribution* by repeatedly reallocating individuals to treatment groups. Fortunately for us, randomization distributions of $\hat{p}_1 - \hat{p}_2$ roughly follow the same rules for shape, center, and variability as sampling distributions of $\hat{p}_1 - \hat{p}_2$.

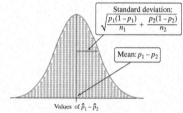

FIGURE 7.11 Select independent SRSs from two populations having proportions of successes p_1 and p_2. The proportions of successes in the two samples are $\hat{p}_1$ and $\hat{p}_2$. When the conditions are met, the sampling distribution of $\hat{p}_1 - \hat{p}_2$ is approximately Normal with mean $p_1 - p_2$ and standard deviation $\sqrt{\dfrac{p_1(1-p_1)}{n_1} + \dfrac{p_2(1-p_2)}{n_2}}$.

Making Connections

In Section 4.3, students analyzed the distracted driving experiment using many simulated random assignments. They created a dotplot of the difference in proportions for each random assignment. This dotplot (page 302) approximates the *randomization distribution*.

ALTERNATE EXAMPLE   Skills 3.B, 3.C

Grade level and gender
Describing the sampling distribution of $\hat{p}_1 - \hat{p}_2$

PROBLEM:
In a very large high school, the junior class has 800 students, 54% of whom are female. The senior class has 700 students, 49% of whom are female. The student council selects a random sample of 40 juniors and a separate random sample of 35 seniors. Let $\hat{p}_{jr} - \hat{p}_{sr}$ be the difference in the sample proportions of females.
(a) What is the shape of the sampling distribution of $\hat{p}_{jr} - \hat{p}_{sr}$? Why?
(b) Find the mean of the sampling distribution.
(c) Calculate and interpret the standard deviation of the sampling distribution.
(d) Estimate the probability that the proportion of females in the sample from the junior class is less than the proportion of females in the sample from the senior class.

SOLUTION:
(a) Approximately Normal, because $n_{jr}p_{jr} = 40(0.54) = 21.6$, $n_{jr}(1 - p_{jr}) = 40(0.46) = 18.4$, $n_{sr}p_{sr} = 35(0.49) = 17.15$, and $n_{sr}(1 - p_{sr}) = 35(0.51) = 17.85$ are all ≥ 10.

(b) $\mu_{\hat{p}_{jr} - \hat{p}_{sr}} = 0.54 - 0.49 = 0.05$

(c) Because $40 < 10\%$ of all juniors and $35 < 10\%$ of all seniors,

---

EXAMPLE

Yummy goldfish!
The sampling distribution of $\hat{p}_1 - \hat{p}_2$

PROBLEM: Your teacher brings two bags of colored goldfish crackers to class. Bag 1 has 25% red crackers and Bag 2 has 35% red crackers. Each bag contains more than 1000 crackers. Using a paper cup, your teacher takes an SRS of 50 crackers from Bag 1 and an independent SRS of 40 crackers from Bag 2. Let $\hat{p}_1 - \hat{p}_2$ be the difference in the sample proportions of red crackers.
(a) What is the shape of the sampling distribution of $\hat{p}_1 - \hat{p}_2$? Why?
(b) Find the mean of the sampling distribution.
(c) Calculate and interpret the standard deviation of the sampling distribution. Verify that the 10% condition is met.
(d) Estimate the probability that the proportion of red crackers in the sample from Bag 1 is greater than the proportion of red crackers in the sample from Bag 2.

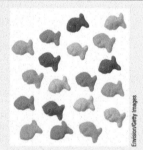
Envision/Getty Images

$$\sigma_{\hat{p}_{jr} - \hat{p}_{sr}} = \sqrt{\frac{0.54(0.46)}{40} + \frac{0.49(0.51)}{35}}$$
$$= 0.1155$$

The difference (Junior – Senior) in the sample proportions of females typically varies by about 0.1155 from the true difference in proportions of 0.05.

(d) We want to find $P(\hat{p}_{jr} < \hat{p}_{sr})$, which is equivalent to $P(\hat{p}_{jr} - \hat{p}_{sr} < 0)$.

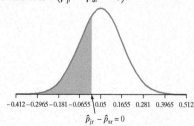

$\hat{p}_{jr} - \hat{p}_{sr} = 0$

$\hat{p}_{jr} - \hat{p}_{sr}$ = Difference in the sample proportions of females in junior class and senior class

(i) $z = \dfrac{0 - 0.05}{0.1155} = -0.43$

Using Table A:
$P(\hat{p}_{jr} - \hat{p}_{sr} < 0) = P(z < -0.43) = 0.3336$
Using Technology: normalcdf(lower:−1000, upper:−0.43, mean:0, SD:1) = 0.3366
(ii) normalcdf(lower:−1000, upper:0, mean:0.05, SD:0.1155) = 0.3325

SOLUTION:

(a) Approximately Normal, because $n_1 p_1 = 50(0.25) = 12.5$, $n_1(1-p_1) = 50(0.75) = 37.5$, $n_2 p_2 = 40(0.35) = 14$, and $n_2(1-p_2) = 40(0.65) = 26$ are all ≥ 10.

> Note that these values are the expected numbers of successes and failures in the two samples.

(b) $\mu_{\hat{p}_1 - \hat{p}_2} = 0.25 - 0.35 = -0.10$

(c) Because $50 < 10\%$ of all crackers in Bag 1 and $40 < 10\%$ of all crackers in Bag 2,

$$\sigma_{\hat{p}_1 - \hat{p}_2} = \sqrt{\frac{0.25(0.75)}{50} + \frac{0.35(0.65)}{40}} = 0.0971$$

> $$\mu_{\hat{p}_1 - \hat{p}_2} = p_1 - p_2$$
>
> $$\sigma_{\hat{p}_1 - \hat{p}_2} = \sqrt{\frac{p_1(1-p_1)}{n_1} + \frac{p_2(1-p_2)}{n_2}}$$

The difference (Bag 1 − Bag 2) in the sample proportions of red goldfish crackers typically varies by about 0.097 from the true difference in proportions of −0.10.

(d) We want to find $P(\hat{p}_1 > \hat{p}_2)$, which is equivalent to $P(\hat{p}_1 - \hat{p}_2 > 0)$.

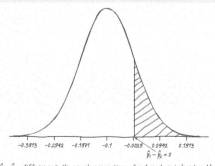

> **1. Draw a Normal distribution.**

$\hat{p}_1 - \hat{p}_2$ = difference in the sample proportions of red crackers in bag 1 and bag 2

(i) $z = \dfrac{0 - (-0.10)}{0.0971} = 1.03$

> **2. Perform calculations.**
> (i) Standardize and use Table A or technology; or
> (ii) Use technology without standardizing.
> Be sure to answer the question that was asked.

Using Table A: $P(\hat{p}_1 - \hat{p}_2 > 0) = P(z > 1.03) = 1 - 0.8485 = 0.1515$

Using technology: normalcdf (lower:1.03, upper:1000, mean:0, SD:1) = 0.1515

(ii) normalcdf(lower:0, upper:1000, mean:−0.10, SD:0.0971) = 0.1515

> **FOR PRACTICE, TRY EXERCISE 49**

Section 7.2 | Summary

- When we want information about the population proportion p of successes, we often take an SRS and use the sample proportion $\hat{p}$ to estimate the unknown parameter p. The **sampling distribution of the sample proportion** $\hat{p}$ describes how the statistic $\hat{p}$ varies in all possible samples of the same size from the population.
  - **Shape:** The sampling distribution of $\hat{p}$ is **approximately Normal** when both $np \geq 10$ and $n(1-p) \geq 10$ (the *Large Counts condition*).

Answers to Section 7.2 Exercises

7.33 (a) $\mu_{\hat{p}} = p = 0.55$
(b) $n = 500$ is less than 10% of the population of all registered voters, so

$$\sigma_{\hat{p}} = \sqrt{\frac{0.55(1 - 0.55)}{500}} = 0.022.$$

In SRSs of size $n = 500$, the sample proportion of registered voters who are Democrats will typically vary by about 0.022 from the true proportion of $p = 0.55$. **(c)** Because $np = 500(0.55) = 275 \geq 10$ and $n(1 - p) = 500(1 - 0.55) = 225 \geq 10$, the sampling distribution of $\hat{p}$ is approximately Normal.

7.34 (a) $\mu_{\hat{p}} = p = 0.59$
(b) $n = 50$ is less than 10% of the population of all married couples with children, so

$$\sigma_{\hat{p}} = \sqrt{\frac{0.59(1 - 0.59)}{50}} = 0.0696.$$

In SRSs of size $n = 50$, the sample proportion of married couples with children in which both parents work outside the home will typically vary by about 0.0696 from the true proportion of $p = 0.59$. **(c)** Because $np = 50(0.59) = 29.5 \geq 10$ and $n(1 - p) = 50(1 - 0.59) = 20.5 \geq 10$, the sampling distribution of $\hat{p}$ is approximately Normal.

7.35 (a) $\mu_{\hat{p}} = p = 0.20$
(b) $n = 30$ is less than 10% of the population of all Skittles, so

$$\sigma_{\hat{p}} = \sqrt{\frac{0.20(1 - 0.20)}{30}} = 0.0730.$$

In SRSs of size $n = 30$, the sample proportion of Skittles that are orange will typically vary by about 0.073 from the true proportion of $p = 0.20$. **(c)** Because $np = 30(0.2) = 6 < 10$, the sampling distribution of $\hat{p}$ is not approximately Normal. Because $p = 0.20$ is closer to 0 than to 1, the sampling distribution of $\hat{p}$ is skewed to the right.

7.36 (a) $\mu_{\hat{p}} = p = 0.90$ **(b)** $n = 15$ is less than 10% of the population of the 3000 workers at the factory, so

$$\sigma_{\hat{p}} = \sqrt{\frac{0.90(1 - 0.90)}{15}} = 0.0775.$$

In SRSs of size $n = 15$, the sample proportion of factory workers who are male will typically vary by about 0.0775 from the true proportion of $p = 0.90$.
(c) $np = 15(0.9) = 13.5 \geq 10$, but $n(1 - p) = 15(1 - 0.9) = 1.5 < 10$, so the sampling distribution of $\hat{p}$ is not

- **Center:** The **mean** of the sampling distribution of $\hat{p}$ is $\mu_{\hat{p}} = p$. So $\hat{p}$ is an unbiased estimator of p.
- **Variability:** The **standard deviation** of the sampling distribution of $\hat{p}$ is approximately $\sigma_{\hat{p}} = \sqrt{p(1 - p)/n}$ for an SRS of size n. This formula can be used if the sample size is less than 10% of the population size (the *10% condition*).

- Choose independent SRSs of size n_1 from Population 1 with proportion of successes p_1 and of size n_2 from Population 2 with proportion of successes p_2. The sampling distribution of $\hat{p}_1 - \hat{p}_2$ has the following properties:
  - **Shape: Approximately Normal** if the Large Counts condition is met: $n_1 p_1$, $n_1(1 - p_1)$, $n_2 p_2$, and $n_2(1 - p_2)$ are all at least 10.
  - **Center:** The **mean** of the sampling distribution is $\mu_{\hat{p}_1 - \hat{p}_2} = p_1 - p_2$.
  - **Variability:** The **standard deviation** of the sampling distribution is approximately $\sigma_{\hat{p}_1 - \hat{p}_2} = \sqrt{\frac{p_1(1 - p_1)}{n_1} + \frac{p_2(1 - p_2)}{n_2}}$ as long as the 10% condition is met: $n_1 < 0.10N_1$ and $n_2 < 0.10N_2$.

Section 7.2 | Exercises

33. Registered voters In a congressional district, 55% of registered voters are Democrats. A polling organization selects a random sample of 500 registered voters from this district. Let $\hat{p}$ = the proportion of Democrats in the sample.

(a) Identify the mean of the sampling distribution of $\hat{p}$.

(b) Calculate and interpret the standard deviation of the sampling distribution of $\hat{p}$. Verify that the 10% condition is met.

(c) Describe the shape of the sampling distribution of $\hat{p}$. Justify your answer.

34. Married with children According to a recent U.S. Bureau of Labor Statistics report, the proportion of married couples with children in which both parents work outside the home is 59%.[6] You select an SRS of 50 married couples with children and let $\hat{p}$ = the sample proportion of couples in which both parents work outside the home.

(a) Identify the mean of the sampling distribution of $\hat{p}$.

(b) Calculate and interpret the standard deviation of the sampling distribution of $\hat{p}$. Verify that the 10% condition is met.

(c) Describe the shape of the sampling distribution of $\hat{p}$. Justify your answer.

35. Orange Skittles® The makers of Skittles claim that 20% of Skittles candies are orange. Suppose this claim is true. You select a random sample of 30 Skittles from a large bag. Let $\hat{p}$ = the proportion of orange Skittles in the sample.

(a) Identify the mean of the sampling distribution of $\hat{p}$.

(b) Calculate and interpret the standard deviation of the sampling distribution of $\hat{p}$. Verify that the 10% condition is met.

(c) Describe the shape of the sampling distribution of $\hat{p}$. Justify your answer.

36. Male workers A factory employs 3000 unionized workers, 90% of whom are male. A random sample of 15 workers is selected for a survey about worker satisfaction. Let $\hat{p}$ = the proportion of males in the sample.

(a) Identify the mean of the sampling distribution of $\hat{p}$.

(b) Calculate and interpret the standard deviation of the sampling distribution of $\hat{p}$. Verify that the 10% condition is met.

(c) Describe the shape of the sampling distribution of $\hat{p}$. Justify your answer.

37. More Skittles® What sample size would be required to reduce the standard deviation of the sampling distribution to one-half the value you found in Exercise 35(b)? Justify your answer.

approximately Normal. Because $p = 0.90$ is closer to 1 than to 0, the sampling distribution of $\hat{p}$ is skewed to the left.

7.37 A sample size of $30(4) = 120$

7.38 A sample size of $15(9) = 135$

7.39 (a) More than 10% of the population $(10/76 = 13\%)$ was selected. **(b)** The actual standard deviation of the sampling distribution of $\hat{p}$ will be smaller than the value provided by the formula.

7.40 (a) More than 10% of the population $(5/30 = 16.7\%)$ was selected. **(b)** The actual standard deviation of the sampling distribution of $\hat{p}$ will be smaller than the value provided by the formula.

7.41 (a) $\mu_{\hat{p}} = p = 0.70$ **(b)** $n = 1012$ is less than 10% of the population of all U.S. adults, so $\sigma_{\hat{p}} = \sqrt{\frac{0.7(1 - 0.7)}{1012}} = 0.0144$.

(c) $np = 1012(0.70) = 708.4 \geq 10$ and $n(1 - p) = 1012(0.30) = 303.6 \geq 10$ **(d)** We want to find $P(\hat{p} \geq 0.67)$.
(i) $z = -2.08$; $P(z \leq -2.08) = 0.0188$
(ii) $P(\hat{p} \geq 0.67) = \text{normalcdf(lower:} -1000, \text{upper: } 0.67, \text{mean: } 0.70, \text{SD: } 0.0144) = 0.0186$ **(e)** Because 0.0186 is a small probability, there is convincing evidence against the claim that $p = 0.70$—it isn't plausible to get a sample proportion this small by chance alone.

38. More workers What sample size would be required to reduce the standard deviation of the sampling distribution to one-third the value you found in Exercise 36(b)? Justify your answer.

39. Airport security The Transportation Security Administration (TSA) is responsible for airport safety. On some flights, TSA officers randomly select passengers for an extra security check before boarding. One such flight had 76 passengers—12 in first class and 64 in coach class. TSA officers selected an SRS of 10 passengers for screening. Let $\hat{p}$ be the proportion of first-class passengers in the sample.

(a) Show that the 10% condition is not met in this case.

(b) What effect does violating the 10% condition have on the standard deviation of the sampling distribution of $\hat{p}$?

40. Don't pick me! Instead of collecting homework from all of her students, Mrs. Friedman randomly selects 5 of her 30 students and collects homework from only those students. Let $\hat{p}$ be the proportion of students in the sample that completed their homework.

(a) Show that the 10% condition is not met in this case.

(b) What effect does violating the 10% condition have on the standard deviation of the sampling distribution of $\hat{p}$?

41. Do you drink the cereal milk? A USA *Today* poll asked a random sample of 1012 U.S. adults what they do with the milk in the bowl after they have eaten the cereal. Let $\hat{p}$ be the proportion of people in the sample who drink the cereal milk. A spokesman for the dairy industry claims that 70% of all U.S. adults drink the cereal milk. Suppose this claim is true.

(a) What is the mean of the sampling distribution of $\hat{p}$?

(b) Find the standard deviation of the sampling distribution of $\hat{p}$. Verify that the 10% condition is met.

(c) Verify that the sampling distribution of $\hat{p}$ is approximately Normal.

(d) Of the poll respondents, 67% said that they drink the cereal milk. Find the probability of obtaining a sample of 1012 adults in which 67% or fewer say they drink the cereal milk, assuming the milk industry spokesman's claim is true.

(e) Does this poll give convincing evidence against the spokesman's claim? Explain your reasoning.

42. Do you go to church? The Gallup Poll asked a random sample of 1785 adults if they attended church during the past week. Let $\hat{p}$ be the proportion of people in the sample who attended church. A newspaper report claims that 40% of all U.S. adults went to church last week. Suppose this claim is true.

(a) What is the mean of the sampling distribution of $\hat{p}$?

(b) Find the standard deviation of the sampling distribution of $\hat{p}$. Verify that the 10% condition is met.

(c) Verify that the sampling distribution of $\hat{p}$ is approximately Normal.

(d) Of the poll respondents, 44% said they did attend church last week. Find the probability of obtaining a sample of 1785 adults in which 44% or more say they attended church last week, assuming the newspaper report's claim is true.

(e) Does this poll give convincing evidence against the newspaper's claim? Explain your reasoning.

43. Students on diets Suppose that 70% of college women have been on a diet within the past 12 months. A sample survey interviews an SRS of 267 college women. What is the probability that 75% or more of the women in the sample have been on a diet?

44. Who owns a Harley? Harley-Davidson motorcycles make up 14% of all the motorcycles registered in the United States. You plan to interview an SRS of 500 motorcycle owners. How likely is your sample to contain 20% or more who own Harleys?

45. On-time shipping A mail-order company advertises that it ships 90% of its orders within three working days. You select an SRS of 100 of the 5000 orders received in the past week for an audit. The audit reveals that 86 of these orders were shipped on time.

(a) If the company really ships 90% of its orders on time, what is the probability that the proportion in an SRS of 100 orders is 0.86 or less?

(b) Based on your answer to part (a), is there convincing evidence that less than 90% of all orders from this company are shipped within three working days? Explain your reasoning.

46. Wait times A hospital claims that 75% of people who come to its emergency room are seen by a doctor within 30 minutes of checking in. To verify this claim, an auditor inspects the medical records of 55 randomly selected patients who checked into the emergency room during the last year. Only 32 (58.2%) of these patients were seen by a doctor within 30 minutes of checking in.

(a) If the wait time is less than 30 minutes for 75% of all patients in the emergency room, what is the probability that the proportion of patients who wait less than 30 minutes is 0.582 or less in a random sample of 55 patients?

(b) Based on your answer to part (a), is there convincing evidence that less than 75% of all patients in the emergency room wait less than 30 minutes? Explain your reasoning.

7.42 (a) $\mu_{\hat{p}} = p = 0.4$ **(b)** $n = 1785$ is less than 10% of the population of all U.S. adults, so $\sigma_{\hat{p}} = \sqrt{\dfrac{0.4(0.6)}{1785}} = 0.0116$.

(c) $np = 1785(0.4) = 714 \geq 10$ and $n(1-p) = 1785(0.6) = 1071 \geq 10$

(d) We want to find $P(\hat{p} \geq 0.44)$.

(i) $z = \dfrac{0.44 - 0.40}{0.0116} = 3.45$; $P(z \geq 3.45) = 0.0003$ (ii) $P(\hat{p} \geq 0.44) =$ normalcdf (lower: 0.44, upper: 1000, mean: 0.40, SD: 0.0116) = 0.0003 **(e)** Because 0.0003 is a small probability, there is convincing evidence against the claim that $p = 0.40$—it isn't plausible to get a sample proportion this large by chance alone.

7.43 $\mu_{\hat{p}} = 0.70$; because 267 is less than 10% of the population of all college women,

$$\sigma_{\hat{p}} = \sqrt{\dfrac{0.7(0.3)}{267}} = 0.0280. \text{ Because}$$

$np = 267(0.7) = 186.9 \geq 10$ and $n(1-p) = 267(0.3) = 80.1 \geq 10$, the sampling distribution of $\hat{p}$ can be approximated by a Normal distribution. We want to find $P(\hat{p} \geq 0.75)$.

(i) $z = 1.79$; $P(z \geq 1.79) = 0.0367$

(ii) $P(\hat{p} \geq 0.75) =$ normalcdf (lower: 0.75, upper: 1000, mean: 0.7, SD: 0.0280) = 0.0371

7.44 $\mu_{\hat{p}} = 0.14$; because 500 is less than 10% of the population of registered motorcycles,

$$\sigma_{\hat{p}} = \sqrt{\dfrac{0.14(0.86)}{500}} = 0.0155$$

Because $np = 500(0.14) = 70 \geq 10$ and $n(1-p) = 500(0.86) = 430 \geq 10$, the sampling distribution of $\hat{p}$ can be approximated by a Normal distribution. We want to find $P(\hat{p} \geq 0.20)$.

(i) $z = \dfrac{0.20 - 0.14}{0.0155} = 3.87$;

$P(z \geq 3.87) \approx 0$ (ii) $P(\hat{p} \geq 0.20) =$ normalcdf (lower: 0.20, upper: 1000, mean: 0.14, SD: 0.0155) = 0.0001

7.45 (a) $\mu_{\hat{p}} = 0.90$; because 100 is less than 10% of the population of orders $(100/5000 = 2\%)$,

$$\sigma_{\hat{p}} = \sqrt{\dfrac{0.90(0.10)}{100}} = 0.03.$$

Because $np = 100(0.90) = 90 \geq 10$ and $n(1-p) = 100(0.10) = 10 \geq 10$, the sampling distribution of $\hat{p}$ can be approximated by a Normal distribution. We want to find $P(\hat{p} \leq 0.86)$.

(i) $z = \dfrac{0.86 - 0.90}{0.03} = -1.33$;

$P(z \leq -1.33) = 0.0918$

(ii) $P(\hat{p} \leq 0.86) =$ normalcdf(lower: -1000, upper: 0.86, mean: 0.90, SD: 0.03) = 0.0912

(b) No; it isn't unusual to get a sample proportion of 0.86 or smaller when selecting an SRS of 100 from a population in which $p = 0.90$. Thus, it is plausible that the 90% claim is correct and that the lower-than-expected percentage is due to chance alone.

7.46 (a) $\mu_{\hat{p}} = 0.75$; because 55 is less than 10% of the population of all people who have been checked into the emergency room during the last year,

$$\sigma_{\hat{p}} = \sqrt{\dfrac{0.75(0.25)}{55}} = 0.0584. \text{ Because}$$

$np = 55(0.75) = 41.25 \geq 10$ and $n(1-p) = 55(0.25) = 13.75 \geq 10$, the sampling distribution of $\hat{p}$ can be approximated by a Normal distribution. We want to find $P(\hat{p} \leq 0.582)$.

(i) $z = \dfrac{0.582 - 0.75}{0.0584} = -2.88$;

$P(z \geq -2.88) \approx 0.0020$

(ii) $P(\hat{p} \leq 0.582) =$ normalcdf(lower: -1000, upper: 0.582, mean: 0.75, SD: 0.0584) = 0.0002

(b) Yes; it is unusual to get a sample proportion of 0.582 or smaller when selecting a random sample of 55 from a population in which $p = 0.75$. Therefore, we have convincing evidence that less than 75% of all patients in the emergency room wait less than 30 minutes.

7.47 (a) $\mu_{\hat{p}_F - \hat{p}_M} = 0.30 - 0.25 = 0.05$
(b) Because $20 < 10\%$ of all females at the large high school and $20 < 10\%$ of all males at the large high school, the standard deviation of $\hat{p}_F - \hat{p}_M$ is approximately

$$\sigma_{\hat{p}_F - \hat{p}_M} = \sqrt{\frac{0.30(0.70)}{20} + \frac{0.25(0.75)}{20}} = $$

0.141.

The difference (Female – Male) in the sample proportions that are enrolled in an AP® class typically varies by about 0.141 from the true difference in proportions of 0.05.

(c) No, $n_F p_F = 20(0.3) = 6$, $n_F(1 - p_F) = 20(0.7) = 14$, $n_M p_M = 20(0.25) = 5$, $n_M(1 - p_M) = 20(0.75) = 15$, which are not all ≥ 10.

7.48 (a) $\mu_{\hat{p}_A - \hat{p}_B} = 0.20 - 0.18 = 0.02$
(b) Because $30 < 10\%$ of all students at high school A and $30 < 10\%$ of all students at high school B, the standard deviation of $\hat{p}_A - \hat{p}_B$ is approximately

$$\sigma_{\hat{p}_A - \hat{p}_B} = \sqrt{\frac{0.20(0.80)}{30} + \frac{0.18(0.82)}{30}} = $$

0.101.

The difference (High school A – High school B) in the sample proportions that participate on a school athletic team typically varies by about 0.101 from the true difference in proportions of 0.02.

(c) No, $n_A p_A = 30(0.2) = 6$, $n_A(1 - p_A) = 30(0.8) = 24$, $n_B p_B = 30(0.18) = 5.4$, $n_B(1 - p_B) = 30(0.82) = 24.6$, which are not all ≥ 10.

7.49 (a) Approximately Normal;
$n_C p_C = 50(0.30) = 15$, $n_C(1 - p_C) = 50(0.7) = 35$, $n_A p_A = 100(0.15) = 15$, and $n_A(1 - p_A) = 100(0.85) = 85$ are all ≥ 10. **(b)** $\mu_{\hat{p}_C - \hat{p}_A} = 0.30 - 0.15 = 0.15$

(c) Because $50 < 10\%$ of the jelly beans in the Child mix and $100 < 10\%$ of the jelly beans in the Adult mix, the standard deviation of $\hat{p}_C - \hat{p}_A$ is approximately

$$\sigma_{\hat{p}_C - \hat{p}_A} = \sqrt{\frac{0.3(0.7)}{50} + \frac{0.15(0.85)}{100}} = $$

0.0740.

The difference (Child mix – Adult mix) in the sample proportions of red jelly beans typically varies by about 0.0740 from the true difference in proportions of 0.15.

(d) We want to find $P(\hat{p}_C - \hat{p}_A \leq 0)$.

(i) $z = \dfrac{0 - 0.15}{0.0740} = -2.03$

$P(z \leq -2.03) = 0.0212$

(ii) $P(\hat{p}_C - \hat{p}_A \leq 0) =$ normalcdf (lower: −1000, upper: 0, mean: 0.15, SD: 0.0740) = 0.0213

47. AP® enrollment Suppose that 30% of female students and 25% of male students at a large high school are enrolled in an AP® class. Independent random samples of 20 females and 20 males are selected and are asked if they are enrolled in an AP® class. Let $\hat{p}_F$ represent the sample proportion of females enrolled in an AP® class and let $\hat{p}_M$ represent the sample proportion of males enrolled in an AP® class.

(a) Find the mean of the sampling distribution of $\hat{p}_F - \hat{p}_M$.

(b) Calculate and interpret the standard deviation of the sampling distribution. Verify that the 10% condition is met.

(c) Is the shape of the sampling distribution approximately Normal? Justify your answer.

48. Athletic participation Suppose that 20% of students at high school A and 18% of students at high school B participate on a school athletic team. Independent random samples of 30 students from each school are selected and are asked if they participate on a school athletic team. Let $\hat{p}_A$ represent the sample proportion of students at school A who participate on a school athletic team and let $\hat{p}_B$ represent the sample proportion of students at school B who participate on a school athletic team.

(a) Find the mean of the sampling distribution of $\hat{p}_A - \hat{p}_B$.

(b) Calculate and interpret the standard deviation of the sampling distribution. Verify that the 10% condition is met.

(c) Is the shape of the sampling distribution approximately Normal? Justify your answer.

49. I want red! A candy maker offers Child and Adult bags
pg 496 of jelly beans with different color mixes. The company claims that the Child mix has 30% red jelly beans, while the Adult mix contains 15% red jelly beans. Assume that the candy maker's claim is true. Suppose we take a random sample of 50 jelly beans from the Child mix and an independent random sample of 100 jelly beans from the Adult mix. Let $\hat{p}_C$ and $\hat{p}_A$ be the sample proportions of red jelly beans from the Child and Adult mixes, respectively.

(a) What is the shape of the sampling distribution of $\hat{p}_C - \hat{p}_A$? Why?

(b) Find the mean of the sampling distribution.

(c) Calculate and interpret the standard deviation of the sampling distribution. Verify that the 10% condition is met.

(d) Find the probability that the proportion of red jelly beans in the Child sample is less than or equal to the proportion of red jelly beans in the Adult sample, assuming that the company's claim is true.

50. Literacy A researcher reports that 80% of high school graduates, but only 40% of high school dropouts, would pass a basic literacy test.[7] Assume that the researcher's claim is true. Suppose we give a basic literacy test to a random sample of 60 high school graduates and an independent random sample of 75 high school dropouts. Let $\hat{p}_G$ and $\hat{p}_D$ be the sample proportions of graduates and dropouts, respectively, who pass the test.

(a) What is the shape of the sampling distribution of $\hat{p}_G - \hat{p}_D$? Why?

(b) Find the mean of the sampling distribution.

(c) Calculate and interpret the standard deviation of the sampling distribution. Verify that the 10% condition is met.

(d) Find the probability that the proportion of graduates who pass the test is at most 0.20 higher than the proportion of dropouts who pass, assuming that the researcher's report is correct.

51. I want red! Refer to Exercise 49. Suppose that the Child and Adult samples contain an equal proportion of red jelly beans. Based on your result in part (d) of Exercise 49, would this give you reason to doubt the company's claim? Explain your reasoning.

52. Literacy Refer to Exercise 50. Suppose that the difference (Graduate – Dropout) in the sample proportions who pass the test is exactly 0.20. Based on your result in part (d) in Exercise 50, would this give you reason to doubt the researcher's claim? Explain your reasoning.

Multiple Choice *Select the best answer for Exercises 53–56.*

Exercises 53–55 refer to the following setting. The magazine Sports Illustrated *asked a random sample of 750 Division I college athletes, "Do you believe performance-enhancing drugs are a problem in college sports?" Suppose that 30% of all Division I athletes think that these drugs are a problem. Let $\hat{p}$ be the sample proportion who say that these drugs are a problem.*

53. Which of the following are the mean and standard deviation of the sampling distribution of the sample proportion $\hat{p}$?

(a) Mean = 0.30, SD = 0.017

(b) Mean = 0.30, SD = 0.55

(c) Mean = 0.30, SD = 0.0003

(d) Mean = 225, SD = 12.5

(e) Mean = 225, SD = 157.5

There is a 0.0213 probability that the proportion of red jelly beans in the Child sample is less than or equal to the proportion of red jelly beans in the Adult sample.

7.50 (a) Approximately Normal;
$n_G p_G = 60(0.8) = 48$,
$n_G(1 - p_G) = 60(0.2) = 12$,
$n_D p_D = 75(0.4) = 30$, and
$n_D(1 - p_D) = 75(0.6) = 45$ are all ≥ 10.

(b) $\mu_{\hat{p}_G - \hat{p}_D} = p_G - p_D = 0.80 - 0.40 = 0.40$

(c) Because $60 < 10\%$ of all high school graduates and $75 < 10\%$ of all high school dropouts, the standard deviation of $\hat{p}_G - \hat{p}_D$ is approximately

$$\sigma_{\hat{p}_G - \hat{p}_D} = \sqrt{\frac{p_G(1 - p_G)}{n_G} + \frac{p_D(1 - p_D)}{n_D}} = $$

$$\sqrt{\frac{0.80(0.20)}{60} + \frac{0.40(0.60)}{75}} = 0.0766.$$

The difference (High school graduates − High school dropouts) in the sample proportions of those who would pass a basic literacy test typically varies by about 0.0766 from the true difference in proportions of 0.40.

(d) We want to find $P(\hat{p}_G - \hat{p}_D \leq 0.2)$.

(i) $z = \dfrac{0.2 - 0.4}{0.0766} = -2.61$

$P(z \leq -2.61) = 0.0045$

(ii) $P(\hat{p}_G - \hat{p}_D \leq 0.2) =$ normalcdf (lower: −1000, upper: 0.2, mean: 0.40, SD: 0.0766) = 0.0045

There is a 0.0045 probability that the proportion of graduates who pass the test is at most 0.20 greater than the proportion of dropouts who pass.

54. Decreasing the sample size from 750 to 375 would multiply the standard deviation by

(a) 2.

(b) $\sqrt{2}$

(c) 1/2.

(d) $1/\sqrt{2}$.

(e) none of these.

55. The sampling distribution of $\hat{p}$ is approximately Normal because

(a) there are at least 7500 Division I college athletes.

(b) $np = 225$ and $n(1 - p) = 525$ are both at least 10.

(c) a random sample was chosen.

(d) the athletes' responses are quantitative.

(e) the sampling distribution of $\hat{p}$ always has this shape.

56. In a congressional district, 55% of the registered voters are Democrats. Which of the following is equivalent to the probability of getting less than 50% Democrats in a random sample of size 100?

(a) $P\left(z < \dfrac{0.50 - 0.55}{100}\right)$

(b) $P\left(z < \dfrac{0.50 - 0.55}{\sqrt{\dfrac{0.55(0.45)}{100}}}\right)$

(c) $P\left(z < \dfrac{0.55 - 0.50}{\sqrt{\dfrac{0.55(0.45)}{100}}}\right)$

(d) $P\left(z < \dfrac{0.50 - 0.55}{\sqrt{100(0.55)(0.45)}}\right)$

(e) $P\left(z < \dfrac{0.55 - 0.50}{\sqrt{100(0.55)(0.45)}}\right)$

Recycle and Review

57. Sharing music online (5.2, 5.3) A sample survey reports that 29% of Internet users download music files online, 21% share music files from their computers, and 12% both download and share music.[8]

(a) Make a two-way table that displays this information.

(b) What percent of Internet users neither download nor share music files?

(c) Given that an Internet user downloads music files online, what is the probability that this person also shares music files?

58. Whole grains (4.2) A series of observational studies revealed that people who typically consume 3 servings of whole grain per day have about a 20% lower risk of dying from heart disease and about a 15% lower risk of dying from stroke or cancer than those who consume no whole grains.[9]

(a) Explain how confounding makes it difficult to establish a cause-and-effect relationship between whole grain consumption and risk of dying from heart disease, stroke, or cancer, based on these studies.

(b) Explain how researchers could establish a cause-and-effect relationship in this context.

Sample Means

LEARNING TARGETS *By the end of the section, you should be able to:*

- Calculate the mean and standard deviation of the sampling distribution of a sample mean $\bar{x}$ and interpret the standard deviation.
- Explain how the shape of the sampling distribution of $\bar{x}$ is affected by the shape of the population distribution and the sample size.
- Calculate the mean and the standard deviation of the sampling distribution of a difference in sample means $\bar{x}_1 - \bar{x}_2$ and interpret the standard deviation.
- Determine if the sampling distribution of $\bar{x}_1 - \bar{x}_2$ is approximately Normal.
- If appropriate, use a Normal distribution to calculate probabilities involving $\bar{x}$ or $\bar{x}_1 - \bar{x}_2$.

7.51 Yes, we might doubt the company's claim. There is only a 2% chance of getting a proportion of red jelly beans in the Child sample less than or equal to the proportion of red jelly beans in the Adult sample if the company's claim is true. This is not very likely.

7.52 Yes, we might doubt the researcher's claim. There is only a 0.45% chance of getting a difference in proportions of at most 0.2 if the researcher's report is correct. This is not very likely.

7.53 a

7.54 b

7.55 b

7.56 b

7.57 (a)

		Share music from computer?		
		Yes	No	Total
Download music files online?	Yes	12	17	29
	No	9	62	71
	Total	21	79	100

(b) $62/100 = 62\%$ **(c)** $12/29 = 0.414$

7.58 (a) The effect of other variables like age, gender, genetics, exercise, or getting enough sleep may be confounded with whole-grain consumption. **(b)** Researchers should randomly assign volunteers to one of two groups. One group will be told to eat 3 servings of whole grain per day and the other group will be told to eat fewer than 3 servings per day. At the end of the study, the researchers will compare the proportion of participants in each group who died from heart disease, stroke, or cancer.

PD **Section 7.3 Overview**

To watch the video overview of Section 7.3 (for teachers), click on the link in the TE-Book, look on the TRFD, or download from the Teacher's Resources on the book's digital platform.

TRM **Section 7.3 Alternate Examples**

You can find the Alternate Examples for this section in Microsoft Word format by clicking on the link in the TE-Book, opening the TRFD, or downloading from the Teacher's Resources on the book's digital platform.

Sample proportions arise most often when we are interested in categorical variables. We then ask questions like "What proportion of U.S. adults has watched *Survivor*?" or "What percent of the adult population attended church last week?" But when we record quantitative variables—household income, lifetime of car brake pads, blood pressure—we are interested in other statistics, such as the median or mean or standard deviation of the variable. The sample mean $\bar{x}$ is the most common statistic computed from quantitative data.

The Sampling Distribution of $\bar{x}$

When Mrs. Gallas's class did the "Penny for your thoughts" activity at the beginning of the chapter, her students produced the "dotplots" in Figure 7.12. These graphs approximate the **sampling distribution of the sample mean** year of penny for samples of size $n = 5$ and for samples of size $n = 20$ from Mrs. Gallas's population of pennies.

> **DEFINITION** Sampling distribution of the sample mean
>
> The **sampling distribution of the sample mean** $\bar{x}$ describes the distribution of values taken by the sample mean $\bar{x}$ in all possible samples of the same size from the same population.

FIGURE 7.12 Approximate sampling distribution of the sample mean year of pennies in 50 samples of size $n = 5$ and 50 samples of size $n = 20$ from a population of pennies.

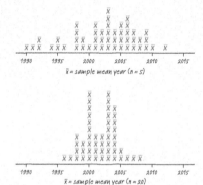

How do these approximate sampling distributions compare?

- *Shape:* The distribution of $\bar{x}$ is slightly skewed to the left when using samples of size $n = 5$ but roughly symmetric when using samples of size $n = 20$.
- *Center:* The distribution of $\bar{x}$ is centered at around 2002 for both sample sizes ($\mu_{\bar{x}} \approx 2002$).
- *Variability:* The distribution of $\bar{x}$ is about half as variable when using samples of size $n = 20$ ($\sigma_{\bar{x}} \approx 2.6$) than with samples of size $n = 5$ ($\sigma_{\bar{x}} \approx 5.2$).

Like the sampling distribution of $\hat{p}$, there are some simple rules that describe the mean and standard deviation of the sampling distribution of $\bar{x}$. Describing the shape of the sampling distribution of $\bar{x}$ is more complicated, so we'll save that for later.

SAMPLING DISTRIBUTION OF $\bar{x}$: MEAN AND STANDARD DEVIATION

Suppose that $\bar{x}$ is the mean of an SRS of size n drawn from a large population with mean μ and standard deviation σ. Then:

- The **mean** of the sampling distribution of $\bar{x}$ is $\mu_{\bar{x}} = \mu$.
- The **standard deviation** of the sampling distribution of $\bar{x}$ is approximately

$$\sigma_{\bar{x}} = \frac{\sigma}{\sqrt{n}}$$

as long as the *10% condition* is satisfied: $n < 0.10N$. The value $\sigma_{\bar{x}}$ measures the typical distance between a sample mean $\bar{x}$ and the population mean μ.

The behavior of $\bar{x}$ in repeated samples is much like that of the sample proportion $\hat{p}$:

- The sample mean $\bar{x}$ is an *unbiased estimator* of the population mean μ.
- The variability of $\bar{x}$ depends on both the variability in the population σ and the sample size n. Values of $\bar{x}$ will be more variable for populations that have more variability. Values of $\bar{x}$ will be less variable for larger samples. *Specifically, multiplying the sample size by 4 cuts the standard deviation in half.*
- When we sample *with* replacement, the standard deviation of the sampling distribution of $\bar{x}$ is exactly $\sigma_{\bar{x}} = \frac{\sigma}{\sqrt{n}}$. When we sample *without* replacement, the observations are not independent and the actual standard deviation of the sampling distribution of $\bar{x}$ is smaller than the value given by the formula. If the sample size is less than 10% of the population size, however, the value given by the formula is nearly correct. This doesn't mean you should aim for a small sample size! Bigger samples provide more information than smaller samples. If the sample size is more than 10% of the population size, we need to use the finite population correction to calculate the standard deviation of the sampling distribution of $\bar{x}$. We'll avoid these situations in this text.

Notice that these facts about the mean and standard deviation of $\bar{x}$ are true *no matter what shape the population distribution has.*

AP® EXAM TIP

Notation matters. The symbols $\hat{p}$, $\bar{x}$, n, p, μ, σ, $\mu_{\hat{p}}$, $\sigma_{\hat{p}}$, $\mu_{\bar{x}}$, and $\sigma_{\bar{x}}$ all have specific and different meanings. Either use notation correctly—or don't use it at all. You can expect to lose credit if you use incorrect notation.

AP® EXAM TIP

Let your students know that the formulas for the mean and standard deviation of the sampling distribution of a sample mean are included on the formula sheet provided to students on both sections of the AP® Statistics exam.

Teaching Tip

Students are sometimes confused by the symbols $\mu_{\bar{x}}$ and $\sigma_{\bar{x}}$. These are simply the mean and standard deviation of the sampling distribution of the sample mean. To illustrate, show your students one of the graphs of sample means you made in the "A penny for your thoughts?" activity and remind them that each of the values plotted on the graph is an $\bar{x}$. Then tell them that the way we denote the mean of the $\bar{x}$ distribution is $\mu_{\bar{x}}$ and the way we denote the standard deviation of the $\bar{x}$ distribution is $\sigma_{\bar{x}}$.

Teaching Tip

The more complicated formula for the standard deviation includes the finite population correction (FPC) given by

$FPC = \sqrt{\dfrac{N-n}{N-1}}$. The standard deviation of the sampling distribution of a sample mean *when sampling without replacement* is $\sigma_{\bar{x}} = \dfrac{\sigma}{\sqrt{n}} \times FPC$. Notice that when n is small relative to N, the FPC is close to 1. Students are not required to know the FPC for the AP® Statistics exam.

Preparing for Inference

As mentioned in Section 7.1, be tenacious about using correct notation. Don't let students get away with using μ when it really should be $\bar{x}$ or p when it should be $\hat{p}$. Being picky about notation now will help students be successful when we get to the formulas for confidence intervals and significance tests in Chapters 8–12.

Don't be absent
Mean and standard deviation of $\bar{x}$

PROBLEM:
The number of absences last year for students at a large high school has a mean of 5.6 days with a standard deviation of 4.1 days. Suppose we take an SRS of 80 of last year's students and calculate the mean number of absences for the members of the sample.
(a) Identify the mean of the sampling distribution of $\bar{x}$.
(b) Calculate and interpret the standard deviation of the sampling distribution of $\bar{x}$. Verify that the 10% condition is met.

SOLUTION:
(a) $\mu_{\bar{x}} = 5.6$ days
(b) Assuming that $n = 80$ is less than 10% of students at the large high school, $\sigma_{\bar{x}} = \dfrac{4.1}{\sqrt{80}} = 0.46$ days. In SRSs of size 80, the sample mean number of absences last year will typically vary by about 0.46 days from the true mean of 5.6 days.

Teaching Tip

Students are expected to recognize that problems involving totals can also be solved with means and vice versa. For example, to find the probability that the total weight of 4 men is greater than 800 pounds, you can find the probability that the mean weight of the 4 men is greater than 200 pounds.

EXAMPLE

Been to the movies recently?
Mean and standard deviation of $\bar{x}$

PROBLEM: The number of movies viewed in the last year by students at a large high school has a mean of 19.3 movies with a standard deviation of 15.8 movies. Suppose we take an SRS of 100 students from this school and calculate $\bar{x}$ = the mean number of movies viewed by the members of the sample.
(a) Identify the mean of the sampling distribution of $\bar{x}$.
(b) Calculate and interpret the standard deviation of the sampling distribution of $\bar{x}$. Verify that the 10% condition is met.

SOLUTION:
(a) $\mu_{\bar{x}} = 19.3$ movies
(b) Assuming that $n = 100$ is less than 10% of students at the large high school, the

standard deviation is approximately $\sigma_{\bar{x}} = \dfrac{15.8}{\sqrt{100}} = 1.58$ movies.

In SRSs of size 100, the sample mean number of movies viewed will typically vary by about 1.58 movies from the true mean of 19.3 movies.

$\mu_{\bar{x}} = \mu$

When $n < 0.10N$,
$\sigma_{\bar{x}} = \dfrac{\sigma}{\sqrt{n}}$

FOR PRACTICE, TRY EXERCISE 59

Think About It

WHERE DO THE FORMULAS FOR THE MEAN AND STANDARD DEVIATION OF $\bar{x}$ COME FROM? Choose an SRS of size n from a population, and measure a quantitative variable X on each individual in the sample. Call the individual measurements $X_1, X_2, \ldots, X_n$. If the population is large relative to the sample, we can think of these X_i's as independent random variables, each with mean μ and standard deviation σ. Because

$$\bar{x} = \frac{1}{n}(X_1 + X_2 + \cdots + X_n)$$

we can use the rules for random variables from Chapter 6 to find the mean and standard deviation of $\bar{x}$. If we let $T = X_1 + X_2 + \cdots + X_n$, then $\bar{x} = \dfrac{1}{n}T$.

Using the addition rules for means and variances, we get

$$\mu_T = \mu_{X_1} + \mu_{X_2} + \cdots + \mu_{X_n} = \mu + \mu + \cdots + \mu = n\mu$$
$$\sigma_T^2 = \sigma_{X_1}^2 + \sigma_{X_2}^2 + \cdots + \sigma_{X_n}^2 = \sigma^2 + \sigma^2 + \cdots + \sigma^2 = n\sigma^2$$
$$\Rightarrow \sigma_T = \sqrt{n\sigma^2} = \sigma\sqrt{n}$$

Because $\bar{x}$ is just a constant multiple of the random variable T,

$$\mu_{\bar{x}} = \frac{1}{n}\mu_T = \frac{1}{n}(n\mu) = \mu$$

$$\sigma_{\bar{x}} = \frac{1}{n}\sigma_T = \frac{1}{n}(\sigma\sqrt{n}) = \sigma\sqrt{\frac{n}{n^2}} = \sigma\sqrt{\frac{1}{n}} = \sigma\frac{1}{\sqrt{n}} = \frac{\sigma}{\sqrt{n}}$$

Teaching Tip

For the AP® Statistics exam, students would not be expected to derive the formulas for the mean and standard deviation of the sampling distribution of the sample mean as shown in this Think About It.

Sampling from a Normal Population

We have described the mean and standard deviation of the sampling distribution of a sample mean $\bar{x}$ but not its shape. That's because the shape of the sampling distribution of $\bar{x}$ depends on the shape of the population distribution. In one important case, there is a simple relationship between the two distributions. The following activity shows what we mean.

ACTIVITY

Exploring the sampling distribution of $\bar{x}$ for a Normal population

Professor David Lane of Rice University has developed a wonderful applet for investigating the sampling distribution of $\bar{x}$. It's dynamic, and it's fun to play with. In this activity, you'll use Professor Lane's applet to explore the shape of the sampling distribution when the population is Normally distributed.

1. Go to http://onlinestatbook.com/stat_sim/sampling_dist/ or search for "online statbook sampling distributions applet" and go to the website. When the BEGIN button appears on the left side of the screen, click on it. You will then see a yellow page entitled "Sampling Distributions" like the one in the screen shot.

2. There are choices for the population distribution: Normal, uniform, skewed, and custom. Keep the default option: Normal. Click the "Animated" button. What happens? Click the button several more times. What do the black boxes represent? What is the blue square that drops down onto the plot below?

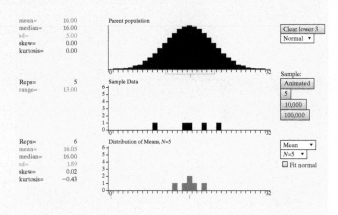

ACTIVITY OVERVIEW

To prepare for using this activity, watch the overview video by clicking on the link in the TE-Book, opening the TRFD, or downloading from the Teacher's Resources on the book's digital platform.

Time: 25 minutes

Materials: Computers with Internet access

Teaching Advice: Use this activity to remind students of the difference between a population distribution, the distribution of sample data, and a sampling distribution. The population distribution is the one labeled "Parent population," the distribution of sample data is the one labeled "Sample Data," and the (approximate) sampling distribution is one labeled "Distribution of Means." Note also the color scheme: the sample "blocks" are the same color as the population, because they are made of the same things. But the $\bar{x}$'s are a different color, because they are a different type of thing.

If the mean of the sampling distribution is not exactly the same as the mean of the population, continue to take more samples until they match. We don't want students to think that the mean of the sampling distribution is *close to* the population mean; it is *exactly equal to* the population mean.

It is important that you show several animated samples before jumping to the simulated sampling distribution with 100,000 samples. Students sometimes fail to grasp a concept if they have to move too quickly into using technology.

Answers:

1–2. Clicking the "animated" button takes a random sample of 5. Each black box represents an individual from the population. The blue square represents the mean of the sample of 5.

3. The shape is approximately Normal. The mean of the sampling distribution is the same as the mean of the population. The standard deviation of the sampling distribution is less than the standard deviation of the population.

4. The shape of both distributions is approximately Normal with a mean equal to the population mean. The distribution of $\bar{x}$ has less variability when $n = 20$ than when $n = 5$.

5. If the population distribution has a Normal shape, the sampling distribution has a Normal shape, no matter the sample size.

3. Click on "Clear lower 3" to start clean. Then click on the "100,000" button under "Sample:" to simulate taking 100,000 SRSs of size $n = 5$ from the population. Answer these questions:

   - Does the simulated sampling distribution of $\bar{x}$ (blue bars) have a recognizable shape? Click the box next to "Fit normal."
   - To the left of each distribution is a set of summary statistics. Compare the mean of the simulated sampling distribution with the mean of the population.
   - How is the standard deviation of the simulated sampling distribution related to the standard deviation of the population?

4. Click "Clear lower 3." Use the drop-down menus to set up the bottom graph to display the mean for samples of size $n = 20$. Then sample 100,000 times. How do the two distributions of $\bar{x}$ compare: shape, center, and variability?

5. What have you learned about the shape of the sampling distribution of $\bar{x}$ when the population has a Normal shape?

As the preceding activity demonstrates, if the population distribution is Normal, then so is the sampling distribution of $\bar{x}$. *This is true no matter what the sample size is.*

> **SAMPLING DISTRIBUTION OF THE SAMPLE MEAN $\bar{x}$ WHEN SAMPLING FROM A NORMAL POPULATION**
>
> Suppose that a population is Normally distributed with mean μ and standard deviation σ. Then the sampling distribution of $\bar{x}$ has the Normal distribution with mean $\mu_{\bar{x}} = \mu$ and standard deviation $\sigma_{\bar{x}} = \sigma/\sqrt{n}$ (provided the 10% condition is met).

We already knew the mean and standard deviation of the sampling distribution. All we have added is the Normal shape. Now we have enough information to calculate probabilities involving $\bar{x}$ when the population distribution is Normal.

Teaching Tip: Using Technology

Use the *Normal Distributions* applet in the Extra Applets on the Student Site at highschool.bfwpub.com/updatedtps6e. Select the *Probability* applet and choose "Normal Distributions" from the pull down menu to do Normal distribution calculations.

EXAMPLE

Young women's heights

Sampling from a Normal population

PROBLEM: The heights of young women follow a Normal distribution with mean $\mu = 64.5$ inches and standard deviation $\sigma = 2.5$ inches.

(a) Find the probability that a randomly selected young woman is taller than 66.5 inches.

(b) Find the probability that the mean height of an SRS of 10 young women exceeds 66.5 inches.

SOLUTION:

(a) Let X = height of a randomly selected young woman.

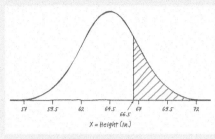

> **1. Draw a Normal distribution.**

X = Height (in.)

(i) $z = \dfrac{66.5 - 64.5}{2.5} = 0.80$

> **2. Perform calculations.**
> (i) Standardize and use Table A or technology
> or
> (ii) Use technology without standardizing. Be sure to answer the question that was asked.

Using Table A: $P(X > 66.5) = P(z > 0.80) = 1 - 0.7881$
$= 0.2119$

Using technology: normalcdf(lower:0.80, upper:1000, mean:0, SD:1) = 0.2119

(ii) normalcdf(lower:66.5, upper:1000, mean:64.5, SD:2.5) = 0.2119

(b) Let $\bar{x}$ = mean height of 10 randomly selected young women.

$\mu_{\bar{x}} = 64.5$

Because $10 < 10\%$ of all young women,

> Calculate the mean and standard deviation of the sampling distribution of $\bar{x}$.
> $$\mu_{\bar{x}} = \mu \qquad \sigma_{\bar{x}} = \dfrac{\sigma}{\sqrt{n}}$$

$\sigma_{\bar{x}} = \dfrac{2.5}{\sqrt{10}} = 0.79$

Because the population of heights is Normal, the distribution of $\bar{x}$ is also Normal.

> Justify that the distribution of $\bar{x}$ is Normal.

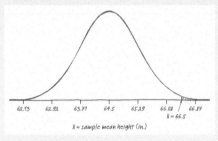

$\bar{x}$ = sample mean height (in.)

(i) $z = \dfrac{66.5 - 64.5}{0.79} = 2.53$

Using Table A: $P(\bar{x} > 66.5) = P(z > 2.53) = 1 - 0.9943 = 0.0057$

Using technology: normalcdf(lower:2.53, upper:1000, mean:0, SD:1) = 0.0057

(ii) normalcdf(lower:66.5, upper:1000, mean:64.5, SD:0.79) = 0.0057

> **FOR PRACTICE, TRY EXERCISE 65**

Teaching Tip

For the "Young women's heights" example above, ask students to interpret the answers in the context of the problem.

(a) There is a 0.2119 probability of randomly selecting one young woman who has a height that exceeds 66.5 inches.

(b) There is a 0.0057 probability of randomly selecting 10 young women whose *mean* height exceeds 66.5 inches.

> **ALTERNATE EXAMPLE**   Skill 4.B
> ### Welcome to the factory
> Sampling from a Normal population

PROBLEM:

Mr. Rose is a new quality-control inspector at a factory that produces axles for a certain model of car. Because the manufacturing process takes several steps, there is some variability in the length of the axles produced at the factory. Specifically, the distribution of axle lengths is Normal with a mean of 597 mm and a standard deviation of 5.3 mm.

(a) Mr. Rose randomly selects one axle. Find the probability that the length of the axle is less than 595 mm.

(b) Mr. Rose randomly selects 20 axles. Find the probability that the mean length of the axles is less than 595 mm.

SOLUTION:

(a) (i) $z = \dfrac{595 - 597}{5.3} = -0.38$

Table A:
$P(X < 595) = P(z < -0.38) = 0.352$
Tech: normalcdf(lower: -1000, upper: -0.38, mean: 0, SD: 1) = 0.352

(ii) normalcdf(lower: -10000, upper: 595, mean: 597, SD: 5.3) = 0.353

(b) $\mu_{\bar{x}} = 597$

Because $20 < 10\%$ of all axles produced at this factory, $\sigma_{\bar{x}} = \dfrac{5.3}{\sqrt{20}} = 1.185$.

Because the population distribution of axle lengths is Normal, the distribution of $\bar{x}$ is also Normal.

(i) $z = \dfrac{595 - 597}{1.185} = -1.69$

Table A:
$P(\bar{x} < 595) = P(z < -1.69) = 0.0455$
Tech: normalcdf(lower: -10000, upper: -1.69, mean: 0, SD: 1) = 0.0455

(ii) normalcdf(lower: -10000, upper: 595, mean: 597, SD: 1.185) = 0.0457

> **AP® EXAM TIP**
>
> Students often forget to divide by the square root of the sample size when finding probabilities involving the sample mean. When reading the example on this page, make sure to discuss the distinction between parts (a) and (b) with this in mind. In part (a), we are asked to find a probability involving X, the height of *one* woman. In part (b), we are asked to find a probability involving $\bar{x}$, the *mean* height of a sample of women.

Figure 7.13 compares the population distribution and the sampling distribution of $\bar{x}$ for the example about young women's heights. It also shows the areas corresponding to the probabilities that we computed. You can see that it is much less likely for the average height of 10 randomly selected young women to exceed 66.5 inches than it is for the height of one randomly selected young woman to exceed 66.5 inches.

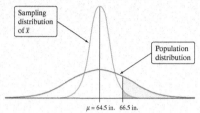

FIGURE 7.13 The sampling distribution of the mean height $\bar{x}$ for SRSs of 10 young women compared with the population distribution of young women's heights.

AP® EXAM TIP

Many students lose credit on probability calculations involving $\bar{x}$ because they forget to divide the population standard deviation by $\sqrt{n}$. Remember that averages are less variable than individual observations!

The fact that averages of several observations are less variable than individual observations is important in many settings. For example, it is common practice in science and medicine to repeat a measurement several times and report the average of the results.

CHECK YOUR UNDERSTANDING

The length of human pregnancies from conception to birth varies according to a distribution that is approximately Normal with mean 266 days and standard deviation 16 days.

1. Find the probability that a randomly chosen pregnant woman has a pregnancy that lasts for more than 270 days.

Suppose we choose an SRS of 6 pregnant women. Let $\bar{x}$ = the mean pregnancy length for the sample.

2. What is the mean of the sampling distribution of $\bar{x}$?

3. Calculate and interpret the standard deviation of the sampling distribution of $\bar{x}$. Verify that the 10% condition is met.

4. Find the probability that the mean pregnancy length for the women in the sample exceeds 270 days.

The Central Limit Theorem

Most population distributions are not Normal. What is the shape of the sampling distribution of $\bar{x}$ when sampling from a non-Normal population? The following activity sheds some light on this question.

Teaching Tip: AP® Connections

Students in AP® science classes have likely collected data from experiments in which they took several measurements and found the mean of those measurements. Because they took more than one measurement, if the initial measurement was an outlier, the other measurements will rein in that value toward the truth.

Answers to CYU

1. We want to find $P(X > 270)$.
(i) $z = 0.25$; $P(z > 0.25) =$
$1 - 0.5987 = 0.4013$
(ii) $P(X > 270) =$ normalcdf(lower: 270, upper: 1000, mean: 266, SD: 16) = 0.4013. There is a 0.4013 probability of selecting a woman whose pregnancy lasts for more than 270 days.

2. $\mu_{\bar{x}} = \mu = 266$ days

3. The sample of size 6 is less than 10% of all pregnant women. Therefore, the standard deviation of the sampling distribution of $\bar{x}$ is approximately $\sigma_{\bar{x}} = \dfrac{\sigma}{\sqrt{n}} = \dfrac{16}{\sqrt{6}} = 6.532$ days. In SRSs of size 6, the sample mean length of human pregnancies from conception to birth will typically vary by about 6.532 days from the true mean of 266 days.

4. The mean length of pregnancy in days follows a Normal distribution with a mean of 266 and a standard deviation of 6.532; we want to find $P(\bar{x} > 270)$.
(i) $z = 0.61$; $P(z > 0.61) = 1 - 0.7291 = 0.2709$
(ii) $P(\bar{x} > 270) =$ normalcdf(lower: 270, upper: 1000, mean: 266, SD: 6.532) = 0.2701 There is a 0.2701 probability of selecting a sample of 6 women whose mean pregnancy length exceeds 270 days.

ACTIVITY

Exploring the sampling distribution of $\bar{x}$ for non-Normal populations

Let's use the sampling distributions applet from the preceding activity to investigate what happens when we start with a non-Normal population distribution.

1. Go to onlinestatbook.com/stat_sim/sampling_dist/ and launch the applet. Select "Skewed" population. Set the bottom two graphs to display the mean—one for samples of size 2 and the other for samples of size 5. Click the "Animated" button a few times to be sure you see what's happening. Then "Clear lower 3" and take 100,000 SRSs. Describe what you see.

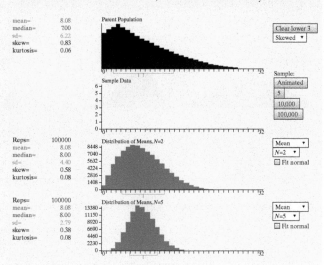

2. Change the sample sizes to $n = 10$ and $n = 16$ and take 100,000 samples. What do you notice?

3. Now change the sample sizes to $n = 20$ and $n = 25$ and take 100,000 more samples. Did this confirm what you saw in Step 2?

4. Clear the page, and select "Custom" distribution. Click on a point on the population graph to insert a bar of that height. Or click on a point on the horizontal axis, and drag up to define a bar. Make a distribution that looks as strange as you can. (Note: You can shorten a bar or get rid of it completely by clicking on the top of the bar and dragging down to the axis.) Then repeat Steps 1 to 3 for your custom distribution. Cool, huh?

5. Summarize what you learned about the shape of the sampling distribution of $\bar{x}$.

ACTIVITY OVERVIEW

To prepare for using this activity, watch the overview video by clicking on the link in the TE-Book, opening the TRFD, or downloading from the Teacher's Resources on the book's digital platform.

Time: 20 minutes

Materials: Computers with Internet access

Teaching Advice: Make connections between the skewed distribution in the applet and the "A penny for your thoughts?" activity from the beginning of the chapter. It is very likely that the shape of the population distribution of year of the pennies was strongly skewed to the left, and the shape of sampling distribution of $\bar{x}$ was getting close to Normal for samples of size 20. When students are creating their own custom distributions, challenge them to create one in which the shape of the sampling distribution when $n = 25$ is not approximately Normal. *Note:* The applet uses capital "N" where we would prefer to see the lowercase "n" to represent the sample size.

Answers:

1. The shape of the sampling distribution of $\bar{x}$ for $n = 2$ is skewed right (but less skewed than the population distribution) and for $n = 5$ is only slightly skewed right. Both distributions have a mean equal to the mean of the population and the variability is less for $n = 5$.

2. As the sample size increases, the shape of the sampling distribution of $\bar{x}$ becomes closer to Normal.

3. Yes.

4. So cool.

5. As the sample size increases, the shape of the sampling distribution of $\bar{x}$ becomes closer to Normal.

The screen shots in Figure 7.14 show the approximate sampling distributions of $\bar{x}$ for samples of size $n = 2$ and samples of size $n = 25$ from three different populations.

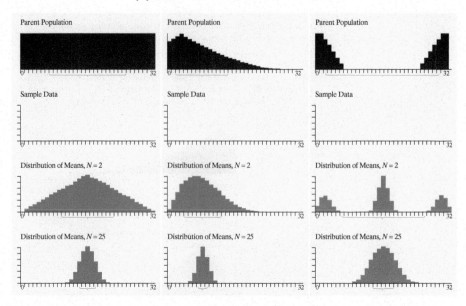

FIGURE 7.14 Approximate sampling distributions of $\bar{x}$ for different population shapes and sample sizes.

It is a remarkable fact that as the sample size increases, the sampling distribution of $\bar{x}$ changes shape: it looks less like that of the population and more like a Normal distribution. When the sample size is large enough, the sampling distribution of $\bar{x}$ is very close to Normal. This is true no matter what shape the population distribution has, as long as the population has a finite mean μ and standard deviation σ, and that the observations in the sample are independent. This important fact of probability theory is called the **central limit theorem** (sometimes abbreviated as CLT).

> **DEFINITION** Central limit theorem (CLT)
>
> Draw an SRS of size n from any population with mean μ and standard deviation σ. The **central limit theorem (CLT)** says that when n is sufficiently large, the sampling distribution of the sample mean $\bar{x}$ is approximately Normal.

How large a sample size n is needed for the sampling distribution of $\bar{x}$ to be close to Normal depends on the population distribution. More observations are required if the shape of the population distribution is far from Normal. In that case, the sampling distribution of $\bar{x}$ will also be very non-Normal if the sample size is small.

AP® EXAM TIP

Students sometimes refer to the central limit theorem when describing how the variability of a sampling distribution decreases as the sample size increases. While this is true, it isn't what the central limit theorem says. The CLT is only about the *shape* of the sampling distribution of the sample mean, not the variability.

As Figure 7.14 illustrates, even when the population distribution is very non-Normal, the sampling distribution of $\bar{x}$ often looks approximately Normal with sample sizes as small as $n = 25$. *To be safe, we'll require that n be at least 30 to invoke the CLT.*

SHAPE OF THE SAMPLING DISTRIBUTION OF THE SAMPLE MEAN $\bar{x}$

- If the population distribution is Normal, the sampling distribution of $\bar{x}$ will also be Normal, no matter what the sample size n is.
- If the population distribution is not Normal, the sampling distribution of $\bar{x}$ will be approximately Normal when the sample size is sufficiently large ($n \geq 30$ in most cases). If the sample size is small and the population distribution is not Normal, the sampling distribution of $\bar{x}$ will retain some characteristics of the population distribution (e.g., skewness).

EXAMPLE

Free oil changes
Calculations using the CLT

PROBLEM: Keith is the manager of an auto-care center. Based on service records from the past year, the time (in hours) that a technician requires to complete a standard oil change and inspection follows a right-skewed distribution with $\mu = 30$ minutes and $\sigma = 20$ minutes. For a promotion, Keith randomly selects 40 current customers and offers them a free oil change and inspection if they redeem the offer during the next month. Keith budgets an average of 35 minutes per customer for a technician to complete the work. Will this be enough?

(a) Calculate the probability that the average time it takes to complete the work exceeds 35 minutes.

(b) How much average time per customer should Keith budget if he wants to be 99% certain that he doesn't go "over budget"?

SOLUTION:

(a) Let $\bar{x}$ = sample mean time to complete work (in minutes).

$\mu_{\bar{x}} = 30$

Assuming $40 < 10\%$ of all current customers,

$\sigma_{\bar{x}} = \dfrac{20}{\sqrt{40}} = 3.16$

> Calculate the mean and standard deviation of the sampling distribution of $\bar{x}$.
> $\mu_{\bar{x}} = \mu \qquad \sigma_{\bar{x}} = \dfrac{\sigma}{\sqrt{n}}$

Because the sample size is large ($40 \geq 30$), the distribution of $\bar{x}$ is approximately Normal.

> Justify that the distribution of $\bar{x}$ is approximately Normal.

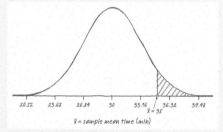

$\bar{x}$ = sample mean time (min)

> 1. Draw a Normal distribution.

ALTERNATE EXAMPLE

Skills 3.C, 4.B

Stock the ATM   Calculations using the CLT

PROBLEM:
A bank in a small town has to stock its single automated teller machine (ATM) with cash at the beginning of each day. Based on thousands of transactions in the past year, they know that the average amount of money requested by a customer is $87.00 with a standard deviation of $52.00. The distribution of money requested is strongly skewed to the right. The bank estimates that there will be 50 customer requests for cash from the ATM each day. The bank is trying to decide how much total cash should be stocked in the machine at the beginning of each day.

(a) Calculate the probability that a random sample of 50 customer requests for cash would require a total of $5000 or more (or an average request of $100 per customer).
(b) How much money should the bank put in the ATM to be 99% certain the machine will have enough money to satisfy a random sample of 50 customer requests?

SOLUTION:
(a) $\mu_{\bar{x}} = 87$ and assuming $50 < 10\%$ of all transactions, $\sigma_{\bar{x}} = \dfrac{52}{\sqrt{50}} = 7.35$. Because the sample size is large ($50 \geq 30$), the central limit theorem states that the distribution of $\bar{x}$ is approximately Normal.

Preparing for Inference

We check the *Normal/Large Sample* condition before doing inference for means using t distributions in Chapters 10–11. We have discussed two ways that this condition can be satisfied:

1. If the population distribution is approximately Normal.
2. If the sample size ≥ 30.

What if the population distribution has unknown shape and the sample size is less than 30? In Chapter 10, we will find that there is a third way to check the Normal/Large Counts condition.
3. A graph of the sample data shows no strong skewness or outliers.

(i) $z = \dfrac{100 - 87}{7.35} = 1.77$

Table A: $P(\bar{x} > 100) = P(z > 1.77) = 1 - 0.9616 = 0.0384$

Tech: normalcdf(lower: 1.77, upper: 1000, mean: 0, SD: 1) = 0.0384

(ii) (lower: 100, upper: 1000, mean: 87, SD: 7.35) = 0.0385

There is only a 3.85% probability that the bank will have to stock $5000 or more to meet the demand of a random sample of 50 customer requests.

(b) (i) *Table A:* 0.99 area to left → $z = 2.33$

Tech: invnorm(area: 0.99, mean: 0, SD: 1) = 2.33

$2.33 = \dfrac{\bar{x} - 87}{7.35} \rightarrow \bar{x} = \$104.13$

(ii) invnorm(area: 0.99, mean: 87, SD: 7.35) = $104.10

To be 99% sure the bank has stocked enough cash for a random sample of 50 customer requests, the bank managers should plan for an average of $104.10 per customer, or $104.10 × 50 = $5205.00 total.

(i) $z = \dfrac{35-30}{3.16} = 1.58$

Using Table A: $P(\bar{x} > 35) = P(z > 1.58) = 1 - 0.9429 = 0.0571$

Using technology:
normalcdf (lower:1.58, upper:1000, mean:0, SD:1) = 0.0571

(ii) normalcdf (lower:35, upper:1000, mean:30, SD:3.16) = 0.0568

There is only a 5.68% probability that Keith hasn't budgeted enough time to complete the work.

> **2. Perform calculations.**
> (i) Standardize and use Table A or technology; or
> (ii) Use technology without standardizing.
> Be sure to answer the question that was asked.

(b)

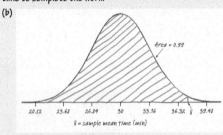

$\bar{x}$ = sample mean time (min)

> **1. Draw a Normal distribution.**

(i) Using Table A: 0.99 area to left $\rightarrow z = 2.33$

Using technology: invnorm(area:0.99, mean:0, SD:1) = 2.33

$2.33 = \dfrac{\bar{x} - 30}{3.16} \rightarrow \bar{x} = 37.4$ minutes

(ii) invnorm(area:0.99, mean:30, SD:3.16) = 37.4 minutes

To be 99% sure he has budgeted enough time, Keith should plan for an average of 37.4 minutes per customer.

> **2. Perform calculations.**
> (i) Use Table A or technology to find the value of z with the indicated area under the standard Normal curve, then "unstandardize" to transform back to the original distribution; or
> (ii) Use technology to find the desired value without standardizing.
> Be sure to answer the question that was asked.

FOR PRACTICE, TRY EXERCISE 73

What if Keith decided to give away only 10 free oil changes? Because the population distribution is skewed to the right and the sample size is small ($10 < 30$), we can't use a Normal distribution to do probability calculations. The sampling distribution of $\bar{x}$ is likely to be skewed to the right—although not as strongly as the population distribution itself.

The Sampling Distribution of a Difference Between Two Means

In the preceding section, we developed methods for comparing two proportions. What if we want to compare the mean of some quantitative variable for the individuals in Population 1 and Population 2, such as the mean income for high school graduates and non-high school graduates? Our parameters of interest are the population means μ_1 and μ_2. Once again, the best approach is to take independent random samples from each population and to compare the sample means $\bar{x}_1$ and $\bar{x}_2$.

Ask the StatsMedic
Which Version of the AP® Exam Was Harder?

This post outlines an activity that can be used to introduce inference for the difference of means. Suppose that two different versions of the AP® Statistics exam were randomly assigned to a group of students. Students who took Version A of the exam had a mean AP® score of 4.20, whereas students who took Version B had a mean score of 4.00. Do the data provide convincing evidence that Version B was harder?

TRM Comparing Pulse Rates Activity

The "Comparing pulse rates" activity is another way to get students thinking about how random chance can affect the difference between two means. Find it in the Teacher's Resource Materials located in the TE-Book, on the TRFD, or in the Teacher's view on the book's digital platform.

Suppose we want to compare the average effectiveness of two headache medications in a randomized experiment. In this case, the parameters μ_1 and μ_2 are the true mean responses for Treatment 1 and Treatment 2, respectively. We use the mean response in the two groups, $\bar{x}_1$ and $\bar{x}_2$, to make the comparison. Here's a table that summarizes these two situations:

Population or treatment	Parameter	Statistic	Sample size
1	μ_1	$\bar{x}_1$	n_1
2	μ_2	$\bar{x}_2$	n_2

We compare the populations or treatments by doing inference about the difference $\mu_1 - \mu_2$ between the parameters. The statistic that estimates this difference is the difference between the two sample means, $\bar{x}_1 - \bar{x}_2$.

To explore the sampling distribution of $\bar{x}_1 - \bar{x}_2$, let's start with two Normally distributed populations having known means and standard deviations. Based on information from the U.S. National Health and Nutrition Examination Survey (NHANES), the heights of 10-year-old girls can be modeled by a Normal distribution with mean $\mu_G = 56.4$ inches and standard deviation $\sigma_G = 2.7$ inches. The heights of 10-year-old boys can be modeled by a Normal distribution with mean $\mu_B = 55.7$ inches and standard deviation $\sigma_B = 3.8$ inches.[10] The table summarizes this information.

Monkey Business Images/
Shutterstock

Population	Shape	Mean	Standard deviation
10-year-old girls	Approximately Normal	$\mu_G = 56.4$ in	$\sigma_G = 2.7$ in
10-year-old boys	Approximately Normal	$\mu_B = 55.7$ in	$\sigma_B = 3.8$ in

Suppose we take independent SRSs of 12 girls and 8 boys of this age and measure their heights. What can we say about the difference $\bar{x}_G - \bar{x}_B$ in the average heights of the sample of girls and the sample of boys?

Earlier in this section, we saw that the sampling distribution of a sample mean $\bar{x}$ has the following properties:

Shape: (1) If the population distribution is Normal, then so is the sampling distribution of $\bar{x}$; (2) If the population distribution isn't Normal, the sampling distribution of $\bar{x}$ will be approximately Normal if the sample size is large enough (say, $n \geq 30$) by the central limit theorem (CLT).

Center: $\mu_{\bar{x}} = \mu$

Variability: $\sigma_{\bar{x}} = \dfrac{\sigma}{\sqrt{n}}$ if $n < 0.10N$

For the sampling distributions of $\bar{x}_G$ and $\bar{x}_B$ in this case:

	Sampling distribution of $\bar{x}_G$	Sampling distribution of $\bar{x}_B$
Shape	Approximately Normal, because the population distribution is approximately Normal	Approximately Normal, because the population distribution is approximately Normal
Center	$\mu_{\bar{x}_G} = \mu_G = 56.4$ inches	$\mu_{\bar{x}_B} = \mu_B = 55.7$ inches
Variability	$\sigma_{\bar{x}_G} = \dfrac{\sigma_G}{\sqrt{n_G}} = \dfrac{2.7}{\sqrt{12}} = 0.78$ inch	$\sigma_{\bar{x}_B} = \dfrac{\sigma_B}{\sqrt{n_B}} = \dfrac{3.8}{\sqrt{8}} = 1.34$ inches
	because $12 < 10\%$ of all 10-year-old girls in the United States.	because $8 < 10\%$ of all 10-year-old boys in the United States.

What about the sampling distribution of $\bar{x}_G - \bar{x}_B$? We used software to take independent SRSs of 12 ten-year-old girls and 8 ten-year-old boys. Our first

Making Connections

Notation matters! A statistic is used to estimate a parameter.

- $\bar{x}$ from the sample is used to estimate the population mean μ.
- s_x from the sample is used to estimate the population standard deviation σ.
- $\bar{x}_1 - \bar{x}_2$ from two samples is used to estimate the difference of population means $\mu_1 - \mu_2$.

Making Connections

Remember that a sampling distribution shows the possible values of a statistic ($\bar{x}_1 - \bar{x}_2$, in this case) and how often those values occur. Also, remind students that we are temporarily looking at populations in which we know the true means so we can investigate the sampling distribution of the difference in sample means. Simulation is a great way to get a sense of what a sampling distribution looks like.

Point at one of the dots in the dotplot and ask students what it represents. The correct response is, "A random sample of twelve 10-year-old girls and a random sample of eight 10-year-old boys and a difference in means calculated from those samples."

Making Connections

In Chapter 6, we showed that "one does not simply add standard deviations" when combining random variables. Rather, we add variances and then take the square root. For more discussion, see the Think About It on the next page.

AP® EXAM TIP

The formula for the standard deviation of the sampling distribution of $\bar{x}_1 - \bar{x}_2$ is provided on the formula sheet for the AP® Statistics exam.

Making Connections

In Chapter 6, we learned that we should add variances when combining random variables, but only if the two random variables are independent. To use the formula for standard deviation on this page, we must be convinced that the two sample means are independent of each other. We also need observations to be independent within each sample, which is satisfied if we are sampling with replacement. If we are sampling without replacement, we must check the 10% condition to be comfortable viewing the observations within each sample as independent.

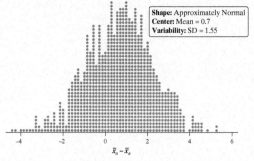

Shape: Approximately Normal
Center: Mean = 0.7
Variability: SD = 1.55

FIGURE 7.15 Simulated sampling distribution of the difference in sample means $\bar{x}_G - \bar{x}_B$ in 1000 SRSs of size $n_G = 12$ from an approximately Normally distributed population with $\mu_G = 56.4$ inches and $\sigma_G = 2.7$ inches and 1000 SRSs of size $n_B = 8$ from an approximately Normally distributed population with $\mu_B = 55.7$ inches and $\sigma_B = 3.8$ inches.

set of samples gave $\bar{x}_G = 56.09$ inches and $\bar{x}_B = 54.68$ inches, resulting in a difference of $\bar{x}_G - \bar{x}_B = 56.09 - 54.68 = 1.41$ inches. A red dot for this value appears in Figure 7.15. The dotplot shows the results of repeating this process 1000 times.

The figure suggests that the sampling distribution of $\bar{x}_G - \bar{x}_B$ has an approximately Normal shape. This makes sense from what you learned in Section 6.2 because we are subtracting two independent random variables, $\bar{x}_G$ and $\bar{x}_B$, that have approximately Normal distributions.

The mean of the sampling distribution is 0.7. The true mean height of all 10-year-old girls is $\mu_G = 56.4$ inches and the true mean height of all 10-year-old boys is $\mu_B = 55.7$ inches. We expect the difference $\bar{x}_G - \bar{x}_B$ to center on the actual difference in the population means, $\mu_G - \mu_B = 56.4 - 55.7 = 0.7$ inch.

The standard deviation of the sampling distribution is 1.55 inches. It can be found using the formula

$$\sqrt{\frac{\sigma_G^2}{n_G} + \frac{\sigma_B^2}{n_B}} = \sqrt{\frac{2.7^2}{12} + \frac{3.8^2}{8}} = 1.55$$

That is, the difference (Girls – Boys) in the sample mean heights typically varies by about 1.55 inches from the true difference in mean heights of 0.7 inch.

THE SAMPLING DISTRIBUTION OF $\bar{x}_1 - \bar{x}_2$

Choose an SRS of size n_1 from Population 1 with mean μ_1 and standard deviation σ_1 and an independent SRS of size n_2 from Population 2 with mean μ_2 and standard deviation σ_2. Then:

- The sampling distribution of $\bar{x}_1 - \bar{x}_2$ is **Normal** if both population distributions are Normal. It is **approximately Normal** if both sample sizes are large ($n_1 \geq 30$ and $n_2 \geq 30$) or if one population is Normally distributed and the other sample size is large.
- The **mean** of the sampling distribution of $\bar{x}_1 - \bar{x}_2$ is $\mu_{\bar{x}_1 - \bar{x}_2} = \mu_1 - \mu_2$.
- The **standard deviation** of the sampling distribution of $\bar{x}_1 - \bar{x}_2$ is approximately

$$\sigma_{\bar{x}_1 - \bar{x}_2} = \sqrt{\frac{\sigma_1^2}{n_1} + \frac{\sigma_2^2}{n_2}}$$

as long as the *10% condition* is met for both samples: $n_1 < 0.10N_1$ and $n_2 < 0.10N_2$.

Note that the formula for the standard deviation is exactly correct only when we have two types of independence:

- Independent samples, so that we can add the variances of $\bar{x}_1$ and $\bar{x}_2$.
- Independent observations within each sample. When sampling without replacement, the actual value of the standard deviation is smaller than the

formula suggests. However, if the 10% condition is met for both samples, the difference is negligible.

The standard deviation of the sampling distribution tells us how much the difference in sample means will typically vary from the difference in the population means if we repeat the random sampling process many times.

Think About It

WHERE DO THE FORMULAS FOR THE MEAN AND STANDARD DEVIATION OF THE SAMPLING DISTRIBUTION OF $\bar{x}_1 - \bar{x}_2$ COME FROM? Both $\bar{x}_1$ and $\bar{x}_2$ are random variables. That is, their values would vary in repeated independent SRSs of size n_1 and n_2. Independent random samples yield independent random variables $\bar{x}_1$ and $\bar{x}_2$. The statistic $\bar{x}_1 - \bar{x}_2$ is the difference of these two independent random variables.

In Chapter 6, we learned that for any two random variables X and Y,

$$\mu_{X-Y} = \mu_X - \mu_Y$$

For the random variables $\bar{x}_1$ and $\bar{x}_2$, we have

$$\mu_{\bar{x}_1 - \bar{x}_2} = \mu_{\bar{x}_1} - \mu_{\bar{x}_2} = \mu_1 - \mu_2$$

We also learned in Chapter 6 that for *independent* random variables X and Y,

$$\sigma_{X-Y}^2 = \sigma_X^2 + \sigma_Y^2$$

For the random variables $\bar{x}_1$ and $\bar{x}_2$, we have

$$\sigma_{\bar{x}_1 - \bar{x}_2}^2 = \sigma_{\bar{x}_1}^2 + \sigma_{\bar{x}_2}^2 = \left(\frac{\sigma_1}{\sqrt{n_1}}\right)^2 + \left(\frac{\sigma_2}{\sqrt{n_2}}\right)^2 = \frac{\sigma_1^2}{n_1} + \frac{\sigma_2^2}{n_2}$$

So $\sigma_{\bar{x}_1 - \bar{x}_2} = \sqrt{\dfrac{\sigma_1^2}{n_1} + \dfrac{\sigma_2^2}{n_2}}$.

When the conditions are met, we can use the Normal density curve shown in Figure 7.16 to model the sampling distribution of $\bar{x}_1 - \bar{x}_2$. Note that this would allow us to calculate probabilities involving $\bar{x}_1 - \bar{x}_2$ with a Normal distribution.

When we analyzed the results of randomized experiments in Section 4.3, we used simulation to create a *randomization distribution* by repeatedly reallocating individuals to treatment groups. Fortunately for us, randomization distributions of $\bar{x}_1 - \bar{x}_2$ roughly follow the same rules for shape, center, and variability as sampling distributions of $\bar{x}_1 - \bar{x}_2$.

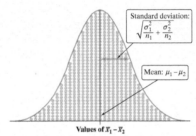

FIGURE 7.16 Select independent SRSs from two populations having means μ_1 and μ_2 and standard deviations σ_1 and σ_2. The two sample means are $\bar{x}_1$ and $\bar{x}_2$. When the conditions are met, the sampling distribution of the difference $\bar{x}_1 - \bar{x}_2$ is approximately Normal with mean $\mu_1 - \mu_2$ and standard deviation $\sqrt{\dfrac{\sigma_1^2}{n_1} + \dfrac{\sigma_2^2}{n_2}}$.

Making Connections

Students shouldn't be expected to derive the formulas for the mean and standard deviation of a difference in sample means, as we do here. However, reading through this Think About It is a great way to review the rules for combining random variables that we learned in Chapter 6 and the properties of the sampling distribution of a sample mean that we learned earlier in this section.

Making Connections

In Section 4.3, students analyzed the caffeine experiment using many simulated random assignments. They created a dotplot of the difference in means for each random assignment. This dotplot (page 301) approximates the randomization distribution.

Who is taller?
Describing the sampling distribution
of $\bar{x}_1 - \bar{x}_2$

PROBLEM:

The distribution of heights for U.S. adult males follows a Normal distribution with mean 69.3 inches and standard deviation 2.94 inches. For adult females, the distribution of heights follows a Normal distribution with a mean of 63.8 inches and standard deviation of 2.80 inches. Suppose that a researcher selects an SRS of 30 males and an SRS of 50 females. Let $\bar{x}_M - \bar{x}_F$ be the difference in the sample mean height between the sample of males and the sample of females.
(a) What is the shape of the sampling distribution of $\bar{x}_M - \bar{x}_F$? Why?
(b) Find the mean of the sampling distribution.
(c) Calculate and interpret the standard deviation of the sampling distribution.
(d) Estimate the probability that the difference in sample means is within 1 inch of the true difference in means.

SOLUTION:
(a) Normal, because both population distributions are Normal.
(b) $\mu_{\bar{x}_M - \bar{x}_F} = 69.3 - 63.8 = 5.5$ inches
(c) Because 30 < 10% of all U.S. adult males and 50 < 10% of all U.S. adult females:

$$\sigma_{\bar{x}_M - \bar{x}_F} = \sqrt{\frac{2.94^2}{30} + \frac{2.80^2}{50}} = 0.667 \text{ inch}$$

The difference (Male − Female) in the sample mean heights typically varies by about 0.667 inch from the true difference in means of 5.5 inches.
(d) We want to find
$P(4.5 \le \bar{x}_M - \bar{x}_F \le 6.5)$.

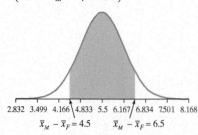

2.832 3.499 4.166 4.833 5.5 6.167 6.834 7.501 8.168

$\bar{x}_M - \bar{x}_F = 4.5$ $\bar{x}_M - \bar{x}_F = 6.5$

$\bar{x}_M - \bar{x}_F$ = Difference in the sample mean
height of males and females

EXAMPLE

Medium or large drink?
Describing the sampling
distribution of $\bar{x}_1 - \bar{x}_2$

RichLegg/Getty Images

PROBLEM: A fast-food restaurant uses an automated filling machine to pour its soft drinks. The machine has different settings for small, medium, and large drink cups. According to the machine's manufacturer, when the large setting is chosen, the amount of liquid L dispensed by the machine follows a Normal distribution with mean 27 ounces and standard deviation 0.8 ounce. When the medium setting is chosen, the amount of liquid M dispensed follows a Normal distribution with mean 17 ounces and standard deviation 0.5 ounce. To test this claim, the manager selects independent random samples of 20 cups filled using the large setting and 25 cups filled using the medium setting during one week. Let $\bar{x}_L - \bar{x}_M$ be the difference in the sample mean amount of liquid under the two settings.
(a) What is the shape of the sampling distribution of $\bar{x}_L - \bar{x}_M$? Why?
(b) Find the mean of the sampling distribution.
(c) Calculate and interpret the standard deviation of the sampling distribution. Verify that the 10% condition is met.
(d) Estimate the probability that the difference in sample means is within 0.25 ounce of the true difference in means.

SOLUTION:
(a) Normal, because both population distributions are Normal.
(b) $\mu_{\bar{x}_L - \bar{x}_M} = 27 - 17 = 10$ ounces
(c) Because 20 < 10% of all large cups of soft drinks and 25 < 10% of all medium cups of soft drinks that week,

$$\sigma_{\bar{x}_L - \bar{x}_M} = \sqrt{\frac{0.80^2}{20} + \frac{0.50^2}{25}} = 0.205 \text{ ounce}$$

The difference (Large cup − Medium cup) in the sample mean amounts of liquid typically varies by about 0.2 ounce from the true difference in means of 10 ounces.
(d) We want to find $P(9.75 \le \bar{x}_L - \bar{x}_M \le 10.25)$.

| $\mu_{\bar{x}_1 - \bar{x}_2} = \mu_1 - \mu_2$ |

| $\sigma_{\bar{x}_1 - \bar{x}_2} = \sqrt{\dfrac{\sigma_1^2}{n_1} + \dfrac{\sigma_2^2}{n_2}}$ |

1. Draw a Normal distribution.

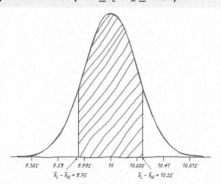

9.385   9.59   9.795   10   10.205   10.41   10.615

$\bar{x}_L - \bar{x}_M = 9.75$ $\bar{x}_L - \bar{x}_M = 10.25$

$\bar{x}_L - \bar{x}_M$ = difference in the sample mean volume of soft drink in large and medium cups

---

(i) $z = \dfrac{4.5 - 5.5}{0.667} = -1.50$ and

$z = \dfrac{6.5 - 5.5}{0.667} = 1.50$

Using Table A: $P(4.5 \le \bar{x}_M - \bar{x}_F \le 6.5) = P(-1.50 \le z \le 1.50) = 0.9332 - 0.0668 = 0.8664$
Using Technology: normalcdf(lower: −1.50, upper: 1.50, mean: 0, SD: 1) = 0.8664
(ii) normalcdf(lower: 4.5, upper: 6.5, mean: 5.5, SD: 0.667) = 0.8662

(i) $z = \dfrac{9.75-10}{0.205} = -1.22$ and $z = \dfrac{10.25-10}{0.205} = 1.22$

Using Table A: $P(9.75 \le \bar{x}_L - \bar{x}_M \le 10.25) = P(-1.22 \le z \le 1.22) = 0.8888 - 0.1112 = 0.7776$.

Using technology: normalcdf(lower: -1.22, upper: 1.22, mean: 0, SD: 1) = 0.7775.

(ii) normalcdf(lower: 9.75, upper: 10.25, mean: 10, SD: 0.205) = 0.7774.

> **2. Perform calculations.**
> (i) Standardize and use Table A or technology; or
> (ii) Use technology without standardizing.
> Be sure to answer the question that was asked.

FOR PRACTICE, TRY EXERCISE 83

Section 7.3 | Summary

- When we want information about the population mean μ for some quantitative variable, we often take an SRS and use the sample mean $\bar{x}$ to estimate the unknown parameter μ. The **sampling distribution of the sample mean** $\bar{x}$ describes how the statistic $\bar{x}$ varies in all possible samples of the same size from the population.
  - **Shape:** If the population distribution is Normal, then so is the sampling distribution of the sample mean $\bar{x}$. If the population distribution is not Normal, the **central limit theorem (CLT)** states that when n is sufficiently large, the sampling distribution of $\bar{x}$ is approximately Normal. For most non-Normal populations, it is safe to use a Normal distribution to calculate probabilities involving $\bar{x}$ when $n \ge 30$.
  - **Center:** The **mean** of the sampling distribution of $\bar{x}$ is $\mu_{\bar{x}} = \mu$, so $\bar{x}$ is an unbiased estimator of μ.
  - **Variability:** The **standard deviation** of the sampling distribution of $\bar{x}$ is approximately $\sigma_{\bar{x}} = \sigma/\sqrt{n}$ for an SRS of size n if the population has standard deviation σ. This formula can be used if the sample size is less than 10% of the population size (*10% condition*).
- Choose independent SRSs of size n_1 from Population 1 with mean μ_1 and standard deviation σ_1 and of size n_2 from Population 2 with mean μ_2 and standard deviation σ_2. The sampling distribution of $\bar{x}_1 - \bar{x}_2$ has the following properties:
  - **Shape:** Normal if both population distributions are Normal; approximately Normal if both sample sizes are large ($n_1 \ge 30$ and $n_2 \ge 30$) or if one population is Normally distributed and the other sample size is large.
  - **Center:** The **mean** of the sampling distribution is $\mu_{\bar{x}_1 - \bar{x}_2} = \mu_1 - \mu_2$.
  - **Variability:** The **standard deviation** of the sampling distribution is approximately $\sigma_{\bar{x}_1 - \bar{x}_2} = \sqrt{\dfrac{\sigma_1^2}{n_1} + \dfrac{\sigma_2^2}{n_2}}$ as long as the 10% condition is met for both samples: $n_1 < 0.10N_1$ and $n_2 < 0.10N_2$.

TRM Section 7.3 Quiz

There is one quiz available for this section in the Teacher's Resource Materials. Click on the link in the TE-Book, look on the TRFD, or download from the Teacher's Resources on the book's digital platform. You can also create your own quiz using the ExamView® Assessment Suite that is part of the TPS6 program. Questions are coded by Learning Target to make it easy to build parallel quizzes.

TRM Full Solutions to Section 7.3 Exercises

Click on the link in the TE-Book, open the TRFD, or download from the Teacher's Resources on the book's digital platform.

Answers to Section 7.3 Exercises

7.59 (a) $\mu_{\bar{x}} = \mu = 225$ seconds
(b) Because the sample size (10) is less than 10% of the population of songs on David's iPod, $\sigma_{\bar{x}} = 18.974$ seconds. In SRSs of size 10, the sample mean play time will typically vary by about 18.974 seconds from the true mean of 225 seconds.

7.60 (a) $\mu_{\bar{x}} = \mu = 40.125$ mm
(b) Assuming the sample size (4) is less than 10% of all axles produced this hour, $\sigma_{\bar{x}} = 0.001$ mm. In SRSs of size 4, the sample mean diameter will typically vary by about 0.001 mm from the true mean of 40.125 mm.

7.61 $10 = \dfrac{60}{\sqrt{n}} \rightarrow 10\sqrt{n} = 60 \rightarrow \sqrt{n} =$
$6 \rightarrow n = 36$

7.62 $0.0005 = \dfrac{0.002}{\sqrt{n}} \rightarrow 0.0005\sqrt{n} =$
$0.002 \rightarrow \sqrt{n} = 4 \rightarrow n = 16$

7.63 (a) 50 is not less than 10% of all students at the small school. The administrators sampled $50/200 = 25\%$ of the students from the school. **(b)** The standard deviation of the sampling distribution of $\bar{x}$ will be smaller than the value provided by the formula $\sigma_{\bar{x}} = \dfrac{\sigma}{\sqrt{n}}$.

7.64 (a) 5 is not less than 10% of all trees planted. The students selected 5 out of the 35, or 14.3% of the trees that were planted. **(b)** The standard deviation of the sampling distribution of $\bar{x}$ will be smaller than the value provided by the formula $\sigma_{\bar{x}} = \dfrac{\sigma}{\sqrt{n}}$.

7.65 (a) We want to find $P(X < 295)$.
(i) $z = \dfrac{295 - 298}{3} = -1$; $P(z < -1) =$
0.1587
(ii) $P(X < 295) = $ normalcdf(lower: -1000, upper: 295, mean: 298, SD: 3) $= 0.1587$
(b) Because X has a Normal distribution, the sampling distribution of $\bar{x}$ has a Normal distribution; $\mu_{\bar{x}} = \mu = 298$ ml. Because 6 is less than 10% of all bottles produced, $\sigma_{\bar{x}} = \dfrac{\sigma}{\sqrt{n}} = \dfrac{3}{\sqrt{6}} = 1.2247$ ml. We want to find $P(\bar{x} < 295)$. (i) $z = \dfrac{295 - 298}{1.2247} =$
-2.45; $P(z < -2.45) = 0.0071$
(ii) $P(\bar{x} < 295) = $ normalcdf(lower: -1000, upper: 295, mean: 298, SD: 1.2247) $= 0.0072$

7.66 (a) We want to find $P(X < 9.65)$.
(i) $z = \dfrac{9.65 - 9.70}{0.03} = -1.67$;
$P(z < -1.67) = 0.0475$

Section 7.3 | Exercises

59. **Songs on an iPod** David's iPod has about 10,000 songs. The distribution of the play times for these songs is heavily skewed to the right with a mean of 225 seconds and a standard deviation of 60 seconds. Suppose we choose an SRS of 10 songs from this population and calculate the mean play time $\bar{x}$ of these songs.

    (a) Identify the mean of the sampling distribution of $\bar{x}$.

    (b) Calculate and interpret the standard deviation of the sampling distribution of $\bar{x}$. Verify that the 10% condition is met.

60. **Making auto parts** A grinding machine in an auto parts plant prepares axles with a target diameter $\mu = 40.125$ millimeters (mm). The machine has some variability, so the standard deviation of the diameters is $\sigma = 0.002$ mm. The machine operator inspects a random sample of 4 axles each hour for quality control purposes and records the sample mean diameter $\bar{x}$. Assume the machine is working properly.

    (a) Identify the mean of the sampling distribution of $\bar{x}$.

    (b) Calculate and interpret the standard deviation of the sampling distribution of $\bar{x}$. Verify that the 10% condition is met.

61. **Songs on an iPod** Refer to Exercise 59. How many songs would you need to sample if you wanted the standard deviation of the sampling distribution of $\bar{x}$ to be 10 seconds? Justify your answer.

62. **Making auto parts** Refer to Exercise 60. How many axles would you need to sample if you wanted the standard deviation of the sampling distribution of $\bar{x}$ to be 0.0005 mm? Justify your answer.

63. **Screen time** Administrators at a small school with 200 students want to estimate the average amount of time students spend looking at a screen (phone, computer, television, and so on) per day. The administrators select a random sample of 50 students from the school to ask.

    (a) Show that the 10% condition is not met in this case.

    (b) What effect does violating the 10% condition have on the standard deviation of the sampling distribution of $\bar{x}$?

64. **Beautiful trees** As part of a school beautification project, students planted 35 trees along a road next to their school. Each month, students in the environmental science class randomly select 5 trees to estimate the average height of all the trees planted for the project.

    (a) Show that the 10% condition is not met in this case.

    (b) What effect does violating the 10% condition have on the standard deviation of the sampling distribution of $\bar{x}$?

65. **Bottling cola** A bottling company uses a filling machine to fill plastic bottles with cola. The bottles are supposed to contain 300 milliliters (ml). In fact, the contents vary according to a Normal distribution with mean $\mu = 298$ ml and standard deviation $\sigma = 3$ ml.

    (a) What is the probability that a randomly selected bottle contains less than 295 ml?

    (b) What is the probability that the mean contents of six randomly selected bottles is less than 295 ml?

66. **Cereal** A company's cereal boxes advertise that each box contains 9.65 ounces of cereal. In fact, the amount of cereal in a randomly selected box follows a Normal distribution with mean $\mu = 9.70$ ounces and standard deviation $\sigma = 0.03$ ounce.

    (a) What is the probability that a randomly selected box of the cereal contains less than 9.65 ounces of cereal?

    (b) Now take an SRS of 5 boxes. What is the probability that the mean amount of cereal in these boxes is less than 9.65 ounces?

67. **Cholesterol** Suppose that the blood cholesterol level of all men aged 20 to 34 follows the Normal distribution with mean $\mu = 188$ milligrams per deciliter (mg/dl) and standard deviation $\sigma = 41$ mg/dl.

    (a) Choose an SRS of 100 men from this population. Describe the sampling distribution of $\bar{x}$.

    (b) Find the probability that $\bar{x}$ estimates μ within ± 3 mg/dl. (This is the probability that $\bar{x}$ takes a value between 185 and 191 mg/dl.)

    (c) Choose an SRS of 1000 men from this population. Now what is the probability that $\bar{x}$ falls within ± 3 mg/dl of μ? In what sense is the larger sample "better"?

68. **Finch beaks** One dimension of bird beaks is "depth" — the height of the beak where it arises from the bird's head. During a research study on one island in the Galapagos archipelago, the beak depth of all Medium Ground Finches on the island was found to be Normally distributed with mean $\mu = 9.5$ millimeters (mm) and standard deviation $\sigma = 1.0$ mm.[11]

    (a) Choose an SRS of 5 Medium Ground Finches from this population. Describe the sampling distribution of $\bar{x}$.

(ii) $P(X < 9.65) = $ normalcdf(lower: -1000, upper: 9.65, mean: 9.70, SD:0.03) $= 0.0478$
(b) Because X has a Normal distribution, the sampling distribution of $\bar{x}$ has a Normal distribution; $\mu_{\bar{x}} = \mu = 9.70$ ounces. Because 5 is less than 10% of all boxes produced,
$\sigma_{\bar{x}} = \dfrac{\sigma}{\sqrt{n}} = \dfrac{0.03}{\sqrt{5}} = 0.0134$ ounces.
We want to find $P(\bar{x} < 9.65)$.
(i) $z = \dfrac{9.65 - 9.70}{0.0134} = -3.73$
Table A: $P(z < -3.73) \approx 0$
(ii) *Tech:* $P(\bar{x} < 9.65) = $ normalcdf(lower: -1000, upper: 9.65, mean: 9.70, SD: 0.0134) $= 0.0001$

7.67 (a) The sampling distribution of $\bar{x}$ is Normal with $\mu_{\bar{x}} = \mu = 188$ mg/dl. Because the sample

---

size (100) is less than 10% of all men aged 20 to 34, $\sigma_{\bar{x}} = \dfrac{\sigma}{\sqrt{n}} = \dfrac{41}{\sqrt{100}} = 4.1$ mg/dl.

(b) We want to find $P(185 \le \bar{x} \le 191)$.
(i) $z = -0.73$ and $z = 0.73$;
$P(-0.73 \le z \le 0.73) =$
$0.7673 - 0.2327 = 0.5346$
(ii) *Tech:* $P(185 \le \bar{x} \le 191) =$
normalcdf(lower: 185, upper: 191, mean: 188, SD: 4.1) $= 0.5357$

(c) $\sigma_{\bar{x}} = \dfrac{\sigma}{\sqrt{n}} = \dfrac{41}{\sqrt{1000}} = 1.30$ mg/dl;
we want to find $P(185 \le \bar{x} \le 191)$.
(i) $z = -2.31$ and $z = 2.31$;
$P(-2.31 \le z \le 2.31) = 0.9896 - 0.0104 =$
0.9792 (ii) $P(185 \le \bar{x} \le 191) = $ normalcdf(lower: 185, upper: 191, mean: 188, SD: 1.30) $= 0.9790$

(b) Find the probability that $\bar{x}$ estimates μ within ± 0.5 mm. (This is the probability that $\bar{x}$ takes a value between 9 and 10 mm.)

(c) Choose an SRS of 50 Medium Ground Finches from this population. Now what is the probability that $\bar{x}$ falls within ± 0.5 mm of μ? In what sense is the larger sample "better"?

69. **Dead battery?** A car company claims that the lifetime of its batteries varies from car to car according to a Normal distribution with mean $\mu = 48$ months and standard deviation $\sigma = 8.2$ months. A consumer organization installs this type of battery in an SRS of 8 cars and calculates $\bar{x} = 42.2$ months.

(a) Find the probability that the sample mean lifetime is 42.2 months or less if the company's claim is true.

(b) Based on your answer to part (a), is there convincing evidence that the company is overstating the average lifetime of its batteries?

70. **Foiled again?** The manufacturer of a certain brand of aluminum foil claims that the amount of foil on each roll follows a Normal distribution with a mean of 250 square feet (ft^2) and a standard deviation of 2 ft^2. To test this claim, a restaurant randomly selects 10 rolls of this aluminum foil and carefully measures the mean area to be $\bar{x} = 249.6 \, ft^2$.

(a) Find the probability that the sample mean area is 249.6 ft^2 or less if the manufacturer's claim is true.

(b) Based on your answer to part (a), is there convincing evidence that the company is overstating the average area of its aluminum foil rolls?

71. **Songs on an iPod** David's iPod has about 10,000 songs. The distribution of the play times for these songs is heavily skewed to the right with a mean of 225 seconds and a standard deviation of 60 seconds.

(a) Describe the shape of the sampling distribution of $\bar{x}$ for SRSs of size $n = 5$ from the population of songs on David's iPod. Justify your answer.

(b) Describe the shape of the sampling distribution of $\bar{x}$ for SRSs of size $n = 100$ from the population of songs on David's iPod. Justify your answer.

72. **High school GPAs** The distribution of grade point average for students at a large high school is skewed to the left with a mean of 3.53 and a standard deviation of 1.02.

(a) Describe the shape of the sampling distribution of $\bar{x}$ for SRSs of size $n = 4$ from the population of students at this high school. Justify your answer.

(b) Describe the shape of the sampling distribution of $\bar{x}$ for SRSs of size $n = 50$ from the population of students at this high school. Justify your answer.

73. **More on insurance** An insurance company claims that in the entire population of homeowners, the mean annual loss from fire is $\mu = \$250$ and the standard deviation of the loss is $\sigma = \$5000$. The distribution of losses is strongly right-skewed: many policies have $0 loss, but a few have large losses. The company hopes to sell 1000 of these policies for $300 each.

(a) Assuming that the company's claim is true, what is the probability that the mean loss from fire is greater than $300 for an SRS of 1000 homeowners?

(b) If the company wants to be 90% certain that the mean loss from fire in an SRS of 1000 homeowners is less than the amount it charges for the policy, how much should the company charge?

74. **Cash grab** At a traveling carnival, a popular game is called the "Cash Grab." In this game, participants step into a sealed booth, a powerful fan turns on, and dollar bills are dropped from the ceiling. A customer has 30 seconds to grab as much cash as possible while the dollar bills swirl around. Over time, the operators of the game have determined that the mean amount grabbed is $13 with a standard deviation of $9. They charge $15 to play the game and expect to have 40 customers at their next carnival.

(a) What is the probability that an SRS of 40 customers grab an average of $15 or more?

(b) How much should the operators charge if they want to be 95% certain that the mean amount grabbed by an SRS of 40 customers is less than what they charge to play the game?

75. **Bad carpet** The number of flaws per square yard in a type of carpet material varies with mean 1.6 flaws per square yard and standard deviation 1.2 flaws per square yard.

(a) Without doing any calculations, explain which event is more likely:

- randomly selecting 1 square yard of material and finding 2 or more flaws

- randomly selecting 50 square yards of material and finding an average of 2 or more flaws

(b) Explain why you cannot use a Normal distribution to calculate the probability of the first event in part (a).

(c) Calculate the probability of the second event in part (a).

The larger sample is better because it is more likely to produce a sample mean within 3 mg/dl of the population mean.

7.68 (a) Normal with $\mu_{\bar{x}} = 9.5$ mm and $\sigma_{\bar{x}} = 1/\sqrt{5} = 0.4472$ mm **(b)** We want to find $P(9 \le \bar{x} \le 10)$.
(i) $z = -1.12$ and $z = 1.12$; $P(-1.12 \le z \le 1.12) = 0.7373$
(ii) $P(9 \le \bar{x} \le 10) = $ normalcdf(lower: 9, upper: 10, mean: 9.5, SD: 0.4472) $= 0.7365$
(c) $\sigma_{\bar{x}} = 1/\sqrt{50} = 0.1414$ mm; we want to find $P(9 \le \bar{x} \le 10)$.
(i) $z = -3.54$ and $z = 3.54$; $P(-3.54 \le z \le 3.54) = 0.9998 - 0.0002 = 0.9996$

(ii) $P(9 \le \bar{x} \le 10) = $ normalcdf(lower: 9, upper: 10, mean: 9.5, SD: 0.1414) $= 0.9996$

The larger sample is better because it is more likely to produce a sample mean within 0.5 mm of the population mean.

7.69 (a) The sampling distribution of $\bar{x}$ is Normal with $\mu_{\bar{x}} = 48$ months and $\sigma_{\bar{x}} = 8.2/\sqrt{8} = 2.899$ months. We want to find $P(\bar{x} \le 42.2)$.
(i) $z = -2.00$; $P(z \le -2.00) = 0.0228$
(ii) $P(\bar{x} \le 42.2) = $ normalcdf(lower: −1000, upper: 42.2, mean: 48, SD: 2.899) $= 0.0227$
(b) Because this probability is very small, there is convincing evidence that the company is overstating the average lifetime.

7.70 (a) The sampling distribution of $\bar{x}$ is Normal with $\mu_{\bar{x}} = 250$ square feet and $\sigma_{\bar{x}} = 2/\sqrt{10} = 0.6325$ square feet. We want to find $P(\bar{x} \le 249.6)$.
(i) $z = -0.63$; $P(z \le -0.63) = 0.2643$
(ii) $P(\bar{x} \le 249.6) = $ normalcdf(lower: −1000, upper: 249.6, mean: 250, SD: 0.6325) $= 0.2636$

(b) Because this probability is not small, there is not convincing evidence that the company is overstating the average area.

7.71 (a) Because $n = 5 < 30$, the sampling distribution of $\bar{x}$ will also be skewed to the right, but not quite as strongly as the population. **(b)** Because $n = 100 \ge 30$, the sampling distribution of $\bar{x}$ is approximately Normal by the central limit theorem.

7.72 (a) Because $n = 4 < 30$, the sampling distribution of $\bar{x}$ will also be skewed to the left, but not quite as strongly as the population. **(b)** Because $n = 50 \ge 30$, the sampling distribution of $\bar{x}$ is approximately Normal by the CLT.

7.73 (a) The sampling distribution of $\bar{x}$ is approximately Normal with $\mu_{\bar{x}} = \$250$ and $\sigma_{\bar{x}} = 5000/\sqrt{1000} = 158.114$. We want to find $P(\bar{x} > 300)$.

(i) $z = 0.32$; $P(z > 0.32) = 1 - 0.6255 = 0.3745$

(ii) $P(\bar{x} > 300) = $ normalcdf(lower: 300, upper: 1000, mean: 250, SD: 158.114) $= 0.3759$

(b) (i) $1.28 = \dfrac{x - 250}{158.114}$, so $x = \$452.39$.
(ii) *Tech:* invNorm(area: 0.90, mean: 250, SD: 158.114) $= \$452.63$

7.74 The sampling distribution of $\bar{x}$ is approximately Normal with $\mu_{\bar{x}} = \$13$ and $\sigma_{\bar{x}} = 9/\sqrt{40} = 1.423$. We want to find $P(\bar{x} \ge 15)$.
(i) $z = 1.41$; $P(z \ge 1.41) = 1 - 0.9207 = 0.0793$
(ii) $P(\bar{x} \ge 15) = $ normalcdf(lower: 15, upper: 1000, mean: 13, SD: 1.423) $= 0.0799$

(b) (i) $1.645 = \dfrac{x - 13}{1.423}$, so $x = \$15.34$.
(ii) *Tech:* invNorm(area: 0.95, mean: 13, SD: 1.423) $= \$15.34$

Answer to 7.75 on next page.

7.75 (a) It is more likely to randomly select 1 square yard of material and find 2 or more flaws. There is more variability from the mean of 1.6 in the number of flaws found in individual square yards of material than in the average number of flaws found in a sample of 50 square yards of material. **(b)** You cannot use a Normal distribution to calculate the probability of the first event in part (a) because the population distribution is not Normal and the sample size is not at least 30. **(c)** Because the sample size is large ($n = 50 \geq 30$), the sampling distribution of $\bar{x}$ is approximately Normal with $\mu_{\bar{x}} = 1.6$. $\sigma_{\bar{x}} = 1.2/\sqrt{50} = 0.1697$. We want to find $P(\bar{x} \geq 2)$.
(i) $z = 2.36$; $P(z \geq 2.36) = 1 - 0.9909 = 0.0091$
(ii) $P(\bar{x} \geq 2) = $ normalcdf(lower: 2, upper: 1000, mean: 1.6, SD: 0.1697) = 0.0092

7.76 (a) It is more likely to randomly select 1 car entering this interchange during rush hours and finding 2 or more people in the car. There is more variability from the mean of 0.75 in the number of passengers found in individual cars than in the average number of passengers found in a sample of 35 cars. **(b)** You cannot use a Normal distribution to calculate the probability of the first event in part (a) because the population distribution is not Normal and the sample size is not at least 30. **(c)** Because the sample size is large ($n = 35 \geq 30$), the sampling distribution of $\bar{x}$ is approximately Normal with $\mu_{\bar{x}} = 1.6$. Because 35 is less than 10% of all cars entering the freeway during rush hour, $\sigma_{\bar{x}} = 0.75/\sqrt{35} = 0.1268$. We want to find $P(\bar{x} \geq 2)$.

(i) $z = 3.15$; $P(z \geq 3.15) = 1 - 0.9992 = 0.0008$
(ii) $P(\bar{x} \geq 2) = $ normalcdf(lower: 2, upper: 1000, mean: 1.6, SD: 0.1268) = 0.0008

7.77 No; the histogram of the sample values will look like the population distribution, whatever it may be. The central limit theorem (CLT) says that the histogram of the sampling distribution of the *sample mean* will look more and more Normal as the sample size increases.

7.78 Although this is a correct statement about the standard deviation of the sampling distribution of $\bar{x}$, the CLT says otherwise. The CLT addresses the *shape* of the sampling distribution, not the variability.

7.79 Because the sample size is large ($n = 30 \geq 30$), the distribution of $\bar{x}$ is approximately Normal with $\mu_{\bar{x}} = \mu = 190$ pounds. Because $n = 30$ is less than 10% of all possible passengers, $\sigma_{\bar{x}} = 35/\sqrt{30} = 6.3901$ pounds.

76. How many people in a car? A study of rush-hour traffic in San Francisco counts the number of people in each car entering a freeway at a suburban interchange. Suppose that this count has mean 1.6 and standard deviation 0.75 in the population of all cars that enter at this interchange during rush hour.

(a) Without doing any calculations, explain which event is more likely:

- randomly selecting 1 car entering this interchange during rush hour and finding 2 or more people in the car
- randomly selecting 35 cars entering this interchange during rush hour and finding an average of 2 or more people in the cars

(b) Explain why you cannot use a Normal distribution to calculate the probability of the first event in part (a).

(c) Calculate the probability of the second event in part (a).

77. **What does the CLT say?** Asked what the central limit theorem says, a student replies, "As you take larger and larger samples from a population, the histogram of the sample values looks more and more Normal." Is the student right? Explain your answer.

78. **What does the CLT say?** Asked what the central limit theorem says, a student replies, "As you take larger and larger samples from a population, the variability of the sampling distribution of the sample mean decreases." Is the student right? Explain your answer.

79. **Airline passengers get heavier** In response to the increasing weight of airline passengers, the Federal Aviation Administration (FAA) told airlines to assume that passengers average 190 pounds in the summer, including clothes and carry-on baggage. But passengers vary, and the FAA did not specify a standard deviation. A reasonable standard deviation is 35 pounds. A commuter plane carries 30 passengers. Find the probability that the total weight of 30 randomly selected passengers exceeds 6000 pounds. (*Hint:* To calculate this probability, restate the problem in terms of the mean weight.)

80. **Lightning strikes** The number of lightning strikes on a square kilometer of open ground in a year has mean 6 and standard deviation 2.4. The National Lightning Detection Network (NLDN) uses automatic sensors to watch for lightning in 1-square-kilometer plots of land. Find the probability that the total number of lightning strikes in a random sample of 50 square-kilometer plots of land is less than 250. (*Hint:* To calculate this

probability, restate the problem in terms of the mean number of strikes.)

81. **House prices** In the northern part of a large city, the distribution of home values is skewed to the right with a mean of \$410,000 and a standard deviation of \$250,000. In the southern part of the city, the distribution of home values is skewed to the right with a mean of \$375,000 and a standard deviation of \$240,000. Independent random samples of 10 houses in each part of the city are selected. Let $\bar{x}_N$ represent the sample mean value of homes in the northern part and let $\bar{x}_S$ represent the sample mean value of homes in the southern part.

(a) Find the mean of the sampling distribution of $\bar{x}_N - \bar{x}_S$.

(b) Calculate and interpret the standard deviation of the sampling distribution. Verify that the 10% condition is met.

(c) Is the shape of the sampling distribution approximately Normal? Justify your answer.

82. **Young players** In the National Football League (NFL), the distribution of age is skewed to the right with a mean of 26.2 years and a standard deviation of 3.24 years. In the National Basketball Association (NBA), the distribution of age is skewed to the right with a mean of 25.8 years and a standard deviation of 4.24 years. Independent random samples of 20 players in each league are selected. Let $\bar{x}_{NFL}$ represent the sample mean age of NFL players and let $\bar{x}_{NBA}$ represent the sample mean age of NBA players.

(a) Find the mean of the sampling distribution of $\bar{x}_{NFL} - \bar{x}_{NBA}$.

(b) Calculate and interpret the standard deviation of the sampling distribution. Verify that the 10% condition is met.

(c) Is the shape of the sampling distribution approximately Normal? Justify your answer.

83. **Cholesterol** The level of cholesterol in the blood for all men aged 20 to 34 follows a Normal distribution with mean $\mu_M = 188$ milligrams per deciliter (mg/dl) and standard deviation $\sigma_M = 41$ mg/dl. For 14-year-old boys, blood cholesterol levels follow a Normal distribution with mean $\mu_B = 170$ mg/dl and standard deviation $\sigma_B = 30$ mg/dl. Suppose we select independent SRSs of 25 men aged 20 to 34 and 36 boys aged 14 and calculate the sample mean cholesterol levels $\bar{x}_M$ and $\bar{x}_B$.

(a) What is the shape of the sampling distribution of $\bar{x}_M - \bar{x}_B$? Why?

(b) Find the mean of the sampling distribution.

We want to find $P(\bar{x} > 200)$. **(i)** $z = 1.56$; $P(z > 1.56) = 1 - 0.9406 = 0.0594$

(ii) $P(\bar{x} > 200) = $ normalcdf(lower: 200, upper: 1000, mean: 190, SD: 6.3901) = 0.0588

7.80 Because the sample size is large ($n = 50 \geq 30$), the distribution of $\bar{x}$ is approximately Normal with $\mu_{\bar{x}} = \mu = 6$ strikes/km². Because 50 is less than 10% of all 1-square-kilometer plots, $\sigma_{\bar{x}} = 2.4/\sqrt{50} = 0.3394$ strikes/km². We want to find $P(\bar{x} < 5)$.
(i) $z = -2.95$; $P(z < -2.95) = 0.0016$
(ii) $P(\bar{x} < 5) = $ normalcdf(lower: -1000, upper: 5, mean: 6, SD: 0.3394) = 0.0016

7.81 (a) $\mu_{\bar{x}_N - \bar{x}_S} = \$410{,}000 - \$375{,}000 = \$35{,}000$
(b) Because $10 < 10\%$ of all homes in the northern part of the large city and $10 < 10\%$ of all homes in the southern part of the large city,

$$\sigma_{\bar{x}_N - \bar{x}_S} = \sqrt{\frac{250{,}000^2}{10} + \frac{240{,}000^2}{10}} = \$109{,}590.15$$

The difference (Northern homes − Southern homes) in the sample mean home value typically varies by about \$109,590.15 from the true difference in means of \$35,000. **(c)** No; the distributions of home value for both populations are skewed to the right and the sample sizes are both small ($n = 10 < 30$).

7.82 (a) $\mu_{\bar{x}_{NFL} - \bar{x}_{NBA}} = 26.2 - 25.8 = 0.4$ years
(b) Because $20 < 10\%$ of all players in the NFL and $20 < 10\%$ of all players in the NBA,

$$\sigma_{\bar{x}_{NFL} - \bar{x}_{NBA}} = \sqrt{\frac{3.24^2}{20} + \frac{4.24^2}{20}} = 1.193 \text{ years.}$$

The difference (NFL − NBA) in the sample mean age typically varies by about 1.193 years from the true difference in means of 0.4 years. **(c)** No; the distributions of age for both populations are skewed to the right and the sample sizes are both small ($n = 20 < 30$).

(c) Calculate and interpret the standard deviation of the sampling distribution. Verify that the 10% condition is met.

(d) Find the probability of getting a difference in sample means $\bar{x}_M - \bar{x}_B$ that's less than 0 mg/dl.

84. **How tall?** The heights of young men follow a Normal distribution with mean $\mu_M = 69.3$ inches and standard deviation $\sigma_M = 2.8$ inches. The heights of young women follow a Normal distribution with mean $\mu_W = 64.5$ inches and standard deviation $\sigma_W = 2.5$ inches. Suppose we select independent SRSs of 16 young men and 9 young women and calculate the sample mean heights $\bar{x}_M$ and $\bar{x}_W$.

(a) What is the shape of the sampling distribution of $\bar{x}_M - \bar{x}_W$? Why?

(b) Find the mean of the sampling distribution.

(c) Calculate and interpret the standard deviation of the sampling distribution. Verify that the 10% condition is met.

(d) Find the probability of getting a difference in sample means $\bar{x}_M - \bar{x}_W$ that's greater than 2 inches.

85. **Cholesterol** Refer to Exercise 83. Should we be surprised if the sample mean cholesterol level for the 14-year-old boys exceeds the sample mean cholesterol level for the men? Explain your answer.

86. **How tall?** Refer to Exercise 84. Should we be surprised if the sample mean height for the young men is at least 2 inches greater than the sample mean height for the young women? Explain your answer.

Multiple Choice: *Select the best answer for Exercises 87–90.*

87. The distribution of scores on the mathematics part of the SAT exam in a recent year was approximately Normal with mean 515 and standard deviation 114. Imagine choosing many SRSs of 100 students who took the exam and finding the average score for each sample. Which of the following are the mean and standard deviation of the sampling distribution of $\bar{x}$?

(a) Mean $= 515$, SD $= 114$

(b) Mean $= 515$, SD $= 114/\sqrt{100}$

(c) Mean $= 515/100$, SD $= 114/100$

(d) Mean $= 515/100$, SD $= 114/\sqrt{100}$

(e) Cannot be determined without knowing the 100 scores.

88. Why is it important to check the 10% condition before calculating probabilities involving $\bar{x}$?

(a) To reduce the variability of the sampling distribution of $\bar{x}$

(b) To ensure that the distribution of $\bar{x}$ is approximately Normal

(c) To ensure that we can generalize the results to a larger population

(d) To ensure that $\bar{x}$ will be an unbiased estimator of μ

(e) To ensure that the observations in the sample are close to independent

89. A machine is designed to fill 16-ounce bottles of shampoo. When the machine is working properly, the amount poured into the bottles follows a Normal distribution with mean 16.05 ounces and standard deviation 0.1 ounce. Assume that the machine is working properly. If 4 bottles are randomly selected and the number of ounces in each bottle is measured, then there is about a 95% probability that the sample mean will fall in which of the following intervals?

(a) 16.05 to 16.15 ounces

(b) 16.00 to 16.10 ounces

(c) 15.95 to 16.15 ounces

(d) 15.90 to 16.20 ounces

(e) 15.85 to 16.25 ounces

90. The number of hours a lightbulb burns before failing varies from bulb to bulb. The population distribution of burnout times is strongly skewed to the right. The central limit theorem says that

(a) as we look at more and more bulbs, their average burnout time gets ever closer to the mean μ for all bulbs of this type.

(b) the average burnout time of a large number of bulbs has a sampling distribution with the same shape (strongly skewed) as the population distribution.

(c) the average burnout time of a large number of bulbs has a sampling distribution with a similar shape but not as extreme (skewed, but not as strongly) as the population distribution.

(d) the average burnout time of a large number of bulbs has a sampling distribution that is close to Normal.

(e) the average burnout time of a large number of bulbs has a sampling distribution that is exactly Normal.

Recycle and Review

Exercises 91 and 92 refer to the following setting. In the language of government statistics, you are "in the labor force" if you are available for work and either working or actively seeking work. The unemployment rate is the proportion of the

(i) $z = \dfrac{2 - 4.8}{1.09} = -2.57$; $P(z > -2.57) = 0.9949$

(ii) $P(\bar{x}_M - \bar{x}_W > 2) = \text{normalcdf(lower: }2, \text{ upper: }1000, \text{ mean: }4.8, \text{ SD: }1.09) = 0.9949$

There is a 0.9949 probability of getting a difference in the sample means $\bar{x}_M - \bar{x}_W$ that is greater than 2 inches.

7.85 Yes; the likelihood that the sample mean cholesterol level of the boys is greater than the sample mean cholesterol level of the men is only 3%.

7.86 No; it is almost certain (a 99.49% chance) that the sample mean height for young men is at least 2 inches greater than the sample mean height for young women.

7.87 b

7.88 e

7.89 c

7.90 d

7.83 (a) Normal because both population distributions are Normal. **(b)** $\mu_{\bar{x}_M - \bar{x}_B} = 188 - 170 = 18$ mg/dl **(c)** Because $25 < 10\%$ of all 20- to 34-year-old males and $36 < 10\%$ of all 14-year-old boys,

$$\sigma_{\bar{x}_M - \bar{x}_B} = \sqrt{\frac{41^2}{25} + \frac{30^2}{36}} = 9.60 \text{ mg/dl. The}$$

difference (20- to 34-year old males − 14-year-old boys) in the sample mean blood cholesterol levels typically varies by about 9.60 mg/dl from the true difference in means of 18 mg/dl.

(d) We want to find $P(\bar{x}_M - \bar{x}_B < 0)$.

(i) $z = \dfrac{0 - 18}{9.60} = -1.88$; $P(z < -1.88) = 0.0304$

(ii) $P(\bar{x}_M - \bar{x}_B < 0) = \text{normalcdf(lower: }-1000, \text{ upper: }0, \text{ mean: }18, \text{ SD: }9.60) = 0.0304$

There is a 0.0304 probability of getting a difference in the sample means $\bar{x}_M - \bar{x}_B$ that is less than zero.

7.84 (a) Normal because both population distributions are Normal.
(b) $\mu_{\bar{x}_M - \bar{x}_W} = 69.3 - 64.5 = 4.8$ inches **(c)** Because $16 < 10\%$ of all young men and $9 < 10\%$ of all young women,

$$\sigma_{\bar{x}_M - \bar{x}_W} = \sqrt{\frac{2.8^2}{16} + \frac{2.5^2}{9}} = 1.09 \text{ inches. The}$$

difference (Men − Women) in the sample mean heights typically varies by about 1.09 inches from the true difference in means of 4.8 inches.

(d) We want to find $P(\bar{x}_M - \bar{x}_W > 2)$.

7.91 The unemployment rates for each level of education are

$P(\text{unemployed} \mid \text{didn't finish HS}) =$
$\dfrac{1062}{12,470} = 0.0852$

$P(\text{unemployed} \mid \text{HS but no college}) =$
$\dfrac{1977}{37,834} = 0.0523$

$P(\text{unemployed} \mid \text{less than bachelor's degree}) =$
$\dfrac{1462}{34,439} = 0.0425$

$P(\text{unemployed} \mid \text{college graduate}) =$
$\dfrac{1097}{40,390} = 0.0272$

There is an association between unemployment rate and education. The unemployment rate decreases with additional education.

7.92 (a) $P(\text{in labor force}) =$
$\dfrac{12,470 + 37,834 + 34,439 + 40,390}{27,669 + 59,860 + 47,556 + 51,582} =$
$\dfrac{125,133}{186,667} = 0.6704.$ There is a 0.6704 probability that a randomly selected person age 25 or older is in the workforce.
(b) $P(\text{in labor force} \mid \text{college graduate}) =$
$\dfrac{40,390}{51,582} = 0.7830.$ Given that a randomly selected person is a college graduate, there is a 0.7830 probability that the person is in the labor force. **(c)** The events "in the labor force" and "college graduate" are not independent, because the probability of being in the labor force (0.6704) does not equal the probability of being in the labor force given that the person is a college graduate (0.7830).

TRM FRAPPY! Materials

Please consult the Teacher's Resource Materials for sample student responses, a scoring rubric, and a printable version of the original question with space for students to write their responses. We present a model solution here.

Answers:

(a) The sampling distribution of the sample mean will be approximately Normal, because the sample size is large ($50 \geq 30$).
(b) $\mu_{\bar{x}} = \mu = 1.1$ and because the sample size is less than 10% of all students

$(50 < 10\% \text{ of } 2500),\ \sigma_{\bar{x}} = \dfrac{\sigma}{\sqrt{n}} =$

$\dfrac{1.4}{\sqrt{50}} = 0.198.$

---

labor force (not of the entire population) that is unemployed. Here are estimates from the Current Population Survey for the civilian population aged 25 years and over in a recent year. The table entries are counts in thousands of people.

Highest education	Total population	In labor force	Employed
Didn't finish high school	27,669	12,470	11,408
High school but no college	59,860	37,834	35,857
Less than bachelor's degree	47,556	34,439	32,977
College graduate	51,582	40,390	39,293

91. **Unemployment** (1.1) Find the unemployment rate for people with each level of education. Is there an association between unemployment rate and education? Explain your answer.

92. **Unemployment** (5.2, 5.3) Suppose that you randomly select one person 25 years of age or older.

(a) What is the probability that a randomly chosen person 25 years of age or older is in the labor force?

(b) If you know that a randomly chosen person 25 years of age or older is a college graduate, what is the probability that he or she is in the labor force?

(c) Are the events "in the labor force" and "college graduate" independent? Justify your answer.

Chapter 7 Wrap-Up

FRAPPY! FREE RESPONSE AP® PROBLEM, YAY!

The following problem is modeled after actual AP® Statistics exam free response questions. Your task is to generate a complete, concise response in 15 minutes.

Directions: Show all your work. Indicate clearly the methods you use, because you will be scored on the correctness of your methods as well as on the accuracy and completeness of your results and explanations.

The principal of a large high school is concerned about the number of absences for students at his school. To investigate, he prints a list showing the number of absences during the last month for each of the 2500 students at the school. For this population of students, the distribution of absences last month is skewed to the right with a mean of $\mu = 1.1$ and a standard deviation of $\sigma = 1.4$.

Suppose that a random sample of 50 students is selected from the list printed by the principal and the sample mean number of absences is calculated.

(a) What is the shape of the sampling distribution of the sample mean? Explain.

(b) What are the mean and standard deviation of the sampling distribution of the sample mean?

(c) What is the probability that the mean number of absences in a random sample of 50 students is less than 1?

(d) Because the population distribution is skewed, the principal is considering using the median number of absences last month instead of the mean number of absences to summarize the distribution. Describe how the principal could use a simulation to estimate the standard deviation of the sampling distribution of the sample median for samples of size 50.

After you finish, you can view two example solutions on the book's website (highschool.bfwpub.com/updatedtps6e). Determine whether you think each solution is "complete," "substantial," "developing," or "minimal." If the solution is not complete, what improvements would you suggest to the student who wrote it? Finally, your teacher will provide you with a scoring rubric. Score your response and note what, if anything, you would do differently to improve your own score.

(c) (i) $z = \dfrac{1 - 1.1}{0.198} = -0.51$

Table A: $P(\bar{x} < 1) = P(z < -0.51) = 0.3050$
Tech: normalcdf(lower: −1000, upper: −0.51, mean: 0, SD: 1) = 0.3050
(ii) normalcdf(lower: −1000, upper: 1, mean: 1.1, SD: 0.198) = 0.3068
There is about a 31% chance that the mean number of absences in a random sample of 50 students is less than 1.
(d) Take many, many random samples of size 50. For each sample, calculate the sample median. Finally, calculate the standard deviation of the distribution of sample medians.

Chapter 7 Review

Section 7.1: What Is a Sampling Distribution?

In this section, you learned the "big ideas" of sampling distributions. The first big idea is the difference between a statistic and a parameter. A parameter is a number that describes some characteristic of a population. A statistic estimates the value of a parameter using a sample from the population. Making the distinction between a statistic and a parameter will be crucial throughout the rest of the course.

The second big idea is that statistics vary. For example, the mean weight in a sample of high school students is a variable that will change from sample to sample. This means that statistics have distributions. The distribution of a statistic in all possible samples of the same size from the same population is called the sampling distribution of the statistic. Knowing the sampling distribution of a statistic tells us how far we can expect a statistic to vary from the parameter value and what values of the statistic should be considered unusual.

The third big idea is the distinction between the distribution of the population, the distribution of a sample, and the sampling distribution of a sample statistic. Reviewing the illustration on page 474 will help you understand the difference between these three distributions. When you are writing your answers, be sure to indicate which distribution you are referring to. Don't make ambiguous statements like "the distribution will become less variable."

The final big idea is how to describe a sampling distribution. To adequately describe a sampling distribution, you need to address shape, center, and variability. If the center (mean) of the sampling distribution is the same as the value of the parameter being estimated, then the statistic is called an unbiased estimator. An estimator is unbiased if it doesn't consistently underestimate or consistently overestimate the parameter in many samples. Ideally, the variability of a sampling distribution will be very small, meaning that the statistic provides precise estimates of the parameter. Larger sample sizes result in sampling distributions with less variability.

Section 7.2: Sample Proportions

In this section, you learned about the shape, center, and variability of the sampling distribution of a sample proportion $\hat{p}$. When the Large Counts condition ($np \geq 10$ and $n(1-p) \geq 10$) is met, the sampling distribution of $\hat{p}$ will be approximately Normal. The mean of the sampling distribution of $\hat{p}$ is $\mu_{\hat{p}} = p$, the population proportion. As a result, the sample proportion $\hat{p}$ is an unbiased estimator of the population proportion p. When the sample size is less than 10% of the population size (the 10% condition), the standard deviation of the sampling distribution of the sample proportion is approximately $\sigma_{\hat{p}} = \sqrt{p(1-p)/n}$. The standard deviation measures how far the sample proportion $\hat{p}$ typically varies from the population proportion p.

In this section, you also learned about the sampling distribution of a difference in sample proportions $\hat{p}_1 - \hat{p}_2$. The shape of the sampling distribution of $\hat{p}_1 - \hat{p}_2$ will be approximately Normal when the Large Counts condition is met for both samples. The center of the sampling distribution of $\hat{p}_1 - \hat{p}_2$ is $\mu_{\hat{p}_1 - \hat{p}_2} = p_1 - p_2$. The standard deviation of the sampling distribution is approximately

$$\sigma_{\hat{p}_1 - \hat{p}_2} = \sqrt{\frac{p_1(1-p_1)}{n_1} + \frac{p_2(1-p_2)}{n_2}}$$ when the 10% condition is met for both samples.

Section 7.3: Sample Means

In this section, you learned about the shape, center, and variability of the sampling distribution of a sample mean $\bar{x}$. When the population is Normally distributed, the sampling distribution of $\bar{x}$ will also be Normal for any sample size. When the population distribution is not Normal and the sample size is small, the sampling distribution of $\bar{x}$ will resemble the population shape. However, the central limit theorem says that the sampling distribution of $\bar{x}$ will become approximately Normal for larger sample sizes (typically when $n \geq 30$), no matter what the population shape. You can use a Normal distribution to calculate probabilities involving the sampling distribution of $\bar{x}$ if the population is Normal or the sample size is at least 30.

The mean of the sampling distribution of $\bar{x}$ is $\mu_{\bar{x}} = \mu$, the population mean. As a result, the sample mean $\bar{x}$ is an unbiased estimator of the population mean μ. When the sample size is less than 10% of the population size (the 10% condition), the standard deviation of the sampling distribution of the sample mean is approximately $\sigma_{\bar{x}} = \sigma/\sqrt{n}$. The standard deviation measures how far the sample mean $\bar{x}$ typically varies from the population mean μ.

In this section, you also learned about the sampling distribution of a difference in sample means $\bar{x}_1 - \bar{x}_2$. The shape of the sampling distribution of $\bar{x}_1 - \bar{x}_2$ will be approximately Normal when both population distributions are Normal, both sample sizes are at least 30, or one population distribution is Normal and the sample size from the other population distribution is at least 30. The center of the sampling distribution of $\bar{x}_1 - \bar{x}_2$ is $\mu_{\bar{x}_1 - \bar{x}_2} = \mu_1 - \mu_2$. The standard deviation of the sampling distribution is approximately

$$\sigma_{\bar{x}_1 - \bar{x}_2} = \sqrt{\frac{\sigma_1^2}{n_1} + \frac{\sigma_2^2}{n_2}}$$ when the 10% condition is met for both samples.

TRM **Chapter 7 Test**

There is one Chapter Test available for this section in the Teacher's Resource Materials. Click on the link in the TE-Book, open in the TRFD, or download from the Teacher's Resources on the book's digital platform. You can also create your own Test using the TPS quiz and test builder (ExamView).

Comparing sampling distributions		
	Sampling distribution of $\hat{p}$	**Sampling distribution of $\bar{x}$**
Center	$\mu_{\hat{p}} = p$	$\mu_{\bar{x}} = \mu$
Variability	$\sigma_{\hat{p}} = \sqrt{\dfrac{p(1-p)}{n}}$ when the 10% condition is met	$\sigma_{\bar{x}} = \dfrac{\sigma}{\sqrt{n}}$ when the 10% condition is met
Shape	Approximately Normal when the Large Counts condition is met: $np \geq 10$ and $n(1-p) \geq 10$	Normal when the population distribution is Normal; approximately Normal if the population distribution is not Normal but the sample size is large ($n \geq 30$)
	Sampling distribution of $\hat{p}_1 - \hat{p}_2$	**Sampling distribution of $\bar{x}_1 - \bar{x}_2$**
Center	$\mu_{\hat{p}_1 - \hat{p}_2} = p_1 - p_2$	$\mu_{\bar{x}_1 - \bar{x}_2} = \mu_1 - \mu_2$
Variability	$\sigma_{\hat{p}_1 - \hat{p}_2} = \sqrt{\dfrac{p_1(1-p_1)}{n_1} + \dfrac{p_2(1-p_2)}{n_2}}$ when the 10% condition is met for both samples	$\sigma_{\bar{x}_1 - \bar{x}_2} = \sqrt{\dfrac{\sigma_1^2}{n_1} + \dfrac{\sigma_2^2}{n_2}}$ when the 10% condition is met for both samples
Shape	Approximately Normal when the Large Counts condition is met for both samples: $n_1 p_1$, $n_1(1-p_1)$, $n_2 p_2$, $n_2(1-p_2)$ all ≥ 10	Normal when both population distributions are Normal; approximately Normal if the population distributions are not Normal but the sample sizes are both at least 30 or one population distribution is Normal and the sample size from the other population is at least 30.

What Did You Learn?

Learning Target	Section	Related Example on Page(s)	Relevant Chapter Review Exercise(s)
Distinguish between a parameter and a statistic.	7.1	470	R7.1
Create a sampling distribution using all possible samples from a small population.	7.1	472	R7.2
Use the sampling distribution of a statistic to evaluate a claim about a parameter.	7.1	473	R7.5, R7.7
Distinguish among the distribution of a population, the distribution of a sample, and the sampling distribution of a statistic.	7.1	Discussed on 474	R7.3
Determine if a statistic is an unbiased estimator of a population parameter.	7.1	477	R7.3
Describe the relationship between sample size and the variability of a statistic.	7.1	Discussed on 479	R7.2
Calculate the mean and standard deviation of the sampling distribution of a sample proportion $\hat{p}$ and interpret the standard deviation.	7.2	490	R7.4, R7.5
Determine if the sampling distribution of $\hat{p}$ is approximately Normal.	7.2	490	R7.4, R7.5
Calculate the mean and the standard deviation of the sampling distribution of a difference in sample proportions $\hat{p}_1 - \hat{p}_2$ and interpret the standard deviation.	7.2	496	R7.8
Determine if the sampling distribution of $\hat{p}_1 - \hat{p}_2$ is approximately Normal.	7.2	496	R7.8

Teaching Tip

At the end of each chapter, a "What Did You Learn?" grid lists all the targets for the chapter. Make sure to discuss this grid with your students. Ask them to read each learning target and self-assess whether or not they can do each one. For each target, the grid shows the section in which it was covered, page references for examples that illustrate the target, and relevant chapter review exercises that students can use to assess their understanding of that target. Encourage your students to use this grid as part of their preparations for the chapter test.

Teaching Tip

Now that you have completed Chapter 7 in this book, students can login to their College Board accounts and take the **Personal Progress Check for Unit 5**.

Learning Target	Section	Related Example on Page(s)	Relevant Chapter Review Exercise(s)
If appropriate, use a Normal distribution to calculate probabilities involving $\hat{p}$ or $\hat{p}_1 - \hat{p}_2$.	7.2	492, 496	R7.4, R7.5, R7.8
Calculate the mean and standard deviation of the sampling distribution of a sample mean $\bar{x}$ and interpret the standard deviation.	7.3	504	R7.6, R7.7
Explain how the shape of the sampling distribution of $\bar{x}$ is affected by the shape of the population distribution and the sample size.	7.3	Discussed on 510–511	R7.6, R7.7
Calculate the mean and standard deviation of the sampling distribution of a difference in sample means $\bar{x}_1 - \bar{x}_2$ and interpret the standard deviation.	7.3	516	R7.9
Determine if the sampling distribution of $\bar{x}_1 - \bar{x}_2$ is approximately Normal.	7.3	516	R7.9
If appropriate, use a Normal distribution to calculate probabilities involving $\bar{x}$ or $\bar{x}_1 - \bar{x}_2$.	7.3	506, 511, 516	R7.6, R7.7

Chapter 7 Review Exercises

These exercises are designed to help you review the important ideas and methods of the chapter.

R7.1 Bad eggs Selling eggs that are contaminated with salmonella can cause food poisoning in consumers. A large egg producer randomly selects 200 eggs from all the eggs shipped in one day. The laboratory reports that 9 of these eggs had salmonella contamination. Identify the population, the parameter, the sample, and the statistic.

R7.2 Five books An author has written 5 children's books. The numbers of pages in these books are 64, 66, 71, 73, and 76.

(a) List all 10 possible SRSs of size $n = 3$, calculate the median number of pages for each sample, and display the sampling distribution of the sample median on a dotplot.

(b) Describe how the variability of the sampling distribution of the sample median would change if the sample size was increased to $n = 4$.

(c) Construct the sampling distribution of the sample median for samples of size $n = 4$. Does this sampling distribution support your answer to part (b)? Explain your reasoning.

R7.3 Birth weights Researchers in Norway analyzed data on the birth weights of 400,000 newborns over a 6-year period. The distribution of birth weights is

approximately Normal with a mean of 3668 grams and a standard deviation of 511 grams.[12]

(a) Sketch a graph that displays the distribution of birth weights for this population.

(b) Sketch a possible graph of the distribution of birth weights for an SRS of size 5. Calculate the range for this sample.

In this population, the range (Maximum − Minimum) of birth weights is 3417 grams. We used technology to take 500 SRSs of size $n = 5$ and calculate the range (Maximum − Minimum) for each sample. The dotplot shows the results.

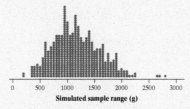

Simulated sample range (g)

(c) In the graph provided, there is a dot at approximately 2800. Explain what this value represents.

(d) Is the sample range an unbiased estimator of the population range? Give evidence from the graph to support your answer.

Answers to Chapter 7 Review Exercises

R7.1 The population is the set of all eggs shipped in one day. The parameter is $p =$ the proportion of eggs shipped that day that had salmonella. The sample consists of the 200 eggs examined. The statistic is $\hat{p} =$ the proportion of eggs in the sample that had salmonella = 9/200 = 0.045.

R7.2 (a) The 10 possible SRSs of size $n = 3$ and the median score of each sample are:

1: 64, 66, 71	median = 66
2: 64, 66, 73	median = 66
3: 64, 66, 76	median = 66
4: 64, 71, 73	median = 71
5: 64, 71, 76	median = 71
6: 64, 73, 76	median = 73
7: 66, 71, 73	median = 71
8: 66, 71, 76	median = 71
9: 66, 73, 76	median = 73
10: 71, 73, 76	median = 73

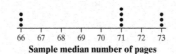

Sample median number of pages

(b) If the sample size was increased to $n = 4$, the variability of the sampling distribution of the sample median would decrease.

(c) The 5 possible SRSs of size $n = 4$ and the median number of pages of each sample are:

1: 64, 66, 71, 73	median = 68.5
2: 64, 66, 71, 76	median = 68.5
3: 64, 66, 73, 76	median = 69.5
4: 64, 71, 73, 76	median = 72
5: 66, 71, 73, 76	median = 72

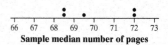

Sample median number of pages

This sampling distribution does support the answer to part **(b)**. The variability of the sampling distribution with $n = 4$ is less than the variability of the sampling distribution with $n = 3$.

R7.3 (a)

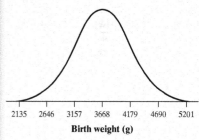

Birth weight (g)

(b) Answers will vary. An example dotplot is given. The range for this sample is 4375 − 2790 = 1585.

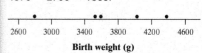

Birth weight (g)

(c) The dot at 2800 represents one SRS of size $n = 5$ from this population where the sample range was 2800 grams.

(d) The sample range is not an unbiased estimator of the population range. If it were unbiased, then the sampling distribution of the sample range would have 3417 (the population range) as its mean.

R7.4 (a) $\mu_{\hat{p}} = p = 0.15$ **(b)** Because the sample size of $n = 1540$ is less than 10% of the population of all adults, the standard deviation of the sampling distribution of $\hat{p}$ is

$$\sigma_{\hat{p}} = \sqrt{\frac{0.15(0.85)}{1540}} = 0.0091.$$

(c) Because $np = 1540(0.15) = 231 \geq 10$ and $n(1-p) = 1540(0.85) = 1309 \geq 10$, the sampling distribution of $\hat{p}$ is approximately Normal. **(d)** We want to find $P(0.13 \leq \hat{p} \leq 0.17)$.

(i) $z = \dfrac{0.13 - 0.15}{0.0091} = -2.20$ and

$z = \dfrac{0.17 - 0.15}{0.0091} = 2.20$;

$P(-2.20 \leq z \leq 2.20) = 0.9861 - 0.0139 = 0.9722$

(ii) $P(0.13 \leq \hat{p} \leq 0.17) =$ normalcdf(lower: 0.13, upper: 0.17, mean: 0.15, SD: 0.0091) = 0.9720 There is a 0.9720 probability of obtaining a sample in which between 13% and 17% are joggers.

R7.5 (a) We have an SRS of size 100 drawn from a population in which the proportion who get a red light is $p = 0.30$, assuming the agents' claim is true. Thus, $\mu_{\hat{p}} = p = 0.30$. Because 100 is less than 10% of the population of travelers, $\sigma_{\hat{p}} = \sqrt{\dfrac{0.30(0.70)}{100}} = 0.0458$.

Because $np = 100(0.30) = 30 \geq 10$ and $n(1 - p) = 100(0.70) = 70 \geq 10$, the sampling distribution of $\hat{p}$ can be approximated by a Normal distribution. We want to find $P(\hat{p} \leq 0.20)$.

(i) $z = \dfrac{0.20 - 0.30}{0.0458} = -2.18$;

$P(z \leq -2.18) = 0.0146$

(ii) $P(\hat{p} \leq 0.20) =$ normalcdf(lower: −1000, upper: 0.20, mean: 0.30, SD: 0.0458) = 0.0145 There is a 0.0145 probability that 20% or fewer of the travelers get a red light.

(b) Because this is a small probability, there is convincing evidence against the agents' claim—it isn't plausible to get a sample proportion this small by chance alone.

R7.6 (a) We want to find $P(X \geq 105)$.

(i) $z = \dfrac{105 - 100}{15} = 0.33$;

$P(z \geq 0.33) = 1 - 0.6293 = 0.3707$

(ii) $P(X \geq 105) =$ normalcdf(lower: 105, upper: 1000, mean: 100, SD: 15) = 0.3694 There is a 0.3694 probability of selecting an individual with a WAIS score of at least 105.

R7.4 Do you jog? The Gallup Poll asked a random sample of 1540 adults, "Do you happen to jog?" Suppose that the true proportion of all adults who jog is $p = 0.15$.

(a) What is the mean of the sampling distribution of $\hat{p}$?

(b) Calculate and interpret the standard deviation of the sampling distribution of $\hat{p}$. Check that the 10% condition is met.

(c) Is the sampling distribution of $\hat{p}$ approximately Normal? Justify your answer.

(d) Find the probability that between 13% and 17% of people jog in a random sample of 1540 adults.

R7.5 Bag check Thousands of travelers pass through the airport in Guadalajara, Mexico, each day. Before leaving the airport, each passenger must pass through the customs inspection area. Customs agents want to be sure that passengers do not bring illegal items into the country. But they do not have time to search every traveler's luggage. Instead, they require each person to press a button. Either a red or a green bulb lights up. If the red light flashes, the passenger will be searched by customs agents. A green light means "Go ahead." Customs agents claim that 30% of all travelers will be stopped (red light), because the light has probability 0.30 of showing red on any push of the button. To test this claim, a concerned citizen watches a random sample of 100 travelers push the button. Only 20 get a red light.

(a) Assume that the customs agents' claim is true. Find the probability that the proportion of travelers who get a red light in a random sample of 100 travelers is less than or equal to the result in this sample.

(b) Based on your results in part (a), is there convincing evidence that less than 30% of all travelers will be stopped? Explain your reasoning.

R7.6 IQ tests The Wechsler Adult Intelligence Scale (WAIS) is a common IQ test for adults. The distribution of WAIS scores for persons over 16 years of age is approximately Normal with mean 100 and standard deviation 15.

(a) What is the probability that a randomly chosen individual has a WAIS score of 105 or greater?

(b) Find the mean and standard deviation of the sampling distribution of the average WAIS score $\bar{x}$ for an SRS of 60 people. Verify that the 10% condition is met. Interpret the standard deviation.

(c) What is the probability that the average WAIS score of an SRS of 60 people is 105 or greater?

(d) Would your answers to any of parts (a), (b), or (c) be affected if the distribution of WAIS scores in the adult population was distinctly non-Normal? Explain your reasoning.

R7.7 Detecting gypsy moths The gypsy moth is a serious threat to oak and aspen trees. A state agriculture department places traps throughout the state to detect the moths. Each month, an SRS of 50 traps is inspected, the number of moths in each trap is recorded, and the mean number of moths is calculated. Based on years of data, the distribution of moth counts is discrete and strongly skewed with a mean of 0.5 and a standard deviation of 0.7.

(a) Explain why it is reasonable to use a Normal distribution to approximate the sampling distribution of $\bar{x}$ for SRSs of size 50.

(b) Estimate the probability that the mean number of moths in a sample of size 50 is greater than or equal to 0.6.

(c) In a recent month, the mean number of moths in an SRS of size 50 was $\bar{x} = 0.6$. Based on this result, is there convincing evidence that the moth population is getting larger in this state? Explain your reasoning.

R7.8 American-made cars Nathan and Kyle both work for the Department of Motor Vehicles (DMV), but they live in different states. In Nathan's state, 80% of the registered cars are made by American manufacturers. In Kyle's state, only 60% of the registered cars are made by American manufacturers. Nathan selects a random sample of 100 cars in his state and Kyle selects a random sample of 70 cars in his state. Let $\hat{p}_N - \hat{p}_K$ be the difference (Nathan's state − Kyle's state) in the sample proportion of cars made by American manufacturers.

(a) What is the shape of the sampling distribution of $\hat{p}_N - \hat{p}_K$? Why?

(b) Find the mean of the sampling distribution.

(c) Calculate and interpret the standard deviation of the sampling distribution. Verify that the 10% condition is met.

(d) What is the probability that the proportion in the sample from Kyle's state exceeds the proportion in the sample from Nathan's state?

R7.9 Candles A company produces candles. Machine 1 makes candles with a mean length of 15 centimeters and a standard deviation of 0.15 centimeter. Machine 2 makes candles with a mean length of 15 centimeters and a standard deviation of 0.10 centimeter. A random sample of 49 candles is taken from each machine. Let $\bar{x}_1 - \bar{x}_2$ be the difference (Machine 1 − Machine 2) in the sample mean length of candles. Describe the shape, center, and variability of the sampling distribution of $\bar{x}_1 - \bar{x}_2$.

(b) The mean of the sampling distribution of $\bar{x}$ is $\mu_{\bar{x}} = \mu = 100$. Because the sample of size 60 is less than 10% of all adults, the standard deviation of the sampling distribution of $\bar{x}$ is

$$\sigma_{\bar{x}} = \frac{\sigma}{\sqrt{n}} = \frac{15}{\sqrt{60}} = 1.9365.$$ The mean WAIS score for a random sample of 60 adults will typically vary from the mean (100) by about 1.9365 points. **(c)** We want to find $P(\bar{x} \geq 105)$.

(i) $z = \dfrac{105 - 100}{1.9365} = 2.58$;

$P(z \geq 2.58) = 1 - 0.9951 = 0.0049$

(ii) $P(\bar{x} \geq 105) =$ normalcdf(lower: 105, upper: 1000, mean: 100, SD: 1.9365) = 0.0049

There is a 0.0049 probability of selecting a sample of 60 adults whose mean WAIS score is at least 105.

(d) The answer to part (a) can differ greatly, depending on the shape of the population distribution. The answer to part (b) would be the same because the mean and standard deviation do not depend on the shape of the population distribution. Because of the large sample size ($60 \geq 30$), the answer we gave for part (c) would still be fairly reliable due to the CLT.

R7.7 (a) Because the sample size is large ($n = 50 \geq 30$), the CLT says that the distribution of $\bar{x}$ will be approximately Normal. **(b)** $\mu_{\bar{x}} = \mu = 0.5$; because 50 is less than 10% of all traps,

$$\sigma_{\bar{x}} = \frac{\sigma}{\sqrt{n}} = \frac{0.7}{\sqrt{50}} = 0.0990.$$ We want to find $P(\bar{x} \geq 0.6)$.

Chapter 7 AP® Statistics Practice Test

Section I: Multiple Choice *Select the best answer for each question.*

T7.1 A study of voting chose 663 registered voters at random shortly after an election. Of these, 72% said they had voted in the election. Election records show that only 56% of registered voters voted in the election. Which of the following statements is true?

(a) 72% is a sample; 56% is a population.

(b) 72% and 56% are both statistics.

(c) 72% is a statistic and 56% is a parameter.

(d) 72% is a parameter and 56% is a statistic.

(e) 72% and 56% are both parameters.

T7.2 The Gallup Poll has decided to increase the size of its random sample of voters from about 1500 people to about 4000 people right before an election. The poll is designed to estimate the proportion of voters who favor a new law banning smoking in public buildings. The effect of this increase is to

(a) reduce the bias of the estimate.

(b) increase the bias of the estimate.

(c) reduce the variability of the estimate.

(d) increase the variability of the estimate.

(e) reduce the bias and variability of the estimate.

T7.3 Suppose we select an SRS of size $n = 100$ from a large population having proportion p of successes. Let $\hat{p}$ be the proportion of successes in the sample. For which value of p would it be safe to use the Normal approximation to the sampling distribution of $\hat{p}$?

(a) 0.01   (d) 0.975

(b) 0.09   (e) 0.999

(c) 0.85

T7.4 The central limit theorem is important in statistics because it allows us to use a Normal distribution to find probabilities involving the sample mean if the

(a) sample size is sufficiently large (for any population).

(b) population is Normally distributed (for any sample size).

(c) population is Normally distributed and the sample size is reasonably large.

(d) population is Normally distributed and the population standard deviation is known (for any sample size).

(e) population size is reasonably large (whether the population distribution is known or not).

T7.5 The number of undergraduates at Johns Hopkins University is approximately 2000, while the number at Ohio State University is approximately 60,000. At both schools, a simple random sample of about 3% of the undergraduates is taken. Each sample is used to estimate the

proportion p of all students at that university who own an iPod. Suppose that, in fact, $p = 0.80$ at both schools. Which of the following is the best conclusion?

(a) We expect that the estimate from Johns Hopkins will be closer to the truth than the estimate from Ohio State because it comes from a smaller population.

(b) We expect that the estimate from Johns Hopkins will be closer to the truth than the estimate from Ohio State because it is based on a smaller sample size.

(c) We expect that the estimate from Ohio State will be closer to the truth than the estimate from Johns Hopkins because it comes from a larger population.

(d) We expect that the estimate from Ohio State will be closer to the truth than the estimate from Johns Hopkins because it is based on a larger sample size.

(e) We expect that the estimate from Johns Hopkins will be about the same distance from the truth as the estimate from Ohio State because both samples are 3% of their populations.

T7.6 A researcher initially plans to take an SRS of size 160 from a certain population and calculate the sample mean $\bar{x}$. Later, the researcher decides to increase the sample size so that the standard deviation of the sampling distribution of $\bar{x}$ will be half as big as when using a sample size of 160. What sample size should the researcher use?

(a) 40   (d) 640

(b) 80   (e) There is not enough information to

(c) 320   determine the sample size.

T7.7 The student newspaper at a large university asks an SRS of 250 undergraduates, "Do you favor eliminating the carnival from the term-end celebration?" All in all, 150 of the 250 are in favor. Suppose that (unknown to you) 55% of all undergraduates favor eliminating the carnival. If you took a very large number of SRSs of size $n = 250$ from this population, the sampling distribution of the sample proportion $\hat{p}$ would be

(a) exactly Normal with mean 0.55 and standard deviation 0.03.

(b) approximately Normal with mean 0.55 and standard deviation 0.03.

(c) exactly Normal with mean 0.60 and standard deviation 0.03.

(d) approximately Normal with mean 0.60 and standard deviation 0.03.

(e) heavily skewed with mean 0.55 and standard deviation 0.03.

(ii) $P(\hat{p}_N - \hat{p}_K < 0) =$ normalcdf (lower: −1000, upper: 0, mean: 0.2, SD: 0.071) = 0.0024

There is a 0.0024 probability that the proportion of registered cars that are made by American manufacturers in Kyle's state exceeds the proportion of registered cars that are made by American manufacturers in Nathan's state.

R7.9 *Shape*: Because $n_1 = 49 \geq 30$ and $n_2 = 49 \geq 30$, the shape of the distribution of $\bar{x}_1 - \bar{x}_2$ is approximately Normal. *Center*: $\mu_{\bar{x}_1 - \bar{x}_2} = 15 - 15 = 0$ cm *Variability*: Because $49 < 10\%$ of all candles made by machine 1 and $49 < 10\%$ of all candles made by machine 2,

$$\sigma_{\bar{x}_1 - \bar{x}_2} = \sqrt{\frac{0.15^2}{49} + \frac{0.10^2}{49}} = 0.026 \text{ cm}.$$

TRM Full Solutions to Chapter 7 AP® Statistics Practice Test

The full solutions can be found by clicking on the link in the TE-Book, opening the TRFD, or downloading from the Teacher's Resources on the book's digital platform.

Answers to Chapter 7 AP® Statistics Practice Test

T7.1 c

T7.2 c

T7.3 c

T7.4 a

T7.5 d

T7.6 d

T7.7 b

(i) $z = \dfrac{0.6 - 0.5}{0.0990} = 1.01$;

$P(z \geq 1.01) = 1 - 0.8438 = 0.1562$

(ii) $P(\bar{x} \geq 0.6) =$ normalcdf(lower: 0.6, upper: 1000, mean: 0.5, SD: 0.0990) = 0.1562 There is a 0.1562 probability that the mean number of moths is greater than or equal to 0.6.

(c) No; because this probability is not small, it is plausible that the sample mean number of moths is this high by chance alone. There is not convincing evidence that the moth population is increasing.

R7.8 (a) Approximately Normal because $n_N p_N = 100(0.80) = 80$, $n_N(1 - p_N) = 100(0.2) = 20$, $n_K p_K = 70(0.6) = 42$, and $n_K(1 - p_K) = 70(0.4) = 28$ are all ≥ 10.

(b) $\mu_{\hat{p}_N - \hat{p}_K} = 0.80 - 0.60 = 0.20$

(c) Because $100 < 10\%$ of all registered cars in Nathan's state and $70 < 10\%$ of all registered cars in Kyle's state,

$$\sigma_{\hat{p}_N - \hat{p}_K} = \sqrt{\frac{0.8(0.2)}{100} + \frac{0.6(0.4)}{70}} = 0.071.$$

The difference (Nathan − Kyle) in the sample proportions of registered cars that are made by American manufacturers typically varies by about 0.071 from the true difference in proportions of 0.20.

(d) We want to find $P(\hat{p}_N - \hat{p}_K < 0)$.

(i) $z = \dfrac{0 - 0.2}{0.071} = -2.82$

$P(z < -2.82) = 0.0024$

T7.8 e

T7.9 b

T7.10 e

T7.11 c

T7.12 Sample statistic A provides the best estimate of the parameter. Both statistics A and B appear to be unbiased, while statistic C appears to be biased because the center of its sampling distribution is smaller than the value of the parameter. In addition, statistic A has less variability than statistic B. In this situation, we want low bias and small variability, so statistic A is the best choice.

T7.13 (a) The probability that a single household pays more than $55 cannot be calculated, because we do not know the shape of the population distribution of monthly fees. **(b)** $\mu_{\bar{x}} = \mu = \$50$; because the sample of size 500 is less than 10% of all households with Internet access, $\sigma_{\bar{x}} = \dfrac{\sigma}{\sqrt{n}} = \dfrac{20}{\sqrt{50}} = \$2.828$.

(c) Because the sample size is large ($n = 50 \geq 30$), the distribution of $\bar{x}$ will be approximately Normal. **(d)** We want to find $P(\bar{x} > 55)$.

(i) $z = \dfrac{55 - 50}{2.828} = 1.77$; $P(z > 1.77) = 1 - 0.9616 = 0.0384$

(ii) $P(\bar{x} > 55) =$ normalcdf(lower: 55, upper: 1000, mean: 50, SD: 2.828) = 0.0385

There is a 0.0385 probability that the mean monthly fee paid by the sample of 50 households exceeds $55.

T7.8 Which of the following statements about the sampling distribution of the sample mean is *incorrect*?

(a) The standard deviation of the sampling distribution will decrease as the sample size increases.

(b) The standard deviation of the sampling distribution measures how far the sample mean typically varies from the population mean.

(c) The sample mean is an unbiased estimator of the population mean.

(d) The sampling distribution shows how the sample mean is distributed around the population mean.

(e) The sampling distribution shows how the sample is distributed around the sample mean.

T7.9 A newborn baby has extremely low birth weight (ELBW) if it weighs less than 1000 grams. A study of the health of such children in later years examined a random sample of 219 children. Their mean weight at birth was $\bar{x} = 810$ grams. This sample mean is an *unbiased estimator* of the mean weight μ in the population of all ELBW babies, which means that

(a) in all possible samples of size 219 from this population, the mean of the values of $\bar{x}$ will equal 810.

(b) in all possible samples of size 219 from this population, the mean of the values of $\bar{x}$ will equal μ.

(c) as we take larger and larger samples from this population, $\bar{x}$ will get closer and closer to μ.

(d) in all possible samples of size 219 from this population, the values of $\bar{x}$ will have a distribution that is close to Normal.

(e) the person measuring the children's weights does so without any error.

T7.10 Suppose that you are a student aide in the library and agree to be paid according to the "random pay" system. Each week, the librarian flips a coin. If the coin comes up heads, your pay for the week is $80. If it comes up tails, your pay for the week is $40. You work for the library for 100 weeks. Suppose we choose an SRS of 2 weeks and calculate your average earnings $\bar{x}$. The shape of the sampling distribution of $\bar{x}$ will be

(a) Normal.

(b) approximately Normal.

(c) right-skewed.

(d) left-skewed.

(e) symmetric but not Normal.

T7.11 An SRS of size 100 is taken from Population A with proportion 0.8 of successes. An independent SRS of size 400 is taken from Population B with proportion 0.5 of successes. The sampling distribution of the difference (A − B) in sample proportions has what mean and standard deviation?

(a) mean = 0.3; standard deviation = 1.3

(b) mean = 0.3; standard deviation = 0.40

(c) mean = 0.3; standard deviation = 0.047

(d) mean = 0.3; standard deviation = 0.0022

(e) mean = 0.3; standard deviation = 0.0002

Section II: Free Response *Show all your work. Indicate clearly the methods you use, because you will be graded on the correctness of your methods as well as on the accuracy and completeness of your results and explanations.*

T7.12 Here are histograms of the values taken by three sample statistics in several hundred samples from the same population. The true value of the population parameter is marked with an arrow on each histogram.

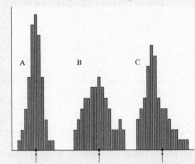

Which statistic would provide the best estimate of the parameter? Justify your answer.

T7.13 The amount that households pay service providers for access to the Internet varies quite a bit, but the mean monthly fee is $50 and the standard deviation is $20. The distribution is not Normal: many households pay a low rate as part of a bundle with phone or television service, but some pay much more for Internet only or for faster connections.[13] A sample survey asks an SRS of 50 households with Internet access how much they pay. Let $\bar{x}$ be the mean amount paid.

(a) Explain why you can't determine the probability that the amount a randomly selected household pays for access to the Internet exceeds $55.

(b) What are the mean and standard deviation of the sampling distribution of $\bar{x}$?

(c) What is the shape of the sampling distribution of $\bar{x}$? Justify your answer.

(d) Find the probability that the average fee paid by the sample of households exceeds $55.

T7.14 According to government data, 22% of American children under the age of 6 live in households with incomes less than the official poverty level. A study of learning in early childhood chooses an SRS of 300 children from one state and finds that $\hat{p} = 0.29$.

(a) Find the probability that at least 29% of the sample are from poverty-level households, assuming that 22% of all children under the age of 6 in this state live in poverty-level households.

(b) Based on your answer to part (a), is there convincing evidence that the percentage of children under the age of 6 living in households with incomes less than the official poverty level in this state is greater than the national value of 22%? Explain your reasoning.

T7.15 In a children's book, the mean word length is 3.7 letters with a standard deviation of 2.1 letters. In a novel aimed at teenagers, the mean word length is 4.3 letters with a standard deviation of 2.5 letters. Both distributions of word length are unimodal and skewed to the right. Independent random samples of 35 words are selected from each book. Let $\bar{x}_C$ represent the sample mean word length in the children's book and let $\bar{x}_T$ represent the sample mean word length in the teen novel.

(a) Find the mean of the sampling distribution of $\bar{x}_C - \bar{x}_T$.

(b) Calculate and interpret the standard deviation of the sampling distribution. Verify that the 10% condition is met.

(c) Justify that the shape of the sampling distribution is approximately Normal.

(d) What is the probability that the sample mean word length is greater in the sample from the children's book than in the sample from the teen novel?

word. **(c)** Approximately Normal because $n_C = 35 > 30$ and $n_T = 35 > 30$.

(d) We want to find $P(\bar{x}_C - \bar{x}_T > 0)$.

(i) $z = \dfrac{0 - (-0.6)}{0.552} = 1.09$

$P(z > 1.09) = 0.1379$

(ii) $P(\bar{x}_C - \bar{x}_T > 0) = $ normalcdf(lower: 0, upper: 1000, mean: -0.6, SD: 0.552) = 0.1385

There is a 0.1385 probability that the sample mean word length is greater in the sample from the children's book than in the sample from the teen novel.

Cumulative AP® Practice Test 2

Section I: Multiple Choice *Choose the best answer for each question.*

AP2.1 The five-number summary for a data set is given by min = 5, $Q_1 = 18$, median = 20, $Q_3 = 40$, max = 75. If you wanted to construct a boxplot for the data set that would show outliers, if any existed, what would be the maximum possible length of the right-side "whisker"?

(a) 33
(b) 35
(c) 45
(d) 53
(e) 55

AP2.2 The probability distribution for the number of heads in four tosses of a coin is given by

Number of heads	0	1	2	3	4
Probability	0.0625	0.2500	0.3750	0.2500	0.0625

The probability of getting at least one *tail* in four tosses of a coin is

(a) 0.2500.
(b) 0.3125.
(c) 0.6875.
(d) 0.9375.
(e) 0.0625.

AP2.3 In a certain large population of adults, the distribution of IQ scores is strongly left-skewed with a mean of 122 and a standard deviation of 5. Suppose 200 adults are randomly selected from this population for a market research study. For SRSs of size 200, the distribution of sample mean IQ score is

(a) left-skewed with mean 122 and standard deviation 0.35.
(b) exactly Normal with mean 122 and standard deviation 5.
(c) exactly Normal with mean 122 and standard deviation 0.35.
(d) approximately Normal with mean 122 and standard deviation 5.
(e) approximately Normal with mean 122 and standard deviation 0.35.

AP2.4 A 10-question multiple-choice exam offers 5 choices for each question. Jason just guesses the answers, so he has probability 1/5 of getting any one answer correct. You want to perform a simulation to determine the number of correct answers that Jason gets. What would be a proper way to use a table of random digits to do this?

(a) One digit from the random digit table simulates one answer, with 5 = correct and all other digits = incorrect. Ten digits from the table simulate 10 answers.

(b) One digit from the random digit table simulates one answer, with 0 or 1 = correct and all other

---

Answers to Cumulative AP® Practice Test 2

AP2.1 a

AP2.2 d

AP2.3 e

AP2.4 b

T7.14 (a) $\mu_{\hat{p}} = p = 0.22$; because 300 is less than 10% of children under the age of 6, $\sigma_{\hat{p}} = \sqrt{\dfrac{0.22(0.78)}{300}} = 0.0239$. Because $np = 300(0.22) = 66$ and $n(1 - p) = 300(0.78) = 234$ are both at least 10, the sampling distribution of $\hat{p}$ can be approximated by a Normal distribution. We want to find $P(\hat{p} > 0.29)$.

(i) $z = \dfrac{0.29 - 0.22}{0.0239} = 2.93$; $P(z \geq 2.93) = 1 - 0.9983 = 0.0017$

(ii) $P(\hat{p} > 0.29) = $ normalcdf (lower: 0.29, upper: 1000, mean: 0.22, SD: 0.0239) = 0.0017

There is a 0.0017 probability that more than 29% of the sample are from poverty-level households.

(b) Because it is unlikely to get a sample proportion of 29% or greater by chance alone, there is convincing evidence that the percentage of children under the age of 6 living in households with incomes less than the official poverty level in this state is greater than the national value of 22%.

T7.15 (a) $\mu_{\bar{x}_C - \bar{x}_T} = 3.7 - 4.3 = -0.6$ words **(b)** Because $35 < 10\%$ of all words in the children's book and $35 < 10\%$ of all words in the novel aimed at teenagers,

$\sigma_{\bar{x}_C - \bar{x}_T} = \sqrt{\dfrac{2.1^2}{35} + \dfrac{2.5^2}{35}} = 0.552$ words. The difference (Child − Teen) in the sample mean word lengths typically varies by about 0.552 word from the true difference in means of −0.6

AP2.5 c

AP2.6 d

AP2.7 c

AP2.8 a

AP2.9 d

digits = incorrect. Ten digits from the table simulate 10 answers.

(c) One digit from the random digit table simulates one answer, with odd = correct and even = incorrect. Ten digits from the table simulate 10 answers.

(d) One digit from the random digit table simulates one answer, with 0 or 1 = correct and all other digits = incorrect, ignoring repeats. Ten digits from the table simulate 10 answers.

(e) Two digits from the random digit table simulate one answer, with 00 to 20 = correct and 21 to 99 = incorrect. Ten pairs of digits from the table simulate 10 answers.

AP2.5 Suppose we roll a fair die four times. What is the probability that a 6 occurs on exactly one of the rolls?

(a) $4\left(\frac{1}{6}\right)^3\left(\frac{5}{6}\right)^1$

(b) $\left(\frac{1}{6}\right)^3\left(\frac{5}{6}\right)^1$

(c) $4\left(\frac{1}{6}\right)^1\left(\frac{5}{6}\right)^3$

(d) $\left(\frac{1}{6}\right)^1\left(\frac{5}{6}\right)^3$

(e) $6\left(\frac{1}{6}\right)^1\left(\frac{5}{6}\right)^3$

AP2.6 On one episode of his show, a radio show host encouraged his listeners to visit his website and vote in a poll about proposed tax increases. Of the 4821 people who vote, 4277 are against the proposed increases. To which of the following populations should the results of this poll be generalized?

(a) All people who have ever listened to this show

(b) All people who listened to this episode of the show

(c) All people who visited the show host's website

(d) All people who voted in the poll

(e) All people who voted against the proposed increases

AP2.7 The number of unbroken charcoal briquets in a 20-pound bag filled at the factory follows a Normal distribution with a mean of 450 briquets and a standard deviation of 20 briquets. The company expects that a certain number of the bags will be underfilled, so the company will replace for free the 5% of bags that have too few briquets. What is the minimum number of unbroken briquets the bag would have to contain for the company to avoid having to replace the bag for free?

(a) 404

(b) 411

(c) 418

(d) 425

(e) 448

AP2.8 You work for an advertising agency that is preparing a new television commercial to appeal to women. You have been asked to design an experiment to compare the effectiveness of three versions of the commercial. Each subject will be shown one of the three versions and then asked to reveal her attitude toward the product. You think there may be large differences in the responses of women who are employed and those who are not. Because of these differences, you should use

(a) a block design, but not a matched pairs design.

(b) a completely randomized design.

(c) a matched pairs design.

(d) a simple random sample.

(e) a stratified random sample.

AP2.9 Suppose that you have torn a tendon and are facing surgery to repair it. The orthopedic surgeon explains the risks to you. Infection occurs in 3% of such operations, the repair fails in 14%, and both infection and failure occur together 1% of the time. What is the probability that the operation is successful for someone who has an operation that is free from infection?

(a) 0.8342

(b) 0.8400

(c) 0.8600

(d) 0.8660

(e) 0.9900

AP2.10 Social scientists are interested in the association between high school graduation rate (HSGR, measured as a percent) and the percent of U.S. families living in poverty (POV). Data were collected from all 50 states and the District of Columbia, and a regression analysis was conducted. The resulting least-squares regression line is given by $\overline{POV} = 59.2 - 0.620(HSGR)$ with $r^2 = 0.802$. Based on the information, which of the following is the best interpretation for the slope of the least-squares regression line?

(a) For each 1% increase in the graduation rate, the percent of families living in poverty is predicted to decrease by approximately 0.896.

(b) For each 1% increase in the graduation rate, the percent of families living in poverty is predicted to decrease by approximately 0.802.

(c) For each 1% increase in the graduation rate, the percent of families living in poverty is predicted to decrease by approximately 0.620.

(d) For each 1% increase in the percent of families living in poverty, the graduation rate is predicted to decrease by approximately 0.802.

(e) For each 1% increase in the percent of families living in poverty, the graduation rate is predicted to decrease by approximately 0.620.

Questions AP2.11–AP2.13 refer to the following graph. Here is a dotplot of the adult literacy rates in 177 countries in a recent year, according to the United Nations. For example, the lowest literacy rate was 23.6%, in the African country of Burkina Faso. Mali had the next lowest literacy rate at 24.0%.

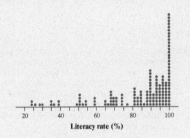

Literacy rate (%)

AP2.11 The overall shape of this distribution is

(a) clearly skewed to the right.

(b) clearly skewed to the left.

(c) roughly symmetric.

(d) approximately uniform.

(e) There is no clear shape.

AP2.12 The mean of this distribution (*don't* try to find it) will be

(a) very close to the median.

(b) greater than the median.

(c) less than the median.

(d) You can't say, because the distribution isn't symmetric.

(e) You can't say, because the distribution isn't Normal.

AP2.13 The country with a literacy rate of 49% is closest to which of the following percentiles?

(a) 6th

(b) 11th

(c) 28th

(d) 49th

(e) There is not enough information to calculate the percentile.

AP2.14 The correlation between the age and height of children under the age of 12 is found to be $r = 0.60$. Suppose we use the age x of a child to predict the height y of the child. What can we conclude?

(a) The height is generally 60% of a child's age.

(b) About 60% of the time, age will accurately predict height.

(c) Thirty-six percent of the variation in height is accounted for by the linear model relating height to age.

(d) For every 1 year older a child is, the regression line predicts an increase of 0.6 foot in height.

(e) Thirty-six percent of the time, the least-squares regression line accurately predicts height from age.

AP2.15 An agronomist wants to test three different types of fertilizer (A, B, and C) on the yield of a new variety of wheat. The yield will be measured in bushels per acre. Six 1-acre plots of land were randomly assigned to each of the three fertilizers. The treatment, experimental unit, and response variable are, respectively,

(a) a specific fertilizer, bushels per acre, a plot of land.

(b) variety of wheat, bushels per acre, a specific fertilizer.

(c) variety of wheat, a plot of land, wheat yield.

(d) a specific fertilizer, a plot of land, wheat yield.

(e) a specific fertilizer, the agronomist, wheat yield.

AP2.16 According to the U.S. Census, the proportion of adults in a certain county who owned their own home was 0.71. An SRS of 100 adults in a certain section of the county found that 65 owned their home. Which one of the following represents the approximate probability of obtaining a sample of 100 adults in which 65 or fewer own their home, assuming that this section of the county has the same overall proportion of adults who own their home as does the entire county?

AP2.10 c

AP2.11 b

AP2.12 c

AP2.13 a

AP2.14 c

AP2.15 d

AP2.16 c

AP2.17 e

AP2.18 a

AP2.19 c

AP2.20 b

(a) $\binom{100}{65}(0.71)^{65}(0.29)^{35}$

(b) $\binom{100}{65}(0.29)^{65}(0.71)^{35}$

(c) $P\left(z \le \dfrac{0.65 - 0.71}{\sqrt{\dfrac{(0.71)(0.29)}{100}}}\right)$

(d) $P\left(z \le \dfrac{0.65 - 0.71}{\sqrt{\dfrac{(0.65)(0.35)}{100}}}\right)$

(e) $P\left(z \le \dfrac{0.65 - 0.71}{\dfrac{(0.71)(0.29)}{\sqrt{100}}}\right)$

AP2.17 Which one of the following would be a correct interpretation if you have a z-score of $+2.0$ on an exam?

(a) It means that you missed two questions on the exam.

(b) It means that you got twice as many questions correct as the average student.

(c) It means that your grade was 2 points higher than the mean grade on this exam.

(d) It means that your grade was in the upper 2% of all grades on this exam.

(e) It means that your grade is 2 standard deviations above the mean for this exam.

AP2.18 Records from a dairy farm yielded the following information on the number of male and female calves born at various times of the day.

		Time of day			
		Day	Evening	Night	Total
Gender	Males	129	15	117	261
	Females	118	18	116	252
	Total	247	33	233	513

What is the probability that a randomly selected calf was born in the night or was a female?

(a) $\dfrac{369}{513}$

(b) $\dfrac{485}{513}$

(c) $\dfrac{116}{513}$

(d) $\dfrac{116}{252}$

(e) $\dfrac{116}{233}$

AP2.19 When people order books from a popular online source, they are shipped in boxes. Suppose that the mean weight of the boxes is 1.5 pounds with a standard deviation of 0.3 pound, the mean weight of the packing material is 0.5 pound with a standard deviation of 0.1 pound, and the mean weight of the books shipped is 12 pounds with a standard deviation of 3 pounds. Assuming that the weights are independent, what is the standard deviation of the total weight of the boxes that are shipped from this source?

(a) 1.84

(b) 2.60

(c) 3.02

(d) 3.40

(e) 9.10

AP2.20 A grocery chain runs a prize game by giving each customer a ticket that may win a prize when the box is scratched off. Printed on the ticket is a dollar value ($500, $100, $25) or the statement "This ticket is not a winner." Monetary prizes can be redeemed for groceries at the store. Here is the probability distribution of the amount won on a randomly selected ticket:

Amount won	$500	$100	$25	$0
Probability	0.01	0.05	0.20	0.74

Which of the following are the mean and standard deviation, respectively, of the winnings?

(a) $15.00, $2900.00

(b) $15.00, $53.85

(c) $15.00, $26.93

(d) $156.25, $53.85

(e) $156.25, $26.93

AP2.21 A large company is interested in improving the efficiency of its customer service and decides to examine the length of the business phone calls made to clients by its sales staff. Here is a cumulative relative frequency graph from data collected over the past year. According to the graph, the shortest 80% of calls will take how long to complete?

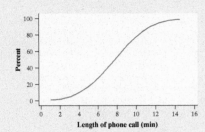

(a) Less than 10 minutes
(b) At least 10 minutes
(c) Exactly 10 minutes
(d) At least 5.5 minutes
(e) Less than 5.5 minutes

Section II: Free Response *Show all your work. Indicate clearly the methods you use, because you will be graded on the correctness of your methods as well as on the accuracy and completeness of your results and explanations.*

AP2.22 A health worker is interested in determining if omega-3 fish oil can help reduce cholesterol in adults. She obtains permission to examine the health records of 200 people in a large medical clinic and classifies them according to whether or not they take omega-3 fish oil. She also obtains their latest cholesterol readings and finds that the mean cholesterol reading for those who are taking omega-3 fish oil is 18 points less than the mean for the group not taking omega-3 fish oil.

(a) Is this an observational study or an experiment? Justify your answer.

(b) Explain the concept of confounding in the context of this study and give one example of a variable that could be confounded with whether or not people take omega-3 fish oil.

(c) Researchers find that the 18-point difference in the mean cholesterol readings of the two groups is statistically significant. Can they conclude that omega-3 fish oil is the cause? Why or why not?

AP2.23 The scatterplot shows the relationship between the number of yards allowed by teams in the National Football League and the number of wins for that team in a recent season, along with the least-squares regression line. Computer output is also provided.

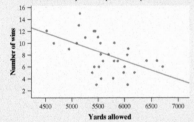

Term	Coef	SE Coef	T-Value	P-Value
Constant	25.66	5.37	4.78	0.000
Yards_allowed	−0.003131	0.000948	−3.30	0.002

S = 2.65358   R-Sq = 26.65%   R-Sq(adj) = 24.21%

(a) State the equation of the least-squares regression line. Define any variables you use.

(b) Calculate and interpret the residual for the Seattle Seahawks, who allowed 4668 yards and won 10 games.

(c) The Carolina Panthers allowed 5167 yards and won 15 games. What effect does the point representing the Panthers have on the equation of the least-squares regression line? Explain.

AP2.24 Every 17 years, swarms of cicadas emerge from the ground in the eastern United States, live for about six weeks, and then die. (There are several different "broods," so we experience cicada eruptions more often than every 17 years.) There are so many cicadas that their dead bodies can serve as fertilizer and increase plant growth. In a study, a researcher added 10 dead cicadas under 39 randomly selected plants in a natural plot of American bellflowers on the forest floor, leaving other plants undisturbed. One of the response variables measured was the size of seeds produced by the plants. Here are the boxplots and summary statistics of seed mass (in milligrams) for 39 cicada plants and 33 undisturbed (control) plants:

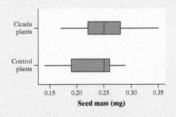

	n	Minimum	Q_1	Median	Q_3	Maximum
Cicada plants	39	0.17	0.22	0.25	0.28	0.35
Control plants	33	0.14	0.19	0.25	0.26	0.29

AP2.21 a

AP2.22 (a) This is an observational study. Subjects were not assigned to take (or not take) fish oil. **(b)** Two variables are confounded when their effects on the cholesterol level cannot be distinguished from one another. For example, people who take omega-3 fish oil might also be more health conscious in general and do other things such as eat more healthfully or exercise more. If eating more healthfully or exercising more lowers cholesterol, researchers would not know whether it was the omega-3 fish oil or the more healthy food consumption or exercise that lowered cholesterol. **(c)** No; this wasn't an experiment, and taking fish oil is possibly confounded with other good habits, such as healthful eating and exercise.

AP2.23 (a) $\hat{y} = 25.66 - 0.003131x$, where $\hat{y}$ is the predicted number of wins and x is the number of yards allowed. **(b)** $\hat{y} = 25.66 - 0.003131(4668) = 11.04$ wins; the residual $= 10 - 11.04 = -1.04$ wins. The actual number of Seattle Seahawks wins was 1.04 less than the number of wins predicted by the regression line with $x = 4668$ yards allowed. **(c)** Because the Carolina Panthers allowed fewer yards than average and also had more wins than average, this point will increase the steepness of the negative slope of the least-squares regression line (make it more negative) and increase the y intercept of the least-squares regression line.

AP2.24 (a) The distribution of seed mass for the cicada plants is roughly symmetric, whereas the distribution of seed mass for the control plants is skewed to the left. Neither group had any outliers. The median seed mass (0.25) is the same for both groups. The cicada plants had a larger range in seed mass, but the control plants had a larger *IQR*. **(b)** The distribution of seed mass for the cicada plants is roughly symmetric, which suggests that the mean should be about the same as the median. However, the distribution of seed mass for the control plants is skewed to the left, which will pull the mean of this distribution below its median toward the smaller values. Because the medians of both distributions are equal, the mean for the cicada plants is likely greater than the mean for the control plants. **(c)** The purpose of the random assignment is to create two groups of plants that are roughly equivalent at the beginning of the experiment. **(d)** A benefit of using only American bellflowers is that the researchers may then control a source of variability. Different types of flowers will have different seed masses, making the response more variable if other types of plants were used. A drawback to using only American bellflowers is that we can't make inferences about the effect of cicadas on other types of plants, because other plants may respond differently to cicadas.

AP2.25 (a) Because the sample size is large ($n = 50 \geq 30$), the distribution of $\bar{x}$ is approximately Normal with $\mu_{\bar{x}} = \mu = 525$ pages. Because $n = 50$ is less than 10% of all novels in the library,

$$\sigma_{\bar{x}} = \frac{\sigma}{\sqrt{n}} = \frac{200}{\sqrt{50}} = 28.28 \text{ pages.}$$

We want to find $P(\bar{x} < 500)$.

(i) $z = \dfrac{500 - 525}{28.28} = -0.88$;

$P(z < -0.88) = 0.1894$

(ii) $P(\bar{x} < 500) = \text{normalcdf(lower: } -1000,$ upper: 500, mean: 525, SD: 28.28) = 0.1883

There is a 0.1883 probability that the average number of pages in the sample is less than 500.

(b) X is a binomial random variable with $n = 50$ and $p = 0.30$. We want to find $P(X \geq 20)$. *Tech:* $P(X \geq 20) = 1 - \text{binomcdf(trials: 50,}$ *p:* 0.30, *x* value: 19) = $1 - 0.9152 = 0.0848$. There is a 0.0848 probability of selecting at least 20 novels that have fewer than 400 pages.

(a) Write a few sentences comparing the distributions of seed mass for the two groups of plants.

(b) Based on the graphical displays, which distribution likely has the larger mean? Justify your answer.

(c) Explain the purpose of the random assignment in this study.

(d) Name one benefit and one drawback of only using American bellflowers in the study.

AP2.25 In a city library, the mean number of pages in a novel is 525 with a standard deviation of 200. Furthermore, 30% of the novels have fewer than 400 pages. Suppose that you randomly select 50 novels from the library.

(a) What is the probability that the average number of pages in the sample is less than 500?

(b) What is the probability that at least 20 of the novels have fewer than 400 pages?

TRM **Chapter 7 Case Study**

The Case Study feature that was found in the previous two editions of TPS has been moved to the Teacher's Resource Materials. Click on the link in the TE-Book, open in the TRFD, or download from the Teacher's Resources on the book's digital platform.

Chapter 8

Chapter 8

Estimating Proportions with Confidence

PD Overview

Chapter 8 marks the beginning of our formal study of statistical inference. Starting in this chapter, we make a major transition from Chapter 7. When we studied sampling distributions in Chapter 7, we pretended to know the value of the population parameter and then asked questions about the distribution of a statistic. For example, "Suppose that 35% of all first-year students attend college within 50 miles of home. What is the probability that a random sample of 1500 students will give a result within 2 percentage points of this true value?" In Chapter 8, we no longer pretend to know the truth. Instead, we use information from a random sample or randomized experiment to create an estimate for the parameter.

In this chapter, we introduce the idea of a confidence interval—an interval of plausible values for a parameter. We also provide the specific details for constructing confidence intervals for a population proportion and confidence intervals for a difference of two proportions. In Chapter 10, we will introduce confidence intervals for a mean and a difference of two means, and later in Chapter 12 confidence intervals for the slope of a least-squares regression line. Fortunately, many of the big ideas of this chapter apply to all confidence intervals. This should make it easier to get through Chapters 9–12, assuming students have a solid understanding of the material in this chapter.

An excellent resource for you to learn more about inference is available for free at the AP® Central website: apcentral .collegeboard.org/pdf/statistics-inference.pdf?course=ap-statistics. A Google search for "AP® Statistics Special Focus: Inference" will reveal the materials that were originally developed for use in College Board workshops and were written by people familiar with the AP® Statistics course.

The Main Ideas

One of the challenges in teaching the AP® Statistics course is keeping students focused on the big picture, not just the details of each section. We outline the main ideas for the chapter here.

Chapter 8 Introduction

We strongly recommend starting this chapter with the "Beads" activity on page 536. It is a great way to get students thinking about estimating with confidence—how to use information from a sample to estimate the value of a population parameter. We also refer back to this activity throughout Section 8.1, so it will definitely help if your students are familiar with it.

Section 8.1 Confidence Intervals: The Basics

We start this section by introducing the idea of a point estimate—a single best guess for the value of a population parameter. Because of sampling variability, however, it is very unlikely that our point estimate will be correct. To increase our chances of getting a correct estimate, we prefer to use an interval of plausible values to estimate the value of the parameter rather than a single point estimate.

To create this interval of plausible values, we rely on our understanding of sampling distributions from Chapter 7. When the distribution of a statistic (such as $\bar{x}$ or $\hat{p}$) is approximately Normal, we know that in about 95% of possible samples, the value of the statistic will be within 2 standard deviations of the population parameter (such as μ or p). This implies that in about 95% of possible samples, the population parameter (μ or p) will be within 2 standard deviations of the statistic ($\bar{x}$ or $\hat{p}$) .

In this section, we have chosen to focus on the big picture and avoid details about the calculations of different types of confidence intervals. For example, we believe that using 2 as the critical value in this section allows students to focus on the most important concepts, such as interpreting a confidence interval, interpreting a confidence level, and making a decision about the value of a parameter. We always interpret a confidence interval as an interval of plausible values for a parameter. Likewise, we always interpret a confidence level as the long-run capture rate of the method we use to compute the interval. Make sure to do the activity on page 542 to reinforce this idea.

The basic structure of a confidence interval is: point estimate ± margin of error. The size of the margin of error is determined mainly by two factors. Increasing the sample size decreases the margin of error, making our intervals narrower. Increasing the confidence level, however, increases the margin of error, making our intervals wider. Finally, the margin of error only accounts for sampling variability—it does not account for other problems in the data collection process that can lead to bias, such as undercoverage or nonresponse.

Section 8.2 Estimating a Population Proportion

In this section, we focus exclusively on confidence intervals for population proportions. We discuss what conditions students should check when calculating a confidence interval for a population proportion. Students have to verify that the data come from a random sample from the population of interest (the Random condition), that the sample is a small fraction of the population when sampling without replacement (the 10% condition), and that the sample size is large enough so the distribution of $\hat{p}$ is approximately Normal (the Large Counts condition).

Although we used a 95% confidence level and a critical value of 2 in Section 8.1, these are not our only choices. In this section, we introduce other common confidence levels and show how to calculate the critical values for them.

The four-step process is ideally suited for confidence interval questions, and we ask students to use this process each time they calculate and interpret a confidence interval. Students should *state* the parameter they are estimating and at what confidence level, *plan* their work by naming the inference method they will use and checking the appropriate conditions, *do* the calculations, and make a *conclusion* in the context of the problem.

Finally, we consider how to choose an appropriate sample size when planning a study. The necessary sample size is based on the confidence level, the proportion of successes, and the desired margin of error.

Section 8.3 Estimating a Difference in Proportions

Here we focus on creating confidence intervals to estimate the difference between two proportions. We know that we can reliably model the sampling distribution of $\hat{p}_1 - \hat{p}_2$ (and perform inference about $p_1 - p_2$) when the Random, 10%, and Large Counts conditions are met.

A confidence interval for a difference between two proportions provides an interval of plausible values for the true difference in proportions. The logic of confidence intervals—including how to interpret a confidence interval, what conditions to check, and how to interpret a confidence level—is the same as it was in Section 8.2.

At the end of this section, we introduce inference for completely randomized experiments. Fortunately, the methods used for analyzing the difference in the proportions from two independent random samples work very well for analyzing the difference in the proportions from two groups in a completely randomized experiment.

Chapter 8: Resources
Teacher's Resource Materials

The following resources, identified by the **TRM** in the annotated student pages, can be found by clicking on the link in the Teacher's e-Book (TE-book), searching by category or chapter on the Teacher's Resource Flash Drive (TRFD), or logging into the book's digital platform highschool.bfwpub.com/updatedtps6e and searching the Teacher's Resources menu (teacher log-in required).

- Alternate Examples: one file per section
- Lecture Presentation Slides: one per section
- "Nerf gun" alternate activity
- Chapter 8 Learning Targets Grid
- How to Construct a Bead Sampler Paddle
- "Hershey's Kisses" alternate activity
- The Plus Four Interval for a Population Proportion
- "Good books" alternate activity
- FRAPPY! Materials
- Chapter 8 Case Study
- Complete solutions for the Check Your Understanding problems, section exercises, review exercises, and AP® practice test
- Quizzes: one per section
- Chapter 8 Test

Free Response Questions from Previous AP® Statistics Exams

Questions can be found on the AP® Central website: apcentral.collegeboard.org/courses/ap-statistics/exam.

Students should be able to answer all the free response questions listed below with material learned in this chapter. Questions that contain content from this chapter but also require content from later chapters are listed in the last chapter required to complete the entire question. This list will be updated after each AP® Statistics exam and will be posted to the Teacher's Resource section of the books' digital platform and to www.statsmedic.com/free-response-questions.

Year	#	Content
2018	2	• Calculate n for a one-sample z interval for a proportion • Explain how bias affects point estimate • Randomized response
2017	2	• One-sample z interval for a proportion • Using interval to estimate total cost
2016	5	• One-sample z interval for proportion • Large Counts condition • Why two-sample interval is not OK
2015	2	• Using confidence intervals to make decisions • Effect of quadrupling sample size on margin of error
2011B	5	• One-sample z interval for a proportion • Using a CI to assess a claim • Determining sample size
2011	6	• One-sample z interval for a proportion • Tree diagrams • Using information from a tree diagram to create a new confidence interval
2010	3	• Interpreting confidence level • Using confidence intervals to make decisions • Determining sample size (CI for a proportion)
2010B	4	• One-sample z interval for a population proportion • Effect of sampling without replacement
2009B	6	• Double blind • Two-sample z interval for difference of proportions • Using relative risk

Year	#	Content
2008	4	• Scatterplots • Describing association • Mean and SE of average of proportions
2006B	2	• Two-sample z interval for difference of proportions • Significant difference
2005	5	• Sources of bias in a survey • Determining sample size (CI for a proportion) • Stratified random sampling
2003	6	• Interpreting a graph • One-sample z interval for a population proportion • Using confidence intervals to make decisions
2003B	6	• One-sample z interval for a population proportion • Interpreting confidence level • Determining sample sizes for different subgroups (CI for a proportion)
2002B	4	• One-sample z interval for a population proportion • Interpreting confidence level • Using confidence intervals to make decisions
2000	6	• One-sample z interval for a population proportion • Combining Normal random variables • Independence • Anticipating patterns in a scatterplot

Applets

- The *Confidence Intervals for Proportions* applet at highschool.bfwpub.com/updatedtps6e allows students to investigate the success rate of confidence intervals for population proportions using different population proportions, confidence levels, and sample sizes.

- The *Simulating Confidence Intervals for Population Parameter* applet at www.rossmanchance.com/applets allows students to investigate the success rate of confidence intervals for population proportions under different circumstances.

Chapter 8

Chapter 8: Pacing Guide, Learning Targets, and Suggested Assignments

This pacing guide is based on a schedule with 110, 50-minute sessions before the AP® Statistics exam. If you have a different number of sessions before the AP® exam, you can modify the pacing guide to suit your needs. If you have additional time, consider incorporating quizzes, released AP® Statistics free response questions, or additional activities. See the Resources section above for suggestions.

The suggested homework assignments list odd-numbered exercises, whenever possible, so students can check their answers against the back-of-book answers. If you would rather students not have access to the answers while doing homework, adding 1 to the exercise numbers usually will do the trick, because the homework exercises typically are paired. For example, Exercises 1 and 2 will generally cover the same topics, but in different contexts. You may also choose to include the Recycle and Review questions at the end of each section, which review topics from previous sections or chapters.

If your school is using the digital platform that accompanies TPS6, you will find these assignments pre-built as online homework assignments for Chapter 8.

Day	Content	Learning Targets: Students will be able to . . .	Suggested Assignment (MC bold)
1	Chapter 8 Introduction, 8.1 The Idea of a Confidence Interval	• Identify an appropriate point estimator and calculate the value of a point estimate. • Interpret a confidence interval in context. • Determine the point estimate and margin of error from a confidence interval. • Use a confidence interval to make a decision about the value of a parameter.	1, 3, 5, 7, 9
2	8.1 Interpreting Confidence Level, What Affects the Margin of Error?	• Interpret a confidence level in context. • Describe how the sample size and confidence level affect the margin of error. • Explain how practical issues like nonresponse, undercoverage, and response bias can affect the interpretation of a confidence interval.	11, 15, 17, 19, 21, **23–26**
3	8.2 Constructing a Confidence Interval for p	• State and check the Random, 10%, and Large Counts conditions for constructing a confidence interval for a population proportion. • Determine the critical value for calculating a $C\%$ confidence interval for a population proportion using a table or technology.	29, 31, 35, 37, 39
4	8.2 Putting It All Together: The Four-Step Process, Choosing the Sample Size	• Construct and interpret a confidence interval for a population proportion. • Determine the sample size required to obtain a $C\%$ confidence interval for a population proportion with a specified margin of error.	41, 45, 49, **55–58**
5	8.3 Confidence Intervals for $p_1 - p_2$, Putting It All Together: Two-Sample z Interval for $p_1 - p_2$	• Determine whether the conditions are met for constructing a confidence interval about a difference between two proportions. • Construct and interpret a confidence interval for a difference between two proportions.	61, 65, 67, 71, **73–75**
6	Chapter 8 Review/FRAPPY!		Chapter 8 Review Exercises
7	Chapter 8 Test		

Chapter 8 Alignment to the College Board's Fall 2019 AP® Statistics Course Framework*

Relationship to College Board Units

Chapter 8 in this book covers Topics 6.1–6.3 and 6.8–6.9 in Unit 6 of the College Board Course Framework. Students will be ready to take the Personal Progress Check for Unit 6 once they have completed Chapters 8 and 9.

Big Ideas and Enduring Understandings

Chapter 8 develops these Big Ideas and related Enduring Understandings outlined in the Course Framework:

- **Big Idea 1: Variation and Distribution (EU: VAR 1):** The distribution of measures for individuals within a sample or population describes variation. The value of a statistic varies from sample to sample. How can we determine whether differences between measures represent random variation or meaningful distinctions? Statistical methods based on probabilistic reasoning provide the basis for shared understandings about variation and about the likelihood that variation between and among measures, samples, and populations is random or meaningful.
- **Big Idea 2: Patterns and Uncertainty (EU: UNC 4):** Statistical tools allow us to represent and describe patterns in data and to classify departures from patterns. Simulation and probabilistic reasoning allow us to anticipate patterns in data and to determine the likelihood of errors in inference.

Course Skills

Chapter 8 helps students to develop the skills identified in the Course Framework.

- **1: Selecting Statistical Models** (1.A, 1.D)
- **3: Using Probability and Simulation** (3.D)
- **4: Statistical Argumentation** (4.A, 4.B, 4.C, 4.D)

Learning Objectives and Essential Knowledge Statements

Section	Learning Objectives	Essential Knowledge Statements
8.1	UNC-4.C, UNC-4.F, UNC-4.G, UNC-4.H	UNC-4.C.2, UNC-4.F.1, UNC-4.F.2, UNC-4.F.3, UNC-4.F.4, UNC-4.G.1, UNC-4.H.2, UNC-4.H.3
8.2	VAR-1.H, UNC-4.A, UNC-4.B, UNC-4.C, UNC-4.D, UNC-4.E, UNC-4.H	VAR-1.H.1, UNC-4.A.1, UNC-4.B.1, UNC-4.B.2, UNC-4.C.1, UNC-4.C.3, UNC-4.C.4, UNC-4.D.1, UNC-4.D.2, UNC-4.E.1, UNC-4.H.1
8.3	UNC-4.I, UNC-4.J, UNC-4.K, UNC-4.L, UNC-4.M, UNC-4.N	UNC-4.I.1, UNC-4.J.1, UNC-4.K.1, UNC-4.L.1, UNC-4.M.1, UNC-4.M.2, UNC-4.N.1

A detailed alignment (The Nitty Gritty Guide) that can be sorted by Course Framework Unit, Topic, Learning Objective, Essential Knowledge Statement, or textbook section, is available on the TRFD and in the Teacher's Resources folder on Sapling Plus. **TRM**

*Should changes be made to the Course Framework in the future, an updated alignment will be placed on our AP® updates page at go.bfwpub.com/ap-course-updates.

Chapter 8

Estimating Proportions with Confidence

Bill Gozansky/Alamy

Teaching Tip

Unit 6 in the College Board Course Framework aligns to Chapters 8 and 9 in this book. Students will be ready to take the Personal Progress Check for Unit 6 once they have completed Chapter 9.

PD Chapter 8 Overview

To watch the video overview of Chapter 8 (for teachers), click on the link in the TE-Book, look on the TRFD, or download from the Teacher's Resources on the book's digital platform.

TRM Lecture Presentation Slides

If you are new to teaching AP® Statistics or are short on time when preparing for class, you may find the Lecture Presentation Slides to be helpful. Experienced AP® Teacher Doug Tyson has created one slide presentation per section. You may use them as is, modify them to fit your needs, or share them with students who miss class. Find them on the TRFD and in the Teacher's Resources on the book's digital platform.

Teaching Tip

Make students aware of the structure of this chapter. Offering a framework at the beginning of the chapter gives them an organized way of storing their learning.

8.1 Confidence Intervals: The Basics
8.2 Estimating a Population Proportion
8.3 Estimating a Difference in Proportions

We will use a parallel structure in Chapter 10 when we first estimate a population mean and then estimate a difference in means.

Preparing for Inference

End of preparation. It's time.

Making Connections

Inference includes confidence intervals and significance tests. A large part of the content from Chapters 1–7 is building knowledge that will be put together in Chapter 8 (confidence intervals for proportions) and Chapter 9 (significance tests for proportions). Each of these chapters will start out with inference about one sample and then move to inference about two samples (or two groups).

TRM Nerf Gun Activity

This alternate activity can be used to introduce students to the idea of a confidence interval. Find it in the Teacher's Resource Materials located in the TE-Book, on the TRFD, or in the Teacher's view on the book's digital platform. Thank you to Steven Malan for this one.

TRM **How to Construct a Bead Sampler Paddle**

This document outlines the process for making a bead sampling tool using pegboard. Find it in the Teacher's Resource Materials located in the TE-Book, on the TRFD, or in the Teacher's view on the book's digital platform.

INTRODUCTION

How long does a battery last on the newest iPhone, on average? What proportion of college undergraduates attended all of their classes last week? How much does the weight of a quarter-pound hamburger at a fast-food restaurant vary after cooking? These are the types of questions we would like to answer.

It wouldn't be practical to determine the lifetime of *every* iPhone battery, to ask *all* undergraduates about their attendance, or to weigh *every* burger after cooking. Instead, we choose a random sample of individuals (batteries, undergraduates, burgers) to represent the population and collect data from those individuals. From what we learned in Chapter 4, if we randomly select the sample, we should be able to generalize our results to the population of interest. However, we cannot be certain that our conclusions are correct—a different sample would likely yield a different estimate. Probability helps us account for the chance variation due to random selection or random assignment.

Chapter 8 begins the formal study of statistical inference—using information from a sample to draw conclusions about a population parameter such as p or μ. This is an important transition from Chapter 7, where you were given information about a population and asked questions about the distribution of a sample statistic, such as the sample proportion $\hat{p}$ or the sample mean $\bar{x}$.

The following activity gives you an idea of what lies ahead.

▶ ACTIVITY OVERVIEW

To prepare for using this activity, watch the overview video by clicking on the link in the TE-Book, opening the TRFD, or downloading from the Teacher's Resources on the book's digital platform.

Time: 20 minutes

Materials: Beads, cup

Teaching Advice: Beads can be bought at a craft store or online. Prepare a population of beads in which the true proportion of the beads that are a certain color is known. Be sure you have a large enough population so that the "scoop" is less than 10% of the population. We suggest a proportion between 0.30 and 0.40. This is an excellent activity for having students create or discover a formula. They have all the pieces to the puzzle, and they just have to put them all together.

- Chapter 2: Empirical rule
- Chapter 7: Details about the sampling distribution of $\hat{p}$ (shape, center, variability)

Allow students time to struggle in groups before revealing the formula (and the true proportion!) at the end of the activity.

Answers:

1. Mix beads thoroughly. Scoop up sample using the cup.

2–5. Answers will vary. Use

$$\hat{p} \pm z^* \sqrt{\frac{\hat{p}(1-\hat{p})}{n}}$$ to construct the

interval.

ACTIVITY   The beads

Before class, your teacher prepared a large population of different-colored beads and put them into a container. In this activity, you and your team will create an interval of plausible values for p = the true proportion of beads in the container that are a particular color (e.g., red).

1. As a class, discuss how to use the cup provided to select a simple random sample of beads from the container.

2. Have one student select an SRS of beads. Separate the beads into two groups: those that are red and those that are not red. Count the number of beads in each group.

3. Calculate $\hat{p}$ = the sample proportion of beads in the container that are red. Do you think this value is equal to the true proportion of red beads in the container? Explain your answer.

4. In teams of 3 or 4 students, determine an interval of plausible (believable) values for the true proportion p using the value of $\hat{p}$ from Step 3 and what you learned in Section 7.2 about the sampling distribution of a sample proportion.

5. Compare your results with those of the other teams in the class. Discuss any problems you encountered and how you dealt with them.

Making Connections

In Chapter 7, we had to know the population parameter to consider what would happen if we took many, many samples. We did this to help students build the concept of a sampling distribution. In practice, we will not know the population parameter. We will collect data from a single sample and use the sample data to make an inference about the population. It is very important that your students recognize this transition.

TRM **Hershey's Kisses Alternate Activity**

For an additional activity, see the "Hershey's Kisses" alternate activity in the Teacher's Resource Materials. Find it in the Teacher's Resource Materials located in the TE-Book, on the TRFD, or in the Teacher's view on the book's digital platform.

In this chapter and the next, we will introduce the two most common types of formal statistical inference. Chapter 8 concerns *confidence intervals* for estimating the value of a parameter. Chapter 9 presents *significance tests,* which assess the evidence for a claim about a parameter. Both types of inference are based on the sampling distributions you studied in Chapter 7.

In this chapter, we start by presenting the idea of a confidence interval in a general way that applies to estimating any unknown parameter. In Section 8.2, we show how to estimate a population proportion using a confidence interval. Section 8.3 focuses on confidence intervals for a difference in proportions.

PD **Section 8.1 Overview**

To watch the video overview of Section 8.1 (for teachers), click on the link in the TE-Book, look on the TRFD, or download from the Teacher's Resources on the book's digital platform.

TRM **Learning Targets Grid**

At the beginning of each section, we present the relevant learning targets. Point these out to your students and refer back to the targets when you cover them in class. There is a PDF version of the grid with an additional column that students can use to keep track of their progress. Find it in the Teacher's Resource Materials located in the TE-Book, on the TRFD, or in the Teacher's view in the book's digital platform.

TRM **Section 8.1 Alternate Examples**

You can find the Alternate Examples for this section in Microsoft Word format by clicking on the link in the TE-Book, opening the TRFD, or downloading from the Teacher's Resources on the book's digital platform.

SECTION 8.1 Confidence Intervals: The Basics

LEARNING TARGETS *By the end of the section, you should be able to:*

- Identify an appropriate point estimator and calculate the value of a point estimate.
- Interpret a confidence interval in context.
- Determine the point estimate and margin of error from a confidence interval.
- Use a confidence interval to make a decision about the value of a parameter.
- Interpret a confidence level in context.
- Describe how the sample size and confidence level affect the margin of error.
- Explain how practical issues like nonresponse, undercoverage, and response bias can affect the interpretation of a confidence interval.

Mr. Buckley's class did "The beads" activity from the Introduction. In their sample of 251 beads, they selected 107 red beads and 144 other beads. If we had to give a single number to estimate p = the true proportion of beads in the container that are red, what would it be? Because the sample proportion $\hat{p}$ is an unbiased estimator of the population proportion p, we use the statistic $\hat{p}$ as a **point estimator** of the parameter p. The best guess for the value of p is $\hat{p} = 107/251 = 0.426$. This value is known as a **point estimate**.

> A statistic is called a *point* estimate because it represents a single point on a number line.

> **DEFINITION** **Point estimator, Point estimate**
>
> A **point estimator** is a statistic that provides an estimate of a population parameter.
>
> The value of that statistic from a sample is called a **point estimate**.

As we saw in Chapter 7, the ideal point estimator will have no bias and little variability. Here's an example involving some of the more common point estimators.

Teaching Tip

Make sure to point out the margin notes in the student edition, such as the one on this page. They include useful information, often from the new Course Framework.

Teaching Tip

Notation matters! A point estimator (statistic) is used to estimate a parameter.

- $\bar{x}$ from the sample is used to estimate the population mean μ.
- $\hat{p}$ from the sample is used to estimate the population proportion p.
- s_x from the sample is used to estimate the population standard deviation σ.

Teaching Tip

The "play" icon at the top of this example (and many others) indicates that students can watch a video presentation of the example by clicking on the link in the e-Book or viewing the videos on the Student Site at highschool.bfwpub.com/updatedtps6e.

ALTERNATE EXAMPLE   Skill 3.B

Do you get enough sleep?
Point estimators

PROBLEM:

Identify the point estimator you would use to estimate the parameter in each of the following settings and calculate the value of the point estimate.

(a) A counselor at a large high school wants to estimate the mean amount of sleep μ that students got the previous night. She selects a random sample of 10 students and asks them to record the number of hours they slept last night. Here are the results:

4   5   5.5   6   6   7   7   7.5   8   10

(b) It is recommended that high school students get 8 hours or more of sleep each night, so the counselor wants to estimate the proportion p of all students at this large high school who got the recommended amount of sleeping time.

(c) The counselor also wants to investigate the variability in sleep times by estimating the population standard deviation σ.

SOLUTION:

(a) Use the sample mean $\bar{x}$ as a point estimator for the population mean μ. The point estimate is

$$\bar{x} = \frac{4+5+5.5+6+6+7+7+7.5+8+10}{10}$$

$$= 6.6 \text{ hours}$$

(b) Use the sample proportion $\hat{p}$ as a point estimator for the population proportion p.

The point estimate is $\hat{p} = \dfrac{2}{10} = 0.20$.

(c) Use the sample standard deviation s_x as a point estimator for the population standard deviation σ. The point estimate is $s_x = 1.696$ hours.

EXAMPLE	From batteries to smoking
	Point estimators

PROBLEM: Identify the point estimator you would use to estimate the parameter in each of the following settings and calculate the value of the point estimate.

(a) Quality control inspectors want to estimate the mean lifetime μ of the AA batteries produced each hour at a factory. They select a random sample of 50 batteries during each hour of production and then drain them under conditions that mimic normal use. Here are the lifetimes (in hours) of the batteries from one such sample:

16.73	15.60	16.31	17.57	16.14	17.28	16.67	17.28	17.27	17.50
15.59	17.54	16.46	15.63	16.82	17.16	16.62	16.71	16.69	17.98
15.99	15.64	17.20	17.24	16.68	16.55	17.48	15.58	17.61	15.98
15.46	16.50	16.19	16.36	17.80	16.61	16.99	16.93	16.01	16.46
17.54	17.41	16.91	16.60	16.78	15.75	17.31	16.50	16.72	17.55

(b) What proportion p of U.S. adults would classify themselves as vegan or vegetarian? A Pew Research Center report surveyed 1473 randomly selected U.S. adults. Of these, 124 said they were vegan or vegetarian.[1]

(c) The quality control inspectors in part (a) want to investigate the variability in battery lifetimes by estimating the population standard deviation σ.

SOLUTION:

(a) Use the sample mean $\bar{x}$ as a point estimator for the population mean μ. The point estimate is

$$\bar{x} = \frac{16.73 + 15.60 + \cdots + 17.55}{50} = 16.718 \text{ hours}.$$

(b) Use the sample proportion $\hat{p}$ as a point estimator for the population proportion p. The point estimate is

$$\hat{p} = \frac{124}{1473} = 0.084.$$

(c) Use the sample standard deviation s_x as a point estimator for the population standard deviation σ. The point estimate is $s_x = 0.664$ hour.

FOR PRACTICE, TRY EXERCISE 1

The Idea of a Confidence Interval

When Mr. Buckley's class did the beads activity, they obtained a sample proportion of $\hat{p} = 107/251 = 0.426$. To account for sampling variability, one team created an interval of plausible values by adding 0.062 to and subtracting 0.062 from $\hat{p} = 0.426$ to get an interval from 0.364 to 0.488. Where did the 0.062 come from? Their reasoning was based on the sampling distribution of the sample proportion from Section 7.2:

- Because the number of successes (107) and the number of failures (144) were both at least 10, the Large Counts condition is met. Therefore, the sampling distribution of $\hat{p}$ is approximately Normal.

Making Connections

We know from Chapter 7 that there will be variability in the sample proportion due to random sampling. Therefore, it doesn't make sense for Mr. Buckley to claim that the proportion is *exactly* 0.426. It seems much more reasonable to instead provide an interval of plausible values.

Making Connections

The interval 0.364 to 0.488 is a 95% confidence interval for a proportion, which we will cover more formally in Section 8.2. At this point, don't worry too much about the calculation. Right now we are working to build up a conceptual understanding of a confidence interval.

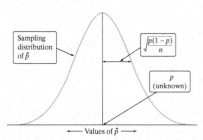

Sampling distribution of $\hat{p}$

$\sqrt{\dfrac{p(1-p)}{n}}$

p (unknown)

Values of $\hat{p}$

- In about 95% of samples, the value of $\hat{p}$ will be within 2 standard deviations $\left(2\sigma_{\hat{p}} \approx 2\sqrt{\dfrac{0.426(1-0.426)}{251}} = 0.062\right)$ of the true proportion p.

- Therefore, in about 95% of samples, the value of the true proportion p will be within 2 standard deviations $\left(2\sigma_{\hat{p}} \approx 2\sqrt{\dfrac{0.426(1-0.426)}{251}} = 0.062\right)$ of $\hat{p}$.

When the estimate of a parameter is reported as an interval of values, it is called a **confidence interval**.

> A confidence interval is called an *interval estimate* because it represents an interval of values on a number line, rather than a single point.

> **DEFINITION   Confidence interval**
>
> A **confidence interval** gives an interval of plausible values for a parameter based on sample data.

Plausible does not mean the same thing as possible. You could argue that just about any value of a parameter is *possible*. *Plausible* means that we shouldn't be surprised if any one of the values in the interval is equal to the value of the parameter. Based on their calculations, the class shouldn't be surprised if Mr. Buckley revealed that the true proportion of red beads in the container is any value from 0.364 to 0.488. However, it would be surprising if the true proportion was less than 0.364 or greater than 0.488.

We use an interval of plausible values rather than a single point estimate to account for sampling variability and increase our confidence that we have a correct value for the parameter. Of course, as the cartoon illustrates, there is a trade-off between the amount of confidence we have that our estimate is correct and how much information the interval provides.

Confidence intervals are constructed so that we know *how much* confidence we should have in the interval. The most common **confidence level** is 95%. You will learn how to interpret confidence levels shortly.

Teaching Tip

When finding the standard deviation of the sample proportion, we use the $\approx$ symbol because we are using the sample proportion $\hat{p}$ to estimate the true proportion p.

Teaching Tip

Emphasize the word *plausible* when describing the values in a confidence interval. Here, *plausible* means "believable," suggesting that we shouldn't be surprised if we found out that any one of the values in the interval is the value of the parameter.

Making Connections

Recall from previous chapters that an outcome was considered unusual or surprising if the probability was less than 5%. In Chapter 9, this value of 5% will be identified as the *significance level* (α). Later, we will see a close connection between the 5% significance level and the 95% confidence level.

Teaching Tip

Use the Garfield cartoon on this page to point out two facts about using an interval estimate:

1. Using an interval of values rather than a point estimate greatly increases the chances of being correct.

2. Using an interval that is overly wide isn't helpful (e.g., the weather report in the cartoon won't help you pick out clothes for the day). This is why statisticians typically prefer 95% confidence intervals: they are wide enough to be correct most of the time, but not wider than they should be. For example, moving from 95% to 99.7% confidence increases our success rate by a little, but increases the width of the interval by 50%.

Teaching Tip

To better understand the idea of "success rate" in the definition of confidence level, see the next activity: the *Confidence Intervals* applet (page 542).

Teaching Tip

The way we interpret a confidence interval is slightly different than you will see in many other books. Our interpretation of Mr. Buckley's interval is "We are 95% confident that the interval from 0.364 to 0.488 captures the true proportion of red beads." Other books might interpret the interval this way: "We are 95% confident that the true proportion of red beads is between 0.364 and 0.488." Both interpretations are fine on the AP® Statistics exam, but we prefer our interpretation because it emphasizes that the interval *captures* the parameter rather than saying that the parameter falls into the interval. After all, the parameter doesn't move—it is a constant. The interval is what varies from sample to sample. We believe this slight change in wording will also help students interpret the confidence level correctly and avoid a very common error on the AP® Statistics exam.

Teaching Tip

When interpreting a confidence interval, it is important that students make it clear they are estimating a parameter for the population, rather than a statistic from a sample. When referring to the parameter in context, use the words *true* and *all* to make it clear you are referring to the population and not the sample. Also, avoid using the past tense, if possible, because this sounds as if you are referring to the sample.

Teaching Tip

This is the "General formula" that works for all the confidence intervals we will calculate in this course:

point estimate ± margin of error

> Some people include the phrase "based on the sample" when interpreting a confidence interval: "Based on the sample, we are *C*% confident that the interval from _____ to _____ captures the [parameter in context]."

> **AP® EXAM TIP**
>
> When interpreting a confidence interval, make sure that you are describing the parameter and not the statistic. It's wrong to say that we are 95% confident the interval from 0.613 to 0.687 captures the proportion of U.S. adults who *admitted* they would experience financial difficulty. The "proportion who *admitted* they would experience financial difficulty" is the sample proportion, which is known to be 0.65. The interval gives plausible values for the proportion who *would admit* to experiencing some financial difficulty if asked.

> **DEFINITION** Confidence level
>
> The **confidence level** *C* gives the overall success rate of the method used to calculate the confidence interval. That is, the interval computed from the sample data will capture the true parameter value in *C*% of all possible samples when the conditions for inference are met.

The Associated Press and the NORC Center for Public Affairs Research recently asked a random sample of U.S. adults how much financial difficulty they would experience if they had to pay an unexpected bill of $1000 right away. Overall, 65% of respondents admitted they would have "a little" or "a lot" of difficulty. A summary of the study reported that the 95% confidence interval for the proportion of U.S. adults who would admit to experiencing some financial difficulty is 0.613 to 0.687. That is, they are 95% confident that the interval from 0.613 to 0.687 captures the true proportion of all U.S. adults who would admit to experiencing some financial difficulty paying an unexpected bill of $1000 right away.

INTERPRETING A CONFIDENCE INTERVAL

To interpret a *C*% confidence interval for an unknown parameter, say, "We are *C*% confident that the interval from _____ to _____ captures the [parameter in context]."

To create an interval of plausible values for a parameter based on data from a sample, we need two components: a point estimate to use as the midpoint of the interval and a **margin of error** to account for sampling variability. The structure of a confidence interval is

point estimate ± margin of error

We can visualize a *C*% confidence interval like this:

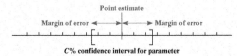

Earlier, we learned that the 95% confidence interval for the proportion of all U.S. adults who would admit to experiencing some financial difficulty paying an unexpected bill of $1000 right away is 0.613 to 0.687. This interval could also be expressed as

$$0.65 \pm 0.037$$

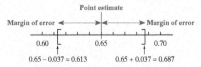

95% confidence interval for p = proportion of all U.S. adults who would admit to experiencing some financial difficulty

---

Making Connections

We will directly assess the evidence for a claim about a parameter in Chapter 9 using a significance test. But a confidence interval can provide us with the same insight about evidence for a claim. In Chapter 9, we will highlight the connection between confidence intervals and significance tests.

Confidence intervals reported in the media are often presented as a point estimate and a margin of error.

> **DEFINITION   Margin of error**
>
> The **margin of error** of an estimate describes how far, at most, we expect the estimate to vary from the true population value. That is, in a $C\%$ confidence interval, the distance between the point estimate and the true parameter value will be less than the margin of error in $C\%$ of all samples.

In addition to estimating a parameter, we can also use confidence intervals to assess claims about a parameter, as in the following example.

EXAMPLE

Who will win the election?
Interpreting a confidence interval

PROBLEM: Two weeks before a presidential election, a polling organization asked a random sample of registered voters the following question: "If the presidential election were held today, would you vote for Candidate A or Candidate B?" Based on this poll, the 95% confidence interval for the population proportion who favor Candidate A is (0.48, 0.54).

(a) Interpret the confidence interval.
(b) What is the point estimate that was used to create the interval? What is the margin of error?
(c) Based on this poll, a political reporter claims that the majority of registered voters favor Candidate A. Use the confidence interval to evaluate this claim.

SOLUTION:

(a) We are 95% confident that the interval from 0.48 to 0.54 captures the true proportion of all registered voters who favor Candidate A in the election.

(b) point estimate $= \dfrac{0.48 + 0.54}{2} = 0.51$

margin of error $= 0.54 - 0.51 = 0.03$

> The point estimate is the midpoint of the interval. The margin of error is the distance from the point estimate to the endpoints of the interval.

(c) Because there are plausible values of p less than or equal to 0.50 in the confidence interval, the interval does not give convincing evidence that a majority (more than 50%) of registered voters favor Candidate A.

> Another way to calculate the margin of error is to divide the width of the confidence interval by 2: $(0.54 - 0.48)/2 = 0.03$.

> Any value from 0.48 to 0.54 is a plausible value for the true proportion who favor Candidate A.

FOR PRACTICE, TRY EXERCISE 5

Teaching Tip

Remind students that the margin of error in a confidence interval accounts for sampling variability due only to random selection. It does not compensate for any bias in the data collection process. For example, if the sample wasn't random or the questions were worded poorly, you shouldn't have much confidence at all!

Teaching Tip

Notice in the example on this page that the confidence interval is written using parentheses (0.48, 0.54), which is considered standard notation in statistics. This is a bit confusing, as 0.48 and 0.54 *are* considered plausible values for the true parameter, which suggests that we should use the formal interval notation with brackets [0.48, 0.54]. On the AP® Statistics exam (and most statistics textbooks), students will see the form using parentheses.

ALTERNATE EXAMPLE   Skills 4.B, 4.D

More problems sleeping   Interpreting a confidence interval

PROBLEM:
A counselor at a large high school wants to estimate the mean amount of sleep μ that students got the previous night. She selects a random sample of 10 students and asks them to record the number of hours they slept last night. Based on this sample, the 95% confidence interval for the population mean amount of sleep for all students at the high school is (5.39, 7.81) hours.
(a) Interpret the confidence interval.
(b) What is the point estimate that was used to create the interval? What is the margin of error?

(c) It is recommended that high school students get 8 hours or more of sleep each night. Based on this survey, the students claim that, on average, students at this high school are not getting the recommended amount of sleep. Use the confidence interval to evaluate this claim.

SOLUTION:
(a) We are 95% confident that the interval from 5.39 to 7.81 hours captures the true mean amount of sleep the previous night for all students at this high school.

(b) point estimate $= \dfrac{5.39 + 7.81}{2}$
$= 6.6$ hours
margin of error $= 7.81 - 6.6 = 1.21$ hours
(c) Because all the plausible values of μ are less than 8 hours in the confidence interval, the interval gives convincing evidence that students at this high school do not get the recommended amount of sleep, on average.

ACTIVITY OVERVIEW

To prepare for using this activity, watch the overview video by clicking on the link in the TE-Book, opening the TRFD, or downloading from the Teacher's Resources on the book's digital platform.

Time: 15 minutes

Materials: Device with Internet access

Teaching Advice: We chose to use $p = 0.6$ and $n = 100$ in the activity because the capture rate ends up being very close to the confidence level. For some combinations of p and n, the capture rate is less than the confidence level, even when the conditions are met, due to the discreteness of the sampling distribution of $\hat{p}$. We don't want this to be a distraction to students, so we tried to choose numbers that work out nicely. FYI: To remedy this problem, statisticians came up with the "Plus 4" interval. See the margin note on page 556. There is some additional explanation of the Plus 4 interval in the TRM.

Consider calling the black confidence intervals the "winners" and the red confidence intervals the "losers." Most samples produce "winners" but some will produce a "loser." Make sure students understand what happened to arrive at a "loser" interval. This random sample from the population must have included far fewer (or far more) successes than expected so that the sample proportion was so small (large) that even after the margin of error was included, the interval still didn't capture the true proportion.

Be sure that students recognize the difference between the sample size and the number of samples. Also help students notice that as the confidence level decreases, the width of the interval decreases, making it less likely to capture the true mean.

Answers:

1–2. Done

3. Answers will vary. The sample size here is 100, while the number of samples is 10.

4. After taking many samples, the "Percent hit" is 95%.

5. About 99% of the intervals capture the true proportion.

6. About 90% of the intervals capture the true proportion.

7. The confidence level matches the "Percent hit" after taking many samples.

Interpreting Confidence Level

What does it mean to be 95% confident? The following activity gives you a chance to explore the meaning of the confidence level.

ACTIVITY   The *Confidence Intervals for Proportions* applet

In this activity, you will use the *Confidence Intervals for Proportions* applet to learn what it means to say that we are "95% confident" that our confidence interval captures the parameter value.

1. Go to highschool.bfwpub.com/updatedtps6e and launch the applet. Change the settings to: Population Proportion (*p*): 0.6, confidence level: 95, and sample size (*n*): 100. The display shows the values from 0.00 to 1.00, with a green line at $p = 0.60$ indicating the value of the true proportion.

2. Click "Sample" to choose an SRS of size $n = 100$ and display the resulting confidence interval. The confidence interval is shown as a horizontal line segment with a dot representing the sample proportion $\hat{p}$ in the middle of the interval.

3. Did the interval capture the population proportion p (what the applet calls a "hit")? Click "Sample" a total of 10 times. How many of the intervals captured the population proportion p? *Note:* So far, you have used the applet to take 10 SRSs, each of size $n = 100$. Be sure you understand the difference between sample size and the number of samples taken.

4. Reset the applet. Click "Sample 25" 40 times to choose 1000 SRSs and display the confidence intervals based on those samples. What percent of the intervals captured the true proportion p?

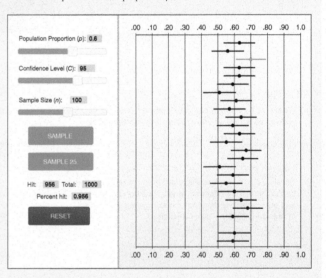

Teaching Tip

When asked to interpret the confidence level, students should picture this applet in their minds. If many samples are taken and 95% confidence intervals are constructed for each of these samples, then about 95% of the intervals will capture the true parameter being estimated. Also, remind students that in real life we usually calculate just one interval. The figure describes what would happen if we were to take many, many samples when the conditions are met.

5. Change the confidence level to 99%. The applet will automatically recalculate all 1000 confidence intervals using a 99% confidence level. What percent of the intervals capture the true proportion p?

6. Repeat Step 5 using a 90% confidence level.

7. Summarize what you have learned about the relationship between confidence level and capture rate (percent hit) after taking many samples.

We will investigate the effect of changing the sample size later.

As the activity confirms, *when the conditions are met and the method is used many times, the capture rate will be very close to the stated confidence level.*

INTERPRETING A CONFIDENCE LEVEL

To interpret a confidence level C, say, "If we were to select many random samples of the same size from the same population and construct a $C\%$ confidence interval using each sample, about $C\%$ of the intervals would capture the [parameter in context]."

Let's revisit the presidential election poll to practice interpreting a confidence level.

EXAMPLE

Another look at the election poll
Interpreting a confidence level

PROBLEM: Two weeks before a presidential election, a polling organization asked a random sample of registered voters the following question: "If the presidential election were held today, would you vote for Candidate A or Candidate B?" Based on this poll, the 95% confidence interval for the population proportion who favor Candidate A is (0.48, 0.54). Interpret the confidence level.

SOLUTION:

If we were to select many random samples of the same size from the population of registered voters and construct a 95% confidence interval using each sample, about 95% of the intervals would capture the true proportion of all registered voters who favor Candidate A in the election.

> Remember that interpretations of confidence level are about the method used to construct the interval—not one particular interval. In fact, we can interpret confidence levels before data are collected!

FOR PRACTICE, TRY EXERCISE 11

AP® EXAM TIP

On a given problem, you may be asked to interpret the confidence interval, the confidence level, or both. Be sure you understand the difference: the confidence interval gives a set of plausible values for the parameter and the confidence level describes the overall capture rate of the method.

In the preceding example, there are only two possibilities:

1. The interval from 0.48 to 0.54 captures the population proportion p. Our random sample was one of the many samples for which the difference between p and $\hat{p}$ is less than the margin of error. When using a 95% confidence level, about 95% of samples result in a confidence interval that captures p.

2. The interval from 0.48 to 0.54 does *not* capture the population proportion p. Our random sample was one of the few samples for which the difference between p and $\hat{p}$ is greater than the margin of error. When using a 95% confidence level, only about 5% of all samples result in a confidence interval that fails to capture p.

Teaching Tip

Notice that we don't need an actual sample to interpret the confidence level. Interpreting the confidence level is about describing the method for calculating a confidence interval, not about interpreting a specific confidence interval calculated from an actual sample.

ALTERNATE EXAMPLE	Skill 4.B

Last sleep
Interpreting a confidence level

PROBLEM:
Based on a random sample of 10 students, a counselor calculates that the 95% confidence interval for the population mean amount of sleep for all students is (5.39, 7.81) hours. Interpret the confidence level.

SOLUTION:
If we were to select many random samples of 10 students at this school and construct a 95% confidence interval using each sample, about 95% of the intervals would capture the true mean amount of sleep the previous night for all students at this school.

 COMMON STUDENT ERROR

Many students confuse interpreting confidence *levels* with interpreting confidence *intervals*. Every time students construct a confidence interval, they are expected to interpret the interval they calculated. They are expected to interpret the confidence level only when they are specifically asked to do so. If students interpret the confidence level unnecessarily, they will not gain any credit and could even *lose* credit on the AP® Statistics exam if the interpretation is incorrect.

Teaching Tip

Many students (and teachers!) struggle to understand why it is inappropriate to use the word *probability* when interpreting a confidence interval. Here are two scenarios that might help:

1. Show students a fair coin. Before tossing, ask them if it is appropriate to say, "There is a 50% *probability*" that the next toss will be heads. Most will agree that this is appropriate.

2. Toss the coin in front of the students, but keep the result hidden under your hand. Ask students if it is appropriate to say, "There is a 50% *probability*" that the coin is heads. This statement is *not* appropriate because the outcome of the random event is already determined (it either is heads or it isn't). If it is heads, then the probability is 1. If it is tails, the probability is 0.

Scenario 1 is analogous to interpreting a confidence *level* (before data are collected), where it is appropriate to use "probability." Scenario 2 is analogous to interpreting a confidence *interval* (after data are collected), where it is not appropriate to use "probability."

Without conducting a census, we cannot know whether our sample is one of the 95% for which the interval captures p or whether it is one of the unlucky 5% that does not. The statement that we are "95% confident" is shorthand for saying, "We got these numbers using a method that gives correct results for 95% of samples."

 The confidence level does not tell us the probability that a particular confidence interval captures the population parameter. Once a particular confidence interval is calculated, its endpoints are fixed. And because the value of a parameter is also a constant, a particular confidence interval either includes the parameter (probability $= 1$) or doesn't include the parameter (probability $= 0$). As Figure 8.1 illustrates, no individual 95% confidence interval has a 95% probability of capturing the true parameter value.

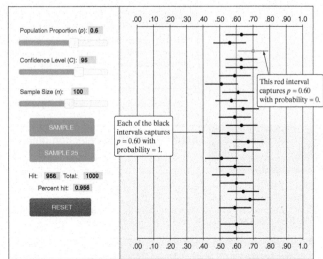

FIGURE 8.1 Image from the *Confidence Intervals for Proportions* applet showing that the probability a particular 95% confidence interval captures the true parameter value is either 0 or 1 (and not 0.95).

CHECK YOUR UNDERSTANDING

The Pew Research Center and *Smithsonian* magazine recently quizzed a random sample of 1006 U.S. adults on their knowledge of science.[2] One of the questions asked, "Which gas makes up most of the Earth's atmosphere: hydrogen, nitrogen, carbon dioxide, or oxygen?" A 95% confidence interval for the proportion who would correctly answer nitrogen is 0.175 to 0.225.

1. Interpret the confidence interval.
2. Interpret the confidence level.
3. Calculate the point estimate and the margin of error.
4. If people guess one of the four choices at random, about 25% should get the answer correct. Does this interval provide convincing evidence that less than 25% of all U.S. adults would answer this question correctly? Explain your reasoning.

✓ **Answers to CYU**

1. We are 95% confident that the interval from 0.175 to 0.225 captures the true proportion of all U.S. adults who would answer the question correctly.

2. If we were to select many random samples of the same size from the population of U.S. adults and construct a 95% confidence interval using each sample, about 95% of the intervals would capture the true proportion of all U.S. adults who would answer the question correctly.

3. The point estimate is $\dfrac{0.175 + 0.225}{2} = 0.20$; the margin of error is $0.225 - 0.20 = 0.025$.

4. All the plausible values in the 95% confidence interval are less than the proportion expected if people were to simply guess from the four choices at random. Therefore, this interval does provide convincing evidence that less than 25% of all U.S. adults would answer this question correctly.

What Affects the Margin of Error?

Why settle for 95% confidence when estimating an unknown parameter? Do larger random samples yield "better" intervals? The *Confidence Intervals for Proportions* applet will shed some light on these questions.

ACTIVITY | Exploring margin of error with the *Confidence Intervals for Proportions* applet

In this activity, you will use the applet to explore the relationship between the confidence level, the sample size, and the margin of error.

Part 1: Adjusting the Confidence Level

1. Go to highschool.bfwpub.com/updatedtps6e and launch the applet. Change the settings to: Population Proportion (p): 0.6, confidence level: 95, and sample size (n): 100. Click "Sample 25" 40 times to select 1000 SRSs and make 1000 confidence intervals.

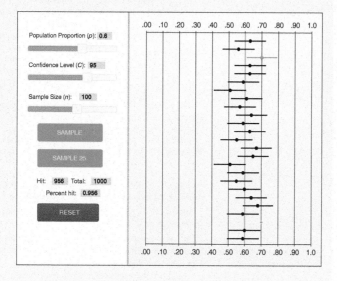

2. Change the confidence level to 99%. What happens to the length of the confidence intervals? What happens to the capture rate (percent hit)? Drag the slider back and forth between 95% and 99% confidence to make sure you see what is happening.
3. Now change the confidence level to 90% and repeat Step 2.
4. Summarize what you learned about the relationship between the confidence level and the margin of error for a fixed sample size.

▶ **ACTIVITY OVERVIEW**

To prepare for using this activity, watch the overview video by clicking on the link in the TE-Book, opening the TRFD, or downloading from the Teacher's Resources on the book's digital platform.

Time: 15 minutes

Materials: Devices with Internet access

Teaching Advice: Be sure that students recognize how margin of error is represented in the applet. The margin of error is the length from the center dot of the confidence interval to the edge of the interval. In part 1, students discover that increasing the confidence level increases the margin of error (with a fixed sample size). In part 2, students discover that increasing the sample size reduces the margin of error (with a fixed confidence level).

Answers:

1. Done

2. The length of the confidence intervals increases. The capture rate (percent hit) increases.

3. The length of the confidence intervals decreases. The capture rate (percent hit) decreases.

4. For a fixed sample size, increasing the confidence level increases the margin of error.

5. Done

6. The length of the confidence intervals decreases.

7. The length of the confidence intervals further decreases.

8. For a fixed confidence level, increasing the sample size reduces the margin of error.

9. No, increasing the sample size does not change the capture rate.

+ Ask the StatsMedic

"Guess My Age" Activity

Consider using the "Guess my age" activity to show students that we should use a wider interval to be more confident that it contains the true parameter.

Making Connections

We know from Chapter 7 that larger sample sizes lead to less variability in the sampling distribution, so it makes sense that the margin of error will decrease.

Making Connections

In this section, we don't want students using formulas to think about ways to decrease the margin of error, because we are trying to build conceptual understanding. Later, these ideas will be justified with formulas. In Sections 8.2 and 10.1, we will see formulas for margin of error for proportions and means:

$$ME = z^* \sqrt{\frac{\hat{p}(1-\hat{p})}{n}} \quad \text{and} \quad ME = t^* \frac{s_x}{\sqrt{n}}$$

Both formulas show that the margin of error will get smaller when z^* or t^* is smaller (smaller confidence level) or the sample size n increases.

Part 2: Adjusting the Sample Size

5. Reset the applet settings to: Population Proportion (p): 0.6, confidence level: 95, and sample size (n): 100. Press "Sample 25" to select 25 SRSs of size $n = 100$ and make 25 confidence intervals.

6. Using the slider, increase the sample size to $n = 500$. Press "Sample 25" to select 25 SRSs of size $n = 500$ and make 25 confidence intervals. What do you notice about the length of the confidence intervals?

7. Using the slider, increase the sample size to $n = 1000$. Press "Sample 25" to select 25 SRSs of size $n = 1000$ and make 25 confidence intervals. What do you notice about the length of the confidence intervals?

8. Summarize what you learned about the relationship between the sample size and the margin of error for a fixed confidence level.

9. Does increasing the sample size increase the capture rate (percent hit)? Use the applet to investigate.

As the activity illustrates, the price we pay for greater confidence is a wider interval. If we're satisfied with 90% confidence, then our interval of plausible values for the parameter will be narrower than if we insist on 95% or 99% confidence. For example, here is a 90% confidence interval and a 99% confidence interval for the proportion of red beads in Mr. Buckley's container based on the class's sample data. Unfortunately, intervals constructed at a 90% confidence level will capture the true value of the parameter less often than intervals that use a 99% confidence level.

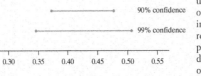

The activity also shows that we can get a more precise estimate of a parameter by increasing the sample size. Larger samples generally yield narrower confidence intervals at any confidence level. *In fact, the width of a confidence interval for a proportion or a mean is proportional to* $1/\sqrt{n}$, *so that quadrupling the sample size cuts the margin of error in half.* However, larger samples don't affect the capture rate and cost more time and money to obtain.

DECREASING THE MARGIN OF ERROR

In general, we prefer an estimate with a small margin of error. The margin of error gets smaller when:

- *The confidence level decreases.* To obtain a smaller margin of error from the same data, you must be willing to accept less confidence.
- *The sample size n increases.* In general, increasing the sample size n reduces the margin of error for any fixed confidence level.

To see why these facts are true, let's look a bit more closely at the method Mr. Buckley's class used to calculate a confidence interval for the true proportion of beads in the container that are red. They started with a point estimate of $\hat{p} = 107/251 = 0.426$. Then they added and subtracted 2 standard deviations to get the interval of plausible values from 0.364 to 0.488.

We could rewrite this interval as

point estimate ± margin of error

$$\hat{p} \pm 2\sigma_{\hat{p}}$$

$$0.426 \pm 2\sqrt{\frac{0.426(1-0.426)}{251}}$$

This leads to the more general formula for a confidence interval:

statistic ± (critical value) · (standard deviation of statistic)

The **critical value** depends on both the confidence level C and the sampling distribution of the statistic. Mr. Buckley's class used a critical value of 2 to be 95% confident. If they wanted to be 99.7% confident, they could have gone 3 standard deviations in each direction. Greater confidence requires a larger critical value.

DEFINITION Critical value

The **critical value** is a multiplier that makes the interval wide enough to have the stated capture rate.

The margin of error also depends on the standard deviation of the statistic. As you learned in Chapter 7, the sampling distribution of a statistic will have a smaller standard deviation when the sample size is larger. This is why the margin of error decreases as you increase the sample size.

WHAT THE MARGIN OF ERROR DOESN'T ACCOUNT FOR When we calculate a confidence interval, we include the margin of error because we expect the value of the point estimate to vary somewhat from the parameter. However, the margin of error accounts for *only* the variability we expect from random sampling. It does not account for practical difficulties, such as undercoverage and nonresponse in a sample survey. These problems can produce estimates that are much farther from the parameter than the margin of error would suggest. Remember this unpleasant fact when reading the results of an opinion poll or other sample survey. The margin of error does *not* account for any sources of bias in the data collection process.

EXAMPLE

What's your GPA?
Factors that affect the margin of error

PROBLEM: As part of a project about response bias, Ellery surveyed a random sample of 25 students from her school. One of the questions in the survey required students to state their GPA aloud. Based on the responses, Ellery said she was 90% confident that the interval from 0.40 to 0.72 captures the proportion of all students at her school with GPAs greater than 3.0.[3]

(a) Explain what would happen to the width of the interval if the confidence level were increased to 99%.

Making Connections

Point out that using a critical value of 1 instead of 2 would result in 68% confidence. Likewise, using a critical value of 3 would result in 99.7% confidence. Don't worry about calculating other critical values in this section, however. We will learn how to do this in Sections 8.2 and 10.1.

Making Connections

This is a different version of the earlier "general formula" we showed for confidence intervals:

point estimate ± margin of error
statistic ± (critical value) ·
(standard error of statistic)

The second formula is given on the AP® Statistics exam. Later, we will develop a "specific formula" for estimating a proportion (Section 8.2) and for estimating a mean (Section 10.1).

Teaching Tip

The word *error* is defined as "a mistake." Unfortunately, the term *margin of error* is a bit misleading. We do not report margin of error for an estimate because there has been a mistake made in the data collection. The margin of error is reported to account for the variability in the sample results that occur due to the fact that we are taking a random sample. The "error" refers to the difference between our estimate and the true value. A better term might be *margin of estimation error* or *margin of sampling error*.

ALTERNATE EXAMPLE

Skill 3.D

Do you brush twice a day? Factors that affect the margin of error

PROBLEM:
As part of a project about response bias, Kody surveyed a random sample of 40 students from his school. Both of Kody's parents are dentists, so many of his survey questions had to do with brushing teeth. One of the questions required students to state aloud whether or not they brush their teeth twice a day. Based on the responses, Kody said he was 90% confident that the interval from 0.79 to 0.96 captures the true proportion of all students at the school who brush their teeth twice a day.
(a) Explain what would happen to the length of the interval if the confidence level were increased to 99%.

(b) Explain what would happen to the length of a 90% confidence interval if the sample size was increased to 100 students.
(c) Describe one potential source of bias in Kody's study that is not accounted for by the margin of error.

SOLUTION:
(a) The confidence interval would be wider because increasing the confidence level increases the margin of error.
(b) The confidence interval would be narrower than the original 90% confidence interval because increasing the sample size decreases the margin of error.

(c) The margin of error doesn't account for the fact that many students might lie about their tooth-brushing habits. They may know that Kody's parents are dentists or they may be embarrassed when having to respond without anonymity. The true proportion of all students who brush twice a day might be even less than 0.79!

Teaching Tip

For an excellent article that explains some of the other factors that the margin of error doesn't account for, check out: www.nytimes.com/2016/10/06/upshot/when-you-hear-the-margin-of-error-is-plus-or-minus-3-percent-think-7-instead.html

TRM **Section 8.1 Quiz**

There is one quiz available for this section in the Teacher's Resource Materials. Click on the link in the TE-Book, look on the TRFD, or download from the Teacher's Resources on the book's digital platform. You can also create your own quiz using the ExamView® Assessment Suite that is part of the TPS6 program. Questions are coded by Learning Target to make it easy to build parallel quizzes.

TRM **Full Solutions to Section 8.1 Exercises**

Click on the link in the TE-Book, open the TRFD, or download from the Teacher's Resources on the book's digital platform.

(b) How would the width of a 90% confidence interval based on a sample of size 100 compare to the original 90% interval, assuming the sample proportion remained the same?

(c) Describe one potential source of bias in Ellery's study that is not accounted for by the margin of error.

SOLUTION:

(a) The confidence interval would be wider because increasing the confidence level increases the margin of error.

> To increase the confidence level (capture rate), we need to use a larger critical value, which increases the margin of error.

(b) The confidence interval would be half as wide because the sample size is 4 times as big.

> Increasing the sample size decreases the standard deviation of the sampling distribution of the sample proportion (assuming the sample proportion doesn't change).

(c) The margin of error doesn't account for the fact that many students might lie about their GPAs when having to respond without anonymity. The proportion of students with GPAs greater than 3.0 might be even less than 0.40!

FOR PRACTICE, TRY EXERCISE 19

Section 8.1 | Summary

- To estimate an unknown population parameter, start with a statistic that will provide a reasonable guess. The chosen statistic is a **point estimator** for the parameter. The specific value of the point estimator that we use gives a **point estimate** for the parameter.

- A **confidence interval** gives an interval of plausible values for an unknown population parameter based on sample data. The interval estimate has the form

  point estimate $\pm$ margin of error

  When calculating a confidence interval, it is common to use the form

  statistic $\pm$ (critical value) · (standard deviation of statistic)

- To interpret a C% confidence interval, say, "We are C% confident that the interval from ____ to ____ captures the [parameter in context]." Be sure that your interpretation describes a parameter and not a statistic.

- The **confidence level** C is the success rate (capture rate) of the method that produces the interval. If you use 95% confidence intervals often, about 95% of your intervals will capture the true parameter value when certain conditions are met. You don't know whether a particular 95% confidence interval calculated from a set of data actually captures the true parameter value.

- Other things being equal, the **margin of error** of a confidence interval gets smaller as:
  - the confidence level C decreases;
  - the sample size n increases.

- Remember that the margin of error for a confidence interval only accounts for chance variation, not other sources of error like nonresponse and undercoverage.

Section 8.1 | Exercises

In Exercises 1–4, identify the point estimator you would use to estimate the parameter and calculate the value of the point estimate.

1. **Got shoes?** How many pairs of shoes, on average, do female teens have? To find out, an AP® Statistics class selected an SRS of 20 female students from their school. Then they recorded the number of pairs of shoes that each student reported having. Here are the data:

50	26	26	31	57	19	24	22	23	38
13	50	13	34	23	30	49	13	15	51

2. **Got shoes?** The class in Exercise 1 wants to estimate the variability in the number of pairs of shoes that female students have by estimating the population standard deviation σ.

3. **Going to the prom** Tonya wants to estimate the proportion of seniors in her school who plan to attend the prom. She interviews an SRS of 50 of the 750 seniors in her school and finds that 36 plan to go to the prom.

4. **Reporting cheating** What proportion of students are willing to report cheating by other students? A student project put this question to an SRS of 172 undergraduates at a large university: "You witness two students cheating on a quiz. Do you go to the professor?" Only 19 answered "Yes."[4]

5. **Prayer in school** A *New York Times*/CBS News Poll asked a random sample of U.S. adults the question "Do you favor an amendment to the Constitution that would permit organized prayer in public schools?" Based on this poll, the 95% confidence interval for the population proportion who favor such an amendment is (0.63, 0.69).

(a) Interpret the confidence interval.

(b) What is the point estimate that was used to create the interval? What is the margin of error?

(c) Based on this poll, a reporter claims that more than two-thirds of U.S. adults favor such an amendment. Use the confidence interval to evaluate this claim.

6. **Losing weight** A Gallup poll asked a random sample of U.S. adults, "Would you like to lose weight?" Based on this poll, the 95% confidence interval for the population proportion who want to lose weight is (0.56, 0.62).[5]

(a) Interpret the confidence interval.

(b) What is the point estimate that was used to create the interval? What is the margin of error?

(c) Based on this poll, Gallup claims that more than half of U.S. adults want to lose weight. Use the confidence interval to evaluate this claim.

7. **Bottling cola** A particular type of diet cola advertises that each can contains 12 ounces of the beverage. Each hour, a supervisor selects 10 cans at random, measures their contents, and computes a 95% confidence interval for the true mean volume. For one particular hour, the 95% confidence interval is 11.97 ounces to 12.05 ounces.

(a) Does the confidence interval provide convincing evidence that the true mean volume is different than 12 ounces? Explain your answer.

(b) Does the confidence interval provide convincing evidence that the true mean volume is 12 ounces? Explain your answer.

8. **Fun size candy** A candy bar manufacturer sells a "fun size" version that is advertised to weigh 17 grams. A hungry teacher selected a random sample of 44 fun size bars and found a 95% confidence interval for the true mean weight to be 16.945 grams to 17.395 grams.

(a) Does the confidence interval provide convincing evidence that the true mean weight is different than 17 grams? Explain your answer.

(b) Does the confidence interval provide convincing evidence that the true mean weight is 17 grams? Explain your answer.

9. **Shoes** The AP® Statistics class in Exercise 1 also asked an SRS of 20 boys at their school how many pairs of shoes they have. A 95% confidence interval for $\mu_G - \mu_B$ = the true difference in the mean number of pairs of shoes for girls and boys is 10.9 to 26.5.

(a) Interpret the confidence interval.

(b) Does the confidence interval give convincing evidence of a difference in the true mean number of pairs of shoes for boys and girls at the school? Explain your answer.

10. **Lying online** Many teens have posted profiles on sites such as Facebook. A sample survey asked random samples of teens with online profiles if they included false information in their profiles. Of 170 younger teens (ages 12 to 14) polled, 117 said "Yes." Of 317 older teens (ages 15 to 17) polled, 152 said "Yes."[6] A 95% confidence interval for $p_Y - p_O$ = the true difference in the proportions of younger teens and older teens who

---

8.5 (a) We are 95% confident that the interval from 0.63 to 0.69 captures the true proportion of all U.S. adults who favor an amendment to the Constitution that would permit organized prayer in public schools. **(b)** The point estimate $= \hat{p} = \dfrac{0.63 + 0.69}{2} = 0.66$; the margin of error $= 0.69 - 0.66 = 0.03$. **(c)** Because the value 2/3 = 0.667 (and values less than 2/3) are in the interval, it is plausible that two-thirds or less of the population favor such an amendment. There is not convincing evidence that more than two-thirds of U.S. adults favor such an amendment.

8.6 (a) We are 95% confident that the interval from 0.56 to 0.62 captures the true proportion of all U.S. adults who would like to lose weight. **(b)** The point estimate $= \hat{p} = \dfrac{0.56 + 0.62}{2} = 0.59$; the margin of error $= 0.62 - 0.59 = 0.03$. **(c)** Because all the plausible values in the interval are greater than 0.5, there is convincing evidence that more than half of U.S. adults want to lose weight.

8.7 (a) Because 12 is one of the plausible values in the 95% confidence interval, there is not convincing evidence that the true mean volume is different from 12 ounces. **(b)** No; although 12 is a plausible value for the true mean volume, there are many other plausible values in the confidence interval. Because any of these values could be the true mean, there is not convincing evidence that the true mean volume is 12 ounces.

8.8 (a) Because 17 is one of the plausible values in the 95% confidence interval, there is not convincing evidence that the true mean weight is different than 17 grams. **(b)** No; although 17 is a plausible value for the true mean weight, there are many other plausible values in the confidence interval. Because any of these values could be the true mean, there is not convincing evidence that the true mean weight is 17 grams.

8.9 (a) We are 95% confident that the interval from 10.9 to 26.5 captures the true difference (Girls – Boys) in the mean number of pairs of shoes at this school. **(b)** Yes; because the 95% confidence interval does not include 0 as a plausible value for the difference in means, there is convincing evidence of a difference in the mean number of shoes for boys and girls.

Teaching Tip

Make sure to point out the two icons next to Exercise 1. The "pg 538" icon reminds students that the example on page 538 is very similar to this exercise. The "play" icon reminds students that there is a video solution available in the student e-Book or at the Student Site.

Answers to Section 8.1 Exercises

8.1 Sample mean, $\bar{x} = 30.35$

8.2 Sample standard deviation, $s_x = 14.24$

8.3 Sample proportion, $\hat{p} = \dfrac{36}{50} = 0.72$

8.4 Sample proportion, $\hat{p} = \dfrac{19}{172} = 0.11$

8.10 (a) We are 95% confident that the interval from 0.120 to 0.297 captures the true difference (Younger – Older) in the proportions of all teens who include false information on their profiles. **(b)** Yes; because the 95% confidence interval does not include 0 as a plausible value for the difference in proportions, there is convincing evidence of a difference in the proportion of younger teens and older teens who include false information on their profiles.

8.11 If we were to select many random samples of the same size from the population of U.S. adults and construct a 95% confidence interval using each sample, about 95% of the intervals from each random sample would capture the true proportion of all U.S. adults who would favor an amendment to the Constitution that would permit organized prayer in public schools.

8.12 If we were to select many random samples of the same size from the population of U.S. adults and construct a 95% confidence interval using each sample, about 95% of the intervals would capture the true proportion of all U.S. adults who want to lose weight.

8.13 (a) The confidence interval is $51,492 − $431 = $51,061 to $51,492 + $431 = $51,923. We are 90% confident that the interval from $51,061 to $51,923 captures the true median household income for all households in Arizona in 2015. **(b)** About 90% of the intervals from the random samples would capture the true median household income for all households in Arizona in 2015.

8.14 (a) The confidence interval is $71,612 to $72,832. We are 90% confident that the interval from $71,612 to $72,832 captures the true median household income for all households in New Jersey in 2015. **(b)** About 90% of the intervals from the random samples would capture the true median household income for all households in New Jersey in 2015.

8.15 84% of the intervals did contain the true parameter, which suggests that these were 80% or 90% confidence intervals.

8.16 100% of the 25 confidence intervals did contain the true mean. This suggests that the confidence level was quite high— probably 99%, but possibly 95%.

8.17 (a) Incorrect; the interval provides plausible values for the *mean* BMI of all women, not plausible values for individual BMI measurements, which will be much

include false information in their profile is 0.120 to 0.297.

(a) Interpret the confidence interval.

(b) Does the confidence interval give convincing evidence of a difference in the true proportions of younger and older teens who include false information in their profiles? Explain your answer.

11. **More prayer in school** Refer to Exercise 5. Interpret
pg 543 the confidence level.

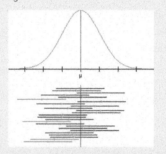

12. **More weight loss** Refer to Exercise 6. Interpret the confidence level.

13. **Household income** The 2015 American Community Survey estimated the median household income for each state. According to ACS, the 90% confidence interval for the 2015 median household income in Arizona is $51,492 ± $431.

(a) Interpret the confidence interval.

(b) Interpret the confidence level.

14. **More income** The 2015 American Community Survey estimated the median household income for each state. According to ACS, the 90% confidence interval for the 2015 median household income in New Jersey is $72,222 ± $610.

(a) Interpret the confidence interval.

(b) Interpret the confidence level.

15. **How confident?** The figure shows the result of taking 25 SRSs from a Normal population and constructing a confidence interval for the population mean using each sample. Which confidence level—80%, 90%, 95%, or 99%—do you think was used? Explain your reasoning.

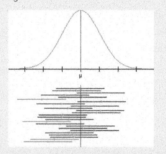

16. **How confident?** The figure shows the result of taking 25 SRSs from a Normal population and constructing

a confidence interval for the population mean using each sample. Which confidence level—80%, 90%, 95%, or 99%—do you think was used? Explain your reasoning.

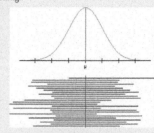

17. **Explaining confidence** A 95% confidence interval for the mean body mass index (BMI) of young American women is 26.8 ± 0.6. Discuss whether each of the following explanations is correct, based on that information.

(a) We are confident that 95% of all young women have BMI between 26.2 and 27.4.

(b) We are 95% confident that future samples of young women will have mean BMI between 26.2 and 27.4.

(c) Any value from 26.2 to 27.4 is believable as the true mean BMI of young American women.

(d) If we take many samples, the population mean BMI will be between 26.2 and 27.4 in about 95% of those samples.

(e) The mean BMI of young American women cannot be 28.

18. **Explaining confidence** The admissions director for a university found that (107.8, 116.2) is a 95% confidence interval for the mean IQ score of all freshmen. Discuss whether each of the following explanations is correct, based on that information.

(a) There is a 95% probability that the interval from 107.8 to 116.2 contains μ.

(b) There is a 95% chance that the interval (107.8, 116.2) contains $\bar{x}$.

(c) This interval was constructed using a method that produces intervals that capture the true mean in 95% of all possible samples.

(d) If we take many samples, about 95% of them will contain the interval (107.8, 116.2).

(e) The probability that the interval (107.8, 116.2) captures μ is either 0 or 1, but we don't know which.

more variable. **(b)** Incorrect; we shouldn't use the results of one sample to predict the results for future samples. **(c)** Correct; a confidence interval provides an interval of plausible values for a parameter.

(d) Incorrect; the population mean always stays the same, regardless of the number of samples taken. **(e)** Incorrect; we are 95% confident that the population mean is between 26.2 and 27.4, but that doesn't rule out any other possibilities.

8.18 (a) Incorrect; the population mean is always the same, so the probability that μ is in a particular interval is either 0 or 1 (but we don't know which). **(b)** Incorrect; the point estimate $\bar{x}$ will always be in the center of the confidence interval, so there is a 100% chance that $\bar{x}$ will be in the interval. **(c)** Correct; this is the meaning of 95% confidence. **(d)** Incorrect; it doesn't make

sense to say that a sample contains an interval. **(e)** Correct; the value of μ is always the same, so it is either always in a particular interval or always not in a particular interval.

19. **Prayer in school** Refer to Exercise 5.

 pg 547

(a) Explain what would happen to the length of the interval if the confidence level were increased to 99%.

(b) How would the width of a 95% confidence interval based on a sample size 4 times as large compare to the original 95% interval, assuming the sample proportion remained the same?

(c) The news article goes on to say: "The theoretical errors do not take into account additional errors resulting from the various practical difficulties in taking any survey of public opinion." List some of the "practical difficulties" that may cause errors which are not included in the ±3 percentage point margin of error.

20. **Losing weight** Refer to Exercise 6.

(a) Explain what would happen to the length of the interval if the confidence level were decreased to 90%.

(b) How would the width of a 95% confidence interval based on a sample size 4 times as large compare to the original 95% interval, assuming the sample proportion remained the same?

(c) As Gallup indicates, the 3 percentage point margin of error for this poll includes only sampling variability (what they call "sampling error"). What other potential sources of error (Gallup calls these "nonsampling errors") could affect the accuracy of the 95% confidence interval?

21. **California's traffic** People love living in California for many reasons, but traffic isn't one of them. Based on a random sample of 572 employed California adults, a 90% confidence interval for the average travel time to work for all employed California adults is 23 minutes to 26 minutes.[7]

(a) Interpret the confidence level.

(b) Name two things you could do to reduce the margin of error. What drawbacks do these actions have?

(c) Describe how nonresponse might lead to bias in this survey. Does the stated margin of error account for this possible bias?

22. **Employment in California** Each month the government releases unemployment statistics. The stated unemployment rate doesn't include people who choose not to be employed, such as retirees. Based on a random sample of 1000 California adults, a 99% confidence interval for the proportion of all California adults employed in the workforce is 0.532 to 0.612.[8]

(a) Interpret the confidence level.

(b) Name two things you could do to reduce the margin of error. What drawbacks do these actions have?

(c) Describe how untruthful answers might lead to bias in this survey. Does the stated margin of error account for this possible bias?

Multiple Choice: *Select the best answer for Exercises 23–26.*

Exercises 23 and 24 refer to the following setting. A researcher plans to use a random sample of houses to estimate the mean size (in square feet) of the houses in a large population.

23. The researcher is deciding between a 95% confidence level and a 99% confidence level. Compared with a 95% confidence interval, a 99% confidence interval will be

(a) narrower and would involve a larger risk of being incorrect.

(b) wider and would involve a smaller risk of being incorrect.

(c) narrower and would involve a smaller risk of being incorrect.

(d) wider and would involve a larger risk of being incorrect.

(e) wider and would have the same risk of being incorrect.

24. After deciding on a 95% confidence level, the researcher is deciding between a sample of size $n = 500$ and a sample of size $n = 1000$. Compared with using a sample size of $n = 500$, a confidence interval based on a sample size of $n = 1000$ will be

(a) narrower and would involve a larger risk of being incorrect.

(b) wider and would involve a smaller risk of being incorrect.

(c) narrower and would involve a smaller risk of being incorrect.

(d) wider and would involve a larger risk of being incorrect.

(e) narrower and would have the same risk of being incorrect.

25. In a poll conducted by phone,

I. Some people refused to answer questions.

II. People without telephones could not be in the sample.

III. Some people never answered the phone in several calls.

Which of these possible sources of bias is included in the ±2% margin of error announced for the poll?

(a) I only

(b) II only

(c) III only

(d) I, II, and III

(e) None of these

---

that our interval will capture the true propoportion. Increase the sample size. *Drawback:* larger samples cost more time and money. **(c)** People ashamed of being unemployed may give untruthful answers on the survey by claiming to be currently employed. This would cause the estimate from the sample to be greater than the true proportion of California adults who are in the workforce. The bias due to untruthful answers is not accounted for by the margin of error because the margin of error accounts for only the variability we expect from random sampling.

8.23 b

8.24 e

8.25 e

8.19 (a) The length would increase. **(b)** The confidence interval would be half as wide. **(c)** One of the practical difficulties would include nonresponse. For example, if people selected but not responding have different views from those responding, the estimated proportion may be off by more than 3 percentage points.

8.20 (a) The length would decrease. **(b)** The confidence interval would be half as wide. **(c)** One of the practical difficulties would include nonresponse. For example, if people selected but not responding have different views from those responding, the estimated proportion may be off by more than 3 percentage points.

8.21 (a) If we constructed a 90% confidence interval using each of the many random samples taken, about 90% of the intervals would capture the true average travel time to work for all employed California adults. **(b)** Decrease confidence level. *Drawback:* less confidence that our interval captures the true average. Increase sample size. *Drawback:* larger samples cost more time and money. **(c)** People who have longer travel times to work may have less time to respond to a survey. This would cause our estimate from the sample to be less than the true mean travel time to work. The bias due to nonresponse is not accounted for by the margin of error because the margin of error accounts for only the variability we expect from random sampling.

8.22 (a) If the government took many random samples and constructed a 99% confidence interval for each sample, about 99% of these intervals would capture the true proportion of all California adults in the workforce. **(b)** Decrease the confidence level. *Drawback:* less confidence

8.26 c

8.27 (a) This was an observational study. Pregnant women and children were not assigned to different amounts of exposure to magnetic fields. **(b)** No; we cannot make any conclusions about cause and effect because this was not an experiment. We can only conclude that there isn't convincing evidence that living near power lines is related to whether children develop cancer.

8.28 (a) The scatterplot shows a moderate, positive linear association between the height of a brother and the height of his sister.

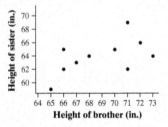

(b) $\hat{y} = 27.64 + 0.5270x$, where $\hat{y} =$ the predicted height of the sister and x is the height of the brother. **(c)** The slope is 0.5270; the predicted height of the sister increases by 0.5270 for each additional 1-inch increase in the brother's height. **(d)** $\hat{y} = 27.64 + 0.5270(71) = 65.057$ inches; the residual = $69 - 65.057 = 3.943$ inches. The actual height of the sister was 3.943 greater than the height predicted by the regression line with $x = 71$ inches.

PD **Section 8.2 Overview**

To watch the video overview of Section 8.2 (for teachers), click on the link in the TE-Book, look on the TRFD, or download from the Teacher's Resources on the book's digital platform.

TRM **Section 8.2 Alternate Examples**

You can find the Alternate Examples for this section in Microsoft Word format by clicking on the link in the TE-Book, opening the TRFD, or downloading from the Teacher's Resources on the book's digital platform

26. You have measured the systolic blood pressure of an SRS of 25 company employees. Based on the sample, a 95% confidence interval for the mean systolic blood pressure for the employees of this company is (122, 138). Which of the following statements is true?

(a) 95% of the sample of employees have a systolic blood pressure between 122 and 138.

(b) 95% of the population of employees have a systolic blood pressure between 122 and 138.

(c) If the procedure were repeated many times, 95% of the resulting confidence intervals would contain the population mean systolic blood pressure.

(d) If the procedure were repeated many times, 95% of the time the population mean systolic blood pressure would be between 122 and 138.

(e) If the procedure were repeated many times, 95% of the time the sample mean systolic blood pressure would be between 122 and 138.

Recycle and Review

27. **Power lines and cancer** (4.2, 4.3) Does living near power lines cause leukemia in children? The National Cancer Institute spent 5 years and $5 million gathering data on this question. The researchers compared 638 children who had leukemia with 620 who did not. They went into the homes and measured the magnetic fields in children's bedrooms, in other rooms, and at the front door. They recorded facts about power lines near the family home and also near the mother's residence when she was pregnant. *Result:* No association between leukemia and exposure to magnetic fields of the kind produced by power lines was found.[9]

(a) Was this an observational study or an experiment? Justify your answer.

(b) Does this study prove that living near power lines doesn't cause cancer? Explain your answer.

28. **Sisters and brothers** (3.1, 3.2) How strongly do physical characteristics of sisters and brothers correlate? Here are data on the heights (in inches) of 11 adult pairs:[10]

Brother	71	68	66	67	70	71	70	73	72	65	66
Sister	69	64	65	63	65	62	65	64	66	59	62

(a) Construct a scatterplot using brother's height as the explanatory variable. Describe what you see.

(b) Use technology to compute the least-squares regression line for predicting sister's height from brother's height.

(c) Interpret the slope in context.

(d) Calculate and interpret the residual for the first pair listed in the table.

SECTION 8.2 **Estimating a Population Proportion**

LEARNING TARGETS *By the end of the section, you should be able to:*

- State and check the Random, 10%, and Large Counts conditions for constructing a confidence interval for a population proportion.

- Determine the critical value for calculating a $C\%$ confidence interval for a population proportion using a table or technology.

- Construct and interpret a confidence interval for a population proportion.

- Determine the sample size required to obtain a $C\%$ confidence interval for a population proportion with a specified margin of error.

In Section 8.1, we saw that a confidence interval can be used to estimate an unknown population parameter. We are often interested in estimating the proportion p of some outcome in a population. Here are some examples:

- What proportion of U.S. adults are unemployed right now?
- What proportion of high school students have cheated on a test?

- What proportion of pine trees in a national park are infested with beetles?
- What proportion of college students pray daily?
- What proportion of a company's laptop batteries last as long as the company claims?

This section shows you how to construct and interpret a confidence interval for a population proportion.

Constructing a Confidence Interval for *p*

When Mr. Buckley's class did "The beads" activity in Section 8.1, the random sample of 251 beads they selected included 107 red beads and 144 other beads. Starting with the general formula for a confidence interval from Section 8.1:

$$\text{point estimate} \pm \text{margin of error}$$
$$= \text{statistic} \pm (\text{critical value}) \cdot (\text{standard deviation of statistic})$$

they determined the values to substitute into the formula using what they learned about the sampling distribution of the sample proportion $\hat{p}$ in Section 7.2.

- *Statistic:* The class decided to use $\hat{p} = 107/251 = 0.426$ because $\hat{p}$ is an unbiased estimator of *p*.
- *Critical value:* The class decided to use critical value $= 2$ based on the empirical rule for Normal distributions.
- *Standard deviation of statistic:* Remembering that the standard deviation of the sampling distribution of $\hat{p}$ is $\sigma_{\hat{p}} = \sqrt{\dfrac{p(1-p)}{n}}$, the class decided to use $\hat{p} = 0.426$ in the formula to get $\sqrt{\dfrac{0.426(1-0.426)}{251}} = 0.031$.

The class's 95% confidence interval is

$$0.426 \pm 2(0.031)$$
$$= 0.426 \pm 0.062 = (0.364, 0.488)$$

The class is 95% confident that the interval from 0.364 to 0.488 captures the true proportion of red beads in Mr. Buckley's container.

The interval constructed by Mr. Buckley's class is nearly correct. Here is the exact formula for a *one-sample z interval for a population proportion.*

ONE-SAMPLE *z* INTERVAL FOR A POPULATION PROPORTION

When the conditions are met, a C% confidence interval for the unknown proportion *p* is

$$\hat{p} \pm z^* \sqrt{\frac{\hat{p}(1-\hat{p})}{n}}$$

where z^* is the critical value for the standard Normal curve with C% of its area between $-z^*$ and z^*.

Teaching Tip

The formula for the standard deviation of the sampling distribution of $\hat{p}$ requires using the true proportion of the whole population (*p*). Because we do not know *p*, we will use our best guess ($\hat{p}$). This estimate of standard deviation of the sampling distribution of $\hat{p}$ is called the standard error.

Teaching Tip

We say "nearly" correct here because the critical value of 2 will soon be more precisely revealed to be 1.96.

➕ Ask the StatsMedic

Interpret the *z*-Score Like It's Your Job

In Chapter 2, we made sure to emphasize the interpretation of the *z*-score. This should help students understand the formula used here. The *z*-score tells us *how many standard deviations* away from the estimate we should go to get the desired level of confidence.

AP® EXAM TIP

The specific formula for a one-sample *z* interval for a population proportion is *not* on the formula sheet provided to students on the AP® Statistics exam. However, the formula sheet does include the general formula for a confidence interval and the formula for the standard error of the statistic $\hat{p}$.

Teaching Tip

Random condition: The data come from a random sample from the population of interest.
So what?
→ so individual observations are independent and we can generalize to the population.

Teaching Tip

10% condition: $n < (0.10)N$
So what?
→ so we can view observations as independent even though we are sampling without replacement.

Teaching Tip

If we sample more than 10% of the population, the formula we use for the standard deviation of $\hat{p}$ should include the finite population correction factor (*FPC*):

$$FPC = \sqrt{1 - \frac{n}{N}}, \text{ and}$$

$$\sigma_{\hat{p}} = \sqrt{\frac{p(1-p)}{n}} \cdot FPC$$

The finite population correction factor is not a topic on the AP® Statistics exam, however. When we don't use the *FPC*, our intervals will be more conservative (wider) than needed.

CONDITIONS FOR ESTIMATING p To make sure the formula for a one-sample z interval for a population proportion is valid, we need to verify that the observations in the sample can be viewed as independent and that the sampling distribution of $\hat{p}$ is approximately Normal. We do this by checking three conditions. Let's discuss them one at a time.

1. **The Random Condition** When our data come from a random sample, we can make an inference about the population from which the sample was selected. If the data come from a convenience sample or voluntary response sample, we should have no confidence that the resulting value of $\hat{p}$ is a good estimate of p. To be sure that $\hat{p}$ is a valid point estimate, we check the *Random condition*: The data come from a random sample from the population of interest.

Random sampling also helps ensure that individual observations in the sample can be viewed as independent. Finally, random sampling introduces chance into the data-production process. We can model random behavior with a probability distribution, like the sampling distributions of Chapter 7. Our method of calculation assumes that the data come from an SRS of size n from the population of interest. Other types of random samples (e.g., stratified or cluster) might be preferable to an SRS in a given setting, but they require more complex calculations than the ones we'll use. When an example, exercise, or AP® Statistics exam item refers to a "random sample" without saying "stratified," "cluster," or "systematic," you can assume the sample is an SRS.

2. **The 10% Condition** As you learned in Chapter 7, the formula for the standard deviation of the sampling distribution of $\hat{p}$ assumes that individual observations are independent. However, when we're sampling without replacement from a (finite) population, the observations are not independent because knowing the outcome of one trial helps us predict the outcome of future trials. Whenever we are sampling without replacement—which is nearly always—we need to check the *10% condition*: $n < 0.10N$, where n is the sample size and N is the population size.

When the 10% condition is met, the standard deviation of the sampling distribution of $\hat{p}$ is approximately

$$\sigma_{\hat{p}} = \sqrt{\frac{p(1-p)}{n}}$$

In practice, of course, we don't know the value of p. If we did, we wouldn't need to construct a confidence interval for it! In large random samples, $\hat{p}$ will be close to p. So we replace p in the formula for the standard deviation of the sample proportion with $\hat{p}$ to get the **standard error (SE)** of the sample proportion $\hat{p}$:

$$SE_{\hat{p}} = \sqrt{\frac{\hat{p}(1-\hat{p})}{n}}$$

The formula sheet provided on the AP® Statistics exam uses the notation $s_{\hat{p}}$ rather than $SE_{\hat{p}}$ for the standard error of the sample proportion $\hat{p}$.

Like the standard deviation, the standard error describes how much the sample proportion $\hat{p}$ typically varies from the population proportion p in repeated SRSs of size n.

> **DEFINITION Standard error**
>
> When the standard deviation of a statistic is estimated from data, the result is called the **standard error** of the statistic.

Teaching Tip

Students sometimes have trouble distinguishing the standard error from the margin of error. The standard error describes how far the sample statistic typically is from the population parameter. The margin of error is a multiple of the standard error, calculated so that the statistic will be within the margin of error of the parameter in $C\%$ of samples.

3. **The Large Counts Condition** When Mr. Buckley's class used the empirical rule to determine the critical value for their confidence interval, they were assuming that the sampling distribution of $\hat{p}$ was approximately Normal. If the distribution of $\hat{p}$ is approximately Normal, then $\hat{p}$ will be within 2 standard deviations of p in about 95% of samples. This means the value of p will be within 2 standard deviations of $\hat{p}$ in about 95% of samples. Thus, using a critical value of 2 will result in approximately 95% confidence.

From what we learned in Chapter 7, we can use the Normal approximation to the sampling distribution of $\hat{p}$ as long as $np \geq 10$ and $n(1-p) \geq 10$. Like the standard error, we replace p with $\hat{p}$ when checking the *Large Counts condition*: $n\hat{p} \geq 10$ and $n(1-\hat{p}) \geq 10$.

When the Large Counts condition is met, we can use a Normal distribution to calculate the critical value z^* for any confidence level. You'll learn how to do this soon.

Here is a summary of the three conditions for constructing a confidence interval for p.

CONDITIONS FOR CONSTRUCTING A CONFIDENCE INTERVAL ABOUT A PROPORTION

- **Random:** The data come from a random sample from the population of interest.
  - **10%:** When sampling without replacement, $n < 0.10N$.
- **Large Counts:** Both $n\hat{p}$ and $n(1-\hat{p})$ are at least 10.

Let's verify that the conditions were met for the interval calculated by Mr. Buckley's class.

EXAMPLE

The beads
Checking conditions

PROBLEM: Mr. Buckley's class wants to construct a confidence interval for $p =$ the true proportion of red beads in the container, which includes 3000 beads. Recall that the class's sample of 251 beads had 107 red beads and 144 other beads. Check if the conditions for constructing a confidence interval for p are met.

SOLUTION:
- Random: The class took a random sample of 251 beads from the container. ✓
  - 10%: 251 beads is less than 10% of 3000. ✓
- Large Counts:

$$n\hat{p} = 251\left(\frac{107}{251}\right) = 107 \geq 10 \quad \text{and}$$

$$n(1-\hat{p}) = 251\left(1 - \frac{107}{251}\right) = 251\left(\frac{144}{251}\right) = 144 \geq 10 ✓$$

FOR PRACTICE, TRY EXERCISE 29

ALTERNATE EXAMPLE   Skill 4.C

Who follows Katy Perry? Checking conditions

PROBLEM:
Katy Perry is one of the most popular accounts on Twitter®, with more than 100 million followers. Mr. Starnes wondered what proportion of Katy Perry's Twitter followers are female, so he selected a random sample of 500 of Katy Perry's followers. In the sample, 308 are female. Check if the conditions for constructing a confidence interval for p are met.

SOLUTION:
- *Random:* Mr. Starnes took a random sample of 500 of Katy Perry's Twitter followers. ✓
  - *10%:* 500 is less than 10% of 100 million. ✓
- *Large Counts:*

$$n\hat{p} = 500\left(\frac{308}{500}\right) = 308 \geq 10 \quad \text{and} \quad n(1-\hat{p}) = 500\left(1 - \frac{308}{500}\right) = 500\left(\frac{192}{500}\right) = 192 \geq 10 ✓$$

Notice that $n\hat{p}$ and $n(1-\hat{p})$ are the number of successes and failures in the sample. In the preceding example, we could address the Large Counts condition simply by saying, "The numbers of successes (107) and failures (144) in the sample are both at least 10."

> Simulation studies have shown that a variation of our method for calculating a 95% confidence interval for p can result in closer to a 95% capture rate in the long run, especially for small sample sizes. This simple adjustment, first suggested by Edwin Bidwell Wilson in 1927, is sometimes called the *plus four* estimate. Just pretend we have four additional observations, two of which are successes and two of which are failures. Then calculate the "plus four interval" using the plus four estimate in place of $\hat{p}$ and sample size $n + 4$ in our usual formula.

WHAT HAPPENS IF ONE OF THE CONDITIONS IS VIOLATED? If the data come from a voluntary response or convenience sample, there's no point in constructing a confidence interval for p. Violation of the Random condition severely limits our ability to make any inference beyond the data at hand.

The figure shows a screen shot from the *Confidence Intervals for Proportions* applet at the book's website, highschool.bfwpub.com/updatedtps6e. We set $n = 20$ and $p = 0.25$. The Large Counts condition is not met because $np = 20(0.25) = 5$ is not at least 10. We used the applet to generate 1000 random samples and construct 1000 95% confidence intervals for p. Only 902 of those 1000 intervals contained $p = 0.25$, a capture rate of 90.2%. When the Large Counts condition is violated, the capture rate will almost always be *less* than the one advertised by the confidence level when the method is used many times.

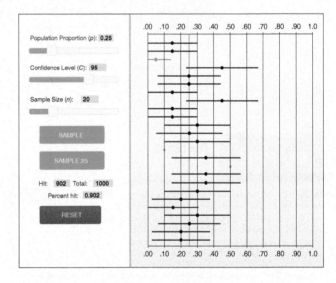

Violating the 10% condition means that we are sampling a large fraction of the population, which should be a good thing! But, as you learned in Section 7.2, the formula we use for the standard deviation of $\hat{p}$ gives a value that is too large when the 10% condition is violated. Confidence intervals based on this formula are wider than they need to be. If many 95% confidence intervals for a population proportion are constructed in this way, more than 95% of them will capture p.

Teaching Tip: Using Technology

Use the applet described on this page (or the *Simulating Confidence Intervals for a Population Parameter* at www.rossmanchance.com/applets) to explore other combinations of n and p. For example, if we take samples of size $n = 30$ from a population where $p = 0.01$ and calculate 95% confidence intervals, only about 25% of the intervals will capture the value of the population parameter!

TRM The Plus Four Interval for a Population Proportion

After reading the margin note on this page, students may wonder why we seek other variations of confidence intervals for a proportion. Studies have shown that when the usual formula for a confidence interval is used, the actual capture rate is often below the stated capture rate, especially when the sample size is small. For example, a "95% confidence interval" may capture the true proportion in only 92% of all possible samples. This discrepancy occurs because the distribution of $\hat{p}$ is discrete, not continuous. Because we use the standard Normal distribution, which is continuous, to calculate the critical values, these critical values are only approximations. If you or your students want to

know more about the "plus four interval," find it in the Teacher's Resource Materials located in the TE-Book, on the TRFD, or in the Teacher's view on the book's digital platform. Just know that the "plus four interval" is not on the AP® Statistics exam, although one of the Investigative Tasks for Unit 6 in the College Board's Question Bank explores the Plus 4 interval.

The actual capture rate is almost always *greater* than the reported confidence level when the 10% condition is violated.

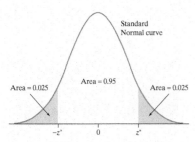

FIGURE 8.2 Finding the critical value z^* for a 95% confidence interval starts by labeling the middle 95% under a standard Normal curve and calculating the area in each tail.

CALCULATING CRITICAL VALUES How do we get the critical value z^* for our confidence interval? If the Large Counts condition is met, we can use the standard Normal curve. For their 95% confidence interval in the beads activity, Mr. Buckley's class used a critical value of 2. Based on the empirical rule for Normal distributions, they figured that $\hat{p}$ will be within 2 standard deviations of p in about 95% of all samples. Thus, p should be within 2 standard deviations of $\hat{p}$ in about 95% of all samples

We can get a more precise critical value from Table A or a calculator. As Figure 8.2 shows, the central 95% of the standard Normal distribution is marked off by 2 points, z^* and $-z^*$. We use the * to remind you that this is a critical value, not a standardized score that has been calculated from data.

Because of the symmetry of the Normal curve, the area in each tail is $0.05/2 = 0.025$. Once you know the tail areas, there are two ways to calculate the value of z^*:

- *Using Table A:* Search the body of Table A to find the point $-z^*$ with area 0.025 to its left. The entry $z = -1.96$ is what we are looking for, so $z^* = 1.96$.

z	.05	.06	.07
−2.0	.0202	.0197	.0192
−1.9	.0256	.0250	.0244
−1.8	.0322	.0314	.0307

- *Using technology:* The command invNorm(area:0.025, mean:0, SD:1) gives $z = -1.960$, so $z^* = 1.960$.

Now we can officially calculate a 95% confidence interval using the data from Mr. Buckley's class:

$$\hat{p} \pm z^* \sqrt{\frac{\hat{p}(1-\hat{p})}{n}} = 0.426 \pm 1.96\sqrt{\frac{0.426(1-0.426)}{251}}$$
$$= 0.426 \pm 0.061$$
$$= (0.365, 0.487)$$

Notice that the margin of error is slightly smaller for this interval than when the class used 2 for the critical value. Mr. Buckley's class is 95% confident that the interval from 0.365 to 0.487 captures the true proportion of red beads in his container. They can also be 95% confident that the interval from $3000(0.365) = 1095$ to $3000(0.487) = 1461$ captures the true *number* of red beads in his container of 3000 beads.

To find a level C confidence interval, we need to catch the central C% under the standard Normal curve. Here's an example that shows how to get the critical value z^* for a different confidence level and use it to calculate a confidence interval.

The ever-popular Katy Perry
Calculating a critical value and
confidence interval

PROBLEM:
Mr. Starnes wondered what proportion of Katy Perry's Twitter® followers are female, so he selected a random sample of 500 of Katy Perry's followers. In the sample, 308 are female. Assume the conditions for inference are met.
(a) Determine the critical value z^* for a 90% confidence interval for a proportion.
(b) Construct a 90% confidence interval for the proportion of all of Katy Perry's Twitter followers who are female.
(c) Interpret the interval from part (b).

SOLUTION:
(a) *Table A:* $z^* = 1.64$ or $z^* = 1.65$;
Tech: The command invNorm
(area: 0.05, mean: 0, SD: 1) gives
$z = -1.645$, so $z^* = 1.645$.
(b) $\hat{p} = 308/500 = 0.616$

$$0.616 \pm 1.645\sqrt{\frac{0.616(1 - 0.616)}{500}}$$

$= 0.616 \pm 0.036 = (0.580, 0.652)$

(c) We are 90% confident that the interval from 0.580 to 0.652 captures the true proportion of Katy Perry's Twitter followers who are female.

Teaching Tip

Consider having students write a general formula and a specific formula with variables before substituting numbers:
General formula:

point estimate $\pm$ margin of error

Specific formula:

$$\hat{p} \pm z^*\sqrt{\frac{\hat{p}(1 - \hat{p})}{n}}$$

One advantage to establishing this expectation is that the general formula will be the same for confidence intervals in later chapters. Only the specific formula changes for each new confidence interval.

EXAMPLE

Read any good books lately?
Calculating a critical value and confidence interval

PROBLEM: According to a 2016 Pew Research Center report, 73% of American adults have read a book in the previous 12 months. This estimate was based on a random sample of 1520 American adults.[11] Assume the conditions for inference are met.
(a) Determine the critical value z^* for a 90% confidence interval for a proportion.
(b) Construct a 90% confidence interval for the proportion of all American adults who have read a book in the previous 12 months.
(c) Interpret the interval from part (b).

SOLUTION:

(a) *Using Table A:* $z^* = 1.64$ or $z^* = 1.65$
Using technology: The command invNorm (area:0.05, mean:0, SD:1) gives $z = -1.645$, so $z^* = 1.645$.

(b) $0.73 \pm 1.645\sqrt{\dfrac{0.73(1 - 0.73)}{1520}}$

$= 0.73 \pm 0.019$
$= (0.711, 0.749)$

(c) We are 90% confident that the interval from 0.711 to 0.749 captures $p =$ the true proportion of American adults who have read a book in the previous 12 months.

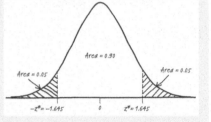

FOR PRACTICE, TRY EXERCISE 35

There are about 250 million U.S. adults. How *many* of them have read a book in the previous year? Using the confidence interval from the preceding example, we can be 90% confident that the interval from $250(0.711) = 177.75$ million to $250(0.749) = 187.25$ million captures the true number of U.S. adults who have read a book in the previous year.

CHECK YOUR UNDERSTANDING

Sleep Awareness Week begins in the spring with the release of the National Sleep Foundation's annual poll of U.S. sleep habits and ends with the beginning of daylight savings time, when most people lose an hour of sleep.[12] In the foundation's random sample of 1029 U.S. adults, 48% reported that they "often or always" got enough sleep during the past 7 nights.

1. Identify the parameter of interest.
2. Check if the conditions for constructing a confidence interval for p are met.
3. Find the critical value for a 99% confidence interval. Then calculate the interval.
4. Interpret the interval in context.

✓ Answers to CYU

1. $p =$ the true proportion of all U.S. adults who "often or always" got enough sleep during the last 7 nights.

2. The Random condition is met because the statement says that the adults were chosen randomly. The 10% condition is met because the sample size $n = 1029$ is less than 10% of all U.S. adults. The Large Counts condition is met because $n\hat{p} = (1029)(0.48) \approx 494 \geq 10$ and $n(1 - \hat{p}) = (1029)(1 - 0.48) \approx 535 \geq 10$.

3. $\dfrac{1 - 0.99}{2} = 0.005$; the closest area is 0.0051 (or 0.0049), corresponding to a critical value of $z^* = 2.57$ (or 2.58). *Tech:* invNorm(area: 0.005, mean: 0, SD: 1) $= -2.576$. So $z^* = 2.576$;

$0.48 \pm 2.576\sqrt{\dfrac{0.48(1 - 0.48)}{1029}} = 0.48 \pm 0.04$

$= (0.44, 0.52)$

4. We are 99% confident that the interval from 0.44 to 0.52 captures $p =$ the true proportion of all U.S. adults who would report that they "often or always" got enough sleep during the last 7 nights.

Putting It All Together: The Four-Step Process

Taken together, the examples about Mr. Buckley's class and "The Beads" activity show you how to get a confidence interval for an unknown population proportion p. Because there are many details to remember when constructing and interpreting a confidence interval, it is helpful to group them into four steps.

CONFIDENCE INTERVALS: A FOUR-STEP PROCESS

State: State the parameter you want to estimate and the confidence level.
Plan: Identify the appropriate inference method and check conditions.
Do: If the conditions are met, perform calculations.
Conclude: Interpret your interval in the context of the problem.

The next example illustrates the four-step process in action.

EXAMPLE

Distracted walking
Constructing and interpreting a confidence interval for p

PROBLEM: A recent poll of 738 randomly selected cell-phone users found that 170 of the respondents admitted to walking into something or someone while talking on their cell phone.[13] Construct and interpret a 95% confidence interval for the proportion of all cell-phone users who would admit to walking into something or someone while talking on their cell phone.

SOLUTION:

STATE: 95% CI for $p =$ the true proportion of all cell-phone users who would admit to walking into something or someone while talking on their cell phone.

PLAN: One-sample z interval for p.

- Random: Random sample of 738 cell-phone users. ✓
  - 10%: It is reasonable to assume that 738 is less than 10% of all cell-phone users. ✓
- Large Counts: The number of successes (170) and the number of failures (738 − 170 = 568) are both at least 10. ✓

> **STATE:** State the parameter you want to estimate and the confidence level.

> **PLAN:** Identify the appropriate inference method and check conditions.

> Remember that $n\hat{p}$ is the number of successes and $n(1-\hat{p})$ is the number of failures in the sample:
>
> $$n\hat{p} = 738\left(\frac{170}{738}\right) = 170$$
>
> $$n(1-\hat{p}) = 738\left(\frac{568}{738}\right) = 568$$

Teaching Tip

It is essential that your students follow this four-step process whenever a problem asks them to calculate a confidence interval. These steps were carefully designed based on the scoring rubrics for confidence interval questions on the AP® Statistics exam. Remind students that **S**tatistics **P**roblems **D**emand **C**onsistency! This would also be a good time to show your students the wealth of information in the back of their books. The Inference Summary repeats the four-step process for confidence intervals (and significance tests) at the top and then presents the details for each of the different inference procedures in a clearly organized grid.

Making Connections

The four-step process will be used repeatedly in Chapters 8–12. Familiarize students with the format and structure, and be consistent in your expectations. High expectations for the four-step process will make Chapters 9–12 much more successful.

AP® EXAM TIP

When naming the inference method, it is acceptable to list the calculator command 1-PropZInt. We prefer that students use the one-sample z interval for p, because it demonstrates deeper understanding and leads to fewer misconceptions. Some students think that a 2-PropZInt can be used in this example because there are two proportions—those who admit and those who don't. See 2016 AP® Exam #5c as an example.

ALTERNATE EXAMPLE   Skills 3.D, 4.B

Apple for college?
Constructing and interpreting a confidence interval for p

PROBLEM:
A recent study asked a random sample of 527 college-age adults (18–24 years old) about their plans for purchasing a new computer. Of the sample, 105 said that they planned to purchase an Apple computer within the next year. Construct and interpret a 95% confidence interval for the proportion of all college-age adults who plan to purchase an Apple computer within the next year.

SOLUTION:
STATE: 95% CI for $p =$ the true proportion of all college-age adults who plan to purchase an Apple computer within the next year.

PLAN: One-sample z interval for p

- *Random:* Random sample of 527 college-age adults. ✓
  - *10%:* It is reasonable to assume that 527 is less than 10% of all college-age adults. ✓
- *Large Counts:* The number of successes (105) and the number of failures (527 − 105 = 422) are both at least 10. ✓

DO: $\hat{p} = 105/527 = 0.199$

$$0.199 \pm 1.96\sqrt{\frac{0.199(1-0.199)}{527}}$$

$= 0.199 \pm 0.034 = (0.165, 0.233)$

CONCLUDE: We are 95% confident that the interval from 0.165 to 0.233 captures $p =$ the true proportion of all college-age adults who plan to purchase an Apple computer within the next year.

Teaching Tip: Using Technology

Remind students that they can watch a video of Technology Corner 18 on the e-Book or at the Student Site.

Teaching Tip: Using Technology

Remind students to look for the option that says "1-PropZInt," not the option that says "ZInterval." The latter option is for calculating a confidence interval for μ when σ is known. The 2-PropZInt option is for the two-sample z interval for a difference in proportions that we will learn about in Section 8.3.

Teaching Tip: Using Technology

When using 1-PropZInt, the value of "x" that students enter into the calculator is the number of successes and must be an integer or else the calculator will return a domain error. If students are given a proportion and a sample size and have to do a calculation to get x, be sure they round this value to the closest integer. Another common error is that students often enter the value of $\hat{p}$ instead of the number of successes. Remind your students that $\hat{p} = x/n$ and that the calculator wants both x and n.

DO: $\hat{p} = 170/738 = 0.230$

$0.230 \pm 1.96 \sqrt{\dfrac{0.230(1 - 0.230)}{738}}$

$= 0.230 \pm 0.030$

$= (0.200, 0.260)$

CONCLUDE: *We are 95% confident that the interval from 0.200 to 0.260 captures p = the true proportion of all cell-phone users who would admit to walking into something or someone while talking on their cell phone.*

DO: If the conditions are met, perform calculations.

CONCLUDE: Interpret your interval in the context of the problem.

Make sure your conclusion is about the population proportion (users who *would admit*) and not the sample proportion (those who *admitted*).

FOR PRACTICE, TRY EXERCISE 41

AP® EXAM TIP

If a free response question asks you to construct and interpret a confidence interval, you are expected to do the entire four-step process. That includes clearly defining the parameter, identifying the procedure, and checking conditions.

 Remember that the margin of error in a confidence interval only accounts for sampling variability! There are other sources of error that are not taken into account. As is the case with many surveys, we are forced to assume that respondents answer truthfully. If they don't, then we shouldn't be 95% confident that our interval captures the truth. Other problems like nonresponse and question wording can also affect the results of a survey. *Lesson:* Sampling beads is much easier than sampling people!

Your calculator will handle the "Do" part of the four-step process, as the following Technology Corner illustrates.

18. Technology Corner   CONSTRUCTING A CONFIDENCE INTERVAL FOR A POPULATION PROPORTION

TI-Nspire and other technology instructions are on the book's website at highschool.bfwpub.com/updatedtps6e.

The TI-83/84 can be used to construct a confidence interval for an unknown population proportion. We'll demonstrate using the previous example. Of $n = 738$ cell-phone users surveyed, $X = 170$ admitted to walking into something or someone while talking on their cell phone. To construct a confidence interval:

- Press $\boxed{\text{STAT}}$, then choose TESTS and 1-PropZInt.

- When the 1-PropZInt screen appears, enter $X = 170$, $n = 738$, and confidence level $= 0.95$. *Note:* The value you enter for X is the number of successes (not the proportion of successes) and must be an integer.

- Highlight "Calculate" and press $\boxed{\text{ENTER}}$. The 95% confidence interval for p is reported, along with the sample proportion $\hat{p}$ and the sample size, as shown here.

Choosing the Sample Size

In planning a study, we may want to choose a sample size that allows us to estimate a population proportion within a given margin of error. The formula for the margin of error (ME) in the confidence interval for p is

$$ME = z^* \sqrt{\frac{\hat{p}(1-\hat{p})}{n}}$$

To calculate the sample size, substitute values for ME, z^*, and $\hat{p}$, and solve for n. Unfortunately, we won't know the value of $\hat{p}$ until *after* the study has been conducted. This means we have to guess the value of $\hat{p}$ when choosing n. Here are two ways to do this:

1. Use a guess for $\hat{p}$ based on a pilot (preliminary) study or past experience with similar studies.
2. Use $\hat{p} = 0.5$ as the guess. The margin of error ME is largest when $\hat{p} = 0.5$, so this guess yields an upper bound for the sample size that will result in a given margin of error. If we get any other $\hat{p}$ when we do our study, the margin of error will be smaller than planned.

Once you have a guess for $\hat{p}$, the formula for the margin of error can be solved to give the required sample size n.

> **SAMPLE SIZE FOR DESIRED MARGIN OF ERROR WHEN ESTIMATING p**
>
> To determine the sample size n that will yield a $C\%$ confidence interval for a population proportion p with a maximum margin of error ME, solve the following inequality for n:
>
> $$z^* \sqrt{\frac{\hat{p}(1-\hat{p})}{n}} \leq ME$$
>
> where $\hat{p}$ is a guessed value for the sample proportion. The margin of error will always be less than or equal to ME if you use $\hat{p} = 0.5$.

Here's an example that shows you how to determine the sample size.

Teaching Tip

Even though it is acceptable to use the calculator's 1-PropZInt feature on the AP® Statistics exam, you can insist that students show additional work in your class. It will help students understand the structure of confidence intervals and prepare them for multiple-choice questions that focus on the way confidence intervals are constructed.

Teaching Tip

Ask students to imagine sitting around a conference room table to plan a survey. We want to determine how many people to interview. Most importantly, we have not yet collected any data.

Teaching Tip

Show students a few quick calculations of $\hat{p}(1 - \hat{p})$ for different values of $\hat{p}$:

$(0.2)(0.8) = 0.16$

$(0.4)(0.6) = 0.24$

$(0.5)(0.5) = 0.25$

$(0.6)(0.4) = 0.24$

$(0.8)(0.2) = 0.16$

It is easy to see that we get the maximum margin of error when $\hat{p} = 0.5$.

AP® EXAM TIP

The specific formula for the margin of error on this page is not included on the formula sheet provided to students on the AP® Statistics exam. However, the general formula for a confidence interval and the formula for the standard error of the statistic $\hat{p}$ are included, so students should be able to put it together.

Who is going to college?
Determining sample size

PROBLEM:

Abbie wonders about college plans for all the students at her large high school (over 3000 students). Specifically, she wants to know the proportion of students who are planning to go to college. Abbie wants her estimate to be within 5 percentage points (0.05) of the true proportion at a 90% confidence level. How many students should she randomly select?

SOLUTION:

$$1.645\sqrt{\frac{0.5(1-0.5)}{n}} \le 0.05 \text{ gives}$$

$n \ge 270.6$. The sample should include at least 271 students. *Note:* If $z^* = 1.64$ is used, the answer is at least 269 students; if $z^* = 1.65$ is used, the answer is at least 273 students.

Teacher Tip

For this example, require students to use exact values from the calculator (2ND ANS) during intermediate steps of solving for sample size. If they start rounding values during the process of solving the inequality, they can obtain widely varying final answers. Students should only round in the final step. Using the inequality $\le$ or $\ge$ will help students understand why they should always round up the final answer if the result is a decimal.

For the "Customer satisfaction" example, students can check their final answer by taking $n = 1068$ and plugging it into the margin of error formula:

$$ME = 1.96\sqrt{\frac{(0.5)(0.5)}{1068}} = 0.029999$$

This is a great way to double-check an answer and a good strategy for multiple-choice items about this content.

EXAMPLE

Customer satisfaction
Determining sample size

PROBLEM: A company has received complaints about its customer service. The managers intend to hire a consultant to carry out a survey of customers. Before contacting the consultant, the company president wants some idea of the sample size that she will be required to pay for. One value of interest is the proportion p of customers who are satisfied with the company's customer service. She decides that she wants the estimate to be within 3 percentage points (0.03) at a 95% confidence level. How large a sample is needed?

SOLUTION:

$$1.96\sqrt{\frac{0.5(1-0.5)}{n}} \le 0.03$$

> We have no idea about the true proportion p of satisfied customers, so we use $\hat{p} = 0.5$ as our guess to be safe.

$$\sqrt{\frac{0.5(1-0.5)}{n}} \le \frac{0.03}{1.96}$$

> Divide both sides by 1.96.

$$\frac{0.5(1-0.5)}{n} \le \left(\frac{0.03}{1.96}\right)^2$$

> Square both sides.

$$0.5(1-0.5) \le n\left(\frac{0.03}{1.96}\right)^2$$

> Multiply both sides by n.

$$\frac{0.5(1-0.5)}{\left(\frac{0.03}{1.96}\right)^2} \le n$$

> Divide both sides by $\left(\frac{0.03}{1.96}\right)^2$.

$$n \ge 1067.111$$

> Make sure to follow the inequality when rounding your answer.

The sample needs to include at least 1068 customers.

FOR PRACTICE, TRY EXERCISE 49

Why not round to the nearest whole number—in this case, 1067? Because a smaller sample size will result in a larger margin of error, possibly more than the desired 3 percentage points for the poll. In general, we round to the next highest integer when solving for sample size to make sure the margin of error is less than or equal to the desired value.

CHECK YOUR UNDERSTANDING

Refer to the preceding example about the company's customer satisfaction survey.

1. In the company's prior-year survey, 80% of customers surveyed said they were satisfied. Using this value as a guess for $\hat{p}$, find the sample size needed for a margin of error of at most 3 percentage points with 95% confidence. How does this compare with the required sample size from the example?

2. What if the company president demands 99% confidence instead of 95% confidence? Would this require a smaller or larger sample size, assuming everything else remains the same? Explain your answer.

✓ Answers to CYU

1. Solving $1.96\sqrt{\frac{0.80(0.20)}{n}} \le 0.03$ for n gives

$$n \ge (0.80)(0.20)\left(\frac{1.96}{0.03}\right)^2 = 682.95.$$ This

means that we should select a sample of at least 683 customers. This sample size is less than the sample size determined in the example using $\hat{p} = 0.5$.

2. If the company president demands 99% confidence instead, the required sample size will be larger because the critical value is larger for 99% confidence (2.576) versus 95% confidence (1.96). The company would need to select at least 1180 customers to have 99% confidence.

Section 8.2 | Summary

- When constructing a confidence interval for a population proportion p, we need to ensure that the observations in the sample can be viewed as independent and that the sampling distribution of $\hat{p}$ is approximately Normal. The required conditions are:
  - **Random:** The data come from a random sample from the population of interest.
    - **10%:** When sampling without replacement, $n < 0.10N$.
  - **Large Counts:** Both $n\hat{p}$ and $n(1 - \hat{p})$ are at least 10. That is, the number of successes and the number of failures in the sample are both at least 10.
- When the conditions are met, the C% **one-sample z interval for p** is

$$\hat{p} \pm z^* \sqrt{\frac{\hat{p}(1 - \hat{p})}{n}}$$

where z^* is the **critical value** for the standard Normal curve with C% of its area between $-z^*$ and z^*.

- When we use the value of $\hat{p}$ to estimate the standard deviation of the sampling distribution of $\hat{p}$, the result is the **standard error** of $\hat{p}$: $SE_{\hat{p}} = \sqrt{\dfrac{\hat{p}(1 - \hat{p})}{n}}$.

- When asked to construct and interpret a confidence interval, follow the four-step process:

  STATE: State the parameter you want to estimate and the confidence level.

  PLAN: Identify the appropriate inference method and check conditions.

  DO: If the conditions are met, perform calculations.

  CONCLUDE: Interpret your interval in the context of the problem.

- The sample size needed to obtain a confidence interval with a maximum margin of error ME for a population proportion involves solving

$$z^* \sqrt{\frac{\hat{p}(1 - \hat{p})}{n}} \leq ME$$

for n, where $\hat{p}$ is a guessed value for the sample proportion, and z^* is the critical value for the confidence level you want. Use $\hat{p} = 0.5$ if you don't have a good idea about the value of $\hat{p}$.

8.2 Technology Corner

TI-Nspire and other technology instructions are on the book's website at highschool.bfwpub.com/updatedtps6e.

18. Constructing a confidence interval for a population proportion Page 560

TRM Section 8.2 Quiz

There is one quiz available for this section in the Teacher's Resource Materials. Click on the link in the TE-Book, look on the TRFD, or download from the Teacher's Resources on the book's digital platform. You can also create your own quiz using the ExamView® Assessment Suite that is part of the TPS6 program. Questions are coded by Learning Target to make it easy to build parallel quizzes.

TRM Full Solutions to Section 8.2 Exercises

Click on the link in the TE-Book, open the TRFD, or download from the Teacher's Resources on the book's digital platform.

8.29 *Random:* Met—SRS. *10%:* Not met because 50 > 10% of seniors in the dormitory. *Large Counts:* Met because $14 \geq 10$ and $36 \geq 10$.

8.30 *Random:* Met—SRS. *10%:* Met because 50 < 10% of students at his college. *Large Counts:* Met because $38 \geq 10$ and $12 \geq 10$.

8.31 *Random:* Met—SRS. *10%:* Met because 25 < 10% of the thousands of bags filled in an hour. *Large Counts:* Not met because 3 bags with too much salt < 10.

8.32 *Random:* May not be met because we do not know if the whelk eggs were a random sample. *10%:* Met because 98 < 10% of all whelk eggs. *Large Counts:* Not met because $n\hat{p} = 9 < 10$.

8.33 (a) To ensure that the observations can be viewed as independent. Otherwise, the formula for the standard error of $\hat{p}$ will overestimate the standard deviation of the sampling distribution of $\hat{p}$.

(b) The capture rate will usually be greater than the specified confidence level.

8.34 (a) So the shape of the sampling distribution of $\hat{p}$ will be approximately Normal, which allows use of a Normal distribution to calculate the critical value z^*. **(b)** The capture rate will almost always be less than the specified confidence level.

8.35 (a) *Table A:* $z^* = 2.33$; *Tech:* $z^* = 2.326$

(b) $0.1996 \pm 2.326\sqrt{\dfrac{0.1996(1 - 0.1996)}{4579}}$

$\rightarrow (0.1859, 0.2133)$

(c) We are 95% confident that the interval from 0.1859 to 0.2133 captures $p =$ the true proportion of American adults who have earned money by selling something online in the previous year.

8.36 (a) *Table A:* $z^* = 2.05$; *Tech:* $z^* = 2.054$

(b) $0.1105 \pm 2.054\sqrt{\dfrac{0.1105(1 - 0.1105)}{172}}$

$\rightarrow (0.0614, 0.1596)$

(c) We are 95% confident that the interval from 0.0614 to 0.1596 captures $p =$ the true proportion of undergraduates at the large university who would go to the professor if they witnessed two students cheating on a quiz.

8.37 $SE_{\hat{p}} = \sqrt{\dfrac{0.1996(1 - 0.1996)}{4579}} =$ 0.0059; in repeated SRSs of size 4579,

---

Section 8.2 | Exercises

For Exercises 29 to 32, check whether each of the conditions is met for calculating a confidence interval for the population proportion p.

29. **Rating school food** Latoya wants to estimate the proportion of the seniors at her boarding school who like the cafeteria food. She interviews an SRS of 50 of the 175 seniors and finds that 14 think the cafeteria food is good.

30. **High tuition costs** Glenn wonders what proportion of the students at his college believe that tuition is too high. He interviews an SRS of 50 of the 2400 students and finds 38 of those interviewed think tuition is too high.

31. **Salty chips** A quality control inspector takes a random sample of 25 bags of potato chips from the thousands of bags filled in an hour. Of the bags selected, 3 had too much salt.

32. **Whelks and mussels** The small round holes you often see in seashells were drilled by other sea creatures, who ate the former dwellers of the shells. Whelks often drill into mussels, but this behavior appears to be more or less common in different locations. Researchers collected whelk eggs from the coast of Oregon, raised the whelks in the laboratory, then put each whelk in a container with some delicious mussels. Only 9 of 98 whelks drilled into a mussel.[14]

33. **The 10% condition** When constructing a confidence interval for a population proportion, we check that the sample size is less than 10% of the population size.

(a) Why is it necessary to check this condition?

(b) What happens to the capture rate if this condition is violated?

34. **The Large Counts condition** When constructing a confidence interval for a population proportion, we check that both $n\hat{p}$ and $n(1 - \hat{p})$ are at least 10.

(a) Why is it necessary to check this condition?

(b) What happens to the capture rate if this condition is violated?

35. **Selling online** According to a recent Pew Research Center report, many American adults have made money by selling something online. In a random sample of 4579 American adults, 914 reported that they earned money by selling something online in the previous year.[15] Assume the conditions for inference are met.

(a) Determine the critical value z^* for a 98% confidence interval for a proportion.

(b) Construct a 98% confidence interval for the proportion of all American adults who would report having earned money by selling something online in the previous year.

(c) Interpret the interval from part (b).

36. **Reporting cheating** What proportion of students are willing to report cheating by other students? A student project put this question to an SRS of 172 undergraduates at a large university: "You witness two students cheating on a quiz. Do you go to the professor?" Only 19 answered "Yes."[16] Assume the conditions for inference are met.

(a) Determine the critical value z^* for a 96% confidence interval for a proportion.

(b) Construct a 96% confidence interval for the proportion of all undergraduates at this university who would go to the professor.

(c) Interpret the interval from part (b).

37. **More online sales** Refer to Exercise 35. Calculate and interpret the standard error of $\hat{p}$ for these data.

38. **More cheating** Refer to Exercise 36. Calculate and interpret the standard error of $\hat{p}$ for these data.

39. **Going to the prom** Tonya wants to estimate what proportion of her school's seniors plan to attend the prom. She interviews an SRS of 50 of the 750 seniors in her school and finds that 36 plan to go to the prom.

(a) Identify the population and parameter of interest.

(b) Check conditions for constructing a confidence interval for the parameter.

(c) Construct a 90% confidence interval for p.

(d) Interpret the interval in context.

40. **Student government** The student body president of a high school claims to know the names of at least 1000 of the 1800 students who attend the school. To test this claim, the student government advisor randomly selects 100 students and asks the president to identify each by name. The president successfully names only 46 of the students.

(a) Identify the population and parameter of interest.

(b) Check conditions for constructing a confidence interval for the parameter.

(c) Construct a 99% confidence interval for p.

(d) Interpret the interval in context.

41. **Video games** A Pew Research Center report on gamers and gaming estimated that 49% of U.S. adults play video games on a computer, TV, game console, or portable device such as a cell phone. This estimate was based on a random sample of 2001 U.S. adults. Construct and interpret a 95% confidence interval for the proportion of all U.S. adults who play video games.[17]

---

the sample proportion of American adults who have earned money by selling something online in the previous year typically varies from the population proportion by about 0.0059.

8.38 $SE_{\hat{p}} = \sqrt{\dfrac{0.1105(1 - 0.1105)}{172}} = 0.0239$;

in repeated SRSs of size 172, the sample proportion of undergraduate students at a large university who would go to the professor if they witness two students cheating on a quiz typically varies from the population proportion by about 0.0239.

8.39 (a) All seniors at Tonya's high school; the true proportion of all seniors who plan to attend the prom. **(b)** *Random:* Random sample. *10%:* 50 < 10% of the population. *Large Counts:* $36 \geq 10$ and $14 \geq 10$.

(c) $0.72 \pm 1.645\sqrt{\dfrac{0.72(0.28)}{50}} = (0.616, 0.824)$

(d) We are 90% confident that the interval from 0.616 to 0.824 captures $p =$ the true proportion of all seniors at Tonya's high school who plan to attend the prom.

8.40 (a) All students at this high school; the true proportion of all students that the student body president knows by name. **(b)** *Random:* Random sample. *10%:* 100 < 10% of the population. *Large Counts:* $46 \geq 10$ and $54 \geq 10$.

(c) $0.46 \pm 2.576\sqrt{\dfrac{0.46(0.54)}{100}} = (0.332, 0.588)$

(d) We are 99% confident that the interval from 0.332 to 0.588 captures $p =$ the true proportion of all students at this high school whom the student body president knows by name.

42. September 11 A recent study asked U.S. adults to name 10 historic events that occurred in their lifetime that have had the greatest impact on the country. The most frequently chosen answer was the September 11, 2001, terrorist attacks, which was included by 76% of the 2025 randomly selected U.S. adults. Construct and interpret a 95% confidence interval for the proportion of all U.S. adults who would include the 9/11 attacks on their list of 10 historic events.

43. Age and video games Refer to Exercise 41. The study also estimated that 67% of adults aged 18–29 play video games, but only 25% of adults aged 65 and older play video games.

(a) Explain why you do not have enough information to give confidence intervals for these two age groups separately.

(b) Do you think a 95% confidence interval for adults aged 18–29 would have a larger or smaller margin of error than the estimate from Exercise 41? Explain your answer.

44. Age and September 11 Refer to Exercise 42. The study also reported that 86% of millennials included 9/11 in their top-10 list and 70% of baby boomers included 9/11.

(a) Explain why you do not have enough information to give confidence intervals for millennials and baby boomers separately.

(b) Do you think a 95% confidence interval for baby boomers would have a larger or smaller margin of error than the estimate from Exercise 42? Explain your answer.

45. Food fight A 2016 survey of 1480 randomly selected U.S. adults found that 55% of respondents agreed with the following statement: "Organic produce is better for health than conventionally grown produce."[18]

(a) Construct and interpret a 99% confidence interval for the proportion of all U.S. adults who think that organic produce is better for health than conventionally grown produce.

(b) Does the interval from part (a) provide convincing evidence that a majority of all U.S. adults think that organic produce is better for health? Explain your answer.

46. Three branches According to a recent study by the Annenberg Foundation, only 36% of adults in the United States could name all three branches of government. This was based on a survey given to a random sample of 1416 U.S. adults.[19]

(a) Construct and interpret a 90% confidence interval for the proportion of all U.S. adults who could name all three branches of government.

(b) Does the interval from part (a) provide convincing evidence that less than half of all U.S. adults could name all three branches of government? Explain your answer.

47. Prom totals Use your interval from Exercise 39 to construct and interpret a 90% confidence interval for the total number of seniors planning to go to the prom.

48. Student body totals Use your interval from Exercise 40 to construct and interpret a 99% confidence interval for the total number of students at the school that the student body president can identify by name. Then use your interval to evaluate the president's claim.

49. School vouchers A small pilot study estimated that
pg 562
44% of all American adults agree that parents should be given vouchers that are good for education at any public or private school of their choice.

(a) How large a random sample is required to obtain a margin of error of at most 0.03 with 99% confidence? Answer this question using the pilot study's result as the guessed value for $\hat{p}$.

(b) Answer the question in part (a) again, but this time use the conservative guess $\hat{p} = 0.5$. By how much do the two sample sizes differ?

50. Can you taste PTC? PTC is a substance that has a strong bitter taste for some people and is tasteless for others. The ability to taste PTC is inherited. About 75% of Italians can taste PTC, for example. You want to estimate the proportion of Americans who have at least one Italian grandparent and who can taste PTC.

(a) How large a sample must you test to estimate the proportion of PTC tasters within 0.04 with 90% confidence? Answer this question using the 75% estimate as the guessed value for $\hat{p}$.

(b) Answer the question in part (a) again, but this time use the conservative guess $\hat{p} = 0.5$. By how much do the two sample sizes differ?

51. Starting a nightclub A college student organization wants to start a nightclub for students under the age of 21. To assess support for this proposal, they will select an SRS of students and ask each respondent if he or she would patronize this type of establishment. What sample size is required to obtain a 90% confidence interval with a margin of error of at most 0.04?

52. Election polling Gloria Chavez and Ronald Flynn are the candidates for mayor in a large city. We want to estimate the proportion p of all registered voters in the city who plan to vote for Chavez with 95% confidence and a margin of error no greater than 0.03. How large a random sample do we need?

53. Teens and their TV sets According to a Gallup Poll report, 64% of teens aged 13 to 17 have TVs in their rooms. Here is part of the footnote to this report:

These results are based on telephone interviews with a randomly selected national sample of 1028 teenagers in the Gallup Poll Panel of households, aged 13 to 17.

8.41 S: p = the true proportion of all U.S. adults who play video games. **P:** One-sample z interval for p. *Random:* Random sample. *10%:* 2001 < 10% of all U.S. adults. *Large Counts:* 980 ≥ 10 and 1021 ≥ 10. **D:** (0.468, 0.512). **C:** We are 95% confident that the interval from 0.468 to 0.512 captures p = the true proportion of U.S. adults who play video games.

8.42 S: p = the true proportion of all U.S. adults who would include the 9/11 attacks on their list of 10 historic events. **P:** One-sample z interval for p. *Random:* Random sample. *10%:* 2025 < 10% of all U.S. adults. *Large Counts:* 1539 ≥ 10 and 486 ≥ 10. **D:** (0.741, 0.779). **C:** We are 95% confident that the interval from 0.741 to 0.779 captures p = the true proportion of U.S. adults who would include the 9/11 attacks on their list of 10 historic events.

8.43 (a) We do not know the size of the sample that was taken from each age group. **(b)** Larger, because this is a smaller sample size than the original group.

8.44 (a) We do not know the size of the sample that was taken from each group. **(b)** Larger, because this is a smaller sample size than the original group.

8.45 (a) S: p = the true proportion of all U.S. adults who think that organic produce is better for health than conventionally grown produce. **P:** One-sample z interval for p. *Random:* Random sample. *10%:* 1480 < 10% of all U.S. adults. *Large Counts:* 814 ≥ 10 and 666 ≥ 10. **D:** (0.517, 0.583). **C:** We are 99% confident that the interval from 0.517 to 0.583 captures p = the true proportion of U.S. adults who would agree with the statement: "Organic produce is better for health than conventionally grown produce." **(b)** Yes; all the plausible values in the interval are greater than 0.5, which provides convincing evidence that a majority of all U.S. adults think that organic produce is better for health than conventionally grown produce.

8.46 (a) S: p = the true proportion of all U.S. adults who could name all three branches of government. **P:** One-sample z interval for p. *Random:* Random sample. *10%:* 1416 < 10% of all U.S. adults. *Large Counts:* 510 ≥ 10 and 906 ≥ 10. **D:** (0.339, 0.381). **C:** We are 90% confident that the interval from 0.339 to 0.381 captures p = the true proportion of U.S. adults who could name all three branches of government. **(b)** Yes; all the plausible values in the interval are less than 0.5, which provides convincing evidence that less than half of all U.S. adults could name all three branches of government.

8.47 We expect between 61.6% and 82.4% of the 750 seniors to attend the prom. We can be 90% confident that the interval from 462 to 618 captures the total number of seniors planning to go to the prom.

8.48 We expect that the student body president knows the name of between 33.2% and 58.8% of the 1800 students who attend the school. We can be 90% confident that the interval from 598 to 1058 captures the total number of students who attend the school whose name is known by the student body president. The president's claim cannot be ruled out.

8.49 (a) Solving $2.576\sqrt{\dfrac{0.44(0.56)}{n}} \le 0.03$ gives $n \ge 1817$. **(b)** Solving $2.576\sqrt{\dfrac{0.5(0.5)}{n}} \le 0.03$ gives $n \ge 1844$. The conservative approach requires 27 more adults.

8.50 (a) Solving $1.645\sqrt{\dfrac{0.75(0.25)}{n}} \le 0.04$ gives $n \ge 318$. **(b)** Solving $1.645\sqrt{\dfrac{0.5(0.5)}{n}} \le 0.04$ gives $n \ge 423$. The conservative approach requires 105 more people.

8.51 Solving $1.645\sqrt{\dfrac{0.5(0.5)}{n}} \le 0.04$ gives $n \ge 423$.

8.52 Solving $1.96\sqrt{\dfrac{0.5(0.5)}{n}} \le 0.03$ gives $n \ge 1068$.

8.53 (a) Solving $0.03 = z^* \sqrt{\dfrac{0.64(0.36)}{1028}}$ gives $z^* = 2.00$. The confidence level is likely 95%. **(b)** Teens are hard to reach and often unwilling to participate in surveys, so nonresponse bias is a "practical difficulty." If teens with TVs in their rooms are less likely to answer the poll, the estimate from the poll would likely be too small.

8.54 (a) Solving $0.01 = z^* \sqrt{\dfrac{0.6341(0.3659)}{5594}}$ gives $z^* = 1.55$. The area between -1.55 and 1.55 under the standard Normal curve is 0.8788. The confidence level is likely 88%. **(b)** We do not know if those who *did* respond can reliably represent those who did not. Because the student athletes might worry about being identified if they have gambled, the estimate from the poll would likely be too small.

8.55 a

8.56 a

8.57 d

8.58 a

8.59 (a) X follows the Normal distribution with a mean of 21.1 and a standard deviation of 1.8. We want $P(X > 22)$.

(i) $z = \dfrac{22 - 21.1}{1.8} = 0.5$; $P(z > 0.5) = 0.3085$

(ii) $P(X > 22) =$ normalcdf(lower: 22, upper: 1000, mean: 21.1, SD: 1.8) $= 0.3085$

There is a 0.3085 probability that a randomly selected orange from this tree has a circumference greater than 22 cm.

(b) Normal distribution; $\mu_{\bar{x}} = \mu = 21.1$ cm. Because 20 is less than 10% of all oranges on the tree,

$\sigma_{\bar{x}} = \dfrac{\sigma}{\sqrt{n}} = \dfrac{1.8}{\sqrt{20}} = 0.4025$ cm.

We want to find $P(\bar{x} > 22)$.

(i) $z = \dfrac{22 - 21.1}{0.4025} = 2.24$; $P(z > 2.24) = 0.0125$

(ii) $P(\bar{x} > 22) =$ normalcdf(lower: 22, upper: 1000, mean: 21.1, SD: 0.4025) $= 0.0127$

There is a 0.0127 probability that the mean circumference of 20 randomly selected oranges from this tree is greater than 22 cm.

For results based on this sample, one can say . . . that the maximum error attributable to sampling and other random effects is ± 3 percentage points. In addition to sampling error, question wording and practical difficulties in conducting surveys can introduce error or bias into the findings of public opinion polls.[20]

(a) We omitted the confidence level from the footnote. Use what you have learned to estimate the confidence level, assuming that Gallup took an SRS.

(b) Give an example of a "practical difficulty" that could lead to bias in this survey.

54. Gambling and the NCAA Gambling is an issue of great concern to those involved in college athletics. Because of this concern, the National Collegiate Athletic Association (NCAA) surveyed randomly selected student athletes concerning their gambling-related behaviors.[21] Of the 5594 Division I male athletes who responded to the survey, 3547 reported participation in some gambling behavior. This includes playing cards, betting on games of skill, buying lottery tickets, betting on sports, and similar activities. A report of this study cited a 1% margin of error.

(a) The confidence level was not stated in the report. Use what you have learned to estimate the confidence level, assuming that the NCAA took an SRS.

(b) The study was designed to protect the anonymity of the student athletes who responded. As a result, it was not possible to calculate the number of students who were asked to respond but did not. How does this fact affect the way that you interpret the results?

Multiple Choice: *Select the best answer for Exercises 55–58.*

55. A Gallup poll found that only 28% of American adults expect to inherit money or valuable possessions from a relative. The poll's margin of error was ± 3 percentage points at a 95% confidence level. This means that

(a) the poll used a method that gets an answer within 3% of the truth about the population 95% of the time.

(b) the percent of all adults who expect an inheritance must be between 25% and 31%.

(c) if Gallup takes another poll on this issue, the results of the second poll will lie between 25% and 31%.

(d) there's a 95% chance that the percent of all adults who expect an inheritance is between 25% and 31%.

(e) Gallup can be 95% confident that between 25% and 31% of the sample expect an inheritance.

56. Refer to Exercise 55. Suppose that Gallup wanted to cut the margin of error in half from 3 percentage points to 1.5 percentage points. How should they adjust their sample size?

(a) Multiply the sample size by 4.

(b) Multiply the sample size by 2.

(c) Multiply the sample size by 1/2.

(d) Multiply the sample size by 1/4.

(e) There is not enough information to answer this question.

57. Most people can roll their tongues, but many can't. The ability to roll the tongue is genetically determined. Suppose we are interested in determining what proportion of students can roll their tongues. We test a simple random sample of 400 students and find that 317 can roll their tongues. The margin of error for a 95% confidence interval for the true proportion of tongue rollers among students is closest to which of the following?

(a) 0.0008　　　(d) 0.04

(b) 0.02　　　(e) 0.05

(c) 0.03

58. A newspaper reporter asked an SRS of 100 residents in a large city for their opinion about the mayor's job performance. Using the results from the sample, the C% confidence interval for the proportion of all residents in the city who approve of the mayor's job performance is 0.565 to 0.695. What is the value of C?

(a) 82　　　(d) 95

(b) 86　　　(e) 99

(c) 90

Review and Recycle

59. **Oranges** (6.1, 7.3) A home gardener likes to grow various kinds of citrus fruit. One of his mandarin orange trees produces oranges whose circumferences follow a Normal distribution with mean 21.1 cm and standard deviation 1.8 cm.

(a) What is the probability that a randomly selected orange from this tree has a circumference greater than 22 cm?

(b) What is the probability that a random sample of 20 oranges from this tree has a mean circumference greater than 22 cm?

60. **More oranges** (1.2, 2.2) The gardener in the previous exercise randomly selects 20 mandarin oranges from the tree and counts the number of seeds in each orange. Here are the data:

3 4 6 6 9 11 11 12 13 13 14 14 16 17 22 23 23 24 28 30

(a) Graph the data using a dotplot.

(b) Based on your graph, is it plausible that the number of seeds from oranges on this tree follows a distribution that is approximately Normal? Explain your answer.

8.60 (a)

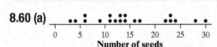

Number of seeds

(b) Based on the dotplot, it is plausible that the number of seeds follows a distribution that is approximately Normal and is fairly symmetric. There is no strong skewness, nor are there any outliers in the data.

<table>
<tr><td>

SECTION 8.3

Estimating a Difference in Proportions

</td></tr>
</table>

LEARNING TARGETS *By the end of the section, you should be able to:*

- Determine whether the conditions are met for constructing a confidence interval about a difference between two proportions.

- Construct and interpret a confidence interval for a difference between two proportions.

In Section 8.2, you learned how to calculate and interpret a confidence interval for a population proportion p. Many interesting statistical questions involve *comparing* the proportion of successes in two populations. What is the difference between the proportion of Democrats and the proportion of Republicans who favor the death penalty? How has the proportion of teenagers with a smartphone changed from 10 years ago? In both of these cases, we want to estimate the value of $p_1 - p_2$, where p_1 and p_2 are the proportions of success in Population 1 and Population 2.

Other statistical questions involve comparing the effectiveness of two treatments in an experiment. For example, how much more effective is a new medication for relieving headaches? What is the difference in the survival rate for two cancer treatments? In these cases, we want to estimate the value of $p_1 - p_2$, where p_1 and p_2 are the true proportions of success for individuals like the ones in the experiment who receive Treatment 1 or Treatment 2.

Confidence Intervals for $p_1 - p_2$

When data come from two independent random samples or two groups in a randomized experiment (the Random condition), the statistic $\hat{p}_1 - \hat{p}_2$ is our best guess for the value of $p_1 - p_2$. The method we use to calculate a confidence interval for $p_1 - p_2$ requires that the sampling distribution of $\hat{p}_1 - \hat{p}_2$ be approximately Normal. In Section 7.2, you learned that this will be true whenever $n_1 p_1$, $n_1(1 - p_1)$, $n_2 p_2$, and $n_2(1 - p_2)$ are all at least 10. Because we don't know the value of p_1 or p_2 when we are estimating their difference, we use $\hat{p}_1$ and $\hat{p}_2$ when checking the Large Counts condition.

CONDITIONS FOR CONSTRUCTING A CONFIDENCE INTERVAL ABOUT A DIFFERENCE IN PROPORTIONS

- **Random:** The data come from two independent random samples or from two groups in a randomized experiment.
  - **10%:** When sampling without replacement, $n_1 < 0.10N_1$ and $n_2 < 0.10N_2$.
- **Large Counts:** The counts of "successes" and "failures" in each sample or group—$n_1\hat{p}_1$, $n_1(1 - \hat{p}_1)$, $n_2\hat{p}_2$, $n_2(1 - \hat{p}_2)$—are all at least 10.

Recall from Chapter 4 that the Random condition is important for determining the scope of inference. Random sampling allows us to generalize our results to the populations of interest; random assignment in an experiment permits us to draw cause-and-effect conclusions.

PD Section 8.3 Overview

To watch the video overview of Section 8.3 (for teachers), click on the link in the TE-Book, look on the TRFD, or download from the Teacher's Resources on the book's digital platform.

TRM Section 8.3 Alternate Examples

You can find the Alternate Examples for this section in Microsoft Word format by clicking on the link in the TE-Book, opening the TRFD, or downloading from the Teacher's Resources on the book's digital platform.

Teaching Tip

Point out that the conditions for doing inference for a difference in proportions are much like the conditions for doing inference for a single proportion. The main difference is that we have to check the conditions for *both* samples/groups. Also, the Random condition now requires that the two samples are independent or that the data come from two groups in a randomized experiment.

Teaching Tip

While it is important that students know how to check each condition, it is equally important that they understand *why* we check the condition. We call this the "*So what?*"

Random condition: So what?
→ so we can generalize to both populations (random samples) *or* so we can show causation (random assignment).
→ "independent" random samples allow us to add variances in the formula for standard deviation of the statistic.

10% condition: So what?
→ so we can view observations as independent in each sample even though we are sampling without replacement.

Large counts condition: So what?
→ so the sampling distribution of $\hat{p}_1 - \hat{p}_2$ will be approximately Normal and we can use z^* to do calculations.

Nike or Adidas?
Checking conditions

PROBLEM:
A student project tried to determine if males and females have the same brand preference between Nike and Adidas. The students took separate random samples of 40 females and 30 males from a large high school and asked them if they prefer Nike or Adidas. Of the females, 55% preferred Nike, compared to 80% of the males who preferred Nike. Let p_F = the true proportion of all females at the school who prefer Nike over Adidas and p_M = the true proportion of all males at the school who prefer Nike over Adidas. Check if the conditions for calculating a confidence interval for $p_F - p_M$ are met.

SOLUTION:
- *Random:* Independent random samples of 40 females and 30 males from the school. ✓
  - *10%:* 40 < 10% of all females at the large school and 30 < 10% of all males at the large school ✓
- *Large Counts:* Because $n_M(1 - \hat{p}_M) = 30(0.20) = 6$ is not ≥ 10, this condition is not satisfied.

EXAMPLE

Do you prefer brand names?
Checking conditions

PROBLEM: A Harris Interactive survey asked independent random samples of adults from the United States and Germany about the importance of brand names when buying clothes. Of the 2309 U.S. adults surveyed, 26% said brand names were important, compared with 22% of the 1058 German adults surveyed. Let p_U = the true proportion of all U.S. adults who think brand names are important when buying clothes and p_G = the true proportion of all German adults who think brand names are important when buying clothes. Check if the conditions for calculating a confidence interval for $p_U - p_G$ are met.

SOLUTION:
- *Random:* Independent random samples of 2309 U.S. adults and 1058 German adults. ✓
  - *10%:* 2309 < 10% of all U.S. adults and 1058 < 10% of all German adults ✓
- *Large Counts?* $n_U \hat{p}_U = 2309(0.26) = 600.34 \to 600$, $n_U(1 - \hat{p}_U) = 2309(0.74) = 1708.66 \to 1709$, $n_G \hat{p}_G = 1058(0.22) = 232.76 \to 233$, $n_G(1 - \hat{p}_G) = 1058(0.78) = 825.24 \to 825$ are all ≥ 10. ✓

> Be sure to mention *independent* random samples from the populations of interest when checking the Random condition.

> We round these values to the nearest whole number because they represent the actual numbers of successes and failures in the two samples.

FOR PRACTICE, TRY EXERCISE 61

If the conditions are met, we can use our familiar formula to calculate a confidence interval for $p_1 - p_2$:

$$\text{statistic} \pm (\text{critical value}) \cdot (\text{standard deviation of statistic})$$
$$= (\hat{p}_1 - \hat{p}_2) \pm z^* \cdot (\text{standard deviation of statistic})$$

In Section 7.2, you also learned that the standard deviation of the sampling distribution of $\hat{p}_1 - \hat{p}_2$ is

$$\sigma_{\hat{p}_1 - \hat{p}_2} = \sqrt{\frac{p_1(1 - p_1)}{n_1} + \frac{p_2(1 - p_2)}{n_2}}$$

when we have two types of independence:

- Independent samples, so we can add the variances of $\hat{p}_1$ and $\hat{p}_2$. This is why we reminded you to mention *independent* random samples when checking the Random condition in the preceding example.
- Independent observations within each sample. When sampling without replacement, the actual value of the standard deviation is smaller than the formula suggests. However, if the 10% condition is met for both samples, the given formula is approximately correct.

Because we don't know the values of the parameters p_1 and p_2, we replace them in the standard deviation formula with the sample proportions. The result is the *standard error* of $\hat{p}_1 - \hat{p}_2$:

$$\text{SE}_{\hat{p}_1 - \hat{p}_2} = \sqrt{\frac{\hat{p}_1(1 - \hat{p}_1)}{n_1} + \frac{\hat{p}_2(1 - \hat{p}_2)}{n_2}}$$

> The formula sheet provided on the AP® Statistics exam uses the notation $s_{\hat{p}_1 - \hat{p}_2}$ rather than $\text{SE}_{\hat{p}_1 - \hat{p}_2}$ for the standard error of the difference in sample proportions $\hat{p}_1 - \hat{p}_2$.

This value estimates how much the difference in sample proportions will typically vary from the difference in the true proportions if we repeat the random sampling or random assignment many times.

When the Large Counts condition is met, we find the critical value z^* for the given confidence level using Table A or technology. Our confidence interval for $p_1 - p_2$ is therefore

$$\text{statistic} \pm (\text{critical value}) \cdot (\text{standard error of statistic})$$

$$= (\hat{p}_1 - \hat{p}_2) \pm z^* \sqrt{\frac{\hat{p}_1(1 - \hat{p}_1)}{n_1} + \frac{\hat{p}_2(1 - \hat{p}_2)}{n_2}}$$

This is often called a *two-sample z interval for a difference between two proportions.*

TWO-SAMPLE *z* INTERVAL FOR A DIFFERENCE BETWEEN TWO PROPORTIONS

When the conditions are met, a C% confidence interval for $p_1 - p_2$ is

$$(\hat{p}_1 - \hat{p}_2) \pm z^* \sqrt{\frac{\hat{p}_1(1 - \hat{p}_1)}{n_1} + \frac{\hat{p}_2(1 - \hat{p}_2)}{n_2}}$$

where z^* is the critical value for the standard Normal curve with C% of the area between $-z^*$ and z^*.

Let's return to the brand names example. Recall that Harris Interactive took independent random samples of 2309 U.S. adults and 1058 German adults, and found that $\hat{p}_U = 0.26$ and $\hat{p}_G = 0.22$. We already confirmed that the conditions are met. A 95% confidence interval for $p_U - p_G$ is

$$(0.26 - 0.22) \pm 1.96 \sqrt{\frac{0.26(1 - 0.26)}{2309} + \frac{0.22(1 - 0.22)}{1058}}$$

$$= 0.04 \pm 0.03$$

$$= (0.01, 0.07)$$

> The interpretation of a confidence interval for a difference in proportions is the same as for a single proportion. Interpreting the confidence *level* is the same as well: If we were to repeat the sampling process many times and compute a 95% confidence interval each time, about 95% of those intervals would capture the difference in the true proportion of all U.S. adults and all German adults who think brand names are important when buying clothes.

Interpretation: We are 95% confident that the interval from 0.01 to 0.07 captures $p_U - p_G =$ the difference in the true proportions of all U.S. adults and all German adults who think brand names are important when buying clothes.

Note that the confidence interval does not include 0 (no difference) as a plausible value for $p_U - p_G$, so we have convincing evidence of a difference between the population proportions. In fact, it is believable that $p_U - p_G$ has any value between 0.01 and 0.07. We can restate this in context as follows: The interval suggests that the importance of brand names when buying clothes for U.S. adults is between 1 and 7 percentage points higher than for German adults.

It would *not* be correct to say that the importance of brand names when buying clothes for U.S. adults is between 1 and 7 *percent* higher than for German adults. To see why, suppose that $p_U = 0.25$ and $p_G = 0.20$. The difference $p_U - p_G = 0.25 - 0.20 = 0.05$, or 5 percentage points. But the proportion of U.S. adults who think brand names are important when buying clothes is $0.05/0.20 = 0.25$, or 25% higher than the corresponding proportion of German adults.

The researchers in the preceding example selected independent random samples from the two populations they wanted to compare. In practice, it's common

AP® EXAM TIP

The formula for the confidence interval for a difference of proportions is *not* included on the AP® Statistics exam formula sheet. However, students are given the general formula for a confidence interval and the formula for standard error of the sampling distribution of $\hat{p}_1 - \hat{p}_2$.

Making Connections

Remind students of the general formula for confidence intervals from Section 8.1 and connect it to the specific formula for difference in proportions.

General formula:
$$\text{estimate} \pm \text{margin of error}$$

Specific formula:
$$(\hat{p}_1 - \hat{p}_2) \pm z^* \sqrt{\frac{\hat{p}_1(1 - \hat{p}_1)}{n_1} + \frac{\hat{p}_2(1 - \hat{p}_2)}{n_2}}$$

Preparing for Inference

For a significance test, the null hypothesis will state that there is no difference between the population proportions ($H_0: p_U - p_G = 0$).

Making Connections

In Chapter 9, we will establish the close connection between confidence intervals and significance tests. A confidence interval gives a set of plausible values for the parameter. If the null hypothesis value is not contained within the interval, we have convincing evidence to reject the null hypothesis.

Making Connections

In this chapter, we do not make a distinction between taking two independent random samples and taking one random sample and separating it into two groups. However, in Chapter 12, this will become an important distinction when we study chi-square tests.

Teaching Tip

The "*So what?*" is different for random sampling and random assignment.
Random sample → so we can generalize to both populations.
Random assignment → so we can show causation.

to take one random sample that includes individuals from both populations of interest and then to separate the chosen individuals into two groups. For instance, a polling company may randomly select 1000 U.S. adults, then separate the Republicans from the Democrats to estimate the difference in the proportion of all people in each party who favor the death penalty. The two-sample z procedures for comparing proportions are still valid in such situations, provided that the two groups can be viewed as independent samples from their respective populations of interest.

INFERENCE FOR EXPERIMENTS So far, we have focused on doing inference using data that were produced by random sampling. However, many important statistical results come from randomized comparative experiments. Fortunately, the formula for calculating a confidence interval for a difference between two proportions is the same whether there are two independent random samples or two randomly assigned groups in an experiment. However, there are differences in the way we define parameters and check conditions.

In an experiment to compare treatments for prostate cancer, 731 men with localized prostate cancer were randomly assigned either to have surgery or to be observed only. After 20 years, $\hat{p}_S = 141/364 = 0.387$ of the men assigned to surgery were still alive and $\hat{p}_O = 122/367 = 0.332$ of the men assigned to observation were still alive.[22] The parameters in this setting are:

p_S = the true proportion of men like the ones in the experiment who would survive 20 years when getting surgery

p_O = the true proportion of men like the ones in the experiment who would survive 20 years when only being observed

> **AP® EXAM TIP**
>
> Many students lose credit when defining parameters in an experiment by describing the sample proportion rather than the true proportion. For example, "the true proportion of the men who *had* surgery and survived 20 years" describes $\hat{p}_S$ not p_S.

Most experiments on people use recruited volunteers as subjects. When subjects are not randomly selected, researchers cannot generalize the results of an experiment to some larger populations of interest. But researchers can draw cause-and-effect conclusions that apply to people like those who took part in the experiment. This same logic applies to experiments on animals or things.

In addition to the difference in the Random condition, there is a change to the 10% condition when analyzing an experiment. Unless the experimental units are randomly selected without replacement from some population, we don't have to check it. Fortunately, there is no change to the Large Counts condition.

Putting It All Together: Two-Sample z Interval for $p_1 - p_2$

The following example shows how to construct and interpret a confidence interval for a difference in proportions. As usual with inference problems, we follow the four-step process.

Making Connections

From Chapter 4, remind students that a voluntary response sample might not be representative of the whole population, whereas a random sample should be representative of the population, allowing us to generalize to the population.

Making Connections

Remind students about the scope of inference table from Chapter 4:

| | | Were individuals randomly assigned to groups? | |
		Yes	No
Were individuals randomly selected?	Yes	Inference about the population: YES Inference about cause and effect: YES	Inference about the population: YES Inference about cause and effect: NO
	No	Inference about the population: NO Inference about cause and effect: YES	Inference about the population: NO Inference about cause and effect: NO

EXAMPLE

Treating lower back pain
Confidence interval for $p_1 - p_2$

4 STEP

PROBLEM: Patients with lower back pain are often given nonsteroidal anti-inflammatory drugs (NSAIDs) like naproxen to help ease their pain. Researchers wondered if taking Valium along with the naproxen would affect pain relief. To find out, they recruited 112 patients with severe lower back pain and randomly assigned them to one of two treatments: naproxen and Valium or naproxen and placebo. After 1 week, 39 of the 57 subjects who took naproxen and Valium reported reduced lower back pain, compared with 43 of the 55 subjects in the naproxen and placebo group.[23]

(a) Construct and interpret a 99% confidence interval for the difference in the proportion of patients like these who would report reduced lower back pain after taking naproxen and Valium versus after taking naproxen and placebo for a week.

(b) Based on the confidence interval in part (a), what conclusion would you make about whether taking Valium along with naproxen affects pain relief? Justify your answer.

SOLUTION:

(a) STATE: 99% CI for $p_1 - p_2$, where $p_1 =$ true proportion of patients like these who would report reduced lower back pain after taking naproxen and Valium for a week and $p_2 =$ true proportion of patients like these who would report reduced lower back pain after taking naproxen and placebo for a week.

> Be sure to indicate the order of subtraction when defining the parameter.

PLAN: Two-sample z interval for $p_1 - p_2$

- Random: Randomly assigned patients to take naproxen and Valium or naproxen and placebo. ✓

> We don't have to check the 10% condition because researchers did not sample patients without replacement from a larger population.

- Large Counts: $39, 57 - 39 = 18, 43,$ and $55 - 43 = 12$ are all ≥ 10. ✓

DO: $\hat{p}_1 = \dfrac{39}{57} = 0.684, \hat{p}_2 = \dfrac{43}{55} = 0.782$

> $(\hat{p}_1 - \hat{p}_2) \pm z^* \sqrt{\dfrac{\hat{p}_1(1 - \hat{p}_1)}{n_1} + \dfrac{\hat{p}_2(1 - \hat{p}_2)}{n_2}}$

$(0.684 - 0.782) \pm 2.576 \sqrt{\dfrac{0.684(0.316)}{57} + \dfrac{0.782(0.218)}{55}}$

$= -0.098 \pm 0.214$

$= (-0.312, 0.116)$

> Refer to the Technology Corner that follows the example. The calculator's 2-PropZInt gives $(-0.3114, 0.1162)$.

CONCLUDE: We are 99% confident that the interval from -0.312 to 0.116 captures $p_1 - p_2 =$ the difference in the true proportions of patients like these who would report reduced pain after taking naproxen and Valium versus after taking naproxen and a placebo for a week.

> The interval suggests that the true proportion of patients like these who would report reduced pain after taking naproxen and Valium is between 31.2 percentage points lower and 11.6 percentage points higher than after taking naproxen and placebo.

(b) Because the interval includes 0 as a plausible value for $p_1 - p_2$, we don't have convincing evidence that taking Valium along with naproxen affects pain relief for patients like these.

FOR PRACTICE, TRY EXERCISE 65

ALTERNATE EXAMPLE

Skills 3.D, 4.D

Approval of the president Confidence interval for $p_1 - p_2$

PROBLEM:
Gallup regularly conducts polls asking U.S. adults if they approve of the job the president is doing. How did President Trump's approval rating change from his first day in office to his 100th day in office? According to a Gallup poll of 1500 randomly selected U.S. adults taken on his first day in office, 45% approved of Trump's job performance. Another Gallup poll of 1500 U.S. adults was taken on his 100th day in office and showed a 38% approval rating.
(a) Construct and interpret a 95% confidence interval for the change in the true proportions of U.S. adults who approved of Trump's job performance from his first day to his 100th day in office.

(b) Based on the confidence interval in part (a), what conclusion would you make about whether Trump's approval rating has changed? Justify your answer.

SOLUTION:
(a) STATE: 95% CI for $p_{100} - p_1$, where $p_{100} =$ true proportion of all U.S. adults who approved of Trump's performance on his 100th day and $p_1 =$ true proportion of all U.S. adults who approved of Trump's performance on his first day in office.

PLAN: Two-sample z interval for $p_{100} - p_1$

- *Random:* Independent random samples of 1500 U.S. adults on first day and 1500 U.S. adults on the 100th day. ✓

- *10%:* $1500 < 10\%$ of all U.S. adults on the first day and $1500 < 10\%$ of all U.S. adults on the 100th day. ✓

- *Large Counts:* The numbers of successes ($1500 \cdot 0.45 = 675$, $1500 \cdot 0.38 = 570$) and failures ($1500 \cdot 0.55 = 825$, $1500 \cdot 0.62 = 930$) in the two samples are all ≥ 10. ✓

DO: $(0.38 - 0.45) \pm$

$1.96\sqrt{\dfrac{0.38(0.62)}{1500} + \dfrac{0.45(0.55)}{1500}}$

$= (-0.1052, -0.0348)$

CONCLUDE: We are 95% confident that the interval -0.1052 to -0.0348 captures $p_{100} - p_1 =$ the change in the true proportions of all U.S. adults who approved of Trump's job performance from his first day to his 100th day in office.
(b) Because the interval does not include 0 as a plausible value for $p_{100} - p_1$, we have convincing evidence that Trump's approval rating changed from the first day to the 100th day. The interval suggests that the true proportion of all U.S. adults who approved of Trump's performance on his 100th day is between 10.5 percentage points lower and 3.5 percentage points lower than on his first day.

Teaching Tip

Here are some important tips for teaching students to use the four-step process for constructing a two-sample z interval for $p_1 - p_2$:

- Inform students they can use the wording of the question to define their parameters and to write the conclusion. We call this "parroting" the stem of the question.
- When naming the inference method, it is acceptable to list the calculator command 2-PropZInt. We prefer that students use the two-sample z interval for $p_1 - p_2$, because it demonstrates deeper understanding and leads to fewer misconceptions.
- Consider having students write the "*So what?*" for each condition.
- Consider having students write a general formula and a specific formula with variables before substituting numbers:

General formula:
estimate $\pm$ margin of error

Specific formula:

$$(\hat{p}_1 - \hat{p}_2) \pm z^*\sqrt{\frac{\hat{p}_1(1 - \hat{p}_1)}{n_1} + \frac{\hat{p}_2(1 - \hat{p}_2)}{n_2}}$$

One advantage to establishing this expectation is that the general formula is the same for all the confidence intervals in this course. Only the specific formula changes for each new confidence interval. It is risky to provide this much work on the AP® Statistics exam because students may lose credit due to a minor notation/substitution error. Consider requiring these formulas during the school year and then relaxing this expectation at the end of the course when preparing for the AP® exam.

We could have subtracted the proportions in the opposite order in part (a) of the example. The resulting 99% confidence interval for $p_2 - p_1$ is

$$(0.782 - 0.684) \pm 2.576\sqrt{\frac{0.782(0.218)}{55} + \frac{0.684(0.316)}{57}}$$
$$= 0.098 \pm 0.214$$
$$= (-0.116, 0.312)$$

Notice that the endpoints of the interval have the same values but opposite signs to the ones in the example. This interval suggests that the true proportion of patients like these who would report reduced pain after taking naproxen and placebo is between 11.6 percentage points lower and 31.2 percentage points higher than after taking naproxen and Valium. That's equivalent to our interpretation of the confidence interval for $p_1 - p_2$ in part (a) of the example.

The fact that 0 is included in a confidence interval for $p_1 - p_2$ means that we don't have convincing evidence of a difference between the true proportions. Keep in mind that 0 is just one of many plausible values for $p_1 - p_2$ based on the sample data. Never suggest that you believe the difference between the true proportions is 0 just because 0 is in the interval!

You can use technology to perform the calculations in the "Do" step. Remember that this comes with potential benefits and risks on the AP® Statistics exam.

---

19. Technology Corner

CONSTRUCTING A CONFIDENCE INTERVAL FOR A DIFFERENCE IN PROPORTIONS

TI-Nspire and other technology instructions are on the book's website at highschool.bfwpub.com/updatedtps6e.

The TI-83/84 can be used to construct a confidence interval for $p_1 - p_2$. We'll demonstrate using the preceding example. Of $n_1 = 57$ subjects who took naproxen and Valium, $X_1 = 39$ reported reduced lower back pain after a week. Of $n_2 = 55$ subjects who took naproxen and placebo, $X_2 = 43$ reported reduced lower back pain after a week. To construct a confidence interval:

- Press STAT, then choose TESTS and 2-PropZInt.
- When the 2-PropZInt screen appears, enter the values shown. Note that the values of x_1, n_1, x_2, and n_2 must be integers or you will get a domain error.
- Highlight "Calculate" and press ENTER.

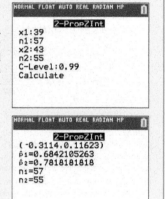

Teaching Tip: Using Technology

Remind students that they can watch videos for each Technology Corner by clicking the link in the e-Book or at the Student Site.

Teaching Tip: Using Technology

Warn students against choosing the wrong procedure on their calculators. The 2-SampZInt is for comparing *means* when the population standard deviations are known. What we want here is the 2-PropZInt.

Teaching Tip: Using Technology

The values of x_1 and x_2 are the numbers of successes, not the proportions of successes. Also, they must be integers! If students are given proportions and sample sizes and have to do calculations to get x_1 and x_2, be sure they round these values to the closest integer.

AP® EXAM TIP

The formula for the two-sample z interval for $p_1 - p_2$ often leads to calculation errors by students. As a result, your teacher may recommend using the calculator's 2-PropZInt feature to compute the confidence interval on the AP® Statistics exam. Be sure to name the procedure (two-sample z interval for $p_1 - p_2$) in the "Plan" step and give the interval $(-0.311, 0.116)$ in the "Do" step.

CHECK YOUR UNDERSTANDING

A Pew Research Center poll asked independent random samples of working women and men how much they value job security. Of the 806 women, 709 said job security was very or extremely important, compared with 802 of the 944 men surveyed. Construct and interpret a 95% confidence interval for the difference in the proportion of all working women and men who consider job security very or extremely important.

Section 8.3 Summary

- Confidence intervals to estimate the difference between the proportions p_1 and p_2 of successes for two populations or treatments are based on the difference $\hat{p}_1 - \hat{p}_2$ between the sample proportions.
- When constructing a confidence interval for a difference in population proportions, we must check for independence in the data collection process and that the sampling distribution of $\hat{p}_1 - \hat{p}_2$ is approximately Normal. The required conditions are:
  - **Random:** The data come from two independent random samples or from two groups in a randomized experiment.
    - **10%:** When sampling without replacement, $n_1 < 0.10N_1$ and $n_2 < 0.10N_2$.
  - **Large Counts:** The counts of "successes" and "failures" in each sample or group—$n_1\hat{p}_1$, $n_1(1-\hat{p}_1)$, $n_2\hat{p}_2$, $n_2(1-\hat{p}_2)$—are all at least 10.
- When conditions are met, a C% confidence interval for $p_1 - p_2$ is

$$(\hat{p}_1 - \hat{p}_2) \pm z^* \sqrt{\frac{\hat{p}_1(1-\hat{p}_1)}{n_1} + \frac{\hat{p}_2(1-\hat{p}_2)}{n_2}}$$

where z^* is the standard Normal critical value with C% of its area between $-z^*$ and z^*. This is called a **two-sample z interval for $p_1 - p_2$.**
- Be sure to follow the four-step process whenever you construct and interpret a confidence interval for the difference between two proportions.
- You can use a confidence interval for a difference in proportions to determine if a claimed value is plausible. For example, if 0 is included in a confidence interval for $p_1 - p_2$, it is plausible that there is no difference between the population proportions.

Teaching Tip

Another option is to require students to show the calculations for their work within this chapter. Then, at the end of the course when preparing for the AP® Statistics exam, relax this expectation and suggest that students instead use the calculator to get the interval.

✓ Answers to CYU

STATE: 95% CI for $p_1 - p_2$, where $p_1 =$ true proportion of working women who would say that job security is very or extremely important and $p_2 =$ true proportion of working men who would say that job security is very or extremely important.

PLAN: Two-sample z interval for $p_1 - p_2$.

- *Random:* The data come from independent random samples of 806 working women and 944 working men. ✓
  - *10%:* $n_1 = 806 < 10\%$ of the population of all working women and $n_2 = 944 < 10\%$ of the population of all working men. ✓

- *Large Counts:* 709, $806 - 709 = 97$, 802, and $944 - 802 = 142$ are all ≥ 10. ✓

DO: $\hat{p}_1 = \dfrac{709}{806} = 0.880$, $\hat{p}_2 = \dfrac{802}{944} = 0.850$;

$(0.88 - 0.85) \pm$

$1.96\sqrt{\dfrac{0.88(0.12)}{806} + \dfrac{0.85(0.15)}{944}} =$

$0.03 \pm 0.032 = (-0.002, 0.062)$

CONCLUDE: We are 95% confident that the interval from -0.002 to 0.062 captures $p_1 - p_2 =$ the difference in the true proportions of all working women and all working men who would say that job security is very or extremely important.

TRM Section 8.3 Quiz

There is one quiz available for this section in the Teacher's Resource Materials. Click on the link in the TE-Book, look on the TRFD, or download from the Teacher's Resources on the book's digital platform. You can also create your own quiz using the ExamView® Assessment Suite that is part of the TPS6 program. Questions are coded by Learning Target to make it easy to build parallel quizzes.

8.3 Technology Corner

TI-Nspire and other technology instructions are on the book's website at highschool.bfwpub.com/updatedtps6e.

19. Constructing a confidence interval for a difference in proportions   Page 572

TRM **Full Solutions to Section 8.3 Exercises**

Click on the link in the TE-Book, open the TRFD, or download from the Teacher's Resources on the book's digital platform.

Answers to Section 8.3 Exercises

8.61 *Random:* Not met because these data do not come from independent random samples or from two groups in a randomized experiment. *10%:* Since no sampling took place, the 10% condition does not apply. *Large Counts:* Not met because there were fewer than 10 successes (3) in the group from the west side of Woburn.

8.62 *Random:* Crackers were randomly assigned to one of the two treatment groups. *10%:* Since no sampling took place, the 10% condition does not apply. *Large Counts:* Not met because there were fewer than 10 successes (0) in the microwave group.

8.63 *Random:* Random assignment. *Large Counts:* 25, 11, 18, 18 $\geq$ 10.

8.64 *Random:* Independent random samples. *10%:* 500 < 10% of all DVDs produced by each machine. *Large Counts:* 480, 20, 484, 16 $\geq$ 10.

8.65 (a) S: p_1 = true proportion of young men who live in their parents' home and p_2 = that of young women . . . **P:** Two-sample z interval for $p_1 - p_2$. *Random:* Independent random samples. *10%:* 2253 < 10% of all young men and 2629 < 10% of all young women. *Large Counts:* 986, 1267, 923, 1706 $\geq$ 10. **D:** (0.051, 0.123) **C:** We are 99% confident that the interval from 0.051 to 0.123 captures $p_1 - p_2$ = the true difference in the proportions of young men and young women who live in their parents' home. **(b)** Because the interval does not contain 0, there is convincing evidence that the true proportion of young men who live at their parents' home is different from the true proportion of young women who do.

Section 8.3 | Exercises

61. Don't drink the water! The movie *A Civil Action* (1998) tells the story of a major legal battle that took place in the small town of Woburn, Massachusetts. A town well that supplied water to east Woburn residents was contaminated by industrial chemicals. During the period that residents drank water from this well, 16 of 414 babies born had birth defects. On the west side of Woburn, 3 of 228 babies born during the same time period had birth defects. Let p_1 = the true proportion of all babies born with birth defects in west Woburn and p_2 = the true proportion of all babies born with birth defects in east Woburn. Check if the conditions for calculating a confidence interval for $p_1 - p_2$ are met.

62. Broken crackers We don't like to find broken crackers when we open the package. How can makers reduce breaking? One idea is to microwave the crackers for 30 seconds right after baking them. Randomly assign 65 newly baked crackers to the microwave and another 65 to a control group that is not microwaved. After 1 day, none of the microwave group were broken and 16 of the control group were broken.[24] Let p_1 = the true proportion of crackers like these that would break if baked in the microwave and p_2 = the true proportion of crackers like these that would break if not microwaved. Check if the conditions for calculating a confidence interval for $p_1 - p_2$ are met.

63. Cockroaches The pesticide diazinon is commonly used to treat infestations of the German cockroach, *Blattella germanica*. A study investigated the persistence of this pesticide on various types of surfaces. Researchers applied a 0.5% emulsion of diazinon to glass and plasterboard. After 14 days, they randomly assigned 72 cockroaches to two groups of 36, placed one group on each surface, and recorded the number that died within 48 hours. On glass, 18 cockroaches died, while on plasterboard, 25 died. If p_1 and p_2 are the true proportions of cockroaches like these that would die within 48 hours on glass treated with diazinon and on plasterboard treated with diazinon, respectively, check if the conditions for calculating a confidence interval for $p_1 - p_2$ are met.

64. Digital video disks A company that records and sells rewritable DVDs wants to compare the reliability of DVD fabricating machines produced by two different manufacturers. They randomly select 500 DVDs produced by each fabricator and find that 484 of the disks produced by the first machine are acceptable and 480 of the disks produced by the second machine are acceptable. If p_1 and p_2 are the proportions of acceptable DVDs produced by the first and second machines, respectively, check if the conditions for calculating a confidence interval for $p_1 - p_2$ are met.

65. Young adults living at home A surprising number of young adults (ages 19 to 25) still live in their parents' homes. The National Institutes of Health surveyed independent random samples of 2253 men and 2629 women in this age group.[25] The survey found that 986 of the men and 923 of the women lived with their parents.

(a) Construct and interpret a 99% confidence interval for the difference in the true proportions of men and women aged 19 to 25 who live in their parents' homes.

(b) Does your interval from part (a) give convincing evidence of a difference between the population proportions? Justify your answer.

66. Where's Egypt? In a Pew Research poll, 287 out of 522 randomly selected U.S. men were able to identify Egypt when it was highlighted on a map of the Middle East. When 520 randomly selected U.S. women were asked, 233 were able to do so.

(a) Construct and interpret a 95% confidence interval for the difference in the true proportions of U.S. men and U.S. women who can identify Egypt on a map.

(b) Based on your interval, is there convincing evidence of a difference in the true proportions of U.S. men and women who can identify Egypt on a map? Justify your answer.

67. **More young adults** Interpret the confidence level for the interval in Exercise 65.

68. **More about Egypt** Interpret the confidence level for the interval in Exercise 66.

69. **Response bias** Does the appearance of the interviewer influence how people respond to a survey question? Ken (white, with blond hair) and Hassan (darker, with Middle Eastern features) conducted an experiment to address this question. They took turns (in a random order) walking up to people on the main street of a small town, identifying themselves as students from a local high school, and asking them, "Do you support President Obama's decision to launch airstrikes in Iraq?" Of the 50 people Hassan spoke to, 11 said "Yes," while 21 of the 44 people Ken spoke to said "Yes." Construct and interpret a 90% confidence interval for the difference in the proportions of people like these who would say they support President Obama's decision when asked by Hassan versus when asked by Ken.

70. **Quit smoking** Nicotine patches are often used to help smokers quit. Does giving medicine to fight depression help? A randomized double-blind experiment assigned 244 smokers to receive nicotine patches and another 245 to receive both a patch and the antidepressant drug bupropion. After a year, 40 subjects in the nicotine patch group had abstained from smoking, as had 87 in the patch-plus-drug group. Construct and interpret a 99% confidence interval for the difference in the true proportion of smokers like these who would abstain when using bupropion and a nicotine patch and the proportion who would abstain when using only a patch.

71. **Ban junk food!** A CBS News poll asked 606 randomly selected women and 442 randomly selected men, "Do you think putting a special tax on junk food would encourage more people to lose weight?" 170 of the women and 102 of the men said "Yes."[26] A 99% confidence interval for the difference (Women – Men) in the true proportion of people in each population who would say "Yes" is −0.020 to 0.120.

(a) Does the confidence interval provide convincing evidence that the two population proportions are different? Explain your answer.

(b) Does the confidence interval provide convincing evidence that the two population proportions are equal? Explain your answer.

72. **Artificial trees?** An association of Christmas tree growers in Indiana wants to know if there is a difference in preference for natural trees between urban and rural households. So the association sponsored a survey of Indiana households that had a Christmas tree last year to find out. In a random sample of 160 rural households, 64 had a natural tree. In a separate random sample of 261 urban households, 89 had a natural tree. A 95% confidence interval for the difference (Rural – Urban) in the true proportion of households in each population that had a natural tree is −0.036 to 0.154.

(a) Does the confidence interval provide convincing evidence that the two population proportions are different? Explain your answer.

(b) Does the confidence interval provide convincing evidence that the two population proportions are equal? Explain your answer.

Multiple Choice: *Select the best answer for Exercises 73–75.*

73. Earlier in this section, you read about an experiment comparing surgery and observation as treatments for men with prostate cancer. After 20 years, $\hat{p}_S = 141/364 = 0.387$ of the men who were assigned to surgery were still alive and $\hat{p}_O = 122/367 = 0.332$ of the men who were assigned to observation were still alive. Which of the following is the 95% confidence interval for $p_S - p_O$?

(a) $(141 - 122) \pm 1.96 \sqrt{\dfrac{141 \cdot 223}{364} + \dfrac{122 \cdot 245}{367}}$

(b) $(141 - 122) \pm 1.96 \left(\sqrt{\dfrac{141 \cdot 223}{364}} + \sqrt{\dfrac{122 \cdot 245}{367}} \right)$

(c) $(0.387 - 0.332) \pm 1.96 \sqrt{\dfrac{0.387 \cdot 0.613}{364} + \dfrac{0.332 \cdot 0.668}{367}}$

(d) $(0.387 - 0.332) \pm 1.96 \left(\sqrt{\dfrac{0.387 \cdot 0.613}{364}} + \sqrt{\dfrac{0.332 \cdot 0.668}{367}} \right)$

(e) $(0.387 - 0.332) \pm 1.96 \sqrt{\dfrac{0.387 \cdot 0.613}{364} - \dfrac{0.332 \cdot 0.668}{367}}$

8.66 S: p_M = true proportion of men who can identify Egypt on a map and p_W = that of women . . . **P:** Two-sample z-interval for $p_M - p_W$. *Random:* Independent random samples. *10%:* 522 < 10% of all men and 520 < 10% of all women. *Large Counts:* 287, 235, 233, 287 $\geq$ 10. **D:** (0.041, 0.162) **C:** We are 95% confident that the interval from 0.041 to 0.162 captures $p_1 - p_2$ = the true difference in the proportions of men and women who can identify Egypt on a map. **(b)** Because the interval does not contain 0, there is convincing evidence that the true proportion of men who can identify Egypt on a map is different from the true proportion of women who can identify Egypt on a map.

8.67 If we were to select many independent random samples of 2253 men and 2629 women from the population of all young adults (ages 19 to 25) and construct a 99% confidence interval for the difference in the true proportions each time, about 99% of the intervals would capture the difference in the true proportions of men and women aged 19 to 25 who live in their parents' homes.

8.68 If we were to select many independent random samples of 522 men and 520 women from the population of all men and women in the U.S. and construct a 95% confidence interval for the difference in the true proportions each time, about 95% of the intervals would capture the difference in the true proportions of U.S. men and U.S. women who can identify Egypt on a map.

8.69 S: p_1 = true proportion of people like the ones in this study who would say they support President Obama's decision when asked by Hassan and p_2 = true proportion . . . when asked by Ken. **P:** Two-sample z-interval for $p_1 - p_2$. *Random:* Random assignment. *Large Counts:* 11, 39, 21, 23 are $\geq$ 10. **D:** $(-0.414, -0.100)$ **C:** We are 90% confident that the interval from -0.414 to -0.100 captures $p_1 - p_2$ = the true difference in the proportions of people like the ones in this study who would say they support President Obama's decision when asked by Hassan or by Ken.

8.70 S: p_1 = true proportion of smokers like these who would abstain when using only a patch and p_2 = true proportion . . . when using both. **P:** Two-sample z-interval for $p_1 - p_2$. *Random:* Random assignment. *Large Counts:* 40, 204, 87, 158 are $\geq$ 10. **D:** $(-0.291, -0.092)$ **C:** We are 99% confident that the interval from $(-0.291, -0.092)$ captures $p_1 - p_2$ = the true difference in the proportion of smokers like these who would abstain when using only a patch and the proportion of smokers like these who would abstain when using bupropion and a nicotine patch.

8.71 (a) Because the interval includes 0 as a plausible value for the difference (Women – Men) in the true proportion of people in each population who would say "Yes," we do not have convincing evidence that the two population proportions are different. **(b)** No; because 0 is captured in the interval, it is *plausible* that the population proportions are equal, but this does not provide *convincing evidence* that the two proportions are equal.

8.72 (a) Because the interval includes 0 as a plausible value for the difference (Rural – Urban) in the true proportion of households that have a natural tree, we do not have convincing evidence that the two population proportions are different. **(b)** No; because 0 is captured in the interval, it is *plausible* that the population proportions are equal, but this does not provide *convincing evidence* that the two proportions are equal.

8.73 c

8.74 d

8.75 d

8.76 (a) The distributions of number of putts for those that did and did not describe their putting technique are skewed to the right. There is one upper outlier in each distribution. The median number of putts for those who described their putting technique (≈ 17 putts) is greater than the median number of putts for those who did not describe their putting technique (≈ 5.5 putts). The distribution of the number of putts for those who did not describe their putting technique is less variable (*IQR* ≈ 15 − 4 = 11 putts) than the distribution of the number of putts for those who did describe their putting technique (*IQR* ≈ 28 − 12 = 16 putts). **(b)** It is better to compare medians because the distributions are both skewed with an outlier. **(c)** Random assignment helps to create two groups of golfers that are roughly equivalent at the beginning of the study. This ensures that the effect of other variables (e.g., putting skill, mental focus, quality of equipment) are spread evenly among the two groups of golfers.

8.77 (a) The difference (Described − Didn't describe) in the median number of putts is positive. This means that those who described their putting technique have a median number of putts that is greater than those who did not describe their putting technique. **(b)** Yes; of the 100 trials in the simulation, none of the simulated differences in the median number of putts required was 11.5 or greater by chance alone. This suggests that it is harder for golfers like these to make putts after being asked to describe their putting technique.

---

74. When constructing a confidence interval for a difference between two population proportions, why is it important to check that the number of successes and the number of failures in each sample is at least 10?

(a) So we can generalize the results to the populations from which the samples were selected.

(b) So we can assume that the two samples are independent.

(c) So we can assume that the observations within each sample are independent.

(d) So we can assume the sampling distribution of $\hat{p}_1 - \hat{p}_2$ is approximately Normal.

(e) So we can assume that population 1 and population 2 are approximately Normal.

75. To estimate the difference in the proportion of students at high school A and high school B who drive themselves to school, a district administrator selected a random sample of 100 students from each school. At school A, 23 of the students said they drive themselves; at school B, 29 of the students said they drive themselves. A 90% confidence interval for $p_A - p_B$ is −0.16 to 0.04. Based on this interval, which conclusion is best?

(a) Because −0.06 is in the interval, there is convincing evidence of a difference in the population proportions.

(b) Because 0 is in the interval, there is convincing evidence of a difference in the population proportions.

(c) Because −0.06 is in the interval, there is not convincing evidence of a difference in the population proportions.

(d) Because 0 is in the interval, there is not convincing evidence of a difference in the population proportions.

(e) Because most of the interval is negative, there is convincing evidence that a greater proportion of students at high school B drive themselves to school.

Review and Recycle

Exercises 76 and 77 refer to the following setting. Athletes often comment that they try not to "overthink it" when competing in their sport. Is it possible to "overthink it"? Or is this just another cliché that athletes use in interviews? To investigate, researchers put some golfers to the test.[27] In the experiment, researchers recruited 40 experienced golfers and allowed them some time to practice their putting. After practicing, they randomly assigned the golfers to one of two groups. Golfers in one group had to write a detailed

---

description of their putting technique (which could lead to "overthinking it"). Golfers in the other group had to do an unrelated verbal task for the same amount of time. After completing their tasks, each golfer was asked to attempt putts from a fixed distance until he or she made 3 putts in a row.

76. **Don't overthink it!** (1.3, 4.2) The boxplots summarize the results of this experiment.

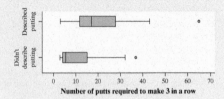

(a) Compare these distributions.

(b) Why might it be better to compare medians rather than means?

(c) What was the purpose of randomly assigning the 40 golfers to the two treatments?

77. **More overthinking it** (4.3) The difference in the medians for the two groups was 17 − 5.5 = 11.5 putts.

(a) Explain how this difference gives some evidence that it is harder for golfers like these to make putts after being asked to describe their putting technique.

(b) To determine if a difference in medians of 11.5 or more could happen simply due to the chance variation in random assignment, 100 trials of a simulation were conducted assuming that the treatment a subject receives does not affect the number of putts required. Each dot in the dotplot represents the difference in medians for one trial. Based on the results of the simulation, is there convincing evidence that it is harder for golfers like these to make putts after being asked to describe their putting technique? Explain your answer.

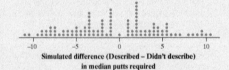

Chapter 8 Wrap-Up

FRAPPY! FREE RESPONSE AP® PROBLEM, YAY!

The following problem is modeled after actual AP® Statistics exam free response questions. Your task is to generate a complete, concise response in 15 minutes.

Directions: Show all your work. Indicate clearly the methods you use, because you will be scored on the correctness of your methods as well as on the accuracy and completeness of your results and explanations.

Members at a popular fitness club currently pay a $40 per month membership fee. The owner of the club wants to raise the fee to $50 but is concerned that some members will leave the gym if the fee increases. To investigate, the owner plans to survey a random sample of the club members and construct a 95% confidence interval for the proportion of all members who would quit if the fee was raised to $50.

(a) Explain the meaning of "95% confidence" in the context of the study.
(b) After the owner conducted the survey, he calculated the confidence interval to be 0.18 ± 0.075. Interpret this interval in the context of the study.

(c) According to the club's accountant, the fee increase will be worthwhile if fewer than 20% of the members quit. According to the interval from part (b), can the owner be confident that the fee increase will be worthwhile? Explain.
(d) One of the conditions for calculating the confidence interval in part (b) is that $n\hat{p} \geq 10$ and $n(1 - \hat{p}) \geq 10$. Explain why it is necessary to check this condition.

After you finish, you can view two example solutions on the book's website (highschool.bfwpub.com/updatedtps6e). Determine whether you think each solution is "complete," "substantial," "developing," or "minimal." If the solution is not complete, what improvements would you suggest to the student who wrote it? Finally, your teacher will provide you with a scoring rubric. Score your response and note what, if anything, you would do differently to improve your own score.

Chapter 8 Review

Section 8.1: Confidence Intervals: The Basics

In this section, you learned that a point estimate is the single best guess for the value of a population parameter. You also learned that a confidence interval provides an interval of plausible values for a parameter based on sample data. To interpret a confidence interval, say, "We are C% confident that the interval from ___ to ___ captures the [parameter in context]," where C is the confidence level of the interval.

The confidence level C describes the percentage of confidence intervals that we expect to capture the value of the parameter in repeated sampling. To interpret a C% confidence level, say, "If we took many samples of the same size from the same population and used them to construct C% confidence intervals, about C% of those intervals would capture the [parameter in context]."

Confidence intervals are formed by including a margin of error on either side of the point estimate. The size of the margin of error is determined by several factors, including the confidence level C and the sample size n. Increasing the sample size n makes the standard deviation of our statistic smaller, decreasing the margin of error. Increasing the confidence level C makes the margin of error larger, to ensure that the capture rate of the interval increases to C%. Remember that the margin of error only accounts for sampling variability—it does not account for any bias in the data collection process.

Section 8.2: Estimating a Population Proportion

In this section, you learned how to construct and interpret confidence intervals for a population proportion. Several important conditions must be met for this type of confidence interval

577

TRM FRAPPY! Materials

Please consult the Teacher's Resource Materials for sample student responses, a scoring rubric, and a printable version of the original question with space for students to write their responses. We present a model solution here.

Answers:

(a) If many samples of members were selected and many 95% confidence intervals were constructed, about 95% of the intervals would capture the true proportion of all members who would quit.
(b) We are 95% confident that the interval from 0.105 to 0.255 captures the true proportion of all members who would quit if the fee was increased.
(c) No, there are plausible values in the interval above 0.20. So the owner cannot be confident that the true proportion is less than 0.20.
(d) We check this condition to make sure $\hat{p}$ has an approximately Normal distribution, which allows us to use a Normal distribution to calculate the z^* critical value used in the interval.

TRM Chapter 8 Test

There is one Chapter Test available for this section in the Teacher's Resource Materials. Click on the link in the TE-Book, open in the TRFD, or download from the Teacher's Resources on the book's digital platform. You can also create your own Test using the TPS quiz and test builder (ExamView).

to be valid. First, the data used to calculate the interval must come from a random sample from the population of interest (the Random condition). When the sample is taken without replacement from the population, the sample size should be less than 10% of the population size (the 10% condition). Finally, the observed number of successes $n\hat{p}$ and observed number of failures $n(1-\hat{p})$ must both be at least 10 (the Large Counts condition).

The formula for calculating a confidence interval for a population proportion is

$$\hat{p} \pm z^* \sqrt{\frac{\hat{p}(1-\hat{p})}{n}}$$

where $\hat{p}$ is the sample proportion, z^* is the critical value, and n is the sample size. The value of z^* is based on the confidence level C. To find z^*, use Table A or technology to determine the values of z^* and $-z^*$ that capture the middle C% of the standard Normal distribution.

The four-step process (State, Plan, Do, Conclude) is perfectly suited for problems that ask you to construct and interpret a confidence interval. You should *state* the parameter you are estimating and the confidence level, *plan* your work by naming the type of interval you will use and checking the appropriate conditions, *do* the calculations, and make a *conclusion* in the context of the problem. You can use technology for the Do step, but make sure that you identify the procedure you are using and type in the values correctly.

Finally, an important part of planning a study is determining the size of the sample to be selected. The necessary sample size is based on the confidence level, the proportion of successes, and the desired margin of error. To calculate the minimum sample size, solve the following inequality for n, where $\hat{p}$ is a guessed value for the sample proportion:

$$z^* \sqrt{\frac{\hat{p}(1-\hat{p})}{n}} \leq ME$$

If you do not have an approximate value of $\hat{p}$ from a previous study or a pilot study use $\hat{p} = 0.5$ to determine the sample size that will yield a value less than or equal to the desired margin of error.

Section 8.3: Estimating a Difference in Proportions

In this section, you learned how to construct confidence intervals for a difference between two proportions. To verify independence in data collection and that the sampling distribution of $\hat{p}_1 - \hat{p}_2$ is approximately Normal, we check three conditions. The Random condition says that the data must be from two independent random samples or two groups in a randomized experiment. The 10% condition says that each sample size should be less than 10% of the corresponding population size when sampling without replacement. The Large Counts condition says that the number of successes and the number of failures from each sample/group should be at least 10. That is, $n_1\hat{p}_1, n_1(1-\hat{p}_1), n_2\hat{p}_2, n_2(1-\hat{p}_2)$ are all ≥ 10.

A confidence interval for a difference between two proportions provides an interval of plausible values for the difference in the true proportions. The formula is

$$(\hat{p}_1 - \hat{p}_2) \pm z^* \sqrt{\frac{\hat{p}_1(1-\hat{p}_1)}{n_1} + \frac{\hat{p}_2(1-\hat{p}_2)}{n_2}}$$

The logic of confidence intervals, including how to interpret the confidence interval and the confidence level, is the same as when estimating a single population proportion. Likewise, you can use a confidence interval for a difference in proportions to evaluate claims about the population proportions. For example, if 0 is not included in a confidence interval for $p_1 - p_2$, there is convincing evidence that the population proportions are different.

Comparing confidence intervals for proportions		
	Confidence interval for p	**Confidence interval for $p_1 - p_2$**
Name (TI-83/84)	One-sample z interval for p (1-PropZInt)	Two-sample z interval for $p_1 - p_2$ (2-PropZInt)
Conditions	• **Random:** The data come from a random sample from the population of interest. ○ **10%:** When sampling without replacement, $n < 0.10N$. • **Large Counts:** Both $n\hat{p}$ and $n(1-\hat{p})$ are at least 10. That is, the number of successes and the number of failures in the sample are both at least 10.	• **Random:** The data come from two independent random samples or from two groups in a randomized experiment. ○ **10%:** When sampling without replacement, $n_1 < 0.10N_1$ and $n_2 < 0.10N_2$. • **Large Counts:** The counts of "successes" and "failures" in each sample or group— $n_1\hat{p}_1, n_1(1-\hat{p}_1), n_2\hat{p}_2, n_2(1-\hat{p}_2)$—are all at least 10.
Formula	$\hat{p} \pm z^* \sqrt{\dfrac{\hat{p}(1-\hat{p})}{n}}$	$(\hat{p}_1 - \hat{p}_2) \pm z^* \sqrt{\dfrac{\hat{p}_1(1-\hat{p}_1)}{n_1} + \dfrac{\hat{p}_2(1-\hat{p}_2)}{n_2}}$

What Did You Learn?

Learning Target	Section	Related Example on Page(s)	Relevant Chapter Review Exercise(s)
Identify an appropriate point estimator and calculate the value of a point estimate.	8.1	538	R8.1
Interpret a confidence interval in context.	8.1	541	R8.4, R8.6
Determine the point estimate and margin of error from a confidence interval.	8.1	541	R8.2
Use a confidence interval to make a decision about the value of a parameter.	8.1	541	R8.2, R8.6
Interpret a confidence level in context.	8.1	543	R8.2
Describe how the sample size and confidence level affect the margin of error.	8.1	547	R8.3
Explain how practical issues like nonresponse, undercoverage, and response bias can affect the interpretation of a confidence interval.	8.1	547	R8.4
State and check the Random, 10%, and Large Counts conditions for constructing a confidence interval for a population proportion.	8.2	555	R8.4
Determine the critical value for calculating a C% confidence interval for a population proportion using a table or technology.	8.2	558	R8.4
Construct and interpret a confidence interval for a population proportion.	8.2	558, 559	R8.4
Determine the sample size required to obtain a C% confidence interval for a population proportion with a specified margin of error.	8.2	562	R8.5
Determine whether the conditions are met for constructing a confidence interval about a difference between two proportions.	8.3	568	R8.6
Construct and interpret a confidence interval for a difference between two proportions.	8.3	571	R8.6

Chapter 8 Review Exercises

These exercises are designed to help you review the important ideas and methods of the chapter.

R8.1 We love football! A Gallup poll conducted telephone interviews with a random sample of adults aged 18 and older. Data were obtained for 1000 people. Of these, 370 said that football is their favorite sport to watch on television.

(a) Define the parameter p in this setting.

(b) What point estimator will you use to estimate p? What is the value of the point estimate?

(c) Do you believe that the value of the point estimate is equal to the value of p? Explain your answer.

R8.2 Sports fans Are you a sports fan? That's the question the Gallup polling organization asked a random sample of 1527 U.S. adults.[28] Gallup reported that a 95% confidence interval for the proportion of all U.S. adults who are sports fans is 0.565 to 0.615.

(a) Calculate the point estimate and the margin of error.

(b) Interpret the confidence level.

(c) Based on the interval, is there convincing evidence that a majority of U.S. adults are sports fans? Explain your answer.

R8.3 (a) The margin of error must get larger to increase the capture rate of the intervals. **(b)** The margin of error will decrease by a factor of 2.

R8.4 (a) S: p = the true proportion of all drivers who have run at least one red light in the last 10 intersections they have entered. **P:** One-sample z interval. *Random:* Random sample. *10%:* 880 < 10% of all drivers. *Large Counts:* 171 ≥ 10 and 709 ≥ 10. **D:** (0.168, 0.220). **C:** We are 95% confident that the interval from 0.168 to 0.220 captures the true proportion of all drivers who have run at least one red light in the last 10 intersections they have entered. **(b)** It is likely that more than 171 respondents have run red lights. We would not expect very many people to claim they *have* run red lights when they have not, but some people will deny running red lights when they have. The margin of error does not account for these sources of bias, only sampling variability.

R8.5 Solving $2.576\sqrt{\dfrac{0.5(0.5)}{n}} \le 0.01$ gives n = 16,590 adults.

R8.6 (a) S: 99% CI for $p_T - p_A$, where p_T = the true proportion of U.S. teens who use Facebook and p_A = the true proportion of U.S. adults who use Facebook. **P:** Two-sample z-interval for $p_T - p_A$. *Random:* The data come from independent random samples of 1060 U.S. teens and 2003 U.S. adults. *Large Counts:* $1060(0.71) \approx 753$, $1060(1 - 0.71) \approx 297$, $2003(0.58) \approx 1162$, and $2003(1 - 0.58) \approx 841$ are all ≥ 10. **D:** $\hat{p}_T$ = 0.71 and $\hat{p}_A$ = 0.58(0.084, 0.176) **C:** We are 99% confident that the interval from 0.084 to 0.176 captures $p_T - p_A$ = the difference in the true proportions of teens and adults who use Facebook. **(b)** Because the interval does not include 0 as a plausible value for the difference in the true proportion of U.S. teens and U.S. adults who use Facebook, we do have convincing evidence that the two population proportions are different.

R8.3 It's about *ME* Explain how each of the following would affect the margin of error of a confidence interval, if all other things remained the same.

(a) Increasing the confidence level

(b) Quadrupling the sample size

R8.4 **Running red lights** A random digit dialing telephone survey of 880 drivers asked, "Recalling the last 10 traffic lights you drove through, how many of them were red when you entered the intersections?" Of the 880 respondents, 171 admitted that at least one light had been red.[29]

(a) Construct and interpret a 95% confidence interval for the population proportion.

(b) Nonresponse is a practical problem for this survey—only 21.6% of calls that reached a live person were completed. Another practical problem is that people may not give truthful answers. What is the likely direction of the bias: Do you think more or fewer than 171 of the 880 respondents really ran a red light? Why? Are these sources of bias included in the margin of error?

R8.5 Do you go to church? The Gallup Poll plans to ask a random sample of adults whether they attended a religious service in the past 7 days. How large a sample would be required to obtain a margin of error of at most 0.01 in a 99% confidence interval for the population proportion who would say that they attended a religious service?

R8.6 **Facebook** As part of the Pew Internet and American Life Project, researchers conducted two surveys. The first survey asked a random sample of 1060 U.S. teens about their use of social media. A second survey posed similar questions to a random sample of 2003 U.S. adults. In these two studies, 71.0% of teens and 58.0% of adults used Facebook.[30] Let p_T = the true proportion of all U.S. teens who use Facebook and p_A = the true proportion of all U.S. adults who use Facebook.

(a) Calculate and interpret a 99% confidence interval for the difference in the true proportions of U.S. teens and adults who use Facebook.

(b) Based on the confidence interval from part (a), is there convincing evidence of a difference in the population proportions? Explain your answer.

Chapter 8 AP® Statistics Practice Test

Section I: Multiple Choice *Select the best answer for each question.*

T8.1 The Gallup Poll interviews 1600 people. Of these, 18% say that they jog regularly. The news report adds: "The poll had a margin of error of plus or minus 3 percentage points at a 95% confidence level." You can safely conclude that

(a) 95% of all Gallup Poll samples like this one give answers within ±3% of the true population value.

(b) the percent of the population who jog is certain to be between 15% and 21%.

(c) 95% of the population jog between 15% and 21% of the time.

(d) we can be 95% confident that the sample proportion is captured by the confidence interval.

(e) if Gallup took many samples, 95% of them would find that 18% of the people in the sample jog.

T8.2 A confidence interval for a difference in proportions is −0.077 to 0.013. What are the point estimate and the margin of error for this interval?

(a) −0.032, 0.045

(b) −0.032, 0.090

(c) −0.032, 0.180

(d) −0.045, 0.032

(e) −0.045, 0.090

T8.3 In a random sample of 100 students from a large high school, 37 regularly bring a reusable water bottle from home. Which of the following gives the correct value and interpretation of the standard error of the sample proportion?

(a) In samples of size 100 from this school, the sample proportion of students who bring a reusable water bottle from home will be at most 0.095 from the true proportion.

(b) In samples of size 100 from this school, the sample proportion of students who bring a reusable water bottle from home will be at most 0.048 from the true proportion.

TRM **Full Solutions to Chapter 8 AP® Statistics Practice Test**

The full solutions can be found by clicking on the link in the TE-Book, opening the TRFD, or downloading from the Teacher's Resources on the book's digital platform.

Answers to Chapter 8 AP® Statistics Practice Test

T8.1 a

T8.2 a

(c) In samples of size 100 from this school, the sample proportion of students who bring a reusable water bottle from home typically varies by about 0.095 from the true proportion.

(d) In samples of size 100 from this school, the sample proportion of students who bring a reusable water bottle from home typically varies by about 0.048 from the true proportion.

(e) There is not enough information to calculate the standard error.

T8.4 Many television viewers express doubts about the validity of certain commercials. In an attempt to answer their critics, Timex Group USA wishes to estimate the true proportion p of all consumers who believe what is shown in Timex television commercials. Which of the following is the smallest number of consumers that Timex can survey to guarantee a margin of error of 0.05 or less at a 99% confidence level?

(a) 550

(b) 600

(c) 650

(d) 700

(e) 750

T8.5 Which of the following is the critical value for calculating a 94% confidence interval for a population proportion?

(a) 1.555

(b) 1.645

(c) 1.881

(d) 1.960

(e) 2.576

T8.6 A radio talk show host with a large audience is interested in the proportion p of adults in his listening area who think the drinking age should be lowered to 18. To find this out, he poses the following question to his listeners: "Do you think that the drinking age should be reduced to 18 in light of the fact that 18-year-olds are eligible for military service?" He asks listeners to go to his website and vote "Yes" if they agree the drinking age should be lowered and "No" if not. Of the 100 people who voted, 70 answered "Yes." Which of the following conditions are violated?

  I. Random   II. 10%   III. Large Counts

(a) I only

(b) II only

(c) III only

(d) I and II only

(e) I, II, and III

T8.7 A marketing assistant for a technology firm plans to randomly select 1000 customers to estimate the proportion who are satisfied with the firm's performance. Based on the results of the survey, the assistant will construct a 95% confidence interval for the proportion of all customers who are satisfied. The marketing manager, however, says that the firm can afford to survey only 250 customers. How will this decrease in sample size affect the margin of error?

(a) The margin of error will be about 4 times larger.

(b) The margin of error will be about 2 times larger.

(c) The margin of error will be about the same size.

(d) The margin of error will be about half as large.

(e) The margin of error will be about one-fourth as large.

T8.8 Thirty-five people from a random sample of 125 workers from Company A admitted to using sick leave when they weren't really ill. Seventeen employees from a random sample of 68 workers from Company B admitted that they had used sick leave when they weren't ill. Which of the following is a 95% confidence interval for the difference in the proportions of workers at the two companies who would admit to using sick leave when they weren't ill?

(a) $0.03 \pm \sqrt{\dfrac{(0.28)(0.72)}{125} + \dfrac{(0.25)(0.75)}{68}}$

(b) $0.03 \pm 1.96 \sqrt{\dfrac{(0.28)(0.72)}{125} + \dfrac{(0.25)(0.75)}{68}}$

(c) $0.03 \pm 1.96 \sqrt{\dfrac{(0.28)(0.72)}{125} - \dfrac{(0.25)(0.75)}{68}}$

(d) $57 \pm 1.96 \sqrt{\dfrac{(0.28)(0.72)}{125} + \dfrac{(0.25)(0.75)}{68}}$

(e) $57 \pm 1.96 \sqrt{\dfrac{(0.28)(0.72)}{125} - \dfrac{(0.25)(0.75)}{68}}$

T8.9 A telephone poll of an SRS of 1234 adults found that 62% are generally satisfied with their lives. The announced margin of error for the poll was 3%. Does the margin of error account for the fact that some adults do not have telephones?

(a) Yes; the margin of error accounts for all sources of error in the poll.

(b) Yes; taking an SRS eliminates any possible bias in estimating the population proportion.

(c) Yes; the margin of error accounts for undercoverage but not nonresponse.

(d) No; the margin of error accounts for nonresponse but not undercoverage.

(e) No; the margin of error only accounts for sampling variability.

T8.3 d

T8.4 d

T8.5 c

T8.6 a

T8.7 b

T8.8 b

T8.9 e

T8.10 c

T8.11 (a) STATE: $p =$ the true proportion of all visitors to Yellowstone who would say they favor the restrictions. **PLAN:** One-sample z interval. *Random:* The visitors were selected randomly. *10%:* $n = 150$ is less than 10% of all visitors to Yellowstone National Park. *Large Counts:* $n\hat{p} = 89 \geq 10$ and $n(1 - \hat{p}) = 61 \geq 10$.

DO: $\hat{p} = \dfrac{89}{150} = 0.593$; $(0.490, 0.696)$.

CONCLUDE: We are 99% confident that the interval from 0.490 to 0.696 captures $p =$ the true proportion of all visitors who would say that they favor the restrictions. **(b)** Because there are values less than 0.50 in the confidence interval, the U.S. Forest Service cannot conclude that more than half of visitors to Yellowstone National Park favor the proposal. It is plausible that only 49% favor the proposal.

T8.12 (a) If we were to select many random samples of the size used in this study from the same population of mesquite trees and construct a 95% confidence interval using each sample, about 95% of the intervals would capture the true proportion of all mesquite trees in this park that are infested with mistletoe. **(b)** Point estimate $= 0.4$ and margin of error $= 0.1753$;

$$0.1753 = 1.96\sqrt{\dfrac{0.4(0.6)}{n}};\ n = 30$$

T8.13 STATE: 90% CI for $p_1 - p_2$, where $p_1 =$ the true proportion of people like these who would say that buying coffee at Starbucks is a waste of money when the girls hold cups from Starbucks and $p_2 =$ the true proportion of people like these who would say that buying coffee at Starbucks is a waste of money when the girls were empty handed. **PLAN:** Two-sample z-interval for $p_1 - p_2$. *Random:* Subjects were randomly assigned to be asked while the girls held Starbucks cups or were empty handed. *Large Counts:* 19, $50 - 19 = 31$, 23, $50 - 23 = 27$ are all ≥ 10. **DO:** $p_1 = \dfrac{19}{50} = 0.38$ and

$p_2 = \dfrac{23}{50} = 0.46$; $(-0.242, 0.082)$.

CONCLUDE: We are 99% confident that the interval from -0.242 to 0.082 captures $p_1 - p_2 =$ the difference (Holding – Empty handed) in the true proportions of people like these who would say that buying coffee at Starbucks is a waste of money.

T8.10 At a baseball game, 42 of 65 randomly selected people own an iPod. At a rock concert occurring at the same time across town, 34 of 52 randomly selected people own an iPod. A researcher wants to test the claim that the proportion of iPod owners at the two venues is different. A 90% confidence interval for the difference (Game – Concert) in population proportions is $(-0.154,\ 0.138)$. Which of the following gives the correct outcome of the researcher's test of the claim?

(a) Because the interval includes 0, the researcher can conclude that the proportion of iPod owners at the two venues is the same.

(b) Because the center of the interval is -0.008, the researcher can conclude that a higher proportion of people at the rock concert own iPods than at the baseball game.

(c) Because the interval includes 0, the researcher cannot conclude that the proportion of iPod owners at the two venues is different.

(d) Because the interval includes -0.008, the researcher cannot conclude that the proportion of iPod owners at the two venues is different.

(e) Because the interval includes more negative than positive values, the researcher can conclude that a higher proportion of people at the rock concert own iPods than at the baseball game.

Section II: Free Response *Show all your work. Indicate clearly the methods you use, because you will be graded on the correctness of your methods as well as on the accuracy and completeness of your results and explanations.*

T8.11 The U.S. Forest Service is considering additional restrictions on the number of vehicles allowed to enter Yellowstone National Park. To assess public reaction, the service asks a random sample of 150 visitors if they favor the proposal. Of these, 89 say "Yes."

(a) Construct and interpret a 99% confidence interval for the proportion of all visitors to Yellowstone who favor the restrictions.

(b) Based on your work in part (a), is there convincing evidence that more than half of all visitors to Yellowstone National Park favor the proposal? Justify your answer.

T8.12 For some people, mistletoe is a symbol of romance. Mesquite trees, however, have no love for the parasitic plant that attaches itself and steals nutrients from the tree. To estimate the proportion of mesquite trees in a desert park that are infested with mistletoe, a random sample of mesquite trees was randomly selected. After inspecting the trees in the sample, the park supervisor calculated a 95% confidence interval of 0.2247 to 0.5753.

(a) Interpret the confidence level.

(b) Calculate the sample size used to create this interval.

T8.13 Do "props" make a difference when researchers interact with their subjects? Emily and Madi asked 100 people if they thought buying coffee at Starbucks was a waste of money.[31] Half of the subjects were asked while Emily and Madi were holding cups from Starbucks, and the other half of the subjects were asked when the girls were empty handed. The choice of holding or not holding the cups was determined at random for each subject. When holding the cups, 19 of 50 subjects agreed that buying coffee at Starbucks was a waste of money. When they weren't holding the cups, 23 of 50 subjects said it was a waste of money. Calculate and interpret a 90% confidence interval for the difference in the proportion of people like the ones in this experiment who would say that buying coffee from Starbucks is a waste of money when asked by interviewers holding or not holding a cup from Starbucks.

TRM Chapter 8 Case Study

The Case Study feature that was found in the previous two editions of *The Practice of Statistics* student edition has been moved to the Teacher's Resource Materials. Click on the link in the TE-Book, open in the TRFD, or download from the Teacher's Resources on the book's digital platform.

Chapter 9

Chapter 9

Testing Claims About Proportions

PD Overview

In the last chapter, we used data from random samples and randomized experiments to estimate population parameters with confidence intervals. In this chapter, we will use data from random samples and randomized experiments to test claims about a population parameter. This type of inference is called a *significance test*.

The logic of significance testing should already be somewhat familiar to students, especially if you did the "Hiring discrimination" activity from Chapter 1 and the "Caffeine experiment" activity from Chapter 4 with your class. In both these activities, we used simulation to investigate whether the observed results gave convincing evidence in support of a claim or could be the result of random chance. Likewise, many of the simulation questions in Section 5.1 asked students to use the logic of significance testing to determine whether an observed result was unusual enough to cast doubt on a claim, such as in the "1 in 6 wins" activity.

As introduced in Chapter 8, students learn how to make decisions using the more formal, 4-step process. We introduce the basics of significance testing in Section 9.1. In Section 9.2, we focus on one-sample tests for a population proportion. In Section 9.3, we focus on two-sample tests for a difference in proportions. By the end of the course, students will have learned eight different significance tests. And while each test varies in details, they all use the same logic and 4-step process.

If you didn't already use this excellent resource in Chapter 8, you can learn more about inference for free at the AP® Central website (apcentral.collegeboard.org/pdf/statistics-inference.pdf?course=ap-statistics). An online search for "AP® Statistics Special Focus: Inference" will reveal the materials that were originally developed for use in College Board workshops and were written by people familiar with the AP® Statistics course.

The Main Ideas

One of the challenges in teaching the AP® Statistics course is keeping students focused on the big picture, not just the details of each section. We outline the main ideas for the chapter here.

Chapter 9 Introduction

The purpose of the "I'm a great free-throw shooter" activity in the Introduction is to get students thinking about how to test a claim, without the formulas and structure of a formal significance test. The basketball player claims to make 80% of his free throws but then makes only 64% of a sample of 50 free throws. Students have to decide whether they think the player is telling the truth and this sample happened purely by chance, or if the results give convincing evidence that the player is exaggerating his skills.

Section 9.1 Significance Tests: The Basics

In this section, we introduce the basic ideas of significance testing. The first step in any significance test is stating the hypotheses that we want to test. The null hypothesis typically is a statement of "no difference," such as "The treatments are equally effective" or "The player's true free-throw percentage is no different from what he claims." The alternative hypothesis describes what we suspect is true and what we are looking for evidence to support—for example, "Treatment A is more effective than Treatment B" or "The player's true free-throw percentage is less than he claims."

At the beginning of each significance test, make sure your students understand that there are two ways to explain why the sample data disagree with the null hypothesis: (1) The null hypothesis is true, and the disagreement is due to chance—sampling variability when using data from a random sample or chance variation in the random assignment when using data from a randomized experiment; or (2) the null hypothesis is false, and the sample data are consistent with an alternative value of the parameter. In a significance test, we always start with the belief that the null hypothesis is true and choose Explanation 2 only when we are convinced that chance is *not* a plausible explanation for the difference between the sample statistic and the hypothesized parameter value.

A *P*-value measures how likely it is that a value of the sample statistic will be at least as extreme as the observed sample

statistic by chance alone, assuming the null hypothesis is true. If the *P*-value is large, it is plausible that the difference between the sample statistic and hypothesized parameter value could have been due to chance alone. However, if the *P*-value is small, we can rule out chance as a plausible explanation for the difference and conclude that the hypothesized parameter value is incorrect. To determine if a *P*-value is "small," we often compare it to a predetermined significance level, such as $\alpha = 0.05$.

Because we are basing our conclusions on sample data, there is always a chance that our conclusions will be in error. We can make two types of errors in a significance test. A Type I error occurs if we find convincing evidence that the alternative hypothesis is true, when in reality it isn't true. A Type II error occurs if we fail to find convincing evidence that the alternative hypothesis is true, when in reality it is true.

Section 9.2 Tests About a Population Proportion

In this section, we present the details for conducting a significance test for a population proportion. Not surprisingly, we suggest that students perform significance tests using the State–Plan–Do–Conclude 4-step process.

In the "State" step, students are expected to state the hypotheses they are testing, state the significance level they are using, and define the parameter they are using in their hypotheses. In the "Plan" step, students are expected to name the procedure they plan to use (e.g., one-sample *z* test for a population proportion) and check appropriate conditions to see if the procedure is reasonable. In the "Do" step, students should calculate the standardized test statistic and *P*-value. The standardized test statistic measures how far the sample statistic is from the hypothesized parameter value, in standardized units. Finally, in the "Conclude" step, students should use the *P*-value to make an appropriate decision about the hypotheses in the context of the problem.

When an alternative hypothesis is two-sided, there is a clear connection between confidence intervals and significance tests. Because confidence intervals provide a range of plausible

values for a population parameter, we can use confidence intervals to reject or fail to reject a null hypothesis.

Finally, we revisit Type I and Type II errors. The probability that we avoid making a Type II error when an alternative value of the parameter is true is called the power of the test. Power is good—if the alternative hypothesis is true, we want to maximize the probability of coming to this conclusion. We can increase the power of a significance test by increasing the sample size, increasing the significance level, and minimizing sources of variability in the data collection process.

Section 9.3 Tests About a Difference in Proportions

In this section, we present the details for conducting a significance test for a difference in proportions. As with significance tests for a single population proportion, we expect students to use the State–Plan–Do–Conclude 4-step process when performing significance tests for a difference in proportions.

There is not much new in this section. The 4-step process is the same as in Section 9.2, and the conditions for conducting a two-sample *z* test for $p_1 - p_2$ are nearly the same as those used in Section 9.2. Make sure to point this out to help your students avoid getting overwhelmed.

For a significance test for a difference between two proportions, we start by assuming that the null hypothesis is true and asking how likely it would be to get results at least as unusual as the results observed in a study by chance alone. If it is plausible that a difference in proportions could be the result of sampling variability or chance variation due to the random assignment, we do not have convincing evidence that the alternative hypothesis is true. However, if the difference is too big to attribute to chance, there is convincing evidence that the alternative hypothesis is true.

At the end of this section, we introduce inference for completely randomized experiments. Fortunately, the methods used for analyzing the difference in the proportions from two independent random samples work very well for analyzing the difference in the proportions from two groups in a completely randomized experiment.

Chapter 9: Resources

Teacher's Resource Materials

The following resources, identified by the **TRM** in the annotated student pages, can be found by clicking on the link in the Teacher's e-Book (TE-Book), searching by category or chapter on the Teacher's Resource Flash Drive (TRFD), or logging into the book's digital platform (highschool.bfwpub.com/updatedtps6e) and searching the Teacher's Resources menu (teacher log-in required).

- Alternate Examples: one file per section
- Lecture Presentation Slides: one per section
- Free-Throw Spinner
- Chapter 9 Learning Targets Grid
- FRAPPY! Materials
- Complete solutions for the Check Your Understanding problems, section exercises, review exercises, and practice test.
- Quizzes: one per section
- Chapter 9 Test

Free Response Questions from Previous AP® Statistics Exams

Questions can be found on the AP® Central website: apcentral .collegeboard.org/courses/ap-statistics/exam.

Students should be able to answer all the free response questions listed below with material learned in this chapter. Questions that contain content from this chapter but also require content from later chapters are listed in the last chapter required to complete the entire question. This list will be updated after each AP® Statistics exam and will be posted to the Teacher's Resource section of the book's digital platform and to www.statsmedic.com/free-response-questions.

Year	#	Content
2019	4	• Two-sample z test for proportions
2015	4	• Two-sample z test for proportions
2013	5	• Scope of inference • Conditions for two-sample z test for proportions • Logic of inference
2012	4	• Two-sample z test (or interval) for proportions
2012	5	• Type II error and consequence • Conclusion to a significance test for a single proportion • Voluntary response bias

2009B	3	• Two-sample z test for proportions
2009B	4	• Random assignment in blocks • Increasing the power of a test
2009	5	• Two-sample z test for proportions • Type I and Type II errors and consequences
2008B	4	• Experimental design • Type I and Type II errors and consequences
2007	5	• Experiment or observational study • Two-sample z test for proportions
2006B	6	• Stating hypotheses • Conditions for a one-sample z test for a proportion • Binomial probability calculations • Significance levels • Calculating P-values and drawing conclusions • Improving a study
2005	4	• One-sample z test for a proportion
2004B	6	• Two-sample z test for proportions • Re-capture technique • Random sampling
2003B	3	• Experiment or observational study • Two-sample z test for proportions
2003	2	• Stating hypotheses • Type I and Type II errors and consequences
2002	6	• One-sample z interval for a proportion • Interpreting confidence level • Two-sample z test for proportions • Pooling
1998	5	• One-sample z test for a proportion • Effect of nonresponse
1997	4	• Two-sample z test for proportions

Applets

- *The Reasoning of a Statistical Test* applet at highschool .bfwpub.com/updatedtps6e allows students to investigate the logic of hypothesis testing by simulating free-throw attempts for a player who claims to make 80% of his shots.
- The *Statistical Power* applet at istats.shinyapps.io/power/ allows students to investigate how the power of a significance test for a population proportion is affected by changes in the sample size, significance level, and alternative parameter value.
- The *Improved Batting Averages (Power)* applet at www .rossmanchance.com/applets estimates the power of

Chapter 9

a one-sample *z* test for a population proportion using simulation. It allows students to investigate how the power of the test is affected by changes in the sample size, significance level, and alternative parameter value.

Chapter 9: Pacing Guide, Learning Targets, and Suggested Assignments

This pacing guide is based on a schedule with 110, 50-minute sessions before the AP® Statistics exam. If you have a different number of sessions before the exam, you can modify the pacing guide to suit your needs. If you have additional time,

consider incorporating quizzes, released AP® Statistics free response questions, or additional activities. See the Resources section above for suggestions.

The suggested homework assignments list odd-numbered exercises, whenever possible, so students can check their answers against the back-of-book answers. If you would rather students not have access to the answers while doing homework, adding 1 to the exercise numbers usually will do the trick, because the homework exercises typically are paired. For example, Exercises 1 and 2 will generally cover the same topics, but in different contexts. You may also choose to include the Recycle and Review questions at the end of each section, which review topics from previous sections or chapters. If your school is using the digital platform that accompanies TPS6, you will find these assignments pre-built as online homework assignments for Chapter 9.

Day	Content	Learning Targets: Students will be able to . . .	Suggested Assignment (MC bold)
1	Chapter 9 Introduction, 9.1 Stating Hypotheses, Interpreting *P*-values, Making Conclusions	• State appropriate hypotheses for a significance test about a population parameter. • Interpret a *P*-value in context. • Make an appropriate conclusion for a significance test.	1, 3, 5, 7, 9, 13, 14, 15, 19
2	9.1 Type I and Type II Errors	• Interpret a Type I and a Type II error in context. Give a consequence of each error in a given setting.	21, 23, 25, 27, **29–32**
3	9.2 Performing a Significance Test About *p*	• State and check the Random, 10%, and Large Counts conditions for performing a significance test about a population proportion. • Calculate the standardized test statistic and *P*-value for a test about a population proportion.	35, 37, 39, 41
4	9.2 Putting It All Together: One-Sample *z* Test for *p*, Two-Sided Tests	• Perform a significance test about a population proportion.	43, 45, 47, 51, 53, 55
5	9.2 The Power of a Test	• Interpret the power of a significance test and describe what factors affect the power of a test.	59, 61, 63, 65, 67, **70–73**
6	9.3 Significance Tests for $p_1 - p_2$	• State appropriate hypotheses for a significance test about a difference between two proportions. • Determine whether the conditions are met for performing a test about a difference between two proportions. • Calculate the standardized test statistic and *P*-value for a test about a difference between two proportions.	77, 79, 83, 85
7	9.3 Putting It All Together: Two-sample *z* Test for $p_1 - p_2$	• Perform a significance test about a difference between two proportions.	87, 89, 93, **95–98**
8	Chapter 9 Review/FRAPPY!		Chapter 9 Review Exercises
9	Chapter 9 Test		

Chapter 9: Alignment to the College Board's Fall 2019 AP® Statistics Course Framework*

Relationship to College Board Units

Chapter 9 in this book covers Topics 6.4–6.7 and 6.10–6.11 in Unit 6 of the College Board Course Framework. Students will be ready to take the Personal Progress Check for Unit 6 once they have completed Chapter 9.

Big Ideas and Enduring Understanding Statements

Chapter 9 develops these Big Ideas and related Enduring Understandings outlined in the Course Framework:

- **Big Idea 1: Variation and Distribution (EU: VAR 6):** The distribution of measures for individuals within a sample or population describes variation. The value of a statistic varies from sample to sample. How can we determine whether differences between measures represent random variation or meaningful distinctions? Statistical methods based on probabilistic reasoning provide the basis for shared understandings about variation and about the likelihood that variation between and among measures, samples, and populations is random or meaningful.

- **Big Idea 2: Patterns and Uncertainty (EU: UNC 3,5):** Statistical tools allow us to represent and describe patterns in data and to classify departures from patterns. Simulation and probabilistic reasoning allow us to anticipate patterns in data and to determine the likelihood of errors in inference.

- **Big Idea 3: Data-Based Predictions, Decisions, and Conclusions (EU: DAT 3):** Data-based regression models describe relationships between variables and are a tool for making predictions for values of a response variable. Collecting data using random sampling or randomized experimental design means that findings may be generalized to the part of the population from which the selection was made. Statistical inference allows us to make data-based decisions.

Course Skills

Chapter 9 helps students to develop the skills identified in the Course Framework.

- **1: Selecting Statistical Models** (1.B, 1.E, 1.F)
- **3: Using Probability and Simulation** (3.A, 3.C, 3.E)
- **4: Statistical Argumentation** (4.A, 4.B, 4.C, 4.E)

Learning Objectives and Essential Knowledge Statements

Section	Learning Objectives	Essential Knowledge Statements
9.1	VAR-6.D, VAR-6.G, UNC-5.A, UNC-5.B, UNC-5.D, DAT-3.A, DAT-3.B	VAR-6.D.1, VAR-6.D.2, VAR-6.D.3, VAR-6.D.4, VAR-6.D.5, VAR-6.G.4, UNC-5.A.1, UNC-5.A.2, UNC-5.B.1, UNC-5.D.1, UNC-5.D.2, DAT-3.A.2, DAT-3.B.1, DAT-3.B.2, DAT-3.B.3, DAT-3.B.4, DAT-3.B.5, DAT-3.B.6, DAT-3.B.7
9.2	VAR-6.E, VAR-6.F, VAR-6.G, UNC-5.B, UNC-5.C, DAT-3.A	VAR-6.E.1, VAR-6.F.1, VAR-6.G.1, VAR-6.G.2, VAR-6.G.3, UNC-5.B.2, UNC-5.B.3, UNC-5.C.1, DAT-3.A.1, DAT-3.B.8, DAT-3.B.9
9.3	VAR-6.H, VAR-6.I, VAR-6.J, VAR-6.K, UNC-3.H, DAT-3.C, DAT-3.D	VAR-6.H.1, VAR-6.H.2, VAR-6.H.3, VAR-6.I.1, VAR-6.J.1, VAR-6.K.1, UNC-3.H.4, DAT-3.C.1, DAT-3.D.1, DAT-3.D.2

A detailed alignment (The Nitty Gritty Guide) that can be sorted by Course Framework Unit, Topic, Learning Objective, Essential Knowledge Statement, or textbook section, is available on the TRFD and in the Teacher's Resources folder on Sapling Plus. **TRM**

*Should changes be made to the Course Framework in the future, an updated alignment will be placed on our AP® updates page at go.bfwpub.com/ap-course-updates.

UNIT 6
Inference for Categorical Data: Proportions

Chapter 9

Testing Claims About Proportions

Bill Gozansky/Alamy

Teaching Tip

Unit 6 in the College Board Course Framework aligns to Chapters 8 and 9 in this book. Students will be ready to take the Personal Progress Check for Unit 6 once they have completed Chapter 9.

PD Chapter 9 Overview

To watch the video overview of Chapter 9 (for teachers), click on the link in the TE-Book, look on the TRFD, or download from the Teacher's Resources on the book's digital platform.

TRM Lecture Presentation Slides

If you are new to teaching AP® Statistics or are short on time when preparing for class, you may find the Lecture Presentation Slides to be helpful. Experienced AP® Teacher Doug Tyson has created one slide presentation per section. You may use them as is, modify them to fit your needs, or share them with students who miss class. Find them on the TRFD and in the Teacher's Resources on the book's digital platform.

Teaching Tip

Make students aware of the structure of this chapter. Giving them a framework at the beginning of the chapter allows them an organized way of storing their learning.

9.1 Significance Tests: The Basics
9.2 Tests About a Population Proportion
9.3 Tests About a Difference in Proportions

We used a parallel structure in Chapter 8.

Teaching Tip

Here is an activity suggested by Roxy Peck that can be used at the start of the chapter to get students thinking about when an outcome is too unlikely to be due to chance. For this activity, use two identical decks of cards to prepare a single deck consisting entirely of red cards. If you can return the newly formed deck to a sealed box and have a student break the seal to open the box, students will be "hooked" from the outset. The best way to do this is to cut the plastic packaging on the bottom of the box and carefully tape it back together.

Invite students to take part in a challenge: draw one card from the shuffled deck. If the card is black, you will give them a cookie. If not, the card will be returned to the deck and the deck shuffled again for the next player. Allow several students to draw.

At what point do students start to suspect that something is amiss? After 1 draw? Probably not. After 2 draws? Probably not. In most classes, by the time the 4th or 5th consecutive red card has been drawn, students are beginning to doubt that they suffer from bad luck. Because the probability of 4 consecutive red cards is 1/16 and the probability of 5 consecutive red cards is 1/32 if the deck is "fair," students naturally gravitate toward the popular 5% significance level.

ACTIVITY OVERVIEW

To prepare for using this activity, watch the overview video by clicking on the link in the TE-Book, opening the TRFD, or downloading from the Teacher's Resources on the book's digital platform.

Time: 20 minutes

Materials: Spinners, paper clips

Teaching Advice: This simulation can also be performed using a single die (1–4 → make, 5 → miss, 6 → ignore), a table of random digits (1–8 → make, 9 and 0 → miss), or a random number generator (1–8 → make, 9 and 10 → miss). Ask students to put sticker dots on a poster board. Post the completed poster board in the classroom so that you can refer to this activity throughout the rest of the chapter and the course.

We are previewing several ideas that will be formally presented later in Chapter 9. The basketball player's claim to make 80% will be called the *null hypothesis,* while our claim that he shoots less than 80% will be called the *alternative hypothesis.* The probability of getting the observed result or less ($\hat{p} \le 0.64$) given that he is an 80% free-throw shooter will later be identified as the *P-value* of the significance test. Recall from previous chapters that an outcome was considered unusual or surprising if the probability was less than 5%, which will be identified as the *significance level* (α).

Use technology to extend this simulation to many, many trials. Go to the Extra Applets on the Student Site at highschool .bfwpub.com/updatedtps6e and choose "One Categorical Variable." Make two categories of "make" (32) and "miss" (18) and begin analysis. Then simulate a sample proportion with a hypothesized proportion of 0.80.

Ask students if they can think of any other ways to calculate the probability that the shooter will make 32 or fewer shots in 50 attempts. At this point in the course, students should recognize that this probability can be calculated with the binomial distribution, the Normal approximation to the binomial distribution, or the sampling distribution of $\hat{p}$.

Answers:

1. Done.

2–4. Answers will vary.

5. Answers will vary. In the dotplot, count the number of dots that are at or below $\hat{p} = 0.64$ and divide it by the total number of dots.

6. Assuming your answer to Question 5 is less than 5%, we have convincing evidence that the player is exaggerating about his free-throw percentage.

INTRODUCTION

Confidence intervals are one of the two most common methods of statistical inference. You can use a confidence interval to estimate parameters, like the proportion p of all U.S. adults who exercise regularly, the true mean amount of time μ that students at a large university spent on social media yesterday, or the difference $p_D - p_C$ in the population proportions of dogs and cats that are frightened by thunder.

What if we want to test a claim about a parameter? For instance, the U.S. Bureau of Labor Statistics claims that the national unemployment rate in March 2017 was 4.5%. A citizens' group suspects that the actual rate was higher. The second common method of inference, called a **significance test**, allows us to weigh the evidence in favor of or against a particular claim.

A significance test is sometimes referred to as a *test of significance,* a *hypothesis test,* or a *test of hypotheses.*

> **DEFINITION   Significance test**
>
> A **significance test** is a formal procedure for using observed data to decide between two competing claims (called *hypotheses*). The claims are usually statements about parameters.

Here is an activity that illustrates the reasoning of a significance test.

ACTIVITY   I'm a great free-throw shooter!

In this activity, you and your classmates will perform a simulation to test a claim about a population proportion.

A basketball player claims to make 80% of the free throws that he attempts. That is, he claims $p = 0.80$, where p is the true proportion of free throws he will make in the long run. We suspect that he is exaggerating and that $p < 0.80$.

Suppose the player shoots 50 free throws and makes 32 of them. His sample proportion of made shots is $\hat{p} = 32 / 50 = 0.64$. This result gives *some* evidence that the player really makes less than 80% of his free throws in the long run. But do we have *convincing* evidence that $p < 0.80$? Or is it plausible that an 80% shooter will have a performance this poor by chance alone? You can use a simulation to find out.

1. Using the pie chart provided by your teacher, label the 80% region "made shot" and the 20% region "missed shot." Straighten out one of the ends of a paper clip so that there is a loop on one side and a pointer on the other. On a flat surface, place a pencil through the loop, and put the tip of the pencil on the center of the pie chart. Then flick the paper clip and see where the pointed end lands: made shot or missed shot.

2. Flick the paper clip a total of 50 times, and count the number of times that the pointed end lands in the "made shot" region.

3. Compute the sample proportion $\hat{p}$ of made shots in your trial of the simulation from Step 2. Plot this value on the class dotplot drawn by your teacher.

4. Repeat Steps 2 and 3 as needed to get at least 40 trials of the simulation for your class.

TRM   Free-Throw Spinner

You will need a spinner with 10 equal spaces to do the "Free-throw shooter" activity. Find it in the Teacher's Resource Materials located in the TE-Book, on the TRFD, or in the Teacher's view on the book's digital platform.

5. Based on the class's simulation results, how likely is it for an 80% shooter to make 64% or less when he shoots 50 free throws?

6. Based on your answer to Question 5, does the observed $\hat{p} = 0.64$ result give convincing evidence that the player is exaggerating? Or is it plausible that an 80% shooter can have a performance this poor by chance alone?

In the activity, the shooter made only 32 of 50 free-throw attempts ($\hat{p} = 0.64$). There are two possible explanations for why he made less than 80% of his shots:

1. The player's claim is true ($p = 0.80$) and his bad performance happened by chance alone.

2. The player's claim is false ($p < 0.80$). That is, the population proportion is less than 0.80 so the sample result is not an unlikely outcome.

If explanation 1 is plausible, then we don't have convincing evidence that the shooter is exaggerating—his poor performance could have occurred purely by chance. However, if it is unlikely for an 80% shooter to get a proportion of 0.64 or less in 50 attempts, then we can rule out explanation 1.

We used software to simulate 400 sets of 50 shots, assuming that the player is really an 80% shooter. Figure 9.1 shows a dotplot of the results. Each dot on the graph represents the sample proportion $\hat{p}$ of made shots in one set of 50 attempts.

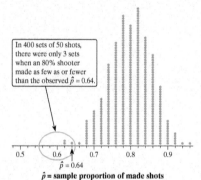

In 400 sets of 50 shots, there were only 3 sets when an 80% shooter made as few as or fewer than the observed $\hat{p} = 0.64$.

$\hat{p} = 0.64$

$\hat{p}$ = sample proportion of made shots

FIGURE 9.1 Dotplot of the simulated sampling distribution of $\hat{p}$, the proportion of free throws made by an 80% shooter in a sample of 50 shots.

The simulation shows that it would be very unlikely for an 80% free-throw shooter to make 32 or fewer out of 50 free throws ($\hat{p} \leq 0.64$) just by chance. This gives us convincing evidence that the player is less than an 80% shooter.

Section 9.1 focuses on the underlying logic of significance tests. Once the foundation is laid, we consider the implications of using these tests to make decisions—about everything from free-throw shooting to the effectiveness of a new drug. In Section 9.2, we present the details of performing a test about a population proportion. Section 9.3 shows how to test a claim about a difference between two proportions. Along the way, we examine the connection between confidence intervals and significance tests. We'll discuss testing claims about means in Chapter 11.

PD **Section 9.1 Overview**

To watch the video overview of Section 9.1 (for teachers), click on the link in the TE-Book, look on the TRFD, or download from the Teacher's Resources on the book's digital platform.

TRM **Learning Targets Grid**

At the beginning of each section, we present the relevant learning targets. Point these out to your students and refer back to the targets when you cover them in class. There is a PDF version of the grid with an additional column that students can use to keep track of their progress. Find it in the Teacher's Resource Materials located in the TE-Book, on the TRFD, or in the Teacher's view in the book's digital platform.

TRM **Section 9.1 Alternate Examples**

You can find the Alternate Examples for this section in Microsoft Word format by clicking on the link in the TE-Book, opening the TRFD, or downloading from the Teacher's Resources on the book's digital platform.

Making Connections

In Section 9.1, make students aware that we will use contexts that involve a proportion and contexts that involve a mean, but without formulas and the 4-step process. Section 9.2 will provide the formulas and 4-step process for a significance test about a proportion. Section 10.1 will provide the formulas and 4-step process for a significance test about a mean.

AP® EXAM TIP

There is almost always one free response question that asks students to perform a significance test. Students will most likely be asked if the data provide convincing evidence for [alternative hypothesis], rather than if the data provide convincing evidence against [null hypothesis].

SECTION 9.1 # Significance Tests: The Basics

LEARNING TARGETS *By the end of the section, you should be able to:*

- State appropriate hypotheses for a significance test about a population parameter.
- Interpret a *P*-value in context.
- Make an appropriate conclusion for a significance test.
- Interpret a Type I error and a Type II error in context. Give a consequence of each error in a given setting.

$\mathbf{A}$s noted in the earlier definition, a significance test starts with a careful statement of the claims we want to compare. Let's take a closer look at how to state these claims.

Stating Hypotheses

> Remember: The null hypothesis is the dull hypothesis!

In the free-throw shooter activity, the player claims that his long-run proportion of made free throws is $p = 0.80$. This is the claim we seek evidence *against*. We call it the **null hypothesis**, abbreviated H_0. Usually, the null hypothesis is a statement of "no difference." For the free-throw shooter, no difference from what he claimed gives H_0: $p = 0.80$.

The claim we hope or suspect to be true instead of the null hypothesis is called the **alternative hypothesis.** We abbreviate the alternative hypothesis as H_a. In this case, we suspect the player might be exaggerating, so our alternative hypothesis is H_a: $p < 0.80$.

> Some people insist that all three possibilities—greater than, less than, and equal to—should be accounted for in the hypotheses. Because the alternative hypothesis in this case is H_a: $p < 0.80$, they would write the null hypothesis as H_0: $p \geq 0.80$. But if we find convincing evidence against $p = 0.80$, our evidence would be even more convincing against $p = 0.81$ or $p = 0.82$. So even if the null hypothesis included many possible values for p (like $p \geq 0.80$), we only have to check the evidence against $p = 0.80$. This is why we prefer the null hypothesis H_0: $p = 0.80$.

DEFINITION Null hypothesis H_0, Alternative hypothesis H_a

The claim that we weigh evidence against in a significance test is called the **null hypothesis (H_0).**

The claim that we are trying to find evidence for is the **alternative hypothesis (H_a).**

In the free-throw shooter example, our hypotheses are

$$H_0: p = 0.80$$
$$H_a: p < 0.80$$

where p is the true proportion of free throws he will make in the long run. The alternative hypothesis is **one-sided** ($p < 0.80$) because we suspect the player makes less than 80% of his free throws. If you suspect that the true value of a parameter may be either greater than or less than the null value, use a **two-sided** alternative hypothesis.

Teaching Tip: AP® Connections

Based on their experiences in AP® science classes, students may be surprised to hear that we state two hypotheses. Students are probably accustomed to stating a single research hypothesis, which is the hypothesis that they hope to find evidence to support (what we call the *alternative hypothesis*).

> **DEFINITION   One-sided and two-sided alternative hypothesis**
>
> The alternative hypothesis is **one-sided** if it states that a parameter is *greater than* the null value or if it states that the parameter is *less than* the null value.
>
> The alternative hypothesis is **two-sided** if it states that the parameter is *different from* the null value (it could be either greater than or less than).

The null hypothesis has the form H_0: parameter = null value. A one-sided alternative hypothesis has one of the forms H_a: parameter < null value or H_a: parameter > null value. A two-sided alternative hypothesis has the form H_a: parameter ≠ null value. To determine the correct form of H_a, read the problem carefully.

It is common to refer to a significance test with a one-sided alternative hypothesis as a *one-sided test* or *one-tailed test* and to a test with a two-sided alternative hypothesis as a *two-sided test* or *two-tailed test*.

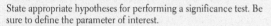

EXAMPLE

Juicy pineapples
Stating hypotheses

PROBLEM: At the Hawaii Pineapple Company, managers are interested in the size of the pineapples grown in the company's fields. Last year, the mean weight of the pineapples harvested from one large field was 31 ounces. A different irrigation system was installed in this field after the growing season. Managers wonder if this change will affect the mean weight of pineapples grown in the field this year.

State appropriate hypotheses for performing a significance test. Be sure to define the parameter of interest.

SOLUTION:

$H_0: \mu = 31$

$H_a: \mu \neq 31$

where μ = the true mean weight (in ounces) of all pineapples grown in the field this year.

> Because managers wonder if the mean weight of this year's pineapples will *differ* from last year's mean weight of 31 ounces, the alternative hypothesis is two-sided.

FOR PRACTICE, TRY EXERCISE 1

 The hypotheses should express the belief or suspicion we have *before* we see the data. It is cheating to look at the data first and then frame the alternative hypothesis to fit what the data show. For example, the data for the pineapple study showed that $\bar{x} = 31.935$ ounces for a random sample of 50 pineapples grown in the field this year. You should *not* change the alternative hypothesis to $H_a: \mu > 31$ after looking at the data.

> **AP® EXAM TIP**
>
> Hypotheses always refer to a population, not to a sample. Be sure to state H_0 and H_a in terms of population parameters. It is *never* correct to write a hypothesis about a sample statistic, such as $H_0: \hat{p} = 0.80$ or $H_a: \bar{x} \neq 31$.

Teaching Tip

Students will be able to tell if a test is one-sided or two-sided by carefully examining the wording used in the question. Ask students, "Which words in the question inform you that this is a one-sided test?" Then have them underline or circle the identified words in the stem of the question. The "Juicy pineapples" example on this page would look like this: "Managers wonder how this change will <u>affect</u> the mean" (two-sided). Students should not use the sample data to form the hypotheses. See the Caution at the bottom of this student page.

Teaching Tip

The "play" icon at the top of this example (and many others) indicates that students can watch a video presentation of the example by clicking on the link in the e-Book or viewing the videos on the Student Site at highschool.bfwpub.com/updatedtps6e.

ALTERNATE EXAMPLE   Skill 1.F

Are you college bound?
Stating hypotheses

PROBLEM:
The U.S. Bureau of Labor Statistics estimates that 69.7% of high school graduates enroll in a college or university. Bernard has great pride in the quality of his large high school. He thinks the proportion of college-bound students is greater for last year's graduating class.

State appropriate hypotheses for performing a significance test. Be sure to define the parameter of interest.

SOLUTION:
$H_0: p = 0.697$

$H_a: p > 0.697$

where p = the true proportion of all last year's graduates at Bernard's school who enroll in a college or university.

Teaching Tip

Notation matters! A statistic is used to estimate a parameter.

- $\bar{x}$ from the sample is used to estimate the population mean μ.
- $\hat{p}$ from the sample is used to estimate the population proportion p.
- s_x from the sample is used to estimate the population standard deviation σ.

✓ Answers to CYU

1. H_0: $p = 0.85$ and H_a: $p \neq 0.85$, where p = proportion of all students at Jannie's high school who sleep fewer than 8 hours a night.

2. H_0: $\mu = 10$ and H_a: $\mu > 10$, where μ = true mean amount of time it takes to complete the census form.

Making Connections

Students have already been exposed to the thinking involved in interpreting *P*-values, thanks to the "Hiring discrimination" activity in Chapter 1, the "1 in 6 wins" activity in Chapter 5, and the "I'm a great free-throw shooter" activity at the start of Chapter 9.

Teaching Tip

The analogy with a criminal trial is useful in discussing the logic of significance testing. For a defendant to be brought to trial, there must be some evidence that the person is guilty. It is the job of the jury to decide if the evidence is *convincing*— that is, to decide that there are no other plausible explanations for how the crime was committed. We do a significance test when the sample statistic provides evidence that the alternative hypothesis is true. To determine if the evidence is convincing, we calculate a *P*-value. If the *P*-value is small, there is convincing evidence in favor of the alternative hypothesis because we can rule out chance as a plausible explanation.

➕ Ask the StatsMedic
The Holy Grail of AP® Statistics

In this post, we discuss the importance of students being able to understand and interpret a *P*-value, rather than memorizing rules for making a decision in a significance test.

CHECK YOUR UNDERSTANDING

For each of the following settings, state appropriate hypotheses for performing a significance test. Be sure to define the parameter of interest.

1. According to the National Sleep Foundation, 85% of teens are getting too little sleep on school nights. Jannie wonders whether this result holds in her large high school. She asks an SRS of 100 students at the school how much sleep they get on a typical night. In all, 75 of the students are getting less than the recommended amount of sleep.

2. As part of its marketing campaign for the 2010 census, the U.S. Census Bureau advertised "10 questions, 10 minutes—that's all it takes." On the census form itself, we read, "The U.S. Census Bureau estimates that, for the average household, this form will take about 10 minutes to complete, including the time for reviewing the instructions and answers." We suspect that the average time it takes to complete the form may be longer than advertised.

Interpreting *P*-values

It may seem strange to you that we state a null hypothesis and then try to find evidence *against* it. Maybe it would help to think about how a criminal trial works in the United States. The defendant is "innocent until proven guilty." That is, the null hypothesis is innocence and the prosecution must offer convincing evidence against this hypothesis and in favor of the alternative hypothesis: guilt. That's exactly how significance tests work, although in statistics we deal with evidence provided by data and use a probability to say how strong the evidence is.

In the free-throw shooter activity at the beginning of the chapter, a player who claimed to make 80% of his free throws made only $\hat{p} = 32/50 = 0.64$ in a random sample of 50 free throws. This is evidence *against* the null hypothesis that $p = 0.80$ and *in favor of* the alternative hypothesis $p < 0.80$. But is the evidence convincing? To answer this question, we want to know how likely it is for an 80% shooter to make 64% or less by chance alone in a random sample of 50 attempts. This probability is called a **P-value**.

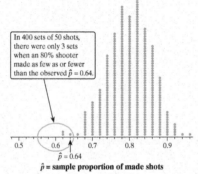

In 400 sets of 50 shots, there were only 3 sets when an 80% shooter made as few as or fewer than the observed $\hat{p} = 0.64$.

$\hat{p} = 0.64$

$\hat{p}$ = **sample proportion of made shots**

> **DEFINITION   P-value**
>
> The **P-value** of a test is the probability of getting evidence for the alternative hypothesis H_a as strong or stronger than the observed evidence when the null hypothesis H_0 is true.

We used simulation to estimate the *P*-value for our free-throw shooter: $3/400 = 0.0075$. How do we interpret this *P*-value? Assuming that the player makes 80% of his free throws in the long run, there is about a 0.0075 probability of getting a sample proportion of 0.64 or less just by chance in a set of 50 shots.

Small *P*-values give convincing evidence for H_a because they say that the observed result is unlikely to occur when H_0 is true. Large *P*-values fail to give convincing evidence for H_a because they say that the observed result is likely to occur by chance alone when H_0 is true.

Making Connections

Later in the chapter, we will use the *P*-value to make a decision about the null hypothesis. Assuming that the player makes 80% of his free throws in the long run, there is about a 0.0075 probability of getting a sample proportion of 0.64 or less just by chance in a set of 50 shots. Because this outcome is unlikely to happen purely by chance, we have convincing evidence against the null hypothesis, and will thus reject H_0.

We'll show you how to calculate *P*-values later. For now, let's focus on interpreting them.

Making Connections

In Section 9.2, we will calculate a *P*-value for a significance test about a proportion. In Section 9.3, we will calculate a *P*-value for a significance test about a difference in proportions.

EXAMPLE

Healthy bones
Interpreting a P-value

PROBLEM: Calcium is a vital nutrient for healthy bones and teeth. The National Institutes of Health (NIH) recommends a calcium intake of 1300 milligrams (mg) per day for teenagers. The NIH is concerned that teenagers aren't getting enough calcium, on average. Is this true? Researchers decide to perform a test of

$$H_0: \mu = 1300$$
$$H_a: \mu < 1300$$

where μ is the true mean daily calcium intake in the population of teenagers. They ask a random sample of 20 teens to record their food and drink consumption for 1 day. The researchers then compute the calcium intake for each student. Data analysis reveals that $\bar{x} = 1198$ mg and $s_x = 411$ mg. Researchers performed a significance test and obtained a *P*-value of 0.1404.

(a) Explain what it would mean for the null hypothesis to be true in this setting.
(b) Interpret the *P*-value.

SOLUTION:

(a) If $H_0: \mu = 1300$ is true, then the mean daily calcium intake in the population of teenagers is 1300 mg.

(b) Assuming that the mean daily calcium intake in the teen population is 1300 mg, there is a 0.1404 probability of getting a sample mean of 1198 mg or less just by chance in a random sample of 20 teens.

FOR PRACTICE, TRY EXERCISE 9

ALTERNATE EXAMPLE Skill 4.B

Are you college bound? Part 2
Interpreting a P-value

PROBLEM:
The U.S. Bureau of Labor Statistics estimates that 69.7% of high school graduates enroll in a college or university. Bernard thinks the proportion of college-bound students is greater for last year's graduates from his large high school. He decides to perform a test of

$$H_0: p = 0.697$$
$$H_a: p > 0.697$$

where p = the true proportion of all last year's graduates at Bernard's school who enroll in a college or university. Bernard asks a random sample of 40 of last year's graduates from his high school if they are enrolled in a college or university, and 34 say "Yes." The sample proportion enrolled in a college or university is

$$\hat{p} = \frac{34}{40} = 0.85.$$ Bernard performed a significance test and obtained a *P*-value of 0.018.

(a) Explain what it would mean for the null hypothesis to be true in this setting.
(b) Interpret the *P*-value.

SOLUTION:
(a) If $H_0: p = 0.697$ is true, then the proportion of all last year's graduates at Bernard's school who enroll in a college or university is 0.697 (the same as the national proportion).

(b) Assuming that the proportion of all last year's graduates at Bernard's school who enroll in a college or university is 0.697, there is a 0.018 probability of getting a sample proportion of 0.85 or greater just by chance in a random sample of 40 graduates.

Remember: A *P*-value measures the strength of evidence for the alternative hypothesis (and against the null hypothesis). In the preceding example, the sample mean was $\bar{x} = 1198$ mg. This result gives some evidence for $H_a: \mu < 1300$ because $1198 < 1300$. The *P*-value is the probability of getting evidence for H_a as strong or stronger than the observed result when H_0 is true. We can write the *P*-value as a conditional probability. For the "Healthy bones" example, the *P*-value = $P(\bar{x} \leq 1198 \mid \mu = 1300) = 0.1404$.

When H_a is two-sided (parameter $\neq$ null value), values of the sample statistic less than or greater than the null value both count as evidence for H_a. Suppose we want to perform a test of $H_0: p = 0.5$ versus $H_a: p \neq 0.5$ based on an SRS with a sample proportion of $\hat{p} = 0.65$. This result gives some evidence for $H_a: p \neq 0.5$ because $0.65 \neq 0.5$. In this case, evidence for H_a as strong or stronger than the observed result includes any value of $\hat{p}$ greater than or equal to 0.65 as well as any value of $\hat{p}$ less than or equal to 0.35. Why? Because $\hat{p} = 0.35$ is just as different from the null value of $p = 0.5$ as $\hat{p} = 0.65$. For this scenario, the *P*-value is equal to the conditional probability $P(\hat{p} \leq 0.35 \text{ or } \hat{p} \geq 0.65 \mid p = 0.5)$.

Making Connections

In Chapter 5, students learned about conditional probabilities. Inform them that a *P*-value is a conditional probability, with the condition being that the null hypothesis is true.

Teaching Tip

Using the conditional probability notation to define a *P*-value will be helpful for some students yet intimidating for others. We suggest exposing students to the notation but not requiring it.

Teaching Tip

At this point, it is tempting to have students memorize a rule or a slogan, such as "If the *P*-value is low, the null must go." Instead, we highly recommend developing students' understanding of the interpretation of the *P*-value, which naturally leads them to the correct conclusion.

Teaching Tip

"Convincing" evidence is the same as "sufficient" evidence. Recent AP® Statistics exams have used the term *convincing*, while the College Board Course and Exam Description uses the term *sufficient*. In this text, we stick with the language of "convincing."

AP® EXAM TIP

When the *P*-value is not small, we "fail to reject H_0." Instead of failing to reject a null hypothesis, many students use language that sounds like they "accept the null hypothesis." Accepting the null hypothesis will always lose credit on the AP® Statistics exam. Further discussion of this idea appears at the top of page 592.

Teaching Tip

When the *P*-value is not small, help students see the "3 **nots**" in the conclusion. For example: "Because the *P*-value of 0.32 is **not** less than $\alpha = 0.05$, we do **not** reject H_0. There is **not** convincing evidence for …"

Making Conclusions

The final step in performing a significance test is to draw a conclusion about the competing claims being tested. We make a decision based on the strength of the evidence in favor of the alternative hypothesis (and against the null hypothesis) as measured by the *P*-value.

- If the observed result is unlikely to occur by chance alone when H_0 is true (small *P*-value), we will "reject H_0."
- If the observed result is not unlikely to occur by chance alone when H_0 is true (large *P*-value), we will "fail to reject H_0."

This wording may seem unusual at first, but it's consistent with what happens in a criminal trial. Once the jury has weighed the evidence against the null hypothesis of innocence, they return one of two verdicts: "guilty" (reject H_0) or "not guilty" (fail to reject H_0). A not-guilty verdict doesn't guarantee that the defendant is innocent, just that there's not convincing evidence of guilt. Likewise, a fail-to-reject H_0 decision in a significance test doesn't guarantee that H_0 is true.

HOW TO MAKE A CONCLUSION IN A SIGNIFICANCE TEST

- If the *P*-value is small, reject H_0 and conclude there is convincing evidence for H_a (in context).
- If the *P*-value is not small, fail to reject H_0 and conclude there is not convincing evidence for H_a (in context).

In the free-throw shooter activity, the estimated *P*-value was 0.0075. Because the *P*-value is small, we reject H_0: $p = 0.80$. We have convincing evidence that the player makes fewer than 80% of his free throws in the long run.

For the teen calcium study, the *P*-value was 0.1404. Because the *P*-value is not small, we fail to reject H_0: $\mu = 1300$. We don't have convincing evidence that teens are getting less than 1300 mg of calcium per day, on average.

How small does a *P*-value have to be for us to reject H_0? In Chapter 4, we suggested that you use a boundary of 5% when determining whether a result is statistically significant. That is equivalent to saying, "View a *P*-value less than 0.05 as small." Choosing this boundary value means we require evidence for H_a so strong that it would happen less than 5% of the time just by chance when H_0 is true.

Sometimes it may be preferable to use a different boundary value—like 0.01 or 0.10—when drawing a conclusion in a significance test. We will explain why shortly. The chosen boundary value is called the **significance level**. We denote it by α, the Greek letter alpha.

DEFINITION   Significance level

The **significance level** α is the value that we use as a boundary for deciding whether an observed result is unlikely to happen by chance alone when the null hypothesis is true.

AP® EXAM TIP

Remind students that the word *significant* has a specific meaning in statistics, so they should not use it in the more casual sense as a synonym for *important*. When Readers score AP® Statistics exams, they will assume that a student who uses this word is using it in the statistical sense. Likewise, students should assume that when a question uses the word *significant*, it means "statistically significant."

Teaching Tip

As with the alternative hypothesis, the significance level should be identified *before* collecting any data.

When we use a fixed significance level α to draw a conclusion in a significance test, here are the two possibilities:

$$P\text{-value} < \alpha \rightarrow \text{reject } H_0 \rightarrow \text{convincing evidence for } H_a (\text{in context})$$

$$P\text{-value} > \alpha \rightarrow \text{fail to reject } H_0 \rightarrow \text{not convincing evidence for } H_a (\text{in context})$$

If the P-value is less than α, we say that the result is "statistically significant at the $\alpha = \underline{\quad}$ level."

Significance at the $\alpha = 0.05$ level is often expressed by the statement "The results were significant $(P < 0.05)$." Here, P stands for the P-value. *The P-value is more informative than a statement of significance* because it describes the strength of evidence for the alternative hypothesis. For example, both an observed result with $P = 0.03$ and an observed result with $P = 0.0003$ are significant at the $\alpha = 0.05$ level. But the P-value of 0.0003 gives much stronger evidence against H_0 and in favor of H_a than the P-value of 0.03.

EXAMPLE

Better batteries
Making conclusions

PROBLEM: A company has developed a new deluxe AAA battery that is supposed to last longer than its regular AAA battery.[1] However, these new batteries are more expensive to produce, so the company would like to be convinced that they really do last longer. Based on years of experience, the company knows that its regular AAA batteries last for 30 hours of continuous use, on average. The company selects an SRS of 15 deluxe AAA batteries and uses them continuously until they are completely drained. The sample mean lifetime is $\bar{x} = 33.93$ hours. A significance test is performed using the hypotheses

$$H_0: \mu = 30$$
$$H_a: \mu > 30$$

where μ is the true mean lifetime (in hours) of the deluxe AAA batteries. The resulting P-value is 0.0717. What conclusion would you make at the $\alpha = 0.05$ level?

SOLUTION:

Because the P-value of 0.0717 > α = 0.05, we fail to reject H_0. We don't have convincing evidence that the true mean lifetime of the company's deluxe AAA batteries is greater than 30 hours.

> What does the P-value of 0.0717 tell us? Assuming $H_0: \mu = 30$ is true, there is a 0.0717 probability of getting a sample mean lifetime of 33.9 hours or more in a random sample of 15 batteries.

FOR PRACTICE, TRY EXERCISE 15

Beginning users of significance tests generally find it easier to compare a P-value to a significance level than to interpret the P-value correctly in context. For that reason, we will include stating a significance level as a required part of every significance test. We'll also ask you to explain what a P-value means in a variety of settings. Just remember that the P-value measures the strength of evidence for the alternative hypothesis and against the null hypothesis.

Teaching Tip

If the results are statistically significant at significance level α, the results are also significant for any level greater than α. For example, results that are significant at the $\alpha = 0.02$ level are also significant at the $\alpha = 0.05$ and $\alpha = 0.10$ levels. We would not be sure if the results were significant at the $\alpha = 0.01$ level unless we knew the exact P-value.

ALTERNATE EXAMPLE   Skill 4.E

Speeding on the highway
Making conclusions

PROBLEM:
The police department in a certain city is trying to determine if it is worth the cost to install a speed sensor and traffic camera on a highway near the city. They will install the speed sensor and traffic camera if convinced that more than 20% of cars are speeding. The police department selects a random sample of 100 cars on the highway, measures their speed, and finds that 28 of the 100 cars are speeding. A significance test is performed using the hypotheses

$$H_0: p = 0.20$$
$$H_a: p > 0.20$$

where p is the true proportion of all cars on this highway that are speeding. The resulting P-value is 0.023. What conclusion would you make at the $\alpha = 0.05$ level?

SOLUTION:
Because the P-value of $0.023 < \alpha = 0.05$, we reject H_0. We have convincing evidence that the true proportion of all cars on this highway that are speeding is greater than 0.20.

Teaching Tip

There are two schools of thought about how to use a P-value to make a conclusion in a significance test. The "strength of evidence" school doesn't "reject" or "fail to reject" the null hypothesis but rather uses the P-value to make statements about how strongly the evidence supports the alternative hypothesis (e.g., "Because the P-value is so small, we have very convincing evidence to support the alternative hypothesis"). The "fixed significance level" school uses a preset boundary (e.g., $\alpha = 0.05$) to decide whether the null hypothesis should be rejected in favor of the alternative hypothesis. To help students avoid mistakes and potentially losing credit on the AP® Statistics exam, we recommend using the fixed significance level approach. We will model this approach in the student text.

COMMON STUDENT ERROR

Some students will want to conclude that the true mean battery lifetime is 30 hours. Help them see that if we were to create a confidence interval for μ, it would include many plausible values, not just $\mu = 30$, so it would not be correct to conclude the true mean is exactly 30 hours.

AP® EXAM TIP

Consider adding the interpretation of the *P*-value to begin the conclusion for a significance test, followed by the two sentences modeled in the text. *Consistent practice with interpreting P-values will lead to more success on the inference questions in the multiple-choice section of the AP® Statistics exam.*

At the end of the year when preparing for the exam, tell students to drop the interpretation of the *P*-value from the conclusion, because it is never required in a conclusion to a significance test and students often make a mistake in their interpretation.

Teaching Tip

Consider going back to the example of a U.S. jury trial. Recall that a person is innocent until "proven" guilty. Start by writing the hypotheses:

H_0: person is innocent
H_a: person is guilty

There are two possible decisions in the end, and each one has the potential to be a mistake.

- Reject H_0: The jury finds convincing evidence that the person is guilty. It is possible that the jury has made a mistake, and the person really is innocent (Type I error).
- Fail to reject H_0: The jury fails to find convincing evidence that the person is guilty. It is possible that the jury has made a mistake, and the person really is guilty (Type II error).

Be careful how you write conclusions when the *P*-value is large. Don't conclude that the null hypothesis is true just because we didn't find convincing evidence for the alternative hypothesis. For example, it would be incorrect to conclude that the company's deluxe AAA batteries last exactly 30 hours, on average. We found *some* evidence that the new batteries last longer, but the evidence wasn't convincing enough to reject H_0. Never "accept H_0" or conclude that H_0 is true! That would be like the jury in a criminal trial declaring the defendant "innocent" when they really mean "not guilty"!

AP® EXAM TIP

We recommend that you follow the two-sentence structure from the example when writing the conclusion to a significance test. The first sentence should give a decision about the null hypothesis—reject H_0 or fail to reject H_0—based on an explicit comparison of the *P*-value to a stated significance level. The second sentence should provide a statement about whether or not there is convincing evidence for H_a in the context of the problem.

In practice, the most commonly used significance level is $\alpha = 0.05$. This is mainly due to Sir Ronald A. Fisher, a famous statistician who worked on agricultural experiments in England during the early twentieth century. Fisher was the first to suggest deliberately using random assignment in an experiment. In a paper published in 1926, Fisher wrote that it is convenient to draw the line at about the level at which we can say: "Either there is something in the treatment, or a coincidence has occurred such as does not occur more than once in twenty trials."[2]

When a researcher plans to draw a conclusion based on a significance level, α should be stated *before* the data are produced. Otherwise, a deceptive user of statistics might choose α *after* the data have been analyzed in an attempt to manipulate the conclusion. This is just as inappropriate as choosing an alternative hypothesis after looking at the data.

Type I and Type II Errors

When we draw a conclusion from a significance test, we hope our conclusion will be correct. But sometimes it will be wrong. There are two types of mistakes we can make: a **Type I error** or a **Type II error**.

> **DEFINITION**   **Type I error, Type II error**
>
> A **Type I error** occurs if we reject H_0 when H_0 is true. That is, the data give convincing evidence that H_a is true when it really isn't.
>
> A **Type II error** occurs if we fail to reject H_0 when H_a is true. That is, the data do not give convincing evidence that H_a is true when it really is.

Here's a helpful reminder to keep the two types of errors straight. "Fail to" goes with Type II.

The possible outcomes of a significance test are summarized in Figure 9.2.

		Truth about the population	
		H_0 true	H_a true
Conclusion based on sample	Reject H_0	Type I error	Correct conclusion
	Fail to reject H_0	Correct conclusion	Type II error

FIGURE 9.2 The two types of errors in significance tests.

Making Connections

The probabilities of a Type I error and of a Type II error are both conditional probabilities:

$P(\text{Type I error}) = P(\text{reject } H_0 \mid H_0 \text{ is true})$

$P(\text{Type II error}) = P(\text{fail to reject } H_0 \mid H_a \text{ is true})$

Making Connections

If H_a is true, the probability that we find convincing evidence that H_a is true (top right box) is called the *power* of the test. Power will be discussed in detail in Section 9.2.

If H_0 is true:

- Our conclusion is correct if we don't find convincing evidence that H_a is true.
- We make a Type I error if we find convincing evidence that H_a is true.

If H_a is true:

- Our conclusion is correct if we find convincing evidence that H_a is true.
- We make a Type II error if we do not find convincing evidence that H_a is true.

Only one error is possible at a time, depending on the conclusion we make.

It is important to be able to describe Type I and Type II errors in the context of a problem. Considering the consequences of each of these types of error is also important, as the following example shows.

EXAMPLE

Perfect potatoes
Type I and Type II errors

Steve Cukrov/Alamy Stock Photo

PROBLEM: A potato chip producer and its main supplier agree that each shipment of potatoes must meet certain quality standards. If the producer finds convincing evidence that more than 8% of the potatoes in the shipment have "blemishes," the truck will be sent away to get another load of potatoes from the supplier. Otherwise, the entire truckload will be used to make potato chips. To make the decision, a supervisor will inspect a random sample of 500 potatoes from the shipment. The producer will then perform a test at the $\alpha = 0.05$ significance level of

$$H_0: p = 0.08$$
$$H_a: p > 0.08$$

where $p =$ the true proportion of potatoes with blemishes in a given truckload. Describe a Type I and a Type II error in this setting, and give a possible consequence of each.

SOLUTION:

Type I error: The producer finds convincing evidence that more than 8% of the potatoes in the shipment have blemishes, when the true proportion is really 0.08.

Consequence: The potato-chip producer sends away the truckload of acceptable potatoes, wasting time and depriving the supplier of money.

Type II error: The producer does not find convincing evidence that more than 8% of the potatoes in the shipment have blemishes, when the true proportion is greater than 0.08.

Consequence: More potato chips are made with blemished potatoes, which may upset customers and lead to decreased sales.

FOR PRACTICE, TRY EXERCISE 23

Which is more serious: a Type I error or a Type II error? That depends on the situation. For the potato-chip producer, a Type II error seems more serious because it may lead to lower-quality potato chips and decreased sales.

Teaching Tip

In the perfect potatoes example, a student may argue that the Type I error is more serious, because the company doesn't want to ruin its relationship with the supplier. A good argument can be made for either error being more serious, so there is not necessarily a "right" answer. Most important is that students give reasoning that is consistent with their decision.

Teaching Tip

When you describe Type II errors, avoid language that suggests you are accepting the null hypothesis. For example, using the potato context from the example on this page, avoid saying, "We conclude that the true proportion of potatoes with blemishes is 0.08, when in reality the true proportion is greater than 0.08." A better response would be to say that "we don't find convincing evidence that the true proportion of potatoes with blemishes is greater than 0.08, when in reality the true proportion is greater than 0.08."

ALTERNATE EXAMPLE   Skill 4.B

Speeding on the highway, Part 2
Type I and Type II errors

PROBLEM:
In the previous alternate example, a police department wants to determine if it is worth the cost to install a speed sensor and traffic camera on the highway near the city. The department will install the speed sensor and traffic camera if more than 20% of the cars on this highway are speeding. The police will select a random sample of 100 cars on the highway and then perform a test at the $\alpha = 0.05$ significance level of

$$H_0: p = 0.20$$
$$H_a: p > 0.20$$

where p is the true proportion of all cars on this highway that are speeding. Describe a Type I error and a Type II error in this setting, and give a possible consequence of each.

SOLUTION:

Type I error: The police department finds convincing evidence that more than 20% of cars on this highway are speeding, when the true proportion is really 0.20.

Consequence: The police department installs the speed sensor and traffic camera on the highway when it isn't needed, wasting money.

Type II error: The police department does not find convincing evidence that more than 20% of cars on this highway are speeding, when the true proportion is greater than 0.20.

Consequence: The police department does not install the speed sensor and traffic camera on the highway when it should. People continue to speed on the highway, putting their safety, and that of others, at risk.

Teaching Tip

On this page, we see that the probability of making a Type I error is the significance level, denoted α. Some textbooks use the notation β for the probability of a Type II error. We will save the use of the notation β for Chapter 12, where it will represent the slope of a population regression line. Notice that the probability of a Type I error and the probability of a Type II error are *not* complementary, even though they are inversely related.

Making Connections

If the conditions for inference are not satisfied, the actual Type I error rate may be higher or lower than the significance level. In Chapter 8, we learned the same lesson with confidence intervals: when the conditions for inference aren't met, the actual capture rate may vary from the stated confidence level.

Teaching Tip

If a Type I error has more serious consequences, consider using a smaller α level. If a Type II error has more serious consequences, consider using a larger α level.

✓ **Answers to CYU**

1. *Type I error:* The manager finds convincing evidence that less than 63% of the drive-thru wait times are longer than 2 minutes, when the true proportion really is 0.63.

Type II error: The manager does not find convincing evidence that less than 63% of the drive-thru wait times are longer than 2 minutes, when the true proportion really is less than 0.63.

2. In this case, a Type I error is more serious because the manager will believe that the additional employee reduces the proportion of drive-thru customers who have to wait longer than 2 minutes to receive their food, when that is not the case.

3. No; if the null hypothesis is true, a significance level of $\alpha = 0.10$ will result in a Type I error 10% of the time just by chance. Since a Type I error is more serious in this case, it would be better to pick a smaller value of α, such as $\alpha = 0.01$.

The most common significance levels are $\alpha = 0.05$, $\alpha = 0.01$, and $\alpha = 0.10$. Which is the best choice for a given significance test? That depends on whether a Type I error or a Type II error is more serious.

In the "Perfect potatoes" example, a Type I error occurs if the true proportion of blemished potatoes in a shipment is $p = 0.08$, but we get a value of the sample proportion $\hat{p}$ large enough to yield a P-value less than $\alpha = 0.05$. When H_0 is true, this will happen 5% of the time just by chance. In other words, $P(\text{Type I error}) = \alpha$.

TYPE I ERROR PROBABILITY

The probability of making a Type I error in a significance test is equal to the significance level α.

We can decrease the probability of making a Type I error in a significance test by using a smaller significance level. For instance, the potato-chip producer could use $\alpha = 0.01$ instead of $\alpha = 0.05$. But there is a trade-off between $P(\text{Type I error})$ and $P(\text{Type II error})$: as one increases, the other decreases, assuming everything else remains the same. If we make it more difficult to reject H_0 by decreasing α, we increase the probability that we will not find convincing evidence for H_a when it is true. That's why it is important to consider the possible consequences of each type of error before choosing a significance level.

 CHECK YOUR UNDERSTANDING

The manager of a fast-food restaurant wants to reduce the proportion of drive-thru customers who have to wait longer than 2 minutes to receive their food after placing an order. Based on store records, the proportion of customers who had to wait longer than 2 minutes was $p = 0.63$. To reduce this proportion, the manager assigns an additional employee to drive-thru orders. During the next month, the manager collects a random sample of 250 drive-thru times and finds that $\hat{p} = \dfrac{144}{250} = 0.576$. The manager then performs a test of the following hypotheses at the $\alpha = 0.10$ significance level:

$$H_0: p = 0.63$$
$$H_a: p < 0.63$$

where p = the true proportion of drive-thru customers who have to wait longer than 2 minutes to receive their food.

1. Describe a Type I error and a Type II error in this setting.
2. Which type of error is more serious in this case? Justify your answer.
3. Based on your answer to Question 2, do you agree with the company's choice of $\alpha = 0.10$? Why or why not?
4. The P-value of the manager's test is 0.0385. Interpret the P-value.

4. Assuming that the true proportion of all drive-thru customers who have to wait longer than 2 minutes to receive their food after placing an order is 0.63, there is a 0.0385 probability of getting a sample proportion of 0.576 or less who have to wait longer than 2 minutes just by chance in a random sample of 250 drive-thru customers.

Section 9.1 | Summary

- A **significance test** is a procedure for using observed data to decide between two competing claims, called hypotheses. The hypotheses are often statements about a parameter, like the population proportion p or the population mean μ.
- The claim that we weigh evidence *against* in a significance test is called the **null hypothesis** (H_0). The null hypothesis has the form H_0: parameter = null value.
- The claim about the population that we are trying to find evidence *for* is the **alternative hypothesis** (H_a).
  - A **one-sided** alternative hypothesis has the form H_a: parameter < null value or H_a: parameter > null value.
  - A **two-sided** alternative hypothesis has the form H_a: parameter ≠ null value.
- Often, H_0 is a statement of no change or no difference. The alternative hypothesis states what we hope or suspect is true.
- The **P-value** of a test is the probability of getting evidence for the alternative hypothesis H_a that is as strong as or stronger than the observed evidence when the null hypothesis H_0 is true.
- Small P-values are evidence against the null hypothesis and for the alternative hypothesis because they say that the observed result is unlikely to occur when H_0 is true. To determine if a P-value should be considered small, we compare it to the **significance level** α.
- We make a conclusion in a significance test based on the P-value.
  - If P-value < α: Reject H_0 and conclude there is convincing evidence for H_a (in context).
  - If P-value > α: Fail to reject H_0 and conclude there is not convincing evidence for H_a (in context).
- When we make a conclusion in a significance test, there are two kinds of mistakes we can make.
  - A **Type I error** occurs if we reject H_0 when it is, in fact, true. In other words, the data give convincing evidence for H_a when the null hypothesis is correct.
  - A **Type II error** occurs if we fail to reject H_0 when H_a is true. In other words, the data don't give convincing evidence for H_a, even though the alternative hypothesis is correct.
- The probability of making a Type I error is equal to the significance level α. There is a trade-off between P(Type I error) and P(Type II error): as one increases, the other decreases. So it is important to consider the possible consequences of each type of error before choosing a significance level.

Section 9.1 | Exercises

In Exercises 1–6, state appropriate hypotheses for performing a significance test. Be sure to define the parameter of interest.

1. **No homework?** Mr. Tabor believes that less than 75% of the students at his school completed their math homework last night. The math teachers inspect the homework assignments from a random sample of 50 students at the school.

   pg 587

2. **Don't argue!** A Gallup poll report revealed that 72% of teens said they seldom or never argue with their friends.[3] Yvonne wonders whether this result holds true in her large high school, so she surveys a random sample of 150 students at her school.

Teaching Tip

Make sure to point out the two icons next to Exercise 1. The "pg 587" icon reminds students that the example on page 587 is much like this exercise. The "play" icon reminds students that there is a video solution available in the student e-Book or at the Student Site.

Answers to Section 9.1 Exercises

9.1 H_0: $p = 0.75$; H_a: $p < 0.75$, where p = the true proportion of the students at Mr. Tabor's school who completed their math homework last night.

9.2 H_0: $p = 0.72$; H_a: $p \neq 0.72$, where p = the true proportion of teens in Yvonne's school who rarely or never argue with their friends.

9.3 H_0: $\mu = 180$; H_a: $\mu \neq 180$, where μ = the true mean volume of liquid dispensed by the machine.

9.4 H_0: $\mu = 115$; H_a: $\mu > 115$, where μ = the true mean score on the SSHA for all students at least 30 years of age at the teacher's college.

9.5 H_0: $\sigma = 3$; H_a: $\sigma > 3$, where σ = the true standard deviation of the temperature in the cabin.

9.6 H_0: $\sigma = 10$; H_a: $\sigma > 10$, where σ = the true standard deviation of the distance jumped by the ski jumpers.

9.7 (a) The null hypothesis is always that there is "no difference" or "no change"; the alternative hypothesis is what we suspect is true. These ideas are reversed in the stated hypotheses. *Correct:* H_0: $p = 0.37$; H_a: $p > 0.37$.
(b) Hypotheses are always about population parameters. However, the stated hypotheses are in terms of the sample statistic. *Correct:* H_0: $\mu = 3000$ grams; H_a: $\mu < 3000$ grams.

9.8 (a) Hypotheses are always about population parameters, but the stated hypotheses are about the sample statistic. *Correct:* H_0: $p = 0.37$; H_a: $p > 0.37$.
(b) Values in both hypotheses must be the same; the null hypothesis should have $=$ and the alternative should have $<$. *Correct:* H_0: $\mu = 3000$ grams; H_a: $\mu < 3000$ grams.

9.9 (a) If H_0: $p = 0.75$ is true, then the proportion of all students at Mr. Tabor's school who completed their homework last night is 0.75. **(b)** Assuming the proportion is 0.75, there is a 0.1265 probability of getting a sample proportion of 0.68 or less by chance in a random sample of 50 students at the school.

9.10 (a) If H_0: $\mu = 115$ is true, then the true mean score is 115. **(b)** Assuming the true mean score is 115, there is a 0.0101 probability of getting a sample mean of 125.7 or greater just by chance in an SRS of 45 older students.

9.11 Assuming the true mean volume of liquid dispensed by the machine is 180 ml, there is a 0.0589 probability of getting a sample mean at least as far from 180 as 179.6 (in either direction) by chance in a random sample of 40 bottles filled by the machine.

9.12 Assuming the true proportion of teens in Yvonne's school who rarely or never argue with their friends is 0.72, there is a 0.0291 probability of getting a sample proportion at least as far from 0.72 as 0.64 (in either direction) by chance in a random sample of 150 students from her school.

3. **How much juice?** One company's bottles of grape-fruit juice are filled by a machine that is set to dispense an average of 180 milliliters (ml) of liquid. A quality-control inspector must check that the machine is working properly. The inspector takes a random sample of 40 bottles and measures the volume of liquid in each bottle.

4. **Attitudes** The Survey of Study Habits and Attitudes (SSHA) is a psychological test that measures students' attitudes toward school and study habits. Scores range from 0 to 200. Higher scores indicate better attitudes and study habits. The mean score for U.S. college students is about 115. A teacher suspects that older students have better attitudes toward school, on average. She gives the SSHA to an SRS of 45 of the over 1000 students at her college who are at least 30 years of age.

5. **Cold cabin?** During the winter months, the temperatures at the Starneses' Colorado cabin can stay well below freezing (32°F or 0°C) for weeks at a time. To prevent the pipes from freezing, Mrs. Starnes sets the thermostat at 50°F. The manufacturer claims that the thermostat allows variation in home temperature of $\sigma = 3$°F. Mrs. Starnes suspects that the manufacturer is overstating the consistency of the thermostat.

6. **Ski jump** When ski jumpers take off, the distance they fly varies considerably depending on their speed, skill, and wind conditions. Event organizers must position the landing area to allow for differences in the distances that the athletes fly. For a particular competition, the organizers estimate that the variation in distance flown by the athletes will be $\sigma = 10$ meters. An experienced jumper thinks that the organizers are underestimating the variation.

In Exercises 7 and 8, explain what's wrong with the stated hypotheses. Then give correct hypotheses.

7. **Stating hypotheses**
(a) A change is made that should improve student satisfaction with the parking situation at a local high school. Before the change, 37% of students approve of the parking that's provided. The null hypothesis H_0: $p > 0.37$ is tested against the alternative H_a: $p = 0.37$.

(b) A researcher suspects that the mean birth weights of babies whose mothers did not see a doctor before delivery is less than 3000 grams. The researcher states the hypotheses as

$$H_0: \bar{x} = 3000 \text{ grams}$$
$$H_a: \bar{x} < 3000 \text{ grams}$$

8. **Stating hypotheses**
(a) A change is made that should improve student satisfaction with the parking situation at your school.

Before the change, 37% of students approve of the parking that's provided. The null hypothesis H_0: $\hat{p} = 0.37$ is tested against the alternative H_a: $\hat{p} > 0.37$.

(b) A researcher suspects that the mean birth weights of babies whose mothers did not see a doctor before delivery is less than 3000 grams. The researcher states the hypotheses as

$$H_0: \mu = 3000 \text{ grams}$$
$$H_a: \mu \leq 2999 \text{ grams}$$

9. **No homework?** Refer to Exercise 1. The math teachers inspect the homework assignments from a random sample of 50 students at the school. Only 68% of the students completed their math homework. A significance test yields a P-value of 0.1265.

(a) Explain what it would mean for the null hypothesis to be true in this setting.

(b) Interpret the P-value.

10. **Attitudes** Refer to Exercise 4. In the study of older students' attitudes, the sample mean SSHA score was 125.7 and the sample standard deviation was 29.8. A significance test yields a P-value of 0.0101.

(a) Explain what it would mean for the null hypothesis to be true in this setting.

(b) Interpret the P-value.

11. **How much juice?** Refer to Exercise 3. The mean amount of liquid in the bottles is 179.6 ml and the standard deviation is 1.3 ml. A significance test yields a P-value of 0.0589. Interpret the P-value.

12. **Don't argue** Refer to Exercise 2. Yvonne finds that 96 of the 150 students (64%) say they rarely or never argue with friends. A significance test yields a P-value of 0.0291. Interpret the P-value.

13. **Interpreting a P-value** A student performs a test of H_0: $\mu = 100$ versus H_a: $\mu > 100$ and gets a P-value of 0.044. The student says, "There is a 0.044 probability of getting the sample result I did by chance alone." Explain why the student's explanation is wrong.

14. **Interpreting a P-value** A student performs a test of H_0: $p = 0.3$ versus H_a: $p < 0.3$ and gets a P-value of 0.22. The student says, "This means there is about a 22% chance that the null hypothesis is true." Explain why the student's explanation is wrong.

15. **No homework** Refer to Exercises 1 and 9. What conclusion would you make at the $\alpha = 0.05$ level?

16. **Attitudes** Refer to Exercises 4 and 10. What conclusion would you make at the $\alpha = 0.05$ level?

9.13 The student forgot to include the conditions and the direction in the interpretation. *Assuming the null hypothesis is true,* there is a 0.044 probability of getting the sample result I did *or one even larger* by chance alone.

9.14 Either H_0 is true or H_0 is false. Assuming the null hypothesis is true, there is a 0.22 probability of getting a sample proportion as small as or smaller than the one observed just by chance.

9.15 Because the P-value of 0.1265 is greater than $\alpha = 0.05$, we fail to reject H_0. We do not have convincing evidence that the true proportion of students at Mr. Tabor's school who completed their math homework last night is less than 0.75.

9.16 Because the P-value of 0.0101 is less than $\alpha = 0.05$, we reject H_0. We have convincing evidence that the true mean score on the SSHA for all students at least 30 years of age at the teacher's college is greater than 115.

17. **How much juice?** Refer to Exercises 3 and 11.

(a) What conclusion would you make at the $\alpha = 0.10$ level?

(b) Would your conclusion from part (a) change if a 5% significance level was used instead? Explain your reasoning.

18. **Don't argue** Refer to Exercises 2 and 12.

(a) What conclusion would you make at the $\alpha = 0.01$ level?

(b) Would your conclusion from part (a) change if a 5% significance level was used instead? Explain your reasoning.

19. **Making conclusions** A student performs a test of $H_0: p = 0.75$ versus $H_a: p < 0.75$ at the $\alpha = 0.05$ significance level and gets a P-value of 0.22. The student writes: "Because the P-value is large, we accept H_0. The data provide convincing evidence that the null hypothesis is true." Explain what is wrong with this conclusion.

20. **Making conclusions** A student performs a test of $H_0: \mu = 12$ versus $H_a: \mu \neq 12$ at the $\alpha = 0.05$ significance level and gets a P-value of 0.01. The student writes: "Because the P-value is small, we reject H_0. The data prove that H_a is true." Explain what is wrong with this conclusion.

21. **Heavy bread?** The mean weight of loaves of bread produced at the bakery where you work is supposed to be 1 pound. You are the supervisor of quality control at the bakery, and you are concerned that new employees are producing loaves that are too light. Suppose you weigh an SRS of bread loaves and find that the mean weight is 0.975 pound.

(a) State appropriate hypotheses for performing a significance test. Be sure to define the parameter of interest.

(b) Explain why there is some evidence for the alternative hypothesis.

(c) The P-value for the test in part (a) is 0.0806. Interpret the P-value.

(d) What conclusion would you make at the $\alpha = 0.01$ significance level?

22. **Philly fanatics?** Nationally, the proportion of red cars on the road is 0.12. A statistically minded fan of the Philadelphia Phillies (whose team color is red) wonders if Phillies fans are more likely to drive red cars. One day during a home game, he takes a random sample of 210 cars parked at Citizens Bank Park (the Phillies' home field), and counts 35 red cars.

(a) State appropriate hypotheses for performing a significance test. Be sure to define the parameter of interest.

(b) Explain why there is some evidence for the alternative hypothesis.

(c) The P-value for the test in (a) is 0.0187. Interpret the P-value.

(d) What conclusion would you make at the $\alpha = 0.05$ significance level?

23. **Opening a restaurant** You are thinking about opening a restaurant and are searching for a good location. From research you have done, you know that the mean income of those living near the restaurant must be over $85,000 to support the type of upscale restaurant you wish to open. You decide to take a simple random sample of 50 people living near one potential location. Based on the mean income of this sample, you will perform a test of

$$H_0: \mu = \$85,000$$
$$H_a: \mu > \$85,000$$

where μ is the true mean income in the population of people who live near the restaurant.[4] Describe a Type I error and a Type II error in this setting, and give a possible consequence of each.

24. **Reality TV** Television networks rely heavily on ratings of TV shows when deciding whether to renew a show for another season. Suppose a network has decided that "Miniature Golf with the Stars" will only be renewed if it can be established that more than 12% of U.S. adults watch the show. A polling company asks a random sample of 2000 U.S. adults if they watch "Miniature Golf with the Stars." The network uses the data to perform a test of

$$H_0: p = 0.12$$
$$H_a: p > 0.12$$

where p is the true proportion of all U.S. adults who watch the show. Describe a Type I error and a Type II error in this setting, and give a possible consequence of each.

25. **Awful accidents** Slow response times by paramedics, firefighters, and policemen can have serious consequences for accident victims. In the case of life-threatening injuries, victims generally need medical attention within 8 minutes of the accident. Several cities have begun to monitor emergency response times. In one such city, emergency personnel took more than 8 minutes to arrive on 22% of all calls involving life-threatening injuries last year. The city manager shares this information and encourages these first responders to "do better." After 6 months, the city manager selects an SRS of 400 calls involving life-threatening injuries and examines the response

---

9.17 (a) Because the P-value of 0.0589 is less than $\alpha = 0.10$, we reject H_0. We have convincing evidence that the true mean volume of liquid dispensed differs from 180 ml. **(b)** Yes; because the P-value of $0.0589 > \alpha = 0.05$, we would fail to reject H_0. We lack convincing evidence that the true mean volume differs from 180 ml.

9.18 (a) Because the P-value of $0.0291 > \alpha = 0.01$, we fail to reject H_0. We lack convincing evidence that the true proportion of teens in Yvonne's school who rarely or never argue differs from 0.72. **(b)** Yes; because the P-value of $0.0291 < \alpha = 0.05$, we would reject H_0. We have convincing evidence that the true proportion differs from 0.72.

9.19 It is never correct to "accept the null hypothesis." If the P-value is large, the data do not provide convincing evidence that the alternative hypothesis is true. Lacking evidence for the alternative hypothesis does not provide convincing evidence that the null hypothesis is true.

9.20 The data never "prove" a hypothesis true, no matter how large or small the P-value. There can always be error.

9.21 (a) $H_0: \mu = 1$, $H_a: \mu < 1$, where $\mu =$ the true mean weight (in pounds) of bread loaves produced. **(b)** Some evidence for the alternative hypothesis exists because the mean weight of an SRS of bread loaves is only 0.975 pound—less than the suspected mean (1 pound).

(c) Assuming the true mean weight of bread loaves is 1 pound, a 0.0806 probability exists of getting a sample mean of 0.975 pound or less by chance in a random sample of loaves. **(d)** Because the P-value of $0.0806 > \alpha = 0.01$, we fail to reject H_0. We do not have convincing evidence that the true mean weight for all loaves of bread produced is less than 1 pound.

9.22 (a) $H_0: p = 0.12$, $H_a: p > 0.12$, where $p =$ the true proportion of all cars parked at Phillies home field that are red. **(b)** Some evidence for the alternative hypothesis exists because the sample proportion of red parked cars is $35/210 = 0.167$, which is greater than the national proportion of red cars (0.12). **(c)** Assuming the true proportion of all red cars parked at Phillies home field is 0.12, a 0.0187 probability exists of getting a sample proportion of $35/210 = 0.167$ or greater by chance in a random sample of 210 cars. **(d)** Because the P-value of $0.0187 < \alpha = 0.05$, we reject H_0. We have convincing evidence that the true proportion of all cars parked at Phillies home field that are red is greater than 0.12.

9.23 *Type I:* You find convincing evidence that the mean income of all residents near the restaurant exceeds $85,000 when in reality it does not. *Consequence:* You open in an undesirable location, so your restaurant may go out of business. *Type II:* You do not find convincing evidence that the mean income of all residents near the restaurant exceeds $85,000 when in reality it does. *Consequence:* You do not open in a desirable location and lose potential income.

9.24 *Type I:* The network finds convincing evidence that the true proportion of U.S. adults who watch "Miniature Golf with the Stars" is greater than 0.12, when it really equals 0.12. *Consequence:* The show will be renewed when interest does not support it. *Type II:* The network doesn't find convincing evidence that the true proportion who watch "Miniature Golf with the Stars" is greater than 0.12, when it really is greater than 0.12. *Consequence:* The show will not renew when there was a market for it.

9.25 (a) A Type I error would be finding convincing evidence that the proportion of all calls in which first responders took more than 8 minutes to arrive had decreased when it really hadn't. A Type II error would be not finding convincing evidence that the proportion of all calls in which first responders took more than 8 minutes to arrive decreased when it really had. **(b)** A Type I error would be worse because the city would overestimate the ability of the emergency personnel to get to the scene quickly and people may end up dying. **(c)** The probability of a Type I error is $\alpha = 0.05$; because the consequence may be the difference between life and death, the manager should use a value for α that is lower, such as $\alpha = 0.01$.

9.26 (a) A Type I error would be finding convincing evidence that the true mean copper content of the water from the new source is greater than 1.3 mg/liter when it really isn't. A Type II error would be *not* finding convincing evidence that the true mean copper content of the water from the new source is greater than 1.3 mg/l when it really is. **(b)** A Type II error would be worse because the water would not be safe for drinking, yet would seem safe. **(c)** Here, we seek to minimize the probability of making a Type II error. As the probability of a Type II error increases, the probability of Type I error decreases; so I would recommend that the company use a value of α greater than 0.05.

9.27 (a) H_0: $p = 0.10$, H_a: $p > 0.10$, where p = the true proportion of all students at Simon's school who are left-handed. **(b)** Based on the simulation results, 24 of the 200 simulated trials yielded a sample proportion of 0.16 or greater, so the P-value is approximately $24/200 = 0.12$. Assuming the true proportion of all students at Simon's school who are left-handed is 0.10, there is a 0.12 probability of getting a sample proportion of 0.16 or greater just by chance in a random sample of 50 students. **(c)** Use $\alpha = 0.05$; because the P-value of $0.12 > \alpha = 0.05$, we fail to reject H_0. We do not have convincing evidence that the true proportion of all students at Simon's school who are left-handed is greater than 0.10.

times. She then performs a test at the $\alpha = 0.05$ level of H_0: $p = 0.22$ versus H_a: $p < 0.22$, where p is the true proportion of calls involving life-threatening injuries during this 6-month period for which emergency personnel took more than 8 minutes to arrive.

(a) Describe a Type I error and a Type II error in this setting.

(b) Which type of error is more serious in this case? Justify your answer.

(c) Based on your answer to part (b), do you agree with the manager's choice of $\alpha = 0.05$? Why or why not?

26. **Clean water** The Environmental Protection Agency (EPA) has determined that safe drinking water should contain at most 1.3 mg/liter of copper, on average. A water supply company is testing water from a new source and collects water in small bottles at each of 30 randomly selected locations. The company performs a test at the $\alpha = 0.05$ significance level of H_0: $\mu = 1.3$ versus H_a: $\mu > 1.3$, where μ is the true mean copper content of the water from the new source.

(a) Describe a Type I error and a Type II error in this setting.

(b) Which type of error is more serious in this case? Justify your answer.

(c) Based on your answer to part (b), do you agree with the company's choice of $\alpha = 0.05$? Why or why not?

27. **More lefties?** In the population of people in the United States, about 10% are left-handed. After bumping elbows at lunch with several left-handed students, Simon wondered if more than 10% of students at his school are left-handed. To investigate, he selected an SRS of 50 students and found 8 lefties ($\hat{p} = 8/50 = 0.16$).

To determine if these data provide convincing evidence that more than 10% of the students at Simon's school are left-handed, 200 trials of a simulation were conducted. Each dot in the graph shows the proportion of students that are left-handed in a random sample of 50 students, assuming that each student has a 10% chance of being left handed.

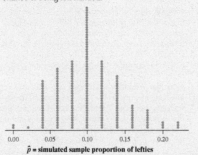

$\hat{p}$ = simulated sample proportion of lefties

(a) State appropriate hypotheses for performing a significance test. Be sure to define the parameter of interest.

(b) Use the simulation results to estimate the P-value of the test in part (a). Interpret the P-value.

(c) What conclusion would you make?

28. **Who wrote this poem?** Statistics can help decide the authorship of literary works. Sonnets by a well-known poet contain an average of $\mu = 6.9$ new words (words not used in the poet's other works) and a standard deviation of $\sigma = 2.7$ words, and the number of new words is approximately Normally distributed. Scholars expect sonnets by other authors to contain more new words than this famous poet's works. A new manuscript has been discovered with many new sonnets, and scholars are debating whether it is this poet's work. They take a random sample of five sonnets from the new manuscript and count the number of new words in each one. The mean number of new words in these five sonnets is $\bar{x} = 9.2$.

The following dotplot shows the results of simulating 200 random samples of size 5 from a Normal distribution with a mean of 6.9 and a standard deviation of 2.7, and calculating the mean for each sample.

4.0 4.4 4.8 5.2 5.6 6.0 6.4 6.8 7.2 7.6 8.0 8.4 8.8 9.2 9.6 10.0
$\bar{x}$ = simulated mean number of new words

(a) State appropriate hypotheses for performing a significance test. Be sure to define the parameter of interest.

(b) Use the simulation results to estimate the P-value of the test in part (a). Interpret the P-value.

(c) What conclusion would you make?

Multiple Choice: *Select the best answer for Exercises 29–32.*

29. Experiments on learning in animals sometimes measure how long it takes mice to find their way through a maze. The mean time is 18 seconds for one particular maze. A researcher thinks that a loud noise will cause the mice to complete the maze faster. She measures how long each of 10 mice takes with a loud noise as stimulus. The appropriate hypotheses for the significance test are

(a) H_0: $\mu = 18$; H_a: $\mu \neq 18$.

(b) H_0: $\mu = 18$; H_a: $\mu > 18$.

9.28 (a) H_0: $\mu = 6.9$, H_a: $\mu > 6.9$, where μ = the true mean number of new words in the newly discovered manuscript. **(b)** Based on the simulation results, 12 of the 200 simulated trials yielded a sample mean of 9.2 or greater, so the P-value is approximately $12/200 = 0.06$. Assuming the true mean number of new words in the manuscript is 6.9, there is a 0.06 probability of getting a sample mean of 9.2 or greater by chance in a random sample of 5 sonnets. **(c)** Use $\alpha = 0.05$; because the P-value of $0.06 > \alpha = 0.05$, we fail to reject H_0. We do not have convincing evidence that the true mean number of new words in the newly discovered manuscript is greater than 6.9.

(c) H_0: $\mu < 18$; H_a: $\mu = 18$.

(d) H_0: $\mu = 18$; H_a: $\mu < 18$.

(e) H_0: $\bar{x} = 18$; H_a: $\bar{x} < 18$.

Exercises 30–32 refer to the following setting. Members of the city council want to know if a majority of city residents support a 1% increase in the sales tax to fund road repairs. To investigate, they survey a random sample of 300 city residents and use the results to test the following hypotheses:

$$H_0: p = 0.50$$
$$H_a: p > 0.50$$

where p is the proportion of all city residents who support a 1% increase in the sales tax to fund road repairs.

30. A Type I error in the context of this study occurs if the city council

(a) finds convincing evidence that a majority of residents supports the tax increase, when in reality there isn't convincing evidence that a majority supports the increase.

(b) finds convincing evidence that a majority of residents supports the tax increase, when in reality at most 50% of city residents support the increase.

(c) finds convincing evidence that a majority of residents supports the tax increase, when in reality more than 50% of city residents do support the increase.

(d) does not find convincing evidence that a majority of residents supports the tax increase, when in reality more than 50% of city residents do support the increase.

(e) does not find convincing evidence that a majority of residents supports the tax increase, when in reality at most 50% of city residents do support the increase.

31. In the sample, $\hat{p} = 158/300 = 0.527$. The resulting P-value is 0.18. What is the correct interpretation of this P-value?

(a) Only 18% of the city residents support the tax increase.

(b) There is an 18% chance that the majority of residents supports the tax increase.

(c) Assuming that 50% of residents support the tax increase, there is an 18% probability that the sample proportion would be 0.527 or greater by chance alone.

(d) Assuming that more than 50% of residents support the tax increase, there is an 18% probability that the sample proportion would be 0.527 or greater by chance alone.

(e) Assuming that 50% of residents support the tax increase, there is an 18% chance that the null hypothesis is true by chance alone.

32. Based on the P-value in Exercise 31, which of the following would be the most appropriate conclusion?

(a) Because the P-value is large, we reject H_0. There is sufficient evidence that more than 50% of city residents support the tax increase.

(b) Because the P-value is large, we fail to reject H_0. There is sufficient evidence that more than 50% of city residents support the tax increase.

(c) Because the P-value is large, we reject H_0. There is sufficient evidence that at most 50% of city residents support the tax increase.

(d) Because the P-value is large, we fail to reject H_0. There is sufficient evidence that at most 50% of city residents support the tax increase.

(e) Because the P-value is large, we fail to reject H_0. There is not sufficient evidence that more than 50% of city residents support the tax increase.

Recycle and Review

33. **Women in math** (5.3) Of the 24,611 degrees in mathematics given by U.S. colleges and universities in a recent year, 70% were bachelor's degrees, 24% were master's degrees, and the rest were doctorates. Moreover, women earned 43% of the bachelor's degrees, 41% of the master's degrees, and 29% of the doctorates.[5]

(a) How many of the mathematics degrees given that year were earned by women? Justify your answer.

(b) Suppose we randomly select a person who earned a mathematics degree in this recent year. Are the events "Degree earned by a woman" and "Degree was a bachelor's degree" independent? Justify your answer using appropriate probabilities.

(c) If you choose 2 of the 24,611 mathematics degrees at random, what is the probability that at least 1 of the 2 degrees was earned by a woman? Show your work.

34. **Explaining confidence** (8.1) Here is an explanation from a newspaper concerning one of its opinion polls. Explain what is wrong with the following statement.

For a poll of 1600 adults, the variation due to sampling error is no more than 3 percentage points either way. The error margin is said to be valid at the 95% confidence level. This means that, if the same questions were repeated in 20 polls, the results of at least 19 surveys would be within 3 percentage points of the results of this survey.

9.29 d

9.30 b

9.31 c

9.32 e

9.33 (a) P(degree earned by a woman) = 0.4168; approximately $(24{,}611)(0.4168) = 10{,}258$ mathematics degrees were awarded to women. **(b)** Not independent; P(woman) = 0.4168, which is not equal to P(woman | bachelors) = 0.43.

(c) P(at least 1 of the 2 degrees earned by a woman) $= 1 - \left(\dfrac{14{,}353}{24{,}611}\right)\left(\dfrac{14{,}352}{24{,}610}\right)$

$= 0.6599$

9.34 One mistake is to say that 95% of other polls would have results within 3 percentage points *of the results of this survey*. It should say that other polls would have results within 3 percentage points *of the true proportion*. Another mistake is to say that "at least 19" of the 20 surveys will be within 3 percentage points when it should say "about 19" of the 20 surveys instead.

Teaching Tip

While it is important that students know how to check each condition, it is equally important that they understand *why* we check the condition. We call this the "*So what?*"

Random condition: So what?
→ so individual observations are independent and we can generalize to the population.

10% condition: So what?
→ so we can view observations as independent even though we are sampling without replacement.

Large Counts condition: So what?
→ so the sampling distribution of $\hat{p}$ will be approximately Normal and we can use z to find a *P*-value.

SECTION 9.2 **Tests About a Population Proportion**

LEARNING TARGETS *By the end of the section, you should be able to:*

- State and check the Random, 10%, and Large Counts conditions for performing a significance test about a population proportion.
- Calculate the standardized test statistic and *P*-value for a test about a population proportion.
- Perform a significance test about a population proportion.
- Interpret the power of a significance test and describe what factors affect the power of a test.

A significance test can be used to test a claim about a population parameter. Section 9.1 presented the reasoning of significance tests, including the idea of a *P*-value. This section describes how to perform a significance test about a population proportion.

Performing a Significance Test About *p*

In Section 9.1, we met a basketball player who claimed to make 80% of his free throws. We thought that he might be exaggerating. In a sample of 50 free throws, the player made only 32. His sample proportion of made free throws was therefore

$$\hat{p} = \frac{32}{50} = 0.64$$

This result is much lower than what he claimed. Does it provide *convincing* evidence against the player's claim? To find out, we need to perform a significance test of

$$H_0: p = 0.80$$
$$H_a: p < 0.80$$

where $p =$ the true proportion of free throws that the shooter makes in the long run.

CHECKING CONDITIONS In Chapter 8, we introduced conditions that should be met before we construct a confidence interval for a population proportion p. We called them Random, 10%, and Large Counts. These same conditions must be verified before carrying out a significance test. Recall that the purpose of these conditions is to ensure that the observations in the sample can be viewed as independent and that the sampling distribution of $\hat{p}$ is approximately Normal.

The Large Counts condition for proportions requires that both np and $n(1-p)$ be at least 10. When constructing a confidence interval for p, we use the sample proportion $\hat{p}$ in place of the unknown p to check this condition. As we discussed in Section 8.2, $n\hat{p} \geq 10$ and $n(1-\hat{p}) \geq 10$ are the symbolic equivalent of saying that the observed counts of successes and failures in the sample are both at least 10. Because we assume H_0 is true when performing a significance test, we use the parameter value specified by the null hypothesis (denoted p_0) when checking the Large Counts condition. In this case, the Large Counts condition says that the *expected* count of successes np_0 and of failures $n(1-p_0)$ are both at least 10.

AP® EXAM TIP

The free response section almost always has a question that asks students to perform a significance test. Students should always check these conditions before performing the test, even if the question doesn't specifically ask for the conditions. Students will not be asked to perform a significance test in a context where the conditions have not been met. There may, however, be a question that focuses on just the conditions. In this case, the conditions may not be met.

> ## CONDITIONS FOR PERFORMING A SIGNIFICANCE TEST ABOUT A PROPORTION
>
> - **Random:** The data come from a random sample from the population of interest.
>   - **10%:** When sampling without replacement, $n < 0.10N$.
> - **Large Counts:** Both np_0 and $n(1 - p_0)$ are at least 10.

If the data come from a convenience sample or a voluntary response sample, there's no point carrying out a significance test for p. The same is true if there are other sources of bias during data collection. If the Large Counts condition is violated, a P-value calculated from a Normal distribution will not be accurate. And if the 10% condition is violated, the value of $\sigma_{\hat{p}}$ that we get from the formula will be greater than the actual standard deviation of the sampling distribution of $\hat{p}$.

Let's check the conditions for performing a significance test of the basketball player's claim that $p = 0.80$.

- *Random:* The 50 shots can be viewed as a random sample from the population of all possible shots that the shooter takes.
  - *10%:* We're not sampling without replacement from a finite population (because the player can keep on shooting), so we don't check the 10% condition.
- *Large Counts:* Assuming H_0 is true, $p = 0.80$. Then $np_0 = (50)(0.80) = 40$ and $n(1 - p_0) = (50)(0.20) = 10$ are both at least 10, so this condition is met.

EXAMPLE	**Get a job!**
	Checking conditions

PROBLEM: According to the U.S. Census Bureau, the proportion of students in high school who have a part-time job is 0.25. An administrator at a large high school suspects that the proportion of students at her school who have a part-time job is less than the national figure. She would like to carry out a test at the $\alpha = 0.05$ significance level of

$$H_0: p = 0.25$$
$$H_a: p < 0.25$$

where $p =$ the true proportion of all students at the school who have a part-time job.

The administrator selects a random sample of 200 students from the school and finds that 39 of them have a part-time job. Check if the conditions for performing the significance test are met.

SOLUTION:

- Random? Random sample of 200 students from the school. ✓
  - 10%: 200 is less than 10% of students at a large high school. ✓
- Large Counts? $np_0 = 200(0.25) = 50 \geq 10$ and $n(1 - p_0) = 200(1 - 0.25) = 150 \geq 10$ ✓

> Be sure to use p_0, not $\hat{p}$, when checking the Large Counts condition!

FOR PRACTICE, TRY EXERCISE 35

Teaching Tip

Consider having students check the condition *and* give the "*So what?*" Here are the "So what?" answers for this example.

Random:
→ so individual observations are independent and we can generalize to the population of all students from the school.

10%:
→ so we can view observations as independent even though we are sampling without replacement.

Large Counts:
→ so the sampling distribution of $\hat{p}$ will be approximately Normal and we can use z to find a P-value.

ALTERNATE EXAMPLE	Skill 4.C

Do you Instagram?
Checking conditions

PROBLEM:
A recent study concluded that 76% of teens aged 13–17 use Instagram. Joseph is an AP® Statistics student who believes the proportion is greater for students at his large high school. He would like to carry out a test at the $\alpha = 0.05$ significance level of

$$H_0: p = 0.76$$
$$H_a: p > 0.76$$

where $p =$ the true proportion of all students at this high school who use Instagram. Joseph selects a random sample of 50 students from his high school and finds that 41 of them use Instagram. Check if the conditions for performing the significance test are met.

SOLUTION:
- *Random:* Random sample of 50 students from the high school. ✓
  - *10%:* 50 is less than 10% of students at a large high school. ✓
- *Large Counts:* $np_0 = 50(0.76) = 38 \geq 10$ and $n(1 - p_0) = 50(1 - 0.76) = 12 \geq 10$ ✓

Teaching Tip

Tell your students that Figure 9.3 on this page is a model for the simulated sampling distribution of $\hat{p}$ in Figure 9.1 on page 585 (or the dotplot poster you saved from the "Free-throw shooter" activity at the beginning of the chapter). That is, it shows the possible values of $\hat{p}$ and how often they will occur, assuming the shooter makes 80% of his shots in the long run. It's never wrong to use a simulation (with a large number of trials) to estimate a *P*-value, but it's more traditional and often easier to estimate a *P*-value using a probability distribution (e.g., Normal) that models the sampling distribution of the statistic.

Making Connections

To help students understand the meaning of the standardized test statistic, remind them of the interpretation of the *z*-score. The *z*-score tells us how many standard deviations from the mean of the null distribution the observed statistic falls, and in what direction. The farther away from the mean in the direction specified by H_a, the smaller the probability of this outcome happening by chance (*P*-value).

AP® EXAM TIP

The formula for a standardized test statistic is included on the formula sheet that is provided for both parts of the AP® Statistics exam. The formula sheet also lists the standard deviation and standard errors of several statistics, including the ones we will use in this chapter, $\sigma_{\hat{p}}$ and $\sigma_{\hat{p}_1 - \hat{p}_2}$.

Teaching Tip

We like referring to *z* as the *standardized test statistic*, rather than simply the *test statistic*, because $\hat{p}$ can also be considered a test statistic.

CALCULATIONS: STANDARDIZED TEST STATISTIC AND *P*-VALUE For the free-throw shooter, the sample proportion of made shots was $\hat{p} = \frac{32}{50} = 0.64$.

Because this result is less than 0.80, there is *some* evidence against H_0: $p = 0.80$ and in favor of H_a: $p < 0.80$. But do we have *convincing* evidence that the player is exaggerating? To answer this question, we have to know how likely it is to get a sample proportion of 0.64 or less by chance alone when the null hypothesis is true. In other words, we are looking for a *P*-value.

Suppose for now that the null hypothesis H_0: $p = 0.80$ is true. Consider the sample proportion $\hat{p}$ of made free throws in a random sample of size $n = 50$. You learned in Section 7.2 that the sampling distribution of $\hat{p}$ will have mean

$$\mu_{\hat{p}} = p = 0.80$$

and standard deviation

$$\sigma_{\hat{p}} = \sqrt{\frac{p(1-p)}{n}} = \sqrt{\frac{0.80(0.20)}{50}} = 0.0566$$

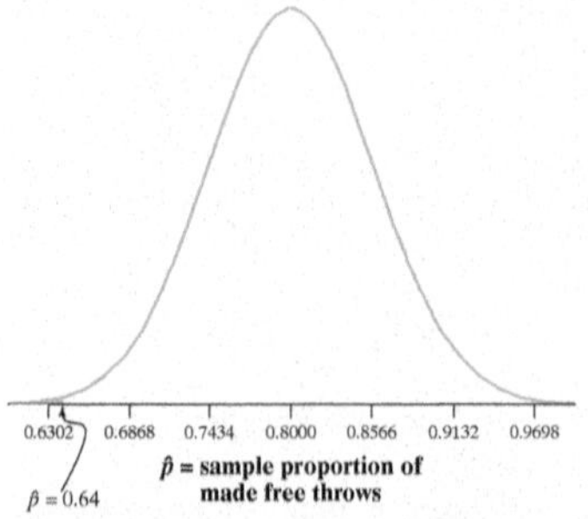

Because the Large Counts condition is met, the sampling distribution of $\hat{p}$ will be approximately Normal. Figure 9.3 displays this distribution. We have added the player's sample result, $\hat{p} = \frac{32}{50} = 0.64$.

0.6302   0.6868   0.7434   0.8000   0.8566   0.9132   0.9698

$\hat{p} = 0.64$

$\hat{p}$ = sample proportion of made free throws

FIGURE 9.3 Normal distribution that models the sampling distribution of the sample proportion $\hat{p}$ of made shots in random samples of 50 free throws by an 80% shooter.

To assess how far the statistic ($\hat{p} = 0.64$) is from the null value of the parameter ($p_0 = 0.80$), we standardize the statistic:

$$z = \frac{\hat{p} - p_0}{\sqrt{\dfrac{p_0(1 - p_0)}{n}}} = \frac{0.64 - 0.80}{0.0566} = -2.83$$

This value is called the **standardized test statistic**.

> Some people refer to the value of *z* as the *test statistic*.

> The formula sheet provided on the AP® Statistics exam has "standard error of statistic" in the denominator of the standardized test statistic formula. However, a note at the bottom of the formula sheet reminds students to use the standard deviation of the statistic when it is assumed to be known. That is the case in a one-sample *z* test for a proportion because we perform calculations assuming that H_0: $p = p_0$ is true.

DEFINITION   Standardized test statistic

A **standardized test statistic** measures how far a sample statistic is from what we would expect if the null hypothesis H_0 were true, in standard deviation units. That is,

$$\text{standardized test statistic} = \frac{\text{statistic} - \text{parameter}}{\text{standard deviation (error) of statistic}}$$

The standardized test statistic says how far the sample result is from the null value, and in what direction, on a standardized scale. In this case, the sample proportion $\hat{p} = 0.64$ of made free throws is 2.83 standard deviations less than the null value of $p = 0.80$.

You can use the standardized test statistic to find the *P*-value for a significance test. In this case, the *P*-value is the probability of getting a sample proportion less than or equal to $\hat{p} = 0.64$ by chance alone when H_0: $p = 0.80$ is true. The shaded area in Figure 9.4(a) shows this probability. Figure 9.4(b) shows the corresponding area to the left of $z = -2.83$ in the standard Normal distribution.

Making Connections

$$\text{standardized test statistic} = \frac{\text{statistic} - \text{parameter}}{\text{standard deviation (error) of statistic}}$$

This is the "general formula" for a standardized test statistic that works for all significance tests in this course (except for chi-square tests). Each specific significance test will then have its own "specific formula." For example, in Section 9.2, we will see this specific formula that works for a one-sample *z*-test for *p*:

$$z = \frac{\hat{p} - p_0}{\sqrt{\dfrac{p_0(1 - p_0)}{n}}}$$

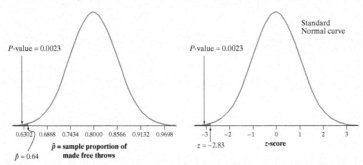

Standard
Normal curve

P-value = 0.0023

P-value = 0.0023

0.6302 0.6868 0.7434 0.8000 0.8566 0.9132 0.9698

$\hat{p}$ = sample proportion of
made free throws

$\hat{p}$ = 0.64

−3   −2   −1   0   1   2   3

z = −2.83   z-score

FIGURE 9.4 The shaded area shows the P-value for the player's sample proportion of made shots (a) on the Normal distribution that models the sampling distribution of $\hat{p}$ from Figure 9.3 and (b) on the standard Normal curve.

Note that the calculated P-value of 0.0023 is even smaller than the estimated P-value of 0.0075 from our earlier simulation.

We can find the P-value using Table A or technology. Table A gives $P(z \le -2.83) = 0.0023$. The TI-83/84 command normalcdf(lower: −1000, upper: −2.83, mean:0, SD:1) also gives a P-value of 0.0023. Remember that P-value calculations are valid only when our probability model is true—that is, when the conditions for inference are met.

If H_0 is true and the player makes 80% of his free throws in the long run, there's only about a 0.0023 probability that he would make 32 or fewer of 50 shots by chance alone. This small probability confirms our earlier decision to reject H_0 and gives convincing evidence that the player is exaggerating.

EXAMPLE

Part-time jobs
Calculating the standardized test statistic and P-value

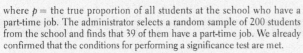

PROBLEM: In the preceding example, an administrator at a large high school decided to perform a test at the $\alpha = 0.05$ significance level of

$$H_0: p = 0.25$$
$$H_a: p < 0.25$$

where $p =$ the true proportion of all students at the school who have a part-time job. The administrator selects a random sample of 200 students from the school and finds that 39 of them have a part-time job. We already confirmed that the conditions for performing a significance test are met.

(a) Explain why the sample result gives some evidence for the alternative hypothesis.
(b) Calculate the standardized test statistic and P-value.
(c) What conclusion would you make?

SOLUTION:

(a) The sample proportion of students at this school with a part-time job is $\hat{p} = \dfrac{39}{200} = 0.195$, which is less than the national proportion of $p = 0.25$ (as suggested by H_a).

ALTERNATE EXAMPLE   Skill 3.E

Do you Instagram? Part 2   Calculating the standardized test statistic and P-value

PROBLEM:
In the preceding alternate example, Joseph decided to perform a test at the $\alpha = 0.05$ significance level of

$$H_0: p = 0.76$$
$$H_a: p > 0.76$$

where $p =$ the true proportion of all students at his large high school who use Instagram. Joseph selects a random sample of 50 students from the school and finds that 41 of them use Instagram. We already confirmed that the conditions for performing a significance test are met.
(a) Explain why the sample result gives some evidence for the alternative hypothesis.

(b) Calculate the standardized test statistic and P-value.
(c) What conclusion would you make?

SOLUTION:
(a) The sample proportion of students at this school who use Instagram is $\hat{p} = \dfrac{41}{50} = 0.82$, which is greater than the national proportion of $p = 0.76$ (supporting H_a).

(b) $z = \dfrac{0.82 - 0.76}{\sqrt{\dfrac{0.76(0.24)}{50}}} = 0.99$

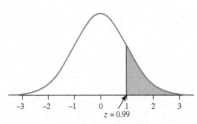

z = 0.99

Table A: $P(z \ge 0.99) = 0.1611$
Tech: normalcdf(lower: 0.99, upper: 1000, mean: 0, SD: 1) = 0.1611
(c) Because the P-value of 0.1611 > $\alpha = 0.05$, we fail to reject H_0. We do not have convincing evidence that the true proportion of all students at this large high school who use Instagram is greater than 0.76.

Teaching Tip

Teaching Tip

During the school year, consider having students write a general formula and a specific formula with variables before substituting numbers:

General formula:

$$\text{standardized test statistic} = \frac{\text{statistic} - \text{parameter}}{\text{standard deviation (error) statistic}}$$

Specific formula:

$$z = \frac{\hat{p} - p_0}{\sqrt{\dfrac{p_0(1 - p_0)}{n}}}$$

One advantage to establishing this expectation is that the general formula will be the same for all significance tests in this course (except for chi-square tests). Only the specific formula changes for each new significance test.

At the end of the year, when preparing for the AP® Statistics exam, tell students they no longer have to include these formulas, because they are not required for full credit on the AP® exam and students often make mistakes in their notation.

(b)

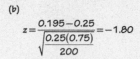

$$z = \frac{0.195 - 0.25}{\sqrt{\dfrac{0.25(0.75)}{200}}} = -1.80$$

$$\text{standardized test statistic} = \frac{\text{statistic} - \text{parameter}}{\text{standard deviation (error) of statistic}}$$

$$z = \frac{\hat{p} - p_0}{\sqrt{\dfrac{p_0(1 - p_0)}{n}}}$$

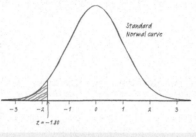

Standard Normal curve

$z = -1.80$

Using Table A: $P(z \leq -1.80) = 0.0359$

Using Technology: normalcdf(lower: -1000, upper: -1.80, mean: 0, SD: 1) $= 0.0359$

(c) Because the P-value of $0.0359 < \alpha = 0.05$, we reject H_0. We have convincing evidence that the true proportion of all students at this large high school who have a part-time job is less than 0.25.

FOR PRACTICE, TRY EXERCISE 39

AP® EXAM TIP

Notice that we did not include an option (ii) to "Use technology to find the desired area without standardizing" when performing the Normal calculation in part (b) of the example. That's because you are always required to give the standardized test statistic and the P-value when performing a significance test on the AP® Statistics exam.

Remember that there are two possible explanations for why the sample proportion of students who have part-time jobs ($\hat{p} = 39/200 = 0.195$) in the example is less than $p = 0.25$. The first explanation is that the true proportion of all students at the school who have part-time jobs is 0.25 and that we got a sample proportion this small due to sampling variability. The second explanation is that the true proportion of students at the school with part-time jobs is less than 0.25. We ruled out the first explanation due to the small P-value (0.0359) and settled on the second explanation. Of course, it is possible that we made a Type I error by rejecting H_0 when it is true.

Putting It All Together: One-Sample *z* Test for *p*

To perform a significance test, we state hypotheses, check conditions, calculate a standardized test statistic and P-value, and draw a conclusion in the context of the problem. The four-step process is ideal for organizing our work.

SIGNIFICANCE TESTS: A FOUR-STEP PROCESS

State: State the hypotheses you want to test and the significance level, and define any parameters you use.

Plan: Identify the appropriate inference method and check conditions.

Do: If the conditions are met, perform calculations.

- Give the sample statistic(s).
- Calculate the standardized test statistic.
- Find the P-value.

Conclude: Make a conclusion about the hypotheses in the context of the problem.

We have shown you how to complete each of the four steps in a test about a population proportion. Let's reflect on how the pieces fit together in a test of $H_0: p = p_0$. When the conditions are met, the sampling distribution of $\hat{p}$ is approximately Normal with

$$\text{mean } \mu_{\hat{p}} = p \quad \text{and} \quad \text{standard deviation } \sigma_{\hat{p}} = \sqrt{\frac{p(1-p)}{n}}$$

For confidence intervals, we substitute $\hat{p}$ for p in the standard deviation formula to obtain the standard error. When performing a significance test, however, we start by assuming that the null hypothesis $H_0: p = p_0$ is true. We use this null value when calculating the standard deviation.

If we standardize the statistic $\hat{p}$ by subtracting its mean and dividing by its standard deviation, we get the standardized test statistic.

$$z = \frac{\hat{p} - p_0}{\sqrt{\dfrac{p_0(1-p_0)}{n}}}$$

There are three conditions that must be met for this formula to be valid.

1. The Random condition allows us to make an inference about the population from which the sample was randomly selected, as you learned in Chapter 4. In addition, random sampling helps ensure that individual observations in the sample can be viewed as independent.

2. The 10% condition allows us to use the familiar formula for the standard deviation of the sampling distribution of $\hat{p}$ (with p_0 replacing p) when we are sampling without replacement from a finite population.

3. The Large Counts condition allows us to use a Normal distribution to model the sampling distribution of $\hat{p}$. When this condition is met and H_0 is true, the standardized test statistic z has approximately the standard Normal distribution. Then we can obtain P-values from this distribution using Table A or technology.

Here is a summary of the Do step for a **one-sample z test for a proportion.**

ONE-SAMPLE z TEST FOR A PROPORTION

Suppose the conditions are met. To perform a test of $H_0: p = p_0$, compute the standardized test statistic

$$z = \frac{\hat{p} - p_0}{\sqrt{\dfrac{p_0(1-p_0)}{n}}}$$

Find the P-value by calculating the probability of getting a z statistic this large or larger in the direction specified by the alternative hypothesis H_a using a standard Normal distribution.

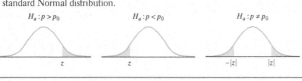

Teaching Tip

When performing significance tests, we always do the calculations assuming that the null hypothesis is true. This is equivalent to assuming that the defendant in a criminal trial is innocent until proven guilty. It is also why we typically state the null hypothesis as an equality, not an inequality such as $H_0: p \geq 0.8$. In the latter case, assuming the null hypothesis is true doesn't tell us what value of p to use in the calculations.

Teaching Tip

This discussion is a more detailed explanation of the "So what?" for each condition and a direct connection of each condition to the formula used to calculate the standardized test statistic.

AP® EXAM TIP

The formula for the standardized test statistic in a one-sample z test for a proportion is not included on the formula sheet provided to students on the AP® Statistics exam. Rather, the formula sheet includes the general formula for a standardized test statistic and lists the different standard deviations and standard errors students may need, including $\sigma_{\hat{p}}$. Students are expected to figure out which standard deviation or standard error to use based on the context of the question.

Teaching Tip

Many older statistics books use a "rejection region" approach to decide what conclusion to make rather than calculating a P-value. To use this approach, a critical value is identified for the standardized test statistic, such as $z^* = 1.645$ when $\alpha = 0.05$ and the test is one-sided. If the observed z test statistic is greater than $z^* = 1.645$, the null hypothesis is rejected. We do not recommend this approach, because knowing the P-value gives us more information than just whether to reject or fail to reject the null hypothesis.

Teaching Tip

Notice in the first two graphs that the direction of the shading matches the direction of the alternative hypothesis.

Does California watch more Netflix?
Significance test for a proportion

PROBLEM:
A recent report states that 55% of U.S. adults use Netflix to stream shows and movies. An advertising company believes the proportion of California residents who use Netflix is greater than the national proportion, because Netflix headquarters is located in Los Gatos, California. The company selects a random sample of 600 adults from California and finds that 360 of them use Netflix. Is there convincing evidence at the $\alpha = 0.05$ level that more than 55% of California residents use Netflix?

SOLUTION:
STATE: We want to test

$$H_0: p = 0.55$$
$$H_a: p > 0.55$$

where p = the true proportion of all California residents who use Netflix using $\alpha = 0.05$.

PLAN: One-sample z test for p

- *Random:* Random sample of 600 residents from California. ✓
  - *10%:* 600 < 10% of all California residents. ✓
- *Large Counts:* 600(0.55) = 330 ≥ 10 and 600(0.45) = 270 ≥ 10 ✓

DO:

- $\hat{p} = \dfrac{360}{600} = 0.60$

- $z = \dfrac{0.60 - 0.55}{\sqrt{\dfrac{0.55(0.45)}{600}}} = 2.46$

- *P*-value:

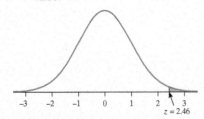

Table A: $P(z \geq 2.46) = 1 - 0.9931 = 0.0069$

Tech: normalcdf(lower: 2.46, upper: 1000, mean: 0, SD: 1) = 0.0069

CONCLUDE: Because our *P*-value of $0.0069 < \alpha = 0.05$, we reject H_0. There is convincing evidence that the true proportion of all California residents who use Netflix is greater than 0.55.

Here is an example of the one-sample z test for p in action.

EXAMPLE

One potato, two potato
Significance test for a proportion

PROBLEM: Recall that the potato-chip producer we met in Section 9.1 and its main supplier agree that each shipment of potatoes must meet certain quality standards. If the producer finds convincing evidence that more than 8% of the potatoes in the shipment have "blemishes," the truck will be sent away to get another load of potatoes from the supplier. Otherwise, the entire truckload will be used to make potato chips. The potato-chip producer has just received a truckload of potatoes from the supplier. A supervisor selects a random sample of 500 potatoes from the truck. An inspection reveals that 47 of the potatoes have blemishes. Is there convincing evidence at the $\alpha = 0.10$ level that more than 8% of the potatoes in the shipment have blemishes?

SOLUTION:
STATE: We want to test

$$H_0: p = 0.08$$
$$H_a: p > 0.08$$

where p = the true proportion of potatoes in this shipment with blemishes, using $\alpha = 0.10$.

> **STATE:** State the hypotheses you want to test and the significance level, and define any parameters you use.

PLAN: One-sample z test for p.

- *Random:* Random sample of 500 potatoes from the shipment. ✓
  - *10%:* It's reasonable to assume that 500 < 10% of all potatoes in the shipment. ✓
- *Large Counts:* 500(0.08) = 40 ≥ 10 and 500(0.92) = 460 ≥ 10 ✓

> **PLAN:** Identify the appropriate inference method and check conditions.

DO:

- $\hat{p} = \dfrac{47}{500} = 0.094$

- $z = \dfrac{0.094 - 0.08}{\sqrt{\dfrac{0.08(0.92)}{500}}} = 1.15$

- *P*-value:

> **DO:** If the conditions are met, perform calculations.
> - Identify the sample statistic(s).
> - Calculate the standardized test statistic.
> - Find the *P*-value.

> The sample result gives *some* evidence in favor of H_a because 0.094 > 0.08.

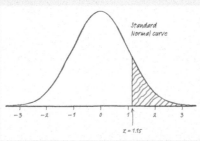

Standard Normal curve

$z = 1.15$

Using Table A: $P(z \geq 1.15) = 1 - 0.8749 = 0.1251$
Using technology: **normalcdf(lower: 1.15, upper: 1000, mean: 0, SD: 1)** = 0.1251

AP® EXAM TIP

When naming the inference method, it is acceptable to list the calculator command 1-PropZTest. However, we prefer that students use the "one-sample z test for p" because it shows deeper understanding and leads to fewer misconceptions. Also, the "one-sample z test for p" is often used as a possible answer in multiple choice questions.

 COMMON STUDENT ERROR

Some students incorrectly believe that a two-sample z test for $p_1 - p_2$ (coming soon in Section 9.3) can be used in this context because there are two proportions—those potatoes with blemishes and those without. Because there is only one sample from one population, this must be a one-sample z test for p. See the 2016 AP® Statistics exam #5C for a similar question.

CONCLUDE: Because our P-value of 0.1251 > α = 0.10, we fail to reject H_0. There is not convincing evidence that the true proportion of blemished potatoes in the shipment is greater than 0.08.

> CONCLUDE: Make a conclusion about the hypotheses in the context of the problem.

FOR PRACTICE, TRY EXERCISE 43

The preceding example reminds us why significance tests are important. The sample proportion of blemished potatoes was $\hat{p} = 47/500 = 0.094$. This result gave some evidence against H_0 and in favor of H_a. To see whether such an outcome is unlikely to occur by chance alone when H_0 is true, we had to carry out a significance test. The P-value told us that a sample proportion this large or larger would occur in about 12.5% of all random samples of 500 potatoes when H_0 is true. So we can't rule out sampling variability as a plausible explanation for getting a sample proportion of $\hat{p} = 0.094$. Of course, we could have made a Type II error in this case by failing to reject H_0 when $H_a: p > 0.08$ is true.

WHAT HAPPENS WHEN THE DATA DON'T SUPPORT H_a? Suppose the supervisor had inspected a random sample of 500 potatoes from the shipment and found 33 with blemishes. This yields a sample proportion of $\hat{p} = 33/500 = 0.066$. This sample doesn't give any evidence to support the alternative hypothesis $H_a: p > 0.08$! Don't continue with the significance test. The conclusion is clear: we should fail to reject $H_0: p = 0.08$. This truckload of potatoes will be used by the potato-chip producer.

If you weren't paying attention, you might end up carrying out the test. Let's see what would happen. The corresponding standardized test statistic is

$$z = \frac{\hat{p} - p_0}{\sqrt{\dfrac{p_0(1 - p_0)}{n}}} = \frac{0.066 - 0.08}{\sqrt{\dfrac{0.08(0.92)}{500}}} = -1.15$$

What's the P-value? It's the probability of getting a z statistic this large or larger in the direction specified by H_a, $P(z \geq -1.15)$. Figure 9.5 shows this

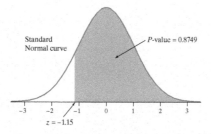

FIGURE 9.5 The P-value for the one-sided test of $H_0: p = 0.08$ versus $H_a: p > 0.08$.

P-value as an area under the standard Normal curve. Using Table A or technology, the *P*-value is $1 - 0.1251 = 0.8749$. There's about an 87.5% chance of getting a sample proportion as large as or larger than $\hat{p} = 0.066$ if $p = 0.08$. As a result, we would fail to reject H_0. Same conclusion, but with lots of unnecessary work!

Always check to see whether the data give evidence against H_0 in the direction specified by H_a before you do calculations.

CHECK YOUR UNDERSTANDING

According to the National Campaign to Prevent Teen and Unplanned Pregnancy, 20% of teens aged 13 to 19 say that they have electronically sent or posted sexually suggestive images of themselves.[6] The counselor at a large high school worries that the actual figure might be higher at her school. To find out, she administers an anonymous survey to a random sample of 250 of the school's 2800 students. All 250 respond, and 63 admit to sending or posting sexual images. Carry out a significance test at the $\alpha = 0.05$ significance level.

Your calculator will handle the "Do" part of the four-step process, as this Technology Corner illustrates. However, be sure to read the AP® Exam Tip that follows.

20. Technology Corner   PERFORMING A ONE-SAMPLE *Z* TEST FOR A PROPORTION

TI-Nspire and other technology instructions are on the book's website at highschool.bfwpub.com/updatedtps6e.

The TI-83/84 can be used to test a claim about a population proportion. We'll demonstrate using the preceding example. In a random sample of size $n = 500$, the supervisor found X = 47 potatoes with blemishes. To perform a significance test:

- Press STAT, then choose TESTS and 1-PropZTest.

- On the 1-PropZTest screen, enter the values shown: $p_0 = 0.08$, $x = 47$, and $n = 500$. Specify the alternative hypothesis as "prop > p_0." *Note:* x is the number of successes and n is the number of trials. Both must be whole numbers!

- If you select "Calculate" and press ENTER, you will see that the standardized test statistic is $z = 1.15$ and the *P*-value is 0.1243.

Answers to CYU

STATE: H_0: $p = 0.20$, H_a: $p > 0.20$, where p is the true proportion of all teens at the school who would say they have electronically sent or posted sexually suggestive images of themselves using $\alpha = 0.05$.

PLAN: One-sample *z* test for *p*. *Random:* We have a random sample of 250 students from the school. *10%:* The sample size (250) is less than 10% of the 2800 students at the school. *Large Counts:* $np_0 = 250(0.2) = 50 \geq 10$ and $n(1 - p_0) = 250(0.8) = 200 \geq 10$.

DO: $\hat{p} = \dfrac{63}{250} = 0.252$

$$z = \frac{0.252 - 0.20}{\sqrt{\dfrac{0.20(0.80)}{250}}} = 2.06$$

P-value $= 0.0197$

CONCLUDE: Because the *P*-value of $0.0197 < \alpha = 0.05$, we reject H_0. We have convincing evidence that more than 20% of the teens in her school would say they have electronically sent or posted sexually suggestive images of themselves.

Teaching Tip: Using Technology

Remind students that they can watch a video of Technology Corner 20 on the e-Book or at the Student Site.

Teaching Tip: Using Technology

The calculator abbreviates the one-sample *z* test for a proportion as "1-Prop*Z*Test." Make sure your students don't accidentally choose the *Z*-Test option, which is for testing a population mean when the population standard deviation σ is known.

Teaching Tip: Using Technology

Remind students using 1-Prop*Z*Test that *x* must be a whole number, which is the number of successes. Students often incorrectly use the value of the sample proportion of successes $\hat{p}$ for *x* instead of the number of successes. Also, if students are given a proportion and a sample size and have to do a calculation to get *x*, be sure they round this value to the closest integer.

- If you select the "Draw" option, you will see the screen shown on the right. Compare these results with those in the example on page 606.

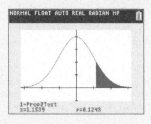

Two-Sided Tests

The free-throw shooter, part-time job, and blemished-potato settings involved one-sided tests. The P-value in a one-sided test about a proportion is the area in one tail of a standard Normal distribution—the tail specified by H_a. In a two-sided test, the alternative hypothesis has the form $H_a: p \neq p_0$. The P-value in such a test is the probability of getting a sample proportion as far as or farther from p_0 in *either direction* than the observed value of $\hat{p}$. As a result, you have to find the area in both tails of a standard Normal distribution to get the P-value. The following example shows how this process works.

EXAMPLE

Nonsmokers
A two-sided test

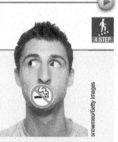

PROBLEM: According to the Centers for Disease Control and Prevention (CDC) website, 68% of high school students have never smoked a cigarette.[7] Yanhong wonders whether this national result holds true in her large, urban high school. For her AP® Statistics class project, Yanhong surveys a simple random sample of 150 students from her school. She gets responses from all 150 students, and 90 say that they have never smoked a cigarette. Is there convincing evidence that the CDC's claim does not hold true at Yanhong's school?

Teaching Tip

Your students may wonder why we have to find the area in both tails when performing a significance test with a two-sided alternative hypothesis. Explain that when we use a two-sided alternative hypothesis, a sample statistic less than or greater than the hypothesized value of the parameter gives evidence against the null hypothesis. Thus, the P-value is the probability of getting a sample statistic at least as extreme (in either direction) as the observed statistic when the null hypothesis is true.

Teaching Tip: Using Technology

When students use the 1-Prop$Z$Test feature and choose the two-sided alternative hypothesis, the calculator automatically finds the two-sided P-value. Many students incorrectly double the P-value provided by the calculator even when they chose the two-sided alternative.

Teaching Tip

Students often use a one-sided alternative hypothesis when H_a should be two-sided. Remind students to use the wording of the question to determine the alternative hypothesis, not the data! In this example, Yanhong wonders if the result will hold true at her school. Because there was no direction specified, she should use a two-sided alternative hypothesis. To recognize that a problem is calling for a two-sided test, look for key questions such as "Has the true value changed?" or "Is there convincing evidence that the true value is different?"

ALTERNATE EXAMPLE Skills 1.E, 4.E

Amazon is prime
A two-sided test

PROBLEM:
Amazon.com is one of the largest Internet-based retailers in the world. The company offers a "prime" membership for a fee, which provides certain benefits for members. A recent report states that 46% of U.S. households have an Amazon "prime" membership. Penelope wants to know if this proportion is different for her large hometown. She selects a random sample of 80 households from her hometown and finds that 43 of them have an Amazon "prime" membership. Is there convincing evidence that the proportion of households in this town differs from the claim in the report?

SOLUTION:
STATE: We want to test

$$H_0: p = 0.46$$
$$H_a: p \neq 0.46$$

where p = the proportion of all households in this town that have an Amazon "prime" membership. We'll use $\alpha = 0.05$.

PLAN: One-sample z test for p

- *Random:* Penelope surveyed a random sample of 80 households from her hometown. ✓
  - *10%:* 80 households < 10% of all households in this town. ✓
- *Large Counts:* 80(0.46) = 36.8 ≥ 10 and 80(0.54) = 43.2 ≥ 10 ✓

DO:

- $\hat{p} = \dfrac{43}{80} = 0.5375$

- $z = \dfrac{0.5375 - 0.46}{\sqrt{\dfrac{0.46(0.54)}{80}}} = 1.39$

- *P*-value

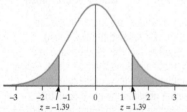

Table A: $P(z \leq -1.39$ or $z \geq 1.39) = 2(0.0823) = 0.1646$

Tech: normalcdf(lower: −1000, upper: −1.39, mean: 0, SD: 1) × 2 = 0.1645

CONCLUDE: Because our *P*-value of 0.1646 > α = 0.05, we fail to reject H_0. We do not have convincing evidence that the proportion of all households in this town that have an Amazon "prime" membership differs from the report's claim of 0.46.

SOLUTION:
STATE: We want to test

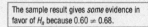
Follow the four-step process!

$H_0: p = 0.68$
$H_a: p \neq 0.68$

where p = the proportion of all students in Yanhong's school who would say they have never smoked a cigarette. We'll use $\alpha = 0.05$.

PLAN: One-sample z test for p.
- *Random:* Yanhong surveyed an SRS of 150 students from her school. ✓
  - 10%: 150 students is less than 10% of all students at a large high school. ✓
- *Large Counts:* 150(0.68) = 102 ≥ 10 and 150(0.32) = 48 ≥ 10 ✓

DO: • $\hat{p} = \dfrac{90}{150} = 0.60$

| The sample result gives *some* evidence in favor of H_a because 0.60 ≠ 0.68. |

• $z = \dfrac{0.60 - 0.68}{\sqrt{\dfrac{0.68(0.32)}{150}}} = -2.10$

• *P*-value:

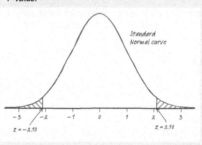

| The TI-83/84's 1-PropZTest gives $z = -2.10$ and *P*-value = 0.0357. |

Using Table A: $P(z \leq -2.10$ or $z \geq 2.10) = 2(0.0179) = 0.0358$
Using technology: normalcdf(lower: − 1000, upper: − 2.10, mean:0, SD:1) × 2 = 0.0357
CONCLUDE: Because our *P*-value of 0.0357 < α = 0.05, we reject H_0. We have convincing evidence that the proportion of all students at Yanhong's school who would say they have never smoked a cigarette differs from the CDC's claim of 0.68.

FOR PRACTICE, TRY EXERCISE 51

AP® EXAM TIP

When making a conclusion in a significance test, be sure that you are describing the parameter and not the statistic. In the preceding example, it's wrong to say that we have convincing evidence that the proportion of students at Yanhong's school who *said* they have never smoked differs from the CDC's claim of 0.68. The "proportion who *said* they have never smoked" is the sample proportion, which is known to be 0.60. The test gives convincing evidence that the proportion of all students at Yanhong's school who *would say* they have never smoked a cigarette differs from 0.68.

Teaching Tip

Tell students to avoid using the past tense of verbs in their conclusions. In most cases, past tense would refer to the results from the sample, rather than making an inference about the population. In some very specific cases, past tense is acceptable. For example: if p = true proportion of all U.S. adults who approved of Obama's performance as president in 2010.

WHY CONFIDENCE INTERVALS GIVE MORE INFORMATION The result of a significance test begins with a decision to reject H_0 or fail to reject H_0. In Yanhong's smoking study, for instance, the data led us to reject H_0: $p = 0.68$ because we found convincing evidence that the proportion of students at her school who would say they have never smoked cigarettes differs from the CDC's claim. We're left wondering what the actual proportion p may be. A confidence interval can shed some light on this issue. You learned how to calculate a confidence interval for a population proportion in Section 8.2.

A 95% confidence interval for p is

$$0.60 \pm 1.96\sqrt{\frac{0.60(1-0.60)}{150}} = 0.60 \pm 0.078 = (0.522, 0.678)$$

This interval gives the values for p that are plausible based on the sample data. We would not be surprised if the proportion of all students at Yanhong's school who would say they have never smoked cigarettes was any value between 0.522 and 0.678. However, we would be surprised if the true proportion was 0.68 because this value is not contained in the confidence interval. Figure 9.6 gives computer output from Minitab software that includes both the results of the significance test and the confidence interval.

```
Session                                              [□][■][X]

Test and CI for One Proportion

Test of p = 0.68 vs p ≠ 0.68


Sample   X    N  Sample p        95% CI      Z-Value  P-Value
1       90  150  0.600000  (0.521601, 0.678399)  -2.10    0.036
```

FIGURE 9.6 Minitab output for the two-sided significance test and a 95% confidence interval for Yanhong's smoking study.

There is a link between confidence intervals and *two-sided* tests. The 95% confidence interval (0.522, 0.678) gives an approximate set of p_0's that should not be rejected by a two-sided test at the $\alpha = 0.05$ significance level. Any p_0 value outside the interval should be rejected as implausible.

With proportions, the link isn't perfect because the standard error used for the confidence interval is based on the sample proportion $\hat{p}$, while the denominator of the standardized test statistic is based on the value p_0 from the null hypothesis.

$$\text{Standardized test statistic: } z = \frac{\hat{p} - p_0}{\sqrt{\frac{p_0(1-p_0)}{n}}}$$

$$\text{Confidence interval: } \hat{p} \pm z^*\sqrt{\frac{\hat{p}(1-\hat{p})}{n}}$$

The big idea is still worth considering: a two-sided test at significance level α and a $100(1-\alpha)\%$ confidence interval (a 95% confidence interval if $\alpha = 0.05$) give similar information about the population parameter. There is a connection between *one-sided* tests and confidence intervals, but it is beyond the scope of this course.

Teaching Tip

Notice in the formula for standard error that we use $\hat{p} = 0.60$, rather than $p = 0.68$, as done for the significance test. Remember when doing a significance test, we assume H_0 to be true, so we know the true parameter $p = 0.68$. For a confidence interval, we don't know the value of the true parameter p, so we use $\hat{p} = 0.60$ as our best guess.

Teaching Tip

Use the smoking study to help students discover the connection between significance tests and confidence intervals (P-value $= 0.036$).

Ask students what decision we would make at

$\alpha = 0.10$ (reject H_0)
$\alpha = 0.05$ (reject H_0)
$\alpha = 0.01$ (fail to reject H_0)

Then use the calculator to find the following:

90% CI: (0.534, 0.666)
Is $p = 0.68$ plausible? No.
95% CI: (0.522, 0.678)
Is $p = 0.68$ plausible? No.
99% CI: (0.497, 0.703)
Is $p = 0.68$ plausible? Yes.

Students should notice that when the null hypothesis value is contained within the confidence interval, we fail to reject H_0 for a significance level α that is the complement of the confidence level.

AP® EXAM TIP

On the AP® Statistics exam, it is acceptable for students to use a confidence interval rather than the standardized test statistic and P-value to address a *two*-sided alternative hypothesis. However, if the alternative hypothesis is one-sided, students will lose credit for using a confidence interval approach unless they explicitly address the imperfect link between one-sided tests and confidence intervals (e.g., by adjusting the confidence level appropriately). **Our recommendation for the AP® Statistics exam is to always stick with a significance test unless a confidence interval is asked for specifically.**

✓ **Answers to CYU**

1. STATE: H_0: $p = 0.75$, H_a: $p \neq 0.75$, where p is the true proportion of all restaurant employees at this chain who would say that work stress has a negative impact on their personal lives using $\alpha = 0.05$.

PLAN: One-sample z test for p. *Random:* We have a random sample of 100 employees from the large restaurant chain. *10%:* Assume the sample size (100) is less than 10% of all employees at this large restaurant chain. *Large Counts:* $np_0 = 100(0.75) = 75 \geq 10$ and $n(1 - p_0) = 100(0.25) = 25 \geq 10$.

DO: $\hat{p} = \dfrac{68}{100} = 0.68$

$z = \dfrac{0.68 - 0.75}{\sqrt{\dfrac{0.75(0.25)}{100}}} = -1.62;$

P-value = 0.1052

CONCLUDE: Because the P-value of $0.1052 > \alpha = 0.10$, we fail to reject H_0. We do not have convincing evidence that the true proportion of all restaurant employees at this large restaurant chain who would say that work stress has a negative impact on their personal lives is different from 0.75.

2. The confidence interval gives the values of p that are plausible based on the sample data. The value 0.75 is plausible because it is within the 90% confidence interval; so based on the confidence interval, we would fail to reject $p = 0.75$ at the $\alpha = 0.10$ significance level. The 90% confidence interval provided gives an approximate set of p_0's that would not be rejected by a two-sided test at the $\alpha = 0.10$ significance level. A two-sided test only allows us to reject (or fail to reject) a hypothesized value for a particular population parameter, whereas a confidence interval provides a set of hypothesized values that would *not* be rejected based on a two-sided test.

Teaching Tip

If the alternative hypothesis H_a: $p < 0.10$ is true, it could be the case that $p = 0.02$ or $p = 0.05$ or $p = 0.08$. In fact, there are infinitely many possibilities. When calculating the power of a significance test, we must consider only *one* specific value for the parameter (like $p = 0.08$).

612   CHAPTER 9   TESTING CLAIMS ABOUT PROPORTIONS

CHECK YOUR UNDERSTANDING

According to the National Institute for Occupational Safety and Health, job stress poses a major threat to the health of workers. A news report claims that 75% of restaurant employees feel that work stress has a negative impact on their personal lives.[8] Managers of a large restaurant chain wonder whether this claim is valid for their employees. A random sample of 100 employees finds that 68 answer "Yes" when asked, "Does work stress have a negative impact on your personal life?"

1. Do these data provide convincing evidence at the $\alpha = 0.10$ significance level that the proportion of all employees in this chain who would say "Yes" differs from 0.75?

2. The figure shows Minitab output from a significance test and confidence interval for the restaurant worker data. Explain how the confidence interval is consistent with, but gives more information than, the test.

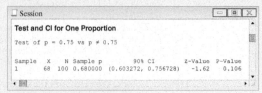

The Power of a Test

Researchers often perform a significance test in hopes of finding convincing evidence *for* the alternative hypothesis. Why? Because H_a states the claim about the population parameter that they believe, suppose, suspect, or worry is true. For instance, a drug manufacturer claims that less than 10% of patients who take its new drug for treating Alzheimer's disease will experience nausea. To test this claim, researchers want to carry out a test of

$$H_0: p = 0.10$$
$$H_a: p < 0.10$$

where $p =$ the true proportion of patients like the ones in the study who would experience nausea when taking the new Alzheimer's drug. They plan to give the new drug to a random sample of Alzheimer's patients whose families have given informed consent for the patients to participate in the study.

Suppose that the true proportion of subjects like the ones in the Alzheimer's study who would experience nausea after taking the new drug is $p = 0.08$. This means that the alternative hypothesis H_a: $p < 0.10$ is true. Researchers would make a Type II error if they failed to find convincing evidence for H_a based on the sample data. How likely is the significance test to *avoid* a Type II error in this case? We refer to this probability as the **power** of the test.

DEFINITION   **Power**

The **power** of a test is the probability that the test will find convincing evidence for H_a when a specific alternative value of the parameter is true.

Teaching Tip

Here are a few Internet resources that can help students with a conceptual understanding of power:

- An Internet search for "The Story of Sad Sack" will reveal a video that gives personalities to Type I errors, Type II errors, and power.
- An Internet search for "Graphpad power analogy" will reveal a website with an analogy that relates a child looking for a tool in a basement to power.

Teaching Tip

Students are not expected to be able to calculate power by hand. Focus on the interpretation and ways to increase power. *Note:* The Investigative Task #6 from the 2018 AP® Statistics exam did ask students to calculate power by hand, which is a reminder that the Investigative Task will always ask questions outside the specified curriculum.

612   Chapter 9   Testing Claims About Proportions

The power of a test is a *conditional* probability: power $= P(\text{reject } H_0 \mid$ parameter $=$ some specific alternative value). In other words, power is the probability that we find convincing evidence the alternative hypothesis is true, given that the alternative hypothesis really is true. To interpret the power of a test in a given setting, just interpret the relevant conditional probability.

Let's return to the Alzheimer's study. Suppose the researchers decide to perform a test of $H_0: p = 0.10$ versus $H_a: p < 0.10$ at the $\alpha = 0.05$ significance level based on data from a random sample of 300 Alzheimer's patients. Advanced calculations reveal that the power of this test to detect $p = 0.08$ is 0.29. *Interpretation:* If the true proportion of Alzheimer's patients like these who would experience side effects when taking the new drug is $p = 0.08$, there is a 0.29 probability that the researchers will find convincing evidence for $H_a: p < 0.10$.

EXAMPLE

Can we tell if the new batteries last longer?
Interpreting the power of a test

PROBLEM: A company has developed a new deluxe AAA battery that is supposed to last longer than its regular AAA battery. Based on years of experience, the company's regular AAA batteries last for 30 hours of continuous use, on average. The company plans to select an SRS of 50 deluxe AAA batteries and use them continuously until they are completely drained. Then it will perform a test at the $\alpha = 0.05$ significance level of

$$H_0: \mu = 30$$
$$H_a: \mu > 30$$

where $\mu =$ the true mean lifetime (in hours) of the deluxe AAA batteries. The new batteries are more expensive to produce, so the company would like to be convinced that they really do last longer. The power of the test to detect that $\mu = 31$ hours is 0.762. Interpret this value.

SOLUTION:

If the true mean lifetime of the company's deluxe AAA batteries is $\mu = 31$ hours, there is a 0.762 probability that the company will find convincing evidence for $H_a: \mu > 30$.

FOR PRACTICE, TRY EXERCISE 59

The company in the example has a good chance (power $= 0.762$) of rejecting H_0 if the true mean lifetime of its deluxe AAA batteries is 31 hours. That is, $P(\text{reject } H_0 \mid \mu = 31) = 0.762$. What's the probability that the company makes a Type II error in this case? It's $P(\text{fail to reject } H_0 \mid \mu = 31) = 1 - 0.762 = 0.238$. We can generalize this relationship between the power of a significance test and the probability of a Type II error.

RELATING POWER AND TYPE II ERROR

The power of a test to detect a specific alternative parameter value is related to the probability of a Type II error for that alternative:

$$\text{Power} = 1 - P(\text{Type II error}) \quad \text{and} \quad P(\text{Type II error}) = 1 - \text{Power}$$

Making Connections

Remind students of this table from the section on Type I and Type II errors in Section 9.1, shown here. The power belongs in the top right box.

		Truth about the population	
		H_0 true	H_a true
Conclusion based on sample	Reject H_0	Type I error	Correct conclusion
	Fail to reject H_0	Correct conclusion	Type II error

Teaching Tip

If H_a is true, there are only two possible outcomes. Either we make a correct decision to reject H_0 (probability of this outcome $=$ power) or we don't make a correct decision (Type II error). It makes sense that these two events are complementary because they are the only two outcomes when H_a is true.

The power of the test in the Alzheimer's study to detect $p = 0.08$ is only 0.29. In other words, researchers have a $1 - 0.29 = 0.71$ probability of making a Type II error by failing to find convincing evidence for H_a: $p < 0.10$ when $p = 0.08$. What can researchers do to decrease the probability of making a Type II error and increase the power of the test?

WHAT AFFECTS THE POWER OF A TEST? Here is an activity that will help you answer this question.

ACTIVITY A great free-throw shooter?

In this activity, we will use an applet to investigate the factors that affect the power of a test. A basketball player claims to make 80% of his free throws. Suppose the player is exaggerating—he really makes less than 80% in the long run. We have the player shoot $n = 50$ shots and record the sample proportion $\hat{p}$ of made free throws. We then use the sample result to perform a test at the $\alpha = 0.05$ significance level of

$$H_0: p = 0.80$$
$$H_a: p < 0.80$$

where p = the true proportion of free throws the shooter makes in the long run.

1. Go to https://istats.shinyapps.io/power/. Enter 0.80 for the null hypothesis value p_0. Choose less for the type of alternative hypothesis. Click the buttons to show Type I error, Type II error, and Power. Set the True value of p to 0.66 (indicating that the player is truly a 66% free-throw shooter). Move the Sample Size slider to $n = 50$ and the Type I Error slider to $\alpha = 0.05$. Confirm that the power of the test is 0.76.

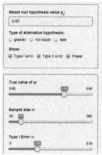 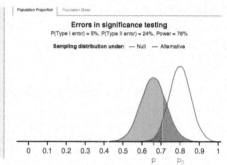

- The curve on the right shows the sampling distribution of the sample proportion $\hat{p}$ for random samples of size $n = 50$ when H_0: $p = 0.80$ is true. We refer to this as the *null distribution*. A value of $\hat{p}$ that falls along the horizontal axis within the darker brownish region is far enough below 0.80 that we should reject H_0: $p = 0.80$. The area of the brownish region is equal to $\alpha = 0.05$.

- The curve on the left shows the sampling distribution of the sample proportion $\hat{p}$ for random samples of size $n = 50$ when $p = 0.66$. We refer to this as the *alternative distribution*. A value of $\hat{p}$ that falls along the horizontal axis within the green region would lead to a correct rejection of H_0: $p = 0.80$. In other words, the green region represents the power of the test. A value of $\hat{p}$ that falls along the horizontal axis within the blue region would lead to a Type II error because it is not far enough below 0.80 to reject H_0: $p = 0.80$.

2. *Sample size*: Change the sample size from $n = 50$ to $n = 100$. What happens to the null and alternative distributions in the applet? Does the power increase or decrease? Explain why this makes sense.

3. *Significance level*: Reset the sample size to $n = 50$.

   (a) Using the slider, change the significance level to $\alpha = 0.01$. How does this affect the probability of a Type I error? How about the probability of a Type II error? Does the power increase or decrease?

   (b) Make a guess about what will happen if you change the significance level to $\alpha = 0.10$. Use the applet to test your conjecture.

   (c) Explain what the results in parts (a) and (b) tell you about the relationship between Type I error probability, Type II error probability, and power.

4. *Difference between null and alternative parameter value*: Reset the sample size to $n = 50$ and the significance level to $\alpha = 0.05$. Will we be more likely to detect that the player is exaggerating if he is really a 60% shooter or if he is really a 70% shooter? Use the slider for the true value of p to test your conjecture.

Increasing the sample size decreases the variability of both the null and alternative distributions. This change decreases the amount of overlap between the two distributions, making it easier to detect a difference between the null and alternative parameter values.

As Step 2 of the activity confirms, we get better information about the true proportion of free throws that the player makes from a random sample of 100 shots than from a random sample of 50 shots. The power of the test to detect that $p = 0.66$ increases from 0.76 to 0.94 when the sample size increases from $n = 50$ to $n = 100$.

Will it be easier to reject H_0 if $\alpha = 0.05$ or if $\alpha = 0.10$? When α is larger, it is easier to reject H_0 because the *P*-value doesn't need to be as small. Step 3 of the activity shows that the power of the test to detect that $p = 0.66$ increases from 0.76 to 0.84 when the significance level increases from $\alpha = 0.05$ to $\alpha = 0.10$. Increasing α means that we are less likely to make a Type II error—failing to reject H_0 when H_a is true. Unfortunately, increasing α also makes it more likely that we make a Type I error—rejecting H_0 when H_0 is true.

Figure 9.7 on the next page illustrates the connection between Type I and Type II error probabilities and the power of a test. Note that $P(\text{Type I error}) = \alpha = 0.10$ and that $P(\text{Type II error}) = 1 - \text{Power} = 1 - 0.16 = 0.84$. You can see that increasing the Type I error probability α would decrease the Type II error probability and increase the power of the test. By the same logic, decreasing the chance of a Type I error results in a higher chance of a Type II error and less power.

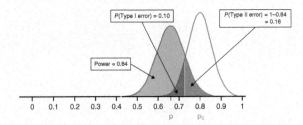

FIGURE 9.7 The connection between Type I error, Type II error, and power. Because $P(\text{Type I error}) = \alpha$, as the significance level α increases: $P(\text{Type I error})$ increases, $P(\text{Type II error})$ decreases, and power increases.

Step 4 of the activity shows that it is easier to detect large differences between the null and alternative parameter values than small differences. When $n = 50$ and $\alpha = 0.05$, the power of the test to detect $p = 0.60$ is 0.94, whereas the power of the test to detect $p = 0.70$ is only 0.54.

The difference between the null parameter value and the specific alternative parameter value of interest is often referred to as the *effect size*. For the basketball player setting with a null value of $p = 0.80$, it is much easier to detect if the true proportion of free throws the player makes is $p = 0.60$ (an effect size of 0.20, or 20 percentage points) than if the true proportion is $p = 0.70$ (an effect size of 0.10, or 10 percentage points).

INCREASING THE POWER OF A SIGNIFICANCE TEST

The power of a significance test to detect an alternative value of the parameter when H_0 is false and H_a is true, based on a random sample of size n and significance level α, will be larger when:

- The sample size n is larger.
- The significance level α is larger.
- The null and alternative parameter values are farther apart.

Making wise choices when collecting data can decrease the standard deviation (standard error) of the null and alternative distributions, resulting in less overlap between the two. That makes it easier to find convincing evidence for the alternative hypothesis when the alternative hypothesis is true.

In addition to these three factors, we can also gain power by making wise choices when collecting data. For example, using blocking in an experiment or stratified random sampling can greatly increase the power of a test in some circumstances. In an experiment, power will increase if you have fewer sources of variability by keeping other variables constant.

Some researchers prefer to think about decreasing the probability of a Type II error rather than increasing the power when performing a test. Because Power $= 1 - P(\text{Type II error})$, these two goals are equivalent. The same factors that lead to increased power also result in decreased Type II error probability.

EXAMPLE

Can we tell if the new drug reduces nausea?
What affects the power of a test

PROBLEM: The researchers in the Alzheimer's experiment want to test the drug manufacturer's claim that fewer than 10% of patients who take its new drug for treating Alzheimer's disease will experience nausea. That is, they want to carry out a test of

$$H_0: p = 0.10$$
$$H_a: p < 0.10$$

where $p =$ the true proportion of patients like the ones in the study who would experience nausea when taking the new Alzheimer's drug. Earlier, we mentioned that the power of the test to detect $p = 0.08$ using a random sample of 300 patients and a significance level of $\alpha = 0.05$ is 0.29.

Determine whether each of the following changes would increase or decrease the power of the test. Explain your answers.

(a) Use $\alpha = 0.01$ instead of $\alpha = 0.05$.

(b) If the true proportion is $p = 0.06$ instead of $p = 0.08$.

(c) Use $n = 200$ instead of $n = 300$.

SOLUTION:

(a) *Decrease; using a smaller significance level makes it harder to reject H_0 when H_a is true.*

(b) *Increase; it is easier to detect a bigger difference between the null and alternative parameter value.*

(c) *Decrease; a smaller sample size gives less information about the true proportion p.*

FOR PRACTICE, TRY EXERCISE 61

In Section 8.2, you learned how to calculate the sample size needed for a desired margin of error in a confidence interval for p. When researchers plan to use a significance test to analyze their results, they will often calculate the sample size needed for a desired power. For a specific study design, the sample size required depends on three factors:

1. *Significance level.* How much risk of a Type I error—rejecting the null hypothesis when H_0 is actually true—are we willing to accept? If a Type I error has serious consequences, we might opt for $\alpha = 0.01$. Otherwise, we should choose $\alpha = 0.05$ or $\alpha = 0.10$. Recall that using a higher significance level would decrease the Type II error probability and increase the power.

2. *Effect size.* How large a difference between the null parameter value and the actual parameter value is important for us to detect?

> Sometimes budget constraints get in the way of achieving high power. It can be expensive to collect data from a large enough sample of individuals to give a significance test the power that researchers desire.

3. *Power.* What probability do we want our study to have to detect a difference of the size we think is important? Most researchers insist on a power of at least 0.80 for their significance tests.

Calculating power by hand is possible but unpleasant. It's better to let technology do the work for you. There are many applets available to help determine sample size and to calculate the power of a test.

ALTERNATE EXAMPLE   Skill 4.A

Powerful ways of preventing ADHD
What affects the power of a test

PROBLEM:

In the preceding alternate example, researchers were testing a new vitamin tablet that claims to lower a child's risk of developing attention deficit/hyperactivity disorder (ADHD). They want to carry out a test of

$$H_0: p = 0.11$$
$$H_a: p < 0.11$$

where $p =$ the true proportion of all children like those in the study who would develop ADHD when given the new vitamin tablet. Earlier, we mentioned that the power of the test to detect $p = 0.05$ using 200 subjects and a significance level of $\alpha = 0.05$ is 0.937.

Determine whether each of the following changes would increase or decrease the power of the test. Explain your answers.

(a) Use $\alpha = 0.10$ instead of $\alpha = 0.05$

(b) If the true proportion is $p = 0.08$ instead of $p = 0.05$

(c) Use $n = 500$ instead of $n = 200$

SOLUTION:

(a) Increase; using a larger significance level makes it easier to reject H_0 when H_a is true.

(b) Decrease; it is harder to detect a smaller difference between the null and alternative parameter value.

(c) Increase; a larger sample size gives more information about the true proportion p.

Teaching Tip

Determining the sample size is an important part of the planning process. Many studies fail because of a lack of power—something a statistician might have predicted before the study began. Lesson: Always consult a statistician before investing time and money in a research project.

✓ Answers to CYU

1. If the true mean percent change in TBBMC during the exercise program is $\mu = 1$, there is a 0.80 probability that the researchers will find convincing evidence for H_a: $\mu > 0$.

2. $P(\text{Type I error}) = \alpha = 0.05$; $P(\text{Type II error}) = 1 - \text{Power} = 1 - 0.80 = 0.20$

3. The researchers could increase the power of the test in Question 1 by increasing the significance level (α) or increasing the sample size (n).

TRM **Section 9.2 Quiz**

There is one quiz available for this section in the Teacher's Resource Materials. Click on the link in the TE-Book, look on the TRFD, or download from the Teacher's Resources on the book's digital platform. You can also create your own quiz using the ExamView® Assessment Suite that is part of the TPS6 program. Questions are coded by Learning Target to make it easy to build parallel quizzes.

CHECK YOUR UNDERSTANDING

Can a six-month exercise program increase the total body bone mineral content (TBBMC) of young women? A team of researchers is planning a study to examine this question. The researchers would like to perform a test of

$$H_0: \mu = 0$$
$$H_a: \mu > 0$$

where μ is the true mean percent change in TBBMC during the exercise program.

1. The power of the test to detect a mean increase in TBBMC of 1% using $\alpha = 0.05$ and $n = 25$ subjects is 0.80. Interpret this value.
2. Find the probability of a Type I error and the probability of a Type II error for the test in Question 1.
3. Describe two ways that researchers could increase the power of the test in Question 1.

Section 9.2 | **Summary**

- To perform a significance test of H_0: $p = p_0$, we need to verify that the observations in the sample can be viewed as independent and that the sampling distribution of $\hat{p}$ is approximately Normal. The required conditions are:
  - **Random:** The data come from a random sample from the population of interest.
    - 10%: When sampling without replacement, $n < 0.10N$.
  - **Large Counts:** Both np_0 and $n(1 - p_0)$ are at least 10.
- The **standardized test statistic** for a *one-sample z test for a proportion* is

$$z = \frac{\hat{p} - p_0}{\sqrt{\dfrac{p_0(1 - p_0)}{n}}}$$

- When the Large Counts condition is met, the standardized test statistic has approximately a standard Normal distribution. You can use Table A or technology to find the *P*-value.
- Follow the four-step process when you perform a significance test:

   State: State the hypotheses you want to test and the significance level, and define any parameters you use.

  Plan: Identify the appropriate inference method and check conditions.

  Do: If the conditions are met, perform calculations:
  - Give the sample statistic(s).
  - Calculate the standardized test statistic.
  - Find the *P*-value.

  Conclude: Make a conclusion about the hypotheses in the context of the problem.

- Confidence intervals provide additional information that significance tests do not — namely, a set of plausible values for the population proportion p. A two-sided test of $H_0: p = p_0$ at significance level α usually gives the same conclusion as a $100(1 - \alpha)\%$ confidence interval.

- The **power** of a test is the probability that the test will find convincing evidence for H_a when a specific alternative value of the parameter is true. In other words, the power of a test is the probability of avoiding a Type II error. For a specific alternative, Power $= 1 - P(\text{Type II error})$.

- We can increase the power of a significance test (decrease the probability of a Type II error) by increasing the sample size, increasing the significance level, or increasing the difference that is important to detect between the null and alternative parameter values (known as the *effect size*). Wise choices when collecting data, such as controlling for other variables and blocking in experiments or stratified random sampling, can also increase power.

- For a specific study design, three factors influence the sample size required for a statistical test: significance level, effect size, and the desired power of the test.

9.2 Technology Corner

TI-Nspire and other technology instructions are on the book's website at highschool.bfwpub.com/updatedtps6e.

20. Performing a one-sample z test for a proportion   Page 608

Section 9.2 | Exercises

35. **Home computers** Jason reads a report that says 80% pg 601 of U.S. high school students have a computer at home. He believes the proportion is smaller than 0.80 at his large rural high school. Jason chooses an SRS of 60 students and finds that 41 have a computer at home. He would like to carry out a test at the $\alpha = 0.05$ significance level of $H_0: p = 0.80$ versus $H_a: p < 0.80$, where $p =$ the true proportion of all students at Jason's high school who have a computer at home. Check if the conditions for performing the significance test are met.

36. **Walking to school** A recent report claimed that 13% of students typically walk to school.[9] DeAnna thinks that the proportion is higher than 0.13 at her large elementary school. She surveys a random sample of 100 students and finds that 17 typically walk to school. DeAnna would like to carry out a test at the $\alpha = 0.05$ significance level of $H_0: p = 0.13$ versus $H_a: p > 0.13$, where $p =$ the true proportion of all students at her elementary school who typically walk to school. Check if the conditions for performing the significance test are met.

37. **The chips project** Zenon decided to investigate whether students at his school prefer name-brand potato chips to generic potato chips. He randomly selected 50 of the 400 students at his school and asked each student to try both types of chips, in a random order. Overall, 41 of the 50 students preferred the name-brand chips. Zenon wants to perform a test of $H_0: p = 0.5$ versus $H_a: p > 0.5$, where $p =$ the true proportion of students at his school who prefer name-brand chips.

(a) Can the observations in the sample be viewed as independent? Justify your answer.

(b) Is the sampling distribution of $\hat{p}$ approximately Normal? Explain your reasoning.

38. **Better to be last?** On TV shows that feature singing competitions, contestants often wonder if there is an advantage in performing last. To investigate, researchers selected a random sample of 100 students from their large university and showed each student the audition video of 12 different singers with similar vocal

TRM **Full Solutions to Section 9.2 Exercises**

Click on the link in the TE-Book, open the TRFD, or download from the Teacher's Resources on the book's digital platform.

Answers to Section 9.2 Exercises

9.35 *Random:* We have an SRS of 60 students from a large rural high school. *10%:* The sample size (60) is less than 10% of all students at this large high school. *Large Counts:* $np_0 = 60(0.80) = 48 \geq 10$ and $n(1 - p_0) = 60(0.20) = 12 \geq 10$.

9.36 *Random:* We have a random sample of 100 students from a large elementary school. *10%:* The sample size (100) is less than 10% of the students at this large elementary school. *Large Counts:* $np_0 = 100(0.13) = 13 \geq 10$ and $n(1 - p_0) = 100(0.87) = 87 \geq 10$.

9.37 (a) No; the students ($n = 50$) are selected from a finite population ($N = 400$) without replacement. Because $n = 50$ is not less than 10% of the population size (400), the observations are not independent. **(b)** Yes; assuming $p = 0.5$ is true, the sampling distribution of $\hat{p}$ will be approximately Normal because $np_0 = 50(0.5) = 25$ and $n(1 - p_0) = 50(1 - 0.5) = 25$ are both ≥ 10.

9.38 (a) Yes; although the students ($n = 100$) are selected from a finite population (a large university) without replacement, $n = 100 < 10\%$ of the population size (all students at the large university). **(b)** No; assuming that $p = 1/12$ is true, the sampling distribution of $\hat{p}$ will not be approximately Normal because $np_0 = 100(1/12) = 8.33$ and $n(1 - p_0) = 100(1 - 1/12) = 91.67$ are not both ≥ 10.

9.39 (a) The sample result gives some evidence for H_a: $p < 0.80$ because

$$\hat{p} = \frac{41}{60} = 0.683,$$ which is less than 0.80.

(b) $z = \dfrac{0.683 - 0.80}{\sqrt{\dfrac{0.80(0.20)}{60}}} = -2.27;$

$P\text{-value} = 0.0116$

(c) Because the P-value of 0.0116 $< \alpha = 0.05$, we reject H_0. We have convincing evidence that the true proportion of all students at this large rural high school who have a computer at home is less than 0.80.

9.40 (a) The sample result gives some evidence for H_a: $p > 0.13$ because

$$\hat{p} = \frac{17}{100} = 0.17,$$ which is greater than 0.13.

(b) $z = \dfrac{0.17 - 0.13}{\sqrt{\dfrac{0.13(0.87)}{100}}} = 1.19;$

$P\text{-value} = 0.1170$

(c) Because the P-value of 0.1170 $> \alpha = 0.05$, we fail to reject H_0. We do not have convincing evidence that the true proportion of all students at this elementary school who typically walk to school is greater than 0.13.

9.41 (a) The P-value is $P(z \geq 2.19) = 0.0143$. Assuming that the true population proportion is 0.5, there is a 0.0143 probability of getting a sample proportion as large as or larger than the one observed just by chance in a random sample of size $n = 200$.
(b) $\alpha = 0.01$: Because the P-value of 0.0143 $> \alpha = 0.01$, we fail to reject H_0. There is not convincing evidence that $p > 0.5$. $\alpha = 0.05$: Yes! The P-value of 0.0143 $< \alpha = 0.05$, so this time we will reject H_0. There is convincing evidence that $p > 0.5$. **(c)** Solve for $\hat{p}$:

$2.19 = \dfrac{\hat{p} - 0.5}{\sqrt{\dfrac{0.5(0.5)}{200}}}$

$\hat{p} = 0.5774$

9.42 (a) P-value $= 0.0375$; if the true population proportion is 0.65, a 0.0375 probability exists of getting a sample proportion as small as or smaller than the one observed by chance in a random sample of size $n = 400$. **(b)** When $\alpha = 0.10$ and $\alpha = 0.05$, reject H_0. **(c)** $\hat{p} = 0.6075$

9.43 S: H_0: $p = 0.75$, H_a: $p > 0.75$, where $p = $ the true proportion of all middle school students who engage in

skills. Each student viewed the videos in a random order. We would expect approximately 1/12 of the students to prefer the last singer seen, assuming order doesn't matter. In this study, 11 of the 100 students preferred the last singer they viewed. The researchers want to perform a test of H_0: $p = 1/12$ versus H_a: $p > 1/12$, where $p = $ the true proportion of students at this university who prefer the singer they last see.

(a) Can the observations in the sample be viewed as independent? Justify your answer.

(b) Is the sampling distribution of $\hat{p}$ approximately Normal? Explain your reasoning.

39. **Home computers** Refer to Exercise 35.

pg 603 (a) Explain why the sample result gives some evidence for the alternative hypothesis.

(b) Calculate the standardized test statistic and P-value.

(c) What conclusion would you make?

40. **Walking to school** Refer to Exercise 36.

(a) Explain why the sample result gives some evidence for the alternative hypothesis.

(b) Calculate the standardized test statistic and P-value.

(c) What conclusion would you make?

41. **Significance tests** A test of H_0: $p = 0.5$ versus H_a: $p > 0.5$ based on a sample of size 200 yields the standardized test statistic $z = 2.19$. Assume that the conditions for performing inference are met.

(a) Find and interpret the P-value.

(b) What conclusion would you make at the $\alpha = 0.01$ significance level? Would your conclusion change if you used $\alpha = 0.05$ instead? Explain your reasoning.

(c) Determine the value of $\hat{p} = $ the sample proportion of successes.

42. **Significance tests** A test of H_0: $p = 0.65$ against H_a: $p < 0.65$ based on a sample of size 400 yields the standardized test statistic $z = -1.78$.

(a) Find and interpret the P-value.

(b) What conclusion would you make at the $\alpha = 0.10$ significance level? Would your conclusion change if you used $\alpha = 0.05$ instead? Explain your reasoning.

(c) Determine the value of $\hat{p} = $ the sample proportion of successes.

43. **Bullies in middle school** A media report claims that

pg 606 more than 75% of middle school students engage in bullying behavior. A University of Illinois study on aggressive behavior surveyed a random sample of 558 middle school students. When asked to describe their

behavior in the last 30 days, 445 students admitted that they had engaged in physical aggression, social ridicule, teasing, name-calling, and issuing threats—all of which would be classified as bullying.[10] Do these data provide convincing evidence at the $\alpha = 0.05$ significance level that the media report's claim is correct?

44. **Watching grass grow** The germination rate of seeds is defined as the proportion of seeds that sprout and grow when properly planted and watered. A certain variety of grass seed usually has a germination rate of 0.80. A company wants to see if spraying the seeds with a chemical that is known to increase germination rates in other species will increase the germination rate of this variety of grass. The company researchers spray a random sample of 400 grass seeds with the chemical, and 339 of the seeds germinate. Do these data provide convincing evidence at the $\alpha = 0.05$ significance level that the chemical is effective for this variety of grass?

45. **Better parking** A local high school makes a change that should improve student satisfaction with the parking situation. Before the change, 37% of the school's students approved of the parking that was provided. After the change, the principal surveys an SRS of 200 from the more than 2500 students at the school. In all, 83 students say that they approve of the new parking arrangement. The principal cites this as evidence that the change was effective.

(a) Describe a Type I error and a Type II error in this setting, and give a possible consequence of each.

(b) Is there convincing evidence that the principal's claim is true?

46. **Raising the sales tax** Members of the city council want to know if a majority of city residents supports a 1% increase in the sales tax to fund road repairs. To investigate, council members survey a random sample of 300 city residents. In the sample, 158 residents say that they are in favor of the sales tax increase.

(a) Describe a Type I error and a Type II error in this setting, and give a possible consequence of each.

(b) Do these data provide convincing evidence that a majority of city residents support the tax increase?

47. **Cell-phone passwords** A consumer organization suspects that less than half of parents know their child's cell-phone password. The Pew Research Center asked a random sample of parents if they knew their child's cell-phone password. Of the 1060 parents surveyed, 551 reported that they knew the password. Explain why it isn't necessary to carry out a significance test in this setting.

48. **Proposition X** A political organization wants to determine if there is convincing evidence that a

bullying behavior using $\alpha = 0.05$. **P:** One-sample z test for p. *Random:* Random sample. *10%:* 558 < 10% of all middle school students. *Large Counts:* 418.5 ≥ 10 and 139.5 ≥ 10. **D:** $z = 2.59$; P-value = 0.0048. **C:** Because the P-value of 0.0048 $< \alpha = 0.05$, we reject H_0. We have convincing evidence that the true proportion of all middle school students who engage in bullying behavior is > 0.75.

9.44 S: H_0: $p = 0.80$, H_a: $p > 0.80$ **P:** *Large Counts:* 320 ≥ 10 and 80 ≥ 10. **D:** $z = 2.38$; P-value = 0.0087. **C:** Reject H_0.

9.45 (a) *Type I:* Finding convincing evidence that more than 37% of students were satisfied with the new arrangement, when in reality only 37% were satisfied. *Consequence:* The principal believes students are satisfied and takes no

further action. *Type II:* Failing to find convincing evidence that more than 37% are satisfied with the new arrangement, when in reality more than 37% are satisfied. *Consequence:* The principal takes further action when none is needed.
(b) S: H_0: $p = 0.37$, H_a: $p > 0.37$, where $p = $ the true proportion of all students who are satisfied with the parking situation after the change using $\alpha = 0.05$. **P:** One-sample z test for p. *Random:* Random sample. *10%:* 200 < 10% of 2500. *Large Counts:* 74 ≥ 10 and 126 ≥ 10. **D:** $z = 1.32$; P-value = 0.0934. **C:** Because the P-value of 0.0934 $> \alpha = 0.05$, we fail to reject H_0. We do not have convincing evidence that the true proportion of all students who are satisfied with the parking situation after the change is greater than 0.37.

majority of registered voters in a large city favor Proposition X. In an SRS of 1000 registered voters, 482 favor the proposition. Explain why it isn't necessary to carry out a significance test in this setting.

49. **Mendel and the peas** Gregor Mendel (1822–1884), an Austrian monk, is considered the father of genetics. Mendel studied the inheritance of various traits in pea plants. One such trait is whether the pea is smooth or wrinkled. Mendel predicted a ratio of 3 smooth peas for every 1 wrinkled pea. In one experiment, he observed 423 smooth and 133 wrinkled peas. Assume that the conditions for inference are met.

(a) State appropriate hypotheses for testing Mendel's claim about the true proportion of smooth peas.

(b) Calculate the standardized test statistic and P-value.

(c) Interpret the P-value. What conclusion would you make?

50. **Spinning heads?** When a fair coin is flipped, we all know that the probability the coin lands on heads is 0.50. However, what if a coin is spun? According to the article "Euro Coin Accused of Unfair Flipping" in the *New Scientist*, two Polish math professors and their students spun a Belgian euro coin 250 times. It landed heads 140 times. One of the professors concluded that the coin was minted asymmetrically. A representative from the Belgian mint indicated the result was just chance. Assume that the conditions for inference are met.

(a) State appropriate hypotheses for testing these competing claims about the true proportion of spins that will land on heads.

(b) Calculate the standardized test statistic and P-value.

(c) Interpret the P-value. What conclusion would you make?

51. **Teen drivers** A state's Division of Motor Vehicles (DMV) claims that 60% of all teens pass their driving test on the first attempt. An investigative reporter examines an SRS of the DMV records for 125 teens; 86 of them passed the test on their first try. Is there convincing evidence at the $\alpha = 0.05$ significance level that the DMV's claim is incorrect?

52. **We want to be rich** In a recent year, 73% of first-year college students responding to a national survey identified "being very well-off financially" as an important personal goal. A state university finds that 132 of an SRS of 200 of its first-year students say that this goal is important. Is there convincing evidence at the $\alpha = 0.05$ significance level that the proportion of all first-year students at this university who think being very well-off is important differs from the national value of 73%?

53. **Teen drivers** Refer to Exercise 51.

(a) Construct and interpret a 95% confidence interval for the true proportion p of all teens in the state who passed their driving test on the first attempt. Assume that the conditions for inference are met.

(b) Explain why the interval in part (a) provides more information than the test in Exercise 51.

54. **We want to be rich** Refer to Exercise 52.

(a) Construct and interpret a 95% confidence interval for the true proportion p of all first-year students at the university who would identify being very well-off as an important personal goal. Assume that the conditions for inference are met.

(b) Explain why the interval in part (a) provides more information than the test in Exercise 52.

55. **Do you Tweet?** The Pew Internet and American Life Project asked a random sample of U.S. adults, "Do you ever . . . use Twitter or another service to share updates about yourself or to see updates about others?" According to Pew, the resulting 95% confidence interval is (0.123, 0.177).[11] Based on the confidence interval, is there convincing evidence that the true proportion of U.S. adults who would say they use Twitter or another service to share updates differs from 0.17? Explain your reasoning.

56. **Losing weight** A Gallup poll found that 59% of the people in its sample said "Yes" when asked, "Would you like to lose weight?" Gallup announced: "For results based on the total sample of national adults, one can say with 95% confidence that the margin of (sampling) error is ±3 percentage points."[12] Based on the confidence interval, is there convincing evidence that the true proportion of U.S. adults who would say they want to lose weight differs from 0.55? Explain your reasoning.

57. **Reporting cheating** What proportion of students are willing to report cheating by other students? A student project put this question to an SRS of 172 undergraduates at a large university: "You witness two students cheating on a quiz. Do you go to the professor?" The Minitab output shows the results of a significance test and a 95% confidence interval based on the survey data.[13]

(a) Define the parameter of interest.

(b) Check that the conditions for performing the significance test are met in this case.

9.46 (a) *Type I error*: Finding convincing evidence that a majority of city residents support a 1% increase in the sales tax to fund road repairs, when in reality at most 50% would. *Consequence:* The sales tax will be increased by 1%, potentially upsetting most city residents.
Type II error: Failing to find convincing evidence that a majority of city residents support a 1% increase in the sales tax to fund road repairs, when in reality a majority would. *Consequence:* The sales tax will not increase and the road repairs will not take place, despite the support of the city residents.
(b) S: H_0: $p = 0.5$, H_a: $p > 0.5$. **P:** *Large Counts:* $150 \geq 10$ and $150 \geq 10$. **D:** $z = 0.92$; P-value $= 0.1788$. **C:** Fail to reject H_0.

9.47 Because $\hat{p} > 0.5$, there is no evidence for H_a: $p < 0.50$.

9.48 Because $\hat{p} < 0.5$, there is no evidence for H_a: $p > 0.50$.

9.49 (a) H_0: $p = 0.75$, H_a: $p \neq 0.75$, where $p =$ the true proportion of peas that will be smooth using $\alpha = 0.05$. **(b)** $z = 0.59$; P-value $= 0.5552$ **(c)** Assuming the true proportion of smooth peas is 0.75, there is a 0.5552 probability of getting a sample proportion as different from 0.75 as 0.761 (in either direction) by chance in a random sample of 556 peas. Because the P-value of $0.5552 > \alpha = 0.05$, we fail to reject H_0. We do not have convincing evidence that the true proportion of peas that are smooth differs from 0.75.

9.50 (a) H_0: $p = 0.50$, H_a: $p \neq 0.50$ **(b)** $z = 1.90$; P-value $= 0.0574$ **(c)** If the true proportion of heads is 0.50, a 0.0574 probability exists of getting a sample proportion as different from 0.50 as 0.56 (in either direction) just by chance in a random sample of 250 spun Belgian euros. Fail to reject H_0.

9.51 S: H_0: $p = 0.60$, H_a: $p \neq 0.60$, where $p =$ the true proportion of teens who pass on the first attempt using $\alpha = 0.05$. **P:** One-sample z test for p. *Random:* Random sample. *10%:* $125 < 10\%$ of population. *Large Counts:* $75 \geq 10$ and $50 \geq 10$. **D:** $z = 2.01$; P-value $= 0.0444$. **C:** Reject H_0. There is convincing evidence that the true proportion of teens who pass on their first attempt differs from 0.60.

9.52 S: H_0: $p = 0.73$, H_a: $p \neq 0.73$ **P:** *Large Counts:* $146 \geq 10$ and $54 \geq 10$. **D:** $z = -2.23$; P-value $= 0.0258$. **C:** Reject H_0.

9.53 (a) S: $p =$ the true proportion of teens who pass on the first attempt. **P:** One-sample z interval for p. **D:** $(0.607, 0.769)$ **C:** We are 95% confident that the interval from 0.607 to 0.769 captures the true proportion of teens who pass on the first attempt. **(b)** The confidence interval gives the values of p that are plausible based on the sample data and not just a reject or fail to reject decision about H_0.

9.54 (a) S: $p =$ the true proportion of first-year students who think being very well-off financially is an important personal goal. **P:** One-sample z interval for p. **D:** $(0.594, 0.726)$ **C:** We are 95% confident that the interval from 0.594 to 0.726 captures the true proportion of first-year students who identify being very well-off as an important personal goal. **(b)** The confidence interval gives the values of p that are plausible based on the sample data and not just a reject or fail to reject decision about H_0.

9.55 No; because the value 0.17 is included in the confidence interval, it is a plausible value for the true proportion of U.S. adults who would say they use Twitter.

9.56 Yes; because 0.55 is not included in the confidence interval from 0.56 to 0.62, it is not a plausible value for the true proportion of U.S. adults who would say that they want to lose weight.

9.57 (a) $p =$ the true proportion of undergraduates willing to report cheating by other students. **(b)** *Random:* Random sample. *10%:* $172 < 10\%$ of all undergrads. *Large Counts:* $25.8 \geq 10$ and $146.2 \geq 10$. **(c)** Assuming the true proportion of undergraduates at this large university who would report cheating is 0.15, there is a 0.146 probability of getting a sample proportion that is at least as different from 0.15 (in either direction) as $\hat{p} = 0.11$. **(d)** No; because the P-value of $0.146 > \alpha = 0.05$, we fail to reject H_0. We do not have convincing evidence that the true proportion of undergraduates who would report cheating differs from 0.15.

9.58 (a) p = the true proportion of U.S. teens aged 13–17 who think young people should wait until marriage to have sex. **(b)** *Random:* Random sample. *10%:* 439 < 10% of population. *Large Counts:* 219.5 ≥ 10 and 219.5 ≥ 10. **(c)** Assuming the true proportion of U.S. teens aged 13–17 who think young people should wait is 0.50, there is a 0.011 probability of getting a sample proportion that is at least as different from 0.5 (in either direction) as $\hat{p}$ = 0.56. **(d)** Yes; because the *P*-value of 0.011 < α = 0.05, we reject H_0. There is convincing evidence that the true proportion of U.S. teens aged 13–17 who think young people should wait differs from 0.5.

9.59 If the true proportion of potatoes with blemishes in a given truckload is p = 0.11, there is a 0.764 probability that the company will find convincing evidence for H_a: p > 0.08.

9.60 If the true mean income in the population of people who live near the restaurant is μ = $86,000, there is a 0.64 probability that I will find convincing evidence for H_a: μ > $85,000.

9.61 (a) Increase; using a larger significance level makes it easier to reject H_0 when H_a is true. **(b)** Decrease; a smaller sample size gives less information about the true proportion p. **(c)** Decrease; it is harder to detect a smaller difference between the null and alternative parameter value.

9.62 (a) Decrease; a smaller sample size gives less information about the true mean μ. **(b)** Decrease; it is harder to detect a smaller difference between the null and alternative parameter value. **(c)** Increase; using a larger significance level makes it easier to reject H_0 when H_a is true.

9.63 (a) The larger significance level will increase the probability of a Type I error. **(b)** The larger sample size would require more time and money.

9.64 (a) The larger significance level will increase the probability of a Type I error. **(b)** The larger sample size would require more time and money.

9.65 (a) If the true proportion of students at the school who are satisfied with the parking situation after the change is p = 0.45 there is a 0.75 probability that the principal will find convincing evidence for H_a: p > 0.37. **(b)** P(Type I error) = α = 0.05 and P(Type II error) = 1 − Power = 1 − 0.75 = 0.25. **(c)** The power would increase by increasing the sample size or using a larger significance level.

(c) Interpret the *P*-value.

(d) Do these data give convincing evidence that the population proportion differs from 0.15? Justify your answer with appropriate evidence.

58. **Teens and sex** The Gallup Youth Survey asked a random sample of U.S. teens aged 13 to 17 whether they thought that young people should wait until marriage to have sex.[14] The Minitab output shows the results of a significance test and a 95% confidence interval based on the survey data.

```
Session                                          _ □ ×

Test and CI for One Proportion

Test of p = 0.5 vs p not = 0.5

Sample   X   N  Sample p        95% CI           Z-Value  P-Value
1      246 439 0.560364 (0.513935, 0.606794)     2.53     0.011
```

(a) Define the parameter of interest.

(b) Check that the conditions for performing the significance test are met in this case.

(c) Interpret the *P*-value.

(d) Do these data give convincing evidence that the actual population proportion differs from 0.5? Justify your answer with appropriate evidence.

59. **Potato chips** A company that makes potato chips
pg 613 requires each shipment of potatoes to meet certain quality standards. If the company finds convincing evidence that more than 8% of the potatoes in the shipment have "blemishes," the truck will be sent back to the supplier to get another load of potatoes. Otherwise, the entire truckload will be used to make potato chips. To make the decision, a supervisor will inspect a random sample of potatoes from the shipment. He will then perform a test of H_0: p = 0.08 versus H_a: p > 0.08, where p is the true proportion of potatoes with blemishes in a given truckload. The power of the test to detect that p = 0.11, based on a random sample of 500 potatoes and significance level α = 0.05, is 0.764. Interpret this value.

60. **Upscale restaurant** You are thinking about opening a restaurant and are searching for a good location. From the research you have done, you know that the mean income of those living near the restaurant must be over $85,000 to support the type of upscale restaurant you wish to open. You decide to take a simple random sample of 50 people living near one potential site. Based on the mean income of this sample, you will perform a test at the α = 0.05 significance level of H_0: μ = $85,000 versus H_a: μ > $85,000, where μ is

the true mean income in the population of people who live near the restaurant. The power of the test to detect that μ = $86,000 is 0.64. Interpret this value.

61. **Powerful potatoes** Refer to Exercise 59. Determine
pg 617 if each of the following changes would increase or decrease the power of the test. Explain your answers.

(a) Change the significance level to α = 0.10.

(b) Take a random sample of 250 potatoes instead of 500 potatoes.

(c) The true proportion is p = 0.10 instead of p = 0.11.

62. **Restaurant power** Refer to Exercise 60. Determine if each of the following changes would increase or decrease the power of the test. Explain your answers.

(a) Use a random sample of 30 people instead of 50 people.

(b) Try to detect that μ = $85,500 instead of μ = $86,000.

(c) Change the significance level to α = 0.10.

63. **Potato power problems** Refer to Exercises 59 and 61.

(a) Explain one disadvantage of using α = 0.10 instead of α = 0.05 when performing the test.

(b) Explain one disadvantage of taking a random sample of 500 potatoes instead of 250 potatoes from the shipment.

64. **Restaurant power problems** Refer to Exercises 60 and 62.

(a) Explain one disadvantage of using α = 0.10 instead of α = 0.05 when performing the test.

(b) Explain one disadvantage of taking a random sample of 50 people instead of 30 people.

65. **Better parking** A local high school makes a change that should improve student satisfaction with the parking situation. Before the change, 37% of the school's students approved of the parking that was provided. After the change, the principal surveys an SRS of students at the school. She would like to perform a test of

$$H_0: p = 0.37$$
$$H_a: p > 0.37$$

where p is the true proportion of students at school who are satisfied with the parking situation after the change.

(a) The power of the test to detect that p = 0.45 based on a random sample of 200 students and a significance level of α = 0.05 is 0.75. Interpret this value.

(b) Find the probability of a Type I error and the probability of a Type II error for the test in part (a).

(c) Describe two ways to increase the power of the test in part (a).

66. Strong chairs? A company that manufactures classroom chairs for high school students claims that the mean breaking strength of the chairs is 300 pounds. One of the chairs collapsed beneath a 220-pound student last week. You suspect that the manufacturer is exaggerating the breaking strength of the chairs, so you would like to perform a test of

$$H_0: \mu = 300$$
$$H_a: \mu < 300$$

where μ is the true mean breaking strength of this company's classroom chairs.

(a) The power of the test to detect that $\mu = 294$ based on a random sample of 30 chairs and a significance level of $\alpha = 0.05$ is 0.71. Interpret this value.

(b) Find the probability of a Type I error and the probability of a Type II error for the test in part (a).

(c) Describe two ways to increase the power of the test in part (a).

67. Error probabilities and power You read that a significance test at the $\alpha = 0.01$ significance level has probability 0.14 of making a Type II error when a specific alternative is true.

(a) What is the power of the test against this alternative?

(b) What's the probability of making a Type I error?

68. Power and error A scientist calculates that a test at the $\alpha = 0.05$ significance level has probability 0.23 of making a Type II error when a specific alternative is true.

(a) What is the power of the test against this alternative?

(b) What's the probability of making a Type I error?

69. Power calculation: Potatoes Refer to Exercise 59.

(a) Suppose that $H_0: p = 0.08$ is true. Describe the shape, center, and variability of the sampling distribution of $\hat{p}$ in random samples of size 500.

(b) Use the sampling distribution from part (a) to find the value of $\hat{p}$ with an area of 0.05 to the right of it. If the supervisor obtains a random sample of 500 potatoes with a sample proportion of blemished potatoes greater than this value of $\hat{p}$, he will reject $H_0: p = 0.08$ at the $\alpha = 0.05$ significance level.

(c) Now suppose that $p = 0.11$. Describe the shape, center, and variability of the sampling distribution of $\hat{p}$ in random samples of size 500.

(d) Use the sampling distribution from part (c) to find the probability of getting a sample proportion greater than the value you found in part (b). This result is the power of the test to detect $p = 0.11$.

Multiple Choice: *Select the best answer for Exercises 70–74.*

70. After once again losing a football game to the archrival, a college's alumni association conducted a survey to see if alumni were in favor of firing the coach. An SRS of 100 alumni from the population of all living alumni was taken, and 64 of the alumni in the sample were in favor of firing the coach. Suppose you wish to see if a majority of all living alumni is in favor of firing the coach. The appropriate standardized test statistic is

(a) $z = \dfrac{0.64 - 0.5}{\sqrt{\dfrac{0.64(0.36)}{100}}}$

(b) $z = \dfrac{0.5 - 0.64}{\sqrt{\dfrac{0.64(0.36)}{100}}}$

(c) $z = \dfrac{0.64 - 0.5}{\sqrt{\dfrac{0.5(0.5)}{100}}}$

(d) $z = \dfrac{0.64 - 0.5}{\sqrt{\dfrac{0.64(0.36)}{64}}}$

(e) $z = \dfrac{0.5 - 0.64}{\sqrt{\dfrac{0.5(0.5)}{100}}}$

71. Which of choices (a) through (d) is *not* a condition for performing a significance test about a population proportion p?

(a) The data should come from a random sample from the population of interest.

(b) Both np_0 and $n(1 - p_0)$ should be at least 10.

(c) If you are sampling without replacement from a finite population, then you should sample less than 10% of the population.

(d) The population distribution should be approximately Normal, unless the sample size is large.

(e) All of the above are conditions for performing a significance test about a population proportion.

72. The standardized test statistic for a test of $H_0: p = 0.4$ versus $H_a: p \neq 0.4$ is $z = 2.43$. This test is

(a) not significant at either $\alpha = 0.05$ or $\alpha = 0.01$.

(b) significant at $\alpha = 0.05$, but not at $\alpha = 0.01$.

(c) significant at $\alpha = 0.01$, but not at $\alpha = 0.05$.

(d) significant at both $\alpha = 0.05$ and $\alpha = 0.01$.

(e) inconclusive because we don't know the value of $\hat{p}$.

73. Which of the following 95% confidence intervals would lead us to reject $H_0: p = 0.30$ in favor of $H_a: p \neq 0.30$ at the 5% significance level?

(a) $(0.19, 0.27)$

(b) $(0.24, 0.30)$

(c) $(0.27, 0.31)$

(d) $(0.29, 0.38)$

(e) None of these

9.66 (a) If the true mean breaking strength of this company's classroom chairs is $\mu = 294$ pounds, there is a 0.71 probability that I will find convincing evidence for $H_a: \mu < 300$. **(b)** P(Type I error) $= \alpha = 0.05$ and P(Type II error) $= 1 - 0.71 = 0.29$ **(c)** The power would increase by increasing the sample size or using a larger significance level.

9.67 (a) Power $= 1 - P$(Type II error) $= 1 - 0.14 = 0.86$ **(b)** P(Type I error) $= \alpha = 0.01$

9.68 (a) Power $= 1 - P$(Type II error) $= 1 - 0.23 = 0.77$ **(b)** P(Type I error) $= \alpha = 0.05$

9.69 (a) *Shape:* Because $np = 500(0.08) = 40 \geq 10$ and $n(1 - p) = 500(0.92) = 460 \geq 10$, the sampling distribution of $\hat{p}$ is approximately Normal. *Center:* The mean of the sampling distribution of $\hat{p}$ is equal to the population proportion. *Variability:* The sample size (500) is less than 10% of the population of all potatoes, so the 10% condition has been met;

$$\sigma_{\hat{p}} = \sqrt{\frac{0.08(0.92)}{500}} = 0.0121.$$

(b) (i) $z = 1.645$; solving $1.645 = \dfrac{\hat{p} - 0.08}{0.0121}$ gives $\hat{p} = 0.0999$.

(ii) *Tech:* invNorm(area: 0.95, mean: 0.08, SD: 0.0121) $= 0.0999$ **(c)** *Shape:* Because $np = 500(0.11) = 55 \geq 10$ and $n(1 - p) = 500(0.89) = 445 \geq 10$, the sampling distribution of $\hat{p}$ is approximately Normal. *Center:* The mean of the sampling distribution of $\hat{p}$ is equal to the population proportion; $\mu_{\hat{p}} = p = 0.11$. *Variability:* The sample size (500) is less than 10% of the population of all potatoes, so the 10% condition has been met;

$$\sigma_{\hat{p}} = \sqrt{\frac{0.11(0.89)}{500}} = 0.0140.$$

(d) (i) $z = \dfrac{0.0999 - 0.11}{0.0140} = -0.72$

$P(z \geq -0.72) = 0.7642$
(ii) $P(\hat{p} \geq 0.0999) =$ normalcdf(lower: 0.0999, upper: 1000, mean: 0.11, SD: 0.014) $= 0.7647$. The power of the test to detect $p = 0.11$ is 0.7647.

9.70 c

9.71 d

9.72 b

9.73 a

9.74 b

9.75 (a) $X - Y$ has a Normal distribution with mean

$\mu_{X-Y} = \mu_X - \mu_Y = 5.3 - 5.26 = 0.04$

and standard deviation

$\sigma_{X-Y} = \sqrt{\sigma_X^2 + \sigma_Y^2} = \sqrt{(0.01)^2 + (0.02)^2}$
$= 0.0224$. $X - Y$ must take on a positive number. **(b)** We want to find $P(X - Y > 0)$.

(i) $z = \dfrac{0 - 0.04}{0.0224} = -1.79$;

$P(z > -1.79) = 0.9633$

(ii) normalcdf(lower: 0, upper: 1000, mean: 0.04, SD: 0.0224) = 0.9629

There is a 0.9629 probability that a randomly selected DVD will fit in a randomly selected case.

(c) $0.9629^{100} = 0.0228$. There is a 0.0228 probability that all 100 DVDs will fit in their cases.

9.76 (a) Take the 10,065 people and randomly assign them to the three treatments, with 3355 people receiving each treatment (bonus for worker, bonus for employer, no bonus). After the experiment, compare the mean time to return to work in the three groups or the proportion of each group who get a job within 11 weeks. **(b)** *Label:* Label the people from 1 to 10,065 in alphabetical order. *Randomize:* Use the command randInt(1, 10065) to select 3355 different integers from 1 to 10,065. Repeat this to identify the individuals to be assigned to each of two different groups. *Select:* The first 3355 individuals selected will be assigned to Treatment 1: $500 bonus for worker; the second 3355 individuals selected will be assigned to Treatment 2: $500 bonus for employer; all remaining individuals will be assigned to Treatment 3: no bonus. **(c)** The purpose of a control group is to give the researchers a baseline for comparison to determine if a financial incentive is better than no incentive at all. If there were no control group, the researchers could only make a conclusion about which type of incentive (to the worker or to the employer) is more effective.

74. A researcher plans to conduct a significance test at the $\alpha = 0.01$ significance level. She designs her study to have a power of 0.90 at a particular alternative value of the parameter of interest. The probability that the researcher will commit a Type II error for the particular alternative value of the parameter she used is

(a) 0.01.

(b) 0.10.

(c) 0.89.

(d) 0.90.

(e) 0.99.

Recycle and Review

75. **Packaging DVDs** (6.2, 5.3) A manufacturer of digital video discs (DVDs) wants to be sure that the DVDs will fit inside the plastic cases used as packaging. Both the cases and the DVDs are circular. According to the supplier, the diameters of the plastic cases vary Normally with mean $\mu = 5.3$ inches and standard deviation $\sigma = 0.01$ inch. The DVD manufacturer produces DVDs with mean diameter $\mu = 5.26$ inches. Their diameters follow a Normal distribution with $\sigma = 0.02$ inch.

(a) Let X = the diameter of a randomly selected case and Y = the diameter of a randomly selected DVD.

Describe the shape, center, and variability of the distribution of the random variable $X - Y$. What is the importance of this random variable to the DVD manufacturer?

(b) Calculate the probability that a randomly selected DVD will fit inside a randomly selected case.

(c) The production process runs in batches of 100 DVDs. If each of these DVDs is paired with a randomly chosen plastic case, find the probability that all the DVDs fit in their cases.

76. **Cash to find work?** (4.2) Will cash bonuses speed the return to work of unemployed people? The Illinois Department of Employment Security designed an experiment to find out. The subjects were 10,065 people aged 20 to 54 who were filing claims for unemployment insurance. Some were offered $500 if they found a job within 11 weeks and held it for at least 4 months. Others could tell potential employers that the state would pay the employer $500 for hiring them. A control group got neither kind of bonus.[15]

(a) Describe a completely randomized design for this experiment.

(b) Explain how you would use a random number generator to assign the treatments.

(c) What is the purpose of the control group in this setting?

SECTION 9.3 **Tests About a Difference in Proportions**

LEARNING TARGETS *By the end of the section, you should be able to:*

- State appropriate hypotheses for a significance test about a difference between two proportions.

- Determine whether the conditions are met for performing a test about a difference between two proportions.

- Calculate the standardized test statistic and *P*-value for a test about a difference between two proportions.

- Perform a significance test about a difference between two proportions.

Which of two often-prescribed drugs—Lipitor or Pravachol—helps lower "bad cholesterol" more? Researchers designed an experiment, called the PROVE-IT Study, to find out. They used about 4000 people with heart disease as subjects. These individuals were randomly assigned to one of two treatment groups: Lipitor or Pravachol. At the end of the study, researchers compared the proportion of

PD **Section 9.3 Overview**

To watch the video overview of Section 9.3 (for teachers), click on the link in the TE-Book, look on the TRFD, or download from the Teacher's Resources on the book's digital platform.

TRM **Section 9.3 Alternate Examples**

You can find the Alternate Examples for this section in Microsoft Word format by clicking on the link in the TE-Book, opening the TRFD, or downloading from the Teacher's Resources on the book's digital platform.

subjects in each group who died, had a heart attack, or suffered other serious consequences within two years. For those using Pravachol, the proportion was 0.263; for those using Lipitor, it was 0.224.[16] Could such a difference have occurred purely by the chance involved in the random assignment?

Comparing two proportions based on random sampling or a randomized experiment is one of the most common situations encountered in statistical practice. In the PROVE-IT experiment, the goal of inference is to determine whether the treatments (Lipitor and Pravachol) *cause* a difference in the proportion of people who experience serious consequences for subjects like the ones in this study. See Exercise 88 (page 637). In other studies, the goal of inference is to compare two population proportions. For example, researchers could survey independent random samples of U.S. adult women and men to see if there is convincing evidence that a higher proportion of females than males have read a book in the past year.

The following activity gives you a taste of what lies ahead in this section.

ACTIVITY   Who likes tattoos?

For their response bias project (page 320), Sarah and Miranda investigated whether the characteristics of the interviewer can affect the response to a survey question. At the Tucson Mall, they asked 60 shoppers the question "Do you like tattoos?" When interviewing 30 of the shoppers, Sarah and Miranda wore long-sleeved tops that covered their tattoos. For the remaining 30 shoppers, the interviewers wore tank tops that revealed their tattoos. The choice of long-sleeve or tank top was determined at random. Sarah and Miranda suspected that more people would answer "Yes" when their tattoos were visible.

What happened in the experiment? The two-way table summarizes the results:[17]

		Clothing worn		
		Tank top	Long sleeves	Total
Like tattoos?	Yes	18	14	32
	No	12	16	28
	Total	30	30	60

The difference (Tank top − Long sleeves) in the proportions of people who said they like tattoos is $18/30 - 14/30 = 0.600 - 0.467 = 0.133$. Does this difference provide convincing evidence that the appearance of the interviewer has an effect on the response, or could the difference be due to chance variation in the random assignment?

In this activity, your class will investigate whether the results of the experiment are statistically significant. Let's see what would happen just by chance if we randomly reassign the 60 people in this experiment to the two treatments (tank top and long sleeves) many times, *assuming the treatment received doesn't affect whether or not a person says they like tattoos.*

1. Using 60 index cards or equally sized pieces of paper, write "Yes" on 32 and "No" on 28. Shuffle the cards and divide them at random into two piles of 30 — one for the tank top treatment group and one for the long sleeves

---

✚ Ask the StatsMedic
Is Yawning Contagious?

This post outlines an alternate activity that can be used to introduce inference for two populations or treatments. In an episode of *MythBusters*, the team performed an experiment to arrive at the conclusion that yawning is contagious. The activity investigates whether or not the MythBusters made a good statistical conclusion. In the end, students will bust the MythBusters.

TRM   Who Likes Tattoos? Template

This printable template has 60 cards labeled "Yes" and "No" for use in this activity. Find it in the Teacher's Resource Materials located in the TE-Book, on the TRFD, or in the Teacher's view on the book's digital platform.

Making Connections

Students were introduced to inference for experiments in Section 4.3. The "Analyzing the caffeine experiment" activity from Section 4.3 is much like the "Who likes tattoos?" activity on this page.

▶ ACTIVITY OVERVIEW

To prepare for using this activity, watch the overview video by clicking on the link in the TE-Book, opening the TRFD, or downloading from the Teacher's Resources on the book's digital platform.

Time: 20 minutes

Materials: Index cards, poster, sticker dots

Teaching Advice: The difference in proportions of 0.133 certainly provides *some evidence* that the appearance of the interviewer has an effect on the response because the difference is greater than 0. The purpose of the activity is to determine if that evidence is *convincing*. We start by assuming the null hypothesis is true, which says that *the treatment received doesn't affect whether or not a person says he or she likes tattoos*. To help students understand how we are using this idea in the activity, hold up an index card that says "Yes" and remind students that this person likes tattoos and will say "Yes" regardless of which treatment group he or she is randomly assigned to.

Ask students to put sticker dots on a poster board. Post the completed poster board in the classroom so that you can refer to this activity throughout the rest of the chapter and the course. This dotplot is an approximation of a *randomization distribution*, which can be closely modeled with a sampling distribution. For more discussion, see the Think About It on page 634. To increase the number of dots on your dotplot, use the *One Categorical Variable* applet at highschool.bfwpub.com/updatedtps6e. Choose multiple groups and input the data from the two-way table. Then "Simulate difference in proportions." Use the applet to count how many times a difference in proportions of 0.133 or greater occurs purely by chance. In the long run, a difference this large or larger should occur about 22% of the time.

OVERVIEW (continued)

Answers:

1–5. Answers will vary. Here is a possible dotplot for a class of 30 students:

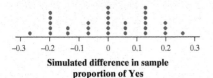

Simulated difference in sample proportion of Yes

According to this dotplot, the difference in proportions was 0.133 or greater on 10 trials. Because a difference this large could easily occur just due to the chance variation in random assignment, the data do not provide convincing evidence that the appearance of the interviewer caused a higher proportion of people to say they like tattoos when the interviewer's tattoos were visible.

Teaching Tip

Sometimes, the hypothesized difference is not 0. For example, suppose that a pharmaceutical company will decide to market a new drug if its success rate is more than 0.05 higher than the currently used drug. In this case, the null hypothesis would be H_0: $p_{new} - p_{current} = 0.05$ and the alternative hypothesis would be H_a: $p_{new} - p_{current} > 0.05$. In this course, we will stick to cases where the hypothesized difference is 0.

Teaching Tip

Make students familiar with both versions of the null hypothesis:

H_0: $p_1 - p_2 = 0$ and H_0: $p_1 = p_2$

We prefer the first version because the P-value calculation uses the sampling distribution of $\hat{p}_1 - \hat{p}_2$.

treatment group. Be sure to determine which pile will represent each group before you deal.

2. Calculate the difference (Tank top – Long sleeves) in the proportions of "Yes" responses for the two groups.
3. Your teacher will draw and label axes for a class dotplot. Plot your result from Step 2 on the graph.
4. Repeat Steps 2 and 3 if needed to get a total of at least 40 trials of the simulation for your class.
5. How often did a difference in proportions of 0.133 or greater occur due only to the chance involved in the random assignment? What conclusion would you make about the effect of the interviewer's appearance?

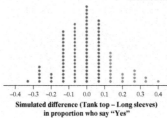

Simulated difference (Tank top – Long sleeves) in proportion who say "Yes"

We used software to perform 100 trials of the simulation described in the activity. In each trial, the computer randomly reassigned the 60 subjects in the experiment to the two treatments, assuming that the treatment received would not affect whether each person says that he or she likes tattoos. Each dot in the graph represents the difference (Tank top – Long sleeves) in the proportion of respondents in the two groups who say they like tattoos for a particular trial. In this simulation, 19 of the 100 trials (in red) produced a difference in proportions of at least 0.133, so the approximate P-value is 0.19. A difference this large could easily occur just due to the chance variation in random assignment! Sarah and Miranda's data do not provide convincing evidence that the appearance of the interviewer causes a higher proportion of people like these to say they like tattoos when the interviewer's tattoos are visible.

Significance Tests for $p_1 - p_2$

An observed difference between two sample proportions $\hat{p}_1$ and $\hat{p}_2$ can reflect an actual difference in the parameters p_1 and p_2, or it may just be due to chance variation in random sampling or random assignment. Significance tests help us decide which explanation makes more sense.

STATING HYPOTHESES In a test for comparing two proportions, the null hypothesis has the general form

$$H_0: p_1 - p_2 = \text{hypothesized value}$$

We're often interested in situations where the hypothesized difference is 0. Then the null hypothesis says that there is no difference between the two parameters:

$$H_0: p_1 - p_2 = 0$$

(You will sometimes see the null hypothesis written in the equivalent form H_0: $p_1 = p_2$.) The alternative hypothesis says what kind of difference we expect.

Here's an example that illustrates how to state hypotheses for a test about a difference between two proportions.

EXAMPLE	Hungry children
	Stating hypotheses

PROBLEM: Researchers designed a study to compare the proportion of children who come to school without eating breakfast in two low-income elementary schools. An SRS of 80 students from School 1 found that 19 missed breakfast today. At School 2, an SRS of 150 students included 26 who missed breakfast today. More than 1500 students attend each school. Do these data give convincing evidence of a difference in the population proportions at the $\alpha = 0.05$ significance level? State appropriate hypotheses for performing a significance test. Be sure to define the parameters of interest.

SOLUTION:

$H_0: p_1 - p_2 = 0$

$H_a: p_1 - p_2 \neq 0$

where p_1 = the true proportion of all students at School 1 who missed breakfast today and p_2 = the true proportion of all students at School 2 who missed breakfast today.

> You can also state the hypotheses as
> $H_0: p_1 = p_2$
> $H_a: p_1 \neq p_2$

FOR PRACTICE, TRY EXERCISE 77

CHECKING CONDITIONS The Random and 10% conditions for performing a significance test about $p_1 - p_2$ are the same as for constructing a confidence interval. However, we check the Large Counts condition differently when performing a test of $H_0: p_1 - p_2 = 0$.

A significance test begins by assuming that the null hypothesis is true. In that case, $p_1 = p_2$. We call the common value of these two parameters p. Unfortunately, we don't know this common value. To estimate p, we combine (or "pool") the data from the two samples as if they came from one larger sample. This combined sample proportion is

> We can also write $\hat{p}_C = \frac{n_1\hat{p}_1 + n_2\hat{p}_2}{n_1 + n_2}$.
> This shows that the combined (pooled) estimate of p is a weighted average of the sample proportions.

$$\hat{p}_C = \frac{\text{number of successes in both samples combined}}{\text{number of individuals in both samples combined}} = \frac{X_1 + X_2}{n_1 + n_2}$$

In other words, $\hat{p}_C$ gives the overall proportion of successes in the combined samples.

Let's return to the "Hungry children" example. The two-way table summarizes the sample data on whether or not children missed breakfast today at the two schools.

		School		
		1	2	Total
Missed breakfast?	Yes	19	26	45
	No	61	124	185
	Total	80	150	230

Making Connections

When we use the combined proportion to calculate the expected counts, we get the same values as we would when checking the expected counts for a chi-square test (Chapter 12) on the same data. Consider the breakfast data from the previous page:

		School		
		1	2	Total
Missed breakfast?	Yes	19	26	45
	No	61	124	185
	Total	80	150	230

- In Section 9.3 when we are doing a two-sample z test for difference of proportions, we get
$$\hat{p}_C = \frac{19 + 26}{80 + 150} = \frac{45}{230} = 0.196,$$
so the expected count of "Yes" for school 1 is $(80)(0.196) = 15.7$.

- In Chapter 12 when we are doing a chi-square test, the expected count of "Yes" for school 1 is
$$\frac{\text{row total} \cdot \text{column total}}{\text{table total}} = \frac{45 \cdot 80}{230}$$
$$= 15.7.$$

Teaching Tip

Some people prefer to use the sample proportions to check the Large Counts condition. This is certainly the right thing to do when the null hypothesis is $H_0: p_1 - p_2 =$ a non-zero value. For AP® Statistics, students will most likely only see tests with a null hypothesis value of 0 (meaning the two proportions are equal), so we will always use the combined (pooled) proportion of success when checking the Large Counts condition.

In this setting, the null hypothesis is $H_0: p_1 - p_2 = 0$ (or, equivalently, $H_0: p_1 = p_2$). So we can estimate the common proportion p of all students at School 1 and all students at School 2 who missed breakfast today using

$$\hat{p}_C = \frac{X_1 + X_2}{n_1 + n_2} = \frac{19 + 26}{80 + 150} = \frac{45}{230} = 0.196$$

The rightmost column in the two-way table makes it easy to see that the overall proportion of successes in the two samples is $\hat{p}_C = 45/230$.

We use the combined (pooled) proportion of successes $\hat{p}_C$ when checking the Large Counts condition for a test of $H_0: p_1 - p_2 = 0$. In that case, $n_1\hat{p}_C$, $n_1(1 - \hat{p}_C)$, $n_2\hat{p}_C$, and $n_2(1 - \hat{p}_C)$ give us the *expected* numbers of successes and failures in the two samples.

> ### CONDITIONS FOR PERFORMING A SIGNIFICANCE TEST ABOUT A DIFFERENCE BETWEEN TWO PROPORTIONS
>
> - **Random:** The data come from two independent random samples or from two groups in a randomized experiment.
>   - **10%:** When sampling without replacement, $n_1 < 0.10N_1$ and $n_2 < 0.10N_2$.
> - **Large Counts:** The expected numbers of successes and failures in each sample or group—$n_1\hat{p}_C$, $n_1(1 - \hat{p}_C)$, $n_2\hat{p}_C$, $n_2(1 - \hat{p}_C)$—are all at least 10.

Some people prefer to use the sample proportions $\hat{p}_1$ and $\hat{p}_2$ to check the Large Counts condition for significance tests, just like we did for confidence intervals. Either way, the purpose of checking this condition is to ensure that the sampling distribution of $\hat{p}_1 - \hat{p}_2$ is approximately Normal. The Random and 10% conditions help us check for independence in the data collection process.

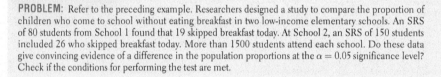

EXAMPLE	**Hungry children**
	Checking conditions

PROBLEM: Refer to the preceding example. Researchers designed a study to compare the proportion of children who come to school without eating breakfast in two low-income elementary schools. An SRS of 80 students from School 1 found that 19 skipped breakfast today. At School 2, an SRS of 150 students included 26 who skipped breakfast today. More than 1500 students attend each school. Do these data give convincing evidence of a difference in the population proportions at the $\alpha = 0.05$ significance level? Check if the conditions for performing the test are met.

SOLUTION:

- Random? Independent random samples of 80 students from School 1 and 150 students from School 2. ✓

  ○ 10%: 80 < 10% of students at School 1; 150 < 10% of students at School 2. ✓

- Large Counts? $\hat{p}_C = \dfrac{19 + 26}{80 + 150} = \dfrac{45}{230} = 0.196$

  $\boxed{\hat{p}_C = \dfrac{X_1 + X_2}{n_1 + n_2}}$

  $n_1 \hat{p}_C = 80(0.196) = 15.68,\ n_1(1 - \hat{p}_C) = 80(0.804) = 64.32,$
  $n_2 \hat{p}_C = 150(0.196) = 29.40,\ n_2(1 - \hat{p}_C) = 150(0.804) = 120.6$
  are all $\geq$ 10. ✓

  The expected numbers of successes (missed breakfast!) and failures (ate breakfast) in both samples are at least 10.

FOR PRACTICE, TRY EXERCISE 79

CALCULATIONS: STANDARDIZED TEST STATISTIC AND *P*-VALUE If the conditions are met, we can proceed with calculations. To do a test of $H_0: p_1 - p_2 = 0$, standardize $\hat{p}_1 - \hat{p}_2$ to get a z statistic:

$$\text{standardized test statistic} = \frac{\text{statistic} - \text{parameter}}{\text{standard deviation (error) of statistic}}$$

$$z = \frac{(\hat{p}_1 - \hat{p}_2) - 0}{\sqrt{\dfrac{p_1(1 - p_1)}{n_1} + \dfrac{p_2(1 - p_2)}{n_2}}}$$

You may be tempted to replace p_1 and p_2 in the denominator with the corresponding sample proportions $\hat{p}_1$ and $\hat{p}_2$. Don't do it!

Our test begins by assuming that $H_0: p_1 - p_2 = 0$ is true. In that case, $p_1 = p_2 = p$. Substitute this common value in place of both p_1 and p_2 in the denominator of the standardized test statistic:

$$z = \frac{(\hat{p}_1 - \hat{p}_2) - 0}{\sqrt{\dfrac{p(1 - p)}{n_1} + \dfrac{p(1 - p)}{n_2}}}$$

With a little factoring, we can rewrite the denominator of the standardized test statistic as

$\sqrt{\hat{p}_C(1 - \hat{p}_C)\left(\dfrac{1}{n_1} + \dfrac{1}{n_2}\right)}$. You'll find this expression on the AP® Statistics exam formula sheet under the heading "Standard Error of Sample Statistic" for a difference of sample proportions in the special case when $p_1 = p_2$. The formula sheet uses the notation $s_{\hat{p}_1 - \hat{p}_2}$ for the standard error.

Now use the combined (pooled) estimate $\hat{p}_C$ in place of p to get the standard error in the denominator:

$$z = \frac{(\hat{p}_1 - \hat{p}_2) - 0}{\sqrt{\dfrac{\hat{p}_C(1 - \hat{p}_C)}{n_1} + \dfrac{\hat{p}_C(1 - \hat{p}_C)}{n_2}}}$$

When the Large Counts condition is met, this z statistic will have approximately the standard Normal distribution. We can find the appropriate P-value using Table A or technology.

ALTERNATE EXAMPLE   Skill 4.C

Who Instagrams?
Checking conditions

PROBLEM:
Refer to the preceding alternate example. The Pew Research Center recently selected a random sample of 269 18- to 29-year-olds and found that 59% of them use Instagram. A separate random sample of 401 30- to 49-year-olds found that 31% of them use Instagram. Do these data give convincing evidence that a greater proportion of 18- to 29-year-olds use Instagram than 30- to 49-year-olds at the $\alpha = 0.05$ significance level? Check if the conditions for performing the test are met.

SOLUTION:
- *Random:* Independent random samples of 269 18- to 29-year-olds and 401 30- to 49-year-olds. ✓
  ○ *10%:* 269 < 10% of all 18- to 29-year-olds; 401 < 10% of all 30- to 49-year-olds. ✓
- *Large Counts:* There are (0.59)(269) = 159 18- to 29-year-olds who use Instagram. There are (0.31)(401) = 124 30- to 49-year-olds who use Instagram.

  $\hat{p}_C = \dfrac{159 + 124}{269 + 401} = \dfrac{283}{670} = 0.422$

  $n_1 \hat{p}_C = 269(0.422) = 113.5$
  $n_1(1 - \hat{p}_C) = 269(0.578) = 155.5$
  $n_2 \hat{p}_C = 401(0.422) = 169.2$
  $n_2(1 - \hat{p}_C) = 401(0.578) = 231.8$ are all ≥ 10. ✓

Teaching Tip

Before revealing the formula for the z test statistic, give students the "Hungry children" example (or the alternate example) and have them work in pairs to write a four-step significance test. Call this an "investigative task" because students are being asked to apply their previous knowledge in a new context. Ask students to put the solution on the board. They will very likely use $\hat{p}_1$ and $\hat{p}_2$ in the denominator of the calculation for the z test statistic. Refer to the null hypothesis, which makes it clear that we shouldn't be plugging in different values for $\hat{p}_1$ and $\hat{p}_2$ (because we are assuming the two proportions are equal!). Use a different color marker to cross off the parts of the solution that are incorrect and make the necessary corrections.

ALTERNATE EXAMPLE   Skill 3.E

Gramming
Calculating the standardized test statistic and *P*-value

PROBLEM:
Refer to the preceding alternate example. The two-way table below summarizes the data from the independent random samples of 18- to 29-year-olds and 30- to 49-year-olds. We already confirmed that the conditions for performing a significance test are met.

		Age (years)		
		18–29	30–49	Total
Use Instagram	Yes	159	124	283
	No	110	277	387
	Total	269	401	670

(a) Explain why the sample results give some evidence for the alternative hypothesis.
(b) Calculate the standardized test statistic and *P*-value.
(c) What conclusion would you make using $\alpha = 0.05$?

SOLUTION:
(a) The observed difference in the sample proportions is $\hat{p}_1 - \hat{p}_2 = \dfrac{159}{269} - \dfrac{124}{401} =$
$0.59 - 0.31 = 0.28$, which gives some evidence in favor of $H_a: p_1 - p_2 > 0$ because $0.28 > 0$.

(b) $\hat{p}_C = \dfrac{159 + 124}{269 + 401} = \dfrac{283}{670} = 0.422$

$$z = \dfrac{(0.59 - 0.31) - 0}{\sqrt{\dfrac{0.422(0.578)}{269} + \dfrac{0.422(0.578)}{401}}}$$

$$= \dfrac{0.28}{0.0389} = 7.20$$

Table A: $P(z > 7.20) \approx 0$
Tech: normalcdf(lower: 7.20, upper: 1000, mean: 0, SD: 1) ≈ 0
(c) Because the *P*-value of $\approx 0 < \alpha = 0.05$, we reject H_0. There is convincing evidence that a greater proportion of 18- to 29-year-olds than 30- to 49-year-olds use Instagram.

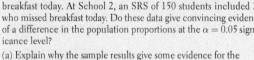

EXAMPLE
Calculating the standardized test statistic and *P*-value
Who eats breakfast?

PROBLEM: Refer to the previous two examples. Researchers designed a study to compare the proportion of children who come to school without eating breakfast in two low-income elementary schools. An SRS of 80 students from School 1 found that 19 missed breakfast today. At School 2, an SRS of 150 students included 26 who missed breakfast today. Do these data give convincing evidence of a difference in the population proportions at the $\alpha = 0.05$ significance level?

(a) Explain why the sample results give some evidence for the alternative hypothesis.
(b) Calculate the standardized test statistic and *P*-value.
(c) What conclusion would you make using $\alpha = 0.05$?

SOLUTION:

(a) The observed difference in the sample proportions is

$$\hat{p}_1 - \hat{p}_2 = \dfrac{19}{80} - \dfrac{26}{150} = 0.2375 - 0.1733 = 0.0642, \text{ which}$$

gives some evidence in favor of $H_a: p_1 - p_2 \neq 0$ because $0.0642 \neq 0$.

(b) $\hat{p}_C = \dfrac{19 + 26}{80 + 150} = \dfrac{45}{230} = 0.196$

$$z = \dfrac{(0.2375 - 0.1733) - 0}{\sqrt{\dfrac{0.196(0.804)}{80} + \dfrac{0.196(0.804)}{150}}} = \dfrac{0.0642}{0.055} = 1.17$$

$$z = \dfrac{(\hat{p}_1 - \hat{p}_2) - 0}{\sqrt{\dfrac{\hat{p}_C(1 - \hat{p}_C)}{n_1} + \dfrac{\hat{p}_C(1 - \hat{p}_C)}{n_2}}}$$

Using Table A: $P(z \leq -1.17 \text{ or } z \geq 1.17)$
$= 2(0.1210) = 0.2420$
Using technology: normalcdf(lower: 1.17, upper: 1000, mean: 0, SD: 1) $\times 2 = 0.2420$

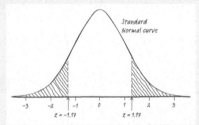

Standard Normal curve

(c) Because the *P*-value of $0.2420 > \alpha = 0.05$, we fail to reject H_0. There is not convincing evidence of a difference in the true proportions of students at School 1 and School 2 who skipped breakfast today.

FOR PRACTICE, TRY EXERCISE 83

⚠ COMMON STUDENT ERROR

When students fail to reject the null hypothesis, they often state that the null hypothesis is true (i.e., they accept the null hypothesis). In this example, it would be incorrect to conclude that the proportion of students who didn't eat breakfast is the *same* at the two schools. We don't have evidence that this is true; we simply don't have convincing evidence that it *isn't* true. To emphasize that 0 is just one of many plausible values for the true difference in proportions, point out the confidence interval shown on the next page. Any of the values from -0.047 to 0.175 could be the true difference in proportions— there is nothing special about 0.

What does the P-value in the example tell us? If there is no difference in the population proportions of students who missed breakfast today at the two schools and we repeated the random sampling process many times, we'd get a difference in sample proportions as large or larger than 0.0642 in either direction about 24% of the time. With such a high probability of getting a result like this just by chance when the null hypothesis is true, we don't have enough evidence to reject H_0.

We can get additional information about the difference between the population proportions who missed breakfast today at School 1 and School 2 with a confidence interval. Technology gives the 95% confidence interval for $p_1 - p_2$ as -0.047 to 0.175. That is, we are 95% confident that the true proportion of students who missed breakfast at School 1 is between 4.7 percentage points lower and 17.5 percentage points higher than at School 2. This is consistent with our "fail to reject H_0" conclusion because 0 is included in the interval of plausible values for $p_1 - p_2$.

21. Technology Corner

PERFORMING A SIGNIFICANCE TEST FOR A DIFFERENCE IN PROPORTIONS

TI-Nspire and other technology instructions are on the book's website at highschool.bfwpub.com/updatedtps6e.

The TI-83/84 can be used to perform significance tests for comparing two proportions when the null hypothesis is no difference. Here, we use the data from the hungry children example. To perform a test of $H_0: p_1 - p_2 = 0$ versus $H_a: p_1 - p_2 \neq 0$:

- Press $\boxed{\text{STAT}}$, then choose TESTS and 2-PropZTest.

- When the 2-PropZTest screen appears, enter the values shown. Specify the alternative hypothesis $p_1 \neq p_2$. Note that the values of x_1, n_1, x_2, and n_2 must be whole numbers or the calculator will report an error.

- If you select "Calculate" and press $\boxed{\text{ENTER}}$, you will see that the standardized test statistic is $z = 1.168$ and the P-value is 0.2427. Do you see the combined proportion of students who didn't eat breakfast? It's the value labeled $\hat{p}$, 0.1957.

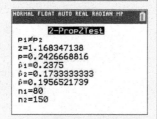

- If you select the "Draw" option, you will see the screen shown here.

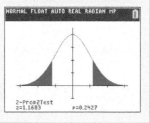

Teaching Tip

Another option is to require students to show the calculations for their work within this chapter. Then at the end of the course, when preparing for the AP® Statistics exam, relax this expectation and suggest that students instead use the calculator to get the standardized test statistic and the *P*-value.

AP® EXAM TIP

The formula for the standardized test statistic for a difference of proportions is *not* included on the AP® Statistics exam formula sheet. However, students are given the general formula for a standardized test statistic and the formula for a standard error of the sampling distribution of $\hat{p}_1 - \hat{p}_2$. Students should remember to calculate the standard error using the combined proportion $\hat{p}_C$ instead of the sample proportions ($\hat{p}_1$ and $\hat{p}_2$) because the null hypothesis will assume that $p_1 = p_2$.

Making Connections

Remind students of the general formula for a standardized test statistic and connect it to the specific formula for a difference in proportions.

General formula:

$$\text{standardized test statistic} = \frac{\text{statistic} - \text{parameter}}{\text{standard error of statistic}}$$

Specific formula:

$$z = \frac{(\hat{p}_1 - \hat{p}_2) - (p_1 - p_2)}{\sqrt{\dfrac{\hat{p}_C(1 - \hat{p}_C)}{n_1} + \dfrac{\hat{p}_C(1 - \hat{p}_C)}{n_2}}}$$

AP® EXAM TIP

The formula for the two-sample *z* statistic for a test about $p_1 - p_2$ often leads to calculation errors by students. As a result, your teacher may recommend using the calculator's 2-PropZTest feature to perform calculations on the AP® Statistics exam. Be sure to name the procedure (two-sample *z* test for $p_1 - p_2$) in the "Plan" step and report the standardized test statistic ($z = 1.17$) and *P*-value (0.2427) in the "Do" step.

Putting It All Together: Two-Sample *z* Test for $p_1 - p_2$

Here is a summary of the Do step for the two-sample *z* test for the difference between two proportions.

TWO-SAMPLE *z* TEST FOR THE DIFFERENCE BETWEEN TWO PROPORTIONS

Suppose the conditions are met. To test the hypothesis $H_0: p_1 - p_2 = 0$, first find the combined (pooled) proportion $\hat{p}_C$ of successes in both samples combined. Then compute the standardized test statistic

$$z = \frac{(\hat{p}_1 - \hat{p}_2) - 0}{\sqrt{\dfrac{\hat{p}_C(1 - \hat{p}_C)}{n_1} + \dfrac{\hat{p}_C(1 - \hat{p}_C)}{n_2}}}$$

Find the *P*-value by calculating the probability of getting a *z* statistic this large or larger in the direction specified by the alternative hypothesis H_a in a standard Normal distribution.

As with any test, be sure to follow the four-step process.

EXAMPLE

Cholesterol and heart attacks

Performing a significance test about $p_1 - p_2$

PROBLEM: High levels of cholesterol in the blood are associated with a higher risk of heart attacks. Will using a drug to lower blood cholesterol reduce heart attacks? The Helsinki Heart Study recruited middle-aged men with high cholesterol but no history of other serious medical problems to investigate this question. The volunteer subjects were assigned at random to one of two treatments: 2051 men took the drug gemfibrozil to reduce their cholesterol levels, and a control group of 2030 men took a placebo. During the next five years, 56 men in the gemfibrozil group and 84 men in the placebo group had heart attacks.

(a) Do the results of this study give convincing evidence at the $\alpha = 0.01$ significance level that gemfibrozil is effective in preventing heart attacks?

(b) Interpret the P-value you got in part (a) in the context of this experiment.

SOLUTION:

(a) STATE: We want to test

$H_0: p_G - p_{PL} = 0$

$H_a: p_G - p_{PL} < 0$

where p_G = the true heart attack rate for middle-aged men like these who take gemfibrozil and p_{PL} = the true heart attack rate for middle-aged men like these who take a placebo using $\alpha = 0.01$.

> We could have subtracted in the opposite order when stating hypotheses
>
> $H_0: p_{PL} - p_G = 0$
>
> $H_a: p_{PL} - p_G > 0$

PLAN: Two-sample z test for $p_G - p_{PL}$

- Random: Volunteer subjects were randomly assigned to gemfibrozil or placebo. ✓

> Note that we do not have to check the 10% condition because the subjects in the experiment were not sampled without replacement from some larger population.

- Large Counts: $\hat{p}_C = \dfrac{56+84}{2051+2030} = \dfrac{140}{4081} = 0.0343$

> The expected numbers of successes (heart attacks!) and failures (no heart attacks) in both groups are at least 10.

$n_1\hat{p}_C = 2051(0.0343) = 70.35, n_1(1-\hat{p}_C) = 2051(0.9657) = 1980.65,$

$n_2\hat{p}_C = 2030(0.0343) = 69.63, n_2(1-\hat{p}_C) = 2030(0.9657) = 1960.37$

are all ≥ 10. ✓

DO:

- $\hat{p}_G = \dfrac{56}{2051} = 0.0273, \hat{p}_{PL} = \dfrac{84}{2030} = 0.0414$

> The sample result gives *some* evidence in favor of $H_a: p_G - p_{PL} < 0$ because $0.0273 - 0.0414 = -0.0141 < 0$.

- $z = \dfrac{(0.0273-0.0414)-0}{\sqrt{\dfrac{0.0343(0.9657)}{2051} + \dfrac{0.0343(0.9657)}{2030}}} = \dfrac{-0.0141}{0.0057} = -2.47$

- P-value

Using Table A: $P(z \leq -2.47) = 0.0068$

Using technology: normalcdf(lower: −1000, upper: −2.47, mean:0, SD:1) = 0.0068

> On the TI-83/84, the 2-PropZTest gives $z = -2.47$ and P-value = 0.0068.

CONCLUDE: Because our P-value of $0.0068 < \alpha = 0.01$, we reject H_0. There is convincing evidence of a lower heart attack rate for middle-aged men like these who take gemfibrozil than for those who take only a placebo.

(b) Assuming $H_0: p_G - p_{PL} = 0$ is true, and that there is no difference in the effectiveness of the two treatments, there is a 0.0068 probability of getting a difference (Gemfibrozil − Placebo) in heart attack rate for the two groups of −0.0141 or less just by the chance involved in the random assignment.

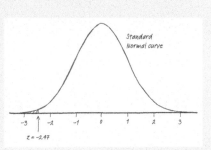

FOR PRACTICE, TRY EXERCISE 85

COMMON STUDENT ERROR

Many students don't use correct language when defining parameters, especially for problems about experiments. In the "Cholesterol and heart attacks" example, it is incorrect to define p_1 as the heart attack rate for men who *took* gemfibrozil. The heart attack rate for men who *took* gemfibrozil is $\hat{p}_1$, not p_1. The parameter should refer to those who *take* gemfibrozil.

- Large Counts:

$\hat{p}_C = \dfrac{18 + 12}{150 + 150} = \dfrac{30}{300} = 0.10$

$n_F\hat{p}_C = 150(0.10) = 15$

$n_F(1 - \hat{p}_C) = 150(0.90) = 135$

$n_T\hat{p}_C = 150(0.10) = 15$

$n_T(1 - \hat{p}_C) = 150(0.90) = 135$ are all ≥ 10. ✓

DO:

- $\hat{p}_F = \dfrac{18}{150} = 0.12, \hat{p}_T = \dfrac{12}{150} = 0.08$

- $z = \dfrac{(0.12 - 0.08) - 0}{\sqrt{\dfrac{0.10(0.90)}{150} + \dfrac{0.10(0.90)}{150}}}$

$= \dfrac{0.04}{0.0346} = 1.16$

- P-value

Using Table A: $P(z \geq 1.16) = 1 - 0.8770 = 0.1230$

Using Tech: normalcdf(lower: 1.16, upper: 1000, mean: 0, SD: 1) = 0.1230

CONCLUDE: Because our P-value of $0.1230 > \alpha = 0.05$, we fail to reject H_0. There is not convincing evidence that a financial incentive helps employees like these quit smoking.

ALTERNATE EXAMPLE

Cash for quitters

Performing a significance test about $p_1 - p_2$

Skills 1.E, 4.E

PROBLEM:

In an effort to reduce health care costs, a large company sponsored a study to help employees stop smoking. Half the subjects were randomly assigned to receive $1000 for quitting smoking for a year, while the other half were simply encouraged to use traditional methods to stop smoking. None of the 300 total volunteers knew of the financial incentive when they first signed up. At the end of one year, 12% of those in the financial rewards group had quit smoking, whereas only 8% in the traditional group had quit smoking. Do the results of this study give convincing evidence that a financial incentive helps employees like these quit smoking?

SOLUTION:

STATE: We want to test

$H_0: p_F - p_T = 0$

$H_a: p_F - p_T > 0$

where p_F = the true quitting rate for employees like these who get a financial incentive to quit smoking and p_T = the true quitting rate for employees like these who are encouraged to use traditional methods to quit smoking using $\alpha = 0.05$.

PLAN: Two-sample z test for $p_F - p_T$

- *Random:* Volunteer subjects were randomly assigned to financial incentive and traditional methods. ✓

Teaching Tip

Here are some important tips for teaching students to use the four-step process for performing a two-sample z test for $p_1 - p_2$:

- Inform students they can use the wording of the question to define their parameters and to write the conclusion. We call this "parroting" the stem of the question.
- Ask students, "Which words in the question inform you this is a one- or two-sided test?"
- When naming the inference method, it is acceptable to list the calculator command 2-PropZTest. We prefer that students use "two-sample z test for $p_1 - p_2$," because it demonstrates deeper understanding and leads to fewer misconceptions.
- Consider having students write the "*So what?*" for each condition.
- Consider having students write a general formula and a specific formula with variables before substituting numbers:

General formula:

$$\text{standardized test statistic} = \frac{\text{statistic} - \text{parameter}}{\text{standard error of statistic}}$$

Specific formula:

$$z = \frac{(\hat{p}_1 - \hat{p}_2) - (p_1 - p_2)}{\sqrt{\dfrac{\hat{p}_C(1 - \hat{p}_C)}{n_1} + \dfrac{\hat{p}_C(1 - \hat{p}_C)}{n_2}}}$$

One advantage to establishing this expectation is that the general formula is the same for all significance tests covered in AP® Statistics (except for chi-square tests). Only the specific formula changes for each new significance test. Realize that it is risky to provide this much work on the AP® Statistics exam because students can lose credit due to a minor notation/substitution error. Consider requiring these formulas during the school year and then relaxing this expectation at the end of the course when preparing for the AP® exam.

We chose $\alpha = 0.01$ in the example to reduce the chance of making a Type I error—finding convincing evidence that gemfibrozil reduces heart attack risk when it really doesn't. This error could have serious consequences if an ineffective drug was given to lots of middle-aged men with high cholesterol!

The random assignment in the Helsinki Heart Study allowed researchers to draw a cause-and-effect conclusion. They could say that gemfibrozil reduces the rate of heart attacks for middle-aged men like those who took part in the experiment. Because the subjects were not randomly selected from a larger population, researchers could not generalize the findings of this study any further. No conclusions could be drawn about the effectiveness of gemfibrozil at preventing heart attacks for all middle-aged men, for men of other ages, or for women.

Think About It

WHY DO THE INFERENCE METHODS FOR RANDOM SAMPLING WORK FOR RANDOMIZED EXPERIMENTS? Confidence intervals and tests for $p_1 - p_2$ are based on the sampling distribution of $\hat{p}_1 - \hat{p}_2$. But in most experiments, researchers don't select subjects at random from any larger populations. They do randomly assign subjects to treatments. We can think about what would happen if the random assignment were repeated many times under the assumption that $H_0: p_1 - p_2 = 0$ is true. That is, we assume that the specific treatment received doesn't affect an individual subject's response.

Let's see what would happen just by chance if we randomly reassign the 4081 subjects in the Helsinki Heart Study to the two groups many times, assuming the drug received *doesn't affect* whether or not each individual has a heart attack. We used software to redo the random assignment 500 times. Figure 9.8 shows the value of $\hat{p}_G - \hat{p}_{PL}$ in the 500 simulated trials. This distribution (sometimes referred to as the *randomization distribution* of $\hat{p}_G - \hat{p}_{PL}$) has an approximately Normal shape with mean 0 and standard deviation 0.0058. This matches well with the distribution we used to perform calculations in the example. Because the Large Counts condition was met and we assumed that $H_0: p_G - p_{PL} = 0$ is true, we used

$$\sqrt{\frac{\hat{p}_C(1 - \hat{p}_C)}{n_1} + \frac{\hat{p}_C(1 - \hat{p}_C)}{n_2}} = \sqrt{\frac{0.0343(1 - 0.0343)}{2051} + \frac{0.0343(1 - 0.0343)}{2030}}$$
$$= 0.0057$$

for the standard error of the statistic.

In the Helsinki Heart Study, the difference in the proportions of subjects who had a heart attack in the gemfibrozil and placebo groups was $0.0273 - 0.0414 = -0.0141$. How likely is it that a difference this large or larger would happen just by chance when H_0 is true? Figure 9.8 provides a rough answer: 5 of the 500 random reassignments yielded a difference in proportions less than or equal to -0.0141. That is, our estimate of the P-value is 0.01. This is quite close to the 0.0068 P-value that we calculated in the preceding example, suggesting that it's OK to use inference methods for random sampling to analyze randomized experiments. Neither of these values is the exact P-value. To get the exact P-value in an experiment, you would have to consider the difference in sample proportions for *all* possible random assignments. This approach is called a *permutation test*.

Teaching Tip

The above claim that "the drug received *doesn't affect* whether or not each individual has a heart attack" is our null hypothesis from the significance test. Assuming this claim is true, a subject who had a heart attack would have the heart attack whether randomly assigned to the gemfibrozil or placebo group.

Teaching Tip: Using Technology

To perform the simulation in Figure 9.8, go to the Extra Applets on the Student Site at highschool .bfwpub.com/updatedtps6e and use the *One Categorical Variable* applet with multiple groups. Enter the data from the gemfibrozil experiment and then "Simulate difference in proportions."

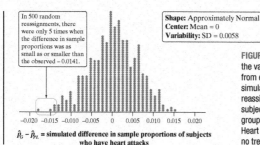

In 500 random reassignments, there were only 5 times when the difference in sample proportions was as small as or smaller than the observed − 0.0141.

Shape: Approximately Normal
Center: Mean = 0
Variability: SD = 0.0058

$\hat{p}_G - \hat{p}_{PL}$ = simulated difference in sample proportions of subjects who have heart attacks

FIGURE 9.8 Dotplot of the values of $\hat{p}_G - \hat{p}_{PL}$ from each of 500 simulated random reassignments of subjects to treatment groups in the Helsinki Heart Study, assuming no treatment effect.

CHECK YOUR UNDERSTANDING

To study the long-term effects of preschool programs for poor children, researchers designed an experiment. They recruited 123 children who had never attended preschool from low-income families in Michigan. Researchers randomly assigned 62 of the children to attend preschool (paid for by the study budget) and the other 61 to serve as a control group who would not go to preschool. One response variable of interest was the need for social services as adults. Over a 10-year period, 38 children in the preschool group and 49 in the control group have needed social services.[18]

1. Do these data provide convincing evidence that preschool reduces the later need for social services for children like the ones in this study? Justify your answer.

2. Based on your conclusion to Question 1, could you have made a Type I error or a Type II error? Explain your reasoning.

3. Should you generalize the result in Question 1 to all children from low-income families who have never attended preschool? Why or why not?

Section 9.3 | Summary

- Significance tests to compare the proportions p_1 and p_2 of successes for two populations or treatments are based on the difference $\hat{p}_1 - \hat{p}_2$ between the sample proportions.

- Significance tests of H_0: $p_1 - p_2 = 0$ use the *combined (pooled) sample proportion* $\hat{p}_C$ to estimate the common value p of parameters p_1 and p_2.

$$\hat{p}_C = \frac{\text{count of successes in both samples combined}}{\text{count of individuals in both samples combined}} = \frac{X_1 + X_2}{n_1 + n_2}$$

- When testing a claim about a difference in proportions, we must check for independence in the data collection process and that the sampling distribution of $\hat{p}_1 - \hat{p}_2$ is approximately Normal. The required conditions are:
  - **Random:** The data come from two independent random samples or from two groups in a randomized experiment.
    - **10%:** When sampling without replacement, $n_1 < 0.10N_1$ and $n_2 < 0.10N_2$.
  - **Large Counts:** The expected counts of "successes" and "failures" in each sample or group— $n_1\hat{p}_C$, $n_1(1-\hat{p}_C)$, $n_2\hat{p}_C$, $n_2(1-\hat{p}_C)$ —are all at least 10.

TRM Section 9.3 Quiz

There is one quiz available for this section in the Teacher's Resource Materials. Click on the link in the TE-Book, open in the TRFD, or download from the Teacher's Resources on the book's digital platform. You can also create your own quiz using the ExamView® Assessment Suite that is part of the TPS6 program. Questions are coded by Learning Target to make it easy to build parallel quizzes.

✓ Answers to CYU

1. STATE: H_0: $p_1 - p_2 = 0$, H_a: $p_1 - p_2 < 0$, where p_1 = the true proportion of children like the ones in the study who attend preschool and use social services later and p_2 = the true proportion of children like the ones in the study who do not attend preschool and use social services later using $\alpha = 0.05$.

PLAN: Two-sample z test for $p_1 - p_2$.
- *Random:* The children were randomly assigned to attend or not attend preschool. ✓
- Large Counts: 38, 62 − 38 = 24, 49, and 61 − 49 = 12 are all ≥ 10. ✓

DO: $\hat{p}_1 = \dfrac{38}{62} = 0.6129$, $\hat{p}_2 = \dfrac{49}{61} = 0.8033$,

and $\hat{p} = \dfrac{38 + 49}{62 + 61} = \dfrac{87}{123} = 0.7073$.

$$z = \frac{(0.6129 - 0.8033) - 0}{\sqrt{\dfrac{0.7073(0.2927)}{62} + \dfrac{0.7073(0.2927)}{61}}}$$

$$= -2.32$$

- *P*-value

Table A: $P(z \leq -2.32) = 0.0102$

Tech: normalcdf(lower: −1000, upper: −2.32, mean: 0, SD: 1) = 0.0102

CONCLUDE: Because the *P*-value of 0.0102 < α = 0.005, we reject H_0. There is convincing evidence that the true proportion of children like the ones in the study who do attend preschool and use social services later is less than the true proportion of children like the ones in the study who do not attend preschool and use social services later. In other words, children like those in this study who participate in preschool are less likely to use social services later in life.

2. Because we rejected H_0, it is possible that we made a Type I error: finding convincing evidence that the true proportion of children like the ones in the study who attend preschool and use social services later is less than the true proportion of children like the ones in the study who do not attend preschool and use social services later, when in reality the true proportions are equal.

3. No; the children who participated in the study were recruited from low-income families in Michigan, not randomly selected, so the results cannot be generalized to all children from low-income families.

- When conditions are met, the **two-sample z test for $p_1 - p_2$** uses the standardized test statistic

$$z = \frac{(\hat{p}_1 - \hat{p}_2) - 0}{\sqrt{\dfrac{\hat{p}_C(1 - \hat{p}_C)}{n_1} + \dfrac{\hat{p}_C(1 - \hat{p}_C)}{n_2}}}$$

with P-values calculated from the standard Normal distribution.

- Be sure to follow the four-step process whenever you perform a significance test about a difference in proportions.

9.3 Technology Corner

TI-Nspire and other technology instructions are on the book's website at highschool.bfwpub.com/updatedtps6e.

21. Performing a significance test for a difference in proportions   Page 631

Answers to Section 9.3 Exercises

9.77 H_0: $p_1 - p_2 = 0$, H_a: $p_1 - p_2 < 0$, where p_1 = the true proportion of 4- to 5-year-olds who would sort correctly and p_2 = the true proportion of 6- to 7-year-olds who would sort correctly.

9.78 H_0: $p_1 - p_2 = 0$, H_a: $p_1 - p_2 \neq 0$, where p_1 = the true proportion of high school freshmen in Illinois who have used anabolic steroids and p_2 = the true proportion of high school seniors in Illinois who have used anabolic steroids.

9.79 *Random:* Independent random samples. *10%:* 50 < 10% of all 4- to 5-year-olds and 53 < 10% of all 6- to 7-year olds. *Large Counts:* $\hat{p}_C = \dfrac{10 + 28}{50 + 53} = 0.369$; 18.45, 31.55, 19.56, 33.44 are all $\geq$ 10.

9.80 *Random:* Independent random samples. *10%:* 1679 < 10% of all high school freshmen in Illinois and 1366 < 10% of all high school seniors in Illinois. *Large Counts:* $\hat{p}_C = \dfrac{34 + 24}{1679 + 1366} = 0.019$; 31.9, 1647.1, 25.95, 1340.05 are $\geq$ 10.

9.81 (a) H_0: $p_1 - p_2 = 0$, H_a: $p_1 - p_2 > 0$, where p_1 = true proportion of shrubs that would resprout after being clipped and burned and p_2 = . . . after being clipped. **(b)** *Random:* Random assignment. *Large Counts:* Not met because all are < 10.

9.82 (a) H_0: $p_1 - p_2 = 0$, H_a: $p_1 - p_2 > 0$, where p_1 = true proportion of mice that are in breeding condition in the area of abundant acorn crop and p_2 = . . . in the untouched area. **(b)** *Random:* Unknown. *Large Counts:* Not met because 4.777 < 10.

Section 9.3 | Exercises

77. Children make choices Many new products introduced into the market are targeted toward children. The choice behavior of children with regard to new products is of particular interest to companies that design marketing strategies for these products. As part of one study, randomly selected children in different age groups were compared on their ability to sort new products into the correct product category (milk or juice).[19] Here are some of the data:

Age group	N	Number who sorted correctly
4- to 5-year-olds	50	10
6- to 7-year-olds	53	28

Researchers want to know if a greater proportion of 6- to 7-year-olds can sort correctly than 4- to 5-year-olds. State appropriate hypotheses for performing a significance test. Be sure to define the parameters of interest.

78. Steroids in high school A study by the National Athletic Trainers Association surveyed random samples of 1679 high school freshmen and 1366 high school seniors in Illinois. Results showed that 34 of the freshmen and 24 of the seniors had used anabolic steroids. Steroids, which are dangerous, are sometimes used in an attempt to improve athletic performance.[20]

Researchers want to know if there is a difference in the proportion of all Illinois high school freshmen and seniors who have used anabolic steroids. State appropriate hypotheses for performing a significance test. Be sure to define the parameters of interest.

79. Children make choices Refer to Exercise 77. Check if the conditions for performing the test are met.

80. Steroids in high school Refer to Exercise 78. Check if the conditions for performing the test are met.

81. Shrubs and fire Fire is a serious threat to shrubs in dry climates. Some shrubs can resprout from their roots after their tops are destroyed. Researchers wondered if fire would help with resprouting. One study of resprouting took place in a dry area of Mexico.[21] The researchers randomly assigned shrubs to treatment and control groups. They clipped the tops of all the shrubs. They then applied a propane torch to the stumps of the treatment group to simulate a fire. All 12 of the shrubs in the treatment group resprouted. Only 8 of the 12 shrubs in the control group resprouted.

(a) State appropriate hypotheses for performing a significance test. Be sure to define the parameters of interest.

(b) Check if the conditions for performing the test are met.

9.83 (a) $\hat{p}_1 - \hat{p}_2 = -0.328 < 0$ **(b)** $z = -3.45$, P-value = 0.0003 **(c)** Because the P-value of $0.0003 < \alpha = 0.05$, we reject H_0. We have convincing evidence that the true proportion of 4- to 5-year-olds who would sort correctly is less than the true proportion of 6- to 7-year-olds who would do so.

9.84 (a) $\hat{p}_1 - \hat{p}_2 = 0.0027 \neq 0$ **(b)** $z = 0.54$, P-value = 0.5904 **(c)** Because the P-value of $0.5904 > \alpha = 0.05$, we fail to reject H_0. We do not have convincing evidence that the true proportion of high school freshmen in Illinois who have used anabolic steroids differs from the true proportion of high school seniors in Illinois who have.

9.85 (a) S: H_0: $p_1 - p_2 = 0$, H_a: $p_1 - p_2 < 0$, where p_1 = true proportion of sophomores who bring a bag lunch and p_2 = . . . seniors . . .

P: Two-sample z test for $p_1 - p_2$. *Random:* Independent random samples. *10%:* 80 < 10% of all sophomores and 104 < 10% of all seniors. *Large Counts:* 56.56, 23.44, 73.528, 30.472 are all $\geq$ 10. **D:** $z = -1.48$, P-value = 0.0699. **C:** Because the P-value of $0.0699 > \alpha = 0.05$, we fail to reject H_0. There is not convincing evidence that the true proportion of sophomores who bring a bag lunch is less than the true proportion of seniors who do. **(b)** If there is no difference in the true proportion of sophomores and seniors who bring a bag lunch, there is a 0.0699 probability of getting a difference in the proportions as large as or larger than the one observed $(0.65 - 0.75 = -0.10)$ by chance alone.

82. Ticks Lyme disease is spread in the northeastern United States by infected ticks. The ticks are infected mainly by feeding on mice, so more mice result in more infected ticks. The mouse population, in turn, rises and falls with the abundance of acorns, their favored food. Experimenters studied two similar forest areas in a year when the acorn crop failed. To see if mice are more likely to breed when there are more acorns, the researchers added hundreds of thousands of acorns to one area to imitate an abundant acorn crop, while leaving the other area untouched. The next spring, 54 of the 72 mice trapped in the first area were in breeding condition, versus 10 of the 17 mice trapped in the second area.[22]

(a) State appropriate hypotheses for performing a significance test. Be sure to define the parameters of interest.

(b) Check if the conditions for performing the test are met.

83. Children make choices Refer to Exercises 77 and 79.

pg 630 (a) Explain why the sample results give some evidence for the alternative hypothesis.

(b) Calculate the standardized test statistic and P-value.

(c) What conclusion would you make?

84. Steroids in high school Refer to Exercises 78 and 80.

(a) Explain why the sample results give some evidence for the alternative hypothesis.

(b) Calculate the standardized test statistic and P-value.

(c) What conclusion would you make?

85. Bag lunch? Phoebe has a hunch that older students
pg 632 at her very large high school are more likely to bring a bag lunch than younger students because they have grown tired of cafeteria food. She takes a simple random sample of 80 sophomores and finds that 52 of them bring a bag lunch. A simple random sample of 104 seniors reveals that 78 of them bring a bag lunch.

(a) Do these data give convincing evidence to support Phoebe's hunch at the $\alpha = 0.05$ significance level?

(b) Interpret the P-value from part (a) in the context of this study.

86. Are teenagers going deaf? In a study of 3000 randomly selected teenagers in 1990, 450 showed some hearing loss. In a similar study of 1800 teenagers reported in 2010, 351 showed some hearing loss.[23]

(a) Do these data give convincing evidence that the proportion of all teens with hearing loss has increased at the $\alpha = 0.01$ significance level?

(b) Interpret the P-value from part (a) in the context of this study.

87. Preventing peanut allergies A recent study of peanut allergies—the LEAP trial—explored whether early exposure to peanuts helps or hurts subsequent development of an allergy to peanuts. Infants (4 to 11 months old) who had shown evidence of other kinds of allergies were randomly assigned to one of two groups. Group 1 consumed a baby-food form of peanut butter. Group 2 avoided peanut butter. At 5 years old, 10 of 307 children in the peanut-consumption group were allergic to peanuts, and 55 of 321 children in the peanut-avoidance group were allergic to peanuts.[24]

(a) Does this study provide convincing evidence of a difference at the $\alpha = 0.05$ significance level in the development of peanut allergies in infants like the ones in this study who consume or avoid peanut butter?

(b) Based on your conclusion in part (a), which mistake—a Type I error or a Type II error—could you have made? Explain your answer.

(c) Should you generalize the result in part (a) to all infants? Why or why not?

(d) A 95% confidence interval for $p_1 - p_2$ is $(-0.185, -0.093)$. Explain how the confidence interval provides more information than the test in part (a).

88. Lowering bad cholesterol Which of two widely prescribed drugs—Lipitor or Pravachol—helps lower "bad cholesterol" more? In an experiment, called the PROVE-IT Study, researchers recruited about 4000 people with heart disease as subjects. These volunteers were randomly assigned to one of two treatment groups: Lipitor or Pravachol. At the end of the study, researchers compared the proportion of subjects in each group who died, had a heart attack, or suffered other serious consequences within two years. For the 2063 subjects using Pravachol, the proportion was 0.263. For the 2099 subjects using Lipitor, the proportion was 0.224.[25]

(a) Does this study provide convincing evidence at the $\alpha = 0.05$ significance level of a difference in the effectiveness of Lipitor and Pravachol for people like the ones in this study?

(b) Based on your conclusion in part (a), which mistake—a Type I error or a Type II error—could you have made? Explain your answer.

(c) Should you generalize the result in part (a) to all people with heart disease? Why or why not?

(d) A 95% confidence interval for $p_{PR} - p_L$ is $(0.013, 0.065)$. Explain how the confidence interval provides more information than the test in part (a).

89. Preventing peanut allergies Refer to Exercise 87. Explain how each of the following changes to the

peanuts at age 5. **(b)** Type I error **(c)** No; when subjects are not randomly selected, we should not generalize the results of an experiment to some larger population. **(d)** The confidence interval does not include 0 as a plausible value for $p_1 - p_2$, which is consistent with the decision to reject H_0: $p_1 - p_2 = 0$ in part (a). The confidence interval tells us that any value between -0.185 and -0.093 is plausible for $p_1 - p_2$ based on the sample data.

9.88 (a) S: H_0: $p_1 - p_2 = 0$, H_a: $p_1 - p_2 \neq 0$, where p_1 = true proportion of subjects like these who would die, have a heart attack, or suffer other serious consequences within 2 years if they took Pravachol and p_2 = true proportion who would suffer serious consequences if they took Lipitor; $\alpha = 0.05$. **P:** Two-sample z test for $p_1 - p_2$. *Random:* Volunteers were randomly assigned to the treatments. *Large Counts:* 510.057, 1588.943, 501.309, 1561.691 are ≥ 10. **D:** $z = -2.95$; P-value $= 0.0031$. **C:** Because the P-value of $0.0031 < \alpha = 0.05$, we reject H_0. There is convincing evidence the true proportion of subjects like these who would die, have a heart attack, or suffer other serious consequences within 2 years if they took Pravachol differs from the true proportion of subjects like these who would die, have a heart attack, or suffer other serious consequences within 2 years if they took Lipitor. **(b)** Because we rejected H_0, it is possible we made a Type I error. **(c)** No; this experiment recruited volunteers as subjects. We should not generalize to some larger population of interest. **(d)** The confidence interval does not include 0 as a plausible value for $p_1 - p_2$, which is consistent with the decision to reject H_0: $p_1 - p_2 = 0$ in part (a). The confidence interval tells us that any value between 0.013 and 0.065 is plausible for $p_1 - p_2$ based on the sample data.

9.86 (a) S: H_0: $p_1 - p_2 = 0$, H_a: $p_1 - p_2 < 0$, where p_1 = true proportion of teenagers in 1990 with some hearing loss and p_2 = . . . in 2010 . . . **P:** Two-sample z test for $p_1 - p_2$. *Random:* Independent random samples. *10%:* 3000 < 10% of all teenagers in 1990 and 1800 < 10% of all teenagers in 2010. *Large Counts:* 501, 2499, 300.6, 1499.4 ≥ 10. **D:** $z = -4.05$, P-value ≈ 0. **C:** Because the P-value of approximately $0 < \alpha = 0.05$, we reject H_0. There is convincing evidence that the true proportion of teenagers in 1990 with some hearing loss is less than the true proportion of teenagers in 2010 with some hearing loss. **(b)** If there is no difference in the true proportion of teenagers in 1990 and 2010 with some hearing loss, there is about a 0 probability of getting a difference in the proportions as large as or larger than the one observed $(0.15 - 0.195 = -0.045)$ by chance alone.

9.87 (a) S: H_0: $p_1 - p_2 = 0$, H_a: $p_1 - p_2 \neq 0$, where p_1 = true proportion of children like the ones in this study who are exposed to peanut butter as infants who are allergic to peanuts at age 5 and p_2 = . . . not exposed . . . **P:** Two-sample z test for $p_1 - p_2$. *Random:* Random assignment. *Large Counts:* 31.928, 275.072, 33.384, 287.616 ≥ 10. **D:** $z = -5.71$, P-value ≈ 0. **C:** Because the P-value of approximately $0 < \alpha = 0.05$, we reject H_0. There is convincing evidence that the true proportion of children like the ones in this study who are exposed to peanut butter as infants who are allergic to peanuts at age 5 differs from the true proportion of children like the ones in this study who are not exposed to peanut butter as infants who are allergic to

9.89 (a) Increasing the sample size will increase the power of the test. Increasing the sample size decreases the variability of both the null and alternative distributions, making it easier to reject the null hypothesis when it is false. A drawback is that an experiment that uses twice as many infants will be more expensive and will require a lot more work in following up with the parents of these infants when they are 5 years old. **(b)** Using $\alpha = 0.10$ instead of $\alpha = 0.05$ will increase the power of the test. When α is larger, it is easier to reject the null hypothesis because the P-value doesn't need to be as small. A drawback to increasing α is that doing so increases the probability of making a Type I error. **(c)** Using male infants only would increase the power of the test by eliminating a source of variability, making it easier to reject the null hypothesis when it is false. A drawback is that this also limits the scope of inference to males only.

9.90 (a) Increasing the sample size will increase the power of the test. Increasing the sample size decreases the variability of both the null and alternative distributions, making it easier to reject the null hypothesis when it is false. A drawback is that an experiment that uses twice as many subjects will be more expensive and will require a lot more work in following up with the subjects at the end of the 2-year long experiment. **(b)** Using $\alpha = 0.10$ instead of $\alpha = 0.05$ will increase the power of the test. When α is larger, it is easier to reject the null hypothesis because the P-value doesn't need to be as small. A drawback to increasing α is that doing so increases the probability of making a Type I error. **(c)** Using only subjects under the age of 60 would increase the power of the test by eliminating a source of variability, making it easier to reject the null hypothesis when it is false. A drawback is that this also limits the scope of inference to those under the age of 60.

9.91 (a) Two-sample z test for $p_1 - p_2$. *Random:* Two groups in a randomized experiment. *Large Counts:* 33.88, 54.12, 31.185, 49.815 are $\geq$ 10. **(b)** If there is no difference in the true pregnancy rates of women prayed for and those who are not, there is a 0.0007 probability of getting a difference in pregnancy rates as large as or larger than the one observed $(0.500 - 0.259 = 0.241)$ by chance alone. **(c)** Because the P-value of 0.0007 is less than $\alpha = 0.05$, we reject H_0. There is convincing evidence the pregnancy rates among women like these who are prayed for is higher than

design of the experiment would affect the power of the test. Then give a drawback of making that change.

(a) Suppose that researchers had recruited twice as many infants for the LEAP trial.

(b) Suppose that researchers had used $\alpha = 0.10$ instead of $\alpha = 0.05$.

(c) Suppose that researchers had used 628 male subjects but no female subjects in the study.

90. **Lowering bad cholesterol** Refer to Exercise 88. Explain how each of the following changes to the design of the experiment would affect the power of the test. Then give a drawback of making that change.

(a) Suppose that researchers had recruited twice as many subjects for the PROVE-IT Study.

(b) Suppose that researchers had used $\alpha = 0.10$ instead of $\alpha = 0.05$.

(c) Suppose that researchers had used 4162 subjects under age 60 but no older subjects in the study.

Exercises 91 and 92 involve the following setting. Some women would like to have children but cannot do so for medical reasons. One option for these women is a procedure called in vitro fertilization (IVF), which involves fertilizing an egg outside the woman's body and implanting it in her uterus.

91. **Prayer and pregnancy** Two hundred women who were about to undergo IVF served as subjects in an experiment. Each subject was randomly assigned to either a treatment group or a control group. Several people (called intercessors) prayed intentionally for the women in the treatment group, although they did not know the women, a process known as intercessory prayer. Prayers continued for 3 weeks following IVF. The intercessors did not pray for the women in the control group. Here are the results: 44 of the 88 women in the treatment group got pregnant, compared to 21 out of 81 in the control group.[26]

Is the pregnancy rate significantly higher for women who received intercessory prayer? To find out, researchers perform a test of $H_0: p_1 = p_2$ versus $H_a: p_1 > p_2$, where p_1 and p_2 are the actual pregnancy rates for women like those in the study who do and don't receive intercessory prayer, respectively.

(a) Name the appropriate test and check that the conditions for carrying out this test are met.

(b) The appropriate test from part (a) yields a P-value of 0.0007. Interpret this P-value in context.

(c) What conclusion should researchers draw at the $\alpha = 0.05$ significance level?

(d) The women in the study did not know whether they were being prayed for. Explain why this is important.

92. **Acupuncture and pregnancy** A study reported in the medical journal *Fertility and Sterility* sought to determine whether the ancient Chinese art of acupuncture could help infertile women become pregnant.[27] A total of 160 healthy women who planned to have IVF were recruited for the study. Half of the subjects (80) were randomly assigned to receive acupuncture 25 minutes before implanting the embryo and again 25 minutes after the implant. The remaining 80 women were assigned to a control group and instructed to lie still for 25 minutes after the embryo transfer. Results are shown in the table.

	Acupuncture group	Control group
Pregnant	34	21
Not pregnant	46	59
Total	80	80

Is the pregnancy rate significantly higher for women who received acupuncture? To find out, researchers perform a test of $H_0: p_1 = p_2$ versus $H_a: p_1 > p_2$, where p_1 and p_2 are the actual pregnancy rates for women like those in the study who do and don't receive acupuncture, respectively.

(a) Name the appropriate test and check that the conditions for carrying out this test are met.

(b) The appropriate test from part (a) yields a P-value of 0.0152. Interpret this P-value in context.

(c) What conclusion should researchers draw at the $\alpha = 0.05$ significance level?

(d) The women in the study knew whether or not they received acupuncture. Explain why this is important.

93. **Texting and driving** Does providing additional information affect responses to a survey question? Two statistics students decided to investigate this issue by asking different versions of a question about texting and driving. Fifty mall shoppers were divided into two groups of 25 at random. The first group was asked version A and the other half were asked version B. The students believed that version A would result in more "Yes" answers. Here are the actual questions:

- *Version A:* A lot of people text and drive. Are you one of them?
- *Version B:* About 6000 deaths occur per year due to texting and driving. Knowing the potential consequences, do you text and drive?

Of the 25 shoppers assigned to version A, 18 admitted to texting and driving. Of the 25 shoppers assigned to version B, only 14 admitted to texting and driving.

the rates for those not prayed for. **(d)** Knowing might have affected their behavior in some way (even unconsciously) that would have affected if they became pregnant. Then we wouldn't know if the prayer or other behaviors caused the higher pregnancy rate.

9.92 (a) Two-sample z test for $p_1 - p_2$. *Random:* Two groups in a randomized experiment. *Large Counts:* 27.52, 52.48, 27.52, 54.12 are $\geq$ 10. **(b)** If there is no difference in pregnancy rates of women who receive acupuncture and those who don't, there is a 0.0152 probability of getting a difference in rates as large as or larger than the one observed $(0.425 - 0.2625 = 0.1625)$ by chance alone. **(c)** Because the P-value of $0.0152 < \alpha = 0.05$, we reject H_0. We have convincing evidence that the pregnancy rate among women like these who receive

acupuncture is higher than the rate for those who do not. **(d)** Knowing whether or not they received acupuncture might have affected the women's behavior in some way (even unconsciously) that would have affected if they became pregnant. Then we wouldn't know if the acupuncture or other behaviors caused the higher rate.

(a) State appropriate hypotheses for performing a significance test. Be sure to define the parameters of interest.

(b) Explain why you should not use the methods of this section to calculate the *P*-value.

(c) We performed 100 trials of a simulation to see what differences (Version A – Version B) in proportions would occur due only to chance variation in the random assignment, assuming that the question asked doesn't matter. A dotplot of the results is shown here. What is the estimated *P*-value?

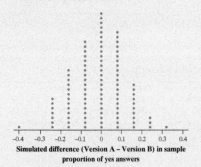

Simulated difference (Version A – Version B) in sample proportion of yes answers

(d) What conclusion would you draw?

94. **Botox benefits?** You may have heard that Botox (botulinum toxin type A) is used for cosmetic surgery, but could it have other beneficial uses? A total of 31 patients who suffered chronic low-back pain were randomly assigned to receive 200 units of either Botox or saline solution through 5 injections at 5 different locations in their backs. The saline injection was not expected to reduce pain but was given as a placebo treatment. Pain relief was defined as a patient's pain level being reduced to less than half of the original pain level after 8 weeks. Here are the results:

	Botox	Saline	Total
Pain relief	10	3	13
No pain relief	5	13	18
Total	15	16	31

(a) State appropriate hypotheses for performing a significance test. Be sure to define the parameters of interest.

(b) Explain why you should not use the methods of this section to calculate the *P*-value.

(c) We performed 100 trials of a simulation to see what differences in proportions would occur due only to chance variation in the random assignment, assuming

that the type of injection (Botox or saline) doesn't matter. A dotplot of the results is shown here. What is the estimated *P*-value?

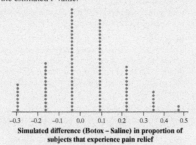

Simulated difference (Botox – Saline) in proportion of subjects that experience pain relief

(d) What conclusion would you draw?

Multiple Choice: *Select the best answer for Exercises 95–98.*

Exercises 95–97 refer to the following setting. A sample survey interviews SRSs of 500 female college students and 550 male college students. Researchers want to determine whether there is a difference in the proportion of male and female college students who worked for pay last summer. In all, 410 of the females and 484 of the males say they worked for pay last summer.

95. Let p_M and p_F be the proportions of all college males and females who worked last summer. The hypotheses to be tested are

(a) $H_0: p_M - p_F = 0$ versus $H_a: p_M - p_F \neq 0$.

(b) $H_0: p_M - p_F = 0$ versus $H_a: p_M - p_F > 0$.

(c) $H_0: p_M - p_F = 0$ versus $H_a: p_M - p_F < 0$.

(d) $H_0: p_M - p_F > 0$ versus $H_a: p_M - p_F = 0$.

(e) $H_0: p_M - p_F \neq 0$ versus $H_a: p_M - p_F = 0$.

96. The researchers report that the results were statistically significant at the 1% level. Which of the following is the most appropriate conclusion?

(a) Because the *P*-value is less than 1%, fail to reject H_0. There is not convincing evidence that the proportion of male college students in the study who worked for pay last summer is different from the proportion of female college students in the study who worked for pay last summer.

(b) Because the *P*-value is less than 1%, fail to reject H_0. There is not convincing evidence that the proportion of all male college students who worked for pay last summer is different from the proportion of all female college students who worked for pay last summer.

9.93 (a) $H_0: p_1 - p_2 = 0$, $H_a: p_1 - p_2 \neq 0$, where p_1 = true proportion of subjects like these who would admit to texting and driving when asked Version A and p_2 = true proportion of subjects like these who would admit it when asked Version B. **(b)** Conditions not satisfied; 9 is less than 10. **(c)** $\hat{p}_1 - \hat{p}_2 = 0.72 - 0.56 = 0.16$; there are 14 trials of the simulation with a difference of proportions greater than or equal to 0.16. Estimated *P*-value = 14/100= 0.14. **(d)** Because the estimated *P*-value of 0.14 > α = 0.05, we fail to reject H_0. We do not have convincing evidence the true proportion of subjects like these who would admit to texting and driving when asked Version A of the question is greater than the true proportion of subjects like these who would admit to texting and driving when asked Version B of the question.

9.94 (a) $H_0: p_1 - p_2 = 0$, $H_a: p_1 - p_2 > 0$, where p_1 = true proportion of subjects like these who would experience pain relief after Botox and p_2 = true proportion who experience relief with saline. **(b)** Conditions not satisfied; all are < 10. **(c)** $\hat{p}_1 - \hat{p}_2 = 0.67 - 0.19 = 0.48$; 2 trials of the simulation have a difference of proportions greater than or equal to 0.48. Estimated *P*-value = 2/100 = 0.02. **(d)** Because the estimated *P*-value of 0.02 < α = 0.05, we reject H_0. We have convincing evidence that the true proportion of subjects like these who would experience pain relief after Botox is greater than the true proportion of subjects like these who would experience pain relief after saline.

9.95 a

9.96 e

9.97 b

9.98 e

9.99 (a) $\hat{y} = -13,832 + 14,954x$, where $\hat{y}$ = the predicted mileage and x = the age in years of the cars. **(b)** The predicted number of miles goes up by 14,954 for each increase of 1 year in age.
(c) $\hat{y} = -13,832 + 14,954(10) = 135,708$; the residual = $110,000 - 135,708 = -25,708$ miles. The student's car had 25,708 fewer miles than predicted, based on its age.

9.100 (a) 77% of the variation in mileage is accounted for by the least-squares regression line with x = age (years).
(b) $\overline{y} = -13,832 + 14,954\overline{x} = 13,832 + 14,954(5) = 60,938$ miles. **(c)** The actual mileage is typically about 22,723 miles away from the amount predicted by the least-squares regression line with x = age (years). **(d)** No, it would not be reasonable to use the least-squares line to predict a car's mileage from its age for a teacher. The least-squares line is based on a sample of cars owned and driven by students, not teachers.

(c) Because the P-value is less than 1%, reject H_0. There is convincing evidence that the proportion of all male college students who worked for pay last summer is the same as the proportion of all female college students who worked for pay last summer.

(d) Because the P-value is less than 1%, reject H_0. There is convincing evidence that the proportion of all male college students in the study who worked for pay last summer is different from the proportion of all female college students in the study who worked for pay last summer.

(e) Because the P-value is less than 1%, reject H_0. There is convincing evidence that the proportion of all male college students who worked for pay last summer is different from the proportion of all female college students who worked for pay last summer.

97. Which of the following is the correct standard error for a test of the hypotheses in Exercise 95?

(a) $\sqrt{\dfrac{0.851(0.149)}{1050}}$

(b) $\sqrt{\dfrac{0.851(0.149)}{550} + \dfrac{0.851(0.149)}{500}}$

(c) $\sqrt{\dfrac{0.880(0.120)}{550} + \dfrac{0.820(0.180)}{500}}$

(d) $\sqrt{\dfrac{0.851(0.149)}{1050} + \dfrac{0.851(0.149)}{1050}}$

(e) $\sqrt{\dfrac{0.880(0.120)}{1050} + \dfrac{0.820(0.180)}{1050}}$

98. In an experiment to learn whether substance M can help restore memory, the brains of 20 rats were treated to damage their memories. First, the rats were trained to run a maze. After a day, 10 rats (determined at random) were given substance M and 7 of them succeeded in the maze. Only 2 of the 10 control rats were successful. The two-sample z test for the difference in the true proportions

(a) gives $z = 2.25, P < 0.02$.

(b) gives $z = 2.60, P < 0.005$.

(c) gives $z = 2.25, P < 0.04$ but not < 0.02.

(d) should not be used because the Random condition is violated.

(e) should not be used because the Large Counts condition is violated.

Recycle and Review

Exercises 99 and 100 refer to the following setting. Thirty randomly selected seniors at Council High School were asked to report the age (in years) and mileage of their main vehicles. Here is a scatterplot of the data:

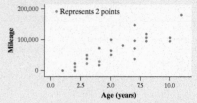

We used Minitab to perform a least-squares regression analysis for these data. Part of the computer output from this regression is shown here.

Predictor	Coef	Stdev	t-ratio	P
Constant	−13832	8773	−1.58	0.126
Age	14954	1546	9.67	0.000
s = 22723	R-sq = 77.0%		R-sq(adj) = 76.1%	

99. **Drive my car** (3.2)

(a) What is the equation of the least-squares regression line? Be sure to define any symbols you use.

(b) Interpret the slope of the least-squares line.

(c) One student reported that her 10-year-old car had 110,000 miles on it. Find and interpret the residual for this data point.

100. **Drive my car** (3.2, 4.3)

(a) Explain what the value of r^2 tells you about how well the least-squares line fits the data.

(b) The mean age of the students' cars in the sample was $\overline{x} = 5$ years. Find the mean mileage of the cars in the sample.

(c) Interpret the value of s.

(d) Would it be reasonable to use the least-squares line to predict a car's mileage from its age for a Council High School teacher? Justify your answer.

Chapter 9 Wrap-Up

FRAPPY! FREE RESPONSE AP® PROBLEM, YAY!

The following problem is modeled after actual AP® Statistics exam free response questions. Your task is to generate a complete, concise response in 15 minutes.

Directions: Show all your work. Indicate clearly the methods you use, because you will be scored on the correctness of your methods as well as on the accuracy and completeness of your results and explanations.

Do employer-sponsored wellness programs work? About 5000 employees at the University of Illinois at Urbana-Champaign volunteered for a study to find out. Researchers randomly assigned 3300 volunteers to a treatment group and the remaining 1534 to a control group. Employees in the treatment group were invited to take paid time off to participate in a wellness program. Those in the control group were not allowed to participate. One measure of the program's effectiveness was employee attrition.

In the 2-year period following the start of the study, 356 of the people in the treatment group left their job for any reason, compared to 184 of the people in the control group.

Do these results provide convincing evidence at the $\alpha = 0.05$ level that offering employees paid time off to participate in a wellness program reduces the proportion who leave their job within 2 years for people similar to the ones in this study?

After you finish, you can view two example solutions on the book's website (highschool.bfwpub.com/updatedtps6e). Determine whether you think each solution is "complete," "substantial," "developing," or "minimal." If the solution is not complete, what improvements would you suggest to the student who wrote it? Finally, your teacher will provide you with a scoring rubric. Score your response and note what, if anything, you would do differently to improve your own score.

Chapter 9 Review

Section 9.1: Significance Tests: The Basics

In this section, you learned the basic ideas of significance testing. Start by stating the hypotheses that you want to test. The null hypothesis (H_0) is typically a statement of "no difference" and the alternative hypothesis (H_a) describes what we suspect is true. Remember that hypotheses are always about population parameters, not sample statistics.

When sample data provide evidence for the alternative hypothesis, there are two possible explanations: (1) The null hypothesis is true, and data supporting the alternative hypothesis occurred just by chance, or (2) the alternative hypothesis is true, and the data are consistent with an alternative value of the parameter. In a significance test, always start with the belief that the null hypothesis is true. If you can rule out chance as a plausible explanation for the observed data, there is *convincing* evidence that the alternative hypothesis is true.

The *P*-value in a significance test measures how likely it is to get evidence for H_a as strong or stronger than the observed evidence, assuming the null hypothesis is true. To determine if the *P*-value is small enough to reject H_0, compare it to a predetermined significance level such as $\alpha = 0.05$. If *P*-value < α, reject H_0—there is convincing evidence that the alternative hypothesis is true. However, if *P*-value > α, fail to reject H_0—there is not convincing evidence that the alternative hypothesis is true.

Because conclusions are based on sample data, there is a possibility that the conclusion will be incorrect. You can make two types of errors in a significance test: a Type I error occurs if you find convincing evidence for the alternative hypothesis when, in reality, the null hypothesis is true. A Type II error occurs when you don't find convincing evidence that the alternative hypothesis is true when, in reality,

641

TRM FRAPPY! Materials

Please consult the Teacher's Resource Materials for sample student responses, a scoring rubric, and a printable version of the original question with space for students to write their responses. We present a model solution here.

TRM Chapter 9 Test

There is one Chapter Test available for this section in the Teacher's Resource Materials. Click on the link in the TE-Book, open in the TRFD, or download from the Teacher's Resources on the book's digital platform. You can also create your own Test using the TPS quiz and test builder (ExamView).

Answers:

STATE: We want to test

$$H_0: p_T - p_{CN} = 0$$
$$H_a: p_T - p_{CN} < 0$$

where p_T = the true proportion of employees similar to the ones in this study who leave their job within two years after being invited to take paid time off to participate in a wellness program and p_{CN} = the true proportion of employees similar to the ones in this study who leave their job within two years after not being invited to participate in a wellness program, using $\alpha = 0.05$.

PLAN: Two-sample z test for a difference in proportions.

- *Random:* The volunteers were randomly assigned to the treatment and control groups.

- *Large Counts:* $\hat{p}_C = \dfrac{356 + 184}{3300 + 1534}$
  $= 0.112$;
  $n_T \hat{p}_C = 3300(0.112) = 369.6$,
  $n_T(1 - \hat{p}_C) = 3300(0.888) = 2930.4$,
  $n_{CN}\hat{p}_C = 1534(0.112) = 171.8$,
  $n_{CN}(1 - \hat{p}_C) = 1534(0.888) = 1362.2$
  are all ≥ 10.

DO:

- $\hat{p}_T = \dfrac{356}{3300} = 0.108$, $\hat{p}_{CN} = \dfrac{184}{1534} = 0.120$

- $z = \dfrac{(0.108 - 0.120) - 0}{\sqrt{\dfrac{(0.112)(0.888)}{3300} + \dfrac{(0.112)(0.888)}{1534}}}$
  $= -1.23$

- *P*-value: $P(z \leq -1.23) = 0.1093$

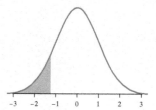

(Technology gives $z = -1.24$, *P*-value = 0.1075)

CONCLUDE: Because the *P*-value of 0.1075 is greater than $\alpha = 0.05$, we fail to reject the null hypothesis. We do not have convincing evidence that offering employees like these paid time off to participate in a wellness program reduces the proportion who leave their job within 2 years.

the alternative hypothesis is true. The probability of making a Type I error is equal to the significance level (α) of the test.

Section 9.2: Tests About a Population Proportion

In this section, you learned the details of performing a significance test about a population proportion p. Whenever you are asked if there is convincing evidence for a claim about a population proportion, you are expected to respond using the familiar four-step process.

STATE: Give the hypotheses you are testing in terms of p, define the parameter p, and state the significance level.

PLAN: Name the procedure you plan to use (one-sample z test for a population proportion) and check the appropriate conditions (Random, 10%, Large Counts) to see if the procedure is appropriate.

- Random: The data come from a random sample from the population of interest.
  - 10%: The sample size is less than 10% of the population when sampling without replacement.
- Large Counts: Both np_0 and $n(1 - p_0)$ are at least 10, where p_0 is the value of p in the null hypothesis.

DO: Calculate the standardized test statistic and P-value. The standardized test statistic z measures how far away the sample statistic is from the hypothesized parameter value in standardized units:

$$z = \frac{\hat{p} - p_0}{\sqrt{\dfrac{p_0(1 - p_0)}{n}}}$$

To calculate the P-value, use Table A or technology.

CONCLUDE: Use the P-value to make an appropriate conclusion about the hypotheses, in context.

Perform a two-sided test when looking for convincing evidence that the true value of the parameter is *different* from the hypothesized value, in either direction. The P-value for a two-sided test is calculated by finding the probability of getting a sample statistic at least as extreme as the observed statistic, in either direction, assuming the null hypothesis is true.

You can also use a confidence interval to make a conclusion for a two-sided test. If the null parameter value is one of the plausible values in the interval, there isn't convincing evidence that the alternative hypothesis is true. However, if the null parameter value is not one of the plausible values in the interval, there is convincing evidence that the alternative hypothesis is true. Besides helping you draw a conclusion, the interval tells you which alternative parameter values are plausible.

The probability that you avoid making a Type II error when an alternative value of the parameter is true is called the power of the test. Power is good—if the alternative hypothesis is true, we want to maximize the probability of finding convincing evidence that it is true. We can increase the power of a significance test by increasing the sample size or by increasing the significance level. The power of a test will also be greater when the alternative value of the parameter is farther away from the null hypothesis value.

Section 9.3: Tests About a Difference in Proportions

In this section, you learned how to perform significance tests for a difference between two proportions. As in any test, you start by stating hypotheses. The null hypothesis is usually $H_0: p_1 - p_2 = 0$ (or, equivalently, $H_0: p_1 = p_2$). The alternative hypothesis can be one-sided ($<$, $>$) or two-sided ($\neq$).

The conditions for significance tests about a difference in proportions are similar to the ones for confidence intervals. The Random condition says that the data must be from two independent random samples or two groups in a randomized experiment. The 10% condition says that each sample size should be less than 10% of the corresponding population size when sampling without replacement. The Large Counts condition says that the *expected* numbers of successes and failures in each sample/group should be at least 10. For a test of $H_0: p_1 - p_2 = 0$, we estimate the common value p of the parameters p_1 and p_2 using the combined (pooled) proportion of successes in the two samples/groups: $\hat{p}_C = \dfrac{X_1 + X_2}{n_1 + n_2}$. So the Large counts condition requires us to check that $n_1\hat{p}_C$, $n_1(1 - \hat{p}_C)$, $n_2\hat{p}_C$, and $n_2(1 - \hat{p}_C)$ are all at least 10.

For a test of $H_0: p_1 - p_2 = 0$, the standardized test statistic is

$$z = \frac{(\hat{p}_1 - \hat{p}_2) - 0}{\sqrt{\dfrac{\hat{p}_C(1 - \hat{p}_C)}{n_1} + \dfrac{\hat{p}_C(1 - \hat{p}_C)}{n_2}}}$$

When conditions are met, P-values can be obtained from the standard Normal distribution.

A significance test for a difference between two proportions uses the same logic as the significance test for one population proportion from Section 9.2. We start by assuming the null hypothesis is true and asking how likely it would be to get evidence for H_a as strong as or stronger than the observed result in a study by chance alone. If it is plausible that a difference in proportions could be the result of sampling variability or the chance variation due to random assignment, we do not have convincing evidence that the alternative hypothesis is true. However, if the difference is too big to attribute to chance, there is convincing evidence that the alternative hypothesis is true.

Teaching Tip

Now that you have completed Chapters 8 and 9 in this book, students can login to their College Board accounts and take the **Personal Progress Check for Unit 6**.

Comparing significance tests for proportions		
	Significance test for p	**Significance test for $p_1 - p_2$**
Name (TI-83/84)	One-sample z test for p (1-PropZTest)	Two-sample z test for $p_1 - p_2$ (2-PropZTest)
Null hypothesis	$H_0: p = p_0$	$H_0: p_1 - p_2 = 0$

Conditions	• **Random:** The data come from a random sample from the population of interest.
	  ○ **10%:** When sampling without replacement, $n < 0.10N$.
	• **Large Counts:** Both np_0 and $n(1-p_0)$ are at least 10. That is, the expected number of successes and the expected number of failures in the sample are both at least 10.
		• **Random:** The data come from two independent random samples or from two groups in a randomized experiment.
		  ○ **10%:** When sampling without replacement, $n_1 < 0.10N_1$ and $n_2 < 0.10N_2$.
		• **Large Counts:** The expected counts of successes and failures in each sample or group—$n_1\hat{p}_C$, $n_1(1-\hat{p}_C)$, $n_2\hat{p}_C$, $n_2(1-\hat{p}_C)$—are all at least 10, where $$\hat{p}_C = \frac{X_1 + X_2}{n_1 + n_2}.$$

Formula	$$z = \frac{\hat{p} - p_0}{\sqrt{\dfrac{p_0(1-p_0)}{n}}}$$ P-value from standard Normal distribution	$$z = \frac{(\hat{p}_1 - \hat{p}_2) - 0}{\sqrt{\dfrac{\hat{p}_C(1-\hat{p}_C)}{n_1} + \dfrac{\hat{p}_C(1-\hat{p}_C)}{n_2}}}$$ P-value from standard Normal distribution

What Did You Learn?

Learning Target	Section	Related Example on Page(s)	Relevant Chapter Review Exercise(s)
State appropriate hypotheses for a significance test about a population parameter.	9.1	587	R9.1, R9.2, R9.3
Interpret a P-value in context.	9.1	589	R9.4
Make an appropriate conclusion for a significance test.	9.1	591	R9.1, R9.3, R9.4
Interpret a Type I and a Type II error in context. Give a consequence of each error in a given setting.	9.1	593	R9.2
State and check the Random, 10%, and Large Counts conditions for performing a significance test about a population proportion.	9.2	601	R9.1, R9.3
Calculate the standardized test statistic and P-value for a test about a population proportion.	9.2	603	R9.1, R9.3
Perform a significance test about a population proportion.	9.2	606, 609	R9.3
Interpret the power of a significance test and describe what factors affect the power of a test.	9.2	613, 617	R9.2
State appropriate hypotheses for a significance test about a difference between two proportions.	9.3	627	R9.5
Determine whether the conditions are met for performing a test about a difference between two proportions.	9.3	628	R9.5
Calculate the standardized test statistic and P-value for a test about a difference between two proportions.	9.3	630	R9.5
Perform a significance test about a difference between two proportions.	9.3	632	R9.5

Teaching Tip

At the end of each chapter, a "What Did You Learn?" grid lists all the targets for the chapter. Make sure to discuss this grid with your students. Ask them to read each learning target and self-assess whether or not they can do each one. For each target, the grid shows the section in which it was covered, page references for examples that illustrate the target, and relevant chapter review exercises that students can use to assess their understanding of that target. Encourage your students to use this grid as part of their preparations for the chapter test.

TRM Full Solutions to Chapter 9 Review Exercises

The full solutions can be found by clicking on the link in the TE-Book, opening the TRFD, or downloading from the Teacher's Resources on the book's digital platform.

Answers to Chapter 9 Review Exercises

R9.1 (a) H_0: $p = 0.25$; H_a: $p > 0.25$, where p = the true proportion of all students at this school who have played/danced in the rain. **(b)** The sample data give some evidence for H_a because $\hat{p} = \dfrac{28}{80} = 0.35 > 0.25$. **(c)** *Appropriate test:* One-sample z test for p. *Random:* We have a random sample of 80 students from this school. *10%:* Assume the sample size (80) is less than 10% of all students at this school. *Normal/Large Sample:* $np_0 = 80(0.25) = 20 \geq 10$ and $n(1 - p_0) = 80(0.75) = 60 \geq 10$. **(d)** $z = 2.07$; *P*-value = 0.0188 *Conclusion:* We will use $\alpha = 0.05$. Because the *P*-value of 0.0118 $< \alpha = 0.05$, we reject H_0. We have convincing evidence that the true proportion of all students at this school who have played/danced in the rain in their lifetime is greater than 0.25.

R9.2 (a) H_0: $p = 0.05$, H_a: $p < 0.05$, where p is the true proportion of adults who will get the flu after using the vaccine. We will use $\alpha = 0.05$. **(b)** *Type I:* Finding convincing evidence that less than 5% of patients would get the flu, when in reality at least 5% would. *Consequence:* The company might get sued for false advertising. *Type II:* Failing to find convincing evidence that less than 5% would get the flu, when in reality less than 5% would. *Consequence:* Loss of potential income. **(c)** Because a Type I error is more serious in this case, and $P(\text{Type I error}) = \alpha$, I would recommend a significance level of $\alpha = 0.01$ to minimize the possibility of making this type of error. **(d)** If the true proportion of adults who will get the flu after using the vaccine is $p = 0.03$, there is a 0.9437 probability that the researchers will find convincing evidence for H_a: $p < 0.05$. **(e)** The power could be increased by increasing the sample size or increasing the significance level α.

R9.3 S: H_0: $p = 0.05$, H_a: $p < 0.05$, where p = the true proportion of adults who will get the flu after using the vaccine. We will use $\alpha = 0.05$. **P:** One-sample z test for p. *Random:* We have a random sample of 1000 adults. *10%:* The sample size (1000) is less than 10% of the population of adults. *Large Counts:* $np_0 = 1000(0.05) = 50 \geq 10$ and $n(1 - p_0) = 1000(0.95) = 950 \geq 10$. **D:** $z = -1.02$; *P*-value = 0.1539. **C:** Because the *P*-value of 0.1539 $> \alpha = 0.05$, we fail to reject H_0. We do not have convincing evidence that the true proportion of adults who will get the flu after using the vaccine is less than 0.05.

Chapter 9 Review Exercises

These exercises are designed to help you review the important ideas and methods of the chapter.

R9.1 **Playing in the rain?** Rob once read that one-quarter of all people have played/danced in the rain at some point in their lives. His friend Justin thinks that the proportion is higher than 0.25 for their high school. To settle their dispute, they ask a random sample of 80 students in their school and find out that 28 have played/danced in the rain.

(a) State the appropriate null and alternative hypotheses.

(b) Explain why the sample data give some evidence for H_a.

(c) Identify the appropriate test to perform and show that the conditions for carrying out the test are met.

(d) Find the standardized test statistic and *P*-value, and make an appropriate conclusion.

R9.2 **Flu vaccine** A drug company has developed a new vaccine for preventing the flu. The company claims that fewer than 5% of adults who use its vaccine will get the flu. To test the claim, researchers give the vaccine to a random sample of 1000 adults.

(a) State appropriate hypotheses for testing the company's claim. Be sure to define your parameter.

(b) Describe a Type I error and a Type II error in this setting, and give the consequences of each.

(c) Would you recommend a significance level of 0.01, 0.05, or 0.10 for this test? Justify your choice.

(d) The power of the test to detect the fact that only 3% of adults who use this vaccine would develop flu using $\alpha = 0.05$ is 0.9437. Interpret this value.

(e) Explain two ways that you could increase the power of the test from part (d).

R9.3 **Flu vaccine** Refer to Exercise R9.2. Of the 1000 adults who were given the vaccine, 43 got the flu. Do these data provide convincing evidence to support the company's claim?

R9.4 **Roulette** An American roulette wheel has 18 red slots among its 38 slots. To test if a particular roulette wheel is fair, you spin the wheel 50 times and the ball lands in a red slot 31 times. The resulting *P*-value is 0.0384.

(a) Interpret the *P*-value.

(b) What conclusion would you make at the $\alpha = 0.05$ level?

(c) The casino manager uses your data to produce a 99% confidence interval for p and gets (0.44, 0.80). He says that this interval provides convincing evidence that the wheel is fair. How do you respond?

R9.5 **Treating AIDS** AZT was the first drug that seemed effective in delaying the onset of AIDS. Evidence of AZT's effectiveness came from a large randomized comparative experiment. The subjects were 870 volunteers who were infected with HIV (the virus that causes AIDS), but who did not yet have AIDS. The study assigned 435 of the subjects at random to take 500 milligrams of AZT each day and another 435 subjects to take a placebo. At the end of the study, 38 of the placebo subjects and 17 of the AZT subjects had developed AIDS.

(a) If the results of the study are statistically significant, is it reasonable to conclude that AZT is the cause of the decrease in the proportion of people like these who will develop AIDS?

(b) Do the data provide convincing evidence at the $\alpha = 0.05$ level that taking AZT lowers the proportion of infected people like the ones in this study who will develop AIDS in a given period of time?

R9.4 (a) Assuming the roulette wheel is fair, there is a 0.0384 probability that we would get a sample proportion of reds ($\hat{p} = 31/50$) at least this different from the expected proportion of reds (18/38) by chance alone. **(b)** Because the *P*-value of 0.0384 $< \alpha = 0.05$, we reject H_0. We have convincing evidence that the true proportion of reds is not equal to 18/38 $= 0.474$. **(c)** Because 18/38 $= 0.474$ is one of the plausible values in the interval, this interval does not provide convincing evidence that the wheel is unfair. But it also does not prove that the wheel is fair; there are many plausible values in the interval that are not equal to 18/38. This conclusion is inconsistent with the conclusion in part (b) because the manager used a 99% CI, which is equivalent to a test using $\alpha = 0.01$. If the manager had used a 95% CI, 18/38 would not be considered a plausible value.

R9.5 (a) Yes, if the results are statistically significant, we have reason to conclude that AZT is the cause of the decrease in the proportion of people like these who will develop AIDS because the subjects were randomly assigned to the treatment and control groups in this randomized comparative experiment. **(b) S:** H_0: $p_1 - p_2 = 0$, H_a: $p_1 - p_2 < 0$, where p_1 = true proportion of patients like these who take AZT and develop AIDS and p_2 = true proportion of patients like these who take placebo and develop AIDS; $\alpha = 0.05$. **P:** Two-sample z test for $p_1 - p_2$. *Random:* The subjects were assigned at random to take AZT or a placebo. *Large Counts:* 27.405, 407.595, 27.405, 407.595 are ≥ 10. **D:** $z = -2.91$; *P*-value = 0.0018. **C:** Because the *P*-value of 0.0018 $< \alpha = 0.05$, we reject H_0. We have convincing evidence that taking AZT lowers the proportion of patients like these who develop AIDS compared to a placebo.

Chapter 9 AP® Statistics Practice Test

TRM Full Solutions to Chapter 9 AP® Statistics Practice Test

The full solutions can be found by clicking on the link in the TE-Book, opening the TRFD, or downloading from the Teacher's Resources on the book's digital platform.

Section I: Multiple Choice *Select the best answer for each question.*

T9.1 An opinion poll asks a random sample of adults whether they favor banning ownership of handguns by private citizens. A commentator believes that more than half of all adults favor such a ban. The null and alternative hypotheses you would use to test this claim are

(a) $H_0: \hat{p} = 0.5; H_a: \hat{p} > 0.5.$
(b) $H_0: p = 0.5; H_a: p > 0.5.$
(c) $H_0: p = 0.5; H_a: p < 0.5.$
(d) $H_0: p = 0.5; H_a: p \neq 0.5.$
(e) $H_0: p > 0.5; H_a: p = 0.5.$

T9.2 The power takeoff driveline on tractors used in agriculture can be a serious hazard to operators of farm equipment. The driveline is covered by a shield in new tractors, but the shield is often missing on older tractors. Two types of shields are the bolt-on and the flip-up. It was believed that the bolt-on shield was perceived as a nuisance by the operators and deliberately removed, but the flip-up shield is easily lifted for inspection and maintenance and may be left in place. In a study by the U.S. National Safety Council, random samples of older tractors with both types of shields were taken to see what proportion of shields were removed. Of 183 tractors designed to have bolt-on shields, 35 had been removed. Of the 136 tractors with flip-up shields, 15 were removed. We wish to perform a test of $H_0: p_B = p_F$ versus $H_a: p_B > p_F$, where p_B and p_F are the proportions of all tractors with the bolt-on and flip-up shields removed, respectively. Which of the following is *not* a condition for performing the significance test?

(a) Both populations are Normally distributed.
(b) The data come from two independent samples.
(c) Both samples were chosen at random.
(d) The expected counts of successes and failures are large enough to use Normal calculations.
(e) Both populations are more than 10 times the corresponding sample sizes.

T9.3 To determine the reliability of experts who interpret lie detector tests in criminal investigations, a random sample of 280 such cases was studied. The results were

Examiner's Decision	Suspect's True Status	
	Innocent	Guilty
"Innocent"	131	15
"Guilty"	9	125

If the hypotheses are H_0: Suspect is innocent versus H_a: Suspect is guilty, which of the following is the best estimate of the probability that an expert commits a Type II error?

(a) 15/280
(b) 9/280
(c) 15/140
(d) 9/140
(e) 15/146

T9.4 A significance test allows you to reject a null hypothesis H_0 in favor of an alternative hypothesis H_a at the 5% significance level. What can you say about significance at the 1% level?

(a) H_0 can be rejected at the 1% significance level.
(b) There is insufficient evidence to reject H_0 at the 1% significance level.
(c) There is sufficient evidence to accept H_0 at the 1% significance level.
(d) H_a can be rejected at the 1% significance level.
(e) The answer can't be determined from the information given.

T9.5 A random sample of 100 likely voters in a small city produced 59 voters in favor of Candidate A. The value of the standardized test statistic for performing a test of $H_0: p = 0.5$ versus $H_a: p > 0.5$ is which of the following?

(a) $z = \dfrac{0.59 - 0.5}{\sqrt{\dfrac{0.59(0.41)}{100}}}$

(b) $z = \dfrac{0.59 - 0.5}{\sqrt{\dfrac{0.5(0.5)}{100}}}$

(c) $z = \dfrac{0.5 - 0.59}{\sqrt{\dfrac{0.59(0.41)}{100}}}$

(d) $z = \dfrac{0.5 - 0.59}{\sqrt{\dfrac{0.5(0.5)}{100}}}$

(e) $z = \dfrac{0.59 - 0.5}{\sqrt{100}}$

T9.6 A researcher claims to have found a drug that causes people to grow taller. The coach of the basketball team at Brandon University has expressed interest but demands evidence. Over 1000 Brandon students volunteer to participate in an experiment to test this new drug. Fifty of the volunteers are randomly selected, their heights are measured, and they are given the drug. Two weeks later, their heights are measured again. The power of the test to detect an average increase in height of 1 inch could be increased by

(a) using only the 12 members of the basketball team in the experiment.

Answers to Chapter 9 AP® Statistics Practice Test

T9.1 b
T9.2 a
T9.3 c
T9.4 e
T9.5 b
T9.6 c

T9.7 e

T9.8 d

T9.9 a

T9.10 c

T9.11 (a) *Type I:* Finding convincing evidence that more than 20% of customers would pay for the upgrade, when in reality they would not. *Type II:* Finding convincing evidence that more than 20% of customers would pay for the upgrade, when in reality more than 20% would. For the company, a Type I error is worse because they would go ahead with the upgrade and lose money. **(b) S:** $H_0: p = 0.20$, $H_a: p \geq 0.20$, where p = the true proportion of customers who would pay \$100 for the upgrade using $\alpha = 0.05$. **P:** One-sample z test for p. *Random:* We have a random sample of 60 customers. *10%:* The sample size (60) is less than 10% of this company's customers. *Large Counts:* $np_0 = 60(0.20) = 12 \geq 10$ and $n(1 - p_0) = 60(0.8) = 48 \geq 10$. **D:** $z = 1.29$; P-value $= 0.0985$. **C:** Because the P-value of $0.0985 > \alpha = 0.05$, we fail to reject H_0. We do not have convincing evidence that the true proportion of customers who would pay \$100 for the upgrade is greater than 0.20.

T9.12 (a) STATE: $H_0: p_1 - p_2 = 0$, $H_a: p_1 - p_2 > 0$, where p_1 = true proportion of cars that have the brake defect in last year's model and p_2 = true proportion of cars with the brake defect in this year's model; $\alpha = 0.05$. PLAN: Two-sample z test for $p_1 - p_2$. *Random:* Independent random samples. *10%:* $n_1 = 100 < 10\%$ of last year's model and $n_2 = 350 < 10\%$ of this year's model. *Large Counts:* 15.6, 84.4, 54.6, and 295.4 are ≥ 10. DO: $z = 1.39$; P-value $= 0.0822$.

CONCLUDE: Because the P-value of $0.0822 > \alpha = 0.05$, we fail to reject H_0. We do not have convincing evidence that the true proportion of brake defects is smaller in this year's model compared to last year's model. **(b)** If the true proportion of cars with the defect has decreased by 0.10 from last year to this year, there is a 0.72 probability that the automaker will find convincing evidence for $H_a: p_1 - p_2 > 0$. **(c)** Besides increasing the sample sizes, the power of the test can be increased by using a larger value for α, the significance level of the test.

(b) using $\alpha = 0.01$ instead of $\alpha = 0.05$.

(c) using $\alpha = 0.05$ instead of $\alpha = 0.01$.

(d) giving the drug to 25 randomly selected students instead of 50.

(e) using a two-sided test instead of a one-sided test.

T9.7 A 95% confidence interval for the proportion of viewers of a certain reality television show who are over 30 years old is (0.26, 0.35). Suppose the show's producers want to test the hypothesis $H_0: p = 0.25$ against $H_a: p \neq 0.25$. Which of the following is an appropriate conclusion for them to draw at the $\alpha = 0.05$ significance level?

(a) Fail to reject H_0; there is convincing evidence that the true proportion of viewers of this reality TV show who are over 30 years old equals 0.25.

(b) Fail to reject H_0; there is not convincing evidence that the true proportion of viewers of this reality TV show who are over 30 years old differs from 0.25.

(c) Reject H_0; there is not convincing evidence that the true proportion of viewers of this reality TV show who are over 30 years old differs from 0.25.

(d) Reject H_0; there is convincing evidence that the true proportion of viewers of this reality TV show who are over 30 years old is greater than 0.25.

(e) Reject H_0; there is convincing evidence that the true proportion of viewers of this reality TV show who are over 30 years old differs from 0.25.

T9.8 In a test of $H_0: p = 0.4$ against $H_a: p \neq 0.4$, a random sample of size 100 yields a standardized test statistic of $z = 1.28$. Which of the following is closest to the P-value for this test?

(a) 0.90   **(d)** 0.20

(b) 0.40   **(e)** 0.10

(c) 0.05

Section II: Free Response *Show all your work. Indicate clearly the methods you use, because you will be graded on the correctness of your methods as well as on the accuracy and completeness of your results and explanations.*

T9.11 A software company is trying to decide whether to produce an upgrade of one of its programs. Customers would have to pay \$100 for the upgrade. For the upgrade to be profitable, the company must sell it to more than 20% of their customers. The company will survey a random sample of 60 customers about this issue.

(a) Which would be a more serious mistake in this setting—a Type I error or a Type II error? Justify your answer.

(b) In the company's survey, 16 customers are willing to pay \$100 each for the upgrade. Do the sample data give convincing evidence that more than 20% of the company's customers are willing to purchase the upgrade? Carry out an appropriate test at the $\alpha = 0.05$ significance level.

T9.9 Bags of a certain brand of tortilla chips claim to have a net weight of 14 ounces. A representative of a consumer advocacy group wishes to see if there is convincing evidence that the mean net weight is less than advertised and so intends to test the hypotheses

$$H_0: \mu = 14$$
$$H_a: \mu < 14$$

A Type I error in this situation would mean concluding that the bags

(a) are being underfilled when they really aren't.

(b) are being underfilled when they really are.

(c) are not being underfilled when they really are.

(d) are not being underfilled when they really aren't.

(e) are being overfilled when they are really underfilled.

T9.10 Conference organizers wondered whether posting a sign that says "Please take only one cookie" would reduce the proportion of conference attendees who take multiple cookies from the snack table during a break. To find out, the organizers randomly assigned 212 attendees to take their break in a room where the snack table had the sign posted, and 189 attendees to take their break in a room where the snack table did not have a sign posted. In the room without the sign posted, 24.3% of attendees took multiple cookies. In the room with the sign posted, 17.0% of attendees took multiple cookies. Is this decrease in proportions statistically significant at the $\alpha = 0.05$ level?

(a) No. The P-value is 0.034.

(b) No. The P-value is 0.068.

(c) Yes. The P-value is 0.034.

(d) Yes. The P-value is 0.068.

(e) Cannot be determined from the information given.

T9.12 A random sample of 100 of last year's model of a certain popular car found that 20 had a specific minor defect in the brakes. The automaker adjusted the production process to try to reduce the proportion of cars with the brake problem. A random sample of 350 of this year's model found that 50 had the minor brake defect.

(a) Was the company's adjustment successful? Carry out an appropriate test using $\alpha = 0.05$ to support your answer.

(b) Suppose that the proportion of cars with the defect was reduced by 0.10 from last year to this year. The power of the test in part (a) to detect this decrease is 0.72. Interpret this value.

(c) Other than increasing the sample sizes, identify one way of increasing the power of the test.

Chapter 10

Chapter 10

Estimating Means with Confidence

PD Overview

In the previous two chapters, we showed how to perform inference for categorical data using confidence intervals for proportions (Chapter 8) and significance tests for proportions (Chapter 9). Both times, we started with inference about a single population proportion and then progressed to inference about a difference in proportions. In the next two chapters, we will follow the same structure in demonstrating how to perform inference for quantitative data. Chapter 10 focuses on confidence intervals for means, and Chapter 11 focuses on significance tests for means.

Consider starting the chapter with the "Confidence interval BINGO!" activity on page 649. In this activity, students discover that the Normal distribution does not work well as a model for calculating confidence intervals for means, which leads to the introduction of the t distribution. Once equipped with critical values from the t distribution, students are ready to construct confidence intervals for a mean and for a difference in means.

At this point, students should be quite comfortable using the four-step process for inference, which will serve them well in this chapter. The structure and reasoning for constructing and interpreting confidence intervals for means is nearly the same as for proportions. Remember that *every* time students are asked to construct and interpret a confidence interval, they should use the four-step process. Rubrics on previous AP® Statistics exams have demanded all four steps, even if the question doesn't specifically ask for them.

At the end of the chapter, students analyze paired data. Using what they learned early in the chapter, students use the four-step process to construct and interpret confidence intervals for the mean difference for paired data.

Throughout this chapter, take advantage of articles and stories in the news. Almost every day, there are summaries of comparative experiments and observational studies that use the inferential methods we learn about in this chapter. Even if the data aren't provided in the article, you can ask your students to consider how the study was designed and to determine what inference procedure would be most appropriate to analyze the results.

The Main Ideas

One of the challenges in teaching the AP® Statistics course is keeping students focused on the big picture, not just the details of each section. We outline the main ideas for the chapter here.

Chapter 10 Introduction

The introduction highlights the types of studies we'll analyze in this chapter. We will show how to estimate a mean, such as the mean number of books read by all American adults in the previous 12 months. We will also demonstrate how to estimate a difference between two means, like the mean diameter of longleaf pines in the northern half and southern half of the Wade Tract Preserve. Finally, we will illustrate how to estimate a mean difference involving paired data, such as the IQ scores of identical twins raised in separate households. In all three cases, we will use confidence intervals to analyze the results of these types of studies.

Section 10.1 Estimating a Population Mean

In this section, we focus exclusively on confidence intervals for a population mean. Although the basic ideas remain the same as confidence intervals for a population proportion, estimating a population mean comes with an extra complication. Because we typically don't know the population standard deviation σ when estimating the population mean μ, we have to use the sample standard deviation s_x instead. As we show in the "Confidence interval BINGO!" activity on page 649, using critical values from the standard Normal distribution when

replacing σ with s_x makes the capture rate too small. To fix the problem, we get critical values from a t distribution rather than the standard Normal distribution. The t distributions are bell-shaped, symmetric, centered at 0, and slightly wider than the standard Normal distribution.

The form of the one-sample t interval for μ is the same as that of the one-sample z interval for p. The way we interpret confidence intervals and confidence levels remains the same as well, as do the Random and 10% conditions. However, the Normal/Large Sample condition replaces the Large Counts condition that we used in Section 8.2.

Section 10.2 Estimating a Difference in Means

In this section, we begin to analyze quantitative data that comes from two independent random samples or from two groups formed by random assignment. The conditions to check here are nearly identical to those from Section 10.1, except that we will need to check each condition for both samples (or both groups) rather than just for a single sample. We provide two different options for calculating the number of degrees of freedom, and then students use the familiar 4-step process to construct and interpret a confidence interval for a difference in means.

Many comparative studies use paired data and should not be analyzed with two-sample inference procedures. Instead of looking at two lists of data, we analyze the single list of differences for each pair. More specifically, we calculate the mean of the differences for the pairs in the sample. Using one-sample t procedures from Section 10.1, we can construct a confidence interval to estimate the true mean of the differences for the population. We will revisit the distinction between two-sample data and paired data in Section 11.2. If students need help understanding the distinction between paired data and two samples while in Section 10.2, consider doing the "Get your heart beating!" activity (or the cholesterol alternate activity) from Section 11.2.

Chapter 10: Resources
Teacher's Resource Materials

The following resources, identified by the **TRM** in the annotated student pages, can be found by clicking on the link in the Teacher's e-Book (TE-book), searching by category or chapter on the Teacher's Resource Flash Drive (TRFD), or logging into the book's digital platform highschool.bfwpub.com/updatedtps6e and searching the Teacher's Resources menu (teacher log-in required).

- Alternate Examples: one file per section
- Lecture Presentation Slides: one per section
- Chapter 10 Learning Targets Grid
- Good Books Alternate Activity
- FRAPPY! Materials
- Chapter 10 Case Study
- Complete solutions for the Check Your Understanding problems, section exercises, review exercises, and practice test.
- Quizzes: one per section
- Chapter 10 Test

Free Response Questions from Previous AP® Statistics Exams

Questions can be found on the AP® Central website: apcentral.collegeboard.org/courses/ap-statistics/exam.

Students should be able to answer all the free response questions listed below with material learned in this chapter. Questions that contain content from this chapter but also require content from later chapters are listed in the last chapter required to complete the entire question. This list will be updated after each AP® Statistics exam and will be posted to the Teacher's Resource section of the book's digital platform and to www.statsmedic.com/free-response-questions.

Year	#	Content
2019	6	• Scope of inference • Mean versus median • Sampling distributions • Bootstrapping a median
2013	1	• Interpreting stemplots • One-sample t interval for a population mean
2012	3	• Comparing histograms • Conditions for two-sample t procedures
2009	4	• Two-sample t interval for the difference between two means • Using a confidence interval to test hypotheses
2008B	3	• Determining sample size (CI for a mean) • Practical constraints
2007	1	• Interpreting standard deviation • Comparing center • Using a confidence interval to test hypotheses
2006	4	• Two-sample t interval for the difference between two means • Using a confidence interval to test hypotheses
2005B	4	• Paired t interval • Using a confidence interval to assess significance
2005	6	• Two-sample t interval for the difference between two means • Constructing and interpreting an interaction plot
2004B	4	• Two-sample t interval for the difference between two means • Two-sample versus paired t interval
2002	1	• Precision of interval estimates • Using confidence intervals to make decisions
2000	2	• Conditions for a one-sample t interval for a population mean

Applets

- The *Confidence Intervals* applet at highschool.bfwpub .com/updatedtps6e allows students to investigate the success rate of confidence intervals for population means using different confidence levels and sample sizes.

- The *One Quantitative Variable* applet in the Extra Applets section can be used to simulate a difference between two means. This applet can also handle confidence intervals for a difference in means.

- The *Simulating Confidence Intervals for Population Parameter* applet at www.rossmanchance.com/applets allows students to investigate the success rate of confidence intervals for population means under different circumstances.

Chapter 10: Pacing Guide, Learning Targets, and Suggested Assignments

This pacing guide is based on a schedule with 110, 50-minute sessions before the AP® Statistics exam. If you have a different number of sessions before the AP® Statistics exam, you can modify the pacing guide to suit your needs. If you have additional time, consider incorporating quizzes, released AP® Statistics free response questions, or additional activities. See the Resources section above for suggestions.

The suggested homework assignments list odd-numbered exercises, whenever possible, so students can check their answers against the back-of-book answers. If you would rather students not have access to the answers while doing homework, adding 1 to the exercise numbers usually will do the trick, because the homework exercises typically are paired. For example, Exercises 1 and 2 will generally cover the same topics, but in different contexts. You may also choose to include the Recycle and Review questions at the end of each section, which review topics from previous sections or chapters. These questions are denoted with a symbol. If your school is using the digital platform that accompanies TPS6, you will find these assignments pre-built as online homework assignments for Chapter 10.

Day	Content	Learning Targets: Students will be able to . . .	Suggested Assignment (MC bold)
1	10.1 The Problem of Unknown σ, Conditions for Estimating μ	• Determine the critical value for calculating a $C\%$ confidence interval for a population mean using a table or technology. • State and check the Random, 10%, and Normal/Large Sample conditions for constructing a confidence interval for a population mean.	1, 3, 5, 7
2	10.1 Constructing a Confidence Interval for μ	• Construct and interpret a confidence interval for a population mean.	9, 13, 15, **21–24**
3	10.2 Confidence Intervals for $\mu_1 - \mu_2$	• Determine whether the conditions are met for constructing a confidence interval for a difference between two means. • Construct and interpret a confidence interval for a difference between two means.	27, 31, 33, 35
4	10.2 Comparing Two Means: Paired Data, Confidence Intervals for μ_{diff}	• Analyze the distribution of differences in a paired data set using graphs and summary statistics. • Construct and interpret a confidence interval for a mean difference.	37, 41, 45, **49–52**
5	Chapter 10 Review/ FRAPPY!		Chapter 10 Review Exercises
6	Chapter 10 Test		

Chapter 10 Alignment to the College Board's Fall 2019 AP® Statistics Course Framework*

Relationship to College Board Units

Chapter 10 in this book covers Topics 7.1–7.3 and 7.6–7.7 in Unit 7 of the College Board Course Framework. Students will be ready to take the Personal Progress Check for Unit 7 once they have completed Chapters 10 and 11.

Big Ideas and Enduring Understandings

Chapter 10 develops these Big Ideas and related Enduring Understandings outlined in the Course Framework:

- **Big Idea 1: Variation and Distribution (EU: VAR 1, 7):** The distribution of measures for individuals within a sample or population describes variation. The value of a statistic varies from sample to sample. How can we determine whether differences between measures represent random variation or meaningful distinctions? Statistical methods based on probabilistic reasoning provide the basis for shared understandings about variation and about the likelihood that variation between and among measures, samples, and populations is random or meaningful.
- **Big Idea 2: Patterns and Uncertainty (EU: UNC 4):** Statistical tools allow us to represent and describe patterns in data and to classify departures from patterns. Simulation and probabilistic reasoning allow us to anticipate patterns in data and to determine the likelihood of errors in inference.

Course Skills

Chapter 10 helps students to develop the skills identified in the Course Framework.

- **1: Selecting Statistical Methods** (1.A, 1.D)
- **3: Using Probability and Simulation** (3.C, 3.D)
- **4: Statistical Argumentation** (4.A, 4.B, 4.C, 4.D)

Learning Objectives and Essential Knowledge Statements

Section	Learning Objectives	Essential Knowledge Statements
10.1	VAR-1.I, VAR-7.A, UNC-4.O, UNC-4.P, UNC-4.Q, UNC-4.R, UNC-4.S, UNC-4.T, UNC-4.U	VAR-1.I.1, VAR-7.A.1, VAR-7.A.2, UNC-4.O.1, UNC-4.O.2, UNC-4.P.1, UNC-4.Q.1, UNC-4.Q.2, UNC-4.Q.3, UNC-4.R.1, UNC-4.R.2, UNC-4.S.1, UNC-4.S.2, UNC-4.S.3, UNC-4.T.1, UNC-4.U.1, UNC-4.U.2, UNC-4.U.3
10.2	UNC-4.O, UNC-4.V, UNC-4.W, UNC-4.X, UNC-4.Y, UNC-4.Z, UNC-4.AA, UNC-4.AB	UNC-4.O.3, UNC-4.V.1, UNC-4.V.2, UNC-4.W.1, UNC-4.X.1, UNC-4.X.2, UNC-4.Y.1, UNC-4.Y.2, UNC-4.Z.1, UNC-4.Z.2, UNC-4.AA.1, UNC-4.AB.1

A detailed alignment (The Nitty Gritty Guide) that can be sorted by Course Framework Unit, Topic, Learning Objective, Essential Knowledge Statement, or textbook section, is available on the TRFD and in the Teacher's Resources folder on Sapling Plus. **TRM**

*Should changes be made to the Course Framework in the future, an updated alignment will be placed on our AP® updates page at go.bfwpub.com/ap-course-updates.

UNIT 7
Inference for Quantitative Data: Means

Chapter **10**

Estimating Means with Confidence

Bill Gozansky/Alamy

Teaching Tip

Unit 7 in the College Board Course Framework aligns to Chapters 10 and 11 in this book. Students will be ready to take the Personal Progress Check for Unit 7 once they have completed Chapter 11.

PD Chapter 10 Overview

To watch the video overview of Chapter 10 (for teachers), click on the link in the TE-Book, look on the TRFD, or download from the Teacher's Resources on the book's digital platform.

TRM Lecture Presentation Slides

If you are new to teaching AP® Statistics or are short on time when preparing for class, you may find the Lecture Presentation Slides to be helpful. Experienced AP® Teacher Doug Tyson has created one slide presentation per section. You may use them as is, modify them to fit your needs, or share them with students who miss class. Find them on the TRFD and in the Teacher's Resources on the book's digital platform.

Teaching Tip

The structure of this chapter will follow the same flow as each of the previous two chapters, which means we will start with analyzing data from one sample (estimating a mean in Section 10.1) and then move to analyzing data from two samples/groups (estimating a difference in means in Section 10.2).

Making Connections

In Chapters 8 and 9, we learned about inference for proportions, which comes from analyzing categorical data. In Chapters 10 and 11, we learn about inference for means, which comes from analyzing quantitative data.

PD **Section 10.1 Overview**

To watch the video overview of Section 10.1 (for teachers), click on the link in the TE-Book, look on the TRFD, or download from the Teacher's Resources on the book's digital platform.

TRM **Learning Targets Grid**

At the beginning of each section, we present the relevant learning targets. Point these out to your students and refer back to the targets when you cover them in class. There is a PDF version of the grid with an additional column that students can use to keep track of their progress. Find it in the Teacher's Resource Materials located in the TE-Book, on the TRFD, or in the Teacher's view on the book's digital platform.

TRM **Section 10.1 Alternate Examples**

You can find the Alternate Examples for this section in Microsoft Word format by clicking on the link in the TE-Book, opening the TRFD, or downloading from the Teacher's Resources on the book's digital platform.

TRM **Good Books Alternate Activity**

A useful activity to introduce this section is the "Good books" activity. Find it in the Teacher's Resource Materials located in the TE-Book, on the TRFD, or in the Teacher's view on the book's digital platform.

INTRODUCTION

What is the mean number of books read by U.S. adults in the previous year? Do Oreo cookies weigh as much, on average, as the packaging suggests? To answer these questions, we select a random sample from the population of interest and calculate the sample mean $\bar{x}$. To account for sampling variability, we use an interval estimate rather than a single point estimate. In other words, we construct a *confidence interval for a population mean*.

How much more do two-bedroom apartments near a university cost than one-bedroom apartments, on average? To answer this question, we select a random sample from each population, calculate the mean of each sample, and construct a *confidence interval for a difference in means*. We can also use a confidence interval to estimate the effect of a treatment in an experiment. How much more effective is red wine than white wine in increasing the amount of heart-healthy polyphenols in the blood? Randomly assign subjects to drink red or white wine and calculate the mean amount of polyphenols for the subjects in each group. Then construct a confidence interval to estimate the true difference in mean polyphenol levels for subjects like the ones in the experiment who drink red versus white wine.

Does listening to music help or hinder performance in math? Ask subjects to take a math test while listening to music and to take a different test without listening to music, in random order. Then calculate the difference in amount of time needed to complete the tests for each subject and the mean of these differences for all subjects. Finally, construct a *confidence interval for a mean difference* to answer the question of interest.

In Chapters 8 and 9, you learned the basics of confidence intervals and significance tests, along with the details of estimating and testing claims about proportions. In Chapters 10 and 11, we revisit these big ideas, but focus on inference for means instead of inference for proportions. In this chapter, you'll learn how to estimate means with confidence.

SECTION 10.1 **Estimating a Population Mean**

LEARNING TARGETS *By the end of the section, you should be able to:*

- Determine the critical value for calculating a $C\%$ confidence interval for a population mean using a table or technology.
- State and check the Random, 10%, and Normal/Large Sample conditions for

- constructing a confidence interval for a population mean.
- Construct and interpret a confidence interval for a population mean.

Inference about a population proportion usually arises when we study *categorical* variables. We learned how to construct and interpret confidence intervals for a population proportion p in Section 8.2. To estimate a population mean, we have

to record values of a *quantitative* variable for a sample of individuals. It makes sense to try to estimate the mean amount of sleep that students at a large high school got last night but not their mean eye color! In this section, we'll examine confidence intervals for a population mean μ.

The Problem of Unknown σ

In Section 8.2, we used this formula for a confidence interval for a population proportion:

$$\hat{p} \pm z^* \sqrt{\frac{\hat{p}(1-\hat{p})}{n}}$$

In more general terms, this is

point estimate $\pm$ margin of error

$=$ statistic $\pm$ (critical value)(standard error of statistic)

A confidence interval for a population mean has a formula with the same structure. Using the sample mean $\bar{x}$ as the point estimate for the population mean μ and $\sigma/\sqrt{n}$ as the standard deviation of the sampling distribution of $\bar{x}$ gives

$$\bar{x} \pm z^* \frac{\sigma}{\sqrt{n}}$$

This interval is called a *one-sample z interval for a population mean*. If we know σ and the conditions are met, then about 95% of random samples of size n will produce 95% confidence intervals that capture the population mean when using this formula. Unfortunately, if we don't know the true value of μ, we rarely know the true value of σ either. We can use s_x as an estimate for σ, but things don't work out as nicely as we might like. Let's explore why this is true.

ACTIVITY | Confidence interval BINGO!

In this activity, you will investigate the problem caused by replacing σ with s_x when calculating a confidence interval for μ, and how to fix it.

A farmer wants to estimate the mean weight (in grams) of all tomatoes grown on her farm. To do so, she will select a random sample of 4 tomatoes, calculate the mean weight (in grams), and use the sample mean $\bar{x}$ to create a 99% confidence interval for the population mean μ. Suppose that the weights of all tomatoes on the farm are approximately Normally distributed with a mean of 100 grams and a standard deviation of 20 grams.

Let's use an applet to simulate taking an SRS of $n = 4$ tomatoes and calculating a 99% confidence interval for μ using three different methods.

Method 1 (assuming σ is known)

$$\bar{x} \pm z^* \frac{\sigma}{\sqrt{n}} = \bar{x} \pm 2.576 \frac{20}{\sqrt{4}}$$

1. Launch the *Simulating Confidence Intervals for a Population Parameter* applet at www.rossmanchance.com/applets. Choose "Means" from the

---

OVERVIEW (continued)

Answers: Method 1

1–3. Done

4. Method 1 worked as intended. In the long run, 99% of the intervals captured the true mean.

drop-down menu and leave the other menus as "Normal" and "z with sigma." Then enter $\mu = 100$, $\sigma = 20$, $n = 4$, and intervals = 1. Set the confidence level = 99%, as shown in the screen shot.

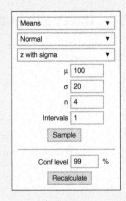

2. Press "Sample." The applet will select an SRS of $n = 4$ tomatoes and calculate a confidence interval for μ. This interval will be displayed as a horizontal line. The vertical line identifies $\mu = 100$. If the interval captures $\mu = 100$, the interval will be green. If the interval misses $\mu = 100$, the interval will be red.

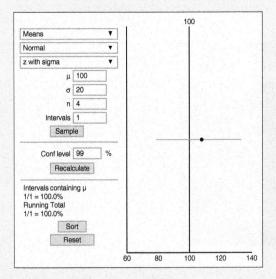

3. Press "Sample" many times, shouting out "BINGO!" whenever you get an interval that misses $\mu = 100$ (i.e., a red interval). Stop when your teacher calls time.

4. How well did Method 1 work? Compare the running total in the lower left corner with the stated confidence level of 99%.

Method 2 (using s_x as an estimate for σ)

$$\bar{x} \pm z^* \frac{s_x}{\sqrt{n}} = \bar{x} \pm 2.576 \frac{s_x}{\sqrt{4}}$$

1. Press the "Reset" button in the lower left. Then, in the third drop-down menu, choose "z with s." This replaces σ with s_x in the formula for the confidence interval. Keep everything else the same.

2. Press "Sample" many times, shouting out "BINGO!" whenever you get an interval that misses $\mu = 100$ (i.e., a red interval). Stop when your teacher calls time.

3. Did Method 2 work as well as Method 1? Discuss with your classmates.

4. Now change the number of intervals to 50 and press "Sample" 20 times, for a total of more than 1000 intervals. Compare the running total in the lower left corner with the stated confidence level of 99%. What do you notice about the length of the intervals that missed?

To increase the capture rate of the intervals to 99%, we need to make the intervals longer. We can do this by using a different critical value, called a t^* critical value. You'll learn how to calculate this number soon.

Method 3 (using s_x as an estimate for σ and a t^* critical value instead of a z^* critical value)

$$\bar{x} \pm t^* \frac{s_x}{\sqrt{n}} = \bar{x} \pm ??? \frac{s_x}{\sqrt{4}}$$

1. Press the "Reset" button in the lower left. Then, in the third drop-down menu, choose "t" and change "Intervals" to 1. Keep everything else the same.

2. Press "Sample" many times, shouting out "BINGO!" whenever you get an interval that misses $\mu = 100$ (i.e., a red interval). Stop when your teacher calls time.

3. Did Method 3 work better than Method 2? How does it compare to Method 1? Discuss with your classmates.

4. Now change the number of intervals to 50 and press "Sample" 20 times, for a total of at least 1000 intervals. Compare the running total in the lower left corner with the stated confidence level of 99%. What do you notice about the length of the intervals compared to Method 2?

Figure 10.1 (on the next page) shows the results of repeatedly constructing confidence intervals using a z^* critical value and the sample standard deviation s_x, as described in Method 2 of the preceding activity. Of the 1000 intervals constructed, only 925 (that's 92.5%) captured the population mean. That's far below our desired 99% confidence level!

67

COMMON STUDENT ERROR

Many students want to say that each interval has a 99% **probability** of capturing the true mean. This is not correct. Once the sample is collected and a confidence interval is created from that sample, the interval either contains the true mean (100% probability) or it does not (0% probability). To help students understand this idea, point to one of the red intervals and ask, "Does this interval contain the true mean?" Then have students look away and look back and then ask the same question again. The interval will not contain the true mean, no matter how many times you ask this question.

Making Connections

From Chapter 7, we know how to standardize values from a sampling distribution that is approximately Normal with a known standard deviation (σ):

$$z = \frac{\bar{x} - \mu}{\frac{\sigma}{\sqrt{n}}}$$

This transformation for all possible values of $\bar{x}$ produces a standardized distribution that is approximately a *Normal distribution*. In Chapter 11, we will standardize values from a sampling distribution that is approximately Normal with an unknown standard deviation. We will use s_x to estimate σ.

$$t = \frac{\bar{x} - \mu}{\frac{s_x}{\sqrt{n}}}$$

Under certain conditions, this transformation for all possible values of $\bar{x}$ produces a standardized distribution that is not Normal but rather a *t distribution*.

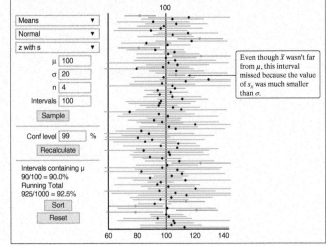

FIGURE 10.1 Screen shot from the *Simulating Confidence Intervals for a Population Parameter* applet at www.rossmanchance.com/applets, showing the capture rate of a "99%" confidence interval when using a z^* critical value with the sample standard deviation s_x.

What went wrong? The intervals that missed (those in red) came from samples with small standard deviations s_x and from samples in which the sample mean $\bar{x}$ was far from the population mean μ. In those cases, multiplying $s_x/\sqrt{4}$ by $z^* = 2.576$ didn't produce long enough intervals to reach $\mu = 100$. To achieve a 99% capture rate, we need to multiply by a larger critical value. But what critical value should we use?

CALCULATING t^* CRITICAL VALUES In the activity, you discovered that Method 3 produces confidence intervals that capture μ at the advertised confidence level. That is, when constructing "99%" confidence intervals using the formula $\bar{x} \pm t^* \frac{s_x}{\sqrt{n}}$, about 99% of the intervals capture the population mean μ. This interval is called a *one-sample t interval for a population mean*.

The critical value in this interval is denoted t^* because it comes from a t distribution, not the standard Normal distribution. The critical value t^* has the same interpretation as z^*: it measures how many standard errors we need to extend from the point estimate to get the desired level of confidence. There is a different t distribution for each sample size. We specify a particular t distribution by giving its *degrees of freedom* (df). When we perform inference about a population mean μ using a t distribution, the appropriate degrees of freedom are found by subtracting 1 from the sample size n, making df $= n - 1$.

Figure 10.2 shows two different t distributions, along with the standard Normal distribution. Because the t distributions have more area in the tails than the standard Normal distribution, t^* critical values will always be larger than z^* critical values for a specified level of confidence. As you learned in the "Bingo" activity, we need to use a critical value larger than z^* to compensate for the variability introduced by using the sample standard deviation s_x as an estimate for the population standard deviation σ. As the degrees of freedom increase, the t distributions

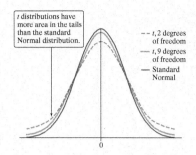

FIGURE 10.2 Density curves for the *t* distributions with 2 and 9 degrees of freedom and the standard Normal distribution. All are symmetric with center 0. The *t* distributions have more variability than the standard Normal distribution, but approach the standard Normal distribution as df increases.

The *t* distribution and the *t* inference procedures were invented by William S. Gosset (1876–1937). Gosset worked for the Guinness brewery, and his goal in life was to make better beer. He used his new *t* procedures to find the best varieties of barley and hops. Gosset's statistical work helped him become head brewer. Because Gosset published under the pen name "Student," you will often see the *t* distribution called "Student's *t*" in his honor.

have less area in the tails and approach the standard Normal distribution. This makes sense because the value of s_x will typically be closer to σ as the sample size increases.

You will learn more about *t* distributions in Chapter 11. For now, we will focus on how to calculate the critical value t^* for various sample sizes and confidence levels.

Table B in the back of the book gives t^* critical values for the *t* distributions. Each row in the table contains critical values for the *t* distribution whose degrees of freedom (df) appear at the left of the row. To make the table easy to use, several of the more common confidence levels C are given at the bottom of the table.

When you use Table B to determine the correct value of t^* for a given confidence interval, all you need to know are the confidence level C and the degrees of freedom (df). In the activity, we calculated 99% confidence intervals with $n = 4$, so df $= 4 - 1 = 3$.

According to the Table B excerpt in the margin, for 99% confidence and 3 degrees of freedom, $t^* = 5.841$. That is, the interval should extend 5.841 standard errors on both sides of the point estimate to have a capture rate of 99%.

Unfortunately, Table B doesn't include every possible df. If the correct df isn't listed, use the greatest df available that is less than the correct df. "Rounding up" to a larger df will result in confidence intervals that are too narrow. The intervals won't be wide enough to include the true population value as often as suggested by the confidence level.

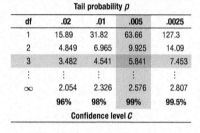

	Tail probability *p*			
df	.02	.01	.005	.0025
1	15.89	31.82	63.66	127.3
2	4.849	6.965	9.925	14.09
3	3.482	4.541	5.841	7.453
⋮	⋮	⋮	⋮	⋮
∞	2.054	2.326	2.576	2.807
	96%	98%	99%	99.5%

Confidence level *C*

EXAMPLE

Finding t^* critical values
Using Table B

PROBLEM: What critical value t^* from Table B should be used in constructing a confidence interval for the population mean in each of the following settings?
(a) A 95% confidence interval based on an SRS of size $n = 12$
(b) A 90% confidence interval from a random sample of 48 observations

---

ALTERNATE EXAMPLE   Skill 3.D

Finding t^* critical values   Using Table B

PROBLEM:
What critical value t^* from Table B should be used to construct a confidence interval for the population mean in each of the following settings?
(a) A 98% confidence interval from a random sample of 20 observations
(b) A 90% confidence interval based on an SRS of size $n = 95$

SOLUTION:
(a) Use df $= 20 - 1 = 19$ and 98% confidence; $t^* = 2.539$.
(b) Use df $= 95 - 1 = 94$ and 90% confidence. There is no df $= 94$, so we use the more conservative df $= 80$; $t^* = 1.664$.

Teaching Tip

In Chapter 6, we learned that transforming a variable by subtracting a constant and then dividing by a constant changes the center and variability of the distribution of that variable but not the shape. However, we are now dividing by a variable (s_x), which changes the shape as well as the center and variability.

Teaching Tip

Students will probably ask you to explain what degrees of freedom are all about. Unfortunately, there is no simple answer. For now, simply explain that the shape and variability of the *t* distributions depend on the degrees of freedom, which depend on the sample size (df $= n - 1$).

Teaching Tip

Because the variability of the *t* distributions is slightly larger than the variability of a standard Normal distribution, we have to go farther out in the *t* distributions than ±1.96 to capture the central 95%. To determine exactly how far, we use Table B (the *t* table) or technology. If we continued to use 1.96, the intervals would be too short and our capture rate would be smaller than 95%, as illustrated in the "Confidence interval BINGO!" activity.

Teaching Tip

Remind students that a t^* value can be interpreted the same as a z^* value. It is the *number of standard deviations* away from the estimate we need to extend on either side to get the desired level of confidence.

Teaching Tip

The "play" icon at the top of this example (and many others) indicates that students can watch a video presentation of the example by clicking on the link in the e-Book or viewing the videos on the Student Site at highschool.bfwpub.com/updatedtps6e.

SOLUTION:

(a) $t^* = 2.201$

	Tail probability p			
df	.05	.025	.02	.01
10	1.812	2.228	2.359	2.764
11	1.796	2.201	2.328	2.718
12	1.782	2.179	2.303	2.681
∞	1.645	1.960	2.054	2.326
	90%	95%	96%	98%
	Confidence level C			

> In Table B, we consult the row corresponding to df = 12 − 1 = 11. We move across that row to the entry that is directly above 95% confidence level on the bottom of the chart.

(b) $t^* = 1.684$

	Tail probability p			
df	.10	.05	.025	.02
30	1.310	1.697	2.042	2.147
40	1.303	1.684	2.021	2.123
50	1.299	1.676	2.009	2.109
∞	1.282	1.645	1.960	2.054
	80%	90%	95%	96%
	Confidence level C			

> With 48 observations, we want to find the t^* critical value for df = 48 − 1 = 47 and 90% confidence. There is no df = 47 row in Table B, so we use the more conservative df = 40.

FOR PRACTICE, TRY EXERCISE 1

Teaching Tip

Once students understand how to use Table B, there is no need to keep using Table A to find z^* values. The z^* value can now be quickly identified at the bottom of Table B. On the AP® Statistics exam, the t table provided to students also includes the critical values for df = ∞ in the last line.

Teaching Tip: Using Technology

If your students have a TI-83 or older TI-84, they may not have the invT command. Installing an upgraded operating system (OS) will usually do the trick. Visit education.ti.com or copy the newer OS directly from another calculator using the Link function.

Because the bottom row of Table B gives z^* critical values, you can use this row when calculating critical values for confidence intervals involving proportions.

The bottom row of Table B gives z^* critical values. That's because t distributions approach the standard Normal distribution as the degrees of freedom approach infinity.

Using t^* rather than z^* solves the problem created by using the sample standard deviation s_x as an estimate for the population standard deviation σ. Because t^* is larger than z^* for a given sample size and confidence level, intervals calculated with t^* will have a higher capture rate than intervals using z^*. Furthermore, when using t^*, the actual percent of intervals that contain the population mean should be equal to the stated confidence level when the conditions are met.

Technology will quickly produce t^* critical values for any sample size.

22. Technology Corner   USING INVERSE t

TI-Nspire and other technology instructions are on the book's website at highschool.bfwpub.com/updatedtps6e.

Most newer TI-84 calculators allow you to find t^* critical values using the inverse t command. Let's use the inverse t command to find the critical values in parts (a) and (b) of the example.

- Press 2nd VARS (DISTR) and choose invT(.

- For part (a), we need to find the critical value for 95% confidence, so we want an area of 0.025 in each tail.

  OS 2.55 or later: In the dialog box, enter these values: area: 0.025, df: 11, choose Paste, and then press ENTER. *Note:* For the inverse *t* command, the area always refers to area *to the left*.

  Older OS: Complete the command invT(0.025,11) and press ENTER.

NORMAL FLOAT AUTO REAL RADIAN MP
invT(0.025,11)
-2.200985143
invT(0.05,47)
-1.677926664

- For part (b), we need 90% confidence, so we want 0.05 in each tail. Use the command invT(0.05,47).

Note that the *t* critical values are the positive values $t^* = 2.201$ and $t^* = 1.678$. The critical value for part (b) is slightly smaller here, because we were able to use df $= 47$ rather than rounding down to df $= 40$.

Now that you know how to calculate a t^* critical value, it's time to make a simple observation. Inference for *proportions* uses z; inference for *means* uses t. That's one reason why distinguishing categorical from quantitative variables is so important.

CHECK YOUR UNDERSTANDING

Use Table B to find the critical value t^* that you would use for a confidence interval for a population mean μ in each of the following settings. If possible, check your answer with technology.

1. A 96% confidence interval based on a random sample of 22 observations

2. A 99% confidence interval from an SRS of 71 observations

Conditions for Estimating μ

As with proportions, you must check that the observations in the sample can be viewed as independent and that the sampling distribution of $\bar{x}$ is approximately Normal before constructing a confidence interval for a population mean. The first two conditions should be familiar by now. The *Random condition* is crucial for doing inference. If the data don't come from a random sample, you can't draw conclusions about a larger population. When sampling without replacement, the *10% condition* allows us to view the observations in the sample as independent, so our formula for the standard deviation is approximately correct. If the 10% condition is violated, our formula will overestimate the standard deviation, making the margin of error larger than it needs to be for the stated confidence level.

The third condition is different, however. When calculating a confidence interval for a population proportion, we check the Large Counts condition to ensure it is appropriate to use the standard Normal distribution to calculate the z^* critical value. When calculating a confidence interval for a population mean, we check the *Normal/Large Sample condition* to ensure it is appropriate to use a t distribution to calculate the t^* critical value. In both cases, violating the third condition usually results in confidence intervals that have a capture rate less than the stated confidence level.

✔ **Answers to CYU**

1. df $= 22 - 1 = 21$, $t^* = 2.189$; *Tech:* invT(area: 0.02, df: 21) $= -2.189$, so $t^* = 2.189$

2. df $= 71 - 1 = 70$, $t^* = 2.660$ (using df $= 60$); *Tech:* invT(area: 0.005, df: 70) $= -2.648$, so $t^* = 2.648$

Teaching Tip

While it is important that students know how to check each condition, it is equally important that they understand *why* we check the condition. We call this the "*So what?*"

Teaching Tip

- **Random condition:** The data come from a random sample from the population of interest.
  So what?
  → so individual observations are independent and we can generalize to the population.

- **10% condition:** $n < (0.10)N$
  So what?
  → so we can view observations as independent even though we are sampling without replacement.

- **Normal/Large Sample condition:** The population has a Normal distribution or the sample size is large ($n \geq 30$). If the population distribution has unknown shape and $n < 30$, use a graph of the sample data to assess the Normality of the population.
  So what?
  → so the sampling distribution of $\bar{x}$ will be approximately Normal and we can use a t distribution to do calculations.

For the t^* critical value to produce confidence intervals with a capture rate that is equal to the confidence level, the population distribution must be Normal. If we are told that the population is Normally distributed, then the condition has been met. Unfortunately, it is rare that we actually know a population distribution is Normal. When the population shape is unknown, there are two ways to satisfy the Normal/Large Sample condition.

- If the sample size is small ($n < 30$), graph the sample data and ask, "Is it plausible that these data came from a Normal population?" If there is no strong skewness or outliers in the data, then the answer is "Yes." Remember to include the graph of sample data when checking the Normal/Large Sample condition in this way.

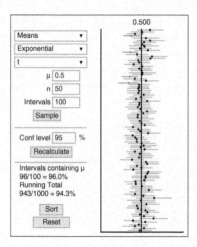

- If the sample size is large ($n \geq 30$), using a t^* critical value will produce confidence intervals with a capture rate that is approximately equal to the confidence level, *even when the population distribution is not Normal.* The large sample size ensures that the sampling distribution of $\bar{x}$ is close to Normal and that the sampling distribution of s_x is close enough to what it would be if the population were Normal.

Here is an image from the *Simulating Confidence Intervals for a Population Parameter* applet at www.rossmanchance.com/applets, showing the capture rate of a 95% one-sample t interval when selecting 1000 samples of size $n = 50$ from a population that is strongly skewed to the right (an exponential distribution).

The capture rate isn't quite 95%, but it's fairly close. The agreement between the stated confidence level and actual capture rate will improve by increasing the sample size or by sampling from a population that is closer to Normal.

CONDITIONS FOR CONSTRUCTING A CONFIDENCE INTERVAL ABOUT A MEAN

- **Random:** The data come from a random sample from the population of interest.
  ◦ **10%:** When sampling without replacement, $n < 0.10N$.
- **Normal/Large Sample:** The population has a Normal distribution or the sample size is large ($n \geq 30$). If the population distribution has unknown shape and $n < 30$, use a graph of the sample data to assess the Normality of the population. Do not use t procedures if the graph shows strong skewness or outliers.

Here is an example of checking the Normal/Large Sample condition in several different contexts.

EXAMPLE

GPAs, wood, and SATs
Checking the Normal/Large Sample condition

PROBLEM: Determine if the Normal/Large Sample condition is met in each of the following settings.

(a) To estimate the average GPA of students at your school, you randomly select 50 students. Here is a histogram of their GPAs:

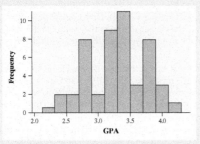

(b) How much force does it take to pull wood apart? The stemplot shows the force (in pounds) required to pull a piece of Douglas fir apart for each of 20 randomly selected pieces.

(c) Suppose you want to estimate the mean SAT Math score at a large high school. Here is a boxplot of the SAT Math scores for a random sample of 20 students at the school:

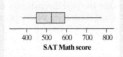

```
23 | 0
24 | 0
25 |
26 | 5
27 |
28 | 7
29 |
30 | 259
31 | 399
32 | 033677
33 | 0236
```

Key: 31|3 = 313 pounds of force required to pull a piece of Douglas fir apart

SOLUTION:

(a) Yes; the sample size is large (50 ≥ 30).

In addition, it is also plausible that the sample came from a Normal population because the histogram doesn't have strong skewness or outliers.

(b) No; the stemplot is strongly skewed to the left with possible low outliers and n = 20 < 30.

It is *not* plausible that this sample came from a Normal population because of the strong skewness and outliers in the stemplot.

(c) Yes; even though n = 20 < 30, the boxplot is only moderately skewed to the right and there are no outliers.

It is plausible that this sample came from a Normal population because the boxplot doesn't show strong skewness or outliers.

FOR PRACTICE, TRY EXERCISE 3

Coffee, height, and homework
Checking the Normal/Large Sample condition

PROBLEM:
Determine if the Normal/Large Sample condition is met in each of the following settings.

(a) We want to estimate the average time (in minutes) to order and receive a regular coffee at a local coffee shop. The times for five randomly selected visits to a local coffee shop are shown in the dotplot.

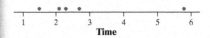

(b) We want to estimate the average height of high school students. The boxplot shows the distribution of height (in centimeters) for 100 randomly selected students.

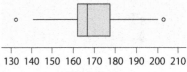

(c) A middle school counselor wants to estimate how many minutes, on average, students spend doing homework. A histogram shows the amount of time a random sample of 20 students spent doing homework the previous evening.

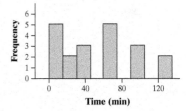

SOLUTION:
(a) No; the sample size is small, and the dotplot shows a possible outlier.
(b) Yes; even though there are outliers, the sample size is large ($n = 100 \geq 30$).
(c) Yes; although the sample size is small ($n = 20 < 30$), there is no strong skewness or outliers.

Making Connections

In Chapter 2, we introduced the Normal probability plot as a graph to assess the Normality of a given set of data. If the Normal probability plot is fairly linear, it is reasonable to believe the data set is approximately Normal.

In the preceding example, we used a histogram, a stemplot, and a boxplot to address the Normal/Large Sample condition. You can also use dotplots and Normal probability plots to assess Normality. Each of these graphs has strengths and weaknesses, so we recommend following the advice of your teacher when choosing a graph to use. In particular, be careful when describing the shape of a distribution based on a boxplot. Boxplots hide modes and gaps, making it impossible to determine if a distribution is approximately Normal based on the boxplot alone. However, because boxplots clearly show skewness and outliers, they can be helpful for identifying important departures from Normality.

In some cases, it is challenging to determine if the skewness in a graph should be considered "strong." One way to judge skewness is to compare the distance from the maximum to the median and from the median to the minimum. Let's take a closer look at the stemplot from part (b) and the boxplot from part (c) in the preceding example.

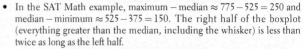
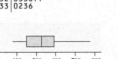

- In the stemplot that records the force required for pulling a piece of wood apart, maximum − median $= 336 - 319.5 = 16.5$ and median − minimum $= 319.5 - 230 = 89.5$. The half of the stemplot with smaller values is more than 5 times as long as the half of the stemplot with larger values.

- In the SAT Math example, maximum − median $\approx 775 - 525 = 250$ and median − minimum $\approx 525 - 375 = 150$. The right half of the boxplot (everything greater than the median, including the whisker) is less than twice as long as the left half.

The stemplot is quite a bit more skewed than the boxplot. Unfortunately, there is no accepted rule of thumb for identifying strong skewness. For that reason, we have chosen the data sets in examples and exercises to avoid borderline cases.

AP® EXAM TIP

If a question on the AP® Statistics exam asks you to construct and interpret a confidence interval, all the conditions should be met. However, you are still required to state the conditions and show evidence that they are met—including a graph if the sample size is small and the data are provided.

Constructing a Confidence Interval for μ

When the conditions are met, a C% confidence interval for the unknown mean μ is

$$\bar{x} \pm t^* \frac{s_x}{\sqrt{n}}$$

The value $\dfrac{s_x}{\sqrt{n}}$ is called the *standard error of the sample mean* $\bar{x}$, or just the standard error of the mean:

$$SE_{\bar{x}} = \frac{s_x}{\sqrt{n}}$$

The formula sheet provided on the AP® Statistics exam uses the notation $s_{\bar{x}}$ rather than $SE_{\bar{x}}$ for the standard error of the sample mean $\bar{x}$.

Like the standard deviation of $\bar{x}$, the standard error of $\bar{x}$ describes how much the sample mean $\bar{x}$ typically varies from the population mean μ in repeated SRSs of size n.

Teaching Tip: AP® Connections

AP® Biology students use the abbreviation SEM for the standard error of the mean. They most often use SEM to draw error bars on bar graphs.

THE ONE-SAMPLE *t* INTERVAL FOR A POPULATION MEAN

When the conditions are met, a C% confidence interval for the unknown mean μ is

$$\bar{x} \pm t^* \frac{s_x}{\sqrt{n}}$$

where t^* is the critical value for the *t* distribution with $n-1$ degrees of freedom (df) with C% of the area between $-t^*$ and t^*.

The following example illustrates how to construct and interpret a confidence interval for a population mean. By now, you should recognize the four-step process.

EXAMPLE

More books
A one-sample *t* interval for μ

PROBLEM: In Chapter 8, you learned that about 73% of American adults claim to have read a book in the previous 12 months. This was based on a 2016 Pew Research Center study that interviewed a random sample of 1520 American adults. The same study also reported that the average number of books read by the members of the sample (including those who reported reading 0 books) was $\bar{x} = 12$ books with a standard deviation of $s_x = 18$ books.[1]

(a) Construct and interpret a 95% confidence interval for the mean number of books read by all American adults in the previous 12 months.

(b) In 2011, Pew reported that American adults read an average of 14 books in the previous 12 months. Does your interval from part (a) provide convincing evidence that the mean for all American adults in 2016 is different than 14 books? Explain your answer.

SOLUTION:

(a) STATE: 95% CI for μ = the true mean number of books read by all American adults in the previous 12 months.

> Make sure to do the four-step process!

PLAN: One-sample *t* interval for μ

- **Random:** Random sample of 1520 American adults ✓
  - **10%:** 1520 is less than 10% of all American adults. ✓
- **Normal/Large Sample:** The sample size is large ($n = 1520 \geq 30$). ✓

DO: *Using Table B:* df = 1000 and $t^* = 1.962$

Using technology: df = 1519 and $t^* = 1.962$

$$12 \pm 1.962\frac{18}{\sqrt{1520}} = 12 \pm 0.91 = (11.09, 12.91)$$

$$\bar{x} \pm t^* \frac{s_x}{\sqrt{n}}$$

ALTERNATE EXAMPLE

Skills 3.D, 4.D

Too much screen time
A one-sample *t* interval for μ

PROBLEM:

Reese believes the students in her large high school spend way too much time on their phones and other devices. To investigate, she estimates the mean amount of daily screen time for all the students in her school by taking a random sample of 50 students. She finds the average daily screen time for the members of the sample is $\bar{x} = 7.1$ hours with a standard deviation of $s_x = 2.4$ hours.

(a) Construct and interpret a 90% confidence interval for the mean amount of daily screen time for all the students in this school.

(b) A recent report stated that U.S. teens have 6.7 hours of daily screen time, on average. Does the interval from part (a) provide convincing evidence that the mean at Reese's school differs from the national average? Explain your answer.

AP® EXAM TIP

The specific formula for a one-sample *t* interval for a population mean is not on the formula sheet provided to students on the AP® Statistics exam. However, the formula sheet does include the general formula for a confidence interval and the formula for the standard error of the statistic $\bar{x}$, $s_{\bar{x}} = \dfrac{s}{\sqrt{n}}$.

Teaching Tip

Consider having students check the condition and give the "*So what?*" answers. Here they are for the "More books" example:

Random:
→ so individual observations are independent and we can generalize to the population of all American adults.

10%:
→ so we can view observations as independent even though we are sampling without replacement.

Normal/Large Sample:
→ so the sampling distribution of $\bar{x}$ will be approximately Normal and we can use a *t* distribution to do calculations.

SOLUTION:

(a) STATE: 90% CI for μ = the true mean amount of daily screen time for all students at this school.

PLAN: One-sample *t* interval for μ

- *Random:* Random sample of 50 students ✓
  - *10%:* 50 is less than 10% of all students at a large high school. ✓
- *Normal/Large Sample:* The sample size is large ($n = 50 \geq 30$). ✓

DO: *Table B:* df = 49 (round down to 40) and $t^* = 1.684$

Tech: df = 49 and $t^* = 1.677$

$$7.1 \pm 1.677\frac{2.4}{\sqrt{50}} = 7.1 \pm 0.57$$

$$= (6.53, 7.67)$$

CONCLUDE: We are 90% confident that the interval from 6.53 hours to 7.67 hours captures μ = the true mean amount of daily screen time for all students at this school.

(b) No; because 6.7 is in the confidence interval, 6.7 is a plausible value for the mean amount of daily screen time for all students at this school.

Students will often forget to use the word *mean* or *average* in their conclusion. In the example on this page, a student might incorrectly conclude "We are 95% confident that the interval from 11.09 books to 12.91 books captures the true number of books read . . ." Remind students that a confidence interval is always trying to estimate a population parameter and the parameter here is the mean.

COMMON STUDENT ERROR

Students will often forget to include units when interpreting a confidence interval for a mean. Remind them that a unit is not needed when estimating a proportion, but a unit is required when estimating a mean.

Preparing for Inference

In Chapter 11, we will use a significance test to answer part (b) in the "More books" example. The null hypothesis is $H_0: \mu = 14$. The P-value is 0.00002, and we reject H_0 and conclude there is convincing evidence that the 2016 mean number of books read differs from 14. This is consistent with our conclusion in part (b).

ALTERNATE EXAMPLE Skills 3.D, 4.B

The G.O.A.T.

Constructing a confidence interval for μ

PROBLEM:

Many people think Michael Jordan is the greatest basketball player of all time, with a career scoring average of 30.1 points per game. Some curious statistics students wondered about his scoring average for all his *home* games. They decided to take a random sample of 15 home games from Michael Jordan's career. Here is the number of points he scored in each of these games:

35	24	29	32
31	25	28	38
35	32	36	27
31	29	26	

Construct and interpret a 95% confidence interval for the mean number of points μ that Michael Jordan scored in all of his home games.

CONCLUDE: We are 95% confident that the interval from 11.09 books to 12.91 books captures μ = the true mean number of books read by all American adults in the previous 12 months.

(b) Yes; because 14 is not in the confidence interval, 14 is not a plausible value for the mean number of books read in the previous 12 months by all American adults in 2016.

> Make sure your conclusion is in context, includes units (books), and is about a population mean. Some people also like to include the phrase "Based on the sample" when interpreting a confidence interval.

FOR PRACTICE, TRY EXERCISE 9

In this example, the sample standard deviation ($s_x = 18$) is greater than the sample mean ($\bar{x} = 12$). If the population were Normally distributed, we would expect the minimum to be about 2 or 3 standard deviations below the mean. However, because the minimum of 0 books is not even 1 standard deviation below the mean, we know the population distribution is skewed to the right. Furthermore, Pew reported that the median number of books read by the members of the sample was 4 books. This is quite a bit less than the mean, again suggesting that the population distribution is skewed to the right. Of course, because of the large sample size, we don't need to worry about the shape of the population distribution when checking the Normal/Large Sample condition.

Here is another example, this time with the actual data values and a smaller sample size.

EXAMPLE

Video screen tension

Constructing a confidence interval for μ

PROBLEM: A manufacturer of high-resolution video terminals must control the tension on the mesh of fine wires that lies behind the surface of the viewing screen. Too much tension will tear the mesh, and too little will allow wrinkles. The tension is measured by an electrical device with output readings in millivolts (mV). Some variation is inherent in the production process. Here are the tension readings from a random sample of 20 screens from a single day's production:

269.5	297.0	269.6	283.3	304.8	280.4	233.5	257.4	317.5	327.4
264.7	307.7	310.0	343.3	328.1	342.6	338.8	340.1	374.6	336.1

Construct and interpret a 90% confidence interval for the mean tension μ of all the screens produced on this day.

SOLUTION:

STATE: 90% CI for μ = the true mean tension of all the video terminals produced this day.

> Remember to do the four-step process!

PLAN: One-sample t interval for μ

- Random: Random sample of 20 screens produced that day ✓
    - 10%: Assume that 20 is less than 10% of all video terminals produced that day. ✓

SOLUTION:
STATE: 95% CI for μ = the true mean number of points that Michael Jordan scored in all his home games.

PLAN: One-sample t interval for μ

- *Random:* Random sample of 15 home games played by Michael Jordan. ✓
    - *10%:* 15 is less than 10% of all home games played by Michael Jordan. ✓
- *Normal/Large Sample:* The dotplot does not show strong skewness or outliers. ✓

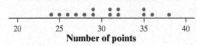

Number of points

DO: $\bar{x} = 30.53$ points and $s_x = 4.21$ points. With df = 14, $t^* = 2.145$:

$$30.53 \pm 2.145\frac{4.21}{\sqrt{15}} = 30.53 \pm 2.33$$
$$= (28.20, 32.86)$$

CONCLUDE: We are 95% confident that the interval from 28.20 points to 32.86 points captures μ = the true mean number of points that Michael Jordan scored in all his home games.

- **Normal/Large Sample:** The dotplot does not show strong skewness or outliers. ✓

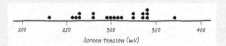

Screen tension (mV)

> Because there is no strong skewness or outliers in the sample, it is plausible that the population distribution of screen tension is Normal.

DO: $\bar{x} = 306.32\,\text{mV}$ and $s_x = 36.21\,\text{mV}$

With $df = 19$, $t^* = 1.729$.

$306.32 \pm 1.729 \dfrac{36.21}{\sqrt{20}}$

$= 306.32 \pm 14.00 = (292.32, 320.32)$

> When raw data are provided, use your calculator to find the mean and standard deviation of the sample.

$$\bar{x} \pm t^* \frac{s_x}{\sqrt{n}}$$

CONCLUDE: We are 90% confident that the interval from 292.32 mV to 320.32 mV captures $\mu =$ the true mean tension in the entire batch of video terminals produced that day.

> Make sure your conclusion is in context, includes units (mV), and is about a population mean.

FOR PRACTICE, TRY EXERCISE 13

■ AP® EXAM TIP

It is not enough just to make a graph of the data on your calculator when assessing Normality. You must *sketch* the graph on your paper to receive credit. You don't have to draw multiple graphs—any appropriate graph will do.

As you probably guessed, your calculator will compute a one-sample t interval for a population mean from sample data or summary statistics.

23. Technology Corner   CONSTRUCTING A CONFIDENCE INTERVAL FOR A POPULATION MEAN

TI-Nspire and other technology instructions are on the book's website at highschool.bfwpub.com/updatedtps6e.

Confidence intervals for a population mean using t distributions can be constructed on the TI-83/84, avoiding the use of Table B. Here is a brief summary of the techniques when you have the actual data values and when you have only numerical summaries:

1. Using summary statistics (see the "More books" example, page 659)
   - Press STAT, arrow over to TESTS, and choose TInterval. ...
   - On the TInterval screen, adjust your settings as shown and choose Calculate.

Teaching Tip

Many students will simply memorize the fact that z^* is used for proportions and t^* is used for means. Be sure that they understand why a t^* is required for means (because we don't know σ and we have to estimate it using s_x).

⚠ COMMON STUDENT ERROR

Many students incorrectly state the Normal/Large Sample condition by saying that the *sample* must have a Normal distribution. This isn't true (or even possible!). Instead, we use the sample data to make an inference about the shape of the population from which the sample was selected.

Teaching Tip: Using Technology

The TI-83/84 and TI-89 also list a ZInterval for calculating a confidence interval for a population mean. This procedure is appropriate only if the population standard deviation is known (which is very rare).

■ AP® EXAM TIP

Many students use the TInterval feature to correctly calculate the confidence interval and then try to "show their work" with an incorrect formula. This will result in a loss of credit because the two attempts are considered "parallel solutions" and students are scored on the worse response. If students want to include a formula in their response, they should make sure it produces the same results as the calculator. If the results aren't the same, students should choose one of the answers and cross the other out.

Teaching Tip

Even though it is acceptable to use the calculator's TInterval feature on the AP® Statistics exam, you can insist that students show additional work in your class. While the formulas are still relatively simple, we recommend this practice. It will help students understand the structure of confidence intervals and prepare them for multiple-choice questions that focus on the way confidence intervals are constructed.

 **Answers to CYU**

STATE: μ = the true mean healing rate

PLAN: One-sample t interval for μ.
Random: The newts were randomly chosen. *10%:* n = 18 is less than 10% of all newts. *Normal/Large Sample:* The histogram below does not show strong skewness or outliers, so this condition is met.

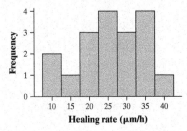

DO: $\bar{x}$ = 25.67 and s_x = 8.32; df = 17 and

t^* = 2.110: $25.67 \pm 2.110 \left(\dfrac{8.32}{\sqrt{18}} \right)$

$= 25.67 \pm 4.14 = (21.53, 29.81)$

CONCLUDE: We are 95% confident that the interval from 21.53 to 29.81 micrometers per hour captures μ = the true mean healing rate for all newts.

---

2. Using raw data (see the "Video screen tension" example, page 660)

- Enter the 20 video screen tension readings in list L1. Proceed to the TInterval screen as in Step 1, but choose Data as the input method. Then adjust your settings as shown and calculate the interval.

 CHECK YOUR UNDERSTANDING

Biologists studying the healing of skin wounds measured the rate at which new cells closed a cut made in the skin of an anesthetized newt. Here are data from a random sample of 18 newts, measured in micrometers (millionths of a meter) per hour:[2]

29 27 34 40 22 28 14 35 26 35 12 30 23 18 11 22 23 33

Calculate and interpret a 95% confidence interval for the mean healing rate μ.

More About the Margin of Error

In Chapter 8, you learned how different factors affect the margin of error when calculating a confidence interval for a population proportion. The lessons are the same in this section:

- The margin of error decreases as the sample size increases, assuming the confidence level and sample standard deviation s_x remain the same.
- The margin of error is proportional to $1/\sqrt{n}$, so quadrupling the sample size will cut the margin of error in half, assuming everything else remains the same.
- The margin of error will be larger for higher confidence levels. This makes sense, as wider intervals will have a greater capture rate than narrower intervals.
- The margin of error doesn't account for bias in the data collection process, only sampling variability.

In Section 8.2, you learned how to calculate the sample size needed to ensure a certain margin of error when estimating a population proportion. Calculating the required sample size is more difficult when estimating a population mean. Consequently, this topic is not included in the AP® Statistics Course Framework. If you're curious, however, read the following Think About It.

Think About It

HOW DO WE CALCULATE THE NECESSARY SAMPLE SIZE WHEN ESTIMATING A POPULATION MEAN? When the population standard deviation σ is unknown and conditions are met, the $C\%$ confidence interval for μ is

$$\bar{x} \pm t^* \frac{s_x}{\sqrt{n}}$$

where t^* is the critical value for confidence level C and degrees of freedom $df = n - 1$. The margin of error (ME) of the confidence interval is

$$ME = t^* \frac{s_x}{\sqrt{n}}$$

To determine the sample size for a desired margin of error, it makes sense to set the expression for ME less than or equal to the specified value and solve the inequality for n. There are two problems with this approach:

1. We don't know the sample standard deviation s_x because we haven't produced the data yet.
2. The critical value t^* depends on the sample size n that we choose.

The second problem is more serious. To get the correct value of t^*, we need to know the sample size. But that's what we're trying to find!

To approximate the sample size needed, start with a reasonable estimate for the *population* standard deviation σ from a similar study that was done in the past or from a small-scale pilot study. By pretending that σ is known, we can use the one-sample z interval for μ:

$$\bar{x} \pm z^* \frac{\sigma}{\sqrt{n}}$$

Using the appropriate standard Normal critical value z^* for confidence level C, we can solve for n using

$$z^* \frac{\sigma}{\sqrt{n}} \leq ME$$

There are more complicated methods of determining sample size that do not require us to use a known value of the population standard deviation σ. Our advice: consult with a statistician when planning a study to estimate a population mean.

Section 10.1 | Summary

- Confidence intervals for the mean μ of a population are based on the sample mean $\bar{x}$. If we know σ (which is very rare), we use a z^* critical value and the standard Normal distribution to calculate a *one-sample z interval for* μ.
- In practice, we usually don't know σ. Replacing the standard deviation of the sampling distribution of $\bar{x}$ ($\sigma_{\bar{x}} = \sigma/\sqrt{n}$) with the **standard error of** $\bar{x}$ ($SE_{\bar{x}} = s_x/\sqrt{n}$) requires use of the t distribution with $n - 1$ degrees of freedom (df) rather than the standard Normal distribution when calculating the critical value for a confidence interval for a population mean.

Teaching Tip

Ask students to imagine sitting around a conference room table to plan a survey. We want to determine how many people to interview. Most importantly, we have not yet collected any data.

AP® EXAM TIP

On the AP® Statistics exam, it is more likely that students will see a question about desired sample size in the context of a proportion rather than a mean.

Teaching Tip

The method we present for estimating the sample size underestimates the sample size needed by using a z^* critical value instead of a slightly larger t^* critical value. However, when the anticipated sample size is large, the difference between the z^* and t^* critical values—and the difference in the sample size calculations—are very small.

TRM Section 10.1 Quiz

There is one quiz available for this section in the Teacher's Resource Materials. Click on the link in the TE-Book, look on the TRFD, or download from the Teacher's Resources on the book's digital platform. You can also create your own quiz using the ExamView® Assessment Suite that is part of the TPS6 program. Questions are coded by Learning Target to make it easy to build parallel quizzes.

- When constructing a confidence interval for a population mean μ, we need to ensure that the observations in the sample can be viewed as independent and that the sampling distribution of $\bar{x}$ is approximately Normal. The required conditions are:
  - **Random:** The data come from a random sample from the population of interest.
    - ○ **10%:** When sampling without replacement, $n < 0.10N$.
  - **Normal/Large Sample:** The population has a Normal distribution or the sample size is large ($n \geq 30$). If the population distribution has unknown shape and $n < 30$, use a graph of the sample data to assess the Normality of the population. Do not use t procedures if the graph shows strong skewness or outliers.
- When conditions are met, a C% confidence interval for the population mean μ is given by the **one-sample t interval**:

$$\bar{x} \pm t^* \frac{s_x}{\sqrt{n}}$$

The critical value t^* is chosen so that the t curve with $n-1$ degrees of freedom has C% of the area between $-t^*$ and t^*. Use Table B or technology to calculate t^*.

- Follow the four-step process—State, Plan, Do, Conclude—whenever you are asked to construct and interpret a confidence interval for a population mean. Remember: inference for proportions uses z; inference for means uses t.
- When estimating a population mean, the **margin of error** will be smaller when the sample size increases and larger when the confidence level increases, assuming that everything else remains the same.

10.1 Technology Corners

TI-Nspire and other technology instructions are on the book's website at highschool.bfwpub.com/updatedtps6e.

22. Using inverse t	Page 654
23. Constructing a confidence interval for a population mean	Page 661

Answers to Section 10.1 Exercises

10.1 (a) df = 9, t^* = 2.262; *Tech:* invT(area: 0.025, df: 9) = −2.262, so t^* = 2.262.

(b) df = 19, t^* = 2.861; *Tech:* invT(area: 0.005, df: 19) = −2.861, so t^* = 2.861.

(c) df = 60, t^* = 1.671; *Tech:* invT(area: 0.05, df: 76) = −1.665, so t^* = 1.665.

10.2 (a) df = 11, t^* = 1.796; *Tech:* invT(area: 0.05, df: 11) = −1.796, so t^* = 1.796.

(b) df = 29, t^* = 2.045; *Tech:* invT(area: 0.025, df: 29) = −2.045, so t^* = 2.045.

(c) df = 50, t^* = 2.678' *Tech:* invT(area: 0.005, df: 57) = −2.665, so t^* = 2.665.

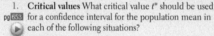
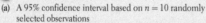

Section 10.1 | Exercises

1. **Critical values** What critical value t^* should be used for a confidence interval for the population mean in each of the following situations?
   (a) A 95% confidence interval based on $n = 10$ randomly selected observations
   (b) A 99% confidence interval from an SRS of 20 observations
   (c) A 90% confidence interval based on a random sample of 77 individuals

2. **Critical values** What critical value t^* should be used for a confidence interval for the population mean in each of the following situations?
   (a) A 90% confidence interval based on $n = 12$ randomly selected observations
   (b) A 95% confidence interval from an SRS of 30 observations
   (c) A 99% confidence interval based on a random sample of size 58

Teaching Tip

Make sure to point out the two icons next to Exercise 1. The "pg 653" icon reminds students that the example on page 653 is very similar to this exercise. The "play" icon reminds students that there is a video solution available in the student e-Book or at the Student Site.

3. Weeds among the corn Velvetleaf is a particularly
annoying weed in cornfields. It produces lots of seeds, pg 657
and the seeds wait in the soil for years until conditions
are right for sprouting. How many seeds do velvetleaf
plants produce? The histogram shows the counts
from a random sample of 28 plants that came up in
a cornfield when no herbicide was used.[3] Determine
if the Normal/Large Sample condition is met in this
context.

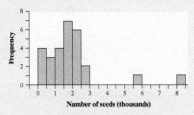

Number of seeds (thousands)

4. Tough read? Judy is interested in the reading level of
a medical journal. She records the length of a random
sample of 100 words. The histogram displays the
distribution of word length for her sample. Determine if
the Normal/Large Sample condition is met in this context.

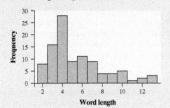

Word length

5. Check them all Determine if the conditions are met
for constructing a confidence interval for the population mean in each of the following settings.

(a) How much time do students at your school spend on
the Internet? You collect data from the 32 members
of your AP® Statistics class and calculate the mean
amount of time that these students spent on the
Internet yesterday.

(b) Is the real-estate market heating up? To estimate the
mean sales price, a realtor in a large city randomly
selected 100 home sales from the previous 6 months
in her city. These sales prices are displayed in the
boxplot.

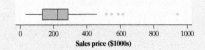

Sales price ($1000s)

6. Check them all Determine if the conditions are
met for constructing a confidence interval for the
population mean in each of the following settings.

(a) We want to estimate the average age at which U.S.
presidents have died. So we obtain a list of all U.S.
presidents who have died and their ages at death.

(b) Do teens text more than they call? To find out, an AP®
Statistics class at a large high school collected data on
the number of text messages and calls sent or received
by each of 25 randomly selected students. The boxplot
displays the difference (Texts – Calls) for each student.

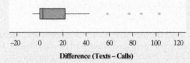

Difference (Texts – Calls)

7. Blood pressure A medical study finds that $\bar{x} = 114.9$
and $s_x = 9.3$ for the seated systolic blood pressure of
27 randomly selected adults. What is the standard error
of the mean? Interpret this value in context.

8. Travel time to work A study of commuting times
reports the travel times to work of a random sample
of 20 employed adults in New York State. The mean
is $\bar{x} = 31.25$ minutes and the standard deviation is
$s_x = 21.88$ minutes. What is the standard error of the
mean? Interpret this value in context.

9. Bone loss by nursing mothers Breast-feeding mothers
secrete calcium into their milk. Some of the calcium pg 659
may come from their bones, so mothers may lose bone
mineral. Researchers measured the percent change
in bone mineral content (BMC) of the spines of 47
randomly selected mothers during three months of
breast-feeding.[4] The mean change in BMC was –3.587%
and the standard deviation was 2.506%.

(a) Construct and interpret a 99% confidence interval
to estimate the mean percent change in BMC in the
population of breast-feeding mothers.

(b) Based on your interval from part (a), do these data give
convincing evidence that nursing mothers lose bone
mineral, on average? Explain your answer.

10. Reading scores in Atlanta The Trial Urban District
Assessment (TUDA) is a government-sponsored study
of student achievement in large urban school districts.
TUDA gives a reading test scored from 0 to 500.
A score of 243 is a "basic" reading level and a score
of 281 is "proficient." Scores for a random sample of
1470 eighth-graders in Atlanta had a mean of 240 with
standard deviation of 42.17.[5]

(a) Construct and interpret a 99% confidence interval for
the mean reading test score of all Atlanta eighth-graders.

10.7 $SE_{\bar{x}} = \dfrac{s_x}{\sqrt{n}} = \dfrac{9.3}{\sqrt{27}} = 1.7898.$
In many random samples of size 27, the
sample mean blood pressure will typically
vary by about 1.7898 from the population
mean blood pressure.

10.8 $SE_{\bar{x}} = \dfrac{s_x}{\sqrt{n}} = \dfrac{21.88}{\sqrt{20}} = 4.8925.$ In
many random samples of size 20, the sample
mean commute time will typically vary by
about 4.8925 from the population mean
commute time.

10.9 (a) S: μ = the true mean percent
change in BMC for breast-feeding
mothers. **P:** One-sample t interval.
Random: The mothers were randomly
selected. *10%:* 47 is less than 10%
of all breast-feeding mothers.
Normal/Large Sample: $n = 47 \geq 30$.
D: df = 40 (−4.575, −2.599); *Tech:*
(−4.569, −2.605) with df = 40. **C:** We
are 99% confident that the interval from
−4.569 to −2.605 captures μ = the
true mean percent change in BMC for
breast-feeding mothers. **(b)** Because all
the plausible values in the interval are
negative (indicating bone loss), the data
give convincing evidence.

10.10 (a) S: μ = the true mean reading
test score for all Atlanta eighth-graders.
P: One-sample t interval. *Random:* Students
were randomly selected. *10%:* 1470 is less
than 10% of eighth-graders in Atlanta.
Normal/Large Sample: $n = 1470 \geq 30$.
D: df = 1469 (237.16, 242.84);
Tech: (237.16, 242.84) with df = 1469.
C: We are 99% confident that the interval
from 237.16 to 242.84 captures μ = the
true mean reading test score for all Atlanta
eighth-graders. **(b)** Because all the plausible
values in the interval are less than 243,
there is convincing evidence.

Teaching Tip

Exercises 6 and 9 on this page have students
working with paired data. Consider going
over these questions with your students as an
introduction to paired data, which will be formally
introduced at the end of Section 10.2.

10.3 Not met; the sample size is small
($n = 28 < 30$) and there are outliers in the
data. We cannot assume that the population is
approximately Normal.

10.4 Met; $n = 100 \geq 30$.

10.5 (a) *Random:* No; the AP® Statistics students
are not a random sample of all students. *10%:* 32
< 10% of all students in the school. (Condition
met if the number of students at the school
is at least 320). *Normal/Large Sample:* Yes;
$n = 32 \geq 30$. **(b)** *Random:* Yes; random sample

of 100 home sales from the previous 6 months
in her city. *10%:* Yes; assume that 100 is less
than 10% of all homes sold during the past
6 months in her city. *Normal/Large Sample:* Yes;
$n = 100 \geq 30$. Although the boxplot shows the
distribution of home sales is strongly right-
skewed with outliers, this condition is still met
because the sample size is large.

10.6 (a) *Random:* No; a list of all U.S. presidents
is not a random sample. *10%:* No; the n
presidents who have died is not less than 10%
of all presidents that have died. *Normal/Large
Sample:* Yes; $n \geq 30$. **(b)** *Random:* Yes; random
sample of 25 students. *10%:* Yes; 25 is less
than 10% of all students at this high school.
Normal/Large Sample: No; $n = 25 < 30$ and
the distribution of differences (Texts − Calls) is
strongly right-skewed with outliers.

10.11 (a) S: μ = the true mean weight of an Oreo cookie. **P:** One-sample t interval. *Random:* The cookies were randomly selected. *10%:* 36 is less than 10% of all Oreo cookies. *Normal/Large Sample:* $n = 36 \geq 30$. **D:** df = 30 (11.369, 11.4152); *Tech:* (11.369, 11.415) with df = 30. **C:** We are 90% confident that the interval from 11.369 to 11.4152 captures μ = the true mean weight for all Oreo cookies. **(b)** If we were to select many random samples of size 36 from the population of all Oreo cookies and construct a 90% confidence interval using each sample, about 90% of the intervals would capture the true mean weight of an Oreo cookie.

10.12 (a) S: μ = the true mean thorax length of a male fruit fly. **P:** One-sample t interval. *Random:* Flies were randomly selected. *10%:* 49 is less than 10% of all male fruit flies. *Normal/Large Sample:* $n = 49 \geq 30$. **D:** df = 48 (0.7816, 0.8192); *Tech:* (0.7817, 0.8191) with df = 48. **C:** We are 90% confident that the interval from 0.7817 to 0.8191 captures μ = the true mean thorax length for all male fruit flies. **(b)** If we were to select many random samples of size 49 from the population of all male fruit flies and construct a 90% confidence interval using each sample, about 90% of the intervals would capture the true mean thorax length.

10.13 S: μ = the true mean number of pepperonis on a large pizza at this restaurant. **P:** One-sample t interval. *Random:* Pizzas were randomly selected. *10%:* 10 is less than 10% of all pepperoni pizzas made at this restaurant. *Normal/ Large Sample:* The dotplot doesn't show any outliers or strong skewness.

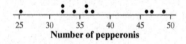

Number of pepperonis

D: $\bar{x} = 37.4$, $s_x = 7.662$, and $n = 10$; df = 9 (31.919, 42.881); *Tech:* (31.919, 42.881) with df = 9. **C:** We are 95% confident that the interval from 31.919 to 42.881 captures μ = the true mean.

10.14 S: μ = the true mean number of crackers in a bag of original goldfish. **P:** One-sample t interval. *Random:* Bags of goldfish were randomly selected. *10%:* 12 is less than 10% of all bags of original-flavored goldfish. *Normal/Large Sample:* The dotplot doesn't show any outliers or strong skewness.

315 320 325 330 335 340 345 350
Number of goldfish crackers

(b) Based on your interval from part (a), is there convincing evidence that the mean reading test score for all Atlanta eighth-graders is less than the basic level? Explain your answer.

11. **America's favorite cookie** Ann and Tori wanted to estimate the average weight of an Oreo cookie to determine if it was less than advertised (34 grams for 3 cookies). They selected a random sample of 36 cookies and found the weight of each cookie (in grams). The mean weight was $\bar{x} = 11.3921$ grams with a standard deviation of $s_x = 0.0817$ grams.[6]

(a) Construct and interpret a 90% confidence interval for the true mean weight of an Oreo cookie.

(b) Interpret the confidence level.

12. **Fruit fly thorax lengths** Fruit flies are used frequently in genetic research because of their quick reproductive cycle. The length of the thorax (in millimeters) was measured for each fly in a random sample of 49 male fruit flies. The mean length was $\bar{x} = 0.8004$ mm, with a standard deviation of $s_x = 0.0782$ mm.[7]

(a) Construct and interpret a 90% confidence interval for the true mean thorax length of a male fruit fly.

(b) Interpret the confidence level.

13. **Pepperoni pizza** Melissa and Madeline love pepperoni
pg 660   pizza, but sometimes they are disappointed with the small number of pepperonis on their pizza. To investigate, they went to their favorite pizza restaurant at 10 random times during the week and ordered a large pepperoni pizza.[8] Here are the number of pepperonis on each pizza:

47	36	25	37	46	36	49	32	32	34

Construct and interpret a 95% confidence interval for the true mean number of pepperonis on a large pizza at this restaurant.

14. **Catching goldfish for school** Carly and Maysem plan to be preschool teachers after they graduate from college. To prepare for snack time, they want to know the mean number of goldfish crackers in a bag of original-flavored goldfish. To estimate this value, they randomly selected 12 bags of original-flavored goldfish and counted the number of crackers in each bag.[9] Here are their data:

317	330	325	323	332	337	324	342	330	349	335	333

Construct and interpret a 95% confidence interval for the true mean number of crackers in a bag of original-flavored goldfish.

15. **A plethora of pepperoni?** Refer to Exercise 13.

(a) Explain why it was necessary to inspect a graph of the sample data when checking the Normal/Large Sample condition.

(b) According to the manager of the restaurant, there should be an average of 40 pepperonis on a large pizza.

Based on the interval, is there convincing evidence that the average number of pepperonis is less than 40? Explain your answer.

(c) Explain two ways that Melissa and Madeline could reduce the margin of error of their estimate. Why might they object to these changes?

16. **A school of fish** Refer to Exercise 14.

(a) Explain why it was necessary to inspect a graph of the sample data when checking the Normal/Large Sample condition.

(b) According to the packaging, there are supposed to be 330 goldfish in each bag of crackers. Based on the interval, is there convincing evidence that the average number of goldfish is less than 330? Explain your answer.

(c) Explain two ways that Carly and Maysem could reduce the margin of error of their estimate. Why might they object to these changes?

17. **Estimating BMI** The body mass index (BMI) of all American young women is believed to follow a Normal distribution with a standard deviation of about 7.5. How large a sample would be needed to estimate the mean BMI μ in this population to within ± 1 with 99% confidence?

18. **The SAT again** High school students who take the SAT Math exam a second time generally score higher than on their first try. Past data suggest that the score increase has a standard deviation of about 50 points. How large a sample of high school students would be needed to estimate the mean change in SAT score to within 2 points with 95% confidence?

19. **Willows in Yellowstone** Writers in some fields summarize data by giving $\bar{x}$ and its standard error rather than $\bar{x}$ and s_x. Biologists studying willow plants in Yellowstone National Park reported their results in a table with columns labeled $\bar{x} \pm$ SE. The table entry for the heights of willow plants (in centimeters) in one region of the park was 61.55 ± 19.03.[10] The researchers measured a total of 23 plants.

(a) Find the sample standard deviation s_x for these measurements.

(b) A hasty reader believes that the interval given in the table is a 95% confidence interval for the mean height of willow plants in this region of the park. Find the actual confidence level for the given interval.

20. **Blink** When two lights that are close together blink alternately, we "see" one light moving back and forth if the time between blinks is short. What is the longest interval of time between blinks that preserves the illusion of motion? Ask subjects to turn a knob that slows the blinking until they "see" two lights rather than one light moving. A report gives the results in the form "mean plus or minus the standard error of the mean."[11] Data for 12 subjects are summarized as 251 ± 45 (in milliseconds).

D: $\bar{x} = 331.417$, $s_x = 8.775$, and $n = 12$; df = 11 (325.842, 336.992); *Tech:* (325.84, 336.99) with df = 11. **C:** We are 95% confident that the interval from 325.84 to 336.99 captures μ = the true mean.

10.15 (a) It was necessary because the sample size was not large (<30). When the sample size is less than 30, we must assume the population is Normally distributed. **(b)** The value 40 is a plausible value found within the confidence interval, so we do not have convincing evidence that the average number of pepperonis is less than 40. **(c)** Melissa and Madeline could increase the sample size or decrease the confidence level. They may not want to increase the sample size because it would be expensive and they also may not want to eat that much pizza. They may not want to decrease the

---

confidence level because it is desirable to have a high degree of confidence that the population parameter has been captured by the interval.

10.16 (a) It was necessary because the sample size was not large (<30). When the sample size is less than 30, we must assume the population is Normally distributed. **(b)** The value 330 is a plausible value found within the confidence interval, so we do not have convincing evidence that the average number of goldfish is less than 330. **(c)** Carly and Maysem could increase the sample size or decrease the confidence level. They may not want to increase the sample size because that would become expensive. They may not want to decrease the confidence level because it is desirable to have a high degree of confidence that the population parameter has been captured by the interval.

(a) Find the sample standard deviation s_x for these measurements.

(b) A hasty reader believes that the interval given in the report is a 95% confidence interval for the population mean. Find the actual confidence level for the given interval.

Multiple Choice: *Select the best answer for Exercises 21–24.*

21. One reason for using a t distribution instead of the standard Normal distribution to find critical values when calculating a level C confidence interval for a population mean is that

(a) z can be used only for large samples.

(b) z requires that you know the population standard deviation σ.

(c) z requires that you can regard your data as an SRS from the population.

(d) z requires that the sample size is less than 10% of the population size.

(e) a z critical value will lead to a wider interval than a t critical value.

22. You have an SRS of 23 observations from a large population. The distribution of sample values is roughly symmetric with no outliers. What critical value would you use to obtain a 98% confidence interval for the mean of the population?

(a) 2.177   (d) 2.500

(b) 2.183   (e) 2.508

(c) 2.326

23. A quality control inspector will measure the salt content (in milligrams) in a random sample of bags of potato chips from an hour of production. Which of the following would result in the smallest margin of error in estimating the mean salt content μ?

(a) 90% confidence; $n = 25$

(b) 90% confidence; $n = 50$

(c) 95% confidence; $n = 25$

(d) 95% confidence; $n = 50$

(e) $n = 100$ at any confidence level

24. Scientists collect data on the blood cholesterol levels (milligrams per deciliter of blood) of a random sample of 24 laboratory rats. A 95% confidence interval for the mean blood cholesterol level μ is 80.2 to 89.8. Which of the following would cause the most worry about the validity of this interval?

(a) There is a clear outlier in the data.

(b) A stemplot of the data shows a mild right skew.

(c) You do not know the population standard deviation σ.

(d) The population distribution is not exactly Normal.

(e) None of these are a problem when using a t interval.

Recycle and Review

25. **Watching TV** (6.1, 7.3) Choose a young person (aged 19 to 25) at random and ask, "In the past seven days, how many days did you watch television?" Call the response X for short. Here is the probability distribution for X:[12]

Days	0	1	2	3	4	5	6	7
Probability	0.04	0.03	0.06	0.08	0.09	0.08	0.05	???

(a) What is the probability that $X = 7$? Justify your answer.

(b) The mean of the random variable X is $\mu_X = 5.44$ and the standard deviation is $\sigma_X = 2.14$. Interpret these values.

(c) Suppose that you asked 100 randomly selected young people (aged 19 to 25) to respond to the question and found that the mean $\bar{x}$ of their responses was 4.96. Would this result surprise you? Justify your answer.

26. **Price cuts** (4.2) Stores advertise price reductions to attract customers. What type of price cut is most attractive? Experiments with more than one factor allow insight into interactions between the factors. A study of the attractiveness of advertised price discounts had two factors: percent of all foods on sale (25%, 50%, 75%, or 100%) and whether the discount was stated precisely (e.g., as in "60% off") or as a range (as in "40% to 70% off"). Subjects rated the attractiveness of the sale on a scale of 1 to 7.

(a) List the treatments for this experiment, assuming researchers will use all combinations of the two factors.

(b) Describe how you would randomly assign 200 volunteer subjects to treatments.

(c) Explain the purpose of the random assignment in part (b).

(d) The figure shows the mean ratings for the eight treatments formed from the two factors.[13] Based on these results, write a careful description of how percent on sale and precise discount versus range of discounts influence the attractiveness of a sale.

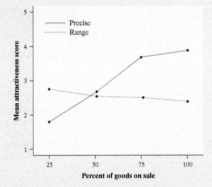

Standard deviation: The number of days of television watched typically varies by about 2.14 days from the mean of 5.44 days. **(c)** Because $n = 100 \geq 30$, we expect the mean number of days $\bar{x}$ for 100 randomly selected young people to be approximately Normally distributed with mean $\mu_{\bar{x}} = \mu = 5.44$. Because the sample size (100) is less than 10% of all young people, $\sigma_{\bar{x}} = \dfrac{\sigma}{\sqrt{n}} = \dfrac{2.14}{\sqrt{100}} = 0.214$. We want $P(\bar{x} \leq 4.96)$.

(i) $z = \dfrac{4.96 - 5.44}{0.214} = -2.24$;

$P(z \leq -2.24) = 0.0125$

(ii) $P(\bar{x} \leq 4.96) = \text{normalcdf(lower: } -1000,$ upper: 4.96, mean: 5.44, SD: 0.214) $=$ 0.0124

There is a 0.0124 probability of getting a sample mean of 4.96 or smaller. Because this probability is small, a sample mean of 4.96 or smaller would be surprising.

10.26 (a) The treatments are:

1: 25% of food on sale, 60% off
2: 50% of food on sale, 60% off
3: 75% of food on sale, 60% off
4: 100% of food on sale, 60% off
5: 25% of food on sale, 40–70% off
6: 50% of food on sale, 40–70% off
7: 75% of food on sale, 40–70% off
8: 100% of food on sale, 40–70% off

(b) Because there are 200 subjects, we label the subjects 001, 002, . . . , 200. Moving from left to right through the table of random digits, select three-digit numbers. The labels 000 and 201 to 999 are not assigned to a subject, so we ignore them. We also ignore any repeats of a label, because that subject has already been assigned to a treatment group. Once we have 25 subjects for the first treatment, we select 25 subjects for the second treatment, and so on, until all subjects have been assigned to a treatment group. **(c)** The purpose of the random assignment in part (b) is to produce treatment groups that are as similar as possible at the beginning of the experiment. **(d)** The range "40% to 70% off" slowly decreases in attractiveness to customers as the percent of food on sale increases. However, the precise "60% off" grows increasingly attractive to customers as the percent of food on sale increases.

10.17 Solving $n \geq \left(\dfrac{2.576(7.5)}{1}\right)^2 = 373.26$.

Select an SRS of 374 women.

10.18 Solving $n \geq \left(\dfrac{1.96(50)}{2}\right)^2 = 2401$. Select a random sample of 2401 students.

10.19 (a) $SE_{\bar{x}} = 19.03 = \dfrac{s_x}{\sqrt{n}} = \dfrac{s_x}{\sqrt{23}}$, so $s_x = 19.03\sqrt{23} = 91.26$. **(b)** Because the researchers are using a critical value of $t^* = 1$. With df $= 23 - 1 = 22$, the area between $t = -1$ and $t = 1$ is approximately tcdf(lower: -1, upper: 1, df: 22) $= 0.67$. So, the confidence level is 67%.

10.20 (a) $SE_{\bar{x}} = 45 = \dfrac{s_x}{\sqrt{n}} = \dfrac{s_x}{\sqrt{12}}$, so $s_x = 45\sqrt{12} = 155.88$. **(b)** Because the researchers are using a critical value of $t^* = 1$. With df $= 12 - 1 = 11$, the area between $t = -1$ and $t = 1$ is approximately tcdf(lower: -1, upper: 1, df: 11) $= 0.66$. So the confidence level is 66%.

10.21 b

10.22 e

10.23 b

10.24 a

10.25 (a) $P(X = 7) = 0.57$ **(b)** *Mean:* If we were to randomly select many young people, the average number of days they watched television in the past 7 days would be about 5.44.

To watch the video overview of Section 10.2 (for teachers), click on the link in the TE-Book, look on the TRFD, or download from the Teacher's Resources on the book's digital platform.

You can find the Alternate Examples for this section in Microsoft Word format by clicking on the link in the TE-Book, opening the TRFD, or downloading from the Teacher's Resources on the book's digital platform.

Teaching Tip

Point out that the conditions for doing inference for a difference in means are much like the conditions for doing inference for a single mean. The main difference is that we have to check the conditions for *both* samples/groups. Also, the Random condition now requires that there be two independent random samples or two groups in a randomized experiment.

SECTION 10.2 # Estimating a Difference in Means

LEARNING TARGETS *By the end of the section, you should be able to:*

- Determine whether the conditions are met for constructing a confidence interval for a difference between two means.
- Construct and interpret a confidence interval for a difference between two means.
- Analyze the distribution of differences in a paired data set using graphs and summary statistics.
- Construct and interpret a confidence interval for a mean difference.

In Section 10.1, you learned how to construct and interpret a confidence interval for a population mean μ. Many interesting statistical questions involve estimating the *difference* between two means based on random samples from the populations of interest. How much larger are the trees in the southern part of a forest compared to trees in the northern part, on average? For wells in a certain region, how much higher is the mean zinc concentration near the bottom of the well than near the top?

Other questions involve estimating the difference in the effectiveness of two treatments in an experiment. How much do pulse rates increase, on average, when drinking cola with caffeine compared to drinking caffeine-free cola? What is the average increase in short-term memory when chewing gum while studying for and taking a test? In this section, you'll learn how to construct confidence intervals to address each of these questions.

Confidence Intervals for $\mu_1 - \mu_2$

When data come from two independent random samples or two groups in a randomized experiment (the Random condition), the statistic $\bar{x}_1 - \bar{x}_2$ is our point estimate for the value of $\mu_1 - \mu_2$. Before constructing a confidence interval for a difference in means, we must check for independence in the data collection process and that the sampling distribution of $\bar{x}_1 - \bar{x}_2$ is approximately Normal.

CONDITIONS FOR CONSTRUCTING A CONFIDENCE INTERVAL ABOUT A DIFFERENCE IN MEANS

- **Random:** The data come from two independent random samples or from two groups in a randomized experiment.
  - **10%:** When sampling without replacement, $n_1 < 0.10N_1$ and $n_2 < 0.10N_2$.
- **Normal/Large Sample:** For *each* sample, the corresponding population distribution (or the true distribution of response to the treatment) is Normal or the sample size is large $(n \geq 30)$. For each sample, if the population (treatment) distribution has unknown shape and $n < 30$, a graph of the sample data shows no strong skewness or outliers.

Teaching Tip

While it is important that students know how to check each condition, it is equally important that they understand *why* we check the condition. We call this the "*So what?*"

Random condition: So what?
→ so we can generalize to both populations (random samples) *or* we can show causation (random assignment).
→ "independent" random samples allow us to add variances in the formula for standard error of the statistic.

10% condition: So what?
→ so we can view observations as independent in each sample even though we are sampling without replacement.

Normal/Large Sample condition: So what?
→ so the sampling distribution of $\bar{x}_1 - \bar{x}_2$ will be approximately Normal and we can use a *t* distribution to do calculations.

Recall from Chapter 4 that the Random condition is important for determining the scope of inference. Random sampling allows us to generalize our results to the populations of interest; random assignment in an experiment permits us to draw cause-and-effect conclusions.

When sampling without replacement, the 10% and Normal/Large Sample conditions are essentially the same as in Section 10.1, except that they need to be checked for *both* samples. In the case of a randomized experiment, the 10% condition doesn't usually apply, but the Normal/Large Sample condition must be checked for *each* treatment group.

EXAMPLE

Do bigger apartments cost more money?
Checking conditions

PROBLEM: A college student wants to compare the cost of one- and two-bedroom apartments near campus. She collects the following data on monthly rents (in dollars) for a random sample of 10 apartments of each type.

1 bedroom	500 650 600 505 450 550 515 495 650 395
2 bedroom	595 500 580 650 675 675 750 500 495 670

Let μ_1 = the true mean monthly rent of all one-bedroom apartments near campus and μ_2 = the true mean monthly rent of all two-bedroom apartments near campus. Check if the conditions for calculating a confidence interval for $\mu_1 - \mu_2$ are met.

SOLUTION:

- **Random:** Independent random samples of 10 one-bedroom apartments and 10 two-bedroom apartments near campus. ✓

  ○ **10%:** We can assume that $10 < 10\%$ of all one-bedroom apartments near campus and that $10 < 10\%$ of all two-bedroom apartments near campus. ✓

- **Normal/Large Sample?** The sample sizes are small, but the dotplots don't show any outliers or strong skewness. ✓

> Be sure to mention *independent* random samples from the populations of interest when checking the Random condition.

> Because there is no strong skewness or outliers in either sample, it is plausible that the population distributions of monthly rent for one-bedroom and two-bedroom apartments near the campus are Normal.

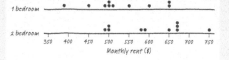

FOR PRACTICE, TRY EXERCISE 27

If the conditions are met, we can use our familiar formula to calculate a confidence interval for $\mu_1 - \mu_2$:

$$\text{statistic} \pm (\text{critical value})(\text{standard error of statistic})$$

Chocolate chip cookies
Checking conditions

PROBLEM:
Ashtyn and Olivia wanted to know if generic chocolate chip cookies have as many chocolate chips as name-brand chocolate chip cookies, on average. To investigate, they randomly selected 10 bags of Chips Ahoy!® cookies and 10 bags of Great Value cookies and randomly selected 1 cookie from each bag. Then they carefully broke apart each cookie and counted the number of chocolate chips in each. Here are their results:

Chips Ahoy: 17 19 21 16 17
                18 20 21 17 18
Great Value: 22 20 14 17 21
                22 15 19 26 18

Let μ_1 = the true mean number of chocolate chips for all Chips Ahoy! chocolate chip cookies and μ_2 = the true mean number of chocolate chips for all Great Value chocolate chip cookies. Check if the conditions for calculating a confidence interval for $\mu_1 - \mu_2$ are met.

SOLUTION:

- *Random:* Independent random samples of 10 Chips Ahoy! chocolate chip cookies and 10 Great Value chocolate chip cookies. ✓

  ○ *10%:* We can assume that $10 < 10\%$ of all Chips Ahoy! chocolate chip cookies and that $10 < 10\%$ of all Great Value chocolate chip cookies. ✓

- *Normal/Large Sample:* The sample sizes are small, but the dotplots don't show any outliers or strong skewness. ✓

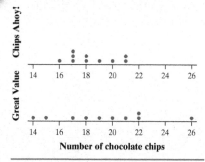

In Section 7.3, you learned that the standard deviation of the sampling distribution of $\bar{x}_1 - \bar{x}_2$ is

$$\sigma_{\bar{x}_1 - \bar{x}_2} = \sqrt{\frac{\sigma_1^2}{n_1} + \frac{\sigma_2^2}{n_2}}$$

when we have two types of independence:

- Independent samples, so we can add the variances of $\bar{x}_1$ and $\bar{x}_2$. This is why we reminded you to mention *independent* random samples when checking the Random condition in the preceding example.
- Independent observations within each sample. When sampling without replacement, the actual value of the standard deviation is smaller than the formula suggests. However, if the 10% condition is met for both samples, the given formula is approximately correct.

Because we usually don't know the values of σ_1 and σ_2, we replace them with the sample standard deviations s_1 and s_2. The result is the *standard error* of $\bar{x}_1 - \bar{x}_2$:

$$SE_{\bar{x}_1 - \bar{x}_2} = \sqrt{\frac{s_1^2}{n_1} + \frac{s_2^2}{n_2}}$$

This value estimates how much the difference in sample means will typically vary from the difference in the true means if we repeat the random sampling or random assignment many times.

When the Normal/Large Sample condition is met, we find the critical value t^* for a given confidence level using Table B or technology. Our confidence interval for $\mu_1 - \mu_2$ is therefore

statistic ± (critical value)(standard error of statistic)

$$= (\bar{x}_1 - \bar{x}_2) \pm t^* \sqrt{\frac{s_1^2}{n_1} + \frac{s_2^2}{n_2}}$$

This is often called a **two-sample t interval for a difference between two means**.

There is just one issue left to resolve: What df should we use to find the t^* critical value? It turns out that there are two practical options.

Option 1 (Technology): Use the t distribution with degrees of freedom calculated by technology using the following formula. Note that the df given by this formula is usually *not* a whole number. This option results in confidence intervals with a margin of error that is approximately correct for the stated confidence level. That is, about 95% of the confidence intervals calculated using Option 1 and a 95% confidence level will capture the true difference in means.

$$df = \frac{\left(\frac{s_1^2}{n_1} + \frac{s_2^2}{n_2}\right)^2}{\frac{1}{n_1-1}\left(\frac{s_1^2}{n_1}\right)^2 + \frac{1}{n_2-1}\left(\frac{s_2^2}{n_2}\right)^2}$$

Option 2 (Conservative): Use the t distribution with degrees of freedom equal to the *smaller* of $n_1 - 1$ and $n_2 - 1$. With this option, the resulting confidence interval has a margin of error *as large as or larger than* is needed for the desired confidence level.

As you can imagine, Option 2 was much more popular in the days when Table B and a four-function calculator were the main calculation tools. We prefer Option 1 because it gives a larger number of degrees of freedom than Option 2. This results in a smaller margin of error for a given confidence level and sample size.

Fortunately, the formula we use for the standard deviation is the same whether we have two independent samples or two groups in a randomized experiment. For more details, see the Think About It on page 726.

The formula sheet provided on the AP® Statistics exam uses the notation $s_{\bar{x}_1 - \bar{x}_2}$ rather than $SE_{\bar{x}_1 - \bar{x}_2}$ for the standard error of the difference in sample means $\bar{x}_1 - \bar{x}_2$.

Statisticians B. L. Welch and F. E. Satterthwaite discovered this fairly remarkable formula in the 1940s.

Teaching Tip

It is not necessary to do a calculation with this formula by hand. Using 2-SampTInt or 2-SampTTest on the calculator will give students the degrees of freedom from this formula. Show students this formula in the book so that they can appreciate what the calculator is giving them. This option is preferred over Option 2 because it gives narrower confidence intervals, smaller *P*-values, and consequently more power.

Teaching Tip

Note that the df from Option 1 is greater than the df from Option 2, but less than the sum of the df for each individual sample:

$(n_1 - 1) + (n_2 - 1) = n_1 + n_2 - 2$

Teaching Tip

Option 2 is considered the conservative approach because it will always result in a "safer" interval. Option 2 will use a smaller number of degrees of freedom than Option 1, leading to a larger t^* value and margin of error in the interval. This will result in a longer interval, which makes it "safer" to assume a C% capture rate.

In the unlikely event that both population standard deviations σ_1 and σ_2 are known, the interval becomes

$$(\bar{x}_1 - \bar{x}_2) \pm z^* \sqrt{\frac{\sigma_1^2}{n_1} + \frac{\sigma_2^2}{n_2}}$$

This interval is known as a *two-sample z interval for a difference in means*. It is rarely used in practice.

> ### TWO-SAMPLE *t* INTERVAL FOR A DIFFERENCE BETWEEN TWO MEANS
>
> When the conditions are met, a C% confidence interval for $\mu_1 - \mu_2$ is
>
> $$(\bar{x}_1 - \bar{x}_2) \pm t^* \sqrt{\frac{s_1^2}{n_1} + \frac{s_2^2}{n_2}}$$
>
> where t^* is the critical value with C% of its area between $-t^*$ and t^* for the *t* distribution with degrees of freedom using either Option 1 (technology) or Option 2 (the smaller of $n_1 - 1$ and $n_2 - 1$).

Let's return to the apartment-rental example. Recall that the college student took independent random samples of $n_1 = 10$ one-bedroom apartments and $n_2 = 10$ two-bedroom apartments. Here are summary statistics on the monthly rents (in dollars) of these apartments:

Group name	n	Mean	SD	Min	Q_1	Med	Q_3	Max
1: 1-bedroom	10	531	82.792	395	495	510.0	600	650
2: 2-bedroom	10	609	89.312	495	500	622.5	675	750

We already confirmed that the conditions are met.

Using Option 1, df = 17.898 and the 90% confidence interval is $(-144.802, -11.198)$. See the Technology Corner after the next example for details on how to obtain this interval on the TI-83/84.

Using Option 2, df = the smaller of $n_1 - 1$ and $n_2 - 1 = 9$. The critical value for a 90% confidence level in a *t* distribution with 9 degrees of freedom is $t^* = 1.833$. So the resulting interval is

$$(531 - 609) \pm 1.833 \sqrt{\frac{82.792^2}{10} + \frac{89.312^2}{10}}$$

$$= -78 \pm 70.591$$

$$= (-148.591, -7.409)$$

Notice that this interval is wider than the one obtained from technology. Because Option 1 uses a *t* distribution with more degrees of freedom, it produces a more precise estimate of $\mu_1 - \mu_2$ than Option 2.

Because Option 1 produces more precise estimates of $\mu_1 - \mu_2$ than Option 2, we will always interpret the confidence interval obtained from technology.

Based on the sample, we are 90% confident that the interval from -144.802 to -11.198 dollars captures $\mu_1 - \mu_2$ = the difference in the true mean monthly rents of one-bedroom and two-bedroom apartments close to the college campus. The interval suggests that the mean monthly rent of one-bedroom apartments is between $11.20 and $144.80 less than the mean monthly rent for two-bedroom apartments near campus.

Putting It All Together: Two-Sample *t* Interval for $\mu_1 - \mu_2$

The following example shows how to construct and interpret a confidence interval for a difference in means. As usual with inference problems, we follow the four-step process.

ALTERNATE EXAMPLE   Skills 3.D, 4.B

Katy Perry or Justin Timberlake?
Confidence interval for $\mu_1 - \mu_2$

PROBLEM:
Students wondered whose Twitter account gets more likes, on average: Katy Perry's or Justin Timberlake's. The students selected a random sample of 15 tweets from Katy Perry and a separate random sample of 15 tweets from Justin Timberlake and recorded the number of likes for each tweet. Here are the dotplots of the data and summary statistics:

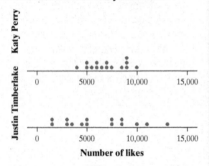

	n	mean	SD
Katy Perry	15	7193	1749
Justin Timberlake	15	6393	3470

(a) Based on the graph and numerical summaries, write a few sentences comparing the number of likes for Katy Perry's tweets with those for Justin Timberlake's tweets.
(b) Construct and interpret a 95% confidence interval for the difference in the mean number of likes for tweets from Katy Perry and Justin Timberlake.

SOLUTION:
(a) *Shape:* The distribution of number of likes for tweets from Katy Perry is fairly symmetric, while the distribution of likes for tweets from Justin Timberlake is slightly skewed to the right.
Outliers: There are no obvious outliers in either graph.
Center: It appears that Katy Perry has more likes for her tweets, on average. The mean number of likes for tweets from Katy Perry is larger than the mean number of likes for tweets from Justin Timberlake.
Variability: There is more variability in the number of likes for tweets from Justin Timberlake. The range and standard deviation of the number of likes are larger for Justin Timberlake.
(b) STATE: 95% CI for $\mu_{KP} - \mu_{JT}$, where μ_{KP} = the true mean number of likes for all tweets from Katy Perry and μ_{JT} = the true mean number of likes for all tweets from Justin Timberlake.

EXAMPLE

Big trees, small trees, short trees, tall trees
Confidence interval for $\mu_1 - \mu_2$

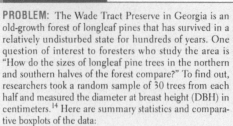

PROBLEM: The Wade Tract Preserve in Georgia is an old-growth forest of longleaf pines that has survived in a relatively undisturbed state for hundreds of years. One question of interest to foresters who study the area is "How do the sizes of longleaf pine trees in the northern and southern halves of the forest compare?" To find out, researchers took a random sample of 30 trees from each half and measured the diameter at breast height (DBH) in centimeters.[14] Here are summary statistics and comparative boxplots of the data:

Descriptive Statistics: North, South

Group	n	Mean	StDev
North	30	23.70	17.50
South	30	34.53	14.26

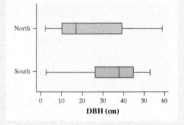

(a) Based on the graph and numerical summaries, write a few sentences comparing the sizes of longleaf pine trees in the two halves of the forest.
(b) Construct and interpret a 90% confidence interval for the difference in the mean DBH of longleaf pines in the northern and southern halves of the Wade Tract Preserve.

SOLUTION:
(a) **Shape:** The distribution of DBH in the northern sample appears skewed to the right, while the distribution of DBH in the southern sample appears skewed to the left.
Outliers: No outliers are present in either sample.
Center: It appears that trees in the southern half of the forest have larger diameters. The mean and median DBH for the southern sample are much larger than the corresponding values for the northern sample.
Variability: There is more variability in the DBH of the northern longleaf pines. The range, IQR, and standard deviation are all larger for the northern sample.
(b) STATE: 90% CI for $\mu_1 - \mu_2$, where μ_1 = the true mean DBH of all trees in the southern half of the forest and μ_2 = the true mean DBH of all trees in the northern half of the forest.

> It would not be correct to say that the northern half of the forest is skewed to the right! Only distributions of quantitative variables (like DBH) can be skewed.

> Furthermore, the boxplots show that more than 75% of the southern trees have diameters that are greater than the northern sample's median.

> Don't forget to include context (the variable of interest) when comparing distributions of quantitative data. In this case, that's DBH.

> Be sure to indicate the order of subtraction when defining the parameter.

PLAN: Two-sample *t* interval for $\mu_{KP} - \mu_{JT}$
- *Random:* Independent random samples of 15 tweets each from Katy Perry and Justin Timberlake. ✓
  - *10%:* 15 < 10% of all Tweets from Katy Perry and 15 < 10% of all tweets from Justin Timberlake. ✓
- *Normal/Large Sample:* The sample sizes are small, but the dotplots don't show any outliers or strong skewness. ✓

DO:
Option 1: 2-SampTInt gives (−1288, 2888.5) using df = 20.682.
Option 2: df = 14, t* = 2.145

$$(7193 - 6393) \pm 2.145\sqrt{\frac{1749^2}{15} + \frac{3470^2}{15}}$$
$$= (-1352.13, 2952.13)$$

CONCLUDE: We are 95% confident that the interval from −1288 to 2888.5 captures $\mu_{KP} - \mu_{JT}$ = the difference in the true mean number of likes for tweets from Katy Perry and Justin Timberlake.

PLAN: Two-sample t interval for $\mu_1 - \mu_2$

- Random: Independent random samples of 30 trees each from the northern and southern halves of the forest. ✓
  - 10%: Assume $30 < 10\%$ of all trees in the northern half of the forest and $30 < 10\%$ of all trees in the southern half of the forest. ✓
- Normal/Large Sample: $n_1 = 30 \geq 30$ and $n_2 = 30 \geq 30$. ✓

DO: $\bar{x}_1 = 34.53, s_1 = 14.26, n_1 = 30,$
$\bar{x}_2 = 23.70, s_2 = 17.50, n_2 = 30$

Option 1: 2-SampTInt gives (3.9362, 17.724) using $df = 55.728$

Refer to Technology Corner 24 for details on how to do calculations for a 2-sample t interval on the TI-83/84.

Option 2: $df = 29, t^* = 1.699$

For Option 2, df = the smaller of $n_1 - 1$ and $n_2 - 1$

$$= (34.53 - 23.70) \pm 1.699\sqrt{\frac{14.26^2}{30} + \frac{17.50^2}{30}}$$

$$= 10.83 \pm 7.00$$

$$= (3.83, 17.83)$$

$(\bar{x}_1 - \bar{x}_2) \pm t^*\sqrt{\dfrac{s_1^2}{n_1} + \dfrac{s_2^2}{n_2}}$

CONCLUDE: We are 90% confident that the interval from 3.9362 to 17.724 centimeters captures $\mu_1 - \mu_2$ = the difference in the true mean DBH of all the southern trees and the true mean DBH of all the northern trees.

FOR PRACTICE, TRY EXERCISE 31

Simulation studies reveal that the two-sample t procedures are most accurate when the sizes of the two samples are equal and the population (treatment) distributions have similar shapes. In planning a two-sample study, choose equal sample sizes whenever possible.

The 90% confidence interval in the example does not include 0. This gives convincing evidence that the difference in the mean diameter of northern and southern trees in the Wade Tract Preserve isn't 0. However, the confidence interval provides more information than a simple reject or fail to reject H_0 conclusion. It gives a set of plausible values for $\mu_1 - \mu_2$. The interval suggests that the mean diameter of the southern trees is between 3.94 and 17.72 cm larger than the mean diameter of the northern trees.

We chose the parameters in the DBH example so that $\bar{x}_1 - \bar{x}_2$ would be positive. What if we had defined μ_1 as the true mean DBH of the northern trees and μ_2 as the true mean DBH of the southern trees? The 90% confidence interval for $\mu_1 - \mu_2$ from technology is $(-17.72, -3.94)$. This interval suggests that the mean diameter of the northern trees is between 3.94 and 17.72 cm smaller than the mean diameter of the southern trees. Changing the order of subtraction doesn't change the result.

Notice that the interval produced by technology is narrower than the one calculated using the conservative method. That's because technology uses the formula on page 670 to obtain a larger (and more accurate) value of df. To further narrow the confidence interval, researchers could increase the sample sizes—although that would cost additional time and money.

Teaching Tip

Suppose you have the work for the "Big trees, small trees. . ." example up on the whiteboard. Ask students, "What parts of this work would change if we switched the order of subtraction?" Using a different color marker, cross off the parts that change, and write in their replacement calculations:

- The $(34.53 - 23.70)$ becomes $(23.70 - 34.53)$.
- The 10.83 becomes -10.83.
- The $(3.83, 17.83)$ becomes $(-17.83, -3.83)$.

Teaching Tip

Remind students that it's OK if the endpoints of the interval in their answers have opposite signs as the endpoints of the interval in the back-of-book answers. Either subtraction order is correct, as long as students clearly identify the order they are using.

Teaching Tip

Here are some important tips for teaching students to use the four-step process for constructing a two-sample t interval for $\mu_1 - \mu_2$:

- Inform students they can use the wording of the question to define their parameters and to write the conclusion. We call this "parroting" the stem of the question.
- When naming the inference method, it is acceptable to list the calculator command "2-SampTInt." We prefer that students use the two-sample t interval for $\mu_1 - \mu_2$, because it demonstrates deeper understanding and leads to fewer misconceptions.
- Consider having students write the "So what?" for each condition.
- Consider having students write a general formula and a specific formula with variables before substituting numbers:

General formula: estimate $\pm$ margin of error

Specific formula: $(\bar{x}_1 - \bar{x}_2) \pm t^*\sqrt{\dfrac{s_1^2}{n_1} + \dfrac{s_2^2}{n_2}}$

One advantage to establishing this expectation is that the general formula is the same for all the confidence intervals in this course. Only the specific formula changes for each new confidence interval. It is risky to provide this much work on the AP® Statistics exam because students may lose credit due to a minor notation/substitution error. Consider requiring these formulas during the school year and then relaxing this expectation at the end of the course when preparing for the AP® exam.

Teaching Tip

We strongly recommend that students use technology for two-sample *t* procedures to ease the computations and take advantage of the larger number of degrees of freedom.

Teaching Tip: Using Technology

For two-sample *t* procedures, we always recommend saying "No" to pooling. We'll discuss why on page 727, but for now, tell your students: "Just say No!"

As with other inference procedures, you can use technology to perform the calculations in the "Do" step. Remember that technology comes with potential benefits and risks on the AP® Statistics exam.

24. Technology Corner | CONSTRUCTING A CONFIDENCE INTERVAL FOR A DIFFERENCE IN MEANS

TI-Nspire and other technology instructions are on the book's website at highschool.bfwpub.com/updatedtps6e.

You can use the two-sample *t* interval option on the TI-83/84 to construct a confidence interval for the difference between two means. Let's confirm the results of the two previous examples with this feature.

1. Do bigger apartments cost more money? (page 669)
   - Enter the one-bedroom monthly rents in L1 and the two-bedroom monthly rents in L2.
   - Press STAT, then choose TESTS and 2-SampTInt....
   - Choose Data as the input method and enter the inputs as shown.
   - Enter the confidence level: C-level: 0.90. For Pooled: choose "No." We'll discuss pooling in Section 11.2.
   - Highlight Calculate and press ENTER.

 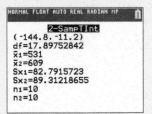

2. Big Trees, Small Trees, Short Trees, Tall Trees (page 672)
   - Press STAT, then choose TESTS and 2-SampTInt....
   - Choose Stats as the input method and enter the summary statistics as shown.
   - Enter the confidence level: C-level: 0.90. For Pooled: choose "No." We'll discuss pooling in Section 11.2.
   - Highlight Calculate and press ENTER.

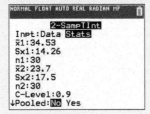 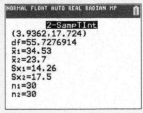

The formula for the two-sample t interval for $\mu_1 - \mu_2$ often leads to calculation errors by students. Also, the interval produced by technology is narrower than the one calculated using the conservative method. As a result, your teacher may recommend using the calculator's 2-SampTInt feature to compute the confidence interval. Be sure to name the procedure (two-sample t interval for $\mu_1 - \mu_2$) in the "Plan" step and give the interval (3.9362, 17.724) and df (55.728) in the "Do" step.

CHECK YOUR UNDERSTANDING

Mr. Wilcox's class performed an experiment to investigate whether drinking a caffeinated beverage would increase pulse rates. Twenty students in the class volunteered to take part in the experiment. All of the students measured their initial pulse rates (in beats per minute). Then Mr. Wilcox randomly assigned the students into two groups of 10. Each student in the first group drank 12 ounces of cola with caffeine. Each student in the second group drank 12 ounces of caffeine-free cola. All students then measured their pulse rates again. The table displays the change in pulse rate for the students in both groups.

	Change in pulse rate (Final pulse rate − Initial pulse rate)										Mean change
Caffeine	8	3	5	1	4	0	6	1	4	0	3.2
No caffeine	3	−2	4	−1	5	5	1	2	−1	4	2.0

1. Construct and interpret a 95% confidence interval for the difference in true mean change in pulse rate for subjects like these who drink caffeine versus who drink no caffeine.
2. What does the interval in Question 1 suggest about whether caffeine increases the average pulse rate of subjects like these? Justify your answer.

Comparing Two Means: Paired Data

Sometimes we want to compare means in a setting that involves measuring a quantitative variable twice for the same individual or for two individuals who are much alike. For instance, a researcher studied a random sample of identical twins who had been separated and adopted at birth. In each case, one twin (Twin A) was adopted by a high-income family and the other (Twin B) by a low-income family. Both twins were given an IQ test as adults. Here are their scores:[15]

Pair	1	2	3	4	5	6	7	8	9	10	11	12
Twin A's IQ (high-income family)	128	104	108	100	116	105	100	100	103	124	114	112
Twin B's IQ (low-income family)	120	99	99	94	111	97	99	94	104	114	113	100

Notice that these two groups of IQ scores did *not* come from independent samples of people who were raised in low-income and high-income families. The

 COMMON STUDENT ERROR

Students often struggle to distinguish between paired data and two-sample data. We decided to put paired data in the same section as two-sample data to help make this distinction clearer for students.

Teaching Tip

Notice that this twin scenario is an observational study, rather than an experiment. The twins were not randomly assigned to the high-income and low-income family. This means that we cannot make any conclusions about cause and effect.

Teaching Tip

Another option is to require students to show the calculations for their work within this chapter. Then at the end of the course, when preparing for the AP® Statistics exam, relax this expectation and suggest that students instead use the calculator to get the interval.

✓ **Answers to CYU**

1. STATE: 95% CI for $\mu_1 - \mu_2$, where $\mu_1 =$ the true mean change in pulse rate for students like these after drinking 12 ounces of cola with caffeine and $\mu_2 =$ the true mean change in pulse rate for students like these after drinking 12 ounces of caffeine-free cola.

PLAN: Two-sample t interval for $\mu_1 - \mu_2$.

- *Random:* The volunteers were randomly assigned to drink either cola with caffeine or caffeine-free cola. ✓
- *Normal/Large Sample:* The sample sizes are small, but the dotplots do not show any outliers or strong skewness. ✓

Caffeine ·· · ·· · ·
No caffeine · · ·· · ·
−2 −1 0 1 2 3 4 5 6 7 8

Change (Final − Initial) in pulse rate (beats/min)

DO: $\bar{x}_1 = 3.2$, $s_1 = 2.70$, $n_1 = 10$, $\bar{x}_2 = 2$, $s_2 = 2.62$, $n_2 = 10$

Option 1: 2-Samp TInt gives $(-1.302, 3.702)$ using df $= 17.986$.

Option 2: df $= 9$, $t^* = 2.262$

$$(3.2 - 2) \pm 2.262 \sqrt{\frac{2.70^2}{10} + \frac{2.62^2}{10}} =$$

$1.2 \pm 2.691 = (-1.491, 3.891)$.

CONCLUDE: We are 99% confident that the interval from -1.491 to 3.891 beats per minute captures $\mu_1 - \mu_2 =$ the difference in the true mean change in pulse rate for all students like these after drinking 12 ounces of cola with caffeine versus without caffeine.

2. The interval in Question 1 does not suggest that caffeine increases the average pulse rate of subjects like these. Because the interval contains 0, it is plausible that the true mean change in pulse rate for all students like these after drinking 12 ounces of cola with caffeine versus without caffeine may equal 0, indicating no difference.

Teaching Tip

Paired data can be produced in observational studies and in experiments. In an experiment, a matched pairs design is a special case of a randomized block design wherein each block has two units. In experiments or observational studies, using paired data can increase power by accounting for a source of variability in the response variable.

Teaching Tip

In fact, the two-sample *t* interval for the difference in mean IQ scores is −1.85 to 13.52, which includes 0 as a plausible value. Note that we shouldn't actually do this calculation, as the samples are not independent (because the data are paired!).

data were obtained from *pairs* of very similar people (identical twins), one living with a low-income family and the other living with a high-income family. This set of IQ scores is an example of **paired data**.

DEFINITION **Paired data**

Paired data result from recording two values of the same quantitative variable for each individual or for each pair of similar individuals.

The remainder of this section focuses on how to analyze paired data and how to perform inference about a true *mean difference*.

ANALYZING PAIRED DATA   The graph in Figure 10.3 shows a parallel dotplot of the IQ scores from the study of identical twins. We can see that the twins raised in high-income households had a higher mean IQ ($\bar{x}_A = 109.5$) than the twins raised in low-income households ($\bar{x}_B = 103.667$). There is a similar amount of variability in IQ scores for these two groups of twins: $s_A = 9.47$ and $s_B = 8.66$. But with so much overlap between the groups, the difference in means does not seem statistically significant.

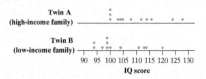

FIGURE 10.3 Parallel dotplots of the IQ scores for pairs of identical twins raised in high-income (Twin A) and low-income (Twin B) households.

The previous analysis ignores the fact that these are *paired* data. Let's look at the difference in IQ scores for each pair of twins.

Pair	1	2	3	4	5	6	7	8	9	10	11	12
Twin A's IQ (high-income family)	128	104	108	100	116	105	100	100	103	124	114	112
Twin B's IQ (low-income family)	120	99	99	94	111	97	99	94	104	114	113	100
Difference (A − B)	8	5	9	6	5	8	1	6	−1	10	1	12

The dotplot in Figure 10.4 displays these differences. Almost all of the differences (11 out of 12) are positive. This graph gives strong evidence that identical twins raised in a high-income household have higher IQ scores as adults, on average, than identical twins raised in a low-income household.

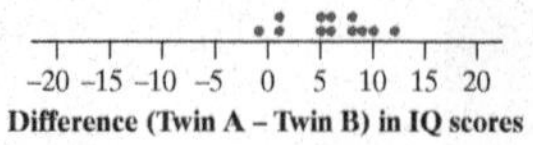

FIGURE 10.4 Dotplot of difference in IQ scores for each pair of twins.

```
        • •  ••  ••• •  •
 -20 -15 -10 -5  0   5  10  15  20
 Difference (Twin A – Twin B) in IQ scores
```

The *mean difference* in IQ scores is

$$\bar{x}_{\text{diff}} = \bar{x}_{\text{A−B}} = \frac{8+5+9+\cdots+1+12}{12} = \frac{70}{12} = 5.833 \text{ points}$$

Making Connections

Consider the scope of inference for the identical twins study. We can generalize to the population of all identical twins reared apart because we had a random sample of such twins. We cannot infer a cause-and-effect relationship between family income and twins' IQ here because there was no random assignment.

Making Connections

Even though we are presented with two lists of values, we are doing analysis with only a single list of values (the differences). Therefore, this is a one-sample scenario and the analysis requires one-sample inference procedures.

This value tells us that the IQ score of the twin in each pair who was raised in a high-income household is 5.83 points higher than the twin who was raised in a low-income household, on average.

The standard deviation of the difference in IQ scores is $s_{\text{diff}} = 3.93$ points. This value is much smaller than the standard deviations we computed earlier when we (incorrectly) viewed the two groups of twins as unrelated: $s_A = 9.47$ points and $s_B = 8.66$ points. Remember: The proper method of analysis depends on how the data are produced.

ANALYZING PAIRED DATA

To analyze paired data, start by computing the difference for each pair. Then make a graph of the differences. Use the mean difference $\bar{x}_{\text{diff}}$ and the standard deviation of the differences s_{diff} as summary statistics.

EXAMPLE

Math and music
Analyzing paired data

gaurav/Getty Images

PROBLEM: Does music help or hinder performance in math? Student researchers Abigail, Carolyn, and Leah designed an experiment using 30 student volunteers to find out. Each subject completed a 50-question single-digit arithmetic test with and without music playing. For each subject, the order of the music and no music treatments was randomly assigned, and the time to complete the test in seconds was recorded for each treatment. Here are the data, along with the difference in time for each subject:

Student	1	2	3	4	5	6	7	8	9	10	11	12	13	14	15
Time with music (sec)	83	119	77	75	64	106	70	69	60	76	47	97	68	77	48
Time without music (sec)	70	106	71	67	59	112	83	69	65	83	38	90	76	68	50
Difference (Music – Without music)	13	13	6	8	5	–6	–13	0	–5	–7	9	7	–8	9	–2
Student	16	17	18	19	20	21	22	23	24	25	26	27	28	29	30
Time with music (sec)	78	113	71	77	37	50	58	52	47	71	146	44	53	57	39
Time without music (sec)	73	93	59	70	39	52	60	54	51	60	141	40	56	53	37
Difference (Music – Without music)	5	20	12	7	–2	–2	–2	–2	–4	11	5	4	–3	4	2

(a) Make a dotplot of the difference (Music – Without music) in time for each subject to complete the test.
(b) Describe what the graph reveals about whether music helps or hinders math performance.
(c) Calculate the mean difference and the standard deviation of the differences. Interpret the mean difference.

(a) Make a dotplot of the difference (Version A – Version B) in scores for each student.
(b) Describe what the graph reveals about whether Version A is harder or easier than Version B.
(c) Calculate the mean difference and the standard deviation of the differences. Interpret the mean difference.

SOLUTION:
(a)

Difference (A – B) in exam score

(b) There is some evidence that Version A was easier than Version B. Of the 20 students, 16 did better on Version A than Version B.

(c) Mean: $\bar{x}_{\text{diff}} = \bar{x}_{A-B} = 3.2$, SD: $s_{\text{diff}} = 3.533$

The score for Version A of the final exam for these 20 students was 3.2 points greater, on average, than their score for Version B of the final exam.

ALTERNATE EXAMPLE

Skill 1.D

Final exam revisited   Analyzing paired data

PROBLEM:

In a previous alternate exercise, Mrs. Gallas hoped to determine if the two versions of her AP® Statistics final exam were equally difficult. Last year, she set up a randomized comparative experiment, and the data showed no statistically significant difference in the scores between the two versions. This year, she selected a random sample of 20 students from the large district and used a matched pairs design, where each student was given both versions of the test. For each student, the test order was randomized. Here are the scores, along with the differences in scores for each student:

Student	1	2	3	4	5	6	7	8	9	10	11	12	13	14	15	16	17	18	19	20
Version A	90	85	77	96	88	83	93	81	84	71	65	98	93	83	85	91	82	79	74	72
Version B	86	84	80	91	83	80	87	83	80	70	60	99	96	75	78	88	81	74	65	66
Difference (A – B)	4	1	–3	5	5	3	6	–2	4	1	5	–1	–3	8	7	3	1	5	9	6

SOLUTION:

(a)

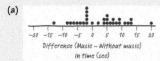

Difference (Music – Without music)
in time (sec)

(b) There is some evidence that music hinders performance on the math test. 17 of the 30 subjects took longer to complete the test when listening to music.

> To get these summary statistics using the TI-83/84, start by typing the difference values into L1. Then do 1-Var Stats.

(c) Mean: $\bar{x}_{diff} = \bar{x}_{Music-Without} = 2.8$ seconds, SD: $s_{diff} = 7.49$ seconds.

The time it took these 30 students to complete the arithmetic quiz with music was 2.8 seconds longer, on average, than the time it took without the music.

FOR PRACTICE, TRY EXERCISE 37

There are two ways that a statistical study involving a single quantitative variable can yield paired data:

1. Researchers can record two values of the variable for each individual.
2. The researcher can form pairs of similar individuals and record the value of the variable once for each individual.

We have seen one example of each method so far. The observational study of identical twins' IQ scores used Method 2, with the pairs consisting of identical twins who were raised separately—one in a high-income household and one in a low-income household. The experiment investigating whether music helps or hinders learning used Method 1, with each subject taking a 50-question single-digit arithmetic test twice—once with music playing and once without—in a random order. We referred to this type of experiment as a *matched pairs* design in Chapter 4. Note that it is also possible to carry out a matched pairs experiment using Method 2 if the researcher forms pairs of similar subjects and randomly assigns each treatment to exactly one member of every pair.

Confidence Intervals for μ_{diff}

When paired data come from a random sample or a randomized experiment, the statistic $\bar{x}_{diff}$ is a point estimate for the true mean difference μ_{diff}. Before constructing a confidence interval for a mean difference, we must check that the conditions for performing inference are met. Aside from the paired data requirement, these conditions are the same as the ones for constructing a confidence interval for μ in Section 10.1.

CONDITIONS FOR CONSTRUCTING A CONFIDENCE INTERVAL ABOUT A MEAN DIFFERENCE

- **Random:** Paired data come from a random sample from the population of interest or from a randomized experiment.
  - **10%:** When sampling without replacement, $n_{\text{diff}} < 0.10N_{\text{diff}}$.
- **Normal/Large Sample:** The population distribution of differences (or the true distribution of differences in response to the treatments) is Normal or the number of differences in the sample is large ($n_{\text{diff}} \geq 30$). If the population (true) distribution of differences has unknown shape and the number of differences in the sample is less than 30, a graph of the sample differences shows no strong skewness or outliers.

The Random condition reminds us that we need paired data to construct a confidence interval for μ_{diff}. Both the 10% and Normal/Large Sample conditions emphasize that the proper method of analyzing paired data is to focus on the differences within each pair.

All of the inference procedures you have learned so far require that individual observations can be viewed as independent. That's why we check the 10% condition when sampling without replacement. For paired data, the two values of the response variable in each pair are generally *not* independent. After all, we are measuring the same quantitative variable twice for one individual or for two very similar individuals. Knowing the value of the variable for one of the measurements should help us predict the value for the other measurement. What is important is that the *difference* values be independent from each other.

You learned how to construct a confidence interval for a population mean μ in Section 10.1. Assuming that the population standard deviation σ is unknown, we can calculate a one-sample t interval for the mean. The appropriate formula is

$$\text{statistic} \pm (\text{critical value})(\text{standard error of statistic})$$
$$= \quad \bar{x} \pm t^* \frac{s_x}{\sqrt{n}}$$

We find the critical value t^* for a given confidence level from a t distribution with df $= n-1$ using Table B or technology.

Now we want to estimate the true mean difference μ_{diff} based on a single sample of differences calculated from paired data. All we have to do is modify the above formula to fit the new setting:

$$\text{statistic} \pm (\text{critical value})(\text{standard error of statistic})$$
$$= \quad \bar{x}_{\text{diff}} \pm t^* \frac{s_{\text{diff}}}{\sqrt{n_{\text{diff}}}}$$

This can be referred to as a **one-sample t interval for a mean difference** or as a **paired t interval for a mean difference**.

Teaching Tip

While it is important that students know how to check each condition, it is equally important that they understand *why* we check the condition. We call this the "*So what?*"

Random condition: So what?
→ so individual differences are independent.
→ so we can generalize to the population of all pairs (random sample) *or* so we can show causation (random assignment).

10% condition: So what?
→ so we can view differences as independent even though we are sampling without replacement.

Normal/Large Sample condition: So what?
→ so the sampling distribution of $\bar{x}_{\text{diff}}$ will be approximately Normal, and we can use a t distribution to do calculations.

 COMMON STUDENT ERROR

Having just done two-sample t intervals, many students incorrectly state "independent random samples" as the check of the random condition when analyzing paired data. This is the opposite of the truth—if the data are paired, the two samples aren't independent!

> ### ONE-SAMPLE t INTERVAL FOR A MEAN DIFFERENCE (PAIRED t INTERVAL FOR A MEAN DIFFERENCE)
>
> When the conditions are met, a C% confidence interval for μ_{diff} is
>
> $$\bar{x}_{\text{diff}} \pm t^* \frac{s_{\text{diff}}}{\sqrt{n_{\text{diff}}}}$$
>
> where t^* is the critical value with C% of its area between $-t^*$ and t^* for the t distribution with $n_{\text{diff}} - 1$ degrees of freedom.

As with any inference procedure, follow the four-step process.

ALTERNATE EXAMPLE Skills 3.D, 4.B

Final exam revisited
Confidence interval for a mean difference

PROBLEM:
In the preceding alternate example, Mrs. Gallas used a matched pairs design to collect data on scores for two different versions of her final exam for 20 students. Recall that $\bar{x}_{\text{diff}} = \bar{x}_{A-B} = 3.2$ and $s_{\text{diff}} = 3.533$. Construct and interpret a 99% confidence interval for the true mean difference (Version A – Version B) in final exam scores for AP® Statistics students in this district.

SOLUTION:
STATE: 99% CI for μ_{diff} = the true mean difference (Version A – Version B) in final exam scores for AP® Statistics students in this district.

PLAN: One-sample t interval for μ_{diff}

- *Random:* Random sample of 20 AP® Statistics students in the district. Also, students were randomly assigned the order of the two versions of the exam. ✓
  - *10%:* Assume 20 < 10% of all AP® Statistics students in the large district. ✓
- *Normal/Large Sample:* The number of differences is small, but the dotplot doesn't show any strong skewness or outliers. ✓

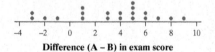

Difference (A – B) in exam score

DO: With 99% confidence and df = 20 − 1 = 19, $t^* = 2.861$.

$$3.2 \pm 2.861 \frac{3.533}{\sqrt{20}} = (0.94, 5.46)$$

CONCLUDE: We are 99% confident that the interval from 0.94 to 5.46 captures the true mean difference (Version A – Version B) in final exam scores for AP® Statistics in this district.

EXAMPLE

Which twin is smarter?
Confidence interval for a mean difference

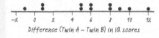

PROBLEM: The data from the random sample of identical twins are shown again in the following table. Construct and interpret a 95% confidence interval for the true mean difference in IQ scores among twins raised in high-income and low-income households.

Pair	1	2	3	4	5	6	7	8	9	10	11	12
Twin A's IQ (high-income family)	128	104	108	100	116	105	100	100	103	124	114	112
Twin B's IQ (low-income family)	120	99	99	94	111	97	99	94	104	114	113	100
Difference (A – B)	8	5	9	6	5	8	1	6	−1	10	1	12

SOLUTION:
STATE: 95% CI for μ_{diff} = the true mean difference (High income – Low income) in IQ scores for pairs of identical twins raised in separate households.

PLAN: One-sample t interval for μ_{diff}

- Random? Random sample of 12 pairs of identical twins, one raised in a high-income household and the other in a low-income household. ✓
  - 10%: Assume 12 < 10% of all pairs of identical twins raised in separate households. ✓
- Normal/Large Sample? The number of differences is small, but the dotplot doesn't show any strong skewness or outliers. ✓

Difference (Twin A – Twin B) in IQ scores

> Follow the four-step process!
>
> Be sure to indicate the order of subtraction when defining the parameter.

⚠ COMMON STUDENT ERROR

In problems involving paired data, students often graph the two original lists of data instead of making a single graph of the differences. If the problem involves paired data, students should use the differences for each step of the four-step process.

DO: $\bar{x}_{diff} = 5.833, s_{diff} = 3.93, n_{diff} = 12$

With 95% confidence and df $= 12 - 1 = 11, t^* = 2.201$.

$5.833 \pm 2.201 \dfrac{3.93}{\sqrt{12}}$

$= 5.833 \pm 2.497$

$= (3.336, 8.330)$

$$\bar{x}_{diff} \pm t^* \frac{s_{diff}}{\sqrt{n_{diff}}}$$

The TInterval function on the TI-83/84 gives (3.338, 8.329).

CONCLUDE: We are 95% confident that the interval from 3.336 to 8.330 captures the true mean difference (High income − Low income) in IQ scores among pairs of identical twins raised in separate households.

FOR PRACTICE, TRY EXERCISE 41

The 95% confidence interval in the example suggests that IQs are between 3.336 and 8.330 points higher, on average, for twins raised in high-income households. However, we can't conclude that household income level *caused* an increase in average IQ score because this was an observational study, not an experiment.

Because the two samples are not independent, it would be inappropriate to construct a two-sample t interval for $\mu_1 - \mu_2$ using the data from the twins study. If data are paired—either two observations of the same variable for each individual or one observation of a variable for each of two similar individuals—you must use a one-sample t interval for μ_{diff}. We will revisit this distinction in Section 11.2 when discussing significance tests about a difference in means.

CHECK YOUR UNDERSTANDING

The data from the matched pairs experiment investigating whether music helps or hinders math performance (page 677) are reproduced here.

Student	1	2	3	4	5	6	7	8	9	10	11	12	13	14	15
Time with music (sec)	83	119	77	75	64	106	70	69	60	76	47	97	68	77	48
Time without music (sec)	70	106	71	67	59	112	83	69	65	83	38	90	76	68	50
Difference (Music − Without music)	13	13	6	8	5	−6	−13	0	−5	−7	9	7	−8	9	−2

Student	16	17	18	19	20	21	22	23	24	25	26	27	28	29	30
Time with music (sec)	78	113	71	77	37	50	58	52	47	71	146	44	53	57	39
Time without music (sec)	73	93	59	70	39	52	60	54	51	60	141	40	56	53	37
Difference (Music − Without music)	5	20	12	7	−2	−2	−2	−2	−4	11	5	4	−3	4	2

1. Construct and interpret a 90% confidence interval for the true mean difference.
2. What does the interval in Question 1 suggest about whether music helps or hinders math performance? Explain your answer.

✔ Answers to CYU

1. STATE: 90% CI for μ_{diff} = the true mean time difference (Music − Without music) for students like the ones in this study to complete the arithmetic test.

PLAN: One-sample t interval for μ_{diff}.

- *Random:* The volunteers were randomly assigned the order of the music and no-music treatments. ✓
- *Normal/Large Sample:* $n_{diff} = 30 \geq 30$. ✓

DO: $\bar{x}_{diff} = 2.8, s_{diff} = 7.490, n_{diff} = 30$. With 90% confidence and df $= 30 - 1 = 29, t^* = 1.699$.

$2.8 \pm 1.699 \dfrac{7.490}{\sqrt{30}} = 2.8 \pm 2.323 =$

$(0.477, 5.123)$

Tech: The TInterval function gives (0.477, 5.123) with df $= 29$.

CONCLUDE: We are 90% confident that the interval from 0.477 to 5.123 captures the true mean time difference (Music − Without music) for students like these to complete the arithmetic test.

2. The 90% confidence interval in Question 1 suggests that students are between 0.477 and 5.123 seconds slower, on average, when completing the arithmetic test while music is playing.

Section 10.2 | Summary

- Confidence intervals for the difference between the means of two populations or the mean responses to two treatments μ_1 and μ_2 are based on the difference $\bar{x}_1 - \bar{x}_2$ between the sample means.

- Before estimating $\mu_1 - \mu_2$, we need to check for independence and that the sampling distribution of $\bar{x}_1 - \bar{x}_2$ is approximately Normal. The required conditions are:

  - **Random:** The data come from two independent random samples or from two groups in a randomized experiment.
    - **10%:** When sampling without replacement, $n_1 < 0.10N_1$ and $n_2 < 0.10N_2$.
  - **Normal/Large Sample:** For each sample, the corresponding population distribution (or the true distribution of response to the treatment) is Normal or the sample size is large ($n \geq 30$). For each sample, if the population (treatment) distribution has unknown shape and $n < 30$, confirm that a graph of the sample data shows no strong skewness or outliers.

- When conditions are met, a C% confidence interval for $\mu_1 - \mu_2$ is

$$(\bar{x}_1 - \bar{x}_2) \pm t^* \sqrt{\frac{s_1^2}{n_1} + \frac{s_2^2}{n_2}}$$

  where t^* is the critical value with C% of its area between $-t^*$ and t^* for the t distribution with degrees of freedom from either Option 1 (technology) or Option 2 (the smaller of $n_1 - 1$ and $n_2 - 1$). This is called a **two-sample t interval for $\mu_1 - \mu_2$**.

- **Paired data** result from recording two values of the same quantitative variable for each individual or for each pair of similar individuals.

- To analyze paired data, start by computing the difference for each pair. Then make a graph of the differences. Use the mean difference $\bar{x}_{\text{diff}}$ and the standard deviation of the differences s_{diff} as summary statistics.

- Before estimating μ_{diff}, we need to check for independence between the differences and that the sampling distribution of $\bar{x}_{\text{diff}}$ is approximately Normal. The required conditions are:

  - **Random:** Paired data come from a random sample from the population of interest or from a randomized experiment.
    - **10%:** When sampling without replacement, $n_{\text{diff}} < 0.10N_{\text{diff}}$.
  - **Normal/Large Sample:** The population distribution of differences (or the true distribution of differences in response to the treatments) is Normal or the number of differences in the sample is large ($n_{\text{diff}} \geq 30$). If the population (true) distribution of differences has unknown shape and the number of differences in the sample is less than 30, a graph of the sample differences shows no strong skewness or outliers.

- When the conditions are met, a C% confidence interval for the true mean difference μ_{diff} is

$$\bar{x}_{\text{diff}} \pm t^* \frac{s_{\text{diff}}}{\sqrt{n_{\text{diff}}}}$$

where t^* is the critical value for a t distribution with df $= n_{diff} - 1$ and $C\%$ of its area between $-t^*$ and t^*. This is called a **one-sample t interval for a mean difference** or a **paired t interval for a mean difference**.

- The proper inference method depends on how the data were produced. For paired data, use a one-sample t interval for μ_{diff}. For quantitative data that come from independent random samples from two populations of interest or from two groups in a randomized experiment, use a two-sample t interval for $\mu_1 - \mu_2$.
- Be sure to follow the four-step process whenever you construct a confidence interval for a difference between two means or for a mean difference.

10.2 Technology Corner

TI-Nspire and other technology instructions are on the book's website at highschool.bfwpub.com/updatedtps6e.

24. Constructing a confidence interval for a difference in means Page 674

Section 10.2 | Exercises

27. Shoes How many pairs of shoes do teenagers have? To pg 669 find out, a group of AP® Statistics students conducted a survey. They selected a random sample of 20 female students and a separate random sample of 20 male students from their school. Then they recorded the number of pairs of shoes that each student reported having. Here are their data:

Males	14	7	6	5	12	38	8	7	10	10
	10	11	4	5	22	7	5	10	35	7
Females	50	26	26	31	57	19	24	22	23	38
	13	50	13	34	23	30	49	13	15	51

Let μ_1 = the true mean number of pairs of shoes that male students at the school have and μ_2 = the true mean number of pairs of shoes that female students at the school have. Check if the conditions for calculating a confidence interval for $\mu_1 - \mu_2$ are met.

28. Who texts more? For their final project, a group of AP® Statistics students wanted to compare the texting habits of males and females. They asked a random sample of students from their school to record the number of text messages sent and received over a 2-day period. Here are their data:

Males	127	44	28	83	0	6	78	6
	5	213	73	20	214	28	11	
Females	112	203	102	54	379	305	179	24
	127	65	41	27	298	6	130	0

Let μ_1 = the true mean number of texts sent by male students at the school and μ_2 = the true mean number of texts sent by female students at the school. Check if the conditions for calculating a confidence interval for $\mu_1 - \mu_2$ are met.

29. Household size How do the numbers of people living in households in the United Kingdom (U.K.) and South Africa compare? To help answer this question, we chose independent random samples of 50 students from each country. Here is a dotplot of the household sizes reported by the students in the survey:

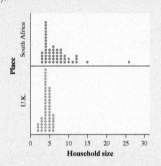

Let μ_{UK} = the true mean number of people living in U.K. households and μ_{SA} = the true mean number of

Answers to Section 10.2 Exercises

10.27 *Random:* Met because these are two independent random samples. *10%:* Met because 20 < 10% of all males at the school and 20 < 10% of all females at the school. *Normal/Large Sample:* Not met; there are fewer than 30 observations in each group and a dotplot for males shows several outliers.

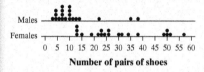

10.28 *Random:* Met; even though the data came from a single random sample, it is reasonable to consider the two samples independent because knowing the response of a male shouldn't predict the response of a female. *10%:* Met because 15 < 10% of all males at the school and 16 < 10% of all females at the school. *Normal/Large Sample:* Not met; there are fewer than 30 observations in each group and a dotplot for males shows several outliers.

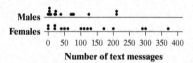

10.29 *Random:* Met because these are two independent random samples. *10%:* Met because 50 < 10% of students in the United Kingdom and 50 < 10% of students in South Africa. *Normal/Large Sample:* Met because 50 $\geq$ 30 and 50 $\geq$ 30, even though there is an outlier in the South African distribution.

10.30 *Random:* Not met because the words chosen from each article were the first words, not a random sample of words. *10%:* Met because 400 < 10% of the words in a medical journal and 100 < 10% of the words in an airline magazine. *Normal/Large Sample:* Met because $n_{AM} = 100 \geq 30$ and $n_{MJ} = 400 \geq 30$.

10.31 (a) The distributions of percent change are both slightly skewed to the left, with no apparent outliers. The centers of the two distributions seem to differ, with red wine drinkers generally having more polyphenols in their blood. The distribution of percent change for the white wine drinkers is slightly more variable. **(b) S:** μ_1 = true mean percent change in polyphenol level in the blood of people like those in the study who drink red wine and μ_2 = true mean percent change in polyphenol level in the blood of people like those in the study who drink white wine. **P:** Two-sample t interval for $\mu_1 - \mu_2$. *Random:* Two groups in a randomized experiment. *Normal/Large Sample:* The dotplots show no strong skewness and no outliers. **D:** $\bar{x}_1 = 5.5$, $s_1 = 2.517$, $n_1 = 9$, $\bar{x}_2 = 0.23$, $s_2 = 3.292$, and $n_2 = 9$. Using df = 14.97, (2.845, 7.689); using df = 8, (2.701, 7.839). **C:** We are 90% confident that the interval from 2.845 to 7.689 captures $\mu_1 - \mu_2$ = true difference in mean percent change in polyphenol level for men like these who drink red wine and men like these who drink white wine.

10.32 (a) The distribution of length of red flowers is skewed to the right, whereas the distribution of length of yellow flowers is roughly symmetric. Neither has any apparent outliers. The center of the distribution of length is much greater for the red flowers than for the yellow. Lengths of the red flowers are more variable than the lengths of the yellow. **(b) S:** μ_1 = true mean length of red flowers and μ_2 = true mean length of yellow flowers. **P:** Two-sample t interval for $\mu_1 - \mu_2$. *Random:* Independent random samples. *10%:* $n_1 = 23 < 10\%$ of all red flowers and $n_2 = 15 < 10\%$ of all yellow flowers. *Normal/Large Sample:* The dotplots show no strong skewness and no outliers. **D:** $\bar{x}_1 = 39.698$, $s_1 = 1.786$, $n_1 = 23$, $\bar{x}_2 = 36.18$, $s_2 = 0.975$, and $n_2 = 15$. Using df = 35.16, (2.606, 4.431); using df = 14, (2.554, 4.482). **C:** We are 95% confident the interval from 2.606 to 4.431 captures $\mu_1 - \mu_2$ = true difference in mean length of red and yellow flowers.

people living in South African households. Check if the conditions for calculating a confidence interval for $\mu_{UK} - \mu_{SA}$ are met.

30. **Long words** Mary was interested in comparing the mean word length in articles from a medical journal and an airline's in-flight magazine. She counted the number of letters in the first 400 words of an article in the medical journal and in the first 100 words of an article in the airline magazine. Mary then used statistical software to produce the histograms shown.

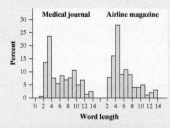

Let μ_{MJ} = the true mean length of all words in the medical journal and μ_{AM} = the true mean length of all words in the airline magazine. Check if the conditions for calculating a confidence interval for $\mu_{MJ} - \mu_{AM}$ are met.

31. **Is red wine better than white wine?** Observational studies suggest that moderate use of alcohol by adults reduces heart attacks and that red wine may have special benefits. One reason may be that red wine contains polyphenols, substances that do good things to cholesterol in the blood and so may reduce the risk of heart attacks. In an experiment, healthy men were assigned at random to drink half a bottle of either red or white wine each day for two weeks. The level of polyphenols in their blood was measured before and after the 2-week period. Here are the percent changes in polyphenols for the subjects in each group:[16]

Red wine	3.5	8.1	7.4	4.0	0.7	4.9	8.4	7.0	5.5
White wine	3.1	0.5	−3.8	4.1	−0.6	2.7	1.9	−5.9	0.1

(a) A dotplot of the data is shown, along with summary statistics. Write a few sentences comparing the distributions.

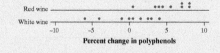

Group name	n	Mean	SD	Min	Q₁	Med	Q₃	Max
Red wine	9	5.500	2.517	0.7	3.75	5.5	7.75	8.4
White wine	9	0.233	3.292	−5.9	−2.20	0.5	2.90	4.1

(b) Construct and interpret a 90% confidence interval for the difference in true mean percent change in polyphenol levels for healthy men like the ones in this study when drinking red wine versus white wine.

32. **Tropical flowers** Different varieties of the tropical flower *Heliconia* are fertilized by different species of hummingbirds. Researchers believe that over time, the lengths of the flowers and the forms of the hummingbirds' beaks have evolved to match each other. Here are data on the lengths in millimeters for random samples of two color varieties of the same species of flower on the island of Dominica:[17]

H. *caribaea* red							
41.90	42.01	41.93	43.09	41.17	41.69	39.78	40.57
39.63	42.18	40.66	37.87	39.16	37.40	38.20	38.07
38.10	37.97	38.79	38.23	38.87	37.78	38.01	

H. *caribaea* yellow							
36.78	37.02	36.52	36.11	36.03	35.45	38.13	37.10
35.17	36.82	36.66	35.68	36.03	34.57	34.63	

(a) A dotplot of the data is shown, along with summary statistics. Write a few sentences comparing the distributions.

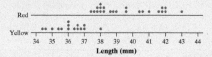

Group	n	Mean	SD	Min	Q₁	Med	Q₃	Max
Red	23	39.698	1.786	37.4	38.07	39.16	41.69	43.09
Yellow	15	36.18	0.975	34.57	35.45	36.11	36.82	38.13

(b) Construct and interpret a 95% confidence interval for the difference in the true mean lengths of these two varieties of flowers.

33. **Paying for college** College financial aid offices expect students to use summer earnings to help pay for college. But how large are these earnings? One large university studied this question by asking a random sample of 1296 students who had summer jobs how much they earned. The financial aid office separated the responses into two groups based on gender, so these can be viewed as independent samples. Here are the data in summary form:[18]

Group	n	$\bar{x}$	s_x
Males	675	$1884.52	$1368.37
Females	621	$1360.39	$1037.46

This interval suggests the true mean length of red flowers is between 2.606 and 4.431 millimeters longer than yellow flowers.

(a) How can you tell from the summary statistics that the distribution of earnings in each group is strongly skewed to the right? The use of two-sample t procedures is still justified. Why?

(b) Construct and interpret a 90% confidence interval for the difference between the true mean summer earnings of male and female students at this university.

(c) Interpret the 90% confidence level in the context of this study.

34. **Beta blockers** In a study of heart surgery, one issue was the effect of drugs called beta blockers on the pulse rate of patients during surgery. The available subjects were randomly assigned into two groups. One group received a beta blocker; the other group received a placebo. The pulse rate of each patient at a critical point during the operation was recorded. Here are the data in summary form:

Group	n	$\bar{x}$	s_x
Beta blocker	30	65.2	7.8
Placebo	30	70.3	8.3

(a) The distribution of pulse rate in each group is not Normal. The use of two-sample t procedures is still justified. Why?

(b) Construct and interpret a 99% confidence interval for the difference in mean pulse rates for patients like these who receive a beta blocker or a placebo.

(c) Interpret the 99% confidence level in the context of this study.

35. **Reaction times** Catherine and Ana wanted to know if student athletes (students on at least one varsity team) have faster reaction times than non-athletes. They took separate random samples of 33 athletes and 30 non-athletes from their school and tested their reaction time using an online reaction test, which measured the time (in seconds) between when a green light went on and the subject pressed a key on the computer keyboard. A 95% confidence interval for the difference (Non-athlete − Athlete) in the mean reaction time was 0.018 ± 0.034 seconds.

(a) Does the interval provide convincing evidence of a difference in the true mean reaction time of athletes and non-athletes? Explain your answer.

(b) Does the interval provide convincing evidence that the true mean reaction time of athletes and non-athletes is the same? Explain your answer.

(c) Identify two ways Catherine and Ana could reduce the width of their interval. Describe any drawbacks to these actions.

36. **Bird eggs** A researcher wants to see if birds that build larger nests lay larger eggs. She selects two random samples of nests: one of small nests and the other of large nests. Then she weighs one egg (chosen at random if there is more than one egg) from each nest. A 95% confidence interval for the difference (Large − Small) between the mean mass (in grams) of eggs in small and large nests is 1.6 ± 2.0.

(a) Does the interval provide convincing evidence of a difference in the true mean egg mass of birds with small nests and birds with large nests? Explain your answer.

(b) Does the interval provide convincing evidence that the true mean egg mass of birds with small nests and birds with large nests is the same? Explain your answer.

(c) Identify two ways the researcher could reduce the width of her interval. Describe any drawbacks to these actions.

37. **Groovy tires** Researchers were interested in comparing two methods for estimating tire wear. The first method used the amount of weight lost by a tire. The second method used the amount of wear in the grooves of the tire. A random sample of 16 tires was obtained. Both methods were used to estimate the total distance traveled by each tire. The table provides the two estimates (in thousands of miles) for each tire.[19]

Tire	Weight	Groove	Diff.	Tire	Weight	Groove	Diff.
1	45.9	35.7	10.2	9	30.4	23.1	7.3
2	41.9	39.2	2.7	10	27.3	23.7	3.6
3	37.5	31.1	6.4	11	20.4	20.9	−0.5
4	33.4	28.1	5.3	12	24.5	16.1	8.4
5	31.0	24.0	7.0	13	20.9	19.9	1.0
6	30.5	28.7	1.8	14	18.9	15.2	3.7
7	30.9	25.9	5.0	15	13.7	11.5	2.2
8	31.9	23.3	8.6	16	11.4	11.2	0.2

(a) Make a dotplot of the difference (Weight − Groove) in the estimate of wear for each tire using the two methods.

(b) Describe what the graph reveals about whether the two methods give similar estimates of tire wear, on average.

(c) Calculate the mean difference and the standard deviation of the differences. Interpret the mean difference.

38. **Well water** Trace metals found in wells affect the taste of drinking water, and high concentrations

---

10.33 (a) Skewed to the right because the earnings cannot be negative, yet the standard deviation is almost as large as the distance between the mean and 0. The use of the two-sample t procedures is justified because the sample sizes are both very large ($675 \geq 30$ and $621 \geq 30$). **(b) S:** $\mu_1 =$ true mean summer earnings of male students and $\mu_2 =$ true mean for female students. **P:** Two-sample t interval for $\mu_1 - \mu_2$. *Random:* It is reasonable to consider the two samples independent because knowing the response of a male shouldn't predict the response of a female. *10%:* $n_1 = 675 < 10\%$ of male students and $n_2 = 621 < 10\%$ of female students. *Normal/Large Sample:* $n_1 = 675 \geq 30$ and $n_2 = 621 \geq 30$. **D:** Using df = 1249.21, (413.62, 634.64); using df = 620, (413.52, 634.72). **C:** We are 90% confident the interval from \$413.62 to \$634.64

captures $\mu_1 - \mu_2 =$ true difference in mean summer earnings of male and female students at this university. **(c)** If we took many random samples of 675 males and 621 females from this university and each time constructed a 90% confidence interval in this same way, about 90% of the resulting intervals would capture the true difference in mean earnings for males and females.

10.34 (a) Because the sample sizes are both large ($30 \geq 30$ and $30 \geq 30$). **(b) S:** $\mu_1 =$ true mean pulse rate of patients like those in the experiment who take beta-blockers and $\mu_2 =$ mean pulse rate for those who do not. **P:** Two-sample t interval for $\mu_1 - \mu_2$. *Random:* Two groups in a randomized experiment. *Normal/Large Sample:* $n_1 = 30 \geq 30$ and $n_2 = 30 \geq 30$. **D:** Using df = 57.78, (−10.64, 0.44); using

df = 29, (−10.83, 0.63). **C:** We are 99% confident the interval from −10.64 to 0.44 captures $\mu_1 - \mu_2 =$ true difference in the mean pulse rate of patients receiving a beta blocker during surgery and those who take a placebo. **(c)** If we took 60 patients and repeated many random assignments of them to the treatments of a beta blocker and a placebo and each time constructed a 99% confidence interval in this same way, about 99% of the resulting intervals would capture the true difference in mean pulse rates for patients given a beta blocker or placebo.

10.35 (a) No, because our interval includes a difference of 0 (no difference) as a plausible value. **(b)** No; instead, we *don't have convincing evidence* the mean reaction times *differ*. Zero is a plausible value for the difference in means, but many other plausible values besides 0 are in the confidence interval. **(c)** They could increase the sample sizes or decrease the confidence level. *Drawbacks:* Increasing the sample sizes would require more work; decreasing the confidence level would give them less certainty that they captured the difference in the true mean reaction time of athletes and non-athletes.

10.36 (a) No, because our interval includes a difference of 0 (no difference) as a plausible value. **(b)** No; instead, we *don't have convincing evidence* the mean reaction times *differ*. Zero is a plausible value for the difference in means, but many other plausible values besides 0 are in the confidence interval. **(c)** The researcher could increase the sample sizes or decrease the confidence level. *Drawbacks:* Increasing the sample sizes could make it difficult to find additional nests with eggs; decreasing the confidence level decreases the certainty of capturing the difference in the true mean egg mass of birds with small nests and birds with large nests.

10.37 (a)

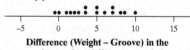

Difference (Weight − Groove) in the estimate of total distance traveled (1000s of mi)

(b) Most of the differences are positive, meaning that the estimates of distance traveled as gauged by the weight tend to be greater than the estimates of distance traveled as gauged by the grooves of the tire. **(c)** $\bar{x}_{diff} = 4.556$, $s_{diff} = 3.226$; the estimate of the number of miles driven using the "weight" method is 4.556 thousand miles greater, on average, than the estimate of the number of miles driven using the "groove" method.

10.38 (a)

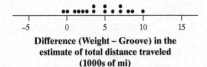

Difference (Bottom − Top) in zinc concentrations (mg/l)

(b) All the differences are positive, meaning that zinc concentrations tend to be greater at the bottom of the well than at the top.
(c) $\bar{x}_{diff} = 0.0804$, $s_{diff} = 0.0523$; the zinc concentration at the bottom of the wells is 0.0804 mg/l greater, on average, than the zinc concentration at the top of the wells.

10.39 (a) The pilot recorded two values for each individual (each of 12 randomly selected days). **(b)** Most of the differences are positive, meaning that the outbound flights (Dubai to Doha) tend to take longer. **(c)** $\bar{x}_{diff} = 10.083$, $s_{diff} = 10.766$; the difference (Outbound − Return) in flight times typically varies by about 10.766 minutes from the mean of 10.083 minutes.

10.40 (a) These are paired data because the table shows two values for each individual (each of 24 largely Muslim nations). **(b)** Most of the differences are positive, meaning that males tend to have higher literacy rates than females in these countries. **(c)** $\bar{x}_{diff} = 9.70\%$, $s_{diff} = 9.751\%$; the difference (Male − Female) in literacy rates typically varies by about 9.751% from the mean of 9.70%.

10.41 S: μ_{diff} = true mean difference in the estimates from these two methods in the population of tires. **P:** One-sample t interval for μ_{diff}. *Random:* Random sample of 16 tires. *10%:* 16 < 10% of all tires. *Normal/Large Sample:* The dotplot in Exercise 75 does not show any strong skewness or outliers.

Difference (Weight − Groove) in the estimate of total distance traveled (1000s of mi)

D: $\bar{x}_{diff} = 4.556$, $s_{diff} = 3.226$, and $n_{diff} = 16$; df = 15; (2.837, 6.275). **C:** We are 95% confident the interval from 2.837 to 6.275 thousands of miles captures the true mean difference in the estimates from these two methods in the population of tires.

---

can pose a health risk. Researchers measured the concentration of zinc (in milligrams/liter) near the top and the bottom of 10 randomly selected wells in a large region. The data are provided in the following table.[20]

Well	1	2	3	4	5	6	7	8	9	10
Bottom	0.430	0.266	0.567	0.531	0.707	0.716	0.651	0.589	0.469	0.723
Top	0.415	0.238	0.390	0.410	0.605	0.609	0.632	0.523	0.411	0.612
Difference	0.015	0.028	0.177	0.121	0.102	0.107	0.019	0.066	0.058	0.111

(a) Make a dotplot of the difference (Bottom − Top) in the zinc concentrations for these 10 wells.

(b) Describe what the graph reveals about whether the amount of zinc at the top and bottom of the wells is the same, on average.

(c) Calculate the mean difference and the standard deviation of the differences. Interpret the mean difference.

39. **Flight times** Emirates Airline offers one outbound flight from Dubai, United Arab Emirates, to Doha, Qatar, and one return flight from Doha to Dubai each day. An experienced Emirates pilot suspects that the Dubai-to-Doha outbound flight typically takes longer. To find out, the pilot collects data about these flights on a random sample of 12 days. The table displays the flight times in minutes.[21]

Day	1	2	3	4	5	6	7	8	9	10	11	12
Outbound (Dubai → Doha)	75	42	62	63	54	46	52	50	42	46	43	52
Return (Doha → Dubai)	42	37	37	44	42	40	41	44	41	42	48	48

(a) Explain why these are paired data.

(b) A dotplot of the difference (Outbound − Return) in flight time for each day is shown. Describe what the graph reveals about whether the outbound or return flight takes longer, on average.

Difference (Outbound − Return) in flight time (min)

(c) Calculate the mean difference and the standard deviation of the differences. Interpret the standard deviation.

40. **Literacy** Do males have higher literacy rates than females, on average, in Islamic countries? The following table shows the percent of men and women who were literate in 24 largely Muslim nations at the time of this writing.[22]

Country	Male literacy (%)	Female literacy (%)
Afghanistan	43.1	12.6
Algeria	86.0	86.0
Azerbaijan	99.9	99.7
Bangladesh	62.0	53.4
Egypt	82.2	65.4
Indonesia	97.0	89.6
Iran	91.2	82.5
Iraq	89.0	73.6
Jordan	96.6	90.2
Kazakhstan	99.8	99.3
Kyrgyzstan	99.3	98.1
Lebanon	93.4	86.0
Libya	98.6	90.7
Malaysia	95.4	90.7
Morocco	76.1	57.6
Pakistan	67.0	42.0
Saudi Arabia	90.4	81.3
Syria	86.0	73.6
Tajikistan	99.8	99.6
Tunisia	95.1	80.3
Turkey	99.3	98.2
Turkmenistan	99.3	98.3
Uzbekistan	99.6	99.0
Yemen	81.2	46.8

(a) Explain why these are paired data.

(b) A dotplot of the difference (Male − Female) in literacy rate for each country is shown. Describe what the graph reveals about whether males have higher literacy rates than females in these countries, on average.

Difference (Male − Female) in literacy rate

(c) Calculate the mean difference and the standard deviation of the differences. Interpret the standard deviation.

41. **Groovy tires** Refer to Exercise 37. Construct and interpret a 95% confidence interval for the true mean difference (Weight − Groove) in the estimates of tire wear using these two methods in the population of tires. pg 680

42. **Well water** Refer to Exercise 38. Construct and interpret a 95% confidence interval for the true mean difference (Bottom − Top) in the zinc concentrations of the wells in this region.

43. **Does playing the piano make you smarter?** Do piano lessons improve the spatial-temporal reasoning of preschool children? A study designed to investigate

---

10.42 S: μ_{diff} = true mean difference in the amount of zinc (mg/l) in the top and bottom of wells in this region. **P:** One-sample t interval for μ_{diff}. *Random:* Random sample of 10 wells. *10%:* Assume 10 < 10% of the population of wells in this large region. *Normal/Large Sample:* The dotplot does not show any strong skewness or outliers.

Difference (Bottom − Top) in zinc concentrations (mg/l)

D: $\bar{x}_{diff} = 0.0804$, $s_{diff} = 0.0523$, and $n_{diff} = 10$; df = 9; (0.0430, 0.1178). **C:** We are 95% confident the interval from 0.0430 to 0.1178 mg/l captures the true mean difference in the amount of zinc between the bottom and the top of the wells in this region.

this question measured the spatial-temporal reasoning of a random sample of 34 preschool children before and after 6 months of piano lessons. The difference (After – Before) in the reasoning scores for each student has mean 3.618 and standard deviation 3.055.[23]

(a) Construct and interpret a 90% confidence interval for the true mean difference.

(b) Based on your interval from part (a), can you conclude that taking 6 months of piano lessons would cause an increase in preschool students' average reasoning scores? Why or why not?

44. **No annual fee?** A bank wonders if omitting the annual credit card fee for customers who charge at least $2400 in a year will increase the amount charged on its credit cards. The bank makes this offer to an SRS of 200 of its credit card customers. It then compares how much these customers charge this year with the amount that they charged last year. The mean increase in the sample is $332, and the standard deviation is $108.

(a) Construct and interpret a 99% confidence interval for the true mean increase.

(b) Based on the interval from (a), can you conclude that dropping the annual fee would cause an increase in the average amount spent by this bank's credit card customers? Why or why not?

45. **Chewing gum** After hearing that students can improve short-term memory by chewing the same flavor of gum while studying for and taking a test, Leila and Valerie designed an experiment to investigate.[24] Using 30 volunteers, they randomly assigned 15 to chew gum while they studied a list of 40 words for 90 seconds. Immediately after the 90-second study period—and while chewing the same gum—the subjects wrote down as many words as they could remember. The remaining 15 followed the same procedure without chewing gum. Two weeks later, each of the 30 subjects did the opposite treatment. The number of words correctly remembered for each test was recorded for each subject.

(a) Explain why these are paired data.

(b) Verify that the conditions for constructing a one-sample t interval for a mean difference are satisfied.

(c) The 95% confidence interval for the true mean difference (Gum – No gum) in number of words remembered is –0.67 to 1.54. Interpret the confidence level.

(d) Based on the interval, is there convincing evidence that chewing gum helps subjects like these with short-term memory? Explain your answer.

46. **Stressful puzzles** Do people get stressed out when other people watch them work? To find out, Sean

and Shelby recruited 30 volunteers to take part in an experiment.[25] Fifteen of the subjects were randomly assigned to complete a word search puzzle while Sean and Shelby stood close by and visibly took notes. The remaining 15 were assigned to complete a word search puzzle while Sean and Shelby stood at a distance. After each subject completed the word search, they completed a second word search under the opposite treatment. The amount of time required to complete each puzzle was recorded for each subject.

(a) Explain why these are paired data.

(b) Verify that the conditions for constructing a one-sample t interval for a mean difference are satisfied.

(c) The 95% confidence interval for the true mean difference (Close by – At a distance) in amount of time needed to complete the puzzle is –12.7 seconds to 119.4 seconds. Interpret the confidence level.

(d) Based on the interval, is there convincing evidence that standing close by causes subjects like these to take longer to complete a word search? Explain your answer.

47. **Flight times** Refer to Exercise 39. Explain why it is not appropriate to construct a paired t interval about the true mean difference (Outbound – Return) in flight times between Dubai and Doha.

48. **Literacy** Refer to Exercise 40. Explain why it is not appropriate to construct a paired t interval about the true mean difference (Male – Female) in literacy rates in Islamic countries.

Multiple Choice: *Select the best answer for Exercises 49–52.*
Exercises 49 and 50 refer to the following setting. Researchers suspect that Variety A tomato plants have a different average yield than Variety B tomato plants. To find out, researchers randomly select 10 Variety A and 10 Variety B tomato plants. Then the researchers divide in half each of 10 small plots of land in different locations. For each plot, a coin toss determines which half of the plot gets a Variety A plant; a Variety B plant goes in the other half. After harvest, they compare the yield in pounds for the plants at each location. The 10 differences (Variety A – Variety B) in yield are recorded. A graph of the differences looks roughly symmetric and unimodal with no outliers. The mean difference is $\bar{x}_{A-B} = 0.34$ and the standard deviation of the differences is $s_{A-B} = 0.83$. Let μ_{A-B} = the true mean difference (Variety A – Variety B) in yield for tomato plants of these two varieties.

49. Which of the following is the best reason to use a one-sample t interval for a mean difference rather than a two-sample t interval for a difference in means to analyze these data?

---

the intervals would capture the true mean difference (Gum – No gum) in the number of words remembered for students like these. **(d)** Because 0 is included in the CI, there is not convincing evidence that chewing gum helps subjects like these with short-term memory.

10.46 (a) They result from recording two values for each individual. **(b)** PLAN: One-sample *t* interval for μ_{diff}. *Random:* The students were randomly assigned a treatment order. *Normal/Large Sample:* $n_{diff} = 30 \geq 30$. **(c)** If they conducted this experiment many times with these 30 volunteers and constructed a 95% CI using the results of each experiment, about 95% of the intervals would capture the true mean difference (Close by – At a distance) in the amount of time needed to complete the puzzle for subjects like these. **(d)** Because 0 is included in the CI, there is not convincing evidence that standing close by causes subjects like these to take longer to complete a word search.

10.47 The dotplot of the distribution of difference (Outbound – Return) in flight times is skewed right with a possible high outlier, violating the Normal/Large Sample condition.

10.48 These data were not collected from a random sample, violating the Random condition. The 24 Islamic countries selected are not less than 10% of all Islamic nations; and the dotplot of difference (Male – Female) in literacy rates is skewed right with a possible high outlier, violating the Normal/Large Sample condition.

10.43 (a) S: μ_{diff} = true mean difference (After – Before) in reasoning scores for all preschool students who take 6 months of piano lessons. **P:** One-sample *t* interval for μ_{diff}. *Random:* Random sample of 34 preschool children. *10%:* 34 < 10% of all preschool children. *Normal/Large Sample:* $n_{diff} = 34 \geq 30$. **D:** $\bar{x}_{diff} = 3.618$, $s_{diff} = 3.055$, and $n_{diff} = 34$; df = 33; (2.731, 4.505). **C:** We are 90% confident the interval from 2.731 to 4.505 captures the true mean difference (After – Before) in reasoning scores for all preschool students who take 6 months of piano lessons. **(b)** No; a randomized experiment is needed to show causation.

10.44 (a) S: μ_{diff} = true mean increase in the amount this bank's credit card customers would spend with no annual fee. **P:** One-sample *t*

interval for μ_{diff}. *Random:* SRS of 200 credit card customers. *10%:* 200 < 10% of the population of all credit card customers. *Normal/Large Sample:* $n_{diff} = 200 \geq 30$. **D:** $\bar{x}_{diff} = 332$, $s_{diff} = 108$, and $n_{diff} = 200$; df = 199; (312.14, 351.86). **C:** We are 99% confident that the interval from $312.14 to $351.86 captures the true mean increase in the amount this bank's credit card customers would spend with no annual fee. **(b)** No; there is no control group to compare results.

10.45 (a) They result from recording two values for each individual. **(b)** PLAN: One-sample *t* interval for μ_{diff}. *Random:* The students were randomly assigned a treatment order. *Normal/Large Sample:* $n_{diff} = 30 \geq 30$. **(c)** If they repeated this experiment many times with these 30 volunteers and constructed a 95% CI using the results of each experiment, about 95% of

10.49 d

10.50 e

10.51 b

10.52 d

10.53 (a) By the empirical rule, about 5% of all observations will fall outside of this interval; P(at least one mean outside interval) $= 1 - (0.95)^2 = 0.0975$. **(b)** By the empirical rule, about 2.5% of all observations will be greater than $\mu_{\bar{x}} + 2\sigma_{\bar{x}}$. Let $X =$ the number of samples that must be taken to observe one greater than $\mu_{\bar{x}} + 2\sigma_{\bar{x}}$. X is a geometric random variable with $p = 0.025$; $P(X = 4) =$ $(1 - 0.025)^3(0.025) = 0.0232$. **(c)** By the empirical rule, the probability of any one observation falling within the interval $\mu_{\bar{x}} - \sigma_{\bar{x}}$ to $\mu_{\bar{x}} + \sigma_{\bar{x}}$ is about 0.68. Let $X =$ the number of sample means out of 5 that fall outside this interval. Assuming that the samples are independent, X is a binomial random variable with $n = 5$ and $p = 0.32$; $P(X \geq 4) = 1 -$ binomcdf(trials: 5, p: 0.32, x value: 3) $= 0.039$. There is a 0.039 probability that at least 4 of the 5 sample means fall outside of this interval. This is a reasonable criterion, because when the process is under control, we would get a "false alarm" only about 4% of the time.

10.54 It is not appropriate to conclude that students who reduce their homework time from 120 minutes to 70 minutes will likely improve their performance on tests such as those used in this study. An experiment is required to show a cause-and-effect relationship between study time and test scores. It is possible that those students who are studying more than 100 minutes per night are really struggling academically compared to those that are studying less (maybe this is why they are studying so much).

(a) The number of plots is the same for Variety A and Variety B plants.

(b) The response variable, yield of tomatoes, is quantitative.

(c) This is an experiment with randomly assigned treatments.

(d) Each plot is given both varieties of tomato plant.

(e) The sample size is less than 30 for both treatments.

50. A 95% confidence interval for μ_{A-B} is given by

(a) $0.34 \pm 1.96(0.83)$

(b) $0.34 \pm 1.96\left(\dfrac{0.83}{\sqrt{10}}\right)$

(c) $0.34 \pm 1.812\left(\dfrac{0.83}{\sqrt{10}}\right)$

(d) $0.34 \pm 2.262(0.83)$

(e) $0.34 \pm 2.262\left(\dfrac{0.83}{\sqrt{10}}\right)$

51. Jordan wondered if the bean burritos at Restaurant A tend to be heavier than the bean burritos at Restaurant B. To investigate, she visited each restaurant at 10 randomly selected times, ordered a bean burrito, and weighed the burrito. The 95% confidence interval for the difference (A − B) in the mean weight of burrito is 0.06 ounce ± 0.20 ounce. Based on the confidence interval, which conclusion is most appropriate?

(a) Because 0 is included in the interval, there is convincing evidence that the mean weight is the same at both restaurants.

(b) Because 0 is included in the interval, there isn't convincing evidence that the mean weight is different at the two restaurants.

(c) Because 0.06 is included in the interval, there is convincing evidence that the mean weight is greater at Restaurant A than Restaurant B.

(d) Because 0.06 is included in the interval, there isn't convincing evidence that the mean weight is greater at Restaurant A than Restaurant B.

(e) Because there are more positive values in the interval than negative values, there is convincing evidence that the mean weight is greater at Restaurant A than Restaurant B.

52. A random sample of 30 words from Jane Austen's *Pride and Prejudice* had a mean length of 4.08 letters with a standard deviation of 2.40. A random sample of 30 words from Henry James's *What Maisie Knew* had a mean length of 3.85 letters with a standard deviation of 2.26. Which of the following is a correct expression for a 95% confidence interval for the difference in mean word length for these two novels?

(a) $(4.08 - 3.85) \pm 2.576\left(\dfrac{2.40}{\sqrt{30}} + \dfrac{2.26}{\sqrt{30}}\right)$

(b) $(4.08 - 3.85) \pm 2.045\left(\dfrac{2.40}{\sqrt{30}} + \dfrac{2.26}{\sqrt{30}}\right)$

(c) $(4.08 - 3.85) \pm 2.045\sqrt{\dfrac{2.40^2}{29} + \dfrac{2.26^2}{29}}$

(d) $(4.08 - 3.85) \pm 2.045\sqrt{\dfrac{2.40^2}{30} + \dfrac{2.26^2}{30}}$

(e) $(4.08 - 3.85) \pm 2.576\sqrt{\dfrac{2.40^2}{30} + \dfrac{2.26^2}{30}}$

Recycle and Review

53. Quality control (2.2, 5.3, 6.3, 7.3) Many manufacturing companies use statistical techniques to ensure that the products they make meet certain standards. One common way to do this is to take a random sample of products at regular intervals throughout the production shift. Assuming that the process is working properly, the mean measurement $\bar{x}$ from a random sample varies according to a Normal distribution with mean $\mu_{\bar{x}}$ and standard deviation $\sigma_{\bar{x}}$. For each question that follows, assume that the process is working properly.

(a) What's the probability that at least one of the next two sample means will fall more than $2\sigma_{\bar{x}}$ from the target mean $\mu_{\bar{x}}$? Show your work.

(b) What's the probability that the first sample mean that is greater than $\mu_{\bar{x}} + 2\sigma_{\bar{x}}$ is the one from the fourth sample taken?

Plant managers are trying to develop a criterion for determining when the process is not working properly. One idea they have is to look at the 5 most recent sample means. If at least 4 of the 5 fall outside the interval $(\mu_{\bar{x}} - \sigma_{\bar{x}}, \mu_{\bar{x}} + \sigma_{\bar{x}})$, they will conclude that the process isn't working.

(c) Find the probability that at least 4 of the 5 most recent sample means fall outside the interval, assuming the process is working properly. Is this a reasonable criterion? Explain your reasoning.

54. Stop doing homework! (4.3) Researchers in Spain interviewed 7725 13-year-olds about their homework habits—how much time they spent per night on homework and whether they got help from their parents or not—and then had them take a test with 24 math questions and 24 science questions. They found that students who spent between 90 and 100 minutes on homework did only a little better on the test than those who spent 60 to 70 minutes on homework. Beyond 100 minutes, students who spent more time did worse than those who spent less time. The researchers concluded that 60 to 70 minutes per night is the optimum time for students to spend on homework.[26] Is it appropriate to conclude that students who reduce their homework time from 120 minutes to 70 minutes will likely improve their performance on tests such as those used in this study? Why or why not?

Chapter 10 Wrap-Up

FRAPPY! FREE RESPONSE AP® PROBLEM, YAY!

The following problem is modeled after actual AP® Statistics exam free response questions. Your task is to generate a complete, concise response in 15 minutes.

Directions: Show all your work. Indicate clearly the methods you use, because you will be scored on the correctness of your methods as well as on the accuracy and completeness of your results and explanations.

Jessica wondered if elementary school students tend to write more when they are given a larger amount of space in which to write.[27] To investigate, she asked 100 fourth and fifth grade students to respond to the following prompt: "Do you like recess? Why or why not?" About half of the students were randomly assigned to use a larger 1/4-sheet of paper to write their response and the remaining students were assigned a smaller 1/6-sheet of paper. After the students finished, Jessica

counted the number of words in each response. Summary statistics for each group are shown in the table.

Group	n	Mean	Standard Deviation
Larger sheet	46	18.3	8.5
Smaller sheet	54	16.9	7.9

(a) Calculate and interpret a 95% confidence interval for the difference in mean number of words written for students like these who are asked to respond to the prompt on either a 1/4-sheet of paper or a 1/6-sheet of paper.
(b) Based only on the confidence interval from part (a), is there convincing evidence that the size of the paper affects the mean number of words written for students like these?

Chapter 10 Review

Section 10.1: Estimating a Population Mean

In this section, you learned how to construct and interpret confidence intervals for a population mean. As with other types of confidence intervals, you should always use the four-step process whenever you are asked to construct and interpret a confidence interval for a population mean. Fortunately, the State and Conclude steps are essentially the same as they were in Chapter 8 when you learned about confidence intervals for proportions. It is only the Plan and Do steps that differ.

Three conditions should be verified before calculating a one-sample t interval for μ. The Random condition says that the data come from a random sample from the population of interest. The 10% condition says that the sample size must be less than 10% of the population size when sampling without replacement. The Normal/Large Sample condition says that the population is Normally distributed or the sample size is at least 30. If the population distribution's shape is unknown and the sample size is less than 30, graph the sample data and check for strong skewness or outliers. If there is

no strong skewness or outliers, it is reasonable to assume that the population distribution is approximately Normal.

The formula for calculating a confidence interval for a population mean is

$$\bar{x} \pm t^* \frac{s_x}{\sqrt{n}}$$

where $\bar{x}$ is the sample mean, t^* is the critical value, s_x is the sample standard deviation, and n is the sample size. We use a t critical value instead of a z critical value when the population standard deviation is unknown—which is almost always the case. The value of t^* is based on the confidence level C and the degrees of freedom (df $= n - 1$). To find t^*, use Table B or technology to determine the values of t^* and $-t^*$ that capture the middle $C\%$ of the appropriate t distribution. As with other intervals, increasing the confidence level leads to a larger margin of error and increasing the sample size leads to a smaller margin of error, assuming other things remain the same.

689

TRM Chapter 10 Test

There is one Chapter Test available for this section in the Teacher's Resource Materials. Click on the link in the TE-Book, open in the TRFD, or download from the Teacher's Resources on the book's digital platform. You can also create your own Test using the TPS quiz and test builder (ExamView).

TRM FRAPPY! Materials

Please consult the Teacher's Resource Materials for sample student responses, a scoring rubric, and a printable version of the original question with space for students to write their responses. We present a model solution here.

Answers:

(a) STATE: 95% CI for $\mu_1 - \mu_2$, where μ_1 = the true mean number of words written for students like these who are asked to respond to the prompt on a 1/4-sheet of paper, and μ_2 = the true mean number of words written for students like these who are asked to respond to the prompt on a 1/6-sheet of paper.

PLAN: Two-sample t interval for $\mu_1 - \mu_2$.

- *Random:* Treatments randomly assigned; this was stated in the stem of the question. ✓
- *Normal/Large Sample:* $n_1 = 46 \geq 30$ and $n_2 = 54 \geq 30$ ✓

DO:

$$(\bar{x}_1 - \bar{x}_2) \pm t^* \sqrt{\frac{s_1^2}{n_1} + \frac{s_2^2}{n_2}} =$$

$$(18.3 - 16.9) \pm 1.986 \sqrt{\frac{8.5^2}{46} + \frac{7.9^2}{54}} =$$

$$1.4 \pm 3.279 = (-1.879, 4.679)$$

with df $= 92.9$

CONCLUDE: We are 95% confident that the interval from -1.879 words to 4.679 words captures the true difference (1/4-sheet $-$ 1/6-sheet) in mean number of words written for students like these who are asked to respond to the prompt on either a 1/4-sheet of paper or a 1/6-sheet of paper.

(b) Because 0 is one of the values in the interval from part (a), it is plausible that there is no difference in the mean number of words that students like these would write on a 1/4-sheet or 1/6-sheet of paper. Therefore, there isn't convincing evidence that the size of the paper affects the mean number of words written for students like these.

Section 10.2: Estimating a Difference in Means

In this section, you learned how to construct confidence intervals for a difference between two means and confidence intervals for a mean difference.

There are three conditions that should be verified before constructing a two-sample t interval for $\mu_1 - \mu_2$. The Random condition says that the data must be from two independent random samples or two groups in a randomized experiment. The 10% condition says that each sample size should be less than 10% of the corresponding population size when sampling without replacement. The Normal/Large Sample condition says that for each sample, the corresponding population distribution (or the true distribution of response to the treatment) is Normal or the sample size is large ($n \geq 30$). If either population (treatment) distribution has unknown shape and $n < 30$, confirm that a graph of the sample data shows no strong skewness or outliers.

Because we rarely know the two population standard deviations, we use a t distribution to determine the critical value t^*. There are two options for calculating the number of degrees of freedom to use. The first option is to use a complicated formula to calculate the degrees of freedom, which is best done with technology. The second option is to use the smaller of $n_1 - 1$ and $n_2 - 1$. The technology option is preferred because it is more accurate and produces a larger number of degrees of freedom, resulting in narrower confidence intervals. If you are using technology, always choose the *unpooled* option when constructing a confidence interval for a difference between two means. The formula is

$$(\bar{x}_1 - \bar{x}_2) \pm t^* \sqrt{\frac{s_1^2}{n_1} + \frac{s_2^2}{n_2}}$$

In this section, you also learned how to analyze paired data, which result from measuring the same quantitative variable twice for one individual or once for each of two very similar individuals. Start by finding the difference between the values in each pair. Then make a graph of the differences. Use the mean difference $\bar{x}_{\text{diff}}$ and the standard deviation of the differences s_{diff} as summary statistics.

There are three conditions that need to be verified before calculating a paired t interval for μ_{diff}. The Random condition says that paired data must come from a random sample from the population of interest or from a randomized experiment. The 10% condition says that the sample size n_{diff} should be less than 10% of the corresponding population of differences when sampling without replacement. The Normal/Large Sample condition says that the population distribution of differences is Normal or that the sample size is large ($n_{\text{diff}} \geq 30$). If the number of differences is small and the population shape is unknown, graph the difference values to make sure there is no strong skewness or outliers.

Once you have computed the differences, calculations proceed as if the data came from a single sample. Using a t distribution with df $= n_{\text{diff}} - 1$, the formula is

$$\bar{x}_{\text{diff}} \pm t^* \frac{s_{\text{diff}}}{\sqrt{n_{\text{diff}}}}$$

To decide whether a two-sample t interval for a difference between two means or paired t interval for a mean difference is appropriate, consider how the data were produced.

	Comparing confidence intervals for means		
	Confidence interval for μ	**Confidence interval for $\mu_1 - \mu_2$**	**Confidence interval for μ_{diff}**
Name (TI-83/84)	One-sample t interval for μ (TInterval)	Two-sample t interval for $\mu_1 - \mu_2$ (2-SampTInt)	Paired t interval for μ_{diff} or One-sample t interval for μ_{diff} (TInterval)

Conditions	• **Random:** The data come from a random sample from the population of interest.
	  ◦ **10%:** When sampling without replacement, $n < 0.10N$.
	• **Normal/Large Sample:** The population distribution is Normal or the sample size is large ($n \geq 30$). If the population distribution has unknown shape and $n < 30$, a graph of the sample data shows no strong skewness or outliers.
		• **Random:** The data come from two independent random samples or from two groups in a randomized experiment.
		  ◦ **10%:** When sampling without replacement, $n_1 < 0.10N_1$ and $n_2 < 0.10N_2$.
		• **Normal/Large Sample:** For *each* sample, the corresponding population distribution (or the true distribution of response to the treatment) is Normal or the sample size is large ($n \geq 30$). If either population (treatment) distribution has unknown shape and $n < 30$, a graph of the corresponding sample data shows no strong skewness or outliers.
			• **Random:** Paired data come from a random sample from the population of interest or from a randomized experiment.
			  ◦ **10%:** When sampling without replacement, $n_{\text{diff}} < 0.10N_{\text{diff}}$.
			• **Normal/Large Sample:** The population distribution of differences (or the true distribution of differences in response to the treatments) is Normal or the number of differences in the sample is large ($n_{\text{diff}} \geq 30$). If the population (true) distribution of differences has unknown shape and $n_{\text{diff}} < 30$, a graph of the sample differences shows no strong skewness or outliers.

Comparing confidence intervals for means		
Confidence interval for μ	Confidence interval for $\mu_1 - \mu_2$	Confidence interval for μ_{diff}
Formula $\bar{x} \pm t^* \dfrac{s_x}{\sqrt{n}}$ $df = n - 1$	$(\bar{x}_1 - \bar{x}_2) \pm t^* \sqrt{\dfrac{s_1^2}{n_1} + \dfrac{s_2^2}{n_2}}$ df from technology or smaller of $n_1 - 1$ and $n_2 - 1$	$\bar{x}_{diff} \pm t^* \dfrac{s_{diff}}{\sqrt{n_{diff}}}$ $df = n_{diff} - 1$

What Did You Learn?

Learning Target	Section	Related Example on Page(s)	Relevant Chapter Review Exercise(s)
Determine the critical value for calculating a $C\%$ confidence interval for a population mean using a table or technology.	10.1	653	R10.1, R10.2
State and check the Random, 10%, and Normal/Large Sample conditions for constructing a confidence interval for a population mean.	10.1	657	R10.3
Construct and interpret a confidence interval for a population mean.	10.1	659, 660	R10.3
Determine whether the conditions are met for constructing a confidence interval for a difference between two means.	10.2	669	R10.4
Construct and interpret a confidence interval for a difference between two means.	10.2	672	R10.4
Analyze the distribution of differences in a paired data set using graphs and summary statistics.	10.2	677	R10.5
Construct and interpret a confidence interval for a mean difference.	10.2	680	R10.5

Chapter 10 Review Exercises

R10.1 It's critical Find the appropriate critical value for constructing a confidence interval in each of the following settings.

(a) Estimating a population mean μ at a 95% confidence level based on an SRS of size 12.

(b) Estimating a true mean difference μ_{diff} at a 99% confidence level based on paired data set with 75 differences.

R10.2 Batteries A company that produces AA batteries tests the lifetime of a random sample of 30 batteries using a special device designed to imitate real-world use. Based on the testing, the company makes the following statement: "Our AA batteries last an average of 430 to 470 minutes, and our confidence in that interval is 95%."[28]

(a) Determine the point estimate, margin of error, standard error, and sample standard deviation.

(b) Explain the phrase "our confidence in that interval is 95%."

(c) Explain two ways the company could reduce the margin of error.

R10.3 Engine parts A random sample of 16 of the more than 200 auto engine crankshafts produced in one

---

Teaching Tip

At the end of each chapter, a "What Did You Learn?" grid lists all the targets for the chapter. Make sure to discuss this grid with your students. Ask them to read each learning target and self-assess whether or not they can do each one. For each target, the grid shows the section in which it was covered, page references for examples that illustrate the target, and relevant chapter review exercises that students can use to assess their understanding of that target. Encourage your students to use this grid as part of their preparations for the chapter test.

TRM Full Solutions to Chapter 10 Review Exercises

The full solutions can be found by clicking on the link in the TE-Book, opening the TRFD, or downloading from the Teacher's Resources on the book's digital platform.

Answers to Chapter 10 Review Exercises

R10.1 (a) df = 11, $t^* = 2.201$ **(b)** df = 74, using Table B and df = 60, $t^* = 2.660$. Using technology and df = 74, $t^* = 2.644$.

R10.2 (a) Point estimate = $(430 + 470)/2 = 450$ minutes; margin of error = $470 - 450 = 20$. SE $= \dfrac{s_x}{\sqrt{30}} = 9.780$ because $20 = 2.045 \dfrac{s_x}{\sqrt{30}}$. Finally, $s_x = 53.57$ because $\dfrac{s_x}{\sqrt{30}} = 9.780$.

(b) If we were to select many samples of 30 batteries from this population and compute a 95% confidence interval for the mean lifetime from each sample, about 95% of these intervals will capture the true mean lifetime of the batteries. **(c)** The company could reduce the margin of error by increasing the sample size or decreasing the confidence level.

R10.3 (a) S: μ = the true mean measurement of the critical dimension for the engine crankshafts produced in one day. **P:** One-sample t interval. *Random:* Random sample. *10%:* 16 < 10% of all crankshafts produced in one day. *Normal/Large Sample:* The histogram shows no strong skewness or outliers.

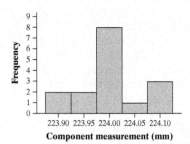

D: $\bar{x}$ = 224.002, s_x = 0.0618, and n = 16. Thus, df = 15 and t^* = 2.131. (223.969, 224.035).

C: We are 95% confident that the interval from 223.969 to 224.035 mm captures μ = the true mean measurement of the critical dimension for engine crankshafts produced on this day. **(b)** Because 224 is in this interval, it is a plausible value for the true mean. We don't have convincing evidence that the process mean has drifted.

R10.4 (a) S: μ_1 = true mean NAEP quantitative skills test score for young men and μ_2 = true mean NAEP quantitative skills test score for young women. **P:** Two-sample t interval for $\mu_1 - \mu_2$. *Random:* It is reasonable to consider the two samples independent because knowing the response of a male shouldn't predict the response of a female. *10%:* n_1 = 840 < 10% of all young men and n_2 = 10.77 < 10% of all young women. *Normal/Large Sample:* n_1 = 840 ≥ 30 and n_2 = 1077 ≥ 30. **D:** $\bar{x}_1$ = 272.40, s_1 = 59.2, n_1 = 840, $\bar{x}_2$ = 274.73, s_2 = 57.5, n_2 = 1077. Using df = 1777.52, (−6.76, 2.10); using df = 839, (−6.76, 2.10). **C:** We are 90% confident that the interval from −6.76 to 2.10 captures $\mu_1 - \mu_2$ = true difference in the mean NAEP quantitative skills test score for young men and the mean test score for young women. **(b)** Because 0 is in the interval, it is plausible the true difference is 0. That is, we do not have convincing evidence of a difference in mean score for male and female young adults.

R10.5 (a)

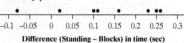

Difference (Standing − Blocks) in time (sec)

---

day was selected. Here are measurements (in millimeters) of a critical component on these crankshafts:

224.120 224.001 224.017 223.982 223.989 223.961 223.960 224.089
223.987 223.976 223.902 223.980 224.098 224.057 223.913 223.999

(a) Construct and interpret a 95% confidence interval for the mean length of this component on all the crankshafts produced on that day.

(b) The mean length is supposed to be μ = 224 mm but can drift away from this target during production. Does your interval from part (a) provide convincing evidence that the mean has drifted from 224 mm? Explain your answer.

R10.4 Men versus women The National Assessment of Educational Progress (NAEP) Young Adult Literacy Assessment Survey interviewed separate random samples of 840 men and 1077 women aged 21 to 25 years.[29] The mean and standard deviation of scores on the NAEP's test of quantitative skills were $\bar{x}_1$ = 272.40 and s_1 = 59.2 for the men in the sample. For the women, the results were $\bar{x}_2$ = 274.73 and s_2 = 57.5. Construct and interpret a 90% confidence interval for the difference in mean score for male and female young adults.

R10.5 Express lanes Is the express lane faster? Statistics students Libby and Kathryn decided to investigate which lane was faster at their local supermarket.[30] To collect their data, they randomly selected 15 times during a week, went to the same store, and bought the same item. Based on a coin flip, one of the students used the express lane and the other used a regular lane. They entered their lanes at the same time, and each recorded the time (in seconds) it took them to complete the transaction. The times are shown in the table.

Visit	Express	Regular
1	337	342
2	226	472
3	502	456
4	408	529
5	151	181
6	284	339
7	150	229
8	357	263
9	349	332
10	257	352
11	321	341
12	383	397
13	565	694
14	363	324
15	85	127

(a) Make a dotplot of the differences (Regular − Express) in wait time for each trip to the store. What does the graph suggest about whether the express lane is faster? Remember that smaller times are better.

(b) Calculate the mean difference and the standard deviation of the differences. Interpret the mean difference.

(c) Calculate and interpret a 99% confidence interval for the true mean difference (Regular − Express) in wait time for shoppers like these at this store.

(d) Based on the interval from part (c), is there convincing evidence that the express lane is faster, on average? Explain your answer.

Chapter 10 AP® Statistics Practice Test

Section I: Multiple Choice *Select the best answer for each question.*

T10.1 Anne claims that a store-brand fertilizer works better than homemade compost as a soil enhancement when growing tomatoes. To test her theory, she plants two tomato plants in each of five planters. One plant in each planter is grown in soil with store-brand fertilizer and the other plant is grown in soil with homemade compost, with the choice of soil determined at random. In three months, she will harvest and weigh the tomatoes from each plant. Which of the following is the correct confidence interval Anne should use to analyze these data?

(a) Two-sample z interval for $\mu_1 - \mu_2$

(b) Paired z interval for μ_{diff}

(c) Two-sample t interval for $\mu_1 - \mu_2$

(d) Paired t interval for μ_{diff}

(e) The correct interval cannot be determined without the data.

(b) $\bar{x}_{diff}$ = 42.667, s_{diff} = 84.019. The wait time for the regular lane was 42.667 seconds greater, on average, than the wait time for the express lane. **(c)** STATE: 99% CI for μ_{diff} = the true mean difference (Regular − Express) in wait time for shoppers like these at this store. PLAN: One-sample t interval for μ_{diff}. *Random:* The students randomly selected times to visit the store. *10%:* 15 < 10% of all possible times that Libby and Kathryn could visit the store. *Normal/Large Sample:* The number of differences is small, but the dotplot does not show any strong skewness or outliers. DO: $\bar{x}_{diff}$ = 42.667, s_{diff} = 84.019, n_{diff} = 15. df = 14, t^* = 2.977. (−21.915, 107.249)

CONCLUDE: We are 99% confident that the interval from −21.915 to 107.249 captures the true mean difference (Regular − Express) in

wait time for shoppers like these at this store. **(d)** No; because 0 is in the interval, we do not have convincing evidence that the express lane is faster, on average.

TRM **Full Solutions to Chapter 10 AP® Statistics Practice Test**

The full solutions can be found by clicking on the link in the TE-Book, opening the TRFD, or downloading from the Teacher's Resources on the book's digital platform.

T10.1 d

T10.2 The weights (in pounds) of three adult males are 160, 215, and 195. What is the standard error of the mean for these data?

(a) 190

(d) 16.07

(b) 27.84

(e) 13.13

(c) 22.73

T10.3 You want to compute a 90% confidence interval for the mean difference in height for mothers and their adult daughters using a random sample of 30 mothers who have an adult daughter. What critical value should you use for this interval?

(a) 1.645

(d) 1.699

(b) 1.671

(e) 1.761

(c) 1.697

T10.4 We want to construct a one-sample t interval for a population mean using data from a population with unknown shape. In which of the following circumstances would it be inappropriate to construct the interval based on an SRS of size 14 from the population?

(a) A stemplot of the data is roughly bell-shaped.

(b) A histogram of the data shows slight skewness.

(c) A boxplot shows that the values above the median are much more variable than the values below the median.

(d) The sample standard deviation is large.

(e) The sample standard deviation is small.

T10.5 A 90% confidence interval for the mean μ of a population is computed from a random sample and is found to be 90 ± 30. Which of the following *could* be the 95% confidence interval based on the same data?

(a) 90 ± 21

(b) 90 ± 30

(c) 90 ± 39

(d) 90 ± 70

(e) Without knowing the sample size, any of the above answers could be the 95% confidence interval.

T10.6 Do high school seniors with part-time jobs spend less time doing homework per week, on average, than seniors without part-time jobs? For a random sample of 45 seniors with part-time jobs, the mean amount of homework time is 4.2 hours with a standard deviation of 3.8 hours. For a random sample of 45 seniors without part time jobs, the mean amount of homework time is 5.8 hours with a standard deviation of 4.9 hours. Assuming the conditions are met, which of the following is the correct standard error for a 95% confidence interval for a difference in the population means?

(a) $\sqrt{\dfrac{4.9^2}{45} - \dfrac{3.8^2}{45}}$

(d) $\sqrt{\dfrac{5.8^2}{45} - \dfrac{4.2^2}{45}}$

(b) $\sqrt{\dfrac{4.9^2}{45} + \dfrac{3.8^2}{45}}$

(e) $\sqrt{\dfrac{5.8^2}{45} + \dfrac{4.2^2}{45}}$

(c) $\dfrac{(4.9 - 3.8)}{\sqrt{45}}$

T10.7 Few people enjoy melted ice cream. Being from the sunny state of Arizona, Megan and Jenna decided to test if generic vanilla ice cream melts faster than Breyers vanilla ice cream.[31] At 10 different times during the day and night, the girls put a single scoop of each type of ice cream in the same location outside and timed how long it took for each scoop to melt completely. When constructing a paired t interval for a mean difference using these data, which of the following distributions should Megan and Jenna check for Normality?

I. The distribution of melt time for the generic ice cream

II. The distribution of melt time for the Breyers ice cream

III. The distribution of difference in melt time

(a) I only

(d) I and II only

(b) II only

(e) I, II, and III

(c) III only

T10.8 A Census Bureau report on the income of Americans says that, with 90% confidence, the median income of all U.S. households in a recent year was $57,005 with a margin of error of $742. Which of the following is the most appropriate conclusion?

(a) 90% of all households had incomes in the interval $57,005 \pm $742.

(b) We can be sure that the median income for all households in the country lies in the interval $57,005 \pm $742.

(c) 90% of the households in the sample interviewed by the Census Bureau had incomes in the interval $57,005 \pm $742.

(d) The Census Bureau got the result $57,005 \pm $742 using a method that will capture the true median income 90% of the time when used repeatedly.

(e) 90% of all possible samples of this same size would result in a sample median that falls within $742 of $57,005.

T10.9 A quiz question gives random samples of $n = 10$ observations from each of two Normally distributed populations. Tom uses a table of t distribution critical values and 9 degrees of freedom to calculate a 95% confidence interval for the difference in the two population means. Janelle uses her calculator's two-sample t interval with 16.87 degrees of freedom to compute the 95% confidence interval. Assume that both students calculate the intervals correctly. Which of the following is true?

T10.2 d

T10.3 d

T10.4 c

T10.5 c

T10.6 b

T10.7 c

T10.8 d

T10.9 a

T10.10 d

T10.11 (a) If we were to select many random samples of the same size from the population of all Tootsie Pops and construct a 95% confidence interval using each sample, about 95% of the intervals would capture the true mean number of licks required to get to the center. **(b)** The point estimate is

$$\frac{317.64 + 394.56}{2} = 356.1.$$ The margin of

error is $394.56 - 356.1 = 38.46$. **(c)** The researcher could reduce the margin of error by decreasing the confidence level. The drawback is that we can't be as confident that our interval will capture the true mean. The researcher could also reduce the margin of error by increasing the sample size. The drawback is that larger samples cost more time and money to obtain.

T10.12 S: μ = the true mean number of bacteria per milliliter in raw milk received at the factory. **P:** One-sample t interval. *Random:* The data come from a random sample. *10%:* $n = 10$ is less than 10% of all 1-milliliter specimens that arrive at the factory. *Normal/Large Sample:* The dotplot shows that there is no strong skewness or outliers.

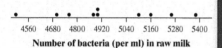

Number of bacteria (per ml) in raw milk

D: $\bar{x} = 4950.0$, $s_x = 268.5$, and $n = 10$; df = 9 and $t^* = 1.833$. (4794.37, 5105.63).

C: We are 90% confident that the interval from 4794.37 to 5105.63 bacteria per ml captures μ = the true mean number of bacteria in the milk received at this factory.

T10.13 (a) S: 95% CI for $\mu_1 - \mu_2$, where μ_1 = the true mean hospital stay for patients like these who get heating blankets during surgery and μ_2 = the true mean hospital stay for patients like these who have core temperatures reduced during surgery. **P:** Two-sample t interval for $\mu_1 - \mu_2$. *Random:* The patients were assigned at random to the normothermic group and the hypothermic group. *Normal/Large Sample:* $n_1 = 104 \geq$ 30 and $n_2 = 96 \geq 30$. **D:** $\bar{x}_1 = 12.1$, $s_1 = 4.4$, $n_1 = 104$, $\bar{x}_2 = 14.7$, $s_2 = 6.5$, $n_2 = 96$

Option 1: 2-Samp TInt gives $(-4.16, -1.04)$ using df = 165.12.

(a) Tom's confidence interval is wider.

(b) Janelle's confidence interval is wider.

(c) Both confidence intervals are the same width.

(d) There is insufficient information to determine which confidence interval is wider.

(e) Janelle made a mistake; degrees of freedom has to be a whole number.

T10.10 The makers of a specialty brand of bottled water claim that their "mini" bottles contain 8 ounces of water. To investigate this claim, a consumer advocate randomly selected a sample of 10 bottles and carefully measured the amount of water in each bottle. The mean volume was 7.98 ounces and the 95% confidence interval for true mean volume is 7.93 to 8.03 ounces. Based on the sample, which of the following conclusions best addresses the makers' claim?

(a) Because 7.98 is in the interval, there is convincing evidence that their claim is correct.

(b) Because 7.98 is in the interval, there is not convincing evidence that their claim is incorrect.

(c) Because 8 is in the interval, there is convincing evidence that their claim is correct.

(d) Because 8 is in the interval, there is not convincing evidence that their claim is incorrect.

(e) Because 0 is not the interval, there is convincing evidence that their claim is incorrect.

Section II: Free Response *Show all your work. Indicate clearly the methods you use, because you will be graded on the correctness of your methods as well as on the accuracy and completeness of your results and explanations.*

T10.11 Many people have asked the question, but few have been patient enough to collect the data. How many licks does it take to get to the center of a Tootsie Pop? After some intense research, a researcher revealed a 95% confidence interval for the mean number of licks to be 317.64 licks to 394.56 licks.[32]

(a) Interpret the confidence level.

(b) Calculate the point estimate and margin of error used to construct the interval.

(c) Name two things the researcher could do to decrease the margin of error. Discuss a drawback of each.

T10.12 A milk processor monitors the number of bacteria per milliliter in raw milk received at the factory. A random sample of 10 one-milliliter specimens of milk supplied by one producer gives the following data:

| 5370 | 4890 | 5100 | 4500 | 5260 | 5150 | 4900 | 4760 | 4700 | 4870 |

Construct and interpret a 90% confidence interval for the population mean μ.

T10.13 Researchers wondered whether maintaining a patient's body temperature close to normal by warming the patient during surgery would affect rates of infection of wounds. Patients were assigned at random to two groups: the normothermic group (core temperatures were maintained at near normal, 36.5°C using heating blankets) and the hypothermic group (core temperatures were allowed to decrease to about 34.5°C). If keeping patients warm during surgery alters the chance of infection, patients in the two groups should show a difference in the average length of their hospital stays. Here are summary statistics on hospital stay (in number of days) for the two groups:

Group	n	$\bar{x}$	s_x
Normothermic	104	12.1	4.4
Hypothermic	96	14.7	6.5

(a) Construct and interpret a 95% confidence interval for the difference in the true mean length of hospital stay for normothermic and hypothermic patients like these.

(b) Does your interval in part (a) suggest that keeping patients warm during surgery affects the average length of patients' hospital stays? Justify your answer.

Option 2: df = 95, $t^* = 1.990$

$$(12.1 - 14.7) \pm 1.990\sqrt{\frac{(4.4)^2}{104} + \frac{(6.5)^2}{96}} =$$

$$-2.6 \pm 1.57 = (-4.17, -1.03).$$

C: We are 95% confident that the interval from -4.16 to -1.04 captures $\mu_1 - \mu_2$ = the true difference in mean length of hospital stay for patients like these who get heating blankets during surgery and those who have their core temperatures reduced during surgery.

(b) Yes; because 0 is not in the interval, we have convincing evidence that the true mean hospital stay for patients like these who get heating blankets during surgery differs from the true mean hospital stay for patients like these who have core temperatures reduced during surgery.

TRM **Chapter 10 Case Study**

The Case Study feature that was found in the previous two editions of *The Practice of Statistics* *student* edition has been moved to the Teacher's Resource Materials. Click on the link in the TE-Book, open in the TRFD, or download from the Teacher's Resources on the book's digital platform.

Chapter 11

Chapter 11

Testing Claims About Means

PD Overview

In Chapter 10, we began inference for quantitative variables by estimating means. We started by constructing and interpreting confidence intervals for a single population mean (Section 10.1) and then progressed to confidence intervals for a difference in means (Section 10.2). In this chapter, we continue inference for quantitative variables by testing claims about means. We follow the same structure as in Chapter 10, first performing significance tests for a single population mean (Section 11.1) and then for a difference in means (Section 11.2).

At this point, students should be quite comfortable using the 4-step process for inference, which will serve them well in this chapter. The structure and reasoning for performing significance tests for means is nearly the same as for proportions. Remember that *every* time students are asked to perform a significance test, they should use the four-step process. Rubrics on previous AP® Statistics exams have demanded all four steps, even if the question doesn't specifically ask for them.

Throughout this chapter, take advantage of articles and stories in the news. Almost every day, there are summaries of comparative experiments and observational studies that use the inferential methods we learn about in this chapter. Even if the data aren't provided in the article, you can ask your students to consider how the study was designed and to determine what inference procedure would be most appropriate to analyze the results.

The Main Ideas

One of the challenges in teaching the AP® Statistics course is keeping students focused on the big picture, not just the details of each section. We outline the main ideas for the chapter here.

Chapter 11 Introduction

The purpose of the "Does polyester decay?" activity in the Introduction is to get students thinking about how to test a claim, without the formulas and structure of a formal significance test. A researcher investigates if polyester strips buried for longer periods of time will have lower breaking strengths, on average. Students will have to decide if there is convincing evidence that polyester decays more, on average, when left in the ground for longer periods of time, or if it is plausible that time buried has no effect and the difference was due to chance alone.

Section 11.1 Tests About a Population Mean

In this section, we present the details for conducting a significance test for a population mean. As with significance tests for proportions, we expect students to use the State–Plan–Do–Conclude 4-step process when performing significance tests for a population mean.

There is not much new in this section. The 4-step process is the same as in Chapter 9, and the conditions for conducting a one-sample t test for μ are the same as those used for constructing a one-sample t interval for μ discussed in Section 10.1. Make sure to point out this similarity to help your students avoid getting overwhelmed.

As with any significance test, always make sure your students can provide the two explanations for why the value of the sample mean differs from the hypothesized population mean. The first explanation is that the hypothesized population mean is correct, so the sample mean differs because of sampling variability. The second explanation is that the sample mean differs because the hypothesized population mean is *incorrect*. We believe the first explanation until we have convincing evidence that the difference did not occur by chance alone.

Finally, we again emphasize the clear connection between confidence intervals and significance tests. Specifically, we show that a 95% confidence interval for μ that does not capture the null value μ_0, will lead us to reject $H_0\colon \mu = \mu_0$ in a two-sided test at the $\alpha = 0.05$ significance level.

Section 11.2 Tests About a Difference in Means

In this section, we move to analyzing quantitative data that comes from *two* independent random samples or from *two* groups formed by random assignment in an experiment. The conditions to check here are nearly identical to those from Section 11.1, except that we will need to check each condition for both samples (or both groups) rather than just for a single sample. Students then use the familiar 4-step process to perform a significance test for a difference in means.

There are some additional details to be concerned with when doing inference for a difference in means. The use of the t statistic requires that we know the proper number of degrees of freedom to use, and the option of pooling creates confusion for some students. We give students clear guidance on both of these issues.

Finally, we return to the analysis of paired data that we started in Section 10.2, but here students perform significance tests for a mean difference instead of constructing a confidence interval. Consider doing the "Get your heart beating!" activity (or the cholesterol alternate activity) to help students understand the distinction between paired data and two-sample data. In this activity, students discover that moving from a completely randomized design to a matched pairs design can account for a source of variability in the response variable, increasing the power of a significance test.

Chapter 11: Resources
Teacher's Resource Materials

The following resources, identified by the **TRM** in the annotated student pages, can be found by clicking on the link in the Teacher's e-Book (TE-book), searching by category or chapter on the Teacher's Resource Flash Drive (TRFD), or logging into the book's digital platform highschool.bfwpub.com /updatedtps6e and searching the Teacher's Resources menu (teacher log-in required).

- Alternate Examples: one file per section
- Lecture Presentation Slides: one per section
- Paper Planes Experiment
- Exercising to Lose Weight
- Cholesterol Activity
- Chapter 11 Learning Targets Grid
- FRAPPY! Materials
- Chapter 11 Project
- Complete solutions for the Check Your Understanding problems, section exercises, review exercises, practice test, and cumulative practice test.
- Quizzes: one per section
- Chapter 11 Test

Free Response Questions from Previous AP® Statistics Exams

Questions can be found on the AP® Central website: apcentral.collegeboard.org/courses/ap-statistics/exam.

Students should be able to answer all the free response questions listed below with material learned in this chapter. Questions that contain content from this chapter but also require content from later chapters are listed in the last chapter required to complete the entire question. This list will be updated after each AP® Statistics exam and will be posted to the Teacher's Resource section of the book's digital platform and to www.statsmedic.com/free-response-questions.

Year	#	Content
2018	4	• Scope of inference for experiment • Two-sample *t* test for the difference between two means
2018	6	• Type II error • Calculating power • How sample size affects power
2014	5	• Paired *t* test
2011	4	• Two-sample *t* test for the difference between two means
2010	5	• Two-sample *t* test for the difference between two means
2009B	5	• One-sample *t* test for a mean • Using simulation to test a standard deviation
2009	6	• Stating hypotheses • Relationship between mean and median • Testing for skewness • Creating a test statistic
2008B	1	• Constructing and comparing dotplots • Logic of hypothesis tests
2008B	6	• Interpreting scatterplots • Paired *t* test • Creating a classification rule
2007B	5	• Two-sample *t* test for the difference between two means
2007	4	• Paired *t* test
2006B	4	• Paired *t* test
2005B	3	• Completely randomized design versus matched pairs design • Two-sample *t* test versus paired *t* test

Year	#	Content
2005B	6	• One-sample *t* test for a mean • Normal probability calculation • Multiplication rule for independent events • Using simulation to estimate a probability
2004B	5	• Boxplots • One-sample *t* interval for a mean (conditions only) • Two-sample *t* test for the difference between two means (conditions only)
2004	6	• One-sample *t* interval for a mean • Relationship between confidence intervals and significance tests • One-sided confidence intervals
2003B	4	• Random assignment • Control groups • Choosing a correct inference procedure • Sources of variability
2003	1	• Constructing boxplots • Using boxplots to compare variability • Stating hypotheses
2002	5	• Stating hypotheses • Two-sample *t* test for the difference between two means
2001	5	• Paired *t* test
2000	4	• Two-sample *t* test for the difference between two means • Inference about cause and effect
1999	6	• One sample *t* test for a mean • Paired *t* test • Displaying relationships with scatterplots
1997	5	• Paired *t* test

Applets

- The *Statistical Power* applet at highschool.bfwpub.com/updatedtps6e allows students to investigate how the power of a significance test for a population mean is affected by changes in the sample size, significance level, and alternative parameter value.

- The *One Quantitative Variable* applet in the Extra Applets section can be used to simulate a difference in two means, as done in the first experiment in the "Get your heart beating!" activity. This applet can also handle confidence intervals and significance tests for difference in means.

Chapter 11

Chapter 11: Pacing Guide, Learning Targets, and Suggested Assignments

This pacing guide is based on a schedule with 110, 50-minute sessions before the AP® Statistics exam. If you have a different number of sessions before the AP® Statistics exam, you can modify the pacing guide to suit your needs. If you have additional time, consider incorporating quizzes, released AP® Statistics free response questions, or additional activities. See the Resources section above for suggestions.

The suggested homework assignments list odd-numbered exercises, whenever possible, so students can check their answers against the back-of-book answers. If you would rather students not have access to the answers while doing homework, adding 1 to the exercise numbers usually will do the trick, because the homework exercises typically are paired. For example, Exercises 1 and 2 will generally cover the same topics, but in different contexts. You may also choose to include the Recycle and Review questions at the end of each section, which review topics from previous sections or chapters. These questions are denoted with a symbol. If your school is using the digital platform that accompanies TPS6, you will find these assignments pre-built as online homework assignments for Chapter 11.

Day	Content	Learning Targets: Students will be able to...	Suggested Assignment (MC bold)
1	11.1 Carrying Out a Significance Test for μ, Putting It All Together: One-Sample t Test for μ	• State and check the Random, 10%, and Normal/Large Sample conditions for performing a significance test about a population mean. • Calculate the standardized test statistic and P-value for a test about a population mean. • Perform a significance test about a population mean.	1, 3, 5, 9, 11
2	11.1 Two-Sided Tests and Confidence Intervals, Using Tests Wisely	• Use a confidence interval to make a conclusion for a two-sided test about a population mean.	13, 15, 17, 19, 21, 23, **27–32**
3	11.2 Significance Tests for $\mu_1 - \mu_2$	• State appropriate hypotheses for a significance test about a difference between two means. • Determine whether the conditions are met for performing a test about a difference between two means. • Calculate the standardized test statistic and P-value for a test about a difference between two means.	35, 37, 39
4	11.2 Putting It All Together: Two-Sample t Test for $\mu_1 - \mu_2$	• Perform a significance test about a difference between two means.	41, 43, 45, 47, 51
5	11.2 Significance Tests for μ_{diff}, Paired Data or Two Samples?	• Perform a significance test about a mean difference. • Determine when it is appropriate to use paired t procedures versus two-sample t procedures.	53, 55, 57, 59, **65–69**
6	Chapter 11 Review/ FRAPPY!		Chapter 11 Review Exercises
7	Chapter 11 Test		Cumulative AP® Practice Test 3

Chapter 11 Alignment to the College Board's Fall 2019 AP® Statistics Course Framework*

Relationship to College Board Units

Chapter 11 in this book covers Topics 7.4–7.5 and 7.8–7.10 in Unit 7 of the College Board Course Framework. Students will be ready to take the Personal Progress Check for Unit 7 once they have completed Chapter 11.

Big Ideas and Enduring Understandings

Chapter 11 develops these Big Ideas and related Enduring Understandings outlined in the Course Framework:

- **Big Idea 1: Variation and Distribution (EU: VAR 7):** The distribution of measures for individuals within a sample or population describes variation. The value of a statistic varies from sample to sample. How can we determine whether differences between measures represent random variation or meaningful distinctions? Statistical methods based on probabilistic reasoning provide the basis for shared understandings about variation and about the likelihood that variation between and among measures, samples, and populations is random or meaningful.

- **Big Idea 3: Data-Based Predictions, Decisions, and Conclusions (EU: DAT 3):** Data-based regression models describe relationships between variables and are a tool for making predictions for values of a response variable. Collecting data using random sampling or randomized experimental design means that findings may be generalized to the part of the population from which the selection was made. Statistical inference allows us to make data-based decisions.

Course Skills

Chapter 11 helps students to develop the skills identified in the Course Framework.

- **1: Selecting Statistical Models** (1.E, 1.F)
- **3: Using Probability and Simulation** (3.E)
- **4: Statistical Argumentation** (4.B, 4.C, 4.E)

Learning Objectives and Essential Knowledge

Section	Learning Objectives	Essential Knowledge Statements
11.1	VAR-7.B, VAR-7.C, VAR-7.D, VAR-7.E, DAT-3.E, DAT-3.F	VAR-7.B.1, VAR-7.C.1, VAR-7.D.1, VAR-7.E.1, DAT-3.E.1, DAT-3.F.1, DAT-3.F.2
11.2	VAR-7.B, VAR-7.C, VAR-7.F, VAR-7.G, VAR-7.H, VAR-7.I, DAT-3.G, DAT-3.H	VAR-7.B.2, VAR-7.C.2, VAR-7.F.1, VAR-7.G.1, VAR-7.H.1, VAR-7.I.1, DAT-3.G.1, DAT-3.H.1, DAT-3.H.2

A detailed alignment (The Nitty Gritty Guide) that can be sorted by Course Framework Unit, Topic, Learning Objective, Essential Knowledge Statement, or textbook section, is available on the TRFD and in the Teacher's Resources folder on Sapling Plus. **TRM**

*Should changes be made to the Course Framework in the future, an updated alignment will be placed on our AP® updates page at go.bfwpub.com/ap-course-updates.

Chapter **11**

Testing Claims About Means

Bill Ozorasky/Alamy

Teaching Tip

Unit 7 in the College Board Course Framework aligns to Chapters 10 and 11 in this book. Students will be ready to take the Personal Progress Check for Unit 7 once they have completed Chapter 11.

PD Chapter 11 Overview

To watch the video overview of Chapter 11 (for teachers), click on the link in the TE Book, look on the TRFD, or download from the Teacher's Resources on the book's digital platform.

TRM Lecture Presentation Slides

If you are new to teaching AP® Statistics or are short on time when preparing for class, you may find the Lecture Presentation Slides to be helpful. Experienced AP® Teacher Doug Tyson has created one slide presentation per section. You may use them as is, modify them to fit your needs, or share them with students who miss class. Find them on the TRFD and in the Teacher's Resources on the book's digital platform.

Teaching Tip

The structure of this chapter will follow the same flow as each of the previous three chapters. We will start with analyzing data from one sample (tests about a population mean in Section 11.1) and then move to analyzing data from two samples/groups (tests about a difference in means in Section 11.2).

Teaching Tip

At the end of Chapter 10, we created confidence intervals for the special case of paired data. In a parallel structure, we will perform significance tests for paired data at the end of Chapter 11.

Making Connections

In Chapters 8 and 9, we learned about inference for proportions, which comes from analyzing categorical data. In Chapters 10 and 11, we learn about inference for means, which comes from analyzing quantitative data.

✛ Ask the StatsMedic

An alternative to the activity on this page is the "Is one form of the AP® exam harder?" activity found on Stats Medic. In this activity, students analyze AP® scores from two different versions of the AP® Statistics exam to determine if there is convincing evidence that one version is harder.

▶ ACTIVITY OVERVIEW

To prepare for using this activity, watch the overview video by clicking on the link in the TE-Book, opening the TRFD, or downloading from the Teacher's Resources on the book's digital platform.

Time: 25 minutes

Materials: Index cards

Teaching Advice: This activity is a great way to reinforce the logic of inference and introduce significance tests for a difference in two means. Remind your students that we always begin a significance test with the assumption that the null hypothesis is true. It is possible that the burial time has no effect on breaking strength and polyester appears to decay over time because the weaker strips happened to be assigned to the 16-week group. To investigate how likely this is, we assume that each individual strip would have the same breaking strength if it was assigned to the other group and see how often we get a random assignment that results in a difference in means at least as extreme as the difference in the actual study.

To extend this simulation using technology, go to the Extra Applets on the Student Site at www.highschool.bfwpub.com/updatedtps6e and use the *One Quantitative Variable* applet with two groups (2 weeks and 16 weeks). Enter the data from the experiment and then "Simulate difference in two means."

INTRODUCTION

Is normal body temperature really 98.6°F? Several years ago, researchers conducted a study to test this commonly accepted value. They used an oral thermometer to measure the temperatures of a random sample of healthy men and women aged 18 to 40. The mean temperature was $\bar{x} = 98.25$°F and the standard deviation was $s_x = 0.73$°F.[1] Do these data provide convincing evidence that the average body temperature in the population of healthy 18- to 40-year-olds is *not* 98.6°F? To answer this question, we need to perform a *test about a population mean*.

Which promotes creativity better: external rewards or internal motivation? Researcher Teresa Amabile carried out an experiment to find out. She recruited experienced creative writers and divided them at random into two groups. One group was given a list of statements about external reasons for writing, like money, praise, or grades. The other group was given a list of statements about internal reasons for writing, like expressing yourself. Both groups were instructed to write a poem about laughter. The subjects' poems were given creativity scores by a panel of poets. Did one group have a significantly higher average score than the other? To answer this question, we need to perform a *test about a difference in means*.

Many people think it's faster to order at the drive-thru than to order inside at fast-food restaurants. Two students decided to investigate. At 10 randomly selected times over a 2-week period, both students went to a local Dunkin' Donuts restaurant. Each time, one student (determined by a coin flip) ordered iced coffee at the drive-thru while the other student ordered iced coffee at the counter inside. Was the average service time significantly less at the drive-thru than inside? To answer this question, we need to perform a *test about a mean difference*.

This chapter focuses on testing claims about means. The following activity gives you a taste of what lies ahead.

ACTIVITY   Does polyester decay?

How quickly do synthetic fabrics such as polyester decay in landfills? A researcher buried polyester strips in the soil for different lengths of time, then dug up the strips and measured the force required to break them. Breaking strength is easy to measure and is a good indicator of decay. Lower strength means the fabric has decayed.

The researcher buried 10 strips of polyester fabric in well-drained soil in the summer. The strips were randomly assigned to two groups: 5 of them were buried for 2 weeks and the other 5 were buried for 16 weeks. Here are the breaking strengths in pounds:[2]

Group 1 (2 weeks)	118	126	126	120	129
Group 2 (16 weeks)	124	98	110	140	110

Do the data give convincing evidence that the mean breaking strength is smaller for polyester strips like these that are buried for 16 weeks rather than 2 weeks?

After concluding that there isn't convincing evidence that polyester decays more in 16 weeks than in 2 weeks, on average, ask your students which type of error could have been committed (Type II). Then ask them what could be done to reduce the probability of a Type II error (increase the sample size for both groups, increase the difference in burial times).

1. The dotplot displays the data from the experiment. Explain why the graph gives some evidence that the mean breaking strength of polyester strips like these is smaller when buried for 16 weeks rather than 2 weeks.

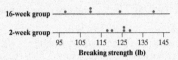

For the 2-week group, the mean breaking strength was $\bar{x}_2 = 123.8$ pounds. For the 16-week group, the mean breaking strength was $\bar{x}_{16} = 116.4$ pounds. The observed difference in average breaking strength for the two groups is $\bar{x}_2 - \bar{x}_{16} = 123.8 - 116.4 = 7.4$ pounds. Is it plausible that this difference is due to the chance involved in the random assignment and not to the treatments themselves? To find out, your class will perform a simulation.

Suppose that the length of time in the ground has no effect on the mean breaking strength of the polyester specimens. (That's our null hypothesis.) Then each specimen would have the same breaking strength regardless of whether it was assigned to the 2-week group or the 16-week group. In that case, we could examine the results of repeated random assignments of the specimens to the two groups, assuming that the null hypothesis of no effect is true.

2. Write each of the 10 breaking-strength measurements on a separate card. Mix the cards well and deal them face down into two piles of 5 cards each. Be sure to decide which pile is the 2-week group and which is the 16-week group before you look at the cards. Calculate the difference in the mean breaking strength (2-week group – 16-week group). Record this value.

3. Your teacher will draw and label axes for a class dotplot. Plot the result you got in Step 2 on the graph.

4. Repeat Steps 2 and 3 if needed to get a total of at least 40 trials of the simulation for your class.

5. Based on the class's simulation results, how surprising would it be to get a difference in means of 7.4 or larger simply due to the chance involved in the random assignment?

6. What conclusion would you draw about whether polyester decays more, on average, when left in the ground for longer periods of time? Explain your answer.

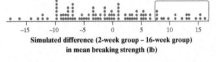

Simulated difference (2-week group – 16-week group) in mean breaking strength (lb)

In this simulation, 17 of the 100 trials produced a difference in means of at least 7.4 pounds, so the approximate P-value is 0.17. It is likely that a difference this big could have happened just due to the chance variation in random assignment. The observed difference is not statistically significant and does not provide convincing evidence that the mean breaking strength of polyester strips like these is smaller when buried for 16 weeks rather than 2 weeks.

SECTION 11.1 # Tests About a Population Mean

LEARNING TARGETS *By the end of the section, you should be able to:*

- State and check the Random, 10%, and Normal/Large Sample conditions for performing a significance test about a population mean.
- Calculate the standardized test statistic and P-value for a test about a population mean.
- Perform a significance test about a population mean.
- Use a confidence interval to make a conclusion for a two-sided test about a population mean.

You learned how to construct a confidence interval for a population mean in Section 10.1. Now we'll examine the details of testing a claim about a population mean μ.

Recall from Chapter 9 that the first step in a significance test is stating hypotheses. When conditions are met, we can calculate the P-value using an appropriate probability distribution. The P-value helps us make a conclusion about the competing claims being tested based on the strength of evidence in favor of the alternative hypothesis H_a and against the null hypothesis H_0.

Carrying Out a Significance Test for μ

In an example from Section 9.1, a company claimed to have developed a deluxe AAA battery that lasts longer than its regular AAA batteries. Based on years of experience, the company knows that its regular AAA batteries last for 30 hours of continuous use, on average. To test the company's claim, we want to perform a test at the $\alpha = 0.05$ significance level of

$$H_0: \mu = 30$$
$$H_a: \mu > 30$$

where μ is the true mean lifetime (in hours) of the deluxe AAA batteries. Our next step is to check that the conditions for performing this significance test are met.

CHECKING CONDITIONS In Chapter 10, we introduced conditions that should be met before we construct a confidence interval for a population mean. We called them Random, 10%, and Normal/Large Sample. These same conditions must be verified before performing a significance test about a population mean. Recall that the purpose of these conditions is to ensure that the observations in the sample can be viewed as independent and that the sampling distribution of $\bar{x}$ is approximately Normal.

AP® EXAM TIP

The free response section almost always has a question that asks students to perform a significance test. Students should always check conditions before performing the test, even if the question doesn't specifically ask for the conditions. Students will not be asked to perform a significance test in a context where the conditions have not been met. There may be, however, a question that focuses on just the conditions. In this case, the conditions may not be met.

CONDITIONS FOR PERFORMING A SIGNIFICANCE TEST ABOUT A MEAN

- **Random:** The data come from a random sample from the population of interest.
  - **10%:** When sampling without replacement, $n < 0.10N$.
- **Normal/Large Sample:** The population has a Normal distribution or the sample size is large ($n \geq 30$). If the population distribution has unknown shape and $n < 30$, use a graph of the sample data to assess the Normality of the population. Do not use t procedures if the graph shows strong skewness or outliers.

If the sample size is large ($n \geq 30$), the Normal/Large Sample condition is met. This condition is more difficult to check if the sample size is small ($n < 30$). In that case, we have to examine a graph of the sample data to see if it is reasonable to believe that the population distribution is Normal.

EXAMPLE

Better batteries
Checking conditions

PROBLEM: Here are the lifetimes (in hours) of the 15 deluxe AAA batteries from the company's simple random sample:

17	32	22	45	30	36	51	27
37	47	35	33	44	22	31	

Check if the conditions for performing the significance test are met.

SOLUTION:
- *Random:* SRS of 15 deluxe AAA batteries. ✓
  - 10%: Assume that 15 is less than 10% of all the company's deluxe AAA batteries. ✓
- Normal/Large Sample: The dotplot does not show strong skewness or outliers. ✓

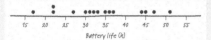
Battery life (h)

> Because the graph shows that there are no outliers or strong skewness in the sample, it is plausible that the population distribution of deluxe AAA battery lifetimes is Normal.

FOR PRACTICE, TRY EXERCISE 1

We used a dotplot to check the Normal/Large Sample condition in the example because it is an easy graph to make by hand. You can also make a stemplot, histogram, boxplot, or Normal probability plot to check this condition. Figure 11.1 shows all four of these graphs for the battery lifetime data. The stemplot, histogram, and boxplot are roughly symmetric and have no outliers. The Normal probability plot is fairly linear, as we would expect if the data came from a Normally distributed population of battery lifetimes.

ALTERNATE EXAMPLE

Skill 4.C

Chips Ahoy! Checking conditions

PROBLEM:

The makers of Chips Ahoy! claim that their chocolate chip cookies weigh 11 grams, on average. Some hungry AP® Statistics students wonder if this claim is true. They took a random sample of 24 Chips Ahoy! chocolate chip cookies. Here are the weights (in grams):

11.6	10.7	11.0	10.4	12.1	11.1
11.3	11.5	11.2	10.4	11.6	11.5
11.1	10.0	11.3	11.2	10.7	10.7
11.8	10.2	10.3	10.6	11.6	11.6

Check if the conditions for performing the significance test are met.

SOLUTION:
- *Random:* Random sample of 24 Chips Ahoy! chocolate chip cookies. ✓
  - *10%:* 24 < 10% of all the Chips Ahoy! chocolate chip cookies. ✓
- *Normal/Large Sample:* The dotplot of sample data does not show strong skewness or outliers. ✓

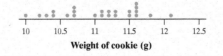
Weight of cookie (g)

Teaching Tip

Emphasize that the Random, 10%, and Normal/Large Sample conditions are exactly the same for significance tests and confidence intervals for a population mean μ.

Teaching Tip

While it is important that students know how to check each condition, it is equally important that they understand *why* we check the condition. We call this the "*So what?*"

Random condition: So what?
→ so individual observations are independent and we can generalize to the population.

10% condition: So what?
→ so we can view observations as independent even though we are sampling without replacement.

Normal/Large Sample condition: So what?
→ so the sampling distribution of $\bar{x}$ will be approximately Normal and we can use a t distribution to do calculations.

Teaching Tip

Here are the "So what?" answers for this example:

Random:
→ so individual observations are independent and we can generalize to the population of all deluxe AAA batteries.

10%:
→ so we can view observations as independent even though we are sampling without replacement.

Normal/Large Sample:
→ so the sampling distribution of $\bar{x}$ will be approximately Normal and we can use a t distribution to do calculations.

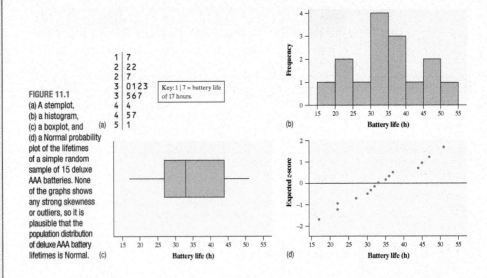

FIGURE 11.1
(a) A stemplot,
(b) a histogram,
(c) a boxplot, and
(d) a Normal probability plot of the lifetimes of a simple random sample of 15 deluxe AAA batteries. None of the graphs shows any strong skewness or outliers, so it is plausible that the population distribution of deluxe AAA battery lifetimes is Normal.

Making Connections

$$\text{standardized test statistic} = \frac{\text{statistic} - \text{parameter}}{\text{standard error of statistic}}$$

This is the general formula for the standardized test statistic that works for the significance tests we will perform in Chapters 9, 11, and 12.

CALCULATIONS: STANDARDIZED TEST STATISTIC AND *P*-VALUE In the "Better batteries" example, the sample mean lifetime for the SRS of 15 deluxe batteries is $\bar{x} = 33.93$ hours and the standard deviation is $s_x = 9.82$ hours. Because the sample mean of 33.93 hours is greater than 30 hours, we have *some* evidence against $H_0: \mu = 30$ and in favor of $H_a: \mu > 30$. But do we have *convincing* evidence that the true mean lifetime of the company's new AAA batteries is greater than 30 hours? To answer this question, we have to know how likely it is to get a sample mean of 33.93 hours or more by chance alone when the null hypothesis is true. As with proportions, we will calculate a standardized test statistic and a *P*-value to find out.

When performing a significance test, we do calculations assuming that the null hypothesis H_0 is true. The standardized test statistic measures how far the sample result diverges from the null parameter value, in standardized units. As before,

$$\text{standardized test statistic} = \frac{\text{statistic} - \text{parameter}}{\text{standard deviation (error) of statistic}}$$

For a test of $H_0: \mu = \mu_0$, our statistic is the sample mean $\bar{x}$. The standard deviation of the sampling distribution of $\bar{x}$ is

$$\sigma_{\bar{x}} = \frac{\sigma}{\sqrt{n}}$$

In an ideal world, our standardized test statistic would be

$$z = \frac{\bar{x} - \mu_0}{\dfrac{\sigma}{\sqrt{n}}}$$

When the Normal/Large Sample condition is met, the standardized test statistic z can be modeled by the standard Normal distribution. We could then use this distribution to find the P-value.

Because the population standard deviation σ is almost always unknown, we use the sample standard deviation s_x in its place. The resulting standardized test statistic has the *standard error* of $\bar{x}$ in the denominator and is denoted by t (you will see why shortly). So the formula becomes

$$t = \frac{\bar{x} - \mu_0}{\frac{s_x}{\sqrt{n}}}$$

The standardized test statistic for the "Better batteries" example is therefore

$$t = \frac{33.93 - 30}{\frac{9.82}{\sqrt{15}}} = 1.55$$

When the Normal/Large Sample condition is met and the null hypothesis is true, the standardized test statistic

$$t = \frac{\bar{x} - \mu_0}{\frac{s_x}{\sqrt{n}}}$$

can be modeled by a t **distribution**. As you learned in Section 10.1, we specify a particular t distribution by giving its *degrees of freedom* (df). When we perform inference about a population mean μ using a t distribution, the appropriate degrees of freedom are found by subtracting 1 from the sample size n, making df $= n - 1$.

Figure 11.2 compares the density curves of the standard Normal distribution and the t distributions with 2 and 9 degrees of freedom. The figure illustrates these facts about the t distributions:

- The t distributions are similar in shape to the standard Normal distribution. They are symmetric about 0, single-peaked, and bell-shaped.

- The t distributions have more variability than the standard Normal distribution. It is more likely to get an extremely large value of t (say, greater than 3) than an extremely large value of z because the t distributions have more area in the tails of the distribution.

- As the degrees of freedom increase, the area in the tails decreases and the t distributions approach the standard Normal distribution.

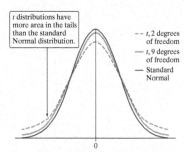

t distributions have more area in the tails than the standard Normal distribution.

- - t, 2 degrees of freedom
— t, 9 degrees of freedom
— Standard Normal

FIGURE 11.2 Density curves for the t distributions with 2 and 9 degrees of freedom and the standard Normal distribution. All are symmetric with center 0. The t distributions have more variability and a slightly different shape than the standard Normal distribution.

> **DEFINITION** t distribution
>
> A t **distribution** is described by a symmetric, single-peaked, bell-shaped density curve. Any t distribution is completely specified by its *degrees of freedom* (df). When performing inference about a population mean based on a random sample of size n using the sample standard deviation s_x to estimate the population standard deviation σ, use a t distribution with df $= n - 1$.

There are only a few real-world situations in which we might know the population standard deviation σ. If we do, then we can calculate P-values using a Normal distribution. The TI-83/84's Z-Test option in the TESTS menu is designed for this special situation.

Recall that the formula sheet provided on the AP® Statistics exam uses the notation $s_{\bar{x}}$ rather than $SE_{\bar{x}}$ for the standard error of the sample mean $\bar{x}$.

Teaching Tip

The specific formula for a one-sample t test for μ is given by

$$t = \frac{\bar{x} - \mu_0}{\frac{s_x}{\sqrt{n}}}$$

> **AP® EXAM TIP**
>
> The specific formula for the standardized test statistic in a one-sample t test for a mean is *not* included on the formula sheet provided to students on the AP® Statistics exam. Rather, the formula sheet includes the general formula for a standardized test statistic and the specific formulas needed to substitute into the general formula.

Teaching Tip

Make sure your students understand what variable follows a t distribution. When a sample is selected from a Normal population, the sample mean $\bar{x}$ follows a Normal distribution, but the standardized $t = \dfrac{\bar{x} - \mu}{\frac{s_x}{\sqrt{n}}}$ statistic follows a t distribution.

Teaching Tip

Remember that the area under any density curve must equal 1. Because the *t* distributions have more variability than the standard Normal distribution, they must also be shorter in height to maintain an area of 1.

We can use Table B to find a *P*-value from the appropriate *t* distribution when performing a test about a population mean. In the "Better batteries" example, we are testing

$$H_0: \mu = 30$$
$$H_a: \mu > 30$$

where μ = the true mean lifetime (in hours) of the company's deluxe AAA batteries. An SRS of $n = 15$ batteries yielded an average lifetime of $\bar{x} = 33.93$ hours and a standard deviation of $s_x = 9.82$ hours. The *P*-value is the probability of getting a result as large as or larger than $\bar{x} = 33.93$ just by chance when $H_0: \mu = 30$ is true. In symbols, *P*-value = $P(\bar{x} \geq 33.93 | \mu = 30)$. Earlier, we calculated the standardized test statistic to be $t = 1.55$. So we estimate the *P*-value by finding $P(t \geq 1.55)$ in a *t* distribution with df = $15 - 1 = 14$. The shaded area in Figure 11.3 shows this probability.

We can find this *P*-value using Table B. Go to the df = 14 row. The *t* statistic falls between the values 1.345 and 1.761. If you look at the top of the corresponding columns in Table B, you'll find that the "Upper-tail probability *p*" is between 0.10 and 0.05. (See the excerpt from Table B.) Because we are looking for $P(t \geq 1.55)$, this is the probability we seek. That is, the *P*-value for this test is between 0.05 and 0.10.

As you can see, Table B gives an interval of possible *P*-values for a significance test. We can still draw a conclusion from the test in much the same way as if we had a single probability. Let's illustrate using the "Better batteries" example. Because the *P*-value of between 0.05 and 0.10 is greater than $\alpha = 0.05$, we fail to reject H_0. We don't have convincing evidence that the true mean lifetime of the company's deluxe AAA batteries is greater than 30 hours.

Table B has two other limitations for finding *P*-values.

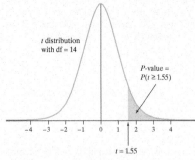

FIGURE 11.3 The shaded area shows the *P*-value for the "Better batteries" example as the area to the right of $t = 1.55$ in a *t* distribution with 14 degrees of freedom.

	Upper-tail probability *p*		
df	0.10	0.05	0.025
13	1.350	1.771	2.160
14	1.345	1.761	2.145
15	1.341	1.753	2.131
	80%	90%	95%
	Confidence level *C*		

- The table shows probabilities for only *positive* values of *t*. To find a *P*-value for a negative value of *t*, we use the symmetry of the *t* distributions. For example, $P(t \geq 1.55) = P(t \leq -1.55)$ when using a *t* distribution with a particular df.

- The table includes probabilities only for *t* distributions with degrees of freedom from 1 to 30 and then skips to df = 40, 50, 60, 80, 100, and 1000. (The bottom row gives probabilities for df = ∞, which corresponds to the standard Normal distribution.) If the df you need isn't provided in Table B, use the next lower df that is available. It's not fair "rounding up" to a larger df, which is like pretending that your sample size is larger than it really is. Doing so would give you a smaller *P*-value than is true and would make you more likely to incorrectly reject H_0 when it's true (i.e., make a Type I error). Of course, "rounding down" to a smaller df will give you a larger *P*-value than is true, which makes you more likely to commit a Type II error!

The next example shows how to deal with both of these issues.

EXAMPLE

Two-sided tests
Calculating the standardized test statistic and *P*-value

PROBLEM: Suppose you want to perform a test of $H_0: \mu = 5$ versus $H_a: \mu \neq 5$ at the $\alpha = 0.01$ significance level. A random sample of size $n = 37$ from the population of interest yields $\bar{x} = 4.81$ and $s_x = 0.365$. Assume that the conditions for carrying out the test are met.

(a) Explain why the sample result gives some evidence for the alternative hypothesis.

(b) Calculate the standardized test statistic and *P*-value.

SOLUTION:

(a) The sample mean is $\bar{x} = 4.81$, which is not equal to 5 (as suggested by H_a).

(b) $t = \dfrac{4.81 - 5}{\dfrac{0.365}{\sqrt{37}}} = -3.17$

$$\text{standardized test statistic} = \frac{\text{statistic} - \text{parameter}}{\text{standard deviation (error) of statistic}}$$

$$t = \frac{\bar{x} - \mu_0}{\dfrac{s_x}{\sqrt{n}}}$$

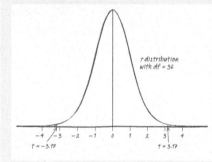

	Upper-tail probability *p*		
df	0.005	0.0025	0.001
29	2.756	3.038	3.396
30	2.750	3.030	3.385
40	2.704	2.971	3.307
	99%	99.5%	99.8%
	Confidence level *C*		

Using Table B: Because df = 37 − 1 = 36 is not available on the table, use df = 30.

$P(t \geq 3.17)$ is between 0.001 and 0.0025.

$P(t \leq -3.17 \text{ or } t \geq 3.17)$ is between $2(0.001) = 0.002$ and $2(0.0025) = 0.005$.

FOR PRACTICE, TRY EXERCISE 5

Given the limitations of Table B, our advice is to use technology to find *P*-values when carrying out a significance test about a population mean. Remember that *P*-value calculations are valid only when our probability model is true—that is, when the conditions for inference are met.

Chips Ahoy! Still hungry
Calculating the standardized test statistic and *P*-value

PROBLEM:
In the preceding alternate example, hungry students collected data on the weights of a random sample of 24 Chips Ahoy! chocolate chip cookies. They want to perform a test of $H_0: \mu = 11$ versus $H_a: \mu \neq 11$ at the $\alpha = 0.05$ significance level, where μ = the true mean weight of all Chips Ahoy! chocolate chip cookies (in grams). The sample of 24 cookies had a mean weight of $\bar{x} = 11.063$ grams and $s_x = 0.562$ grams. Assume that the conditions for carrying out the test are met.

(a) Explain why the sample result gives some evidence for the alternative hypothesis.

(b) Calculate the standardized test statistic and *P*-value.

SOLUTION:
(a) The sample mean is $\bar{x} = 11.063$ grams, which is not equal to 11 grams (supporting H_a).

(b) $t = \dfrac{11.063 - 11}{\dfrac{0.562}{\sqrt{24}}} = 0.55$

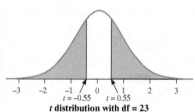

t distribution with df = 23

Table B: Using df = 24 − 1 = 23, $P(t \geq 0.55)$ is greater than 0.25.

$P(t \leq -0.55 \text{ or } t \geq 0.55)$ is greater than $2(0.25) = 0.50$.

Teaching Tip

During the school year, consider having students write a general formula and a specific formula with variables before substituting numbers:

General formula:

$$\frac{\text{standardized}}{\text{test statistic}} = \frac{\text{statistic} - \text{parameter}}{\text{standard error of statistic}}$$

Specific formula:

$$t = \frac{\bar{x} - \mu_0}{\dfrac{s_x}{\sqrt{n}}}$$

One advantage to establishing this expectation is that the general formula will be the same for significance tests in Chapters 9, 11, and 12. Only the specific formula changes for each new significance test.

At the end of the year, when preparing for the AP® Statistics exam, tell students they no longer have to include these formulas, because they are not required for full credit on the AP® exam and students often make mistakes in their notation.

Teaching Tip: Using Technology

Remind students that they can watch a video of Technology Corner 25 on the e-Book or at the Student Site.

Teaching Tip: Using Technology

Your students may wonder why we use the value 1000 or −1000 when using the tcdf command. Technically, we would like to use ∞ or −∞, but the TI-84 calculator doesn't have an ∞ button. Instead, we use a value that is extremely large. Choosing a value 1000 standard deviations above the mean seems plenty large to us!

Teaching Tip: Using Technology

If your students are using a TI-84 with OS 2.55 or later and aren't given dialog boxes when they use the tcdf command, have them turn on the Stat Wizards by pressing "Mode" and scrolling down to the second page. If students are using a TI-83 Plus or TI-84 with an older operating system, they can run an app called *Catalog Help* (ctlghelp) that will remind them what values to enter for certain commands. Press the APPS button to see if it is already loaded. If it is, press "Enter" and read the directions. If not, download the app from education.ti.com or copy it by linking from another calculator.

25. Technology Corner COMPUTING *P*-VALUES FROM *t* DISTRIBUTIONS

TI-Nspire and other technology instructions are on the book's website at highschool.bfwpub.com/updatedtps6e.

You can use the tcdf command on the TI-83/84 to calculate areas under a *t* distribution curve. The syntax is tcdf(lower bound, upper bound, df).

Let's use the tcdf command to compute the *P*-values from the last two examples.

Better batteries: To find $P(t \geq 1.55)$,

- Press [2nd] [VARS] (DISTR) and choose tcdf(.

  OS 2.55 or later: In the dialog box, enter these values: lower: 1.55, upper: 1000, df:14, choose Paste, and then press [ENTER].

  Older OS: Complete the command tcdf(1.55,1000,14) and press [ENTER].

Two-sided test: To find $P(t \leq -3.17$ or $t \geq 3.17)$,

- Press [2nd] [VARS] (DISTR) and choose tcdf(.

  OS 2.55 or later: In the dialog box, enter these values: lower: −1000, upper: −3.17, df:36, choose Paste, multiply by 2, and then press [ENTER].

  Older OS: Complete the command tcdf(−1000, −3.17, 36)∗2 and press [ENTER].

```
NORMAL FLOAT AUTO REAL RADIAN MP
tcdf(1.55,1000,14)
                0.0717235647
```

```
NORMAL FLOAT AUTO REAL RADIAN MP
tcdf(-1000,-3.17,36)*2
                0.0031080065
```

CHECK YOUR UNDERSTANDING

The makers of Aspro brand aspirin want to be sure that their tablets contain the right amount of active ingredient (acetylsalicylic acid). So they inspect a random sample of 30 tablets from a batch in production. When the production process is working properly, Aspro tablets contain an average of $\mu = 320$ milligrams (mg) of active ingredient. The amount of active ingredient in the 30 selected tablets has mean 319 mg and standard deviation 3 mg.

1. State appropriate hypotheses for a significance test in this setting.
2. Check that the conditions are met for carrying out the test.
3. Calculate the standardized test statistic and *P*-value.
4. What conclusion would you make?

✓ **Answers to CYU**

1. H_0: $\mu = 320$, H_a: $\mu \neq 320$, where μ = the true mean amount of active ingredient (mg) in Aspro tablets from this batch of production using $\alpha = 0.05$.

2. *Random:* We have a random sample of 30 tablets. *10%:* Assume the sample of size 30 is less than 10% of the population of all tablets in this batch. *Normal/Large Sample:* $n = 30 \geq 30$. All conditions are met.

3. $t = \dfrac{319 - 320}{\dfrac{3}{\sqrt{30}}} = -1.83$

4. Because the *P*-value of $0.0775 > \alpha = 0.05$, we fail to reject H_0. There is not convincing evidence that the true mean amount of the active ingredient in Aspro tablets from this batch of production differs from 320 mg.

Putting It All Together: One-Sample *t* Test for μ

We have shown you how to complete each of the four steps in a test about a population mean. Let's reflect on how the pieces fit together in a test of $H_0: \mu = \mu_0$. We start by assuming that the null hypothesis is true. Because we usually don't know the population standard deviation σ, we use the sample standard deviation s_x to estimate it. If we standardize the statistic $\bar{x}$ by subtracting its mean and dividing by its standard error, we get the standardized test statistic.

$$t = \frac{\bar{x} - \mu_0}{\frac{s_x}{\sqrt{n}}}$$

There are three conditions that must be met for this formula to be valid.

1. The Random condition allows us to make an inference about the population from which the sample was randomly selected, as you learned in Chapter 4. Random sampling also helps ensure that individual observations in the sample are independent.

2. The 10% condition allows us to view the observations in the sample as independent when we are sampling without replacement from a finite population. Then we can use the familiar formula for the standard error of $\bar{x}$ in the denominator of the standardized test statistic. If the 10% condition is violated, our formula will overestimate the standard error, making the *t* statistic smaller and the *P*-value larger than they should be.

3. The Normal/Large Sample condition allows us to model the distribution of the standardized test statistic *t* using a *t* distribution with $n - 1$ degrees of freedom. Then we can obtain *P*-values from this distribution using Table B or technology. If the Normal/Large Sample condition is violated, the *P*-value calculated from a *t* distribution with $n - 1$ degrees of freedom will not be accurate.

Here is a summary of the Do step for a **one-sample *t* test for a mean**.

ONE-SAMPLE *t* TEST FOR A MEAN

Suppose the conditions are met. To test the hypothesis $H_0: \mu = \mu_0$, compute the standardized test statistic

$$t = \frac{\bar{x} - \mu_0}{\frac{s_x}{\sqrt{n}}}$$

Find the *P*-value by calculating the probability of getting a *t* statistic this large or larger in the direction specified by the alternative hypothesis H_a in a *t* distribution with df $= n - 1$.

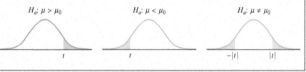

$H_a: \mu > \mu_0$ $H_a: \mu < \mu_0$ $H_a: \mu \neq \mu_0$

IQ scores for AP® teachers
Performing a significance test about μ

PROBLEM:
Human intelligence can be estimated using an intelligence quotient (IQ) score. The distribution of IQ scores for the adult population is approximately Normal with a mean of 100. Some students hypothesize that AP® teachers have a higher mean IQ score than the general adult population. They select a random sample of 10 AP® teachers and record their IQ scores. Here are the results:

| 106 | 114 | 121 | 95 | 125 |
| 102 | 110 | 103 | 99 | 108 |

(a) Do the data provide convincing evidence at the $\alpha = 0.05$ significance level that AP® teachers have a higher mean IQ score than the general adult population?
(b) Given your conclusion in part (a), which kind of mistake—a Type I error or a Type II error—could you have made? Explain what this mistake would mean in context.

SOLUTION:
(a) STATE: We want to test

$H_0: \mu = 100$
$H_a: \mu > 100$

where μ = the true mean IQ score for all AP® teachers using $\alpha = 0.05$.

PLAN: One-sample t test for μ

- *Random:* Random sample of 10 AP® teachers. ✓
  - *10%:* 10 < 10% of all AP® teachers. ✓
- *Normal/Large Sample:* The population distribution of IQ scores is approximately Normal. ✓

DO:

- $\bar{x} = 108.3$, $s_x = 9.499$

- $t = \dfrac{108.3 - 100}{\dfrac{9.499}{\sqrt{10}}} = 2.76$

- P-value

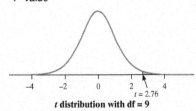

$t = 2.76$
***t* distribution with df = 9**

df = 10 − 1 = 9

Table B: P-value is between 0.01 and 0.02.

Tech: tcdf(lower: 2.76, upper: 1000, df: 9) = 0.011

Now we are ready to test a claim about a population mean. Once again, we follow the four-step process.

| **EXAMPLE** | **Healthy streams**
Performing a significance test about μ |

PROBLEM: The level of dissolved oxygen (DO) in a stream or river is an important indicator of the water's ability to support aquatic life. A researcher measures the DO level at 15 randomly chosen locations along a stream. Here are the results in milligrams per liter (mg/l):

| 4.53 | 5.04 | 3.29 | 5.23 | 4.13 | 5.50 | 4.83 | 4.40 |
| 5.42 | 6.38 | 4.01 | 4.66 | 2.87 | 5.73 | 5.55 | |

An average dissolved oxygen level below 5 mg/l puts aquatic life at risk.
(a) Do the data provide convincing evidence at the $\alpha = 0.05$ significance level that aquatic life in this stream is at risk?
(b) Given your conclusion in part (a), which kind of mistake—a Type I error or a Type II error—could you have made? Explain what this mistake would mean in context.

SOLUTION:
(a) STATE: We want to test

$H_0: \mu = 5$
$H_a: \mu < 5$

where μ = the true mean dissolved oxygen (DO) level in the stream, using $\alpha = 0.05$.

PLAN: One-sample t test for μ.

- *Random:* The researcher measured the DO level at 15 randomly chosen locations. ✓
- *Normal/Large Sample:* The histogram looks roughly symmetric and shows no outliers. ✓

> Follow the four-step process!

> There are an infinite number of possible locations along the stream, so it isn't necessary to check the 10% condition.

> Because the histogram shows no strong skewness or outliers in the sample, it is plausible that the population distribution of dissolved oxygen levels in the stream is Normal.

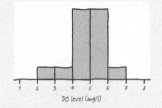

DO level (mg/l)

CONCLUDE: Because the P-value of 0.011 < α = 0.05, we reject H_0. We have convincing evidence that the true mean IQ score for all AP® teachers is greater than the general adult population mean IQ score of 100.

(b) Because we rejected H_0 in part (a), we could have made a Type I error (rejecting H_0 when H_0 is true). If we did, the true mean IQ score for all AP® teachers is 100, but we found convincing evidence that it was greater than 100 with our significance test. This would mean that AP® teachers aren't really more intelligent than the rest of the population, but we incorrectly concluded that they are more intelligent.

Teaching Tip

Ask students, "Which words in the question inform you this is a one-sided test?"
Answer: Oxygen level <u>below</u> 5 mg/l puts aquatic life at risk (one-sided).

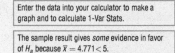

DO: • $\bar{x} = 4.771, s_x = 0.9396$

• $t = \dfrac{4.771 - 5}{\dfrac{0.9396}{\sqrt{15}}} = -0.94$

> Enter the data into your calculator to make a graph and to calculate 1-Var Stats.

> The sample result gives *some* evidence in favor of H_a because $\bar{x} = 4.771 < 5$.

• P-value: df = 15 − 1 = 14

Using Table B: P-value is between 0.15 and 0.20.

Using technology: tcdf(lower: −1000, upper: − 0.94, df: 14) = 0.1816

CONCLUDE: *Because the P-value of 0.1816 > α = 0.05, we fail to reject H_0. We don't have convincing evidence that the true mean DO level in the stream is less than 5 mg/l.*

(b) *Because we failed to reject H_0 in part (a), we could have made a Type II error (failing to reject H_0 when H_a is true). If we did, then the true mean dissolved oxygen level μ in the stream is less than 5 mg/l, but we didn't find convincing evidence with our significance test. That would imply aquatic life in this stream is at risk, but we weren't able to detect that fact.*

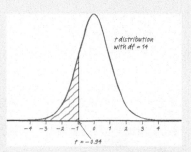

FOR PRACTICE, TRY EXERCISE 9

To reduce the chance of making a Type II error, the researcher in the preceding example could have taken a larger random sample of stream locations. Doing so would yield a more precise estimate of the stream's true mean dissolved oxygen level μ. That is, a larger sample size increases the power of the test to detect if $H_a: \mu < 5$ is true. Another way to increase the power of the test would be to use a higher significance level, like $\alpha = 0.10$. This change makes it easier to reject $H_0: \mu = 5$ when the average DO level in the stream is below 5 mg/l. But increasing the significance level also increases the probability of making a Type I error—finding convincing evidence that the stream is unhealthy when it really isn't—from 0.05 to 0.10.

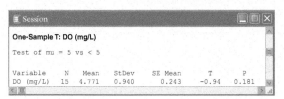

FIGURE 11.4 Minitab output for the one-sample *t* test from the dissolved oxygen example.

Because the *t* procedures are so common, all statistical software packages will do the calculations for you. Figure 11.4 shows the output from Minitab for the one-sample *t* test in the preceding example. Note that the results match!

You can also use your calculator to carry out a one-sample *t* test. But be sure to read the AP® Exam Tip at the end of the following Technology Corner.

AP® EXAM TIP

When naming the inference method, it is acceptable to list the calculator command "*t* test." However, we prefer that students use the "one-sample *t* test for μ," because it demonstrates deeper understanding and leads to fewer misconceptions. Also, we want to differentiate this significance test from the *t* test for slope that is coming up in Chapter 12.

Teaching Tip

In the "Conclude" step of the example on this page, remind students that it is incorrect to conclude that "the mean dissolved oxygen level *is 5 mg/l*." This is equivalent to "accepting the null."

Teaching Tip: Using Technology

When computer output provides the "standard error of the mean," we can use this value in the denominator of the *t* statistic instead of $s_x/\sqrt{n}$. Using this Minitab output, show your students that $0.940/\sqrt{15} = 0.243$, so $t = \dfrac{4.771 - 5}{0.243}$.

26. Technology Corner PERFORMING A ONE-SAMPLE *t* TEST FOR A MEAN

TI-Nspire and other technology instructions are on the book's website at highschool.bfwpub.com/updatedtps6e

You can perform a one-sample *t* test using either raw data or summary statistics on the TI-83/84. Let's use the calculator to carry out the test of $H_0: \mu = 5$ versus $H_a: \mu < 5$ from the dissolved oxygen example. Start by entering the sample data in L1. Then, to do the test:

- Press STAT, choose TESTS and T-Test.
- Adjust your settings as shown.

If you select "Calculate," the screen below left appears.

The standardized test statistic is $t = -0.94$ and the *P*-value is 0.1809.

If you specify "Draw," you see a *t* distribution curve (df = 14) with the lower tail shaded, along with the standardized test statistic and *P*-value.

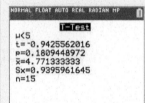

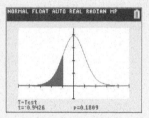

Note: If you are given summary statistics instead of the original data, you would select the input option "Stats" instead of "Data" in the first line and enter the summary statistics.

AP® EXAM TIP

Remember: If you give just calculator results with no work, and one or more values are wrong, you probably won't get any credit for the "Do" step. If you opt for the calculator-only method, name the procedure (one-sample *t* test for μ) and report the standardized test statistic ($t = -0.94$), degrees of freedom (df = 15), and *P*-value (0.1809).

CHECK YOUR UNDERSTANDING

A teacher suspects that students at his school are getting less than the recommended 8 hours of sleep a night, on average. To test his belief, the teacher asks a random sample of 28 students, "How much sleep did you get last night?" Here are the data (in hours):

| 9 | 6 | 8 | 6 | 8 | 8 | 6 | 6.5 | 6 | 7 | 9 | 4 | 3 | 4 |
| 5 | 6 | 11 | 6 | 3 | 6 | 6 | 10 | 7 | 8 | 4.5 | 9 | 7 | 7 |

Do these data provide convincing evidence at the $\alpha = 0.05$ significance level in support of the teacher's suspicion?

✓ **Answers to CYU**

STATE: $H_0: \mu = 8$, $H_a: \mu < 8$, where μ = the true mean amount of sleep that students at the teacher's school get each night using $\alpha = 0.05$.

PLAN: One-sample *t* test for μ. *Random:* The teacher selected a random sample of 28 students. *10%:* Assume the sample size (28) is less than 10% of the population of students at this school. *Normal/Large Sample:* There were only 28 students, so we need to examine the sample data. The histogram indicates that there is not much skewness and no outliers, so it is reasonable to use a *t* procedure.

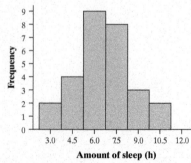

DO: $t = \dfrac{6.643 - 8}{\dfrac{1.981}{\sqrt{28}}} = -3.625$;

P-value = 0.0006

CONCLUDE: Because the *P*-value of 0.0006 < $\alpha = 0.05$, we reject H_0. There is convincing evidence that students at this school get less than 8 hours of sleep, on average.

Two-Sided Tests and Confidence Intervals

You learned in Section 9.2 that a confidence interval gives more information than a significance test does—it provides the entire set of plausible values for the parameter based on the data. The connection between two-sided tests and confidence intervals is even stronger for means than it was for proportions. That's because both inference methods for means use the standard error of $\bar{x}$ in the calculations:

$$\text{standardized test statistic: } t = \frac{\bar{x} - \mu_0}{\frac{s_x}{\sqrt{n}}} \qquad \text{confidence interval: } \bar{x} \pm t^* \frac{s_x}{\sqrt{n}}$$

The link between two-sided tests and confidence intervals for a population mean allows us to make a conclusion directly from a confidence interval.

- If a 95% confidence interval for μ does not capture the null value μ_0, we can reject $H_0: \mu = \mu_0$ in a two-sided test at the $\alpha = 0.05$ significance level.
- If a 95% confidence interval for μ captures the null value μ_0, then we should fail to reject $H_0: \mu = \mu_0$ in a two-sided test at the $\alpha = 0.05$ significance level.

The same logic applies for other confidence levels, but *only* for a two-sided test.

EXAMPLE

Juicy pineapples
Confidence intervals and two-sided tests

PROBLEM: At the Hawaii Pineapple Company, the mean weight of the pineapples harvested from one large field was 31 ounces last year. A different irrigation system was installed in this field after the growing season. Managers wonder if this change will affect the mean weight of future pineapples grown in the field. To find out, they select and weigh a random sample of 50 pineapples from this year's crop.

(a) State an appropriate pair of hypotheses for a significance test in this setting. Be sure to define the parameter of interest.

(b) Check conditions for performing the test in part (a).

(c) A 95% confidence interval for the mean weight of all pineapples grown in the field this year is (31.255, 32.616). Based on this interval, what conclusion would you make for a test of the hypotheses in part (a) at the $\alpha = 0.05$ significance level?

(d) Can we conclude that the different irrigation system caused a change in the mean weight of pineapples produced? Explain your answer.

SOLUTION:

(a) We want to test
$H_0: \mu = 31$
$H_a: \mu \neq 31$
where μ = the true mean weight (in ounces) of all pineapples grown in the field this year.

(b) • Random: Random sample of 50 pineapples from this year's crop. ✓
  ○ 10%: It is safe to assume that $50 < 10\%$ of all pineapples in a large field. ✓
• Normal/Large Sample: $n = 50 \geq 30$. ✓

Maks Narodenko/ Shutterstock.com

ALTERNATE EXAMPLE

Skill 4.D

Are radio stations honest?   Confidence intervals and two-sided tests

PROBLEM:
A classic-rock radio station claims to play an average of 50 minutes of music every hour. To investigate the station's claim, you randomly select 12 different hours during the next week and record what the radio station plays in each of the 12 hours. Here is how much music (in minutes) was played during each of these hours:

48	49	50	51	49	53
49	47	47	50	46	48

(a) State an appropriate pair of hypotheses for a significance test in this setting. Be sure to define the parameter of interest.

(b) Check conditions for performing the test in part (a).

(c) A 95% confidence interval for the mean play time (in minutes) of all hours this week is (47.691, 50.142). Based on this interval, what conclusion would you make for a test of the hypotheses in part (a) at the $\alpha = 0.05$ significance level?

(d) Can we generalize our conclusion for this radio station for the whole year? Explain your answer.

Teaching Tip

When students are given a confidence interval from which to make a decision for a significance test, ask the following questions:

1. What is the confidence level of the interval? What α value does this correspond to?

2. According to the confidence interval, is the null hypothesis value a plausible value?

3. Based on your answer to Question 2, should we reject the null hypothesis or fail to reject the null hypothesis?

4. Make a proper conclusion that includes whether or not we have convincing evidence for the alternative hypothesis.

SOLUTION:

(a) We want to test
$H_0: \mu = 50$
$H_a: \mu \neq 50$

where μ = the true mean play time (in minutes) of music in all hours this week at this radio station.

(b)
- *Random:* Random sample of 12 hours from this radio station for this week. ✓
  ○ *10%:* $12 < 10\%$ of all hours during the week. ✓
- *Normal/Large Sample:* The dotplot of sample data does not show strong skewness or outliers. ✓

Time playing music (min)

(c) The 95% CI does include 50 as a plausible value, so we would fail to reject H_0. We do not have convincing evidence that the true mean hourly play time at this radio station is not 50 minutes.

(d) No; we can only generalize our conclusions to the population from which we took our random sample. Because we took a random sample of hours for this one week, we can only generalize our results to this one week. To generalize to the whole year, we must take a random sample of hours from the whole year.

(c) The 95% confidence interval does not include 31 as a plausible value, so we would reject H_0. We have convincing evidence that the true mean weight of all pineapples grown this year is not 31 ounces.

(d) No; this was not a randomized comparative experiment, so we cannot infer causation. It is possible that other things besides the irrigation system changed from last year's growing season. Maybe the weather was different this year, and that's why the pineapples have a different mean weight than last year.

> Recall from Chapter 4 that only well-designed experiments allow us to establish cause-and-effect conclusions.

FOR PRACTICE, TRY EXERCISE 17

Minitab output for a significance test and confidence interval based on the pineapple data is shown. Notice that the weights of the 50 randomly selected pineapples from this year's crop had a mean of $\bar{x} = 31.935$ ounces and a standard deviation of $s_x = 2.394$ ounces. The standardized test statistic is

$$t = \frac{\bar{x} - \mu_0}{\frac{s_x}{\sqrt{n}}} = \frac{31.935 - 31}{\frac{2.394}{\sqrt{50}}} = 2.76$$

The corresponding P-value from a t distribution with 49 degrees of freedom is 0.008. Because the P-value of $0.008 < \alpha = 0.05$, we would reject H_0. This is consistent with our decision based on the 95% confidence interval in the example.

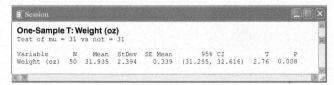

```
Session                                                    [_][□][X]

One-Sample T: Weight (oz)
Test of mu = 31 vs not = 31

Variable      N    Mean   StDev   SE Mean      95% CI         T      P
Weight (oz)  50   31.935  2.394    0.339   (31.255, 32.616)  2.76  0.008
```

Would a 99% confidence interval for μ include 31 ounces as a plausible value for the parameter? Only if a two-sided test would fail to reject $H_0: \mu = 31$ at a 1% significance level. Because the P-value of $0.008 < \alpha = 0.01$, we would reject H_0. We once again have convincing evidence that the mean weight of all pineapples produced this year is not 31 ounces. So the 99% confidence interval would not contain 31. You can check that the interval is (31.028, 32.842).

Think About It

IS THERE A CONNECTION BETWEEN *ONE*-SIDED TESTS AND CONFIDENCE INTERVALS FOR A POPULATION MEAN? As you might expect, the answer is yes. But the link is more complicated. Consider a one-sided test of $H_0: \mu = 10$ versus $H_a: \mu > 10$ based on an SRS of 30 observations. With df $= 30 - 1 = 29$, Table B says that the test will reject H_0 at $\alpha = 0.05$ if the standardized test statistic t is greater than 1.699. For this to happen, the sample mean $\bar{x}$ would have to exceed $\mu_0 = 10$ by more than 1.699 standardized units.

Table B also shows that $t^* = 1.699$ is the critical value for a 90% confidence interval. That is, a 90% confidence interval will extend 1.699 standardized units on either side of the sample mean $\bar{x}$. If $\bar{x}$ exceeds 10 by more than 1.699 standardized units, the resulting interval will not include 10. And the one-sided test will reject $H_0: \mu = 10$. There's the link: our one-sided test at $\alpha = 0.05$ gives the same conclusion about H_0 as a 90% confidence interval for μ.

Teaching Tip

Notice the null hypothesis value of 31 is just barely outside the 99% confidence interval of (31.028, 32.842). This means the observed result is just barely significant at the $\alpha = 0.01$ level, which we can confirm with the P-value $= 0.008$ being just barely less than 0.01.

Teaching Tip

The Investigative Task (Question 6) from the 2004 AP® Statistics exam explores the link between one-sided tests and confidence intervals and introduces students to one-sided confidence intervals. Although one-sided confidence intervals are not on the AP® Statistics Topic Outline, discussing this question is a handy way to explore the relationship between confidence intervals and significance tests. Remember that the Investigative Task expects students to apply knowledge in new contexts or nonroutine ways, including areas not on the Topic Outline.

CHECK YOUR UNDERSTANDING

According to the National Center for Health Statistics, the mean systolic blood pressure for males 35 to 44 years of age is 128. The health director of a large company wonders if this national average holds for the company's middle-aged male employees. So the director examines the medical records of a random sample of 72 male employees in this age group and records each of their systolic blood pressure readings.

1. State an appropriate pair of hypotheses for a significance test in this setting. Be sure to define the parameter of interest.
2. Check conditions for performing the test in Question 1.
3. A 95% confidence interval for the mean systolic blood pressure of all 35- to 44-year-old male employees at this company is (126.43, 133.43). Based on this interval, what conclusion would you make for a test of the hypotheses in Question 1 at the $\alpha = 0.05$ significance level?

Using Tests Wisely

Significance tests are widely used in reporting the results of research in many fields. New drugs require significant evidence of effectiveness and safety. Courts ask about statistical significance in hearing discrimination cases. Marketers want to know whether a new ad campaign significantly outperforms the old one, and medical researchers want to know whether a new therapy performs significantly better. In all these uses, statistical significance is valued because it points to an effect that is unlikely to occur simply by chance.

Carrying out a significance test is often quite simple, especially if you use technology. Using tests wisely is not so simple. Here are some points to keep in mind when using or interpreting significance tests.

STATISTICAL SIGNIFICANCE AND PRACTICAL IMPORTANCE When a null hypothesis of no effect or no difference can be rejected at the usual significance levels ($\alpha = 0.10$ or $\alpha = 0.05$ or $\alpha = 0.01$), there is convincing evidence of a difference. But that difference may be very small. When large samples are used, even tiny deviations from the null hypothesis will be significant.

Suppose we're testing a new antibacterial cream, "Formulation NS," on a small cut made on the inner forearm. We know from previous research that with no medication, the mean healing time (defined as the time for the scab to fall off) is 7.6 days. The claim we want to test here is that Formulation NS speeds healing. So our hypotheses are

$$H_0: \mu = 7.6$$
$$H_a: \mu < 7.6$$

where $\mu =$ the true mean healing time (in days) in the population of college students whose cuts are treated with Formulation NS. We will use a 5% significance level.

A random sample of 250 college students give informed consent to participate in a study and apply Formulation NS to their wounds. The mean healing time for these subjects is $\overline{x} = 7.5$ days and the standard deviation is $s_x = 0.9$ day. Note that the conditions for performing a one-sample t test for μ are met. We carry out

✓ Answers to CYU

1. $H_0: \mu = 128$, $H_a: \mu \neq 128$, where $\mu =$ the true mean systolic blood pressure for the company's middle-aged male employees.

2. *Random:* The director examines the medical records of a random sample of 72 male employees in this age group. *10%:* The sample size (72) is less than 10% of the population of middle-aged male employees at this large company. *Normal/Large Sample:* $n = 72 \geq 30$

3. The 95% confidence interval does include 128 as a plausible value, so we fail to reject H_0 at the $\alpha = 0.05$ significance level. We do not have convincing evidence that the true mean systolic blood pressure for the company's middle-aged male employees is different from 128.

Teaching Tip

The CBS News program *60 Minutes* aired an excellent segment about the placebo effect called "Treating Depression: Is There a Placebo Effect?" The 13-minute video addresses several of the issues we discuss in this section, including the difference between statistical significance and practical (clinical) importance and the use of multiple tests. To find the video, search YouTube for "Treating Depression: Is There a Placebo Effect?"

Some people say "not clinically significant" when they mean not practically important.

the test and find that $t = -1.76$ and *P*-value $= 0.04$ with df $= 249$. Because 0.04 is less than $\alpha = 0.05$, we reject H_0. Formulation NS "significantly" reduces the average healing time. However, this result is not practically important. Having your scab fall off one-tenth of a day sooner is no big deal!

Remember the wise saying: **Statistical significance is not the same thing as practical importance.** The remedy for attaching too much importance to statistical significance is to pay attention to the data as well as to the *P*-value. The foolish user of statistics who feeds the data to a calculator or computer without exploratory analysis will often be embarrassed. Plot your data, and examine them carefully. Is the difference you are seeking visible in your graphs? If not, ask yourself whether the difference is large enough to be practically important. Are there outliers or other departures from a consistent pattern? A few outlying observations can produce highly significant results if you blindly apply common significance tests. Outliers can also destroy the significance of otherwise convincing data.

To help evaluate practical importance, give a confidence interval for the parameter in which you are interested. A confidence interval provides a set of plausible values for the parameter, rather than simply asking if the observed result is too surprising to occur by chance alone when H_0 is true. Confidence intervals are not used as often as they should be, whereas significance tests are perhaps overused.

BEWARE OF MULTIPLE ANALYSES Statistical significance ought to mean that you have found a difference that you were looking for. The reasoning behind statistical significance works well if you decide what difference you are seeking, design a study to search for it, and use a significance test to weigh the evidence you gather. In other settings, significance may have little meaning. Here's one such example.

Might the radiation from cell phones be harmful to users? Many studies have found little or no connection between using cell phones and various illnesses. Here is part of a news account of one study:

> A hospital study that compared brain cancer patients and a similar group without brain cancer found no statistically significant difference in cell-phone use for the two groups. But when 20 distinct types of brain cancer were considered separately, a significant difference in cell-phone use was found for one rare type. Puzzlingly, however, this risk appeared to decrease rather than increase with greater mobile phone use.[3]

Think for a moment. Suppose that the 20 null hypotheses for these 20 significance tests are all true. Then each test has a 5% chance of being significant at the 5% level. That's what $\alpha = 0.05$ means: results this extreme occur only 5% of the time just by chance when the null hypothesis is true. We expect about 1 of 20 tests to give a significant result just by chance. Running one test and reaching the $\alpha = 0.05$ level is reasonably good evidence that you have found something; running 20 tests and reaching that level only once is not.

Searching data for patterns is certainly legitimate. Performing every conceivable significance test on a data set with many variables until you obtain a statistically significant result is not. This unfortunate practice is known by many names, including data dredging and *P*-hacking.

For more on the pitfalls of multiple analyses, do an Internet search for the XKCD comic about jelly beans causing acne.

Section 11.1 | Summary

- To perform a significance test of $H_0: \mu = \mu_0$, we need to ensure that the observations in the sample can be viewed as independent and that the sampling distribution of $\bar{x}$ is approximately Normal. The required conditions are:
  - **Random:** The data come from a random sample from the population of interest.
    - ◦ **10%:** When sampling without replacement, $n < 0.10N$.
  - **Normal/Large Sample:** The population has a Normal distribution or the sample size is large ($n \geq 30$). If the population distribution has unknown shape and $n < 30$, use a graph of the sample data to assess the Normality of the population. Do not use t procedures if the graph shows strong skewness or outliers.

- The standardized test statistic for a **one-sample t test for a mean** is

$$t = \frac{\bar{x} - \mu_0}{\frac{s_x}{\sqrt{n}}}$$

- When the Normal/Large Sample condition is met, the distribution of this standardized test statistic can be modeled by a t **distribution** with $n - 1$ degrees of freedom. You can use Table B or technology to find the P-value.

- Confidence intervals provide additional information that significance tests do not—namely, a set of plausible values for the parameter μ. A 95% confidence interval for μ gives consistent results with a two-sided test of $H_0: \mu = \mu_0$ at the $\alpha = 0.05$ significance level.

- Very small differences can be highly significant (small P-value) when a test is based on a large sample. A statistically significant result may not be practically important.

- Many tests run at once will likely produce some significant results by chance alone, even if all the null hypotheses are true. Beware of P-hacking.

11.1 Technology Corners

TI-Nspire and other technology instructions are on the book's website at highschool.bfwpub.com/updatedtps6e.

25. Computing P-values from t distributions	Page 704
26. Performing a one-sample t test for a mean	Page 708

TRM Section 11.1 Quiz

There is one quiz available for this section in the Teacher's Resource Materials. Click on the link in the TE-Book, look on the TRFD, or download from the Teacher's Resources on the book's digital platform. You can also create your own quiz using the ExamView® Assessment Suite that is part of the TPS6 program. Questions are coded by Learning Target to make it easy to build parallel quizzes.

Teaching Tip

Make sure to point out the two icons next to Exercise 1. The "pg 699" icon reminds students that the example on page 699 is very similar to this exercise. The "play" icon reminds students that there is a video solution available in the student e-Book or at the Student Site.

Answers to Section 9.3 Exercises

11.1 *Random:* Random sample of students. *10%:* $45 < 10\%$ of 1000. *Normal/Large Sample:* $n = 45 \geq 30$.

11.2 *Random:* Random sample of bags. *10%:* Assume $75 < 10\%$ of population. *Normal/Large Sample:* $n = 75 \geq 30$.

11.3 (a) $H_0: \mu = 11.5$, $H_a: \mu < 11.5$, where $\mu =$ the true mean battery life when playing videos for all tablets. **(b)** *Random:* Random sample. *10%:* $20 < 10\%$ of population. *Normal/Large Sample:* Not met because the sample size is less than 30 and the dotplot of the distribution of battery life is strongly skewed to the right.

11.4 (a) $H_0: \mu = 50$, $H_a: \mu > 50$, where $\mu =$ the true mean percent of purchases for which an alternative supplier offered lower prices. **(b)** *Random:* Random sample. *10%:* $25 < 10\%$ of population. *Normal/Large Sample:* Not met because the sample size is less than 30 and the histogram of the data is strongly skewed to the left.

11.5 (a) $\bar{x} = 62.8 \neq 64$

(b) $t = \dfrac{62.8 - 64}{\dfrac{5.36}{\sqrt{25}}} = -1.12$;

P-value: df $= 24$

Table B: Between 0.20 and 0.30; *Tech:* 0.2738

11.6 (a) $\bar{x} = 4.7 < 5$

(b) $t = \dfrac{4.7 - 5}{\dfrac{0.74}{\sqrt{20}}} = -1.81$;

P-value: df $= 19$

Table B: Between 0.025 and 0.05; *Tech:* 0.0431

Section 11.1   Exercises

1. **Attitudes** The Survey of Study Habits and Attitudes
pg 699 (SSHA) is a psychological test with scores that range from 0 to 200. The mean score for U.S. college students is 115. A teacher suspects that older students have better attitudes toward school. She gives the SSHA to an SRS of 45 students from the more than 1000 students at her college who are at least 30 years of age. The teacher wants to perform a test at the $\alpha = 0.05$ significance level of

$$H_0: \mu = 115$$
$$H_a: \mu > 115$$

where $\mu =$ the mean SSHA score in the population of students at her college who are at least 30 years old. Check if the conditions for performing the test are met.

2. **Candy!** A machine is supposed to fill bags with an average of 19.2 ounces of candy. The manager of the candy factory wants to be sure that the machine does not consistently underfill or overfill the bags. So the manager plans to conduct a significance test at the $\alpha = 0.10$ significance level of

$$H_0: \mu = 19.2$$
$$H_a: \mu \neq 19.2$$

where $\mu =$ the true mean amount of candy (in ounces) that the machine put in all bags filled that day. The manager takes a random sample of 75 bags of candy produced that day and weighs each bag. Check if the conditions for performing the test are met.

3. **Battery life** A tablet computer manufacturer claims that its batteries last an average of 11.5 hours when playing videos. The quality-control department randomly selects 20 tablets from each day's production and tests the fully charged batteries by playing a video repeatedly until the battery dies. The quality-control department will discard the batteries from that day's production run if they find convincing evidence that the mean battery life is less than 11.5 hours. Here are a dotplot and summary statistics of the data from one day:

10 10.5 11 11.5 12 12.5 13 13.5 14
Battery life (hours)

n	Mean	SD	Min	Q_1	Med	Q_3	Max
20	11.07	1.097	10	10.3	10.6	11.85	13.9

(a) State appropriate hypotheses for the quality-control department to test. Be sure to define your parameter.

(b) Check if the conditions for performing the test in part (a) are met.

4. **Paying high prices?** A retailer entered into an exclusive agreement with a supplier who guaranteed to provide all products at competitive prices. To be sure the supplier honored the terms of the agreement, the retailer had an audit performed on a random sample of 25 invoices. The percent of purchases on each invoice for which an alternative supplier offered a lower price than the original supplier was recorded.[4] For example, a data value of 38 means that the price would be lower with a different supplier for 38% of the items on the invoice. A histogram and some numerical summaries of the data are shown here. The retailer would like to determine if there is convincing evidence that the mean percent of purchases for which an alternative supplier offered lower prices is greater than 50% in the population of this company's invoices.

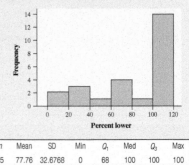

Percent lower

n	Mean	SD	Min	Q_1	Med	Q_3	Max
25	77.76	32.6768	0	68	100	100	100

(a) State appropriate hypotheses for the retailer's test. Be sure to define your parameter.

(b) Check if the conditions for performing the test in part (a) are met.

5. **Two-sided test** Suppose you want to perform a test of
pg 703 $H_0: \mu = 64$ versus $H_a: \mu \neq 64$ at the $\alpha = 0.05$ significance level. A random sample of size $n = 25$ from the population of interest yields $\bar{x} = 62.8$ and $s_x = 5.36$. Assume that the conditions for carrying out the test are met.

(a) Explain why the sample result gives some evidence for the alternative hypothesis.

(b) Calculate the standardized test statistic and *P*-value.

11.7 (a) $t = \dfrac{125.7 - 115}{\dfrac{29.8}{\sqrt{45}}} = 2.41$

(b) *P*-value: df $= 44$

Table B: Between 0.01 and 0.02; *Tech:* 0.0101. Assuming that the true mean SSHA score for older students is 115, there is a 0.0101 probability of getting a sample mean of at least 125.7 by chance alone.

(c) Because the *P*-value of $0.0101 < \alpha = 0.05$, we reject H_0. We have convincing evidence that the true mean SSHA score in the population of students at her college who are at least 30 years old is greater than 115.

11.8 (a) $t = \dfrac{19.28 - 19.2}{\dfrac{0.81}{\sqrt{75}}} = 0.86$

(b) *P*-value: df $= 74$

Table B: Using df $= 60$, between $2(0.15) = 0.30$ and $2(0.20) = 0.40$; *Tech:* 0.3926. Assuming that the true mean weight of candy bags is 19.2 ounces, there is a 0.3926 probability of getting a sample mean as different from 19.2 as 19.28 by chance alone.

(c) Because the *P*-value of $0.3926 > \alpha = 0.10$, we fail to reject H_0. We do not have convincing evidence that the true mean amount of candy (in ounces) that the machine put in all bags filled that day is different from 19.2.

6. **One-sided test** Suppose you want to perform a test of $H_0: \mu = 5$ versus $H_a: \mu < 5$ at the $\alpha = 0.05$ significance level. A random sample of size $n = 20$ from the population of interest yields $\bar{x} = 4.7$ and $s_x = 0.74$. Assume that the conditions for carrying out the test are met.

(a) Explain why the sample result gives some evidence for the alternative hypothesis.

(b) Calculate the standardized test statistic and P-value.

7. **Attitudes** In the study of older students' attitudes from Exercise 1, the sample mean SSHA score was 125.7 and the sample standard deviation was 29.8.

(a) Calculate the standardized test statistic.

(b) Find and interpret the P-value.

(c) What conclusion would you make?

8. **Candy!** In the study of the candy machine from Exercise 2, the sample mean weight for the bags of candy was 19.28 ounces and the sample standard deviation was 0.81 ounce.

(a) Calculate the standardized test statistic.

(b) Find and interpret the P-value.

(c) What conclusion would you make?

9. **Construction zones** Every road has one at some
pg 706 point—construction zones that have much lower speed limits. To see if drivers obey these lower speed limits, a police officer uses a radar gun to measure the speed (in miles per hour, or mph) of a random sample of 10 drivers in a 25 mph construction zone. Here are the data:

| 27 | 33 | 32 | 21 | 30 | 30 | 29 | 25 | 27 | 34 |

(a) Is there convincing evidence at the $\alpha = 0.01$ significance level that the average speed of drivers in this construction zone is greater than the posted speed limit?

(b) Given your conclusion in part (a), which kind of mistake—a Type I error or a Type II error—could you have made? Explain what this mistake would mean in context.

10. **Ending insomnia** A study was carried out with a random sample of 10 patients who suffer from insomnia to investigate the effectiveness of a drug designed to increase sleep time. The following data show the number of additional hours of sleep per night gained by each subject after taking the drug.[5] A negative value indicates that the subject got less sleep after taking the drug.

| 1.9 | 0.8 | 1.1 | 0.1 | −0.1 | 4.4 | 5.5 | 1.6 | 4.6 | 3.4 |

(a) Is there convincing evidence at the $\alpha = 0.01$ significance level that the average sleep increase is greater than 0 for insomnia patients when taking this drug?

(b) Given your conclusion in part (a), which kind of mistake—a Type I error or a Type II error—could you have made? Explain what this mistake would mean in context.

11. **Reading level** A school librarian purchases a novel for her library. The publisher claims that the book is written at a fifth-grade reading level, but the librarian suspects that the reading level is lower than that. The librarian selects a random sample of 40 pages and uses a standard readability test to assess the reading level of each page. The mean reading level of these pages is 4.8 with a standard deviation of 0.8. Do these data give convincing evidence at the $\alpha = 0.05$ significance level that the average reading level of this novel is less than 5?

12. **How much juice?** One company's bottles of grapefruit juice are filled by a machine that is set to dispense an average of 180 milliliters (ml) of liquid. The company has been getting negative feedback from customers about underfilled bottles. To investigate, a quality-control inspector takes a random sample of 40 bottles and measures the volume of liquid in each bottle. The mean amount of liquid in the bottles is 179.6 ml and the standard deviation is 1.3 ml. Do these data provide convincing evidence at the $\alpha = 0.05$ significance level that the machine is underfilling the bottles, on average?

13. **Pressing pills** A drug manufacturer forms tablets by compressing a granular material that contains the active ingredient and various fillers. The hardness of a sample from each batch of tablets produced is measured to control the compression process. The target value for the hardness is $\mu = 11.5$. The hardness data for a random sample of 20 tablets from one large batch are

11.627	11.613	11.493	11.602	11.360
11.374	11.592	11.458	11.552	11.463
11.383	11.715	11.485	11.509	11.429
11.477	11.570	11.623	11.472	11.531

Is there convincing evidence at the 5% level that the mean hardness of the tablets in this batch differs from the target value?

14. **Jump around** Student researchers Haley, Jeff, and Nathan saw an article on the Internet claiming that the average vertical jump for teens was 15 inches. They wondered if the average vertical jump of students at their school differed from 15 inches, so they obtained a list of student names and selected a random sample

11.9 (a) S: $H_0: \mu = 25$, $H_a: \mu > 25$, where μ = the true mean speed of all drivers in a construction zone using $\alpha = 0.01$. **P:** One-sample t test for μ. *Random:* Random sample. *10%:* 10 < 10% of population. *Normal/Large Sample:* There is no strong skewness or outliers in the sample.

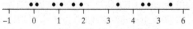

Speed (mph)

D: $\bar{x} = 28.8$, $s_x = 3.94$, $t = 3.05$, df = 9, P-value = 0.0069. **C:** Because the P-value of 0.0069 < $\alpha = 0.01$, we reject H_0. We have convincing evidence that the true mean speed of all drivers in the construction zone is greater than 25 mph. **(b)** Type I error: Finding convincing

evidence that the true mean speed is greater than 25 mph when it really isn't.

11.10 (a) S: $H_0: \mu = 0$, $H_a: \mu > 0$, where μ = the true mean number of additional hours of sleep per night gained by using the drug for all people who would take it using $\alpha = 0.01$. **P:** One-sample t test for μ. *Random:* Random sample. *10%:* 10 < 10% of population. *Normal/Large Sample:* The dotplot doesn't show any outliers or strong skewness.

Number of additional hours of sleep

D: $\bar{x} = 2.33$, $s_x = 2.002$, $t = 3.68$, df = 9, P-value = 0.00254. **C:** Because the P-value of 0.00254 < $\alpha = 0.05$, we reject H_0. There is convincing evidence that the drug is effective at

increasing the average sleep time for patients who suffer from insomnia.
(b) Type I error: Finding convincing evidence that the true mean sleep increase is greater than 0 when it really isn't.

11.11 S: $H_0: \mu = 5$, $H_a: \mu < 5$, where μ = the true mean reading level of all pages in this novel using $\alpha = 0.05$. **P:** One-sample t test for μ. *Random:* Random sample. *10%:* Assume 40 < 10% of population. *Normal/Large Sample:* $n = 40 \geq 30$. **D:** $\bar{x} = 4.8$, $s_x = 0.8$, $t = -1.58$, df = 39, and P-value = 0.0610. **C:** Because the P-value of 0.0610 > $\alpha = 0.05$, we fail to reject H_0. There is not convincing evidence that the true mean reading level for this novel is less than 5.

11.12 S: $H_0: \mu = 180$, $H_a: \mu < 180$, where μ = the true mean amount (ml) of grapefruit juice dispensed using $\alpha = 0.05$. **P:** One-sample t test for μ. *Random:* Random sample. *10%:* Assume 40 < 10% of population. *Normal/Large Sample:* $n = 40 \geq 30$. **D:** $\bar{x} = 179.6$, $s_x = 1.3$, $t = -1.95$, df = 39, P-value = 0.0292. **C:** Because the P-value of 0.0292 < $\alpha = 0.05$, we reject H_0. There is convincing evidence that the true mean amount of grapefruit juice dispensed is less than 180 ml.

11.13 S: $H_0: \mu = 11.5$, $H_a: \mu \neq 11.5$, where μ = the true mean hardness of the tablets using $\alpha = 0.05$. **P:** One-sample t test for μ. *Random:* Random sample. *10%:* 20 < 10% of population. *Normal/Large Sample:* There is no strong skewness or outliers in the sample.

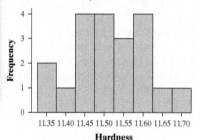

Hardness

D: $\bar{x} = 11.5164$, $s_x = 0.095$, $t = 0.77$, df = 19, and P-value = 0.4494. **C:** Because the P-value of 0.4494 > $\alpha = 0.05$, we fail to reject H_0. We do not have convincing evidence that the true mean hardness of these tablets is different from 11.5.

11.14 S: $H_0: \mu = 15$, $H_a: \mu \neq 15$, where μ = the true mean vertical jump of all students at this school using $\alpha = 0.10$.
P: One-sample t test for μ. *Random:* Random sample. *10%:* $20 < 10\%$ of population. *Normal/Large Sample:* There is no strong skewness or outliers in the sample, so it is reasonable to use a t procedure.

Vertical jump (in.)

D: $\bar{x} = 17$, $s_x = 5.368$, $t = 1.67$, df = 19, and P-value = 0.1121.
C: Because the P-value of $0.1121 > \alpha = 0.10$, we fail to reject H_0. We do not have convincing evidence that the true mean vertical jump distance for all students at this school is different from 15 inches.

11.15 (a) The test in Exercise 13 only allowed us to fail to reject $H_0: \mu = 11.5$. The confidence interval tells us that any value of μ between 11.472 and 11.561 is plausible based on the sample data. This is consistent with, but gives more information than, the test in Exercise 13.
(b) If the true mean hardness of the tablets is = 11.55, there is a 0.61 probability that the drug manufacturer will find convincing evidence for $H_a: \mu \neq 11.5$.
(c) $1 - 0.61 = 0.39$ **(d)** Increase the sample size or use a larger significance level.

11.16 (a) The test in Exercise 14 only allowed us to fail to reject $H_0: \mu = 15$. The confidence interval tells us that any value of μ between 14.472 and 19.076 is plausible based on the sample data. This is consistent with, but gives more information than, the test in Exercise 14.
(b) If the true mean vertical jump of the students at this school is $\mu = 17$ inches, there is a 0.49 probability that the student researchers will find convincing evidence for $H_a: \mu \neq 15$. **(c)** $1 - 0.49 = 0.51$
(d) Increase the sample size or use a larger significance level.

11.17 (a) $H_0: \mu = 200$, $H_a: \mu \neq 200$, where μ = the true mean response time of European servers (in milliseconds).
(b) *Random:* Random sample. *10%:* $14 < 10\%$ of population. *Normal/Large Sample:* A graph of the data reveals no strong skewness or outliers. **(c)** Because the 95% confidence interval does not contain 200, we reject H_0 at the $\alpha = 0.05$ significance level. We have convincing evidence that the mean response time of European servers is different from 200 milliseconds. **(d)** No! We cannot draw any conclusions about the United States

of 20 students. After contacting these students several times, they finally convinced them to allow their vertical jumps to be measured. Here are the data (in inches):

| 11.0 | 11.5 | 12.5 | 26.5 | 15.0 | 12.5 | 22.0 | 15.0 | 13.5 | 12.0 |
| 23.0 | 19.0 | 15.5 | 21.0 | 12.5 | 23.0 | 20.0 | 8.5 | 25.5 | 20.5 |

Do these data provide convincing evidence at the $\alpha = 0.10$ level that the average vertical jump of students at this school differs from 15 inches?

15. **Pressing pills** Refer to Exercise 13.
(a) A 95% confidence interval for the true mean hardness μ is (11.472, 11.561). Explain how this interval gives more information than the test in Exercise 13.
(b) The power of the test to detect that $\mu = 11.55$ is 0.61. Interpret this value.
(c) Find the probability of a Type II error if $\mu = 11.55$.
(d) Describe two ways to decrease the probability in part (c).

16. **Jump around** Refer to Exercise 14.
(a) A 90% confidence interval for the true mean vertical jump μ is (14.924, 19.076). Explain how this interval gives more information than the test in Exercise 14.
(b) The power of the test to detect that $\mu = 17$ inches is 0.49. Interpret this value
(c) Find the probability of a Type II error if $\mu = 17$.
(d) Describe two ways to decrease the probability in part (c).

17. **Fast connection?** How long does it take for a chunk of information to travel from one server to another and back on the Internet? According to the site internettrafficreport.com, the average response time is 200 milliseconds (about one-fifth of a second). Researchers wonder if this claim is true, so they collect data on response times (in milliseconds) for a random sample of 14 servers in Europe. A graph of the data reveals no strong skewness or outliers.
(a) State an appropriate pair of hypotheses for a significance test in this setting. Be sure to define the parameter of interest.
(b) Check conditions for performing the test in part (a).
(c) The 95% confidence interval for the mean response time is 158.22 to 189.64 milliseconds. Based on this interval, what conclusion would you make for a test of the hypotheses in part (a) at the 5% significance level?
(d) Do we have convincing evidence that the mean response time of servers in the United States is different from 200 milliseconds? Justify your answer.

18. **Water!** A blogger claims that U.S. adults drink an average of 40 ounces (that's five 8-ounce glasses) of water per day. Researchers wonder if this claim is true, so they ask a random sample of 24 U.S. adults about their daily water intake. A graph of the data shows a roughly symmetric shape with no outliers.
(a) State an appropriate pair of hypotheses for a significance test in this setting. Be sure to define the parameter of interest.
(b) Check conditions for performing the test in part (a).
(c) The 90% confidence interval for the mean daily water intake is 30.35 to 36.92 ounces. Based on this interval, what conclusion would you make for a test of the hypotheses in part (a) at the 10% significance level?
(d) Do we have convincing evidence that the amount of water U.S. children drink per day differs from 40 ounces? Justify your answer.

19. **Tests and confidence intervals** The P-value for a two-sided test of the null hypothesis $H_0: \mu = 10$ is 0.06.
(a) Does the 95% confidence interval for μ include 10? Why or why not?
(b) Does the 90% confidence interval for μ include 10? Why or why not?

20. **Tests and confidence intervals** The P-value for a two-sided test of the null hypothesis $H_0: \mu = 15$ is 0.03.
(a) Does the 99% confidence interval for μ include 15? Why or why not?
(b) Does the 95% confidence interval for μ include 15? Why or why not?

21. **Do you have ESP?** A researcher looking for evidence of extrasensory perception (ESP) tests 500 subjects. Four of these subjects do significantly better ($P < 0.01$) than random guessing.
(a) Is it proper to conclude that these four people have ESP? Explain your answer.
(b) What should the researcher now do to test whether any of these four subjects has ESP?

22. **Preventing colds** A medical experiment investigated whether taking the herb echinacea could help prevent colds. The study measured 50 different response variables usually associated with colds, such as low-grade fever, congestion, frequency of coughing, and so on. At the end of the study, those taking echinacea displayed significantly better responses at the $\alpha = 0.05$ level than those taking a placebo for 3 of the 50 response variables studied. Should we be convinced that echinacea helps prevent colds? Why or why not?

because we only collected information from a random sample of servers in Europe.

11.18 (a) $H_0: \mu = 5$, $H_a: \mu \neq 5$, where μ = the true mean number of 8-ounce glasses of water that a U.S. adult drinks per day, or $H_0: \mu = 40$ vs. $H_a: \mu \neq 40$, where μ = the true mean daily water intake (in ounces) for all U.S. adults.
(b) *Random:* Random sample. *10%:* $24 < 10\%$ of population. *Normal/Large Sample:* A graph shows a roughly symmetric shape with no outliers. **(c)** Because the 90% confidence interval does not contain 40 ounces, we reject H_0 at the $\alpha = 0.10$ significance level. We have convincing evidence that the true mean daily water intake for U.S. adults is different from 40 ounces.
(d) No! We cannot draw any conclusions about the mean amount of water U.S. children drink

per day because we only collected information from a random sample of U.S. adults.

11.19 (a) Yes; because the P-value of $0.06 > \alpha = 0.05$, we fail to reject $H_0: \mu = 10$ at the 5% level of significance. The 95% confidence interval will include 10. **(b)** No; because the P-value of $0.06 < \alpha = 0.10$, we reject $H_0: \mu = 10$ at the 10% level of significance. The 90% confidence interval would not include 10 as a plausible value.

11.20 (a) Yes; because the P-value of $0.03 > \alpha = 0.01$, we fail to reject $H_0: \mu = 15$ at the 1% level of significance. The 99% confidence interval will include 15. **(b)** No; because the P-value of $0.03 < \alpha = 0.05$, we reject $H_0: \mu = 15$ at the 5% level of significance. The 95% confidence interval would not include 15 as a plausible value.

23. **Improving SAT scores** A national chain of SAT-preparation schools wants to know if using a smartphone app in addition to its regular program will help increase student scores more than using just the regular program. On average, the students in the regular program increase their scores by 128 points during the 3-month class. To investigate using the smartphone app, the prep schools have 5000 students use the app along with the regular program and measure their improvement. Then the schools will test the following hypotheses: $H_0: \mu = 128$ versus $H_a: \mu > 128$, where μ is the true mean improvement in the SAT score for students who attend these prep schools. After 3 months, the average improvement was $\bar{x} = 130$ with a standard deviation of $s_x = 65$. The standardized test statistic is $t = 2.18$ with a P-value of 0.0148. Explain why this result is statistically significant, but not practically important.

24. **Music and mazes** A researcher wishes to determine if people are able to complete a certain pencil and paper maze more quickly while listening to classical music. Suppose previous research has established that the mean time needed for people to complete a certain maze (without music) is 40 seconds. The researcher decides to test the hypotheses $H_0: \mu = 40$ versus $H_a: \mu < 40$, where $\mu =$ the average time in seconds to complete the maze while listening to classical music. To do so, the researcher has 10,000 people complete the maze with classical music playing. The mean time for these people is $\bar{x} = 39.92$ seconds, and the P-value of his significance test is 0.0002. Explain why this result is statistically significant, but not practically important.

25. **Sampling shoppers** A marketing consultant observes 50 consecutive shoppers at a supermarket, recording how much each shopper spends in the store. Explain why it would not be wise to use these data to carry out a significance test about the mean amount spent by all shoppers at this supermarket.

26. **Ages of presidents** Joe is writing a report on the backgrounds of American presidents. He looks up the ages of all the presidents when they entered office. Because Joe took a statistics course, he uses these numbers to perform a significance test about the mean age of all U.S. presidents. Explain why this makes no sense.

Multiple Choice: *Select the best answer for Exercises 27–32.*

27. The reason we use t procedures instead of z procedures when carrying out a test about a population mean is that

(a) z requires that the sample size be large.

(b) z requires that you know the population standard deviation σ.

(c) z requires that the data come from a random sample.

(d) z requires that the population distribution be Normal.

(e) z can only be used for proportions.

28. You are testing $H_0: \mu = 75$ against $H_a: \mu < 75$ based on an SRS of 20 observations from a Normal population. The t statistic is $t = -2.25$. The P-value

(a) falls between 0.01 and 0.02.

(b) falls between 0.02 and 0.04.

(c) falls between 0.04 and 0.05.

(d) falls between 0.05 and 0.25.

(e) is greater than 0.25.

29. You are testing $H_0: \mu = 10$ against $H_a: \mu \neq 10$ based on an SRS of 15 observations from a Normal population. What values of the t statistic are statistically significant at the $\alpha = 0.005$ level?

(a) $t > 3.326$

(b) $t > 3.286$

(c) $t > 2.977$

(d) $t < -3.326$ or $t > 3.326$

(e) $t < -3.286$ or $t > 3.286$

30. After checking that conditions are met, you perform a significance test of $H_0: \mu = 1$ versus $H_a: \mu \neq 1$. You obtain a P-value of 0.022. Which of the following must be true?

(a) A 95% confidence interval for μ will include the value 1.

(b) A 95% confidence interval for μ will include the value 0.022.

(c) A 99% confidence interval for μ will include the value 1.

(d) A 99% confidence interval for μ will include the value 0.022.

(e) None of these is necessarily true.

31. The most important condition for making an inference about a population mean from a significance test is that

(a) the data come from a random sample.

(b) the population distribution is exactly Normal.

(c) the data contain no outliers.

(d) the sample size is less than 10% of the population size.

(e) the sample size is at least 30.

32. Vigorous exercise helps people live several years longer (on average). Whether mild activities like slow walking extend life is not clear. Suppose that the added life expectancy from regular slow walking is just 2 months.

11.25 It would not be wise to use these data to carry out a significance test about the mean amount spent by all shoppers at the supermarket because this sample wasn't randomly selected—it was a convenience sample. Depending on the time of day or the day of the week, certain types of shoppers may be underrepresented.

11.26 It makes no sense to perform a significance test about the mean age of all U.S. presidents because Joe looked up the ages of every president when he entered office. This is not information taken from a sample. We have information about all presidents—the whole population of interest—so Joe can calculate the exact value of μ.

11.27 b

11.28 a

11.29 d

11.30 c

11.31 a

11.21 (a) No; in a sample of size $n = 500$, we expect to see about $(500)(0.01) = 5$ people who do better than random guessing, with a significance level of 0.01. These four might have ESP, or they may simply be among the "lucky" ones we expect to see just by chance. **(b)** The researcher should repeat the procedure on these four people to see if they again perform well.

11.22 At the $\alpha = 0.05$ level, we expect to have a Type I error (getting a significant result when the null hypothesis is true) 5% of the time. If 50 tests are performed at this significance level and all 50 null hypotheses were true, we would expect, on average, $(50)(0.05) = 2.5$ Type I errors. The three significant results in this situation may be such errors.

11.23 Although the hypothesis test shows that the results are statistically significant (we have convincing evidence that $\mu > 128$), this is not practically significant. A test with such a large sample size will often produce a significant result for a very small departure from the null value. There is little practical significance to an increase in average SAT score of only 2 points.

11.24 Although the hypothesis test shows that the results are statistically significant (we have convincing evidence that $\mu < 40$), this is not practically significant. A test with such a large sample size will often produce a significant result for a very small departure from the null value. There is little practical significance to a change in average finish time of 0.08 second.

11.32 a

11.33 (a) Not included; the margin of error does not account for undercoverage. **(b)** Not included; the margin of error does not account for nonresponse. **(c)** Included; the margin of error does account for sampling variability.

11.34 (a) P(at least one apple) $=$ $1 - P$(no apples) $= 1 - (0.80)^5 =$ 0.67232. There is a 0.67232 probability of getting at least one apple in one pull of the lever. **(b)** We assume that all pulls of the lever are independent. The probability of no apples on a given pull of the lever is $1 - 0.67232 = 0.32768$. So the probability of no apples on any of 5 pulls is $(0.32768)^5 = 0.00378$ and the probability of at least 1 apple in 5 pulls of the lever is $1 - 0.00378 = 0.99622$.

PD **Section 11.2 Overview**

To watch the video overview of Section 11.2 (for teachers), click on the link in the TE-Book, look on the TRFD, or download from the Teacher's Resources on the book's digital platform.

TRM **Section 11.2 Alternate Examples**

You can find the Alternate Examples for this section in Microsoft Word format by clicking on the link in the TE-Book, opening the TRFD, or downloading from the Teacher's Resources on the book's digital platform.

A significance test is more likely to find a significant increase in mean life expectancy with regular slow walking if

(a) it is based on a very large random sample and a 5% significance level is used.

(b) it is based on a very large random sample and a 1% significance level is used.

(c) it is based on a very small random sample and a 5% significance level is used.

(d) it is based on a very small random sample and a 1% significance level is used.

(e) the size of the sample doesn't have any effect on the significance of the test.

Recycle and Review

33. **Is your food safe?** (8.1) "Do you feel confident or not confident that the food available at most grocery stores is safe to eat?" When a Gallup poll asked this question, 87% of the sample said they were confident.[6] Gallup announced the poll's margin of error for 95% confidence as ±3 percentage points. Which of the following sources of error are included in this margin of error? Explain your answer.

(a) Gallup dialed landline telephone numbers at random and so missed all people without landline phones, including people whose only phone is a cell phone.

(b) Some people whose numbers were chosen never answered the phone in several calls or answered but refused to participate in the poll.

(c) There is chance variation in the random selection of telephone numbers.

34. **Spinning for apples** (5.3 or 6.3) In the "Ask Marilyn" column of *Parade* magazine, a reader posed this question: "Say that a slot machine has five wheels, and each wheel has five symbols: an apple, a grape, a peach, a pear, and a plum. I pull the lever five times. What are the chances that I'll get at least one apple?" Suppose that the wheels spin independently and that the five symbols are equally likely to appear on each wheel in a given spin.

(a) Find the probability that the slot player gets at least one apple in one pull of the lever.

(b) Now answer the reader's question.

SECTION 11.2 **Tests About a Difference in Means**

LEARNING TARGETS *By the end of the section, you should be able to:*

- State appropriate hypotheses for a significance test about a difference between two means.
- Determine whether the conditions are met for performing a test about a difference between two means.
- Calculate the standardized test statistic and P-value for a test about a difference between two means.
- Perform a significance test about a difference between two means.
- Perform a significance test about a mean difference.
- Determine when it is appropriate to use paired t procedures versus two-sample t procedures.

In Section 11.1, you learned how to perform a significance test about a population mean μ. Many interesting statistical questions involve comparing means μ_1 and μ_2 for two populations or treatments. Has the mean number of hours worked per week by Americans changed from what it was 30 years ago? Which of two chemotherapy regimens results in longer average survival for early-stage

pancreatic cancer patients? In each of these cases, we want to test a claim about the *difference in means* $\mu_1 - \mu_2$.

Other questions involve comparing means for paired data. Does the mean price per gallon of unleaded gasoline and of diesel fuel differ at Colorado gas stations that sell both? Are people's standing pulse rates higher, on average, than their sitting pulse rates? In each of these cases, we want to test a claim about the *mean difference* μ_{diff}.

This section shows you how to perform significance tests for comparing means in both of these settings.

Significance Tests for $\mu_1 - \mu_2$

An observed difference between two sample means $\bar{x}_1$ and $\bar{x}_2$ can reflect an actual difference in the parameters μ_1 and μ_2, or it may just be due to chance variation in random sampling or random assignment. Significance tests help us decide which explanation makes more sense.

STATING HYPOTHESES AND CHECKING CONDITIONS In a test for comparing two means, the null hypothesis has the general form

$$H_0: \mu_1 - \mu_2 = \text{hypothesized value}$$

We're often interested in situations in which the hypothesized difference is 0. Then the null hypothesis says that there is no difference between the two parameters:

$$H_0: \mu_1 - \mu_2 = 0$$

(You will sometimes see the null hypothesis written in the equivalent form $H_0: \mu_1 = \mu_2$.) The alternative hypothesis says what kind of difference we expect.

The conditions for performing a significance test about $\mu_1 - \mu_2$ are the same as for constructing a confidence interval. Recall that the purpose of these conditions is to check for independence in the data collection process and that the sampling distribution of $\bar{x}_1 - \bar{x}_2$ is approximately Normal.

CONDITIONS FOR PERFORMING A SIGNIFICANCE TEST ABOUT A DIFFERENCE IN MEANS

- **Random:** The data come from two independent random samples or from two groups in a randomized experiment.
  - **10%:** When sampling without replacement, $n_1 < 0.10N_1$ and $n_2 < 0.10N_2$.
- **Normal/Large Sample:** For *each* sample, the corresponding population distribution (or the true distribution of response to the treatment) is Normal or the sample size is large ($n \geq 30$). For each sample, if the population (treatment) distribution has unknown shape and $n < 30$, a graph of the sample data shows no strong skewness or outliers.

Teaching Tip

Familiarize students with both versions of the null hypothesis:

$$H_0: \mu_1 - \mu_2 = 0 \quad \text{and} \quad H_0: \mu_1 = \mu_2$$

We prefer the first version because it makes it easy to see why we plug in 0 for the parameter in the calculation of the standardized test statistic:

$$t = \frac{(\bar{X}_1 - \bar{X}_2) - (\mu_1 - \mu_2)}{\sqrt{\dfrac{s_1^2}{n_1} + \dfrac{s_2^2}{n_2}}}$$

Teaching Tip

This significance test can be performed in exactly the same way for a null hypothesis with a difference of means other than 0, but in this course, we will generally stick to cases where the difference is 0. In Exercises 49 and 50, students can investigate a null hypothesis with a difference of means that is non-zero.

Teaching Tip

The Random, 10%, and Normal/Large Sample conditions are exactly the same for significance tests for a difference in means as they were for a confidence interval for a difference in means. WooHoo!

Here's an example that illustrates how to state hypotheses and check conditions.

ALTERNATE EXAMPLE Skills 1.F, 4.C

Which version of the exam is harder?
Stating hypotheses and checking conditions

PROBLEM:

Mrs. Gallas has created two different versions of a final exam. She wonders if the difficulty is the same for each version. To find out, she randomly assigns 80 student volunteers to two groups: 40 students take Version A, and 40 students take Version B. Here are the results:

Version	Mean	SD
A	84.2	8.9
B	79.9	12.3

Do these data give convincing evidence at the $\alpha = 0.05$ significance level of a difference in the true mean score on the final exam for Version A and Version B for students like the ones in the study?
(a) State appropriate hypotheses for performing a significance test. Be sure to define the parameters of interest.
(b) Check if the conditions for performing the test are met.

SOLUTION:

(a) $H_0: \mu_A - \mu_B = 0$
$H_a: \mu_A - \mu_B \neq 0$

where μ_A = the true mean score on Version A of the final exam for students like the ones in the study and μ_B = the true mean score on Version B of the final exam for students like the ones in the study.
(b) • *Random:* The 80 subjects were randomly assigned to Version A or Version B. ✓
• *Normal/Large Sample:* $n_A = 40 \geq 30$ and $n_B = 40 \geq 30$. ✓

EXAMPLE

A longer work week?
Stating hypotheses and checking conditions

PROBLEM: Has the mean number of hours Americans work in a week changed? One of the questions on the General Social Survey (GSS) asked respondents how many hours they work each week. Responses from random samples of employed Americans were recorded for 1975 and 2014—they are summarized here:

Year	Sample size	Mean	SD
1975	764	38.97 hours	13.13 hours
2014	1501	41.91 hours	14.35 hours

Do these data give convincing evidence at the $\alpha = 0.05$ significance level of a difference in the true mean number of work hours per week for employed Americans in 1975 and 2014?

(a) State appropriate hypotheses for performing a significance test. Be sure to define the parameters of interest.

(b) Check if the conditions for performing the test are met.

SOLUTION:

(a) $H_0: \mu_{2014} - \mu_{1975} = 0$
$H_a: \mu_{2014} - \mu_{1975} \neq 0$

where μ_{1975} = **the true mean hours worked per week by employed Americans in 1975** and μ_{2014} = **the true mean hours worked per week by employed Americans in 2014.**

> You could also state the hypotheses as
> $H_0: \mu_{2014} = \mu_{1975}$
> $H_a: \mu_{2014} \neq \mu_{1975}$

(b) • **Random? Independent random samples of 764 employed Americans in 1975 and 1501 employed Americans in 2014.** ✓
  ○ **10%: 764 < 10% of all employed Americans in 1975; 1501 < 10% of all employed Americans in 2014.** ✓
• **Normal/Large Sample?** $n_{1975} = 764 \geq 30$ **and** $n_{2014} = 1501 \geq 30.$ ✓

> Be sure to mention *independent* random samples from the populations of interest when checking the Random condition.

FOR PRACTICE, TRY EXERCISE 35

CALCULATIONS: STANDARDIZED TEST STATISTIC AND *P*-VALUE If the conditions are met, we can proceed with calculations. To do a test of $H_0: \mu_1 - \mu_2 = 0$, start by standardizing $\bar{x}_1 - \bar{x}_2$:

$$\text{standardized test statistic} = \frac{\text{statistic} - \text{parameter}}{\text{standard deviation (error) of statistic}}$$

In the unlikely event that both population standard deviations σ_1 and σ_2 are known, the standardized test statistic becomes

$$z = \frac{(\overline{x}_1 - \overline{x}_2) - 0}{\sqrt{\dfrac{\sigma_1^2}{n_1} + \dfrac{\sigma_2^2}{n_2}}}$$

We could then use the standard Normal distribution to find the P-value. This is known as a *two-sample z test for a difference in means.* It is rarely used in practice.

Because we seldom know the population standard deviations σ_1 and σ_2, we use the standard error of $\overline{x}_1 - \overline{x}_2$ in the denominator of the standardized test statistic. If the hypothesized difference in means is 0, the standardized test statistic is

$$t = \frac{(\overline{x}_1 - \overline{x}_2) - 0}{\sqrt{\dfrac{s_1^2}{n_1} + \dfrac{s_2^2}{n_2}}}$$

When the Normal/Large Sample condition is met, we can find the P-value using the t distribution with degrees of freedom given by Option 1 (technology) or Option 2 (df = smaller of $n_1 - 1$ and $n_2 - 1$). See page 670 for details about the degrees of freedom used by technology.

Teaching Tip

Before revealing the formula for the t test statistic, assign students the preceding "Longer work week" example (or the alternate example) and have them work in pairs to write a four-step solution. Call this an "investigative task" because students are being asked to apply their previous knowledge in a new context.

Teaching Tip

Our two-sample t statistic doesn't quite follow a t distribution, even when the conditions are met. However, when an additional condition is met (equal population standard deviations), the *pooled* version of the two-sample t statistic does follow a t distribution with df = $n_1 - 1 + n_2 - 1$. We don't recommend using the pooled version for two reasons: (1) The equal population standard deviation condition is seldom met, and (2) the approximation using our (unpooled) two-sample t statistic is quite accurate. For more on pooling when doing inference about a difference in means, see page 727.

EXAMPLE

A longer work week?
Calculating the standardized test statistic and P-value

David Papazian/Getty Images

PROBLEM: Refer to the previous example. The table summarizes data on hours worked per week from the independent random samples of employed Americans in 1975 and 2014.

Year	Sample size	Mean	SD
1975	764	38.97 hours	13.13 hours
2014	1501	41.91 hours	14.35 hours

We already confirmed that the conditions for performing a significance test are met.
(a) Explain why the sample results give some evidence for the alternative hypothesis.
(b) Calculate the standardized test statistic and P-value.
(c) What conclusion would you make?

SOLUTION:

(a) The observed difference in the sample means is $\overline{x}_{2014} - \overline{x}_{1975} = 41.91 - 38.97 = 2.94$, which gives some evidence in favor of H_a: $\mu_{2014} - \mu_{1975} \neq 0$ because $2.94 \neq 0$.

(b) $t = \dfrac{(41.91 - 38.97) - 0}{\sqrt{\dfrac{14.35^2}{1501} + \dfrac{13.13^2}{764}}} = \dfrac{2.94}{0.602} = 4.88$

$$t = \frac{(\overline{x}_1 - \overline{x}_2) - 0}{\sqrt{\dfrac{s_1^2}{n_1} + \dfrac{s_2^2}{n_2}}}$$

P-value:

Option 1: 2-SampTTest gives $t = 4.88$ and P-value = 0.00000115 using df = 1660.58.

> Refer to Technology Corner 27 on page 724 for details on how to do calculations for a 2-sample t test on the TI-83/84.

Option 2: df = smaller of $764 - 1$ and $1501 - 1 = 763$
Using Table B: df = 100 gives P-value < $2(0.0005) = 0.001$

ALTERNATE EXAMPLE

Skill 3.E

Which version of the exam is harder? Calculating the standardized test statistic and P-value

PROBLEM:
Refer to the preceding alternate example. The table summarizes data on the scores for two different versions of the final exam.

Version	Number of students	Mean	SD
A	40	84.2	8.9
B	40	79.9	12.3

We already confirmed that the conditions for performing a significance test are met.
(a) Explain why the sample results give some evidence for the alternative hypothesis.

(b) Calculate the standardized test statistic and P-value.
(c) What conclusion would you make?

SOLUTION:
(a) The observed difference in the sample means is $\overline{x}_A - \overline{x}_B = 84.2 - 79.9 = 4.3$, which gives some evidence in favor of H_a: $\mu_A - \mu_B \neq 0$ because $4.3 \neq 0$.

(b) $t = \dfrac{(84.2 - 79.9) - 0}{\sqrt{\dfrac{8.9^2}{40} + \dfrac{12.3^2}{40}}} = \dfrac{4.3}{2.40} = 1.79$

P-value:
Option 1: 2-SampTTest gives $t = 1.79$ and P-value = 0.078 using df = 71.05.

Option 2: df = 39
Table B: df = 30 gives $2(0.025) <$ P-value < $2(0.05)$ or $0.05 <$ P-value < 0.10
Tech: tcdf(lower: 1.79, upper: 1000, df: 39) × 2 = 0.081

(c) Because the P-value of $0.078 > \alpha = 0.05$, we fail to reject H_0. There is not convincing evidence of a difference in the true mean score on Version A and Version B of the final exam for students like the ones in the study.

Using technology: **tcdf(lower: 4.88, upper: 1000, df:763)×2=0.000001292.**

(c) Because the P-value of 0.00000115 < α = 0.05, we reject H_0. There is convincing evidence of a difference in the true mean hours worked by employed Americans in 1975 and 2014.

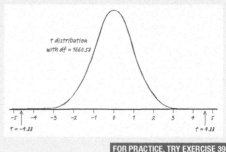

t distribution with df = 1660.58

$t = -4.88$ $t = 4.88$

FOR PRACTICE, TRY EXERCISE 39

What does the P-value in the example tell us? If there is no difference in the true mean hours worked by employed Americans in 1975 and in 2014, there is a 0.00000115 probability of getting a difference in sample means as large as or larger than 2.94 hours in either direction purely by the chance involved in random sampling. With such a small probability of getting a result like this just by chance when the null hypothesis is true, we have convincing evidence to reject H_0.

We can get more information about the difference between the population mean hours worked by employed Americans in 1975 and 2014 with a confidence interval. Technology gives the 95% confidence interval for $\mu_{2014} - \mu_{1975}$ as 1.76 to 4.12 hours. That is, we are 95% confident that the true mean hours worked per week by employed Americans is between 1.76 hours and 4.12 hours larger in 2014 than in 1975. This is consistent with our "reject H_0" conclusion because 0 is not included in the interval of plausible values for $\mu_{2014} - \mu_{1975}$.

Putting It All Together: Two-Sample *t* Test for $\mu_1 - \mu_2$

Here is a summary of the Do step for the **two-sample *t* test for a difference between two means**.

Recall that the formula sheet provided on the AP® Statistics exam uses the notation $s_{\bar{x}_1 - \bar{x}_2}$ rather than $SE_{\bar{x}_1 - \bar{x}_2}$ for the standard error of the difference in sample means $\bar{x}_1 - \bar{x}_2$.

> ### TWO-SAMPLE *t* TEST FOR A DIFFERENCE BETWEEN TWO MEANS
>
> Suppose the conditions are met. To test the hypothesis $H_0: \mu_1 - \mu_2 = 0$, compute the standardized test statistic
>
> $$t = \frac{(\bar{x}_1 - \bar{x}_2) - 0}{\sqrt{\dfrac{s_1^2}{n_1} + \dfrac{s_2^2}{n_2}}}$$
>
> Find the P-value by calculating the probability of getting a *t* statistic this large or larger in the direction specified by the alternative hypothesis H_a. Use the *t* distribution with degrees of freedom approximated by Option 1 (technology) or Option 2 (the smaller of $n_1 - 1$ and $n_2 - 1$).

AP® EXAM TIP

The specific formula for the standardized test statistic for a difference between two means is *not* included on the AP® Statistics exam formula sheet. However, students are given the general formula for a standardized test statistic and the specific standard error formula needed to substitute into the general formula.

Making Connections

Remind students of the general formula for a standardized test statistic from Chapter 9 and connect it to the specific formula for difference between two means.

General formula:

$$\text{standardized test statistic} = \frac{\text{statistic} - \text{parameter}}{\text{standard error of statistic}}$$

Specific formula:

$$t = \frac{(\bar{x}_1 - \bar{x}_2) - (\mu_1 - \mu_2)}{\sqrt{\dfrac{s_1^2}{n_1} + \dfrac{s_2^2}{n_2}}}$$

Here's an example that shows how to perform a two-sample t test for a difference between two means in a randomized experiment.

EXAMPLE

Calcium and blood pressure
Significance test for a difference between two means

PROBLEM: Does increasing the amount of calcium in our diet reduce blood pressure? Examination of a large sample of people revealed a relationship between calcium intake and blood pressure. Such observational studies do not establish causation. Researchers therefore designed a randomized comparative experiment.

The subjects were 21 healthy men who volunteered to take part in the experiment. They were randomly assigned to two groups: 10 of the men received a calcium supplement for 12 weeks, while the control group of 11 men received a placebo pill that looked identical. The experiment was double-blind. The response variable is the decrease in systolic (top number) blood pressure for a subject after 12 weeks, in millimeters of mercury. An increase appears as a negative number.[7] Here are the data:

Group 1 (calcium)	7	−4	18	17	−3	−5	1	10	11	−2	
Group 2 (placebo)	−1	12	−1	−3	3	−5	5	2	−11	−1	−3

(a) Do the data provide convincing evidence that a calcium supplement reduces blood pressure more than a placebo, on average, for subjects like the ones in this study?

(b) Interpret the P-value you got in part (a) in the context of this experiment.

SOLUTION:

(a) STATE: $H_0: \mu_C - \mu_P = 0$

$H_a: \mu_C - \mu_P > 0$

> Follow the four-step process!

where μ_C = the true mean decrease in systolic blood pressure for healthy men like the ones in this study who take a calcium supplement and μ_P = the true mean decrease in systolic blood pressure for healthy men like the ones in this study who take a placebo. No significance level was given, so we'll use $\alpha = 0.05$.

PLAN: Two-sample t test for $\mu_C - \mu_P$

- Random: The 21 subjects were randomly assigned to the calcium or placebo treatments. ✓

> Note that we did not have to check the 10% condition because the subjects in the experiment were not sampled without replacement from some larger population.

- Normal/Large Sample: The sample sizes are small, but the dotplots show no strong skewness and no outliers. ✓

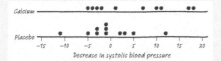

> Based on these graphs, it's reasonable to believe that the true distributions of decrease in systolic blood pressure are Normal for subjects like these when taking calcium or the placebo.

SOLUTION:

(a) STATE: $H_0: \mu_{AP} - \mu_N = 0$

$H_a: \mu_{AP} - \mu_N > 0$

where μ_{AP} = the true mean IQ score for all high school students in this district who take AP® Statistics and μ_N = the true mean IQ score for all high school students in this district who don't take AP® Statistics. No significance level was given, so we'll use $\alpha = 0.05$.

PLAN: Two-sample t test for $\mu_{AP} - \mu_N$

- *Random:* Independent random samples of 10 students in the district who take AP® Statistics and 10 students in the district who don't take AP® Statistics. ✓
  - *10%:* Assume 10 < 10% of all students in this large district who take AP® Statistics; 10 < 10% of all students in this large district who don't take AP® Statistics. ✓

- *Normal/Large Sample:* The sample sizes are small, but the dotplots show no strong skewness and no outliers. ✓

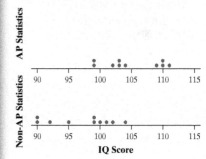

DO:

- $t = \dfrac{(105 - 97.2) - 0}{\sqrt{\dfrac{4.619^2}{10} + \dfrac{5.095^2}{10}}} = \dfrac{7.8}{2.17} = 3.59$

- *P-value*

Option 1: 2-SampTTest gives $t = 3.59$ and P-value = 0.001 using df = 17.83.

Option 2: df = 9

Table B: P-value is between 0.0025 and 0.005.

Tech: tcdf(lower: 3.59, upper: 1000, df: 9) = 0.003

CONCLUDE: Because the P-value of $0.001 < \alpha = 0.05$, we reject H_0. There is convincing evidence that high school students in this district who take AP® Statistics have a greater mean IQ score than high school students in this district who don't take AP® Statistics.

(b) Assuming $H_0: \mu_{AP} - \mu_N = 0$ is true, there is a 0.001 probability of getting a difference (AP® Statistics students − Students not taking AP® Statistics) in mean IQ scores from the samples of 7.8 or greater just by the chance involved in the random sampling.

Skills 4.B, 4.E

ALTERNATE EXAMPLE

AP® Statistics is smart   Significance test for a difference between two means

PROBLEM:

Some students in a very large school district wondered if students who take AP® Statistics in high school have greater IQ scores, on average, than students who don't take AP® Statistics. To investigate, they took a random sample of 10 AP® Statistics high school students in the district, and a separate random sample of 10 non-AP® Statistics high school students in the district, and had each of the students take an IQ test. Here are the results:

AP® Statistics	103	110	99	103	109	111	99	102	104	110
Non-AP® Statistics	102	99	100	104	95	92	99	101	90	90

(a) Do the data provide convincing evidence that high school students in this district who take AP® Statistics have a greater mean IQ score than high school students in this district who don't take AP® Statistics?

(b) Interpret the P-value you got in part (a) in the context of this study.

Teaching Tip

Here are some important tips for teaching students to use the four-step process for performing a two-sample t test for $\mu_1 - \mu_2$:

- Inform students they can use the wording of the question to define their parameters and to write the conclusion. We call this "parroting" the stem of the question.
- Ask students, "Which words in the question inform you this is a one- or two-sided test?"
- When naming the inference method, it is acceptable to list the calculator command "2-SampTTest." We prefer that students use the "two-sample t test for $\mu_1 - \mu_2$," because it demonstrates deeper understanding and leads to fewer misconceptions.
- Consider having students write the "*So what?*" for each condition.
- Consider having students write a general formula and a specific formula with variables before substituting numbers:

General formula:

$$\text{standardized test statistic} = \frac{\text{statistic} - \text{parameter}}{\text{standard error of statistic}}$$

Specific formula:

$$t = \frac{(\bar{x}_1 - \bar{x}_2) - (\mu_1 - \mu_2)}{\sqrt{\dfrac{s_1^2}{n_1} + \dfrac{s_2^2}{n_2}}}$$

One advantage to establishing this expectation is that the general formula is the same for significance tests in Chapter 9. Only the specific formula changes for each new significance test. It is risky to provide this much work on the AP® Statistics exam because students could lose credit due to a minor notation/substitution error. Consider requiring these formulas during the school year and then relaxing this expectation at the end of the course, when preparing for the AP® exam.

DO:
- $\bar{x}_C = 5.000$, $s_C = 8.743$, $n_C = 10$; $\bar{x}_P = -0.273$, $s_P = 5.901$, $n_P = 11$

- $t = \dfrac{[5.000 - (-0.273)] - 0}{\sqrt{\dfrac{8.743^2}{10} + \dfrac{5.901^2}{11}}} = \dfrac{5.273}{3.2878} = 1.604$

> The sample results give *some* evidence in favor of H_a: $\mu_C - \mu_P > 0$ because $5.000 - (-0.273) = 5.273 > 0$.

- *P*-value

Option 1: 2-SampTTest gives $t = 1.604$ and *P*-value $= 0.0644$ using $df = 15.59$.

Option 2: $df = 9$

Using Table B: $0.05 < P\text{-value} < 0.10$

Using technology: tcdf(lower:1.604, upper:1000, df:9) $= 0.0716$

> Refer to Technology Corner 27 below for details on how to do calculations for a two-sample t test on the TI-83/84.

CONCLUDE: Because the *P*-value of $0.0644 > \alpha = 0.05$, we fail to reject H_0. The experiment does not provide convincing evidence that the true mean decrease in systolic blood pressure is higher for men like these who take calcium than for men like these who take a placebo.

(b) Assuming H_0: $\mu_1 - \mu_2 = 0$ is true, there is a 0.0644 probability of getting a difference (Calcium − Placebo) in mean blood pressure reduction for the two groups of 5.273 or greater just by the chance involved in the random assignment.

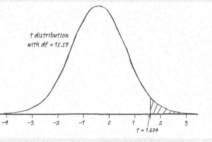

t distribution with df = 15.59

FOR PRACTICE, TRY EXERCISE 41

Notice that technology gives smaller, more accurate *P*-values for two-sample t tests than the conservative method. That's because calculators and software use the more complicated formula on page 670 to obtain a larger number of degrees of freedom. The degrees of freedom calculated by technology will always fall between the df from Option 2 and $(n_1 - 1) + (n_2 - 1) = n_1 + n_2 - 2$.

Why didn't researchers find a significant difference in the calcium and blood pressure experiment? The difference in mean systolic blood pressures for the two groups was 5.273 millimeters of mercury. This seems like a fairly large difference. With the small group sizes of 10 and 11, however, this difference wasn't large enough to reject H_0: $\mu_C - \mu_P = 0$ in favor of the one-sided alternative. We suspect that larger groups might show a similar difference in mean blood pressure reduction, which would indicate that calcium has a significant effect. If so, then the researchers in this experiment made a Type II error—failing to reject a false H_0. In fact, later analysis of data from an experiment with more subjects resulted in a *P*-value of 0.008. *Sample size strongly affects the power of a test.* It is easier to detect a difference in the effectiveness of two treatments if both are applied to large numbers of subjects.

27. Technology Corner · PERFORMING A SIGNIFICANCE TEST FOR A DIFFERENCE IN MEANS

TI-Nspire and other technology instructions are on the book's website at highschool.bfwpub.com/updatedtps6e.

You can use the two-sample t test option on the TI-83/84 to do the calculations for a significance test about the difference between two means. Let's confirm the results of the two previous examples with this feature.

Teaching Tip

When planning a study that will result in a two-sample t test for a difference in means, it is helpful to keep the two sample sizes about the same. This makes the *P*-value calculations more accurate and decreases the standard error of the difference in means, making the power of the test higher (assuming the standard deviations for both populations are roughly the same). For example, suppose you have 60 subjects for an experiment and the standard deviation of the response variable in both groups is 10. Trying different values for n_1 in the formula $\sqrt{\dfrac{10^2}{n_1} + \dfrac{10^2}{60 - n_1}}$ shows that the standard error will be smallest when the sample sizes are the same.

1. **A longer work week? (page 721)**
   - Press STAT, then choose TESTS and 2-SampTTest.
   - In the 2-SampTTest screen, specify "Stats" and adjust your other settings as shown. For Pooled: choose "No." We'll discuss pooling shortly.
   - Highlight "Calculate" and press ENTER.

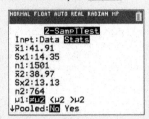 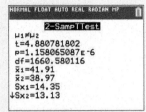

2. **Calcium and blood pressure (page 723)**
   - Enter the Group 1 (calcium) data in L1 and the Group 2 (placebo) data in L2.
   - Press STAT, then choose TESTS and 2-SampTTest.
   - In the 2-SampTTest screen, specify "Data" and adjust your other settings as shown. For Pooled: choose "No." We'll discuss pooling shortly.
   - Highlight "Calculate" and press ENTER.

 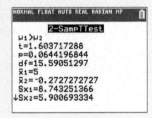

Note: If you select "Draw" instead of "Calculate," the appropriate t distribution will be displayed, showing the standardized test statistic and the shaded area corresponding to the P-value.

AP® EXAM TIP

The formula for the two-sample t statistic for a test about $\mu_1 - \mu_2$ often leads to calculation errors by students. Also, the P-value from technology is smaller and more accurate than the one obtained using the conservative method. As a result, your teacher may recommend using the calculator's 2-SampTTest feature to perform calculations. Be sure to name the procedure (two-sample t test for $\mu_1 - \mu_2$) in the "Plan" step and to report the standardized test statistic ($t = 1.60$), P-value (0.0644), and df (15.59) in the "Do" step.

Teaching Tip

We strongly recommend that students use technology for two-sample t procedures to ease the computations and take advantage of the larger number of degrees of freedom.

Teaching Tip: Using Technology

For two-sample t procedures, we always recommend saying "No" to pooling. We'll discuss why on page 727, but for now, tell your students: "Just say No!"

Teaching Tip

Another option is to require students to show the calculations for their work within this chapter. Then at the end of the course, when preparing for the AP® Statistics exam, relax this expectation and suggest that students instead use the calculator to get the standardized test statistic and the P-value.

TRM Paper Airplanes Experiment

In this activity, students test two different models of paper airplane to see if one model flies significantly farther than the other. Find it in the Teacher's Resource Materials located in the TE-Book, on the TRFD, or in the Teacher's view on the book's digital platform. Find templates for different models by searching Google for "fun paper airplanes."

TRM Exercising to Lose Weight

The authors of research studies often use specialized notation when reporting their results in academic journals. This problem provides some details about a weight-loss experiment and asks students to discuss aspects of the experimental design and confirm the results of the study. Find it in the Teacher's Resource Materials located in the TE-Book, on the TRFD, or in the Teacher's view on the book's digital platform.

Teaching Tip

The claim that "the drug received *doesn't affect* each individual's change in systolic blood pressure" is our null hypothesis from the significance test. Assuming this claim is true, the subject who had a change in systolic blood pressure of 7 would have the same change whether randomly assigned to the calcium or placebo group.

Teaching Tip: Using Technology

To perform the simulation on this page, go to the Extra Applets on the Student Site at highschool.bfwpub.com/updatedtps6e and use the *One Quantitative Variable* applet with two groups. Enter the data from the calcium experiment and then "Simulate difference in two means."

WHY DO THE INFERENCE METHODS FOR RANDOM SAMPLING WORK FOR RANDOMIZED EXPERIMENTS? Confidence intervals and tests for $\mu_1 - \mu_2$ are based on the sampling distribution of $\bar{x}_1 - \bar{x}_2$. But in experiments, we aren't sampling at random from any larger populations. We can think about what would happen if the random assignment were repeated many times under the assumption that $H_0: \mu_1 - \mu_2 = 0$ is true. That is, we assume that the specific treatment received doesn't affect an individual subject's response.

Let's see what would happen just by chance if we randomly reassign the 21 subjects in the calcium and blood pressure experiment to the two groups many times, assuming the drug received *doesn't affect* each individual's change in systolic blood pressure. We used software to redo the random assignment 1000 times. Figure 11.5 shows the value of $\bar{x}_C - \bar{x}_P$ in each of the 1000 simulation trials. This distribution (sometimes referred to as the *randomization distribution* of $\bar{x}_C - \bar{x}_P$) has an approximately Normal shape with mean 0 (no difference) and standard deviation 3.42. This matches fairly well with the distribution we used to perform calculations in the example.

In the actual experiment, the difference in the mean change in blood pressure in the calcium and placebo groups was $5.000 - (-0.273) = 5.273$. How likely is it that a difference this large or larger would happen just by chance when H_0 is true? Figure 11.5 provides a rough answer: 64 of the 1000 random reassignments (indicated by the red dots) yielded a difference in means greater than or equal to 5.273. That is, our estimate of the *P*-value is 0.064. This is quite close to the 0.0644 *P*-value that we obtained in the Technology Corner, suggesting that it's OK to use inference methods for random sampling to analyze randomized experiments. Neither of these values is the exact *P*-value. To get the exact *P*-value in an experiment, you would consider the difference in sample means for *all* possible random assignments. This approach is called a *permutation test*.

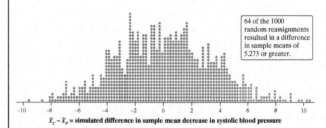

64 of the 1000 random reassignments resulted in a difference in sample means of 5.273 or greater.

$\bar{x}_C - \bar{x}_P$ = simulated difference in sample mean decrease in systolic blood pressure

FIGURE 11.5 Dotplot of the values of $\bar{x}_C - \bar{x}_P$ from each of 1000 simulated random reassignments of subjects to treatment groups in the calcium and blood pressure experiment, assuming no treatment effect.

CHECK YOUR UNDERSTANDING

For their final statistics project, two students performed an experiment to determine whether plants grow better if they are exposed to classical music or to metal music. Ten bean seeds were selected and each was planted in a Styrofoam cup. Half of these cups were randomly assigned to be exposed to metal music each night, while the other half were exposed to classical music each night. The amount of growth, in millimeters, was recorded for each plant after 2 weeks. Here are the data.[5]

Metal	22	36	73	57	3
Classical	87	78	124	121	19

Do these data give convincing evidence at the $\alpha = 0.05$ significance level of a difference in the mean growth of plants like these that are exposed to classical music or to metal music?

THE POOLED TWO-SAMPLE t PROCEDURES (DON'T USE THEM!) Most software offers a choice of two-sample t procedures. One is often labeled "unequal" variances; the other, "equal" variances. The unequal variance procedure uses our formula for the two-sample t interval and test, with df calculated as shown on page 670. *This test is valid whether or not the population variances are equal.*

The other choice is a special version of the two-sample t procedures that assumes the two population distributions have the same variance. This procedure combines (the statistical term is *pools*) the two sample variances to estimate the common population variance. The resulting statistic is called the *pooled two-sample t statistic.*

The pooled t statistic has exactly the t distribution with $n_1 + n_2 - 2$ degrees of freedom *if* the two population variances really are equal and the population distributions are exactly Normal. This method offers more degrees of freedom than Option 1 (Technology), which leads to narrower confidence intervals and smaller P-values. The pooled t procedures were in common use before software made it easy to use Option 1 for our two-sample t procedures.

In the real world, distributions are not exactly Normal, and population variances are not exactly equal. In practice, the Option 1 two-sample t procedures are almost always more accurate than the pooled procedures. Our advice: *Never use the pooled t procedures if you have technology that will carry out Option 1.*

> Recall that the variance is the square of the standard deviation.

> Remember, we always use the pooled sample proportion $\hat{p}_C = \dfrac{X_1 + X_2}{n_1 + n_2}$ in the standard error when performing a test of $H_0: p_1 - p_2 = 0$. That's because p_1 and p_2 are equal if the null hypothesis is true. We estimate this common value using the overall proportion $\hat{p}_C$ of successes in the two samples combined.

Significance Tests for μ_{diff}

What if we want to compare means in a setting that involves *paired data*? Recall from Section 10.2 that paired data result from recording two values of the same quantitative variable for each individual or for each pair of similar individuals. To analyze paired data, start by finding the difference within each pair. Be sure to indicate the order of subtraction. Then calculate $\bar{x}_{\text{diff}}$ and s_{diff}.

STATE: We want to test $H_0: \mu_1 - \mu_2 = 0$, $H_a: \mu_1 - \mu_2 \neq 0$, where $\mu_1 =$ the true mean growth (in millimeters) after 2 weeks for bean plants like these that are exposed to metal music each night and $\mu_2 =$ the true mean growth (in millimeters) after 2 weeks for bean plants like these that are exposed to classical music each night.

PLAN: Two-sample t test for $\mu_1 - \mu_2$.

- *Random:* The bean seeds were randomly assigned to be exposed to either metal or classical music. ✓
- *Normal/Large Sample:* The sample sizes are small, but the dotplots show no strong skewness and no outliers. ✓

DO: $\bar{x}_1 = 38.2$, $s_1 = 27.707$, $n_1 = 5$, $\bar{x}_2 = 85.8$, $s_2 = 42.494$, and $n_2 = 5$ 2-SampTTest gives $t = -2.098$ and P-value $= 0.0748$ using df $= 6.884$.

CONCLUDE: Because the P-value of $0.0748 > \alpha = 0.05$, we fail to reject H_0. We do not have convincing evidence that the true mean growth of plants like these that are exposed to metal music at night differs from the true mean growth of plants like these that are exposed to classical music at night.

Teaching Tip

Notice that df $= n_1 + n_2 - 2$ for the pooled t statistic is the sum of the degrees of freedom from each sample:

$$(n_1 - 1) + (n_2 - 1) = n_1 + n_2 - 2$$

Teaching Tip

The "Get your heart beating!" activity and the alternate "Cholesterol" activity on page 732 both provide an excellent transition between two-sample and paired data. Consider using one of these activities here as an introduction to significance tests about a mean difference.

Teaching Tip

After emphasizing that we do not recommend pooling for two-sample t procedures, many students will ask why we pool for a two-sample z test for a difference in proportions. The difference is in the standard errors.

In a two-sample z test for a difference in proportions, we assume that the two population proportions are the same when doing the calculations. Because the formula for the standard error includes these proportions, we are required to assume they are the same when calculating the standard error. We pool the sample proportions to get a better estimate of the true proportion for this calculation.

In a two-sample t test for a difference in means, we assume that the two population means are the same when doing calculations. However, the formula for the standard error does not include these means, so assuming that the null hypothesis is true has no effect on the standard error calculations.

When paired data come from a random sample or a randomized experiment, we may want to perform a significance test about the true mean difference μ_{diff}. The null hypothesis has the general form

$$H_0\text{: } \mu_{\text{diff}} = \text{hypothesized value}$$

We'll focus on situations where the hypothesized value is 0. Then the null hypothesis says that the true mean difference is 0:

$$H_0\text{: } \mu_{\text{diff}} = 0$$

The alternative hypothesis says what kind of difference we expect.

The conditions for performing a significance test about μ_{diff} are the same as the ones for constructing a confidence interval for a mean difference.

CONDITIONS FOR PERFORMING A SIGNIFICANCE TEST ABOUT A MEAN DIFFERENCE

- **Random:** Paired data come from a random sample from the population of interest or from a randomized experiment.
  - **10%:** When sampling without replacement, $n_{\text{diff}} < 0.10 N_{\text{diff}}$.
- **Normal/Large Sample:** The population distribution of differences (or the true distribution of differences in response to the treatments) is Normal or the number of differences in the sample is large ($n_{\text{diff}} \geq 30$). If the population (true) distribution of differences has unknown shape and the number of differences in the sample is less than 30, a graph of the sample differences shows no strong skewness or outliers.

When conditions are met, we can carry out a **one-sample t test for a mean difference** (also known as a **paired t test for a mean difference**). The standardized test statistic is

$$t = \frac{\bar{x}_{\text{diff}} - 0}{\frac{s_{\text{diff}}}{\sqrt{n_{\text{diff}}}}}$$

We can use Table B or technology to find the P-value from the t distribution with df $= n_{\text{diff}} - 1$.

ONE-SAMPLE t TEST FOR A MEAN DIFFERENCE (PAIRED t TEST FOR A MEAN DIFFERENCE)

Suppose the conditions are met. To test the hypothesis H_0: $\mu_{\text{diff}} = 0$, compute the standardized test statistic

$$t = \frac{\bar{x}_{\text{diff}} - 0}{\frac{s_{\text{diff}}}{\sqrt{n_{\text{diff}}}}}$$

Find the P-value by calculating the probability of getting a t statistic this large or larger in the direction specified by the alternative hypothesis H_a. Use the t distribution with $n_{\text{diff}} - 1$ degrees of freedom.

Making Connections

The formula for the t test statistic shown here is exactly the same formula for the t test statistic for a one-sample t test for μ from Section 11.1. It now includes the subscript "diff."

As with any inference procedure, be sure to follow the four-step process when performing a test about a mean difference.

EXAMPLE

Is caffeine dependence real?
Significance test about a mean difference

4 STEP

PROBLEM: Researchers designed an experiment to study the effects of caffeine withdrawal. They recruited 11 volunteers who were diagnosed as being caffeine dependent to serve as subjects. Each subject was barred from coffee, colas, and other substances with caffeine for the duration of the experiment. During one 2-day period, subjects took capsules containing their normal caffeine intake. During another 2-day period, they took placebo capsules. The order in which subjects took caffeine and the placebo was randomized. At the end of each 2-day period, a test for depression was given to all 11 subjects. Researchers wanted to know whether being deprived of caffeine would lead to an increase in depression.[9]

The table displays data on the subjects' depression test scores. Higher scores show more symptoms of depression.

Subject	1	2	3	4	5	6	7	8	9	10	11
Depression (caffeine)	5	5	4	3	8	5	0	0	2	11	1
Depression (placebo)	16	23	5	7	14	24	6	3	15	12	0

Do the data provide convincing evidence at the $\alpha = 0.05$ significance level that caffeine withdrawal increases depression score, on average, for subjects like the ones in this experiment?

SOLUTION:

STATE: $H_0: \mu_{diff} = 0$

$H_a: \mu_{diff} > 0$

where μ_{diff} = the true mean difference (Placebo − Caffeine) in depression test score for subjects like these. Because no significance level is given, we'll use $\alpha = 0.05$.

PLAN: Paired t test for μ_{diff}

- Random: Researchers randomly assigned the treatments—placebo then caffeine, caffeine then placebo—to the subjects. ✓

- Normal/Large Sample: The sample size is small, but the histogram of differences doesn't show any outliers or strong skewness. ✓

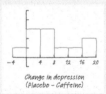

Change in depression
(Placebo − Caffeine)

> Follow the four-step process!

> Start by calculating the difference in depression test scores for each subject. We chose the order placebo − caffeine for subtraction so that most values would be positive. Always state the order of subtraction when defining the parameter!

> Note that we do not have to check the 10% condition because we are not sampling without replacement from a finite population.

> Because there is no strong skewness or outliers, it is plausible that the true distribution of difference (Placebo − Caffeine) in depression test scores for subjects like these is Normal.

- Normal/Large Sample: The number of differences is small, but the dotplot doesn't show any strong skewness or outliers. ✓

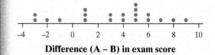

Difference (A − B) in exam score

DO:

- $t = \dfrac{3.2 - 0}{\dfrac{3.533}{\sqrt{20}}} = 4.05$

- P-value: df = 20 − 1 = 19

Table B: P-value is less than 2(0.0005) = 0.001.

Tech: tcdf(lower: 4.05, upper: 1000, df: 19) = 0.00034 × 2 = 0.00068

CONCLUDE: Because the P-value of 0.00068 < α = 0.05, we reject H_0. We have convincing evidence of a difference in the mean test score between Version A and Version B of the final exam for AP® Statistics students in this district. This conclusion is consistent with the confidence interval calculated in the previous alternate example (0.94, 5.46) because the 99% confidence interval did not capture the null hypothesis value of μ_{diff} = 0.

Notice that when Mrs. Gallas used a randomized comparative experiment (Alternate Example on page 721) and found a difference of means of $\bar{x}_A - \bar{x}_B$ = 4.3, the two-sample t test did not show this to be a statistically significant difference (P-value = 0.078). However, when Mrs. Gallas used a matched pairs design and found a mean difference of $\bar{x}_{diff}$ = 3.2, the paired t test showed this to be very statistically significant (P-value = 0.00068). The matched pairs results were significant even with a smaller sample size of 20, rather than 80.

ALTERNATE EXAMPLE

Skills 1.E, 4.E

Final exam solved Significance test about a mean difference

PROBLEM:
In an alternate example from Chapter 10, Mrs. Gallas used a matched pairs design to collect data on scores for two different versions of her final exam for a random sample of 20 AP® Statistics students from her district. Each student took both Version A and Version B of the exam and the order was randomized. Recall that $\bar{x}_{diff} = \bar{x}_{A-B}$ = 3.2 and s_{diff} = 3.533. Do the data provide convincing evidence at the $\alpha = 0.01$ significance level that there is a difference in the average test scores between Version A and Version B of the final exam for AP® Statistics students in this district?

SOLUTION:
STATE:

$H_0: \mu_{diff} = 0$
$H_a: \mu_{diff} \neq 0$

where μ_{diff} = the true mean difference (Version A − Version B) in final exam scores for AP® Statistics students in this district using $\alpha = 0.01$.

PLAN: Paired t test for μ_{diff}
- Random: Random sample of 20 AP® Statistics students in the district. ✓
  - 10%: Assume 20 < 10% of all AP® Statistics students in the large district. ✓

Teaching Tip

Some students may want to do a two-sample t test in this example. Here is a strategy for distinguishing between settings that call for a paired t test and settings that call for a two-sample t test. List the data in two columns and ask: "Is there a label that would apply to the two values in each row, and only the two values in that row?" For example, in the calcium and blood pressure example from earlier in this section, no label would apply to the two values in the first row and only those two values (because there were two randomized groups):

Label?	Calcium	Placebo
?	7	−1
?	−4	12
...	...	...

However, in this caffeine deprivation example, a label does apply to the two values in the first row, and only those two values. In this case, both numbers came from Subject 1, and all of Subject 1's numbers appear in the first row:

Label?	Depression (caffeine)	Depression (placebo)
Subject 1	5	16
Subject 2	5	23
...	...	...

Making Connections

In Chapter 4, we learned that a variable that is associated with the explanatory variable and is influencing the response variable is called a confounding variable. In this context, the weather could be a confounding variable.

DO: $\bar{x}_{\text{diff}} = 7.364$, $s_{\text{diff}} = 6.918$, $n_{\text{diff}} = 11$

- $t = \dfrac{7.364 - 0}{\dfrac{6.918}{\sqrt{11}}} = 3.53$

- P-value df $= 11 - 1 = 10$

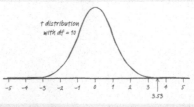

Using Table B: P-value is between 0.0025 and 0.005.

Using Technology: tcdf(lower: 3.53, upper: 1000, df: 10) = 0.0027

CONCLUDE: Because the P-value of 0.0027 < α = 0.05, we reject H_0. We have convincing evidence that caffeine withdrawal increases depression test score, on average, for subjects like the ones in this experiment.

> The sample result gives some evidence in favor of H_a: $\mu_{\text{diff}} > 0$ because $\bar{x}_{\text{diff}} = 7.364 > 0$.

$$t = \frac{\bar{x}_{\text{diff}} - 0}{\dfrac{s_{\text{diff}}}{\sqrt{n_{\text{diff}}}}}$$

> The TTest function on the TI-83/84 gives $t = 3.53$ and P-value = 0.0027.

> Be sure to include "on average" (or equivalent) in the conclusion to indicate that the test is about the true mean difference.

FOR PRACTICE, TRY EXERCISE 53

Why did the researchers randomly assign the order in which subjects received placebo and caffeine in the example? Researchers want to be able to conclude that any statistically significant change in depression score is due to the treatments themselves and not to some other variable. One obvious concern is the order of the treatments. Suppose that caffeine were given to all the subjects during the first 2-day period. What if the weather were nicer on these 2 days than during the second 2-day period when all subjects were given a placebo? Researchers wouldn't be able to tell if a large increase in the mean depression score is due to the difference in weather or due to the treatments. Random assignment of the caffeine and placebo to the two time periods in the experiment should help ensure that no other variable (like the weather) is systematically affecting subjects' responses.

Because researchers randomly assigned the treatments, they can make an inference about cause and effect. The data from this experiment provide convincing evidence that depriving caffeine-dependent subjects like these of caffeine causes an increase in depression scores, on average.

The significance test in the example led to a simple decision: reject H_0. We know from past experience that a confidence interval gives more information than a test—it provides the entire set of plausible values for the parameter based on the data. A 90% confidence interval for μ_{diff} is

> Notice that we used a 90% confidence level here, which corresponds to the one-sided significance test with $\alpha = 0.05$.

$$\bar{x}_{\text{diff}} \pm t^* \frac{s_{\text{diff}}}{\sqrt{n_{\text{diff}}}} = 7.364 \pm 1.812 \frac{6.918}{\sqrt{11}} = 7.364 \pm 3.780 = (3.584, 11.144)$$

We are 90% confident that the interval from 3.584 to 11.144 captures the true mean difference (Placebo − Caffeine) in depression test score for caffeine-dependent individuals like the ones in this study. The interval suggests that caffeine deprivation results in an average increase in depression test score of between 3.584 and 11.144 points for subjects like these.

Teaching Tip

Here are some important tips for teaching students to use the four-step process for performing a one-sample t test for μ_{diff}:

- Inform students they can use the wording of the question to define their parameters and to write the conclusion. We call this "parroting" the stem of the question.
- Ask students, "Which words in the question inform you this is a one- or two-sided test?"
- When naming the inference method, the calculator command "TTest" may be accepted, but it isn't a guarantee because there are other tests that use a t test statistic. We prefer that students use the "paired t test for μ_{diff}," because it demonstrates deeper understanding and leads to fewer misconceptions.

- Consider having students write the "*So what?*" for each condition.
- Consider having students write a general formula and a specific formula with variables before substituting numbers:

General formula:

$$\frac{\text{standardized}}{\text{test statistic}} = \frac{\text{statistic} - \text{parameter}}{\text{standard error of statistic}}$$

Specific formula: $t = \dfrac{\bar{x}_{\text{diff}} - \mu_{\text{diff}}}{\dfrac{s_{\text{diff}}}{\sqrt{n_{\text{diff}}}}}$

CHECK YOUR UNDERSTANDING

Consumers Union designed an experiment to test whether nitrogen-filled tires would maintain pressure better than air-filled tires. They obtained two tires from each of several brands and then randomly assigned one tire in each pair to be filled with air and the other to be filled with nitrogen. All tires were inflated to the same pressure and placed outside for a year. At the end of the year, Consumers Union measured the pressure in each tire. The pressure loss (in pounds per square inch) during the year for the tires of each brand is shown in the table.[10]

Brand	Air	Nitrogen	Brand	Air	Nitrogen
BF Goodrich Traction T/A HR	7.6	7.2	Pirelli P6 Four Seasons	4.4	4.2
Bridgestone HP50 (Sears)	3.8	2.5	Sumitomo HTR H4	1.4	2.1
Bridgestone Potenza G009	3.7	1.6	Yokohama Avid H4S	4.3	3.0
Bridgestone Potenza RE950	4.7	1.5	BF Goodrich Traction T/A V	5.5	3.4
Bridgestone Potenza EL400	2.1	1.0	Bridgestone Potenza RE950	4.1	2.8
Continental Premier Contact H	4.9	3.1	Continental ContiExtreme Contact	5.0	3.4
Cooper Lifeliner Touring SLE	5.2	3.5	Continental ContiProContact	4.8	3.3
Dayton Daytona HR	3.4	3.2	Cooper Lifeliner Touring SLE	3.2	2.5
Falken Ziex ZE-512	4.1	3.3	General Exclaim UHP	6.8	2.7
Fuzion Hrl	2.7	2.2	Hankook Ventus V4 H105	3.1	1.4
General Exclaim	3.1	3.4	Michelin Energy MXV4 Plus	2.5	1.5
Goodyear Assurance Tripletred	3.8	3.2	Michelin Pilot Exalto A/S	6.6	2.2
Hankook Optimo H418	3.0	0.9	Michelin Pilot HX MXM4	2.2	2.0
Kumho Solus KH16	6.2	3.4	Pirelli P6 Four Seasons	2.5	2.7
Michelin Energy MXV4	2.0	1.8	Sumitomo HTR$^+$	4.4	3.7
Michelin Pilot XGT H4	1.1	0.7			

Do the data give convincing evidence at the $\alpha = 0.05$ significance level that air-filled tires lose more pressure, on average, than nitrogen-filled tires for brands like these?

Paired Data or Two Samples?

Earlier in this section, we used two-sample t procedures to compare the mean change in blood pressure for healthy men who take calcium and for healthy men who take a placebo. These methods require data that come from *independent random samples* from the two populations of interest or from *two groups in a randomized experiment* (as was the case in the example on page 723). When the conditions are met, we can perform inference about the difference $\mu_1 - \mu_2$ in the true means.

In the preceding example, we used paired t procedures to compare the mean change in depression scores for caffeine-dependent individuals when taking caffeine versus when taking a placebo. These methods require *paired data* that come from a random sample from the population of interest or from a randomized experiment (as was the case in this example because the same 11 subjects received both treatments). When the conditions are met, we can perform inference about the true mean difference μ_{diff}. The proper inference method depends on how the data were produced.

Answers to CYU

STATE: $H_0: \mu_{diff} = 0$, $H_a: \mu_{diff} > 0$, where μ_{diff} = the true mean difference (Air – Nitrogen) in pressure lost using $\alpha = 0.05$

PLAN: Paired t test for μ_{diff}.

- *Random:* Tires in each pair are randomly assigned to be filled with air or nitrogen. ✓

- *Normal/Large Sample:* $n_{diff} = 31 \geq 30$. ✓

DO: $\bar{x}_{diff} = 1.252$, $s_{diff} = 1.202$, and $n_{diff} = 31$

$$t = \frac{1.252 - 0}{\frac{1.202}{\sqrt{31}}} = 5.80$$

df = 30
P-value ≈ 0

CONCLUDE: Because the *P*-value of approximately $0 < \alpha = 0.05$, we reject H_0. We have convincing evidence that the true mean difference (Air – Nitrogen) in pressure lost is greater than 0. In other words, we have convincing evidence that tires lose less pressure when filled with nitrogen than when filled with air, on average.

AP® EXAM TIP

There are many AP® Statistics exam questions that can help students practice inference procedures for paired data. See 2014 #5, 2008B #6, 2007 #4, 2006B #4, 2005B #4, 2001 #5, 1999 #6, or 1997 #5.

Teaching Tip

Many students have trouble deciding when it is appropriate to use a two-sample t test and when it is appropriate to use a paired t test. The key distinction is how the data were produced. In an experiment, if groups were formed using a completely randomized design, a two-sample t test is the right choice. However, if subjects were paired and then split at random into the two treatment groups, or if each subject received both treatments, a paired t test is appropriate. Paired data can also arise when the data are produced by random sampling. For example, if wells are selected at random and the quality of water is measured at the top and bottom of each well, we should treat these measurements as paired data.

Luke's taco shop
Two samples or paired data?

PROBLEM:
In each of the following settings, decide whether you should use two-sample *t* procedures to perform inference about a difference in means or paired *t* procedures to perform inference about a mean difference. Explain your choice.

(a) Luke's taco shop is considering a switch to a new tortilla that supposedly has a larger diameter. To test this claim, Luke takes a random sample of 50 of the old tortillas and 50 of the new tortillas and records the diameter of each.

(b) Luke's taco shop wants to be sure that the new tortillas taste better than the old tortillas. Luke selects a random sample of 20 regular customers. Each customer is asked to try both tortillas and then record a "taste" score for each. The order in which the customers try the two tortillas is randomized.

(c) Luke's taco shop is not sure whether to cook the tortillas in the oven or on the grill. The chefs want tortillas to cook as quickly as possible. Luke sets up an experiment taking a batch of 50 tortillas and randomly assigning half of them to be cooked one at a time in the oven and half of them to be cooked one at a time on the grill. The time it takes until ready to serve is recorded for each tortilla.

SOLUTION:

(a) Two-sample *t* procedures; the data come from independent random samples of the old and new tortillas.

(b) Paired *t* procedures; the data come from two measurements of the same variable ("taste" score) for each regular customer.

(c) Two-sample *t* procedures; the data come from two groups in a randomized experiment, with each group consisting of 25 tortillas in which method of cooking (oven or grill) was randomly assigned.

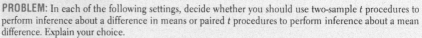

EXAMPLE

Are you all wet?
Two samples or paired data?

PROBLEM: In each of the following settings, decide whether you should use two-sample *t* procedures to perform inference about a difference in means or paired *t* procedures to perform inference about a mean difference. Explain your choice.

(a) Before exiting the water, scuba divers remove their fins. A maker of scuba equipment advertises a new style of fins that is supposed to be faster to remove. A consumer advocacy group suspects that the time to remove the new fins may be no different than the time required to remove old fins, on average. Twenty experienced scuba divers are recruited to test the new fins. Each diver flips a coin to determine if they wear the new fin on the left foot and the old fin on the right foot, or vice versa. The time to remove each type of fin is recorded for every diver.

(b) To study the health of aquatic life, scientists gathered a random sample of 60 White Piranha fish from a tributary of the Amazon River during one year. The average length of these fish was compared to a random sample of 82 White Piranha from the same tributary a decade ago.

(c) Can a wetsuit deter shark attacks? A researcher has designed a new wetsuit with color variations that are suspected to deter shark attacks. To test this idea, she fills two identical drums with bait and covers one in the standard black neoprene wetsuit and the other in the new suit. Over a period of one week, she selects 16 two-hour time periods and randomly assigns 8 of them to the drum in the black wetsuit. The other 8 are assigned to the drum with the new suit. During each time period, the appropriate drum is submerged in waters that sharks frequent, and the number of times a shark bites the drum is recorded.

SOLUTION:

(a) Paired *t* procedures. The data come from two measurements of the same variable (time to remove fin) for each diver.

(b) Two-sample *t* procedures. The data come from independent random samples of White Piranha in two different years.

> If the sample sizes are different, it can't be paired data.

(c) Two-sample *t* procedures. The data come from two groups in a randomized experiment, with each group consisting of 8 time periods in which a drum with a specific wetsuit (standard or new) was randomly assigned to be submerged.

FOR PRACTICE, TRY EXERCISE 59

When designing an experiment to compare two means, a completely randomized design may not be the best option. A matched pairs design might be a better choice, as the following activity shows.

ACTIVITY   Get your heart beating!

Are standing pulse rates higher, on average, than sitting pulse rates? In this activity, you will perform two experiments to try to answer this question.

Experiment #1: Completely randomized design

1. Your teacher will randomly assign half of the students in your class to stand and the other half to sit. Once the two treatment groups have been formed, students should stand or sit as required. Then they should measure their

pulses for 1 minute. Have the subjects in each group record their data on the board.

2. Analyze the data for the completely randomized design. Make parallel dotplots and calculate the mean pulse rate for each group. Is there *some* evidence that standing pulse rates are higher, on average? Explain your answer.

Experiment #2: Matched pairs design

3. To produce paired data in this setting, each student should receive both treatments in a random order. Because you already sat or stood in Step 1, do the opposite now. As before, everyone should measure his or her pulse for 1 minute after the treatment is imposed (i.e., once everyone is standing or sitting). Then each subject should calculate his or her difference (Standing − Sitting) in pulse rate and record this value on the board.

4. Analyze the data for the matched pairs design. Make a dotplot of these differences and calculate their mean. Is there *some* evidence that standing pulse rates are higher, on average? Explain your answer.

5. Which design provides more convincing evidence that standing pulse rates are higher, on average, than sitting pulse rates? Justify your answer.

A statistics class with 24 students performed the "Get your heart beating" activity. Figure 11.6 shows a dotplot of the pulse rates for their completely randomized design. The mean pulse rate for the standing group is $\bar{x}_1 = 74.83$; the mean for the sitting group is $\bar{x}_2 = 68.33$. So the average pulse rate is 6.5 beats per minute higher in the standing group. However, the variability in pulse rates for the two groups creates a lot of overlap in the dotplots. A two-sample t test of $H_0: \mu_1 - \mu_2 = 0$ versus $H_a: \mu_1 - \mu_2 > 0$ yields $t = 1.42$ and a P-value of 0.09. These data do not provide convincing evidence that standing pulse rates are higher, on average, than sitting pulse rates for people like the students in this class.

What about the class's matched pairs design? Figure 11.7 shows a dotplot of the difference (Standing − Sitting) in pulse rate for each of the 24 students. We can see that 21 of the 24 students recorded a positive difference, indicating that their standing pulse rate was higher. The mean difference is $\bar{x}_{\text{diff}} = 6.83$ beats per minute. A one-sample t test of $H_0: \mu_{\text{diff}} = 0$ versus $H_a: \mu_{\text{diff}} > 0$ gives $t = 6.483$ and a P-value of approximately 0. These data provide *very* convincing evidence that standing pulse rates are higher, on average, than sitting pulse rates for people like the students in this class.

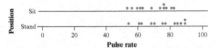

FIGURE 11.6 Parallel dotplots of the pulse rates for the standing and sitting groups in a statistics class's completely randomized design.

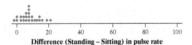

FIGURE 11.7 Dotplot of the difference (Standing − Sitting) in pulse rate for each student in a statistics class's matched pairs design.

TRM **Cholesterol Activity**

This activity can be used as an alternate to the "Get your heart beating!" activity on this page. Find it in the Teacher's Resource Materials located in the TE-Book, on the TRFD, or in the Teacher's view on the book's digital platform.

Teaching Tip

As a second alternate activity, consider an experiment to determine if students can grab more candy (as in Chapter 3) with their dominant hand. Start by assigning half to use the dominant hand, the other half to use the non-dominant hand. Then have each person do the opposite and look at the differences for each person.

ACTIVITY OVERVIEW

To prepare for using this activity, watch the overview video by clicking on the link in the TE-Book, opening the TRFD, or downloading from the Teacher's Resources on the book's digital platform.

Time: 20 minutes

Materials: Posters, sticker dots

Teaching Advice: Give students explicit instructions on the procedure for measuring their pulse. Standardizing the procedure will reduce variability in the results. Ask students to put sticker dots on a poster board. Post the completed poster board in the classroom so that you can refer to this activity throughout the rest of the course.

After completing Experiment #1, have students do a two-sample t test (if conditions are met) to determine if there is convincing evidence that standing pulse rates are higher, on average. Use the same scale for the parallel dotplots in Experiment #1 and the dotplot of differences for Experiment #2. We want students to clearly see that the matched pairs design accounted for a source of variability.

After completing Experiment #2, ask students to do a paired t test for a mean difference (if conditions are met) to determine if there is convincing evidence that standing pulse rates are higher, on average. Students should notice a much smaller P-value for the matched pairs design over the completely randomized design.

Answers:

1. Done.

2. If the mean pulse rate is greater for the standing group than for the sitting group, there is some evidence that standing pulse rates are higher, on average. If so, perform a two-sample t test for the difference of means.

3. Done.

4. If the mean difference in pulse rate (Standing − Sitting) is greater than 0, there is some evidence that standing pulse rates are higher, on average. If so, perform a paired t test for the mean difference.

5. Compare the P-value for Experiment #1 and Experiment #2. It is very likely that the matched pairs design has a lesser P-value and, therefore, provides more convincing evidence that standing pulse rates are higher, on average, than sitting pulse rates than does the completely randomized design.

Making Connections

In Chapter 9, we discovered three factors that can increase the power of a significance test:

1. Increase the sample size.
2. Increase α.
3. Increase effect size.

We now have a fourth factor: Use a sampling method or experimental design that accounts for a source of the variability in the response variable (stratified random sampling, block design, matched pairs, controlling other variables in an experiment).

TRM **Section 11.2 Quiz**

There is one quiz available for this section in the Teacher's Resource Materials. Click on the link in the TE-Book, open in the TRFD, or download from the Teacher's Resources on the book's digital platform. You can also create your own quiz using the ExamView® Assessment Suite that is part of the TPS6 program. Many questions are coded by Learning Target to make it easy to build parallel quizzes.

Teaching Tip: Using Technology

We have now covered many different inference procedures. Students have to be able to choose the correct inference procedure for different settings. Search the Internet for "categorizing statistics problems" to find the Itcconline page that allows students to practice choosing the correct inference procedure. Uncheck the procedures we don't know yet (prediction intervals, the three chi-square tests, and 1-way ANOVA), press Submit, and have fun!

Let's take one more look at the two figures. Notice that we used the same scale for both graphs. The matched pairs design reduced the variability in the response variable by accounting for a big source of variability—the differences between individual students. That made it easier to detect the fact that standing causes an increase in the average pulse rate. In other words, using a paired design resulted in more power. With the large amount of variability in the completely randomized design, we could not draw such a conclusion.

Section 11.2 | Summary

- Tests for the difference $\mu_1 - \mu_2$ between the means of two populations or the mean responses to two treatments are based on the difference $\bar{x}_1 - \bar{x}_2$ between the sample means.
- Before testing a claim about $\mu_1 - \mu_2$, we need to check for independence in the data collection process and that the sampling distribution of $\bar{x}_1 - \bar{x}_2$ is approximately Normal. The required conditions are:
  - **Random:** The data come from two independent random samples or from two groups in a randomized experiment.
    - **10%:** When sampling without replacement, $n_1 < 0.10N_1$ and $n_2 < 0.10N_2$.
  - **Normal/Large Sample:** For each sample, the corresponding population distribution (or the true distribution of response to the treatment) is Normal or the sample size is large ($n \geq 30$). For each sample, if the population (treatment) distribution has unknown shape and $n < 30$, a graph of the sample data shows no strong skewness or outliers.
- To test $H_0: \mu_1 - \mu_2 = 0$, use a **two-sample t test for $\mu_1 - \mu_2$**. The standardized test statistic is

$$t = \frac{(\bar{x}_1 - \bar{x}_2) - 0}{\sqrt{\dfrac{s_1^2}{n_1} + \dfrac{s_2^2}{n_2}}}$$

  P-values are calculated using the t distribution with degrees of freedom from either Option 1 (technology) or Option 2 (the smaller of $n_1 - 1$ and $n_2 - 1$).
- Paired data result from recording two values of the same quantitative variable for each individual or for each pair of similar individuals. To compare means in a setting that involves paired data, start by computing the difference for each pair. Be sure to indicate the order of subtraction. Use the mean difference $\bar{x}_{\text{diff}}$ and the standard deviation of the differences s_{diff} as summary statistics.
- Before testing a claim about μ_{diff}, we need to check for independence of the differences and that the sampling distribution of $\bar{x}_{\text{diff}}$ is approximately Normal. The required conditions are:
  - **Random:** Paired data come from a random sample from the population of interest or from a randomized experiment.
    - **10%:** When sampling without replacement, $n_{\text{diff}} < 0.10N_{\text{diff}}$.
  - **Normal/Large Sample:** The population distribution of differences (or the true distribution of differences in response to the treatments) is Normal or the number of differences in the sample is large ($n_{\text{diff}} \geq 30$). If the

population (true) distribution of differences has unknown shape and the number of differences in the sample is less than 30, a graph of the sample differences shows no strong skewness or outliers.

- A significance test of $H_0: \mu_{\text{diff}} = 0$ is called a **one-sample t test for a mean difference** or a **paired t test for a mean difference**. The standardized test statistic is

$$t = \frac{\bar{x}_{\text{diff}} - 0}{\dfrac{s_{\text{diff}}}{\sqrt{n_{\text{diff}}}}}$$

When the Normal/Large Sample condition is met, find the P-value using the t distribution with degrees of freedom equal to $n_{\text{diff}} - 1$.
- The proper inference method for comparing two means depends on how the data were produced. For paired data, use one-sample t procedures for μ_{diff}. For quantitative data that come from independent random samples from two populations of interest or from two groups in a randomized experiment, use two-sample t procedures for $\mu_1 - \mu_2$.
- Be sure to follow the four-step process whenever you perform a significance test for comparing two means.

11.2 Technology Corner

TI-Nspire and other technology instructions are on the book's website at *highschool.bfwpub.com/updatedtps6e.*

27. Performing a significance test for a difference in means   Page 724

TRM **Full Solutions to Section 11.2 Exercises**

Click on the link in the TE-Book, open the TRFD, or download from the Teacher's Resources on the book's digital platform.

Section 11.2 | Exercises

35. **Sorting the music** Student researchers Adam, Edward, and Kian wondered if music would affect performance for certain tasks. To find out, they had student volunteers sort a shuffled set of 26 playing cards by face value and by color. Nineteen of the 38 volunteers were randomly assigned to listen to music during the sorting, while the others listened to no music. Here are parallel boxplots of the time in seconds that it took to sort the cards for the students in each group:

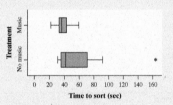

Time to sort (sec)

Do these data give convincing evidence of a difference in the true mean sorting times at the $\alpha = 0.10$ significance level?

(a) State appropriate hypotheses for performing a significance test. Be sure to define the parameters of interest.

(b) Check if the conditions for performing the test are met.

36. **Ice cream** For a statistics class project, Jonathan and Crystal held an ice-cream-eating contest. They randomly selected 29 males and 35 females from their large high school to participate. Each student was given a small cup of ice cream and instructed to eat it as fast as possible. Jonathan and Crystal then recorded each contestant's gender and time (in seconds), as shown in the dotplots.

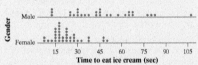

Time to eat ice cream (sec)

Do these data give convincing evidence of a difference in the population means at the $\alpha = 0.10$ significance level?

Answers to Section 11.2 Exercises

11.35 (a) $H_0: \mu_1 - \mu_2 = 0$, $H_a: \mu_1 - \mu_2 \neq 0$, where $\mu_1 = $ true mean time needed to sort the cards for students like these while listening to music and $\mu_2 = $ true mean time . . . not listening to music. **(b)** *Random:* Two groups in a randomized experiment. *Normal/Large Sample:* Not met; the samples are small and the distribution of time to sort for those who listen to no music is skewed right with one upper outlier.

11.36 (a) H_0: $\mu_1 - \mu_2 = 0$, H_a: $\mu_1 - \mu_2 \neq 0$, where μ_1 = true mean time it takes male students to eat a small cup of ice cream and μ_2 = true mean time . . . for female students. **(b)** *Random:* Independent random samples. *10%:* 29 < 10% of all males and 35 < 10% of all females. *Normal/Large Sample:* The dotplots show no outliers or strong skewness.

11.37 (a) H_0: $\mu_1 - \mu_2 = 0$, H_a: $\mu_1 - \mu_2 \neq 0$, where μ_1 = true mean reliability rating of Anglo customers and μ_2 = true mean reliability for Hispanic customers. **(b)** *Random:* Independent random samples. *10%:* 92 < 10% of Anglo customers and 86 < 10% of Hispanic customers. *Normal/Large Sample:* 92 ≥ 30 and 86 ≥ 30.

11.38 (a) H_0: $\mu_1 - \mu_2 = 0$, H_a: $\mu_1 - \mu_2 \neq 0$, where μ_1 = true mean number of turns required to finish the memory game for students like these while listening to music and μ_2 = true mean number . . . not listening to music. **(b)** *Random:* Random assignment. *Normal/Large Sample:* 42 ≥ 30 and 42 ≥ 30.

11.39 (a) $\bar{x}_1 - \bar{x}_2 = 6.37 - 5.91 = 0.46 \neq 0$ **(b)** $t = 3.89$; using df = 143.69, P-value = 0.0002. Using df = 85, P-value = 0.0002. **(c)** Because the P-value of 0.0002 < α = 0.05, we reject H_0. We have convincing evidence the true mean reliability rating for all Hispanic customers differs from the true mean reliability rating for all Anglo customers.

11.40 (a) $\bar{x}_1 - \bar{x}_2 = 15.833 - 13.714 = 2.119 \neq 0$ **(b)** $t = 2.59$; using df = 81.108, P-value = 0.0114. Using df = 41, P-value = 0.0132. **(c) C:** Because the P-value of 0.0114 < α = 0.05, we reject H_0. We have convincing evidence the true mean number of turns required to finish the memory game for students like these while listening to music differs from the true mean number of turns required to finish the memory game for students like these while not listening to music.

11.41 (a) S: H_0: $\mu_1 - \mu_2 = 0$, H_a: $\mu_1 - \mu_2 > 0$, where μ_1 = true mean reduction in blood pressure for subjects like these who take fish oil and μ_2 = true mean reduction . . . regular oil. **P:** Two-sample t test for $\mu_1 - \mu_2$. *Random:* Randomized assignment. *Normal/Large Sample:* The dotplots show no strong skewness and no outliers.

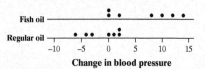

Fish oil

Regular oil

-10 -5 0 5 10 15

Change in blood pressure

(a) State appropriate hypotheses for performing a significance test. Be sure to define the parameters of interest.

(b) Check if the conditions for performing the test are met.

37. Happy customers As the Hispanic population in the United States has grown, businesses have tried to understand what Hispanics like. One study interviewed separate random samples of Hispanic and Anglo customers leaving a bank. Customers were classified as Hispanic if they preferred to be interviewed in Spanish or as Anglo if they preferred English. Each customer rated the importance of several aspects of bank service on a 10-point scale.[11] Here are summary results for the importance of "reliability" (the accuracy of account records, etc.):

Group	n	$\bar{x}$	s_x
Anglo	92	6.37	0.60
Hispanic	86	5.91	0.93

Researchers want to know if there is a difference in the mean reliability ratings of all Anglo and Hispanic bank customers.

(a) State appropriate hypotheses for performing a significance test. Be sure to define the parameters of interest.

(b) Check that the conditions for performing the test are met.

38. Does music help or hinder memory? Many students at Matt's school claim they can think more clearly while listening to their favorite kind of music. Matt believes that music interferes with thinking clearly. To find out which is true, Matt recruits 84 volunteers and randomly assigns them to two groups. The "Music" group listens to their favorite music while playing a "match the animals" memory game. The "No Music" group plays the same game in silence. Here are some descriptive statistics for the number of turns it took the subjects in each group to complete the game (fewer turns indicate a better performance):

Group	Sample size	Mean	SD
Music	42	15.833	3.944
No music	42	13.714	3.550

Matt wants to know if listening to music affects the average number of turns required to finish the memory game for students like these.

(a) State appropriate hypotheses for performing a significance test. Be sure to define the parameters of interest.

(b) Check if the conditions for performing the test are met.

39. Happy customers Refer to Exercise 37.

pg 721

(a) Explain why the sample results give some evidence for the alternative hypothesis.

(b) Calculate the standardized test statistic and P-value.

(c) What conclusion would you make?

40. Does music help or hinder memory? Refer to Exercise 38.

(a) Explain why the sample results give some evidence for the alternative hypothesis.

(b) Calculate the standardized test statistic and P-value.

(c) What conclusion would you make?

41. Fish oil To see if fish oil can help reduce blood pressure, males with high blood pressure were recruited and randomly assigned to different treatments. Seven of the men were assigned to a 4-week diet that included fish oil. Seven other men were assigned to a 4-week diet that included a mixture of oils that approximated the types of fat in a typical diet. Each man's blood pressure was measured at the beginning of the study. At the end of the 4 weeks, each volunteer's blood pressure was measured again and the reduction in diastolic blood pressure was recorded. These differences are shown in the table. Note that a negative value means that the subject's blood pressure *increased*.[12]

pg 723

Fish oil	8	12	10	14	2	0	0
Regular oil	-6	0	1	2	-3	-4	2

(a) Do these data provide convincing evidence that fish oil helps reduce blood pressure more, on average, than regular oil for men like these?

(b) Interpret the P-value from part (a) in the context of this study.

42. Baby birds Do birds learn to time their breeding? Blue titmice eat caterpillars. The birds would like lots of caterpillars around when they have young to feed, but they must breed much earlier. Do the birds learn from one year's experience when to time their breeding next year? Researchers randomly assigned 7 pairs of birds to have the natural caterpillar supply supplemented while feeding their young and another 6 pairs to serve as a control group relying on natural food supply. The next year, they measured how many days after the caterpillar peak the birds produced their nestlings.[13] The investigators expected the control group to adjust their breeding date the following year, whereas the well-fed supplemented group had no reason to change. Here are the data (days after caterpillar peak):

Control	4.6	2.3	7.7	6.0	4.6	-1.2	
Supplemented	15.5	11.3	5.4	16.5	11.3	11.4	7.7

D: $\bar{x}_1 = 6.571$, $s_1 = 5.855$, $n_1 = 7$, $\bar{x}_2 = -1.143$, $s_2 = 3.185$, $n_2 = 7$, and $t = 3.06$. Using df = 9.264, P-value = 0.0065; using df = 6, P-value = 0.0111. **C:** Because the P-value of 0.0065 < α = 0.05, we reject H_0. We have convincing evidence fish oil helps reduce blood pressure more, on average, than regular oil for subjects like these. **(b)** Assuming the true mean reduction in blood pressure is the same regardless of whether the individual takes fish oil or regular oil, there is a 0.0065 probability that we would observe a difference in sample means of 7.714 or greater by chance alone.

11.42 (a) S: H_0: $\mu_1 - \mu_2 = 0$, H_a: $\mu_1 - \mu_2 < 0$, where μ_1 = true mean time to breeding for birds relying on natural food supply and μ_2 = true

mean . . . food supplements. **P:** Two-sample t test for $\mu_1 - \mu_2$. *Random:* Random assignment. *Normal/Large Sample:* The dotplots show no strong skewness or outliers.

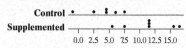

Control

Supplemented

0.0 2.5 5.0 7.5 10.0 12.5 15.0

Time to breeding (days behind caterpillar peak)

D: $\bar{x}_1 = 4.0$, $s_1 = 3.11$, $n_1 = 6$, $\bar{x}_2 = 11.3$, $s_2 = 3.93$, $n_2 = 7$, and $t = -3.74$. Using df = 10.955, P-value = 0.0016; using df = 5, P-value = 0.0067. **C:** Because the P-value of 0.0016 < α = 0.05, we reject H_0. We have convincing evidence the true mean time to breeding is less for birds relying on natural food supply than for birds with food supplements.

(a) Do the data provide convincing evidence that birds like these that have to rely on the natural food supply produce their nestlings closer to the caterpillar peak, on average, than birds like these that have the caterpillar supply supplemented?

(b) Interpret the P-value from part (a) in the context of this study.

43. **Who talks more—men or women?** Researchers equipped random samples of 56 male and 56 female students from a large university with a small device that secretly records sound for a random 30 seconds during each 12.5-minute period over 2 days. Then they counted the number of words spoken by each subject during each recording period and, from this, estimated how many words per day each subject speaks. The female estimates had a mean of 16,177 words per day with a standard deviation of 7520 words per day. For the male estimates, the mean was 16,569 and the standard deviation was 9108. Do these data provide convincing evidence at the $\alpha = 0.05$ significance level of a difference in the average number of words spoken in a day by all male and all female students at this university?

44. **Gray squirrel** In many parts of the northern United States, two color variants of the Eastern Gray Squirrel—gray and black—are found in the same habitats. A scientist studying squirrels in a large forest wonders if there is a difference in the sizes of the two color variants. He collects random samples of 40 squirrels of each color from a large forest and weighs them. The 40 black squirrels have a mean weight of 20.3 ounces and a standard deviation of 2.1 ounces. The 40 gray squirrels have a mean weight of 19.2 ounces and a standard deviation of 1.9 ounces. Do these data provide convincing evidence at the $\alpha = 0.01$ significance level of a difference in the mean weights of all gray and black Eastern Gray Squirrels in this forest?

45. **Who talks more—men or women?** Refer to Exercise 43.

(a) Construct and interpret a 95% confidence interval for the difference between the true means. If you already defined parameters and checked conditions in Exercise 43, you don't need to do so again here.

(b) Explain how the confidence interval provides more information than the test in Exercise 43.

46. **Gray squirrel** Refer to Exercise 44.

(a) Construct and interpret a 99% confidence interval for the difference between the true means. If you already defined parameters and checked conditions in Exercise 44, you don't need to do so again here.

(b) Explain how the confidence interval provides more information than the test in Exercise 44.

47. **Teaching reading** An educator believes that new reading activities in the classroom will help elementary school pupils improve their reading ability. She recruits 44 third-grade students and randomly assigns them into two groups. One group of 21 students does these new activities for an 8-week period. A control group of 23 third-graders follows the same curriculum without the activities. At the end of the 8 weeks, all students are given the Degree of Reading Power (DRP) test, which measures the aspects of reading ability that the treatment is designed to improve. Here are parallel boxplots of the data:[14]

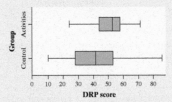

(a) Write a few sentences comparing the DRP scores for the two groups.

After checking that the conditions for inference are met, the educator performs a test of $H_0: \mu_A - \mu_C = 0$ versus $H_a: \mu_A - \mu_C > 0$, where μ_A = the true mean DRP score of third-graders like these who do the new reading activities and μ_C = the true mean DRP score of third-graders like these who follow the same curriculum without the activities. Computer output from the test is shown.

```
Two-sample T for DRP score

Group        N    Mean   StDev   SE Mean
Activities   21   51.5   11.0    2.4
Control      23   41.5   17.1    3.6

T-Value = 2.31   P-Value = 0.013   DF = 37
```

(b) What conclusion should the educator make at the $\alpha = 0.05$ significance level?

(c) Can we conclude that the new reading activities caused an increase in the mean DRP score? Explain your answer.

(d) Based on your conclusion in part (b), which type of error—a Type I error or a Type II error—could you have made? Explain your answer.

48. **Does breast-feeding weaken bones?** Breast-feeding mothers secrete calcium into their milk. Some of the calcium may come from their bones, so mothers may lose bone mineral. Researchers compared a random sample of 47 breast-feeding women with a random sample of 22 women of similar age who were neither

(b) Assuming the true mean time to breeding is the same for birds relying on natural food supply and birds with food supplements, there is a 0.0016 probability that we would observe a difference in sample means of -7.3 or smaller by chance alone.

11.43 S: $H_0: \mu_1 - \mu_2 = 0$, $H_a: \mu_1 - \mu_2 \neq 0$, where μ_1 = true mean number of words spoken per day by female students and μ_2 = true mean number . . . by male students. **P:** Two-sample t test for $\mu_1 - \mu_2$. *Random:* Independent random samples. *10%:* $n_1 = 56 < 10\%$ of females at a large university and $n_2 = 56 < 10\%$ of males at a large university. *Normal/Large Sample:* $n_1 = 56 \geq 30$ and $n_2 = 56 \geq 30$. **D:** $t = -0.25$; using df = 106.195, P-value = 0.8043. Using df = 55, P-value = 0.8035. **C:** Because the P-value of $0.8043 > \alpha = 0.05$, we fail to reject H_0.

We do not have convincing evidence the true mean number of words spoken per day by female students differs from the true mean number of words spoken per day by male students at this university.

11.44 S: $H_0: \mu_1 - \mu_2 = 0$, $H_a: \mu_1 - \mu_2 \neq 0$, where μ_1 = true mean weight of black squirrels and μ_2 = that of gray squirrels. **P:** Two-sample t test for $\mu_1 - \mu_2$. *Random:* Independent random samples. *10%:* $n_1 = 40 < 10\%$ of black squirrels and $n_2 = 40 < 10\%$ of gray squirrels. *Normal/Large Sample:* $n_1 = 40 \geq 30$ and $n_2 = 40 \geq 30$. **D:** $t = 2.46$; using df = 77.232, P-value = 0.0163. Using df = 39, P-value = 0.0184. **C:** Because the P-value of $0.0163 > \alpha = 0.01$, we fail to reject H_0. We do not have convincing evidence the true mean weight of black squirrels differs from the true mean weight of gray squirrels.

11.45 (a) D: $\bar{x}_1 = 16{,}177$, $s_1 = 7520$, $n_1 = 56$, $\bar{x}_2 = 16{,}569$, $s_2 = 9108$, $n_2 = 56$. Using df = 106.2, $(-3521, 2737)$; using df = 55, $(-3555, 2771)$. **C:** We are 95% confident that the interval from -3521 to 2737 words captures $\mu_1 - \mu_2$ = true difference in mean number of words spoken per day by female students versus male students at this university. **(b)** The two-sided test only allows us to reject (or fail to reject) a difference of 0, whereas a confidence interval provides a set of plausible values for the true difference in means.

11.46 (a) D: $\bar{x}_1 = 20.3$, $s_1 = 2.1$, $n_1 = 40$, $\bar{x}_2 = 19.2$, $s_2 = 1.9$, $n_2 = 40$. Using df = 77.23, $(-0.0826, 2.2826)$; using df = 30, $(-0.131, 2.331)$. **C:** We are 99% confident that the interval from -0.0826 to 2.2826 ounces captures $\mu_1 - \mu_2$ = true difference in mean weight of all black and gray squirrels. **(b)** The two-sided test only allows us to reject (or fail to reject) a difference of 0, whereas a confidence interval provides a set of plausible values for the true difference in mean weight of all black and gray squirrels.

11.47 (a) The score distribution for the activities group is slightly skewed to the left, while the score distribution for the control group is slightly skewed to the right. Neither distribution has any clear outliers. The center of the activities group is higher than the center of the control group. The scores in the activities group are less variable than the scores in the control group. Overall, it appears that scores are typically higher for students in the activities group. **(b)** Because the P-value of $0.013 < \alpha = 0.05$, we reject H_0. We have convincing evidence that the true mean DRP score for third-grade students like the ones in the experiment who do the activities is greater than the true mean DRP score for third-grade students like the ones in the experiment who don't do the activities. **(c)** Yes, because this was a randomized controlled experiment. **(d)** Because we rejected the null hypothesis, we could have committed a Type I error.

11.48 (a) Both distributions of BMC change are slightly skewed to the right. Neither distribution has any clear outliers. The center of the BMC change distribution is much smaller for breast-feeding mothers than for women who are not pregnant or lactating. The BMC changes are much more variable for breast-feeding mothers than for women who are not pregnant or lactating. Overall, it appears that breast-feeding mothers do lose bone mineral. **(b)** Because the P-value of approximately 0 is less than $\alpha = 0.05$, we reject H_0. We have convincing evidence that breast-feeding women have a greater mean percent bone-mineral loss than women who are neither pregnant nor lactating. **(c)** No, this was not a randomized controlled experiment. **(d)** Because we rejected the null hypothesis, we could have committed a Type I error.

11.49 (a) $t = 0.98$, P-value $= 0.1675$ using df $= 26.96$. **(b)** Assuming that the true mean cholesterol reduction for subjects who take the new drug is 10 mg/dl more than subjects who take the old drug, there is a 0.1675 probability that we would observe a difference in sample means that is at least 4.6 mg/dl or greater beyond a difference of 10 mg/dl by chance alone. Because the P-value of $0.1675 > \alpha = 0.05$, we fail to reject H_0. We do not have convincing evidence that the true mean cholesterol reduction is more than 10 mg/dl greater for the new drug than for the current drug. **(c)** To increase the power of this test, we could increase the sample sizes or the significance level.

11.50 (a) $t = 0.80$, P-value $= 0.2131$ using df $= 48.46$. **(b)** Assuming that the true mean amount of water used by the new toilets is 0.5 gallon less than the current toilets, there is a 0.2131 probability that we would observe a difference in sample means that is at least 0.05 gallon or greater beyond a difference of 0.5 gallon by chance alone. Because the P-value of $0.2131 > \alpha = 0.05$, we fail to reject H_0. We do not have convincing evidence that the new model toilet reduces the amount of water used by greater than 0.50 gallon per flush than the current-model toilet, on average. **(c)** To increase the power of this test, we could increase the sample sizes or the significance level.

pregnant nor lactating. They measured the percent change in the bone mineral content (BMC) of the women's spines over 3 months. Here are comparative boxplots of the data:[15]

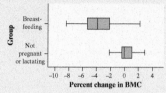

(a) Write a few sentences comparing the percent changes in BMC for the two groups.

After checking that the conditions for inference are met, the researchers perform a test of $H_0: \mu_{BF} - \mu_{NP} = 0$ versus $H_a: \mu_{BF} - \mu_{NP} < 0$, where $\mu_{BF} =$ the true mean percent change in BMC for breast-feeding women and $\mu_{NP} =$ the true mean percent change in BMC for women who are not pregnant or lactating. Computer output from the test is shown.

```
Two-sample T for BMC change

Group        N    Mean   StDev   SE Mean
Breastfeed   47   -3.59   2.51    0.37
Notpregnant  22    0.31   1.30    0.28

T-Value = -8.50   P-Value = 0.000   DF = 66
```

(b) What conclusion should the researchers make at the $\alpha = 0.05$ significance level?

(c) Can we conclude that breast-feeding causes a mother's bones to lose bone mineral? Why or why not?

(d) Based on your conclusion in part (b), which type of error—a Type I error or a Type II error—could you have made? Explain your answer.

49. **A better drug?** In a pilot study, a company's new cholesterol-reducing drug outperforms the currently available drug. If the data provide convincing evidence that the mean cholesterol reduction with the new drug is more than 10 milligrams per deciliter of blood (mg/dl) greater than with the current drug, the company will begin the expensive process of mass-producing the new drug. For the 14 subjects who were assigned at random to the current drug, the mean cholesterol reduction was 54.1 mg/dl with a standard deviation of 11.93 mg/dl. For the 15 subjects who were randomly assigned to the new drug, the mean cholesterol reduction was 68.7 mg/dl with a standard deviation of 13.3 mg/dl. Graphs of the data reveal no outliers or strong skewness.

Researchers want to perform a test of
$H_0: \mu_{new} - \mu_{cur} = 10$ versus $H_a: \mu_{new} - \mu_{cur} > 10$.

(a) Calculate the standardized test statistic and P-value.

(b) Interpret the P-value. What conclusion would you make?

(c) Describe two ways to increase the power of the test.

50. **Down the toilet** A company that makes hotel toilets claims that its new pressure-assisted toilet reduces the average amount of water used by more than 0.5 gallon per flush when compared to its current model. To test this claim, the company randomly selects 30 toilets of each type and measures the amount of water that is used when each toilet is flushed once. For the current-model toilets, the mean amount of water used is 1.64 gal with a standard deviation of 0.29 gal. For the new toilets, the mean amount of water used is 1.09 gal with a standard deviation of 0.18 gal.

Researchers want to perform a test of:
$H_0: \mu_{cur} - \mu_{new} = 0.5$ versus $H_a: \mu_{cur} - \mu_{new} > 0.5$.

(a) Calculate the standardized test statistic and P-value.

(b) Interpret the P-value. What conclusion would you make?

(c) Describe two ways to increase the power of the test.

51. **Rewards and creativity** Do external rewards—things like money, praise, fame, and grades—promote creativity? Researcher Teresa Amabile suspected that the answer is no, and that internal motivation enhances creativity. To find out, she recruited 47 experienced creative writers who were college students and divided them at random into two groups. The students in one group were given a list of statements about extrinsic reasons (E) for writing, such as public recognition, making money, or pleasing their parents. Students in the other group were given a list of statements about intrinsic reasons (I) for writing, such as expressing yourself and enjoying playing with words. Both groups were then instructed to write a poem about laughter. Each student's poem was rated separately by 12 different poets using a creativity scale.[16] These ratings were averaged to obtain an overall creativity score for each poem. The table shows summary statistics for the two groups.

Group name	n	Mean	SD
Intrinsic	24	19.883	4.440
Extrinsic	23	15.739	5.253

We used software to randomly reassign the 47 subjects to the two groups 100 times, assuming the treatment received doesn't affect each individual's creativity rating. A dotplot of the simulated difference (Intrinsic – Extrinsic) in mean creativity rating is shown.

11.51 (a) The researchers randomly assigned the subjects to create two groups that were roughly equivalent at the beginning of the experiment. **(b)** *Type I error:* The researcher finds convincing evidence that the true mean rating for students like these who are provided with internal reasons is higher than the true mean rating for students like these who are provided with external reasons, when the true mean rating is the same. *Type II error:* The manager does not find convincing evidence that the true mean rating for students like these who are provided with internal reasons is higher than the true mean rating for students like these who are

provided with external reasons, when this truly is the case. **(c)** The difference in the sample means is 4.14. Based on the dotplot, only 1 of the 100 differences were that great, meaning the P-value is approximately 0.01. Assuming the true difference (Intrinsic – Extrinsic) in mean creativity rating is 0, there is a 0.01 probability that we would observe a difference in sample means of 4.14 or greater by chance alone. **(d)** Because the P-value of $0.01 < \alpha = 0.05$, we have convincing evidence the true mean rating for students like these with internal reasons is higher than the true mean rating for students like these with external reasons.

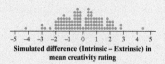

Simulated difference (Intrinsic – Extrinsic) in mean creativity rating

(a) Why did researchers randomly assign the subjects to the two treatment groups?

(b) Describe a Type I error and a Type II error in this setting.

(c) Use the results of the simulation to estimate and interpret the P-value.

(d) What conclusion would you make?

52. **Sleep deprivation** Does sleep deprivation linger for more than a day? Researchers designed a study using 21 volunteer subjects between the ages of 18 and 25. All 21 participants took a computer-based visual discrimination test at the start of the study. Then the subjects were randomly assigned into two groups. The 11 subjects in one group were deprived of sleep for an entire night in a laboratory setting. The 10 subjects in the other group were allowed unrestricted sleep for the night. Both groups were allowed as much sleep as they wanted for the next two nights. On Day 4, all the subjects took the same visual discrimination test on the computer. Researchers recorded the improvement in time (measured in milliseconds) from Day 1 to Day 4 on the test for each subject.[17] The table shows summary statistics for the two groups.

Group	Mean	SD
Unrestricted sleep	19.82	12.17
Sleep-deprived	3.90	14.73

We used software to randomly reassign the 21 subjects to the two groups 100 times, assuming the treatment received doesn't affect each individual's time improvement on the test. A dotplot of the simulated difference (Unrestricted – Sleep-deprived) in mean time improvement is shown.

Simulated difference (Unrestricted – Sleep-deprived) in mean time improvement

(a) Explain why the researchers didn't let the subjects choose whether to be in the sleep-deprivation group or the unrestricted sleep group.

(b) Describe a Type I error and a Type II error in this setting.

(c) Use the results of the simulation to estimate and interpret the P-value.

(d) What conclusion would you make?

53. **Drive-thru or go inside?** Many people think it's faster to order at the drive-thru than to order inside at fast-food restaurants. To find out, Patrick and William used a random number generator to select 10 times over a 2-week period to visit a local Dunkin' Donuts restaurant. At each of these times, one boy ordered an iced coffee at the drive-thru and the other ordered an iced coffee at the counter inside. A coin flip determined who went inside and who went to the drive-thru. The table shows the times, in seconds, that it took for each boy to receive his iced coffee after he placed the order.[18]

Visit	Inside time	Drive-thru time
1	62	55
2	63	50
3	325	321
4	105	110
5	135	124
6	55	54
7	92	90
8	75	69
9	203	200
10	100	103

Do these data provide convincing evidence at the $\alpha = 0.05$ level of a true mean difference in service time inside and at the drive-thru for this Dunkin' Donuts restaurant?

54. **Better barley** Does drying barley seeds in a kiln increase the yield of barley? A famous experiment by William S. Gosset (who discovered the t distributions) investigated this question. Eleven pairs of adjacent plots were marked out in a large field. For each pair, regular barley seeds were planted in one plot and kiln-dried seeds were planted in the other. A coin flip was used to determine which plot in each pair got the regular barley seed and which got the kiln-dried seed. The following table displays the data on barley yield (pound per acre) for each plot.[19]

Plot	Regular	Kiln
1	1903	2009
2	1935	1915
3	1910	2011
4	2496	2463
5	2108	2180
6	1961	1925
7	2060	2122
8	1444	1482
9	1612	1542
10	1316	1443
11	1511	1535

11.53 S: H_0: $\mu_{diff} = 0$, H_a: $\mu_{diff} \neq 0$, where μ_{diff} = true mean difference (Inside – Drive-thru) in time it would take to receive an iced coffee at this Dunkin' Donuts restaurant after placing the order; $\alpha = 0.05$. **P:** Paired *t* test for μ_{diff}. *Random:* The times were randomly selected. *Normal/Large Sample:* The dotplot does not show any strong skewness or outliers.

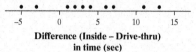

Difference (Inside – Drive-thru) in time (sec)

D: $\bar{x}_{diff} = 3.9$ seconds, $s_{diff} = 5.646$, and $n_{diff} = 10$; $t = 2.18$; df = 9; P-value = 0.0568. **C:** Because the P-value of $0.0568 > \alpha = 0.05$, we fail to reject H_0. We do not have convincing evidence the true mean difference (Inside – Drive-thru) in the time it would take to receive an iced coffee at this Dunkin' Donuts restaurant after placing the order is not 0.

11.54 S: H_0: $\mu_{diff} = 0$, H_a: $\mu_{diff} < 0$, where μ_{diff} = true mean difference (Regular – Kiln) in yield between regular barley seeds and kiln-dried barley seeds; $\alpha = 0.05$. **P:** Paired *t* test for μ_{diff}. *Random:* The treatments were randomly assigned. *Normal/Large Sample:* The dotplot does not show any strong skewness or outliers.

Difference (Regular – Kiln) in yield (lb/acre)

D: $\bar{x}_{diff} = -33.7$, $s_{diff} = 66.2$, and $n_{diff} = 11$; $t = -1.69$; df = 10; P-value = 0.0609. **C:** Because the P-value of $0.0609 > \alpha = 0.05$, we fail to reject H_0. We do not have convincing evidence that the true mean difference (Regular – Kiln) in yield is less than 0.

11.52 (a) If people were allowed to choose, it is likely that all those who choose one particular treatment (e.g., sleep deprivation) might be systematically different from those who choose to be in the other treatment group. Then we wouldn't know if the difference in characteristics or difference in sleep was the cause of a difference in mean time improvement. **(b)** *Type I error:* The researcher finds convincing evidence that the mean increase in score is higher for subjects like these who are allowed to sleep than for subjects like these who are sleep deprived, when the true mean increase in score is the same. *Type II error:* The manager does not find convincing evidence that the mean increase in score is higher for subjects like these who are allowed to sleep than for subjects like these who are sleep deprived, when this truly is the case. **(c)** The difference in the sample means is 15.92. Based on the dotplot, only 2 of the 100 differences were that great, meaning the P-value is approximately 0.02. Assuming the true difference (Unrestricted – Sleep deprived) in mean time improvement is 0, there is a 0.02 probability that we would observe a difference in sample means of 15.92 or greater by chance alone. **(d)** Because the P-value of $0.02 < \alpha = 0.05$, we have convincing evidence that the mean increase in score is higher for subjects like these who are allowed to sleep than for subjects like these who are sleep deprived. **(e)** Because we found convincing evidence that the mean is higher for subjects with unrestricted sleep when it is possible that there is no difference in the means, we could have made a Type I error.

11.55 (a) Yes! The data arose from a randomized comparative experiment, so if the result of this study is statistically significant, we can conclude that the difference in the ability to memorize words was caused by whether students were performing the task in silence or with music playing. **(b) S:** H_0: $\mu_{diff} = 0$, H_a: $\mu_{diff} \neq 0$, where μ_{diff} = true mean difference (Music − Silence) in the average number of words recalled by students at this school; $\alpha = 0.01$. **P:** Paired t test for μ_{diff}. *Random:* The treatments were assigned in a random order. *Normal/Large Sample:* $n_{diff} = 30 \geq 30$. **D:** $\bar{x}_{diff} = 1.57$, $s_{diff} = 2.70$, and $n_{diff} = 30$; $t = 3.18$; df = 29; P-value = 0.0034. **C:** Because the P-value of 0.0034 < α = 0.01, we reject H_0. We have convincing evidence that the true mean difference (Music − Silence) in the average number of words recalled by students at this school is different from 0. **(c)** Because we rejected H_0, it is possible that we made a Type I error.

11.56 (a) No; this is an observational study, not a controlled experiment. So even if the result of this study is statistically significant, we cannot conclude that the difference in shopping behavior is due to the effect of Friday the 13th. **(b) S:** H_0: $\mu_{diff} = 0$, H_a: $\mu_{diff} \neq 0$, where μ_{diff} = true mean difference (6th − 13th) in the mean number of shoppers at grocery stores; $\alpha = 0.10$. **P:** Paired t test for μ_{diff}. *Random:* Paired data; random sample of 45 grocery stores. *10%:* 45 < 10% of all grocery stores. *Normal/Large Sample:* $n_{diff} = 45 \geq 30$. **D:** $\bar{x}_{diff} = -46.5$, $s_{diff} = 178.0$, and $n_{diff} = 45$; $t = -1.75$; df = 44; P-value = 0.0867. **C:** Because the P-value of 0.0867 < α = 0.10, we reject H_0. We have convincing evidence that the true mean difference (6th − 13th) in the number of shoppers at grocery stores on these two days differs from 0. **(c)** Because we rejected H_a, it is possible that we made a Type I error.

11.57 (a) S: μ_{diff} = true mean difference (Music − Silence) in the average number of words recalled by students at this school. **P:** One-sample t interval for μ_{diff}. *Random:* The treatments were assigned in a random order. *Normal/Large Sample:* $n_{diff} = 30 \geq 30$. **D:** $\bar{x}_{diff} = 1.57$, $s_{diff} = 2.70$, $n_{diff} = 30$; df = 29, (0.211, 2.929). **C:** We are 99% confident the interval from 0.211 to 2.929 captures the true mean difference (Music − Silence) in the number of words recalled by students at this school. **(b)** The two-sided test

Do these data provide convincing evidence at the $\alpha = 0.05$ level that drying barley seeds in a kiln increases the yield of barley, on average?

55. **Music and memory** Does listening to music while studying help or hinder students' learning? Two statistics students designed an experiment to find out. They selected a random sample of 30 students from their medium-sized high school to participate. Each subject was given 10 minutes to memorize two different lists of 20 words, once while listening to music and once in silence. The order of the two word lists was determined at random; so was the order of the treatments. The difference (Silence − Music) in the number of words recalled was recorded for each subject. The mean difference was 1.57 and the standard deviation of the differences was 2.70.

(a) If the result of this study is statistically significant, can you conclude that the difference in the ability to memorize words was caused by whether students were performing the task in silence or with music playing? Why or why not?

(b) Do the data provide convincing evidence at the $\alpha = 0.01$ significance level that the number of words recalled in silence or when listening to music differs, on average, for students at this school?

(c) Based on your conclusion in part (a), which type of error—a Type I error or a Type II error—could you have made? Explain your answer.

56. **Friday the 13th** Do people behave differently on Friday the 13th? Researchers collected data on the number of shoppers at a random sample of 45 grocery stores on Friday the 6th and Friday the 13th in the same month. Then they calculated the difference (subtracting in the order 6th minus 13th) in the number of shoppers at each store on these 2 days. The mean difference is −46.5 and the standard deviation of the differences is 178.0.[20]

(a) If the result of this study is statistically significant, can you conclude that the difference in shopping behavior is due to the effect of Friday the 13th on people's behavior? Why or why not?

(b) Do these data provide convincing evidence at the $\alpha = 0.05$ level that the number of shoppers at grocery stores on these 2 days differs, on average?

(c) Based on your conclusion in part (a), which type of error—a Type I error or a Type II error—could you have made? Explain your answer.

57. **Music and memory** Refer to Exercise 55.

(a) Construct and interpret a 99% confidence interval for the true mean difference. If you already defined the

parameter and checked conditions in Exercise 55, you don't need to do so again here.

(b) Explain how the confidence interval provides more information than the test in Exercise 55.

58. **Friday the 13th** Refer to Exercise 56.

(a) Construct and interpret a 90% confidence interval for the true mean difference. If you already defined the parameter and checked conditions in Exercise 56, you don't need to do so again here.

(b) Explain how the confidence interval provides more information than the test in Exercise 56.

59. **Two samples or paired data?** In each of the following settings, decide whether you should use two-sample t procedures to perform inference about a difference in means or paired t procedures to perform inference about a mean difference. Explain your choice.[21]

(a) To test the wear characteristics of two tire brands, A and B, each of 50 cars of the same make and model is randomly assigned Brand A tires or Brand B tires.

(b) To test the effect of background music on productivity, factory workers are observed. For one month, each subject works without music. For another month, the subject works while listening to music on an MP3 player. The month in which each subject listens to music is determined by a coin toss.

(c) How do young adults look back on adolescent romance? Investigators interviewed a random sample of 40 couples in their mid-twenties. The female and male partners were interviewed separately. Each was asked about his or her current relationship and also about a romantic relationship that lasted at least 2 months when they were aged 15 or 16. One response variable was a measure on a numerical scale of how much the attractiveness of the adolescent partner mattered. You want to find out how much men and women differ on this measure.

60. **Two samples or paired data?** In each of the following settings, decide whether you should use two-sample t procedures to perform inference about a difference in means or paired t procedures to perform inference about a mean difference. Explain your choice.[22]

(a) To compare the average weight gain of pigs fed two different diets, nine pairs of pigs were used. The pigs in each pair were littermates. A coin toss was used to decide which pig in each pair got Diet A and which got Diet B.

only allows us to reject a difference of 0, where the confidence interval provided a set of plausible values (0.211, 2.929) for the true mean difference (Music − Silence) in the number of words recalled by students at this school.

11.58 (a) S: μ_{diff} = true mean difference (6th − 13th) in the number of shoppers at grocery stores. **P:** One-sample t interval for μ_{diff}. *Random:* Paired data; random sample of 45 grocery stores. *10%:* 45 < 10% of all grocery stores. *Normal/Large Sample:* $n_{diff} = 45 \geq 30$. **D:** $\bar{x}_{diff} = -46.5$, $s_{diff} = 178.0$, $n_{diff} = 45$; using df = 40, (−90.284, −0.916). Using df = 44, (−91.08, −1.92). **C:** We are 90% confident that the interval from −91.08 to −1.92 captures the true mean difference (6th − 13th) in the number of shoppers at

grocery stores. **(b)** The two-sided test only allowed us to reject a difference of zero, where the confidence interval provided a set of plausible values (−91.08, −1.92) for true mean difference (6th − 13th) in the number of shoppers at grocery stores.

11.59 (a) Two-sample t test; the data are being produced using two distinct groups of cars in a randomized experiment. **(b)** Paired t test; this is a matched pairs experimental design in which both treatments are applied to each subject in a random order. **(c)** Paired t test; the data were collected from the male and female partners in 40 couples.

(b) Separate random samples of male and female college professors are taken. We wish to compare the average salaries of male and female teachers.

(c) To test the effects of a new fertilizer, 100 plots are treated with the new fertilizer, and 100 plots are treated with another fertilizer. A computer's random number generator is used to determine which plots get which fertilizer.

61. **Have a ball!** Can students throw a baseball farther than a softball? To find out, researchers conducted a study involving 24 randomly selected students from a large high school. After warming up, each student threw a baseball as far as he or she could and threw a softball as far as he or she could, in a random order. The distance in yards for each throw was recorded. Here are the data, along with the difference (Baseball – Softball) in distance thrown, for each student:

Student	Baseball	Softball	Difference (Baseball – Softball)
1	65	57	8
2	90	58	32
3	75	66	9
4	73	61	12
5	79	65	14
6	68	56	12
7	58	53	5
8	41	41	0
9	56	44	12
10	70	65	5
11	64	57	7
12	62	60	2
13	73	55	18
14	50	53	-3
15	63	54	9
16	48	42	6
17	34	32	2
18	49	48	1
19	48	45	3
20	68	67	1
21	30	27	3
22	26	25	1
23	28	25	3
24	26	31	-5

(a) Explain why these are paired data.

(b) A boxplot of the differences is shown. Explain how the graph gives *some* evidence that students like these can throw a baseball farther than a softball.

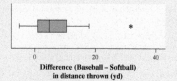

Difference (Baseball – Softball)
in distance thrown (yd)

(c) State appropriate hypotheses for performing a test about the true mean difference. Be sure to define any parameter(s) you use.

(d) Explain why the Normal/Large Sample condition is not met in this case.

The mean difference (Baseball – Softball) in distance thrown for these 24 students is $\bar{x}_{diff} = 6.54$ yards. Is this a surprisingly large result if the null hypothesis is true? To find out, we can perform a simulation assuming that students have the same ability to throw a baseball and a softball. For each student, write the two distances thrown on different note cards. Shuffle the two cards and designate one distance to baseball and one distance to softball. Then subtract the two distances (Baseball – Softball). Do this for all the students and find the simulated mean difference. Repeat many times. Here are the results of 100 trials of this simulation:

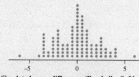

Simulated mean difference (Baseball – Softball)
in distance thrown (yd)

(e) Use the results of the simulation to estimate the P-value. What conclusion would you draw?

62. **Flight times** Emirates Airlines offers one outbound flight from Dubai, United Arab Emirates, to Doha, Qatar, and one return flight from Doha to Dubai each day. An experienced Emirates pilot suspects that the Dubai-to-Doha outbound flight typically takes longer. To find out, the pilot collects data about these flights on a random sample of 12 days.

If the pilot's suspicion is correct, the difference (Outbound – Return) in flight times will be positive on more days than it is negative. What if either flight is equally likely to take longer? Then we can model the outcome on a randomly selected day with a coin toss. Heads means the Dubai-to-Doha outbound flight lasts longer; tails means the Doha-to-Dubai return flight lasts longer. To imitate a random sample of 12 days, imagine tossing a fair coin 12 times.

11.60 (a) Paired *t* test; this is a matched pairs design using pairs of pigs that were littermates. One pig in each pair received one treatment and the other pig in the pair received the other treatment. **(b)** Two-sample *t* test; the data come from independent random samples of male and female college professors. **(c)** Two-sample *t* test; the data are being produced using two distinct groups of plots in a randomized experiment.

11.61 (a) The table shows two values for each individual (each student). **(b)** Most of the differences (Baseball – Softball) are positive, meaning the students tend to throw the baseball farther than the softball. **(c)** H_0: $\mu_{diff} = 0$, H_a: $\mu_{diff} > 0$, where μ_{diff} = the true mean difference (Baseball – Softball) in distance thrown. **(d)** $n_{diff} = 24 < 30$ and the boxplot of the differences has an outlier. **(e)** A mean difference of 6.54 or larger happened in 0 of the 100 simulation trials, so the estimated P-value is 0. Because the P-value $< \alpha = 0.05$, we reject H_0. These data provide convincing evidence that students like these can throw the baseball farther, on average, than the softball.

11.62 (a) X = number of heads has a binomial distribution with $n = 12$ and $p = 0.50$; $P(X \geq 11) = 1 -$ binomcdf(trials: 12, p: 0.50, x value: 10) = 0.0032. There is a 0.32% probability of getting 11 or more heads in 12 tosses of a fair coin. **(b)** Because the probability $0.0032 < 0.05$, we have convincing evidence that the Emirates pilot's suspicion is correct. We have reason to believe that the Dubai-to-Doha flight typically takes longer.

11.63 (a) Paired t test; we have paired data (two scores for each student). **(b) S:** H_0: $\mu_{diff} = 0$, H_a: $\mu_{diff} > 0$, where μ_{diff} = the true mean increase in SAT verbal scores of students who were coached; $\alpha = 0.05$. **P:** Paired t test for μ_{diff}. *Random:* Random sample of 427 students. *10%:* $n_{diff} = 427$ is less than 10% of students who are coached. *Normal/Large Sample:* $n_{diff} = 427 \geq 30$. **D:** $t = 10.16$; df $= 426$; P-value ≈ 0. **C:** Because the P-value of approximately $0 < \alpha = 0.05$, we reject H_0. There is convincing evidence that students who are coached increase their scores on the SAT verbal test, on average.

11.64 (a) S: We want to test H_0: $\mu_C - \mu_U = 0$, H_a: $\mu_C - \mu_U > 0$, where μ_C = the true mean gain in SAT verbal score for all coached students and μ_U = the true mean gain in SAT verbal score for all uncoached students using $\alpha = 0.05$. **P:** Two-sample t test for $\mu_C - \mu_U$. *Random:* These data come from independent random samples of 427 coached students and 2733 uncoached students. *10%:* $n_C = 427 < 10\%$ of all coached students and $n_U = 2733 < 10\%$ of all uncoached students. *Normal/Large Sample:* $n_C = 427 \geq 30$ and $n_U = 2733 \geq 30$. **D:** 2-SampTTest gives $t = 2.646$ and P-value $= 0.0042$ using df $= 534.45$. **C:** Because the P-value of $0.0041 < \alpha = 0.05$, we reject H_0. We have convincing evidence that coached students improve more than uncoached students, on average. **(b)** The interval gives a range of plausible values for the mean gain in SAT verbal score that coached students might expect to receive beyond that of uncoached students. **(c)** The average amount of points gained (3.02 to 12.98) is not very large. It does not seem like the money spent on coaching is worth it.

11.65 d

11.66 b

11.67 d

(a) Find the probability of getting 11 or more heads in 12 tosses of a fair coin.

(b) The outbound flight took longer on 11 of the 12 days. Based on your result in part (a), what conclusion would you make about the Emirates pilot's suspicion?

Exercises 63 and 64 refer to the following setting. Coaching companies claim that their courses can raise the SAT scores of high school students. Of course, students who retake the SAT without paying for coaching generally raise their scores. A random sample of students who took the SAT twice found 427 who were coached and 2733 who were uncoached.[23] Starting with their Verbal scores on the first and second tries, we have these summary statistics:

		Try 1		Try 2		Gain	
	n	$\bar{x}$	s_x	$\bar{x}$	s_x	$\bar{x}$	s_x
Coached	427	500	92	529	97	29	59
Uncoached	2733	506	101	527	101	21	52

63. **Coaching and SAT scores** Let's first ask if students who are coached increased their scores significantly, on average.

(a) You could use the information on the Coached line to carry out either a two-sample t test comparing Try 1 with Try 2 or a paired t test using Gain. Which is the correct test? Why?

(b) Carry out the proper test. What do you conclude?

64. **Coaching and SAT scores** What we really want to know is whether coached students improve more than uncoached students, on average, and whether any advantage is large enough to be worth paying for.

(a) Carry out an appropriate test at the $\alpha = 0.05$ significance level to determine if there is convincing evidence that coached students improve more than uncoached students, on average.

(b) A 90% confidence interval for $\mu_{coached} - \mu_{uncoached}$ is (3.02, 12.98). Explain how this interval gives more information than the test in part (a).

(c) Based on your work, what is your opinion: Do you think coaching courses are worth paying for?

Multiple Choice: *Select the best possible answer for Exercises 65–69.*

Exercises 65–67 refer to the following setting. A study of road rage asked random samples of 596 men and 523 women about their behavior while driving. Based on their answers, each person was assigned a road rage score on a scale of 0 to 20. The participants were chosen by random digit dialing of phone numbers. The researchers performed a test of the following hypotheses: H_0: $\mu_M = \mu_F$ versus H_a: $\mu_M \neq \mu_F$.

65. Which of the following describes a Type II error in the context of this study?

(a) Finding convincing evidence that the true means are different for males and females, when in reality the true means are the same

(b) Finding convincing evidence that the true means are different for males and females, when in reality the true means are different

(c) Not finding convincing evidence that the true means are different for males and females, when in reality the true means are the same

(d) Not finding convincing evidence that the true means are different for males and females, when in reality the true means are different

(e) Not finding convincing evidence that the true means are different for males and females, when in reality there is convincing evidence that the true means are different

66. The P-value for the stated hypotheses is 0.002. Interpret this value in the context of this study.

(a) Assuming that the true mean road rage score is the same for males and females, there is a 0.002 probability of getting a difference in sample means equal to the one observed in this study.

(b) Assuming that the true mean road rage score is the same for males and females, there is a 0.002 probability of getting a difference in sample means at least as large in either direction as the one observed in this study.

(c) Assuming that the true mean road rage score is different for males and females, there is a 0.002 probability of getting a difference in sample means at least as large in either direction as the one observed in this study.

(d) Assuming that the true mean road rage score is the same for males and females, there is a 0.002 probability that the null hypothesis is true.

(e) Assuming that the true mean road rage score is the same for males and females, there is a 0.002 probability that the alternative hypothesis is true.

67. Based on the P-value in Exercise 66, which of the following must be true?

(a) A 90% confidence interval for $\mu_M - \mu_F$ will contain 0.

(b) A 95% confidence interval for $\mu_M - \mu_F$ will contain 0.

(c) A 99% confidence interval for $\mu_M - \mu_F$ will contain 0.

(d) A 99.9% confidence interval for $\mu_M - \mu_F$ will contain 0.

(e) It is impossible to determine whether any of these statements is true based only on the P-value.

68. A study of the impact of caffeine consumption on reaction time was designed to correct for the impact of subjects' prior sleep deprivation by dividing the 24 subjects into 12 pairs on the basis of the average hours of sleep they had had for the previous 5 nights. That is, the two with the highest average sleep were a pair, then the two with the next highest average sleep, and so on. One randomly assigned member of each pair drank 2 cups of caffeinated coffee, and the other drank 2 cups of decaf. Each subject's performance on a standard reaction-time test was recorded. Which of the following is the correct check of the "Normal/Large Sample" condition for this significance test?

I. Confirm graphically that the scores of the caffeine drinkers could have come from a Normal distribution.

II. Confirm graphically that the scores of the decaf drinkers could have come from a Normal distribution.

III. Confirm graphically that the differences in scores within each pair of subjects could have come from a Normal distribution.

(a) I only

(b) II only

(c) III only

(d) I and II only

(e) I, I, and III

69. There are two common methods for measuring the concentration of a pollutant in fish tissue. Do the two methods differ, on average? You apply both methods to each fish in a random sample of 18 carp and use

(a) the paired t test for μ_{diff}.

(b) the one-sample z test for p.

(c) the two-sample t test for $\mu_1 - \mu_2$.

(d) the two-sample z test for $p_1 - p_2$.

(e) none of these.

Recycle and Review

In each part of Exercises 70 and 71, state which inference procedure from Chapter 8, 9, 10, or 11 you would use. Be specific. For example, you might say, "Two-sample z test for the difference between two proportions." You do not have to carry out any procedures.

70. **Which inference method?**

(a) Drowning in bathtubs is a major cause of death in children less than 5 years old. A random sample of parents was asked many questions related to bathtub safety. Overall, 85% of the sample said they used baby bathtubs for infants. Estimate the percent of all parents of young children who use baby bathtubs.

(b) How seriously do people view speeding in comparison with other annoying behaviors? A large random sample of adults was asked to rate a number of behaviors on a scale of 1 (no problem at all) to 5 (very severe problem). Do speeding drivers get a higher average rating than noisy neighbors?

(c) You have data from interviews with a random sample of students who failed to graduate from a particular college in 7 years and also from a random sample of students who entered at the same time and did graduate within 7 years. You will use these data to estimate the difference in the percents of students from rural backgrounds among dropouts and graduates.

(d) Do experienced computer-game players earn higher scores when they play with someone present to cheer them on or when they play alone? Fifty teenagers with experience playing a particular computer game have volunteered for a study. We randomly assign 25 of them to play the game alone and the other 25 to play the game with a supporter present. Each player's score is recorded.

71. **Which inference method?**

(a) A city planner wants to determine if there is convincing evidence of a difference in the average number of cars passing through two different intersections. He randomly selects 12 times between 6:00 A.M. and 10:00 P.M., and he and his assistant count the number of cars passing through each intersection during the 10-minute interval that begins at that time.

(b) Are more than 75% of Toyota owners generally satisfied with their vehicles? Let's design a study to find out. We'll select a random sample of 400 Toyota owners. Then we'll ask each individual in the sample, "Would you say that you are generally satisfied with your Toyota vehicle?"

(c) Are male college students more likely to binge drink than female college students? The Harvard School of Public Health surveys random samples of male and female undergraduates at four-year colleges and universities about whether they have engaged in binge drinking.

(d) A bank wants to know which of two incentive plans will most increase the use of its credit cards and by how much. It offers each incentive to a group of current credit card customers, determined at random, and compares the amount charged during the following 6 months.

11.68 c

11.69 a

11.70 (a) One-sample z interval for a proportion. **(b)** Paired t test for the mean difference. **(c)** Two-sample z interval for the difference in proportions. **(d)** Two-sample t test for a difference in means.

11.71 (a) Paired t test for the mean difference. **(b)** One-sample z test for a proportion. **(c)** Two-sample z test for a difference in proportions. **(d)** Two-sample t interval for a difference in means.

TRM Chapter 11 FRAPPY! Materials

Please consult the Teacher's Resource Materials for sample student responses, a scoring rubric, and a printable version of the original question with space for students to write their responses. We present a model solution here.

Answers:

STATE: H_0: $\mu_N - \mu_S = 0$

H_a: $\mu_N - \mu_S > 0$

where μ_N = the true mean percentage of popped kernels for name-brand microwave popcorn and μ_S = the true mean percentage of popped kernels for store-brand microwave popcorn. Use $\alpha = 0.05$.

PLAN: Two-sample t test for $\mu_N - \mu_S$.

• *Random:* The data come from independent random samples.
  ○ *10%:* There are more than $10(10) = 100$ bags of name-brand microwave popcorn and more than $10(10) = 100$ bags of store-brand microwave popcorn

• *Normal/Large Sample:* The sample sizes are small, but the dotplots show no strong skewness or outliers

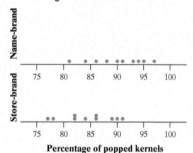

Percentage of popped kernels

DO:

$\bar{x}_N = 89.9$, $s_N = 5.13$, $n_N = 10$; $\bar{x}_S = 84.5$, and $s_S = 4.81$, $n_S = 10$

$$t = \frac{(89.9 - 84.5) - 0}{\sqrt{\dfrac{5.13^2}{10} + \dfrac{4.81^2}{10}}} = 2.43$$

P-value:

Option 1: 2-SampTTest gives $t = 2.43$ and *P*-value $= 0.013$, using df $= 17.93$.

Option 2: df $= 9$

Table B: $0.01 < $ *P*-value < 0.02

Tech: tcdf(lower: 2.43, upper: 1000, df: 9) $= 0.019$

CONCLUDE: Because the *P*-value of $0.013 < \alpha = 0.05$, we reject H_0. We have convincing evidence that the true mean percentage of popped kernels is greater for name-brand microwave popcorn than for store-brand microwave popcorn.

FRAPPY! FREE RESPONSE AP® PROBLEM, YAY!

The following problem is modeled after actual AP® Statistics exam free response questions. Your task is to generate a complete, concise response in 15 minutes.

Directions: Show all your work. Indicate clearly the methods you use, because you will be scored on the correctness of your methods as well as on the accuracy and completeness of your results and explanations.

Will using name-brand microwave popcorn result in a greater percentage of popped kernels than using store-brand microwave popcorn? To find out, Briana and Maggie randomly selected 10 bags of name-brand microwave popcorn and 10 bags of store-brand microwave popcorn. The chosen bags were arranged in a random order. Then each bag was popped for 3.5 minutes, and the percentage of popped kernels was calculated. The results are displayed in the following table.

Name-brand	95	88	84	94	81	90	97	93	91	86
Store-brand	91	89	82	82	77	78	84	86	86	90

Do the data provide convincing evidence that using name-brand microwave popcorn will result in a greater mean percentage of popped kernels?

After you finish, you can view two example solutions on the book's website (highschool.bfwpub.com/updatedtps6e). Determine whether you think each solution is "complete," "substantial," "developing," or "minimal." If the solution is not complete, what improvements would you suggest to the student who wrote it? Finally, your teacher will provide you with a scoring rubric. Score your response and note what, if anything, you would do differently to improve your own score.

Chapter 11 Review

Section 11.1: Tests About a Population Mean

In this section, you learned the details of performing a significance test about a population mean. Whenever you are asked if there is convincing evidence for a claim about a population mean, you are expected to respond using the familiar four-step process.

STATE: Give the hypotheses you are testing in terms of μ, define the parameter μ, and state the significance level.

PLAN: Name the procedure you are using (one-sample t test for a population mean), and check the conditions to see if the procedure is appropriate.

• Random: The data come from a random sample from the population of interest.
  ○ 10%: The sample size is less than 10% of the population size when sampling without replacement.

• Normal/Large Sample: The population distribution is Normal or the sample size is large ($n \geq 30$). If the sample size is small and the population shape is unknown, a graph of the sample data shows no strong skewness or outliers that would suggest a non-Normal population.

DO: Calculate the standardized test statistic and *P*-value. The standardized test statistic measures how far away the sample statistic is from the hypothesized parameter value in standardized units:

$$t = \frac{\bar{x} - \mu_0}{\dfrac{s_x}{\sqrt{n}}}$$

To calculate the *P*-value using a t distribution, determine the degrees of freedom (df $= n - 1$) and use Table B or technology.

TRM Chapter 11 Test

There is one Chapter Test available for this section in the Teacher's Resource Materials. Click on the link in the TE-Book, open in the TRFD, or download from the Teacher's Resources on the book's digital platform. You can also create your own Test using the TPS quiz and test builder (ExamView).

Teaching Tip

Now that you have completed Chapter 11 in this book, students can login to their College Board accounts and take the **Personal Progress Check for Unit 7**.

CONCLUDE: Use the *P*-value to make an appropriate conclusion about the hypotheses, in context.

Remember to use significance tests wisely. When planning a study, use a large enough sample size so the test will have adequate power. Also, remember that statistically significant results aren't always practically important. Finally, be aware that the probability of making at least one Type I error goes up dramatically when conducting multiple tests.

Section 11.2 Tests About a Difference in Means

In this section, you learned how to perform significance tests for a difference between two means. As in any test, you start by stating hypotheses. The null hypothesis is usually $H_0: \mu_1 - \mu_2 = 0$ (or equivalently, $H_0: \mu_1 = \mu_2$). The alternative hypothesis can be one-sided ($<$ or $>$) or two-sided ($\neq$).

The conditions for inference about a difference in means are the same for significance tests as for confidence intervals. The Random condition says that the data must be from two independent random samples or two groups in a randomized experiment. The 10% condition says that each sample size should be less than 10% of the corresponding population size when sampling without replacement. The Normal/Large Sample condition says that for each sample, the corresponding population distribution (or the true distribution of response to the treatment) is Normal or the sample size is large ($n \geq 30$). If either population (treatment) distribution has unknown shape and $n < 30$, a graph of the corresponding sample data shows no strong skewness or outliers.

For a test of $H_0: \mu_1 - \mu_2 = 0$, the standardized test statistic is

$$t = \frac{(\overline{x}_1 - \overline{x}_2) - 0}{\sqrt{\dfrac{s_1^2}{n_1} + \dfrac{s_2^2}{n_2}}}$$

When conditions are met, *P*-values can be obtained using a *t* distribution with degrees of freedom from technology (Option 1) or the smaller of $n_1 - 1$ and $n_2 - 1$ (Option 2).

In this section, you also learned how to compare two means in the special case of paired data. When two values of the same quantitative variable are recorded for each individual or for each of two similar individuals, start by finding the difference of the values in each pair. Be sure to indicate the order of subtraction. Use the mean difference $\overline{x}_{\text{diff}}$ and the standard deviation of the differences s_{diff} as summary statistics.

To perform a significance test about the true mean difference μ_{diff}, start by stating hypotheses. The null hypothesis is usually $H_0: \mu_{\text{diff}} = 0$. The conditions are the same for significance tests as for confidence intervals. The Random condition says that paired data must come from a random sample from the population of interest or from a randomized experiment. The 10% condition says that the sample size n_{diff} should be less than 10% of the corresponding population of differences when sampling without replacement. The Normal/Large Sample condition says that the population distribution of differences is Normal or that the sample size is large ($n_{\text{diff}} \geq 30$). If the sample size is small and the population shape is unknown, a graph of the differences shows no strong skewness or outliers.

For a test of $H_0: \mu_{\text{diff}} = 0$, the standardized test statistic is

$$t = \frac{\overline{x}_{\text{diff}} - 0}{\dfrac{s_{\text{diff}}}{\sqrt{n}}}$$

When the conditions are met, the *P*-value can be obtained from a *t* distribution with df $= n_{\text{diff}} - 1$.

To decide whether two-sample *t* procedures for a difference between two means or paired *t* procedures for a mean difference are appropriate, consider how the data were produced.

	Comparing significance tests for means		
	Significance test for μ	**Significance test for $\mu_1 - \mu_2$**	**Significance test for μ_{diff}**
Name (TI-83/84)	One-sample *t* test for μ (T-Test)	Two-sample *t* test for $\mu_1 - \mu_2$ (2-SampTTest)	Paired *t* test for μ_{diff} or One-sample *t* test for μ_{diff} (T-Test)
Null hypothesis	$H_0: \mu = \mu_0$	$H_0: \mu_1 - \mu_2 = 0$	$H_0: \mu_{\text{diff}} = 0$
Conditions	• **Random:** The data come from a random sample from the population of interest. ○ **10%:** When sampling without replacement, $n < 0.10N$.	• **Random:** The data come from two independent random samples or from two groups in a randomized experiment. ○ **10%:** When sampling without replacement, $n_1 < 0.10N_1$ and $n_2 < 0.10N_2$.	• **Random:** Paired data come from a random sample from the population of interest or from a randomized experiment. ○ **10%:** When sampling without replacement, $n_{\text{diff}} < 0.10N_{\text{diff}}$.

	Comparing significance tests for means		
	Significance test for μ	**Significance test for $\mu_1 - \mu_2$**	**Significance test for μ_{diff}**
Conditions (continued)	• **Normal/Large Sample**: The population distribution is Normal or the sample size is large ($n \geq 30$). If the population distribution has unknown shape and $n < 30$, a graph of the sample data shows no strong skewness or outliers.	• **Normal/Large Sample**: For each sample, the corresponding population distribution (or the true distribution of response to the treatment) is Normal or the sample size is large ($n \geq 30$). If either population (treatment) distribution has unknown shape and $n < 30$, a graph of the corresponding sample data shows no strong skewness or outliers.	• **Normal/Large Sample**: The population distribution of differences (or the true distribution of differences in response to the treatments) is Normal or the number of differences in the sample is large ($n_{diff} \geq 30$). If the population (true) distribution of differences has unknown shape and $n_{diff} < 30$, a graph of the sample differences shows no strong skewness or outliers.
Formula	$$t = \frac{\bar{x} - \mu_0}{\frac{s_x}{\sqrt{n}}}$$ P-value from t distribution with df $= n - 1$	$$t = \frac{(\bar{x}_1 - \bar{x}_2) - 0}{\sqrt{\frac{s_1^2}{n_1} + \frac{s_2^2}{n_2}}}$$ P-value from t distribution with df from technology or smaller of $n_1 - 1$ and $n_2 - 1$	$$t = \frac{\bar{x}_{diff} - 0}{\frac{s_{diff}}{\sqrt{n_{diff}}}}$$ P-value from t distribution with df $= n_{diff} - 1$

Teaching Tip

At the end of each chapter, a "What Did You Learn?" grid lists all the targets for the chapter. Make sure to discuss this grid with your students. Ask them to read each learning target and self-assess whether or not they can do each one. For each target, the grid shows the section in which it was covered, page references for examples that illustrate the target, and relevant chapter review exercises that students can use to assess their understanding of that target. Encourage your students to use this grid as part of their preparations for the chapter test.

What Did You Learn?

Learning Target	Section	Related Example on Page(s)	Relevant Chapter Review Exercise(s)
State and check the Random, 10%, and Normal/Large Sample conditions for performing a significance test about a population mean.	11.1	699	R11.1, R11.2
Calculate the standardized test statistic and P-value for a test about a population mean.	11.1	703	R11.1, R11.2
Perform a significance test about a population mean.	11.1	706	R11.2
Use a confidence interval to make a conclusion for a two-sided test about a population mean.	11.1	709	R11.3
State appropriate hypotheses for a significance test about a difference in means.	11.2	720	R11.4
Determine whether the conditions are met for performing a test about a difference between two means.	11.2	720	R11.4
Calculate the standardized test statistic and P-value for a test about a difference between two means.	11.2	721	R11.4
Perform a significance test about a difference between two means.	11.2	723	R11.4
Perform a significance test about a mean difference.	11.2	729	R11.5
Determine when it is appropriate to use paired t procedures versus two-sample t procedures.	11.2	732	R11.4, R11.5

Chapter 11 Review Exercises

These exercises are designed to help you review the important ideas and methods of the chapter.

R11.1 Graduating to new heights? The average height of 18-year-old American women is 64.2 inches. You wonder whether the mean height of this year's female graduates from a large local high school differs from the national average. You measure an SRS of 48 female graduates and find that $\bar{x} = 63.5$ inches and $s_x = 3.7$ inches.

(a) State the appropriate null and alternative hypotheses for performing a significance test.

(b) Explain why the sample data give some evidence for H_a.

(c) Identify the appropriate test and show that the conditions for carrying out the test are met.

(d) Find the standardized test statistic and P-value.

(e) Interpret the P-value. What conclusion would you make?

R11.2 Fonts and reading ease Does the use of fancy type fonts slow down the reading of text on a computer screen? Adults can read four paragraphs of a certain text in the common Times New Roman font in an average of 22 seconds. Researchers asked a random sample of 24 adults to read this text in the ornate font named *Gigi*. Here are their times (in seconds).

23.2 21.2 28.9 27.7 29.1 27.3 16.1 22.6 25.6 34.2 23.9 26.8
20.5 34.3 21.4 32.6 26.2 34.1 31.5 24.6 23.0 28.6 24.4 28.1

Do these data provide convincing evidence that it takes adults longer than 22 seconds, on average, to read these four paragraphs in Gigi font?

R11.3 Icebreaker? In the children's game Don't Break the Ice, small plastic ice cubes are squeezed into a square frame. Each child takes turns tapping out a cube of "ice" with a plastic hammer, hoping that the remaining cubes don't collapse. For the game to work correctly, the cubes must be big enough so that they hold each other in place in the plastic frame, but not so big that they are too difficult to tap out. The machine that produces the plastic cubes is designed to make cubes that are 25.4 millimeters (mm) wide, but the width varies a little. To ensure that the machine is working well, a supervisor inspects a random sample of 50 cubes every hour and measures their width. If the sample provides convincing evidence at the $\alpha = 0.05$ significance level that the true mean width

μ of the cubes produced that hour differs from 25.4 mm, the supervisor will discard all of the cubes.

(a) Describe a Type II error in this setting.

(b) Identify one benefit and one drawback of changing the significance level to $\alpha = 0.10$.

(c) The data from a sample taken during 1 hour result in a 95% confidence interval of (25.39, 25.441). What conclusion should the supervisor make?

R11.4 Each day I am getting better in math A "subliminal" message is below our threshold of awareness but may nonetheless influence us. Can subliminal messages help students learn math? A group of 18 students who had failed the mathematics part of the City University of New York Skills Assessment Test agreed to participate in a study to find out. All received a daily subliminal message, flashed on a screen too rapidly to be consciously read. The treatment group of 10 students (assigned at random) was exposed to "Each day I am getting better in math." The control group of 8 students was exposed to a neutral message, "People are walking on the street." All 18 students participated in a summer program designed to improve their math skills, and all took the assessment test again at the end of the program. The following table gives data on the subjects' scores before and after the program.[24]

Treatment group			Control group		
Pretest	Posttest	Difference	Pretest	Posttest	Difference
18	24	6	18	29	11
18	25	7	24	29	5
21	33	12	20	24	4
18	29	11	18	26	8
18	33	15	24	38	14
20	36	16	22	27	5
23	34	11	15	22	7
23	36	13	19	31	12
21	34	13			
17	27	10			

(a) Explain why a two-sample t test and not a paired t test is the appropriate inference procedure in this setting.

(b) The following boxplots display the differences in pretest and posttest scores for the students in the control (C) and treatment (T) groups. Write a few sentences comparing the performance of these two groups.

TRM Full Solutions to Chapter 11 Review Exercises

The full solutions can be found by clicking on the link in the TE-Book, opening the TRFD, or downloading from the Teacher's Resources on the book's digital platform.

Answers to Chapter 11 Review Exercises

R11.1 (a) H_0: $\mu = 64.2$, H_a: $\mu \neq 64.2$, where μ = the true mean height of all female graduates from the large local high school this year. **(b)** The observed sample mean is $\bar{x} = 63.5$ inches, which does not equal 64.2 inches. **(c)** One-sample t test for a population mean. *Random:* Random sample of 48 female

graduates. *10%:* $n = 48 < 10\%$ of all female graduates from this large high school. *Normal/ Large Sample:* $n = 48 \geq 30$ **(d)** $t = -1.31$, df $= 47$, P-value $= 0.1966$ **(e)** P-value interpretation: Assuming that the true mean height of all female graduates is 64.2 inches, there is a 0.1966 probability of getting a sample mean as different as or more different than 63.5 inches (in either direction) by chance alone. *Conclusion:* Because the P-value of $0.1966 > \alpha = 0.05$, we fail to reject H_0. We do not have convincing evidence that the true mean height of all female graduates from this high school is different from the national average of 64.2 inches.

R11.2 S: H_0: $\mu = 22$, H_a: $\mu > 22$, where μ = the true mean amount of time (in seconds) it takes adults to read four paragraphs of text in the ornate font Gigi. We will use $\alpha = 0.05$. **P:** One-sample t test for μ. *Random:* We have a random sample of 24 adults. *10%:* The sample size (24) is less than 10% of the population of adults. *Normal/ Large Sample:* Because the sample size is small, we need to graph the sample data. The histogram shows that the distribution is roughly symmetric with no outliers, so using a t procedure is appropriate.

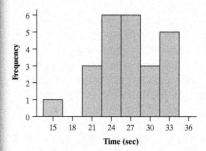

D: $\bar{x} = 26.496$, $s_x = 4.728$, $t = 4.66$, df $= 23$, and P-value $= 0.000054$. **C:** Because the P-value of $0.000054 < \alpha = 0.05$, we reject H_0. There is convincing evidence that the true mean amount of time it takes adults to read four paragraphs of text in the ornate font Gigi is greater than 22 seconds.

R11.3 (a) *Type II error:* Failing to find convincing evidence that the true mean width of the cubes differs from 25.4 mm, when the true mean width of the cubes differs from 25.4 mm. **(b)** This change will reduce the probability of making a Type II error. A drawback to this change is an increased chance of rejecting batches of cubes that are of the appropriate width. **(c)** Because $\mu = 25.4$ mm is one of the plausible values in this 95% confidence interval, the supervisor should not reject this batch as being significantly different from the intended mean width of 25.4 mm.

R11.4 (a) The difference data come from two groups in a randomized experiment. **(b)** The distribution of differences for the control group is slightly skewed to the right, while the distribution of differences for the treatment group is roughly symmetric. Neither distribution has any clear outliers. The center for the treatment group is greater than the center for the control group. The differences in the control group are more variable (based on the *IQR*) than the differences in the treatment group. Overall, it appears that students in the treatment group had bigger improvements, on average. **(c) S:** $H_0: \mu_1 - \mu_2 = 0$, $H_a: \mu_1 - \mu_2 > 0$, where μ_1 = true mean difference in test scores for students like these who get the treatment message and μ_2 = true mean difference in test scores for students like these who get the neutral message, using $\alpha = 0.05$. **P:** Two-sample t test for $\mu_1 - \mu_2$. *Random:* The students were assigned at random to the treatment and control groups. *Normal/Large Sample:* The boxplots show no strong skewness and no outliers. **D:** $\bar{x}_1 = 11.4$, $s_1 = 3.169$, $n_1 = 10$, $\bar{x}_2 = 8.25$, $s_2 = 3.69$, $n_2 = 8$; $t = 1.91$. Using df = 13.919, P-value = 0.0382; using df = 7, P-value = 0.0489. **C:** Because the P-value of $0.0382 > \alpha = 0.01$, we fail to reject H_0. There is not convincing evidence that the true mean difference in test scores for students like these who get the treatment message is greater than the true mean difference in test scores for students like these who get the neutral message. **(d)** We cannot generalize to all students who failed because our sample was not a random sample of all students who failed. It was a group of students who agreed to participate in the experiment.

R11.5 S: $H_0: \mu_{\text{diff}} = 0$, $H_a: \mu_{\text{diff}} > 0$, where μ_{diff} = true mean difference (Standing − Blocks) in 50-meter run time; $\alpha = 0.05$. **P:** Paired t test for μ_{diff}. *Random:* The order was randomized. *Normal/Large Sample:* The dotplot shows no strong skewness and no outliers.

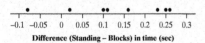

Difference (Standing – Blocks) in time (sec)

D: $\bar{x}_{\text{diff}} = 0.131$, $s_{\text{diff}} = 0.119$, and $n_{\text{diff}} = 8$; $t = 3.11$. Using df = 7, P-value = 0.0085. **C:** Because the P-value of $0.0085 < \alpha = 0.05$, we reject H_0. We have convincing evidence that the true mean difference (Standing − Blocks) in 50-meter run time for sprinters like these is greater than 0.

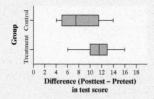

Group

(c) Do the data provide convincing evidence at the $\alpha = 0.01$ significance level that subliminal messages help students like the ones in this study learn math, on average?

(d) Can we generalize these results to the population of all students who failed the mathematics part of the City University of New York Skills Assessment Test? Why or why not?

R11.5 On your mark In track, sprinters typically use starting blocks because they think it will help them run a faster race. To test this belief, an experiment was designed where each sprinter on a track team ran a 50-meter dash two times, once using starting blocks and once with a standing start. The order of the two different types of starts was determined at random for each sprinter. The times (in seconds) for 8 different sprinters are shown in the table.

Sprinter	With blocks	Standing start
1	6.12	6.38
2	6.42	6.52
3	5.98	6.09
4	6.80	6.72
5	5.73	5.98
6	6.04	6.27
7	6.55	6.71
8	6.78	6.80

Do these data provide convincing evidence that sprinters like these run a faster race when using starting blocks, on average?

Chapter 11 AP® Statistics Practice Test

Section I: Multiple Choice *Select the best answer for each question.*

T11.1 Experiments on learning in animals sometimes measure how long it takes mice to find their way through a maze. The mean time is 18 seconds for one particular maze. A researcher thinks that a loud noise will cause the mice to complete the maze faster. She measures how long each of 10 mice takes with a noise stimulus. What are the null and alternative hypotheses for the appropriate significance test?

(a) $H_0: \mu = 18$ vs. $H_a: \mu \neq 18$

(b) $H_0: \mu = 18$ vs. $H_a: \mu < 18$

(c) $H_0: \mu = 18$ vs. $H_a: \mu > 18$

(d) $H_0: \mu < 18$ vs. $H_a: \mu = 18$

(e) $H_0: \mu \neq 18$ vs. $H_a: \mu = 18$

T11.2 You are thinking of conducting a one-sample t test about a population mean μ using a 0.05 significance level. Which of the following statements is correct?

(a) You should not carry out the test if the sample does not have a Normal distribution.

(b) You can safely carry out the test if there are no outliers, regardless of the sample size.

(c) You can carry out the test if a graph of the data shows no strong skewness, regardless of the sample size.

(d) You can carry out the test only if the population standard deviation is known.

(e) You can safely carry out the test if your sample size is at least 30.

TRM **Full Solutions to Chapter 11 AP® Statistics Practice Test**

The full solutions can be found by clicking on the link in the TE-Book, opening the TRFD, or downloading from the Teacher's Resources on the book's digital platform.

Answers to Chapter 10 AP® Statistics Practice Test

T11.1 b

T11.2 e

T11.3 A 95% confidence interval for μ based on $n = 15$ observations from a Normal population is $(-0.73, 1.92)$. If we use this confidence interval to test the hypothesis $H_0: \mu = 0$ against $H_a: \mu \neq 0$, which of the following is the most appropriate conclusion?

(a) Reject H_0 at the $\alpha = 0.05$ level of significance.

(b) Fail to reject H_0 at the $\alpha = 0.05$ level of significance.

(c) Reject H_0 at the $\alpha = 0.10$ level of significance.

(d) Fail to reject H_0 at the $\alpha = 0.10$ level of significance.

(e) We cannot perform the required test since we do not know the value of the standardized test statistic.

T11.4 Which of the following has the smallest probability?

(a) $P(t > 2)$ if t has 5 degrees of freedom.

(b) $P(t > 2)$ if t has 2 degrees of freedom.

(c) $P(z > 2)$ if z is a standard Normal random variable.

(d) $P(t < 2)$ if t has 5 degrees of freedom.

(e) $P(z < 2)$ if z is a standard Normal random variable.

T11.5 A significance test was performed to test $H_0: \mu = 2$ versus the alternative $H_a: \mu \neq 2$. A sample of size 28 produced a standardized test statistic of $t = 2.051$. Assuming all conditions for inference were met, which of the following intervals contains the P-value for this test?

(a) $0.01 < P < 0.02$

(b) $0.02 < P < 0.025$

(c) $0.025 < P < 0.05$

(d) $0.05 < P < 0.10$

(e) $P > 0.10$

T11.6 A study of road rage asked separate random samples of 596 men and 523 women about their behavior while driving. Based on their answers, each respondent was assigned a road rage score on a scale of 0 to 20. Are the conditions for performing a two-sample t test satisfied?

(a) Maybe; we have independent random samples, but we should look at the data to check Normality.

(b) No; road rage scores on a scale from 0 to 20 can't be Normal.

(c) No; we don't know the population standard deviations.

(d) Yes; the large sample sizes guarantee that the corresponding population distributions will be Normal.

(e) Yes; we have two independent random samples and large sample sizes.

Exercises T11.7 and T11.8 refer to the following setting. A researcher wished to compare the average amount of time spent in extracurricular activities by high school students in a suburban school district with the average time spent by students in a large city school district. The researcher obtained an SRS of 60 high school students in a large suburban school district and found the mean time spent in extracurricular activities per week to be 6 hours with a standard deviation of 3 hours. The researcher also obtained an independent SRS of 40 high school students in a large city school district and found the mean time spent in extracurricular activities per week to be 5 hours with a standard deviation of 2 hours. Suppose that the researcher decides to carry out a significance test of $H_0: \mu_{\text{suburban}} = \mu_{\text{city}}$ versus a two-sided alternative.

T11.7 Which is the correct standardized test statistic?

(a) $z = \dfrac{(6-5)-0}{\sqrt{\dfrac{3}{60} + \dfrac{2}{40}}}$

(b) $z = \dfrac{(6-5)-0}{\sqrt{\dfrac{3^2}{60} + \dfrac{2^2}{40}}}$

(c) $t = \dfrac{(6-5)-0}{\dfrac{3}{\sqrt{60}} + \dfrac{2}{\sqrt{40}}}$

(d) $t = \dfrac{(6-5)-0}{\sqrt{\dfrac{3}{60} + \dfrac{2}{40}}}$

(e) $t = \dfrac{(6-5)-0}{\sqrt{\dfrac{3^2}{60} + \dfrac{2^2}{40}}}$

T11.8 The P-value for the test is 0.048. A correct conclusion is to

(a) fail to reject H_0 because $0.048 < \alpha = 0.05$. There is convincing evidence of a difference in the average time spent on extracurricular activities by students in the suburban and city school districts.

(b) fail to reject H_0 because $0.048 < \alpha = 0.05$. There is not convincing evidence of a difference in the average time spent on extracurricular activities by students in the suburban and city school districts.

(c) fail to reject H_0 because $0.048 < \alpha = 0.05$. There is convincing evidence that the average time spent on extracurricular activities by students in the suburban and city school districts is the same.

(d) reject H_0 because $0.048 < \alpha = 0.05$. There is not convincing evidence of a difference in the average time spent on extracurricular activities by students in the suburban and city school districts.

(e) reject H_0 because $0.048 < \alpha = 0.05$. There is convincing evidence of a difference in the average time spent on extracurricular activities by students in the suburban and city school districts.

T11.3 b

T11.4 c

T11.5 d

T11.6 e

T11.7 e

T11.8 e

T11.9 c

T11.10 a

T11.11 (a) We want to test $H_0: \mu = \$158$, $H_a: \mu \neq \$158$, where $\mu =$ the true mean amount spent on food by households in this city. We will perform the test at the $\alpha = 0.05$ significance level. **(b)** The Normal/Large sample condition is met because $n = 50 \geq 30$. **(c)** Assuming that the true mean amount of money spent on food per household in this city is $158, there is a 0.1283 probability of getting a sample mean as different as or more different than $165 (in either direction) by chance alone. *Conclusion:* Because the *P*-value of $0.128 > \alpha = 0.05$, we fail to reject H_0. We do not have convincing evidence that the true mean amount spent on food per household in this city is different from the national average of $158.

T11.12 (a) S: We want to test $H_0: \mu_1 - \mu_2 = 0$, $H_a: \mu_1 - \mu_2 > 0$, where $\mu_1 =$ the true mean number of "errors" found when the author is believed to be a non-native English speaker and $\mu_2 =$ the true mean number of "errors" found when nothing is said about the author using $\alpha = 0.05$. **P:** Two-sample *t* test for $\mu_1 - \mu_2$. *Random:* The volunteers were assigned at random to the two groups. *Normal/Large Sample:* $n_1 = 30 \geq 30$ and $n_2 = 30 \geq 30$. **D:** $\bar{x}_1 = 3.15$, $s_1 = 2.58$, $n_1 = 30$, $\bar{x}_2 = 0.85$, $s_2 = 1.09$, and $n_2 = 30$. 2-SampTTest gives $t = 4.498$ and *P*-value ≈ 0 using df $= 39.033$. **D:** Because the *P*-value of approximately $0 < \alpha = 0.05$, we reject H_0. There is convincing evidence that the true mean number of "errors" found by volunteers like these who believe the author is a non-native English speaker is greater than the true mean number of "errors" found by volunteers like these when nothing is said about the author. **(b)** Based on this conclusion, it is possible that we made a Type I error, meaning we may have found convincing evidence for the alternative hypothesis when the null hypothesis was actually true.

T11.9 Are TV commercials louder than their surrounding programs? To find out, researchers collected data on 50 randomly selected commercials in a given week. With the television's volume at a fixed setting, they measured the maximum loudness of each commercial and the maximum loudness in the first 30 seconds of regular programming that followed. Assuming conditions for inference are met, the most appropriate method for answering the question of interest is

(a) a two-sample *t* test for a difference in means.

(b) a two-sample *t* interval for a difference in means.

(c) a paired *t* test for a mean difference.

(d) a paired *t* interval for a mean difference.

(e) a two-sample *z* test for a difference in proportions.

T11.10 Researchers want to evaluate the effect of a natural product on reducing blood pressure. They plan to carry out a randomized experiment to compare the mean reduction in blood pressure of a treatment (natural product) group and a placebo group. Then they will use the data to perform a test of $H_0: \mu_T - \mu_P = 0$ versus $H_a: \mu_T - \mu_P > 0$, where $\mu_T =$ the true mean reduction in blood pressure when taking the natural product and $\mu_P =$ the true mean reduction in blood pressure when taking a placebo for subjects like the ones in the experiment. The researchers would like to detect whether the natural product reduces blood pressure by at least 7 points more, on average, than the placebo. If groups of size 50 are used in the experiment, a two-sample *t* test using $\alpha = 0.01$ will have a power of 80% to detect a 7-point difference in mean blood pressure reduction. If the researchers want to be able to detect a 5-point difference instead, then the power of the test

(a) would be less than 80%.

(b) would be greater than 80%.

(c) would still be 80%.

(d) could be either less than or greater than 80%.

(e) would vary depending on the values of μ_T and μ_P.

Section II: Free Response *Show all your work. Indicate clearly the methods you use, because you will be graded on the correctness of your methods as well as on the accuracy and completeness of your results and explanations.*

T11.11 A government report says that the average amount of money spent per U.S. household per week on food is about $158. A random sample of 50 households in a small city is selected, and their weekly spending on food is recorded. The sample data have a mean of $165 and a standard deviation of $32. Is there convincing evidence that the mean weekly spending on food in this city differs from the national figure of $158?

(a) State appropriate hypotheses for performing a significance test in this setting. Be sure to define the parameter of interest.

(b) The distribution of household spending in this small city is heavily skewed to the right. Explain why the Normal/Large Sample condition is met in this case.

(c) The *P*-value of the test is 0.128. Interpret this value. What conclusion would you make?

T11.12 As a non-native English speaker, Sanda is convinced that people find more grammar and spelling mistakes in essays when they think the writer is a non-native English speaker. To test this, she randomly sorts a group of 60 volunteers into two groups of 30. Both groups are given the same paragraph to read. One group is told that the author of the paragraph is someone whose native language is not English. The other group is told nothing about the author. The subjects are asked to count the number of spelling and grammar mistakes in the paragraph. While the two groups found about the same number of real mistakes in the passage, the number of things that were incorrectly identified as mistakes was more interesting. Some numerical summaries of the data follow.

Group	n	Mean	SD
Non-native	30	3.15	2.58
No info	30	0.85	1.09

(a) Is there convincing evidence at the $\alpha = 0.05$ significance level that the mean number of "mistakes" found will be greater when people like these volunteers are told that the author is a non-native speaker?

(b) Based on your conclusion in part (a), which type of error—a Type I error or a Type II error—could you have made? Explain your answer.

T11.13 "I can't get through my day without coffee" is a common statement from many college students. They assume that the benefits of coffee include staying awake during lectures and remaining more alert during exams and tests. Students in a statistics class designed an experiment to measure memory retention with and without drinking a cup of coffee before a test. This experiment took place on two different days in the same week (Monday and Wednesday). Ten students were used. Each student received no coffee or one cup of coffee 1 hour before the test on a particular day. The test consisted of a series of words flashed on a screen, after which the student had to write down as many of the words as possible. On the other day, each student received a different amount of coffee (none or one cup).

(a) One of the researchers suggested that all the subjects in the experiment drink no coffee before Monday's test and one cup of coffee before Wednesday's test. Explain to the researcher why this is a bad idea *and* suggest a better method of deciding when each subject receives the two treatments.

(b) The researchers actually used the better method of deciding when each subject receives the two treatments that you identified in part (a). For each subject, the number of words recalled when drinking no coffee and when drinking one cup of coffee is recorded in the table. Carry out an appropriate test to determine whether there is convincing evidence that drinking coffee improves memory, on average, for students like the ones in this study.

Student	No cup	One cup
1	24	25
2	30	31
3	22	23
4	24	24
5	26	27
6	23	25
7	26	28
8	20	20
9	27	27
10	28	30

Chapter 11 Project   Which Costs More: Diesel or Unleaded?

When buying or leasing a new car, one of the factors that customers consider is the type of fuel it uses. Some people prefer vehicles that use diesel fuel, while others favor vehicles that use regular unleaded gasoline. Which of these two types of fuel costs more at the pump, on average? Does the answer to this question depend on the state where you live?

Researchers collected data on the price per gallon (in dollars) of diesel fuel and regular unleaded gasoline from a random sample of gas stations in six states: Colorado, Illinois, Indiana, Kansas, Missouri, and Ohio. The file gas prices ch 11 project.xls, which can be accessed from the book's website at highschool.bfwpub.com/updatedtps6e, contains data from a total of 82 gas stations. Download the file to a computer for further analysis using the application specified by your teacher. Use the file provided to answer the following questions.

1. Start by calculating the difference (Diesel − Unleaded) in gas prices for all 82 stations, and store these values in a new column titled Difference.

2. Make a graph to display the distribution of difference in gas prices. Describe the shape, center, and variability of the distribution. Are there any outliers?

3. Construct and interpret a 95% confidence interval for the true mean difference. Does the interval provide convincing evidence of a difference in the mean price per gallon of diesel fuel and regular unleaded gasoline?

4. Make a graph that compares the difference (Diesel − Unleaded) in gas prices for the six states from which the data were collected. Write a few sentences comparing the distributions.

5. We might expect the mean difference (Diesel − Unleaded) in gas prices to be the same for adjacent states. Choose two adjacent states from the data set. Then carry out an appropriate test to see if the data provide convincing evidence to contradict this expectation.

T11.13 (a) Students may improve from Monday to Wednesday just because they have already done the task once. Then we wouldn't know if the experience with the test or the caffeine is the cause of the difference in scores. A better way to run the experiment would be to randomly assign half the students to receive 1 cup of coffee on Monday and the other half to get no coffee on Monday. Then have the opposite treatment given to each person on Wednesday. **(b)** STATE: H_0: $\mu_{diff} = 0$, H_a: $\mu_{diff} < 0$, where $\mu_{diff} =$ the true mean difference (No coffee − Coffee) in the number of words recalled by students like the ones in this study who received no coffee and received 1 cup; $\alpha = 0.05$. PLAN: Paired *t* test for μ_{diff}. *Random:* The treatments were assigned in a random order. *Normal/Large Sample:* The dotplot shows no strong skewness and no outliers.

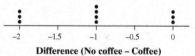

Difference (No coffee − Coffee) in number of words recalled

DO: $\bar{x}_{diff} = -1$, $s_{diff} = 0.816$, and $n_{diff} = 10$; $t = -3.87$; df = 9; *P*-value = 0.0019. CONCLUDE: Because the *P*-value of 0.0019 $< \alpha = 0.05$, we reject H_0. We have convincing evidence that the true mean difference (No coffee − Coffee) in word recall for students like the ones in this study is less than 0.

Answers to Cumulative AP® Practice Test 3

AP3.1 e

AP3.2 b

AP3.3 d

AP3.4 c

AP3.5 d

AP3.6 d

AP3.7 c

Cumulative AP® Practice Test 3

Section I: Multiple Choice *Choose the best answer.*

AP3.1 Suppose the probability that a softball player gets a hit in any single at-bat is 0.300. Assuming that her chance of getting a hit on a particular time at bat is independent of her other times at bat, what is the probability that she will not get a hit until her fourth time at bat in a game?

(a) $\binom{4}{3}(0.3)^1(0.7)^3$

(b) $\binom{4}{3}(0.3)^3(0.7)^1$

(c) $\binom{4}{1}(0.3)^3(0.7)^1$

(d) $(0.3)^3(0.7)^1$

(e) $(0.3)^1(0.7)^3$

AP3.2 A survey asked a random sample of U.S. adults about their political party affiliation and how long they thought they would survive compared to most people in their community if an apocalyptic disaster were to strike. The responses are summarized in the following two-way table.

		Political affiliation			
		Democrat	Independent	Republican	Total
Expected survival length	Longer	79	134	101	314
	About as long	169	163	84	416
	Not as long	43	49	23	115
	Not sure	69	58	26	153
	Total	360	404	234	998

Suppose we select one of the survey respondents at random. Which of the following probabilities is the largest?

(a) P(Independent and Longer)

(b) P(Independent or Not as long)

(c) P(Democrat | Not as long)

(d) P(About as long | Democrat)

(e) P(About as long)

AP3.3 *Sports Illustrated* planned to ask a random sample of Division I college athletes, "Do you believe performance-enhancing drugs are a problem in college sports?" Which of the following is the *smallest* number of athletes that must be interviewed to estimate the true proportion who believe performance-enhancing drugs are a problem within ±2% with 90% confidence?

(a) 17

(b) 21

(c) 1680

(d) 1702

(e) 2401

AP3.4 The distribution of grade point averages (GPAs) for a certain college is approximately Normal with a mean of 2.5 and a standard deviation of 0.6. The minimum possible GPA is 0.0 and the maximum possible GPA is 4.33. Any student with a GPA less than 1.0 is put on probation, while any student with a GPA of 3.5 or higher is on the dean's list. About what percent of students at the college are on probation or on the dean's list?

(a) 0.6

(b) 4.7

(c) 5.4

(d) 94.6

(e) 95.3

AP3.5 Which of the following will increase the power of a significance test?

(a) Increase the Type II error probability.

(b) Decrease the sample size.

(c) Reject the null hypothesis only if the *P*-value is less than the significance level.

(d) Increase the significance level α.

(e) Select a value for the alternative hypothesis closer to the value of the null hypothesis.

AP3.6 You can find some interesting polls online. Anyone can become part of the sample just by clicking on a response. One such poll asked, "Do you prefer watching first-run movies at a movie theater, or waiting until they are available to watch at home or on a digital device?" In all, 8896 people responded, with only 12% (1118 people) saying they preferred theaters. You can conclude that

(a) American adults strongly prefer watching movies at home or on their digital devices.

(b) the high nonresponse rate prevents us from drawing a conclusion.

(c) the sample is too small to draw any conclusion.

(d) the poll uses voluntary response, so the results tell us little about all American adults.

(e) American adults strongly prefer seeing movies at a movie theater.

AP3.7 A certain candy has different wrappers for various holidays. During Holiday 1, the candy wrappers are 30% silver, 30% red, and 40% pink. During Holiday 2, the wrappers are 50% silver and 50% blue. In separate random samples of 40 candies on Holiday 1 and 40 candies on Holiday 2, what are the mean and standard deviation of the total number of silver wrappers?

(a) 32, 18.4

(b) 32, 6.06

(c) 32, 4.29

(d) 80, 18.4

(e) 80, 4.29

AP3.8 A beef rancher randomly sampled 42 cattle from her large herd to obtain a 95% confidence interval for the mean weight (in pounds) of the cattle in the herd. The interval obtained was (1010, 1321). If the

rancher had used a 98% confidence interval instead, the interval would have been

(a) wider with less precision than the original estimate.

(b) wider with more precision than the original estimate.

(c) wider with the same precision as the original estimate.

(d) narrower with less precision than the original estimate.

(e) narrower with more precision than the original estimate.

AP3.9 School A has 400 students and School B has 2700 students. A local newspaper wants to compare the distributions of SAT scores for the two schools. Which of the following would be the most useful for making this comparison?

(a) Back-to-back stemplots for A and B

(b) A scatterplot of A versus B

(c) Two dotplots for A and B drawn on the same scale

(d) Two relative frequency histograms of A and B drawn on the same scale

(e) Two bar graphs for A and B drawn on the same scale

AP3.10 Let X represent the outcome when a fair six-sided die is rolled. For this random variable, $\mu_X = 3.5$ and $\sigma_X = 1.71$. If the die is rolled 100 times, what is the approximate probability that the sum is at least 375?

(a) 0.0000   (b) 0.0017   (c) 0.0721

(d) 0.4420   (e) 0.9279

AP3.11 An agricultural station is testing the yields for six different varieties of seed corn. The station has four large fields available, located in four distinctly different parts of the county. The agricultural researchers consider the climatic and soil conditions in the four parts of the county as being quite different, but are reasonably confident that the conditions within each field are fairly similar throughout. The researchers divide each field into six sections and then randomly assign one variety of corn seed to each section in that field. This procedure is done for each field. At the end of the growing season, the corn will be harvested, and the yield (measured in tons per acre) will be compared. Which one of the following statements about the design is correct?

(a) This is an observational study because the researchers are watching the corn grow.

(b) This a randomized block design with fields as blocks and seed types as treatments.

(c) This is a randomized block design with seed types as blocks and fields as treatments.

(d) This is a completely randomized design because the six seed types were randomly assigned to the four fields.

(e) This is a completely randomized design with 24 treatments—6 seed types and 4 fields.

AP3.12 The correlation between the heights of fathers and the heights of their grownup sons, both measured in inches, is $r = 0.52$. If fathers' heights were measured in feet instead, the correlation between heights of fathers and heights of sons would be

(a) much smaller than 0.52.

(b) slightly smaller than 0.52.

(c) unchanged; equal to 0.52.

(d) slightly larger than 0.52.

(e) much larger than 0.52.

AP3.13 A random sample of 200 New York State voters included 88 Republicans, while a random sample of 300 California voters produced 141 Republicans. Which of the following represents the 95% confidence interval for the true difference in the proportion of Republicans in New York State and California?

(a) $(0.44 - 0.47) \pm 1.96 \left(\dfrac{(0.44)(0.56) + (0.47)(0.53)}{\sqrt{200 + 300}} \right)$

(b) $(0.44 - 0.47) \pm 1.96 \left(\dfrac{(0.44)(0.56)}{\sqrt{200}} + \dfrac{(0.47)(0.53)}{\sqrt{300}} \right)$

(c) $(0.44 - 0.47) \pm 1.96 \sqrt{\dfrac{(0.44)(0.56)}{200} + \dfrac{(0.47)(0.53)}{300}}$

(d) $(0.44 - 0.47) \pm 1.96 \sqrt{\dfrac{(0.44)(0.56) + (0.47)(0.53)}{200 + 300}}$

(e) $(0.44 - 0.47) \pm 1.96 \sqrt{\dfrac{(0.45)(0.55)}{200} + \dfrac{(0.45)(0.55)}{300}}$

AP3.14 Which of the following is *not* a property of a binomial setting?

(a) Outcomes of different trials are independent.

(b) The chance process consists of a fixed number of trials, n.

(c) The probability of success is the same for each trial.

(d) Trials are repeated until a success occurs.

(e) Each trial can result in either a success or a failure.

AP3.15 Mrs. Woods and Mrs. Bryan are avid vegetable gardeners. They use different fertilizers, and each claims that hers is the best fertilizer to use when growing tomatoes. Both agree to do a study using the weight of their tomatoes as the response variable. Each planted the same varieties of tomatoes on the same day and fertilized the plants on the same schedule throughout the growing season. At harvest time, each randomly selects 15 tomatoes from her garden and weighs them. After performing a two-sample t test on the difference in mean weights of tomatoes, they get $t = 5.24$ and $P = 0.0008$. Can the gardener with the larger

AP3.8 a

AP3.9 d

AP3.10 c

AP3.11 b

AP3.12 c

AP3.13 c

AP3.14 d

AP3.15 d

AP3.16 c

AP3.17 b

AP3.18 b

AP3.19 e

AP3.20 c

mean claim that her fertilizer caused her tomatoes to be heavier?

(a) Yes, because a different fertilizer was used on each garden.

(b) Yes, because random samples were taken from each garden.

(c) Yes, because the *P*-value is so small.

(d) No, because the condition of the soil in the two gardens is a potential confounding variable.

(e) No, because $15 < 30$.

AP3.16 The Environmental Protection Agency (EPA) is charged with monitoring industrial emissions that pollute the atmosphere and water. So long as emission levels stay within specified guidelines, the EPA does not take action against the polluter. If the polluter violates regulations, the offender can be fined, forced to clean up the problem, or possibly closed. Suppose that for a particular industry the acceptable emission level has been set at no more than 5 parts per million (5 ppm). The null and alternative hypotheses are $H_0: \mu = 5$ versus $H_a: \mu > 5$. Which of the following describes a Type II error?

(a) The EPA fails to find convincing evidence that emissions exceed acceptable limits when, in fact, they are within acceptable limits.

(b) The EPA finds convincing evidence that emissions exceed acceptable limits when, in fact, they are within acceptable limits.

(c) The EPA fails to find convincing evidence that emissions exceed acceptable limits when, in fact, they do exceed acceptable limits.

(d) The EPA finds convincing evidence that emissions exceed acceptable limits when, in fact, they do exceed acceptable limits.

(e) The EPA fails to find convincing evidence that emissions exceed acceptable limits when, in fact, there is convincing evidence.

AP3.17 Which of the following statements is *false*?

(a) A measure of center alone does not completely summarize a distribution of quantitative data.

(b) If the original measurements are in inches, converting them to centimeters will not change the mean or standard deviation.

(c) One of the disadvantages of a histogram is that it doesn't show each data value.

(d) In a quantitative data set, adding a new data value equal to the mean will decrease the standard deviation.

(e) If a distribution of quantitative data is strongly skewed, the median and interquartile range should be reported rather than the mean and standard deviation.

AP3.18 A 96% confidence interval for the proportion of the labor force that is unemployed in a certain city is (0.07, 0.10). Which of the following statements is true?

(a) The probability is 0.96 that between 7% and 10% of the labor force is unemployed.

(b) About 96% of the intervals constructed by this method will contain the true proportion of the labor force that is unemployed in the city.

(c) In repeated samples of the same size, there is a 96% chance that the sample proportion will fall between 0.07 and 0.10.

(d) The true rate of unemployment in the labor force lies within this interval 96% of the time.

(e) Between 7% and 10% of the labor force is unemployed 96% of the time.

AP3.19 A large toy company introduces many new toys to its product line each year. The company wants to predict the demand as measured by *y*, first-year sales (in millions of dollars) using *x*, awareness of the product (as measured by the percent of customers who had heard of the product by the end of the second month after its introduction). A random sample of 65 new products was taken, and a correlation of 0.96 was computed. Which of the following is true?

(a) The least-squares regression line accurately predicts first-year sales 96% of the time.

(b) About 92% of the time, the percent of people who have heard of the product by the end of the second month will correctly predict first-year sales.

(c) About 92% of first-year sales can be accounted for by the percent of people who have heard of the product by the end of the second month.

(d) For each increase of 1% in awareness of the new product, the predicted sales will go up by 0.96 million dollars.

(e) About 92% of the variation in first-year sales can be accounted for by the least-squares regression line with the percent of people who have heard of the product by the end of the second month as the explanatory variable.

AP3.20 Final grades for a class are approximately Normally distributed with a mean of 76 and a standard deviation of 8. A professor says that the top 10% of the class will receive an A, the next 20% a B, the next 40% a C, the next 20% a D, and the bottom 10% an F. What is the approximate maximum grade a student could attain and still receive an F for the course?

(a) 70   (b) 69.27   (c) 65.75

(d) 62.84   (e) 57

AP3.21 National Park rangers keep data on the bears that inhabit their park. Here is a histogram of the weights of 143 bears measured in a recent year:

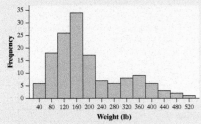

Weight (lb)

Which of the following statements is correct?

(a) The median will lie in the interval (140, 180), and the mean will lie in the interval (180, 220).

(b) The median will lie in the interval (140, 180), and the mean will lie in the interval (260, 300).

(c) The median will lie in the interval (100, 140), and the mean will lie in the interval (180, 220).

(d) The mean will lie in the interval (140, 180), and the median will lie in the interval (260, 300).

(e) The mean will lie in the interval (100, 140), and the median will lie in the interval (180, 220).

AP3.22 A random sample of size n will be selected from a population, and the proportion $\hat{p}$ of those in the sample who have a Facebook page will be calculated. How would the margin of error for a 95% confidence interval be affected if the sample size were increased from 50 to 200 and the sample proportion of people who have a Facebook page is unchanged?

(a) It remains the same. (d) It is divided by 2.

(b) It is multiplied by 2. (e) It is divided by 4.

(c) It is multiplied by 4.

AP3.23 A scatterplot and a least-squares regression line are shown in the figure. What effect does point P have on the slope of the regression line and the correlation?

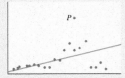

(a) Point P increases the slope and increases the correlation.

(b) Point P increases the slope and decreases the correlation.

(c) Point P decreases the slope and decreases the correlation.

(d) Point P decreases the slope and increases the correlation.

(e) Point P increases the slope but does not change the correlation.

AP3.24 The following dotplots show the average high temperatures (in degrees Celsius) for a sample of tourist cities from around the world. Both the January and July average high temperatures are shown. What is one statement that can be made with certainty from an analysis of the graphical display?

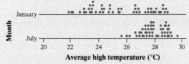

Average high temperature (°C)

(a) Every city has a larger average high temperature in July than in January.

(b) The distribution of temperatures in July is skewed right, while the distribution of temperatures in January is skewed left.

(c) The median average high temperature for January is higher than the median average high temperature for July.

(d) There appear to be outliers in the average high temperatures for January and July.

(e) There is more variability in average high temperatures in January than in July.

AP3.25 Suppose the null and alternative hypotheses for a significance test are defined as
$$H_0: \mu = 40$$
$$H_a: \mu < 40$$
Which of the following specific values for H_a will give the highest power?

(a) $\mu = 38$ (b) $\mu = 39$ (c) $\mu = 41$

(d) $\mu = 42$ (e) $\mu = 43$

AP3.26 A large university is considering the establishment of a schoolwide recycling program. To gauge interest in the program by means of a questionnaire, the university takes separate random samples of undergraduate students, graduate students, faculty, and staff. This is an example of what type of sampling design?

(a) Simple random sample

(b) Stratified random sample

(c) Convenience sample

(d) Cluster sample

(e) Systematic random sample

AP3.27 Suppose the true proportion of people who use public transportation to get to work in the Washington, D.C., area is 0.45. In a simple random

AP3.21 a

AP3.22 d

AP3.23 b

AP3.24 e

AP3.25 a

AP3.26 b

AP3.27 c

AP3.28 d

AP3.29 a

AP3.30 b

AP3.31 STATE: H_0: $\mu_{\text{diff}} = 0$, H_a: $\mu_{\text{diff}} < 0$, where μ_{diff} = true mean change in weight (After – Before) in pounds for people like these who follow a 5-week crash diet; $\alpha = 0.05$. PLAN: Paired *t* test for μ_{diff}. *Random:* Random sample of dieters. *10%:* 15 < 10% of all dieters. *Normal/Large Sample:* The boxplot shows no strong skewness and no outliers.

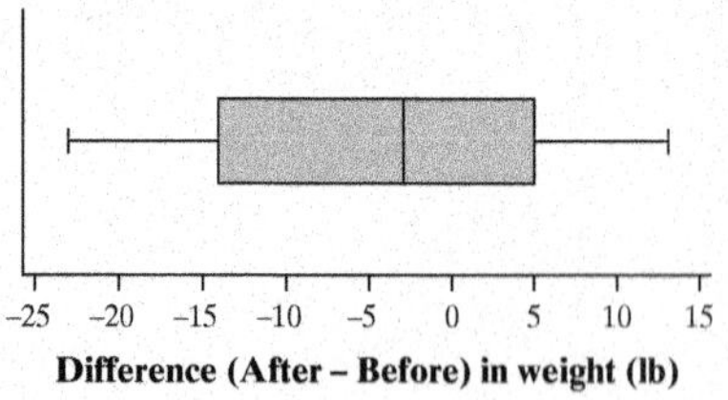

DO: $\bar{x}_{\text{diff}} = -3.6$, $s_{\text{diff}} = 11.53$, $n_{\text{diff}} = 15$; $t = -1.21$; df = 14; *P*-value = 0.1232. CONCLUDE: Because the *P*-value of $0.1232 > \alpha = 0.05$, we fail to reject H_0. We do not have convincing evidence that the true mean change in weight (After – Before) for people like these who follow a 5-week crash diet is less than 0.

AP3.32 (a) This is an observational study; no treatments were imposed. **(b)** H_0: $p_1 - p_2 = 0$, H_a: $p_1 - p_2 < 0$, where p_1 = the true proportion of VLBW babies who graduate from high school by age 20 and p_2 = the true proportion of non-VLBW babies who graduate from high school by age 20, using $\alpha = 0.05$. PLAN: Two-sample *z* test for $p_1 - p_2$. *Random:* Independent random samples. *10%:* $n_1 = 242$ < 10% of all VLBW babies, and $n_2 = 233$ < 10% of all non-VLBW babies. *Large Counts:* 189.486, 52.514, 182.439, and 50.561 are ≥ 10. DO: $z = -2.34$; *P*-value = 0.0095. CONCLUDE: Because the *P*-value of $0.0095 < \alpha = 0.05$, we reject H_0. We have convincing evidence that the true proportion of VLBW babies who graduate from high school by age 20 is less than the true proportion of non-VLBW babies who graduate from high school by age 20.

sample of 250 people who work in Washington, about how far do you expect the sample proportion to be from the true proportion?

(a) 0.4975 (b) 0.2475 (c) 0.0315
(d) 0.0009 (e) 0

Questions 28 and 29 refer to the following setting. According to sleep researchers, if you are between the ages of 12 and 18 years old, you need 9 hours of sleep to function well. A simple random sample of 28 students was chosen from a large high school, and these students were asked how much sleep they got the previous night. The mean of the responses was 7.9 hours with a standard deviation of 2.1 hours.

AP3.28 If we are interested in whether students at this high school are getting too little sleep, on average, which of the following represents the appropriate null and alternative hypotheses?

(a) $H_0: \mu = 7.9$ and $H_a: \mu < 7.9$
(b) $H_0: \mu = 7.9$ and $H_a: \mu \neq 7.9$
(c) $H_0: \mu = 9$ and $H_a: \mu \neq 9$
(d) $H_0: \mu = 9$ and $H_a: \mu < 9$
(e) $H_0: \mu \leq 9$ and $H_a: \mu \geq 9$

AP3.29 Which of the following is the standardized test statistic for the hypothesis test?

(a) $t = \dfrac{7.9 - 9}{\frac{2.1}{\sqrt{28}}}$
(b) $t = \dfrac{9 - 7.9}{\frac{2.1}{\sqrt{28}}}$
(c) $t = \dfrac{7.9 - 9}{\sqrt{\frac{2.1}{28}}}$
(d) $t = \dfrac{7.9 - 9}{\frac{2.1}{\sqrt{27}}}$
(e) $t = \dfrac{9 - 7.9}{\frac{2.1}{\sqrt{27}}}$

AP3.30 Shortly before the 2012 presidential election, a survey was taken by the school newspaper at a very large state university. Randomly selected students were asked, "Whom do you plan to vote for in the upcoming presidential election?" Here is a two-way table of the responses by political persuasion for 1850 students:

		Political persuasion			
		Democrat	Republican	Independent	Total
Candidate of choice	Obama	925	78	26	1029
	Romney	78	598	19	695
	Other	2	8	11	21
	Undecided	32	28	45	105
	Total	1037	712	101	1850

Which of the following statements about these data is true?

(a) The percent of Republicans among the respondents is 41%.
(b) The marginal relative frequencies for the variable choice of candidate are given by Obama: 55.6%; Romney: 37.6%; Other: 1.1%; Undecided: 5.7%.
(c) About 11.2% of Democrats reported that they planned to vote for Romney.
(d) About 44.6% of those who are undecided are Independents.
(e) The distribution of political persuasion among those for whom Romney is the candidate of choice is Democrat: 7.5%; Republican: 84.0%; Independent: 18.8%.

Section II: Free Response *Show all your work. Indicate clearly the methods you use, because you will be graded on the correctness of your methods as well as on the accuracy and completeness of your results and explanations.*

AP3.31 A researcher wants to determine whether or not a 5-week crash diet is effective over a long period of time. A random sample of 15 five-week crash dieters is selected. Each person's weight (in pounds) is recorded before starting the diet and 1 year after it is concluded. Do the data provide convincing evidence that 5-week crash dieters weigh less, on average, 1 year after finishing the diet?

Dieter	1	2	3	4	5	6	7	8
Before	158	185	176	172	164	234	258	200
After	163	182	188	150	161	220	235	191

Dieter	9	10	11	12	13	14	15
Before	228	246	198	221	236	255	231
After	228	237	209	220	222	268	234

AP3.32 Starting in the 1970s, medical technology has enabled babies with very low birth weight (VLBW, less than 1500 grams, or about 3.3 pounds) to survive without major handicaps. It was noticed that these children nonetheless had difficulties in school and as adults. A long-term study has followed 242 randomly selected VLBW babies to age 20 years, along with a control group of 233 randomly selected babies from the same population who had normal birth weight.[25]

(a) Is this an experiment or an observational study? Why?
(b) At age 20, 179 of the VLBW group and 193 of the control group had graduated from high school. Do these data provide convincing evidence at the

$\alpha = 0.05$ significance level that the graduation rate among VLBW babies is less than for normal-birth-weight babies?

AP3.33 A nuclear power plant releases water into a nearby lake every afternoon at 4:51 P.M. Environmental researchers are concerned that fish are being driven away from the area around the plant. They believe that the temperature of the water discharged may be a factor. The scatterplot shows the temperature of the water (in degrees Celsius) released by the plant and the measured distance (in meters) from the outflow pipe of the plant to the nearest fish found in the water on eight randomly chosen afternoons.

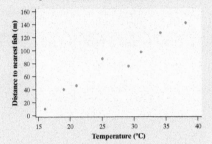

Here are computer output from a least-squares regression analysis on these data and a residual plot:

Predictor	Coef	SE Coef	T	P
Constant	−73.64	15.48	−4.76	0.003
Temperature	5.7188	0.5612	10.19	0.000

$S = 11.4175$ R-Sq = 94.5%   R-Sq(adj) = 93.6%

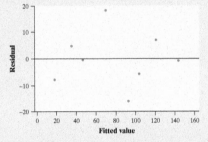

(a) Explain why a linear model is appropriate for describing the relationship between temperature and distance to the nearest fish.

(b) Write the equation of the least-squares regression line. Define any variables you use.
(c) Interpret the slope of the regression line.
(d) Compute the residual for the point (29, 78). Interpret this residual.

AP3.34 The Candy Shoppe assembles gift boxes that contain 8 chocolate truffles and 2 handmade caramel nougats. The truffles have a mean weight of 2 ounces with a standard deviation of 0.5 ounce, and the nougats have a mean weight of 4 ounces with a standard deviation of 1 ounce. The empty boxes have mean weight 3 ounces with a standard deviation of 0.2 ounce.

(a) Assuming that the weights of the truffles, nougats, and boxes are independent, what are the mean and standard deviation of the weight of a box of candy?
(b) Assuming that the weights of the truffles, nougats, and boxes are approximately Normally distributed, what is the probability that a randomly selected box of candy will weigh more than 30 ounces?
(c) If five gift boxes are randomly selected, what is the probability that at least one of them will weigh more than 30 ounces?
(d) If five gift boxes are randomly selected, what is the probability that the mean weight of the five boxes will be more than 30 ounces?

AP3.35 An investor is comparing two stocks, A and B. She wants to know if over the long run, there is a significant difference in the return on investment as measured by the percent increase or decrease in the price of the stock from its date of purchase. The investor takes a random sample of 50 annualized daily returns over the past 5 years for each stock. The data are summarized in the table.

Stock	Mean return	Standard deviation
A	11.8%	12.9%
B	7.1%	9.6%

(a) The investor uses the data to perform a two-sample t test of $H_0: \mu_A - \mu_B = 0$ versus $H_a: \mu_A - \mu_B \neq 0$, where μ_A = the true mean annualized daily return for Stock A and μ_B = the true mean annualized daily return for Stock B. The resulting P-value is 0.042. Interpret this value in context. What conclusion would you make?
(b) The investor believes that although the return on investment for Stock A usually exceeds that of Stock B, Stock A represents a riskier investment, where the risk is measured by the price volatility of the stock. The sample variance s_x^2 is a statistical measure of the price

(i) $z = \dfrac{30 - 27}{0.899} = 3.34$

$P(z > 3.34) = 0.0004$

(ii) $P(\overline{W} > 30)$ = normalcdf(lower: 30, upper: 1000, mean: 27, SD: 0.899) = 0.0004 There is a 0.0004 probability of randomly selecting 5 boxes that have a mean weight of more than 30 ounces.

AP3.35 (a) Assuming that the true difference in the mean annualized daily return for Stock A and Stock B is 0, there is a 0.042 probability that we would observe a difference in sample means of 4.7 or greater by chance alone. CONCLUDE: Because the P-value of 0.042 < α = 0.05, we reject H_0. We have convincing evidence that the true mean annualized return for Stock A is different from the true mean annualized return for Stock B.

AP3.33 (a) Because there is no leftover curved pattern in the residual plot, the residuals look randomly scattered around the residual = 0 line. **(b)** $\hat{y} = -73.64 + 5.7188x$, where $\hat{y}$ = predicted distance and x = temperature (°C). **(c)** The predicted distance from the nearest fish to the outflow pipe increases by 5.7188 meters for each additional 1-degree-Celsius increase in water discharge temperature. **(d)** $\hat{y} = -73.64 + 5.7188(29) = 92.21$ meters; the residual = 78 − 92.21 = −14.21 meters. The actual distance from the nearest fish to the outflow pipe on this afternoon was 14.21 meters closer than the distance predicted by the regression line with $x = 29$°C.

AP3.34 (a) Define W = the weight of a randomly selected gift box. Then $\mu_W = 27$ and $\sigma_W = 2.01$.

(b) The distribution of W will also be Normal. We want to find $P(W > 30)$.

(i) $z = \dfrac{30 - 27}{2.01} = 1.49$

$P(z > 1.49) = 0.0681$

(ii) $P(W > 30)$ = normalcdf(lower: 30, upper: 1000, mean: 27, SD: 2.01) = 0.0678 There is a 0.0678 probability of randomly selecting a box that weighs more than 30 ounces. **(c)** P(at least one box is greater than 30 ounces) = $1 - (1 - 0.0678)^5 = 0.2960$. There is a 0.2960 probability of selecting a random sample of 5 boxes and having at least one box weigh more than 30 ounces. **(d)** The distribution of $\overline{W}$ will also be Normal, with $\mu_{\overline{W}} = \mu_W = 27$ and

$\sigma_{\overline{W}} = \dfrac{2.01}{\sqrt{5}} = 0.899$. We want to find $P(\overline{W} > 30)$.

AP3.35 (b) $H_0: \sigma_A^2 - \sigma_B^2 = 0$, $H_a: \sigma_A^2 - \sigma_B^2 > 0$, where $\sigma_A^2 =$ the true variance of returns for Stock A and $\sigma_B^2 =$ the true variance of returns for Stock B.

(c) $F = \dfrac{(12.9)^2}{(9.6)^2} = 1.806$; values of F that are greater than 1 indicate that the price volatility for Stock A is higher than that for Stock B. The statistic provides some evidence for the alternative hypothesis because $1.806 > 1$.

(d) A test statistic of 1.806 or greater occurred in only 6 out of the 200 trials. Thus, the approximate P-value is $6/200 = 0.03$. Because $0.03 < \alpha = 0.05$, we reject H_0. There is convincing evidence that the true variance of returns for Stock A is greater than the true variance of returns for Stock B.

volatility and indicates how much an investment's actual performance during a specified period varies from its average performance over a longer period. Do the price fluctuations in Stock A significantly exceed those of Stock B, as measured by their variances? State an appropriate set of hypotheses that the investor is interested in testing.

(c) To measure this, we will construct a test statistic defined as

$$F = \frac{\text{larger sample variance}}{\text{smaller sample variance}}$$

Calculate the value of the F statistic using the information given in the table. Explain how the value of the statistic provides some evidence for the alternative hypothesis you stated in part (b).

(d) Two hundred simulated values of this test statistic, F, were calculated assuming that the two stocks have the same variance in daily price. The results

of the simulation are displayed in the following dotplot.

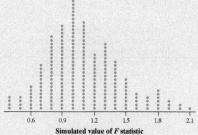

Simulated value of F statistic

Use these simulated values and the test statistic that you calculated in part (c) to determine whether the observed data provide convincing evidence that Stock A is a riskier investment than Stock B. Explain your reasoning.

Chapter 12

Chapter 12

Inference for Distributions and Relationships

PD Overview

In this chapter, we learn three different significance tests for distributions of categorical data and one significance test for relationships between two quantitative variables. The chi-square test for goodness of fit is used to determine whether the distribution of a single categorical variable differs from a hypothesized distribution in some population. The chi-square test for homogeneity is used to compare the distribution of a single categorical variable for two or more populations or treatments. The chi-square test for independence is used to investigate the relationship between two categorical variables in a single population. Finally, the t test for slope is used to determine if there is a linear association between two quantitative variables. Students often have a hard time deciding which test to use, so remind them continually of these distinctions.

Chi-square tests differ in several ways from the other significance tests we have learned about. First, we usually state the hypotheses in words rather than symbols. Second, there are no one-sided alternative hypotheses. Essentially, the alternative hypothesis is always "the null hypothesis isn't true." Students are often happy that they don't have to think as much about how to state the alternative hypothesis or calculate the P-value. Also, chi-square tests do not have corresponding confidence intervals like the other significance tests do. For example, although a two-sample z interval can accompany a two-sample z test for a difference between two proportions, there is no confidence interval that corresponds to a chi-square test for homogeneity.

In the final section, we introduce confidence intervals and significance tests for the slope of a least-squares regression line. Confidence intervals for the slope give a range of plausible values for the slope. Significance tests for a slope allow us to choose between two competing claims about the true value of the slope. We also take the opportunity to review many other topics in regression that your students may have forgotten since Chapter 3.

The "Inference" special-focus materials on the AP® Central website (apcentral.collegeboard.org/pdf/statistics-inference.pdf) are a great resource for helping you understand the logic of significance tests, why we need to check conditions, and how the tests in this chapter fit in the overall scheme of inference.

If you haven't read it recently, we recommend that you reread the section "Preparing Students for the AP® Statistics Exam" on page ATE-xv of this Annotated Teacher's Edition. Also, make sure to review with students the "About the AP® Exam and AP® Exam Tips" appendix on pages A-1 through A-5, near the end of the Student Edition. Finally, make sure to use the inference review questions and the set of 120 flash cards provided in the Teacher's Resource Materials.

We also strongly recommend that students do a final project to pull together the four major content areas of the course. Ask them to design a study, collect the data, analyze the data, and perform inference to draw a conclusion. A project description and sample rubric are included in the Teacher's Resource Materials.

The Main Ideas

One of the challenges of teaching the AP® Statistics course is keeping students focused on the big picture, not just the details of each section. We outline the main ideas for the chapter here.

Chapter 12 Introduction

We strongly recommend using "The candy man can" activity to open this chapter. It is a great way to introduce chi-square tests, and students always enjoy getting a tasty treat in class. Also, we refer back to this activity in Section 12.1 as we develop the chi-square test for goodness of fit. Some more teaching suggestions are listed in the Activity Overview on page 760.

Section 12.1 Chi-Square Tests for Goodness of Fit

This section begins by developing the chi-square statistic as a way to measure the difference between an observed distribution and a hypothesized distribution of categorical data. Like any other statistic, the chi-square statistic has its own sampling distribution that can be estimated by simulation. Fortunately, when certain conditions are satisfied, we can accurately model the sampling distribution of the chi-square statistic using a density curve, making it easier to calculate P-values. Not coincidentally, the density curves come from a family of distributions known as chi-square distributions. This is much like what we learned in Chapter 10—the sampling distribution of the t statistic can be modeled with a t distribution when certain conditions are satisfied.

Chi-square tests for goodness of fit always compare the observed distribution of a categorical variable with a hypothesized distribution. In the M&M'S® example, we compare the observed distribution of color with the distribution specified by the company. Other examples test if the distribution of outcome for a die is fair or if the distribution of NHL player birthdays is uniform throughout the year.

Finally, we suggest that students do a follow-up analysis when the results of a chi-square test are significant. This involves looking at which categories of the variable had the largest contributions to the chi-square statistic and whether the observed values in those categories were larger or smaller than expected.

Section 12.2 Inference for Two-Way Tables

In previous chapters, we learned how to do one-sample significance tests for a single population proportion and a single population mean. In both cases, we compared the observed value of a statistic to a hypothesized value, such as $p = 0.5$ or $\mu = 10$. For two-sample significance tests, we compared the proportions or means from two populations or treatments to each other, rather than to some hypothesized value. Similarly, while a chi-square test for goodness of fit compares an observed distribution to a hypothesized distribution, a chi-square test for homogeneity compares the observed distributions of a categorical variable to each other for two or more populations or treatments. Fortunately, the chi-square statistic and chi-square distributions we use to estimate P-values are the same for both tests.

The chi-square test for independence is the third chi-square test discussed in this chapter and is used to investigate the relationship between two categorical variables in one population. While there are some important differences in how we state

hypotheses and check conditions, many components of the test for independence are the same as those for the test for homogeneity. In fact, many statisticians don't even make a distinction between the two tests. However, because the tests are listed separately in the College Board Course Framework, we recommend that your students know the difference.

Section 12.3 Inference for Slope

In this section, we introduce students to the idea of performing inference about the relationship between two *quantitative* variables. If there is a linear association between two variables and the data come from a random sample or a randomized experiment, the least-squares regression line we calculate is just an estimate of the true least-squares regression line. We have to use inference to create an interval estimate for, or to make decisions about, the true value of the slope.

If you have time, we highly recommend doing the "Sampling from Old Faithful" activity as an introduction to this section. It will help students review some important ideas from Chapter 3, while also preparing them for the inference to come.

We start by developing the sampling distribution of the sample slope b so we can create confidence intervals and conduct significance tests for the population slope β. Although it is possible to do inference for other regression parameters, the College Board Course Framework includes only inference for slope, so that is where we will focus our efforts.

As with any other sampling distribution, we are interested in the shape, center, and variability of the sampling distribution of b. When certain conditions are met, the sampling distribution of b is approximately Normal and centered at the value of the true slope β, with a standard deviation of $\sigma_b = \dfrac{\sigma}{\sigma_x\sqrt{n}}$.

Unfortunately, the conditions are more complicated than in previous chapters. Students should know how to check if the relationship is *linear*, if the observations are *independent*, if the residuals are *Normally distributed*, if the residuals have an *equal standard deviation* at each value of x, and if the data were produced at *random*. Hopefully, the LINER acronym will help students remember all five conditions!

Once we are satisfied that the conditions have been met, performing inference should be fairly routine. As always, we expect students to follow the 4-step process. Other than the conditions, the remaining steps are much like those given in previous chapters.

On previous AP® Statistics exams, students have almost always been provided with computer output when asked questions about inference for regression. It is very important that students know how to obtain the equation of the least-squares regression line from the computer output, as well as the standard error of

the slope, the standardized test statistic, *P*-value, standard deviation of the residuals, and r^2. If you are very short on time, it would be wise to spend it teaching students how to use computer output to perform inference for the slope.

Chapter 12: Resources
Teacher's Resource Materials

The following resources, identified by the **TRM** in the annotated student pages, can be found by clicking on the link in the Teacher's e-Book (TE-book), searching by category or chapter on the Teacher's Resource Flash Drive (TRFD), or logging into the book's digital platform and searching the Teacher's Resources menu (teacher log-in required).

- Alternate Examples: one file per section
- Lecture Presentation Slides: one per section
- "Sampling from Old Faithful" activity
- "The helicopter experiment" activity and template
- Does Seat Location Matter?
- FRAPPY! Materials
- Flash Cards for AP® Statistics exam review
- Inference Review
- Chapter 12 Learning Targets Grid
- Chapter 12 Case Study
- Chapter 12 Final Project
- Complete solutions for the Check Your Understanding problems, section exercises, chapter review exercises, chapter practice test, and cumulative AP® practice test
- Quizzes: one per section
- Chapter 12 Test

Free Response Questions from Previous AP® Statistics Exams

Questions can be found on the AP® Central website: apcentral.collegeboard.org/courses/ap-statistics/exam.

Students should be able to answer all the free response questions listed below with material learned in this chapter. This list will be updated after each AP® Statistics exam and will be posted to the Teacher's Resource section of the book's digital platform and to www.statsmedic.com/free-response-questions.

Sections 12.1–12.2: Inference for Categorical Data: Chi-Square

Year	#	Content
2017	5	• Chi-square test for independence
2016	2	• Chi-square test for homogeneity • Follow-up analysis
2014	1	• Conditional relative frequency • Association between categorical variables • Chi-square test for independence
2013	4	• Chi-square test for independence
2011B	4	• Chi-square test for independence • Type I and Type II errors
2010B	5	• General addition rule • Conditional probability • Independence • Chi-square test for independence
2010	6	• Graphing and comparing distributions • Evaluating and using an unfamiliar test statistic
2009	1	• Graphing categorical data • Describing an association between categorical variables • Choosing a correct inference procedure • Stating hypotheses
2008	5	• Chi-square test for goodness of fit • Follow-up analysis
2006	6	• Stating hypotheses • Calculating a test statistic and *P*-value • Rejection regions • Identifying simulated distributions of a test statistic
2004	5	• Chi-square test for independence • Scope of inference
2003	5	• Chi-square test for independence
2003B	5	• Multiplication rule for independent events • Expected value • Chi-square test for goodness of fit
2002B	6	• Two-sample *t* test • Chi-square test for homogeneity • Comparing distributions using graphs
1999	2	• Chi-square test for independence
1998	3	• Methods of random assignment • Choosing the correct inference procedure

Section 12.3: Inference for Quantitative Data: Slopes

Year	#	Content
2011	5	• Regression output • Interpreting slope • Meaning of r^2 • t test for slope (conclusion only)
2011B	6	• Interpreting slope • Extrapolation • Sampling distribution of $\bar{y}$ in a regression context • Optimal design for estimating slope
2010B	6	• Interpreting slope • Interpreting a residual • Using the residuals to estimate an effect • Testing for a difference between two slopes using a confidence interval • Using two different least-squares regression lines to estimate an effect
2008	6	• Two-sample t test for a difference in means • Stating the equation of a least-squares regression line from computer output • Interpreting slope • t test for a slope • Comparing inference methods
2007	6	• Interpreting slope • Using a model with no constant term, t test for slope with H_0: $\beta = 1$ • Graphing a multiple regression model with an indicator variable • Interpreting the coefficients of a multiple regression model
2007B	6	• Two-sample z test for a difference of proportions • Confidence interval for slope • Using a confidence interval to make a decision • Using transformed data and a least-squares regression line to make predictions
2006	2	• Stating the equation of a least-squares regression line from computer output • Interpreting the standard deviation of the residuals • Interpreting the standard error of the slope
2005B	5	• Stating the equation of a least-squares regression line from computer output • Interpreting the slope and y intercept • Confidence interval for slope
2001	6	• Making graphs and comparing two distributions • t test for slope • Classifying a new observation

Applets

- The *One Categorical Variable* applet in the Extra Applets section of the Student Site at highschool.bfwpub.com/updatedtps6e can be used to perform a chi-square test for data from a one-way or two-way table.

- The *Sampling Regression Lines* applet at www.rossmanchance.com/applets allows students to define a population of bivariate data, take random samples from the population, and record the value of the sample slope and intercept.

- The *Two Quantitative Variables* applet in the Extra Applets section of the Student Site at highschool.bfwpub.com/updatedtps6e allows students to construct scatterplots, residual plots, and dotplots of residuals. Students can also perform inference for bivariate data.

Chapter 12: Pacing Guide, Learning Targets, and Suggested Assignments

This pacing guide is based on a schedule with 110, 50-minute sessions before the AP® Statistics exam. If you have a different number of sessions before the AP® exam, you can modify the pacing guide to suit your needs. If you have additional time, consider incorporating quizzes, released AP® Statistics free response questions, or additional activities. See the Resources section for suggestions.

The suggested homework assignments list odd-numbered exercises, whenever possible, so students can check their answers against the back-of-book answers. If you would rather students not have access to the answers while doing homework, adding 1 to the exercise numbers usually will do the trick, because the homework exercises typically are paired. For example, Exercises 1 and 2 will generally cover the same topics, but in different contexts. You may also choose to include the Recycle and Review questions at the end of each section, which review topics from previous sections or chapters. If your school is using the digital platform that accompanies TPS6, you will find these assignments pre-built as online homework assignments for Chapter 12.

Day	Content	Learning Targets: Students will be able to...	Suggested Assignment (MC bold)
1	Chapter 12 Introduction, 12.1 Stating Hypotheses, Comparing Observed and Expected Counts: The Chi-Square Test Statistic, The Chi-Square Distributions and P-Values	• State appropriate hypotheses and compute the expected counts and chi-square test statistic for a chi-square test for goodness of fit. • State and check the Random, 10%, and Large Counts conditions for performing a chi-square test for goodness of fit. • Calculate the degrees of freedom and P-value for a chi-square test for goodness of fit.	1, 3, 5, 7
2	12.1 Carrying Out a Test	• Perform a chi-square test for goodness of fit. • Conduct a follow-up analysis when the results of a chi-square test are statistically significant.	9, 13, **19–22**
3	12.2 Tests for Homogeneity: Stating Hypotheses, Expected Counts and the Chi-Square Test Statistic, Conditions and P-Values; The Chi-Square Test for Homogeneity	• State appropriate hypotheses and compute the expected counts and chi-square test statistic for a chi-square test based on data in a two-way table. • State and check the Random, 10%, and Large Counts conditions for a chi-square test based on data in a two-way table. • Calculate the degrees of freedom and P-value for a chi-square test based on data in a two-way table. • Perform a chi-square test for homogeneity.	27, 29, 31, 33, 35
4	12.2 Relationships Between Two Categorical Variables, The Chi-Square Test for Independence, Using Chi-Square Tests Wisely	• Perform a chi-square test for independence. • Choose the appropriate chi-square test in a given setting.	41, 43, 47, 49, 51, **55–60**
5	12.3 Sampling Distribution of b, Conditions for Regression Inference	• Check the conditions for performing inference about the slope β of the population (true) regression line.	65, 67, 69
6	12.3 Estimating the Parameters, Constructing a Confidence Interval for the Slope	• Interpret the values of a, b, s, and SE_b in context, and determine these values from computer output. • Construct and interpret a confidence interval for the slope β of the population (true) regression line.	71, 73, 75
7	12.3 Performing a Significance Test for the Slope	• Perform a significance test about the slope β of the population (true) regression line.	79, 83, **87–92**
8	Chapter 12 Review/ FRAPPY!		Chapter 12 Review Exercises
9	Chapter 12 Test		Cumulative AP® Practice Test 4

Chapter 12 Alignment to the College Board's Fall 2019 AP® Statistics Course Framework*

Relationship to College Board Units

Chapter 12 in this book covers Topics 8.1–8.7 and 9.1–9.6 in Units 8 and 9 of the College Board Course Framework. Students will be ready to take the Personal Progress Check for Unit 8 after Section 12.2 and Unit 9 after Section 12.3.

Big Ideas and Enduring Understandings

Chapter 12 develops these Big Ideas and related Enduring Understandings outlined in the Course Framework:

- **Big Idea 1: Variation and Distribution (EU: VAR 1, 7, 8):** The distribution of measures for individuals within a sample or population describes variation. The value of a statistic varies from sample to sample. How can we determine whether differences between measures represent random variation or meaningful distinctions? Statistical methods based on probabilistic reasoning provide the basis for shared understandings about variation and about the likelihood that variation between and among measures, samples, and populations is random or meaningful.
- **Big Idea 2: Patterns and Uncertainty (EU: UNC 4):** Statistical tools allow us to represent and describe patterns in data and to classify departures from patterns. Simulation and probabilistic reasoning allow us to anticipate patterns in data and to determine the likelihood of errors in inference.
- **Big Idea 3: Data-Based Predictions, Decisions, and Conclusions (EU: DAT 3):** Data-based regression models describe relationships between variables and are a tool for making predictions for values of a response variable. Collecting data using random sampling or randomized experimental design means that findings may be generalized to the part of the population from which the selection was made. Statistical inference allows us to make data-based decisions.

Course Skills

Chapter 12 helps students to develop the skills identified in the Course Framework.

- **1: Selecting Statistical Models** (1.A, 1.D, 1.E, 1.F)
- **3: Using Probability and Simulation** (3.A, 3.C, 3.D, 3.E)
- **4: Statistical Argumentation** (4.A, 4.B, 4.C, 4.D, 4.E)

Learning Objectives and Essential Knowledge

Section	Learning Objectives	Essential Knowledge Statements
12.1	VAR-1.J, VAR-8.A, VAR-8.B, VAR-8.C, VAR-8.D, VAR-8.E, VAR-8.F, VAR-8.G, DAT-3.I, DAT-3.J	VAR-1.J.1, VAR-8.A.1, VAR-8.A.2, VAR-8.A.3, VAR-8.B.1, VAR-8.C.1, VAR-8.D.1, VAR-8.E.1, VAR-8.F.1, VAR-8.F.2, VAR-8.G.1, DAT-3.I.1, DAT-3.J.1, DAT-3.J.2
12.2	VAR-8.H, VAR-8.I, VAR-8.J, VAR-8.K, VAR-8.L, VAR-8.M, DAT-3.K, DAT-3.L	VAR-8.H.1, VAR-8.I.1, VAR-8.I.2, VAR-8.J.1, VAR-8.J.2, VAR-8.K.1, VAR-8.L.1, VAR-8.M.1, VAR-8.M.2, DAT-3.K.1, DAT-3.L.1, DAT-3.L.2
12.3	VAR-1.K, UNC-4.AC, UNC-4.AD, UNC-4.AE, UNC-4.AF, UNC-4.AG, UNC-4.AH, UNC-4.AI, VAR-7.J, VAR-7.K, VAR-7.L, VAR-7.M, DAT-3.M, DAT-3.N	VAR-1.K.1, UNC-4.AC.1, UNC-4.AC.2, UNC-4.AC.3, UNC-4.AD.1, UNC-4.AE.1, UNC-4.AE.2, UNC-4.AF.1, UNC-4.AF.2, UNC-4.AG.1, UNC-4.AG.2, UNC-4.AH.1, UNC-4.AI.1, VAR-7.J.1, VAR-7.K.1, VAR-7.L.1, VAR-7.M.1, VAR-7.M.2, DAT-3.M.1, DAT-3.N.1, DAT-3.N.2

A detailed alignment (The Nitty Gritty Guide) that can be sorted by Course Framework Unit, Topic, Learning Objective, Essential Knowledge Statement, or textbook section, is available on the TRFD and in the Teacher's Resources folder on Sapling Plus. **TRM**

*Should changes be made to the Course Framework in the future, an updated alignment will be placed on our AP® updates page at go.bfwpub.com/ap-course-updates.

Notes

Chapter 12

Inference for Distributions and Relationships

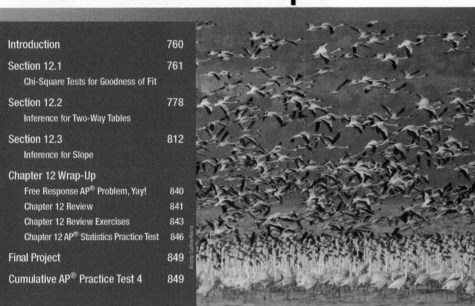

Antrey Gudkov/Alamy

Teaching Tip

Units 8 and 9 in the College Board Course Framework align to Chapter 12 in this book. Students will be ready to take the Personal Progress Check for Unit 8 after Section 12.2 and Unit 9 after Section 12.3.

Teaching Tip

Chapter 12 is the only chapter that covers two College Board Units (8 and 9). We decided to combine them into a single chapter because of their very low relative weight in the College Board's Fall 2019 CED. Each of these Units account for 2% to 5% of the multiple choice part of the exam, so together they account for 4% to 10%.

PD Chapter 12 Overview

To watch the video overview of Chapter 12 (for teachers), click on the link in the TE-Book, look on the TRFD, or download from the Teacher's Resources on the book's digital platform.

TRM Lecture Presentation Slides

If you are new to teaching AP® Statistics or are short on time when preparing for class, you may find the Lecture Presentation Slides to be helpful. Experienced AP® Teacher Doug Tyson has created one slide presentation per section. You may use them as is, modify them to fit your needs, or share them with students who miss class. Find them on the TRFD and in the Teacher's Resources on the book's digital platform.

Teaching Tip

Make students aware that there will be four different significance tests studied in this chapter:

1. Chi-square test for goodness of fit (Section 12.1)
2. Chi-square test for homogeneity (Section 12.2)
3. Chi-square test for independence (Section 12.2)
4. t test for slope (Section 12.3); in addition, students will learn about t intervals for the slope.

Teaching Tip

Each of the three chi-square tests will be introduced within a different context. Students will use these contexts to organize their thinking about the three chi-square tests, enabling them to read a new context and decide which chi-square test is appropriate.

1. M&M'S® or NHL birthdays → Chi-square test for goodness of fit
2. Music and purchases → Chi-square test for homogeneity
3. Anger level and heart disease → Chi-square test for independence

Teaching Tip

Emphasize that Sections 12.1 and 12.2 are all about categorical data by noting that each of the variables in the examples (M&M'S color, birth month, type of music, type of purchase, anger level, heart disease status) is categorical.

Teaching Tip

M&M'S® are produced at different factories. Each factory has its own claimed distribution. Here are two examples:

Hackettstown, NJ
Brown: 12.5%, *Red:* 12.5%, *Yellow:* 12.5%, *Green:* 12.5%, *Orange:* 25%, *Blue:* 25%

Cleveland, OH
Brown: 12.4%, *Red:* 13.1%, *Yellow:* 13.5%, *Green:* 19.8%, *Orange:* 20.5%, *Blue:* 20.7%

To find out which factory your package of M&Ms came from, check the serial code: a code that includes "HKP" was produced at the Hackettstown factory; a code that includes "CLV" was produced at the Cleveland factory. See the Teaching Tip on page 778 if you are using a different type of M&M'S product (Almond, Crispy, Peanut, etc.).

INTRODUCTION

In Section 9.2, we discussed tests for the proportion of successes in a single population. These tests were based on a single categorical variable with values that were divided into two categories: success and failure. Sometimes we want to perform a test for the distribution of a categorical variable with two *or more* categories. The *chi-square test for goodness of fit* allows us to determine whether a hypothesized distribution seems valid. This test is useful in a field like genetics, where the laws of probability give the expected proportion of outcomes in each category.

In Section 9.3, we discussed tests for a difference in proportions for two populations or treatments. Sometimes we'd like to compare the distribution of a categorical variable for two *or more* populations or treatments, where the variable can have two *or more* categories. We can decide whether the distribution of a categorical variable differs for two or more populations or treatments using a *chi-square test for homogeneity*. This test will help us answer the question: Does background music influence customer purchases?

Tests for homogeneity use data summarized in a two-way table. It is also possible to use the information in a two-way table to study the relationship between two categorical variables. The *chi-square test for independence* allows us to determine if there is convincing evidence of an association between two categorical variables in a population, such as anger level and heart disease status.

If we want to know if there is convincing evidence of a linear association between two quantitative variables, such as the drop height and flight time for paper helicopters, we can use a *t test for the slope of a least-squares regression line*. We can also estimate the slope of a population regression line using a *t interval for the slope*.

Here's an activity that gives you a taste (pun intended) of what lies ahead.

ACTIVITY The candy man can

Ramón Rivera Moret

Mars, Inc., is famous for its milk chocolate candies. Here's what the company's Consumer Affairs Department says about the distribution of color for M&M'S® Milk Chocolate Candies produced at its Hackettstown, New Jersey, factory:

Brown: 12.5%   Red: 12.5%   Yellow: 12.5%
Green: 12.5%   Orange: 25%   Blue: 25%

The purpose of this activity is to investigate if the distribution of color in a large bag of M&M'S Milk Chocolate Candies differs from the claimed distribution.

1. Your class will take a random sample of 60 M&M'S Milk Chocolate Candies from a large bag and count the number of candies of each color. Make a table on the board that summarizes these *observed counts*.
2. How can you tell if the sample data give convincing evidence against the company's claim? Each team of 3 or 4 students should discuss this question and devise a formula for a test statistic that measures the difference between the observed and expected color distributions. The test statistic should yield a single number when the observed and expected values are plugged in. Also, larger differences between the observed and expected distributions should result in a larger value for the statistic.

▶ ACTIVITY OVERVIEW

To prepare for this activity, watch the overview video by clicking on the link in the TE-Book, opening the TRFD, or downloading from the Teacher's Resources on the book's digital platform.

Time: 30 minutes

Materials: Large bag of M&M'S Milk Chocolate Candies

Teaching Advice: Tell students that you emailed the company to get the claimed distribution of colors. Today, we will test their claim, and if we find convincing evidence against it, we will write an angry letter to Mars, Inc.

We are using one sample for the whole class instead of a different sample for each student because we want everyone to be working with the same set of numbers (to make the conversations easier).

Here are some potential guiding questions to help students brainstorm their own formulas for the chi-square test statistic:

- Should we look at the difference between the observed and expected *proportions* in each color category or between the observed and expected *counts* in each category?
- Should we use the differences themselves, the absolute value of the differences, or the square of the differences?
- Should we divide each difference value by the sample size, expected count, or nothing at all?

When calculating the actual chi-square test statistic for the class sample, consider formatting a table with the following five columns:

$$O,\ E,\ (O - E),\ (O - E)^2,\ \frac{(O - E)^2}{E}.$$

At the end of the activity, calculate the proportion of dots greater than or equal to the chi-square test statistic calculated for your class sample. This is the estimated *P*-value, which can be used to make a conclusion about the claimed distribution of colors.

If the estimated *P*-value is less than 5%, there is convincing evidence that the distribution of color in the large bag of M&M'S Milk Chocolate Candies differs from the company's claimed distribution. In this case, ask students to do a follow-up analysis to identify which color provided the strongest evidence for this conclusion.

Answers are on the next page.

3. Each team will share its proposed test statistic with the class. Your teacher will then reveal how the *chi-square test statistic* χ^2 is calculated.

4. Discuss as a class: If your sample is consistent with the company's claim, will the value of χ^2 be large or small? If your sample is not consistent with the company's claim, will the value of χ^2 be large or small?

5. Compute the value of the chi-square test statistic for the class's data.

We can use simulation to determine if your class's chi-square test statistic is large enough to provide convincing evidence that the distribution of colors in the large bag differs from the company's claim. To conduct the simulation, 100 random samples of size 60 were selected from a population of M&M'S Milk Chocolate Candies that matches the company's claim. For each random sample, the value of the χ^2 test statistic was calculated and plotted on the dotplot.

6. There is one dot at $\chi^2 = 16$. Explain what this dot represents.

7. Where does your class's value of χ^2 fall relative to the other dots on the dotplot? What conclusion can you make about the distribution of colors in the large bag?

0 2 4 6 8 10 12 14 16 18 20

Simulated χ^2 test statistic

The dotplot at the end of the activity shows the values of the chi-square test statistic that are likely to occur by chance alone when sampling from the company's claimed M&M'S Milk Chocolate Candies color distribution. You may have noticed that the shape of the distribution *isn't* approximately Normal. Will it always look like this? You will learn more about the sampling distribution of the chi-square test statistic shortly.

SECTION 12.1   Chi-Square Tests for Goodness of Fit

LEARNING TARGETS *By the end of the section, you should be able to:*

- State appropriate hypotheses and compute the expected counts and chi-square test statistic for a chi-square test for goodness of fit.
- State and check the Random, 10%, and Large Counts conditions for performing a chi-square test for goodness of fit.
- Calculate the degrees of freedom and *P*-value for a chi-square test for goodness of fit.
- Perform a chi-square test for goodness of fit.
- Conduct a follow-up analysis when the results of a chi-square test are statistically significant.

Jerome's class did the "Candy man can" activity using a bag from the Hackettstown factory. The one-way table summarizes the data from the class's sample of M&M'S Milk Chocolate Candies.

Color	Brown	Red	Yellow	Green	Orange	Blue	Total
Count	12	3	7	9	9	20	60

PD **Section 12.1 Overview**

To watch the video overview of Section 12.1 (for teachers), click on the link in the TE-Book, look on the TRFD, or download from the Teacher's Resources on the book's digital platform.

TRM **Section 12.1 Alternate Examples**

You can find the Alternate Examples for this section in Microsoft Word format by clicking on the link in the TE-Book, opening the TRFD, or downloading from the Teacher's Resources on the book's digital platform.

TRM **Learning Targets Grid**

At the beginning of each section, we present the relevant learning targets. Point these out to your students and refer back to the targets when you cover them in class. There is a PDF version of the grid with an additional column that students can use to keep track of their progress. Find it in the Teacher's Resource Materials located in the TE-Book, on the TRFD, or in the Teacher's view on the book's digital platform.

The sample proportion of brown candies is $\hat{p} = \dfrac{12}{60} = 0.20$. Because the company claims that 12.5% of M&M'S® Milk Chocolate Candies are brown, Jerome might believe that something fishy is going on. He could use the one-sample z test for a proportion from Chapter 9 to test the hypotheses

$$H_0: p = 0.125$$
$$H_a: p \neq 0.125$$

where p is the true proportion of M&M'S Milk Chocolate Candies in the large bag that are brown. He could then perform additional significance tests for each of the remaining colors.

Besides being fairly inefficient, this method would also lead to the problem of multiple tests, which we discussed in Section 11.1. More important, this approach wouldn't tell us how likely it is to get a random sample of 60 candies with a color *distribution* that differs as much from the one claimed by the company as the class's sample does, taking all the colors into consideration at one time. For that, we use a new kind of significance test, called a *chi-square test for goodness of fit*.

> Note that the correct alternative hypothesis H_a is two-sided. A sample proportion of brown candies much higher or much lower than 0.125 would give Jerome reason to be suspicious about the company's claim. It's not appropriate to adjust H_a after looking at the sample data!

Stating Hypotheses

As with any significance test, we begin by stating hypotheses. The null hypothesis in a chi-square test for goodness of fit should state a claim about the distribution of a single categorical variable in the population of interest. In the "Candy man can" activity, the categorical variable we're measuring is color and the population of interest is all the M&M'S Milk Chocolate Candies in the large bag. The appropriate null hypothesis is:

H_0: The distribution of color in the large bag of M&M'S Milk Chocolate Candies is the same as the claimed distribution.

The alternative hypothesis in a chi-square test for goodness of fit is that the categorical variable does not have the specified distribution. For the "Candy man can" activity, our alternative hypothesis is

H_a: The distribution of color in the large bag of M&M'S Milk Chocolate Candies is *not* the same as the claimed distribution.

Although we usually write the hypotheses in words, we can also write them in symbols. For example, here are the hypotheses for the "Candy man can" activity:

H_0: $p_{brown} = 0.125$, $p_{red} = 0.125$, $p_{yellow} = 0.125$, $p_{green} = 0.125$, $p_{orange} = 0.25$, $p_{blue} = 0.25$

H_a: At least two of these proportions differ from the values stated by the null hypothesis

where p_{color} = the true proportion of M&M'S Milk Chocolate Candies of that color in the large bag.

Why don't we write the alternative hypothesis as "H_a: At least one of these proportions differs from the values stated by the null hypothesis" instead? If the stated proportion in one category is wrong, then the stated proportion in at least one other category must be wrong because the sum of the proportions must be 1.

Don't state the alternative hypothesis in a way that suggests that *all* the proportions in the null hypothesis are wrong. For instance, it would be *incorrect* to write

$$H_a: p_{brown} \neq 0.125, p_{red} \neq 0.125, p_{yellow} \neq 0.125, p_{green} \neq 0.125,$$
$$p_{orange} \neq 0.25, p_{blue} \neq 0.25$$

Comparing Observed and Expected Counts: The Chi-Square Test Statistic

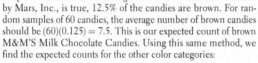

The idea of the chi-square test for goodness of fit is this: we compare the observed counts from our sample with the counts that would be expected if H_0 is true. (*Remember:* we always assume that H_0 is true when performing a significance test.) The more the observed counts differ from the expected counts, the more evidence we have against the null hypothesis and for the alternative hypothesis.

Recall that Jerome's class collected data from a random sample of M&M'S® Milk Chocolate Candies. How many candies of each color should they expect to find in their sample of 60 candies? Assuming that the color distribution stated by Mars, Inc., is true, 12.5% of the candies are brown. For random samples of 60 candies, the average number of brown candies should be $(60)(0.125) = 7.5$. This is our expected count of brown M&M'S Milk Chocolate Candies. Using this same method, we find the expected counts for the other color categories:

Red: $(60)(0.125) = 7.5$ Orange: $(60)(0.25) = 15$

Yellow: $(60)(0.125) = 7.5$ Blue: $(60)(0.25) = 15$

Green: $(60)(0.125) = 7.5$

> ### CALCULATING EXPECTED COUNTS IN A CHI-SQUARE TEST FOR GOODNESS OF FIT
>
> The expected count for category i in the distribution of a categorical variable is
>
> $$np_i$$
>
> where p_i is the proportion for category i specified by the null hypothesis.

Did you notice that the expected count sounds a lot like the expected value of a random variable from Chapter 6? That's no coincidence. The number of M&M'S Milk Chocolate Candies of a specific color in a random sample of 60 candies is a binomial random variable. Its expected value is np, the average number of candies of this color in many samples of 60 M&M'S Milk Chocolate Candies. *The expected count is not likely to be a whole number and shouldn't be rounded to a whole number.*

To see if the data give convincing evidence for the alternative hypothesis, we compare the observed counts from our sample with the expected counts. If the observed counts are far from the expected counts, that's the evidence we were

Teaching Tip

Walking into a room full of statisticians and referring to the "chai" square test or the "chee" square test statistic will result in immediate loss of credibility. "Chai" should only be used when ordering a drink at Starbucks®. Please help your AP® Biology teachers with this one.

> **AP® EXAM TIP**
>
> Some students mistakenly round the expected counts, believing that the counts must be integers. Students will lose points for this on the AP® Statistics exam because it shows a misunderstanding of what expected counts represent—the average number of observations in a given category in many, many random samples.

Making Connections

In the M&M'S® example, the distribution of the count of each color is a binomial distribution. Back in Chapter 6, we calculated the expected value for a binomial distribution as $\mu = np$, so the formula for expected counts on this page is not really a new one.

Teaching Tip

Here are some common student questions about the formula for the chi-square test statistic:

Why square the difference between observed count and expected count?
Sometimes the difference will be positive, and sometimes it will be negative. Squaring the difference ensures we have all positive values. Remind students that we used a similar approach in the formula for standard deviation.

Why divide by expected count?
We are interested in how far away the observed count is from the expected count *relative to the expected count*. See the Think About It on page 765 for more details.

AP® EXAM TIP

If students encounter a chi-square test on the AP® Statistics exam, be sure they include the summation symbol Σ when writing the formula for the test statistic.

Making Connections

For Jerome's M&M'S® data, six "components" are added together to get the chi-square test statistic. These components give us insight about each of the M&M'S colors. The larger components correspond to colors in which the observed count was relatively farther from the expected count. We will use these components later to do a follow-up analysis.

Teaching Tip: AP® Connections

The formula sheet for the AP® Biology exam presents the chi-square formula as:

$$\chi^2 = \sum \frac{(o - e)^2}{e}$$

AP® Biology students are also given a small chi-square table of critical values.

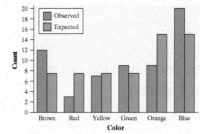

FIGURE 12.1 Bar graph comparing observed and expected counts for Jerome's class sample of 60 M&M'S® Milk Chocolate Candies.

AP® EXAM TIP

The formula for the chi-square test statistic is included on the formula sheet that is provided on the AP® Statistics exam. However, it doesn't include the word *count*:

$$\chi^2 = \sum \frac{(\text{Observed} - \text{Expected})^2}{\text{Expected}}$$

We included the word *count* to emphasize that you must use the observed and expected counts—not the observed and expected proportions—when calculating the chi-square test statistic.

seeking. The table gives the observed and expected counts for the sample of 60 candies from Jerome's class. Figure 12.1 shows these counts as a side-by-side bar graph.

Color	Observed	Expected
Brown	12	7.5
Red	3	7.5
Yellow	7	7.5
Green	9	7.5
Orange	9	15.0
Blue	20	15.0

We see some fairly large differences between the observed and expected counts in several color categories. How likely is it that differences this large or larger would occur just by chance in random samples of size 60 from the population distribution claimed by Mars, Inc.? To answer this question, we calculate a statistic that measures how far apart the observed and expected counts are, relative to the expected counts. The statistic we use to make the comparison is the **chi-square test statistic** χ^2. (The symbol χ is the lowercase Greek letter chi, pronounced "kye" like "rye.")

DEFINITION Chi-square test statistic

The **chi-square test statistic** is a measure of how far the observed counts are from the expected counts, relative to the expected counts. The formula for the statistic is

$$\chi^2 = \sum \frac{(\text{Observed count} - \text{Expected count})^2}{\text{Expected count}}$$

where the sum is over all possible values of the categorical variable.

For Jerome's data, we add six terms—one for each color category:

$$\chi^2 = \frac{(12-7.5)^2}{7.5} + \frac{(3-7.5)^2}{7.5} + \frac{(7-7.5)^2}{7.5} + \frac{(9-7.5)^2}{7.5} + \frac{(9-15)^2}{15} + \frac{(20-15)^2}{15}$$

$$= 2.7 + 2.7 + 0.03 + 0.30 + 2.4 + 1.67 = 9.8$$

Here's an example to help you practice what you have learned so far.

EXAMPLE

A fair die
Hypotheses, expected counts, and the chi-square statistic

PROBLEM: Carrie made a 6-sided die in her ceramics class and rolled it 90 times to test if each side was equally likely to show up. The table summarizes the outcomes of her 90 rolls.

(a) State the hypotheses that Carrie should test.
(b) Calculate the expected count for each of the possible outcomes.
(c) Calculate the value of the chi-square test statistic.

Outcome of roll	1	2	3	4	5	6	Total
Frequency	12	28	12	13	10	15	90

ALTERNATE EXAMPLE   Skills 1.F, 3.A

The color of Reese's Pieces®   Hypotheses, expected counts, and the chi-square test statistic

PROBLEM:
The Hershey Company makes Reese's Pieces and claims the distribution of colors is as follows: 50% orange, 25% brown, and 25% yellow. Skeptical of this claim, Trey purchases a very large bag of Reese's Pieces and selects a random sample of 80 pieces. Here are the results:

Color	Orange	Brown	Yellow
Count	31	22	27

(a) State the hypotheses that Trey should test.
(b) Calculate the expected count for each of the possible outcomes.
(c) Calculate the value of the chi-square test statistic.

(continues)

SOLUTION:

(a) H_0: The sides of Carrie's die are equally likely to show up.

H_a: The sides of Carrie's die are not equally likely to show up.

(b) If H_0 is true, each of the 6 sides should show up 1/6 of the time. The expected count is $90(1/6) = 15$ for each side.

Outcome of roll	1	2	3	4	5	6	Total
Expected count	15	15	15	15	15	15	90

You can also state the null hypothesis as:

H_0: Carrie's die is fair.

or

H_0: The distribution of outcome for Carrie's die is uniform.

or

H_0: $p_1 = p_2 = p_3 = p_4 = p_5 = p_6 = 1/6$

(c) $\chi^2 = \dfrac{(12-15)^2}{15} + \dfrac{(28-15)^2}{15} + \dfrac{(12-15)^2}{15} + \dfrac{(13-15)^2}{15} + \dfrac{(10-15)^2}{15} + \dfrac{(15-15)^2}{15}$

$= 0.6 + 11.27 + 0.6 + 0.27 + 1.67 + 0$

$= 14.41$

$\chi^2 = \sum \dfrac{(\text{Observed count} - \text{Expected count})^2}{\text{Expected count}}$

FOR PRACTICE, TRY EXERCISE 1

Ramón Rivera Moret

Think About It

WHY DO WE DIVIDE BY THE EXPECTED COUNT WHEN CALCULATING THE CHI-SQUARE TEST STATISTIC? In Jerome's class sample, they got 4.5 more browns than expected $(12 - 7.5)$ and 5 more blues than expected $(20 - 15)$. Which of these is more surprising?

In both cases, the number of M&M'S® Milk Chocolate Candies in the sample exceeds the expected count by about the same amount. But it's much more surprising to be off by 4.5 out of an expected 7.5 brown candies (a 60% discrepancy) than to be off by 5 out of an expected 15 blue candies (a 33% discrepancy). For that reason, we want the category with a larger *relative* difference to contribute more heavily to the evidence against H_0 and in favor of H_a measured by the χ^2 test statistic.

If we just computed (Observed Count − Expected Count)^2 for each category instead, the contributions of these two color categories would be about the same (with the contribution from blue being slightly larger):

Brown: $(12 - 7.5)^2 = 20.25$ Blue: $(20 - 15)^2 = 25$

By using (Observed count − Expected count)^2/Expected count, we guarantee that the color category with the larger relative difference will contribute more heavily to the total:

Brown: $\dfrac{(12 - 7.5)^2}{7.5} = 2.7$ Blue: $\dfrac{(20 - 15)^2}{15} = 1.67$

SOLUTION:

(a) H_0: Hershey's claimed distribution of color for Reese's Pieces is correct for this bag.

H_a: Hershey's claimed distribution of color for Reese's Pieces is incorrect for this bag.

(b) If H_0 is true, 50% of the Reese's Pieces should be orange, 25% should be brown, and 25% should be yellow. The expected count is $80(0.50) = 40$ for orange, $80(0.25) = 20$ for brown, and $80(0.25) = 20$ for yellow.

Color	Orange	Brown	Yellow
Expected count	40	20	20

(c) $\chi^2 = \dfrac{(31 - 40)^2}{40} + \dfrac{(22 - 20)^2}{20}$

$+ \dfrac{(27 - 20)^2}{20}$

$= 2.025 + 0.2 + 2.45 = 4.675$

AP® EXAM TIP

On the AP® Statistics exam, students do not have to show the calculation of every term. Showing something like

$\chi^2 = \dfrac{(12 - 15)^2}{15} + \dfrac{(28 - 15)^2}{15} + \cdots$

$= 14.41$ is sufficient.

 COMMON STUDENT ERROR

Some students mistakenly use proportions when calculating the chi-square test statistic. Even though it is possible to state the hypotheses for a chi-square test using proportions, the chi-square test statistic is always calculated using counts.

Teaching Tip

Why do we divide by the expected count when calculating the chi-square test statistic? Here is another way to approach this idea with students:

Scenario 1: The expected number of red M&M'S® is 6 and we get 16 red M&M'S.

Scenario 2: The expected number of red M&M'S is 500 and we get 510 M&M'S.

Which scenario provides more convincing evidence against the company's claim? In both scenarios, the observed value is 10 away from the expected. But Scenario 1 provides much more convincing evidence. The important idea is how far away the observed count is from the expected count *relative to the expected count.*

✓ **Answers to CYU**

1. H_0: The company's claimed color distribution for its Peanut M&M'S® is correct.

H_a: The company's claimed color distribution is not correct.

2. There were 65 candies in the sample from the bag. The expected count of blue, orange, green, and yellow candies is $65(0.20) = 13$, and the expected count of red and brown is $65(0.10) = 6.5$.

3.

$$\chi^2 = \frac{(14-13)^2}{13} + \frac{(9-13)^2}{13} + \cdots$$
$$= 2.3847$$

Teaching Tip

There are two possible explanations for why Jerome's class got a chi-square statistic as large as 9.8:

- The claimed distribution is correct, and the differences they observed were due to sampling variability.
- The claimed distribution is not correct.

To decide which of these explanations is more plausible, we need to know how likely it is to get a chi-square statistic at least as large as 9.80 by chance, assuming the claimed distribution is correct. In other words, we want a *P*-value!

CHECK YOUR UNDERSTANDING

Mars, Inc., reports that the M&M'S® Peanut Chocolate Candies from their Cleveland factory have the following distribution of color: 20% blue, 20% orange, 20% green, 20% yellow, 10% red, and 10% brown. Joey bought a large bag of them and selected a random sample of 65 candies. He found 14 blue, 9 orange, 15 green, 14 yellow, 5 red, and 8 brown.

1. State appropriate hypotheses for testing the company's claim about the color distribution of M&M'S Peanut Chocolate Candies in Joey's large bag.
2. Calculate the expected count for each color.
3. Calculate the chi-square test statistic for Joey's sample.

The Chi-Square Distributions and *P*-Values

Think of χ^2 as a measure of the relative distance the observed counts are from the expected counts. Like any distance, it is always zero or positive, and it is zero only when the observed counts are exactly equal to the expected counts. Large values of χ^2 are stronger evidence for H_a because they say that the observed counts are far from what we would expect if H_0 were true. Small values of χ^2 suggest that the data are consistent with the null hypothesis. Is the value from Jerome's class, $\chi^2 = 9.8$, a large value? You know the drill: compare the observed value 9.8 against the sampling distribution that shows how χ^2 would vary in repeated random sampling if the null hypothesis were true.

We used software to simulate taking 1000 random samples of size 60 from the population distribution of M&M'S Milk Chocolate Candies given by Mars, Inc. Figure 12.2 shows a dotplot of the values of the chi-square test statistic for these 1000 samples.

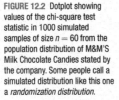

FIGURE 12.2 Dotplot showing values of the chi-square test statistic in 1000 simulated samples of size $n = 60$ from the population distribution of M&M'S Milk Chocolate Candies stated by the company. Some people call a simulated distribution like this one a *randomization distribution*.

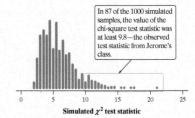

Recall that larger values of χ^2 give more convincing evidence against H_0 and in favor of H_a. According to the dotplot, 87 of the 1000 simulated samples resulted in a chi-square test statistic of 9.8 or higher. Our estimated *P*-value is $87/1000 = 0.087$. Because the *P*-value exceeds the default $\alpha = 0.05$ significance level, we fail to reject H_0. We do not have convincing evidence that the color distribution in Jerome's bag is different from the distribution claimed by the company.

As Figure 12.2 suggests, the sampling distribution of the chi-square test statistic is *not* a Normal distribution. It is a right-skewed distribution that allows only non-negative values because χ^2 can never be negative.

In Section 7.2, you learned that the sampling distribution of a sample proportion $\hat{p}$ is modeled well by a Normal distribution when the Large Counts condition ($np \geq 10$ and $n(1-p) \geq 10$) is met. There is a similar Large Counts condition for chi-square tests: when the expected counts are all at least 5, the sampling distribution of the χ^2 test statistic is modeled well by a **chi-square distribution** with degrees of freedom (df) equal to the number of categories minus 1. As with the t distributions, there is a different chi-square distribution for each possible value of df.

> **DEFINITION** **Chi-square distribution**
>
> A **chi-square distribution** is defined by a density curve that takes only nonnegative values and is skewed to the right. A particular chi-square distribution is specified by its degrees of freedom.

Figure 12.3 shows the density curves for three members of the chi-square family of distributions. As the degrees of freedom (df) increase, the density curves become less skewed, and larger values become more probable.

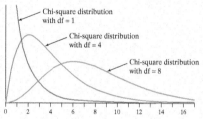

FIGURE 12.3 The density curves for three members of the chi-square family of distributions.

Here are two other interesting facts about the chi-square distributions:

- The mean of a particular chi-square distribution is equal to its degrees of freedom.
- For df > 2, the mode (peak) of the chi-square density curve is at df − 2.

For example, when df = 8, the chi-square distribution has a mean of 8 and a mode of 6.

To get *P*-values from a chi-square distribution, we can use technology or Table C in the back of the book. For Jerome's class data, $\chi^2 = 9.8$. Because all the expected counts are at least 5, the χ^2 test statistic will be modeled well by a chi-square distribution when H_0 is true. There are 6 color categories for M&M'S® Milk Chocolate Candies, so df $= 6 - 1 = 5$.

The *P*-value is the probability of getting a value of χ^2 as large as or larger than 9.8 when H_0 is true. Figure 12.4 shows this probability as an area under the chi-square density curve with 5 degrees of freedom.

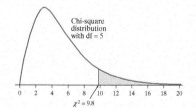

FIGURE 12.4 The *P*-value for a chi-square test for goodness of fit using Jerome's M&M'S Milk Chocolate Candies class data.

Teaching Tip: Using Technology

Remind students that they can watch a video of Technology Corner 28 on the e-Book or at the Student Site.

Teaching Tip: Using Technology

You can also use the *One Categorical Variable* applet to find the *P*-value. Go to the Extra Applets on the Student Site at highschool.bfwpub.com/updatedtps6e. This applet requires the user to input the original observed counts. Then choose "Chi-square goodness-of-fit test" and input the expected values.

Making Connections

We now have three calculator functions that can help us find areas for different continuous probability distributions: normalcdf, tcdf, and χ^2cdf.

AP® EXAM TIP

When stating conclusions for chi-square tests, students often accept the null hypothesis. Points will always be deducted on the AP® Statistics exam for this error. For Jerome's M&M'S® data, even though the evidence is not convincing, there *is* some evidence that the company's claim is incorrect. Consequently, it would be wrong to conclude that the company's claim is correct. Remind your students about the criminal trial analogy: a verdict of "not guilty" doesn't mean the defendant is innocent!

		P	
df	.15	.10	.05
4	6.74	7.78	9.49
5	8.12	9.24	11.07
6	9.45	10.64	12.59

To find the *P*-value using Table C, look in the df = 5 row. The value $\chi^2 = 9.8$ falls between the critical values 9.24 and 11.07. The corresponding areas in the right tail of the chi-square distribution with 5 degrees of freedom are 0.10 and 0.05. So the *P*-value for a test based on Jerome's data is between 0.05 and 0.10. Now let's look at how to find the *P*-value with your calculator.

28. Technology Corner   FINDING *P*-VALUES FOR CHI-SQUARE TESTS

TI-Nspire and other technology instructions are on the book's website at highschool.bfwpub.com/updatedtps6e.

To find the *P*-value in the M&M'S® Milk Chocolate Candies example with your calculator, use the χ^2cdf command. We ask for the area between $\chi^2 = 9.8$ and a very large number (we'll use 10,000) under the chi-square density curve with 5 degrees of freedom.

Press [2nd] [VARS] (DISTR) and choose χ^2cdf(.

OS 2.55 or later: In the dialog box, enter these values: lower:9.8, upper:10000, df:5, choose Paste, and then press [ENTER].

Older OS: Complete the command χ^2cdf(9.8,10000,5) and press [ENTER].

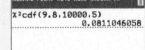

```
NORMAL FLOAT AUTO REAL RADIAN MP
χ²cdf(9.8,10000,5)
                       0.0811046058
```

Table C gives us an interval in which the *P*-value falls. The calculator's χ^2cdf command gives a result that is consistent with Table C but more precise. For that reason, we recommend using your calculator to compute *P*-values from a chi-square distribution.

How do we interpret the *P*-value from Jerome's test? The same way we interpret other *P*-values: Assuming that the claimed distribution of color is correct, there is a 0.081 probability of getting a chi-square statistic of 9.8 or larger by chance alone. The way we make conclusions is the same as well: Because our *P*-value of 0.081 is greater than $\alpha = 0.05$, we fail to reject H_0. We don't have convincing evidence that the distribution of color in the large bag of M&M'S Milk Chocolate Candies differs from the claimed distribution. This is consistent with the results of the simulation from the activity.

 Failing to reject H_0 does not mean that the null hypothesis is true! That is, we can't conclude that the bag has the color distribution claimed by Mars, Inc. All we can say is that the sample data did not provide convincing evidence to reject H_0.

EXAMPLE   Return of the fair die
Finding a *P*-value

PROBLEM: Carrie made a 6-sided die in her ceramics class and rolled it 90 times to test if each side was equally likely to show up. The table summarizes the observed and expected counts.

Outcome of roll	1	2	3	4	5	6	Total
Observed count	12	28	12	13	10	15	90
Expected count	15	15	15	15	15	15	90

In the preceding example, we calculated $\chi^2 = 14.41$. Find the *P*-value using Table C. Then calculate a more precise value using technology. Assume the conditions for inference are met.

ALTERNATE EXAMPLE   Skill 3.E

Return of Reese's Pieces®   Finding a *P*-value

PROBLEM:

The Hershey Company makes Reese's Pieces and claims the distribution of colors is as follows: 50% orange, 25% brown, and 25% yellow. Trey bought a large bag of Reese's Pieces and takes a random sample of 80 pieces. Here are the results:

Color	Orange	Brown	Yellow
Count	31	22	27

In the preceding alternate example, we calculated $\chi^2 = 4.675$. Find the *P*-value using Table C, then calculate a more precise value using technology. Assume the conditions for inference are met.

SOLUTION:

df = 3 − 1 = 2

Using Table C: The *P*-value is between 0.05 and 0.10.

Using technology: χ^2cdf (lower: 4.675, upper: 10000, df: 2) = 0.0966.

SOLUTION:

df = 6 − 1 = 5

Using Table C: **the P-value is between 0.01 and 0.02.**

Using technology:
χ^2cdf(lower:14.41, upper:10000, df:5)
= 0.0132.

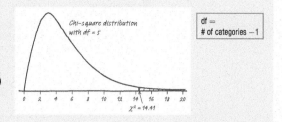

df =
# of categories − 1

FOR PRACTICE, TRY EXERCISE 3

Assuming that Carrie's die is fair, there is only a 0.0132 probability of getting a chi-square statistic of 14.41 or larger by chance alone. Because the P-value of 0.0132 is less than $\alpha = 0.05$, we should reject H_0. There is convincing evidence that Carrie's die is unfair.

CHECK YOUR UNDERSTANDING

Let's continue our analysis of Joey's sample of M&M'S® Peanut Chocolate Candies from the preceding Check Your Understanding (page 766).

1. Confirm that the expected counts are large enough to use a chi-square distribution to calculate the P-value. Which df should you use?
2. Use Table C to find the P-value. Then use technology.
3. Interpret the P-value.
4. What conclusion should you draw about the company's claimed color distribution for M&M'S Peanut Chocolate Candies?

Carrying Out a Test

Like our test for a population proportion, the chi-square test for goodness of fit uses an approximation that becomes more accurate as we take larger random samples. As discussed in the previous subsection, check that the expected counts are all at least 5 (the Large Counts condition) to ensure the sample size is large enough. In addition to the Large Counts condition, we need to check the Random and 10% conditions. Together, these conditions ensure that the observations in the sample can be viewed as independent and that the sampling distribution of the χ^2 test statistic can be modeled by a chi-square distribution.

CONDITIONS FOR PERFORMING A CHI-SQUARE TEST FOR GOODNESS OF FIT

- **Random:** The data come from a random sample from the population of interest.
  - **10%:** When sampling without replacement, $n < 0.10N$.
- **Large Counts:** All *expected* counts are at least 5.

Teaching Tip

Continue to ask students to interpret the P-value. Understanding the meaning of the P-value here (0.0132) is what leads to the correct conclusion (reject H_0).

✓ Answers to CYU

1. The expected counts (13, 13, 13, 13, 6.5, 6.5) are all at least 5. We should use df = 6 − 1 = 5.

2. From Table C, the P-value is greater than 0.25. From the calculator, P-value = χ^2cdf(lower: 2.3847, upper: 10000, df: 5) = 0.7938.

3. Because the P-value of 0.7938 > $\alpha = 0.05$, we fail to reject H_0. There is not convincing evidence that the color distribution of M&M'S® Peanut Chocolate Candies differs from what the company claims.

Teaching Tip

While it is important that students know how to check each condition, it is equally important that they understand *why* we check the condition. We call this the *"So what?"*

Teaching Tip

Here is the *"So what?"* answer for each condition:

Random:
→ so individual observations are independent and we can generalize to the population.

10%:
→ so we can view observations as independent even though we are sampling without replacement.

Large Counts:
→ so the sampling distribution is approximately a chi-square distribution and we can use χ^2 to find a P-value.

Teaching Tip

The Large Counts condition for this test serves the same purpose as the Large Counts condition for tests for proportions: to ensure that the probability distribution we use to calculate the P-value is a good model for the actual sampling distribution of the test statistic we are using.

Teaching Tip

What happens when we violate the 10% condition? When dealing with means or proportions, violating the 10% condition means our formulas for the standard deviation (error) of the statistic give values that are larger than the truth, leading to an overestimate of the true P-value. In a chi-square test for goodness of fit, violating the 10% condition produces a sampling distribution of the χ^2 test statistic with a smaller mean and smaller standard deviation than expected based on the degrees of freedom for the test. Therefore, our P-value estimates from a chi-square distribution will be an overestimate of the truth. In all cases, violating the 10% conditions means our conclusions will be too conservative.

When we want to compare the distribution of a categorical variable in one population to a claimed distribution, we use a **chi-square test for goodness of fit**.

THE CHI-SQUARE TEST FOR GOODNESS OF FIT

Suppose the conditions are met. To perform a test of

H_0: The stated distribution of a categorical variable in the population of interest is correct

compute the chi-square test statistic:

$$\chi^2 = \sum \frac{(\text{Observed count} - \text{Expected count})^2}{\text{Expected count}}$$

where the sum is over all categories. The *P*-value is the area to the right of χ^2 under the chi-square density curve with degrees of freedom = number of categories -1.

The next example shows the chi-square test for goodness of fit in action. As always, we follow the four-step process when performing inference.

EXAMPLE

Birthdays in hockey
A test for equal proportions

PROBLEM: In his book *Outliers*, Malcolm Gladwell suggests that a hockey player's birth month has a big influence on his chance to make it to the highest levels of the game. Specifically, because January 1 is the cut-off date for youth leagues in Canada [where many National Hockey League (NHL) players come from], players born in January will be competing against players up to 12 months younger. The older players tend to be bigger, stronger, and more coordinated and hence get more playing time, more coaching, and have a better chance of being successful.

To see if birth date is related to success (judged by whether a player makes it into the NHL), a random sample of 80 NHL players from a recent season was selected and their birthdays were recorded. The one-way table summarizes the data on birthdays for these 80 players.

Birthday	Jan–Mar	Apr–Jun	Jul–Sep	Oct–Dec
Number of players	32	20	16	12

Do these data provide convincing evidence that the birthdays of NHL players are not uniformly distributed across the four quarters of the year?

ALTERNATE EXAMPLE   Skills 1.E, 4.E

Equal animal crackers
A test for equal proportions

PROBLEM:
Nathan wondered if the different types of animals used for animal crackers were in equal proportions. He took a random sample of 686 animal crackers and got the distribution of animals shown in the table. Do these data provide convincing evidence that the type of animal used for animal crackers is not uniformly distributed?

Animal	Cow	Horse	Buffalo	Moose	Elephant	Camel	Goat	Polar Bear	Donkey	Cat (tail)	Cat (no tail)	Lion	Wombat
Number	40	64	42	52	60	61	44	51	58	48	56	55	55

SOLUTION:
STATE:
H_0: The type of animal used for animal crackers is uniformly distributed.

H_a: The type of animal used for animal crackers is not uniformly distributed.

We'll use $\alpha = 0.05$.

PLAN: Chi-square test for goodness of fit

- *Random:* The data came from a random sample of animal crackers. ✓
  - *10%:* We must assume that 686 is less than 10% of all animal crackers. ✓

- *Large Counts:* All expected counts = $686(1/13) = 52.77 \geq 5$. ✓

DO:
- Test statistic:
$$\chi^2 = \frac{(40 - 52.77)^2}{52.77} + \cdots + \frac{(55 - 52.77)^2}{52.77}$$
$$= 3.09 + 2.39 + 2.20 + 0.01 + 0.99$$
$$+ 1.28 + 1.46 + 0.06 + 0.52 + 0.43$$
$$+ 0.20 + 0.09 + 0.09$$
$$= 12.82$$

- *P*-value: df = 13 − 1 = 12

Table C: The *P*-value is greater than 0.25.

Tech: χ^2cdf (lower: 12.816, upper: 10000, df: 12) = 0.383

CONCLUDE: Because the *P*-value of 0.383 > α = 0.05, we fail to reject H_0. We do not have convincing evidence that the type of animal used for animal crackers is not uniformly distributed.

SOLUTION:

STATE:

H_0: The birthdays of all NHL players are uniformly distributed across the four quarters of the year.

H_a: The birthdays of all NHL players are not uniformly distributed across the four quarters of the year.

We'll use $\alpha = 0.05$.

PLAN: Chi-square test for goodness of fit

- Random: The data came from a random sample of NHL players. ✓
  - 10%: We must assume that 80 is less than 10% of all NHL players. ✓
- Large Counts: All expected counts = $80(1/4) = 20 \geq 5$. ✓

DO:

- Test statistic:

$$\chi^2 = \frac{(32-20)^2}{20} + \frac{(20-20)^2}{20} + \frac{(16-20)^2}{20} + \frac{(12-20)^2}{20}$$

$$= 7.2 + 0 + 0.8 + 3.2 = 11.2$$

- P-value: df = $4 - 1 = 3$

Using Table C: The P-value is between 0.01 and 0.02.

Using technology: χ^2cdf(lower: 11.2, upper: 10000, df: 3)

$$= 0.011$$

CONCLUDE: Because the P-value of $0.011 < \alpha = 0.05$, we reject H_0. We have convincing evidence that the birthdays of NHL players are not uniformly distributed across the four quarters of the year.

Follow the four-step process!

We could write the hypotheses in symbols as
H_0: $p_{\text{Jan-Mar}} = p_{\text{Apr-Jun}} = p_{\text{Jul-Sep}} = p_{\text{Oct-Dec}} = 1/4$
H_a: At least two of the proportions are not 1/4

There were 879 NHL players in the population from which we selected our sample.

There is *some* evidence in favor of H_a because the observed counts differ from the expected counts.

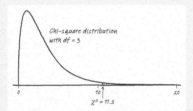

Chi-square distribution with df = 3

$\chi^2 = 11.2$

FOR PRACTICE, TRY EXERCISE 9

You can use your calculator to carry out the "Do" step for a chi-square test for goodness of fit. Remember that using your calculator comes with potential benefits and risks on the AP® Statistics exam.

---

29. Technology Corner   PERFORMING A CHI-SQUARE TEST FOR GOODNESS OF FIT

TI-Nspire and other technology instructions are on the book's website at highschool.bfwpub.com/updatedtps6e.

You can use the TI-84 to perform the calculations for a chi-square test for goodness of fit. We'll use the data from the hockey and birthdays example to illustrate the steps.

1. Enter the observed counts in L1 and the expected counts in L2.

Birthday	Observed	Expected
Jan–Mar	32	20
Apr–Jun	20	20
Jul–Sep	16	20
Oct–Dec	12	20

Teaching Tip: Using Technology

The TI-84 calculator has two different chi-square tests available. For a chi-square test for goodness of fit (Section 12.1), students will use lists and χ^2GOF-Test. For a chi-square test for homogeneity or chi-square test for independence (Section 12.2), students will use matrices and χ^2-Test.

Teaching Tip

Consider having students check the condition *and* give the "*So what?*" Here is the "*So what?*" answer for each condition in this example:

Random:
→ so individual observations are independent and we can generalize to the population of all NHL hockey players in a recent season.

10%:
→ so we can view observations as independent even though we are sampling without replacement.

Large Counts:
→ so the sampling distribution is approximately a chi-square distribution and we can use χ^2 to find a P-value.

⚠ COMMON STUDENT ERROR

Students are familiar with the degrees of freedom formula $n - 1$ from the t distributions, where n is the sample size. The formula looks almost the same for the chi-square test for goodness of fit (number of categories $- 1$). Students will sometimes mistakenly use the sample size when calculating the degrees of freedom, so check in this example that students use df = $4 - 1$, not df = $80 - 1$.

Teaching Tip

Consider having students write a follow-up analysis in their conclusion. See the Teaching Tip at the bottom of page 772.

Making Connections

Consistently ask your students to think about Type I and Type II errors. It is important that students understand these errors are possible in any significance test, not just the ones in Chapter 9, where these errors were introduced. In the NHL example, it is possible we made a Type I error—finding convincing evidence that the birthdays of NHL players are not uniformly distributed when they really are.

Teaching Tip: Using Technology

There are two ways to upgrade the operating system on the TI-84. The easiest is to link an older calculator with an upgraded calculator using a USB or calculator cable. Then press the "LINK" button on both machines. On the receiving machine, arrow over to "RECEIVE" and press "ENTER." On the sending machine, arrow down to "SendOS" and press "ENTER." The other method is to use free connection software and OS updates from education.ti.com.

Teaching Tip

Consider requiring formulas and work during the school year and then relaxing this expectation at the end of the course, when preparing for the AP® Statistics exam.

Teaching Tip

A follow-up analysis is only suggested if the data are statistically significant. If the data are not statistically significant, there is no reason to do further analysis.

AP® EXAM TIP

Over the years, several free response questions have required students to perform a chi-square significance test. The follow-up analysis is not required for full credit when performing the test, but the thinking involved in the follow-up analysis may help students succeed in part (b) of the question. See 2008 #5 and 2016 #2.

Teaching Tip

When doing a follow-up analysis, don't focus only on the size of the contribution—be sure to also discuss the direction of the difference. For example, it's good to say that the component for January to March is the largest, but it is also important to point out that the observed number of players is greater than expected during that time period.

---

2. Press [STAT], arrow over to TESTS, and choose χ^2GOF–Test... .

Note: TI-83s and some older TI-84s don't have this test. TI-84 users can get this functionality by upgrading their operating systems.

3. Enter the inputs shown. If you choose Calculate, you'll get a screen with the test statistic, *P*-value, and df. If you choose the Draw option, you'll get a picture of the appropriate chi-square distribution with the test statistic marked and shaded area corresponding to the *P*-value.

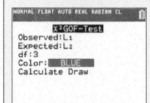  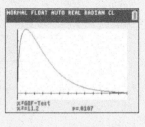

We'll discuss the CNTRB results shortly.

AP® EXAM TIP

You can use your calculator to carry out the mechanics of a significance test on the AP® Statistics exam. But there's a risk involved. If you just give the calculator answer with no work, and one or more of your values is incorrect, you will likely get no credit for the "Do" step. We recommend writing out the first few terms of the chi-square calculation followed by "...". This approach might help you earn partial credit if you enter a number incorrectly. Be sure to name the procedure (chi-square test for goodness of fit) and to report the test statistic ($\chi^2 = 11.2$), degrees of freedom (df = 3), and *P*-value (0.011).

FOLLOW-UP ANALYSIS In the chi-square test for goodness of fit, we test the null hypothesis that a categorical variable has a specified distribution in the population of interest. If the sample data lead to a statistically significant result, we can conclude that our variable has a distribution different from the one stated. To investigate *how* the distribution is different, start by identifying the categories that contribute the most to the chi-square statistic. Then describe how the observed and expected counts differ in those categories, noting the direction of the difference.

Let's return to the hockey and birthdays example. The table of observed and expected counts for the 80 randomly selected NHL players is repeated on the following page. The last column shows the contributions (also called components) of the chi-square test statistic. The two biggest contributions to the chi-square statistic came from Jan–Mar and Oct–Dec. In January through March, 12 *more* players were born than expected. In October through December, 8 *fewer* players were born than expected. These results support Malcolm Gladwell's claim that NHL players are more likely to be born early in the year.

Teaching Tip

Consider having students add a follow-up analysis sentence at the end of the conclusion for the significance test:

CONCLUDE: Because the *P*-value of $0.011 < \alpha = 0.05$, we reject H_0. We have convincing evidence that the birthdays of NHL players are not uniformly distributed across the four quarters of the year. *The biggest components of the χ^2 test statistic are 7.2 and 3.2 because in January through March, 12 more players than expected were born, and in October through December, 8 fewer players than expected were born.*

Note: The follow-up analysis does not have to provide discussion of the *two* largest contributors to the χ^2 test statistic. Discussion of *one* is often enough.

Consider requiring students to do a follow-up analysis during the school year and then relaxing this expectation at the end of the course, when preparing for the AP® Statistics exam.

Birthday	Observed	Expected	O − E	$(O − E)^2/E$
Jan–Mar	32	20	12	7.2
Apr–Jun	20	20	0	0.0
Jul–Sep	16	20	−4	0.8
Oct–Dec	12	20	−8	3.2

Note: When we ran the chi-square test for goodness of fit on the calculator, a list of these individual components was produced and stored in the list menu. On the TI-84, the list is called CNTRB (for contribution).

CHECK YOUR UNDERSTANDING

Does the warm, sunny weather in Arizona affect a driver's choice of car color? Cass thinks that Arizona drivers might opt for a lighter color with the hope that it will reflect some of the heat from the sun. To see if the distribution of car colors in Oro Valley, near Tucson, is different from the distribution of car colors across North America, she selected a random sample of 300 cars in Oro Valley. The table shows the distribution of car color for Cass's sample in Oro Valley and the distribution of car color in North America, according to www.ppg.com.[1]

Color	White	Black	Gray	Silver	Red	Blue	Green	Other	Total
Oro Valley sample	84	38	31	46	27	29	6	39	300
North America	23%	18%	16%	15%	10%	9%	2%	7%	100%

1. Do these data provide convincing evidence that the distribution of car color in Oro Valley differs from the North American distribution?
2. If there is convincing evidence of a difference in the distribution of car color, perform a follow-up analysis.

Section 12.1 | Summary

- The **chi-square test for goodness of fit** tests the null hypothesis that a categorical variable has a specified distribution in the population of interest. The alternative hypothesis is that the variable does not have the specified distribution in the population of interest.
- This test compares the **observed count** in each category with the counts that would be expected if H_0 were true. The **expected count** for any category is found by multiplying the sample size by the proportion in each category according to the null hypothesis.
- The **chi-square test statistic** is

$$\chi^2 = \sum \frac{(\text{Observed count} - \text{Expected count})^2}{\text{Expected count}}$$

where the sum is over all possible categories.

✓ Answers to CYU

1. STATE: H_0: The distribution of car colors in Oro Valley is the same as the distribution of car colors across North America.

H_a: The distribution of car colors in Oro Valley is not the same as the distribution of car colors across North America. We'll use $\alpha = 0.05$.

PLAN: Chi-square test for goodness of fit. *Random:* The data come from a random sample of 300 cars in Oro Valley. *10%:* $n = 300$ is less than 10% of all cars in Oro Valley. *Large Counts:* All expected counts (69, 54, 48, 45, 30, 27, 6, 21) are at least 5.

DO: Test statistic:

$$\chi^2 = \frac{(84 - 69)^2}{69} + \frac{(38 - 54)^2}{54} + \cdots$$

$$= 29.921$$

P-value: df = 8 − 1 = 7

Table C: P-value is less than 0.0005.

Tech: χ^2cdf(lower: 29.921, upper: 10000, df: 7) ≈ 0

CONCLUDE: Because the *P*-value of approximately $0 < \alpha = 0.05$, we reject H_0. We have convincing evidence that the distribution of car colors in Oro Valley is not the same as the distribution of car colors across North America.

TRM Section 12.1 Quiz

There is one quiz available for this section in the Teacher's Resource Materials. Click on the link in the TE-Book, look on the TRFD, or download from the Teacher's Resources on the book's digital platform. You can also create your own quiz using the ExamView® Assessment Suite that is part of the TPS6 program. Questions are coded by Learning Target to make it easy to build parallel quizzes.

2. The table of chi-square contributions is shown below.

Color	Observed	Expected	O − E	$(O − E)^2/E$
White	84	69	15	3.2609
Black	38	54	−16	4.7407
Gray	31	48	−17	6.0208
Silver	46	45	1	0.0222
Red	27	30	−3	0.03
Blue	29	27	2	0.1482
Green	6	6	0	0
Other	39	21	18	15.429

The two biggest contributions to the chi-square statistic came from gray and other colored cars. There were fewer gray cars than expected and more other-colored cars than expected. As for Cass's question: It does seem that drivers in Oro Valley prefer lighter-colored cars as there were more white cars than expected and fewer black cars than expected, so it seems Cass might be onto something!

Teaching Tip

Make sure to point out the two icons next to Exercise 1. The "pg 764" icon reminds students that the example on page 764 is much like this exercise. The "play" icon reminds students that there is a video solution available in the student e-Book or at the Student Site.

Teaching Tip

Many solutions provided in the Annotated Teacher's Edition have been shortened to fit in the margin.

Answers to Section 12.1 Exercises

12.1 (a) H_0: The company's claimed distribution for its deluxe mixed nuts is correct. H_a: The company's claimed distribution is not correct. **(b)** Cashews: $150(0.52) = 78$; Almonds: $150(0.27) = 40.5$; Macadamia nuts: $150(0.13) = 19.5$; and Brazil nuts: $150(0.08) = 12$.

(c)

$$\chi^2 = \frac{(83 - 78)^2}{78} + \frac{(29 - 40.5)^2}{40.5} + \cdots$$
$$= 6.599$$

12.2 (a) H_0: The distribution of outcomes is what it should be (the wheel is fair). H_a: The distribution of outcomes is not what it should be (the wheel is not fair). Or

(b) Red: $200\left(\dfrac{18}{38}\right) = 94.74$

Black: $200\left(\dfrac{18}{38}\right) = 94.74$

Green: $200\left(\dfrac{2}{38}\right) = 10.53$

(c) $\chi^2 = \dfrac{(85 - 94.74)^2}{94.74} + \dfrac{(99 - 94.74)^2}{94.74}$

$+ \dfrac{(16 - 10.53)^2}{10.53} = 4.034$

12.3 (a) The P-value (0.061) is between 0.05 and 0.10. **(b)** The P-value (0.0003) is less than 0.0005.

- To ensure the observations in the sample can be viewed as independent and that the sampling distribution of the χ^2 test statistic can be modeled by a chi-square distribution, the required conditions are:
  - **Random:** The data come from a random sample from the population of interest.
    - ◦ **10%:** When sampling without replacement, $n < 0.10N$.
  - **Large Counts:** All *expected* counts are at least 5.
- Large values of χ^2 are evidence against H_0 and in favor of H_a. The P-value is the area to the right of χ^2 under the chi-square density curve with degrees of freedom df = number of categories $- 1$.
- If the test finds a statistically significant result, consider doing a **follow-up analysis** that looks for the largest contributions to the chi-square test statistic and compares the observed and expected counts.

12.1 Technology Corners

TI-Nspire and other technology instructions are on the book's website at highschool.bfwpub.com/updatedtps6e.

28. Finding P-values for chi-square tests — page 768
29. Performing a chi-square test for goodness of fit — page 771

Section 12.1   Exercises

1. **Aw, nuts!** A company claims that each batch of its deluxe mixed nuts contains 52% cashews, 27% almonds, 13% macadamia nuts, and 8% Brazil nuts. To test this claim, a quality-control inspector takes a random sample of 150 nuts from the latest batch. The table displays the sample data.

Type of nut	Cashew	Almond	Macadamia	Brazil
Count	83	29	20	18

(a) State appropriate hypotheses for performing a test of the company's claim.

(b) Calculate the expected count for each type of nut.

(c) Calculate the value of the chi-square test statistic.

2. **Roulette** Casinos are required to verify that their games operate as advertised. American roulette wheels have 38 slots—18 red, 18 black, and 2 green. In one casino, managers record data from a random sample of 200 spins of one of their American roulette wheels. The table displays the results.

Color	Red	Black	Green
Count	85	99	16

(a) State appropriate hypotheses for testing whether these data give convincing evidence that the

distribution of outcomes on this wheel is not what it should be.

(b) Calculate the expected count for each color.

(c) Calculate the value of the chi-square test statistic.

3. **P-values** For each of the following, find the P-value using Table C. Then calculate a more precise value using technology.

(a) $\chi^2 = 19.03$, df $= 11$ (b) $\chi^2 = 19.03$, df $= 3$

4. **More P-values** For each of the following, find the P-value using Table C. Then calculate a more precise value using technology.

(a) $\chi^2 = 4.49$, df $= 5$ (b) $\chi^2 = 4.49$, df $= 1$

5. **Aw, nuts!** Refer to Exercise 1.

(a) Confirm that the expected counts are large enough to use a chi-square distribution to calculate the P-value. What degrees of freedom should you use?

(b) Use Table C to find the P-value. Then use your calculator's χ^2cdf command.

(c) Interpret the P-value.

(d) What conclusion would you draw about the company's claimed distribution for its deluxe mixed nuts?

12.4 (a) The P-value (0.481) is greater than 0.25. **(b)** The P-value (0.034) is between 0.025 and 0.05.

12.5 (a) All expected counts (see Exercise 12.1) are ≥ 5; df $= 3$. **(b)** The P-value (0.0858) is between 0.05 and 0.10. **(c)** Assuming the company's claimed distribution for its deluxe mixed nuts is correct, there is a 0.0858 probability of getting a chi-square statistic of 6.599 or larger by chance alone. **(d)** Because the P-value of $0.0858 > \alpha = 0.05$, we fail to reject H_0. We do not have convincing evidence that the company's claimed distribution for its deluxe mixed nuts is not correct.

6. **Roulette** Refer to Exercise 2.

(a) Confirm that the expected counts are large enough to use a chi-square distribution to calculate the *P*-value. What degrees of freedom should you use?

(b) Use Table C to find the *P*-value. Then use your calculator's χ^2cdf command.

(c) Interpret the *P*-value.

(d) What conclusion would you draw about whether or not the roulette wheel is operating correctly?

7. **No chi-square** A school's principal wants to know if students spend about the same amount of time on homework each night of the week. She asks a random sample of 50 students to keep track of their homework time for a week. The following table displays the average amount of time (in minutes) students reported per night.

Night	Sunday	Monday	Tuesday	Wednesday	Thursday	Friday	Saturday
Average time	130	108	115	104	99	37	62

Explain carefully why it would *not* be appropriate to perform a chi-square test for goodness of fit using these data.

8. **No chi-square** The principal in Exercise 7 also asked the random sample of students to record whether they did all of the homework that was assigned on each of the five school days that week. Here are the data:

School day	Monday	Tuesday	Wednesday	Thursday	Friday
Number who did all homework	34	29	32	28	19

Explain carefully why it would *not* be appropriate to perform a chi-square test for goodness of fit using these data.

9. **Munching Froot Loops** Kellogg's Froot Loops cereal comes in six colors: orange, yellow, purple, red, blue, and green. Charise randomly selected 120 loops and noted the color of each. Here are her data:

pg 770

Color	Orange	Yellow	Purple	Red	Blue	Green
Count	28	21	16	25	14	16

Do these data provide convincing evidence at the 5% significance level that Kellogg's Froot Loops do not contain an equal proportion of each color?

10. **What's your sign?** The University of Chicago's General Social Survey (GSS) is the nation's most important social science sample survey. For reasons known only to social scientists, the GSS regularly asks a random sample of people their astrological sign. Here are the counts of responses from a recent GSS of 4344 people:

Sign	Aries	Taurus	Gemini	Cancer	Leo	Virgo
Count	321	360	367	374	383	402
Sign	Libra	Scorpio	Sagittarius	Capricorn	Aquarius	Pisces
Count	392	329	331	354	376	355

If births are spread uniformly across the year, we expect all 12 signs to be equally likely. Do these data provide convincing evidence at the 1% significance level that all 12 signs are *not* equally likely?

11. **Fruit flies** Biologists wish to mate pairs of fruit flies having genetic makeup RrCc, indicating that each has one dominant gene (R) and one recessive gene (r) for eye color, along with one dominant (C) and one recessive (c) gene for wing type. Each offspring will receive one gene for each of the two traits from each parent, so the biologists predict that the following phenotypes should occur in a ratio of 9:3:3:1.

Phenotype	Red eyes and straight wings	Red eyes and curly wings	White eyes and straight wings	White eyes and curly wings
Frequency	99	42	49	10

Assume that the conditions for inference are met. Carry out a test at the $\alpha = 0.05$ significance level of the proposed genetic model.

12. **You say tomato** The paper "Linkage Studies of the Tomato" (*Transactions of the Canadian Institute*, 1931) reported the following data on phenotypes resulting from crossing tall cut-leaf tomatoes with dwarf potato-leaf tomatoes. We wish to investigate whether the following frequencies are consistent with genetic laws, which state that the phenotypes should occur in the ratio 9:3:3:1.

Phenotype	Tall cut	Tall potato	Dwarf cut	Dwarf potato
Frequency	926	288	293	104

Assume that the conditions for inference are met. Carry out a test at the $\alpha = 0.05$ significance level of the proposed genetic model.

13. **Birds in the trees** Researchers studied the behavior of birds that were searching for seeds and insects in an Oregon forest. In this forest, 54% of the trees are Douglas firs, 40% are ponderosa pines, and 6% are other types of trees. At a randomly selected time during the day, the researchers observed 156 red-breasted nuthatches: 70 were seen in Douglas firs, 79 in ponderosa pines, and 7 in other types of trees.[2]

(a) Do these data provide convincing evidence that nuthatches prefer particular types of trees when they're searching for seeds and insects?

(b) Relative to the proportion of each tree type in the forest, which type of trees do the nuthatches seem to prefer the most? The least?

12.6 (a) All are ≥ 5; df = 2.

(b)

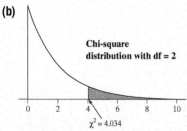

Chi-square distribution with df = 2

$\chi^2 = 4.034$

The *P*-value (0.133) is between 0.10 and 0.15. **(c)** Assuming the distribution of outcomes is what it should be (the wheel is fair), there is a 0.133 probability of getting a chi-square statistic of 4.034 or larger by chance alone. **(d)** Because the *P*-value of 0.133 $> \alpha = 0.05$, we fail to reject H_0. We do not have convincing evidence that the distribution of outcomes is not what it should be.

12.7 Time spent doing homework is quantitative. Chi-square tests for goodness of fit should be used only for distributions of categorical data.

12.8 The data do not describe the distribution of a single categorical variable. Homework completion status (yes or no) is being recorded for each student in the sample on 5 different days of the week.

12.9 S: H_0: Froot Loops contain an equal proportion of each color. H_a: Froot Loops do not contain an equal proportion of each color; $\alpha = 0.05$. **P:** Chi-square test for goodness of fit. *Random:* Random sample of Froot Loops. *10%:* $n = 120 < 10\%$ of all Froot Loops. *Large Counts:* All expected counts $= 120(1/6) = 20 \geq 5$. **D:** $\chi^2 = 7.9$; df = 5. The *P*-value (0.1618) is between 0.15 and 0.20. **C:** Because the *P*-value of 0.1618 $> \alpha = 0.05$, we fail to reject H_0. We do not have convincing evidence Froot Loops do not contain an equal proportion of each color.

12.10 S: H_0: All 12 astrological signs are equally likely. H_a: All 12 astrological signs are not equally likely; $\alpha = 0.01$. **P:** Chi-square test for goodness of fit. *Random:* Random sample of people. *10%:* $n = 4344 < 10\%$ of all people. *Large Counts:* All expected counts $=$
$$4344\left(\frac{1}{12}\right) = 362 \geq 5.$$

D: $\chi^2 = 19.76$; df = 11. The *P*-value (0.0487) is between 0.025 and 0.05. **C:** Because the *P*-value of 0.0487 $> \alpha = 0.01$, we fail to reject H_0. There is not convincing evidence the 12 astrological signs are not equally likely.

12.11 S: H_0: The distribution of eye color and wing shape is the same as what the biologists predict. H_a: The distribution of eye color and wing shape is not the same as what the biologists predict; $\alpha = 0.05$. **P:** Chi-square test for goodness of fit. The conditions are met. **D:** $\chi^2 = 6.187$; df = 3. The *P*-value (0.1029) is between 0.10 and 0.15. **C:** Because the *P*-value of 0.1029 $> \alpha = 0.05$, we fail to reject H_0. We do not have convincing evidence that the distribution of eye color and wing shape differs from what the biologists predict.

12.12 S: H_0: The proposed 9:3:3:1 genetic model is correct. H_a: The proposed 9:3:3:1 genetic model is not correct; $\alpha = 0.05$. **P:** Chi-square test for goodness of fit. The conditions are met. **D:** $\chi^2 = 1.469$; df = 3. The *P*-value (0.6895) is greater than 0.25. **C:** Because the *P*-value of 0.6895 $> \alpha = 0.05$, we fail to reject H_0. We do not have convincing evidence that the proposed 9:3:3:1 genetic model is not correct.

12.13 (a) S: H_0: Nuthatches do not prefer particular types of trees when searching for seeds and insects. H_a: Nuthatches do prefer particular types; $\alpha = 0.05$. **P:** Chi-square test for goodness of fit. *Random:* Random sample of 156 red-breasted nuthatches. *10%:* $n = 156 < 10\%$ of all nuthatches. *Large Counts:* 84.24, 62.4, and 9.36 are all ≥ 5. **D:** $\chi^2 = 7.418$; df = 2. The *P*-value (0.0245) is between 0.02 and 0.025. **C:** Because the *P*-value of 0.0245 $< \alpha = 0.05$, we reject H_0. There is convincing evidence that nuthatches prefer particular types of trees when searching for seeds and insects. **(b)** The breakdown of the chi-square statistic is $\chi^2 = 2.407 + 4.416 + 0.595$. The largest contributors are Douglas firs and ponderosa pines. Fewer red-breasted nuthatches are observed in Douglas firs $(70 - 84.24 = -14.24)$ and more in ponderosa pines $(79 - 62.4 = 16.6)$ than expected, so the nuthatches seem to prefer ponderosa pines the most and Douglas firs the least.

12.14 (a) S: H_0: Seagulls do not have a preference for where they land. H_a: Seagulls do have a preference for where they land; $\alpha = 0.05$. **P:** Chi-square test for goodness of fit. *Random:* Random sample of 200 seagulls. *10%:* $n = 200 < 10\%$ of all seagulls. *Large Counts:* 112, 58, and 30 are all ≥ 5. **D:** $\chi^2 = 14.474$; df = 2. The *P*-value (0.0007) is between 0.0005 and 0.001. **C:** Because the *P*-value of $0.0007 < \alpha = 0.05$, we reject H_0. There is convincing evidence that seagulls have a preference where they land. **(b)** The breakdown of the chi-square statistic is $\chi^2 = 2.286 + 0.155 + 12.033$. The largest contributors are sand and rocks. More seagulls are observed landing on sand $(128 - 112 = 16)$ and fewer land on rocks $(11 - 30 = -19)$ than expected, so the seagulls seem to prefer the sand most and the rocks least.

12.15 (a) S: H_0: Mendel's 3:1 genetic model is correct. H_a: Mendel's 3:1 genetic model is not correct; at $\alpha = 0.05$. **P:** Chi-square test for goodness of fit. The conditions are met. **D:** $\chi^2 = 0.3453$; df = 1. The *P*-value (0.5568) is greater than 0.25. **C:** Because the *P*-value of $0.5568 > \alpha = 0.05$, we fail to reject H_0. We do not have convincing evidence that Mendel's 3:1 genetic model is not correct. **(b)** H_0: $p = 0.75$, H_a: $p \neq 0.75$, where p = the true proportion of peas that will be smooth; at $\alpha = 0.05$; $z = 0.5876$; *P*-value = 0.5568. **C:** Because the *P*-value of $0.5568 > \alpha = 0.05$, we fail to reject H_0. We do not have convincing evidence that the true proportion of peas that are smooth differs from 0.75. In both cases, we fail to reject the null hypothesis. The *P*-value for each test has the same value, and $z^2 = \chi^2$!

12.16 (a) S: H_0: Heads and tails are equally likely when the Belgian euro coin is flipped. H_a: Heads and tails are not equally likely when the Belgian euro coin is flipped; at $\alpha = 0.05$. **P:** Chi-square test for goodness of fit. The conditions are met. **D:** $\chi^2 = 3.6$; df = 1. The *P*-value (0.0578) is between 0.05 and 0.10. **C:** Because the *P*-value of $0.0578 > \alpha = 0.05$, we fail to reject H_0. We do not have convincing evidence that heads and tails are not equally likely when the coin is flipped. **(b)** H_0: $p = 0.50$, H_a: $p \neq 0.50$, where p = true proportion of flipped Belgian euro coins that will land heads up; at $\alpha = 0.05$; $z = 1.897$; *P*-value = 0.0578. **C:** Because the *P*-value of $0.0578 > \alpha = 0.05$, we fail to reject H_0. We do not have convincing evidence the true proportion of flipped Belgian euro coins landing heads up differs from 0.50. In both cases, we fail to reject the null hypothesis. The *P*-value for each test has the same value, and $z^2 = \chi^2$!

14. **Seagulls by the seashore** Do seagulls show a preference for where they land? To answer this question, biologists conducted a study in an enclosed outdoor space with a piece of shore whose area was made up of 56% sand, 29% mud, and 15% rocks. The biologists chose 200 seagulls at random. Each seagull was released into the outdoor space on its own and observed until it landed somewhere on the piece of shore. In all, 128 seagulls landed on the sand, 61 landed in the mud, and 11 landed on the rocks.

(a) Do these data provide convincing evidence that seagulls show a preference for where they land?

(b) Relative to the proportion of each ground type on the shore, which type of ground do the seagulls seem to prefer the most? The least?

15. **Mendel and the peas** Gregor Mendel (1822–1884), an Austrian monk, is considered the father of genetics. Mendel studied the inheritance of various traits in pea plants. One such trait is whether the pea is smooth or wrinkled. Mendel predicted a ratio of 3 smooth peas for every 1 wrinkled pea. In one experiment, he observed 423 smooth and 133 wrinkled peas. Assume that the conditions for inference are met.

(a) Carry out a chi-square test for goodness of fit for the genetic model that Mendel predicted.

(b) In Chapter 9, Exercise 49, you tested Mendel's prediction using a one-sample z test for a proportion. The hypotheses were H_0: $p = 0.75$ and H_a: $p \neq 0.75$, where p = the true proportion of smooth peas. Calculate the z statistic and *P*-value for this test. How do these values compare to the values from part (a)?

16. **Spinning heads?** When a fair coin is flipped, we all know that the probability the coin lands on heads is 0.50. However, what if a coin is spun? According to the article "Euro Coin Accused of Unfair Flipping" in the *New Scientist*, two Polish math professors and their students spun a Belgian euro coin 250 times. It landed heads 140 times. One of the professors concluded that the coin was minted asymmetrically. A representative from the Belgian mint indicated the result was just chance. Assume that the conditions for inference are met.

(a) Carry out a chi-square test for goodness of fit to test if heads and tails are equally likely when a euro coin is spun.

(b) In Chapter 9, Exercise 50, you analyzed these data with a one-sample z test for a proportion. The hypotheses were H_0: $p = 0.5$ and H_a: $p \neq 0.5$, where p = the true proportion of heads. Calculate the z statistic and *P*-value for this test. How do these values compare to the values from part (a)?

17. **Is your random number generator working?** Use your calculator's RandInt function to generate 200 digits from 0 to 9 and store them in a list.

(a) State appropriate hypotheses for a chi-square test for goodness of fit to determine whether your calculator's random number generator gives each digit an equal chance of being generated.

(b) Carry out a test at the $\alpha = 0.05$ significance level. *Hint:* To obtain the observed counts, make a histogram of the list containing the 200 random digits, and use the trace feature to see how many of each digit were generated. You may have to adjust your window to go from -0.5 to 9.5 with an increment of 1.

(c) Assuming that a student's calculator is working properly, what is the probability that the student will make a Type I error in part (b)?

(d) Suppose that 25 students in an AP® Statistics class independently do this exercise for homework and that all of their calculators are working properly. Find the probability that at least one of them makes a Type I error.

18. **Skittles®** Statistics teacher Jason Molesky contacted Mars, Inc., to ask about the color distribution for Skittles candies. Here is an excerpt from the response he received: "The original flavor blend for the Skittles Bite Size Candies is lemon, green apple, orange, strawberry and grape. They were chosen as a result of consumer preference tests we conducted. The flavor blend is 20 percent of each flavor."

(a) State appropriate hypotheses for a significance test of the company's claim.

(b) Find the expected counts for a random sample of 60 candies.

(c) How large a χ^2 test statistic would you need to have significant evidence against the company's claim at the $\alpha = 0.05$ level? At the $\alpha = 0.01$ level?

(d) Create a set of observed counts for a random sample of 60 candies that gives a *P*-value between 0.01 and 0.05. Show the calculation of your chi-square test statistic.

Multiple Choice: *Select the best answer for Exercises 19–22.*

Exercises 19–21 refer to the following setting. The manager of a high school cafeteria is planning to offer several new types of food for student lunches in the new school year. She wants to know if each type of food will be equally popular so she can start ordering supplies and making other plans. To find out, she selects a random sample of 100 students and asks them, "Which type of food do you prefer: Ramen, tacos, pizza, or hamburgers?" Here are her data:

Type of food	Ramen	Tacos	Pizza	Hamburgers
Count	18	22	39	21

12.17 (a) H_0: Each of the digits 0–9 from my random number generator is equally likely. H_a: Each of the digits 0–9 from my random number generator is not equally likely. **(b) P:** Chi-square test for goodness of fit. *Random:* Random sample. *10%:* Not needed, because we are not sampling without replacement. *Large Counts:* Each expected count = $200(0.10) = 20 \geq 5$. **D:** In one test of the calculator, we got 18 zeroes, 22 ones, 23 twos, 21 threes, 21 fours, 21 fives, 17 sixes, 14 sevens, 21 eights, and 22 nines; $\chi^2 = 3.5$; df = 9. The *P*-value (0.9411) is greater than 0.25. **C:** Because the *P*-value of $0.9411 > \alpha = 0.05$, we fail to reject H_0. We don't have convincing evidence that each of the digits 0–9 from my random number generator is not equally likely. **(c)** Because we are using $\alpha = 0.05$, there is a 0.05 probability of committing a Type I error, assuming the random number generator is working properly. **(d)** P(at least one Type I error) = $1 - (0.95)^{25} = 0.723$; there is a 0.723 probability that at least one student makes a Type I error, assuming all random number generators are working properly.

12.18 (a) H_0: The true distribution of flavors for Skittles® candies is the same as the company's claim. H_a: The true distribution of flavors for Skittles candies is not the same as the company's claim. **(b)** Each expected count = $60(0.2) = 12 \geq 5$. **(c)** Using df = 4 and Table C, the value for $\alpha = 0.05$ is 9.49 and for $\alpha = 0.01$ is 13.28. So χ^2 statistics greater than 9.49 would provide significant evidence at the $\alpha = 0.05$ level and χ^2 values greater than 13.28 would provide significant evidence at the $\alpha = 0.01$ level. **(d)** One possibility is 6 lemon, 6 green apple, 16 orange, 16 strawberry, and 16 grape; $\chi^2 = 10$, which is between 9.49 and 13.28.

19. An appropriate null hypothesis to test whether the food choices are equally popular is

(a) $H_0: \mu = 25$, where μ = the mean number of students that prefer each type of food.

(b) $H_0: p = 0.25$, where p = the proportion of all students who prefer ramen.

(c) $H_0: n_R = n_T = n_P = n_H = 25$, where n_R is the number of students in the school who would choose ramen, and so on.

(d) $H_0: p_R = p_T = p_P = p_H = 0.25$, where p_R is the proportion of students in the school who would choose ramen, and so on.

(e) $H_0: \hat{p}_R = \hat{p}_T = \hat{p}_P = \hat{p}_H = 0.25$, where $\hat{p}_R$ is the proportion of students in the sample who chose ramen, and so on.

20. The value of the chi-square test statistic is

(a) $\dfrac{(18-25)^2}{25} + \dfrac{(22-25)^2}{25} + \dfrac{(39-25)^2}{25} + \dfrac{(21-25)^2}{25}$

(b) $\dfrac{(25-18)^2}{18} + \dfrac{(25-22)^2}{22} + \dfrac{(25-39)^2}{39} + \dfrac{(25-21)^2}{21}$

(c) $\dfrac{(18-25)}{25} + \dfrac{(22-25)}{25} + \dfrac{(39-25)}{25} + \dfrac{(21-25)}{25}$

(d) $\dfrac{(18-25)^2}{100} + \dfrac{(22-25)^2}{100} + \dfrac{(39-25)^2}{100} + \dfrac{(21-25)^2}{100}$

(e) $\dfrac{(0.18-0.25)^2}{0.25} + \dfrac{(0.22-0.25)^2}{0.25} + \dfrac{(0.39-0.25)^2}{0.25}$
$+ \dfrac{(0.21-0.25)^2}{0.25}$

21. The P-value for a chi-square test for goodness of fit is 0.0129. Which of the following is the most appropriate conclusion at a significance level of 0.05?

(a) Because 0.0129 is less than $\alpha = 0.05$, reject H_0. There is convincing evidence that the food choices are equally popular.

(b) Because 0.0129 is less than $\alpha = 0.05$, reject H_0. There is not convincing evidence that the food choices are equally popular.

(c) Because 0.0129 is less than $\alpha = 0.05$, reject H_0. There is convincing evidence that the food choices are not equally popular.

(d) Because 0.0129 is less than $\alpha = 0.05$, fail to reject H_0. There is not convincing evidence that the food choices are equally popular.

(e) Because 0.0129 is less than $\alpha = 0.05$, fail to reject H_0. There is convincing evidence that the food choices are equally popular.

22. Which of the following is *false*?

(a) A chi-square distribution with k degrees of freedom is more right-skewed than a chi-square distribution with $k+1$ degrees of freedom.

(b) A chi-square distribution never takes negative values.

(c) The degrees of freedom for a chi-square test are determined by the sample size.

(d) $P(\chi^2 > 10)$ is greater when df $= k+1$ than when df $= k$.

(e) The area under a chi-square density curve is always equal to 1.

Recycle and Review

23. **Video games** (1.1) To determine if there is a relationship between age and playing video games, the Pew Research Center asked randomly selected adults for their age and if they "ever play video games on a computer, TV, game console, or portable device like a cell phone."[3] Here is a two-way table summarizing the results of the study:

		\multicolumn{5}{c}{Age group}				
		18–29	30–49	50–64	65+	Total
Play video games	Yes	887	1217	650	279	3033
	No	429	872	985	840	3126
	Total	1316	2089	1635	1119	6159

(a) Construct a segmented bar graph to display the relationship between age group and response to the question about video games.

(b) Describe the association shown in the segmented bar graph in part (a).

Exercises 24–26 refer to the following setting. Do students who read more books for pleasure tend to earn higher grades in English? The boxplots show data from a simple random sample of 79 students at a large high school. Students were classified as light readers if they read fewer than 3 books for pleasure per year. Otherwise, they were classified as heavy readers. Each student's average English grade for the previous two marking periods was converted to a GPA scale, where A = 4.0, A− = 3.7, B+ = 3.3, and so on.

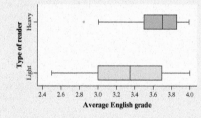

12.19 d

12.20 a

12.21 c

12.22 c

12.23 (a)

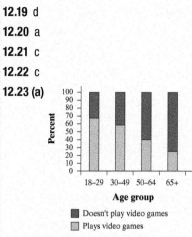

(b) There is an association between age group and playing video games for the subjects in the study. As age increases, the proportion of adults who play video games decreases.

12.24 The distribution of English grades for the heavy readers is skewed to the left, while for the light readers, it is roughly symmetric. There is one low outlier in the heavy reading group but no outliers in the light reading group. The center of the distribution of English grades is greater for the heavy readers than for the light readers, indicating that heavy readers typically get higher grades in English.

12.25 (a) Conditions are met. *Random:* Independent, because knowing the response of a heavy reader shouldn't help us predict the response of a light reader. *10%:* $n_1 = 47 < 10\%$ of heavy readers and $n_2 = 32 < 10\%$ of light readers at this large school. *Normal/Large Sample:* $n_1 = 47 \geq 30$ and $n_2 = 32 \geq 30$. **(b) S:** μ_1 = true mean English grade of heavy readers and μ_2 = true mean English grade of light readers. **P:** Two-sample t interval for $\mu_1 - \mu_2$. **D:** df = 31; using Table B and df = 30, (0.1163, 0.4517). *Tech:* (0.1197, 0.4483) with df = 59.46. **C:** We are 95% confident that the interval from 0.1197 to 0.4483 captures the true difference in the mean English grade of heavy and light readers. **(c)** No; even though 0 is not in the confidence interval, this was an observational study, so no conclusion of cause and effect can be made.

12.26 (a) *Slope:* 0.024; the predicted average English grade goes up by 0.024 point for each increase of 1 book read. *y intercept:* 3.42; the predicted English grade is about 3.42 for a student who has read 0 books. **(b)** $\hat{y} = 3.42 + 0.024(17) = 3.828$; the residual = $2.85 - 3.828 = -0.978$. The actual English grade for this student was 0.978 point less than predicted by the regression line with $x = 17$ books read. **(c)** About 8.3% of the variability in English grades is accounted for by the least-squares regression line with x = number of books read.

PD **Section 12.2 Overview**

TRM **Section 12.2 Alternate Examples**

Making Connections

For both the chi-square test for homogeneity and the chi-square test for independence, the data will be presented in a two-way table. The formulas and calculations for both tests are identical. The difference between the two tests lies in the way the data are collected, which will affect the conclusions we can make at the end of each test.

24. **Reading and grades** (1.3) Write a few sentences comparing the distributions of average English grade for light and heavy readers.

25. **Reading and grades** (10.2) Summary statistics for the two groups from Minitab are provided.

Type of reader	n	Mean	StDev	SE Mean
Heavy	47	3.640	0.324	0.047
Light	32	3.356	0.380	0.067

(a) Explain why it is acceptable to use two-sample t procedures in this setting.

(b) Construct and interpret a 95% confidence interval for the difference in the mean English grade for light and heavy readers.

(c) Does the interval in part (b) provide convincing evidence that reading more causes a difference in students' English grades? Justify your answer.

26. **Reading and grades** (3.2) The scatterplot shows the number of books read and the English grade for all 79 students in the study. The least-squares regression line $\hat{y} = 3.42 + 0.024x$ has been added to the graph.

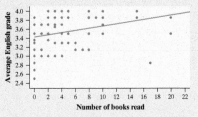

(a) Interpret the slope and y intercept.

(b) The student who reported reading 17 books for pleasure had an English GPA of 2.85. Calculate and interpret this student's residual.

(c) For this linear model, $r^2 = 0.083$. Interpret this value.

SECTION 12.2 **Inference for Two-Way Tables**

LEARNING TARGETS *By the end of the section, you should be able to:*

- State appropriate hypotheses and compute the expected counts and chi-square test statistic for a chi-square test based on data in a two-way table.

- State and check the Random, 10%, and Large Counts conditions for a chi-square test based on data in a two-way table.

- Calculate the degrees of freedom and *P*-value for a chi-square test based on data in a two-way table.

- Perform a chi-square test for homogeneity.

- Perform a chi-square test for independence.

- Choose the appropriate chi-square test in a given setting.

The two-sample z tests in Section 9.3 allow us to compare the proportions of successes in two populations or for two treatments. What if we want to compare more than two samples or groups? More generally, what if we want to compare the distributions of a single categorical variable across several populations or treatments? We rely on a new significance test, called a *chi-square test for homogeneity*.

The test for homogeneity starts by presenting the data in a two-way table. However, two-way tables have other uses than comparing distributions of a single categorical variable. As we saw in Section 1.1, they can also be used to summarize relationships between two categorical variables. To determine if there is convincing evidence of an association between two categorical variables in a single population, we perform a *chi-square test for independence*.

Teaching Tip

Here is a way to introduce the chi-square test for homogeneity. Ask students if they think the distribution of color is the same for different types of M&M'S® candies. Take a random sample of two or more types of M&M'S and use the chi-square test for homogeneity to analyze the data. *Note:* The following percents are NOT needed to do a chi-square test for homogeneity, as students would compare the distribution of color for one sample to the distribution of color for another sample. They are simply provided here as a reference.

M&M'S Milk and dark chocolate HKP: 25% cyan blue, 25% orange, 12.5% green, 12.5% bright yellow, 12.5% red, 12.5% brown

M&M'S Milk and dark chocolate CLV: 20.7% cyan blue, 20.5% orange, 19.8% green, 13.5% bright yellow, 13.1% red, 12.4% brown

M&M'S Milk and dark chocolate peanut HKP: 22.2% cyan blue, 22.2% orange, 16.7% green, 16.7% bright yellow, 11.1% red, 11.1% brown

M&M'S Milk and dark chocolate peanut CLV: 20% cyan blue, 20% orange, 20% green, 20% bright yellow, 10% red, 10% brown

M&M'S Almond: 19% cyan blue, 19% orange, 19% green, 19% bright yellow, 12% brown, 12% red

M&M'S Crispy: 14.3% cyan blue, 28.5% orange, 14.3% green, 14.3% bright yellow, 14.3% brown, 14.3% red

M&M'S Minis: 25% cyan blue, 25% orange, 12% green, 13% bright yellow, 12% red, 13% brown

M&M'S Peanut butter and almond: 20% cyan blue, 20% orange, 20% green, 20% bright yellow, 10% red, 10% brown

M&M'S Pretzel: 28.5% blue, 14.3% each of yellow, orange, green, brown, red

Tests for Homogeneity: Stating Hypotheses

Does background music influence what customers buy? One experiment in a European restaurant compared three randomly assigned treatments: no music, French accordion music, and Italian string music. Under each condition, the researchers recorded the number of customers who ordered French, Italian, and other entrées.[4] The null hypothesis in this example is

H_0: There is no difference in the true distributions of entrées ordered at this restaurant when no music, French accordion music, or Italian string music is played.

In general, the null hypothesis in a chi-square test for homogeneity says that there is *no difference* in the true distribution of a categorical variable in the populations of interest or for the treatments in an experiment.

The alternative hypothesis says that there *is* a difference in the distributions but does not specify the nature of that difference. In the restaurant example, the alternative hypothesis is

H_a: There is a difference in the true distributions of entrées ordered at this restaurant when no music, French accordion music, or Italian string music is played.

 The alternative hypothesis does not state that all of the distributions are different. Instead, the alternative hypothesis will be true even if just one of the true distributions is different from the others. Consequently, any difference among the three observed distributions of entrées ordered is evidence against the null hypothesis and for the alternative hypothesis.

So how did the experiment turn out? The two-way table summarizes the data and Figure 12.5 shows the conditional relative frequencies of entrée ordered for each of the three treatments.

		Type of background music			
		None	French	Italian	Total
Entrée ordered	French	30	39	30	99
	Italian	11	1	19	31
	Other	43	35	35	113
	Total	84	75	84	243

The type of entrée that customers order seems to differ considerably across the three music treatments. Orders of Italian entrées are very low (1.3%) when French music is playing, but are higher when Italian music (22.6%) or no music (13.1%) is playing. French entrées seem popular in this restaurant, as they are ordered often under all music conditions—but notably more often when French music is playing. For all three music treatments, the percent of Other entrées ordered was similar.

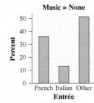

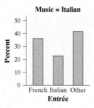

FIGURE 12.5 Relative frequency bar graphs comparing the distributions of entrées ordered for different music conditions.

Making Connections

Be sure to point out that this chi-square test for homogeneity has 1 categorical variable (entrée choice) and 3 treatments (no music, French music, and Italian music). This will help students make the distinction among the three types of chi-square tests.

1. Chi-square test for goodness of fit: 1 variable, 1 population.
2. Chi-square test for homogeneity: 1 variable, 2+ populations/ treatments.
3. Chi-square test for independence: 2 variables, 1 population.

Making Connections

Recall from Section 9.3 that we preferred the null hypothesis H_0: $p_1 - p_2 = 0$ (no difference) over H_0: $p_1 = p_2$ (the same) because our calculation of the P-value uses the sampling distribution of $\hat{p}_1 - \hat{p}_2$. Here, we also prefer the hypothesis that states "no difference."

Teaching Tip

We used relative frequencies (percents) instead of counts because the group sizes were not the same for each treatment (84 for None, 75 for French, and 84 for Italian).

Teaching Tip

We can use a chi-square test for homogeneity to compare the distribution of a single *categorical* variable for 2 or more populations or treatments. To compare the mean of a *quantitative* variable for more than 2 populations or treatments, we use a significance test called ANOVA. This test is not included in the AP® Topic Outline, but it is covered in online Chapter 13, available on the Student Site at highschool.bfwpub.com /updatedtps6e.

Making Connections

In Chapter 1, we introduced conditional relative frequencies. These three bar charts are the three conditional relative frequencies of entrée ordered for each treatment (heard no music, heard French music, and heard Italian music). What would these graphs looks like if the null hypothesis were true? *Answer:* The three graphs would be identical.

Do the differences in these distributions provide convincing evidence that background music affects customer behavior at this restaurant? Or is it plausible that the background music has no effect on customer behavior and that these differences are due to the chance involved in the random assignment of treatments? To decide, we have to know how likely it is to get differences this big or bigger when the null hypothesis is true. In other words, we need a *P*-value!

With only the methods we already know, we might start by comparing the proportions of French entrées ordered when no music ($\hat{p} = 30/84 = 0.357$) and when French accordion music are played ($\hat{p} = 39/75 = 0.52$) using a two-sample *z* test. We could similarly compare other pairs of proportions, ending up with many tests and many *P*-values. This is a bad idea. Performing multiple tests on the same data increases the probability that we make a Type I error in at least one of the tests.

Because of the increased probability of a false positive, it's cheating to pick out one large difference from the two-way table and then perform a significance test as if it were the only comparison we had in mind. Statisticians even have a name for this unethical practice: *P*-hacking.

For example, a test comparing the proportions of French entrées ordered under the no music and French accordion music treatments shows that the difference is statistically significant ($z = 2.06$, $P = 0.039$). However, the proportions of Italian entrées ordered for the no music and Italian string music treatments do not differ significantly ($z = 1.61$, $P = 0.107$). Reporting only the results of the first test wouldn't be telling the whole story.

The problem of how to do many comparisons at once without increasing the overall probability of a Type I error is common in statistics. Statistical methods for dealing with multiple comparisons usually have two steps:

1. Perform an *overall test* to see if there is convincing evidence of any differences among the parameters that we want to compare.
2. When the overall test shows there is convincing evidence of a difference, perform a detailed *follow-up analysis* to decide which of the parameters differ and to estimate how large the differences are.

When we want to compare the distribution of a categorical variable for several populations or treatments, the overall test uses the familiar chi-square test statistic.

Tests for Homogeneity: Expected Counts and the Chi-Square Test Statistic

A chi-square test for homogeneity begins with the hypotheses

H_0: There is no difference in the distribution of a categorical variable for several populations or treatments.

H_a: There is a difference in the distribution of a categorical variable for several populations or treatments.

To perform the test, we compare the observed counts in a two-way table with the counts we would expect if H_0 were true. Calculating the expected counts isn't difficult, but we calculate them slightly differently than in the chi-square test for goodness of fit.

Ingram Publishing/Alamy

The null hypothesis in the restaurant experiment is that there's no difference in the distributions of entrées ordered when no music, French accordion music, or Italian string music is played. To find the expected counts, we start by assuming that H_0 is true. We can see from the two-way table that 99 of the 243 entrées ordered during the study were French.

		Type of background music			
		None	French	Italian	Total
Entrée ordered	French	30	39	30	99
	Italian	11	1	19	31
	Other	43	35	35	113
	Total	84	75	84	243

If the specific type of music that's playing has no effect on entrée orders, the proportion of French entrées ordered under each music condition should be $99/243 = 0.4074$. Because there were 84 total entrées ordered when no music was playing, we would expect

$$84 \cdot \frac{99}{243} = 84(0.4074) = 34.22$$

of those entrées to be French, on average. The expected counts of French entrées ordered under the other two music conditions can be found in a similar way:

French music: $75(0.4074) = 30.56$

Italian music: $84(0.4074) = 34.22$

> **AP® EXAM TIP**
>
> As with chi-square tests for goodness of fit, the expected counts should *not* be rounded to the nearest whole number. While an observed count of entrées ordered must be a whole number, an expected count need not be a whole number. The expected count gives the average number of entrées ordered if H_0 is true and the random assignment process is repeated many times.

> As with a test for goodness of fit, the expected count is a sample size times a proportion specified by the null hypothesis.

We repeat the process to find the expected counts for the other two types of entrées. The overall proportion of Italian entrées ordered during the study was $31/243 = 0.1276$. So the expected counts of Italian entrées ordered under each treatment are

No music: $84(0.1276) = 10.72$

French music: $75(0.1276) = 9.57$

Italian music: $84(0.1276) = 10.72$

The overall proportion of Other entrées ordered during the experiment was $113/243 = 0.465$. So the expected counts of Other entrées ordered for each treatment are

No music: $84(0.465) = 39.06$

French music: $75(0.465) = 34.88$

Italian music: $84(0.465) = 39.06$

Teaching Tip

Consider having students record the expected values in parentheses directly next to or below the observed values in the two-way table. Students should then also provide a key indicating that the values in parentheses are expected counts. Students lost credit on the 2017 AP® Exam #5 if they did not include a key or label for the expected counts.

		Type of background music			
		None	French	Italian	Total
Entrée ordered	French	30 (34.22)	39 (30.56)	30 (34.22)	99
	Italian	11 (10.72)	1 (9.57)	19 (10.72)	31
	Other	43 (39.06)	35 (34.88)	35 (39.06)	113
	Total	84	75	84	243

KEY: (expected counts)

The following table summarizes the expected counts for all three treatments. Note that the values for no music and Italian music are the same because 84 total entrées were ordered under each condition, and we expect the distributions of entrée choice to be the same.

If we were to make relative frequency bar charts for each treatment based on these expected counts, all three graphs would look exactly the same. This is because the expected counts are calculated assuming that the null hypothesis—no difference in the distribution of entrée choice—is true.

Expected counts				
	Type of background music			
	None	French	Italian	Total
Entrée ordered French	34.22	30.56	34.22	99
Italian	10.72	9.57	10.72	31
Other	39.06	34.88	39.06	113
Total	84	75	84	243

We can check our work by adding the expected counts to obtain the row and column totals, as in the table. These should be the same as those in the table of observed counts except for small roundoff errors, such as 75.01 rather than 75 for the total number of entrées ordered when French music was playing.

Let's take a look at the two-way table from the restaurant study one more time. In this context, we found the expected count of French entrées ordered when no music was playing as follows:

$$84 \cdot \frac{99}{243} = 34.22$$

Observed counts				
	Type of background music			
	None	French	Italian	Total
Entrée ordered French	30	39	30	99
Italian	11	1	19	31
Other	43	35	35	113
Total	84	75	84	243

We've marked in the table the three numbers used in this calculation. These values are the row total for French entrées ordered, the column total for entrées ordered when no music was playing, and the table total of entrées ordered during the experiment. We can rewrite the original calculation as

$$\frac{84 \cdot 99}{243} = \frac{99 \cdot 84}{243} = 34.22$$

This suggests a more general formula for the expected count in any cell of a two-way table.

CALCULATING EXPECTED COUNTS FOR A CHI-SQUARE TEST BASED ON DATA IN A TWO-WAY TABLE

When H_0 is true, the expected count in any cell of a two-way table is

$$\text{expected count} = \frac{\text{row total} \cdot \text{column total}}{\text{table total}}$$

Teaching Tip

Recall that in Section 9.3, we used the combined (pooled) sample proportion $\hat{p}_C$ to check the Large Counts condition for a two-sample z test for $p_1 - p_2$. The expected count calculation for a chi-square test based on data in a two-way table also uses the combined (pooled) proportion.

$$\text{expected count} = \frac{\text{row total} \cdot \text{column total}}{\text{table total}}$$

$$\text{expected count} = \frac{\text{row total}}{\text{table total}} \cdot \text{column total}$$

$$\text{expected count} = \hat{p}_C \cdot \text{column total}$$

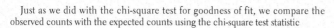

Just as we did with the chi-square test for goodness of fit, we compare the observed counts with the expected counts using the chi-square test statistic

$$\chi^2 = \sum \frac{(\text{Observed count} - \text{Expected count})^2}{\text{Expected count}}$$

This time, the sum is over all cells (not including the totals!) in the two-way table.

For French entrées with no music, the observed count is 30 orders and the expected count is 34.22. The contribution to the χ^2 test statistic for this cell is

$$\frac{(\text{Observed count} - \text{Expected count})^2}{\text{Expected count}} = \frac{(30 - 34.22)^2}{34.22} = 0.52$$

The χ^2 test statistic is the sum of nine such terms:

$$\chi^2 = \frac{(30 - 34.22)^2}{34.22} + \frac{(39 - 30.56)^2}{30.56} + \cdots + \frac{(35 - 39.06)^2}{39.06}$$

$$= 0.52 + 2.33 + \cdots + 0.42 = 18.28$$

> **AP® EXAM TIP**
>
> In the "Do" step, you aren't required to show every term in the chi-square test statistic. Writing the first few terms of the sum followed by "..." is considered as "showing work." We suggest that you do this and then let your calculator tackle the computations.

Here is an example to practice what you have learned so far.

| EXAMPLE | **Would you vote for a female president?**
Hypotheses, expected counts, and the chi-square test statistic |

PROBLEM: For a class project, Abby and Mia wanted to know if the gender of an interviewer could affect the responses to a survey question. The subjects in their experiment were 100 males from their school. Half of the males were randomly assigned to be asked, "Would you vote for a female president?" by a female interviewer. The other half of the males were asked the same question by a male interviewer.[5] The table shows the results.

(a) State the appropriate null and alternative hypotheses.

(b) Show the calculation for the expected count in the Male/Yes cell. Then provide a complete table of expected counts.

(c) Calculate the value of the chi-square test statistic.

		Gender of interviewer		
		Male	Female	Total
Response to question	Yes	30	39	69
	No	8	3	11
	Maybe	12	8	20
	Total	50	50	100

Teaching Tip: Using Technology

To calculate the chi-square statistic using your calculator, follow the instructions from the Technology Corner on page 787 or use lists. Simply enter the observed counts in L1, the expected counts in L2, and the following formula in L3: (L1 − L2)^2/L2. Then find the sum of the values in L3 using 1-Var Stats L3. The advantage to this approach is that L3 now stores all the "components" of the chi-square statistic, which will help if a follow-up analysis is needed.

> **⚠ COMMON STUDENT ERROR**
>
> Students might calculate expected values for the "Total" row and "Total" column and then use these values as part of the calculation for the chi-square test statistic. This is incorrect—expected values should only be calculated for the individual counts, not the totals.

Teaching Tip

In this example, Abby and Mia are measuring one categorical variable (response to question) for two treatment groups (people interviewed by a male and people interviewed by a female), making this a test for homogeneity. In fact, all significance tests for two-way tables created from experiments will use a test for homogeneity.

| ALTERNATE EXAMPLE | Skills 1.F, 3.A |

Are you going into STEM? Hypotheses, expected counts, and the chi-square test statistic

PROBLEM:
Every year, Mrs. Gallas surveys a random sample of 50 seniors at her large high school to find out what career they intend to pursue after high school. She wonders if the distribution of intended career choice has changed in the last 5 years. The two-way table summarizes the data from the two samples.

(a) State the appropriate null and alternative hypotheses.
(b) Show the calculation for the expected count in the Art/Humanities in 2014 cell. Then provide a complete table of expected counts.
(c) Calculate the value of the chi-square test statistic.

		Graduating class		
		2014	2019	Total
Intended career	Arts/Humanities	22	10	32
	Social sciences	13	13	26
	STEM	10	20	30
	Other/Undecided	5	7	12
	Total	50	50	100

(continues)

ALTERNATE EXAMPLE **(continued)**

SOLUTION:

(a) H_0: There is no difference in the true distributions of intended careers for the graduating class of 2014 and the graduating class of 2019 at Mrs. Gallas's high school.

H_a: There is a difference in the true distributions of intended careers for the graduating class of 2014 and the graduating class of 2019 at Mrs. Gallas's high school.

(b) The expected count for the Art/Humanities in 2014 cell is $\dfrac{32 \cdot 50}{100} = 16$.

The rest of the expected counts are shown in the table below:

		Graduating class		
		2014	2019	Total
	Arts/Humanities	16	16	32
Intended career	Social sciences	13	13	26
	STEM	15	15	30
	Other/Undecided	6	6	12
	Total	50	50	100

(c) $\chi^2 = \dfrac{(22 - 16)^2}{16} + \dfrac{(10 - 16)^2}{16} + \cdots$

$= 8.167$

✓ **Answers to CYU**

1. H_0: There is no difference in the true distributions of superpower preference for all survey takers from the United Kingdom and from the United States.

H_a: There is a difference in the true distribution of superpower preference . . .

2. The expected count for the U.K./Fly cell is $\dfrac{(99)(200)}{415} = 47.711$. The rest of the expected counts are shown in the table below:

		Country		
		U.K.	U.S.	Total
	Fly	47.711	51.289	99
	Freeze time	46.265	49.735	96
Superpower preference	Invisibility	32.289	34.711	67
	Super strength	20.723	22.277	43
	Telepathy	53.012	56.988	110
	Total	200	215	415

SOLUTION:

(a) H_0: There is no difference in the true distributions of response to this question when asked by a male interviewer and when asked by a female interviewer for subjects like these.

H_a: There is a difference in the true distributions of response to this question when asked by a male interviewer and when asked by a female interviewer for subjects like these.

(b) The expected count for the Male/Yes cell is $\dfrac{69 \cdot 50}{100} = 34.5$. The rest of the expected counts are shown in the table:

$$\text{expected count} = \frac{\text{row total} \cdot \text{column total}}{\text{table total}}$$

		Gender of interviewer		
		Male	Female	Total
	Yes	34.5	34.5	69
Response to question	No	5.5	5.5	11
	Maybe	10.0	10.0	20
	Total	50	50	100

(c) $\chi^2 = \dfrac{(30 - 34.5)^2}{34.5} + \dfrac{(39 - 34.5)^2}{34.5} + \cdots$

$= 4.25$

$$\chi^2 = \sum \frac{(\text{Observed count} - \text{Expected count})^2}{\text{Expected count}}$$

FOR PRACTICE, TRY EXERCISE 29

CHECK YOUR UNDERSTANDING

Separate random samples of children from the United Kingdom and the United States who completed a survey in a recent year were selected. For each student, we recorded the superpower he or she would most like to have: the ability to fly, ability to freeze time, invisibility, super strength, or telepathy (ability to read minds). Is there convincing evidence that the distributions of superpower preference are different for survey takers in the two countries? The data are summarized in the two-way table.[6]

		Country		
		U.K.	U.S.	Total
	Fly	54	45	99
	Freeze time	52	44	96
Superpower preference	Invisibility	30	37	67
	Super strength	20	23	43
	Telepathy	44	66	110
	Total	200	215	415

1. State the appropriate null and alternative hypotheses.
2. Show the calculation for the expected count in the U.K./Fly cell. Then provide a complete table of expected counts.
3. Calculate the value of the chi-square test statistic.

3. $\chi^2 = \dfrac{(54 - 47.711)^2}{47.711} + \dfrac{(45 - 51.289)^2}{51.289} + \cdots = 6.29$

Tests for Homogeneity: Conditions and *P*-values

Like every other significance test, there are conditions that must be met to justify our calculations and conclusions. The Random and 10% conditions are the same as they were for the two-sample *z* test for $p_1 - p_2$. And the Large Counts condition is the same as it was for the chi-square test for goodness of fit. Making sure that the expected counts are all at least 5 ensures that the sampling distribution of the χ^2 test statistic can be well modeled by a chi-square distribution.

Because a stratified random sample consists of a simple random sample from each stratum and these samples are independent, you could also use a test for homogeneity to compare the distribution of a categorical variable among the different strata.

> ### CONDITIONS FOR PERFORMING A CHI-SQUARE TEST FOR HOMOGENEITY
>
> - **Random:** The data come from independent random samples or from groups in a randomized experiment.
>   - **10%:** When sampling without replacement, $n < 0.10N$ for each sample.
> - **Large Counts:** All *expected* counts are at least 5.

Because the three treatments were assigned at random and all expected counts are at least 5, the conditions are met for the restaurant experiment. We don't have to check the 10% condition because the researchers were not sampling without replacement from some population of interest. They performed an experiment using customers who happened to be in the restaurant at the time.

Observed counts

Type of background music

		None	French	Italian	Total
Entrée ordered	French	30	39	30	99
	Italian	11	1	19	31
	Other	43	35	35	113
	Total	84	75	84	243

Expected counts

Type of background music

		None	French	Italian	Total
Entrée ordered	French	34.22	30.56	34.22	99
	Italian	10.72	9.57	10.72	31
	Other	39.06	34.88	39.06	113
	Total	84	75	84	243

As in the test for goodness of fit, you should think of the chi-square test statistic χ^2 as a measure of how much the observed counts deviate from the expected counts, relative to the expected counts. Once again, large values of χ^2 are evidence against H_0 and in favor of H_a. The *P*-value measures the strength of this evidence. When the conditions are met, *P*-values for a chi-square test for homogeneity come from a chi-square distribution with

$$\text{df} = (\text{number of rows} - 1) \times (\text{number of columns} - 1)$$

For the restaurant experiment, $\chi^2 = 18.28$. Because there are 3 rows and 3 columns in the two-way table (not including the totals),

$$\text{df} = (3-1) \times (3-1) = 4$$

We can find the *P*-value using Table C or technology.

Teaching Tip

While it is important that students know how to check each condition, it is equally important that they understand *why* we check the conditions. We call this the "*So what?*"

Teaching Tip

Here is the "*So what?*" answer for each condition.

Random:
→ so individual observations are independent and we can generalize to the populations (random sample).

AND/OR

→ so we can show causation (random assignment).

10%:
→ so we can view observations as independent even though we are sampling without replacement.

Large Counts:
→ so the sampling distribution is approximately a chi-square distribution and we can use χ^2 to find a *P*-value.

 COMMON STUDENT ERROR

When checking conditions, many students mistakenly examine the observed counts rather than the expected counts. Also, many students forget to list the expected counts, especially when doing the calculations on their calculators. Simply stating that the expected counts are at least 5 is not sufficient—students must list the expected counts to prove they have checked them.

Teaching Tip

Here is one way to explain why df = (number of rows − 1) × (number of columns − 1). Using the music and entrée example, take students through the expected value calculations for the 4 cells in the top left of the two-way table. Without any further expected value calculations, we know the rest of the expected values because we have row and column totals. For example, once we know that None, French has an expected value of 34.22 and None, Italian has an expected value of 10.72, the expected value for None, Other is already determined: 84 − (34.22 + 10.72) = 39.06. Once totals are known, then only 4 of the cells provide new information (df = 4) about the distributions. The remaining cells are constrained (not free) because the totals are known.

	No music	French music	Italian music	Total
French entrée	34.22	30.56	34.22	99
Italian entrée	10.72	9.57	10.72	31
Other entrée	39.06	34.88	39.06	113
Total	84	75	84	243

Teaching Tip: Using Technology

You can also use the *One Categorical Variable* applet to find the *P*-value. Go to the Extra Applets on the Student Site at highschool.bfwpub.com/updatedtps6e. This applet requires the user to input the original observed counts. Be sure to use Groups: "Multiple," then choose "Chi-square test for homogeneity."

ALTERNATE EXAMPLE Skills 4.C, 3.E

Are you going into STEM? Part 2
Conditions, *P*-value, and conclusion

PROBLEM:
In the preceding alternate example, you read about an annual survey done at a large high school to determine students' intended career choices for life after high school. Earlier, we calculated $\chi^2 = 8.167$.

(a) Verify that the conditions for inference are met.
(b) Use Table C to find the *P*-value, then use your calculator's χ^2cdf command.
(c) Interpret the *P*-value from the calculator in context.
(d) What conclusion would you draw? Justify your answer.

SOLUTION:
(a)
- *Random:* Independent random samples of seniors from the graduating classes of 2014 and 2019. ✓
  - *10%:* 50 < 10% of all seniors in the graduating class of 2014; 50 < 10% of all seniors in the graduating class of 2019. ✓

Large counts: All expected counts ≥ 5 (see preceding Alternate Example). ✓

(b) df = (4 − 1)(2 − 1) = 3
- *Table C:* The *P*-value is between 0.025 and 0.05.
- *Tech:* χ^2cdf (lower: 8.167, upper: 10000, df: 3) = 0.043

(c) Assuming no difference in the true distributions of intended careers for the graduating class of 2014 and the graduating class of 2019 at Mrs. Gallas's high school, there is a 0.043 probability of observing differences in the distributions of responses as large as or larger than the ones in this study by chance alone.

(d) Because the *P*-value of 0.043 < α = 0.05, we reject H_0. There is convincing evidence of a difference in the true distributions of intended careers for the graduating class of 2014 and the graduating class of 2019 at Mrs. Gallas's high school.

---

df	P	
	.0025	.001
4	16.42	18.47

- *Using Table C:* Look at the df = 4 row in Table C. The calculated value $\chi^2 = 18.28$ lies between the critical values 16.42 and 18.47. The corresponding *P*-value is between 0.001 and 0.0025.

Chi-square distribution with df = 4

$\chi^2 = 18.28$

- *Using technology:* The command χ^2cdf(lower:18.28, upper:10000, df:4) gives 0.0011.

Assuming background music has no effect on what entrée a customer orders, there is a 0.0011 probability of getting a χ^2 test statistic of 18.28 or greater by chance alone. Because the *P*-value of 0.0011 < α = 0.05, we reject H_0. There is convincing evidence that there is a difference in the true distributions of entrées ordered at this restaurant when no music, French accordion music, or Italian string music is played.

EXAMPLE

Does the gender of an interviewer matter?
Conditions, *P*-value, and conclusion

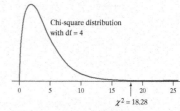

sturti/Getty Images

PROBLEM: In the preceding example, you read about an experiment to determine if the gender of an interviewer affects responses to the question "Would you vote for a female president?" Here are tables showing the observed and expected counts:

Observed counts				
		Gender of interviewer		
		Male	Female	Total
Response to question	Yes	30	39	69
	No	8	3	11
	Maybe	12	8	20
	Total	50	50	100

Expected counts				
		Gender of interviewer		
		Male	Female	Total
Response to question	Yes	34.5	34.5	69
	No	5.5	5.5	11
	Maybe	10.0	10.0	20
	Total	50	50	100

Earlier, we calculated $\chi^2 = 4.25$.

(a) Verify that the conditions for inference are met.
(b) Use Table C to find the *P*-value. Then use your calculator's χ^2cdf command.
(c) Interpret the *P*-value from the calculator.
(d) What conclusion would you draw?

SOLUTION:

(a) **Random:** Treatments were randomly assigned. ✓

Large Counts: All expected counts ≥ 5 (see table of expected counts). ✓

We don't check the 10% condition because Abby and Mia didn't randomly select subjects from some population.

(b) $df = (3 − 1)(2 − 1) = 2$

Using Table C: The P-value is between 0.10 and 0.15.

Using technology: χ^2cdf(lower: 4.25, upper: 10000, df: 2) = 0.119

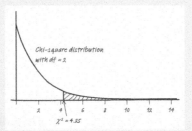

Chi-square distribution with df = 2

$\chi^2 = 4.25$

(c) Assuming that the gender of the interviewer doesn't affect responses to this question, there is a 0.119 probability of observing differences in the distributions of responses as large as or larger than those in this study by chance alone.

(d) Because the P-value of 0.119 > $\alpha = 0.05$, we fail to reject H_0. There is not convincing evidence of a difference in the true distributions of response to this question when asked by a male interviewer and when asked by a female interviewer for subjects like these.

FOR PRACTICE, TRY EXERCISE 31

Calculating the expected counts and then the chi-square test statistic by hand is a bit time-consuming. As usual, technology saves time and gets the arithmetic right.

30. Technology Corner

PERFORMING CHI-SQUARE TESTS FOR TWO-WAY TABLES

TI-Nspire and other technology instructions are on the book's website at highschool.bfwpub.com/updatedtps6e.

You can use the TI-83/84 to perform calculations for a chi-square test for homogeneity. We'll use the data from the restaurant study to illustrate the process.

1. Enter the observed counts in matrix [A].

- Press 2nd X⁻¹ (MATRIX), arrow to EDIT, and choose A.

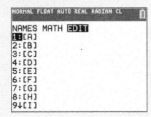

- Enter the dimensions of the matrix: 3 × 3.

Teaching Tip

Consider having students check the condition *and* give the "*So what?*" Here is the "*So what?*" answer for each condition in this example:

Random:
→ so individual observations are independent and we can say the gender of the interviewer causes changes in the responses.

Large Counts:
→ so the sampling distribution is approximately a chi-square distribution and we can use χ^2 to find a P-value.

Teaching Tip: Using Technology

Using this χ^2 test will not work for a chi-square test for goodness of fit. For this test to run properly, there must be at least two rows and two columns of data. In a test for goodness of fit, there is only one row (or one column) of data. Remind students that they can watch videos for each Technology Corner by clicking the link in the e-Book or at the Student Site.

- Enter the observed counts from the two-way table in the same locations in the matrix.

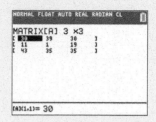

2. Press STAT, arrow to TESTS, and choose χ^2-Test. Adjust your settings as shown.

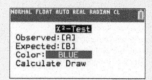

Note: You do not have to enter the expected counts in matrix [B]. Once you have run the test, the expected counts will be stored in matrix [B].

3. Choose "Calculate" or "Draw" to carry out the test. If you choose "Calculate," you should get the test statistic, *P*-value, and df shown here. If you specify "Draw," the chi-square distribution with 4 degrees of freedom will be drawn, the area in the tail will be shaded, and the *P*-value will be displayed.

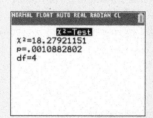 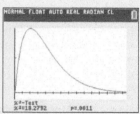

4. To see the expected counts, Press 2nd X⁻¹ (MATRIX), arrow to EDIT, and choose [B].

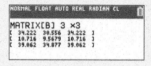

CHECK YOUR UNDERSTANDING

In the preceding Check Your Understanding (page 784), we presented data about superpower preferences for random samples of children from the United Kingdom and the United States. Here are the data once again:

Superpower preference	Country		
	U.K.	U.S.	Total
Fly	54	45	99
Freeze time	52	44	96
Invisibility	30	37	67
Super strength	20	23	43
Telepathy	44	66	110
Total	200	215	415

1. Verify that the conditions for inference are met.
2. Use Table C to find the P-value. Then use your calculator's χ^2cdf command.
3. Interpret the P-value from the calculator.
4. What conclusion would you draw?

Putting It All Together: The Chi-Square Test for Homogeneity

In Section 12.1, we used a chi-square test for goodness of fit to test a hypothesized model for the distribution of a categorical variable. When we want to compare the distribution of a categorical variable in several populations or for several treatments, we use a **chi-square test for homogeneity.**

This test is also known as a chi-square test for homogeneity of proportions. We prefer the simpler name.

CHI-SQUARE TEST FOR HOMOGENEITY

Suppose the conditions are met. To perform a test of

H_0: There is no difference in the distribution of a categorical variable for several populations or treatments

compute the chi-square test statistic

$$\chi^2 = \sum \frac{(\text{Observed count} - \text{Expected count})^2}{\text{Expected count}}$$

where the sum is over all cells (not including totals) in the two-way table. The P-value is the area to the right of χ^2 under the chi-square density curve with degrees of freedom $=$ (number of rows $- 1$)(number of columns $- 1$).

Teaching Tip

Consider requiring formulas and work during the school year and then relaxing this expectation at the end of the course, when preparing for the AP® Statistics exam.

✓ Answers to CYU

1. *Random:* Independent random samples of 200 survey takers from the U.K. and 215 survey takers from the U.S. *10%:* $n = 200$ is less than 10% of all U.K. survey takers and $n = 215$ is less than 10% of all U.S. survey takers. *Large counts:* All expected counts (47.7, 51.3, 46.3, 49.7, 32.3, 34.7, 20.7, 22.3, 53.0, 57.0) are at least 5.

2. The test statistic is $\chi^2 = 6.29$; P-value: df $= (5 - 1)(2 - 1) = 4$.

Table C: The P-value is between 0.15 and 0.20.

Tech: χ^2cdf(lower: 6.29, upper: 10000, df: 4) $= 0.1785$

3. Assuming the distribution of superpower preference is the same for all U.K. and U.S. survey takers, there is a 0.1785 probability of observing differences in the distributions as large as or larger than the ones in this study by chance alone.

4. Because the P-value of 0.1785 $> \alpha = 0.05$, we fail to reject H_0. We do not have convincing evidence of a difference in the true distributions of superpower preference for all survey takers from the United Kingdom and the United States.

What is your music preference?
The chi-square test for homogeneity

PROBLEM:
Do high school students in Michigan and California have the same music preferences? We used the Census at School® website to select separate random samples of 100 high school students from Michigan and 100 high school students from California. Students were asked, "What is your favorite music genre?" The two-way table summarizes their responses.

		State		
		Michigan	California	Total
Favorite music genre	Country	12	4	16
	Pop	15	14	29
	Rap	21	22	43
	Rock	7	10	17
	Other	45	50	95
	Total	100	100	200

Do these data provide convincing evidence at the $\alpha = 0.05$ level that the distributions of favorite music genre differ for high school students in Michigan and California?

SOLUTION:
STATE:
H_0: There is no difference in the distributions of favorite music genre for high school students in Michigan and California.

H_a: There is a difference in the distributions of favorite music genre for high school students in Michigan and California.

We'll use $\alpha = 0.05$.

PLAN: Chi-square test for homogeneity
- *Random:* Independent random samples of students from Michigan and California. ✓
  - *10%:* 100 < 10% of all high school students in Michigan, 100 < 10% of all high school students in California. ✓
- *Large Counts:* All expected counts are ≥ 5 (see table). ✓

		State		
		Michigan	California	Total
Favorite music genre	Country	8	8	16
	Pop	14.5	14.5	29
	Rap	21.5	21.5	43
	Rock	8.5	8.5	17
	Other	47.5	47.5	95
	Total	100	100	200

Let's look at an example of a chi-square test for homogeneity from start to finish. As usual, we follow the four-step process when performing a significance test.

EXAMPLE

Speaking English
The chi-square test for homogeneity

PROBLEM: The Pew Research Center conducts surveys about a variety of topics in many different countries. In one survey, it wanted to investigate how residents of different countries feel about the importance of speaking the national language. Separate random samples of residents of Australia, the United Kingdom, and the United States were asked many questions, including the following: "Some people say that the following things are important for being truly [survey country nationality]. Others say they are not important. How important do you think it is to be able to speak English?" The two-way table summarizes the responses to this question.

		Country			
		Australia	U.K.	U.S.	Total
Opinion about speaking English	Very important	690	1177	702	2569
	Somewhat important	250	242	221	713
	Not very important	40	28	50	118
	Not at all important	20	13	30	63
	Total	1000	1460	1003	3463

Do these data provide convincing evidence at the $\alpha = 0.05$ level that the distributions of opinion about speaking English differ for residents of Australia, the U.K., and the U.S.?

SOLUTION:

> Use the four-step process!

STATE: H_0: There is no difference in the true distributions of opinion about speaking English for residents of Australia, the U.K., and the U.S.

H_a: There is a difference in the true distributions of opinion about speaking English for residents of Australia, the U.K., and the U.S.

We'll use $\alpha = 0.05$.

PLAN: Chi-square test for homogeneity.
- Random: Independent random samples of residents from the three countries. ✓
  - 10%: 1000 is <10% of all Australian residents, 1460 is <10% of all U.K. residents, and 1003 is <10% of all U.S. residents. ✓
- Large Counts: All expected counts are ≥5 (see table below). ✓

		Expected counts Country			
		Australia	U.K.	U.S.	Total
Opinion about speaking English	Very important	741.8	1083.1	744.1	2569
	Somewhat important	205.9	300.6	206.5	713
	Not very important	34.1	49.7	34.2	118
	Not at all important	18.2	26.6	18.2	63
	Total	1000	1460	1003	3463

DO:
- *Test statistic:*

$$\chi^2 = \frac{(12-8)^2}{8} + \frac{(4-8)^2}{8} + \frac{(15-14.5)^2}{14.5} \cdots = 4.85$$

- *P-value:* df = $(5-1)(2-1) = 4$

Table C: P-value > 0.25

Tech: χ^2cdf(lower: 4.85, upper: 10000, df: 4) = 0.303

CONCLUDE: Because the *P*-value of 0.303 > $\alpha = 0.05$, we fail to reject H_0. There is not convincing evidence of a difference in the distributions of favorite music genre for high school students in Michigan and California.

DO:

- *Test statistic:*

$$\chi^2 = \frac{(690-741.8)^2}{741.8} + \frac{(1177-1083.1)^2}{1083.1} + \cdots = 68.57$$

> Because the observed counts differ from the expected counts, there is *some* evidence for H_a.

- *P-value:* $df = (4-1)(3-1) = 6$

Using Table C: $P\text{-value} < 0.0005$

Using technology:
$\chi^2 cdf(\text{lower}:68.57, \text{upper}:10000, \text{df}:6) \approx 0$

CONCLUDE: Because the *P*-value of approximately $0 < \alpha = 0.05$, we reject H_0. There is convincing evidence that there is a difference in the true distributions of opinion about speaking English for residents of Australia, the U.K., and the U.S.

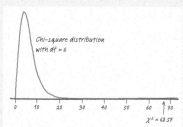

Chi-square distribution with df = 6

$\chi^2 = 68.57$

> If you want to know *how* the distributions differ, do a follow-up analysis.

FOR PRACTICE, TRY EXERCISE 35

> **AP® EXAM TIP**
>
> Many students lose credit on the AP® Statistics exam because they don't write down and label the expected counts in their response. It isn't enough to claim that all the expected counts are at least 5. You must provide clear evidence.

What if we want to compare several proportions? Many studies involve comparing the proportion of successes for each of several populations or treatments. The two-sample z test from Section 9.3 allows us to test the null hypothesis $H_0: p_1 = p_2$, where p_1 and p_2 are the true proportions of successes for the two populations or treatments. The chi-square test for homogeneity allows us to test $H_0: p_1 = p_2 = \ldots = p_k$. This null hypothesis says that there is no difference in the proportions of successes for the k populations or treatments. The alternative hypothesis is H_a: At least two of the proportions are different. Many students *incorrectly state H_a* as "all the proportions are different." Think about it this way: the opposite of "all the proportions are equal" is "some of the proportions are not equal."

FOLLOW-UP ANALYSIS The chi-square test for homogeneity allows us to compare the distribution of a categorical variable for any number of populations or treatments. If the test allows us to reject the null hypothesis of no difference, we may want to do a follow-up analysis that examines the differences in detail. As with the chi-square test for goodness of fit, start by identifying the cells that contribute the most to the chi-square statistic. Then describe how the observed and expected counts differ in those categories, noting the direction of the difference.

Our earlier restaurant study found significant differences among the distributions of entrées ordered under each of the three music conditions. We entered the two-way table for the study into Minitab software and requested a chi-square

> **AP® EXAM TIP**
>
> Another method that suffices for checking the Large Counts condition is to state that the *smallest* expected count of (give value) is at least 5.

> **AP® EXAM TIP**
>
> Students should specify which of the chi-square tests they are using. Just saying "chi-square test" is unlikely to earn credit on the AP® Statistics exam (even though in the past it has on occasion).

Teaching Tip

As with the chi-square test for goodness of fit, students are not expected to do a follow-up analysis on the AP® Statistics exam unless they are specifically asked to. If asked to do a follow-up analysis, students should address the size of the contribution *and* the direction of the difference (e.g., there were *fewer* Italian entrées ordered than expected when French music was playing).

AP® EXAM TIP

Carefully point out the information provided in the computer output on this page. On the AP® Statistics exam, students will be expected to understand how to read computer output. For computer analyses of two-way tables, a "key" usually appears somewhere in the output. Here, the second and third lines describe where to find the observed and expected counts and the chi-square contributions.

Teaching Tip

Consider having students add a follow-up sentence at the end of the conclusion for the significance test:

CONCLUDE: Because the *P*-value of $0.001 < \alpha = 0.05$, we reject H_0. We have convincing evidence of a difference among the true distributions of entrées ordered under each of the three music conditions. *The largest components of the χ^2 test statistic are 7.672 and 6.404 because the orders of Italian entrées are far below what we expect when French music is playing and far above what we expect when Italian music is playing.*

Note: The follow-up analysis does not have to provide discussion of the *two* largest contributors to the χ^2 test statistic. Discussion of *one* is often enough.

Consider requiring students to do a follow-up analysis during the school year and then relaxing this expectation at the end of the course, when preparing for the AP® Statistics exam.

```
Chi-Square Test: None, French, Italian

Expected counts are printed below observed counts

Chi-Square contributions are printed below expected counts

          None   French   Italian   Total
French      30       39        30      99
entrée   34.22    30.56     34.22
         0.521    2.334     0.521
Italian     11        1        19      31
entrée   10.72     9.57     10.72
         0.008    7.672     6.404
Other       43       35        35     113
entrée   39.06    34.88     39.06
         0.397    0.000     0.422
Total       84       75        84     243

Chi-Sq = 18.279, DF = 4, P-Value = 0.001
```

FIGURE 12.6 Minitab output for the two-way table in the restaurant study. The output gives the observed counts, the expected counts, and the individual components of the chi-square test statistic.

test. The output appears in Figure 12.6. Minitab repeats the two-way table of observed counts and puts the expected count for each cell below the observed count, followed by the nine individual components that contribute to the χ^2 test statistic.

Looking at the output, we see that just two of the nine components contribute about 14 (almost 77%) of the total $\chi^2 = 18.28$. Comparing the observed and expected counts in these two cells, we see that orders of Italian entrées are far below what we expect when French music is playing and far above what we expect when Italian music is playing. We are led to a specific conclusion: orders of Italian entrées are strongly affected by Italian and French music. More advanced methods provide tests and confidence intervals that make this follow-up analysis more complete.

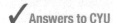

CHECK YOUR UNDERSTANDING

Canada has universal health care. The United States does not but often offers more elaborate treatment to patients who have access. How do the two systems compare in treating heart attacks? Researchers compared random samples of U.S. and Canadian heart attack patients. One key outcome was the patients' own assessment of their quality of life relative to what it had been before the heart attack. Here are the data for the patients who survived a year:

Quality of life	Canada	United States
Much better	75	541
Somewhat better	71	498
About the same	96	779
Somewhat worse	50	282
Much worse	19	65
Total	**311**	**2165**

Is there a significant difference between the two distributions of quality-of-life ratings? Carry out an appropriate test at the $\alpha = 0.01$ level.

Relationships Between Two Categorical Variables

Two-way tables can summarize data from different types of studies. The restaurant experiment compared entrées ordered for three music treatments. The observational study about speaking English compared independent random samples from three different populations. In both cases, we are comparing the distribution of a categorical variable for several populations or treatments. We use the chi-square test for homogeneity to perform inference in such settings.

Answers to CYU

STATE: H_0: There is no difference in the distribution of quality of life for patients who have suffered a heart attack in Canada and the U.S.

H_a: There is a difference in the distribution of quality of life . . .

We will use $\alpha = 0.01$.

PLAN: Chi-square test for homogeneity. *Random:* The data came from independent random samples from Canada and the United States. *10%:* $n_1 = 311$ is less than 10% of all Canadian heart attack patients and $n_2 = 2165$ is less than 10% of all U.S. heart attack patients. *Large Counts:* All expected counts (77.37, 538.63, 71.47, 497.53, 109.91, 765.09, 41.70, 290.30, 10.55, 73.45) are at least 5.

DO: $\chi^2 = \dfrac{(75-77.37)^2}{77.37} + \cdots + \dfrac{(65-73.45)^2}{73.45}$

$= 11.725$

P-value: df $= (5-1)(2-1) = 4$

Table C: The *P*-value is between 0.01 and 0.02.

Tech: χ^2cdf(lower: 11.725, upper: 10000, df: 4) = 0.0195

CONCLUDE: Because the *P*-value of $0.0195 > \alpha = 0.01$, we fail to reject H_0. There is not convincing evidence of a difference in the distribution of quality of life for heart attack patients in Canada and the United States.

Making Connections

Inform students that we will learn how to test for a relationship between two *quantitative* variables in Section 12.3.

Francesco Carta fotografo/Getty Images

Another common situation that leads to a two-way table is when a *single* random sample of individuals is chosen from a *single* population and then classified based on *two* categorical variables. In that case, our goal is to analyze the relationship between the variables.

Are people who are prone to sudden anger more likely to develop heart disease? A prospective observational study followed a random sample of 8474 people with normal blood pressure for about four years.[7] All the individuals were free of heart disease at the beginning of the study. Each person took the Spielberger Trait Anger Scale test, which measures how prone a person is to sudden anger. Researchers also recorded whether each individual developed coronary heart disease (CHD). This includes people who had heart attacks, as well as those who needed medical treatment for heart disease. Here is a two-way table that summarizes the data:

		Anger level			
		Low	Moderate	High	Total
CHD status	Yes	53	110	27	190
	No	3057	4621	606	8284
	Total	3110	4731	633	8474

FIGURE 12.7 Bar graph comparing the percents of people in each anger category who developed coronary heart disease (CHD).

The bar graph in Figure 12.7 shows the percent of people in each of the three anger categories who developed CHD. There is a clear trend: as the anger score increases, so does the percent who suffer heart disease. A much higher percent of people in the high anger category developed CHD (4.27%) than in the moderate (2.33%) and low (1.70%) anger categories.

Do these data provide convincing evidence of an association between the variables in the larger population? Or is it plausible that there is no association between the variables in the population and that we observed an association in the sample by chance alone? To answer that question, we rely on a new significance test.

Tests for Independence: Stating Hypotheses

When we gather data from a single random sample and measure two categorical variables, we are often interested in whether the sample data provide convincing evidence that the variables have an association in the population. That is, does knowing the value of one variable help predict the value of the other variable for individuals in the population? To determine if evidence from the sample is convincing, we perform a *chi-square test for independence*.

In this test, our null hypothesis is that there is *no association* between the two categorical variables in the population of interest. The alternative hypothesis is that there *is* an association between the variables. For the observational study of anger level and coronary heart disease, we want to test the hypotheses

H_0: There is no association between anger level and heart-disease status in the population of people with normal blood pressure.

H_a: There is an association between anger level and heart-disease status in the population of people with normal blood pressure.

No association between two variables means that knowing the value of one variable does not help us predict the value of the other. That is, the variables are *independent*. An equivalent way to state the hypotheses is

H_0: Anger and heart-disease status are independent in the population of people with normal blood pressure.

H_a: Anger and heart-disease status are not independent in the population of people with normal blood pressure.

Tests for Independence: Expected Counts

As with the two previous types of chi-square tests, we begin by comparing the observed counts with the expected counts if H_0 is true. In the anger study, the null hypothesis is that there is no association between anger level and heart-disease status in the population of interest. If we assume that H_0 is true, then anger level and CHD status are independent. We can find the expected cell counts in the two-way table using the definition of independent events from Chapter 5: $P(A|B) = P(A)$. The chance process here is randomly selecting a person and recording his or her anger level and CHD status.

		Anger level			
		Low	Moderate	High	Total
CHD status	Yes	53	110	27	190
	No	3057	4621	606	8284
	Total	3110	4731	633	8474

Let's start by considering the events "Yes" and "Low anger." We see from the two-way table that 190 of the 8474 people in the study had CHD. If we imagine choosing one member of the sample at random, $P(\text{Yes}) = 190/8474$. If the null hypothesis is true and anger level and CHD status are independent, knowing that the selected individual is low anger does not change the probability that this person develops CHD. That is to say, $P(\text{Yes}|\text{Low anger}) = P(\text{Yes}) = 190/8474 = 0.02242$. Of the 3110 low-anger people in the study, we'd expect

$$3110 \cdot \frac{190}{8474} = 3110(0.02242) = 69.73$$

to get CHD. You can see that the general formula we developed earlier for a test for homogeneity applies in this situation also:

$$\text{expected count} = \frac{\text{row total} \cdot \text{column total}}{\text{table total}} = \frac{190 \cdot 3110}{8474} = 69.73$$

You can complete the table of expected counts by using the formula for each cell or by using the formula for some cells and subtracting to find the remaining cells. Here is the completed table of expected counts:

		Expected counts			
		Anger level			
		Low	Moderate	High	Total
CHD status	Yes	69.73	106.08	14.19	190
	No	3040.27	4624.92	618.81	8284
	Total	3110	4731	633	8474

Making Connections

This is a good context to review a few ideas from Chapter 5, including the definition of independent events and conditional probability.

Making Connections

To calculate expected counts here, we assume the two variables are independent. Point out that $P(\text{CHD}|\text{low anger}) = P(\text{CHD}|\text{moderate anger}) = P(\text{CHD}|\text{high anger}) = P(\text{CHD})$ if there is no association between anger and heart disease.

Teaching Tip

How many expected values should we calculate with the formula before we can simply subtract values from the totals to get the remaining expected values? The answer is 2, if you pick them strategically, which is exactly the number of degrees of freedom.

AP® EXAM TIP

Remind students that it is not appropriate to round the expected counts to the nearest integer.

Tests for Independence: Conditions and Calculations

The 10% and Large Counts conditions for the chi-square test for independence are the same as for the test for homogeneity. There is a slight difference in the Random condition for the two tests: a test for independence uses data from a single random sample, but a test for homogeneity uses data from two or more independent random samples or from two or more groups in a randomized experiment. Together, the three conditions ensure that the observations in the sample can be viewed as independent and that the sampling distribution of the χ^2 test statistic can be modeled by a chi-square distribution.

> ### CONDITIONS FOR PERFORMING A CHI-SQUARE TEST FOR INDEPENDENCE
>
> - **Random:** The data come from a random sample from the population of interest.
>   - **10%:** When sampling without replacement, $n < 0.10N$.
> - **Large Counts:** All *expected* counts are at least 5.

The conditions are met in the anger and heart disease study because the data came from a random sample of people with normal blood pressure, 8474 is less than 10% of all people with normal blood pressure, and all the expected counts are at least 5 (see table on previous page).

When the conditions are met, we use the familiar χ^2 test statistic to measure the strength of the association between the variables in the sample. P-values for this test come from a chi-square distribution with

$$\text{df} = (\text{number of rows} - 1) \times (\text{number of columns} - 1)$$

For the anger and heart disease study,

- Test statistic: $\chi^2 = \dfrac{(53 - 69.73)^2}{69.73} + \dfrac{(110 - 106.08)^2}{106.08} + \cdots = 16.077$

- *Using technology*: With df $= (2-1)(3-1) = 2$, P-value $= 0.00032$

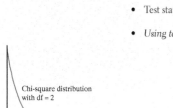

Chi-square distribution with df = 2

$\chi^2 = 16.077$

Assuming that there is no association between anger level and heart-disease status in the population of people with normal blood pressure, there is a 0.00032 probability of getting a χ^2 value of 16.077 or larger by chance alone. Because the P-value of $0.00032 < \alpha = 0.05$, we reject H_0. We have convincing evidence of an association between anger level and heart-disease status in the population of people with normal blood pressure.

A follow-up analysis reveals that two cells contribute most of the chi-square test statistic: Low anger, Yes (4.014) and High anger, Yes (11.564). Many fewer low-anger people developed CHD than expected. And many more high-anger people got CHD than expected.

Can we conclude that proneness to anger *causes* heart disease? No. The anger and heart-disease study is an observational study, not an experiment. It isn't surprising that some other variables are confounded with anger level. For example, people prone to anger are more likely than others to be men

---

Teaching Tip

In the CHD status and anger level context, there was a single random sample from the population of people with normal blood pressure. If instead they had taken 3 independent random samples (a random sample from the population of people with low anger level, a random sample from the population of people with moderate anger level, and a random sample of people with high anger level), we would use a chi-square test for homogeneity.

Teaching Tip

While it is important that students know how to check each condition, it is equally important that they understand *why* we check the conditions. We call this the "*So what?*"

Teaching Tip

Here is the "*So what?*" answer for each condition.

Random:
→ so individual observations are independent and we can generalize to the population.

10%:
→ so we can view observations as independent even though we are sampling without replacement.

Large Counts:
→ so the sampling distribution is approximately a chi-square distribution and we can use χ^2 to find a P-value.

Teaching Tip

Continue to ask students to interpret the P-value. The new College Board Course Framework has an increased emphasis on this learning objective. Additionally, understanding the P-value helps students make the correct conclusion at the end of a significance test.

who drink and smoke. We don't know whether the increased rate of heart disease among those with higher anger levels in the study is due to their anger or perhaps to their drinking and smoking or maybe even to gender.

Putting It All Together: The Chi-Square Test for Independence

When we want to test for an association between two categorical variables in a population, we use a **chi-square test for independence.** Here are the key details.

> The chi-square test for independence is also known as the chi-square test for association.

CHI-SQUARE TEST FOR INDEPENDENCE

Suppose the conditions are met. To perform a test of

H_0: There is no association between two categorical variables in the population of interest

compute the chi-square test statistic

$$\chi^2 = \sum \frac{(\text{Observed count} - \text{Expected count})^2}{\text{Expected count}}$$

where the sum is over all cells in the two-way table (not including the totals). The *P*-value is the area to the right of χ^2 under the chi-square density curve with degrees of freedom = (number of rows − 1)(number of columns − 1).

We're now ready to perform a complete test for independence.

EXAMPLE

Snowmobiles in Yellowstone
Chi-square test for independence

PROBLEM: In Chapter 1, you read about a random sample of winter visitors to Yellowstone National Park. Each of the 1526 visitors in the sample was asked two questions:

1. Do you belong to an environmental club (like the Sierra Club)?
2. What is your experience with a snowmobile: own, rent, or never used?

The two-way table summarizes the results.

		Environmental club status		
		Not a member	Member	Total
Snowmobile experience	Never used	445	212	657
	Renter	497	77	574
	Owner	279	16	295
	Total	1221	305	1526

Do the data provide convincing evidence of an association between environmental club status and type of snowmobile use in the population of winter visitors to Yellowstone National Park?

Teaching Tip

Although the Do step is the same for a chi-square test for homogeneity and a chi-square test for independence, the State, Plan, and Conclude steps are each a little different. In a test for homogeneity, we are comparing the distribution of one categorical variable in *two or more* populations or treatments. However, in a test for independence, we are investigating the relationship between two categorical variables in *one* population. Make sure to highlight these differences as you introduce the test for independence.

Making Connections

In Chapter 1, we looked at conditional relative frequencies of type of snowmobile use and concluded that there is an association between environmental club status and type of snowmobile use *for the sample.* We are now trying to assess if we have convincing evidence of an association *for the population.* Making an inference about the relationship between two categorical variables in a population based on data from a sample requires that we do a significance test to account for the variability due to random sampling.

ALTERNATE EXAMPLE

Do you use Facebook? The chi-square test for independence

PROBLEM:
Mark Z. is interested in finding out if there is an association between age group and Facebook use among U.S. adults. He takes a random sample of 250 U.S. adults aged 18 or older and records their age and whether or not they use Facebook. The two-way table summarizes the results.

	Age (years)				
		18–29	30–49	50–64	65+	Total
Facebook user?	Yes	60	64	46	21	191
	No	8	20	22	9	59
	Total	68	84	68	30	250

Do the data provide convincing evidence of an association between age group and whether or not someone uses Facebook for all U.S. adults?

SOLUTION:
STATE: H_0: There is no association between age group and whether or not someone uses Facebook for all U.S. adults.

H_a: There is an association between age group and whether or not someone uses Facebook for all U.S. adults.

We'll use $\alpha = 0.05$.

PLAN: Chi-square test for independence.

- *Random:* Random sample of 250 U.S. adults. ✓
  - *10%:* 250 < 10% of all U.S. adults. ✓
- *Large Counts:* All the expected counts are at least 5 (see table). ✓

(continues)

SOLUTION:

STATE:

H_0: There is no association between environmental club status and type of snowmobile use in the population of winter visitors to Yellowstone.

H_a: There is an association between environmental club status and type of snowmobile use in the population of winter visitors to Yellowstone.

We'll use $\alpha = 0.05$.

PLAN: Chi-square test for independence.

- Random: Random sample of 1526 winter visitors to Yellowstone. ✓
  - 10%: It is reasonable to assume that 1526 < 10% of all winter visitors to Yellowstone. ✓
- Large Counts: All the expected counts are at least 5 (see table below). ✓

		Expected counts Environmental club status		
		Not a member	Member	Total
Snowmobile experience	Never used	525.7	131.3	657
	Renter	459.3	114.7	574
	Owner	236.0	59.0	295
	Total	1221	305	1526

DO:

- Test statistic: $\chi^2 = \dfrac{(445 - 525.7)^2}{525.7} + \dfrac{(212 - 131.3)^2}{131.3} + \cdots$

  $= 116.6$

- P-value: df $= (3 - 1)(2 - 1) = 2$

Using Table C: P-value < 0.0005

Using technology: The calculator's χ^2-Test gives $\chi^2 = 116.6$ and P-value $= 4.82 \times 10^{-26}$ using df $= 2$.

CONCLUDE: Because the P-value of approximately $0 < \alpha = 0.05$, we reject H_0. We have convincing evidence of an association between environmental club status and type of snowmobile use in the population of winter visitors to Yellowstone National Park.

> Use the four-step process!

> You could also say H_0: Environmental club status and type of snowmobile use are independent in the population of winter visitors to Yellowstone.

> If no significance level is provided, use $\alpha = 0.05$.

> We know this is a test for independence and not homogeneity because there was only one random sample.

> Because the observed counts differ from the expected counts, there is *some* evidence for H_a.

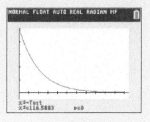

NORMAL FLOAT AUTO REAL RADIAN MP

χ^2-Test
χ^2=116.5883   p≈0

FOR PRACTICE, TRY EXERCISE 43

AP® EXAM TIP

When the P-value is very small, the calculator will report it using scientific notation. Remember that P-values are probabilities and must be between 0 and 1. If your calculator reports the P-value with a number that appears to be greater than 1, look to the right, and you will see that the P-value is being expressed in scientific notation. If you claim that the P-value is 4.82, you will certainly lose credit.

Teaching Tip

Consider having students add a follow-up sentence at the end of the conclusion for the significance test. For this example:

CONCLUDE: Because the P-value of approximately $0 < \alpha = 0.05$, we reject H_0. We have convincing evidence of an association between environmental club status and type of snowmobile use in the population of winter visitors to Yellowstone National Park. *The largest components of the χ^2 test statistic are 49.6 and 31.3 because the members who never used a snowmobile was far above what we expect and the members who owned a snowmobile was far below what we expect if there was no association between these variables in the population.*

ALTERNATE EXAMPLE (continued)

		Expected counts Age (years)				
		18–29	30–49	50–64	65+	Total
Facebook user?	Yes	51.952	64.176	51.952	22.92	191
	No	16.048	19.824	16.048	7.08	59
	Total	68	84	68	30	250

DO: • Test statistic:

$$\chi^2 = \frac{(60 - 51.952)^2}{51.952} + \frac{(64 - 64.176)^2}{64.176} + \cdots$$

$$= 8.856$$

- P-value: df $= (2 - 1)(4 - 1) = 3$

Table C: P-value is between 0.025 and 0.05.

Tech: The calculator's χ^2-Test gives $\chi^2 = 8.856$ and P-value $= 0.031$ using df $= 3$.

CONCLUDE: Because the P-value of $0.031 < \alpha = 0.05$, we reject H_0. We have convincing evidence of an association between age group and whether or not someone uses Facebook for all U.S. adults.

✔ **Answers to CYU**

STATE: H_0: There is no association between gender and perceived body image in the population of U.S. college students.

H_a: There is an association between gender and perceived body image in the population of U.S. college students.

We will use $\alpha = 0.01$.

PLAN: Chi-square test for independence. *Random:* Random sample of 1200 U.S. college students. *10%:* $n = 1200$ is less than 10% of the population of all college students. *Large Counts:* All expected counts (541.50, 313.50, 148.83, 86.17, 69.67, 40.33) are at least 5.

DO:

$$\chi^2 = \frac{(560 - 541.5)^2}{541.5} + \frac{(295 - 313.50)^2}{313.50}$$

$$= 47.176$$

P-value: df $= (3 - 1)(2 - 1) = 2$

Table C: The *P*-value is less than 0.0005.

Tech: χ^2cdf(lower: 47.176, upper: 10000, df: 2) ≈ 0

CONCLUDE: Because the *P*-value of approximately $0 < \alpha = 0.01$, we reject H_0. There is convincing evidence of an association between gender and perceived body image in the population of U.S. college students.

CHECK YOUR UNDERSTANDING

A random sample of 1200 U.S. college students was asked, "What is your perception of your own body? Do you feel that you are overweight, underweight, or about right?" The two-way table summarizes the data on perceived body image by gender.[5]

		Gender		
		Female	Male	Total
Body image	About right	560	295	855
	Overweight	163	72	235
	Underweight	37	73	110
	Total	760	440	1200

Do these data provide convincing evidence at the $\alpha = 0.01$ level of an association between gender and perceived body image in the population of U.S. college students?

Using Chi-Square Tests Wisely

Both the chi-square test for homogeneity and the chi-square test for independence start with a two-way table of observed counts. They even calculate the test statistic, degrees of freedom, and *P*-value in the same way. *The questions that these two tests answer are different, however.* A chi-square test for homogeneity tests whether the distribution of a categorical variable is the same for each of several populations or treatments. The chi-square test for independence tests whether two categorical variables are associated in some population of interest.

One way to help you distinguish these two tests is to consider the two sets of totals in the two-way table. In tests for homogeneity, one set of totals is known by the researchers *before the data are collected.* For example, in the experiment to determine if the gender of the interviewer affects the response to a question about a female president, Abby and Mia decided in advance to randomly assign 50 subjects to each treatment.

		Gender of interviewer		
		Male	Female	Total
Response to question	Yes	?	?	?
	No	?	?	?
	Maybe	?	?	?
	Total	50	50	100

Likewise, in the observational study comparing opinions about speaking English, researchers knew in advance that they would survey 1000 people from Australia, 1460 from the U.K., and 1003 from the U.S. In both cases, only one set of totals was left to vary. This is consistent with the design of the study: select independent random samples (or randomly assign treatments) and compare the distribution of a single categorical variable.

However, in a test for independence, neither set of totals is known in advance. In the observational study about snowmobile use in Yellowstone, the researchers didn't know anything about either variable ahead of time—they only knew that they would survey 1526 visitors. This is consistent with the design of the study: select one sample and record the values of two variables for each member.

		Environmental club status		
		Not a member	Member	Total
Snowmobile experience	Never used	?	?	?
	Renter	?	?	?
	Owner	?	?	?
	Total	?	?	1526

Unfortunately, it is quite common to see questions about association when a test for homogeneity applies; when a test for independence applies, questions about differences between proportions or the distribution of a variable are quite common. Many people avoid the distinction altogether and pose questions about the "relationship" between two variables.

Instead of focusing on the question asked, the best plan is to consider *how the data were produced.* If the data come from two or more independent random samples or treatment groups in a randomized experiment, then do a chi-square test for homogeneity. If the data come from a single random sample, with the individuals classified according to two categorical variables, use a chi-square test for independence.

Scary movies and fear
Choosing the right type of chi-square test

PROBLEM: Are men and women equally likely to suffer lingering fear from watching scary movies as children? Researchers asked a random sample of 117 college students to write narrative accounts of their exposure to scary movies before the age of 13. More than one-fourth of the students said that some of the fright symptoms are still present when they are awake.[9] The following table breaks down these results by gender.

		Gender		
		Male	Female	Total
Fright symptoms?	Yes	7	29	36
	No	31	50	81
	Total	38	79	117

Assume that the conditions for performing inference are met. Minitab output for a chi-square test using these data is shown.

ALTERNATE EXAMPLE

Are you headed for college? Choosing the right type of chi-square test

Skills 1.E, 4.E

PROBLEM:

A curious AP® Statistics student wondered if students in her high school would respond differently to a teacher than to a student when asked about their intention to go to college after high school. One hundred students from her high school volunteered to participate in an experiment, and they were randomly assigned to have either a student or a teacher interview them. This two-way table summarizes the data.

		Interviewer		
		Student	Teacher	Total
Going to college?	Yes	14	27	41
	No	36	23	59
	Total	50	50	100

Assume that the conditions for performing inference are met. Minitab output for a chi-square test using these data is shown.

Teaching Tip

Chi-square test for goodness of fit: 1 variable, 1 population.

Chi-square test for homogeneity: 1 variable, 2+ populations/treatments.

Chi-square test for independence: 2 variables, 1 population.

Teaching Tip

To help students remember the difference between the 3 chi-square tests, urge them to recall the context that you used in class to introduce the test. Here are the contexts that we used in the text:

1. M&M'S® or NHL birthdays → Chi-square test for goodness of fit
2. Music and purchases → Chi-square test for homogeneity
3. Anger and heart disease → Chi-square test for independence

```
Chi-Square Test: Student,
Teacher

Expected counts are printed
below observed counts

Chi-Square contributions are
printed below expected counts

          Student   Teacher   Total
Yes          14        27        41
           20.5      20.5
          2.061     2.061

No           36        23        59
           29.5      29.5
          1.432     1.432

Total        50        50       100

Chi-Sq = 6.986, DF = 1,
P-Value = 0.008
```

(a) Should a chi-square test for independence or a chi-square test for homogeneity be used in this setting? Explain your reasoning.
(b) State an appropriate pair of hypotheses for researchers to test in this setting.
(c) Which cell contributes most to the chi-square test statistic? In what way does this cell differ from what the null hypothesis suggests?
(d) Interpret the *P*-value in context. What conclusion would you draw at $\alpha = 0.01$?

(continues)

SOLUTION:
(a) Chi-square test for homogeneity. The data were produced using two randomly assigned groups of students, who were then classified according to one variable: whether or not they say they are going to college. The chi-square test for independence requires one random sample from one population and classification according to two variables.
(b) H_0: There is no difference in the true distribution of whether or not students say they are going to college when interviewed by a student and by a teacher.

H_a: There is a difference in the true distribution of whether or not students say they are going to college when interviewed by a student and by a teacher.
(c) There are two cells with the same contribution. Subjects interviewed by a student who said "Yes" to going to college (2.061) and students interviewed by a teacher who said "Yes" to going to college (2.061). Far fewer students from the group interviewed by the student say they are going to college (14) than we would expect if H_0 were true (20.5) and far more students from the group interviewed by the teacher say they are going to college (27) than we would expect if H_0 were true (20.5).
(d) If there is no difference in the true distribution of whether or not students say they are going to college when interviewed by a student and by a teacher, there is a 0.008 probability of observing differences in the distributions of responses as large as or larger than the ones in this experiment. Because the P-value of $0.008 < \alpha = 0.01$, we reject H_0. We have convincing evidence that the distribution of whether or not students say they are going to college differs when interviewed by a student and by a teacher.

```
Chi-Square Test: Male, Female
Expected counts are printed below observed counts
Chi-Square contributions are printed below expected counts

                Male        Female       Total
Yes              7            29           36
              11.69         24.31
              1.883         0.906
No              31           50           81
              26.31         54.69
              0.837         0.403
Total           38           79          117
Chi-Sq = 4.028,      DF = 1,      P-Value = 0.045
```

(a) Should a chi-square test for independence or a chi-square test for homogeneity be used in this setting? Explain your reasoning.
(b) State an appropriate pair of hypotheses for researchers to test in this setting.
(c) Which cell contributes most to the chi-square test statistic? In what way does this cell differ from what the null hypothesis suggests?
(d) Interpret the P-value. What conclusion would you draw at $\alpha = 0.01$?

SOLUTION:

(a) Chi-square test for independence. The data were produced using a single random sample of college students, who were then classified according to two variables: gender and whether or not they had lingering fright symptoms. The chi-square test for homogeneity requires independent random samples from each population.

> In this setting, the researchers wouldn't know either set of totals until the data were collected. They would know only the overall total (117) in advance. Thus, a test for independence is the correct choice.

(b) H_0: There is no association between gender and whether or not college students have lingering fright symptoms.

H_a: There is an association between gender and whether or not college students have lingering fright symptoms.

(c) Men who admit to having lingering fright symptoms account for the largest component of the chi-square test statistic (1.883). Far fewer men in the sample admitted to lingering fright symptoms (7) than we would expect if H_0 were true (11.69).

(d) If there is no association between gender and whether or not college students have lingering fright symptoms, there is a 0.045 probability of obtaining an association as strong as or stronger than the one observed in the random sample of 117 students. Because the P-value of $0.045 > \alpha = 0.01$, we fail to reject H_0. We do not have convincing evidence that there is an association between gender and whether or not college students have lingering fright symptoms.

FOR PRACTICE, TRY EXERCISE 49

WHAT IF WE WANT TO COMPARE TWO PROPORTIONS? Shopping at secondhand stores is becoming more popular and has even attracted the attention of business schools. A study of customers' attitudes toward secondhand stores interviewed separate random samples of shoppers at two secondhand stores of the same chain in different cities. The two-way table shows the breakdown of respondents by gender.[10]

AP® EXAM TIP

The remaining topics in this section are interesting and deepen our understanding of concepts, but they are not part of the College Board Course Framework. If you are short on time, these ideas can be skipped.

		Store		
		A	B	Total
Gender	Male	38	68	106
	Female	203	150	353
	Total	241	218	459

Do the data provide convincing evidence of a difference in the distributions of gender for shoppers at these two stores? To answer this question, we could perform a chi-square test for homogeneity. Our hypotheses are

H_0: There is no difference in the distributions of gender for shoppers at these two stores.

H_a: There is a difference in the distributions of gender for shoppers at these two stores.

A difference in distribution of gender would mean that there is a difference in the true proportion of female shoppers at the two stores. So we could also use a two-sample z test from Section 9.3 to compare two proportions. The hypotheses for this test are

$$H_0: p_A - p_B = 0$$
$$H_a: p_A - p_B \neq 0$$

where p_A and p_B are the true proportions of female shoppers at Store A and Store B, respectively.

The TI-84 screen shots show the results from a two-sample z test for $p_A - p_B$ and from a chi-square test for homogeneity. (We verified that the Random, 10%, and Large Counts conditions were met before carrying out the calculations.)

Note that the P-values from the two tests are the same except for rounding errors. You can also check that the chi-square test statistic is the square of the two-sample z statistic: $(3.915...)^2 = 15.334$.

As the preceding example suggests, the chi-square test for homogeneity based on a 2×2 table is equivalent to the two-sample z test for $p_1 - p_2$ with a two-sided alternative hypothesis. However, there are other settings where only one of the options is valid:

- If the two-way table is larger than 2×2, the only option is a chi-square test for homogeneity.
- If the table is 2×2 and the alternative hypothesis is one-sided, use a two-sample z test for a difference in proportions rather than a chi-square test.

✚ Ask the StatsMedic

$\chi^2 = (z)^2$: A Student Proof

This post outlines the journey one AP® Statistics student took to prove that the chi-square test statistic is the square of the two-sample z statistic.

Teaching Tip

To illustrate that comparing the distributions can provide more information than comparing means, consider the following two distributions of scores on the AP® Statistics exam. Even though both schools had a mean score of 3, the distributions are very different!

Score	School A	School B
5	10	1
4	5	5
3	1	10
2	5	5
1	10	1

However, be careful not to use too few categories when converting a quantitative variable into a categorical variable. For example, knowing the mean score might be more informative than simply reporting the percent of students who scored 3 or higher.

Teaching Tip

Remind students that we require the expected counts to be at least 5 so that the chi-square probability distribution is a good model for the sampling distribution of the chi-square statistic. A distribution isn't necessarily a good model for the sampling distribution of a statistic just because it has the same name!

- If the table is 2×2 and you want to construct a confidence interval for a difference in proportions, the only option is a two-sample z interval.

Because of the points made in the last two bullets, we recommend the methods in Sections 8.3 and 9.3 for comparing two proportions whenever you are given a choice.

GROUPING QUANTITATIVE DATA INTO CATEGORIES As we mentioned in Chapter 1, it is possible to convert a quantitative variable to a categorical variable by grouping together intervals of values. Here's an example. Researchers surveyed independent random samples of shoppers at two secondhand stores of the same chain in different cities. The two-way table summarizes data on the incomes of the shoppers in the two samples.

		Store		
		A	B	Total
Income	Under $10,000	70	62	132
	$10,000 to $19,999	52	63	115
	$20,000 to $24,999	69	50	119
	$25,000 to $34,999	22	19	41
	$35,000 or more	28	24	52
	Total	241	218	459

Personal income is a quantitative variable. But by grouping the values of this variable, we create a categorical variable. We could use these data to carry out a chi-square test for homogeneity because the data came from independent random samples of shoppers at the two stores. Comparing the distributions of income for shoppers at the two stores would give more information than simply comparing their mean incomes.

WHAT IF SOME OF THE EXPECTED CELL COUNTS ARE LESS THAN 5? Let's look at a situation where this is the case. A sample survey asked a random sample of young adults, "Where do you live now? That is, where do you stay most often?" Here is a two-way table of all 2984 people in the sample (both men and women) classified by their age and by where they lived.[11] Living arrangement is a categorical variable. Even though age is quantitative, the two-way table treats age as categorical by dividing the young adults into four categories. The table gives the observed counts for all 20 combinations of age and living arrangement.

		Age (years)				
		19	20	21	22	Total
Living arrangement	Parents' home	324	378	337	318	1357
	Another person's home	37	47	40	38	162
	Your own place	116	279	372	487	1254
	Group quarters	58	60	49	25	192
	Other	5	2	3	9	19
	Total	540	766	801	877	2984

Our null hypothesis is H_0: There is no association between age and living arrangement in the population of young adults. The following table shows the expected counts assuming H_0 is true. We can see that two of the expected counts (circled in red) are less than 5. This violates the Large Counts condition.

		Age (years)				
		19	20	21	22	Total
	Parents' home	245.57	348.35	364.26	398.82	1357
	Another person's home	29.32	41.59	43.49	47.61	162
Living arrangement	Your own place	226.93	321.90	336.61	368.55	1254
	Group quarters	34.75	49.29	51.54	56.43	192
	Other	3.44	4.88	5.10	5.58	19
	Total	540	766	801	877	2984

To make all of the expected counts 5 or more, a clever strategy is to "collapse" the table by combining two or more rows or columns. In this case, it might make sense to combine the "Group quarters" and "Other" living arrangements. Doing so and then running a chi-square test in Minitab gives the following output. Notice that the Large Counts condition is now met.

```
Chi-Square Test: 19, 20, 21, 22

Expected counts are printed below observed counts

Chi-Square contributions are printed below expected counts

                   19       20       21        22    Total
Parent's home     324      378      337       318     1357
               245.57   348.35   364.26    398.82
               25.049    2.525     2.04    16.379

Another home       37       47       40        38      162
                29.32    41.59    43.49     47.61
                2.014    0.705    0.279      1.94

Own place         116      279      372       487     1254
               226.93    321.9   336.61    368.55
               54.226    5.719     3.72    38.068

Other              63       62       52        34      211
                38.18    54.16    56.64     62.01
               16.129    1.134     0.38    12.654

Total             540      766      801       877     2984

Chi-Sq = 182.961,          DF = 9,          P-Value = 0.000
```

Teaching Tip

An alternative to collapsing rows or columns is to use technology to simulate the distribution of the chi-square test statistic, assuming the null hypothesis is true. Then use the resulting randomization distribution to estimate the P-value. Note that this is what we did at the end of the M&M'S® activity in the Introduction to Chapter 12.

Teaching Tip: Using Technology

We have now covered many different inference procedures. Students have to be able to choose the correct inference procedure for different settings. Search the Internet for "categorizing statistics problems" to find the Itcconline page that allows students to practice choosing the correct inference procedure. Uncheck the procedures we don't know yet (prediction intervals, and 1-way ANOVA), press "Submit," and have fun!

Section 12.2 | Summary

- We use the **chi-square test for homogeneity** to compare the distribution of a single categorical variable for each of several populations or treatments. The null hypothesis is that there is no difference in the distribution of the categorical variable for each of the populations or treatments.
- When performing a chi-square test for homogeneity, we need to check for independence and that the sampling distribution of the χ^2 test statistic can be modeled by a chi-square distribution. The required conditions are:
  - **Random:** The data come from independent random samples or groups in a randomized experiment.
    - **10%:** When sampling without replacement, $n < 0.10N$ for each sample.
  - **Large Counts:** All expected counts must be at least 5.
- We use the **chi-square test for independence** to test the association between two categorical variables. The null hypothesis is that there is no association between the two categorical variables in the population of interest. Another way to state the null hypothesis is that the two categorical variables are independent in the population of interest.
- When performing a chi-square test for independence, we need to check that the observations in the sample can be viewed as independent and that the sampling distribution of the χ^2 test statistic can be modeled by a chi-square distribution. The required conditions are:
  - **Random:** The data come from a random sample from the population of interest.
    - **10%:** When sampling without replacement, $n < 0.10N$.
  - **Large Counts:** All expected counts must be at least 5.
- The **expected count** in any cell of a two-way table when H_0 is true is
$$\text{expected count} = \frac{\text{row total} \cdot \text{column total}}{\text{table total}}$$
- The **chi-square test statistic** is
$$\chi^2 = \sum \frac{(\text{Observed count} - \text{Expected count})^2}{\text{Expected count}}$$
where the sum is over all cells in the two-way table (not including the totals).
- Calculate the P-value by finding the area to the right of χ^2 under the chi-square density curve with (number of rows -1)(number of columns -1) **degrees of freedom.**
- If the test finds a statistically significant result, consider doing a **follow-up analysis** that looks for the largest components of the chi-square test statistic and compares the observed and expected counts in the corresponding cells.

12.2 Technology Corner

TI-Nspire and other technology instructions are on the book's website at highschool.bfwpub.com/updatedtps6e.

30. Performing chi-square tests for two-way tables   Page 787

Teaching Tip

Now that you have completed Sections 12.1 and 12.2 in this book, you have completed the content for Unit 8. Students can login to their College Board accounts and take the **Personal Progress Check for Unit 8**.

Section 12.2 | Exercises

27. The color of candy Inspired by the example about how background music influences choice of entrée at a restaurant, a statistics student decided to investigate other ways to influence a person's behavior. Using 60 volunteers, she randomly assigned 20 volunteers to get a "red" survey, 20 volunteers to get a "blue" survey, and 20 volunteers to get a control survey. The first three questions on each survey were the same, but the fourth and fifth questions were different. For example, the fourth question on the "red" survey was "When you think of the color red, what do you think about?" On the blue survey, the question replaced *red* with *blue*. On the control survey, the last two questions were not about color. As a reward, each volunteer was allowed to choose a chocolate candy in a red wrapper or a chocolate candy in a blue wrapper. Here are segmented bar graphs showing the results of the experiment. Describe what you see.

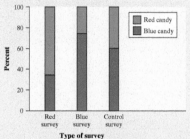

28. Python eggs How is the hatching of water python eggs influenced by the temperature of the snake's nest? Researchers randomly assigned newly laid eggs to one of three water temperatures: hot, neutral, or cold. Hot duplicates the extra warmth provided by the mother python, and cold duplicates the absence of the mother.[12] Here are segmented bar graphs showing the results of the experiment. Describe what you see.

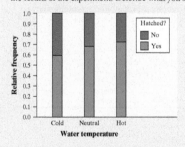

29. More candy The two-way table shows the results of the experiment described in Exercise 27. pg 783

		Survey type			
		Red	Blue	Control	Total
Color of candy	Red	13	5	8	26
	Blue	7	15	12	34
	Total	20	20	20	60

(a) State the appropriate null and alternative hypotheses.

(b) Show the calculation for the expected count in the Red/Red cell. Then provide a complete table of expected counts.

(c) Calculate the value of the chi-square test statistic.

30. More pythons The two-way table shows the results of the experiment described in Exercise 28.

		Water temperature			
		Cold	Neutral	Hot	Total
Hatching status	Yes	16	38	75	129
	No	11	18	29	58
	Total	27	56	104	187

(a) State the appropriate null and alternative hypotheses.

(b) Show the calculation for the expected count in the Cold/Yes cell. Then provide a complete table of expected counts.

(c) Calculate the value of the chi-square test statistic.

31. Last candy Refer to Exercises 27 and 29. pg 786

(a) Verify that the conditions for inference are met.

(b) Use Table C to find the *P*-value. Then use your calculator's χ^2cdf command.

(c) Interpret the *P*-value from the calculator.

(d) What conclusion would you draw using $\alpha = 0.01$?

32. Last python Refer to Exercises 28 and 30.

(a) Verify that the conditions for inference are met.

(b) Use Table C to find the *P*-value. Then use your calculator's χ^2cdf command.

(c) Interpret the *P*-value from the calculator.

(d) What conclusion would you draw using $\alpha = 0.10$?

Answers to Section 12.2 Exercises

12.27 Those who completed a survey on red paper were more likely to choose a chocolate candy in a red wrapper; those who completed a survey on blue paper were more likely to choose a chocolate candy in a blue wrapper; and those who completed the "control survey" had a slight preference for chocolate candy in a blue wrapper.

12.28 As the temperature warms up from cold to neutral to hot, the proportion of eggs that hatch increases.

12.29 (a) H_0: The distribution of candy choice is the same for subjects like these who receive the red survey, the blue survey, and the control survey. H_a: The distribution of candy choice is not the same . . . **(b)** Expected count = $(20)(26)/60 = 8.67$; remaining expected counts are 8.67, 8.67, 11.33, 11.33, 11.33. **(c)** $\chi^2 = 6.65$

12.30 (a) H_0: The distribution of hatching status is the same for eggs like these that are placed in cold water, neutral water, or hot water. H_a: There is a difference in the distribution of hatching status. **(b)** Expected counts: 18.63, 38.63, 71.74, 8.37, 17.37, and 32.26 are all ≥ 5. **(c)** $\chi^2 = 1.703$

12.31 (a) *Random:* Surveys were randomly assigned. *10%:* Not needed because the volunteers were not randomly selected from some population. *Large Counts:* All expected counts 8.67, 8.67, 8.67, 11.33, 11.33, and 11.33 are ≥ 5. **(b)** $\chi^2 = 6.65$; using df = $(2 - 1)(3 - 1) = 2$, the *P*-value (0.0359) is between 0.025 and 0.05. **(c)** Assuming the distribution of candy choice is the same for subjects like these who receive the red survey, the blue survey, or the control survey, there is a 0.0359 probability of observing differences in the distributions as large as or larger than the ones in this study by chance alone. **(d)** Because the *P*-value of $0.0359 > \alpha = 0.01$, we fail to reject H_0. We do not have convincing evidence that the distribution of candy choice differs for subjects like these who receive the red survey, the blue survey, or the control survey.

12.32 (a) *Random:* The data came from 3 groups in a randomized experiment. *10%:* The 10% condition is not needed because the eggs were not randomly selected from some population. *Large Counts:* All expected counts 18.63, 38.63, 71.74, 8.37, 17.37, and 32.26 are ≥ 5. **(b)** $\chi^2 = 1.703$; using df = $(2 - 1)(3 - 1) = 2$, the *P*-value (0.4267) is greater than 0.25. **(c)** Assuming the distribution of hatching status is the same for eggs like these that are placed in cold water, neutral water, or hot water, there is a 0.4267 probability of observing differences in distribution as large as or larger than the ones in this study by chance alone. **(d)** Because the *P*-value of $0.4267 > \alpha = 0.10$, we fail to reject H_0. We do not have convincing evidence of a difference in the true proportion of eggs like these that hatch in cold, neutral, or hot water.

12.33 We do not have the actual counts of the travelers in each category.

12.34 These data are quantitative, not categorical.

12.35 S: H_0: There is no difference in the distribution of color for name-brand and store-brand gummy bears. **P:** Chi-square test for homogeneity. *Random:* Independent random samples. *10%:* 37 < 10% of all name-brand bears and 622 < 10% of store-brand bears. *Large Counts:* 130.83, 218.17, 58.86, 98.14, 50.61, 84.39, 77.97, 130.03, 54.73, and 91.27 are ≥ 5. **D:** $\chi^2 = 1.81$, df = 4, *P*-value > 0.25 (0.7698) **C:** Because the *P*-value of 0.7698 > $\alpha = 0.05$, we fail to reject H_0. There is not convincing evidence of a difference in the distribution of color for name-brand and store-brand gummy bears.

12.36 S: H_0: There is no difference in the distribution of opinions about how high schools are doing among black, Hispanic, and white parents. **P:** Chi-square test for homogeneity. *Random:* Independent random samples. *10%:* 202 < 10% of all black parents, 202 < 10% of all Hispanic parents, and 201 < 10% of all white parents. *Large Counts:* All expected counts are ≥ 5. **D:** $\chi^2 = 22.426$, df = 8, 0.0025 < *P*-value < 0.005 (0.0042) **C:** Because the *P*-value of 0.0042 < $\alpha = 0.05$, we reject H_0. There is convincing evidence of a difference in the distribution of opinions about how high schools are doing among black, Hispanic, and white parents.

12.37 (a)

	Nicotine patch	Drug	Patch plus drug	Placebo	Total
Success	40	74	87	25	226
Failure	204	170	158	135	667
Total	244	244	245	160	893

(b) S: H_0: The true proportions of smokers like these who quit for a year are the same for the four treatments. **P:** Chi-square test for homogeneity. *Random:* 4 groups in a randomized experiment. *Large Counts:* 61.75, 61.75, 62, 40.49, 182.25, 182.25, 183, and 119.51 are ≥ 5. **D:** $\chi^2 = 34.937$, df = 3, *P*-value < 0.0005 (≈ 0) **C:** Because the *P*-value of approximately 0 < $\alpha = 0.05$, we reject H_0. There is convincing evidence that the true proportions of smokers like these who are able to quit for a year are not the same for the four treatments.

33. **Sorry, no chi-square** How do U.S. residents who travel overseas for leisure differ from those who travel for business? The following is the breakdown by occupation.[13]

Occupation	Leisure travelers (%)	Business travelers (%)
Professional/technical	36	39
Manager/executive	23	48
Retired	14	3
Student	7	3
Other	20	7
Total	**100**	**100**

Explain why we can't use a chi-square test to learn whether these two distributions differ significantly.

34. **Going nuts** The UR Nuts Company sells Deluxe and Premium nut mixes, both of which contain only cashews, Brazil nuts, almonds, and peanuts. The Premium nuts are much more expensive than the Deluxe nuts. A consumer group suspects that the two nut mixes are really the same. To find out, the group took separate random samples of 20 pounds of each nut mix and recorded the weights of each type of nut in the sample. Here are the data:[14]

	Type of mix	
Type of nut	Premium	Deluxe
Cashew	6 lb	5 lb
Brazil nut	3 lb	4 lb
Almond	5 lb	6 lb
Peanut	6 lb	5 lb

Explain why we can't use a chi-square test to determine whether these two distributions differ significantly.

35. **Gummy bears** Courtney and Lexi wondered if the distribution of color was the same for name-brand gummy bears (Haribo Gold) and store-brand gummy bears (Great Value). To investigate, they randomly selected 6 bags of each type and counted the number of gummy bears of each color.[15] Here are the data:

pg 790

		Brand		
		Name	Store	Total
	Red	137	212	349
	Green	53	104	157
Color	Yellow	50	85	135
	Orange	81	127	208
	White	52	94	146
	Total	373	622	995

Do these data provide convincing evidence that the distributions of color differ for name-brand gummy bears and store-brand gummy bears?

12.38 (a)

	Placebo	Aspirin	Dipyridamole	Both	Total
Stroke	250	206	211	157	824
No stroke	1399	1443	1443	1493	5778
Total	1649	1649	1654	1650	6602

(b) S: H_0: The true proportions of patients like these who have a stroke are the same for the four treatments. **P:** Chi-square test for homogeneity. *Random:* 4 groups in a randomized experiment. *Large Counts:* 205.81, 205.81, 206.44, 205.94, 1443.19, 1443.19, 1447.56, and 1444.06 ≥ 5.

36. **How are schools doing?** The nonprofit group Public Agenda conducted telephone interviews with three randomly selected groups of parents of high school children. There were 202 black parents, 202 Hispanic parents, and 201 white parents. One question asked, "Are the high schools in your state doing an excellent, good, fair, or poor job, or don't you know enough to say?" Here are the survey results:[16]

		Parents' race/ethnicity			
		Black	Hispanic	White	Total
	Excellent	12	34	22	68
Opinion about high schools	Good	69	55	81	205
	Fair	75	61	60	196
	Poor	24	24	24	72
	Don't know	22	28	14	64
	Total	202	202	201	605

Do these data provide convincing evidence that the distributions of opinion about high schools differ for the three populations of parents?

37. **How to quit smoking** It's hard for smokers to quit. Perhaps prescribing a drug to fight depression will work as well as the usual nicotine patch. Perhaps combining the patch and the drug will work better than either treatment alone. Here are data from a randomized, double-blind trial that compared four treatments.[17] A "success" means that the subject did not smoke for a year following the start of the study.

Group	Treatment	Subjects	Successes
1	Nicotine patch	244	40
2	Drug	244	74
3	Patch plus drug	245	87
4	Placebo	160	25

(a) Summarize these data in a two-way table.

(b) Do the data provide convincing evidence of a difference in the effectiveness of the four treatments at the $\alpha = 0.05$ significance level?

38. **Preventing strokes** Aspirin prevents blood from clotting and so helps prevent strokes. The Second European Stroke Prevention Study asked whether adding another anticlotting drug named dipyridamole would be more effective for patients who had already had a stroke. Here are the data on strokes during the two years of the study:[18]

Group	Treatment	Number of patients	Number who had a stroke
1	Placebo	1649	250
2	Aspirin	1649	206
3	Dipyridamole	1654	211
4	Both	1650	157

D: $\chi^2 = 24.243$, df = 3, *P*-value < 0.0005 (≈ 0) **C:** Because the *P*-value of approximately 0 < $\alpha = 0.05$, we reject H_0. There is convincing evidence that the true proportions of patients like these who have a stroke are not the same for the four treatments.

(a) Summarize these data in a two-way table.

(b) Do the data provide convincing evidence of a difference in the effectiveness of the four treatments at the $\alpha = 0.05$ significance level?

39. **How to quit smoking** Refer to Exercise 37. Which treatment seems to be most effective? Least effective? Justify your choices.

40. **Preventing strokes** Refer to Exercise 38. Which treatment seems to be most effective? Least effective? Justify your choices.

41. **Relaxing in the sauna** Researchers followed a random sample of 2315 middle-aged men from eastern Finland for up to 30 years. They recorded how often each man went to a sauna and whether or not he suffered sudden cardiac death (SCD). The two-way table shows the data from the study.[19]

		Weekly sauna frequency			
		1 or fewer	2–3	4 or more	Total
SCD	Yes	61	119	10	190
	No	540	1394	191	2125
	Total	601	1513	201	2315

(a) State appropriate hypotheses for performing a chi-square test for independence in this setting.

(b) Compute the expected counts assuming that H_0 is true.

(c) Calculate the chi-square test statistic, df, and P-value.

(d) What conclusion would you draw?

42. **Is astrology scientific?** The General Social Survey (GSS) asked a random sample of adults their opinion about whether astrology is very scientific, sort of scientific, or not at all scientific. Here is a two-way table of counts for people in the sample who had three levels of higher education:[20]

		Degree held			
		Associate's	Bachelor's	Master's	Total
Opinion about astrology	Not at all scientific	169	256	114	539
	Very or sort of scientific	65	65	18	148
	Total	234	321	132	687

(a) State appropriate hypotheses for performing a chi-square test for independence in this setting.

(b) Compute the expected counts assuming that H_0 is true.

(c) Calculate the chi-square test statistic, df, and P-value.

(d) What conclusion would you draw?

43. **Finger length** Is your index finger longer than your ring finger? Or is it the other way around? It isn't the same for everyone. To investigate if there is a relationship between gender and relative finger length, we selected a random sample of 460 U.S. high school students who completed a survey. The two-way table shows the results.

		Gender		
		Female	Male	Total
Relative finger length	Index longer	85	73	158
	Same length	42	44	86
	Ring longer	100	116	216
	Total	227	233	460

Do these data provide convincing evidence at the $\alpha = 0.10$ level of an association between gender and relative finger length in the population of students who completed the survey?

44. **Regulating guns** The National Gun Policy Survey asked a random sample of adults, "Do you think there should be a law that would ban possession of handguns except for the police and other authorized persons?" Here are the responses, broken down by the respondent's level of education:[21]

		Education					
		Less than HS	HS grad	Some college	College grad	Postgrad degree	Total
Opinion about handgun ban	Yes	58	84	169	98	77	486
	No	58	129	294	135	99	715
	Total	116	213	463	233	176	1201

Do these data provide convincing evidence at the $\alpha = 0.05$ level of an association between education level and opinion about a handgun ban in the adult population?

45. **Tuition bills** A random sample of U.S. adults was recently asked, "Would you support or oppose major new spending by the federal government that would help undergraduates pay tuition at public colleges without needing loans?" The two-way table shows the responses, grouped by age.[22]

		Age				
		18–34	35–49	50–64	65+	Total
Response	Support	91	161	272	332	856
	Oppose	25	74	211	255	565
	Don't know	4	13	20	51	88
	Total	120	248	503	638	1509

Do these data provide convincing evidence of an association between age and opinion about loan-free tuition in the population of U.S. adults?

12.39 *Most effective:* Far more people than expected were successful using both $(87 - 62 = 25)$. *Least effective:* Far fewer people than expected were successful using just the patch $(40 - 61.75 = -21.75)$.

12.40 *Most effective:* Far fewer people than expected had strokes while on both $(157 - 205.94 = -48.94)$. *Least effective:* Far more people than expected had strokes while on placebo $(250 - 205.81 = 44.19)$.

12.41 (a) H_0: There is no association between weekly sauna frequency and suffering from sudden cardiac death in the population of middle-aged men from eastern Finland. **(b)** 49.326, 124.177, 16.497, 551.674, 1388.823, 184.503

(c) $\chi^2 = 6.032$; using df = 2, $0.025 < P$-value $(0.0490) < 0.05$ **(d)** Because the P-value of $0.0490 < \alpha = 0.05$, we reject H_0. There is convincing evidence of an association between weekly sauna frequency and suffering from sudden cardiac death in the population of middle-aged men from eastern Finland.

12.42 (a) H_0: There is no association between level of education and opinion about astrology in the population of adults with some higher education. **(b)** 183.59, 251.85, 103.56, 50.41, 69.15, and 28.44 **(c)** $\chi^2 = 10.582$; using df = 2, $0.005 < P$-value $(0.0050) < 0.01$ **(d)** Because the P-value of $0.0050 < \alpha = 0.05$, we reject H_0. There is convincing evidence of an

association between level of education and opinion about astrology in the population of adults with some higher education.

12.43 S: H_0: There is no association between gender and relative finger length in the population of U.S. high school students who completed the survey. **P:** Chi-square test for independence. *Random:* Random sample of U.S. high school students. *10%:* $460 < 10\%$ of the population of all U.S. high school students who completed the survey. *Large Counts:* 77.97, 80.03, 42.44, 43.56, 106.59, and 109.41 are ≥ 5. **D:** $\chi^2 = 2.065$ and P-value $= 0.3561$ using df = 2. **C:** Because the P-value of $0.3561 > \alpha = 0.10$, we fail to reject H_0. There is not convincing evidence of an association between gender and relative finger length in the population of U.S. high school students who completed the survey.

12.44 S: H_0: There is no association between education level and opinion about a handgun ban in the adult population. **P:** Chi-square test for independence. *Random:* Random sample of adults. *10%:* $1201 < 10\%$ of all adults. *Large Counts:* 46.94, 86.19, 187.36, 94.29, 71.22, 69.06, 126.81, 275.64, 138.71, 104.78 are ≥ 5. **D:** $\chi^2 = 8.525$ and P-value $= 0.0741$ using df = 4. **C:** Because the P-value of $0.0741 > \alpha = 0.05$, we fail to reject H_0. We do not have convincing evidence of an association between educational level and opinion about a handgun ban in the adult population.

12.45 S: H_0: There is no association between age and opinion about loan-free tuition in the population of U.S. adults. **P:** Chi-square test for independence. *Random:* Random sample of U.S. adults. *10%:* $1509 < 10\%$ of all U.S. adults. *Large Counts:* All expected counts (68.07, 140.68, 285.33, 361.91, 44.93, 92.86, 188.33, 238.88, 7.00, 14.46, 29.33, 37.21) are ≥ 5. **D:** $\chi^2 = 39.755$ and P-value ≈ 0 using df = 6. **C:** Because the P-value of approximately $0 < \alpha = 0.05$, we reject H_0. We have convincing evidence of an association between age and opinion about loan-free tuition in the population of U.S. adults.

12.46 STATE: H_0: There is no association between age and use of online banking in the population of Internet users. H_a: There is an association . . . We'll use $\alpha = 0.05$. PLAN: Chi-square test for independence. *Random:* Random sample of Internet users. *10%:* $n = 1846 < 10\%$ of all Internet users. *Large Counts:* All expected counts 232.81, 319.45, 325.93, 209.82, 162.19, 222.55, 227.07, 146.18 ≥ 5. DO: The calculator's χ^2-Test gives $\chi^2 = 43.797$ and P-value ≈ 0 using df $= 3$. CONCLUDE: Because the P-value of approximately $0 < \alpha = 0.05$, we reject H_0. We have convincing evidence of an association between age and use of online banking in the population of Internet users.

12.47 (a) Chi-square test for homogeneity. The data came from two independent random samples. **(b)** Chi-square test for independence. The data came from a single random sample ($n = 1480$ U.S. adults), with the individuals classified in a random sample of 1480 U.S. adults according to their age group and whether or not they are vegan/vegetarian.

12.48 (a) Chi-square test for independence. The data came from a single random sample ($n = 1169$ people who had suffered heart attacks), with the individuals classified according to two categorical variables (chocolate consumption in the previous year and whether or not they die within 8 years). **(b)** Chi-square test for homogeneity. The data came from two independent random samples.

12.49 (a) Chi-square test for independence. The data came from a single random sample ($n = 4854$ young adults aged 19 to 25 years), with the individuals classified according to two categorical variables (gender and "Where do you live now?"). **(b)** H_0: There is no association between gender and where people live in the population of young adults. H_a: There is an association . . . **(c)** *Random:* Random sample of young adults. *10%:* $n = 4854 < 10\%$ of all young adults. *Large Counts:* The expected counts in the Minitab output are all ≥ 5. **(d)** *Interpretation:* Assuming there is no association between gender and where people live in the population of young adults, there is a 0.012 probability of getting a random sample of 4854 young adults with an association as strong as or stronger than the one found in this study by chance alone. *Conclusion:* Because the P-value of $0.012 < \alpha = 0.05$, we reject H_0. There is convincing evidence of an association between gender and where people live in the population of young adults.

46. **Online banking** A recent poll conducted by the Pew Research Center asked a random sample of 1846 Internet users if they do any of their banking online. The table summarizes their responses by age.[23] Is there convincing evidence of an association between age and use of online banking for Internet users?

Online banking		Age				
		18–29	30–49	50–64	65+	Total
	Yes	265	352	304	167	1088
	No	130	190	249	189	758
	Total	395	542	553	356	1846

47. **Which test?** Determine which chi-square test is appropriate in each of the following settings. Explain your reasoning.

(a) With many babies being delivered by planned cesarean section, Mrs. McDonald's statistics class hypothesized that there would be fewer younger people born on the weekend. To investigate, they selected a random sample of people born before 1980 and a separate random sample of people born after 1993. In addition to year of birth, they also recorded the day of the week on which each person was born.

(b) Are younger people more likely to be vegan/vegetarian? To investigate, the Pew Research Center asked a random sample of 1480 U.S. adults for their age and whether or not they are vegan/vegetarian.

48. **Which test?** Determine which chi-square test is appropriate in each of the following settings. Explain your reasoning.

(a) Does chocolate help heart-attack victims live longer? Researchers in Sweden randomly selected 1169 people who had suffered heart attacks and asked them about their consumption of chocolate in the previous year. Then the researchers followed these people and recorded whether or not they had died within 8 years.[24]

(b) Random-digit-dialing telephone surveys used to exclude cell-phone numbers. If the opinions of people who have only cell phones differ from those of people who have landline service, the poll results may not represent the entire adult population. The Pew Research Center interviewed separate random samples of cell-only and landline telephone users who were less than 30 years old and asked them to describe their political party affiliation.[25]

49. **Where do young adults live?** A survey by the National Institutes of Health asked a random sample of young adults (aged 19 to 25 years), "Where do you live now? That is, where do you stay most often?" Here is the full two-way table (omitting a few who refused to answer and one who reported being homeless):[26]

Living location		Gender		
		Female	Male	Total
	Parents' home	923	986	1909
	Another person's home	144	132	276
	Own place	1294	1129	2423
	Group quarters	127	119	246
	Total	2488	2366	4854

(a) Should we use a chi-square test for homogeneity or a chi-square test for independence in this setting? Justify your answer.

(b) State appropriate hypotheses for performing the type of test you chose in part (a).

Here is Minitab output from a chi-square test.

```
Chi-Square Test: Female, Male
Expected counts are printed below observed
counts
Chi-Square contributions are printed below
expected counts
                    Female      Male    Total
Parents' home          923       986     1909
                    978.49    930.51
                     3.147     3.309
Another home           144       132      276
                    141.47    134.53
                     0.045     0.048
Own place             1294      1129     2423
                   1241.95   1181.05
                     2.181     2.294
Group                  127       119      246
                    126.09    119.91
                     0.007     0.007
Total                 2488      2366     4854
Chi-Sq = 11.038,     DF = 3,   P-Value = 0.012
```

(c) Check that the conditions for carrying out the test are met.

(d) Interpret the P-value. What conclusion would you draw?

50. **Distance from home** A study of first-year college students asked separate random samples of students from private and public universities the following question: "How many miles is this university from your permanent home?" Students had to choose from the following options: 5 or fewer, 6 to 10, 11 to 50, 51 to

100, 101 to 500, or more than 500.[27] Here is the two-way table summarizing the responses:

		Type of university		
		Public	Private	Total
	5 or fewer	1951	1028	2979
	6 to 10	2688	1285	3973
Distance from home (miles)	11 to 50	10,971	5527	16,498
	51 to 100	6765	2211	8976
	101 to 500	15,177	6195	21,372
	Over 500	5811	9486	15,297
	Total	43,363	25,732	69,095

(a) Should we use a chi-square test for homogeneity or a chi-square test for independence in this setting? Justify your answer.

(b) State appropriate hypotheses for performing the type of test you chose in part (a).

Here is Minitab output from a chi-square test.

```
Chi-Square Test: Public, Private
Expected counts are printed below observed
counts
Chi-Square contributions are printed below
expected counts
              Public    Private    Total
5 or less       1951       1028     2979
              1869.6     1109.4
                3.54       5.97
6 to 10         2688       1285     3973
              2493.4     1479.6
               15.19      25.59
11 to 50       10971       5527    16498
              10354      6144.1
               36.77      61.98
51 to 100       6765       2211     8976
              5633.2     3342.8
              227.4       383.2
101 to 500     15177       6195    21372
              13413      7959.2
              232         391
Over 500        5811       9486    15297
              9600.2     5696.8
              1496       2520.4
Total          43363      25732    69095
Chi-Sq = 5398.7,   DF = 5,   P-Value = 0.0000
```

(c) Check that the conditions for carrying out the test are met.

(d) Interpret the P-value. What conclusion would you draw?

51. **Where do you live?** Conduct a follow-up analysis for the test in Exercise 49.

52. **How far away do you live?** Conduct a follow-up analysis for the test in Exercise 50.

53. **Treating ulcers** Gastric freezing was once a recommended treatment for ulcers in the upper intestine. Use of gastric freezing stopped after experiments showed it had no effect. One randomized comparative experiment found that 28 of the 82 gastric-freezing patients improved, while 30 of the 78 patients in the placebo group improved.[28] We can test the hypothesis of "no difference" in the effectiveness of the treatments in two ways: with a two-sample z test or with a chi-square test.

(a) State appropriate hypotheses for a chi-square test.

(b) Here is Minitab output for a chi-square test. Interpret the P-value. What conclusion would you draw?

```
Chi-Square Test: Gastric freezing,
Placebo
Expected counts are printed below
observed counts
Chi-Square contributions are printed
below expected counts
                 Gastric
              freezing   Placebo    Total
Improved           28        30       58
                29.73     28.27
                 0.1       0.105
Didn't improve     54        48      102
                52.27     49.73
                 0.057     0.06
Total              82        78      160
Chi-Sq = 0.322,     DF = 1,    P-Value = 0.570
```

(c) Here is Minitab output for a two-sample z test. Explain how these results are consistent with the test in part (a).

```
Test for Two Proportions
Sample      X         N      Sample p
1          28        82      0.341463
2          30        78      0.384615
Difference = p(1) - p(2)
Estimate for difference: -0.0431520
Test for difference = 0 (vs not = 0):
Z = -0.57  P-Value = 0.570
```

54. **Opinions about the death penalty** The General Social Survey (GSS) asked separate random samples of people with only a high school degree and people with a bachelor's degree, "Do you favor or oppose the death penalty for persons convicted of murder?" Of the 1379

12.52 $\chi^2 = 3.54 + 5.97 + 15.19 + 25.59 + 36.77 + 61.97 + 227.4 + 383.2 + 232 + 391 + 1496 + 2520.4 = 5398.7$; the largest component comes from private school students who attend school more than 500 miles from home. There were more individuals in this category than would have been expected ($9486 - 5696.8 = 3789.2$). The next largest component comes from public school students who attend school more than 500 miles from home. There were fewer individuals in this category than would have been expected ($5811 - 9600.2 = -3789.2$).

12.53 (a) H_0: There is no difference in the improvement rates for patients like these who receive gastric freezing and those who receive the placebo. H_a: There is a difference in improvement rates . . . **(b)** Assuming there is no difference in the improvement rates between patients like these who receive gastric freezing and those who receive the placebo, there is a 0.570 probability of observing a difference in improvement rates as large as or larger than the difference observed in the study by chance alone. *Conclude:* Because the P-value of $0.570 > \alpha = 0.05$, we fail to reject H_0. There is not convincing evidence of a difference in the improvement rates for patients like these who receive gastric freezing and those who receive the placebo. **(c)** The P-value for this test is identical to the P-value for the test in part (a). Also, $z^2 = (-0.57)^2 = 0.3249 \approx \chi^2 = 0.322$.

12.50 (a) Chi-square test for homogeneity. The data came from two independent random samples. **(b)** H_0: There is no difference in the distribution of "distance from home" among first-year private and public university students. H_a: There is a difference . . . **(c)** *Random:* Two independent random samples. *10%:* $n_1 = 43,363 < 10\%$ of all public university students and $n_2 = 25,732 < 10\%$ of all private university students. *Large Counts:* The expected counts are all ≥ 5. **(d)** Assuming the distribution of "distance from home" is the same for all first-year private and public university students, there is approximately 0 probability of observing differences in the distributions as large as or larger than the ones in this study by

chance alone. *Conclude:* Because the P-value of approximately $0 < \alpha = 0.05$, we reject H_0. There is convincing evidence of a difference in the distribution of "distance from home" among first-year private and public university students.

12.51 $\chi^2 = 3.147 + 3.309 + 0.045 + 0.048 + 2.181 + 2.294 + 0.007 + 0.007 = 11.038$; the largest component comes from males who currently live in their parents' homes. There were more males living in their parents' homes than would have been expected ($986 - 930.51 = 55.49$). The next largest component comes from females who currently live in their parents' homes. There were fewer females living in their parents' homes than would have been expected ($923 - 978.49 = -55.49$).

12.54 (a) H_0: There is no difference in opinion about the death penalty among all people who have only a high school diploma and all those who have a bachelor's degree. H_a: There is a difference in opinion . . . **(b)** Assuming there is no difference in opinion about the death penalty among all people who have only a high school diploma and all those who have a bachelor's degree, there is approximately 0 probability of observing a difference in opinion as large as or larger than the difference observed in the study by chance alone. *Conclude:* Because the P-value of approximately $0 < \alpha = 0.05$, we reject H_0. There is convincing evidence of a difference in opinion about the death penalty among all people who have only a high school diploma and all those who have a bachelor's degree. **(c)** The P-value for this test is identical to the P-value for the test in part (a). Also, $z^2 = (4.19)^2 = 17.556 \approx \chi^2 = 17.590$.

12.55 e

12.56 d

12.57 a

12.58 a

12.59 c

people with only a high school degree, 1010 favored the death penalty, while 319 of the 504 people with a bachelor's degree favored the death penalty. We can test the hypothesis of "no difference" in support for the death penalty among people in these educational categories in two ways: with a two-sample z test or with a chi-square test.

(a) State appropriate hypotheses for a chi-square test.

(b) Here is Minitab output for a chi-square test. Interpret the P-value. What conclusion would you draw?

```
Chi-Square Test: HS, Bachelor
Expected counts are printed below observed
counts
Chi-Square contributions are printed below
expected counts

                HS      Bachelor     Total
Favor         1010          319       1329
             973.28       355.72
              1.385        3.790

Oppose         369          185        554
             405.72       148.28
              3.323        9.092

Total         1379          504       1883
Chi-Sq = 17.590,    DF = 1,    P-Value = 0.000
```

(c) Here is Minitab output for a two-sample z test. Explain how these results are consistent with the test in part (a).

```
Test for Two Proportions
Sample       X        N       Sample p
1         1010      1379      0.732415
2          319       504      0.632937

Difference = p(1) - p(2)
Estimate for difference: 0.0994783
Test for difference = 0 (vs not = 0):
Z = 4.19  P-Value = 0.000
```

Multiple Choice: *Select the best answer for Exercises 55–60.*

Exercises 55–58 refer to the following setting. The National Longitudinal Study of Adolescent Health interviewed a random sample of 4877 teens (grades 7 to 12). One question asked, "What do you think are the chances you will be married in the next 10 years?" Here is a two-way table of the responses by gender:[29]

		Gender		
		Female	Male	Total
Opinion about marriage	Almost no chance	119	103	222
	Some chance, but probably not	150	171	321
	A 50–50 chance	447	512	959
	A good chance	735	710	1445
	Almost certain	1174	756	1930
	Total	2625	2252	4877

55. Which of the following is the appropriate null hypothesis for performing a chi-square test?

(a) Equal proportions of female and male teenagers are almost certain they will be married in 10 years.

(b) There is no difference between the distributions of female and male teenagers' opinions about marriage in this sample.

(c) There is no difference between the distributions of female and male teenagers' opinions about marriage in the population.

(d) There is no association between gender and opinion about marriage in the sample.

(e) There is no association between gender and opinion about marriage in the population.

56. Which of the following is the expected count of females who respond "Almost certain"?

(a) 487.7   (d) 1038.8

(b) 525   (e) 1174

(c) 965

57. Which of the following is the correct number of degrees of freedom for the chi-square test using these data?

(a) 4   (d) 20

(b) 8   (e) 4876

(c) 10

58. For these data, $\chi^2 = 69.8$ with a P-value of approximately 0. Assuming that the researchers used a significance level of 0.05, which of the following is true?

(a) A Type I error is possible.

(b) A Type II error is possible.

(c) Both a Type I and a Type II error are possible.

(d) There is no chance of making a Type I or Type II error because the P-value is approximately 0.

(e) There is no chance of making a Type I or Type II error because the calculations are correct.

59. When analyzing survey results from a two-way table, the main distinction between a test for independence and a test for homogeneity is

(a) how the degrees of freedom are calculated.

(b) how the expected counts are calculated.

(c) the number of samples obtained.

(d) the number of rows in the two-way table.

(e) the number of columns in the two-way table.

60. Cocaine addicts need cocaine to feel any pleasure, so perhaps giving them an antidepressant drug will help. A 3-year study with 72 chronic cocaine users compared an antidepressant drug called desipramine with lithium (a standard drug to treat cocaine addiction) and a placebo. One-third of the subjects were randomly assigned to receive each treatment. At the end of the study, researchers recorded whether or not the subjects relapsed.[30] Which of the following conditions must be satisfied to perform the appropriate chi-square test using the data from this study?

I.   The population distribution is approximately Normal.

II. The treatments were randomly assigned.

III. The observed counts are all at least 5.

(a) I only   (d) II and III

(b) II only   (e) I, II, and III

(c) III only

Recycle and Review

For Exercises 61 and 62, you may find the inference summary chart near the back cover to be helpful.

61. **Inference recap** (8.1 to 12.2) In each of the following settings, state which inference procedure from Chapter 8, 9, 10, 11, or 12 you would use. Be specific. For example, you might answer, "Two-sample z test for the difference between two proportions." You do not have to carry out any procedures.[31]

(a) What is the average voter turnout during an election? A random sample of 38 cities was asked to report the percent of registered voters who voted in the most recent election.

(b) Are blondes more likely to have a boyfriend than the rest of the single world? Independent random samples of 300 blondes and 300 nonblondes were asked whether they have a boyfriend.

62. **Inference recap** (8.1 to 12.2) In each of the following settings, state which inference procedure from Chapter 8, 9, 10, 11, or 12 you would use. Be specific. For example, you might answer, "Two-sample z test for the difference between two proportions." You do not have to carry out any procedures.[32]

(a) Is there a relationship between attendance at religious services and alcohol consumption? A random sample of 1000 adults was asked whether they regularly attend religious services and whether they drink alcohol daily.

(b) Separate random samples of 75 college students and 75 high school students were asked how much time, on average, they spend watching television each week. We want to estimate the difference in the average amount of TV watched by high school and college students.

Exercises 63 and 64 refer to the following setting. For their final project, a group of AP® Statistics students investigated the following question: "Will changing the rating scale on a survey affect how people answer the question?" To find out, the group took an SRS of 50 students from an alphabetical roster of the school's just over 1000 students. The first 22 students chosen were asked to rate the cafeteria food on a scale of 1 (terrible) to 5 (excellent). The remaining 28 students were asked to rate the cafeteria food on a scale of 0 (terrible) to 4 (excellent). Here are the data:

	1 to 5 scale				
Rating	1	2	3	4	5
Frequency	2	3	1	13	3

	0 to 4 scale				
Rating	0	1	2	3	4
Frequency	0	0	2	18	8

63. **Design and analysis** (4.2, 12.2)

(a) Was this an observational study or an experiment? Justify your answer.

(b) Explain why it would *not* be appropriate to perform a chi-square test in this setting.

64. **Average ratings** (1.3, 2.1, 11.2) The students decided to compare the average ratings of the cafeteria food on the two scales.

(a) Find the mean and standard deviation of the ratings for the students who were given the 1-to-5 scale.

(b) For the students who were given the 0-to-4 scale, the ratings have a mean of 3.21 and a standard deviation of 0.568. Since the scales differ by one point, the group decided to add 1 to each of these ratings. What are the mean and standard deviation of the adjusted ratings?

(c) Would it be appropriate to compare the means from parts (a) and (b) using a two-sample t test? Justify your answer.

12.60 b

12.61 (a) One-sample t interval for μ
(b) Two-sample z test for the difference between two proportions

12.62 (a) Chi-square test for independence **(b)** Two-sample t interval for the difference between two means

12.63 (a) This was an experiment because a treatment (type of rating scale) was deliberately imposed on the students who took part in the study. **(b)** Several of the expected counts are less than 5.

12.64 (a) The mean for the 1–5 scale is 3.545; the standard deviation is 1.184. **(b)** The new mean for the 0–4 scale is $3.21 + 1 = 4.21$. Adding 1 to each of the values does not change the standard deviation, so it remains 0.568. **(c)** No, because the Normal/Large Sample condition is not met. The sample sizes are both less than 30 and there are two low outliers in the distribution of responses for students who were given the 1–5 scale.

Teaching Tip

Consider giving students a quiz over the first two sections of Chapter 12 or assigning the Unit 8 Personal Progress Check before moving to Section 12.3. Section 12.3 makes a big shift from inference for categorical variables (chi-square) to inference for the relationship between two quantitative variables (inference for slope).

PD **Section 12.3 Overview**

To watch the video overview of Section 12.3 (for teachers), click on the link in the TE-Book, look on the TRFD, or download from the Teacher's Resources on the book's digital platform.

TRM **Section 12.3 Alternate Examples**

You can find the Alternate Examples for this section in Microsoft Word format by clicking on the link in the TE-Book, opening the TRFD, or downloading from the Teacher's Resources on the book's digital platform.

 ACTIVITY OVERVIEW

To prepare for using this activity, watch the overview video by clicking on the link in the TE-Book, opening the TRFD, or downloading from the Teacher's Resources on the book's digital platform.

Time: 20 minutes

Materials: 263 cards with duration on one side and interval on the other

Teaching Advice: Prepare the 263 cards before the activity by printing the "Sampling from Old Faithful" document in the Teacher's Resource Materials and cutting out the cards. Each group of students will be sampling from this same population. If you have a very large class, create multiple populations to make the data collection go quickly. Use a different color for each population set of cards so students know which bag to put their cards back into at the end of the activity.

Be sure that students understand the distinction between the population (all 263 eruptions) and the sample (the 15 eruptions randomly selected). Also be sure they understand that the parameter (the true slope of the least-squares regression line for all 263 eruptions) is being estimated using a statistic (the slope of the least-squares regression line for the sample of 15 eruptions).

Students can use their calculators or the *Two Quantitative Variables* applet available in the Extra Applets on the Student Site at highschool.bfwpub.com/updatedtps6e to calculate the slope of the least-squares regression line for their random sample of 15 eruptions. You could also use this applet to reveal the true slope of the population regression line at the end of the activity.

LEARNING TARGETS *By the end of the section, you should be able to:*

- Check the conditions for performing inference about the slope β of the population (true) regression line.
- Interpret the values of a, b, s, and SE_b in context, and determine these values from computer output.
- Construct and interpret a confidence interval for the slope β of the population (true) regression line.
- Perform a significance test about the slope β of the population (true) regression line.

When a scatterplot shows a linear relationship between a quantitative explanatory variable x and a quantitative response variable y, we can use the least-squares line calculated from the data to predict y for a given value of x. If the data are a random sample from a larger population, we use statistical inference to answer questions like these:

- Is there really a linear relationship between x and y in the population, or is it plausible that the pattern we see in the scatterplot happened by chance alone?
- In the population, how much will the predicted value of y change for each increase of 1 unit in x? What's the margin of error for this estimate?

If the data come from a randomized experiment, the values of the explanatory variable correspond to the levels of some factor that is being manipulated by the researchers. For instance, malaria researchers might want to investigate how temperature affects the life span of mosquitoes. They could set up several tanks at each of several different temperatures and then randomly assign hundreds of mosquitoes to each of the tanks. The response variable of interest is the time (in days) from hatching to death. Suppose that a scatterplot of average life span versus temperature has a linear form. We use statistical inference to decide if these data provide convincing evidence that changes in temperature cause changes in life span.

In this section, we will show you how to estimate and test claims about the slope of the population (true) regression line. Before you perform inference about a slope, you should understand the sampling distribution of the sample slope. The following activity gets you started.

> It is conventional to refer to a scatterplot of the points (x,y) as a graph of y versus x. So a scatterplot of life span versus temperature uses life span as the response variable and temperature as the explanatory variable.

ACTIVITY   **Sampling from Old Faithful**

In Chapter 7, you learned about the sampling distribution of a sample proportion $\hat{p}$ and the sampling distribution of a sample mean $\bar{x}$. In this activity, you will explore the sampling distribution of the sample slope b when calculating least-squares regression lines from random samples.

Old Faithful geyser is one of the most popular attractions in Yellowstone National Park. As you saw in Chapter 3, it is possible to use a least-squares regression line to predict y = the interval of time (in minutes) until the next eruption from x = duration (in minutes) of an eruption. In one particular month, Old Faithful erupted 263 times. Your teacher has printed out this population on 263 cards. Each card gives the duration of a particular eruption on one side and the interval to the next eruption on the other side.

Most values of the sample slopes should be between 5 and 20, with the center of the distribution around 13. Be sure that the class takes enough samples so the dotplot begins to reveal the shape, center, and variability of the true sampling distribution. We suggest at least 20 samples. Post the completed poster board in the classroom so that you can refer to this activity throughout the rest of the chapter.

The shape of the simulated sampling distribution should be approximately Normal. The center should be close to the true slope of the least-squares regression line for the whole population (13.29). The standard deviation should be close to 1.42, which will later be denoted σ_b.

TRM **Sampling from Old Faithful**

There is a PDF of the data from the 263 eruptions of Old Faithful that can be printed and cut out to create the cards needed for this activity. Also included are Excel and Fathom files with all the data. Find these in the Teacher's Resource Materials located in the TE-Book, on the TRFD, or in the Teacher's Resources in the book's digital platform.

1. Form teams of 2 or 3 students.

2. Have one student from each team select a random sample of 15 cards from the population and return to the team. This student should read aloud the duration (x) and interval (y) values for each eruption while the remaining team members enter these values into their calculators (or other technology). Replace the 15 cards in the population.

3. After each team has recorded the values for its 15 randomly selected eruptions, the team should calculate the sample least-squares regression line for its data. Record the value of the sample slope b.

4. Your teacher will prepare an axis for a dotplot to display the distribution of the sample slope. Add the value you calculated in Step 3 to this dotplot, using the symbol "b" instead of a dot. Repeat Steps 2–4 as necessary to get at least 20 sample slopes.

5. As a class, describe this simulated sampling distribution. Remember to discuss shape, center, and variability.

6. What do you think would happen to the simulated sampling distribution if the sample size was increased from 15 to 50? Discuss as a class.

Figure 12.8 shows the relationship between x = duration (in minutes) of an eruption and y = the interval of time (in minutes) until the next eruption for all 263 eruptions of the Old Faithful geyser during a particular month. The least-squares regression line is also shown.

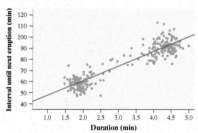

FIGURE 12.8 Scatterplot of the duration and interval between eruptions of Old Faithful for all 263 eruptions in a single month. The population least-squares line is shown in blue.

Because the scatterplot includes all the eruptions in a particular month, the least-squares regression line is called a **population regression line** (or true regression line). In most cases, we don't have data from the entire population, so we use the **sample regression line** (or estimated regression line) to estimate the population regression line.

Making Connections

Remind students that a statistic from a sample is used to estimate a parameter for a population. In previous chapters, we used a sample proportion ($\hat{p}$) to estimate a population proportion (p), a sample mean ($\bar{x}$) to estimate a population mean (μ), a difference of sample proportions ($\hat{p}_1 - \hat{p}_2$) to estimate a difference in population proportions ($p_1 - p_2$), and a difference of sample means ($\bar{x}_1 - \bar{x}_2$) to estimate a difference of population means ($\mu_1 - \mu_2$). In this section, we will use a sample regression line ($\hat{y} = a + bx$) to estimate a population regression line ($\mu_y = \alpha + \beta x$).

Teaching Tip

Notation matters! A statistic is used to estimate a parameter.

- a from the sample regression line is used to estimate the population y intercept α.
- b from the sample regression line is used to estimate the population slope β.
- s from the sample regression line is used to estimate the population standard deviation σ (coming later).

Teaching Tip

While there are several different parameters being estimated with the sample regression line, inform students that in this chapter we focus on estimating and testing claims about the slope β.

Teaching Tip

Make sure that students understand the graphs in Figure 12.9. In each graph, a different SRS of size 15 was selected and the sample least-squares regression line was calculated. Emphasize that different samples will produce different lines—and different slopes. This means that the sample slope is a random variable. The sampling distribution of the sample slope describes the possible values of b and how likely they are to occur.

Note that the symbols α and β here refer to the intercept and slope of the population regression line. They are in no way related to the probabilities of Type I and Type II errors, which are also designated by the Greek letters α and β. To prevent this confusion, some books use the form $\mu_y = \beta_0 + \beta_1 x$ for the population regression line.

> **DEFINITION** **Population regression line, Sample regression line**
>
> A regression line calculated from every value in the population is called a **population regression line** (true regression line). The equation of a population regression line is $\mu_y = \alpha + \beta x$ where
>
> - μ_y is the mean y-value for a given value of x.
> - α is the population y intercept.
> - β is the population slope.
>
> A regression line calculated from a sample is called a **sample regression line** (estimated regression line). The equation of a sample regression line is $\hat{y} = a + bx$ where
>
> - $\hat{y}$ is the estimated mean y-value for a given value of x.
> - a is the sample y intercept.
> - b is the sample slope.

In Chapter 3, we interpreted $\hat{y}$ as the predicted value of y for a given value of x. That interpretation is still valid. What if we want to use the sample regression line to make an inference about the population regression line $\mu_y = \alpha + \beta x$? Now we can think of $\hat{y}$ as being an estimate for μ_y, the mean value of y for all individuals in the population with the given value of x.

How does the slope of the sample regression line b relate to the slope of the population regression line β? To find out, we'll learn about the sampling distribution of the sample slope.

Sampling Distribution of b

Confidence intervals and significance tests about the slope of the population regression line are based on the sampling distribution of b, the slope of the sample regression line. Figure 12.9 shows the results of taking three different SRSs of 15 Old Faithful eruptions from the population described earlier. Each graph displays the selected points and the least-squares regression line for that sample (in green). The population regression line ($\mu_y = 33.35 + 13.29x$) is also shown (in blue).

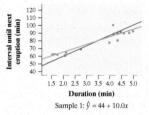 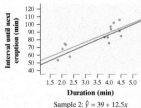 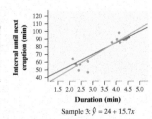

Sample 1: $\hat{y} = 44 + 10.0x$ Sample 2: $\hat{y} = 39 + 12.5x$ Sample 3: $\hat{y} = 24 + 15.7x$

FIGURE 12.9 Scatterplots and least-squares regression lines (in green) for three different SRSs of 15 Old Faithful eruptions, along with the population regression line (in blue).

Notice that the slopes of the sample regression lines ($b = 10.0$, $b = 12.5$, and $b = 15.7$) vary quite a bit from the slope of the population regression line, $\beta = 13.29$. The pattern of variation in the sample slope b is described by its sampling distribution.

To get a better picture of this variation, we used technology to simulate choosing 1000 SRSs of $n = 15$ points from the Old Faithful data, each time calculating the sample regression line $\hat{y} = a + bx$. Figure 12.10 displays the values of the slope b for the 1000 sample regression lines. We have added a vertical line (in blue) at 13.29 corresponding to the slope of the population regression line β.

Let's describe this simulated sampling distribution of b.

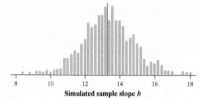

FIGURE 12.10 Dotplot of the sample slope b of the least-squares regression line in 1000 simulated SRSs of $n = 15$ eruptions. The population slope (13.29) is marked with a blue vertical line.

Shape: We can see that the distribution of b-values is roughly symmetric and single peaked. Figure 12.11(a) is a Normal probability plot of these sample regression line slopes. The strong linear pattern in the graph tells us that the simulated sampling distribution of b is close to Normal.

Center: The mean of the 1000 b-values is 13.34. This value is quite close to the slope of the population (true) regression line, 13.29.

Variability: The standard deviation of the 1000 b-values is 1.40. We will soon see that the standard deviation of the sampling distribution of b is actually 1.42.

Figure 12.11(b) is a histogram of the b-values from the 1000 simulated SRSs. We have superimposed the density curve for a Normal distribution with mean 13.29 and standard deviation 1.42. This curve models the approximate sampling distribution of the slope quite well.

Let's do a quick recap. For all 263 eruptions of Old Faithful in a single month, the population regression line is $\mu_y = 33.35 + 13.29x$. We use the symbols $\alpha = 33.35$ and $\beta = 13.29$ to represent the y intercept and slope parameters. The standard deviation of the residuals for this line is the parameter $\sigma = 6.47$.

Figure 12.11(b) shows the approximate sampling distribution of the slope b of the sample regression line for samples of 15 eruptions. If we take *all* possible

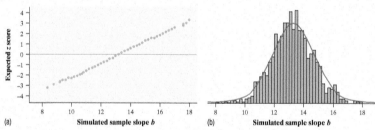

FIGURE 12.11 (a) Normal probability plot and (b) histogram of the 1000 sample regression line slopes from Figure 12.10. The blue density curve in Figure 12.11(b) is for a Normal distribution with mean 13.29 and standard deviation 1.42.

SRSs of size $n = 15$ from this population, we get the sampling distribution of b. Can you guess its shape, center, and variability?

Shape: Approximately Normal

Center: $\mu_b = \beta = 13.29$ (b is an unbiased estimator of β.)

Variability: $\sigma_b = \dfrac{\sigma}{\sigma_x \sqrt{n}} = \dfrac{6.47}{1.18\sqrt{15}} = 1.42$, where σ_x is the standard deviation of duration for the 263 eruptions.

We interpret σ_b just like any other standard deviation: the slopes of the sample regression lines typically vary from the slope of the population regression line by about 1.42.

Here's a summary of the important facts about the sampling distribution of b.

SAMPLING DISTRIBUTION OF A SLOPE

Choose an SRS of n observations (x, y) from a population of size N with least-squares regression line

$$\mu_y = \alpha + \beta x$$

Let b be the slope of the sample regression line. Assuming the conditions are met:

- The **mean** of the sampling distribution of b is $\mu_b = \beta$.
- The **standard deviation** of the sampling distribution of b is approximately

$$\sigma_b = \frac{\sigma}{\sigma_x \sqrt{n}}$$

- The **shape** of the sampling distribution of b is approximately Normal.

We'll say more about the conditions in a moment.

Think About It

WHAT'S WITH THAT FORMULA FOR σ_b? Three factors affect the standard deviation of the sampling distribution of b:

1. σ, the standard deviation of the residuals for the population regression line. Because σ is in the numerator of the formula, when σ is larger, so is σ_b. When the points vary more from the population (true) regression line, we should expect more variability in the slopes b of sample regression lines from repeated random sampling or random assignment.

2. σ_x, the standard deviation of the explanatory variable. Because σ_x is in the denominator of the formula, when σ_x is larger, σ_b is smaller. More variability in the values of the explanatory variable leads to a more precise estimate of the slope of the true regression line.

3. n, the sample size. As with every other formula for the standard deviation of a statistic, the variability of the statistic gets smaller as the sample size increases. A larger sample size will lead to a more precise estimate of the true slope.

Teaching Tip

The guidelines for determining the shape, center, and variability of the *sampling* distribution of b also apply to the *randomization* distribution of b. This makes it possible to use the same inference procedures to analyze the results of random samples and randomized experiments.

Making Connections

To calculate the standard deviation of the sampling distribution of the sample slope σ_b, we need the data from all individuals in the population. Most often, we will only have a sample from the population. Using the sample, we will calculate SE_b, which is an estimate of σ_b. The discussion of SE_b is coming soon.

Teaching Tip

To illustrate how the standard deviation of x affects the variability of the slope, hold a meter stick with one hand on each end. Then "wiggle" the meter stick by moving your hands up and down by an inch in opposite directions. Ask your students to comment on how much the "slope" of the meter stick changes. Next, move your hands toward the middle of the meter stick and wiggle the meter stick by moving your hands up and down by an inch in opposite directions. The "slope" of the meter stick is now much more variable.

Teaching Tip

Question 6 on the 2011 AP® Statistics exam (Form B) led students to discover that the sampling distribution of the slope will be less variable when the values of x are more spread out. In an experiment, researchers can reduce the standard error of the slope (and gain more power) by spreading out the levels of the explanatory variable as much as possible so that the standard deviation of x is larger.

Conditions for Regression Inference

We can fit a least-squares line to any data relating two quantitative variables, but the results are useful only if the scatterplot shows a linear pattern. Inference about regression involves more detailed conditions. Figure 12.12 shows the regression model in picture form *when the conditions are met*.

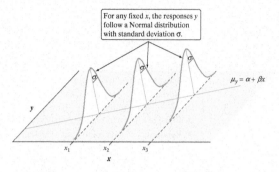

For any fixed x, the responses y follow a Normal distribution with standard deviation σ.

$\mu_y = \alpha + \beta x$

FIGURE 12.12 The regression model when the conditions for inference are met. The line is the population (true) regression line, which shows how the mean response μ_y changes as the explanatory variable x changes. For any fixed value of x, the observed response y varies according to a Normal distribution having mean μ_y and standard deviation σ.

Shape: For each possible value of the explanatory variable x, the values of the response variable y follow a Normal distribution.

Center: For each possible value of the explanatory variable x, the mean value of the response variable μ_y falls on the population (true) regression line $\mu_y = \alpha + \beta x$. Because the regression line goes through the mean value of y for each value of x, we can interpret the slope as the change in the *average* value of y for each 1-unit increase in x.

Variability: For each possible value of the explanatory variable x, the values of the response variable y have the same standard deviation σ.

Consider the population of all eruptions of the Old Faithful geyser in a given year. For each eruption, let x be the duration (in minutes) and y be the interval of time (in minutes) until the next eruption. Suppose that the conditions for regression inference are met, the population regression line is $\mu_y = 34 + 13x$, and the variability around the line is given by $\sigma = 6$.

Let's focus on the eruptions that lasted $x = 2$ minutes. For this "subpopulation":

- The average amount of time until the next eruption is $\mu_y = 34 + 13(2) = 60$ minutes.

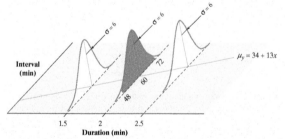

Interval (min)

$\sigma = 6$ $\sigma = 6$ $\sigma = 6$

$\mu_y = 34 + 13x$

72

60

48

1.5   2   2.5

Duration (min)

- The amount of time until the next eruption follows a Normal distribution with mean 60 minutes and standard deviation 6 minutes.
- For about 95% of these eruptions, the amount of time y until the next eruption is between $60 - 2(6) = 48$ minutes and $60 + 2(6) = 72$ minutes. That is, if the previous eruption lasted 2 minutes, 95% of the time the next eruption will occur in 48 to 72 minutes.

Teaching Tip

Make sure to spend time carefully explaining Figure 12.12 and how it illustrates why we need the upcoming Linear, Normal, and Equal SD conditions.

Teaching Tip

Tell students there is a mountain range of Normal distributions. Each mountain has the same variability (σ), and the least-squares regression line is used to predict the value at the center of each mountain.

Teaching Tip

Students often struggle with the concept of μ_y. Help them understand that there is a distribution of y-values for each x-value. In the context of the Old Faithful eruptions, there is a distribution of wait times for each duration. See the discussion below.

Teaching Tip

You could also consider using the acronym LINEaR to help students remember the conditions for regression inference, where the "a" stands for "and": Linear, Independent, Normal, Equal SD, and Random.

Teaching Tip

Remind students that if the conditions for inference aren't met, the stated confidence level and significance levels may not be correct. For example, if the conditions aren't met, "95%" confidence intervals for the slope may capture the true slope less than 95% of the time.

Here are the conditions for performing inference about the linear regression model. The acronym LINER should help you remember them!

CONDITIONS FOR REGRESSION INFERENCE

Suppose we have n observations on a quantitative explanatory variable x and a quantitative response variable y. Our goal is to study or predict the behavior of y for given values of x.

- **Linear:** The true relationship between x and y is linear. For any particular value of x, the mean response μ_y falls on the population (true) regression line $\mu_y = \alpha + \beta x$.
- **Independent:** Individual observations are independent of each other. When sampling without replacement, check the *10% condition*.
- **Normal:** For any particular value of x, the response y varies according to a Normal distribution.
- **Equal SD:** The standard deviation of y (call it σ) is the same for all values of x.
- **Random:** The data come from a random sample from the population of interest or a randomized experiment.

Although the conditions for regression inference are a bit complicated, it is not hard to check for major violations. Most of the conditions involve the population (true) regression line and the deviations of responses from this line. We usually can't observe the population line, but the sample regression line estimates it. The residuals from the sample regression line estimate the deviations from the population line. We can check several of the conditions for regression inference by looking at graphs of the residuals. Start by making a residual plot and a histogram, dotplot, stemplot, boxplot, or Normal probability plot of the residuals.

Here's a summary of how to check the conditions one by one.

- **Linear:** Examine the scatterplot to see if the overall pattern is roughly linear. Make sure there are no leftover curved patterns in the residual plot.

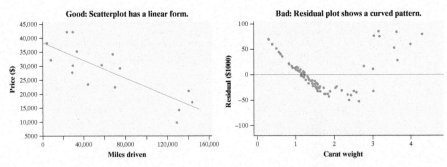

- **Independent:** Knowing the value of the response variable for one individual shouldn't help predict the value of the response variable for other individuals. If sampling is done without replacement, remember to check that the

Teaching Tip

Here are some additional comments about checking the conditions for regression inference:

Linear: If there is a leftover curved pattern in the residual plot (such as the above U shape), this condition is not met. One way to address this violation is to transform the data so the relationship becomes roughly linear. This is the topic of Section 3.3.

Independent: There are methods for doing inference for time-series data, but they are beyond the scope of this book.

Normal: We check the Normal condition in the same manner as in earlier chapters. Make a graph of the residuals (not a residual plot) and look for strong skewness or outliers. If the graph

of the residuals shows strong skew or outliers, this condition can still be satisfied if the sample size is large ($n \geq 30$).

Equal SD: Fan-shaped patterns in the residual plot indicate a violation of this condition.

Random: Remind students about the scope of inference introduced in Chapter 4. If the data don't come from a random sample, we shouldn't make an inference about a larger population. Likewise, if the data do not come from a randomized experiment, we shouldn't make inferences about cause and effect.

sample size is less than 10% of the population size (*10% condition*). There are also other issues that can lead to a lack of independence. One example is measuring the same variable over time, yielding what is known as *time-series data*. Knowing that a young girl's height at age 6 is 48 inches would definitely give you additional information about her height at age 7. We will avoid doing inference for time-series data in this course.

> If the distribution of residuals has strong skewness or outliers, inference procedures will still be reasonably accurate if the sample size is large (e.g., $n \geq 30$).

- **Normal:** Make a histogram, dotplot, stemplot, boxplot, or Normal probability plot of the residuals and check for strong skewness or outliers. Ideally, we would check the Normality of the residuals at each *x*-value. Because we rarely have enough observations at each *x*-value, however, we make one graph of all the residuals to check for Normality.

- **Equal SD:** Look at the scatter of the residuals above and below the "residual = 0" line in the residual plot. The variability of the residuals in the vertical direction should be roughly the same from the smallest to the largest *x*-value.

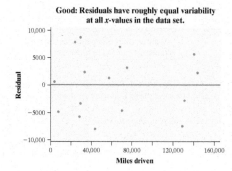

Good: Residuals have roughly equal variability at all *x*-values in the data set.

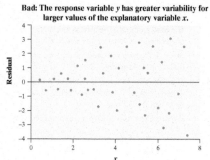

Bad: The response variable *y* has greater variability for larger values of the explanatory variable *x*.

- **Random:** See if the data came from a random sample from the population of interest or a randomized experiment. If not, we can't make inferences about a larger population or about cause and effect.

Let's look at an example that illustrates the process of checking conditions.

EXAMPLE

The helicopter experiment
Checking conditions

PROBLEM: Mrs. Barrett's class did a fun experiment using paper helicopters. After making 70 helicopters using the same template, students randomly assigned 14 helicopters to each of five drop heights: 152 centimeters (cm), 203 cm, 254 cm, 307 cm, and 442 cm. Teams of students released the 70 helicopters in a random order and measured the flight times in seconds. The class used computer software to carry out a least-squares regression analysis for these data. Here are a scatterplot, residual plot, and histogram of the residuals.

Teaching Tip: Using Technology

Use the *Two Quantitative Variables* applet in the Extra Applets on the Student Site at highschool.bfwpub.com/updatedtps6e to check conditions for regression inference. After inputting the sample data, click "Calculate least-squares regression line." The applet will make a scatterplot, a residual plot, and offer the option to create a dotplot of residuals.

Teaching Tip

Here are the quick summaries for satisfying each condition:

Linear: Scatterplot of sample data is fairly linear.

Independent: 10% condition if sampling without replacement.

Normal: Distribution of residuals is approximately Normal.

Equal SD: No > or < shape in the residual plot.

Random: Random sample or random assignment.

ALTERNATE EXAMPLE   Skill 4.C

Math or English?
Checking conditions

PROBLEM:

In the state of Michigan, all high school juniors are required to take the SAT, which is made up of a Math section and a Reading and Writing section. A random sample of 50 school districts in Michigan is selected, and the mean Math score and mean Reading and Writing score for a recent year are recorded. A scatterplot, residual plot, and dotplot of the residuals are shown.

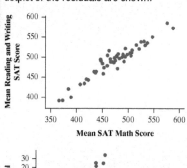

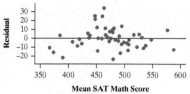

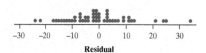

(continues)

Check whether the conditions for performing inference about the regression model are met.

SOLUTION:

- *Linear:* The scatterplot shows a clear linear form and there is no leftover curved pattern in the residual plot. ✓
- *Independent:* Assume that 50 < 10% of all school districts in Michigan. ✓
- *Normal:* There is no strong skewness or outliers in the histogram of the residuals. ✓
- *Equal SD:* The residual plot shows a similar amount of scatter about the residual = 0 line for each mean SAT Math score. ✓
- *Random:* Random sample of 50 school districts in the state of Michigan. ✓

Teaching Tip

The standard deviation σ measures the variability of *y* for a given value of *x*. The "Equal SD" condition says that this standard deviation σ is the same for each value of *x*. Refer to the "mountain range of Normal distributions" picture on page 817.

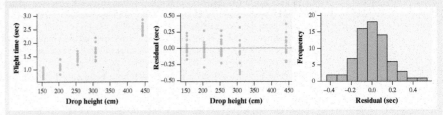

Check whether the conditions for performing inference about the regression model are met.

SOLUTION:

- Linear: The scatterplot shows a clear linear form and there is no leftover curved pattern in the residual plot. ✓
- Independent: Because the helicopters were released in a random order and no helicopter was used twice, knowing the result of one observation should not help us predict the value of another observation. ✓
- Normal: There is no strong skewness or outliers in the histogram of the residuals. ✓
- Equal SD: The residual plot shows a similar amount of scatter about the residual = 0 line for each drop height. However, flight times seem to vary a little more for the helicopters that were dropped from a height of 307 cm. ✓
- Random: The helicopters were randomly assigned to the five possible drop heights. ✓

> Use the LINER acronym!

> Although there are "stacks" in the residual plot, the residuals are centered on the horizontal line at 0 for each drop height used in the experiment.

> Note that we do not have to check the 10% condition here because there was no random sampling.

FOR PRACTICE, TRY EXERCISE 69

You will always see some irregularity when you look for Normality and equal standard deviation in the residuals, especially when you have few observations. Don't overreact to minor issues in the graphs when checking the Normal and Equal SD conditions.

Estimating the Parameters

When the conditions are met, we can do inference about the regression model $\mu_y = \alpha + \beta x$. The first step is to estimate the unknown parameters. If we calculate the sample regression line $\hat{y} = a + bx$, the sample slope b is an unbiased estimator of the true slope β, and the sample y intercept a is an unbiased estimator of the true y intercept α. The remaining parameter is the standard deviation σ, which describes the variability of the response y about the population (true) regression line.

The least-squares regression line computed from the sample data estimates the population (true) regression line. So the residuals estimate how much y varies about the population line. Because σ is the standard deviation of responses about the population (true) regression line, we estimate it with the standard deviation of the residuals s:

$$s = \sqrt{\frac{\sum \text{residuals}^2}{n-2}} = \sqrt{\frac{\sum (y_i - \hat{y}_i)^2}{n-2}}$$

Making Connections

The concept and formula for the standard deviation of the residuals *s* were first introduced in Chapter 3. It is more important that students be able to interpret the value of *s* rather than calculate its value.

Making Connections

With two-variable data, we will soon find that the number of degrees of freedom is $n - 2$, which is what you see in the formula for *s*. If students are curious about where the $n - 2$ comes from, here is one way to explain it: If you had only two points, you could calculate a least-squares regression line but would have no way to estimate the amount of variability from the line. Only when you have three or more points can you begin to estimate variability from the line.

Recall from Chapter 3 that s describes the size of a "typical" prediction error. Because s is estimated from data, it is sometimes called the *regression standard error* or the *root mean squared error*.

It is possible to do inference about any of the three parameters in the regression model: α, β, or σ. However, the slope β of the population (true) regression line is usually the most important parameter in a regression problem. So we'll restrict our attention to inference about the slope.

When the conditions are met, the sampling distribution of the slope b is approximately Normal with mean $\mu_b = \beta$ and standard deviation

$$\sigma_b = \frac{\sigma}{\sigma_x \sqrt{n}}$$

In practice, we don't know σ for the true regression line. So we estimate it with the standard deviation of the residuals, s. We also don't know the standard deviation σ_x for the population of x-values. (For reasons beyond the scope of this text, we replace the denominator with $s_x\sqrt{n-1}$.) So we estimate the variability of the sampling distribution of b with the *standard error of the slope*

> The formula sheet provided on the AP® Statistics exam uses the notation s_b rather than SE_b for the standard error of the sample slope b.

$$SE_b = \frac{s}{s_x\sqrt{n-1}}$$

This standard error has the same interpretation as the standard error of $\bar{x}$ or the standard error of $\hat{p}$. It measures how far our estimate typically varies from the truth. In this case, it measures how far the sample slope typically varies from the population (true) slope if we repeat the data production process many times.

Although we give the formula for the standard error of b, you should rarely have to calculate it by hand. Computer output gives the standard error SE_b immediately to the right of the sample slope b.

EXAMPLE

The helicopter experiment, part 2
Estimating the parameters

PROBLEM: Earlier, Mrs. Barrett's class used computer software to perform a least-squares regression analysis of their helicopter data. Recall that the data came from dropping 70 paper helicopters from various heights (in inches) and measuring the flight times (in seconds). Some output from this regression analysis is shown here. We checked conditions for performing inference earlier.

```
Regression Analysis: Flight time versus Drop height
Predictor              Coef      SE Coef      T       P
Constant             −0.03761    0.05838    −0.64   0.522
Drop height (cm)    0.0057244   0.0002018   28.37   0.000
S = 0.168181      R-Sq = 92.2%        R-Sq(adj) = 92.1%
```

(a) What is the estimate for α? Interpret this value.
(b) What is the estimate for β? Interpret this value.
(c) What is the estimate for σ? Interpret this value.
(d) Give the standard error of the slope SE_b. Interpret this value.

ALTERNATE EXAMPLE

Skill 4.B

Math or English? Part 2 Estimating parameters

PROBLEM:
In the preceding alternate example, a random sample of 50 school districts in Michigan was selected, and the mean SAT Math score and mean SAT Reading and Writing score were recorded for each district. Some output from this regression analysis is shown below. We checked conditions for performing inference earlier.

```
Regression Analysis: Mean Reading and Writing
     SAT score versus Mean Math SAT score

Predictor    Coef    SE Coef      T       P
Constant     96.5    16.9       5.71    0.000
Mean SAT    0.8312   0.0355    23.40    0.000
  Math score

S = 11.8045   R-Sq = 91.94%   R-Sq(adj) = 91.77%
```

(a) What is the estimate for α? Interpret this value.
(b) What is the estimate for β? Interpret this value.
(c) What is the estimate for σ? Interpret this value.
(d) Give the standard error of the slope SE_b. Interpret this value.

(continues)

Teaching Tip

Ask students to explain why s is sometimes called the "root mean squared error." The word *error* is used as a synonym for *residual*. We "square" the errors so that they don't add up to 0. We find the *mean* by dividing by $n - 2$, and we take the square "root" to return to the original units of the response variable.

AP® EXAM TIP

The formula for SE_b is provided, but it is very unlikely that students will have to use this formula on the AP® Statistics exam. Most often, students will be provided the computer output with this value. It is most important that they are able to use this value to construct a confidence interval or perform a significance test for slope.

Teaching Tip

Make sure that students know how to interpret the standard error of the slope. Like other standard errors, it measures how far we expect our estimate to be from the truth in repeated random sampling or repeated random assignment.

SOLUTION:

(a) a = 96.5; for school districts with a mean SAT Math score of 0, we predict their mean SAT Reading and Writing score to be 96.5. This estimate is an extrapolation and doesn't make sense for the SAT—scores must be between 200 and 800.

(b) b = 0.8312; for each increase of 1 point in the mean SAT Math score, the predicted mean SAT Reading and Writing score increases by 0.8312 point.

(c) s = 11.8045; the actual mean SAT Reading and Writing scores typically vary by about 11.8045 from the mean scores predicted with the least-squares regression line using x = mean SAT Math score.

(d) SE_b = 0.0355; if we repeated the random sampling many times, the slope of the sample regression line would typically vary by about 0.0355 from the slope of the true regression line for predicting the mean SAT Reading and Writing score from the mean SAT Math score.

Making Connections

In Chapters 8 and 10, we suggested asking students to write a general formula and a specific formula for each confidence interval. This structure should help students come up with the formula on this page:

General formula:
    point estimate ± margin of error

Specific formula:
    $b \pm t^*SE_b$

Teaching Tip

Students might wonder why we use a t distribution rather than a Normal distribution. The reason is the same for slopes as it is for means: even though $\bar{x}$ and b have approximately Normal distributions when certain conditions are met, when standardizing these statistics, we use an estimate of the standard deviation. Because this estimate is a variable, not a constant, the shape of the distribution of the standardized test statistic is no longer Normal.

SOLUTION:

(a) $a = -0.03761$; if a helicopter is dropped from 0 cm, it will take -0.03761 second to land, on average.

(b) $b = 0.0057244$; for each increase of 1 cm in drop height, the average flight time increases by 0.0057244 second.

(c) $s = 0.168181$; the actual flight times typically vary by about 0.168181 second from the times predicted with the least-squares regression line using x = drop height.

(d) $SE_b = 0.0002018$; if we repeated the random assignment many times, the slope of the sample regression line would typically vary by about 0.0002018 from the slope of the true regression line for predicting flight time from drop height.

> A negative flight time is clearly unrealistic and is likely due to extrapolating to $x = 0$ from the data in the experiment.

> Remember that the slope and y intercept describe *average* flight times, not the actual flight times. You could also describe the slope as the increase in *predicted* flight time for each increase of 1 cm in drop height.

> The standard error of the slope is just to the right of the slope, in the column called "SE Coef."

FOR PRACTICE, TRY EXERCISE 71

When we compute the least-squares regression line based on a random sample of data, we can think about doing inference for the *population* regression line. When our least-squares regression line is based on data from a randomized experiment, as in the helicopter example, the resulting inference is about the *true* regression line relating the explanatory and response variables. From now on, we'll use the term *population regression line* in sampling situations and the term *true regression line* when describing experiments.

Constructing a Confidence Interval for the Slope

In a regression setting, we often want to estimate the slope β of the population (true) regression line. The slope b of the sample regression line is our point estimate for β. An interval estimate is more useful than the point estimate because it gives a set of plausible values for β based on the sample data.

The confidence interval for β has the familiar form

$$\text{statistic} \pm (\text{critical value}) \cdot (\text{standard error of statistic})$$

We use the statistic b as our point estimate and SE_b as the standard error of the statistic. Because we don't know the true standard deviation σ_b, we must use a t^* critical value rather than a z^* critical value. Get the t^* critical value from a t distribution with $n - 2$ degrees of freedom. (The explanation of why df $= n - 2$ is beyond the scope of this book.) Here is the formula:

$$b \pm t^*SE_b$$

We call this a t **interval for the slope.** Here are the details.

t INTERVAL FOR THE SLOPE

When the conditions are met, a $C\%$ confidence interval for the unknown slope β of the population (true) regression line is

$$b \pm t^*SE_b$$

where t^* is the critical value for the t distribution with $n - 2$ degrees of freedom and $C\%$ of the area between $-t^*$ and t^*.

AP® EXAM TIP

The general formula for a confidence interval is included on the formula sheet provided to students on both parts of the AP® Statistics exam. However, the specific formula for a confidence interval for the slope is *not* included on the formula sheet.

The values of t given in the computer regression output are not the critical values for a confidence interval. They come from carrying out a significance test about the y intercept or slope of the population (true) regression line. We'll discuss tests in more detail shortly.

You can find a confidence interval for the y intercept α of the population (true) regression line in the same way, using a and SE_a from the "Constant" row of the computer output. However, we are usually interested only in the point estimate for α that's provided in the output.

Here is an example using a familiar context that illustrates the four-step process for calculating and interpreting a confidence interval for the slope.

EXAMPLE

How much is that truck worth?
Confidence interval for a slope

PROBLEM: Everyone knows that cars and trucks lose value the more they are driven. Can we predict the price of a used Ford F-150 SuperCrew 4 × 4 if we know how many miles it has on the odometer? A random sample of 16 used Ford F-150 SuperCrew 4 × 4s was selected from among those listed for sale on autotrader.com. The number of miles driven and price (in dollars) were recorded for each of the trucks. Here are the data:

Miles driven	70,583	129,484	29,932	29,953	24,495	75,678	8359	4447
Price ($)	21,994	9500	29,875	41,995	41,995	28,986	31,891	37,991
Miles driven	34,077	58,023	44,447	68,474	144,162	140,776	29,397	131,385
Price ($)	34,995	29,988	22,896	33,961	16,883	20,897	27,495	13,997

Here is some computer output from a least-squares regression analysis of these data. Construct and interpret a 90% confidence interval for the slope of the population regression line.

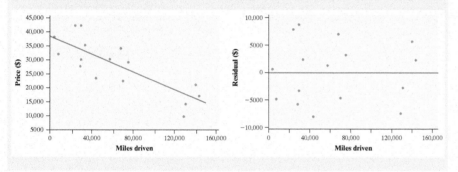

Making Connections

In Chapter 3, we used this F-150 data set to introduce least-squares regression lines for making predictions, interpreting residuals, constructing a residual plot, interpreting slope and r^2, and interpreting computer output.

Here are graphs and computer output from a least-squares regression analysis for these data. Construct and interpret a 95% confidence interval for the slope of the population regression line.

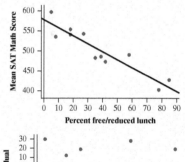

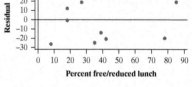

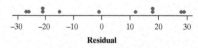

ALTERNATE EXAMPLE

Lunch and SAT Math scores   Confidence interval for a slope

Skills 1.D, 3.D

PROBLEM:
A random sample of 11 high schools was selected from Michigan. The percent of students who participate in the free/reduced lunch program and the mean SAT Math score of each high school in the sample were recorded for each high school. Here are the data:

	Percent free/reduced	Mean SAT Math
East Kentwood High School	58	490.4
Rockford High School	8	535.5
Caledonia High School	18	541.3
Cedar Springs High School	39	485.9
Muskegon High School	85	427.3
Comstock Park High School	42	473.2
Sparta High School	35	483.1
Lowell High School	27	542.7
Spring Lake High School	18	554.1
Ottawa Hills High	78	402.3
Northville High School	5	597.6

```
Regression Analysis: Mean SAT Math score
    versus Percent free/reduced lunch

Predictor         Coef    SE Coef      T      P
Constant         577.9      12.5   46.16  0.000
Percent free/   -1.993     0.276   -7.22  0.000
  reduced lunch

S = 23.3168  R-Sq = 85.29%  R-Sq(adj) = 83.66%
```

(continues)

ALTERNATE EXAMPLE (continued)

SOLUTION:

STATE: 95% CI for β = the slope of the population regression line relating y = mean SAT Math score to x = percent free/reduced lunch for high schools in the state of Michigan.

PLAN: t interval for the slope

Linear: The scatterplot shows a clear linear pattern. Also, the residual plot shows no leftover curved patterns. ✓

Independent: Assume that 11 < 10% of all high schools in the state of Michigan. ✓

Normal: There is no strong skewness or outliers in the dotplot of residuals. ✓

Equal SD: The scatter of points around the residual = 0 line appears to be about the same at all x-values. ✓

Random: Random sample of 11 high schools in the state of Michigan. ✓

DO: With df = 11 − 2 = 9, t^* = 2.262.

$-1.993 \pm 2.262(0.276)$
$= -1.993 \pm 0.624$
$= (-2.617, -1.369)$

CONCLUDE: We are 95% confident that the interval from −2.617 to −1.369 captures the slope of the population regression line relating y = mean SAT Math score to x = percent free/reduced lunch for high schools in the state of Michigan.

Teaching Tip

In the "Do" step of the example on this page, consider having students write a general formula and a specific formula with variables before substituting numbers:

General formula:
    point estimate ± margin of error

Specific formula:
    $b \pm t^*SE_b$

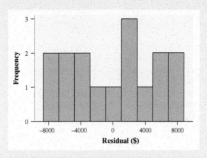

Regression Analysis: Price ($) versus Miles driven				
Predictor	Coef	SE Coef	T	P
Constant	38257	2446	15.64	0.000
Miles driven	−0.16292	0.03096	−5.26	0.000
S = 5740.13	R-Sq = 66.4%		R-Sq(adj) = 64.0%	

SOLUTION:

STATE: 90% CI for β = the slope of the population regression line relating y = price to x = miles driven for used Ford F-150 Super-Crew 4 × 4s listed for sale on autotrader.com.

PLAN: t interval for the slope.

Linear: The scatterplot shows a clear linear pattern. Also, the residual plot shows no leftover curved patterns. ✓

Independent: Assume that 16 < 10% of all used Ford F-150 SuperCrew 4 × 4s. ✓

Normal: There is no strong skewness or outliers in the histogram of residuals. ✓

Equal SD: The scatter of points around the residual = 0 line appears to be about the same at all x-values. ✓

Random: Random sample of 16 used Ford F-150 SuperCrew 4 × 4s. ✓

DO: With df = 16 − 2 = 14, t^* = 1.761.
  $-0.16292 \pm 1.761(0.03096)$
$= -0.16292 \pm 0.05452$
$= (-0.21744, -0.10840)$

CONCLUDE: We are 90% confident that the interval from −0.2174 to −0.1084 captures the slope of the population regression line relating y = price to x = miles driven for used Ford F-150 SuperCrew 4 × 4s listed for sale on autotrader.com.

> Follow the four-step process!

> Remember to use the acronym LINER when checking the conditions.

> $b \pm t^* SE_b$

> Refer to Technology Corner 31 for details on how to do calculations for a t interval for slope. The calculator gives (−0.2173, −0.1084) using df = 14.

> This means we are 90% confident that the average price of a used Ford F-150 goes down between $0.1084 and $0.2174 per mile.

FOR PRACTICE, TRY EXERCISE 75

The change in average price of a used Ford F-150 is quite small for a 1-mile increase in miles driven. What if miles driven increased by 1000 miles? We can just multiply both endpoints of the confidence interval in the example by 1000 to get a 90% confidence interval for the corresponding change in average price. The resulting interval is (− 217.44, −108.40). That is, based on this sample, we are 90% confident that the average price of a used Ford F-150 will decrease between $108.40 and $217.44 for every additional 1000 miles driven. If we want to reduce the width of our confidence interval, we can increase the size of our sample.

Preparing for Inference

In the example above, the confidence interval contains only negative numbers as plausible values for the slope. Because this interval does not contain 0, we have convincing evidence that there is a linear relationship between miles driven and price ($) for all used F-150 SuperCrew 4 × 4s listed for sale on www.autotrader.com.

AP® EXAM TIP

This idea of multiplying a confidence interval by a constant to create a new confidence interval was part of Free Response Question #2 on the 2017 AP® Statistics exam.

So far, we have used computer regression output when performing inference about the slope of a population (true) regression line. The TI-84 and other calculators can do the calculations for inference when the sample data are provided.

31. Technology Corner   CONSTRUCTING A CONFIDENCE INTERVAL FOR SLOPE

TI-Nspire and other technology instructions are on the book's website at highschool.bfwpub.com/updatedtps6e.

Let's use the data from the preceding example to construct a confidence interval for the slope of a population (true) regression line on the TI-84. *Note:* The TI-83 and older operating systems for the TI-84 do not include this option.

- Enter the *x*-values (miles driven) into L1 and the *y*-values (price) into L2.

- Press STAT, then choose TESTS and LinRegTInt. . . .

- In the LinRegTInt screen, adjust the inputs as shown. Then highlight "Calculate" and press ENTER.

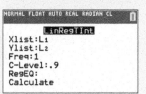

- The linear regression *t* interval results are shown here. Note that *s* is the standard deviation of the residuals, *not* the standard error of the slope.

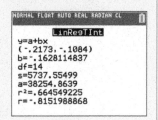

AP® EXAM TIP

When you see a list of data values on an exam question, wait a moment before typing the data into your calculator. Read the question through first. Often, information is provided that makes it unnecessary for you to enter the data at all. This can save you valuable time on the AP® Statistics exam.

CHECK YOUR UNDERSTANDING

Many people believe that students learn better if they sit closer to the front of the classroom. Does sitting closer cause higher achievement, or do better students simply choose to sit in the front? To investigate, a statistics teacher randomly assigned students to seat locations in the classroom for a particular chapter and recorded the test score for each student at the end of the chapter. Here are the data, along with output from a regression analysis. Construct and interpret a 95% confidence interval for the slope of the true regression line.

Teaching Tip: Using Technology

Remind students that they can watch videos of each Technology Corner by clicking the link in the e-Book or at the Student Site.

Teaching Tip: Using Technology

Students with TI-83s or older TI-84s may not have an option for calculating a confidence interval for a slope. However, they can still use their calculators to find the slope *b* and the standard error of the slope SE_b. If they use the LinRegTTest option (see page 829), b = slope and

$$t = \frac{b - 0}{SE_b}, \text{ so } SE_b = \frac{b}{t}.$$

Teaching Tip: Using Technology

Notice that the "RegEQ" is left blank. The other option is to put Y_1 (VARS: Y-VARS: Function: Y_1) in this position. Using this option, the calculator will automatically store the least-squares regression line into Y_1.

Answers to CYU

STATE: β = the slope of the true regression line relating y = score to x = row for students like these.

PLAN: *t* interval for the slope.
Linear: The scatterplot shows a linear pattern. Also, the residual plot shows no leftover curved patterns.
Independent: Knowing the score for one student shouldn't help predict the score of another student.
Normal: There is no strong skewness or outliers in the dotplot of the residuals.
Equal SD: The amount of scatter of points around the residual = 0 line varies somewhat, but it is still plausible that σ is the same for all x-values and that the differences are due to sampling variability.
Random: Students were randomly assigned a seat location within the classroom.

DO: With df = 30 − 2 = 28, t^* = 2.048; −1.117 ± 2.048(0.947) = (−3.056, 0.822)

Tech: The calculator gives (−3.057, 0.823) using df = 28.

CONCLUDE: We are 95% confident that the interval from −3.056 to 0.822 captures the slope of the true regression line relating y = score to x = row for students like these.

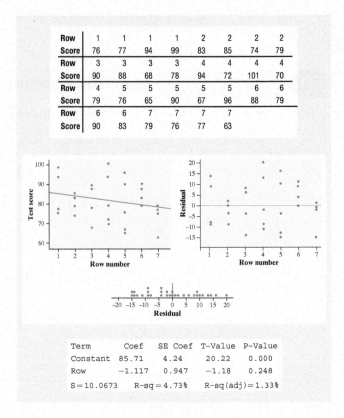

Term	Coef	SE Coef	T-Value	P-Value
Constant	85.71	4.24	20.22	0.000
Row	−1.117	0.947	−1.18	0.248

S=10.0673 R-sq=4.73% R-sq(adj)=1.33%

Performing a Significance Test for the Slope

When the conditions for inference are met, we can use the slope b of the sample regression line to construct a confidence interval for the slope β of the population (true) regression line. We can also perform a significance test to determine whether a specified value of β is plausible. The null hypothesis has the general form $H_0: \beta = \beta_0$, where β_0 is the null hypothesis value of the slope. To do a test, calculate the standardized test statistic:

$$\text{standardized test statistic} = \frac{\text{statistic} - \text{parameter}}{\text{standard deviation (error) of statistic}}$$

$$t = \frac{b - \beta_0}{SE_b}$$

To find the *P*-value, use a *t* distribution with $n - 2$ degrees of freedom. Here is a summary of the Do step for the **t test for the slope**.

Making Connections

In earlier chapters, we suggested asking students to write a general formula and a specific formula for each significance test. This structure should help students come up with the formula on this page:

General formula:
$$\text{standardized test statistic} = \frac{\text{statistic} - \text{parameter}}{\text{standard error of statistic}}$$

Specific formula:
$$t = \frac{b - \beta_0}{SE_b}$$

Teaching Tip

See 2007 #6 for an example of a regression model that passes through (0,0).

In some contexts, it makes sense to fit a regression model that goes through the point (0,0). Because there is only one parameter to estimate in this case (the slope), the degrees of freedom would be df = $n - 1$.

t TEST FOR THE SLOPE

Suppose the conditions are met. To perform a test of the hypothesis $H_0: \beta = \beta_0$, compute the standardized test statistic

$$t = \frac{b - \beta_0}{SE_b}$$

Find the *P*-value by calculating the probability of getting a *t* statistic this large or larger in the direction specified by the alternative hypothesis H_a in a *t* distribution with df = $n - 2$.

AP® EXAM TIP

The general formula for a test statistic is included on the formula sheet provided to students on both parts of the AP® Statistics exam. However, the specific formula for the test statistic for the slope is *not* included on the formula sheet.

If sample data suggest a linear relationship between two variables, how can we determine whether this happened just by chance or whether there is actually a linear association between *x* and *y* in the population? By performing a test of $H_0: \beta = 0$. A regression line with slope 0 is horizontal. That is, the mean of *y* does not change at all when *x* changes. So $H_0: \beta = 0$ says that there is *no linear association* between *x* and *y* in the population. Put another way, H_0 says that *linear regression of y on x is no better for predicting y than the mean of the response variable* $\bar{y}$.

Does a null hypothesis of "no association" sound familiar? It should. In Section 12.2, you learned about chi-square tests for independence, which analyze the relationship between two *categorical* variables. The *t* test for slope looks for an association between two *quantitative* variables.

Regression output from statistical software usually gives the value of *t* for a test of $H_0: \beta = 0$. If you want to test a null hypothesis other than $H_0: \beta = 0$, get the slope and standard error from the output and use the formula to calculate the *t* statistic.

Computer software also reports the *P*-value for a *two-sided* test of $H_0: \beta = 0$. For a one-sided test in the proper direction, just divide the *P*-value in the output by 2. The following example shows what we mean.

Making Connections

The null hypothesis $\beta = 0$ means the linear regression of *y* on *x* is no better for predicting *y* than the mean of the response variable $\bar{y}$. This sounds exactly like the interpretation of r^2 from Chapter 3. Therefore, the null hypothesis is also stating that $r^2 = 0$.

EXAMPLE

Crying and IQ
Significance test for β

PROBLEM: Infants who cry easily may be more easily stimulated than others. This may be a sign of higher IQ. Child development researchers explored the relationship between the crying of infants 4 to 10 days old and their later IQ test scores. A snap of a rubber band on the sole of the foot caused the infants to cry. The researchers recorded the crying and measured its intensity by the number of peaks in the most active 20 seconds. They later measured the children's IQ at age three years using the Stanford–Binet IQ test. The table contains data from a random sample of 38 infants.[33]

Crycount	IQ	Crycount	IQ	Crycount	IQ	Crycount	IQ
10	87	20	90	17	94	12	94
12	97	16	100	19	103	12	103
9	103	23	103	13	104	14	106
16	106	27	108	18	109	10	109
18	109	15	112	18	112	23	113
15	114	21	114	16	118	9	119
12	119	12	120	19	120	16	124
20	132	15	133	22	135	31	135
16	136	17	141	30	155	22	157
33	159	13	162				

Teaching Tip

Make sure that students understand why we perform a significance test for a slope. When data from a random sample or a randomized experiment suggest that a linear association exists between two variables, there are two possible explanations for why the slope differs from 0. The first explanation is that there really is no association between the variables, and we got a non-zero slope due to sampling variability or the chance variation due to random assignment. The second explanation is that there really is an association between the two variables. We do a significance test to decide which explanation is more plausible.

ALTERNATE EXAMPLE

Skills 1.E, 4.E

Tipping at a buffet   Significance test for β

PROBLEM:
Do customers who stay longer at buffets give larger tips? Charlotte, an AP® Statistics student who worked at an Asian-style buffet restaurant, decided to investigate this question for her second-semester project. During her job as a hostess, she obtained a random sample of receipts, which included the length of time (in minutes) a party stayed at the restaurant and the tip amount (in dollars) those customers left. Here are the data:

Time (min)	Tip ($)
23	5.00
39	2.75
44	7.75
55	5.00
61	7.00
65	8.88
67	9.01
70	5.00
74	7.29
85	7.50
90	6.00
99	6.50

(continues)

Here is computer output from a least-squares regression analysis of these data. Do the data provide convincing evidence at the $\alpha = 0.05$ level of a positive linear relationship between the amount of time and amount of tip for customers at this Asian buffet?

Predictor	Coef	SE Coef	T	P
Constant	4.535	1.657	2.74	0.021
Time(minutes)	0.03013	0.02448	1.23	0.247

S = 1.77931   R-Sq = 13.2%   R-Sq(adj) = 4.5%

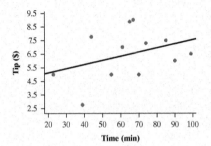

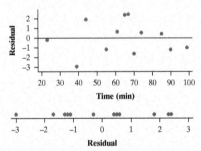

SOLUTION:

STATE: We want to test

$H_0 : \beta = 0$

$H_a : \beta > 0$

where β is the slope of the regression line relating y = amount of tip to x = amount of time. Use $\alpha = 0.05$.

PLAN: t test for the slope

Linear: The scatterplot shows a linear relationship between amount of time (minutes) and amount of tip (dollars) and there are no leftover curved patterns in the residual plot. ✓

Independent: 12 is less than 10% of all receipts at the restaurant. ✓

Normal: The dotplot of residuals does not show strong skewness or clear outliers. ✓

Equal SD: The residual plot shows a fairly equal amount of scatter around the horizontal line at 0 for all x-values. ✓

Random: Random sample of 12 receipts. ✓

DO:
- $t = 1.23$
- P-value: $0.247/2 = 0.1235$

CONCLUDE: Because the P-value of $0.1235 > \alpha = 0.05$, we fail to reject H_0. There is not convincing evidence of a positive linear relationship between the amount of time and amount of tip for customers at this Asian buffet.

Here is computer output from a least-squares regression analysis of these data. Do these data provide convincing evidence at the $\alpha = 0.05$ level of a positive linear relationship between count of crying peaks and IQ in the population of infants?

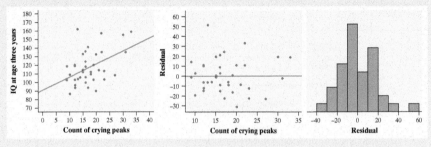

Regression Analysis: IQ versus Crycount

Predictor	Coef	SE Coef	T	P
Constant	91.268	8.934	10.22	0.000
Crycount	1.4929	0.4870	3.07	0.004

S = 17.50   R-Sq = 20.7%   R-Sq(adj) = 18.5%

SOLUTION:

STATE: We want to test

$H_0 : \beta = 0$

$H_a : \beta > 0$

where β = the slope of the regression line relating y = IQ score to x = count of crying peaks in the population of infants. Use $\alpha = 0.05$.

PLAN: t test for the slope.

Linear: The scatterplot shows a linear relationship between crying peaks and IQ, and there are no leftover curved patterns in the residual plot. ✓

Independent: 38 is less than 10% of all infants. ✓

Normal: The histogram of residuals does not show strong skewness or clear outliers. ✓

Equal SD: The residual plot shows a fairly equal amount of scatter around the horizontal line at 0 for all x-values. ✓

Random: Random sample of 38 infants. ✓

DO: df = 36
- $t = 3.07$
- P-value = $0.004/2 = 0.002$

CONCLUDE: Because the P-value of $0.002 < \alpha = 0.05$, we reject H_0. There is convincing evidence of a positive linear relationship between the count of crying peaks and IQ score in the population of infants.

> Follow the four-step process!

> There is some evidence for H_a because $b = 1.4929 > 0$.

> The t statistic and P-value are found in the row for "Crycount" under the corresponding headings. Because the P-value is for a two-sided test and there is evidence for H_a, we cut the P-value in half for the one-sided test.

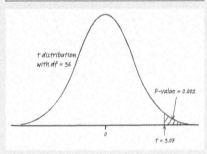

> Refer to Technology Corner 32 for details on how to do calculations for a t test for slope. The calculator gives $t = 3.065$ and P-value $= 0.002$ using df = 36.

FOR PRACTICE, TRY EXERCISE 79

Teaching Tip

In the "Do" step for the example on this page, ask students to write a general formula and a specific formula with variables before substituting numbers:

General formula:

$$\text{standardized test statistic} = \frac{\text{statistic} - \text{parameter}}{\text{standard error of statistic}}$$

Specific formula:

$$t = \frac{b - \beta_0}{SE_b} = \frac{1.4920 - 0}{0.4870} = 3.07$$

Based on the results of the crying and IQ study, should we ask doctors and parents to make infants cry more so that they'll be smarter later in life? Hardly. This observational study gives statistically significant evidence of a positive linear relationship between the two variables. However, we can't conclude that more intense crying as an infant *causes* an increase in IQ. Maybe infants who cry more are more alert to begin with and tend to score higher on intelligence tests.

32. Technology Corner PERFORMING A SIGNIFICANCE TEST FOR SLOPE

TI-Nspire and other technology instructions are on the book's website at highschool.bfwpub.com/updatedtps6e.

Let's use the data from the crying and IQ study to perform a significance test for the slope of the population regression line on the TI-83/84.

- Enter the *x*-values (crying count) into L1 and the *y*-values (IQ score) into L2.

- Press STAT, then choose TESTS and LinRegTTest. . . .

- In the LinRegTTest screen, adjust the inputs as shown. Then highlight "Calculate" and press ENTER.

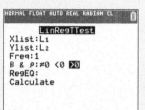

- The linear regression *t* test results take two screens to present. We show only the first screen. The value of *s* is the standard deviation of the residuals, not the standard error of the slope.

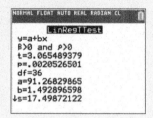

Think About It

WHAT'S WITH THAT $\rho > 0$ IN THE LinRegTTest SCREEN? The slope b of the least-squares regression line is closely related to the correlation r between the explanatory and response variables x and y (recall that $b = r\dfrac{s_y}{s_x}$). In the same way, the slope β of the population regression line is closely related to the correlation ρ (the lowercase Greek letter rho) between x and y in the population. In particular, the slope is 0 when the correlation is 0.

Testing the null hypothesis $H_0: \beta = 0$ is exactly the same as testing that there is *no correlation* between x and y in the population from which we drew our data. You can use the test for zero slope to test the hypothesis $H_0: \rho = 0$ of zero correlation between any two quantitative variables. That's a useful trick. Because correlation also makes sense when there is no explanatory–response distinction, it is handy to be able to test correlation without doing regression.

Teaching Tip: Using Technology

Notice that the "RegEQ" is left blank. The other option is to put Y₁ (VARS: Y-VARS: Function: Y₁) in this position. Using this option, the calculator will automatically store the least-squares regression line into Y₁.

Teaching Tip

While this test for a zero correlation is interesting, it isn't part of the AP® Statistics curriculum (although it could show up on an Investigative Task!). If your students are interested, the test statistic for the correlation is

$$t = \frac{r - \rho}{\sqrt{\dfrac{1 - r^2}{n - 2}}},$$ where ρ is the

hypothesized value of the correlation and df $= n - 2$.

Teaching Tip: Using Technology

We have now covered all the inference procedures for the course. Students must be able to choose the correct inference procedure for different settings. Search the Internet for "categorizing statistics problems" to find the ltcconline page that allows students to practice choosing the correct inference procedure. Uncheck the procedures we don't know yet (prediction intervals and 1-way ANOVA), press Submit, and have fun!

✚ Ask the StatsMedic
Student Created Inference Test

In this post, we outline an assignment that empowers students to create their own context and inference questions, develop rubrics for each, and then use their rubrics to grade another student's work.

✓ Answers to CYU

1. STATE: We want to test $H_0: \beta = 0$, $H_a: \beta < 0$, where β = the slope of the true regression line relating y = score to x = row for students like these, using $\alpha = 0.05$.

PLAN: t test for the slope; the conditions for regression inference are met.

DO: According to the output, the standardized test statistic is $t = -1.18$. Because the computer output includes a two-sided P-value and the sample slope is consistent with H_a, P-value $= 0.248/2 = 0.124$.

CONCLUDE: Because the P-value of $0.124 > \alpha = 0.05$, we fail to reject H_0. There is not convincing evidence of a negative linear relationship between row and score for students like these.

2. *Type I error:* The teacher finds convincing evidence of a negative linear relationship between row and score, when the true slope is 0.

Type II error: The teacher does not find convincing evidence of a negative linear relationship between row and score, when the true slope is < 0.

Because we failed to reject the null hypothesis, it is possible that we made a Type II error.

TRM **Section 12.3 Quiz**

There is one quiz available for this section in the Teacher's Resource Materials. Click on the link in the TE-Book, look on the TRFD, or download from the Teacher's Resources on the book's digital platform. You can also create your own quiz using the ExamView® Assessment Suite that is part of the TPS6 program. Questions are coded by Learning Target to make it easy to build parallel quizzes.

CHECK YOUR UNDERSTANDING

The preceding Check Your Understanding (page 825) described some results from an experiment about seat locations and test scores. Here again is the output from a least-squares regression analysis for these data:

Term	Coef	SE Coef	T-Value	P-Value
Constant	85.71	4.24	20.22	0.000
Row	−1.117	0.947	−1.18	0.248
S = 10.0673		R-sq = 4.73%	R-sq(adj) = 1.33%	

1. Do these data provide convincing evidence at the $\alpha = 0.05$ significance level of a negative linear relationship between row and test score for students like those in the experiment? Assume that the conditions for regression inference are met.
2. Describe a Type I error and a Type II error in this context. Which error is possible based on your conclusion? Explain your reasoning.

Section 12.3 | Summary

- When an association between two quantitative variables is linear, use a least-squares regression line to model the relationship between the explanatory variable x and the response variable y.
- Inference in this setting uses the **sample regression line** $\hat{y} = a + bx$ to estimate or test a claim about the **population (true) regression line** $\mu_y = \alpha + \beta x$.
- The conditions for regression inference are
  - **Linear:** The true relationship between x and y is linear. For any particular value of x, the mean response μ_y falls on the population (true) regression line $\mu_y = \alpha + \beta x$.
  - **Independent:** Individual observations are independent of each other. When sampling without replacement, check the *10% condition*.
  - **Normal:** For any particular value of x, the response y varies according to a Normal distribution.
  - **Equal SD:** The standard deviation of y (call it σ) is the same for all values of x.
  - **Random:** The data come from a random sample from the population of interest or a randomized experiment.
- When the conditions for inference are met, the **sampling distribution of the sample slope** b is approximately Normal with mean $\mu_b = \beta$ and standard deviation $\sigma_b = \dfrac{\sigma}{\sigma_x \sqrt{n}}$.
- The slope b and intercept a of the sample regression line estimate the slope β and intercept α of the population (true) regression line. Use the standard deviation of the residuals s to estimate σ and the standard error of the slope $\mathrm{SE}_b = \dfrac{s}{s_x \sqrt{n-1}}$ to estimate σ_b.

Teaching Tip

Now that you have completed Section 12.3 in this book, students can login to their College Board accounts and take the **Personal Progress Check for Unit 9**.

- Confidence intervals and significance tests for the slope β of the population regression line are based on a t distribution with $n-2$ degrees of freedom.
- The t **interval for the slope** β is $b \pm t^* \, SE_b$.
- To test the null hypothesis $H_0: \beta = \beta_0$, carry out a t **test for the slope.** This test uses the standardized test statistic

$$t = \frac{b - \beta_0}{SE_b}$$

The most common null hypothesis is $H_0: \beta = 0$, which says that there is no linear relationship between x and y in the population.

12.3 Technology Corners

TI-Nspire and other technology instructions are on the book's website at highschool.bfwpub.com/updatedtps6e.

31. Constructing a confidence interval for slope	Page 825
32. Performing a significance test for slope	Page 829

Section 12.3 | Exercises

65. **Predicting height** Using the health records of every student at a high school, the school nurse created a scatterplot relating y = height (in centimeters) to x = age (in years). After verifying that the conditions for the regression model were met, the nurse calculated the equation of the population regression line to be $\mu_y = 105 + 4.2x$ with $\sigma = 7$ cm.

(a) According to the population regression line, what is the average height of 15-year-old students at this high school?

(b) About what percent of 15-year-old students at this school are taller than 180 cm?

(c) If the nurse used a random sample of 50 students from the school to calculate the regression line instead of using all the students, would the slope of the sample regression line be exactly 4.2? Explain your answer.

66. **Predicting high temperatures** Using the daily high and low temperature readings at Chicago's O'Hare International Airport for an entire year, a meteorologist made a scatterplot relating y = high temperature to x = low temperature, both in degrees Fahrenheit.[34] After verifying that the conditions for the regression model were met, the meteorologist calculated the equation of the population regression line to be $\mu_y = 16.6 + 1.02x$ with $\sigma = 6.64°F$.

(a) According to the population regression line, what is the average high temperature on days when the low temperature is 40°F?

(b) About what percent of days with a low temperature of 40°F have a high temperature greater than 70°F?

(c) If the meteorologist used a random sample of 10 days to calculate the regression line instead of using all the days in the year, would the slope of the sample regression line be exactly 1.02? Explain your answer.

67. **Oil and residuals** Researchers examined data on the depth of small defects in the Trans-Alaska Oil Pipeline. The researchers compared the results of measurements on 100 defects made in the field with measurements of the same defects made in the laboratory.[35] The figure shows a residual plot for the least-squares regression line based on these data. Explain why the conditions for performing inference about the slope β of the population regression line are *not* met.

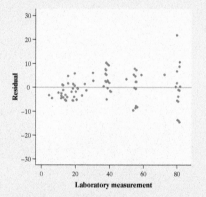

TRM **Full Solutions to Section 12.3 Exercises**

Click on the link in the TE-Book, open the TRFD, or download from the Teacher's Resources on the book's digital platform.

Answers to Section 12.3 Exercises

12.65 (a) $\mu_y = 105 + 4.2(15) = 168$ cm **(b)** Because the conditions were met, the distribution of heights for 15-year-old students follow a Normal distribution with a mean of 168 cm and a standard deviation of 7 cm. Let X = the height of 15-year-old students at this school. We want to find $P(X > 180)$.

(i) $z = \dfrac{180 - 168}{7} = 1.71$; the proportion of z-scores above 1.71 is 0.0436.

(ii) normalcdf(lower: 180, upper: 1000, mean: 168, SD: 7) = 0.0432

About 4.32% of 15-year-old students are taller than 180 cm. **(c)** Probably not. The slope of the sample regression line would almost certainly differ from 4.2 due to sampling variability. We would, however, expect the slope of the sample regression line to be close to 4.2.

12.66 (a) $\mu_y = 16.6 + 1.02(40) = 57.4°F$ **(b)** Because the conditions were met, the distribution of temperature on days when the low temperature is 40°F follows a Normal distribution with a mean of 57.4°F and a standard deviation of 6.64°F. Let X = the high temperature. We want to find $P(X > 70)$.

(i) $z = \dfrac{70 - 57.4}{6.64} = 1.90$; the proportion of z-scores above 1.90 is 0.0287.

(ii) normalcdf(lower: 70, upper: 1000, mean: 57.4, SD: 6.64) = 0.0289

About 2.89% of days with a low temperature of 40°F have a high temperature greater than 70 degrees. **(c)** Probably not. If the meteorologist used a random sample of 10 days to calculate the regression line instead of using all the days in the year, the slope of the sample regression line would almost certainly differ from 1.02 due to sampling variability. We would, however, expect the slope of the sample regression line to be close to 1.02.

12.67 The Equal SD condition is not met because the standard deviation of the residuals clearly increases as the laboratory measurement (x) increases.

12.68 The Linear condition is not met. There is clear curvature in the residual plot, which suggests that the relationship between mean SAT score and percent taking is not linear.

12.69 *Linear:* There is no leftover curved pattern in the residual plot, indicating that a linear model is appropriate. *Independent:* Knowing the BAC for one subject should not help us predict the BAC for another subject. *Normal:* The histogram of the residuals shows no strong skewness or outliers. *Equal SD:* The residual plot shows a similar amount of scatter about the residual = 0 line for each *x* = number of beers. *Random:* The students were randomly assigned a certain number of cans of beer to drink.

12.70 *Linear:* There is no leftover curved pattern in the residual plot, indicating that a linear model is appropriate. *Independent:* Knowing the proportion of perch killed in one pen should not help us predict the proportion of perch killed in another pen. *Normal:* The histogram of the residuals shows no strong skewness or outliers. *Equal SD:* The residual plot shows a similar amount of scatter about the residual = 0 line for each *x* = number of perch. *Random:* The perch were randomly assigned to the 4 pens.

68. **SAT Math scores** Is there a relationship between the percent of high school graduates in each state who took the SAT and the state's mean SAT Math score? Here is a residual plot from a linear regression analysis that used data from all 50 states in a recent year. Explain why the conditions for performing inference about the slope β of the population regression line are *not* met.

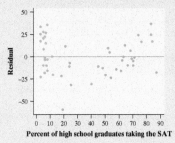

69. **Beer and BAC** How well does the number of beers a person drinks predict his or her blood alcohol content (BAC)? Sixteen volunteers aged 21 or older with an initial BAC of 0 took part in a study to find out. Each volunteer drank a randomly assigned number of cans of beer. Thirty minutes later, a police officer measured their BAC. A least-squares regression analysis was performed on the data using x = number of beers and y = BAC. Here is a residual plot and a histogram of the residuals. Check whether the conditions for performing inference about the regression model are met.

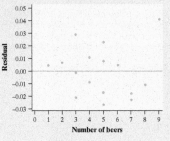

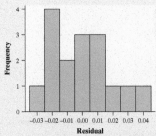

70. **Prey attracts predators** Here is one way in which nature regulates the size of animal populations: high population density attracts predators, which remove a higher proportion of the population than when the density of the prey is low. One study looked at kelp perch and their common predator, the kelp bass. On each of four occasions, the researcher set up four large circular pens on sandy ocean bottoms off the coast of southern California. He randomly assigned young perch to 1 of 4 pens so that one pen had 10 perch, one pen had 20 perch, one pen had 40 perch, and the final pen had 60 perch. Then he dropped the nets protecting the pens, allowing bass to swarm in, and counted the number of perch killed after two hours.[36] A regression analysis was performed on the 16 data points using x = number of perch in pen and y = proportion of perch killed. Here is a residual plot and a histogram of the residuals. Check whether the conditions for performing inference about the regression model are met.

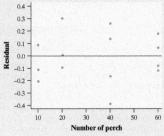

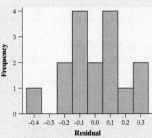

71. **Beer and BAC** Refer to Exercise 69. Here is computer output from the least-squares regression analysis of the beer and blood alcohol data.

```
Dependent variable is:   BAC
No Selector
R squared = 80.0%  R squared (adjusted) = 78.6%
s = 0.0204 with 16 − 2 = 14 degrees of freedom
```

Variable	Coefficient	s.e. of Coeff	t-ratio	prob
Constant	−0.012701	0.0126	−1.00	0.3320
Beers	0.017964	0.0024	7.84	≤ 0.0001

(a) What is the estimate for α? Interpret this value.

(b) What is the estimate for β? Interpret this value.

(c) What is the estimate for σ? Interpret this value.

(d) Give the standard error of the slope SE_b. Interpret this value.

72. **Prey attracts predators** Refer to Exercise 70. Here is computer output from the least-squares regression analysis of the perch data.

```
Predictor   Coef      Stdev.    t-ratio   p
Constant    0.12049   0.09269   1.30      0.215
Perch       0.008569  0.002456  3.49      0.004
S=0.1886    R-Sq=46.5%    R-Sq(adj)=42.7%
```

(a) What is the estimate for α? Interpret this value.

(b) What is the estimate for β? Interpret this value.

(c) What is the estimate for σ? Interpret this value.

(d) Give the standard error of the slope SE_b. Interpret this value.

73. **Beer and BAC** Refer to Exercises 69 and 71.

(a) Find the critical value for a 99% confidence interval for the slope of the true regression line. Then calculate the confidence interval.

(b) Interpret the interval from part (a).

(c) Explain the meaning of "99% confident" in this context.

(d) Based on the interval from part (a), is there convincing evidence of a linear association between number of beers and BAC? Explain your answer.

74. **Prey attracts predators** Refer to Exercises 70 and 72.

(a) Find the critical value for a 90% confidence interval for the slope of the true regression line. Then calculate the confidence interval.

(b) Interpret the interval from part (a).

(c) Explain the meaning of "90% confident" in this context.

(d) Based on the interval from part (a), is there convincing evidence of a linear association between number of perch and percent killed? Explain your answer.

75. **Less mess?** Kerry and Danielle wanted to investigate if tapping on a can of soda would reduce the amount of soda expelled after the can has been shaken. For their experiment, they vigorously shook 40 cans of soda and randomly assigned each can to

be tapped for 0 seconds, 4 seconds, 8 seconds, or 12 seconds. After opening the cans and waiting for the fizzing to stop, they measured the amount expelled (in milliliters) by subtracting the amount remaining from the original amount in the can.[37] Here are their data:

Amount expelled (mL)			
(0 sec)	(4 sec)	(8 sec)	(12 sec)
110	95	88	80
100	105	84	75
105	105	87	80
105	105	85	75
105	95	79	70
110	90	100	65
107	88	85	71
105	95	85	77
104	94	80	76
106	96	80	75

Here and on the next page is some computer output from a least-squares regression analysis of these data. Construct and interpret a 95% confidence interval for the slope of the true regression line.

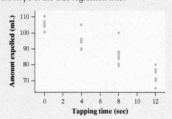

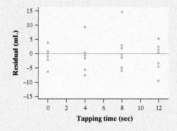

12.71 (a) $\alpha = -0.012701$; if a student drinks 0 beers, the BAC will be -0.012701, on average. Of course, with 0 beers, BAC should be 0. **(b)** $b = 0.017964$; for each increase of 1 beer consumed, the average BAC increases by about 0.017964. **(c)** $s = 0.0204$; the actual BAC amounts typically vary by about 0.0204 from the amounts predicted, with the least-squares regression line using $x =$ number of beers. **(d)** $SE_b = 0.0024$. If we repeated the random assignment many times, the slope of the sample regression line would typically vary by about 0.0024 from the slope of the true regression line for predicting BAC from number of beers.

12.72 (a) $\alpha = 0.12049$; if a pen has 0 perch, the proportion killed will be 0.12049, on

average. Of course, with 0 perch, the proportion killed should be 0. The P-value is 0.215, which suggests 0 is a plausible value for the true y intercept. **(b)** $b = 0.008569$; for each increase of 1 perch, the average proportion killed increases by 0.008569. **(c)** $s = 0.1886$; the actual proportion of perch killed typically varies by about 0.1886 from the proportion predicted with the least-squares regression line using $x =$ number of perch. **(d)** $SE_b = 0.002456$; if we repeated the random assignment many times, the slope of the sample regression line would typically vary by about 0.002456 from the slope of the true regression line for predicting proportion of perch killed from number of perch.

12.73 (a) df $= 14$, $t^* = 2.977$; (0.011, 0.025) **(b)** We are 99% confident that the interval from 0.011 to 0.025 captures the slope of the true regression line relating $y =$ BAC to $x =$ number of beers consumed. **(c)** If we repeated the experiment many times and computed a confidence interval for the slope each time, about 99% of the resulting intervals would contain the slope of the true regression line relating $y =$ BAC to $x =$ number of beers consumed. **(d)** Based on the interval from part (a), there is convincing evidence of a linear association between number of beers and BAC because 0 is not a plausible value of the slope of the true regression line.

12.74 (a) df $= 14$, $t^* = 1.761$; (0.00424, 0.0129) **(b)** We are 90% confident that the interval from 0.00424 to 0.0129 captures the slope of the true regression line relating $y =$ the proportion of perch killed to $x =$ the number of perch in the pen. **(c)** If we repeated the experiment many times and computed a confidence interval for the slope each time, about 90% of the resulting intervals would contain the slope of the true regression line relating $y =$ the proportion of perch killed to $x =$ the number of perch in the pen. **(d)** Based on the interval from part (a), there is convincing evidence of a linear association between the proportion of perch killed and the number of perch in the pen because 0 is not a plausible value of the slope of the true regression line.

12.75 STATE: $\beta =$ the slope of the true regression line relating $y =$ amount expelled to $x =$ tapping time. PLAN: t interval for the slope. *Linear:* The scatterplot shows a linear pattern, and the residual plot shows no leftover curved patterns. *Independent:* Knowing the amount expelled by one can does not help us predict the amount expelled by another can. *Normal:* There is no strong skewness or outliers in the histogram of the residuals. *Equal SD:* The amount of scatter of points around the residual $= 0$ line appears to be about the same at all x-values. *Random:* Cans were randomly assigned a tapping time. DO: df $= 38$; using df $= 30$, $t^* = 2.042$; $(-2.9962, -2.2738)$ CONCLUDE: We are 95% confident that the interval from -2.9962 to -2.2738 captures the slope of the true regression line relating $y =$ amount expelled to $x =$ tapping time.

12.76 STATE: β = the slope of the true regression line relating y = amount expelled to x = number of Mentos. PLAN: t interval for the slope. *Linear:* The residual plot shows no leftover curved patterns. *Independent:* Knowing the amount expelled by one can does not help us predict the amount expelled by another can. *Normal:* There is no strong skewness or outliers in the histogram of the residuals. *Equal SD:* The amount of scatter of points around the residual = 0 line appears to be about the same at all x-values. *Random:* Cans were randomly assigned 2, 3, 4, or 5 Mentos. DO: df = 22, $t^* = 2.074$; (0.0453, 0.0963) CONCLUDE: Line relating y = amount expelled to x = number of Mentos.

12.77 (a) STATE: β = the slope of the population regression line relating y = number of clusters of beetle larvae to x = number of stumps. PLAN: t interval for the slope. The conditions are met. DO: df = 21, $t^* = 2.831$; (8.678, 15.11) CONCLUDE: We are 99% confident that the interval from 8.678 to 15.11 captures the slope of the population regression line relating y = number of clusters of beetle larvae to x = number of stumps. **(b)** The researchers could increase the sample size or decrease the confidence level. They may not want to increase the sample size because that would be expensive and time consuming. They may not want to decrease the confidence level because it is desirable to have a high degree of confidence that the population parameter has been captured by the interval.

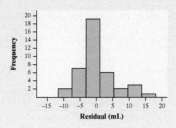

Predictor	Coef	SE Coef	T	P
Constant	106.36	1.3238	80.345	0.000
Tapping time	−2.6350	0.1769	−14.895	0.000
S = 5.00347		R-Sq = 85.4%	R-Sq(adj) = 85.0%	

76. More mess? When Mentos are dropped into a newly opened bottle of Diet Coke, carbon dioxide is released from the Diet Coke very rapidly, causing the Diet Coke to be expelled from the bottle. To see if using more Mentos causes more Diet Coke to be expelled, Brittany and Allie used twenty-four 2-cup bottles of Diet Coke and randomly assigned each bottle to receive either 2, 3, 4, or 5 Mentos. After waiting for the fizzing to stop, they measured the amount expelled (in cups) by subtracting the amount remaining from the original amount in the bottle.[38] Here are their data:

Amount expelled (cups)			
(2 Mentos)	(3 Mentos)	(4 Mentos)	(5 Mentos)
1.125	1.1875	1.25	1.25
1.25	1.125	1.3125	1.4375
1.0625	1.25	1.25	1.3125
1.25	1.1875	1.375	1.3125
1.125	1.3125	1.3125	1.375
1.0625	1.1875	1.25	1.4375

Here is computer output from a least-squares regression analysis of these data. Construct and interpret a 95% confidence interval for the slope of the true regression line.

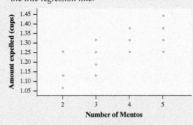

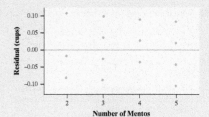

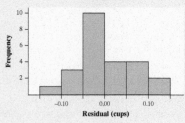

Predictor	Coef	SE Coef	T	P
Constant	1.0021	0.0451	22.215	0.000
Mentos	0.0708	0.0123	5.770	0.000
S = 0.06724		R-Sq = 60.2%	R-Sq(adj) = 58.4%	

77. Beavers and beetles Do beavers benefit beetles? Researchers laid out 23 circular plots, each 4 meters in diameter, at random in an area where beavers were cutting down cottonwood trees. In each plot, they counted the number of stumps from trees cut by beavers and the number of clusters of beetle larvae. Ecologists think that the new sprouts from stumps are more tender than other cottonwood growth so that beetles prefer them. If so, more stumps should produce more beetle larvae.[39]

Here is computer output for a regression analysis of these data.

Predictor	Coef	SE Coef	T	P
Constant	−1.286	2.853	−0.45	0.657
Stumps	11.894	1.136	10.47	0.000
S = 6.41939		R-Sq = 83.9%	R-Sq(adj) = 83.1%	

(a) Construct and interpret a 99% confidence interval for the slope of the population regression line. Assume that the conditions for performing inference are met.

(b) Describe two ways the researchers could reduce the width of the confidence interval in part (a). Explain any drawbacks to these actions.

78. Ideal proportions The students in Mr. Shenk's class measured the arm spans and heights (in inches) of a random sample of 18 students from their large high school. Here is computer output from a least-squares regression analysis of these data.

```
Predictor    Coef     Stdev    t-ratio    P
Constant    11.547     5.600     2.06    0.056
Armspan     0.84042   0.08091   10.39    0.000
S=1.613    R-Sq=87.1%    R-Sq(adj)=86.3%
```

(a) Construct and interpret a 90% confidence interval for the slope of the population regression line. Assume that the conditions for performing inference are met.

(b) Describe two ways Mr. Shenk's students could reduce the width of the confidence interval in part (a). Explain any drawbacks to these actions.

79. Weeds among the corn Lamb's quarters is a common weed that interferes with the growth of corn. An agriculture researcher planted corn at the same rate in 16 small plots of ground and then weeded the plots by hand to allow a fixed number of lamb's quarters plants to grow in each meter of corn row. The decision on how many of these plants to leave in each plot was made at random. No other weeds were allowed to grow. Here are the yields of corn (bushels per acre) in each of the plots:[40]

Weeds per meter	Yield	Weeds per meter	Yield
0	166.7	3	158.6
0	172.2	3	176.4
0	165.0	3	153.1
0	176.9	3	156.0
1	166.2	9	162.8
1	157.3	9	142.4
1	166.7	9	162.8
1	161.1	9	162.4

Here is some computer output from a least-squares regression analysis of these data. Do these data provide convincing evidence at the $\alpha = 0.05$ level that more lamb's quarters reduce corn yield?

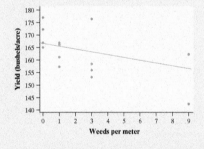

```
Predictor          Coef      SE Coef    T       P
Constant         166.483     2.725     61.11   0.000
Weeds per meter  -1.0987     0.5712    -1.92   0.075
S=7.97665    R-Sq=20.9%    R-Sq(adj)=15.3%
```

80. Time at the table Does how long young children remain at the lunch table help predict how much they eat? Here are data on a random sample of 20 toddlers observed over several months.[41] "Time" is the average number of minutes a child spent at the table when lunch was served. "Calories" is the average number of calories the child consumed during lunch, calculated from careful observation of what the child ate each day.

Time	Calories	Time	Calories
21.4	472	42.4	450
30.8	498	43.1	410
37.7	465	29.2	504
33.5	456	31.3	437
32.8	423	28.6	489
39.5	437	32.9	436
22.8	508	30.6	480
34.1	431	35.1	439
33.9	479	33.0	444
43.8	454	43.7	408

Some computer output from a least-squares regression analysis of these data is shown on the next page. Do these data provide convincing evidence at the $\alpha = 0.01$ level of a linear relationship between time at the table and calories consumed in the population of toddlers?

meter is negative. In other words, we have convincing evidence that having more weeds reduces corn yield.

12.80 STATE: $H_0: \beta = 0$, $H_a: \beta \neq 0$, where $\beta =$ the slope of the population regression line relating $y =$ calorie consumption to $x =$ time at the table in the population of toddlers; $\alpha = 0.01$. PLAN: t test for slope. *Linear:* There is no leftover curved pattern in the residual plot. *Independent:* $20 < 10\%$ of all toddlers. *Normal:* The dotplot of the residuals shows no strong skewness or outliers. *Equal SD:* The residual plot shows a similar amount of scatter about the residual = 0 line for all x-values. *Random:* The 20 toddlers were randomly selected. DO: $t = -3.62$ and the two-sided P-value is 0.002. CONCLUDE: Because the P-value of $0.002 < \alpha = 0.01$, we reject H_0. There is convincing evidence of a linear relationship between time at the table and calorie consumption in the population of toddlers.

12.78 (a) STATE: $\beta =$ the slope of the population regression line relating $y =$ height to $x =$ arm span for students in this high school. PLAN: t interval for the slope. The conditions are met. DO: df $= 16$, $t^* = 1.746$; (0.6992, 0.9817) CONCLUDE: We are 90% confident that the interval from 0.6992 to 0.9817 captures the slope of the population regression line relating $y =$ height to $x =$ arm span for students in this high school. **(b)** Mr. Shenk's students could increase the sample size or decrease the confidence level. They may not want to increase the sample size because that would be time consuming. They may not want to decrease the confidence level because it is desirable to have a high degree of confidence that the population parameter has been captured by the interval.

12.79 STATE: $H_0: \beta = 0$, $H_a: \beta < 0$, where $\beta =$ the slope of the true regression line relating $y =$ corn yield to $x =$ weeds per meter; $\alpha = 0.05$. PLAN: t test for slope. *Linear:* There is no leftover curved pattern in the residual plot. *Independent:* Knowing the yield for one meter does not help us predict the yield for another meter. *Normal:* The dotplot of the residuals shows no strong skewness or outliers. *Equal SD:* The residual plot shows a similar amount of scatter about the residual = 0 line for all x-values. *Random:* The plots were randomly assigned how many weeds were allowed to grow. DO: $t = -1.92$; P-value $= 0.075/2 = 0.0375$ CONCLUDE: Because the P-value of $0.0375 < \alpha = 0.05$, we reject H_0. There is convincing evidence that the slope of the true regression line relating corn yield to weeds per

12.81 STATE: H_0: $\beta = 0$, H_a: $\beta < 0$, where β = the slope of the population regression line relating y = heart disease death rate to x = wine consumption in the population of countries; $\alpha = 0.05$. PLAN: t test for the slope. *Linear:* There is no leftover curved pattern in the residual plot. *Independent:* The sample size $n = 19 < 10\%$ of all countries. *Normal:* The histogram of residuals shows no strong skewness or outliers. *Equal SD:* The residual plot shows that the standard deviation of the death rates might be a little smaller for large values of wine consumption x, but it is hard to tell with so few data values. *Random:* The data come from a random sample. Because the Equal SD condition is questionable, we will proceed with caution.

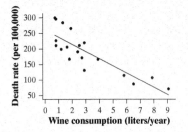

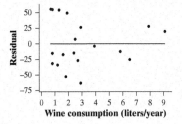

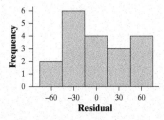

DO: $t = -6.46$, df = 17, and P-value ≈ 0. CONCLUDE: Because the P-value of approximately $0 < \alpha = 0.05$, we reject H_0. There is convincing evidence of a negative linear relationship between wine consumption and heart disease death rate in the population of countries.

12.82 STATE: H_0: $\beta = 0$, H_a: $\beta < 0$, where β = slope of the population regression line relating y = pulse rate to x = swim time for Professor Moore; $\alpha = 0.05$. PLAN: t test for the slope. *Linear:* There is no leftover curved pattern in the residual plot. *Independent:* The sample size $n = 23 < 10\%$ of all days when Professor Moore swims 2000 yards. *Normal:* The histogram of residuals

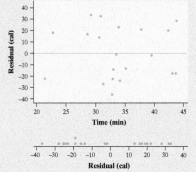

Predictor Coef   SE Coef   T   P
Constant   560.65 29.37   19.09 0.000
Time   -3.0771 0.8498   -3.62 0.002
S = 23.3980   R-Sq = 42.1%   R-Sq(adj) = 38.9%

81. Is wine good for your heart? A researcher from the University of California, San Diego, collected data on average per capita wine consumption and heart disease death rate in a random sample of 19 countries for which data were available. The following table displays the data.[42]

Alcohol from wine (liters/year)	Heart disease death rate (per 100,000)	Alcohol from wine (liters/year)	Heart disease death rate (per 100,000)
2.5	211	7.9	107
3.9	167	1.8	167
2.9	131	1.9	266
2.4	191	0.8	227
2.9	220	6.5	86
0.8	297	1.6	207
9.1	71	5.8	115
2.7	172	1.3	285
0.8	211	1.2	199
0.7	300		

(shown at bottom right) shows no strong skewness or outliers. *Equal SD:* The residual plot shows that the standard deviation of the pulse rates might be a little smaller for small values of x, but it is hard to tell with so few data values. *Random:* The data come from a random sample. Because the Equal SD condition is questionable, we will proceed with caution.

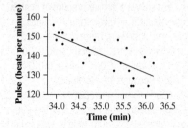

Is there convincing evidence of a negative linear relationship between wine consumption and heart disease deaths in the population of countries?

82. The professor swims Here are data on the time (in minutes) Professor Moore takes to swim 2000 yards and his pulse rate (beats per minute) after swimming on a random sample of 23 days:

Time	34.12	35.72	34.72	34.05	34.13	35.72
Pulse	152	124	140	152	146	128
Time	36.17	35.57	35.37	35.57	35.43	36.05
Pulse	136	144	148	144	136	124
Time	34.85	34.70	34.75	33.93	34.60	34.00
Pulse	148	144	140	156	136	148
Time	34.35	35.62	35.68	35.28	35.97	
Pulse	148	132	124	132	139	

Is there convincing evidence of a negative linear relationship between Professor Moore's swim time and his pulse rate in the population of days on which he swims 2000 yards?

83. Turn up the volume? Nicole and Elena wanted to know if listening to music at a louder volume negatively impacts test performance. To investigate, they recruited 30 volunteers and randomly assigned 10 volunteers to listen to music at 30 decibels, 10 volunteers to listen to music at 60 decibels, and 10 volunteers to listen to music at 90 decibels. While listening to the music, each student took a 10-question math test. Here is computer output from a least-squares regression analysis using x = volume and y = number correct:[43]

Predictor Coef   SE Coef   T   P
Constant   9.9000   0.7525   13.156 0.0000
Volume   -0.0483 0.0116   -4.163 0.0003
S = 1.55781   R-Sq = 38.2%   R-Sq(adj) = 36.0%

(a) Is there convincing evidence that listening to music at a louder volume hurts test performance? Assume the conditions for inference are met.

(b) Interpret the P-value from part (a).

84. Pencils and GPA Is there a relationship between a student's GPA and the number of pencils in his or her backpack? Jordynn and Angie decided to find out by selecting a random sample of students from their high school. Here is computer output from a least-squares regression analysis using x = number of pencils and y = GPA:[44]

Predictor Coef   SE Coef   T   P
Constant   3.2413   0.1809   17.920   0.0000
Pencils   -0.0423 0.0631   -0.670   0.5062
S = 0.738533   R-Sq = 0.9%   R-Sq(adj) = 0.0%

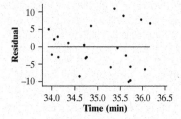

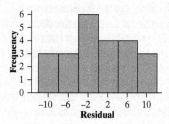

(a) Is there convincing evidence of a linear relationship between GPA and number of pencils for students at this high school? Assume the conditions for inference are met.

(b) Interpret the P-value from part (a).

85. **Stats teachers' cars** A random sample of 21 AP® Statistics teachers was asked to report the age (in years) and mileage of their primary vehicles. Here is a scatterplot of the data:

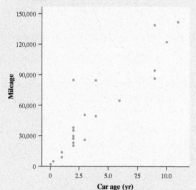

Car age (yr)

Here is some computer output from a least-squares regression analysis of these data. Assume that the conditions for regression inference are met.

Variable	Coef	SE Coef	t-ratio	prob
Constant	7288.54	6591	1.11	0.2826
Car age	11630.6	1249	9.31	<0.0001
S=19280	R-Sq=82.0%		R-Sq(adj)=81.1%	

(a) Verify that the 95% confidence interval for the slope of the population regression line is (9016.4, 14,244.8).

(b) A national automotive group claims that the typical driver puts 15,000 miles per year on his or her main vehicle. We want to test whether AP® Statistics teachers are typical drivers. Explain why an appropriate pair of hypotheses for this test is $H_0: \beta = 15,000$ versus $H_a: \beta \neq 15,000$.

(c) Compute the standardized test statistic and P-value for the test in part (b). What conclusion would you draw at the $\alpha = 0.05$ significance level?

(d) Does the confidence interval in part (a) lead to the same conclusion as the test in part (c)? Explain your answer.

86. **Paired tires** Exercise 37 in Chapter 10 (page 685) compared two methods for estimating tire wear. The first method used the amount of weight lost by a tire. The second method used the amount of wear in the grooves of the tire. A random sample of 16 tires was obtained. Both methods were used to estimate the total distance traveled by each tire. The following scatterplot displays the two estimates (in thousands of miles) for each tire.[45]

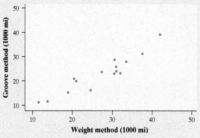

Weight method (1000 mi)

Here is some computer output from a least-squares regression analysis of these data. Assume that the conditions for regression inference are met.

Predictor	Coef	SE Coef	T	P
Constant	1.351	2.105	0.64	0.531
Weight	0.79021	0.07104	11.12	0.000
S=2.62078	R-Sq=89.8%	R-Sq(adj)=89.1%		

(a) Verify that the 99% confidence interval for the slope of the population regression line is (0.5787, 1.0017).

(b) Researchers want to test whether there is a difference in the two methods of estimating tire wear. Explain why the researchers might want to test the hypotheses $H_0: \beta = 1$ versus $H_a: \beta \neq 1$.

(c) Compute the standardized test statistic and P-value for the test in part (b). What conclusion would you draw at the $\alpha = 0.01$ significance level?

(d) Does the confidence interval in part (a) lead to the same conclusion as the test in part (c)? Explain your answer.

DO: $t = -5.13$, df $= 23 - 2 = 21$, and P-value ≈ 0. CONCLUDE: Because the P-value of approximately $0 < \alpha = 0.05$, we reject H_0. There is convincing evidence of a negative linear relationship between swim time and pulse rate in the population of days when Professor Moore swims 2000 yards.

12.83 (a) STATE: $H_0: \beta = 0$, $H_a: \beta < 0$, where β = slope of the true regression line relating y = performance on the 10-question math test to x = volume of the music for students like these; $\alpha = 0.05$. PLAN: t test for the slope. The conditions are met. DO: $t = -4.163$; P-value $= 0.0003/2 = 0.00015$. CONCLUDE: Because the P-value of $0.00015 < \alpha = 0.05$, we reject H_0. There is convincing evidence of a negative linear relationship relating y = performance on the 10-question math

test to x = volume of the music for students like these. **(b)** Assuming the slope of the true regression line relating y = performance on the 10-question math test to x = volume of the music is zero, there is a 0.00015 probability of getting a sample slope of -0.0483 or smaller by chance alone.

12.84 (a) STATE: $H_0: \beta = 0$, $H_a: \beta \neq 0$, where β = slope of the population regression line relating y = GPA to x = number of pencils in his or her backpack in the population of students at Jordynn and Angie's high school; $\alpha = 0.05$. PLAN: t test for the slope. The conditions are met. DO: $t = -0.670$ and the two-sided P-value is 0.5062. CONCLUDE: Because the P-value of $0.5062 > \alpha = 0.01$, we fail to reject H_0. There is not convincing evidence of a linear relationship between y = GPA and x = number of pencils

in his or her backpack in the population of students at Jordynn and Angie's high school. **(b)** Assuming the slope of the true regression line relating y = GPA to x = number of pencils in a student's backpack is zero, there is a 0.5062 probability of getting a sample slope at least as extreme as -0.0423 by chance alone.

12.85 (a) With df = 19, $t^* = 2.093$; (9016.4, 14,244.8) **(b)** Because the automotive group claims that people drive 15,000 miles per year, it says that for every increase of 1 year, the mileage would increase by 15,000 miles.
(c) $t = \dfrac{11{,}630.6 - 15{,}000}{1249} = -2.70$;
df = 19, the P-value is between $2(0.005) = 0.01$ and $2(0.01) = 0.02$. *Tech:* P-value $= 0.0142$. Because the P-value of $0.0142 < \alpha = 0.05$, we reject H_0. We have convincing evidence that the slope of the population regression line relating miles to years is not equal to 15,000. **(d)** Yes; because the interval in part (a) does not include the value 15,000, the interval also provides convincing evidence that the slope of the population regression line relating miles to years is not equal to 15,000.

12.86 (a) With df = 14, $t^* = 2.977$; (0.5787, 1.0017) **(b)** Both variables are measuring tire wear in the same units. If one method of measuring wear gives an increase in wear of 1 unit, we expect the other way of measuring wear to also give an increase in wear of 1 unit. This translates into a slope of $1/1 = 1$.
(c) $t = \dfrac{0.79021 - 1}{0.07104} = -2.95$; df = 14, the P-value is between $2(0.005) = 0.01$ and $2(0.01) = 0.02$. *Tech:* P-value $= 0.0105$. Because the P-value of $0.0105 > \alpha = 0.01$, we fail to reject H_0. There is not convincing evidence that the slope of the population line relating wear measurements using the groove method to wear measurements using the weight method differs from 1. **(d)** Yes; because the interval in part (a) includes the value 1, the interval also does not provide convincing evidence that the slope of the population line relating wear measurements using the groove method to wear measurements using the weight method differs from 1. However, the values might not be the same even if the slope is 1.

12.87 c

12.88 c

12.89 a

12.90 e

12.91 b

12.92 a

Multiple Choice: *Select the best answer for Exercises 87–92.*
Exercises 87–92 refer to the following setting. To see if students with longer feet tend to be taller, a random sample of 25 students was selected from a large high school. For each student, x = foot length (cm) and y = height (cm) were recorded. We checked that the conditions for inference about the slope of the population regression line are met. Here is a portion of the computer output from a least-squares regression analysis using these data:

```
Predictor      Coef    SE Coef    T       P
Constant     91.9766   10.2204   8.999   0.000
Foot length   3.0867    0.4117   7.498   0.000
S = 6.47044    R-Sq = 72.8%    R-Sq(adj) = 71.5%
```

87. Which of the following is the equation of the least-squares regression line for predicting height from foot length?

(a) $\widehat{\text{height}} = 10.2204 + 0.4117 \,(\text{foot length})$

(b) $\widehat{\text{height}} = 0.4117 + 3.0867 \,(\text{foot length})$

(c) $\widehat{\text{height}} = 91.9766 + 3.0867 \,(\text{foot length})$

(d) $\widehat{\text{height}} = 91.9766 + 6.47044 \,(\text{foot length})$

(e) $\widehat{\text{height}} = 3.0867 + 6.47044 \,(\text{foot length})$

88. The slope β of the population regression line describes

(a) the exact increase in height (cm) for students at this high school when foot length increases by 1 cm.

(b) the average increase in foot length (cm) for students at this high school when height increases by 1 cm.

(c) the average increase in height (cm) for students at this high school when foot length increases by 1 cm.

(d) the average increase in foot length (cm) for students in the sample when height increases by 1 cm.

(e) the average increase in height (cm) for students in the sample when foot length increases by 1 cm.

89. Is there convincing evidence that height increases as foot length increases? To answer this question, test the hypotheses

(a) $H_0: \beta = 0$ versus $H_a: \beta > 0$.

(b) $H_0: \beta = 0$ versus $H_a: \beta < 0$.

(c) $H_0: \beta = 0$ versus $H_a: \beta \neq 0$.

(d) $H_0: \beta > 0$ versus $H_a: \beta = 0$.

(e) $H_0: \beta = 1$ versus $H_a: \beta > 1$.

90. Which of the following is the best interpretation of the value 0.4117 in the computer output?

(a) For each increase of 1 cm in foot length, the average height increases by about 0.4117 cm.

(b) When using this model to predict height, the predictions will typically be off by about 0.4117 cm.

(c) The linear relationship between foot length and height accounts for 41.17% of the variation in height.

(d) The linear relationship between foot length and height is moderate and positive.

(e) In repeated samples of size 25, the slope of the sample regression line for predicting height from foot length will typically vary from the population slope by about 0.4117.

91. Which of the following is a 95% confidence interval for the population slope β?

(a) 3.0867 ± 0.4117

(b) 3.0867 ± 0.8518

(c) 3.0867 ± 0.8069

(d) 3.0867 ± 0.8497

(e) 3.0867 ± 0.8481

92. Which of the following would have resulted in a violation of the conditions for inference?

(a) If the entire sample was selected from one classroom

(b) If the sample size was 15 instead of 25

(c) If the scatterplot of x = foot length and y = height did not show a perfect linear relationship

(d) If the histogram of heights had an outlier

(e) If the standard deviation of foot length was different from the standard deviation of height

Recycle and Review

Exercises 93–95 refer to the following setting. Does the color in which words are printed affect your ability to read them? Do the words themselves affect your ability to name the color in which they are printed? Mr. Starnes designed a study to investigate these questions using the 16 students in his AP® Statistics class as subjects. Each student performed the following two tasks in random order while a partner timed his or her performance: (1) Read 32 words aloud as quickly as possible, and (2) say the color in which each of 32 words is printed as quickly as possible. Try both tasks for yourself using the word list given.

BROWN	RED	BLUE	GREEN
RED	GREEN	BROWN	BROWN
GREEN	RED	BLUE	BLUE
BROWN	BLUE	GREEN	RED
BLUE	BROWN	RED	RED
RED	BLUE	BROWN	GREEN
BLUE	GREEN	GREEN	BLUE
GREEN	BROWN	RED	BROWN

93. **Color words** (4.2) Let's review the design of the study.

(a) Explain why this was an experiment and not an observational study.

(b) Did Mr. Starnes use a completely randomized design or randomized block design? Why do you think he chose this experimental design?

(c) Explain the purpose of the random assignment in the context of the study.

Here are the data from Mr. Starnes's experiment. For each subject, the time to perform the two tasks is given to the nearest second.

Subject	Words	Colors	Subject	Words	Colors
1	13	20	9	10	16
2	10	21	10	9	13
3	15	22	11	11	11
4	12	25	12	17	26
5	13	17	13	15	20
6	11	13	14	15	15
7	14	32	15	12	18
8	16	21	16	10	18

94. **Color words** (10.2) Now let's analyze the data.

(a) Calculate the difference (Colors – Words) for each subject and summarize the distribution of differences with a boxplot. Does the graph provide evidence of a difference in the average time required to perform the two tasks? Explain your answer.

(b) Explain why it is not safe to use paired t procedures to do inference about the mean difference in time to complete the two tasks.

95. **Color words** (3.1, 3.2, 12.3) Can we use a student's word task time to predict his or her color task time?

(a) Make an appropriate scatterplot to help answer this question. Describe what you see.

(b) Use technology to find the equation of the least-squares regression line. Define any variables you use.

(c) Find and interpret the residual for the student who completed the word task in 9 seconds.

(d) Assume that the conditions for performing inference about the slope of the true regression line are met. The P-value for a test of $H_0 : \beta = 0$ versus $H_a : \beta > 0$ is 0.0215. Interpret this value.

Note: John Ridley Stroop is often credited with the discovery in 1935 of the fact that the color in which "color words" are printed interferes with people's ability to identify the color. The paper outlining the so-called Stroop effect, though, was originally published by German researchers in 1929.

96. **Yahtzee** (5.3, 6.3) In the game of Yahtzee, 5 six-sided dice are rolled simultaneously. To get a Yahtzee, the player must get the same number on all 5 dice.

(a) Luis says that the probability of getting a Yahtzee in one roll of the dice is $\left(\dfrac{1}{6}\right)^5$. Explain why Luis is wrong.

(b) Nassir decides to keep rolling all 5 dice until he gets a Yahtzee. He is surprised when he still hasn't gotten a Yahtzee after 25 rolls. Should he be? Calculate an appropriate probability to support your answer.

12.95 (a) There appears to be a moderately strong, positive linear association between the length of time to read the word and length of time to identify the color.

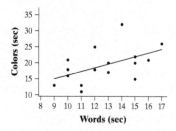

(b) $\hat{y} = 4.887 + 1.1321x$, where $\hat{y} =$ the predicted time to identify the color and $x =$ the amount of time to read the words. **(c)** $\hat{y} = 4.887 + 1.1321(9) = 15.076$ seconds, so the residual is $y - \hat{y} = 13 - 15.076 = -2.076$ seconds. The actual time to complete the color task was 2.076 seconds less than the time predicted by the regression line, with $x = 9$ seconds to complete the word task. **(d)** If the true slope of the regression line relating time to identify the colors and time to read the words is 0 and this experiment were repeated many times, there is a 0.0215 probability of getting an observed slope of 1.1321 or larger by chance alone.

12.96 (a) Luis has calculated the probability of getting 5 of one particular number (e.g., getting all 6's). What he has not taken into account are the 6 different ways to get a Yahtzee—one for each number on the die. **(b)** No; the probability of getting a Yahtzee on any roll is

$6\left(\dfrac{1}{6}\right)^5 = 0.000772$. The probability of

getting no Yahtzee on any roll is $1 - 0.000772 = 0.999228$. The probability of getting 25 rolls with no Yahtzee is very large: $(0.999228)^{25} = 0.9809$.

12.93 (a) This was an experiment because the two treatments were deliberately assigned to the students. **(b)** To help account for the different abilities of students to read the words or to say the color the words were printed in, he used a randomized block design where each student was a block. **(c)** Random assignment helped to average out the effects of the order in which people did the two treatments. For example, if every subject said the color of the printed word first and was frustrated by this task, the times for the second treatment (reading the word) might be worse. Then we wouldn't know why the times were longer for the second treatment—because of frustration or because the second method actually takes longer.

12.94 (a) Yes, the graph provides evidence of a difference in the average time required to perform the two tasks. For all students, the time it took to identify the color was the same as or longer than the time needed to read the words.

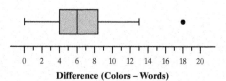

Difference (Colors – Words)

(b) It is not safe to use a paired t procedure because there are few differences ($n_d = 16 < 30$) and there is an outlier.

Please consult the Teacher's Resource Materials for sample student responses, a scoring rubric, and a printable version of the original question with space for students to write their responses. We present a model solution here.

Answers:

(a) H_0: The true distribution of responses is the same for the two questions *versus* H_a: The true distribution of responses is not the same for the two questions.

(b) *Type I:* Finding convincing evidence that the true distribution of responses is different for the two questions, when the true distributions really are the same.

Type II: Not finding convincing evidence that the true distribution of responses is different for the two questions, when the true distributions really are not the same.

(c) We should not use a chi-square distribution to estimate the *P*-value because the Large Counts condition is not met. All the expected counts are less than 5:

	A	B
Very important	4.5	4.5
Important	3.5	3.5
Somewhat important	2.5	2.5
Not that important	1.5	1.5
Not at all important	3	3

(d) Because 24 of the 100 trials resulted in a χ^2-value of 6.12 or more, the estimated *P*-value is 0.24. Because the approximate *P*-value of 0.24 is greater than $\alpha = 0.05$, we fail to reject H_0. We do not have convincing evidence that the true distribution of responses is different for the two questions.

Chapter 12 Wrap-Up

FRAPPY! FREE RESPONSE AP® PROBLEM, YAY!

The following problem is modeled after actual AP® Statistics exam free response questions. Your task is to generate a complete, concise response in 15 minutes.

Directions: Show all your work. Indicate clearly the methods you use, because you will be scored on the correctness of your methods as well as on the accuracy and completeness of your results and explanations.

Two statistics students wanted to know if including additional information in a survey question would change the distribution of responses. To find out, they randomly selected 30 teenagers and asked them one of the following two questions. Fifteen of the teenagers were randomly assigned to answer Question A, and the other 15 students were assigned to answer Question B.

Question A: When choosing a college, how important is a good athletic program: very important, important, somewhat important, not that important, or not important at all?

Question B: It's sad that some people choose a college based on its athletic program. When choosing a college, how important is a good athletic program: very important, important, somewhat important, not that important, or not important at all?

The table below summarizes the responses to both questions. For these data, the chi-square test statistic is $\chi^2 = 6.12$.

		Question		
		A	B	Total
	Very important	7	2	9
	Important	4	3	7
Importance of a good athletic program	Somewhat important	2	3	5
	Not that important	1	2	3
	Not important at all	1	5	6
	Total	15	15	30

(a) State the hypotheses that the students are interested in testing.

(b) Describe a Type I error and a Type II error in the context of the hypotheses stated in part (a).

(c) For these data, explain why it would *not* be appropriate to use a chi-square distribution to calculate the *P*-value.

(d) To estimate the *P*-value, 100 trials of a simulation were conducted, assuming that the additional information didn't have an effect on the response to the question. In each trial of the simulation, the value of the chi-square test statistic was calculated. These simulated chi-square test statistics are displayed in the dotplot shown here.

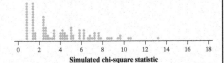

Simulated chi-square statistic

Based on the results of the simulation, what conclusion would you make about the hypotheses stated in part (a)?

After you finish, you can view two example solutions on the book's website (highschool.bfwpub.com/updatedtps6e). Determine whether you think each solution is "complete," "substantial," "developing," or "minimal." If the solution is not complete, what improvements would you suggest to the student who wrote it? Finally, your teacher will provide you with a scoring rubric. Score your response and note what, if anything, you would do differently to improve your own score.

840

Chapter 12 Review

Section 12.1: Chi-Square Tests for Goodness of Fit

In this section, you learned the details for performing a chi-square test for goodness of fit. The null hypothesis is that a single categorical variable follows a specified distribution in a population of interest. The alternative hypothesis is that the variable does not follow the specified distribution in the population of interest.

The chi-square test statistic measures the difference between the observed distribution of a categorical variable and its hypothesized distribution. To calculate the chi-square test statistic, use the following formula that involves the observed and expected counts for each value of the categorical variable:

$$\chi^2 = \sum \frac{(\text{Observed count} - \text{Expected count})^2}{\text{Expected count}}$$

To calculate the expected counts, multiply the total sample size by the proportion specified by the null hypothesis for each category. Larger values of the chi-square test statistic provide more convincing evidence that the distribution of the categorical variable differs from the hypothesized distribution in the population of interest.

When the Random, 10%, and Large Counts conditions are satisfied, we can accurately model the sampling distribution of the chi-square test statistic using a chi-square distribution (density curve). The Random condition says that the data are from a random sample from the population of interest. The 10% condition says that the sample size should be less than 10% of the population size when sampling without replacement. The Large Counts condition says that the *expected* counts for each category must be at least 5. In a test for goodness of fit, use a chi-square distribution with degrees of freedom = number of categories − 1.

When the results of a test for goodness of fit are significant, consider doing a follow-up analysis. Identify which categories of the variable had the largest contributions to the chi-square test statistic and whether the observed values in those categories were larger or smaller than expected.

Section 12.2: Inference for Two-Way Tables

In this section, you learned two different tests to analyze categorical data that are summarized in a two-way table. A test for homogeneity compares the distribution of a single categorical variable for two or more populations or treatments. A test for independence looks for an association between two categorical variables in a single population.

In a chi-square test for homogeneity, the null hypothesis is that there is no difference in the true distribution of a categorical variable for two or more populations or treatments. The alternative hypothesis is that there is a difference in the distributions. The Random condition is that the data come from independent random samples or groups in a randomized experiment. The 10% condition applies for each sample when sampling without replacement, but not for experiments that lack random selection. Finally, the Large Counts condition remains the same — the expected counts must be at least 5 in each cell of the two-way table.

To calculate the expected counts for a test for homogeneity, use the following formula:

$$\text{expected count} = \frac{\text{row total} \cdot \text{column total}}{\text{table total}}$$

To calculate the P-value, compute the chi-square test statistic and use a chi-square distribution with degrees of freedom = (number of rows − 1)(number of columns − 1).

In a chi-square test for independence, the null hypothesis is that there is no association between two categorical variables in one population (or that the two variables are independent in the population). The alternative hypothesis is that there is an association between the two variables (or that the two variables are not independent). For this test, the Random condition says that the data must come from a single random sample. The 10% condition applies when sampling without replacement. The Large Counts condition is still the same — the expected counts must all be at least 5. The method for calculating expected counts, the chi-square test statistic, the degrees of freedom, and the P-value are exactly the same in a test for independence and a test for homogeneity.

As with tests for goodness of fit, when the results of a test for homogeneity or independence are significant, consider doing a follow-up analysis. Identify which cells in the two-way table had the largest contributions to the chi-square test statistic and whether the observed counts in those cells were larger or smaller than expected.

Section 12.3: Inference for Slope

In this section, you learned that the sample regression line $\hat{y} = a + bx$ estimates the population (true) regression line $\mu_y = \alpha + \beta x$. The sampling distribution of the sample slope b is the foundation for doing inference about the population (true) slope β. When the conditions are met, the sampling distribution of b has an approximately Normal distribution with mean $\mu_b = \beta$ and standard deviation $\sigma_b = \dfrac{\sigma}{\sigma_x \sqrt{n}}$.

There are five conditions for performing inference about the slope of a population (true) least-squares regression line. Remember them with the acronym LINER.

- The **linear** condition says that the mean value of the response variable μ_y falls on the population (true) regression line $\mu_y = \alpha + \beta x$. To check the linear condition, verify that there are no leftover curved patterns in the residual plot.

- The **independent** condition says that individual observations are independent of each other. Be sure to verify that the sample size is less than 10% of the population size when sampling without replacement from a population (the 10% condition).

- The **Normal** condition says that the distribution of y-values is approximately Normal for each value of x. To check the Normal condition, graph a dotplot, histogram, stemplot, boxplot, or Normal probability plot of the residuals and verify that there are no outliers or strong skewness.

- The **equal SD** condition says that for each value of x, the distribution of y should have the same standard deviation. To check the equal SD condition, verify that the residuals have roughly the same amount of scatter around the residual $= 0$ line for each value of x on the residual plot.

- The **random** condition says that the data are from a random sample or a randomized experiment. To check the

random condition, verify that randomness was properly used in the data collection process.

To construct and interpret a confidence interval for the slope of the population (true) least-squares regression line, follow the familiar four-step process. The formula for the confidence interval is $b \pm t^* \, SE_b$, where t^* is the t critical value with df $= n - 2$. The standard error of the slope SE_b describes how far the sample slope typically varies from the population (true) slope in repeated random samples or random assignments. The formula for the standard error of the slope is $SE_b = \dfrac{s}{s_x \sqrt{n-1}}$. The standard error of the slope is typically provided with standard computer output for least-squares regression.

When you conduct a significance test for the slope of the population (true) least-squares regression line, use the standardized test statistic $t = \dfrac{b - \beta_0}{SE_b}$ with df $= n - 2$. In most cases, the null hypothesis is $H_0: \beta = 0$. This hypothesis says that a straight-line relationship between x and y is no better at predicting y than using the mean value $\bar{y}$. The value of the standardized test statistic for a test of $H_0: \beta = 0$, along with a two-sided P-value, is typically provided with standard computer output for least-squares regression.

Comparing the Three Chi-Square Tests			
	Goodness of fit	**Homogeneity**	**Independence**
Number of samples/ treatments	1	2 or more	1
Number of variables	1	1	2
Null hypothesis	The stated distribution of a categorical variable in the population of interest is correct.	There is no difference in the distribution of a categorical variable for several populations or treatments.	There is no association between two categorical variables in the population of interest.
Random condition	The data come from a random sample from the population of interest.	The data come from independent random samples or groups in a randomized experiment.	The data come from a random sample from the population of interest.
10% condition	When sampling without replacement, $n < 0.10N$ for each sample.		
Large Counts condition	All expected counts ≥ 5		
Expected counts	$\begin{pmatrix} \text{sample} \\ \text{size} \end{pmatrix}\begin{pmatrix} \text{expected} \\ \text{proportion} \end{pmatrix}$	$\dfrac{\text{row total} \cdot \text{column total}}{\text{table total}}$	
Formula for test statistic	$\chi^2 = \sum \dfrac{(\text{Observed count} - \text{Expected count})^2}{\text{Expected count}}$		
Degrees of freedom	(# categories $-$ 1)	(# rows $-$ 1)(# columns $-$ 1)	
TI-83/84 name	χ^2 GOF-test	χ^2-test	

What Did You Learn?

Learning Target	Section	Related Example on Page(s)	Relevant Chapter Review Exercise(s)
State appropriate hypotheses and compute the expected counts and chi-square test statistic for a chi-square test for goodness of fit.	12.1	764	R12.1
State and check the Random, 10%, and Large Counts conditions for performing a chi-square test for goodness of fit.	12.1	770	R12.1
Calculate the degrees of freedom and P-value for a chi-square test for goodness of fit.	12.1	768	R12.1
Perform a chi-square test for goodness of fit.	12.1	770	R12.1
Conduct a follow-up analysis when the results of a chi-square test are statistically significant.	12.1, 12.2	Discussion on 772, 791	R12.3
State appropriate hypotheses and compute the expected counts and chi-square test statistic for a chi-square test based on data in a two-way table.	12.2	783	R12.4, R12.5
State and check the Random, 10%, and Large Counts conditions for a chi-square test based on data in a two-way table.	12.2	786	R12.2, R12.4, R12.5
Calculate the degrees of freedom and P-value for a chi-square test based on data in a two-way table.	12.2	786	R12.4, R12.5
Perform a chi-square test for homogeneity.	12.2	790	R12.4
Perform a chi-square test for independence.	12.2	796	R12.5
Choose the appropriate chi-square test in a given setting.	12.2	799	R12.3
Check the conditions for performing inference about the slope β of the population (true) regression line.	12.3	819	R12.7
Interpret the values of a, b, s, and SE_b in context, and determine these values from computer output.	12.3	821	R12.6
Construct and interpret a confidence interval for the slope β of the population (true) regression line.	12.3	823	R12.9
Perform a significance test about the slope β of the population (true) regression line.	12.3	827	R12.8

Teaching Tip

At the end of each chapter, a "What Did You Learn?" grid lists all the targets for the chapter. Make sure to discuss this grid with your students. Ask them to read each learning target and self-assess whether or not they can do each one. For each target, the grid shows the section in which it was covered, page references for examples that illustrate the target, and relevant chapter review exercises that students can use to assess their understanding of that target. Encourage your students to use this grid as part of their preparations for the chapter test.

Chapter 12 Review Exercises

These exercises are designed to help you review the important ideas and methods of the chapter.

R12.1 **Testing a genetic model** Biologists wish to cross pairs of tobacco plants having genetic makeup Gg, indicating that each plant has one dominant gene (G) and one recessive gene (g) for color. Each offspring plant will receive one gene for color from each parent. The Punnett square shows the possible combinations of genes received by the offspring.

		Parent 2 passes on:	
		G	g
Parent 1 passes on:	G	GG	Gg
	g	Gg	gg

The Punnett square suggests that the expected ratio of green (GG) to yellow-green (Gg) to albino (gg) tobacco plants should be 1:2:1. In other words, the biologists predict that 25% of the offspring will be

TRM **Full Solutions to Chapter 12 Review Exercises**

The full solutions can be found by clicking on the link in the TE-Book, opening the TRFD, or downloading from the Teacher's Resources on the book's digital platform.

Answers to Chapter 12 Review Exercises

R12.1 STATE: H_0: The proposed 1:2:1 genetic model is correct. H_a: The proposed 1:2:1 genetic model is not correct; at $\alpha = 0.01$. **PLAN:** Chi-square test for goodness of fit. *Random:* Random sample of 84 pairs of yellow-green parent plants. *10%:* $n = 84 < 10\%$ of all yellow-green parent plants. *Large Counts:* The expected counts 21, 42, and 21 are all ≥ 5. **DO:** $\chi^2 = 6.476$; using df $= 3 - 1 = 2$, the P-value (0.0392) is between 0.025 and 0.05. **CONCLUDE:** Because the P-value of $0.0392 > \alpha = 0.01$, we fail to reject H_0. We do not have convincing evidence that the proposed 1:2:1 genetic model is not correct.

R12.2 The result is not valid because several of the expected counts are less than 5. The expected counts are 12.5, 0.5, 4.5, 1.5, 5, 12.5, 0.5, 4.5, 1.5, and 5.

R12.3 (a) Chi-square test for homogeneity. The data came from two independent random samples. **(b)** The value 252 is the expected count for the 2009/None cell. It is calculated $(366)(14{,}557)/21{,}177$. The value 14.113 is the component of the chi-square test statistic that comes from the cell "2009/None" and is calculated as $\dfrac{(192 - 251.587)^2}{251.587} = 14.113$. **(c)** The three cells that contribute most to the chi-square test statistic are: "2015/None," which had more households than expected with no TV $(174 - 114 = 60)$; "2015/One," which had more households than expected with 1 TV $(1680 - 1494 = 186)$; and "2015/Five or more," which had fewer households than expected with 5 or more TVs $(392 - 511 = -119)$.

R12.4 (a)

Treatment

		Stress management	Exercise	Usual care	Total
Cardiac event?	Yes	3	7	12	22
	No	30	27	28	85
	Total	33	34	40	107

(b) The success rate was highest for stress management $(30/33 = 0.909)$, followed by exercise $(27/34 = 0.794)$ and usual care $(28/40 = 0.70)$.

(c) STATE: H_0: The true success rates for patients like these are the same for all three treatments. H_a: The true success rates . . . are not the same . . .

---

green, 50% will be yellow-green, and 25% will be albino. To test their hypothesis about the distribution of offspring, the biologists mate 84 randomly selected pairs of yellow-green parent plants. Of 84 offspring, 23 plants were green, 50 were yellow-green, and 11 were albino. Do the data provide convincing evidence at the $\alpha = 0.01$ level that the true distribution of color of offspring is different from what the biologists predict?

R12.2 Sorry, no chi-square We would prefer to learn from teachers who know their subject. Perhaps even preschool children are affected by how knowledgeable they think teachers are. Assign 48 three- and four-year-olds at random to be taught the name of a new toy by either an adult who claims to know about the toy or an adult who claims not to know about it. Then ask the children to pick out a picture of the new toy from a set of pictures of other toys and say its name. The response variable is the count of right answers in four tries. Here are the data:[46]

Count of correct answers

		0	1	2	3	4	Total
Knowledge of teacher	Knowledgeable	5	1	6	3	9	24
	Ignorant	20	0	3	0	1	24
	Total	25	1	9	3	10	48

The researchers report that children who were taught by the teacher claiming to be knowledgeable did significantly better ($\chi^2 = 20.4$, $P < 0.05$). Explain why this result isn't valid.

R12.3 Fewer TVs? The United States Energy Information Administration periodically surveys a random sample of U.S. households to determine how they use energy.[47] One of the variables they track is how many TVs are in a household (None, 1, 2, 3, 4, or 5 or more). The computer output compares the distribution of number of TVs for households in 2009 and 2015.

	2009	2015	All
None	192	174	366
	252	114	
	14.113	31.034	
1	3098	1680	4778
	3284	1494	
	10.577	23.258	
2	4801	2190	6991
	4806	2185	
	0.004	0.010	
3	3406	1495	4901
	3369	1532	
	0.408	0.897	
4	1818	689	2507
	1723	784	
	5.204	11.442	
	2009	2015	All
5 or more	1242	392	1634
	1123	511	
	12.564	27.628	
All	14557	6620	21177

Cell Contents: Count
                Expected count
                Contribution to Chi-square

Chi-Square $= 137.137$, DF $= 5$, P-Value $= 0.000$

(a) Which chi-square test is appropriate to analyze these data? Explain your answer.

(b) Show how the numbers 252 and 14.113 were obtained for the 2009/None cell.

(c) Which 3 cells contribute most to the chi-square test statistic? How do the observed and expected counts compare for these cells?

R12.4 Stress and heart attacks You read a newspaper article that describes a study of whether stress management can help reduce heart attacks. The 107 subjects all had reduced blood flow to the heart and so were at risk of a heart attack. They were assigned at random to three groups. The article goes on to say:

> One group took a four-month stress management program, another underwent a four-month exercise program, and the third received usual heart care from their personal physicians. In the next three years, only 3 of the 33 people in the stress management group suffered "cardiac events," defined as a fatal or non-fatal heart attack or a surgical procedure such as a bypass or angioplasty. In the same period, 7 of the 34 people in the exercise group and 12 out of the 40 patients in usual care suffered such events.[48]

(a) Use the information in the news article to make a two-way table that describes the study results.

(b) Compare the success rates of the three treatments in preventing cardiac events.

(c) Do the data provide convincing evidence at the $\alpha = 0.05$ level that the true success rates for patients like these are not the same for the three treatments?

---

at $\alpha = 0.05$. **PLAN:** Chi-square test for homogeneity. *Random:* 3 groups in a randomized experiment. *Large Counts:* The expected counts (6.79, 6.99, 8.22, 26.21, 27.01, 31.78) are all ≥ 5. **DO:** $\chi^2 = 4.84$; using df $= (2 - 1)(3 - 1) = 2$, the P-value (0.0889) is between 0.05 and 0.10. **CONCLUDE:** Because the P-value of $0.0889 > \alpha = 0.05$, we fail to reject H_0. We do not have convincing evidence that the true success rates for patients like these are not the same for all three treatments.

R12.5 Popular kids Who were the popular kids at your elementary school? Did they get good grades or have good looks? Were they good at sports? A study was performed in Michigan to examine the factors that determine social status for children in grades 4, 5, and 6. Researchers administered a questionnaire to a random sample of 478 students in these grades. One of the questions asked, "What would you most like to do at school: make good grades, be good at sports, or be popular?" The two-way table summarizes the students' responses.[49] Is there convincing evidence of an association between gender and goal for students in grades 4, 5, and 6?

		Gender		
		Female	Male	Total
Goal	Grades	130	117	247
	Popular	91	50	141
	Sports	30	60	90
	Total	251	227	478

Exercises R12.6–R12.9 refer to the following setting. Do taller students require fewer steps to walk a fixed distance? The scatterplot shows the relationship between x = height (in inches) and y = number of steps required to walk the length of a school hallway for a random sample of 36 students at a high school.

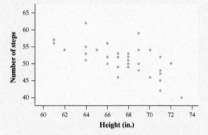

A least-squares regression analysis was performed on the data. Here is some computer output from the analysis:

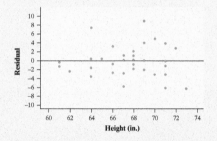

```
Predictor    Coef    SE Coef    T      P
Constant    113.57   13.085   8.679  0.000
Height      -0.9211   0.1938   ***    ***
S=3.50429  R-Sq=39.9%  R-Sq(adj)=38.1%
```

R12.6 Long legs

(a) Describe what the scatterplot tells you about the relationship between height and number of steps.

(b) What is the equation of the least-squares regression line? Define any variables you use.

(c) Identify the value of each of the following from the computer output. Then provide an interpretation of each value.
   (i) a
   (ii) b
   (iii) s
   (iv) SE_b

R12.7 Long legs Verify that the conditions for inference about the slope of the least-squares regression line are met in this context.

R12.8 Long legs Do these data provide convincing evidence at the $\alpha = 0.05$ level that taller students at this school require fewer steps to walk a fixed distance? Assume that the conditions for inference are met.

R12.9 Long legs Construct and interpret a 95% confidence interval for the slope of the population regression line. Assume that the conditions for inference are met. Explain how the interval provides more information than the test in Exercise R12.8.

line using x = height. (iv) $SE_b = 0.1938$; if we repeated the random selection many times, the slope of the sample regression line would typically vary by about 0.1938 from the slope of the population regression line for predicting number of steps from height.

R12.7 *Linear:* The residual plot shows no leftover curved pattern. *Independent:* Knowing the number of steps needed for one student should not help us predict the number of steps needed for another student. Also, the sample size $n = 36 < 10\%$ of all students at the high school. *Normal:* The histogram of the residuals does not show strong skewness or outliers. *Equal SD:* The residual plot shows roughly equal amounts of scatter for all x-values. *Random:* The data come from a random sample.

R12.8 STATE: H_0: $\beta = 0$, H_a: $\beta < 0$, where β = slope of the population regression line relating y = number of steps to x = height. PLAN: t test for the slope. The conditions are met. DO: $t = -4.75$; with df = 34, using df = 30, the P-value (0) is less than 0.0005. CONCLUDE: Because the P-value of approximately $0 < \alpha = 0.05$, we reject H_0. There is convincing evidence that the slope of the true regression line relating height to number of steps is negative. In other words, we have convincing evidence that taller students require fewer steps to walk the length of the hallway.

R12.9 STATE: β = the slope of the population regression line relating y = number of steps to x = height. PLAN: t interval for the slope. The conditions are met. DO: With df = 34, using df = 30, $t^* = 2.042$; $(-1.317, -0.525)$. CONCLUDE: We are 95% confident that the interval from -1.317 to -0.525 captures the slope of the population regression line relating y = number of steps to x = height. The interval provides more information because it gives an interval of plausible values for the population slope.

R12.5 STATE: H_0: There is no association between gender and goals for 4th, 5th, and 6th grade students. H_a: There is an association between gender and goals . . . at $\alpha = 0.05$. PLAN: Chi-square test for independence. *Random:* Random sample of students. *10%:* $n = 478 < 10\%$ of all 4th, 5th, and 6th grade students in Michigan. *Large Counts:* All expected counts 129.7, 117.3, 74.04, 66.96, 47.26, and 42.74 ≥ 5. DO: The calculator's χ^2-Test gives $\chi^2 = 21.455$ and P-value ≈ 0.00002 using df = 2. CONCLUDE: Because the P-value of $0.00002 < \alpha = 0.05$, we reject H_0. There is convincing evidence of an association between gender and goals for 4th, 5th, and 6th grade students in Michigan.

R12.6 (a) The scatterplot reveals a negative, moderately strong linear relationship between height (inches) and number of steps.
(b) $\hat{y} = 113.57 - 0.9211x$, where $\hat{y}$ is the predicted number of steps and x is the height of the student (in inches).
(c) (i) $a = 113.57$; if a student has a height of 0 inches, the number of steps to walk the length of the school hallway is 113.57 steps, on average. Of course, a student with a height of 0 will not be able to walk anywhere, so this is unreasonable.
(ii) $b = -0.9211$; for each 1-inch increase in height, the average number of steps to walk the length of the school hallway decreases by 0.9211 steps. (iii) $s = 3.50429$; the actual number of steps needed to walk the length of the school hallway typically varies by about 3.50429 from the predicted value, with the least-squares regression

Answers to Chapter 12 AP® Statistics Practice Test

T12.1 c

T12.2 e

T12.3 d

T12.4 c

T12.5 d

T12.6 d

Chapter 12 AP® Statistics Practice Test

Section I: Multiple Choice *Select the best answer for each question.*

Exercises T12.1 and T12.2 refer to the following setting. Recent revenue shortfalls in a midwestern state led to a reduction in the state budget for higher education. To offset the reduction, the largest state university proposed a 25% tuition increase. It was determined that such an increase was needed simply to compensate for the lost support from the state. Separate random samples of 50 freshmen, 50 sophomores, 50 juniors, and 50 seniors from the university were asked whether they were strongly opposed to the increase, given that it was the minimum increase necessary to maintain the university's budget at current levels. Here are the results:

		Year				
		Freshman	Sophomore	Junior	Senior	Total
Strongly opposed?	Yes	39	36	29	18	122
	No	11	14	21	32	78
	Total	50	50	50	50	200

T12.1 Which null hypothesis would be appropriate for performing a chi-square test?

(a) The closer students get to graduation, the less likely they are to be opposed to tuition increases.

(b) The mean number of students who are strongly opposed is the same for each of the 4 years.

(c) The distribution of student opinion about the proposed tuition increase is the same for each of the 4 years at this university.

(d) Year in school and student opinion about the tuition increase are independent in the sample.

(e) There is an association between year in school and opinion about the tuition increase at this university.

T12.2 The conditions for carrying out the chi-square test in Exercise T12.1 are:

I.   Independent random samples from the populations of interest.

II. All sample sizes are less than 10% of the populations of interest.

III. All expected counts are at least 5.

Which of the conditions is (are) satisfied in this case?

(a) I only   (d) II and III only

(b) II only   (e) I, II, and III

(c) I and III only

Exercises T12.3–T12.5 refer to the following setting. A random sample of traffic tickets given to motorists in a large city is examined. The tickets are classified according to the race or ethnicity of the driver. The results are summarized in the following table.

Race/Ethnicity	White	Black	Hispanic	Other
Number of tickets	69	52	18	9

The proportion of this city's population in each of the racial/ethnic categories listed is as follows.

Race/Ethnicity	White	Black	Hispanic	Other
Proportion	0.55	0.30	0.08	0.07

We wish to test H_0: The racial/ethnic distribution of traffic tickets in the city is the same as the racial/ethnic distribution of the city's population.

T12.3 Assuming H_0 is true, what is the expected number of Hispanic drivers who would receive a ticket?

(a) 8   (d) 11.84

(b) 10.36   (e) 12

(c) 11

T12.4 We compute the value of the χ^2 test statistic to be 6.57. Assuming that the conditions for inference are met, which of the following is correct?

(a) P-value > 0.20

(b) $0.10 < P$-value < 0.20

(c) $0.05 < P$-value < 0.10

(d) $0.01 < P$-value < 0.05

(e) P-value < 0.01

T12.5 The category that contributes the largest component to the χ^2 test statistic is

(a) White, with 12.4 fewer tickets than expected.

(b) White, with 12.4 more tickets than expected.

(c) Hispanic, with 6.16 fewer tickets than expected.

(d) Hispanic, with 6.16 more tickets than expected.

(e) Other, with 1.36 fewer tickets than expected.

T12.6 Which of the following statements about chi-square distributions are true?

I.   For all chi-square distributions, $P(\chi^2 \geq 0) = 1$.

II. A chi-square distribution with fewer than 10 degrees of freedom is roughly symmetric.

III. The more degrees of freedom a chi-square distribution has, the larger the mean of the distribution.

(a) I only   (d) I and III

(b) II only   (e) I, II, and III

(c) III only

T12.7 Which of the following is *not* one of the conditions that must be satisfied in order to perform inference about the slope of a least-squares regression line?

(a) For each value of x, the population of y-values is Normally distributed.

(b) The standard deviation σ of the population of y-values corresponding to a given value of x is always the same, regardless of the specific value of x.

(c) The sample size—that is, the number of paired observations (x, y)—exceeds 30.

(d) There exists a straight line such that, for each value of x, the mean μ_y of the corresponding population of y-values lies on that straight line.

(e) The data come from a random sample or a randomized experiment.

T12.8 Inference about the slope β of a least-squares regression line is based on which of the following distributions?

(a) The t distribution with $n-1$ degrees of freedom

(b) The standard Normal distribution

(c) The chi-square distribution with $n-1$ degrees of freedom

(d) The t distribution with $n-2$ degrees of freedom

(e) The Normal distribution with mean μ and standard deviation σ

Exercises T12.9–T12.11 refer to the following setting. An old saying in golf is "You drive for show and you putt for dough." The point is that good putting is more important than long driving for shooting low scores and hence winning money. To see if this is the case, data from a random sample of 69 of the nearly 1000 players on the PGA Tour's world money list are examined. The average number of putts per hole (fewer is better) and the player's total winnings for the previous season are recorded and a least-squares regression line was fitted to the data. Assume the conditions for inference about the slope are met. Here is computer output from the regression analysis:

```
Predictor    Coef     SE Coef     T      P
Constant     7897179  3023782     6.86   0.000
Avg. putts   -4139198 1698371     ****   ****
S = 281777   R-Sq = 8.1%   R-Sq(adj) = 7.8%
```

T12.9 Suppose that the researchers test the hypotheses $H_0: \beta = 0$ versus $H_a: \beta < 0$. Which of the following is the value of the t statistic for this test?

(a) 2.61   (d) −2.44

(b) 2.44   (e) −20.24

(c) 0.081

T12.10 The P-value for the test in Exercise T12.9 is 0.0087. Which of the following is a correct interpretation of this result?

(a) The probability there is no linear relationship between average number of putts per hole and total winnings for these 69 players is 0.0087.

(b) The probability there is no linear relationship between average number of putts per hole and total winnings for all players on the PGA Tour's world money list is 0.0087.

(c) If there is no linear relationship between average number of putts per hole and total winnings for the players in the sample, the probability of getting a random sample of 69 players that yields a least-squares regression line with a slope of −4,139,198 or less is 0.0087.

(d) If there is no linear relationship between average number of putts per hole and total winnings for the players on the PGA Tour's world money list, the probability of getting a random sample of 69 players that yields a least-squares regression line with a slope of −4,139,198 or less is 0.0087.

(e) The probability of making a Type I error is 0.0087.

T12.11 Which of the following would make the P-value in Exercise T12.10 invalid?

(a) If the scatterplot of the sample data wasn't perfectly linear

(b) If the distribution of earnings has an outlier

(c) If the distribution of earnings wasn't approximately Normal

(d) If the earnings for golfers with small putting averages was much more variable than the earnings for golfers with large putting averages

(e) If the standard deviation of earnings is much larger than the standard deviation of putting average

T12.7 c

T12.8 d

T12.9 d

T12.10 d

T12.11 d

T12.12 (a) Random assignment was used to create three roughly equivalent groups at the beginning of the study. **(b)** H_0: The true proportion of spouse abusers like the ones in the study who will be arrested within 6 months is the same for all three police responses. H_a: The true proportion like these who will be arrested in 6 months is not the same . . . **(c)** If the true proportion of spouse abusers like the ones in the study who will be arrested within 6 months is the same for all three police responses, there is a 0.0796 probability of getting differences between the 3 groups as large as or larger than the ones observed by chance alone. **(d)** Because the P-value of 0.0796 is larger than $\alpha = 0.05$, we fail to reject H_0. There is not convincing evidence that the true proportion of spouse abusers like the ones in the study who will be arrested within 6 months is not the same for all three police responses.

T12.13 STATE: H_0: There is no association between smoking status and educational level among French men aged 20–60 years. H_a: There is an association . . . at $\alpha = 0.05$. PLAN: Chi-square test for independence. *Random:* Random sample of French men. *10%:* $n = 459 < 10\%$ of all French men aged 20–60 years. *Large Counts:* The expected counts are all ≥ 5. DO: The calculator's χ^2-Test gives $\chi^2 = 13.305$ and P-value $= 0.0384$ using df $= 6$. CONCLUDE: Because the P-value of $0.0384 < \alpha = 0.05$, we reject H_0. There is convincing evidence of an association between smoking status and educational level among French men aged 20–60 years.

T12.14 (a) (i) $b = 4.8323$; for each increase of 1 mg in growth hormone, the average weight gain increases by 4.8323 ounces. (ii) $a = 4.5459$; for a chicken given no growth hormone ($x = 0$), the predicted weight gain is 4.546 ounces, on average. (iii) $s = 3.135$; the actual weight gain will typically vary by about 3.135 ounces from the weight gain predicted with the least-squares regression line using $x =$ dose of growth hormone. (iv) $SE_b = 1.0164$; if we repeated the random assignment many times, the slope of the sample regression line would typically vary by about 1.0164 from the slope of the true regression line for predicting weight gain from dose of growth hormone. **(b)** STATE: H_0: $\beta = 0$, H_a: $\beta \neq 0$, where $\beta =$ slope of the true regression line relating $y =$ weight gain to $x =$ dose of growth hormone; we use $\alpha = 0.05$. PLAN: t test for the slope. The conditions are met. DO: $t = 4.75$ and

Section II: Free Response *Show all your work. Indicate clearly the methods you use, because you will be graded on the correctness of your methods as well as on the accuracy and completeness of your results and explanations.*

T12.12 A study conducted in Charlotte, North Carolina, tested the effectiveness of three police responses to spouse abuse: (1) advise and possibly separate the couple, (2) issue a citation to the offender, and (3) arrest the offender. Police officers were trained to recognize eligible cases. When presented with an eligible case, a police officer called the dispatcher, who would randomly assign one of the three available treatments to be administered. There were a total of 650 cases in the study. Each case was classified according to whether the abuser was arrested within 6 months of the original incident.[30]

		Police response			
		Advise and separate	Citation	Arrest	Total
Subsequent arrest?	No	187	181	175	543
	Yes	25	43	39	107
	Total	212	224	214	650

(a) Explain the purpose of the random assignment in the design of this study.

(b) State an appropriate pair of hypotheses for performing a chi-square test in this setting.

(c) Assume that the conditions for performing the test in part (b) are met. The test yields $\chi^2 = 5.063$ and a P-value of 0.0796. Interpret this P-value.

(d) What conclusion should we draw from the study?

T12.13 In the United States, there is a strong relationship between education and smoking: well-educated people are less likely to smoke. Does a similar relationship hold in France? To find out, researchers recorded the level of education and smoking status of a random sample of 459 French men aged 20 to 60 years.[51] The two-way table displays the data.

		Education			
		Primary school	Secondary school	University	Total
Smoking status	Nonsmoker	56	37	53	146
	Former	54	43	28	125
	Moderate	41	27	36	104
	Heavy	36	32	16	84
	Total	187	139	133	459

Is there convincing evidence of an association between smoking status and educational level among French men aged 20 to 60 years?

T12.14 Growth hormones are often used to increase the weight gain of chickens. In an experiment using 15 chickens, 3 chickens were randomly assigned to each of 5 different doses of growth hormone (0, 0.2, 0.4, 0.8, and 1.0 milligrams). The subsequent weight gain (in ounces) was recorded for each chicken. A researcher plots the data and finds that a linear relationship appears to hold. Here is computer output from a least-squares regression analysis of these data. Assume that the conditions for performing inference about the slope β of the true regression line are met.

Predictor	Coef	SE Coef	T	P
Constant	4.5459	0.6166	7.37	<0.0001
Dose	4.8323	1.0164	4.75	0.0004

$S = 3.135$ $R-Sq = 38.4\%$ $R-Sq(adj) = 37.7\%$

(a) Interpret each of the following in context:
   (i) The slope
   (ii) The y intercept
   (iii) The standard deviation of the residuals
   (iv) The standard error of the slope

(b) Do the data provide convincing evidence of a linear relationship between dose and weight gain? Carry out a significance test at the $\alpha = 0.05$ level.

(c) Construct and interpret a 95% confidence interval for the slope parameter.

P-value $= 0.0004$. CONCLUDE: Because the P-value of $0.0004 < \alpha = 0.05$, we reject H_0. There is convincing evidence of a linear relationship between the dose of growth hormone and weight gain for chickens like these. **(c)** STATE: $\beta =$ the slope of the true regression line relating $y =$ weight gain to $x =$ dose of growth hormone. PLAN: t interval for the slope. The conditions are met. DO: df $= 13$, (2.6369, 7.0277) CONCLUDE: We are 95% confident that the interval from 2.6369 to 7.0277 captures the slope of the true regression line relating $y =$ weight gain to $x =$ dose of growth hormone for chickens like these.

Final Project

In this project, your team will formulate a statistical question, design a study to answer the question, conduct the study, collect the data, analyze the data, and use statistical inference to answer the question. You may do your study on any topic, but you must be able to include the six steps listed above.

1. Write a proposal describing the design of your study. Make sure to include the following:

   - Describe the statistical question you are trying to answer.

   - List the explanatory and response variables (or just the response variable when appropriate).

   - State the null and alternative hypothesis you will be testing, along with the name of the test you will use to analyze the results.

   - Describe how you will collect the data, so the conditions for inference will be satisfied.

   - Explain how your study will be safe and ethical if you are using human subjects.

2. Once your teacher has approved your proposal, carry out the study.

3. Create a poster to summarize your project. Make sure to include the following:

   - *Title* (in the form of a question).

   - *Introduction.* The introduction should discuss what question you are trying to answer, why you chose this topic, what your hypotheses are, and how you will analyze your data.

   - *Data Collection.* In this section, you will describe how you obtained your data. Be specific.

   - *Graphs and Summary Statistics, Including Raw Data.* Begin by providing the raw data. If the data are quantitative, list them in a table. If the data are categorical, summarize them in a two-way table. Then make graphs that are well labeled and easy to compare, and list appropriate summary statistics. Use the graphs and summary statistics to describe the evidence for the alternative hypothesis.

   - *Analysis and Conclusion.* Identify the inference procedure you used and discuss the conditions for inference. Give the (standardized) test statistic and *P*-value (with interpretation), along with the appropriate conclusion. Then provide the corresponding confidence interval (with interpretation) or follow-up analysis (for chi-square tests).

   - *Reflections.* In this section, you should discuss any possible errors (e.g., Type I or Type II), limitations to your conclusion, what you could do to improve the study the next time, and any other critical reflections.

   - The key to a good statistical poster is communication and organization. Make sure all components of the poster are focused on answering the question of interest and that statistical vocabulary is used correctly. Include live action pictures of your data collection in progress.

Teaching Tip

If you use the final project, plan ahead to give students several days (or weeks) to complete it. Consider assigning the project at the beginning of the chapter and setting the due date for the completion of the chapter. This project can easily be completed using a TI-83/84 calculator or an online applet, such as the ones found on the Student Site.

TRM **Chapter 12 Final Project**

The supporting resources for this project can be found by clicking on the link in the TE-Book, opening the TRFD, or downloading from the Teacher's Resources on the book's digital platform.

TRM **Chapter 12 Case Study**

The Case Study feature that was found in the previous two editions of *The Practice of Statistics* student edition has been moved to the Teacher's Resource Materials. Click on the link in the TE-Book, open in the TRFD, or download from the Teacher's Resources on the book's digital platform.

✚ Ask the StatsMedic

How Do I Help Students Review for the AP® Statistics Exam?

Planning your review for the AP® Statistics exam can make a significant (!) difference in student scores. In this post, we give some ideas for high-leverage activities you can use to get students ready.

TRM **Full Solutions to Cumulative AP® Practice Test 4**

The full solutions can be found by clicking on the link in the TE-Book, opening the TRFD, or downloading from the Teacher's Resources on the book's digital platform.

AP4.1 e

AP4.2 c

AP4.3 d

AP4.4 a

AP4.5 b

AP4.6 e

Cumulative AP® Practice Test 4

Section I: Multiple Choice *Choose the best answer for Questions AP4.1–AP4.40.*

AP4.1 A major agricultural company is testing a new variety of wheat to determine whether it is more resistant to certain insects than the current wheat variety. The proportion of a current wheat crop lost to insects is 0.04. Thus, the company wishes to test the following hypotheses:

$$H_0: p = 0.04$$
$$H_a: p < 0.04$$

Which of the following significance levels and sample sizes would lead to the highest power for this test?

(a) $n = 200$ and $\alpha = 0.01$ (d) $n = 500$ and $\alpha = 0.01$
(b) $n = 400$ and $\alpha = 0.05$ (e) $n = 500$ and $\alpha = 0.05$
(c) $n = 400$ and $\alpha = 0.01$

AP4.2 If $P(A) = 0.24$ and $P(B) = 0.52$ and events A and B are independent, what is $P(A \text{ or } B)$?

(a) 0.1248
(b) 0.28
(c) 0.6352
(d) 0.76
(e) The answer cannot be determined from the given information.

AP4.3 Sam has determined that the weights of unpeeled bananas from his local store have a mean of 116 grams with a standard deviation of 9 grams. Assuming that the distribution of weight is approximately Normal, to the nearest gram, the heaviest 30% of these bananas weigh at least how much?

(a) 107 g   (d) 121 g
(b) 111 g   (e) 125 g
(c) 116 g

AP4.4 The school board in a certain school district obtained a random sample of 200 residents and asked if they were in favor of raising property taxes to fund the hiring of more statistics teachers. The resulting confidence interval for the true proportion of residents in favor of raising taxes was (0.183, 0.257). Which of the following is the margin of error for this confidence interval?

(a) 0.037   (d) 0.220
(b) 0.074   (e) 0.257
(c) 0.183

AP4.5 After a name-brand drug has been sold for several years, the Food and Drug Administration (FDA) will allow other companies to produce a generic equivalent. The FDA will permit the generic drug to be sold as long as there isn't convincing evidence that it is less effective than the name-brand drug. For a proposed generic drug intended to lower blood pressure, the following hypotheses will be used:

$$H_0: \mu_G = \mu_N \text{ versus } H_a: \mu_G < \mu_N$$

where

μ_G = true mean reduction in blood pressure using the generic drug

μ_N = true mean reduction in blood pressure using the name-brand drug

In the context of this situation, which of the following describes a Type I error?

(a) The FDA finds convincing evidence that the generic drug is less effective, when in reality it is less effective.
(b) The FDA finds convincing evidence that the generic drug is less effective, when in reality it is equally effective.
(c) The FDA finds convincing evidence that the generic drug is equally effective, when in reality it is less effective.
(d) The FDA fails to find convincing evidence that the generic drug is less effective, when in reality it is less effective.
(e) The FDA fails to find convincing evidence that the generic drug is less effective, when in reality it is equally effective.

AP4.6 The town council wants to estimate the proportion of all adults in their medium-sized town who favor a tax increase to support the local school system. Which of the following sampling plans is most appropriate for estimating this proportion?

(a) A random sample of 250 names from the local phone book
(b) A random sample of 200 parents whose children attend one of the local schools
(c) A sample consisting of 500 people from the city who take an online survey about the issue
(d) A random sample of 300 homeowners in the town
(e) A random sample of 100 people from an alphabetical list of all adults who live in the town

AP4.7 Which of the following is a categorical variable?

(a) The weight of an automobile
(b) The time required to complete the Olympic marathon
(c) The fuel efficiency (in miles per gallon) of a hybrid car

(d) The brand of shampoo purchased by shoppers in a grocery store

(e) The closing price of a particular stock on the New York Stock Exchange

AP4.8 A large machine is filled with thousands of small pieces of candy, 40% of which are orange. When money is deposited, the machine dispenses 60 randomly selected pieces of candy. Which of the following expressions best approximates the probability of getting fewer than 18 orange candies in a random sample of 60 candies, assuming that z follows a standard Normal distribution?

(a) $P\left(z < \dfrac{0.3 - 0.4}{\sqrt{\dfrac{(0.4)(0.6)}{60}}}\right)$

(b) $P\left(z < \dfrac{0.3 - 0.4}{\sqrt{\dfrac{(0.3)(0.7)}{60}}}\right)$

(c) $P\left(z < \dfrac{0.3 - 0.4}{\dfrac{\sqrt{(0.4)(0.6)}}{60}}\right)$

(d) $P\left(z < \dfrac{0.3 - 0.4}{\dfrac{(0.4)(0.6)}{\sqrt{60}}}\right)$

(e) $P\left(z < \dfrac{0.4 - 0.3}{\sqrt{\dfrac{(0.3)(0.7)}{60}}}\right)$

AP4.9 A random sample of 900 students at a very large university was asked which social networking site they used most often during a typical week. Their responses are shown in the table.

	Gender		
	Male	Female	Total
Facebook	221	283	504
Twitter	42	38	80
LinkedIn	108	87	195
Pinterest	23	26	49
Snapchat	29	43	72
Total	423	477	900

(Networking site labels the row group Facebook/Twitter/LinkedIn/Pinterest/Snapchat/Total)

If gender and preferred networking site were independent, how many females would you expect to choose LinkedIn?

(a) 87.00
(b) 90.00
(c) 95.40
(d) 97.50
(e) 103.35

AP4.10 Insurance adjusters are always concerned about being overcharged for accident repairs. The adjusters suspect that Repair Shop 1 quotes higher estimates than Repair Shop 2. To check their suspicion, the adjusters randomly select 12 cars that were recently involved in an accident and then take each of the cars to both repair shops to obtain separate estimates of the cost to fix the vehicle. The estimates are given in hundreds of dollars.

Car	1	2	3	4	5	6
Shop 1	21.2	25.2	39.0	11.3	15.0	18.1
Shop 2	21.3	24.1	36.8	11.5	13.7	17.6
Car	7	8	9	10	11	12
Shop 1	25.3	23.2	12.4	42.6	27.6	12.9
Shop 2	24.8	21.3	12.1	42.0	26.7	12.5

Assuming that the conditions for inference are met, which of the following significance tests should be used to determine whether the adjusters' suspicion is correct?

(a) A paired t test
(b) A two-sample t test
(c) A t test to see if the slope of the population regression line is 0
(d) A chi-square test for homogeneity
(e) A chi-square test for goodness of fit

AP4.11 A survey firm wants to ask a random sample of adults in Ohio if they support an increase in the state sales tax from 5.75% to 6%, with the additional revenue going to education. Let $\hat{p}$ denote the proportion in the sample who say that they support the increase. Suppose that 40% of all adults in Ohio support the increase. If the survey firm wants the standard deviation of the sampling distribution of $\hat{p}$ to equal 0.01, how large a sample size is needed?

(a) 1500
(b) 2400
(c) 2401
(d) 2500
(e) 9220

AP4.12 A set of 10 cards consists of 5 red cards and 5 black cards. The cards are shuffled thoroughly, and you choose one at random, observe its color, and replace it in the set. The cards are thoroughly reshuffled, and you again choose a card at random, observe its color, and replace it in the set. This is done a total of four times. Let X be the number of red cards observed in these four trials. The random variable X has which of the following probability distributions?

(a) The Normal distribution with mean 2 and standard deviation 1
(b) The binomial distribution with $n = 10$ and $p = 0.5$
(c) The binomial distribution with $n = 5$ and $p = 0.5$
(d) The binomial distribution with $n = 4$ and $p = 0.5$
(e) The geometric distribution with $p = 0.5$

AP4.7 d

AP4.8 a

AP4.9 e

AP4.10 a

AP4.11 b

AP4.12 d

AP4.13 e

AP4.14 d

AP4.15 b

AP4.16 a

AP4.13 Do students who are on an athletic team spend less time doing homework, on average, compared to students who are not on an athletic team? To investigate, the athletic director at a large high school randomly selected 32 students who were on an athletic team and randomly selected 48 students who were not on an athletic team. He asked them how many hours they typically spend on homework per week. Are the conditions for inference about a difference in means satisfied?

(a) Maybe; the data came from independent random samples, but we should examine the data to check for Normality.

(b) No, the sample sizes are not the same.

(c) No; a paired t test should be used in this case.

(d) Yes; the large sample sizes guarantee that the corresponding population distributions will be Normal.

(e) Yes, assuming that there are at least 320 students on athletic teams and at least 480 students not on athletic teams.

AP4.14 Do hummingbirds prefer store-bought food made from concentrate or a simple mixture of sugar and water? To find out, a researcher obtains 10 identical hummingbird feeders and fills 5, chosen at random, with store-bought food from concentrate and the other 5 with a mixture of sugar and water. The feeders are then randomly assigned to 10 possible hanging locations in the researcher's yard. Which inference procedure should you use to test whether hummingbirds show a preference for store-bought food based on amount consumed?

(a) A one-sample z test for a proportion

(b) A two-sample z test for a difference in proportions

(c) A chi-square test for independence

(d) A two-sample t test

(e) A paired t test

AP4.15 A Harris poll found that 54% of American adults don't think that human beings developed from earlier species. The poll's margin of error for 95% confidence was 3%. This means that

(a) there is a 95% chance the interval (51%, 57%) contains the true percent of American adults who do not think that human beings developed from earlier species.

(b) the poll used a method that provides an estimate within 3% of the truth about the population in 95% of samples.

(c) if Harris conducts another poll using the same method, the results of the second poll will lie between 51% and 57%.

(d) there is a 3% chance that the interval is incorrect.

(e) the poll used a method that would result in an interval that contains 54% in 95% of all possible samples of the same size from this population.

AP4.16 Two six-sided dice are rolled and the sum of the faces showing is recorded after each roll. Let X = the number of rolls required to obtain a sum greater than 7.

If 100 trials are conducted, which of the following is most likely to be the result of the simulation?

(a)

Number of rolls X	Number of trials
1	34
2	20
3	16
4	10
5	6
6	6
7	3
8	2
9	1
10	0
11	1
12	0
13	1

(b)

Number of rolls X	Number of trials
0	34
1	20
2	16
3	10
4	6
5	6
6	3
7	2
8	1
9	0
10	1
11	0
12	1

(c)

Number of rolls X	Number of trials
1	18
2	23
3	26
4	15
5	9
6	6
7	1
8	0
9	1
10	0
11	0
12	0
13	1

(d)

Number of rolls X	Number of trials
1	10
2	9
3	10
4	12
5	7
6	13
7	10
8	7
9	9
10	10
11	2
12	1

(e)

Number of rolls X	Number of trials
1	2
2	2
3	5
4	10
5	11
6	15
7	22
8	17
9	9
10	4
11	2
12	0
13	1

AP4.17 Women who are severely overweight suffer economic consequences, a study has shown. They have household incomes that are $6710 less than other women, on average. The findings are from an eight-year observational study of 10,039 randomly selected women who were 16 to 24 years old when the research began. If the difference in average incomes is statistically significant, does this study give convincing evidence that being severely overweight causes a woman to have a lower income?

(a) Yes; the study included both women who were severely overweight and women who were not.

(b) Yes; the subjects in the study were selected at random.

(c) Yes, because the difference in average incomes is larger than would be expected by chance alone.

(d) No; the study showed that there is no connection between income and being severely overweight.

(e) No; the study suggests an association between income and being severely overweight, but we can't draw a cause-and-effect conclusion.

Questions AP4.18 and AP4.19 refer to the following situation. Could mud wrestling be the cause of a rash contracted by University of Washington students? Two physicians at the university's student health center wondered about this when one male and six female students complained of rashes after participating in a mud-wrestling event. Questionnaires were sent to a random sample of students who participated in the event. The results, by gender, are summarized in the following table.

Gender

		Male	Female	Total
Developed rash?	Yes	12	12	24
	No	38	12	50
	Total	50	24	74

Here is some computer output for the preceding table. The output includes the observed counts, the expected counts, and the chi-square statistic.

```
Expected counts are printed below
observed counts
           MALE      FEMALE      Total
Yes         12         12         24
          16.22       7.78
No          38         12         50
          33.78      16.22
Total       50         24         74
ChiSq = 5.002
```

AP4.18 The cell that contributes most to the chi-square statistic is

(a) men who developed a rash.

(b) men who did not develop a rash.

(c) women who developed a rash.

(d) women who did not develop a rash.

(e) both (a) and (d).

AP4.19 From the chi-square test performed in this study, we may conclude that

(a) there is convincing evidence of an association between the gender of an individual participating in the event and development of a rash.

(b) mud wrestling causes a rash, especially for women.

(c) there is absolutely no evidence of any relationship between the gender of an individual participating in the event and the subsequent development of a rash.

(d) development of a rash is a real possibility if you participate in mud wrestling, especially if you do so regularly.

(e) the gender of the individual participating in the event and the development of a rash are independent.

AP4.20 Random assignment is part of a well-designed comparative experiment because

(a) it is more fair to the subjects.

(b) it helps create roughly equivalent groups before treatments are imposed on the subjects.

(c) it allows researchers to generalize the results of their experiment to a larger population.

(d) it helps eliminate any possibility of bias in the experiment.

(e) it prevents the placebo effect from occurring.

AP4.21 The following back-to-back stemplots compare the ages of players from two minor-league hockey teams ($1 \mid 7 = 17$ years).

Team A		Team B
98777	1	788889
44333221	2	00123444
7766555	2	556679
521	3	023
86	3	55

Which of the following *cannot* be justified from the plots?

(a) Team A has the same number of players in their 30s as does Team B.

(b) The median age of both teams is the same.

(c) Both age distributions are skewed to the right.

(d) The range of age is greater for Team A

(e) There are no outliers by the 1.5 *IQR* rule in either distribution.

AP4.17 e

AP4.18 c

AP4.19 a

AP4.20 b

AP4.21 e

AP4.22 c

AP4.23 b

AP4.24 c

AP4.25 d

AP4.26 e

AP4.22 A distribution that represents the number of cars X parked in a randomly selected residential driveway on any night is given by

x_i	0	1	2	3	4
p_i	0.10	0.20	0.35	0.20	0.15

Given that there is at least 1 car parked on a randomly selected residential driveway on a particular night, which of the following is closest to the probability that exactly 4 cars are parked on that driveway?

(a) 0.10 (b) 0.15 (c) 0.17

(d) 0.75 (e) 0.90

AP4.23 Which sampling method was used in each of the following settings, in order from I to IV?

   I. A student chooses to survey the first 20 students to arrive at school.

   II. The name of each student in a school is written on a card, the cards are well mixed, and 10 names are drawn.

   III. A state agency randomly selects 50 people from each of the state's senatorial districts.

   IV. A city council randomly selects eight city blocks and then surveys all the voting-age residents on those blocks.

(a) Voluntary response, SRS, stratified, cluster

(b) Convenience, SRS, stratified, cluster

(c) Convenience, cluster, SRS, stratified

(d) Convenience, SRS, cluster, stratified

(e) Cluster, SRS, stratified, convenience

AP4.24 Western lowland gorillas, whose main habitat is in central Africa, have a mean weight of 275 pounds with a standard deviation of 40 pounds. Capuchin monkeys, whose main habitat is Brazil and other parts of Latin America, have a mean weight of 6 pounds with a standard deviation of 1.1 pounds. Both distributions of weight are approximately Normally distributed. If a particular western lowland gorilla is known to weigh 345 pounds, approximately how much would a capuchin monkey have to weigh, in pounds, to have the same standardized weight as the gorilla?

(a) 4.08

(b) 7.27

(c) 7.93

(d) 8.20

(e) There is not enough information to determine the weight of a capuchin monkey.

AP4.25 Suppose that the mean weight of a certain breed of pig is 280 pounds with a standard deviation of 80 pounds. The distribution of weight for these pigs tends to be somewhat skewed to the right. A random sample of 100 pigs is taken. Which of the following statements about the sampling distribution of the sample mean weight $\bar{x}$ is true?

(a) It will be Normally distributed with a mean of 280 pounds and a standard deviation of 80 pounds.

(b) It will be Normally distributed with a mean of 280 pounds and a standard deviation of 8 pounds.

(c) It will be approximately Normally distributed with a mean of 280 pounds and a standard deviation of 80 pounds.

(d) It will be approximately Normally distributed with a mean of 280 pounds and a standard deviation of 8 pounds.

(e) There is not enough information to determine the mean and standard deviation of the sampling distribution.

AP4.26 Which of the following statements about the t distribution with degrees of freedom df is (are) true?

   I. It is symmetric.

   II. It has more variability than the t distribution with df + 1 degrees of freedom.

   III. As df increases, the t distribution approaches the standard Normal distribution.

(a) I only

(b) II only

(c) III only

(d) I and III

(e) I, II, and III

Questions AP4.27–AP4.29 refer to the following situation. Park rangers are interested in estimating the weight of the bears that inhabit their state. The rangers have data on weight (in pounds) and neck girth (distance around the neck in inches) for 10 randomly selected bears. Here is some regression output for these data:

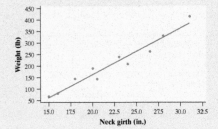

Predictor	Coef	SE Coef	T	P
Constant	−241.70	38.57	−6.27	0.000
Neck girth	20.230	1.695	11.93	0.000

S = 26.7565   R-Sq = 94.7%

AP4.27 Which of the following is the correct value of the correlation and its corresponding interpretation?

(a) The correlation is 0.947, and 94.7% of the variability in a bear's weight can be accounted for by the least-squares regression line using neck girth as the explanatory variable.

(b) The correlation is 0.947. The linear association between a bear's neck girth and its weight is strong and positive.

(c) The correlation is 0.973, and 97.3% of the variability in a bear's weight can be accounted for by the least-squares regression line using neck girth as the explanatory variable.

(d) The correlation is 0.973. The linear association between a bear's neck girth and its weight is strong and positive.

(e) The correlation cannot be calculated without the data.

AP4.28 Which of the following represents a 95% confidence interval for the slope of the population least-squares regression line relating the weight of a bear and its neck girth?

(a) 20.230 ± 1.695

(b) 20.230 ± 3.83

(c) 20.230 ± 3.91

(d) 20.230 ± 20.22

(e) 26.7565 ± 3.83

AP4.29 A bear was recently captured whose neck girth was 35 inches and whose weight was 466.35 pounds. If this bear were added to the data set, what would be the effect on the value of s?

(a) It would decrease the value of s because the added point is an outlier.

(b) It would decrease the value of s because the added point lies on the least-squares regression line.

(c) It would increase the value of s because the added point is an outlier.

(d) It would increase the value of s because the added point lies on the least-squares regression line.

(e) It would have no effect on the value of s because the added point lies on the least-squares regression line.

AP4.30 An experimenter wishes to test if one of two types of fish food (a standard fish food and a new product) is better for producing fish of equal weight after a two-month feeding program. The experimenter has two identical fish tanks (1 and 2) and is considering how to assign 40 fish, each of which has a numbered tag, to the tanks. The best way to do this would be to

(a) put all the odd-numbered fish in Tank 1 and the even-numbered fish in Tank 2. Give the standard food to Tank 1 and the new product to Tank 2.

(b) obtain pairs of fish whose weights are roughly equal at the start of the experiment and randomly assign one of the pair to Tank 1 and the other to Tank 2. Give the standard food to Tank 1 and the new product to Tank 2.

(c) proceed as in option (b), but put the heavier of each pair into Tank 2. Give the standard food to Tank 1 and the new product to Tank 2.

(d) assign the fish completely at random to the two tanks using a coin flip: heads means Tank 1 and tails means Tank 2. Give the standard food to Tank 1 and the new product to Tank 2.

(e) divide the 40 fish into two groups, with the 20 heaviest fish in one group. Randomly choose which tank to assign the heaviest fish and assign the lightest fish to the other tank. Give the standard food to Tank 1 and the new product to Tank 2.

AP4.31 A city wants to conduct a poll of taxpayers to determine the level of support for constructing a new city-owned baseball stadium. Which of the following is the main reason for using a large sample size in constructing a confidence interval to estimate the proportion of city taxpayers who would support such a project?

(a) To increase the confidence level

(b) To eliminate any confounding variables

(c) To reduce nonresponse bias

(d) To increase the precision of the estimate

(e) To reduce undercoverage

AP4.32 A standard deck of playing cards contains 52 cards, of which 4 are aces and 13 are hearts. You are offered a choice of the following two wagers:

I. Draw one card at random from the deck. You win $10 if the card drawn is an ace. Otherwise, you lose $1.

II. Draw one card at random from the deck. If the card drawn is a heart, you win $2. Otherwise, you lose $1.

Which of the two wagers should you prefer if you plan to play many times?

(a) Wager 1, because it has a greater expected value

(b) Wager 2, because it has a greater expected value

(c) Wager 1, because it has a greater probability of winning

(d) Wager 2, because it has a greater probability of winning

(e) Both wagers are equally favorable.

AP4.27 d

AP4.28 c

AP4.29 b

AP4.30 b

AP4.31 d

AP4.32 a

AP4.33 e

AP4.34 b

AP4.35 a

AP4.36 b

AP4.37 d

AP4.33 Here are boxplots of SAT Critical Reading and Math scores for a randomly selected group of female juniors at a highly competitive suburban school:

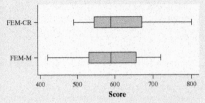

Which of the following *cannot* be justified by the plots?

(a) The maximum Critical Reading score is greater than the maximum Math score.

(b) Critical Reading scores are skewed to the right, whereas Math scores are somewhat skewed to the left.

(c) The median Critical Reading score and the median Math score for females are about the same.

(d) There appear to be no outliers in the distributions of SAT Critical Reading score.

(e) The mean Critical Reading score and the mean Math score for females are about the same.

AP4.34 A distribution of exam scores has mean 60 and standard deviation 18. If each score is doubled, and then 5 is subtracted from that result, what will the mean and standard deviation of the new scores be?

(a) mean = 115 and standard deviation = 31

(b) mean = 115 and standard deviation = 36

(c) mean = 120 and standard deviation = 6

(d) mean = 120 and standard deviation = 31

(e) mean = 120 and standard deviation = 36

AP4.35 In a clinical trial, 30 patients with a certain blood disease are randomly assigned to two groups. One group is then randomly assigned the currently marketed medicine, and the other group receives the experimental medicine. Every week, patients report to the clinic where blood tests are conducted. The clinic technician is unaware of the kind of medicine each patient is taking, and the patient is also unaware of which medicine he or she has been given. This design can be described as

(a) a double-blind, completely randomized experiment, with the currently marketed medicine and the experimental medicine as the two treatments.

(b) a single-blind, completely randomized experiment, with the currently marketed medicine and the experimental medicine as the two treatments.

(c) a double-blind, matched pairs design, with the currently marketed medicine and the experimental medicine forming a pair.

(d) a double-blind, block design that is not a matched pairs design, with the currently marketed medicine and the experimental medicine as the two blocks.

(e) a double-blind, randomized observational study.

AP4.36 A local investment club that meets monthly has 200 members ranging in age from 27 to 81. A cumulative relative frequency graph of age is shown. Approximately how many members of the club are more than 60 years of age?

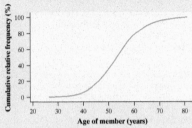

(a) 20   (b) 44   (c) 78

(d) 90   (e) 110

AP4.37 A manufacturer of electronic components is testing the durability of a newly designed integrated circuit to determine whether its life span is longer than that of the earlier model, which has a mean life span of 58 months. The company takes a simple random sample of 120 integrated circuits and simulates typical use until they stop working. The null and alternative hypotheses used for the significance test are $H_0: \mu = 58$ and $H_a: \mu > 58$. The P-value for the resulting one-sample t test is 0.035. Which of the following best describes what the P-value measures?

(a) The probability that the new integrated circuit has the same life span as the current model is 0.035.

(b) The probability that the test correctly rejects the null hypothesis in favor of the alternative hypothesis is 0.035.

(c) The probability that a single new integrated circuit will not last as long as one of the earlier circuits is 0.035.

(d) The probability of getting a sample mean greater than or equal to the observed sample mean if there really is no difference between the new and old circuits is 0.035.

(e) The probability of getting a sample mean greater than or equal to 58 if there really is no difference between the new and old circuits is 0.035.

Questions AP4.38 and AP4.39 refer to the following situation. Do children's fear levels change over time and, if so, in what ways? Little research has been done on the prevalence and persistence of fears in children. Several years ago, two researchers surveyed a randomly selected group of 94 third- and fourth-grade children, asking them to rate their level of fearfulness about a variety of situations. Two years later, the children again completed the same survey. The researchers computed the overall fear rating for each child in both years and were interested in the relationship between these ratings. They then assumed that the true regression line was

$$\mu_{\text{later rating}} = \alpha + \beta \,(\text{initial rating})$$

and that the conditions for regression inference were satisfied. This model was fitted to the data using least-squares regression. The following results were obtained from statistical software.

```
Predictor        Coefficient   St. Dev.
Constant           0.877917     0.1184
Initial rating     0.397911     0.0676
S = 0.2374         R-Sq = 0.274
```

Here is a scatterplot of the later ratings versus the initial ratings and a plot of the residuals versus the initial ratings:

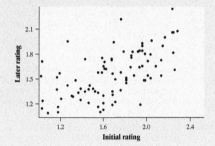

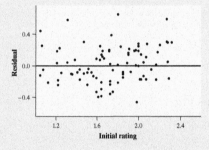

AP4.38 Which of the following statements is supported by these plots?

(a) The abundance of outliers and influential observations in the plots means that the conditions for regression are clearly violated.

(b) These plots contain dramatic evidence that the standard deviation of the response about the true regression line is not approximately the same for each x-value.

(c) These plots call into question the validity of the condition that the later ratings vary Normally about the least-squares line for each value of the initial ratings.

(d) A linear model isn't appropriate here because the residual plot shows no association.

(e) There is no striking evidence that the conditions for regression inference are violated.

AP4.39 George's initial fear rating was 0.2 higher than Jonny's. What does the model predict about their later fear ratings?

(a) George's will be about 0.96 higher than Jonny's.

(b) George's will be about 0.40 higher than Jonny's.

(c) George's will be about 0.20 higher than Jonny's

(d) George's will be about 0.08 higher than Jonny's.

(e) George's will be about the same as Jonny's.

AP4.40 The table provides data on the political affiliation and opinion about the death penalty of 850 randomly selected voters from a congressional district.

		Opinion about death penalty		
		Favor	Oppose	Total
Political party	Republican	299	98	397
	Democrat	77	171	248
	Other	118	87	205
	Total	494	356	850

Which of the following does *not* support the conclusion that being a Republican and favoring the death penalty are not independent?

(a) $\dfrac{299}{494} \neq \dfrac{98}{356}$

(b) $\dfrac{299}{494} \neq \dfrac{397}{850}$

(c) $\dfrac{494}{850} \neq \dfrac{299}{397}$

(d) $\dfrac{494}{850} \neq \dfrac{397}{850}$

(e) $\dfrac{(397)(494)}{850} \neq 299$

AP4.38 e

AP4.39 d

AP4.40 d

AP4.41 STATE: $H_0: \mu_1 - \mu_2 = 0$, $H_a: \mu_1 - \mu_2 \neq 0$, where $\mu_1 =$ true mean difference in electrical potential for diabetic mice and $\mu_2 =$ true mean difference for normal mice; $\alpha = 0.05$. **PLAN:** Two-sample t test for $\mu_1 - \mu_2$. *Random:* Independent random samples. *10%:* $n_1 = 24 < 10\%$ of all diabetic mice and $n_2 = 18 < 10\%$ of all normal mice. *Normal/Large Sample:* The graphs of the data reveal no outliers or strong skewness, so a two-sample t procedure is appropriate. **DO:** $t = 2.55$; using df $= 38.46$, P-value $= 0.0149$. **CONCLUDE:** Because the P-value of $0.0149 < \alpha = 0.05$, we reject H_0. There is convincing evidence that the true mean electric potential for diabetic mice differs from that for normal mice.

AP4.42 (a) $H_0: p_1 - p_2 = 0$, $H_a: p_1 - p_2 < 0$, where $p_1 =$ true proportion of women like the ones in the study who were physically active as teens and would suffer a cognitive decline and $p_2 =$ the true proportion not physically active as teens and would suffer a cognitive decline. **(b)** A two-sample z test for $p_1 - p_2$. **(c)** No; because participants were mostly white women from only four states, the findings may not be generalizable to women in other racial and ethnic groups or those who live in other states. **(d)** Two variables are confounded when their effects on the response variable (measure of cognitive decline) cannot be distinguished from one another. For example, women who were physically active as teens might have also done other things differently, such as eating a healthier diet. If healthier diets lead to less cognitive decline, we would be unable to determine if it was the women's physically active youth or their healthier diet that slowed their cognitive decline.

AP4.43 (a) Because the first question called it a "fat tax," people may have reacted negatively because they believe this is a tax on those who are overweight. The second question provides extra information that gets people thinking about the obesity problem in the U.S. and the increased health care that could be provided with the tax money, which might make them respond more positively to the proposed tax. The question should be worded more straightforwardly. **(b)** This method samples only people at fast-food restaurants who may go there because they like sugary drinks and wouldn't want to pay a tax on their favorite beverages. The proportion of those who would

---

Section II: Free Response *Show all your work. Indicate clearly the methods you use, because you will be graded on the correctness of your methods as well as on the accuracy and completeness of your results and explanations.*

AP4.41 The body's natural electrical field helps wounds heal. If diabetes changes this field, it might explain why people with diabetes heal more slowly. A study of this idea compared randomly selected normal mice and randomly selected mice bred to spontaneously develop diabetes. The investigators attached sensors to the right hip and front feet of the mice and measured the difference in electrical potential (in millivolts) between these locations. Graphs of the data for each group reveal no outliers or strong skewness. The following computer output provides numerical summaries of the data.[52]

Variable	N	Mean	StDev
Diabetic mice	24	13.090	4.839
Normal mice	18	10.022	2.915

Minimum	Q1	Median	Q3	Maximum
1.050	10.038	12.650	17.038	22.600
4.950	8.238	9.250	12.375	16.100

Is there convincing evidence at the $\alpha = 0.05$ level that the mean electrical potential differs for normal mice and mice with diabetes?

AP4.42 Can physical activity in youth lead to mental sharpness in old age? A 2010 study investigating this question involved 9344 randomly selected, mostly white women over age 65 from four U.S. states. These women were asked about their levels of physical activity during their teenage years, 30s, 50s, and later years. Those who reported being physically active as teens enjoyed the lowest level of cognitive decline—only 8.5% had cognitive impairment—compared with 16.7% of women who reported not being physically active at that time.

(a) State an appropriate pair of hypotheses that the researchers could use to test whether the proportion of women who suffered a cognitive decline was significantly smaller for women who were physically active in their youth than for women who were not physically active at that time. Be sure to define any parameters you use.

(b) Assuming the conditions for performing inference are met, what inference method would you use to test the hypotheses you identified in part (a)? Do *not* carry out the test.

(c) Suppose the test in part (b) shows that the proportion of women who suffered a cognitive decline was significantly smaller for women who were physically active in their youth than for women who were not physically active at

that time. Can we generalize the results of this study to all women aged 65 and older? Justify your answer.

(d) We cannot conclude that being physically active as a teen *causes* a lower level of cognitive decline for women over 65, due to possible confounding with other variables. Explain the concept of confounding and give an example of a potential confounding variable in this study.

AP4.43 In a recent poll, randomly selected New York State residents at various fast-food restaurants were asked if they supported or opposed a "fat tax" on sugared soda. Thirty-one percent said that they were in favor of such a tax and 66% were opposed. But when asked if they would support such a tax if the money raised were used to fund health care given the high incidence of obesity in the United States, 48% said that they were in favor and 49% were opposed.

(a) In this situation, explain how bias may have been introduced based on the way the questions were worded *and* suggest a way that the questions could have been worded differently in order to avoid this bias.

(b) In this situation, explain how bias may have been introduced based on the way the sample was taken *and* suggest a way that the sample could have been obtained in order to avoid this bias.

(c) This poll was conducted only in New York State. Suppose the pollsters wanted to ensure that estimates for the proportion of people who would support a tax on sugared soda were available for each state as well as an overall estimate for the nation as a whole. Identify a sampling method that would achieve this goal *and* briefly describe how the sample would be taken.

AP4.44 Each morning, coffee is brewed in the school workroom by one of three faculty members, depending on who arrives first at work. Mr. Worcester arrives first 10% of the time, Dr. Currier arrives first 50% of the time, and Mr. Legacy arrives first on the remaining mornings. The probability that the coffee is strong when brewed by Dr. Currier is 0.1, while the corresponding probabilities when it is brewed by Mr. Legacy and Mr. Worcester are 0.2 and 0.3, respectively. Mr. Worcester likes strong coffee!

(a) What is the probability that on a randomly selected morning the coffee will be strong?

(b) If the coffee is strong on a randomly selected morning, what is the probability that it was brewed by Dr. Currier?

---

oppose such a tax is likely to be overestimated with this method. A random sample of all New York State residents should be taken to provide a better estimate of the level of support for such a tax. **(c)** Use a stratified random sampling method in which each state is a stratum. Select a random sample from each state to obtain results for each state and combine the random samples to obtain an overall estimate for the nation as a whole.

AP4.44 Let $W =$ Mr. Worcester arrives first, $L =$ Mr. Legacy arrives first, $C =$ Dr. Currier arrives first, and $S =$ the coffee is strong. The tree diagram below organizes the given information.

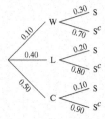

(a) $P(S) = 0.16$; there is a 0.16 probability that the coffee will be strong on a randomly selected morning.

(b) $P(C \mid S) = \dfrac{0.05}{0.16} = 0.3125$; given that the coffee is strong, there is a 0.3125 probability that it was brewed by Dr. Currier.

AP4.45 The following table gives data on the mean number of seeds produced in a year by several common tree species and the mean weight (in milligrams) of the seeds produced. Two species appear twice because their seeds were counted in two locations. We might expect that trees with heavy seeds produce fewer of them, but what mathematical model best describes the relationship?[53]

Tree species	Seed count	Seed weight (mg)
Paper birch	27,239	0.6
Yellow birch	12,158	1.6
White spruce	7202	2.0
Engelmann spruce	3671	3.3
Red spruce	5051	3.4
Tulip tree	13,509	9.1
Ponderosa pine	2667	37.7
White fir	5196	40.0
Sugar maple	1751	48.0
Sugar pine	1159	216.0
American beech	463	247.0
American beech	1892	247.0
Black oak	93	1851.0
Scarlet oak	525	1930.0
Red oak	411	2475.0
Red oak	253	2475.0
Pignut hickory	40	3423.0
White oak	184	3669.0
Chestnut oak	107	4535.0

(a) Describe the association between seed count and seed weight shown in the scatterplot.

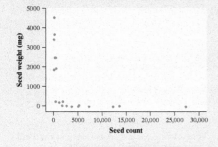

(b) Two alternative models based on transforming the original data are proposed to predict the seed weight from the seed count. Here are graphs and computer output from a least-squares regression analysis of the transformed data.

Model A:

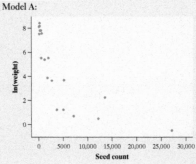

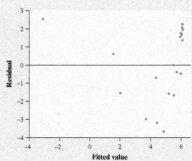

```
Predictor      Coef        SE Coef        T      P
Constant     6.1394       0.5726      10.72  0.000
Seed count  -0.00033869  0.00007187   -4.71  0.000
S = 2.08100    R-Sq = 56.6%     R-Sq(adj) = 54.1%
```

AP4.45 (a) The scatterplot reveals a strong, negative curved relationship between seed count and seed weight.
(b) Model B is better because the scatterplot shows a much more linear pattern and its residual plot shows no leftover curved pattern. The scatterplot for Model A still has a curved pattern and the residual plot has a leftover U-shaped pattern.
(c) $\overline{\ln(\text{weight})} = 15.491 - 1.5222 \ln(3700) = 2.984$,
so $\sum \text{weight} = e^{2.984} = 19.77$ mg.

AP4.46 (a) Let $X =$ diameter of a randomly selected lid. *Shape:* Normal distribution. *Center:* $\mu_{\bar{x}} = \mu = 4$ inches. *Variability:* Because 25 is less than 10% of all lids produced that hour, $\sigma_{\bar{x}} = 0.004$ inch. **(b)** We want to find $P(\bar{x} < 3.99$ or $\bar{x} > 4.01)$.

(i) $z = \dfrac{3.99 - 4}{0.004} = -2.50$ and

$z = \dfrac{4.01 - 4}{0.004} = 2.50$; the proportion of z-scores less than -2.50 or greater than 2.50 is $0.0062 + 0.0062 = 0.0124$.

(ii) $1 - $normalcdf(lower: 3.99, upper: 4.01, mean: 4, SD: 0.004) $= 0.0124$; assuming the machine is working properly, there is a 0.0124 probability that the mean diameter of a sample of 25 lids is less than 3.99 inches or greater than 4.01 inches.

(c) We want to find $P(4 < \bar{x} < 4.01)$.

(i) $z = \dfrac{4 - 4}{0.004} = 0$ and $z = \dfrac{4.01 - 4}{0.004} = 2.50$; the proportion of z-scores between 0 and 2.50 is $0.9938 - 0.5000 = 0.4938$.

(ii) normalcdf(lower: 4, upper: 4.01, mean: 4, SD: 0.004) $= 0.4938$; assuming the machine is working properly, there is a 0.4938 probability that the mean diameter of a sample of 25 lids is between 4.00 and 4.01 inches. **(d)** Let $Y =$ the number of samples (out of 5) in which the sample mean is between 4.00 and 4.01. The random variable Y has a binomial distribution with $n = 5$ and $p = 0.4938$. *Tech:* $P(X \geq 4) = 1 - $binomcdf(trials: 5, p: 0.4938, x-value: 3) $= 0.1798$; assuming the manufacturing process is working correctly, there is a 0.1798 probability that in 5 consecutive samples, 4 or 5 of the sample means will be above the desired mean of 4.00 but below the upper boundary of 4.01. **(e)** Because the probability found in part (b) is less than the probability found in part (d), getting a sample mean below 3.99 or above 4.01 is more convincing evidence that the machine should be shut down. This event is much less likely to happen by chance when the machine is working correctly.

Model B:

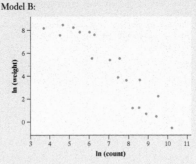

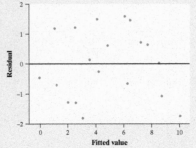

Predictor	Coef	SE Coef	T	P
Constant	15.491	1.081	14.33	0.000
ln(count)	-1.5222	0.1470	-10.35	0.000
S=1.16932		R-Sq=86.3%	R-Sq(adj)=85.5%	

Which model, A or B, is more appropriate for predicting seed weight from seed count? Justify your answer.

(c) Using the model you chose in part (b), predict the seed weight if the seed count is 3700.

AP4.46 A company manufactures plastic lids for disposable coffee cups. When the manufacturing process is working correctly, the diameters of the lids are approximately Normally distributed with a mean diameter of 4 inches and a standard deviation of 0.02 inch. To make sure the machine is not producing lids that are too big or too small, each hour a random sample of 25 lids is selected and the sample mean $\bar{x}$ is calculated.

(a) Describe the shape, center, and variability of the sampling distribution of the sample mean diameter, assuming the machine is working properly.

The company decides that it will shut down the machine if the sample mean diameter is less than 3.99 inches or greater than 4.01 inches, because this indicates that some lids will be too small or too large for the cups. If the sample mean is less than 3.99 or greater than 4.01, all the lids manufactured that hour are thrown away because the company does not want to sell bad products.

(b) Assuming that the machine is working properly, what is the probability that a random sample of 25 lids will have a mean diameter less than 3.99 inches or greater than 4.01 inches?

Also, to identify any trends, each hour the company records the value of the sample mean on a chart, like the one given here.

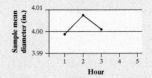

One benefit of using this type of chart is that out-of-control production trends can be noticed before it is too late and lids have to be thrown away. For example, if the sample mean is consistently greater than 4 (but less than 4.01), this would suggest that something might be wrong with the machine. If such a trend is noticed before the sample mean gets larger than 4.01, then the machine can be fixed without having to throw away any lids.

(c) Assuming that the manufacturing process is working correctly, what is the probability that the sample mean diameter will be above the desired mean of 4.00 but below the upper boundary of 4.01?

(d) Assuming that the manufacturing process is working correctly, what is the probability that in 5 consecutive samples, 4 or 5 of the sample means will be above the desired mean of 4.00 but below the upper boundary of 4.01?

(e) Which of the following results gives more convincing evidence that the machine needs to be shut down? Explain your answer.

1. Getting a single sample mean below 3.99 or above 4.01

or

2. Taking 5 consecutive samples and having at least 4 of the sample means be between 4.00 and 4.01

About the AP® Exam and AP® Exam Tips

The AP® Statistics exam consists of two distinct sections: Multiple Choice and Free Response. Specific details about the composition and scoring of the exam are provided below.

AP® Statistics Exam Composition

Section I: Multiple Choice　　　90 minutes　　　50% of exam score

40 multiple choice questions, each with 5 answer choices

The composition of the Multiple Choice section is based on both content and skills. Here are the numbers of questions for each unit and skill category.

Unit	Number of questions
Unit 1: Exploring One-Variable Data	6–9
Unit 2: Exploring Two-Variable Data	2–3
Unit 3: Collecting Data	5–6
Unit 4: Probability, Random Variables, and Probability Distributions	4–8
Unit 5: Sampling Distributions	3–5
Unit 6: Inference for Categorical Data: Proportions	5–6
Unit 7: Inference for Quantitative Data: Means	4–7
Unit 8: Inference for Categorical Data: Chi-Square	1–2
Unit 9: Inference for Quantitative Data: Slopes	1–2

Skill Category	Number of questions
Skill 1: Selecting Statistical Methods	6–9
Skill 2: Data Analysis	6–9
Skill 3: Using Probability and Simulation	12–16
Skill 4: Statistical Argumentation	10–14

Section II: Free Response　　　90 minutes　　　50% of exam score

Part A: Questions 1–5	65 minutes	37.5% of exam score
Part B: Question 6 (Investigative task)	25 minutes	12.5% of exam score

The composition of the Free Response section is based on skills. The first 5 free-response questions (Part A) include:

• One multi-part question focusing primarily on Collecting Data (Skill Category 1)

• One multi-part question focusing primarily on Exploring Data (Skill Category 2)

• One multi-part question focusing primarily on Probability and Sampling Distributions (Skill Category 3)

• One question focusing primarily on Inference (Skill Categories 1, 3, 4)

• One question focusing on two or more skill categories

The sixth free-response question (Part B) is the Investigative Task, which assesses multiple skill categories and content areas, focusing on applying the skills and content in new contexts or non-routine ways.

Formulas, Tables, and Calculator Use

Formulas and tables (like the ones near the back of the book) are provided on both sections of the exam. The formulas are at the beginning of the test booklets and the tables are at the end. You may use your calculator throughout the exam.

AP® Statistics Exam Scoring

Section I: Multiple Choice The score is based only on the number of questions answered correctly. So don't leave any questions unanswered!

Weighted Section I Score = Number of correct answers × 1.25

Section II: Free Response Each free-response question is scored holistically on a 0 to 4 scale. The score categories represent different levels of quality in a student's response across two dimensions: statistical knowledge and communication.

Weighted Section II Score = (Sum of scores on Questions 1 – 5) × 1.875 + Question 6 score × 3.125

Composite Score = Weighted Section I Score + Weighted Section II Score

Composite scores (on a 100-point scale) are converted to AP® Scores (on a 1 to 5 scale) using cutoffs determined each year based on statistical analysis of overall student performance on the exam. In a recent year, it took 33 points to earn a 2, 44 points to earn a 3, 57 points to earn a 4, and 70 points to earn a 5.

AP® EXAM TIPS

Chapter 1

- If you learn to distinguish categorical from quantitative variables now, it will pay big rewards later. You will be expected to analyze categorical and quantitative variables correctly on the AP® Statistics exam.

- When comparing groups of different sizes, be sure to use relative frequencies (percents or proportions) instead of frequencies (counts) when analyzing categorical data. Comparing only the frequencies can be misleading, as in this setting (page 18). There are many more people who never use snowmobiles among the non-environmental club members in the sample (445) than among the environmental club members (212). However, the *percentage* of environmental club members who never use snowmobiles is much higher (69.5% to 36.4%). Finally, make sure to avoid statements like "More club members never use snowmobiles" when you mean "A greater percentage of club members never use snowmobiles."

- Always be sure to include context when you are asked to describe a distribution. This means using the variable name, not just the units the variable is measured in.

- When comparing distributions of quantitative data, it's not enough just to list values for the center and variability of each distribution. You have to explicitly *compare* these values, using words like "greater than," "less than," or "about the same as."

- If you're asked to make a graph on a free response question, be sure to label and scale your axes. Unless your calculator shows labels and scaling, don't just transfer a calculator screenshot to your paper.

- The formula sheet provided with the AP® Statistics exam also gives the sample standard deviation in the equivalent form

$$s_x = \sqrt{\frac{1}{n-1}\sum(x_i - \bar{x})^2}.$$

- You may be asked to determine whether a quantitative data set has any outliers. Be prepared to state and use the rule for identifying outliers.

- Use statistical terms carefully and correctly on the AP® Statistics exam. Don't say "mean" if you really mean "median." Range is a single number; so are Q_1, Q_3, and IQR. Avoid poor use of

language, like "the outlier *skews* the mean" or "the median is in the middle of the IQR." Skewed is a shape and the IQR is a single number, not a region. If you misuse a term, expect to lose some credit.

Chapter 2

- Students often do not get full credit on the AP® Statistics exam because they only use option (ii) with "calculator-speak" to show their work on Normal calculation questions—for example, normal cdf(–1000,6,6.84,1.55). This is *not* considered clear communication. To get full credit, follow the two-step process (page 122), making sure to carefully label each of the inputs in the calculator command if you use technology in Step 2: normalcdf(lower: – 1000, upper: 6, mean: 6.84, SD: 1.55).

- As noted previously, to make sure that you get full credit on the AP® Statistics exam, do not use "calculator-speak" alone—for example, invNorm(0.90,6.84,1.55). This is *not* considered clear communication. To get full credit, follow the two-step process (page 129), making sure to carefully label each of the inputs in the calculator command if you use technology in Step 2: invNorm (area: 0.90, mean: 6.84, SD:1.55).

- Never say that a distribution of quantitative data *is* Normal. Real-world data always show at least slight departures from a Normal distribution. The most you can say is that the distribution is "approximately Normal."

- Normal probability plots are not included on the AP® Statistics course framework. However, these graphs are very useful for assessing Normality. You may use them on the AP® exam if you wish—just be sure that you know what you're looking for.

Chapter 3

- When you are asked to *describe* the association shown in a scatterplot, you are expected to discuss the direction, form, and strength of the association, along with any unusual features, *in the context of the problem*. This means that you need to use both variable names in your description.

- If you are asked to make a scatterplot, be sure to label and scale both axes. *Don't* just copy an unlabeled calculator graph directly onto your paper.

- When asked to interpret the slope or *y* intercept, it is very important to include the word *predicted* (or equivalent) in your response. Otherwise, it might appear that you believe the regression equation provides actual values of *y*.

- When displaying the equation of a least-squares regression line, the calculator will report the slope and intercept with much more precision than we need. There is no firm rule for how many decimal places to show for answers on the AP® Statistics exam. Our advice: decide how much to round based on the context of the problem you are working on.

Chapter 4

- If you're asked to describe how the design of a sample survey leads to bias, you're expected to do two things: (1) describe how the members of the sample might respond differently from the rest of the population, and (2) explain how this difference would lead to an underestimate or overestimate. Suppose you were asked to explain how using your statistics class as a sample to estimate the proportion of all high school students who own a graphing calculator could result in bias. You might respond, "This is a convenience sample. It would probably include a much higher proportion of students with a graphing calculator than in the population at large because a graphing calculator is required for the statistics class. So this method would probably lead to an overestimate of the actual population proportion."

- If you are asked to identify a possible confounding variable in a given setting, you are expected to explain how the variable you choose (1) is associated with the explanatory variable and (2) is associated with the response variable.

- If you are asked to describe a completely randomized design, stay away from flipping coins. For example, suppose we ask each student in the caffeine experiment (page 279) to toss a coin. If it's heads, then the student will drink the cola with caffeine. If it's tails, then the student will drink the caffeine-free cola. As long as all 20 students toss a coin, this is still a completely randomized design. Of course, the two groups are unlikely to contain exactly 10 students because it is unlikely that 20 coin tosses will result in a perfect 50-50 split between heads and tails.

  The problem arises if we try to force the two groups to have equal sizes. Suppose we continue to have students toss coins until one of the groups has 10 students and then place the remaining students in the other group. In this case, the last two students in line are very likely to end up in the same group. However, in a completely randomized design, the last two subjects should only have a 50% chance of ending up in the same group.

- Don't mix the language of experiments and the language of sample surveys or other observational studies. You will lose credit for saying things like "use a randomized block design to select the sample for this survey" or "this experiment suffers from nonresponse because some subjects dropped out during the study."

Chapter 5

- On the AP® Statistics exam, you may be asked to describe how to perform a simulation using rows of random digits. If so, provide a clear enough description of your process for the reader to get the same results from *only* your written explanation. Remember that every label needs to be the same length. In the golden ticket lottery example (page 334), the labels should be 01 to 95 (all two digits),

not 1 to 95. When sampling without replacement, be sure to mention that repeated numbers should be ignored.

- Many probability problems involve simple computations that you can do on your calculator. It may be tempting to just write down your final answer without showing the supporting work. Don't do it! A "naked answer," even if it's correct, will usually be penalized on a free response question.

- You can write statements like $P(A|B)$ if events A and B are clearly defined in a problem. Otherwise, it's probably easier to use contextual labels, like $P(I|F)$ in the preceding example (page 361). Or you can just use words: $P(\text{Instagram}|\text{Facebook})$.

Chapter 6

- If the mean of a random variable has a non-integer value but you report it as an integer, your answer will not get full credit.

- If you are asked to calculate the mean or standard deviation of a discrete random variable on a free response question, you must show numerical values substituted into the appropriate formula, as in the previous two examples (pages 395 and 397). Feel free to use ellipses (...) if there are many terms in the summation, as we did. You may then use the method described in Technology Corner 13 to perform the calculation with 1-Var Stats. Writing only 1-Var Stats L1, L2 and then giving the correct values of the mean and standard deviation will *not* earn credit for showing work. Also, be sure to avoid incorrect notation when labeling these parameters.

- Students often do not get full credit on the AP® Statistics exam because they only use option (ii) with "calculator-speak" to show their work on Normal calculation questions—for example, normal cdf(68,70,64,2.7). This is not considered clear communication. To get full credit, follow the two-step process (page 401), making sure to carefully label each of the inputs in the calculator command if you use technology in Step 2: normalcdf(lower:68, upper:70, mean:64, SD: 2.7).

- Don't rely on "calculator speak" when showing your work on free response questions. Writing binompdf (5, 0.25, 3) = 0.08789 will *not* earn you full credit for a binomial probability calculation. At the very least, you must indicate what each of those calculator inputs represents. For example, "binompdf(trials:5,p:0.25,x value:3) = 0.08789."

Chapter 7

- Many students lose credit on the AP® Statistics exam when defining parameters because their description refers to the sample instead of the population or because the description isn't clear about which group of individuals the parameter is describing. When defining a parameter, we suggest including the word *all* or the word *true* in your description to make it clear that you aren't referring to a sample statistic.

- Terminology matters. Never just say "the distribution." Always say "the distribution of [blank]," being careful to distinguish the distribution of the population, the distribution of sample data, and the sampling distribution of a statistic. Likewise, don't use ambiguous terms like "sample distribution," which could refer to the distribution of sample data or to the sampling distribution of a statistic. You will lose credit on free response questions for misusing statistical terms.

- Make sure to understand the difference between accuracy and precision when writing responses on the AP® Statistics exam. Many

students use "accurate" when they really mean "precise." For example, a response that says "increasing the sample size will make an estimate more accurate" is incorrect. It should say that increasing the sample size will make an estimate more precise. If you can't remember which term to use, don't use either of them. Instead, explain what you mean without using statistical vocabulary.

• Notation matters. The symbols $\hat{p}$, $\bar{x}$, n, p, μ, σ, $\mu_{\hat{p}}$, $\sigma_{\hat{p}}$, $\mu_{\bar{x}}$, and $\sigma_{\bar{x}}$ all have specific and different meanings. Either use notation correctly—or don't use it at all. You can expect to lose credit if you use incorrect notation.

• Many students lose credit on probability calculations involving $\bar{x}$ because they forget to divide the population standard deviation by $\sqrt{n}$. Remember that averages are less variable than individual observations!

Chapter 8

• When interpreting a confidence interval, make sure that you are describing the parameter and not the statistic. It's wrong to say that we are 95% confident the interval from 0.613 to 0.687 captures the proportion of U.S. adults who *admitted* they would experience financial difficulty. The "proportion who *admitted* they would experience financial difficulty" is the sample proportion, which is known to be 0.65. The interval gives plausible values for the proportion who *would admit* to experiencing some financial difficulty if asked.

• On a given problem, you may be asked to interpret the confidence interval, the confidence level, or both. Be sure you understand the difference: the confidence interval gives a set of plausible values for the parameter and the confidence level describes the overall capture rate of the method.

• If a free response question asks you to construct and interpret a confidence interval, you are expected to do the entire four-step process. That includes clearly defining the parameter, identifying the procedure, and checking conditions.

• You may use your calculator to compute a confidence interval on the AP® Statistics exam. But there's a risk involved. If you just give the calculator answer with no work, you'll get either full credit for the "Do" step (if the interval is correct) or no credit (if it's wrong). If you opt for the calculator-only method, be sure to complete the other three steps, including identifying the procedure (e.g., one-sample z interval for p) and give the interval in the Do step (e.g., 0.19997 to 0.26073).

• Many students lose credit when defining parameters in an experiment by describing the sample proportion rather than the true proportion. For example, "the true proportion of the men who *had* surgery and survived 20 years" describes $\hat{p}_S$, not p_S.

• The formula for the two-sample z interval for $p_1 - p_2$ often leads to calculation errors by students. As a result, your teacher may recommend using the calculator's 2-PropZInt feature to compute the confidence interval on the AP® Statistics exam. Be sure to name the procedure (two-sample z interval for $p_1 - p_2$) in the "Plan" step and give the interval $(-0.311, 0.116)$ in the "Do" step.

Chapter 9

• Hypotheses always refer to a population, not a sample. Be sure to state H_0 and H_a in terms of population parameters. It is *never* correct to write a hypothesis about a sample statistic, such as $H_0: \hat{p} = 0.80$ or $H_a: \bar{x} \neq 31$.

• We recommend that you follow the two-sentence structure from the example (page 591) when writing the conclusion to a significance test. The first sentence should give a decision about the null hypothesis—reject H_0 or fail to reject H_0—based on an explicit comparison of the P-value to a stated significance level. The second sentence should provide a statement about whether or not there is convincing evidence for H_a in the context of the problem.

• Notice that we did not include an option (ii) to "Use technology to find the desired area without standardizing" when performing the Normal calculation in part (b) of the example. That's because you are always required to give the standardized test statistic and the P-value when performing a significance test on the AP® Statistics exam.

• When a significance test leads to a fail to reject H_0 decision, as in the preceding example (page 606), be sure to interpret the results as "We don't have convincing evidence for H_a." Saying anything that sounds like you believe H_0 is (or might be) true will lead to a loss of credit. For instance, it would be *wrong* to conclude, "There is convincing evidence that the true proportion of blemished potatoes is 0.08." And don't write responses as text messages, like "FTR the H_0."

• You can use your calculator to carry out the mechanics of a significance test on the AP® Statistics exam. But there's a risk involved. If you give just the calculator answer with no work, and one or more of your values are incorrect, you will probably get no credit for the "Do" step. If you opt for the calculator-only method, be sure to name the procedure (one-sample z test for a proportion) and to report the test statistic ($z = 1.15$) and P-value (0.1243).

• When making a conclusion in a significance test, be sure that you are describing the parameter and not the statistic. In the preceding example (page 609), it's wrong to say that we have convincing evidence that the proportion of students at Yanhong's school who *said* they have never smoked differs from the CDC's claim of 0.68. The "proportion who *said* they have never smoked" is the sample proportion, which is known to be 0.60. The test gives convincing evidence that the proportion of all students at Yanhong's school who *would say* they have never smoked a cigarette differs from 0.68.

• The formula for the two-sample z statistic for a test about $p_1 - p_2$ often leads to calculation errors by students. As a result, your teacher may recommend using the calculator's 2-PropZTest feature to perform calculations on the AP® Statistics exam. Be sure to name the procedure (two-sample z interval for $p_1 - p_2$) in the "Plan" step and report the standardized test statistic ($z = 1.17$) and P-value (0.2427) in the "Do" step.

Chapter 10

• If a question on the AP® Statistics exam asks you to construct and interpret a confidence interval, all the conditions should be met. However, you are still required to state the conditions and show evidence that they are met—including a graph if the sample size is small and the data are provided.

• It is not enough just to make a graph of the data on your calculator when assessing Normality. You must *sketch* the graph on your paper to receive credit. You don't have to draw multiple graphs—any appropriate graph will do.

• The formula for the two-sample t interval for $\mu_1 - \mu_2$ often leads to calculation errors by students. Also, the interval produced by technology is narrower than the one calculated using the conservative method. As a result, your teacher may recommend using the calculator's 2-SampTInt feature to compute the confidence

interval. Be sure to name the procedure (two-sample t interval for $\mu_1 - \mu_2$) in the "Plan" step and give the interval (3.9362, 17.724) and df (55.728) in the "Do" step.

Chapter 11

• It is not enough just to make a graph of the data on your calculator when assessing Normality. You must *sketch* the graph on your paper and make an appropriate comment about it to receive credit. You don't have to draw more than one graph—a single appropriate graph will do.

• Remember: If you give just calculator results with no work, and one or more values are wrong, you probably won't get any credit for the "Do" step. If you opt for the calculator-only method, name the procedure (one-sample t test for μ) and report the test statistic ($t = -0.94$), degrees of freedom (df = 14), and P-value (0.1809).

• The formula for the two-sample t statistic for a test about $\mu_1 - \mu_2$ often leads to calculation errors by students. Also, the P-value from technology is smaller and more accurate than the one obtained using the conservative method. As a result, your teacher may recommend using the calculator's 2-SampTTest feature to perform calculations. Be sure to name the procedure (two-sample t test for $\mu_1 - \mu_2$) in the "Plan" step and to report the standardized test statistic ($t = 1.60$), P-value (0.0644), and df (15.59) in the "Do" step.

Chapter 12

• The formula for the chi-square test statistic is included on the formula sheet that is provided on the AP® Statistics exam. However, it doesn't include the word *count*:

$$\chi^2 = \sum \frac{(\text{Observed} - \text{Expected})^2}{\text{Expected}}$$

We included the word *count* to emphasize that you must use the observed and expected counts—not the observed and expected proportions—when calculating the chi-square test statistic.

• When checking the Large Counts condition, be sure to examine the *expected* counts, not the observed counts. Make sure to write and label the expected counts on your paper or you won't receive credit. And never round them to the nearest integer!

• You can use your calculator to carry out the mechanics of a significance test on the AP® Statistics exam. But there's a risk involved. If you just give the calculator answer with no work, and one or more of your values is incorrect, you will likely get no credit for

the "Do" step. We recommend writing out the first few terms of the chi-square calculation followed by "...". This approach might help you earn partial credit if you enter a number incorrectly. Be sure to name the procedure (chi-square test for goodness of fit) and to report the test statistic ($\chi^2 = 11.2$), degrees of freedom (df = 3), and P-value (0.011).

• As with chi-square tests for goodness of fit, the expected counts should *not* be rounded to the nearest whole number. While an observed count of entrées ordered must be a whole number, an expected count need not be a whole number. The expected count gives the average number of entrées ordered if H_0 is true and the random assignment process is repeated many times.

• In the "Do" step, you aren't required to show every term in the chi-square test statistic. Writing the first few terms of the sum followed by "..." is considered as "showing work." We suggest that you do this and then let your calculator tackle the computations.

• You can use your calculator to carry out the mechanics of a significance test on the AP® Statistics exam. But there's a risk involved. If you just give the calculator answer without showing work, and one or more of your entries is incorrect, you will likely get no credit for the "Do" step. We recommend writing out the first few terms of the chi-square calculation followed by "...". This approach may help you earn partial credit if you enter a number incorrectly. Be sure to name the procedure (χ^2 test for homogeneity) and to report the test statistic ($\chi^2 = 18.279$), degrees of freedom (df = 4), and P-value (0.0011).

• Many students lose credit on the AP® Statistics exam because they don't write down and label the expected counts in their response. It isn't enough to claim that all the expected counts are at least 5. You must provide clear evidence.

• When the P-value is very small, the calculator will report it using scientific notation. Remember that P-values are probabilities and must be between 0 and 1. If your calculator reports the P-value with a number that appears to be greater than 1, look to the right, and you will see that the P-value is being expressed in scientific notation. If you claim that the P-value is 4.82, you will certainly lose credit.

• When you see a list of data values on an exam question, wait a moment before typing the data into your calculator. Read the question through first. Often, information is provided that makes it unnecessary for you to enter the data at all. This can save you valuable time on the AP® Statistics exam.

Formulas for the AP® Statistics Exam

Students are provided with the following formulas on both the multiple choice and free-response sections of the AP® Statistics exam.

I. Descriptive Statistics

$$\bar{x} = \frac{1}{n}\sum x_i = \frac{\sum x_i}{n}$$

$$s_x = \sqrt{\frac{1}{n-1}\sum(x_i - \bar{x})^2} = \sqrt{\frac{\sum(x_i - \bar{x})^2}{n-1}}$$

$$\hat{y} = a + bx$$

$$\bar{y} = a + b\bar{x}$$

$$r = \frac{1}{n-1}\sum\left(\frac{x_i - \bar{x}}{s_x}\right)\left(\frac{y_i - \bar{y}}{s_y}\right)$$

$$b = r\frac{s_y}{s_x}$$

II. Probability and Distributions

$$P(A \cup B) = P(A) + P(B) - P(A \cap B)$$

$$P(A|B) = \frac{P(A \cap B)}{P(B)}$$

Probability Distribution	Mean	Standard Deviation
Discrete random variable, X	$\mu_X = E(X) = \sum x_i \cdot P(x_i)$	$\sigma_X = \sqrt{\sum(x_i - \mu_X)^2 \cdot P(x_i)}$
If X has a **binomial** distribution with parameters n and p, then: $$P(X = x) = \binom{n}{x}p^x(1-p)^{n-x}$$ where $x = 0, 1, 2, 3, \ldots, n$	$\mu_X = np$	$\sigma_X = \sqrt{np(1-p)}$
If X has a **geometric** distribution with parameter p, then: $$P(X = x) = (1-p)^{x-1}p$$ where $x = 1, 2, 3, \ldots$	$\mu_X = \frac{1}{p}$	$\sigma_X = \frac{\sqrt{1-p}}{p}$

III. Sampling Distributions and Inferential Statistics

Standardized test statistic: $\dfrac{\text{statistic} - \text{parameter}}{\text{standard error of statistic}}$

Confidence interval: statistic $\pm$ (critical value) (standard error of statistic)

Chi-square statistic: $\chi^2 = \sum\dfrac{(\text{observed} - \text{expected})^2}{\text{expected}}$

III. Sampling Distributions and Inferential Statistics *(continued)*

Sampling Distributions for Proportions

Random Variable	Parameters of Sampling Distribution		Standard Error* of Sample Statistic
For one population $\hat{p}$	$\mu_{\hat{p}} = p$	$\sigma_{\hat{p}} = \sqrt{\dfrac{p(1-p)}{n}}$	$s_{\hat{p}} = \sqrt{\dfrac{\hat{p}(1-\hat{p})}{n}}$
For two populations: $\hat{p}_1 - \hat{p}_2$	$\mu_{\hat{p}_1 - \hat{p}_2} = p_1 - p_2$	$\sigma_{\hat{p}_1 - \hat{p}_2} = \sqrt{\dfrac{p_1(1-p_1)}{n_1} + \dfrac{p_2(1-p_2)}{n_2}}$	$s_{\hat{p}_1 - \hat{p}_2} = \sqrt{\dfrac{\hat{p}_1(1-\hat{p}_1)}{n_1} + \dfrac{\hat{p}_2(1-\hat{p}_2)}{n_2}}$ When $p_1 = p_2$ is assumed: $s_{\hat{p}_1 - \hat{p}_2} = \sqrt{\hat{p}_c(1-\hat{p}_c)\left(\dfrac{1}{n_1} + \dfrac{1}{n_2}\right)}$ where $\hat{p}_c = \dfrac{X_1 + X_2}{n_1 + n_2}$

Sampling Distributions for Means

Random Variable	Parameters of Sampling Distribution		Standard Error* of Sample Statistic
For one population $\overline{x}$	$\mu_{\overline{x}} = \mu$	$\sigma_{\overline{x}} = \dfrac{\sigma}{\sqrt{n}}$	$s_{\overline{x}} = \dfrac{s}{\sqrt{n}}$
For two populations: $\overline{x}_1 - \overline{x}_2$	$\mu_{\overline{x}_1 - \overline{x}_2} = \mu_1 - \mu_2$	$\sigma_{\overline{x}_1 - \overline{x}_2} = \sqrt{\dfrac{\sigma_1^2}{n_1} + \dfrac{\sigma_2^2}{n_2}}$	$s_{\overline{x}_1 - \overline{x}_2} = \sqrt{\dfrac{s_1^2}{n_1} + \dfrac{s_2^2}{n_2}}$

Sampling Distributions for Simple Linear Regression

Random Variable	Parameters of Sampling Distribution		Standard Error* of Sample Statistic
For slope: b	$\mu_b = \beta$	$\sigma_b = \dfrac{\sigma}{\sigma_x \sqrt{n}}$ where $\sigma_x = \sqrt{\dfrac{\sum(x_i - \mu)^2}{n}}$	$s_b = \dfrac{s}{s_x \sqrt{n-1}}$ where $s = \sqrt{\dfrac{\sum(y_i - \hat{y}_i)^2}{n-2}}$ and $s_x = \sqrt{\dfrac{\sum(x_i - \overline{x})^2}{n-1}}$

*Standard deviation is a measurement of variability from the theoretical population. Standard error is the estimate of the standard deviation. If the standard deviation of the statistic is assumed to be known, then the standard deviation should be used instead of the standard error.

Notes and Data Sources

Overview: What Is Statistics?

1. Patrizia Frei et al., "Use of mobile phones and risk of brain tumours: Update of Danish cohort study," *BMJ*, 343 (2011), d6387 (Published 20 October 2011).
2. Nikhil Swaminathan, "Gender jabber: Do women talk more than men?" *Scientific American*, July 6, 2007.
3. Stephen Moss, "Do women really talk more?" *The Guardian*, November 26, 2006.
4. Data from gapminder.org, accessed June 15, 2017.
5. Data from the Josephson Institute's "2012 Report Card of American Youth," accessed online at charactercounts.org, March 16, 2013.

Chapter 1

1. Roller coaster data for 2015 from www.rcdb.com.
2. *Radio Today 2013*, Executive Summary, downloaded from www.arbitron.com.
3. Data obtained using the Random Sampler tool on the American Statistical Association's Census at School website: www.amstat.org/censusatschool/index.cfm.
4. We got the idea for this example from David Lane's case study "Who is buying iMacs?" which we found online at onlinestatbook .com/case_studies_rvls/.
5. April 10, 2002, "Effect of hypericum perforatum (St John's Wort) in major depressive disorder," *Journal of the American Medical Association (JAMA)*, 287(14), 2002, pp. 1807–1814.
6. *Nutrition Action*, "Weighing the options: Do extra pounds mean extra years?" (March 2013).
7. Centers for Disease Control and Prevention, *Births: Final Data for 2005*, National Vital Statistics Reports, 56, No. 6, 2007, at www.cdc.gov.
8. Data from the article "New York's elevators define the city," by Oliver Roeder, May 4, 2016 at fivethirtyeight.com.
9. *Global Automotive 2014 Color Popularity Report* by Axalta Coating Systems, LLC.
10. Found at spam-filter-review.toptenreviews.com, which claims to have compiled data "from a number of different reputable sources."
11. *The Hispanic Population: 2010*, at www.census.gov, based on the data collected from the 2010 U.S. Census.
12. Data for 2010 from the 2012 *Statistical Abstract of the United States* at the Census Bureau website, www.census.gov.
13. This exercise is based on information from www .minerandcostudio.com.
14. Elizabeth F. Loftus and John C. Palmer, "Reconstruction of automobile destruction: An example of the interaction between language and memory," *Journal of Verbal Learning and Verbal Behavior*, 13 (1974), pp. 585–589, www.researchgate .net/publication/222307973_Reconstruction_of_Automobile _Destruction_An_Example_of_the_Interaction_Between _Language_and_Memory.

15. The U.K. data were obtained using the Random Sampler tool at www.censusatschool.com. The U.S. data were obtained using the Random Sampler tool at www.amstat.org/censusatschool.
16. Data from the University of Florida Biostatistics Open Learning Textbook at bolt.mph.ufl.edu/.
17. R. Shine et al., "The influence of nest temperatures and maternal brooding on hatchling phenotypes in water pythons," *Ecology*, 78 (1997), pp. 1713–1721.
18. K. Eagan, J. B. Lozano, S. Hurtado, and M. H. Case. *The American Freshman: National Norms, Fall 2013*. (2013) Los Angeles: Higher Education Research Institute, UCLA.
19. Data from Axalta Coating Systems, "Global Automotive 2015 Color Popularity Report."
20. Pew Research Center American Trends Panel Survey, 2014.
21. *Statistical Abstract of the United States*, 2004–2005, Table 1233.
22. Siem Oppe and Frank De Charro, "The effect of medical care by a helicopter trauma team on the probability of survival and the quality of life of hospitalized victims," *Accident Analysis and Prevention*, 33 (2001), pp. 129–138. The authors give the data in this example as a "theoretical example" to illustrate the need for more elaborate analysis of actual data using severity scores for each victim.
23. Data obtained from taxfoundation.org/article/state-and -local-sales-tax-rates-2016.
24. James T. Fleming, "The measurement of children's perception of difficulty in reading materials," *Research in the Teaching of English*, 1 (1967), pp. 136–156.
25. Data for 2010 from the 2012 *Statistical Abstract of the United States* at the Census Bureau website, www.census.gov.
26. The original paper is T. M. Amabile, "Motivation and creativity: Effects of motivational orientation on creative writers," *Journal of Personality and Social Psychology*, 48, No. 2 (February 1985), pp. 393–399. The data for Exercise 57 came from Fred L. Ramsey and Daniel W. Schafer, *The Statistical Sleuth*, 3rd ed., Brooks/ Cole Cengage Learning, 2013.
27. The cereal data came from the Data and Story Library, lib.stat.cmu.edu/DASL/.
28. USDA National Nutrient Database for Standard Reference 26 Software v.1.4.
29. From the Electronic Encyclopedia of Statistics Examples and Exercises (EESEE) story, "Acorn Size and Oak Tree Range."
30. CO_2 emissions data from the World Bank's website: data.worldbank.org/indicator/EN.ATM.CO2E.PC.
31. From the American Community Survey, at factfinder2 .census.gov.
32. Maribeth Cassidy Schmitt, "The effects of an elaborated directed reading activity on the metacomprehension skills of third graders," PhD thesis, Purdue University, 1987.
33. Monthly stock returns from the website of Professor Kenneth French of Dartmouth, mba.tuck.dartmouth.edu/pages/faculty /ken.french. A fine point: the data are the "excess returns" on stocks, the actual returns less the small monthly returns on Treasury bills.

34. The cereal data came from the Data and Story Library, www.stat.cmu.edu/StatDat/.

35. Advanced Placement exam results from AP® Central website, apcentral.collegeboard.com.

36. H. Lindberg, H. Roos, and P. Gardsell, "Prevalence of coxarthritis in former soccer players," *Acta Orthopedica Scandinavica*, 64 (1993), pp. 165–167.

37. Data on PVC pipe length from a related example in Minitab statistical software.

38. Data from pirate.shu.edu/~wachsmut/Teaching/MATH1101/Descriptives/variability.html.

39. Tablet ratings from www.consumerreports.org/cro/electronics-computers/computers-internet/tablets/tablet-ratings/ratings-overview/selector.htm.

40. Data from www.realtor.org.

41. Tom Lloyd et al., "Fruit consumption, fitness, and cardio-vascular health in female adolescents: The Penn State Young Women's Health Study," *American Journal of Clinical Nutrition*, 67 (1998), pp. 624–630.

42. C. B. Williams, *Style and Vocabulary: Numerological Studies*, Griffin, 1970.

43. Data from the most recent Annual Demographic Supplement can be found at www.census.gov/cps.

44. The idea for this exercise came from the Plotly website: help.plot.ly.

45. 2007 CIRP Freshman Survey; data from the Higher Education Research Institute's report "The American freshman: National norms for fall 2007," published January 2008.

46. The report "Cell phones key to teens' social lives, 47% can text with eyes closed" is published online by Harris Interactive, July 2008, www.marketingcharts.com.

47. A. Karpinski and A. Duberstein, "A description of Facebook use and academic performance among undergraduate and graduate students," paper presented at the American Educational Research Association annual meeting, April 2009. Thanks to Aryn Karpinski for providing us with some original data from the study.

48. S. M. Stigler, "Do robust estimators work with real data?" *Annals of Statistics*, 5 (1977), pp. 1055–1078.

49. T. Bjerkedal, "Acquisition of resistance in guinea pigs infected with different doses of virulent tubercle bacilli," *American Journal of Hygiene*, 72 (1960), pp. 130–148.

50. Data from the Bureau of Labor Statistics, Annual Demographic Supplement, www.bls.census.gov.

51. Data from the report "Is our tuna family-safe?" prepared by Defenders of Wildlife, 2006.

Chapter 2

1. Data obtained from Rex Boggs.

2. www.census.gov/topics/income-poverty/income.html

3. Information on bone density in the reference populations was found at www.courses.washington.edu/bonephys/opbmd.html.

4. Stephen Jay Gould, "Entropic homogeneity isn't why no one hits .400 anymore," *Discover*, August 1986, pp. 60–66. Gould does not standardize but gives a speculative discussion instead.

5. Data from Gary Community School Corporation, courtesy of Celeste Foster, Department of Education, Purdue University.

6. Detailed data appear in P. S. Levy et al., *Total Serum Cholesterol Values for Youths 12–17 Years*, Vital and Health Statistics, Series 11, No. 155, National Center for Health Statistics, 1976.

7. The cereal data came from the Data and Story Library, www.stat.cmu.edu/StatDat/.

8. See Chapter 1, Note 24.

9. See Chapter 1, Note 49.

10. Kevin Quealy, Amanda Cox, and Josh Katz, "At Chipotle, how many calories do people really eat?" *New York Times*, February 17, 2015.

11. *Model Year 2016 Fuel Economy Guide*, from the website www.fueleconomy.gov.

12. Pew Hispanic Center tabulations of 2011 American Community Survey.

13. Data provided by Chris Olsen, who found the information in *Scuba News* and *Skin Diver* magazines.

14. The data set was constructed based on information provided in P. D. Wood et al., "Plasma lipoprotein distributions in male and female runners," in P. Milvey (ed.), *The Marathon: Physiological, Medical, Epidemiological, and Psychological Studies*, New York Academy of Sciences, 1977.

15. Data found online at www.earthtrends.wri.org.

16. See Chapter 1, Note 48.

17. We found the information on birth weights of Norwegian children on the National Institute of Environmental Health Sciences website. The relevant article can be accessed here: www.ncbi.nlm.nih.gov/pubmed/1536353.

18. Thomas K. Cureton et al., *Endurance of Young Men*, Monographs of the Society for Research in Child Development, Vol. 10, No. 1, 1945.

Chapter 3

1. Thanks to Paul Myers for sharing this idea.

2. Data from mlb.com and spoctrac.com. Thanks to Jeff Eicher for sharing on the AP® Statistics Teacher's Community.

3. From the random sampler at ww2.amstat.org/CensusAtSchool/.

4. www.gapminder.org

5. *Nutrition Action*, December 2009.

6. www.tylervigen.com/

7. Data from Aaron Waggoner.

8. www.nejm.org/doi/full/10.1056/NEJMon1211064

9. www.stat.columbia.edu/~gelman/research/published/golf.pdf

10. Data from www.teamusa.org/road-to-rio-2016/team-usa/athletes. Thanks to Jeff Eicher for sharing on the AP® Statistics Teacher's Community.

11. www.starbucks.com/promo/nutrition

12. Based on T. N. Lam, "Estimating fuel consumption from engine size," *Journal of Transportation Engineering*, 111 (1985), pp. 339–357. The data for 10 to 50 km/h are measured; those for 60 and higher are calculated from a model given in the paper and are therefore smoothed.

13. Here are the data on boat registrations: www.flhsmv.gov/dmv/vslfacts.html. And here on manatee deaths: myfwc.com/research/manatee/rescue-mortality-response/mortality-statistics/yearly/.

14. www.ncei.noaa.gov/

15. www.basketball-reference.com

16. Samuel Karelitz et al., "Relation of crying activity in early infancy to speech and intellectual development at age three years," *Child Development*, 35 (1964), pp. 769–777.

17. *Consumer Reports*, June 1986, pp. 366–367.

18. G. A. Sacher and E. F. Staffelt, "Relation of gestation time to brain weight for placental mammals: Implications for the theory of vertebrate growth," *American Naturalist*, 108 (1974), pp. 593–613. We found the data in Fred L. Ramsey and Daniel W. Schafer, *The Statistical Sleuth: A Course in Methods of Data Analysis*, Duxbury, 1997, p. 228.

19. M. A. Houck et al., "Allometric scaling in the earliest fossil bird, Archaeopteryx lithographica," *Science*, 247 (1990), pp. 195–198. The authors conclude from a variety of evidence that all specimens represent the same species.

20. Table 1 of E. Thomassot et al., "Methane-related diamond crystallization in the earth's mantle: Stable isotopes evidence from a single diamond-bearing xenolith," *Earth and Planetary Science Letters*, 257 (2007), pp. 362–371.

21. From a graph in L. Partridge and M. Farquhar, "Sexual activity reduces lifespan of male fruit flies," *Nature*, 294 (1981), pp. 580–582. Provided by Brigitte Baldi.

22. Data on used car prices from autotrader.com, September 8, 2012. We searched for F-150 44's on sale within 50 miles of College Station, Texas.

23. www.lumeradiamonds.com/diamonds/results?price =1082-1038045&carat=0.30-16.03&shapes=B%20&cut =EX&clarity=FL,IF&color=D,E,F#

24. Frank J. Anscombe, "Graphs in statistical analysis," *American Statistician*, 27 (1973), pp. 17–21.

25. N. R. Draper and J. A. John, "Influential observations and outliers in regression," *Technometrics*, 23 (1981), pp. 21–26.

26. P. Goldblatt (ed.), Longitudinal study: Mortality and social organization, Her Majesty's stationery office, 1990. At least, so claims Richard Conniff, in *The Natural History of the Rich*, Norton, 2002, p. 45. We have not been able to access the Goldblatt report.

27. Data from George W. Pierce, *The Songs of Insects*, Cambridge, MA, Harvard University Press, 1949, pp. 12–21.

28. i.nbcolympics.com/figure-skating/resultsandschedules /event=FSW010000/index.html

29. Data from Brittany Foley and Allie Dutson, Canyon del Oro High School.

30. Data from Kerry Lane and Danielle Neal, Canyon del Oro High School.

31. From a graph in G. D. Martinsen, E. M. Driebe, and T. G. Whitham, "Indirect interactions mediated by changing plant chemistry: Beaver browsing benefits beetles," *Ecology*, 79 (1998), pp. 192–200.

32. Gary Smith, "Do statistics test scores regress toward the mean?" *Chance*, 10, No. 4 (1997), pp. 42–45.

33. www.baseball-reference.com

34. Data from Haley Vaughn, Nate Trona, and Jeff Green, Central York High School.

35. *Consumer Reports*, November 2005.

36. Debora L. Arsenau, "Comparison of diet management instruction for patients with non–insulin dependent diabetes mellitus: Learning activity package vs. group instruction," MS thesis, Purdue University, 1993.

37. www.pro-football-reference.com/teams/jax/2011.htm

38. A. K. Yousafzai et al., "Comparison of armspan, arm length and tibia length as predictors of actual height of disabled and nondisabled children in Dharavi, Mumbai, India," *European Journal of Clinical Nutrition*, 57 (2003), pp. 1230–1234. In fact, $r^2 = 0.93$.

39. David M. Fergusson and L. John Horwood, "Cannabis use and traffic accidents in a birth cohort of young adults," *Accident Analysis and Prevention*, 33 (2001), pp. 703–711.

40. Information about the sources used to obtain the data can be found under "Documentation" at the Gapminder website, www .gapminder.org.

41. Gordon L. Swartzman and Stephen P. Kaluzny, *Ecological Simulation Primer*, Macmillan, 1987, p. 98.

42. en.wikipedia.org/wiki/Transistor_count

43. From Discount Tire advertisement (Tandy Engineering and Associates), November 28, 2015.

44. Sample Pennsylvania female rates provided by Life Quotes, Inc., in *USA Today*, December 20, 2004.

45. G. A. Sacher and E. F. Staffelt, "Relation of gestation time to brain weight for placental mammals: Implications for the theory of vertebrate growth," *American Naturalist*, 108 (1974), pp. 593–613. We found these data in Fred L. Ramsey and Daniel W. Schafer, *The Statistical Sleuth: A Course in Methods of Data Analysis*, Duxbury, 1997.

46. Jérôme Chave, Bernard Riéra, and Marc-A. Dubois, "Estimation of biomass in a neotropical forest of French Guiana: Spatial and temporal variability," *Journal of Tropical Ecology*, 17 (2001), pp. 79–96.

47. S. Chatterjee and B. Price, Regression Analysis by Example, Wiley, 1977.

48. www.stat.columbia.edu/~gelman/research/published/golf.pdf

49. Chris Carbone and John L. Gittleman, "A common rule for the scaling of carnivore density," *Science*, 295 (2002), pp. 2273–2276.

50. Data originally from A. J. Clark, *Comparative Physiology of the Heart*, Macmillan, 1927, p. 84. Obtained from Frank R. Giordano and Maurice D. Weir, *A First Course in Mathematical Modeling*, Brooks/Cole, 1985, p. 56.

51. www.advancedwebranking.com/cloud/ctrstudy/

52. Thanks to Jeff Eicher for collecting and cleaning these data from www.espn.com.

53. The World Almanac and Book of Facts (2009).

54. G. L. Kooyman et al., "Diving behavior and energetics during foraging cycles in king penguins," *Ecological Monographs*, 62 (1992), pp. 143–163.

55. We found the data on cherry blossoms in the paper "Linear equations and data analysis," which was posted on the North Carolina School of Science and Mathematics website, www .ncssm.edu.

56. www.lumeradiamonds.com/diamonds/results?price =1082-1038045&carat=0.30-16.03&shapes=B%20&cut =EX&clarity=FL,IF&color=D,E,F#

57. From a graph in Craig Packer et al., "Ecological change, group territoriality, and population dynamics in Serengeti lions," *Science*, 307 (2005), pp. 390–393.

58. www.baseball-reference.com

Chapter 4

1. www.nhtsa.gov/press-releases/seat-belt-use-us-reaches -historic-90-percent

2. www.commonsensemedia.org/sites/default/files/uploads /research/census_researchreport.pdf

3. Sheldon Cohen, William J. Doyle, Cuneyt M. Alper, Denise Janicki-Deverts, and Ronald B. Turner, "Sleep habits and susceptibility to the common cold," *Archives of Internal Medicine*, 169, No. 1 (2009), pp. 62–67.

4. Rebecca Smith and Alison Wessner, The Lawrenceville School, "Does listening to music while studying affect performance on learning assessments?" May 2009.

5. Frederick Mosteller and David L. Wallace, Inference and Disputed Authorship: The Federalist. Addison-Wesley, Reading, MA, 1964 & according to the website en.wikipedia.org/wiki/Federalist_papers.

6. Excerpt obtained from www.constitution.org/fed/federa51.htm.

7. *Arizona Daily Star*, 4-18-16.

8. fivethirtyeight.com/features/isthepollingindustry instasisorincrisis/

9. For information on the American Community Survey of households (there is a separate sample of group quarters), go to www.census.gov/acs.

10. Quinnipiac University Poll, November 5, 2015.

11. www.cleaninginstitute.org/assets/1/AssetManager/2010%20 Hand%20Washing%20Findings.pdf

12. Cynthia Crossen, "Margin of error: Studies galore support products and positions, but are they reliable?" *Wall Street Journal*, November 14, 1991.

13. Robert C. Parker and Patrick A. Glass, "Preliminary results of double-sample forest inventory of pine and mixed stands with high- and low-density LiDAR," in Kristina F. Connoe (ed.), Proceedings of the 12th Biennial Southern Silvicultural Research Conference, U.S. Department of Agriculture, Forest Service, Southern Research Station, 2004. The researchers actually sampled every tenth plot. This is a systematic sample.

14. Gary S. Foster and Craig M. Eckert, "Up from the grave: A socio-historical reconstruction of an African American community from cemetery data in the rural Midwest," *Journal of Black Studies*, 33 (2003), pp. 468–489.

15. Bryan E. Porter and Thomas D. Berry, "A nationwide survey of self-reported red light running: Measuring prevalence, predictors, and perceived consequences," *Accident Analysis and Prevention*, 33 (2001), pp. 735–741.

16. Mario A. Parada et al., "The validity of self-reported seatbelt use: Hispanic and non-Hispanic drivers in El Paso," *Accident Analysis and Prevention*, 33 (2001), pp. 139–143.

17. Data from Marcos Chavez-Martinez, Canyon del Oro High School.

18. fivethirtyeight.com/features/donttakeyourvitamins/

19. *Arizona Daily Star*, 3-6-12 "Study links Vitamin D, stronger bones in girls."

20. www.nbcwashington.com/news/health/ADHD_Linked_To _Lead_and_Mom_s_Smoking.html.

21. *J. Clin. Endocrinol. Metab.* 2016. doi:10.1210/jc.2015-4013. Cited in Nutrition Action, April 2016.

22. www.sciencedirect.com/science/article/pii /S0747563214001563

23. tucson.com/news/science/healthmedfit /dietersinstudyshedmorepoundsifcashison /article_4bf8a176052d57f1bc2d2a5972713bae.html

24. "Family dinner linked to better grades for teens: Survey finds regular meal time yields additional benefits," written by John Mackenzie for ABC News's *World News Tonight*, September 13, 2005.

25. www.plosmedicine.org/article/info%3Adoi%2F10 .1371%2Fjournal.pmed.1001595

26. aem.asm.org/content/early/2016/08/15/AEM.01838-16 .abstract?sid=61679ac7-4522-4b65-9284-04d7bfbbcdab

27. The placebo effect examples are from Sandra Blakeslee, "Placebos prove so powerful even experts are surprised," *New York Times*, October 13, 1998.

28. The "three-quarters" estimate is cited by Martin Enserink, "Can the placebo be the cure?" *Science*, 284 (1999), pp. 238–240. An extended treatment is Anne Harrington (ed.), The Placebo Effect: An Interdisciplinary Exploration, Harvard University Press, 1997.

29. Carlos Vallbona et al., "Response of pain to static magnetic fields in postpolio patients, a double blind pilot study," *Archives of Physical Medicine and Rehabilitation*, 78 (1997), pp. 1200–1203.

30. *Arizona Daily Star*, 9-16-2012, "New analysis casts doubt on omega-3 heart health benefits."

31. Steering Committee of the Physicians' Health Study Research Group, "Final report on the aspirin component of the ongoing Physicians' Health Study," *New England Journal of Medicine*, 321 (1989), pp. 129–135.

32. The flu trial quotation is from Kristin L. Nichol et al., "Effectiveness of live, attenuated intranasal influenza virus vaccine in healthy, working adults," *Journal of the American Medical Association*, 282 (1999), pp. 137–144.

33. Nutrition Action, April 2016.

34. Nutrition Action, July/August 2008.

35. *Time*, December 9, 2010, www.tinyurl.com/2dt8hjj.

36. archinte.jamanetwork.com/article.aspx?articleid=1899554

37. National Institute of Child Health and Human Development, Study of Early Child Care and Youth Development. The article appears in the July 2003 issue of *Child Development*. The quotation is from the summary on the NICHD website, www.nichd.nih.gov.

38. *Early Human Development*, 76, No. 2 (February 2004), pp. 139–145.

39. www.ncbi.nlm.nih.gov/pubmed/16304443

40. foodpsychology.cornell.edu/OP/buffet_pricing

41. Nutrition Action, October 2013 (*Archives of Physical Medicine and Rehabilitation*, 93, p. 1269).

42. Marielle H. Emmelot-Vonk et al., "Effect of testosterone supplementation on functional mobility, cognition, and other parameters in older men," *Journal of the American Medical Association*, 299 (2008), pp. 39–52.

43. Naomi D. L. Fisher, Meghan Hughes, Marie Gerhard-Herman, and Norman K. Hollenberg, "Flavonol-rich cocoa induces nitricoxide-dependent vasodilation in healthy humans," *Journal of Hypertension*, 21, No. 12 (2003), pp. 2281–2286.

44. Joel Brockner et al., "Layoffs, equity theory, and work performance: Further evidence of the impact of survivor guilt," *Academy of Management Journal*, 29 (1986), pp. 373–384.

45. Mason, M. F., et al., Precise offers are potent anchors: Conciliatory counteroffers and attributions of knowledge in negotiations, *Journal of Experimental Social Psychology* (2013), dx.doi.org/10.1016/j.jesp.2013.02.012.

46. *Arizona Daily Star*, February 15, 2016. "Blood boosters may help preemies develop."

47. Details of the Carolina Abecedarian Project, including references to published work, can be found online at abc.fpg.unc.edu.

48. Christopher Anderson, "Measuring what works in health care," *Science*, 263 (1994), pp. 1080–1082.

49. *Arizona Daily Star*, May 10, 2016. "Diet soda, pregnancy: Mix may fuel childhood obesity, study says."

50. Mary O. Mundinger et al., "Primary care outcomes in patients treated by nurse practitioners or physicians," *Journal of the American Medical Association*, 238 (2000), pp. 59–68.

51. Study conducted by cardiologists at Athens Medical School, Greece, and announced at a European cardiology conference in February 2004.

52. rainfall.weatherdb.com/compare/28-21240/Tucson-Arizona-vs-Princeton-New-Jersey

53. The sleep deprivation study is described in R. Stickgold, L. James, and J. Hobson, "Visual discrimination learning requires post-training sleep," *Nature Neuroscience*, 2000, pp. 1237–1238. We obtained the data from Allan Rossman, who got it courtesy of the authors.

54. David L. Strayer, Frank A. Drews, and William A. Johnston, "Cell phone–induced failures of visual attention during simulated driving," *Journal of Experimental Psychology: Applied*, 9 (2003), pp. 23–32.

55. The idea for this chart came from Fred L. Ramsey and Daniel W. Schafer, *The Statistical Sleuth: A Course in Methods of Data Analysis*, 2nd edition, Duxbury Press, 2002.

56. journals.sagepub.com/doi/pdf/10.3102/0002831213488818

57. pediatrics.aappublications.org/content/pediatrics/early/2016/08/25/peds.2016-0910.full.pdf

58. The Health Consequences of Smoking: 1983, U.S. Health Service, 1983.

59. See the details on the website of the Office for Human Research Protections of the Department of Health and Human Services, www.hhs.gov/ohrp.

60. *San Gabriel Valley Tribune* (February 13, 2003).

61. Data from Michael Khawam, Canyon del Oro High School.

62. www.nejm.org/doi/full/10.1056/NEJMoa0905471

63. W. E. Paulus et al., "Influence of acupuncture on the pregnancy rate in patients who undergo assisted reproductive therapy," *Fertility and Sterility*, 77, No. 4 (2002), pp. 721–724.

64. Linda Stern et al., "The effects of low-carbohydrate versus conventional weight loss diets in severely obese adults: One-year follow up of a randomized trial," *Annals of Internal Medicine*, 140, No. 10 (May 2004), pp. 778–785.

65. Charles A. Nelson III et al., "Cognitive recovery in socially deprived young children: The Bucharest Early Intervention Project," *Science*, 318 (2007), pp. 1937–1940.

66. Marilyn Ellis, "Attending church found factor in longer life," *USA Today*, August 9, 1999.

67. jamanetwork.com/journals/jamasurgery/fullarticle/2601320

68. *Nutrition Action*, March 2013 (Circulation 127, p. 188), www.ncbi.nlm.nih.gov/pubmed/23319811.

69. *Nutrition Action*, December 2014 (*Acta Psychologica*, 153, p. 13).

70. www.pnas.org/content/111/24/8788.full

71. Scott DeCarlo with Michael Schubach and Vladimir Naumovski, "A decade of new issues," *Forbes*, March 5, 2001, www.forbes.com.

72. "Antibiotics no better than placebo for sinus infections," in.news.yahoo.com/antibiotics-no-better-placebo-sinusinfections-074240018.html.

73. www.gallup.com/poll/180260/americans-ratenurses-highest-honesty-ethical-standards.aspx

74. L. E. Moses and F. Mosteller, "Safety of anesthetics," in J. M. Tanur et al. (eds.), *Statistics: A Guide to the Unknown*, 3rd ed., Wadsworth, 1989, pp. 15–24.

75. *Arizona Daily Star*, 10-29-2008.

76. R. C. Shelton et al., "Effectiveness of St. John's wort in major depression," *Journal of the American Medical Association*, 285 (2001), pp. 1978–1986.

77. *Arizona Daily Star*, 2-20-2010, "Scientists: Act happy; it might help your heart."

78. health.usnews.com/health-news/news/articles/2015/09/22/txt-msgs-may-lead-to-broad-heart-linked-benefits-study-says

79. Based on a news item "Bee off with you," *Economist*, November 2, 2002, p. 78.

80. services.google.com/fh/files/misc/images-of-computer-science-report.pdf

81. From the Electronic Encyclopedia of Statistical Examples and Exercises (EESEE) case study "Is caffeine dependence real?"

82. This project is based on an activity suggested in Richard L. Schaeffer, Ann Watkins, Mrudulla Gnanadesikan, and Jeffrey A. Witmer, Activity-Based Statistics, Springer, 1996.

Chapter 5

1. R. Vallone and A. Tversky, "The hot hand in basketball: On the misperception of random sequences," *Cognitive Psychology*, 17 (1985), pp. 295–314.

2. Gur Yaari and Shmuel Eisenmann, "The hot (invisible?) hand: Can time sequence patterns of success/failure in sports be modeled as repeated random independent trials?" *PLoS ONE* 6(10), 2011, p. e24532. doi:10.1371/journal.pone.0024532.

3. www.commonsensemedia.org/technology-addiction-concern-controversy-and-finding-balance-infographic

4. www.prweek.com/article/1247089/paul-holmes-aarp-fooling-itself-misrepresenting-facts-gop-medicare-bill-members

5. Data from the Pew Research Center, November 2016, "Social Media Update 2016."

6. Data from the website of Statistics Canada, www.statcan.ca.

7. 2009 National Household Travel Survey, nhts.ornl.gov/2009/pub/stt.pdf.

8. Gail Burrill, "Two-way tables: Introducing probability using real data," paper presented at the Mathematics Education into the Twenty-first Century Project, Czech Republic, September 2003. Burrill cites as her source H. Kranendonk, P. Hopfensperger, and R. Scheaffer, *Exploring Probability*, Dale Seymour Publications, 1999.

9. From the EESEE story "What Makes a Pre-teen Popular?"

10. From the EESEE story "Is It Tough to Crawl in March?"

11. Pierre J. Meunier et al., "The effects of strontium ranelate on the risk of vertebral fracture in women with postmenopausal osteoporosis," *New England Journal of Medicine*, 350 (2004), pp. 459–468.

12. www.ncaa.org/about/resources/research/probability-competing-beyond-high-school

13. Lenhart, A., Smith, A., and Anderson, M. "Teens, technology and romantic relationships," Pew Research Center, October 2015.

14. Thanks to Michael Legacy for suggesting the context of this problem.

15. This is one of several tests discussed in Bernard M. Branson, "Rapid HIV testing: 2005 update," a presentation by the Centers for Disease Control and Prevention, at www.cdc.gov. The Malawi clinic result is reported by Bernard M. Branson, "Point-of-care rapid tests for HIV antibody," *Journal of Laboratory Medicine*, 27 (2003), pp. 288–295.

16. The National Longitudinal Study of Adolescent Health interviewed a stratified random sample of 27,000 adolescents, then reinterviewed many of the subjects six years later, when most were aged 19 to 25. These data are from the Wave III reinterviews in

2000 and 2001, found at the website of the Carolina Population Center, www.cpc.unc.edu.

17. Data from the University of Florida Biostatistics Open Learning Textbook at bolt.mph.ufl.edu/.

18. From the EESEE story "What Makes a Pre-teen Popular?"

19. Information about Internet users comes from sample surveys carried out by the Pew Internet and American Life Project, found online at www.pewinternet.org.

20. We got these data from the Energy Information Administration on their website at www.eia.gov.

21. www.kff.org/entmedia/mh012010pkg.cfm

22. www.atpworldtour.com/en/players/andy-murray/mc10/player-stats?year=2016&surfaceType=all

23. From the National Institutes of Health's National Digestive Diseases Information Clearinghouse, found at digestive.niddk.nih.gov/.

24. Probabilities from trials with 2897 people known to be free of HIV antibodies and 673 people known to be infected are reported in J. Richard George, "Alternative specimen sources: Methods for confirming positives," 1998 Conference on the Laboratory Science of HIV, found online at the Centers for Disease Control and Prevention, www.cdc.gov.

25. The probabilities given are realistic, according to the fundraising firm SCM Associates, at scmassoc.com.

26. Robert P. Dellavalle et al., "Going, going, gone: Lost Internet references," *Science*, 302 (2003), pp. 787–788.

27. Thanks to Corey Andreasen for suggesting the idea for this exercise.

28. Margaret A. McDowell et al., "Anthropometric reference data for children and adults: U.S. population, 1999–2002," *National Center for Health Statistics, Advance Data from Vital and Health Statistics*, No. 361 (2005), at www.cdc.gov/nchs.

29. The probability distribution was based on data found at online.wsj.com/mdc/public/page/2_3022-autosales.html.

30. www.cbsnews.com/stories/2010/06/01/health/webmd/main6537635.shtml

31. Thanks to Tim Brown, The Lawrenceville School, for providing the idea for this exercise.

Chapter 6

1. The Apgar score data came from *National Center for Health Statistics*, Monthly Vital Statistics Reports, Vol. 30, No. 1, Supplement, May 6, 1981.

2. The probability distribution is based on data obtained from gradedistribution.registrar.indiana.edu/.

3. The mean of a continuous random variable X with density function can be found by integration:

$$\mu_X = \int x f(x)\, dx$$

This integral is a kind of weighted average, analogous to the discrete-case mean

$$\mu_X = \sum x_i p_i$$

The variance of a continuous random variable X is the average squared deviation of the values of X from their mean, found by the integral

$$\sigma_X^2 = \int (x - \mu)^2 f(x)\, dx$$

4. You can find a mathematical explanation of Benford's law in Ted Hill, "The first-digit phenomenon," *American Scientist*, 86 (1996), pp. 358–363; and Ted Hill, "The difficulty of faking data," *Chance*, 12, No. 3 (1999), pp. 27–31. Applications in fraud detection are discussed in the second paper by Hill and in Mark A. Nigrini, "I've got your number," *Journal of Accountancy*, May 1999, available online at www.journalofaccountancy.com/issues/1999/may/nigrini.

5. The National Longitudinal Study of Adolescent Health interviewed a stratified random sample of 27,000 adolescents, then reinterviewed many of the subjects six years later, when most were aged 19 to 25. These data are from the Wave III reinterviews in 2000 and 2001, found at the website of the Carolina Population Center, www.cpc.unc.edu.

6. Data from the Census Bureau's American Housing Survey.

7. Thomas K. Cureton et al., Endurance of Young Men, *Monographs of the Society for Research in Child Development*, Vol. 10, No. 1, 1945.

8. Some results of the survey are described at www.nclnet.org/personal-finance/66-teensand-money/120-ncl-survey-teens-andfinancial-education.

9. Ed O'Brien and Phoebe C. Ellsworth, "Saving the last for best: A positivity bias for end experiences," *Psychological Science*, 23, No. 2 (September 2011), pp. 163–165.

10. We got the 9% figure from The Pew Research Center's "Assessing the Representativeness of Public Opinion Surveys," published on May 15, 2012.

11. Office of Technology Assessment, Scientific Validity of Polygraph Testing: A Research Review and Evaluation, Government Printing Office, 1983.

Chapter 7

1. This activity is based on a similar activity suggested in Richard L. Schaeffer, Ann Watkins, Mrudulla Gnanadesikan, and Jeffrey A. Witmer, Activity-Based Statistics, Springer, 1996.

2. www.census.gov/data/tables/time-series/demo/income-poverty/cps-pinc/pinc-01.html

3. This and similar results of Gallup polls are from the Gallup Organization website, www.gallup.com.

4. From a graph in Stan Boutin et al., "Anticipatory reproduction and population growth in seed predators," *Science*, 314 (2006), pp. 1928–1930.

5. See Note 3.

6. www.bls.gov/news.release/pdf/famee.pdf

7. The idea for this exercise was inspired by an example in David M. Lane's Hyperstat Online textbook at davidmlane.com/hyperstat.

8. Amanda Lenhart and Mary Madden, "Music downloading, file-sharing and copyright," Pew Internet and American Life Project, 2003, at www.pewinternet.org.

9. *Nutrition Action Healthletter*, September 2016.

10. We obtained the National Health and Nutrition Examination Survey data from the Centers for Disease Control and Prevention website at www.cdc.gov/nchs/nhanes.htm.

11. Based on a figure in Peter R. Grant, *Ecology and Evolution of Darwin's Finches*, Princeton University Press, 1986.

12. We found the information on birth weights of Norwegian children on the National Institute of Environmental Health Sciences

website: The relevant article can be accessed here: www.ncbi.nlm
.nih.gov/pubmed/1536353.

13. time.com/money/3712480/pay-less-internet-cut-cable/

Chapter 8

1. assets.pewresearch.org/wp-content/uploads/sites/14/2016/12
/19170147/PS_2016.12.01_Food-Science_FINAL.pdf

2. www.smithsonianmag.com/ideas-innovations/How-Much
-Do-Americans-Know-About-Science.html

3. Data from Ellery Page, Canyon del Oro High School.

4. Michele L. Head, "Examining college students' ethical
values," Consumer Science and Retailing honors project, Purdue
University, 2003.

5. This and similar results of Gallup polls are from the Gallup
Organization website, www.gallup.com.

6. Amanda Lenhart and Mary Madden, "Teens, privacy and
online social networks," Pew Internet and American Life Project,
2007, at www.pewinternet.org.

7. Data from 2013 Current Population Survey, found at www
.eeps.com/zoo/acs/source/index.php.

8. See Note 7.

9. E. W. Campion, "Editorial: Power lines, cancer, and fear,"
New England Journal of Medicine, 337, No. 1 (1997), pp. 44–46.
The study report is M. S. Linet et al., "Residential exposure to
magnetic fields and acute lymphoblastic leukemia in children,"
pp. 1–8 in the same issue. See also G. Taubes, "Magnetic field–
cancer link: will it rest in peace?" Science, 277 (1997), p. 29.

10. Karl Pearson and A. Lee, "On the laws of inheritance in
man," Biometrika, 2 (1902), p. 357. These data also appear in D.
J. Hand et al., A Handbook of Small Data Sets, Chapman & Hall,
1994. This book offers more than 500 data sets that can be used in
statistical exercises.

11. Pew Research Center, September 2016, "Book Reading
2016."

12. sleepfoundation.org/sleep-polls-data/2015-sleepand-pain

13. www.usatoday.com/picture-gallery/news/2015/04/07/usa
-today-snapshots/6340793/

14. Eric Sanford et al., "Local selection and latitudinal variation
in a marine predator–prey interaction," Science, 300 (2003), pp.
1135–1137.

15. Pew Research Center, November 2016, "Gig Work, Online
Selling and Home Sharing."

16. See Note 4.

17. Duggan, Maeve, "Gaming and Gamers." Pew Research
Center, December 2015.

18. Pew Research Center, December 2016, "The New Food
Fights: U.S. Public Divides Over Food Science."

19. cdn.annenbergpublicpolicycenter.org/wp-content
/uploads/Civics-survey-press-release-09-17-2014-for-PRNewswire.pdf

20. Linda Lyons, "Most teens have in-room entertainment,"
February 22, 2005, www.gallup.com/poll/14989/Most-Teens
-InRoom-Entertainment.aspx.

21. Based on information in "NCAA 2003 national study of
collegiate sports wagering and associated health risks," which can
be found on the NCAA website: www.ncaa.org.

22. www.nejm.org/doi/full/10.1056/NEJMoa1615869

23. Benjamin W. Friedman et al., "Diazepam is no better than
placebo when added to naproxen for acute low back pain," Annals
of Emergency Medicine, February 7, 2017.

24. Saiyad S. Ahmed, "Effects of microwave drying on checking
and mechanical strength of low-moisture baked products," MS
thesis, Purdue University, 1994.

25. The National Longitudinal Study of Adolescent Health
interviewed a stratified random sample of 27,000 adolescents,
then reinterviewed many of the subjects six years later, when most
were aged 19 to 25. These data are from the Wave III reinterviews,
found at the website of the Carolina Population Center: www.cpc
.unc.edu.

26. www.cbsnews.com/htdocs/pdf/poll_whereamericastands
_obesity_010710.pdf

27. Psychonomic Bulletin & Review, 2008, 15 (5), 927-932, doi:
10.3758/PBR.15.5.927. Thanks to Kristin Flegal for sharing the
data from this study.

28. news.gallup.com/poll/183689/industry-grows-percentage
-sports-fans-steady.aspx

29. Bryan E. Porter and Thomas D. Berry, "A nationwide
survey of self-reported red light running: Measuring prevalence,
predictors, and perceived consequences," Accident Analysis and
Prevention, 33 (2001), pp. 735–741.

30. Amanda Lenhart, "Teens, social media & technology over-
view 2015," Pew Research Center, April 9, 2015; and Maeve
Duggan, Nicole B. Ellison, Cliff Lampe, Amanda Lenhart, and
Mary Madden, "Social Media Update 2014," Pew Research
Center, January 9, 2015.

31. Emily Cohen and Madi McDole, Canyon del Oro High School.

Chapter 9

1. Thanks to Josh Tabor for suggesting the idea for this example.

2. R. A. Fisher, "The arrangement of field experiments," Journal of
the Ministry of Agriculture of Great Britain, 33 (1926), p. 504, quoted in
Leonard J. Savage, "On rereading R. A. Fisher," Annals of Statistics, 4
(1976), p. 471. Fisher's work is described in a biography by his daughter: Joan Fisher Box, R. A. Fisher: The Life of a Scientist, Wiley, 1978.

3. Julie Ray, "Few teens clash with friends," May 3, 2005, on the
Gallup Organization website, www.gallup.com.

4. The idea for this exercise was provided by Michael Legacy
and Susan McGann.

5. Projections from the 2011 Digest of Education Statistics,
found online at nces.ed.gov.

6. From the report "Sex and tech: Results from a study of teens
and young adults," published by the National Campaign to Prevent Teen and Unplanned Pregnancy, www.thenationalcampaign
.org/sextech.

7. www.cdc.gov/healthyyouth/data/yrbs/pdf/2015/2015_us
_tobacco.pdf

8. National Institute for Occupational Safety and Health, Stress
at Work, 2000, available online at www.cdc.gov/niosh/docs/99101/.
Results of this survey were reported in Restaurant Business,
September 15, 1999, pp. 45–49.

9. Thanks to DeAnna McDonald for allowing us some creative
license with her teaching assignment!

10. Dorothy Espelage et al., "Factors associated with bullying
behavior in middle school students," Journal of Early Adolescence,
19, No. 3 (August 1999), pp. 341–362.

11. Aaron Smith and Joanna Brenner, "Twitter Use 2012," published
by the Pew Internet and American Life Project, May 31, 2012.

12. This and similar results of Gallup polls are from the Gallup
Organization website, www.gallup.com.

13. Michele L. Head, "Examining college students' ethical values," Consumer Science and Retailing honors project, Purdue University, 2003.

14. Linda Lyons, "Teens: Sex can wait," December 14, 2004, from the Gallup Organization website, www.gallup.com.

15. Based on Stephen A. Woodbury and Robert G. Spiegelman, "Bonuses to workers and employers to reduce unemployment: randomized trials in Illinois," American Economic Review, 77 (1987), pp. 513–530.

16. C. P. Cannon et al., "Intensive versus moderate lipid lowering with statins after acute coronary syndromes," New England Journal of Medicine, 350 (2004), pp. 1495–1504.

17. Data from student project by Miranda Edwards and Sarah Juarez, Canyon del Oro High School.

18. The study is reported in William Celis III, "Study suggests Head Start helps beyond school," New York Times, April 20, 1993. See www.highscope.org.

19. Based on Deborah Roedder John and Ramnath Lakshmi-Ratan, "Age differences in children's choice behavior: The impact of available alternatives," Journal of Marketing Research, 29 (1992), pp. 216–226.

20. National Athletic Trainers Association, press release, dated September 30, 1994.

21. Francisco Lloret et al., "Fire and resprouting in Mediterranean ecosystems: Insights from an external biogeographical region, the Mexican shrubland," American Journal of Botany, 88 (1999), pp. 1655–1661.

22. Clive G. Jones et al., "Chain reactions linking acorns to gypsy moth outbreaks and Lyme disease risk," Science, 279 (1998), pp. 1023–1026.

23. Arizona Daily Star, August 18, 2010.

24. George Du Toit, M. B., B. Ch., et al., "Randomized Trial of Peanut Consumption in Infants at Risk for Peanut Allergy," New England Journal of Medicine, 372 (February 2015), pp. 803–813.

25. See Note 16.

26. Kwang Y. Cha, Daniel P. Wirth, and Rogerio A. Lobo, "Does prayer influence the success of in vitro fertilization–embryo transfer?" Journal of Reproductive Medicine, 46 (2001), pp. 781–787.

27. W. E. Paulus et al., "Influence of acupuncture on the pregnancy rate in patients who undergo assisted reproductive therapy," Fertility and Sterility, 77, No. 4 (2002), pp. 721–724.

Chapter 10

1. Pew Research Center, September 2016, "Book Reading 2016." Standard deviation estimated from frequency table provided in the report.

2. Data provided by Drina Iglesia, Purdue University. The data are part of a larger study reported in D. D. S. Iglesia, E. J. Cragoe, Jr., and J. W. Vanable, "Electric field strength and epithelization in the newt (Notophthalmus viridescens)," Journal of Experimental Zoology, 274 (1996), pp. 56–62.

3. Harry B. Meyers, "Investigations of the life history of the velvetleaf seed beetle, Althaeus folkertsi Kingsolver," MS thesis, Purdue University, 1996.

4. M. Ann Laskey et al., "Bone changes after 3 mo of lactation: Influence of calcium intake, breast-milk output, and vitamin D–receptor genotype," American Journal of Clinical Nutrition, 67 (1998), pp. 685–692.

5. TUDA results for 2003 from the National Center for Education Statistics, at nces.ed.gov/nationsreportcard.

6. Data from Tori Heimink and Ann Perry, Canyon del Oro High School.

7. From a graph in L. Partridge and M. Farquhar, "Sexual activity reduces lifespan of male fruit flies," Nature, 294 (1981), pp. 580–582. Provided by Brigitte Baldi.

8. Data from Melissa Silva and Madeline Dunlap, Canyon del Oro High School.

9. Data from Carly Myers and Maysem Ahmad, Canyon del Oro High School.

10. Lisa M. Baril et al., "Willow-bird relationships on Yellowstone's northern range," Yellowstone Science, 17, No. 3 (2009), pp. 19–26.

11. Helen E. Staal and D. C. Donderi, "The effect of sound on visual apparent movement," American Journal of Psychology, 96 (1983), pp. 95–105.

12. Based on interviews in 2000 and 2001 by the National Longitudinal Study of Adolescent Health. Found at the website of the Carolina Population Center, www.cpc.unc.edu.

13. Simplified from Sanjay K. Dhar, Claudia Gonzalez-Vallejo, and Dilip Soman, "Modeling the effects of advertised price claims: Tensile versus precise pricing," Marketing Science, 18 (1999), pp. 154–177.

14. Data for this example from Noel Cressie, Statistics for Spatial Data, Wiley, 1993.

15. Niels Juel-Nielsen, Individual and Environment: Monozygotic Twins Reared Apart, International Universities Press, 1980.

16. Shailija V. Nigdikar et al., "Consumption of red wine polyphenols reduces the susceptibility of low-density lipoproteins to oxidation in vivo," American Journal of Clinical Nutrition, 68 (1998), pp. 258–265.

17. Ethan J. Temeles and W. John Kress, "Adaptation in a plant–hummingbird association," Science, 300 (2003), pp. 630–633. We thank Ethan J. Temeles for providing the data.

18. Data provided by Marvin Schlatter, Division of Financial Aid, Purdue University.

19. R. D. Stichler, G. G. Richey, and J. Mandel, "Measurement of treadware of commercial tires," Rubber Age, 73, No. 2 (May 1953).

20. Data from Pennsylvania State University Stat 500 Applied Statistics online course, onlinecourses.science.psu.edu/stat500/.

21. Data obtained from flightaware.com on November 27, 2015.

22. United Nations data on literacy were found at en.openei.org/wiki/WRI-Earth_Trends_Data.

23. F. H. Rauscher et al., "Music training causes long-term enhancement of preschool children's spatial-temporal reasoning," Neurological Research, 19 (1997), pp. 2–8.

24. Leila El-Ali and Valerie Pederson, Canyon del Oro High School

25. Sean Leader and Shelby Zismann, Canyon del Oro High School

26. blogs.edweek.org/edweek/curriculum/2015/03/homework_math_science_study.html from this study: www.apa.org/pubs/journals/releases/edu-0000032.pdf.

27. Jessica Sheldon, Canyon del Oro High School

28. From program 19, "Confidence Intervals," in the Against All Odds video series.

29. Francisco L. Rivera-Batiz, "Quantitative literacy and the likelihood of employment among young adults," *Journal of Human Resources*, 27 (1992), pp. 313–328.

30. Libby Foulk and Kathryn Hilton, Canyon del Oro High School

31. Megan Zeeb and Jenna Wilson, Canyon del Oro High School

32. Cory Heid, "Tootsie pops: How many licks to the chocolate?" *Significance Magazine*, October 2013, page 47.

Chapter 11

1. P. A. Mackowiak, S. S. Wasserman, and M. M. Levine, "A critical appraisal of 98.6 degrees F, the upper limit of the normal body temperature, and other legacies of Carl Reinhold August Wunderlich," *Journal of the American Medical Association*, 268 (1992), pp. 1578–1580.

2. Sapna Aneja, "Biodeterioration of textile fibers in soil," MS thesis, Purdue University, 1994.

3. Warren E. Leary, "Cell phones: Questions but no answers," *New York Times*, October 26, 1999.

4. This exercise is based on events that are real. The data and details have been altered to protect the privacy of the individuals involved.

5. W. S. Gosset, "The probable error of a mean," *Biometrika*, 6 (1908), pp. 1–25. We obtained the sleep data from the Data and Story Library (DASL) website, lib.stat.cmu.edu/DASL/. They cite as a reference R. A. Fisher, *The Design of Experiments*, 3rd ed., Oliver and Boyd, 1942, p. 27.

6. From the Gallup Organization website: www.gallup.com.

7. This study is reported in Roseann M. Lyle et al., "Blood pressure and metabolic effects of calcium supplementation in normotensive white and black men," *Journal of the American Medical Association*, 257 (1987), pp. 1772–1776. The data were provided by Dr. Lyle.

8. Data from student project, Canyon del Oro High School.

9. E. C. Strain et al., "Caffeine dependence syndrome: Evidence from case histories and experimental evaluation," *Journal of the American Medical Association*, 272 (1994), pp. 1604–1607.

10. We obtained the tire pressure loss data from the *Consumer Reports* website: www.consumerreports.org/cro/news/2007/10/tires-nitrogen-air-loss-study/index.htm.

11. Gabriela S. Castellani, "The effect of cultural values on Hispanics' expectations about service quality," MS thesis, Purdue University, 2000.

12. *New England Journal of Medicine*, 320 (1989), pp. 1037–1043; cited in Fred Ramsey and Daniel Schafer, *The Statistical Sleuth*, Pacific Grove, CA: Duxbury Press, 2002, p. 23.

13. From a graph in Fabrizio Grieco, Arie J. van Noordwijk, and Marcel E. Visser, "Evidence for the effect of learning on timing of reproduction in blue tits," *Science*, 296 (2002), pp. 136–138.

14. Adapted from Maribeth Cassidy Schmitt, "The effects of an elaborated directed reading activity on the metacomprehension skills of third graders," PhD dissertation, Purdue University, 1987.

15. M. Ann Laskey et al., "Bone changes after 3 months of lactation: Influence of calcium intake, breast-milk output, and vitamin D–receptor genotype," *American Journal of Clinical Nutrition*, 67 (1998), pp. 685–692.

16. The original paper is T. M. Amabile, "Motivation and creativity: Effects of motivational orientation on creative writers," *Journal of Personality and Social Psychology*, 48, No. 2 (February 1985), pp. 393–399. The data for Exercise 43 came from Fred L. Ramsey and Daniel W. Schafer, *The Statistical Sleuth*, 3rd ed., Brooks/Cole Cengage Learning, 2013.

17. The data for this exercise came from Rossman, Cobb, Chance, and Holcomb's National Science Foundation project shared at JMM 2008 in San Diego. Their original source was Robert Stickgold, LaTanya James, and J. Allan Hobson, "Visual discrimination learning requires sleep after training," *Nature Neuroscience*, 3 (2000), pp. 1237–1238.

18. Data from a student project by Patrick Baker and William Manheim, Canyon del Oro High School.

19. W. S. Gosset, "The probable error of a mean," *Biometrika*, 6 (1908), 1–25.

20. From the story "Friday the 13th," at the Data and Story Library, lib.stat.cmu.edu/DASL.

21. The idea for this exercise was provided by Robert Hayden.

22. See Note 21.

23. Wayne J. Camera and Donald Powers, "Coaching and the SAT I," TIP (online journal at www.siop.org/tip), July 1999.

24. Data provided by Warren Page, New York City Technical College, from a study done by John Hudesman.

25. Maureen Hack et al., "Outcomes in young adulthood for very low-birth-weight infants," *New England Journal of Medicine*, 346 (2002), pp. 149–157. Exercise AP3.32 is simplified, in that the measures reported in this paper have been statistically adjusted for "sociodemographic status."

Chapter 12

1. Data from Cassandra Randal-Greene, Canyon del Oro High School, and newsroom.ppg.com/getmedia/ef974015-f89b4211-97ab-18d5ff3e7303/2014-NA-PPG.jpg.aspx

2. R. W. Mannan and E. C. Meslow, "Bird populations and vegetation characteristics in managed and old-growth forests, northwestern Oregon," *Journal of Wildlife Management*, 48 (1984), pp. 1219–1238.

3. www.pewinternet.org/2015/12/15/gaming-and-gamers/

4. The context of this example was inspired by C. M. Ryan et al., "The effect of in-store music on consumer choice of wine," *Proceedings of the Nutrition Society*, 57 (1998), p. 1069A.

5. Data from Abigail Gentzler and Mia Sapone, Canyon del Oro High School.

6. ww2.amstat.org/CensusAtSchool/

7. Janice E. Williams et al., "Anger proneness predicts coronary heart disease risk," *Circulation*, 101 (2000), pp. 63–95.

8. Data from the University of Florida Biostatistics Open Learning Textbook at bolt.mph.ufl.edu/.

9. J. Cantor, "Long-term memories of frightening media often include lingering trauma symptoms," poster paper presented at the Association for Psychological Science Convention, New York, May 26, 2006.

10. William D. Darley, "Store-choice behavior for preowned merchandise," *Journal of Business Research*, 27 (1993), pp. 17–31.

11. The National Longitudinal Study of Adolescent Health interviewed a stratified random sample of 27,000 adolescents, then

reinterviewed many of the subjects six years later, when most were aged 19 to 25. These data are from the Wave III reinterviews in 2000 and 2001, found at www.cpc.unc.edu.

12. R. Shine et al., "The influence of nest temperatures and maternal brooding on hatchling phenotypes in water pythons," *Ecology*, 78 (1997), pp. 1713–1721.

13. U.S. Department of Commerce, Office of Travel and Tourism Industries, in-flight survey, 2007, at tinet.ita.doc.gov.

14. The idea for this exercise came from Bob Hayden.

15. Data from Lexi Epperson and Courtney Johnson, Canyon del Oro High School.

16. Data compiled from a table of percents in "Americans view higher education as key to the American dream," press release by the National Center for Public Policy and Higher Education, www.highereducation.org, May 3, 2000.

17. Douglas E. Jorenby et al., "A controlled trial of sustained -release bupropion, a nicotine patch, or both for smoking cessation," *New England Journal of Medicine*, 340 (1990), pp. 685–691.

18. Martin Enserink, "Fraud and ethics charges hit stroke drug trial," *Science*, 274 (1996), pp. 2004–2005.

19. archinte.jamanetwork.com/article.aspx?articleid =2130724#tab1

20. All General Social Survey exercises in this chapter present tables constructed using the search function at the GSS archive, sda.berkeley.edu/archive.htm.

21. Based closely on Susan B. Sorenson, "Regulating firearms as a consumer product," *Science*, 286 (1999), pp. 1481–1482. Because the results in the paper were "weighted to the U.S. population," we have changed some counts slightly for consistency.

22. www.quinnipiac.edu/news-and-events/quinnipiacuniversity -poll/national/release-detail?ReleaseID=2275

23. www.pewinternet.org/files/old-media//Files/Reports/2013 /PIP_OnlineBanking.pdf

24. *Journal of Internal Medicine*, 266 (2009), pp. 248–257.

25. Pew Research Center for the People and the Press, "The cell phone challenge to survey research," news release for May 15, 2006, at www.people-press.org.

26. See Note 11.

27. K. Eagan, J. B. Lozano, S. Hurtado, & M. H. Case. The American Freshman: National Norms, Fall 2013. (2013) Los Angeles: Higher Education Research Institute, UCLA.

28. Lillian Lin Miao, "Gastric freezing: An example of the evaluation of medical therapy by randomized clinical trials," in John P. Bunker, Benjamin A. Barnes, and Frederick Mosteller (eds.), *Costs, Risks, and Benefits of Surgery*, Oxford University Press, 1977, pp. 198–211.

29. See Note 11.

30. D. M. Barnes, "Breaking the cycle of addiction," *Science*, 241 (1988), pp. 1029–1030.

31. Thanks to Larry Green, Lake Tahoe Community College, for giving us permission to use several of the contexts from his website at www.ltcconline.net/greenl/java/Statistics/catStatProb /categorizingStatProblems12.html.

32. See Note 31.

33. Samuel Karelitz et al., "Relation of crying activity in early infancy to speech and intellectual development at age three years," *Child Development*, 35 (1964), pp. 769–777.

34. www.ncei.noaa.gov

35. Data from National Institute of Standards and Technology, Engineering Statistics Handbook, www.itl.nist.gov/div898 /handbook. The analysis there does not comment on the bias of field measurements.

36. Todd W. Anderson, "Predator responses, prey refuges, and density-dependent mortality of a marine fish," *Ecology*, 81 (2001), pp. 245–257.

37. Data from Kerry Lane and Danielle Neal, Canyon del Oro High School.

38. Data from Brittany Foley and Allie Dutson, Canyon del Oro High School.

39. Based on a plot in G. D. Martinsen, E. M. Driebe, and T. G. Whitham, "Indirect interactions mediated by changing plant chemistry: Beaver browsing benefits beetles," *Ecology*, 79 (1998), pp. 192–200.

40. Data provided by Samuel Phillips, Purdue University.

41. Based on Marion E. Dunshee, "A study of factors affecting the amount and kind of food eaten by nursery school children," *Child Development*, 2 (1931), pp. 163–183. This article gives the means, standard deviations, and correlation for 37 children but does not give the actual data.

42. M. H. Criqui, University of California, San Diego, reported in the *New York Times*, December 28, 1994.

43. Data from Nicole Enos and Elena Tesluk, Canyon del Oro High School.

44. Data from Jordynn Watson and Angelica Valenzuela, Canyon del Oro High School.

45. R. D. Stichler, G. G. Richey, and J. Mandel, "Measurement of treadware of commercial tires," *Rubber Age*, 73, No. 2 (May 1953).

46. Mark A. Sabbagh and Dare A. Baldwin, "Learning words from knowledgeable versus ignorant speakers: Links between preschoolers' theory of mind and semantic development," *Child Development*, 72 (2001), pp. 1054–1070. Many statistical software packages offer "exact tests" that are valid even when there are small expected counts.

47. www.eia.gov/consumption/residential/index.php

48. Brenda C. Coleman, "Study: Heart attack risk cut 74% by stress management," Associated Press dispatch appearing in the *Lafayette* (Ind.) *Journal and Courier*, October 20, 1997.

49. Based on the EESEE story "What Makes Pre-teens Popular?"

50. Based on the EESEE story "Domestic Violence."

51. Karine Marangon et al., "Diet, antioxidant status, and smoking habits in French men," *American Journal of Clinical Nutrition*, 67 (1998), pp. 231–239.

52. Data provided by Corinne Lim, Purdue University, from a student project supervised by Professor Joseph Vanable.

53. Data from many studies compiled in D. F. Greene and E. A. Johnson, "Estimating the mean annual seed production of trees," *Ecology*, 75 (1994), pp. 642–647.

Glossary/Glosario

English	Español
1.5 × IQR rule for outliers An observation is called an outlier if it falls more than $1.5 \times IQR$ above the third quartile or below the first quartile. (p. 66)	**regla 1.5 × la gama entre cuartiles para valores atípicos** Se le dice valor atípico a una observación si cae a más de $1.5 \times$ la gama entre cuartiles por encima del tercer cuartil o por debajo del primer cuartil. (p. 66)
10% condition When taking a random sample of size n without replacement from a population of size N, we can view individual observations as independent when performing calculations as long as $n < 0.10N$. (p. 447)	**condición del 10%** Cuando se toma una muestra aleatoria de tamaño n sin reposición de una población de tamaño N, es posible considerar las observaciones individuales de modo independiente al realizar cálculos en tanto $n < 0.10N$. (p. 447)

A

English	Español
accurate An estimator is accurate if it is unbiased. (p.481)	**preciso** Un estimador se considera preciso si no presenta sesgo. (p. 481)
addition rule for mutually exclusive events If A and B are mutually exclusive events, $P(A \text{ or } B) = P(A) + P(B)$. (p. 345)	**regla de suma para eventos que se excluyen mutuamente** Si A y B son eventos que se excluyen entre sí, $P(A \text{ o } B) = P(A) + P(B)$. (p. 345)
alternative hypothesis H_a The claim that we are trying to find evidence *for* in a significance test. (p. 586)	**hipótesis H_a alternativa** La proposición de que en una prueba de significancia estadística estamos tratando de hallar evidencia que esté *a favor*. (p. 586)
anonymity The names of individuals participating in a study are not known even to the director of the study. (p. 307)	**anonimato** Cuando se desconocen los nombres de las personas que participan en un estudio; incluso el director del estudio los ignora. (p. 307)
approximate sampling distribution The distribution of a statistic in many samples (but not all possible samples) of the same size from the same population. (p. 472)	**distribución aproximada del muestreo** Distitución de una estadística entre muchas muestras (aunque no entre todas las muestras posibles) del mismo tamaño de la misma población. (p. 472)
association A relationship between two variables in which knowing the value of one variable helps predict the value of the other. If knowing the value of one variable does not help predict the value of the other, there is no association between the variables. (p. 20)	**asociación** Relación entre dos variables en la cual saber el valor de una variable facilita la predicción del valor de la otra. Si saber el valor de una variable no facilita la predicción del valor de la otra, entonces no existe ninguna asociación entre las variables. (p. 20)

B

English	Español
back-to-back stemplot (*also called* **back-to-back stem-and-leaf plot**) Plot used to compare the distribution of a quantitative variable for two groups. Each observation in both groups is separated into a stem, consisting of all but the final digit, and a leaf, the final digit. The stems are arranged in a vertical column with the smallest at the top. The values from one group are plotted on the left side of the stem and the values from the other group are plotted on the right side of the stem. Each leaf is written in the row next to its stem, with the leaves arranged in increasing order out from the stem. (p. 39)	**diagrama de tallos contiguos** (también se le dice **diagrama de tallos y hojas contiguos**) Se utiliza para comparar la distribución de una variable cuantitativa en dos grupos. Cada observación efectuada en ambos grupos se separa en un tallo, que consiste de todos los dígitos salvo el último, y una hoja, que consta del último dígito. Los tallos se organizan en una columna vertical con las cifras más pequeñas arriba. Los valores de un grupo se diagraman al lado izquierdo del tallo y los valores del otro grupo se diagraman al lado derecho del tallo. Cada hoja se coloca en el renglón que está al lado de su tallo, y las hojas dispuestas en orden ascendiente extendiéndose hacia fuera a partir del tallo. (p. 39)

bar graph Graph used to display the distribution of a categorical variable or to compare the sizes of different quantities. The horizontal axis of a bar graph identifies the categories or quantities being compared. The heights of the bars show the frequency or relative frequency for each value of the categorical variable. The graph is drawn with blank spaces between the bars to separate the items being compared. (p. 10)	**gráfico de barras** Se usa para ilustrar la distribución de una variable categorizada o para comparar el tamaño de diferentes cantidades. El eje horizontal del gráfico de barras identifica las categorías o las cantidades que se han de comparar. La altura de las barras muestra la frecuencia o la frecuencia relativa de cada valor de la variable categórica. Se puede dibujar con espacios en blanco entre las barras a fin de separar las diversas categorías que se desea comparar. (p. 10)
bias The design of a statistical study shows bias if it is very likely to underestimate or very likely to overestimate the value you want to know. (p. 252)	**sesgo** El diseño de un estudio estadístico refleja un sesgo si existe una alta probabilidad de subestimar o sobreestimar el valor que se busca. (p. 252)
biased estimator A statistic used to estimate a parameter is biased if the mean of its sampling distribution is not equal to the value of the parameter being estimated. (p. 478)	**calculador sesgado** La estadística que se usa para computar un parámetro está sesgada si la media de la distribución de su muestreo no equivale al valor del parámetro que se está computando. (p. 478)
bimodal A graph of quantitative data with two clear peaks. (p. 33)	**bimodal** Gráfico de datos cuantitativos con dos picos bien definidos. (p. 33)
binomial coefficient The number of ways to arrange x successes among n trials is given by the binomial coefficient $\binom{n}{x} = \dfrac{n!}{x!(n-x)!}$ for $x = 0,1,2,\ldots,n$ where $n! = n(n-1)(n-2)\cdots 3\cdot 2\cdot 1$ and $0! = 1$. (p. 435)	**coeficiente binomial** La cantidad de maneras de organizar x aciertos entre n ensayos se representa con el coeficiente binomial $\binom{n}{x} = \dfrac{n!}{x!(n-x)!}$ para $x = 0,1,2,\ldots,n$ en el que $n! = n(n-1)(n-2)\cdots 3\cdot 2\cdot 1$ y $0! = 1$. (p. 435)
binomial distribution In a binomial setting, suppose we let $X =$ the number of successes. The probability distribution of X is a binomial distribution with parameters n and p, where n is the number of trials of the random process and p is the probability of a success on each trial. (p. 434)	**distribución binomial** En un entorno binomial, supongamos que se permite que $X =$ la cantidad de aciertos. La distribución de la probabilidad de X es una distribución binomial con los parámetros n y p, en la que n es la cantidad de ensayos del proceso aleatorio y p es la probabilidad de un acierto en cualquiera de los ensayos. (p. 434)
binomial probability formula If X has the binomial distribution with n trials and probability p of success on each trial, the probability of getting exactly x successes in n trials $(x = 0, 1, 2, \ldots, n)$ is $P(X = x) = \binom{n}{x}p^x(1-p)^{n-x}$. (p. 436)	**fórmula de probabilidad binomial** Si X tiene la distribución binomial con n ensayos y la probabilidad p de acierto en cada ensayo, la probabilidad de obtener exactamente x aciertos en n ensayos $(x = 0, 1, 2, \ldots, n)$ es $P(X = x) = \binom{n}{x}p^x(1-p)^{n-x}$. (p. 436)
binomial random variable The count X of successes in a binomial setting. The possible values of X are 0, 1, 2, ..., n. (p. 434)	**variable aleatoria binomial** La cuenta X de aciertos en un entorno binomial. Los valores posibles de X son 0, 1, 2, ..., n. (p. 434)
binomial setting Arises when we perform n independent trials of the same random process and record the number of times that a particular outcome (called a success) occurs. The four conditions for a binomial setting are: • Binary? The possible outcomes of each trial can be classified as "success" or "failure." • Independent? Trials must be independent; that is, knowing the outcome of one trial must not tell us anything about the outcome of any other trial. • Number? The number of trials n of the random process must be fixed in advance. • Same probability? There is the same probability p of success on each trial. (p. 432)	**entorno binomial** Surge cuando se realizan n ensayos independientes del mismo proceso aleatorio y se anota la cantidad de veces que se produce un resultado dado. Las cuatro condiciones que definen un entorno binomial son: • ¿Binario? Los resultados posibles de cada ensayo se pueden clasificar como "acierto" o "fracaso". • ¿Independiente? Los ensayos han de ser independientes; es decir, saber el resultado de un ensayo no debe indicar nada acerca del resultado de otro ensayo. • ¿Número? El número de ensayos de n en el proceso aleatorio se tiene que fijar con anticipación. • ¿Misma probabilidad? Existe la misma probabilidad p de lograr un acierto en cada ensayo. (p. 432)
bivariate data A data set that describes the relationship between two variables. (p. 153)	**datos bivariados** Grupo de datos que describen la relación entre dos variables. (p. 153)

block Group of experimental units that are known before the experiment to be similar in some way that is expected to affect the response to the treatments. (p. 285)	**bloque** Grupo de unidades experimentales que antes del experimento se sabe son similares de alguna manera previsible que afecte la respuesta a los tratamientos. (p. 285)
boxplot A visual representation of the five-number summary. The box spans the quartiles and shows the variability of the central half of the distribution. The median is marked within the box. Lines extend from the box to the smallest and largest observations that are not outliers. Outliers are marked with a special symbol such as an asterisk (*). (p. 68)	**diagrama de caja y bigotes** Representación visual del resumen de cinco cifras. La caja abarca los cuartiles y muestra la variabilidad de la mitad central de la distribución. Dentro de la caja se marca la media. Las líneas se extienden a partir de la caja a las observaciones más pequeña y más grande que no son valores atípicos. Los valores atípicos se marcan con un símbolo especial tal como un asterisco (*). (p. 68)

C

categorical variable A variable that assigns labels that place each individual into a particular group, called a category. (p. 3)	**variable categorizada** Variable que asigna una etiqueta a cada individuo para colocarlo dentro un grupo particular, conocido como categoría. (p. 3)
census Study that collects data from every individual in the population. (p. 249)	**censo** Un estudio en el que recogen datos acerca de cada individuo en la población. (p. 249)
central limit theorem (CLT) In an SRS of size n from any population with mean μ and finite standard deviation σ, when n is sufficiently large, the sampling distribution of the sample mean $\bar{x}$ is approximately Normal. (p. 510)	**teorema del límite central** Traza una muestra aleatoria sencilla de tamaño n a partir de una población con la media μ y una desviación estándar finita de σ. El teorema del límite central manifiesta que cuando n es lo suficientemente grande, la distribución de muestreo de la media de la muestra $\bar{x}$ es aproximadamente normal. (p. 510)
Chebyshev's inequality In any distribution, the proportion of observations falling within k standard deviations of the mean is at least $1 - \dfrac{1}{k^2}$. (p. 119)	**desigualdad de Chebychov** En cualquier distribución, la proporción de observaciones que yacen dentro de k desviaciones estándar de la media es al menos $1 - \dfrac{1}{k^2}$. (p. 119)
chi-square distribution A distribution that is defined by a density curve that takes only non-negative values and is skewed to the right. A particular chi-square distribution is specified by giving its degrees of freedom. (p. 767)	**distribución de ji cuadrado** Distribución que se define por una curva de densidad que sólo acepta valores no negativos y está sesgada hacia la derecha. Se especifica una distribución de ji cuadrado dada citando sus grados de libertad. (p. 767)
chi-square test statistic Measure of how far the observed counts are from the expected counts relative to the expected counts. The formula is $$\chi^2 = \sum \frac{\left(\text{Observed Count} - \text{Expected Count}\right)^2}{\text{Expected Count}}$$ where the sum is over all possible values of the categorical variable or all cells in the two-way table. (p. 764)	**prueba estadística de ji cuadrado** Una medición de la distancia entre las cuentas observadas y las cuentas previstas en relación con las cuentas previstas. La fórmula es $$\chi^2 = \sum \frac{\left(\text{Cuenta Observadas} - \text{Cuenta Previstas}\right)^2}{\text{Cuenta Previstas}}$$ en la que la suma está sobre todos los valores posibles de la variable categorizada o sobre todas las celdas en la tabla de doble vía. (p. 764)
chi-square test for goodness of fit A test of the null hypothesis that a categorical variable has a specified distribution in the population of interest. (p. 770) For more details, see the inference summary at the back of the book.	**prueba de ji cuadrado para confirmar la bondad de ajuste** Prueba de la hipótesis nula en la cual la variable categórica tiene una distribución especificada en la población de interés. (p. 770) Para más información, ver el resumen de inferencia del final del libro.
chi-square test for homogeneity A test of the null hypothesis that the distribution of a categorical variable is the same for two or more populations/treatments. (p. 789) For more details, see the inference summary at the back of the book.	**prueba de ji cuadrado de homogeneidad** Prueba de la hipótesis nula en la cual la distribución de una variable categórica por la misma en dos o más poblaciones/tratamientos. (p. 789) Para más información, ver el resumen de inferencia del final del libro.

chi-square test for independence A test of the null hypothesis that there is no association between two categorical variables in the population of interest. (p. 796) For more details, see the inference summary at the back of the book.

prueba de ji cuadrado de independencia Prueba de la hipótesis nula en la cual no hay asociación entre dos variables categóricas dentro de la población de interés. (p. 796) Para mayor información, ver el resumen de inferencia del final del libro.

cluster A group of individuals in the population that are located near each other. (p. 258)

clúster Grupo de individuos de una población que se ubican cerca uno del otro. (p. 258)

cluster sampling Method of sampling that divides the population into non-overlapping groups (*clusters*) of individuals that are located near each other, randomly chooses clusters, and includes each member of the selected clusters in the sample. (p. 258)

muestra de clúster Método de muestreo por el que se divide a la población en grupos de individuos (clústers) no superpuestos, se seleccionan aleatoriamente clústers y se incluye a cada miembro de los clústers seleccionados en la muestra. (p. 258)

coefficient of determination r^2 A measure of the percent reduction in the sum of squared residuals when using the least-squares regression line to make predictions, rather than the mean value of y. In other words, r^2 measures the percent of the variability in the response variable that is accounted for by the least-squares regression line. (p. 190)

coeficiente de determinación r^2 Medida del porcentaje de reducción de la suma de cuadrados residuales cuando se hacen predicciones por medio de una línea de regresión de mínimos cuadrados, en lugar de usar el valor medio de y. Es decir, r^2 mide el porcentaje de variabilidad en la variable de respuesta que se calcula a través de la línea de regresión de mínimos cuadrados. (p. 190)

comparison Experimental design principle. Use a design that compares two or more treatments. (p. 281)

comparación Principio de diseño experimental. Se usa un diseño que compara dos o más tratamientos. (p. 281)

complement The complement of event A, written as A^C, is the event that A does not occur. (p. 344)

complemento El complemento del evento A, escrito como A^C, es el evento en el que A no ocurre. (p. 344)

complement rule The probability that an event does not occur is 1 minus the probability that the event does occur. In symbols, $P(A^C) = 1 - P(A)$. (p. 344)

regla del complemento La probabilidad de que no suceda un evento es 1 menos la probabilidad de que el evento sí suceda. En representación simbólica, $P(A^C) = 1 - P(A)$. (p. 344)

completely randomized design Design in which the experimental units are assigned to the treatments completely at random. (p. 283)

diseño completamente aleatorizado Cuando las unidades experimentales se les asignan a los tratamientos de manera completamente aleatoria. (p. 283)

components Individual terms that are added together to produce the chi-square test statistic. Also called *contributions*:

$$\text{component} = \frac{(\text{Observed Count} - \text{Expected Count})^2}{\text{Expected Count}}$$

(p. 772)

componentes Los términos individuales que se suman para producir la estadística de prueba de ji cuadrado. También se denominan *contribuciones*:

$$\text{component} = \frac{(\text{Cuenta Observadas} - \text{Cuenta Previstas})^2}{\text{Cuenta Previstas}}$$

(p. 772)

conditional distribution Describes how the values of one variable vary among individuals who have a specific value of another variable. There is a separate conditional distribution for each value of the other variable. (p. 18)

distribución condicional Describe cómo varían los valores de una variable entre individuos que tienen un valor específico de otra variable. Hay una distribución condicional separada para cada valor de la otra variable. (p. 18)

conditional probability Probability that one event happens given that another event is already known to have happened. The probability that event A happens given that event B has happened is denoted by $P(A|B)$. To find the conditional probability $P(A|B)$, use the formula

$$P(A|B) = \frac{P(A \cap B)}{P(B)}$$
$$= \frac{P(\text{both events occur})}{P(\text{given event occurs})}$$

(p. 359)

probabilidad condicional La probabilidad de que un evento suceda a la luz de que se sabe que otro evento ya sucedió. La probabilidad de que el evento A suceda, dado que el evento B ya sucedió, se denota con $P(A|B)$. Para hallar la probabilidad condicional $P(A|B)$, se usa la fórmula

$$P(A|B) = \frac{P(A \cap B)}{P(B)}$$
$$= \frac{P(\text{both events occur})}{P(\text{given even occurs})}$$

(p. 359)

conditional relative frequency Gives the percent or proportion of individuals that have a specific value for one categorical variable among individuals who share the same value of another categorical variable (the condition). (p. 17)

frecuencia relativa condicional Ofrece el porcentaje o proporción de individuos que tienen un valor específico para una variable categórica entre individuos que comparten el mismo valor de otra variable categórica (condición). (p. 17)

confidence interval Gives an interval of plausible values for a parameter. The interval is calculated from sample data and has the form

$$\text{point estimate} \pm \text{margin of error}$$

or, alternatively,

$$\text{statistic} \pm (\text{critical value}) \cdot (\text{standard error of statistic})$$

(p. 539)

intervalo de confianza Ofrece un intervalo de valores plausibles para un parámetro. El intervalo se computa a partir de los datos muestrales y tiene la forma

$$\text{Estimado de punto} \pm \text{margen de error}$$

o alternativamente,

$$\text{estadística} \pm (\text{valor crítico}) \cdot (\text{desviación estándar de la estadística})$$

(p. 539)

confidence level C Overall success rate of the method used to calculate the confidence interval. In C% of all possible samples, the method would yield an interval that captures the true parameter value when the conditions for inference are met. (p. 540)

nivel de confianza C La tasa general de aciertos del método con el que se computa el intervalo de confianza. En el C% de todas las muestras posibles, el método produciría un intervalo que capta el valor verdadero del parámetro cuando se cumplen las condiciones de inferencia. (p. 540)

confidential A basic principle of data ethics that requires that an individual's data be kept private. Only statistical summaries for groups of subjects may be made public. (p. 307)

confidencial Principio básico de la ética de la gestión de datos. Requiere que los datos de un individuo se mantengan en reserva. Solo pueden hacerse públicos los resúmenes estadísticos de grupos de individuos. (p. 307)

confounding When two variables are associated in such a way that their effects on a response variable cannot be distinguished from each other. (p. 271)

confuso Cuando dos variables se asocian de tal manera que sus efectos en una variable de respuesta no se pueden distinguir el uno del otro. (p. 271)

continuous random variable Variable that can take any value in an interval on the number line. The probability of any event is the area under a density curve and above the values of the variable that make up the event. (p. 399)

variable aleatoria continua Emplea cualquier valor en un intervalo de cifras. La probabilidad de cualquier evento en el área debajo de una curva de densidad y encima de los valores de la variable que componen el evento. (p. 399)

continuous variable A quantitative variable that can take any value in an interval on the number line (p. 4)

variable continua Variabe cuantitativa que puede tomar cualquier valor en un intervalo de cifras. (p. 4)

control Experimental design principle. Keeping variables (other than the explanatory variable) the same for all groups, especially variables that are likely to affect the response variable. Helps avoid confounding and reduces variability in the response variable. (pp. 280–281).

control Principio del diseño experimental. Se mantienen las mismas variables (con excepción de la variable explicativa) para todos los grupos, en especial las variables que podrían afectar la variable de respuesta. Permite evitar la confusión y reduce la variabilidad en la variable de respuesta. (págs. 280–281)

control group Experimental group whose primary purpose is to provide a baseline for comparing the effects of the other treatments. Depending on the purpose of the experiment, a control group may be given an inactive treatment (placebo), an active treatment, or no treatment at all. (p. 276)

grupo de control Grupo experimental cuyo fin primario es establecer una línea base mediante la cual se comparan los efectos de otros tratamientos. Según el objeto del experimento, a un grupo de control se le puede administrar un tratamiento inactivo (placebo), un tratamiento activo, o ningún tratamiento. (p. 276)

convenience sampling Method of sampling that selects individuals from the population who are easy to reach. (p. 251)

muestreo de conveniencia Método de muestreo por el cual se seleccionan individuos de la población con quienes es fácil hacer contacto. (p. 251)

correlation r (*also called* **correlation coefficient**) Measures the direction and strength of the linear relationship between two quantitative variables. We can calculate r using the formula

$$r = \frac{1}{n-1} \sum \left(\frac{x_i - \bar{x}}{s_x} \right) \left(\frac{y_i - \bar{y}}{s_y} \right).$$ (pp. 160, 166)

correlación r (*también conocida como* **coeficiente de correlación**) Mide el sentido y la fuerza de la relación lineal entre dos variables cuantitativas. La r puede calcularse con la fórmula

$$r = \frac{1}{n-1} \sum \left(\frac{x_i - \bar{x}}{s_x} \right) \left(\frac{y_i - \bar{y}}{s_y} \right).$$ (págs. 160, 166)

critical value Multiplier that makes a confidence interval wide enough to have the stated capture rate. The critical value depends on both the confidence level C and the sampling distribution of the statistic. (p. 547)

valor crítico Multiplicador que amplía el intervalo de confianza lo suficiente para retener la tasa de captación indicada. El valor crítico depende de tanto el nivel de confianza C como de la distribución de muestreo de la estadística. (p. 547)

cumulative probability distribution Gives the percentile corresponding to each possible value of the variable. (p. 395)

distribución de probabilidad acumulativa Indica el percentil correspondiente a cada valor posible de la variable. (p. 395)

cumulative relative frequency graph A cumulative relative frequency graph plots a point corresponding to the percentile of a given value in a distribution of quantitative data. Consecutive points are then connected with a line segment to form the graph. (p. 93)

gráfico de la frecuencia relativa acumulada El gráfico de frecuencia relativa acumulada traza un punto correspondiente al percentil de un valor dado en una distribución de datos cuantitativos. A partir de ahí, los puntos consecutivos se conectan con un segmento lineal para formar el gráfico. (p. 93)

D

density curve Models the distribution of a quantitative variable with a curve that (a) is always on or above the horizontal axis and (b) has area exactly 1 underneath it. The area under the curve and above any interval of values on the horizontal axis estimates the proportion of all observations that fall in that interval. (p. 110)

curva de densidad Modelos de distribución de una variable cuantitativa con una curva que (a) siempre está sobre o por encima del eje horizontal y (b) tiene 1 área exactamente debajo. El área debajo de la curva y por encima de todo intervalo de valores en el eje horizontal estima la proporción de todas las observaciones que caen en dicho intervalo. (p. 110)

descriptive statistics Process of describing data using graphs and numerical summaries. (p. 5)

estadística descriptiva Proceso que describe los datos haciendo uso de gráficos y resúmenes numéricos. (p. 5)

discrete random variable Variable that takes a fixed set of possible values with gaps between them. The probability of any event is the sum of the probabilities for the values of the variable that make up the event. (p. 390)

variable aleatoria discreta Variable que emplea un conjunto fijo de valores posibles entre los cuales hay brechas. La probabilidad de cualquier evento es la suma de las probabilidades de los valores de la variable que compone el evento. (p. 390)

discrete variable A quantitative variable that takes a fixed set of possible values with gaps between them. (p. 4)

variable discreta Variable cuantitativa que toma un conjunto fijo de valores posibles entre los cuales hay brechas. (p. 4)

distribution Tells what values a variable takes and how often it takes these values. (p. 5)

distribución Indica qué valores adopta una variable y con qué frecuencia adopta dichos valores. (p. 5)

distribution of sample data Tells what values a variable takes for all individuals in a particular sample. (p. 474)

distribución de valores muestrales Indica qué valores asume una variable para todos los individuos en una muestra particular. (p. 474)

dotplot A graph that displays the distribution of a quantitative variable by plotting each data value as a dot above its location on a number line. (p. 30)

gráfico de puntos Gráfico que muestra la distribución de una variable cuantitativa trazando el valor de cada dato encima de su ubicación a lo largo de una línea de cifras. (p. 30)

double-blind An experiment in which neither the subjects nor those who interact with them and measure the response variable know which treatment a subject is receiving. (p. 277)

doble ciego Experimento en el que ninguno de los sujetos ni aquellos que interactúan con los sujetos y que miden la variable de repuesta saben qué tratamiento recibe el sujeto. (p. 277)

E

empirical probability Estimated probability of a specific outcome of a random process (like getting a head when tossing a fair coin) obtained by actually performing many trials of the random process. (p. 329)

probabilidad empírica Probabilidad estimada de obtener un resultado específico en un proceso aleatorio (como lanzar una moneda al aire y salga cara) luego de realizar muchos ensayos del proceso aleatorio. (p. 329)

empirical rule (*also known as the* **68–95–99.7 rule**) In a Normal distribution with mean μ and standard deviation σ, (a) approximately 68% of the observations fall within σ of the mean μ, (b) approximately 95% of the observations fall within 2σ of μ, and (c) approximately 99.7% of the observations fall within 3σ of μ. (p. 117)

regla empírica (*también conocida como* **regla 68–95–99.7**) En una distribución normal, com media μ y desviación estándar σ, (a) el 68% de las observaciones aproximadamente caen dentro de σ de la media μ, (b) el 95% de las observaciones aproximadamente caen dentro de 2σ de la media μ y (c) el 99,7% de las observaciones aproximadamente caen dentro de 3σ de la media μ. (p. 117)

event Any collection of outcomes from some random process. Events are usually designated by capital letters, like A, B, C, and so on. (p. 343)	**evento** Cualquier colección de los resultados de un proceso aleatorio. Los eventos generalmente se designan en letra mayúscula, como A, B, C, y así sucesivamente. (p. 343)
expected counts Numbers of individuals in the sample that would fall in each cell of the one-way or two-way table if H_0 were true. (pp. 763, 782)	**cuentas previstas** Cantidades de individuos en la muestra que caerían en cada celda en la tabla, sea de una vía o de dos vías, si H_0 fuera verdad. (págs. 763, 782)
experiment A study in which researchers deliberately impose treatments (conditions) on individuals to measure their responses. (p. 271)	**experimento** Estudio en el que los investigadores deliberadamente les imponen tratamientos (condiciones) a individuos con el fin de medir sus respuestas. (p. 271)
experimental unit The object to which a treatment is randomly assigned. When the experimental units are human beings, they are often called subjects. (p. 273)	**unidad experimental** Objeto al cual se le asigna un tratamiento aleatorio. Cuando las unidades experimentales son seres humanos, generalmente se se refiere a ellos como sujetos. (p. 273)
explanatory variable Variable that may help predict or explain changes in a response variable. (pp. 153, 270)	**variable explicativa** Variable que puede ayudar a predecir o explicar cambios en una variable de respuesta. (págs. 153, 270)
exponential model Relationship of the form $y = ab^x$. If the relationship between two variables follows an exponential model and we plot the logarithm (base 10 or base e) of y against x, we should observe a straight-line pattern in the transformed data. (p. 221)	**modelo exponencial** Relación de la forma $y = ab^x$. Si la relación entre dos variables se ajusta a un modelo exponencial, y trazamos el logaritmo (de base 10 o de base e) de y con respecto a x, se debe observar un patrón en línea recta en los datos transformados. (p. 221)
extrapolation Use of a regression model for prediction outside the interval of x values used to obtain the model. The further we extrapolate, the less reliable the predictions. (p. 178)	**extrapolación** Uso de un modelo de regresión para hacer predicciones por fuera del intervalo de valores x que se utiliza para obtener el modelo. Cuanto mayor sea la extrapolación, menos confiable serán las predicciones. (p. 178)

F

factor Explanatory variable in an experiment that is manipulated and may cause a change in the response variable. (p. 274)	**factor** La variable explicativa en un experimento que se manipula y puede causar un cambio en la variable de respuesta. (p. 274)
factorial For any positive whole number n, its factorial $n!$ is $n! = n(n-1)(n-2) \cdot \ldots \cdot 3 \cdot 2 \cdot 1$ In addition, we define $0! = 1$. (p. 435)	**factorial** Para cualquier número entero positivo n, su factorial $n!$ es $n! = n(n-1)(n-2) \cdot \ldots \cdot 3 \cdot 2 \cdot 1$ Además, definimos $0! = 1$. (p. 435)
fail to reject H_0 If the observed result is not unlikely to occur when the null hypothesis is true, we should fail to reject H_0 and say that we do not have convincing evidence for H_a. (p. 590)	**no rechazar H_0** Si no es improbable que el resultado observado suceda cuando es verdad la hipótesis nula, no se debe rechazar H_0 y se ha de indicar que no contamos con evidencia convincente de H_a (p. 590)
first quartile Q_1 If the observations in a data set are ordered from smallest to largest, the first quartile Q_1 is the median of the data values that are to the left of the median in the ordered list (p. 64)	**primer cuartil Q_1** Si las observaciones del conjunto de datos se ordenan de menor a mayor, el primer cuartil Q_1 es la media de los valores de los datos ubicados a la izquierda de la media en la lista ordenada. (p. 64)
five-number summary The minimum, first quartile Q_1, median, third quartile Q_3, and maximum of a distribution of quantitative data. (p. 68)	**resumen de cinco cifras** El mínimo, el primer cuartil Q_1, la media, el tercer cuartil Q_3, y el máximo de una distribución de datos cuantitativos. (p. 68)
frequency table Table that displays the number of individuals having each value (p. 9)	**tabla de frecuencias** Tabla que muestra el número de individuos que tiene cada valor. (p. 9)

G

general addition rule If A and B are two events resulting from some random process, then the probability that event A or event B (or both) occur is $P(A \text{ or } B) = P(A \cup B) = P(A) + P(B) - P(A \cap B)$. (p. 348)	**regla general de adición** Si A y B son dos eventos resultantes de algún proceso aleatorio, la probabilidad de que el evento A o el evento B (o ambos) suceda es $P(A \circ B) = P(A \cup B) = P(A) + P(B) - P(A \cap B)$ (p. 348)

general multiplication rule The probability that events A and B both occur can be found using the formula $P(A \text{ and } B) = P(A \cap B) = P(A) \cdot P(B \mid A)$ (p. 366)	**regla general de multiplicación** La probabilidad de que sucedan los eventos A y B se puede determinar utilizando la fórmula $P(A \text{ y } B) = P(A \cap B) = P(A) \cdot P(B \mid A)$ (p. 366)
geometric distribution In a geometric setting, suppose we let $X =$ the number of trials it takes to get a success. The probability distribution of X is a geometric distribution with parameter p, the probability of a success on any trial. The possible values of X are $1, 2, 3, \ldots$. (p. 451)	**distribución geométrica** En un entorno geométrico, supongamos que se permite que $X =$ la cantidad de ensayos que se precisan para lograr un acierto. La distribución de la probabilidad de X es una distribución geométrica con el parámetro p, la probabilidad de lograr un acierto en cualquier ensayo. Los valores posibles de X son $1, 2, 3, \ldots$. (p. 451)
geometric probability formula If X has the geometric distribution with probability p of success on each trial, the probability of getting the first success on the xth trial $(x = 1, 2, 3, \ldots)$ is $P(X = x) = (1-p)^{x-1}p$. (p. 451)	**fórmula de probabilidad geométrica** Si X tiene una distribución geometría con la probabilidad p de acierto en cada ensayo, la probabilidade de acertar por primera vez en el ensayo número x $(x = 1, 2, 3, \ldots)$ es $P(X = x) = (1-p)^{x-1}p$. (p. 451)
geometric random variable The number of trials X that it takes to get a success in a geometric setting. (p. 451)	**variable aleatoria geométrica** La cantidad de ensayos X que se precisan para lograr un acierto en un entorno geométrico. (p. 451)
geometric setting Arises when we perform independent trials of the same random process and record the number of trials it takes to get one success. On each trial, the probability p of success must be the same. (p. 450)	**entorno geométrico** Surge un entorno geométrico cuando se realizan ensayos independientes del mismo proceso aleatorio y se graban la cantidad de ensayos que se precisan para lograr un acierto. En cada ensayo, la probabilidad p de lograr un acierto tiene que ser la misma. (p. 450)

H

heterogeneous When the individuals in a group differ considerably with respect to the variable of interest. Ideally, the individuals in clusters are heterogeneous, and each cluster mirrors the variability in the population. (p. 258)	**heterogéneo** Condición que se presenta cuando los individuos del un grupo difieren considerablemente respecto de la variable de interés. Idealmente, los individuos de los grupos son heterogéneos y cada grupo refleja la variabilidad de la población. (p. 258)
high leverage Points that have much larger or much smaller x values than the other points in a bivariate quantitative data set. (p. 200)	**apalancamiento alto** Puntos con valores x mucho mayores o mucho menores que los otros puntos de un conjunto de datos cuantitativos bivariados. (p. 200)
histogram Graph that displays the distribution of a quantitative variable by showing each interval of values as a bar. The heights of the bars show the frequencies or relative frequencies of values in each interval. (p. 40)	**histograma** Muestra la distribución de una variable cuantitativa que expresa cada intervalo de valores en forma de barra. La altura de las barras muestra las frecuencias o las frecuencias relativas de los valores en cada intervalo. (p. 40)
homogeneous When the individuals in a group are quite similar with respect to the variable of interest. Ideally, the individuals in strata are homogeneous. (p. 257)	**homogéneo** Condición que se presenta cuando los individuos de un grupo son bastante parecidos respecto de la variable de interés. Idealmente, los individuos en estratos son homogéneos. (p. 257)

I

independent events Two events are independent if knowing whether or not one event has occurred does not change the probability that the other event will happen. In other words, events A and B are independent if $$P(A \mid B) = P(A \mid B^C) = P(A)$$ and $$P(B \mid A) = P(B \mid A^C) = P(B)$$ (p. 363)	**eventos independientes** Dos eventos son independientes cuando el hecho de saber o desconocer si un evento ha ocurrido no cambia la probabilidad de que el otro suceda. Es decir, los eventos A y B son independientes si $$P(A \mid B) = P(A \mid B^C) = P(A)$$ y $$P(B \mid A) = P(B \mid A^C) = P(B)$$ (p. 363)
independent random variables If knowing the value of X does not help us predict the value of Y, then X and Y are independent random variables. In other words, two random variables are independent if knowing the value of one variable does not change the probability distribution of the other variable. (p. 418)	**variables aleatorias independientes** Si conocer los valores de X no nos sirve para predecir el valor de Y, entonces X e Y son variables aleatorias independientes. Es decir, dos variables aleatorias son independientes si conocer el valor de una variable no cambia la probabilidad de distribución de la otra variable. (p. 418)

individual An object described by a set of data. Individuals can be people, animals, or things. (p. 3)	**individuo** Un objeto descrito por un conjunto de datos. Los individuos pueden ser personas, animales o cosas. (p. 3)
inferential statistics Drawing conclusions that go beyond the data at hand. (p. 6)	**estadística inferencial** Llegar a conclusiones que van más allá de los datos que están a la mano. (p. 6)
inference about cause and effect Conclusion from the results of an experiment that the treatments caused the difference in responses. Requires a well-designed experiment in which the treatments are randomly assigned to the experimental units and statistically significant results. (p. 303)	**inferencia sobre causa y efecto** Uso de los resultados de un experimento para llegar a la conclusión de que son los tratamientos los que marcan la diferencia en las respuestas. Exige un experimento bien diseñado en el que los tratamientos se asignan de manera aleatoria a las unidades experimentales y resultados estadísticamente significativos. (p. 303)
inference about a population Conclusion about the larger population based on sample data. Requires that the individuals taking part in a study be randomly selected from the population of interest. (p. 303)	**inferencia sobre una población** Conclusión sobre una población en general con base en datos muestrales. Se precisa que los participantes del estudio sean escogidos de manera aleatoria a partir de la población de interés. (p. 303)
influential point Any point that, if removed, substantially changes the slope, y intercept, correlation, coefficient of determination, or standard deviation of the residuals. (p. 200)	**punto influyente** Cualquier punto que, al quitarse, cambia sustancialmente la pendiente, la interceptación y, el coeficiente de determinación o la desviación estándar de las residuales. (p. 200)
informed consent Basic principle of data ethics that states that individuals must be informed in advance about the nature of a study and any risk of harm it may bring. Participating individuals must then consent in writing. (p. 307)	**consentimiento informado** Principio básico de ética en la gestión de los datos. A los individuos se les ha de informar, con antelación, de la naturaleza de un estudio y de los riesgos o perjuicios que podría conllevar. Entonces los participantes tendrán que dar autorización por escrito. (p. 307)
institutional review board Board charged with protecting the safety and well-being of the participants in advance of a planned study and with monitoring the study itself. (p. 307)	**junta de revisión institucional** Junta encargada de salvaguardar la seguridad y el bienestar de los participantes, anticipándose a un estudio planeado, además de supervisar el estudio mismo. (p. 307)
interquartile range The distance between the first and third quartiles of a distribution. In symbols, $IQR = Q_3 - Q_1$. (p. 64)	**gama entre cuartiles** Distancia entre el primer cuartil y el tercer cuartil de una distribución. En símbolos, $IQR = Q_3 - Q_1$. (p. 64)
intersection The event "A and B" is called the intersection of events A and B. It consists of all outcomes that are common to both events, and is denoted by $A \cap B$. (p. 351)	**intersección** Al evento "A y B" se le conoce como la intersección de los eventos A y B. Consiste en todos los resultados que son comunes para ambos eventos, y se expresa $A \cap B$. (p. 351)

J

joint probability The probability P(A and B) that events A and B both occur. (p. 348)	**probabilidad conjunta** Probabilidad P(A y B) de que ocurran ambos eventos A y B. (p. 348)
joint relative frequency Gives the percent or proportion of individuals that have a specific value for one categorical variable and a specific value for another categorical variable. (p. 15)	**frecuencia relativa conjunta** Ofrece el porcentaje o proporción de individuos que tienen un valor específico para una variable categórica y un valor específico para otra variable categórica. (p. 15)

L

Large Counts condition Suppose X is the number of successes and $\hat{p}$ is the proportion of successes in a binomial setting with n trials and success probability p. The Large Counts condition says that the distribution of X and the distribution of $\hat{p}$ will be approximately Normal if $np \geq 10$ and $n(1-p) \geq 10$. That is, the expected numbers (counts) of successes and failures are both at least 10. (pp. 448, 490)	**condición de cuentas grandes** Supongamos que X representa el número de aciertos y $\hat{p}$ es la proporción de aciertos en un entorno binomial con n ensayos y una probabilidad de aciertos p. La condición de cuentas grandes establece que la distribución de X y la distribución de $\hat{p}$ será aproximadamente normal si $np \geq 10$ y $n(1-p) \geq 10$. Es decir, el número (cuenta) de aciertos y fracasos es, por lo menos, 10. (págs. 448, 490)
Large Counts condition for a chi-square test It is safe to use a chi-square distribution to perform P-value calculations if all expected counts are at least 5. (p. 769)	**condición de cuentas grandes en la prueba de ji cuadrado** Se puede utilizar sin problemas la distribución de ji cuadrado para realizar cómputos de valor P si todas las cuentas previstas son de al menos 5. (p. 769)

law of large numbers If we observe more and more trials of any random process, the proportion of times that a specific outcome occurs approaches its probability. (p. 329)

ley de las cifras grandes Si se observan más y más ensayos en cualquier proceso aleatorio, la proporción de veces que se da un resultado específico se aproxima a su probabilidad. (p. 329)

least-squares regression line The line that makes the sum of the squared residuals as small as possible. (p. 183)

línea de regresión de mínimos cuadrados La línea que reduce al mínimo posible la suma de los cuadrados residuales. (p. 183)

level Specific value of an explanatory variable (factor) in an experiment. (p. 274)

nivel Valor específico de una variable explicativa (factor) en un experimento. (p. 274)

linear combination A linear combination of two random variables X and Y can be written in the form $aX + bY$, where a and b are constants. (p. 418)

combinación lineal La combinación lineal de dos variables aleatorias X e Y puede representarse con la fórmula $aX + bY$, en la cual a y b son constantes. (p. 418)

linear transformation A linear transformation of the random variable X involves multiplying the values of the variable by a constant and/or adding a constant. It can be written in the form $Y = a + bX$. (p. 414)

transformación lineal La transformación lineal de la variable aleatoria X implica multiplicar los valores de la variable por una constante o sumar una constante. Puede representarse con la fórmula $Y = a + bX$. (p. 414)

M

margin of error Describes how far, at most, we expect the estimate to vary from the true population value. That is, the difference between the point estimate and the true parameter value will be less than the margin of error in C% of all samples, where C is the confidence level. (pp. 299, 541)

margen de error Describe hasta qué punto máximo puede esperarse una variación del valor estimado respecto del valor verdadero de la población. Esto es, que la diferencia entre el estimado del punto y el valor real del parámetro será menor que el margen de error en C% de todas las muestras, donde C es el nivel de confianza. (págs. 299, 541)

marginal relative frequency Gives the percent or proportion of individuals that have a specific value for one categorical variable. (p. 14)

frecuencia relativa marginal Ofrece el porcentaje o proporción de individuos que tienen un valor específico para una variable categórica. (p. 14)

matched pairs design Common experimental design for comparing two treatments that uses blocks of size 2. In some matched pairs designs, each subject receives both treatments in a random order. In others, two very similar experimental units are paired and the two treatments are randomly assigned within each pair. (p. 288)

diseño de pares coincidentes Diseño experimental común que emplea bloques de tamaño 2 para comparar dos tratamientos. En algunos diseños de pares coincidentes, cada sujeto se somete a ambos tratamientos en un orden aleatorio. En otros, se emparejan dos unidades experimentales muy similares y los tratamientos se asignan aleatoriamente a los sujetos de cada par. (p. 288)

mean The average of all the individual data values in a distribution of quantitative data. To find the mean, add all the values and divide by the total number of data values. In symbols, the sample mean $\bar{x}$ is given by

$$\bar{x} = \frac{\sum x_i}{n}$$

Use μ to denote the population mean. (p. 54)

media El promedio de todos los valores de datos individuales en una distribución de datos cuantitativos. Para obtener la media, se suman todos los valores y se divide el resultado entre el número total de observaciones. En símbolos, la media de la muestra $\bar{x}$ se determina de la siguiente manera:

$$\bar{x} = \frac{\sum x_i}{n}$$

Para referirse a la media de la población debe usarse μ. (p. 54)

mean of a density curve Point at which a density curve would balance if made of solid material. (p. 112)

media de una curva de densidad El punto en el cual la curva se equilibraría si estuviera elaborada de un material macizo. (p. 112)

mean (expected value) of a discrete random variable Describes the variable's long-run average value over many, many trials of the same random process. To find the mean (expected value) of X, multiply each possible value by its probability, then add all the products:

$$\mu_X = E(X) = x_1 p_1 + x_2 p_2 + x_3 p_3 + \dots = \sum x_i p_i$$

(p. 394)

media (valor previsto) de una variable aleatoria discreta Describe el valor medio de la variable en el largo plazo a partir de muchos ensayos de un proceso aleatorio. Para hallar la media (un valor previsto) de X, se multiplica cada valor posible por su probabilidad y luego se suman todos los productos:

$$\mu_X = E(X) = x_1 p_1 + x_2 p_2 + x_3 p_3 + \dots = \sum x_i p_i$$

(p. 394)

median The midpoint of a distribution; the number such that about half the observations are smaller and about half are larger. To find the median, arrange the data values from smallest to largest. If the number n of data values is odd, the median is the middle value in the ordered list. If the number n of data values is even, use the average of the two middle values in the ordered list as the median. (p. 57)

mediana El punto intermedio de una distribución, con una cifra tal que aproximadamente la mitad de las observaciones son más pequeñas y la mitad son más grandes. Para hallar la mediana, los datos deben organizarse en orden ascedente, es decir de menor a mayor valor. Si el número n valores es impar, la mediana es el valor del medio de la lista ordenada. Si el número n de valores es par, la mediana es el promedio de los dos valores del medio de la lista ordenada. (p. 57)

median of a density curve Equal-areas point, the point that divides the area under the curve in half. (p. 112)

media de una curva de densidad Es el punto que divide en dos mitades iguales la zona ubicada por debajo de la curva. (p. 112)

mode Value in a distribution having the greatest frequency. (p. 54)

modo En una distribución, el valor que tiene la mayor frecuencia. (p. 54)

mosaic plot A modified segmented bar graph in which the width of each rectangle is proportional to the number of individuals in the corresponding category. (p. 19)

Gráfico de mosaico Variación del gráfico de barras segmentadas en el cual el ancho de cada rectángulo es proporcional al número de individuos de la categoría correspondiente. (p. 19)

multiplication rule for independent events If A and B are independent events, then the probability that A and B both occur is $P(A \cap B) = P(A) \cdot P(B)$. (p. 372)

regla de multiplicación de eventos independientes Si A y B son eventos independientes, la probabilidad de que sucedan ambos, tanto A como B es $P(A \cap B) = P(A) \cdot P(B)$. (p. 372)

mutually exclusive (disjoint) Two events A and B that have no outcomes in common and so can never occur together. That is, $P(A \text{ and } B) = 0$. (p. 345)

exclusivos mutuamente (desencajamiento) Dos eventos A y B que no tienen resultados en común y por lo tanto nunca pueden suceder a la vez. Es decir, $P(A \text{ y } B) = 0$. (p. 345)

N

negative association When values of one variable tend to decrease as the values of the other variable increase. (p. 157)

asociación negativa Se presenta cuando los valores de una variable tienden a disminuir al tiempo que los de la otra variable aumentan. (p. 157)

no association Knowing the value of one variable does not help predict the value of another variable. (p. 157)

sin asociación Conocer el valor de una variable no sirve para anticipar el valor de otra. (p. 157)

nonresponse Occurs when an individual chosen for the sample can't be contacted or refuses to participate. (p. 261)

no respondió Sucede cuando a un individuo escogido para la muestra no se le puede contactar o el sujeto se niega a participar. (p. 261)

Normal approximation to a binomial distribution Suppose that a count X of successes has the binomial distribution with n trials and success probability p. When n is large, the distribution of X is approximately Normal with mean np and standard deviation $\sqrt{np(1-p)}$. We use this Normal approximation when $np \geq 10$ and $n(1-p) \geq 10$. (p. 448)

aproximación Normal hacia una distribución binomial Supongamos que una cuenta X de aciertos tiene la distribución binomial con n ensayos y una probabilidad de acierto p. Cuando n es grande, la distribución de X es aproximadamente Normal con media np y desviación estándar $\sqrt{np(1-p)}$. Se hace uso de esta aproximación Normal cuando $np \geq 10$ y $n(1-p) \geq 10$. (p. 448)

Normal curve Important kind of density curve that is symmetric, single-peaked, and bell-shaped. (p. 114)

curva Normal Tipo importante de curva de densidad que está simétrica, de un solo pico y con la forma de curva de campana. (p. 114)

Normal distribution Distribution described by a Normal curve. Any Normal distribution is completely specified by two parameters, its mean μ and standard deviation σ. The mean of a Normal distribution is at the center of the symmetric Normal curve. The standard deviation is the distance from the center to the change-of-curvature points on either side. (p. 114)

distribución Normal Según la describe una curva Normal. Cualquier distribución Normal se especifica completamente con dos parámetros, su media μ y la desviación estándar σ. La media de una distribución Normal yace en el centro de la curva Normal simétrica. La desviación estándar es la distancia del centro a los puntos a ambos lados en los que cambia la curva. (p. 114)

Normal/Large Sample condition A condition for performing inference about a mean, which requires that the data come from a Normally distributed population or that the sample size is large ($n \geq 30$). When the sample size is small and the shape of the population distribution is unknown, a graph of the sample data shows no strong skewness or outliers. When performing inference about a difference between two means, check that this condition is met for both samples. (p. 655)

condición de muestra Normal/Grande Condición para hacer una inferencia sobre una media, que requiere que los datos procedan de una población con distribución Normal o que el tamaño de la muestra sea grande ($n \geq 30$) Si la muestra es pequeña y se desconoce la forma de la distribución de la población, el gráfico de la muestra no tendrá asimetría ni valores atípicos. Al hacer una inferencia sobre la diferencia entre dos medias, revisa que ambas muestras cumplan con esta condición. (p. 655)

Normal probability plot A scatterplot of the ordered pair (data value, expected z-score) for each of the individuals in a quantitative data set. That is, the x-coordinate of each point is the actual data value and the y-coordinate is the expected z-score corresponding to the percentile of that data value in a standard Normal distribution. If the points on a Normal probability plot lie close to a straight line, the data are approximately Normally distributed. A nonlinear form in a Normal probability plot indicates a non-Normal distribution. (p. 133)

gráfico de probabilidad Normal Gráfico de dispersión del par ordenado (valor, puntuación z esperada) para cada individuo del conjunto de datos cuantitativos. La coordenada x de cada punto es el valor real mientras que la coordenada y constituye la puntuación z esperada correspondiente al percentil del valor de los datos en una distribución normal estándar. Si los puntos en un gráfico de probabilidad Normal yacen cerca de una línea recta, los datos son distribuidos Normalmente de manera aproximada. En un gráfico de probabilidad normal, un formato no lineal es signo de distribución no normal. (p. 133)

null hypothesis H_0 Claim we weigh evidence against in a significance test. Often the null hypothesis is a statement of "no difference." (p. 586)

hipótesis nula H_0 Contrapeso de la evidencia en una prueba de significancia. A menudo la hipótesis nula es una declaración de que "no hay diferencia." (p. 586)

O

observational study Study that observes individuals and measures variables of interest but does not attempt to influence the responses. (p. 270)

estudio de observación Se observan los individuos y se miden las variables de interés pero no se trata de influir en las respuestas. (p. 270)

observed counts Actual numbers of individuals in the sample that fall in each cell of the one-way or two-way table. (p. 760)

cuentas observadas Las cifras reales que corresponden a individuos en la muestra que caen en cada celda de la tabla de una vía o en la de dos vías. (p. 760)

one-sample t interval for a mean Confidence interval used to estimate a population mean. (p. 659) For more details, see the inference summary at the back of the book.

intervalo t de una sola muestra para una media Intervalo de confianza que sirve para estimar la media de una población (p. 659) Para más información, ver el resumen de inferencia del final del libro.

one-sample t test for a mean A test of the null hypothesis that a population mean is equal to a specified value. (p. 705) For more details, see the inference summary at the back of the book.

intervalo t de una sola muestra para una media Prueba de la hipótesis nula en la cual la media de la población es igual a un valor específicado. (p. 705) Para más información, ver el resumen de inferencia del final del libro.

one-sample z interval for a proportion Confidence interval used to estimate a population proportion. (p. 553) For more details, see the inference summary at the back of the book.

intervalo z de una sola muestra para una proporción Intervalo de confianza que se usa para estimar la proporción de una población. (p. 553) Para más información, ver el resumen de inferencia del final del libro.

one-sample z test for a proportion A test of the null hypothesis that a population proportion is equal to a specified value. (p. 605) For more details, see the inference summary at the back of the book.

prueba z de una sola muestra para una proporción Prueba de la hipótesis nula en la cual la proporción de una población es igual a un valor especificado. (p. 605) Para más información, ver el resumen de inferencia del final del libro.

one-sided alternative hypothesis An alternative hypothesis is one-sided if it states that a parameter is greater than the null value or if it states that the parameter is less than the null value. (p. 587)

hipótesis alternativa unilateral Una hipótesis alternativa es unilateral si indica que un parámetro es mayor que el valor nulo o si indica que el parámetro es más pequeño que el valor nulo. (p. 587)

one-way table Table used to display the distribution of a single categorical variable. (p. 761)

tabla de una vía Se usa para mostrar la distribución de una sola variable categorizada. (p. 761)

outlier Individual value that falls outside the overall pattern of a distribution. Call an observation an outlier if it falls more than 1.5 IQR above the third quartile or more than 1.5 IQR below the first quartile. (pp. 34, 66)

valor atípico Valor individual que queda por fuera del patrón general de la distribución. La observación es un valor atípico si queda más de 1.5 IQR por encima del tercer cuartil o más de 1.5 IQR por debajo del primer cuartil. (págs. 34, 66)

outlier in regression A point that does not follow the pattern of the data and has a large residual. (p. 200)

valor atípico en regresión Punto que no sigue el patrón de los datos y cuenta con una residual grande. (p. 200)

P

P-value The probability of getting evidence for the alternative hypothesis H_a as strong as or stronger than the observed evidence when the null hypothesis H_0 is true. The smaller the P-value, the stronger the evidence against H_0 and in favor of H_a provided by the data. (p. 588)

valor P La probabilidad de obtener evidencia para la hipótesis alternativa H_a es tan fuerte o más fuerte que la evidencia observada cuando la hipótesis H_0 es verdadera. Cuanto menor sea el valor P, más fuerte será la evidencia contra H_0 y en favor de H_a que proporcionan los datos. (p. 588)

paired data The result of recording two values of the same quantitative variable for each individual or for each pair of similar individuals. To analyze paired data, start by computing the difference for each pair. Then make a graph of the differences. Use the mean difference and the standard deviation of the differences as summary statistics. (p. 676)

datos apareados El resultado de registrar dos valores de la misma variable cuantitativa por cada individuo o por cada par de individuos similares. Para analizar datos apareados, calcula primero la diferencia entre cada par. Entonces, haz un gráfico de las diferencias. Usa la diferencia media y la desviación estándar de las diferencias como estadísticas de resumen. (p. 676)

paired *t* interval for a mean difference (*also called a* **one-sample *t* interval for a mean difference**) Confidence interval used to estimate a population (true) mean difference. (p. 680) For more details, see the inference summary at the back of the book.

intervalo *t* apareado para la diferencia media (*también llamado* **intervalo *t* para una diferencia media**) Intervalo de confianza que sirve para estimar la diferencia media (real) de una población. (p. 680) Para más información, ver el resumen de inferencia del final del libro.

paired *t* test for a mean difference (*also called a* **one-sample *t* test for a mean difference**) A test of the null hypothesis that a population (true) mean difference is equal to a specified value, usually 0. (p. 728) For more details, see the inference summary at the back of the book.

prueba *t* apareada para la diferencia media (*también llamada* **prueba *t* de una sola muestra para la diferencia media**) Prueba de la hipótesis nula en la cual la diferencia media (real) de una población es igual a un valor especificado, generalmente 0. (p. 728) Para más información, ver el resumen de inferencia del final del libro.

parameter A number that describes some characteristic of a population. (pp. 56, 470)

parámetro Número que describe ciertas características de una población. (págs. 56, 470)

percentile The pth percentile of a distribution is the value with $p\%$ of observations less than or equal to it. (p. 91)

percentil El percentil pth de una distribución es el valor cuyo porcentaje de las observaciones es menor o igual que la cifra. (p. 91)

pie chart Chart that shows the distribution of a categorical variable as a "pie" whose slices have areas proportional to the category frequencies or relative frequencies. A pie chart must include all the categories that make up a whole. (p. 10)

gráfico circular Muestra la distribución de una variable categorizada con la forma de un círculo subdividido en porciones cuyo tamaño es proporcional a las frecuencias de categoría o frecuencias relativas. El gráfico circular tiene que incluir todas las categorías que componen la totalidad. (p. 10)

placebo A treatment that has no active ingredient but is otherwise like other treatments. (p. 272)

placebo Tratamiento que, más allá de no tener un ingrediente activo, es como todos los tratamientos. (p. 272)

placebo effect Describes the fact that some subjects in an experiment will respond favorably to any treatment, even an inactive one (placebo). (p. 277)

efecto placebo Describe el hecho de que algunos sujetos del experimento responden de manera favorable a cualquier tratamiento, incluso uno inactivo (con placebo). (p. 277)

point estimate Specific value of a point estimator from a sample. (p. 537)

estimado de punto El valor específico de un estimador de punto tomado de una muestra. (p. 537)

point estimator Statistic that provides an estimate of a population parameter. (p. 537)

estimador de punto Estadística que nos da un estimado de un parámetro de la población. (p. 537)

pooled or combined sample proportion The overall proportion of successes in the two samples is

$$\hat{p}_C = \frac{\text{number of successes in both samples combined}}{\text{number of individuals in both samples combined}}$$
$$= \frac{X_1 + X_2}{n_1 + n_2}$$

(p. 627)

proporción combinada de la muestra La proporción total de aciertos en las dos muestras es

$$\hat{p}_C = \frac{\text{cuenta de aciertos en ambas muestras combinadas}}{\text{cuenta de individuos en ambas muestras combinadas}} = \frac{X_1 + X_2}{n_1 + n_2}$$

(p. 627)

population In a statistical study, the entire group of individuals we want information about. (p. 249)	**población** En un estudio estadístico, la población es el grupo completo de individuos sobre el cual deseamos contar con información. (p. 249)
population distribution The distribution of a variable for all individuals in the population. (pp. 392, 474)	**distribución de la población** Distribución de una variable para todos los individuos de la población. (págs. 392, 474)
population (true) regression line Regression line $\mu_y = \alpha + \beta x$ calculated from every value in the population. (pp. 813–814)	**línea de regresión (real) de la población** La línea de regresión $\mu_y = \alpha + \beta x$ calculada a partir de cada valor de la población. (págs. 813–814)
positive association When values of one variable tend to increase as the values of the other variable increase. (p. 157)	**asociación positiva** Cuando los valores de una variable tienden a aumentar a medida que los otra variable aumentan. (p. 157)
power The probability that a test will find convincing evidence for H_a when a specific alternative value of the parameter is true. The power of a test against any alternative is 1 minus the probability of a Type II error for that alternative; that is, power = $1 - P$(Type II error). (p. 612)	**poder** La probabilidad de que en una prueba halle evidencia convincente de H_a cuando un valor alternativo especificado del parámetro es verdadero. El poder de una prueba con respecto a cualquier alternativa es 1 menos la probabilidad de un error Tipo II para dicha alternativa; es decir, poder = $1 - P$(error Tipo II). (p. 612)
power model Relationship of the form $y = ax^p$. When experience or theory suggests that the relationship between two variables is described by a power model, you can transform the data to achieve linearity in two ways: (1) raise the values of the explanatory variable x to the p power and plot the points (x^p, y), or (2) take the pth root of the values of the response variable y and plot the points $(x, \sqrt[p]{y})$. If you don't know what power to use, taking the logarithms of both variables should produce a linear pattern. (p. 214)	**modelo de poder** Relación de la forma $y = ax^p$. Cuando la experiencia o una teoría sugiere que la relación entre dos variables la describe un modelo de poder, se pueden transformar los datos para que logren la linealidad de dos maneras: (1) elevar los valores de la variable explicativa x a la potencia p y trazar los puntos (x^p, y), o (2) tomar la raíz p de los valores de la variable de respuesta y y trazar los puntos $(x, \sqrt[p]{y})$. Si no se sabe qué potencia se ha de usar, se toman los logaritmos de ambas variables para producir un patrón lineal. (p. 214)
predicted value In regression, $\hat{y}$ (read "y hat") is the predicted value of the response variable y for a given value of the explanatory variable x. (p. 177)	**valor proyectado** $\hat{y}$ es el valor proyectado de la variable de respuesta y para un valor dado de la viariable explicativa x. (p. 177)
probability A number between 0 and 1 that describes the proportion of times an outcome of a random process would occur in a very long series of trials. (p. 329)	**probabilidad** Cifra entre 0 y 1 que describe la proporción de veces que un resultado de un proceso aleatorio sucedería en una serie muy prolongada de repeticiones. (p. 329)
probability distribution Gives the possible values of a random variable and their probabilities. (p. 390)	**distribución de la probabilidad** Presenta los valores posibles de una variable aleatoria y (p. 390)
probability model Description of some random process that consists of two parts: a list of all possible outcomes and a probability for each outcome. (p. 342)	**modelo de probabilidad** Descripción de un proceso aleatorio que consta de dos partes: una lista de todos los resultados posibles y una probabilidad para cada resultado. (p. 342)
prospective observational study An observational study that tracks individuals into the future. (p. 270)	**estudio de observación prospectivo** Estudio de observación que sigue la trayectoria futura de los individuos. (p. 270)

Q

quantitative variable Variable that takes number values that are quantities—counts or measurements. (p. 3)	**variable cuantitativa** Variable que toma valores numéricos que representan cantidades: cuentas o medidas. (p. 3)
quartiles The quartiles of a distribution divide the ordered data set into four groups having roughly the same number of values. (p. 64)	**cuartiles** Los cuartiles de una distribución dividen un conjunto de datos ordenados en cuatro grupos que tienen aproximadamente el mismo número de valores. (p. 64)

R

random assignment Experimental design principle. Use a chance process to assign experimental units to treatments (or treatments to experimental units). Doing so helps create roughly equivalent groups of experimental units by balancing the effects of other variables among the treatment groups. (pp. 279, 281)	**asignación aleatoria** Principio de diseño experimental. Se usa el azar para asignar unidades experimentales a los tratamientos (o tratamientos a las unidades experimentales) de manera tal que se formen grupos de unidades experimentales más o menos equivalentes al equilibrar los efectos de otras variables entre los grupos de tratamiento. (págs. 279, 281)

random condition A condition for performing inference, which requires that the data come from a random sample from the population of interest or from a randomized experiment. When comparing two or more populations or treatments, check that the data come from independent random samples from the populations of interest or from groups in a randomized experiment. (p. 554)

condición aleatoria Condición para hacer inferencias que requiere que los datos hayan sido tomados de una muestra aleatoria de la población de interés o de un experimento aleatorio. Al comparar dos o más poblaciones o tratamientos, verifica que los datos hayan sido tomados por muestras aleatorias independientes de la población de interés o de grupos de un experimento aleatorio. (p. 554)

random process Generates outcomes that are determined purely by chance. (p. 329)

proceso aleatorio Proceso que genera resultados determinados puramente al azar. (p. 329)

random sampling Using a chance process to determine which members of a population are chosen for the sample. (p. 253)

muestreo aleatorio Uso de un proceso de probabilidad para determinar qué miembros de la población son elegidos para la muestra. (p. 253)

random variable Variable that takes numerical values that describe the outcomes of a random process. (p. 390)

variable aleatoria Toma valores numéricos que describen los resultados de un proceso aleatorio. (p. 390)

randomization distribution Distribution of a statistic (like $\hat{p}_1 - \hat{p}_2$ or $\overline{x}_1 - \overline{x}_2$) in repeated random assignments of experimental units to treatment groups, assuming that the specific treatment received doesn't affect individual responses. When the conditions are met, usual inference procedures based on the sampling distribution of the statistic will be approximately correct. (pp. 496, 515, 634, 726)

distribución de la aleatoriedad La distribución de una estadística (como $\hat{p}_1 - \hat{p}_2$ or $\overline{x}_1 - \overline{x}_2$) en designaciones aleatorizadas reiteradas de unidades experimentales a grupos de tratamiento, asumiendo que el tratamiento específico no afecte las respuestas individuales. Cuando se cumplan las condiciones, los procedimientos de inferencia corrientes que se basan en la distribución del muestreo de la estadística serán aproximadamente correctos. (págs. 496, 515, 634, 726)

randomized block design Experimental design that forms groups (*blocks*) consisting of individuals that are similar in some way that is expected to affect the response to the treatments and randomly assigns experimental units to treatments separately within each block. (p. 285)

diseño de bloques aleatorios Diseño experimental en el que se forman grupos (*bloques*) compuestos por individuos que se parecen en algo que supuestamente afectará la respuesta a los tratamientos y aleatoriamente asigna unidades experimentales a tratamientos de manera separada dentro de cada bloque. (p. 285)

range A measure of variability equal to the distance between the minimum value and the maximum value of a distribution. That is, range = maximum − minimum. (p. 60)

gama Medida de variabilidad que equivale a la distancia entre el valor mínimo y el valor máximo de una distribución. Es decir, gama = máximo − mínimo. (p. 60)

regression line (*also called a* **simple linear regression model**) Line that models how a response variable *y* changes as an explanatory variable *x* changes. Regression lines are expressed in the form $\hat{y} = a + bx$, where $\hat{y}$ is the predicted value of *y* for a given value of *x*. (p. 176)

línea de regresión (*también llamada* **modelo de regresión lineal simple**) Línea que describe cómo una variable de respuesta *y* cambia a medida que cambia una variable explicativa *x*. Las líneas de regresión se expresan mediante la fórmula $\hat{y} = a + bx$, en la que $\hat{y}$ es el valor anticipado de *y* para un valor dado de *x*. (p. 176)

reject H_0 If the observed result is too unlikely to occur by chance alone when the null hypothesis is true, we can reject H_0 and say that there is convincing evidence for H_a. (p. 590)

rechazar H_0 Si es demasiado improbable que el resultado observado suceda solo por simple casualidad cuando la hipótesis nula es verdad, se puede rechazar H_0 y decir que existe evidencia convincente a favor de H_a. (p. 590)

relative frequency table Table that shows the proportion or percentage of individuals having each value. (p. 9)

tabla de frecuencia relativa Tabla que muestra la proporción o porcentaje de individuos que tiene cada valor. (p. 9)

replication Experimental design principle. Giving each treatment to enough experimental units so that a difference in the effects of the treatments can be distinguished from chance variation due to the random assignment. (p. 281)

replicación Principio de diseño experimental. Cada tratamiento se administra a suficientes unidades experimentales con el propósito de que cualquier diferencia en los efectos de los tratamientos pueda distinguirse de una variación casual debida a la asignación aletaoria. (p. 281)

residual Difference between an actual value of the response variable and the value predicted by the regression line:

$$\text{residual} = \text{actual } y - \text{predicted } y = y - \hat{y}$$

(p. 179)

residual La diferencia entre un valor real de la variable de respuesta y el valor proyectado por la línea de regresión:

$$\text{residual} = \text{actual } y - \text{predicted } y = y - \hat{y}$$

(p. 179)

residual plot A scatterplot that displays the residuals on the vertical axis and the explanatory variable on the horizontal axis. Residual plots help us assess whether a regression model is appropriate. (p. 185)

gráfico residual Gráfico de dispersión que muestra los residuales sobre el eje vertical y la variable explicativa sobre el eje horizonal. Los trazados residuales nos permiten evaluar si el modelo de regresión es apropiado. (p. 185)

resistant A statistical measure that isn't sensitive to extreme values. (p. 56)	**resistente** Medida estadística no sensible a valores extremos. (p. 56)
response bias Occurs when there is a systematic pattern of inaccurate responses to a survey question. Includes bias due to question wording. (p. 262)	**sesgo de la respuesta** Sucede cuando existe un patrón sistemático de respuestas imprecisas a una pregunta de encuesta. Se incluye el sesgo dada la formulación de la pregunta. (p. 262)
response variable Variable that measures the outcome of a study. (pp. 153, 270)	**variable de respuesta** Variable que mide el resultado de un estudio. (págs. 153, 270)
retrospective observational study An observational study that uses existing data for a sample of individuals. (p. 270)	**estudio de observación retrospectiva** Estudio de observación que usa los datos existentes para las muestras de individuos. (p. 270)

S

sample Subset of individuals in the population from which we collect data. (p. 249)	**muestra** Subconjunto de individuos en la población a partir de la cual se recogen datos. (p. 249)
sample regression line (estimated regression line) Least-squares regression line $\hat{y} = a + bx$ computed from the sample data. (p. 813)	**línea de regresión de la muestra (línea de regresión estimada)** La línea de regresión de mínimos cuadrados $\hat{y} = a + bx$ computada a partir de los datos de la muestra. (p. 813)
sample space List of all possible outcomes of a random process. (p. 342)	**espacios de la muestra** Lista de todos los resultados posibles de un proceso aleatorio. (p. 342)
sample survey Study that collects data from a sample to learn about the population from which the sample was selected. (p. 250)	**valoración de la muestra** Estudio que reúne datos de una muestra para conocer a la población de la cual se tomó la muestra. (p. 250)
sampling distribution The distribution of values taken by a statistic in all possible samples of the same size from the same population. (p. 472)	**distribución del muestreo** La distribución de valores tomados por la estadística en todas las muestras posibles del mismo tamaño obtenidas de una misma población. (p. 472)
sampling distribution of a sample mean $\bar{x}$ The distribution of values taken by the sample mean $\bar{x}$ in all possible samples of the same size from the same population. (p. 502)	**distribución de muestreo de la media de una muestra $\bar{x}$** Distribución de valores tomados de la media muestral $\bar{x}$ en todas las posibles muestras del mismo tamaño obtenidas de la misma población. (p. 502)
sampling distribution of a sample proportion $\hat{p}$ The distribution of values taken by the sample proportion $\hat{p}$ in all possible samples of the same size from the same population. (p. 487)	**distribución del muestreo de una proporción de la muestra $\hat{p}$** Distribución de valores tomados de la proporción muestral $\hat{p}$ en todas las posibles muestras del mismo tamaño de la misma población. (p. 487)
sampling distribution of a slope b The distribution of values taken by the sample slope b in all possible samples of the same size from the same population. (p. 816)	**distribución del muestreo de una pendiente b** Distribución de valores tomados de la pendiente muestral b en todas las posibles muestras del mismo tamaño de la misma población. (p. 816)
sampling distribution of $\hat{p}_1 - \hat{p}_2$ The distribution of values taken by the statistic $\hat{p}_1 - \hat{p}_2$ in all possible samples of size n_1 from population 1 and all possible samples of size n_2 from population 2. (p. 495)	**distribución del muestreo $\hat{p}_1 - \hat{p}_2$** Distribución de valores tomados por la estadística $\hat{p}_1 - \hat{p}_2$ de todas las muestras posibles de tamaño n_1 de la población 1 y de todas las muestras posibles del tamaño n_2 de la población 2. (p. 495)
sampling distribution of $\bar{x}_1 - \bar{x}_2$ The distribution of values taken by the statistic $\bar{x}_1 - \bar{x}_2$ in all possible samples of size n_1 from population 1 and all possible samples of size n_2 from population 2. (p. 514)	**distribución de muestreo de $\bar{x}_1 - \bar{x}_2$** Distribución de valores tomados por la estadística $\bar{x}_1 - \bar{x}_2$ de todas las muestras posibles de tamaño n_1 de la población 1 y de todas las muestras posibles del tamaño n_2 de la población 2. (p. 514)
sampling frame A list of individuals in a population. (p. 261)	**marco de muestreo** Lista de individuos de una población. (p. 261)
sampling variability The fact that different random samples of the same size from the same population produce different estimates. (pp. 298, 471)	**variabilidad del muestreo** El hecho de que muestras aleatorias de un mismo tamaño tomadas de la misma población producen estimaciones distintas. (págs. 298, 471)
sampling with replacement When an individual from a population can be selected more than once when choosing a sample. (p. 254)	**muestreo con reemplazo** Muestreo en el cual un individuo puede elegirse más de una vez cuando se selecciona una muestra. (p. 254)

sampling without replacement When an individual from a population can be selected only once when choosing a sample. (p. 254)

muestreo sin reemplazo Muestreo en el cual al seleccionar una muestra cada individuo de la población puede elegirse solo una vez. (p. 254)

scatterplot Plot that shows the relationship between two quantitative variables measured on the same individuals. The values of one variable appear on the horizontal axis, and the values of the other variable appear on the vertical axis. Each individual in the data appears as a point in the graph. (p. 154)

gráfico de dispersión Permite apreciar la relación entre dos variables cuantitativas midiendo los mismos individuos. Los valores de una variable figuran en el eje horizontal y los valores de la otra variable figuran en el eje vertical. Cada individuo en los datos figura como un punto en el gráfico. (p. 154)

segmented bar graph Graph that displays the distribution of a categorical variable as segments of a rectangle, with the area of each segment proportional to the percent of individuals in the corresponding category. (p. 19)

gráfico de barras segmentado Gráfico que muestra la distribución de una variable categorizada como segmentos de un rectángulo cuyo tamaño es proporcional al porcentaje de individuos de la categoría correspondiente. (p. 19)

side-by-side bar graph Graph used to compare the distribution of a categorical variable in each of several groups. For each value of the categorical variable, there is a bar corresponding to each group. The height of each bar is determined by the count or percent of individuals in the group with that value. (p. 19)

gráfico de barras contiguas Se usa para comparar la distribución de una variable categorizada en cada uno de varios grupos. Para cada valor de la variable categorizada, hay una barra que corresponde a cada grupo. La altura de la barra la determina el conteo o el porcentaje de individuos en el grupo que tengan ese valor. (p. 19)

significance level Fixed value α that we use as a boundary for deciding whether an observed result is too unlikely to happen by chance alone when the null hypothesis is true. The significance level gives the probability of a Type I error. (p. 590)

nivel de significancia Valor fijo α que se usa como límite para decidir si un resultado observado es demasiado improbable que suceda solo al azar cuando la hipótesis nula es verdad. El nivel de significancia nos da la probabilidad de un error Tipo 1. (p. 590)

significance test Formal procedure for using observed data to decide between two competing claims (the null hypothesis and the alternative hypothesis). The claims are usually statements about parameters. (p. 584)

prueba de significancia Procedimiento formal en el que se usan datos observados para decidir entre dos opciones que compiten entre sí (la hipótesis nula y la hipótesis alternativa). Las opciones comúnmente son enunciados acerca de un parámetro. (p. 584)

simple random sample (SRS) Sample chosen in such a way that every group of n individuals in the population has an equal chance to be selected as the sample. (p. 254)

muestra aleatoria sencilla Muestra tomada de tal manera que cada grupo de n individuos en la población tenga la misma oportunidad de ser escogido como la muestra. (p. 254)

simulation Imitation of a random process in such a way that simulated outcomes are consistent with real-world outcomes. (p. 332)

simulación Imitación de un proceso aleatorio de manera que los resultados simulados sean consecuentes con los resultados reales. (p. 332)

single-blind An experiment in which either the subjects or the people who interact with them and measure the response don't know which treatment a subject is receiving. (p. 277)

ciego sencillo Experimento en el que ya sea los sujetos, o bien las personas que interactúan con los sujetos y miden la variable de respuesta, desconocen qué tratamiento recibe un sujeto. (p. 277)

skewed A distribution of quantitative data is *skewed to the right* if the right side of the graph (containing the half of the observations with larger values) is much longer than the left side. It is *skewed to the left* if the left side of the graph is much longer than the right side. (p. 32)

asimétrica Distribución de datos cuantitativos que está *sesgada hacia la derecha* si la derecha del gráfico (que contiene la mitad de las observaciones con valores más grandes) es mucho más larga que el lado izquierdo. Está *sesgada hacia la izquierda* si el lado izquierdo del gráfico es mucho más largo que el derecho. (p. 32)

slope In the regression equation $\hat{y} = a + bx$, the slope b is the amount by which the predicted value of y changes when x increases by 1 unit. (p. 181)

pendiente En la ecuación de regresión $\hat{y} = a + bx$, la pendiente b es la cantidad por la cual el valor anticipado de y cambia si x aumenta 1 unidad. (p. 181)

standard deviation Measures the typical distance of the values in a distribution from the mean. It is calculated by finding an "average" of the squared deviations from the mean and then taking the square root. In symbols, the sample standard deviation s_x is given by

$$s_x = \sqrt{\frac{\sum (x_i - \overline{x})^2}{n-1}}$$

Use σ_x to denote the population standard deviation. (p. 61)

desviación estándar Mide la distancia típica de los valores en una distribución a partir de la media. Se computa hallando un "promedio" de las desviaciones al cuadrado a las que luego se les computa la raíz cuadrada. En símbolos, la desviación estándar de la muestra s_x se representa de la siguiente manera:

$$s_x = \sqrt{\frac{\sum (x_i - \overline{x})^2}{n-1}}$$

Para referirse a la media de la población debe usarse σ_x. (p. 61)

standard deviation of a discrete random variable Square root of the variance σ_X^2 of a discrete random variable. The standard deviation measures how much the values of the variable typically vary from the mean in many, many trials of the random process. In symbols,

$$\sigma_x = \sqrt{\sum (x_i - \mu_x)^2 p_i}$$

(p. 396)

desviación estándar de una variable aleatoria discreta La raíz cuadrada de la variación de una variable aleatoria discreta σ_X^2. La desviación estándar mide cuánto varían normalmente los valores de la variable respecto de la media repitiendo una gran cantidad de veces los ensayos del proceso aleatorio. En símbolos se representa de la siguiente manera:

$$\sigma_x = \sqrt{\sum (x_i - \mu_x)^2 p_i}$$

(p. 396)

---

standard deviation of the residuals (s) If we use a least-squares line to predict the values of a response variable y from an explanatory variable x, the standard deviation of the residuals (s) is given by

$$s = \sqrt{\frac{\sum \text{residuals}^2}{n-2}} = \sqrt{\frac{\sum (y_i - \hat{y}_i)^2}{n-2}}$$

This value measures the size of a typical residual. That is, s measures the typical distance between the actual y values and the predicted y values. (p. 189)

desviación estándar de las residuales (s) Si se usa la línea de cuadrados mínimos para predecir los valores de una variable de respuesta y a partir de una variable explicativa x, la desviación estándar de las residuales (s) la da

$$s = \sqrt{\frac{\sum \text{residuals}^2}{n-2}} = \sqrt{\frac{\sum (y_i - \hat{y}_i)^2}{n-2}}$$

Este valor mide el tamaño de una residual típica. Es decir, s mide la distancia típica entre los valores de y reales y los valores de y esperados. (p. 189)

---

standard error When the standard deviation of a statistic is estimated from data, the result is the standard error of the statistic. The standard error estimates how far the value of the statistic typically varies from the value it is trying to estimate. (p. 554)

error estándar Cuando se computa la desviación estándar de una estadística a partir de datos, el resultado es el error estándar de la estadística. El error estándar estima cuánto varía típicamente el valor de la estadística del valor que se busca estimar. (p. 554)

---

standard Normal distribution Normal distribution with mean 0 and standard deviation 1. (p. 120)

distribución normal estándar La distribución normal con media de 0 y desviación estándar 1. (p. 120)

---

standardized score (z-score) For an individual value in a distribution, the standardized score (z-score) tells us how many standard deviations from the mean the value falls, and in what direction. To find the standardized score (z-score), compute

$$z = \frac{\text{value} - \text{mean}}{\text{standard deviation}}$$

(p. 95)

puntuación estandarizada (puntuación z) Para un valor individual en una distribución, la puntuación estandarizada (puntuación z) expresa cuántas desviaciones estándar se producen respecto de la media y en qué dirección ocurren. Para determinar la puntuación estandarizada (puntuación z) debe aplicarse la siguiente fórmula:

$$z = \frac{\text{valor} - \text{media}}{\text{desviación estándar}}$$

(p. 95)

---

standardized test statistic Value that measures how far a sample statistic is from what we would expect if the null hypothesis H_0 were true, in standard deviation units. That is,

$$\text{standardized test statistic} = \frac{\text{statistic} - \text{parameter}}{\text{standard deviation (error) of statistic}}$$

(p. 602)

estadística de prueba estandarizada Valor que mide la divergencia entre la estadística de muestra y lo que esperaríamos si la hipótesis nula H_0 fuera cierta, en unidades de desviación estándar. Esto es:

estadística de prueba estandarizada

$$= \frac{\text{estadística} - \text{parámetro}}{\text{(error de) desviación estándar de la estadística}}$$

(p. 602)

---

statistic Number that describes some characteristic of a sample. (pp. 56, 470)

dato estadístico Número que describe alguna característica de una muestra. (págs. 56, 470)

---

statistically significant (1) When the observed results of a study are too unusual to be explained by chance alone, the results are called statistically significant. (p. 300)
(2) If the P-value is less than alpha, we say that the results of a statistical study are significant at level α. In that case, we reject the null hypothesis H_0 and conclude that there is convincing evidence for the alternative hypothesis H_a. (p. 590)

estadísticamente significativo (1) Cuando los resultados observados de un estudio son demasiado atípicos para poder decir que son consecuencia únicamente del azar, a estos se les llama estadísticamente significativos. (p. 300)
(2) Si el valor P es menor que alfa, se dice que los resultados de un estudio estadístico son significativos al nivel α. En tal caso, se rechaza la hipótesis nula H_0 y se concluye que hay evidencia convincente para la hipótesis H_a alternativa. (p. 590)

---

statistics The science and art of collecting, analyzing, and drawing conclusions from data. (p. 2)

estadística La ciencia y arte de coleccionar, analizar y llegar a conclusiones a partir de un conjunto de datos. (p. 2)

stemplot (*also called* **stem-and-leaf plot**) Simple graphical display for fairly small quantitative data sets that gives a quick picture of the shape of a distribution while including the actual numerical values in the graph. Each data value is separated into two parts: a *stem*, which consists of all but the final digit, and a *leaf*, the final digit. The stems are ordered from lowest to highest and arranged in a vertical column. The leaves are arranged in increasing order out from the appropriate stems. (p. 37)

gráfico de tallos al que (*también se le dice* **gráfico de tallos y hojas**) Representación gráfica sencilla de conjuntos de datos cuantitativos relativamente pequeños que dan una imagen rápida de la forma de una distribución, al tiempo que incluyen los valores numéricos mismos en el gráfico. Cada valor se separa en dos partes: un *tallo*, compuesto por todos los dígitos excepto el último, y una *hoja*, que es ese último dígito. Los tallos se ordenan de menor a mayor en una columna. Las hojas, en orden ascendente partiendo de los tallos aproximados. (p. 37)

strata Groups of individuals in a population who share characteristics thought to be associated with the variables being measured in a study. (p. 257)

estratos Grupos de individuos de una población que comparten características supuestamente asociadas con las medidas en un estudio. (p. 257)

stratified random sampling Method of sampling that divides the population into non-overlapping groups (*strata*) of individuals who share characteristics thought to be associated with the variables being measured in a study, selects an SRS from each stratum, and combines the SRSs into one overall sample. (p. 257)

muestreo aleatorio estratificado Método de muestreo que divide a la población en grupos no superpuestos (*estratos*) de individuos que comparten características parecidas supuestamente asociadas con las variables medidas en el estudio, selecciona una muestra aleatoria sencilla de cada estrato y combina las muestras aleatorias sencillas en una muestra general. (p. 257)

subjects Experimental units that are human beings. (p. 273)

sujetos Unidades experimentales que son seres humanos. (p. 273)

symmetric A graph of quantitative data in which the right and left sides are approximately mirror images of each other is roughly symmetric. (p. 32)

simétrico Si los lados derecho e izquierdo de un gráfico de datos cuantitativos son reflejos aproximados uno del otro, se dice que son más o menos simétricos. (p. 32)

systematic random sampling Method of sampling that chooses individuals from an ordered arrangement of the population by randomly selecting one of the first k individuals and choosing every kth individual thereafter. (p. 260)

muestreo sistemático aleatorio Método de muestreo que elige individuos de un grupo ordenado de la población de manera aleatoria mediante la elección de uno de los primeros individuos k y cada individuo k-ésimo de ahí en adelante. (p. 260)

T

***t* distribution** A distribution described by a symmetric, single-peaked, bell-shaped density curve that is completely specified by its degrees of freedom (df). When performing inference about a population mean based on a random sample of size n using the sample standard deviation s_x to estimate the population standard deviation σ, use a t distribution with df $= n - 1$. (pp. 652, 701)

distribución *t* Distribución descrita por una curva de densidad simétrica, con un solo vértice y forma de campana, completamente especificada por sus grados de libertad (df). Al realizar la inferenecia de una población basada en una muestra aleatoria de tamaño n usando la desviación estándar de la muestra s_x para estimar la desviación estándar de población, se debe usar t distribución t con df $= n - 1$. (págs. 652, 701)

***t* interval for the slope** Confidence interval used to estimate the slope of a population (true) regression line. (p. 822) For more details, see the inference summary in the back of the book.

intervalo *t* para la pendiente Intervalo de confianza que se usa para estimar la pendiente de la línea de regresión de una población (real). (p. 822) Para más información, ver el resumen de inferencia del final del libro.

***t* test for the slope** A test of the null hypothesis that there is no linear association between two quantitative variables. (p. 827) For more details, see the inference summary in the back of the book.

prueba *t* para la pendiente Prueba de la hipótesis nula en la cual no hay asociación lineal entre dos variables cuantitativas. (p. 827) Para más información, ver el resumen de inferencia del final del libro.

third quartile Q_3 If the observations in a data set are ordered from smallest to largest, the third quartile Q_3 is the median of the data values that are to the right of the median in the ordered list. (p. 64)

tercer cuartil Q_3 Si las observaciones en un conjunto de datos se organizan de menor a mayor, el tercer cuartil Q_3 es la media de los valores ubicados a la derecha de la media de la lista ordenada. (p. 64)

transforming data Changing the scale of measurement for a quantitative variable. Applying a linear, power, or logarithmic function to a quantitative variable are common ways of transforming data. (p. 214)

transformación de datos Conlleva cambiar la escala de medición para una variable cuantitativa. Aplicar una función lineal, de fuerza o logarítmica a una variable cuantitativa son maneras comunes de transformar los datos. (p. 214)

treatment Specific condition applied to the individuals in an experiment. If an experiment has several explanatory variables, a treatment is a combination of specific values of these variables. (p. 273)	**tratamiento** Una condición específica que se les aplica a los individuos en un experimento. Si un experimento tiene varias variables explicativas, el tratamiento es una combinación de los valores específicos de estas variables. (p. 273)
tree diagram A diagram that shows the sample space of a random process involving multiple stages. The probability of each outcome is shown on the corresponding branch of the tree. All probabilities after the first stage are conditional probabilities. (p. 367)	**diagrama de árbol** Diagrama que presenta el espacio de muestra de un proceso aleatorio de múltiples etapas. La probabilidad de cada resultado se muestra en la rama correspondiente del árbol. Todas las probabilidades que están después de la primera etapa son probabilidades condicionales. (p. 367)
trial One repetition of a random process. (p. 329)	**ensayo** Repetición de un proceso aleatorio. (p. 329)
two-sample t interval for a difference between two means Confidence interval used to estimate a difference in the means of two populations/treatments. (p. 671) For more details, see the inference summary in the back of the book.	**intervalo t de dos muestras para obtener la diferencia entre dos medias** Intervalo de confianza que se usa para estimar la diferencia entre los medios de dos poblaciones/tratamientos. (p. 671) Para más información, ver el resumen de inferencia del final del libro.
two-sample t test for the difference between two means A test of the null hypothesis that the difference in the means of two populations/treatments is equal to a specified value (usually 0). (p. 722) For more details, see the inference summary in the back of the book.	**prueba t de dos muestras para obtener la diferencia entre dos medias** Prueba de la hipótesis nula en la cual la diferencia entre las medias de dos poblaciones/tratamientos es igual a un valor especificado (normalmente 0). (p. 722) Para más información, ver el resumen de inferencia del final del libro.
two-sample z interval for a difference between two proportions Confidence interval used to estimate a difference in the proportions of successes in two populations/treatments. (p. 569) For more details, see the inference summary in the back of the book.	**intervalo z de dos muestras para obtener la diferencia entre dos proporciones** Intervalo de confianza que se usa para estimar la diferencia entre las proporciones de éxitos en dos poblaciones/tratamientos. (p. 569) Para más información, ver el resumen de inferencia del final del libro.
two-sample z test for the difference between two proportions A test of the null hypothesis that the difference in the proportions of successes in two populations/treatments is equal to a specified value (usually 0). (p. 632) For more details, see the inference summary in the back of the book.	**prueba z de dos muestras para obtener la diferencia entre dos proporciones** Prueba de la hipótesis nula en la cual diferencia en la proporción de aciertos entre dos poblaciones/tratamientos es igual a un valor especificado (normalmente 0). (p. 632) Para más información, ver el resumen de inferencia del final del libro.
two-sided alternative hypothesis The alternative hypothesis is two-sided if it states that the parameter is different from the null value (it could be either smaller or larger). (p. 587)	**hipótesis alternativa bilateral** La hipótesis alternativa es bilateral si indica que el parámetro es diferente del valor nulo (podría ser más pequeño o más grande). (p. 587)
two-way table Table of counts that summarizes data on the relationship between two categorical variables for some group of individuals. (p. 14)	**tabla de doble vía** Una tabla de cuentas que resume los datos sobre la relación entre dos variables categorizadas para un grupo de individuos. (p. 14)
Type I error An error that occurs if we reject H_0 when H_0 is true. That is, the data give convincing evidence that H_a is true when it really isn't. (p. 592)	**error Tipo I** Error que sucede si se rechaza H_0 aunque H_0 sea verdadera. Es decir, la prueba ofrece evidencia contundente de que H_a es verdadera cuando en realidad no lo es. (p. 592)
Type II error An error that occurs if we fail to reject H_0 when H_a is true. That is, the data do not give convincing evidence that H_a is true when it really is. (p. 592)	**error Tipo II** Error que sucede si no se rechaza H_0 incluso cuando H_a es verdadera. Es decir, los datos no ofrecen evidencia contundente de que H_a sea verdadera cuando sí lo es. (p. 592)

U

unbiased estimator A statistic used for estimating a parameter is unbiased if the mean of its sampling distribution is equal to the value of the parameter being estimated. (p. 476)	**estimador sin sesgo** La estadística que se usa para computar un parámetro es un estimador sin sesgo si la media de distribución de su muestreo equivale al valor del parámetro que se está computando. (p. 476)
undercoverage Occurs when some members of the population are less likely to be chosen or cannot be chosen in a sample. (p. 261)	**subcobertura** Sucede cuando algunos miembros de la población tienen menor probablidad de ser elegidos o no pueden ser elegidos para una muestra. (p. 261)

uniform A distribution in which the frequency (relative frequency) of each possible value is about the same is *approximately uniform*. (p. 33)	**uniforme** Una distribución en la cual la frecuencia (frecuencia relativa) de cada valor posible es casi la misma se considera *aproximandamente uniforme*. (p. 33)
unimodal A graph of quantitative data with a single peak. (p. 33)	**unimodal** Gráfico de datos cuantitativos con un solo pico. (p. 33)
union The event "A or B" is called the union of events A and B. It consists of all outcomes in A or B or both, and is denoted by $A \cup B$. (p. 351)	**unión** El evento "A or B" "se denomina la unión de los eventos A y B. Abarca todos los resultados en A o B, o ambos y se expresa con $A \cup B$. (p. 351)
univariate data A one-variable data set. (p.153)	**datos univariados** Conjunto de datos de una sola variable. (p.153)

V

variability of a statistic Describes the variation in a statistic's sampling distribution. Statistics from larger samples have less variability. (p. 479)	**variabilidad de una estadística** Describe la variabilidad del muestreo de una estadística. Las estadísticas de las muestras grandes tienen menos variabilidad. (p. 479)
variable Any attribute of an individual. A variable can take different values for different individuals. (p. 3)	**variable** Cualquier atributo de un individuo. Una variable puede tener distintos valores según cada individuo. (p. 3)
variance of a discrete random variable σ_X^2 Weighted average of the squared deviations of the values of the variable from their mean. In symbols, $$\sigma_X^2 = \sum(x_i - \mu_X)^2 p_i$$ (p. 396)	**variación de una variable aleatoria discreta** σ_X^2 Promedio sopesado de las desviaciones cuadráticas de los valores de la variable a partir de su media. En símbolos se expresa $$\sigma_X^2 = \sum(x_i - \mu_X)^2 p_i$$ (p. 396)
variance s_x^2 "Average" squared deviation of the values in a distribution from their mean. In symbols, the sample variance s_x^2 is given by $$s_x^2 = \frac{\sum(x_i - \overline{x})^2}{n-1}$$ (p. 61)	**variación** s_x^2 Desviación cuadrática "promedio" en que se producen los valores de una distribución con respecto a la media. En símbolos se expresa de la siguiente manera: $$s_x^2 = \frac{\sum(x_i - \overline{x})^2}{n-1}$$ (p. 61)
Venn diagrams A diagram that consists of one or more circles surrounded by a rectangle. Each circle represents an event. The region inside the rectangle represents the sample space of the random process. (p. 350)	**diagramas Venn** Diagrama que consiste en uno o más círculos colocados dentro de un rectángulo. Cada círculo representa un evento. El interior del rectángulo representa el espacio muestral del proceso aleatorio. (p. 322)
voluntary response sampling Method of sampling that allows people to choose to be in the sample by responding to a general invitation. (p. 252)	**muestreo de respuesta voluntaria** Método de muestreo que permite que alguien elija ser parte de la muestra al aceptar una invitación general. (p. 252)

W

wording of questions An important influence on the answers given in a survey. Confusing or leading questions can introduce strong bias, and changes in wording can greatly change a survey's outcome. Even the order in which questions are asked matters. (p. 262)	**terminología de las preguntas** Influencia importante sobre las respuestas que se dan en un sondeo. Las preguntas confusas o capciosas puede introducir un sesgo marcado, y los cambios de terminología pueden modificar con mucho los resultados de tal sondeo. Incluso el orden en que se hacen las preguntas tiene importancia. (p. 262)

Y

y intercept In the regression equation $\hat{y} = a + bx$, the y intercept a is the predicted value of y when $x = 0$. (p. 181)	**interceptación y** En la ecuación de regresión $\hat{y} = a + bx$, la interceptación y a es el valor anticipado de y cuando $x = 0$. (p. 181)

Z

z-score See *standardized score*.	**puntaje z** Ver *puntajes estandarizados*.

Index

Note: Page numbers followed by f indicate figures.

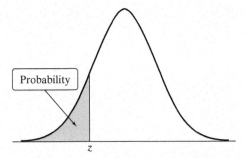

Table entry for z is the area under the standard Normal curve to the left of z.

| Table A Standard Normal probabilities | | | | | | | | | |
z	.00	.01	.02	.03	.04	.05	.06	.07	.08	.09
−3.4	.0003	.0003	.0003	.0003	.0003	.0003	.0003	.0003	.0003	.0002
−3.3	.0005	.0005	.0005	.0004	.0004	.0004	.0004	.0004	.0004	.0003
−3.2	.0007	.0007	.0006	.0006	.0006	.0006	.0006	.0005	.0005	.0005
−3.1	.0010	.0009	.0009	.0009	.0008	.0008	.0008	.0008	.0007	.0007
−3.0	.0013	.0013	.0013	.0012	.0012	.0011	.0011	.0011	.0010	.0010
−2.9	.0019	.0018	.0018	.0017	.0016	.0016	.0015	.0015	.0014	.0014
−2.8	.0026	.0025	.0024	.0023	.0023	.0022	.0021	.0021	.0020	.0019
−2.7	.0035	.0034	.0033	.0032	.0031	.0030	.0029	.0028	.0027	.0026
−2.6	.0047	.0045	.0044	.0043	.0041	.0040	.0039	.0038	.0037	.0036
−2.5	.0062	.0060	.0059	.0057	.0055	.0054	.0052	.0051	.0049	.0048
−2.4	.0082	.0080	.0078	.0075	.0073	.0071	.0069	.0068	.0066	.0064
−2.3	.0107	.0104	.0102	.0099	.0096	.0094	.0091	.0089	.0087	.0084
−2.2	.0139	.0136	.0132	.0129	.0125	.0122	.0119	.0116	.0113	.0110
−2.1	.0179	.0174	.0170	.0166	.0162	.0158	.0154	.0150	.0146	.0143
−2.0	.0228	.0222	.0217	.0212	.0207	.0202	.0197	.0192	.0188	.0183
−1.9	.0287	.0281	.0274	.0268	.0262	.0256	.0250	.0244	.0239	.0233
−1.8	.0359	.0351	.0344	.0336	.0329	.0322	.0314	.0307	.0301	.0294
−1.7	.0446	.0436	.0427	.0418	.0409	.0401	.0392	.0384	.0375	.0367
−1.6	.0548	.0537	.0526	.0516	.0505	.0495	.0485	.0475	.0465	.0455
−1.5	.0668	.0655	.0643	.0630	.0618	.0606	.0594	.0582	.0571	.0559
−1.4	.0808	.0793	.0778	.0764	.0749	.0735	.0721	.0708	.0694	.0681
−1.3	.0968	.0951	.0934	.0918	.0901	.0885	.0869	.0853	.0838	.0823
−1.2	.1151	.1131	.1112	.1093	.1075	.1056	.1038	.1020	.1003	.0985
−1.1	.1357	.1335	.1314	.1292	.1271	.1251	.1230	.1210	.1190	.1170
−1.0	.1587	.1562	.1539	.1515	.1492	.1469	.1446	.1423	.1401	.1379
−0.9	.1841	.1814	.1788	.1762	.1736	.1711	.1685	.1660	.1635	.1611
−0.8	.2119	.2090	.2061	.2033	.2005	.1977	.1949	.1922	.1894	.1867
−0.7	.2420	.2389	.2358	.2327	.2296	.2266	.2236	.2206	.2177	.2148
−0.6	.2743	.2709	.2676	.2643	.2611	.2578	.2546	.2514	.2483	.2451
−0.5	.3085	.3050	.3015	.2981	.2946	.2912	.2877	.2843	.2810	.2776
−0.4	.3446	.3409	.3372	.3336	.3300	.3264	.3228	.3192	.3156	.3121
−0.3	.3821	.3783	.3745	.3707	.3669	.3632	.3594	.3557	.3520	.3483
−0.2	.4207	.4168	.4129	.4090	.4052	.4013	.3974	.3936	.3897	.3859
−0.1	.4602	.4562	.4522	.4483	.4443	.4404	.4364	.4325	.4286	.4247
−0.0	.5000	.4960	.4920	.4880	.4840	.4801	.4761	.4721	.4681	.4641

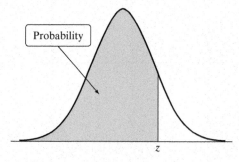

Table entry for *z* is the area under the standard Normal curve to the left of *z*.

Table A Standard Normal probabilities (continued)									
z	**.00**	**.01**	**.02**	**.03**	**.04**	**.05**	**.06**	**.07**	**.08**	**.09**
0.0	.5000	.5040	.5080	.5120	.5160	.5199	.5239	.5279	.5319	.5359
0.1	.5398	.5438	.5478	.5517	.5557	.5596	.5636	.5675	.5714	.5753
0.2	.5793	.5832	.5871	.5910	.5948	.5987	.6026	.6064	.6103	.6141
0.3	.6179	.6217	.6255	.6293	.6331	.6368	.6406	.6443	.6480	.6517
0.4	.6554	.6591	.6628	.6664	.6700	.6736	.6772	.6808	.6844	.6879
0.5	.6915	.6950	.6985	.7019	.7054	.7088	.7123	.7157	.7190	.7224
0.6	.7257	.7291	.7324	.7357	.7389	.7422	.7454	.7486	.7517	.7549
0.7	.7580	.7611	.7642	.7673	.7704	.7734	.7764	.7794	.7823	.7852
0.8	.7881	.7910	.7939	.7967	.7995	.8023	.8051	.8078	.8106	.8133
0.9	.8159	.8186	.8212	.8238	.8264	.8289	.8315	.8340	.8365	.8389
1.0	.8413	.8438	.8461	.8485	.8508	.8531	.8554	.8577	.8599	.8621
1.1	.8643	.8665	.8686	.8708	.8729	.8749	.8770	.8790	.8810	.8830
1.2	.8849	.8869	.8888	.8907	.8925	.8944	.8962	.8980	.8997	.9015
1.3	.9032	.9049	.9066	.9082	.9099	.9115	.9131	.9147	.9162	.9177
1.4	.9192	.9207	.9222	.9236	.9251	.9265	.9279	.9292	.9306	.9319
1.5	.9332	.9345	.9357	.9370	.9382	.9394	.9406	.9418	.9429	.9441
1.6	.9452	.9463	.9474	.9484	.9495	.9505	.9515	.9525	.9535	.9545
1.7	.9554	.9564	.9573	.9582	.9591	.9599	.9608	.9616	.9625	.9633
1.8	.9641	.9649	.9656	.9664	.9671	.9678	.9686	.9693	.9699	.9706
1.9	.9713	.9719	.9726	.9732	.9738	.9744	.9750	.9756	.9761	.9767
2.0	.9772	.9778	.9783	.9788	.9793	.9798	.9803	.9808	.9812	.9817
2.1	.9821	.9826	.9830	.9834	.9838	.9842	.9846	.9850	.9854	.9857
2.2	.9861	.9864	.9868	.9871	.9875	.9878	.9881	.9884	.9887	.9890
2.3	.9893	.9896	.9898	.9901	.9904	.9906	.9909	.9911	.9913	.9916
2.4	.9918	.9920	.9922	.9925	.9927	.9929	.9931	.9932	.9934	.9936
2.5	.9938	.9940	.9941	.9943	.9945	.9946	.9948	.9949	.9951	.9952
2.6	.9953	.9955	.9956	.9957	.9959	.9960	.9961	.9962	.9963	.9964
2.7	.9965	.9966	.9967	.9968	.9969	.9970	.9971	.9972	.9973	.9974
2.8	.9974	.9975	.9976	.9977	.9977	.9978	.9979	.9979	.9980	.9981
2.9	.9981	.9982	.9982	.9983	.9984	.9984	.9985	.9985	.9986	.9986
3.0	.9987	.9987	.9987	.9988	.9988	.9989	.9989	.9989	.9990	.9990
3.1	.9990	.9991	.9991	.9991	.9992	.9992	.9992	.9992	.9993	.9993
3.2	.9993	.9993	.9994	.9994	.9994	.9994	.9994	.9995	.9995	.9995
3.3	.9995	.9995	.9995	.9996	.9996	.9996	.9996	.9996	.9996	.9997
3.4	.9997	.9997	.9997	.9997	.9997	.9997	.9997	.9997	.9997	.9998

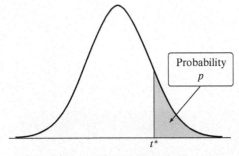

Probability
p

Table entry for *p* and *C* is the point *t** with probability *p* lying to its right
and probability *C* lying between −*t** and *t**.

Table B   *t* distribution critical values

df	.25	.20	.15	.10	.05	.025	.02	.01	.005	.0025	.001	.0005
1	1.000	1.376	1.963	3.078	6.314	12.71	15.89	31.82	63.66	127.3	318.3	636.6
2	0.816	1.061	1.386	1.886	2.920	4.303	4.849	6.965	9.925	14.09	22.33	31.60
3	0.765	0.978	1.250	1.638	2.353	3.182	3.482	4.541	5.841	7.453	10.21	12.92
4	0.741	0.941	1.190	1.533	2.132	2.776	2.999	3.747	4.604	5.598	7.173	8.610
5	0.727	0.920	1.156	1.476	2.015	2.571	2.757	3.365	4.032	4.773	5.893	6.869
6	0.718	0.906	1.134	1.440	1.943	2.447	2.612	3.143	3.707	4.317	5.208	5.959
7	0.711	0.896	1.119	1.415	1.895	2.365	2.517	2.998	3.499	4.029	4.785	5.408
8	0.706	0.889	1.108	1.397	1.860	2.306	2.449	2.896	3.355	3.833	4.501	5.041
9	0.703	0.883	1.100	1.383	1.833	2.262	2.398	2.821	3.250	3.690	4.297	4.781
10	0.700	0.879	1.093	1.372	1.812	2.228	2.359	2.764	3.169	3.581	4.144	4.587
11	0.697	0.876	1.088	1.363	1.796	2.201	2.328	2.718	3.106	3.497	4.025	4.437
12	0.695	0.873	1.083	1.356	1.782	2.179	2.303	2.681	3.055	3.428	3.930	4.318
13	0.694	0.870	1.079	1.350	1.771	2.160	2.282	2.650	3.012	3.372	3.852	4.221
14	0.692	0.868	1.076	1.345	1.761	2.145	2.264	2.624	2.977	3.326	3.787	4.140
15	0.691	0.866	1.074	1.341	1.753	2.131	2.249	2.602	2.947	3.286	3.733	4.073
16	0.690	0.865	1.071	1.337	1.746	2.120	2.235	2.583	2.921	3.252	3.686	4.015
17	0.689	0.863	1.069	1.333	1.740	2.110	2.224	2.567	2.898	3.222	3.646	3.965
18	0.688	0.862	1.067	1.330	1.734	2.101	2.214	2.552	2.878	3.197	3.611	3.922
19	0.688	0.861	1.066	1.328	1.729	2.093	2.205	2.539	2.861	3.174	3.579	3.883
20	0.687	0.860	1.064	1.325	1.725	2.086	2.197	2.528	2.845	3.153	3.552	3.850
21	0.686	0.859	1.063	1.323	1.721	2.080	2.189	2.518	2.831	3.135	3.527	3.819
22	0.686	0.858	1.061	1.321	1.717	2.074	2.183	2.508	2.819	3.119	3.505	3.792
23	0.685	0.858	1.060	1.319	1.714	2.069	2.177	2.500	2.807	3.104	3.485	3.768
24	0.685	0.857	1.059	1.318	1.711	2.064	2.172	2.492	2.797	3.091	3.467	3.745
25	0.684	0.856	1.058	1.316	1.708	2.060	2.167	2.485	2.787	3.078	3.450	3.725
26	0.684	0.856	1.058	1.315	1.706	2.056	2.162	2.479	2.779	3.067	3.435	3.707
27	0.684	0.855	1.057	1.314	1.703	2.052	2.158	2.473	2.771	3.057	3.421	3.690
28	0.683	0.855	1.056	1.313	1.701	2.048	2.154	2.467	2.763	3.047	3.408	3.674
29	0.683	0.854	1.055	1.311	1.699	2.045	2.150	2.462	2.756	3.038	3.396	3.659
30	0.683	0.854	1.055	1.310	1.697	2.042	2.147	2.457	2.750	3.030	3.385	3.646
40	0.681	0.851	1.050	1.303	1.684	2.021	2.123	2.423	2.704	2.971	3.307	3.551
50	0.679	0.849	1.047	1.299	1.676	2.009	2.109	2.403	2.678	2.937	3.261	3.496
60	0.679	0.848	1.045	1.296	1.671	2.000	2.099	2.390	2.660	2.915	3.232	3.460
80	0.678	0.846	1.043	1.292	1.664	1.990	2.088	2.374	2.639	2.887	3.195	3.416
100	0.677	0.845	1.042	1.290	1.660	1.984	2.081	2.364	2.626	2.871	3.174	3.390
1000	0.675	0.842	1.037	1.282	1.646	1.962	2.056	2.330	2.581	2.813	3.098	3.300
∞	0.674	0.841	1.036	1.282	1.645	1.960	2.054	2.326	2.576	2.807	3.091	3.291
	50%	60%	70%	80%	90%	95%	96%	98%	99%	99.5%	99.8%	99.9%

Confidence level *C*

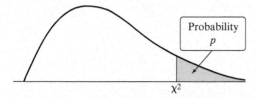

Table entry for p is the point χ^2 with probability p lying to its right.

Table C Chi–square distribution critical values											
					Tail probability p						
df	.25	.20	.15	.10	.05	.025	.02	.01	.005	.0025	.001	.0005
1	1.32	1.64	2.07	2.71	3.84	5.02	5.41	6.63	7.88	9.14	10.83	12.12
2	2.77	3.22	3.79	4.61	5.99	7.38	7.82	9.21	10.60	11.98	13.82	15.20
3	4.11	4.64	5.32	6.25	7.81	9.35	9.84	11.34	12.84	14.32	16.27	17.73
4	5.39	5.99	6.74	7.78	9.49	11.14	11.67	13.28	14.86	16.42	18.47	20.00
5	6.63	7.29	8.12	9.24	11.07	12.83	13.39	15.09	16.75	18.39	20.51	22.11
6	7.84	8.56	9.45	10.64	12.59	14.45	15.03	16.81	18.55	20.25	22.46	24.10
7	9.04	9.80	10.75	12.02	14.07	16.01	16.62	18.48	20.28	22.04	24.32	26.02
8	10.22	11.03	12.03	13.36	15.51	17.53	18.17	20.09	21.95	23.77	26.12	27.87
9	11.39	12.24	13.29	14.68	16.92	19.02	19.68	21.67	23.59	25.46	27.88	29.67
10	12.55	13.44	14.53	15.99	18.31	20.48	21.16	23.21	25.19	27.11	29.59	31.42
11	13.70	14.63	15.77	17.28	19.68	21.92	22.62	24.72	26.76	28.73	31.26	33.14
12	14.85	15.81	16.99	18.55	21.03	23.34	24.05	26.22	28.30	30.32	32.91	34.82
13	15.98	16.98	18.20	19.81	22.36	24.74	25.47	27.69	29.82	31.88	34.53	36.48
14	17.12	18.15	19.41	21.06	23.68	26.12	26.87	29.14	31.32	33.43	36.12	38.11
15	18.25	19.31	20.60	22.31	25.00	27.49	28.26	30.58	32.80	34.95	37.70	39.72
16	19.37	20.47	21.79	23.54	26.30	28.85	29.63	32.00	34.27	36.46	39.25	41.31
17	20.49	21.61	22.98	24.77	27.59	30.19	31.00	33.41	35.72	37.95	40.79	42.88
18	21.60	22.76	24.16	25.99	28.87	31.53	32.35	34.81	37.16	39.42	42.31	44.43
19	22.72	23.90	25.33	27.20	30.14	32.85	33.69	36.19	38.58	40.88	43.82	45.97
20	23.83	25.04	26.50	28.41	31.41	34.17	35.02	37.57	40.00	42.34	45.31	47.50
21	24.93	26.17	27.66	29.62	32.67	35.48	36.34	38.93	41.40	43.78	46.80	49.01
22	26.04	27.30	28.82	30.81	33.92	36.78	37.66	40.29	42.80	45.20	48.27	50.51
23	27.14	28.43	29.98	32.01	35.17	38.08	38.97	41.64	44.18	46.62	49.73	52.00
24	28.24	29.55	31.13	33.20	36.42	39.36	40.27	42.98	45.56	48.03	51.18	53.48
25	29.34	30.68	32.28	34.38	37.65	40.65	41.57	44.31	46.93	49.44	52.62	54.95
26	30.43	31.79	33.43	35.56	38.89	41.92	42.86	45.64	48.29	50.83	54.05	56.41
27	31.53	32.91	34.57	36.74	40.11	43.19	44.14	46.96	49.64	52.22	55.48	57.86
28	32.62	34.03	35.71	37.92	41.34	44.46	45.42	48.28	50.99	53.59	56.89	59.30
29	33.71	35.14	36.85	39.09	42.56	45.72	46.69	49.59	52.34	54.97	58.30	60.73
30	34.80	36.25	37.99	40.26	43.77	46.98	47.96	50.89	53.67	56.33	59.70	62.16
40	45.62	47.27	49.24	51.81	55.76	59.34	60.44	63.69	66.77	69.70	73.40	76.09
50	56.33	58.16	60.35	63.17	67.50	71.42	72.61	76.15	79.49	82.66	86.66	89.56
60	66.98	68.97	71.34	74.40	79.08	83.30	84.58	88.38	91.95	95.34	99.61	102.7
80	88.13	90.41	93.11	96.58	101.9	106.6	108.1	112.3	116.3	120.1	124.8	128.3
100	109.1	111.7	114.7	118.5	124.3	129.6	131.1	135.8	140.2	144.3	149.4	153.2

Table D	Random digits						

Line							
101	19223	95034	05756	28713	96409	12531	42544	82853
102	73676	47150	99400	01927	27754	42648	82425	36290
103	45467	71709	77558	00095	32863	29485	82226	90056
104	52711	38889	93074	60227	40011	85848	48767	52573
105	95592	94007	69971	91481	60779	53791	17297	59335
106	68417	35013	15529	72765	85089	57067	50211	47487
107	82739	57890	20807	47511	81676	55300	94383	14893
108	60940	72024	17868	24943	61790	90656	87964	18883
109	36009	19365	15412	39638	85453	46816	83485	41979
110	38448	48789	18338	24697	39364	42006	76688	08708
111	81486	69487	60513	09297	00412	71238	27649	39950
112	59636	88804	04634	71197	19352	73089	84898	45785
113	62568	70206	40325	03699	71080	22553	11486	11776
114	45149	32992	75730	66280	03819	56202	02938	70915
115	61041	77684	94322	24709	73698	14526	31893	32592
116	14459	26056	31424	80371	65103	62253	50490	61181
117	38167	98532	62183	70632	23417	26185	41448	75532
118	73190	32533	04470	29669	84407	90785	65956	86382
119	95857	07118	87664	92099	58806	66979	98624	84826
120	35476	55972	39421	65850	04266	35435	43742	11937
121	71487	09984	29077	14863	61683	47052	62224	51025
122	13873	81598	95052	90908	73592	75186	87136	95761
123	54580	81507	27102	56027	55892	33063	41842	81868
124	71035	09001	43367	49497	72719	96758	27611	91596
125	96746	12149	37823	71868	18442	35119	62103	39244
126	96927	19931	36809	74192	77567	88741	48409	41903
127	43909	99477	25330	64359	40085	16925	85117	36071
128	15689	14227	06565	14374	13352	49367	81982	87209
129	36759	58984	68288	22913	18638	54303	00795	08727
130	69051	64817	87174	09517	84534	06489	87201	97245
131	05007	16632	81194	14873	04197	85576	45195	96565
132	68732	55259	84292	08796	43165	93739	31685	97150
133	45740	41807	65561	33302	07051	93623	18132	09547
134	27816	78416	18329	21337	35213	37741	04312	68508
135	66925	55658	39100	78458	11206	19876	87151	31260
136	08421	44753	77377	28744	75592	08563	79140	92454
137	53645	66812	61421	47836	12609	15373	98481	14592
138	66831	68908	40772	21558	47781	33586	79177	06928
139	55588	99404	70708	41098	43563	56934	48394	51719
140	12975	13258	13048	45144	72321	81940	00360	02428
141	96767	35964	23822	96012	94591	65194	50842	53372
142	72829	50232	97892	63408	77919	44575	24870	04178
143	88565	42628	17797	49376	61762	16953	88604	12724
144	62964	88145	83083	69453	46109	59505	69680	00900
145	19687	12633	57857	95806	09931	02150	43163	58636
146	37609	59057	66967	83401	60705	02384	90597	93600
147	54973	86278	88737	74351	47500	84552	19909	67181
148	00694	05977	19664	65441	20903	62371	22725	53340
149	71546	05233	53946	68743	72460	27601	45403	88692
150	07511	88915	41267	16853	84569	79367	32337	03316

Inference Summary

How to Organize an Inference Problem: The Four-Step Process	
Confidence intervals	**Significance tests**
STATE: State the parameter you want to estimate and the confidence level.	State the hypotheses you want to test and the significance level, and define any parameters you use.
PLAN: Identify the appropriate inference method and check conditions.	Identify the appropriate inference method and check conditions.
DO: If the conditions are met, perform calculations.	If the conditions are met, perform calculations. • Give the sample statistic(s). • Calculate the standardized test statistic. • Find the *P*-value.
CONCLUDE: Interpret your interval in the context of the problem.	Make a conclusion about the hypotheses in the context of the problem.

$$\text{Confidence interval} = \text{statistic} \pm (\text{critical value}) \cdot (\text{standard error of statistic})$$

$$\text{Standardized test statistic} = \frac{\text{statistic} - \text{parameter}}{\text{standard deviation (error) of statistic}}$$

Inference about	Number of samples/ groups	Interval or test (Section)	Name of procedure (TI-83/84 name) Formula	Conditions
Proportions	1	Interval (8.2)	One-sample z interval for p (1-PropZInt) $$\hat{p} \pm z^* \sqrt{\frac{\hat{p}(1-\hat{p})}{n}}$$	**Random:** The data come from a random sample from the population of interest. ○ **10%:** When sampling without replacement, $n < 0.10N$. **Large Counts:** *Interval:* Both $n\hat{p}$ and $n(1-\hat{p}) \geq 10$. *Test:* Both np_0 and $n(1-p_0) \geq 10$.
		Test (9.2)	One-sample z test for p (1-PropZTest) $$z = \frac{\hat{p} - p_0}{\sqrt{\frac{p_0(1-p_0)}{n}}}$$	
	2	Interval (8.3)	Two-sample z interval for $p_1 - p_2$ (2-PropZInt) $$(\hat{p}_1 - \hat{p}_2) \pm z^* \sqrt{\frac{\hat{p}_1(1-\hat{p}_1)}{n_1} + \frac{\hat{p}_2(1-\hat{p}_2)}{n_2}}$$	**Random:** The data come from two independent random samples or from two groups in a randomized experiment. ○ **10%:** When sampling without replacement, $n_1 < 0.10N_1$ and $n_2 < 0.10N_2$. **Large Counts:** *Interval:* The counts of successes and failures in each sample or group—$n_1\hat{p}_1$, $n_1(1-\hat{p}_1)$, $n_2\hat{p}_2$ and $n_2(1-\hat{p}_2)$—are all ≥ 10. *Test:* The expected counts of successes and failures in each sample or group—$n_1\hat{p}_C$, $n_1(1-\hat{p}_C)$, $n_2\hat{p}_C$, $n_2(1-\hat{p}_C)$—are all ≥ 10.
		Test (9.3)	Two-sample z test for $p_1 - p_2$ (2-PropZTest) $$z = \frac{(\hat{p}_1 - \hat{p}_2) - 0}{\sqrt{\frac{\hat{p}_C(1-\hat{p}_C)}{n_1} + \frac{\hat{p}_C(1-\hat{p}_C)}{n_2}}}$$ where $\hat{p}_C = \frac{\text{total successes}}{\text{total sample size}} = \frac{X_1 + X_2}{n_1 + n_2}$	

Inference about	Number of samples/ groups	Interval or test (Section)	Name of procedure (TI-83/84 name) Formula	Conditions
Means	1	Interval (10.1)	One-sample t interval for μ (TInterval) $$\bar{x} \pm t^* \frac{s_x}{\sqrt{n}} \quad df = n - 1$$	**Random:** The data come from a random sample from the population of interest. ○ **10%:** When sampling without replacement, $n < 0.10N$.
		Test (11.1)	One-sample t test for μ (T-Test) $$t = \frac{\bar{x} - \mu_0}{\frac{s_x}{\sqrt{n}}} \quad df = n - 1$$	**Normal/Large Sample:** The population has a Normal distribution or the sample size is large ($n \geq 30$). If the population distribution has unknown shape and $n < 30$, a graph of the sample data shows no strong skewness or outliers.
	2	Interval (10.2)	Two-sample t interval for $\mu_1 - \mu_2$ (2-SampTInt) $$(\bar{x}_1 - \bar{x}_2) \pm t^* \sqrt{\frac{s_1^2}{n_1} + \frac{s_2^2}{n_2}}$$ df from technology or smaller of $n_1 - 1$, $n_2 - 1$	**Random:** The data come from two independent random samples or from two groups in a randomized experiment. ○ **10%:** When sampling without replacement, $n_1 < 0.10N_1$ and $n_2 < 0.10N_2$. **Normal/Large Sample:** For each sample, the corresponding population distribution (or the true distribution of response to the treatment) is Normal or the sample size is large ($n \geq 30$). For each sample, if the population (treatment) distribution has unknown shape and $n < 30$, a graph of the sample data shows no strong skewness or outliers.
		Test (11.2)	Two-sample t test for $\mu_1 - \mu_2$ (2-SampTTest) $$t = \frac{(\bar{x}_1 - \bar{x}_2) - 0}{\sqrt{\frac{s_1^2}{n_1} + \frac{s_2^2}{n_2}}}$$ df from technology or smaller of $n_1 - 1$, $n_2 - 1$	
	Paired data	Interval (10.2)	Paired t interval for μ_{diff} (TInterval) $$\bar{x}_{\text{diff}} \pm t^* \frac{s_{\text{diff}}}{\sqrt{n_{\text{diff}}}} \quad df = n_{\text{diff}} - 1$$	**Random:** Paired data come from a random sample from the population of interest or from a randomized experiment. ○ **10%:** When sampling without replacement, $n_{\text{diff}} < 0.10N_{\text{diff}}$.
		Test (11.2)	Paired t test for μ_{diff} (T-Test) $$t = \frac{\bar{x}_{\text{diff}} - \mu_0}{\frac{s_{\text{diff}}}{\sqrt{n_{\text{diff}}}}} \quad df = n_{\text{diff}} - 1$$	**Normal/Large Sample:** The population distribution of differences (or the true distribution of differences in response to the treatments) is Normal or the number of differences in the sample is large ($n_{\text{diff}} \geq 30$). If the population (treatment) distribution of differences has unknown shape and $n_{\text{diff}} < 30$, a graph of the sample differences shows no strong skewness or outliers.

Inference about	Number of samples/ groups	Interval or test (Section)	Name of procedure (TI-83/84 name) Formula	Conditions
Distribution of a categorical variable	1	Test (12.1)	Chi-square test for goodness of fit (χ^2GOF-Test) $$\chi^2 = \sum \frac{(\text{observed count} - \text{expected count})^2}{\text{expected count}}$$ df = number of categories − 1	**Random:** The data come from a random sample from the population of interest. ○ **10%:** When sampling without replacement, $n < 0.10N$. **Large Counts:** All expected counts at least 5
	2 or more	Test (12.2)	Chi-square test for homogeneity (χ^2-Test) $$\chi^2 = \sum \frac{(\text{observed count} - \text{expected count})^2}{\text{expected count}}$$ df = (# of rows − 1)(# of columns − 1)	**Random:** Data from independent random samples or from groups in a randomized experiment. ○ **10%:** When sampling without replacement, $n_1 < 0.10N_1$, $n_2 < 0.10N_2$, and so on. **Large Counts:** All expected counts at least 5
Relationship between 2 categorical variables	1	Test (12.2)	Chi-square test for independence (χ^2-Test) $$\chi^2 = \sum \frac{(\text{observed count} - \text{expected count})^2}{\text{expected count}}$$ df = (# of rows − 1)(# of columns − 1)	**Random:** The data come from a random sample from the population of interest. ○ **10%:** When sampling without replacement, $n < 0.10N$. **Large Counts:** All expected counts at least 5
Relationship between 2 quantitative variables (slope)	1	Interval (12.3)	t interval for the slope (LinRegTInt) $b \pm t^*(\text{SE}_b)$ with df $= n - 2$	**Linear:** The actual relationship between x and y is linear. For any particular value of x, the mean response μ_y falls on the population (true) regression line $\mu_y = \alpha + \beta x$. **Independent:** Individual observations are independent of each other. When sampling without replacement, check the *10% condition*, $n < 0.10N$.
		Test (12.3)	t test for the slope (LinRegTTest) $$t = \frac{b - \beta_0}{\text{SE}_b}$$ with df $= n - 2$	**Normal:** For any particular value of x, the response y varies according to a Normal distribution. **Equal SD:** The standard deviation of y (call it σ) is the same for all values of x. **Random:** The data come from a random sample from the population of interest or a randomized experiment.

Technology Corner References

TI-Nspire and other technology instructions are on the website at highschool.bfwpub.com/updatedtps6e.